Springer Collected Works in Mathematics

For further volumes:
http://www.springer.com/series/11104

Goro Shimura

Collected Papers III

1978 – 1988

Reprint of the 2003 Edition

 Springer

Goro Shimura
Mathematics Department
Princeton University
Princeton, NJ 08544-1000
USA

ISSN 2194-9875
ISBN 978-1-4939-1833-1   (Softcover)
        978-0-387-95417-2   (Hardcover)
DOI 10.1007/978-1-4612-2060-2
Springer New York Heidelberg Dordrecht London

Library of Congress Control Number: 2012954381

Printed on acid-free paper

Springer is part of Springer Science+Business Media (www.springer.com)

Contents

List of Articles   **vii**

[78a] On certain reciprocity-laws for theta functions and modular forms   1
[78b] The arithmetic of automorphic forms with respect to a unitary group . . .   38
[78c] The special values of the zeta functions associated with Hilbert
       modular forms   75
[79a] Automorphic forms and the periods of abelian varieties   115
[79b] On some problems of algebraicity   147
[80]   The arithmetic of certain zeta functions and automorphic forms on
       orthogonal groups   154
[81a] The critical values of certain zeta functions associated with modular
       forms of half-integral weight   217
[81b] On certain zeta functions attached to two Hilbert modular forms:
       I. The case of Hecke characters   241
[81c] On certain zeta functions attached to two Hilbert modular forms:
       II. The case of automorphic forms on a quaternion algebra   279
[81d] Arithmetic of differential operators on symmetric domains   318
[82a] Models of an abelian variety with complex multiplication over
       small fields   349
[82b] The periods of certain automorphic forms of arithmetic type   360
[82c] Confluent hypergeometric functions on tube domains   388
[83a] Algebraic relations between critical values of zeta functions and
       inner products   422
[83b] On Eisenstein series   455
[84a] Differential operators and the singular values of Eisenstein series   515
[84b] On differential operators attached to certain representations of
       classical groups   584

[85a] On Eisenstein series of half-integral weight 610

[85b] On the Eisenstein series of Hilbert modular groups 644

[86] On a class of nearly holomorphic automorphic forms 686

[87a] Nearly holomorphic functions on hermitian symmetric spaces 746

[87b] On Hilbert modular forms of half-integral weight 774

[88] On the critical values of certain Dirichlet series and the periods of
automorphic forms 848

Notes II 909

List of Articles

The numbers in parentheses are the articles not included in this collection.

Volume I

[(52)] On a certain ideal of the center of a Frobeniusean algebra, Scientific Papers of the College of General Education, University of Tokyo, 2 (1952), 117–124.

[54] A note on the normalization-theorem of an integral domain, Scientific Papers of the College of General Education, University of Tokyo, 4 (1954), 1–8.

[55] Reduction of algebraic varieties with respect to a discrete valuation of the basic field, American Journal of Mathematics, 77 (1955), 134–176.

[56] On complex multiplications, Proceedings of the International Symposium on Algebraic Number Theory, Tokyo-Nikko, 1955, Science Council of Japan, Tokyo (1956), 23–30.

[57a] La fonction ζ du corps des fonctions modulaires elliptiques, Comptes rendus des séances de l'Académie des Sciences, 244 (1957), 2127–2130.

[(57b)] Kindai-teki Seisu-ron (Modern Number Theory, in Japanese, with Yutaka Taniyama), Kyoritsu Shuppan, 1957.

[58a] Correspondances modulaires et les fonctions zeta de courbes algébriques, Journal of the Mathematical Society of Japan, 10 (1958), 1–28.

[58b] Modules des variétés abéliennes polarisées et fonctions modulaires, Séminaire Henri Cartan, École Normale Supérieure, 1957/58, *Fonctions Automorphes*, Exposé 18–20 (1958).

[(58c)] Fonctions automorphes et variétés abéliennes, Séminaire Bourbaki, 1957/58, Exposé 167 (1958).

[59a] Fonctions automorphes et correspondances modulaires, Proceedings of the International Congress of Mathematicians, Edinburgh, 1958 (1959), 330–338.

[59b] On the theory of automorphic functions, Annals of Mathematics, 70 (1959), 101–144.

[59c] Sur les intégrales attachées aux formes automorphes, Journal of the Mathematical Society of Japan, 11 (1959), 291–311.

[59d] On specializations of abelian varieties (with Shoji Koizumi), Scientific Papers of the College of General Education, University of Tokyo, 9 (1959), 187–211.

[60a] On vector differential forms attached to automorphic forms (with Michio Kuga), Journal of the Mathematical Society of Japan, 12 (1960), 258–270.

[(60b)] Automorphic functions and number theory: I (in Japanese), Sugaku, 11 (1960), 193–205.

[(61a)] Complex multiplication of abelian varieties and its applications to number theory (with Yutaka Taniyama), Publications of the Mathematical Society of Japan, No. 6 (1961).

[61b] On the zeta functions of the algebraic curves uniformized by certain automorphic functions, Journal of the Mathematical Society of Japan, 13 (1961), 275–331.

[(61c)] Automorphic functions and number theory: II (in Japanese), Sugaku, 13 (1961), 65–80.

[62a] On Dirichlet series and abelian varieties attached to automorphic forms, Annals of Mathematics, 76 (1962), 237–294.

[62b] On the class-fields obtained by complex multiplication of abelian varieties, Osaka Mathematical Journal, 14 (1962), 33-44.

[63a] Arithmetic of alternating forms and quaternion hermitian forms, Journal of the Mathematical Society of Japan, 15 (1963), 33–65.

[63b] On analytic families of polarized abelian varieties and automorphic functions, Annals of Mathematics, 78 (1963), 149–192.

[63c] On the cohomology groups attached to certain vector valued differential forms on the product of the upper half planes (with Yozo Matsushima), Annals of Mathematics, 78 (1963), 417–449.

[63d] On modular correspondences for $Sp(n, \mathbf{Z})$ and their congruence relations, Proceedings of the National Academy of Sciences, 49 (1963), 824–828.

[63e] On the fields of definition for fields of automorphic functions, Proceedings of the 1963 Number Theory Conference, Boulder, Colorado, August 5–24, 1963, 26–32.

[64a] Arithmetic of unitary groups, Annals of Mathematics, 79 (1964), 369–409.

[64b] On the field of definition for a field of automorphic functions, Annals of Mathematics, 80 (1964), 160–189.

[64c] Class-fields and automorphic functions, Annals of Mathematics, 80 (1964), 444–463.

[64d] On purely transcendental fields of automorphic functions of several variables, Osaka Journal of Mathematics, 1 (1964), 1–14.

[64e] The zeta function of an algebraic variety and automorphic functions, Summer Research Institute on Algebraic Geometry, Woods Hole, Massachusetts, July 6–31, 1964 (mimeographed notes of lectures).

[65a] On the field of definition for a field of automorphic functions: II, Annals of Mathematics, 81 (1965), 124–165.

[65b] On the zeta function of a fibre variety whose fibres are abelian varieties (with Michio Kuga), Annals of Mathematics, 82 (1965), 478–539.

[66a] A reciprocity law in non-solvable extensions, Journal für die reine und angewandte Mathematik, 221 (1966), 209–220.

[66b] Moduli and fibre systems of abelian varieties, Annals of Mathematics, 83 (1966), 294–338.

[66c] On the field of definition for a field of automorphic functions: III, Annals of Mathematics, 83 (1966), 377–385.

[66d] Moduli of abelian varieties and number theory, Proceedings of the Symposium on Pure Mathematics, 9, *Algebraic groups and discontinuous subgroups*, 1965, American Mathematical Society, (1966), 312–332.

Volume II

[67a] Discontinuous groups and abelian varieties, Mathematische Annalen, 168 (1967), 171–199.

[67b] Construction of class fields and zeta functions of algebraic curves, Annals of Mathematics, 85 (1967), 58–159.

[67c] Number fields and zeta functions associated with discontinuous groups and algebraic varieties, Proceedings of the International Congress of Mathematicians, Moscow, 1966, 100–107 (1967).

[67d] Algebraic number fields and symplectic discontinuous groups, Annals of Mathematics, 86 (1967), 503–592.

[68a] Algebraic varieties without deformation and the Chow variety, Journal of the Mathematical Society of Japan, 20 (1968), 336–341.

[(68b)] Automorphic functions and number theory, Lecture notes in mathematics 54, Springer, 1968.

[68c] An ℓ-adic method in the theory of automorphic forms, the text of a lecture at the conference *Automorphic functions for arithmetically defined groups*, Oberwolfach, Germany, July 28–August 3, 1968.

[69] Local representations of Galois groups, Annals of Mathematics, 89 (1969), 99–124.

[70a] On canonical models of arithmetic quotients of bounded symmetric domains, Annals of Mathematics, 91 (1970), 144–222.

[70b] On canonical models of arithmetic quotients of bounded symmetric domains: II, Annals of Mathematics, 92 (1970), 528–549.

[71a] On arithmetic automorphic functions, Proceedings of the International Congress of Mathematicians, Nice, 1970, vol. 2, 343–348 (1971).

[71b] On the zeta-function of an abelian variety with complex multiplication, Annals of Mathematics, 94 (1971), 504–533.

[71c] Class fields over real quadratic fields in the theory of modular functions, in *Several Complex Variables* II, Maryland 1970, Lecture notes in mathematics 185 (1971), 169–188.

[(71d)] Introduction to the arithmetic theory of automorphic functions, Publications of the Mathematical Society of Japan, No. 11, Iwanami Shoten and Princeton University Press, 1971.

[71e] On elliptic curves with complex multiplication as factors of the Jacobians of modular function fields, Nagoya Mathematical Journal, 43 (1971), 199–208.

[72a] On the field of rationality for an abelian variety, Nagoya Mathematical Journal, 45 (1972), 167–178.

[72b] Class fields over real quadratic fields and Hecke operators, Annals of Mathematics, 95 (1972), 130–190.

[73a] On modular forms of half integral weight, Annals of Mathematics, 97 (1973), 440–481.

[(73b)] Complex multiplication, Proceedings of the International Summer School of Modular Functions of One Variable, Antwerp, 1972, Lecure notes in mathematics, 320 (1973), 37–56.

[(73c)] Modular forms of half integral weight, Proceedings of the International Summer School of Modular Functions of One Variable, Antwerp, 1972, Lecure notes in mathematics, 320 (1973), 57–74.

[73d] On the factors of the jacobian variety of a modular function field, Journal of the Mathematical Society of Japan, 25 (1973), 523–544.

[74] On the trace formula for Hecke operators, Acta mathematica, 132 (1974), 245–281.

[75a] On the holomorphy of certain Dirichlet series, Proceedings of the London Mathematical Society, 3rd ser. 31 (1975), 79–98.

[75b] On the real points of an arithmetic quotient of a bounded symmetric domain, Mathematische Annalen, 215 (1975), 135–164.

[75c] On some arithmetic properties of modular forms of one and several variables, Annals of Mathematics, 102 (1975), 491–515.

[75d] On the Fourier coefficients of modular forms of several variables, Nachrichten der Akademie der Wissenschaften in Göttingen, Mathematisch-Physikalische Klasse, 1975, 261–268.

[76a] Theta functions with complex multiplication, Duke Mathematical Journal, 43 (1976), 673–696.

[76b] The special values of the zeta functions associated with cusp forms, Communications on pure and applied Mathematics, 29 (1976), 783–804.

[77a] On abelian varieties with complex multiplication, Proceedings of the London Mathematical Society, 3rd ser. 34 (1977), 65–86.

[77b] Unitary groups and theta functions, Proceedings of the International Symposium on Algebraic Number Theory, Kyoto, 1976 (1977), 195–200.

[77c] On the derivatives of theta functions and modular forms, Duke Mathematical Journal, 44 (1977), 365–387.

[77d] On the periods of modular forms, Mathematische Annalen, 229 (1977), 211–221.

Volume III

[78a] On certain reciprocity-laws for theta functions and modular forms, Acta mathematica, 141 (1978), 35–71.

[78b] The arithmetic of automorphic forms with respect to a unitary group, Annals of Mathematics, 107 (1978), 569–605.

[78c] The special values of the zeta functions associated with Hilbert modular forms, Duke Mathematical Journal, 45 (1978), 637–679.

[79a] Automorphic forms and the periods of abelian varieties, Journal of the Mathematical Society of Japan, 31 (1979), 561–592.

[79b] On some problems of algebraicity, Proceedings of the International Congress of Mathematicians, Helsinki, 1978 (1979), 373–379.

[80] The arithmetic of certain zeta functions and automorphic forms on orthogonal groups, Annals of Mathematics, 111 (1980), 313–375.

[81a] The critical values of certain zeta functions associated with modular forms of half-integral weight, Journal of the Mathematical Society of Japan, 33 (1981), 649–672.

[81b] On certain zeta functions attached to two Hilbert modular forms: I. The case of Hecke characters, Annals of Mathematics, 114 (1981), 127–164.

[81c] On certain zeta functions attached to two Hilbert modular forms: II. The case of automorphic forms on a quaternion algebra, Annals of Mathematics, 114 (1981), 569–607.

[81d]   Arithmetic of differential operators on symmetric domains, Duke Mathematical Journal, 48 (1981), 813–843.

[(81e)]   Corrections to "The special values of the zeta functions associated with Hilbert modular forms," vol 45 (1978), 637–679, Duke Mathematical Journal, 48 (1981), 697.

[82a]   Models of an abelian variety with complex multiplication over small fields, Journal of Number Theory, 15 (1982), 25–35.

[82b]   The periods of certain automorphic forms of arithmetic type, Journal of the Faculty of Science, University of Tokyo, Sec. IA, 28 (1982), 605–632.

[82c]   Confluent hypergeometric functions on tube domains, Mathematische Annalen, 260 (1982), 269–302.

[83a]   Algebraic relations between critical values of zeta functions and inner products, American Journal of Mathematics, 105 (1983), 253–285.

[83b]   On Eisenstein series, Duke Mathematical Journal, 50 (1983), 417–476.

[84a]   Differential operators and the singular values of Eisenstein series, Duke Mathematical Journal, 51 (1984), 261–329.

[84b]   On differential operators attached to certain representations of classical groups, Inventiones mathematicae, 77 (1984), 463–488.

[85a]   On Eisenstein series of half-integral weight, Duke Mathematical Journal, 52 (1985), 281–314.

[85b]   On the Eisenstein series of Hilbert modular groups, Revista Matemática Iberoamericana, 1 (1985), 1–42.

[86]   On a class of nearly holomorphic automorphic forms, Annals of Mathematics, 123 (1986), 347–406.

[87a]   Nearly holomorphic functions on hermitian symmetric spaces, Mathematische Annalen, 278 (1987), 1–28.

[87b]   On Hilbert modular forms of half-integral weight, Duke Mathematical Journal, 55 (1987), 765–838.

[88]   On the critical values of certain Dirichlet series and the periods of automorphic forms, Inventiones mathematicae, 94 (1988), 245–305.

Volume IV

[89a]   Yutaka Taniyama and his time, Bulletin of the London Mathematical Society, 21 (1989), 186–196.

[89b]   *L*-functions and eigenvalue problems, *Algebraic analysis, geometry, and number theory*, Proceedings of the JAMI Inaugural Conference 1988, Supplement to the American Journal of Mathematics, 1989, 341–396.

[90a] Invariant differential operators on hermitian symmetric spaces, Annals of Mathematics, 132 (1990), 237–272.

[90b] On the fundamental periods of automorphic forms of arithmetic type, Inventiones mathematicae, 102 (1990), 399–428.

[(90c)] Some old and recent arithmetical results concerning modular forms and related zeta functions, University of Istanbul, Faculty of Science, Journal of Mathematics, 49 (1990), 45–56.

[91] The critical values of certain Dirichlet series attached to Hilbert modular forms, Duke Mathematical Journal, 63 (1991), 557–613.

[(93a)] Arithmeticity of the special values of various zeta functions and the periods of abelian integrals (in Japanese), Sugaku, 45 (1993), 111–127; English translation by Toshitsune Miyake, Sugaku expositions, 8 (1995), 17–38.

[93b] On the transformation formulas of theta series, American Journal of Mathematics, 115 (1993), 1011–1052.

[93c] On the Fourier coefficients of Hilbert modular forms of half-integral weight, Duke Mathematical Journal, 71 (1993), 501–557.

[94a] Fractional and trigonometric expressions for matrices, American Mathematical Monthly, 101 (1994), 744–758.

[94b] Euler products and Fourier coefficients of automorphic forms on symplectic groups, Inventiones mathematicae, 116 (1994), 531–576.

[94c] Differential operators, holomorphic projection, and singular forms, Duke Mathematical Journal, 76 (1994), 141–173.

[95a] Eisenstein series and zeta functions on symplectic groups, Inventiones mathematicae, 119 (1995), 539–584.

[95b] Zeta functions and Eisenstein series on metaplectic groups, Inventiones mathematicae, 121 (1995), 21–60.

[96a] Convergence of zeta functions on symplectic and metaplectic groups, Duke Mathematical Journal, 82 (1996), 327–347.

[96b] Response, Notices of the American Mathematical Society, vol. 43, No. 11 (November 1996), 1344–1347.

[(97a)] Euler Products and Eisenstein series, CBMS Regional Conference Series in Mathematics, No. 93, American Mathematical Society, 1997.

[97b] Zeta functions and Eisenstein series on classical groups, Proceedings of the National Academy of Sciences, 94, 11133–11137 (1997).

[(98)] Abelian varieties with complex multiplication and modular functions, Princeton University Press, 1998.

[99a] An exact mass formula for orthogonal groups, Duke Mathematical Journal, 97 (1999), 1–66.

[99b] The number of representations of an integer by a quadratic form, Duke Mathematical Journal, 100 (1999), 59–92.

[99c] Generalized Bessel functions on symmetric spaces, Journal für die reine und angewandte Mathematik, 509 (1999), 35–66.

[99d] Some exact formulas on quaternion unitary groups, Journal für die reine und angewandte Mathematik, 509 (1999), 67–102.

[99e] André Weil as I knew him, Notices of the American Mathematical Society, vol. 46, No. 4 (April 1999), 428–433

[(00)] Arithmeticity in the theory of automorphic forms, Mathematical Surveys and Monographs, vol. 82, American Mathematical Society, 2000.

[(01a)] Letter to the editor, Notices of the American Mathematical Society, vol. 48, No. 7 (August 2001), 678.

[01b] Arithmeticity of Dirichlet series and automorphic forms on unitary groups, unpublished.

[01c] The relative regulator of an algebraic number field, unpublished.

78a

On certain reciprocity-laws for theta functions and modular forms

Acta mathematica, 141 (1978), 35-71

The present paper has several objectives. The first apparent theme is an investigation of holomorphic functions $f(u, z)$ on $\mathbf{C}_s^n \times \mathfrak{H}_n$ which generalize the classical theta function

$$\theta(u, z) = \sum_{x \in \mathbf{Z}^n} \exp\left(\pi i({}^t x z x + 2 \cdot {}^t x u)\right) \quad (u \in \mathbf{C}_1^n,\ z \in \mathfrak{H}_n),$$

where $\mathfrak{H}_n$ is the Siegel upper space of degree n, and $\mathbf{C}_s^n$ the vector space of all $n \times s$ complex matrices. As is well known, θ satisfies a transformation formula under $Sp(n, \mathbf{Z})$, and also another formula under a translation $u \mapsto u + za + b$ with a, b in $\mathbf{Z}^n$. Generalizations of these two formulas are the conditions we impose on f. Multiplying f by a certain exponential factor, we associate with f a non-holomorphic function $f_*(u, z)$ whose value at a point $u = zp + q$ is a holomorphic modular form on $\mathfrak{H}_n$ if p and q belong to $\mathbf{Q}_s^n$, the set of $\mathbf{Q}$-rational elements of $\mathbf{C}_s^n$. In the special case $f = \theta$, we have

$$\theta_*(u, z) = \exp\left(\pi i \cdot {}^t(u - \bar{u})(z - \bar{z})^{-1} u\right)\theta(u, z).$$

Now we consider the group G of similitudes of an alternating form and its restricted adelization $G_{\mathbf{A}+}$, the restriction to the identity component being made at the archimedean place. In order to deal with the modular forms of half integral weight, we introduce a certain covering $\mathfrak{G}$ of $G_{\mathbf{A}+}$, which is modelled on the metaplectic group of Weil [10]. Then we define the action of every element of $\mathfrak{G}$ both on modular forms on $\mathfrak{H}_n$ and on the functions $f(u, z)$ with cyclotomic Fourier coefficients so that a reciprocity-law

$$(1) \qquad f_*(\Omega_z v, z)^\nu = (f^\nu)_*(\Omega_z \cdot {}^t x v, z)$$

holds for all $y \in \mathfrak{G}$ and all $v \in \mathbf{Q}_s^{2n}$, where $\Omega_z = (z\ 1_n)$ and x is the projection of y to $G_{\mathbf{A}+}$; x can replace y if the forms are of integral weight (Theorem 3.10). The action of $G_{\mathbf{A}+}$ or $\mathfrak{G}$

---

[1] Supported by NSF Grant MCS 76-11376.

on modular forms is consistent with the concept of canonical models in the sense of [4]. Formula (1) may be considered a "generic form" of the main theorem of complex multiplication of abelian varieties or of Siegel modular functions, which was originally proved without theta functions, and then formulated in terms of theta functions in our recent paper [7]. In fact, (1) enables us to give a simplified proof of the main theorem of [7].

The first two sections of the present paper are devoted to the action of $G_{\mathbf{A}+}$ and $\mathfrak{G}$ on modular forms on $\mathfrak{H}_n$, which may be of independent interest, though this part is preliminary to the rest of the paper. Formula (1) and its "specializations" will be proved in § 3. The next two sections are of technical nature; the proofs of some statements of the previous sections will be completed there. In § 6, we first observe that if F is a modular form on $\mathfrak{H}_{n+s}$ and if a point Z of $\mathfrak{H}_{n+s}$ is expressed in the form

$$Z = \begin{pmatrix} z & u \\ {}^t u & w \end{pmatrix}$$

with $u \in \mathbf{C}_s^n$, $z \in \mathfrak{H}_n$, and $w \in \mathfrak{H}_s$, then F has a Fourier expansion

$$(2) \qquad\qquad F(Z) = \sum_{\xi} f_{\xi}(u, z) \exp(\pi i \cdot \mathrm{tr}\,(\xi z))$$

whose Fourier coefficients f_{ξ} belong to the functions of the above type, where ξ runs over non-negative rational symmetric matrices of degree n. We shall then prove another reciprocity-law concerning the action of $\mathfrak{G}$ and its counterpart $\mathfrak{G}'$ of degree $n+s$ on f_{ξ} and F (Theorem 6.2).

We shall present all these for holomorphic modular forms with a rather general automorphic factor $\det(cz+d)^{k/2}\varrho(cz+d)$, where $k \in \mathbf{Z}$ and ϱ is an arbitrary rational representation of GL_n. The "theta functions" f will also be defined relative to a representation τ of GL_{n+s}. The consideration of an arbitrary ϱ or τ is made not merely for the sake of generality, but because there are good reasons for believing that such is natural and even necessary for the future development of the arithmetic theory of modular forms and zeta functions. It should be pointed out, however, that the nature of the reciprocity-laws is essentially revealed in the case of trivial ϱ and trivial τ, and therefore if the reader wishes to have a quick grasp of the ideas, he may be advised to assume throughout ϱ and τ to be trivial. In fact we have stated (1) in that special case.

While our theory may be accepted on its own merits, it has one significant and hidden aspect. Expansion (2) is actually an example of Fourier–Jacobi series in the sense of Pyatetskii-Shapiro [3]. Such a series occurs naturally as a Fourier expansion of an automorphic form on a Siegel domain of the third kind. When the discontinuous group is arithmetically defined, one can ask whether there is a natural class of "arithmetic automorphic forms" which is characterized by some properties of Fourier coefficients and which plays the role similar to that of the elliptic modular forms with algebraic or cyclotomic Fourier coefficients. Now our results, especially the above two laws of reciprocity, seem to suggest an affirmative answer to this question. A detailed explanation of the ideas about this is another of our main purposes of this paper, and will be given in the last section, to which the preceding sections may serve as a long introduction.

Notation. For a ring X with an identity element, the set of all $r \times s$ matrices with coefficients in X is denoted by X_s^r, and simply by X^r if $s=1$; the identity element of X_s^s is denoted by 1_s. Further $X^\times$ denotes the group of all invertible elements of X. The diagonal matrix with diagonal elements $d_1, ..., d_n$ is denoted by

$$\mathrm{diag}\,[d_1, ..., d_n].$$

For $T \in \mathbf{C}_s^s$, we put

$$e_s(T) = \exp\,(2\pi i \cdot \mathrm{tr}\,(T)),$$

and especially $e(u) = e_1(u) = e^{2\pi i u}$ for $u \in \mathbf{C}$. If T is a hermitian matrix, we write $T \geqslant 0$ or $T > 0$ according as T is non-negative or positive definite. The Siegel upper space of degree n is denoted by $\mathfrak{H}_n$, thus

$$\mathfrak{H}_n = \{z \in \mathbf{C}_n^n \,|\, {}^t z = z,\, \mathrm{Im}\,(z) > 0\}.$$

We put $\Omega_z = (z\ \ 1_n)$ for $z \in \mathfrak{H}_n$.

If K is an algebraic number field, $K_\mathbf{A}$ denotes the ring of adeles of K, $K_\mathbf{A}^\times$ the group of ideles of K, and K_{ab} the maximal abelian extension of K. By class field theory, every element y of $K_\mathbf{A}^\times$ acts on K_{ab} as an automorphism. We denote by a^y the image of $a \in K_{\mathrm{ab}}$ under y. In particular, if $K = \mathbf{Q}$, we put, as usual, $\mathbf{Q}_\mathbf{A} = \mathbf{A}$ and $\mathbf{Q}_\mathbf{A}^\times = \mathbf{A}^\times$; further, we denote by $\mathbf{A}_f$ the non-archimedean part of $\mathbf{A}$, and by $\mathbf{A}_+^\times$ the subgroup of $\mathbf{A}^\times$ consisting of the elements whose archimedean components are positive. For $0 < N \in \mathbf{Z}$ and $x,\, y \in \mathbf{A}_s^s$, we write $x \equiv y \pmod{N}$ if $x_p - y_p \in N(\mathbf{Z}_p)_s^s$ for all primes p, where x_p and y_p are the p-components of x and y, respectively. For $b \in \mathbf{A}^\times$, we denote by $|b|$ the positive rational number such that $|b|\mathbf{Z}_p = b_p \mathbf{Z}_p$ for all p. We define a compact subgroup $\mathbf{Z}_f^\times$ of $\mathbf{A}^\times$ by $\mathbf{Z}_f^\times = \prod_p \mathbf{Z}_p^\times$, where the product is taken over all primes p.

1. The action of $G_{\mathbf{A}+}$ on modular forms of integral weight

Throughout the first six sections, we let G denote the algebraic subgroup of GL_{2n} defined over $\mathbf{Q}$ such that

$$G_\mathbf{Q} = \{\alpha \in GL_{2n}(\mathbf{Q}) \,|\, {}^t\alpha J \alpha = \nu(\alpha)J \quad \text{with } \nu(\alpha) \in \mathbf{Q}\},$$

where

$$\text{(1.1)} \qquad J = \begin{pmatrix} 0 & -1_n \\ 1_n & 0 \end{pmatrix},$$

and $G_\mathbf{A}$ its adelization. The map $\nu: G_\mathbf{Q} \to \mathbf{Q}^\times$ can be naturally extended to a continuous map of $G_\mathbf{A}$ into $\mathbf{A}$, which is still denoted by ν. We denote by G_f, G_∞, and $G_{\infty+}$ the non-archimedean part of $G_\mathbf{A}$, the archimedean part of $G_\mathbf{A}$, and the identity component of G_∞, respectively. We then put $G_{\mathbf{A}+} = G_f G_{\infty+}$ and

$$G_{\mathbf{Q}+} = G_{\mathbf{A}+} \cap G_\mathbf{Q} = \{\alpha \in G_\mathbf{Q} \,|\, \nu(\alpha) > 0\}.$$

Every element $\alpha = \begin{pmatrix} a & b \\ c & d \end{pmatrix}$ of $G_{\mathbf{Q}+}$ acts on $\mathfrak{H}_n$ by the rule $\alpha(z) = (az+b)(cz+d)^{-1}$ for $z \in \mathfrak{H}_n$. Now we take an arbitrary rational representation

$$\varrho: GL_n(\mathbf{Q}) \to GL_m(\mathbf{Q}),$$

and extend it to a holomorphic representation $GL_n(\mathbf{C}) \to GL_m(\mathbf{C})$, which is again denoted by ϱ. For $\alpha = \begin{pmatrix} a & b \\ c & d \end{pmatrix} \in G_{\mathbf{Q}+}$ and a $\mathbf{C}^m$-valued function f on $\mathfrak{H}_n$, we define a function $f|_\varrho\alpha$ on $\mathfrak{H}_n$ by

$$\text{(1.2)} \qquad (f|_\varrho\alpha)(z) = \varrho(cz+d)^{-1}f(\alpha(z)) \quad (z \in \mathfrak{H}_n).$$

Given a congruence subgroup Γ of $G_{\mathbf{Q}+}$, we denote by $\mathcal{M}_\varrho(\Gamma)$ the vector space of all $\mathbf{C}^m$-valued holomorphic functions f on $\mathfrak{H}_n$ which satisfy $f|_\varrho\gamma = f$ for all $\gamma \in \Gamma$ and which are finite at cusps. Such f may be called *modular forms of weight* ϱ. If $f \in \mathcal{M}_\varrho(\Gamma)$, f has a Fourier expansion

$$\text{(1.3)} \qquad f(z) = \sum_\xi c(\xi) e_n(\xi z)$$

with $c(\xi) \in \mathbf{C}^m$, where ξ runs over positive semi-definite symmetric elements of $\mathbf{Q}_n^n$. If $n > 1$, every holomorphic f satisfying $f|_\varrho\gamma = f$ for $\gamma \in \Gamma$ has such an expansion, and hence belongs to $\mathcal{M}_\varrho(\Gamma)$ (see for example [1]). For a subfield $\mathfrak{R}$ of $\mathbf{C}$, we denote by $\mathcal{M}_\varrho(\Gamma, \mathfrak{R})$ the set of all f in $\mathcal{M}_\varrho(\Gamma)$ with $c(\xi)$ in $\mathfrak{R}^m$, and by $\mathcal{M}_\varrho(\mathfrak{R})$ the union of $\mathcal{M}_\varrho(\Gamma, \mathfrak{R})$ for all congruence subgroups Γ. Further we denote by $\mathcal{A}_\varrho(\mathfrak{R})$ the set of all quotients $g^{-1}h$ with $h \in \mathcal{M}_\tau(\mathfrak{R})$ and $0 \neq g \in \mathcal{M}_\omega(\mathfrak{R})$, where $\omega(X) = \det(X)^k$, $\tau = \omega\varrho$ with some integer k. Then we put

$$\mathcal{A}_\varrho(\Gamma, \mathfrak{R}) = \{f \in \mathcal{A}_\varrho(\mathfrak{R}) \,|\quad f|_\varrho\gamma = f \quad \text{for all } \gamma \in \Gamma\}.$$

It can easily be shown that $\mathcal{M}_\varrho(\Gamma, \mathfrak{R})$ consists of all the holomorphic elements of $\mathcal{A}_\varrho(\Gamma, \mathfrak{R})$ finite at cusps. If $\varrho(X) = \det(X)^k$ with $k \in \mathbf{Z}$, we write $f|_k\alpha$, $\mathcal{M}_k$, $\mathcal{A}_k$ for $f|_\varrho\alpha$, $\mathcal{M}_\varrho$, $\mathcal{A}_\varrho$.

Let σ be a field-automorphism of $\mathbf{C}$. Define the action of σ on $\mathbf{C}^m$ component-wise. For $f \in \mathcal{M}_\varrho(\mathbf{C})$ with expansion (1.3), we define f^σ by

$$(1.4) \qquad f^\sigma(z) = \sum_\xi c(\xi)^\sigma e_n(\xi z).$$

If $\varrho(X) = \det(X)^k$, f^σ is actually an element of $\mathcal{M}_k(\mathbf{C})$; moreover σ can be naturally extended to an automorphism of $\mathcal{A}_k(\mathbf{C})$ (see [6, § 4]). We shall prove in § 5 that these hold for an arbitrary ϱ. The notation f^σ is meaningful if σ is an isomorphism of a subfield $\mathfrak{R}$ of $\mathbf{C}$ onto another subfield and $f \in \mathcal{A}_\varrho(\mathfrak{R})$; in particular f^t is meaningful for $t \in \mathbf{A}^\times$ if $f \in \mathcal{A}_\varrho(\mathbf{Q}_{ab})$.

Now we define an injection $\iota\colon \mathbf{A}_f^\times \to G_{\mathbf{A}+}$ by

$$(1.5) \qquad \iota(t) = \begin{pmatrix} 1_n & 0 \\ 0 & t^{-1}1_n \end{pmatrix} \qquad (t \in \mathbf{A}_f^\times).$$

In our treatment, we shall often need the strong approximation theorem in $G_\mathbf{A}$, which can be given as

LEMMA 1.1. *Let T be an open subgroup of $\{x \in G_\mathbf{A} \mid v(x) = 1\} G_{\infty+}$. Then $G_{\mathbf{A}+}$ is the product of $G_{\mathbf{Q}+}$, $\iota(\mathbf{Z}_f^\times)$, and T in an arbitrary order, i.e., $G_{\mathbf{A}+} = G_{\mathbf{Q}+}\iota(\mathbf{Z}_f^\times)T = T\iota(\mathbf{Z}_f^\times)G_{\mathbf{Q}+} = \cdots$.*

Cf. [4, I, 3.4; II, (3.10.3)].

We are going to define the action of $G_{\mathbf{A}+}$ on $\mathcal{A}_\varrho(\mathbf{Q}_{ab})$. First we recall that $G_{\mathbf{A}+}$ acts on the field $\mathcal{A}_0(\mathbf{Q}_{ab})$ as a group of automorphisms, and for every $t \in \mathbf{A}_f^\times$, the action of $\iota(t)$ on $\mathbf{Q}_{ab}$ is the same as that of t (see [4], [5], [6]).

THEOREM 1.2. *There is an action of $G_{\mathbf{A}+}$ on $\mathcal{A}_\varrho(\mathbf{Q}_{ab})$, written as $(x, f) \mapsto f^x$ for $x \in G_{\mathbf{A}+}$ and $f \in \mathcal{A}_\varrho(\mathbf{Q}_{ab})$, with the following properties:*

(i) *$\mathcal{M}_\varrho(\mathbf{Q}_{ab})$ is stable under the action;*

(ii) *the action of $G_{\mathbf{A}+}$ on $\mathcal{A}_0(\mathbf{Q}_{ab})$ is the same as that mentioned above;*

(iii) *the action is associative, i.e., $f^{xy} = (f^x)^y$;*

(iv) *$(f \oplus g)^x = f^x \oplus g^x$, $(f \otimes g)^x = f^x \otimes g^x$, where $f \oplus g$ and $f \otimes g$ are naturally defined as elements of $\mathcal{A}_{\varrho \oplus \tau}(\mathbf{Q}_{ab})$ and $\mathcal{A}_{\varrho \otimes \tau}(\mathbf{Q}_{ab})$ for $f \in \mathcal{A}_\varrho(\mathbf{Q}_{ab})$ and $g \in \mathcal{A}_\tau(\mathbf{Q}_{ab})$;*

(v) *$f^\alpha = f|_\varrho \alpha$ if $\alpha \in G_{\mathbf{Q}+}$;*

(vi) *$f^{\iota(t)} = f^t$ for $t \in \mathbf{Z}_f^\times$;*

(vii) *$f^x = f$ if $x \in G_{\infty+}$;*

(viii) *for each $f \in \mathcal{A}_\varrho(\mathbf{Q}_{ab})$, the element x of $G_{\mathbf{A}+}$ such that $f^x = f$ form an open subset of $G_{\mathbf{A}+}$.*

This extends the previous result [6, Theorem 5], which concerns the action of $G_{\mathbf{A}+}$ on $\mathcal{A}_k(\mathbf{Q}_{ab})$ with $k \in 2\mathbf{Z}$. In the previous papers, we defined $f|_k \alpha$ for $\alpha \in G_{\mathbf{A}+}$ with a scalar

5

factor $\nu(\alpha)^{k/2}$, which is different from our present definition (1.2). This change is necessary to guarantee (iii) in the general case. Before proving our theorem, we first state the uniqueness of the action as

PROPOSITION 1.3. *The action of $G_{\mathbf{A}+}$ on $\mathcal{A}_0(\mathbf{Q}_{ab})$ is uniquely determined by* (iii), (v), (vi), (viii).

Proof. Let $U = \{x \in G_{\mathbf{A}+} \,|\, f^x = f\}$ with any given f. By Lemma 1.1, if U is open, each element x of $G_{\mathbf{A}+}$ can be written as $x = u\iota(t)\alpha$ with $u \in U$, $t \in \mathbf{Z}_{\mathfrak{f}}^{\times}$ and $\alpha \in G_{\mathbf{Q}+}$. Therefore, assuming (iii), (v), (vi), we have $f^x = f^t|_\varrho \alpha$, which proves our assertion.

In this section, we prove our theorem only for $\mathcal{A}_k(\mathbf{Q}_{ab})$ with $k \in \mathbf{Z}$. The general case will be discussed in § 5. First put

$$(1.6) \qquad \theta(z) = \sum_{v \in \mathbf{Z}^n} e({}^t vzv/2) \qquad (z \in \mathfrak{H}_n),$$

and for each positive integer N,

$$(1.7_{\mathrm{a}}) \qquad \Gamma_N = \{\gamma \in G_{\mathbf{Q}} \cap SL_{2n}(\mathbf{Z}) \,|\, \gamma \equiv 1_{2n}(\mathrm{mod}\ N)\},$$

$$(1.7_{\mathrm{b}}) \qquad S_1 = G_{\mathbf{A}+} \cap \{G_{\infty+} \prod_p GL_{2n}(\mathbf{Z}_p)\},$$

$$(1.7_{\mathrm{c}}) \qquad S_N' = \left\{x \in S_1 \,\Big|\, x \equiv \begin{pmatrix} 1_n & 0 \\ 0 & t1_n \end{pmatrix} (\mathrm{mod}\ N) \text{ with } t \in \mathbf{Z}_{\mathfrak{f}}^{\times}\right\},$$

$$(1.7_{\mathrm{d}}) \qquad S_N = \{x \in S_1 \,|\, x \equiv 1_{2n}(\mathrm{mod}\ N)\}.$$

Obviously it is sufficient to consider the case $k > 0$. Fix a positive integer k, and put $h(z) = \theta(z)^{2k}$. As shown in [7], there is a positive integer M such that $h \in \mathcal{M}_k(\Gamma_M)$. Take any such M that is a multiple of 4, and put $T = S_M'$. By Lemma 1.1, every element x of $G_{\mathbf{A}+}$ can be written as $x = u\alpha$ with $u \in T$ and $\alpha \in G_{\mathbf{Q}+}$. Now we have $\mathcal{A}_k(\mathbf{Q}_{ab}) = h \cdot \mathcal{A}_0(\mathbf{Q}_{ab})$. Define the action of x on $\mathcal{A}_k(\mathbf{Q}_{ab})$ by $g^x = (g/h)^x(h|_k\alpha)$ for $g \in \mathcal{A}_k(\mathbf{Q}_{ab})$. By [6, Th. 4] or [7, Prop. 1.5], $h|_k\alpha \in \mathcal{M}_k(\mathbf{Q}_{ab})$, so that $g^x \in \mathcal{A}_k(\mathbf{Q}_{ab})$. We see easily that this does not depend on the choice of u and α. Also we have obviously $g^\alpha = g|_k\alpha$, $g^{\alpha\beta} = (g^\alpha)^\beta$, $g^{uv} = (g^u)^v$, $g^{u\alpha} = (g^u)^\alpha$ for $\alpha, \beta \in G_{\mathbf{Q}+}$ and $u, v \in T$. Assume that

$$(1.8) \qquad (g^\alpha)^y = g^{\alpha y} \quad \text{for } g \in \mathcal{A}_k(\mathbf{Q}_{ab}),\ \alpha \in G_{\mathbf{Q}+},\ y \in G_{\mathbf{A}+}$$

holds. Given x and y of $G_{\mathbf{A}+}$, put $x = u\alpha$ as above and $\alpha y = v\beta$ with $v \in T$ and $\beta \in G_{\mathbf{Q}+}$. Then, by (1.8), $(g^x)^y = (g^{u\alpha})^y = (g^u)^{\alpha y} = (g^u)^{v\beta} = (g^{uv})^\beta = g^{uv\beta} = g^{xy}$. Thus the proof of (iii) can be reduced to (1.8). Observe that if (1.8) is true for some fixed α, y, and $g \neq 0$, then it is true for all g in $\mathcal{A}_k(\mathbf{Q}_{ab})$ and for the same α and y; if (1.8) is true for α, then so is for α^{-1}. Therefore it is sufficient to prove (1.8) for $g = h$ and for α belonging to a set of generators, say B, of $G_{\mathbf{Q}+}$. Given $\alpha \in B$ and $y \in G_{\mathbf{A}+}$, put $y = v\beta$ with $v \in T$ and $\beta \in G_{\mathbf{Q}+}$, and $\alpha v = u\gamma$ with $u \in T$ and $\gamma \in G_{\mathbf{Q}+}$. Suppose $(h^\alpha)^v = h^{\alpha v}$. Then $(h^\alpha)^y = ((h^\alpha)^v)^\beta = (h^{u\gamma})^\beta = ((h^u)^\gamma)^\beta = h^{u\gamma\beta} = h^{\alpha y}$. Thus it is sufficient to prove

$$(1.9) \qquad (h^\alpha)^v = h^{\alpha v} \quad \text{for } \alpha \in B \quad \text{and } v \in T.$$

Now by [7, Lemma 2], we can take B to be the set consisting of J and the elements of the form $\begin{pmatrix} a & b \\ 0 & d \end{pmatrix}$ contained in $\mathbf{Z}_{2n}^{2n}$. First consider the case $\alpha = J$. If $v \in T$, we have $v \equiv \begin{pmatrix} 1 & 0 \\ 0 & t \end{pmatrix} \pmod{M}$ with $t \in \mathbf{Z}_t^\times$. Take $\begin{pmatrix} p & q \\ r & s \end{pmatrix} \in SL_2(\mathbf{Z})$ so that $s > 0$, $\begin{pmatrix} p & q \\ r & s \end{pmatrix} \equiv \begin{pmatrix} t & 0 \\ 0 & t^{-1} \end{pmatrix} \pmod{M}$, and put $\beta = \begin{pmatrix} p1_n & q1_n \\ r1_n & s1_n \end{pmatrix}$, $\alpha v = w\beta\alpha$. Then $w \equiv v \pmod{M}$, so that $w \in T$. Hence $h^{\alpha v} = h^{\beta\alpha}$. By Prop. A.2 of the Appendix, we have $h^\beta = \left(\dfrac{-1}{s}\right)^{kn} h$. On the other hand, $h^\alpha = (-i)^{kn} h$ by [7, (16')] and $i^v = \left(\dfrac{-1}{s}\right) i$, so that $(h^\alpha)^v = \left(\dfrac{-1}{s}\right)^{kn} (-i)^{kn} h = h^{\beta\alpha} = h^{\alpha v}$, which proves (1.9) for $\alpha = J$. Next assume $\alpha = \begin{pmatrix} a & b \\ 0 & d \end{pmatrix} \in G_{\mathbf{Q}+} \cap \mathbf{Z}_{2n}^{2n}$. Let $N = M \cdot \det(\alpha)$ and put $v = w\beta$ with $w \in S_N'$ and $\beta \in G_{\mathbf{Q}+}$. Then $w \equiv \iota(t) \pmod{N}$ with $t \in \mathbf{Z}_t^\times$. Let s be a positive integer such that $t^{-1} \equiv s \pmod{N}$, and put $\gamma = \begin{pmatrix} a & sb \\ 0 & d \end{pmatrix}$, $u = \alpha w\gamma^{-1}$. Then $u \in S_M'$. Therefore $h^{\alpha v} = h^{u\gamma\beta} = h^{\gamma\beta}$. On the other hand $(h^\alpha)^v = (h^\alpha)^{w\beta} = (h^\alpha/h)^{w\beta} h^\beta$. Let $h(z) = \sum_\xi c(\xi) e_n(\xi z)$ be the Fourier expansion of h. Then

$$h^\alpha(z) = \det(d)^{-k} \sum_\xi c(\xi) e_n(\xi b d^{-1}) e_n(d^{-1}\xi a z),$$

$$h^\gamma(z) = \det(d)^{-k} \sum_\xi c(\xi) e_n(s\xi b d^{-1}) e_n(d^{-1}\xi a z).$$

Since $2\xi \in \mathbf{Z}_n^n$, we see that $h^\alpha \in \mathfrak{M}_k(\Gamma_N, \mathbf{Q}(e(1/N)))$, and hence $(h^\alpha)^t = h^\gamma$. By [6, Th. 2, (ii); Th. 3, (i)], we have $(h^\alpha/h)^w = (h^\alpha)^t/h = h^\gamma/h$. Therefore $(h^\alpha)^v = h^{\gamma\beta} = h^{\alpha v}$. Thus (1.9) is true for all $\alpha = \begin{pmatrix} a & b \\ 0 & d \end{pmatrix}$ and $\alpha = J$. This completes the proof of (iii). Now assertions (ii), (v), (vi), (vii), (viii) are obvious from our definition of the action and [6, Th. 2, (ii)]. If $f \in \mathfrak{M}_k(\mathbf{Q}_{ab})$ and $x \in G_{\mathbf{A}+}$, $f^x = (f^t)^\alpha$ with suitable t and α as shown in the proof of Prop. 1.3. This together with [6, Th. 2, Th. 4] proves (i).

The above property (vi) can be generalized as follows:

$$(1.10) \qquad \left(\sum_\xi c(\xi) e_n(\xi z)\right)^{\iota(b)} = \varrho(|b|1_n) \sum_\xi c(\xi)^b e_n(|b|\xi z)$$

if $b \in \mathbf{A}_+^\times$ and $\sum_\xi c(\xi) e_n(\xi z) \in \mathcal{M}_\varrho(\mathbf{Q}_{ab})$. This can be shown by decomposing b as $b = stu$ with $s = |b|$, $t \in \mathbf{Z}_t^\times$, and $u \in \mathbf{Q}_{\infty+}^\times$.

Let S_0 be an open compact subgroup of G_t, and let $S = S_0 G_{\infty+}$, $\Gamma_S \cap G_\mathbf{Q} = S$. Further let k_S be the subfield of $\mathbf{Q}_{ab}$ corresponding to the subgroup $\mathbf{Q}^\times \nu(S)$ of $\mathbf{A}^\times$. Suppose that the following condition is satisfied:

$$(1.11) \qquad \iota(t) \in S \quad \textit{if } t \in \mathbf{Z}_t^\times \textit{ and } t \textit{ gives the identity map on } k_S.$$

For example, the above S_N and S_N' satisfy this.

PROPOSITION 1.4. *The notation being as above, if (1.11) is satisfied, then* $A_\varrho(\Gamma_S, k_S) = \{f \in A_\varrho(\mathbf{Q}_{ab}) \mid f^x = f \text{ for all } x \in S\}$.

Proof. Given $f \in A_\varrho(\Gamma_S, k_S)$ and $x \in S$, put $U = \{y \in S \mid f^y = f\}$ and $x = u\iota(t)\alpha$ with $u \in U$, $t \in \mathbf{Z}_t^\times$ and $\alpha \in G_{\mathbf{Q}+}$. Then $\iota(t) \in S$ by (1.11), so that $\alpha \in \Gamma_S$. Therefore f is invariant under u, $\iota(t)$, and α, and hence $f^x = f$. Conversely, if $f^x = f$ for all $x \in S$, f is invariant under Γ_S and $\iota(\mathbf{Z}_t^\times) \cap S$, so that $f \in A_\varrho(\Gamma_S, k_S)$, Q.E.D.

2. The forms of half-integral weight

The purpose of this section is to define a group $\mathfrak{G}$ acting on the modular forms of half-integral weight in exactly the same fashion as $G_{\mathbf{A}+}$ on $A_\varrho(\mathbf{Q}_{ab})$. Let ϱ be as in § 1. With $k \in 2^{-1}\mathbf{Z}$ and a subfield $\mathfrak{R}$ of $\mathbf{C}$, we define $A_{\varrho,k}(\mathfrak{R})$ to be the set of all functions of the form $\theta^{2k} f$ with $f \in A_\varrho(\mathfrak{R})$, where θ is defined by (1.6). Also we denote by $\mathcal{M}_{\varrho,k}(\mathfrak{R})$ the set of all holomorphic elements of $A_{\varrho,k}(\mathfrak{R})$ that are finite at cusps. If $k - l \in \mathbf{Z}$ and $\tau(X) = \det(X)^{k-l}\varrho(X)$, then $A_{\varrho,k} = A_{\tau,l}$ and $\mathcal{M}_{\varrho,k} = \mathcal{M}_{\tau,l}$. If ϱ is trivial, we write $A_{\varrho,k}$ and $\mathcal{M}_{\varrho,k}$ simply as A_k and $\mathcal{M}_k$. The action of an isomorphism σ of $\mathfrak{R}$ onto a subfield of $\mathbf{C}$ can be defined on $A_{\varrho,k}(\mathfrak{R})$ by $f^\sigma = (f/\theta^{2k})^\sigma \theta^{2k}$ for $f \in A_{\varrho,k}(\mathfrak{R})$. We shall prove in § 5 that $\mathcal{M}_{\varrho,k}(\mathbf{C}) = \mathcal{M}_{\varrho,k}(\mathbf{Q}) \otimes_\mathbf{Q} \mathbf{C}$ so that σ maps $\mathcal{M}_{\varrho,k}(\mathbf{C})$ into itself; its action is defined again by (1.4).

Let $\mathfrak{E}$ denote the group of all $\mathbf{Q}$-linear automorphisms of the module $A_{1/2}(\mathbf{Q}_{ab})$. For any finite subset F of $A_{1/2}(\mathbf{Q}_{ab})$, take

$$\{v \in \mathfrak{E} \mid f^v = f \quad \text{for all } f \in F\}$$

as a neighborhood of the identity element of $\mathfrak{E}$. Then $\mathfrak{E}$ becomes a topological group. Let W be the group of all roots of unity. Now we define a subgroup $\mathfrak{G}$ of $G_{\mathbf{A}+} \times \mathfrak{E}$ with induced topology by

$$(2.1) \qquad \mathfrak{G} = \{(x, v) \in G_{\mathbf{A}+} \times \mathfrak{E} \mid (f^v)^2 = \zeta(f^2)^x \quad \text{for all } f \in A_{1/2}(\mathbf{Q}_{ab})$$
$$\text{with } \zeta \in W \text{ independent of } f\}.$$

The root of unity ζ being determined by $(f^v)^2 = \zeta(f^2)^x$, we shall write (x, v) also as (x, ζ, v) to emphasize ζ. If $(x, \zeta, v) \in \mathfrak{G}$ and $f, g \in \mathcal{A}_{1/2}(\mathbf{Q}_{ab})$, we have $((f+g)^v)^2 = \zeta((f+g)^2)^x$, so that $f^v g^v = \zeta(fg)^x$, and hence

$$(2.2) \qquad g^v/f^v = (g/f)^x \quad \text{if } (x, v) \in \mathfrak{G} \quad \text{and } f, g \in \mathcal{A}_{1/2}(\mathbf{Q}_{ab}).$$

For every $\zeta \in W$, we can define an element of $\mathfrak{E}$ by $f \mapsto \zeta f$, which is denoted again by ζ. Then $(1, \zeta^2, \zeta)$ is an element of $\mathfrak{G}$. Let W_1 denote the set of all such elements. Then we see easily that

$$(2.3) \qquad 1 \to W_1 \to \mathfrak{G} \to G_{\mathbf{A}+} \to 1$$

is exact, with a natural projection map of $\mathfrak{G}$ into $G_{\mathbf{A}+}$. In fact, to see the surjectivity, let $U = \{y \in G_{\mathbf{A}+} \mid (\theta^2)^y = \theta^2\}$. Given $x \in G_{\mathbf{A}+}$, we have $x = y\iota(t)\alpha$ with $y \in U$, $t \in \mathbf{Z}_{\mathbf{f}}^\times$, and $\alpha = \begin{pmatrix} a & b \\ c & d \end{pmatrix} \in G_{\mathbf{Q}+}$ by Lemma 1.1. Define $v \in \mathfrak{E}$ by

$$f^v = (f/\theta)^x \theta(\alpha(z)) \det (cz+d)^{-1/2}$$

with any choice of $\det (cz+d)^{-1/2}$. Then $(x, 1, v) \in \mathfrak{G}$.

Now we define the action of $\mathfrak{G}$ on $\mathcal{A}_{\varrho.k}(\mathbf{Q}_{ab})$ as follows. For $(x, v) \in \mathfrak{G}$ and $f \in \mathcal{A}_{\varrho.k}(\mathbf{Q}_{ab})$, we put

$$(2.4) \qquad f^{(x, v)} = (f/h^{2k})^x (h^v)^{2k}$$

with any non-zero $h \in \mathcal{A}_{1/2}(\mathbf{Q}_{ab})$. This is independent of the choice of h. Moreover $f^{(x, v)} = f^v$ if ϱ is trivial and $k = 1/2$; further, if $(x, \zeta, v) \in \mathfrak{G}$ and $k \in \mathbf{Z}$, we have $f^{(x, \zeta, v)} = \zeta^k f^x$. As mentioned above, $\mathcal{A}_{\varrho.k}(\mathbf{Q}_{ab}) = \mathcal{A}_{\tau, l}(\mathbf{Q}_{ab})$ if $\tau(X) = \det (X)^{k-l}\varrho(X)$. But the action of (x, v) on this same set depends on (ϱ, k). To avoid a complicated notation, we hereafter understand that if the set is denoted by $\mathcal{A}_{\varrho.k}(\mathbf{Q}_{ab})$, the action of $\mathfrak{G}$ is defined with the same ϱ and k. Notice that the action of $(x, 1, v)$ is independent of ϱ and k.

Let $\mathfrak{G}_{\mathbf{Q}}$ denote the set consisting of all pairs (α, ψ) formed by an element $\alpha = \begin{pmatrix} a & b \\ c & d \end{pmatrix}$ of $G_{\mathbf{Q}+}$ and a holomorphic function ψ on $\mathfrak{H}_n$ such that $\psi(z)^2 = \zeta \cdot \det (cz+d)$ with $\zeta \in W$. A law of composition

$$(\alpha, \psi(z))(\alpha', \psi'(z)) = (\alpha\alpha', \psi(\alpha'(z))\psi'(z))$$

makes $\mathfrak{G}_{\mathbf{Q}}$ a group. For a function f on $\mathfrak{H}_n$ and $\beta = (\alpha, \psi) \in \mathfrak{G}_{\mathbf{Q}}$, we define a function $f|_{\varrho.k}\beta$ by

$$(2.5) \qquad f|_{\varrho.k}\beta = (f|_\varrho \alpha)\psi^{-2k}.$$

By [7, Prop. 1.5], we see easily that $\mathcal{A}_{\varrho,k}(\mathbf{Q}_{ab})$ is stable under this action, and hence, in particular, it defines an element of $\mathfrak{E}$. Then

$$(\alpha, \psi) \mapsto (\alpha, (\alpha, \psi))$$

gives an embedding of $\mathfrak{G}_{\mathbf{Q}}$ into $\mathfrak{G}$. Identify $\mathfrak{G}_{\mathbf{Q}}$ with its image in $\mathfrak{G}$. Then $\mathfrak{G}_{\mathbf{Q}}$ is the inverse image of $G_{\mathbf{Q}+}$, and thus

$$1 \to W_1 \to \mathfrak{G}_{\mathbf{Q}} \to G_{\mathbf{Q}+} \to 1$$

is exact. We see easily that $f^{\beta} = f|_{\varrho,k}\beta$ if $\beta \in \mathfrak{G}_{\mathbf{Q}}$ and $f \in \mathcal{A}_{\varrho,k}(\mathbf{Q}_{ab})$.

Next, for $t \in \mathbf{Z}_{\mathfrak{l}}^{\times}$, denote by t' the element of $\mathfrak{E}$ given by $f \mapsto f^t$ for $f \in \mathcal{A}_{1/2}(\mathbf{Q}_{ab})$. Then $(\iota(t), t') \in \mathfrak{G}$. We see easily that the map $t \mapsto (\iota(t), t')$ can be extended to an injection ι_1 of $\mathbf{A}_{+}^{\times}$ into $\mathfrak{G}$ so that

$$(2.6) \qquad \left(\sum_{\xi} c(\xi) e_n(\xi z)\right)^{\iota_1(b)} = |b|^{kn} \varrho(|b| 1_n) \sum_{\xi} c(\xi)^b e_n(|b| \xi z)$$

if $\sum_{\xi} c(\xi) e_n(\xi z) \in \mathcal{M}_{\varrho,k}(\mathbf{Q}_{ab})$ and $b \in \mathbf{A}_{+}^{\times}$.

Now we can state the properties of the action of $\mathfrak{G}$ on $\mathcal{A}_{\varrho,k}(\mathbf{Q}_{ab})$ in exactly the same fashion as in Theorem 1.2 with $\mathfrak{G}_{\mathbf{Q}}$ and ι_1 in place of $G_{\mathbf{Q}+}$ and ι. In substance, this is already done in the above discussion. (See also Prop. 2.2 below.) We note that if Γ_{θ} is defined by (a.1) of the Appendix, then

$$(2.7) \qquad \gamma \mapsto (\gamma, (\theta \circ \gamma)/\theta)$$

defines an injection of Γ_{θ} into $\mathfrak{G}_{\mathbf{Q}}$. By a *congruence subgroup* of $\mathfrak{G}_{\mathbf{Q}}$, we understand a subgroup Δ of $\mathfrak{G}_{\mathbf{Q}}$ such that:

(2.8_a) *the projection map of $\mathfrak{G}_{\mathbf{Q}}$ to $G_{\mathbf{Q}+}$ gives a one-to-one map of Δ onto a subgroup Γ of $\{\alpha \in G_{\mathbf{Q}} \mid \nu(\alpha) = 1\}$ which has Γ_N as a subgroup of finite index for some N;*

(2.8_b) *the inverse of the projection map coincides with (2.7) on Γ_N for sufficiently large N.*

For such a Δ and a subfield $\mathfrak{R}$ of $\mathbf{C}$, we define $\mathcal{M}_{\varrho,k}(\Delta, \mathfrak{R})$ (resp. $\mathcal{A}_{\varrho,k}(\Delta, \mathfrak{R})$) to be the set of all elements f of $\mathcal{M}_{\varrho,k}(\mathfrak{R})$ (resp. $\mathcal{A}_{\varrho,k}(\mathfrak{R})$) such that $f^{\beta} = f$ for all $\beta \in \Delta$. If $n > 1$, (2.8_b) is always satisfied by virtue of the congruence subgroup property of $Sp(n, \mathbf{Z})$.

PROPOSITION 2.1. *For every $k \in 2^{-1}\mathbf{Z}$, > 0, we have $\mathcal{M}_k(\mathbf{C}) = \mathcal{M}_k(\mathbf{Q}) \otimes_{\mathbf{Q}} \mathbf{C}$. Moreover, for each even positive integer N, let Δ_N be the image of Γ_N under (2.7); then $\mathcal{M}_k(\Delta_N, \mathbf{C}) = \mathcal{M}_k(\Delta_N, \mathbf{Q}) \otimes_{\mathbf{Q}} \mathbf{C}$.*

Proof. As shown in [5] and [6, § 3], there is a model V defined over $\mathbf{Q}$ for the compactification of $\mathfrak{H}_n/\Gamma_N$ whose function field can be identified with $\mathcal{A}_0(\Gamma_N, \mathbf{Q})$; moreover, there is a non-zero element g of $\mathcal{A}_h(\Gamma_N, \mathbf{Q})$, for some positive integer h, whose divisor on V is rational over $\mathbf{Q}$. We can take h so that $\theta^{2h} \in \mathcal{M}_h(\Gamma_2, \mathbf{Q})$. Then $\theta^{2h}/g \in \mathcal{A}_0(\Gamma_N, \mathbf{Q})$. It follows that the divisor of θ on V is rational over $\mathbf{Q}$. Therefore, identifying $\mathcal{M}_k(\Delta_N, \mathbf{Q})$ with a linear system on V rational over $\mathbf{Q}$ in the same manner as in the proof of [5, Th. 6], we obtain the second assertion, from which the first one follows immediately.

PROPOSITION 2.2. *The projection map of $\mathfrak{G}$ onto G_{A+} is open and continuous. Moreover, $\mathfrak{G}$ is locally compact, and $\iota_1(\mathbf{Z}_\mathfrak{f}^{\times})\mathfrak{G}_\mathbf{Q} G_{\infty+}$ is dense in $\mathfrak{G}$, where we embed $G_{\infty+}$ in $\mathfrak{G}$ by identifying x with $(x, 1)$ for $x \in G_{\infty+}$.*

Proof. With an open compact subgroup S of $G_\mathfrak{f}$, a neighborhood Y of the identity in $G_{\infty+}$, and a finite subset F of $\mathcal{A}_{1/2}(\mathbf{Q_{ab}})$ containing θ, we put

$$(2.9) \qquad U(SY, F) = \{(x, v) \in \mathfrak{G} \mid x \in SY,\ h^v = h \quad \text{for all } h \in F\}.$$

The sets of this type form a basis of neighborhoods of the identity element of $\mathfrak{G}$. Given such a set, let

$$T = \{x \in S \mid (h/\theta)^x = h/\theta,\ (h^2)^x = h^2 \quad \text{for all } h \in F\}.$$

Then T is an open compact subgroup of $G_\mathfrak{f}$, and $U(SY, F) \supset U(TY, F)$. For every $x \in TY$, define an element v of $\mathfrak{E}$ by $f^v = (f/\theta)^x \theta$. Then $(x, v) \in U(TY, F)$. Thus we see that the projection of $\mathfrak{G}$ onto G_{A+} gives a one-to-one map of $U(TY, F)$ onto TY. Our assertions except the last one follow from this fact. Let (y, w) be an arbitrary element of $\mathfrak{G}$. Given S and F, define T as above, and put $y = x\iota(t)\alpha$ with $x \in TG_{\infty+}$, $t \in \mathbf{Z}_\mathfrak{f}^{\times}$, and $\alpha \in G_{\mathbf{Q}+}$ by Lemma 1.1. Define again v for x as above. Then $(y, w)^{-1}(x, v)\iota_1(t)$ is an element of $\mathfrak{G}$ whose projection to G_{A+} is α^{-1}, and hence it belongs to $\mathfrak{G}_\mathbf{Q}$. This proves the last assertion.

Given F, S, and T as above, take an even positive integer N so that $S_N \subset TG_\infty$. Obviously

$$(2.10) \qquad U(S_N, \{\theta\}) = U(S_N, F) \subset U(TG_{\infty+}, F).$$

Thus the sets $U(S_N, \{\theta\})$ with $0 < N \in \mathbf{Z}$ form a basis of neighborhoods of $\mathfrak{G}$ modulo $G_{\infty+}$.

PROPOSITION 2.3. *For every even positive integer N, we have*

$$(2.11) \qquad U(S_N, \{\theta\}) \cap \mathfrak{G}_\mathbf{Q} = \Delta_N,$$

$$(2.12) \qquad f^v = f \quad \text{if } y \in U(S_N, \{\theta\}) \quad \text{and } f \in \mathcal{A}_{\varrho,k}(\Delta_N, \mathbf{Q}(e(1/N)).$$

Proof. Equality (2.11) is obvious. To show (2.12), let $y \in U(S_N, \{\theta\})$, $f \in \mathcal{A}_{\varrho,k}(\Delta_N, \mathbf{Q}(e(1/N)))$, and $V = \{u \in U(S_N, \{\theta\}) \mid (f/\theta^k)^u = f/\theta^k\}$. By Prop. 2.2, $y = u\iota_1(t)\beta$ with $u \in V$, $t \in \mathbf{Z}_\mathfrak{f}^{\times}$ and $\beta \in \mathfrak{G}_\mathbf{Q}$. We see that $t \equiv 1 \pmod{N}$, and hence $\iota_1(t) \in U(S_N, \{\theta\})$, so that $\beta \in \Delta_N$ by (2.11). Therefore $f^v = (f^t)^\beta = f$.

3. Generalization of theta functions

In this section, we consider a certain class of holomorphic functions on $\mathbf{C}_s^n \times \mathfrak{H}_n$ which includes classical theta functions as special cases, where s and n are arbitrary positive integers. First we define the action of an element $\alpha = \begin{pmatrix} a & b \\ c & d \end{pmatrix}$ of $G_{\mathbf{Q}+}$ on $\mathbf{C}_s^n \times \mathfrak{H}_n$ by

$$(3.1) \qquad \alpha(u, z) = ({}^t(cz + d)^{-1} u, \, \alpha(z)) \quad (u \in \mathbf{C}_s^n, z \in \mathfrak{H}_n).$$

We also define the action of an element $\beta = (\alpha, \psi)$ of $\mathfrak{G}_{\mathbf{Q}}$ on $\mathfrak{H}_n$ and on $\mathbf{C}_s^n \times \mathfrak{H}_n$ to be the same as that of α, and put $\nu(\beta) = \nu(\alpha)$. Throughout this section, k will denote a non-negative element of $2^{-1}\mathbf{Z}$, and ξ a non-negative symmetric element of $\mathbf{Q}_s^s$. Also we fix a polynomial representation

$$(3.2) \qquad \tau: GL_{n+s}(\mathbf{Q}) \to GL_m(\mathbf{Q}),$$

and use the same notation τ for its natural extension to $GL_{n+s}(\mathbf{C})$. For a $\mathbf{C}^m$-valued function $f(u, z)$ on $\mathbf{C}_s^n \times \mathfrak{H}_n$ and $\beta = (\alpha, \psi) \in \mathfrak{G}_{\mathbf{Q}}$ with $\alpha = \begin{pmatrix} a & b \\ c & d \end{pmatrix}$, we define functions $f|_{\tau, k}\beta$ and $f|_{\tau, k, \xi}\beta$ on $\mathbf{C}_s^n \times \mathfrak{H}_n$ by

$$(3.3) \qquad (f|_{\tau, k}\beta)(u, z) = \psi(z)^{-2k} \tau \begin{pmatrix} cz + d & 0 \\ 0 & \nu(\alpha)\,1_s \end{pmatrix}^{-1} f(\alpha(u, z)),$$

$$(3.4) \qquad (f|_{\tau, k, \xi}\beta)(u, z) = \psi(z)^{-2k} e_s(-\tfrac{1}{2}\nu(\alpha)^{-1}\xi \cdot {}^t u(cz + d)^{-1} cu) \, \tau \begin{pmatrix} cz + d & cu \\ 0 & \nu(\alpha)\,1_s \end{pmatrix}^{-1} f(\alpha(u, z)).$$

When $k \in \mathbf{Z}$ and $\psi(z)^2 = \det(cz + d)$, we denote $f|_{\tau, k}\beta$ and $f|_{\tau, k, \xi}\beta$ also by $f|_{\tau, k}\alpha$ and $f|_{\tau, k, \xi}\alpha$.

Let us now associate with a given f two $\mathbf{C}^m$-valued functions $P^{\tau, \xi}f$ and $P_{\tau, \xi}f$ defined by

$$(3.5) \qquad (P^{\tau, \xi}f)(u, z) = e_s(\tfrac{1}{2}\xi \cdot {}^t u(z - \bar{z})^{-1} u) \, \tau \begin{pmatrix} 1_n & (z - \bar{z})^{-1} u \\ 0 & 1_s \end{pmatrix} f(u, z),$$

$$(3.6) \qquad (P_{\tau, \xi}f)(u, z) = e_s(\tfrac{1}{2}\xi \cdot {}^t(u - \bar{u})(z - \bar{z})^{-1} u) \, \tau \begin{pmatrix} 1_n & (z - \bar{z})^{-1}(u - \bar{u}) \\ 0 & 1_s \end{pmatrix} f(u, z).$$

It can easily be verified that for every $\beta \in \mathfrak{G}_{\mathbf{Q}}$,

$$(3.7) \qquad P^{\tau, \eta}(f|_{\tau, k, \xi}\beta) = (P^{\tau, \xi}f)|_{\tau, k}\beta,$$

$$(3.8) \qquad P_{\tau, \eta}(f|_{\tau, k, \xi}\beta) = (P_{\tau, \xi}f)|_{\tau, k}\beta \quad (\eta = \nu(\beta)^{-1}\xi).$$

From this and an obvious relation $f|_{\tau, k}(\beta\delta) = (f|_{\tau, k}\beta)|_{\tau, k}\delta$, we obtain

$$(3.9) \qquad (f|_{\tau, k, \xi}\beta)|_{\tau, k, \eta}\delta = f|_{\tau, k, \xi}(\beta\delta) \quad \text{if } \eta = \nu(\beta)^{-1}\xi.$$

12

Let Δ be a congruence subgroup of $\mathfrak{G}_{\mathbf{Q}}$ as defined in § 2, and Λ a lattice in $\mathbf{Q}_s^{2n}$. Our main object of study in this section is a holomorphic $\mathbf{C}^m$-valued function f on $\mathbf{C}_s^n \times \mathfrak{H}_n$ satisfying the following two conditions:

$$(3.10) \qquad f|_{\tau,k,\xi}\gamma = f \quad \text{for all } \gamma \in \Delta,$$

$$(3.11) \qquad f(u+zp+q, z) = e_s(-\xi(\tfrac{1}{2} \cdot {}^tpzp + {}^tpu))\,\tau \begin{pmatrix} 1_n & -p \\ 0 & 1_s \end{pmatrix} f(u, z)$$

$$\text{for all } \begin{pmatrix} p \\ q \end{pmatrix} \in \Lambda \text{ with } p \text{ and } q \text{ in } \mathbf{Q}_s^n.$$

Obviously such an f is periodic as a function in the real parts of u and z, so that it has a Fourier expansion of the form

$$(3.12) \qquad f(u, z) = \sum_{\eta \in Y} \sum_{\lambda \in L} c(\eta, \lambda)\, e_n(\eta z + u\lambda)$$

with $c(\eta, \lambda) \in \mathbf{C}^m$, where L is a lattice in $\mathbf{Q}_n^s$, and Y is a lattice in the vector space $\{\eta \in \mathbf{Q}_n^n \mid {}^t\eta = \eta\}$. If $n > 1$, we denote by $T_{\tau,k,\xi}(\Delta, \Lambda)$ the vector space of all holomorphic f satisfying (3.10) and (3.11). In the case $n = 1$, we define $T_{\tau,k,\xi}(\Delta, \Lambda)$ under a certain additional condition (3.23) below. For the moment, let us assume either $n > 1$, or as-if the definition for $n = 1$ is already given.

For a subfield $\mathfrak{R}$ of $\mathbf{C}$, we denote by $T_{\tau,k,\xi}(\Delta, \Lambda; \mathfrak{R})$ the set of all $f \in T_{\tau,k,\xi}(\Delta, \Lambda)$ with expansion (3.12) whose coefficients $c(\eta, \lambda)$ have components in $\mathfrak{R}$. Further we denote by $T_{\tau,k,\xi}(\mathfrak{R})$ the union of $T_{\tau,k,\xi}(\Delta, \Lambda; \mathfrak{R})$ for all possible Δ and Λ, and put $T_{\tau,k,\xi} = T_{\tau,k,\xi}(\mathbf{C})$. If $k \in \mathbf{Z}$, $T_{\tau,k,\xi}(\Gamma, \Lambda)$ and $T_{\tau,k,\xi}(\Gamma, \Lambda; \mathfrak{R})$ can be defined for a congruence subgroup Γ of $G_{\mathbf{Q}+}$ in a similar way. To simplify our notation, we write hereafter f^*, f_*, and f^β for $P^{\tau,\xi}f$, $P_{\tau,\xi}f$, and $f|_{\tau,k,\xi}\beta$, when $f \in T_{\tau,k,\xi}$ or more generally when f satisfies (3.10) and (3.11); further, if τ is trivial we write $T_{k,\xi}$ for $T_{\tau,k,\xi}$.

For each $z \in \mathfrak{H}_n$, define a hermitian form $H_{\xi,z}$ and an alternating form $E_{\xi,z}$ by

$$(3.13) \qquad H_{\xi,z}(u, v) = 2i \cdot \operatorname{tr}(\xi \cdot {}^t\bar{u}(z-\bar{z})^{-1}v),$$

$$(3.14) \qquad 2i \cdot E_{\xi,z}(u, v) = H_{\xi,z}(u, v) - H_{\xi,z}(v, u) \quad (u, v \in \mathbf{C}_s^n).$$

Put $\Omega_z = (z \quad 1_n)$. Then

$$(3.15) \qquad E_{\xi,z}(\Omega_z a, \Omega_z b) = \operatorname{tr}(\xi \cdot {}^t a J b) \quad (a, b \in \mathbf{R}_s^{2n}).$$

If $f \in T_{\tau,k,\xi}(\Delta, \Lambda)$ and $l = zp + q$ with $\begin{pmatrix} p \\ q \end{pmatrix} \in \Lambda$, we have

$$(3.16) \qquad f^*(u+l, z) = e_s(\tfrac{1}{2}\xi \cdot {}^tpq)\, e\!\left(\frac{1}{2i}\, H_{\xi,z}(l, u + \tfrac{1}{2}l)\right)\tau \begin{pmatrix} 1_n & (z-\bar{z})^{-1}l \\ 0 & 1_s \end{pmatrix} f^*(u, z)$$

$$(3.17) \qquad f_*(u+l, z) = e_s(\tfrac{1}{2}\xi \cdot {}^tpq)\, e(E_{\xi,z}(l, u)/2)\, f_*(u, z).$$

Now, for $f \in T_{\tau,k,\xi}$ and elements r and r' of $\mathbf{R}_s^n$, we have

$$(3.18) \qquad f_*(zr + r', z) = e_s(\tfrac{1}{2}\xi \cdot {}^t r(zr + r'))\, \tau \begin{pmatrix} 1_n & r \\ 0 & 1_s \end{pmatrix} f(zr + r', z).$$

This is obviously holomorphic in z. Now for $\beta = (\alpha, \psi) \in \mathfrak{G}_{\mathbf{Q}}$ with $\alpha = \begin{pmatrix} a & b \\ c & d \end{pmatrix}$ and $v \in \mathbf{R}_s^{2n}$, we have

$$(3.19) \qquad f_*(\Omega_{\alpha(z)} v, \alpha(z)) = \psi(z)^{2k}\, \tau \begin{pmatrix} cz + d & 0 \\ 0 & \nu(\alpha)\,1_s \end{pmatrix} (f^\beta)_*(\Omega_z \cdot {}^t \alpha v, z).$$

This follows immediately from (3.8). Therefore, if we define a representation ϱ of GL_n by

$$(3.20) \qquad \varrho(X) = \tau \begin{pmatrix} X & 0 \\ 0 & 1_s \end{pmatrix},$$

and put $h_v(z) = f_*(\Omega_z v, z)$ and $h_v'(z) = (f^\beta)_*(\Omega_z v, z)$, then

$$(3.21) \qquad h_v|_{\varrho,k}\beta = \tau \begin{pmatrix} 1_n & 0 \\ 0 & \nu(\alpha)\,1_s \end{pmatrix} h_w', \quad \text{where } w = {}^t\alpha v.$$

On the other hand, put $g(v) = f_*(\Omega_z v, z)$ for $v \in \mathbf{Q}_s^{2n}$ with a fixed z. Then (3.17) together with (3.15) shows that, given any lattice L in $\mathbf{Q}_s^{2n}$, we can find another lattice $L' \subset L$ so that g defines a function on L/L'. Therefore, if $\mathbf{A}_f$ denotes the non-archimedean part of $\mathbf{A}$, then g can be uniquely extended to a continuous function on $(\mathbf{A}_f)_s^{2n}$. (Cf. [7, pp. 683–4].) Thus $f_*(\Omega_z v, z)$ for $v \in (\mathbf{A}_f)_s^{2n}$ is meaningful. Coming back to the above h_v, (3.21) implies

(3.22) $h_v|_{\varrho,k}\gamma = h_v$ if $v \in \mathbf{Q}_s^{2n}$ and γ belongs to a sufficiently small congruence subgroup of $\mathfrak{G}_{\mathbf{Q}}$ depending on v.

Therefore if $n > 1$, we see that

(3.23) $f_*(\Omega_z v, z)$ for every fixed $v \in \mathbf{Q}_s^{2n}$ belongs, as a function of z, to $\mathfrak{M}_{\varrho,k}(\mathbf{C})$, where ϱ is defined by (3.20).

Now in the case $n = 1$, we define $T_{\tau,k,\xi}(\Delta, \Lambda)$ to be the set of all holomorphic f satisfying (3.10), (3.11), and (3.23). Condition (3.23) is essentially a condition on the Fourier coefficients of f (at all cusps) as shown by the following proposition and its proof.

PROPOSITION 3.1. *If f is an element of $T_{\tau,k,\xi}$ with expansion (3.12), then $c(\eta, \lambda) \neq 0$ only when $\eta \geq 0$, and moreover, for a fixed η, there are only finitely many λ such that $c(\eta, \lambda) \neq 0$.*

Proof. Observe that if $p, q \in \mathbf{Q}_s^n$, we have

$$f_*(zp + q, z) = \tau \begin{pmatrix} 1_n & p \\ 0 & 1_s \end{pmatrix} \sum_\zeta a(\zeta) e_n(\zeta z)$$

with

(3.24)
$$a(\zeta) = e_s(\xi \cdot {}^t pq/2) \sum_{\lambda, \eta} e_s(\lambda q) c(\eta, \lambda),$$

where the sum is taken under the condition

(3.24')
$$2\zeta = 2\eta + p\xi \cdot {}^t p + p\lambda + {}^t\lambda \cdot {}^t p.$$

If ζ is not non-negative, (3.23) implies that (3.24) must be 0 for all $q \in \mathbf{Q}_s^n$. Therefore

(3.25) $c(\eta, \lambda) \neq 0$ *only if* $2\eta + p\xi \cdot {}^t p + p\lambda + {}^t\lambda \cdot {}^t p \geqslant 0$ *for all* $p \in \mathbf{Q}_s^n$.

Taking p to be 0, we see that $c(\eta, \lambda) = 0$ unless $\eta \geqslant 0$. For a fixed η, there are only finitely many λ in a given lattice satisfying the inequality of (3.25) for all $p \in \mathbf{Q}_s^n$. This completes the proof.

Note that the sum of (3.24) is a finite sum. In fact, if $c(\eta, \lambda) \neq 0$, we have $2\eta + 4p\xi \cdot {}^t p + 2p\lambda + 2 \cdot {}^t\lambda \cdot {}^t p \geqslant 0$ by (3.25), so that (3.24') shows $\eta \leqslant 2\zeta + p\xi \cdot {}^t p$. There are only finitely many such $\eta \geqslant 0$ in a lattice for fixed p and ζ, which proves the desired finiteness.

PROPOSITION 3.2. *Let $\Re$ be a subfield of $\mathbf{C}$ containing $\mathbf{Q}_{ab}$, and let $f \in T_{\tau, k, \xi}$. Then* (i) $f \in T_{\tau, k, \xi}(\Re)$ *if and only if* $f_*(\Omega_2 v, z) \in \mathfrak{M}_{\varrho, k}(\Re)$ *for all* $v \in \mathbf{Q}_s^{2n}$; (ii) *if* $f \in T_{\tau, k, \xi}(\Re)$ *and* $\beta \in \mathfrak{G}_\mathbf{Q}$, *then* $f^\beta \in T_{\tau, k, \zeta}(\Re)$ *with* $\zeta = \nu(\beta)^{-1}\xi$.

Proof. Consider expansion (3.12) for f. If $f \in T_{\tau, k, \xi}(\Re)$, then obviously $f_*(\Omega_z v, z) \in \mathfrak{M}_{\varrho, k}(\Re)$ for all $v \in \mathbf{Q}_s^{2n}$. Conversely suppose $f_*(r, z)$ has Fourier coefficients in $\Re^m$ for every $r \in \mathbf{Q}_s^n$. This implies that

$$\sum_\lambda c(\eta, \lambda) e_s(\lambda r) \in \Re^m \quad \text{for all } r \in \mathbf{Q}_s^n,$$

and hence $c(\eta, \lambda) \in \Re^m$. To prove assertion (ii), first assume $\Re = \mathbf{C}$. If $n > 1$, the desired conclusion follows easily from our definition of $T_{\tau, k, \xi}$; if $n = 1$, we need (3.21). This result together with assertion (i), (3.21), and the fact that $\mathfrak{M}_{\varrho, k}(\Re)^\beta = \mathfrak{M}_{\varrho, k}(\Re)$ proves assertion (ii) in the general case.

Typical examples of functions of $T_{k, \xi}$ with $s = 1$ are provided by

(3.26)
$$\theta(u, z; p, q) = \sum_{x - p \in \mathbf{Z}^n} e(\tfrac{1}{2} \cdot {}^t xzx + {}^t x(u + q)),$$

(3.27)
$$\varphi(u, z; p, q) = e(\tfrac{1}{2} \cdot {}^t u(z - \bar{z})^{-1} u) \theta(u, z; p, q),$$

(3.28)
$$\varphi'(u, z; p, q) = e(\tfrac{1}{2} \cdot {}^t(u - \bar{u})(z - \bar{z})^{-1} u) \theta(u, z; p, q),$$

4 − 782901 *Acta mathematica* 141. Imprimé le 1 Septembre 1978

15

where $u \in \mathbf{C}^n$ and $p, q \in \mathbf{R}^n$. In fact, if $f(u, z) = \theta(\lambda u, \mu z; p, q)$ with $p, q \in \mathbf{Q}^n$ and positive rational numbers λ, μ, then $f \in T_{1/2, \nu}(\mathbf{Q}_{ab})$ with $\nu = \lambda^2/\mu$, and

$$f^*(u, z) = \varphi(\lambda u, \mu z; p, q), \quad f_*(u, z) = \varphi'(\lambda u, \mu z; p, q)$$

(cf. Appendix and [7, § 1]). To give an explicit example of (3.21) or (3.22), put

$$(3.29) \qquad \chi(z; v, w; p, q) = \varphi'(zv + w, z; p, q) \quad (v, w, p, q \in \mathbf{R}^n),$$
$$\psi_\gamma = (\theta \circ \gamma)/\theta \qquad\qquad (\gamma \in \Gamma_\theta).$$

We obtain, from (a.3) of the Appendix,

$$(3.30) \qquad \chi(\gamma(z); v, w; p, q) = e(({}^t pq - {}^t p^* q^*)/2)\psi_\gamma(z)\, \chi(z; v^*, w^*; p^*, q^*)$$

for every $\gamma \in \Gamma_\theta$, where

$$\begin{pmatrix} v^* \\ w^* \end{pmatrix} = {}^t\gamma \begin{pmatrix} v \\ w \end{pmatrix}, \quad \begin{pmatrix} p^* \\ q^* \end{pmatrix} = {}^t\gamma \begin{pmatrix} p \\ q \end{pmatrix}.$$

Also we note that

$$(3.31) \qquad \chi(z; v, w; p, q) = e(-\tfrac{1}{2} \cdot {}^t vw - {}^t vq)\theta(0, z; v + p, w + q).$$

This follows from [7, (11)] and was actually stated as [7, (27)].

For $f \in T_{\tau, k, \xi}(\Delta, \Lambda; \mathfrak{R})$ and $\varepsilon \in GL_s(\mathbf{Q})$, put

$$g(u, z) = \tau \begin{pmatrix} 1_n & 0 \\ 0 & \varepsilon \end{pmatrix} f(u\varepsilon, z), \ \zeta = \varepsilon\xi \cdot {}^t\varepsilon.$$

Then $g \in T_{\tau, k, \zeta}(\Delta, \Lambda\varepsilon^{-1}; \mathfrak{R})$. Therefore, to discuss the nature of f, we can simplify our discussion by assuming that

$$(3.32) \qquad \xi = \text{diag}\,[\xi_1, ..., \xi_r, 0, ..., 0], \quad r = \text{rank}\,(\xi),$$

with positive integers ξ_i, and also that $\Lambda \supset \mathbf{Z}_s^{2n}$.

PROPOSITION 3.3. *Suppose that ξ is given by (3.32), and write the variable u on $\mathbf{C}_s^n$ in the form $u = (v, w)$ with $v \in \mathbf{C}_r^n$ and $w \in \mathbf{C}_{s-r}^n$. Define a representation ω of GL_{n+} by*

$$\omega(X) = \tau \begin{pmatrix} X & 0 \\ 0 & 1_{s-r} \end{pmatrix} \quad (X \in GL_{n+r}).$$

Then, for each $f \in T_{\tau, k, \xi}$, there is an element g of $T_{\omega, k, \eta}$ such that $f(v, w, z) = g(v, z)$, where $\eta = \text{diag}\,[\xi_1, ..., \xi_r]$. Moreover, $f_*(v, w, z) = g_*(v, z)$ and

$$f^\beta(v, w, z) = \tau \begin{pmatrix} 1_{n+r} & 0 \\ 0 & \nu(\beta)^{-1}1_{s-r} \end{pmatrix} g^\beta(v, z) \ \text{for every } \beta \in \mathfrak{G}_\mathbf{Q}.$$

Proof. For any fixed v and z, let $h(w)$ be a component of the vector $f_*(v, w, z)$. Then (3.17) implies that

$$h(w + zp + q) = h(w) \quad \text{for all } p, q \in L, \tag{3.33}$$

where L is a lattice in $\mathbf{Q}^n_{s-r}$. Now (3.6) shows that h is a finite sum $\sum_\nu A_\nu(w) B_\nu(x)$ with holomorphic functions A_ν and monomials B_ν of the components, say x_{ij}, of $x = \mathrm{Im}\,(w)$. Let us consider this as a polynomial in x_{ij}, and prove, by induction on the degree of the polynomial, that it is a constant, involving neither w nor x. First suppose that it is of degree 0, i.e., h is holomorphic in w. Then (3.33) implies that h is a holomorphic function on a complex torus, which must be a constant. In the general case, applying the induction assumption to $\partial h / \partial \bar{w}_{ij}$, we see that $h(w) = A_0(w) + \sum_{i,j} a_{ij} \bar{w}_{ij}$ with a holomorphic A_0 and $a_{ij} \in \mathbf{C}$. Then (3.33) shows that $\partial A_0 / \partial w_{ij}$ is invariant under $w \mapsto w + zp + q$, and hence is a constant. It follows that $h(w) = b + \mathrm{tr}\,(Dw + E\bar{w})$ with $b \in \mathbf{C}$ and $D, E \in C_n^{s-r}$. Again from (3.33), we obtain

$$\mathrm{tr}\,(D(zp + q) + E(\bar{z}p + q)) = 0 \quad \text{for all } p, q \in L,$$

so that $D = E = 0$. Thus we have proved that $f_*(v, w, z)$ does not depend on w. This together with (3.6) shows that $f(v, w, z)$, for fixed v and z, is a polynomial function in $w - \bar{w}$. Being holomorphic in w, it must be a constant. Thus we can put $f(v, w, z) = g(v, z)$. Then it is straightforward to verify all our assertions.

Before proceeding further, we prove an easy

LEMMA 3.4. *Let A be an arbitrary set, and B a domain in $\mathbf{C}^n$, and let $f_1(u, z), \ldots, f_m(u, z)$, $g(u, z)$ be complex valued functions on $A \times B$, holomorphic in z, where $u \in A$ and $z \in B$. Suppose that the functions $f_1(u, z_0), \ldots, f_m(u, z_0)$ on A are linearly independent over $\mathbf{C}$ for every $z_0 \in B$, and $g(u, z) = \sum_{k=1}^m h_k(z) f_k(u, z)$ with functions h_k on B. Then h_k is holomorphic on B for every k.*

Proof. For fixed u and z, put

$$X(u, z) = \{(c_1, \ldots, c_m) \in \mathbf{C}^1_m \mid \sum_k c_k f_k(u, z) = 0\}.$$

For any fixed $z_0 \in B$, we have $\bigcap_{u \in A} X(u, z_0) = \{0\}$, so that there exist m points $u_1, \ldots, u_m$ of A such that $\bigcap_{j=1}^m X(u_j, z_0) = \{0\}$. This means that $\det\,(f_k(u_j, z_0))_{j,k} \neq 0$. Solving the equations

$$g(u_j, z) = \sum_{k=1}^m h_k(z) f_k(u_j, z) \quad (j = 1, \ldots, m),$$

we see that h_k is holomorphic in a neighborhood of z_0, QED.

PROPOSITION 3.5. *If $n > 1$, every f of $T_{k,\xi}(\mathfrak{R})$ has an expression*

$$(3.34) \qquad f(u, z) = \sum_{h \in H/K} c_h(z) \sum_{x - h \in K} e_s(\xi(\tfrac{1}{2} \cdot {}^t x z x + {}^t x u))$$

with $c_h \in \mathcal{M}_{k-(r/2)}(\mathfrak{R})$, where H and K are lattices in the vector space

$$\mathbf{Q}_s^n / \{ y \in \mathbf{Q}_s^n \mid y\xi = 0 \},$$

and $r = \mathrm{rank}\,(\xi)$. Conversely, any such expression defines an element of $T_{k,\xi}(\mathfrak{R})$; this is so even if $n = 1$.

Proof. Suppose $f \in T_{k,\xi}(\Delta, \Lambda)$. Take $\varepsilon \in GL_s(\mathbf{Q})$ so that $\mathbf{Z}_s^{2n} \subset \Lambda \varepsilon^{-1}$ and $\varepsilon \xi \cdot {}^t \varepsilon = \mathrm{diag}\,[\zeta_1, \ldots, \zeta_r, 0, \ldots, 0]$ with positive integers ζ_ν. Put $\zeta = \varepsilon \xi \cdot {}^t \varepsilon$ and $g(u, z) = f(u\varepsilon, z)$. Then $g \in T_{k,\zeta}(\Delta, \mathbf{Z}_s^{2n})$. Let u_ν be the ν-th column of u. By Prop. 3.3, g depends only on $u_1, \ldots, u_r$, and if $\nu \leqslant r$, we have

$$g(\ldots, u_\nu + zp + q, \ldots, z) = e(-\zeta_\nu(\tfrac{1}{2} \cdot {}^t p z p + {}^t p u_\nu)) g(\ldots, u_\nu, \ldots, z)$$

for $p, q \in \mathbf{Z}^n$. Therefore, as a function of u_ν, it is a linear combination of $\theta(\zeta_\nu u_\nu, \zeta_\nu z; h, 0)$ with h in a set of representatives for $\zeta_\nu^{-1} \mathbf{Z}^n / \mathbf{Z}^n$. Consequently we obtain, in view of Lemma 3.4,

$$(3.35) \qquad g(u, z) = \sum_h c_h(z) \prod_{\nu=1}^r \theta(\zeta_\nu u_\nu, \zeta_\nu z; h_\nu, 0) \quad (h = (h_1, \ldots, h_r))$$

$$= \sum_{h \in M/L} c_h(z) \sum_{x - h \in L} e_r(\zeta'(\tfrac{1}{2} \cdot {}^t x z x + {}^t x u'))$$

with holomorphic functions c_h on $\mathfrak{H}_n$, where $\zeta' = \mathrm{diag}\,[\zeta_1, \ldots, \zeta_r]$, $u' = (u_1, \ldots, u_r)$, $L = \mathbf{Z}_r^n$, and $M = \mathbf{Z}_r^n \zeta'^{-1}$. Since g satisfies (3.10), we can easily verify that $c_h \in \mathcal{M}_{k-(r/2)}(\mathfrak{R})$ if $n > 1$. Consider L and M as submodules of $\mathbf{Q}_s^n$ in an obvious way, and let H and K be their images under ε. Then we obtain our first assertion. The converse part follows immediately from the fact that $\theta(\lambda u, \mu z; h, 0)$ defines an element of $T_{1/2,\nu}(\mathbf{Q})$ with $\nu = \lambda^2/\mu$.

LEMMA 3.6. *For $f \in T_{k,\xi}(\mathfrak{R})$, put*

$$(3.36) \qquad d_\lambda f = \begin{bmatrix} \dfrac{1}{2\pi i} \dfrac{\partial f}{\partial u_{1\lambda}} \\ \vdots \\ \dfrac{1}{2\pi i} \dfrac{\partial f}{\partial u_{n\lambda}} \\ \xi_{1\lambda} f \\ \vdots \\ \xi_{s\lambda} f \end{bmatrix} \quad (\lambda = 1, \ldots, s).$$

Then $d_\lambda f \in T_{\tau,k,\xi}(\mathfrak{R})$ with $\tau(X) = X$, and

$$(3.36') \qquad d_\lambda(f^\beta) = \begin{pmatrix} 1_n & 0 \\ 0 & \nu(\beta)\, 1_s \end{pmatrix} (d_\lambda f)^\beta \quad \text{for every } \beta \in \mathfrak{G}_{\mathbf{Q}}.$$

This can be verified in a straightforward way, except the fact that $d_\lambda(f)$ satisfies (3.23) when $n = 1$, which can be shown as follows. Since f^β satisfies (3.25) for every $\beta \in \mathfrak{G}_{\mathbf{Q}}$, we see easily that $d_\lambda(f^\beta)$ satisfies the same condition. Therefore, by (3.36') and (3.21), $(d_\lambda f)_* (\Omega_z v, z)^\beta$ is finite at $i\infty$ for every $v \in \mathbf{Q}_s^2$ and every $\beta \in \mathfrak{G}_{\mathbf{Q}}$. Q.E.D.

PROPOSITION 3.7. *Suppose that ξ is non-degenerate. Then, for an arbitrary point (u_0, z_0) of $\mathbf{C}_s^n \times \mathfrak{H}_n$, there exists a function A on $\mathbf{C}_s^n \times \mathfrak{H}_n$ with values in $\mathbf{C}_m^n$ such that*

(i) *the columns of A belong to $T_{\tau, j, \xi}(\mathbf{Q})$, where j is a positive element of $2^{-1}\mathbf{Z}$ depending only on τ;*

(ii) $\det(A(u_0, z_0)) \neq 0$.

Proof. If τ is trivial, the function f defined by

$$f(u\varepsilon, z) = c(z) \prod_{\nu=1}^{s} \theta(\zeta_\nu u_\nu, \zeta_\nu z; h_\nu, 0)$$

with suitable $c \in \mathcal{M}_{j-(s/2)}(\mathbf{Q})$ and h_ν can be taken as A, where ε and ζ_ν are as in the proof of Prop. 3.5. In this case j may be arbitrarily chosen under the condition $j \geqslant s/2$. Next assume $\tau(X) = X$. Let $k \geqslant s/2$ and take $f \in T_{k,\xi}(\mathbf{Q})$ so that $f(u_0, z_0) \neq 0$. We are going to define A by

$$A(u, z) = \begin{bmatrix} \dfrac{1}{2\pi i}\dfrac{\partial f}{\partial u_{11}} & \cdots & \dfrac{1}{2\pi i}\dfrac{\partial f}{\partial u_{1s}} & b_{11} & \cdots & b_{1n} \\ \cdots & \cdots & \cdots & \cdots & \cdots & \cdots \\ \dfrac{1}{2\pi i}\dfrac{\partial f}{\partial u_{n1}} & \cdots & \dfrac{1}{2\pi i}\dfrac{\partial f}{\partial u_{ns}} & b_{n1} & \cdots & b_{nn} \\ \xi_{11} f & \cdots & \xi_{1s} f & 0 & \cdots & 0 \\ \cdots & \cdots & \cdots & \cdots & \cdots & \cdots \\ \xi_{s1} f & \cdots & \xi_{ss} f & 0 & \cdots & 0 \end{bmatrix}$$

with a suitable $B = (b_{\lambda\mu}(u, z))$. Let ϱ be the identity representation of $GL_n(\mathbf{C})$ onto itself. By [8, Prop. 1.2] and its proof, we can find an element P of $\mathcal{M}_{\varrho, 1/2}(\mathbf{Q})$ such that $\det(P(z_0)) \neq 0$. Take $g \in T_{k-(1/2), \xi}(\mathbf{Q})$ so that $g(u_0, z_0) \neq 0$, and put $B(u, z) = g(u, z)P(z)$. By virtue of Lemma 3.6, A has the required property. In this case, j must be $> s/2$. Let $\omega(X) = X \otimes \ldots \otimes X$ and $C = A \otimes \ldots \otimes A$ (both t copies). Then $\det(C(u_0, z_0)) \neq 0$, and the columns of C belong to $T_{\omega, tk, t\xi}(\mathbf{Q})$. Since an arbitrary irreducible τ is a $\mathbf{Q}$-rational component of such ω, we obtain our assertion.

PROPOSITION 3.8. $T_{\tau,k,\xi} = T_{\tau,k,\xi}(\mathbf{Q}) \otimes_{\mathbf{Q}} \mathbf{C}$.

This will be proved in the next section.

We are going to define an action of $\mathfrak{G}$ on $T_{\tau,k,\xi}(\mathbf{Q}_{ab})$. First, for $f \in T_{\tau,k,\xi}(\mathfrak{R})$ with expansion (3.12) and an injection σ of $\mathfrak{R}$ into $\mathbf{C}$, we define f^{σ} by

$$(3.37) \qquad f^{\sigma}(u, z) = \sum_{\eta, \lambda} c(\eta, \lambda)^{\sigma} e_n(\eta z + u\lambda).$$

By virtue of Prop. 3.8, f^{σ} actually defines an element of $T_{\tau,k,\xi}$. In particular, if $f \in T_{\tau,k,\xi}(\mathbf{Q}_{ab})$, f^b is meaningful for $b \in \mathbf{A}^{\times}$.

As already shown, $f_*(\Omega_z v, z)$ is meaningful for $v \in (\mathbf{A}_f)_s^{2n}$, and defines, as a function of z, an element of $\mathcal{M}_{\varrho,k}(\mathbf{C})$, where ϱ is given by (3.20). Now we let $G_{\mathbf{A}}$ act on $(\mathbf{A}_f)_s^{2n}$ by left matrix multiplication, of course ignoring the archimedean part.

PROPOSITION 3.9. Let $f \in T_{\tau,k,\xi}$, $v \in (\mathbf{A}_f)_s^{2n}$, $r \in \mathbf{Z}_f^{\times}$, and let σ be an automorphism of $\mathbf{C}$ that coincides with the action of r on $\mathbf{Q}_{ab}$. Then

$$f_*(\Omega_z v, z)^{\sigma} = (f^{\sigma})_*(\Omega_z \iota(r) v, z),$$

where the left-hand side is defined by the action of σ on $\mathcal{M}_{\varrho,k}(\mathbf{C})$.

Proof. If $v = \begin{pmatrix} p \\ q \end{pmatrix}$ with p and q in $\mathbf{Q}_s^n$, and if f has expansion (3.12), then, as shown in the proof of Prop. 3.1, we have

$$f_*(\Omega_z v, z) = \tau \begin{pmatrix} 1_n & p \\ 0 & 1_s \end{pmatrix} \sum_{\zeta} a(\zeta) e_n(\zeta z)$$

with $a(\zeta)$ given by (3.24). Since the sum expressing $a(\zeta)$ is a finite sum, we have

$$a(\zeta)^{\sigma} = e_s(\xi \cdot {}^t pq'/2) \sum_{\lambda} e_s(\lambda q') c(\eta, \lambda)^{\sigma},$$

where q' is an element of $\mathbf{Q}_s^n$ sufficiently close to $r^{-1}q$. This proves our proposition.

We are now ready to state our first main result:

THEOREM 3.10. Given $f \in T_{\tau,k,\xi}(\mathbf{Q}_{ab})$ and $y \in \mathfrak{G}$, there is a unique element f^y of $T_{\tau,k,\eta}(\mathbf{Q}_{ab})$ with $\eta = |\nu(x)|^{-1}\xi$ such that

$$(3.38) \qquad f_*(\Omega_z v, z)^y = \tau \begin{pmatrix} 1_n & 0 \\ 0 & |\nu(x)| 1_s \end{pmatrix} (f^y)_*(\Omega_z \cdot {}^t xv, z) \quad \text{for all } v \in (\mathbf{A}_f)_s^{2n},$$

where x is the projection of y to $G_{\mathbf{A}+}$, and $f_*(\Omega_z v, z)^y$ is the image of $f_*(\Omega_z v, z)$ as an element of $\mathcal{M}_{\varrho,k}(\mathbf{Q}_{ab})$ under y as defined in § 2. If $k \in \mathbf{Z}$, the assertion holds with $G_{\mathbf{A}+}$ and x in place of $\mathfrak{G}$ and y.

Proof. Define Δ_N as in Prop. 2.1. In view of Prop. 3.8, we can find a positive integer N and a lattice Λ such that $f^t \in T_{\tau,k,\xi}(\Delta_N, \Lambda; \mathbf{Q}(e(1/N)))$ for all $t \in \mathbf{Z}_t^\times$. Define $U(S_N, \{\theta\})$ by (2.9). By Prop. 2.2, we have $y = y' \iota_1(t) \beta$ with $y' \in U(S_N, \{\theta\})$, $t \in \mathbf{Z}_t^\times$ and $\beta \in \mathfrak{G}_\mathbf{Q}$. Now we define f^y to be $(f^t)^\beta$. By our choice of N, we see easily that this is independent of the choice of y', t, β, in view of Prop. 2.3. Moreover, $f^y \in T_{\tau,k,\eta}(\mathbf{Q}_{\mathrm{ab}})$ by Prop. 3.2, since $\nu(\beta) = |\nu(x)|$. To show (3.38), given $v \in \mathbf{Q}_s^{2n}$, take a multiple M of N so that $Mv \in \mathbf{Z}_s^{2n}$ and $f_*(\Omega_z v, z) \in \mathcal{M}_{\varrho,k}(\Delta_M, \mathbf{Q}(e(1/M)))$. Changing y', t, and β if necessary, we may assume that $y' \in U(S_M, \{\theta\})$. By Prop. 2.3, (2.5), (2.6), (3.19), and Prop. 3.9, we have

$$
f_*(\Omega_z v, z)^y = (f_*(\Omega_z v, z)^t)^\beta = (f^t)_*(\Omega_z \iota(t) v, z)^\beta
$$
$$
= \tau \begin{pmatrix} 1_n & 0 \\ 0 & \nu(\alpha)\, 1_s \end{pmatrix} (f^y)_*(\Omega_z \cdot {}^t\alpha\iota(t)\, v, z),
$$

where α is the projection of β to $G_{\mathbf{Q}+}$. This proves (3.38) and completes the proof, since it is obvious that f^y is uniquely determined by (3.38).

PROPOSITION 3.11. *The action of the elements of $\mathfrak{G}$ on $T_{\tau,k,\xi}(\mathbf{Q}_{\mathrm{ab}})$ has the following properties:*

(i) $(af + bg)^y = a^y f^y + b^y g^y$ *for* $a, b \in \mathbf{Q}_{\mathrm{ab}}$ *and* $f, g \in T_{\tau,k,\xi}(\mathbf{Q}_{\mathrm{ab}})$;

(ii) $(f^x)^y = f^{xy}$ *for* $x, y \in \mathfrak{G}$;

(iii) $f^\beta = f|_{\tau,k,\xi}\beta$ *if* $\beta \in \mathfrak{G}_\mathbf{Q}$;

(iv) $f^{\iota_1(r)} = f^r$ *if* $r \in \mathbf{Z}_t^\times$;

(v) $f^y = f$ *if* $f \in T_{\tau,k,\xi}(\Delta_N, \Lambda; \mathbf{Q}(e(1/N)))$ *and* $y \in U(S_N, \{\theta\})$.

Proof. The first four properties follow immediately from the above proof and (3.38). The last one can be proved in exactly the same fashion as Prop. 2.3, since $\{y \in \mathfrak{G} \,|\, f^y = f\}$ is an open subset of $\mathfrak{G}$, as our definition of f^y in the above proof shows.

Notice that Theorem 3.10, together with Proposition 3.11, gives an analogue of the main theorem of [7], which concerns the action of the idele group of an algebraic number field on the theta functions with complex multiplication. Let us now state a consequence of Theorem 3.10 as a theorem in which an abelian variety and its division points are more conspicuous.

Let δ be a diagonal matrix whose diagonal elements are positive integers $\delta_1, ..., \delta_n$ such that $3 \leqslant \delta_1$, $\delta_\nu \,|\, \delta_{\nu+1}$, and let $L(z, \delta) = z\mathbf{Z}^n + \delta\mathbf{Z}^n$ for each $z \in \mathfrak{H}_n$. We define a projective embedding Θ_z of $\mathbf{C}^n/L(z, \delta)$ by

$$
(3.39) \qquad\qquad u \mapsto \Theta_z(u) = (\theta(u, z; j, 0))_{j \in \mathfrak{J}} \quad (u \in \mathbf{C}^n),
$$

where $\mathfrak{J}$ is a complete set of representatives for $\delta^{-1}\mathbf{Z}^n/\mathbf{Z}^n$. We denote by $A(z, \delta)$ the image variety. Put, for simplicity, $\mathfrak{R} = \mathcal{A}_0(\mathbf{Q}_{ab})$. Fix any point $\mathfrak{z}$ of $\mathfrak{H}_n$, and let $B_\mathfrak{z}$ be the ring consisting of all elements of $\mathfrak{R}$ holomorphic at $\mathfrak{z}$, and $\mathfrak{R}_\mathfrak{z}$ the field generated over $\mathbf{Q}$ by the values $F(\mathfrak{z})$ for all $F \in B_\mathfrak{z}$. The set of all F in $B_\mathfrak{z}$ such that $F(\mathfrak{z}) = 0$ form a unique maximal ideal $P_\mathfrak{z}$ of $B_\mathfrak{z}$, and $F \mapsto F(\mathfrak{z})$ gives an isomorphism of $B_\mathfrak{z}/P_\mathfrak{z}$ onto $\mathfrak{R}_\mathfrak{z}$. Now let x be an element of $G_{\mathbf{A}+}$ such that

$$(3.40) \qquad\qquad (B_\mathfrak{z})^x = B_\mathfrak{z}.$$

Obviously $(P_\mathfrak{z})^x = P_\mathfrak{z}$, so that $F(\mathfrak{z}) \mapsto F^x(\mathfrak{z})$ (with $F \in B_\mathfrak{z}$) gives an automorphism of $\mathfrak{R}_\mathfrak{z}$. For example, if $\mathfrak{z}$ is "generic" for the functions of $\mathfrak{R}$, (3.40) is satisfied for all $x \in G_{\mathbf{A}+}$. Another extreme case is the points $\mathfrak{z}$ with complex multiplication, which we shall discuss afterwards. Now we have the following theorem which generalizes [8, Th. 2.4].

THEOREM 3.12. *Given an element x of $G_{\mathbf{A}+}$ and a point $\mathfrak{z}$ of $\mathfrak{H}_n$ satisfying (3.40), let σ denote the automorphism of $\mathfrak{R}_\mathfrak{z}$ obtained from x as above, and let R_δ be the subgroup of $G_{\mathbf{A}}$ defined by [8, (1.11)]. Put $x = q\alpha$ with $q \in R_\delta$ and $\alpha \in G_{\mathbf{Q}+}$ as guaranteed by Lemma 1.1. Put also $\alpha = \begin{pmatrix} a & b \\ c & d \end{pmatrix}$ and $\lambda = {}^t(c\mathfrak{z} + d)^{-1}$ with any such α and q. Then*

    (i) $A(\mathfrak{z}, \delta)^\sigma = A(\alpha(\mathfrak{z}), \delta)$,

    (ii) $\Theta_\mathfrak{z}(\Omega_\mathfrak{z} v)^\sigma = \Theta_{\alpha(\mathfrak{z})}(\lambda\Omega_\mathfrak{z} \cdot {}^t xv) = \Theta_{\alpha(\mathfrak{z})}(\Omega_{\alpha(\mathfrak{z})} \cdot {}^t qv)$ *for all* $v \in (\mathbf{A}_f)^{2n}$.

Proof. Put $f_j(u, z) = \theta(u, z; j, 0)$ for each $j \in \mathfrak{J}$. As noted above, $f_j \in T_{1/2, 1}(\mathbf{Q}_{ab})$. Let y and r be elements of $\mathfrak{G}$ lying above x and q, respectively. By Th. 3.10, we have

$$(3.41) \qquad (f_j)_*(\Omega_z v, z)^y = (f_j^y)_*(\Omega_z \cdot {}^t xv, z) \quad (v \in (\mathbf{A}_f)^{2n},\ j \in \mathfrak{J}).$$

Put $\Gamma_\delta = R_\delta \cap G_{\mathbf{Q}}$. This is the group defined in [8, (1.9)]. Let $r = r' \iota_1(t) \beta$ with $r' \in U(S_N, \{\theta\})$, $t \in \mathbf{Z}_f^\times$, and $\beta \in \mathfrak{G}_{\mathbf{Q}}$ as in the proof of Theorem 3.10 with a suitable N. Then $f_j = f_j^\beta$, and β has an element γ of Γ_δ as its projection to $G_{\mathbf{Q}+}$. Therefore, by (a.2) of the Appendix, we have $f_h/f_j = f_h^\beta/f_j^\beta = (f_h/f_j)^\gamma = f_h/f_j$ for h, $j \in \mathfrak{J}$. Consequently $(f_h^y)_*/(f_j^y)_* = f_h^y/f_j^y = (f_h/f_j) \circ \alpha = (f_h \circ \alpha)/(f_j \circ \alpha)$. Therefore the values of (3.41) are proportional to

$$f_j(\alpha(\Omega_z \cdot {}^t xv, z)) = f_j(\lambda\Omega_z \cdot {}^t xv, \alpha(z)) = f_j(\Omega_{\alpha(z)} \cdot {}^t qv, \alpha(z)).$$

This proves (ii). Since the points $\Theta_\mathfrak{z}(\Omega_\mathfrak{z} v)$ with $v \in (\mathbf{A}_f)^{2n}$ are dense in $A(\mathfrak{z}, \delta)$, we obtain (i) from (ii).

Equality (ii) can also be expressed by the commutative diagram

(3.42)

$$\begin{array}{ccc} \mathbf{Q}^{2n}/L & \xrightarrow{\ \omega\ } & A_{\mathfrak{z}} \\[2pt] {}^t x \downarrow & & \downarrow \sigma \\[2pt] \mathbf{Q}^{2n}/{}^t\alpha L & \xrightarrow{\ \omega'\ } & A_{\alpha(\mathfrak{z})} \end{array}$$

where $L = \begin{pmatrix} 1 & 0 \\ 0 & \delta \end{pmatrix} \mathbf{Z}^{2n}$, $\omega(v) = \Theta_{\mathfrak{z}}(\Omega_{\mathfrak{z}} v)$, and $\omega'(v) = \Theta_{\alpha(\mathfrak{z})}(\Omega_{\alpha(\mathfrak{z})} \cdot {}^t\alpha^{-1} v)$. Notice that ${}^t x L = {}^t\alpha L$; ω and ω' are bijective maps onto the groups of all points of finite order on $A_{\mathfrak{z}}$ and $A_{\alpha(\mathfrak{z})}$, respectively.

We insert here a proposition concerning a non-vanishing of an element of $\mathcal{A}_k(\mathbf{Q}_{ab})$ at a point $\mathfrak{z}$ of the above type, which generalizes [8, Lemma 2.5] and is similar to [5, Prop. 10].

PROPOSITION 3.13. *Suppose that a point $\mathfrak{z}$ of $\mathfrak{H}_n$ and an element x of G_{A+} satisfy (3.40). If $f \in \mathcal{A}_k(\mathbf{Q}_{ab})$, $k \in \mathbf{Z}$, f is finite at $\mathfrak{z}$, and $f(\mathfrak{z}) \neq 0$, then f^x is finite at $\mathfrak{z}$ and $f^x(\mathfrak{z}) \neq 0$.*

Proof. This is obvious if $k = 0$. The case of negative k can be reduced to positive k by taking f^{-1}. Assuming $k > 0$, take any non-zero element g of $\mathcal{M}_k(\mathbf{Q}_{ab})$. Since $G_{\mathbf{Q}+}$ is dense in $G_{\mathbf{R}+}$, we can find an element α of $G_{\mathbf{Q}+}$ so that $g(\alpha(\mathfrak{z})) \neq 0$. Put $h = g^{\alpha x^{-1}}$. Then $h^x(\mathfrak{z}) \neq 0$, $h/f \in \mathcal{A}_0(\mathbf{Q}_{ab})$, and $h/f \in B_{\mathfrak{z}}$. By (3.40), $h^x/f^x = (h/f)^x \in B_{\mathfrak{z}}$, and hence $f^x(\mathfrak{z}) \neq 0$. Applying this result to h^x and x^{-1}, we see that $h(\mathfrak{z}) \neq 0$. Thus $f/h \in B_{\mathfrak{z}}$, so that $f^x/h^x \in B_{\mathfrak{z}}$, which proves that f^x is finite at $\mathfrak{z}$ and $f^x(\mathfrak{z}) \neq 0$.

Let us now briefly show that the main theorem of [7] can actually be derived from Th. 3.10. Let $\mathfrak{z}$ be a point of $\mathfrak{H}_n$ such that $A(\mathfrak{z}, \delta)$ has many complex multiplications in the sense of [7, §2]. Define Y and K' as in [7, p. 684], and let $\Phi: Y \to \mathbf{C}_n^n$ be the anti-representation which gives the action of Y on $A(\mathfrak{z}, \delta)$ as the endomorphism algebra. Define an injection ε of Y into $\mathbf{Q}_{2n}^{2n}$ by $\Phi(a)\Omega_{\mathfrak{z}} = \Omega_{\mathfrak{z}} \cdot {}^t\varepsilon(a)$ for $a \in Y$, and a map $\eta: K_A'^{\times} \to Y_A^{\times}$ by [7, (25)]. Then $\varepsilon \circ \eta$ maps $K_A'^{\times}$ into G_{A+}. Given any $r \in K_A'^{\times}$, put $x = \varepsilon(\eta(r))^{-1}$. By [4, I, (2.7.3); II, (6.2.3)], we have $(B_{\mathfrak{z}})^x = B_{\mathfrak{z}}$, and

(3.43) $$F(\mathfrak{z}) \in K_{ab}' \quad \text{and} \quad F(\mathfrak{z})^r = F^x(\mathfrak{z}) \quad \text{for all } F \in B_{\mathfrak{z}}.$$

Now let $f \in T_{k, \xi}(\mathbf{Q}_{ab})$. Choose any h of $\mathcal{M}_k(\mathbf{Q}_{ab})$ so that $h(\mathfrak{z}) \neq 0$, and put $g(u) = f_*(u, \mathfrak{z})/h(\mathfrak{z})$ for $u \in \mathbf{C}_s^n$. Then

$$g(\Omega_{\mathfrak{z}} v) = f_*(\Omega_{\mathfrak{z}} v, \mathfrak{z})/h(\mathfrak{z}) \in K_{ab}'$$

58   G. SHIMURA

for all $v \in \mathbf{Q}_s^{2n}$ by (3.43), since $f_*(\Omega_z v, z) \in \mathcal{M}_k(\mathbf{Q}_{ab})$. Take an element y of $\mathfrak{G}$ lying above x, and put $g'(u) = (f^y)_*(u, \mathfrak{z})/h^y(\mathfrak{z})$. This is meaningful since $h^y(\mathfrak{z}) \neq 0$ by Prop. 3.13. By (3.43) and Th. 3.10, we have

$$g(\Omega_\mathfrak{z} v)^r = [f_*(\Omega_z v, z)/h(z)]^x \big|_{z=\mathfrak{z}} = (f^y)_*(\Omega_\mathfrak{z} \cdot {}^t x v, \mathfrak{z})/h^y(\mathfrak{z})$$
$$= g'(\Omega_\mathfrak{z} \cdot {}^t x v) = g'(\Phi(\eta(r)^{-1})\Omega_\mathfrak{z} v),$$

and thus

(3.44) $\qquad g(u)^r = g'(\Phi(\eta(r)^{-1})u) \quad$ for all $u \in \Omega_\mathfrak{z} \mathbf{Q}_s^{2n} \quad$ and all $r \in K_\mathbf{A}'^\times$.

When $s = 1$ and $k = 1/2$, this is exactly the fundamental relation needed for the proof of the main theorem of [7]. Since the functions $h(\mathfrak{z})^{-1}f^*$ span the space of "arithmetic theta functions" as shown by [7, Prop. 2.5], we obtain the theorem.

4. Proof of Proposition 3.8

Given a lattice Y and an element ζ in the vector space $\{\eta \in \mathbf{Q}_n^n \mid {}^t\eta = \eta\}$ and a lattice L in $\mathbf{Q}_n^s$, consider all formal series of the form

(4.1) $\qquad p(u, z) = \sum_{\zeta \leqslant \eta \in Y} \sum_{\lambda \in L} a(\eta, \lambda) e_n(\eta z + u\lambda) \quad (u \in \mathbf{C}_s^n, z \in \mathfrak{H}_n)$

with complex coefficients $a(\eta, \lambda)$ satisfying the following condition:

(4.2) $\qquad$ *For a fixed η, there are only finitely many λ such that $a(\eta, \lambda) \neq 0$.*

As proved in Prop. 3.1, the components of an element of $T_{\tau, k, \xi}$ satisfy (4.2). Let Φ denote the set of all such formal series, with all choices of Y and L. Obviously Φ forms a ring. Take linearly independent positive definite symmetric elements $\zeta_1, \ldots, \zeta_M$ of $\mathbf{Q}_n^n$, where $M = n(n+1)/2$. To each (η, λ) we can assign its "coordinates"

$$(\mathrm{tr}\,(\zeta_1\eta), \ldots, \mathrm{tr}\,(\zeta_M\eta), \lambda_{11}, \lambda_{12}, \ldots, \lambda_{sn}).$$

By means of these coordinates, we introduce a lexicographic order into the set of all (η, λ). Then, for $0 \neq p \in \Phi$, the "first" non-vanishing coefficient of p is meaningful. Therefore Φ is an integral domain.

LEMMA 4.1. *Suppose $p, q, r \in \Phi$, $pq = r$, $p \neq 0$, p has coefficients in a subfield F of $\mathbf{C}$, and r has coefficients in a vector space W over F. Then q has coefficients in W.*

This can be easily proved by means of the above lexicographic order.

Our proof of Prop. 3.8 needs special care when $n = 1$, since the holomorphy on $\mathfrak{H}_1$ does not guarantee the finiteness at cusps. To avoid the difficulty, let $S_{k, \xi}(\Delta, \Lambda)$ denote the set of all holomorphic f satisfying (3.10) and (3.11) with trivial τ such that $f_*(\Omega_z v, z)$ belongs, as a function of z, to $\mathcal{A}_k(\mathbf{C})$ for all $v \in Q_s^{2n}$. Then we define $S_{k,\xi}(\mathfrak{R})$ for a subfield $\mathfrak{R}$ of $\mathbf{C}$ to be the set of all such f, with any possible Δ and Λ, and with Fourier coefficients (i.e., $c(\eta, \lambda)$ of (3.12)) in $\mathfrak{R}$. Obviously $S = T$ if $n > 1$.

LEMMA 4.2. *Let $\mathcal{L}_k(\mathfrak{R})$ denote the set of all elements of $\mathcal{A}_k(\mathfrak{R})$ holomorphic on $\mathfrak{H}_1$ (but not necessarily so at cusps). Then $\mathcal{L}_k(\mathbf{C}) = \mathcal{L}_k(\mathbf{Q}) \otimes_\mathbf{Q} \mathbf{C}$.*

Proof. Given $f \in \mathcal{L}_k(\mathbf{C})$, consider a cusp form

$$\eta(z) = e(z/24) \prod_{n=1}^{\infty} (1 - e(nz)),$$

and observe that $\eta^{2h} f \in \mathcal{M}_{k+h}(\mathbf{C})$ for a suitable positive integer h. Therefore $\eta^{2h} f$ is a finite $\mathbf{C}$-linear combination of elements of $\mathcal{M}_{k+h}(\mathbf{Q})$, Dividing this by η^{2h}, we obtain our lemma.

LEMMA 4.3. *If $n = 1$ and $f \in S_{k,\xi}(\mathbf{C})$, then*

$$(4.3) \qquad f(u, z) = \sum_{j=1}^{N} c_j(z) F_j(u, z)$$

with $c_j \in \mathcal{L}_{k-(r/2)}(\mathbf{C})$ and $F_j \in T_{r/2,\xi}(\mathbf{Q})$, where $r = \mathrm{rank}\,(\xi)$, and hence f has a Fourier expansion of type (4.1).

Proof. The proof of Prop. 3.5 shows at least an expression (4.3) for f with c_j holomorphic on $\mathfrak{H}_1$ and functions $F_j \in T_{r/2,\xi}(\mathbf{Q})$ such that $F_j(u, z_0)$ for $j = 1, \ldots, N$ are linearly independent over $\mathbf{C}$ for every z_0. Put, for each $v \in Q_s^2$ and for any fixed z_0,

$$X(v) = \left\{ (b_1, \ldots, b_N) \in \mathbf{C}_N^1 \,\middle|\, \sum_{j=1}^{N} b_j F_j(\Omega_{z_0} v, z_0) = 0 \right\}.$$

The intersection of all such $X(v)$ is $\{0\}$, and hence we can find N elements $v_1, \ldots, v_N$ of Q_s^2 such that $\bigcap_{i=1}^{N} X(v_i) = \{0\}$. This implies that

$$\det \left((F_j)_*(\Omega_z v_i, z) \right)_{i,j} \neq 0.$$

Denote this determinant by $D(z)$. The equations

$$f_*(\Omega_z v_i, z) = \sum_{j=1}^{N} c_j(z) (F_j)_*(\Omega_z v_i, z) \quad (i = 1, \ldots, N)$$

show that $Dc_j \in \mathcal{A}_l(\mathbf{C})$ with $l = k + (N-1)r/2$. Since $D \in \mathcal{M}_{rN/2}(\mathbf{C})$, we obtain the desired conclusion.

Now, to prove Prop. 3.8, we may assume, in view of Prop. 3.3, that ξ is non-degenerate. Obviously any elements of $T_{\tau,k,\xi}(\mathbf{Q})$ linearly independent over $\mathbf{Q}$ are also linearly independent over $\mathbf{C}$. Therefore our aim is to show that $T_{\tau,k,\xi}$ is spanned by $T_{\tau,k,\xi}(\mathbf{Q})$ over $\mathbf{C}$. Let $f \in T_{\tau,k,\xi}$ and let W_f be the vector space spanned over $\mathbf{Q}$ by the Fourier coefficients of the components of f. Let us first prove

$$(4.4) \qquad W_f \text{ is finite-dimensional over } \mathbf{Q};$$

$$(4.5) \quad \text{If } \{r_\nu\} \text{ is a basis of } W_f \text{ over } \mathbf{Q} \text{ and } \tau \text{ is trivial, then } f = \sum_\nu r_\nu g_\nu \text{ with } g_\nu \in S_{k,\xi}(\mathbf{Q}).$$

If τ is trivial, we see, from Prop. 3.5 and Prop. 2.1 when $n > 1$ and from Lemmas 4.2 and 4.3 when $n = 1$, that f is a finite sum $\sum t_\varkappa f_\varkappa$ with $t_\varkappa \in \mathbf{C}$ and $f_\varkappa \in S_{k,\xi}(\mathbf{Q})$. Therefore W_f is contained in $\sum \mathbf{Q} t_\varkappa$ and hence finite-dimensional. Given a basis $\{r_\nu\}$ of W_f, take a basis of $\sum \mathbf{Q} t_\varkappa$ of the form $\{r_\nu\} \cup \{r'_\mu\}$ and express each $t_\varkappa$ as a $\mathbf{Q}$-linear combination of r_ν and r'_μ. Then $f = \sum r_\nu g_\nu + \sum r'_\mu g'_\mu$ with g_ν and g'_μ in $S_{k,\xi}(\mathbf{Q})$. Since f has coefficients in W_f, comparison of coefficients shows that g'_μ must be 0, and thus $f = \sum r_\nu g_\nu$, which proves (4.5).

In the general case with an arbitrary τ, take A as in Prop. 3.7 with any point (u_0, z_0), and put $d = \det(A)$, $B = d \cdot {}^t A^{-1}$, and $g = {}^t B f$ for a given $f \in T_{\tau,k,\xi}$. Then the components of g belong to $T_{l,m\xi}$ with some l, and $W_g \subset W_f$. By Lemma 4.1, the relation $df = Ag$ shows that $W_f \subset W_g$, so that $W_f = W_g$. Since (4.4) is true for trivial τ, we have (4.4) for an arbitrary τ. Let $\{r_\nu\}$ be a basis of W_f over $\mathbf{Q}$. By (4.5), $g = \sum r_\nu h_\nu$ with $h_\nu \in S_{l,m\xi}(\mathbf{Q})^m$, so that $f = \sum r_\nu d^{-1} A h_\nu$. Note that $d(u_0, z_0) \neq 0$. Given another point (u_1, z_1), we take A_1, B_1, and d_1 similarly so that $d_1(u_1, z_1) \neq 0$, and obtain an expression $f = \sum r_\nu d_1^{-1} A_1 h_{1\nu}$ with $h_{1\nu} \in S_{l,m\xi}(\mathbf{Q})^m$ and with the same r_ν. Obviously

$$\sum_\nu r_\nu (dA_1 h_{1\nu} - d_1 A h_\nu) = 0.$$

Now $dA_1 h_{1\nu}$ and $d_1 A h_\nu$ have Fourier expansions of type (4.1). Since the components of A, A_1, d, and d_1 satisfy (4.2), each sum expressing a Fourier coefficient of $dA_1 h_{1\nu}$ or $d_1 A h_\nu$ in terms of those of d, d_1, A, A_1, h_ν, and $h_{1\nu}$ is a finite sum, so that $dA_1 h_{1\nu}$ and $d_1 A h_\nu$ have coefficients in $\mathbf{Q}^m$. Since $\{r_\nu\}$ are linearly independent over $\mathbf{Q}$, we have $dA_1 h_{1\nu} = d_1 A h_\nu$, and so $d^{-1} A h_\nu = d_1^{-1} A_1 h_{1\nu}$. This shows that $d^{-1} A h_\nu$ is holomorphic on the whole $\mathbf{C}_s^n \times \mathfrak{H}_n$. Obviously it satisfies (3.10) and (3.11). For every automorphism σ of $\mathbf{C}$, define a holomorphic function f_σ by $f_\sigma = \sum_\nu r_\nu^\sigma d^{-1} A h_\nu$. Both df and $A h_\nu$ have Fourier expansions of type (3.12). Define the action of σ on any such series by (3.37), which, in general, is only a formal series. Since $A h_\nu$ has coefficients in $\mathbf{Q}^m$ for the same reason as above, we have $(df)^\sigma = \sum r_\nu^\sigma A h_\nu = df_\sigma$, and thus $(df)^\sigma$ is meaningful as a function. Let t be an element of

$\mathbf{Z}_l^{\times}$ whose action on $\mathbf{Q}_{ab}$ coincides with σ. Then the proof of Prop. 3.9 is applicable to both d and df, so that for $v \in \mathbf{Q}_s^{2n}$, we have

$$d_{*}(\Omega_z \iota(t) v, z) = d_{*}(\Omega_z v, z)^{\sigma},$$

$$(df_{\sigma})_{*}(\Omega_z \iota(t) v, z) = ((df)^{\sigma})_{*}(\Omega_z \iota(t) v, z) = (df)_{*}(\Omega_z v, z)^{\sigma}.$$

Since $(df)_{*} = d_{*} f_{*}$ and $(df_{\sigma})_{*} = d_{*}(f_{\sigma})_{*}$, we have

(4.6) $$(f_{\sigma})_{*}(\Omega_z \iota(t) v, z) = f_{*}(\Omega_z v, z)^{\sigma}$$

if $d_{*}(\Omega_z v, z) \neq 0$. For every $v \in \mathbf{Q}_s^{2n}$, we can take A so that $d_{*}(\Omega_z v, z) \neq 0$, and so we have (4.6) for all $v \in \mathbf{Q}_s^{2n}$. (Note that f_{σ} is independent of the choice of A.) This shows that f_{σ} satisfies (3.23), i.e., $f_{\sigma} \in T_{\tau, k, \xi}$. Now, by Lemma 4.4 below, we can find a set of automorphisms $\{\sigma\}$ such that $\det(r_{\nu}^{\sigma})_{\sigma, \nu} \neq 0$. Then the relations $f_{\sigma} = \sum r_{\nu}^{\sigma} d^{-1} A h_{\nu}$ show that $d^{-1} A h_{\nu} \in T_{\tau, k, \xi}$. By Lemma 4.1, we obtain $d^{-1} A h_{\nu} \in T_{\tau, k, \xi}(\mathbf{Q})$, which completes the proof.

LEMMA 4.4. *If $r_1, \ldots, r_N$ are N complex numbers linearly independent over $\mathbf{Q}$, then there is a set $\{\sigma\}$ of N automorphisms of $\mathbf{C}$ such that $\det(r_{\nu}^{\sigma})_{\sigma, \nu} \neq 0$.*

Proof. Let H denote the linear subspace of $\mathbf{C}_N^1$ consisting of all $(x_1, \ldots, x_N)$ of $\mathbf{C}_N^1$ such that $\sum_{\nu=1}^{N} r_{\nu}^{\sigma} x_{\nu} = 0$ for all automorphisms σ of $\mathbf{C}$. Obviously H is stable under the map $(x_1, \ldots, x_N) \mapsto (x_1^{\sigma}, \ldots, x_N^{\sigma})$ for all σ. Therefore H is defined over $\mathbf{Q}$. If $H \neq \{0\}$, H contains a non-zero vector of $\mathbf{Q}_N^1$, which contradicts the linear independence of $\{r_{\nu}\}$. Hence $H = \{0\}$, from which our assertion follows.

5. Proof of Theorem 1.2

We shall now complete the proof of Theorem 1.2 in the case of non-trivial ϱ. Observe that the theory of § 3 concerning $T_{k, \xi}$ is independent of Theorem 1.2 with non-trivial ϱ. Consider $T_{k, \xi}$ with $s = 1$ and $0 < \xi \in \mathbf{Q}$. Given an element $f(u, z)$ of $T_{k, \xi}$, define a $\mathbf{C}^n$-valued function Df on $\mathfrak{H}_n$ by

(5.1) $$(Df)(z) = \frac{1}{2\pi i} \begin{pmatrix} (\partial f/\partial u_1)(0, z) \\ \vdots \\ (\partial f/\partial u_n)(0, z) \end{pmatrix} \qquad (z \in \mathfrak{H}_n).$$

We see easily that, for every $\beta = (\alpha, \psi) \in \mathfrak{G}_{\mathbf{Q}}$ with $\alpha = \begin{pmatrix} a & b \\ c & d \end{pmatrix}$,

(5.2) $$(Df)|_k \beta = (cz + d) D(f^{\beta}),$$

and hence $Df \in \mathfrak{M}_{\varrho, k}(\mathbf{C})$, where $\varrho(X) = X$. If σ is an automorphism of $\mathbf{C}$, we have obviously

(5.3) $$(Df)^{\sigma} = D(f^{\sigma}).$$

Given n elements $f_1, ..., f_n$ of $T_{k,\xi}$, we consider a $\mathbf{C}_n^1$-valued function $\mathfrak{f} = (f_1, ..., f_n)$ on $\mathbf{C}_1^n \times \mathfrak{H}_n$, and put

$$(5.4) \qquad\qquad D\mathfrak{f} = (Df_1, ..., Df_n).$$

Formulas (5.2) and (5.3) hold also with $\mathfrak{f}$ in place of f. We call $\mathfrak{f}$ *regular* if $\det(D\mathfrak{f})$ is not identically equal to 0.

PROPOSITION 5.1. (i) *For every* $z_0 \in \mathfrak{H}_n$, *there exists a regular* $\mathfrak{f} = (f_1, ..., f_n)$ *with* f_ν *in* $T_{1/2,1}(\mathbf{Q})$ *such that* $\det(D\mathfrak{f})(z_0) \neq 0$.

(ii) *Suppose that* $f_\nu \in T_{k,\xi}(\mathbf{Q})$ *and* $\mathfrak{f} = (f_1, ..., f_n)$ *is regular. Then* $\mathfrak{f}^y = (f_1^y, ..., f_n^y)$ *is regular for every* $y \in \mathfrak{G}$.

(iii) *If* $\mathfrak{f}$ *is as in* (ii) *and* $\mathfrak{h} = (h_1, ..., h_n)$ *with* $h_\nu \in T_{k,\xi}(\mathbf{Q}_{ab})$, *then* $D(\mathfrak{f}^y)^{-1} D(\mathfrak{h}^y) = ((D\mathfrak{f})^{-1} D\mathfrak{h})^y$ *for every* $y \in \mathfrak{G}$.

(*In the last relation,* $(D\mathfrak{f})^{-1} D\mathfrak{h}$ *has components in* $\mathcal{A}_0(\mathbf{Q}_{ab})$, *so that* $((D\mathfrak{f})^{-1} D\mathfrak{h})^y$ *is meaningful.*)

Proof. Assertion (i) follows immediately from [8, Prop. 1.2] and its proof. Assertion (ii) is obvious, since $\mathfrak{f}^y = \mathfrak{f}^\beta$ for some $\beta \in \mathfrak{G}_\mathbf{Q}$, as shown in the proof of Th. 3.10. To prove (iii), put $E = (D\mathfrak{f})^{-1} D\mathfrak{h}$. We can find $\sigma \in \mathrm{Gal}(\mathbf{Q}_{ab}/\mathbf{Q})$ and $\beta \in \mathfrak{G}_\mathbf{Q}$ such that $\mathfrak{f}^y = \mathfrak{f}^\beta$, $\mathfrak{h}^y = (\mathfrak{h}^\sigma)^\beta$, and $E^y = (E^\sigma)^\beta$. Then

$$((D\mathfrak{f})^{-1} D\mathfrak{h})^y = (((D\mathfrak{f})^{-1} D\mathfrak{h})^\sigma)^\beta = (((D\mathfrak{f})^\sigma)^\beta)^{-1} ((D\mathfrak{h})^\sigma)^\beta$$
$$= D((\mathfrak{f}^\sigma)^\beta)^{-1} D((\mathfrak{h}^\sigma)^\beta) = D(\mathfrak{f}^y)^{-1} D(\mathfrak{h}^y)$$

by (5.2) and (5.3). Q.E.D.

PROPOSITION 5.2. *For every rational representation* ϱ *of* $GL_n(\mathbf{Q})$ *and* $k \in 2^{-1}\mathbf{Z}$, *we have* $\mathcal{M}_{\varrho,k}(\mathbf{C}) = \mathcal{M}_{\varrho,k}(\mathbf{Q}) \otimes_\mathbf{Q} \mathbf{C}$.

Proof. Let m be the degree of ϱ. For every $z_0 \in \mathfrak{H}_n$, there exist a positive integer j and a $\mathbf{C}_m^m$-valued function A on $\mathfrak{H}_n$ such that:

(i) the columns of A belong to $\mathcal{M}_{\varrho,j}(\mathbf{Q})$;

(ii) $\det(A(z_0)) \neq 0$.

This follows either from Prop. 3.7 or from (i) of Prop. 5.1. With any such A and given $f \in \mathcal{M}_{\varrho,k}(\mathbf{C})$, put $d = \det(A)$, $B = d \cdot {}^t A^{-1}$, and $g = {}^t Bf$. Then the components of g belong to $\mathcal{M}_t(\mathbf{C})$ for some t. Since $\mathcal{M}_t(\mathbf{C}) = \mathcal{M}_t(\mathbf{Q}) \otimes_\mathbf{Q} \mathbf{C}$ by Prop. 2.1, we obtain our assertion by the same type of reasoning as in the proof of Prop. 3.8.

This proposition shows that the action of an automorphism σ of C on $\mathcal{M}_{\varrho,k}(\mathbf{C})$ defined by (1.4) actually maps $\mathcal{M}_{\varrho,k}(\mathbf{C})$ onto itself.

To define the action of $G_{\mathbf{A}+}$ on $\mathcal{A}_{\varrho}(\mathbf{Q}_{ab})$, first assume $\varrho(X)=X$. Choose any regular $\mathfrak{f}=(f_1, ..., f_n)$ with f_ν in $T_{1/2,1}(\mathbf{Q})$. Given $g \in \mathcal{A}_{\varrho}(\mathbf{Q}_{ab})$, observe that the components of $(D\mathfrak{f})^{-1}g$ belong to $\mathcal{A}_{1/2}(\mathbf{Q}_{ab})$, so that $((D\mathfrak{f})^{-1}g)^y$ is meaningful for $y \in \mathfrak{G}$. Now for $x \in G_{\mathbf{A}+}$, define g^x by $g^x = D(\mathfrak{f}^y)((D\mathfrak{f})^{-1}g)^y$ with any element y of $\mathfrak{G}$ with projection x on $G_{\mathbf{A}+}$ of the form $y=(x, 1, v)$. This is independent of the choice of y, and also of the choice of $\mathfrak{f}$ because of (iii) of Prop. 5.1. If $\varrho(X) = X \otimes ... \otimes X$, then considering $D\mathfrak{f} \otimes ... \otimes D\mathfrak{f}$ instead of $D\mathfrak{f}$, we can similarly define the action of $G_{\mathbf{A}+}$ on $\mathcal{A}_{\varrho}(\mathbf{Q}_{ab})$. Since every irreducible rational representation of $GL_n(\mathbf{Q})$ is obtained as a suitable power of determinant times a $\mathbf{Q}$-rational constituent of such a tensor representation, it is now easy to define the action of $G_{\mathbf{A}+}$ on $\mathcal{A}_{\varrho}(\mathbf{Q}_{ab})$ with an arbitrary ϱ. Then the properties (i–viii) of Theorem 1.2 can be verified in a straightforward way. In particular, to prove the associativity (iii), we need (iii) of Prop. 5.1.

6. Partial Fourier expansions of modular forms

Let n and s be positive integers. We shall now show that a certain Fourier expansion of a modular form on $\mathfrak{H}_{n+s}$ yields naturally some elements of $T_{\tau,k,\xi}$. First we write the variable point Z of $\mathfrak{H}_{n+s}$ in the form

$$(6.1) \qquad Z = \begin{pmatrix} z & u \\ {}^t u & w \end{pmatrix} \quad \text{with } z \in \mathfrak{H}_n,\ w \in \mathfrak{H}_s,\ u \in \mathbf{C}_s^n.$$

We define the objects G, $\mathfrak{G}$, $\mathcal{A}_{\varrho,k}$, $\mathcal{M}_{\varrho,k}$, Γ_N, etc. with degree $n+s$ instead of n, and denote them by G', $\mathfrak{G}'$, $\mathcal{A}'_{\varrho,k}$, $\mathcal{M}'_{\varrho,k}$, Γ'_N, etc. Let us now consider the elements of $G'_{\mathbf{Q}}$ of the form

$$(6.2) \qquad \gamma = \begin{pmatrix} 1_n & 0 & 0 & q \\ {}^t p & 1_s & {}^t q & r \\ 0 & 0 & 1_n & -p \\ 0 & 0 & 0 & 1_s \end{pmatrix}, \quad p \in \mathbf{Q}_s^n,\ q \in \mathbf{Q}_s^n,\ r \in \mathbf{Q}_s^s.$$

Such a γ belongs to $G'_{\mathbf{Q}}$ if and only if ${}^t qp + r$ is symmetric. For Z as in (6.1), we have

$$(6.3) \qquad \gamma(Z) = \begin{pmatrix} z & u + zp + q \\ {}^t u + {}^t pz + {}^t q & w + {}^t pzp + {}^t pu + {}^t up + {}^t qp + r \end{pmatrix}.$$

Let τ be a polynomial representation $GL_{n+s}(\mathbf{Q}) \to GL_m(\mathbf{Q})$ as in § 3, and let $F(Z) = F(z, u, w)$ be an element of $\mathcal{M}'_{\tau,k}(\mathbf{C})$ with $0 \leqslant k \in 2^{-1}\mathbf{Z}$. Then we have a Fourier expansion of the form

$$(6.4) \qquad F(z, u, w) = \sum_{\xi} f_{\xi}(u, z)\, e_s(\tfrac{1}{2}\xi w)$$

with holomorphic $\mathbf{C}^m$-valued functions $f_\xi(u, z)$ on $\mathbf{C}^n_s \times \mathfrak{H}_n$, where ξ runs over non-negative elements of a lattice in $\{\xi \in \mathbf{Q}^s_s \mid {}^t\xi = \xi\}$. Notice that f_ξ is defined on the whole $\mathbf{C}^n_s \times \mathfrak{H}_n$, though $F(Z)$ is defined only for $\mathrm{Im}\,(Z) > 0$. Given $\alpha = \begin{pmatrix} a & b \\ c & d \end{pmatrix} \in G$, define an element α' of G' by

$$(6.5) \qquad \alpha' = \begin{pmatrix} a & 0 & b & 0 \\ 0 & 1_s & 0 & 0 \\ c & 0 & d & 0 \\ 0 & 0 & 0 & \nu(\alpha)\,1_s \end{pmatrix}.$$

The map $\alpha \mapsto \alpha'$ is a $\mathbf{Q}$-rational injection, so that it can be extended to a continuous map of $G_\mathbf{A}$ into $G'_\mathbf{A}$. With $\alpha \in G_{\mathbf{Q}+}$ and α' as in (6.5), we have

$$(6.6) \qquad \alpha'(Z) = \begin{pmatrix} \alpha(z) & {}^t(cz + d)^{-1} u \\ {}^tu(cz + d)^{-1} & \nu(\alpha)^{-1} w - \nu(\alpha)^{-1} \cdot {}^tu(cz + d)^{-1} cu \end{pmatrix},$$

$$(6.7) \qquad \begin{pmatrix} c & 0 \\ 0 & 0 \end{pmatrix} Z + \begin{pmatrix} d & 0 \\ 0 & \nu(\alpha)\,1_s \end{pmatrix} = \begin{pmatrix} cz + d & cu \\ 0 & \nu(\alpha)\,1_s \end{pmatrix},$$

$$(6.8) \qquad F(\alpha'(Z)) = \sum_\xi f_\xi(\alpha(u, z))\, e_s(-\tfrac{1}{2}\nu(\alpha)^{-1}\xi \cdot {}^tu(cz + d)^{-1} cu)\, e_s(\tfrac{1}{2}\nu(\alpha)^{-1}\xi w).$$

If ι' is the map of $\mathbf{A}^\times_+$ into $G'_{\mathbf{A}+}$ corresponding to ι, we see that

$$(6.9) \qquad \iota(t)' = \iota'(t) \quad (t \in \mathbf{A}^\times_+).$$

Let $(\alpha, \psi) \in \mathfrak{G}_\mathbf{Q}$ with $\alpha = \begin{pmatrix} a & b \\ c & d \end{pmatrix}$. Define a holomorphic function ψ' on $\mathfrak{H}_{n+s}$ by

$$(6.10) \qquad \psi'\begin{pmatrix} z & u \\ {}^tu & w \end{pmatrix} = \nu(\alpha)^{s/2}\psi(z).$$

From (6.7), we know that $(\alpha', \psi') \in \mathfrak{G}'_\mathbf{Q}$, and thus obtain an injection

$$(6.11) \qquad (\alpha, \psi) \mapsto (\alpha', \psi')$$

of $G_\mathbf{Q}$ into $G'_\mathbf{Q}$.

A good example of Fourier expansion (6.4) can be obtained from theta functions as follows. Let θ' denote the function defined by (3.26) with $n + s$ instead of n, and put $\Phi(Z) = \theta'(0, Z; p, q)$ with

$$p = \begin{pmatrix} p_1 \\ p_2 \end{pmatrix}, \; q = \begin{pmatrix} q_1 \\ q_2 \end{pmatrix} \quad (p_1, q_1 \in \mathbf{Q}^n; \; p_2, q_2 \in \mathbf{Q}^s).$$

Then

$$\Phi(Z) = \sum_{y - p_2 \in \mathbf{Z}^s} e({}^tyq_2)\, \theta(uy, z; p_1, q_1)\, e(\tfrac{1}{2} \cdot {}^tywy).$$

Thus the coefficients f_ξ for Φ are linear combinations of $\theta(uy, z; p_1, q_1)$. In particular, if we put $\theta'(Z) = \theta'(0, Z; 0, 0)$, then

$$(6.12) \qquad \theta'(Z) = \sum_{y \in Z^s} \theta(uy, z; 0, 0)\, e(\tfrac{1}{2} \cdot {}^t y w y).$$

Define Γ_θ and $\lambda(\gamma)$ for $\gamma \in \Gamma_\theta$ by (a.1) and (a.2) of the Appendix, and similarly a subgroup Γ'_θ of $G'_{\mathbf{Q}}$ and $\lambda'(\beta)$ for $\beta \in \Gamma'_\theta$. From (6.8), (6.12), and (a.2), we see easily that $\lambda(\gamma) = \lambda'(\gamma')$ for every $\gamma \in \Gamma_\theta$, or more precisely,

$$(6.13) \qquad \theta'(\gamma'(Z))/\theta'(Z) = \theta(\gamma(z))/\theta(z) \quad \text{for every } \gamma \in \Gamma_\theta.$$

Therefore, if Δ_N and Δ'_N are the images of Γ_N and Γ'_N under the map (2.7) (respectively with n and $n+s$), then the map (6.11) of $\mathfrak{G}_{\mathbf{Q}}$ into $\mathfrak{G}'_{\mathbf{Q}}$ sends Δ_N into Δ'_N.

PROPOSITION 6.1. *Suppose* $F \in \mathfrak{M}'_{\tau,k}(\Gamma'_N, \mathfrak{R})$ *or* $F \in \mathfrak{M}'_{\tau,k}(\Delta'_N, \mathfrak{R})$ *according as* $2k$ *is even or odd, where* $\mathfrak{R}$ *is a subfield of* $\mathbf{C}$. *Then the Fourier coefficient* f_ξ *of* F *defined by* (6.4) *belongs to* $T_{\tau,k,\xi}(\Gamma_N, \Lambda; \mathfrak{R})$ *or* $T_{\tau,k,\xi}(\Delta_N, \Lambda; \mathfrak{R})$ *accordingly, with a suitable* Λ.

Proof. Let $F \in \mathfrak{M}'_{\tau,k}(\Delta'_N, \mathfrak{R})$. If $\delta = (\gamma, 1) \in \mathfrak{G}_{\mathbf{Q}}$ with γ defined by (6.2) and ${}^t q p + r = 0$, formula (6.3) shows that

$$F^\delta(z, u, w) = \tau \begin{pmatrix} 1_n & p \\ 0 & 1_s \end{pmatrix} \sum_\xi e_s(\xi(\tfrac{1}{2} \cdot {}^t p z p + {}^t p u)) f_\xi(u + zp + q, z) e_s(\tfrac{1}{2} \xi w).$$

If $\beta = (\alpha, \psi) \in \mathfrak{G}_{\mathbf{Q}}$ and $\beta' = (\alpha', \psi') \in \mathfrak{G}'_{\mathbf{Q}}$ with α' of (6.5) as above, we obtain, from (6.8),

$$(6.14) \qquad F^{\beta'}(z, u, w) = \nu(\alpha)^{-ks} \sum_\xi (f_\xi)^\beta (u, z) e_s(\tfrac{1}{2} \nu(\alpha)^{-1} \xi w).$$

Therefore f_ξ satisfies (3.10) and (3.11) with a suitable Λ and $\Delta = \Delta_N$. Obviously f_ξ has Fourier coefficients in $\mathfrak{R}$, and hence $f_\xi \in T_{\tau,k,\xi}(\Delta_N, \Lambda; \mathfrak{R})$ if $n > 1$. Suppose $n = 1$. By (3.18) and (3.19), we have

$$F^{\beta'\delta}(z, 0, w) = \nu(\alpha)^{-ks} \tau \begin{pmatrix} 1_n & 0 \\ 0 & \nu(\beta) 1_s \end{pmatrix}^{-1} \sum_\xi e_s(-\tfrac{1}{2}\nu(\alpha)^{-1}\xi \cdot {}^t pq) (f_\xi)_*(\Omega_z v, z)^\beta e_s(\tfrac{1}{2}\nu(\alpha)^{-1}\xi w)$$

where ${}^t \alpha v = \begin{pmatrix} p \\ q \end{pmatrix}$. Since $F^{\beta'\delta} \in \mathfrak{M}'_{\tau,k}$, this shows that $(f_\xi)_*(\Omega_z v, z)^\beta$ is finite at $i\infty$ for every $\beta \in \mathfrak{G}_{\mathbf{Q}}$ and every $v \in \mathbf{Q}_s^2$. Therefore $f_\xi \in T_{\tau,k,\xi}$.

THEOREM 6.2. *There is a unique continuous injective homomorphism* $y \mapsto y'$ *of* $\mathfrak{G}$ *into* $\mathfrak{G}'$ *which coincides with* $\iota_1(t) \mapsto \iota'_1(t)$ *on* $\iota_1(\mathbf{A}_+^\times)$ *and with* (6.11) *on* $\mathfrak{G}_{\mathbf{Q}}$, *and which makes the diagram*

5 − 782901 *Acta mathematica* 141. Imprimé le 1 Septembre 1978

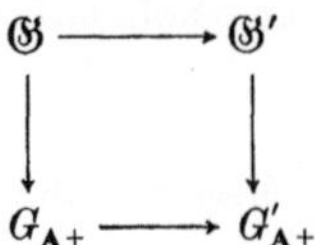

commutative. Moreover, if F is an element of $\mathcal{M}'_{\tau,k}(\mathbf{Q}_{ab})$ with expansion (6.4), then, for every $y \in \mathfrak{G}$, we have

$$(6.15) \qquad F^{y'}(z, u, w) = |\nu(y)|^{-ks} \sum_{\xi} (f_{\xi})^y (u, z)\, e_s(\tfrac{1}{2} |\nu(y)|^{-1} \xi w),$$

where $(f_\xi)^y$ is defined as in Theorem 3.10. If $k \in \mathbf{Z}$, (6.15) holds with $y \in G_{\mathbf{A}+}$ and $y' \in G'_{\mathbf{A}+}$.

Proof. If $y = (\alpha, \psi)$ and $y' = (\alpha', \psi')$ as in (6.11), we obtain (6.15) from (6.14). Next (6.15) is obvious if $y = \iota_1(t)$ and $y' = \iota'_1(t)$ with $t \in \mathbf{Z}_{\mathfrak{f}}^\times$. Now, given $y = (x, v) \in \mathfrak{G}$, an element y' of $\mathfrak{G}'$ with projection x' on $G'_{\mathbf{A}+}$ is uniquely determined by (6.15), if it exists. Let $\mathfrak{P}$ be the subgroup of $\mathfrak{G}$ generated by $\iota_1(\mathbf{Z}_{\mathfrak{f}}^\times)$, $G_{\infty+}$, and $\mathfrak{G}_{\mathbf{Q}}$. Then we can define an injective homomorphism $y \mapsto y'$ of $\mathfrak{P}$ into $\mathfrak{G}'$ which satisfies (6.15) and makes the diagram

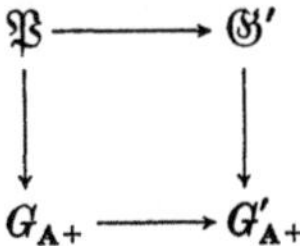

commutative. This homomorphism is continuous, since a basis of neighborhoods of the identity element of $\mathfrak{G}$ can be given by (2.9). By Prop. 2.2, $\mathfrak{P}$ is dense in $\mathfrak{G}$, and $\mathfrak{G}'$ is locally compact, and hence the map can be extended to the whole $\mathfrak{G}$ as desired.

In the above, we assumed $n > 0$, but the ordinary Fourier expansion $F(w) = \sum_\xi f_\xi e_s(\xi w)$ of a modular form F on $\mathfrak{H}_s$ with $f_\xi \in \mathbf{C}$ may be regarded as the extreme case $n = 0$. Then (2.6) or (1.10) may be considered the extreme case of (6.15) with the map ι of $\mathbf{A}^\times$ into $G'_{\mathbf{A}}$ in place of the injection $G_{\mathbf{A}} \to G'_{\mathbf{A}}$.

7. Concluding remarks

A more informative title of this section may be "What Theorems 3.10 and 6.2 suggest in the general theory of arithmetic automorphic forms". Let G be a $\mathbf{Q}$-simple algebraic group such that $G_{\mathbf{R}}$ modulo a maximal compact subgroup is a symmetric domain $\mathfrak{S}$, and let Γ be a congruence subgroup of $G_{\mathbf{Q}}$. Our main interest here is in the nature of the Fourier expansion of an automorphic form on $\mathfrak{S}$ with respect to Γ, when $\mathfrak{S}/\Gamma$ is not compact. If $\mathfrak{S}$ is a tube domain and G has a sufficiently large subgroup acting on $\mathfrak{S}$ as a group of translations, an automorphic form has a Fourier expansion similar to the expansion (1.3) of a Siegel modular form. In general, however, $\mathfrak{S}$ may not be a tube domain; even if $\mathfrak{S}$ is a tube domain, the group of translations may not be large enough to guarantee such an expansion. In order to study this point in detail, Pyatetskii-Shapiro introduced in [3] the notion of Siegel domains of the first, second, and third kinds, and discussed a certain Fourier expansion for an automorphic form, which he called a Fourier–Jacobi series. When $\mathfrak{S}$ is represented as a domain of the third kind, the series has the form

$$(7.1) \qquad\qquad \sum_{\lambda \in L} g_\lambda(u, z)\, e(\langle \lambda, w \rangle),$$

where $\langle\,,\,\rangle$ is a C-bilinear form in a certain complex vector space, $g_\lambda(u, z)$ is a theta function in $u \in \mathbf{C}^m$ with a parameter z, and λ runs over a lattice L. Expansion (6.4) is actually an example of (7.1). If $\mathfrak{S}$ is of the first or second kind, $g_\lambda(u, z)$ is just a constant, or a function only in u without z. Now suppose that we can construct canonical models with respect to G, or rather, with respect to a reductive group containing G, and therefore can speak of arithmetic automorphic functions with respect to Γ. Then we may naturally ask the following questions.

(I) *Can one define "arithmetic automorphic forms" consistent with the notion of canonical models and arithmetic automorphic functions?*

This question can be asked even in the case of compact quotient.

(II) *Assuming the quotient non-compact, can one characterize such arithmetic forms in terms of the properties of the Fourier coefficients g_λ of (7.1)?*

(III) *Are holomorphic Eisenstein series, up to constant factors, arithmetic?*

The answers to all these questions seem to be in the affirmative. For example, the results of our present and previous papers give affirmative answers to (I) and (II) for the symplectic groups. The works of Siegel, Klingen, Baily and others answer (III) positively for many groups Γ when the Fourier coefficients are constants. In general, if the Fourier coefficients are constants, it is reasonable to call (7.1) arithmetic when the g_λ belong to $\mathbf{Q}_{\mathrm{ab}}$.

Now, in the case in which $\mathfrak{S}$ is of the third kind, the answers to (I), (II) will probably be given in the following way. First, the parameter z belongs to a symmetric domain $\mathfrak{S}_0$ on which an algebraic group G_0 acts. There is an injection of G_0 into a symplectic group which induces a holomorphic embedding ε of $\mathfrak{S}_0$ into $\mathfrak{H}_n$ for some n. Then $g_\lambda(u, z)$, possibly modified by a suitable factor, is a function on $\mathbf{C}^m \times \mathfrak{S}_0$ whose behavior is similar to $h(u, \varepsilon(z))$ with an element h of $T_{k, \xi}$ of § 3. In other words, if z is a "generic point" of $\mathfrak{S}_0$ and A is an abelian variety associated with the point $\varepsilon(z)$, then g_λ as a function in u is a theta function of A, behaving like $\theta(u, \varepsilon(z); r, s)$. Define $(g_\lambda)_*$ in the same manner as in

§ 3. Then we call (7.1) *arithmetic* if the values $(g_\lambda)_*(v, z)$ for all λ and all v commensurable with the periods are arithmetic automorphic forms in $z \in \mathfrak{S}_0$. Thus question (II) for G can be reduced to question (I) for G_0, which is comparatively easy. The quotient of $\mathfrak{S}_0$ by a discontinuous subgroup of G_0 may be compact. For example, let G be the unitary group of a hermitian form over a totally indefinite quaternion algebra whose center is totally real. Then a natural choice of a parabolic subgroup yields Fourier expansions in a domain of the first or third kind according as the degree of the hermitian form is even or odd. A recent work of Garrett [2] gives affirmative answers to questions (I) and (II) in either case, endorsing what we said above.

In general, it is expected that A is a generic member of a family of abelian varieties characterized by their endomorphism algebras, or the product of several copies of such. Then the action of the adelization of G_0 on g_λ is essentially obtained as a specialization of Theorem 3.10, in a fashion similar to that explained at the end of § 3. Also, one should be able to prove an analogue of Theorem 6.2 with $\mathfrak{S}$ and $\mathfrak{S}_0$ in place of $\mathfrak{H}_{n+s}$ and $\mathfrak{H}_n$.

The domain $\mathfrak{S}$ of the second kind may be viewed as an extreme case where $\mathfrak{S}_0$ consists of a single point; the behavior of g_λ in this case is similar to h^* with $h \in T_{k,\xi}$. A typical example of G with such an $\mathfrak{S}$ is provided by the unitary group of a hermitian form over an imaginary quadratic field, which we treated in [9]. The answers to questions (I) and (II) can be given exactly in the above described fashion. The domain of the second kind occurs also when G is the unitary group of a skew-hermitian form over a definite quaternion algebra. As for question (III), we indicated in the same article [9], if without details, that the answer is affirmative at least if $\mathfrak{S}$ is a complex unit ball

$$\left\{ (z_1, \ldots, z_n) \in \mathbf{C}_n^1 \,\middle|\, \sum_{\nu=1}^{n} |z_\nu|^2 < 1 \right\}.$$

It is very likely that the same holds in a more general case.

Appendix

Our purpose here is to determine the constant factor in the transformation formula of the function $\theta(u, z; r, s)$ defined by (3.26). To simplify our notation, for every symmetric matrix S, let $\{S\}$ denote the column vector consisting of the diagonal elements of S; also put $S[X] = {}^t\!XSX$ for a matrix X such that SX can be defined. Let a subgroup Γ_θ of Γ_1 be defined by

$$(\mathrm{a.1}) \qquad \Gamma_\theta = \left\{ \begin{pmatrix} a & b \\ c & d \end{pmatrix} \in \Gamma_1 \,\middle|\, \{{}^t\!ac\} \equiv \{{}^t\!bd\} \equiv 0 \pmod{2\mathbf{Z}^n} \right\} \qquad (\text{cf. } [8, \S 1]).$$

For every $\gamma = \begin{pmatrix} a & b \\ c & d \end{pmatrix} \in \Gamma_\theta$, we have

(a.2)

$$\theta({}^t(cz+d)^{-1}u, \gamma(z); g, h) = \lambda(\gamma)e(({}^tgh - {}^tg^*h^*)/2)\det(cz+d)^{1/2}e(\tfrac{1}{2}\cdot {}^tu(cz+d)^{-1}cu)\theta(u, z; g^*, h^*),$$

where $\lambda(\gamma)$ is a constant depending only on γ, and

$$\begin{pmatrix} g^* \\ h^* \end{pmatrix} = {}^t\gamma \begin{pmatrix} g \\ h \end{pmatrix}.$$

This is classical; for a short proof, see [7]. Define φ and φ' by (3.27) and (3.28). Then formula (a.2) is equivalent with

(a.3) $$\varphi'({}^t(cz+d)^{-1}u, \gamma(z); g, h) = \lambda(\gamma)e(({}^tgh - {}^tg^*h^*)/2)\det(cz+d)^{1/2}\varphi'(u, z; g^*, h^*).$$

This holds also with φ instead of φ'.

PROPOSITION A.1. *If* $\gamma = \begin{pmatrix} a & b \\ c & d \end{pmatrix} \in \Gamma_\theta$, $\det(d) > 0$, *and* $\det(cz+d)^{1/2}$ *is chosen so as to become positive when* $z = iy$ *with real* y *tends to* 0, *then* $\lambda(\gamma) = \det(d)^{-1/2}\sum_{v \in V} e(bd^{-1}[v]/2)$, *where* V *is a complete set of representatives for* $\mathbf{Z}^n/d\mathbf{Z}^n$.

Proof. Put $\theta(z) = \theta(0, z; 0, 0)$, $w = {}^td^{-1}z(cz+d)^{-1}$, $f = bd^{-1}$. Then $\gamma(z) = w + f$, so that

(a.4) $$\lambda(\gamma)\det(cz+d)^{1/2}\theta(z) = \theta(\gamma(z)) = \theta(w+f) = \sum_{x \in \mathbf{Z}^n} e(\tfrac{1}{2}(f+w)[x]).$$

With V as above, put $x = v + ds$ with $v \in V$ and $s \in \mathbf{Z}^n$. Then

$$\theta(w+f) = \sum_v \sum_s e(\tfrac{1}{2}f[v+ds] + \tfrac{1}{2}w[v+ds])$$

$$= \sum_v e(\tfrac{1}{2}f[v])\theta(0, z(cz+d)^{-1}d; d^{-1}v, 0),$$

since $f[v+ds]/2 \equiv f[v]/2 \pmod{\mathbf{Z}}$. Now, from [7, (16')], we obtain

$$\det(-iz)^{1/2}\theta(0, z; g, h) = e({}^tgh)\theta(0, -z^{-1}; -h, g).$$

Put $z = i\tau 1_n$ with $0 < \tau \in \mathbf{R}$, and observe that

(a.5) $$\lim_{\tau \to 0} \tau^{n/2}\theta(0, i\tau 1_n; g, h) = e({}^tgh)\delta(h),$$

where $\delta(h) = 1$ or 0 according as $h \in \mathbf{Z}^n$ or $h \notin \mathbf{Z}^n$. Taking the limit of $\tau^{n/2}$ times (a.4) when τ tends to 0, we obtain the desired formula for $\lambda(\gamma)$.

PROPOSITION A.2. *Let σ be a symmetric element of $\mathbf{Z}_n^n \cap GL_n(\mathbf{Q})$, and let $\begin{pmatrix} p & q \\ r & s \end{pmatrix} \in SL_2(\mathbf{Z})$.*

Suppose that $0 < s \equiv 1 \pmod 2$, $r\sigma^{-1} \in \mathbf{Z}_n^n$, $r \neq 0$, and $\{pr\sigma^{-1}\} \equiv \{qs\sigma\} \equiv 0 \pmod{2\mathbf{Z}^n}$. Then

$$\lambda \begin{pmatrix} p1_n & q\sigma \\ r\sigma^{-1} & s1_n \end{pmatrix} = \varepsilon_s^n \left(\frac{-2r}{s} \right)^n \left(\frac{\det(\sigma)}{s} \right),$$

where $\varepsilon_s = 1$ or i according as $s \equiv 1$ or $-1 \pmod 4$, and $\left(\dfrac{\;}{\;} \right)$ is the quadratic residue symbol.

Proof. Substituting $z + h$ for z in (a.2) with $h = {}^t h \in \mathbf{Z}_n^n$ such that $\{h\} \in 2\mathbf{Z}^n$, we find

$$(\text{a.6}) \qquad \lambda \begin{pmatrix} a & b + ah \\ c & d + ch \end{pmatrix} = \lambda \begin{pmatrix} a & b \\ c & d \end{pmatrix}.$$

Since $(2r, s) = 1$, there is a prime l, not dividing $\det(\sigma)$, of the form $l = 8mr + s$ with $0 < m \in \mathbf{Z}$. Put $k = 8mp + q$. By (a.6) and Prop. A.1, we have

$$(\text{a.7}) \qquad \lambda \begin{pmatrix} p1_n & q\sigma \\ r\sigma^{-1} & s1_n \end{pmatrix} = \lambda \begin{pmatrix} p1_n & k\sigma \\ r\sigma^{-1} & l1_n \end{pmatrix} = l^{-n/2} \sum_{v \in V} e\left(\frac{k}{2l} \sigma[v] \right),$$

where V is a complete set of representatives for $\mathbf{Z}^n / l\mathbf{Z}^n$. Now we can find an element α of $GL_n(\mathbf{Q}) \cap \mathbf{Z}_n^n$ such that l does not divide $\det(\alpha)$ and ${}^t \alpha \sigma \alpha$ is congruent to a diagonal matrix modulo l. Let $\xi_1, \ldots, \xi_n$ be the diagonal elements of ${}^t \alpha \sigma \alpha$. Then (a.7) is equal to

$$l^{-n/2} \prod_{v=1}^{n} \left\{ \sum_{x=1}^{l} e\left(\frac{k\xi_v}{2l} x^2 \right) \right\} = \prod_{v=1}^{n} \varepsilon_l \left(\frac{2k\xi_v}{l} \right) = \varepsilon_l^n \left(\frac{2k}{l} \right)^n \left(\frac{\det(\sigma)}{l} \right).$$

Since $\begin{pmatrix} p & k \\ r & l \end{pmatrix} \in SL_2(\mathbf{Z})$, we have $kr \equiv -1 \pmod l$. Also we have $l \equiv s \pmod{8r}$, so that $\varepsilon_l = \varepsilon_s$, and $\left(\dfrac{2k}{l} \right) = \left(\dfrac{-2r}{l} \right) = \left(\dfrac{-2r}{s} \right)$. Since $r^n \det(\sigma)^{-1}$ is an integer, $\det(\sigma)$ is a square times a divisor of r, and hence $\left(\dfrac{\det(\sigma)}{l} \right) = \left(\dfrac{\det(\sigma)}{s} \right)$, which completes the proof.

References

[1]. CARTAN, H., Formes modulaires. Séminaire H. Cartan 1957/58, exp. 4.

[2]. GARRETT, P., Arithmetic automorphic forms for quaternion unitary groups. Thesis, Princeton University, 1977.

[3]. PYATETSKII-SHAPIRO, I. I., *Geometry of classical domains and the theory of automorphic functions.* Moscow, 1961 (Russian).

[4]. SHIMURA, G., On canonical models of arithmetic quotients of bounded symmetric domains, I, II. *Ann. of Math.*, 91 (1970), 144–222; 92 (1970), 528–549.

[5]. —— On some arithmetic properties of modular forms of one and several variables. *Ann. of Math.*, 102 (1975), 491–515.

[6]. —— On the Fourier coefficients of modular forms of several variables. *Göttingen Nachr. Akad. Wiss.*, 1975, 261–268.

[7]. —— Theta functions with complex multiplication. *Duke Math. J.*, 43 (1976), 673–696.

[8]. —— On the derivatives of theta functions and modular forms. *Duke Math. J.*, 44 (1977), 365–387.

[9]. —— Unitary groups and theta functions. *Proceedings of Int. Symp. on Alg. N. Th. Kyoto 1976* (1977), 195–200.

[10]. WEIL, A., Sur certains groupes d'opérateurs unitaires. *Acta Math.*, 111 (1964), 143–211.

Received June 20, 1977

The arithmetic of automorphic forms
with respect to a unitary group

Annals of Mathematics, 107 (1978), 569-605

Introduction

The Fourier expansion of an elliptic modular form has been fruitfully utilized in various arithmetical problems as well as in the study of the analytic properties of the form itself. The same can be said also for the Hilbert and Siegel modular forms. One expresses a given modular form F as a function of complex variables $u_1, \cdots, u_m$ with an expansion

$$(0.1) \qquad F(u_1, \cdots, u_m) = \sum_x c_x \exp(2\pi i \cdot \sum_{\nu=1}^m x_\nu u_\nu) ,$$

where the coefficients c_x are complex numbers and x runs over a lattice. Especially important are those F for which all c_x are algebraic, or more restrictedly, cyclotomic. They form a distinguishable class which is stable under the transformation by the elements of the algebraic group in question. The algebraicity of c_x is also indispensable if the value of a modular function at a special point is the problem, as in the theory of complex multiplication. In general, an expansion of type (0.1) exists for an automorphic form if it is defined on a tube domain and the group contains sufficiently many translations. There are, however, cases in which no such expansion is available. A typical example is provided by the symmetric domain

$$(0.2) \qquad \mathfrak{Z} = \{(w, z) \in \mathbf{C}_q^n \times \mathbf{C}_q^q \mid i({}^t\overline{w}Sw + {}^t\overline{z} - z) > 0\}$$

with a skew-hermitian matrix S of size n such that $-iS > 0$. Here A_q^n, for a ring A, denotes the set of all $n \times q$-matrices with entries in A; we write $h > 0$ if a hermitian matrix h is positive definite. Assuming that S has coefficients in an imaginary quadratic field K, we can define a natural action on $\mathfrak{Z}$ of the unitary group

$$U = \{\alpha \in \mathrm{GL}_{n+2q}(K) \mid {}^t\overline{\alpha}R\alpha = R\} , \qquad R = \begin{bmatrix} S & 0 & 0 \\ 0 & 0 & 1_q \\ 0 & -1_q & 0 \end{bmatrix} .$$

0003-486X/78/0107-0001 $ 01.85

© 1978 by Princeton University Press

For copying information, see inside back cover.

[1] Supported by NSF Grant MCS76-1176.

In this case, an automorphic form $f(w, z)$ on $\mathfrak{Z}$ with respect to a congruence subgroup of U (for example, with a power of jacobian as the factor of automorphy) has a Fourier expansion of the form

$$(0.3) \qquad f(w, z) = \sum_{\tau} g_{\tau}(w)\exp(2\pi i \cdot \mathrm{tr}(\tau z)) \,,$$

where τ runs over non-negative hermitian matrices belonging to a lattice in K_q^q. The Fourier coefficients g_{τ} are theta functions on C_q^n relative to a certain lattice L contained in K_q^n. Now we can ask the following questions:

(I) *Can one define the notion of arithmeticity of $f(w, z)$ by means of some properties of g_{τ} similar to the algebraicity of c_x?*

(II) *Supposing there is such a notion, are (holomorphic) Eisenstein series arithmetic?*

The main purpose of this paper is to give affirmative answers to these questions, the second one being restricted to the case $q = 1$. We first observe that C_q^n/L is isogenous to a product of nq elliptic curves with K as the field of complex multiplications. Therefore we can apply to the present situation the theory of arithmetic theta functions in our previous paper [9], which has been developed, in fact, with this application in view. The desired class of "arithmetic forms" is defined to consist of all those f whose coefficients g_{τ} are arithmetic in the sense of [9]; here g_{τ} is called arithmetic if a certain exponential factor times g_{τ} takes values in the maximal abelian extension K_{ab} of K at all points of K_q^n. Now the theory of canonical models as developed in [3] and [6] enables us to speak of the field of arithmetic automorphic functions, say $\mathfrak{K}$, with respect to our group. Our first main result asserts that these two definitions of arithmeticity are compatible; namely, the composite $K_{\mathrm{ab}}\mathfrak{K}$ coincides with the field of all quotients of arithmetic automorphic forms of the same weight.

Since the isotropy subgroup of $U_{\mathbf{R}}$ at a point of $\mathfrak{Z}$ has two simple factors, we are naturally faced with two factors of automorphy taking values in $\mathrm{GL}_{n+q}(C)$ and $\mathrm{GL}_q(C)$. Therefore we shall consider the automorphic forms that are vector-valued holomorphic functions on $\mathfrak{Z}$ with an automorphic property expressed in terms of an arbitrary K-rational representation of $\mathrm{GL}_{n+q} \times \mathrm{GL}_q$. They have Fourier expansions similar to (0.3) whose coefficients are generalized theta functions of the type studied in [12]. We shall prove that the arithmetic forms, which can be defined by the same ideas as above, are stable as a whole under the transformation by the group elements, and of course compatible with canonical models.

The second question will be answered only in the case $q = 1$, though our methods are probably applicable to the general case. The explicit form of our Eisenstein series is

$$E_{j,k}(w, z) = \sum (aw + bz + c)^{-k} \begin{bmatrix} \bar{b}w - S^{-1} \cdot {}^t\bar{a} \\ \bar{b}z + \bar{c} \end{bmatrix}^{(j)} \qquad ((w, z) \in \mathfrak{B}) ,$$

where $0 \leqq j \in \mathbf{Z}$, $0 < k \in \mathbf{Z}$, $x^{(j)} = x \otimes \cdots \otimes x$ with j copies of $x \in \mathbf{C}^{n+1}$, and (a, b, c) runs over all non-vanishing elements of $K_n^1 \times K \times K$ satisfying some congruence conditions and an equality

$$aS^{-1} \cdot {}^t\bar{a} = b\bar{c} - c\bar{b} .$$

If $k > 2n + j + 2$, this is convergent and defines an automorphic form relative to a representation ρ of $\mathrm{GL}_{n+1} \times \mathrm{GL}_1$ defined by

$$\rho(X, Y) = Y^k \cdot {}^tX^{-1} \otimes \cdots \otimes {}^tX^{-1} \qquad (j \text{ copies of } {}^tX^{-1}; \ X \in \mathrm{GL}_{n+1}, \ Y \in \mathrm{GL}_1) .$$

It will be shown, as the second main result of the paper, that $E_{j,k}$, multiplied by a certain well-defined constant, is arithmetic. This fact will be applied, in a subsequent paper, to some questions concerning the special values of a certain type of zeta function. Actually this application is one of the main motives of our investigation. Since the proof of arithmeticity is rather intricate, we mention at this point only that the main theorem of [7] concerning a non-holomorphic derivative of an elliptic modular form is crucial in the analysis of $E_{j,k}$ in the case $j > 0$. This is not accidental; in fact, even the non-holomorphic derivative of a theta function discussed in [10] plays an essential role in the proof of the compatibility of holomorphic differentiation on $\mathfrak{B}$ with the arithmeticity. To conclude the introduction, we note that an essential part of our results was announced in [11].

Notation. For a ring A, we define A_q^n as above, and denote A_1^n simply by A^n; 1_n or simply 1 denotes the identity matrix of degree n. The transpose of a matrix X is denoted by ${}^tX$. For a complex $n \times n$-matrix Z, we put $e_n(Z) = \exp(2\pi i \cdot \mathrm{tr}(Z))$, where $\mathrm{tr}(Z)$ is the trace of Z, and in particular $e(z) = e_1(z) = e^{2\pi i z}$ for $z \in \mathbf{C}$. The bar will always mean the complex conjugate. If h is a complex hermitian matrix, we write $h > 0$ or $h \geqq 0$ according as h is positive definite or non-negative. If ρ is a finite dimensional representation of a group, $d(\rho)$ denotes the dimension of the representation space. In most cases, $\mathbf{C}^{d(\rho)}$ is taken as the representation space of ρ. If K is a number field, K_{ab} denotes the maximal abelian extension of K.

1. Unitary groups and factors of automorphy

Throughout the paper, we denote by K an imaginary quadratic field embedded in C. Our main interest is in the automorphic forms with respect to the unitary group of an indefinite hermitian form over K. For some technical reasons, we take a skew-hermitian form instead of a hermitian form. Thus, with an element R of $\mathrm{GL}_m(K)$ such that ${}^t\bar{R} = -R$, we consider an algebraic group G defined over $\mathbf{Q}$ such that

$$G_{\mathbf{Q}} = \{\alpha \in \mathrm{GL}_m(K) \mid {}^t\bar{\alpha}R\alpha = \nu(\alpha)R \quad \text{with } \nu(\alpha) \in \mathbf{Q}\} ,$$
$$G_{\mathbf{R}} = \{\alpha \in \mathrm{GL}_m(\mathbf{C}) \mid {}^t\bar{\alpha}R\alpha = \nu(\alpha)R \quad \text{with } \nu(\alpha) \in \mathbf{R}\} .$$

Then we put

$$G_{\mathbf{Q}+} = \{\alpha \in G_{\mathbf{Q}} \mid \nu(\alpha) > 0\} , \quad G_{\mathbf{R}+} = \{\alpha \in G_{\mathbf{R}} \mid \nu(\alpha) > 0\} .$$

Let q be the number of negative eigenvalues of $-iR$. We may assume, changing R for $-R$ if necessary, that $2q \leq m$. Put $m = n + 2q$. In this paper, we are interested only in the case where R is indefinite, i.e., $q > 0$. Also we assume $n > 0$ in the first five sections. The case $n = 0$ will be discussed in the last section. By the Hasse principle, the equivalence class of R is determined by m, q, and the class of $\det(R)$ modulo $N_{K/\mathbf{Q}}(K^{\times})$. Therefore, if $n > 0$, we may assume, without losing generality, that R has the form

$$(1.1) \qquad R = \begin{bmatrix} S & 0 & 0 \\ 0 & 0 & 1_q \\ 0 & -1_q & 0 \end{bmatrix} , \qquad -{}^t\bar{S} = S \in \mathrm{GL}_n(K), \quad -iS > 0 .$$

We may even assume that S is diagonal, although this is unnecessary in the first two sections. Our automorphic forms will be considered on the space $\mathfrak{Z}$ defined by (0.2). The variable point (w, z) of $\mathfrak{Z}$ will often be denoted by a single letter $\mathfrak{z}$. Let $\mathfrak{Y}$ denote the set of all elements Y of $\mathrm{GL}_m(\mathbf{C})$ such that

$$(1.2) \qquad i \cdot {}^t\bar{Y}RY = \begin{bmatrix} -A & 0 \\ 0 & B \end{bmatrix}$$

with positive definite hermitian matrices A and B of size $n + q$ and q, respectively. For $\mathfrak{z} = (w, z) \in \mathfrak{Z}$, put

$$(1.3) \qquad P(\mathfrak{z}) = \begin{bmatrix} S^{-1} & 0 & w \\ {}^t\bar{w} & {}^t\bar{z} & z \\ 0 & 1_q & 1_q \end{bmatrix} .$$

A direct calculation shows that

$$(1.4) \qquad i \cdot {}^t\overline{P(\mathfrak{z})}RP(\mathfrak{z}) = \begin{bmatrix} -\xi(\mathfrak{z}) & 0 \\ 0 & \eta(\mathfrak{z}) \end{bmatrix} ,$$

with

$$(1.5) \qquad \xi(w, z) = i \begin{bmatrix} S^{-1} & -w \\ {}^t\bar{w} & {}^t\bar{z} - z \end{bmatrix} ,$$

$$(1.6) \qquad \eta(w, z) = i({}^t\bar{w}Sw + {}^t\bar{z} - z) .$$

Since $\eta(\mathfrak{z}) > 0$ and iR has q positive eigenvalues, we see that $\xi(\mathfrak{z}) > 0$, and hence $P(\mathfrak{z}) \in \mathfrak{Y}$. This proves a part of

PROPOSITION 1.1. *The map*

$$(1.7) \qquad (\mathfrak{z}, \ C, \ D) \longmapsto P(\mathfrak{z}) \begin{bmatrix} C & 0 \\ 0 & D \end{bmatrix}$$

gives a bijection of $\mathfrak{Z} \times \mathrm{GL}_{n+q}(\mathbf{C}) \times \mathrm{GL}_q(\mathbf{C})$ *onto* $\mathfrak{Y}$.

Proof. The only non-trivial point is the surjectivity. Let $Y = (V, X) \in \mathfrak{Y}$ with $V \in \mathbf{C}^m_{n+q}$ and $X \in \mathbf{C}^m_q$, and put ${}^t X = ({}^t x_1 \ {}^t x_2 \ {}^t x_3)$ with x_2 and x_3 in $\mathbf{C}^q_q$. Then $i({}^t\bar{x}_1 S x_1 + {}^t\bar{x}_2 x_3 - {}^t\bar{x}_3 x_2) > 0$. Since $-iS > 0$, we see that x_3 is invertible. Put $w = x_1 x_3^{-1}$ and $z = x_2 x_3^{-1}$. Then $(w, z) \in \mathfrak{Z}$. Since $Y \in \mathfrak{Y}$, we have $i \cdot {}^t\bar{V}RV < 0$. Let V_1 be the first $n + q$ rows of V, and suppose $V_1 u = 0$ with $u \in \mathbf{C}^{n+q}$. Since R has the form (1.1), we have $i \cdot {}^t\bar{u} \cdot {}^t\bar{V}RVu = 0$, and hence $u = 0$. Thus V_1 is invertible. Therefore, choosing suitable (C, D) of $\mathrm{GL}_{n+q}(\mathbf{C}) \times \mathrm{GL}_q(\mathbf{C})$, we can put Y in the form

$$\begin{bmatrix} S^{-1} & 0 & w \\ {}^t\bar{w} & {}^t\bar{z} & z \\ r & s & 1_q \end{bmatrix} \begin{bmatrix} C & 0 \\ 0 & D \end{bmatrix} .$$

Since $Y \in \mathfrak{Y}$, we can easily verify that $r = 0$ and $s = 1$, which completes the proof.

If $\alpha \in G_{\mathbf{R}+}$ and $Y \in \mathfrak{Y}$, then obviously $\alpha Y \in \mathfrak{Y}$. Therefore, if $\mathfrak{z} = (w, z) \in \mathfrak{Z}$ we can define a point $\alpha(\mathfrak{z})$ of $\mathfrak{Z}$ by

$$(1.8) \qquad \alpha P(\mathfrak{z}) = P(\alpha(\mathfrak{z})) \begin{bmatrix} \overline{\kappa(\alpha, \ \mathfrak{z})} & 0 \\ 0 & \mu(\alpha, \ \mathfrak{z}) \end{bmatrix}$$

with $\kappa(\alpha, \ \mathfrak{z}) \in \mathrm{GL}_{n+q}(\mathbf{C})$ and $\mu(\alpha, \ \mathfrak{z}) \in \mathrm{GL}_q(\mathbf{C})$. More explicitly, if

$$(1.9) \qquad \alpha = \begin{bmatrix} a_1 & b_1 & c_1 \\ a_2 & b_2 & c_2 \\ a_3 & b_3 & c_3 \end{bmatrix}$$

with blocks corresponding to those of R, then

$$(1.10) \qquad \begin{aligned} \alpha(w, z) = &((a_1 w + b_1 z + c_1)(a_3 w + b_3 z + c_3)^{-1} , \\ &(a_2 w + b_2 z + c_2)(a_3 w + b_3 z + c_3)^{-1}) , \end{aligned}$$

$$(1.11) \qquad \kappa(\alpha, \ \mathfrak{z}) = \begin{bmatrix} \bar{S}\bar{a}_1\bar{S}^{-1} + \bar{S}\bar{b}_1 \cdot {}^t w & \bar{S}\bar{c}_1 + \bar{S}\bar{b}_1 \cdot {}^t z \\ \bar{a}_3\bar{S}^{-1} + \bar{b}_3 \cdot {}^t w & \bar{c}_3 + \bar{b}_3 \cdot {}^t z \end{bmatrix} ,$$

$$(1.12) \qquad \mu(\alpha, \ \mathfrak{z}) = a_3 w + b_3 z + c_3 .$$

The matrices κ and μ are obviously factors of automorphy holomorphic in

$\mathfrak{z}$. It can easily be seen that $\mathfrak{Z}$ is isomorphic to the quotient of the simple part of $G_{\mathbf{R}+}$ by its maximal compact subgroup. To find the relationship between κ, μ, and the jacobian of α, put

$$(1.13) \qquad Y(\mathfrak{z}, \mathfrak{z}_1) = {}^t\overline{P(\mathfrak{z})}RP(\mathfrak{z}_1) \qquad\qquad (\mathfrak{z}, \mathfrak{z}_1 \in \mathfrak{Z}) .$$

Obviously

$$(1.14) \qquad \nu(\alpha)\, Y(\mathfrak{z}, \mathfrak{z}_1) = \begin{bmatrix} {}^t\kappa(\alpha, \mathfrak{z}) & 0 \\ 0 & {}^t\mu(\alpha, \mathfrak{z}) \end{bmatrix} Y\big(\alpha(\mathfrak{z}), \alpha(\mathfrak{z}_1)\big) \begin{bmatrix} \overline{\kappa(\alpha, \mathfrak{z}_1)} & 0 \\ 0 & \mu(\alpha, \mathfrak{z}_1) \end{bmatrix}$$

for every $\alpha \in G_{\mathbf{R}+}$. Let $\delta(\mathfrak{z}, \mathfrak{z}_1)$ denote the $(n+q)\times q$-matrix which is an upper right block of $Y(\mathfrak{z}, \mathfrak{z}_1)$. If $\mathfrak{z} = (w, z)$ and $\mathfrak{z}_1 = (w_1, z_1)$, a simple computation shows that

$$\delta(\mathfrak{z}, \mathfrak{z}_1) = \begin{bmatrix} w & -w_1 \\ z & -z_1 \end{bmatrix} .$$

Relation (1.14) shows that

$$\nu(\alpha)\delta(\mathfrak{z}, \mathfrak{z}_1) = {}^t\kappa(\alpha, \mathfrak{z})\delta\big(\alpha(z), \alpha(\mathfrak{z}_1)\big)\mu(\alpha, \mathfrak{z}_1) ,$$

and hence

$$(1.15) \qquad \begin{bmatrix} dw \\ dz \end{bmatrix} \circ \alpha = \nu(\alpha) \cdot {}^t\kappa(\alpha, \mathfrak{z})^{-1} \begin{bmatrix} dw \\ dz \end{bmatrix} \mu(\alpha, \mathfrak{z})^{-1} .$$

Again from (1.14), we obtain

$$(1.16) \qquad \nu(\alpha)\xi(\mathfrak{z}) = {}^t\kappa(\alpha, \mathfrak{z})\xi\big(\alpha(\mathfrak{z})\big)\overline{\kappa(\alpha, \mathfrak{z})} ,$$
$$(1.17) \qquad \nu(\alpha)\eta(\mathfrak{z}) = {}^t\overline{\mu(\alpha, \mathfrak{z})}\eta\big(\alpha(\mathfrak{z})\big)\mu(\alpha, \mathfrak{z}) \qquad\qquad (\alpha \in G_{\mathbf{R}+}) .$$

Noting that

$$\begin{bmatrix} S^{-1} & 0 & w \\ {}^t\overline{w} & {}^t\overline{z} & z \\ 0 & 1_q & 1_q \end{bmatrix} \begin{bmatrix} 1_n & 0 & -Sw \\ 0 & 1_q & -1_q \\ 0 & 0 & 1_q \end{bmatrix} = \begin{bmatrix} S^{-1} & 0 & 0 \\ {}^t\overline{w} & {}^t\overline{z} & i\eta(\mathfrak{z}) \\ 0 & 1_q & 0 \end{bmatrix} ,$$

we obtain

$$(1.18) \qquad \det\big(P(\mathfrak{z})\big) = \det(S^{-1})\det\big(-i\eta(\mathfrak{z})\big) .$$

Taking the determinant of (1.8), we obtain, from (1.17) and (1.18),

$$(1.19) \qquad \det\big(\kappa(\alpha, \mathfrak{z})\big) = \det(\alpha)^{-1}\nu(\alpha)^{n+q}\det\big(\mu(\alpha, \mathfrak{z})\big) .$$

Then we find, from (1.15), that the jacobian of α at $\mathfrak{z}$ is equal to

$$(1.20) \qquad \det(\alpha)^q\det\big(\mu(\alpha, \mathfrak{z})\big)^{-m} .$$

To study the "parabolic elements" of $G_{\mathbf{R}+}$, we first note that if $b_3 = 0$ for $\alpha \in G_{\mathbf{R}+}$ as in (1.9), then $b_1 = 0$ and $a_3 = 0$. Further if, in addition, the diagonal blocks a_1, b_2, c_3 are all identities, α has the form

$$(1.21) \qquad \alpha = \begin{bmatrix} 1_n & 0 & x \\ {}^t\bar{x}S & 1_q & y + \dfrac{1}{2}{}^t\bar{x}Sx \\ 0 & 0 & 1_q \end{bmatrix}, \qquad\qquad x \in \mathbf{C}_q^n, \quad {}^t\bar{y} = y \in \mathbf{C}_q^q.$$

We denote by P_Q the subgroup of G_{Q+} consisting of all such α with K-rational x and y.

2. Automorphic forms

For a lattice Λ in the vector space K_m^1, put

$$\Gamma_\Lambda = \{\gamma \in G_Q \mid \det(\gamma) = \nu(\gamma) = 1, \quad \Lambda\gamma = \Lambda\}.$$

By a *congruence subgroup* of G_{Q+}, we understand a subgroup Γ of G_{Q+} which contains

$$\{\gamma \in \Gamma_\Lambda \mid \Lambda(\gamma - 1) \subset N\Lambda\}$$

for some positive integer N and such that $\Gamma/(\Gamma \cap K^\times)$ is commensurable with $\Gamma_\Lambda/(\Gamma_\Lambda \cap K^\times)$. Throughout the paper, Γ will denote such a group. To define automorphic forms, let us consider a K-rational representation

$$(2.1) \qquad\qquad \rho\colon \mathrm{GL}_{n+q} \times \mathrm{GL}_q \longrightarrow \mathrm{GL}_{d(\rho)},$$

where $d(\rho)$ is a positive integer which is the degree of ρ. For $\alpha \in G_{R+}$ and a $\mathbf{C}^{d(\rho)}$-valued function f on $\mathfrak{Z}$, we define a $\mathbf{C}^{d(\rho)}$-valued function $f|_\rho\alpha$ on $\mathfrak{Z}$ by

$$(2.2) \qquad (f|_\rho\alpha)(\mathfrak{z}) = \rho(\kappa(\alpha, \mathfrak{z}), \mu(\alpha, \mathfrak{z}))^{-1} f(\alpha(\mathfrak{z})) \qquad\qquad (\mathfrak{z} \in \mathfrak{Z}).$$

We denote by $\mathfrak{M}_\rho(\Gamma)$, resp. $\mathfrak{A}_\rho(\Gamma)$, the set of all holomorphic, resp., meromorphic $\mathbf{C}^{d(\rho)}$-valued functions f on $\mathfrak{Z}$ such that $f|_\rho\gamma = f$ for all $\gamma \in \Gamma$. An element of $\mathfrak{M}_\rho(\Gamma)$ is called a Γ-*automorphic form of weight* ρ. In particular, if $\rho(X, Y) = \det(Y)^k$ with $k \in \mathbf{Z}$, we write $f|_k\alpha$, $\mathfrak{M}_k$, and $\mathfrak{A}_k$ for $f|_\rho\alpha$, $\mathfrak{M}_\rho$, and $\mathfrak{A}_\rho$. Notice that $\mathfrak{A}_0(\Gamma)$ is the field of all Γ-invariant meromorphic functions on $\mathfrak{Z}$.

Given a congruence subgroup Γ, we can find a lattice L in K_q^n and a lattice J' in the vector space

$$(2.3) \qquad\qquad \mathfrak{K}_q = \{y \in K_q^q \mid {}^t\bar{y} = y\}$$

such that $\Gamma \cap P_Q$ contains all elements of the form (1.21) with $x \in L$ and $y \in J'$. Therefore, if $f \in \mathfrak{M}_\rho(\Gamma)$, we have $f(w, z + y) = f(w, z)$ for all $y \in J'$, and hence $f(w, z)$ has an expansion of the form

$$(2.4) \qquad\qquad f(w, z) = \sum_{r \in J} g_r(w)e_q(rz),$$

where J is the lattice in $\mathfrak{K}_q$ determined by

$$J = \{r \in \mathfrak{K}_q \mid \mathrm{tr}(rJ') \subset \mathbf{Z}\},$$

and the g_r are holomorphic functions on $\mathbf{C}_q^n$. Now observe that if α is of the form (1.21), then $\mu(\alpha,\ \mathfrak{z}) = 1_q$ and

$$\kappa(\alpha,\ \mathfrak{z}) = \begin{vmatrix} 1_n & \bar{S}\bar{x} \\ 0 & 1_q \end{vmatrix}.$$

Therefore we have

$$(2.5) \qquad f\Big(w + x,\ z + {}^t\bar{x}Sw + \frac{1}{2}\,{}^t\bar{x}Sx\Big) = \rho\Big(\begin{pmatrix} 1 & \bar{S}\bar{x} \\ 0 & 1 \end{pmatrix},\ 1_q\Big)f(w,\ z)$$

$$\text{for all}\quad x \in L\ .$$

Define a hermitian form H_r for each $r \in \mathcal{H}_q$ by

$$(2.6) \qquad\qquad H_r(u,\ v) = -2i \cdot \mathrm{tr}(r \cdot {}^t\bar{u}Sv) \qquad\qquad ((u,\ v) \in \mathbf{C}_q^n \times \mathbf{C}_q^n)\ ,$$

and a representation τ of GL_{n+q} by

$$(2.7) \qquad\qquad\qquad \tau(X) = \rho(X,\ 1_q) \qquad\qquad\qquad (X \in \mathrm{GL}_{n+q})\ .$$

Then (2.5) holds if and only if g_r belongs to the class of functions g satisfying

$$(2.8) \quad g(w + x) = e\Big(\frac{1}{2i}H_r\Big(x,\ w + \frac{1}{2}x\Big)\Big)\tau\begin{bmatrix} 1_n & \bar{S}\bar{x} \\ 0 & 1_q \end{bmatrix}g(w) \qquad \text{for all}\quad x \in L\ .$$

Given an arbitrary K-rational representation τ of GL_{n+q}, an element r of $\mathcal{H}_q$, and a skew-hermitian matrix S of K_n^n, we understand by a *theta function of type* $(\tau,\ r,\ S,\ L)$ a $\mathbf{C}^{d(\tau)}$-valued holomorphic function g on $\mathbf{C}_q^n$ satisfying (2.8). The set of all such g is denoted by $\mathfrak{T}_{\tau,r}(L)$ when S is fixed. We then denote by $\mathfrak{T}_{\tau,r}$ the union of $\mathfrak{T}_{\tau,r}(L)$ for all lattices L in K_q^n.

LEMMA 2.1.   (i) $\mathfrak{T}_{\tau,r} \neq \{0\}$ *only if* $r \geq 0$.
(ii) *Every element of* $\mathfrak{T}_{\tau,0}$ *is a constant.*
(iii) *If* $g \in \mathfrak{T}_{\tau,r}$, $s = br \cdot {}^t\bar{b}$ *with* $b \in \mathrm{GL}_q(K)$, *and*

$$h(w) = \tau\begin{bmatrix} 1_n & 0 \\ 0 & \bar{b} \end{bmatrix}g(wb) \qquad\qquad (w \in \mathbf{C}_q^n)\ ,$$

then $h \in \mathfrak{T}_{\tau,s}$.
  (iv) *If* $g \in \mathfrak{T}_{\tau,r}$ *and* $r = \begin{pmatrix} s & 0 \\ 0 & 0 \end{pmatrix}$ *with* $s \in \mathrm{GL}_p(K)$, *then* $g(u,\ v) = f(u)$ *with* $f \in \mathfrak{T}_{\tau,s}$, *where* $u \in \mathbf{C}_p^n$ *and* $v \in \mathbf{C}_{q-p}^n$.

Proof. Since $-iS > 0$, H_r is non-negative if and only if $r \geq 0$. Thus (i) implies that expansion (2.4) is extended only over $r \geq 0$. Now the first two assertions follow from the classical theory of theta functions if τ is trivial. In the general case, the matrices $\tau\begin{pmatrix} 1 & \bar{S}\bar{x} \\ 0 & 1 \end{pmatrix}$ for $x \in L$ form a commutative unipotent group, and hence we may assume, changing the coordinate system f necessary, that they are upper triangular matrices with 1 as all the diagonal elements. Let $0 \neq g \in \mathfrak{X}_{\tau,r}$. If h is the last non-vanishing component of g, then h satisfies (2.8) with trivial τ, so that H_r must be non-negative. Next, supposing $r = 0$, let p be the last non-constant component of g. Then $p(w + x) = p(w) + c(\bar{x})$ with a constant independent of w that is a rational function of $\bar{x}$. Then the derivatives of p define holomorphic functions on $\mathbf{C}_q^n/L$, which must be constants. Therefore p is a polynomial in w of degree $\leqq 1$. This implies that $c(\bar{x})$ is linear in x, and hence must be 0. Therefore p is a constant, which proves (ii). The third assertion can be verified in a straightforward way. If r and g are as in (iv), we can apply (ii) to $g(u, v)$ as a function of v and obtain the desired conclusion.

For each $g \in \mathfrak{X}_{\tau,r}$, we define a function g_* on $\mathbf{C}_q^n$ (which may be non-holomorphic) by

$$(2.9) \qquad g_*(w) = e\left(\frac{i}{4} H_r(w, w)\right) \tau \begin{bmatrix} 1_n & -\bar{S}\bar{w} \\ 0 & 1_q \end{bmatrix} g(w) \qquad (w \in \mathbf{C}_q^n) .$$

It can easily be verified that if $g \in \mathfrak{X}_{\tau,r}(L)$, we have

$$(2.10) \qquad g_*(w + x) = e\big(\mathrm{Re}[\mathrm{tr}(r \cdot {}^t\bar{w} S x)]\big) g_*(w) \qquad \text{for all } x \in L .$$

If $r = 0$, we see that $g_* = g$. Further if g and h are as in (iii) of Lemma 2.1, we have

$$h_*(w) = \tau \begin{bmatrix} 1_n & 0 \\ 0 & \bar{b} \end{bmatrix} g_*(wb) .$$

The function g_* occurs naturally in the following way. Define α by (1.21) with $x \in K_q^n$ and $y = 0$; let $f = \sum_r g_r(w)e_q(rz)$ and $(f|_\rho\alpha) = \sum_r h_r(w)e_q(rz)$. Then we see easily that

$$(2.11) \qquad h_r(0) = (g_r)_*(x) .$$

Now observe that $\mathbf{C}_q^n/L$ is isogenous to the product of qn copies of an elliptic curve with complex multiplications in K. Therefore we can apply the theory of arithmetic theta functions developed in [9] to the present situation if τ is trivial. The essential ideas about this were explained in [11], and are actually suggested by formula (2.11). To define the arithmeticity in the case of an arbitrary representation τ, it is necessary to introduce Siegel modular forms. Let $G^{(n)}$ be the algebraic subgroup of GL_{2n} defined over $\mathbf{Q}$ such that

$$(2.12) \qquad G_{\mathbf{Q}}^{(n)} = \{\beta \in \mathrm{GL}_{2n}(\mathbf{Q}) \mid {}^t\beta J_n\beta = \nu(\beta)J_n \text{ with } \nu(\beta) \in \mathbf{Q}\} ,$$

where

$$(2.12') \qquad\qquad J_n = \begin{bmatrix} 0 & -1_n \\ 1_n & 0 \end{bmatrix}.$$

We put $G_{Q+}^{(n)} = \{\beta \in G_Q^{(n)} \mid \nu(\beta) > 0\}$, and

$$(2.13) \qquad\qquad \mathfrak{H}_n = \{z \in \mathbf{C}_n^n \mid {}^t z = z,\ \mathrm{Im}(z) > 0\}.$$

The action of $\beta = \begin{pmatrix} a & b \\ c & d \end{pmatrix} \in G_{Q+}^{(n)}$ on $\mathfrak{H}_n$ is defined as usual by $\beta(z) = (az + b)(cz + d)^{-1}$. Let ρ be a rational representation of GL_n. For a $\mathbf{C}^{d(\rho)}$-valued function f on $\mathfrak{H}_n$ and $\beta = \begin{pmatrix} a & b \\ c & d \end{pmatrix} \in G_{Q+}^{(n)}$, we define a $\mathbf{C}^{d(\rho)}$-valued function $f|_\rho \beta$ on $\mathfrak{H}_n$ by

$$(2.14) \qquad\qquad (f|_\rho \beta)(z) = \rho(cz + d)^{-1} f(\beta(z)) \qquad\qquad (z \in \mathfrak{H}_n).$$

Given a congruence subgroup Γ of $\mathrm{Sp}(n, \mathbf{Z})$ and a subfield $\mathfrak{R}$ of $\mathbf{C}$, we define $\mathfrak{M}_\rho^{(n)}(\Gamma, \mathfrak{R})$ to be the module of all holomorphic $\mathbf{C}^{d(\rho)}$-valued functions f on $\mathfrak{H}_n$ such that:

(i) $f|_\rho \gamma = f$ for all $\gamma \in \Gamma$;

(ii) f is finite at every cusp;

(iii) the Fourier coefficients of f have components in $\mathfrak{R}$.

For details, the reader is referred to [12]. We denote by $\mathfrak{M}_\rho^{(n)}(\mathfrak{R})$ the union of $\mathfrak{M}_\rho^{(n)}(\Gamma, \mathfrak{R})$ for all congruence subgroups Γ, and by $\mathcal{Q}_\rho^{(n)}(\mathfrak{R})$ the set of all quotients $g^{-1}h$ with $h \in \mathfrak{M}_\sigma^{(n)}(\mathfrak{R})$ and $0 \neq g \in \mathfrak{M}_\tau^{(n)}(\mathfrak{R})$, where $\tau(X) = \det(X)^k$ and $\sigma = \tau\rho$ with some integer k. We write $f|_k \beta$, $\mathfrak{M}_k^{(n)}$, and $\mathcal{Q}_k^{(n)}$ for $f|_\tau \beta$, $\mathfrak{M}_\tau^{(n)}$, and $\mathcal{Q}_\tau^{(n)}$, if $\tau(X) = \det(X)^k$ with $k \in \mathbf{Z}$.

Given a point v of $\mathfrak{H}_n$, we understand by an $\mathfrak{R}$-*rational ρ-uniformizer* at v, a $\mathbf{C}_{d(\rho)}^{d(\rho)}$-valued meromorphic function E on $\mathfrak{H}_n$ finite at v whose columns belong to $\mathcal{Q}_\rho^{(n)}(\mathfrak{R})$ and such that $\det(E(v)) \neq 0$. If ρ is $\mathbf{Q}$-rational, the existence of a $\mathbf{Q}$-rational ρ-uniformizer at any point of $\mathfrak{H}_n$ follows from [12, Prop. 3.7], which also shows that for every $\mathbf{Q}$-rational representation ρ, there is a positive integer k depending only on ρ such that: (i) $\sigma(X) = \det(X)^k \rho(X)$ is a polynomial representation, and (ii) a σ-uniformizer with columns in $\mathfrak{M}_\sigma^{(n)}(\mathbf{Q})$ exists at any point of $\mathfrak{H}_n$. (As a consequence of this fact, we know that if $n > 1$, $\mathcal{Q}_\rho^{(n)}(\mathbf{C})$ is the set of all meromorphic functions f on $\mathfrak{H}_n$ such that $f|_\rho \gamma = f$ for all γ in a congruence subgroup of $\mathrm{Sp}(n, \mathbf{Z})$.)

Let us consider in particular a point v of $K_n^n \cap \mathfrak{H}_n$. Observe that $\mathbf{C}^n/(v\mathbf{Z}^n + \mathbf{Z}^n)$ is isogenous to the product of n elliptic curves with elements of K as complex multiplications. Therefore $g(v) \in K_{ab}$ for every $g \in \mathcal{Q}_0^{(n)}(K_{ab})$ finite at v. Now, for a K-rational representation ρ of GL_n, an element B of $\mathrm{GL}_{d(\rho)}(\mathbf{C})$ is called a ρ-*regulator* at v if $Bf(v)$ has components in K_{ab} for every $f \in \mathcal{Q}_\rho^{(n)}(K_{ab})$ finite at v. Such a B always exists. In fact, if E is a K-rational ρ-uniformizer at v, $E^{-1}f$ has components in $\mathcal{Q}_0^{(n)}(K_{ab})$ for every $f \in \mathcal{Q}_\rho^{(n)}(K_{ab})$;

hence $CE(v)^{-1}$ is a ρ-regulator at v for every $C \in \mathrm{GL}_{d(\rho)}(K_{\mathrm{ab}})$. Conversely, if B is a ρ-regulator at v, $BE(v)$ has components in K_{ab}. Thus we see that all the ρ-regulators at v form a coset $\mathrm{GL}_{d(\rho)}(K_{\mathrm{ab}})E(v)^{-1}$.

PROPOSITION 2.2. *Let* $h \in \mathfrak{C}_1^{(1)}(K_{\mathrm{ab}})$, $w_0 \in K \cap \mathfrak{H}_1$, *and* $\rho(X) = \det(X)^k \sigma(X)$ *for* $X \in \mathrm{GL}_n$ *with* $k \in \mathbf{Z}$ *and a* **Q**-*rational component* σ *of the representation*

$$(2.15) \qquad\qquad X \longmapsto X \otimes \cdots \otimes X \qquad\qquad (s \text{ copies of } X) .$$

If $h(w_0)$ *is finite and* $\neq 0$, *then* $h(w_0)^{-s-kn} \cdot 1_{d(\rho)}$ *is a* ρ-*regulator at every point of* $K_n^n \cap \mathfrak{H}_n$.

Proof. Put $M = w_0 \mathbf{Z} + \mathbf{Z}$ and $L = v\mathbf{Z}^n + \mathbf{Z}^n$ with any $v \in K_n^n \cap \mathfrak{H}_n$. Let f and g be the embeddings of $\mathbf{C}/M$ and $\mathbf{C}^n/L$ into projective spaces by theta functions as defined in [10, (1.5$_\mathrm{b}$)], and let E and A be their image varieties. Then we obtain a diagram

$$\begin{array}{ccc}
\mathbf{C}^n/M^n & \xrightarrow{\ f^n\ } & E^n \\
{\scriptstyle \alpha}\downarrow & & \downarrow{\scriptstyle \beta} \\
\mathbf{C}^n/L & \xrightarrow{\ g\ } & A
\end{array}$$

where $f^n(x_1, \cdots, x_n) = \big(f(x_1), \cdots, f(x_n)\big)$ for $x_j \in \mathbf{C}/M$, $\alpha(u \bmod M^n) = pu \bmod L$ for $u \in \mathbf{C}^n$ with a positive integer p such that $pM^n \subset L$, and β is an isogeny which makes the diagram commutative. Since all the torsion points of E and A are K_{ab}-rational [10, Theorem 1.1], we see that β is K_{ab}-rational. Let $u_1, \cdots, u_n$ be the standard variables on $\mathbf{C}^n$, and let $\rho(X) = X$ for $X \in \mathrm{GL}_n$. If B is a ρ-regulator at v, [10, Th. 1.5] shows that the derivations $\mathfrak{d}_1, \cdots, \mathfrak{d}_n$ defined by

$$\begin{bmatrix} \mathfrak{d}_1 \\ \vdots \\ \mathfrak{d}_n \end{bmatrix} = (2\pi i)^{-1} B \begin{bmatrix} \partial/\partial u_1 \\ \vdots \\ \partial/\partial u_n \end{bmatrix}$$

are K_{ab}-rational derivations on A. The same theorem applied to $\mathbf{C}/M$ shows that $\big(2\pi i \cdot h(w_0)\big)^{-1}\partial/\partial u_j$ are also K_{ab}-rational. Therefore B is $h(w_0)^{-1}$ times an element of $\mathrm{GL}_n(K_{\mathrm{ab}})$. Our assertion can be derived from this fact by considering $\det(B)^k B \otimes \cdots \otimes B$.

This proposition shows that the ρ-regulators depend only on K and ρ, and are independent of the choice of points of $K_n^n \cap \mathfrak{H}_n$. Therefore we hereafter call them simply *ρ-regulators* (when K is fixed, as in the present paper).

To define the arithmeticity of the elements of $\mathfrak{T}_{\tau, r}$ and $\mathfrak{M}_\rho(\Gamma)$, let τ_0 be a representation of GL_n defined by

$$(2.18) \qquad \tau_0(X) = \tau\begin{bmatrix} X & 0 \\ 0 & 1_q \end{bmatrix} \qquad (X \in \mathrm{GL}_n).$$

Take a τ_0-regulator B. We call an element g of $\mathfrak{X}_{\tau,r}$ *arithmetic* if $Bg_*(w)$ has components in K_{ab} for all $w \in K_q^n$, where g_* is defined by (2.9). Also we call an element f of $\mathfrak{M}_\rho(\Gamma)$ *arithmetic* if its Fourier coefficients g_r determined by (2.4) are all arithmetic. More generally, we consider an arbitrary subfield Ψ of C containing K_{ab}, and denote by $\mathfrak{X}_{\tau,r}(L, \Psi)$ (resp. $\mathfrak{X}_{\tau,r}(\Psi)$) the set of all $g \in \mathfrak{X}_{\tau,r}(L)$ (resp. $\mathfrak{X}_{\tau,r}$) such that $Bg_*(w) \in \Psi^{d(\tau)}$ for all $w \in K_q^n$, and by $\mathfrak{M}_\rho(\Gamma, \Psi)$ the set of all $f \in \mathfrak{M}_\rho(\Gamma)$ whose Fourier coefficients g_r belong to $\mathfrak{X}_{\tau,r}(\Psi)$ for all r, where τ is defined by (2.7). Further we denote by $\mathfrak{M}_\rho(\Psi)$ the union of $\mathfrak{M}_\rho(\Gamma, \Psi)$ for all congruence subgroups Γ, and by $\mathfrak{A}_\rho(\Psi)$ the set of all quotients f_1/f_2 with $f_1 \in \mathfrak{M}_\sigma(\Psi)$ and $0 \neq f_2 \in \mathfrak{M}_k(\Psi)$, where k is any integer and $\sigma(X, Y) = \det(Y)^k \rho(X, Y)$. Then we put

$$(2.19) \qquad \mathfrak{A}_\rho(\Gamma, \Psi) = \{ f \in \mathfrak{A}_\rho(\Psi) \mid f\!\mid_\rho\!\gamma = f \text{ for all } \gamma \in \Gamma \}.$$

Remark. If σ and τ are K-rational representations of GL_{n+q} such that $\sigma\begin{pmatrix} 1 & Y \\ 0 & 1 \end{pmatrix} = \tau\begin{pmatrix} 1 & Y \\ 0 & 1 \end{pmatrix}$ for $Y \in K_q^n$, then obviously $\mathfrak{X}_{\sigma,r} = \mathfrak{X}_{\tau,r}$. However, $\mathfrak{X}_{\sigma,r}(\Psi)$ does not necessarily coincide with $\mathfrak{X}_{\tau,r}(\Psi)$. For example, if $\sigma(X) = \det(X)^k \tau(X)$ with $k \in \mathbf{Z}$, we see easily that $\mathfrak{X}_{\sigma,r}(\Psi) = h(w_0)^{kn} \mathfrak{X}_{\tau,r}(\Psi)$ with the number $h(w_0)$ of Proposition 2.2. Similarly, if $\pi(X, Y) = \big(\det(X)/\det(Y)\big)^k \rho(X, Y)$ with $k \in \mathbf{Z}$ for $X \in \mathrm{GL}_{n+q}$ and $Y \in \mathrm{GL}_q$, then $\mathfrak{M}_\pi(\mathbf{C}) = \mathfrak{M}_\rho(\mathbf{C})$, but $\mathfrak{M}_\pi(\Psi) = h(w_0)^{kn} \mathfrak{M}_\rho(\Psi)$. This may look awkward, but causes no inconvenience.

To see that our definition of arithmeticity is natural and meaningful, let us now give an example of an element of $\mathfrak{X}_{\tau,r}(K_{\mathrm{ab}})$ with non-trivial τ. For $g \in \mathfrak{X}_{1,r}(L)$, where the subscript 1 denotes the trivial representation of GL_{n+q}, define $d_{\nu,r}g$ for $\nu = 1, \cdots, q$ by

$$(2.20) \qquad d_{\nu,r}g = \begin{bmatrix} (2\pi i)^{-1} \partial g/\partial w_{1\nu} \\ \vdots \\ (2\pi i)^{-1} \partial g/\partial w_{n\nu} \\ \bar{r}_{1\nu} g \\ \vdots \\ \bar{r}_{n\nu} g \end{bmatrix}.$$

It can easily be verified that $d_{\nu,r}g \in \mathfrak{X}_{\tau,r}(L)$, where $\tau(X) = X$. Moreover, we have

PROPOSITION 2.3. *Let* $g \in \mathfrak{X}_{1,r}(\Psi)$ *with a subfield* Ψ *of* C *containing* K_{ab}. *Then* $d_{\nu,r}g \in \mathfrak{X}_{\tau,r}(\Psi)$, *where* $\tau(X) = X$.

Proof. Put $b = h(w_0)^{-1}$ with the number $h(w_0)$ of Proposition 2.2, and define nq derivations $\mathfrak{d}_{\mu\nu}$ by $\mathfrak{d}_{\mu\nu} = (2\pi i)^{-1} b \cdot \partial/\partial w_{\mu\nu}$. As shown in the proof of Proposition 2.2, these $\mathfrak{d}_{\mu\nu}$ form a basis of K_{ab}-rational derivations of the field of K_{ab}-rational functions on C_q^n/L. Define $\mathfrak{d}_H$ by [10, (2.3)] with $H = H_r$. Then

$$\begin{bmatrix} (\mathfrak{d}_{1\nu})_H g \\ \vdots \\ \vdots \\ (\mathfrak{d}_{n\nu})_H g \\ \bar{r} g \end{bmatrix} = B \begin{bmatrix} 1_n & \bar{S}\bar{w} \\ 0 & 1_q \end{bmatrix} d_{\nu,r} g, \text{ where } B = \begin{bmatrix} b1_n & 0 \\ 0 & 1_q \end{bmatrix}.$$

Now [10, Th. 2.2] shows that if $g \in \mathfrak{T}_{1,r}(\Psi)$, the left-hand side times $e(i \cdot H(w, w)/4)$ is Ψ-rational for every $w \in K_q^n$. Since B is a τ_0-regulator, it follows that $d_{\nu,r} g \in \mathfrak{T}_{r,r}(\Psi)$.

PROPOSITION 2.4. *The set $\mathfrak{M}_\rho(\Psi)$ consists of the holomorphic elements of $\mathfrak{A}_\rho(\Psi)$ for every subfield Ψ of $\mathbf{C}$ containing K_{ab}.*

Proof. Let $h = f/g$ with $f \in \mathfrak{M}_\sigma(\Psi)$ and $0 \neq g \in \mathfrak{M}_k(\Psi)$, where $\sigma(X, Y) = \det(Y)^k \rho(X, Y)$. Suppose h holomorphic. Consider the Fourier expansions

$$f(w, z) = \sum_r a_r(w) e_q(rz), \ g(w, z) = \sum_r b_r(w) e_q(rz), \ h(w, z) = \sum_r c_r(w) e_q(rz),$$

where r runs over non-negative elements of a lattice J in $\mathfrak{K}_q$. Then

$$a_r(w) = \sum_{s+t=r} b_s(w) c_t(w).$$

Take a basis $\{p_\nu\}$ of $\mathfrak{K}_q$ over $\mathbf{Q}$ consisting of positive definite elements. To each $r \in \mathfrak{K}_q$, we can assign its "coordinates" $\mathrm{tr}(p_\nu r)$ and put the elements of J in a lexicographic order according to the coordinates. If b_u is the "first" non-zero coefficient of g, we see that $b_u(w)^\lambda c_t(w)$, with a suitable positive integer λ, is a linear combination $\sum \varphi_r a_r$ with polynomials φ_r in b_s with coefficients in $\mathbf{Q}$. Therefore, if B is a τ_0-regulator, where $\tau_0(X) = \rho\left(\begin{pmatrix} X & 0 \\ 0 & 1 \end{pmatrix}, 1\right)$, we see that $B(c_t)_*(w)$ has components in Ψ for every w in K_q^n such that $b_u(w) \neq 0$. This is sufficient to guarantee that $h \in \mathfrak{M}_\rho(\Psi)$.

3. The embedding of $\mathfrak{Z}$ into $\mathfrak{H}_m$

The space $\mathfrak{Z}$ modulo a congruence subgroup of $G_{\mathbf{Q}+}$ is the variety of moduli of a certain family of abelian varieties which can be obtained as follows. First define a diagonal representation Ξ of K by

$$(3.1) \qquad \Xi(a) = \begin{bmatrix} a1_{n+q} & 0 \\ 0 & \bar{a}1_q \end{bmatrix} \qquad (a \in K).$$

Let $(A, \mathcal{C}, \iota)$ be a structure formed by an abelian variety A of dimension m defined over $\mathbf{C}$, a polarization $\mathcal{C}$ of A, and an injection ι of K into $\mathrm{End}(A) \otimes \mathbf{Q}$

such that: (i) the representation of K through ι on the tangent space of A at the origin is equivalent to Ξ; (ii) the involution of $\mathrm{End}(A)\otimes\mathbf{Q}$ determined by $\mathcal{C}$ maps $\iota(a)$ onto $\iota(\bar{a})$ for every $a \in K$. Under these conditions, A can be given as a complex torus $\mathbf{C}^m/D$ with a lattice D in $\mathbf{C}^m$, and $\iota(a)$ is represented by $\Xi(a)$. Then K acts, through Ξ, on $\mathbf{Q}D$, and hence there is an isomorphism p of K_m^1 onto $\mathbf{Q}D$ such that $p(ax) = \Xi(a)p(x)$ for $a \in K$ and $x \in K_m^1$. Put $M = p^{-1}(D)$. Then M is a lattice in K_m^1. Let $E(x, y)$ be the Riemann (alternating) form of a basic divisor in $\mathcal{C}$. Then there is a skew-hermitian element T of K_m^m such that

$$(3.2) \qquad E\big(p(u), p(v)\big) = \mathrm{Tr}_{K/\mathbf{Q}}(u\,T\cdot{}^t\bar{v}) \qquad\qquad \big((u, v) \in K_m^1 \times K_m^1\big) .$$

In this situation we say that $(A, \mathcal{C}, \iota)$ is of type $(K, \Xi; T, M)$. (For a more detailed discussion, see [4].) Now let $\{e_1, \cdots, e_m\}$ be the standard basis of K_m^1, and put $p(e_\nu) = \begin{bmatrix} u_\nu \\ u_\nu' \end{bmatrix}$ with $u_\nu \in \mathbf{C}^{n+q}$ and $u_\nu' \in \mathbf{C}^q$. Define an element X of $\mathbf{C}_m^m$ by

$$X = \begin{bmatrix} u_1 \cdots u_m \\ \bar{u}_1' \cdots \bar{u}_m' \end{bmatrix} .$$

As shown in [4], we have $i\bar{X}T^{-1}\cdot{}^tX = \begin{pmatrix} P & 0 \\ 0 & -Q \end{pmatrix}$ with positive definite hermitian matrices P and Q of size $n + q$ and q, respectively.

Let us now assume that $T = (2\bar{R})^{-1}$ with the matrix R of the form (1.1). Then ${}^t\bar{X} \in \mathfrak{Y}$, and hence ${}^t\bar{X}$ corresponds to a point $\mathfrak{z}$ of $\mathfrak{Z}$ through the map (1.7). Conversely, for every $\mathfrak{z} = (w, z) \in \mathfrak{Z}$, put

$$(3.3) \qquad {}^t\overline{P(\mathfrak{z})} = \begin{bmatrix} u_1 \cdots u_m \\ \bar{u}_1' \cdots \bar{u}_m' \end{bmatrix}$$

with $u_\nu \in \mathbf{C}^{n+q}$ and $u_\nu' \in \mathbf{C}^q$, and define $p_\mathfrak{z}: K_m^1 \to \mathbf{C}^m$ by

$$p_\mathfrak{z}\big(\textstyle\sum_{\nu=1}^m a_\nu e_\nu\big) = \sum_{\nu=1}^m \Xi(a_\nu) \begin{bmatrix} u_\nu \\ u_\nu' \end{bmatrix} \qquad\qquad (a_\nu \in K) .$$

Also define an $\mathbf{R}$-valued alternating form $E_\mathfrak{z}(u, v)$ on $\mathbf{C}^m$ so that

$$E_\mathfrak{z}\big(p_\mathfrak{z}(x), p_\mathfrak{z}(y)\big) = \mathrm{Tr}_{K/\mathbf{Q}}(x\,T\cdot{}^t\bar{y}) \qquad\qquad \big((x, y) \in K_m^1 \times K_m^1\big) .$$

Then $\mathbf{C}^m/p_\mathfrak{z}(M)$, together with $E_\mathfrak{z}$, gives a structure of type $(K, \Xi; T, M)$, if M is chosen so that $E_\mathfrak{z}\big(p_\mathfrak{z}(M), p_\mathfrak{z}(M)\big) = \mathbf{Z}$.

Now for each $\alpha \in G_{\mathbf{Q}+}$, put

$$(3.4) \qquad \Lambda(\alpha, \mathfrak{z}) = \begin{bmatrix} {}^t\kappa(\alpha, \mathfrak{z}) & 0 \\ 0 & {}^t\mu(\alpha, \mathfrak{z}) \end{bmatrix} \qquad\qquad (\mathfrak{z} \in \mathfrak{Z}) .$$

Writing $p(x, \mathfrak{z})$ for $p_\mathfrak{z}(x)$, we obtain, from (1.8),

$$(3.5) \qquad p(x\bar{\alpha}, \mathfrak{z}) = \Lambda(\alpha, \mathfrak{z})p(x, \alpha(\mathfrak{z})) \qquad\qquad (x \in K_m^1, \mathfrak{z} \in \mathfrak{Z}) .$$

Given a structure $(A, \mathcal{C}, \iota)$ of the above type corresponding to a point $\mathfrak{z} = (w, z)$ of $\mathfrak{Z}$, we have a polarized abelian variety $(A, \mathcal{C})$, which determines a point Z of $\mathfrak{H}_m$. Let us now derive an explicit expression for Z in terms of (w, z). We now assume (without losing generality) that S is diagonal. This assumption will be retained until the end of Section 5. Take a generating element δ of K such that $\bar{\delta} = -\delta$. Observe that

$$\{e_1, \cdots, e_n, e_{n+1}, \cdots, e_{n+q}, \delta e_{n+1}, \cdots, \delta e_{n+q},$$
$$e_1\bar{R}, \cdots, e_n\bar{R}, e_{n+q+1}, \cdots, e_m, -\delta^{-1}e_{n+q+1}, \cdots, -\delta^{-1}e_m\}$$

is a basis of K_m^1 over $\mathbf{Q}$. Moreover, if we denote these elements by $g_1, \cdots, g_{2m}$, then

$$\left(\mathrm{Tr}_{K/\mathbf{Q}}(g_i T \cdot {}^t\bar{g}_j)\right)_{i,j=1,\cdots,2m} = J_m .$$

The period matrix of A, or rather of the map $p_\mathfrak{z}$, with respect to the basis $\{g_k\}$ is $(\omega_1 \ \omega_2)$ with

$$\omega_1 = \left(p_\mathfrak{z}(g_1) \cdots p_\mathfrak{z}(g_m)\right), \qquad \omega_2 = \left(p_\mathfrak{z}(g_{m+1}) \cdots p_\mathfrak{z}(g_{2m})\right) .$$

Put ${}^t\overline{P(\mathfrak{z})}$ in the form

$$\overline{{}^tP(\mathfrak{z})} = \begin{bmatrix} u & v & x \\ \bar{u}' & \bar{v}' & \bar{x}' \end{bmatrix}, \quad u \in \mathbf{C}_n^{n+q}, \quad v, x \in \mathbf{C}_q^{n+q} .$$

Since $\left(p_\mathfrak{z}(e_1) \cdots p_\mathfrak{z}(e_m)\right) = \begin{pmatrix} u & v & x \\ u' & v' & x' \end{pmatrix}$, we have

$$\omega_1 = \begin{bmatrix} u & v & \delta v \\ u' & v' & -\delta v' \end{bmatrix}, \qquad \omega_2 = \begin{bmatrix} -uS & x & -\delta^{-1}x \\ u'S & x' & \delta^{-1}x' \end{bmatrix},$$

or more explicitly

$$(3.6) \qquad \omega_1 = \begin{bmatrix} -S^{-1} & w & \delta w \\ 0 & z & \delta z \\ {}^tw & {}^tz & -\delta \cdot {}^tz \end{bmatrix}, \qquad \omega_2 = \begin{bmatrix} 1_n & 0 & 0 \\ 0 & 1_q & -\delta^{-1}1_q \\ {}^twS & 1_q & \delta^{-1}1_q \end{bmatrix} .$$

Then $Z = \omega_2^{-1}\omega_1$ is the point of $\mathfrak{H}_m$ corresponding to $(A, \mathcal{C})$ relative to $\{g_k\}$. We have actually

$$(3.7) \quad Z = \omega_2^{-1}\omega_1 = \begin{bmatrix} -S^{-1} & w & \delta w \\[2mm] {}^tw & \dfrac{1}{2}(z + {}^tz - {}^twSw) & \dfrac{\delta}{2}(z - {}^tz - {}^twSw) \\[2mm] \delta \cdot {}^tw & \dfrac{\delta}{2}({}^tz - z - {}^twSw) & -\dfrac{\delta^2}{2}(z + {}^tz + {}^twSw) \end{bmatrix} .$$

Put $Z = \varepsilon(w, z)$. Then ε defines an embedding of $\mathfrak{Z}$ into $\mathfrak{H}_m$ compatible with the group action in the following sense. To each $\alpha \in K_m^m$, assign an element $\alpha^* = (a_{jk})$ of $\mathbf{Q}_{2m}^{2m}$ by

$$(3.8) \qquad g_j \bar{\alpha} = \sum_{k=1}^{2m} a_{jk} g_k \qquad (j = 1, \cdots, 2m) .$$

Then $\alpha \mapsto \alpha^*$ defines an embedding of $G_{\mathbf{Q}}$ into $G_{\mathbf{Q}}^{(m)}$ satisfying $\nu(\alpha) = \nu(\alpha^*)$ and

$$(3.9) \qquad \varepsilon \circ \alpha = \alpha^* \circ \varepsilon \qquad (\alpha \in G_{\mathbf{Q}+}) .$$

Indeed, put $p'(x) = p(x, \alpha(\mathfrak{z}))$. From (3.5), we obtain

$$(3.10) \qquad p'(g_j) = \Delta(\alpha, \mathfrak{z})^{-1} p\left(\sum_{k=1}^{2m} a_{jk} g_k, \mathfrak{z}\right) .$$

Viewing ω_j for $j = 1, 2$ as functions of $\mathfrak{z}$, put $\omega_j = \omega_j(\mathfrak{z})$. Then (3.10) implies that

$$\big(\omega_1(\alpha(\mathfrak{z})) \ \ \omega_2(\alpha(\mathfrak{z}))\big) = \Delta(\alpha, \mathfrak{z})^{-1}\big(\omega_1(\mathfrak{z}) \ \ \omega_2(\mathfrak{z})\big) \cdot {}^t\alpha^* ,$$

and therefore $\varepsilon(\alpha(\mathfrak{z})) = \alpha^*(\varepsilon(\mathfrak{z}))$. Furthermore, if $\alpha^* = \begin{pmatrix} A & B \\ C & D \end{pmatrix}$ and $Z = \varepsilon(\mathfrak{z})$, then

$$(3.11) \qquad CZ + D = {}^t\omega_2(\alpha(\mathfrak{z})) \cdot {}^t\Delta(\alpha, \mathfrak{z}) \cdot {}^t\omega_2(\mathfrak{z})^{-1} .$$

Define a $\mathbf{C}_m^m$-valued function χ on $\mathfrak{B}$ by

$$(3.12) \qquad \chi(\mathfrak{z}) = \begin{bmatrix} 1_{n+q} & 0 \\ 0 & \dfrac{\delta}{2} 1_q \end{bmatrix} \omega_2(\mathfrak{z}) = \begin{bmatrix} 1_n & 0 & 0 \\ 0 & 1_q & -\delta^{-1} 1_q \\ \dfrac{\delta}{2} {}^t w S & \dfrac{\delta}{2} 1_q & \dfrac{1}{2} 1_q \end{bmatrix} \qquad (\mathfrak{z} = (w, z) \in \mathfrak{B}) .$$

From (3.11), we obtain

$$(3.13) \qquad CZ + D = {}^t\chi(\alpha(\mathfrak{z})) \begin{bmatrix} \kappa(\alpha, \mathfrak{z}) & 0 \\ 0 & \mu(\alpha, \mathfrak{z}) \end{bmatrix} \cdot {}^t\chi(\mathfrak{z})^{-1} ,$$

and hence, by (1.19),

$$(3.14) \qquad \det(CZ + D) = \det(\alpha)^{-1} \nu(\alpha)^{n+q} \det\big(\mu(\alpha, \mathfrak{z})\big)^2 .$$

We now study the pull-back of theta functions and Siegel modular forms. Define $T_{\sigma,k,\xi}$ as in [12, § 3] with a $\mathbf{Q}$-rational polynomial representation σ of GL_m, a non-negative integer k, and a non-negative symmetric element ξ of $\mathbf{Q}_{2q}^{2q}$. Write ξ in the form

$$(3.15) \qquad \xi = \begin{bmatrix} \xi_1 & \xi_2 \\ {}^t\xi_2 & \xi_3 \end{bmatrix}$$

with ξ_1, ξ_2, ξ_3 of size q. Recall that an element f of $T_{\sigma,k,\xi}$ is a $\mathbf{C}^{d(\sigma)}$-valued function on $\mathbf{C}_{2q}^n \times \mathfrak{H}_n$, written $f(u; z)$ with $u \in \mathbf{C}_{2q}^n$ and $z \in \mathfrak{H}_n$.

PROPOSITION 3.1. *Given* $f \in T_{\sigma,k,\xi}(\Psi)$ *with* σ, k, ξ *as above and with a subfield* Ψ *of* $\mathbf{C}$ *containing* K_{ab}, *put*

$$(3.16) \qquad h(w) = \sigma(C)(P^{\sigma,\varepsilon}f)(w,\, \delta w;\, -S^{-1}) \qquad\qquad (w \in \mathbf{C}_q^{\times}),$$

where $P^{\sigma,\varepsilon}f$ is defined by [12, (3.5)], and

$$(3.17) \qquad C = \begin{bmatrix} 1_n & 0 & 0 \\ 0 & \dfrac{1}{2}1_q & \dfrac{-\delta}{2}1_q \\ 0 & \delta^{-1}1_q & 1_q \end{bmatrix}.$$

Then $h \in \mathfrak{T}_{\tau,r}(\Psi)$, and

$$(3.18) \qquad h_*(w) = \sigma(C)(P_{\sigma,\varepsilon}f)(w,\, \delta w;\, -S^{-1}),$$

where $P_{\sigma,\varepsilon}f$ is defined by [12, (3.6)], and

$$\tau(X) = \sigma\begin{bmatrix} X & 0 \\ 0 & 1_q \end{bmatrix} \qquad\qquad (X \in \mathrm{GL}_{n+q}),$$

$$r = \frac{1}{2}\left(\xi_1 - \delta^2\xi_3 + ({}^t\xi_2 - \xi_2)\right).$$

Proof. That $h \in \mathfrak{T}_{\tau,r}$ and formula (3.18) can be verified in a straight-forward way. Now, by [12, (3.23)], for every a, b in $\mathbf{Q}_q^n$, $(P_{\sigma,\varepsilon}f)(za + b;\, z)$ as a function of z belongs to $\mathfrak{M}_{\pi}^{(n)}(\Psi)$, where

$$\pi(Y) = \sigma\begin{bmatrix} Y & 0 \\ 0 & 1_{2q} \end{bmatrix} \qquad\qquad (Y \in \mathrm{GL}_n).$$

Since $\sigma(C)$ commutes with $\pi(Y)$, $\sigma(C)(P_{\sigma,\varepsilon}f)(za + b;\, z)$ belongs to $\mathfrak{M}_{\pi}^{(n)}(\Psi)$. Therefore if B is a π-regulator, $Bh_*(w)$ is Ψ-rational for every $w \in K_q^n$, and hence $h \in \mathfrak{T}_{\tau,r}(\Psi)$.

Let $F \in \mathfrak{A}_{\sigma}^{(m)}(\mathbf{C})$ with a $\mathbf{Q}$-rational (not necessarily polynomial) represen-tation σ of GL_m. Assuming $F \circ \varepsilon$ is meaningful, define a meromorphic func-tion F^{ε} on $\mathfrak{Z}$ by

$$(3.19) \qquad F^{\varepsilon}(\mathfrak{z}) = \sigma\big({}^t\chi(\mathfrak{z})^{-1}\big)F\big(\varepsilon(\mathfrak{z})\big) \qquad\qquad (\mathfrak{z} \in \mathfrak{Z}),$$

and define also a representation ρ of $\mathrm{GL}_{n+q} \times \mathrm{GL}_q$ by

$$(3.20) \qquad \rho(X,\, Y) = \sigma\begin{bmatrix} X & 0 \\ 0 & Y \end{bmatrix} \qquad\qquad (X \in \mathrm{GL}_{n+q},\ Y \in \mathrm{GL}_q).$$

By (3.9) and (3.13), we have

$$(3.21) \qquad F^{\varepsilon}|_{\rho}\alpha = (F|_{\sigma}\alpha^*)^{\varepsilon} \qquad \text{for every} \quad \alpha \in G_{\mathbf{Q}+}.$$

Therefore $F^{\varepsilon} \in \mathfrak{A}_{\rho}(\mathbf{C})$. Obviously $F^{\varepsilon} \in \mathfrak{M}_{\rho}(\mathbf{C})$ if $F \in \mathfrak{M}_{\sigma}^{(m)}(\mathbf{C})$. In particular, if $F \in \mathfrak{M}_k^{(m)}(\mathbf{C})$ with $k \in \mathbf{Z}$, $F^{\varepsilon} \in \mathfrak{M}_{2k}(\mathbf{C})$ by virtue of (3.14).

PROPOSITION 3.2. *The notation being as above, let Ψ be a subfield of $\mathbf{C}$ containing K_{ab}. Then $F^{\varepsilon} \in \mathfrak{A}_{\rho}(\Psi)$ if $F \in \mathfrak{A}_{\sigma}^{(m)}(\Psi)$; $F^{\varepsilon} \in \mathfrak{M}_{\rho}(\Psi)$ if $F \in \mathfrak{M}_{\sigma}^{(m)}(\Psi)$.*

Proof. First suppose that σ is a polynomial representation, and $F \in \mathfrak{M}_\sigma^{(m)}(\Psi)$. Write the variable Z on $\mathfrak{H}_m$ in the form

$$Z = \begin{bmatrix} Z' & U \\ {}^t U & W \end{bmatrix}, \quad U \in \mathbf{C}_{2q}^n, \quad Z' \in \mathfrak{H}_n, \quad W \in \mathfrak{H}_{2q}.$$

As observed in [12, § 6], F has a Fourier expansion

$$F(Z) = \sum_\xi \varphi_\xi(U; Z') e_{2q}\left(\frac{1}{2} \xi W\right)$$

with $\varphi_\xi \in T_{\sigma,0,\xi}(\Psi)$. Substituting expression (3.7) for Z, we obtain

$$F\big(\varepsilon(w, z)\big) = \sum_r \psi_r(w) e_q(rz),$$

$$\psi_r(w) = \sum e_{2q}\left(\frac{-1}{4} \xi \cdot {}^t(w \ \delta w) S(w \ \delta w)\right) \varphi_\xi(w, \delta w; -S^{-1}),$$

where the last sum is extended over all ξ of the form (3.15) such that $r = (1/2)(\xi_1 - \delta^2 \xi_3 + ({}^t\xi_2 - \xi_2))$. Obviously this is a finite sum. Now we have

$$
{}^t\chi(\mathfrak{z})^{-1} = C \begin{bmatrix} 1_n & \dfrac{1}{2}\bar{S}w & \dfrac{\delta}{2}\bar{S}w \\ 0 & 1_q & 0 \\ 0 & 0 & 1_q \end{bmatrix},
$$

and hence $F^\varepsilon(\mathfrak{z}) = \sum_r g_r(w) e_q(rz)$, where

$$g_r(w) = \sigma(C) \sum_\xi (P^{\sigma,\xi}\varphi_\xi)(w, \delta w; -S^{-1}).$$

This together with Proposition 3.1 shows that $F^\varepsilon \in \mathfrak{M}_\rho(\Psi)$. Now let σ be an arbitrary **Q**-rational representation of GL_m. For $F \in \mathfrak{A}_\sigma^{(m)}(\Psi)$, we have $F = h^{-1}F_1$ with $h \in \mathfrak{M}_k^{(m)}(\Psi)$ and $F_1 \in \mathfrak{M}_\tau^{(m)}(\Psi)$, where $\tau(X) = \det(X)^k \sigma(X)$. By Lemma 3.3 below, we may assume that $k > 0$, τ is a polynomial representation, and $h \circ \varepsilon \neq 0$. Applying our result to h and F_1, we see that $F^\varepsilon = (h^\varepsilon)^{-1} F_1^\varepsilon \in \mathfrak{A}_\rho(\Psi)$ as desired. If $F \in \mathfrak{M}_\sigma^{(m)}(\Psi)$, we have $F^\varepsilon \in \mathfrak{M}_\rho(\Psi)$ by virtue of Proposition 2.4. In this proof we needed

LEMMA 3.3. *Suppose an element f of $\mathfrak{A}_\sigma^{(m)}(\Psi)$ with a subfield Ψ of* **C** *is finite at a point Z_0 of $\mathfrak{H}_m$. Then there exist a positive integer k and an element h of $\mathfrak{M}_k^{(m)}(\Psi)$ such that $\tau(X) = \det(X)^k \sigma(X)$ is a polynomial representation, $h(Z_0) \neq 0$, and $hf \in \mathfrak{M}_\tau^{(m)}(\Psi)$.*

Proof. As already mentioned, there is a polynomial representation ρ of GL_m of the form $\rho(X) = \det(X)^s \sigma(X)$ with $0 < s \in \mathbf{Z}$ such that a ρ-uniformizer, say B, at Z_0 whose columns belong to $\mathfrak{M}_\rho^{(m)}(\mathbf{Q})$ exists. Put $p = B^{-1}f$. Then the components of p belong to $\mathfrak{A}_{-s}^{(m)}(\Psi)$. Take $q \in \mathfrak{A}_s^{(m)}(\mathbf{Q})$ so that $q(Z_0) \neq 0$. (For example, take $q(Z) = \theta\,(0, Z; r, 0)$ with a suitable $r \in \mathbf{Q}^m$, where θ is the classical theta function defined by [9, (10)].) Then $qp \in \mathcal{A}_0^{(m)}(\Psi)$. Since $\mathfrak{M}_k^{(m)}(\Psi) = \mathfrak{M}_k^{(m)}(\mathbf{Q}) \otimes_{\mathbf{Q}} \Psi$ as shown in [8, (9)] or [12, Prop. 2.1], we see that qp belongs to $\Psi \cdot \mathcal{A}_0^{(m)}(\Gamma_N, \mathbf{Q})$ with a positive integer N, where

$$\Gamma_N = \{\gamma \in \mathrm{Sp}(m, \mathbf{Z}) \mid \gamma \equiv 1_{2m} \ (\mathrm{mod} \ N)\} \, .$$

It is well known that a basis of $\mathfrak{M}_t^{(m)}(\Gamma_N)$, for sufficiently large t, gives a projective embedding of $\Gamma_N \backslash \mathfrak{H}_m$. (This is a special case of the theorem of Baily-Borel.) Again by [8, (9)], the basis, say $\{b_i\}$, can be taken from $\mathfrak{M}_t^{(m)}(\Gamma_N, \mathbf{Q})$. This gives a $\mathbf{Q}$-rational model of $\Gamma_N \backslash \mathfrak{H}_m$ whose function field over $\mathbf{Q}$ is $\mathcal{A}_0^{(m)}(\Gamma_N, \mathbf{Q})$. Therefore, since qp is finite at Z_0, every component of qp can be expressed as a quotient u/v with homogeneous polynomials u and v of $\{b_i\}$ with coefficients in Ψ such that $v(Z_0) \neq 0$. Let w be the product of all such v chosen for each component of qp, and let $h = wq$. Then $h(Z_0) \neq 0$, $h \in \mathfrak{M}_k^{(m)}(\Psi)$ for some $k > s$, the components of wqp belong to $\mathfrak{M}_{k-s}^{(m)}(\Psi)$, and $hf = Bwqp \in \mathfrak{M}_r^{(m)}(\Psi)$ with $\tau(X) = \det(X)^k \sigma(X)$ as desired.

4. Arithmetic automorphic forms in the context of canonical models

We now consider the canonical models, in the sense of [3] and [6], of $\mathfrak{Z}$ modulo congruence subgroups of $G_{\mathbf{Q}+}$. Let $\mathfrak{K}$ denote the union of the function fields of all such canonical models, i.e., the field $\mathfrak{L}_{j_0}$ of [3, 4.2]; $\mathfrak{K}$ may be called the field of arithmetic automorphic functions on $\mathfrak{Z}$ with respect to G. Similarly, let $\mathfrak{K}^{(m)}$ be the field of all arithmetic automorphic functions on $\mathfrak{H}_m$ with respect to $G^{(m)}$. As shown in [7] and [8], $\mathfrak{K}^{(m)} = \mathcal{A}_0^{(m)}(\mathbf{Q}_{\mathrm{ab}})$. The corresponding result holds for G, namely

THEOREM 4.1. *For every subfield Ψ of $\mathbf{C}$ containing K_{ab}, we have* $\mathfrak{A}_0(\Psi) = \Psi\mathfrak{K}$.

To prove this, we first note

PROPOSITION 4.2. *If $f \in K_{\mathrm{ab}}\mathfrak{K}^{(m)}$ and $f \circ \varepsilon$ is defined, then $f \circ \varepsilon \in K_{\mathrm{ab}}\mathfrak{K}$. Moreover, every element of $K_{\mathrm{ab}}\mathfrak{K}$ can be obtained as such $f \circ \varepsilon$.*

This follows from a general principle of the embedding of canonical models of a group into those of another group, and can be proved by the same argument as in [6, I, § 8]. See also Deligne [2, Cor. 5.7] for a general theorem in a somewhat different formulation.

Since $\Psi\mathfrak{K}^{(m)} = \mathcal{A}_0^{(m)}(\Psi)$ as shown in [7] and [8], Propositions 3.2 and 4.2 imply that $\Psi\mathfrak{K} = \Psi\mathfrak{K}^{(m)} \circ \varepsilon \subset \mathfrak{A}_0(\Psi)$. To prove the opposite inclusion, we note that $\mathfrak{A}_0(\Psi)$ and $\mathbf{C}$ are linearly disjoint over Ψ. In fact, let $c_1, \cdots, c_s$ be elements of $\mathbf{C}$ linearly independent over Ψ, and suppose $\sum_\nu c_\nu f_\nu / g_\nu = 0$ with f_ν and g_ν in $\mathfrak{M}_{k_\nu}(\Psi)$. Multiplying this with $g_1 \cdots g_s$, we obtain $\sum_\nu c_\nu h_\nu = 0$

with $h_\nu \in \mathfrak{M}_k(\Psi)$, where $k = \sum_\nu k_\nu$. Put $h_\nu(w, z) = \sum_r h_{\nu r}(w)e_q(rz)$. Then $\sum_\nu c_\nu(h_{\nu r})_*(w) = 0$ for all r. Now $(h_{\nu r})_*(w) \in \Psi$ for all $w \in K_q^n$. Therefore the linear independence of $\{c_\nu\}$ implies that $(h_{\nu r})_*(w) = 0$ for all ν, all r, and all $w \in K_q^n$, and hence $h_\nu = 0$ for all ν, which proves the desired linear disjointness. Let $f \in \mathfrak{A}_0(\Psi)$. Since $\mathfrak{A}_0(\mathbf{C}) = \mathbf{C}\mathfrak{R}$, we have $f = (\sum_\lambda b_\lambda g_\lambda)/(\sum_\lambda b_\lambda h_\lambda)$ with finitely many elements g_λ, h_λ of $\Psi\mathfrak{R}$ and $b_\lambda \in \mathbf{C}$. Changing $\{b_\lambda\}$ for its suitable subset, we may assume that the b_λ are linearly independent over Ψ. We have $\sum_\lambda b_\lambda(g_\lambda - fh_\lambda) = 0$ and $g_\lambda - fh_\lambda \in \mathfrak{A}_0(\Psi)$, so that the linear independence of b_λ over $\mathfrak{A}_0(\Psi)$ implies that $g_\lambda = fh_\lambda$ for all λ, and hence $f \in \Psi\mathfrak{R}$, which completes the proof of Theorem 4.1.

PROPOSITION 4.3. *For every K-rational representation ρ of $\mathrm{GL}_{n+q} \times \mathrm{GL}_q$ and every point $\mathfrak{z}_0$ of $\mathfrak{Z}$, there exists a $\mathbf{C}_{d(\rho)}^{d(\rho)}$-valued meromorphic function A on $\mathfrak{Z}$ such that*

(i) *A is finite at $\mathfrak{z}_0$ and $\det(A(\mathfrak{z}_0)) \neq 0$;*

(ii) *the columns of $A|_\rho \alpha$ belong to $\mathfrak{A}_\rho(K_{ab})$ for all $\alpha \in G_{\mathbf{Q}+}$.*

Proof. Take a $\mathbf{Q}$-rational π-uniformizer F at $\varepsilon(\mathfrak{z}_0)$, where $\pi(U) = U$ for $U \in \mathrm{GL}_m$. Put $E = F^\varepsilon$. By (3.21) and Proposition 3.2, the columns of $E|_{\rho_1}\alpha$ belong to $\mathfrak{A}_{\rho_1}(K_{ab})$ for every $\alpha \in G_{\mathbf{Q}+}$, where $\rho_1(X, Y) = \begin{pmatrix} X & 0 \\ 0 & Y \end{pmatrix}$ for $X \in \mathrm{GL}_{n+q}$ and $Y \in \mathrm{GL}_q$. Obviously $\det(E(\mathfrak{z}_0)) \neq 0$. We can take an $(n + q) \times (n + q)$-matrix B from the first $n + q$ rows of E so that $\det(B(\mathfrak{z}_0)) \neq 0$, and a $(q \times q)$-matrix C from the last q rows of E so that $\det(C(\mathfrak{z}_0)) \neq 0$. Then the columns of $B|_\sigma\alpha$ and $C|_\tau\alpha$ belong to $\mathfrak{A}_\sigma(K_{ab})$ and $\mathfrak{A}_\tau(K_{ab})$ respectively, for all $\alpha \in G_{\mathbf{Q}+}$, where

$$\sigma(X, Y) = X, \quad \tau(X, Y) = Y \qquad (X \in \mathrm{GL}_{n+q}, \ Y \in \mathrm{GL}_q).$$

Now every K-rational irreducible representation of $\mathrm{GL}_{n+q} \times \mathrm{GL}_q$ can be obtained as a K-rational component of a representation of the type

$$(X, Y) \longmapsto \det(X)^h \det(Y)^k X \otimes \cdots \otimes X \otimes Y \otimes \cdots \otimes Y$$

with $h, k \in \mathbf{Z}$. Therefore considering a suitable "submatrix" of

$$\det(B)^h \det(C)^k B \otimes \cdots \otimes B \otimes C \otimes \cdots \otimes C,$$

we obtain our asssertion.

THEOREM 4.4. *If ρ is a K-rational representation of $\mathrm{GL}_{n+q} \times \mathrm{GL}_q$, $f \in \mathfrak{A}_\rho(K_{ab})$, and $\alpha \in G_{\mathbf{Q}+}$, then $f|_\rho\alpha \in \mathfrak{A}_\rho(K_{ab})$; $f|_\rho\alpha \in \mathfrak{M}_\rho(K_{ab})$ if $f \in \mathfrak{M}_\rho(K_{ab})$.*

This implies in particular that our definition of the arithmeticity of automorphic forms does not depend on the choice of "cusp."

Proof. Take A as in Proposition 4.3 with any $\mathfrak{z}_0$. If $f \in \mathfrak{A}_\rho(K_{ab})$, we see that $A^{-1}f$ has components in $\mathfrak{A}_0(K_{ab})$. Now $\mathfrak{A}_0(K_{ab}) = K_{ab}\mathfrak{R}$, and $\mathfrak{R}$ is stable under G_{Q+} as proved in [3]. Therefore $f|_\rho\alpha = (A|_\rho\alpha)(A^{-1}f)\circ\alpha \in \mathfrak{A}_\rho(K_{ab})$. If f is holomorphic, so is $f|_\rho\alpha$, and hence $f|_\rho\alpha \in \mathfrak{M}_\rho(K_{ab})$ by Proposition 2.4.

Define now an operator D which assigns to a function f on $\mathfrak{Z}$ a $\mathbf{C}_q^{n+q}$-valued function Df on $\mathfrak{Z}$ by

$$(4.1) \qquad Df = \begin{bmatrix} \partial f/\partial w_{11} & \cdots & \partial f/\partial w_{1q} \\ \cdots & \cdots & \cdots \\ \partial f/\partial w_{n1} & \cdots & \partial f/\partial w_{nq} \\ \partial f/\partial z_{11} & \cdots & \partial f/\partial z_{1q} \\ \cdots & \cdots & \cdots \\ \partial f/\partial z_{q1} & \cdots & \partial f/\partial z_{qq} \end{bmatrix}.$$

By (1.15), we have

$$(4.2) \qquad (Df)\circ\alpha = \nu(\alpha)^{-1}\kappa(\alpha, \mathfrak{z})D(f\circ\alpha)\cdot{}^t\mu(\alpha, \mathfrak{z}) \qquad (\alpha \in G_{\mathbf{R}+}).$$

Let σ_1 be a representation of $\mathrm{GL}_{n+q}\times\mathrm{GL}_q$ on $\mathbf{C}_q^{n+q}$ defined by

$$(4.3) \qquad \sigma_1(X, Y)U = XU\cdot{}^tY \qquad (X\in\mathrm{GL}_{n+q}(\mathbf{C}),\ Y\in\mathrm{GL}_q(\mathbf{C}),\ U\in\mathbf{C}_q^{n+q}).$$

Then (4.2) shows that $Df \in \mathfrak{A}_{\sigma_1}(\mathbf{C})$ if $f \in \mathfrak{A}_0(\mathbf{C})$.

PROPOSITION 4.5. $(2\pi i)^{-1}Df \in \mathfrak{A}_{\sigma_1}(K_{ab})$ if $f \in \mathfrak{A}_0(K_{ab})$.

Proof. Write $f = g/h$ with $g, h \in \mathfrak{M}_k(K_{ab}), k \in \mathbf{Z}$. Then $Df = h^{-2}(hDg - gDh)$. Consider Fourier expansions

$$g(w, z) = \sum_r g_r(w)e_q(rz), \quad h(w, z) = \sum_r h_r(w)e_q(rz).$$

Then $h^2Df = \sum_t p_t(w)e_q(tz)$, where p_t is the $(n + q) \times q$-matrix whose ν^{th} column is

$$2\pi i \sum_{r+s=t} (h_s\cdot d_{\nu,r}g_r - g_s\cdot d_{\nu,r}h_r),$$

where $d_{\nu,r}$ is defined by (2.20). Therefore our assertion follows from Proposition 2.3.

THEOREM 4.6. *For every K-rational representation ρ of $\mathrm{GL}_{n+q} \times \mathrm{GL}_q$, we have $\mathfrak{M}_\rho(\mathbf{C}) = \mathfrak{M}_\rho(K_{ab}) \otimes_{K_{ab}}\mathbf{C}$.*

Proof. For simplicity, put $\Psi = K_{ab}$. We first consider the case of $\mathfrak{M}_k$ with $k \in \mathbf{Z}$. Given a congruence subgroup Γ of G_{Q+}, $\mathfrak{Z}/\Gamma$ has a canonical model V rational over a subfield of Ψ, and the function field of V over Ψ can be identified with the set of all Γ-invariant elements of $\mathfrak{A}_0(\Psi)$, namely $\mathfrak{A}_0(\Gamma, \Psi)$. If $0\neq g \in \mathfrak{A}_k(\Gamma, \Psi)$, the divisor of g on V, denoted $\mathrm{div}(g)$, is meaningful. Let us now show that $\mathrm{div}(g)$ is Ψ-rational. Let $N = (n + q)q$, and take N functions $\{f_\nu\}$ from $\mathfrak{A}_0(\Gamma, \Psi)$ that are algebraically independent. Let F be $(2\pi i)^{-N}$ times the jacobian of $\{f_\nu\}$ with respect to the variables $w_{\lambda\nu}$ and

$z_{\mu\nu}$; F may be expressed as

$$F = \det\left((2\pi i)^{-1}Df_1, \cdots, (2\pi i)^{-1}Df_N\right).$$

We see easily from Proposition 4.5 that $F \in \mathfrak{A}_\sigma(\Psi)$, where $\sigma(X, Y)=\det(X)^q \cdot \det(Y)^{n+q}$. Now $\mathrm{div}(F)$ is the divisor of the N-form $df_1 \wedge \cdots \wedge df_N$ on V, and hence is Ψ-rational. Since $\mathfrak{A}_\sigma(\Psi) = c \cdot \mathfrak{A}_m(\Psi)$ with a constant c, we see that $c^k g^m/F^k \in \mathfrak{A}_0(\Psi)$. Therefore $m \cdot \mathrm{div}(g)$ is Ψ-rational, so that $\mathrm{div}(g)$ itself is Ψ-rational. Once this rationality is established, we can show that

$$\mathfrak{M}_k(\Gamma, \mathbf{C}) = \mathfrak{M}_k(\Gamma, \Psi) \otimes_\Psi \mathbf{C}$$

by identifying $\mathfrak{M}_k(\Gamma, \mathbf{C})$ with a linear system of a Ψ-rational divisor on V, in exactly the same fashion as in the proof of [7, Th. 6]. (We note that a compactified canonical model of $\mathfrak{Z}/\Gamma$ exists and is Ψ-rational, since the proof of [7, Th. 5] is applicable to the present case.) Therefore, to prove our assertion for $\mathfrak{M}_k$, it is sufficient to show that $\mathfrak{A}_k(\Psi) \neq \{0\}$. But this is guaranteed by Proposition 4.3.

In the case of an arbitrary K-rational representation ρ, we take, for any $\mathfrak{z}_0 \in \mathfrak{Z}$, a holomorphic $\mathbf{C}_{d(\rho)}^{d(\rho)}$-valued function T on $\mathfrak{Z}$ such that $\det(T(\mathfrak{z}_0)) \neq 0$ and the columns of T belong to $\mathfrak{M}_\sigma(\Psi)$, where $\sigma(X, Y)=\det(X)^s \det(Y)^s \rho(X, Y)$ with a positive integer s. This can be proved by the same technique as in Proposition 4.3. We only have to start with a holomorphic π-uniformizer, where $\pi(U) = \det(U)^s U$ with a suitable $s > 0$. The existence of such a uniformizer is guaranteed by [12, Prop. 3.7]. Notice that s is independent of $\mathfrak{z}_0$. Now, given $f \in \mathfrak{M}_\rho(\mathbf{C})$, put $g = \det(T)T^{-1}f$. Then $g \in \mathfrak{M}_\tau(\mathbf{C})$, where $\tau(X, Y) = \det(X)^a \det(Y)^b$ with some a and b. Take a basis $\{c_\nu\}$ of $\mathbf{C}$ over Ψ. Recall that $\mathfrak{M}_\tau(\mathbf{C}) = \mathfrak{M}_{a+b}(\mathbf{C})$ and $\mathfrak{M}_\tau(\Psi) = \eta \mathfrak{M}_{a+b}(\Psi)$ with a constant η. Therefore, by the above result, g is a finite sum $\sum c_\nu h_\nu$ with $h_\nu \in \mathfrak{M}_\tau(\Psi)$, so that $f = \sum c_\nu p_\nu$ with $p_\nu = \det(T)^{-1}Th_\nu$. Observe that $p_\nu \in \mathfrak{A}_\rho(\Psi)$. For every point $\mathfrak{z}'$ of $\mathfrak{Z}$, we can find another T' such that $\det(T'(\mathfrak{z}')) \neq 0$. We then obtain another expression $f = \sum c_\nu p'_\nu$ with $p'_\nu \in \mathfrak{A}_\rho(\Psi)$ holomorphic at $\mathfrak{z}'$, and $\sum c_\nu T^{-1}(p_\nu - p'_\nu) = 0$. As shown in the proof of Theorem 4.1, $\mathfrak{A}_0(\Psi)$ and $\mathbf{C}$ are linearly disjoint over Ψ, and hence $p_\nu = p'_\nu$. This shows that p_ν is holomorphic everywhere, so that $p_\nu \in \mathfrak{M}_\rho(\Psi)$ by Proposition 2.4. This completes the proof.

PROPOSITION 4.7. *For every K-rational representation τ of GL_{n+q}, we have $\mathfrak{T}_{\tau,r}(\mathbf{C}) = \mathfrak{T}_{\tau,r}(K_{\mathrm{ab}}) \otimes_{K_{\mathrm{ab}}} \mathbf{C}$.*

We omit the proof, since it is similar to the argument in the last part of the above proof.

So far we defined only the rationality of automorphic forms over K_{ab}.

In order to speak of the rationality over subfields of K_{ab}, say over K, we have to study the behavior of the elements of $\mathfrak{A}_\rho(K_{ab})$ under a certain adelized group. To be more explicit, if we define a subgroup $\mathfrak{G}$ of $G_{A+} \times K_A^\times$ by

$$(4.4) \qquad \mathfrak{G} = \{(x, a) \in G_{A+} \times K_A^\times \mid \det(x)\bar{a}^{n+q}a^q \in K^\times K_\infty^\times, \ \nu(x)\bar{a}a \in \mathbf{Q}^\times \mathbf{Q}_{\infty+}^\times\} \ ,$$

then we can define a natural action of $\mathfrak{G}$ on $\mathfrak{A}_\rho(K_{ab})$ similar to that of $G_{A+}^{(n)}$ on $\mathcal{C}_\sigma^{(n)}(\mathbf{Q}_{ab})$ stated in [12, Th. 1.2]. In the present paper, however, we shall not go into details of this topic. We only mention that the analogue of formula [12, (1.10)], or rather of [12, (6.15)] can be given in terms of the action of $K_A^\times$ on $\mathfrak{T}_{\tau,r}(K_{ab})$, which was discussed in [9] when τ is trivial.

5. Eisenstein series on $\mathfrak{Z}$

With the matrix R of (1.1), define a subset $\mathfrak{X}$ of K_m^q by

$$(5.1) \qquad \mathfrak{X} = \{x \in K_m^q \mid xR^{-1}\cdot{}^t\bar{x} = 0 \ , \ \mathrm{rank}(x) = q\} \ .$$

Let G_Q' be the subgroup of G_Q consisting of all α such that $\nu(\alpha) = 1$, and P_Q' the subgroup of G_Q' consisting of α of the form

$$(5.2) \qquad \alpha = \begin{bmatrix} * & 0 & * \\ * & 1_q & * \\ 0 & 0 & 1_q \end{bmatrix}.$$

For each $\alpha \in G_Q$, let $\mathfrak{x}(\alpha)$ denote the last q rows of α. We observe that $\mathfrak{x}(\alpha\beta) = \mathfrak{x}(\alpha)\beta$, and $\mathfrak{x}(\alpha) \in \mathfrak{X}$ since $\alpha R^{-1}\cdot{}^t\bar{\alpha} = \nu(\alpha)R^{-1}$.

PROPOSITION 5.1. *The map $\alpha \mapsto \mathfrak{x}(\alpha)$ gives a bijection of $P_Q'\backslash G_Q'$ onto $\mathfrak{X}$.*

Proof. Suppose $\mathfrak{x}(\alpha) = \mathfrak{x}(\beta)$ for $\alpha, \beta \in G_Q'$. Let v, w, and x be the first n rows, the second q rows, and the last q rows of α, respectively. We have

$$\beta\alpha^{-1} = \beta R^{-1}\cdot{}^t\bar{\alpha}R = \begin{bmatrix} * & * & * \\ xR^{-1}\cdot{}^t\bar{v} & xR^{-1}\cdot{}^t\bar{w} & xR^{-1}\cdot{}^t\bar{x} \end{bmatrix} R \ .$$

Since $\alpha R^{-1}\cdot{}^t\bar{\alpha} = R^{-1}$, this is equal to

$$\begin{bmatrix} * & * & * \\ * & * & * \\ 0 & 1 & 0 \end{bmatrix}\begin{bmatrix} S & 0 & 0 \\ 0 & 0 & 1 \\ 0 & -1 & 0 \end{bmatrix} = \begin{bmatrix} * & * & * \\ * & * & * \\ 0 & 0 & 1 \end{bmatrix},$$

and hence $\beta\alpha^{-1} \in P_Q'$. To prove the surjectivity, let $x \in \mathfrak{X}$. Since $\mathrm{rank}(x) = q$, we can find $a \in K_m^q$ such that $xR^{-1}\cdot{}^t\bar{a} = 1_q$. Put $w = a - (1/2)aR^{-1}\cdot{}^t\bar{a}x$. Then $xR^{-1}\cdot{}^t\bar{w} = 1_q$, and $wR^{-1}\cdot{}^t\bar{w} = 0$. Let

$$U = \{u \in K_m^1 \mid uR^{-1}\cdot{}^t\bar{w} = uR^{-1}\cdot{}^t\bar{x} = 0\} \ .$$

Then U has dimension n over K. Let β be an element of K_m^m whose first n

rows form a basis of U over K, second q rows form w, and last q rows form x. Then

$$\beta R^{-1} \cdot {}^t\bar{\beta} = \begin{bmatrix} T & 0 & 0 \\ 0 & 0 & -1 \\ 0 & 1 & 0 \end{bmatrix} \quad \text{with } {}^t\bar{T} = -T \in \mathrm{GL}_n(K).$$

By the Witt theorem, T must be equivalent to S^{-1}, namely $S^{-1} = BT \cdot {}^t\bar{B}$ for some $B \in \mathrm{GL}_n(K)$. Put

$$\alpha = \begin{bmatrix} B & 0 \\ 0 & 1_{2q} \end{bmatrix}\beta.$$

Then $\alpha \in G'_Q$ and $\mathfrak{x}(\alpha) = x$.   Q.E.D.

Remark. The assertion of the above proposition holds even if G'_Q and P'_Q are replaced by their subgroups consisting of the elements of determinant 1. In fact, let $c = \det(\alpha)$ with $\alpha \in G'_Q$. Since $c\bar{c} = 1$, we can easily find an element D of $\mathrm{GL}_n(K)$ such that $DS^{-1} \cdot {}^t\bar{D} = S^{-1}$ and $\det(D) = c^{-1}$. For example, take S^{-1} in the diagonal form, and let $D = \mathrm{diag}[c^{-1}, 1, \cdots, 1]$. Put

$$\gamma = \begin{bmatrix} D & 0 \\ 0 & 1_{2q} \end{bmatrix}\alpha.$$

Then $\gamma \in G'_Q$, $\det(\gamma) = 1$, and $\mathfrak{x}(\gamma) = \mathfrak{x}(\alpha)$.

Let us now define Eisenstein series on $\mathfrak{Z}$ associated with a K-rational representation ρ of $\mathrm{GL}_{n+q} \times \mathrm{GL}_q$. Take a vector v of $K^{d(\rho)}$ (or more generally, of $\mathbf{C}^{d(\rho)}$ if the arithmeticity is not the problem) such that

$$(5.3) \qquad \rho\left(\begin{pmatrix} * & * \\ 0 & 1_q \end{pmatrix}, \; 1_q\right)v = v$$

for all elements of GL_{n+q} of the form $\begin{pmatrix} * & * \\ 0 & 1_q \end{pmatrix}$, and put

$$(5.4) \qquad \varphi(x, \mathfrak{z}) = \rho\big(\kappa(\alpha, \mathfrak{z}), \mu(\alpha, \mathfrak{z})\big)^{-1}v \qquad\qquad (x \in \mathfrak{X}, \; \mathfrak{z} \in \mathfrak{Z})$$

with an element α of G'_Q such that $x = \mathfrak{x}(\alpha)$. In view of Proposition 5.1, this is well-defined, and

$$(5.5) \qquad \varphi(x\gamma, \mathfrak{z}) = \rho\big(\kappa(\gamma, \mathfrak{z}), \mu(\gamma, \mathfrak{z})\big)^{-1}\varphi(x, \gamma(\mathfrak{z})) \qquad \text{for every } \gamma \in G'_Q.$$

Take a subset X of $\mathfrak{X}$, and put

$$(5.6) \qquad E_{X,\varphi}(\mathfrak{z}) = \sum_{x \in X} \varphi(x, \mathfrak{z}).$$

Assume that this is uniformly convergent in every compact subset of $\mathfrak{Z}$. Then (5.5) implies that

(5.7) $$E_{x,\varphi}|_\rho\gamma = E_{X\gamma,\varphi} \qquad \text{for every} \quad \gamma \in G'_\mathbf{Q},$$

where $X\gamma = \{x\gamma \mid x \in X\}$. Therefore, if $\{\gamma \in G'_\mathbf{Q} \mid X\gamma = X\}$ contains a congruence subgroup Γ, $E_{x,\varphi}$ defines an element of $\mathfrak{M}_\rho(\Gamma)$.

To define $E_{x,\varphi}$ more explicitly, let us hereafter assume $q = 1$. Every K-rational irreducible representation of $\mathrm{GL}_{n+1} \times \mathrm{GL}_1$ is a K-rational component of the representation

(5.8) $$\rho(U,\, V) = \det(U)^l V^k \cdot {}^t U^{-1} \otimes \cdots \otimes {}^t U^{-1}$$
$$(p \text{ copies of } {}^t U^{-1}; \ U \in \mathrm{GL}_{n+1},\ V \in \mathrm{GL}_1)$$

with $0 \leq p \in \mathbf{Z}$, $k \in \mathbf{Z}$, and $l \in \mathbf{Z}$. Obviously a non-trivial vector v satisfying (5.3) for this ρ exists only if $l \geq 0$. So assume $l \geq 0$, and take v in the form

(5.9) $$v = \underbrace{\begin{bmatrix} 0 \\ \vdots \\ \vdots \\ 0 \\ 1 \end{bmatrix} \otimes \cdots \otimes \begin{bmatrix} 0 \\ \vdots \\ \vdots \\ 0 \\ 1 \end{bmatrix}}_{p-(n+1)l} \otimes u, \qquad \underbrace{u \in K^{n+1} \otimes \cdots \otimes K^{n+1}}_{(n+1)l},$$

with a vector u such that $(A \otimes \cdots \otimes A)u = \det(A)^l u$ for all $A \in \mathrm{GL}_{n+1}$. Then v satisfies (5.3) for the representation (5.8). (It seems that all vectors satisfying (5.3) are essentially of this type, up to linear combinations and permutations of factors of the tensor product, though the author has no complete proof of this fact. Notice also that such a non-vanishing u, and hence v, exists for any $l \geq 0$ such that $(n + 1)l \leq p$.) Define $\varphi(x, \mathfrak{z})$ by (5.4) with this choice of v. Putting $j = p - (n + 1)l$ and $x = (a, b, c)$ with $a \in K^1_n$ and $b, c \in K$, we obtain, from (1.11),

(5.10) $$\varphi(a,\, b,\, c;\, w,\, z) = (aw + bz + c)^{-k} \begin{bmatrix} \bar{b}w - S^{-1} \cdot {}^t\bar{a} \\ \bar{b}z + \bar{c} \end{bmatrix}^{(j)} \otimes u,$$

where $V^{(j)} = V \otimes \cdots \otimes V$ with j copies of V.

Next, take an arbitrary lattice $\mathfrak{m}$ in K^1_m and an element y of $\mathfrak{X}$, and put

(5.11) $$X = X(\mathfrak{m},\, y) = \{x \in \mathfrak{X} \mid x - y \in \mathfrak{m}\},$$

(5.12) $$\Gamma(\mathfrak{m},\, y) = \{\gamma \in G'_\mathbf{Q} \mid \mathfrak{m}\gamma = \mathfrak{m},\ y\gamma - y \in \mathfrak{m}\}.$$

Obviously $X\gamma = X$ for $\gamma \in \Gamma(\mathfrak{m}, y)$, and $\Gamma(\mathfrak{m}, y)$ is a congruence subgroup of $G_{\mathbf{Q}+}$. The series $E_{x,\varphi}$ defined with these φ and X can be written as $E_{j,k} \otimes u$ with

(5.13) $$E_{j,k} = E_{j,k}(w,\, z;\, \mathfrak{m},\, y)$$
$$= \sum (aw + bz + c)^{-k} \begin{bmatrix} \bar{b}w - S^{-1} \cdot {}^t\bar{a} \\ \bar{b}z + \bar{c} \end{bmatrix}^{(j)},$$

where the sum is taken over all $(a, b, c) \in X(\mathfrak{m}, y)$. Thus, starting with a vector v of (5.9) with $l \geq 0$, we can reduce the series to the case where $l = 0$.

PROPOSITION 5.2. *If* $k > 2n + j + 2$ *and* $j \geq 0$, *the series* $E_{j,k}$ *is absolutely and uniformly convergent in every compact subset of* $\mathfrak{Z}$.

Thus, if $k > 2n + j + 2$, $E_{j,k}$ is an element of $\mathfrak{M}_\rho(\Gamma(\mathfrak{m}, y))$, where

$$(5.14) \qquad \rho(U, V) = V^k \cdot {}^t U^{-1} \otimes \cdots \otimes {}^t U^{-1}$$
$$(j \text{ copies of } {}^t U^{-1}; \ U \in \mathrm{GL}_{n+1}, \ V \in \mathrm{GL}_1) .$$

Now our main result about Eisenstein series can be stated as

THEOREM 5.3. *Let* $p \in \mathfrak{G}_1^{(1)}(K_{ab})$ *and* $\tau_0 \in K \cap \mathfrak{H}_1$; *suppose* p *is finite at* τ_0 *and* $p(\tau_0) \neq 0$. *Then* $\pi^{-k} p(\tau_0)^{-j-k} E_{j,k}$ *belongs to* $\mathfrak{M}_\rho(K_{ab})$.

The convergence of $E_{j,k}$ is perhaps a special case of the result of [1] about the convergence of an Eisenstein series of an algebraic group of a more general type. Here we prove it in an elementary way, since it gives an interpretation of the series which will be employed in our discussion in a subsequent paper.

Let $Q = {}^t Q \in \mathrm{GL}_n(\mathbf{Q})$ and $0 < P = {}^t P \in \mathrm{GL}_n(\mathbf{R})$. Let κ and λ be the numbers of positive and negative eigenvalues of Q. We consider an infinite series

$$(5.15) \qquad \sum_{\substack{Q[x]=0 \\ 0 \neq x \equiv y \bmod L}} f(x) P[x]^{-s} ,$$

where f is a homogeneous polynomial function of degree d on $\mathbf{R}^n$, L is a lattice in $\mathbf{Q}^n$, y is a fixed element of $\mathbf{Q}^n$, $P[x] = {}^t x P x$, $Q[x] = {}^t x Q x$, and s is a complex variable.

PROPOSITION 5.4. *The series* (5.15) *is absolutely convergent for* $\mathrm{Re}(s) > \sigma_0$ *where*

$$\sigma_0 = \begin{cases} (n + d - 2)/2 & \text{if } \kappa > 1 \text{ and } \lambda > 1 , \\ (n + d - 1)/2 & \text{if } \kappa > 1 \text{ and } \lambda = 1 , \\ (n + d)/2 & \text{if } \kappa = \lambda = 1 . \end{cases}$$

The convergence is uniform when s, P, *and the coefficients of* f *stay in compact sets.*

Proof. We may assume that $Q = \begin{pmatrix} A & 0 \\ 0 & -B \end{pmatrix}$ with positive definite symmetric matrices A and B of degree κ and λ, respectively. Put $R = \begin{pmatrix} A & 0 \\ 0 & B \end{pmatrix}$. Then there exists a positive number h such that $P > hR$. Therefore it is sufficient to consider the convergence of $\sum f(x) R[x]^{-s}$. Also the problem can easily be reduced to the case where $Q \in \mathbf{Z}_n^n$, $L = \mathbf{Z}^n$, and $y = 0$. Put

$x = \begin{pmatrix} u \\ v \end{pmatrix}$ with $u \in \mathbf{R}^{\kappa}$ and $v \in \mathbf{R}^{\lambda}$. Then

$$(5.16) \qquad \sum_{\substack{Q[x]=0 \\ 0 \neq x \in L}} |f(x)| P[x]^{-s} = 2^{-s} \sum_{m=1}^{\infty} c_m m^{-s} \,,$$

where c_m is the sum of $\left| f\begin{pmatrix} u \\ v \end{pmatrix} \right|$ for all $u \in \mathbf{Z}^{\kappa}$ and $v \in \mathbf{Z}^{\lambda}$ such that $A[u] = B[v] = m$. Let a_m and b_m be the numbers of all such vectors u and v, respectively. Then $c_m \leq E m^{d/2} a_m b_m$ with a positive constant E depending on f. If $\kappa > 1$ and $\lambda > 1$, we have

$$a_m = O(m^{(\kappa/2)-1+\varepsilon}) \,, \quad b_m = O(m^{(\lambda/2)-1+\varepsilon})$$

with any $\varepsilon > 0$. Therefore (5.16) is convergent for $\mathrm{Re}(s) > (n + d - 2)/2$. If $\lambda = 1$, we have $b_m = O(1)$, which proves the remaining cases. The uniformity of the convergence is obvious from our argument.

We now apply this to $E_{j,k}$. For each $x \in \mathfrak{X}$, put $xP(\mathfrak{z}) = (\mathfrak{a}, \mathfrak{b})$ with $\mathfrak{a} \in \mathbf{C}^q_{n+q}$ and $\mathfrak{b} \in \mathbf{C}^q_q$. Also put

$$(5.17) \qquad U = P(\mathfrak{z}) \begin{bmatrix} \xi(\mathfrak{z})^{-1} & 0 \\ 0 & \eta(\mathfrak{z})^{-1} \end{bmatrix} {}^t\overline{P(\mathfrak{z})}$$

with P, ξ, η defined by (1.3) and (1.4). We have $\mathfrak{a}\xi(\mathfrak{z})^{-1} \cdot {}^t\overline{\mathfrak{a}} = \mathfrak{b}\eta(\mathfrak{z})^{-1} \cdot {}^t\overline{\mathfrak{b}}$ in view of (1.4), since $xR^{-1} \cdot {}^t\overline{x} = 0$. Therefore $xU \cdot {}^t\overline{x} = 2\mathfrak{b}\eta(\mathfrak{z})^{-1} \cdot {}^t\overline{\mathfrak{b}}$. Now assume $q = 1$ and put $x = (a, b, c)$ as in (5.10). Then $xU \cdot {}^t\overline{x} = 2\eta(\mathfrak{z})^{-1}|aw + bz + c|^2$. If f is a polynomial function in the real and imaginary parts of a, b, c, we have

$$\sum_{x \in X} f(x)|aw + bz + c|^{-2s} = |\eta(\mathfrak{z})/2|^s \sum_{x \in X} f(x)|xU \cdot {}^t\overline{x}|^{-s} \,,$$

which is a series of type (5.15) with $\kappa = 2n + 2$ and $\lambda = 2$. Therefore, if the degree of f is d, it is convergent for $\mathrm{Re}(s) > (2m + d - 2)/2 = (2n + d + 2)/2$. This proves Proposition 5.2.

To prove Theorem 5.3, let us consider the Fourier expansion

$$(5.18) \qquad \pi^{-k}p(\tau_0)^{-j-k}E_{j,k}(w, z; \mathfrak{m}, y) = \sum_r g_r(w; \mathfrak{m}, y)e(rz) \,.$$

Since $q = 1$, r runs over non-negative elements of a lattice in $\mathbf{Q}$. Let α be defined by (1.21) with $x \in K^n_q$ and $y = 0$. By (5.7) and (2.11), we have

$$(5.19) \qquad (g_r)_*(x; \mathfrak{m}, y) = g_r(0; \mathfrak{m}\alpha, y\alpha) \,.$$

Let σ be the representation of GL_n defined by

$$(5.20) \qquad \sigma(W) = \begin{bmatrix} {}^tW^{-1} & 0 \\ 0 & 1 \end{bmatrix} \otimes \cdots \otimes \begin{bmatrix} {}^tW^{-1} & 0 \\ 0 & 1 \end{bmatrix} \quad (j \text{ copies}, W \in \mathrm{GL}_n) \,.$$

By Proposition 2.2, a matrix C of the form

$$(5.21) \qquad C = \begin{bmatrix} p(\tau_0)1_n & 0 \\ 0 & 1 \end{bmatrix} \otimes \cdots \otimes \begin{bmatrix} p(\tau_0)1_n & 0 \\ 0 & 1 \end{bmatrix} \qquad (j \text{ copies})$$

is a σ-regulator. Our aim is to show that $C(g_r)_*(x; \mathfrak{m}, y)$ has components in K_{ab} for all $x \in K_q^n$. By virtue of (5.19), it is sufficient to show that $Cg_r(0; \mathfrak{m}, y)$ is K_{ab}-rational for all $\mathfrak{m}$ and y. We can even restrict $\mathfrak{m}$ and y to those of a special type. In fact, let $\mathfrak{o} = \mathbf{Z}\tau_0 + \mathbf{Z}$ with an arbitrary $\tau_0 \in \mathfrak{H}_1 \cap K$, and take a positive integer q so that $q\mathfrak{m}$ and qy are contained in $\mathfrak{o}_\mathfrak{m}^1$; further take a positive integer N so that $N\mathfrak{o}_\mathfrak{m}^1 \subset q\mathfrak{m}$. Then

$$E_{j,k}(\mathfrak{z}; \mathfrak{m}, y) = q^{k-j} E_{j,k}(\mathfrak{z}; q\mathfrak{m}, qy) = q^{k-j} \sum_\nu E_{j,k}(\mathfrak{z}; N\mathfrak{o}_\mathfrak{m}^1, y_\nu)$$

with finitely many y_ν such that $y_\nu \equiv qy \pmod{q\mathfrak{m}}$. Thus our task is to show that

$$(5.22) \qquad Cg_r(0; N\mathfrak{o}_\mathfrak{m}^1, y) \in (K_{ab})^{d(\rho)} \qquad \text{for all} \quad y \in \mathfrak{X} \cap \mathfrak{o}_\mathfrak{m}^1 \quad \text{and} \quad 0 < N \in \mathbf{Z}.$$

For this purpose, we first decompose (5.11) into its "components"

$$(5.23) \quad \sum (aw + bz + c)^{-k}(\bar{b}z + \bar{c})^h(\bar{b}w - S^{-1} \cdot {}^t\bar{a})^{(l)} \quad (h = 0, \cdots, j; l = j - h),$$

where $x^{(l)} = x \otimes \cdots \otimes x$ with l copies of $x \in \mathbf{C}^n$, and (a, b, c) runs over $X(N\mathfrak{o}_\mathfrak{m}^1, y)$. We are going to establish the Fourier expansion of (5.23) in the form $\sum g_r(w)e(rz)$ with g_r given explicitly as certain infinite series. Since C is given by (5.21), we see that (5.22) amounts to

$$(5.24) \qquad \pi^{-k} p(\tau_0)^{-h-k} g_r(0) \quad \text{has components in} \quad K_{ab}.$$

Our computation will require

LEMMA 5.5. *Let* $0 < N \in \mathbf{Z}$, $0 \leq h \in \mathbf{Z}$, $0 < k \in \mathbf{Z}$, *and*

$$F_{h,k}(z, z') = \sum_{m \in \mathbf{Z}} (z' + Nm)^h (z + Nm)^{-k} \qquad (z \in \mathfrak{H}_1, z' \in \mathbf{C}).$$

If $k - h > 1$, *this is absolutely and uniformly convergent in every compact subset of* $\mathfrak{H}_1 \times \mathbf{C}$, *and*

$$F_{h,k}(z, z') = \sum_{\nu=0}^h \binom{h}{\nu} \frac{(-2\pi i)^{k-\nu}}{(k - \nu - 1)!} N^{\nu-k}(z' - z)^{h-\nu} \sum_{\lambda=1}^\infty \lambda^{k-\nu-1} e(\lambda z/N).$$

Proof. It is well known that

$$\sum_{m \in \mathbf{Z}} (z + Nm)^{-k} = \frac{(-2\pi i)^k}{(k - 1)!} N^{-k} \sum_{\lambda=1}^\infty \lambda^{k-1} e(\lambda z/N) \qquad (1 < k \in \mathbf{Z}).$$

Calling this sum $f_k(z)$, we have

$$\begin{aligned} F_{h,k}(z, z') &= \sum_m [(z' - z) + (z + Nm)]^h (z + Nm)^{-k} \\ &= \sum_\nu \binom{h}{\nu} (z' - z)^{h-\nu} f_{k-\nu}(z), \end{aligned}$$

which proves our lemma.

With $\tau_0 \in K \cap \mathfrak{H}_1$ and $\mathfrak{o} = \mathbf{Z}\tau_0 + \mathbf{Z}$ as above, put

$$(5.25) \qquad \delta = \tau_0 - \bar{\tau}_0 ,$$

$$(5.26) \qquad \Phi(a) = a \cdot (-\delta S)^{-1} \cdot {}^t\bar{a} \qquad\qquad (a \in K_n^1) .$$

Further, for two elements $b = b_1\tau_0 + b_2$ and $c = c_1\tau_0 + c_2$ of K with b_ν and c_ν in $\mathbf{Q}$, we put

$$(5.27) \qquad [c, b] = \begin{bmatrix} c_1 & c_2 \\ b_1 & b_2 \end{bmatrix} .$$

Then $\det[c, b] = (c\bar{b} - \bar{c}b)/\delta$. For $0 \neq x = (a, b, c) \in K_m^1$, we see that $x \in \mathfrak{X}$ if and only if $\det[c, b] = \Phi(a)$. Writing α for $[c, b]$, we have

$$bz + c = (1 \quad z)\alpha \begin{bmatrix} \tau_0 \\ 1 \end{bmatrix} .$$

If $b = 0$ and $\det[c, b] = \Phi(a)$, then $a = 0$, since Φ is definite. Thus (5.23) can be written as

$$(5.28) \qquad \sum_{\alpha,a} \left\{ (1 \quad z)\alpha \begin{bmatrix} \bar{\tau}_0 \\ 1 \end{bmatrix} \right\}^h \left\{ (1 \quad z)\alpha \begin{bmatrix} \tau_0 \\ 1 \end{bmatrix} + aw \right\}^{-k} (j(\alpha, \bar{\tau}_0)w - S^{-1} \cdot {}^t\bar{a})^{(l)}$$

$$+ \begin{cases} 0 & \text{if } l \neq 0 , \\ \sum_c \bar{c}^h/c^k & \text{if } l = 0 , \end{cases}$$

where we put

$$(5.29) \qquad j(\alpha, \tau) = b_1\tau + b_2 \quad \text{for } \alpha = \begin{bmatrix} * & * \\ b_1 & b_2 \end{bmatrix} \text{ and } \tau \in \mathfrak{H}_1 .$$

The sums are taken over all $c \in K$ and all $(\alpha, a) \in \mathbf{Z}_2^2 \times K_n^1$ under the conditions

$$0 \neq c \equiv c_0 \pmod{N\mathfrak{o}} ,$$
$$\alpha \equiv \alpha_0 \pmod{N\mathbf{Z}_2^2}, \text{ the second row of } \alpha \neq 0 ,$$
$$a \equiv a_0 \pmod{N\mathfrak{o}_n^1}, \quad \Phi(a) = \det(\alpha) .$$

Here c_0, α_0, and a_0 are determined modulo N by y in an obvious way. The arithmeticity of the term $\sum \bar{c}^h/c^k$ is relatively simple. Indeed, it is the value at τ_0 of the series

$$(5.30) \qquad \sum (c_1\bar{\tau} + c_2)^h(c_1\tau + c_2)^{-k} \qquad\qquad (\tau \in \mathfrak{H}_1) ,$$

where the sum is taken over all integers $(c_1, c_2) \neq (0, 0)$ congruent to some fixed integers modulo N. Therefore $\pi^{-k}p(\tau_0)^{-h-k} \sum \bar{c}^h/c^k$ belongs to K_{ab} by virtue of [7, Th. 1] (or by the earlier results mentioned in the paper).

The first term of (5.28) is far more difficult. We decompose it into more manageable sums by using

LEMMA 5.6. *Let A be the set of all elements of $\mathbf{Z}_2^2$ with non-vanishing* *second rows, and let R_N be a complete set of representatives for $\Delta_{N\infty}\backslash\Delta_N$, where*

$$\Delta_N = \{\gamma \in \mathrm{SL}_2(\mathbf{Z}) \mid \gamma \equiv 1 \pmod{N}\},$$
$$\Delta_{N\infty} = \left\{\begin{pmatrix} 1 & mN \\ 0 & 1 \end{pmatrix} \Big| m \in \mathbf{Z}\right\}.$$

Then the map $(r, s, t; \gamma) \mapsto \begin{pmatrix} r & s \\ 0 & t \end{pmatrix}\gamma$ gives a bijection of $\{(r, s, t; \gamma) \in \mathbf{Z}^3 \times R_1 \mid t > 0\}$ onto A. Moreover, let C be a complete set of representatives for $\Delta_{1\infty}\backslash\Delta_1/\Delta_N$, and let $\alpha_0 \in \mathbf{Z}_2^2$. Then there is a subset B of C determined by α_0 and there exist integers r_β, s_β, t_β determined by each $\beta \in B$ such that $\{\alpha \in A \mid \alpha \equiv \alpha_0 \pmod{N}\}$ is a disjoint union

$$\bigcup_{\beta \in B} \left\{\begin{pmatrix} r & s \\ 0 & t \end{pmatrix}\gamma \mid \gamma \in R_N\beta,\ r \equiv r_\beta,\ s \equiv s_\beta,\ 0 < t \equiv t_\beta \pmod{N}\right\}.$$

We omit the proof, which is elementary and straightforward. Notice that B is a finite set, since C is finite.

If $\alpha = \begin{pmatrix} r & s \\ 0 & t \end{pmatrix}\gamma$, we have

$$(1\quad z)\alpha\begin{bmatrix} \tau_0 \\ 1 \end{bmatrix} = (r\cdot\gamma(\tau_0) + tz + s)j(\gamma, \tau_0).$$

Therefore, by means of Lemma 5.6, the first sum of (5.28) can be written as a finite sum $\sum_{\beta \in B} L_\beta$ with

(5.31)　　$L_\beta = \sum_\gamma j(\gamma, \bar{\tau}_0)^h j(\gamma, \tau_0)^{-k} \sum_{t,a} (j(\gamma, \bar{\tau}_0)tw - S^{-1}\cdot{}^t\bar{a})^{(l)} F_{t,a}$,

(5.32)　　$F_{t,a} = \sum_s (r\cdot\gamma(\bar{\tau}_0) + tz + s)^h (r\cdot\gamma(\tau_0) + tz + j(\gamma, \tau_0)^{-1}aw + s)^{-k}$,

where the sums are taken under the conditions

$$\gamma \in R_N\beta,$$
$$0 < t \equiv t_\beta,\ s \equiv s_\beta,\ r \equiv r_\beta \pmod{N},$$
$$rt = \Phi(a),\ a \equiv a_0 \pmod{N\mathfrak{o}_n^1}.$$

Before proceeding further, we note that $\mathrm{Im}(b^{-1}aw + z + b^{-1}c) > 0$ if $(w, z) \in \mathfrak{Z}$, $(a, b, c) \in \mathfrak{X}$, and $b \neq 0$. In fact, put $u = b^{-1}a$, $v = b^{-1}c$, and

$$\alpha = \begin{bmatrix} 1_n & 0 & -S^{-1}\cdot{}^t\bar{u} \\ u & 1 & v \\ 0 & 0 & 1 \end{bmatrix}.$$

Then $\alpha \in G'_{\mathbf{Q}}$ and $\alpha(w, z) = (w - S^{-1}\cdot{}^t\bar{u}, uw + z + v)$. Since $\alpha(w, z) \in \mathfrak{Z}$, we have $\mathrm{Im}(uw + z + v) > 0$ as expected. Therefore, putting $s = s_\beta + Nm$ with $m \in \mathbf{Z}$ in (5.32), we can apply Lemma 5.5 to $F_{t,a}$ to find

$$F_{t,a} = \sum_{\nu=0}^{h} \binom{h}{\nu} \frac{(-2\pi i)^{k-\nu}}{(k-\nu-1)!} N^{\nu-k}[r \cdot \gamma(\overline{\tau}_0) - r \cdot \gamma(\tau_0) - j(\gamma,\tau_0)^{-1}aw]^{h-\nu}$$

$$\times \sum_{\lambda=1}^{\infty} \lambda^{k-\nu-1} e(N^{-1}\lambda[r \cdot \gamma(\tau_0) + tz + j(\gamma,\tau_0)^{-1}aw + s_\beta]) \, .$$

Thus we can write the Fourier expansion of (5.28) in the form $\sum_{m=0}^{\infty} g_m(w)e(mz/N)$ with

$$(5.33) \qquad g_m(w) = \sum_{\beta \in B} \sum_{t\lambda=m} e(\lambda s_\beta/N) g_{t,\lambda}(w) \qquad \text{for} \quad m > 0 \, ,$$

where $0 < t \equiv t_\beta \pmod N$, $0 < \lambda \in \mathbf{Z}$, and

$$(5.34) \quad g_{t,\lambda}(w) = \sum_{\nu=0}^{h} \binom{h}{\nu} \frac{(-2\pi i)^{k-\nu}}{(k-\nu-1)!} N^{\nu-k}\lambda^{k-\nu-1} \sum_{\gamma} \sum_{a} j(\gamma,\overline{\tau}_0)^h j(\gamma,\tau_0)^{-k}$$

$$\times (j(\gamma,\overline{\tau}_0)tw - S^{-1}\cdot {}^t\overline{a})^{(l)}\{t^{-1}\Phi(a)[\gamma(\overline{\tau}_0) - \gamma(\tau_0)] - j(\gamma,\tau_0)^{-1}aw\}^{h-\nu}$$

$$\times e(N^{-1}\lambda[t^{-1}\Phi(a)\cdot\gamma(\tau_0) + j(\gamma,\tau_0)^{-1}aw]) \, ,$$

where $\gamma \in R_N\beta$ and a runs under the conditions

$$a \equiv a_0 \pmod{N\mathfrak{o}_n^1}, \quad \Phi(a)/t \equiv r_\beta \pmod N \, .$$

The set of all such a is a disjoint union

$$\bigcup_{y \in Y}\{a \in \mathfrak{o}_n^1 \mid a \equiv y \pmod{P\mathfrak{o}_n^1}\}$$

with a multiple P of N and a finite subset Y of $\mathfrak{o}_n^1$ depending on Φ, t, and r_β. For each $y \in Y$, define a vector-valued function f_y on $\mathfrak{H}_1$ by

$$(5.35) \qquad f_y(\tau) = \sum_{a \equiv y \bmod P}(S^{-1}\cdot {}^t\overline{a})^{(l)}e(N^{-1}t^{-1}\lambda\Phi(a)\tau) \qquad (\tau \in \mathfrak{H}_1) \, .$$

Observe that, if $a_1, \cdots, a_n$ are the components of a, every homogeneous polynomial in $\overline{a}_1, \cdots, \overline{a}_n$ is a spherical function of the quadratic form which assigns $\Phi(a)$ to $(\mathrm{Re}(a), \mathrm{Im}(a))$. Therefore the components of f_y belong to $\mathfrak{M}_{l+n}^{(1)}(\Delta_M, K)$ with a suitable multiple M of N. If $j = 0$ (and hence $h = l = 0$), this fact leads easily to (5.24). Indeed, we have

$$(5.36) \qquad g_{t,\lambda}(0) = \frac{(-2\pi i)^k}{(k-1)!} N^{-k}\lambda^{k-1} \sum_{y \in Y} \sum_{\gamma \in R_N\beta} j(\gamma,\tau_0)^{-k}f_y(\gamma(\tau_0)) \, .$$

Observe that we can change R_N without changing B, r_β, s_β, t_β in Lemma 5.6. Therefore f_y is independent of the choice of R_N. Now we can take $\bigcup_{\varepsilon \in E} R_M \varepsilon$ as R_N with a suitable finite subset E of Δ_N. This replacement of R_N may change $g_{t,\lambda}$, but of course does not change (5.33) as a whole. Take M so that $f_y \in \mathfrak{M}_{l+n}^{(1)}(\Delta_M)$ for all f_y necessary for expressing g_m with a fixed m. (There are only finitely many such f_y.) Put, for $k > 2$,

$$(5.37) \qquad \varepsilon_k^M(\tau) = \sum_{\gamma \in R_M} j(\gamma,\tau)^{-k} \qquad (\tau \in \mathfrak{H}_1) \, .$$

Since $k > n + 2$, we have

$$(5.38) \qquad \sum_{\gamma \in R_{N\beta}} j(\gamma, \tau)^{-k} f_\nu(\gamma(\tau))$$
$$= \sum_{\zeta \in E\beta} \sum_{\gamma \in R_M} j(\gamma, \zeta(\tau))^{n-k} j(\zeta, \tau)^{-k} f_\nu(\zeta(\tau))$$
$$= \sum_{\zeta \in E\beta} j(\zeta, \tau)^{-k} \varepsilon_{k-n}^M(\zeta(\tau)) f_\nu(\zeta(\tau)) \; .$$

It is well known that $\varepsilon_{k-n}^M \in \mathfrak{M}_{k-n}^{(1)}(\mathbf{Q}_{\mathrm{ab}})$, and hence $f_\nu \varepsilon_{k-n}^M \in \mathfrak{M}_k^{(1)}(K_{\mathrm{ab}})$. Let φ_ν denote the sum of (5.38). Since $\mathfrak{M}_k^{(1)}(K_{\mathrm{ab}})$ is stable under the action of $G_{\mathbf{Q}+}^{(1)}$ as shown by [7, Prop. 8] or [12, Th. 1.2], we have $\varphi_\nu \in \mathfrak{M}_k^{(1)}(K_{\mathrm{ab}})$, so that $p(\tau_0)^{-k} \varphi_\nu(\tau_0) \in K_{\mathrm{ab}}$, and so $\pi^{-k} p(\tau_0)^{-k} g_{t,\lambda}(0) \in K_{\mathrm{ab}}$. This is the desired result for the first term of (5.28) in the case $j = 0$.

In the general case, we have to introduce differential operators D_r and D_r^h by

$$(5.39) \qquad D_r = \frac{1}{2\pi i}\left(\frac{r}{\tau - \bar\tau} + \frac{\partial}{\partial \tau}\right) \qquad\qquad (0 < r \in \mathbf{Z}) \, ,$$

$$(5.40) \qquad D_r^h = \begin{cases} \text{the identity operator if} \quad h = 0 \, , \\ D_{r+2h-2} \cdots D_{r+2} D_r \quad \text{if} \quad 0 < h \in \mathbf{Z} \, . \end{cases}$$

We can easily verify that

$$(5.41) \qquad D_r^h f = (2\pi i)^{-h}(r + h - 1)! \sum_{\nu=0}^h \binom{h}{\nu} \frac{(\tau - \bar\tau)^{\nu-h}}{(r + \nu - 1)!} \partial^\nu f/\partial \tau^\nu \, ,$$

$$(5.42) \quad D_r^h(f|_r \gamma) = (D_r^h f)|_{r+2h} \gamma \text{ for every } \gamma \in \mathrm{SL}_2(\mathbf{R}) \, .$$

In particular, if $f \in \mathfrak{M}_r^{(1)}(\Delta_M)$, then $D_r^h f$ has the same automorphic property as the elements of $\mathfrak{M}_{r+2h}^{(1)}(\Delta_M)$. The same reasoning as in (5.38) shows that

$$\sum_{\gamma \in R_{N\beta}} f_\nu|_{k-h}\gamma = \sum_{\zeta \in E\beta} (f_\nu \varepsilon_{k-h-l-n}^M)|_{k-h}\zeta \, .$$

Call this sum s_ν. Then s_ν has components in $\mathfrak{M}_{k-h}^{(1)}(K_{\mathrm{ab}})$. By (5.42), we have

$$D_{k-h}^h s_\nu = \sum_{\gamma \in R_{N\beta}} (D_{k-h}^h f_\nu)|_{k+h}\gamma \, .$$

Substituting expression (5.41) with f_ν as f into this and comparing the result with (5.34), we find that

$$(5.43) \quad g_{t,\lambda}(0) = \frac{(2\pi i)^k}{(k-1)!}(-1)^{h+k+l} N^{h-k} \lambda^{k-h-l}(\tau_0 - \bar\tau_0)^h \sum_{\nu \in \mathrm{Y}} (D_{k-h}^h s_\nu)(\tau_0) \, .$$

Now the main theorem II of [7] asserts that if $s \in \mathfrak{M}_r^{(1)}(K_{\mathrm{ab}})$, the value $p(\tau_0)^{-r-2h}(D_r^h s)(\tau_0)$ belongs to K_{ab}. Applying this to (5.43), we obtain (5.24), which completes the proof of Theorem 5.3.

The series (5.30) under a suitable congruence condition is a non-vanishing function. Therefore $E_{j,k}(w, z; \mathfrak{m}, y)$ has a non-vanishing constant term at least for a suitable choice of $\mathfrak{m}$ and y. This shows the existence of non-vanishing $E_{j,k}$ for every j and k under the convergence condition.

So far we restricted our Eisenstein series to the case $q = 1$. If $q > 1$, we can define an Eisenstein series E_k by

$$(5.49) \qquad E_k(w, z; \mathfrak{m}, y) = \sum_{a,b,c} \det(aw + bz + c)^{-k} \,,$$

where (a, b, c) runs over a complete set of representatives for $X(\mathfrak{m}, y)$ modulo

$$\{u \in \mathrm{GL}_q(K) \mid u\mathfrak{m} = \mathfrak{m},\ uy - y \in \mathfrak{m}\} \,;$$

$X(\mathfrak{m}, y)$ is defined by (5.11) with a lattice $\mathfrak{m}$ of K_n^q and $y \in \mathfrak{X}$. There is no obvious analogue of $E_{j,k}$ with $j > 0$ in this case, although we can perhaps define similar objects by introducing a convergence factor or taking a different parabolic subgroup.

Let us now come back to the case $q = 1$. In the above proof we specialized w to 0 and observed that the value in question is the value of a (holomorphic or non-holomorphic) modular form at τ_0. We can actually discuss it without specialization. Although this inevitably makes things more complicated, there are some interesting points worth mentioning. For example, in the simplest case $h = l = 0$, define a function $q_y(w, \tau)$ by

$$(5.44) \qquad q_y(w, \tau) = \sum_{a \equiv y(P)} e\big(N^{-1}\lambda[t^{-1}\Phi(a)\tau + aw]\big) \qquad (w \in \mathbf{C}^n,\ \tau \in \mathfrak{H}_1) \,.$$

Then we have

$$(5.45) \quad g_{t,\lambda}(w) = \frac{(-2\pi i)^k}{(k - 1)!} N^{-k}\lambda^{k-1} \sum_{y \in Y} \sum_{\tau \in R_N\beta} j(\gamma, \tau_0)^{-k} q_y\big(j(\gamma, \tau_0)^{-1}w, \gamma(\tau_0)\big) \,.$$

Now q_y has an automorphic property

$$(5.46) \qquad q_y\big(j(\gamma, \tau)^{-1}w, \gamma(\tau)\big) = j(\gamma, \tau)^n q_y(w, \tau) \quad \text{for all } \gamma \in \Delta_M \,,$$

with a sufficiently large M. Therefore the same technique as in (5.38) is applicable to q_y. This connects $(g_{t,\lambda})_*(w)$ with the values of the "intrinsic theta functions" of [9, § 4]. Without going into details, we only note here that (5.46) is a special case of

PROPOSITION 5.7. *Let r and n be positive integers. With a positive definite hermitian matrix H of K_r^r, a lattice L in K_n^r, and an element ν of K_n^r, put*

$$f(w, z) = \sum_{x - \nu \in L} e_n\left(\frac{1}{2} \cdot {}^t\overline{x}Hxz + wx\right) \qquad (w \in \mathbf{C}_r^n,\ z \in \mathfrak{H}_n) \,.$$

Then

$$(5.47) \qquad f\big({}^t(cz + d)^{-1}w, \gamma(z)\big) = \det(cz + d)^r f(w, z)$$

for every $\gamma = \begin{pmatrix} a & b \\ c & d \end{pmatrix}$ in a sufficiently small congruence subgroup of $\mathrm{Sp}(n, \mathbf{Z})$.

This can be obtained by specialization of a classical theta function. It is noteworthy that the exponential factor of the type

$$e\left(\frac{1}{2}\cdot{}^{t}u(cz + d)^{-1}cu\right),$$

which occurs in the transformation of classical theta functions, does not appear in (5.47), even though f is holomorphic both in w and z.

If $h > 0$ or $l > 0$, the analogue of $q_\nu(w, \tau)$ involves the antiholomorphic variable $\bar{\tau}$. As the above proof shows, when w is specialized to 0, the nature of this value is explained by the operators D_r^h. If we want to study it without specializing w, we have to introduce a differential operator involving both w and τ. To be more explicit, let $f(u, z)$ be a function on $\mathbf{C}_s^n \times \mathfrak{H}_n$ belonging to $T_{k,\xi}$ with $0 \leq k \in \mathbf{Z}$ and ${}^{t}\xi = \xi \in \mathbf{Q}_s^s$ as defined in [12, § 3]. Define an operator $D_{k,\xi}$ which assigns to f a $\mathbf{C}_{n+s}^{n+s}$-valued function $D_{k,\xi}f$ by

$$(5.48) \qquad D_{k,\xi}f = \begin{bmatrix} \dfrac{1}{\pi i}\Delta f + \dfrac{k}{\pi i}(z - \bar{z})^{-1}f & \dfrac{1}{2\pi i}\dfrac{\partial f}{\partial u} \\[2ex] \dfrac{1}{2\pi i}{}^{t}\!\left(\dfrac{\partial f}{\partial u}\right) & \xi f \end{bmatrix},$$

$$\Delta f = \left(\frac{1+\delta_{\lambda\mu}}{2}\frac{\partial f}{\partial z_{\lambda\mu}}\right)_{\lambda\,\mu=1,\cdots,n}, \qquad \frac{\partial f}{\partial u} = \left(\frac{\partial f}{\partial u_{\lambda\nu}}\right)_{\substack{\lambda=1,\cdots,n \\ \nu=1,\cdots,s}}.$$

Then $D_{k,\xi}f$ behaves like an element of $T_{\sigma,k,\xi}$ with a representation σ of GL_{n+s} on the space of symmetric matrices defined by

$$\sigma(X)Y = XY\cdot{}^{t}X \qquad\qquad (X \in \mathrm{GL}_{n+s},\ Y = {}^{t}Y).$$

The functions $g_{t,\lambda}$ in the case $j = 1$ are given in terms of such $D_{k,\xi}f$. It will be an interesting problem to study the case $j > 1$ in a similar vein.

Finally we add that a non-holomorphic Eisenstein series on $\mathfrak{Z}$, similar to (5.30), can be defined by replacing the factor

$$\begin{bmatrix} \bar{b}w - S^{-1}\cdot{}^{t}\bar{a} \\ \bar{b}z + \bar{c} \end{bmatrix}^{(j)}$$

of (5.13) with its complex conjugate. The arithmetic significance of the values of such series as well as those of (5.13) at special points of $\mathfrak{Z}$ will be the topic of a subsequent paper.

6. The case $n = 0$

If iR has the same number of positive and negative eigenvalues, R is equivalent over K to

$$(6.1) \qquad \begin{bmatrix} 0 & 1_q \\ -1_q & 0 \end{bmatrix} \text{ or } \begin{bmatrix} s & 0 & 0 & 0 \\ 0 & t & 0 & 0 \\ 0 & 0 & 0 & 1_{q-1} \\ 0 & 0 & -1_{q-1} & 0 \end{bmatrix},$$

where s and t are numbers of K such that $is < 0$, $it > 0$, and $st \notin N_{K/Q}(K)$. Therefore let us first take R to be the first matrix of (6.1). In this case, we can make all the formulas of Section 1 valid, by disregarding the first n rows and n columns of matrices. For example, we understand that

$$(6.2) \qquad \mathfrak{Z} = \{z \in C_q^q \mid i({}^t\bar{z} - z) > 0\},$$
$$\mu(\alpha, z) = b_3 z + c_3, \quad \kappa(\alpha, z) = \bar{b}_3 \cdot {}^t z + \bar{c}_3$$

$$\text{if } \alpha = \begin{bmatrix} b_2 & c_2 \\ b_3 & c_3 \end{bmatrix} \in G_{R+}.$$

The group P_Q consists of all matrices of the form

$$(6.3) \qquad \begin{bmatrix} 1_q & y \\ 0 & 1_q \end{bmatrix}, \quad {}^t\bar{y} = y \in K_q^q.$$

Define $\mathfrak{M}_\rho(\Gamma)$ and $\mathfrak{A}_\rho(\Gamma)$ as in Section 2 with a K-rational representation ρ of $GL_q \times GL_q$. If $f \in \mathfrak{M}_\rho(\Gamma)$, we have a Fourier expansion

$$(6.4) \qquad f(z) = \sum_{r \in J} c_r e_q(rz)$$

with constant coefficients c_r, where J is a lattice in $\mathcal{H}_q$. Therefore for every subfield $\mathfrak{R}$ of C, we define $\mathfrak{M}_\rho(\Gamma, \mathfrak{R})$ to be the set of all such f with c_r in $\mathfrak{R}^{d(\rho)}$. Then we define $\mathfrak{M}_\rho(\mathfrak{R})$ and $\mathfrak{A}_\rho(\mathfrak{R})$ in exactly the same fashion as we did for Siegel modular forms. Now all the formulas of Section 3 are meaningful in the present case. In particular, we have an embedding ε of $\mathfrak{Z}$ into $\mathfrak{H}_{2q}$ defined by (3.7), where S and w should be ignored. Also observe that $\omega(z) = z$ and

$$(6.5) \qquad {}^t\chi(z)^{-1} = \begin{bmatrix} \dfrac{1}{2} 1_q & \dfrac{-\delta}{2} 1_q \\ \delta^{-1} 1_q & 1_q \end{bmatrix} = C$$

in the present case. For $F \in \mathfrak{A}_\sigma^{(m)}(C)$, define F^ε by $F^\varepsilon = \sigma(C) F \circ \varepsilon$. Then $F^\varepsilon \in \mathfrak{A}_\rho(C)$ with ρ defined by (3.20). If Ψ is a subfield of C containing K, and $F \in \mathfrak{A}_\sigma^{(m)}(\Psi)$, then $F^\varepsilon \in \mathfrak{A}_\rho(\Psi)$. Moreover, since $\det(C) = 1$, if $F \in \mathfrak{A}_k^{(m)}(\Psi)$ with $k \in Z$ and with an arbitrary subfield Ψ of C not necessarily containing K, we have $F^\varepsilon \in \mathfrak{A}_{2k}(\Psi)$. Define the field $\mathfrak{R}$ of arithmetic automorphic functions on $\mathfrak{Z}$ with respect to G. Since $n = 0$, the constant field of $\mathfrak{R}$ is Q_{ab} as shown in [3] and [5]. Therefore the analogue of Theorem 4.1 can be stated as follows:

THEOREM 6.1. *Suppose R is the first matrix of (6.1). Then, for every subfield Ψ of $\mathbf{C}$ containing $\mathbf{Q}_{ab}$, we have $\mathfrak{A}_0(\Psi) = \Psi\mathfrak{R}$.*

The proof can be given in the same manner.

The second case of (6.1) is more complicated, and in a sense more interesting. If $q = 1$ and R is equivalent to the second matrix of (6.1), that is, to $\begin{pmatrix} s & 0 \\ 0 & t \end{pmatrix}$, then the quotient of the symmetric domain modulo a congruence subgroup is compact, so that there is no Fourier expansion. To discuss the case $q>1$, take more generally R in the form

$$(6.6) \qquad R = \begin{bmatrix} S & 0 & 0 & 0 \\ 0 & 0 & 0 & 1_p \\ 0 & 0 & T & 0 \\ 0 & -1_p & 0 & 0 \end{bmatrix}, \qquad \begin{array}{l} -{}^t\bar{S} = S \in \mathrm{GL}_l(K),\ iS < 0, \\ -{}^t\bar{T} = T \in \mathrm{GL}_h(K),\ iT > 0. \end{array}$$

We assume $l \geqq h$ and $p > 0$. Note that a matrix of type (1.1) is equivalent to a matrix of type (6.6) if $q = p + h$ and $n = l - h$. Define a domain $\mathfrak{Z}'$ by

$$(6.7) \qquad \mathfrak{Z}' = \left\{ Z \in \mathbf{C}^{l+p}_{h+p} \,\middle|\, i({}^t\bar{Z} \quad 1_{h+p})R \begin{bmatrix} Z \\ 1_{h+p} \end{bmatrix} > 0 \right\}.$$

The action of an element $\alpha = \begin{pmatrix} A & B \\ C & D \end{pmatrix}$ of $G_{\mathbf{R}+}$ with $A \in \mathbf{C}^{l+p}_{l+p}$ on $\mathfrak{Z}'$ can be defined by $\alpha(Z) = (AZ + B)(CZ + D)^{-1}$. (The action of $G_{\mathbf{R}+}$ on $\mathfrak{Z}$ in Section 1 is the extreme case $h = 0$.) We can define automorphic forms on $\mathfrak{Z}'$ with the same ideas as in Section 2; for example, $\det(CZ + D)^k$ is a factor of automorphy. In this case, an automophic form f as a function of

$$(6.8) \qquad Z = \begin{pmatrix} u & w \\ v & z \end{pmatrix} \quad \text{with} \quad u \in \mathbf{C}^l_h,\ v \in \mathbf{C}^p_h,\ w \in \mathbf{C}^l_p,\ z \in \mathbf{C}^p_p,$$

has an expansion

$$(6.9) \qquad f(Z) = \sum_{r \in J} g_r(u;\, v,\, w)e_p(rz),$$

where J is a lattice in $\mathcal{H}_p$. Now it can be shown that g_r is a theta function of $(v,\, w) \in \mathbf{C}^p_h \times \mathbf{C}^l_p$; u belongs to a symmetric domain

$$(6.10) \qquad \{ u \in \mathbf{C}^l_h \mid i({}^t\bar{u}Su + T) > 0 \}.$$

Moreover, g_r has a certain automorphic property, similar to [12, (3.10)], under the elements of an arithmetic subgroup Δ of the unitary group of $\begin{pmatrix} S & 0 \\ 0 & T \end{pmatrix}$. Thus g_r is similar to the functions of $T_{k,\varepsilon}$ studied in [12]. If $l=h=1$, (6.10) is one-dimensional, and Δ is essentially a Fuchsian group obtained from an indefinite quaternion division algebra over $\mathbf{Q}$. In any case, we can define the arithmeticity of such g_r as well as that of f by means of the ideas explained in [12, § 7]. If $n = l - h > 0$ and $q = p + h$, this coincides with the previous definition of arithmeticity of automorphic forms in Section 3.

PRINCETON UNIVERSITY, PRINCETON, N.J.

REFERENCES

[1] A. BOREL, Introduction to automorphic forms, *Proc. Symp. Pure Math.* IX, *Algebraic Groups and Discontinuous Groups*, A.M.S. 1966, 199–210.

[2] P. DELIGNE, Travaux de Shimura, Sém. Bourbaki 1971, exp. 389, Lecture Notes in Math. 244, Springer.

[3] K. MIYAKE, Models of certain automorphic function fields, Acta Math. **126** (1971), 245–307.

[4] G. SHIMURA, On anlytic families of polarized abelian varieties and automorphic functions, Ann. of Math. **78** (1963), 149–192.

[5] ————, On the field of definition for a field of automorphic functions, I, II, III, Ann. of Math. **80** (1964), 160–189; **81** (1965), 124–165; **83** (1966), 377–385.

[6] ————, On canonical models of arithmetic quotients of bounded symmetric domains, I, II, Ann. of Math. **91** (1970), 144–222; **92** (1970), 528–549.

[7] ————, On some arithmetic properties of modular forms of one and several variables, Ann. of Math. **102** (1975), 491–515.

[8] ————, On the Fourier coefficients of modular forms of several variables, Göttingen Nachr. Akad. Wiss. 1975, 261–268.

[9] ————, Theta functions with complex multiplication, Duke Math. J. **43** (1976), 673–696.

[10] ————, On the derivatives of theta functions and modular forms, Duke Math. J. **44** (1977), 365–387.

[11] ————, Unitary groups and theta functions, *Proc. Int. Symp. on Alg. N. Th.* Kyoto 1976 (1977), 195–200.

[12] ————, On certain reciprocity-laws for theta functions and modular forms, to appear.

(Received November 9, 1977)

The special values of the zeta functions associated with Hilbert modular forms

Duke Mathematical Journal, 45 (1978), 637-679*

The zeta functions to be studied are of the following two types:

(I) $$D(s, \mathbf{f}, \mathbf{g}) = \sum C(\mathfrak{n}, \mathbf{f}) C(\mathfrak{n}, \mathbf{g}) N(\mathfrak{n})^{-s},$$

(II) $$D(s, \mathbf{f}, \omega) = \sum C(\mathfrak{n}, \mathbf{f}) \omega(\mathfrak{n}) N(\mathfrak{n})^{-s}.$$

Here $\mathbf{f}$ and $\mathbf{g}$ are Hilbert modular forms defined relative to a totally real alge-braic number filed F of degree n; ω is a Hecke ideal character of F of finite order; $\mathfrak{n}$ runs over all integral ideals of F; $C(\mathfrak{n}, \mathbf{f})$ and $C(\mathfrak{n}, \mathbf{g})$ are the "Fourier coefficients" of $\mathbf{f}$ and $\mathbf{g}$. The theorems in Section 4 will assert certain alge-braicity properties of the values of $D(s, \mathbf{f}, \mathbf{g})$ and $D(s, \mathbf{f}, \omega)$ at an integer or a half-integer, when $\mathbf{f}$ is a primitive cusp form. This generalizes our previous results in the elliptic modular case ([12], [14]). Our methods of proof are the same as in that case. In particular, it is indispensable to investigate the behavior of a modular form under the action of automorphisms on its Fourier coefficients.

If F has degree n, Hilbert modular forms are defined with respect to a factor of automorphy of the form

$$\prod_{\nu=1}^{n} (c_\nu z_\nu + d_\nu)^{k_\nu},$$

where $k_1, \ldots, k_n$ are positive integers. The modular forms with arbitrary $k_1, \ldots, k_n$ are as natural and important as those in the special case $k_1 = \cdots = k_n$. In fact, such "multiple weights" present many interesting features which do not exist in the one-dimensional case. This is one of the reasons why we have taken up the present work. Another motive is that there is an application of our results to the theory of arithmetic automorphic forms, which the author hopes to discuss on some future accasion.

One final remark may be added. The series of type (I) are significant on their own merits; they should not be regarded as auxiliary objects to the study of the series of type (II), as the introduction of the previous paper [12] might have misleadingly suggested.

Notation. If R is an associative ring with identity element, $R^\times$ denotes the group of all invertible elements of R. A diagonal matrix with diagonal entries $c_1, \ldots, c_m$ is denoted by $\mathrm{diag}[c_1, \ldots, c_m]$. We put $\mathbf{e}(z) = e^{2\pi i z}$ for $z \in \mathbf{C}$ and $\mathrm{sgn}(x) = x/|x|$ for $x \in \mathbf{R}^\times$. $\mathrm{Aut}(\mathbf{C})$ stands for the group of all ring-automorphisms of $\mathbf{C}$ (without continuity). If K is an algebraic number field, K_{ab} denotes the maximal abelian extension of K. The ring of p-adic integers is denoted by $\mathbf{Z}_p$. We denote by $\overline{\mathbf{Q}}$ the algebraic closure of $\mathbf{Q}$ in $\mathbf{C}$. We use $\bigsqcup$ to indicate a disjoint union.

Throughout the paper, F will denote a totally real algebraic number field of degree n, $\mathfrak{r}$ the maximal order of F, $F_{\mathbf{A}}$ the ring of adeles of F, $F_{\mathbf{A}}^\times$ the group of ideles of F, D_F the discriminant of F, R_F the regulator of F, and $\mathfrak{P}_\infty$

1

the product of all archimedean primes of F. The norm of an ideal $\mathfrak{a}$ (resp. an element ξ) of F is denoted by $N(\mathfrak{a})$ (resp. $N(\xi)$). By a Hecke character of $F_{\mathbf{A}}^{\times}$ we understand a continuous homomorphism η of $F_{\mathbf{A}}^{\times}$ into $\mathbf{C}^{\times}$ such that $|\eta| = 1$ and $\eta(F^{\times}) = 1$. For such an η we denote by η^{*} the ideal character associated with η.

1. Automorphisms on the space of Hilbert modular forms

Given F as above, we view $GL_2(F)$ as the group $G_{\mathbf{Q}}$ of $\mathbf{Q}$-rational points of a $\mathbf{Q}$-rational algebraic subgroup G of $GL(2n)$. Then the adelization $G_{\mathbf{A}}$ of G can be identified with $GL_2(F_{\mathbf{A}})$. We put

$$(1.1\text{a}) \qquad \mathfrak{H} = \{\, z \in \mathbf{C} \,|\, \mathrm{Im}(z) > 0 \,\},$$

$$(1.1\text{b}) \qquad GL_2^{+}(\mathbf{R}) = \{\, \alpha \in GL_2(\mathbf{R}) \,|\, \det(\alpha) > 0 \,\}.$$

We are going to define the action of certain elements of $G_{\mathbf{A}}$ on the Hilbert modular forms defined on $\mathfrak{H}^n$. We first let every element $\alpha = (\alpha_1, \ldots, \alpha_n)$ of $GL_2^{+}(\mathbf{R})^n$ with $\alpha_\nu = \begin{bmatrix} a_\nu & b_\nu \\ c_\nu & d_\nu \end{bmatrix}$ act on $\mathfrak{H}^n$ by

$$\alpha(z_1, \ldots, z_n) = (\alpha_1(z_1), \ldots, \alpha_n(z_n)), \quad \alpha_\nu(z_\nu) = (a_\nu z_\nu + b_\nu)/(c_\nu z_\nu + d_\nu).$$

For $k = (k_1, \ldots, k_n) \in \mathbf{Z}^n$ and $z = (z_1, \ldots, z_n) \in \mathbf{C}^n$, we write

$$(1.2\text{a}) \qquad z^k = \prod_{\nu=1}^{n} z_\nu^{k_\nu},$$

$$(1.2\text{b}) \qquad \{k\} = \sum_{\nu=1}^{n} k_\nu, \qquad \{z\} = \sum_{\nu=1}^{n} z_\nu,$$

$$(1.2\text{c}) \qquad \mathbf{e}_F(z) = \mathbf{e}(\{z\}) = \exp\left(2\pi i \sum_{\nu=1}^{n} z_\nu\right).$$

If $z_1, \ldots, z_n$ are real positive, we define real positive z^k also for $k \in \mathbf{Q}^n$ in the same manner. For a complex-valued function f on $\mathfrak{H}^n$ and $\alpha = \begin{bmatrix} * & * \\ c & d \end{bmatrix} \in GL_2^{+}(\mathbf{R})^n$ as above, we define functions $f_k|\alpha$ and $f\|_k\alpha$ by

$$(1.3\text{a}) \qquad (f|_k\alpha)(z) = (cz + d)^{-k} f(\alpha(z)),$$

$$(1.3\text{b}) \qquad (f\|_k\alpha)(z) = \det(\alpha)^{k/2}(f|_k\alpha)(z),$$

where abbreviation (1.2a) is used, with the understanding that

$$cz + d = (c_1 z_1 + d_1, \ldots, c_n z_n + d_n), \quad \det(\alpha) = \big(\det(\alpha_1), \ldots, \det(\alpha_n)\big).$$

We fix an embedding of F into $\mathbf{R}^n$ given by $a \mapsto (a^{\tau_1}, \ldots, a^{\tau_n})$, where $\tau_1, \ldots, \tau_n$ are all the ring-injections of F into $\mathbf{R}$. This identifies $\mathbf{R}^n$ and $GL_2(\mathbf{R})^n$ with the archimedean factors of $F_{\mathbf{A}}$ and $G_{\mathbf{A}}$, which we denote by F_∞ and G_∞, respectively. Then we put $G_{\infty+} = GL_2^{+}(\mathbf{R})^n$,

$$G_{\mathbf{A}+} = \{\, x \in G_{\mathbf{A}} \,|\, x_\infty \in G_{\infty+} \,\}, \qquad G_{\mathbf{Q}+} = G_{\mathbf{Q}} \cap G_{\mathbf{A}+},$$

where x_∞ denotes the archimedean part of x. We call a subgroup Γ of $G_{\mathbf{Q}+}$ a *congruence subgroup* if it contains

$$(1.4) \qquad \Gamma_N = \{\, \gamma \in SL_2(\mathfrak{r}) \,|\, \gamma - 1 \in N \cdot M_2(\mathfrak{r}) \,\}$$

for some poisitve integer N, and if $\Gamma/(\Gamma\cap F)$ is commensurable with $SL_2(\mathfrak{r})/\{\pm 1\}$. For such a Γ, we denote by $\mathcal{M}_k(\Gamma)$ the **C**-module of all holomorphic functions f on $\mathfrak{H}^n$ satisfying $f\|_k\gamma = f$ for all $\gamma \in \Gamma$, that are holomorphic at every cusp. (The holomorphy at a cusp is automatically satisfied if $F \neq \mathbf{Q}$.) Such an f has a Fourier expansion

$$(1.5) \qquad f(z) = \sum_\xi c(\xi)\mathbf{e}_F(\xi z) \qquad (z \in \mathfrak{H}^n)$$

with $c(\xi) \in \mathbf{C}$, where ξ runs over 0 and all totally positive elements of a lattice in F, and

$$\xi z = (\xi^{\tau_1} z_1, \ldots, \xi^{\tau_n} z_n).$$

An element of $\mathcal{M}_k(\Gamma)$ is called a *Hilbert modular form of weight* k *with respect to* Γ. We hereafter identify $\mathbf{Z}^n$ with the free module $\sum_{\nu=1}^n \mathbf{Z}\tau_\nu$ by putting $(k_1, \ldots, k_n) = \sum_{\nu=1}^n k_\nu\tau_\nu$, which makes the notation ξ^k for $\xi \in F$ natural and consistent. Also we put

$$(1.6) \qquad \mathbf{1} = (1, \ldots, 1) = \sum_{\nu=1}^n \tau_\nu.$$

Thus the elements of $\mathcal{M}_{\kappa\mathbf{1}}(\Gamma)$ with $0 < \kappa \in \mathbf{Z}$ are the Hilbert modular forms of weight κ in the traditional sense. Now let $\mathcal{M}_k$ denote the union of $\mathcal{M}_k(\Gamma_N)$ for all positive integers N. For a subfield $\mathfrak{R}$ of $\mathbf{C}$, we denote by $\mathcal{M}_k(\Gamma, \mathfrak{R})$ (resp. $\mathcal{M}_k(\mathfrak{R})$) the set of all $f \in \mathcal{M}_k(\Gamma)$ (resp. $f \in \mathcal{M}_k$) whose Fourier coefficients $c(\xi)$ determined by (1.5) belong to $\mathfrak{R}$, and by $\mathcal{A}_k(\mathfrak{R})$ the set of all meromorphic functions of the form f/g with $f \in \mathcal{M}_{h+k}(\mathfrak{R})$ and $0 \neq g \in \mathcal{M}_h(\mathfrak{R})$ (with any $h \in \mathbf{Z}^n$); further we put

$$(1.7a) \qquad \mathcal{A}_k(\Gamma, \mathfrak{R}) = \{\, f \in \mathcal{A}_k(\mathfrak{R}) \mid f\|_k\gamma = f \quad \text{for all} \quad \gamma \in \Gamma \,\},$$

$$(1.7b) \qquad \mathcal{A}_k(\Gamma) = \mathcal{A}_k(\Gamma, \mathbf{C}), \quad \mathcal{A}_k = \mathcal{A}_k(\mathbf{C}).$$

It can easily be seen that $\mathcal{M}_k(\mathfrak{R}) = \mathcal{A}_k(\mathfrak{R}) \cap \mathcal{M}_k$. If $f \in \mathcal{A}_k$, we write simply $f|\alpha$ and $f\|\alpha$ for $f|_k\alpha$ and $f\|_k\alpha$.

Every congruence subgroup Γ contains the elements $\mathrm{diag}[\varepsilon, \varepsilon^{-1}]$ with ε in a subgroup of $\mathfrak{r}^\times$ of finite index. If $f \in \mathcal{M}_k(\Gamma)$ and f has expansion (1.5), we have obviously

$$(1.8a) \qquad c(\varepsilon^2\xi) = \varepsilon^k c(\xi) \quad \text{if} \quad \mathrm{diag}[\varepsilon, \varepsilon^{-1}] \in \Gamma,$$

$$(1.8b) \qquad c(\varepsilon\xi) = \varepsilon^{k/2} c(\xi) \quad \text{if} \quad \mathrm{diag}[\varepsilon, 1] \in \Gamma.$$

An element f of $\mathcal{M}_k$ is called a *cusp form* if the constant term of the Fourier expansion of $f|_k\alpha$ is 0 for every $\alpha \in G_{\mathbf{Q}+}$. We denote by $\mathcal{S}_k$, $\mathcal{S}_k(\Gamma)$, $\mathcal{S}_k(\mathfrak{R})$, etc. the sets of cusp forms of $\mathcal{M}_k$, $\mathcal{M}_k(\Gamma)$, $\mathcal{M}_k(\mathfrak{R})$, etc. Relation (1.8a) shows that $\mathcal{M}_k = \mathcal{S}_k$ unless $k_1 = \cdots = k_n$.

Proposition 1.1. $\mathcal{M}_k \neq \{0\}$ *only if* $k_1 = \cdots = k_n \geq 0$ *or all* k_ν *are positive.*

PROOF. Given $0 \neq f \in \mathcal{M}_k$, put $g(w) = (f|_k\alpha)(w, \ldots, w)$ for $w \in \mathfrak{H}$ with any $\alpha \in G_{\mathbf{Q}+}$. Suppose $k_1 = \cdots = k_n$; then g is an elliptic modular form of weight nk_1. We can take α so that $g \neq 0$ (cf. [9, pp. 501-2]). Hence $k_1 \geq 0$. Next suppose that the k_ν are not all equal and contain nonpositive ones. The vanishing of $\mathcal{M}_k$ for such k in the compact quotient case was proved in [3, Theorem 3.1]. The same reasoning applies to the present case, since $\mathcal{M}_k = \mathcal{S}_k$ for such k. Thus we obtain our proposition.

Now we consider the ring Ψ of all formal power series of the form (1.5) with $c(\xi) \in \mathbf{C}$; we allow ξ to run over 0 and all totally positive elements of a lattice in F; the lattice may depend on each series. Since Ψ is an integral domain, we can form its field of quotients, say Ω, in which $\mathcal{A}_k$ can be embedded in a natural way. Let $\sigma \in \mathrm{Aut}(\mathbf{C})$. For each $f \in \Psi$ with expansion (1.5), we define f^σ by $f^\sigma = \sum_\xi c(\xi)^\sigma \mathbf{e}_F(\xi z)$. This action of σ can be uniquely extended to an automorphism of Ω, which is denoted also by $g \mapsto g^\sigma$ for $g \in \Omega$. Under our identification $\mathbf{Z}^n = \sum_{\nu=1}^n \mathbf{Z}\tau_\nu$, we can let σ act on $\mathbf{Z}^n$ by
$$\left(\textstyle\sum_{\nu=1}^n k_\nu \tau_\nu\right)^\sigma = \sum_{\nu=1}^n k_\nu \tau_\nu \sigma.$$

Proposition 1.2. *For every* $\sigma \in \mathrm{Aut}(\mathbf{C})$, *we have* $(\mathcal{A}_k)^\sigma = \mathcal{A}_l$ *and* $(\mathcal{M}_k)^\sigma = \mathcal{M}_l$ *with* $l = k^\sigma$.

PROOF. This was shown in [9] when $k_1 \equiv \cdots \equiv k_n \pmod 2$. To prove the general case, we may naturally assume $n > 1$. The main idea is to consider the pull-back of Siegel modular forms. Take a basis $\{\beta_1, \ldots, \beta_n\}$ of F over $\mathbf{Q}$ and put
$$B = \begin{bmatrix} \beta_1^{\tau_1} & \cdots & \beta_n^{\tau_1} \\ \cdots & \cdots & \cdots \\ \beta_1^{\tau_n} & \cdots & \beta_n^{\tau_n} \end{bmatrix},$$
$$W_B(z) = {}^t B \cdot \mathrm{diag}[z_1, \ldots, z_n] B \qquad (z = (z_1, \ldots, z_n) \in \mathfrak{H}^n).$$

Then $z \mapsto W_B(z)$ gives an embedding of $\mathfrak{H}^n$ into $\mathfrak{H}_n$, the Siegel upper space of degree n. This is compatible with the injection I of $SL_2(F)$ into $Sp(n, \mathbf{Q})$ defined by
$$I\left(\begin{bmatrix} a & b \\ c & d \end{bmatrix}\right) = \begin{bmatrix} {}^t B & 0 \\ 0 & B^{-1} \end{bmatrix} \begin{bmatrix} \Delta(a) & \Delta(b) \\ \Delta(c) & \Delta(d) \end{bmatrix} \begin{bmatrix} {}^t B^{-1} & 0 \\ 0 & B \end{bmatrix},$$
$$\Delta(a) = \mathrm{diag}[a^{\tau_1}, \ldots, a^{\tau_n}] \qquad (a \in F).$$

Let us now consider an $M_n(\mathbf{C})$-valued holomorphic function T on $\mathfrak{H}_n$ such that
$$T\big(\alpha(Z)\big) = (CZ + D)T(Z)\det(CZ + D) \quad \text{for all} \quad \alpha = \begin{bmatrix} * & * \\ C & D \end{bmatrix} \in \Gamma'$$
with a congruence subgroup Γ' of $Sp(n, \mathbf{Z})$. For every point Z_0 of $\mathfrak{H}_n$, we can find such a T with rational Fourier coefficients under the condition $\det\big(T(Z_0)\big) \neq 0$. Indeed, we obtained in [13] an $M_n(\mathbf{C})$-valued meromorphic function P on $\mathfrak{H}_n$ that is "$\mathbf{Q}$-rational" and satisfies
$$P\big(\alpha(Z)\big) = (CZ + D)P(Z) \quad \text{for all} \quad \alpha = \begin{bmatrix} * & * \\ C & D \end{bmatrix} \in \Gamma'.$$

With the notation of [13, p. 370, (1.19)], it is sufficient to take $\theta_j(0, Z)^2 P(Z)$ as $T(Z)$. Now we take such P and T so that they are finite and have non-zero determinant at least at a point of $W_B(\mathfrak{H}^n)$, and put, with any non-zero $q \in \mathbf{Q}^n$,

(1.9) $\qquad h(z) = B \cdot P\big(W_B(z)\big)q, \quad g(z) = B \cdot T\big(W_B(z)\big)q \qquad (z \in \mathfrak{H}^n).$

Let $h_1, \ldots, h_n$ and $g_1, \ldots, g_n$ be the components of h and g. Then $h_\nu \in \mathcal{A}_{\tau_\nu}(F^{\tau_\nu})$ and $g_\nu \in \mathcal{M}_{\tau_\nu+1}(F^{\tau_\nu})$, where $\{\tau_1, \ldots, \tau_n\}$ is considered a basis of $\mathbf{Z}^n$; moreover, $h_\nu^\sigma = h_\mu$ and $g_\nu^\sigma = g_\mu$ if $\sigma \in \mathrm{Aut}(\mathbf{C})$ and $\tau_\nu \sigma = \tau_\mu$ (cf. [13, Proposition 3.1]) ; none of the h_ν and g_ν is 0. Given $k \in \mathbf{Z}^n$ and $f \in \mathcal{A}_k$, let m be a positive integer greater than all k_ν, and put $\varphi = f \prod_{\nu=1}^n g_\nu^{m-k_\nu}$. Then $\varphi \in \mathcal{A}_{t\mathbf{1}}$ with $t = (n+1)m - \{k\}$, and $f^\sigma = \varphi^\sigma \prod_{\nu=1}^n (g_\nu^\sigma)^{k_\nu - m}$. Since $\varphi \in \mathcal{A}_{t\mathbf{1}}$, we have $\varphi^\sigma \in \mathcal{A}_{t\mathbf{1}}$ by [9, Proposition 7]. Therefore $f^\sigma \in \mathcal{A}_l$ with $l = k^\sigma$. Suppose $f \in \mathcal{M}_k$. For every $z_0 \in \mathfrak{H}^n$, we can choose the above T and q so that $g_\nu(z_0) \neq 0$ for all ν. Therefore f^σ is holomorphic at z_0, and hence $f^\sigma \in \mathcal{M}_l$. This completes the proof.

Let f and g be elements of $\mathcal{M}_{\kappa\mathbf{1}}$ $(0 < \kappa \in \mathbf{Z})$ with expansions

$$f(z) = \sum_\xi c(\xi)\mathbf{e}_F(\xi z), \qquad g(z) = \sum_\xi b(\xi)\mathbf{e}_F(\xi z).$$

Suppose $c(\xi)^\sigma = b(\xi)$ for all $\xi \neq 0$ with an element σ of $\mathrm{Aut}(\mathbf{C})$. Then $c(0)^\sigma - b(0) = f^\sigma - g \in \mathcal{M}_{\kappa\mathbf{1}}$, and hence $c(0)^\sigma = b(0)$. In particular, if $c(\xi)^\sigma = c(\xi)$ for all $\xi \neq 0$, we have $c(0)^\sigma = c(0)$. This proves the first assertion of the following proposition, since $\mathcal{M}_k = \mathcal{S}_k$ if $k \notin \mathbf{Z1}$.

Proposition 1.3. *Let* $f(z) = \sum_\xi c(\xi)\mathbf{e}_F(\xi z) \in \mathcal{M}_k$ *with* $k \neq 0$, *and let* K_f *be the field generated over* $\mathbf{Q}$ *by the* $c(\xi)$ *for* $\xi \neq 0$. *Then* $c(0) \in K_f$. *Moreover* K_f *is finitely generated over* $\mathbf{Q}$.

(That $c(0) \in K_f$ was proved by Klingen in [1] when $k = \kappa\mathbf{1}$ with $\kappa > 1$.)

PROOF. If $k = \kappa\mathbf{1}$, we have $\mathcal{M}_{\kappa\mathbf{1}} = \mathcal{M}_{\kappa\mathbf{1}}(\mathbf{Q}) \otimes_{\mathbf{Q}} \mathbf{C}$ by [9, Theorem 7], which proves that K_f is finitely generated over $\mathbf{Q}$. In the general case, given $f \in \mathcal{M}_k$, define φ as in the proof of Proposition 1.2. Then $f = \varphi \prod_{\nu=1}^n g_\nu^{k_\nu - m}$. Since g_ν has coefficients in F^{τ_ν}, the desired result for K_f follows from that for K_φ, which is already established.

Proposition 1.4. *Let* F' *denote the composite field* $F^{\tau_1} \cdots F^{\tau_n}$. *For each* $k \in \mathbf{Z}^n$, *let* Φ_k *be the subfield of* F' *such that*

$$\mathrm{Gal}(F'/\Phi_k) = \{\, \sigma \in \mathrm{Gal}(F'/\mathbf{Q}) \mid k^\sigma = k \,\}.$$

Let $f \in \mathcal{A}_k(\Lambda)$ *with a subfield* Λ *of* $\mathbf{C}$ *containing both* $\mathbf{Q}_{\mathrm{ab}}$ *and* Φ_k. *Then* $f|_k\alpha \in \mathcal{A}_k(\Lambda)$ *for every* $\alpha \in G_{\mathbf{Q}+}$; $f|_k\alpha \in \mathcal{M}_k(\Lambda)$ *if* $f \in \mathcal{M}_k(\Lambda)$.

PROOF. It is sufficient to prove our assertions for $\alpha = \mathrm{diag}[a, 1]$ and for $\alpha \in SL_2(F)$. The former case is obvious. Let $\alpha \in SL_2(F)$. If h is defined by (1.9) and $\beta = I(\alpha) = \begin{bmatrix} * & * \\ C & D \end{bmatrix}$, we have $h|\alpha = B \cdot S(W_B(z))q$ with $S(Z) = (CZ + D)^{-1}P(\beta(Z))$. By [13, Proposition 1.4] or [15, Theorem 1.2], S is $\mathbf{Q}_{\mathrm{ab}}$-rational, and hence $h_\nu|\alpha \in \mathcal{A}_{\tau_\nu}(F^{\tau_\nu}\mathbf{Q}_{\mathrm{ab}})$, and $(h_\nu|\alpha)^\sigma = h_\mu|\alpha$ if $\sigma = \mathrm{id}$. on $\mathbf{Q}_{\mathrm{ab}}$ and $\tau_\nu \sigma = \tau_\mu$. Given $f \in \mathcal{A}_k(\Lambda)$, put $\psi = f \prod_{\nu=1}^n h_\nu^{-k_\nu}$. Then $\psi \in \mathcal{A}_0(\Lambda)$ and $f|\alpha = (\psi|\alpha) \prod_{\nu=1}^n (h_\nu|\alpha)^{k_\nu}$. Since $\psi|\alpha \in \mathcal{A}_0(\Lambda)$, we obtain the first assertion. The second assertion is obvious, since $\mathcal{M}_k(\Lambda) = \mathcal{A}_k(\Lambda) \cap \mathcal{M}_k$.

It is now necessary to define several subgroups of $G_{\mathbf{A}}$ as follows:

$$D_+ = \left\{\, x \in G_{\mathbf{A}+} \,\middle|\, \det(x) \in \mathbf{Q}_{\mathbf{A}}^{\times} \,\right\}, \quad D_{\mathbf{Q}+} = \left\{\, x \in G_{\mathbf{Q}+} \,\middle|\, \det(x) \in \mathbf{Q}^{\times} \,\right\},$$
$$\mathcal{G}_+ = D_+ G_{\mathbf{Q}+} G_{\infty+} = \left\{\, x \in G_{\mathbf{A}+} \,\middle|\, \det(x) \in \mathbf{Q}_{\mathbf{A}}^{\times} F^{\times} F_{\infty+}^{\times} \,\right\} \quad (\text{ cf. } [6, (2.2.1)]),$$
$$R_N = G_{\infty+} \cdot \left\{\, x \in \textstyle\prod_{\mathfrak{p}} GL_2(\mathfrak{r}_{\mathfrak{p}}) \,\middle|\, x_{\mathfrak{p}} \equiv \mathrm{diag}[1, a_{\mathfrak{p}}] \pmod{N \cdot M_2(\mathfrak{r}_{\mathfrak{p}})} \right.$$
$$\left. \text{with } a_{\mathfrak{p}} \in \mathfrak{r}_{\mathfrak{p}}^{\times} \text{ for every finite place } \mathfrak{p} \,\right\},$$
$$S_N = G_{\infty+} \cdot \left\{\, x \in R_N \,\middle|\, \det(x) \in \mathbf{Q}_{\mathbf{A}}^{\times} \,\right\}.$$

Here and henceforth $F_{\mathfrak{p}}$ and $\mathfrak{r}_{\mathfrak{p}}$ denote the $\mathfrak{p}$-completions of F and $\mathfrak{r}$ for a finite place $\mathfrak{p}$; $x_{\mathfrak{p}}$ denotes the $\mathfrak{p}$-component of x. Given $x \in \mathcal{G}_+$, take $c \in \mathbf{Q}_{\mathbf{A}}^{\times}$ so that $\det(x) \in cF^{\times} F_{\infty+}^{\times}$. We then obtain a homomorphism ρ of $\mathcal{G}_+$ onto $\mathrm{Gal}(\mathbf{Q}_{\mathrm{ab}}/\mathbf{Q})$ by defining $\rho(x)$ to be the action of c^{-1} on $\mathbf{Q}_{\mathrm{ab}}$. (This is a special case of $[6, (2.3.7)]$.) Now we define a subgroup $\mathfrak{G}$ of $\mathcal{G}_+ \times \mathrm{Gal}(\overline{\mathbf{Q}}/\mathbf{Q})$ by

$$\mathfrak{G} = \left\{\, (x, \sigma) \in \mathcal{G}_+ \times \mathrm{Gal}(\overline{\mathbf{Q}}/\mathbf{Q}) \,\middle|\, \rho(x) = \sigma \ \text{ on } \ \mathbf{Q}_{\mathrm{ab}} \,\right\}.$$

Theorem 1.5. *There is an action of $\mathfrak{G}$ on the graded algebra $\sum_k \mathcal{A}_k(\overline{\mathbf{Q}})$, written $((x, \sigma), f) \mapsto f^{(x,\sigma)}$ for $(x, \sigma) \in \mathfrak{G}$ and $f \in \mathcal{A}_k(\overline{\mathbf{Q}})$, with the following properties:*

(i) $(f + g)^{(x,\sigma)} = f^{(x,\sigma)} + g^{(x,\sigma)}$, $(fg)^{(x,\sigma)} = f^{(x,\sigma)} g^{(x,\sigma)}$;

(ii) $(f^{(x,\sigma)})^{(y,\tau)} = f^{(xy,\sigma\tau)}$;

(iii) $f^{(\alpha,1)} = f|\alpha$ if $\alpha \in G_{\mathbf{Q}+}$;

(iv) $f^{(x,\sigma)} = f^{\sigma}$ if $x = \mathrm{diag}[1, t]$ with $t \in \prod_p \mathbf{Z}_p^{\times}$;

(v) $\mathcal{A}_k(\overline{\mathbf{Q}})^{(x,\sigma)} = \mathcal{A}_l(\overline{\mathbf{Q}})$ and $\mathcal{M}_k(\overline{\mathbf{Q}})^{(x,\sigma)} = \mathcal{M}_l(\overline{\mathbf{Q}})$ if $l = k^{\sigma}$;

(vi) *The action of (x, σ) on $\mathcal{A}_0(\mathbf{Q}_{\mathrm{ab}})$ is the same as that of x defined in the framework of canonical models (see $[6]$, $[9]$).*

PROOF. We first note that $\Gamma_N = G_{\mathbf{Q}+} \cap S_N$ and $\mathcal{G}_+ = G_{\mathbf{Q}+} S_N = S_N G_{\mathbf{Q}+}$ for every N; the latter follows easily from the strong approximation theorem; see $[6, \text{Proposition } 3.4]$. Given $f \in \mathcal{A}_k(\overline{\mathbf{Q}})$, the images f^{σ} under $\sigma \in \mathrm{Gal}(\overline{\mathbf{Q}}/\mathbf{Q})$ form a finite set, by virtue of Proposition 1.3. Therefore we can find N so that $f^{\sigma} \in \mathcal{A}_*(\Gamma_N)$ for all $\sigma \in \mathrm{Gal}(\overline{\mathbf{Q}}/\mathbf{Q})$. Let $(x, \sigma) \in \mathfrak{G}$. We have $x = s\alpha$ with $s \in S_N$ and $\alpha \in G_{\mathbf{Q}+}$. Then we define $f^{(x,\sigma)}$ to be $f^{\sigma}|\alpha$. This does not depend on the choice of s and α, and has the properties of our theorem. The verification is easy except the associativity (ii). If $k \in 2\mathbf{Z}^n$, the associativity follows from $[6, \text{Theorem } 8]$, which, in essence, is a special case of the present theorem; the only difference is that we previously defined $f^{(\alpha,1)}$ to be $f\|_k\alpha$. This change causes no difficulty so long as $k \in 2\mathbf{Z}^n$. To prove the associativity in the general case, we first observe that it follows from the special case

$$(1.10) \qquad (f^{(\beta,1)})^{(x,\sigma)} = f^{(\beta x,\sigma)} \quad \text{if} \ \beta \in G_{\mathbf{Q}+} \text{ and } (x, \sigma) \in \mathfrak{G}.$$

Now take $h = {}^t(h_1, \ldots, h_n)$ defined by (1.9), and suppose

$$(1.11) \qquad (h^{(\beta,1)})^{(x,\sigma)} = h^{(\beta x,\sigma)}$$

is true for one choice of $\{h, \beta, (x, \sigma)\}$. Then the expression $f = \psi \prod h_{\nu}^{k_{\nu}}$ in the proof of Proposition 1.4 shows that (1.10) is true for the same choice of β and (x, σ), and for all $f \in \mathcal{A}_k(\overline{\mathbf{Q}})$ with any k. Now every element of $G_{\mathbf{Q}+}$ is a product of an element of $SL_2(F)$ and an element of the form $\mathrm{diag}[a, 1]$. Therefore it is sufficient to show (1.11) with these two types of elements as β. Let $\beta = \mathrm{diag}[a, 1]$; take N so that $h|\gamma = h$ for $\gamma \in \Gamma_N$. Further take a multiple M of N so that $\beta S_M \beta^{-1} \subset S_N$. Let $x = s\alpha$ with $s \in S_M$ and $\alpha \in G_{\mathbf{Q}+}$. Since $h^{(\beta,1)}(z) = h(az)$, we have $(h^{(\beta,1)})^{(x,\sigma)} = h^\sigma | \beta\alpha$. Now $(\beta x, \sigma) = (\beta s \beta^{-1}, \sigma)(\beta\alpha, 1)$, so that $h^{(\beta x, \sigma)} = h^\sigma | \beta\alpha$, which proves the case $\beta = \mathrm{diag}[a, 1]$. Next let $\beta \in SL_2(F)$. Take a multiple M of N so that $(h|\beta)^\sigma$ is Γ_M-invariant for all $\sigma \in \mathrm{Gal}(\overline{\mathbf{Q}}/\mathbf{Q})$. Given (x, σ), let $x = s\alpha$ with $s \in S_M$, $\alpha \in G_{\mathbf{Q}+}$, and let $\beta s = u\gamma$ with $u \in S_N$ and $\gamma \in G_{\mathbf{Q}+}$. Suppose $(h^{(\beta,1)})^{(s,\sigma)} = h^{(\beta s, \sigma)}$. Then

$$(h^{(\beta,1)})^{(x,\sigma)} = (h^{(\beta s,\sigma)})^{(\alpha,1)} = \left((h^{(u,\sigma)})^{(\gamma,1)}\right)^{(\alpha,1)} = h^{(\beta x, \sigma)}.$$

Therefore it is sufficient to prove

$$(1.12) \qquad (h^{(\beta,1)})^{(s,\sigma)} = h^{(\beta s, \sigma)} \quad \text{for} \quad \beta \in SL_2(F) \text{ and } s \in S_M.$$

Since the action of $G_{\infty+}$ is trivial, we may assume that $\det(s) \in \mathbf{Q}_{\mathbf{A}}^\times$. Now the injection I of $SL_2(F)$ into $Sp(n, \mathbf{Q})$ can be extended to an injection of D_+ into $G_{\mathbf{A}+}^{(n)}$, where $G_{\mathbf{A}+}^{(n)}$ is the (restricted) adelization of the group of similitudes of an allternating form considered in [6], [10], and [15]. If $x \in D_+$, we easily see that $h^{(x,\sigma)} = B^\sigma P^{I(x)}(W_B(z))q$. Therefore (1.12) follows from the associativity of the action of $G_{\mathbf{A}+}^{(n)}$, which is guaranteed by [15, Theorem 1.2]. This completes the proof.

Let us now consider a congruence subgroup Γ defined by

$$(1.13) \qquad \Gamma = \Gamma(\mathfrak{b}, \mathfrak{c})$$

$$= \left\{ \begin{bmatrix} a & b \\ c & d \end{bmatrix} \in G_{\mathbf{Q}+} \;\middle|\; a \in \mathfrak{r},\ b \in \mathfrak{b}^{-1},\ c \in \mathfrak{bc},\ d \in \mathfrak{r},\ ad - bc \in \mathfrak{r}^\times \right\}$$

with a fractional ideal $\mathfrak{b}$ and an integral ideal $\mathfrak{c}$ in F. Let $\mathfrak{r}_+^\times$ denote the group of all totally positive units of $\mathfrak{r}$. With a character χ_0 of $(\mathfrak{r}/\mathfrak{c})^\times \times (\mathfrak{r}/\mathfrak{c})^\times$ and a character χ_1 of $\mathfrak{r}_+^\times$, both of finite order, define a character χ of Γ by

$$\chi\left(\begin{bmatrix} a & b \\ c & d \end{bmatrix}\right) = \chi_1(ad - bc)\chi_0\big(a \ (\mathrm{mod}\ \mathfrak{c}),\ d \ (\mathrm{mod}\ \mathfrak{c})\big),$$

and set

$$(1.14) \qquad \mathcal{M}_k(\Gamma, \chi) = \left\{ f \in \mathcal{M}_k \;\middle|\; f\|\gamma = \chi(\gamma)f \text{ for all } \gamma \in \Gamma \right\}.$$

Proposition 1.6. *If $f \in \mathcal{M}_k(\Gamma, \chi)$ and $\sigma \in \mathrm{Aut}(\mathbf{C})$, then $f^\sigma \in \mathcal{M}_l(\Gamma, \chi_\sigma)$, where $l = k^\sigma$ and $\chi_\sigma(\gamma) = \chi(\gamma)^\sigma \det(\gamma)^{l/2}/\big(\det(\gamma)^{k/2}\big)^\sigma$.*

PROOF. Let $f \in \mathcal{M}_k(\Gamma, \chi)$ and $\gamma \in \Gamma$. It is sufficient to show that $f^\sigma\|\gamma = \chi_\sigma(\gamma)f^\sigma$ in the cases either $\gamma = \mathrm{diag}[a, 1]$ or $\det(\gamma) = 1$. The former case can be verified in a straightforward way. Let $\gamma = \begin{bmatrix} a & b \\ c & d \end{bmatrix} \in \Gamma \cap SL_2(F)$. To prove this case, we assume $f \in \mathcal{M}_k(\overline{\mathbf{Q}})$. The general case will be discussed later. Take $t \in \prod_p \mathbf{Z}_p^\times$ so that σ coincides on $\mathbf{Q}_{\mathrm{ab}}$ with the action of t^{-1}, and put $s = \mathrm{diag}[1, t]$. By strong approximation, we have $s\gamma s^{-1} = \beta u$ with $\beta \in SL_2(F)$

and $u \in S_N \cap SL_2(F_\mathbf{A})$, where N is any positive integer. Put $\beta = \begin{bmatrix} a' & b' \\ c' & d' \end{bmatrix}$. With a sufficiently large N, we have $a' \equiv a$, $d' \equiv d \pmod{\mathfrak{c}}$ and $\beta \in \Gamma$. Then

$$f^\sigma \| \gamma = f^{(s\gamma, \sigma)} = f^{(\beta u s, \sigma)} = (\chi(\beta) f)^\sigma = \chi(\beta)^\sigma f^\sigma.$$

This is the desired result, since $\chi(\beta) = \chi(\gamma)$.

Proposition 1.7. *Let Φ_k be the field defined in Proposition 1.4. Then $\mathcal{M}_k = \mathcal{M}_k(\Phi_k) \otimes_{\Phi_k} \mathbf{C}$.*

PROOF. Define $g = {}^t(g_1, \ldots, g_n)$ by (1.9) and put $r = g_1 \cdots g_n$. Then $r \in \mathcal{M}_{(n+1)\mathbf{1}}(\mathbf{Q})$. Take N so that $\mathcal{M}_\mathbf{1}(\Gamma_N, \mathbf{Q}) \neq \{0\}$ and g is Γ_N-invariant; further take a canonical model V of $\Gamma_N \backslash \mathfrak{H}^n$ rational over $\mathbf{Q}$ as considered in the proof of [9, Theorem 6]. (V is denoted as V_T there.) Every $f \in \mathcal{A}_k(\Gamma_N)$, $\neq 0$, defines its divisor on V, written $\mathrm{div}(f)$, which is an algebraic divisor. The proof of [9, Theorem 5] established an element p of $\mathcal{M}_{t\mathbf{1}}(\Gamma_N, \mathbf{Q})$ with $0 < t \in \mathbf{Z}$ such that $\mathrm{div}(p)$ is $\mathbf{Q}$-rational. Since r^t/p^{n+1} defines a $\mathbf{Q}$-rational function on V, we see that $\mathrm{div}(r)$ is $\mathbf{Q}$-rational. Obviously $\mathrm{div}(r) = \mathrm{div}(g_1) + \cdots + \mathrm{div}(g_n)$, and $\mathrm{div}(g_\nu)$ is a positive divisor since g_ν is holomorphic; consequently $\mathrm{div}(g_\nu)$ is $\overline{\mathbf{Q}}$-rational for every ν. Take a non-zero element φ of $\mathcal{M}_\mathbf{1}(\Gamma_N, \mathbf{Q})$. Given $k \in \mathbf{Z}^n$, put $\psi = \varphi^{-\{k\}} \prod_{\nu=1}^n g_\nu^{k_\nu}$. Then $\psi \in \mathcal{A}_k(\Gamma_N, \Phi_k)$, and $\mathrm{div}(\psi)$ is $\overline{\mathbf{Q}}$-rational. Using the isomorphism between $\mathcal{M}_k(\Gamma_N)$ and the linear ssystem of $\mathrm{div}(\psi)$ on V_N, we can prove

$$(1.15) \qquad \mathcal{M}_k(\Gamma_N) = \mathcal{M}_k(\Gamma_N, \overline{\mathbf{Q}}) \otimes_{\overline{\mathbf{Q}}} \mathbf{C}$$

in exactly the same fashion as in the proof of [9, Theorem 6]. Now

$$(1.16) \qquad \mathcal{M}_k(\Gamma_N, \overline{\mathbf{Q}})^\sigma = \mathcal{M}_l(\Gamma_N, \overline{\mathbf{Q}}) \quad \text{if} \quad \sigma \in \mathrm{Gal}(\overline{\mathbf{Q}}/\mathbf{Q}) \quad \text{and} \quad l = k^\sigma.$$

This can be shown by the same and simpler argument as in the proof of Proposition 1.6. Let $f \in \mathcal{M}_k(\Gamma_N, \overline{\mathbf{Q}})$ and let K be a finite Galois extension of Φ_k containing the Fourier coefficients of f. Then $\sum_\sigma b^\sigma f^\sigma \in \mathcal{M}_k(\Gamma_N, \Phi_k)$ for every $b \in K$, where σ runs over $\mathrm{Gal}(K/\Phi_k)$. This shows that

$$(1.17) \qquad \mathcal{M}_k(\Gamma_N, \overline{\mathbf{Q}}) = \mathcal{M}_k(\Gamma_N, \Phi_k) \otimes_{\Phi_k} \overline{\mathbf{Q}},$$

which, combined with (1.15), completes the proof.

We can now complete the proof of Proposition 1.6. Let $f \in \mathcal{M}_k(\Gamma, \chi)$. By Proposition 1.7, f is a finite sum $\sum c_\lambda g_\lambda$ with $c_\lambda \in \mathbf{C}$ and $g_\lambda \in \mathcal{M}_k(\overline{\mathbf{Q}})$. Here we may assume that the c_λ are linearly independent over $\overline{\mathbf{Q}}$. For every $\gamma \in \Gamma$, we have $\sum c_\lambda (g_\lambda \| \gamma - \chi(\gamma) g_\lambda) = 0$. Since $g_\lambda \| \gamma - \chi(\gamma) g_\lambda \in \mathcal{M}_k(\overline{\mathbf{Q}})$, we see that $g_\lambda \| \gamma - \chi(\gamma) g_\lambda = 0$ for all λ, that is, $g_\lambda \in \mathcal{M}_k(\Gamma, \chi)$. This proves

$$(1.18) \qquad \mathcal{M}_k(\Gamma, \chi) = [\mathcal{M}_k(\Gamma, \chi) \cap \mathcal{M}_k(\overline{\mathbf{Q}})] \otimes_{\overline{\mathbf{Q}}} \mathbf{C},$$

as well as Proposition 1.6, since the assertion has been shown for such g_λ.

Proposition 1.8. *The space of cusp forms $\mathcal{S}_k(\overline{\mathbf{Q}})$ is stable under the action of $\mathfrak{G}$.*

PROOF. Since $\mathcal{S}_k(\overline{\mathbf{Q}})$ is stable under $G_{\mathbf{Q}+}$, it is sufficient to show that it is stable under every σ of $\mathrm{Gal}(\overline{\mathbf{Q}}/\mathbf{Q})$. Given σ, take t and s as in the proof of

Proposition 1.6. Let $f \in \mathcal{S}_k(\overline{\mathbf{Q}})$ and $\alpha \in G_{\mathbf{Q}+}$. Observe that $\{ g^{(v,\tau)} \mid (v, \tau) \in \mathfrak{G},\ v \in S_1 \}$ is a finite set for every $g \in \mathcal{M}_k(\overline{\mathbf{Q}})$. Therefore we can find a positive integer N so that $(f^\sigma|\alpha)^{(v,\tau)} \in \mathcal{M}_*(\Gamma_N)$ for all $(v, \tau) \in \mathfrak{G}$ with $v \in S_1$. Let $s\alpha = \beta w$ with $\beta \in G_{\mathbf{Q}+}$ and $w \in S_N$. Then $(f|\beta)^\tau \in \mathcal{M}_*(\Gamma_N)$ for every $\tau \in \mathrm{Gal}(\overline{\mathbf{Q}}/\mathbf{Q})$, so that $f^\sigma|\alpha = (f|\beta)^\sigma$. Since $f|\beta \in \mathcal{S}_k$, the first Fourier coefficient of $f^\sigma|\alpha$ is 0, Q.E.D.

In the proof of Proposition 1.6 we have seen that if $0 \neq f \in \mathcal{A}_{t1}(\Gamma_N, \mathbf{Q})$, then $\mathrm{div}(f)$ is $\mathbf{Q}$-rational. One can naturally ask the following question:

For $0 \neq f \in \mathcal{A}_k(\Gamma_N, \Lambda)$ with a field Λ containing Φ_k, is $\mathrm{div}(f)$ rational over Λ?

The answer seems affirmative.

2. Hecke operators and Dirichlet series

In this section we first introduce Hecke operators on the space of Hilbert modular forms, and associate Dirichlet series with them. We follow the formulation of our previous paper [5]. In the latter part of this section, we shall discuss the behavior of the eigenfunctions of Hecke operators under $\mathrm{Gal}(\overline{\mathbf{Q}}/\mathbf{Q})$.

For a fractional ideal $\mathfrak{a}$ of F and a prime ideal $\mathfrak{p}$ of F, we define the localization $\mathfrak{a}_\mathfrak{p}$ of $\mathfrak{a}$ as usual as a submodule of $F_\mathfrak{p}$. Let $\mathfrak{d}$ denote the different of F relative to $\mathbf{Q}$, and $\mathfrak{c}$ an arbitrarily fixed integral ideal of F. We define subsets $W = W(\mathfrak{c})$ and $Y = Y(\mathfrak{c})$ of $G_\mathbf{A}$ by

$$(2.1a) \qquad W = G_{\infty+} \cdot \prod_\mathfrak{p} W_\mathfrak{p}, \qquad Y = G_\mathbf{A} \cap \left(G_{\infty+} \cdot \prod_\mathfrak{p} Y_\mathfrak{p} \right),$$

where the product is taken over all prime ideals $\mathfrak{p}$ of F, and

$$(2.1b) \quad Y_\mathfrak{p} = \left\{ \begin{bmatrix} a & b \\ c & d \end{bmatrix} \in GL_2(F_\mathfrak{p}) \,\middle|\, a\mathfrak{r}_\mathfrak{p} + \mathfrak{c}_\mathfrak{p} = \mathfrak{r}_\mathfrak{p},\ b \in \mathfrak{d}_\mathfrak{p}^{-1},\ c \in \mathfrak{c}_\mathfrak{p}\mathfrak{d}_\mathfrak{p},\ d \in \mathfrak{r}_\mathfrak{p} \right\},$$

$$(2.1c) \qquad\qquad W_\mathfrak{p} = \left\{ x \in Y_\mathfrak{p} \,\middle|\, \det(x) \in \mathfrak{r}_\mathfrak{p}^\times \right\}.$$

Obviously Y is a semi-group and W is a subgroup of $G_{\mathbf{A}+}$. For every $t \in F_\mathbf{A}^\times$, we denote by $t\mathfrak{r}$ the fractional ideal of F associated with t; if $\mathfrak{a}$ is another fractional ideal, we put $t\mathfrak{a} = t\mathfrak{r} \cdot \mathfrak{a}$. Let h be the number of ideal classes modulo $\mathfrak{P}_\infty$. Take h elements $t_1, \ldots, t_h$ of $F_\mathbf{A}^\times$ so that $(t_\lambda)_\infty = 1$ and $t_1\mathfrak{r}, \ldots, t_h\mathfrak{r}$ form a complete set of representatives for such ideal classes, and put

$$(2.2) \qquad\qquad x_\lambda = \mathrm{diag}[1, t_\lambda], \qquad x_\lambda^{-\iota} = \mathrm{diag}[t_\lambda^{-1}, 1].$$

Here and henceforth, ι denotes the main involution of $M_2(F)$ and its extension to $M_2(F_\mathbf{A})$. We easily see that $G_\mathbf{A}$ can be expressed as a disjoint union

$$(2.3) \qquad\qquad G_\mathbf{A} = \bigsqcup_{\lambda=1}^h G_\mathbf{Q} x_\lambda W = \bigsqcup_{\lambda=1}^h G_\mathbf{Q} x_\lambda^{-\iota} W.$$

Put

$$\Gamma_\lambda = \Gamma_\lambda(\mathfrak{c}) = x_\lambda W x_\lambda^{-1} \cap G_\mathbf{Q}.$$

With the notation of (1.13), we have $\Gamma_\lambda = \Gamma(t_\lambda\mathfrak{d}, \mathfrak{c})$. Now take a character ψ_0 of $(\mathfrak{r}/\mathfrak{c})^\times$, and define a homomorphism $\psi_Y : Y \to \mathbf{C}^\times$ by

$$(2.4) \qquad \psi_Y \left(\begin{bmatrix} a & b \\ c & d \end{bmatrix} \right) = \psi_0(a_{\mathfrak{c}} \bmod \mathfrak{c}),$$

where $a_{\mathfrak{c}}$ denotes "the $\mathfrak{c}$-part" of a. Given ψ_0 and $k = (k_1, \ldots, k_n) \in \mathbf{Z}^n$, we put

$$(2.5a) \qquad \mathcal{M}_k(\Gamma_\lambda, \psi_0) = \{\, f \in \mathcal{M}_k \mid f\|_k\gamma = \psi_Y(x_\lambda^{-1}\gamma x_\lambda)f \ \text{ for all } \ \gamma \in \Gamma_\lambda \,\},$$

$$(2.5b) \qquad \mathcal{S}_k(\Gamma_\lambda, \psi_0) = \mathcal{M}_k(\Gamma_\lambda, \psi_0) \cap \mathcal{S}_k,$$

$$(2.5c) \qquad \mathcal{M}_k(\mathfrak{c}, \psi_0) = \prod_{\lambda=1}^h \mathcal{M}_k(\Gamma_\lambda, \psi_0), \qquad \mathcal{S}_k(\mathfrak{c}, \psi_0) = \prod_{\lambda=1}^h \mathcal{S}_k(\Gamma_\lambda, \psi_0).$$

Obviously $\mathcal{M}_k(\mathfrak{c}, \psi_0) = \{0\}$ unless ψ_0 satisfies

$$(2.6) \qquad \psi_0(\varepsilon) = \operatorname{sgn}(\varepsilon^k) \ \text{ for every } \ \varepsilon \in \mathfrak{r}^\times.$$

Therefore we hereafter impose this condtion on ψ_0. Then

$$\psi_Y(x_\lambda^{-1}\gamma^\iota x_\lambda) = \psi_Y(x_\lambda^{-1}\gamma x_\lambda)^{-1}$$

for every $\gamma \in \Gamma_\lambda$. Let $(f_1, \ldots, f_h) \in \mathcal{M}_k(\mathfrak{c}, \psi_0)$ with $f_\lambda \in \mathcal{M}_k(\Gamma_\lambda, \psi_0)$. We now associate with $(f_1, \ldots, f_h)$ a $\mathbf{C}$-valued function $\mathbf{f}$ on $G_\mathbf{A}$ by

$$(2.7) \qquad \mathbf{f}(\alpha x_\lambda^{-\iota} w) = \psi_Y(w^\iota)(f_\lambda\|_k w_\infty)(\mathbf{i}) \ \text{ for } \ \alpha \in G_\mathbf{Q} \ \text{ and } \ w \in W,$$

where $\mathbf{i} = (i, \ldots, i)$. Since $\mathbf{f}$ determines $f_1, \ldots, f_h$, we shall write $\mathbf{f} = (f_1, \ldots, f_h)$, and regard $\mathcal{M}_k(\mathfrak{c}, \psi_0)$ also as the set of all such functions $\mathbf{f}$ on $G_\mathbf{A}$. Notice that $\mathbf{f}(\beta x v) = \psi_Y(v^\iota)\mathbf{f}(x)$ for $\beta \in G_\mathbf{Q}$ and $v \in W$, $v_\infty = 1$.

We now introduce the algebra $R(W, Y)$ of all finite sums $\sum c_y WyW$ with $y \in Y$ and $c_y \in \mathbf{C}$, and also the algebra $\{R_{\lambda\mu}\}$, where $R_{\lambda\mu}$ consists of all finite sums $\sum_\alpha \Gamma_\lambda \alpha \Gamma_\mu$ with $\alpha \in G_\mathbf{Q} \cap x_\lambda Y x_\mu^{-1}$ and $c_\alpha \in \mathbf{C}$. The law of multiplication is defined as in [5, §2] and [7, §3.2]. Put $W_\lambda = x_\lambda W x_\lambda^{-1}$. Then we have $\Gamma_\lambda = G_\mathbf{Q} \cap W_\lambda$ and

$$(2.8a) \qquad \alpha \in G_\mathbf{Q} \cap x_\lambda Y x_\mu^{-1} \implies W_\lambda \alpha W_\mu = W_\lambda \alpha \Gamma_\mu = \Gamma_\lambda \alpha W_\mu,$$

$$(2.8b) \qquad \Gamma_\lambda \alpha \Gamma_\mu = \bigsqcup_j \Gamma_\lambda \alpha_j \implies W x_\lambda^{-1} \alpha x_\mu W = \bigsqcup_j W x_\lambda^{-1} \alpha_j x_\mu.$$

To prove these, for such an α, put $x_\lambda^{-1}\alpha x_\mu = arb$ with $a, b \in W$ and a diagonal element r of $G_\mathbf{A}$, and $U = \alpha^{-1}W_\lambda\alpha \cap W_\mu$. Then $U = x_\mu b^{-1}(r^{-1}Wr \cap W)bx_\mu^{-1}$; hence $\det(U) = \det(r^{-1}Wr \cap W)$, which equals $\det(W) = \det(W_\mu)$, as can easily be seen. Thus, given $w \in W_\mu$, we can find $u \in U$ such that $\det(w) = \det(u)$. By strong approximation, $u^{-1}w \in UG_\mathbf{Q}$, which shows that $W_\mu \subset UG_\mathbf{Q}$. Since $U \subset W_\mu$ and $\Gamma_\mu = W_\mu \cap G_\mathbf{Q}$, we obtain $W_\mu \subset U\Gamma_\mu$; hence $W_\lambda\alpha W_\mu \subset W_\lambda\alpha U\Gamma_\mu = W_\lambda\alpha\Gamma_\mu$. Similarly we can show that $W_\lambda\alpha W_\mu \subset \Gamma_\lambda\alpha W_\mu$. Thus we obtain (2.8a), from which (2.8b) follows immediately, since $W x_\lambda^{-1}\alpha x_\mu W = x_\lambda^{-1}W_\lambda\alpha W_\mu x_\mu = x_\lambda^{-1}W_\lambda\alpha\Gamma_\mu x_\mu$.

Now given $\alpha \in G_\mathbf{Q} \cap x_\lambda Y x_\mu^{-1}$, take a coset decomposition

$$\Gamma_\lambda \alpha \Gamma_\mu = \bigsqcup_j \Gamma_\lambda \alpha_j$$

with finitely many α_j. For $f \in \mathcal{M}_k(\Gamma_\lambda, \psi_0)$, we can define an element $f|\Gamma_\lambda\alpha\Gamma_\mu$ of $\mathcal{M}_k(\Gamma_\mu, \psi_0)$ by

$$(2.9) \qquad f|\Gamma_\lambda\alpha\Gamma_\mu = \sum_j \psi_Y(x_\lambda^{-1}\alpha_j x_\mu)^{-1} f\|\alpha_j.$$

Next, given $y \in Y$, let $WyW = \bigsqcup_j Wy_j$ with elements y_j such that $(y_j)_\infty = 1$. For $\mathbf{f} \in \mathcal{M}_k(\mathfrak{c}, \psi_0)$ we define a function $\mathbf{f}|WyW$ on $G_\mathbf{A}$, which is an element of $\mathcal{M}_k(\mathfrak{c}, \psi_0)$, by

$$(2.10) \qquad (\mathbf{f}|WyW)(x) = \sum_j \psi_Y(y_j)^{-1}\mathbf{f}(xy_j^\iota) \qquad (x \in G_\mathbf{A}).$$

For each fixed λ, in view of (2.3), we can find an element α_λ of $G_\mathbf{Q}$ and μ so that $x_\lambda y \in \alpha_\lambda x_\mu W$. Then $\alpha_\lambda \in x_\lambda Y x_\mu^{-1} \cap G_\mathbf{Q}$ and $WyW = Wx_\lambda^{-1}\alpha_\lambda x_\mu W$; μ is uniquely determined under the condition that $\det(y)t_\lambda t_\mu^{-1}\mathfrak{r}$ belongs to the principal class modulo $\mathfrak{P}_\infty$. If $\mathbf{f} = (f_1, \ldots, f_h)$, we have

$$(2.11) \qquad \mathbf{f}|WyW = (g_1, \ldots, g_h) \quad \text{with} \quad g_\mu = f_\lambda | \Gamma_\lambda \alpha_\lambda \Gamma_\mu.$$

In fact, if $\Gamma_\lambda \alpha_\lambda \Gamma_\mu = \bigsqcup_j \Gamma_\lambda \gamma_j$, we have $WyW = \bigsqcup_j Wx_\lambda^{-1}\gamma_j x_\mu$ by (2.8b); hence we can take $\{x_\lambda^{-1}(\gamma_j)_0 x_\mu\}$ to be $\{y_j\}$, where the subscript 0 indicates the projection to the non-archimedean factor of $G_\mathbf{A}$. Then, for $w \in G_{\infty+}$ we have $(g_\mu\|w)(\mathbf{i})$ $= (\mathbf{f}|WyW)(x_\mu^{-\iota}w) = \sum_j \psi_Y(x_\lambda^{-1}\gamma_j x_\mu)\mathbf{f}(x_\mu^{-\iota}wx_\mu^\iota(\gamma_j^\iota)_0 x_\lambda^{-\iota}) = \sum_j \psi_Y(x_\lambda^{-1}\gamma_j x_\mu)$ $\cdot\mathbf{f}((\gamma_j^\iota)_0 x_\lambda^{-\iota}w)$. For $\beta \in G_{\mathbf{Q}+}$ let β_∞ be its projection to G_∞. Then $\mathbf{f}(\beta_0 x) = \mathbf{f}(\beta\beta_\infty^{-1}x) = \mathbf{f}(\beta_\infty^{-1}x)$, so that $\mathbf{f}((\gamma_j^\iota)_0 x_\lambda^{-\iota}w) = \mathbf{f}(x_\lambda^{-\iota}(\gamma_j^\iota)_\infty^{-1}w) = ((f_\lambda\|\gamma_j)\|w)(\mathbf{i})$. Thus we obtain (2.11).

For each integral ideal $\mathfrak{m}$ of F, let $T_\mathfrak{c}(\mathfrak{m})$ denote the element of $R(W, Y)$ which is the sum of all different WyW with $y \in Y$ such that $\det(y)\mathfrak{r} = \mathfrak{m}$; further we define $S_\mathfrak{c}(\mathfrak{m})$ as follows: $S_\mathfrak{c}(\mathfrak{m}) = WaW$ with an element a of $F_\mathbf{A}^\times$ such that $\mathfrak{m} = a\mathfrak{r}$ if $\mathfrak{m}$ is prime to $\mathfrak{c}$; $S_\mathfrak{c}(\mathfrak{m}) = 0$ if $\mathfrak{m}$ is not prime to $\mathfrak{c}$. Then $R(W, Y)$ is a commutative algebra generated by $T_\mathfrak{c}(\mathfrak{p})$ and $S_\mathfrak{c}(\mathfrak{p})$ for all prime ideals $\mathfrak{p}$, and

$$(2.12) \qquad T_\mathfrak{c}(\mathfrak{m})T_\mathfrak{c}(\mathfrak{n}) = \sum_{\mathfrak{m}+\mathfrak{n}\subset\mathfrak{a}} N(\mathfrak{a})S_\mathfrak{c}(\mathfrak{a})T_\mathfrak{c}(\mathfrak{a}^{-2}\mathfrak{m}\mathfrak{n});$$

we have also a formal Euler product

$$(2.13) \qquad \sum_\mathfrak{m} T_\mathfrak{c}(\mathfrak{m})N(\mathfrak{m})^{-s} = \prod_\mathfrak{p} \left[1 - T_\mathfrak{c}(\mathfrak{p})N(\mathfrak{p})^{-s} + S_\mathfrak{c}(\mathfrak{p})N(\mathfrak{p})^{1-2s}\right]^{-1}$$

(cf. [5, Proposition 2.6] and [7, §§3.3, 3.4]). We shall write simply $T(\mathfrak{m})$ and $S(\mathfrak{m})$ for $T_\mathfrak{c}(\mathfrak{m})$ and $S_\mathfrak{c}(\mathfrak{m})$ when there is no fear of confusion.

If $\mathfrak{p} \nmid \mathfrak{c}$ and $\pi_\mathfrak{p}$ is a prime element of $\mathfrak{r}_\mathfrak{p}$, we have

$$(2.14) \qquad (\mathbf{f}|S(\mathfrak{p}))(x) = \mathbf{f}(\pi_\mathfrak{p}x) \qquad (\mathfrak{p} \nmid \mathfrak{c}, \; x \in G_\mathbf{A}).$$

Proposition 2.1. *If a non-zero element $\mathbf{f}$ of $\mathcal{M}_k(\mathfrak{c}, \psi_0)$ is a common eigenfunction of $S(\mathfrak{p})$ for all prime ideals $\mathfrak{p}$ not dividing $\mathfrak{c}$, then there is a Hecke character ψ of $F_\mathbf{A}^\times$ of finite order such that $\mathbf{f}(vx) = \psi(v)\mathbf{f}(x)$ for all $v \in F_\mathbf{A}^\times$. Moreover, put $\mathfrak{U} = F_{\infty+}^\times \prod_\mathfrak{p} \mathfrak{r}_\mathfrak{p}^\times$; then $\psi(u) = \mathrm{sgn}(u_\infty)^k\psi_0(u_\mathfrak{c} \bmod \mathfrak{c})$ for every $u \in \mathfrak{U}$.*

PROOF. From (2.7) we obtain $\mathbf{f}(bux) = \mathrm{sgn}(u_\infty)^k\psi_0(u_\mathfrak{c} \bmod \mathfrak{c})\mathbf{f}(x)$ for $b \in F^\times$ and $u \in \mathfrak{U}$. Given $t \in F_\mathbf{A}^\times$, take a prime ideal $\mathfrak{p}$ prime to $\mathfrak{c}$ belonging to the ideal class of $t\mathfrak{r}$ modulo $\mathfrak{P}_\infty$. Then $t \in \pi_\mathfrak{p}F^\times\mathfrak{U}$. Therefore, if $\mathbf{f}$ is an eigenfunction of $S(\mathfrak{p})$, then $\mathbf{f}(tx)$ is a constant times $\mathbf{f}(x)$. Thus we can define a character ψ of $F_\mathbf{A}^\times$ by $\mathbf{f}(vx) = \psi(v)\mathbf{f}(x)$ for $v \in F_\mathbf{A}^\times$. Since $\psi(bu) =$

$\text{sgn}(u_\infty)^k \psi_0(u_{\mathfrak{c}} \bmod \mathfrak{c})$ for $b \in F^\times$ and $u \in \mathfrak{U}$, ψ is a Hecke character of finite order, as expected.

For fixed ψ_0 and k, consider the set of all Hecke characters ψ of $F_{\mathbf{A}}^\times$ such that $\psi(u) = \text{sgn}(u_\infty)^k \psi_0(u_{\mathfrak{c}} \bmod \mathfrak{c})$ for every $u \in \mathfrak{U}$. Clearly the conductor of such a ψ divides $\mathfrak{c}$. Now, for each such ψ, let $\mathcal{M}_k(\mathfrak{c}, \psi)$ denote the set of all $\mathbf{f} \in \mathcal{M}_k(\mathfrak{c}, \psi_0)$ such that $\mathbf{f}(v x) = \psi(v)\mathbf{f}(x)$ for every $v \in F_{\mathbf{A}}^\times$. Then $\mathcal{M}_k(\mathfrak{c}, \psi_0)$ is the direct sum of the subspaces $\mathcal{M}_k(\mathfrak{c}, \psi)$ for all such ψ. The subspace of $\mathcal{M}_k(\mathfrak{c}, \psi)$ consisting of cusp forms is denoted by $\mathcal{S}_k(\mathfrak{c}, \psi)$. It can easily be seen that $\mathcal{M}_k(\mathfrak{c}, \psi)$ consists of all $\mathbf{C}$-valued functions $\mathbf{f}$ on $G_{\mathbf{A}}$ satisfying the following conditions.

(2.15a) $\mathbf{f}(s\alpha x w) = \psi(s)\psi_Y(w^\iota)\mathbf{f}(x)$ for $s \in F_{\mathbf{A}}^\times$, $\alpha \in G_{\mathbf{Q}}$, $w \in W(\mathfrak{c})$, $w_\infty = 1$.

(2.15b) For every $x \in G_{\mathbf{A}}$ with $x_\infty = 1$, there exists an element g_x of $\mathcal{M}_k$ such that $\mathbf{f}(xv) = (g_x \| v)(\mathbf{i})$ for all $v \in G_{\infty+}$.

The latter condition can be replaced by a weaker one:

(2.15c) For each λ, there exists an element f_λ of $\mathcal{M}_k$ such that $\mathbf{f}(x_\lambda^{-\iota} v) = (f_\lambda \| v)(\mathbf{i})$ for all $v \in G_{\infty+}$.

Let us now consider the Fourier expansion of an element of $\mathcal{M}_k(\mathfrak{c}, \psi_0)$ and examine the effect of the operators $T(\mathfrak{m})$ on it. Let $\mathbf{f} = (f_1, \ldots, f_h) \in \mathcal{M}_k(\mathfrak{c}, \psi_0)$ with $f_\lambda \in \mathcal{M}_k(\Gamma_\lambda, \psi_0)$. Since Γ_λ contains the elements of the form $\begin{bmatrix} 1 & b \\ 0 & 1 \end{bmatrix}$ with $b \in (t_\lambda \mathfrak{d})^{-1}$, f_λ has a Fourier expansion

$$(2.16) \qquad f_\lambda(z) = \sum_\xi a_\lambda(\xi) e_F(\xi z) \qquad (\xi = 0 \ \text{ or } \ 0 \ll \xi \in t_\lambda),$$

where we put $t_\lambda = t_\lambda \mathfrak{r}$, and write $0 \ll \xi$ if ξ is totally positive. Since $f_\lambda(\varepsilon z)\varepsilon^{k/2} = f_\lambda(z)$ for $0 \ll \varepsilon \in \mathfrak{r}^\times$, we see that $a_\lambda(\xi)\xi^{-k/2}$ for $\xi \neq 0$ depends only on the ideal $\xi \mathfrak{r}$. Now every integral ideal of F can be written as ξt_λ^{-1} with a unique λ and a totally positive $\xi \in t_\lambda$. Thus, define $c(\mathfrak{m}, \mathbf{f})$ for every fractional ideal $\mathfrak{m}$ of F by

$$(2.17) \qquad c(\mathfrak{m}, \mathbf{f}) = \begin{cases} a_\lambda(\xi)\xi^{-k/2} & \text{if } \ \mathfrak{m} = \xi t_\lambda^{-1} \subset \mathfrak{r}, \ 0 \ll \xi, \\ 0 & \text{if } \ \mathfrak{m} \not\subset \mathfrak{r}. \end{cases}$$

Then we can show that $\mathbf{f}$ has the following expansion:

$$(2.18) \qquad \mathbf{f}\left(\begin{bmatrix} y & x \\ 0 & 1 \end{bmatrix}\right) = \sum_{0 \ll \zeta \in F} c(\zeta y \mathfrak{r}, \mathbf{f})(\zeta y_\infty)^{k/2} e_F(i\zeta y_\infty)\chi_F(\zeta x)$$

$$+ \begin{cases} c_0(y\mathfrak{r})|y|^{k_1/2} & \text{if } \ k_1 = \cdots = k_n, \\ 0 & \text{otherwise}. \end{cases}$$

Here $x \in F_{\mathbf{A}}$, $y \in F_{\mathbf{A}}^\times$, $y_\infty \gg 0$; χ_F is the continuous character of the additive group $F_{\mathbf{A}}/F$ such that $\chi_F(x_\infty) = e_F(x_\infty)$ for $x_\infty \in F_\infty$; $|y| = N(y\mathfrak{r})^{-1}|y_\infty|^{\mathbf{1}}$; c_0 is the function on the ideal group of F defined by

$$(2.19) \qquad c_0(\eta t_\lambda^{-1}) = a_\lambda(0)N(t_\lambda)^{-k_1/2} \quad \text{for} \quad 0 \ll \eta \in F.$$

To prove (2.18), we first observe that $F_{\mathbf{A}}^{\times} = \bigcup_{\lambda=1}^{h} F^{\times}\mathfrak{U}t_{\lambda}^{-1}$. Given $y \in F_{\mathbf{A}}^{\times}$ and $x \in F_{\mathbf{A}}$, take $p \in F^{\times}$, $u \in \mathfrak{U}$, λ, and $q \in F$ so that $y = put_{\lambda}^{-1}$ and $x - q = pt_{\lambda}^{-1}v$ with $v \in F_{\infty} \times \prod_{\mathfrak{p}} \mathfrak{o}_{\mathfrak{p}}^{-1}$. Then $\begin{bmatrix} y & x \\ 0 & 1 \end{bmatrix} = \alpha x_{\lambda}^{-\iota}w$ with $\alpha = \begin{bmatrix} p & q \\ 0 & 1 \end{bmatrix}$ and $w = \begin{bmatrix} u & v \\ 0 & 1 \end{bmatrix}$; hence

$$\mathbf{f}\left(\begin{bmatrix} y & x \\ 0 & 1 \end{bmatrix}\right) = (f_{\lambda}\|w_{\infty})(\mathbf{i}) = u_{\infty}^{k/2}f_{\lambda}(v_{\infty} + iu_{\infty})$$

$$= (p^{-1}y)_{\infty}^{k/2}f_{\lambda}\left(p^{-1}(x_{\infty} + iy_{\infty} - q_{\infty})\right)$$

$$= (p^{-1}y)_{\infty}^{k/2}\sum_{\xi}a_{\lambda}(\xi)\mathbf{e}_{F}\left(p^{-1}\xi(x_{\infty} + iy_{\infty} - q_{\infty})\right)$$

$$= a_{\lambda}(0)(p^{-1}y)_{\infty}^{k/2} + \sum_{\zeta}c(\zeta y\mathfrak{r}, \mathbf{f})(\zeta y)_{\infty}^{k/2}\mathbf{e}_{F}(i\zeta y_{\infty})\mathbf{e}_{F}(\zeta x_{\infty} - \zeta q_{\infty}),$$

where we put $\zeta = p^{-1}\xi$; $a_{\lambda}(0)$ occurs only when $k_1 = \cdots = k_n$. Put $\kappa = k_1$ in such a case. Then $(p^{-1}y)_{\infty}^{\kappa \mathbf{1}} = |N(p)|^{-\kappa}|y_{\infty}|^{\kappa \mathbf{1}} = N(yt_{\lambda})^{-\kappa}|y_{\infty}|^{\kappa \mathbf{1}} = N(t_{\lambda})^{-\kappa}|y|^{\kappa}$; also, $\chi_{F}(\zeta x) = \chi_{F}((\zeta x)_{\infty})\chi_{F}((\zeta x)_0) = \mathbf{e}_{F}(\zeta x_{\infty})\chi_{F}((\zeta x)_0)$ and $\chi_{F}((\zeta x)_0) = \chi_{F}((p^{-1}\xi x)_0) = \chi_{F}((\xi t_{\lambda}^{-1}v + \xi p^{-1}q)_0) = \chi_{F}((\xi p^{-1}q)_0) = \chi_{F}(-(\zeta q)_{\infty}) = \mathbf{e}_{F}(-\zeta q)$, where we used the fact that $\chi_{F}((\xi t_{\lambda}^{-1}v)_0) = 1$ since $(\xi t_{\lambda}^{-1}v)_{\mathfrak{p}} \in \mathfrak{o}_{\mathfrak{p}}^{-1}$ for every $\mathfrak{p}$. Thus we obtain (2.18).

Define an ideal character ψ^{*} modulo $\mathfrak{c}\mathfrak{P}_{\infty}$ by $\psi^{*}(\mathfrak{p}) = \psi(\pi_{\mathfrak{p}})$ for every prime ideal $\mathfrak{p}$ not dividing $\mathfrak{c}$ with a prime element $\pi_{\mathfrak{p}}$ of $F_{\mathfrak{p}}$; we understand that $\psi^{*}(\mathfrak{a}) = 0$ if $\mathfrak{a}$ is not prime to $\mathfrak{c}$. Notice that $\psi^{*}(\xi\mathfrak{r}) = \psi_0(\xi)^{-1}\mathrm{sgn}(\xi)^{k}$ for ξ prime to $\mathfrak{c}$. Suppose now that $\mathbf{f} \in \mathcal{M}_k(\mathfrak{c}, \psi)$. Then we have

$$(2.20) \qquad c(\mathfrak{m}, \mathbf{f}|T_{\mathfrak{c}}(\mathfrak{n})) = \sum_{\mathfrak{m}+\mathfrak{n}\subset\mathfrak{a}} \psi^{*}(\mathfrak{a})N(\mathfrak{a}^{-1}\mathfrak{n})c(\mathfrak{a}^{-2}\mathfrak{m}\mathfrak{n}, \mathbf{f}).$$

To prove this, we first consider the set of all integral ideals $\mathfrak{a}$ dviding $\mathfrak{n}$ and prime to $\mathfrak{c}$. For each $\mathfrak{a}$ we fix elements a and d of $F_{\mathbf{A}}^{\times}$ such that $a_{\mathfrak{c}} = 1$, $a_{\infty} = d_{\infty} = 1$, $a\mathfrak{r} = \mathfrak{a}$, $ad\mathfrak{r} = \mathfrak{n}$; take an element δ of $F_{\mathbf{A}}^{\times}$ such that $\delta_{\infty} = 1$ and $\delta\mathfrak{r} = \mathfrak{o}$. We then observe that $T(\mathfrak{n})$ viewed as a union of double cosets WyW is a disjoint union of the cosets Wv, $v = \begin{bmatrix} a & -\delta^{-1}b_0 \\ 0 & d \end{bmatrix}$ with such a, d, and $b \in \mathfrak{r}/\mathfrak{d}\mathfrak{r}$. For such a v, we have $\psi_{Y}(v) = 1$ and $\begin{bmatrix} y & x \\ 0 & 1 \end{bmatrix}v^{\iota} = \begin{bmatrix} a & 0 \\ 0 & a \end{bmatrix}\begin{bmatrix} a^{-1}dy & x + a^{-1}\delta^{-1}b_0y \\ 0 & 1 \end{bmatrix}$, where x, y are as in (2.18). To simplify the matter, let us assume that the constant term is 0. By (2.10) and (2.18) we have

$$(\mathbf{f}|T(\mathfrak{n}))\left(\begin{bmatrix} y & x \\ 0 & 1 \end{bmatrix}\right) = \sum_{a,d,b}\psi(a)\mathbf{f}\left(\begin{bmatrix} a^{-1}dy & x + a^{-1}\delta^{-1}b_0y \\ 0 & 1 \end{bmatrix}\right)$$

$$= \sum_{a,d,b,\zeta}\psi(a)c(\zeta a^{-1}dy\mathfrak{r}, \mathbf{f})(\zeta y_{\infty})^{k/2}\mathbf{e}_{F}(i\zeta y_{\infty})\chi_{F}(\zeta x)\chi_{F}(\zeta a^{-1}b_0\delta^{-1}y).$$

Now $c(\zeta a^{-1}dy\mathfrak{r}, \mathbf{f}) = 0$ unless $\zeta a^{-1}dy\mathfrak{r} \subset \mathfrak{r}$; hence, assuming the inclusion, we have $\sum_{b}\chi_{F}(\zeta a^{-1}b_0\delta^{-1}y) = \sum_{b}\chi_{F}\left(\delta^{-1}\cdot\zeta a^{-1}dy\cdot(d^{-1}b_0)\right)$, which is 0 unless $\zeta a^{-1}y\mathfrak{r} \subset \mathfrak{r}$, in which case the sum if $N(d\mathfrak{r})$. Thus the sum $\sum_{a,d,b,\zeta}$ becomes

$$\sum_{a,d,\zeta} \psi(a)N(d\mathfrak{r})c(\zeta a^{-1}dy\mathfrak{r}, \mathbf{f})(\zeta y_\infty)^{k/2}\mathbf{e}_F(i\zeta y_\infty)\chi_F(\zeta x)$$

$$= \sum_\zeta \left\{ \sum_a \psi^*(a)N(a^{-1}\mathfrak{n})c(a^{-2}\mathfrak{n}\zeta y\mathfrak{r}, \mathbf{f}) \right\} (\zeta y_\infty)^{k/2}\mathbf{e}_F(i\zeta y_\infty)\chi_F(\zeta x)$$

with a running under the condition $\mathfrak{n} + \zeta y\mathfrak{r} \subset a$. Putting $\mathfrak{m} = \zeta y\mathfrak{r}$, we obtain (2.20) if there is no constant term. Clearly the result is valid even if the constant term is nontrivial.

Let $k_0 = \mathrm{Max}(k_1, \ldots, k_n)$; put

$$(2.21) \qquad\qquad T'(\mathfrak{n}) = T'_c(\mathfrak{n}) = N(\mathfrak{n})^{(k_0-2)/2}T_c(\mathfrak{n}),$$

$$(2.22) \qquad\qquad C(\mathfrak{m}, \mathbf{f}) = N(\mathfrak{m})^{k_0/2}c(\mathfrak{m}, \mathbf{f}).$$

Then (2.20) can be rewritten as

$$(2.23) \qquad C(\mathfrak{m}, \mathbf{f}|T'(\mathfrak{n})) = \sum_{\mathfrak{m}+\mathfrak{n}\subset a} \psi^*(a)N(a)^{k_0-1}C(a^{-2}\mathfrak{m}\mathfrak{n}, \mathbf{f}).$$

If $\mathfrak{m} = \xi t_\lambda^{-1}$ with $0 \ll \xi \in F$, we have

$$(2.24) \qquad\qquad C(\mathfrak{m}, \mathbf{f}) = N(t_\lambda)^{-k_0/2}a_\lambda(\xi)\xi^{(k_0 \mathbf{1}-k)/2}.$$

We now attach to $\mathbf{f}$ a Dirichlet series

$$(2.25) \qquad\qquad D(s, \mathbf{f}) = \sum_{\mathfrak{m}} C(\mathfrak{m}, \mathbf{f})N(\mathfrak{m})^{-s},$$

where $\mathfrak{m}$ runs over all integral ideals of F. As will be shown later, this converges for sufficiently large $\mathrm{Re}(s)$, and can be continued to a meromorphic function on the whole s-plane. If $\mathbf{f}$ is a common eigenfunction of $T'(\mathfrak{m})$ for all $\mathfrak{m}$ and $\mathbf{f}|T'(\mathfrak{m}) = \lambda(\mathfrak{m})\mathbf{f}$, then $C(\mathfrak{m}, \mathbf{f}) = \lambda(\mathfrak{m})C(\mathfrak{r}, \mathbf{f})$, and we have

$$(2.26) \qquad \sum_{\mathfrak{m}} \lambda(\mathfrak{m})N(\mathfrak{m})^{-s} = \prod_{\mathfrak{p}} \left[1 - \lambda(\mathfrak{p})N(\mathfrak{p})^{-s} + \psi^*(\mathfrak{p})N(\mathfrak{p})^{k_0-1-2s}\right]^{-1}.$$

We call $\mathbf{f}$ *normalized* if $c(\mathfrak{r}, \mathbf{f}) = 1$, in which case $C(\mathfrak{r}, \mathbf{f}) = 1$ and $C(\mathfrak{m}, \mathbf{f}) = \lambda(\mathfrak{m})$.

Proposition 2.2. *The eigenvalues of $T(\mathfrak{m})$ are algebraic numbers. Moeoveer, if $k_1 = \cdots = k_n$, then the eigenvalues of $T'(\mathfrak{m})$ are algebraic integers.*

PROOF. Let $\mathcal{M}_k(\mathfrak{c}, \psi, \overline{\mathbf{Q}})$ denote the set of all $\mathbf{f} = (f_1, \ldots, f_h)$ in $\mathcal{M}_k(\mathfrak{c}, \psi)$ such that $f_\lambda \in \mathcal{M}_k(\overline{\mathbf{Q}})$ for all λ. Then $T(\mathfrak{m})$ maps $\mathcal{M}_k(\mathfrak{c}, \psi, \overline{\mathbf{Q}})$ into itself, as can be seen either from Proposition 1.4 or from (2.20). Therefore we obtain our first assertion in view of (1.18). If $k_1 = \cdots = k_n$, $\mathcal{M}_k(\mathfrak{c}, \psi, \overline{\mathbf{Q}})$ is spanned by the forms $\mathbf{f}$ for which $C(\mathfrak{m}, \mathbf{f})$ are algebraic integers. This follows from [10] and [11, pp. 682–683]. They are stable under $T'(\mathfrak{m})$ by virtue of (2.23), which proves the second assertion.

Let us now introduce an inner product in $\mathcal{M}_k$ as follows. For f and g in $\mathcal{M}_k$, take a congruence subgroup Γ so that $\mathcal{M}_k(\Gamma)$ contains both f and g, and put

$$(2.27) \qquad\qquad \langle f, g \rangle = \mu(\Gamma\backslash\mathfrak{H}^n)^{-1} \int_{\Gamma\backslash\mathfrak{H}^n} \overline{f(z)}g(z)y^k d\mu(z),$$

where $d\mu(z) = \prod_{\nu=1}^{n} y_\nu^{-2} dx_\nu dy_\nu$ with $z_\nu = x_\nu + iy_\nu$ and $\mu(\Gamma \backslash \mathfrak{H}^n) = \int_{\Gamma \backslash \mathfrak{H}^n} d\mu(z)$. The inner product is meaningful if fg is a cusp form, and does not depend on the choice of Γ. We use the notation $\langle f, g \rangle$ also for C^∞-functions f and g with the same property of automorphy as the elements of $\mathcal{M}_k$, whenever the integral is convergent. Next, for $\mathbf{f} = (f_1, \ldots, f_h)$ and $\mathbf{g} = (g_1, \ldots, g_h)$ of $\mathcal{M}_k(\mathfrak{c}, \psi_0)$, we define their inner product by

$$(2.28) \qquad \langle \mathbf{f}, \mathbf{g} \rangle = \sum_{\lambda=1}^{h} \langle f_\lambda, g_\lambda \rangle,$$

which is meaningful if $f_\lambda g_\lambda$ is a cusp form for every λ.

Let ζ_F denote the Dedekind zeta function of F. Then we have

$$(2.29) \qquad \mu\big(SL_2(\mathfrak{r}) \backslash \mathfrak{H}^n\big) = 2\pi^{-n} D_F^{3/2} \zeta_F(2)$$

(cf. Siegel [16]). To simplify our later computation, let us put

$$(2.30) \qquad c_F = 2\pi^{-2n} D_F^{1/2} \zeta_F(2).$$

It is well-known that this is a rational number. (In fact, Proposition 3.1 below includes this fact.) We can easily derive from (2.29) that

$$(2.31) \qquad \mu\big(\Gamma_\lambda(\mathfrak{c}) \backslash \mathfrak{H}^n\big) = \pi^n c_F D_F [\mathfrak{r}_+^\times : \mathfrak{r}^{\times 2}]^{-1} N(\mathfrak{c}) \prod_{\mathfrak{p} | \mathfrak{c}} \big(1 + N(\mathfrak{p})^{-1}\big),$$

where $\mathfrak{r}_+^\times = \big\{a \in \mathfrak{r}^\times \, | \, a \gg 0\big\}$ and $\mathfrak{r}^{\times 2} = \big\{a^2 \, | \, a \in \mathfrak{r}^\times \big\}$.

Proposition 2.3. *For every integral ideal* $\mathfrak{q}$ *and every* $\mathbf{f} \in \mathcal{M}_k(\mathfrak{c}, \psi)$, *there exists a unique element of* $\mathcal{M}_k(\mathfrak{qc}, \psi)$, *written* $\mathbf{f}|\mathfrak{q}$, *such that*

$$(2.32) \qquad C(\mathbf{m}, \mathbf{f}|\mathfrak{q}) = C(\mathfrak{q}^{-1}\mathbf{m}, \mathbf{f}).$$

PROOF. Let $s = \mathrm{diag}[q, 1]$ with an element $q \in F_\mathbf{A}^\times$ such that $q\mathfrak{r} = \mathfrak{q}$ and $q_\infty = 1$. Given $\mathbf{f} = (f_1, \ldots, f_h) \in \mathcal{M}_k(\mathfrak{c}, \psi)$, put $\mathbf{f}'(x) = \mathbf{f}(xs^{-1})$ for $x \in G_\mathbf{A}$. For each λ, we can find an index μ and a totally positive element a_μ of F such that $q\mathfrak{t}_\lambda = a_\mu \mathfrak{t}_\mu$. Then $\mathbf{f}' = (f_1', \ldots, f_h')$ with

$$(2.33) \qquad f_\lambda' = f_\mu \| \mathrm{diag}[a_\mu, 1].$$

This shows that $\mathbf{f}' \in \mathcal{M}_k(\mathfrak{qc}, \psi)$. Then $N(\mathfrak{q})^{-k_0/2} \mathbf{f}'$. gives the desired $\mathbf{f}|\mathfrak{q}$.

Now, let $\mathcal{S}_k'(\mathfrak{c}, \psi)$ denote the subspace of $\mathcal{S}_k(\mathfrak{c}, \psi)$ spanned by all $\mathbf{g}|\mathfrak{q}$ obtained from $\mathbf{g} \in \mathcal{S}_k(\mathfrak{b}, \psi)$ for all divisors $\mathfrak{b}$ of $\mathfrak{c}$, $\neq \mathfrak{c}$, that are divisible by the conductor of ψ; $\mathfrak{q}$ runs over all divisors of $\mathfrak{b}^{-1}\mathfrak{c}$. Obviously $\mathcal{S}_k'(\mathfrak{c}, \psi)$ is stable under $T_\mathfrak{c}(\mathbf{m})$ for all $\mathbf{m}$ prime to $\mathfrak{c}$. We denote by $\mathcal{S}_k^0(\mathfrak{c}, \psi)$ the orthogonal complement of $\mathcal{S}_k'(\mathfrak{c}, \psi)$ in $\mathcal{S}_k(\mathfrak{c}, \psi)$ with respect to (2.28). This is also stable under $T_\mathfrak{c}(\mathbf{m})$ for all $\mathbf{m}$ prime to $\mathfrak{c}$, by virtue of Proposition 2.4 below. Moreover, an element $\mathbf{f}$ of $\mathcal{S}_k^0(\mathfrak{c}, \psi)$ is a common eigenfunction of $T_\mathfrak{c}(\mathbf{m})$ for all $\mathbf{m}$ if that is so for almost all prime ideals $\mathbf{m}$. This is due to Miyake [4], though our formulation is somewhat different from [4]. By a *primitive (cusp form) of type* (k, ψ), we understand an element $\mathbf{f}$ of $\mathcal{S}_k^0(\mathfrak{c}, \psi)$, with some $\mathfrak{c}$, that is a normalized common eigenfunction of $T_\mathfrak{c}(\mathbf{m})$ for all $\mathbf{m}$; we call $\mathfrak{c}$ the *conductor* of $\mathbf{f}$.

Proposition 2.4. *If* $y \in Y$ *and* $\det(y)\mathfrak{r}$ *is prime to* $\mathfrak{c}$, *then*

$$\psi^*\big(\det(y)\mathfrak{r}\big)\langle \mathbf{f}|WyW, \mathbf{g}\rangle = \langle \mathbf{f}, \mathbf{g}|WyW\rangle$$

for all $\mathbf{f}$ *and* $\mathbf{g}$ *of* $\mathcal{S}_k(\mathfrak{c}, \psi)$.

PROOF. As already noted, given y and λ, we can find an index μ and $\alpha \in x_\lambda Y x_\mu^{-1}\cap G_{\mathbf{Q}}$ such that $WyW = WzW$ with $z = x_\lambda^{-1}\alpha x_\mu$. By [7, Lemma 3.5 and Proposition 3.6] we can find β_j such that $\Gamma_\mu \alpha^\iota \Gamma_\lambda = \bigsqcup_j \Gamma_\mu \beta_j = \bigsqcup_j \beta_j \Gamma_\lambda$. Then $\Gamma_\lambda \alpha \Gamma_\mu = \bigsqcup_j \Gamma_\lambda \beta_j^\iota$. Define a map $[\Gamma_\mu \alpha^\iota \Gamma_\lambda]'$ of $\mathcal{M}_k(\Gamma_\mu, \psi_0)$ into $\mathcal{M}_k(\Gamma_\lambda, \psi_0)$ by

$$(2.34) \qquad g|[\Gamma_\mu \alpha^\iota \Gamma_\lambda]' = \sum_j \psi_Y(x_\lambda^{-1}\beta_j^\iota x_\mu)g\|\beta_j \qquad (g \in \mathcal{M}_k(\Gamma_\mu, \psi_0)).$$

It can easily be shown, for cusp forms f and g, that

$$(2.35) \qquad \langle f|\Gamma_\lambda \alpha \Gamma_\mu, g\rangle = \langle f, g|[\Gamma_\mu \alpha^\iota \Gamma_\lambda]'\rangle.$$

We have $W^\iota = W$, $WzW = Wz^\iota W$, and $z^\iota = t_\lambda^{-1}t_\mu x_\mu^{-1}\alpha^\iota x_\lambda$; hence $WyW = \bigsqcup_j Wt_\lambda^{-1}t_\mu x_\mu^{-1}\beta_j x_\lambda$ by (2.8b). For $\mathbf{g} = (g_1, \ldots, g_h) \in \mathcal{S}_k(\mathfrak{c}, \psi)$ put $\mathbf{g}|WyW = (l_1, \ldots, l_h)$. Then for $w \in G_{\infty+}$ we have, by (2.10) and (2.15c),

$$(l_\lambda\|w)(\mathbf{i}) = \sum_j \psi_Y(t_\lambda^{-1}t_\mu x_\mu^{-1}\beta_j x_\lambda)^{-1}\mathbf{g}\big(x_\lambda^{-\iota}w t_\lambda^{-1}t_\mu x_\lambda^\iota(\beta_j^\iota)_0 x_\mu^{-\iota}\big)$$

$$= \sum_j \psi_Y\big(t_\lambda t_\mu^{-1}\det(\beta_j)^{-1}\big)\psi_Y(x_\lambda^{-1}\beta_j^\iota x_\mu)\psi(t_\lambda^{-1}t_\mu)\mathbf{g}\big((\beta_j^\iota)_0 w x_\mu^{-\iota}\big).$$

Clearly $\det(y)\mathfrak{r} = t_\lambda^{-1}t_\mu \det(\beta_j)\mathfrak{r}$; hence we easily see that $\psi_Y\big(t_\lambda t_\mu^{-1}\det(\beta_j)^{-1}\big)$ $\cdot\psi(t_\lambda^{-1}t_\mu) = \psi^*\big(\det(y)\mathfrak{r}\big)$; also $\mathbf{g}\big((\beta_j^\iota)_0 w x_\mu^{-\iota}\big) = \mathbf{g}\big((\beta_j^\iota)_\infty^{-1} w x_\mu^{-\iota}\big) = (g_\mu\|\beta_j w)(\mathbf{i})$. Thus $l_\lambda = \psi^*\big(\det(y)\mathfrak{r}\big)g_\mu|[\Gamma_\mu \alpha^\iota \Gamma_\lambda]'$. This, combined with (2.35) and (2.28), gives our proposition.

Proposition 2.5. *If* $\mathbf{f} \in \mathcal{S}_k(\mathfrak{c}, \psi)$ *and* $\mathbf{f}|T'(\mathfrak{m}) = \lambda(\mathfrak{m})\mathbf{f}$ *with* $\mathfrak{m}$ *prime to* $\mathfrak{c}$, *then* $\lambda(\mathfrak{m}) = \psi^*(\mathfrak{m})\overline{\lambda(\mathfrak{m})}$.

This follows immediately from the above proposition.

Proposition 2.6. *For* $\sigma \in \mathrm{Gal}(\overline{\mathbf{Q}}/\mathbf{Q})$ *and* $\mathbf{f} = (f_1, \ldots, f_h) \in \mathcal{M}_k(\mathfrak{c}, \psi_0)$ *with* $f_\lambda \in \mathcal{M}_k(\overline{\mathbf{Q}})$, *define* $\mathbf{f}^\sigma$ *by*

$$(2.36) \qquad \mathbf{f}^\sigma = (f_1^{[\sigma]}, \ldots, f_h^{[\sigma]}), \quad f_\lambda^{[\sigma]} = f_\lambda^\sigma \cdot \big(N(\mathfrak{t}_\lambda)^{k_0/2}\big)^\sigma N(\mathfrak{t}_\lambda)^{-k_0/2},$$

where $k_0 = \mathrm{Max}(k_1, \ldots, k_n)$. *Further let* $m = k^\sigma$, *and suppose*

$$(2.37) \qquad \big\{\xi^{(k_0 1 - k)/2}\big\}^\sigma = \xi^{(k_0 1 - m)/2} \text{ for all totally positive } \xi \in F.$$

Then $\mathbf{f}^\sigma \in \mathcal{M}_m(\mathfrak{c}, \psi_0^\sigma)$ *and* $C(\mathfrak{m}, \mathbf{f}^\sigma) = C(\mathfrak{m}, \mathbf{f})^\sigma$; $\mathbf{f}^\sigma \in \mathcal{M}_m(\mathfrak{c}, \psi^\sigma)$ *if* $\mathbf{f} \in \mathcal{M}_k(\mathfrak{c}, \psi)$. *Moreover, if* $\mathbf{f}$ *is primitive, so is* $\mathbf{f}^\sigma$.

Obviously (2.37) holds for every $\sigma \in \mathrm{Gal}(\overline{\mathbf{Q}}/\mathbf{Q})$ if

$$(2.38) \qquad k_1 \equiv \cdots \equiv k_n \pmod 2.$$

Therefore our assertions hold for every $\sigma \in \mathrm{Gal}(\overline{\mathbf{Q}}/\mathbf{Q})$ under (2.38).

PROOF. By Proposition 1.6, (2.37) implies that $f_\lambda^\sigma \in \mathcal{M}_m(\Gamma_\lambda, \psi_0^\sigma)$, and hence $\mathbf{f}^\sigma \in \mathcal{M}_m(\mathfrak{c}, \psi_0^\sigma)$; obviously $C(\mathfrak{m}, \mathbf{f}^\sigma) = C(\mathfrak{m}, \mathbf{f})^\sigma$. Let us now show that

$$(2.39) \qquad \big(\mathbf{f}|S(\mathfrak{m})\big)^\sigma = \mathbf{f}^\sigma|S(\mathfrak{m}), \qquad \big(\mathbf{f}|T'(\mathfrak{m})\big)^\sigma = \mathbf{f}^\sigma|T'(\mathfrak{m}).$$

Let $\alpha \in x_\lambda Y x_\mu^{-1} \cap G_{\mathbf{Q}}$ and $\Gamma_\lambda \alpha \Gamma_\mu = \bigcup_j \Gamma_\lambda \alpha_j$ (disjoint). Choosing suitable α_j, we may assume that $\det(\alpha_j) = \det(\alpha)$. Let t be an element of $\prod_p \mathbf{Z}_p^\times$ whose action on $\mathbf{Q}_{\mathrm{ab}}$ coincides with σ, and put $s = \mathrm{diag}[1,\, t^{-1}]$. Now $W_\lambda \alpha W_\mu = \bigcup_j W_\lambda \alpha_j$, so that, for each j, we have $\alpha_j s = r_j \alpha_i$ with a unique i and $r_j \in W_\lambda$. We now consider the action of the group $\mathfrak{G}$ on $\mathcal{M}_k$ defined in §1. Given $f_\lambda \in \mathcal{M}_k(\Gamma_\lambda, \psi_0)$, there is a positive integer N such that $f_\lambda^{(v,\tau)} = f_\lambda^\tau$ for every $v \in S_N$. Take N so large that $S_N \subset W_\lambda$ for all λ. Since $\det(s^{-1} r_j) = 1$, we have $s^{-1} r_j = q_j \gamma_j$ with $q_j \in S_N$ and $\gamma_j \in SL_2(F)$. Then $\det(q_j) = 1$, $\gamma_j \in \Gamma_\lambda$, and $(f_\lambda | \alpha_j)^\sigma = (f_\lambda | \alpha_j)^{(s,\sigma)} = f_\lambda^\sigma | \gamma_j \alpha_i$. Putting $k' = k_0 \mathbf{1} - k$ and $m' = k_0 \mathbf{1} - m$, we obtain

$$\left[N\big(\det(\alpha)\big)^{(k_0 - 2)/2} f_\lambda | \Gamma_\lambda \alpha \Gamma_\mu \right]^\sigma$$

$$= N\big(\det(\alpha)\big)^{k_0 - 1} \sum_j \overline{\psi}_Y (x_\lambda^{-1} \alpha_j x_\mu)^\sigma \big(\det(\alpha)^{-k'/2}\big)^\sigma (f_\lambda | \alpha_j)^\sigma$$

$$= N\big(\det(\alpha)\big)^{k_0 - 1} \sum_j \overline{\psi}_Y (x_\lambda^{-1} \alpha_j x_\mu)^\sigma \det(\alpha)^{-m'/2} f_\lambda^\sigma | \gamma_j \alpha_i$$

$$= N\big(\det(\alpha)\big)^{k_0 - 1} \sum_j \overline{\psi}_Y (x_\lambda^{-1} \alpha_j x_\mu)^\sigma \det(\alpha)^{-m'/2} f_\lambda^\sigma | \alpha_i$$

$$= N\big(\det(\alpha)\big)^{(k_0 - 2)/2} f_\lambda^\sigma | \Gamma_\lambda \alpha \Gamma_\mu.$$

It follows that if $y \in Y$ and $\mathfrak{m} = \det(y)\mathfrak{r}$, then

$$(2.40) \qquad \left[N(\mathfrak{m})^{(k_0 - 2)/2} \mathbf{f} | WyW \right]^\sigma = N(\mathfrak{m})^{(k_0 - 2)/2} \mathbf{f}^\sigma | WyW,$$

which implies (2.39). Therefore we see that $\mathbf{f}^\sigma \in \mathcal{M}_m(\mathfrak{c}, \psi^\sigma)$ if $\mathbf{f} \in \mathcal{M}_k(\mathfrak{c}, \psi)$. Moreover, if $\mathbf{f}$ is a common eigenfunction of the operators $T'(\mathfrak{m})$, so is $\mathbf{f}^\sigma$. Obviously σ maps $\mathcal{S}'_k(\mathfrak{c}, \psi)$ onto $\mathcal{S}'_m(\mathfrak{c}, \psi^\sigma)$. Suppose $\mathbf{f}$ is a primitive form of conductor $\mathfrak{c}$, and let $\mathbf{f} | T'(\mathfrak{p}) = \lambda(\mathfrak{p})\mathbf{f}$. By Miyake's result in [4], we can find finitely many prime ideals $\mathfrak{p}_1, \ldots, \mathfrak{p}_r$ not dividing $\mathfrak{c}$ such that the set of eigenvalues $\{\lambda(\mathfrak{p}_1), \ldots, \lambda(\mathfrak{p}_r)\}$ does not occur in the subspace $\mathcal{S}'_k(\mathfrak{c}, \psi)$. Then the set $\{\lambda(\mathfrak{p}_1)^\sigma, \ldots, \lambda(\mathfrak{p}_r)^\sigma\}$ does not occur in $\mathcal{S}'_m(\mathfrak{c}, \psi^\sigma)$. It follows from this and Proposition 2.4 that $\mathbf{f}^\sigma$ is orthogonal to $\mathcal{S}'_m(\mathfrak{c}, \psi^\sigma)$, which proves the last assertion.

Remark 2.7. We are assuming that $\psi^*(\xi\mathfrak{r}) = \mathrm{sgn}(\xi)^k$ for $\xi \equiv 1 \pmod{^\times \mathfrak{c}}$. Now (2.37) implies that $(|\xi|^k)^\sigma = |\xi|^m$ for every $\xi \in F^\times$, so that $\psi^*(\xi\mathfrak{r})^\sigma = \mathrm{sgn}(\xi)^m$ for $\xi \equiv 1 \pmod{^\times \mathfrak{c}}$. In other words, (2.37) implies that $k \equiv k^\sigma \pmod{2\mathbf{Z}^n}$.

Proposition 2.8. *Suppose $\mathbf{f}$ is a primitive form of type (k, ψ) with k satisfying (2.38), and let $\mathbf{Q}(\mathbf{f})$ denote the field generated over $\mathbf{Q}$ by $C(\mathfrak{m}, \mathbf{f})$ for all $\mathfrak{m}$. Then the following assertions hold:*

(i) *$\mathbf{Q}(\mathbf{f})$ contains the values $\psi^*(\mathfrak{m})$. (This holds even without (2.38).)*

(ii) *$\mathbf{Q}(\mathbf{f})$ is generated by $C(\mathfrak{p}, \mathbf{f})$ for almost all prime ideals $\mathfrak{p}$.*

(iii) *$\mathbf{Q}(\mathbf{f})$ has finite degree over $\mathbf{Q}$.*

(iv) *$\mathbf{Q}(\mathbf{f})$ is either totally real or a totally imaginary quadratic extension of a totally real algebraic number field.*

PROOF. By (2.23), we have $C(\mathfrak{p}, \mathbf{f})^2 = C(\mathfrak{p}^2, \mathbf{f}) + \psi^*(\mathfrak{p})N(\mathfrak{p})^{k_0-1}$, which proves (i). Suppose $\sigma \in \mathrm{Aut}(\mathbf{C})$ and $C(\mathfrak{p}, \mathbf{f})^\sigma = C(\mathfrak{p}, \mathbf{f})$ for almost all $\mathfrak{p}$. By Proposition 2.6, $\mathbf{f}^\sigma$ is primitive, and $C(\mathfrak{p}, \mathbf{f}^\sigma) = C(\mathfrak{p}, \mathbf{f})^\sigma$. By the result of Miyake [4] we have $\mathbf{f}^\sigma = \mathbf{f}$, so that $C(\mathfrak{m}, \mathbf{f})^\sigma = C(\mathfrak{m}, \mathbf{f})$ for all $\mathfrak{m}$, which proves (ii). Assertion (iii) follows immediately from (2.24), Propositions 1.3 and 2.2. To prove (iv), let ρ denote complex conjugation. By Proposition 2.5 we have, for $\mathfrak{m}$ prime to $\mathfrak{c}$,

$$\psi^*(\mathfrak{m})^\sigma C(\mathfrak{m}, \mathbf{f})^{\rho\sigma} = C(\mathfrak{m}, \mathbf{f})^\sigma = C(\mathfrak{m}, \mathbf{f}^\sigma) = \psi^*(\mathfrak{m})^\sigma C(\mathfrak{m}, \mathbf{f}^\sigma)^\rho.$$

This shows that $\rho\sigma = \sigma\rho$ on $\mathbf{Q}(\mathbf{f})$ for all $\sigma \in \mathrm{Aut}(\mathbf{C})$, which proves (iv).

Remark 2.9. The coefficients $C(\mathfrak{m}, \mathbf{f})$ do not necessarily generate a field of finite degree over $\mathbf{Q}$ unless (2.38) is satisfied. In fact, a counter-example can be obtained by taking the form $\mathbf{f}$ such that $D(s, \mathbf{f})$ is the L-function of a Hecke character of a totally imaginary quadratic extension of F.

Given $f(z) = \sum_\xi a(\xi)\mathbf{e}_F(\xi z) \in \mathcal{M}_k$, let U be a subgroup of $\mathfrak{r}_+^\times$ of finite index such that $a(\varepsilon\xi) = \varepsilon^{k/2}a(\xi)$ for $\varepsilon \in U$. (Such a U always exists; see (1.8a, b).) We then set

$$(2.41a) \qquad D(s, f) = [\mathfrak{r}_+^\times : U]^{-1} \sum_{\xi \in F^\times/U} a(\xi)\xi^{(k_0\mathbf{1}-k)/2}N(\xi)^{-s},$$

$$(2.41b) \qquad R(s, f) = (2\pi)^{-ns} \prod_{\nu=1}^n \Gamma\big(s - (k_0 - k_\nu)/2\big) \cdot D(s, f).$$

This is convergent for sufficiently large $\mathrm{Re}(s)$ and does not depend on the choice of U. Furthermore, $R(s, f)$ can be continued to a meromorphic function on the whole plane that is holomorphic except for possible simple poles; the poles may occur at $s = k_0$ and $s = 0$ only when $k = k_0\mathbf{1}$. If $\beta = \begin{bmatrix} 0 & 1 \\ -p & 0 \end{bmatrix}$ with $0 \ll p \in F$, we have

$$(2.42) \qquad R(s, f) = i^{\{k\}} N(p)^{(k_0/2)-s} R(k_0 - s, f\|\beta^{-1}).$$

The residue of $R(s, f)$ at $s = 0$ (when $k = k_0\mathbf{1}$) is $-2^{-1}[\mathfrak{r}^\times : \mathfrak{r}_+^\times]R_F \cdot a(0)$; the residue at $s = k_0$ can be similarly obtained from the constant term of $f\|\beta^{-1}$.

In particular, if $f \in \mathcal{M}_k(\Gamma_\lambda, \psi_0)$, we take an index κ and a totally positive element q_λ of F so that $\mathfrak{t}_\lambda\mathfrak{t}_\kappa\mathfrak{c}\mathfrak{d}^2 = q_\lambda\mathfrak{r}$, and put

$$(2.43) \qquad \beta_\lambda = \begin{bmatrix} 0 & 1 \\ -q_\lambda & 0 \end{bmatrix}.$$

Obviously κ is uniquely determined by λ and $\mathfrak{c}$; we have $\beta_\lambda\Gamma_\lambda\beta_\lambda^{-1} = \Gamma_\kappa$ and

$$f\|\beta_\lambda^{-1} = (-1)^{\{k\}} f\|\beta_\lambda \in \mathcal{M}_k(\Gamma_\kappa, \psi_0^{-1}).$$

To obtain a "global" result, define an element b of $G_\mathbf{A}$ by

$$(2.44) \qquad b_0 = \begin{bmatrix} 0 & 1 \\ m_0 & 0 \end{bmatrix}, \qquad b_\infty = 1$$

with an element m of $F_\mathbf{A}^\times$ such that $m\mathfrak{r} = \mathfrak{c}\mathfrak{d}^2$, where the subscript 0 indicates the projection to the nonarchimedean part. With λ, κ, and β_λ as above, we have

$$(2.45) \qquad \mathfrak{c}\mathfrak{d}^2\mathfrak{t}_\kappa = q_\lambda\mathfrak{t}_\lambda^{-1} \quad \text{and} \quad b = x_\kappa^{-1}(\beta_\lambda)_0 x_\lambda^{-\iota} \cdot \mathrm{diag}[u, 1]$$

with $u \in \prod_{\mathfrak{p}} \mathfrak{r}_{\mathfrak{p}}^{\times}$; also $bWb^{-1} = W$. For $\mathbf{f} \in \mathcal{M}_k(\mathfrak{c}, \psi)$, define $\mathbf{f}|J_{\mathfrak{c}}$ by

$$(2.46) \qquad (\mathbf{f}|J_{\mathfrak{c}})(x) = \psi\big(\det(x)\big)^{-1}\mathbf{f}(xb) \qquad (x \in G_{\mathbf{A}}).$$

We easily see that $\mathbf{f}|J_{\mathfrak{c}} \in \mathcal{M}_k(\mathfrak{c}, \psi^{-1})$ and $(\mathbf{f}|J_{\mathfrak{c}})(x) = \mathbf{f}(x^{-\iota}b)$. Also, using (2.45) and (2.15a, c), we find that $\mathbf{f}|J_{\mathfrak{c}} = (f_1', \dots, f_h')$ with $f_\kappa' = f_\lambda \|\beta_\lambda^{-1}$. Now define

$$(2.47) \qquad R(s, \mathbf{f}) = N(\mathfrak{c}\mathfrak{d}^2)^{s/2}(2\pi)^{-ns} \prod_{\nu=1}^{n} \Gamma\big(s - (k_0 - k_\nu)/2\big) \cdot D(s, \mathbf{f}).$$

Since $D(s, \mathbf{f}) = \sum_{\lambda=1}^{h} N(\mathfrak{t}_\lambda)^{s - k_0/2} D(s, f_\lambda)$, from (2.42) we obtain

$$(2.48) \qquad R(s, \mathbf{f}) = i^{\{k\}} R(k_0 - s, \mathbf{f}|J_{\mathfrak{c}}).$$

This is meromophic on the whole plane. The poles are at most simple and may occur only at $s = k_0$ and $s = 0$ when $k = k_0\mathbf{1}$. The residue at $s = 0$ is $-2^{-1}[\mathfrak{r}^{\times} : \mathfrak{r}_{+}^{\times}]R_F \sum_{\lambda=1}^{h} c_0(\mathfrak{t}_\lambda)$ with the "constant term" c_0 of the expansion (2.18) of $\mathbf{f}$ (see also (2.19)); the residue at $s = k_0$ can be similarly obtained from the constant term of $\mathbf{f}|J_{\mathfrak{c}}$. Thus $R(s, \mathbf{f})$ is entire if $\mathbf{f}$ is a cusp form.

Proposition 2.10. *If $\mathbf{f}$ is primitive and has conductor $\mathfrak{c}$, then $\mathbf{f}|J_{\mathfrak{c}}$ is a constant times $\mathbf{f}^\rho$, where ρ denotes complex conjugation.*

PROOF. For $\mathbf{f} \in \mathcal{M}_k(\mathfrak{c}, \psi)$ and $y \in Y$, $y_\infty = y_{\mathfrak{c}} = 1$ we have

$$(2.49) \qquad (\mathbf{f}|J_{\mathfrak{c}})|WyW = \psi\big(\det(y)\big)^{-1}(\mathbf{f}|WyW)|J_{\mathfrak{c}}$$

Indeed, we may assume that y is diagonal; then we have $WyW = Wbyb^{-1}W$. Let $WyW = \bigsqcup_j Wy_j$, $(y_j)_\infty = (y_j)_{\mathfrak{c}} = 1$. Since $bWb^{-1} = W$, we have $WyW = \bigsqcup_j Wby_jb^{-1}$. Put $\mathbf{g} = \mathbf{f}|WyW$ and $\mathbf{h} = \mathbf{f}|J_{\mathfrak{c}}$. Observe that $(axa^{-1})^\iota = ax^\iota a^{-1}$ for every $a, x \in G_{\mathbf{A}}$. Thus

$$(\mathbf{g}|J_{\mathfrak{c}})(x) = \mathbf{g}(x^{-\iota}b) = \sum_j \mathbf{f}(x^{-\iota}by_j^\iota) = \sum_j \mathbf{f}\big((xby_j^{-1}b^{-1})^{-\iota}b\big)$$

$$= \sum_j \mathbf{h}(xby_j^{-1}b^{-1}) = \sum_j \mathbf{h}\big(\det(y_j^{-1})x(by_jb^{-1})^\iota\big) = \psi\big(\det(y)\big)^{-1}(\mathbf{h}|WyW)(x),$$

which proves (2.49). This combined with Proposition 2.5 yields

$$(2.50) \qquad (\mathbf{f}|J_{\mathfrak{c}})|T'(\mathfrak{m}) = C(\mathfrak{m}, \mathbf{f})^\rho \mathbf{f}|J_{\mathfrak{c}} \quad \text{for } \mathfrak{m} \text{ prime to } \mathfrak{c},$$

which proves our proposition.

3. Eisenstein series

Let $\mathfrak{a}$ and $\mathfrak{b}$ be fractional ideals of F, a_0 and b_0 be elements of F, $0 \leq \kappa \in \mathbf{Z}$, $r = (r_1, \dots, r_n) \in \mathbf{Z}^n$ with $r_\nu \geq 0$ for all ν, and let U be a subgroup of $\mathfrak{r}^{\times}$ of finite index such that $N(u)^\kappa = 1$, $ua_0 - a_0 \in \mathfrak{a}$, $ub_0 - b_0 \in \mathfrak{b}$ for all $u \in U$. We consider an infinite series

$$(3.1) \qquad E_{\kappa,U}^r(z, s; a_0, b_0; \mathfrak{a}, \mathfrak{b}) = D_F^{1/2} N(\mathfrak{b})(-1)^{\kappa n}(2\pi i)^{-\kappa n - \{r\}}(z - \bar{z})^{-r}$$

$$\cdot \sum_{(a, b)U} \left(\frac{a\bar{z} + b}{az + b}\right)^r (az + b)^{-\kappa \mathbf{1}}|az + b|^{-2s\mathbf{1}}.$$

Here $z \in \mathfrak{H}^n$ and $s \in \mathbf{C}$; the sum is taken over all $(a, b) \neq (0, 0)$ in $\{(a, b) \in F \times F \mid a - a_0 \in \mathfrak{a}, b - b_0 \in \mathfrak{b}\}/U$, where the action of U is defined by $(a, b)u = (au, bu)$ for $u \in U$. The series is convergent for $\mathrm{Re}(2s + \kappa) > 2$. Set

$$(3.2) \qquad \Phi^r_{\kappa,U} = E^r_{\kappa,U} \prod_{\nu=1}^n \Gamma(s + \kappa + r_\nu).$$

Then $\Phi^r_{\kappa,U}$ can be continued as a meromorphic function of s to the whole plane, and is entire if $\kappa > 0$ or $r_\nu > 0$ for some ν. If $\kappa = 0$ and $r = 0$, $\Phi^r_{\kappa,U}$ is holomorphic except at a simple pole at $s = 1$ and at a possible simple pole at $s = 0$; the pole at $s = 0$ occurs if and only if $a_0 \in \mathfrak{a}$ and $b_0 \in \mathfrak{b}$. The residue at $s = 1$ is

$$(3.3) \qquad 2^{n-2}\pi^n[\mathfrak{r}^\times : U]D_F^{-1/2}N(\mathfrak{a})^{-1}R_F(y_1 \cdots y_n)^{-1},$$

where $y_\nu = \operatorname{Im}(z_\nu)$. If s stays in a compact set S on which $\Phi^r_{\kappa,U}$ is holomorphic, we have

$$(3.4) \qquad \Phi^r_{\kappa,U} = O\big((y_1 \cdots y_n)^A\big) \quad \text{when} \quad (y_1, \ldots, y_n) \to (\infty, \ldots, \infty)$$

with a real constant A depending only on S. All these facts can easily be established by considering, for example, the Fourier expansion of $E^r_{\kappa,U}$ in terms of confluent hypergeometric functions.

Now suppose $\kappa > 0$ and put

$$(3.5) \qquad E_{\kappa,U}(z) = E_{\kappa,U}(z; a_0, b_0; \mathfrak{a}, \mathfrak{b}) = E^0_{\kappa,U}(z, 0; a_0, b_0; \mathfrak{a}, \mathfrak{b}).$$

This is holomorphic in z except when $n = 1$ and $\kappa = 2$. To make our discussion simpler, we exclude this special case, assuming always $n > 1$ or $\kappa \neq 2$. (That case can easily be included by some extra calculations. Anyway, our main problems in this paper are in the case $n > 1$.) Then $E_{\kappa,U}(z)$ belongs to $\mathcal{M}_{\kappa\mathbf{1}}$. The Fourier coefficients of $E_{\kappa,U}$ can be determined by simply putting $s = 0$ in the above-mentioned Fourier expansion and given as follows (see [1] and [2], for example):

$$(3.6) \quad E_{\kappa,U}(z) = A(0) + \Gamma(\kappa)^{-n} \sum_{cU \equiv a_0 \ (\mathrm{mod}\ \mathfrak{a})} \sum_{\substack{b \in \mathfrak{b}^* \\ bc \gg 0}} |N(b)|^{-1}N(b)^\kappa \mathbf{e}_F(bcz + bb_0),$$

where $\mathfrak{b}^* = \mathfrak{b}^{-1}\mathfrak{d}^{-1}$ and $A(0)$ is the value at $s = 0$ of the function

$$(3.7) \quad A(s) = \delta(a_0, \mathfrak{a})(-2\pi i)^{-\kappa n}N(\mathfrak{b})D_F^{1/2} \sum_{dU \equiv b_0 \ (\mathrm{mod}\ \mathfrak{b})} \operatorname{sgn}\big(N(d)\big)^\kappa|N(d)|^{-\kappa-2s}$$

$$+ \begin{cases} 0 & \text{if } \kappa > 1, \\ 2^{-n} \displaystyle\sum_{cU \equiv a_0 \ (\mathrm{mod}\ \mathfrak{a})} \operatorname{sgn}\big(N(c)\big)|N(c)|^{-2s} & \text{if } \kappa = 1. \end{cases}$$

Here $\delta(a_0, \mathfrak{a}) = 1$ or 0 according as $a_0 \in \mathfrak{a}$ or $a_0 \notin \mathfrak{a}$; the sums are taken over all cosets dU such that $0 \neq d \equiv b_0 \pmod{\mathfrak{b}}$ and all cosets cU such that $0 \neq c \equiv a_0 \pmod{\mathfrak{a}}$. By Proposition 1.3 we have $A(0) \in \mathbf{Q}_{\mathrm{ab}}$; thus $E_{\kappa,U} \in \mathcal{M}_{\kappa\mathbf{1}}(\mathbf{Q}_{\mathrm{ab}})$.

Let us now consider a character η of the ideal class group of F modulo the product $\mathfrak{c}\mathfrak{P}_\infty$ of an integral ideal $\mathfrak{c}$ and $\mathfrak{P}_\infty$. We put $\eta(\mathfrak{a}) = 0$ for any ideal $\mathfrak{a}$ not prime to $\mathfrak{c}$. Let $r = (r_1, \ldots, r_n) \in \mathbf{Z}^n$ be such that

$$(3.8) \qquad \eta(x\mathfrak{r}) = \operatorname{sgn}(x^r) \quad \text{if } x \equiv 1 \pmod{\mathfrak{c}}.$$

It can easily be seen that for every fractional ideal $\mathfrak{a}$, $\operatorname{sgn}(x)^r\eta(x\mathfrak{a}^{-1})$ as a function of $x \in \mathfrak{a}$ depends only on x modulo $\mathfrak{c}\mathfrak{a}$. Suppose that η is primitive, i.e., $\mathfrak{c}$ is exactly the conductor of η. Then we define the Gauss sum $\tau(\eta)$ of η by

$$(3.9) \qquad \tau(\eta) = \sum_x \mathrm{sgn}(x^r)\eta(x\mathfrak{c}\mathfrak{d})\mathbf{e}_F(x),$$

where x runs over $\mathfrak{c}^{-1}\mathfrak{d}^{-1}/\mathfrak{d}^{-1}$. We understand that $\mathrm{sgn}(x^r)\eta(x\mathfrak{a})$ for $x = 0$ and an ideal $\mathfrak{a}$ means $\eta(\mathfrak{a})$ or 0 according as $\mathfrak{c} = \mathfrak{r}$ or $\mathfrak{c} \neq \mathfrak{r}$. It is well known that $|\tau(\eta)|^2 = N(\mathfrak{c})$. If η is imprimitive, we consider the primitive character η' associated with η, and define $\tau(\eta)$ to be the same as $\tau(\eta')$. If $\sigma \in \mathrm{Gal}(\mathbf{Q}_{\mathrm{ab}}/\mathbf{Q})$ and $\mathbf{e}\big(1/N(\mathfrak{c})\big)^\sigma = \mathbf{e}\big(q/N(\mathfrak{c})\big)$ with $0 < q \in \mathbf{Z}$, then

$$(3.10) \qquad \tau(\eta^\sigma) = \eta(q\mathfrak{r})^\sigma \tau(\eta)^\sigma.$$

If η is primitive, $\mathfrak{a}$ is a fractional ideeal, and $b \in \mathfrak{a}^{-1}\mathfrak{c}^{-1}\mathfrak{d}^{-1}$, then we have

$$(3.11) \qquad \sum_{x \in \mathfrak{a}/\mathfrak{a}\mathfrak{c}} \mathrm{sgn}(x^r)\eta(x\mathfrak{a}^{-1})\mathbf{e}_F(bx) = \mathrm{sgn}(b)^r \overline{\eta}(b\mathfrak{a}\mathfrak{c}\mathfrak{d})\tau(\eta).$$

With a primitive or an imprimitive character η defined modulo $\mathfrak{c}\mathfrak{P}_\infty$ as above, we define an L-function $L(s, \eta)$ as usual by

$$(3.12) \qquad L(s, \eta) = L_{\mathfrak{c}}(s, \eta) = \sum_{\mathfrak{a}+\mathfrak{c}=\mathfrak{r}} \eta(\mathfrak{a})N(\mathfrak{a})^{-s}.$$

We put the subscript $\mathfrak{c}$ in order to emphasize the exclusion of the ideals not prime to $\mathfrak{c}$.

Proposition 3.1. *Suppose* $\eta(x\mathfrak{r}) = \mathrm{sgn}\big(N(x)^\kappa\big)$ *for* $x \equiv 1 \pmod{\mathfrak{c}}$ *with a positive integer* κ, *and put*

$$P_{\mathfrak{c}}(\kappa, \eta) = \tau(\eta)^{-1}(2\pi i)^{-\kappa n} D_F^{1/2} L_{\mathfrak{c}}(\kappa, \eta).$$

Then $P_{\mathfrak{c}}(\kappa, \eta)$ *belongs to the field* $\mathbf{Q}(\eta)$ *generated over* $\mathbf{Q}$ *by the values of* η; *moreover* $P_{\mathfrak{c}}(\kappa, \eta)^\sigma = P_{\mathfrak{c}}(\kappa, \eta^\sigma)$ *for every* $\sigma \in \mathrm{Gal}\big(\mathbf{Q}(\eta)/\mathbf{Q}\big)$.

PROOF. This is a refinement of the result of Klingen [1]. Obviously it is sufficient to prove the case in which η is primitive. Let $\{\mathfrak{a}_1, \ldots, \mathfrak{a}_m\}$ be a complete set of representatives for the ideal class group modulo the group $\big\{ x\mathfrak{r} \,\big|\, x \in F, N(x) > 0 \big\}$. Take a sum

$$(3.13) \quad \tau(\eta)^{-1} \sum_{\lambda=1}^{m} N(\mathfrak{a}_\lambda)^{\kappa-1} \sum_{x \in \mathfrak{a}_\lambda/\mathfrak{a}_\lambda\mathfrak{c}} \mathrm{sgn}\big(N(x)^\kappa\big)\eta(x\mathfrak{a}_\lambda^{-1})E_{\kappa,U}(z; 0, x; \mathfrak{r}, \mathfrak{a}_\lambda\mathfrak{c}),$$

where $U = \big\{ u \in \mathfrak{r}^\times \,\big|\, N(u)^\kappa = 1, u \equiv 1 \pmod{\mathfrak{c}} \big\}$. Using (3.7), we find that the constant term of (3.13) is

$$2(-1)^{\kappa n}N(\mathfrak{c})P_{\mathfrak{c}}(\kappa, \eta)[\mathfrak{r}^\times : U]\big[\mathfrak{r}^\times : \{u \in \mathfrak{r}^\times \,|\, N(u) > 0\}\big]^{-1}.$$

Notice that this is so even when $\kappa = 1$. From (3.11) and (3.6), we easily see that the Fourier coefficients of the non-constant terms belong to $\mathbf{Q}(\eta)$; moreover, the action of σ on them is given by replacing η by η^σ. Therefore we obtain our assertions by Proposition 1.3 and the reasoning preceding it.

Proposition 3.2. *Let* $\mathfrak{c}$ *be an integral ideal,* $\mathfrak{b}$ *a fractional ideal,* κ *a nonnegative integer, and* $r \in \mathbf{Z}^n$ *with* $r_\nu \geq 0$ *for every* ν. *Further let* ω *be a character of* $(\mathfrak{r}/\mathfrak{c})^\times$ *such that* $\omega(u) = \mathrm{sgn}\big(N(u)\big)^\kappa$ *for all* $u \in \mathfrak{r}^\times$. *Put*

$$(3.14) \quad H(z, s, \omega) = (2\pi i)^{-\{r\}}(z-\bar{z})^{-r}\sum_{c,d}\omega(d)\left(\frac{c\bar{z}+d}{cz+d}\right)^r (cz+d)^{-\kappa 1}|cz+d|^{-2s1},$$

where the sum is taken over all $(c, d) \neq (0, 0)$ *in* $\mathfrak{r}^{\times} \setminus \{(c, d) \in \mathfrak{b}\mathfrak{c} \times \mathfrak{r} \mid cb^{-1} + d\mathfrak{r} = \mathfrak{r}\}$; *we put* $\omega(d) = 0$ *if* $d\mathfrak{r} + \mathfrak{c} \neq \mathfrak{r}$. *Then* H *as a meromorphic function of* s *can be continued to the whole plane. Moreover suppose* $\kappa > 0$ *and* $r = 0$; *suppose further* $n > 1$, *or* $\kappa \neq 2$, *or* ω *is nontrivial. Then* H *is holomorphic at* $s = 0$. *If we put*

$$H(z, \omega) = H(z, 0, \omega),$$

then $H(z, \omega) \in \mathcal{M}_{\kappa 1}(\mathbf{Q}(\omega))$, *where* $\mathbf{Q}(\omega)$ *is the field generated over* $\mathbf{Q}$ *by the values of* ω. *Furthermore* $H(z, \omega)^{\sigma} = H(z, \omega^{\sigma})$ *for every* $\sigma \in \mathrm{Gal}(\mathbf{Q}(\omega)/\mathbf{Q})$.

PROOF. Since the assertions in the case $n = 1$ are either well-known or proved in [12, Lemma 5], let us assume $n > 1$. Let μ denote the Möbius function of the ideal class group of F. Then the sum $\sum_{c,d}$ of (3.14) can be written $\sum_{c,d} \sum_{\mathfrak{a}} \mu(\mathfrak{a})$ with $(0, 0) \neq (c, d) \in (\mathfrak{b}\mathfrak{c} \times \mathfrak{r})/\mathfrak{r}^{\times}$ and $\mathfrak{a} \supset c\mathfrak{b}^{-1} + d\mathfrak{r}$. Since $\omega(d) = 0$ if $d\mathfrak{r} + \mathfrak{c} \neq \mathfrak{r}$, we may assume that $\mathfrak{a} + \mathfrak{c} = \mathfrak{r}$. Therefore we have

$$(3.15) \qquad (2\pi i)^{\{r\}}(z - \bar{z})^r H(z, s, \omega)$$
$$= \sum_{\mathfrak{a}} \mu(\mathfrak{a}) \sum_{c,d} \omega(d) \left(\frac{c\bar{z} + d}{cz + d} \right)^r (cz + d)^{-\kappa 1} |cz + d|^{-2s1},$$

where $\mathfrak{a}$ runs over all integral ideals prime to $\mathfrak{c}$, and $(0, 0) \neq (c, d) \in (\mathfrak{a}\mathfrak{b}\mathfrak{c} \times \mathfrak{a})/\mathfrak{r}^{\times}$. Let X denote the set of all ideal characters χ defined modulo $\mathfrak{c}\mathfrak{P}_{\infty}$ such that $\chi(y\mathfrak{r}) = \omega(y)\mathrm{sgn}(N(y)^{\kappa})$ for $y \in F^{\times}$ prime to $\mathfrak{c}$. For each $\chi \in X$ and an ideal class A of F, put

$$(3.16a) \qquad S_A(s, \chi) = \sum_{\mathfrak{r} \in A,\, \mathfrak{r} + \mathfrak{c} = \mathfrak{r}} \mu(\mathfrak{r})\chi(\mathfrak{r})N(\mathfrak{r})^{-\kappa - 2s},$$

$$(3.16b) \qquad J_A(z, s, \chi) = (2\pi i)^{-\{r\}}(z - \bar{z})^{-r} N(\mathfrak{a})^{\kappa + 2s}$$
$$\cdot \sum_{c,d} \mathrm{sgn}(N(d)^{\kappa})\chi(d\mathfrak{a}^{-1}) \left(\frac{c\bar{z} + d}{cz + d} \right)^r (cz + d)^{-\kappa 1} |cz + d|^{-2s1},$$

where $\mathfrak{a}$ is any fixed ideal in A prime to $\mathfrak{c}$ and $(0, 0) \neq (c, d) \in (\mathfrak{a}\mathfrak{b}\mathfrak{c} \times \mathfrak{a})/\mathfrak{r}^{\times}$. Clearly (3.16b) does not depend on the choice of $\mathfrak{a}$. Now, for any fixed χ we have

$$(3.17a) \qquad H(z, s, \omega) = \sum_A S_A(s, \chi) J_A(z, s, \chi),$$

$$(3.17b) \qquad \sum_A (\eta\chi^{-1})(A) S_A(s, \chi) = \sum_{\mathfrak{a}} \mu(\mathfrak{a})\eta(\mathfrak{a}) N(\mathfrak{a})^{-\kappa - 2s} = L_{\mathfrak{c}}(2s + \kappa, \eta)^{-1},$$

$$(3.17c) \qquad h_0 S_A(s, \chi) = \sum_{\eta \in X} (\eta^{-1}\chi)(A) L_{\mathfrak{c}}(2s + \kappa, \eta)^{-1},$$

where $\sum_A$ is the sum over all ideal classes A of F, and h_0 is the class number of F. Thus

$$(3.18) \qquad H(z, s, \omega) = h_0^{-1} \sum_A \sum_{\eta \in X} (\eta^{-1}\chi)(A) L_{\mathfrak{c}}(2s + \kappa, \eta)^{-1} J_A(z, s, \chi)$$

with any fixed $\chi \in X$. Since J_A is a linear combination of $E_{\kappa,U}^r$, we see that $H(z, s, \omega)$ is meromorphic on the whole s-plane. Suppose $\kappa > 0$ and $r = 0$. Then (3.18) shows that $H(z, s, \omega)$ is holomorphic at $s = 0$. Put

$$K_A(z, \chi) = (-2\pi i)^{-\kappa n} D_F^{1/2} J_A(z, 0, \chi),$$

$$T_A(\chi) = \sum_{\eta \in X} (\eta^{-1}\chi)(A)\tau(\eta)^{-1}P_{\mathfrak{c}}(\kappa, \eta)^{-1}.$$

Then

$$H(z, \omega) = (-1)^{\kappa n}h_0^{-1}\sum_A T_A(\chi)K_A(z, \chi),$$

$$K_A(z, \chi) = [\mathfrak{r}^\times : U]^{-1}N(\mathfrak{a})^{\kappa-1}\chi(\mathfrak{a}^{-1})\sum_{d \in \mathfrak{a}/\mathfrak{a}\mathfrak{c}}\omega(d)E_{\kappa, U}(z; 0, d; \mathfrak{a}\mathfrak{b}\mathfrak{c}, \mathfrak{a}\mathfrak{c})$$

with any $\mathfrak{a}$ in A prime to $\mathfrak{c}$. Since $E_{\kappa, U}(z; \cdots)$ belongs to $\mathcal{M}_{\kappa\mathbf{1}}(\mathbf{Q}_{\mathrm{ab}})$, the same is true for $K_A(z, \chi)$ and $H(z, \omega)$. From (3.6) we see that if $\sigma \in \mathrm{Gal}(\mathbf{Q}_{\mathrm{ab}}/\mathbf{Q})$ and $\mathbf{e}(1/N(\mathfrak{c}))^\sigma = \mathbf{e}(q/N(\mathfrak{c}))$ with $0 < q \in \mathbf{Z}$, then $\omega(q)^\sigma K_A(z, \chi)^\sigma = K_A(z, \chi^\sigma)$. Proposition 3.1 together with (3.10) implies that $T_A(\chi)^\sigma = \omega(q)^\sigma T_A(\chi^\sigma)$, and hence $H(z, \omega)^\sigma = H(z, \omega^\sigma)$, which implies that $H(z, \omega) \in \mathcal{M}_{\kappa\mathbf{1}}(\mathbf{Q}_{\mathrm{ab}})$. This completes the proof.

Proposition 3.3. *The function*

$$H^*(z, s, \omega) = H(z, s, \omega)\prod_{\nu=1}^n \Gamma(s + \kappa + r_\nu)$$

is holomorphic in s at least in the following regions:
  (1) $\mathrm{Re}(s) \geq 0$ *if* $\kappa > 0$;
  (2) $\mathrm{Re}(s) > 1$ *if* $\kappa = 0$ *and* $r = 0$;
  (3) $\mathrm{Re}(s) > 1/2$ *if* $\kappa = 0$ *and* $r \neq 0$.
Moreover, if s stays in a compact subset S of any of these regions, then

$$(3.19) \qquad H^*\big(\alpha(z), s, \omega\big)\left(\frac{c\bar{z} + d}{cz + d}\right)^r (cz + d)^{-\kappa\mathbf{1}}|cz + d|^{-2s\mathbf{1}}$$

$$= O\big((y_1 \cdots y_n)^B\big) \quad \text{when} \quad (y_1, \ldots, y_n) \longrightarrow (\infty, \ldots, \infty)$$

for every $\alpha = \begin{bmatrix} * & * \\ c & d \end{bmatrix} \in G_{\mathbf{Q}+}$, where B is a positive constant depending on S, α, κ, and r. Furthermore, suppose $\kappa = 0$, $r = 0$, and ω is the trivial character of $(\mathfrak{r}/\mathfrak{c})^\times$; then $H(z, s, \omega)$ has a simple pole at $s = 1$ with residue

$$(3.20) \qquad 2^{n-2}\pi^n\zeta_F(2)^{-1}(y_1 \cdots y_n)^{-1}D_F^{-1}R_F N(\mathfrak{b}\mathfrak{c})^{-1}\prod_{\mathfrak{p}|\mathfrak{c}}\big(1 + N(\mathfrak{p})^{-1}\big)^{-1}.$$

PROOF. The holomorphy in s and (3.19) follow from (3.18) and the known properties of $\Phi^r_{\kappa, U}$. If $\kappa = 0$, $r = 0$, and ω is trivial, the residue at $s = 1$ can easily be obtained from (3.3).

Proposition 3.4. *Let $\mathfrak{a}$ and $\mathfrak{b}$ be integral ideals and let η (resp. χ) be a (primitive or an imprimitive) character of the ideal class group modulo $\mathfrak{a}\mathfrak{P}_\infty$ (resp. $\mathfrak{b}\mathfrak{P}_\infty$) such that*

$$\eta(v\mathfrak{r}) = \mathrm{sgn}(v^q) \quad \text{if} \quad v \equiv 1 \pmod{\mathfrak{a}}, \quad \text{and} \quad \chi(v\mathfrak{r}) = \mathrm{sgn}(v^r) \quad \text{if} \quad v \equiv 1 \pmod{\mathfrak{b}}$$

with $q, r \in \mathbf{Z}^n$. Suppose $n > 1$ and $q + r \equiv \kappa\mathbf{1} \pmod{2\mathbf{Z}^n}$ with a positive integer κ. Then there exists an element $\mathbf{g}$ of $\mathcal{M}_{\kappa\mathbf{1}}(\mathfrak{a}\mathfrak{b}, \psi)$ such that

$$(3.21) \qquad D(s, \mathbf{g}) = L_{\mathfrak{a}}(s, \eta)L_{\mathfrak{b}}(s + 1 - \kappa, \chi),$$

where ψ is the Hecke character such that $\psi^*(\mathfrak{r}) = (\eta\chi)(\mathfrak{r})$ for $\mathfrak{r}$ prime to $\mathfrak{a}\mathfrak{b}$.

PROOF. With fractional ideals $\mathfrak{r}$, $\mathfrak{y}$, and $a \in F$, $b \in \mathfrak{y}$, put

$$G(a, b, \mathfrak{r}, \mathfrak{y}) = \Gamma(\kappa)^n N(\mathfrak{b})^{-1}\sum_t \mathbf{e}_F(-tb)E_{\kappa, U}(z; a, t; \mathfrak{r}, \mathfrak{d}^{-1}\mathfrak{y}^{-1}),$$

$$H(\mathfrak{x}, \mathfrak{y}) = \sum_{a,b} \operatorname{sgn}(a^q)\eta(a\mathfrak{x}^{-1})\operatorname{sgn}(b^r)\chi(b\mathfrak{y}^{-1})N(\mathfrak{y})^{1-\kappa}G(a, b; \mathfrak{x}a, \mathfrak{y}),$$

where $t \in \mathfrak{b}^{-1}\mathfrak{d}^{-1}\mathfrak{y}^{-1}/\mathfrak{d}^{-1}\mathfrak{y}^{-1}$, $a \in \mathfrak{x}/\mathfrak{x}a$, and $b \in \mathfrak{y}/\mathfrak{y}\mathfrak{b}$; U should be sufficiently small. Then the desired $\mathbf{g} = (g_1, \ldots, g_h)$ can be given by

$$g_\lambda = N(\mathfrak{t}_\lambda)^{\kappa/2}K(\mathfrak{t}_\lambda), \quad K(\mathfrak{t}) = [\mathfrak{r}^\times : U]^{-1}\sum_{\mathfrak{x}} H(\mathfrak{x}, \mathfrak{t}\mathfrak{x}^{-1}), \quad (1 \le \lambda \le h),$$

where $\mathfrak{x}$ runs over a complete set of representatives for the ideal classes of F. To see this, let us write $f \doteq g$ when $f - g$ is a constant. From (3.6) we easily obtain

$$G(a, b, \mathfrak{x}, \mathfrak{y}) \doteq \sum_{\beta \equiv b \ (\mathrm{mod}\ \mathfrak{b}\mathfrak{y})} \sum_{cU, c\beta \gg 0} |N(\beta)|^{-1}N(\beta)^\kappa \mathbf{e}_F(c\beta z),$$

where $0 \ne c \equiv a \ (\mathrm{mod}\ \mathfrak{x})$. Then a direct calculation shows that

$$H(\mathfrak{x}, \mathfrak{y}) \doteq \sum_{0 \ll \xi \in F} \mathbf{e}_F(\xi z) \sum_{\alpha U} \eta(\alpha\mathfrak{x}^{-1})\chi(\alpha^{-1}\xi\mathfrak{y}^{-1})N(\alpha^{-1}\xi\mathfrak{y}^{-1})^{\kappa-1},$$

where $0 \ne \alpha \in \mathfrak{x}$ and $\alpha^{-1}\xi \in \mathfrak{y}$. Putting $\mathfrak{m} = \alpha\mathfrak{x}^{-1}$ and $\mathfrak{n} = \xi(\mathfrak{m}\mathfrak{x}\mathfrak{y})^{-1}$, we obtain

$$H(\mathfrak{x}, \mathfrak{y}) \doteq [\mathfrak{r}^\times : U] \sum_{0 \ll \xi \in F} \mathbf{e}_F(\xi z) \sum_{\mathfrak{m},\mathfrak{n}} \eta(\mathfrak{m})\chi(\mathfrak{n})N(\mathfrak{n})^{\kappa-1},$$

where $\mathfrak{m}$ and $\mathfrak{n}$ are integral ideals such that $\mathfrak{m}\mathfrak{n} = \xi\mathfrak{x}^{-1}\mathfrak{y}^{-1}$ and $\mathfrak{m}\mathfrak{x}$ is a principal ideal. Then

$$K(\mathfrak{t}) \doteq \sum_{0 \ll \xi \in F} \mathbf{e}_F(\xi z) \sum_{\mathfrak{m},\mathfrak{n}} \eta(\mathfrak{m})\chi(\mathfrak{n})N(\mathfrak{n})^{\kappa-1},$$

where $\mathfrak{m}$ and $\mathfrak{n}$ are integral ideals such that $\mathfrak{t}\mathfrak{m}\mathfrak{n} = \xi\mathfrak{t}$. From this we easily obtain $C(\mathfrak{z}, \mathbf{g}) = \sum_{\mathfrak{m}\mathfrak{n}=\mathfrak{z}} \eta(\mathfrak{m})\chi(\mathfrak{n})N(\mathfrak{n})^{\kappa-1}$, which is equivalent to (3.21). To show that $\mathbf{g} \in \mathcal{M}_{\kappa\mathbf{1}}(\mathfrak{a}\mathfrak{b}, \psi)$, we have to verify the behavior of the functions G, H, and K under the group elements, which, if tedious, can be done in a straightforward way.

4. Main theorems

Our main objectives are several properties of rationality of the special values of Dirichlet series of the following two types:

$$(4.1\mathrm{a}) \qquad D(s, \mathbf{f}, \mathbf{g}) = \sum C(\mathfrak{m}, \mathbf{f})C(\mathfrak{m}, \mathbf{g})N(\mathfrak{m})^{-s},$$

$$(4.1\mathrm{b}) \qquad D(s, \mathbf{f}, \chi) = \sum C(\mathfrak{m}, \mathbf{f})\chi^*(\mathfrak{m})N(\mathfrak{m})^{-s}.$$

Here $\mathbf{f} \in \mathcal{S}_k(\mathfrak{c}, \psi)$ and $\mathbf{g} \in \mathcal{M}_l(\mathfrak{c}, \varphi)$; χ, ψ, and φ are Hecke characters of $F_{\mathbf{A}}^\times$ of finite order; χ^* denotes the ideal character associated with χ; $\mathfrak{m}$ runs over all integral ideals of F. We put

$$(4.2\mathrm{a}) \qquad k_0 = \mathrm{Max}(k_1, \ldots, k_n), \quad l_0 = \mathrm{Max}(l_1, \ldots, l_n),$$

$$(4.2\mathrm{b}) \quad k' = (k_1', \ldots, k_n'), \ k_\nu' = k_0 - k_\nu; \quad l' = (l_1', \ldots, l_n'), \ l_\nu' = l_0 - l_\nu.$$

For the reason explained in §2, we assume

$$(4.2\mathrm{c}) \qquad \psi(u) = \operatorname{sgn}(u^k) \quad \text{and} \quad \psi(u) = \operatorname{sgn}(u^l) \quad \text{for} \quad u \in F_\infty^\times.$$

For every subfield Λ of $\mathbf{C}$, we denote by $\Lambda(\mathbf{g})$ (resp. $\Lambda(\chi)$) the field generated over Λ by $C(\mathfrak{m}, \mathbf{g})$ (resp. $\chi^*(\mathfrak{m})$) for all $\mathfrak{m}$; also we set $\Lambda(\mathbf{f}, \mathbf{g}) = \Lambda(\mathbf{f})(\mathbf{g})$ and $\Lambda(\mathbf{f}, \chi) = \Lambda(\mathbf{f})(\chi)$. Now our main results can be stated as the following three theorems.

Theorem 4.1. *Suppose $\mathbf{f}$ is primitive, and*

(4.3a) $$l_\nu < k_\nu \quad \text{for every} \ \nu,$$

(4.3b) $$k_1 - l_1 \equiv \cdots \equiv k_n - l_n \pmod 2.$$

For an integer or a half-integer m such that

(4.4a) $$(k_0 + l_0 - 2)/2 < m < (k_0 + k_\nu + l_0 - l_\nu)/2 \quad \text{for every} \ \nu,$$

(4.4b) $$2m \equiv k_0 + l_0 - k_\nu - l_\nu \pmod 2 \quad \text{for every} \ \nu,$$

put

(4.5) $$S(m, \mathbf{f}, \mathbf{g}) = D_F^{-1/2} \pi^{-\{k\}} D(m, \mathbf{f}, \mathbf{g}) / \langle \mathbf{f}, \mathbf{f} \rangle.$$

where $\{k\} = \sum_{\nu=1}^n k_\nu$. Then the number $S(m, \mathbf{f}, \mathbf{g})$ belongs to $\overline{\mathbf{Q}}(\mathbf{g})$. Moreover

(4.6) $$S(m, \mathbf{f}, \mathbf{g})^\sigma = S(m, \mathbf{f}^\sigma, \mathbf{g}^\sigma)$$

for every $\sigma \in \mathrm{Aut}(\mathbf{C})$ such that

(4.7) $$\left[\prod_{\nu=1}^n (\xi^{\tau_\nu})^{k'_\nu/2} \right]^\sigma = \prod_{\nu=1}^n (\xi^{\tau_\nu \sigma})^{k'_\nu/2} \quad \text{and} \quad \left[\prod_{\nu=1}^n (\xi^{\tau_\nu})^{l'_\nu/2} \right]^\sigma = \prod_{\nu=1}^n (\xi^{\tau_\nu \sigma})^{l'_\nu/2}$$

for all totally positive $\xi \in F$.

Here the conductor of $\mathbf{f}$ is not necessarily $\mathfrak{c}$; $\mathbf{f}^\sigma$ and $\mathbf{g}^\sigma$ are defined as in Proposition 2.6; they are meaningful under (4.7). If $k_1 \equiv \cdots \equiv k_n \pmod 2$, we have $l_1 \equiv \cdots \equiv l_n \pmod 2$ by (4.3b); in this case (4.7) holds for all $\sigma \in \mathrm{Aut}(\mathbf{C})$, m is an integr, and $S(m, \mathbf{f}, \mathbf{g})$ belongs to $\mathbf{Q}(\mathbf{f}, \mathbf{g})$.

For a Hecke character ψ of $F_\mathbf{A}^\times$ as above, we define the Gauss sum $\tau(\psi)$ and the L-function $L_{\mathfrak{c}}(s, \psi)$ to be the same as $\tau(\psi^*)$ and $L_{\mathfrak{c}}(s, \psi^*)$ (see (3.9) and (3.12)). We then put

(4.8) $$\mathcal{D}_{\mathfrak{c}}(s, \mathbf{f}, \mathbf{g}) = L_{\mathfrak{c}}(2s + 2 - k_0 - l_0, \psi\varphi) D(s, \mathbf{f}, \mathbf{g}),$$

(4.9) $$E(\mathbf{f}) = (2\pi i)^{n(1-k_0)} \pi^{\{k\}} \tau(\psi) \langle \mathbf{f}, \mathbf{f} \rangle.$$

We shall prove that $\mathcal{D}_{\mathfrak{c}}(s, \mathbf{f}, \mathbf{g})$ is an entire function if $k \neq l$ (Proposition 4.13). Now a stronger version of Theorem 4.1 can be stated as

Theorem 4.2. *Suppose $\mathbf{f}$ is primitive and conditions (4.3a, b) are satisfied. Put*

(4.10) $$T(m, \mathbf{f}, \mathbf{g}) = (2\pi i)^{n(l_0 - 1 - 2m)} \tau(\varphi)^{-1} \mathcal{D}_{\mathfrak{c}}(m, \mathbf{f}, \mathbf{g}) / E(\mathbf{f})$$

for an integer or a half integer m such that

(4.11a) $$(k_0 + l_0 - k_\nu + l_\nu - 2)/2 < m < (k_0 + l_0 + k_\nu - l_\nu)/2 \quad \text{for every} \ \nu,$$

(4.11b) $$2m \equiv k_0 + l_0 - k_\nu - l_\nu \pmod 2 \quad \text{for every} \ \nu.$$

Then the number $T(m, \mathbf{f}, \mathbf{g})$ belongs to $\overline{\mathbf{Q}}(\mathbf{g})$. Moreover we have

(4.12) $T(m, \mathbf{f}, \mathbf{g})^\sigma = T(m, \mathbf{f}^\sigma, \mathbf{g}^\sigma)$ *for every $\sigma \in \mathrm{Aut}(\mathbf{C})$ satisfying (4.7).*

Theorem 4.3. *Let $\mathbf{f}$ be a primitive cusp form of type (k, ψ) with k such that*

(4.13) $$k_1 \equiv \cdots \equiv k_n \pmod 2.$$

Let $k^0 = \text{Min}(k_1, \dots, k_n)$. For a Hecke character χ of $F_{\mathbf{A}}^{\times}$ of finite order and an integer m such that

$$(4.14) \qquad (k_0 - k^0)/2 < m < (k_0 + k^0)/2,$$

put

$$(4.15) \qquad A(m, \mathbf{f}, \chi) = (2\pi i)^{-mn}\tau(\chi)^{-1}D(m, \mathbf{f}, \chi).$$

If $k_\nu \geq 3$ for all ν, then there exist nonzero complex numbers $u(r, \mathbf{f}^\sigma)$, defined for each $r \in \mathbf{Z}^n/2\mathbf{Z}^n$ and for $\mathbf{f}^\sigma$ with $\sigma \in \text{Gal}(\overline{\mathbf{Q}}/\mathbf{Q})$, with the following properties:

(I) If $\chi(a) = \text{sgn}\big[a^r N(a)^m\big]$ for $a \in F_\infty^\times$, we have

$$A(m, \mathbf{f}, \chi)/u(r, \mathbf{f}) \in \mathbf{Q}(\mathbf{f}, \chi),$$

and

$$\big[A(m, \mathbf{f}, \chi)/u(r, \mathbf{f})\big]^\sigma = A(m, \mathbf{f}^\sigma, \chi^\sigma)/u(r, \mathbf{f}^\sigma)$$

for every $\sigma \in \text{Gal}(\overline{\mathbf{Q}}/\mathbf{Q})$.

(II) If $q, r \in \mathbf{Z}^n/2\mathbf{Z}^n$ and $q_\nu + r_\nu \equiv 1 \pmod 2$ for all ν, then

$$u(q, \mathbf{f})u(r, \mathbf{f})/E(\mathbf{f}) \in \mathbf{Q}(\mathbf{f}),$$

and

$$\big[u(q, \mathbf{f})u(r, \mathbf{f})/E(\mathbf{f})\big]^\sigma = u(q, \mathbf{f}^\sigma)u(r, \mathbf{f}^\sigma)/E(\mathbf{f}^\sigma)$$

for every $\sigma \in \text{Gal}(\overline{\mathbf{Q}}/\mathbf{Q})$.

Remark. (A) If $k^0 = 1$, then there is no integer m satisfying (4.14). If $k^0 = 2$, then k_0 must be even under (4.13), and $m = k_0/2$ is the unique integer satisfying (4.14). The above theorem is stated under the assumption $k^0 \geq 3$. It is conjecurable that (I, II) are true also in the case $k^0 = 2$; this is so at least when $n = 1$, as shown in [14]. In fact, we shall prove, for all $k^0 \geq 2$, the following assertion:

(4.16) *Let m and m' be two integers satisfying (4.14), and χ and η be two Hecke characters of $F_{\mathbf{A}}^{\times}$ of finite order such that $(\chi\eta)(v) = \text{sgn}\big(N(v)\big)^{m-m'-1}$ for $v \in F_\infty^\times$. Then*
$$A(m, \mathbf{f}, \chi)A(m', \mathbf{f}, \eta)/E(\mathbf{f}) \in \mathbf{Q}(\mathbf{f}, \chi, \eta).$$

Moreover, for every $\sigma \in \text{Gal}(\overline{\mathbf{Q}}/\mathbf{Q})$, we have

$$\big[A(m, \mathbf{f}, \chi)A(m', \mathbf{f}, \eta)/E(\mathbf{f})\big]^\sigma = A(m, \mathbf{f}^\sigma, \chi^\sigma)A(m', \mathbf{f}^\sigma, \eta^\sigma)/E(\mathbf{f}^\sigma).$$

If $k^0 \geq 3$, this is equivalent to (II).

(B) As will be shown in Proposition 4.5 below, $D(m, \mathbf{f}, \chi) = D(m, \mathbf{g})$ with a cusp form $\mathbf{g}$. Since $R(s, \mathbf{g})$ defined by (2.47) is entire, the existence of the $\Gamma(s - k_\nu'/2)$ implies that $D(m, \mathbf{f}, \chi) = 0$ if $m \leq k_\nu'/2$ for some ν. Thus it is natural to impose (4.14) on m. Conditions (4.11a, b) are also natural for a similar reason, in view of Proposition 4.13 below. (For a discussion in the case $n = 1$, see [14, p. 219].)

(C) Theorems 4.1 and 4.2 deal with the case of forms with different weights. If $k = l$, it is more natural to take the Euler product whose Euler p-factor is of degree 3 in $N(\mathfrak{p})^{-s}$ (see [14, pp. 219-220]). In fact, Sturm [17] proved a result similar to these theorems for such an Euler product constructed from an elliptic modular form of an arbitrary level. There is no doubt that a corresponding theorem holds in the Hilbert modular case.

The ideas of the proofs of the above theorems are the same as in [12] and [14], in which the one-dimensional case was treated. We begin with

Proposition 4.4. *Let χ^* be a primitive ideal character whose conductor has finite part $\mathfrak{q}$, and $f(z) = \sum_\xi a(\xi)\mathbf{e}_F(\xi z) \in \mathcal{S}_k\big(\Gamma_\lambda(\mathfrak{c}), \psi_0\big)$; let $\mathfrak{c}_0$ be the conductor of ψ_0 and $\mathfrak{a}$ the least common multiple of $\mathfrak{c}$, $\mathfrak{a}^2$, and $\mathfrak{q}\mathfrak{c}_0$. Then*

$$\sum_\xi \chi^*(\xi t_\lambda^{-1})a(\xi)\mathbf{e}_F(\xi z)$$

belongs to $\mathcal{S}_k\big(\Gamma_\lambda(\mathfrak{a}), \omega\big)$, where $\omega(t) = \psi_0(t)\chi^*(t\mathfrak{r})^{-2}$ for $(t, \mathfrak{a}) = 1$.

PROOF. Suppose $\chi^*(v\mathfrak{r}) = \mathrm{sgn}(v^r)$ for $v \equiv 1 \pmod{\mathfrak{q}}$ with $r \in \mathbf{Z}^n$. Put

$$h(z) = \sum_{u \in R} \mathrm{sgn}(u^r)\overline{\chi}^*(u\mathfrak{q}t_\lambda\mathfrak{d})f(z + u) \quad \text{with} \quad R = \mathfrak{q}^{-1}t_\lambda^{-1}\mathfrak{d}^{-1}/t_\lambda^{-1}\mathfrak{d}^{-1}.$$

From (3.11) we obtain $h(z) = \tau(\overline{\chi}^*)\sum_\xi \chi^*(\xi t_\lambda^{-1})a(\xi)\mathbf{e}_F(\xi z)$. If $\gamma = \begin{bmatrix} a & b \\ c & d \end{bmatrix} \in \Gamma_\lambda(\mathfrak{a})$ and $\det(\gamma) = 1$, we can verify that $h\|\gamma = \omega(a)h$ by the same argument as in [7, Proposition 3.64]. Since $\varepsilon^{k/2}h(\varepsilon z) = h(z)$ for all $\varepsilon \in \mathfrak{r}_+^\times$, we obtain our assertion.

Proposition 4.5. *Let χ and ψ be Hecke characters of $F_\mathbf{A}^\times$ of finite order and let $\mathbf{f} \in \mathcal{S}_k(\mathfrak{c}, \psi)$. Define $\mathfrak{a}$ as in Proposition 4.4 with the ideal character associated with χ as χ^* and the "$\mathfrak{c}$-factor" of ψ as ψ_0. Then there exists an element $\mathbf{g}$ of $\mathcal{S}_k(\mathfrak{a}, \chi^2\psi)$ such that $D(s, \mathbf{g}) = D(s, \mathbf{f}, \chi)$. Moreover, if $\mathfrak{q}$ is prime to $\mathfrak{c}$, we have*

$$D(s, \mathbf{g}|J_\mathfrak{a}) = \psi^*(\mathfrak{q})\chi^*(\mathfrak{c})\tau(\chi)^2 N(\mathfrak{c})^{-1}D(s, \mathbf{f}|J_\mathfrak{c}, \overline{\chi}),$$

where $J_\mathfrak{c}$ is defined by (2.46).

PROOF. Given $\mathbf{f} \in \mathcal{S}_k(\mathfrak{c}, \psi)$, define a function $\mathbf{g}$ on $G_\mathbf{A}$ by

$$\mathbf{g}(x) = \tau(\overline{\chi})^{-1}\chi\big(\det(x)\big)\sum_v \overline{\chi}^*(v\mathfrak{q}\mathfrak{d})\mathrm{sgn}(v^r)\mathbf{f}\left(x \begin{bmatrix} 1 & -v \\ 0 & 1 \end{bmatrix}_0\right)$$

for $x \in G_\mathbf{A}$, where v runs over $\mathfrak{q}^{-1}\mathfrak{d}^{-1}/\mathfrak{d}^{-1}$ and r is defined as in the above proof; the subscript 0 indicates the projection to the non-archimedean factor. Then $\mathbf{g} \in \mathcal{S}_k(\mathfrak{a}, \chi^2\psi)$ in view of (2.15a, b). That $C(\mathfrak{m}, \mathbf{g}) = \chi^*(\mathfrak{m})C(\mathfrak{m}, \mathbf{f})$ can be verified by means of Proposition 4.4. The last assertion can be proved by the same method as in [7, Proposition 3.65].

Let a^ρ denote the complex conjugate of $a \in \mathbf{C}$. For $\mathbf{f} \in \mathcal{M}_k(\mathfrak{c}, \psi)$, $\mathbf{f}^\rho$ is a well-defined element of $\mathcal{M}_k(\mathfrak{c}, \psi^{-1})$; see Proposition 2.6. For some technical reason, we shall write also $\mathbf{f}_\rho$ for $\mathbf{f}^\rho$; similarly if $f(z) = \sum_\xi a(\xi)\mathbf{e}_F(\xi z) \in \mathcal{M}_k(\Gamma_\lambda, \psi_0)$, we define $f_\rho \in \mathcal{M}_k(\Gamma_\lambda, \psi_0^{-1})$ by

$$f_\rho(z) = \sum_\xi a(\xi)^\rho \mathbf{e}_F(\xi z).$$

Notice that

$$\overline{f_\rho(z)} = \sum_\xi a(\xi)\mathbf{e}_F(-\xi\overline{z}).$$

Proposition 4.6. *If* **f** *is primitive,* $\sigma \in \mathrm{Aut}(\mathbf{C})$, *and* σ *satisfies* (2.37), *then* $\mathbf{f}^{\rho\sigma} = \mathbf{f}^{\sigma\rho}$.

PROOF. By Proposition 2.5 we can show that $C(\mathfrak{m}, \mathbf{f}^{\rho\sigma}) = C(\mathfrak{m}, \mathbf{f}^{\sigma\rho})$ for every $\mathfrak{m}$ prime to $\mathfrak{c}$ by the same argument as in the proof of Proposition 2.8. Since both $\mathbf{f}^{\rho\sigma}$ and $\mathbf{f}^{\sigma\rho}$ are primitive, we obtain the desired equality.

Let ψ_0 and φ_0 be characters of $(\mathfrak{r}/\mathfrak{c})^{\times}$, and let $f \in \mathcal{S}_k(\Gamma_\lambda, \psi_0)$ and $g \in \mathcal{M}_l(\Gamma_\lambda, \varphi_0)$ with Fourier expansions

$$(4.18) \qquad f(z) = \sum_\xi a(\xi)\mathbf{e}_F(\xi z), \quad g(z) = \sum_\xi b(\xi)\mathbf{e}_F(\xi z) \qquad (\xi \in \mathfrak{t}_\lambda).$$

Putting $L = \mathbf{R}^n / \mathfrak{t}_\lambda^{-1}\mathfrak{d}^{-1}$, we have

$$D_F^{1/2} N(\mathfrak{t}_\lambda) \int_L \overline{f_\rho(z)}g(z)dx = \sum_\xi a(\xi)b(\xi)\exp\left(-4\pi\{\xi y\}\right),$$

where $z_\nu = x_\nu + iy_\nu$. With k' and l' defined by (4.2b), put

$$(4.19) \qquad a'(\xi) = a(\xi)\xi^{k'/2}, \qquad b'(\xi) = b(\xi)\xi^{l'/2}.$$

Let U denote the group of all totally positive units of F. By (1.8b) we have $a'(\varepsilon\xi) = a'(\xi)$ and $b'(\varepsilon\xi) = b'(\xi)$ for all $\varepsilon \in U$. Let $W = \{y \in \mathbf{R}^n \mid y_\nu > 0 \text{ for all } \nu\}$ and $M = W/U$. Then we have

$$(4.20) \qquad D_F^{1/2} N(\mathfrak{t}_\lambda) \int_M \left\{ \int_L \overline{f_\rho(z)}g(z)dx \right\} y^{(s-1)\mathbf{1}-(k'+l')/2}\,dy$$

$$= \int_M \sum_{\varepsilon \in U}\sum_{\xi U} a'(\xi)b'(\xi)\xi^{-(k'+l')/2}(\varepsilon y)^{(s-1)\mathbf{1}-(k'+l')/2}\exp\left(-4\pi\{\xi\varepsilon y\}\right)dy$$

$$= \int_W \sum_{\xi U} a'(\xi)b'(\xi)\xi^{-(k'+l')/2}y^{(s-1)\mathbf{1}-(k'+l')/2}\exp\left(-4\pi\{\xi y\}\right)dy$$

$$= (4\pi)^{\{k'+l'\}/2-ns}\Gamma\left(s, -(k'+l')/2\right)D(s, f, g),$$

where we put

$$(4.21) \qquad \Gamma(s, r) = \prod_{\nu=1}^n \Gamma(s + r_\nu) \quad \text{for} \quad r = (r_1, \ldots, r_n) \in \mathbf{R}^n,$$

$$(4.22) \qquad D(s, f, g) = \sum_{\xi U} a'(\xi)b'(\xi)N(\xi)^{-s};$$

the sum is taken over all cosets ξU with totally positive ξ in $\mathfrak{t}_\lambda$. Obviously $L \times M$ gives $\Gamma_{\lambda\infty}\backslash\mathfrak{H}^n$, where $\Gamma_{\lambda\infty} = \left\{ \begin{bmatrix} * & * \\ 0 & * \end{bmatrix} \in \Gamma_\lambda \right\}$. Let $R = \Gamma_{\lambda\infty}\backslash\Gamma_\lambda$ and $\Phi_\lambda = \Gamma_\lambda\backslash\mathfrak{H}^n$. Then we obtain

$$D_F^{-1/2} N(\mathfrak{t}_\lambda)^{-1}(4\pi)^{\{k'+l'\}/2-ns}\Gamma\left(s, -(k'+l')/2\right)D(s, f, g)$$

$$= \int_{\Phi_\lambda} \left\{ \sum_{\gamma \in R} [\overline{f_\rho(z)}g(z)y^{(s+1)\mathbf{1}-(k'+l')/2}] \circ \gamma \right\} y^{-2\mathbf{1}}dxdy.$$

Observe that for every $\gamma = \begin{bmatrix} * & * \\ c & d \end{bmatrix} \in \Gamma_\lambda$ we have

$$[\overline{f_\rho(z)}g(z)y^{(s+1)\mathbf{1}-(k'+l')/2}] \circ \gamma$$

$$= (\psi_0\varphi_0)(d)^{-1}(c\overline{z} + d)^k(cz + d)^l|cz + d|^{-2(s+1)\mathbf{1}+k'+l'}\overline{f_\rho(z)}g(z)y^{(s+1)\mathbf{1}-(k'+l')/2}.$$

Let us now impose the following set of conditions on k and l :

(4.23a) $$k_\nu \geq l_\nu \quad \text{for every} \quad \nu,$$

(4.23b) $$k_1 - l_1 \equiv \cdots \equiv k_n - l_n \pmod 2.$$

Let κ be a nonnegative integer such that

(4.24a) $$0 \leq k_\nu - l_\nu - \kappa \in 2\mathbf{Z} \quad \text{for every} \quad \nu,$$

and let

(4.24b) $$r = (k - l - \kappa\mathbf{1})/2 \qquad m = (k_0 + l_0 + \kappa - 2)/2.$$

Let $H^r_{\kappa,\lambda}(z, s, \omega)$ be the function defined by (3.14) with $\mathfrak{b} = \mathfrak{t}_\lambda\mathfrak{d}$. Then

$$(4.25) \quad (-1)^{\{r\}}(4\pi)^{n(m+1-s)-\{k\}}D_F^{-1/2}N(\mathfrak{t}_\lambda)^{-1}\Gamma\big(s, -(k'+l')/2\big)D(s, f, g)$$

$$= \int_{\Phi_\lambda} \overline{f_\rho(z)}g(z)H^r_{\kappa,\lambda}(z, s-m, \overline{\psi}_0\overline{\varphi}_0)y^{(s-m-2)\mathbf{1}+k}\,dx\,dy.$$

Substituting (3.18) for H, we observe, in view of (3.4), that $D(s, f, g)$ can be continued to a meromorphic function on the whole plane. As to its holomorphy, we have

Proposition 4.7. *Under assumptions (4.23a, b), the product*

$$D(s, f, g) \prod_{\nu=1}^n \Gamma\big(s + 1 + (k_\nu - l_\nu - k_0 - l_0)/2\big)\Gamma\big(s - (k'_\nu + l'_\nu)/2\big)$$

is holomorphic for $\mathrm{Re}(s) > k_0$ *or* $\mathrm{Re}(s) > (k_0 + l_0 - 1)/2$ *according as* $k = l$ *or* $k \neq l$. *Moreover, if* κ *is a positive integer satisfying (4.24a), then (4.25) is holomorphic at* $s = m = (k_0 + l_0 + \kappa - 2)/2$.

Proof. If $k = l$, we have $\kappa = 0$ and $r = 0$. If $k \neq l$, choose any possible κ. (We can always take $\kappa = 0$ or $\kappa = 1$.) Our assertions then follow from Proposition 3.3.

Assuming now $\kappa > 0$, we put

(4.26) $$H_{\kappa,\lambda}(z, \omega) = H^0_{\kappa,\lambda}(z, 0, \omega).$$

If $n > 1$ or $\kappa \neq 2$, this belongs to $\mathcal{M}_{\kappa\mathbf{1}}\big(\mathbf{Q}(\omega)\big)$ by Proposition 3.2. Given k and $r \in \mathbf{Z}^n$ with $k_\nu \geq 0$ and $r_\nu \geq 0$ for every ν, define a differential operator δ^r_k of order $\{r\}$ by

(4.27) $$\delta^r_k = (2\pi i)^{-\{r\}} \prod_{\nu=1}^n \prod_{j=1}^{r_\nu} \left(\frac{\kappa + 2j - 2}{2iy_\nu} + \frac{\partial}{\partial z_\nu}\right).$$

Here we understand that δ^0_k is the identity operator. (For basic properties of δ^r_k, see [9].) We put $\delta^r_\kappa = \delta^r_{\kappa\mathbf{1}}$ for $0 \leq \kappa \in \mathbf{Z}$. We can easily verify

(4.28) $$\delta^r_\kappa H_{\kappa,\lambda}(z, \omega) = \prod_{\nu=1}^n \big[\Gamma(\kappa + r_\nu)/\Gamma(\kappa)\big] \cdot H^r_{\kappa,\lambda}(z, 0, \omega).$$

Putting $s = m$ in (4.25), we obtain

Proposition 4.8. *Under conditions (4.3a, b), let* m *be an integer or a half-integer satisfying (4.4a, b). Then*

$$D(m, f, g) = D_F^{1/2} N(\mathfrak{t}_\lambda) \pi^{\{k\}} \alpha(k, l, m, \mathfrak{c}) \langle f_\rho, \, g \cdot \delta_\kappa^r H_{\kappa,\lambda}(z, \overline{\psi}_0\overline{\varphi}_0) \rangle,$$

where κ *and* r *are determined by (4.24b), and*

$$\alpha(k, l, m, \mathfrak{c}) = (-1)^{\{r\}} 4^{\{k\}-n} \Gamma(\kappa)^n \cdot c_F D_F [\mathfrak{t}_+^\times : \mathfrak{t}^{\times 2}]^{-1} N(\mathfrak{c}) \prod_{\mathfrak{p}|\mathfrak{c}} \left[1 + N(\mathfrak{p})^{-1} \right]$$

$$\cdot \prod_{\nu=1}^n \Gamma\big(m + 1 + (k_\nu - l_\nu - k_0 - l_0)/2\big)^{-1} \Gamma\big(m + (k_\nu + l_\nu - k_0 - l_0)/2\big)^{-1}.$$

Here c_F is the number of (2.30). Notice that $\alpha(k, l, m, \mathfrak{c})$ is a rational number.

Proposition 4.9. *Suppose* $k = l$ *and* $\overline{\psi}_0 = \varphi_0$. *Then* $D(s, f, g)$ *has a possible simple pole at* $s = k_0$ *with residue*

$$2^{n-1} (4\pi)^{\{k\}} \prod_{\nu=1}^n \Gamma(k_\nu)^{-1} \cdot R_F[\mathfrak{t}_+^\times : \mathfrak{t}^{\times 2}]^{-1} \langle f_\rho, g \rangle.$$

This can be obtained from (3.20) by taking $\kappa = 0$, $r = 0$, and $m = k_0 - 1$ in (4.25).

For two elements q and r of Z^n, let us write $q \leq r$ if $q_\nu \leq r_\nu$ for all ν. Given $0 \leq k$, $r \in \mathbf{Z}^n$ and a congruence subgroup Γ, let $A_k^r(\Gamma, \omega)$ denote the set of all functions of the form $f(z) = \sum_{0 \leq q \leq r} y^{-q} h_q(z)$ with holomorphic functions h_q on $\mathfrak{H}^n$, such that $f\|_k \gamma = \omega(\gamma) f$ for all $\gamma \in \Gamma$, where ω is a character of Γ.

Proposition 4.10. *Suppose* $n > 1$, $k_\nu > 2r_\nu$ *for all* ν, *and let* $f \in A_k^r(\Gamma, \omega)$. *Then* $f = \sum_{0 \leq t \leq r} \delta_{k-2t}^t g_t$ *with elements* g_t *of* $\mathcal{M}_{k-2t}(\Gamma, \omega)$, *which are uniquely determined by* f.

This can be proved in exactly the same fashion as in [12, Lemma 7], which deals with the case $n = 1$.

Lemma 4.11. *Let* $f \in \mathcal{S}_k(\Gamma_\lambda, \psi_0)$, $g \in \mathcal{M}_l(\Gamma_\lambda, \overline{\psi}_0)$ *with* $k = l + 2r$. *Suppose* $k \neq l$ *and* $r_\nu \geq 0$ *for all* ν. *Then* $\langle f_\rho, \delta_l^r g \rangle = 0$.

PROOF. We have obviously $\delta_l^r g = (2\pi i)^{-\{r\}} \sum_{0 \leq p \leq r} c_p (2iy)^{p-r} (\partial/\partial z)^p g$ with $c_p \in \mathbf{Z}$. If $b'(\xi)$ is defined by (4.18) and (4.19), then

$$\delta_l^r g = \sum_p c_p(-4\pi y)^{p-r} \sum_\xi b'(\xi) \xi^{p-(l'/2)} \mathbf{e}_F(\xi z),$$

and hence

$$D_F^{1/2} N(\mathfrak{t}_\lambda) \int_{M \times L} \overline{f_\rho(z)} (\delta_l^r g)(z) y^{(s-1)1-k'} \, dxdy$$

$$= (4\pi)^{\{k'\}-ns} D\big(s - (k_0 - l_0)/2, f, g\big) \sum_{0 \leq p \leq r} (-1)^{\{p-r\}} c_p \Gamma(s, p - r - k').$$

The left-hand side can be transformed to an integral of the same type as (4.25), which has, at $s = k_0$, residue $\langle f_\rho, \delta_l^r g \rangle$ times a nonzero constant. But the right-hand side is holomorphic at $s = k_0$ by Proposition 4.7. Hence we obtain our lemma.

Proposition 4.12. *For Hecke characters* $\chi_1, \ldots, \chi_t$ *of* $F_\mathbf{A}^\times$ *of finite order and for every* $\sigma \in \mathrm{Gal}(\overline{\mathbf{Q}}/\mathbf{Q})$ *we have*

$$\left[\tau(\chi_1) \cdots \tau(\chi_t)/\tau(\chi_1 \cdots \chi_t) \right]^\sigma = \tau(\chi_1^\sigma) \cdots \tau(\chi_t^\sigma)/\tau(\chi_1^\sigma \cdots \chi_t^\sigma).$$

This follows immediately from (3.10).

Let us now take $\mathbf{f} = (f_1, \ldots, f_h) \in \mathcal{S}_k(\mathfrak{c}, \psi)$ and $\mathbf{g} = (g_1, \ldots, g_h) \in \mathcal{M}_l(\mathfrak{c}, \varphi)$. From (2.24) we obtain

$$D(s, \mathbf{f}, \mathbf{g}) = \sum_{\lambda=1}^h N(\mathfrak{t}_\lambda)^{s-(k_0+l_0)/2} D(s, f_\lambda, g_\lambda).$$

Observe that the measure of Φ_λ depends only on $\mathfrak{c}$, but not on λ. Call it $M_{\mathfrak{c}}$. From (4.25) we obtain

(4.29)
$$B(s)D(s, \mathbf{f}, \mathbf{g})$$
$$= \sum_{\lambda=1}^h N(\mathfrak{t}_\lambda)^{s-m+(\kappa/2)} \left\langle f_{\lambda\rho}, g_\lambda y^{(s-m)\mathbf{1}} H^r_{\kappa,\lambda}(z, s-m, \overline{\psi}_0\overline{\varphi}_0) \right\rangle,$$

where

(4.30)
$$B(s) = (-1)^{\{r\}} (4\pi)^{n(m+1-s)-\{k\}} D_F^{-1/2} M_{\mathfrak{c}}^{-1} \Gamma\big(s, -(k'+l')/2\big).$$

Proposition 4.13. *Under (4.23a, b), the product*

$$\mathcal{D}_{\mathfrak{c}}(s, \mathbf{f}, \mathbf{g}) \prod_{\nu=1}^n \Gamma\big(s+1+(k_\nu - l_\nu - k_0 - l_0)/2\big) \Gamma\big(s - (k'_\nu + l'_\nu)/2\big)$$

can be continued to a meromorphic function on the whole plane, which is holomorphic except for possible simple poles at $s = k_0$ and $s = k_0 - 1$. The poles may occur only when $k = l$. If $k = l$, then the residue of $D(s, \mathbf{f}, \mathbf{g})$ at $s = k_0$ is

$$2^{n-1}(4\pi)^{\{k\}} \prod_{\nu=1}^n \Gamma(k_\nu)^{-1} \cdot R_F[\mathfrak{r}_+^\times : \mathfrak{r}^{\times 2}]^{-1} \langle \mathbf{f}_\rho, \mathbf{g} \rangle.$$

PROOF. Take $\kappa\,(\geq 0)$, r, and m as in (4.24a, b). (There may be many choices of κ, but the result will be the same with any choice.) Let ψ_0 and φ_0 denote the "$\mathfrak{c}$-factors" of ψ and φ. Put $\omega = (\psi_0\varphi_0)^{-1}$, $\chi = \psi^*\varphi^*$, and

$$d_\lambda(z, s) = N(\mathfrak{t}_\lambda)^{s+(\kappa/2)} g_\lambda(z) y^{s\mathbf{1}} H^r_{\kappa,\lambda}(z, s, \omega),$$
$$v_{\mathfrak{a},\lambda}(z, s) = N(\mathfrak{t}_\lambda)^{s+(\kappa/2)} g_\lambda(z) y^{s\mathbf{1}} J_{\mathfrak{a},\lambda}(z, s, \chi),$$

where $J_{\mathfrak{a},\lambda}$ denotes the function J_λ defined by (3.16) with $\mathfrak{b} = \mathfrak{t}_\lambda\mathfrak{d}$ and any ideal $\mathfrak{a}$ prime to $\mathfrak{c}$ belonging to A. Observe that d_λ and $v_{\mathfrak{a},\lambda}$ have the same property of automorphy as the elements of $\mathcal{M}_k(\mathfrak{c}, \psi_0^{-1})$. Now, for every prime ideal $\mathfrak{p}$ prime to $\mathfrak{c}$ and an index λ, determine μ by $\mathfrak{t}_\lambda\mathfrak{p}^2 = \mathfrak{t}_\mu q$ with $0 \ll q \in F$. Let $\pi_{\mathfrak{p}}$ be a prime element of $\mathfrak{r}_{\mathfrak{p}}^\times$. Then $t_\lambda\pi_{\mathfrak{p}}^2 = t_\mu q u$ with $u \in F_{\infty+}^\times \prod_{\mathfrak{p}} \mathfrak{r}_{\mathfrak{p}}^\times$ and

$$x_\lambda \cdot \mathrm{diag}[\pi_{\mathfrak{p}}, \pi_{\mathfrak{p}}u^{-1}]x_\mu^{-1} = \mathrm{diag}[\pi_{\mathfrak{p}}, \pi_{\mathfrak{p}}^{-1}q] = \alpha_\lambda w$$

with $\alpha_\lambda \in G_{\mathbf{Q}+}$ and $w \in x_\mu W x_\mu^{-1}$. We can choose α_λ so that w is "very close" to 1 in the nonarchimedean factor of $G_{\mathbf{A}}$. Then $\mathbf{f}|S(\mathfrak{p}) = (f_1', \ldots, f_h')$ with $f_\mu' = f_\lambda\|\alpha_\lambda$. For such an α_λ, we can easily verify that

(4.31a)
$$v_{\mathfrak{p}\mathfrak{a},\lambda}\|_k \alpha_\lambda = (\chi^{-1}\varphi^*)(\mathfrak{p})v_{\mathfrak{a},\mu}.$$

Put $\mathbf{d} = (d_1, \ldots, d_h)$ and $\mathbf{v}_{\mathfrak{a}} = (v_{\mathfrak{a},1}, \ldots, v_{\mathfrak{a},h})$. Then (4.31a) can be written as

(4.31b)
$$\chi(\mathfrak{p})\mathbf{v}_{\mathfrak{p}\mathfrak{a}} = \varphi^*(\mathfrak{p})\mathbf{v}_{\mathfrak{a}}|S(\mathfrak{p})^{-1}.$$

By this and (3.15) (or (3.17a)), we have

$$\mathbf{d} = \sum_{\mathfrak{a}} \mu(\mathfrak{a})N(\mathfrak{a})^{-\kappa-2s}\chi(\mathfrak{a})\mathbf{v}_{\mathfrak{a}} = \sum_{\mathfrak{a}} \mu(\mathfrak{a})N(\mathfrak{a})^{-\kappa-2s}\varphi^*(\mathfrak{a})\mathbf{v}_{\mathfrak{r}}|S(\mathfrak{a})^{-1},$$

where $\mathfrak{a}$ runs over all integral ideals prime to $\mathfrak{c}$. Therefore from (4.29) we obtain

$$B(s+m)D(s+m, \mathbf{f}, \mathbf{g}) = \sum_{\lambda=1}^{h} \langle (f_\lambda)_\rho(z), d_\lambda(z, s) \rangle$$
$$= \langle \mathbf{f}_\rho, \mathbf{d} \rangle = \sum_{\mathfrak{a}} \mu(\mathfrak{a})\varphi^*(\mathfrak{a})N(\mathfrak{a})^{-\kappa-2s} \langle \mathbf{f}_\rho, \mathbf{v}_{\mathfrak{r}}|S(\mathfrak{a})^{-1} \rangle.$$

Using the above α_λ, we see that $\alpha_\lambda^{-1}\Gamma_\lambda\alpha_\lambda = \Gamma_\mu$ and

$$\langle (f_\lambda)_\rho, v_{\mathfrak{r},\mu}\|\alpha_\lambda^{-1} \rangle = \langle (f_\lambda)_\rho\|\alpha_\lambda, v_{\mathfrak{r},\mu} \rangle;$$

hence

$$B(s+m)D(s+m, \mathbf{f}, \mathbf{g}) = \sum_{\mathfrak{a}} \mu(\mathfrak{a})\varphi^*(\mathfrak{a})N(\mathfrak{a})^{-\kappa-2s} \langle \mathbf{f}_\rho|S(\mathfrak{a}), \mathbf{v}_{\mathfrak{r}} \rangle$$
$$= \sum_{\mathfrak{a}} \mu(\mathfrak{a})(\varphi\psi)^*(\mathfrak{a})N(\mathfrak{a})^{-\kappa-2s} \langle \mathbf{f}_\rho, \mathbf{v}_{\mathfrak{r}} \rangle,$$

which implies

$$(4.32) \qquad B(s+m)L_{\mathfrak{c}}(\kappa + 2s, \varphi\psi)D(s+m, \mathbf{f}, \mathbf{g}) = \sum_{\lambda=1}^{h} \langle (f_\lambda)_\rho, v_{\mathfrak{r},\mu} \rangle$$

$$= M_{\mathfrak{c}}^{-1} \sum_{\lambda=1}^{h} N(\mathfrak{t}_\lambda)^{s+(\kappa/2)} \int_{\Phi_\lambda} \overline{(f_\lambda)_\rho(z)}g_\lambda(z)J_{\mathfrak{r},\lambda}(z, s, \psi^*\varphi^*)y^{k+(s-2)\mathbf{1}} \, dxdy.$$

Now $J_{\mathfrak{r},\lambda}$ is a linear combination of $E^r_{\kappa,U}$ defined by (3.1), and therefore (4.32) multiplied by $\Gamma(s, \lambda\mathbf{1}+r)$ yields a function with the desired analytic properties. The residue at $s = k_0$ can be determined by Proposition 4.7.

Proposition 4.14. *Let $\mathbf{f}$ be a primitive cusp form, and let $\mathfrak{a}$ and $\mathfrak{b}$ be integral ideals of F. Then $\langle \mathbf{f}|\mathfrak{a}, \mathbf{f}|\mathfrak{b} \rangle/\langle \mathbf{f}, \mathbf{f} \rangle$ belongs to $\mathbf{Q}(\mathbf{f})$, where $\mathbf{f}|\mathfrak{a}$ is defined by (2.32). Moreover $[\langle \mathbf{f}|\mathfrak{a}, \mathbf{f}|\mathfrak{b} \rangle/\langle \mathbf{f}, \mathbf{f} \rangle]^\sigma = \langle \mathbf{f}^\sigma|\mathfrak{a}, \mathbf{f}^\sigma|\mathfrak{b} \rangle/\langle \mathbf{f}^\sigma, \mathbf{f}^\sigma \rangle$ for every $\sigma \in \text{Aut}(\mathbf{C})$ satisfying (2.37).*

PROOF. Proposition 4.13 shows that

$$\langle \mathbf{f}|\mathfrak{a}, \mathbf{f}|\mathfrak{b} \rangle/\langle \mathbf{f}, \mathbf{f} \rangle = \big[D(s, \mathbf{f}_\rho|\mathfrak{a}, \mathbf{f}|\mathfrak{b})/D(s, \mathbf{f}_\rho, \mathbf{f})\big]_{s=k_0}.$$

Now both $D(s, \mathbf{f}_\rho, \mathbf{f})$ and $D(s, \mathbf{f}_\rho|\mathfrak{a}, \mathbf{f}|\mathfrak{b})$ can be expressed as Euler products of the type discussed in [12, Lemma 1]. Therefore we obtain our assertions by the same argument as in [12, p. 792, Proof of Proposition 1, and Lemma 3].

Proposition 4.15. *Let $\mathbf{f}$ and $\mathbf{g}$ be elements of $\mathcal{M}_k(\mathfrak{c}, \psi_0)$. Suppose $k \neq \mathbf{1}$ and $\mathbf{f}$ is a primitive cusp form. Then*

$$\big(\langle \mathbf{f}, \mathbf{g} \rangle/\langle \mathbf{f}, \mathbf{f} \rangle\big)^\sigma = \langle \mathbf{f}^\sigma, \mathbf{g}^\sigma \rangle/\langle \mathbf{f}^\sigma, \mathbf{f}^\sigma \rangle$$

for every $\sigma \in \text{Aut}(\mathbf{C})$ satisfying (2.37). Moreover, under (4.13), $\langle \mathbf{f}, \mathbf{g} \rangle/\langle \mathbf{f}, \mathbf{f} \rangle$ belongs to $\mathbf{Q}(\mathbf{f}, \mathbf{g})$.

PROOF. If $k \neq \mathbf{1}$, every form is a sum of a cusp form and an Eisenstein series (see [2]). Therefore our asserions can be proved in exactly the same manner as in [12, Lemma 4]. Notice that Miyake's rsult in [4] as well as Propositions 2.6 and 4.14 are essential.

PROOF OF THEOREM 4.1. Since the case $n = 1$ was proved in [12], we assume $n > 1$. (We shall do so also in the proofs of Theorems 4.2 and 4.3.) First we observe that $\langle \mathbf{f}^\rho, \mathbf{f}^\rho \rangle = \langle \mathbf{f}, \mathbf{f} \rangle$ by virtue of Proposition 4.13, since $D(s, \mathbf{f}, \mathbf{f}^\rho) = D(s, \mathbf{f}^\rho, \mathbf{f})$. By Proposition 4.8 we have

$$S(m, \mathbf{f}, \mathbf{g}) = \alpha(k, l, m, \mathfrak{c})\langle \mathbf{f}^\rho, \mathbf{p} \rangle/\langle \mathbf{f}^\rho, \mathbf{f}^\rho \rangle$$

with $\mathbf{p} = (p_1, \ldots, p_h)$, $p_\lambda = N(\mathfrak{t}_\lambda)^{\kappa/2} g_\lambda \cdot \delta_\kappa^r H_{\kappa,\lambda}(z, \omega)$, $\omega = (\psi_0 \varphi_0)^{-1}$. Applying Lemma 4.10 to p_λ, we have $p_\lambda = \sum_{0 \leq t \leq r} \delta_{k-2t}^t q_{t\lambda}$ with $q_{t\lambda} \in \mathcal{M}_{k-2t}(\Gamma_\lambda, \psi_0^\rho)$. Put $\mathbf{q} = (q_{01}, \ldots, q_{0h})$. Then $\mathbf{q} \in \mathcal{M}_k(\mathfrak{c}, \psi_0^\rho)$. By Lemma 4.11, we have $\langle \mathbf{f}^\rho, \mathbf{p} \rangle = \langle \mathbf{f}^\rho, \mathbf{q} \rangle$, and hence

$$S(m, \mathbf{f}, \mathbf{g}) = \alpha(k, l, m, \mathfrak{c}) \langle \mathbf{f}^\rho, \mathbf{q} \rangle / \langle \mathbf{f}^\rho, \mathbf{f}^\rho \rangle.$$

Given $\sigma \in \mathrm{Aut}(\mathbf{C})$, put $j = k^\sigma$, $s = r^\sigma$, and $\mathbf{p}' = (p_1', \ldots, p_h')$ with $p_\lambda' = \big(N(\mathfrak{t}_\lambda)^{\kappa/2} g_\lambda\big)^\sigma \delta_\kappa^s H_{\kappa,\lambda}(z, \omega^\sigma)$. We easily see that $p_\lambda' = \sum_{0 \leq u \leq s} \delta_{j-2u}^u q_{t\lambda}^\sigma$ in view of Proposition 3.2. Suppose σ satisfies (4.7). Then $p_\lambda' \in A_j^s(\Gamma_\lambda, \psi_0^{\rho\sigma})$ and $q_{0\lambda}^\sigma \in \mathcal{M}_j(\Gamma_\lambda, \psi_0^{\rho\sigma})$. Put

$$\big(N(\mathfrak{t}_\lambda)^{k_0/2}\big)^\sigma = \zeta_\lambda N(\mathfrak{t}_\lambda)^{k_0/2}, \qquad \big(N(\mathfrak{t}_\lambda)^{l_0/2}\big)^\sigma = \eta_\lambda N(\mathfrak{t}_\lambda)^{l_0/2}.$$

Then $\mathbf{f}^\sigma = (\zeta_1 f_1^\sigma, \ldots, \zeta_h f_h^\sigma)$ and $\mathbf{g}^\sigma = (\eta_1 g_1^\sigma, \ldots, \eta_h g_h^\sigma)$ according to our definition in Proposition 2.6. Since $\kappa - k_0 - l_0 \in 2\mathbf{Z}$, we have $\big(N(\mathfrak{t}_\lambda)^{\kappa/2}\big)^\sigma = \eta_\lambda \zeta_\lambda N(\mathfrak{t}_\lambda)^{\kappa/2}$. Thus

$$\langle \mathbf{f}^{\sigma\rho}, \mathbf{q}^\sigma \rangle = \sum_{\lambda=1}^h \langle \zeta_\lambda f_\lambda^{\sigma\rho}, \zeta_\lambda q_{0\lambda}^\sigma \rangle = \sum_{\lambda=1}^h \langle \zeta_\lambda f_\lambda^{\sigma\rho}, \zeta_\lambda p_\lambda' \rangle$$

$$= \sum_{\lambda=1}^h \langle \zeta_\lambda f_\lambda^{\sigma\rho}, \eta_\lambda N(\mathfrak{t}_\lambda)^{\kappa/2} g_\lambda^\sigma \cdot \delta_\kappa^s H_{\kappa,\lambda}(z, \omega^\sigma) \rangle.$$

This shows that

$$S(m, \mathbf{f}^\sigma, \mathbf{g}^\sigma) = \alpha(k^\sigma, l^\sigma, m, \mathfrak{c}) \langle \mathbf{f}^{\sigma\rho}, \mathbf{q}^\sigma \rangle / \langle \mathbf{f}^{\sigma\rho}, \mathbf{f}^{\sigma\rho} \rangle.$$

Since $\alpha(k, l, m, \mathfrak{c}) = \alpha(k^\sigma, l^\sigma, m, \mathfrak{c})$ and $\mathbf{f}^{\sigma\rho} = \mathbf{f}^{\sigma\rho}$, we obtain (4.6) from Proposition 4.15. If $\sigma = \mathrm{id.}$ on $\overline{\mathbf{Q}}(\mathbf{g})$, we have $S(m, \mathbf{f}, \mathbf{g})^\sigma = S(m, \mathbf{f}, \mathbf{g})$, which completes the proof.

PROOF OF THEOREM 4.2. If $m > (k_0 + l_0 - 2)/2$, our assertions follow immediately from Proposition 3.1, Lemma 4.12, and Theorem 4.1. The remaining case is concerned with an integer or a half-integer m' satisfying (4.11a, b) and such that $m' \leq (k_0 + l_0 - 1)/2$. Put $m = k_0 + l_0 - 1 - m'$. Then m satisfies (4.4a, b). Define κ and r by (4.24b) with this m, and consider $J_{\mathfrak{a},\lambda}(z, s, \chi)$ in the proof of Proposition 4.13. Let $\mathfrak{e}$ be the conductor of χ $(= \psi^* \varphi^*)$, $\mathfrak{f} = \mathfrak{e}^{-1}\mathfrak{c}$, and let η be the primitive character associated with χ. Define functions Δ_κ^r, K_κ^r, and L_κ^r by

$$\Delta_\kappa^r(z, s) = \pi^{-ns} y^{(\kappa+s)\mathbf{1}} \Gamma(s, \kappa\mathbf{1} + r),$$

$$(2\pi i)^{\{r\}} (z - \bar{z})^r K_\kappa^r(z, s; \mathfrak{a}, \mathfrak{b}; \eta)$$

$$= \sum_{c,d} \mathrm{sgn}(N(d)^\kappa) \eta(d\mathfrak{b}^{-1}) \left(\frac{c\bar{z} + d}{cz + d}\right)^r (cz + d)^{-\kappa\mathbf{1}} |cz + d|^{-2s\mathbf{1}},$$

$$(2\pi i)^{\{r\}} (z - \bar{z})^r L_\kappa^r(z, s; \mathfrak{a}, \mathfrak{b}; \bar{\eta})$$

$$= \sum_{c,d} \mathrm{sgn}(N(c)^\kappa) \bar{\eta}(c\mathfrak{a}^{-1}) \left(\frac{c\bar{z} + d}{cz + d}\right)^r (cz + d)^{-\kappa\mathbf{1}} |cz + d|^{-2s\mathbf{1}},$$

where $(0, 0) \neq (c, d) \in (\mathfrak{a} \times \mathfrak{b})/\mathfrak{r}^\times$, and $\eta(\mathfrak{x}) = \bar{\eta}(\mathfrak{x}) = 0$ if $\mathfrak{x} + \mathfrak{e} \neq \mathfrak{r}$. Then

$$\Delta_\kappa^r(z, 1 - \kappa - s) K_\kappa^r(z, 1 - \kappa - s; \mathfrak{a}, \mathfrak{b}; \eta)$$

$$= D_F^{-1} \tau(\chi) N(\mathfrak{a}\mathfrak{b}\mathfrak{e})^{-1} \Delta_\kappa^r(z, s) L_\kappa^r(z, s; (\partial\mathfrak{b}\mathfrak{e})^{-1}, (\partial\mathfrak{a})^{-1}; \bar{\eta}).$$

This can be proved in exactly the same fashion as in [8, Lemma 3.5]. Now we have

$$J_{\mathfrak{r},\lambda}(z,\, s,\, \chi) = \sum_{\mathfrak{b}|\mathfrak{f}} \mu(\mathfrak{b})\eta(\mathfrak{b})K_\kappa^r(z,\, s;\, \mathfrak{t}_\lambda\mathfrak{c}\mathfrak{d},\, \mathfrak{b};\, \eta).$$

Therefore, substituting $1 - \kappa - s$ for s in (4.32), we obtain

$$D_F \cdot \tau(\chi)^{-1} M_{\mathfrak{c}}\pi^{n(2s+\kappa-1)}\Gamma(-s,\, \mathbf{1}+r)B(m'-s)\mathcal{D}_{\mathfrak{c}}(m'-s,\, \mathbf{f},\, \mathbf{g})$$

$$= \Gamma(s,\, \kappa\mathbf{1}+r)\sum_{\lambda=1}^{h}\sum_{\mathfrak{b}|\mathfrak{f}}\mu(\mathfrak{b})\eta(\mathfrak{b})N(\mathfrak{b}\mathfrak{c}\mathfrak{d}\mathfrak{e})^{-1}N(\mathfrak{t}_\lambda)^{-s-(\kappa/2)}$$

$$\cdot \int_{\Phi_\lambda} \overline{(f_\lambda)_\rho(z)}g_\lambda(z)L_\kappa^r\big(z,\, s;\, (\mathfrak{b}\mathfrak{d}\mathfrak{e})^{-1},\, (\mathfrak{c}\mathfrak{d}^2\mathfrak{t}_\lambda)^{-1};\, \overline{\eta}\big)y^{k+(s-2)\mathbf{1}}\, dxdy.$$

Put

$$L_{\mathfrak{b},\lambda}(z,\, \overline{\eta}) = D_F^{1/2}(2\pi i)^{-\kappa n}L_\kappa^0\big(z,\, 0;\, (\mathfrak{b}\mathfrak{d}\mathfrak{e})^{-1},\, (\mathfrak{c}\mathfrak{d}^2\mathfrak{t}_\lambda)^{-1};\, \overline{\eta}\big).$$

Then we have a formula similar to (4.28) for L_κ^r, and putting $s = 0$, obtain

$$T(m',\, \mathbf{f},\, \mathbf{g}) = \tau(\chi)\tau(\varphi)^{-1}\tau(\psi)^{-1}\beta(k,\, l,\, m',\, \mathfrak{c})$$

$$\cdot \sum_{\mathfrak{b}|\mathfrak{f}}\mu(\mathfrak{b})\eta(\mathfrak{b})N(\mathfrak{b}\mathfrak{c}\mathfrak{d}\mathfrak{e})^{-1}\langle\mathbf{f}_\rho,\, \mathbf{p}_\mathfrak{b}\rangle/\langle\mathbf{f},\, \mathbf{f}\rangle$$

with a rational number $\beta(k,\, l,\, m',\, \mathfrak{c})$ similar to $\alpha(k,\, l,\, m,\, \mathfrak{c})$ of Proposition 4.8, and $\mathbf{p}_\mathfrak{b} = (p_{\mathfrak{b}1},\, \ldots,\, p_{\mathfrak{b}h})$, $p_{\mathfrak{b}\lambda} = N(\mathfrak{t}_\lambda)^{-\kappa/2}g_\lambda \cdot \delta_\kappa^r L_{\mathfrak{b},\lambda}(z,\, \eta^\rho)$. Now $L_{\mathfrak{b},\lambda}$ is a linear combination of $E_{\kappa,U}$ defined by (3.5), and its Fourier expansion can be obtained from (3.6). Then we easily see that $L_{\mathfrak{b},\lambda}(z,\, \eta^\rho)^\sigma = L_{\mathfrak{b},\lambda}(z,\, \eta^{\rho\sigma})$ for every $\sigma \in \mathrm{Gal}(\overline{\mathbf{Q}}/\mathbf{Q})$. Therefore, repeating the same reasoning as in the proof of Theorem 4.1 and using Lemma 4.12, we obtain Theorem 4.2.

Proposition 4.16. *Let* $\mathbf{f}$ *be an element of* $\mathcal{S}_k(\mathfrak{c},\, \psi)$ *that is a normalized eigenfunction of* $T(\mathfrak{m})$ *for all* $\mathfrak{m}$. *Then* $D(s,\, \mathbf{f}) \neq 0$ *for* $\mathrm{Re}(s) \geq (k_0 + 1)/2$.

PROOF. Put $\lambda(\mathfrak{m}) = C(\mathfrak{m},\, \mathbf{f})$. In general, if $f(z) = \sum a(\xi)\mathbf{e}_F(\xi z) \in \mathcal{S}_k$, then $y^{k/2}f$ is a bounded function on $\mathfrak{H}^n$, so that a well-known argument shows that $a(\xi) = O(\xi^{k/2})$. Thus $\lambda(\mathfrak{m}) = O\big(N(\mathfrak{m})^{k_0/2}\big)$. For each prime ideal $\mathfrak{p}$ of F, put

$$1 - \lambda(\mathfrak{p})N(\mathfrak{p})^{-s} + \psi^*(\mathfrak{p})N(\mathfrak{p})^{k_0-1-2s} = \big(1 - \alpha_\mathfrak{p}N(\mathfrak{p})^{-s}\big)\big(1 - \beta_\mathfrak{p}N(\mathfrak{p})^{-s}\big)$$

with $\alpha_\mathfrak{p},\, \beta_\mathfrak{p} \in \mathbf{C}$. (Either or both of $\alpha_\mathfrak{p}$ and $\beta_\mathfrak{p}$ may be 0 if $\mathfrak{p}|\mathfrak{c}$.) Since

$$(\alpha_\mathfrak{p}^{n+1} - \beta_\mathfrak{p}^{n+1})/(\alpha_\mathfrak{p} - \beta_\mathfrak{p}) = \lambda(\mathfrak{p}^n) = O\big(N(\mathfrak{p})^{nk_0/2}\big),$$

we easily see that $\mathrm{Max}\big(|\alpha_\mathfrak{p}|,\, |\beta_\mathfrak{p}|\big) \leq N(\mathfrak{p})^{k_0/2}$. Put

$$D(s) = \prod_{\mathfrak{p}\nmid\mathfrak{c}} \big[\big(1 - \alpha_\mathfrak{p}N(\mathfrak{p})^{-s}\big)\big(1 - \beta_\mathfrak{p}N(\mathfrak{p})^{-s}\big)\big]^{-1},$$

$$D_\rho(s) = \prod_{\mathfrak{p}\nmid\mathfrak{c}} \big[\big(1 - \overline{\alpha}_\mathfrak{p}N(\mathfrak{p})^{-s}\big)\big(1 - \overline{\beta}_\mathfrak{p}N(\mathfrak{p})^{-s}\big)\big]^{-1},$$

$$B(s) =$$
$$\prod_{\mathfrak{p}\nmid\mathfrak{c}} \big[\big(1 - \alpha_\mathfrak{p}\overline{\alpha}_\mathfrak{p}N(\mathfrak{p})^{-s}\big)\big(1 - \alpha_\mathfrak{p}\overline{\beta}_\mathfrak{p}N(\mathfrak{p})^{-s}\big)\big(1 - \beta_\mathfrak{p}\overline{\alpha}_\mathfrak{p}N(\mathfrak{p})^{-s}\big)\big(1 - \beta_\mathfrak{p}\overline{\beta}_\mathfrak{p}N(\mathfrak{p})^{-s}\big)\big]^{-1}.$$

The first two products are convergent for $\mathrm{Re}(s) > (k_0/2) + 1$; B is convergent for $\mathrm{Re}(s) > k_0 + 1$. $D(s)$ is entire, since it is $D(s,\, \mathbf{f})$ times an entire function; the same holds for D_ρ. As for B, it is an entire function times $\mathcal{D}_{\mathfrak{c}}(s,\, \mathbf{f},\, \mathbf{f}_\rho)$, by virtue of [12, Lemma 1]. By Proposition 4.13, B is holomorphic on the whole plane except for a possible simple pole at $s = k_0$. (Since $\Gamma(s + 1 - k_0)$ is among the Γ-factors of the product in Proposition 4.13, $\mathcal{D}_{\mathfrak{c}}$ has no pole at $s = k_0 - 1$.) Put $r = it + (k_0 - 1)/2$ with real t, $\xi_{\mathfrak{p}} = \alpha_{\mathfrak{p}} N(\mathfrak{p})^{-r}$, $\eta_{\mathfrak{p}} = \beta_{\mathfrak{p}} N(\mathfrak{p})^{-r}$, and

$$A(s) = D(s + r)D_\rho(s + \bar{r})B(s + r + \bar{r})\prod_{\mathfrak{p} \nmid \mathfrak{c}}\left(1 - N(\mathfrak{p})^{-s}\right)^{-1}.$$

Then A is holomorphic for $\mathrm{Re}(s) > 1$, and

$$(4.33) \qquad \log A(s) = \sum_{m=1}^{\infty}\sum_{\mathfrak{p}\nmid\mathfrak{c}} m^{-1}N(\mathfrak{p})^{-ms}\left|1 + \xi_{\mathfrak{p}}^m + \eta_{\mathfrak{p}}^m\right|^2$$

for $\mathrm{Re}(s) > 2$. Let $\mathrm{Re}(s) = \sigma_0$ be the line of convergence of the right-hand side of (4.33). Suppose $\sigma_0 > 1$. Then, for real $s > \sigma_0$ we have $\log A(s) \geq 0$, and hence $A(s) = \exp\left(\log A(s)\right) \geq 1$, so that $A(\sigma_0) \geq 1$. Since A is holomorphic at σ_0, this means that $\log A(s)$ can be continued holomorphically in a neighborhood of σ_0, which contradicts the well-known fact that a Dirichlet series with non-negative coefficients is not holomorphic at the real point on the line of convergence. Therefore $\sigma_0 \leq 1$. This implies in particular that if $\mathrm{Re}(s) > 1$, then $A(s) = \exp\left[\log A(s)\right] \neq 0$, and hence $D(s) \neq 0$ for $\mathrm{Re}(s) > (k_0 + 1)/2$. Suppose $D(1 + r) = 0$. Then $D_\rho(1 + \bar{r}) = 0$, so that $A(s)$ is entire. Suppose σ_0 is a finite value, i.e., $\sigma_0 \neq -\infty$. By means of the same argument as above, we see that $A(\sigma_0) \geq 1$ and $\log A(s)$ is holomorphic at σ_0, which is again a contradiction. Therefore $\sigma_0 = -\infty$, which means that $A(s) = \exp\left[\log A(s)\right] \neq 0$ everywhere. But the factor $\prod_{\mathfrak{p}\nmid\mathfrak{c}}\left(1 - N(\mathfrak{p})^{-s}\right)^{-1}$, being $\zeta_F(s)$ times an entire function, has many zeros for $\mathrm{Re}(s) < 0$, a contradiction. Thus $D(1 + r) \neq 0$. Since $D(s, \mathbf{f})$ is $D(s)$ times a nonvanishing factor, this completes the proof.

PROOF OF THEOREM 4.3. Let m and m' be two integers satisfying (4.14). Assume $m' \leq m$. Take two Hecke characters χ and η of $F_{\mathbf{A}}^\times$ of finite order. Let $t = m - m' + 1$; suppose $(\chi\eta)(v) = \mathrm{sgn}(N(v)^t)$ for $v \in F_\infty^\times$. Let $\mathfrak{a}$ and $\mathfrak{b}$ be the conductors of χ and η, respectively, and $\mathfrak{c}$ the least common multiple of $\mathfrak{a}$, $\mathfrak{b}$, and the conductor of $\mathbf{f}$. By Proposition 3.4, there is an element $\mathbf{g}$ of $\mathcal{M}_{t1}(\mathfrak{c}, \chi\eta)$ such that

$$D(s, \mathbf{g}) = L_{\mathfrak{a}}(s, \chi)L_{\mathfrak{b}}(s + 1 - t, \eta).$$

By [12, Lemma 1], we have

$$\begin{aligned}\mathcal{D}_{\mathfrak{c}}(s, \mathbf{f}, \mathbf{g}) &= L_{\mathfrak{c}}(2s + 2 - k_0 - t, \psi\chi\eta)D(s, \mathbf{f}, \mathbf{g}) \\ &= D(s, \mathbf{f}, \chi)D(s + 1 - t, \mathbf{f}, \eta),\end{aligned}$$

and hence

$$A(m, \mathbf{f}, \chi)A(m', \mathbf{f}, \eta)/E(\mathbf{f}) = T(m, \mathbf{f}, \mathbf{g})\tau(\chi\eta)/\left[\tau(\chi)\tau(\eta)\right].$$

Therefore, observing that (4.11a, b) are satified, we obtain (4.16) from Theorem 4.2 and Proposition 4.12, since (4.7) holds for every $\sigma \in \mathrm{Gal}(\overline{\mathbf{Q}}/\mathbf{Q})$ in the present case. Suppose $k^0 \geq 3$; take $m' = (k_0 + k^0 - 2)/2$ in (4.16). By Propositions 4.16 and 4.5 we have $A(m', \mathbf{f}, \eta) \neq 0$. Now, for each $r \in \mathbf{Z}^n/2\mathbf{Z}^n$, we can find a Hecke character η of $F_{\mathbf{A}}^\times$ of order 2 such that $\eta(a) = \mathrm{sgn}[a^r N(a)^{m'+1}]$ for $a \in F_\infty^\times$. For each $\sigma \in \mathrm{Gal}(\overline{\mathbf{Q}}/\mathbf{Q})$ put

$$(4.34) \qquad u(r, \mathbf{f}^\sigma) = E(\mathbf{f}^\sigma)/A(m', \mathbf{f}^\sigma, \eta)$$

with any choice of such η. Then we obtain (I, II) of Theorem 4.3 immediately from (4.16).

5. Cusp forms associated with Hecke characters of a CM-field

Let K be a totally imaginary quadratic extension of F. Take n injective homomorphisms $\varphi_1, \ldots, \varphi_n$ of K into $\mathbf{C}$ such that $\varphi_\nu = \tau_\nu$ on F for every ν, and put, for $t = (t_1, \ldots, t_n) \in \mathbf{Z}^n$ with $t_\nu \geq 0$ and $a \in K$,

$$(5.1\mathrm{a}) \qquad \varphi = \sum_{\nu=1}^n t_\nu \varphi_\nu,$$

$$(5.1\mathrm{b}) \qquad a^\varphi = \prod_{\nu=1}^n (a^{\varphi_\nu})^{t_\nu}.$$

The purpose of this section is to reformulate Theorems 4.1 and 4.3 in the case where $\mathbf{f}$ is obtained from the forms of the type

$$(5.2) \qquad \sum_{a-c_0 \in \mathfrak{e}} a^\varphi \mathbf{e}_F(aa^\rho bz),$$

where $c_0 \in K$, $\mathfrak{e}$ is a fractional ideal of K, ρ denotes complex conjugation, and $0 \ll b \in F$. We first make simple observations about an arbitrary element f of $\mathcal{M}_k$. If U is a subgroup of $\mathfrak{r}_+^\times$ of finite index, we call f U-invariant if $\varepsilon^{k/2} f(\varepsilon z) = f(z)$ for every $\varepsilon \in U$. Given two U-invariant elements

$$(5.3) \qquad f(z) = \sum_\xi a(\xi) \mathbf{e}_F(\xi z) \quad \text{and} \quad g(z) = \sum_\xi b(\xi) \mathbf{e}_F(\xi z)$$

of $\mathcal{M}_k$ and $\mathcal{M}_l$, respectively, we put

$$(5.4) \qquad D_0(s, f, g) = [\mathfrak{r}_+^\times : U]^{-1} \sum_{\xi U} a(\xi) b(\xi) \xi^{(k'+l')/2} N(\xi)^{-s},$$

where k' and l' are defined by (4.2a, b), and the sum is taken over all different ξU with $0 \ll \xi \in F$. This does not depend on the choice of U, and has the same property of holomorhy as stated in Proposition 4.7.

Lemma 5.1. *Suppose f is U-invariant. Let $V = \mathfrak{r}_+^\times$ and*

$$h(z) = [V : U]^{-1} \sum_{\varepsilon \in V/U} \varepsilon^{k/2} f(\varepsilon z),$$

where ε runs over a complete set of representatives for V/U. Then h is V-invariant, $D(s, h) = D(s, f)$, and $D_0(s, h, g) = D_0(s, f, g)$ for every V-invariant g. ($D(s, f)$ has been defined by (2.41a).)

This can be verified in a straightforward way. Notice that $D(s, f) = 0$ if $\varepsilon^{k/2} f(\varepsilon z) = \chi(\varepsilon) f$ with a nontrivial character χ of $\mathfrak{r}_+^\times$.

Lemma 5.2. *Let $f \in \mathcal{M}_k(\overline{\mathbf{Q}})$, $\sigma \in \mathrm{Gal}(\overline{\mathbf{Q}}/\mathbf{Q}_{\mathrm{ab}})$, and $\alpha \in G_{\mathbf{Q}+}$. If $k_1 \equiv \cdots \equiv k_n \pmod{2}$, then $(f|\alpha)^\sigma = f^\sigma|\alpha$ and $(f\|\alpha)^\sigma = f^\sigma\|\alpha$.*

PROOF. We take the group $\mathfrak{G}$ of Theorem 1.5. Since $(1, \sigma)(\alpha, 1) = (\alpha, 1)(1, \sigma)$ in $\mathfrak{G}$, we have $(f|\alpha)^\sigma = f^\sigma|\alpha$. Put $k^\sigma = l$, $m = -k'/2$, and $m^\sigma = q$. (Observe that $m \in \mathbf{Z}^n$.) If $0 \ll b \in F$, we have $b^{k/2} = b^m N(b)^{k_0/2}$. Therefore, if $\sigma = $ id. on $\mathbf{Q}_{\mathrm{ab}}$, then $(b^{k/2})^\sigma = b^q N(b)^{k_0/2} = b^{l/2}$, which combined with the first equality proves the second equality.

Given K and $\varphi = \sum_{\nu=1}^n t_\nu \varphi_\nu$ as above, let $\mathcal{L}(K, \varphi)$ denote the set of all finite $\mathbf{C}$-linear combinations of forms of type (5.2).

Proposition 5.3. *Put $k = t + 1$; then $\mathcal{L}(K, \varphi) \subset \mathcal{M}_k$. Moreover, put*

(5.5)
$$\mathcal{L}(K, \varphi, \overline{\mathbf{Q}}) = \mathcal{L}(K, \varphi) \cap \mathcal{M}_k(\overline{\mathbf{Q}}).$$

Then $\mathcal{L}(K, \varphi, \overline{\mathbf{Q}})$ is stable under the action of $\|_k \alpha$ for every $\alpha \in G_{\mathbf{Q}+}$; furthermore $\mathcal{L}(K, \varphi, \overline{\mathbf{Q}})^\sigma = \mathcal{L}(K, \varphi\sigma, \overline{\mathbf{Q}})$ for every $\sigma \in \mathrm{Gal}(\overline{\mathbf{Q}}/\mathbf{Q})$.

PROOF. Since the forms of type (5.2) are generalized theta series, we can show that $\mathcal{L}(K, \varphi) \subset \mathcal{M}_k$ in a standard way (for example, by using the pull-back of classical theta functions as in [13, §3]). The last assertion is obvious. The stability of $\mathcal{L}(K, \varphi, \overline{\mathbf{Q}})$ under $\|_k \alpha$ is easy if $\alpha = \begin{bmatrix} a & b \\ 0 & d \end{bmatrix}$. If $\alpha = \begin{bmatrix} 0 & -1 \\ 1 & 0 \end{bmatrix}$, it follows from the Poisson summation. Since $G_{\mathbf{Q}+}$ is generated by these two types of elements, we obtain the desired result.

Let us now consider a point w of $\mathfrak{H}^n$ of the form

(5.6)
$$w = (w_0^{\varphi_1}, \ldots, w_0^{\varphi_n}) \quad \text{with} \quad w_0 \in K.$$

Let K' be the field generated over $\mathbf{Q}$ by the numbers $\sum_{\nu=1}^n x^{\varphi_\nu}$ for all $x \in K$. A basic theorem of canonical models asserts that $h(w) \in K'_{\mathrm{ab}}$ for every $h \in \mathcal{A}_0(\mathbf{Q}_{\mathrm{ab}})$ finite at w; see [6], [9]. Given elements $f \in \mathcal{M}_k, g \in \mathcal{M}_l, \ldots$, and a subfield Λ of $\mathbf{C}$, we denote by $\Lambda(f, g, \ldots)$ the field generated over Λ by the Fourier coefficients of $f, g, \ldots$.

Theorem 5.4. *Given K and $\varphi = \sum_{\nu=1}^n t_\nu \varphi_\nu$ as above, put $k = t + 1$ and $t_0 = \mathrm{Max}(t_1, \ldots, t_n)$. Suppose*

(5.7)
$$t_1 \equiv \cdots \equiv t_n \pmod 2.$$

For $f \in \mathcal{L}(K, \varphi, \overline{\mathbf{Q}})$, $p \in \mathcal{M}_l(\overline{\mathbf{Q}})$, and a point w of type (5.6) such that $p(w) \neq 0$, put

(5.8)
$$X(m, f, p, w) = \pi^{\{t'\}/2 - mn} D(m, f)/p(w),$$

where $t' = t_0 1 - t$ and m is an integer satisfying (4.14). Then $X(m, f, p, w) \in K'_{\mathrm{ab}}(f, p)$. Moreover, for every $\sigma \in \mathrm{Gal}(\overline{\mathbf{Q}}/K'_{\mathrm{ab}})$, we have

(5.9)
$$X(m, f, p, w)^\sigma = X(m, f^\sigma, p^\sigma, w).$$

PROOF. First assume $m > t_0/2$. It is sufficient to prove the case where f is the function of (5.2). Change $(c_0, \mathfrak{e})$ for $(c_0^\rho, \mathfrak{e}^\rho)$ and put $r = (t_0 1 - 2m1 + t)/2$ and $\kappa = 2m - t_0$. We easily see that

$$b^{-t'/2} N(b)^{m+s} D_F^{1/2} (2\pi i)^{-\kappa n - \{r\}} (w - \overline{w})^{-r} D(m + s, f)$$

is a $\mathbf{Q}$-linear combination of the series $E_{r,U}^\kappa$ of type (3.1) evaluated at $z = w$. Given $\sigma \in \mathrm{Gal}(\overline{\mathbf{Q}}/\mathbf{Q})$, put $t^\sigma = u$ and $r^\sigma = q$. If $\sigma = \mathrm{id.}$ on K', then $\varphi\sigma = \sum u_\nu \varphi_\nu$ and $\sum r_\nu \varphi_\nu \sigma = \sum q_\nu \varphi_\nu$, so that $[(w - \overline{w})^r]^\sigma = (w - \overline{w})^q$. Therefore, putting $s = 0$, we obtain our assertions from [9, Theorem 1 and Main Theorem III]. If $m \leq t_0/2$, put $m' = t_0 + 1 - m$. Then $m' > t_0/2$, and the functional equation (2.42) settles our problem, in view of Lemma 5.2.

Let η be a primitive or an imprimitive Hecke ideal character of K defined modulo an integral ideal $\mathfrak{f}$ of K such that

$$(5.10) \qquad \eta(a\mathfrak{r}_K) = \prod_{\nu=1}^{n} \left(a^{\varphi_\nu}/|a^{\varphi_\nu}|\right)^{t_\nu} \quad \text{for} \quad a \equiv 1 \pmod{\mathfrak{f}},$$

where $\mathfrak{r}_K$ denotes the maximal order of K. Define the L-function of η as ususal by

$$L(s, \eta) = \sum_{\mathfrak{a}} \eta(\mathfrak{a}) N(\mathfrak{a})^{-s},$$

where $\mathfrak{a}$ runs over all integral ideals of K prime to $\mathfrak{f}$. If $t_0 = \mathrm{Max}(t_1, \ldots, t_n)$, $t_\nu \geq 0$, and $k = t + 1$, then we can easily verify that

$$(5.11) \qquad L(s, \eta) = D\big(s + (t_0/2), \mathfrak{f}\big)$$

with an element $\mathbf{f}$ of $\mathcal{M}_k\big(\mathfrak{f}\mathfrak{f}^\rho D(K/F), \psi\big)$, where $D(K/F)$ is the relative discriminant of K/F, and ψ is the Hecke idele character such that

$$(5.12) \qquad \psi^*(\mathfrak{a}) = \psi_{K/F}(\mathfrak{a})\eta(\mathfrak{a}\mathfrak{r}_K),$$

$\psi_{K/F}$ denoting the ideal character of F corresponding to the extension K/F. Moreover, $\mathbf{f}$ is primitive if and only if η is primitive. Obviously the components of $\mathbf{f}$ belong to $\mathcal{L}(K, \varphi)$. Conversely, every primitive cusp form whose components belong to $\mathcal{L}(K, \varphi)$ must be of this type.

Theorem 5.5. *Suppose $t_\nu > 0$ for all ν; let $\mathbf{f}$ be a primitive form whose components belong to $\mathcal{L}(K, \varphi)$. For every $q \in \mathcal{M}_{2t}(\overline{\mathbf{Q}})$ such that $q(w) \neq 0$, put*

$$E(\mathbf{f}, q, w) = \pi^n \langle \mathbf{f}, \mathbf{f} \rangle / q(w),$$

where w is as in (5.6). Then $E(\mathbf{f}, q, w) \in \overline{\mathbf{Q}}$. Moreover, if $t_1 \equiv \cdots \equiv t_n$ (mod 2), then $E(\mathbf{f}, q, w) \in K'_{\mathrm{ab}}(\mathbf{f}, q)$, and

$$E(\mathbf{f}, q, w)^\sigma = E(\mathbf{f}^\sigma, q^\sigma, w) \quad \text{for every} \quad \sigma \in \mathrm{Gal}(\overline{\mathbf{Q}}/K'_{\mathrm{ab}}).$$

PROOF. Let $\mathfrak{d}(K/F)$ denote the relative different of K over F. Take a primitive character η defined modulo $\mathfrak{f}$ as in (5.11), and define another character ω modulo $\mathfrak{d}(K/F)\mathfrak{f}\mathfrak{f}^\rho$ by $\omega(\mathfrak{a}) = \eta(\mathfrak{a})\eta(\mathfrak{a}^\rho)^\rho$. Then $L(s, \omega) = D(s + t_0, \mathbf{g})$ with an element $\mathbf{g}$ of $\mathcal{M}_{2t+1}(\mathfrak{c}^2, \psi_{K/F})$, where $\mathfrak{c} = D(K/F)\mathfrak{f}\mathfrak{f}^\rho$. We can easily verify that

$$\mathcal{D}_{\mathfrak{c}}(s, \mathbf{f}_\rho, \mathbf{f}) = D(s, \mathbf{g})\zeta_K(s - t_0) \prod_{\mathfrak{q}|\mathfrak{f}} \left(1 - N(\mathfrak{q})^{t_0 - s}\right).$$

Taking the residue at $s = k_0 \, (= t_0 + 1)$, we find, by Proposition 4.13,

$$(5.13) \qquad \langle \mathbf{f}, \mathbf{f} \rangle = A \cdot |D_F/D_K|^{1/2} \pi^{-n-\{k\}} D(k_0, \mathbf{g})$$

with a rational constant A which depends only on t, K/F, and $\mathfrak{f}$. Now $D(k_0, \mathbf{g}) = \sum_{\lambda=1}^{h} N(\mathfrak{t}_\lambda)^{1/2} D(k_0, g_\lambda)$ if $\mathbf{g} = (g_\lambda)_{\lambda=1}^{h}$. Applying Theorem 5.4 to $D(k_0, g_\lambda)$, we obtain our assertions. We also note that

$$(5.14) \qquad E(\mathbf{f}, q, w) \in K'_{\mathrm{ab}}(\mathbf{g}, q).$$

Theorem 5.6. *Given $f \in \mathcal{L}(K, \varphi, \overline{\mathbf{Q}})$, $g \in \mathcal{M}_l(\overline{\mathbf{Q}})$, w as in (5.6), and $q \in \mathcal{M}_{2t}(\overline{\mathbf{Q}})$ such that $q(w) \neq 0$, put*

$$(5.15) \qquad S(m, f, g, q, w) = \pi^{-\{t\}} D_0(m, f, g)/q(w)$$

for an integer m satisfying (4.4a, b). Suppose that at least one of f and g is $\mathfrak{r}_+^\times$-invariant, and (4.3a, b) are satisfied. Then $S(m, f, g, q, w) \in \overline{\mathbf{Q}}$. Moreover, if $t_1 \equiv \cdots \equiv t_n \pmod 2$, then $S(m, f, g, q, w) \in K'_{\mathrm{ab}}(f, g, q)$, and

$$S(m, f, g, q, w)^\sigma = S(m, f^\sigma, g^\sigma, q^\sigma, w) \quad \text{for every} \quad \sigma \in \mathrm{Gal}(\overline{\mathbf{Q}}/K'_{\mathrm{ab}}).$$

PROOF. Define $S(m, \mathbf{f}, \mathbf{g}, q, w)$ for $\mathbf{f} \in \mathcal{M}_k(\mathfrak{c}, \psi)$ and $\mathbf{g} \in \mathcal{M}_l(\mathfrak{c}, \psi')$ by (5.15) with $D(m, \mathbf{f}, \mathbf{g})$ in place of $D_0(m, f, g)$. If $\mathbf{f}$ and $\mathbf{g}$ have Fourier coefficients in $\overline{\mathbf{Q}}$, then we can express $\mathbf{f}$ and $\mathbf{g}$ in the forms $\mathbf{f} = \sum_{\mathbf{h},\mathfrak{a}} c(\mathbf{h}, \mathfrak{a})\mathbf{h}|\mathfrak{a}$ and $\mathbf{g} = \sum_{\mathbf{k},\mathfrak{b}} d(\mathbf{k}, \mathfrak{b})\mathbf{k}|\mathfrak{b}$ with primitive $\mathbf{h}$ and $\mathbf{k}$, integral ideals $\mathfrak{a}$ and $\mathfrak{b}$, and $c(\mathbf{h}, \mathfrak{a})$ and $d(\mathbf{k}, \mathfrak{b})$ in $\overline{\mathbf{Q}}$. Since $S(m, \mathbf{h}|\mathfrak{a}, \mathbf{k}|\mathfrak{b}, q, w)$ is the product of $S(m, \mathbf{h}, \mathbf{k}, q, w)$ and finitely many Euler factors as shown in [12, Lemma 1], we obtain our assertions with $\mathbf{f}$ and $\mathbf{g}$ in place of f and g from Theorems 4.1 and 5.5. Let us now consider the case of f and g given as above. By Lemmas 5.1 and 5.2 we may assume that both f and g are $\mathfrak{r}_+^\times$-invariant. Take N so that $f \in \mathcal{M}_k(\Gamma_N)$ and $g \in \mathcal{M}_l(\Gamma_N)$, and take $\{t_\lambda\}$ as in §2. We can find a positive integer c such that $ct_1^{-1}\mathfrak{d}^{-1} \subset N\mathfrak{r}$. Put $u(z) = f(cz)$ and $v(z) = g(cz)$. Then u and v are $\mathfrak{r}_+^\times$-invariant. There is an integral ideal $\mathfrak{c}$ such that $u\|\alpha = u$ if

$$\alpha \equiv \begin{bmatrix} 1 & * \\ 0 & 1 \end{bmatrix} \pmod{\mathfrak{c}}.$$

Therefore $u = \sum_j u_j$ with $u_j \in \mathcal{M}_k(\Gamma_1(\mathfrak{c}), \psi_{j0})$ with characters ψ_{j0} of $(\mathfrak{r}/\mathfrak{c})^\times$. By Proposition 5.3 we see that $u_j \in \mathcal{L}(K, \varphi, \overline{\mathbf{Q}})$. A similar decomposition holds for v. Thus we can reduce our problem to the elements $f \in \mathcal{M}_k(\Gamma_1, \psi_0)$ and $g \in \mathcal{M}_l(\Gamma_1, \psi'_0)$. Now consider the element $(f, 0, \ldots, 0)$ of $\mathcal{M}_k(\mathfrak{c}, \psi_0)$, and decompose it as $(f, 0, \ldots, 0) = \sum_i \mathbf{f}_i$ with $\mathbf{f}_i \in \mathcal{M}_k(\mathfrak{c}, \psi_i)$ with Hecke characters ψ_i of $F_{\mathbf{A}}^\times$. Again the components of $\mathbf{f}_i$ belong to $\mathcal{L}(K, \varphi, \overline{\mathbf{Q}})$. Decomosing $(g, 0, \ldots, 0)$ in a similar way, we can reduce the question to the case we have already proved. This completes the proof.

It is conjecturable that Theorem 5.6 is true without the assumption of $\mathfrak{r}_+^\times$-invariance of f or g. Also, the above theorems concern only the rationality with K'_{ab} as the basic field. We can actually refine them by analyzing the behavior of the quantities X, E, and S under automorhisms over K'. In this paper, however, we have contented ourselves with weaker results in order to simplify the treatment.

REFERENCES

1. H. KLINGEN, Über den arithmetischen Charakter der Fourier-koeffizienten von Modulformen, Math. Ann. **147** (1962), 176–188.
2. H. D. KLOOSTERMANN, Theorie der Eisensteinschen Reihen von mehreren Veränderlichen, Abh. Math. Sem. Hamburg **6** (1928), 163–188.
3. Y. MATSUSIMA AND G. SHIMURA, On the cohomology groups attached to certain vector valued differential forms on the product of the upper half planes, Ann. of Math. **78** (1963), 417–449.
4. T. MIYAKE, On automorphic forms on GL_2 and Hecke operators, Ann. of Math. **94** (1971), 174–189.
5. G. SHIMURA, On Dirichlet series and abelian varieties attached to automor-

phic forms, Ann. of Math. **76** (1962), 237–294.

6. _____ , On canonical models of arithmetic quotients of bounded symmetric domains, Ann. of Math. **91** (1970), 144–222.

7. _____ , Introduction to the arithmetic theory of automorphic functions, Publ. Math. Soc. Japan, No.11, Iwanami Shoten and Princeton Univ. Press, 1971.

8. _____ , On modular forms of half integral weight, Ann. of Math. **97** (1973), 440–481.

9. _____ , On some arithmetic properties of modular forms of one and several variables, Ann. of Math. **102** (1975), 491–515.

10. _____ , On the Fourier coefficients of modular forms of several variables, Göttingen Nachr. Akad. Wiss. Math.-Phys. Klasse, 1975, 261–268.

11. _____ , Theta functions with complex multiplication, Duke Math. J. **43** (1976), 673–696.

12. _____ , The special values of the zeta functions associated with cusp forms, Comm. pure appl. Math. **29** (1976), 783–804.

13. _____ , On the derivatives of theta functions and modular forms, Duke Math. J. **44** (1977), 365–387.

14. _____ , On the periods of modular forms, Math. Ann. **229** (1977), 211–221.

15. _____ , On certain reciprocity–laws for theta functions and modular forms, Acta math. **141** (1978), 35–71.

16. C. L. SIEGEL, The volume of the fundamental domain for some infinite groups, Trans. Amer. Math. Soc. **39** (1936), 209–218

17. J. STURM, Special values of zeta functions and Eisenstein series of half integral weight, Thesis, Princeton University 1977, to be published.

Automorphic forms and the periods
of abelian varieties

Journal of the Mathematical Society of Japan, 31 (1979), 561-592

There are three interrelated topics to be treated in this paper:

I. Monomial relations between the periods of abelian varieties with complex multiplication;

II. The derivatives of automorphic forms of arithmetic type;

III. The non-vanishing of the first cohomology group of a discrete subgroup of $SU(n, 1)$.

To describe our results, let A be an abelian variety of dimension g defined over $\bar{Q}$, whose endomorphism-algebra contains a totally imaginary quadratic extension K of a totally real algebraic number field F such that $[F:Q]=g$. Throughout the paper, we denote by $\bar{Q}$ the algebraic closure of the rational number field Q embedded in the complex number field C. Let Φ be the representation of K on the space of holomorphic 1-forms on A. Then Φ consists of g injections τ of K into C, and for each τ, there is a $\bar{Q}$-rational 1-form ω_τ on A on which an element x of K acts as a scalar x^τ. As shown in [13], there is a constant $p(\tau, \Phi)$ such that

$$(0.1) \qquad \int_c \omega_\tau \sim \pi \cdot p(\tau, \Phi) \qquad \text{for all 1-cycles } c \text{ on } A.$$

Here and henceforth, we write $a \sim b$ for two complex numbers a and b if $a/b \in \bar{Q}$. The first principal aim of this paper is to prove monomial relations between $p(\tau, \Phi)$ for various Φ with the same K, as well as relations between such "periods" for a given K and those for an extension of K. The constant $p(\tau, \Phi)$ can be obtained from a Hilbert modular function f with respect to a congruence subgroup of $GL_2(F)$ as follows. Take the variable $z=(z_1, \cdots, z_g)$ on the product $\mathfrak{H}^g$ of g copies of the upper half plane $\mathfrak{H}_1$. Let $\Phi=\sum_{\nu=1}^g \tau_\nu$ and $w_0=(w^{\tau_1}, \cdots, w^{\tau_g})$ with an element w of K such that $\mathrm{Im}(w^{\tau_\nu})>0$ for all ν. Then

$$(0.2) \qquad (\partial f/\partial z_\nu)(w_0) \sim \pi \cdot p(\tau_\nu, \Phi)^2 \qquad (\nu=1, \cdots, g),$$

if f is $\bar{Q}$-rational, i.e., if f is the quotient of two Hilbert modular forms with

---

* Supported by NSF Grant MCS76-11376.

Fourier coefficients in $\bar{Q}$ (see [13]). Now we can speak of $\bar{Q}$-rational automorphic functions on $\mathfrak{H}_1^g$ for a more general type of algebraic groups of which $GL_2(F)$ is a special case. The generalization of (0.2) to such functions will be our second main subject.

As the third topic, which is a sort of by-product, we show the existence of a discrete subgroup $\varGamma$ of $SU(n, 1)$ such that $H^1(\varGamma\backslash\mathfrak{D}_n, \boldsymbol{R})\neq\{0\}$, where

$$\mathfrak{D}_n=\{(z_1, \cdots, z_n)\in C^n\,|\,\textstyle\sum_{k=1}^n|z_k|^2<1\}.$$

Such a non-vanishing was first proved by Kazhdan in [3]. Subsequently Wallach obtained more general results for $SU(p, q)$ in the case of compact quotient [1, Ch. VIII].[1]

Our results will be obtained by considering a family of abelian varieties of dimension mg with the above K as a subalgebra of the endomorphism-algebra. This is parametrized by the points of a symmetric domain $\mathfrak{D}$ of which $\mathfrak{D}_n$ is a special case. We embed $\mathfrak{D}$ into the Siegel upper half space $\mathfrak{H}_{mg}$ of degree mg, and pull back a certain modular form on $\mathfrak{H}_{mg}$ to $\mathfrak{D}$, which gives rise to a non-vanishing holomorphic 1-form on $\varGamma\backslash\mathfrak{D}_{m-1}$ when $\mathfrak{D}=\mathfrak{D}_{m-1}$. The study of the first two topics are made by analysing the pull-back of another type of (meromorphic) modular form at CM-points. This combined with our previous results concerning 1-forms on A in [13] yields the main theorems on the monomial relations of $p(\tau, \varPhi)$ as well as those on the generalization of (0.2).

Obviously our investigation of the second topic admits of a further extension to the algebraic groups of a more general type; even for the groups in the present paper, no effort is expended to make the sharpest statement. The author believes, however, that our theorems and propositions are given to such an extent as to explain the basic ideas and indicate what their generalizations or analogues in other cases should be.

1. The main theorems on monomial relations between the periods.

We let an isomorphism σ of a field act on the right; thus x^σ denotes the image under σ of an element x of the field. Throughout the paper, we denote by ρ the complex conjugation. By a *CM-field*, we understand an algebraic number field K of finite degree with an automorphism σ of order 2 such that $\sigma\tau=\tau\rho$ for every injection τ of K into C. Such a σ is obviously unique for K; therefore we use the same letter ρ for σ; we also denote by K_0 the fixed subfield of K by ρ. Then K_0 is totally real and K is a totally

---

1) The reader who is interested only in this topic may dispense with §§ 2, 5, 6, and 7.

imaginary quadratic extension of K_0. Conversely, such an extension is a *CM*-field.

Given an algebraic number field K of finite degree, we denote by I_K (resp. $I_K(Q)$) the module of all formal linear combinations $\Phi=\sum_\tau c_\tau \tau$ with $c_\tau \in Z$ (resp. $c_\tau \in Q$) of the injections τ of K into C. If $c_\tau \geq 0$ for all τ, Φ may be considered an equivalence class of representations of K by complex matrices; we then put $\operatorname{tr}\Phi(a)=\sum_\tau c_\tau a^\tau$ for $a\in K$. We call (K, Φ) a *CM-type* if K is a *CM*-field and the sum of Φ and its complex conjugate $\Phi\rho$ is the class of regular representations of K over Q. We shall also say that Φ is a *CM-type of K*.

Let A be an abelian variety defined over C, and ι an injection of K into $\operatorname{End}(A)\otimes Q$. We say that (A, ι) *is of type* (K, Φ) if the representation of K through ι on the space of holomorphic 1-forms on A belongs to Φ. Suppose (K, Φ) is a *CM*-type and (A, ι) is of type (K, Φ) and defined over $\bar{Q}$. Let $[K:Q]=2g$ and $\Phi=\sum_{\nu=1}^{g}\tau_\nu$. For each ν, there is a $\bar{Q}$-rational holomorphic 1-form ω_ν on A satisfying $\omega_\nu \circ \iota(a)=a^{\tau_\nu}\omega_\nu$ for all $a\in K$ such that $\iota(a)\in\operatorname{End}(A)$; ω_ν is unique up to algebraic factors. In [13, Remark 3.4], we have proved:

PROPOSITION 1.1. *There exists a non-zero complex number $p(\tau_\nu, \Phi)$ depending only on K, Φ, and τ_ν such that $\int_c \omega_\nu \sim \pi \cdot p(\tau_\nu, \Phi)$ for all 1-cycles c on A.*

The constant $p(\tau_\nu, \Phi)$ is determined up to algebraic factors. If the multiplicity of τ in Φ is 0, (i. e., if $\tau\notin\{\tau_1, \cdots, \tau_g\}$,) we simply define $p(\tau, \Phi)$ to be 1. Now our first main result, in a (seemingly) weaker form, can be stated as

THEOREM 1.2. *Let $\Phi_1, \cdots, \Phi_m$ be CM-types of a CM-field K, and τ an injection of K into C. Then the number*

(1.1) $$\prod_{j=1}^{m}[p(\tau, \Phi_j)/p(\tau\rho, \Phi_j)],$$

up to algebraic factors, depends only on τ and $\Phi_1+\cdots+\Phi_m$. Moreover, let (L, Ψ) be a CM-type such that $K\subset L$ and the restriction of Ψ to K is $\Phi_1+\cdots+\Phi_m$. Let $q_K(\tau, \Psi)$ denote the product of $p(\sigma, \Psi)$ for all injections σ of L into C which coincide with τ on K. Then $q_K(\tau, \Psi)/q_K(\tau\rho, \Psi)$ differs from (1.1) by an algebraic factor.

The proof will be given in § 5.

Let I_K^0 (resp. $I_K^0(Q)$) denote the submodule of I_K (resp. $I_K(Q)$) consisting of all $\sum_\tau c_\tau \tau$ such that $c_\tau + c_{\tau\rho}$ does not depend on τ. Notice that if Ψ is a *CM*-type of an extension of K, the restriction of Ψ to K belongs to I_K^0. Now we state the stronger form of the above theorem as

THEOREM 1.3. *Let $\Phi_1, \cdots, \Phi_m$ be CM-types of K and suppose $\sum_{i=1}^{m}r_i\Phi_i=0$ with $r_i\in Q$. Then, for every injection τ of K into C, we have*

$$\prod_{i=1}^{m}[p(\tau, \Phi_i)/p(\tau\rho, \Phi_i)]^{r_i} \sim 1 .$$

Moreover, let Ψ be a CM-type of an extension L of K. Suppose the restriction of Ψ to K can be written $\sum_{i=1}^{m} s_i \Phi_i$ with $s_i \in Q$. Then, for every injection τ of K into C, we have

$$q_K(\tau, \Psi)/q_K(\tau\rho, \Psi) \sim \prod_{i=1}^{m}[p(\tau, \Phi_i)/p(\tau\rho, \Phi_i)]^{s_i} ,$$

where $q_K(\tau, \Psi)$ is defined as in Theorem 1.2.

PROOF. The first assertion can be derived from Theorem 1.2 by expressing the equality $\sum_i r_i \Phi_i = 0$ as $\sum_i a_i \Phi_i = \sum_i b_i \Phi_i$ with positive integers a_i and b_i. To prove the second assertion, observe that the restriction of Ψ to K can be written $\sum_{k=1}^{n} \Phi_k'$ with CM-types Φ_k' of K. Take positive integers f, c_i and d_i so that $fs_i = c_i - d_i$. Then $f \sum_k \Phi_k' + \sum_i d_i \Phi_i = \sum_i c_i \Phi_i$, so that the desired result follows from Theorem 1.2.

Let us now insert here a few propositions which can easily be derived from our definition.

PROPOSITION 1.4. *Let (K, Φ) and (L, Ψ) be CM-types such that $K \subset L$ and the restriction of Ψ to K is $[L : K]$ times Φ. Then $p(\tau, \Phi) \sim p(\sigma, \Psi)$ if τ is the restriction of σ to K.*

PROOF. Let (A, ι) and (A', ι') be of type (K, Φ) and (L, Ψ), respectively. By [7, Proposition 14], A' is isogenous to the product of $[L : K]$ copies of A. Our assertion follows immediately from this fact.

PROPOSITION 1.5. *Let (K, Φ) be a CM-type, and σ an isomorphism of a field K_1 onto K. Then $p(\sigma\tau, \sigma\Phi) \sim p(\tau, \Phi)$, where we let σ act on I_K on the left in an obvious way.*

This is obvious.

The nature of the periods of the anti-holomorphic 1-forms $\bar{\omega}_\nu$ can be seen from

PROPOSITION 1.6. $p(\tau, \Phi)^\rho \sim p(\tau, \Phi)$.

PROOF. As shown in [13, pp. 381–383], the number $p(\tau, \Phi)$ can be obtained as the value $h(w_0)$ of a quotient $h = f_1/f_2$ of two Hilbert modular forms f_1 and f_2, with Fourier coefficients in K_0^r, at a point $w_0 = (w^{\tau_1}, \cdots, w^{\tau_g})$ with $w \in K$ such that $\mathrm{Im}(w^{\tau_\nu}) > 0$ for all τ_ν. We can take w to be pure imaginary. Then $h(w_0)$ is real.

We have so far been interested in $p(\tau, \Phi)$ only up to algebraic factors. Let us now refine this point by considering them and their products up to factors belonging to smaller fields. For simplicity, let us assume that all number fields in the following discussion are subfields of C, and denote by K_{ab} the maximal abelian extension of a number field K. For each $\Phi \in I_K$, let K^Φ denote the field generated over Q by $\mathrm{tr}\, \Phi(a)$ for all $a \in K$. Let M be the Galois closure of K over Q, and let $G = \mathrm{Gal}(M/Q)$ and $H = \mathrm{Gal}(M/K)$. Then I_K

can be identified with the $\mathbf{Z}$-module generated by the cosets $H\alpha$ with $\alpha\in G$; G acts on I_K on the right. It is easy to see that

$$\mathrm{Gal}(M/K^{\Phi})=\{\gamma\in G\,|\,\Phi\gamma=\Phi\}\qquad(\text{cf. [15, §8.3]}).$$

CONJECTURE 1.7. *Let τ be an injection of a CM-field K into $\mathbf{C}$, and $\Phi\in I_K^0$. Then we can specify a non-zero complex number $P(\tau,\Phi)$, up to factors belonging to $(K^{\Phi})_{ab}K_0^{\tau}$, with the property that if $\Phi=\sum_{i=1}^{m}s_i\Phi_i$ with CM-types Φ_i of K and $s_i\in\mathbf{Z}$, then*

$$P(\tau,\Phi)\sim\prod_{i=1}^{m}[p(\tau,\Phi_i)/p(\tau\rho,\Phi_i)]^{s_i}.$$

We now show that this conjecture can be reduced to a certain special case. Let $[K:\mathbf{Q}]=2g$ and let Ω denote the sum of all injections of K into $\mathbf{C}$. Every element Φ of I_K^0 can be expressed as $\Phi=a\Omega+\Xi$ with $a\in\mathbf{Z}$ and

$$\Xi=\sum_{\nu=1}^{g}[a_{\nu}\tau_{\nu}+(n-a_{\nu})\tau_{\nu}\rho],$$

where the τ_{ν} are g injections which form a CM-type of K, and the a_{ν} are non-negative integers such that $n-a_g\leq a_g\leq\cdots\leq a_1=n$. Such n is uniquely determined by Φ. In view of Proposition 1.5, we can put $P(\tau,\Phi)=P(\tau,\Xi)$. Let us prove:

(1.2) *Conjecture 1.7 is true if either n is odd or $n=0$;*

(1.3) *Conjecture 1.7 is true if it is true for $n=2$.*

First if $n=0$, we can put $P(\tau,\Xi)=1$. If $n=1$, Ξ is a CM-type, and $P(\tau,\Xi)$ must be $\sim p(\tau_{\nu},\Xi)$ or $p(\tau_{\nu},\Xi)^{-1}$ according as $\tau=\tau_{\nu}$ or $\tau=\tau_{\nu}\rho$. Now $p(\tau_{\nu},\Xi)$ can be given as the value $h(w)$ as explained in the proof of Proposition 1.6, which settles the problem, since the quotient of two such values belong to $(K^{\Xi})_{ab}K_0^{\tau\nu}$. If $n=2$, Ξ has the form

$$(1.4)\qquad\Xi=\sum_{\nu=1}^{r}2\tau_{\nu}+\sum_{\nu=r+1}^{g}(\tau_{\nu}+\tau_{\nu}\rho)$$

with $0<r\leq g$. Let us assume that one can specify $P(\tau,\Xi)$ for such Ξ as in our conjecture. Suppose $n>2$. If $a_g>n/2$ (which is the case when n is odd), Ξ can be written

$$\Xi=\sum_{j=1}^{t}[b_j\xi_j+(n-b_j)\xi_j\rho]$$

with partial sums $\xi_1,\cdots,\xi_t$ of $\sum_{\nu=1}^{g}\tau_{\nu}$ such that $\sum_{j=1}^{t}\xi_j=\sum_{\nu=1}^{g}\tau_{\nu}$ and positive integers b_j such that $n-b_t<b_t<\cdots<b_1=n$. Then we can find CM-types $\Phi_1,\cdots,\Phi_n$ such that $\Xi=\sum_{i=1}^{n}\Phi_i$ and each Φ_i has an expression $\Phi_i=\xi_1+\cdots+\xi_s+\xi_{s+1}\rho+\cdots+\xi_t\rho$. Observe that if $\Xi\gamma=\Xi$ for an element $\gamma\in G$, then $\xi_j\gamma=\xi_j$ for every j, and hence $\Phi_i\gamma=\Phi_i$ for every i. This implies that $K^{\Phi_i}\subset K^{\Xi}$. Therefore we can put $P(\tau,\Xi)=\prod_{i=1}^{n}P(\tau,\Phi_i)$. Next suppose $a_g=n/2$; then $n\geq 4$. We are going to define $P(\tau,\Xi)$ by induction on n. We observe that

$$\Xi=\sum_{\nu=1}^{r}[a_{\nu}\tau_{\nu}+(n-a_{\nu})\tau_{\nu}\rho]+(n/2)\sum_{\nu=r+1}^{g}(\tau_{\nu}+\tau_{\nu}\rho)$$

566   G. Shimura

with the integer r such that $a_{r-1}>a_r=\cdots=a_g$. Put

$$\Psi=\sum_{\nu=1}^{r}2\tau_\nu+\sum_{\nu=r+1}^{g}(\tau_\nu+\tau_\nu\rho)$$

and $\varXi=\Psi+\varXi'$. Then $\varXi\gamma=\varXi$ implies $\Psi\gamma=\Psi$ and $\varXi'\gamma=\varXi'$, and hence $K^\Psi K^{\varXi'}$ $\subset K^\varXi$. By induction, $P(\tau,\Psi)$ and $P(\tau,\varXi')$ are meaningful. Therefore we can put $P(\tau,\varXi)=P(\tau,\Psi)P(\tau,\varXi')$.

We shall prove in §7 that *if $\varXi$ is given by (1.4), $P(\tau_\nu,\varXi)$ can be specified as in our conjecture at least for $\nu\leq r$.* Thus the conjecture is reduced to the problem of finding $P(\tau_\nu,\varXi)$ for such $\varXi$ and for $\nu>r$.

2. Applications and examples.

The above theorems imply various interesting relations between the periods $p(\tau,\Phi)$. For simplicity, we assume in this section that all fields are subfields of C. Let us begin with the simplest case where K is an imaginary quadratic field. There are two *CM*-types (K,φ) and (K,ρ) with the identity map of K as φ. The period in these cases is given by means of an elliptic modular form h of weight 1 with algebraic Fourier coefficients. In fact, if w is an element of K such that $\mathrm{Im}(w)>0$ and $h(w)\neq0$, we have $p(\varphi,\varphi)\sim$ $p(\rho,\rho)\sim h(w)$ by [13, Remark 3.4]. As a special case of Theorem 1.2, we obtain

PROPOSITION 2.1. *Let (L,Ψ) be a CM-type. Suppose that L contains an imaginary quadratic field K and the restriction of Ψ to K is $r\varphi+s\rho$ with non-negative integers r and s, where φ denotes the identity map of K. Then $q_K(\varphi,\Psi)/q_K(\rho,\Psi)\sim h(w)^{r-s}$.*

A similar but somewhat different result has been obtained by Gross [2] with no mention of $p(\tau,\Phi)$ in the higher-dimensional case.

EXAMPLE 2.2. Suppose in particular $r=s$ in Proposition 2.1. Then we have $q_K(\varphi,\Psi)\sim q_K(\rho,\Psi)$ which is a non-trivial relation between the periods; this happens even when (L,Ψ) is primitive. For instance, let L_0 be a totally real extension of Q of degree 4 with no subfields of degree 2, and let $L=L_0K$ with an imaginary quadratic field K. We can define a *CM*-type $\Psi=\sum_{\nu=1}^{4}\tau_\nu$ of L so that $\tau_1=\tau_2=\varphi$ and $\tau_3=\tau_4=\rho$ on K. Then (L,Ψ) is primitive and

$$(2.1)\qquad\qquad p(\tau_1,\Psi)p(\tau_2,\Psi)\sim p(\tau_3,\Psi)p(\tau_4,\Psi).$$

Similar examples can be obtained for fields of higher degree.

We now study the periods in connection with the reflex of a given *CM*-type. First, for a fixed *CM*-field K of degree $2g$, take a pair of *CM*-types Φ_0 and Φ_1 of the forms

$$\Phi_0=\tau_1+\tau_2+\cdots+\tau_g,\qquad\Phi_1=\tau_1+(\tau_2+\cdots+\tau_g)\rho,$$

where τ_1 is the identity map of K, and for each injection τ of K into C, define an element $p_K(\tau)$ of $C^\times/\bar{Q}^\times$ by

$$(2.2) \qquad p_K(\tau) \sim \begin{cases} p(\tau_1, \Phi_0)p(\tau_1, \Phi_1) & \text{if } \tau = \tau_1, \\ 1/[p[\tau_1, \Phi_0)p(\tau_1, \Phi_1)] & \text{if } \tau = \rho, \\ p(\tau_i, \Phi_0)/p(\tau_i\rho, \Phi_1) & \text{if } \tau = \tau_i, \\ p(\tau_i\rho, \Phi_1)/p(\tau_i, \Phi_0) & \text{if } \tau = \tau_i\rho. \end{cases}$$

By Theorem 1.2, this definition does not depend on the choice of $\tau_2, \cdots, \tau_g$. Let Φ be an arbitrary CM-type of K, and (J, Ψ) the reflex of (K, Φ) in the sense of [11, 1.3] (i.e., *the dual* defined in [9, 5.1] and [15, 8.3]). Then $J = K^\Phi$. Let M be the Galois closure of K over Q and let $G = \mathrm{Gal}(M/Q)$. Consider Φ and Ψ as the sets of injections of K and J into M. For each $\sigma \in \Psi$ take $\gamma \in G$ which gives σ on J. Observe that $\Phi\gamma$ is a CM-type of K depending only on σ, since $\Phi\alpha = \Phi$ for $\alpha \in \mathrm{Gal}(M/J)$; therefore we write $\Phi\gamma$ also $\Phi\sigma$.

THEOREM 2.3. *If (J, Ψ) is the reflex of (K, Φ) and $\sigma \in \Psi$, we have*

$$p(\sigma, \Psi)^2 \sim \Pi_{\tau \in \Phi\sigma} p_K(\tau).$$

PROOF. Observe that $(J^\sigma, \sigma^{-1}\Psi)$ is the reflex of $(K, \Phi\sigma)$. By Proposition 1.5, we have $p(\sigma, \Psi) \sim p(1, \sigma^{-1}\Psi)$. Therefore, taking $\Phi\sigma$ and $\sigma^{-1}\Psi$ in place of Φ and Ψ, it is sufficient to prove

$$(2.3) \qquad p(1, \Psi)^2 \sim \Pi_{\tau \in \Phi} p_K(\tau)$$

when $1 \in \Phi$ and $1 \in \Psi$. Since M is a CM-field, we can define $p_M(\alpha)$ for each $\alpha \in G$. We consider ρ an element of G, which is actually contained in the center. For a subset X of G, we use the same letter X for the element of I_M which is the sum of the elements of X. Let us now prove

$$(2.4) \qquad \Pi_{\alpha \in X} p_M(\alpha) \sim p(1, 1 + X^{-1} + Y)/p(1, 1 + X^{-1}\rho + Y)$$

if X and Y are subsets of G such that $1 + X^{-1} + Y$ is a CM-type of M, where $X^{-1} = \{\gamma^{-1} \mid \gamma \in X\}$. This is obvious if X is empty. Assume (2.4) is true for X and let $W = X \cup \{\beta\}$; suppose $1 + W^{-1} + Z$ is a CM-type. Then we have

$$p_M(\beta) \sim p(\beta, 1 + \beta + \beta X^{-1} + \beta Z)/p(\beta\rho, 1 + \beta\rho + \beta X^{-1}\rho + \beta Z\rho)$$

$$\sim p(1, 1 + W^{-1} + Z)/p(1, 1 + X^{-1} + \beta^{-1}\rho + Z).$$

This multiplied by (2.4) with $Y = Z + \beta^{-1}\rho$ proves the case of $X \cup \{\beta\}$; thus the proof of (2.4) is completed by induction. Let $H = \mathrm{Gal}(M/K)$. Then $p_K(H\beta)$ for $\beta \in G$ is meaningful. Now we have

$$(2.5) \qquad p_K(H\beta) \sim \Pi_{\alpha \in H\beta} p_M(\alpha).$$

In fact, if $\beta \notin H$, we take a CM-type $\sum_{\nu=1}^{g} \tau_\nu$ of K so that $\tau_1 = \mathrm{id}_K$ and τ_2 is represented by $H\beta$. Let $H\beta_\nu$ represent τ_ν for $\nu > 1$ with $\beta = \beta_2$. Then

$$p_K(\tau_2) \sim p(\tau_2, \tau_1 + \tau_2 + \cdots + \tau_g) / p(\tau_2\rho, \tau_1 + \tau_2\rho + \cdots + \tau_g\rho)$$

$$\sim p(\beta, H + H\beta_2 + \cdots + H\beta_g) / p(\beta\rho, H + H\beta_2\rho + \cdots + H\beta_g\rho)$$

by Proposition 1.4. Define Y by $\beta Y + \beta = H\beta_2 + \cdots + H\beta_g$. Then

$$p_K(\tau_2) \sim p(\beta, H + \beta + \beta Y) / p(\beta\rho, H + \beta\rho + \beta Y\rho)$$

$$\sim p(1, 1 + \beta^{-1}H + Y) / p(1, 1 + \beta^{-1}H\rho + Y) \sim \prod_{\alpha \in H\beta} p_M(\alpha)$$

by (2.4); this proves (2.5) in the case $\beta \notin H$. To prove the case $H\beta = H$, put $A = H - 1$ and $B = H\beta_2 + \cdots + H\beta_g$. Then

$$\prod_{\alpha \in H} p_M(\alpha) \sim p_M(1) \prod_{\alpha \in A} p_M(\alpha)$$

$$\sim p(1, 1 + A + B\rho) p(1, 1 + A\rho + B) p(1, 1 + A + B) / p(1, 1 + A\rho + B) \quad \text{(by (2.4))}$$

$$\sim p(1, H + B) p(1, H + B\rho)$$

$$\sim p(\tau_1, \tau_1 + \cdots + \tau_g) p(\tau_1, \tau_1 + (\tau_2 + \cdots + \tau_g)\rho) \quad \text{(by Proposition 1.4)}$$

$$\sim p_K(\tau_1).$$

This completes the proof of (2.5). Now let Φ be represented by $H\gamma_1, \cdots, H\gamma_g$, and let $S = H\gamma_1 + \cdots + H\gamma_g = 1 + X$. By (2.4) and (2.5),

$$\prod_{\tau \in \Phi} p_K(\tau) \sim \prod_{\alpha \in S} p_M(\alpha) \sim p_M(1) \prod_{\alpha \in X} p_M(\alpha)$$

$$\sim p(1, 1 + X^{-1}) p(1, 1 + X^{-1}\rho) p(1, 1 + X^{-1}) / p(1, 1 + X^{-1}\rho)$$

$$\sim p(1, 1 + X^{-1})^2 \sim p(1, S^{-1})^2.$$

Now the restriction of S^{-1} to J is exactly $[M:J]\Psi$ according to the definition of the reflex. Thus $p(1, S^{-1}) \sim p(1, \Psi)$ by Proposition 1.4, and we obtain (2.2). This completes the proof.

REMARK 2.4. Let $\mathfrak{E}_K$ denote the set of all CM-types of K. Then $\mathfrak{E}_K$ has 2^g elements and G acts on $\mathfrak{E}_K$ on the right. Let $\{\Phi_1, \cdots, \Phi_s\}$ be a complete set of representatives for $\mathfrak{E}_K/G$, and let (J_i, Ψ_i) be the reflex of (K, Φ_i). Then $\{\Phi_i\gamma \mid \gamma \in G\}$ contains exactly $[J_i : Q]$ elements, so that $\sum_{i=1}^{s}[J_i : Q] = 2^g$. Since the reflex of $(K, \Phi_i\gamma)$ is $(J_i^\gamma, \gamma^{-1}\Psi_i)$, the CM-types (J_i, Ψ_i), for $i = 1, \cdots, s$, "represents" the reflexes of all the CM-types of K. Observe that $p_K(\tau\rho) = p_K(\tau)^{-1}$. Therefore the above theorem shows that the 2^{g-1} numbers $p(\sigma, \Psi_i)^2$ are monomials of g numbers $p_K(\tau_1), \cdots, p_K(\tau_g)$ and their inverses. As shown in [11, 1.9, 1.10], it can happen that $s = 1$ and $[J_1 : Q] = 2^g$. In any case, at most g of the numbers $p(\sigma, \Psi_i)$ can be algebraically independent. We can refine this point as follows.

THEOREM 2.5. *Let $\Psi=\sum_{\lambda=1}^{h}\sigma_\lambda$ be a CM-type of a CM-field J of degree $2h$, and M the Galois closure of J over $\mathbf{Q}$. Further let $t(\Psi)$ be the dimension of the svbspace of $I_J^0(\mathbf{Q})$ generated over $\mathbf{Q}$ by $\Psi\gamma$ for all $\gamma\in\mathrm{Gal}(M/\mathbf{Q})$, and let $\delta(\Psi)=h+1-t(\Psi)$. Then the module*

$$\{(x_1,\cdots,x_h)\in\mathbf{Z}^h\,|\,\textstyle\prod_{\lambda=1}^{h}p(\sigma_\lambda,\Psi)^{x_\lambda}\sim1\}$$

has rank at least $\delta(\Psi)$.

PROOF. We note that $\delta(\Psi)\geqq0$, since $I_J^0(\mathbf{Q})$ has dimension $h+1$. Let (K,Φ) be the reflex of (J,Ψ), and (J',Ψ') the reflex of (K,Φ). Then the restriction of Ψ to J' is $[J:J']\Psi'$. Obviously $t(\Psi)=t(\Psi')$; therefore Proposition 1.4 reduces the problem to (J',Ψ'); so we assume $J=J'$. Now by Kubota [17], we have

$$(2.6)\qquad\qquad\qquad t(\Psi)=t(\Phi).$$

(This is non-trivial, but can be proved in an elementary way.) Changing Φ for $\Phi\rho$ if necessary, we may assume that $\rho\notin\Phi$. Let W be the subspace of $I_K^0(\mathbf{Q})$ generated by $\Phi\sigma$ for all $\sigma\in\Psi$. Since $\Phi\gamma+\Phi\gamma\rho=\Phi+\Phi\rho$, we see that $\Phi\gamma\in W+\mathbf{Q}\Phi\rho$ for all $\gamma\in G$. Observe that $\Phi\rho\notin W$. Therefore W has dimension $t(\Phi)-1$. If $\sum_{\lambda=1}^{h}x_\lambda\Phi\sigma_\lambda=0$, then Theorem 2.3 implies that

$$\textstyle\prod_{\lambda=1}^{h}p(\sigma_\lambda,\Psi)^{2x_\lambda}\sim1.$$

This proves our theorem.

REMARK 2.6. (a) Since $t(\Psi)=t(\Psi')\leqq\frac{1}{2}[J':\mathbf{Q}]+1$, we have $\delta(\Psi)=0$ only when Ψ is primitive. The converse is not true, however.

(b) If J is a Galois extension of $\mathbf{Q}$, it can easily be seen that $t(\Psi)$ is the dimension of the space generated over $\mathbf{Q}$ by $\gamma\Psi$ for all $\gamma\in\mathrm{Gal}(J/\mathbf{Q})$.

PROPOSITION 2.7. *Let (K,Φ) and (L,Ψ) be CM-types. Suppose that $K\subset L$, $\delta(\Psi)=0$, and L is a Galois extension of $\mathbf{Q}$. Then, for each injection τ of K into $\mathbf{C}$, we have*

$$(2.7)\qquad\qquad p(\tau,\Phi)\sim\textstyle\prod_{\alpha\in G}p(\alpha,\Psi)^{b_\alpha}$$

with $b_\alpha\in\mathbf{Q}$, where $G=\mathrm{Gal}(L/\mathbf{Q})$.

PROOF. There is a CM-type Ψ_1 of L whose restriction to K is $[L:K]\Phi$. Take $\beta\in G$ which gives τ on K. If $\delta(\Psi)=0$, $I_L^0(\mathbf{Q})$ is generated by $\alpha\Psi$ for all $\alpha\in G$, so that $\Psi_1=\sum_{\alpha\in G}c_\alpha\alpha\Psi$ with $c_\alpha\in\mathbf{Q}$. Then, for $\tau\in\Phi$,

$$p(\tau,\Phi)\sim p(\beta,\Psi_1)\sim\textstyle\prod_{\alpha\in G}[p(\beta,\alpha\Psi)/p(\beta\rho,\alpha\Psi)]^{c_\alpha}$$

$$\sim\textstyle\prod_{\alpha\in G}[p(\alpha^{-1}\beta,\Psi)/p(\alpha^{-1}\beta\rho,\Psi)]^{c_\alpha}$$

by Theorem 1.3 and Proposition 1.5. This proves our assertion.

REMARK 2.8. (a) It often happens that I_L^0 is generated over $\mathbf{Z}$ by $\alpha\Psi$ for all $\alpha \in G$. In such a case, we have (2.7) with $b_\alpha \in Z$.

(b) The above proposition includes the case $K=L$. Therefore, when L is normal over Q and $\delta(\Psi)=0$, the quantities $p_L(\alpha)$ are algebraically dependent on $p(\alpha, \Psi)$ and vice versa.

Let us now assume that K is an imaginary cyclic extension of Q of degree $2g$. It can easily be seen that if $\Phi=\sum_{\nu=0}^{g-1}\sigma^\nu$ with a generator σ of $\mathrm{Gal}(K/Q)$, we have $\delta(\Phi)=0$, and $\sigma^\nu\Phi$ for $\nu=0, 1, \cdots, g$ form a basis of I_K^0 over Z. Let (L, Ψ) be a CM-type such that $K \subset L$. Then the restriction of Ψ to K belongs to I_K^0, and hence can be written $\sum_{\nu=0}^{g}s_\nu\sigma^\nu\Phi$ with $s_\nu \in Z$. By Theorem 1.3 and Proposition 1.5, we have

$$(2.8) \qquad q_K(\sigma^\lambda, \Psi)/q_K(\sigma^{\lambda+g}, \Psi) \sim \prod_{\nu=0}^{g}[p(\sigma^{\lambda-\nu}, \Phi)/p(\sigma^{\lambda+g-\nu}, \Phi)]^{s_\nu}$$

$$(\lambda=0, 1, \cdots, g-1).$$

If the restriction of Ψ to K is $\sum_{\nu=0}^{2g-1}r_\nu\sigma^\nu$ with $0 \leq r_\nu \in Z$, we have $s_0=r_0$, $s_1=r_1-r_0, \cdots, s_{g-1}=r_{g-1}-r_{g-2}$, and $s_g=r_g-r_{g-1}+r_0$. Consider (2.8) as "linear equations" in the variables $p(\sigma^\mu, \Phi)$ for $\mu=0, 1, \cdots, g-1$. The determinant of the coefficients can be given in terms of r_ν; if it is non-vanishing, we can express $p(\sigma^\mu, \Phi)$ by means of $q_K(\sigma^\lambda, \Psi)$. For example, if $g=2$, we obtain

PROPOSITION 2.9. *Let K be an imaginary cyclic extension of Q of degree 4, and $\Phi=1+\sigma$ with a generator σ of $\mathrm{Gal}(K/Q)$. Let (L, Ψ) be a CM-type such that $K \subset L$, and let the restriction of Ψ to K be $\sum_{\nu=0}^{3}r_\nu\sigma^\nu$ with $0 \leq r_\nu \in Z$. Suppose $r_0 \neq r_1$ or $r_1 \neq r_2$, and let*

$$a=(r_1-r_2)/[(r_1-r_2)^2+(r_0-r_1)^2], \quad b=(r_0-r_1)/[(r_1-r_2)^2+(r_0-r_1)^2].$$

Then we have

$$p(1, \Phi) \sim q_K(1, \Psi)^a q_K(\sigma^2, \Psi)^{-a} q_K(\sigma, \Psi)^{-b} q_K(\sigma^3, \Psi)^b,$$

$$p(\sigma, \Phi) \sim q_K(1, \Psi)^b q_K(\sigma^2, \Psi)^{-b} q_K(\sigma, \Psi)^a q_K(\sigma^3, \Psi)^{-a}.$$

Observe that $r_0+r_2=r_1+r_3=[L:K]$. Therefore if $[L:K]$ is odd, we have $r_0 \neq r_1$ or $r_1 \neq r_2$.

Coming back to the case of an arbitrary degree, if equations (2.8) are not independent, we obtain non-trivial relations among $p(\tau, \Psi)$.

EXAMPLE 2.10. Suppose that L is a cyclic extension of Q of degree 18, $[L:K]=3$, and τ a generator of $\mathrm{Gal}(L/Q)$ which coincides with σ on K. Put $\Psi=1+\tau^{10}+\sum_{\nu=2}^{8}\tau^\nu$. Then (L, Ψ) is primitive. In this case, the right-hand sides of (2.8) for $\lambda=0$ and $\lambda=2$ are the same; for $\lambda=1$, we obtain its inverse. Therefore

$$p(1, \Psi)p(\tau^6, \Psi)/p(\tau^3, \Psi) \sim p(\tau^4, \Psi)p(\tau^{10}, \Psi)/p(\tau^7, \Psi)$$

$$\sim p(\tau^2, \Psi)p(\tau^8, \Psi)/p(\tau^5, \Psi) \sim p(1, \Phi)p(\sigma^2, \Phi)/p(\sigma, \Phi).$$

In fact, we can easily verify that $\delta(\Psi)=2$, and these relations are those guaranteed by Theorem 2.5 and its proof.

The relations given by Proposition 2.7 can be used to express the periods in terms of the values of the gamma-function. For example, let $L=Q(\zeta)$, $\zeta=\exp(2\pi i/l)$ with an odd prime l. As shown by Weil [16], if Ψ is the CM-type of L obtained from a curve of the form $y^l=x^a(1-x)$, $p(\alpha, \Psi)$ is the product of an algebraic number, a power of π, and a monomial of $\Gamma(t/l)$ for $t=1, \cdots, l-1$. It was shown by Kubota [17] that $\delta(\Psi)=0$ for $a=1$. Therefore Proposition 2.7 guarantees an expression for $p(\tau, \Phi)$ in terms of $\Gamma(t/l)$ for every CM-type Φ of every imaginary subfield of $Q(\zeta)$.

More CM-types of cyclotomic fields can be obtained from the factors of the jacobian of the curve $x^d+y^e=1$ with arbitrary positive integers d and e. The periods of abelian integrals in the case $d=e$ have been computed by Rohrlich in [2, Appendix]. It seems that Theorem 1.3 and the idea in the proof of Proposition 2.7, together with a careful analysis of these CM-types, will lead to a general result about an expression of $p(\tau, \Phi)$ as a monomial of the values of Γ for an arbitrary CM-type Φ of an arbitrary imaginary cyclotomic field, which will give a generalization of the formula of Chowla and Selberg in [6]. In the cases in which the periods are expressed by $\Gamma(t/l)$, our monomial relations can often be checked directly by means of the known relations of Γ such as $\Gamma(s)\Gamma(1-s)\sim\pi$ for $s\in Q$. Probably this is always so in such cyclotomic cases. In other words, it is unlikely that our relations produce new algebraic relations between the values of Γ.

Finally we mention that $p(\tau, \Phi)$ is closely connected with the values of L-functions of a CM-field with Hecke characters, as shown by [12, Theorem 2].

3. The symmetric domain of a unitary group.

To simplify our notation, we denote by $\mathrm{diag}[X_1, \cdots, X_m]$ the square matrix with square matrices $X_1, \cdots, X_m$ on the diagonal blocks and with 0 in all other blocks. If S is a ring, we let S_q^p denote the module of all $p\times q$ matrices with entries in S; we put $S^p=S_1^p$ and $M_p(S)=S_p^p$. For a complex hermitian matrix H, we write $H>0$ if H is positive definite. Given two non-negative integers r and s, put $J_{r,s}=\mathrm{diag}[1_r, -1_s]$, $m=r+s$, where 1_r denotes the identity matrix of degree r, and define a unitary group $U(r, s)$ by

$$(3.1) \qquad U(r, s)=\{\alpha\in GL_m(C)\,|\,{}^t\bar{\alpha}J_{r,s}\alpha=J_{r,s}\} ;$$

we understand that $J_{r,s}=1_m$ if $rs=0$. The quotient of $U(r, s)$ modulo a maximal compact subgroup is isomorphic to a bounded symmetric domain $\mathfrak{D}(r, s)$ defined by

$$(3.2) \qquad \mathfrak{D}(r, s) = \{ z \in C^r_s \mid 1_r - z \cdot {}^t\bar{z} > 0 \} \, .$$

To define the action of $U(r, s)$ on $\mathfrak{D}(r, s)$, let $\mathfrak{X}$ be the set of all elements Y of $GL_m(C)$ such that ${}^t\bar{Y} J_{r,s} Y = \mathrm{diag}[A, -B]$ with $0 < A = {}^t\bar{A} \in C^r_r$ and $0 < B = {}^t\bar{B} \in C^s_s$. Put

$$(3.3) \qquad Y(z) = \begin{bmatrix} 1_r & z \\ {}^t\bar{z} & 1_s \end{bmatrix} \qquad (z \in \mathfrak{D}(r, s)) \, .$$

It can easily be seen that

$$(z, u, v) \mapsto Y(z) \begin{pmatrix} u & 0 \\ 0 & v \end{pmatrix}$$

gives a bijective map of $\mathfrak{D}(r, s) \times GL_r(C) \times GL_s(C)$ onto $\mathfrak{X}$. Obviously $\alpha Y \in \mathfrak{X}$ if $\alpha \in U(r, s)$ and $Y \in \mathfrak{X}$. Therefore, if $z \in \mathfrak{D}(r, s)$, we can define a point $\alpha(z)$ of $\mathfrak{D}(r, s)$ by

$$(3.4) \qquad \alpha \cdot Y(z) = Y(\alpha(z)) \cdot \mathrm{diag}[\overline{\lambda(\alpha, z)}, \mu(\alpha, z)]$$

with $\lambda(\alpha, z) \in GL_r(C)$ and $\mu(\alpha, z) \in GL_s(C)$. Notice that both λ and μ are holomorphic factors of automorphy. Put

$$(3.5) \qquad R(z, w) = {}^t\overline{Y(z)} J_{r,s} Y(w) = \begin{bmatrix} 1_r - z \cdot {}^t\bar{w} & w - z \\ {}^t\bar{z} - {}^t\bar{w} & {}^t\bar{z}w - 1_s \end{bmatrix} \, .$$

Substituting $\alpha(z)$ and $\alpha(w)$ for z and w, we find

$$(3.6) \qquad R(z, w) = \begin{bmatrix} {}^t\lambda(\alpha, z) & 0 \\ 0 & {}^t\overline{\mu(\alpha, z)} \end{bmatrix} R(\alpha(z), \alpha(w)) \begin{bmatrix} \overline{\lambda(\alpha, w)} & 0 \\ 0 & \mu(\alpha, w) \end{bmatrix} \, .$$

Comparing the upper right blocks of both sides, we obtain

$$(3.7) \qquad dz \circ \alpha = {}^t\lambda(\alpha, z)^{-1} dz \, \mu(\alpha, z)^{-1} \qquad (\alpha \in U(r, s)) \, .$$

Put $\xi(z) = 1_r - z \cdot {}^t\bar{z}$ and $\eta(z) = 1_s - {}^t\bar{z}z$. Since $R(z, z) = \mathrm{diag}[\xi(z), -\eta(z)]$, (3.6) yields

$$(3.8) \qquad \xi(z) = {}^t\lambda(\alpha, z) \xi(\alpha(z)) \overline{\lambda(\alpha, z)}, \quad \eta(z) = {}^t\overline{\mu(\alpha, z)} \eta(\alpha(z)) \mu(\alpha, z) \, .$$

Now we can easily verify that $\det(Y(z)) = \det(\xi(z)) = \det(\eta(z))$. Taking the determinant of (3.4) and using (3.8), we find

$$(3.9) \qquad \det(\mu(\alpha, z)) = \det(\alpha) \det(\lambda(\alpha, z)) \qquad (\alpha \in U(r, s)) \, .$$

If $r = 0$ or $s = 0$, the group $U(r, s)$ is compact. In this case, we denote by $\mathfrak{D}(r, s)$ the space consisting of a single point, say z_0, and put $Y(z_0) = 1_m$; the action of $U(r, s)$ on $\mathfrak{D}(r, s)$ is trivial.

4. A family of abelian varieties.

Let K be a CM-field, and Φ an element of I_K which represents a class of representations of K. We consider a structure (A, ι, C) such that: (i) (A, ι) is of type (K, Φ); (ii) C is a polarization of A; and (iii) the involution of $\mathrm{End}(A) \otimes Q$ determined by C sends $\iota(a)$ to $\iota(a^\rho)$. Take an isomorphism of A onto a complex torus C^n/D with a lattice D in C^n so that $\Phi(a)$ corresponds to $\iota(a)$, where we use Φ for a matrix representation in the given class. Then K acts on QD through Φ, and hence there is an isomorphism p of K_m^1 onto QD such that $p(ax)=\Phi(a)p(x)$ for $a \in K$ and $x \in K_m^1$, where $m=2n/[K:Q]$. Put $\mathfrak{M}=p^{-1}(D)$. Then $\mathfrak{M}$ is a lattice in K_m^1. Let $E(x, y)$ be the Riemann (alternating) form of a basic divisor in C. As shown in [7], there is an element T of $GL_m(K)$ such that $^t T^\rho = -T$ and

$$E(p(x), p(y))=\mathrm{Tr}_{K/Q}(xT \cdot {}^t y^\rho) \qquad ((x, y) \in K_m^1 \times K_m^1).$$

In this situation, we say that $\langle A, \iota, C \rangle$ *is of type* $(K, \Phi, T, \mathfrak{M})$, or simply *of type* (K, Φ, T) if only the isogeny-class is the question. Let $[K:Q]=2g$ and take g injections $\tau_1, \cdots, \tau_g$ of K into C which form a CM-type of K; put $\Phi=\sum_{\nu=1}^g(r_\nu\tau_\nu+s_\nu\tau_\nu\rho)$ with integers r_ν and s_ν. Then a structure of type (K, Φ, T) exists if and only if $-iT^{\tau_\nu}$ has signature (r_ν, s_ν) for every ν (see [7]); especially $r_1+s_1=\cdots=r_g+s_g=m$, and hence $\Phi \in I_K^0$. Suppose in particular Φ is a CM-type and $\Phi=\sum_{\nu=1}^g\tau_\nu$. Then $m=1$ and T is an element of K such that $T^\rho=-T$ and $-iT^{\tau_\nu}>0$ for all ν. If $m>1$, the structures of a given type can be parametrized by the points of a symmetric domain $\mathfrak{D}$, as proved in [7]. Let us now recall this explicit parametrization.

With a fixed element T of $GL_m(K)$ such that $^t T^\rho=-T$, define a group G_Q by

$$(4.1) \qquad G_Q=\{\alpha \in GL_m(K) \mid \alpha T \cdot {}^t\alpha^\rho=T\}.$$

Assuming the signature of $-iT^{\tau_\nu}$ to be (r_ν, s_ν), take an element Q_ν of $GL_m(\bar{Q})$ so that

$$(4.2) \qquad -iT^{\tau_\nu}=Q_\nu^\rho J_{r_\nu, s_\nu} \cdot {}^t Q_\nu.$$

Then G_Q can be embedded into $\prod_{\nu=1}^g U(r_\nu, s_\nu)$ by the map

$$(4.3) \qquad \alpha \mapsto (Q_1^{-1}\alpha^{\tau_1\rho}Q_1, \cdots, Q_g^{-1}\alpha^{\tau_g\rho}Q_g).$$

We let G_Q act on the domain

$$(4.4) \qquad \mathfrak{D}=\mathfrak{D}(r_1, s_1) \times \cdots \times \mathfrak{D}(r_g, s_g)$$

through the map (4.3). Notice that $r_\nu s_\nu=0$ may happen for some ν. For $\alpha \in G_Q$ and $z=(z_1, \cdots, z_g) \in \mathfrak{D}$ with $z_\nu \in \mathfrak{D}(r_\nu, s_\nu)$, we put $\lambda_\nu(\alpha, z)=\lambda(Q_\nu^{-1}\alpha^{\tau_\nu\rho}Q_\nu, z_\nu)$,

$$\mu_\nu(\alpha,\ z)=\mu(Q_\nu^{-1}\alpha^{-\nu\rho}Q_\nu,\ z_\nu),$$

$$Y_\nu(z)=Y(z_\nu)=\begin{cases} \begin{bmatrix} 1_{r_\nu} & z_\nu \\ {}^t\bar{z}_\nu & 1_{s_\nu} \end{bmatrix} & (r_\nu s_\nu>0),\\[2em] 1_m & (r_\nu s_\nu=0). \end{cases}$$

Each point z of $\mathfrak{D}$ determines a structure of type $(K,\ \Phi,\ T,\ \mathfrak{M})$ as follows. First put

$$Y_\nu(z)\cdot{}^tQ_\nu^\rho=\begin{bmatrix} u_1^\nu \cdots u_m^\nu \\ \bar{v}_1^\nu \cdots \bar{v}_m^\nu \end{bmatrix} \qquad (\nu=1,\ \cdots,\ g)$$

with $u_k^\nu\in C^{r_\nu}$ and $v_k^\nu\in C^{s_\nu}$, and define $\mathfrak{x}_k(z)\in C^{mg}$ by

$${}^t\mathfrak{x}_k(z)=({}^tu_k^1,\ {}^tv_k^1,\ \cdots,\ {}^tu_k^g,\ {}^tv_k^g).$$

Put $n=mg$, and define an n-dimensional diagonal representation Φ of K by

$$\Phi(a)=\mathrm{diag}[a^{\tau_1}1_{r_1},\ a^{\tau_1\rho}1_{s_1},\ \cdots,\ a^{\tau_g}1_{r_g},\ a^{\tau_g\rho}1_{s_g}] \qquad (a\in K).$$

For $a=(a_1,\ \cdots,\ a_m)\in K_m^1$ and $z\in\mathfrak{D}$, put

$$p_z(a)=p(a,\ z)=\textstyle\sum_{k=1}^m\Phi(a_k)\mathfrak{x}_k(z).$$

Then p_z can be extended to an R-linear isomorphism of $(K\otimes_Q R)_m^1$ onto C^n. Therefore, if $\mathfrak{M}$ is a lattice in K_m^1, $C^n/p_z(\mathfrak{M})$ is a complex torus. Define an R-valued alternating form E_z on C^n by

$$E_z(p_z(a),\ p_z(b))=\mathrm{Tr}_{K/Q}(aT\cdot{}^tb^\rho) \qquad (a,\ b\in K_m^1).$$

This defines a Riemann form on C^n, so that the torus has a structure of abelian variety; the action of $\Phi(a)$ defines an endomorphism. In this way we obtain a structure $(A_z,\ \iota_z,\ C_z)$ of type $(K,\ \Phi,\ T,\ \mathfrak{M})$ for each $z\in\mathfrak{D}$. Conversely, every structure of type $(K,\ \Phi,\ T,\ \mathfrak{M})$ can be obtained in this fashion. For details, see [7].

We now define an embedding ε of $\mathfrak{D}$ into the Siegel upper space

$$\mathfrak{H}_n=\{z\in C_n^n\mid {}^tz=z,\ \mathrm{Im}(z)>0\}.$$

First take a basis $\{h_1,\ \cdots,\ h_{2n}\}$ of K_m^1 over Q so that

$$(4.5) \qquad (\mathrm{Tr}_{K/Q}(h_iT\cdot{}^th_j^\rho))_{i,\,j=1,\cdots,2n}=\begin{bmatrix} 0 & -1_n \\ 1_n & 0 \end{bmatrix},$$

and define the period matrix $(\omega_1(z),\ \omega_2(z))$ by

$$(4.6) \qquad \omega_1(z)=(p_z(h_1)\cdots p_z(h_n)),\quad \omega_2(z)=(p_z(h_{n+1})\cdots p_z(h_{2n})).$$

As is well known, $\omega_2(z)$ is invertible, and $\omega_2(z)^{-1}\omega_1(z)\in\mathfrak{H}_n$. Thus we obtain an embedding

$$(4.7) \qquad \varepsilon : \mathfrak{D} \to \mathfrak{H}_n , \quad \varepsilon(z)=\omega_2(z)^{-1}\omega_1(z) \qquad (z\in\mathfrak{D}) .$$

This is holomorphic, since $\mathfrak{x}_k(z)$ is holomorphic in z. Every $\beta=\begin{pmatrix} A & B \\ C & D \end{pmatrix}\in\mathrm{Sp}(n, \boldsymbol{R})$ acts on $\mathfrak{H}_n$ as usual under the rule $\beta(Z)=(AZ+B)(CZ+D)^{-1}$ for $Z\in\mathfrak{H}_n$. Now, for every $\alpha\in G_{\boldsymbol{Q}}$, define an element $\alpha^*=(a_{ij})$ of $GL_{2n}(\boldsymbol{Q})$ by $h_i\alpha=\sum_{j=1}^{2n}a_{ij}h_j$. Obviously $\alpha^*\in\mathrm{Sp}(n, \boldsymbol{Q})$. Moreover we have

$$(4.8) \qquad \varepsilon\circ\alpha=\alpha^*\circ\varepsilon .$$

To show this, we first observe that, for $\alpha\in G_{\boldsymbol{Q}}$,

$$(4.9) \qquad Y_\nu(z)\cdot{}^t Q_\nu^\rho\cdot{}^t\alpha^{\div}_\nu=\varLambda'_\nu(\alpha, z)Y_\nu(\alpha(z))\cdot{}^t Q_\nu^\rho ,$$

where

$$\varLambda'_\nu(\alpha, z)=\begin{cases} \mathrm{diag}[{}^t\lambda_\nu(\alpha, z), {}^t\mu_\nu(\alpha, z)^\rho] & (r_\nu s_\nu>0) , \\ {}^t(Q_\nu^{-1}\alpha^{\div}_\nu{}^\rho Q_\nu)^\rho & (r_\nu s_\nu=0) . \end{cases}$$

Define $\varLambda(\alpha, z)$ and $\varLambda_\nu(\alpha, z)$ by

$$(4.10_{\mathrm{a}}) \qquad \varLambda(\alpha, z)=\mathrm{diag}[\varLambda_1(\alpha, z), \cdots, \varLambda_g(\alpha, z)] ,$$

$$(4.10_{\mathrm{b}}) \qquad \varLambda_\nu(\alpha, z)=\begin{cases} \mathrm{diag}[\lambda_\nu(\alpha, z), \mu_\nu(\alpha, z)] & (r_\nu s_\nu>0) , \\ (Q_\nu^{-1}\alpha^{\div}_\nu{}^\rho Q_\nu)^\rho & (s_\nu=0) , \\ Q_\nu^{-1}\alpha^{\div}_\nu{}^\rho Q_\nu & (r_\nu=0) . \end{cases}$$

Then (4.9) implies that

$$(4.11) \qquad p(x\alpha, z)={}^t\varLambda(\alpha, z)p(x, \alpha(z)) \qquad (x\in K_m^1, z\in\mathfrak{D}) .$$

Therefore we have

$$(\omega_1(z), \omega_2(z))\cdot{}^t\alpha^*={}^t\varLambda(\alpha, z)(\omega_1(\alpha(z)), \omega_2(\alpha(z))) ,$$

which proves (4.8). At the same time we see that

$$(4.12) \qquad {}^t\omega_2(\alpha(z))\varLambda(\alpha, z)\cdot{}^t\omega_2(z)^{-1}=C\cdot\varepsilon(z)+D \qquad \text{if} \quad \alpha^*=\begin{pmatrix} A & B \\ C & D \end{pmatrix} .$$

Let $\eta : GL_n(\boldsymbol{C})\to GL_p(\boldsymbol{C})$ be a rational representation. Given a C_q^p-valued function f on $\mathfrak{H}_n$, an element $\beta=\begin{pmatrix} A & B \\ C & D \end{pmatrix}$ of $\mathrm{Sp}(n, \boldsymbol{Q})$, and $k\in\boldsymbol{Z}$, define a C_q^p-valued function $f|_{\eta, k}\beta$ by

$$(f|_{\eta, k}\beta)(Z)=\det(CZ+D)^{-k}\eta(CZ+D)^{-1}f(\beta(Z)) \qquad (Z\in\mathfrak{H}_n) ,$$

and also a function f^ε on $\mathfrak{D}$ by

$$(4.13) \qquad f^\varepsilon(z)=\det(\omega_2(z))^{-k}\eta({}^t\omega_2(z)^{-1})f(\varepsilon(z)) \qquad (z\in\mathfrak{D}) .$$

For every $\alpha\in G_{\boldsymbol{Q}}$, we have, by (4.11),

$$f^s(\alpha(z))=\det(\Lambda(\alpha, z))^k \eta(\Lambda(\alpha, z))(f|_{\tau, k}\alpha^*)^s(z).$$

Now define a congruence subgroup Γ_N of G_Q by

(4.14) $$\Gamma_N=\{\gamma\in G_Q\,|\,\det(\gamma)=1,\ \mathfrak{m}\gamma=\mathfrak{m},\ \mathfrak{m}(1-\gamma)\subset N\mathfrak{m}\},$$

where N is a positive integer and $\mathfrak{m}=\sum_{j=1}^{2n}Zh_j$. The map $\alpha\mapsto\alpha^*$ sends Γ_N into the group

(4.15) $$\varDelta_N=\{\beta\in\mathrm{Sp}(n, Z)\,|\,\beta\equiv 1_{2n}(\mathrm{mod}\ N)\}.$$

Therefore, if $f|_{\tau, k}\beta=f$ for all $\beta\in\varDelta_N$, then f^s satisfies

(4.16) $$f^s(\gamma(z))=\det(\Lambda(\gamma, z))^k \eta(\Lambda(\gamma, z))f^s(z)\quad\text{for all }\gamma\in\Gamma_N.$$

The members of our family with complex multiplication can be obtained as follows (cf. Miyake [5]). Let

(4.17) $$Y=M_{n_1}(L_1)\oplus\cdots\oplus M_{n_t}(L_t)$$

with CM-fields L_i containing K, and let δ be a positive involution of Y. Suppose $m=\sum_{i=1}^t n_i\,[L_i:K]$ and there is a K-linear injection h of Y into $M_m(K)$ such that $h(x^\delta)=T\cdot {}^t h(x)^\rho T^{-1}$ for all $x\in Y$. Put

(4.18) $$Y[\delta]=\{c\in Y\,|\,cc^\delta=1\}.$$

Then $h(Y[\delta])$ is contained in G_Q and has a unique common fixed point z_0 on $\mathfrak{D}$. There is a Q-linear injection $\varXi_\nu$ of Y into $M_m(C)$, for each ν, such that $\varXi_\nu(c)={}^t\Lambda_\nu(h(c), z_0)$ for every $c\in Y[\delta]$. Moreover, there is a CM-type $\varPsi_i$ of L_i such that $\varPsi_i=\sum_{\nu=1}^g\varPsi_{i\nu}$ and $n_i\,\mathrm{tr}(\varPsi_{i\nu}(a))=\mathrm{tr}(\varXi_\nu(a))$ for every $a\in L_i$. If (A_i, ι_i) is of type $(L_i, \varPsi_i)$, the member A_{z_0} is isogenous to $A_1^{n_1}\times\cdots\times A_t^{n_t}$. Conversely, let $(L_i, \varPsi_i)$ be CM-types such that $K\subset L_i$, and let (A_i, ι_i) be of type $(L_i, \varPsi_i)$, Suppose the sum of the restrictions of $n_1\varPsi_1, \cdots, n_t\varPsi_t$ to K is $\varPhi$. Then $A=A_1^{n_1}\times\cdots\times A_t^{n_t}$ has a polarization which defines an involution of $\mathrm{End}(A)\otimes Q$ whose restriction to the image of K corresponds to ρ. As shown in [7], A can be obtained as a member of a family of our type with suitable T and $\mathfrak{M}$. We state here two facts about such special members as lemmas.

LEMMA 4.1. *Let $\varPhi_1, \cdots, \varPhi_m$ be CM-types of K, and let (A_i, ι_i) be of type $(K, \varPhi_i)$. If $\varPhi=\sum_{i=1}^m\varPhi_i$, the above family (without changing T or $\mathfrak{M}$) contains a member isogenous to $A_1\times\cdots\times A_m$.*

PROOF. We can easily find (by induction on m, for example) elements $\zeta_1, \cdots, \zeta_m$ of K such that T is equivalent to $\mathrm{diag}[\zeta_1, \cdots, \zeta_m]$ and $\mathrm{Im}(\zeta_i^\tau)>0$ for τ belonging to $\varPhi_i$, for every i. Then there is a structure (A_i', ι_i', C_i') of type $(K, \varPhi_i, \zeta_i)$ (see [15] or [7]), and $A_1'\times\cdots\times A_t'$ is isogenous to a member of our family.

LEMMA 4.2. *Suppose the member A_z is isogenous to a product of abelian* *varieties belonging to CM-types. Then the coordinates of $\omega_1(z)$ and $\omega_2(z)$ are algebraic.*

PROOF. Observe that the action of G_q on $\mathfrak{D}$ is $\bar{Q}$-rational. As explained above, z is obtained as a unique fixed point of $Y[\delta]$. Therefore the coordinates of z are algebraic, so that the vector $p_z(a)$ is algebraic for every $a \in K^1_m$, which proves our lemma.

5. Proof of Theorem 1.2.

The notation being as in Theorem 1.2, put $\Phi = \Phi_1 + \cdots + \Phi_m$, and consider the family of §4 with this Φ and any choice of T and $\mathfrak{M}$. If $\sum_{\nu=1}^{g} r_\nu s_\nu = 0$, we have $\Phi_1 = \cdots = \Phi_m$. In this case, Theorem 1.2 follows from Proposition 1.4. Therefore we may assume that $\mathfrak{D}$ is a domain of positive dimension.

Suppose first that either $\Gamma_N \backslash \mathfrak{D}$ is compact or $mg > 2$. We denote then by $\mathfrak{A}_0(\Gamma_N)$ the field of all Γ_N-invariant meromorphic functions on $\mathfrak{D}$, and by $\mathfrak{A}_0$ the union of $\mathfrak{A}_0(\Gamma_N)$ for all positive integers N. If $\Gamma_N \backslash \mathfrak{D}$ is not compact and $mg = 2$, Γ_N is conjugate to a congruence subgroup of $SL_2(Z)$; so we can similarly define $\mathfrak{A}_0(\Gamma_N)$ and $\mathfrak{A}_0$ under the usual cusp-condition. In either case, we can speak of canonical models, due to Miyake [5] (cf. also [8], [11]). Therefore $\bar{Q}$-rational elements of $\mathfrak{A}_0$ are meaningful. We denote by $\mathfrak{A}_0(\bar{Q})$ the subfield of $\mathfrak{A}_0$ consisting of all $\bar{Q}$-rational elements. We note that if f is a $\bar{Q}$-rational Siegel modular function on $\mathfrak{H}_n$ and $f \circ \varepsilon$ is meaningful, then $f \circ \varepsilon \in \mathfrak{A}_0(\bar{Q})$.

Let $P(Z)$ be a C^n_n-valued meromorphic function on $\mathfrak{H}_n$ such that

$$(5.1) \qquad P(\beta(Z)) = (CZ + D)P(Z) \qquad \text{for every } \beta = \begin{pmatrix} * & * \\ C & D \end{pmatrix} \in \Delta_M$$

with a positive integer M. We can find such a P which is "Q-rational", holomorphic at $\varepsilon(z)$, and with $\det(P(\varepsilon(z))) \neq 0$, for any given point z of $\mathfrak{D}$, as proved in [13, Proposition 1.2]. Take a generator b of K over Q such that $bb^\rho = 1$; put $\alpha = b1_m$, $\alpha^* = \begin{pmatrix} * & * \\ C & D \end{pmatrix}$, and

$$V(Z) = P(Z)^{-1}(CZ + D)^{-1}P(\alpha^*(Z)) \qquad (Z \in \mathfrak{H}_n),$$

$$U(z) = V(\varepsilon(z)) \qquad\qquad\qquad (z \in \mathfrak{D}).$$

We see easily that the entries of V are Δ_N-invariant for some N. Moreover, they are $\bar{Q}$-rational by [13, Proposition 1.4]; hence the entries of U belong to $\mathfrak{A}_0(\bar{Q})$. By (4.12), we have

$$(5.2) \qquad\qquad {}^t\omega_2(z)\Phi(b) \cdot {}^t\omega_2(z)^{-1} = C \cdot \varepsilon(z) + D,$$

and hence $P(\varepsilon(z))^{-1} \cdot {}^t\omega_2(z)\Phi(b) \cdot {}^t\omega_2(z)^{-1}P(\varepsilon(z)) = U(z)^{-1}$. Since K is generated by

b, we see that

$$a \longmapsto P(\varepsilon(z))^{-1} \cdot {}^t\omega_2(z)\Phi(a) \cdot {}^t\omega_2(z)^{-1}P(\varepsilon(z))$$

is an $\mathfrak{A}_0(\bar{\boldsymbol{Q}})$-rational representation of K, so that there is an element Y of $GL_n(\mathfrak{A}_0(\bar{\boldsymbol{Q}}))$ such that

$$P(\varepsilon(z))^{-1} \cdot {}^t\omega_2(z)\Phi(a) \cdot {}^t\omega_2(z)^{-1}P(\varepsilon(z)) = Y^{-1}\Phi(a)Y$$

for every $a \in K$. In view of our definition of Φ in the diagonal form, we have

$$(5.3) \qquad\qquad {}^t\omega_2(z)^{-1}P(\varepsilon(z)) = \mathrm{diag}[R_1,\, S_1,\, \cdots,\, R_g,\, S_g]Y$$

with square matrices R_ν and S_ν of size r_ν and s_ν, whose entries are meromorphic functions on $\mathfrak{D}$. From (4.16), we see that

$$(5.4_a) \qquad\qquad R_\nu(\gamma(z)) = \begin{cases} \lambda_\nu(\gamma,\, z)R_\nu(z) & (r_\nu s_\nu > 0), \\ (Q_\nu^{-1}\gamma^{\tau_\nu \rho}Q_\nu)^\rho R_\nu(z) & (s_\nu = 0), \end{cases}$$

$$(5.4_b) \qquad\qquad S_\nu(\gamma(z)) = \begin{cases} \mu_\nu(\gamma,\, z)S_\nu(z) & (r_\nu s_\nu > 0), \\ Q_\nu^{-1}\gamma^{\tau_\nu \rho}Q_\nu S_\nu(z) & (r_\nu = 0), \end{cases}$$

if $\gamma \in \Gamma_{N'}$ for sufficiently large N'.

Let us now consider the special member at the fixed point z_0 of $h(Y'[\delta])$ as in §4. For simplicity, we take $t = n_1 = 1$ and drop the subscript 1; thus the member at z_0 defines a structure (A, ι) of type (L, Ψ) and $[L : K] = m$. (This is sufficient for our proof of Theorem 1.2. The following argument with obvious modifications applies to the general case.) Put $W = \varepsilon(z_0)$. Then $\boldsymbol{C}^n/(W\boldsymbol{Z}^n + \boldsymbol{Z}^n)$ is isogenous to A. Now take the projective embedding

$$(5.5_a) \qquad\qquad \Theta : \boldsymbol{C}^n/(W\boldsymbol{Z}^n + \boldsymbol{Z}^n) \to A_0$$

of the torus onto an abelian variety A_0 by classical theta functions as defined in [13, (1.5$_b$)]. Then A_0 is isogenous to A, and by [13, Theorem 1.1], is defined over $\bar{\boldsymbol{Q}}$. Define an isomorphism

$$(5.5_b) \qquad\qquad \Theta' : \boldsymbol{C}^n/[\omega_1(z_0)\boldsymbol{Z}^n + \omega_2(z_0)\boldsymbol{Z}^n] \to A_0$$

by $\Theta'(x) = \Theta(\omega_2(z_0)^{-1}x)$. Let $x_1, \cdots, x_n$ be the coordinate functions on $\boldsymbol{C}^n$. By [13, Theorem 1.5], there are $\bar{\boldsymbol{Q}}$-rational 1-forms $\xi_1, \cdots, \xi_n$ on A_0 such that

$$(5.6) \qquad\qquad \begin{bmatrix} \xi_1 \\ \vdots \\ \xi_n \end{bmatrix} \cdot \Theta' = \pi \cdot {}^t P(\varepsilon(z_0))\omega_2(z_0)^{-1} \begin{bmatrix} dx_1 \\ \vdots \\ dx_n \end{bmatrix}.$$

We have a direct sum decomposition of $\boldsymbol{C}^n$ into subspaces V_ν, V_ν' such that $\Phi(a)$ acts as a^{τ_ν} on V_ν and as $a^{\tau_\nu \rho}$ on V_ν'. Let $x_1^\nu, \cdots, x_{r_\nu}^\nu$ and $y_1^\nu, \cdots, y_{s_\nu}^\nu$ be the coordinate functions on V_ν and V'_ν, respectively, which are renamings of some of $x_1, \cdots, x_n$. Multiplying (5.6) with a $\bar{Q}$-rational matrix ${}^t Y(z_0)^{-1}$, we find that

$$(5.7) \qquad \pi \cdot {}^t R_\nu(z_0) \begin{bmatrix} dx_1^\nu \\ \vdots \\ dx_r^\nu \end{bmatrix} \quad \text{and} \quad \pi \cdot {}^t S_\nu(z_0) \begin{bmatrix} dy_1^\nu \\ \vdots \\ dy_s^\nu \end{bmatrix} \qquad (r=r_\nu, \ s=s_\nu)$$

correspond to $\bar{Q}$-rational 1-forms on A_0.

Define diagonal representations Ψ_ν^+ and Ψ_ν^- of L by

$$\Psi_\nu^+ = \mathrm{diag}[\psi_{\nu 1}, \cdots, \psi_{\nu r}], \quad \Psi_\nu^- = \mathrm{diag}[\psi'_{\nu 1}, \cdots, \psi'_{\nu s}]$$

with injections $\psi_{\nu j}$ and $\psi'_{\nu j}$ of L into C belonging to Ψ which give τ_ν and $\tau_\nu \rho$ on K, respectively. Then $r=r_\nu$, $s=s_\nu$, and $\Psi=\sum_{\nu=1}^{g}(\Psi_\nu^+ + \Psi_\nu^-)$. Take a generator c of L over Q such that $cc^\rho = 1$. As explained in § 4, $c \mapsto \Lambda_\nu(h(c), z_0)$ is essentially $\mathrm{diag}[\Psi_\nu^+(c), \Psi_\nu^-(c)]$; moreover, (if $r_\nu s_\nu > 0$,) $\Psi_\nu^+(c)$ corresponds to $\lambda_\nu(h(c), z_0)$ and $\Psi_\nu^-(c)$ to $\mu_\nu(h(c), z_0)$. Therefore we can find invertible $\bar{Q}$-rational matrices C_ν and C'_ν such that

$$(5.8_\mathrm{a}) \qquad C_\nu^{-1} \Psi_\nu^+(c) C_\nu = \begin{cases} \lambda_\nu(h(c), z_0) & (r_\nu s_\nu > 0), \\ (Q_\nu^{-1} h(c)^{\tau_\nu \rho} Q_\nu)^\rho & (s_\nu = 0), \end{cases}$$

$$(5.8_\mathrm{b}) \qquad C_\nu'^{-1} \Psi_\nu^-(c) C'_\nu = \begin{cases} \mu_\nu(h(c), z_0) & (r_\nu s_\nu > 0), \\ Q_\nu^{-1} h(c)^{\tau_\nu \rho} Q_\nu & (r_\nu = 0). \end{cases}$$

Since ${}^t\lambda_\nu(h(c), z_0)$ or ${}^t(Q_\nu^{-1} h(c)^{\tau_\nu \rho} Q_\nu)^\rho$ represents $\iota(c)$ on the subspace V_ν, if we put

$$\begin{bmatrix} du_1^\nu \\ \vdots \\ du_r^\nu \end{bmatrix} = {}^t C_\nu^{-1} \begin{bmatrix} dx_1^\nu \\ \vdots \\ dx_r^\nu \end{bmatrix} \qquad (r=r_\nu),$$

then $du_j^\nu \circ \iota(c) = c^{\psi_{\nu j}} du_j^\nu$. By Lemma 4.2, du_j^ν, considered a 1-form on A_0, has algebraic periods. By Proposition 1.1, $\pi \cdot p(\psi_{\nu j}, \Psi) du_j$ is $\bar{Q}$-rational, and hence

$$(5.9_\mathrm{a}) \qquad \mathrm{diag}[p(\psi_{\nu 1}, \Psi), \cdots, p(\psi_{\nu r}, \Psi)]^{-1} C_\nu R_\nu(z_0) \in GL_r(\bar{Q}),$$

and similarly

$$(5.9_\mathrm{b}) \qquad \mathrm{diag}[p(\psi'_{\nu 1}, \Psi), \cdots, p(\psi'_{\nu s}, \Psi)]^{-1} C'_\nu S_\nu(z_0) \in GL_s(\bar{Q}).$$

Put $f_\nu(z) = \det(R_\nu(z))$ and $g_\nu(z) = \det(S_\nu(z))$. Then we obtain

$$(5.10) \qquad f_\nu(z_0) \sim q_K(\tau_\nu, \Psi), \quad g_\nu(z_0) \sim q_K(\tau_\nu \rho, \Psi).$$

We can apply the whole procedure to a point of $\mathfrak{D}$ corresponding to a member of a somewhat different type. In fact, take CM-types $\Phi_1, \cdots, \Phi_m$ of K so that $\Phi = \Phi_1 + \cdots + \Phi_m$. Let (A_i, ι_i) be of type (K, Φ_i). Lemma 4.1 guarantees a member of our family isogenous to $A_1 \times \cdots \times A_t$. Let v be a point of $\mathfrak{D}$ corresponding to this member. Since v can be replaced by any point in the set $\{\alpha(v) \mid \alpha \in G_\mathbf{Q}\}$ which is dense in $\mathfrak{D}$, we may assume that the functions of (5.3) are all holomorphic at v and $\det(P(\varepsilon(v))) \neq 0$. For each injection τ of K into C, put $q(\tau) = \prod_{i=1}^{m} p(\tau, \Phi_i)$. Repeating the above reasoning with the direct sum of m copies of K in place of L, we find that $f_\nu(v) \sim q(\tau_\nu)$ and $g_\nu(v) \sim q(\tau_\nu \rho)$.

Suppose $r_\nu s_\nu > 0$. By virtue of $(5.4_{\mathrm{a,b}})$ and (3.9), we see that $f_\nu/g_\nu \in \mathfrak{A}_0$, and $q(\tau_\nu)^{-1} q(\tau_\nu \rho)(f_\nu/g_\nu)(v) \in \bar{\mathbf{Q}}$. Now v can be replaced by $\alpha(v)$ with $\alpha \in G_\mathbf{Q}$ if the functions of (5.3) as well as $(P \circ \varepsilon)^{-1}$ are holomorphic at $\alpha(v)$. Since such points $\alpha(v)$ form a dense subset of $\mathfrak{D}$, we see that $q(\tau_\nu)^{-1} q(\tau_\nu \rho) f_\nu/g_\nu \in \mathfrak{A}_0(\bar{\mathbf{Q}})$. Now $q(\tau_\nu)/q(\tau_\nu \rho)$ depends only on $\Phi_1, \cdots, \Phi_m$ and τ_ν. On the other hand, f_ν/g_ν depends only on ν, Φ, T, and $\mathfrak{M}$. Therefore $q(\tau_\nu)/q(\tau_\nu \rho)$ depends only on Φ and τ_ν, which proves the first assertion of Theorem 1.2 in the case $r_\nu s_\nu > 0$. Moreover, we have $q(\tau_\nu)^{-1} q(\tau_\nu \rho)(f_\nu/g_\nu)(z_0) \in \bar{\mathbf{Q}}$, and hence

$$q(\tau_\nu)/q(\tau_\nu \rho) \sim f_\nu(z_0)/g_\nu(z_0) \sim q_K(\tau_\nu, \Psi)/q_K(\tau_\nu \rho, \Psi),$$

which is the second assertion. If $s_\nu = 0$, we see from (5.4_{a}) that $f_\nu \in \mathfrak{A}_0$, and conclude that $q(\tau_\nu)^{-1} f_\nu \in \mathfrak{A}_0(\bar{\mathbf{Q}})$ by the same reasoning as above. Therefore the above conclusions hold with $g_\nu = 1$ and $q(\tau_\nu \rho) = q_K(\tau_\nu \rho, \Psi) = 1$. The case $r_\nu = 0$ can be treated in a similar way. This completes the proof.

6. The derivatives of the elements of $\mathfrak{A}_0(\bar{Q})$.

The functions R_ν and S_ν obtained in §5 are (meromorphic) automorphic forms on $\mathfrak{D}$ with arithmetic properties $(5.9_{\mathrm{a,b}})$. Let us now study the relationship between R_ν, S_ν, and the derivatives of the elements of $\mathfrak{A}_0(\bar{Q})$; namely, we prove:

THEOREM 6.1. *Suppose $r_\nu s_\nu > 0$ and let z_{jk}^ν be the (j, k)-entry of the variable z_ν on $\mathfrak{D}(r_\nu, s_\nu)$; define for $f \in \mathfrak{A}_0(\bar{Q})$ a matrix-valued meromorphic function $\Delta_\nu f$ on $\mathfrak{D}$ by*

$$(6.1) \qquad \Delta_\nu f = (\partial f/\partial z_{jk}^\nu) \qquad (1 \leq j \leq r_\nu,\ 1 \leq k \leq s_\nu).$$

Then there is a matrix W with entries in $\mathfrak{A}_0(\bar{Q})$ such that $\pi^{-1} \Delta_\nu f = R_\nu W \cdot {}^t S_\nu$.

REMARK. (1) Our assertion with $\mathfrak{A}_0$ instead of $\mathfrak{A}_0(\bar{Q})$ is obvious from (3.7) and $(5.4_{\mathrm{a,b}})$.

(2) Combining $(5.9_{\mathrm{a,b}})$ with the theorem, we obtain a description of the value of $\Delta_\nu f$ at a CM-point z_0 in terms of $p(\psi_{\nu j}, \Psi)$, similar to (0.2).

PROOF. Let $\kappa = \sum_{\nu=1}^{g} r_\nu s_\nu$. We can find $\bar{Q}$-rational Siegel modular functions $\mathfrak{f}_1, \cdots, \mathfrak{f}_\kappa$ on $\mathfrak{H}_n$ so that $\mathfrak{f}_1 \circ \varepsilon, \cdots, \mathfrak{f}_\kappa \circ \varepsilon$ are algebraically independent. Put

$k_j=\mathfrak{f}_j\cdot\varepsilon$. Then $\partial/\partial h_1, \cdots, \partial/\partial h_\kappa$ are well-defined derivations of $\mathfrak{A}_0(\bar{Q})$ and

$$\Delta_\nu f=\sum_{j=1}^\kappa(\partial f/\partial h_j)\Delta_\nu h_j\,.$$

Therefore it is sufficient to prove our theorem in the case where $f=\mathfrak{f}\cdot\varepsilon$ with a $\bar{Q}$-rational Siegel modular function $\mathfrak{f}$. To simplify the notation, we put $r_1=r$, $s_1=s$ and $\tau_1=\tau$, and prove our assertion for $\nu=1$. We may put T in the form $T=\mathrm{diag}[\zeta_1, \cdots, \zeta_m]$ with $\zeta_\lambda\in K$ such that $i\zeta_\lambda^2<0$ for $\lambda\leq r$ and $i\zeta_\lambda^2>0$ for $\lambda>r$. Take real numbers $q_1, \cdots, q_m$ so that $q_\lambda^2=-i\zeta_\lambda^2$ for $\lambda\leq r$ and $q_\lambda^2=i\zeta_\lambda^2$ for $\lambda>r$. Then we can take $Q=\mathrm{diag}[q_1, \cdots, q_m]$ as the matrix Q_1 of (4.2). For each λ, choose bases $\{a_{\lambda 1}, \cdots, a_{\lambda g}\}$ and $\{a'_{\lambda 1}, \cdots, a'_{\lambda g}\}$ of K_0 over Q so that

$$(6.2) \qquad \mathrm{Tr}_{K_0/Q}(\zeta_\lambda^2 a_{\lambda j} a'_{\lambda k})=\delta_{jk}/2\,.$$

Let $\{e_1, \cdots, e_m\}$ be the standard basis of K_m^1, and let $h_{\lambda j}=a_{\lambda j}e_\lambda$ and $h'_{\lambda j}=\zeta_\lambda a'_{\lambda j}e_\lambda$. Then the elements

$$h_{11}, \cdots, h_{1g}, \cdots, h_{m1}, \cdots, h_{mg}, h'_{11}, \cdots, h'_{1g}, \cdots, h'_{m1}, \cdots, h'_{mg}$$

can be taken as $h_1, \cdots, h_{2mg}$ of (4.5). Focusing our attention on the first m rows of matrices, we have

$$\omega_1(z)=\begin{pmatrix} b & z_1c \\ {}^t z_1 b & c \\ * & * \end{pmatrix}, \quad \omega_2(z)=\begin{pmatrix} b' & z_1c' \\ -{}^t z_1 b' & -c' \\ * & * \end{pmatrix}$$

with $b, b'\in C_{rg}^r$ and $c, c'\in C_{sg}^s$. Consider the functions on $\mathfrak{D}$ as functions in z_1, keeping the variables $z_2, \cdots, z_g$ constant; thus $df=\sum_{j=1}^r\sum_{k=1}^s(\partial f/\partial z'_{jk})dz'_{jk}=\mathrm{tr}(\Delta_1 f\cdot{}^t dz_1)$. Since $d\varepsilon=\omega_2^{-1}(d\omega_1-d\omega_2\varepsilon)$, we have

$$\omega_2 d\varepsilon\cdot{}^t\omega_2=d\omega_1\cdot{}^t\omega_2-d\omega_2\cdot{}^t\omega_1$$

$$=\begin{pmatrix} dz_1(c\cdot{}^tc'-c'\cdot{}^tc){}^tz_1 & -dz_1(c\cdot{}^tc'+c'\cdot{}^tc) & * \\ {}^t dz_1(b\cdot{}^tb'+b'\cdot{}^tb) & {}^t dz_1(b'\cdot{}^tb-b\cdot{}^tb')z & * \\ 0 & 0 & 0 \end{pmatrix}.$$

This must be symmetric. The entries of b, c, b', c' can be given in terms of $q_\lambda a'_{\lambda j}$ and $q_\lambda\zeta_\lambda a'_{\lambda j}$. By virtue of (6.2), we find that

$$\omega_2 d\varepsilon\cdot{}^t\omega_2=-i\begin{pmatrix} 0 & dz_1 & 0 \\ {}^t dz_1 & 0 & 0 \\ 0 & 0 & 0 \end{pmatrix}.$$

For a Q-rational Siegel modular function $\mathfrak{f}$ on $\mathfrak{H}_n$, define a C_n^n-valued function $\varDelta\mathfrak{f}$ on $\mathfrak{H}_n$ by

$$\varDelta\mathfrak{f}=\left(\frac{1+\delta_{jk}}{2}\frac{\partial\mathfrak{f}}{\partial Z_{jk}}\right) \qquad (Z=(Z_{jk})\in\mathfrak{H}_n).$$

Then $d\mathfrak{f}=\mathrm{tr}(\varDelta\mathfrak{f}\cdot dZ)$, so that $d(\mathfrak{f}\circ\varepsilon)=\mathrm{tr}([(\varDelta\mathfrak{f})\circ\varepsilon]d\varepsilon)$. Put $X=\pi^{-1}P^{-1}\varDelta\mathfrak{f}\cdot{}^tP^{-1}$ with the function P of (5.1). Then the entries of X are $\bar{Q}$-rational modular functions on $\mathfrak{H}_n$. Consider R_ν, S_ν, and Y of (5.3); put $V=Y\cdot(X\circ\varepsilon)\cdot{}^tY$ and $U=\mathrm{diag}[R_1, S_1, \cdots, R_g, S_g]$. Then

$$\pi^{-1}d(\mathfrak{f}\circ\varepsilon)=\mathrm{tr}(V\cdot{}^tU\omega_2 d\varepsilon\cdot{}^t\omega_2 U)=-2i\cdot\mathrm{tr}(R_1 v_{12}\cdot{}^tS_1\cdot{}^tdz_1),$$

where v_{12} is the $(1, 2)$-block of V. Therefore $\pi^{-1}\varDelta_1(\mathfrak{f}\circ\varepsilon)=-2iR_1 v_{12}\cdot{}^tS_1$, which completes the proof, since the entries of v_{12} belong to $\mathfrak{A}_0(\bar{Q})$.

7. The derivatives of automorphic forms on $\mathfrak{H}_1^r$.

Let F be a totally real algebraic number field of degree g, and B a quaternion algebra óver F. Let $\theta_1, \cdots, \theta_g$ denote the injections of F into R, and suppose that B is unramified at $\theta_1, \cdots, \theta_r$, and ramified at $\theta_{r+1}, \cdots \theta_g$. Then there is an R-linear isomorphism

$$(7.1) \qquad B\otimes_{\mathfrak{q}}R\cong M_2(R)^r\times H^{g-r}$$

where H denotes the Hamilton quaternions. Let B^+ denote the multiplicative group of all elements α of B with totally positive reduced norm to F. We fix an isomorphism (7.1) so that B is embedded in $M_2(\bar{Q}\cap R)^r$; then B^+ can be embedded into $GL_2^+(R)^r$, where

$$GL_2^+(R)=\{\alpha\in GL_2(R)\,|\,\det(\alpha)>0\},$$

and thus acts on the product $\mathfrak{H}_1^r$ of r copies of the upper half plane $\mathfrak{H}_1$. With an arbitrarily fixed lattice $\mathfrak{a}$ in B and a positive integer N, let

$$(7.2) \qquad E_N=\{\gamma\in B^+\,|\,\gamma\gamma^\iota=1,\ \gamma\mathfrak{a}=\mathfrak{a},\ (\gamma-1)\mathfrak{a}\subset N\mathfrak{a}\},$$

where $\gamma\mapsto\gamma^\iota$ is the main involution of B. For $\alpha\in B$, let $(\alpha_1, \cdots, \alpha_r)$ denote the image of α in $M_2(R)^r$. Given $k=(k_1, \cdots, k_r)\in Z^r$, $\alpha\in B^+$ with $\alpha_\nu=\begin{pmatrix}a_\nu & b_\nu\\ c_\nu & d_\nu\end{pmatrix}$, and a function f on $\mathfrak{H}_1^r$, we define a function $f|_k\alpha$ by

$$(7.3) \qquad (f|_k\alpha)(z)=f(\alpha(z))\prod_{\nu=1}^r(c_\nu z_\nu+d_\nu)^{-k_\nu}.$$

We denote by $\mathcal{A}_k(E_N)$ the set of all meromorphic functions f, meromorphic even at cusps, such that $f|_k\gamma=f$ for all $\gamma\in E_N$; the cusp-condition is necessary only when $B=M_2(Q)$. Further we denote by $\mathcal{A}_k$ the union of $\mathcal{A}_k(E_N)$ for all N. Let F' denote the field generated over Q by $\sum_{\nu=1}^r a^{\theta_\nu}$ for all $a\in F$.

The results of [10] and [11] enable us to speak of (F'_{ab})-rational elements of $\mathcal{A}_0$. More generally, if Λ is a subfield of C containing F'_{ab}, we can define the field $\mathcal{A}_0(\Lambda)$ consisting of Λ-rational elements of $\mathcal{A}_0$.

Let J be a totally imaginary quadratic extension of F embeddable in B, and h an F-linear injection of J into B. Then $h(J^\times)$ has a unique fixed point $w=(w_1, \cdots, w_r)$ on $\mathfrak{H}^r$. Each θ_ν can be extended to two injections of J into C. If $\nu \leq r$, we can specify one of them, say ξ_ν, by

$$(7.4) \qquad h(a)_\nu \begin{bmatrix} w_\nu \\ 1 \end{bmatrix} = \begin{bmatrix} w_\nu \\ 1 \end{bmatrix} a^{\xi_\nu} \qquad \text{for every } a \in J.$$

(Cf. [9, 2.6, 2.7].) Thus (J, h) determines (J, Ξ_0) with $\Xi_0 = \sum_{\nu=1}^r \xi_\nu$; this is a *CM*-type if $r=g$. Let J' denote the field generated over Q by $\mathrm{tr}(\Xi_0(b))$ for all $b \in J$. The main theorems of [10] and [11] assert

$$(7.5) \qquad f(w) \in J'_{ab} \quad \text{if} \quad f \in \mathcal{A}_0(F'_{ab}) \ \text{ and } f \text{ is finite at } w.$$

Now the generalization of (0.2) can be stated as

THEOREM 7.1. *The notation J, w, and $\xi_1, \cdots, \xi_r$ being as above, take injections $\xi_{r+1}, \cdots, \xi_g$ of J into C which extend $\theta_{r+1}, \cdots, \theta_g$, and put $\Xi_1 = \sum_{\nu=1}^g \xi_\nu$ and $\Xi_2 = \sum_{\nu=1}^r \xi_\nu + \sum_{\nu=r+1}^g \xi_\nu \rho$. Let f be an element of $\mathcal{A}_0(\bar{Q})$ holomorphic at w. Then, for $1 \leq \nu_1 < \cdots < \nu_s \leq r$, we have*

$$(\partial^s f / \partial z_{\nu_1} \cdots \partial z_{\nu_s})(w) \sim \pi^s \prod_{i=1}^s [p(\xi_{\nu_i}, \Xi_1) p(\xi_{\nu_i}, \Xi_2)] .$$

Notice that $p(\xi_\nu, \Xi_1) p(\xi_\nu, \Xi_2)$ for $\nu \leq r$ depends only on $\xi_1, \cdots, \xi_r$, and is independent of the choice of $\xi_{r+1}, \cdots, \xi_g$, by virtue of Theorem 1.2. Before proving the theorem, we make some observations. First note that $\partial f / \partial z_\nu \neq 0$ for every $\nu \leq r$ if f is a non-constant element of $\mathcal{A}_0$.

PROPOSITION 7.2. *Let $q_\mu = (\partial u / \partial z_\mu)^{-1} \partial v / \partial z_\mu$ with non-constant functions u and v of $\mathcal{A}_0(F'_{ab})$. Then $q_\mu \in \mathcal{A}_0(\bar{Q})$. Moreover, $q_\mu(w) \in J'_{ab} F^{\theta_\mu}$ for the fixed point w of $h(J^\times)$ as above, if q_μ is holomorphic at w.*

PROOF. Take a non-constant element f of $\mathcal{A}_0(F'_{ab})$. Given w, we may assume, changing f for $f \circ \beta$ with a suitable $\beta \in B^+$, that $\partial f / \partial z_\mu(w) \neq 0$ for every $\mu \leq r$. Take $a \in J$ so that $J = Q(a^\rho/a)$; put $\alpha = h(a)$, $c_\lambda = (a^\rho/a)^\lambda$, and $s_\lambda = f \circ \alpha^\lambda$ for $\lambda = 1, \cdots, r$. Then

$$\frac{\partial(s_1, \cdots, s_r)}{\partial(z_1, \cdots, z_r)}(w) = \prod_{\mu=1}^r \frac{\partial f}{\partial z_\mu}(w) \cdot \det(c_\lambda^{\xi_\nu}) \neq 0 .$$

Therefore $s_1, \cdots, s_r$ are algebraically independent, and hence $\partial/\partial s_1, \cdots, \partial/\partial s_r$ are well-defined derivations of $\mathcal{A}_0(F'_{ab})$. Then $\partial u / \partial z_\mu = \sum_\lambda (\partial u / \partial s_\lambda)(\partial s_\lambda / \partial z_\mu)$, so that

$$(\partial f / \partial z_\mu)^{-1}(\partial u / \partial z_\mu)(w) = \sum_\lambda (\partial u / \partial s_\lambda)(w) c_\lambda^{\xi_\mu} \in J'_{ab} J^{\xi_\mu} = J'_{ab} F^{\theta_\mu} ,$$

if u is finite at w. Thus $q_\mu(w) \in J'_{ab} F^{\theta\mu}$ whenever both u and v are finite at w and $(\partial u/\partial z_\mu)(w) \neq 0$. Since such points w with the same J and $\{\xi_1, \cdots, \xi_r\}$ form a dense subset of $\mathfrak{H}_1^r$, we see the $q_\mu \in \mathcal{A}_0(J'_{ab} F^{\theta\mu})$, which implies our assertions.

The above proof tempts us to make the following

CONJECTURE 7.3. (i) $(\partial u/\partial z_\mu)^{-1}(\partial v/\partial z_\mu) \in \mathcal{A}_0(F'_{ab} F^{\theta\mu})$ if u, $v \in \mathcal{A}_0(F'_{ab})$.

(ii) *If u is a non-constant function in $\mathcal{A}_0(E_N) \cap \mathcal{A}_0(F'_{ab})$, the divisor of $\partial u/\partial z_\mu$ on a canonical model of $E_N \backslash \mathfrak{H}_1^r$ is rational over $F'_{ab} F^{\theta\mu}$.*

We now prove Theorem 7.1 in the case $s=1$, i.e.,

$$(7.6) \qquad \pi^{-1}(\partial f/\partial z_\nu)(w) \sim p(\xi_\nu, \varXi_1) p(\xi_\nu, \varXi_2) \qquad \text{for} \quad \nu=1, \cdots, r.$$

In view of Proposition 7.2, it is sufficient to prove this for one particular f in $\mathcal{A}_0(Q)$. Take a totally imaginary quadratic extension K of F, not isomorphic to J, which splits B. Identify $M_2(K)$ with $B \otimes_F K$. We can find an element T of $GL_2(K)$ such that ${}^t T^\rho = -T$ and

$$(7.7) \qquad B = \{\alpha \in M_2(K) \mid \alpha^\iota T = T \cdot {}^t \alpha^\rho\},$$

where ι denotes the main involution of $M_2(K)$. To show this, let γ be the involution of $M_2(K)$ which coincides with ρ on K and with the main involution on B. Then $x^\gamma = u \cdot {}^t x^\rho u^{-1}$ with a hermitian element u of $GL_2(K)$. Putting $T = \zeta u$ with a pure imaginary element ζ of K, we obtain (7.7). We can easily verify that $i T^{-\nu}$ is indefinite or definite according as $\nu \leq r$ or $\nu > r$ (cf. [9, 7.2], [10, 6.2]). Define G_Q of (4.1) with this T. The group $\{\alpha \in B \mid \alpha\alpha^\iota = 1\}$ is contained in G_Q, and there is a holomorphic isomorphism j of $\mathfrak{H}_1^r$ onto $\mathfrak{D}_1^r$ which makes the action of α on $\mathfrak{H}_1^r$ and on $\mathfrak{D}_1^r$ compatible (cf. [9, (7.3.1)], [10, 6.6]). Moreover we have $\mathfrak{A}_0(\bar{Q}) \circ j = \mathcal{A}_0(\bar{Q})$, as our construction of canonical models in [10] and [11] shows; we note that j is given by fractional linear transformations with algebraic coefficients.

Let $f \in \mathfrak{A}_0(\bar{Q})$. If w is the fixed point of $h(J^\times)$, we have, by Theorem 6.1,

$$(7.8) \qquad [\partial(\mathfrak{f} \circ j)/\partial z_1](w) \sim \pi \cdot R_1(j(w)) S_1(j(w)),$$

since the derivatives of j at w are algebraic. Note that $m=2$ and $r_1=s_1=1$ in the present case. Now take injections $\tau_1, \cdots, \tau_g$ of K into C so that $\tau_\nu = \theta_\nu$ on F and $-i T^{-\nu} > 0$ for $\nu > r$. Then

$$\varPhi = \sum_{\nu=1}^r (\tau_\nu + \tau_\nu \rho) + \sum_{\nu=r+1}^g 2\tau_\nu$$

in the present case. The point $j(w)$ of $\mathfrak{D}$ corresponds to an abelian variety of type $(L, \varPsi)$ with $L = JK$ and a CM-type $\varPsi = \sum_{\nu=1}^g (\alpha_\nu + \beta_\nu)$ of L defined by the table

<table>
<tr><td rowspan="2">(7.9)</td><td></td><td colspan="2">$\nu \leq r$</td><td colspan="2">$\nu > r$</td></tr>
<tr><td></td><td>on J</td><td>on K</td><td>on J</td><td>on K</td></tr>
<tr><td>$\alpha_\nu =$</td><td>ξ_ν</td><td>τ_ν</td><td>ξ_ν</td><td>τ_ν</td></tr>
<tr><td>$\beta_\nu =$</td><td>ξ_ν</td><td>$\tau_\nu \rho$</td><td>$\xi_\nu \rho$</td><td>τ_ν</td></tr>
</table>

as shown in [9, 6.5] and [10, 6.8]. By (5.10) and Theorem 1.2, we have

$$R_1(j(w))S_1(j(w)) \sim p(\alpha_1, \Psi)p(\beta_1, \Psi) \sim p(\xi_1, \Xi_1)p(\xi_1, \Xi_2),$$

which combined with (7.8) proves (7.6).

Let us now prove the statement at the end of §1. Put $\Xi = \Xi_1 + \Xi_2$ and observe that $J' = J^\Xi$. By Proposition 7.2, the left-hand side of (7.6) is determined up to factors belonging to $J'_{ab}F^0_\nu$. Therefore we can put $P(\tau_\nu, \Xi) = \pi^{-1}(\partial f/\partial z_\nu)(w)$ with non-constant $f \in \mathcal{A}_0(F'_{ab})$ for each $\nu \leq r$, which has the desired property of Conjecture 1.7. It should be noted that $(\partial f/\partial z_\nu)(w)$ depends on the choice of the algebra B in which J is embedded and also of the isomorphism (7.1).

The existence of "$\bar{Q}$-rational forms of weight 1" can be shown by

PROPOSITION 7.4. *For each $\mu \leq r$, there is a non-zero element t of $\mathcal{A}_k$ such that $\pi t^2 = u \cdot \partial f/\partial z_\mu$ with u and f in $\mathcal{A}_0(\bar{Q})$, where k is determined by μ in an obvious way.*

PROOF. The notation being as in the proof of (7.6), we have proved in §5 that $\delta R_1/S_1 \in \mathfrak{A}_0(\bar{Q})$ with $\delta = q(\tau_1\rho)/q(\tau_1)$. As already remarked, we have $j(z_1, \cdots, z_r) = ((a_1 z_1 + b_1)/(c_1 z_1 + d_1), \cdots)$ with $\begin{pmatrix} a_1 & b_1 \\ c_1 & d_1 \end{pmatrix}$ in $SL_2(\bar{Q})$. Then $t = \delta^{1/2}(R_1 \circ j)/(c_1 z_1 + d_1)$ has the required property for $\mu = 1$, by virtue of Theorem 6.1.

Let $\mathcal{A}_k(\bar{Q})$ denote, for $k = (k_1, \cdots, k_r) \in \mathbf{Z}^r$, the set of elements φ of $\mathcal{A}_k$ such that $\varphi^2 = \pi^{-\{k\}} f \prod_{\nu=1}^r (\partial u/\partial z_\nu)^{k_\nu}$ with a fixed non-constant element u of $\mathcal{A}_0(F'_{ab})$ and an arbitrary f of $\mathcal{A}_0(\bar{Q})$, where $\{k\} = \sum_{\nu=1}^r k_\nu$. Proposition 7.4 shows that $\mathcal{A}_k(\bar{Q}) \neq \{0\}$ for every k; obviously $\mathcal{A}_k(\bar{Q})$ is a one-dimensional vector space over $\mathcal{A}_0(\bar{Q})$. By Proposition 7.2, $\mathcal{A}_k(\bar{Q})$ does not depend on the choice of u. This implies in particular

PROPOSITION 7.5. *If $v \in \mathcal{A}_k(\bar{Q})$ and $\alpha \in B^+$, then $v|_k \alpha \in \mathcal{A}_k(\bar{Q})$.*

For $k = (k_1, \cdots, k_r)$ and $e = (e_1, \cdots, e_r)$ in $\mathbf{Z}^r$ with $e_\nu \geq 0$, define a differential operator d_k^e of degree $\{e\}$ acting on the functions on $\mathfrak{H}_1^r$ by

$$d_k^e = (2\pi i)^{-\{e\}} \prod_{\nu=1}^r \left(\frac{k_\nu + 2e_\nu - 2}{z_\nu - \bar{z}_\nu} + \frac{\partial}{\partial z_\nu} \right) \cdots \left(\frac{k_\nu + 2}{z_\nu - \bar{z}_\nu} + \frac{\partial}{\partial z_\nu} \right) \left(\frac{k_\nu}{z_\nu - \bar{z}_\nu} + \frac{\partial}{\partial z_\nu} \right),$$

where $\{e\} = \sum_{\nu=1}^r e_\nu$; we understand that d_k^0 is the identity operator (cf. [12,

§ 1]).

THEOREM 7.6. *Let $f\in\mathcal{A}_k(\bar{Q})$ and $v\in\mathcal{A}_{k+2e}(\bar{Q})$ with $e_\nu\geq 0$ for all ν. Suppose that both f and v^{-1} are holomorphic at the fixed point w of $h(J^\times)$ as above. Then $(v^{-1}d_k^e f)(w)\in\bar{Q}$.*

This is a generalization of the main theorem I of [12] and can be proved in exactly the same fashion.

COROLLARY 7.7. *If $f\in\mathcal{A}_0(\bar{Q})$ and $0<\nu_1<\cdots<\nu_s\leq r$, then*

$$\pi^{-s}\partial^s f/\partial z_{\nu_1}\cdots\partial z_{\nu_s}\in\mathcal{A}_k(\bar{Q}),$$

where k is determined by $\nu_1,\cdots,\nu_s$ in an obvious way.

This follows immediately from Theorem 7.6, and proves, combined with (7.6), the general case of Theorem 7.1.

The above results concern the archimedean primes of F unramified in B. Let us now prove an analogue of Proposition 7.4 for the remaining primes. Let χ_ν for $\nu=r+1,\cdots,g$ be injections of B into $M_2(\bar{Q})$ such that $\chi_\nu(a)=a^{\theta_\nu}1_2$ for $a\in F$. We consider $M_2(C)$-valued meromorphic functions $\mathfrak{i}$ on $\mathfrak{H}_1^r$ such that

$$(7.10)\qquad\qquad \mathfrak{i}(\gamma(z))=\chi_\nu(\gamma)\mathfrak{i}(z)\qquad\text{for every }\gamma\in E_N$$

with some N. The existence of such $\mathfrak{i}$ can be shown as follows. Identifying B with a subalgebra of $M_2(K)$, we can find an element U_ν of $GL_2(\bar{Q})$ such that $U_\nu^{-1}\chi_\nu(a)U_\nu=(Q_\nu^{-1}a^{\tau_\nu\rho}Q_\nu)^\rho$ for all $a\in B$. Put $\mathfrak{i}=U_\nu R_\nu\circ j$ with the function R_ν of (5.3) for a fixed $\nu>r$. Then (5.4_a) implies (7.10). Moreover, for any point v of $\mathfrak{H}_1^r$, we can take $\mathfrak{i}$ so that $\mathfrak{i}$ is holomorphic at v and $\det(\mathfrak{i}(v))\neq 0$.

To define the $\bar{Q}$-rationality of $\mathfrak{i}$, we consider the fixed point w of $h(J^\times)$ as before, and define ξ_ν, Ξ_1 and Ξ_2 as in Theorem 7.1. Then there is an element D_ν of $GL_2(\bar{Q})$ such that

$$(7.11)\qquad\qquad D_\nu\chi_\nu(h(x))D_\nu^{-1}=\mathrm{diag}[x^{\xi_\nu\rho},x^{\xi_\nu}]\qquad\text{for all }x\in J.$$

Put $q_\nu=[p(\xi_\nu,\Xi_1)/p(\xi_\nu\rho,\Xi_2)]^{1/2}$. We now consider the following property of $\mathfrak{i}$.

$$(7.12)\qquad\qquad \mathrm{diag}[q_\nu,q_\nu^{-1}]D_\nu\mathfrak{i}(w)\in M_2(\bar{Q}).$$

Observe that this property does not depend on the choice of D_ν.

PROPOSITION 7.8. *For each $\nu>r$, there exists an $M_2(C)$-valued meromorphic function $\mathfrak{i}$ on $\mathfrak{H}_1^r$ whose determinant is not identically equal to 0, and which satisfies (7.10) for some N and (7.12) at every fixed point w of the above type where $\mathfrak{i}$ is holomorphic.*

PROOF. The notation being as in the proof of (7.6), put $z_0=j(w)$ with the fixed point w of $h(J^\times)$. Define $\Psi=\sum_{\nu=1}^g(\alpha_\nu+\beta_\nu)$ by (7.9). Since

$$D_\nu U_\nu (Q_\nu^{-1} h(x)^{-\nu\rho} Q_\nu)^\rho U_\nu^{-1} D_\nu^{-1} = \mathrm{diag}[x^{\xi_\nu\rho}, \, x^{\xi_\nu}] \qquad \text{for} \quad x \in J,$$

we can take $D_\nu U_\nu$ as C_ν of (5.8_a); hence we have, by (5.9_a),

$$\mathrm{diag}[p(\beta_\nu, \Psi), \, p(\alpha_\nu, \Psi)]^{-1} D_\nu U_\nu R_\nu(z_0) \in GL_2(\bar{Q}).$$

By Theorem 1.2, (if $\nu > r$,) we have

$$p(\alpha_\nu, \Psi)/p(\beta_\nu, \Psi) \sim p(\xi_\nu, \Xi_1)/p(\xi_\nu\rho, \Xi_2),$$

$$p(\alpha_\nu, \Psi)p(\beta_\nu, \Psi) \sim p(\tau_\nu, \Phi_1)p(\tau_\nu, \Phi_2),$$

where $\Phi_1 = \sum_{\nu=1}^g \tau_\nu$ and $\Phi_2 = \sum_{\nu=1}^r \tau_\nu\rho + \sum_{\nu=r+1}^g \tau_\nu$. Therefore, if we put $b = [p(\tau_\nu, \Phi_1)p(\tau_\nu, \Phi_2)]^{1/2}$ and $\mathfrak{t} = b^{-1} U_\nu R_\nu \circ j$, then $\mathfrak{t}$ satisfies (7.12) at w. Let w^* be the fixed point of $h^*(J^{*\times})$ with some other h^* and J^*. Since b and U are independent of h and J, $\mathfrak{t}$ satisfies (7.12) at w^*, whenever the functions of (5.3) as well as $(P \circ \varepsilon)^{-1}$ are holomorphic at $j(w^*)$. Now we can find a function $\mathfrak{t}^*$ which satisfies (7.12) at w^* and such that $\det(\mathfrak{t}^*(w^*)) \neq 0$. Both $\mathfrak{t}$ and $\mathfrak{t}^*$ satisfy (7.12) at the points in a dense subset S of $\mathfrak{H}_1^r$. Put $\mathfrak{f} = \mathfrak{t}^{*-1}\mathfrak{t}$. We see that the entries of $\mathfrak{f}$ belong to $\mathcal{A}_0$ and take algebraic values on S, and hence belong to $\mathcal{A}_0(\bar{Q})$. Since $\mathfrak{t}$ is holomorphic at w^* if and only if $\mathfrak{f}$ is holomorphic at w^*, we see that $\mathfrak{t}$ satisfies (7.12) at w^* whenever $\mathfrak{t}$ is holomorphic at w^*. This completes the proof.

One can prove the analogues of the above theorems for the derivatives of the functions $\mathfrak{t}$ satisfying (7.10) and (7.12). Also, the field $\bar{Q}$ can be replaced by smaller fields depending on the weight. We have not pursued these points to their full extent in order not to obscure the ideas and to keep the paper short. In fact, almost all of the results in this section can be generalized (at least) to the group

$$\{\alpha \in GL_n(B) \mid {}^t\alpha \iota \alpha \in F\}$$

acting on $\mathfrak{H}_n^r$, which was treated in [10] and [11]. It should be noted that $(5.9_{a,b})$, (7.12), or their generalizations can be utilized for defining the arithmeticity of automorphic forms, especially in the case of compact quotient.

As a final remark, we give here a paraphrase of Proposition 2.1 in terms of modular forms. Let f be a (meromorphic) Hilbert modular form of weight k, i.e., an element of $\mathcal{A}_k$ with $B = M_2(F)$, and let q be an element of F such that $q^{\theta_\nu} > 0$ or < 0 according as $\nu \leq r$ or $> r$. Put

$$(7.13) \qquad u(z) = f(q^{\theta_1}z, \cdots, q^{\theta_r}z, q^{\theta_{r+1}}\bar{z}, \cdots, q^{\theta_g}\bar{z}) \qquad (z \in \mathfrak{H}_1),$$

and suppose $k_1 = \cdots = k_r = -k_{r+1} = \cdots = -k_g = \lambda$. Let $v(z)$ be a (meromorphic) modular form of weight $\lambda(2r-g)$ with respect to a congruence subgroup of $SL_2(Z)$. Observe that

$$(u/v)(\gamma(z))=(u/v)(z)[(cz+d)/(c\bar{z}+d)]^{\lambda(g-r)}$$

for $\gamma=\begin{pmatrix} * & * \\ c & d \end{pmatrix}$ in a congruence subgroup of $SL_2(\mathbf{Z})$.

PROPOSITION 7.9. *Let z_0 be a point of $\mathfrak{H}_1$ which generates an imaginary quadratic field K, and suppose both f and v are $\bar{\mathbf{Q}}$-rational. Then $(u/v)(z_0)\in\bar{\mathbf{Q}}$.*

PROOF. Put $L=FK$ and define a CM-type $\Psi=\sum_{\nu=1}^{g}\sigma_\nu$ so that $\sigma_\nu=\theta_\nu$ on F, $z_0^{\sigma_\nu}=z_0$ or $\bar{z}_0$ according as $\nu\leq r$ or $>r$. Then

$$u(z_0)\sim\prod_{\nu=1}^{r}p(\sigma_\nu,\,\Psi)^\lambda\cdot\prod_{\nu=r+1}^{g}p(\sigma_\nu,\,\Psi)^{-\lambda}\sim v(z_0)$$

by Proposition 2.1.

8. The non-vanishing of $H^1(\Gamma\backslash\mathfrak{D}_r,\,R)$.

Let $\mathfrak{S}$ be an irreducible bounded symmetric domain, and Γ a fixed-point-free discontinuous group of holomorphic automorphisms of $\mathfrak{S}$ such that $\Gamma\backslash\mathfrak{S}$ is compact. It was shown by Matsushima [4] that the first cohomology group of $\Gamma\backslash\mathfrak{S}$ over R vanishes unless $\mathfrak{S}$ is isomorphic to

$$(8.1) \qquad \mathfrak{D}_r=\mathfrak{D}(r,\,1)=\{(z_1,\,\cdots,\,z_r)\in C^r\mid\textstyle\sum_{k=1}^{r}|z_k|^2<1\}\,.$$

The purpose of this section is to show that the last condition is actually meaningful. In other words, for each r, we can produce an example of Γ acting on $\mathfrak{D}_r$ such that there is a holomorphic closed 1-form on $\Gamma\backslash\mathfrak{D}_r$ whose cohomology class is not 0. The paper [3] by Kazhdan and a forthcoming work [1, Ch. VIII] by Wallach should be mentioned as previous investigations on this topic. Our result can be stated as

THEOREM 8.1. *Let K, T, and $\tau_1,\,\cdots,\,\tau_g$ be as in §4. Define G_Q by (4.1) and a subgroup Γ_1 of G_Q by (4.14) with $N=1$ and with an arbitrary lattice $\mathfrak{m}$ of K_m^1. Let $(r_\nu,\,s_\nu)$ be the signature of $-iT^{\tau_\nu}$. Suppose $s_1=1$ and $s_2=\cdots=s_g=0$ (so that $\mathfrak{D}=\mathfrak{D}_{m-1}$). Then there is a subgroup Γ of Γ_1 of finite index such that $\Gamma\backslash\mathfrak{D}$ has a closed holomorphic 1-form whose cohomology class is not 0. Moreover, we can find $m-1$ such 1-forms $\xi_1,\,\cdots,\,\xi_{m-1}$ so that $\xi_1\wedge\cdots\wedge\xi_{m-1}\neq0$.*

This applies to both compact and non-compact quotients. Notice that, under the assumption on s_ν in the theorem, $\Gamma\backslash\mathfrak{D}$ is compact if $g>1$; it is not compact if $g=1$ and $r_1>1$.

To prove our theorem, we introduce the classical theta function

$$\theta(u,\,Z;\,a,\,b)=\sum_{x\in Z^n}e((1/2)\cdot{}^t(x+a)Z(x+a)+{}^t(x+a)(u+b))$$

$$(u\in C^n,\,Z\in\mathfrak{H}_n,\,a\in R^n,\,b\in R^n),$$

where $e(v)=e^{2\pi iv}$. It is well known that, for every $\gamma=\begin{pmatrix} * & * \\ C & D \end{pmatrix}\in\Delta_2$, one has

(8.2) $\quad \theta({}^{t}(CZ+D)^{-1}u,\ \gamma(Z)\ ;\ a,\ b)$

$$=\lambda_{\gamma}e(({}^{t}ab-{}^{t}a'b')/2)\det(CZ+D)^{1/2}e((1/2)\cdot{}^{t}u(CZ+D)^{-1}Cu)\theta(u,\ Z\ ;\ a',\ b'),$$

where $({}^{t}a',\ {}^{t}b')=({}^{t}a,\ {}^{t}b)\gamma$, and λ_{γ} is a fourth root of unity depending only on γ. With fixed $a_{1},\ \cdots,\ a_{n}$ in $\boldsymbol{Q}^{n}$, define an $M_{n}(\boldsymbol{C})$-valued function V on $\mathfrak{H}_{n}$ by

$$V(Z)=(v_{jk}(Z)),\quad v_{jk}(Z)=(\partial\theta/\partial u_{j})(0,\ Z\ ;\ a_{k},\ 0).$$

We see from (8.2) that

$$V(\gamma(Z))=\lambda_{\gamma}\det(CZ+D)^{1/2}\cdot(CZ+D)V(Z)\qquad\text{for every } \gamma=\begin{pmatrix}* & *\\ C & D\end{pmatrix}\in\Delta_{N},$$

where N is a suitably large even positive integer depending on $a_{1},\ \cdots,\ a_{n}$. Moreover, as shown in [13, Proposition 1.2], given a point Z_{0} of $\mathfrak{H}_{n}$, we can choose $a_{1},\ \cdots,\ a_{n}$ so that $\det(V(Z_{0}))\neq 0$. Now define a function V^{ϵ} on $\mathfrak{D}$ by

$$V^{\epsilon}(z)=\det(\omega_{2}(z))^{-1/2}\cdot{}^{t}\omega_{2}(z)^{-1}V(\epsilon(z))\qquad(z\in\mathfrak{D})$$

with any choice of $\det(\omega_{2}(z))^{-1/2}$, where ϵ and ω_{2} are as in §4. By (4.16), (4.10$_{a,b}$), and (3.9), we have

(8.3) $\qquad V^{\epsilon}(\alpha(z))=t(\alpha)\prod_{\nu}'\det(\mu_{\nu}(\alpha,\ z))\cdot\Lambda(\alpha,\ z)V^{\epsilon}(z)\qquad\text{for } \alpha\in\Gamma_{N}$

with a fourth root of unity $t(\alpha)$, where $\prod_{\nu}'$ is the product over all ν such that $r_{\nu}s_{\nu}\neq 0$. Obviously t is a character of Γ_{N}. Choose the a_{j} so that $\det(V^{\epsilon})$ does not vanish identically on $\mathfrak{D}$.

Now assume that $r_{1}=m-1$, $s_{1}=1$, and $s_{2}=\cdots=s_{g}=0$. Then $\mathfrak{D}=\mathfrak{D}_{m-1}$. If R is the first $m-1$ rows of V^{ϵ}, (8.3) implies that

(8.4) $\qquad R(\alpha(z))=t(\alpha)\mu_{1}(\alpha,\ z)\lambda_{1}(\alpha,\ z)R(z)\qquad\text{for } \alpha\in\Gamma_{N}.$

Take an arbitrary non-vanishing column of $R(z)$ and call it $q(z)$, and its components $q_{1},\ \cdots,\ q_{m-1}$. Let $z_{1},\ \cdots,\ z_{m-1}$ be the components of the variable point z on $\mathfrak{D}$. Define a 1-form ξ on $\mathfrak{D}$ by

(8.5) $\qquad \xi=\sum_{k=1}^{m-1}q_{k}(z)dz_{k}={}^{t}q(z)dz.$

From (8.4) and (3.7), we see that $\xi\circ\alpha=t(\alpha)\xi$ for all $\alpha\in\Gamma_{N}$. Therefore if $\Gamma=\mathrm{Ker}(t)$, then $[\Gamma_{N}:\Gamma]\leq 4$, and ξ defines a non-vanishing holomorphic 1-form on $\Gamma\backslash\mathfrak{D}$. Since R has rank $m-1$, we can choose $m-1$ columns of $R(z)$ which produce 1-forms $\xi_{1},\ \cdots,\ \xi_{m-1}$ such that $\xi_{1}\wedge\cdots\wedge\xi_{m-1}\neq 0$.

If $\Gamma\backslash\mathfrak{D}$ is compact, ξ is closed and the cohomology class of ξ is automatically non-vanishing; our proof is completed in that case. In the non-compact case, a further reasoning is necessary. As mentioned above, if $\mathfrak{D}=\mathfrak{D}_{m-1}$, the non-compact quotient occurs only when $g=1$. We are naturally interested only in the case $m>2$. First, to show that ξ is closed, we use the unbounded domain

$$\mathfrak{Z}=\{(w, z)\in C^{m-2}\times C \mid i({}^t\overline{w}Sw+\overline{z}-z)>0\}\ ,$$

which is isomorphic to $\mathfrak{D}_{m-1}$ and was introduced in [14]. Here S is a diagonal skew-hermitian element of $GL_{m-2}(K)$ determined by T. We defined in [14, §3] an embedding ε' of $\mathfrak{Z}$ into $\mathfrak{H}_m$ similar to ε; so we can define everything with respect to $\mathfrak{Z}$ and ε' instead of $\mathfrak{D}$ and ε. The function on $\mathfrak{Z}$ corresponding to V^ε is defined by

$$(8.6) \qquad\qquad X(w, z)={}^t\mathfrak{X}(w, z)^{-1}V(\varepsilon'(w, z)) \qquad ((w, z)\in\mathfrak{Z})$$

with $\mathfrak{X}$ of [14, (3.12)] (cf. (3.19) and Proposition 3.2 of [14]). Then q corresponds to the first $m-1$ elements of a suitable column of $X(w, z)$, which can be given as

$$\begin{bmatrix} x_1 \\ \vdots \\ x_m \end{bmatrix}={}^t\mathfrak{X}(w, z)^{-1}\begin{bmatrix} y_1\circ\varepsilon' \\ \vdots \\ y_m\circ\varepsilon' \end{bmatrix}, \qquad y_k=(\partial\theta/\partial u_k)(0, Z;\ a, 0)$$

with a suitable $a\in Q^m$. Thus ξ corresponds to a 1-form $\eta=\sum_{k=1}^{m-2}x_k dw_k+x_{m-1}dz$ on $\mathfrak{Z}$. Using the explicit form of $\mathfrak{X}$ in [14, (3.12)], we find

$$x_k=y_k\circ\varepsilon'-(s_kw_k/2)y_{m-1}\circ\varepsilon'-(\delta s_kw_k/2)y_m\circ\varepsilon' \qquad \text{if}\ \ k<m-1,$$

$$x_{m-1}=(y_{m-1}\circ\varepsilon'-\delta y_m\circ\varepsilon')/2,$$

where s_k is the k-th diagonal element of S and δ is a generator of K used in [14, §3]. Now [14, (3.7)] shows that

$$d\varepsilon'(w, z)=\begin{pmatrix} 0 & dw & \delta dw \\ {}^t dw & dz-{}^t wSdw & -\delta\cdot{}^t wSdw \\ \delta\cdot{}^t dw & -\delta\cdot{}^t wSdw & -\delta^2(dz+{}^t wSdw) \end{pmatrix}.$$

Combining this with the well known relation

$$(\partial^2\theta/\partial u_j\partial u_k)(u, Z;\ a, b)=2\pi i(1+\delta_{jk})(\partial\theta/\partial Z_{jk})(u, Z;\ a, b),$$

we can show by a direct calculation that $\partial x_{m-1}/\partial w_k=\partial x_k/\partial z$ and $\partial x_j/\partial w_k=\partial x_k/\partial w_j$ if $j<m-1$ and $k<m-1$, which proves the desired closedness.

To show the non-vanishing of the cohomology-class, we go back to $\mathfrak{D}$ and take T in the diagonal form $T=\mathrm{diag}[\zeta_1, \cdots, \zeta_m]$. Since the equivalence class of T can be determined by its signature and the class of $\det(T)$ modulo $N_{K/Q}(K^\times)$, we can choose $\zeta_1, \cdots, \zeta_m$ so that $\zeta_{m-1}\zeta_m>0$ and $\zeta_{m-1}\zeta_m\notin N_{K/Q}(K^\times)$. (We are naturally assuming $m>2$.) Put

$$G'_Q=\{\alpha\in GL_2(K)\mid\alpha T'\cdot{}^t\alpha=T'\}\ , \quad T'=\mathrm{diag}[\zeta_{m-1}, \zeta_m]\ .$$

Then G'_Q acts on the disc $\mathfrak{D}_1$. Moreover, $\alpha\mapsto\mathrm{diag}[1_{m-1}, \alpha]$ gives an injection of G'_Q into G_Q, which is compatible with the embedding $z\mapsto\begin{pmatrix}0\\z\end{pmatrix}$ of $\mathfrak{D}_1$ into

$\mathfrak{D}_{m-1}$. Take a congruence subgroup Γ' of G'_Q so that Γ' is mapped into Γ_N. Then Γ' has a subgroup of finite index Γ'' which is mapped into Γ. We can construct the above ξ of (8.5) so that its pull-back to $\mathfrak{D}_1$, say ξ', is not 0. Our choice of T' implies that $\Gamma''\backslash\mathfrak{D}_1$ is compact, so that there is an element γ of Γ'' and a point w of $\mathfrak{D}_1$ such that $\int_w^{\gamma(w)} \xi' \neq 0$. This shows the non-vanishing of the cohomology-class of ξ on $\Gamma\backslash\mathfrak{D}$, and completes the proof.

Instead of taking the first $m-1$ columns, we can consider the m columns W_ν of V^ε corresponding to the ν-th diagonal block $\Lambda_\nu(\alpha, z)$ of $\Lambda(\alpha, z)$ for $\nu>1$, when $g>1$. Then

$$W_\nu(\alpha(z))=t(\alpha)\mu_1(\alpha, z)\Lambda_\nu(\alpha)W_\nu(z) \qquad \text{for} \quad \alpha\in\Gamma_N ,$$

where $\Lambda_\nu(\alpha)=(Q_\nu^{-1}\alpha^{\varepsilon\nu\rho}Q_\nu)^\eta$. Obviously Λ_ν is an injection of G_Q into $U(m)$. Thus W_ν gives a rather interesting type of automorphic form.

References

[1] A. Borel and N. Wallach, Continuous cohomology, discrete subgroups and representations of reductive groups, to appear in Annals of Math. studies.

[2] B. Gross, On the periods of abelian integrals and a formula of Chowla and Selberg, Invent. Math. 45 (1978), 193-211.

[3] Kazhdan, Some applications of the Weil representation, preprint, 1976.

[4] Y. Matsushima, On the first Betti number of compact quotient spaces of higher dimensional symmetric spaces, Ann. of Math., 75 (1962), 312-330.

[5] K. Miyake, Models of certain automorphic function fields, Acta Math., 126 (1971), 245-307.

[6] A. Selberg and S. Chowla, On Epstein's zeta-function, J. reine angew. Math., 227 (1967), 86-110.

[7] G. Shimura, On analytic families of polarized abelian varieties and automorphic functions, Ann. of Math., 78 (1963), 149-192.

[8] G. Shimura, On the field of definition for a field of automorphic functions, I, II, III, Ann. of Math., 80 (1964), 160-189; 81 (1965), 124-165; 83 (1966), 377-385.

[9] G. Shimura, Construction of class fields and zeta functions of algebraic curves, Ann. of Math., 85 (1967), 58-159.

[10] G. Shimura, Algebraic number fields and symplectic discontinuous groups, Ann. of Math., 86 (1967), 503-592.

[11] G. Shimura, On canonical models of arithmetic quotients of bounded symmetric domains, I, II, Ann. of Math., 91 (1970) 144-222; 92 (1970), 528-549.

[12] G. Shimura, On some arithmetic properties of modular forms of one and several variables, Ann. of Math., 102 (1975), 491-515.

[13] G. Shimura, On the derivatives of theta functions and modular forms, Duke Math. J., 44 (1977), 365-387.

[14] G. Shimura, The arithmetic of automorphic forms with respect to a unitary group, Ann. of Math., 107 (1978), 569-605.

[15] G. Shimura and Y. Taniyama, Complex multiplication of abelian varieties and its applications to number theory, Publ. Math. Soc. Japan, No. 6, 1961.

[16] A. Weil, Sur les périodes des intégrales abéliennes, Comm. Pure Appl. Math., 29 (1976), 813–819.

[17] T. Kubota, On the field extension by complex multiplication, Trans. Amer. Math. Soc., 118 (1965), 113–122.

Goro SHIMURA

Department of Mathematics
Princeton University
Princeton, New Jersey 08540
U.S.A.

On some problems of algebraicity

Proceedings of the International Congress of Mathematicians,
Helsinki, 1978 (1979), 373-379

1. As every mathematician knows, the parallelism between algebraic number theory and transcendental number theory exists only in their appellations and not in their contents. Indeed, the aim of the latter theory is to prove the transcendence of a given number, while algebraic numbers are there from the beginning in the former. Therefore, if one proves the algebraicity of an analytically defined number, it cannot be viewed as a theorem of either theory. It belongs to a new area of investigation for which this lecture is intended and to which I can give no good designation. Although the problem of algebraicity has not attracted much attention until recently, there are at least two classical and well-known examples.

(1) *The values of Riemann's zeta function.* If $\zeta(s)=\sum_{n=1}^{\infty}n^{-s}$, then $\pi^{-m}\zeta(m)$ for an even positive integer m is a rational number.

(2) *Complex multiplication of elliptic modular functions.* If f is an elliptic modular function with rational Fourier coefficients, then the value $f(z)$ at an imaginary quadratic irrational z is algebraic.

In each example, one can go beyond the algebraicity. In (2), for example, one can actually prove that $f(z)$ generates an abelian extension of $Q(z)$; also in both cases, the p-adic nature of the numbers is an interesting topic. In this lecture, however, I will concentrate only in the algebraicity, without touching on more delicate questions. It should be noted that the question of algebraicity can be naturally asked for various mathematical objects. For instance, one can ask whether a given complex manifold V is an algebraic variety. Supposing it is so, one may still ask whether V has a model defined over an algebraic number field. This question, when V is an arithmetic quotient of a bounded symmetric domain, is fundamental in our investigation. However, as I talked on this in a previous Congress ([6], see also [2], [4], [5]), I take the algebraicity of such V to be a starting point rather than a goal in the following discussion.

Now what kind of numbers should be offered for their algebraicity? As a generalization of example (1), a natural choice falls upon the values of an L-function of a number field F with, say, an abelian character χ. If χ is of finite order, the problem is reduced, for some natural reasons, to the case where F is totally real and χ is totally ramified or unramified. The results, due to Siegel and Klingen, being well known, will need no detailed description here. Beyond this, one may attempt to study the values of zeta functions obtained from various algebraic groups, say $GL_n(F)$. If F is totally real and $n=2$, we can again obtain a certain algebraicity theorem for the values of the Mellin transform of a holomorphic Hilbert cusp form [11]. Some results have been obtained also by Sturm [13] for the zeta functions of $GL_3(Q)$ related to cusp forms of $GL_2(Q)$. These results suggest some general principles of algebraicity, but without making any speculation, let us narrow down our subject to the case in which the values of some zeta functions occur as the special values of Eisenstein series. The simplest example is the zeta function of an imaginary quadratic field with a Hecke character of infinite order. That such a coincidence occurs in this case is rather obvious, but its nontrivial generalization (still in the quadratic case) was given by Damerell [1]. Subsequently a different approach was taken by Weil [14]; these results have been generalized to the case related to the Eisenstein series of the Hilbert modular groups [7]. We now propose to extend this to a more general framework, by considering automorphic forms of a Q-rational reductive algebraic group G. We assume that the semi-simple part of G_R modulo a maximal compact subgroup is a bounded symmetric domain S, and take, for simplicity, a power of the jacobian as the factor of automorphy. We then lay out our program by asking the following series of questions:

(I) *Can one define the notion of arithmetic automorphic functions?*

(II) *Can one define the notion of arithmetic automorphic forms?*

(III) *Are (holomorphic) Eisenstein series arithmetic?*

(IV) *Is there any explicit way to construct arithmetic automorphic forms, similar to Eisenstein series, in the case of compact quotient?*

(V) *Supposing the answers to these questions are affirmative, is there any interpretation of the values of such explicit arithmetic automorphic forms at CM-points as the values of zeta functions?*

The arithmeticity should be defined relative to certain number fields. But without seeking the sharper results, we simply take the algebraic closure $\overline{Q}$ of the rational number field as the basic field; so *arithmetic* forms may be called *$\overline{Q}$-rational* forms. The first question is essentially the same as the problem of finding the (canonical) $\overline{Q}$-rational models of the quotients S/Γ for congruence subgroups Γ of G_Q. As we said above, we start by assuming their existence. This means that we can speak of $\bar{Q}$-rational automorphic functions, whose values at CM-points are algebraic. As to (II), if S/Γ is not compact, we can consider Fourier expansions of automorphic forms with respect to a group of translations T contained in Γ. In general, the Fourier coefficients are theta functions; they become constants if T is sufficiently large. The characterization of the $\bar{Q}$-rationality by means of some properties of such Fourier coefficients are treated in [9], [10] and Garrett [3]; so let us now consider a more general method applicable even to the case of compact quotient. The idea is: first define a certain constant $p(w)$ at each CM-point w on S; then call an automorphic form f $\bar{Q}$-rational if $f(w)/p(w) \in \bar{Q}$ for every CM-point w. The constant $p(w)$ is given as a period of an abelian variety. Let us first discuss its definition and basic properties.

2. Let K be a totally imaginary quadratic extension of a totally real algebraic number field F of finite degree. We call such a K a *CM-field*. Let I_K denote the $\mathbf{Z}$-module of all formal sums $\sum_\tau c_\tau \tau$ of embeddings τ of K into $\mathbf{C}$ with $c_\tau \in \mathbf{Z}$. Such a sum with $c_\tau \geqslant 0$ defines an equivalence class of representations of K by complex matrices. An element Φ of I_K is called a *CM-type* of K if $\Phi + \Phi\varrho$ represents the regular representation of K over $\mathbf{Q}$. Here and henceforth ϱ denotes the complex conjugation. Given a CM-type Φ of K, there is an abelian variety A, defined over $\bar{Q}$, such that: (i) $2\dim(A)=[K:Q]$; (ii) there is an injection ι of K into $\mathrm{End}(A) \otimes Q$; (iii) Φ is the class of representation of K on the space of holomorphic 1-forms on A. Let $[K:Q]=2g$ and $\Phi = \sum_{v=1}^{g} \tau_v$. For each v, there exists a $\bar{Q}$-rational holomorphic 1-form ω_v, $\neq 0$, on A satisfying $\omega_v \circ \iota(a) = a^{\tau_v} \omega_v$ for all $a \in K$ such that $\iota(a) \in \mathrm{End}(A)$.

PROPOSITION 1. *There exists a nonzero complex number $p(\tau_v, \Phi)$ depending only on K, Φ, and τ_v such that $\int_c \omega_v / [\pi \cdot p(\tau_v, \Phi)] \in \bar{Q}$ for all 1-cycles c on A (i.e. elements c of $H_1(A, \mathbf{Z})$).*

For the proof, see [8]. We put $p(\tau, \Phi) = p(\tau\varrho, \Phi)^{-1}$ for an embedding τ of K into $\mathbf{C}$ not belonging to $\{\tau_1, ..., \tau_g\}$.

THEOREM 1. *Let $\Phi_1, ..., \Phi_m$ be CM-types of K, and τ an embedding of K into $\mathbf{C}$. Then*

$$(2.1) \qquad \prod_{i=1}^{m} p(\tau, \Phi_i)^{r_i}$$

with $r_i \in \mathbf{Z}$ depends, up to algebraic factors, only on $\sum_{i=1}^{m} r_i \Phi_i$ and τ. Moreover, let Ψ be a CM-type of a CM-field L containing K, and suppose the restriction of Ψ to K is $\sum_i r_i \Phi_i$. Then the product of $p(\sigma, \Psi)$ for all embeddings σ of L into $\mathbf{C}$ which coincide with τ on K differs from (2.1) by an algebraic factor.

We denote by $p(\tau, \sum_i r_i \Phi_i)$ the number (2.1) which is determined up to algebraic factors. Assuming $K \subset \mathbf{C}$, put $\Phi_K = \Omega_K + \varepsilon - \varepsilon\varrho$, where ε is the identity embedding of K into $\mathbf{C}$ and Ω_K the sum of all embeddings of K into $\mathbf{C}$. Then $p(\tau, \Phi_K)$ is meaningful for every embedding τ of K into $\mathbf{C}$.

THEOREM 2. *Let Φ be a CM-type of K and (K', Φ') the reflex of (K, Φ). Then, for every embedding σ of K' into C, $p(\sigma, \Phi')^2$ is an algebraic number times the product of $p(\tau, \Phi_K)$ for all τ belonging to $\Phi\sigma$.*

Note that $\Phi\sigma$ is a well-defined *CM*-type of K. This relation yields an upper bound for the number of algebraically independent elements among $p(\sigma, \Phi')$. The details of these results will be given in [12].

3. We now use $p(\tau, \Phi)$ for the definition of arithmetic automorphic forms. Though the idea is expected to work in a more general case, we consider here the case of arithmetic subgroups of $\mathrm{Sp}\,(n, R)^r$. Let B be a quaternion algebra over a totally real algebraic number field F of degree g. Define a Q-rational algebraic group G so that

$$G_Q = \{\alpha \in \mathrm{GL}_n(B) \,|\, {}^t\alpha^\iota\alpha = \nu(\alpha)1_n \quad \text{with} \quad \nu(\alpha)\in F\},$$

where ι denotes the main involution of B. Let $\theta_1, ..., \theta_n$ be the embeddings of F into R, and suppose that B is unramified at $\theta_1, ..., \theta_r$ and ramified at $\theta_{r+1}, ..., \theta_g$. Then there is an isomorphism

$$(3.1) \qquad\qquad\qquad G_R \simeq \prod_{\nu=1}^{g} G_\nu,$$

$$G_\nu = \begin{cases} \{\alpha \in \mathrm{GL}_{2n}(R) \,|\, {}^t\alpha J\alpha = \nu(\alpha)J \quad \text{with} \quad \nu(\alpha)\in R\} \ (\nu = 1, ..., r), \\ \{\alpha \in \mathrm{GL}_n(H) \,|\, {}^t\alpha^\iota\alpha = \nu(\alpha)1_n \quad \text{with} \quad \nu(\alpha)\in R\} \ (\nu = r+1, ..., g), \end{cases}$$

where H denotes the Hamilton quaternions, ι the main involution of H, and

$$J = \begin{bmatrix} 0 & -1_n \\ 1_n & 0 \end{bmatrix}.$$

Let G_{Q+} denote the subgroup of G_Q consisting of all α such that $\nu(\alpha)$ is totally positive. For each $\alpha\in G_Q$, let $(\alpha_1, ..., \alpha_g)$ be the corresponding element of $\prod_{\nu=1}^{g} G_\nu$. We fix an isomorphism (3.1) so that $\alpha_1, ..., \alpha_r$ have algebraic entries for every $\alpha\in G_Q$. If $\alpha\in G_{Q+}$, we define the action of α on the product $\mathfrak{H}_n^r$ of r copies of

$$\mathfrak{H}_n = \{z \in M_n(C) \,|\, {}^tz = z, \ \mathrm{Im}\,(z) > 0\}$$

by $\alpha(z_1, ..., z_r) = (\alpha_1(z_1), ..., \alpha_r(z_r))$, $\alpha_\nu(z_\nu) = (a_\nu z_\nu + b_\nu)(c_\nu z_\nu + d_\nu)^{-1}$ where

$$\alpha_\nu = \begin{bmatrix} a_\nu & b_\nu \\ c_\nu & d_\nu \end{bmatrix}.$$

A factor of automorphy $\chi_\nu(\alpha, z)$ can be defined by

$$\chi_\nu(\alpha, z) = c_\nu z_\nu + d_\nu \quad (z = (z_1, ..., z_r)\in\mathfrak{H}_n^r; \ \nu = 1, ..., r).$$

Also we consider a Q-linear embedding $\chi_\nu: M_n(B) \to M_{2n}(\bar{Q})$ such that $\chi_\nu(a1_n) = a^{\theta_\nu}1_{2n}$ for $\nu = r+1, ..., g$.

As shown in [5], we can construct canonical models of $\mathfrak{H}_n^r/\Gamma$ for congruence subgroups Γ of G_{Q+}, so that the field of $\bar{Q}$-rational Γ-automorphic functions is meaningful. We denote by $\mathfrak{A}_0(\bar{Q})$ the union of such fields for all Γ. Our task is to study the $\bar{Q}$-rationality of automorphic forms defined relative to $\chi_1, ..., \chi_g$

in connection with their values at the *CM*-points of G_{Q+} on $\mathfrak{H}_n^r$, which can be obtained as follows. Let $Y = Y_1 \oplus \ldots \oplus Y_t$ with simple *F*-algebras Y_i; suppose Y has a positive involution δ and there is an *F*-linear embedding h of Y into $M_n(B)$ such that $h(a)^\delta = {}^t h(a)^t$. Let L_i be the center of Y_i and put $L = L_1 \oplus \ldots \oplus L_t$, $q_i^2 = [Y_i : L_i]$, $2m_i = [L_i : F]$, $Y[\delta] = \{x \in Y | x^\delta x \in F\}$, $L[\delta] = Y[\delta] \cap L$. Then $h(Y[\delta]) \subset G_{Q+}$. Suppose $n = \sum_{i=1}^t q_i m_i$. Then $h(Y[\delta])$ has a unique common fixed point $w = (w_1, \ldots, w_r)$ on $\mathfrak{H}_n^r$. It is this type of fixed point where we examine the values of automorphic forms. For each $v \leqslant r$, we can define a representation $\Xi_v : Y \to M_n(C)$ such that $\Xi_v(a) = \chi_v(h(a), w)$ for all $a \in Y[\delta]$. Then m_i embeddings $\sigma_{i1}^v, \ldots, \sigma_{im_i}^v$ of L_i into C can be determined by $\mathrm{tr}\,(\Xi_v(a)) = q_i \sum_j \sigma_{ij}^v(a)$ for $a \in L_i$. For $v > r$, take m_i embeddings $\sigma_{i1}^v, \ldots, \sigma_{im_i}^v$ of L_i into C which extend θ_v so that $\sum_{v=1}^g \sum_{j=1}^{m_i} \sigma_{ij}^v$ is a *CM*-type of L_i. Consider these σ_{ij}^v homomorphisms of L into C, and call Ψ_v the diagonal representation of L of degree n composed of

$$\mathrm{diag}\,[\sigma_{i1}^v, \ldots, \sigma_{im_i}^v] \otimes 1_{q_i}$$

for $i = 1, \ldots, t$. Define also an element Φ_i of I_{L_i} by

$$\Phi_i = \sum_{j=1}^{m_i} \left\{ \sum_{v=1}^r 2\sigma_{ij}^v + \sum_{v=r+1}^g (\sigma_{ij}^v + \sigma_{ij}^v \varrho) \right\}.$$

Then $p(\sigma_{ij}^v, \Phi_i)$ is meaningful. We can find $E_v \in \mathrm{GL}_n(\bar{Q})$ for $v \leqslant r$ and $E_v \in \mathrm{GL}_{2n}(\bar{Q})$ for $v > r$ so that

$$E_v \chi_v(h(a), w) E_v^{-1} = \Psi_v(a) \quad \text{for all} \quad a \in L[\delta] \quad (v = 1, \ldots, r),$$

$$E_v \chi_v(h(a)) E_v^{-1} = \mathrm{diag}\,[\Psi_v(a)^\varrho, \Psi_v(a)] \quad \text{for all} \quad a \in L \quad (v = r+1, \ldots, g).$$

Put $p_{ij}^v = p(\sigma_{ij}^v, \Phi_i)^{1/2}$ and define diagonal matrices $P_v(w)$ by

$$P_v(w) = \begin{cases} \mathrm{diag}\,[P_{v1} \otimes 1_{q_1}, \ldots, P_{vt} \otimes 1_{q_t}] & (v \leqslant r), \\ \mathrm{diag}\,[P_{v1} \otimes 1_{q_1}, \ldots, P_{vt} \otimes 1_{q_t}, P_{v1}^{-1} \otimes 1_{q_1}, \ldots, P_{vt}^{-1} \otimes 1_{q_t}] & (v > r), \end{cases}$$

$$P_{vi} = \mathrm{diag}\,[p_{i1}^v, \ldots, p_{im_i}^v].$$

Here we denote by $\mathrm{diag}\,[P_1, \ldots, P_s]$ the square matrix with square matrices $P_1, \ldots, P_s$ in the diagonal blocks and 0 in all other blocks.

THEOREM 3. *Given an arbitrary point* z_0 *of* $\mathfrak{H}_n^r$, *there exists, for each* $v \leqslant g$, *a meromorphic function* S_v *on* $\mathfrak{H}_n^r$, *with values in* $M_n(C)$ *or in* $M_{2n}(C)$ *according as* $v \leqslant r$ *or* $v > r$, *such that:* (i) S_v *is holomorphic at* z_0 *and* $\det(S_v(z_0)) \neq 0$; (ii) $S_v(\gamma(z)) = \chi_v(\gamma, z) S_v(z)$ *for all* γ *in a congruence subgroup of* G_{Q+}, *where we understand that* $\chi_v(\gamma, z) = \chi_v(\gamma)$ *if* $v > r$; (iii) *if* w *is the fixed point of* $h(Y[\delta])$ *as above and* S_v *is holomorphic at* w, *then* $P_v(w)^{-1} E_v S_v(w)$ *is a* $\bar{Q}$-*rational matrix.*

This theorem enables us to define $\bar{Q}$-rational automorphic forms of a general type. Take a $\bar{Q}$-rational representation $\omega : (\mathrm{GL}_n)^r \times (\mathrm{GL}_{2n})^{g-r} \to \mathrm{GL}_d$, and consider a C^d-valued meromorphic function f on $\mathfrak{H}_n^r$. For every $\alpha \in G_{Q+}$ and $z \in \mathfrak{H}_n^r$, put

$$\omega(\alpha, z) = \omega(\chi_1(\alpha, z), \ldots, \chi_r(\alpha, z), \chi_{r+1}(\alpha), \ldots, \chi_g(\alpha)),$$

$$(f|_\omega \alpha)(z) = \omega(\alpha, z)^{-1} f(\alpha(z)).$$

We denote by $\mathfrak{A}_\omega$ the set of all C^d-valued meromorphic functions f such that $f|_\omega \gamma = f$ for all γ in a congruence subgroup of G_Q. Taking S_ν as in Theorem 3, we denote by $\mathfrak{A}_\omega(\bar{Q})$ the set of all elements f of $\mathfrak{A}_\omega$ such that $\omega(S_1, \ldots, S_g)^{-1} f$ has components in $\mathfrak{A}_0(\bar{Q})$. This does not depend on the choice of S_ν. Moreover, if $f \in \mathfrak{A}_\omega(\bar{Q})$ and $\alpha \in G_{Q+}$, then $f|_\omega \alpha \in \mathfrak{A}_\omega(\bar{Q})$. The derivatives of elements of $\mathfrak{A}_0(\bar{Q})$ are $\bar{Q}$-rational in the sense described by

THEOREM 4. *For a meromorphic function f on $\mathfrak{H}_n^r$, define an $M_n(C)$-valued function $\Delta_\nu f$ on $\mathfrak{H}_n^r$ by*

$$(\Delta_\nu f)(z_1, \ldots, z_r) = \left(\frac{1 + \delta_{ij}}{2} \frac{\partial f}{\partial z_{\nu ij}} \right)_{i,j=1,\ldots,n} \qquad (\nu = 1, \ldots, r),$$

where $z_{\nu ij}$ is the (i, j)-entry of the matrix variable z_ν on $\mathfrak{H}_n$. If $f \in \mathfrak{A}_0(\bar{Q})$ and S_ν is a function of Theorem 3, then the entries of $\pi^{-1} S_\nu^{-1} \Delta_\nu f \cdot {}^t S_\nu^{-1}$ belong to $\mathfrak{A}_0(\bar{Q})$.

The proofs of these theorems will be given in **[12]** in the case $n=1$. The general case can be proved in exactly the same fashion.

4. Finally we give new examples of affirmative answers to questions (IV) and (V). The notation being as in § 3, let E be a subfield of F such that $[E:Q]=r$ and the restrictions of $\theta_1, \ldots, \theta_r$ to E are all different. Denote by R_E the maximal order of E and by U the group of all totally positive units of R_E. Take a Hilbert modular form

$$\varphi(z) = \sum_a c(a) \exp\left(2\pi i \sum_{\nu=1}^r a^{\theta_\nu} z_\nu \right) \quad (z = (z_1, \ldots, z_r) \in \mathfrak{H}_1^r)$$

of weight $(h_1, \ldots, h_r) \ (\in 2^{-1} Z^r)$ with respect to a congruence subgroup of $SL_2(E)$. Specialize the group G_Q of § 3 to the case $n=1$. Then $G_Q = B^\times$ and $G_\nu = GL_2(R)$ for $\nu \leqslant r$. For $k = (k_1, \ldots, k_r) \in Z^r$, $\alpha \in B$ and $z = (z_1, \ldots, z_r) \in \mathfrak{H}_1^r$, put $T_k(\alpha, z) = \prod_{\nu=1}^r \mathrm{tr}\,([z_\nu] \alpha_\nu)^{-k_\nu}$, where

$$[b] = \begin{bmatrix} b & -b^2 \\ 1 & -b \end{bmatrix} \quad \text{for } b \in C.$$

Given a totally negative element ξ of F and an R_E-lattice Y of $\{\beta \in B | \beta' = -\beta\}$, define a function f on $\mathfrak{H}_1^r$ by

$$(4.1) \qquad\qquad f(z) = \sum_{0 \neq \alpha \in Y/U} c\big(\mathrm{Tr}_{F/E}(\xi \alpha \alpha') \big) T_k(\alpha, z),$$

where the sum is extended over all different cosets αU with $0 \neq \alpha \in Y$.

THEOREM 5. *Suppose $k_1 - h_1 = \ldots = k_r - h_r = t$ with an element t of $2^{-1} Z$ greater than $3[F:E]$ and φ is either a constant or a cusp form. Then (4.1) is convergent and defines an automorphic form of weight $(2k_1, \ldots, 2k_r)$. Moreover, if $2t \equiv [F:E] \ (\mathrm{mod}\ 2)$ and the coefficients $c(a)$ are algebraic, $\pi^{-\{k\}} f$ is $\bar{Q}$-rational in the sense of § 3, where $\{k\} = \sum_{\nu=1}^r k_\nu$.*

This result holds even when φ is not a cusp form, provided that t is sufficiently large. Series of the same type can be defined also for orthogonal groups and unitary groups; they include Eisenstein series as special cases. Closely related to these are the series of type

$$(4.2) \qquad W(s) = \sum_{0 \neq x \in X/U} c\big(\mathrm{Tr}_{F/E}(yxx^{\varrho})\big) \prod_{v=1}^{r} (x^{\sigma_v})^{-k_v} |x^{\sigma_v}|^{-s} \quad (s \in C),$$

where X is an R_E-lattice of a totally imaginary quadratic extension K of F; y is an element of F such that $y^{\sigma_v} > 0$ or < 0 according as $v \leqslant r$ or $> r$; $\sigma_1, \ldots, \sigma_r$ are embeddings of K into C such that $\sigma_v = \theta_v$ on F. Now W can be continued to a meromorphic function on the whole plane. If $F = E$ and φ is an Eisenstein series, W is essentially the product of two L-functions of K with Hecke characters.

THEOREM 6. *Suppose $h \in Z^r$ and $k_1 - h_1 = \ldots = k_r - h_r > [K : E]$. Define $\Psi \in I_K$ by $\Psi = \Omega_K + \sum_{v=1}^{r} (\sigma_v - \sigma_v \varrho)$, where Ω_K is the sum of all embeddings of K into C. Then $\pi^{-(k)} W(0) \prod_{v=1}^{r} p(\sigma_v, \Psi)^{-k_v} \in \bar{Q}$.*

The subject of this section will be treated in more detail in a forthcoming paper.

References

1. R. M. Damerell, *L-functions of elliptic curves with complex multiplication.* I, II, Acta Arith. **17** (1970), 287—301; **19** (1971), 311—317.

2. P. Deligne, *Travaux de Shimura*, Sém. Bourbaki 1971, exp. 389, Lecture Notes in Math. vol. 244, Springer-Verlag, Berlin and New York.

3. P. Garrett, *Arithmetic properties of Fourier-Jacobi expansions of automorphic forms in several variables* (to appear).

4. K.-y. Shih, *Existence of certain canonical models*, Duke Math. J. **45** (1978), 63—66.

5. G. Shimura, *On canonical models of arithmetic quotients of bounded symmetric domains.* I, II, Ann. of Math. (2) **91** (1970), 144—222; **92** (1970), 528—549.

6. ———, *On arithmetic automorphic functions*, Proc. Internat Congress of Math. 1970, vol. II, 1971, pp. 343—348.

7. ———, *On some arithmetic properties of modular forms of one and several variables*, Ann. of Math. (2) **102** (1975), 491—515.

8. ———, *On the derivatives of theta functions and modular forms*, Duke Math. J. **44** (1977). 365—387.

9. ———, *The arithmetic of automorphic forms with respect to a unitary group*, Ann. of Math. (2) **107** (1978), 569—605.

10. ———, *On certain reciprocity-laws for theta functions and modular forms*, Acta Math. **141** (1978), 35—71.

11. ———, *The special values of the zeta functions associated with Hilbert modular forms*, Duke Math. J. **45** (1978), 637—679.

12. ———, *Automorphic forms and the periods of abelian varieties*, J. Math. Soc. Japan (to appear).

13. J. Sturm, *Special values of zeta functions and Eisenstein series of half integral weight* (to appear).

14. A. Weil, *Elliptic functions according to Eisenstein and Kronecker*, Ergebnisse der Math. und ihrer Grenzgebiete, vol. **88**, Springer-Verlag, Berlin and New York, 1976.

PRINCETON UNIVERSITY
PRINCETON, NEW JERSEY 08540, U.S.A.

The arithmetic of certain zeta functions
and automorphic forms on orthogonal groups

Annals of Mathematics, 111 (1980), 313-375

One of the two principal objects of study in this paper is a Dirichlet series which, in a special case, has the form

$$(1) \qquad \mathfrak{D}(s) = \sum_{0 \neq x \equiv a(L)} \omega\big(\mathrm{Tr}_{K/\mathbf{Q}}(\eta x x^{\rho})\big) x^{\Phi}(x^{\psi})^{-k} |x^{\psi}|^{-2s} \qquad (s \in \mathbf{C}).$$

Here K is a totally imaginary quadratic extension of a totally real algebraic number field K_0; $a \in K$ and L is a lattice in K; ω denotes the Fourier coefficients of an elliptic modular form

$$\Omega(z) = \sum_{\xi \in \mathbf{Q}} \omega(\xi) e^{2\pi i \xi z}$$

of an integral weight ≥ 0; k is a positive integer; ψ is an embedding of K into C; η is an element of K_0 such that η^{ψ} is its only positive conjugate; $\Phi = b(\psi + \psi\rho) + \sum_{\nu} c_{\nu}\varphi_{\nu}$, where b and c_{ν} are non-negative integers, ρ is the complex conjugation, and $\{\varphi_{\nu}\}$ is the set of all embeddings of K into C other than ψ and $\psi\rho$. It will be shown that the series is convergent for sufficiently large $\mathrm{Re}(s)$ and can be continued to a meromorphic function on the whole plane. Now one of our main results will assert that the values $\mathfrak{D}(\mu)$ for certain integers μ are algebraic numbers times

$$\pi^{k} p_{K}(\psi, \psi)^{2k} \prod_{\nu} p_{K}(\varphi_{\nu}, \psi)^{-2c_{\nu}}$$

where $p_{K}(\varphi, \psi)$ is a complex number depending only on φ and ψ. We shall actually prove such an algebraicity for the series defined in a similar way with a Hilbert modular form in place of Ω (Theorem 9.2).

If $[K: \mathbf{Q}] = 2$, $\mathfrak{D}$ is essentially of the type

$$(2) \qquad \sum_{0 < \xi \in \mathbf{Q}} \omega(\xi)\lambda(\xi)\xi^{-s},$$

where the $\lambda(\xi)$ are the Fourier coefficients of another modular form $\sum_{\xi} \lambda(\xi) e^{2\pi i \xi z}$ whose Mellin transform is an L-function of K with a Hecke character. Thus $\mathfrak{D}$ may be viewed as a generalization of (2). It should be noted however that in general no rational power of $|x^{\psi}|$ is a rational number if $[K: F] > 2$, and hence $\mathfrak{D}$ is quite different from the zeta functions hitherto

0003-486X/80/0111-2/0313/063 $ 03.15/1

© 1980 by Princeton University Mathematics Department

For copying information, see inside back cover.

* Supported by NSF Grant MCS 78-02781.

investigated. Also, our formulation includes the case where Ω is a constant. Then the series is extended over all x such that $\mathrm{Tr}_{K/\mathbf{Q}}(\eta x x^{\rho}) = 0.$[1]

The series of the above type occurs as a specialization of an explicitly defined automorphic form on an orthogonal group

$$(3) \qquad \{A \in \mathrm{GL}_{m+2}(F) \mid {}^{t}ASA = S\} ,$$

where F is a totally real algebraic number field and S is a symmetric element of $\mathrm{GL}_{m+2}(F)$ whose conjugates have signatures $(m, 2)$ or $(m + 2, 0)$. The construction of such forms characterized by a certain arithmetic property is another main topic of our investigation. The characterization is made by means of the quantities of type $p_{K}(\varphi, \psi)$. Thus our program begins with the definition of $p_{K}(\varphi, \psi)$ and their fundamental properties (Theorem 1.1), which may be of independent interest. We then introduce the automorphic forms with respect to (3) and call them arithmetic or $\overline{\mathbf{Q}}$-rational, if their values at CM-points are algebraic numbers times products of several $p_{K}(\varphi, \psi)$ depending on the weight and the point. This can be done in two ways: one in terms of the orthogonal group and the other relative to the spinor group which covers (3). The formulation of our theorem on the values of $\mathfrak{D}$ makes the former indispensable; the latter is necessary for the existence of $\overline{\mathbf{Q}}$-rational automorphic forms, which we prove by using the embedding of the spinor group into a group of symplectic type obtained by Shih [6]. The consistency of both definitions is crucial in our theory and follows from the algebraic relations between various $p_{K}(\varphi, \psi)$, which we call "the reflex principle" (Theorem 1.2). The reader will find in Section 6 an intimate connection between the reflex of a CM-type and the spin representations. After a preliminary section on Hilbert modular forms of half-integral weight, we introduce in Section 8 certain infinite series f which include Eisenstein series as a special case. They can be defined for both compact and non-compact quotients of (3) (see (8.3) of the text for the explicit form). The arithmeticity of f is one of our main results (Theorem 8.4).

Our proofs are made in the following steps. (A) The arithmeticity of the values of $\mathfrak{D}$, in the cases more general than (1), implies that of f. (B) In a special case, the values of $\mathfrak{D}$ are known to be arithmetic, as proved in

---

[1] The zeta function of this type was mentioned in [13, pp. 198-199]. Actually one of the main motives of the theory developed in [13] and [16] was its application to this zeta function. However, the author realized that a more general result could be obtained by considering orthogonal groups instead of unitary groups. The present work may therefore be regarded as a natural sequel to these papers and also as the fulfilment of the promises made in the introductions of [16] and [17]. Some of the basic ideas are also explained in [19].

a previous paper [17]; this shows the arithmeticity of f in a special case. (C) The values of the derivatives of f (in that special case) can be examined by the methods of [10], and give the arithmeticity of the values of $\mathfrak{D}$ in the most general case. The ideas work only with the use of all the derivatives of Hilbert modular forms in place of Ω, which seem natural and unavoidable.

It is naturally expected that the series of the same nature as f can be defined for groups of other types. We shall show in Section 14 that this is so at least for the unitary group over a CM-field which becomes, over $\mathbf{R}$, the product of copies of $U(m, 1)$ and $U(m + 1)$. In the present paper, we have not tried to obtain the best possible results. For example, we have been content only with the rationality over the field of all algebraic numbers. The rationality over smaller fields, say over cyclotomic fields, could have been treated without great difficulties for "totally indefinite" orthogonal groups, though this would have obscured the main ideas and made the paper unnecessarily long. Other comments and open problems will be discussed in the last section.

Contents

Introduction 313
Notation and terminology...315
  1. The fundamental quantities p_K 316
  2. Orthogonal groups over $\mathbf{R}$ and factors of automorphy 319
  3. Orthogonal groups over F and automorphic forms..............323
  4. Arithmetic automorphic forms on $\mathfrak{H}_n^r$ 324
  5. The CM-points of $G(S)$ on $\mathfrak{Z}^r$ 326
  6. Existence of $\overline{\mathbf{Q}}$-rational forms on $\mathfrak{Z}^r$............................330
  7. Hilbert modular forms of half-integral weight 337
  8. Explicit construction of $\overline{\mathbf{Q}}$-rational automorphic forms on $\mathfrak{Z}^r$....344
  9. The special values of certain zeta functions....................348
 10. Proofs of the main theorems in special cases...................356
 11. Non-holomorphic derivatives of automorphic forms 358
 12. Proof of Theorem 9.2......................................362
 13. Analytic continuation 365
 14. The case of unitary groups 370
 15. Concluding remarks and open problems 372

Notation and terminology. Besides the standard symbols $\mathbf{Z}$, $\mathbf{Q}$, $\mathbf{R}$, and $\mathbf{C}$, we denote by $\overline{\mathbf{Q}}$, $\mathbf{T}$, and $\mathbf{H}$ the algebraic closure of $\mathbf{Q}$ in $\mathbf{C}$, the group of all complex numbers of absolute value 1, and the division ring of Hamilton quaternions. For an associative ring A with identity element, we denote, when there is no fear of confusion, by A_q^p the module of $p \times q$ matrices with entries in A and by $A^\times$ the group of all invertible elements of A. We also put $A_1^p = A^p$, $M_p(A) = A_p^p$ and $\mathrm{GL}_p(A) = M_p(A)^\times$. If V is an A-module, $\mathrm{End}(V, A)$ denotes the ring of all A-linear endomorphisms of V, and

$\mathrm{GL}(V, A) = \mathrm{End}(V, A)^{\times}$. If A is a semi-simple algebra over a field F, $\mathrm{Tr}_{A/F}$ denotes the reduced trace of A to F. For a symmetric bilinear form $S(x, y)$, we put $S[x] = S(x, x)$. This applies also to matrices; namely $S[X] = {}^{t}XSX$ for a symmetric matrix S and a matrix X for which the product is meaningful; the upper left t means the transpose. We denote by $\mathrm{diag}[X_1, \cdots, X_k]$ the square matrix with $X_1, \cdots, X_k$ on the diagonal blocks and 0 in all other blocks. The identity matrix of degree m is denoted by 1_m. For a complex hermitian matrix h, we write $h > 0$ or $h \geq 0$ according as h is positive definite or non-negative. We write $x \gg 0$ for an element x of a totally real algebraic number field if x is totally positive. For $a, b \in \mathbf{C}$, we write $a \sim b$ if $b \neq 0$ and $a/b \in \overline{\mathbf{Q}}$. An *embedding* of a field or a ring means an injective homomorphism, which we often denote by a letter on the upper right; thus x^{τ} denotes the image of x under an embedding τ. In particular we denote by x^{ρ} (and also by $\bar{x}$) the complex conjugate of x. For $x \in \mathbf{C}$, we put $e(x) = \exp(2\pi i x)$. By a CM-*field*, we understand a totally imaginary quadratic extension of a totally real algebraic number field.

1. The fundamental quantities p_K

The numbers $p_K(\varphi, \psi)$ mentioned in the introduction are given by the special values of certain Hilbert modular forms, or equivalently, by the periods of abelian varieties with complex multiplication. Some of their basic properties have been treated in [14] and [18]. The purpose of this section is to give a reformulation of the main results of [18, §§ 1, 2] in a simplified and somewhat sharper form.

For a commutative semi-simple algebra Y over $\mathbf{Q}$, we denote by $J(Y)$ or simply J_Y the set of all non-trivial homomorphisms of Y into $\mathbf{C}$, and by $I(Y)$ or simply I_Y the free $\mathbf{Z}$-module generated by the elements of J_Y. For $x \in Y^{\times}$ and $\alpha = \sum_{\tau} a_{\tau}\tau \in I_Y$ with $a_{\tau} \in \mathbf{Z}$ and $\tau \in J_Y$, we put $x^{\alpha} = \prod_{\tau}(x^{\tau})^{a_{\tau}}$ and consider α a "character" of $Y^{\times}$.

Let K be a CM-field of degree $2n$, and let ρ denote the complex conjugation. Denote by I_K^0 the submodule of I_K consisting of all α such that $\alpha + \alpha\rho$ contains every element of J_K with the same multiplicity. An element of I_K^0 which is the sum of exactly n elements of J_K is called a CM-*type* of K. Let F be the maximal real subfield of K. Take Hilbert modular forms f and g with respect to a congruence subgroup of $\mathrm{SL}_2(F)$ of weight

$$(k_1, \cdots, k_{r-1}, k_r + 1, k_{r+1}, \cdots, k_n) \quad \text{and} \quad (k_1, \cdots, k_n),$$

respectively. Given a CM-type $\varphi = \sum_{i=1}^{n} \tau_i$ of K, we can find an element w of K such that $\mathrm{Im}(w^{\tau_i}) > 0$ for all i. We can also find f and g so that both

f and g have Fourier coefficients in $\overline{\mathbf{Q}}$ and $f(w_0)g(w_0) \neq 0$, where $w_0 = (w^{\tau_1}, \cdots, w^{\tau_n})$. We then define $p_K(\tau_r, \varphi)$ to be the element of $\mathbf{C}^\times/\overline{\mathbf{Q}}^\times$ represented by $f(w_0)/g(w_0)$, which depends only on K, τ_r and φ. The quantity $p_K(\tau_r, \varphi)$ can also be obtained from a period of an abelian variety with K as the field of complex multiplication; for details, the reader is referred to [14, pp. 381–383] and [18, §1].

Let L be a CM-field containing K. For every $\alpha \in J_K$ and $\beta \in J_L$, we denote by $\mathrm{Inf}_{L/K}(\alpha)$ the sum of all the elements γ of J_L which coincide with α on K, and by $\mathrm{Res}_{L/K}(\beta)$ the restriction of β to K. These can be extended to two additive maps

$$\mathrm{Inf}_{L/K}\colon I_K \longrightarrow I_L\,, \qquad \mathrm{Res}_{L/K}\colon I_L \longrightarrow I_K\,.$$

THEOREM 1.1. *There exists a bilinear map p_K of $I_K \times I_K$ into $\mathbf{C}^\times/\overline{\mathbf{Q}}^\times$, defined for each CM-field K, with the following properties:*

  (1) $p_K(\tau_r, \varphi)$ *is defined as above if* $\varphi = \sum_{i=1}^{n} \tau_i$ *is a CM-type of* K;

  (2) $p_K(\xi\rho, \eta) = p_K(\xi, \eta\rho) = p_K(\xi, \eta)^{-1}$ *for every* $\xi, \eta \in I_K$;

  (3) $p_K(\xi, \mathrm{Res}_{L/K}(\zeta)) = p_L(\mathrm{Inf}_{L/K}(\xi), \zeta)$ *if* $\xi \in I_K$, $\zeta \in I_L$, *and* $K \subset L$;

  (4) $p_K(\mathrm{Res}_{L/K}(\zeta), \xi) = p_L(\zeta, \mathrm{Inf}_{L/K}(\xi))$ *if* $\xi \in I_K$, $\zeta \in I_L$, *and* $K \subset L$;

  (5) $p_{K'}(\gamma\xi, \gamma\eta) = p_K(\xi, \eta)$ *if* γ *is an isomorphism of* K' *onto* K.

We shall denote $p_K(\xi, \eta)$ also by $p(\xi, \eta; K)$.

Proof. Put $p_K(\tau_r\rho, \varphi) = p_K(\tau_r, \varphi)^{-1}$ if $\varphi = \sum_{r=1}^{n} \tau_r$ is a CM-type of K. (This is different from the definition $p(\tau_r\rho, \varphi) = 1$ of [18, §1], but coincides with that of [19].) Theorems 1.2 and 1.3 of [18] show that we can put $p_K(\alpha, \sum_i s_i\varphi_i) = \prod_i p_K(\alpha, \varphi_i)^{s_i}$ for $\alpha \in J_K$ and $\sum_i s_i\varphi_i \in I_K^0$ with $s_i \in \mathbf{Z}$ and CM-types φ_i of K without any conflict. Extend this to $I_K \times I_K^0$ linearly. Then [18, Th. 1.3, Prop. 1.4, and Prop. 1.5] imply properties (3), (4), and (5) for $p_K(\xi, \eta)$ when $\eta \in I_K^0$. Observing that $\zeta - \zeta\rho \in I_K^0$ for every $\zeta \in I_K$, we put $p_K(\xi, \zeta) = p_K(\xi, \zeta - \zeta\rho)^{1/2}$. Then all the properties of p_K on $I_K \times I_K$ can be verified in a straightforward way.

Let M be a CM-field contained in $\mathbf{C}$ which is a Galois extension of $\mathbf{Q}$, and let $G = \mathrm{Gal}(M/\mathbf{Q})$. Every $\delta \in I_M$ can be written $\delta = \sum_{\gamma \in G} c_\gamma \gamma$ with $c_\gamma \in \mathbf{Z}$. We then put $\delta^* = \sum_{\gamma \in G} c_\gamma \gamma^{-1}$. Now let K and L be CM-fields contained in $\mathbf{C}$, and let $\xi \in I_K$ and $\eta \in I_L$. We call (K, ξ) and (L, η) a *reflexive pair* if $\mathrm{Inf}_{M/K}(\xi)^* = \mathrm{Inf}_{M/L}(\eta)$ for a Galois extension M of $\mathbf{Q}$ which is a CM-field and contains both K and L. (If this is so for *one* M, it holds for *every* such M.) Let $\mathrm{Gal}(M/\mathbf{Q})$ act on I_K on the right in a natural way. If (K, ξ) and (L, η) form a reflexive pair, we see that $\eta\beta$ for $\beta \in \mathrm{Gal}(M/\mathbf{Q})$ depends only on η and $\mathrm{Res}_{M/K}(\beta)$. Therefore we write $\eta\alpha$ for $\eta\beta$ if $\alpha = \mathrm{Res}_{M/K}(\beta)$. Given $\xi \in I_K$, there is a unique CM-field L_0 and a unique $\eta_0 \in I(L_0)$ with the property that (K, ξ) and (L, η) form a reflexive pair if and only if $L_0 \subset L$ and $\eta = \mathrm{Inf}_{L/L_0}(\eta_0)$. We call (L_0, η_0) the *reflex* of (K, ξ) (cf. [9, §1.3]).

THEOREM 1.2. *If (K, ξ) and (L, η) form a reflexive pair, we have* $p_K(\alpha, \xi) = p_L(\eta\alpha, \mathrm{id}_L)$ *for every* $\alpha \in J_K$.

Proof. Take M as above. By (5) of Theorem 1.1, we have $p_M(\beta, \gamma) = p_M(\gamma^{-1}\beta, \mathrm{id}_M)$ for $\beta, \gamma \in \mathrm{Gal}(M/\mathbf{Q})$, so that $p_M(\beta, \zeta) = p_M(\zeta^*\beta, \mathrm{id}_M)$ for every $\zeta \in I_M$. If $\alpha = \mathrm{Res}_{M/K}(\beta)$, we obtain, by (3) and (4) of Theorem 1.1,

$$p_K(\alpha, \xi) = p_M\big(\beta, \mathrm{Inf}_{M/K}(\xi)\big) = p_M\big(\mathrm{Inf}_{M/K}(\xi)^*\beta, \mathrm{id}_M\big)$$
$$= p_M\big(\mathrm{Inf}_{M/L}(\eta)\beta, \mathrm{id}_M\big) = p_L(\eta\alpha, \mathrm{id}_L) , \qquad \text{Q.E.D.}$$

This theorem may be called *the reflex principle*, and includes [18, Theorem 2.3] as a special case.

Given a CM-field K, take a Galois extension M of $\mathbf{Q}$ which is a CM-field and contains K. For $\xi \in I_K$, let $T(\xi)$ denote the $\mathbf{Z}$-submodule of I_K generated by $\xi\gamma$ for all $\gamma \in G = \mathrm{Gal}(M/\mathbf{Q})$, and $t(\xi)$ the rank of $T(\xi)$.

LEMMA 1.3. *If (K, ξ) and (L, η) form a reflexive pair, we have* $t(\xi) = t(\eta)$.

Proof. Take M containing both K and L. Since $t(\xi) = t\big(\mathrm{Inf}_{M/K}(\xi)\big)$, it is sufficient to prove $t(\zeta) = t(\zeta^*)$ for $\zeta \in I_M$. Let

$$\zeta = \sum_{\gamma \in G} c(\gamma)\gamma, \quad \zeta\alpha = \sum_{\beta \in G} m(\alpha, \beta)\beta, \quad \text{and} \quad \zeta^*\beta = \sum_{\alpha \in G} m^*(\alpha, \beta)\alpha .$$

Then $m(\alpha, \beta) = c(\beta\alpha^{-1}) = m^*(\alpha, \beta)$, so that

$$t(\zeta) = \mathrm{rank}\big(m(\alpha, \beta)\big)_{\alpha,\beta} = \mathrm{rank}\big(m^*(\alpha, \beta)\big)_{\alpha,\beta} = t(\zeta^*) , \qquad \text{Q.E.D.}$$

PROPOSITION 1.4. *Let $[K:\mathbf{Q}] = 2n$ and let $\tau_1, \cdots, \tau_n$ be n elements of J_K which form a CM-type of K. Then, for every $\xi \in I_K$, the module*

$$\{(x_1, \cdots, x_n) \in \mathbf{Z}^n \mid \textstyle\prod_{i=1}^n p_K(\tau_i, \xi)^{x_i} = 1\}$$

has rank at least $n - t(\xi - \xi\rho)$.

Proof. Let (L, η) be the reflex of (K, ξ). Then $(L, \eta - \eta\rho)$ and $(K, \xi - \xi\rho)$ form a reflexive pair. Observe that $T(\eta - \eta\rho)$ is generated by $(\eta - \eta\rho)\tau_i$ for $i = 1, \cdots, n$. Hence $t(\eta - \eta\rho) \leq n$. By Theorem 1.2, we have

$$p_K(\tau_i, \xi)^2 = p_K(\tau_i, \xi - \xi\rho) = p_L\big((\eta - \eta\rho)\tau_i, \mathrm{id}_L\big) .$$

Therefore if $\sum_{i=1}^n x_i(\eta - \eta\rho)\tau_i = 0$, we have $\prod_{i=1}^n p_K(\tau_i, \xi)^{x_i} = 1$. This proves our proposition, since $t(\xi - \xi\rho) = t(\eta - \eta\rho)$ by Lemma 1.3.

If φ is a CM-type of K, we can easily verify that $t(\varphi) = 1 + t(\varphi - \varphi\rho)$. Therefore the above proposition includes [18, Theorem 2.5] as a special case. The notation being as in Proposition 1.4, we may naturally ask a question:

are n quantities $p_K(\tau_i, \mathrm{id}_K)$ for $i = 1, \cdots, n$ algebraically independent?

Let us now consider the direct sum Y of CM-fields $K_1, \cdots, K_t$. Obviously I_Y can be identified with the direct sum of $I(K_1), \cdots, I(K_t)$. Given $\alpha = (\alpha_1, \cdots, \alpha_t)$ and $\beta = (\beta_1, \cdots, \beta_t)$ of I_Y with α_i and β_i in $I(K_i)$, we put

$$(1.1) \qquad p_Y(\alpha, \beta) = \prod_{i=1}^{t} p(\alpha_i, \beta_i; K_i) \,.$$

Let $\theta \colon Y^\times \to \mathrm{GL}_d(\overline{\mathbf{Q}})$ be a homomorphism which can be diagonalized in the form

$$T\theta(x)T^{-1} = \mathrm{diag}[x^{\gamma_1}, \cdots, x^{\gamma_d}] \qquad\qquad (x \in Y^\times)$$

with $\gamma_1, \cdots, \gamma_d \in I_Y$ and $T \in \mathrm{GL}_d(\overline{\mathbf{Q}})$. Then we put, for every $\beta \in I_Y$,

$$(1.2) \qquad P_Y(\theta, \beta) = T^{-1}\mathrm{diag}[p_Y(\gamma_1, \beta), \cdots, p_Y(\gamma_d, \beta)]T \,.$$

Here and henceforth, we use $p_Y(\alpha, \beta)$ to indicate any complex number which represents it. Thus $P_Y(\theta, \beta)$ is a complex matrix which represents a certain well-defined "coset". Property (2) of Theorem 1.1 shows that $P_Y(\theta, \beta)$ depends only on the restriction of θ to the group

$$(1.3) \qquad Y^u = \{(y_1, \cdots, y_t) \in Y \mid y_i y_i^\rho = 1 \text{ for all } i\} \,.$$

Viewing Y^u as the group of $\mathbf{Q}$-rational points of a $\mathbf{Q}$-rational algebraic group which we denote by the same symbol Y^u, we can identify $(Y^u)_{\mathbf{C}}$ with a group

$$\{(x_\alpha)_{\alpha \in J(Y)} \in \mathbf{C}^{J(Y)} \mid x_\alpha x_{\alpha\rho} = 1 \text{ for all } \alpha\} \,,$$

where we identify $y \in Y^u$ with $(y^\alpha)_{\alpha \in J(Y)}$. For $\beta \in I_Y$, let $\mathfrak{p}_Y(\beta)$ denote an element of $(Y^u)_{\mathbf{C}}$ whose α-component represents $p_Y(\alpha, \beta)$ for all $\alpha \in J_Y$. Extend θ to a continuous homomorphism of $(Y^u)_{\mathbf{C}}$ into $\mathrm{GL}_d(\mathbf{C})$. Then we can also define $P_Y(\theta, \beta)$ by

$$(1.4) \qquad P_Y(\theta, \beta) = \theta\bigl(\mathfrak{p}_Y(\beta)\bigr) \,.$$

2. Orthogonal groups over R and factors of automorphy

Let V be a vector space over $\mathbf{R}$ of dimension $m + 2$, and $S(x, y)$ an $\mathbf{R}$-valued $\mathbf{R}$-bilinear symmetric form on V of signature $(m, 2)$. We extend S to a $\mathbf{C}$-valued symmetric form on $V_{\mathbf{C}} = V \otimes_{\mathbf{R}} \mathbf{C}$. For every $v \in V_{\mathbf{C}}$, its complex conjugate $\bar{v}$ is meaningful. We consider the set

$$(2.1) \qquad \mathfrak{Y}(S) = \{v \in V_{\mathbf{C}} \mid S[v] = 0, \, S(v, \bar{v}) < 0\} \,.$$

If $v \in \mathfrak{Y}(S)$, there is a unique two-dimensional subspace W of V such that $v \in W \otimes_{\mathbf{R}} \mathbf{C}$; moreover S is negative on such W. In fact, W is spanned by $v + \bar{v}$ and $iv - i\bar{v}$. Let W' be the orthogonal complement of W in V relative to S. Define a positive symmetric form P_v on V so that $P_v = S$ on W',

$P_v = -S$ on W, and $P_v(W, W') = 0$. Now every $x \in V$ can be expressed as a sum $x = cv + \bar{c}\bar{v} + z$ with $z \in W'$ and $c \in \mathbf{C}$. Then

$$(2.2) \qquad P_v[x] - S[x] = -4|c|^2 S(v, \bar{v}) = -4S(v, \bar{v})^{-1}|S(x, v)|^2 \geqq 0 .$$

This will be used in Section 8.

PROPOSITION 2.1. *There exists a* $\mathbf{C}$-*linear isomorphism* A *of* V_C *onto* $\mathbf{C}^{m+2}$ *such that*

$$S(x, y) = {}^t(Ax)RAy , \qquad S(\bar{x}, y) = {}^t\overline{(Ax)}QAy ,$$

where R *and* Q *are the elements of* $\mathrm{GL}_{m+2}(\mathbf{R})$ *defined by*

$$(2.3) \qquad R = \mathrm{diag}\left[1_m, \begin{pmatrix} 0 & -1 \\ -1 & 0 \end{pmatrix}\right] , \quad Q = \mathrm{diag}[1_m, -1_2] .$$

Proof. Since S is "equivalent" to Q over $\mathbf{R}$, it is sufficient to find an element B of $\mathrm{GL}_{m+2}(\mathbf{C})$ such that ${}^t\bar{B}QB = Q$ and ${}^tBQB = R$. Such a B is given, for example, by

$$B = \mathrm{diag}\left[1_m, 2^{-1/2}\begin{pmatrix} 1 & 1 \\ -i & i \end{pmatrix}\right] .$$

Now define groups $G(S)$ and $G(Q, R)$ by

$$(2.4a) \qquad G(S) = \{\alpha \in \mathrm{GL}(V, \mathbf{R}) \,|\, S(\alpha x, \alpha y) = \nu(\alpha)S(x, y) \text{ with } \nu(\alpha) \in \mathbf{R}\} ,$$

$$(2.4b) \quad G(Q, R) = \{\alpha \in \mathrm{GL}_{m+2}(\mathbf{C}) \,|\, {}^t\alpha R\alpha = \nu(\alpha)R, {}^t\bar{\alpha}Q\alpha = \nu(\alpha)Q \text{ with } \nu(\alpha) \in \mathbf{R}\} .$$

Then $\alpha \mapsto A\alpha A^{-1}$ gives an isomorphism of $G(S)$ onto $G(Q, R)$, and $v \mapsto Av$ maps $\mathfrak{Y}(S)$ onto

$$(2.5) \qquad \mathfrak{Y}(Q, R) = \{u \in \mathbf{C}^{m+2} \,|\, {}^tuRu = 0, {}^t\bar{u}Qu < 0\} .$$

If $u_1, \cdots, u_{m+2}$ are the components of $u \in \mathfrak{Y}(Q, R)$, we have $\sum_{i=1}^{m} u_i^2 = 2u_{m+1}u_{m+2}$ and $\sum_{i=1}^{m}|u_i|^2 < |u_{m+1}|^2 + |u_{m+2}|^2$, so that $|u_{m+1}| \neq |u_{m+2}|$. Thus $\mathfrak{Y}(Q, R)$ has two components defined by $|u_{m+1}| \gtrless |u_{m+2}|$. Suppose $|u_{m+2}| > |u_{m+1}|$ and put $z_i = u_i/u_{m+2}$ for $i \leqq m$, $z = {}^t(z_1, \cdots, z_m)$, and $w = u_{m+1}/u_{m+2}$. Then $w = {}^tzz/2$, and z belongs to the (bounded symmetric) domain

$$(2.6) \qquad \mathfrak{Z}_m = \left\{z \in \mathbf{C}^m \,\Big|\, {}^t\bar{z}z < 1 + \frac{1}{4}|{}^tzz|^2 < 2\right\} .$$

We write $\mathfrak{Z}_m$ simply $\mathfrak{Z}$ when m is fixed. For each $z \in \mathfrak{Z}$, we put

$$(2.7) \qquad p(z) = \begin{bmatrix} z \\ w \\ 1 \end{bmatrix} , \quad w = {}^tzz/2 .$$

The above argument shows that $(c, z) \mapsto c \cdot p(z)$ gives a bijective map of $\mathbf{C}^{\times} \times \mathfrak{Z}$ onto a component of $\mathfrak{Y}(Q, R)$. If $\alpha \in G(Q, R)$ and $\nu(\alpha) > 0$, then α

acts on $\mathfrak{Y}(Q, R)$. Let $G_+(Q, R)$ denote the group of all α with $\nu(\alpha)>0$ under which the components of $\mathfrak{Y}(Q, R)$ are stable. If $\alpha \in G_+(Q, R)$ and $z \in \mathfrak{Z}$, we can define a point $\alpha(z)$ of $\mathfrak{Z}$ and a scalar $\mu(\alpha, z) \in \mathbf{C}^\times$ by

$$(2.8) \qquad \alpha p(z) = p\big(\alpha(z)\big)\mu(\alpha, z) \qquad\qquad \big(\alpha \in G_+(Q, R)\big).$$

Obviously μ is a holomorphic factor of automorphy. If $\alpha \in G_+(Q, R)$, we can easily verify that α has the form

$$(2.9) \qquad \alpha = \begin{bmatrix} a & b & \bar{b} \\ \bar{c} & \bar{e} & \bar{d} \\ c & d & e \end{bmatrix}, \qquad\qquad a \in M_m(\mathbf{R}) .$$

Then $\mu(\alpha, z) = cz + dw + e$ with $w = {}^tzz/2$. Therefore if we put

$$(2.10) \qquad p'(z) = \begin{bmatrix} \bar{z} \\ 1 \\ \bar{w} \end{bmatrix}, \qquad\qquad w = {}^tzz/2 ,$$

then a straightforward calculation shows that

$$(2.11) \qquad \alpha p'(z) = p'\big(\alpha(z)\big)\overline{\mu(\alpha, z)} .$$

Now we define a $GL_m(\mathbf{C})$-valued holomorphic factor of automorphy $\lambda(\alpha, z)$ and an $\mathbf{R}$-valued function η on $\mathfrak{Z}$ by

$$(2.12) \qquad dz \circ \alpha = {}^t\lambda(\alpha, z)^{-1}\mu(\alpha, z)^{-1}dz ,$$

$$(2.13) \qquad \eta(z) = -{}^t\overline{p(z)}Qp(z) = 1 + |w|^2 - {}^t\bar{z}z \qquad (w = {}^tzz/2) .$$

From (2.8), we can easily derive

$$(2.14) \qquad \eta\big(\alpha(z)\big) = \nu(\alpha)|\mu(\alpha, z)|^{-2}\eta(z) .$$

We introduce differential operators D and D_s for $s \in \mathbf{C}$, which assign $\mathbf{C}_m^n$-valued functions Df and $D_s f$ to a $\mathbf{C}^n$-valued function f, by

$$(2.15) \qquad D_s f = \eta^{-s}D(\eta^s f) = Df + s\eta^{-1}f D\eta ,$$

$$(2.16) \qquad Df = (\partial f/\partial z_1, \cdots, \partial f/\partial z_m) ,$$

where $z_1, \cdots, z_m$ are the components of $z \in \mathfrak{Z}$. Notice that

$$(2.17) \qquad (Df) \circ \alpha = \mu(\alpha, z)D(f \circ \alpha) \cdot {}^t\lambda(\alpha, z) .$$

Define a $\mathbf{C}_m^{m+2}$-valued function $q(z)$ on $\mathfrak{Z}$ by $q(z) = D_{-1}\eta$. Then (2.8) together with (2.14) and (2.17) shows that $\alpha q(z) = q\big(\alpha(z)\big) \cdot {}^t\lambda(\alpha, z)^{-1}$. Therefore if we put

$$B(z) = \big(q(z) \quad p'(z) \quad p(z)\big) ,$$

we obtain

$$(2.18) \qquad \alpha B(z) = B\big(\alpha(z)\big) \cdot \mathrm{diag}\big[{}^t\lambda(\alpha, z)^{-1}, \overline{\mu(\alpha, z)}, \mu(\alpha, z)\big] .$$

Computing q explicitly, we find

(2.19) $$\det\left(B(z)\right) = \eta(z) \;,$$

(2.20) $$'B(z)RB(z) = \begin{bmatrix} 1_m & 0 & 0 \\ 0 & 0 & -\eta(z) \\ 0 & -\eta(z) & 0 \end{bmatrix} \;,$$

(2.21) $$'\overline{B(z)}QB(z) = \operatorname{diag}[\xi(z),\, -\eta(z),\, -\eta(z)]$$

with

$$\xi(z) = 1_m - \bar{z}\cdot{}'z + \eta(z)^{-1}(z - w\bar{z})\cdot{}'(\bar{z} - \bar{w}z),\; w = {}'zz/2 \;.$$

Observe that ξ is hermitian and positive. Substitution of $\alpha(z)$ for z in (2.20) and (2.21) yields

(2.22) $$\lambda(\alpha,\, z)\cdot{}'\lambda(\alpha,\, z) = \nu(\alpha)^{-1}1_m \;,$$

(2.23) $$\xi(\alpha(z)) = \nu(\alpha)\overline{\lambda(\alpha,\, z)}\xi(z)\cdot{}'\lambda(\alpha,\, z) = \overline{\lambda(\alpha,\, z)}\xi(z)\lambda(\alpha,\, z)^{-1} \;.$$

Taking the determinant of (2.18) and (2.21), we find

(2.24) $$\det\left(\lambda(\alpha,\, z)\right) = \nu(\alpha)\det(\alpha)^{-1} \;,$$

(2.25) $$\det\left(\xi(z)\right) = 1 \;.$$

From (2.12), we obtain:

(2.26) The jacobian of α at z is $\nu(\alpha)^{-1}\det(\alpha)\mu(\alpha,\, z)^{-m}$.

The point $z = 0$ may be regarded as "the origin" of $\mathfrak{Z}$. Putting $z = 0$ in (2.18), we find that $\alpha(0) = 0$ if and only if $\alpha = c\cdot\operatorname{diag}[a,\, \bar{e},\, e]$ with $0 < c \in \mathbf{R}$, $a \in O(m,\, \mathbf{R})$, and $|e| = 1$. For such α, we have $\alpha(z) = e^{-1}az$. Therefore $z = 0$ is the unique fixed point of α if and only if ce is not an eigenvalue of ca. Since $ca = {}'\lambda(\alpha,\, 0)^{-1}$ and $ce = \mu(\alpha,\, 0)$, we can rephrase this fact as

PROPOSITION 2.2. *Let* $\beta \in G_+(Q,\, R)$ *and suppose* β *has a fixed point* w *on* $\mathfrak{Z}$. *Then* w *is the only fixed point of* β *if and only if* $\mu(\beta,\, w)$ *is not an eigenvalue of* $\lambda(\beta,\, w)^{-1}$.

We conclude this section by introducing the notion of spherical functions. Identify V with $\mathbf{R}^{m+2}$ and use S for the matrix representing the form S; thus $S[x] = {}'xSx$ for $x \in \mathbf{R}^{m+2}$. A complex-valued C^∞-function φ on $\mathbf{R}^{m+2}$ or its subset is called *S-harmonic* if

(2.27) $$\sum_{i,j=1}^{m+2} s_{ij}^*\,\partial^2\varphi/\partial x_i\partial x_j = 0 \;,$$

where $(s_{ij}^*) = S^{-1}$. We call an *S*-harmonic homogeneous polynomial function *S-spherical*. These definitions apply also to any non-degenerate S with no assumption on its signature.

3. Orthogonal groups over F and automorphic forms

Throughout the rest of the paper, we denote by F a totally real algebraic number field of degree n, and by $\tau_1, \cdots, \tau_n$ the embeddings of F into $\mathbf{R}$. We fix a vector space V over F of dimension $m + 2$ and an F-valued F-bilinear symmetric form $S(x, y)$ on V. We define the τ_i-completion V_i of V to be the tensor product $V \otimes_F \mathbf{R}$ with respect to the embedding τ_i, and extend S to an $\mathbf{R}$-valued $\mathbf{R}$-bilinear symmetric form S_i on V_i. We assume

(3.1) $S_1, \cdots, S_r$ *have signature* $(m, 2)$ *and* $S_{r+1}, \cdots, S_n$ *are positive definite;*
$\quad\quad 0 < r \leqq n.$

Define a group $G(S)$ by

(3.2) $\quad\quad G(S) = \{\alpha \in \mathrm{GL}(V, F) \mid S[\alpha x] = \nu(\alpha)S[x] \quad \text{with} \quad \nu(\alpha) \in F\}\,.$

We may consider $G(S)$ the group $G_{\mathbf{Q}}$ of $\mathbf{Q}$-rational points of a $\mathbf{Q}$-rational algebraic group G. Then $G_{\mathbf{R}}$ can be identified with $\prod_{i=1}^{n} G(S_i)$, where $G(S_i)$ is defined by (2.4a) with V_i in place of V. For $\alpha \in G_{\mathbf{R}}$ and $v \in V \otimes_{\mathbf{Q}} \mathbf{R}$, we denote by α_i and v_i their projections to $G(S_i)$ and V_i. By Proposition 2.1, we can find, for each $i \leqq r$, an isomorphism A_i of $(V_i)_{\mathbf{C}}$ onto $\mathbf{C}^{m+2}$ such that

(3.3) $\quad\quad S_i(x, y) = {}^t(A_i x)R A_i y\,, \quad S_i(\bar{x}, y) = {}^t\overline{(A_i x)}Q A_i y\,.$

We choose and fix such A_i so that $A_i x_i$ has algebraic components for every $x \in V$. Now we have an isomorphism:

(3.4)
$$\prod_{i=1}^{r} G(S_i) \longrightarrow G(Q, R)^r$$
$$\underset{\omega}{} \quad\quad\quad\quad \underset{\omega}{}$$
$$(\alpha_1, \cdots, \alpha_r) \longmapsto (A_1\alpha_1 A_1^{-1}, \cdots, A_r\alpha_r A_r^{-1})\,.$$

If $\alpha \in G(S)$, $A_i\alpha_i A_i^{-1}$ is $\overline{\mathbf{Q}}$-rational for every i. We denote by $G_+(S)$ the subgroup of $G(S)$ consisting of all α with totally positive $\nu(\alpha)$ whose image under (3.4) belongs to $G_+(Q, R)^r$, and let $G_+(S)$ act on the product $\mathfrak{Z}^r$ of r copies of $\mathfrak{Z}$ through (3.4). Also we put

(3.5) $\quad\quad \lambda_i(\alpha, z) = \lambda(A_i\alpha_i A_i^{-1}, z_i)\,, \quad \mu_i(\alpha, z) = \mu(A_i\alpha_i A_i^{-1}, z_i)$
$$(\alpha \in G_+(S), z = (z_1, \cdots, z_r) \in \mathfrak{Z}^r).$$

Now take a $\overline{\mathbf{Q}}$-rational representation

(3.6) $\quad\quad \sigma: (\mathrm{GL}_1)^r \times (\mathrm{GL}_m)^r \times \prod_{i=r+1}^{n} G(S_i) \longrightarrow \mathrm{GL}(W, \mathbf{C})\,,$

where W is a finite-dimensional complex vector space with $\overline{\mathbf{Q}}$-rational structure, and put, for $\alpha \in G_+(S)$ and $z \in \mathfrak{Z}^r$,

(3.7) $\quad \Lambda_\sigma(\alpha, z) = \sigma\big(\mu_1(\alpha, z), \cdots, \mu_r(\alpha, z), \lambda_1(\alpha, z), \cdots, \lambda_r(\alpha, z), \alpha_{r+1}, \cdots, \alpha_n\big)\,.$

To a W-valued function f on $\mathfrak{Z}^r$, we assign a W-valued function $f\,|_\sigma \alpha$ by

(3.8) $\quad\quad\quad\quad (f\,|_\sigma \alpha)(z) = \Lambda_\sigma(\alpha, z)^{-1} f\big(\alpha(z)\big)\,.$

By a *congruence subgroup* of $G_+(S)$, we understand a subgroup which contains a group of the form

$$\{\gamma \in G_+(S) \mid \nu(\gamma) = 1, (\gamma - 1)L \subset NL\}$$

as a subgroup of finite index, where L is a lattice in V and N is a positive integer. If Γ is a congruence subgroup of $G_+(S)$, we denote by $\mathfrak{A}_o(\Gamma)$ the set of all f meromorphic on $\mathfrak{Z}^r$ and at the cusps such that $f|_o \gamma = f$ for all $\gamma \in \Gamma$, by $\mathfrak{A}_o$ the union of $\mathfrak{A}_o(\Gamma)$ for all congruence subgroups Γ, and by $\mathfrak{M}_o$ the set of all holomorphic elements of $\mathfrak{A}_o$ which are finite at cusps; each cusp condition is necessary only when $\Gamma \backslash \mathfrak{Z}^r$ is not compact and $m \leq 2$.

4. Arithmetic automorphic forms on $\mathfrak{H}_n^r$

In this section, we consider automorphic forms with respect to subgroups of $\mathrm{Sp}(n, \mathbf{R})^r$. Their pull-backs will be employed in Section 6 to obtain arithmetic automorphic forms on $\mathfrak{Z}^r$.

Let B be a quaternion algebra over F unramified at $\tau_1, \cdots, \tau_r$ and ramified at $\tau_{r+1}, \cdots, \tau_n$. Define, with a positive integer q, a **Q**-rational algebraic group $\mathfrak{G}$ such that

$$(4.1) \qquad \mathfrak{G}_\mathbf{Q} = \{\alpha \in \mathrm{GL}_q(B) \mid {}^t\alpha^\iota \alpha = \nu(\alpha)1_q \quad \text{with} \quad \nu(\alpha) \in F\},$$

where ι denotes the main involution of B. Put

$$\mathfrak{G}_i = \begin{cases} \{\alpha \in \mathrm{GL}_{2q}(\mathbf{R}) \mid {}^t\alpha J\alpha = \nu(\alpha)J \text{ with } \nu(\alpha) \in \mathbf{R}\} & (i = 1, \cdots, r), \\ \{\alpha \in \mathrm{GL}_q(\mathbf{H}) \mid {}^t\alpha^\iota \alpha = \nu(\alpha)1_q \text{ with } \nu(\alpha) \in \mathbf{R}\} & (i = r + 1, \cdots, n), \end{cases}$$

where ι denotes the main involution of $\mathbf{H}$ and $J = \begin{pmatrix} 0 & 1_q \\ -1_q & 0 \end{pmatrix}$. As shown in [8, §4.4], there is an isomorphism

$$(4.2) \qquad \mathfrak{G}_\mathbf{R} \cong \prod_{i=1}^n \mathfrak{G}_i .$$

For each $\alpha \in \mathfrak{G}_\mathbf{R}$, let $(\alpha_1, \cdots, \alpha_n)$ be the corresponding elements of $\prod_i \mathfrak{G}_i$. Examining the reasoning of [8, §4.4], we see that an isomorphism (4.2) can be chosen so that $\alpha_1, \cdots, \alpha_r$ have algebraic entries for every $\alpha \in \mathfrak{G}_\mathbf{Q}$; we fix such an isomorphism in the following treatment. Let $\mathfrak{G}_{\mathbf{Q}+}$ denote the subgroup of $\mathfrak{G}_\mathbf{Q}$ consisting of all α such that $\nu(\alpha) \gg 0$. We define the action of such an α on the product $\mathfrak{H}_q^r$ of r copies of

$$(4.3) \qquad \mathfrak{H}_q = \{z \in M_q(\mathbf{C}) \mid {}^t z = z, \mathrm{Im}(z) > 0\}$$

by $\alpha(z_1, \cdots, z_r) = (\alpha_1(z_1), \cdots, \alpha_r(z_r))$, $\alpha_i(z_i) = (a_i z_i + b_i)(c_i z_i + d_i)^{-1}$ where $\alpha_i = \begin{pmatrix} a_i & b_i \\ c_i & d_i \end{pmatrix}$. A factor of automorphy ω_i can be defined by

$$(4.4) \qquad \omega_i(\alpha, z) = c_i z_i + d_i \qquad (z = (z_1, \cdots, z_r) \in \mathfrak{H}_q^r; i = 1, \cdots, r).$$

We also consider a **Q**-linear embedding Ω_i of $M_q(B)$ into $M_{2q}(\overline{\mathbf{Q}})$ such that

$\Omega_i(a1_q) = a^{r_i}1_{2q}$ for each i.

As shown in [9], we can construct canonical models of $\mathfrak{H}_q^r/\Gamma$ for congruence subgroups Γ of $\mathfrak{G}_{Q+}$, so that the field of $\overline{\mathbf{Q}}$-rational Γ-automorphic functions is meaningful. Denote by $\mathcal{A}_0(\overline{\mathbf{Q}})$ the union of such fields for all Γ. We are going to introduce the $\overline{\mathbf{Q}}$-rationality of automorphic forms in connection with their values at the CM-points on $\mathfrak{H}_q^r$. Let $Z = Z_1 \oplus \cdots \oplus Z_t$ and $L = L_1 \oplus \cdots \oplus L_t$ with simple F-algebras Z_k and their centers L_k which are CM-fields; suppose Z has a positive involution δ and there is an F-linear embedding h of Z into $M_q(B)$ such that $h(a^\delta) = {}^t h(a)^\iota$. Put

$$s_k^2 = [Z_k: L_k],\ 2m_k = [L_k: F],\ L^u = \{x \in L \,|\, xx^\delta = 1\}\ .$$

Then $h(L^u) \subset \mathfrak{G}_{Q+}$. Suppose $q = \sum_{k=1}^t s_k m_k$. Then $h(L^u)$ has a unique common fixed point w on $\mathfrak{H}_q^r$ (for details, see [8, 4.7]). Now we can define, for each $i \leq r$, a representation $\xi_i: L \to M_q(\overline{\mathbf{Q}})$ such that $\xi_i(x) = \omega_i(h(x), w)$ for all $x \in L^u$, and also a representation $\Xi_j: L \to M_{2q}(\overline{\mathbf{Q}})$ by $\Xi_j = \Omega_j \circ h$. It can be seen that the restriction of ξ_i to L_k is represented by s_k times an element η_{ik} of $I(L_k)$. Putting $\eta = \sum_{i=1}^r \sum_{k=1}^t \eta_{ik}$, we can speak of $P_L(\xi_i, \eta)$ and $P_L(\Xi_j, \eta)$ as defined in Section 1.

We now consider automorphic forms relative to a $\overline{\mathbf{Q}}$-rational representation

$$(4.5) \qquad \zeta: (GL_q)^r \times (GL_{2q})^{n-r} \longrightarrow GL_d\ .$$

For $\alpha \in G_q$, $z \in \mathfrak{H}_q^r$ and a $\mathbf{C}^d$-valued function f on $\mathfrak{H}_q^r$, we put

$$(4.6) \qquad (f|_\zeta \alpha)(z) = \Lambda_\zeta(\alpha, z)^{-1} f\big(\alpha(z)\big)\ ,$$
$$\Lambda_\zeta(\alpha, z) = \zeta\big(\omega_1(\alpha, z), \cdots, \omega_r(\alpha, z), \Omega_{r+1}(\alpha), \cdots, \Omega_n(\alpha)\big)\ .$$

We denote by $\mathcal{A}_\zeta$ the set of all $\mathbf{C}^d$-valued meromorphic functions f such that $f|_\zeta \gamma = f$ for all γ in a congruence subgroup of $\mathfrak{G}_Q$ (and which are meromorphic at cusps, when $q = 1$ and $B = M_2(\mathbf{Q})$). For each fixed point w of $h(L^u)$ as above, we put

$$(4.7) \qquad P_\zeta(w) = \zeta\big(P_L(\xi_1, \eta), \cdots, P_L(\xi_r, \eta), P_L(\Xi_{r+1}, \eta), \cdots, P_L(\Xi_n, \eta)\big)\ ,$$

and call an element f of $\mathcal{A}_\zeta$ *arithmetic* or $\overline{\mathbf{Q}}$-*rational* if $P_\zeta(w)^{-1} f(w)$ has components in $\overline{\mathbf{Q}}$ for every fixed point w of the above type where f is holomorphic. We denote by $\mathcal{A}_\zeta(\overline{\mathbf{Q}})$ the set of all such f.

PROPOSITION 4.1. *Given a representation ζ as in* (4.5) *and an arbitrary point z_0 of $\mathfrak{H}_q^r$, there exists an $M_d(\mathbf{C})$-valued meromorphic function U on $\mathfrak{H}_q^r$ such that:*

(i) *T is holomorphic at z_0 and $\det\big(U(z_0)\big) \neq 0$;*

(ii) *the columns of U belong to $\mathcal{A}_\zeta(\overline{\mathbf{Q}})$.*

Proof. This was stated in [19, Theorem 3] when Λ_ζ is ω_i or Ω_j. (The detailed proof was given in [18] in the case $q = 1$. The general case can be proved in the same manner.) Therefore we obtain the desired T by putting $U = \zeta(S_1, \cdots, S_n)$ with the functions S_i of that theorem.

We call a function U with the above properties (i), (ii) a $\mathbf{Q}$-*rational* ζ-*uniformizer at* z_0.

5. The CM-points of $G(S)$ on $\mathfrak{Z}^r$

Let us now study the isolated fixed points of $G_+(S)$ on $\mathfrak{Z}^r$. Given $w \in \mathfrak{Z}^r$, let Y_w be the F-linear span of

$$(5.1) \qquad \{\alpha \in G_+(S) \,|\, \nu(\alpha) = 1, \, \alpha(w) = w\} \,.$$

It can be shown that Y_w has a positive involution δ such that $S(\beta x, y) = S(x, \beta^\delta y)$ for $\beta \in Y_w$ (see the proof of Proposition 5.3 below). It does not seem difficult to classify all possible types of Y_w for which w is the only fixed point of (5.1). For our purpose, however, it is sufficient to consider a somewhat special type. Namely, we take an F-algebra Y of dimension $m + 2$ which is a direct sum of CM-fields and several copies of F, and suppose that there is an F-linear embedding h of Y into $\mathrm{End}(V, F)$ which maps the identity element to the identity element and which satisfies

$$(5.2) \qquad S\big(h(a)x, y\big) = S(x, h(a^\rho)y) \,,$$

where ρ is the unique positive involution of Y. Put

$$(5.3) \qquad Y^* = \{y \in Y \,|\, yy^\rho = 1\} \,.$$

Then $h(Y^*) \subset G(S)$. Now we assume

$$(5.4) \qquad h(Y^*) \text{ is contained in } G_+(S) \,.$$

Then, $h(Y^*)$, its closure being compact, has a common fixed point w on $\mathfrak{Z}^r$. We call w a CM-*point* of $G_+(S)$. Let $w = (w_1, \cdots, w_r)$ with $w_i \in \mathfrak{Z}$. By (2.18) and (2.22), we have

$$(5.5) \qquad B(w_i)^{-1} A_i h(a)_i A_i^{-1} B(w_i)$$
$$= \mathrm{diag}\big[\lambda_i(h(a), w), \, \overline{\mu_i(h(a), w)}, \, \mu_i(h(a), w)\big] \qquad (a \in Y^*) \,.$$

Since Y^* spans Y over F, we can define representations

$$\varphi_i\colon Y \longrightarrow M_m(\mathbf{C}) \,, \qquad \psi_i\colon Y \longrightarrow \mathbf{C} \qquad (i = 1, \cdots, r)$$

so that

$$(5.6\mathrm{a}) \quad B(w_i)^{-1} A_i h(a)_i A_i^{-1} B(w_i) = \mathrm{diag}\big[\varphi_i(a), \, \overline{\psi_i(a)}, \, \psi_i(a)\big] \qquad (i \leq r, \, a \in Y) \,,$$

$$(5.6\mathrm{b}) \qquad \varphi_i(a) = \lambda_i(h(a), w) \,, \qquad \psi_i(a) = \mu_i(h(a), w) \qquad (i \leq r, \, a \in Y^*) \,.$$

From (2.22), we see that ${}^t\varphi_i(a)^{-1} = \varphi_i(a)$ for all $a \in Y^*$.

PROPOSITION 5.1. *Under* (5.4), *w is the unique common fixed point on $\mathfrak{Z}^r$, and has algebraic coordinates. Moreover, let*

$$Y_0 = \{x \in Y \mid \psi_1(x) = \cdots = \psi_r(x) = 0\} \ .$$

Then $Y = K \oplus Y_0$ with a direct sum K of CM-fields.

Proof. Since $[Y : F] = m + 2$, h is a regular representation of Y over F. This composed with τ_i is equivalent to the sum of φ_i, $\bar{\psi}_i$, and ψ_i. Therefore $\bar{\psi}_i \neq \psi_i$, and hence ψ_i is non-trivial on exactly one CM-field which is a direct summand of Y. Thus $Y = K \oplus Y_0$ as desired. Since φ_i does not contain ψ_i, we see that w is a unique common fixed point of $h(Y^u)$, by virtue of Proposition 2.2. The algebraicity follows from the fact that $A_i \alpha_i A_i^{-1}$ is algebraic for $\alpha \in G_+(S)$.

PROPOSITION 5.2. *Condition* (5.4) *is satisfied if Y has only one or no copy of F as a direct summand.*

Proof. In such a case, $h(Y^u)$ has either 1 or 2 connected components. Let Y_+^u be the identity component of Y^u. Observe that Y_+^u spans Y over F, and that $h(Y_+^u)$, its closure being compact and connected, is contained in $G_+(S)$ and has a fixed point w. We can then define φ_i and ψ_i by (5.6a); we have (5.6b) for $a \in Y_+^u$. Repeating the proof of Prop. 5.1, we see that w is a unique common fixed point of $h(Y_+^u)$, and ψ_i is trivial on the possible direct summand F of Y. This shows that $h(Y^u)$ is contained in $G_+(S)$ and has w as a common fixed point.

PROPOSITION 5.3. *The notation being as in Proposition 5.1, $h(K)$ is completely determined by w, and is independent of the choice of Y and h.*

Proof. Consider Y a subalgebra of the F-linear span Y_w of (5.1). Obviously φ_i and ψ_i can be extended to Y_w. We can define an F-linear involution σ of Y_w by $S(bx, y) = S(x, b^\sigma y)$ for $b \in Y_w$. Identify V_i with $\mathbf{R}^{m+2}$ and use the letter S_i for the matrix such that $S_i(x, y) = {}^t x S_i y$. Put $C_i = A_i^{-1} B(w_i)$ and $E = \mathrm{End}(V, F)$. If $i \leq r$ and $0 \neq b \in Y_w$, we have

$$\mathrm{Tr}_{E/F}(bb^\sigma)^{\tau_i} = \mathrm{Tr}\,(b_i S_i^{-1} \cdot {}^t b_i S_i)$$
$$= \mathrm{Tr}\,(C_i^{-1} b_i C_i C_i^{-1} S_i^{-1} \cdot {}^t \bar{C}_i^{-1} \cdot {}^t \bar{C}_i \cdot {}^t b_i \cdot {}^t \bar{C}_i^{-1} \cdot {}^t \bar{C}_i S_i C_i)$$

$$= \mathrm{Tr}\left(\begin{bmatrix} \varphi_i(b) & & \\ & \overline{\psi_i(b)} & \\ & & \psi_i(b) \end{bmatrix} \begin{bmatrix} \xi_i^{-1} & & \\ & -\eta_i^{-1} & \\ & & -\eta_i^{-1} \end{bmatrix} \begin{bmatrix} {}^t\overline{\varphi_i(b)} & & \\ & \psi_i(b) & \\ & & \overline{\psi_i(b)} \end{bmatrix} \begin{bmatrix} \xi_i & & \\ & -\eta_i & \\ & & -\eta_i \end{bmatrix}\right) > 0 \ ,$$

where $\xi_i = \xi(w_i)$ and $\eta_i = \eta(w_i)$ as in (2.21). If $i > r$, S_i is positive, so that $\mathrm{Tr}\,(b_i S_i^{-1} \cdot {}^t b_i S_i) > 0$. This shows that σ is a positive involution. Put $Z = \{x \in Y_w \mid \psi_1(x) = \cdots = \psi_r(x) = 0\}$. Since Y_w is semi-simple, $Y_w = L \oplus Z$ with a commutative F-algebra L. Obviously $Y_0 \subset Z$, and hence $L \oplus Y_0$ is a commutative semi-simple subalgebra of E, and so $[L \oplus Y_0 : F] = m + 2$. On the other hand $K \cap Z = \{0\}$, so that K is isomorphic to a subalgebra of L. Since $[K \oplus Y_0 : F] = m + 2$, we have $[K : F] = [L : F]$. Now $K \oplus Z \subset L \oplus Z$, so that $K = L$. Thus K, or more precisely $h(K)$, is completely determined by w.

PROPOSITION 5.4. *The notation being as in Propositions 5.1 and 5.3, there exists an element v of V such that $V = h(Y)v$. Moreover, for any choice of such v, there exists a unique element δ of Y such that $S(h(a)v, h(b)v) = \mathrm{Tr}_{Y/F}(\delta a b^\rho)$ for all $(a, b) \in Y \times Y$. The element δ has the properties: $\delta = \delta^\rho$; $\delta^{\psi_i} < 0$ for $i = 1, \cdots, r$; $\delta^\alpha > 0$ for every $\alpha \in J_Y$ different from $\psi_1, \psi_1 \rho, \cdots, \psi_r, \psi_r \rho$.*

Proof. The existence of v and δ is obvious; clearly $\delta^\rho = \delta$. Put $u_i = A_i^{-1} p(w_i)$ with the map p of (2.7) for each $i \leq r$. Then

$$(5.7) \qquad h(a)_i u_i = a^{\psi_i} u_i \qquad (a \in Y).$$

Fixing one index i, let H be the set of all elements α of J_Y which induce τ_i on F, and take a disjoint decomposition $H = B \cup B\rho \cup C$ where C consists of all "real" elements of H in an obvious sense. Then

$$(5.8) \qquad \mathrm{Tr}_{Y/F}(\delta a b^\rho)^{\tau_i} = \sum_{\beta \in B} \delta^\beta (a^\beta b^{\beta \rho} + a^{\beta \rho} b^\beta) + \sum_{\gamma \in C} \delta^\gamma a^\gamma b^\gamma.$$

We can define a C-linear isomorphism

$$(5.9) \qquad \chi : \mathbf{C}^H \longrightarrow (V_i)_\mathbf{C}$$

such that $\chi((a^\alpha)_{\alpha \in H}) = h(a)_i v_i$ for all $a \in Y$. Let $\{e_\alpha\}_{\alpha \in H}$ be the standard basis of $\mathbf{C}^H$. Then we see that

$$S_i\left(\chi\left(\sum_{\alpha \in H} x_\alpha e_\alpha\right), \chi\left(\sum_{\alpha \in H} x'_\alpha e_\alpha\right)\right) = \sum_{\beta \in B} \delta^\beta (x_\beta x'_{\beta \rho} + x_{\beta \rho} x'_\beta) + \sum_{\gamma \in C} \delta^\gamma x_\gamma x'_\gamma,$$

and $\overline{\chi(e_\beta)} = \chi(e_{\beta \rho})$ if $\beta \in B$. Moreover, (5.7) implies that u_i is a scalar times $\chi(e_{\psi_i})$. Since $S_i(u_i, \bar{u}_i) < 0$, we have $\delta^{\psi_i} < 0$. The inequality $\delta^\alpha > 0$ for α other than ψ_i, $\psi_i \rho$ is a consequence of our assumption on the signature of S_i.

Using the notation of the above proof, we notice:

$$(5.10) \qquad S_i[\chi(e_\alpha)] = 0 \text{ if } \alpha \neq \alpha \rho; \quad S_i(\chi(e_\alpha), \chi(e_\gamma)) = 0 \text{ if } \alpha \neq \gamma \text{ and } \alpha \neq \gamma \rho.$$

Let $K^u = \{x \in K \mid xx^\rho = 1\}$. We define a map

$$(5.11) \qquad h^u : K^u \longrightarrow G_+(S)$$

as the composite of h with the natural injection $K^u \to Y^u$, i.e., $h^u(a) = h(a + \mathrm{id}_{Y_0})$. We now define the arithmetic elements of $\mathfrak{A}_\sigma$ in exactly the same fashion as for $\mathfrak{A}_\zeta$ of Section 4. Given a $\overline{\mathbf{Q}}$-rational representation σ as in (3.6) and Y, h, w, K as above, define a representation $\xi_\sigma\colon K^\times \to \mathrm{GL}(W, \mathbf{C})$ by

$$(5.12) \qquad \xi_\sigma(a) = \Lambda_\sigma\big(h^\times(a), w\big) \qquad\qquad (a \in K^\times)$$

with Λ_σ of (3.7). We then put

$$(5.13) \qquad P_\sigma(w) = P_K\big(\xi_\sigma, 2\textstyle\sum_{i=1}^r \psi_i\big) \ .$$

By Proposition 5.3, this depends only on σ and w. It can easily be seen that

$$(5.14) \qquad P_\sigma(\alpha(w)) = \Lambda_\sigma(\alpha, w)P_\sigma(w)\Lambda_\sigma(\alpha, w)^{-1} \qquad\qquad (\alpha \in G_+(S)).$$

Note that $\Lambda_\sigma(\alpha, w)$ is algebraic for every $\alpha \in G_+(S)$. Now we call an element f of $\mathfrak{A}_\sigma$ *arithmetic* (or $\overline{\mathbf{Q}}$-*rational*) if $P_\sigma(w)^{-1}f(w)$ is $\overline{\mathbf{Q}}$-rational for every CM-point w where f is holomorphic. We denote by $\mathfrak{A}_\sigma(\overline{\mathbf{Q}})$ the set of all such f. Relation (5.14) shows that if f is arithmetic, so is $f|_\sigma\alpha$ for every $\alpha \in G_+(S)$. If σ is a trivial representation and $\dim(W) = 1$, an element of $\mathfrak{A}_\sigma$ is an *automorphic function*; in this case we denote $\mathfrak{A}_\sigma$ by $\mathfrak{A}_0$. Then $\mathfrak{A}_0(\overline{\mathbf{Q}})$ is the field of all automorphic functions whose values at CM-points are algebraic.

Suppose now that $V = V' \oplus U$ with subspaces V' and U over F of dimension $m' + 2$ and l respectively, which are orthogonal relative to S, and that S is totally positive definite on U. Let S' and T be the restrictions of S to V' and U. Define $G(S')$, V_i', S_i', A_i', etc. with respect to S'; define also the τ_i-completion U_i of U and the extension of T_i of T to U_i. Take a $\overline{\mathbf{Q}}$-rational $\mathbf{R}$-linear isomorphism B_i of U_i onto $\mathbf{R}^l$ so that $T_i(x, y) = {}^t(B_i x)B_i y$; denote by the same letter B_i its $\mathbf{C}$-linear extension. Then we obtain an isomorphism A_i^* of $(V_i)_{\mathbf{C}}$ onto $\mathbf{C}^{m+2}$ by $A_i^*(v' + u) = \begin{pmatrix} B_i u \\ A_i' v' \end{pmatrix}$ for $v' \in V'$ and $u \in U_i$. This has the same property as A_i of (3.3), so that $A_i^* A_i^{-1} \in G(Q, R)$. Now $G_+(S')$ is naturally embedded into $G_+(S)$, and there is a commutative diagram

$$
\begin{array}{ccc}
G_+(S') & \longrightarrow & G(Q', R')^r \times \prod_{i>r} G(S_i') \\
\Big\downarrow & & {\scriptstyle\omega}\Big\downarrow \\
G_+(S) & \longrightarrow & G(Q, R)^r \times \prod_{i>r} G(S_i)
\end{array}
$$

where horizontal arrows are defined by (3.4) with A_i' and A_i; Q' and R' are defined by (2.3) with m' in place of m; $\omega(\alpha) = \gamma^{-1}\,\mathrm{diag}\,[1_l, \alpha]\gamma$, $\gamma = (\gamma_1, \cdots, \gamma_r)$, $\gamma_i = A_i^* A_i^{-1}$ with an obvious interpretation of notation. There is also an injection $z' \mapsto \begin{pmatrix} 0 \\ z' \end{pmatrix}$ of $\mathfrak{Z}_{m'}$ into $\mathfrak{Z}_m$. Therefore we can define an embedding ε of $\mathfrak{Z}_{m'}^r$ into $\mathfrak{Z}_m^r$ by

$$
\varepsilon(z_1', \cdots, z_r') = (z_1, \cdots, z_r), \quad p(z_i)c_i(z') = A_i A_i^{*-1} p\begin{pmatrix} 0 \\ z_i' \end{pmatrix}
$$

with a scalar factor $c_i(z')$. If $\alpha' \in G_+(S')$ is mapped to $\alpha \in G_+(S)$, then obviously $\varepsilon(\alpha'(z')) = \alpha(\varepsilon(z'))$ and

(5.15a)
$$\mu_i(\alpha, \varepsilon(z')) = c_i(\alpha'(z'))\mu_i(\alpha', z')c_i(z')^{-1} .$$

Similarly, putting $C_i(z') = \lambda\left(A_i A_i^{*-1}, \begin{pmatrix} 0 \\ z_i' \end{pmatrix}\right)$, we find

(5.15b)
$$\lambda_i(\alpha, \varepsilon(z')) = C_i(\alpha'(z')) \cdot \mathrm{diag}[1_l, \lambda_i(\alpha', z')]C_i(z')^{-1} .$$

Given σ as in (3.6), define $\sigma'\colon (\mathrm{GL}_1)^r \times (\mathrm{GL}_{m'})^r \times \prod_{i>r} G(S_i') \to \mathrm{GL}(W, \mathbf{C})$ by

$$\sigma'(\beta_1, \cdots, \beta_r, \gamma_1, \cdots, \gamma_r, \delta_{r+1}, \cdots, \delta_n)$$
$$= \sigma(\beta_1, \cdots, \beta_r, \mathrm{diag}[1_l, \gamma_1], \cdots, \mathrm{diag}[1_l, \gamma_r], \delta_{r+1}^*, \cdots, \delta_n^*) ,$$

where δ_i^* is the image of δ_i under the natural injection. Take $f \in \mathfrak{A}_\sigma$ such that $f \circ \varepsilon$ is meaningful, and put $f^\iota(z') = \mathfrak{C}_\sigma(z')^{-1}f(\varepsilon(z'))$, where

$$\mathfrak{C}_\sigma(z') = \sigma(c_1(z'), \cdots, c_r(z'), C_1(z'), \cdots, C_r(z'), 1_{m+2}, \cdots, 1_{m+2}) .$$

We see easily that $f^\iota \in \mathfrak{A}_{\sigma'}$.

PROPOSITION 5.5. *If $f \in \mathfrak{A}_\sigma(\overline{\mathbf{Q}})$ and $f \circ \varepsilon$ is meaningful, then $f^\iota \in \mathfrak{A}_{\sigma'}(\overline{\mathbf{Q}})$.*

Proof. Let w' be a CM-point of $G_+(S')$ on $\mathfrak{Z}_{m'}^r$ fixed by $h'(Y'^u)$. Put $w = \varepsilon(w')$ and $Y = Y' \oplus F \oplus \cdots \oplus F$ with l copies of F. Take a coordinate system of U so that T is represented by a diagonal matrix. Define $h\colon Y \to M_{m+2}(F) = \mathrm{End}(V, F)$ by

$$h(y', b_1, \cdots, b_l) = \mathrm{diag}[b_1, \cdots, b_l, h'(y')]$$

for $y' \in Y'$ and $b_j \in F$. Then $h(Y^u)$ is contained in $G_+(S)$ and has w as a fixed point. Define φ_i, ψ_i, K for Y and similarly φ_i', ψ_i', K' for Y'. We can easily verify that $K = K'$ and $\psi_i = \psi_i'$. From (5.15a, b), we obtain $P_{\sigma'}(w') = X^{-1}P_\sigma(w)X$ with $X = \mathfrak{C}_\sigma(w')$. Our assertion follows immediately from this relation.

6. Existence of $\overline{\mathbf{Q}}$-rational forms on $\mathfrak{Z}^r$

We now consider the Clifford algebra C of S and its subalgebra D consisting of the even elements. Define a $\mathbf{Q}$-rational algebraic group $\mathfrak{F}$ such that

(6.1)
$$\mathfrak{F}_\mathbf{Q} = \{x \in D^\times \mid x V x^{-1} = V\} .$$

Denote by ι the main involution of C and put $\nu(x) = xx^\iota$ for $x \in \mathfrak{F}_\mathbf{Q}$. Then $\nu(x) \in \mathfrak{F}$. There is a natural map

(6.2)
$$\pi\colon \mathfrak{F}_\mathbf{Q} \longrightarrow G(S)$$

defined by $\pi(x)v = xvx^{-1}$ for $x \in V$. It is well known that

$$\pi(\mathfrak{F}_\mathbf{Q}) = \{\alpha \in G(S) \mid \nu(\alpha) = \det(\alpha) = 1\} .$$

Let $\mathfrak{F}_{Q+}$ denote the subgroup of $\mathfrak{F}_Q$ consisting of all x such that $\nu(x) \gg 0$. Then $\pi(\mathfrak{F}_{Q+}) \subset G_+(S)$. We define the action of $\beta \in \mathfrak{F}_{Q+}$ on $\mathfrak{Z}^r$ to be the same as that of $\pi(\beta)$.

Put $m + 2 = 2p$ or $m + 2 = 2p + 1$ according as m is even or odd. By virtue of the well-known structure of D, we can define a representation

$$(6.3) \qquad\qquad \theta_i: D \longrightarrow M_s(\overline{\mathbf{Q}}) \qquad\qquad (s = 2^p)$$

so that $\mathrm{Tr}\big(\theta_i(a)\big) = \mathrm{Tr}_{D/F}(a)^{\tau_i}$. If m is even, θ_i is equivalent to the direct sum of two irreducible representations

$$(6.4) \qquad\qquad \theta_{i1}, \theta_{i2}: D \longrightarrow M_{s/2}(\overline{\mathbf{Q}}) \ .$$

Let Y, h, K, and Y_0 be as in Proposition 5.1. Express K in the form

$$(6.5) \qquad\qquad K = K_1 \oplus \cdots \oplus K_t$$

with CM-fields K_ν. To make our treatment smooth, we view these K_ν as subfields of $\mathbf{C}$ by fixed embeddings which coincide on F. Let r_ν be the element of $I(K_\nu)$ which represents the norm map of K_ν to F, and let $\mathfrak{X}_\nu = \{u \in I(K_\nu) | u + u\rho = r_\nu\}$ and $\mathfrak{X} = \prod_{\nu=1}^{t} \mathfrak{X}_\nu$. Further let M be the Galois closure of $K_1 \cdots K_t$ over $\mathbf{Q}$. We let $\mathrm{Gal}(M/\mathbf{Q})$ act on $I(K)$ on the right; $\mathfrak{X}$ is stable under $\mathrm{Gal}(M/F)$. Let R be a complete set of representatives for $\mathfrak{X}/\mathrm{Gal}(M/F)$. For each $x \in \mathfrak{X}$, let

$$(6.6) \qquad\qquad H_x = \{\gamma \in \mathrm{Gal}(M/\mathbf{Q}) | x\gamma = x\}$$

and define a subfield L_x of M by $H_x = \mathrm{Gal}(M/L_x)$. These L_x are CM-fields containing F. Let L denote the direct sum of L_x for $x \in R$. Put $[K: F] = 2l$. Then $\mathfrak{X}$ has 2^l elements and consists of $x\gamma$ for all $x \in R$ and all $\gamma \in \mathrm{Gal}(M/F)$. Therefore

$$2^l = \sum_{x \in R}[\mathrm{Gal}(M/F): H_x] = \sum_{x \in R}[L_x: F] = [L: F] \ .$$

We can define a homomorphism χ of $K^\times$ into $L^\times$ by

$$(6.7) \qquad\qquad \chi(a) = (a^x)_{x \in R} \qquad\qquad (a \in K^\times) \ ,$$

since $a^x \in L_x$ for every $a \in K^\times$. Define K^u and L^u by (1.3). Viewing K^u and L^u as (the $\mathbf{Q}$-rational points of) $\mathbf{Q}$-rational algebraic groups, we can define $K_{\mathbf{R}}^u$ $K_{\mathbf{C}}^u$, $L_{\mathbf{R}}^u$, and $L_{\mathbf{C}}^u$; extend χ to a homomorphism of $K_{\mathbf{C}}^u$ into $L_{\mathbf{C}}^u$.

LEMMA 6.1. *There exists an element a of K^u such that $\chi(a)$ generates L over $\mathbf{Q}$ and $\{a^q | q \in \mathbf{Z}\}$ is dense in $K_{\mathbf{R}}^u$.*

Proof. Take t different rational primes $p_1, \cdots, p_t$ which split completely in M. Let $\mathfrak{p}_\nu$ be a prime ideal of K_ν dividing p_ν. Take a positive integer e so that $\mathfrak{p}_\nu^e = (c_\nu)$ with $c_\nu \in K_\nu$. Put $a_\nu = c_\nu/c_\nu^\rho$ and $a = (a_1, \cdots, a_t)$. For $x \in \mathfrak{X}$ and for $\gamma \in \mathrm{Gal}(M/\mathbf{Q})$, we have $(a^z)^\gamma = a^z$ if and only if $\gamma \in H_z$. Thus a^z generates L_x over $\mathbf{Q}$ for each x. Also, for two different x and y of R, a^z is not conjugate to a^v over $\mathbf{Q}$. Hence the minimum polynomial of $\chi(a)$ over $\mathbf{Q}$ has degree $[L:\mathbf{Q}]$, so that $\chi(a)$ generates L over $\mathbf{Q}$. Now $K_{\mathbf{R}}^u$ can be identified with $\mathbf{T}^{l n}$. Since the components of a in $\mathbf{T}^{l n}$ are multiplicatively independent (over $\mathbf{Z}$), the points a^q with $q \in \mathbf{Z}$ are uniformly distributed in $\mathbf{T}^{l n}$,   Q.E.D.

LEMMA 6.2. *There exists a unique F-linear embedding g of L into D such that $g(\chi(K^u)) \subset \mathfrak{F}_{\mathbf{Q}+}$, $g(y^\rho) = g(y)^\iota$ for every $y \in L$, and $\pi(g(\chi(b))) = h^u(b)^2$ for every $b \in K^u$ with h^u of (5.11).*

Proof. Take a as in Lemma 6.1 and put $\alpha = h^u(a)$. Take $\delta \in \mathfrak{F}_{\mathbf{Q}}$ so that $\pi(\delta) = \alpha$ and put $\beta = \nu(\delta)^{-1}\delta^2$. Then $\beta\beta' = 1$ and $\pi(\beta) = \alpha^2$. A basic fact of the spin representations tells that if the eigenvalues of α_i are $\{\zeta_1, \zeta_1^{-1}, \cdots, \zeta_p, \zeta_p^{-1}\}$ or $\{\zeta_1, \zeta_1^{-1}, \cdots, \zeta_p, \zeta_p^{-1}, 1\}$ according as m is even or odd, then $\theta_i(\beta)$ has 2^p eigenvalues $\zeta_1^{\pm 1} \cdots \zeta_p^{\pm 1}$ with all combinations of exponents ± 1. In other words, the eigenvalues of $\theta_i(\beta)$ are 2^{p-l} times

(6.8) $$\{a^{x\sigma} \mid x \in R,\ \sigma \in J(L_x),\ \sigma = \tau_i \text{ on } F\} .$$

Since $\chi(a)$ generates L over $\mathbf{Q}$, we can define an F-linear embedding g of L into D by $g(\chi(a)) = \beta$. We have $g(y^\rho) = g(y)^\iota$ for every $y \in L$, since $\chi(a)^\rho = \chi(a)^{-1}$. Now $g \circ \chi$ maps K^u into $D^\times$ and the points a^q with $q \in \mathbf{Z}$ into $\mathfrak{F}_{\mathbf{Q}}$. Since these points are dense in $K_{\mathbf{R}}^u$, we see that $g \circ \chi$ maps K^u into $\mathfrak{F}_{\mathbf{Q}}$, and hence $\pi \circ g \circ \chi$ is meaningful. We have $\pi(g(\chi(a^q))) = h^u(a^q)^2$ for all $q \in \mathbf{Z}$, so that $\pi(g(\chi(b))) = h^u(b)^2$ for all $b \in K^u$. If g' is another map with the same properties as g, we have $g'(\chi(b)) = \pm g(\chi(b))$, so that $g'(\chi(b^2)) = g(\chi(b^2))$ for all $b \in K^u$. This implies $g = g'$, since L is generated by such $\chi(b^2)$.

Now, following Shih [6], [7], we embed $\mathfrak{F}$ into the group $\mathfrak{G}$ of Section 4 as follows. Take a subspace W of V of dimension 3 over F so that S is totally positive definite on the orthogonal complement of W. Let B be the even Clifford algebra of the restriction of S to W. Considering B a subalgebra of D, we view D as a left B-module. Then there is a B-valued ι-hermitian form Φ on D such that

(6.9) $$\mathrm{Tr}_{B/F}(\Phi(x, y)) = \mathrm{Tr}_{D/F}(xy') .$$

Define a $\mathbf{Q}$-rational algebraic group $\mathfrak{G}'$ by

(6.10) $$\mathfrak{G}'_{\mathbf{Q}} = \{\alpha \in \mathrm{GL}(D, B) \mid \Phi(x\alpha, y\alpha) = \nu(\alpha)\Phi(x, y) \text{ with } \nu(\alpha) \in F\} .$$

It can easily be seen that Φ is "definite" at $\tau_{r+1}, \cdots, \tau_n$, so that $\mathfrak{G}'$ can be identified with the group $\mathfrak{G}$ of Section 4 with $q = 2^{m-1}$; we fix such an identification. Right multiplication by an element β of D defines an element

$e(\beta)$ of $\mathrm{End}(D, B)$. In this way, we obtain a **Q**-rational embedding

$$(6.11) \qquad e: \mathfrak{F} \longrightarrow \mathfrak{G} \,,$$

which is compatible with a holomorphic embedding

$$(6.12) \qquad \varepsilon: \mathfrak{Z}^r \longrightarrow \mathfrak{H}_q^r \,;$$

moreover, $f \circ \varepsilon \in \mathfrak{A}_0(\overline{\mathbf{Q}})$ for every $f \in \mathfrak{A}_0(\overline{\mathbf{Q}})$ for which $f \circ \varepsilon$ is meaningful. (For details, see Shih [6], [7].) It can easily be shown that if $m > 1$, the pull-back of the factor of automorphy $\det(\omega_i)$ is equivalent to a power of μ_i in the sense that

$$(6.13) \qquad \det \omega_i\big(e(\alpha),\, \varepsilon(z)\big) = v_i\big(\alpha(z)\big)^{-1}\mu_i\big(\pi(\alpha),\, z\big)^{q/2} v_i(z) \quad \text{for} \quad \alpha \in \mathfrak{F}_{\mathbf{R}}$$

with a holomorphic function v_i on $\mathfrak{Z}^r$, which is unique up to constant factors. If $m = 1$, a similar relation exists between $\det(\omega_i)^2$ and μ_i. If $z = \alpha(z_0)$, we have

$$v_i(z) = \mu_i\big(\pi(\alpha),\, z_0\big)^{q/2} v_i(z_0) \det \omega_i\big(e(\alpha),\, \varepsilon(z_0)\big)^{-1} \,.$$

Therefore, taking $v_i(z_0) = 1$ at one CM-point z_0, we may assume that $v_i(z)$ is algebraic for every CM-point z and even for every z with algebraic coordinates. (This principle applies to factors of automorphy of a more general type.)

We now consider $L, \mathcal{X}, g$ of Lemma 6.2 obtained from Y, h, K; let w be the fixed point of $h(Y^u)$ on $\mathfrak{Z}^r$. Observe that $e \circ g$ embeds L into $M_q(B)$ and L^u into $\mathfrak{G}_Q$; moreover $e\big(g(L^u)\big)$ has $\varepsilon(w)$ as its unique fixed point on $\mathfrak{H}_q^r$. By (6.13), we have

$$(6.14) \qquad \det \omega_i\big(e(\alpha),\, \varepsilon(w)\big)^2 = \mu_i\big(\pi(\alpha),\, w^q\big) \quad \text{for} \quad \alpha \in g(L^u) \,.$$

Define φ_i and ψ_i by (5.6a, b). Take $a \in K^u$ as in Lemma 6.1. Let $\zeta_1, \zeta_1^{-1}, \cdots,$ ζ_p, ζ_p^{-1} (and 1 if m is odd) be the eigenvalues of $h(a)_i$ with a fixed $i < r$. We arrange ζ_j so that $\zeta_1 = \psi_i(a)$. We can easily verify that the eigenvalues of $\omega_i\big(eg(\mathcal{X}(a)),\, \varepsilon(w)\big)$ are exactly N times 2^{p-1} numbers $\zeta_1\zeta_2^{\pm 1}\cdots\zeta_p^{\pm 1}$ where $N = 2^{p-2}$ or 2^{p-1} according as m is even or odd. We see also that the commutor of $e\big(g(L)\big)$ in $M_q(B)$ is an algebra of the type discussed in Section 4.

Let us now determine the representations ξ_i and Ξ_j of L^u defined by $\xi_i(c) = \omega_i\big(e(g(c)),\, \varepsilon(w)\big)$ and $\Xi_j(c) = \Omega_j\big(e(g(c))\big)$ (see §4). For each $i \leq r$, there is a unique $K_{\nu(i)}$ on which ψ_i is non-trivial. Let $H = \mathrm{Gal}(M/K_{\nu(i)})$ with a fixed i. Every element of $J(K_{\nu(i)})$ can be represented by a coset $H\gamma$ with $\gamma \in \mathrm{Gal}(M/\mathbf{Q})$. For $x \in R$, let x_ν be its νth component, and put $x_{\nu(i)} = \sum_j H\gamma_j$ with $\gamma_j \in \mathrm{Gal}(M/\mathbf{Q})$. By (6.6), we have $\bigcup_j H\gamma_j = \bigcup_k \delta_k^{-1}H_x$ with $\{\delta_k\} \subset \mathrm{Gal}(M/\mathbf{Q})$. Define $y_{xi} \in I(L_x)$ by $y_{xi} = \sum_k H_x\delta_k$. Then $(K_{\nu(i)}, x_{\nu(i)})$ and (L_x, y_{xi}) form a reflexive pair. Put $b = \mathcal{X}(a)$. Then the eigenvalues of $\xi_i(b)$ consist of $2^{p-1}N$

times $a^{x\sigma}$ for all $x \in R$ and all $\sigma \in J(L_x)$ such that ψ_i occurs in $x_{\nu(i)}\sigma$. Take any $\alpha_i \in \mathrm{Gal}(M/\mathbf{Q})$ so that $\alpha_i = \psi_i$ on $K_{\nu(i)}$. Then ψ_i occurs in $x_{\nu(i)}\sigma$ if and only if σ occurs in $y_{xi}\alpha_i$. Thus the eigenvalues of $\xi_i(b)$ consist of $2^{p-1}N$ times $a^{x\sigma}$ for all $x \in R$ and all $\sigma \in J(L_x)$ such that σ occurs in $y_{xi}\alpha_i$. Therefore the element of I_L representing $\det(\xi_i)$ is

$$(2^{p-1}Ny_{xi}\alpha_i)_{x \in R} \ .$$

Let η be the element of $I_L = \prod_{x \in R} I(L_x)$ whose x-component is $\eta_x = \sum_{i=1}^{r} y_{xi}\alpha_i$. This is exactly the element η defined at the fixed point $\varepsilon(w)$ in Section 4.

PROPOSITION 6.3. *Define* $\mathfrak{p}_L(\eta)$ *and* $\mathfrak{p}_K(\psi)$ *as in Section 1 with* $\psi = \sum_{i=1}^{r} \psi_i \in I_K$. *Then* $\mathfrak{p}_L(\eta) = \chi(\mathfrak{p}_K(\psi))$.

Proof. For every element of $L_{\mathbf{C}}^u$, we can speak of its (x, σ)-component with $x \in R$ and $\sigma \in J(L_x)$. The (x, σ)-component of $\mathfrak{p}_L(\eta)$ is $p(\sigma, \eta_x; L_x)$. If $a \in K^u$, the (x, σ)-component of $\chi(a)$ is $a^{x\sigma}$. Therefore, the (x, σ)-component of $\chi(\mathfrak{p}_K(\psi))$ is $\prod_{\mu=1}^{t} p(x_\mu\sigma, \sum_{\nu(i)=\mu}\psi_i; K_\mu)$. This is equal to

$$\prod_{\mu=1}^{t}\prod_{\nu(i)=\mu} p(x_\mu\sigma, \alpha_i; K_{\nu(i)})$$
$$= \prod_{\mu=1}^{t}\prod_{\nu(i)=\mu} p(\alpha_i^{-1}x_{\nu(i)}\sigma, 1; K_{\nu(i)}^{\alpha_i}) \qquad \text{(by Theorem 1.1, (5))}$$
$$= \prod_{\mu=1}^{t}\prod_{\nu(i)=\mu} p(\sigma, y_{xi}\alpha_i; L_x) \qquad \text{(by Theorem 1.2)}$$
$$= p(\sigma, \eta_x; L_x) = \text{the } (x, \sigma)\text{-component of } \mathfrak{p}_L(\eta), \qquad \text{Q.E.D.}$$

COROLLARY 6.4. *Define* $\xi_i: L^u \to \mathrm{GL}_q$ *by* $\xi_i(b) = \omega_i(e(g(b)), \varepsilon(w))$. *Then* $\det(P_L(\xi_i, \eta)) = p_K(2^{m-1}\psi_i, \psi)$.

Proof. By (6.14), we have $\det \xi_i(b)^2 = \mu_i(\pi(g(b)), w)^q$, and hence $\det(\xi_i(\chi(a))) = \psi_i(a)^q$ for $a \in K^u$. Substituting $\mathfrak{p}_K(\psi)$ for a, we obtain the desired formula from Prop. 6.3.

THEOREM 6.5. *For every* $\overline{\mathbf{Q}}$-*rational representation* σ *as in* (3.6) *with* $W = \mathbf{C}^d$ *and an arbitrary point* z_0 *of* $\mathfrak{Z}^r$, *there exists an* $M_d(\mathbf{C})$-*valued meromorphic function* T *on* $\mathfrak{Z}^r$ *such that*

(i) T *is holomorphic at* z_0 *and* $\det(T(z_0)) \neq 0$;

(ii) *the columns of* T *belong to* $\mathfrak{A}_\sigma(\overline{\mathbf{Q}})$.

We call such T a $\overline{\mathbf{Q}}$-*rational* σ-*uniformizer at* z_0.

Proof. It is sufficient to prove the existence of an $M_d(\mathbf{C})$-valued meromorphic function T on $\mathfrak{Z}^r$ satisfying (i) and

(iii) *There exist meromorphic functions* $f_1, \cdots, f_\kappa$ *on* $\mathfrak{Z}^r$ *holomorphic at* z_0 *such that* $P_\sigma(w)^{-1}T(w)$ *is* $\overline{\mathbf{Q}}$-*rational for every* CM-*point* w *where* T *and* $f_1, \cdots, f_\kappa$ *are holomorphic; in particular,* $P_\sigma(z_0)^{-1}T(z_0)$ *is* $\overline{\mathbf{Q}}$-*rational if* z_0 *is a* CM-*point*.

In fact, suppose such a T always exists for given σ and z_0. Let g be an element of $\mathfrak{A}_\sigma$ such that $P_\sigma(w)^{-1}g(w)$ is $\overline{\mathbf{Q}}$-rational for every w in a set $\mathfrak{W}$ of CM-points which is dense in $\mathfrak{Z}^r$. Let us show that $g \in \mathfrak{A}_\sigma(\overline{\mathbf{Q}})$. Let z_0 be a CM-point where g is holomorphic. Take T satisfying (i) and (iii) for this z_0. Put $g' = T^{-1}g$. Then g' has components in $\mathfrak{A}_0$, and $g'(w)$ is $\overline{\mathbf{Q}}$-rational for every w in $\mathfrak{W}$ where $f_1, \cdots, f_\kappa$ are holomorphic and $\det(T(w)) \neq 0$. Such points w form a dense subset of $\mathfrak{Z}^r$, so that g' is $\overline{\mathbf{Q}}$-rational. Now $g = Tg'$ and g' is holomorphic at z_0, and hence $P_\sigma(z_0)^{-1}g(z_0)$ is $\overline{\mathbf{Q}}$-rational as desired. Applying this result to the columns of T, we find that T satisfies (ii).

It is also sufficient to prove (i), (iii) for the representations σ_i^1, σ_i^2, and σ_i^3 defined by

(6.15a) $$\sigma_i^1(\alpha_1, \cdots, \alpha_r, \beta_1, \cdots, \beta_r, \gamma_{r+1}, \cdots, \gamma_n) = \alpha_i \,,$$

(6.15b) $$\sigma_i^2(\alpha_1, \cdots, \alpha_r, \beta_1, \cdots, \beta_r, \gamma_{r+1}, \cdots, \gamma_n) = \beta_i \,,$$

(6.15c) $$\sigma_i^3(\alpha_1, \cdots, \alpha_r, \beta_1, \cdots, \beta_r, \gamma_{r+1}, \cdots, \gamma_n) = \gamma_i \,.$$

Indeed, given z_0, let T_i^k be a function satisfying (i), (iii) for σ_i^k. Now every irreducible representation of a general linear group or an orthogonal group is an irreducible constituent of a tensor representation. ($G(S_i)$ is not an orthogonal group, but obviously $\mathfrak{A}_\sigma(\overline{\mathbf{Q}})$ depends only on the restriction of σ to the subgroup defined by $\nu(\alpha_{r+1}) = \cdots = \nu(\alpha_n) = 1$.) Therefore a desired function for a given σ can be obtained as a "submatrix" of a suitable tensor product of T_i^1, T_i^2, and T_i^3. We also note that

(6.16) $$\det(P_\sigma(w)) = 1 \text{ if } \sigma = \sigma_i^2 \text{ or } \sigma = \sigma_i^3 \,,$$

since the representation ξ_σ of (5.12) in these cases contains α and $\alpha\rho$ with the same multiplicity for every $\alpha \in J_K$.

Let us first consider the case $\sigma = \sigma_i^3$ with $i > r$. Suppose m is even. Let U be a $\overline{\mathbf{Q}}$-rational Ω_i-uniformizer at $\varepsilon(z_0)$. Observe that $\Omega_i \circ e$, viewed as a representation of D, is equivalent to 2^{p-2} copies of θ_{i1} and θ_{i2}. Thus

$$E_i \Omega_i(e(d))E_i^{-1} = \operatorname{diag}[\theta_{i1}(d), \theta_{i2}(d), \cdots] \qquad (d \in D)$$

with $E_i \in \mathrm{GL}_{2q}(\overline{\mathbf{Q}})$. Choosing suitable columns and rows of the matrix $E_i U \circ \varepsilon$, we find $M_{s/2}(\mathbf{C})$-valued meromorphic functions T_j on $\mathfrak{Z}^r$ for $j = 1, 2$ such that $\det(T_j(z_0)) \neq 0$ and $T_j \circ \gamma = \theta_{ij}(\gamma)T_j$ for all γ in a congruence subgroup of $\mathfrak{F}_Q$, where $s = 2^p$. Considering V a subset of C, take a vector subspace W of C over F so that $C = V \oplus W$ and define a representation π' of $D^\times$ into $\mathrm{GL}(C, F)$ by $\pi'(d)x = dxd^{-1}$ for $d \in D^\times$ and $x \in C$. Then

$$\pi'(d) = \begin{bmatrix} \pi(d) & 0 \\ * & * \end{bmatrix} \text{ if } d \in \mathfrak{F}_Q \,.$$

Fixing a basis of V and W over F, we can speak of the images of $\pi(d)$ and $\pi'(d)$ under τ_i, which we denote by $\pi_i(d)$ and $\pi_i'(d)$. Now considering $D \otimes_Q \overline{Q}$, we find a $\overline{Q}$-rational representation

$$\rho_i : \mathrm{GL}_{s/2} \times \mathrm{GL}_{s/2} \longrightarrow \mathrm{GL}_u \qquad\qquad (u = 2^{m+2})$$

such that $\pi_i'(d) = \rho_i(\theta_{i1}(d), \theta_{i2}(d))$. Let $T'(z)$ be the first $m + 2$ rows of $\rho_i(T_1(z), T_2(z))$. Then $T'(\gamma(z)) = \gamma_i T'(z)$ for all γ in a congruence subgroup of $G_+(S)$. To show that T' is arithmetic, take the fixed point w of $h(Y^u)$; define L, g, $\mathfrak{p}_L(\eta)$, etc., and put $\mathfrak{P} = g(\mathfrak{p}_L(\eta))$. Then $\Omega_i(e(\mathfrak{P}))^{-1}U(\varepsilon(w))$ is $\overline{Q}$-rational (if $U \circ \varepsilon$ is finite at w), so that $\theta_{ij}(\mathfrak{P})^{-1}T_j(w)$ are $\overline{Q}$-rational for $j = 1, 2$, and hence $\pi_i'(\mathfrak{P})\rho_i(T_1(w), T_2(w))$ is $\overline{Q}$-rational. By Proposition 6.3 and Lemma 6.2, $\mathfrak{P} = g(\mathcal{X}(\mathfrak{p}_K(\psi))) \in \mathfrak{F}_C$ and $\pi_i(\mathfrak{P}) = h^u(\mathfrak{p}_K(\psi))_i^2$. Therefore $h^u(\mathfrak{p}_K(\psi))_i^{-2}T'(w)$ is $\overline{Q}$-rational, whenever $U \circ \varepsilon$ is finite at w and $\det(T_1(w)T_2(w)) \neq 0$. We can form a matrix T from suitable $m + 2$ columns of T' such that $\det(T(z_0)) \neq 0$. Then T has the required property (iii) for σ_i^3. If m is odd, we consider θ_i instead of θ_{i1} and θ_{i2}, and obtain the desired result in the same (and simpler) way.

Next let us focus our attention on σ_i^1, assuming $m > 1$. (If $m = 1$, our theorem follows directly from Proposition 4.1.) Take an element f of $\mathcal{C}_\zeta(\overline{Q})$ so that $f(\varepsilon(z_0)) \neq 0$ with a representation

$$\zeta(\omega_1, \cdots, \omega_r, \Omega_{r+1}, \cdots, \Omega_n) = \det(\omega_i), \qquad\qquad i \leqq r .$$

Put $t(z) = v_i(z)f(\varepsilon(z))$ with v_i of (6.13). Then $t \in \mathfrak{A}_\sigma$ with $\sigma = (\sigma_i^1)^{q/2}$. Corollary 6.4 shows that $P_\sigma(w) = P_\zeta(\varepsilon(w))$, and hence $t \in \mathfrak{A}_\sigma(\overline{Q})$. This is weaker than the desired result for σ_i^1 itself. Before proving a stronger result, we look at σ_i^2. For $\beta \in \mathfrak{G}_{Q+}$, let $\delta_i(\beta, Z)$ denote the jacobian matrix of the action of β on the ith factor $\mathfrak{H}_q$ of $\mathfrak{H}_q^r$, with respect to the entries of the variable matrix on $\mathfrak{H}_q$. Then

$${}^t\delta_i(\beta, Z)^{-1} = \Theta_i(\omega_i(\beta, Z)) \qquad\qquad (Z \in \mathfrak{H}_q^r)$$

with a representation Θ_i of GL_q of degree $N = q(q + 1)/2$. If $C_i(z)$ denotes the $N \times m$ matrix of the partial derivatives of the coordinates of $\varepsilon(z)_i$, we have

$$(6.17) \qquad \delta_i(e(\alpha), \varepsilon(z))C_i(z) = C_i(\alpha(z)) \cdot {}^t\lambda_i(\pi(\alpha), z)^{-1}\mu_i(\pi(\alpha), z)^{-1}$$

for $\alpha \in \mathfrak{F}_Q$ and $z \in \mathfrak{Z}^r$. Obviously $C_i(z)$ has rank m everywhere on $\mathfrak{Z}^r$ and is algebraic at CM-points. Let R be a $\overline{Q}$-rational Θ_i-uniformizer at $\varepsilon(z_0)$, and let $T'(z) = {}^tC_i(z)R(\varepsilon(z))$. Then $T'(\gamma(z)) = \mu_i(\gamma, z)\lambda_i(\gamma, z)T'(z)$ for every γ in a congruence subgroup of $G(S)$. Let $\kappa(b) = \Theta_i(\omega_i(e(g(b)), \varepsilon(w)))$ for $b \in L^u$. Then $\kappa(\mathfrak{p}_L(\eta))^{-1}R(\varepsilon(w))$ is $\overline{Q}$-rational if $R \circ \varepsilon$ is finite at w, and (6.17)

shows that

$$'C_i(w)\kappa(\chi(a)) = \psi_i(h^u(a)^2)\varphi_i(h^u(a)^2) \cdot {}'C_i(w) \qquad \text{for } a \in K^u .$$

Substituting $\mathfrak{p}_K(\psi)$ for a, we obtain, by virtue of Proposition 6.3,

$$'C_i(w)\kappa(\mathfrak{p}_L(\eta)) = \psi_i(h^u(\mathfrak{p}_K(\psi)^2))\varphi_i(h^u(\mathfrak{p}_K(\psi))^2) \cdot {}'C_i(w) .$$

Choosing suitable m columns of T', we find a matrix T satisfying (i) and (iii) with $\sigma = \sigma_i^1\sigma_i^2$. Then $\det(T)$ satisfies (i) and (iii) with $\sigma = (\sigma_i^1)^m$ by (6.16). If m is odd, this proves the case $\sigma = \sigma_i^1$. In fact, we have obtained an element t of $\mathfrak{A}_\sigma(\overline{\mathbf{Q}})$ with $\sigma = (\sigma_i^1)^{q/2}$; therefore $t^u \det(T)^v$ with integers u and v such that $(uq/2) + vm = 1$ is the desired function for σ_i^1. Dividing T by this function, we can prove the case $\sigma = \sigma_i^2$. Thus it remains to prove the case where $\sigma = \sigma_i^1$ and m is even. For this we use the embedding considered in Proposition 5.5. By that proposition, we can pull back an arithmetic form on $(\mathfrak{Z}_{m+1})^r$ to that on $\mathfrak{Z}_m^r$. Therefore our result for odd m produces the desired form for even m.

For a function f on $\mathfrak{Z}^r$, define a column vector $D_i f$ by

$$(6.18) \qquad D_i f = {}^t(\partial f/\partial z_{i1}, \cdots, \partial f/\partial z_{im}) \qquad (i = 1, \cdots, r)$$

with the standard variables $z_{i1}, \cdots, z_{im}$ on the ith factor $\mathfrak{Z}$ of $\mathfrak{Z}^r$.

PROPOSITION 6.6. *Let* $f \in \mathfrak{A}_0(\overline{\mathbf{Q}})$ *and* $\sigma = \sigma_i^1\sigma_i^2$ *with* σ_i^k *of* (6.15). *Then* $\pi^{-1}D_i f \in \mathfrak{A}_\sigma(\overline{\mathbf{Q}})$.

Proof. That $D_i f \in \mathfrak{A}_\sigma$ is obvious from (2.17). Define a similar operator Δ_i on $\mathfrak{H}_q^r$ by

$$(6.19) \qquad \Delta_i g = {}^t(\partial g/\partial Z_{i1}, \cdots, \partial g/\partial Z_{iN}) \qquad (i \leq r,\ N = q(q + 1)/2),$$

with an arrangement of the entries $Z_{i1}, \cdots, Z_{iN}$ of the ith matrix variable on $\mathfrak{H}_q^r$. Then we have $D_i(g \circ \varepsilon) = {}^tC_i \cdot (\Delta_i g) \circ \varepsilon$ with C_i of (6.17). Now $\pi^{-1}\Delta_i g$ is $\overline{\mathbf{Q}}$-rational by virtue of [19, Theorem 4] (cf. also [18, Cor. 7.7]). Hence, taking $\Delta_i g$ in place of R in the proof of Theorem 6.5, we find that $\pi^{-1}D_i(g \circ \varepsilon) \in \mathfrak{A}_\sigma(\overline{\mathbf{Q}})$. Now we can find mr elements $g_1, \cdots, g_{mr}$ of $\mathcal{C}_0(\overline{\mathbf{Q}})$ such that $g_1 \circ \varepsilon, \cdots, g_{mr} \circ \varepsilon$ are algebraically independent. Put $f_j = g_j \circ \varepsilon$. Then $\partial/\partial f_1, \cdots, \partial/\partial f_{mr}$ are well-defined derivations of $\mathfrak{A}_0(\overline{\mathbf{Q}})$. For every $f \in \mathfrak{A}_0(\overline{\mathbf{Q}})$, we have $D_i f = \sum_{j=1}^{mr}(\partial f/\partial f_j)D_i f_j$, which proves our assertion.

7. Hilbert modular forms of half-integral weight

The purpose of this section is to introduce Hilbert modular forms of half-integral weight as well as non-holomorphic theta series which will be employed in next sections. We denote by α_ν for $\alpha \in M_q(F)$ the image of α under τ_ν, and embed $M_q(F)$ into $M_q(\mathbf{R})^n$ through the map $\alpha \mapsto (\alpha_1, \cdots, \alpha_n)$;

here q is an arbitrary positive integer. We put also

$$(7.1) \qquad \mathrm{GL}_q^+(F) = \{\alpha \in \mathrm{GL}_q(F) \,|\, \det(\alpha) \gg 0\}$$

and define the action of $\mathrm{GL}_2^+(F)$ on $\mathfrak{H}_1^n$ component by component as in Section 4. To simplify our notation, for $z = (z_1, \cdots, z_n) \in \mathbf{C}^n$ and $\lambda = (\lambda_1, \cdots, \lambda_n) \in \mathbf{R}^n$, we put

$$(7.2\mathrm{a}) \qquad z^\lambda = z_1^{\lambda_1} \cdots \lambda_n^{\lambda_n},$$

$$(7.2\mathrm{b}) \qquad |\lambda| = |\lambda_1| + \cdots + |\lambda_n| \qquad \text{(this occurs only as exponent)},$$

$$(7.2\mathrm{c}) \qquad \mu 1 = (\mu, \cdots, \mu) \quad \text{for} \quad \mu \in \mathbf{R}.$$

Thus $z^{\mu 1} = (z_1 \cdots z_n)^\mu$. We use these symbols also for $\lambda \in \mathbf{C}^n$ or $\mu \in \mathbf{C}$ when they are meaningful.

Let S be a symmetric element of $\mathrm{GL}_q(F)$ and let (λ_ν, μ_ν) be the signature of S_ν. Take n elements $P_1, \cdots, P_n$ of $\mathrm{GL}_q(\mathbf{R})$ such that ${}^t P_\nu = P_\nu > 0$ and $P_\nu S_\nu^{-1} P_\nu = S_\nu$. Then we put

$$(7.3) \qquad R_\nu(z_\nu) = x_\nu S_\nu + i y_\nu P_\nu \qquad \text{for} \quad z_\nu = x_\nu + i y_\nu \in \mathfrak{H}_1.$$

We consider a "spherical function" χ on $\mathbf{R}^{qn}$ relative to S_ν and P_ν as follows. Decompose $\mathbf{R}^{qn}$ into the direct sum of n copies of $\mathbf{R}^q$ and write the variable w in $\mathbf{R}^{qn}$ in the form $w = (w_1, \cdots, w_n)$ with $w_\nu \in \mathbf{R}^q$. Then our function χ has an expression

$$(7.4) \qquad \chi(w) = \prod_{\nu=1}^n ({}^t r_\nu S_\nu w_\nu)^{l_\nu} ({}^t r_\nu' S_\nu w_\nu)^{m_\nu}$$

with $0 \le l_\nu \in \mathbf{Z}$, $0 \le m_\nu \in \mathbf{Z}$, and elements $r_1, \cdots, r_n, r_1', \cdots, r_n'$ of $\mathbf{C}^q$ such that

$$(7.5) \qquad S_\nu[r_\nu] = S_\nu[r_\nu'] = 0, \quad P_\nu r_\nu = S_\nu r_\nu, \quad P_\nu r_\nu' = -S_\nu r_\nu'.$$

Now we define a theta series of the form

$$(7.6) \qquad f(z; a, g, h, L) = \sum_{w-g \in L} \chi(w) e\left(\sum_{\nu=1}^n \left\{ \frac{1}{2} a_\nu R_\nu(z_\nu)[w_\nu] + {}^t w_\nu h_\nu \right\} \right),$$

where $z = (z_1, \cdots, z_n) \in \mathfrak{H}_1^n$, $0 \ll a \in F$, $g, h \in F^q$, L is a lattice in F^q, and $e(v) = \exp(2\pi i v)$ for $v \in \mathbf{C}$.

PROPOSITION 7.1. *Let $\mathcal{T}(S, \{P_\nu\}, \chi)$ denote the vector space spanned over* $\mathbf{C}$ *by all functions of the form* (7.6) *with fixed S, P_ν, χ and with arbitrary a, g, h, L. Then the following assertions hold:*

(7.7) *If $f \in \mathcal{T}(S, \{P_\nu\}, \chi)$ and $\alpha = \begin{pmatrix} * & * \\ c & d \end{pmatrix} \in \mathrm{GL}_2^+(F)$, then*

$$f(\alpha(z))(cz + d)^{-(\lambda/2)-l}(c\bar{z} + d)^{-(\mu/2)-m}$$

belongs to $\mathcal{T}(S, \{P_\nu\}, \chi)$, where abbreviation (7.2a) is used.

(7.8) *If $f \in \mathcal{T}(S, \{P_\nu\}, \chi)$ and $\gamma = \begin{pmatrix} * & * \\ c & d \end{pmatrix} \in \Gamma$, $d \gg 0$, then*

$$f(\gamma(z)) = f(z)\Big(\frac{2c}{d\mathfrak{r}}\Big)^q (cz + d)^{(\lambda/2)+l}(c\bar{z} + d)^{(\mu/2)+m} \, ,$$

where Γ is a congruence subgroup of $\mathrm{SL}_2(\mathfrak{r})$ depending on f, $(-)$ is the quadratic residue symbol in F; the branches of $(cz + d)^\alpha$ and $(c\bar{z} + d)^\beta$ are chosen so as to become positive as z tends to 0; $\mathfrak{r}$ denotes the maximal order of F.

This is well known at least in various special cases (see Siegel [21] and Shintani [20], for example). For the reader's convenience, we sketch here a proof, since the result in this form may not be found in the existing literature. First we observe that $\mathcal{T}(S, \{P_\nu\}, \chi) = \mathcal{T}(cS, \{c_\nu P_\nu\}, \chi)$ for $0 \ll c \in F$. Hence, changing S and P_ν for cS and $c_\nu P_\nu$ with a suitable c, we may assume:
(7.9) *The entries of S are all algebraic integers.*

Let $\{\beta_1, \cdots, \beta_n\}$ be a **Z**-basis of a fractional ideal $\mathfrak{b}$ in F, and let

$$(7.10a) \qquad B = \begin{bmatrix} \beta_1^{r_1} 1_q \cdots \beta_n^{r_1} 1_q \\ \cdots\cdots\cdots \\ \beta_1^{r_n} 1_q \cdots \beta_n^{r_n} 1_q \end{bmatrix},$$

$$(7.10b) \qquad W(z) = {}^tB \cdot \mathrm{diag}[R_1(z_1), \cdots, R_n(z_n)]B \qquad (z = (z_1, \cdots, z_n) \in \mathfrak{H}_1^n).$$

Also we put

$$(7.11) \qquad \Phi(T) = \mathrm{diag}[T^{r_1}, \cdots, T^{r_n}] \qquad (T \in M_q(F)),$$

$$(7.12) \qquad \sigma_r = \begin{bmatrix} {}^tB\Phi(a1_q) \cdot {}^tB^{-1} & {}^tB\Phi(bS)B \\ B^{-1}\Phi(cS^{-1}) \cdot {}^tB^{-1} & B^{-1}\Phi(d1_q)B \end{bmatrix} \quad \text{for} \quad \gamma = \begin{pmatrix} a & b \\ c & d \end{pmatrix} \in \mathrm{SL}_2(F).$$

It can easily be verified that $W(z) \in \mathfrak{H}_{nq}$, $\sigma_r \in \mathrm{Sp}(nq, \mathbf{Q})$, $\sigma_r(W(z)) = W(\gamma(z))$, and

$$\det\{B^{-1}\Phi(cS^{-1}) \cdot {}^tB^{-1}W(z) + B^{-1}\Phi(d1_q)B\} = (cz + d)^\lambda (c\bar{z} + d)^\mu \, .$$

Now consider the classical theta function

$$(7.13) \qquad \theta(u, Z; r, s) = \sum_{g-r \in \mathbf{Z}^n} e\Big(\frac{1}{2} Z[g] + {}^tg(u + s)\Big)$$

with $u \in \mathbf{C}^{nq}$, $Z \in \mathfrak{H}_{nq}$, $r \in \mathbf{Q}^{nq}$, $s \in \mathbf{Q}^{nq}$, and also the "modified" theta function

$$(7.14) \qquad \varphi(u, Z; r, s) = e\Big(\frac{1}{2} \cdot {}^tu(Z - \bar{Z})^{-1}u \Big)\theta(u, z; r, s)$$

introduced in [11] and [14]. It is known that

$$(7.15) \qquad \varphi({}^t(CZ + D)^{-1}u, \sigma(Z); r, s)$$
$$= \kappa(\sigma)e(({}^trs - {}^tr's')/2)\det(CZ + D)^{1/2}\varphi(u, Z; r', s')$$

for every $\sigma = \begin{pmatrix} * & * \\ C & D \end{pmatrix} \in \Gamma_\theta$, where $({}^tr', {}^ts') = ({}^tr, {}^ts)\gamma$, Γ_θ is a subgroup of

$\mathrm{Sp}(nq, \mathbf{Z})$ of level 2 defined by [14, (1.7)], and $\kappa(\sigma)$ is a root of unity depending only on σ. Define a function $\mathfrak{f}$ by

$$\mathfrak{f}(v, z) = \mathfrak{f}(v, z; g, h, \mathfrak{b}) = \varphi({}^{t}Bv, W(z); B^{-1}g, {}^{t}Bh)$$

for $v \in \mathbf{C}^{nq}$ and $z \in \mathfrak{H}_{1}^{n}$. A direct calculation shows that

(7.16)
$$\mathfrak{f}(v, z) = e\left(\sum_{\nu=1}^{n}(4iy_{\nu}P_{\nu})^{-1}[v_{\nu}]\right)\sum_{x-g\in\mathfrak{b}q}e\left(\sum_{\nu=1}^{n}\left\{\frac{1}{2}R_{\nu}(z_{\nu})[x_{\nu}] + {}^{t}x_{\nu}(v_{\nu} + h_{\nu})\right\}\right).$$

Therefore if $\gamma = \begin{pmatrix} a & b \\ c & d \end{pmatrix} \in \mathrm{SL}_{2}(F)$ and $\sigma_{\gamma} = \begin{pmatrix} * & * \\ C & D \end{pmatrix} \in \Gamma_{\theta}$, we have

(7.17)
$$\mathfrak{f}(\xi(\gamma, z)\tilde{v}, \gamma(z); g, h, \mathfrak{b})$$
$$= \kappa(\sigma_{\gamma})e(X)(cz + d)^{\lambda/2}(c\bar{z} + d)^{\mu/2}\mathfrak{f}(v, z; g', h', \mathfrak{b}),$$

where $\xi(\gamma, z) = B(C \cdot W(z) + D)B^{-1}$ and

(7.18)
$$X = \frac{1}{2}\mathrm{Tr}_{F/Q}({}^{t}gh - {}^{t}g'h'), \qquad \begin{pmatrix} g' \\ h' \end{pmatrix} = \begin{pmatrix} a1_{q} & cS^{-1} \\ bS & d1_{q} \end{pmatrix}\begin{pmatrix} g \\ h \end{pmatrix}.$$

Let $r_{\nu 1}, \cdots, r_{\nu q}$ and $v_{\nu 1}, \cdots, v_{\nu q}$ be the components of the vectors r_{ν} and v_{ν} of $\mathbf{R}^{q}$. Apply a differential operator

(7.19)
$$(2\pi i)^{-|l+m|}\prod_{\nu=1}^{n}\left(\sum_{j=1}^{q}r_{\nu j}\partial/\partial v_{\nu j}\right)^{l_{\nu}}\left(\sum_{j=1}^{q}r'_{\nu j}\partial/\partial v_{\nu j}\right)^{m_{\nu}}$$

to (7.16) and (7.17), and put $v = 0$. Then we obtain the function of (7.6) with $L = \mathfrak{b}^{q}$ and $a = 1$, and

(7.20)
$$f(\gamma(z); 1, g, h, \mathfrak{b}^{q})$$
$$= \kappa(\gamma)e(X)(cz + d)^{l+(\lambda/2)}(c\bar{z} + d)^{m+(\mu/2)}f(z; 1, g', h', \mathfrak{b}^{q})$$

with g', h' and X of (7.18) and $\kappa(\gamma) = \kappa(\sigma_{\gamma})$.

Now $\mathcal{T}(S, \{P_{\nu}\}, \chi)$ is stable under $z \mapsto az$ with $0 \ll a \in F$ and $z \mapsto z + b$ with $b \in F$. Since $\mathrm{GL}_{2}^{+}(F)$ is generated by the elements of the forms $\begin{pmatrix} a & b \\ 0 & d \end{pmatrix}$ and $\begin{pmatrix} 0 & -c \\ c^{-1} & 0 \end{pmatrix}$, it is sufficient to prove that the space is stable under $\gamma = \begin{pmatrix} 0 & -c \\ c^{-1} & 0 \end{pmatrix}$. Observe that $\sigma_{\gamma} = \beta\alpha$ with

$$\beta = \begin{bmatrix} 0 & -1_{qn} \\ 1_{qn} & 0 \end{bmatrix}, \quad \alpha = \begin{bmatrix} A & 0 \\ 0 & {}^{t}A^{-1} \end{bmatrix}, \quad A = B^{-1}\Phi(cS)^{-1}\cdot{}^{t}B^{-1}.$$

Therefore the question can be reduced to the transformation of φ under α and β, which is well known. Applying operator (7.19) to the transformation formulas, we obtain (7.7). In particular we have

(7.21)
$$f(-z^{-1}; 1, S^{-1}g, Sh, S^{-1}c^{q})$$
$$= e\left(\mathrm{Tr}_{F/Q}({}^{t}gh)\right)D_{F}^{q/2}N(\mathfrak{b})^{q}|N_{F/Q}(\det(S))|^{1/2}$$
$$\times (-iz)^{\lambda/2}(i\bar{z})^{\mu/2}z^{l}\bar{z}^{m}f(z; 1, h, -g, \mathfrak{b}^{q}),$$

where D_F is the discriminant of F and $\mathfrak{c} = \{x \in F \mid \mathrm{Tr}_{F/\mathbf{Q}}(x\mathfrak{b}) \subset \mathbf{Z}\}$. To prove (7.8), we first assume $L = \mathfrak{r}^q$. By [15, p. 70, Prop. A.1], we have

$$(7.22) \qquad \kappa(\gamma) = \kappa(\sigma_\gamma) = N_{F/\mathbf{Q}}(d)^{-q/2} \sum_{v \in L/dL} e\left(\frac{1}{2}\mathrm{Tr}_{F/\mathbf{Q}}\left(bd^{-1}S[v]\right)\right)$$

if $\gamma = \begin{pmatrix} a & b \\ c & d \end{pmatrix} \in \mathrm{SL}_2(\mathfrak{r})$, $d \gg 0$ and $\sigma_\gamma \in \Gamma_\theta$.

LEMMA 7.2. *Suppose* $\gamma = \begin{pmatrix} a & b \\ c & d \end{pmatrix} \in \mathrm{SL}_2(\mathfrak{r})$, $S \in M_q(\mathfrak{r})$, $\{bS\} \equiv 0 \ (\mathrm{mod}\ 2\mathfrak{b}^{-1})$, $cS^{-1} \equiv 0 \ (\mathrm{mod}\ \mathfrak{b})$, $\{cS^{-1}\} \equiv 0 \ (\mathrm{mod}\ 2\mathfrak{b})$ *and* $0 \ll d \equiv 1 \ (\mathrm{mod}\ 4\mathfrak{b})$, *where* $\mathfrak{b}$ *denotes the different of* F *over* $\mathbf{Q}$, *and* $\{Y\}$ *the vector consisting of the diagonal elements of* Y. *Then* (7.20), *when* $\mathfrak{b} = \mathfrak{r}$, *holds with*

$$\kappa(\gamma) = \left(\frac{2c}{d\mathfrak{r}}\right)^q \left(\frac{\det(S)}{d\mathfrak{r}}\right),$$

where we understand that $\left(\dfrac{0}{d\mathfrak{r}}\right) = 1$.

This can be derived from (7.22) by the same technique as in [15, Prop. A. 2], except that we need some results on Gauss sums in F, which we state here as

LEMMA 7.3. *Let* $\mathfrak{p}$ *be a prime ideal in* F *not dividing* 2, *and* c *an element of* F *such that* $c\mathfrak{p}\mathfrak{b}$ *is an integral ideal prime to* $\mathfrak{p}$. *Then*

$$(7.23) \qquad \sum_{x \bmod \mathfrak{p}} e\left(\mathrm{Tr}_{F/\mathbf{Q}}(cx^2)\right) = \sum_{y \bmod \mathfrak{p}} \left(\frac{y}{\mathfrak{p}}\right) e\left(\mathrm{Tr}_{F/\mathbf{Q}}(cy)\right).$$

Moreover, suppose that $\mathfrak{p} \nmid \mathfrak{b}$, $\mathfrak{p} = \pi\mathfrak{r}$ *with* $\pi \in \mathfrak{r}$, $\pi \equiv 1 \ (\mathrm{mod}\ 4\mathfrak{r})$. *Then*

$$(7.24) \qquad \sum_{x \bmod \mathfrak{p}} \left(\frac{x}{\mathfrak{p}}\right) e\left(\mathrm{Tr}_{F/\mathbf{Q}}(x/\pi)\right) = (-i)^k \psi(\mathfrak{b}) N(\mathfrak{p})^{1/2},$$

where k *is the number of archimedean primes of* F *for which* π *is negative, and* ψ *is the character modulo* $\mathfrak{p}$ *associated with the quadratic extension* $F(\pi^{1/2})$ *of* F; $\psi(\mathfrak{b}) = 1$ *if* $\pi \equiv 1 \ (\mathrm{mod}\ 4\mathfrak{b})$.

Proof. The first formula is due to Hecke [1, p. 223, (171)]. The second one follows from a result of Hecke that the constant factor in the functional equation of $L(s, \psi)$ is 1. If $\pi \equiv 1 \ (\mathrm{mod}\ 4\mathfrak{b})$, every prime factor of $\mathfrak{b}$ splits in $F(\pi^{1/2})$, and hence $\psi(\mathfrak{b}) = 1$.

Thus we have proved (7.8) for the function f of (7.6) with $a = 1$ and $L = \mathfrak{r}^q$. The case with an arbitrary a and $L = \mathfrak{r}^q$ can be reduced to the case $a = 1$ in an obvious way. Finally we observe that $\mathcal{T}(S, \{P_\nu\}, \mathcal{X})$ is spanned by those f with $L = \mathfrak{r}^q$ as can be verified in a purely formal way. This completes the proof of Proposition 7.1.

The condition $d \gg 0$ imposed in (7.8) and Lemma 7.2 is not so restrictive as it may look, because of

LEMMA 7.4. *There exists a congruence subgroup Γ of $SL_2(\mathfrak{r})$ such that every congruence subgroup of Γ is generated by its elements $\begin{pmatrix} a & b \\ c & d \end{pmatrix}$ with $d \gg 0$.*

The proof is elementary and therefore may be left to the reader.

Let us now consider the special case $q = 1$, $S = 1$, and $P_\nu = 1$ for all ν. Put

$$(7.25) \qquad \theta(z) = \sum_{x \in \mathfrak{r}} e\left(\frac{1}{2}\sum_{\nu=1}^{n} x_\nu^2 z_\nu\right),$$

$$(7.26) \qquad \Gamma_0 = \left\{\begin{pmatrix} a & b \\ c & d \end{pmatrix} \in SL_2(\mathfrak{r}) \,\middle|\, b \in 2\mathfrak{d}^{-1}, \, c \in 2\mathfrak{d}\right\}.$$

Then we have

$$(7.27) \qquad \theta(\gamma(z)) = j(\gamma, z)\theta(z), \quad j(\gamma, z) = \zeta_\gamma(cz + d)^{1/2}$$

with a root of unity ζ_γ for every $\gamma = \begin{pmatrix} a & b \\ c & d \end{pmatrix} \in \Gamma_0$,

$$(7.28) \qquad j\left(\begin{pmatrix} * & * \\ c & d \end{pmatrix}, z\right) = \left(\frac{2c}{d\mathfrak{r}}\right)(cz + d)^{1/2} \quad \text{if} \quad 0 \ll d \equiv 1 \,(\mathrm{mod}\, 4\mathfrak{d}).$$

Let $l = (l_1, \cdots, l_n)$ be an element of $2^{-1}\mathbf{Z}^n$ satisfying

$$(7.29) \qquad 2l_1 \equiv \cdots \equiv 2l_n \,(\mathrm{mod}\, 2), \, l_\nu \geqq 0 \text{ for all } \nu,$$

and let Γ be a congruence subgroup of Γ_0. We denote by $\mathfrak{M}_l(\Gamma)$ the set of all Hilbert modular forms with respect to Γ, which can be defined as follows. If $l \in \mathbf{Z}^n$, they are defined as usual (see [10], [17], for example). If $l \notin \mathbf{Z}^n$, we have $l_\nu = m_\nu + (1/2)$ with $0 \leq m_\nu \in \mathbf{Z}$. Then an element of $\mathfrak{M}_l(\Gamma)$ is a holomorphic function f on $\mathfrak{H}_1^n$ which satisfies

$$f(\gamma(z)) = j(\gamma, z)(cz + d)^m f(z) \quad \text{for all} \quad \gamma = \begin{pmatrix} * & * \\ c & d \end{pmatrix} \in \Gamma,$$

and which is finite at cusps. We denote by $\mathfrak{M}_{l,F}$ or simply by $\mathfrak{M}_l$ the union of $\mathfrak{M}_l(\Gamma)$ for all congruence subgroups of Γ_0. Every element f of $\mathfrak{M}_l$ has a Fourier expansion of the form

$$(7.30) \qquad f(z) = \sum_{\xi \in L} a(\xi)e_F(\xi z), \; e_F(\xi z) = \exp\left(2\pi i \sum_{\nu=1}^{n} \xi_\nu z_\nu\right)$$

with $a(\xi) \in \mathbf{C}$, where L is a lattice in F, and ξ runs over 0 and all totally positive elements of L. We also write $f(z) = \sum_{\xi \in F} a(\xi) e_F(\xi z)$ and understand that $a(\xi) = 0$ for $\xi \notin L$. Given a subfield K of $\mathbf{C}$, we denote by $\mathfrak{M}_l(K)$ the set of all $f \in \mathfrak{M}_l$ whose Fourier coefficients $a(\xi)$ defined by (7.30) belong to K.

PROPOSITION 7.5. *Let $f \in \mathfrak{M}_l$ and $\alpha = \begin{pmatrix} a & b \\ c & d \end{pmatrix} \in \mathrm{GL}_2^+(F)$. Then the function* $g(z) = f(\alpha(z))(cz + d)^{-l}$ *belongs to* $\mathfrak{M}_l$. *Moreover* $g \in \mathfrak{M}_l(\overline{\mathbf{Q}})$ *if* $f \in \mathfrak{M}_l(\overline{\mathbf{Q}})$.

Proof. If $l \in \mathbf{Z}^n$, this is included in [17, Proposition 1.4]. Suppose $l \notin \mathbf{Z}^n$. That $g \in \mathfrak{M}_l$ follows from Proposition 7.1 and Lemma 7.4. Suppose $f \in \mathfrak{M}_l(\overline{\mathbf{Q}})$ and let $f/\theta = h$, $l_\nu = m_\nu + (1/2)$. That $g \in \mathfrak{M}_l(\overline{\mathbf{Q}})$ is obvious if $c = 0$. Since $\mathrm{GL}_2^+(F)$ is generated by the elements $\begin{pmatrix} a & b \\ 0 & d \end{pmatrix}$ and $\begin{pmatrix} 0 & -1 \\ 1 & 0 \end{pmatrix}$, it is sufficient to show that $f(-1/z)z^{-l} \in \mathfrak{M}_l(\overline{\mathbf{Q}})$. By [17, Proposition 1.4], $h(-1/z)z^{-m}$ is $\overline{\mathbf{Q}}$-rational; (7.21) shows that $\theta(-1/z)z^{-1/2}$ is $\overline{\mathbf{Q}}$-rational; hence the desired conclusion follows.

With $l \in 2^{-1}\mathbf{Z}^n$ as in (7.29) and $r \in \mathbf{Z}^n$, we denote by $\mathfrak{M}_{l,F}^r$ or simply by $\mathfrak{M}_l^r$ the set of all functions of the form

$$(7.31) \qquad \sum_{0 \leq q \leq r} (2\pi i)^{-|q|}(\partial/\partial z)^q f_q \quad \text{with} \quad f_q \in \mathfrak{M}_{l+2r-2q}$$

and by $\mathfrak{M}_l^r(\overline{\mathbf{Q}})$ the set of all such functions with $f_q \in \mathfrak{M}_{l+2r-2q}(\overline{\mathbf{Q}})$. Here we write $q \leqq r$ if $q_\nu \leqq r_\nu$ for all ν.

PROPOSITION 7.6. *If* $f \in \mathfrak{M}_k^r(\mathbf{Q})$ *and* $g \in \mathfrak{M}_l^s(\overline{\mathbf{Q}})$, *then* $fg \in \mathfrak{M}_{k+l}^{r+s}(\overline{\mathbf{Q}})$.

Proof. It is sufficient to show that

$$(7.32) \qquad (2\pi i)^{-|p+q|}(\partial/\partial z)^p f \cdot (\partial/\partial z)^q g \in \mathfrak{M}_{k+l}^{p+q}(\overline{\mathbf{Q}})$$

if $f \in \mathfrak{M}_k(\overline{\mathbf{Q}})$ and $g \in \mathfrak{M}_l(\overline{\mathbf{Q}})$. Define a differential operator δ_k^r by

$$(7.33) \qquad \delta_k^r = (2\pi i)^{-|r|} \prod_{\nu=1}^n \delta_\nu(k_\nu + 2r_\nu - 2) \cdots \delta_\nu(k_\nu + 2)\delta_\nu(k_\nu) ,$$
$$\delta_\nu(\kappa) = \kappa(z_\nu - \bar{z}_\nu)^{-1} + (\partial/\partial z_\nu) \qquad\qquad (\kappa \in \mathbf{R}) .$$

Assertion (7.32) is obvious if $l = 0$. Assuming $l \neq 0$, we have

$$(7.34) \qquad \delta_k^r f \cdot \delta_l^s g = \sum_{0 \leq t \leq r+s} \delta_{m-2t}^t h_t , \qquad\qquad m = k + l + 2r + 2s,$$

with unique $h_t \in \mathfrak{M}_{m-2t}$. If both k and l are contained in $\mathbf{Z}^n$, this follows from [17, Lemma 4.10] and [12, Lemma 7]. The same proofs apply to the general case. Obviously h_t is $\overline{\mathbf{Q}}$-rational. Express both sides of (7.34) as polynomials in $\mathrm{Im}(z_1)^{-1}, \cdots, \mathrm{Im}(z_n)^{-1}$, and compare the constant terms (i.e., the holomorphic terms). Then we find

$$(2\pi i)^{-|r+s|}(\partial/\partial z)^r f \cdot (\partial/\partial z)^s g = \sum_t (2\pi i)^{-|t|}(\partial/\partial z)^t h_t$$

as desired.

PROPOSITION 7.7. *Suppose $S_\nu > 0$ for all ν, and let $\varphi(w)$ be a polynomial function on $\mathbf{R}^{qn}$ which is homogeneous of degree r_ν in the νth variable w_ν on $\mathbf{R}^q$ for $\nu = 1, \cdots, n$. Put*

$$f(z) = \sum_{w - \rho \in L} \varphi(w) e_F(S[w]z) \qquad\qquad (z \in \mathfrak{H}_1^n)$$

with a lattice L in F^q and $g \in F^q$. Then $f \in \mathfrak{M}_k^s$, where s_ν is the integral part of $r_\nu/2$ (i.e., $s_\nu = [r_\nu/2]$) and $k_\nu = (q/2) + r_\nu - 2s_\nu$.

Proof. A well known theorem on spherical functions asserts that

$$\varphi(w) = \sum_{0 \le t \le r/2} S[w]^t h_t(w)$$

with functions h_t which are S_ν-spherical of degree $r_\nu - 2t_\nu$ in w_ν. Therefore

$$f(z) = \sum_t (2\pi i)^{-|t|} (\partial/\partial z)^t f_t(z), \quad f_t(z) = \sum_{w - g \in L} h_t(w) e_F(S[w]z) .$$

By Proposition 7.1, $f_t \in \mathfrak{M}_l$ with $l_\nu = (q/2) + r_\nu - 2t_\nu$. This proves our assertion.

PROPOSITION 7.8. *Let h be a C^∞-function on $\mathfrak{H}_1^n$; let $0 \le k \in 2^{-1}\mathbf{Z}^n$ and $0 \le p \in \mathbf{Z}^n$. Then $(2\pi i)^{-|p|}(\partial/\partial z)^p h = \sum_{0 \le r \le p} c_r(-4\pi y)^{r-p} \delta_k^r h$ with $c_r \in \mathbf{Q}$, where $y_\nu = \mathrm{Im}(z_\nu)$.*

This can easily be shown by induction on $|p|$.

Let U_F denote the group of all totally positive units of F. If $f \in \mathfrak{M}_l^r$, there is a subgroup U of U_F of finite index such that $f(\varepsilon z)\varepsilon^{(l/2)+r} = f(z)$ for all $\varepsilon \in U$. We say then that f is *U-invariant*. If (7.30) is the Fourier expansion of f, the U-invariance of f is equivalent to

(7.35) $a(\varepsilon\xi) = a(\xi)\varepsilon^{(l/2)+r}$ for all $\xi \in F$ and all $\varepsilon \in U$.

8. Explicit construction of $\overline{\mathbf{Q}}$-rational automorphic forms on $\mathcal{B}^r$

Let V be a finite-dimensional vector space over $\mathbf{Q}$. By a *congruence subset of V*, we understand a finite union

$$\bigcup_{k=1}^{N} \{x \in V \mid x - u_k \in L_k\}$$

with $u_1, \cdots, u_N \in V$ and lattices $L_1, \cdots, L_N$ in V. A *congruence subset of $V - \{0\}$* means a congruence subset of V with 0 excluded.

We now go back to the vector space V over F, the symmetric form S on V, and the completions V_i of V at τ_i as in Section 3. Let $\mathcal{P}_\kappa$, for $\kappa = (\kappa_{r+1}, \cdots, \kappa_n) \in \mathbf{Z}^{n-r}$, $\kappa_i \ge 0$, be the vector space of C-valued polynomial functions on $V_{r+1} \times \cdots \times V_n$, which are homogeneous of degree κ_i in the variable on V_i for each $i > r$. We call an element t of $\mathcal{P}_\kappa$ $\overline{\mathbf{Q}}$-*rational* if $t(x) \in \overline{\mathbf{Q}}$ for every $x \in V$, where we embed V into $V_{r+1} \times \cdots \times V_n$ in an obvious way.

Let us now assume that F has a subfield E satisfying the following two conditions:

(8.1a) $[E: \mathbf{Q}] = r$;

(8.1b) *the restrictions of $\tau_1, \cdots \tau_r$ to E are all different.*

We put $q = n/r$ and, for each $i \leq r$, denote by $\tau_{i1}, \cdots, \tau_{iq}$ the elements of J_F which coincide with τ_i on E; we assume $\tau_i = \tau_{i1}$. Thus

$$\{\tau_{11}, \cdots, \tau_{1q}, \cdots, \tau_{r1}, \cdots, \tau_{rq}\}$$

is another arrangement of $\tau_1, \cdots, \tau_n$. Let U_E denote the group of all totally positive units of E. For a subgroup U of U_E of finite index, we put

$$(8.2) \qquad M_U = [U_E: U]^{-1} .$$

We now define a C-valued function f on $\mathcal{Z}^r \times \mathbf{C}$ by

$$(8.3) \quad f(z, s; t, \xi, X) = M_U \sum_{x \in X/U} \omega(\mathrm{Tr}_{F/E}(\xi S[x]))t(x)$$
$$\times S(x, A^{-1}p(z))^{-k} | S(x, A^{-1}p(z)) |^{-2s} \quad (z \in \mathcal{Z}^r, s \in \mathbf{C})$$

with a given set of data $\{\Omega, t, \xi, X, k\}$. Here $t \in \mathcal{S}_\kappa$; ξ is a totally negative element of F; the $\omega(a)$ for $a \in E$ are the Fourier coefficients of an element

$$(8.4) \qquad \Omega(z) = \sum_{a \in E} \omega(a)e_E(az) \qquad (z \in \mathfrak{H}_1^r)$$

of $\mathfrak{M}_{l,E}^{\delta}$ with $0 \leq l \in 2^{-1}\mathbf{Z}^r$ and $\delta \in \mathbf{Z}^r$; X is a congruence subset of $V - \{0\}$; U is a subgroup of U_E of finite index such that Ω is U-invariant and $uX = X$ for all $u \in U$; $k = (k_1, \cdots, k_r) \in \mathbf{Z}^r, k_i > 0;$

$$(8.5)$$
$$S(x, A^{-1}p(z))^{-k} | S(x, A^{-1}p(z)) |^{-2s} = \prod_{i=1}^{r} S_i(x_i, A_i^{-1}p(z_i))^{-k_i} | S_i(x_i, A_i^{-1}p(z_i)) |^{-2s}$$

with A_i of (3.3) and p of (2.7); X/U denotes a complete set of representatives for X modulo scalar multiplication by the elements of U.

The U-invariance of Ω implies

$$(8.6) \qquad \omega(ua) = \omega(a)u^{\delta+(l/2)} \qquad \text{for all} \qquad u \in U.$$

To make each term of (8.3) dependent only on Ux, we have to assume

(8.7) *the numbers $k_i - l_i - 2\delta_i - \sum_{j=2}^{q}\kappa_{ij}$ for $i = 1, \cdots, r$ are the same,*

where we put $\kappa_{ij} = \kappa_\nu$ when $\tau_{ij} = \tau_\nu$. The factor M_U makes the sum independent of the choice of U. Each term of (8.3) is meaningful because of

$(8.8) \qquad S_i(x_i, w_i) \neq 0$ *for every $w_i \in \mathcal{Y}(S_i)$ and every $x \in X$ such that*
$$\omega(\mathrm{Tr}_{F/E}(\xi S[x])) \neq 0 .$$

In fact, since $\omega(a) \neq 0$ only if $a = 0$ or $a \gg 0$, we are considering only the elements x of X such that $\mathrm{Tr}_{F/E}(\xi S[x])^{\tau_i} \geq 0$ for all $i \leq r$. For such x, we have

$$(8.9) \qquad S[x]^{\tau_i} \leq -\sum_{j=2}^{q}(\xi^{\tau_{ij}}/\xi^{\tau_i})S[x]^{\tau_{ij}} \leq 0 \qquad (i = 1, \cdots, r) ,$$

since ξ is totally negative. This together with (2.2) implies (8.8).

We include in the above definition the case in which Ω is a constant. If $\Omega = 1$, our sum is taken over all $x \in X/U$ such that $\mathrm{Tr}_{F/E}(\xi S[x]) = 0$.

To prove the convergence, we need

LEMMA 8.1. *Let $T_1, \cdots, T_r$ be positive quadratic forms on $\mathbf{R}^N$, and $g_1, \cdots, g_r$ be homogeneous polynomial functions on R^N of degree $d_1, \cdots, d_r$, respectively. Further, let U be a subgroup of finite index of U_E, and X a congruence subset of $E^N - \{0\}$ such that $uX = X$ for $u \in U$. Then*

$$\sum_{x \in X/U} \prod_{i=1}^{r} g_i(x_i) T_i[x_i]^{-s-d_i/2} \qquad (s \in \mathbf{C})$$

is absolutely convergent for $\mathrm{Re}(s) > N/2$, where $x_i = x^{\tau_i}$. The convergence is uniform if s, T_i, and the coefficients of g_i stay in compact sets.

Proof. There is a constant A such that $|g_i(x)| \leq A T_i[x]^{d_i/2}$ for all i and all $x \in \mathbf{R}^N$. Therefore our problem is reduced to the case where $g_1, \cdots, g_r$ are constants. Given r positive numbers $p_1, \cdots, p_r$, we can find an element u of U such that $u_i^2 p_i \leq c(p_1 \cdots p_r)^{1/r}$ for all i, where c is a positive constant depending only on U. Applying this to $p_i = T_i[x_i]$ with $x \in X$, we have

$$\sum_{i=1}^{r} T_i[u_i x_i] \leq rc \prod_{i=1}^{r} T_i[x_i]^{1/r} .$$

Define a positive quadratic form $P[y]$ on $\mathbf{R}^{Nr}$ by

$$P[(y_1, \cdots, y_r)] = \sum_{i=1}^{r} T_i[y_i] \qquad (y_i \in \mathbf{R}^N) .$$

It is well known that $\sum_{x \in X} P[x]^{-s}$ is convergent for $\mathrm{Re}(s) > Nr/2$. This proves the convergence; the uniformity is obvious from our reasoning.

PROPOSITION 8.2. *Let θ be a real number $\geq -1/2$ such that*

$$(8.10) \qquad \omega(a) \prod_{i=1}^{r} (a^{\tau_i})^{-\delta_i - (l_i/2)} = O(N_{E/\mathbf{Q}}(a)^\theta) \quad for \quad a \neq 0.$$

Then (8.3) is absolutely convergent if

$$(8.11) \quad \mathrm{Re}(2s) + k_i - l_i - 2\delta_i - \sum_{j=2}^{q} \kappa_{ij} > 2\theta + (m+2)q \quad for\ all \quad i \leq r.$$

Suppose in particular Ω is a constant (so that $l = \delta = 0$). Then the series is absolutely convergent if

$$(8.12) \qquad \mathrm{Re}(2s) + k_i - \sum_{j=2}^{q} \kappa_{ij} > (m+2)q - 2 \quad for\ all \quad i \leq r.$$

The convergence is uniform if z, s, and the coefficients of t stay in compact sets.

Proof. Fix $z \in \mathfrak{Z}^r$ and define a positive form P_v on V_i as in Section 2 with $v = A_i^{-1} p(z_i)$; call it P_i. Then (2.2) holds with P_i, S_i, and $p(z_i)$ in place of P_v, S and v. If $\tau_{ij} = \tau_\nu$ on E, denote S_ν and V_ν by S_{ij} and V_{ij}. Let $W_i = V \otimes_E E_i$ with the τ_i-completion E_i of E. Then W_i can be identified with $\prod_{j=1}^{q} V_{ij}$. Define a positive form R_i on W_i by

$$(8.13) \qquad R_i[x_i] = -\xi^{\tau_{i1}} P_i[x_{i1}] - \sum_{j=2}^{q} \xi^{\tau_{ij}} S_{ij}[x_{ij}]$$

for $x_i = (x_{i1}, \cdots, x_{iq}) \in W_i$ with $x_{ij} \in V_{ij}$. If $x \in V$, we have

$$\mathrm{Tr}_{F/E}\big(\xi S[x]\big)^{r_i} = \sum_{j=1}^{q}\xi^{r_{ij}}S_{ij}[x] \leqq R_i[x_i] \, .$$

If $\mathrm{Tr}_{F/E}\big(\xi S[x]\big) \neq 0$, we have

$$\big|\omega\big(\mathrm{Tr}_{F/E}\big(\xi S[x]\big)\big)\big| \leqq B\prod_{i=1}^{r}R_i[x_i]^{\theta+\delta_i+(l_i/2)}$$

with a positive constant B independent of x. Moreover

$$(8.14) \qquad R_i[x_i] \leqq R_i[x_i] + \mathrm{Tr}_{F/E}\big(\xi S[x]\big)^{r_i} = -\xi^{r_i}\big(P_i[x_{i1}] - S_i[x_{i1}]\big)$$
$$= 4\,|\,\xi^{r_i}S_i(v_i,\,\bar v_i)^{-1}S_i(x_{i1},\,v_i)^2\,| \, , \qquad\qquad v_i = A_i^{-1}p(z_i) \, ,$$

by (2.2), and hence

$$\big|\omega\big(\mathrm{Tr}_{F/E}\big(\xi S[x]\big)\big)\big|\,\big|\,S(x,\,A^{-1}p(z))\big|^{-k-\sigma} \leqq C\prod_{i=1}^{r}R_i[x_i]^{\theta+\delta_i+(l_i-k_i-\sigma)/2}$$

with a positive constant C depending on z, but independent of x (if $\sigma \geqq -k_i$ for all $i \leqq r$). Thus the terms with $\mathrm{Tr}_{F/E}\big(\xi S[x]\big) \neq 0$ make a convergent series under (8.11) by virtue of Lemma 8.1. The terms with $\mathrm{Tr}_{F/E}\big(\xi S[x]\big) = 0$ make a series of the form

$$\sum_{x\in X/U,\,Q[x]=0}\prod_{i=1}^{r}g_i(x_i)T_i[x_i]^{-s-d_i/2}$$

with the same notation as in Lemma 8.1, where $Q[x] = \mathrm{Tr}_{F/E}\big(\xi S[x]\big)$. The condition of the convergence of such a series, when $E = \mathbf{Q}$, is given in [16, Proposition 5.4]. This can easily be generalized and yields our condition (8.12), which follows from (8.11) since $\theta \geqq -1/2$. If Ω is a constant, only (8.12) is necessary.

Remark 8.2A. If $E = \mathbf{Q}$, $l \in \mathbf{Z}$ and Ω is a derivative of a cusp form, we can take any $\theta > -1/2$ in (8.10) by virtue of Deligne's result. In the general case, it is well known that (8.10) holds with $\theta = 0$ if Ω is a derivative of a cusp form. Anyway, (8.10) does not hold with $\theta < -1/2$ unless $\Omega = 0$. This is why we have assumed $\theta \geqq -1/2$.

THEOREM 8.3. *Suppose*

$$(8.15) \qquad\qquad l_1 \equiv \cdots \equiv l_r \equiv mq/2 \qquad (\mathrm{mod}\,\mathbf{Z}) \, .$$

Then series (8.3) *as a function of* s *can be continued to a meromorphic function on the whole plane.*

This will be proved in Section 13.

Let $\alpha \in G_+(S)$. We see easily that

$$(8.16) \qquad f\big(\alpha(z),\,s;\,t,\,\nu(\alpha)^{-1}\xi,\,\alpha X\big)$$
$$= f(z,\,s;\,t\circ\alpha,\,\xi,\,X)\prod_{i=1}^{r}\nu(\alpha_i)^{-k_i}\mu_i(\alpha,\,z)^{k_i}|\nu(\alpha_i)^{-1}\mu_i(\alpha,\,z)|^{2s} \, .$$

Let $t^{(1)}, \cdots, t^{(d)}$ be a $\overline{\mathbf{Q}}$-rational basis of $\mathscr{P}_{\mathfrak{k}}$ over $\mathbf{C}$, and define a representation $\sigma_0\colon \prod_{i>r}G(S_i) \to \mathrm{GL}_d(\mathbf{C})$ by

$$(8.17) \qquad \begin{bmatrix} t^{(1)} \circ \alpha \\ \vdots \\ t^{(d)} \circ \alpha \end{bmatrix} = \sigma_0(\alpha_{r+1}, \cdots, \alpha_n) \begin{bmatrix} t^{(1)} \\ \vdots \\ t^{(d)} \end{bmatrix} .$$

Assuming that (8.3) is meaningful at $s = 0$, put

$$(8.18) \qquad \mathfrak{f}(z) = {}^t\!\big(f(z, 0; t^{(1)}, \xi, X), \cdots, f(z, 0; t^{(d)}, \xi, X)\big) .$$

Then (8.16) shows that $\mathfrak{f}|_\sigma \gamma = \mathfrak{f}$ for every γ in a congruence subgroup of $G_+(S)$ where

$$(8.19) \qquad \sigma(a_1, \cdots, a_r, \beta_1, \cdots, \beta_r, \gamma_{r+1}, \cdots, \gamma_n) = a_1^{k_1} \cdots a_r^{k_r} \sigma_0(\gamma_{r+1}, \cdots, \gamma_n) .$$

To make $\mathfrak{f}(z)$ meaningful, we consider the following two cases:

$$(8.20a) \qquad k_i - l_i - 2\delta_i - \sum_{j=2}^q \kappa_{ij} > (m + 2)q + 2\theta \text{ for all } i \text{ with } \theta(\geq -1/2)$$
$$\text{satisfying (8.10);}$$

$$(8.20b) \qquad \Omega = 1 \text{ and } k_i - \sum_{j=2}^q \kappa_{ij} > (m + 2)q - 2 \text{ for all } i.$$

By Proposition 8.2, $\mathfrak{f}(z)$ in these cases is well-defined and holomorphic in z; therefore $\mathfrak{f} \in \mathfrak{M}_\sigma$, since the cusp condition, when $\mathrm{GL}_2(\mathbf{Q})$ is involved, can easily be verified. We shall show in Section 13 that the value of (8.3) at $s = 0$ defines an automorphic form holomorphic in z under a condition weaker than (8.20a, b) provided that $l_1 = \cdots = l_r$, $\delta = 0$, and t is spherical. However, we shall consider here $\mathfrak{f}$ only under (8.20a, b). Now the first main result of this paper can be stated as

Theorem 8.4. *Suppose the Fourier coefficients $\omega(a)$ of Ω are all algebraic and (8.15) is satisfied. Then $\pi^{-|k|}\mathfrak{f}$ is $\overline{\mathbf{Q}}$-rational in each case of (8.20a, b), where $|k| = \sum_{i=1}^r k_i$.*

To put it differently, let w be the fixed point of $h(Y^u)$ with $Y = K \oplus Y_0$ as in Proposition 5.1, and t a $\overline{\mathbf{Q}}$-rational element of $\mathscr{P}_r$ such that $t(h^u(a)x) = a^\Phi t(x)$ for every $a \in K^u$ with $\Phi \in I_K$. Then the $\overline{\mathbf{Q}}$-rationality of $\mathfrak{f}$ is equivalent to

$$(8.21) \qquad f(w, 0; t, \xi, X) \sim \pi^{|k|} p_K\big(\Phi + \sum_{i=1}^r k_i \psi_i, \; 2\sum_{i=1}^r \psi_i\big)$$

with ψ_i defined by (5.6a, b). Theorem 8.4 will be proved in Section 10.

9. The special values of certain zeta functions

To state our second main theorem, let K be a CM-field containing F, not necessarily quadratic over F. Choose and fix n elements $\psi_1, \cdots, \psi_n$ of J_K whose restrictions to F are $\tau_1, \cdots, \tau_n$, respectively. Let $\varphi_{i1}, \cdots, \varphi_{i\lambda}$ be the elements of J_K, other than ψ_i and $\psi_i\rho$, whose restrictions to F coincide with τ_i, where $[K : F] = \lambda + 2$ with $\lambda \geq 0$. (Thus $\lambda = 0$ and $\sum_{i=1}^n \psi_i$ is a

CM-type if and only if $[K:F] = 2$.) Take an element

$$(9.1) \qquad \Omega(z) = \sum_{a \in F} \omega(a) e_F(az) \qquad (z \in \mathfrak{H}_1^n)$$

of $\mathfrak{M}_{l,F}^0$ and define a series

$$(9.2) \qquad \mathfrak{D}(s) = M_U \sum_{x \in X/U} \omega(\mathrm{Tr}_{K/F}(\eta x x^\rho)) x^{\Phi - \rho\Delta} |x^\psi|^{-2s} \qquad (s \in \mathbf{C}) .$$

Here $\eta \in K$ and Φ, Δ, $\psi \in I_K$; we impose the following conditions on them:

$$(9.3) \qquad \eta = \eta^\rho, \, \eta^{\psi_i} > 0 \text{ for } i = 1, \cdots, n \text{ and } \eta^\beta < 0 \text{ for all } \beta \text{ of}$$
$$J_K \text{ other than } \psi_i \text{ and } \psi_i \rho;$$

$$(9.4a) \qquad \psi = \sum_{i=1}^n \psi_i, \, \Delta = \sum_{i=1}^n d_i \psi_i \text{ with positive integers } d_i;$$

$$(9.4b) \qquad \Phi = \sum_{i=1}^n \{ b_i(\psi_i + \psi_i \rho) + \sum_{j=1}^\lambda c_{ij} \varphi_{ij} \} \text{ with non-negative integers}$$
$$b_i \text{ and } c_{ij};$$

X is a congruence subset of $K - \{0\}$; U is a subgroup of U_F of finite index such that Ω is U-invariant and $uX = X$ for all $u \in U$. To make the sum meaningful, we assume, for the same reason as in (8.3),

$$(9.5) \qquad d_i - l_i - 2\delta_i - 2b_i - \sum_{j=1}^\lambda c_{ij} \text{ is independent of } i; \text{ we call this}$$
$$\text{number } \alpha.$$

PROPOSITION 9.1. *Series (9.2) is absolutely convergent for*

$$(9.6a) \qquad \mathrm{Re}(s) > \theta + \frac{1}{2}[K:F] - \frac{\alpha}{2} ,$$

where θ is a real number $\geq -1/2$ such that (8.10) holds with F and n in place of E and r. If Ω is a constant, the series is absolutely convergent for

$$(9.6b) \qquad \mathrm{Re}(s) > \frac{1}{2}[K:F] - \frac{\alpha}{2} - 1 .$$

Proof. Considering K a vector space over its maximal real subfield F' and calling it V, define an F'-valued symmetric form S on V by $S(x, y) = -\mathrm{Tr}_{K/F'}(\eta x y^\rho)$. Then we can apply the technique of the proof of Proposition 8.2 to the present case. In fact, we consider a positive form R_i on V_i by

$$R_i[x] = 2(\eta x x^\rho)^{\psi_i} - \sum_{j=1}^\lambda (\eta x x^\rho)^{\varphi_{ij}} \qquad (x \in V) .$$

The fields E and F there correspond to F and F' here. Then we obtain our results in the same manner.

Now our second main theorem can be stated as follows.

THEOREM 9.2. *Suppose $l \in \mathbf{Z}^n$. Then $\mathfrak{D}(s)$ of (9.2) can be continued to a meromorphic function on the whole plane. Suppose moreover $\omega(a) \in \overline{\mathbf{Q}}$ for all $a \in F$. Let μ be an integer such that*

(9.7a)
$$\frac{1}{2}[K:F] - \frac{\alpha}{2} + \theta < \mu \leqq b_i \quad \textit{for all } i$$

with θ as in Proposition 9.1. Then $\mathfrak{D}(\mu) \sim \pi^{|d|} p_K(\Phi + \Delta, 2\psi)$. If Ω is a constant, the same conclusion holds under

(9.7b)
$$\frac{1}{2}[K;F] - \frac{\alpha}{2} - 1 < \mu \leqq b_i \quad \textit{for all } i \; .$$

The proof will be completed in Sections 12, 13.

Remark 9.3. (A) An integer μ satisfying (9.7a) exists only if

(9.8)
$$d_i - l_i - 2\delta_i - \sum_{j=1}^{\lambda} c_{ij} > [K:F] + 2\theta \quad \text{for all } i.$$

Thus our assertion about $\mathfrak{D}(\mu)$ is vacuous unless this is satisfied. A similar remark applies to (9.7b).

(B) If $l = \delta = 0$ and $\Omega = 1$, we have
$$\mathfrak{D}(s) = \sum_{x \in Y/U} x^{\mathfrak{d} - \rho\Delta} |x^\psi|^{-2s} \; ,$$

where $Y = \{x \in X \,|\, \mathrm{Tr}_{K/F}(\eta x x^\rho) = 0\}$. This is nonempty only when $[K:F] > 2$.

(C) Put $b_i' = b_i + (\alpha - \varepsilon)/2$ with $\varepsilon = 0$ or 1 according as α is even or odd. Further put
$$\Phi' = \sum_{i=1}^{n} \{b_i'(\psi_i + \psi_i \rho) + \sum_{j=1}^{\lambda} c_{ij} \mathcal{P}_{ij}\} \; .$$

Let $\mathfrak{D}'$ denote the series (9.2) defined with Φ' instead of Φ. Then $\mathfrak{D}(s) = \mathfrak{D}'(s + (\alpha - \varepsilon)/2)$. Therefore we may assume $\alpha = 0$ or 1 without losing the essential content of our theorem, though this assumption is sometimes inconvenient.

The purpose of the remaining part of this section is to derive the theorem in the special case $[K:F] = 2$ from a previous result in [17]. We start by considering $f \in \mathfrak{M}_{k,F}$ and $g \in \mathfrak{M}_{l,F}$ with both k and l in $\mathbf{Z}^n$ and with Fourier expansions
$$f(z) = \sum_{\xi} a(\xi) e_F(\xi z) \; , \qquad g(z) = \sum_{\xi} b(\xi) e_F(\xi z) \; .$$

Assume that f is a cusp form; take a subgroup U of U_F so that both f and g are U-invariant, and put

(9.9)
$$L(s; f, g) = M_U \sum_{\xi U} a(\xi) b(\xi) \xi^{(-k-l)/2} N(\xi)^{-s} \qquad (s \in \mathbf{C}) ,$$

where the sum is taken over all different cosets ξU with $\xi \neq 0$. This does not depend on the choice of U, and can be continued to a meromorphic function on the whole plane (cf. [17, Prop. 4.3]). We studied in [17] the values of (9.9) at certain integers or half-integers. Series (9.2), when $[K:F] = 2$, is essentially the same as (9.9) with f obtained from a Hecke character of a CM-field. To be more explicit, let K, ψ, and Δ be the same as above and suppose $[K:F] = 2$; then ψ is a CM-type. With a congruence subset X of K and $0 \ll \eta \in F$, consider

$$(9.10) \qquad h(z) = \sum_{x \in X} x^{\Delta} e_F(\eta x x^{\rho} z) .$$

As observed in [17, §5], h belongs to $\mathfrak{M}_k(\overline{\mathbf{Q}})$ with $k_i = d_i + 1$. Let $\mathfrak{L}(K, \Delta, \overline{\mathbf{Q}})$ denote the set of all $\overline{\mathbf{Q}}$-linear combinations of such h.

Theorem 9.4. *Let* $f \in \mathfrak{L}(K, \Delta, \overline{\mathbf{Q}})$ *and* $g \in \mathfrak{M}_l(\overline{\mathbf{Q}})$. *Suppose* $0 < l_i \leqq d_i$ *for all* i *and*

$$(9.11) \qquad d_1 - l_1 \equiv \cdots \equiv d_n - l_n \pmod{2} .$$

Let μ *be an integer or a half-integer such that*

$$(9.12) \qquad -1 < \mu < (d_i - l_i + 1)/2 , \quad 2\mu \equiv d_i + 1 - l_i \pmod{2} \quad \text{for all } i .$$

Then $L(\mu; f, g) \sim \pi^{|d|} p_K(\Delta, 2\psi)$.

Before proving this, we give a generalization which will be necessary for our later proof of Theorem 9.2. With $\Omega \in \mathfrak{M}_l^0(\overline{\mathbf{Q}})$ as in (9.1), put

$$(9.13) \qquad D(s; \Omega, \Delta) = M_U \sum_{x \in X/U} \omega(\eta x x^{\rho}) x^{\Theta - \rho\Delta} N(x)^{-s}$$

with $0 \ll \eta \in F$, a congruence subset X of K, and

$$\Theta = \sum_{i=1}^{n} \frac{1}{2} (d_i - l_i - 2\delta_i - \varepsilon)(\psi_i + \psi_i \rho) ,$$

where $\varepsilon = 0$ or 1 according as $d_i - l_i$ is even or odd. It can easily be verified that

$$(9.14) \qquad D\big(s; (2\pi i)^{-|t|}(\partial/\partial z)^t f, \Delta\big) = \eta^t D(s; f, \Delta) \quad \text{if} \quad f \in \mathfrak{M}_{l+2\delta-2t} .$$

Proposition 9.5. *Suppose* $d_i \geqq l_i + 2\delta_i > 0$ *for all* i *and* (9.11) *is satisfied. Let* κ *be an integer such that*

$$(9.15) \qquad 0 \leqq \kappa \leqq (d_i - l_i - 2\delta_i - \varepsilon)/2 \quad \text{for all } i .$$

Then $D(\kappa; \Omega, \Delta) \sim \pi^{|d|} p_K(\Delta, 2\psi)$ *for every* $\Omega \in \mathfrak{M}_l^0(\overline{\mathbf{Q}})$.

Proof. If $\delta = 0$ and h is defined by (9.11), we have

$$L(s; h, \Omega) = 2^{1-n} \eta^{(-d-l)/2} N(\eta)^{-s-(1/2)} D\big(s + (1 - \varepsilon)/2; \Omega, \Delta\big)$$

and therefore our assertion follows from Theorem 9.4. The general case can be reduced to this special case by (9.14).

This proposition together with Remark 9.3, (C) implies Theorem 9.2 when $[K:F] = 2$.

Theorem 9.4 was proved in [17, Theorem 5.6] under the assumption that either f or g is U_F-invariant. Our task is to remove that assumption. In fact, the theory of Hecke operators in [17] was developed only for the

U_F-invariant forms. Let us now generalize the theory of [17] to the forms of a more general type. We use, for the most part, the same notation as in [17]. Thus $\mathfrak{r}$ denotes the maximal order of F, $\mathfrak{d}$ the different of F over $\mathbf{Q}$, $F_A^\times$ the idele group of F, G_A the adelization of $\mathrm{GL}_2(F)$, and $\mathfrak{S}_k$ the set of cusp forms belonging to $\mathfrak{M}_k$. We fix a complete set of representatives $\{\mathfrak{t}_1, \cdots, \mathfrak{t}_h\}$ of the ideal classes of F modulo the product of all archimedean primes. Fixing an integral ideal $\mathfrak{c}$ of F and a character φ_0 of $(\mathfrak{r}/\mathfrak{c})^\times$, we define groups $W = W(\mathfrak{c})$, $Y = Y(\mathfrak{c})$, $\Gamma_\lambda = \Gamma_\lambda(\mathfrak{c})$ and a function φ_Y as in [17, §2]. (We used ψ in place of φ in [17].) We fix a complex-valued character θ of U_F of finite order and consider elements f of $\mathfrak{M}_k$ such that

$$(9.16) \qquad f \|_k \gamma = \varphi_0(a) \theta(\det(\gamma)) f \quad \text{for all} \quad \gamma = \begin{pmatrix} a & * \\ * & * \end{pmatrix} \in \Gamma_\lambda ,$$

where $f \|_k \gamma$ is defined by [17, (1.3a, b)]. Let $\mathfrak{M}_k(\Gamma_\lambda, \varphi_0, \theta)$ denote the set of all such f. If θ is trivial, this coincides with $\mathfrak{M}_k(\Gamma_\lambda, \varphi_0)$ of [17, (2.5a)]. We always assume

$$(9.17) \qquad \varphi_0(\varepsilon) \theta(\varepsilon^2) = \mathrm{sgn}(\varepsilon)^k \quad \text{for every} \quad \varepsilon \in \mathfrak{r}^\times ,$$

since $\mathfrak{M}_k(\Gamma_\lambda, \varphi_0, \theta) = \{0\}$ if this is not satisfied. Then (9.16) implies

$$(9.18) \qquad f(\varepsilon z) \varepsilon^{k/2} = \theta(\varepsilon)^{-1} f(z) \quad \text{for every} \quad \varepsilon \in U_F .$$

We put $\mathfrak{S}_k(\Gamma_\lambda, \varphi_0, \theta) = \mathfrak{M}_k(\Gamma_\lambda, \varphi_0, \theta) \cap \mathfrak{S}_k$ and

$$\mathfrak{M}_k(\mathfrak{c}, \varphi_0, \theta) = \prod_{\lambda=1}^h \mathfrak{M}_k(\Gamma_\lambda, \varphi_0, \theta), \quad \mathfrak{S}_k(\mathfrak{c}, \varphi_0, \theta) = \prod_{\lambda=1}^h \mathfrak{S}_k(\Gamma_\lambda, \varphi_0, \theta) .$$

Now we can find $m = (m_1, \cdots, m_n) \in \mathbf{R}^n$ such that

$$(9.19) \qquad \theta(\varepsilon) = \varepsilon^{im} \quad \text{for every} \quad \varepsilon \in U_F ,$$

where $i = \sqrt{-1}$. Such an m is not unique, but we fix one such m in the following discussion. Take $x_1, \cdots, x_h$ as in [17, (2.2)] and consider the coset decomposition [17, (2.3)]. Given $(f_1, \cdots, f_h) \in \mathfrak{M}_k(\mathfrak{c}, \varphi_0, \theta)$, we define a function f on G_A by

$$(9.20) \qquad f(\alpha x^{-\iota} w) = \varphi_Y(w^\iota) \det(w_\infty)^{im} (f_\lambda \| w_\infty)(i) \quad \text{for} \quad \alpha \in G_\mathbf{Q} \text{ and } w \in W,$$

where $i = (i, \cdots, i) \in \mathfrak{H}_1^n$, and ι is the main involution of $M_2(F)$. This f has the following properties:

$$(9.21a) \qquad f(\alpha x w) = \varphi_Y(w^\iota) f(x) \quad for \quad \alpha \in G_\mathbf{Q} \text{ and } w \in W, \ w_\infty = 1 ;$$

$$(9.21b) \qquad \textit{For every } x \in G_A \text{ with } x_\infty = 1, \text{ there is an element } g_x \text{ of } \mathfrak{M}_k \text{ such}$$
$$\textit{that } f(xy) = \det(y)^{im} (g_x \| y)(i) \text{ for all } y \in G_{\infty+}.$$

We denote by $\mathfrak{M}_k(\mathfrak{c}, \varphi_0, m)$ the set of all functions on G_A satisfying (9.21a, b). It can easily be shown that every element f of $\mathfrak{M}_k(\mathfrak{c}, \varphi_0, m)$ can be obtained from $(f_1, \cdots, f_h) \in \mathfrak{M}_k(\mathfrak{c}, \varphi_0, \theta)$ by (9.20). We shall therefore write $f =$

$(f_1, \cdots, f_h; m)$ or simply $f = (f_1, \cdots, f_h)$ when m is fixed. For $s \in F_A^\times$ and $f \in \mathfrak{M}_k(\mathfrak{c}, \varphi_0, m)$, define f^s by $f^s(x) = f(sx)$. Then $f^s \in \mathfrak{M}_k(\mathfrak{c}, \varphi_0, m)$ and $f^s = \varphi_0(s_\mathfrak{c}) \operatorname{sgn}(s_\infty)^k |s_\infty|^{2im} f$ if $s \in F_\infty^\times \prod_\mathfrak{p} \mathfrak{r}_\mathfrak{p}^\times$, where $s_\mathfrak{c}$ denotes the $\mathfrak{c}$-part of s. In view of (9.17), there exists a Hecke character φ of $F_A^\times$ such that

$$(9.22) \qquad \varphi(s) = \varphi_0(s_\mathfrak{c}) \operatorname{sgn}(s_\infty)^k |s_\infty|^{2im} \quad \text{for} \quad s \in F_\infty^\times \prod_\mathfrak{p} \mathfrak{r}_\mathfrak{p}^\times \, .$$

Putting

$$(9.23\mathrm{a}) \qquad \mathfrak{M}_k(\mathfrak{c}, \varphi) = \{ f \in \mathfrak{M}_k(\mathfrak{c}, \varphi_0, m) \mid f(sx) = \varphi(s)f(x) \text{ for all } s \in F_A^\times \} \, ,$$

$$(9.23\mathrm{b}) \qquad \mathfrak{S}_k(\mathfrak{c}, \varphi) = \{ f \in \mathfrak{M}_k(\mathfrak{c}, \varphi) \mid f = (f_1, \cdots, f_h) \text{ with } f_\lambda \in \mathfrak{S}_k \} \, ,$$

we observe that $\mathfrak{M}_k(\mathfrak{c}, \varphi_0, m)$ (resp. $\mathfrak{S}_k(\mathfrak{c}, \varphi_0, m)$) is the direct sum of $\mathfrak{M}_k(\mathfrak{c}, \varphi)$ (resp. $\mathfrak{S}_k(\mathfrak{c}, \varphi)$) for all Hecke characters φ of $F_A^\times$ satisfying (9.22).

Now we consider the algebra of $Wy\,W$ with $y \in Y$ and define the action of $Wy\,W$ on $\mathfrak{M}_k(\mathfrak{c}, \varphi_0, m)$ by [17, (2.10)]. Given $y \in Y$ and an index λ, there exist, as observed in [17, p. 648], an element α of G_{Q+} and an index μ such that $Wy\,W = Wx_\lambda^{-1} \alpha x_\mu W$. Let $\Gamma_\lambda \alpha \Gamma_\mu = \bigcup_j \Gamma_\lambda \alpha_j$ (disjoint). If $f = (f_1, \cdots, f_h) \in \mathfrak{M}_k(\mathfrak{c}, \varphi_0, m)$ and $f \mid Wy\,W = (g_1, \cdots, g_h)$, then

$$(9.24) \qquad g_\mu = \sum_j \varphi_Y(x_\lambda^{-1} \alpha_j x_\mu)^{-1} \det(\alpha_j)^{-im} f_\lambda \|\alpha_j \, .$$

Now every f_λ of $\mathfrak{M}_k(\Gamma_\lambda, \varphi_0, \theta)$ has a Fourier expansion

$$(9.25) \qquad f_\lambda(z) = \sum_\xi a_\lambda(\xi)\, e_F(\xi z) \qquad\qquad (\xi = 0 \text{ or } 0 \ll \xi \in \mathfrak{t}_\lambda) \, ,$$

and (9.18) implies that

$$(9.26) \qquad a_\lambda(\xi\varepsilon) = a_\lambda(\xi)\varepsilon^{(k/2)+im} \quad \text{for} \quad \varepsilon \in U_F \, ,$$

and hence $a_\lambda(\xi)\xi^{-(k/2)-im}$ depends only on $\xi\mathfrak{r}$; we see also that $a_\lambda(0) = 0$ if θ is non-trivial. For a fractional ideal $\mathfrak{m}$ in F, we put

$$(9.27) \qquad c(\mathfrak{m}, f) = \begin{cases} a_\lambda(\xi)\xi^{-(k/2)-im} & \text{if } \mathfrak{m} = \xi\mathfrak{t}_\lambda^{-1} \text{ and } \mathfrak{m} \text{ is integral} \, , \\ 0 & \text{if } \mathfrak{m} \text{ is not integral} \, , \end{cases}$$

and find that if $a_\lambda(0) = 0$ for all λ, f has expansion

$$(9.28) \qquad f\left(\begin{pmatrix} y & x \\ 0 & 1 \end{pmatrix} \right) = \sum_{0 \ll \zeta \in F} c(\zeta y \mathfrak{r}, f)(\zeta y_\infty)^{(k/2)+im} e_F(\zeta i y_\infty)\chi_F(\zeta x) \qquad (y_\infty \gg 0)$$

which is similar to [17, (2.18)]. Further define $T(\mathfrak{n})$ by [17, (2.21)]. If $f \in \mathfrak{M}_k(\mathfrak{c}, \varphi)$, we have

$$(9.29) \qquad c(\mathfrak{m}, f \mid T(\mathfrak{n})) = \sum_{\mathfrak{m}+\mathfrak{n} \subset \mathfrak{a}, \mathfrak{a}+\mathfrak{c}=\mathfrak{r}} \varphi^*(\mathfrak{a}) N(\mathfrak{a}^{-1}\mathfrak{n}) c(\mathfrak{a}^{-2}\mathfrak{m}\mathfrak{n}, f) \, ,$$

where φ^* is the ideal character associated with φ.

For each prime ideal $\mathfrak{p}$ of F, we have $\mathfrak{p}^h = (x)$ with $0 \ll x \in F$. With any choice of such x, we put $\eta(\mathfrak{p}) = x^{im/h}$ and extend η to a character of the whole ideal group, which is not necessarily a "Grössen-character".

PROPOSITION 9.6. *Let $\mathfrak{N}$ be the set of all $\boldsymbol{f} = (f_1, \cdots, f_h)$ in $\mathfrak{M}_k(\mathfrak{c}, \varphi_0, m)$ such that $\eta(\mathfrak{t}_\lambda)^{-1}f_\lambda \in \mathfrak{M}_k(\overline{\mathbf{Q}})$ for every λ. Then $\mathfrak{M}_k(\mathfrak{c}, \varphi_0, m) = \mathfrak{N} \otimes_{\overline{\mathbf{Q}}} \mathbf{C}$, and $\mathfrak{N}$ is stable under $\eta(\det(y)\mathfrak{r}) Wy W$ for every $y \in Y$.*

The first assertion follows immediately from [17, Proposition 1.7], and the second one from (9.24) and [17, Proposition 1.4].

Let $\mathcal{X}$ be a Hecke character of $F_A^\times$ of conductor $\mathfrak{q}$, and let $\mathcal{X}_\infty$ be the infinite part of $\mathcal{X}$. Here and henceforth, the conductor always refers to the finite part. Now we have

$$(9.30) \qquad \mathcal{X}_\infty(y) = \mathrm{sgn}(y)^r |y|^{i\kappa} \quad \text{for} \quad y \in F_\infty^\times$$

with $r \in \mathbf{Z}^a$ and $\kappa \in \mathbf{R}^a$. Let $\mathcal{X}^*$ denote the ideal character associated with $\mathcal{X}$; we put $\mathcal{X}^*(\mathfrak{a}) = 0$ for integral ideals $\mathfrak{a}$ prime to $\mathfrak{q}$. For every fractional ideal $\mathfrak{b}$, $\mathcal{X}_\infty(x)\mathcal{X}^*(x\mathfrak{b}^{-1})$ as a function of $x \in \mathfrak{b}$ depends only on x modulo $\mathfrak{b}\mathfrak{c}$. We define the Gauss sum $\tau(\mathcal{X})$ of $\mathcal{X}$ by

$$(9.31) \qquad \tau(\mathcal{X}) = \tau(\mathcal{X}^*) = \textstyle\sum_{x \in \mathfrak{q}^{-1}\mathfrak{b}^{-1}/\mathfrak{b}^{-1}} \mathcal{X}_\infty(x)\mathcal{X}^*(x\mathfrak{q}\mathfrak{b})e_F(x) .$$

If $\mathfrak{a}$ is a fractional ideal and $b \in (\mathfrak{a}\mathfrak{b}\mathfrak{q})^{-1}$, then

$$(9.32) \qquad \textstyle\sum_{x \in \mathfrak{a}/\mathfrak{a}\mathfrak{q}} \mathcal{X}_\infty(x)\mathcal{X}^*(x\mathfrak{a}^{-1})e_F(bx) = \bar{\mathcal{X}}_\infty(b)\bar{\mathcal{X}}^*(b\mathfrak{a}\mathfrak{q}\mathfrak{b})\tau(\mathcal{X}) .$$

PROPOSITION 9.7. *Let $\mathcal{X}$, $\mathfrak{q}$, r, κ, and $\mathcal{X}^*$ be as above, and let*

$$f(z) = \textstyle\sum_\xi a(\xi)e_F(\xi z) \in \mathfrak{S}_k(\Gamma_\lambda(\mathfrak{c}), \varphi_0, \theta) .$$

Further let $\mathfrak{c}_0$ be the conductor of φ_0, and $\mathfrak{a}$ the least common multiple of $\mathfrak{c}$, $\mathfrak{q}^2$, and $\mathfrak{q}\mathfrak{c}_0$. Then

$$(9.33) \qquad \textstyle\sum_\xi \mathcal{X}_\infty(\xi)\mathcal{X}^*(\xi\mathfrak{t}_\lambda^{-1})a(\xi)e_F(\xi z) \in \mathfrak{S}_k(\Gamma_\lambda(\mathfrak{a}), \varphi_0', \theta') ,$$

where $\varphi_0'(t) = \varphi_0(t)\mathcal{X}_\infty(t)^{-2}\mathcal{X}^*(t\mathfrak{r})^{-2}$ for $(t, \mathfrak{a}) = 1$ and $\theta'(\varepsilon) = \theta(\varepsilon)\mathcal{X}_\infty(\varepsilon)$ for $\varepsilon \in U_F$. Moreover, for every $\boldsymbol{f} \in \mathfrak{M}_k(\mathfrak{c}, \varphi)$, there exists an element $\boldsymbol{g}$ of $\mathfrak{M}_k(\mathfrak{a}, \varphi\mathcal{X}^2)$ such that $c(\mathfrak{m}, \boldsymbol{g}) = \mathcal{X}^*(\mathfrak{m})c(\mathfrak{m}, \boldsymbol{f})$ for all $\mathfrak{m}$.

This can be proved by the same technique as for [17, Prop. 4.4 and Prop. 4.5].

Proof of Theorem 9.4. We may assume $\mathfrak{t}_1 = \mathfrak{r}$. Replacing f and g by $f(cz)$ and $g(cz)$ with a suitable $c \in F$, we may assume that $f\|\gamma = f$ and $g\|\gamma = g$ for all $\gamma = \begin{pmatrix} a & b \\ c & d \end{pmatrix} \in \Gamma_1(\mathfrak{c})$ such that $a \equiv b \equiv 1 \pmod{\mathfrak{c}}$ with a suitable integral ideal $\mathfrak{c}$. Now the vector space of such f is the direct sum of $\mathfrak{M}_k(\Gamma_1(\mathfrak{c}), \varphi_0, \theta)$ for all possible characters φ_0 of $(\mathfrak{r}/\mathfrak{c})^\times$ and characters θ of U_F satisfying

$$(9.34) \qquad \theta(\varepsilon) = 1 \quad \text{for} \quad U_F \ni \varepsilon \equiv 1 \pmod{\mathfrak{c}} .$$

Replacing $\mathfrak{c}$ by its suitable multiple if necessary, we may assume

(9.35) $\mathfrak{c} = c_0 \mathfrak{r}$ with a rational integer $c_0 > 1$.

By virtue of [17, Prop. 5.3], the projection to each direct summand maps an element of $\mathfrak{L}(K, \Delta, \overline{\mathbf{Q}})$ to another element of the same set. Therefore our problem can be reduced to the case where $f \in \mathfrak{M}_k(\Gamma_1(\mathfrak{c}), \varphi_0, \theta)$ and $g \in \mathfrak{M}_l(\Gamma_1(\mathfrak{c}), \varphi_0', \theta')$ with some characters $\varphi_0, \varphi_0', \theta, \theta'$. We may assume $\theta\theta' = 1$, since $L(s; f, g) = 0$ if $\theta\theta' \neq 1$. If $\boldsymbol{p} = (p_1, \cdots, p_h; m) \in \mathfrak{S}_k(\mathfrak{c}, \varphi_0, m)$ and $\boldsymbol{q} = (q_1, \cdots, q_h; -m) \in \mathfrak{M}_l(\mathfrak{c}, \varphi_0', -m)$, we have

$$\sum_{\mathfrak{n}} c(\mathfrak{n}, \boldsymbol{p}) c(\mathfrak{n}, \boldsymbol{q}) N(\mathfrak{n})^{-s} = \sum_{\lambda=1}^{h} N(\mathfrak{t}_\lambda)^s L(s; p_\lambda, q_\lambda) .$$

Call this sum $L(s; \boldsymbol{p}, \boldsymbol{q})$. If $\boldsymbol{f} = (f, 0, \cdots, 0; m)$ and $\boldsymbol{g} = (g, 0, \cdots, 0; -m)$, we have $L(s; \boldsymbol{f}, \boldsymbol{g}) = L(s, f, g)$. Define a subspace $\mathfrak{N}$ of $\mathfrak{M}_k(\mathfrak{c}, \varphi_0, m)$ as in Proposition 9.6, and similarly a subspace $\mathfrak{N}'$ of $\mathfrak{M}_l(\mathfrak{c}, \varphi_0', -m)$ by using η^{-1}. Let $\mathfrak{N}^*$(resp. $\mathfrak{N}^*(\varphi)$) denote the elements $\boldsymbol{f} = (f_1, \cdots, f_h)$ on $\mathfrak{N}$ (resp. $\mathfrak{N} \cap \mathfrak{M}_k(\mathfrak{c}, \varphi)$) such that $\eta(\mathfrak{t}_\lambda)^{-1} f_\lambda \in \mathfrak{L}(K, \Delta, \overline{\mathbf{Q}})$. By Proposition 9.6 and [17, Prop. 5.3], $\mathfrak{N}^*$ is stable under $S(m)$ and $T(m)$; it is the direct sum of the subspaces $\mathfrak{N}^*(\varphi)$. Therefore it is sufficient to prove Theorem 9.4 for $L(s; \boldsymbol{f}, \boldsymbol{g})$ with $\boldsymbol{f} \in \mathfrak{N}^*(\varphi)$ and $\boldsymbol{g} \in \mathfrak{N}'$. Take a basis $\{\boldsymbol{f}_i\}$ of $\mathfrak{N}^*(\varphi)$ over $\overline{\mathbf{Q}}$ and a basis $\{\boldsymbol{g}_j\}$ of $\mathfrak{N}'$ over $\overline{\mathbf{Q}}$, and let

$$\boldsymbol{f}_i \,|\, T(\mathfrak{n}) = \sum_p u_{ip}(\mathfrak{n}) \boldsymbol{f}_p , \qquad \boldsymbol{g}_j \,|\, T(\mathfrak{n}) = \sum_q v_{jq}(\mathfrak{n}) \boldsymbol{g}_q$$

with $u_{ip}(\mathfrak{n})$ and $v_{jq}(\mathfrak{n})$ in $\overline{\mathbf{Q}}$. Decompose each integral ideal $\mathfrak{n}$ into the product $\mathfrak{n} = \mathfrak{ab}$ with $(\mathfrak{b}, \mathfrak{c}) = 1$ and a divisor $\mathfrak{a}$ of a power of $\mathfrak{c}$. Then $N(\mathfrak{a}) c(\mathfrak{ab}, \boldsymbol{f}) = c(\mathfrak{b}, \boldsymbol{f} \,|\, T'(\mathfrak{a}))$ by (9.29), and hence

$$\begin{aligned}
L(s; \boldsymbol{f}_i, \boldsymbol{g}_j) &= \sum_{\mathfrak{a}, \mathfrak{b}} c(\mathfrak{ab}, \boldsymbol{f}_i) c(\mathfrak{ab}, \boldsymbol{g}_j) N(\mathfrak{ab})^{-s} \\
&= \sum_{\mathfrak{a}, \mathfrak{b}} c(\mathfrak{b}, \boldsymbol{f}_i \,|\, T(\mathfrak{a})) c(\mathfrak{b}, \boldsymbol{g}_j \,|\, T(\mathfrak{a})) N(\mathfrak{a})^{-2-s} N(\mathfrak{b})^{-s} \\
&= \sum_{p, q} \sum_{\mathfrak{b}} c(\mathfrak{b}, \boldsymbol{f}_p) c(\mathfrak{b}, \boldsymbol{g}_q) N(\mathfrak{b})^{-s} \sum_{\mathfrak{a}} u_{ip}(\mathfrak{a}) v_{jq}(\mathfrak{a}) N(\mathfrak{a})^{-2-s} .
\end{aligned}$$

Now we have

(9.36) $$\sum_{\mathfrak{a}} \big(u_{ip}(\mathfrak{a}) \big) \otimes \big(v_{jq}(\mathfrak{a}) \big) N(\mathfrak{a})^{-2-s}$$
$$= \prod_{\mathfrak{p} | \mathfrak{c}} [1 - N(\mathfrak{p})^{-2-s} \big(u_{ip}(\mathfrak{p}) \big) \otimes \big(v_{jq}(\mathfrak{p}) \big)]^{-1} .$$

In view of (9.34), we can find a Hecke character χ' of $F_{\mathbf{A}}^{\times}$ whose conductor is a divisor of $\mathfrak{c}$ and such that $\chi_\infty'(a) = a^{im}$. Assumption (9.35) guarantees a Hecke character χ'' of finite order whose conductor $\mathfrak{e}$ satisfies $\mathfrak{c}^N \subset \mathfrak{e} \subset \mathfrak{c}^2$ with a positive integer N. Put $\chi = \chi' \chi''$. Fix a pair of indices (p, q). By Proposition 9.7, there exists an element $\boldsymbol{f}_p'$ of $\mathfrak{M}_k(\mathfrak{e}^2, \varphi \overline{\chi}^2)$ and an element $\boldsymbol{g}_q'$ of $\mathfrak{M}_l(\mathfrak{e}^2, \varphi' \chi^2)$ such that

$$c(\mathfrak{n}, \boldsymbol{f}_p') = \overline{\chi}^*(\mathfrak{n}) c(\mathfrak{n}, \boldsymbol{f}_p) , \qquad c(\mathfrak{n}, \boldsymbol{g}_q') = \chi^*(\mathfrak{n}) c(\mathfrak{n}, \boldsymbol{g}_q)$$

for all $\mathfrak{n}$. Observe that $c(\mathfrak{n}, \boldsymbol{f}_p')$ and $c(\mathfrak{n}, \boldsymbol{g}_q')$ are algebraic and the components of $\boldsymbol{f}_p'$ belong to $\mathfrak{L}(K, \Delta, \overline{\mathbf{Q}})$; we have

$$\sum_{(\mathfrak{b},\mathfrak{c})=1} c(\mathfrak{b}, \boldsymbol{f}_p)c(\mathfrak{b}, \boldsymbol{g}_q)N(\mathfrak{b})^{-s} = L(s; \boldsymbol{f}_p', \boldsymbol{g}_q') \ .$$

Since $\varphi\bar{\chi}^2$ and $\varphi'\chi^2$ are characters of finite order, we can apply [17, Theorem 5.6] to $L(s; \boldsymbol{f}_p', \boldsymbol{g}_q')$. Now

$$L(s; \boldsymbol{f}_i, \boldsymbol{g}_j) = \sum_{p,q} L(s; \boldsymbol{f}_p', \boldsymbol{g}_q')[(p, q)\text{-component of } (9.36)] \ .$$

Evaluating this at $s = \mu$, we obtain the desired conclusion.

THEOREM 9.8. *Suppose $f \in \mathfrak{L}(K, \Delta, \overline{\mathbf{Q}})$, $g \in \mathfrak{M}_k(\overline{\mathbf{Q}})$, $k_\nu = d_\nu + 1 > 1$ for all ν. Then $\langle f, g \rangle \sim \pi^{-n} p_K(\Delta, 2\psi)$, where $\langle f, g \rangle$ is the inner product defined by [17, (2.27)].*

Proof. Writing g as the sum of Eisenstein series and a cusp form, we can reduce the problem to the case where g is a cusp form. For $\boldsymbol{f} = (f_1, \cdots, f_h; m)$ and $\boldsymbol{g} = (g_1, \cdots, g_h; m)$ in $\mathfrak{S}_k(\mathfrak{c}, \varphi)$, put

$$\langle \boldsymbol{f}, \boldsymbol{g} \rangle = \sum_{\lambda=1}^h \langle f_\lambda, g_\lambda \rangle \ .$$

For the same reason as in the above proof, it is sufficient to prove our assertion for $\langle \boldsymbol{f}, \boldsymbol{g} \rangle$ in place of $\langle f, g \rangle$ with $\boldsymbol{f} \in \mathfrak{M}^*(\varphi)$ and $\boldsymbol{g} \in \mathfrak{M}$. By [17, Prop. 4.9], we know that the residue of $L(s; \boldsymbol{f}_\rho, \boldsymbol{g})$ at $s = 0$ is $\pi^{|k|} R_F \langle \boldsymbol{f}, \boldsymbol{g} \rangle$ times a rational number, where R_F is the regulator of F, and $\boldsymbol{f}_\rho$ is the element of $\mathfrak{M}^*(\bar{\varphi})$ such that $c(\mathfrak{n}, \boldsymbol{f}_\rho) = c(\mathfrak{n}, \boldsymbol{f})^\rho$. Taking bases of $\mathfrak{M}^*(\bar{\varphi})$ and $\mathfrak{M}$ over $\overline{\mathbf{Q}}$ and "twisting" them by a character χ of $F_\mathbf{A}^\times$ as in the above proof, we can reduce the problem to the case where $m = 0$. Assuming $m = 0$, express $\boldsymbol{f}$ and $\boldsymbol{g}$ in the form

$$\boldsymbol{f} = \sum_{h,\mathfrak{a}} u(\boldsymbol{h}, \mathfrak{a})\boldsymbol{h} | \mathfrak{a} \ , \qquad \boldsymbol{g} = \sum_{k,\mathfrak{b}} v(\boldsymbol{k}, \mathfrak{b})\boldsymbol{k} | \mathfrak{b}$$

with primitive $\boldsymbol{h}$ and $\boldsymbol{k}$, $u(\boldsymbol{h}, \mathfrak{a})$ and $v(\boldsymbol{k}, \mathfrak{b})$ in $\overline{\mathbf{Q}}$, and integral ideals $\mathfrak{a}$ and $\mathfrak{b}$, where $\boldsymbol{h} | \mathfrak{a}$ is defined as in [17, Prop. 2.3]. This together with [17, Prop. 4.14] reduces the problem to the case where $\boldsymbol{f}$ is primitive. If $\boldsymbol{f}$ is primitive, our assertion follows directly from [17, Theorem 5.5].

10. Proofs of the main theorems in special cases

LEMMA 10.1. *There exists a CM-point of $\mathfrak{Z}^r$ fixed by $h(Y^\times)$ with $Y = K \oplus Y_0$, $[K:F] = 2$, $Y_0 = F^m$ in the notation of Proposition 5.1.*

Proof. Take a coordinate system of V over F so that S is represented by $\mathrm{diag}[s_1, \cdots, s_{m+2}]$ with totally positive $s_1, \cdots, s_m$, and let $Y = K \oplus F^m$, $K = F(\zeta)$, $\zeta^2 = -s_{m+1}/s_{m+2}$. Define $h: Y \to M_{m+2}(F)$ by

$$h(a + b\zeta, c_1, \cdots, c_m) = \mathrm{diag}\left[c_1, \cdots, c_m, \begin{bmatrix} a & b \\ \zeta^2 b & a \end{bmatrix} \right] \qquad (a, b, c_1, \cdots, c_m \in F) \ .$$

Then $Sh(x) = {}^t h(x^\rho)S$. For each $i \le r$, extend τ_i to an element σ_i of J_K.

Then the vectors $(0, \cdots, 0, 1, \pm\zeta^{\sigma_i})$ are eigenvectors under $h(Y^u)_i$ and belong to $\mathfrak{D}(S_i)$. Therefore $h(Y^u)$ has a fixed point on $\mathfrak{Z}^r$, and (5.4) is satisfied; hence we obtain our assertion.

To prove Theorem 8.4, it is sufficient to show that $\pi^{-|k|} P_o(w)^{-1} \mathfrak{f}(w)$ is $\overline{\mathbf{Q}}$-rational for every w in a set of CM-points $\mathfrak{W}$ which is dense in $\mathfrak{Z}^r$, as shown in the first part of the proof of Theorem 6.5. Choosing any CM-point w_0 on $\mathfrak{Z}^r$ (as in the above lemma, for example) we can take $\{\alpha(w_0) \mid \alpha \in G_+(S)\}$ as $\mathfrak{W}$. In view of (5.14) and (8.16), our task is to show that $\pi^{-|k|} P_o(w_0)^{-1} \mathfrak{f}(w_0)$ is $\overline{\mathbf{Q}}$-rational for $\mathfrak{f}$ defined with an arbitrary congruence set X and an arbitrary ξ, but with a *special* w_0. We do this in this section in the case $F = E$.

We first consider f without assuming $F = E$. Let w be the fixed point of $h(Y^u)$ with $Y = K \oplus Y_0$ as in Proposition 5.1, without assuming $[K : F] = 2$. Take a vector v of V so that $V = h(Y)v$ as in Proposition 5.4, and let $W = h(Y_0)v$. If S' denotes the restriction of S to W, we have

$$(10.2) \quad S(h(a)v + x, h(b)v + y) = \mathrm{Tr}_{K/F}(\varepsilon ab^\rho) + S'(x, y) \quad (a, b \in K; x, y \in W)$$

with a "real" element ε of K which can be obtained as the projection of the element δ of Proposition 5.4 to K. Let $u_i = A_i^{-1} p(w_i)$ with the ith component w_i of w. Define ψ_i by (5.6a, b) and view it as an element of J_K. Then $h(a)_i u_i = a^{\psi_i} u_i$ for $a \in K^u$; u_i is an algebraic vector characterized by this property up to algebraic factors. By (5.10), we have $S_i(W, u_i) = 0$, and hence

$$(10.3) \quad S_i(h(b)v + y, u_i) = S_i(v_i, h(b^\rho)_i u_i) = b^{\rho\psi_i} S_i(v_i, u_i) \quad (b \in K, y \in W) \, .$$

Putting $\Psi = \sum_{i=1}^{r} k_i \psi_i$ and $\psi = \sum_{i=1}^{r} \psi_i$, we obtain

$$S(h(b)v + y, A^{-1}p(w))^{-k} |S(h(b)v + y, A^{-1}p(w))|^{-2s} = d'|d|^{-s} b^{-\rho\Psi} |b^\psi|^{-2s}$$

with nonzero algebraic numbers d and d' independent of b and y. Thus

$$(10.4) \quad M_U^{-1} f(w, s; t, \xi, X) = d'|d|^{-s} \sum_{h(b)v + y \in X/U} b^{-\rho\Psi} |b^\psi|^{-2s}$$
$$\times t(h(b)v + y)\omega(\mathrm{Tr}_{K/E}(\varepsilon\xi bb^\rho) + \mathrm{Tr}_{F/E}(\xi S'[y])) \, .$$

Let B be a congruence subset of K, and C a congruence subset of W. Then

$$(10.5) \quad \{h(b)v + y \mid (0, 0) \neq (b, y) \in B \times C\}$$

is a congruence subset of $V - \{0\}$. Since every congruence subset of $V - \{0\}$ is a finite disjoint union of such sets, it is sufficient to prove (8.21) when X has form (10.5). We take U so that B and C are stable under U. Since (8.3) is extended over x satisfying (8.9), $x = h(0)v + y \neq 0$ does not occur in (10.4). Hence the sum is extended over $b \in B'/U$ and $y \in C$, where $B' = B - \{0\}$.

Let us now show that Theorem 8.4 follows from Theorem 9.2. We now assume $[K:F] = 2$. Take a $\overline{\mathbf{Q}}$-rational $t \in \mathscr{P}_k$ such that $t(h^u(a)x) = a^\Phi t(x)$ for $a \in K^u$ with $\Phi \in I_K$. Such a t is a $\overline{\mathbf{Q}}$-linear combination of functions t' on $\prod_{i>r} V_i$ of the form

$$(10.6) \qquad t'(h(b)v + y) = b^\Theta g(y) \qquad (b \in K, y \in W)$$

with a $\overline{\mathbf{Q}}$-rational homogeneous function g on $\prod_{i>r} W_i$ and an element Θ of I_K such that $\Theta \equiv \Phi \pmod{(1+\rho)I_K}$. Let α_{ij} be the degree of g in the variable on V_{ij}; put $\alpha_i = \sum_{j=2}^{q} \alpha_{ij}$ and $\Theta = \sum_{i=1}^{r}\sum_{i=1}^{\lambda} c_{ij}\mathscr{P}_{ij}$ with $c_{ij} \geqq 0$, where $\mathscr{P}_{i1}, \cdots, \mathscr{P}_{i\lambda}$ are the elements of I_K, other than ψ_i and $\psi_i\rho$, which coincide with τ_i on E. Put

$$H(z) = \sum_{y \in C} g(y) e_E\big(-\mathrm{Tr}_{F/E}(\xi S'[y])z\big) \qquad (z \in \mathfrak{H}_1^r)\ ,$$
$$\Omega'(z) = \Omega(z)H(z) = \sum_{a \in E} \omega'(a) e_E(az)\ .$$

Since $-\xi S'$ is totally positive, $\Omega' \in \mathfrak{M}_{l',E}$ by virtue of Propositions 7.6 and 7.7, where

$$l_i' = l_i + (mq/2) + \varepsilon_i\ , \qquad e_i = \delta_i + (\alpha_i - \varepsilon_i)/2 \qquad (i = 1, \cdots, r)\ ;$$

ε_i is 0 or 1 according as α_i is even or odd. Condition (8.15) implies that $l' \in \mathbf{Z}^r$. We have obviously

$$\omega'(a) = \sum_{y \in C} g(y)\omega\big(a + \mathrm{Tr}_{F/E}(\xi S'[y])\big)\ .$$

Therefore (10.4) at $s = 0$ becomes

$$f(w, 0; t, \xi, X) = M_U d' \sum_{b \in B'/U} \omega'\big(\mathrm{Tr}_{K/E}(\varepsilon\xi bb^\rho)\big)b^{\Theta - \rho\Psi}\ .$$

This is the value of (9.2) at $s = 0$ with $E, \varepsilon\xi, \Theta, k$ and Ω' in place of F, η, Φ, d and Ω. Observe that (9.3) follows from Proposition 5.4. We have

$$k_i - l_i' - 2e_i - \sum_{j=2}^{q} c_{ij} = k_i - l_i - 2\delta_i - \sum_{j=2}^{q}\kappa_{ij} - (mq/2) = \sigma$$

with an integer σ independent of i by virtue of (8.7). Therefore we obtain (8.21) from Theorem 9.2 as promised.

Suppose $F = E$ in Theorem 8.4. Then what we need in the above proof is Theorem 9.2 in the case $[K:F] = 2$. Since this case was already proved in Section 9, we have thus established Theorem 8.4 when $F = E$.

11. Non-holomorphic derivatives of automorphic forms

Let $\mathfrak{Z}_{(i)}$ denote the ith factor $\mathfrak{Z}$ of $\mathfrak{Z}^r$ and T_i the (global) complex tangent space of $\mathfrak{Z}_{(i)}$, which can be written $\sum_{j=1}^{m} \mathbf{C}\,\partial/\partial z_{ij}$ with the coordinate functions $z_{i1}, \cdots, z_{im}$ on $\mathfrak{Z}_{(i)}$. The choice of "algebraic" A_i in Section 3 implies that T_i has a $\overline{\mathbf{Q}}$-rational structure. For a C^∞-function on $\mathfrak{Z}^r$ or on its subset with values in a finite-dimensional complex vector space W, we define a function

$D_i f$ on $\mathfrak{Z}^r$ with values in $\mathrm{Hom}\,(T_i,\,W)$ by

$$(11.1) \qquad D_i f(\textstyle\sum_{j=1}^m u_j \partial/\partial z_{ij}) = \sum_{j=1}^m u_j \partial f/\partial z_{ij} \qquad (u_j \in \mathbf{C})\,.$$

Now $\mu_i(\alpha,\,z)^{-1}\cdot{}^t\lambda_i(\alpha,\,z)^{-1}$ is the jacobian matrix of α on $\mathfrak{Z}_{(i)}$, which can be viewed as an element of $\mathrm{GL}\,(T_i)$; as already seen in (2.17), we have

$$(11.2) \qquad (D_i f)\circ \alpha = D_i(f\circ\alpha)\mu_i(\alpha,\,z)\cdot{}^t\lambda_i(\alpha,\,z) \qquad (\alpha \in G_+(S))\,.$$

Suppose W has a $\overline{\mathbf{Q}}$-rational structure and let

$$(11.3) \qquad \sigma\colon (\mathrm{GL}_1)^r \times (\mathrm{GL}_m)^r \times \textstyle\prod_{i>r} G(S_i) \longrightarrow \mathrm{GL}\,(W)$$

be a $\overline{\mathbf{Q}}$-rational representation as considered in Section 3; here $(\mathrm{GL}_m)^r$ actually stands for $\prod_{i=1}^r \mathrm{GL}\,(T_i^*)$ with the dual space T_i^* of T_i. We then define a representation τ of the same group on the space $\mathrm{Hom}\,(T_i,\,W)$, which we write $\tau = \sigma + (i)$, by

$$(11.4) \qquad \tau(A)U = \sigma(A)\circ U\circ{}^t(a_i b_i) \qquad \text{for } U \in \mathrm{Hom}\,(T_i,\,W) \text{ and}$$
$$A = (a_1,\,\cdots,\,a_r,\,b_1,\,\cdots,\,b_r,\,c_{r+1},\,\cdots,\,c_n)\,.$$

Obviously $\tau(A)$ is equivalent to $\sigma(A)\otimes(a_i b_i)$. Let ξ_i and η_i be the functions on $\mathfrak{Z}_{(i)}$ defined by (2.13) and (2.21), and put

$$(11.5) \qquad \Xi_{i,\sigma}(z) = \sigma(\cdots,\,\eta_i(z),\,\cdots,\,\xi_i(z),\,\cdots)\,,$$

where the unwritten components are all equal to the identity. From (2.14) and (2.23), we see that if $\alpha \in G_+(S)$ and $\nu(\alpha) = 1$, then

$$(11.6) \qquad \Xi_{i,\sigma}\big(\alpha(z)\big)\sigma(\cdots,\,\mu_i(\alpha,\,z),\,\cdots,\,\lambda_i(\alpha,\,z),\,\cdots)$$
$$= \sigma(\cdots,\,\overline{\mu_i(\alpha,\,z)^{-1}},\,\cdots,\,\overline{\lambda_i(\alpha,\,z)},\,\cdots)\Xi_{i,\sigma}(z)\,.$$

We now define a differential operator $D_{i,\sigma}$ by

$$(11.7) \qquad (D_{i,\sigma}f)(z) = \Xi_{i,\sigma}(z)^{-1}D_i\big(\Xi_{i,\sigma}(z)f(z)\big) \qquad (z \in \mathfrak{Z}^r;\ i \leqq r)\,.$$

Here f is W-valued and $D_{i,\sigma}f$ is $\mathrm{Hom}\,(T_i,\,W)$-valued. If $\tau = \sigma + (i)$, we can easily verify

$$(11.8) \qquad (D_{i,\sigma}f)|_\tau\alpha = D_{i,\sigma}(f\,|_\sigma\alpha) \qquad \text{for }\ \alpha \in G_+(S),\ \nu(\alpha) = 1\,.$$

In particular, if $f \in \mathfrak{A}_\sigma(\Gamma)$, we have $(D_{i,\sigma}f)|_\tau\gamma = D_{i,\sigma}f$ for $\gamma \in \Gamma$, though $D_{i,\sigma}f$ may not be meromorphic. If σ is a trivial representation, we have $D_{i,\sigma} = D_i$, so that $D_{i,\sigma}f \in \mathfrak{A}_\tau$ if $f \in \mathfrak{A}_0$. Now successive application produces an operator of the type

$$(11.9) \qquad D_{p,\omega}\cdots D_{k,\rho}D_{j,\tau}D_{i,\sigma}\,,$$

where $\tau = \sigma + (i)$, $\rho = \tau + (j)$, etc. This sends a W-valued function to a function with values in the vector space

$$(11.10) \qquad \mathrm{Hom}\,(T_p,\,\mathrm{Hom}\,(\cdots,\,\mathrm{Hom}\,(T_j,\,\mathrm{Hom}\,(T_i,\,W))\cdots))\,,$$

which we can identify with the space of multilinear maps

$$(11.11) \qquad T_i \times T_j \times \cdots \times T_p \longrightarrow W .$$

Let e_i be the number of factors $D_{i,*}$ with index i in (11.9); put $e=(e_1, \cdots, e_r)$. We then denote (11.9) simply by D_σ^e, though strictly speaking (11.9) depends not only on σ and e, but also on the order of i, j, k, etc. If $f \in \mathfrak{A}_\sigma(\Gamma)$, we have $(D_\sigma^e f)|_\chi \gamma = D_\sigma^e f$ for $\gamma \in \Gamma$, where $\chi = \sigma + (i) + (j) + \cdots + (p)$; obviously χ is equivalent with the representation

$$(11.12) \qquad A = (a_1, \cdots, a_r, b_1, \cdots, b_r, c_{r+1}, \cdots, c_n)$$
$$\longmapsto \sigma(A) \otimes (a_1 b_1)^{(e_1)} \otimes \cdots \otimes (a_r b_r)^{(e_r)} ,$$

where $y^{(s)} = y \otimes \cdots \otimes y$ with s copies of y.

LEMMA 11.1. *Let* $0 = (0, \cdots, 0)$ *denote the "origin" of* $\mathfrak{Z}^r$ *(see* §1*). Then* $(D_\sigma^e f)(0) = (D_p \cdots D_j D_i f)(0)$.

Proof. We first observe that $\Xi_{i,\sigma}$ has the form

$$(11.13) \qquad \Xi_{i,\sigma}(z) = 1_W + \sum_h \varphi_h(\bar{z}_{i1}, \cdots, \bar{z}_{im}) z_{i1}^{h_1} \cdots z_{im}^{h_m}$$

with functions φ_h holomorphic in $\bar{z}_{i1}, \cdots, \bar{z}_{im}$ which vanish at 0. In fact, (2.13) and (2.21) show that this is so for η_i and ξ_i. Since σ is a rational representation, we obtain (11.13). This means that $\partial^{|h|} \Xi_{i,\sigma}/(\partial z)^h(0) = 0$ for all $h \neq 0$. Our lemma is an immediate consequence of this fact.

THEOREM 11.2. *Let* $f \in \mathfrak{A}_\sigma(\bar{\mathbf{Q}})$ *and let* $D_\sigma^e f$ *and* χ *be defined as above. Then* $\pi^{-|e|} P_\chi(w)^{-1}(D_\sigma^e f)(w)$ *is* $\bar{\mathbf{Q}}$-*rational for every* CM-*point* w *on* $\mathfrak{Z}^r$ *if* f *is holomorphic at* w, *where* $|e| = e_1 + \cdots + e_r$.

Proof. This is similar to [10, Main Theorem I] and can be proved in the same fashion with a slight modification. We first observe that if $g \in \mathfrak{A}_\sigma(\bar{\mathbf{Q}})$, $\pi^{-1} D_i g$ is arithmetic by virtue of Proposition 6.6. Given $f \in \mathfrak{A}_\sigma(\bar{\mathbf{Q}})$ and a CM-point w where f is holomorphic, take a $\bar{\mathbf{Q}}$-rational σ-uniformizer H at w as guaranteed by Theorem 6.5. Let U be the square matrix whose first column is f and other columns are 0. (Here we identify W with $\mathbf{C}^d$.) Put $H' = U + cH$ with $c \in \mathbf{Q}$. Then H' is holomorphic at w and $\det(H'(w)) \neq 0$ for a suitable choice of c. Since $U = H' - cH$, it is sufficient to prove our assertion for H and H', i.e., for every $\bar{\mathbf{Q}}$-rational σ-uniformizer at w. (Strictly speaking, σ must be replaced by a sum of its copies.) Since w has algebraic coordinates, we can find an element β of $G(Q, R)^r$ with algebraic coefficients such that $\beta(w) = 0$. Transforming everything by β, we may assume that $w = 0$. Take $Y = K + Y_0$ and $h: Y \to \operatorname{End}(V, F)$ as in Proposition 5.1 so that 0 is fixed by $h(Y^u)$. Let $\alpha = h^u(a)$ with $a \in K^u$ and put $M = H^{-1}(H|_\sigma \alpha)$. Then M has components in $\mathfrak{A}_0(\bar{\mathbf{Q}})$, $H|_\sigma \alpha = HM$, and

$M(0) = H(0)^{-1}\Lambda_\sigma(\alpha, 0)^{-1}H(0)$. In view of Lemma 11.1, our task is to show that $\pi^{-|e|}P_\chi(0)^{-1}(D_p \cdots D_i H)(0)$ is algebraic. We do this by induction on $|e|$. The case $|e| = 1$ has been settled when H has components in $\mathfrak{A}_0(\overline{\mathbf{Q}})$. If $\tau = \sigma + (i)$, we have

$$(11.14) \qquad (D_{i,\sigma}H)|_\tau\alpha = D_{i,\sigma}(HM) .$$

Observe that $P_\sigma(0)$ commutes with $\Lambda_\sigma(\alpha, 0)$. Since H is arithmetic, we may assume, multiplying by a suitable algebraic linear transformation on the right, that $H(0) = P_\sigma(0)$; then $M(0) = \Lambda_\sigma(\alpha, 0)^{-1}$. Evaluating (11.14) at 0, we obtain

$$(11.15) \qquad M_0(D_i H)_0\kappa_i u = \big((D_i H)_0 u\big)M_0 + H_0(D_i M)_0 u \qquad (u \in T_i) ,$$

where the subscript 0 indicates the value at $z = 0$, and $\kappa_i = \mu_i(\alpha, 0)^{-1} \cdot {}^t\lambda_i(\alpha, 0)^{-1}$. Let $q_i = P_\theta(0)$ with $\theta = {}^t(\sigma_i^1\sigma_i^2)^{-1}$, where σ_i^k is defined by (6.15). Put $X = \pi^{-1}P_\sigma(0)^{-1}(D_i H)_0 q_i$. Our immediate task is to show that Xu is algebraic for every algebraic $u \in T_i$. Substituting $q_i u$ for u in (11.15), we obtain

$$(11.16) \qquad M_0 X\kappa_i u = (Xu)M_0 + \pi^{-1}(D_i M)_0 q_i u .$$

Since $\pi^{-1}D_i M$ is $\overline{\mathbf{Q}}$-rational as observed above, $\pi^{-1}(D_i M)_0 q_i$ is $\overline{\mathbf{Q}}$-rational. Now we could have assumed at the beginning that

$$\sigma(b_1, \cdots, b_r, c_1, \cdots, c_r, d_{r+1}, \cdots, d_n) = b_1^{k_1} \cdots b_r^{k_r} \sigma'(c_1, \cdots, c_r, d_{r+1}, \cdots, d_n)$$

with $(k_1, \cdots, k_r) \in \mathbf{Z}^r$ and a representation σ' of $(\mathrm{GL}_m)^r \times \prod_{i>r} G(S_i)$. (An arbitrary σ is a direct sum of such σ with various k's.) Under this assumption, put $\Psi = -\sum_{i=1}^r k_i\psi_i$ and

$$N = \sigma'\big(\varphi_1(a), \cdots, \varphi_r(a), h^u(a)_{r+1}, \cdots, h^u(a)_n\big)^{-1}$$

with ψ_i and φ_i of (5.6). Then $\kappa_i = \psi_i(a)^{-1}\varphi_i(a)$ and $M_0 = a^\Psi N$. Therefore (11.16) implies that

$$(11.17) \qquad NX\kappa_i u - XuN \text{ is algebraic for algebraic } u \in T_i.$$

We can choose $a \in K^u$ so that κ_i and $N \otimes N^{-1}$ have no common eigenvalues. Then (11.17) shows that X is algebraic. This proves the case $|e| = 1$. If $|e| > 1$, we have

$$(D_{p,\omega} \cdots D_{j,\tau}D_{i,\sigma}H)|_\chi\alpha = D_{p,\omega} \cdots D_{j,\tau}D_{i,\sigma}(HM) .$$

By Lemma 11.1, this can be written as

$$\begin{aligned}
M_0(D_p &\cdots D_j D_i H)_0(\kappa_i u_i, \kappa_j u_j, \cdots, \kappa_p u_p) \\
&- (D_p \cdots D_j D_i H)_0(u_i, u_j, \cdots, u_p)M_0 \\
&= (D_p \cdots D_j H)_0(u_j, \cdots, u_p)(D_i M)_0 u_i \\
&\quad + \cdots + H_0(D_p \cdots D_j D_i M)_0(u_i, u_j, \cdots, u_p)
\end{aligned}$$

for $u_i \in T_i, \cdots, u_p \in T_p$, where we consider the "derivatives" as multilinear maps of type (11.11) with $\mathrm{End}(W)$ instead of W. Substitute $q_i u_i, \cdots, q_p u_p$ for $u_i, \cdots, u_p$, and take $a \in K^{\times}$ so that $\kappa_i \otimes \cdots \otimes \kappa_p$ and $N \otimes N^{-1}$ have no common eigenvalues. This is feasible since $\psi_1(a), \cdots, \psi_n(a)$ are not involved in N. Applying induction to the right-hand side, we find that

$$\pi^{-|e|} P_\sigma(0)^{-1}(D_p \cdots D_j D_i H)_0(q_i u_i, q_j u_j, \cdots, q_p u_p)$$

is algebraic for every algebraic $u_i, \cdots, u_p$. This completes the proof.

12. Proof of Theorem 9.2

Let w be the fixed point of $h(Y^{\times})$ with $Y = K + Y_0$ as in Proposition 5.1. Given $k, e \in \mathbf{Z}^r$ with $k_i > 0$ and $e_i \geq 0$, we consider

$$(12.1) \qquad L_k^e(x) = [D_\sigma^e S(x, A^{-1}p(z))^{-k}]_{z=w} \qquad (x \in V_1 \times \cdots \times V_r)$$

with $D_\sigma^e = D_{p,\omega} \cdots D_{i,\sigma}$ as in Section 11, where

$$(12.2) \qquad \sigma(a_1, \cdots, a_{r}, \cdots) = a_1^{k_1} \cdots a_r^{k_r} .$$

As explained in Section 11, $L_k^e(x)$ can be viewed as a C-valued multilinear function on $T_i \times \cdots \times T_p$; thus $L_k^e(x)(u_i, \cdots, u_p)$ for $u_i \in T_i, \cdots, u_p \in T_p$ is meaningful. Each T_i has a fixed identification with $\mathbf{C}^m$. Put

$$C_i = \mathrm{diag}[1_{m+1}, \eta(w_i)] B(w_i)^{-1} A_i \qquad (i = 1, \cdots, r) .$$

Then (2.20) shows that $R[C_i x] = S_i[x]$ for $x \in (V_i)_{\mathbf{C}}$ and

$$C_i h(a)_i C_i^{-1} = \mathrm{diag}[\varphi_i(a), \overline{\psi_i(a)}, \psi_i(a)] \qquad (a \in Y)$$

with φ_i and ψ_i of (5.6). Let

$$(12.3) \qquad \Theta_i = \{\gamma \in J_Y \,|\, \gamma = \tau_i \text{ on } F, \gamma \neq \psi_i, \gamma \neq \psi_i \rho\} .$$

For each $\gamma \in \Theta_i$, there exists a nonzero $\overline{\mathbf{Q}}$-rational vector q_γ of $(V_i)_{\mathbf{C}}$, unique up to constant factors, such that $h(a)_i q_\gamma = a^\gamma q_\gamma$ for all $a \in Y$. We see that $C_i q_\gamma = {}^t({}^t u_\gamma, 0, 0)$ with $u_\gamma \in \mathbf{C}^m = T_i$ and $\varphi_i(a)u_\gamma = a^\gamma u_\gamma$ for $a \in Y$.

Let Φ be an element of I_K which contains ψ_i and $\psi_i \rho$ with the same multiplicity for each i; let $\mathcal{Q}_e(\Phi)$ for $e = (e_1, \cdots, e_r) \in \mathbf{Z}^r$ denote the set of all polynomial functions g on $\prod_{i=1}^r V_i$ which are homogeneous of degree e_i in the variable x_i on V_i and satisfy $g(h^{\times}(a)x) = a^\Phi g(x)$ for all $a \in K^{\times}$. Put $A_i x_i = {}^t({}^t v_i, y_i', y_i)$ for $x_i \in V_i$ with $v_i \in \mathbf{C}^m$ and $y_i, y_i' \in \mathbf{C}$. We are now interested in the function $\mathfrak{g}$ of the form

$$\mathfrak{g}(x) = \prod_{i=1}^r (y_i')^{-d_i} g(x)$$

with $\mathfrak{g} \in \mathcal{Q}_e(\Phi)$ and integers $d_i > e_i$.

LEMMA 12.1. *Suppose g is $\overline{\mathbf{Q}}$-valued on V and condition*

$$(12.4) \qquad d_i - e_i > (m - 2)/2 \qquad \text{for all} \quad i \leq r$$

is satisfied. Then $\mathfrak{g}$ *is a* $\overline{\mathbf{Q}}$-*linear combination of functions of the form*

$$S[x]^p L_{d-e+2p}^{e-2p}(u_\alpha, \cdots, u_\zeta), \ 0 \leq p \leq e/2$$

with elements $\alpha, \cdots, \zeta$ *of* $\bigcup_{i=1}^r \Theta_i$ *such that*

$$(12.5) \qquad \alpha + \cdots + \zeta \equiv \Phi\rho \mod (1+\rho)I_K + I(Y_0) .$$

Proof. For the same reason as in the proof of Theorem 11.2, we may assume that $w = 0$. Obviously $L_k^e(x)$ is the product of

$$[(D_o^e)_i S_i(x_i, A_i^{-1}p(z_i))^{-k_i}]_{z=0} \qquad (i = 1, \cdots, r) ,$$

where $(D_o^e)_i$ denotes "the i-part" of D_o^e. We may also assume that $g(x) = \prod_{i=1}^r g_i(x_i)$ with homogeneous functions g_i on V_i such that $g_i(h^u(a)_i x_i) = a^{\Phi_i} g_i(x_i)$ for $a \in K^u$, where Φ_i is "the i-part" of Φ. Therefore it is sufficient to prove our assertion for each i-factor. To do this, we drop, for simplicity, the subscript i, treating things as if $r = 1$. By (3.3) and (2.3), we have

$$S(x, A^{-1}p(z)) = {}^t vz - y' - y \cdot {}^t zz/2 .$$

A direct calculation shows that, if $k = d - e$,

$$(12.6) \quad (-1)^e(-y')^d L_k^e(x)(u_1, \cdots, u_e)$$
$$= \sum_{j=0}^{[e/2]} (-1)^j k(k+1) \cdots (k+e-j-1)(yy')^j H_j(v; u_1, \cdots, u_e)$$

with functions H_j homogeneous in v of degree $e - 2j$. In particular

$$(12.7) \qquad H_0(v; u_1, \cdots, u_e) = {}^t vu_1 \cdots {}^t vu_e .$$

Let $\mathcal{H}_{d,e}(\Phi)$ denote the set of all S-harmonic functions on V in the sense of (2.27) of the form $y'^{-d}s(x)$ with $s \in \mathcal{Q}_e(\Phi)$. Observe that $Ah(a)x = {}^t({}^t(\varphi(a)v), a^{\psi\rho}y', a^\psi y)$. Since ψ and $\psi\rho$ appear in Φ with the same multiplicity, we have

$$(12.8) \qquad s(x) = \sum_{0 \leq j \leq e/2} (yy')^j r_j(v)$$

with $r_j \in \mathcal{Q}_{e-2j}(\Phi)$. The harmonicity implies that

$$(12.9) \qquad \Delta_v r_{j-1} = 2j(j-d)r_j \qquad (1 \leq j \leq e/2) ,$$

where $\Delta_v = \sum_{\lambda=1}^m \partial^2/\partial v_\lambda^2$ with the coordinates $v_1, \cdots, v_m$ of v. Thus s is completely determined by r_0. Since $r_0 \in \mathcal{Q}_e(\Phi)$, r_0 is a linear combination of functions of the form

$$(12.10) \qquad r(v) = {}^t vu_\alpha \cdots {}^t vu_\zeta$$

with $\alpha, \cdots, \zeta \in J_Y$ satisfying (12.5). We see that if $r_0 \in \mathcal{Q}_e(\Phi)$ and s is defined by (12.8) and (12.9), then $s \in \mathcal{H}_{d,e}(\Phi)$. Since D_o^e commutes with the Laplacian with respect to S, we observe that $L_{d-e}^e(x)(u_\alpha, \cdots, u_\zeta)$ is S-harmonic. Therefore (12.6) and (12.7) show that $L_{d-e}^e(x)(u_\alpha, \cdots, u_\zeta)$ is a constant times the element s of $\mathcal{H}_{d,e}(\Phi)$ whose first term is (12.10). Thus $\mathcal{H}_{d,e}(\Phi)$ is spanned by $L_{d-e}^{e}(x)(u_\alpha, \cdots, u_\zeta)$. Observe that $\dim(\mathcal{H}_{d,e}(\Phi))$ does not depend on d. Let $g \in \mathcal{Q}_e(\Phi)$. A well known fact on spherical functions tells that $g(x) = \sum_p S[x]^p \sigma_p(x)$ with spherical σ_p of degree $e - 2p$. The uniqueness of σ_p implies that $\sigma_p \in \mathcal{H}_{0,e}(\Phi)$. This shows that

$$(12.11) \quad \dim(\mathcal{Q}_e(\Phi)) = \sum_{0 \leq p \leq e/2} \dim(\mathcal{H}_{0,e-2p}(\Phi)) = \sum_{0 \leq p \leq e/2} \dim(\mathcal{H}_{d,e-2p}(\Phi)) .$$

If $r_p \in \mathcal{H}_{d,e-2p}(\Phi)$ and $\sum_p S[x]^p r_p(x) = 0$, then $r_p = 0$ for all p. This can easily be shown by induction on e by applying the Laplacian to the equation. (Here we need (12.4).) Therefore (12.11) shows that

$$(y')^{-d} g(x) = \sum_p S[x]^p r_p(x) \quad \text{with} \quad r_p \in \mathcal{H}_{d,e-2p}(\Phi) ,$$

which establishes our lemma with the base field $\mathbf{C}$ instead of $\overline{\mathbf{Q}}$. Since our problem concerns rational functions, it is clear that the desired fact over $\overline{\mathbf{Q}}$ follows from this result.

Let us now prove the part of Theorem 9.2 concerning $\mathcal{D}(\mu)$. We assume $[K: F] > 2$, since the case $[K: F] = 2$ was proved in Section 9. Let Ω, η, ψ, and Δ be as in (9.1), (9.3), and (9.4a). Put

$$(12.12) \qquad\qquad S(x, y) = -\mathrm{Tr}_{K/F}(\eta x y^\rho) \qquad\qquad (x, y \in K) .$$

Viewing K as a vector space over F and writing it V, we find that S satisfies (3.1) with $r = n$. Now the regular representation of K defines an embedding h of K into $\mathrm{End}(V, F)$ satisfying (5.2). By Proposition 5.2, $h(K^u)$ has a unique common fixed point w on $\mathfrak{Z}^n$. Thus the setting is the same as in Proposition 5.4 with $Y = K$ and $v = 1$. We are going to apply the above lemma to this w with $n = r$ and $F = E$. Given Φ as in (9.4b) and an integer μ as in (9.7a, b), put

$$e_i = 2b_i - 2\mu + \sum_{j=1}^{\lambda} c_{ij} \qquad\qquad (i = 1, \cdots, n) ,$$
$$\Xi = \sum_{i=1}^{n} \{(b_i - \mu)(\psi_i + \psi_i \rho) + \sum_{j=1}^{\lambda} c_{ij} \mathcal{P}_{ij}\} .$$

(Note that $\lambda = m$ in the present case.) Define a polynomial function g on $\prod_{i=1}^{n} V_i$ so that $g(x) = x^\Xi$ for $x \in K = V$. Observe that (12.4) follows from (9.7a, b). Therefore, by the above lemma, $x^{\Xi - \rho\Delta}$ is a $\overline{\mathbf{Q}}$-linear combination of

$$S[x]^p L_{d-e+2p}^{e-2p}(x)(u_\alpha, \cdots, u_\zeta) \qquad\qquad (0 \leq p \leq e/2)$$

with $\alpha, \cdots, \zeta \in J_K$ such that $\alpha + \cdots + \zeta \equiv \rho\Xi \bmod (1 + \rho)I_K$. We now take "the pth derivative" Ω_p of Ω:

$$\Omega_p(z) = (2\pi i)^{-|p|}(\partial/\partial z)^p \Omega(z) = \sum_{a \in F} a^p \omega(a) e_F(az) .$$

This belongs to $\mathfrak{M}_l^{d+p}$. Define series (8.3) with $F = E$, $t = 1$, $\xi = -1$, $k = d - e + 2p$, and with Ω_p in place of Ω. Call $f_p(z)$ its value at $s = 0$. The convergence of f_p follows from (9.7a, b). Now we have

$$(D_\sigma^{e-2p} f_p)(w) = M_U \sum_{x \in X/U} (-S[x])^p \omega(-S[x]) L_{d-e+2p}^{e-2p}(x) ,$$

where σ is defined by (12.2). Therefore $\mathfrak{D}(\mu)$ is a $\overline{\mathbf{Q}}$-linear combination of

$$(12.13) \qquad (D_\sigma^{e-2p} f_p)(w)(u_\alpha, \cdots, u_\zeta)$$

with $\alpha, \cdots, \zeta$ as above. Now Theorem 8.4 is true if $F = E$, as proved at the end of Section 10. Therefore $\pi^{-|k|} f_p$ is $\overline{\mathbf{Q}}$-rational, and hence $\pi^{-|d|} P_\chi(w)^{-1}(D_\sigma^{e-2p} f_p)(w)$ is $\overline{\mathbf{Q}}$-rational by Theorem 11.2. The definition of $P_\chi(w)$ implies that

$$(D_\sigma^{e-2p} f_p)(w)(u_\alpha, \cdots, u_\zeta) \sim \pi^{|d|} p_K(\Delta - \alpha - \cdots - \zeta, 2\psi) ,$$

which together with (2) of Theorem 1.1 proves the assertion of Theorem 9.2 about $\mathfrak{D}(\mu)$.

13. Analytic continuation

The symbols $\mathscr{P}_\iota$, X, U, ξ, t, etc. being the same as in Section 8 (especially as in (8.3)), we define a function g^* on $\mathfrak{H}_1^n \times \mathscr{B}^r$ by

$$(13.1)$$

$$g^*(z^*, w; t) = \sum_{v \in X} t(v)\overline{S(v, A^{-1}p(w))^k} e\left(\sum_{\mu=1}^n - \xi_\mu\{x_\mu^* S_\mu[v] + iy_\mu^* P_\mu[v]\}\right) ,$$

where $w \in \mathscr{B}^r$, $z^* = (z_1^*, \cdots, z_n^*)$, $z_\mu^* = x_\mu^* + iy_\mu^* \in \mathfrak{H}_1$, $P_\mu = S_\mu$ for $\mu > r$; if $\mu \leq r$, we take P_μ to be the positive form corresponding to $A_\mu^{-1}p(w_\mu)$ considered in the proof of Proposition 8.2 (where we wrote P_i instead of P_μ). Notice that we can include the trivial term $v = 0$ in (13.1). Let $g(z, w; t)$ denote the function on $\mathfrak{H}_1^r \times \mathscr{B}^r$ obtained from $g^*(z^*, w; t)$ by substituting $z_\mu^* = z_\nu$ for $\tau_\mu = \tau_\nu$ on E (i.e., via the embedding $\mathfrak{H}_1^r \to \mathfrak{H}_1^n$ obtained from $E \to F$). Then

$$(13.2)$$

$$g(z, w; t) = \sum_{v \in X} t(v)\overline{S(v, A^{-1}p(w))^k} e\left(\sum_{\nu=1}^r \{x_\nu \mathrm{Tr}_{F/E}(-\xi S[v])^{\tau_\nu} + iy_\nu R_\nu[v]\}\right) ,$$

where $z = (z_1, \cdots, z_r)$, $z_\nu = x_\nu + iy_\nu \in \mathfrak{H}_1$, and R_ν is defined by (8.13). Take $\Omega \in \mathfrak{M}_{l,E}^\delta$ as in (8.4). Neither Ω nor g is an "automorphic form", unless $\delta = 0$ or some conditions on t are imposed. However, they are at least invariant under $z \mapsto z + b$ for every b in a fractional ideal $\mathfrak{b}$ of E. If L denotes a fundamental domain for $\mathbf{R}^r/\mathfrak{b}$ and D_E the discriminant of E, we have

$$(13.3) \qquad D_E^{-1/2} N(\mathfrak{b})^{-1} \int_L \Omega(x + iy)g(x + iy, w; t)dx = \sum_{v \in X} \omega(\mathrm{Tr}_{F/E}(\xi S[v]))$$

$$\times t(v)\overline{S(v, A^{-1}p(w))^k} e\left(\sum_{\nu=1}^r iy_\nu\{R_\nu[v] + \mathrm{Tr}_{F/E}(\xi S[v])^{\tau_\nu}\}\right) .$$

By (2.2), (2.13), and (8.14), we have

$$R_\nu[v] + \mathrm{Tr}_{F/E}(\xi S[v])^{\tau_\nu} = \left| 4\xi^{\tau_\nu}\eta(w_\nu)^{-1}S_\nu(v_\nu, A_\nu^{-1}p(w_\nu))^2 \right| .$$

Let M be a fundamental domain for

$$\{(y_1, \cdots, y_r) \in \mathbf{R}^r \mid y_\nu > 0\}/\{\varepsilon^2 \mid \varepsilon \in U\} .$$

In view of (8.6) and (8.7), we obtain

$$(13.4) \quad D_E^{-1/2} N(\mathfrak{b})^{-1} \int_M \int_L \Omega(x + iy)g(x + iy, w; t)y^{k+(s-1)\mathbf{1}}\,dx\,dy$$

$$= (8\pi)^{-|k|-rs}N(\xi)^{-s}\prod_{\nu=1}^r \xi_\nu^{-k_\nu \tau_\nu}\eta(w_\nu)^{s+k_\nu}\Gamma(s + k_\nu)f(w, s; t, \xi, X) .$$

Thus our task is to show the analytic continuation of this integral. We reduce the question to the case where Ω and g are "automorphic forms" as follows. Let $\mathscr{P}'_\kappa$ denote the elements of $\mathscr{P}_\kappa$ which are S_μ-spherical on V_μ for each $\mu > r$. Every element t of $\mathscr{P}_\kappa$ can be written in the form $\sum_a S[v]^a \sigma_a(v)$ with $\sigma_a \in \mathscr{P}'_{\kappa-2a}$, $a = (a_{r+1}, \cdots, a_n) \in \mathbf{Z}^{n-r}$. Since $A_\nu^{-1}p(w_\nu)$ is a vector of type (7.5), $g^*(z^*, w; \sigma_a)$ is a function of type (7.6). By Proposition 7.1, there is a congruence subgroup Γ^* of $\mathrm{SL}_2(F)$ such that

$$g^*\big(\gamma(z^*), w; \sigma_a\big) = g^*(z^*, w; \sigma_a)\left(\frac{2c}{d\mathfrak{r}_F}\right)^m (cz^* + d)^h(c\bar{z}^* + d)^{k+1}$$

$$\text{for} \quad \gamma = \begin{pmatrix} * & * \\ c & d \end{pmatrix} \in \Gamma^* ,$$

where $k + 1 = (k_1 + 1, \cdots, k_r + 1, 0, \cdots, 0)$, $h_\mu = m/2$ for $\mu \leqq r$, and $h_\mu = \kappa_\mu - 2a_\mu + (m + 2)/2$ for $\mu > r$. Observe that

$$g^*\big(z^*, w; S[v]^a\sigma_a(v)\big) = (-2\pi i\xi)^{-a}(\partial/\partial z^*)^a g^*(z^*, w; \sigma_a) .$$

By Proposition 7.8, the right-hand side is a linear combination of $(y^*)^{b-a}\delta_h^b g^*(z^*, w; \sigma_a)$, where δ_h^λ is defined with respect to the variables $z^*_{r+1}, \cdots, z^*_n$. Similarly Ω is a linear combination of $y^{t-u}\delta_p^t\Lambda$ with $\Lambda \in \mathfrak{M}_{p,E}$, $p = l + 2\delta - 2u$. Let $G(z, w; \sigma_a)$ be the function obtained from $\delta_h^b g^*(z^*, w; \sigma_a)$ by substituting $z_\mu^* = z_\nu$ for $\tau_\mu = \tau_\nu$ on E. Thus it is sufficient to prove the analytic continuation of (13.4) with Ω and g replaced by $y^{t-u}\delta_p^t\Lambda$ and $y^{b-a}G$, that is,

$$(13.5) \qquad\qquad \int_M \int_L H(z, w, s)y^{-2\mathbf{1}}\,dx\,dy ,$$

$$H(z, w, s) = y^{k-\beta+(s+1)\mathbf{1}}(\delta_p^t\Lambda)(z)G(z, w; \sigma_a)$$

with $0 \leqq \beta \in \mathbf{Z}^r$. Observe that

$$H\big(\gamma(z), w, s\big) = H(z, w, s)(cz + d)^{-2\mathbf{1}}\left(\frac{c\bar{z} + d}{cz + d}\right)^\beta |cz + d|^{-2s\mathbf{1}}$$

$$\text{for all} \quad \gamma = \begin{pmatrix} * & * \\ c & d \end{pmatrix} \in \Gamma ,$$

where Γ is a congruence subgroup of $\mathrm{SL}_2(E)$ and

$$\lambda = k_\nu - l_\nu - 2\delta_\nu - \sum_{j=2}^{q}\kappa_{\nu j} - (mq/2) - q + 2 .$$

By (8.7) and (8.15), this is an integer independent of ν. We may assume that

$$\Gamma = \left\{ \begin{pmatrix} a & b \\ c & d \end{pmatrix} \in \mathrm{SL}_2(E) \,|\, a \equiv d \equiv 1 \,(\mathrm{mod}\,\mathfrak{c}),\ b \in \mathfrak{b},\ c \in \mathfrak{b}^{-1}\mathfrak{c} \right\}$$

with $\mathfrak{b}$ as above and an integral ideal $\mathfrak{c}$. Choosing a suitable $\mathfrak{c}$ and changing U for a subgroup, we may assume that

$$U = \{ u \in \mathfrak{r}^\times \,|\, u \equiv 1 \,(\mathrm{mod}\,\mathfrak{c}) \} ,$$

where $\mathfrak{r}$ denotes the maximal order of E. Now $L \times M$ is a fundamental domain for $\Gamma_\infty \backslash \mathfrak{H}_1^r$, where $\Gamma_\infty = \left\{ \begin{pmatrix} * & * \\ 0 & * \end{pmatrix} \in \Gamma \right\}$. Let Y be the set of all pairs (c, d) such that $c \in \mathfrak{b}^{-1}\mathfrak{c}$, $d \equiv 1 \,(\mathrm{mod}\,\mathfrak{c})$, $c\mathfrak{b} + d\mathfrak{r} = \mathfrak{r}$. Let $u \in U$ act on Y by $(c, d)u = (cu, du)$. Then $\Gamma_\infty \begin{pmatrix} a & b \\ c & d \end{pmatrix} \mapsto (c, d)U$ gives a bijection of $\Gamma_\infty \backslash \Gamma$ onto Y/U. Put

$$\mathfrak{E}(z, s) = \sum_{(c,d) \in Y/U} (cz + d)^{-\lambda}\left(\frac{c\bar{z} + d}{cz + d}\right)^{\beta} |cz + d|^{-2s_1} .$$

Let Φ be a fundamental domain for $\Gamma \backslash \mathfrak{H}_1^r$. Then (13.5) is equal to

$$(13.6) \qquad \int_{\Phi} H(z, w, s)\mathfrak{E}(z, s)y^{-\lambda 1}\,dx\,dy .$$

It is well known that $\mathfrak{E}$, as a function of s, can be continued to a meromorphic function on the whole plane. Moreover, the above integral converges absolutely whenever $\mathfrak{E}$ is holomorphic in s, and defines a meromorphic function on the whole plane. This follows from the expansion of $\mathfrak{E}$ in terms of confluent hypergeometric functions and its behavior at cusps. (Cf. [17, Prop. 3.2].) Thus we obtain Theorem 8.3. (In [17, § 3], we discussed $\mathfrak{E}$ assuming $\lambda \geqq 0$ and $\beta \geqq 0$. The function $\mathfrak{E}$ with arbitrary λ and β can be transformed to that case by substituting $s - (\lambda/2)$ for s or by changing w_ν for $\bar{w}_\nu$.)

We now consider the special case where $\Omega \in \mathfrak{M}_{l,E}$ and $t \in \mathscr{P}'_r$. Then our integral multiplied by a constant and $\eta(w)^{-k-s1}$ is

$$(13.7a) \qquad \eta(w)^{-k-s1}\int_{\Phi} \Omega(z)\mathfrak{E}(z, s)g(z, w; t)y^{k+(s-1)1}\,dx\,dy ,$$

$$(13.7b) \qquad \mathfrak{E}(z, s) = \sum_{(c,d) \in Y/U} (cz + d)^{-\lambda 1}|cz + d|^{-2s1} .$$

Suppose $\lambda > 0$. Then $\mathfrak{E}$ is holomorphic at $s = 0$; moreover $\mathfrak{E}(z, 0) \in \mathfrak{M}_{\lambda 1, E}$ if either $\lambda \neq 2$ or $r > 1$. Anyway, the value of (13.7a) at $s = 0$ is $f(w, 0; t, \xi, X)$ times a nonzero constant. Let us now show that this is a holomorphic modular form in w, at least when Ω is a cusp form and

(13.8) $l_1 = \cdots = l_r$, $\lambda > 0$ and $k_\nu - \sum_{j=2}^q \kappa_{\nu j} > (m + 2)q$; $\lambda \neq 2$ or $r > 1$.
Assume these and put $\mu = \lambda + l_\nu$. Then $\Omega(z)\mathfrak{E}(z, 0) \in \mathfrak{M}_{\mu_1}(\Gamma)$. The desired result is included in

THEOREM 13.1. *Let* $k \in \mathbf{Z}^r$, $\mu \in 2^{-1}\mathbf{Z}$, *and* $t \in \mathscr{P}'_\kappa$; *suppose*

(13.10a) $\mu = k_\nu - \sum_{j=2}^q \kappa_{\nu j} - (mq/2) - q + 2$ $(\nu = 1, \cdots, r)$,

(13.10b) $k_\nu - \sum_{j=2}^q \kappa_{\nu j} > (m + 2)q$ $(\nu = 1, \cdots, r)$.

For a cusp form $\varphi \in \mathfrak{M}_{\mu_1, E}(\Gamma)$, *put*

(13.11) $f(w; t) = \eta(w)^{-k} \int_\Phi \varphi(z)g(z, w; t)y^{k-1}dxdy$ $(w \in \mathfrak{Z}^r,\ \Phi = \Gamma\backslash\mathfrak{H}_1^r)$.

With a $\overline{\mathbf{Q}}$*-rational basis* $\{t_1, \cdots, t_d\}$ *of* $\mathscr{P}'_\kappa$ *over* $\mathbf{C}$, *put*

$${}^t\mathfrak{f}(w) = \left(f(w; t_1),\ \cdots,\ f(w; t_d)\right) ,$$

and define a representation $\sigma': \prod_{\nu > r} G(S_\nu) \longrightarrow \mathrm{GL}_d(\mathbf{C})$ *by*

$$(t_1 \circ \alpha,\ \cdots,\ t_d \circ \alpha) = (t_1,\ \cdots,\ t_d) \cdot {}^t\sigma'(\alpha_{r+1},\ \cdots,\ \alpha_n) .$$

Then $f \in \mathfrak{M}_\sigma$ *with a representation*

$$\sigma(a_1,\ \cdots,\ a_r,\ \beta_1,\ \cdots,\ \beta_r,\ \gamma_{r+1},\ \cdots,\ \gamma_n) = a^k\sigma'(\gamma_{r+1},\ \cdots,\ \gamma_n) .$$

Moreover, if $F = E$ *and* φ *is* $\overline{\mathbf{Q}}$*-rational, so is* $f(w; 1)$.

Proof. For $0 \ll a \in \mathfrak{b}^{-1}\mathfrak{b}^{-1}$, define a Poincaré series $P_a(z)$ by

$$P_a(z) = \sum_{\gamma \in \Gamma_\infty / \Gamma} T_a\big(\gamma(z)\big)J(\gamma, z)^{-1},\ \ T_a(z) = \sum_{u \in U} e_E(u^2 az) ,$$

where J is the factor of automorphy for the elements of $\mathfrak{M}_{\mu_1}(\Gamma)$. Since $\mu > 2$, P_a is a cusp form of $\mathfrak{M}_{\mu_1}(\Gamma)$, and every cusp form of $\mathfrak{M}_{\mu_1}(\Gamma)$ is a finite linear combination of such P_a (see Maass [3]). Therefore, to prove the holomorphy of $f(w, t)$ in w, we may assume that $\varphi = P_a$. Put

$$f_a(w; t) = \sum_v t(v)S(v, A^{-1}p(w))^{-k} ,$$

where the sum is taken over all $v \in X$ such that $\mathrm{Tr}_{F/E}(\xi S[v]) = a$. The proof of Proposition 8.2 shows that this is convergent under (13.10b); hence f_a is holomorphic in w. Now taking $e_E(az)$ in place of Ω in (13.3) and (13.4), we find

$$\int_M \int_L T_a(z)g(z, w; t)y^{k-1}dxdy = c\pi^{-|k|}\eta(w)^k f_a(w; t)$$

with $0 \neq c \in \overline{\mathbf{Q}}$. Since $g(z, w; t)$ is a function of type (7.6), the left side can be transformed into

$$\int_\Phi P_a(z)g(z, w; t)y^{k-1}dxdy .$$

This proves the desired holomorphy. Obviously a formula similar to (8.16) holds for $f_a(w; t)$, and hence $\mathfrak{f} \in \mathfrak{M}_\sigma$. To show that $f(w; 1)$ is arithmetic when $F = E$, specialize w to the fixed point w_0 of $h(Y^u)$ with $Y = K + F^m$ and $[K : F] = 2$ as in Lemma 10.1. For the reason explained in Section 10, our problem is reduced to the assertion that $f(w_0; 1) \sim p_K(\Psi, 2\psi)$ with $\Psi = \sum_{i=1}^n k_i \psi_i$ when X has form (10.5). To see this, with ε, S', B, and C as in (10.4) and (10.5), put

$$\alpha(z) = \sum_{b \in B} b^{\rho\Psi} e_F(2\varepsilon\xi bb^\rho z) , \qquad \beta(z) = \sum_{c \in C} e_F(-\xi S'[c]z) \qquad (z \in \mathfrak{H}_1^n) .$$

By (10.3), we have $g(z, w_0; 1) = \zeta \cdot \overline{\alpha(z)}\beta(z)$ with $\zeta \in \overline{\mathbf{Q}}$. Observe that $\beta \in \mathfrak{M}_{m1/2}(\overline{\mathbf{Q}})$ and $\alpha \in \mathfrak{L}(K, \Psi, \overline{\mathbf{Q}})$. Now we have

$$f(w_0; 1) = \zeta \cdot \eta(w_0)^{-k} \int_\Phi \varphi(z)\beta(z)\overline{\alpha(z)} y^{k-1} dx dy = \zeta'\pi^n \langle \alpha, \varphi\beta \rangle$$

with $\zeta' \in \overline{\mathbf{Q}}$. Therefore the desired result follows from Theorem 9.8.

Remark 13.2. That the integral of type (13.11) produces a holomorphic automorphic form with respect to an orthogonal group, when $F = E = \mathbf{Q}$, has been shown by Shintani, Niwa, Zagier, Oda, Rallis and Schiffman; the reader may find more complete references in [4] and [5]. In particular, it is proved in these papers that the $\mathbf{Q}_{ab}$-rationality of (13.11) follows from that of φ when $F = E = \mathbf{Q}$, S represents 0, and some conditions on the weight are imposed. Actually the $\mathbf{Q}_{ab}$-rationality when $F = E = \mathbf{Q}$ can be derived by our methods, even if S is anisotropic. In fact, the above argument shows the K_{ab}-rationality of $f(w; 1)$ for an imaginary quadratic field K. Changing K for another imaginary quadratic field, the technique used for establishing canonical models allows us to obtain the desired rationality over $\mathbf{Q}_{ab}$.

It remains to prove the analytic continuation of (9.2). Let K be a CM-field such that $[K : F] = 2$, and η an element of F such that $\eta^{\tau_\nu} > 0$ or < 0 according as $\nu \leq r$ or $\nu > r$. Put $V = K$ and consider this a vector space over F. Define an F-valued form S on V by

$$S(a, b) = -\operatorname{Tr}_{K/F}(\eta a b^\rho) \qquad (a, b \in V = K).$$

We can now apply the above methods to this case by simply putting $m = 0$. To be more explicit, let $\psi_1, \cdots, \psi_r$ be extensions of $\tau_1, \cdots, \tau_r$ to K. Put

$$\psi = \sum_{\nu=1}^r \psi_\nu, \ \Delta = \sum_{\nu=1}^r k_\nu \psi_\nu , \qquad \Theta = \sum_{\nu=1}^r b_\nu(\psi_\nu + \psi_\nu\rho) ,$$

and

$$g(z^*; t) = \sum_{v \in X} t(v) v^{\Theta + \Delta} e\left(-2\sum_{\mu=1}^n v_\mu v_\mu^\rho(x_\mu^* \eta_\mu + i y_\mu^* \eta_\mu')\right) ,$$

where $\eta_\mu' = -\eta_\mu$ or η_μ according as $\mu \leq r$ or $\mu > r$. Composing this with the embedding of $\mathfrak{H}_1^r$ into $\mathfrak{H}_1^n$, define a function $g(z; t)$ on $\mathfrak{H}_1^r$. Then we find, for $\Omega \in \mathfrak{M}_{l, E}$,

$$(13.12) \qquad D_E^{-1/2} N(\mathfrak{b})^{-1} \int_M \int_L \Omega(z) g(z;t) y^{k+(s-1)\mathbf{1}} dx dy$$

$$= (8\pi)^{-|k|-rs} \prod_{\nu=1}^r \eta_\nu^{-k_\nu - s} \Gamma(k_\nu + s) \sum_{v \in X/U} \omega\left(\mathrm{Tr}_{K/E}(\eta v v^\rho)\right) t(v) v^{\Theta - \rho \Delta} |v^\psi|^{-2s} .$$

By the same methods as those used for (13.4), we find that (13.12) can be continued to a meromorphic function on the whole plane. (We need $\partial/\partial \bar{z}_\mu^*$ for $\mu \leqq r$ to obtain v^ϑ.) Now $\mathcal{D}(s)$ of (9.2) is exactly the series of (13.12), though we have k, E, and $t(v)v^\Theta$ instead of d, F, and v^Φ; the present F corresponds to the maximal real subfield of K there. This completes the proof of the first part of Theorem 9.2.

14. The case of unitary groups

Our theorems 8.3 and 8.4 have natural analogues for arithmetic subgroups of $U(m, 1)^r \times U(m+1)^{n-r}$, which we shall now discuss briefly. Let K be a CM-field such that $[K:F]=2$. We choose and fix, throughout this section, n elements $\sigma_1, \cdots, \sigma_n$ of J_K which are extensions of $\tau_1, \cdots, \tau_n$. With an element $H = {}^t H^\rho \in \mathrm{GL}_{m+1}(K)$, we define a group $G(H)$ and its subgroup $G_+(H)$ by

$$(14.1\text{a}) \qquad G(H) = \{\alpha \in \mathrm{GL}_{m+1}(K) \,|\, \alpha H \cdot {}^t\alpha^\rho = \nu(a) H \text{ with } \nu(\alpha) \in F\} ,$$

$$(14.1\text{b}) \qquad G_+(H) = \{\alpha \in G(H) \,|\, \nu(\alpha) \gg 0\} .$$

Take an element Q_ν of $\mathrm{GL}_{m+1}(\overline{\mathbf{Q}})$ so that

$$(14.2) \qquad Q_\nu^\rho \cdot \mathrm{diag}\,[1_{p_\nu}, -1_{q_\nu}]\,{}^t Q_\nu = H^{\sigma_\nu} ,$$

where (p_ν, q_ν) is the signature of H^{σ_ν}. Then the map

$$(14.3) \qquad \alpha \longmapsto (Q_\nu^{-1} \alpha^{\sigma_\nu \rho} Q_\nu)_{\nu=1,\cdots,n}$$

gives an injection of $G(H)$ into $\prod_{\nu=1}^n G(p_\nu, q_\nu)$, where

$$G(p, q) = \{\beta \in \mathrm{GL}_{p+q}(\mathbf{C}) \,|\, {}^t\bar{\beta} J_{p,q} \beta = \nu(\beta) J_{p,q} \text{ with } \nu(\beta) \in \mathbf{R}\} ,$$

$$J_{p,q} = \mathrm{diag}\,[1_p, -1_q].$$

Define bounded symmetric domains $\mathcal{D}$ and $D(p, q)$ by

$$(14.4) \qquad \mathcal{D} = \prod_{\nu=1}^n D(p_\nu, q_\nu), \quad D(p, q) = \{z \in \mathbf{C}_q^p \,|\, 1_q - {}^t\bar{z}z > 0\} .$$

For $\beta \in G(p, q)$, $\nu(\beta) > 0$ and $z \in D(p, q)$, we can define $\beta(z)$ by

$$(14.5) \qquad \beta \begin{bmatrix} 1_p & z \\ {}^t\bar{z} & 1_q \end{bmatrix} = \begin{bmatrix} 1_p & \beta(z) \\ {}^t\overline{\beta(z)} & 1_q \end{bmatrix} \begin{bmatrix} \overline{\lambda(\beta, z)} & 0 \\ 0 & \mu(\beta, z) \end{bmatrix} ,$$

with $\lambda(\beta, z) \in \mathrm{GL}_p(\mathbf{C})$ and $\mu(\beta, z) \in \mathrm{GL}_q(\mathbf{C})$. (For details, the reader is referred to [18, §§ 3, 4].) Obviously λ and μ are holomorphic factors of automorphy. We can make this meaningful even in the case $pq = 0$ by putting

$$(14.6\text{a}) \qquad \lambda(\beta, z) = \bar{\beta} \qquad \text{if} \quad q = 0 ,$$

(14.6b) $$\mu(\beta, z) = \beta \qquad \text{if} \quad p = 0 ;$$

we understand that $D(p, q)$ consists of a single point if $pq = 0$, and let $G(p, q)$ act on it trivially. Now we define the action of $G_+(H)$ on $\mathfrak{D}$ through (14.3). For $\alpha \in G_+(H)$ and $z = (z_1, \cdots, z_n) \in \mathfrak{D}$, we put

(14.7) $$\lambda_\nu(\alpha, z) = \lambda(Q_\nu^{-1} \alpha^{\sigma_\nu \rho} Q_\nu, z_\nu) , \qquad \mu_\nu(\alpha, z) = \mu(Q_\nu^{-1} \alpha^{\sigma_\nu \rho} Q_\nu, z_\nu) .$$

Take a $\overline{\mathbf{Q}}$-rational representation $\zeta \colon \prod_{\nu=1}^{n}(\mathrm{GL}(p_\nu) \times \mathrm{GL}(q_\nu)) \to \mathrm{GL}(W)$, where W is a finite dimensional complex vector space with $\overline{\mathbf{Q}}$-structure, and put

(14.8) $$\Lambda_\zeta(\alpha, z) = \zeta(\lambda_1(\alpha, z), \mu_1(\alpha, z), \cdots, \lambda_n(a, z), \mu_n(\alpha, z)) .$$

Denote by $\mathfrak{M}_\zeta$ the set of all W-valued functions f holomorphic on $\mathfrak{D}$ (and at the cusps) such that $f(\gamma(z)) = \Lambda_\zeta(\gamma, z)f(z)$ for all γ in a congruence subgroup of $G_+(H)$.

To define the $\overline{\mathbf{Q}}$-rationality, let $Y = M_{n_1}(L_1) \oplus \cdots \oplus M_{n_s}(L_s)$ with CM-fields L_i containing K, and let δ be a positive involution of Y. Suppose there is a K-linear embedding h of Y into $M_{m+1}(K)$ such that $h(1_Y) = 1_{m+1}$ and $h(y^\delta)H = H \cdot {}^t h(y)^\rho$ for all $y \in Y$. Put

(14.9) $$L = L_1 \oplus \cdots \oplus L_s , \qquad L^\mu = \{x \in L \,|\, xx^\delta = 1\} ,$$

and suppose $m + 1 = \sum_{i=1}^{s} n_i[L_i \colon K]$. Then $h(L^\mu)$ is contained in $G_+(H)$ and has a unique common fixed point, say w, on $\mathfrak{D}$. Now we can define CM-types ψ_i of L_i so that

$$\prod_{i=1}^{s} a^{n_i \psi_i} = \prod_{\nu=1}^{n} \det(\lambda_\nu(h(a), w)) \det(\mu_\nu(h(a), w)) \quad \text{for} \quad a \in L^\mu .$$

(See [18, §4] for details.) Let ψ be the element of I_L whose L_i-component is ψ_i; then $\mathfrak{p}_L(\psi)$ is meaningful (see §1). We call an element f of $\mathfrak{M}_\zeta$ $\overline{\mathbf{Q}}$-rational if $\Lambda_\zeta(h(\mathfrak{p}_L(\psi)), w)^{-1}f(w)$ is $\overline{\mathbf{Q}}$-rational for every such point w.

To construct an analogue of (8.3) in the present setting, let us now assume that $q_\nu = 1$ for $\nu \leq r$ and $q_\nu = 0$ for $\nu > r$, with a positive integer $r \leq n$; also assume the existence of a subfield E of F satisfying (8.1a, b). For $x \in K_{m+1}^1$ put

(14.10a) $$x^{\sigma_\nu \rho} Q_\nu \begin{bmatrix} 1_m & z_\nu \\ {}^t \bar{z}_\nu & 1 \end{bmatrix} = \overline{(\lambda_\nu(x, z), \mu_\nu(x, z))} \qquad (\nu \leq r),$$

(14.10b) $$x^{\sigma_\nu} Q_\nu^\circ = \lambda_\nu(x, z) \qquad (\nu > r)$$

with $\mu_\nu(x, z) \in \mathbf{C}$ and $\lambda_\nu(x, z) \in \mathbf{C}_m^1$ or $\mathbf{C}_{m+1}^1$ according as $\nu \leq r$ or $\nu > r$. Then

$$\lambda_\nu(x\alpha, z) = \lambda_\nu(x, \alpha(z))\lambda_\nu(\alpha, z), \quad \mu_\nu(x\alpha, z) = \mu_\nu(x, \alpha(z))\mu_\nu(\alpha, z) .$$

Now we define a column vector $f(z, s)$ by

(14.11) $${}^t f(z, s) = M_U \sum_{x \in X/U} \omega(\mathrm{Tr}_{F/E}(\xi x H \cdot {}^t x^\rho))\lambda(x, z)^k \mu(x, z)^{-k} |\mu(x, z)|^{-2s} .$$

Here $z \in \mathfrak{D} = D(m, 1)^r$, $s \in \mathbb{C}$; X is a congruence subset of $K_{m+1}^1 - \{0\}$; U, M_U, ξ and ω are the same as in (8.3); $k = (k_1, \cdots, k_r) \in \mathbb{Z}^r$, $h = (h_1, \cdots, h_n) \in \mathbb{Z}^n$, $k_\nu > 0$, $h_\nu \geqq 0$;

$$\mu(x, z)^{-k} |\mu(x, z)|^{-2s} = \prod_{\nu=1}^{r} \mu_\nu(x, z)^{-k_\nu} |\mu_\nu(x, z)|^{-2s} ,$$

$$\lambda(x, z)^h = \underbrace{\lambda_1(x, z) \otimes \cdots \otimes \lambda_1(x, z)}_{h_1 \text{ copies}} \otimes \cdots \otimes \underbrace{\lambda_n(x, z) \otimes \cdots \otimes \lambda_n(x, z)}_{h_n \text{ copies}} .$$

Define $\tau_{\nu j}$ for $\nu \leqq r$ and $j = 1, \cdots, q$ as in Section 8; put $h_{\nu j} = h_\kappa$ if $\tau_{\nu j} = \tau_\kappa$ on E. The series is well-defined if

(14.12) $\qquad\qquad k_\nu - l_\nu - 2\delta_\nu - \sum_{j=1}^{q} h_{\nu j}$ is independent of ν.

Call this number β. The series is absolutely convergent at $s = 0$ under

(14.13) $\qquad\qquad \beta > \begin{cases} 2q(m + 1) + 2\theta & \text{if } \omega \text{ satisfies (8.10)}, \\ 2q(m + 1) - 2 & \text{if } \Omega = 1. \end{cases}$

Assume this and put $\mathfrak{f}(z) = f(z, 0)$. We see that $\mathfrak{f} \in \mathfrak{M}_\zeta$ where

$$\zeta(a_1, b_1, \cdots, a_r, b_r, a_{r+1}, \cdots, a_n)$$
$$= b_1^{k_1} \cdots b_r^{k_r} \cdot \underbrace{{}^t a_1^{-1} \otimes \cdots \otimes {}^t a_1^{-1}}_{h_1 \text{ copies}} \otimes \cdots \otimes \underbrace{{}^t a_n^{-1} \otimes \cdots \otimes {}^t a_n^{-1}}_{h_n \text{ copies}} .$$

THEOREM 14.1. *Suppose* $l \in \mathbb{Z}^r$. *Then* $f(z, s)$ *as a function of* s *can be continued to a meromorphic function on the whole plane. Moreover, if the coefficients* $\omega(a)$ *are algebraic, then, under* (14.12) *and* (14.13),

$$\pi^{-|k|} p_K\left(\textstyle\sum_{\nu=1}^{r} k_\nu \sigma_\nu + \sum_{\nu=1}^{n} h_\nu \sigma_\nu, -\sum_{\nu=1}^{n} \sigma_\nu\right)\mathfrak{f}$$

is $\overline{\mathbb{Q}}$*-rational.*

We can derive the $\overline{\mathbb{Q}}$-rationality from Theorem 9.2 by the same methods as in Section 10. The analytic continuation as well as an analogue of Proposition 13.1 can be proved in exactly the same fashion as in Section 13. Since no essentially new ideas are necessary, we leave the details to the reader.

15. Concluding remarks and open problems

1. Series (8.3) and (14.11) include certain "Eisenstein series" as special cases. In fact, suppose $F = E$ and H is isotropic, and take $\Omega = 1$. Then (14.11) becomes the sum extended over all $x \in X/U$ satisfying $xH \cdot {}^t x^\rho = 0$. For the reason explained in [16, Prop. 5.1], this may be called an Eisenstein series. The same interpretation can be made for (8.3) when $F = E$ and S is isotropic. Now in [16], we proved by a different method the K_{ab}-rationality of $f(z, 0)$ when $n = 1$, $m > 1$ and $\Omega = 1$. The same technique is applicable

to a more general case if H is isotropic and also to the orthogonal case under a certain condition on S.

Langlands proved in [2] the analytic continuation as well as the functional equation of Eisenstein series in a very general setting. If $\Omega = 1$, $F = E$, and S and H are isotropic, the assertions of Theorems 8.3 and 14.1 concerning the analytic continuation seem to follow from his results.

2. Our main theorems are not stated in the best possible form. First of all, the value of (8.3) at $s = 0$ can define a $\overline{\mathbf{Q}}$-rational holomorphic modular form on $\mathscr{Z}^r$ under a weaker condition than (8.20a, b), as proved in Proposition 13.1 in a special case. The same type of results hold undoubtedly for both orthogonal and unitary groups in a more general case. Similarly one should be able to find a smaller lower bound for the integers μ where $\mathfrak{D}(s)$ is evaluated in Theorem 9.2. This problem obviously requires a more careful analysis of $\mathfrak{D}$. In fact, we have proved only the analytic continuation of (8.3) and (9.2). Their functional equations can be "theoretically" derived from that of $\mathfrak{E}(z, s)$ in (13.6), but seem rather complicated. It is desirable to establish them in transparent forms.

3. Condition (8.15) which we needed in Theorems 8.3 and 8.4 is not absolutely necessary. In fact, one can prove the arithmeticity without it, at least when $E = \mathbf{Q}$. To do this, one first has to study the special values of $L(s; f, g)$ defined by (9.9) when g has a half-integral weight. The algebraicity of such special values was obtained by Sturm [22] in the elliptic modular case. Therefore, if $E = \mathbf{Q}$, we can obtain the analogues of Theorems 9.2 and 14.1 when $l \notin \mathbf{Z}^r$ as well as those of Theorems 8.3 and 8.4 when $2l_\nu + 1 \equiv mq \pmod 2$. There is no doubt that the same holds for an arbitrary E.

4. Our proof of analytic continuation in Section 13 suggests the following generalization. With E as in (8.1a, b), let f and h be elements of $\mathfrak{M}_{k,F}$ and $\mathfrak{M}_{l,E}$ with expansions

$$f(z^*) = \sum_{\xi \in F} a(\xi) e_F(\xi z^*), \quad h(z) = \sum_{\eta \in E} b(\eta) e_E(\eta z) \qquad (z^* \in \mathfrak{H}_1^n, \ z \in \mathfrak{H}_1^r)$$

respectively, and let δ be an element of F such that $\delta^{\tau_\nu} > 0$ if and only if $\nu \leqq r$. Put

$$g(z_1, \cdots, z_r) = f(\delta^{\tau_1} z_1^*, \cdots, \delta^{\tau_n} z_n^*),$$

where

$$z_\lambda^* = \begin{cases} z_\lambda & \text{if } \lambda \leqq r, \\ \bar{z}_\nu & \text{if } \tau_\lambda = \tau_\nu \text{ on } E \text{ and } \lambda > r. \end{cases}$$

Taking h and $\bar{g}$ in place of Ω and g in (13.12), we can easily prove that a series $\mathfrak{L}$ defined by

$$(15.1) \qquad \mathfrak{L}(s) = \sum_{\xi \in F/U} b\big(\mathrm{Tr}_{F/E}(\delta\xi)\big)\,\overline{a(\xi)}\prod_{\nu=1}^{r}(\xi^{\tau_\nu})^{-m_\nu - s}$$

can be continued to a meromorphic function on the whole plane, where $2m_\nu = l_\nu + \sum_{j=1}^{g} k_{\nu j}$ and U is a subgroup of U_E of finite index. Series (9.2) is actually a special case of (15.1) if $\Omega \in \mathfrak{M}_l$, $b_i = 1$, and Φ contains no pair of complex conjugate embeddings. It is an open question whether the special values of (15.1) have a property of algebraicity.

PRINCETON UNIVERSITY, PRINCETON, NEW JERSEY

REFERENCES

[1] E. HECKE, *Vorlesungen über die Theorie der algebraischen Zahlen*, Leipzig, 1923.

[2] R. LANGLANDS, On the functional equations satisfied by Eisenstein series, Lecture notes in mathematics 544, Springer, 1976.

[3] H. MAASS, Zur Theorie der automorphen Funktionen von n Veränderlichen, Math. Ann. **117** (1940), 538-578.

[4] T. ODA, On modular forms associated with indefinite quadratic forms of signature $(2, n-2)$, Math. Ann. **231** (1977), 97-144.

[5] S. RALLIS and G. SCHIFFMANN, On a relation between $\widetilde{SL}_2$ cusp forms and cusp forms on tube domains associated to orthogonal groups, preprint, 1977.

[6] K. SHIH, Existence of certain canonical models, Duke Math. J. **45** (1978), 63-66.

[7] ————, Construction of arithmetic automorphic functions for special Clifford groups, to appear.

[8] G. SHIMURA, Algebraic number fields and symplectic discontinuous groups, Ann. of Math. **86** (1967), 503-592.

[9] ————, On canonical models of arithmetic quotients of bounded symmetric domains, Ann. of Math. **91** (1970), 144-222; II, **92** (1970), 528-549.

[10] ————, On some arithmetic properties of modular forms of one and several variables, Ann. of Math. **102** (1975), 491-515.

[11] ————, Theta functions with complex multiplication, Duke Math. J. **43** (1976), 673-696.

[12] ————, The special values of the zeta functions associated with cusp forms, Commun. Pure and App. Math. **29** (1976), 783-804.

[13] ————, Unitary groups and theta functions, Proc. Int. Symp. on Alg. N. Th. Kyoto 1976 (1977), 195-200.

[14] ————, On the derivatives of theta functions and modular forms, Duke Math. J. **44** (1977), 365-387.

[15] ————, On certain reciprocity laws for theta functions and modular forms, Acta Math. **141** (1978), 35-71.

[16] ————, The arithmetic of automorphic forms with respect to a unitary group, Ann. of Math. **107** (1978), 569-605.

[17] ————, The special values of the zeta functions associated with Hilbert modular forms. Duke Math. J. **45** (1978), 637-679.

[18] ————, Automorphic forms and the periods of abelian varieties, J. Math. Soc. Japan, **31** (1979), 561-592.

[19] ————, On some problems of algebraicity, Proc. Int. Congress of Math. Helsinki 1978.

[20] T. SHINTANI, On construction of holomorphic cusp forms of half integral weight, Nagoya Math. J. **58** (1975), 83-126.

[21] C. SIEGEL, Indefinite quadratische Formen und Funktionentheorie I, II, Math. Ann. **124** (1951), 17-54; **124** (1952), 364-387.

[22] J. STURM, Special values of zeta functions and Eisenstein series of half integral weight, to appear in Amer. J. Math.

(Received March 27, 1979)

The critical values of certain zeta functions associated with modular forms of half-integral weight

Journal of the Mathematical Society of Japan, 33 (1981), 649-672

Introduction.

The main theme of this paper is the algebraicity of two types of numbers occurring in connection with modular forms. One is the values at certain integers or half-integers of a Dirichlet series

$$D(s, f, g)=\sum_{n=1}^{\infty} a(n)b(n)n^{-s}$$

obtained from two modular forms

$$f(z)=\sum_{n=1}^{\infty} a(n)e^{2\pi i n z} \quad \text{and} \quad g(z)=\sum_{n=0}^{\infty} b(n)e^{2\pi i n z},$$

and the other is the inner product $\langle f, g \rangle$ of f and g, when they have the same weight. These have been treated in our previous papers [12], [13], and [14] for the forms f and g of integral weight. Therefore the present investigation concerns the cases in which either or both of f and g have half-integral weight.

To describe the nature of the problems as well as of the results, we let m' and m denote the weights of f and g, respectively, which are positive elements of $2^{-1}Z$. If $m=m'$, there is a well known relation between $\langle f, g \rangle$ and the residue of $D(s, f, g)$ at $s=m$, and therefore the second object has a character similar to the first one. Setting this residue aside, we restrict, for some natural reasons, the study of the values of $D(s, f, g)$ to the case $m<m'$. If we exclude the case in which both m and m' are integers, there are three cases:

I. $m \in Z$ and $m' \in Z$,
II. $m \in Z$ and $m' \notin Z$,
III. $m \notin Z$ and $m' \notin Z$.

The first case was treated by Sturm [17], [18]. Therefore we consider here the remaining two cases. Let $m'=k/2$ with an odd integer k, and assume that f

* Supported by NSF Grant MCS 7903631.

is a cusp form which is an eigenfunction of the operators $T(p^2)$ introduced in [9] for almost all primes p; put $f|T(p^2)=c_p f$. The main theorem of [9] shows that there is a modular form F of integral weight $k-1$ such that $F|T(p)=c_p F$ for all such p, where $T(p)$ is the ordinary Hecke operator of degree p and of weight $k-1$. To obtain the algebraicity, we naturally assume that $a(n)$ and $b(n)$ are all algebraic numbers. We shall then show that both $\langle f, g \rangle$ (when $m=m'=k/2$) and the values of $D(s/2, f, g)$ (when $m<m'$) for certain integers s are algebraic numbers times $\pi^{n/2}u_-(F)$, where n is an integer depending on s, m, m', and $u_-(F)$ is a "period" of F. Actually we prove a stronger result which tells the behavior of the algebraic numbers under automorphisms.

The methods of proof are, for the most part, the same as in our previous papers [12], [13], [14] and also in Sturm [17]. However, we are confronted by a new difficulty, since no satisfactory theory of "primitive forms" has been developed in the case of half-integral weight. Such a theory in the case of integral weight was needed in order to obtain the most general result without any artificial conditions. Therefore one of the main technical objectives of this work is to show that the concept of "primitive form" is unnecessary, at least for our present purposes. In fact, we shall obtain our results for an *arbitrary eigenfunction f of $T(p^2)$ for almost all p*. (We can assume, however, that F is primitive.) The dispensability of the concept of primitive form is significant, because it will make the study of the critical values of the zeta functions associated with modular forms of other types practicable. In fact, our technique in the proof of Theorem 1 seems applicable, for example, to the Siegel modular forms. As another technical necessity dictated by our methods, we have to analyse carefully the behavior of modular forms under the action of automorphisms of C on the Fourier coefficients.

Since both inner products and the values of zeta functions are natural objects of arithmetic interest, their investigation hardly needs justification. It may be added, however, that there is an interesting application of our results to the theory of algebraic curves uniformized by automorphic functions of arithmetic type, which the author hopes to treat in a subsequent paper. Though the lemmas are of course preliminaries to the main theorems, some of them (Lemmas 9 and 10, for example) may be of independent interest.

1. Statement of the main results.

We denote by $GL_2^+(R)$ the subgroup of $GL_2(R)$ consisting of the elements with positive determinant, and let every element $\alpha=\begin{pmatrix} a & b \\ c & d \end{pmatrix}$ of $GL_2^+(R)$ act on the upper half plane

$$H = \{z \in C \,|\, \mathrm{Im}(z) > 0\}$$

by the rule $\alpha(z) = (az+b)/(cz+d)$. For every positive integer N, we define congruence subgroups $\Gamma_0(N)$, $\Gamma_1(N)$, and $\Gamma(N)$ of $SL_2(Z)$ by

$$\Gamma_0(N) = \left\{ \begin{pmatrix} a & b \\ c & d \end{pmatrix} \in SL_2(Z) \,\middle|\, c \equiv 0 \;(\mathrm{mod}\; N) \right\},$$

$$\Gamma_1(N) = \left\{ \begin{pmatrix} a & b \\ c & d \end{pmatrix} \in \Gamma_0(N) \,\middle|\, a \equiv d \equiv 1 \;(\mathrm{mod}\; N) \right\},$$

$$\Gamma(N) = \left\{ \begin{pmatrix} a & b \\ c & d \end{pmatrix} \in \Gamma_1(N) \,\middle|\, b \equiv 0 \;(\mathrm{mod}\; N) \right\}.$$

We introduce a factor of automorphy $j(\gamma, z)$ of weight $1/2$ for $\gamma \in \Gamma_0(4)$ by

$$(1.1a) \qquad \theta(\gamma(z)) = j(\gamma, z)\theta(z) \qquad (\gamma \in \Gamma_0(4)),$$

$$(1.1b) \qquad \theta(z) = \sum_{n=-\infty}^{\infty} e(n^2 z).$$

Here and throughout the paper, we put

$$e(z) = \exp(2\pi i z) \qquad (z \in C).$$

We recall that

$$j\left(\begin{pmatrix} a & b \\ c & d \end{pmatrix}, z\right)^2 = \left(\frac{-1}{d}\right)(cz+d) \qquad (\text{see [9]}).$$

Given a non-negative element λ of $2^{-1}Z$, an element $\gamma = \begin{pmatrix} * & * \\ c & d \end{pmatrix}$ of $SL_2(Z)$, and a function f on H, we define a function $f|_\lambda \gamma$ on H by

$$(1.2) \qquad (f|_\lambda \gamma)(z) = \begin{cases} f(\gamma z)(cz+d)^{-\lambda} & \text{if } \lambda \in Z, \\ f(\gamma z)j(\gamma, z)^{-2\lambda} & \text{if } \lambda \notin Z. \end{cases}$$

We naturally assume that $\gamma \in \Gamma_0(4)$ when $\lambda \notin Z$. For a congruence subgroup Γ of $SL_2(Z)$, which we assume to be contained in $\Gamma_0(4)$ if $\lambda \notin Z$, we denote by $\mathcal{M}(\lambda, \Gamma)$ the set of all modular forms of weight λ with respect to Γ, that is, holomorphic functions f on H such that

$$(1.3a) \qquad f|_\lambda \gamma = f \quad \text{for all } \gamma \in \Gamma,$$

$$(1.3b) \qquad f \text{ is holomorphic at every cusp.}$$

We call such an f a *cusp form* if f vanishes at every cusp, and denote by $S(\lambda, \Gamma)$ the set of all cusp forms of $\mathcal{M}(\lambda, \Gamma)$. The reader is referred to [9] for the precise meaning of these cusp conditions. We then put

$$\mathcal{M}(\lambda, N) = \mathcal{M}(\lambda, \Gamma_1(N)), \quad \mathcal{S}(\lambda, N) = \mathcal{S}(\lambda, \Gamma_1(N)),$$

$$\mathcal{M}(\lambda) = \bigcup_{N=1}^{\infty} \mathcal{M}(\lambda, \Gamma(N)), \quad \mathcal{S}(\lambda) = \bigcup_{N=1}^{\infty} \mathcal{S}(\lambda, \Gamma(N))$$

$(4|N$ if $\lambda \in \mathbf{Z})$. Given a character χ modulo N, we put

$$\mathcal{M}(\lambda, N, \chi) = \{f \in \mathcal{M}(\lambda, N) \mid \chi(d_\gamma)f = f|_\lambda\gamma \text{ for every } \gamma \in \Gamma_0(N)\},$$

$$\mathcal{S}(\lambda, N, \chi) = \mathcal{S}(\lambda) \cap \mathcal{M}(\lambda, N, \chi).$$

Here and throughout the paper, d_γ denotes the lower right entry of γ. We always assume

(1.4a) $$\chi(-1) = (-1)^\lambda \quad \text{if } \lambda \in \mathbf{Z},$$

(1.4b) $$\chi(-1) = 1 \quad \text{and} \quad 4|N \quad \text{if } \lambda \notin \mathbf{Z},$$

because $\mathcal{M}(\lambda, N, \chi) = \{0\}$ otherwise.

Introducing an invariant measure $d\mu(z)$ on H by

$$d\mu(z) = y^{-2}dx\,dy, \quad z = x + iy,$$

we define, for two elements f and g of $\mathcal{M}(\lambda)$, their inner product $\langle f, g \rangle$ by

(1.5) $$\langle f, g \rangle = \mu(\Phi)^{-1} \int_\Phi \overline{f(z)}g(z)y^\lambda d\mu(z),$$

$$\mu(\Phi) = \int_\Phi d\mu(z), \quad \Phi = \Gamma(N) \backslash H,$$

where we take N so that both f and g belong to $\mathcal{M}(\lambda, \Gamma(N))$. To guarantee the convergence of the integral, we assume that either f or g is a cusp form. We note that

(1.6) $$\mu(\Gamma(N) \backslash H) = (\pi/3)[\Gamma(1) : \Gamma(N)\{\pm 1\}].$$

We also define the symbol $\langle f, g \rangle$ by (1.5) for continuous functions f and g such that $f|_\lambda\gamma = f$ and $g|_\lambda\gamma = g$ for all $\gamma \in \Gamma(N)$, whenever the integral is convergent.

Every element f of $\mathcal{M}(\lambda, \Gamma(N))$ has a Fourier expansion

$$f(z) = \sum_{n=0}^{\infty} a(n)e(nz/N).$$

Let $\mathrm{Aut}(C)$ denote the group of all ring-automorphisms of the complex number field C, and let $\sigma \in \mathrm{Aut}(C)$. Then we can define an element f^σ of $\mathcal{M}(\lambda)$ by

$$f^\sigma(z) = \sum_{n=0}^{\infty} a(n)^\sigma e(nz/N).$$

Such an action can be defined even for Siegel modular forms of integral and half-integral weights (see [15]). If f is a cusp form, so is f^σ. This is proved in [14, Proposition 1.8] when $\lambda \in Z$. The case of half-integral weight can be reduced to that of integral weight by taking the square of a given form. In particular, we denote by ρ the complex conjugation, and we put thus

$$(1.7) \qquad f^\rho(z) = \sum_{n=0}^\infty \overline{a(n)} e(nz/N) = \overline{f(-\bar{z})}\,.$$

We see easily that

$$(1.8) \qquad \langle f^\rho, g \rangle = \langle g^\rho, f \rangle\,.$$

We say that f is *rational over* a subfield L of C, or f is *L-rational*, if L contains all the coefficients $a(n)$. This is so if and only if $f^\sigma = f$ for all σ of $\mathrm{Aut}(C)$ which give the identity map on L. Given modular forms $f, g, \cdots$ and Dirichlet characters $\chi, \psi, \cdots$, we denote by $K(f, g, \cdots, \chi, \psi, \cdots)$ the field generated over Q by the Fourier coefficients of $f, g, \cdots$ (at $i\infty$) and the values of $\chi, \psi, \cdots$.

Suppose $\lambda \in Z$; let

$$F(z) = \sum_{n=1}^\infty c(n) e(nz) \in \bigcup_{N=1}^\infty S(\lambda, N)$$

and let ψ be a primitive character modulo a positive integer r. Then we put

$$(1.9a) \qquad D(s, F) = \sum_{n=1}^\infty c(n) n^{-s}\,,$$

$$(1.9b) \qquad D(s, F, \psi) = \sum_{n=1}^\infty \psi(n) c(n) n^{-s}\,.$$

We call F *primitive* if the following conditions are satisfied:

(1.10a)   *F is a common eigenfunction of Hecke operators $T(p)$ for almost all primes p (that is, for all except a finite number of p);*

(1.10b) $c(1) = 1$;

(1.10c)   *There is a positive integer N such that $F \in S(\lambda, N)$ and that*
         $\langle F, h(tz) \rangle = 0$ *for all $h \in S(\lambda, M)$, $tM \mid N$, $M < N$.*

The reader is referred to Miyake [3] for the basic results on primitive forms. We recall in particular that such a form F is an eigenfunction of Hecke operators $T(p)$ of level N for all primes p; $F \mid T(p) = c(p)F$; F is uniquely determined by giving the eigenvalues $c(p)$ for almost all p. Moreover, if F is a primitive element belonging to $S(\lambda, N, \chi)$, then F^σ is primitive and belongs to $S(\lambda, N, \chi^\sigma)$ (see [14, Proposition 2.6]).

Now let k be an odd positive integer. In [9], we defined a certain operator

$T_{k,\chi}^N(p^2)$ on $\mathcal{M}(k/2, N, \chi)$, which we simply write $T_N(p^2)$ or $T(p^2)$ when there is no fear of confusion. Suppose $k \geq 3$ and an element g of $S(k/2, N, \chi)$ is an eigenfunction of $T_N(p^2)$ for all primes p, and let $g|T(p^2)=b(p)g$. Then the main theorem of [9] shows that there is an element h of $\mathcal{M}(k-1, N', \chi^2)$ with some N' such that

$$(1.11) \qquad D(s, h)=\prod_p [1-b(p)p^{-s}+\chi(p)^2 p^{k-2-2s}]^{-1}.$$

Moreover, h is a cusp form if $k \geq 5$. With this result in mind, we introduce a new concept as follows.

Suppose $k \geq 3$. Fix a primitive element F of $S(k-1)$ with eigenvalues $c(p)$ as above, or rather, fix a system of eigenvalues $\{c(p)\}$. We say that an element f of $S(k/2, N, \chi)$ *corresponds to* F, or to $\{c(p)\}$, if $f|T(p^2)=c(p)f$ for almost all p. Then we denote by $S(k/2, N, \chi, F)$ the vector subspace of $S(k/2, N, \chi)$ consisting of all the elements corresponding to F. In §2, we shall prove that this subspace is spanned by $K(\chi, F)$-rational elements, and sent to $S(k/2, N, \chi^\sigma, F^\sigma)$ by σ.

Given an arbitrary primitive element h of $S(\lambda)$ and a primitive character ψ, we treated in our previous papers the values of $D(s, h, \psi)$ for certain integers s. To recall the result, put

$$A(m, h, \psi)=(2\pi i)^{-m}\tau(\psi)^{-1}D(m, h, \psi)$$

for every integer m in the range $0<m<\lambda$, where $\tau(\psi)$ is the Gauss sum defined as usual by

$$\tau(\psi)=\sum_{n=1}^{r}\psi(n)e(n/r) \qquad (r=\text{conductor of } \psi).$$

If ψ' is an imprimitive character associated with ψ, we define its Gauss sum $\tau(\psi')$ to be the same as $\tau(\psi)$. Now notice that the number of all different h^σ with $\sigma \in \mathrm{Aut}(C)$ is $[K(h):Q]$. We proved in [13] (see also [12], [14]) that there exist $2[K(h):Q]$ complex numbers $u_\pm(h^\sigma)$ depending only on h^σ with the following properties:

$(1.12a) \qquad A(m, h, \psi)/u_+(h) \in K(h, \psi) \quad if \ \psi(-1)=(-1)^m,$
$\qquad\qquad A(m, h, \psi)/u_-(h) \in K(h, \psi) \quad if \ \psi(-1)=(-1)^{m-1}$
$\qquad\qquad for \ every \ positive \ integer \ m<\lambda.$

$(1.12b) \qquad [A(m, h, \psi)/u_\pm(h)]^\sigma=A(m, h^\sigma, \psi^\sigma)/u_\pm(h^\sigma) \ for \ every \ \sigma \in \mathrm{Aut}(C),$
$\qquad\qquad where \ the \ choice \ between \ u_+ \ and \ u_- \ is \ given \ as \ in \ (1.12a).$

$(1.12c) \qquad i^{\lambda-1}\pi\tau(\psi)\langle h, h\rangle/[u_+(h)u_-(h)] \in K(h).$

After these preliminaries, our first main result can be stated as follows.

THEOREM 1. *Suppose k is an odd integer ≥ 3. For $f \in S(k/2, N, \chi, F)$ and $g \in S(k/2, N)$, put*

$$(1.13) \qquad\qquad I(f, g) = \pi i \tau(\chi) \langle f, g \rangle.$$

Then $I(f, g)/u_-(F) \in K(f^\rho, g)$. Moreover, for every $\sigma \in \mathrm{Aut}(C)$, we have

$$[I(f, g)/u_-(F)]^\sigma = I(f^{\rho \sigma \rho}, g^\sigma)/u_-(F^\sigma).$$

In particular, if f is $K(F, \chi)$-rational, then $I(f, g)/u_-(F)$ belongs to $K(F, \chi, g)$, and we have, for every $\sigma \in \mathrm{Aut}(C)$,

$$[I(f, g)/u_-(F)]^\sigma = I(f^\sigma, g^\sigma)/u_-(F^\sigma).$$

This will be proved in §4.

To state the second main result, we define, for two modular forms

$$f(z) = \sum_{n=1}^{\infty} a(n)e(nz) \in S(m', N, \chi),$$

$$g(z) = \sum_{n=0}^{\infty} b(n)e(nz) \in \mathcal{M}(m, N, \psi),$$

a Dirichlet series $D(s, f, g)$ by

$$(1.14) \qquad\qquad D(s, f, g) = \sum_{n=1}^{\infty} a(n)b(n)n^{-s}.$$

This is convergent for sufficiently large $\mathrm{Re}(s)$, and can be continued to a meromorphic function on the whole plane. In [12] and [13], we examined the critical values of $D(s, f, g)$ when both m and m' are integers and $m < m'$. Sturm treated the case where $m < m' \in Z$, $m \notin Z$ (see [18]). We consider here the following cases:

   (I) $m' = k/2$ and $m = l/2$ with odd integers k and l; $k > l$, $k \geq 5$.

   (II) $m' = k/2$, $m \in Z$, and $m \leq (k-3)/2$.

Let us now put, in Case (I),

$$\mathcal{D}_N(s, f, g) = L_N(2s + 2 - (k+l)/2, \omega)D(s, f, g),$$

where $\omega(d) = \left(\dfrac{-1}{d}\right)^\lambda \chi(d)\psi(d)$, $\lambda = (k-l)/2$, and

$$(1.15) \qquad\qquad L_N(s, \omega) = \sum_{n>0,\ (n, N)=1} \omega(n)n^{-s};$$

we put subscript N to emphasize the condition $(n, N) = 1$. This notation L_N applies to an arbitrary primitive or imprimitive character ω whose conductor may or may not divide N.

THEOREM 2. *Under assumption* (I), $\Gamma(s)\Gamma(s+1-(l/2))\mathfrak{D}_N(s, f, g)$ *is an entire function. Suppose moreover that* $f \in S(k/2, N, \chi, F)$; *let* n *be an odd integer such that*

$$(1.16) \qquad l \leqq n \leqq k-2,$$

and let

$$B(n, f, g) = i\tau(\psi)^{-1}\pi^{(l-1-2n)/2}\mathfrak{D}_N(n/2, f, g)/u_-(F).$$

Then $B(n, f, g) \in K(f, g)$. *Moreover, for every* $\sigma \in \mathrm{Aut}(C)$, *we have*

$$B(n, f, g)^\sigma = B(n, f^\sigma, g^\sigma).$$

The existence of the factor $\Gamma(s+1-(l/2))$ shows that $\mathfrak{D}_N(n/2, f, g) = 0$ for every odd integer n less than l.

THEOREM 3. *Let* $f \in S(k/2, N, \chi, F)$ *and* $g \in \mathfrak{M}(m, N, \psi)$ *with an odd integer* k *and an integer* $m < k/2$. *Put* $\mu = (k+2m-1)/4$ *and*

$$\mathfrak{D}_N(s, f, g) = L_N(4s-4\mu+2, \chi^2\psi^2)D(s, f, g).$$

Then

$$(s-\mu)\Gamma(s)\Gamma(s+1-m)\Gamma(s-\mu+(1+\varepsilon)/2)\mathfrak{D}_N(s, f, g)$$

is an entire function, where ε *is* 0 *or* 1 *according as* 2μ *is even or odd. The factor* $s-\mu$ *is unnecessary if* 2μ *is odd or if* $\chi^2\psi^2$ *is nontrivial. Moreover, put*

$$J_1 = \begin{cases} \{n \in \mathbf{Z} \mid m+(k-1)/2 \leqq n \leqq k-2,\ n \equiv 1 \pmod 2\} & \textit{if } \chi^2\psi^2 \textit{ is nontrivial} \\ \{n \in \mathbf{Z} \mid m+(k-1)/2 < n \leqq k-2,\ n \equiv 1 \pmod 2\} & \textit{if } \chi^2\psi^2 \textit{ is trivial,} \end{cases}$$

$$J_2 = \{n \in \mathbf{Z} \mid 2m \leqq n \leqq m+(k-3)/2,\ n \equiv 0 \pmod 2\},$$

$$C(n, f, g) = \begin{cases} i\tau(\chi\psi^2)^{-1}\pi^{2m-2n+(k-3)/2}\mathfrak{D}_N(n/2, f, g)/u_-(F) & \textit{for } n \in J_1, \\ i^{m-1}\tau(\psi)^{-1}\pi^{m-n-1}\mathfrak{D}_N(n/2, f, g)/u_-(F) & \textit{for } n \in J_2. \end{cases}$$

Then $C(n, f, g)^\sigma = C(n, f^\sigma, g^\sigma)$ *for every* $\sigma \in \mathrm{Aut}(C)$.

We see from the nature of the gamma factors that

$$\mathfrak{D}_N(n/2, f, g) = 0 \quad \text{if} \quad n \equiv 1 \pmod 2 \quad \text{and} \quad n \leqq m+(k-3)/2,$$

$$\text{or if} \quad n \equiv 0 \pmod 2 \quad \text{and} \quad n \leqq 2m-2.$$

Obviously $J_1 \cup J_2$ is empty unless $k-2m \geqq 3$. These two theorems will be proved in §5.

2. The behavior of modular forms under $\mathrm{Aut}(C)$.

Throughout the rest of the paper, we denote by $\bar{\mathbf{Q}}$ the algebraic closure of $\mathbf{Q}$ and by $\mathbf{Q}_{\mathrm{ab}}$ the maximal abelian extension of $\mathbf{Q}$, both embedded in C; A will denote the adeles of $\mathbf{Q}$.

We introduce a group $\mathcal{G}$ consisting of all couples $(\alpha, \varphi(z))$ where

$$\alpha = \begin{pmatrix} a & b \\ c & d \end{pmatrix} \in GL_2^+(\mathbf{R})$$

and φ is a holomorphic function on H such that $\varphi(z)^2 = t \cdot \det(\alpha)^{-1/2}(cz+d)$ with a complex number t of absolute value 1. The law of multiplication in $\mathcal{G}$ is defined by

$$(\alpha, \varphi(z))(\beta, \psi(z)) = (\alpha\beta, \varphi(\beta(z))\psi(z)).$$

We denote by P the projection map of $\mathcal{G}$ onto $GL_2^+(\mathbf{R})$ which sends (α, φ) to α. For $\xi = (\alpha, \varphi) \in \mathcal{G}$, a function f on H, and an odd integer k, we put

$$f|_{k/2}\xi = f(\alpha(z))\varphi(z)^{-k}.$$

If $f \in \mathcal{M}(k/2)$, we shall often write $f|_{k/2}\xi$ simply $f|\xi$. We can define an injection $\gamma \mapsto \gamma^*$ of $\Gamma_0(4)$ into $\mathcal{G}$ by

$$(2.1) \qquad\qquad \gamma^* = (\gamma, j(\gamma, z)) \qquad (\gamma \in \Gamma_0(4)).$$

Our notation (1.2) shows that $f|_{k/2}\gamma = f|_{k/2}\gamma^*$. Another type of injection can be defined by using the well known function

$$(2.2) \qquad\qquad \eta(z) = e(z/24)\prod_{n=1}^{\infty}(1 - e(nz)).$$

In fact, if we put

$$(2.3a) \qquad\qquad \eta(\gamma(z)) = h(\gamma, z)\eta(z),$$

$$(2.3b) \qquad\qquad \gamma_* = (\gamma, h(\gamma, z)) \qquad (\gamma \in \Gamma(1)),$$

then $\gamma \mapsto \gamma_*$ embeds $\Gamma(1)$ into $\mathcal{G}$. We see that $\gamma_* = \gamma^*$ for $\gamma \in \Gamma(24)$, since $\eta(z)$ belongs to $S(1/2, \Gamma(24))$ (see [9, p. 457]).

LEMMA 1. *Let N be a multiple of 24 and let $\sigma \in \mathrm{Aut}(C)$. Then there is an automorphism $\gamma \mapsto \gamma^\sigma$ of $\Gamma(1)/\Gamma(N)$ such that: (i) $\Gamma_0(M)/\Gamma(N)$ is stable under σ for every divisor M of N; (ii) $f^\sigma|(\gamma^\sigma)_* = (f|\gamma_*)^\sigma$ and $f^\sigma|(\gamma^\sigma)^* = (f|\gamma^*)^\sigma$ for all $f \in \mathcal{M}(k/2, \Gamma(N))$; (iii) γ and γ^σ have the same diagonal entries modulo N.*

PROOF. As shown in [15, Proposition 2.1], $\mathcal{M}(k/2, \Gamma(N))$ is spanned by $\mathbf{Q}$-rational elements. Therefore it is sufficient to prove the existence of an automorphism with the required property (ii) for all $\mathbf{Q}$-rational f. For this purpose we consider the action of the adele group G_{A+} (which is the "positive part" of $GL_2(A)$) on $\mathbf{Q}_{\mathrm{ab}}$-rational modular functions defined in [8, Chapter 6] (see also [14], [15]). Given σ, take $t \in \prod_p Z_p^\times$ so that the action of t on $\mathbf{Q}_{\mathrm{ab}}$ coincides with σ. By the strong approximation theorem, we have

$$(2.4) \qquad \begin{pmatrix} 1 & 0 \\ 0 & t \end{pmatrix} \gamma \begin{pmatrix} 1 & 0 \\ 0 & t^{-1} \end{pmatrix} = u\alpha$$

with $\alpha \in SL_2(Q)$ and $u \in SL_2(A)$ such that $u_p \equiv 1 \pmod{NZ_p}$ for all primes p. Obviously $\alpha \in \Gamma(1)$. Given a Q-rational f of $\mathcal{M}(k/2, \Gamma(N))$, put $g = f/\eta^k$. Then g is a Q-rational modular function of level N, so that $g^u = g$. Since $\begin{pmatrix} 1 & 0 \\ 0 & t \end{pmatrix}$ acts as σ^{-1} on modular forms, we have $(g|\gamma)^\sigma = g^\sigma|\alpha$, and hence $(f|\gamma_*)^\sigma = f^\sigma|\alpha_*$. Similarly, considering f/θ^k instead of f/η^k, we find $(f|\gamma^*)^\sigma = f^\sigma|\alpha^*$ when $\gamma \in \Gamma_0(4)$. Obviously α modulo N is uniquely determined for γ and t by (2.4). Writing γ^σ for α, we obtain our lemma.

The above technique is also applicable to the elements of $\mathcal{G}$ whose projections do not necessarily belong to $\Gamma(1)$. For example, we can prove

LEMMA 2. *For a positive integer M, define an operator ξ_M acting on $\mathcal{M}(\lambda)$ by*

$$f|\xi_M = \begin{cases} f(-1/Mz)(Mz)^{-\lambda} & \text{if } \lambda \in Z, \\ f(-1/Mz)(-iMz)^{-\lambda} & \text{if } \lambda \notin Z, \end{cases}$$

where $(-iMz)^{-\lambda}$ is chosen to be positive on the positive imaginary axis. Let $\sigma \in \mathrm{Aut}(C)$. Then, for every $f \in \mathcal{M}(\lambda, N, \chi)$, we have

$$f^\sigma|\xi_M = \chi(s)^\sigma \left(\frac{2}{s}\right)^{2\lambda} (f|\xi_M)^\sigma,$$

where s is an integer such that $e(1/2N)^\sigma = e(s/2N)$, and $\left(\dfrac{2}{s}\right)$ is the quadratic residue symbol.

Here M may or may not be equal to N. We omit the proof, as this will be necessary only for an alternative proof of certain facts in our proof of Theorem 3.

LEMMA 3. *Let X be a vector space over a subfield K of C with countably many elements, and let $U = X \otimes_K C$. Let $\mathrm{Aut}(C/K)$ denote the group of all automorphisms of C over K, and let $\mathrm{Aut}(C/K)$ act on U by $(\sum_n c_n x_n)^\sigma = \sum_n c_n^\sigma x_n$ for $\sigma \in \mathrm{Aut}(C/K)$, $c_n \in C$, and $x_n \in X$. Suppose V is a finite-dimensional C-linear subspace of U stable under $\mathrm{Aut}(C/K)$. Then $V = (X \cap V) \otimes_K C$.*

PROOF. We can find a finite-dimensional subspace Y of X such that $V \subset Y \otimes_K C$. Then our assertion becomes a special case of a well-known fact that every affine variety stable under $\mathrm{Aut}(C/K)$ is K-rational. One can also prove our lemma in an elementary way.

LEMMA 4. *For $0 < \lambda \in 2^{-1}Z$, the following assertions hold.*

(1) *Let $f \in \mathcal{M}(\lambda)$, $\alpha = \begin{pmatrix} a & b \\ c & d \end{pmatrix} \in GL_2(Q)$, $\det(\alpha) > 0$, and let $g(z) = f(\alpha(z))(cz+d)^{-\lambda}$. Then $K(g) \subset Q_{ab}K(f)$.*

(2) $\mathcal{M}(\lambda, \Gamma)$ *and* $\mathcal{S}(\lambda, \Gamma)$ *are spanned by* **Q**-*rational elements, if* Γ *is either* $\Gamma_1(N)$ *or* $\Gamma(N)$.

(3) $\mathcal{M}(\lambda, N, \chi)$ *and* $\mathcal{S}(\lambda, N, \chi)$ *are spanned by* $K(\chi)$-*rational elements.*

(4) $\mathcal{M}(\lambda, N, \chi)^{\sigma} = \mathcal{M}(\lambda, N, \chi^{\sigma})$ *and* $\mathcal{S}(\lambda, N, \chi)^{\sigma} = \mathcal{S}(\lambda, N, \chi^{\sigma})$ *for every* $\sigma \in \mathrm{Aut}(C)$.

PROOF. If $\lambda \in \mathbf{Z}$, assertions (1) and (4) are included in [14, Propositions 1.4, 1.6, 1.8]. Assertion (1) for $\lambda \notin \mathbf{Z}$ is proved in [11, Proposition 1.5] when $K(f) \subset \mathbf{Q}_{ab}$; the same proof is obviously valid even if $K(f) \not\subset \mathbf{Q}_{ab}$. Assertion (2) for $\mathcal{M}(\lambda, \Gamma)$ is included in [15, Proposition 2.1] if $\Gamma = \Gamma(N)$. The same proof is obviously valid for $\Gamma_1(N)$ also. Since $\mathcal{S}(\lambda)$ is stable under $\mathrm{Aut}(C)$ as mentioned in §1, we obtain assertion (2) for $\mathcal{S}(\lambda, \Gamma)$ from Lemma 3. Assertion (4) for $\lambda = k/2$ can be proved in a straightforward way by means of Lemma 1. Then assertion (3) follows from Lemma 3.

We now recall the definition of $T(p^2)$. Fixing a multiple N of 4, put $\Gamma = \Gamma_0(N)$ and denote by Δ the image of Γ under the map $\gamma \mapsto \gamma^*$. Further we consider an element ξ of $\mathcal{G}$ defined by

$$\xi = (\alpha, p^{1/2}), \quad \alpha = \begin{pmatrix} 1 & 0 \\ 0 & p^2 \end{pmatrix}$$

with a prime p and then a coset decomposition

$$\Gamma \alpha \Gamma = \amalg_{\nu} \Gamma \alpha_{\nu}, \quad \alpha_{\nu} = \begin{pmatrix} a_{\nu} & b_{\nu} \\ c_{\nu} & d_{\nu} \end{pmatrix}.$$

As shown in [9], we have $\Delta \xi \Delta = \amalg_{\nu} \Delta \xi_{\nu}$ with elements ξ_{ν} such that $P(\xi_{\nu}) = \alpha_{\nu}$. Now, for $f \in \mathcal{M}(k/2, N, \chi)$, we put

$$(2.5) \qquad f \,|\, T_N(p^2) = p^{(k/2)-2} \sum_{\nu} \chi(a_{\nu}) f \,|\, \xi_{\nu}.$$

LEMMA 5. *Let* p *be a prime not dividing* N. *Then*

$$\chi(p^2) \langle f \,|\, T(p^2), g \rangle = \langle f, g \,|\, T(p^2) \rangle$$

for every $f, g \in \mathcal{S}(k/2, N, \chi)$.

PROOF. Observe that

$$(2.6) \qquad \langle f \,|\, \delta, g \rangle = \langle f, g \,|\, \delta^{-1} \rangle \qquad \text{for every } \delta \in \mathcal{G}.$$

Therefore

$$\langle f \,|\, T(p^2), g \rangle = p^{(k/2)-2} \langle \sum_{\nu} \chi(a_{\nu}) f \,|\, \xi_{\nu}, g \rangle = p^{(k/2)-2} \langle f, \sum_{\nu} \chi(a_{\nu})^{-1} g \,|\, \xi_{\nu}^{-1} \rangle.$$

By [8, Lemma 3.5], we can choose α_{ν} so that $\Gamma \alpha \Gamma = \amalg_{\nu} \Gamma \alpha_{\nu} = \amalg_{\nu} \alpha_{\nu} \Gamma$. Then $\Delta \xi^{-1} \Delta = \amalg_{\nu} \Delta \xi_{\nu}^{-1}$. Put

$$\eta = \left(\begin{pmatrix} p^2 & 0 \\ 0 & 1 \end{pmatrix}, p^{-1/2} \right), \quad \zeta = \left(p^2 \begin{pmatrix} 1 & 0 \\ 0 & 1 \end{pmatrix}, 1 \right).$$

As shown in the proof of [9, Theorem 1.7], we have $\xi^{-1}\zeta=\eta\in\Delta\xi\Delta$, and hence $\Delta\xi\Delta=\Delta\xi^{-1}\zeta\Delta=\amalg_\nu\Delta\xi_\nu^{-1}\zeta$. Since $P(\xi_\nu^{-1}\zeta)=\begin{pmatrix} d_\nu & * \\ * & a_\nu \end{pmatrix}$, we have

$$g\,|\,T(p^2)=p^{(k/2)-2}\sum_\nu\chi(d_\nu)g\,|\,\xi_\nu^{-1}=p^{(k/2)-2}\chi(p^2)\sum_\nu\chi(a_\nu)^{-1}g\,|\,\xi_\nu^{-1},$$

which proves our assertion.

This shows that the operators $\chi(p)^{-1}T(p^2)$ for p prime to N are hermitian on $S(k/2, N, \chi)$; since they commute, we can find an orthogonal basis of $S(k/2, N, \chi)$ consisting of eigenforms for all p prime to N. We also observe that $S(k/2, N, \chi, F)$ is orthogonal to $S(k/2, N, \chi, F')$ if $F\neq F'$.

LEMMA 6. *Suppose k is an odd integer >1. If $S(k/2, N, \chi, F)\neq \{0\}$ with a primitive F, then F belongs to $S(k-1, N')$ with a divisor N' of a power of N. Moreover, if $F\,|\,T(p)=c(p)F$ for $p\nmid N$, then $f\,|\,T(p^2)=c(p)f$ for all $f\in S(k/2, N, \chi, F)$ and all $p\nmid N$.*

PROOF. As remarked above, $S(k/2, N, \chi)$ has an orthogonal basis $\{g_1, \cdots, g_n\}$ consisting of eigenfunctions of $T_N(p^2)$ for all p prime to N. We take it so that $g_1, \cdots, g_m\in S(k/2, N, \chi, F)$ and $g_{m+1}, \cdots, g_n\notin S(k/2, N, \chi, F)$. We see easily that no nontrivial linear combination of $g_{m+1}, \cdots, g_n$ belongs to $S(k/2, N, \chi, F)$. Therefore $g_1, \cdots, g_m$ span $S(k/2, N, \chi, F)$. Let $g_1\,|\,T(p^2)=b(p)g_1$ for $p\nmid N$, and let S_1 be the set of all elements f of $S(k/2, N, \chi)$ such that $f\,|\,T(p^2)=b(p)f$ for all $p\nmid N$. Since S_1 is stable under $T_N(p^2)$ for all p, it contains an eigenfunction f_0 of $T_N(p^2)$ for all p; let $f_0\,|\,T_N(p^2)=b(p)f_0$ for every p (including the prime factors of N). The main result of [9] shows that there is an element h of $\mathcal{M}(k-1, N', \chi^2)$ such that $D(s, h)$ is given by (1.11), where N' can be taken to be a multiple of N which divides a power of N; $h\in S(k-1)$ if $k>3$. Since $D(s, h)$ has an Euler product, h is an eigenfunction of all Hecke operators of level N'; in particular $h\,|\,T(p)=b(p)h$ for all $p\nmid N$. Since $F\,|\,T(p)=b(p)F$ for almost all p, we see that h is a cusp form even when $k=3$. Therefore the "exact level" of F divides N', and hence $F\,|\,T(p)=b(p)F$ for all $p\nmid N$. Applying this result also to $g_2, \cdots, g_m$, we obtain our assertion. Notice that if $k>3$, the same reasoning shows that each of $g_{m+1}, \cdots, g_n$ belongs to $S(k/2, N, \chi, F')$ with some F'.

We mention here Shintani [16], Niwa [4], Gelbart and Piatetski-Shapiro [2], and Flicker [1] as investigations related to the correspondence between $S(k/2)$ and $S(k-1)$.

LEMMA 7. (1) *$S(k/2, N, \chi, F)$ is spanned by $K(\chi, F)$-rational elements.*

(2) *$S(k/2, N, \chi, F)^\sigma=S(k/2, N, \chi^\sigma, F^\sigma)$ for every $\sigma\in\mathrm{Aut}(C)$.*

(3) *Let $\mathcal{T}(k/2, N, \chi, F)$ be the orthogonal complement of $S(k/2, N, \chi, F)$ in $S(k/2)$. Then $\mathcal{T}(k/2, N, \chi, F)^\sigma=\mathcal{T}(k/2, N, \chi^\sigma, F^\sigma)$.*

(4) *If $k\geq 5$, $S(k/2, N, \chi)$ is the direct sum of subspaces $S(k/2, N, \chi, F)$ for*

finitely many F's.

PROOF. We first observe that

$$(2.7) \qquad\qquad (f|T(p^2))^\sigma = f^\sigma|T(p^2).$$

This follows from [9, Theorem 1.7] and Lemma 4, (4). Now take $g_1, \cdots, g_n$ as in the proof of Lemma 6 for a fixed F. Put

$$\mathcal{S}_0(k/2, N, \chi, F) = \mathcal{S}(k/2, N, \chi) \cap \mathcal{S}(k/2, N, \chi, F).$$

Then $\mathcal{S}_0(k/2, N, \chi, F)$ is spanned by $g_{m+1}, \cdots, g_n$. By (2.7), $g_1^\sigma, \cdots, g_n^\sigma$ are eigenfunctions of $T(p^2)$ for all p prime to N in $\mathcal{S}(k/2, N, \chi^\sigma)$, and obviously $g_1^\sigma, \cdots, g_m^\sigma$ belong to $\mathcal{S}(k/2, N, \chi^\sigma, F^\sigma)$. Let $F|T(p)=c(p)F$. If $i>m$, we have $g_i|T(p^2)=c'(p)g_i$ with $c'(p)\neq c(p)$ for infinitely many p. Then, by (2.7), g_i^σ does not belong to $\mathcal{S}(k/2, N, \chi^\sigma, F^\sigma)$; moreover, g_i^σ is orthogonal to $\mathcal{S}(k/2, N, \chi^\sigma, F^\sigma)$ by virtue of Lemma 5. Therefore $g_{m+1}^\sigma, \cdots, g_n^\sigma$ span $\mathcal{S}_0(k/2, N, \chi^\sigma, F^\sigma)$. This proves (2) and $\mathcal{S}_0(k/2, N, \chi, F)^\sigma = \mathcal{S}_0(k/2, N, \chi^\sigma, F^\sigma)$. Then (1) follows from Lemma 3. Assertion (4) follows from the remark at the end of the proof of Lemma 6. To complete the proof of (3), denote by $\mathcal{U}(k/2, M, \chi)$ the orthogonal complement of $\mathcal{S}(k/2, N, \chi)$ in $\mathcal{S}(k/2, \Gamma(M))$ for every multiple M of N. Obviously $\mathcal{S}(k/2, N, \chi, F)$ is the sum of $\mathcal{S}_0(k/2, N, \chi, F)$ and $\mathcal{U}(k/2, M, \chi)$ for all multiples M of N. Thus our task is to show that $\mathcal{U}(k/2, M, \chi)^\sigma = \mathcal{U}(k/2, M, \chi^\sigma)$. Observe that $\mathcal{U}(k/2, M, \chi)$ is stable under $\Gamma_0(N)$. Taking a basis $\{f_1, \cdots, f_m\}$ of $\mathcal{U}(k/2, M, \chi)$, we can define a representation $(a_{ij}(\gamma))$ of $\Gamma_0(N)$ (or rather, of $\Gamma_0(N)/\Gamma(M)$) by $f_i|\gamma = \sum_j a_{ij}(\gamma)f_j$. Since this does not contain the representation $\gamma \mapsto \chi(d_\gamma)$, we have

$$\sum_{\gamma \in R} \chi(d_\gamma)^\rho a_{ij}(\gamma) = 0 \qquad (i, j = 1, \cdots, m),$$

where $R = \Gamma_0(N)/\Gamma(M)$, and ρ denotes the complex conjugation. Given $\sigma \in \mathrm{Aut}(C)$, let $\gamma \mapsto \gamma^\sigma$ be the automorphism of $\Gamma_0(N)/\Gamma(M)$ of Lemma 1. If $h \in \mathcal{S}(k/2, N, \chi^\sigma)$, we have

$$[\Gamma_0(N):\Gamma(M)]\langle h, f_i^\sigma\rangle = \sum_{\gamma \in R} \chi(d_\gamma)^{\sigma\rho}\chi(d_\gamma)^\sigma \langle h, f_i^\sigma\rangle$$

$$= \sum_{\gamma \in R} \chi(d_\gamma)^{\sigma\rho}\langle h|(\gamma^\sigma)^{-1}, f_i^\sigma\rangle = \sum_{\gamma \in R} \chi(d_\gamma)^{\rho\sigma}\langle h, f_i^\sigma|\gamma^\sigma\rangle$$

$$= \sum_j \sum_{\gamma \in R} \chi(d_\gamma)^{\rho\sigma} a_{ij}(\gamma)^\sigma \langle h, f_j^\sigma\rangle = 0.$$

Therefore $f_1^\sigma, \cdots, f_m^\sigma$ belong to $\mathcal{U}(k/2, M, \chi^\sigma)$. This completes the proof.

LEMMA 8. *Suppose* $0 \neq f \in \mathcal{S}(k/2, N, \chi)$. *Then* $K(\chi) \subset K(f)$. *If moreover* $f \in \mathcal{S}(k/2, N, \chi, F)$, *then* $K(\chi, F) \subset K(f)$.

This follows from (4) of Lemma 4 and (2) of Lemma 7.

3. Eisenstein series and differential operators.

LEMMA 9. *Let g be a C-valued continuous function on H such that $g|_{k/2}\gamma = g$ for every $\gamma \in \Gamma$ with a congruence subgroup Γ of $\Gamma_0(4)$. Suppose that $\langle f, g\rangle$ defined by (1.5) is meaningful for all $f \in S(k/2)$. Then there is an element h of $S(k/2, \Gamma)$ uniquely determined by g, such that $\langle f, g\rangle = \langle f, h\rangle$ for all $f \in S(k/2, \Gamma)$. Moreover this h satisfies $\langle f, g\rangle = \langle f, h\rangle$ for all $f \in S(k/2)$.*

We call h *the essential part of* g.

PROOF. The first assertion is obvious. Take N so that $\Gamma(N) \subset \Gamma$. Then there exists an element p of $S(k/2, \Gamma(N))$ such that $\langle f, g\rangle = \langle f, p\rangle$ for all $f \in S(k/2, \Gamma(N))$. Let $\gamma \in \Gamma$. Then $S(k/2, \Gamma(N))$ is stable under $f \mapsto f|\gamma$, and therefore

$$\langle f, p|\gamma\rangle = \langle f|\gamma^{-1}, p\rangle = \langle f|\gamma^{-1}, g\rangle = \langle f, g|\gamma\rangle = \langle f, g\rangle = \langle f, p\rangle$$

for all $f \in S(k/2, \Gamma(N))$. Hence $p|\gamma = p$, i.e., $p \in S(k/2, \Gamma)$. By the uniqueness of h, we have $h = p$, which proves our lemma.

REMARK. The same type of result obviously holds for the forms of integral weight, and even for modular forms on a higher-dimensional domain. In fact, it is a purely algebraic statement concerning a bilinear form on a vector space on which a system of groups acts under some conditions.

For a congruence subgroup Γ of $\Gamma_0(4)$, define now a subspace $\mathcal{E}(k/2, \Gamma)$ of $\mathcal{M}(k/2, \Gamma)$ by

$$\mathcal{E}(k/2, \Gamma) = \{g \in \mathcal{M}(k/2, \Gamma) \mid \langle g, f\rangle = 0 \text{ for all } f \in S(k/2, \Gamma)\}.$$

Obviously $\mathcal{M}(k/2, \Gamma) = S(k/2, \Gamma) \oplus \mathcal{E}(k/2, \Gamma)$. Moreover, taking $h = 0$ and g in $\mathcal{E}(k/2, \Gamma)$ in the above lemma, we find that

$$(3.1) \qquad \mathcal{E}(k/2, \Gamma) = \mathcal{M}(k/2, \Gamma) \cap \mathcal{E}(k/2, \Gamma') \qquad \text{if } \Gamma' \subset \Gamma.$$

LEMMA 10. *Suppose $k \neq 3$. Then $\mathcal{E}(k/2, \Gamma_1(N))$ is spanned by its $\mathbf{Q}$-rational elements.*

PROOF. If $k = 1$, this is a direct consequence of a result of Serre and Stark [7, Theorem 6]. Suppose $k > 3$. Let us first prove our assertion for $\Gamma(N)$ instead of $\Gamma_1(N)$. Fix a positive integer $N > 2$ and denote by R a complete set of representatives of $\Gamma_{N,\infty} \backslash \Gamma(N)$, where

$$(3.2) \qquad \Gamma_{N,\infty} = \left\{ \begin{pmatrix} 1 & mN \\ 0 & 1 \end{pmatrix} \,\middle|\, m \in \mathbf{Z} \right\}.$$

Define $h(\gamma, z)$ by (2.3) and put

$$G(z) = \sum_{\gamma \in R} h(\gamma, z)^{-k},$$

$$G_\alpha(z) = G(\alpha(z))h(\alpha, z)^{-k} \qquad (\alpha \in \Gamma(1)).$$

Also denote by $\mathcal{M}'_N$ the set of all elements f of $\mathcal{M}(k/2)$ such that $f(\gamma(z)) = h(\gamma, z)^k f$ for every $\gamma \in \Gamma(N)$, and put $S'_N = \mathcal{M}'_N \cap S(k/2)$. Now a standard argument (originally due to Petersson) shows that S'_N is orthogonal to G, and hence to G_α for all $\alpha \in \Gamma(1)$. Moreover G_α vanishes at all cusps of $\Gamma(N)$ except at $\alpha^{-1}(\infty)$; therefore every element of $\mathcal{M}'_N$ is a linear combination of G_α plus a cusp form. (For details, see Petersson [5, Satz 2], [6].) Therefore if we put

$$\mathcal{E}'_N = \{f \in \mathcal{M}'_N \mid \langle f, S'_N \rangle = 0\},$$

then $\mathcal{E}'_N$ is spanned by G_α for all $\alpha \in \Gamma(1)$. Suppose now $24 \mid N$, and put, for every character ω modulo N,

$$(3.3a) \qquad G(z, \omega) = \sum_{\gamma \in W} \omega(d_\gamma) j(\gamma, z)^{-k},$$

$$(3.3b) \qquad G^*(z, \omega) = G(-1/z, \omega)(-iz)^{-k/2},$$

where W is a complete set of representatives of $\Gamma_{1,\infty} \backslash \Gamma_0(N)$. Obviously (3.3a) vanishes if $\omega(-1) = -1$; so we consider only those characters ω satisfying $\omega(-1) = 1$. Observe that the sum of $G(z, \omega)$ for all such ω is a positive integer times G. Now the Fourier expansion of G^* was studied in [10]. In fact,

$$G^*(z, \omega) = 2e(k/8)N^{-k/2}E^*(z/N, 0, -k, \omega)$$

with the function E^* of [10, p. 86]. As shown on [10, p. 89], if we put

$$G^*(z, \omega) = \sum_{n=1}^{\infty} B(n, \omega)e(nz/N)$$

and $n = tm^2$ with a square-free t, then

$$N^{k/2}\Gamma(k/2)L_N(k-1, \omega^2)B(n, \omega)$$
$$= 2n^{(k/2)-1}(2\pi)^{k/2}L_N(\lambda, \omega_t)\sum_{a,b}\mu(a)\omega_t(a)\omega(b)^2 a^{-\lambda}b^{2-k},$$

where $\lambda = (k-1)/2$, μ is the Moebius function, ω_t is a primitive character such that $\omega_t(a) = \left(\dfrac{-1}{a}\right)^\lambda\left(\dfrac{tN}{a}\right)\omega(a)$ for $(a, tN) = 1$, and the sum $\sum_{a,b}$ is extended over all positive integers a, b prime to N such that ab divides m. For Dirichlet characters $\chi_1, \cdots, \chi_n$, put

$$(3.4) \qquad [\chi_1, \cdots, \chi_n] = \tau(\chi_1) \cdots \tau(\chi_n)/\tau(\chi_1 \cdots \chi_n),$$

$$(3.5) \qquad P_N(n, \chi_1) = (2\pi i)^{-n}\tau(\chi_1)^{-1}L_N(n, \chi_1) \text{ for } 0 < n \in \mathbf{Z}, \chi_1(-1) = (-1)^n.$$

It can easily be shown that, for every $\sigma \in \mathrm{Aut}(C)$,

$$(3.6a) \qquad [\chi_1, \cdots, \chi_m]^\sigma = [\chi_1^\sigma, \cdots, \chi_m^\sigma],$$

(3.6b)
$$P_N(n, \chi_1)^\sigma = P_N(n, \chi_1^\sigma)$$

(see [14, Lemma 4.12, Proposition 3.1]). Putting $\varphi_t(a) = \left(\dfrac{-1}{a}\right)^\lambda \left(\dfrac{tN}{a}\right)$, we see that

$$\sqrt{2}\,\tau(\omega)P_N(2\lambda, \omega^2)B(n, \omega)$$

$$= A(k, N, n)[\omega, \omega][\varphi_t, \omega]^{-1}P_N(\lambda, \omega_t)\sum_{a,b}\mu(a)\omega_t(a)\omega(b)^2 a^{-\lambda}b^{2-k}$$

with a rational number $A(k, N, n)$ depending only on k, N, n. This combined with (3.6a, b) shows that

(3.7)
$$[\sqrt{2}\,\tau(\omega)G^*(z, \omega)]^\sigma = \sqrt{2}\,\tau(\omega^\sigma)G^*(z, \omega^\sigma)$$

for every $\sigma \in \mathrm{Aut}(C)$. Now $G(z, \omega)$ belongs to $\mathcal{M}(k/2, N)$; again a standard argument shows that it is orthogonal to $\mathcal{S}(k/2, N, \chi)$ for every χ, and hence it belongs to $\mathcal{E}(k/2, \Gamma_1(N))$. Therefore $G^*(z, \omega)$ belongs to $\mathcal{E}(k/2, \Gamma(N))$. Since $\mathcal{E}'_N = \mathcal{E}(k/2, \Gamma(N))$ if $24|N$, this shows that $\mathcal{E}(k/2, \Gamma(N))$ is generated by $G^*(\gamma(z), \omega)h(\gamma, z)^{-k}$ for all $\gamma \in \Gamma(1)$ and all ω. This combined with Lemma 1 and (3.7) shows that $\mathcal{E}(k/2, \Gamma(N))$ is stable under $\mathrm{Aut}(C)$. So far we have assumed $24|N$. Given an arbitrary multiple M of 4, let N be a common multiple of 24 and M. By (3.1), we have

$$\mathcal{E}(k/2, \Gamma_1(M)) = \mathcal{M}(k/2, \Gamma_1(M)) \cap \mathcal{E}(k/2, \Gamma(N)),$$

and hence $\mathcal{E}(k/2, \Gamma_1(M))$ is stable under $\mathrm{Aut}(C)$. Therefore it has a **Q**-rational basis by virtue of Lemma 3.

The corresponding fact for all integral weights $\geqq 1$ can be derived by the same (and actually easier) argument from the well known results of Hecke on Eisenstein series. It is conjecturable that the above lemma is true also for $k = 3$.

We now introduce a differential operator δ_t and δ_t^r for $t \in C$ and $0 \leqq r \in Z$ by

(3.8a)
$$\delta_t^r = \delta_{t+2r-2} \cdots \delta_{t+2}\delta_t,$$

(3.8b)
$$\delta_t = (2\pi i)^{-1}[t(2iy)^{-1} + (\partial/\partial z)] \qquad \text{(cf. [12], [14]).}$$

We see easily that

$$\delta_\lambda^r(f|_\lambda\gamma) = (\delta_\lambda^r f)|_{\lambda+2r}\gamma . \qquad (\lambda \in 2^{-1}Z, \ \gamma \in \Gamma_0(4)).$$

Let $\mathcal{A}_\lambda$ denote the set of all functions g on H of the form

(3.9)
$$g = \sum_{0 \leqq r \leqq \lambda/2} \delta_{\lambda-2r}^r g_r \qquad \text{with} \quad g_r \in \bigcup_{N=1}^\infty \mathcal{M}(\lambda-2r, N).$$

Then the following facts can easily be proved (see [12]):

(3.10)
$$\textit{The } g_r \textit{ are unique for } g.$$

(3.11)
$$\textit{If } f \in \mathcal{A}_\lambda \textit{ and } g \in \mathcal{A}_\mu, \textit{ then } fg \in \mathcal{A}_{\lambda+\mu}$$

For $\sigma \in \mathrm{Aut}(C)$ and g as in (3.9), we define g^σ by

$$(3.12) \qquad g^\sigma = \sum_r \delta_{\lambda-2r}^r g_r^\sigma .$$

This can be defined also as follows: g has an expression

$$g(z) = \sum_{0 \leq s \leq \lambda/2} (-4\pi y)^{-s} \sum_{n=0}^\infty c_s(n)e(nz)$$

with $c_s(n) \in C$; then

$$(3.13) \qquad g^\sigma(z) = \sum_s (-4\pi y)^{-s} \sum_{n=0}^\infty c_s(n)^\sigma e(nz) .$$

Therefore we have $(fg)^\sigma = f^\sigma g^\sigma$ for two such functions f and g.

LEMMA 11. *Let $f \in S(\lambda)$ and $g \in \mathcal{M}(\lambda-2r)$ with $0 < r \in Z$. Then $\langle f, \delta_{\lambda-2r}^r g \rangle = 0$.*

PROOF. It is sufficient to prove the case where $f \in S(\lambda, N, \chi)$ and $g \in \mathcal{M}(\lambda-2r, N, \chi)$. If $\lambda \in Z$, this was proved in [12]. The case of half-integral weight can be proved in exactly the same fashion.

LEMMA 12. *For every $g \in \mathcal{A}_\lambda$, let $[g]$ denote the essential part of g in the sense of Lemma 9. Suppose $\lambda \neq 3/2$. Then $[g]^\sigma = [g^\sigma]$ for every $\sigma \in \mathrm{Aut}(C)$.*

PROOF. Write g in the form (3.9), and let $g_0 = p+q$ with $p \in S(\lambda, N)$ and $q \in \mathcal{E}(\lambda, \Gamma_1(N))$ for some N. Then $p = [g]$ by virtue of Lemma 11. If we change g for g^σ, then naturally g_0 is changed into $g_0^\sigma = p^\sigma + q^\sigma$. We know that $p^\sigma \in S(\lambda)$, and also, by Lemma 10, that $q^\sigma \in \mathcal{E}(\lambda, \Gamma_1(N))$ (though, strictly speaking, we proved only the case $\lambda \notin Z$). Therefore we obtain our assertion.

4. Proof of Theorem 1.

Let k and l be odd integers such that $0 < l < k$, and let $f \in S(k/2, M, \chi)$ and $g \in \mathcal{M}(l/2, M', \psi)$ with Fourier expansions

$$f(z) = \sum_{n=1}^\infty a(n)e(nz), \quad g(z) = \sum_{n=0}^\infty b(n)e(nz) .$$

For a typographical reason, let us denote the function f^ρ defined by (1.7) also by f_ρ. By Lemma 4, $f_\rho \in S(k/2, M, \chi^{-1})$. Now a straightforward calculation shows that

$$(4.1) \qquad \int_0^\infty y^{s-1} \int_0^1 \overline{f_\rho(z)} g(z)\,dx\,dy = \Gamma(s)(4\pi)^{-s} D(s, f, g)$$

for sufficiently large $\mathrm{Re}(s)$. The left-hand side can be transformed, by a well known principle (see [9], [12], for example), to the form

$$(4.2) \qquad \int_\Phi \bar{f}_\rho g \cdot E_{\lambda, N}(z, s+1-(k/2), \omega) y^{s+1} d\mu(z) ,$$

where $\Phi=\Gamma_0(N)\backslash H$, N is any common multiple of M and M', $\lambda=(k-l)/2$, and

$$E_{\lambda,N}(z,\,s,\,\omega)=\sum_{c,\,d}\omega(d)(cz+d)^{-\lambda}|cz+d|^{-2s},$$

$$\omega(d)=\left(\frac{-1}{d}\right)^{\lambda}\chi(d)\psi(d);$$

$(c,\,d)$ runs over the lower entries of the elements of a complete set of representatives of $[\{\pm1\}\Gamma_{1,\infty}]\backslash\Gamma_0(N)$; $\Gamma_{1,\infty}$ is defined by (3.2). It is well known that

$$\Gamma(s+\lambda)L_N(2s+\lambda,\,\omega)E_{\lambda,N}(z,\,s,\,\omega)$$

can be continued to an entire function on the whole plane. Moreover, integral (4.2) multiplied by $\Gamma(s+\lambda)L_N(2s+\lambda,\,\omega)$ is convergent by virtue of [9, Lemma 3.3], and therefore defines an entire function. This proves the first part of Theorem 2.

　　To prove Theorem 1, we specialize the above consideration to the case where $l=1$ and

$$g(z)=(1/2)\sum_{n=-\infty}^{\infty}\varphi(n)e(tn^2z)$$

with a primitive character φ modulo r such that $\varphi(-1)=1$ and a positive integer t. By [9, Propositions 1.3 and 2.2], g belongs to $\mathcal{M}(1/2,\,4tr^2,\,\psi)$ with $\psi(m)=\varphi(m)\left(\dfrac{t}{m}\right)$. The above computation specialized to this case shows that

$$(4.3)\qquad\Gamma(s)(4\pi)^{-s}t^{-s}\sum_{n=1}^{\infty}\varphi(n)a(tn^2)n^{-2s}$$

$$=\int_{\Phi}\bar{f}_\rho g\cdot E_{\lambda,N}(z,\,s+1-(k/2),\,\omega)y^{s+1}d\mu(z),$$

$$\lambda=(k-1)/2,\quad\omega=\varphi_t\chi\varphi,\quad\varphi_t(m)=\left(\frac{-1}{m}\right)^{\lambda}\left(\frac{t}{m}\right).$$

　　Suppose $S(k/2,\,M,\,\chi,\,F)\neq\{0\}$ and $F|T(p)=c(p)F$ for each prime p. By Lemma 6, we have $f|T(p^2)=c(p)f$ for all $f\in S(k/2,\,M,\,\chi,\,F)$ and for all $p\nmid M$. We can also find the above character φ so that

$(4.4a)\qquad\qquad\qquad\textit{every prime factor of } M \textit{ divides } r,$

$(4.4b)\qquad\qquad\qquad\textit{g is a cusp form.}$

For example, let $p_1,\cdots,p_n$ be the odd prime factors of M, and put

$$\varphi(d)=\left(\frac{r}{d}\right),\quad r=8p_1\cdots p_nq$$

with an odd prime q such that $q\equiv3\pmod 4$ and $q\nmid M$. Then φ is not totally even in the sense of Serre and Stark [7, p. 36], and therefore g is a cusp form by virtue of [7, Theorem B]. Now take t under the condition

(4.5) $\qquad\qquad\qquad$ *t has no square factor prime to r.*

Let $0 \neq f \in S(k/2, M, \chi, F)$. With a fixed t, let N be the least common multiple of M and tr^2. Then it follows from [9, Corollary 1.8, Theorem 1.9] that

(4.6) $\qquad\qquad \sum_{n=1}^{\infty} \varphi(n)a(tn^2)n^{-2s} = a(t)L_N(2s+1-\lambda, \omega)^{-1}D(2s, F, \varphi).$

Put $m = k-2$ and evaluate (4.3) at $s = m/2$. Then we have

$$a(t)D(m, F, \varphi) = R\pi^{(m+1)/2}t^{1/2}L_N(\lambda, \omega)\langle f^\rho, g \cdot E_{\lambda, N}(z, \omega)\rangle,$$

where R is a positive rational number which depends only on m, N, and t, and is independent of f, and

(4.7) $\qquad\qquad\qquad E_{\lambda, N}(z, \omega) = E_{\lambda, N}(z, 0, \omega).$

If $\lambda \neq 2$ or ω is non-trivial, $E_{\lambda, N}(z, \omega)$ belongs to $\mathcal{M}(\lambda, N, \omega^{-1})$ and therefore $g \cdot E_{\lambda, N}(z, \omega) \in S(k/2, N, \chi^{-1})$. If $\lambda = 2$ (and hence $k = 5$) and ω is trivial, $g \cdot E_{2, N}(z, \omega)$ belongs to $\mathcal{A}_{5/2}$ for the same reason as on [12, p. 795], and therefore

$$g \cdot E_{2, N}(z, \omega) = g_0 + \delta_{1/2}g_1$$

with $g_0 \in \mathcal{M}(5/2, N, \chi^{-1})$ and $g_1 \in \mathcal{M}(1/2, N, \chi^{-1})$. Since g is a cusp form, we see that both g_0 and g_1 are cusp forms. (Actually g_1 is a constant multiple of g.) Put $g_0 = g \cdot E_{\lambda, N}(z, \omega)$ if $\lambda \neq 2$ or ω is nontrivial, and let $g_0 = h + h'$ with

$$h \in S(k/2, M, \chi^\rho, F^\rho), \quad h' \in \mathcal{T}(k/2, M, \chi^\rho, F^\rho),$$

where $\mathcal{T}$ is defined in Lemma 7. Observing that $i^\lambda t^{1/2}\tau(\varphi_t)$ is a rational number, we obtain, in view of (1.8),

(4.8) $\qquad R'a(t)A(m, F, \varphi) = [\varphi_t, \chi, \varphi]^{-1}P_N(\lambda, \omega)\pi i \cdot \tau(\chi)\langle h^\rho, f\rangle,$

where R' is a nonzero rational number independent of f.

$\qquad$ Since $\lambda \geq 1$, we have $P_N(\lambda, \omega) \neq 0$. Suppose now $k \geq 5$. Then we have $m \geq k/2$, and hence $D(m, F, \varphi) \neq 0$ by [14, Proposition 4.16] (cf. also [12, Proposition 2]). Write $a(t, f)$ for $a(t)$ in order to emphasize the dependence on f. Then the above relation shows that

(4.9) $\qquad\qquad a(t, f) = 0 \quad \text{if and only if} \quad \langle h^\rho, f\rangle = 0.$

Let $\sigma \in \mathrm{Aut}(C)$, and take f^σ, F^σ, and φ^σ instead of f, F, and φ, but with the same t. By [12, Lemma 5], we have $E_{\lambda, N}(z, \omega)^\sigma = E_{\lambda, N}(z, \omega^\sigma)$ if $\lambda \neq 2$ or ω is nontrivial. This is so even when $\lambda = 2$ and ω is trivial, if the action of σ is defined by (3.12). (In fact, it is Q-rational.) Therefore we see from (3) of Lemma 7 that h^σ is exactly the form which plays the role of h for f^σ, F^σ, and φ^σ. Thus we have

$$R'a(t, f^\sigma)A(m, F^\sigma, \varphi^\sigma)=[\varphi_t, \chi^\sigma, \varphi^\sigma]^{-1}P_N(\lambda, \omega^\sigma)\pi i\tau(\chi^\sigma)\langle h^{\sigma\rho}, f^\sigma\rangle.$$

Define $I(f, g)$ by (1.13); divide (4.8) by $u_-(F)$ and apply σ. Then we find, in view of (1.12b) and (3.6a, b), that

$$(4.10) \qquad\qquad [I(h^\rho, f)/u_-(F)]^\sigma=I(h^{\sigma\rho}, f^\sigma)/u_-(F^\sigma).$$

We have obtained h for each fixed t satisfying (4.5); now write h_t for h^ρ. Observe that every positive integer can be written as tn^2 with t satisfying (4.5) and a positive integer n prime to r. Therefore (4.6) and (4.9) show that $f=0$ if $\langle h_t, f\rangle=0$ for all t satisfying (4.5). It follows that the h_t span $S(k/2, M, \chi, F)$.

Theorem 1 for $k\geq 5$ now follows easily from this result. Changing the notation in order to avoid confusion, let $p\in S(k/2, M, \chi, F)$ and $q\in S(k/2, M')$. Then we have $p=\sum_{s\in V}b_s h_s$ and $q=r+\sum_{t\in V}c_t h_t$ with b_s and c_t in C, r in $\mathcal{T}(k/2, M, \chi, F)$ of Lemma 7, and with a finite set V of positive integers satisfying (4.5). Then

$$I(p, q)=\sum_{s, t\in V}\bar{b}_s c_t I(h_s, h_t).$$

By (4.10) and (3) of Lemma 7, we have

$$(4.11) \qquad\qquad [I(p, q)/u_-(F)]^\sigma=I(p^{\rho\sigma\rho}, q^\sigma)/u_-(F^\sigma)$$

for every $\sigma\in\mathrm{Aut}(C)$. Now $K(F)$ is either totally real or is a CM-field. Therefore $K(F, \chi)=K(F^\rho, \chi^\rho)\subset K(p^\rho)$ by Lemma 8, and hence (4.11) shows that $I(p, q)/u_-(F)$ belongs to $K(p^\rho, q)$. If p is $K(F, \chi)$-rational, we have $K(p)=K(p^\rho)=K(F, \chi)$ and $\rho\sigma\rho=\sigma$ on $K(p)$. This proves Theorem 1 when $k>3$.

Next let us assume $k=3$. In this case we have $m=1$ and $D(1, F, \varphi)$ may be 0, and therefore it is necessary to modify the above reasoning. We first choose a character φ_1 under conditions (4.4a, b). If $D(1, F, \varphi_1)\neq 0$, then the whole preceding argument is valid. (Since $\lambda=1$, $g\cdot E_{\lambda, N}(z, \omega)$ is always a cusp form.) Suppose $D(1, F, \varphi_1)=0$, and put

$$F(z)=\sum_{n=1}^{\infty}c(n)e(nz), \quad F^*(z)=\sum_{n=1}^{\infty}\varphi_1(n)c(n)e(nz).$$

Applying [13, Theorem 2] to F^*, we find a character ξ whose conductor is a prime not dividing Mtr^2, and such that $\xi(-1)=1$, $D(1, F^*, \xi)\neq 0$. Put $\varphi=\xi\varphi_1$. Then

$$D(1, F, \varphi)=D(1, F^*, \xi)\neq 0.$$

Therefore the above reasoning for $k>3$ is valid also in the case $k=3$ with this choice of φ. This completes the proof of Theorem 1.

REMARK. In the case $k>3$, we have employed a character φ such that $\varphi^2=1$. Therefore the forms h_t in the above proof are $K(\chi, F)$-rational.

5. Proof of Theorems 2 and 3.

We now go back to (4.1) and (4.2) with $k>3$. A direct calculation shows that

$$(5.1) \qquad E_{\lambda, N}(z, -r, \omega)=\frac{\Gamma(\lambda-2r)}{\Gamma(\lambda-r)}(-4\pi y)^r \delta_{\lambda-2r}^r E_{\lambda-2r, N}(z, \omega) \qquad (0\leqq 2r<\lambda, r\in Z).$$

Let n be an odd integer such that

$$(5.2) \qquad (k+l-4)/2<n\leqq k-2,$$

and let $r=(k-n-2)/2$. Then $0\leqq 2r<\lambda$ and $r\in Z$. Substituting $n/2$ for s in (4.1) and (4.2), we find that

$$(5.3) \qquad D(n/2, f, g)=J\pi^{(k-1)/2}\langle f^\rho, g\delta_{\lambda-2r}^r E_{\lambda-2r, N}(z, \omega)\rangle,$$

where J is a rational number depending only on k, l, n, and N. Now assume that $f\in S(k/2, M, \chi, F)$. Let h be the essential part of $g\delta_{\lambda-2r}^r E_{\lambda-2r, N}(z, \omega)$ in the sense of Lemma 9. Observing that $h\in S(k/2, N, \chi^\rho)$, decompose h into the sum $h=p+q$ with p in $S(k/2, M, \chi^\rho, F^\rho)$ and q in $\mathcal{T}(k/2, M, \chi^\rho, F^\rho)$ of Lemma 7. We can now state a preliminary version to Theorem 2 as

THEOREM 4. *Let* $f\in S(k/2, M, \chi, F)$ *and* $g\in \mathcal{M}(l/2, M')$ *with odd integers* k *and* l *such that* $0<l<k$ *and* $k\geqq 5$. *For every positive integer n satisfying* (5.2), *put*

$$B'(n, f, g)=\pi^{(3-k)/2}i\tau(\chi)D(n/2, f, g)/u_-(F).$$

Then $B'(n, f, g)^\sigma=B'(n, f^\sigma, g^\sigma)$ *for every* $\sigma\in \mathrm{Aut}(C)$.

To prove this, we may assume that $g\in \mathcal{M}(l/2, M', \psi)$ with a character ψ modulo M'. Then (5.3) shows that

$$B'(n, f, g)=J\cdot I(p^\rho, f)/u_-(F).$$

If we change f and g for f^σ and g^σ, then p is changed into p^σ. This follows from Lemma 12 and (3) of Lemma 7. Thus

$$B'(n, f^\sigma, g^\sigma)=J\cdot I(p^{\sigma\rho}, f^\sigma)/u_-(F^\sigma).$$

Hence our assertion follows immediately from Theorem 1.

The assertion of Theorem 2 for n satisfying (5.2) is merely a restatement of this result. In fact, put $\varphi(d)=\left(\dfrac{-1}{d}\right)^\lambda$ and observe that $i^\lambda\tau(\varphi)$ is a rational number, and that $0 < n+2-(k+l)/2 \equiv \lambda \pmod 2$. Then

$$B(n, f, g) = J'[\chi, \phi, \varphi]^{-1} P_N(n+2-(k+l)/2, \chi\psi\varphi) I(p^\rho, f)/u_-(F)$$

with a rational number J' depending only on k, l, n, and N, and hence we obtain the desired conclusion from Theorem 1 and (3.6a, b).

Now let n' be an odd integer such that $l \leq n' < (k+l)/2$. The transformation $s \mapsto (1/2)(k+l)-1-s$ sends $n'/2$ to $n/2$ with an odd integer n satisfying (5.2). Therefore our result for $B(n', f, g)$ can be obtained by using the functional equation of $E_{\lambda, N}(z, s, \omega)$. The computation is exactly the same as what was done on [13, pp. 217-218]; we only have to take $k/2$, $l/2$, and $\chi\psi\varphi$ in place of k, l, and $\chi\psi$ there.

Finally, to prove Theorem 3, we take g from $\mathcal{M}(m, M', \psi)$ with an integer $m < k/2$, and still f in $S(k/2, M, \chi, F)$. In this case, a computation similar to that of §4 shows that

$$(4\pi)^{-s}\Gamma(s)D(s, f, g) = \int_\Phi \bar{f}_\rho g G_{\kappa, N}(z, 2s+2-k, \omega) y^{k/2} d\mu(z)$$

where

$$G_{\kappa, N}(z, s, \omega) = y^{s/2} \sum_{\gamma \in W} \omega(d_\gamma) j(\gamma, z)^{-\kappa} |j(\gamma, z)|^{-2s},$$

$$\kappa = k - 2m, \quad \omega(d) = \left(\frac{-1}{d}\right)^m \chi(d)\psi(d),$$

and W is a complete set of representatives of $[\{\pm 1\}\Gamma_{1,\infty}]\backslash\Gamma_0(N)$. Now $G_{\kappa, N}(z, s, \omega)$ coincides with $E(z, s, -\kappa, \omega)$ of [10]; moreover, [10, Proposition 3] shows that

$$(s+\alpha-1)\Gamma\left(\frac{s+\kappa}{2}\right)\Gamma\left(\frac{s+\alpha+\varepsilon}{2}\right)L_N(2s+2\alpha, \omega^2)E(z, s, -\kappa, \omega)$$

is entire, where $\alpha = (\kappa-1)/2$, and ε is 0 or 1 according as α is even or odd; the factor $s+\alpha-1$ is necessary only when ω^2 is trivial and $\kappa \equiv 1 \pmod 4$. Therefore the first assertion of Theorem 3 follows from [10, Proposition 4] in exactly the same fashion as in [10, § 5].

Now put

$$T_1 = \begin{cases} \{r \in 2Z \mid 0 \leq r < (\kappa-3)/2\} & \text{if } \omega^2 \text{ is trivial,} \\ \{r \in 2Z \mid 0 \leq r \leq (\kappa-3)/2\} & \text{otherwise,} \end{cases}$$

$$T_2 = \{r \in Z \mid (\kappa-1)/2 \leq r < \kappa, \ r \equiv 1 \pmod 2\},$$

$$H(z, r, \omega) = C_r \{L_N(2s-1+\kappa, \omega^2)G_{\kappa, N}(z, s, \omega)\}_{s=-r},$$

$$C_r = \begin{cases} \tau(\omega^2)^{-1}\pi^{1-\varepsilon+(3r/2)} & \text{if } r \in T_1, \\ \tau(\omega)^{-1}\pi^{(r-\kappa)/2} & \text{if } r \in T_2. \end{cases}$$

In [17], Sturm showed that $H(z, r, \omega)$ belongs to $\mathcal{A}_\kappa$, and moreover that $H(z, r, \omega)^\sigma = H(z, r, \omega^\sigma)$. These results can also be obtained by examining the Fourier expansion of the function $E^*(z, s, -\kappa, \omega)$ of [10] at $s = -r$ and applying Lemma 2 to it. This is similar to but somewhat more complicated than the analysis of $G^*(z, \omega)$ in the proof of Lemma 10. Anyhow, the arithmetic part of Theorem 3 can now be proved in the same fashion as for Theorem 2 by virtue of these properties of $H(z, r, \omega)$.

ADDENDUM. Some results of Reference [14] are quoted at several points in the present paper. We inform the reader that more than 20 misprints are in that article, though they affect neither the main theorems nor the results employed here. A full list of corrections has been sent to the Duke Mathematical Journal for publication.

References

[1] Y. Flicker, Automorphic forms on covering groups of $GL(2)$, preprint, 1979.

[2] S. Gelbart and I. Piatetski-Shapiro, On Shimura's correspondence for modular forms of half-integral weight, Proc. Int. Coll., Bombay, 1979.

[3] T. Miyake, On automorphic forms on GL_2 and Hecke operators, Ann. of Math., **94** (1971), 174-189.

[4] S. Niwa, Modular forms of half integral weight and the integral of certain theta-functions, Nagoya Math. J., **56** (1975), 147-161.

[5] H. Petersson, Über die Entwicklungskoeffizienten der ganzen Modulformen und ihre Bedeutung für die Zahlentheorie, Abh. Math. Sem. Univ. Hamburg, **8** (1931), 215-242.

[6] H. Petersson, Über die systematische Bedeutung der Eisensteinschen Reihen, Abh. Math. Sem. Univ. Hamburg, **16** (1949), 104-130.

[7] J.-P. Serre and H. Stark, Modular forms of weight 1/2, Modular functions of one variable VI, Lecture Notes in Math., **627** (1977), Springer, 27-67.

[8] G. Shimura, Introduction to the arithmetic theory of automorphic functions, Publ. Math. Soc. Japan, No. 11, Iwanami Shoten and Princeton Univ. Press, 1971.

[9] G. Shimura, On modular forms of half integral weight, Ann. of Math., **97** (1973), 440-481.

[10] G. Shimura, On the holomorphy of certain Dirichlet series, Proc. London Math. Soc., **31** (1975), 79-98.

[11] G. Shimura, Theta functions with complex multiplication, Duke Math. J., **43** (1976), 673-696.

[12] G. Shimura, The special values of the zeta functions associated with cusp forms, Comm. Pure Appl. Math., **29** (1976), 783-804.

[13] G. Shimura, On the periods of modular forms, Math. Ann., **229** (1977), 211-221.

[14] G. Shimura, The special values of the zeta functions associated with Hilbert modular forms, Duke Math. J., **45** (1978), 637-679.

[15] G. Shimura, On certain reciprocity-laws for theta functions and modular forms, Acta Math., **141** (1978), 35-71.

[16] T. Shintani, On construction of holomorphic cusp forms of half integral weight, Nagoya Math. J., **58** (1975), 83-126.

[17] J. Sturm, Special values of zeta functions and Eisenstein series of half integral
      weight, Amer. J. Math., 102 (1980), 219-240.
[18] J. Sturm, Addendum to special values of zeta functions, Amer. J. Math., 102
      (1980), 781-783.

Goro SHIMURA

Department of Mathematics
Princeton University
Princeton, New Jersey 08544
U. S. A.

On certain zeta functions attached to two Hilbert modular forms:
I. The case of Hecke characters

Annals of Mathematics, 114 (1981), 127-164

Introduction

To make our exposition smooth, let us first introduce some notational conventions. For an algebraic number field K of finite degree, we denote by J_K the set of all embeddings of K into $\mathbf{C}$, and by I_K the free $\mathbf{Z}$-module generated by the elements of J_K. We then put $\mathbf{R}I_K = I_K \otimes \mathbf{R}$ and $\mathbf{C}I_K = I_K \otimes \mathbf{C}$. If $p = \Sigma_\sigma p_\sigma \sigma \in I_K$ with $\sigma \in J_K$ and $p_\sigma \in \mathbf{Z}$, we put $x^p = \prod_\sigma (x^\sigma)^{p_\sigma}$ for $0 \neq x \in K$; this is meaningful for $p \in \mathbf{C}I_K$ if x^σ are all real and positive.

Throughout the paper, we denote by $\mathfrak{H}$ the complex upper half plane and by F a totally real algebraic number field of degree n. Now the zeta function to be studied in this paper, when suitably specialized, has the form

$$(1) \qquad D(s) = \sum_{0 \ll b \in \mathfrak{y}} \omega\big(\mathrm{Tr}_{F/\mathbf{Q}}(\zeta b)\big)\chi(b\mathfrak{x}^{-1})b^{p-s\tau} \qquad (s \in \mathbf{C}).$$

Here $\mathfrak{y}$ is the residue class of an element of F modulo a fractional ideal of F; we write $0 \ll b$ to indicate that b is totally positive; ω denotes the Fourier coefficients of an elliptic modular form

$$(2) \qquad \Omega(z) = \sum_{a \in \mathbf{Q}} \omega(a)e^{2\pi i a z} \qquad (z \in \mathfrak{H})$$

of an integral weight $l \geq 0$; $\zeta \in F, \tau \in J_F, p \in 2^{-1}I_F$; $\mathfrak{x}$ is a fractional ideal of F; $\{\chi(\mathfrak{a})\}$ is a system of eigenvalues of Hecke operators in the sense that there is a holomorphic cusp form $\mathbf{f}$ with respect to a congruence subgroup of $\mathrm{GL}_2(F)$ such that $\mathbf{f}|T(\mathfrak{a}) = \chi(\mathfrak{a})\mathbf{f}$ for all integral ideals $\mathfrak{a}$ of F, where $T(\mathfrak{a})$ is the Hecke operator of degree $\mathfrak{a}$. The eigenvalues $\chi(b\mathfrak{x}^{-1})$ with a fixed $\mathfrak{x}$ can be interpreted as the Fourier coefficients of a Hilbert modular form. In fact, if we put

*Supported by NSF Grant MCS78-02781.

$\xi(b) = b^q \chi(b\mathfrak{x}^{-1})$ with a suitable $q \in 2^{-1}I_F$, then

$$(3) \qquad\qquad f(z) = \sum_{b \in \mathfrak{x}} \xi(b) e^{2\pi i \cdot \mathrm{Tr}(bz)} \qquad\qquad (z \in \mathfrak{H}^n)$$

is a Hilbert modular form, so that D can be expressed as

$$(4) \qquad\qquad D(s) = \sum_{0 \ll b \in \mathfrak{y}} \omega\big(\mathrm{Tr}_{F/\mathbf{Q}}(\zeta b)\big)\xi(b)b^{d-s\tau}$$

with $d = p - q$. We impose the following conditions on ζ and d: (i) $\zeta^\tau > 0$ and $\zeta^\sigma < 0$ for $\sigma \neq \tau \in J_F$; (ii) $d \in I_F$, $d = \Sigma_\sigma d_\sigma \sigma$ with $d_\sigma \geqq 0$ for all $\sigma \neq \tau$. It will be shown as a preliminary step that (4) is convergent for sufficiently large $\mathrm{Re}(s)$ and can be continued to a meromorphic function on the whole plane. Now the main aim of this paper is to prove a theorem of algebraicity which can be stated as follows.

(A) *Under certain conditions on* χ, *the values* $D(\mu)$ *for certain integers* μ *are algebraic numbers times a constant* $C(\chi, \tau)$ *which depends only on* χ *and* τ, *and is independent of* Ω, $\mathfrak{x}$, $\mathfrak{y}$, ζ, p, *and* μ, *provided the coefficients* $\omega(\mathfrak{a})$ *are all algebraic.*

Other theorems will be obtained as various consequences, complements, or analogues of this central principle.

The study of D was made in a previous paper [6] when $\Sigma\chi(\mathfrak{a})N(\mathfrak{a})^{-s}$ is a Hecke L-function of a totally imaginary quadratic extension K of F. This amounts to the assumption that f has the form

$$(5) \qquad\qquad f(z) = \sum_{c \in Y} c^\Psi \exp(2\pi i \cdot \mathrm{Tr}(c\bar{c}z)) \qquad\qquad (z \in \mathfrak{H}^n)$$

with a suitable $\Psi \in I_K$ and a residue class Y in K modulo an ideal. In this case, we have

$$(6) \qquad\qquad D(s) = \sum_{c \in Y} \omega\big(\mathrm{Tr}_{F/\mathbf{Q}}(\zeta c\bar{c})\big)c^\Phi |c^\psi|^{-2s}$$

with $\Phi \in I_K$ and $\psi \in J_K$. Thus the present work is a natural continuation of [6].

If $F = \mathbf{Q}$, a suitable normalization reduces (4) to a Dirichlet series $\Sigma_{n=1}^\infty \omega(n)\xi(n)n^{-s}$ obtained from two elliptic modular forms $\Sigma\omega(n)e^{2\pi i n z}$ and $\Sigma\xi(n)e^{2\pi i n z}$, which we investigated in [3] and [4]. As already mentioned in [6], however, in general no rational power of b^τ in (1) is a rational number, and hence D is not the type of zeta function which ordinarily occurs in number theory. This fact makes the study of D at least technically interesting from an analytic viewpoint, but we should emphasize that D is a very natural object, which reflects some algebro-geometric and number-theoretical characteristics of arithmetic quotients of $\mathfrak{H}^r$ for $r \leqq n$. Before touching on this point, let us first lay out our program.

Because of the length of the paper and the diversity of our methods, we divide the whole paper into two parts. In Part I, we shall principally treat the case where f has form (5). The constant $C(\chi, \tau)$ in this case will be given as a power of π times a "period" $p_K(\Psi, 2\psi)$ introduced in [6]. This fact was already proved in [6] for the values $D(\mu)$ for which series (6) is convergent. Now we carefully examine the analytic continuation of D and show the arithmeticity of $D(\mu)$ for some more integers μ where D is holomorphic, but as a series is divergent. Our second theme of Part I is, as was so in [6], an automorphic form on an orthogonal group which has the following form in the simplest case:

$$(7) \qquad \varphi(w, s) = \sum_{0 \neq x \in X} \omega(-S(x, x))S(x, w)^{-k}|S(x, w)|^{-s}$$

$$(s \in \mathbf{C}, w \in \mathfrak{Z}).$$

Here $S(x, y)$ is a $\mathbf{Q}$-valued symmetric bilinear form of signature $(m, 2)$ on $\mathbf{Q}^{m+2}$; $\mathfrak{Z}$ is a symmetric domain contained in $\mathbf{C}^{m+2}$; we extend S to $\mathbf{C}^{m+2}$ in a natural way; $0 < k \in \mathbf{Z}$; X is a residue class of $\mathbf{Q}^{m+2}$ modulo a lattice. In [6], we showed that this can be continued to a meromorphic function on the whole s-plane and that if the series is convergent at $s = 0$, then $\pi^{-k}\varphi(w, 0)$ is an automorphic form on $\mathfrak{Z}$ of arithmetic type, provided the $\omega(a)$ are all algebraic. The fact that the values $D(\mu)$ occur as the values of $\varphi(w, 0)$ at a CM-point w ties both series inseparably. The convergence at $s = 0$ requires that $k > l + m + 1$ if Ω is a cusp form, for example. Now in the present paper, we show that $\varphi(w, 0)$ is meaningful and holomorphic in w, and even arithmetic, if $k > l + (m/2) + 1$ or $k = l + (m/2)$. Similar results hold for Ω which is not necessarily a cusp form, and especially for $\Omega = 1$; in the last case $\varphi(w, 0)$ may be called an Eisenstein series.

The holomorphy of $\varphi(w, 0)$ follows from a simple principle which includes as a special case the fact that an inner product of the form $\langle p, \delta q \rangle$ vanishes if p is holomorphic and $\delta = (\lambda/2iy) + (\partial/\partial z)$ for some $\lambda \in \mathbf{Q}$. Such a vanishing was proved in our papers [3] and [5] when both p and q are holomorphic. The present generalization is applicable to a very wide class of functions q which are not necessarily eigenfunctions of invariant differential operators. It may be worth noting that the holomorphy as well as the arithmeticity of $\varphi(w, 0)$ can be established even when S is anisotropic.

The proof of the arithmeticity of $\varphi(w, 0)$ and $D(\mu)$ (when f has form (5)) is the same as in [6]. In addition to this arithmeticity, we shall prove that the residue of (6) at a critical point, under some conditions on l and Φ, has the same nature as $D(\mu)$, and also that under a condition on l and k, the residue of (7) at $s = 1$ multiplied by elementary factors is a holomorphic automorphic form on $\mathfrak{Z}$ which is arithmetic.

Essential to our theory is a certain non-holomorphic pull-back of a Hilbert modular form, which, applied to the above f, is given as

$$(8) \qquad g(z) = f(\zeta_1 z, \zeta_2 \bar{z}, \ldots, \zeta_n \bar{z}) \qquad (z \in \mathfrak{H}),$$

where $\zeta_1, \ldots, \zeta_n$ are the conjugates of ζ; we take $\zeta_1 = \zeta^\tau$. If $(k_1, \ldots, k_n)$ is the weight of f and $\lambda = k_2 + \cdots + k_n$, then $\mathrm{Im}(z)^\lambda g(z)$ behaves like a modular form of weight $k_1 - \lambda$ under a congruence subgroup of $\mathrm{SL}_2(\mathbf{Z})$. Our series (4) in a special case can actually be obtained as an integral of the Rankin-Selberg type for Ω and g. The residue of D just mentioned includes as a special case the inner product $\langle h, \mathrm{Im}(z)^\lambda g \rangle$ with a $\overline{\mathbf{Q}}$-rational elliptic modular form h of weight $k_1 - \lambda$. Thus such an inner product has the same property of algebraicity as $D(\mu)$.

In the subsequent Part II, we shall investigate D in the case where f is not necessarily of type (5). More precisely, we shall prove statement (A) when the eigenvalues $\chi(\mathfrak{a})$ occur in the space $\mathfrak{S}(B)$ of cusp forms with respect to a quaternion algebra B over F, which (when D is specialized to (1)) is unramified at τ and ramified at all other archimedean primes. If $\mathfrak{q}$ is a $\overline{\mathbf{Q}}$-rational eigenform in $\mathfrak{S}(B)$ with eigenvalues $\chi(\mathfrak{a})$, then the constant $C(\chi, \tau)$ is given as the inner product $\langle \mathfrak{q}, \mathfrak{q} \rangle$ times a power of π. The result about the residue can be obtained also in this case. These facts seem to imply that the non-holomorphic form g of (8) represents at least some arithmetic nature of $\mathfrak{q}$, although a new philosophy is necessary for the full understanding of this phenomenon. The proof of (A) in this quaternion case relies heavily on the arithmeticity of (7), which in turn relies on that of $D(\mu)$ of type (6). Thus the conclusion of Part II will be reached only through Part I. We shall actually investigate $D(s)$ and $\varphi(w, s)$ with a subfield E of F as a basic field instead of $\mathbf{Q}$; namely we take Ω to be a Hilbert modular form with respect to a subgroup of $\mathrm{GL}_2(E)$. The quaternion algebra B in this case is unramified at exactly $[E:\mathbf{Q}]$ archimedean primes of F.

There are some interesting applications of our results to the arithmetic of the varieties obtained from B. For instance, we shall obtain new information about the critical values of a zeta function with an Euler product attached to an algebraic surface of that type. More details about this will be given in Part II.

Notation. The symbols J_K, I_K, $\mathbf{R}I_K$, $\mathbf{C}I_K$, F, $\ll$, and $\mathfrak{H}$ will have the same meaning as above throughout the paper. We denote by R_F the regulator of F. We write $p \leqq q$ for two elements $p = \Sigma_\sigma p_\sigma \sigma$ and $q = \Sigma_\sigma q_\sigma \sigma$ of $\mathbf{R}I_K$ if $p_\sigma \leqq q_\sigma$ for all $\sigma \in J_K$. If L is a subfield of K, we put $\mathrm{Res}_{K/L}(p) = \Sigma_\sigma p_\sigma \mathrm{Res}_{K/L}(\sigma)$, where $\mathrm{Res}_{K/L}(\sigma)$ denotes the restriction of σ to L. The algebraic closure of $\mathbf{Q}$ in $\mathbf{C}$ is denoted by $\overline{\mathbf{Q}}$. For two elements a and b of $\mathbf{C}$, we write $a \sim b$ if $b \neq 0$ and $a/b \in \overline{\mathbf{Q}}$. We put $\mathbf{e}(z) = \exp(2\pi i z)$ for $z \in \mathbf{C}$. If $S(x, y)$ is a bilinear form, we put $S[x] = S(x, x)$. Other symbols will be introduced in the first few paragraphs of each section, especially in Section 1.

1. Automorphic forms on $\mathfrak{H}^n$

With F as in the introduction, we denote by $\tau_1, \ldots, \tau_n$ the elements of J_F, and by $\mathfrak{r}_F$, or simply by $\mathfrak{r}$, the maximal order of F. We shall consider holomorphic as well as non-holomorphic automorphic forms on $\mathfrak{H}^n$ with respect to congruence subgroups of $\mathrm{SL}_2(\mathfrak{r})$. We embed F into $\mathbf{R}^n$ by the map $a \mapsto (a^{\tau_1}, \ldots, a^{\tau_n})$ and identify $F \otimes_{\mathbf{Q}} \mathbf{R}$ and $F \otimes_{\mathbf{Q}} \mathbf{C}$ with $\mathbf{R}^n$ and $\mathbf{C}^n$ through the map. Then the notation z^p can be extended to the case of $z \in \mathbf{C}^n$; namely, if $z = (z_1, \ldots, z_n) \in \mathbf{C}^n$ $(= F \otimes_{\mathbf{Q}} \mathbf{C})$ and $p = \sum_{i=1}^n p_i \tau_i \in \mathbf{C} I_K$, we have

$$(1.1\mathrm{a}) \qquad z^p = \prod_{i=1}^n z_i^{p_i};$$

we usually assume that either $p \in I_K$ or z_i are all real and positive. We put also

$$(1.1\mathrm{b}) \qquad |p| = p_1 + \cdots + p_n \qquad\qquad (p \in \mathbf{R} I_K),$$

$$(1.2) \qquad \mathbf{e}_F(az) = \mathbf{e}\left(\sum_{\nu=1}^n a^{\tau_\nu} z_\nu \right) \qquad\qquad (a \in F, z \in \mathbf{C}^n).$$

When there is no fear of confusion, we use the letter F also for the element $\sum_{i=1}^n \tau_i$ of I_F; thus

$$(1.3) \qquad z^F = z_1 \cdots z_n.$$

In particular, $a^F = N_{F/\mathbf{Q}}(a)$ for $a \in F$; similarly $|z|^{sF} = |z_1 \cdots z_n|^s$ for $s \in \mathbf{C}$.

Let $k = \sum_i k_i \tau_i$ be an element of $2^{-1} I_F$ such that

$$(1.4) \qquad 2k_1 \equiv \cdots \equiv 2k_n (\mathrm{mod}\ 2).$$

For a congruence subgroup Γ of $\mathrm{SL}_2(F)$, we denote by $\mathcal{C}_k(\Gamma)$ the vector space of all C^∞-functions g on $\mathfrak{H}^n$ such that

$$g(\gamma(z)) = \begin{cases} g(z)(cz+d)^k & \text{if } k \in I_F \\ g(z)j(\gamma, z)(cz+d)^{k-(F/2)} & \text{if } k \notin I_F \end{cases} \quad \text{for all } \gamma = \left(\begin{smallmatrix} * & * \\ c & d \end{smallmatrix}\right) \in \Gamma,$$

where $j(\gamma, z)$ is defined by [6, (7.28)]; if $k \notin I_F$, we assume that Γ is a congruence subgroup of

$$(1.5) \qquad \left\{ \begin{pmatrix} a & b \\ c & d \end{pmatrix} \in \mathrm{SL}_2(\mathfrak{r}) \,\middle|\, b \in 2\mathfrak{d}^{-1}, c \in 2\mathfrak{d} \right\},$$

where $\mathfrak{d}$ is the different of F relative to $\mathbf{Q}$. We denote by $\mathfrak{M}_k(\Gamma)$ the set of all (holomorphic) Hilbert modular forms belonging to $\mathcal{C}_k(\Gamma)$, and by $\mathcal{C}_k$ (resp. $\mathfrak{M}_k$) the union of $\mathcal{C}_k(\Gamma)$ (resp. $\mathfrak{M}_k(\Gamma)$) for all possible Γ. We have $\mathfrak{M}_k(\Gamma) \neq \{0\}$ only when $k \geq 0$, but $\mathcal{C}_k(\Gamma)$ can be nontrivial without this condition on k. If it is necessary to emphasize the basic field F, we write it $\mathcal{C}_{k,F}$ (resp. $\mathfrak{M}_{k,F}$). We introduce a measure $d\mu(z)$ on $\mathfrak{H}^n$ by

$$(1.6) \qquad d\mu(z) = \prod_{\nu=1}^{n} y_\nu^{-2} \, dx_\nu \, dy_\nu \qquad\qquad (z_\nu = x_\nu + iy_\nu),$$

and define the inner product $\langle f, g \rangle$ of two elements f and g of $\mathcal{C}_k$ by

$$(1.7) \qquad \langle f, g \rangle = \mu(\Phi)^{-1} \int_\Phi \overline{f(z)} \, g(z) \operatorname{Im}(z)^k \, d\mu(z)$$

whenever the integral is absolutely convergent, where Φ is a fundamental domain of $\Gamma \setminus \mathfrak{H}^n$ and $\mu(\Phi)$ is its measure; naturally we take Γ so that both f and g belong to $\mathcal{C}_k(\Gamma)$.

In our later discussion, Eisenstein series are essential. To define them, take Γ in the form

$$(1.8) \quad \Gamma = \left\{ \begin{pmatrix} a & b \\ c & d \end{pmatrix} \in \mathrm{SL}_2(F) \,\middle|\, a \equiv d \equiv 1 \ (\mathrm{mod}\ \mathfrak{c}), \, b \in \mathfrak{b}, \, c \in \mathfrak{b}^{-1}\mathfrak{c} \right\}$$

with a fractional ideal $\mathfrak{b}$ and an integral ideal $\mathfrak{c}$ of F. Let U_F denote the group of all totally positive units of $\mathfrak{r}_F$. Take a subgroup U of U_F of finite index so that $u \equiv 1 \ (\mathrm{mod}\ \mathfrak{c})$ for all $u \in U$, and put

$$(1.9) \qquad \Gamma_\infty = \Gamma_\infty(U, \mathfrak{b}) = \left\{ \begin{pmatrix} u & b \\ 0 & u^{-1} \end{pmatrix} \,\middle|\, u \in U, \, b \in \mathfrak{b} \right\},$$

$$(1.10) \qquad R = \left\{ (c,d) \in \mathfrak{b}^{-1}\mathfrak{c} \times \mathfrak{r} \,\middle|\, d \equiv 1 \ (\mathrm{mod}\ \mathfrak{c}), \, c\mathfrak{b} + d\mathfrak{r} = \mathfrak{r} \right\}.$$

We let U act on R by $u(c,d) = (uc, ud)$ for $u \in U$. Then

$$\Gamma_\infty \begin{pmatrix} a & b \\ c & d \end{pmatrix} \mapsto U(c,d)$$

gives a one-to-one map of $\Gamma_\infty \backslash \Gamma$ onto R/U. Now we consider a sum

$$(1.11) \quad E_{\kappa,p}(z,s) = E_{\kappa,p}(z,s; \mathfrak{b}, \mathfrak{c})$$

$$= [U] \sum_{(c,d) \in R/U} \left(\frac{c\bar{z} + d}{cz + d} \right)^p (cz + d)^{-\kappa F} |cz + d|^{-2sF}$$

$$(z \in \mathfrak{H}^n, \, s \in \mathbf{C}),$$

where $\kappa \in \mathbf{Z}$, $p = \Sigma p_i \tau_i \in I_F$, and

$$(1.12) \qquad\qquad [U] = [U_F : U]^{-1};$$

the sum is taken over a complete set of representatives of R/U. This is absolutely convergent if $\operatorname{Re}(2s + \kappa) > 2$, and can be continued to a meromorphic function on the whole s-plane; the factor $[U]$ makes it independent of the choice of U.

PROPOSITION 1.1. (i) $E_{\kappa,p}(z,s)$ *is holomorphic for* $\mathrm{Re}(s) \geqq (1-\kappa)/2$ *if* $\kappa F + 2p \neq 0$.

(ii) *If* $\kappa > 0$, $E_{\kappa,0}(z,0)$ *is an element of* $\mathfrak{M}_{\kappa F}(\Gamma)$ *except for the special case in which* $F = \mathbf{Q}$ *and* $\kappa = 2$.

(iii) $E_{0,0}(z,s)$ *is holomorphic for* $\mathrm{Re}(s) > 1$, *and has a simple pole at* $s = 1$; *the residue at* $s = 1$ *is a rational number times* $\pi^{-n}D_F^{1/2}R_F\mathrm{Im}(z)^{-F}$, *where* D_F *is the discriminant of* F, *and* R_F *the regulator of* F.

Proof. Let us first assume that $\kappa \geqq 0$ and $p \geqq 0$. For every character ω of $(\mathfrak{r}/\mathfrak{c})^\times$ such that $\omega(u) = N(u)^\kappa$ for all $u \in \mathfrak{r}^\times$, put

$$(1.13) \quad E(z,s,\omega) = [U]\sum_{c,d}\omega(d)\left(\frac{c\bar{z}+d}{cz+d}\right)^p(cz+d)^{-\kappa F}|cz+d|^{-2sF},$$

where the sum is taken over all U-equivalence classes of $(c,d) \neq (0,0)$ such that $c \in \mathfrak{b}^{-1}\mathfrak{c}$ and $c\mathfrak{b} + d\mathfrak{r} = \mathfrak{r}$. This is $[\mathfrak{r}^\times : U_F](-4\pi y)^{-p}$ times the function $H(z,s,\omega)$ of [5, (3.14)] (with p and $\mathfrak{b}$ instead of r and $\mathfrak{b}^{-1}$ there). It can easily be seen that the sum of $E(z,s,\omega)$ for all such ω is equal to a positive integer times $E_{\kappa,p}(z,s)$. Now [5,(3.18)] shows that $E(z,s,\omega)$ is holomorphic for $\mathrm{Re}(2s+\kappa) \geqq 1$ if $\kappa > 0$ or $p \neq 0$ (and $p \geqq 0$), since the Hecke L-functions are non-vanishing for $\mathrm{Re}(s) \geqq 1$. Hence we obtain (i) in the case $\kappa \geqq 0$ and $p \geqq 0$. If $\kappa = 0$ and $p = 0$, the holomorphy for $\mathrm{Re}(s) > 1$ is guaranteed by the convergence of (1.11); $E(z,s,\omega)$ is holomorphic even at $s = 1$ if ω is nontrivial, as shown by [5, (3.16), (3.18)]. If ω is trivial, it has a simple pole at $s = 1$, and its residue is given by [5, (3.20)]. This proves (iii). Assertion (ii) follows from [5, Prop. 3.2]. Now given arbitrary $p \in I_F$ and $\kappa \in \mathbf{Z}$, let $\lambda = \mathrm{Min}_\nu|\kappa + 2p_\nu|$ and put $\lambda + 2q_\nu = |\kappa + 2p_\nu|$. Then $0 \leqq q_\nu \in \mathbf{Z}$. Take an element h of F so that $h^{\tau_\nu} < 0$ if $\kappa + 2p_\nu < 0$ and $h^{\tau_\nu} > 0$ if $\kappa + 2p_\nu \geqq 0$. Put $w_\nu = h^{\tau_\nu}z_\nu$ or $= h^{\tau_\nu}\bar{z}_\nu$, according as $h^{\tau_\nu} > 0$ or < 0. Then we see that

$$E_{\kappa,p}(w, s - (\kappa/2); \mathfrak{b}, \mathfrak{c}) = E_{\lambda,q}(z, s - (\lambda/2); h^{-1}\mathfrak{b}, \mathfrak{c}),$$

which proves assertion (i) in the general case.

Our technique requires some differential operators acting on the functions on $\mathfrak{H}^n$. They are defined by

$$(1.14a) \qquad \delta_k^p = \prod_{\nu=1}^n \delta_\nu(k_\nu + 2p_\nu - 2)\cdots\delta_\nu(k_\nu + 2)\delta_\nu(k_\nu),$$

$$(1.14b) \qquad \delta_\nu(k_\nu) = (2\pi i)^{-1}\left[k_\nu(z_\nu - \bar{z}_\nu)^{-1} + (\partial/\partial z_\nu)\right],$$

where $k \in Cl_F$, $0 \leqq p \in I_F$, and $z_1,\ldots,z_n$ are the variables on $\mathfrak{H}^n$. We put also

$$(1.15) \qquad \mathbf{d}^p = (2\pi i)^{-|p|}(\partial/\partial z_1)^{p_1}\cdots(\partial/\partial z_n)^{p_n}.$$

Then there are two formal identities between δ_k^p and $\mathbf{d}^p$:

$$(1.16a) \qquad \mathbf{d}^p = \sum_{0 \leq q \leq p} \binom{p}{q} \frac{\Gamma_n(k+p)}{\Gamma_n(k+q)} (4\pi y)^{q-p} \delta_k^q,$$

$$(1.16b) \qquad \delta_k^p = \sum_{0 \leq q \leq p} \binom{p}{q} \frac{\Gamma_n(k+p)}{\Gamma_n(k+q)} (-4\pi y)^{q-p} \mathbf{d}^q,$$

where $y = \mathrm{Im}(z)$, $\binom{p}{q} = \prod_{\nu=1}^{n} \binom{p_\nu}{q_\nu}$, and $\Gamma_n(k) = \prod_{\nu=1}^{n} \Gamma(k_\nu)$. If $n = 1$, these can be proved by induction on p; the n-dimensional case is merely the "product" of n such one-dimensional formulas.

2. Lemmas on holomorphy

Let Γ, U, $\mathfrak{b}$, $\mathfrak{c}$, and Γ_∞ be as in (1.8) and (1.9). Then every element g of $\mathcal{C}_l(\Gamma)$ has a Fourier expansion of the form

$$(2.1) \qquad g(x + iy) = \sum_{a \in \mathfrak{a}} \lambda(a, y) e_F(ax) \qquad (x, y \in \mathbf{R}^n)$$

with C^∞-functions $\lambda(a, y)$ of $y \in \mathbf{R}_+^n$, where e_F is defined by (1.2), and

$$(2.2) \qquad \mathfrak{a} = \{ a \in F \,|\, \mathrm{Tr}_{F/\mathbf{Q}}(a\mathfrak{b}) \subset \mathbf{Z} \},$$

$$(2.3) \qquad \mathbf{R}_+ = \{ y \in \mathbf{R} \,|\, y > 0 \}.$$

Take any subgroup V of $\{ u^2 \,|\, u \in U,\; u \equiv 1 \pmod{4\mathfrak{b}} \}$ of finite index. Then we see that $g(vz) = v^{-l/2} g(z)$ for every $v \in V$ (cf. [6, (7.28)]) and hence

$$(2.4) \qquad \lambda(va, y) = v^{l/2} \lambda(a, vy) \text{ for every } v \in V.$$

We say then that g is V-*invariant*. Let $f \in \mathfrak{M}_k(\Gamma)$ with $k \in 2^{-1} I_F$. Then f has a Fourier expansion

$$(2.5) \qquad f(z) = \sum_{a \in \mathfrak{a}} \xi(a) e_F(az) \qquad (a = 0 \quad \text{or} \quad a \gg 0).$$

We observe that the V-invariance of f means

$$(2.6) \qquad \xi(va) = v^{k/2} \xi(a) \quad \text{for every } v \in V.$$

It is necessary for our later purposes to treat $\mathbf{d}^p f$ or $\mathbf{d}^p g$ which are not necessarily automorphic forms. Therefore we consider the set $\mathcal{C}_l'(V, \mathfrak{b})$ of all C^∞-functions on $\mathfrak{H}^n$ of the form (2.1) whose coefficients satisfy (2.4). We also denote by $\mathfrak{M}_k'(V, \mathfrak{b})$ the set of all functions f on $\mathfrak{H}^n$ of the form

$$(2.7) \qquad f(z) = \sum_{p \in P} \sum_{a \in \mathfrak{a}} y^{-p} \xi_p(a) e_F(az) \qquad (a = 0 \quad \text{or} \quad a \gg 0),$$

$$\xi_p(va) = v^{(k/2)-p} \xi_p(a) \quad \text{for every } v \in V,$$

where P is a finite subset of $2^{-1}I_F$. Notice that $\mathbf{d}^p$, δ_k^p, and multiplication by y^{-p} send $\mathfrak{M}_k'(V, \mathfrak{b})$ and $\mathcal{C}_k'(V, \mathfrak{b})$ into $\mathfrak{M}_{k+2p}'(V, \mathfrak{b})$ and $\mathcal{C}_{k+2p}'(V, \mathfrak{b})$.

For $f \in \mathcal{C}_k'(V, \mathfrak{b})$ and $g \in \mathcal{C}_l'(V, \mathfrak{b})$, we define a function $D_{k,l}(s, f, g)$ of $s \in \mathbf{C}$ by

$$(2.8) \quad D_{k,l}(s, f, g) = [V] N(\mathfrak{b})^{-1} D_F^{-1/2} \int_{L \times M} \overline{f(z)}\, g(z) y^{sF+(k+l)/2}\, dx\, d^{\times}y.$$

Here D_F is the discriminant of F; $L = \mathbf{R}^n/\mathfrak{b}$, $M = \mathbf{R}_+^n/V$; $z = x + iy$; $x \in L$ and $y \in M$; $d^{\times}y$ is the multiplicative measure. If f and g are given by (2.7) and (2.1), we have formally

$$(2.9) \quad D_{k,l}(s, f, g)$$

$$= [V] \sum_{p \in P} \sum_{a \in \mathfrak{a}} \overline{\xi_p(a)} \int_M \lambda(a, y) e_F(iay) y^{sF-p+(k+l)/2} d^{\times}y$$

$$= [V] \sum_{p \in P} \overline{\xi_p(0)} \int_M \lambda(0, y) y^{sF-p+(k+l)/2} d^{\times}y$$

$$+ [V] \sum_{p \in P} \sum_{0 \ll a \in \mathfrak{a}/V} \overline{\xi_p(a)} \int_{\mathbf{R}_+^n} \lambda(a, y) e_F(iay) y^{sF-p+(k+l)/2} d^{\times}y,$$

where the last sum is taken over all different aV with $0 \ll a \in \mathfrak{a}$. To make this meaningful, we assume that each integral of (2.8) and (2.9) as well as the whole series on the right-hand side of (2.9) is absolutely convergent for sufficiently large $\mathrm{Re}(s)$, which will always be the case in our later applications. We obtain, directly from our definition,

$$(2.10) \qquad D_{k,l}(s, f, g) = D_{k-2p, l-2q}(s, y^p f, y^q g).$$

The right-hand side of (2.8) is convergent for sufficiently large $\mathrm{Re}(s)$ if the following condition is satisfied.

(2.11) *For every positive number c, there exists a positive number A such that $|y^{(k+l)/2} fg| \leq A y^{-cF}$ for all $y \in \mathbf{R}_+^n$.*

LEMMA 2.1. *Let $f \in \mathcal{C}_k'(V, b)$ and $g \in \mathcal{C}_l'(V, \mathfrak{b})$. Fix an index $v \leq n$ and suppose that (2.11) is satisfied and also that $\partial f/\partial x_v \cdot g, \partial f/\partial y_v \cdot g, f \cdot \partial g/\partial x_v, f \cdot \partial g/\partial y_v$ satisfy (2.11) with $k + l + 2\tau_v$ in place of $k + l$, where $z_v = x_v + iy_v$. Put $(k_v + l_v)/2 = t + u$ with complex numbers t and u. Then, for sufficiently large $\mathrm{Re}(s)$, we have*

$$(2.12) \qquad 4\pi D_{k, l+2\tau_v}(s, f, \delta_v(t)g) - 2i D_{k-2\tau_v, l}\left(s, y_v^2 \cdot \partial f/\partial \bar{z}_v, g\right)$$

$$= (s + u) D_{k,l}(s, f, g).$$

In particular, if f is holomorphic, we have

$$(2.13) \qquad 4\pi D_{k,\,l+2\tau_\nu}(s,f,\delta_\nu(t)g) = (s+u)D_{k,\,l}(s,f,g).$$

Notice that $y_\nu^2 \partial f/\partial \bar{z}_\nu \in \mathcal{C}'_{k-2\tau_\nu}(V,\mathfrak{b})$, and moreover $y_\nu^2 \partial f/\partial \bar{z}_\nu \in \mathcal{C}_{k-2\tau_\nu}$ if $f \in \mathcal{C}_k$.

Proof. Take for simplicity ν to be n, and put $\alpha = (k+l)/2$. Observe that

$$(\partial/\partial z_n)(\bar{f}gy^{sF+\alpha}) = (\partial\bar{f}/\partial z_n)gy^{sF+\alpha}$$
$$+ 2\pi i \bar{f}\delta_n(t)gy^{sF+\alpha} - (i/2)(s+u)\bar{f}gy^{sF+\alpha-\tau_n}.$$

Therefore, with V, L, and M as in (2.8), we have

$$-2i[V]N(\mathfrak{b})^{-1}D_F^{-1/2}\int_{L\times M}(\partial/\partial z_n)(\bar{f}gy^{sF+\alpha})y_n\,dx\,d^\times y$$
$$= -2iD_{k-2\tau_n,\,l}(s,y_n^2\partial f/\partial\bar{z}_n,g) + 4\pi D_{k,\,l+2\tau_n}(s,f,\delta_n(t)g)$$
$$-(s+u)D_{k,\,l}(s,f,g).$$

Therefore, to prove (2.12), it is sufficient to show

$$\int_L(\partial/\partial x_n)(\bar{f}gy^{sF+\alpha})\,dx = 0, \qquad \int_M(\partial/\partial y_n)(\bar{f}gy^{sF+\alpha})y_n\,d^\times y = 0$$

for sufficiently large $\mathrm{Re}(s)$. The first integral is the constant term of the Fourier expansion of $(\partial/\partial x_n)(\bar{f}gy^{sF+\alpha})$, which is obviously 0. To deal with the second one, define the action of V on $\mathbf{R}_+^{n-1}$ by taking the projection to the first $n-1$ coordinates, and take a fundamental domain K of $\mathbf{R}_+^{n-1}/V$. Then we can put $M = K \times \mathbf{R}_+$. Keeping the variable x constant, put

$$h(y_n,s) = \int_K \bar{f}gy^{sF+\alpha}d^\times(y_1,\ldots,y_{n-1}).$$

Then the integral in question is

$$\int_0^\infty (\partial h/\partial y_n)(y_n,s)\,dy_n.$$

From (2.11), we see easily that

$$\lim_{y_n\to\infty} h(y_n,s) = \lim_{y_n\to 0} h(y_n,s) = 0 \quad \text{if} \quad \mathrm{Re}(s) > 1,$$

which proves the desired vanishing.

Relation (2.13) will play a key role in our proof of holomorphy of automorphic forms. As an application of (2.13), we obtain

$$D_{k,\,l+2h}(s,f,\mathbf{d}^h g) = (4\pi)^{-|h|}\frac{\Gamma_n(s+h+(k+l)/2)}{\Gamma_n(s+(k+l)/2)}D_{k,\,l}(s,f,g)$$

for every $h \in I_F$, ≥ 0, provided that f is holomorphic and the derivatives of f and g satisfy (2.11). In fact, if $|h| = 1$, this is (2.13) with $t = 0$; the general case can be obtained by induction.

Let us now consider the case where $f \in \mathcal{C}_k$ and $g \in \mathcal{C}_l$. We can find a group Γ of type (1.8) such that $f \in \mathcal{C}_k(\Gamma)$ and $g \in \mathcal{C}_l(\Gamma)$. Take $V = \{u^2 \mid u \in U\}$ with

$$U = \{u \in U_F \mid u \equiv 1 \pmod{\mathfrak{c}}\}.$$

Then $L \times M$ is a fundamental domain of $\Gamma_\infty(U, \mathfrak{b}) \setminus \mathfrak{H}^n$. Suppose that

$$(2.14) \qquad k_1 - l_1 \equiv \cdots \equiv k_n - l_n \equiv \kappa \pmod{2} \text{ with } \kappa \in \mathbf{Z}.$$

Then we can write $k - l = \kappa F + 2t$ with $t \in I_F$. We assume $\mathfrak{c} \subset 4\mathfrak{d}$ if $k \notin I_F$. Then (2.8) can be transformed into the form

$$(2.15) \quad 2^{n-1} N(\mathfrak{b}) D_F^{1/2} D_{k,l}(s, f, g)$$

$$= \int_\Phi \overline{f(z)} g(z) E_{\kappa,t}(z, s + 1 - (\kappa/2)) y^{(s+1)F+(k+l)/2} \, d\mu(z)$$

$$= \int_\Phi \overline{f(z)} g(z) \overline{E_{-\kappa,-t}(z, \bar{s} + 1 + (\kappa/2))} \, y^{(s+1)F+(k+l)/2} \, d\mu(z),$$

where $E_{\kappa,t}$ is defined by (1.11) and $\Phi = \Gamma \setminus \mathfrak{H}^n$.

We now introduce a condition under which the convergence is guaranteed. Given $g \in \mathcal{C}_l$ and $\alpha = \left(\begin{smallmatrix} * & * \\ c & d \end{smallmatrix}\right) \in \mathrm{SL}_2(F)$, put $g_\alpha(z) = (cz + d)^{-l} g(z)$ with any choice of a branch of $(cz + d)^{-l}$. Then $g_\alpha \in \mathcal{C}_l$. We say that g is *rapidly decreasing* if, for every α and every $c \in \mathbf{R}$, there exist positive real numbers A and B (depending on α and c) such that

$$(2.16) \qquad |y^{l/2} g_\alpha(x + iy)| \leq A y^{cF} \quad \text{if} \quad y^F > B.$$

It is well known that a cusp form is rapidly decreasing. We say that g is *slowly increasing* if, for every α, there exist positive numbers A, B, c satisfying (2.16). We can easily show that if g is slowly increasing, then

$$(2.17) \qquad |y^{l/2} g(x + iy)| \leq B(y^{cF} + y^{-cF}) \quad \text{for all } y \in \mathbf{R}_+^n$$

with some positive constants B and c. If $f \in \mathfrak{M}_k$, then $\delta_k^p f$ is slowly increasing for every p.

LEMMA 2.2. *Let $f \in \mathcal{C}_k$ and $g \in \mathcal{C}_l$. Suppose fg is rapidly decreasing. Then the right-hand side of (2.8) is absolutely convergent and defines $D_{k,l}(s, f, g)$ as a holomorphic function in s for sufficiently large* $\mathrm{Re}(s)$. *Moreover, suppose (2.14) is satisfied. Then:*

(i) $D_{k,l}(s, f, g)$ can be continued to a meromorphic function on the whole s-plane;

(ii) if $k \neq l$, $D_{k,l}(s, f, g)$ is holomorphic for $\mathrm{Re}(s) \geq -1/2$;

(iii) *if $k = l$, $D_{k,k}(s, f, g)$ is holomorphic for* $\mathrm{Re}(s) > 0$, *and has at most a simple pole at* $s = 0$; *the residue at* $s = 0$ *is a nonzero rational number times* $R_F\langle f, g \rangle$.

Proof. That the right-hand side of (2.8) is convergent for large $\mathrm{Re}(s)$ follows from (2.16) and (2.17); in fact, fg satisfies (2.11). Assertion (i) follows from (2.15) and the well known behavior of $E_{\kappa, l}$ at cusps. Assertions (ii) and (iii) follow from Proposition 1.1 and (2.15) combined with the fact that $\mu(\Gamma \setminus \mathfrak{H}^n)$ is a rational number times π^n.

LEMMA 2.3. *Let $f \in \mathcal{C}_k$ and $g \in \mathcal{C}_l$. Suppose that $k = l + 2\tau_\nu$ for some $\nu \leq n$ and $fg, y_\nu^2 \partial f / \partial \bar{z}_\nu \cdot g, f\delta_\nu(l_\nu)g$ are rapidly decreasing; suppose also that the assumptions of Lemma 2.1 concerning the derivatives of f and g are satisfied. Then*

$$(2.18) \qquad - 2\pi i \langle f, \delta_\nu(l_\nu)g \rangle = \langle y_\nu^2 \partial f / \partial \bar{z}_\nu, g \rangle.$$

In particular, if f is holomorphic,

$$(2.19) \qquad \langle f, \delta_\nu(l_\nu)g \rangle = 0.$$

Proof. Taking $t = l_\nu$ and $u = 1$ in (2.12), we obtain

$$4\pi D_{k,k}(s, f, \delta_\nu(l_\nu)g) - 2iD_{l,l}(s, y_\nu^2 \partial f / \partial \bar{z}_\nu, g) = (s + 1)D_{k,l}(s, f, g).$$

By Lemma 2.2, the right-hand side is holomorphic at $s = 0$; the residue of each term of the left-hand side at $s = 0$ is a scalar times each inner product of (2.18). Since these scalar factors for both terms are the same, as can be seen from (2.15), we obtain (2.18).

In an earlier version of the present paper, the author proved Lemmas 2.1 and 2.3 only for holomorphic f. The consideration of $\partial f / \partial \bar{z}_\nu$ for non-holomorphic f was suggested by R. Indik; formula (2.18) is due to him.

3. Analytic continuation of zeta functions

To define the zeta function mentioned in the introduction, we consider a subfield E of F and denote by $\tau'_1, \ldots, \tau'_r$ the elements of J_E, where $r = [E: \mathbf{Q}]$. Changing the order of $\tau_1, \ldots, \tau_n$ if necessary, we assume $\tau'_i = \mathrm{Res}_{F/E}(\tau_i)$ for $i = 1, \ldots, r$. We shall often identify $\Sigma_{i=1}^r \mathbf{Z}\tau_i$ with $I_E = \Sigma_{i=1}^r \mathbf{Z}\tau'_i$ in an obvious way, and put $x^E = x^{\tau_1} \cdots x^{\tau_n}$ even for $x \in F$ if there is no fear of confusion.

Let f and Ω be elements of $\mathfrak{M}_{m, F}$ and $\mathfrak{M}_{l, E}$ with $0 \neq m \in 2^{-1}I_F$ and $0 \leq l \in 2^{-1}I_E$ and with expansions

$$(3.0a) \qquad f(w) = \sum_{b \in F} \xi(b)e_F(bw) \qquad (w \in \mathfrak{H}^n),$$

$$(3.0b) \qquad \Omega(z) = \sum_{a \in E} \omega(a)e_E(az) \qquad (z \in \mathfrak{H}^r).$$

Let U be a subgroup of U_E of finite index such that both f and Ω are U-invariant. Fix an element ζ of F such that

$$(3.1) \qquad \zeta^{\tau_i} > 0 \text{ for } i \leq r \text{ and } \zeta^{\tau_i} < 0 \quad \text{for } i > r.$$

Let $\omega_p(a) = a^p \omega(a)$ with an arbitrary $p \in I_E, \geq 0$. Observe that

$$\mathbf{d}^p \Omega(z) = \sum_{a \in E} \omega_p(a) \mathbf{e}_E(z).$$

Now our object of study is an infinite series

$$(3.2) \qquad \Lambda(s) = [U] \sum_{0 \ll b \in F/U} \xi(b) \omega_p\big(\mathrm{Tr}_{F/E}(\zeta b)\big) b^{d - sE} \qquad (s \in \mathbf{C}).$$

Here $d = \sum_{i=1}^n d_i \tau_i \in I_F$ and the sum is extended over all totally positive elements b of F modulo multiplication by the elements of U. To make each term of (3.2) dependent only on Ub, we assume

$$(3.3) \qquad l + 2p + \mathrm{Res}_{F/E}(m + 2d) = \alpha E \quad \text{with } \alpha \in 2^{-1}\mathbf{Z}.$$

The factor $[U]$ $(= [U_E : U]^{-1})$ makes the sum independent of the choice of U. We also assume

$$(3.4) \qquad d_i \geq 0 \quad \text{for } i > r.$$

We can express $(\mathrm{Tr}(\zeta b))^p b^d$ as a finite sum $\sum \eta_x b^x$ with $x \in I_F$ and algebraic numbers η_x independent of b. Therefore Λ is a finite $\overline{\mathbf{Q}}$-linear combination of the series of the same type with $p = 0$ but with exponents x, instead of d, such that $\mathrm{Res}_{F/E}(x - d) = p$. The invariant α determined for the new exponent x is the same as that obtained from p and d. Thus the consideration of $\mathbf{d}^p \Omega$ in addition to Ω is inessential, but will make our later treatment easier.

PROPOSITION 3.1. *Let θ be -1 if Ω is a constant; otherwise let θ be a real number $\geq -1/2$ such that*

$$\omega(a) a^{-l/2} = O\big(N_{E/\mathbf{Q}}(a)^\theta\big) \qquad (0 \neq a \in E).$$

Similarly, let σ be a real number $\geq -1/2$ such that

$$\xi(b) b^{-m/2} = O\big(N_{F/\mathbf{Q}}(b)^\sigma\big) \qquad (0 \neq b \in F).$$

Then (3.2) is absolutely convergent if

$$(3.5) \quad \mathrm{Re}(s) > (\alpha/2) + (1 + \sigma)[F : E] + \theta \quad \textit{with the number } \alpha \textit{ of } (3.3).$$

If f is a cusp form, we can take $\sigma = 0$.

THEOREM 3.2. *Suppose that f is a cusp form and α is an integer. Then Λ can be continued to a meromorphic function on the whole plane which is holomorphic for $\mathrm{Re}(s) > \alpha/2$ and has at most a simple pole at $s = \alpha/2$.*

Moreover, if $2(m_i + d_i) > \alpha$ *for at least one* $i \leqq r$, *then* Λ *is holomorphic for* $\mathrm{Re}(s) \geqq (\alpha - 1)/2$.

Proof of Proposition 3.1. Changing s for $s + c$ with a suitable integer c, we may assume that $d_i \geqq 0$ for all i; also, for the reason explained above, it is sufficient to consider the case $p = 0$. Let $\tau_{i1}, \ldots, \tau_{iq}$ for each $i \leqq r$ denote the elements of J_F whose restrictions to E are τ_i'; we put $\tau_i = \tau_{i1}$. Define a quadratic form $T[x]$ on $\mathbf{R}^q$ by $T[x] = \sum_{j=1}^q x_j^2$ for $x = (x_1, \ldots, x_q) \in \mathbf{R}^q$, and also an embedding $b \mapsto b^{(i)}$ of F into $\mathbf{R}^q$ by $b^{(i)} = (b^{\tau_{i1}}, \ldots, b^{\tau_{iq}})$ for each $i \leqq r$. Now sum (3.2) is taken under the condition $b \gg 0$ and $\mathrm{Tr}_{F/E}(\zeta b)^{\tau_i} \geqq 0$ for all i. Therefore, by (3.1), $|(\zeta b)^{\tau_{ij}}| \leqq (\zeta b)^{\tau_i}$ for all $j \leqq q$ and all $i \leqq r$. Hence $T[b^{(i)}] \leqq A_i b^{2\tau_i}$ for such b with a positive constant A_i independent of b, and hence

$$|b^{-sE}| \leqq C_1 \prod_{i=1}^r |T[b^{(i)}]^{-s/2}| \qquad\qquad (\mathrm{Re}(s) > 0).$$

Here and henceforth $C_1, C_2, \ldots$ mean positive constants independent of b. Since $\mathrm{Tr}_{F/E}(\zeta b)^{\tau_i} \leqq (\zeta b)^{\tau_i} \leqq \zeta^{\tau_i} T[b^{(i)}]^{1/2}$, we have

$$|\omega(\mathrm{Tr}_{F/E}(\zeta b))| \leqq C_2 \prod_{i=1}^r T[b^{(i)}]^{(2\theta + l_i)/4}$$

if $\mathrm{Tr}_{F/E}(\zeta b) \neq 0$. Similarly we have

$$|\xi(b)| \leqq C_3 b^{(m/2) + \sigma F} \leqq C_4 \prod_{i=1}^r b^{k_i \tau_i} \leqq C_4 \prod_{i=1}^r T[b^{(i)}]^{k_i/2},$$

where $\sum_{i=1}^r k_i \tau_i' = \mathrm{Res}_{F/E}(m/2) + \sigma q E$, and

$$b^d \leqq C_5 \prod_{i=1}^r T[b^{(i)}]^{e_i/2},$$

where $\sum_{i=1}^r e_i \tau_i' = \mathrm{Res}_{F/E}(d)$. Thus each term of (3.2) such that $\mathrm{Tr}_{F/E}(\zeta b) \neq 0$ can be majorized by

$$(*) \qquad\qquad C_6 \prod_{i=1}^r \left| T[b^{(i)}]^{(\alpha + 2\theta - 2s + 2\sigma q)/4} \right|.$$

The sum of such terms is convergent if $\mathrm{Re}(s) > (\alpha/2) + \theta + (1 + \sigma)q$ by virtue of [6, Lemma 8.1]. Each term with $\mathrm{Tr}_{F/E}(\zeta b) = 0$ can be majorized by (*) with $\alpha - l_i$ in place of $\alpha + 2\theta$. (If $\omega(0) \neq 0$, we have $l_1 = \cdots = l_r$.) Such b's belong to a $(q - 1)$-dimensional space over E. Therefore, again by [6, Lemma 8.1], the convergence condition is

$$\mathrm{Re}(s) > (\alpha - l_i)/2 + (1 + \sigma)q - 1.$$

This completes the proof.

Before proving Theorem 3.2, we introduce a certain pull-back of a C^∞-function g on $\mathfrak{H}^n$ to $\mathfrak{H}^r$ relative to an element ζ of F satisfying (3.1). Namely, we define a function g^ς on $\mathfrak{H}^r$ by

$$(3.6) \qquad g^\varsigma(z_1,\ldots,z_r) = g(z_1^*,\ldots,z_n^*),$$

$$z_\lambda^* = \begin{cases} \varsigma^{\tau_\lambda} z_\lambda & \text{if } \lambda \leq r, \\ \varsigma^{\tau_\lambda} \bar{z}_\nu & \text{if } \lambda > r \text{ and } \text{Res}_{F/E}(\tau_\lambda) = \tau_\nu'. \end{cases}$$

LEMMA 3.3. *Suppose* $g \in \mathcal{C}_{m,F}$. *Let* $\mu = \text{Res}_{F/E}(\Sigma_{i>r} m_i \tau_i) = \Sigma_{i=1}^r \mu_i \tau_i'$ *and* $\kappa = \Sigma_{i=1}^r (m_i - \mu_i)\tau_i'$. *Then* $\text{Im}(z)^\mu g^\varsigma(z)$ *belongs to* $\mathcal{C}_{\kappa,E}$.

Proof. This is straightforward if $m \in I_F$. Our problem is the consistency of the factors of automorphy in the case $m \notin I_F$. To settle this point, we consider the theta function

$$\theta(w) = \sum_{b \in \mathfrak{r}} \mathbf{e}_F(b^2 w/2) \qquad\qquad (w \in \mathfrak{H}^n).$$

We have

$$\theta^\varsigma(z) = \sum_{b \in \mathfrak{r}} \mathbf{e}\left(\sum_{\nu=1}^r x_\nu \text{Tr}_{F/E}(\varsigma b^2/2)^{\tau_\nu} + \sum_{\nu=1}^r iy_\nu \big[\varsigma b^2 - \text{Tr}_{F/E}(\varsigma b^2/2)\big]^{\tau_\nu} \right).$$

Observe that this is a special case of [6, (7.6)] with E as the basic field. Therefore, by [6, Prop. 7.1],

$$\theta^\varsigma(\gamma(z)) = \left(\frac{2c}{d\mathfrak{r}_E} \right)^q (cz+d)^{E/2}(c\bar{z}+d)^{(q-1)E/2} \theta^\varsigma(z)$$

$$\text{for every } \gamma = \left(\begin{smallmatrix} * & * \\ c & d \end{smallmatrix}\right) \in \Gamma, \quad d \gg 0,$$

where $q = [F:E]$ and Γ is a congruence subgroup of $\text{SL}_2(\mathfrak{r}_E)$. This shows, by virtue of [6, Lemma 7.4], that $\text{Im}(z)^{(q-1)E/2}\theta^\varsigma(z)$ belongs to $\mathcal{C}_{(2-q)E/2}$. If $g \in \mathcal{C}_m$ with $m \notin I_F$, we have $m = l + (1/2)F$ with $l \in I_F$, and therefore the desired consistency follows from the above fact about θ^ς.

Proof of Theorem 3.2. We may again assume $p = 0$ for the reason explained before Proposition 3.1. We also observe, for $0 \leq h \in I_F$, that

$$(3.7) \qquad (\mathbf{d}^h f)^\varsigma = \sum_b \xi(b) b^h \mathbf{e}_E\big(x\text{Tr}_{F/E}(\varsigma b) + iy[2\varsigma b - \text{Tr}_{F/E}(\varsigma b)]\big).$$

Put

$$(3.8a) \qquad \Omega_\rho(z) = \sum_a \overline{\omega(a)}\, \mathbf{e}_E(az),$$

$$(3.8b) \qquad k = \sum_{i=1}^r m_i \tau_i', \qquad \mu = \text{Res}_{F/E}\left(\sum_{i>r} m_i \tau_i \right), \qquad e = \sum_{i=1}^r d_i \tau_i',$$

$$(3.8c) \qquad q = \sum_{i>r} d_i \tau_i, \qquad q' = \text{Res}_{F/E}(q).$$

142   GORO SHIMURA

Now we have, by (1.16a),

$$(3.9) \qquad (\mathbf{d}^q f)^{\zeta} = \sum_{0 \leq u \leq q} a_u (\pi y)^{u' - q'} (\delta_m^u f)^{\zeta}$$

with $a_u \in \overline{\mathbf{Q}}$, where $u' = \mathrm{Res}_{F/E}(u)$. By Lemma 3.3, we see that

$$y^{\mu + 2u'} (\delta_m^u f)^{\zeta} \in \mathcal{C}_{k - \mu - 2u', E}.$$

From (2.9) and (3.7), we obtain, by an easy calculation,

(3.10)

$$D_{l, k - \mu - 2q'}\!\left(s - (\alpha/2), \Omega_\rho, y^{\mu + 2q'} (\mathbf{d}^q f)^{\zeta} \right) = (4\pi\zeta)^{e - sE} \prod_{i=1}^{r} \Gamma(s - d_i) \Lambda(s).$$

On the other hand, by (2.10) and (3.9), this is equal to

$$(3.11) \qquad \sum_u a_u \pi^{u' - q'} D_{l, k - \mu - 2u'}\!\left(s - (\alpha/2), \Omega_\rho, y^{\mu + 2u'} (\delta_m^u f)^{\zeta} \right).$$

We can now apply Lemma 2.2 to each term, since f is a cusp form and the assumption that $\alpha \in \mathbf{Z}$ implies (2.14). Thus we obtain the first assertion of Theorem 3.2. Suppose $2(m_i + d_i) > \alpha$ for at least one $i \leq r$. Then $l \neq k - \mu - 2u'$ for all u. Therefore, by Lemma 2.2, each term of (3.11) is holomorphic for $\mathrm{Re}(s) \geq (\alpha - 1)/2$. This completes the proof.

THEOREM 3.4. *The notation and the assumption being the same as in Theorem 3.2, suppose that $2(m_i + d_i) = \alpha$ for every $i \leq r$. Then the residue of Λ at $s = \alpha/2$ can be expressed in the form*

$$\prod_{i=1}^{r} \Gamma(m_i)^{-1} \pi^{|k|} R_E \langle \Omega_\rho, \sum_v c_v y^{\mu + 2v'} (\delta_m^v f)^{\zeta} \rangle$$

with $c_v \in \overline{\mathbf{Q}}$. Here $\sum_v$ is a finite sum over the elements v of I_F of the form $v = \sum_{i > r} v_i \tau_i$ and $v' = \mathrm{Res}_{F/E}(v)$; Ω_ρ, μ, and k are defined by (3.8a, b) and the superscript ζ is defined by (3.6); v must satisfy $l = k - \mu - 2v'$.

Proof. We first assume $p = 0$, and use the same notation as in the above proof. Our condition $2(m_i + d_i) = \alpha$ implies that $l = k - \mu - 2q'$. Therefore each term of (3.11) is holomorphic at $s = \alpha/2$ unless $q = u$. Thus, by (iii) of Lemma 2.2, the residue of (3.11) at $s = \alpha/2$ is a nonzero rational number times $a_q R_E \langle \Omega_\rho, y^{\mu + 2q'} (\delta_m^q f)^{\zeta} \rangle$. Since a_q is an algebraic number, we obtain our assertion in the case $p = 0$. To prove the general case, call $\Lambda(p, d)$ the given series. As observed before, we have $\Lambda(p, d) = \sum_x \eta_x \Lambda(0, x)$ with $\eta_x \in \overline{\mathbf{Q}}$ and $x \in I_F$ such that $d \leq x$. Each term is either holomorphic at $s = \alpha/2$ or has the residue of the desired form, since $2(m_i + x_i) \geqq 2(m_i + d_i) = \alpha$ for every $i \leqq r$. This completes the proof.

4. Main theorems on zeta functions

Let K be a totally imaginary quadratic extension of F. Throughout the rest of the paper, we denote by ρ the complex conjugation, and often consider ρ also an element of J_K. Let E be a subfield of F of degree r as in Section 3. We fix r elements $\psi_1, \ldots, \psi_r$ whose restrictions to F are $\tau_1, \ldots, \tau_r$, respectively. Take an element

$$(4.1) \qquad \Omega(z) = \sum_{a \in E} \omega(a) \mathbf{e}_E(az) \qquad\qquad (z \in \mathfrak{H}^r)$$

of $\mathfrak{M}_{l, E}$ with $0 \leqq l \in I_E$ and let $\omega_p(a) = \omega(a)a^p$ for $0 \leqq p \in I_E$. We now define a series

$$(4.2) \qquad D(s) = [U] \sum_{x \in X/U} \omega_p\big(\mathrm{Tr}_{F/E}(\zeta x x^\rho)\big) x^\Phi |x^\psi|^{-2s} \qquad (s \in \mathbf{C}).$$

Here X is a congruence subset of $K - \{0\}$ in the sense of [6, §8]; U is a subgroup of U_E of finite index such that Ω is U-invariant and $uX = X$ for all $u \in U$; ζ is an element of F satisfying (3.1); $\psi = \sum_{i=1}^r \psi_i$; Φ is an element of I_K satisfying

$$(4.3) \quad \Phi = \sum_{\gamma \in J_K} c(\gamma)\gamma \text{ with } c(\gamma) \geqq 0 \quad \text{for } \gamma \notin \{\psi_1, \ldots, \psi_r, \psi_1\rho, \ldots, \psi_r\rho\};$$

the sum is taken over all different xU with $x \in X$. To make each term dependent only on xU, we assume

$$(4.4) \qquad \mathrm{Res}_{K/E}(\Phi) + l + 2p = \beta E \quad \text{with } \beta \in \mathbf{Z}.$$

PROPOSITION 4.1. *Let θ be the same as in Proposition 3.1. Then (4.2) is absolutely convergent if*

$$(4.5) \qquad \mathrm{Re}(s) > (\beta/2) + [F:E] + \theta.$$

This is a restatement of [6, Proposition 9.1].

Changing ψ_i for $\psi_i\rho$ if necessary, we may assume

$$(4.6) \qquad c(\psi_i) \geqq c(\psi_i\rho) \quad \text{for } i = 1, \ldots, r.$$

THEOREM 4.2. *Suppose $\Phi \notin (1 + \rho)I_K$. Then $D(s)$ can be continued to a meromorphic function on the whole plane which is holomorphic for $\mathrm{Re}(s) > (\beta + [F:E])/2$, and has at most a simple pole at $s = (\beta + [F:E])/2$. Moreover, suppose*

$$(4.7) \qquad 2c(\psi_i) > \beta + [F:E] - 2 \quad \text{for at least one } i \leqq r.$$

Then D is holomorphic for $\mathrm{Re}(s) \geqq (\beta + [F:E] - 1)/2$.

This is actually a special case of Theorem 3.2 as will be shown below.

THEOREM 4.3. *Suppose* $\omega(a) \in \overline{\mathbf{Q}}$ *for all* $a \in E$. *Let* μ *be an integer satisfying*

$$(4.8\text{a}) \qquad (\beta + [F:E] - 1)/2 \leqq \mu \leqq c(\psi_i) \quad \text{for } i = 1,\ldots,r.$$

$$(4.8\text{b}) \qquad \beta + [F:E] \neq 2\mu \quad \text{if } E = \mathbf{Q} \neq F.$$

Then D is holomorphic at μ, and $D(\mu) \sim \pi^e p_K(\Phi, 2\psi)$, where $e = \Sigma_{i=1}^r\{c(\psi_i) - c(\psi_i\rho)\}$ and p_k is the symbol of [6, §1].

It is conjecturable that condition (4.8b) is unnecessary.

Remark 4.4. The existence of μ satisfying (4.8a) implies (4.7) and also that

$$c(\psi_i) - c(\psi_i\rho) \geqq l_i + [F:E] - 1 \quad \text{for } i = 1,\ldots,r.$$

The right-hand side is 0 only if Ω is a constant and $F = E$; $D = 0$ in such a case. Therefore if this trivial case is excluded, we have $c(\psi_i) > c(\psi_i\rho)$ for every $i \leqq r$, so that $\Phi \notin (1 + \rho)I_K$.

Remark 4.5. To emphasize the exponents p and Φ, denote by $D(p, \Phi)$ the above series with fixed X and Ω. For the reason explained in Section 3, $D(p, \Phi)$ is a finite $\overline{\mathbf{Q}}$-linear combination of $D(0, \Phi + (1 + \rho)\Xi)$ with $0 \leqq \Xi \in I_K$ such that $\text{Res}_{K/E}(\Xi) = p$. The invariants β defined for $D(p, \Phi)$ and $D(0, \Phi + (1 + \rho)\Xi)$ are the same; $p_K(\Phi + (1 + \rho)\Xi, 2\psi) = p_K(\Phi, 2\psi)$ by [6, Theorem 1.1, (2)]; condition (4.7) or (4.8) for $D(p, \Phi)$ implies that for $D(0, \Phi + (1 + \rho)\Xi)$. Thus Theorems 4.2 and 4.3 can be reduced to the case $p = 0$.

The above theorems provide a stronger version of [6, Theorem 9.2]; the symbols K_0, F, n, η there correspond to F, E, r, ζ here.

We now show that (4.2) is a special case of (3.2). Take $\psi_{r+1}, \ldots, \psi_n \in J_K$ so that $\text{Res}_{K/F}(\psi_i) = \tau_i$; let $\Psi = \Sigma_{i=1}^n (\lambda_i - 1)\psi_i \in I_K$ and

$$(4.9) \qquad f(z) = \sum_{b \in X} b^\Psi e_F(bb^\rho z)$$

with $\lambda_i > 0$ and an arbitrary congruence subset X of K. Then $f \in \mathfrak{M}_{\lambda, F}$ with $\lambda = \Sigma_{i=1}^n \lambda_i \tau_i$ by [6, Prop. 7.1] (cf. also [5, §5]). Given $d = \Sigma_{i=1}^n d_i \tau_i \in I_F$ ($d_i \geqq 0$ for $i \geqq 0$), define the above D with

$$(4.10) \qquad \Phi = \Psi + \sum_{i=1}^n d_i(\psi_i + \psi_i\rho),$$

and also Λ of (3.2) with the present f. Then we see easily that

$$(4.11) \qquad D(s) = 2^{r-1}\Lambda(s).$$

Condition (3.3) is equivalent to (4.4); we have $\alpha = \beta + [F:E]$. Conversely, given $\Phi = \Sigma_\gamma c(\gamma) \in I_K$ as in (4.3), put $\lambda_i = c(\psi_i) - c(\psi_i\rho) + 1$ and $d_i = c(\psi_i\rho)$. Here we assume that $c(\psi_i) \geqq c(\psi_i\rho)$ for all i, changing ψ_i for $\psi_i\rho$ if necessary. Then we have (4.10) and (4.11); f is a cusp form if $\Phi \notin (1 + \rho)I_K$. Thus (4.2) is a special case of (3.2). Therefore Theorem 4.2 follows from Theorem 3.2. Notice also that the quantity $\pi^\nu p_K(\Phi, 2\psi)$ depends only on f, since $p_K(\Phi, 2\psi) = p_k(\Psi, 2\psi)$.

THEOREM 4.6. *Suppose* $\omega(a) \in \overline{\mathbf{Q}}$ *for all* $a \in E$; *suppose further that*

$$(4.12a) \qquad\qquad l \geqq (3 - [F:E])E,$$

$$(4.12b) \qquad c(\psi_i\rho) < c(\psi_i) = (\beta + [F:E] - 2)/2 \quad \text{for every } i \leqq r.$$

Then the residue of D *at* $s = (\beta + [F:E])/2$ *is* $R_E \pi^e p_K(\Phi, 2\psi)$ *times an algebraic number, where* R_E *is the regulator of* E *and* e *is the same as in Theorem 4.3.*

THEOREM 4.7. *Let* $\psi_1, \ldots, \psi_n$ *be the elements of* J_K *such that* $\mathrm{Res}_{K/F}(\psi_i) = \tau_i$. *Let* $\lambda = \Sigma_{i=1}^n \lambda_i \tau_i \in I_F, 0 \leqq q = \Sigma_{i>r} q_i \tau_i \in I_F, q' = \mathrm{Res}_{F/E}(q), \mu = \mathrm{Res}_{F/E}(\Sigma_{i>r}\lambda_i\tau_i)$, *and* $l = \Sigma_{i=1}^r \lambda_i\tau_i' - \mu - 2q'$. *Suppose* $\lambda_i > 0$ *for all* $i \leqq n$ *and* $\lambda_i > 1$ *for all* $i \leqq r$; *suppose further that* $l \geqq E$ *if* $[F:E] = 2$. *Define* f *by* (4.9) *with* $\Psi = \Sigma_{i=1}^n(\lambda_i - 1)\psi_i$. *Then, for every* $\overline{\mathbf{Q}}$-*rational element* g *of* $\mathfrak{M}_{l,E}$, *we have*

$$\langle g, y^{\mu + 2q'}(\delta_\lambda^q f)^\varsigma \rangle \sim \pi^{-r} p_K(\Psi, 2\Sigma_{i=1}^r \psi_i),$$

where ( )$^\varsigma$ *is defined by* (3.6).

Let us now show that Theorem 4.6 is equivalent to Theorem 4.7. Given Φ, define λ_i, d_i and Λ as above so that (4.11) holds. Observe that (4.12b) implies $2(\lambda_i + d_i) = \alpha$ for every $i \leqq r$. Now the residue of Λ is given by Theorem 3.4. Therefore we obtain Theorem 4.6 from Theorem 4.7. Conversely, the notation being as in Theorem 4.7, put $d_i = 1 - \lambda_i$ for $i \leqq r$ and $d_i = q_i$ for $i > r$. Then we have again (4.11) with Φ of (4.10); notice that $\alpha = 2$. As shown in the proof of Theorem 3.4, the residue of (3.10) at $s = \alpha/2$ is a nonzero algebraic number times

$$R_E\langle \Omega_\rho, y^{\mu + 2q'}(\delta_\lambda^q f)^\varsigma \rangle.$$

(Notice that $p = 0$ under the assumptions of Theorem 4.7.) Therefore Theorem 4.6 together with (4.11) implies Theorem 4.7.

Proof of Theorems 4.3 and 4.7 in the case $F = E$. If $F = E$, we see easily that $2^{1-n}D(s + (\beta + 1)/2)$ coincides with $L(s, f, g)$ of [6, (9.9)], where f is defined by (4.10) and $g(z) = \Omega(\zeta^{-1}z)$. Hence Theorem 4.3 in the case $F = E$ is equivalent to [6, Theorem 9.4]. As for Theorem 4.7, we have $\mu = q' = 0$, $q = 0$, and $f^{\zeta} = f(\zeta z)$ if $F = E$. In this case, our assertion is the same as [6, Theorem 9.8].

The proof of Theorems 4.3 and 4.7 in the general case will be given in Sections 8 and 9.

5. Construction of automorphic forms of arithmetic type

As in [6], we fix a vector space V of dimension $m + 2$, and an F-valued F-bilinear form $S(x, y)$ on V. Let V_i and S_i be the localization of V and S at the archimedean prime τ_i. Then we assume that $S_1, \ldots, S_r$ have signature $(m, 2)$, and $S_{r+1}, \ldots, S_n$ are positive definite. We define the group $G_+(S)$, its action on the space $\mathfrak{Z}^r$, and the automorphic forms on $\mathfrak{Z}^r$ as in [6, §3]. We assume also that F has a subfield E of degree r, and denote by $\tau_1', \ldots, \tau_r'$ the elements of J_E as in Section 3; for simplicity, we put

$$(5.1) \qquad\qquad q = [F : E] \; (= n/r).$$

For $0 \leqq \kappa = \sum_{i > r} \kappa_i \tau_i \in I_F$, we denote by $\mathcal{P}_\kappa(S)$ the vector space of $\mathbf{C}$-valued polynomial functions on $V_{r+1} \times \cdots \times V_n$ which are S_i-harmonic in the sense of [6, §2] and homogeneous of degree κ_i in the variable on V_i for each $i > r$. Now our second main object of study is an infinite series f defined by

$$(5.2) \quad f(w, s) = f(w, s; t, X, \omega)$$

$$= [U] \sum_{x \in X/U} \omega\big(\mathrm{Tr}_{F/E}(-S[x])\big) t(x) S[x]^i S\big(x, A^{-1}p(w)\big)^{-k}$$

$$\times \big| S\big(x, A^{-1}p(w)\big) \big|^{-2sE}$$

Here $w \in \mathfrak{Z}^r$, $s \in \mathbf{C}$, $t \in \mathcal{P}_\kappa(S)$; X is a congruence subset of $V - \{0\}$ in the sense of [6, §8]; the $\omega(a)$ are the Fourier coefficients of an element

$$\Omega(z) = \sum_{a \in E} \omega(a) e_E(az) \qquad\qquad (z \in \mathfrak{H}^r)$$

of $\mathfrak{M}_{l, E}$ with $0 \leqq l \in 2^{-1} I_E$; U is a subgroup of U_E of finite index such that Ω is U-invariant and $uX = X$ for all $u \in U$; $0 \leqq j \in I_F$, $k = \sum_{i=1}^r k_i \tau_i' \in I_E$ and $k_i > 0$ for all $i \leqq r$; X/U denotes a complete set of representatives of X modulo scalar multiplication by the elements of U;

$$S\big(x, A^{-1}p(w)\big)^{-k} \big| S\big(x, A^{-1}p(w)\big) \big|^{-2sE}$$

$$= \prod_{i=1}^r S_i\big(x, A_i^{-1}p(w_i)\big)^{-k_i} \big| S_i\big(x, A_i^{-1}p(w_i)\big) \big|^{-2s},$$

where A_i is as in [6, (3.3)] and $p(w)$ is defined by [6, (2.7)].

Our sum is essentially the same as [6, (8.3)]. Here we are imposing the conditions that Ω is a modular form and t is harmonic. These restrictions are compensated by the factor $S[x]^j$ (see Remark 5.5 below). To make each term of (5.2) dependent only on Ux, we assume

$$(5.3) \qquad k - l - \operatorname{Res}_{F/E}(\kappa + 2j) = \alpha'E \quad \text{with} \quad \alpha' \in 2^{-1}\mathbf{Z}.$$

PROPOSITION 5.1. *Let θ be the same as in Proposition 3.1. Then (5.2) is absolutely convergent if*

$$\operatorname{Re}(2s) > (m + 2)q - \alpha' + 2\theta.$$

This is a restatement of [6, Proposition 8.2].

THEOREM 5.2. *Put $\alpha = \alpha' + 2 - q(m + 2)/2$, and suppose*

$$(5.4) \qquad l_1 \equiv \cdots \equiv l_r \equiv mq/2 \ (\mathrm{mod}\,\mathbf{Z}).$$

Then (5.2) as a function of s can be continued to a meromorphic function on the whole plane, which is holomorphic at $s = 0$ if $\alpha > 0$. Moreover, when $\alpha > 0$, $f(w,0)$ is a holomorphic function on $\mathfrak{Z}^r$ provided either $\alpha \neq 2$ or $E \neq \mathbf{Q}$.

Notice that α is an integer under (5.4) and also that the condition $\alpha > 0$ is considerably better than the convergence of $f(w, s)$ at $s = 0$ guaranteed by the above proposition. The theorem will be proved in Section 7.

The notation and the assumption being the same as in Theorem 5.2, let $\{t_1, \ldots, t_d\}$ be a $\overline{\mathbf{Q}}$-rational basis of $\mathscr{P}_\kappa(S)$ over $\mathbf{C}$, and define a representation $\sigma' : \Pi_{i>r}G(S_i) \to \mathrm{GL}_d(\mathbf{C})$ by

$$(5.5) \qquad (t_1 \circ \alpha, \ldots, t_d \circ \alpha) = (t_1, \ldots, t_d) \cdot {}^t\sigma'(\alpha_{r+1}, \ldots, \alpha_n)$$

and also a representation $\sigma : (\mathrm{GL}_1)^r \times (\mathrm{GL}_m)^r \times \Pi_{i>r}G(S_i) \to \mathrm{GL}_d(\mathbf{C})$ by

$$(5.6) \qquad \sigma(a_1, \ldots, a_r, \beta_1, \ldots, \beta_r, \gamma_{r+1}, \ldots, \gamma_n) = a^k \sigma'(\gamma_{r+1}, \ldots, \gamma_n).$$

Assuming $\alpha > 0$, define a $\mathbf{C}^d$-valued function $\mathfrak{f}$ on $\mathfrak{Z}^r$ by

$$(5.7) \qquad \mathfrak{f}(w) = {}^t(f(w,0;t_1, X, \Omega), \ldots, f(w,0;t_d, X, \Omega))$$

with fixed X and Ω. Then Theorem 5.2 together with [6, (8.16)] shows that $\mathfrak{f}$ belongs to the space $\mathfrak{M}_\sigma$ of automorphic forms on $\mathfrak{Z}^r$ defined in [6, §3]. Now we can state a stronger version of [6, Theorem 8.4] as

THEOREM 5.3. *Suppose that $\omega(a) \in \overline{\mathbf{Q}}$ for all $a \in E$, (5.4) is satisfied, and $\alpha > 0$; suppose further that $\alpha \neq 2$ or $E \neq \mathbf{Q}$. Then $\pi^{-|k|}\mathfrak{f}$ is arithmetic in the sense of [6, p. 329], where $|k| = \Sigma_{i=1}^r k_i$.*

THEOREM 5.4. *Suppose that (5.4) is satisfied, $\alpha = 0$, and $j_1 = \cdots = j_r = 0$. Then $f(w, s)$ is holomorphic for $\operatorname{Re}(s) > 1$ and has at most a simple pole at*

$s = 1$. *With $\{t_i\}$ as above, put*

$$g(w) = \eta(w)^E \lim_{s \to 1} (s-1) \cdot {}^t(f(w, s; t_1, X, \Omega), \ldots, f(w, s; t_d, X, \Omega)).$$

Then $g(w)$ is holomorphic on $\mathfrak{Z}^r$ and defines an element of $\mathfrak{M}_\sigma$. Moreover, if $\omega(a)$ is algebraic for every $a \in E$, $R_E^{-1}\pi^{-|k|}g$ is arithmetic, where R_E is the regulator of E.

These theorems will be proved in Sections 7 and 8.

Remark 5.5. We observe that a^i for $a \in F$ is a finite $\mathbf{Q}$-linear combination of $\mathrm{Tr}_{F/E}(a)^p a^h$ with $0 \leqq p \in I_E$ and $0 \leqq h = \Sigma_{i>r} h_i \tau_i \in I_F$. Therefore, if we put $\omega_p(a) = \omega(a)a^p$, then (5.2) is a finite sum of series of the form

$$(5.8) \qquad f'(w, s; t) = [U] \sum_{x \in X/U} \omega_p\big(\mathrm{Tr}_{F/E}(-S[x])\big)t(x)S[x]^h$$

$$\times S\big(x, A^{-1}p(w)\big)^{-k}|S\big(x, A^{-1}p(w)\big)|^{-2sE}.$$

This is practically the same as the sum considered in [6, (8.3)]. Conversely, (5.8) is a sum of finitely many series of type (5.2). Notice that $\mathrm{Res}_{F/E}(i) = p + \mathrm{Res}_{F/E}(h)$. Therefore, in order to prove the above theorems, it is sufficient to consider series of type (5.8) instead of (5.2).

Remark 5.6. Suppose $m \leqq 2$, $E = F = \mathbf{Q}$, and S is isotropic. In order to assert that $f(w, 0)$ or $g(w)$ in this case is a modular form in the ordinary sense, we have to verify the "cusp condition". This can be achieved by making an obvious estimate of theta functions (see (6.6) below) in the proof of Theorems 5.2 and 5.4, and by showing that f or g is slowly increasing. We shall not give details, since the whole procedure is straightforward. The same remark should be made on Theorem 6.4 below, but actually will not be repeated.

6. Theta functions

The proof of Theorems 5.2 and 5.4 requires certain theta functions which were employed in [6, §13]. To define them in a somewhat different form, we start with an explicit form of the "majorant" of S_ν. For $w = (w_1, \ldots, w_r) \in \mathfrak{Z}^r$, let $P_\nu(x, y; w)$ denote the positive symmetric $\mathbf{R}$-bilinear form in $(x, y) \in V_\nu \times V_\nu$ corresponding to the point $v_\nu = A_\nu^{-1}p(w_\nu)$ of $\mathfrak{Y}(S_\nu)$ defined in the first paragraph of [6, §2]. We then put

$$(6.1) \qquad P_\nu[x; w] = P_\nu(x, x; w) \qquad (x \in V_\nu).$$

Formula [6, (2.2)] shows that

$$(6.2\text{a}) \qquad P_\nu[x; w] = S_\nu[x] + 4\eta(w_\nu)^{-1}|S_\nu(x, v_\nu)|^2,$$

$$\left(v_\nu = A_\nu^{-1}p(w_\nu), x \in V_\nu\right),$$

(6.2b)

$$P_\nu(x, y; w) = S_\nu(x, y) + 2\eta(w_\nu)^{-1}\{S_\nu(x, v_\nu)S_\nu(y, \bar{v}_\nu) + S_\nu(x, \bar{v}_\nu)S_\nu(y, v_\nu)\},$$

where η is defined by [6, (2.13)]; namely

$$(6.3) \qquad \eta(w_\nu) = -S_\nu(v_\nu, \bar{v}_\nu), \; v_\nu = A_\nu^{-1}p(w_\nu).$$

We see easily that

$$(6.4\text{a}) \qquad P_\nu(v_\nu, x; w) = -S_\nu(v_\nu, x),$$

$$(6.4\text{b}) \qquad P_\nu[\alpha x; \alpha w] = \nu(\alpha)^{\tau_\nu}P_\nu[x; w] \quad \text{for every } \alpha \in G_+(S).$$

Recalling that the space $\mathfrak{Z}$ is a domain embedded in $\mathbf{C}^m$, let $w_{\nu 1}, \ldots, w_{\nu m}$ be the coordinate functions on the ν-th factor $\mathfrak{Z}$ of $\mathfrak{Z}^r$.

LEMMA 6.1. *The notation being as above, let* $u_{\nu j} = (\partial/\partial w_{\nu j})A_\nu^{-1}p(w_\nu)$ *and* $c_{\nu j}(w) = S_\nu(\bar{u}_{\nu j}, v_\nu)\bar{v}_\nu + \eta(w_\nu)\bar{u}_{\nu j}$ *with* $v_\nu = A_\nu^{-1}p(w_\nu)$. *Then*

$$(6.5\text{a}) \qquad P_\nu(c_{\nu j}(w), x; w) = S_\nu(c_{\nu j}(w), x),$$

$$(6.5\text{b}) \qquad (\partial/\partial \bar{w}_{\nu j})P_\nu[x; w] = 4\eta(w_\nu)^{-2}S_\nu(x, c_{\nu j}(w))S_\nu(x, v_\nu) \qquad (x \in V_\nu).$$

This can be verified in a straightforward way.

We now define a theta function $\theta(z, w)$ on $\mathfrak{H}^n \times \mathfrak{Z}^r$ by

$$(6.6) \quad \theta(z, w) = \theta(z, w; t, X, a) = \prod_{\nu=1}^{r} y_\nu^{m/2}\eta(w_\nu)^{-k_\nu}$$

$$\times \sum_{u \in X} t(u)S[u]^h \overline{S(u, A^{-1}p(w))}^k \, e\left(\sum_{\nu=1}^{n} a^{\tau_\nu}R_\nu[u; z, w]\right),$$

$$(6.7) \qquad R_\nu[u; z, w] = \begin{cases} -x_\nu S_\nu[u] + iy_\nu P_\nu[u; w] & (\nu \le r), \\ -z_\nu S_\nu[u] & (\nu > r). \end{cases}$$

Here $z = (z_1, \ldots, z_n)$, $z_\nu = x_\nu + iy_\nu \in \mathfrak{H}$, $w \in \mathfrak{Z}^r$, $t \in \mathscr{P}_\kappa(S)$: X is a congruence subset of V; a is an element of F such that $a^{\tau_\nu} > 0$ or $a^{\tau_\nu} < 0$ according as $\nu \le r$ or $\nu > r$; $0 \le h = \Sigma_{\nu > r}h_\nu \tau_\nu \in I_F$, $k = \Sigma_{\nu=1}^r k_\nu \tau_\nu \in I_F$, $k_\nu > 0$ for all $\nu \le r$. From (6.2a), we obtain

$$(6.8) \qquad -R_\nu[u; z, w] = \bar{z}_\nu S_\nu[u] - 4iy_\nu\eta(w_\nu)^{-1}|S_\nu(u, A_\nu^{-1}p(w_\nu))|^2 \; (\nu \le r),$$

and also from (6.4b),

$$(6.9) \qquad R_\nu[\alpha u; z, \alpha w] = \nu(\alpha)^{\tau_\nu} R_\nu[u; w] \qquad (\alpha \in G_+(S), \nu \leqq n).$$

A direct calculation shows that

$$(6.10) \qquad \theta(z, \alpha w; t, \alpha X, a) = \nu(\alpha)^h \mu(\alpha, w)^k \theta(z, w; t \circ \alpha, X, \nu(\alpha)a)$$

$$(\alpha \in G_+(S)).$$

To emphasize h, denote the above series by $\theta_h(z, w)$. Obviously $(-a)^h \theta_h(z, w) = \mathbf{d}^h \theta_0(z, w)$ with $\mathbf{d}^h$ of (1.15). We observe, in view of (6.4a), that our series θ_0, up to an elementary factor, is a special case of [6, (7.6)], and therefore, by [6, Prop. 7.1], belongs to $\mathcal{C}_{\lambda, F}$ as a function of z, where

$$(6.11) \qquad \lambda_\nu = \begin{cases} k_\nu + 1 - (m/2) & (\nu \leqq r), \\ \kappa_\nu + 1 + (m/2) & (\nu > r). \end{cases}$$

THEOREM 6.2. *Given* $g \in \mathcal{M}_{\lambda, F}$, *put*

$$f(w) = \int_\Psi \overline{g(z)}\, \theta_0(z, w)\, \mathrm{Im}(z)^\lambda\, d\mu(z) \qquad (\Psi = \Gamma \backslash \mathfrak{H}^n)$$

with a suitably small congruence subgroup Γ *of* $\mathrm{SL}_2(F)$ *such that both* g *and* $\theta_0(z, w)$ *belong to* $\mathcal{C}_\lambda(\Gamma)$. (*The integral is the inner product of* (1.7) *up to a factor* $\mu(\Psi)$.) *Then the integral is convergent, and* f *is holomorphic in* w.

This and the next theorem give a stronger version of [6, Theorem 13.1]. In the present theorem, it is unnecessary to assume the existence of the subfield E; also g need not be a cusp form; g can even be a constant if $\lambda = 0$. As mentioned in [6, Remark 13.2], results of this type were obtained by Rallis and Schiffman [2], and by Oda [1] in the case $E = F = \mathbf{Q}$, generalizing an earlier result of Niwa. However, their results apply only to the cusp forms g of a sufficiently large weight λ for which Poincaré series and certain infinite series are convergent; in fact, [1, Theorem 2] is proved under the assumption $\lambda > 3 + (m/2)$. Our theorem has no such restriction. The proof will be given after another theorem.

Now assuming the existence of a subfield E of F as in Section 3, we define a function $\Theta(z, w)$ on $\mathfrak{H}^r \times \mathfrak{Z}^r$ by

$$(6.12) \qquad \Theta(z, w) = \Theta(z, w; t, X) = y^{mE/2} \eta(w)^{-k}$$

$$\times \sum_{u \in X} t(u) S[u]^h \overline{S(u, A^{-1}p(w))}^k \mathbf{e}\left(\sum_{\nu=1}^r \left\{ x_\nu \mathrm{Tr}_{F/E}(-S[u])^{\tau_\nu} + iy_\nu P_\nu^*[u; w] \right\} \right),$$

where t, h, and k are the same as in (6.6), and

$$P_\nu^*[u; w] = P_\nu[u; w] + \mathrm{Tr}_{F/E}(S[u])^{\tau_\nu} - S[u]^{\tau_\nu}.$$

It can easily be seen that Θ is a pull-back of θ in the sense that

$$\Theta(z, w) = a^{mE/2}\theta(z^*, w) \qquad (z \in \mathfrak{H}^r),$$

where z^* is defined by

$$(6.13) \qquad z_\nu^* = \begin{cases} a^{\tau_\nu} z_\nu, & \text{if } \nu \leqq r, \\ a^{\tau_\nu} \bar{z}_\lambda, & \text{if } \nu > r \text{ and } \mathrm{Res}_{F/E}(\tau_\nu) = \tau_\lambda'. \end{cases}$$

Obviously $\Theta(z, w)$ satisfies the same type of transformation formula as (6.10). Let Θ_0 denote the special case of Θ with $h = 0$, and let

$$(6.14\text{a}) \qquad \Theta_0'(z, w) = \Theta_0'(z, w; t, X) = y^\sigma \Theta_0(z, w; t, X);$$

$$(6.14\text{b}) \qquad \sigma = \mathrm{Res}_{F/E}\left(\sum_{i>r} (\kappa_i + 1 + (m/2))\tau_i \right).$$

By Lemma 3.3, Θ_0' as a function of $z \in \mathfrak{H}^r$ belongs to $\mathcal{C}_{\mu, E}$ with

$$(6.15) \qquad \mu = k - \mathrm{Res}_{F/E}(\kappa) + \{2 - [F:E](m + 2)/2\}E.$$

(Actually Θ_0' is, up to an elementary factor, a special case of [6, (7.6)], so that [6, Prop. 7.1] is applicable to it.)

THEOREM 6.3. *Given* $\mathrm{g} \in \mathfrak{M}_{\mu, E}$, *put*

$$f(w) = \int_\Phi \overline{\mathrm{g}(z)}\, \Theta_0'(z, w)\mathrm{Im}(z)^\mu\, d\mu(z) \qquad (\Phi = \Gamma \backslash \mathfrak{H}^r)$$

with a congruence subgroup Γ *of* $\mathrm{SL}_2(E)$ *such that both* g *and* Θ_0' *belong to* $\mathcal{C}_\mu(\Gamma)$. *Then the integral is convergent, and f is holomorphic in w.*

Proof of Theorems 6.2 and 6.3. The convergence is obvious, since both θ_0 and Θ_0 are rapidly decreasing. Fixing our attention on the variable $w_{\nu j}$ as in Lemma 6.1, we are going to show that $\partial f / \partial \bar{w}_{\nu j} = 0$. For this purpose, we put

$$(6.16\text{a}) \quad \xi_{\nu j}(z, w) = y_\nu \eta(w_\nu)^{-1} \prod_{\lambda=1}^r y_\lambda^{m/2} \eta(w_\lambda)^{-k_\lambda} \sum_{u \in X} t(u) \overline{S(u, A^{-1}p(w))}^l$$

$$\times S_\nu(u, c_{\nu j}(w))\mathbf{e}\left(\sum_{\nu=1}^n a^{\tau_\nu} R_\nu[u; z, w] \right) \qquad (z \in \mathfrak{H}^n),$$

$$(6.16\text{b}) \qquad \Xi_{\nu j}(z, w) = y^\sigma \xi_{\nu j}(z^*, w) \qquad (z \in \mathfrak{H}^r),$$

where $c_{\nu j}$ is as in Lemma 6.1, z^* is defined by (6.13), and $l = k - \tau_\nu$. Formula (6.5a) shows that these are special cases of [6, (7.6)] up to elementary factors, and therefore, by [6, Prop. 7.1], $\xi_{\nu j} \in \mathcal{C}_{\lambda - 2\tau_\nu, F}$ and $\Xi_{\nu j} \in \mathcal{C}_{\mu - 2\tau_\nu, E}$; also they are rapidly decreasing. (It should be noted that condition $S_\nu[\tau_\nu] = 0$ of [6, (7.5)] is unnecessary if l_ν of [6, (7.4)] is 1.) Moreover, a direct calculation together with (6.5b)

shows that

(6.17a) $$- 4\pi\delta_\nu(\lambda_\nu - 2)\xi_{\nu j}(z, w) = \partial\theta_0/\partial\overline{w}_{\nu j},$$

(6.17b) $$- 4\pi a^{\tau_\nu + (mE/2)}\delta_\nu(\mu_\nu - 2)\Xi_{\nu j}(z, w) = \partial\Theta_0'/\partial\overline{w}_{\nu j},$$

where $\delta_\nu(\alpha)$ is defined by (1.14b). Therefore, by (2.19), we see that $\langle g, \partial\theta_0/\partial\overline{w}_{\nu j}\rangle = 0$ in Theorem 6.2 and $\langle g, \partial\Theta_0'/\partial\overline{w}_{\nu j}\rangle = 0$ in Theorem 6.3, which proves the desired holomorphy.

With a basis $\{t_1,\ldots, t_d\}$ of $\mathscr{P}_\kappa(S)$, define a representation σ by (5.5) and (5.6). Now, given $g \in \mathfrak{M}_{\mu, E}$, define a $\mathbf{C}^d$-valued function $\mathfrak{f}$ on $\mathfrak{Z}^r$ by

(6.18) $${}^t\mathfrak{f}(w) = (f(w, t_1),\ldots, f(w, t_d)) \qquad (w \in \mathfrak{Z}^r),$$

$$f(w, t) = \int_\Phi \overline{g(z)}\,\Theta_0'(z, w; t, X)\mathrm{Im}(z)^\mu \, d\mu(z)$$

with Φ as in Theorem 6.3. By that theorem, we see that $\mathfrak{f}$ belongs to the space $\mathfrak{M}_\sigma$ of automorphic forms defined in [6, §3]. Now we have

THEOREM 6.4. *Suppose that* g *is* $\overline{\mathbf{Q}}$*-rational. Then* $\mathfrak{f}$ *is* $\overline{\mathbf{Q}}$*-rational.*

This is actually a special case of Theorem 5.4, as will be shown later. If $F = E$, Theorem 6.3 is essentially the same as Theorem 6.2. It seems, however, that there is no good constant c with which the arithmeticity of cf follows from that of g in Theorem 6.2, except when $F = E$.

7. Proof of Theorem 5.2 and the analytic part of Theorem 5.4

For the reason explained in Remark 5.5, we consider functions $f'(w, s; t)$ of (5.8) with $0 \leqq h = \Sigma_{i>r}h_i\tau_i \in I_F$ instead of (5.2), and prove assertions of Theorems 5.2 and 5.4 for them. As to the constant α of Theorem 5.2, we have

(7.1) $$k - l - \mathrm{Res}_{F/E}(\kappa + 2h) - 2p = \{\alpha - 2 + q(m + 2)/2\}E$$
$$(q = [F : E]).$$

Now we put

$$\Omega_\rho(z) = \sum_{a \in E} \overline{\omega(a)}\, e_E(az) \qquad (z \in \mathfrak{H}^r).$$

This belongs to $\mathfrak{M}_{l, E}$ and $\mathbf{d}^p\Omega_\rho = \Sigma\omega_p(a)e_E(az)$. We consider then

(7.2) $$D_{l+2p, \mu+2\sigma+2h'}\big(s - 1 + (\alpha/2), \mathbf{d}^p\Omega_\rho, \Theta\big),$$

where Θ, μ, σ are defined by (6.12), (6.15), (6.14b), and $h' = \mathrm{Res}_{F/E}(h)$. Observe that

(7.3) $$l + 2p + \mu + 2\sigma + 2h' = 2k - (\alpha + m - 2)E.$$

Use (2.9) with $V = \{u^2 | u \in U\}$ to compute (7.2). Then we find that (7.2) is equal to

$$[V] \sum_{x \in X/U} \omega_p\big(\mathrm{Tr}_{F/E}(-S[x])\big)t(x)S[x]^h \overline{S\big(x, A^{-1}p(w)\big)}^k \eta(w)^{-k}$$

$$\times \int_{\mathbf{R}^r_+} \mathbf{e}\left(\sum_{\nu=1}^r iy_\nu \{P_\nu[x; w] - S_\nu[x]\} \right) y^{k+sE} d^\times y.$$

Using (6.2a), we find that this becomes

$$(7.4) \qquad 2^{1-r}(8\pi)^{-|k|-rs}\Gamma_r(k + sE)\eta(w)^{sE}f'(w, s; t).$$

Assign to every function ξ on $\mathfrak{H}^n$ a function ξ^* on $\mathfrak{H}^r$ by $\xi^*(z) = \xi(z^*)$ with z^* of (6.13). If ξ is holomorphic in $z_{r+1}, \ldots, z_n$, we have

$$(7.5) \qquad a^{\tau_\nu}\delta_\nu(\alpha)\xi^* = (\delta_\nu(\alpha)\xi)^* \qquad\qquad (\nu \leqq r, \alpha \in \mathbf{C}).$$

By (1.16a), we have

$$\mathbf{d}^p\Omega_\rho = \sum_{0 \leqq t \leqq p} b_t y^{t-p}\delta_l^t\Omega_\rho,$$

$$\Theta(z, w) = |a|^{(mE/2)-h}\mathbf{d}^h\theta_0(z^*, w) = \sum_{0 \leqq u \leqq h} c_u y^{u'-h}(\delta_\lambda^u\theta_0)^*$$

with real numbers b_t and c_u, where $u' = \mathrm{Res}_{F/E}(u)$ for every $u \in I_F$. Therefore (7.2) (which is equal to (7.4)) is a linear combination of

$$D_{l+2p,\, \mu+2\sigma+2h'}\big(s - 1 + (\alpha/2), y^{t-p}\delta_l^t\Omega_\rho, y^{u'-h'}(\delta_\lambda^u\theta_0)^*\big).$$

By (2.10), this is equal to

$$(7.6) \qquad D_{l+2t,\, \mu-2u'}\big(s - 1 + (\alpha/2), \delta_l^t\Omega_\rho, y^{\sigma+2u'}(\delta_\lambda^u\theta_0)^*\big).$$

Observe that $\delta_l^t\Omega_\rho \in \mathcal{C}_{l+2t}$ and $y^{\sigma+2u'}(\delta_\lambda^u\theta_0)^* \in \mathcal{C}_{\mu-2u'}$. Now $\mu - l = \alpha E + 2h' + 2p$, so that (2.14) is satisfied. Therefore, by Lemma 2.2, (7.4) can be continued to a meromorphic function on the whole s-plane. Moreover, we have $\mu - 2u' - l - 2t = \alpha E + 2(p - t + h' - u')$. Therefore, if $\alpha > 0$, (ii) of Lemma 2.2 shows that (7.4) is holomorphic for $\mathrm{Re}(s) \geqq (1 - \alpha)/2$, especially at $s = 0$.

To prove that $f'(w, 0; t)$ is holomorphic in w, we fix our attention on the variable $w_{\nu j}$ for each $\nu \leqq r$ and $j \leqq m$ as in Lemma 6.1 and show that $(\partial f'/\partial \overline{w}_{\nu j})(w, 0; t) = 0$, or equivalently, that

$$(7.7) \quad D_{l+2p,\, \mu+2\sigma+2h'}\big(s - 1 + (\alpha/2), \mathbf{d}^p\Omega_\rho, \partial\Theta/\partial\overline{w}_{\nu j}\big) = 0 \text{ at } s = 0.$$

For this purpose, we take the function $\xi_{\nu j}$ defined by (6.16a). Put $\varepsilon = \delta_\nu(\lambda_\nu - 2)$

with λ_ν of (6.11). By (6.17a) and (7.5), we have

$$|a|^{h-(mE/2)}\partial\Theta/\partial\overline{w}_{\nu j} = (\partial/\partial\overline{w}_{\nu j})(\mathbf{d}^h\theta_0)^* = (\mathbf{d}^h(\partial\theta_0/\partial\overline{w}_{\nu j}))^*$$

$$= -4\pi(\mathbf{d}^h\varepsilon\xi_{\nu j})^* = -4\pi(\varepsilon\mathbf{d}^h\xi_{\nu j})^* = -4\pi a^{\tau_\nu}\varepsilon((\mathbf{d}^h\xi_{\nu j})^*),$$

since $\nu \leqq r$ and $\mathbf{d}^h$ involves $\partial/\partial z_\nu$ only for $\nu > r$. Thus our task is to show that

$$(7.8) \qquad D_{l+2p,\,\mu+2\sigma+2h'}\big(s - 1 + (\alpha/2), \mathbf{d}^p\Omega_\rho, \varepsilon((\mathbf{d}^h\xi_{\nu j})^*)\big)$$

vanishes at $s = 0$. Recall (7.3) and apply (2.13) to the present case with $t = \lambda_\nu - 2$ and $u = 1 - (\alpha/2)$ to find that (7.8) is equal to $s/(4\pi)$ times

$$(7.9) \qquad D_{l+2p,\,\mu+2\sigma+2h'-2\tau_\nu}\big(s - 1 + (\alpha/2), \mathbf{d}^p\Omega_\rho, (\mathbf{d}^h\xi_{\nu j})^*\big).$$

Therefore (7.8) vanishes at $s = 0$, if (7.9) is finite at $s = 0$. Put $\gamma = \lambda - 2\tau_\nu$ and $\xi = \xi_{\nu j}$ for simplicity. Then (7.9) is a finite linear combination of

$$D_{l+2p,\,\mu+2\sigma+2h'-2\tau_\nu}\big(s - 1 + (\alpha/2), y^{t-p}\delta_l^t\Omega_\rho, (y^{u-h}\delta_\gamma^u\xi)^*\big)$$

with $0 \leqq t \leqq p$ and $0 \leqq u \leqq h$. (We have $\delta_\lambda^u = \delta_\gamma^u$, as h does not involve τ_ν.) By (2.10), this is equal to

$$D_{l+2t,\,\mu-2u'-2\tau_\nu}\big(s - 1 + (\alpha/2), \delta_l^t\Omega_\rho, y^{\sigma+2u'}(\delta_\gamma^u\xi)^*\big).$$

Now $\mu - 2u' - 2\tau_\nu' - l - 2t = \alpha E + 2(p - t + h' - u' - \tau_\nu')$. This is 0 only if $E = \mathbf{Q}$, $\alpha = 2$, $p = t$, and $h = u$. Otherwise it is holomorphic for $\mathrm{Re}(s - 1 + (\alpha/2)) \geqq -1/2$ by (ii) of Lemma 2.2. This completes the proof of Theorem 5.2.

To prove Theorem 5.4, assume $\alpha = 0$ and $j_1 = \cdots = j_r = 0$. Then we can put $p = 0$ and $j = h$ in the above consideration, so that f' coincides with f of (5.2). We are interested in the residue of (7.2) at $s = 1$, which is a constant times the residue of $\eta(w)^{sE}f(w, s)$ at $s = 1$. We again consider (7.6) and observe that it has a nontrivial residue at $s = 1$ only when $h = u$. Thus our problem is the residue of

$$D_{l,\,l}\big(s, \Omega_\rho, y^{\sigma+2h'}(\delta_\lambda^h\theta_0)^*\big)$$

at $s = 0$, which is a nonzero rational number times

$$R_E\langle\Omega_\rho, y^{\sigma+2h'}(\delta_\lambda^h\theta_0)^*\rangle$$

by virtue of Lemma 2.2. This is at least real analytic in w, since $(\delta_\lambda^h\theta_0)^*$ is rapidly decreasing. Observe that

$$\partial/\partial\overline{w}_{\nu j}\big[y^{\sigma+2h'}(\delta_\lambda^h\theta_0)^*\big] = y^{\sigma+2h'}\big(\delta_\lambda^h(\partial\theta_0/\partial\overline{w}_{\nu j})\big)^* = -4\pi y^{\sigma+2h'}\big(\delta_\lambda^h\varepsilon\xi_{\nu j}\big)^*$$

$$= -4\pi y^{\sigma+2h'}\big(\varepsilon\delta_\gamma^h\xi_{\nu j}\big)^* = -4\pi a^{\tau_\nu}y^{\sigma+2h'}\varepsilon\big(\delta_\gamma^h\xi_{\nu j}\big)^*$$

$$= -4\pi a^{\tau_\nu}\varepsilon'\big(y^{\sigma+2h'}(\delta_\gamma^h\xi_{\nu j})^*\big),$$

where $\varepsilon' = \delta_\nu(\lambda_\nu - 2 - \sigma_\nu - 2h'_\nu)$, and also that $\lambda_\nu - 2 - \sigma_\nu - 2h'_\nu = l_\nu - 2$. Therefore $\partial/\partial \overline{w}_{\nu j}$ of our residue is a constant times $\langle \Omega_\rho, \delta_\nu(l_\nu - 2)g \rangle$ with a rapidly decreasing function g. This is 0 by (2.19). Thus our residue is a holomorphic function in w. This proves the analytic part of Theorem 5.4.

This result includes Theorem 6.3 as a special case. In fact, putting $h = 0$, our result asserts that the residue of $D_{l,l}(s, \Omega_\rho, y^\sigma \Theta_0)$ at $s = 0$, and hence $\langle \Omega_\rho, y^\sigma \Theta_0 \rangle$, is holomorphic in w, which is Theorem 6.3. We also note that Theorem 6.4 is a special case of the arithmetic part of Theorem 5.4 for exactly the same reason. Notice that if μ is defined by (6.15), then (5.4) is satisfied.

8. Proof of Theorems 4.3, 5.3, and the arithmetic part of Theorem 5.4

Let w_0 be the fixed point of $h(Y^u)$ as studied in [6, §5]. Here Y is an F-algebra of rank $m + 2$, and h is an F-linear embedding of Y into $\mathrm{End}(V, F)$ satisfying certain conditions; Y^u is defined by [6, (5.3)]. As in Propositions 5.1 and 5.4 of [6], we put $Y = K \oplus Y_0$ and $V = h(Y)v$ with an element v of V. Here we assume that K is a CM-field, but not necessarily $[K : F] = 2$. Define $\psi_1, \ldots, \psi_r$ by [6, (5.6a)]. They are viewed as elements of J_K. If we put $K^u = \{a \in K \,|\, aa^\rho = 1\}$, we have $\mu_i(h^u(a), w_0) = a^{\psi_i}$ for $a \in K^u$, where $h^u(a) = h(a + id_{Y_0})$. Let $W = h(Y_0)v$; denote by S' the restriction of S to W. By [6, (10.2)], we have

$$(8.1) \qquad S[h(b)v + c] = \mathrm{Tr}_{K/F}(\varepsilon bb^\rho) + S'[c] \qquad (b \in K, c \in W)$$

with $\varepsilon \in K$, $\varepsilon^\rho = \varepsilon$. This ε can be obtained as the projection to K of the element δ of [6, Prop. 5.4]. Therefore $\varepsilon^{\psi_i} < 0$ for $i \leq r$ and $\varepsilon^\alpha > 0$ for $\alpha \in J_K - \{\psi_1, \ldots, \psi_r, \psi_1\rho, \ldots, \psi_r\rho\}$. Let $u_i = A_i^{-1}p(w_i)$ with the i-th component w_i of w_0. By [6, (10.3)], we have

$$(8.2) \qquad S_i(h(b)v + c, u_i) = b^{\rho\psi_i}S_i(v_i, u_i) \qquad (b \in K, c \in W, i \leq r).$$

Now, to prove Theorems 5.3 and 5.4, we assume $[K : F] = 2$ and take a $\overline{\mathbf{Q}}$-rational element t of $\mathscr{P}_\kappa(S)$ such that $t(h^u(a)x) = a^\Theta t(x)$ for every $a \in K^u$ with $\Theta \in I_K$. Our desired arithmeticity of Theorem 5.3 is equivalent to

$$(8.3) \qquad f(w_0, 0; t, X, \Omega) \sim \pi^{|k|}p_K(\Theta + \Psi, 2\psi),$$

where $\Psi = \Sigma_{i=1}^r k_i \psi_i$ and $\psi = \Sigma_{i=1}^r \psi_i$, for the reason explained in [6, (8.21), §10]. As for Theorem 5.4, we put

$$(8.4) \qquad g(w) = \eta(w)^E \lim_{s \to 1}(s - 1)f(w, s; t, X, \Omega).$$

Then our task is to show that

$$(8.5) \qquad g(w_0) \sim R_E \pi^{|k|}p_K(\Theta + \Psi, 2\psi).$$

We may assume, for the reason explained in [6, §10], that X has the form

$$(8.6) \qquad X = \{h(b)v + c \,|\, (0,0) \neq (b, c) \in B \times C\}$$

with congruence subsets B of k and C of W. Now t is a $\overline{\mathbf{Q}}$-linear combination of functions t' on $\prod_{i>r}V_i$ of the form

$$t'(h(b)v + c) = b^{\Theta'}g'(c) \qquad\qquad (b \in K, c \in W)$$

with $\overline{\mathbf{Q}}$-rational homogeneous functions g' on $\prod_{i>r}W_i$ and elements Θ' of I_K such that $0 \leqq \Theta' \equiv \Theta \ (\mathrm{mod}(1 + \rho)I_K)$. We observe also that

$$S[h(b)v + c]^i = \{2\varepsilon bb^\rho + S'[c]\}^i,$$

and this is a $\overline{\mathbf{Q}}$-linear combination of $b^{(1+\rho)\gamma}S'[c]^\lambda$ with $0 \leqq \gamma \in I_F$ and $\lambda \in I_F$. Therefore $t(h(b)v + c)S[h(b)v + c]^i$ is a $\overline{\mathbf{Q}}$-linear combination of $b^\Xi g(c)$ with a $\overline{\mathbf{Q}}$-rational homogeneous function g on $\prod_{i>r}W_i$ and an element Ξ of I_K such that $\Xi \equiv \Theta \ (\mathrm{mod}(1 + \rho)I_K)$. Put

$$H(z) = \sum_{c \in C} g(c)\mathbf{e}_E\big(\mathrm{Tr}_{F/E}(S'[c]z)\big) \qquad\qquad (z \in \mathfrak{H}^r),$$

$$\Omega'(z) = \Omega(z)H(z) = \sum_{a \in E} \omega'(a)\mathbf{e}_E(az).$$

Then, in the same fashion as observed in [6, §10], $f(w_0, s; t, X, \Omega)$ is a finite $\overline{\mathbf{Q}}$-linear combination of

$$(8.7) \qquad\qquad A'A^s \sum_{0 \neq b \in B/U} \omega'\big(\mathrm{Tr}_{K/E}(-\varepsilon bb^\rho)\big) b^{\Xi - \rho\Psi}|b^\Psi|^{-2s}$$

with algebraic numbers A and A'. By [6, Prop. 7.6 and 7.7], Ω' is a sum of derivatives of $\overline{\mathbf{Q}}$-rational elements of $\mathfrak{M}_{l',E}$ with various l'. Therefore (8.7) is a sum of several series of type (4.2). A straightforward verification shows that the constant β defined by (4.4) for these series is $2 - q - \alpha$. Thus, if $\alpha > 0$, condition (4.8a) is satisfied with $\mu = 0$; (4.8b) is satisfied if $\alpha \neq 2$ or $E \neq \mathbf{Q}$. Notice that $\Xi \geqq 0$, $c(\psi_i) - c(\psi_i\rho) = k_i$ and $c(\psi_i) \geqq 0$ for $i \leqq r$. Hence the desired (8.3) follows from Theorem 4.3.

Next suppose $\alpha = 0$ and $j_1 = \cdots = j_r = 0$. Then we see that Ξ in the above consideration does not involve $\psi_1, \ldots, \psi_r$. Therefore we have

$$c(\psi_1) = \cdots = c(\psi_r) = 0 = (\beta + q - 2)/2,$$

so that (4.12b) is satisfied. Since $l' \geqq l + (mq/2)E$ and $\alpha = 0$, l' satisfies (4.12a). Thus (8.5) follows from Theorem 4.6, since $\eta(w_0)^E$ is algebraic.

We have shown in Section 4 that Theorems 4.3 and 4.7 are true when $F = E$. Therefore Theorems 5.3, 5.4, and 6.4 are true when $F = E$.

Our remaining task is to prove Theorem 4.3 in the most general case. This is basically the same as [6, §12]; we take, however, E as the basic field. The symbols K, ζ, etc. being the same as in Section 4, put $[K:E] = m + 2$ and

$$(8.8) \qquad\qquad S(x, y) = -\mathrm{Tr}_{K/E}(\zeta xy^\rho) \qquad\qquad (x, y \in K).$$

We may assume that $m > 0$, since the case $[K:E] = 2$ is already proved. Viewing K as a vector space over E and writing it V, we define the group $G_+(S)$, the domain $\mathfrak{Z}^r$, and other symbols relative to this S with E as the basic field instead of F. The regular representation of K over E defines an embedding of K into $\mathrm{End}(V, E)$, and $h(K'')$ has a unique common fixed point w_0 on $\mathfrak{Z}^r$. Take $\psi_1, \ldots, \psi_n \in J_K$ so that $\Sigma_{i=1}^n \psi_i$ is a CM-type of K. Now, given Φ as in (4.3) and an integer μ satisfying (4.8a,b), put $b_i = c(\psi_i)$, $b_i' = c(\psi_i \rho)$ for $i = 1, \ldots, n$ and

$$\Xi = \sum_{i=1}^r (b_i - \mu)(\psi_i + \psi_i \rho) + \sum_{i=r+1}^n (b_i \psi_i + b_i' \psi_i \rho),$$

$$\mathrm{Res}_{K/E}(\Xi) = \sum_{i=1}^r e_i \tau_i', \qquad d_i = b_i - b_i', \qquad \Delta = \sum_{i=1}^r d_i \psi_i, \qquad \psi = \sum_{i=1}^r \psi_i.$$

Then we have $\Phi - \mu(\psi + \psi\rho) = \Xi - \rho\Delta$. We may assume that $\psi_1, \ldots, \psi_r$ are defined by [6, (5.6b)]. Now we can repeat the argument of [6, pp. 364–365] with these Ξ and Δ, always with E instead of F. The only necessary care in the present case is that we should always keep the factor $|S(x, A^{-1}p(w))|^{-2s}$ in the calculation and put $s = 0$ at the end. This procedure can be justified, because (7.4) is given as (7.2). It should also be noted that if we define the constant α for the series of type (5.2) in this proof, we have $\alpha = 2\mu + 2 - \beta - [F:E]$, so that (4.8a) implies that $\alpha > 0$; if $E = \mathbf{Q} \neq F$, (4.8b) implies that $\alpha \neq 2$. We have also

$$\sum_{i=1}^r (d_i - e_i)\tau_i' = 2\mu E - \mathrm{Res}_{K/E}(\Phi) \geqq (2\mu - \beta)E \geqq (mE)/2,$$

so that condition [6, (12.4)] is satisfied. Thus we obtain the desired conclusion to complete the proof of Theorem 4.3.

9. Proof of Theorems 4.6 and 4.7

Let Θ_0 be the function Θ of (6.12) with $h = 0$. Define this with $F = E$, $t = 1$, $k = \Sigma_{i=1}^r k_i \tau_i'$, $k_i > 0$ for all i. We take S to be defined by (8.8) with a given CM-field K. Thus, in this section, $G_+(S)$ is again defined with E as the basic field. We are going to study $\Theta_0(z, w_0)$ with the fixed point w_0 in the above proof of Theorem 4.3. Let v and u_i be as in Section 8. Using the notation of the proof of [6, Prop. 5.4], we find that $u_i = d_i \chi(e_{\psi_i})$ with $0 \neq d_i \in \overline{\mathbf{Q}}$, and therefore

$$\eta(w_i) = -S_i(\bar{u}_i, u_i) = |d_i|^2 \zeta^{\psi_i} \qquad (i \leqq r).$$

From this relation combined with (6.2a) and (8.2), we obtain

$$(9.1) \qquad P_i[h(b)v; w_0] = -\mathrm{Tr}_{K/E}(\zeta bb^\rho)^{\tau_i} + 4(\zeta bb^\rho)^{\psi_i} \qquad (b \in K).$$

Take a $\overline{\mathbf{Q}}$-rational element g of $\mathfrak{M}_{l, E}$ with $l = k + (2 - m)E/2$, and put

$$(9.1') \qquad \mathbf{f}(w) = \langle g(z), \Theta_0(z, w) \rangle \qquad (w \in \mathfrak{Z}^r).$$

Since Theorem 6.3 is a special case of Theorem 5.4 which is true if $F = E$, we know that $\pi^r \mathbf{f}$ is arithmetic. To prove Theorem 4.7, we apply the differential operator D_σ^e defined in [6, §11] to $\mathbf{f}$.

Let T_i denote the global complex tangent space of the i-th factor $\mathfrak{Z}$ of $\mathfrak{Z}^r$ as considered in [6, §11]. Define $\psi_1, \ldots, \psi_n$ as in Section 8 and also representations $\varphi_1, \ldots, \varphi_r$ of K by [6, (5.6a)]. Let $\gamma \in J_K$ and $\mathrm{Res}_{K/E}(\gamma) = \tau_i'$; suppose $\gamma \notin \{\psi_1, \ldots, \psi_r, \psi_1 \rho, \ldots, \psi_r \rho\}$. As shown in [6, §12], there is a nonzero $\overline{\mathbf{Q}}$-rational element u_ζ of T_i such that $\varphi_i(a)u_\gamma = a^\gamma u_\gamma$ for $a \in K$. Define a representation σ of $(\mathrm{GL}_1)^r \times (\mathrm{GL}_m)^r$ by

$$\sigma(a_1, \ldots, a_r, \ldots) = a^k \qquad\qquad (a_i \in \mathrm{GL}_1).$$

Then $(D_\sigma^e \mathbf{f})(w_0)$ with $e = \sum_{i=1}^r e_i \tau_i' \in I_E$ is a multilinear function on $T_i \times T_i \times \cdots \times T_p$ as explained in [6, §11]. By [6, Theorem 11.2], we have

(9.2)

$$D_\sigma^e \mathbf{f}(w_0)(u_\alpha, u_\beta, \ldots, u_\xi) \sim \pi^{|e| - r} p_K\left(\sum_{i=1}^r (k_i + e_i)\psi_i - \alpha - \beta - \cdots - \xi, 2\psi \right).$$

We have $D_\sigma^e \mathbf{f}(w) = \langle g(z), D_\sigma^e \Theta_0(z, w) \rangle$, and so we are going to study $D_\sigma^e \Theta_0(z, w)$ at $w = w_0$. For the reason explained in the proof of [6, Theorem 11.2], we may assume w_0 to be "the origin" 0 of $\mathfrak{Z}^r$. By [6, Lemma 11.1], the effect of D_σ^e at $w = 0$ is the same as that of a holomorphic partial differentiation of order e. (We can also ignore the factor $\eta(w)^{-k}$ for the reason explained in the lemma.) Now put

(9.3)
$$\varepsilon(a, w) = \mathbf{e}\left(\sum_{\nu=1}^r \{ -S_\nu[a]x_\nu + iy_\nu P_\nu[a; w] \} \right) \qquad (a \in V).$$

From (9.1), we obtain, for every $b \in K$,

(9.4)
$$\varepsilon(h(b)v, w_0) = \mathbf{e}\left(\sum_{\nu=1}^r \bar{z}_\nu \mathrm{Tr}_{K/E}(\zeta bb^\rho)^{\tau_\nu'} + \sum_{\nu=1}^r 4iy_\nu(\zeta bb^\rho)^{\psi_\nu} \right)$$
$$= \mathbf{e}_F(2bb^\rho z)^\zeta,$$

where the superscript ζ is defined by (3.6). Fix an index $i \leqq r$, and put, for $a \in V$,

$$A_i a = \begin{pmatrix} c_i'' \\ c_i' \\ c_i \end{pmatrix} \qquad (c_i, c_i' \in \mathbf{C}, c_i'' \in \mathbf{C}^m = T_i).$$

By [6, (2.3), (3.3)], we have

$$S_i\left(a, A_i^{-1} p(w_i)\right) = {}^t c_i'' w_i - c_i' - c_i \cdot {}^t w_i w_i / 2.$$

The explicit form of $P_i[a; w]$ in terms of c_i, c_i', c_i'' can then be given by (6.2a).

Therefore a straightforward calculation shows that

$$\left[D_\sigma^e \varepsilon(a, w)\right]_{w=0} = \varepsilon(a, 0) \otimes_{i=1}^r H_i(a)$$

with multilinear functions H_i on $T_i \times \cdots \times T_i$ (e_i copies) such that

$$H_i(a)(u_1, \ldots, u_{e_i}) = \sum_{j=0}^{[e_i/2]} \alpha_j (\pi y_i \bar{c}_i')^{e_i - i} c_i^j H_{ij}(c_i''; u_1, \ldots, u_{e_i})$$

for elements $u_1, u_2, \ldots$ of T_i with $\alpha_j \in \mathbf{Q}$ and

$$H_{i0}(c_i''; u_1, \ldots, u_{e_i}) = {}^t c_i'' u_1 \cdot {}^t c_i'' u_2 \cdots {}^t c_i'' u_{e_i},$$

$$H_{i1}(c_i''; u_1, \ldots, u_{e_i}) = {}^t u_1 u_2 \cdot {}^t c_i'' u_3 \cdots {}^t c_i'' u_{e_i} + {}^t u_1 u_3 \cdot {}^t c_i'' u_2 \cdots {}^t c_i'' u_{e_i} + \cdots,$$

$$\cdots \quad \cdots \quad \cdots \qquad\qquad (u_1, u_2, \cdots \in T_i).$$

Notice that H_{ij} is a homogeneous function in c_i'' of degree $e_i - 2j$. Let $\alpha, \ldots, \xi$ be elements of J_K such that $e = \mathrm{Res}_{K/E}(\alpha + \cdots + \xi)$. Then, for $b \in K$, we find that

$$\left[D_\sigma^e \varepsilon(h(b)v, w)\right]_{w=0}(u_\alpha, \ldots, u_\xi) = \varepsilon(h(b)v, 0) \sum_\Delta C_\Delta (\pi y)^{e-t} b^{\Xi + \Delta\rho},$$

where $C_\Delta \in \overline{\mathbf{Q}}$, $2t = e - \mathrm{Res}_{K/E}(\Delta)$, $\Xi = \Sigma_{i=1}^r e_i \psi_i$, and Δ runs over the elements of I_K such that $0 \leq \Delta \leq \alpha + \cdots + \xi$. From this and (9.4), we obtain

$$D_\sigma^e \Theta_0(z, w)_{w=0}(u_\alpha, \ldots, u_\xi) = y^{mE/2} \sum_\Delta (\pi y)^{e-t} C_\Delta' q_\Delta^\varsigma(z),$$

where $C_\Delta' \in \overline{\mathbf{Q}}$ and $q_\Delta(z) = \Sigma_{b \in X} b^{\Phi + \Delta\rho} \mathbf{e}_F(2bb^\rho z)$, $\Phi = \Sigma_{i=1}^r (k_i + e_i)\psi_i$. Notice that $C_\Delta' \neq 0$ if $\Delta = \alpha + \cdots + \xi$. Thus

(9.5) $$R_E(D_\sigma^e \mathbf{f})(w_0)(u_\alpha, \ldots, u_\xi)$$

is a nonzero algebraic number times the residue at $s = 0$ of

$$\sum_\Delta C_\Delta' D_{l,l}\left(s, g, (\pi y)^{e-t} y^{mE/2} q_\Delta^\varsigma\right).$$

Fix one Δ, and write $\Phi + \Delta\rho$ in the form

$$\Phi + \Delta\rho = \Psi + \sum_{i>r} p_i(\psi_i + \psi_i\rho), \quad \Psi = \sum_{i=1}^n (\lambda_i - 1)\psi_i, \lambda_i \geq 1, p_i \geq 0.$$

We have $\lambda_i - 1 = k_i + e_i$ for $i \leq r$, since $\Delta\rho$ does not involve $\psi_1, \ldots, \psi_r$, $\psi_1\rho, \ldots, \psi_r\rho$; also, for $i > r$, we change ψ_i for $\psi_i\rho$ if necessary. Put

$$h_\Delta(z) = \sum_{b \in X} b^\Psi \mathbf{e}_F(2bb^\rho z).$$

Then $2^{|p|} q_\Delta = \mathbf{d}^p h_\Delta$, and hence our problem is the residue of

(9.6) $$D_{l,l}\left(s, g, y^{e-t+(mE/2)}(\mathbf{d}^p h_\Delta)^\varsigma\right).$$

Put $\mu = \mathrm{Res}_{F/E}(\Sigma_{i>r} \lambda_i \tau_i)$ and $p' = \mathrm{Res}_{F/E}(p)$. Recall that $[F:E] = (m+2)/2$

in the present situation. Now we have

$$(9.7) \qquad e - 2t = \mathrm{Res}_{K/E}(\Delta\rho) = \mu - (mE/2) + 2p',$$

$$(9.8) \qquad \sum_{i=1}^{r} \lambda_i \tau_i' - \mu - 2p' = k + E - (mE/2) + 2t = l + 2t.$$

By (2.10) and (9.7), we see that (9.6) is equal to

$$D_{l,\, l+2t}\Big(s, g, y^{\mu+2p'}(\mathbf{d}^p h_\Delta)^\zeta\Big).$$

The decomposition of $(\mathbf{d}^p h_\Delta)^\zeta$ of type (3.9) gives this as a linear combination of

$$D_{l,\, l+2t+2p'-2u'}\Big(s, g, y^{\mu+2u'}(\delta_\lambda^u h_\Delta)^\zeta\Big),$$

which are holomorphic at $s = 0$ unless $t = 0$ and $u = p$. Therefore (9.5) is a nonzero algebraic number times the residue of

$$\pi^{|e|} D_{l,\, l}\Big(s, g, y^{\mu+2p'}(\delta_\lambda^p h_\Delta)^\zeta\Big)$$

at $s = 0$, where $\Delta = \alpha + \cdots + \xi$.

We can now derive Theorem 4.7 from this fact. Given λ, q, μ, and l as in Theorem 4.7, put

$$\Delta\rho = \sum_{i>r} \{(\lambda_i - 1 + q_i)\psi_i + q_i\psi_i\rho\}, \quad e = \sum_{i=1}^{r} e_i\tau_i' = \mathrm{Res}_{K/E}(\Delta).$$

Then $e = \mu + 2q' - (mE/2)$. Put $k_i = \lambda_i - 1 - e_i$ for $i \leqq r$. Since $\Sigma_{i=1}^{r}\lambda_i\tau_i' - \mu - 2q' = l \geqq (2 - (m/2))E$, we have $k_i > 0$ for all $i \leqq r$. Taking these k, e, and Δ to be the above k, e, and $\alpha + \cdots + \xi$, we see that $h_\Delta(z/2)$ is exactly the function f of Theorem 4.7 and also that $p = q$. Changing ζ for $\zeta/2$, we obtain the assertion of Theorem 4.7 from (9.2).

Nonvanishing examples of $D(\mu)$ and inner products. Let us now show that the values $D(\mu)$ of Theorem 4.3 and the inner products of Theorem 4.7 are in general nonvanishing in the case $E \neq F$. (If $E = F$, this fact is easy.) For simplicity, we consider here only the case $E = \mathbf{Q}$. Let F, K, Ω, and other symbols be the same as in (4.2); assume $n > 1$ and $E = \mathbf{Q}$. Since $r = 1$, we have $\psi = \psi_1$. Take $\psi_1, \ldots, \psi_n \in J_K$ so that $\mathrm{Res}_{K/F}(\psi_i) = \tau_i$, and take Φ in the form

$$\Phi = -k\psi_1\rho + \Delta, \quad \Delta = \sum_{i=2}^{n} \{(\lambda_i - 1 + q_i)\psi_i + q_i\psi_i\rho\}$$

with integers $k > 0$, $\lambda_i > 0$, $q_i \geqq 0$. Identify K with a subfield of $\mathbf{C}$ via $\psi_1\rho$. Then $D(s)$ with $p = 0$ can be written in the form

$$D(s) = \sum_{x \in X} \omega\big(\mathrm{Tr}_{F/\mathbf{Q}}(\zeta x x^\rho)\big) x^\Delta x^{-k}|x|^{-2s}.$$

Fixing ζ, X, Ω, and Δ, call this D_k. Put $\mu = \Sigma_{i=2}^{n}\lambda_i$, $q' = \Sigma_{i=2}^{n}q_i$. Then $\beta = l - k + \mu + 2q' - n + 1$. Therefore D_k is convergent at $s = 0$ if $k > l + \mu + 2q' + n - 1 + 2\theta$ with θ of Proposition 3.1. Let N be the smallest of such integers k, and let

$$g(t) = \sum_{k=N}^{\infty} D_k(0)t^k \qquad\qquad (t \in \mathbf{C}).$$

We can easily see that this is convergent for sufficiently small $|t|$ and

$$g(t) = \sum_{x \in X} \omega\big(\mathrm{Tr}_{F/\mathbf{Q}}(\zeta x x^\rho)\big)x^\Delta(x^{-1}t)^N / (1 - x^{-1}t).$$

Decompose this into two sums according to whether $|x^{-1}t| \geq 1/2$ or $|x^{-1}t| < 1/2$. The former is a finite sum; the latter is majorized by

$$2|t|^N \sum_{x \in X} |\omega\big(\mathrm{Tr}_{F/\mathbf{Q}}(\zeta x x^\rho)\big)x^\Delta x^{-N}|,$$

which is convergent. Thus g is meromorphic on the whole plane and has a pole at $t = x$ if $\omega(\mathrm{Tr}_{F/\mathbf{Q}}(\zeta x x^\rho)) \neq 0$, which can happen with a suitable choice of $\{\Omega, \zeta, X\}$. Then $D_k(0) \neq 0$ for infinitely many k. This shows also the nontriviality of functions of type (5.2), since the values of (5.2) at CM-points are given by such $D_k(0)$.

Now put $q = 0$. By (4.11) and (3.10), we have

$$(4\pi\zeta)^{-s-k}\Gamma(s + k)D_k(s) = D_{l,k+1-\mu}\big(s - (\alpha/2), \Omega_\rho, y^\mu h^\zeta\big),$$

where $\alpha = \beta + n$, $h(z) = \Sigma_{b \in X} b^\Psi e_F(bb^\rho z)$, $\Psi = k\psi_1 + \Sigma_{i=2}^{n}(\lambda_i - 1)\psi_i$. By (2.15), this is, up to a nonzero constant factor, equal to

$$(9.9) \qquad\qquad \langle y^s E_{\kappa,0}(z, s)\Omega_\rho, y^\mu h^\zeta \rangle,$$

where $\kappa = k + 1 - \mu - l$. Take k so that $\kappa \neq 2$. Then (9.9) at $s = 0$ gives a nonvanishing example of quantity of Theorem 4.7 with $q = 0$.

Coming back to the above proof of Theorem 4.7, take $E = \mathbf{Q}$ and observe that $\Theta_0(z, w_0)$ is of the type h^ζ with $\Psi = k\psi_1$. Therefore we obtain an example of nonvanishing $\mathbf{f}$ defined by (9.1'). Then $D_\sigma^e \mathbf{f}$ is also nonvanishing. The points $\gamma(w_0)$ for all $\gamma \in G_+(S)$ form a dense subset of $\mathfrak{Z}^r$. Thus $(D_\sigma^e \mathbf{f})(\gamma(w_0)) \neq 0$ for some γ. As our proof shows, this gives a nonvanishing example of $\langle g, y^{\mu+2q'}(\delta_\lambda^q f)^\zeta \rangle$ with $q \neq 0$.

10. Automorphic forms on unitary groups

Our main theorems of Sections 5 and 6 have analogues for unitary groups over CM-fields, as already mentioned in [6, §14]. Let us now briefly indicate how they can be stated without going into details of proof. Take a CM-field K such that $[K:F] = 2$ and take $\sigma_1, \ldots, \sigma_n$ of J_K so that $\mathrm{Res}_{K/F}(\sigma_i) = \tau_i$. With a hermitian element H of $\mathrm{GL}_{m+1}(K)$, we define groups $G(H), G_+(H)$, a domain $\mathfrak{D}$, and the action of $G_+(H)$ on $\mathfrak{D}$ as in [6, §14]. We assume that H^{σ_i} has signature $(m,1)$ for $i \leqq r$ and $(m+1,0)$ for $i > r$. Therefore our domain is given by

$$(10.1) \qquad \mathfrak{D} = D(m,1)^r, \ D(m,1) = \{w \in \mathbf{C}^m \mid {}^t\overline{w}w < 1\}.$$

We denote by K_{m+1} the vector space of $(m+1)$-dimensional row vectors with components in K, and put $H[v] = vH \cdot {}^t v^\rho$ for $v \in K_{m+1}$. Assuming the existence of a subfield E of F as in Section 3, we define a column vector $f(w,s)$ by

$$(10.2)$$

$${}^t f(w,s) = [U] \sum_{v \in X/U} \omega\big(-\mathrm{Tr}_{F/E}(H[v])\big) H[v]^t \lambda(v,w)^h \mu(v,w)^{-k} \big|\mu(v,w)^E\big|^{-2s}.$$

Here $w \in \mathfrak{D}, s \in \mathbf{C}$; X is a congruence subset of $K_{m+1} - \{0\}$; ω denotes the Fourier coefficients of an element Ω of $\mathfrak{M}_{l,E}$ with $l \in I_E$ as in (4.1); U is a subgroup of U_E of finite index such that Ω is U-invariant and $uX = X$ for all $u \in U$; $0 \leqq j \in I_F$, $0 \leqq h \in I_F$, $k = \Sigma_{i=1}^r k_i \tau_i \in I_F$, $k_i > 0$ for all $i \leqq r$; $\lambda(v,w)^h, \mu(v,w)^{-k}, |\mu(v,w)|^{-2s}$ are defined as in [6, (14.11)]. In fact, our series is practically the same as [6, (14.11)]. For the same reason as in (4.2) and (5.2), we assume

$$(10.3) \qquad k - l - \mathrm{Res}_{F/E}(h + 2j) = \alpha'E \text{ with } \alpha' \in \mathbf{Z}.$$

If θ is as in Proposition 3.1, $f(w,s)$ is convergent for

$$(10.4) \qquad \mathrm{Re}(s) > (m+1)[F:E] - (\alpha'/2) + \theta.$$

THEOREM 10.1. *Put* $\alpha = \alpha' + 2 - (m+1)[F:E]$. *Then* $f(w,s)$ *as a function of* s *can be continued to a meromorphic function on the whole plane. Moreover suppose* $\alpha > 0$. *Then* $f(w,s)$ *is holomorphic at* $s = 0$; *if* $\alpha \neq 2$ *or* $E \neq \mathbf{Q}$, $f(w,0)$ *is a holomorphic function on* $\mathfrak{D}$ *which belongs to* $\mathfrak{M}_\zeta$ *with a representation* ζ *defined in* [6, p. 372]. *Furthermore, if the coefficients* $\omega(a)$ *are all algebraic,* $c \cdot f(w,0)$ *is arithmetic, where*

$$c = \pi^{-|k|} p_K\left(\sum_{i=1}^r k_i \sigma_i + \sum_{i=1}^n h_i \sigma_i, -\sum_{i=1}^n \sigma_i\right).$$

THEOREM 10.2. *Suppose* $\alpha = 0$ *and* $j_1 = \cdots = j_r = 0$. *Then* $f(w,s)$ *is holomorphic for* $\mathrm{Re}(s) > 1$ *and has at most a simple pole at* $s = 1$. *Put*

$$g(w) = \prod_{i=1}^r \eta(w_i) \cdot \lim_{s \to 1}(s-1)f(w,s),$$

$$\eta(w_i) = 1 - {}^t\overline{w}_i w_i \qquad\qquad (w_i \in D(m,1)).$$

Then $g(w)$ is holomorphic on $\mathfrak{D}$ and defines an element of $\mathfrak{M}_{\zeta}$. Moreover, if $\omega(a)$ are all algebraic, $R_E^{-1}cg$ is arithmetic.

These can be proved in exactly the same fashion as Theorems 5.3 and 5.4. The proof requires a (column-vector-valued) function θ on $\mathfrak{H}^n \times \mathfrak{D}$ defined by

$$'\theta(z, w) = \prod_{\nu=1}^{r} y_{\nu}^{m+h_{\nu}} \eta(w_{\nu})^{k_{\nu}}$$

$$\times \sum_{v \in X} \lambda(v, w)^h \overline{\mu(v, w)}^k \mathbf{e}\left(- \sum_{\nu=1}^{n} a^{\tau_{\nu}} \left\{ x_{\nu} H[v]^{\tau_{\nu}} - i y_{\nu} R_{\nu}[v; w] \right\} \right).$$

Here $z_{\nu} = x_{\nu} + i y_{\nu} \in \mathfrak{H}$, $w = (w_1, \dots, w_r) \in \mathfrak{D}$, $a \in F$, $a^{\tau_{\nu}} > 0$ for $\nu \leqq r$ and $a^{\tau_{\nu}} < 0$ for $\nu > r$,

$$R_{\nu}[v; w] = \begin{cases} H[v]^{\tau_{\nu}} + 2\eta(w_{\nu})^{-1}|\mu_{\nu}(v, w)|^2 & (\nu \leqq r), \\ -H[v]^{\tau_{\nu}} & (\nu > r). \end{cases}$$

This is again a special case of [6, (7.6)], and belongs, as a function of z, to $\mathcal{C}_{\lambda, F}$, where

$$\lambda_{\nu} = \begin{cases} k_{\nu} + 1 - m - h_{\nu} & (\nu \leqq r), \\ m + h_{\nu} + 1 & (\nu > r). \end{cases}$$

As a function of w, it behaves like an element of $\mathfrak{M}_{\zeta}$.

THEOREM 10.3. *Given $g \in \mathfrak{M}_{\lambda, F}$, put*

$$f(w) = \int_{\Psi} \overline{g(z)}\, \theta(z, w) \mathrm{Im}(z)^{\lambda} d\mu(z) \qquad (\Psi = \Gamma \backslash \mathfrak{H}^n)$$

with a suitably small congruence subgroup Γ of $\mathrm{SL}_2(F)$ such that both g and θ belong to $\mathcal{C}_{\lambda}(\Gamma)$. Then the integral is convergent and f is holomorphic in w and defines an element of $\mathfrak{M}_{\zeta}$.

This can be proved by the same technique as for Theorem 6.2. We can also prove the analogues of Theorems 6.3 and 6.4, but do not state them here, since they are actually special cases of Theorem 10.2.

PRINCETON UNIVERSITY, PRINCETON, NEW JERSEY

REFERENCES

[1]   T. ODA, On modular forms associated with indefinite quadratic forms of signature $(2, n-2)$, Math. Ann. **231** (1977), 97–144.
[2]   S. RALLIS and G. SCHIFFMANN, Automorphic forms constructed from the Weil representation: holomorphic case, Amer. J. of Math. **100** (1978), 1049–1122.

[3]   G. Shimura, The special values of the zeta functions associated with cusp forms, Commun. Pure and App. Math. **29** (1976), 783–804.

[4]   ______, On the periods of modular forms, Math. Ann. **229** (1977), 211–221.

[5]   ______, The special values of the zeta functions associated with Hilbert modular forms, Duke Math. J. **45** (1978), 637–679.

[6]   ______, The arithmetic of certain zeta functions and automorphic forms on orthogonal groups, Ann. of Math. **111** (1980), 313–375.

(Received May 27, 1980)

Corrections to [6]

Page 321, line 6 from bottom: $D_{-1}\eta$ should be $D_{-1}p$.

Page 325, line 21: G_q should be $\mathfrak{G}_{\mathbf{Q}+}$.

Page 343, Proposition 7.6: $\mathbf{Q}$ should be $\overline{\mathbf{Q}}$.

Page 358, lines 19, 20: Insert "a finite $\overline{\mathbf{Q}}$-linear combination of functions of the form" at the end of line 19, and delete "$f(w, 0; t, \xi, X) = $" from line 20.

On certain zeta functions attached to two Hilbert modular forms:
II. The case of automorphic forms on a quaternion algebra

Annals of Mathematics, 114 (1981), 569-607

Introduction

As explained in Part I, our main theme of the whole paper is the critical values of a Dirichlet series which in the simplest case has the form

$$(1) \qquad D(s) = \Sigma_{0 \prec b \in \mathfrak{y}} \omega\big(\mathrm{Tr}_{F/\mathbf{Q}}(\zeta b)\big)\chi(b\mathfrak{x}^{-1})b^{p-s}.$$

As to the meaning of the symbols, the reader is referred to the introduction of Part I. We recall here only that F is a totally real algebraic number field of degree n, and ω denotes the Fourier coefficients of an elliptic modular form $\Omega(z) = \Sigma_{a \in \mathbf{Q}}\omega(a)e^{2\pi i a z}$. In Part I, we investigated D assuming that $\Sigma_a \chi(\mathfrak{a})N(\mathfrak{a})^{-s}$ is an L-function of a CM-field with an algebraic-valued Hecke character. In the present Part II, we treat the case where $\chi(\mathfrak{a})$ can be obtained as eigenvalues of Hecke operators $T(\mathfrak{a})$ on a quaternion algebra B over F. Thus we begin our study by recalling the theory of automorphic forms on the product $\mathfrak{H}^r$ of r copies of the upper half plane $\mathfrak{H}$ with respect to congruence subgroups of $B^\times$. Here r is the number of archimedean primes of F unramified in B. Then we introduce in Section 2 the notion of arithmeticity (or $\overline{\mathbf{Q}}$-rationality) of such automorphic forms and show that the space of automorphic forms can be spanned by the arithmetic ones. Our main theorems will be stated in Section 3 and proved in Section 6; Sections 4 and 5 are preliminaries to the proof. The central result, specialized to the case $r = 1$, asserts that if $\mathbf{g}$ is a $\overline{\mathbf{Q}}$-rational eigenform such that $\mathbf{g}|T(\mathfrak{a}) = \chi(\mathfrak{a})\mathbf{g}$ for all integral ideals $\mathfrak{a}$ of F, then $D(s_0)$ for certain integers s_0 are algebraic numbers times $\pi^d\langle \mathbf{g}, \mathbf{g}\rangle$, where d is an integer determined by the weight of $\mathbf{g}$, and $\langle *, *\rangle$ is the Petersson inner product.

---

*Supported by NSF Grant MCS 7903631.

We shall actually prove a more general result (Theorem 3.3 and Theorem 3.7, (i)) of the same nature in the case where r is not necessarily equal to 1. We assume however that F has a subfield E of degree r over $\mathbf{Q}$ and the archimedean primes of F unramified in B have different restrictions to E; we take a Hilbert modular form on $\mathrm{GL}_2(E)$ in place of Ω.

As was shown by Shimizu [7] and Jacquet and Langlands [2] (and as will be proved also in this paper), there is a Hilbert modular form $\mathbf{f}$ such that $\mathbf{f}|T(\mathfrak{a}) = \chi(\mathfrak{a})\mathbf{f}$ for all $\mathfrak{a}$. We shall prove as an application of the main result that $\langle \mathbf{f}, \mathbf{f} \rangle / \langle \mathbf{g}, \mathbf{g} \rangle$ is algebraic if B is totally indefinite (that is, if $F = E$) and if $\mathbf{f}$ is $\overline{\mathbf{Q}}$-rational (Theorem 3.8). It should be noted that our series D can be defined with no reference to B, since $\chi(\mathfrak{a})$ can be obtained from Hilbert modular forms. It seems impossible, however, to replace the constant $\pi^d \langle \mathbf{g}, \mathbf{g} \rangle$ by a quantity of the form $\pi^{d'} \langle \mathbf{f}, \mathbf{f} \rangle$ except when $F = E$. There is a natural question whether the periods of $\mathbf{f}$ are algebraically related to those of $\mathbf{g}$. This, if true, is stronger than the algebraicity of $\langle \mathbf{f}, \mathbf{f} \rangle / \langle \mathbf{g}, \mathbf{g} \rangle$, and can actually be proved in certain cases; the author hopes to treat this problem in a subsequent paper.

Although in general no Euler products seem to be associated with series D, there are two exceptional cases in which natural Euler products can be found; namely the cases $F = E$ and $[F : E] = 2$. If $F = E$, our series is practically the same as a series of the form

$$\Sigma_{\mathfrak{a}} \chi(\mathfrak{a}) \chi'(\mathfrak{a}) N(\mathfrak{a})^{-s},$$

whose critical values we studied in [11]. In the case $[F : E] = 2$, we obtain, as a special case of our series, a sum

$$\Sigma_{\mathfrak{b}} \chi(\mathfrak{b} \mathfrak{r}_F) N(\mathfrak{b})^{-s},$$

where $\mathfrak{b}$ runs over all integral ideals of E, and $\mathfrak{r}_F$ denotes the maximal order of F. This multiplied by a certain L-function of F has an Euler product of the form

$$C(s) = \prod_{\mathfrak{q}} C_{\mathfrak{q}}(N(\mathfrak{q})^{-s})^{-1}$$

with a polynomial $C_{\mathfrak{q}}$ of degree ≤ 4 defined for each prime ideal $\mathfrak{q}$ of E. We can then show that $C(\mu/2)$ for certain integers μ are algebraic numbers times $\pi^e \langle \mathbf{g}, \mathbf{g} \rangle$ with an integer e depending on μ (Theorem 3.11). It should be noted that C is a zeta function attached to an arithmetic quotient of $\mathfrak{H}^2$.

Let us now briefly explain how to prove the algebraicity of $D(s_0)/[\pi^d \langle \mathbf{g}, \mathbf{g} \rangle]$. In our previous paper [13] and also in Part I, we constructed automorphic forms of arithmetic type on $\mathrm{SO}(m,2)^r \times \mathrm{SO}(m+2)^{n-r}$. We specialize this to the case $m = 2$ by taking the norm form of B to be the quadratic form. This gives an automorphic form $H(w, w')$ on $\mathfrak{H}^r \times \mathfrak{H}^r$ with respect to $\Gamma \times \Gamma$, where Γ is a congruence subgroup of $B^\times$. One of the main theorems of Part I shows that $\pi^{-d}H$ is arithmetic, and hence it can be shown that

$$H(w, w') = \sum_i f_i(w)g_i(w')$$

with finitely many $\overline{\mathbf{Q}}$-rational Γ-automorphic forms f_i and g_i on $\mathfrak{H}^r$. The desired algebraicity will be obtained by analysing the nature of the inner product

$$\langle g(w), H(w, w')\rangle = \sum_i \langle g, f_i\rangle g_i(w')$$

with a $\overline{\mathbf{Q}}$-rational Γ-automorphic form g. As an auxiliary result, we shall give a simple proof for the fact that

$$(2) \qquad \sum_{0 \ll b \in \mathfrak{y}} \chi(b\mathfrak{x}^{-1})b^q \mathbf{e}_F(bz)$$

with a suitable $q \in 2^{-1}I_F$ is a Hilbert modular form (Theorem 3.1, Propositions 5.1 and 5.4), which may be of independent interest.

We shall also show that under certain conditions on the weights of g and Ω, the residues of D and C at critical points divided by the regulator of E have the same nature of algebraicity as $D(s_0)$ (Theorem 3.5, Theorem 3.7, (ii), and Theorem 3.11, (iii)). This result can be expressed also as a property of the inner product of a certain non-holomorphic pull-back of (2) with a Hilbert modular form on $GL_2(E)$ (Theorem 3.6 and Theorem 3.7, (iv)), as already mentioned in the introduction of Part I.

Notation. The symbols I_K, J_K, F, $\ll$, R_F, $\mathfrak{H}$, $\overline{\mathbf{Q}}$, $\sim$, and $\mathbf{e}(z)$ introduced in Part I will be employed also in Part II. In addition to these, we denote by $\mathbf{H}$ the division ring of Hamilton quaternions, and by $\mathbf{A}$ the ring of adeles of $\mathbf{Q}$. If R is an associative ring with identity element, $R^\times$ denotes the group of all invertible elements of R, and $M_n(R)$ the ring of all matrices of size n with entries in R. We put $F_{\mathbf{A}} = F \otimes_{\mathbf{Q}} \mathbf{A}$; then $F_{\mathbf{A}}^\times$ denotes the group of ideles of F. For a prime ideal $\mathfrak{p}$ and a fractional ideal $\mathfrak{a}$ of F, we denote by $F_{\mathfrak{p}}$ the $\mathfrak{p}$-completion of F and by $\mathfrak{a}_{\mathfrak{p}}$ the closure of $\mathfrak{a}$ in $F_{\mathfrak{p}}$. The norm of $\mathfrak{a}$ is denoted by $N(\mathfrak{a})$. For $t \in F_{\mathbf{A}}^\times$, we denote by $t\mathfrak{a}$ the fractional ideal of F determined by $(t\mathfrak{a})_{\mathfrak{p}} = t_{\mathfrak{p}}\mathfrak{a}_{\mathfrak{p}}$ for all $\mathfrak{p}$, where $t_{\mathfrak{p}}$ is the $\mathfrak{p}$-component of t. The group of all totally positive units of F is denoted by U_F. The complex conjugate of a complex matrix or a complex-valued function X is denoted by $\overline{X}$ and also by X^ρ in Section 4. The disjoint union of sets S_a for $a \in A$ is denoted by $\amalg_{a \in A} S_a$.

The sections, formulas, theorems, etc. of Part I are referred to, for example, as Section I.7, (I.3.10), Theorem I.5.3, etc., which mean Section 7, (3.10), Theorem 5.3, etc. of Part I.

1. Automorphic forms and Hecke operators
of a quaternion algebra

Our exposition will follow for the most part the formulation of our previous paper [9] with some modifications (cf. also Jacquet and Langlands [2]). We first recall that F denotes a totally real algebraic number field of degree n, and $\tau_1, \ldots, \tau_n$ are the embeddings of F into $\mathbf{R}$. Throughout Part II, we denote by B a quaternion algebra over F unramified at $\tau_1, \ldots, \tau_r$ and ramified at $\tau_{r+1}, \ldots, \tau_n$; we assume $0 < r \le n$. The matrix algebra $M_2(F)$ is included as a special case of B. Now there is an isomorphism

$$(1.1) \qquad\qquad B \otimes_{\mathbf{Q}} \mathbf{R} \to M_2(\mathbf{R})^r \times \mathbf{H}^{n-r}$$

whose restriction to F maps a to $(a^{\tau_1}, \ldots, a^{\tau_n})$. We let ι denote the main involution of B and its extension to $B_{\mathbf{A}}$ $(= B \otimes_{\mathbf{Q}} \mathbf{A})$ and put $N(\alpha) = \alpha\alpha^\iota$, $\mathrm{Tr}(\alpha) = \alpha + \alpha^\iota$ for $\alpha \in B_{\mathbf{A}}$. We view $B^\times$ as the $\mathbf{Q}$-rational points $G_{\mathbf{Q}}$ of a $\mathbf{Q}$-rational algebraic group G and define its adelization $G_{\mathbf{A}} = (B_{\mathbf{A}})^\times$. The infinite part of $G_{\mathbf{A}}$ is denoted by G_∞ and identified with $\mathrm{GL}_2(\mathbf{R})^r \times (\mathbf{H}^\times)^{n-r}$; the identity component of G_∞ is denoted by $G_{\infty+}$. For an element x of $G_{\mathbf{A}}$, we denote by x_∞ its infinite part, and by $x^{(i)}$ for $1 \le i \le n$ the i-th component of x_∞. Then we put

$$G_{\mathbf{A}+} = \{x \in G_{\mathbf{A}} \mid x_\infty \in G_{\infty+}\}, \quad G_{\mathbf{Q}+} = G_{\mathbf{Q}} \cap G_{\mathbf{A}+}.$$

If $x \in G_{\mathbf{A}+}$, we let x act on $\mathfrak{H}^r$ by

$$x(z_1, \ldots, z_r) = \left(x^{(1)}(z_1), \ldots, x^{(r)}(z_r)\right), x^{(i)}(z_i) = (a_i z_i + b_i)/(c_i z_i + d_i),$$

where $x^{(i)} = \begin{pmatrix} a_i & b_i \\ c_i & d_i \end{pmatrix}$; we can also define $x(z)$ for arbitrary $x \in G_{\mathbf{A}}$ and $z \in \mathbf{C}^r$ by the same formula.

For every nonnegative integer m, we define an $\mathbf{R}$-rational polynomial representation $\sigma_m : \mathbf{H}^\times \to \mathrm{GL}_{m+1}(\mathbf{C})$ by composing an injection of $\mathbf{H}$ into $M_2(\mathbf{C})$ with the representation of GL_2 by symmetric tensors of degree m. (We understand that σ_0 is the trivial representation.) Changing it for an equivalent representation and taking a suitable isomorphism (1.1), we may assume:

(1.2a) $\sigma_m(x^\iota) = {}^t\overline{\sigma_m(x)}$ *for all* $x \in \mathbf{H}^\times$;

(1.2b) $\sigma_m(\alpha^{(i)})$ *has algebraic entries for every* $\alpha \in G_{\mathbf{Q}}$, *every* $i > r$, *and every* $m \ge 0$;

(1.3) $\alpha^{(i)}$ *has algebraic entries for every* $\alpha \in G_{\mathbf{Q}}$ *and every* $i \le r$.

Let $k = \Sigma_{i=1}^r k_i \tau_i \in I_F$ and $\kappa = \Sigma_{i=r+1}^n \kappa_i \tau_i \in I_F$ with $\kappa_i \ge 0$, and let $d = d(\kappa) = \prod_{i>r}(\kappa_i + 1)$; thus $d = 1$ if $r = n$. We then define a representation $\sigma_\kappa : G_{\mathbf{A}} \to \mathrm{GL}_d(\mathbf{C})$ by

$$(1.4) \qquad\qquad \sigma_\kappa(x) = \otimes_{i>r}\sigma_{\kappa_i}(x^{(i)}) \qquad\qquad (x \in G_{\mathbf{A}})$$

and also put

$$(1.5) \quad j(x, z)^k = \prod_{i=1}^{r} (c_i z_i + d_i)^{k_i} |N(x^{(i)})|^{-k_i/2} \quad \left(x \in G_{\mathbf{A}}, x^{(i)} = \begin{pmatrix} * & * \\ c_i & d_i \end{pmatrix} \right)$$

for $z = (z_1, \ldots, z_r) \in \mathbf{C}^r$, $\mathrm{Im}(z_i) \neq 0$. For a $\mathbf{C}^d$-valued function f on $\mathfrak{H}^r$ and $x \in G_{\mathbf{A}+}$, we define a $\mathbf{C}^d$-valued function $f\|_{k,\kappa}x$ on $\mathfrak{H}^r$ by

$$(1.6) \qquad (f\|_{k,\kappa}x)(z) = \prod_{i>r} N(x^{(i)})^{\kappa_i/2} \cdot j(x, z)^{-k} \sigma_\kappa(x)^{-1} f(x(z)).$$

For every positive integer N, put

$$(1.7) \qquad \Gamma_N = \{ \gamma \in \mathfrak{o} \,|\, N(\gamma) = 1, \gamma - 1 \in N\mathfrak{o} \}$$

with a fixed maximal order $\mathfrak{o}$. By a *congruence subgroup of* $G_{\mathbf{Q}+}$, we understand (by abuse of language) a subgroup Γ of $G_{\mathbf{Q}+}$ such that $\Gamma_N \subset \Gamma$ and $[\Gamma \mathfrak{r}^\times : \Gamma_N \mathfrak{r}^\times] < \infty$ for some N, where $\mathfrak{r}$ denotes the maximal order of F. For such a Γ, we denote by $\mathfrak{M}_{k,\kappa}(\Gamma)$ (resp. $\mathcal{C}_{k,\kappa}(\Gamma)$) the set of all $\mathbf{C}^d$-valued holomorphic (resp. meromorphic) functions f which are holomorphic (resp. meromorphic) also at cusps and which satisfy $f\|_{k,\kappa}\gamma = f$ for all $\gamma \in \Gamma$; the cusp condition is necessary only when $B = M_2(\mathbf{Q})$. The union of $\mathfrak{M}_{k,\kappa}(\Gamma)$ (resp. $\mathcal{C}_{k,\kappa}(\Gamma)$) for all congruence subgroups Γ will be denoted by $\mathfrak{M}_{k,\kappa}$ (resp. $\mathcal{C}_{k,\kappa}$). We always assume $k_i \geq 0$ for all $i \leq r$ whenever we speak of $\mathfrak{M}_{k,\kappa}$, since $\mathfrak{M}_{k,\kappa} = \{0\}$ otherwise. We shall simply write $f\|x$ for $f\|_{k,\kappa}x$ when $f \in \mathcal{C}_{k,\kappa}$. If φ is a $\mathbf{C}^\times$-valued character of Γ of finite order, we put

$$(1.8) \qquad \mathfrak{M}_{k,\kappa}(\Gamma, \varphi) = \{ f \in \mathfrak{M}_{k,\kappa} \,|\, \varphi(\gamma)f = f\|\gamma \text{ for all } \gamma \in \Gamma \}.$$

We denote by $\mathfrak{S}_{k,\kappa}$ (resp. $\mathfrak{S}_{k,\kappa}(\Gamma), \mathfrak{S}_{k,\kappa}(\Gamma, \varphi)$) the set of elements of $\mathfrak{M}_{k,\kappa}$ (resp. $\mathfrak{M}_{k,\kappa}(\Gamma), \mathfrak{M}_{k,\kappa}(\Gamma, \varphi)$) which vanish at every cusp (if any); thus $\mathfrak{S}_{k,\kappa} = \mathfrak{M}_{k,\kappa}$ if B is a division algebra. We define a measure μ on $\mathfrak{H}^r$ by

$$(1.9\mathrm{a}) \qquad d\mu(z) = \prod_{\nu=1}^{r} y_\nu^{-2} dx_\nu dy_\nu \qquad\qquad (z_\nu = x_\nu + iy_\nu),$$

and the inner product $\langle f, g \rangle$ of elements f and g of $\mathfrak{S}_{k,\kappa}(\Gamma)$ by

$$(1.9\mathrm{b}) \qquad \langle f, g \rangle = \mu(D)^{-1} \int_D {}^t\overline{f(z)}\, g(z) \mathrm{Im}(z)^k \, d\mu(z), \qquad D = \Gamma \backslash \mathfrak{H}^r,$$

where $\mathrm{Im}(z)^k = \prod_{\nu=1}^{r} y_\nu^{k_\nu}$.

To consider automorphic forms as functions on $G_{\mathbf{A}}$, we take a subgroup W of $G_{\mathbf{A}}$ of the form $W = W_0 G_{\infty+}$ with an open compact subgroup W_0 of the finite part of $G_{\mathbf{A}}$. Then we have a finite coset decomposition

$$(1.10) \qquad G_{\mathbf{A}} = \amalg_{\lambda=1}^{h} G_{\mathbf{Q}} x_\lambda W = \amalg_{\lambda=1}^{h} W x_\lambda G_{\mathbf{Q}}$$

with $h = [F_\mathbf{A}^\times : F^\times N(W)]$; in fact, (1.10) holds if and only if $F_\mathbf{A}^\times = \amalg_{\lambda=1}^h N(x_\lambda) F^\times N(W)$. We take $x_{\lambda\infty} = 1$. Given a character Φ of W whose kernel is a group of the same type as W, we denote by $\mathfrak{M}_{k,\kappa}(W, \Phi)$ the set of all $\mathbf{C}^d$-valued functions $\mathbf{f}$ on $G_\mathbf{A}$ satisfying the following two conditions.

(1.11a) $\mathbf{f}(\alpha x w) = \Phi(w)\mathbf{f}(x)$ *for* $\alpha \in G_\mathbf{Q}, x \in G_\mathbf{A}, w \in W, w_\infty = 1$;

(1.11b) *For every* $x \in G_\mathbf{A}$ *with* $x_\infty = 1$, *there is an element* $\mathbf{g}_x$ *of* $\mathfrak{M}_{k,\kappa}$ *such that* $\mathbf{f}(xy) = (\mathbf{g}_x \Vert y)(\mathbf{i})$ *for all* $y \in G_{\infty+}$.

Here $\mathbf{i} = (\sqrt{-1}, \ldots, \sqrt{-1})$ $(\in \mathfrak{H}^r)$. Put $W_\lambda = x_\lambda W x_\lambda^{-1}$, $\Gamma_\lambda = W_\lambda \cap G_\mathbf{Q}$, and $\varphi_\lambda(\gamma) = \Phi(x_\lambda^{-1}\gamma x_\lambda)^{-1}$ for $\gamma \in \Gamma_\lambda$. Now, given $(f_1, \ldots, f_h) \in \Pi_{\lambda=1}^h \mathfrak{M}_{k,\kappa}(\Gamma_\lambda, \varphi_\lambda)$, define a function $\mathbf{f}$ on $G_\mathbf{A}$ by $\mathbf{f}(\alpha x_\lambda^{-\iota} w) = \Phi(w)(f_\lambda \Vert w)(\mathbf{i})$ for $\alpha \in G_\mathbf{Q}, w \in W, 1 \le \lambda \le h$, where $x^{-\iota} = (x^\iota)^{-1} = N(x)^{-1}x$. (Notice that $G_\mathbf{A} = \amalg_{\lambda=1}^h G_\mathbf{Q} x_\lambda^{-\iota} W$.) It can easily be seen that $\mathbf{f} \in \mathfrak{M}_{k,\kappa}(W, \Phi)$, and every element of $\mathfrak{M}_{k,\kappa}(W, \Phi)$ can be obtained in this manner. Therefore we write $\mathbf{f} = (f_1, \ldots, f_h)$ and identify $\mathfrak{M}_{k,\kappa}(W, \Phi)$ with $\Pi_{\lambda=1}^h \mathfrak{M}_{k,\kappa}(\Gamma_\lambda, \varphi_\lambda)$; we denote by $\mathfrak{S}_{k,\kappa}(W, \Phi)$ the set of all $(f_1, \ldots, f_h)$ such that $f_\lambda \in \mathfrak{S}_{k,\kappa}$ for all λ. For $\mathbf{f} = (f_1, \ldots, f_h)$ and $\mathbf{g} = (g_1, \ldots, g_h)$ in $\mathfrak{S}_{k,\kappa}(W, \Phi)$, we put

$$(1.12) \qquad \langle \mathbf{f}, \mathbf{g} \rangle = h^{-1} \sum_{\lambda=1}^h \langle f_\lambda, g_\lambda \rangle.$$

It can easily be seen that this inner product is independent of the choice of x_λ and is consistently defined on the union of $\mathfrak{S}_{k,\kappa}(W, \Phi)$ for all possible W and Φ.

We now consider (W, Φ) of a specific type. Let $\mathfrak{r}$ denote the maximal order of F, and $\mathfrak{e}$ the product of all prime ideals of F ramified in B. For each prime ideal $\mathfrak{p}$ of F, let $B_\mathfrak{p} = B \otimes_F F_\mathfrak{p}$. If $\mathfrak{a}$ is an $\mathfrak{r}$-submodule of B, $\mathfrak{a}_\mathfrak{p}$ denotes its localization at $\mathfrak{p}$. We fix a maximal order $\mathfrak{o}$ of B, and identify $\mathfrak{o}_\mathfrak{p}$ with $M_2(\mathfrak{r}_\mathfrak{p})$ for each $\mathfrak{p}$ prime to $\mathfrak{e}$. Now given an integral ideal $\mathfrak{c}$ of F, we define an order $\mathfrak{o}^\mathfrak{c}$ of "level $\mathfrak{c}$" as an $\mathfrak{r}$-submodule of $\mathfrak{o}$ by the condition

$$(1.13) \qquad (\mathfrak{o}^\mathfrak{c})_\mathfrak{p} = \begin{cases} \left\{ \begin{pmatrix} a & b \\ c & d \end{pmatrix} \in M_2(\mathfrak{r}_\mathfrak{p}) \,\middle|\, c \in \mathfrak{c}_\mathfrak{p} \right\} & \text{if } \mathfrak{p} \not\supset \mathfrak{e}, \\ \mathfrak{r}_\mathfrak{p} + \mathfrak{c}_\mathfrak{p}\mathfrak{o}_\mathfrak{p} & \text{if } \mathfrak{p} \supset \mathfrak{e}. \end{cases}$$

Then we put

$$(1.14) \qquad W = G_{\infty+} \prod_\mathfrak{p} (\mathfrak{o}^\mathfrak{c})_\mathfrak{p}^\times$$

and consider coset decomposition (1.10); we choose x_λ so that their $\mathfrak{p}$-components are 1 for every prime factor $\mathfrak{p}$ of $\mathfrak{c}$ and $x_{\lambda\infty} = 1$. We then define W_λ and Γ_λ as above. We also define a semi-group Y contained in $G_\mathbf{A}$ by $Y = G_\mathbf{A} \cap (G_{\infty+} \Pi_\mathfrak{p} Y_\mathfrak{p})$

with

$$(1.15) \quad Y_{\mathfrak{p}} = \begin{cases} (\mathfrak{o}^{\mathfrak{c}})_{\mathfrak{p}}^{\times} & \text{if } \mathfrak{p} \supset \mathfrak{e} + \mathfrak{c}, \\[2mm] \mathfrak{o}_{\mathfrak{p}} \cap B_{\mathfrak{p}}^{\times} & \text{if } \mathfrak{p} \not\supset \mathfrak{c}, \\[2mm] \left\{ \begin{pmatrix} a & b \\ c & d \end{pmatrix} \in (\mathfrak{o}^{\mathfrak{c}})_{\mathfrak{v}} \cap B_{\mathfrak{p}}^{\times} \mid a \in \mathfrak{r}_{\mathfrak{p}}^{\times} \right\} & \text{if } \mathfrak{p} \supset \mathfrak{c}, \mathfrak{p} \not\supset \mathfrak{e}. \end{cases}$$

Given a character ψ_0 of $(\mathfrak{r}/\mathfrak{c})^{\times}$, we define a multiplicative map ψ_Y of Y into $\mathbf{C}^{\times}$ and a character φ_{λ} of Γ_{λ} by

$(1.16a)$ $\psi_Y\left(\begin{pmatrix} a & b \\ c & d \end{pmatrix} \right) = \psi_{0\mathfrak{p}}(a)$ for $\begin{pmatrix} a & b \\ c & d \end{pmatrix} \in Y_{\mathfrak{p}}$ if $\mathfrak{p} \supset \mathfrak{c}$ and $\mathfrak{p} \not\supset \mathfrak{e}$,

$(1.16b)$ $\psi_Y(x) = \psi_{0\mathfrak{p}}(x \bmod \mathfrak{c}_{\mathfrak{p}}\mathfrak{o}_{\mathfrak{p}})$ for $x \in (\mathfrak{o}^{\mathfrak{c}})_{\mathfrak{p}}^{\times}$ if $\mathfrak{p} \supset \mathfrak{e} + \mathfrak{c}$,

$(1.16c)$ $\psi_Y(Y_{\mathfrak{p}}) = 1$ if $\mathfrak{p} \not\supset \mathfrak{c}$,

$(1.16d)$ $\psi_Y(G_{\infty +}) = 1$,

$$(1.17) \qquad \varphi_{\lambda}(\gamma) = \psi_Y(x_{\lambda}^{-1} \gamma x_{\lambda}) \text{ for } \gamma \in \Gamma_{\lambda},$$

where $\psi_{0\mathfrak{p}}$ is the $\mathfrak{p}$-part of ψ_0. Now we put

$$(1.18) \quad \mathfrak{M}_{k,\kappa}(\mathfrak{c}, \psi_0) = \prod_{\lambda=1}^{h} \mathfrak{M}_{k,\kappa}(\Gamma_{\lambda}, \varphi_{\lambda}), \quad \mathfrak{S}_{k,\kappa}(\mathfrak{c}, \psi_0) = \prod_{\lambda=1}^{h} \mathfrak{S}_{k,\kappa}(\Gamma_{\lambda}, \varphi_{\lambda}).$$

We always assume a condition

$$(1.19) \qquad \psi_0(u) = \prod_{i \le r} \operatorname{sgn}(u^{\tau_i})^{k_i} \prod_{i > r} \operatorname{sgn}(u^{\tau_i})^{\kappa_i} \text{ for all } u \in \mathfrak{r}^{\times},$$

since $\mathfrak{M}_{k,\kappa}(\Gamma_{\lambda}, \varphi_{\lambda}) = \{0\}$ unless this is satisfied. The spaces of (1.18) are the same as $\mathfrak{M}_{k,\kappa}(W, \Phi)$ and $\mathfrak{S}_{k,\kappa}(W, \Phi)$ with $\Phi(w) = \psi_Y(w^{\iota})$. Let ψ be a Hecke character of $F_{\mathbf{A}}^{\times}$ such that

(1.20) $\psi(v) = \prod_{\mathfrak{p}} \psi_{0\mathfrak{p}}(v_{\mathfrak{p}}) \prod_{i \le r} \operatorname{sgn}(v^{(i)})^{k_i} \prod_{i > r} \operatorname{sgn}(v^{(i)})^{\kappa_i}$ if $v \in F_{\mathbf{A}}^{\times}$ and $v_{\mathfrak{p}} \in \mathfrak{r}_{\mathfrak{p}}^{\times}$ for all p.

For such a character ψ, let $\mathfrak{M}_{k,\kappa}(\mathfrak{c}, \psi)$ denote the vector space of all $\mathbf{C}^d$-valued functions $\mathbf{f}$ on $G_{\mathbf{A}}$ satisfying (1.11a,b) with $\Phi(w) = \psi_Y(w^{\iota})$ and also

(1.21) $\mathbf{f}(sx) = \psi(s)\mathbf{f}(x)$ *for all* $s \in F_{\mathbf{A}}^{\times}$.

Then we can easily verify that $\mathfrak{M}_{k,\kappa}(\mathfrak{c}, \psi_0)$ is the direct sum of the spaces $\mathfrak{M}_{k,\kappa}(\mathfrak{c}, \psi)$ for all Hecke characters ψ satisfying (1.20).

To define Hecke operators, we first prove two preliminary lemmas.

LEMMA 1.1. *Let U_1 and U_2 be open subgroups of $G_{\mathbf{A}}$, and y an element of $G_{\mathbf{A}}$; further let $U' = y^{-1}U_1 y \cap U_2$ and $\Gamma = U_2 \cap G_{\mathbf{Q}}^1$, where $G_{\mathbf{Q}}^1 = \{\alpha \in B \mid N(\alpha) = 1\}$. Suppose $N(U_2) = N(U')$. Then $U_2 = U'\Gamma$ and $U_1 y U_2 = U_1 y \Gamma$.*

Proof. Given $u \in U_2$, we have $N(u) = N(v)$ with $v \in U'$. By virtue of strong approximation, we have $v^{-1}u = w\gamma$ with $w \in U'$ and $\gamma \in G_{\mathbf{Q}}^1$. Then

$\gamma \in G_{\mathbf{Q}}^1 \cap U_2 = \Gamma$, and hence $u \in U'\Gamma$, which proves that $U_2 = U'\Gamma$. Therefore $U_1 y U_2 = U_1 y U'\Gamma = U_1 y\Gamma$.

LEMMA 1.2. *If* $\alpha \in G_{\mathbf{Q}} \cap x_\lambda Y x_\mu^{-1}$, *then*

$$N(\alpha^{-1} W_\lambda \alpha \cap W_\mu) = N(W_\mu), \, W_\lambda \alpha W_\mu = W_\lambda \alpha \Gamma_\mu', \, and \, \Gamma_\lambda \alpha \Gamma_\mu = \Gamma_\lambda \alpha \Gamma_\mu',$$

where $\Gamma_\mu' = \{\alpha \in \Gamma_\mu \,|\, N(\alpha) = 1\}$.

Proof. We first observe that $N(y^{-1}Wy \cap W) = N(W)$ for every $y \in Y$. Let $y = x_\lambda^{-1}\alpha x_\mu$ and $W' = y^{-1}Wy \cap W$. Then $x_\mu W' x_\mu^{-1} = \alpha^{-1} W_\lambda \alpha \cap W_\mu$, and hence $N(\alpha^{-1} W_\lambda \alpha \cap W_\mu) = N(W') = N(W) = N(W_\mu)$. By Lemma 1.1, we have $W_\lambda \alpha W_\mu = W_\lambda \alpha \Gamma_\mu'$. Let $\beta \in \Gamma_\lambda \alpha \Gamma_\mu$. Since $\Gamma_\lambda \alpha \Gamma_\mu \subset W_\lambda \alpha W_\mu$, we have $\beta = v\delta$ with $v \in W_\lambda$ and $\delta \in \alpha \Gamma_\mu'$. Then $v \in W_\lambda \cap G_{\mathbf{Q}} = \Gamma_\lambda$, and hence $\beta \in \Gamma_\lambda \alpha \Gamma_\mu'$, which proves that $\Gamma_\lambda \alpha \Gamma_\mu = \Gamma_\lambda \alpha \Gamma_\mu'$.

We now introduce the algebra $R(W, Y)$ of all formal finite sums $\Sigma c_y W y W$ with $y \in Y$ and $c_y \in \mathbf{C}$, as well as the algebra $\{R_{\lambda\mu}\}$ where $R_{\lambda\mu}$ consists of all formal finite sums $\Sigma c_\alpha \Gamma_\lambda \alpha \Gamma_\mu$ with $\alpha \in G_{\mathbf{Q}} \cap x_\lambda Y x_\mu^{-1}$ and $c_\alpha \in \mathbf{C}$. The law of multiplication is defined as in [9, §2] and [10, §3.2]. If $\alpha \in G_{\mathbf{Q}} \cap x_\lambda Y x_\mu^{-1}$, we have $\Gamma_\lambda \alpha \Gamma_\mu = \amalg_i \Gamma_\lambda \alpha_i$ with finitely many α_i. For $f \in \mathfrak{M}_{k,\kappa}(\Gamma_\lambda, \varphi_\lambda)$, we can define an element $f | \Gamma_\lambda \alpha \Gamma_\mu$ of $\mathfrak{M}_{k,\kappa}(\Gamma_\mu, \varphi_\mu)$ by

$$(1.22) \qquad f | \Gamma_\lambda \alpha \Gamma_\mu = \sum_i \psi_Y \left(x_\lambda^{-1} \alpha_i x_\mu \right)^{-1} f \| \alpha_i.$$

On the other hand, given $y \in Y$, let $WyW = \amalg_i Wy_i$ with $y_{i\infty} = 1$. Then we can define the action of WyW on $\mathfrak{M}_{k,\kappa}(\mathfrak{c}, \psi_0)$ by

$$(1.23) \qquad (\mathbf{f} | WyW)(x) = \sum_i \psi_Y(y_i)^{-1} \mathbf{f}(xy_i^\iota) \qquad (x \in G_{\mathbf{A}}).$$

For every index λ, we can find an index μ and an element α of $G_{\mathbf{Q}} \cap x_\lambda Y x_\mu^{-1}$ such that $WyW = W x_\lambda^{-1} \alpha x_\mu W$; μ is uniquely determined by WyW and λ. If $\mathbf{f} = (f_1, \ldots, f_h) \in \mathfrak{M}_{k,\kappa}(\mathfrak{c}, \psi_0)$ and $\mathbf{f} | WyW = (g_1, \ldots, g_h)$ with f_λ and g_λ in $\mathfrak{M}_{k,\kappa}(\Gamma_\lambda, \psi_0)$, we can easily verify that $g_\mu = f_\lambda | \Gamma_\lambda \alpha \Gamma_\mu$.

For each integral ideal $\mathfrak{m}$ of F, let $T_\mathfrak{c}(\mathfrak{m})$, or simply $T(\mathfrak{m})$, denote the sum of all different WyW with $y \in Y$ such that $N(y)\mathfrak{r} = \mathfrak{m}$. Further we let $S_\mathfrak{c}(\mathfrak{m}) = WaW$ with an element a of $F_{\mathbf{A}}^\times$ such that $a\mathfrak{r} = \mathfrak{m}$ if $\mathfrak{m}$ is prime to $\mathfrak{c}$; we put $S_\mathfrak{c}(\mathfrak{m}) = 0$ if $\mathfrak{m}$ is not prime to $\mathfrak{c}$. An element $\mathbf{f}$ of $\mathfrak{M}_{k,\kappa}(\mathfrak{c}, \psi_0)$ belongs to $\mathfrak{M}_{k,\kappa}(\mathfrak{c}, \psi)$ with a Hecke character ψ if and only if $\mathbf{f}$ is a common eigenfunction of $S_\mathfrak{c}(\mathfrak{m})$ for all $\mathfrak{m}$. We have also formally

$$(1.24) \quad \sum_{\mathfrak{m}} T_\mathfrak{c}(\mathfrak{m}) N(\mathfrak{m})^{-s} = \prod_{\mathfrak{p}} \left[1 - T_\mathfrak{c}(\mathfrak{p}) N(\mathfrak{p})^{-s} + \delta_\mathfrak{p} S_\mathfrak{c}(\mathfrak{p}) N(\mathfrak{p})^{1-2s} \right]^{-1},$$

where $\delta_{\mathfrak{p}} = 0$ or 1 according as $\mathfrak{p}$ divides $\mathfrak{e}$ or not. Notice that $T_{\mathfrak{e}}(\mathfrak{p}) = S_{\mathfrak{e}}(\mathfrak{p}) = 0$ if $\mathfrak{p} \supset \mathfrak{e} + \mathfrak{c}$.

LEMMA 1.3. *If $y \in Y$ and $N(y)\mathfrak{r}$ is prime to $\mathfrak{c}$, then $WyW = Wy'W$ and $\psi^*(N(y)\mathfrak{r})\langle \mathbf{f} | WyW, \mathbf{g} \rangle = \langle \mathbf{f}, \mathbf{g} | WyW \rangle$ for $\mathbf{f} \in \mathfrak{S}_{k,\kappa}(\mathfrak{c}, \psi_0)$ and $\mathbf{g} \in \mathfrak{S}_{k,\kappa}(\mathfrak{c}, \psi)$, where ψ^* is the ideal-character associated with ψ.*

This can be proved in exactly the same fashion as [11, Prop. 2.4]. ($[\Gamma_\mu \alpha' \Gamma_\mu]'$ and $[\Gamma_\mu \alpha \Gamma_\lambda]$ of [11, p. 652, lines 20 and 22] should be $[\Gamma_\mu \alpha' \Gamma_\lambda]'$.) Take in particular $\mathbf{f}$ in $\mathfrak{S}_{k,\kappa}(\mathfrak{c}, \psi')$ with $\psi' \neq \psi$ and $y \in F_{\mathbf{A}}^\times$. Then we see that $\mathfrak{S}_{k,\kappa}(\mathfrak{c}, \psi)$ is orthogonal to $\mathfrak{S}_{k,\kappa}(\mathfrak{c}, \psi')$ if $\psi \neq \psi'$.

2. Automorphic forms of arithmetic type

Let us first recall the definition of the arithmeticity of the elements of $\mathcal{C}_{k,\kappa}$ defined in [13]. Let K be a totally imaginary quadratic extension of F, and h an F-linear embedding of K into B. Then $h(K^\times)$ is contained in $G_{\mathbf{Q}+}$ and has a unique common fixed point $w = (w_1, \ldots, w_r)$ on $\mathfrak{H}^r$, which we call a CM-*point*. Moreover, we can define embeddings $\varphi_1, \ldots, \varphi_r$ of K into $\mathbf{C}$ by

$$(2.1) \qquad h(x)^{(i)} \begin{bmatrix} w_i \\ 1 \end{bmatrix} = x^{\varphi_i} \begin{bmatrix} w_i \\ 1 \end{bmatrix} \qquad (x \in K; i = 1, \ldots, r).$$

We consider then an element $\psi = \Sigma_{i=1}^r \varphi_i$ of I_K and put

$$(2.2) \qquad P_{k,\kappa}(w) = p_K(\Sigma_{i=1}^r k_i \varphi_i, \psi) \sigma_\kappa(h(\mathfrak{p}_K(\psi)))$$

with the symbols p_K and $\mathfrak{p}_K$ introduced in [13, §1]. An element f of $\mathcal{C}_{k,\kappa}$ is called *arithmetic* or $\overline{\mathbf{Q}}$-*rational* if the components of the vector $P_{k,\kappa}(w)^{-1}f(w)$ are all algebraic for every CM-point w where f is holomorphic. An element $\mathbf{f} = (f_1, \ldots, f_h)$ of $\mathfrak{M}_{k,\kappa}(\mathfrak{c}, \psi_0)$ is called *arithmetic* or $\overline{\mathbf{Q}}$-*rational* if the f_λ are all arithmetic. This is so if and only if g_x of (1.11b) is arithmetic for every $x \in G_{\mathbf{A}}$, $x_\infty = 1$. We indicate the set of all arithmetic elements of a given set by adding the symbol $\overline{\mathbf{Q}}$; thus $\mathcal{C}_{k,\kappa}(\overline{\mathbf{Q}})$, $\mathfrak{M}_{k,\kappa}(\Gamma, \overline{\mathbf{Q}})$, $\mathfrak{S}_{k,\kappa}(\mathfrak{c}, \psi, \overline{\mathbf{Q}})$, etc. denote respectively the sets of arithmetic elements of $\mathcal{C}_{k,\kappa}$, $\mathfrak{M}_{k,\kappa}(\Gamma)$, $\mathfrak{S}_{k,\kappa}(\mathfrak{c}, \psi)$, etc. We can easily verify that both $\mathcal{C}_{k,\kappa}(\overline{\mathbf{Q}})$ and $\mathfrak{M}_{k,\kappa}(\overline{\mathbf{Q}})$ are stable under the map $f \mapsto f \| \alpha$ for all $\alpha \in G_{\mathbf{Q}+}$.

To simplify our notation, we put $\mathcal{C}_k = \mathcal{C}_{k,0}$, $\mathfrak{M}_k = \mathfrak{M}_{k,0}$, and $\mathfrak{S}_k = \mathfrak{S}_{k,0}$. Then $\mathcal{C}_0(\Gamma)$ is the field of Γ-invariant automorphic functions, which can be identified with the function field of the algebraic variety $\Gamma \backslash \mathfrak{H}^r$. The theory of canonical models of $\Gamma \backslash \mathfrak{H}^r$ shows that $\mathcal{C}_0(\Gamma)$ is the composite of $\mathcal{C}_0(\Gamma, \overline{\mathbf{Q}})$ and $\mathbf{C}$; moreover $\mathcal{C}_0(\Gamma, \overline{\mathbf{Q}})$ and $\mathbf{C}$ are linearly disjoint over $\overline{\mathbf{Q}}$.

THEOREM 2.1. $\mathfrak{M}_{k,\kappa} = \mathfrak{M}_{k,\kappa}(\overline{\mathbf{Q}}) \otimes_{\overline{\mathbf{Q}}} \mathbf{C}$, $\mathfrak{S}_{k,\kappa} = \mathfrak{S}_{k,\kappa}(\overline{\mathbf{Q}}) \otimes_{\overline{\mathbf{Q}}} \mathbf{C}$.

Proof. The first equality was proved in [11, Prop. 1.7] when $B = M_2(F)$. To derive the second one from this result, take a basis $\{c_i\}$ of C over $\overline{\mathbf{Q}}$ and let $f \in \mathfrak{S}_k$. Then $f = \Sigma_i c_i f_i$ with finitely many $f_i \in \mathfrak{M}_k(\overline{\mathbf{Q}})$. Given $\alpha \in G_{\mathbf{Q}+}$, compare the constant terms of the Fourier expansions of both sides of $f \| \alpha = \Sigma_i c_i f_i \| \alpha$. Then we find that the constant terms of $f_i \| \alpha$ must be 0, and hence $f_i \in \mathfrak{S}_k(\overline{\mathbf{Q}})$ as desired. Assuming now that B is a division algebra, we first treat the case in which $\kappa = 0$ and $k = te$ with $t \in \mathbf{Z}$, where $e = \Sigma_{i=1}^r \tau_i$. For every positive integer N, define Γ_N by (1.7). Let V_N be a $\overline{\mathbf{Q}}$-rational algebraic variety isomorphic to $\Gamma_N \backslash \mathfrak{H}^r$, and p_N the projection map of $\mathfrak{H}^r$ onto V_N. For the reason explained above, we can choose V_N so that its function field over $\overline{\mathbf{Q}}$ is $\mathcal{Q}_0(\Gamma_N, \overline{\mathbf{Q}}) \circ p_N^{-1}$. Now $\mathcal{Q}_e(\overline{\mathbf{Q}})$ contains a nonzero element, say q, as shown in [12, Prop. 7.4]. Take a positive integer M so that $q \in \mathcal{Q}_e(\Gamma_M)$ and Γ_M has no elements of finite order other than 1. We now prove

$$(2.3) \qquad \mathfrak{M}_{te}(\Gamma_N) = \mathfrak{M}_{te}(\Gamma_N, \overline{\mathbf{Q}}) \otimes_{\overline{\mathbf{Q}}} C \qquad (0 < t \in \mathbf{Z})$$

for every multiple N of M. Fix N. Every $h \in \mathcal{Q}_k(\Gamma_N), \neq 0$, defines its divisor on V_N, denoted by $\mathrm{div}(h)$. Take r elements $f_1, \ldots, f_r$ of $\mathcal{Q}_0(\Gamma_N, \overline{\mathbf{Q}})$ algebraically independent over C, and let

$$h = \pi^{-r} \partial(f_1, \ldots, f_r)/\partial(z_1, \ldots, z_r).$$

By [12, Cor. 7.7], h belongs to $\mathcal{Q}_{2e}(\overline{\mathbf{Q}})$; moreover, $\mathrm{div}(h)$ is $\overline{\mathbf{Q}}$-rational, since it is the divisor of a $\overline{\mathbf{Q}}$-rational differential form $df_1 \wedge \cdots \wedge df_r$ on V_N. Since $q^{-2}h \in \mathcal{Q}_0(\Gamma_N, \overline{\mathbf{Q}})$, $2 \, \mathrm{div}(q)$ is $\overline{\mathbf{Q}}$-rational, and hence $\mathrm{div}(q)$ is $\overline{\mathbf{Q}}$-rational. Now

$$\mathfrak{M}_{te}(\Gamma_N) = \left\{ fq^t \mid f \in L(t \cdot \mathrm{div}(q)) \right\},$$

where $L(X) = \{ f \in \mathcal{Q}_0(\Gamma_N) \mid \mathrm{div}(f) \geq -X \}$ for a divisor X on V_N. It is well known that $L(X)$ is spanned by $\overline{\mathbf{Q}}$-rational elements if X is $\overline{\mathbf{Q}}$-rational. Therefore we obtain (2.3) for every multiple N of M, which implies our theorem in the case where $\kappa = 0$ and $k = te$. Before proving the general case, we make several observations. A basis of $\mathfrak{M}_{te}(\Gamma_N, \overline{\mathbf{Q}})$ over $\overline{\mathbf{Q}}$ for sufficiently large t defines a $\overline{\mathbf{Q}}$-rational projective embedding of V_N. This implies that every element of $\mathcal{Q}_0(\Gamma_N, \overline{\mathbf{Q}})$ is a quotient of two elements of $\mathfrak{M}_{se}(\Gamma_N, \overline{\mathbf{Q}})$ with a suitable positive integer s. By [13, Prop. 4.1] (see also [12, §7]), there exists, for every $z_0 \in \mathfrak{H}^r$, an $M_d(C)$-valued meromorphic function T such that

    (2.4a) *every column of T belongs to $\mathcal{Q}_{k,\kappa}(\overline{\mathbf{Q}})$;*

    (2.4b) *T is holomorphic at z_0 and $\det(T(z_0)) \neq 0$.*

We call such a T an *arithmetic uniformizer of weight (k, κ) at z_0.*

LEMMA 2.2. *For every (k, κ) and every $z_0 \in \mathfrak{H}^r$, there exists an arithmetic uniformizer of weight $(k + le, \kappa)$ at z_0 whose columns belong to $\mathfrak{M}_{k+le, \kappa}$, where l is a positive integer depending only on (k, κ).*

Proof. This is well known if $B = M_2(F)$. (Such a modular form can easily be obtained by a theta series, for example.) Therefore suppose B is a division algebra. Given $z_0 \in \mathfrak{H}^r$, take an arithmetic uniformizer T of weight (k, κ) at z_0; take a multiple N of M so that every column of T belongs to $\mathcal{Q}_{k, \kappa}(\Gamma_N)$. Let Δ be a complete set of representatives of Γ_M / Γ_N. If $\delta \in \Delta$, we observe that $T^{-1}(T \| \delta)$ has entries in $\mathcal{Q}_0(\Gamma_N, \overline{\mathbf{Q}})$. Expressing every entry as a quotient of two elements of $\mathfrak{M}_{se}(\Gamma_N, \overline{\mathbf{Q}})$ with various s, we find a nonzero element g_δ of $\mathfrak{M}_{be}(\Gamma_N, \overline{\mathbf{Q}})$ with $0 < b \in \mathbf{Z}$ such that $g_\delta T^{-1}(T \| \delta)$ is holomorphic on $\mathfrak{H}^r$. Then $g_\delta T \| \delta$ is holomorphic at z_0, and hence $(g_\delta \| \delta^{-1})T$ is holomorphic at $\delta(z_0)$. Putting $h = \Pi_{\delta \in \Delta} g_\delta \| \delta^{-1}$, we find that hT is holomorphic at every point of $\Gamma_{M_0^*}$. Obviously $0 \neq h \in \mathfrak{M}_{te}(\Gamma_N, \overline{\mathbf{Q}})$ with $0 < t \in \mathbf{Z}$. Now the singularities of the entries of hT form a divisor of V_N whose projection to V_M, say W, is a divisor of V_M. Our construction shows that $p_M(z_0) \notin W$. We can find in this way a finite covering $\{Z_\alpha\}_{\alpha \in A}$ of V_M by Zariski open subsets Z_α, nonzero elements h_α of $\mathfrak{M}_{te}(\overline{\mathbf{Q}})$, and arithmetic uniformizers T_α of weight (k, κ) such that $h_\alpha T_\alpha$ is holomorphic on $p_M^{-1}(Z_\alpha)$. Now given again z_0 and T as above, the entries of $T_\alpha^{-1}T$ belong to $\mathcal{Q}_0(\overline{\mathbf{Q}})$, and hence there exists a nonzero element f_α of $\mathfrak{M}_{s_\alpha e}(\overline{\mathbf{Q}})$ with $0 < s_\alpha \in \mathbf{Z}$ such that $f_\alpha T_\alpha^{-1}T$ is holomorphic on $\mathfrak{H}^r$. Then $h_\alpha f_\alpha T$ is holomorphic on $p_M^{-1}(Z_\alpha)$. Thus if we put $S = (\Pi_{\alpha \in A} h_\alpha f_\alpha) \cdot T$, S is holomorphic on the whole $\mathfrak{H}^r$. Every column of S belongs to $\mathfrak{M}_{k+le, \kappa}(\overline{\mathbf{Q}})$ with $0 < l \in \mathbf{Z}$, and $\det(S)$ is a nonvanishing function on $\mathfrak{H}^r$. Since the points $\beta(z_0)$ for all $\beta \in G_{\mathbf{Q}+}$ are dense on $\mathfrak{H}^r$, we see that $S \| \beta$ with a suitable $\beta \in G_{\mathbf{Q}+}$ is a desired uniformizer.

We are now ready to prove Theorem 2.1 in the general case. We fix a basis $\{c_i\}$ of $\mathbf{C}$ over $\overline{\mathbf{Q}}$. If $h = \Sigma_i c_i h_i \in \mathcal{Q}_{k, \kappa}$ with $h_i \in \mathcal{Q}_{k, \kappa}(\overline{\mathbf{Q}})$, then the h_i are unique for h. In fact, suppose $\Sigma_i c_i h_i = 0$. Take an arithmetic uniformizer T of weight (k, κ). Then $\Sigma_i c_i T^{-1} h_i = 0$. Since $\mathcal{Q}_0(\overline{\mathbf{Q}})$ and $\mathbf{C}$ are linearly disjoint over $\overline{\mathbf{Q}}$, we find that $T^{-1} h_i = 0$, and hence $h_i = 0$ for all i, which proves the uniqueness. Now given $f \in \mathfrak{M}_{k, \kappa}$ and a point z_0 of $\mathfrak{H}^r$, take an arithmetic uniformizer S of weight $(k + le, \kappa)$ as in Lemma 2.2. Let $q = \det(S)$. Then $qS^{-1}f$ has components in $\mathfrak{M}_{se}$ with some $s \in \mathbf{Z}$. By Lemma 2.2, we can find an element u of $\mathfrak{M}_{(t-s)e}(\overline{\mathbf{Q}})$ such that $u(z_0) \neq 0$, with a suitable integer $t > s$. Then $uqS^{-1}f$ has components in $\mathfrak{M}_{te}$, and hence $uqS^{-1}f = \Sigma_i c_i g_i$ with vectors g_i whose components belong to $\mathfrak{M}_{te}(\overline{\mathbf{Q}})$. Thus $f = \Sigma_i c_i f_i$ if we put $f_i = (uq)^{-1} S g_i$. Obviously $f_i \in \mathcal{Q}_{k, \kappa}(\overline{\mathbf{Q}})$. These f_i are unique for f as shown above, and are holomorphic at z_0. Since z_0 is an arbitrary point of $\mathfrak{H}^r$, we have $f_i \in \mathfrak{M}_{k, \kappa}(\overline{\mathbf{Q}})$, which completes the proof.

LEMMA 2.3. *For every congruence subgroup Γ and its character φ, one has $\mathfrak{M}_{k,\kappa}(\Gamma, \varphi) = \mathfrak{M}_{k,\kappa}(\Gamma, \varphi, \overline{\mathbf{Q}}) \otimes_{\overline{\mathbf{Q}}} \mathbf{C}$ and $\mathfrak{S}_{k,\kappa}(\Gamma, \varphi) = \mathfrak{S}_{k,\kappa}(\Gamma, \varphi, \overline{\mathbf{Q}}) \otimes_{\overline{\mathbf{Q}}} \mathbf{C}$.*

Proof. Let $f \in \mathfrak{M}_{k,\kappa}(\Gamma, \varphi)$. By Theorem 2.1, $f = \Sigma_i c_i g_i$ with $g_i \in \mathfrak{M}_{k,\kappa}(\overline{\mathbf{Q}})$, where $\{c_i\}$ is a basis of $\mathbf{C}$ over $\overline{\mathbf{Q}}$. Then

$$\sum_i c_i \varphi(\gamma) g_i = \varphi(\gamma) f = f \| \gamma = \sum_i c_i g_i \| \gamma$$

for every $\gamma \in \Gamma$. Since $g_i \| \gamma$ is arithmetic, we have $g_i \| \gamma = \varphi(\gamma) g_i$ by virtue of the uniqueness in the above proof, and hence $g_i \in \mathfrak{M}_{k,\kappa}(\Gamma, \varphi, \overline{\mathbf{Q}})$. The case of $\mathfrak{S}_{k,\kappa}$ can be proved in a similar way.

LEMMA 2.4. *With Γ and φ as in Lemma 2.3, let Γ' be a congruence subgroup of Γ and φ' the restriction of φ to Γ'. Then the orthogonal complement of $\mathfrak{S}_{k,\kappa}(\Gamma, \varphi)$ in $\mathfrak{S}_{k,\kappa}(\Gamma', \varphi')$ has a $\overline{\mathbf{Q}}$-rational basis.*

Proof. Put $U = \mathfrak{r}^{\times}$. Replacing Γ' by $U\Gamma' \cap \Gamma$, we may assume that Γ/Γ' is finite. Let $\Gamma = \amalg_{i=1}^m \Gamma' \varepsilon_i$. For $f \in \mathfrak{S}_{k,\kappa}(\Gamma', \varphi')$, put $P(f) = m^{-1} \Sigma_{i=1}^m \varphi(\varepsilon_i)^{-1} f \| \varepsilon_i$. Then $P(f) \in \mathfrak{S}_{k,\kappa}(\Gamma, \varphi)$ and $f - P(f)$ is orthogonal to $\mathfrak{S}_{k,\kappa}(\Gamma, \varphi)$. If f is arithmetic, so is $f - P(f)$. Our assertion follows immediately from this fact.

3. Main theorems

We have so far treated automorphic forms with a fixed B. In this section, however, various B's need to be considered. To indicate that our spaces of automorphic forms are defined with respect to B, we write $\mathfrak{S}_{k,\kappa}(B)$ and $\mathfrak{S}_{k,\kappa}(\mathfrak{c}, \psi, B)$ for $\mathfrak{S}_{k,\kappa}$ and $\mathfrak{S}_{k,\kappa}(\mathfrak{c}, \psi)$. We drop the subscript κ if $\kappa = 0$. For example, $\mathfrak{S}_k(M_2(F))$ is the space of cusp forms of weight k with respect to congruence subgroups of $\mathrm{GL}_2(F)$.

Let η be a $\mathbf{C}$-valued function defined on a finite-dimensional vector space V over $\mathbf{Q}$. We call η *locally constant* if there are two lattices L and M in V such that $\eta(x) = 0$ for $x \notin L$, and $\eta(x)$ depends only on x modulo M. If V is a vector space over F, we can always find a subgroup U of $\mathfrak{r}^{\times}$ of finite index such that $\eta(ux) = \eta(x)$ for all $u \in U$; we then call η *U-invariant*.

Let $\{g_p\}_{p \in P}$ be a basis of $\mathfrak{S}_{k,\kappa}(\mathfrak{c}, \psi)$ over $\mathbf{C}$, and let

$$(3.1) \qquad g_p | T_{\mathfrak{c}}(\mathfrak{a}) = \sum_{q \in P} \chi_{pq}(\mathfrak{a}) g_q$$

with $\chi_{pq}(\mathfrak{a}) \in \mathbf{C}$ for each fractional ideal $\mathfrak{a}$ of F; we understand that $T_{\mathfrak{c}}(\mathfrak{a}) = 0$ if $\mathfrak{a}$ is not integral. We first state a preliminary result.

THEOREM 3.1. *Let $\mathfrak{x}$ be a fractional ideal of F and η a locally constant function on F. Put, for each fixed p and q,*

$$(3.2) \qquad A_{pq}(z) = \sum_{0 \ll b \in F} \eta(b)\chi_{pq}(b\mathfrak{x})b^{(m/2)-F}\mathbf{e}_F(bz) \qquad (z \in \mathfrak{H}^n),$$

where $\mathbf{e}_F(bz) = \mathbf{e}(\sum_{\nu=1}^n b^{\tau_\nu}z_\nu)$, the superscript F means $\sum_{\nu=1}^n \tau_\nu$, and $m = \sum_{\nu=1}^n m_\nu\tau_\nu \in I_F$ with

$$(3.3) \qquad m_\nu = \begin{cases} k_\nu & (\nu \le r), \\ \kappa_\nu + 2 & (\nu > r). \end{cases}$$

Then $A_{pq}(z)$ is an element of $\mathfrak{S}_m(M_2(F))$.

This will be proved in Section 6. A result of the same nature in terms of the traces of Hecke operators was first obtained by Eichler when $F = \mathbf{Q}$. More general cases were treated by Shimizu [7], [8], and by Jacquet and Langlands [2, §§14, 16]. Though the essential content of our theorem is included in their results, the above formulation seems new.

Let us now assume that F has a subfield E satisfying the following condition.

(3.4) $[E:\mathbf{Q}] = r$ *and the restrictions of $\tau_1, \ldots, \tau_r$ to E are all different.*

We put $\tau_i' = \mathrm{Res}_{F/E}(\tau_i)$ for $i \le r$ and denote by U_E the group of all totally positive units of E. Taking an element

$$(3.5) \qquad \Omega(z) = \sum_{a \in E} \omega(a)\mathbf{e}_E(az) \qquad (z \in \mathfrak{H}^r)$$

of $\mathfrak{M}_l(M_2(E))$ with $0 \le l \in I_E$ and putting $\omega_u(a) = a^u\omega(a)$ with $0 \le u \in I_E$, we define an infinite series

$$(3.6) \qquad \Lambda_{pq}(s) = [U] \sum_{0 \ll b \in F/U} \eta(b)\chi_{pq}(b\mathfrak{x})\omega_u\left(\mathrm{Tr}_{F/E}(\zeta b)\right)b^{c-sE} \qquad (s \in \mathbf{C}).$$

Here η, χ_{pq}, and $\mathfrak{x}$ are the same as in Theorem 3.1; $\zeta \in F$, $\zeta^{\tau_i} > 0$ for $i \le r$ and $\zeta^{\tau_i} < 0$ for $i > r$; U is a subgroup of U_E of finite index such that η and Ω are U-invariant; $c = \sum_{i=1}^n c_i\tau_i \in 2^{-1}I_F$; $[U] = [U_E:U]^{-1}$; the sum is taken over all different cosets bU with $0 \ll b \in F$; $b^{sE} = (b^{\tau_1}\cdots b^{\tau_r})^s$ (see Section I.3). To make each term of (3.6) dependent only on bU, we assume

$$(3.7) \qquad l + 2u + 2\,\mathrm{Res}_{F/E}(c) = \gamma E \text{ with } \gamma \in \mathbf{Z}.$$

If we put $\xi_{pq}(b) = \eta(b)\chi_{pq}(b\mathfrak{x})b^{(m/2)-F}$ and $d = c - (m/2) + F$, then

$$(3.8\text{a}) \qquad A_{pq}(z) = \sum_{0 \ll b \in F} \xi_{pq}(b)\mathbf{e}_F(bz),$$

$$(3.8\text{b}) \qquad \Lambda_{pq}(s) = [U] \sum_{0 \ll b \in F/U} \xi_{pq}(b)\omega_u\left(\mathrm{Tr}_{F/E}(\zeta b)\right)b^{d-sE}.$$

Therefore Λ_{pq} is a series of type (I.3.2). Since we assumed (I.3.4) and also that $d \in I_F$, we have to impose the following conditions:

(3.9a) $$2c_i \equiv k_i \pmod{2} \text{ for } i \leq r,$$

(3.9b) $$\kappa_i \leq 2c_i \equiv \kappa_i \pmod{2} \text{ for } i > r.$$

By Proposition I.3.1, our series is convergent for sufficiently large $\mathrm{Re}(s)$. In fact, we have $\alpha = \gamma + 2[F:E]$ for the invariant α of (I.3.3). The analytic continuation follows from Theorem I.3.2 and can be stated as follows.

PROPOSITION 3.2. *Under* (3.7) *and* (3.9a, b), $\Lambda_{pq}(s)$ *can be continued to a meromorphic function on the whole plane which is holomorphic for* $\mathrm{Re}(s) > (\gamma/2) + [F:E]$ *and has at most a simple pole at* $s = (\gamma/2) + [F:E]$. *Moreover, if*

(3.10) $$2c_i + k_i + 2 > \gamma + 2[F:E] \text{ for at least one } i \leq r,$$

then Λ_{pq} *is holomorphic for* $\mathrm{Re}(s) \geq (\gamma - 1)/2 + [F:E]$.

We are now ready to present our main theorems of arithmeticity, in which we naturally assume that

(3.11) $$\Omega \text{ and } \{g_p\} \text{ are } \overline{\mathbf{Q}}\text{-rational and } \eta \text{ is } \overline{\mathbf{Q}}\text{-valued}.$$

THEOREM 3.3. *Let* $\Lambda(s)$ *and* Ξ *be the matrices of size* P *with* $\Lambda_{pq}(s)$ *and* $\langle g_p, g_q \rangle$ *as the* (p, q)-*entries, respectively. Then, under* (3.7), (3.9a, b), *and* (3.11), *the entries of the matrix* $\pi^{-|k|} \cdot {}^t\Xi^{-1}\Lambda(s_0)$ *are all algebraic for every integer* s_0 *satisfying*

(3.12a) $$(\gamma/2) - 1 + [F:E] < s_0 \leq (k_i/2) + c_i \text{ for all } i \leq r,$$

(3.12b) $$2s_0 \neq \gamma + 2[F:E] \text{ if } E = \mathbf{Q}.$$

Remark 3.4. (i) Put $\kappa' = \mathrm{Res}_{F/E}(\Sigma_{i>r}\kappa_i\tau_i) = \Sigma_{i=1}^{r}\kappa_i'\tau_i'$. We see easily that, given k, l, u, and κ, there exists $c = \Sigma_{i=1}^{n}c_i\tau_i \in 2^{-1}I_F$ satisfying (3.7) and (3.9a, b) if and only if

$$k_1 - l_1 - \kappa_1' \equiv \cdots \equiv k_r - l_r - \kappa_r' \pmod{2}.$$

(ii) The existence of an integer s_0 satisfying (3.12a) implies (3.10). Hence Λ is finite at s_0 by virtue of Proposition 3.2.

(iii) The consideration of $\omega_u(a)$ in addition to $\omega(a)$ is made only for technical reasons. The series with nontrivial u can easily be reduced to the case $u = 0$, as explained in Section I.3 and Remark I.4.5.

(iv) Condition (3.12b) is probably unnecessary.

THEOREM 3.5. *Suppose, besides* (3.7), (3.9a,b) *and* (3.11), *that*

$$(3.13) \qquad 2c_i + k_i + 2 = \gamma + 2[F:E] \text{ for all } i \leq r.$$

Define Λ *and* Ξ *as in Theorem* 3.3. *Then the residue of* $\pi^{-|k|} \cdot {}^t\Xi^{-1}\Lambda$ *at* $s = (\gamma/2) + [F:E]$ *is* R_E *times an algebraic matrix, where* R_E *is the regulator of* E.

To state the next theorem, we define, for a function f on $\mathfrak{H}^n$, a function f^ς on $\mathfrak{H}^r$ by $f^\varsigma(z) = f(z^*)$ for $z \in \mathfrak{H}^r$, where

$$(3.14) \qquad z_\nu^* = \begin{cases} \zeta^{\tau_\nu} z_\nu & \text{if } \nu \leq r, \\ \zeta^{\tau_\nu} \bar{z}_\lambda & \text{if } \nu > r \text{ and } \operatorname{Res}_{F/E}(\tau_\nu) = \tau_\lambda'. \end{cases}$$

THEOREM 3.6. *Let* $0 \leq v = \Sigma_{i>r} v_i \tau_i \in I_F$, $v' = \operatorname{Res}_{F/E}(v)$, $\mu = \operatorname{Res}_{F/E}(\Sigma_{i>r} m_i \tau_i)$, *and* $l = \Sigma_{i=1}^r k_i \tau_i' - \mu - 2v'$, *where* m_i *is defined by* (3.3). *For every* $h \in \mathfrak{M}_l(M_2(E))$, *let* $\langle h, y^{\mu+2v'}(\delta_m^v A)^\varsigma \rangle$ *denote the matrix whose* (p,q)-*entry is* $\langle h, y^{\mu+2v'}(\delta_m^v A_{pq})^\varsigma \rangle$ *with* A_{pq} *of* (3.2), *where* δ_m^v *is defined by* (I.1.14a). *Then all the entries of the matrix* ${}^t\Xi^{-1}\langle h, y^{\mu+2v'}(\delta_m^v A)^\varsigma \rangle$ *are algebraic, provided that* h *is* $\overline{\mathbf{Q}}$-*rational.*

Theorems 3.3 and 3.5 will be proved in Section 6. Here let us derive Theorem 3.6 from Theorem 3.5. Given the data as in Theorem 3.6, we consider (3.8b) with $u = 0$ and $d = v - k$. Then $c = v - k + (m/2) - F$; (3.7) and (3.13) are satisfied with $\gamma = -2[F:E]$. Therefore Theorem I.3.4, or rather its proof in the case $p = 0$, shows that the residue of Λ_{pq} at $s = 0$ is $C\pi^{|k|}R_E$ $\langle \Omega_\rho, y^{\mu+2v'}(\delta_m^v A_{pq})^\varsigma \rangle$ with a nonzero algebraic number C independent of (p,q). Thus the desired result follows from Theorem 3.5.

The above theorems are given in terms of a fixed $\mathbf{Q}$-rational basis of $\mathfrak{S}_{k,\kappa}(\mathfrak{c}, \psi)$. Let us now derive some results concerning an eigenfunction of Hecke operators. Let $\{\chi(\mathfrak{a})\}$ be a system of eigenvalues of $T_\mathfrak{c}(\mathfrak{a})$ in the sense that there is a nonzero element $\mathbf{f}$ of $\mathfrak{S}_{k,\kappa}(\mathfrak{c}, \psi)$ such that $\mathbf{f}|T_\mathfrak{c}(\mathfrak{a}) = \chi(\mathfrak{a})\mathbf{f}$ for all $\mathfrak{a}$. Then we obtain, from (1.24),

$$(3.15) \qquad \sum_{\mathfrak{a}} \chi(\mathfrak{a})N(\mathfrak{a})^{-s} = \prod_{\mathfrak{p}} \left[1 - \chi(\mathfrak{p})N(\mathfrak{p})^{-s} + \delta_{\mathfrak{p}}\psi^*(\mathfrak{p})N(\mathfrak{p})^{1-2s} \right]^{-1},$$

where ψ^* is the ideal-character attached to ψ; we put $\psi^*(\mathfrak{p}) = 0$ if $\mathfrak{p} \supset \mathfrak{c}$. Observe that the $\chi(\mathfrak{a})$ are algebraic, as easily shown by Lemma 2.3. Given χ, let $Z(\chi)$ (resp. $Z(\chi, \overline{\mathbf{Q}})$) denote the set of all such $\mathbf{f}$ (resp. such arithmetic $\mathbf{f}$). Now let $Z'(\chi)$ (resp. $Z'(\chi, \overline{\mathbf{Q}})$) denote the set of all $\mathbf{f}$ (resp. arithmetic $\mathbf{f}$) of $\mathfrak{S}_{k,\kappa}(\mathfrak{c}, \psi)$ such that $\mathbf{f}|T_\mathfrak{c}(\mathfrak{a}) = \chi(\mathfrak{a})\mathbf{f}$ for all $\mathfrak{a}$ prime to $\mathfrak{c}$. From Lemma 2.3, we obtain

$$Z(\chi) = Z(\chi, \overline{\mathbf{Q}}) \otimes_{\overline{\mathbf{Q}}} \mathbf{C}, \quad Z'(\chi) = Z'(\chi, \overline{\mathbf{Q}}) \otimes_{\overline{\mathbf{Q}}} \mathbf{C}.$$

We say that χ is *regular* if $Z(\chi) = Z'(\chi)$. This is so if and only if $Z(\chi, \overline{\mathbf{Q}}) = Z'(\chi, \overline{\mathbf{Q}})$.

THEOREM 3.7. *Suppose χ regular; let $0 \neq \mathbf{g} \in Z(\chi, \overline{\mathbf{Q}})$. With η, ζ, and $\mathfrak{x}$ as in (3.6), put*

$$(3.16a) \qquad L(s) = [U] \sum_{0 \ll b \in F/U} \eta(b)\chi(b\mathfrak{x})\omega_u(\mathrm{Tr}_{F/E}(\zeta b))b^{c-sE},$$

$$(3.16b) \qquad f(z) = \sum_{0 \ll b \in F} \eta(b)\chi(b\mathfrak{x})b^{(m/2)-F}\mathbf{e}_F(bz) \qquad\qquad (z \in \mathfrak{H}^n).$$

Then, under (3.7), (3.9a,b), and (3.11), the following assertions hold.

(i) $L(s_0) \sim \pi^{|k|}\langle \mathbf{g}, \mathbf{g}\rangle$ *for every integer s_0 satisfying (3.12a,b).*

(ii) *Under (3.13), the residue of L at $s = (\gamma/2) + [F:E]$ is an algebraic number times $R_E\pi^{|k|}\langle \mathbf{g}, \mathbf{g}\rangle$.*

(iii) $f \in \mathcal{S}_m(M_2(F))$.

(iv) *The notation being as in Theorem 3.6, one has*

$$\langle h, y^{\mu+2v'}(\delta_m^v f)^\zeta\rangle \sim \langle \mathbf{g}, \mathbf{g}\rangle$$

for every $\overline{\mathbf{Q}}$-rational element h of $\mathfrak{M}_l(M_2(F))$.

Proof. Assertion (iii) follows directly from Theorem 3.1. By Lemma 1.3, $\mathcal{S}_{k,\kappa}(\mathfrak{c}, \psi)$ has an orthogonal basis composed of common eigenfunctions of $T_{\mathfrak{c}}(\mathfrak{a})$ for all $\mathfrak{a}$ prime to $\mathfrak{c}$. Since the eigenvalues are algebraic, $\mathcal{S}_{k,\kappa}(\mathfrak{c}, \psi, \overline{\mathbf{Q}})$ is the direct sum of $Z'(\chi, \overline{\mathbf{Q}})$ and its orthogonal complement, say W. Let $\{\mathbf{f}_i\}$ be a $\overline{\mathbf{Q}}$-rational basis of $Z'(\chi, \overline{\mathbf{Q}})$ and let $X = \pi^{-|k|}L(s_0)$. Assertion (i) is trivially true if $X = 0$; so assume $X \neq 0$. Applying Theorem 3.3 to $\{\mathbf{f}_i\}$ and a $\overline{\mathbf{Q}}$-rational basis of W, we find that $X^{-1}\langle \mathbf{f}_i, \mathbf{f}_j\rangle$ is algebraic for every i and j, which proves (i). The remaining assertions follow from Theorems 3.5 and 3.6 in the same fashion.

To indicate that χ occurs at "level $\mathfrak{c}$", we write $\{\chi, \mathfrak{c}\}$ for the system of eigenvalues as above. Let $\{\chi', \mathfrak{c}'\}$ be another such system of the same weight (k, κ). We say that $\{\chi, \mathfrak{c}\}$ and $\{\chi', \mathfrak{c}'\}$ are *equivalent* if $\chi(\mathfrak{p}) = \chi'(\mathfrak{p})$ for all but a finite number of prime ideals $\mathfrak{p}$. Now, from the theory of Jacquet-Langlands [2, §§14, 16] (which sharpens the preceding result of Shimizu [7]) combined with the result of Miyake [5] (cf. also [8]), one can derive the following facts:

(I) A given equivalence class of systems of eigenvalues which occur in $\mathcal{S}_{k,\kappa}(B)$ contains a unique $\{\chi_0, \mathfrak{c}_0\}$ such that $\mathfrak{c}_0$ divides $\mathfrak{c}$ for all other $\{\chi, \mathfrak{c}\}$ in the class. We call $\{\chi_0, \mathfrak{c}_0\}$ *primitive.*

(II) A primitive $\{\chi_0, \mathfrak{c}_0\}$ obtained from $\mathcal{S}_{k,\kappa}(\mathfrak{c}_0, \psi, B)$ always occurs as a primitive system of eigenvalues on $\mathcal{S}_m(\mathfrak{b}, \psi, M_2(F))$ with m as in (3.3) and an integral ideal $\mathfrak{b}$ which has the same set of prime factors as $\mathfrak{c}_0\mathfrak{e}$. (This fact without primitiveness is essentially included in Theorem 3.7, (iii).)

(III) A primitive system is regular.

It should be noted that when $e \neq r$, $Z(\chi)$ is not necessarily one-dimensional even if χ is primitive. (To obtain one-dimensional spaces, one has to decompose $\mathfrak{S}_{k,\kappa}(\mathfrak{c}, \psi)$ according to the representations of $\prod_{\mathfrak{p}|\mathfrak{e}} B_{\mathfrak{p}}^{\times}$.) Anyway, if χ is primitive, Theorem 3.7 may be viewed as a collection of results concerning the eigenvalues of Hecke operators on $\mathfrak{S}_m(\mathfrak{b}, \psi, M_2(F))$ (and another Hilbert modular form Ω on $\mathfrak{H}'$), though the constant $\langle \mathbf{g}, \mathbf{g} \rangle$ can be obtained from the Hilbert modular form f rather indirectly through property (iv). We may then ask a natural question: *Given F, $(\tau_1, \ldots, \tau_r)$, and a primitive system $\{\chi, \mathfrak{b}\}$ defined on $\mathfrak{S}_m(M_2(F))$, can one find a quaternion algebra B over F with the following properties?*

(i) *B is unramified at $\tau_1, \ldots, \tau_r$ and ramified at $\tau_{r+1}, \ldots, \tau_n$;*

(ii) *χ occurs in $\mathfrak{S}_{k,\kappa}(B)$ with k and κ as in* (3.3).

An obvious necessary condition is that $m_i \geq 2$ for $i > r$. This is not a sufficient condition; in general, χ cannot necessarily be obtained from a quaternion algebra B with a specified r. However, if $n - r$ is even, the answer is affirmative, provided $m_i \geq 2$ for $i > r$. In fact, if $n - r$ is even, we can take B to be the quaternion algebra ramified exactly at $\tau_{r+1}, \ldots, \tau_n$ and with $e = r$. This follows again from [7], [8] and [2].

THEOREM 3.8. *Suppose B is totally indefinite, that is, $r = n$. Let $\mathbf{f}$ be a nonzero $\overline{\mathbf{Q}}$-rational element of $\mathfrak{S}_k(\mathfrak{b}, \psi, M_2(F))$ such that $\mathbf{f} \mid T(\mathfrak{a}) = \chi(\mathfrak{a})\mathbf{f}$ for all $\mathfrak{a}$. Suppose that $\{\chi, \mathfrak{b}\}$ is primitive and χ occurs in $\mathfrak{S}_k(B)$. Let $\mathbf{g}$ and $\mathbf{h}$ be $\overline{\mathbf{Q}}$-rational elements of $\mathfrak{S}_k(\mathfrak{c}, \psi, B)$; suppose $\mathbf{g} \mid T(\mathfrak{p}) = \chi(\mathfrak{p})\mathbf{g}$ for all except a finite number of prime ideals $\mathfrak{p}$ of F. Then $\langle \mathbf{g}, \mathbf{h} \rangle \sim \langle \mathbf{f}, \mathbf{f} \rangle$; in particular $\langle \mathbf{g}, \mathbf{g} \rangle \sim \langle \mathbf{f}, \mathbf{f} \rangle$.*

Proof. Changing $\mathfrak{c}$ for its suitable multiple, we may assume that $\mathbf{g} \mid T(\mathfrak{p}) = \chi(\mathfrak{p})\mathbf{g}$ for all $\mathfrak{p}$ prime to $\mathfrak{c}$. Let X be the set of all elements $\mathbf{q}$ of $\mathfrak{S}_k(\mathfrak{c}, \psi, B)$ such that $\mathbf{q} \mid T(\mathfrak{p}) = \chi(\mathfrak{p})\mathbf{q}$ for all $\mathfrak{p}$ prime to $\mathfrak{c}$, and Y the orthogonal complement of X in $\mathfrak{S}_k(\mathfrak{c}, \psi)$. For the same reason as in the proof of Theorem 3.7, X and Y have $\overline{\mathbf{Q}}$-rational bases, say $\{\mathbf{g}_1, \ldots, \mathbf{g}_t\}$ and $\{\mathbf{g}_{t+1}, \ldots, \mathbf{g}_u\}$. Now there is a nonzero $\overline{\mathbf{Q}}$-rational element $\mathbf{f}'$ of $\mathfrak{S}_k(\mathfrak{b}', \psi', M_2(F))$ with some $\mathfrak{b}'$ and ψ' such that $\mathbf{f}' \mid T(\mathfrak{a}) = \varepsilon(\mathfrak{a})\chi(\mathfrak{a})\mathbf{f}'$ for all $\mathfrak{a}$ with an ideal-character ε whose conductor is a multiple of $\mathfrak{c}$ (see [11, Prop. 4.5]). We then have $\langle \mathbf{f}', \mathbf{f}' \rangle \sim \langle \mathbf{f}, \mathbf{f} \rangle$. This follows from the fact that $\langle \mathbf{f}, \mathbf{f} \rangle$ is the residue of a series $D_{\mathfrak{c}}(s, \mathbf{f}_{\rho}, \mathbf{f})$ at a critical point, up to an elementary factor (see [11, Prop. 4.13]; it also follows from [11, (4.16)]). Now $\langle \mathbf{f}', \mathbf{f}' \rangle$ is a finite sum $\sum_\lambda \langle f_\lambda, f_\lambda \rangle$ with elements f_λ of $\mathfrak{S}_k(M_2(F))$ of the forms

$$f_\lambda(z) = a_\lambda \sum_{0 \ll b \in \mathfrak{t}_\lambda} \varepsilon(bt_\lambda^{-1})\chi(bt_\lambda^{-1})b^{(k/2)-F}\mathbf{e}_F(bz)$$

with ideals $\mathfrak{t}_\lambda$ prime to $\mathfrak{c}$ and $a_\lambda \in \overline{\mathbf{Q}}$ (see [11, (2.16), (2.17), (2.24), (2.25)]). We now define χ_{pq}, A_{pq}, Λ_{pq} and Ξ with respect to $\{\mathbf{g}_1,\ldots,\mathbf{g}_u\}$; we fix one λ and take $\mathfrak{x} = \mathfrak{t}_\lambda^{-1}$, $\eta(b) = a_\lambda \varepsilon(b\mathfrak{t}_\lambda^{-1})$ for $b \in \mathfrak{t}_\lambda$ and $\eta(b) = 0$ for $b \notin \mathfrak{t}_\lambda$. Then we have $A_{11} = \cdots = A_{tt} = f_\lambda$, $A_{ij} = A_{ji} = 0$ for $i \neq j \leq t$. Taking $v = 0$ in Theorem 3.6, we see that ${}^t\Xi^{-1}\langle f_\lambda, A \rangle$ is algebraic. If $f_\lambda \neq 0$, we see that $\langle \mathbf{g}_i, \mathbf{g}_j \rangle \sim \langle f_\lambda, f_\lambda \rangle$ for $i \leq t$ or $j \leq t$. In particular, $\langle \mathbf{g}, \mathbf{g} \rangle \sim \langle f_\lambda, f_\lambda \rangle$, and hence $\langle \mathbf{g}, \mathbf{g} \rangle \sim \langle \mathbf{f}', \mathbf{f}' \rangle \sim \langle \mathbf{f}, \mathbf{f} \rangle$ and $\langle \mathbf{g}, \mathbf{g}_i \rangle \sim \langle \mathbf{f}, \mathbf{f} \rangle$ for all i, which completes the proof.

As will be explained below, the assertion of Theorem 3.8 is probably false if B is not totally indefinite. We can make, however, the following

CONJECTURE 3.9. *Let the notation be the same as in Theorem 3.8. Remove the assumption that B is totally indefinite. Let B' be another quaternion algebra over F unramified at $\tau_1,\ldots,\tau_r$ and ramified at $\tau_{r+1},\ldots,\tau_n$. Further let $\mathbf{j}$ be a nonzero $\overline{\mathbf{Q}}$-rational element of $\mathcal{S}_{k,\kappa}(\mathfrak{c}',\psi',B')$ with some $\mathfrak{c}'$ and ψ' such that $\mathbf{j} \mid T(\mathfrak{p}) = \chi(\mathfrak{p})\mathbf{j}$ for almost all $\mathfrak{p}$. Then $\langle \mathbf{g}, \mathbf{h} \rangle \sim \langle \mathbf{j}, \mathbf{j} \rangle$.*

Assuming that F has a subfield E satisfying (3.4), we would be able to prove this by the reasoning of the proof of Theorem 3.7, once we could show that $L(s_0) \neq 0$ for some choice of η, Ω, ζ, $\mathfrak{c}$, and s_0.

In Part I, we treated the case where $\Sigma\chi(\mathfrak{a})N(\mathfrak{a})^{-s}$ is an L-function of a CM-field. To explain this connection, take a totally imaginary quadratic extension K of F and elements $\varepsilon_1,\ldots,\varepsilon_n$ of J_K so that $\mathrm{Res}_{K/F}(\varepsilon_i) = \tau_i$. Let $\Psi = \Sigma_{i=1}^n(m_i - 1)\varepsilon_i$ with positive integers m_i, and let ξ be a Hecke ideal-character of K defined modulo an integral ideal $\mathfrak{C}$ such that $\xi(a\mathfrak{r}_K) = a^\Psi/|a^\Psi|$ for $a \equiv 1 \pmod{\mathfrak{C}}$, where $\mathfrak{r}_K$ is the maximal order of K. Then we put as usual $L(s,\xi) = \Sigma\xi(\mathfrak{A})N(\mathfrak{A})^{-s}$, where $\mathfrak{A}$ runs over all integral ideals of K prime to $\mathfrak{C}$. Now there is always a χ occurring in $\mathcal{S}_m(M_2(F))$ with $m = \Sigma_{i=1}^n m_i\tau_i$ such that $L(s-(1/2),\xi) = \Sigma\chi(\mathfrak{a})N(\mathfrak{a})^{-s}$ (see [11, §5], for example). Let $\mathbf{f}$ be a nonzero $\overline{\mathbf{Q}}$-rational element of $\mathcal{S}_m(M_2(F))$ such that $\mathbf{f} \mid T(\mathfrak{p}) = \chi(\mathfrak{p})\mathbf{f}$ for almost all $\mathfrak{p}$. Then

$$(3.18) \qquad \pi^n\langle \mathbf{f}, \mathbf{f} \rangle \sim p_K\left(\Psi, 2\sum_{i=1}^n \varepsilon_i\right) \text{ if } m_i > 1 \text{ for all } i \leq n.$$

by [13, Theorem 9.8] (cf. also [11, Theorem 5.5]). Suppose that χ occurs in $\mathcal{S}_{k,\kappa}(B)$, and let $\mathbf{g}$ be a nonzero $\overline{\mathbf{Q}}$-rational element of $\mathcal{S}_{k,\kappa}(\mathfrak{c},\psi,B)$ such that $\mathbf{g} \mid T(\mathfrak{p}) = \chi(\mathfrak{p})\mathbf{g}$ for almost all $\mathfrak{p}$. Then we are tempted to make

CONJECTURE 3.10. $\pi^r\langle \mathbf{g}, \mathbf{g} \rangle \sim p_K(\Psi, 2\Sigma_{i=1}^r\varepsilon_i)$ *if F has a subfield E satisfying* (3.4) *and $k_i > 1$ for all $i \leq r$.*

Theorem 3.8 combined with (3.18) shows that this is true if B is totally indefinite. Without assuming that B is totally indefinite, we can prove the conjecture if $L(s_0)$ of Theorem 3.7, (i) is nonvanishing for some choice of η, Ω, ζ, c, and s_0. In fact, as explained in Section I.4, series (3.16a) in the present case is the same as (I.4.2), and therefore the desired conclusion follows from (i) of Theorem 3.7 and Theorem I.4.3. It should be noted that there is no evidence supporting Conjectures 3.9 and 3.10 when F has no subfield E satisfying (3.4), though they may possibly be true in a more general case without E. On the other hand, there are nonvanishing examples of $L(s_0)$, at least in the case $E = \mathbf{Q} \neq F$, as shown in Section I.9. These give examples for which Conjecture 3.10 is true with $r < n$. Then the assertion of Theorem 3.8 extended to the case $r < n$, if true, produces some new algebraic relations among the periods $p_K(\alpha, \beta)$, which are highly improbable. Thus the assumption $r = n$ is unlikely to be dropped from Theorem 3.8.

It seems that one cannot construct Euler products out of series of type (3.6) except in some special cases. An obvious special case is when $E = F$. We conclude this section by treating a less obvious example. Suppose $[F:E] = 2$, $\zeta^2 \in E$, $u = 0$, and $\Omega = 1$. Then sum (3.16a) is taken under the condition $0 \ll b \in E/U$. Arrange the elements of J_F so that $\tau_i = \tau_{r+i} = \tau_i'$ on E for $i = 1, \ldots, r$. Since $l = u = 0$, condition (3.7) implies that $c_i + c_{r+i} = \gamma/2$ for all $i \leq r$. Therefore, given k and κ, we can find an element $c = \sum_{i=1}^{n} c_i \tau_i$ of I_F satisfying (3.7) and (3.9a,b) if and only if

$$(3.19) \qquad k_1 - \kappa_{r+1} \equiv \cdots \equiv k_r - \kappa_{2r} \ (\mathrm{mod}\ 2).$$

We have to choose γ so that $\gamma \equiv k_i - \kappa_{r+i} (\mathrm{mod}\ 2)$. Thus series (3.16a) can be written in the form

$$(3.20) \qquad L(s) = [U] \sum_{0 \ll b \in E/U} \eta(b) \chi(b\zeta) N_{E/\mathbf{Q}}(b)^{(\gamma/2)-s}$$

This is independent of the choice of c. To make the range of s_0 of (3.12a) maximum, we put $c_{r+i} = \kappa_{r+i}/2$ for $i = 1, \ldots, r$. Then (3.12a,b) take the forms

$$(3.21\mathrm{a}) \qquad (\gamma/2) + 1 < s_0 \leq (k_i - \kappa_{r+i} + \gamma)/2,$$

$$(3.21\mathrm{b}) \qquad 2s_0 \neq \gamma + 4 \ \text{if}\ E = \mathbf{Q}.$$

To obtain an Euler product from (3.20), we first decompose each Euler $\mathfrak{p}$-factor of (3.15) into linear factors by putting

$$1 - \chi(\mathfrak{p})X + \delta_\mathfrak{p} \psi^*(\mathfrak{p})N(\mathfrak{p})X^2 = (1 - \alpha_\mathfrak{p} X)(1 - \beta_\mathfrak{p} X)$$

with complex numbers $\alpha_\mathfrak{p}$ and $\beta_\mathfrak{p}$, where X is an indeterminate. We then define a polynomial $C_\mathfrak{q}$ for each prime ideal $\mathfrak{q}$ of E as follows.

(I) If $\mathfrak{q}\mathfrak{r}_F = \mathfrak{p}\mathfrak{p}'$ with different prime ideals $\mathfrak{p}$ and $\mathfrak{p}'$ in F, then

$$C_\mathfrak{q}(X) = (1 - \alpha_\mathfrak{p}\alpha_{\mathfrak{p}'}X)(1 - \alpha_\mathfrak{p}\beta_{\mathfrak{p}'}X)(1 - \beta_\mathfrak{p}\alpha_{\mathfrak{p}'}X)(1 - \beta_\mathfrak{p}\beta_{\mathfrak{p}'}X).$$

(II) If $\mathfrak{q}\mathfrak{r}_F = \mathfrak{p}$ with a prime ideal $\mathfrak{p}$ of F, then

$$C_{\mathfrak{q}}(X) = \left[1 - \chi(\mathfrak{p})X + \delta_{\mathfrak{p}}\psi^*(\mathfrak{p})N(\mathfrak{p})X^2\right]\left[1 - \delta_{\mathfrak{p}}\psi^*(\mathfrak{p})N(\mathfrak{p})X^2\right].$$

(III) If $\mathfrak{q}\mathfrak{r}_F = \mathfrak{p}^2$ with a prime ideal $\mathfrak{p}$ of F, then

$$C_{\mathfrak{q}}(X) = \left(1 - \alpha_{\mathfrak{p}}^2 X\right)\left(1 - \alpha_{\mathfrak{v}}\beta_{\mathfrak{v}}X\right)\left(1 - \beta_{\mathfrak{v}}^2 X\right).$$

Now let φ be a character of finite order of the group of all ideals of E prime to an integral ideal, say $\mathfrak{f}$. Then we consider

$$(3.22) \qquad\qquad C(s) = \prod_{\mathfrak{q}} C_{\mathfrak{q}}\left(\varphi(\mathfrak{q})N(\mathfrak{q})^{-s}\right)^{-1},$$

where $\mathfrak{q}$ runs over all prime ideals of E prime to $\mathfrak{f}$. We can easily verify that $C(s) = L'(2s - 2, \xi)D(s)$ with

$$(3.23) \qquad D(s) = \sum_{\mathfrak{a}}\varphi(\mathfrak{a})\chi(\mathfrak{a}\mathfrak{r}_F)N(\mathfrak{a})^{-s}, \quad L'(s,\xi) = \sum_{\mathfrak{b}}\xi(\mathfrak{b})N(\mathfrak{b})^{-s},$$

$$\xi(\mathfrak{b}) = \varphi(\mathfrak{b})^2\psi^*(\mathfrak{b}\mathfrak{r}_F),$$

where $\mathfrak{a}$ (resp. $\mathfrak{b}$) runs over all integral ideals of E prime to $\mathfrak{f}$ (resp. $\mathfrak{c}\mathfrak{e}\mathfrak{f}$).

THEOREM 3.11. *The notation being the same as in Theorem 3.7, suppose that $[F:E] = 2$ and condition (3.19) is satisfied; define C and D by (3.22) and (3.23). Then the following assertions hold.*

(i) C and D can be continued to meromorphic functions on the whole plane; D is holomorphic for $\mathrm{Re}(s) > 2$ and has at most a simple pole at $s = 2$; if $k_i - \kappa_{r+i} > 2$ for at least one $i \le r$, D is holomorphic for $\mathrm{Re}(s) \ge 3/2$.

(ii) $C(\mu/2) \sim \pi^{(\mu-2)r+|k|}\langle \mathbf{g},\mathbf{g}\rangle$ and $D(\mu/2) \sim \pi^{|k|}\langle \mathbf{g},\mathbf{g}\rangle$ for every integer μ such that $2 < \mu \le k_i - \kappa_{r+i}, \mu \equiv k_i - \kappa_{r+i}(\mathrm{mod}\ 2)$ for all $i \le r$, and $\mu \ne 4$ if $E = \mathbf{Q}$.

(iii) If $k_i - \kappa_{r+i} = 2$ for all i, the residue of C (resp. D) at $s = 2$ is an algebraic number times $R_E\pi^{2r+|k|}\langle \mathbf{g},\mathbf{g}\rangle$ (resp. $R_E\pi^{|k|}\langle \mathbf{g},\mathbf{g}\rangle$).

Proof. Observe that $L'(h, \xi) \sim \pi^{hr}$ for $0 < h \in \mathbf{Z}, h \equiv \gamma$ (mod 2), since $\psi^*(\mathfrak{a}\mathfrak{r}_F) = \mathrm{sgn}(N_{E/\mathbf{Q}}(a))^{\gamma}$ for $a \in E, a \equiv 1$ (mod $\mathfrak{c}$). Let $\{\mathfrak{y}\}$ be a complete set of representatives of the ideal classes of E modulo the product of all archimedean primes. Then we have $D(s) = \sum_{\mathfrak{y}} N(\mathfrak{y})^s L_{\mathfrak{y}}(s + (\gamma/2))$ with

$$L_{\mathfrak{y}}(s) = \sum_{0 \ll b \in \mathfrak{y}/U} \varphi(b\mathfrak{y}^{-1})\chi(b\mathfrak{y}^{-1}\mathfrak{r}_F)N_{E/\mathbf{Q}}(b)^{(\gamma/2)-s},$$

where $U = U_E$. Since each $L_{\mathfrak{y}}$ is a series of type (3.20), our assertions follow from Proposition 3.2 and Theorem 3.7.

The condition $\mu \neq 4$ when $E = \mathbf{Q}$ is probably unnecessary.

Specializing χ to the case of Hecke characters of a CM-field, we can derive from Theorem 3.11 that Conjecture 3.10 is true if $[F : E] = 2$, $E \neq \mathbf{Q}$, and $k_i - \kappa_{r+i} \geq 4$ for all $i \leq r$, or if $[F : E] = 2$, $E = \mathbf{Q}$, and $k_1 - \kappa_2 > 4$. The verification is more or less straightforward except that we need [13, Theorem 1.1], and therefore may be left to the reader.

Now coming back to the general case, we observe that χ occurs in $\mathbb{S}_m(M_2(F))$ by (iii) of Theorem 3.7. Let σ be the automorphism of F over E of order 2. Then there is a system χ' of eigenvalues occurring in $\tilde{\mathbb{S}}_{\sigma m}(M_2(F))$ such that $\chi'(\mathfrak{x}) = \chi(\mathfrak{x}^\sigma)$. This can easily be seen by changing variables z_i for z_{r+i} on $\mathfrak{H}^n$. Define a Dirichlet series Φ by

$$(3.24) \qquad \Phi(s) = \sum_{\mathfrak{x}} \chi(\mathfrak{x}) \chi'(\mathfrak{x}) N(\mathfrak{x})^{-s} \sum_{\mathfrak{y}} \psi^*(\mathfrak{y}\mathfrak{y}^\sigma) N(\mathfrak{y})^{2-2s},$$

where $\mathfrak{x}$ runs over all integral ideals of F and $\mathfrak{y}$ over those prime to $\mathfrak{c}\mathfrak{c}^\sigma\mathfrak{e}\mathfrak{e}^\sigma$. Then we see that Φ has an Euler product $\Phi(s) = \prod_{\mathfrak{p}} \Phi_{\mathfrak{p}}(N(\mathfrak{p})^{-s})^{-1}$ with polynomials $\Phi_{\mathfrak{p}}$ of degree ≤ 4 defined for the prime ideals $\mathfrak{p}$ of F. Let C_0 and C_1 denote the above C defined with trivial φ and also with $\varphi(\mathfrak{a}) = \left(\dfrac{F/E}{\mathfrak{a}} \right)$. Then we can easily verify that

$$(3.25) \qquad \Phi(s) = C_0(s) C_1(s) \prod_{\mathfrak{p} \mid \mathfrak{d}(F/E)} \left(1 - \delta_v \psi^*(\mathfrak{p}) N(\mathfrak{p})^{1-s} \right)^{-1},$$

where $\mathfrak{d}(F/E)$ is the different of F over E. Therefore we obtain from Theorem 3.11 a certain result about the critical values of Φ. This is not covered by Theorem 4.2 of [11] which concerns similar series, because the weights m and σm do not satisfy [11, (4.3a)].

Some fifteen years ago, the author found that the Euler product of type (3.22) has an analytic continuation to the whole plane, and a few years later verified that the Euler factor C_q when $E = \mathbf{Q}$ is a factor of the zeta function of a surface $\Gamma \backslash \mathfrak{H}^2$ (over $\mathbf{Q}$) with a congruence subgroup Γ of a totally indefinite quaternion algebra over F. (Cf. Asai [1] for the analytic continuation and relation (3.25) in the case $E = \mathbf{Q}$.) Recently Langlands [4] obtained a more complete result about such a coincidence (but not about analytic continuation) for the basic field F of an arbitrary degree. Though the zeta function of $\Gamma \backslash \mathfrak{H}^2$ in the strict sense corresponds to the case $k_1 = \kappa_2 + 2 = 2$, the Euler products in other cases have also a similar algebro-geometric meaning. Thus the above theorem may be viewed in such a geometric perspective.

4. Automorphic forms on $\mathfrak{H}^r \times \mathfrak{H}^r$

In this section, we denote by X^ρ and occasionally also by $\bar{X}$ the complex conjugate of a complex matrix X with no consistent principle. If X is invertible, we put $X^{-\rho} = (X^{-1})^\rho$. It is necessary for our purpose to consider $M_d(\mathbf{C})$-valued functions on $\mathfrak{H}^r \times \mathfrak{H}^r$, where $d = d(\kappa)$ as in Section 1. Given such a function $\mathfrak{f}(z, w)$ and $(\alpha, \beta) \in G_{\mathbf{Q}+} \times G_{\mathbf{Q}+}$, we define another $M_d(\mathbf{C})$-valued function $\mathfrak{f}\|_{k,\kappa}(\alpha, \beta)$ by

$$(4.1) \qquad (\mathfrak{f}\|_{k,\kappa}(\alpha, \beta))(z, w) = \prod_{i > r} N(\alpha\beta^{-1})^{\kappa_i \tau_i/2} j(\alpha, z)^{-k} j(\beta, w)^{-k}$$
$$\times \sigma_\kappa(\alpha)^{-1} \mathfrak{f}(\alpha z, \beta w) \sigma_\kappa(\beta) \qquad (z, w \in \mathfrak{H}^r).$$

We denote by $\mathfrak{M}^*_{k,\kappa}(\Gamma)$ the vector space of such functions $\mathfrak{f}$ holomorphic on $\mathfrak{H}^r \times \mathfrak{H}^r$ and at cusps satisfying $\mathfrak{f}\|_{k,\kappa}(\alpha, \beta) = \mathfrak{f}$ for all $(\alpha, \beta) \in \Gamma \times \Gamma$, and by $\mathfrak{M}^*_{k,\kappa}$ the union of $\mathfrak{M}^*_{k,\kappa}(\Gamma)$ for all congruence subgroups Γ of $G_{\mathbf{Q}+}$. Our representation σ_κ has the property that

$$(4.2) \qquad \sigma_\kappa(x)^\rho = {}^t\sigma_\kappa(x^\iota) = A_\kappa \sigma_\kappa(x) A_\kappa^{-1} \text{ for all } x \in G_{\mathbf{Q}}$$

with an element A_κ of $\mathrm{GL}_d(\overline{\mathbf{Q}})$, which is unique up to an algebraic scalar factor. Substituting x^ι for x, we see easily that ${}^tA_\kappa^{-\rho}$ has the same property as A_κ. Therefore, changing A_κ for its suitable algebraic scalar multiple, we may assume that A_κ *is unitary*. Now we call an element $\mathfrak{f}$ of $\mathfrak{M}^*_{k,\kappa}$ *arithmetic* if all the entries of the matrix $P_{k,\kappa}(w)^{-1}\mathfrak{f}(w, w')A_\kappa^{-1} \cdot {}^tP_{k,\kappa}(w')^{-1}$ are algebraic for every pair (w, w') of CM-points, where $P_{k,\kappa}(w)$ is defined by (2.2). Given elements g_λ and h_λ of $M_{k,\kappa}$, if we put

$$(4.3) \qquad \mathfrak{f}(z, w) = \sum_{\lambda=1}^{m} g_\lambda(z) \cdot {}^t h_\lambda(w) A_\kappa,$$

then obviously $\mathfrak{f} \in \mathfrak{M}^*_{k,\kappa}$. Moreover we have

LEMMA 4.1. (1) *Every $\mathfrak{f}$ of $\mathfrak{M}^*_{k,\kappa}$ can be expressed in the form (4.3) with g_λ and h_λ in $\mathfrak{M}_{k,\kappa}$.*

(2) *If $\mathfrak{f}$ is arithmetic, g_λ and h_λ can be chosen to be arithmetic.*

(3) *Suppose that g_λ and $\mathfrak{f}$ are arithmetic in expression (4.3), and $g_1, \ldots, g_m$ are linearly independent over $\mathbf{C}$. Then $h_1, \ldots, h_m$ are arithmetic.*

Proof. Suppose $\mathfrak{f} \in \mathfrak{M}^*_{k,\kappa}(\Gamma)$; let $\{g_\lambda\}_{\lambda=1}^m$ be a basis of $\mathfrak{M}_{k,\kappa}(\Gamma)$, and f_ν the ν-th column of $\mathfrak{f}$. Then $f_\nu(z, w) = \sum_{\lambda=1}^m p_{\lambda\nu}(w) g_\lambda(z)$ with $\mathbf{C}$-valued functions $p_{\lambda\nu}$, which are uniquely determined by f_ν and $\{g_\lambda\}$. Now we have

$$\{0\} = \bigcap_z \left\{ (c_1, \ldots, c_m) \in \mathbf{C}^m \,\middle|\, \sum_{\lambda=1}^m c_\lambda g_\lambda(z) = 0 \right\},$$

so that we can find finitely many points $v_1, \ldots, v_\mu$ of $\mathfrak{H}^r$ such that

$$\{0\} = \bigcap_{i=1}^{\mu} \left\{ (c_1, \ldots, c_m) \in \mathbf{C}^m \,\middle|\, \sum_{\lambda=1}^{m} c_\lambda g_\lambda(v_i) = 0 \right\}.$$

This means that $(c_\lambda) \mapsto (\sum_{\lambda=1}^{m} c_\lambda g_\lambda(v_i))_{i=1}^{\mu}$ is a $\mathbf{C}$-linear injection of $\mathbf{C}^m$ into $\mathbf{C}^{d\mu}$. Let S denote the image space of this map, and T the inverse map of S onto $\mathbf{C}^m$. We see that T maps $(f_i(v_i, w))_{i=1}^{\mu}$ onto $(p_{\lambda\nu}(w))$, and hence $p_{\lambda\nu}$ is holomorphic. Thus we can express $\mathfrak{f}$ in the form (4.3) with holomorphic functions h_λ. Considering the behavior of $\mathfrak{f}$ under Γ (and also the Fourier expansion of $\mathfrak{f} \,\|(1, \alpha)$ for $\alpha \in \mathrm{SL}_2(\mathbf{Z})$ when $B = M_2(\mathbf{Q})$), we see easily that $h_\lambda \in \mathfrak{M}_{k, \kappa}$. Next let us prove (3); then (2) will follow from Lemma 2.3. Let $\mathfrak{W}$ be the set of all CM-points. For every $g \in \mathfrak{M}_{k, \kappa}$, define a function g^0 on $\mathfrak{W}$ by $g^0(v) = P_{k, \kappa}(v)^{-1} g(v)$. Then the arithmeticity of $\mathfrak{f}$ implies that $\sum_\lambda g_\lambda^0(v) \cdot {}^t h_\lambda^0(v')$ is algebraic for every $(v, v') \in \mathfrak{W} \times \mathfrak{W}$. Since $\mathfrak{W}$ is dense in $\mathfrak{H}^r$, the functions $g_1^0, \ldots, g_m^0$ are linearly independent over $\mathbf{C}$. Therefore we can find finitely many points $u_1, \ldots, u_l$ of $\mathfrak{W}$ such that

$$\{0\} = \bigcap_{i=1}^{l} \left\{ (c_\lambda) \in \mathbf{C}^m \,\middle|\, \sum_\lambda c_\lambda g_\lambda^0(u_i) = 0 \right\}.$$

Let $\eta_{j\lambda}$ be the j-th component of h_λ^0. Then $\sum \eta_{j\lambda}(v) g_\lambda^0(u_i)$ is algebraic for every i and every $v \in \mathfrak{W}$. Since $g_\lambda^0(u_i)$ are algebraic, we see that $\eta_{j\lambda}(v)$ is algebraic. This completes the proof.

LEMMA 4.2. *Let* $f \in \mathfrak{M}_{k, \kappa}$ *and* $g(z) = j(\delta, z)^{-k} A_\kappa^{-1} \cdot {}^t \sigma_\kappa(\delta) \overline{f(\delta \bar z)}$ *with an element* δ *of* B *such that* $N(\delta)^{\tau_i} < 0$ *for all* $i \leq r$. *Then* $g \in \mathfrak{M}_{k, \kappa}$; *moreover* f *is arithmetic if and only if* g *is arithmetic.*

Proof. That $g \in \mathfrak{M}_{k, \kappa}$ can be verified in a straightforward way. To prove the second assertion, let w be the fixed point of $h(K^\times)$ for an embedding h of K into B as in Section 2. Define φ_i by (2.1) and put $\psi = \sum_{i=1}^{r} \varphi_i$. Define another embedding h' of K into B by $h'(x) = \delta h(x^\rho)\delta^{-1}$ for $x \in K$. Then $h'(K^\times)$ has $w' = \delta \bar w$ as its fixed point; moreover, h' and w' determine the same φ_i. Now there is an element ζ of $\mathrm{GL}_d(\overline{\mathbf{Q}})$ such that

$$\sigma_\kappa(h'(x)) = \zeta^{-1} \mathrm{diag}[x^{\alpha_1}, \ldots, x^{\alpha_d}] \zeta \qquad (x \in K^\times)$$

with $\alpha_\lambda \in I_K$, where $\mathrm{diag}[y_1, \ldots, y_d]$ denotes the diagonal matrix with $y_1, \ldots, y_d$ as the diagonal elements. By [13, (1.2)], we have

$$(4.4) \qquad P_{k, \kappa}(w') = \zeta^{-1} \mathrm{diag}[q_1, \ldots, q_d] \zeta, \quad q_\lambda = p_K\left(\sum_{i=1}^{r} k_i \varphi_i + \alpha_\lambda, \psi \right).$$

From (4.2), we obtain

$$\sigma_\kappa(h(x)) = A_\kappa^{-1}\sigma_\kappa(h(x))^\rho A_\kappa = \eta^{-1}\mathrm{diag}[x^{\alpha_1},\ldots,x^{\alpha_d}]\eta \quad (x \in K^\times)$$

with $\eta = \zeta^\rho\sigma_\kappa(\delta)^\rho A_\kappa$. By [12, Prop. 1.6], we can take q_λ to be real. Therefore, by [13, (1.2)] and (4.4), we have

$$P_{k,\kappa}(w)^\rho = A_\kappa^{-\rho}\sigma_\kappa(\delta)^{-1}P_{k,\kappa}(w')\sigma_\kappa(\delta)A_\kappa^\rho,$$

and hence

$$\left[j(\delta,w)^k P_{k,\kappa}(w)^{-1}g(w)\right]^\rho = A_\kappa^{-\rho}\sigma_\kappa(\delta)^{-1}P_{k,\kappa}(w')^{-1}\sigma_\kappa(\delta\delta^\iota)f(w').$$

The desired result follows from this relation.

After these preliminaries, we now define an F-valued F-bilinear form S on B by

$$(4.5) \qquad\qquad S(x,y) = xy^\iota + yx^\iota \qquad\qquad (x,y \in B).$$

Then S has signature $(2,2)$ at $\tau_1,\ldots,\tau_r$ and $(4,0)$ at $\tau_{r+1},\ldots,\tau_n$. We are going to apply the results of Section I.5 to this form S, viewing B as the vector space V in that section. The completion of $V(=B)$ at τ_i, denoted by V_i, can be identified with $M_2(\mathbf{R})$ if $i \leq r$ and with $\mathbf{H}$ if $i > r$; the space $\mathfrak{B}_m$ of [13, (2.6)] in the present case is isomorphic to $\mathfrak{H}^2$. We therefore define a map $p: \mathbf{C}^2 \to M_2(\mathbf{C})$ by

$$(4.6) \qquad\qquad p(z,w) = \begin{pmatrix} z & -wz \\ 1 & -w \end{pmatrix} \qquad\qquad ((z,w) \in \mathbf{C}^2).$$

The restriction of p to $\mathfrak{H}^2$ is essentially the same as the map of [13, (2.7)]. If S_i denotes the $\mathbf{C}$-bilinear extension of S to $V_i \otimes_\mathbf{R} \mathbf{C}$ $(= M_2(\mathbf{C}))$ for $i \leq r$, we have

$$(4.7) \quad S_i(\alpha, p(z,w)) = (z \quad 1)\varepsilon\alpha\begin{pmatrix} w \\ 1 \end{pmatrix}, \quad \varepsilon = \begin{pmatrix} 0 & 1 \\ -1 & 0 \end{pmatrix} \qquad (\alpha \in M_2(\mathbf{C})).$$

To make our notation simpler, we put

$$(4.8a) \quad [\alpha,z,w] = (z \quad 1)\varepsilon\alpha\begin{pmatrix} w \\ 1 \end{pmatrix} \qquad\qquad (\alpha \in M_2(\mathbf{C}),(z,w) \in \mathbf{C}^2).$$

Further, given $k = \sum_{i=1}^r k_i\tau_i \in I_F$, $\alpha \in B$, $z = (z_1,\ldots,z_r)$ and $w = (w_1,\ldots,w_r)$ in $\mathbf{C}^r$, we put

$$(4.8b) \qquad\qquad [\alpha,z,w]^k = \prod_{i=1}^r [\alpha^{(i)},z_i,w_i]^k.$$

Now define the groups $G(S)$ and $G_+(S)$ of [13, §3] for the present S. Let an element (β,γ) of $G_\mathbf{Q} \times G_\mathbf{Q}$ act on $V = B$ by $(\beta,\gamma)x = \beta x\gamma^\iota$ for $x \in B$. This gives a homomorphism of $G_\mathbf{Q} \times G_\mathbf{Q}$ into $G(S)$ which maps $G_{\mathbf{Q}+} \times G_{\mathbf{Q}+}$ into $G_+(S)$. Moreover, this action localized at τ_i for $i \leq r$ gives

$$(4.9) \qquad (\beta,\gamma)p(z,w) = p(\beta z, \gamma w)j(\beta,z)j(\gamma,w)|N(\beta\gamma)|^{1/2};$$

hence the action of the image of (β, γ) in $G_+(S)$ on $\mathfrak{H}^r \times \mathfrak{H}^r$ is the same as that of (β, γ). We have also

$$(4.10) \qquad [\beta'\alpha\gamma, z, w]^k = |N(\beta\gamma)|^{k/2} j(\beta, z)^k j(\gamma, w)^k [\alpha, \beta z, \gamma w]^k$$

$$(\alpha \in B, (\beta, \gamma) \in G_Q \times G_Q).$$

With E, Ω, and ω as in (3.4) and (3.5), we define an $M_d(\mathbf{C})$-valued infinite series H by

$$(4.11) \quad H(z, w, s) = H(z, w, s; \xi, \Omega)$$

$$= [U] \sum_{0 \neq \alpha \in B/U} \xi(\alpha) \sigma_\kappa(\alpha) \omega_u(\mathrm{Tr}_{F/E}(-N(\alpha))) N(\alpha)^v$$

$$\times [\alpha, z, w]^{-k} |[\alpha, z, w]|^{-2sE}.$$

Here $(z, w) \in \mathfrak{H}^r \times \mathfrak{H}^r, s \in \mathbf{C}$; ξ is a locally constant function on B; k and κ are the same as in Section 1; $k_i > 0$ for all $i \leq r$; $0 \leq u \in I_E, 0 \leq v \in I_F$; $\omega_u(a) = a^u \omega(a)$; U is a subgroup of U_E of finite index such that ξ and Ω are U-invariant. To make each term of (4.11) dependent only on $U\alpha$, we assume

$$(4.12) \qquad k - l - 2u - \mathrm{Res}_{F/E}(2v + \kappa) = \beta'E \text{ with } \beta' \in \mathbf{Z}.$$

Now we observe that the entries of $\sigma_\kappa(x)$ as functions on $\prod_{i>r} V_i$ are S_i-harmonic on V_i. Therefore each entry of matrix (4.11) is a finite linear combination of series of type (I.5.2). Condition (I.5.4) is satisfied since $l \in I_E$. Thus, by Proposition I.5.1 and Theorem I.5.2, H is absolutely convergent for sufficiently large $\mathrm{Re}(s)$, and can be continued to a meromorphic function on the whole s-plane. Moreover, suppose

$$(4.13) \qquad \beta' > 2[F: E] - 2; \beta' \neq 2[F: E] \text{ if } E = \mathbf{Q}.$$

Then Theorem I.5.2 asserts that $H(z, w, 0)$ is meaningful and holomorphic on $\mathfrak{H}^r \times \mathfrak{H}^r$. Relation (4.10) shows easily that $H(z, w, 0)$ belongs to $\mathfrak{M}^*_{k, \kappa}$.

PROPOSITION 4.3. *Suppose that ξ is $\overline{\mathbf{Q}}$-valued, Ω is $\overline{\mathbf{Q}}$-rational, and (4.13) is satisfied. Then $\pi^{-|k|} H(z, w, 0)$ is arithmetic.*

PROPOSITION 4.4. *Suppose that $\beta' = 2[F: E] - 2, u = 0$, and $v_1 = \cdots = v_r = 0$. Then $H(z, w, s)$ is holomorphic in s for $\mathrm{Re}(s) > 1$ and has at most a simple pole at $s = 1$. Moreover, put*

$$\mathfrak{g}(z, w) = \prod_{\nu=1}^{r} \mathrm{Im}(z_\nu) \mathrm{Im}(w_\nu) \cdot \lim_{s \to 1} (s - 1) H(z, w, s).$$

Then $\mathfrak{g}$ is holomorphic on $\mathfrak{H}^r \times \mathfrak{H}^r$ and belongs to $\mathfrak{M}^*_{k, \kappa}$. Furthermore, if ξ is $\overline{\mathbf{Q}}$-valued and Ω is $\overline{\mathbf{Q}}$-rational, $R_E^{-1} \pi^{-|k|} \mathfrak{g}$ is arithmetic.

These propositions follow immediately from Theorems I.5.3 and I.5.4, since
the arithmeticity defined at the beginning of this section coincides with that of
[13, §5], which can be shown as follows. Let K and K' be non-isomorphic
CM-fields quadratic over F, and let $L = K \otimes_F K'$. Take F-linear embeddings h
and h' of K and K' into B, respectively; let w and w' be the fixed points of
$h(K^\times)$ and $h'(K'^\times)$ on $\mathfrak{H}^r$. We can define an F-linear embedding η of L into
$\mathrm{End}_F(B)$ by $\eta(a \otimes b)x = h(a)xh'(b)^\iota$ for $a \in K$, $b \in K'$, and $x \in B$. If φ_i is
determined by h at w_i by (2.1) and similarly φ_i' by h' at w_i', then we see easily
that

$$\eta(c)p(w_i, w_i') = c^{\psi_i}p(w_i, w_i') \qquad (c \in L, cc^\rho = 1)$$

with an element ψ_i of J_L such that $\mathrm{Res}_{L/K}(\psi_i) = \varphi_i$ and $\mathrm{Res}_{L/K'}(\psi_i) = \varphi_i'$. Now
let ω_i be the element of J_L such that $\mathrm{Res}_{L/K}(\omega_i) = \varphi_i$ and $\mathrm{Res}_{L/K'}(\omega_i) = \varphi_i'\rho$.
Then, for every $\alpha \in J_L$, we have

$$p_L(\alpha, 2\psi_i) = p_L(\alpha, \psi_i + \omega_i)p_L(\alpha, \psi_i + \omega_i\rho)$$
$$= p_K(\mathrm{Res}_{L/K}(\alpha), \varphi_i)p_{K'}(\mathrm{Res}_{L/K'}(\alpha), \varphi_i')$$

by [13, Theorem 1.1, (4)], and hence $\mathfrak{p}_L(\Sigma_{i=1}^r 2\psi_i) = \mathfrak{p}_K(\Sigma_{i=1}^r \varphi_i) \otimes \mathfrak{p}_{K'}(\Sigma_{i=1}^r \varphi_i')$
for the symbol $\mathfrak{p}$ of [13, p. 319]. It follows easily from this fact that the present
arithmeticity of the elements of $\mathfrak{M}^*_{k,\kappa}$ is consistent with that of [13, §5].

5. Certain integrals involving theta functions

We now specialize the function θ_0 of (I.6.6) to the present case. Namely we
define an $M_d(\mathbf{C})$-valued function θ on $\mathfrak{H}^n \times \mathfrak{H}^r \times \mathfrak{H}^r$ by

$$(5.1) \quad \theta(z, w, w') = \theta(z, w, w'; \xi, a) = \prod_{\nu=1}^r \mathrm{Im}(z_\nu)\mathrm{Im}(w_\nu)^{-k_\nu}\mathrm{Im}(w_\nu')^{-k_\nu}$$

$$\times \sum_{\alpha \in B} \xi(\alpha)\sigma_\kappa(\alpha)[\alpha, \overline{w}, \overline{w}']^k e\left(\sum_{\nu=1}^n a^\tau R_\nu[\alpha, z, w, w'] \right),$$

$$(5.2) \quad R_\nu[\alpha, z, w, w'] = \begin{cases} -N(\alpha^{(\nu)})\bar{z}_\nu + iy_\nu \dfrac{\|[\alpha^{(\nu)}, w_\nu, w_\nu']\|^2}{2\mathrm{Im}(w_\nu)\mathrm{Im}(w_\nu')} & (\nu \leq r), \\[12pt] -N(\alpha^{(\nu)})z_\nu & (\nu > r). \end{cases}$$

Here $z \in \mathfrak{H}^n$, $w \in \mathfrak{H}^r$, $w' \in \mathfrak{H}^r$, $y_\nu = \mathrm{Im}(z_\nu)$; ξ is a locally constant function on
B; a is an element of F such that $a^{\tau_\nu} > 0$ if and only if $\nu \leq r$. Since the entries of
$\sigma_\kappa(\alpha)$ are S_ν-harmonic on B_ν for each $\nu > r$, we have, by [13, Prop. 7.1],

$$(5.3) \quad \theta(\alpha z, w, w') = (cz + d)^m \theta(z, w, w') \text{ for } \alpha = \begin{pmatrix} * & * \\ c & d \end{pmatrix} \in \Delta,$$

with a congruence subgroup Δ of $SL_2(\mathfrak{r})$, where m is defined by (3.3). A direct calculation shows also that

$$(5.4) \quad \theta(z, \beta w, \gamma w'; \xi, a) = N(\beta\gamma)^{k/2} j(\beta, w)^k j(\gamma, w')^k$$

$$\times \sigma_\kappa(\beta)\theta(z, w, w'; \xi', N(\beta\gamma)a)\sigma_\kappa(\gamma') \qquad (\beta, \gamma \in G_Q.),$$

where $\xi'(\alpha) = \xi(\beta\alpha\gamma')$. This is a special case of (I.6.10).

Now taking an element δ of B such that $N(\delta)^{\mathfrak{r}_\nu} < 0$ for all $\nu \leq r$, put $\varphi(\alpha) = \xi(\alpha\delta^{-1})$ and

$$(5.5) \quad \theta_1(z, w, w') = \theta_1(z, w, w'; \varphi)$$

$$= |N(\delta)|^{k/2}\theta(z, w, \delta\overline{w}'; \xi, -N(\delta))j(\delta, \overline{w}')^{k}\sigma_\kappa(\delta).$$

This is given as an infinite series by

$$(5.6) \quad \theta_1(z, w, w') = \prod_{\nu=1}^{r} \mathrm{Im}(z_\nu)\mathrm{Im}(w_\nu)^{k_\nu}\mathrm{Im}(w'_\nu)^{-k_\nu}$$

$$\times \sum_{\alpha \in B} \varphi(\alpha)\sigma_\kappa(\alpha)[\alpha, \overline{w}, w']^k \mathbf{e}\left(\sum_{\nu=1}^{n} - R_\nu[\alpha, z, \overline{w}, w']\right),$$

where we define $R_\nu[\alpha, z, \overline{w}, w']$ by (5.2) with $\overline{w}$ in place of w (though $\overline{w} \notin \mathfrak{H}^r$). Obviously θ_1 satisfies a formula of type (5.3). We have also

$$(5.7) \quad \theta_1(z, \beta w, \gamma w'; \varphi) = j(\beta, w)^k j(\gamma, \overline{w}')^k \sigma_\kappa(\beta)\theta_1(z, w, w'; \varphi')\sigma_\kappa(\gamma)^{1}$$

for $\beta, \gamma \in B, \nu(\beta) = \nu(\gamma) = 1, \varphi'(\alpha) = \varphi(\beta\alpha\gamma^{-1})$.

Take a congruence subgroup Γ of G_Q so that

$$-1 \notin \Gamma \subset \{\gamma \in B \mid \nu(\gamma) = 1, \varphi(\gamma\alpha) = \varphi(\alpha\gamma) = \varphi(\alpha) \text{ for all } \alpha \in B\}.$$

We now consider an integral of the form

$$(5.8) \qquad \int_\Phi{}' \overline{f(w)}\,\theta_1(z, w, w')\mathrm{Im}(w)^k \, d\mu(w) \qquad (\Phi = \Gamma \backslash \mathfrak{H}^r)$$

with $f \in \mathcal{S}_{k, \kappa}(\Gamma)$. Since f is a cusp form and θ_1 is slowly increasing as a function of w in the sense of Section I.2, the integral is convergent. The main aim of this section is to prove the following proposition which will play a key role in our proof of the main theorems. (The existence of E satisfying (3.4) is not assumed here.)

PROPOSITION 5.1. *Let R be a complete set of representatives of $\Gamma \backslash G_{Q+}$. Then (5.8) is equal to*

$$(5.9) \qquad (-2i)^{|k|} \sum_{\alpha \in R} \varphi(\alpha)N(\alpha)^{(m/2)-F} \cdot {}'\overline{(f\|\alpha)(w')}\,\mathbf{e}_F(N(\alpha)z).$$

To prove this, we first consider the terms of θ_1 for the elements α in $B^\times$. Let R' be a complete set of representatives of $\Gamma \setminus B^\times$. Then such terms form a sum

$$y_1 \cdots y_r \operatorname{Im}(w')^{-k} \sum_{\alpha \in R'} \varphi(\alpha)$$

$$\times \sum_{\gamma \in \Gamma} \int_\Phi \mathbf{e}\left(\sum_{\nu=1}^{n} - R_\nu[\gamma\alpha, z, \overline{w}, w'] \right) [\gamma\alpha, \overline{w}, w']^k \cdot {}^t\overline{f(w)}\, \sigma_\kappa(\gamma\alpha)\, d\mu(w).$$

Since $R_\nu[\gamma\alpha, z, \overline{w}, w'] = R_\nu[\alpha, z, \gamma^{-1}\overline{w}, w']$ and f is Γ-invariant, the last sum over Γ is equal to

$$\int_{\mathfrak{H}'} \mathbf{e}\left(\sum_{\nu=1}^{n} - R_\nu[\alpha, z, \overline{w}, w'] \right) [\alpha, \overline{w}, w']^k \cdot {}^t\overline{f(w)}\, \sigma_\kappa(\alpha)\, d\mu(w).$$

This can be written also in the form

$$(5.10) \quad \mathbf{e}_F(N(\alpha)z)\mathbf{e}\left(-2i \sum_{\nu=1}^{r} y_\nu N(\alpha^{(\nu)}) \right)$$

$$\times \int_{\mathfrak{H}'} \exp\left(-\pi \sum_{\nu=1}^{r} y_\nu \frac{|[\alpha^{(\nu)}, \overline{w}_\nu, w'_\nu]|^2}{\operatorname{Im}(w_\nu)\operatorname{Im}(w'_\nu)} \right) [\alpha, \overline{w}, w']^k \cdot {}^t\overline{f(w)}\, \sigma_\kappa(\alpha)\, d\mu(w).$$

Suppose $N(\alpha) \gg 0$ and put $f_\alpha = f \| \alpha$. Transforming w into αw, we find that the last integral becomes

$$(5.11) \quad N(\alpha)^{(k+\kappa)/2} \int_{\mathfrak{H}'} \exp\left(-\pi \sum_{\nu=1}^{r} N(\alpha^{(\nu)})y_\nu \frac{|[1, \overline{w}_\nu, w'_\nu]|^2}{\operatorname{Im}(w_\nu)\operatorname{Im}(w'_\nu)} \right)$$

$$\times [1, \overline{w}, w']^k \cdot {}^t\overline{f_\alpha(w)}\, d\mu(w).$$

Observe that $[1, \overline{w}, w']^k = (\overline{w} - w')^k$. Now we need

LEMMA 5.2. *Put*

$$P(w, w') = \exp\left(- \sum_{\nu=1}^{r} c_\nu \operatorname{Im}(w_\nu)^{-1}\operatorname{Im}(w'_\nu)^{-1}|\overline{w}_\nu - w'_\nu|^2 \right) \qquad (w, w' \in \mathfrak{H}')$$

with $0 < c_\nu \in \mathbf{R}$. *Let* $h(w)$ *be a* C^∞-*function on* $\mathfrak{H}'$ *such that*

$$L_\nu h = \left(s_\nu^2 - (1/4) \right)h \qquad\qquad (\nu = 1,\ldots,r)$$

with $s_\nu \in \mathbf{C}$, *where* $L_\nu = v_\nu^2(\partial^2/\partial u_\nu^2 + \partial^2/\partial v_\nu^2)$, $w_\nu = u_\nu + iv_\nu$. *Then*

$$\int_{\mathfrak{H}'} P(w, w')h(w)\, d\mu(w) = \prod_{\nu=1}^{r} (\pi/c_\nu)^{1/2} e^{-2c_\nu} K(c_\nu, s_\nu) \cdot h(w'),$$

provided that the integral is convergent, where

$$K(c, s) = \int_0^\infty \exp\left(-c(x + x^{-1})\right)x^{s-1}\,dx \qquad (s \in \mathbf{C}, 0 < c \in \mathbf{R}).$$

Proof. Observe that $P(\gamma w, \gamma w') = P(w, w')$ for $\gamma \in \mathrm{SL}_2(\mathbf{R})^r$. Therefore, by a well-known principle (see Selberg [6]), we have

$$(5.12) \qquad \int_{\mathfrak{H}'} P(w, w')h(w)\,d\mu(w) = K(t_1, \ldots, t_r)h(w')$$

for every h satisfying $L_\nu h = t_\nu h$, where $K(t_1, \ldots, t_r)$ is a scalar depending only on $t_1, \ldots, t_r$. To determine K, we take

$$h(w_1, \ldots, w_r) = \prod_{\nu=1}^r \mathrm{Im}(w_\nu)^{s_\nu + (1/2)}.$$

Then $L_\nu h = (s_\nu^2 - (1/4))h$. If $w' = (i, \ldots, i)$, the left-hand side of (5.12) is

$$\prod_{\nu=1}^r \int_0^\infty v^{s_\nu - (3/2)} \int_{-\infty}^\infty \exp\left(-c_\nu(u^2 + v^2 + 2v + 1)/v\right)du\,dv.$$

Thus a direct calculation gives the desired form of K.

We now apply this result to (5.11) taking $h(w) = (\overline{w} - a)^k \cdot {}^t\overline{f_a(w)}$ with $a \in \mathfrak{H}'$. Since h is anti-holomorphic in w, we have $s_\nu = 1/2$. Recalling that $K(c, 1/2) = (\pi/c)^{1/2}e^{-2c}$, we find that (5.10) is equal to

$$(-2i)^{|k|}\mathbf{e}_F(N(\alpha)z)N(\alpha)^{(m/2)-F}(y_1 \cdots y_r)^{-1}\mathrm{Im}(w')^k \cdot {}^t\overline{f_a(w')}.$$

Therefore the terms for $N(\alpha) \gg 0$ give (5.9). It remains to show that other terms vanish. Suppose $N(\alpha)^{\tau_\lambda} < 0$ for some $\lambda \le r$. Define a map η of $\mathbf{C}^r$ onto itself by $\eta(w) = (\eta_1(w_1), \ldots, \eta_r(w_r))$ with $\eta_\nu(w_\nu) = w_\nu$ or $\overline{w}_\nu$ according as $N(\alpha)^{\tau_\nu} > 0$ or < 0. Put $w^* = \eta(\alpha w')$ and observe that

$$\frac{\left|[\alpha^{(\nu)}, \overline{w}_\nu, w'_\nu]\right|^2}{\mathrm{Im}(w_\nu)\mathrm{Im}(w'_\nu)} = |N(\alpha^{(\nu)})|\frac{|\overline{w}_\nu - \eta_\nu(w_\nu^*)|^2}{\mathrm{Im}(w_\nu)\mathrm{Im}(w_\nu^*)}.$$

Hence we again obtain an integral of the type

$$(5.13) \qquad \int_{\mathfrak{H}'} P'(w, w^*)[\alpha, \overline{w}, w']^k \cdot {}^t\overline{f(w)}\,\sigma_\kappa(\alpha)\,d\mu(w)$$

where

$$P'(w, w^*) = \exp\left(\sum_{\nu=1}^r c_\nu \mathrm{Im}(w_\nu)^{-1}\mathrm{Im}(w_\nu^*)^{-1}|\overline{w}_\nu - \eta_\nu(w_\nu^*)|^2\right)$$

with $0 < c_\nu \in \mathbf{R}$. Since $P'(\gamma w, \gamma w^*) = P'(w, w^*)$ for $\gamma \in \mathrm{SL}_2(\mathbf{R})^r$, (5.13) is an eigenvalue times $[\alpha, \overline{w}^*, w']^k \cdot {}^t\overline{f(w^*)}\sigma_\kappa(\alpha)$, for the same reason as in the proof of Lemma 5.2. Since $N(\alpha)^{\tau_\lambda} < 0$, we have $[\alpha^{(\lambda)}, \overline{w}_\lambda^*, w'_\lambda] = 0$; hence (5.13) is 0.

Finally suppose $N(\alpha) = 0$. The factor $[\alpha,\ldots]^k$ is 0 if $\alpha = 0$; thus we may assume $\alpha \neq 0$. This can happen only when $B = M_2(F)$. To prove the desired vanishing, we may assume that $\alpha = \begin{pmatrix} a & 0 \\ 0 & 0 \end{pmatrix}$. Then

$$\{\gamma \in \Gamma \mid \gamma\alpha = \alpha\} = \left\{ \begin{pmatrix} 1 & b \\ 0 & 1 \end{pmatrix} \middle| b \in \mathfrak{b} \right\}$$

with a lattice $\mathfrak{b}$ in F. Call this group T. Then the sum of the terms for the elements of $\Gamma\alpha$ become

$$\int_D \mathbf{e}\left(\sum_{\nu=1}^{n} - R_\nu[\alpha, z, \overline{w}, w'] \right)[\alpha, \overline{w}, w']^k \overline{f(w)}\, d\mu(w) \qquad (D = T \setminus \mathfrak{H}^n).$$

Since $[\alpha, \overline{w}, w']^k = (-aw')^k$, this involves an integral of the form

$$\int_{\mathbf{R}^n/\mathfrak{b}} \overline{f(u + iv)}\, du,$$

which is 0, as f is a cusp form. This completes the proof of Proposition 5.1.

REMARK 5.3. Integral (5.8) is essentially the same as that of [9, §4.5]. In fact, the kernel function $\exp\{-\pi \cdot \operatorname{tr}(A_\nu^{-1} X_\nu)\}$ of [9, (20)] is practically the same as P of Lemma 5.2, though the symmetric space of [9, (20)] is the product of $\mathfrak{H}^r$ with a Euclidean space.

As an immediate consequence of Proposition 5.1, we obtain

$$(5.14) \qquad \int_{\Phi \times \Phi} {}'\overline{f(w)}\, \theta_1(z, w, w') \mathrm{g}(w') \operatorname{Im}(w)^k \operatorname{Im}(w')^k\, d\mu(w)\, d\mu(w')$$

$$= (-2i)^{|k|} \mu(\Phi) \sum_{\alpha \in R} \varphi(\alpha) N(\alpha)^{(m/2)-F} \langle f \| \alpha, \mathrm{g} \rangle \mathbf{e}_F(N(\alpha)z)$$

for $f, \mathrm{g} \in S_{k,\kappa}(\Gamma)$. The left-hand side is similar to the integral investigated by Shintani [16]. We can actually derive (5.14) also by a computation of his type, which is somewhat longer than the above proof, however,

PROPOSITION 5.4. *Series (5.9) and (5.14) as functions of z are cusp forms belonging to $\mathfrak{S}_m(\Delta, M_2(F))$, where Δ is a group of (5.3).*

Proof. Obviously the functions in question are holomorphic and satisfy the same transformation formula as (5.3). Now, from [13, Proposition 7.1], we see that $(cz + d)^{-m}\theta_1(\alpha z, w, w'; \varphi)$, for every $\alpha = \begin{pmatrix} * & * \\ c & d \end{pmatrix} \in \mathrm{GL}_2^+(F)$, is a finite linear combination of functions of the form $\theta_1(bz, w, w'; \varphi')$ with $0 \ll b \in F$ and locally constant functions φ' on B. Therefore (5.9) or (5.14) transformed by α

is a finite linear combination of integrals of the above type, and hence has a Fourier expansion of type (5.9) or (5.14), whose constant term is 0. This proves our assertion.

6. Proof of Theorems 3.1, 3.3, and 3.5

We first prove a few lemmas on $G_\mathbf{A}$. The letters x_λ, $\mathfrak{o}^{\mathfrak{c}}$, W, Y, and Γ_λ being the same as in Section 1, let $X_{\lambda\mu}$ denote the set of all elements α of the lattice $x_\lambda \mathfrak{o}^{\mathfrak{c}} x_\mu^{-1}$ (that is, the lattice in B whose $\mathfrak{p}$-closure is $x_{\lambda\mathfrak{p}}(\mathfrak{o}^{\mathfrak{c}})_\mathfrak{p} x_{\mu\mathfrak{p}}^{-1}$ for every $\mathfrak{p}$) which, as elements of $B_\mathfrak{p}$, belong to $Y_\mathfrak{p}$ for every $\mathfrak{p}$ dividing $\mathfrak{c}$. We put then $X_{\lambda\mu}^+ = \{\alpha \in X_{\lambda\mu} \,|\, N(\alpha) \gg 0\}$. Notice that $X_{\lambda\mu}^+ = x_\lambda Y x_\mu^{-1} \cap B$. Further we put $\mathfrak{t}_{\lambda\mu} = N(x_\lambda x_\mu^{-1})\mathfrak{r}$, and denote by D the set of all totally positive elements d of F such that $d \in N(W_\mathfrak{p})$ for every $\mathfrak{p}$ dividing $\mathfrak{c} + \mathfrak{e}$.

LEMMA 6.1. (i) $N(X_{\lambda\mu}^+) = D \cap \mathfrak{t}_{\lambda\mu}$; (ii) $N(\Gamma_\lambda) = D \cap U_F$.

Proof. That the left-hand side of (i) or (ii) is contained in the right-hand side is obvious. To prove the opposite inclusion, let $b \in D \cap \mathfrak{t}_{\lambda\mu}$. By our definition of Y and D, we have $b = N(y)$ with $y \in x_\lambda Y x_\mu^{-1}$. Since $b \gg 0$, we have also $b = N(\beta)$ with $\beta \in B$. Then $N(\beta y^{-1}) = 1$. By the strong approximation theorem, $\beta y^{-1} = \alpha z$ with $z \in x_\lambda W x_\lambda^{-1}$ and $\alpha \in G_\mathbf{Q}$, $N(\alpha) = 1$. Then $\alpha^{-1}\beta = zy \in x_\lambda Y x_\mu^{-1} \cap G_\mathbf{Q} = X_{\lambda\mu}^+$, which proves (i). If $b \in U_F \cap D$, we can take y from $x_\lambda W x_\lambda^{-1}$. Then $\alpha^{-1}\beta = zy \in x_\lambda W x_\lambda^{-1} \cap G_\mathbf{Q} = \Gamma_\lambda$, so that $b = N(\alpha^{-1}\beta) \in N(\Gamma_\lambda)$, which proves (ii).

LEMMA 6.2. *Let $\mathfrak{m}$ be an ideal prime to $\mathfrak{e} + \mathfrak{c}$. Then, for every index λ, there is an element b of D and an index μ such that $\mathfrak{m} = b\mathfrak{t}_{\mu\lambda}$. Moreover μ is unique for λ and $\mathfrak{m}$.*

Proof. Let $\mathfrak{m} = y\mathfrak{r}$ with $y \in F_\mathbf{A}^\times$ with $y_\infty = 1$ and $y_\mathfrak{p} = 1$ for $\mathfrak{p} \supset \mathfrak{e} + \mathfrak{c}$. Since $F_\mathbf{A}^\times = \amalg_{\mu=1}^h N(x_\lambda x_\mu^{-1}) F^\times N(W)$, we have $y = N(x_\lambda x_\mu^{-1}) bz$ with some μ, $b \in F^\times$, and $z \in N(W)$. Then we see that $b \in D$ and $\mathfrak{m} = b\mathfrak{t}_{\lambda\mu}$. Conversely, if $\mathfrak{m} = b\mathfrak{t}_{\lambda\mu}$ with $b \in D$, we have $y = N(x_\lambda x_\mu^{-1}) bz$ with $z \in F_\infty^\times \prod_\mathfrak{p} \mathfrak{r}_\mathfrak{p}^\times$. Then $z_\mathfrak{p} \in N(W_\mathfrak{p})$ for every $\mathfrak{p} \supset \mathfrak{e} + \mathfrak{c}$, and hence $z \in N(W)$. This proves the uniqueness of μ.

For every fractional ideal $\mathfrak{m}$, let $T_\mathfrak{c}^{\lambda\mu}(\mathfrak{m})$ denote the sum of all different $\Gamma_\lambda \alpha \Gamma_\mu$ such that $\alpha \in X_{\lambda\mu}^+$ and $\mathfrak{m} = N(\alpha)\mathfrak{t}_{\mu\lambda}$. This is not 0 only if $\mathfrak{m}$ is integral and prime to $\mathfrak{e} + \mathfrak{c}$. By the above lemma, μ is unique for λ and $\mathfrak{m}$. The definition of $T_\mathfrak{c}(\mathfrak{m})$ in Section 1 shows that $T_\mathfrak{c}^{\lambda\mu}(\mathfrak{m})$ is the restriction of $T_\mathfrak{c}(\mathfrak{m})$ to $\mathcal{S}_{k,\kappa}(\Gamma_\lambda, \varphi_\lambda)$. We see easily that, for $0 \ll b \in F$, $T_\mathfrak{c}^{\lambda\mu}(b\mathfrak{t}_{\mu\lambda}) \neq 0$ only if $b \in N(X_{\lambda\mu}^+)U_F$.

LEMMA 6.3. *Let* $\Gamma'_\lambda = \{\gamma \in \Gamma_\lambda \mid N(\gamma) = 1\}$. *For each* $b \in N(X_{\lambda\mu}')$, *let* $Z_b = \{\alpha \in X_{\lambda\mu} \mid N(\alpha) = b\}$, *and let* A_b *be a complete set of representatives of* $\Gamma'_\lambda \backslash Z_b / \Gamma'_\mu$. *Then* $T_c^{\lambda\mu}(bt_{\mu\lambda})$ *is the sum of all* $\Gamma_\lambda \alpha \Gamma_\mu$ *with* $\alpha \in A_b$.

Proof. Observe that $T_c^{\lambda\mu}(bt_{\mu\lambda})$ is the sum of all different $\Gamma_\lambda \alpha \Gamma_\mu$ such that $\alpha \in X_{\lambda\mu}$ and $N(\alpha) \in bU_F$. Since b and $N(\alpha)$ belong to D, we have $b^{-1}N(\alpha) \in D \cap U_F = N(\Gamma_\lambda)$ by Lemma 6.1. Therefore we can choose α so that $N(\alpha) = b$. If $N(\alpha) = N(\beta) = b$ for α and β in $X_{\lambda\mu}$, we have $\Gamma_\lambda \alpha \Gamma_\mu = \Gamma_\lambda \beta \Gamma_\mu$ if and only if $\Gamma'_\lambda \alpha \Gamma'_\mu = \Gamma'_\lambda \beta \Gamma'_\mu$ by virtue of Lemma 1.2. Therefore we obtain our assertion.

Now, given c and s_0 as in Theorem 3.3, put

$$c' = c - s_0 E + (k - \kappa)/2.$$

Conditions (3.9a, b) and (3.12a) imply that $0 \ll c' \in I_F$. Observe that $b^{c'}$ (for $b \in F$) is a finite $\overline{\mathbf{Q}}$-linear combination of $b^t \mathrm{Tr}_{F/E}(\zeta b)^u$ with $0 \le u \in I_E$ and $0 \le v \in I_F$, $v = \sum_{i>r} v_i \tau_i$. This means that, to prove Theorem 3.3, we may assume that $s_0 = 0$ and c has the form

$$(6.1) \qquad c = v + (\kappa - k)/2 \text{ with } 0 \le v = \sum_{i>r} v_i \tau_i \in I_F.$$

We may also assume that $\mathfrak{x}$ is prime to $\mathfrak{c}$, since the general case can easily be reduced to this case. It is also sufficient to prove the assertion of Theorem 3.3 for one particular $\overline{\mathbf{Q}}$-rational basis of $\mathfrak{S}_{k,\kappa}(\mathfrak{c}, \psi_0)$ instead of an arbitrary basis of $\mathfrak{S}_{k,\kappa}(\mathfrak{c}, \psi)$. Thus we take the basis to be the union of $\overline{\mathbf{Q}}$-rational bases of $\mathfrak{S}_{k,\kappa}(\Gamma_\lambda, \varphi_\lambda)$ for $1 \le \lambda \le h$.

Now take U of (3.6) so that $U \subset D$, and write each term of (3.6) simply $X(b)$. Then (3.6) can be decomposed into the sum

$$[U] \sum_{0 \ll \beta \in F/D} \sum_{b \in D/U} X(\beta b).$$

Since $T(\mathfrak{m}) \neq 0$ only if $\mathfrak{m}$ is prime to $\mathfrak{c} + \mathfrak{e}$, we may assume that β is prime to $\mathfrak{c} + \mathfrak{e}$. Now D is defined by a congruence condition modulo a power of $\mathfrak{c} + \mathfrak{e}$, and hence we see that $\sum_{0 \ll \beta \in F/D}$ is actually a finite sum. Therefore we can take $D\beta$ instead of F in (3.6). Fixing one λ, denote by S a $\overline{\mathbf{Q}}$-rational basis of $\mathfrak{S}_{k,\kappa}(\Gamma_\lambda, \varphi_\lambda)$. Let $(d_{fg})_{f,g \in S}$ be the inverse of the matrix $(\langle f, g \rangle)_{f,g \in S}$. Then our task is to show that

$$(6.2) \qquad \pi^{-|k|} \sum_{b \in D/U} \eta(b)\omega_u\left(\mathrm{Tr}_{F/E}(\zeta b)\right) b^{c-sE} \sum_{f \in S} d_{fg} f \mid T_c(b\beta\mathfrak{x})$$

is arithmetic for every $g \in S$ at $s = 0$. By Lemma 6.2, $\beta\mathfrak{x} = et_{\mu\lambda}$ with $e \in D$ and some μ. Changing η and ζ suitably, we can transform (6.2) into

$$(6.3) \qquad \pi^{-|k|} \sum_{b \in D/U} \eta(b)\omega_u\left(\mathrm{Tr}_{F/E}(\zeta b)\right) b^{c-sE} \sum_{f \in S} d_{fg} f \mid T_c^{\lambda\mu}(bt_{\mu\lambda}).$$

We can obviously replace D by $D \cap t_{\lambda\mu}$, which is $N(X_{\lambda\mu}')$ by Lemma 6.1. Thus our problem is to show that (6.3) is arithmetic for every $g \in S$ at $s = 0$.

Now, for every function h on $\mathfrak{H}^n$, define a function h^* on $\mathfrak{H}'$ by $h^*(z) = h(z^*)$ with

$$z_\nu^* = \begin{cases} \zeta^{\tau_\nu} z_\nu, & \text{if } \nu \le r, \\ \zeta^{\tau_\nu} \bar{z}_\mu, & \text{if } \nu > r \text{ and } \mathrm{Res}_{F/E}(\tau_\nu) = \tau_\mu'. \end{cases}$$

Define θ by (5.1) with $\zeta = a^{-1}$ and $0 \le v = \Sigma_{i > r} v_i \tau_i \in I_F$; put

$$\Theta(z, w, w') = |\zeta|^{r - E} (\mathbf{d}^v \theta)(z^*, w, w'; \xi, \zeta^{-1}) \quad (z, w, w' \in \mathfrak{H}'),$$

where $\mathbf{d}^v$ is defined by (I.1.15) with respect to $z_{r+1}, \ldots, z_n$. Define also H as in (4.11) with v of (6.1). Then (4.12) follows from (3.7) and (6.1); in fact, the constant γ of (3.7) coincides with $-\beta'$ with β' of (4.12). Since $s_0 = 0$ satisfies (3.12a, b), we obtain (4.13). To express H as an integral, we consider

$$(6.5a) \qquad D_{l + 2u, \varepsilon}\big(s - 1 + (\beta/2), \mathbf{d}^u \Omega_\rho, \Theta\big)$$

with $\varepsilon = k + \mathrm{Res}_{F/E}(\kappa + 2v) + 2([F : E] - 1)E$, $\beta = \beta' + 2 - 2[F : E]$, and $\Omega_\rho(z) = \Sigma_{a \in E} \omega(a) e_E(az)$. This is a special case of (I.7.2), and can be written as

$$(6.5b) \quad [V] N(\mathfrak{b})^{-1} D_E^{-1/2} \int_{L \times M} \overline{\mathbf{d}^u \Omega_\rho(z)}\, \Theta(z, w, w') y^{k + (s-1)E}\, dx\, d^\times y,$$

where $L = \mathbf{R}'/\mathfrak{b}$, $M = \mathbf{R}'_+/V$ with a suitable ideal $\mathfrak{b}$ of E and a subgroup V of U_E of finite index (see (I.2.8)). Now the computation of Section I.7 specialized to this case shows that (6.5b) is equal to

$$(6.6) \qquad 2^{1-r} \pi^{-|k|-rs} \Gamma_r(k + sE) \mathrm{Im}(w)^{sE} \mathrm{Im}(w')^{sE} H(w, w', s).$$

Suppose ξ is $\overline{\mathbf{Q}}$-valued and Ω is $\overline{\mathbf{Q}}$-rational. By Proposition 4.3, $\pi^{-|k|} H(w, w', 0)$ is arithmetic. Take an element δ of B so that $N(\delta) = -\zeta^{-1}$, and put

$$H_1(w, w', s) = \pi^{-|k|} H(w, \delta \bar{w}', s) j(\delta, \bar{w}')^{-k} |j(\delta, w')|^{-2s} \sigma_\kappa(\delta).$$

By Lemma 4.1, (1) and Lemma 4.2, we have

$$(6.7a) \qquad H_1(w, w', 0) = \sum_i f_i(w) \cdot {}^t\overline{g_i(w')}$$

with arithmetic elements f_i and g_i of $\mathfrak{M}_{k, \kappa}$. Taking a sufficiently small congruence subgroup Γ of B, we may assume $\{f_i\}$ to be a $\overline{\mathbf{Q}}$-rational basis of $\mathfrak{M}_{k, \kappa}(\Gamma)$. We have $\mathfrak{M}_{k, \kappa} \ne \mathfrak{S}_{k, \kappa}$ only when $B = M_2(F)$ and $k_1 = \cdots = k_n$. In this special case, we observe that $\Lambda_{pq} \ne 0$ only if $l \ne 0$, and hence (3.12a) implies that $k_i > 1$. Therefore $\mathfrak{M}_k(\Gamma)$ can be decomposed into the direct sum of $\mathfrak{S}_k(\Gamma)$ and the space of Eisenstein series (see Kloosterman [3]). Since this is done

$\overline{\mathbf{Q}}$-rationally, we have

$$(6.7\text{b}) \qquad H_1(w, w', 0) = \sum_i f_i(w) \cdot {}^t\overline{g_i(w')} + \sum_i e_i(w) \cdot {}^t\overline{h_i(w')}$$

with a $\overline{\mathbf{Q}}$-rational basis $\{f_i\}$ of $\mathcal{S}_k(\Gamma)$, Eisenstein series e_i, and $\overline{\mathbf{Q}}$-rational elements g_i and h_i of $\mathcal{M}_k$. Thus we use (6.7a) or (6.7b) according as B is a division algebra or not.

Define θ_1 by (5.5). Then

$$[V]N(\mathfrak{b})^{-1}D_E^{-1/2}\int_{L \times M} \overline{\mathbf{d}^u\Omega_\rho(z)}\,(\mathbf{d}^v\theta_1)(z^*, w, w')y^{k+(s-1)E}\,dx\,d^\times y$$
$$= A(w, w', s)H_1(w, w', s),$$
$$A(w, w', s) = 2^{1-r}|\zeta|^{(1-s)E-v-(k/2)}\pi^{-rs}\Gamma_r(k + sE)\mathrm{Im}(w)^{sE}\mathrm{Im}(w')^{sE}.$$

For every $f \in \mathcal{S}_{k,\kappa}(\Gamma)$, put

$$K(z, w'; f) = \int_\Phi {}^t\overline{f(w)}\,\theta_1(z, w, w')\mathrm{Im}(w)^k\,d\mu(w)$$

with $\Phi = \Gamma \setminus \mathfrak{H}^r$. Then we have

$$(6.8) \quad \int_\Phi {}^t\overline{f(w)}\,A(w, w', s)H_1(w, w', s)\mathrm{Im}(w)^k\,d\mu(w)$$
$$= [V]N(\mathfrak{b})^{-1}D_E^{-1/2}\int_{L \times M} \overline{\mathbf{d}^u\Omega_\rho(z)}\,(\mathbf{d}^vK)(z^*, w'; f)y^{k+(s-1)E}\,dx\,d^\times y.$$

This holds at least for sufficiently large $\mathrm{Re}(s)$, because θ_1 is slowly increasing (at cusps, if any) as a function of (w, w'). By Proposition 5.1, we know that

$$K(z, w'; f) = (-2i)^{|k|}\sum_{0 \ll b \in F} b^{(m/2)-F}\mathbf{e}_F(bz)\sum_{\alpha \in R_b}\varphi(\alpha) \cdot {}^t\overline{(f\|\alpha)(w')},$$

where $\varphi(\alpha) = \xi(\alpha\delta^{-1})$ and $R_b = \Gamma \setminus \{\alpha \in B \mid N(\alpha) = b\}$. Therefore the right-hand side of (6.8) is a function of type (I.3.10) and equal to

$$(6.9) \quad [V](-2i)^{|k|}(4\pi\zeta)^{(1-s)E-k}\Gamma_r(k + (s-1)E)$$
$$\times \sum_{0 \ll b \in F/V}\omega_u(\mathrm{Tr}_{F/E}(\zeta b))b^{c-sE}\sum_{\alpha \in R_b}\varphi(\alpha) \cdot {}^t\overline{(f\|\alpha)(w')}.$$

Here we have used (6.1). We can of course replace V by U. This evaluated at $s = 0$ is equal to the left-hand side of (6.8) at $s = 0$, which is

$$C\mu(\Phi)\sum_i \langle f, f_i\rangle \cdot {}^t\overline{g_i(w')}$$

with a nonzero algebraic number C independent of f. Recall that $\mu(\Phi)$ is π^r

times an algebraic number; also note that $k_i > 1$ for all $i \leq r$ by virtue of (3.12a). Let (d_{ij}) be the inverse of the matrix $(\langle f_i, f_j \rangle)_{i,j}$. Then we find that ${}'g_i(w')$ is equal to a nonzero algebraic number times the value at $s = 0$ of

$$\pi^{-|k|} \sum_{b U} \omega_u \left(\mathrm{Tr}_{F/E}(\zeta b) \right) b^{c-sE} \sum_{\alpha \in R_b} \varphi(\alpha) \sum_j d_{ij} \cdot \overline{{}'(f_j \| \alpha)(w')}.$$

Change φ and Ω for $\bar{\varphi}$ and Ω_ρ, and take the complex conjugate of the whole series. Since g_i is arithmetic, we see that

$$\pi^{-|k|} \sum_{bU} \omega_u \left(\mathrm{Tr}_{F/E}(\zeta b) \right) b^{c-sE} \sum_{\alpha \in R_b} \varphi(\alpha) \sum_i d_{ji} f_i \| \alpha$$

is arithmetic at $s = 0$.

Given our space $\mathfrak{S}_{k,\kappa}(\Gamma_\lambda, \varphi_\lambda)$, we can take the above Γ so that $\Gamma \subset \Gamma'_\lambda$ and $\varphi_\lambda(\Gamma) = 1$ with Γ'_λ of Lemma 6.3. We have chosen a basis S of $\mathfrak{S}_{k,\kappa}(\Gamma_\lambda, \varphi_\lambda)$ in (6.3). By Lemma 2.4, we may assume that $\{f_i\}$ consists of S and some functions orthogonal to S. Then we find that

$$(6.10) \qquad \pi^{-|k|} \sum_{bU} \omega_u \left(\mathrm{Tr}_{F/E}(\zeta b) \right) b^{c-sE} \sum_{\alpha \in R_b} \varphi(\alpha) \sum_{f \in S} d_{fg} f \| \alpha$$

at $s = 0$ is arithmetic for every $g \in S$. Now with $X_{\lambda\mu}$ as in the beginning of this section, put $\varphi_{\lambda\mu}(\alpha) = \psi_\gamma(x_\lambda^{-1} \alpha x_\mu)$ for $\alpha \in X_{\lambda\mu}^+$ with ψ_γ of (1.16). Given a locally constant function η on F, we now consider a locally constant function φ on B such that

$$(6.11) \qquad \varphi(\alpha) = \begin{cases} \eta(N(\alpha)) \varphi_{\lambda\mu}(\alpha)^{-1} & \text{for } \alpha \in X_{\lambda\mu}^+, \\ 0 & \text{for } \alpha \notin X_{\lambda\mu}. \end{cases}$$

The existence of such a φ is obvious, since $\varphi_{\lambda\mu}$ depends only on α modulo a lattice. Here we fix (λ, μ). Let Z_b and A_b be the same as in Lemma 6.3; let W_b be a complete set of representatives of $\Gamma \backslash Z_b$. Then (6.10) with φ of (6.11) can be written in the form

$$(6.12) \qquad \pi^{-|k|} \sum_{bU} \eta(b) \omega_u \left(\mathrm{Tr}_{F/E}(\zeta b) \right) b^{c-sE} \sum_{f \in S} d_{fg} \sum_{\alpha \in W_b} \varphi_{\lambda\mu}(\alpha)^{-1} f \| \alpha,$$

where b runs over $N(X_{\lambda\mu}^+)/U$. If we change W_b for $\Gamma'_\lambda \backslash Z_b$, the whole sum is multiplied by $[\Gamma'_\lambda : \Gamma]^{-1}$. Therefore (6.12) is $[\Gamma'_\lambda : \Gamma]$ times

$$(6.13) \qquad \pi^{-|k|} \sum_{bU} \eta(b) \omega_u \left(\mathrm{Tr}_{F/E}(\zeta b) \right) b^{c-sE} \sum_{f \in S} d_{fg} \sum_{\alpha \in A_b} f | \Gamma'_\lambda \alpha \Gamma'_\mu,$$

where we put $f | \Gamma'_\lambda \alpha \Gamma'_\mu = \sum_i \varphi_{\lambda\mu}(\alpha_i)^{-1} f \| \alpha_i$ for $\Gamma'_\lambda \alpha \Gamma'_\mu = \sqcup_i \Gamma'_\lambda \alpha_i$. By Lemma 6.3

and Lemma 1.2, we have

$$f \mid T_c^{\lambda\mu}(bt_{\mu\lambda}) = \sum_{\alpha \in \Lambda_b} f \mid \Gamma_\lambda \alpha \Gamma_\mu = \sum_{\alpha \in \Lambda_b} f \mid \Gamma'_\lambda \alpha \Gamma'_\mu,$$

and hence (6.13) is exactly (6.3). This completes the proof of Theorem 3.3.

Proof of Theorem 3.1. For the same reason as in the above proof, it is sufficient to show that

$$(6.14) \qquad \sum_{b \in D} \eta(b) b^{(m/2)}\,{}^F e_F(bz) f \mid T_c^{\lambda\mu}(bt_{\mu\lambda})$$

as a function of z belongs to $\mathfrak{S}_m(M_2(F))$ for every $f \in S$. This sum is extended over b in $N(X_{\lambda\mu})$. The same reasoning as in the last part of the above proof shows that (6.14) is

$$[\Gamma'_\lambda : \Gamma]^{-1} \sum_{0 \ll b \in F} b^{(m/2)-F} e_F(bz) \sum_{\alpha \in R_b} \varphi(\alpha) f \| \alpha$$

with a suitable locally constant function φ on B. Therefore the desired result follows from Proposition 5.4.

Proof of Theorem 3.5. Denote series (3.6) by $\Lambda_{pq}(u, c)$ to emphasize exponents u and c. Since $\mathrm{Tr}_{F/E}(\zeta b)^u b^c$ is a finite sum $\Sigma_x \xi_x b^x$ with $\xi_x \in \overline{\mathbf{Q}}$ and $c \leq x \in 2^{-1} I_F$, we have $\Lambda_{pq}(u, c) = \Sigma_x \xi_x \Lambda_{pq}(0, x)$. By Proposition 3.2, $\Lambda_{pq}(0, x)$ is holomorphic at $(\gamma/2) + [F:E]$ unless $x_i = c_i$ for all $i \leq r$. This means that it is sufficient to prove the case $u = 0$. Put $s_0 = (\gamma/2) + [F:E] - 1$ and $v = c - s_0 E + (k - \kappa)/2$. Conditions (3.9b) and (3.13) imply that $v = \Sigma_{i>r} v_i \tau_i$ with $0 \leq v_i \in \mathbf{Z}$. Thus, changing s for $s + s_0$, we may assume that c has form (6.1). Then our question is the residue of Λ_{pq} at $s = 1$. The notation being the same as in the proof of Theorem 3.3, our task is to show that the residue of (6.3) with $u = 0$ at $s = 1$ multiplied by R_E^{-1} is arithmetic for every $g \in S$. For this purpose, we again use (6.8) and (6.9). Let $\mathfrak{g}_1(w, w')$ be the residue of $\mathrm{Im}(w)^{sF} \mathrm{Im}(w')^{sF} H_1(w, w', s)$ at $s = 1$. Suppose B is a division algebra. By Proposition 4.4, Lemmas 4.1 and 4.2, we have

$$\mathfrak{g}_1(w, w') = R_E \sum_i f_i(w) \cdot {}^t \overline{h_i(w')}$$

with f_i as in the proof of Theorem 3.3 and arithmetic elements h_i of $\mathfrak{M}_{k,\kappa}$. Since (6.8) is equal to (6.9), we see that the residue of (6.9) at $s = 1$ is

$$C' \pi^{-r} \mu(\Phi) R_E \sum_i \langle f, f_i \rangle \cdot {}^t \overline{h_i(w')}$$

with a nonzero algebraic number C'. Therefore we obtain the desired conclusion by exactly the same procedure as in the last part of the proof of Theorem 3.3.

If $B = M_2(F)$, condition (3.10) implies that $k_i = l_i$ for all $i \leq n$. If $k_i > 1$ for some i, we have (6.7b), so that the above proof is valid. If $k_1 = \cdots = k_n = 1$, we cannot use (6.7b), since the decomposition of $\mathfrak{M}_k$ into the direct sum of $\mathfrak{S}_k$ and the space of Eisenstein series is an open question, though it is perhaps true. However, if $B = M_2(F)$, Theorem 3.5 can easily be derived from [11, Proposition 4.3]; or rather, the former is a mere paraphrase of the latter. Therefore this case needs no further discussion.

7. The case of half-integral l

As already mentioned on [13, p. 373], assumption (I.5.4) is not absolutely necessary for the validity of the assertions of Theorem I.5.2, and moreover, Theorem I.4.3 may be extended to the case $l \notin I_E$. There is no doubt that the same is true for the present Theorem 3.3. We can actually prove all these at least when $E = \mathbf{Q}$. Though the general case does not seem to present great difficulties, our current state of knowledge on this problem allows us only to make some conjectural statements, which we now briefly formulate in order. Each statement has its counterpart which we have proved in this paper; we note it in parentheses.

C.7.1 (Theorem I.3.2). *Define Λ by (I.3.2); suppose that f is a cusp form and (I.3.3) holds with $2\alpha \equiv 1 \pmod 2$. Then Λ can be continued to the whole plane as a meromorphic function which is holomorphic for $\mathrm{Re}(s) > (2\alpha - 1)/4$.*

C.7.2 (Theorems I.4.2, I.4.3). *Define D by (I.4.2) with $\Omega \in \mathfrak{M}_{l,E}$, $l \notin I_E$; suppose $\Phi \notin (1 + \rho)I_K$ and $\mathrm{Res}_{K/E}(\Phi) + l + 2p = \beta E$ with $\beta \in 2^{-1}\mathbf{Z}$. Then D can be continued to the whole plane as a meromorphic function which is holomorphic for $\mathrm{Re}(s) > (2\beta + 2[F\colon E] - 1)/4$. Moreover suppose Ω is $\overline{\mathbf{Q}}$-rational; let μ be an integer satisfying*

$$(7.1) \qquad \beta + [F\colon E] - (1/2) < 2\mu \leq 2c(\psi_i) \text{ for } i = 1, \ldots, r.$$

Then D is holomorphic at μ, and $D(\mu) \sim \pi^e p_K(\Phi, 2\psi)$, where $e = \sum_{i=1}^{r} \{c(\psi_i) - c(\psi_i \rho)\}$.

C.7.3 (Theorems I.5.2, I.5.3). *Define $f(w, s)$ by (I.5.2); suppose (I.5.3) holds and put $\alpha = \alpha' + 2 - (m + 2)[F\colon E]/2$. Then f can be continued to the whole s-plane as a meromorphic function. Moreover, if $3 < 2\alpha \equiv 1 \pmod 2$, f is holomorphic in s at $s = 0$, and $f(w, 0)$ is holomorphic in w. Furthermore, if Ω is $\overline{\mathbf{Q}}$-rational and $\mathfrak{f}$ is defined by (I.5.7), then $\pi^{-k}\mathfrak{f}$ is arithmetic.*

C.7.4 (Theorem I.10.1). *Define $f(w, s)$ by (I.10.2) with $\Omega \in \mathfrak{M}_{l,E}$, $l \notin I_E$; suppose the equality of (I.10.3) holds with $\alpha' \in 2^{-1}\mathbf{Z}$; define α and e as in Theorem I.10.1. Then f can be continued to the whole s-plane as a meromorphic* *function. Moreover, if $\alpha > 3/2$, f is holomorphic in s at $s = 0$, and $f(w,0)$ is holomorphic in w; furthermore $cf(w,0)$ is arithmetic if Ω is $\overline{\mathbf{Q}}$-rational.*

C.7.5 (Theorems 3.3 and 3.7). *Define Λ_{pq} by (3.6) and L by (3.16a) with $\Omega \in \mathfrak{M}_{l,E}$, $l \notin I_E$; further define Λ and Ξ as in Theorem 3.3. Suppose $l + 2u + 2\,\mathrm{Res}_{F/E}(c) = \gamma E$ with $\gamma \in 2^{-1}\mathbf{Z}$. Then, under (3.9a,b) and (3.11), the entries of the matrix $\pi^{-|k|}\cdot {}^t\Xi^{-1}\Lambda(s_0)$ are all algebraic and $L(S_0) \sim \pi^{|k|}\langle \mathbf{g},\mathbf{g}\rangle$ for every integer s_0 satisfying*

$$(7.2) \qquad (\gamma/2) - (1/4) + [F\colon E] < s_0 \le (k_i/2) + c_i \ \text{ for all } \ i \le r.$$

These statements are true if $E = \mathbf{Q}$. There are two basic facts necessary for the proof. The first one is the Eisenstein series of half-integral weight which we treated in [15]. In particular, the holomorphy properties in s of Λ, D, f, and Λ_{pq} follow from [15, Proposition 3]. The second one is the algebraicity property of the critical values of

$$(7.3) \qquad \sum_{n=1}^{\infty} \omega(n)\xi(n)n^{-s}$$

for $\Omega(z) = \Sigma\omega(n)e(nz) \in \mathfrak{M}_l(M_2(\mathbf{Q}))$ and $g(z) = \Sigma\xi(n)e(nz) \in \mathcal{S}_m(M_2(\mathbf{Q}))$ when $2l \equiv 1 \pmod 2$ and $m \in \mathbf{Z}$. This is due to Sturm [17]. Combining his result with [11, Theorem 5.3], we obtain the arithmetic part of C.7.2 when $F = E = \mathbf{Q}$. Once this is done, the remaining part of the above statements in the case $E = \mathbf{Q}$ can be derived in exactly the same fashion as for their counterparts. Similarly, the only necessary tools for the proof in the case of an arbitrary basic field E are the Eisenstein series of half-integral weight on $\mathrm{GL}_2(E)$ and the nature of the critical values of certain Dirichlet series which are generalizations of (7.3) with Ω and g replaced by Hilbert modular forms on $\mathrm{GL}_2(E)$ respectively of half-integral and integral weights.

PRINCETON UNIVERSITY, PRINCETON, NEW JERSEY

REFERENCES

[1] T. ASAI, On certain Dirichlet series associated with Hilbert modular forms and Rankin's method, Math. Ann. **226** (1977), 81–94.

[2] H. JACQUET and R. P. LANGLANDS, *Automorphic Forms on* GL(2), Lecture notes in mathematics 114, Springer, 1970.

[3] H. D. KLOOSTERMAN, Theorie der Eisensteinschen Reihen von mehreren Veränderlichen, Abh. Math. Sem. Hamb. **6** (1928), 163–188.

[4] R. P. LANGLANDS, On the zeta-functions of some simple Shimura varieties, Canad. J. of Math. **31** (1979), 1121–1216.

[5] T. MIYAKE, On automorphic forms on GL_2 and Hecke operators, Ann. of Math. 94 (1971), 174–189.

[6] A. SELBERG, Harmonic analysis and discontinuous groups in weakly symmetric Riemannian spaces with applications to Dirichlet series, J. Indian Math. Soc. 20 (1956), 47–87.

[7] H. SHIMIZU, On zeta functions of quaternion algebras, Ann. of Math. 81 (1965), 166–193.

[8] ———, Theta series and automorphic forms on GL_2, J. Math. Soc. Japan, 24 (1972), 638–683.

[9] G. SHIMURA, On Dirichlet series and abelian varieties attached to automorphic forms, Ann. of Math. 76 (1962), 237–294.

[10] ———, *Introduction to the Arithmetic Theory of Automorphic Functions*, Iwanami Shoten and Princeton Univ. Press, 1971.

[11] ———, The special values of the zeta functions associated with Hilbert modular forms, Duke Math. J. 45 (1978), 637–679.

[12] ———, Automorphic forms and the periods of abelian varieties, J. Math. Soc. Japan, 31 (1979), 561–592.

[13] ———, The arithmetic of certain zeta functions and automorphic forms on orthogonal groups, Ann. of Math. 111 (1980), 313–375.

[14] ———, On certain zeta functions attached to two Hilbert modular forms: I. The case of Hecke characters, Ann. of Math. 114 (1981), 127–164.

[15] ———, On the holomorphy of certain Dirichlet series, Proc. London Math. Soc. 31 (1975), 79–98.

[16] T. SHINTANI, On construction of holomorphic cusp forms of half integral weight, Nagoya Math. J. 58 (1975), 83–126.

[17] J. STURM, Addendum to special values of zeta functions, Amer. J. of Math. 102 (1980), 781–783.

(Received September 11, 1980)

Arithmetic of differential operators on symmetric domains

Duke Mathematical Journal, 48 (1981), 813-843

Introduction. Throughout the paper, we denote by $\mathfrak{K}$ the Siegel upper half space of degree m, which consists of all complex symmetric matrices of size m with positive definite imaginary part. If $m = 1$, our operators have the forms

$$D_r^k = D_{r+2k-2} \cdots D_{r+2} D_r, \qquad D_r = (z - \bar{z})^{-1} r + \partial/\partial z \qquad (r \in \mathbf{Z}, z \in \mathfrak{K}).$$

Operator D_r^k sends a (holomorphic or nonholomorphic) modular form of weight r to a form of weight $r + 2k$. Moreover, if f and g are elliptic modular forms of weight r and $r + 2k$ respectively, and if they have algebraic Fourier coefficients, then $\pi^{-k} g^{-1} D_r^k f$ takes an algebraic value at every imaginary quadratic point, as proved in [8]. Now the purpose of the present paper is to define some operators with similar properties on $\mathfrak{K}$ with $m \geqslant 1$ and also on other domains, and to study the arithmetic nature of the values of certain nonholomorphic Eisenstein series at CM-points on $\mathfrak{K}$ by means of the operators. We have already shown in our previous paper [14] how these ends are attained in the case of the orthogonal group $SO(m, 2)$. Although the same ideas are applicable to the symplectic and other groups, there are several interesting aspects of the theory of such operators which were not dealt with in [14]. Therefore we treat here the symplectic case anew, considering in particular the features not covered by what we did in [14].

The definition of a generalization of D_r in the case $m \geqslant 1$ is relatively simple. In fact, fixing a rational representation

$$\rho : GL_m(\mathbf{C}) \to GL(V)$$

with a finite-dimensional complex vector space V, we take $\rho(cz + d)$ as the factor of automorphy with the standard meaning of $cz + d$. Then, for a V-valued function f on $\mathfrak{K}$, we put

$$D_\rho f = \rho(z - \bar{z})^{-1} D(\rho(z - \bar{z})f),$$

where $D = ((1/2)(1 + \delta_{ij})\partial/\partial z_{ij})$; we understand that $D_\rho f$ has values in $\mathrm{Hom}(\mathfrak{S}, V)$, where $\mathfrak{S}$ is the global complex tangent space of $\mathfrak{K}$. This operator has a certain property of commutativity with the action of the symplectic group. It is important, however, to know the nature of iterated operators of type $D_\omega \cdots D_\sigma D_\rho$, which are generalizations of the above D_r^k and considerably more

Received May 26, 1981. Supported by NSF Grant MCS 78-02781.

complicated than a single operator D_ρ. Thus our effort in this paper is expended on the task of treating such operators with as little combinatorial computation as possible, so that, for example, their effect on Eisenstein series can easily be found. In particular, we define in a rather simple way an operator $\Delta_\rho^{(k)}$ such that $\Delta_\rho^{(k)}f$ is still V-valued but belongs to the factor $\det(cz + d)^{2k}\rho(cz + d)$; this is another generalization of D_r^k. If V is one-dimensional and $k = 1$, $\Delta_\rho^{(1)}$ differs from the operator M_α of Maass [3] only by the factor $\det(z - \bar{z})$.

The generalization of the algebraicity of the values of $\pi^{-k}g^{-1}D_r^k f$ will be given in two versions: one in terms of $D_\omega \cdots D_\sigma D_\rho$ and the other in terms of $\Delta_\rho^{(k)}$ (Theorems 3.1 and 3.2). It should be noted that operators D_ρ are indispensable for the algebraicity-theorem concerning $\Delta_\rho^{(k)}$, though the latter can be defined without the former, as will be shown in Section 7. We apply these results of algebraicity to an Eisenstein series of the form

$$E(z) = \sum_{c,\,d} \det(cz + d)^{-q}\rho\big((cz + d)^{-1}(c\bar{z} + d)(z - \bar{z})^{-1}\big).$$

Here $\{c, d\}$ runs over all representatives of coprime symmetric pairs in the usual sense, $m + 1 < q \in 2\mathbf{Z}$, and ρ is an arbitrary polynomial representation of GL_m. We shall show that the value of E at a CM-point w of $\mathfrak{K}$ can be expressed in terms of the symbol p_K which was introduced in [14] and which can be given as several products of the periods of abelian varieties with complex multiplication (Theorems 5.1 and 5.2). In Section 6, we give an interpretation of $E(w)$ as an infinite series constructed directly from a given CM-type with no reference to the factors of automorphy, nor to $\mathfrak{K}$. Up to Section 7, our exposition will be restricted to the symplectic case. In the final section, however, we shall briefly describe how the analogues of D_ρ and $\Delta_\rho^{(k)}$ can be defined on all other three major types of bounded symmetric domains, excluding exceptional ones. The detailed proofs will not be given there, since they can be treated in exactly the same fashion as in the symplectic case with no essentially new ideas.

Notation. In addition to the standard symbols $\mathbf{Z}$, $\mathbf{Q}$, $\mathbf{R}$, and $\mathbf{C}$, we employ $\overline{\mathbf{Q}}$ to denote the algebraic closure of $\mathbf{Q}$ in $\mathbf{C}$. We denote by $M_n(R)$ and $GL_n(R)$ the ring of all square matrices of size n over a ring R and the group of its invertible elements; 1_n denotes the identity element of $M_n(R)$. For a complex vector space V, we denote by $GL(V)$ the group of all $\mathbf{C}$-linear automorphisms of V. If X is a (complex) matrix, ^tX denotes its transpose, and $\bar{X}$ its complex conjugate. If X is hermitian, we write $X > 0$ to indicate that it is positive definite. Further $\mathrm{diag}[X_1, \ldots, X_k]$ denotes the square matrix with square matrices $X_1, \ldots, X_k$ in the diagonal blocks and 0 in all other blocks.

1. Some elementary formulas on symplectic groups. We define a $\mathbf{Q}$-rational algebraic group G so that

$$(1.1) \qquad G_{\mathbf{Q}} = \big\{ \alpha \in GL_{2m}(\mathbf{Q}) \,|\, {}^t\alpha J\alpha = \nu(\alpha)J \quad \text{with} \quad \nu(\alpha) \in \mathbf{Q} \big\}$$

with a positive integer m, where

$$(1.2) \qquad J = J_m = \begin{pmatrix} 0 & -1_m \\ 1_m & 0 \end{pmatrix}.$$

Then $G_{\mathbf{R}}$ and $G_{\mathbf{C}}$ can be defined in the usual way; ν can be extended to a homomorphism of $G_{\mathbf{C}}$ into $GL_1(\mathbf{C})$, which we denote again by ν. Let $\mathcal{S}$ denote the space of all complex symmetric matrices of size m. We put then

$$(1.3) \qquad \mathcal{K} = \{ z \in \mathcal{S} \mid i(\bar{z} - z) > 0 \}.$$

If

$$\alpha = \begin{pmatrix} a & b \\ c & d \end{pmatrix} \in G_{\mathbf{R}}$$

and $\nu(\alpha) > 0$, then we can define as usual the action of α on $\mathcal{K}$ by

$$(1.4) \qquad \alpha z = \alpha(z) = (az + b)(cz + d)^{-1}.$$

For

$$\begin{pmatrix} a & b \\ c & d \end{pmatrix} \in G_{\mathbf{C}}$$

and $z \in \mathcal{S}$, let $p = az + b$ and $q = cz + d$. Then we see easily that ${}^t pq = {}^t qp$, and hence $pq^{-1} \in \mathcal{S}$ if q is invertible. Therefore we can take (1.4) to be the action of an arbitrary element

$$\alpha = \begin{pmatrix} a & b \\ c & d \end{pmatrix}$$

of $G_{\mathbf{C}}$ on $\mathcal{S}$, since $cz + d$ is invertible at least for some z of $\mathcal{S}$ as will be shown below. We understand that the action is a birational map of $\mathcal{S}$ onto itself defined at least at "generic points" but not necessarily at an arbitrary point of $\mathcal{S}$. We also put

$$(1.5) \qquad \mu_\alpha(z) = \mu(\alpha, z) = cz + d$$

$$(1.6) \qquad j_\alpha(z) = j(\alpha, z) = \det(cz + d) \qquad \left(\alpha = \begin{pmatrix} a & b \\ c & d \end{pmatrix} \in G_{\mathbf{C}}, z \in \mathcal{S} \right).$$

Then we have

$$(1.7) \qquad \mu_{\alpha\beta}(z) = \mu_\alpha(\beta z)\mu_\beta(z), \qquad j_{\alpha\beta}(z) = j_\alpha(\beta z)j_\beta(z),$$

$$(1.8) \qquad \alpha \begin{pmatrix} z & z' \\ 1_m & 1_m \end{pmatrix} = \begin{pmatrix} \alpha z & \alpha z' \\ 1_m & 1_m \end{pmatrix} \begin{pmatrix} \mu_\alpha(z) & 0 \\ 0 & \mu_\alpha(z') \end{pmatrix} \qquad (z, z' \in \mathcal{S}).$$

Calling the last product W and forming ${}^t WJW$, we find

$$(1.9) \qquad \nu(\alpha)(z - z') = {}^t \mu_\alpha(z)(\alpha z - \alpha z')\mu_\alpha(z').$$

That $j_\alpha(z) \neq 0$ for some $z \in \mathfrak{S}$ can be shown as follows. This is obvious if c is invertible. Let A be the set of all α with invertible c. Then A is a nonempty open subset of $G_{\mathbf{C}}$ stable under the map $\alpha \mapsto \alpha^{-1}$. Since $G_{\mathbf{C}}$ is connected, every element of $G_{\mathbf{C}}$ is a product of finitely many elements of A. (Actually two elements are sufficient.) Observe that if αz is meaningful for some $z \in \mathfrak{S}$, then $z \mapsto \alpha z$ gives a one-to-one map of an open dense subset of $\mathfrak{S}$ onto another such set. Now (1.7) is true if βz is meaningful. Therefore, if both j_α and j_β are generically nonvanishing, so is $j_{\alpha\beta}$, Q.E.D. It should also be noted that $(\alpha\beta)z = \alpha(\beta z)$.

If $dz = (dz_{ij})$ denotes the differential of the variable matrix $z = (z_{ij})$ on $\mathfrak{S}$, we have

$$(1.10) \qquad \nu(\alpha)\,dz = {}^t \mu_\alpha(z)\,d(\alpha z)\,\mu_\alpha(z),$$

which follows easily from (1.9). We now define differential operators ∂_{ij} and Δ on $\mathfrak{S}$ by

$$(1.11) \qquad \partial_{ij} = \begin{cases} \partial/\partial z_{ij} & \text{if } i = j, \\ (1/2)\partial/\partial z_{ij} & \text{if } i \neq j, \end{cases}$$

$$(1.12) \qquad \Delta = \det(\partial_{ij}) = \sum_\sigma \operatorname{sgn}(\sigma)\partial_{1\sigma(1)} \cdots \partial_{m\sigma(m)},$$

where σ runs over all permutations of $\{1, \ldots, m\}$. (We of course take z_{ij} for $i \leqslant j$ to be the independent variables on $\mathfrak{S}$.) For every integer n, we denote by Δ^n the n-th iteration of Δ. Recall a well known formula

$$(1.13) \qquad \Delta(\det(z)^s) = \prod_{k=0}^{m-1} (s + (k/2)) \cdot \det(z)^{s-1} \qquad (s \in \mathbf{C}).$$

For the reader's convenience, we give a proof here. First we note another well known formula

$$(1.14) \qquad \int_P e^{-\operatorname{tr}(uz)} \det(u)^{s-p}\,du = \Gamma_m(s)\det(z)^{-s}$$

$$(p = (m + 1)/2, z \in \mathfrak{S}, \operatorname{Re}(z) > 0, \operatorname{Re}(s) > m - 1),$$

due to Siegel [16, (110)], where $P = \{u \in \mathfrak{S} \mid \bar{u} = u > 0\}$, $du = \prod_{i \leqslant j} du_{ij}$, and $\Gamma_m(s) = \pi^{m(m-1)/4}\prod_{k=0}^{m-1}\Gamma(s - (k/2))$. Applying Δ to (1.14), we obtain (1.13) when $\operatorname{Re}(z) > 0$ and $\operatorname{Re}(s) < 1 - m$. This is sufficient, because of the purely algebraic nature of the desired formula.

If (c, d) is the lower half of an element of $G_{\mathbf{C}}$, we obtain, as an easy consequence of (1.13), a formula

$$(1.15) \quad \Delta^n\big(\det(cz + d)^s\big) = \epsilon_n(s)\det(c)^n \det(cz + d)^{s - n},$$

$$\epsilon_n(s) = \prod_{k=0}^{m-1} \{\Gamma(s + 1 + (k/2))/\Gamma(s + 1 + (k/2) - n)\}.$$

In addition to $\mathcal{H}$ and $G_{\mathbf{R}}$, we consider a bounded domain $\mathcal{B}$ and a group G' defined by

$$(1.16) \qquad\qquad \mathcal{B} = \{z \in \mathcal{S} \mid 1_m - \bar{z}z > 0\},$$

$$(1.17) \quad G' = \big\{\alpha \in G_{\mathbf{C}} \mid {}^t\bar{\alpha} I_{m,m}\alpha = \nu(\alpha)I_{m,m}\big\}, \qquad I_{m,m} = \mathrm{diag}\big[\,1_m, -1_m\big].$$

This group coincides with the set of all elements of $G_{\mathbf{C}}$ of the form $\left(\begin{smallmatrix} a & b \\ \bar{b} & \bar{a} \end{smallmatrix}\right)$. Put

$$(1.18) \qquad\qquad \beta_0 = \begin{pmatrix} u1_m & u1_m \\ -\bar{u}1_m & \bar{u}1_m \end{pmatrix}, \qquad u = 2^{-1/2}e^{\pi i/4}.$$

Then we see that $\beta_0 \in G_{\mathbf{C}}$, $\nu(\beta_0) = 1$, and $\beta_0 G'\beta_0^{-1} = G_{\mathbf{R}}$. Moreover, it is well known that G' acts on $\mathcal{B}$ and $w \mapsto \beta_0(w) = i(w + 1)(1 - w)^{-1}$ maps $\mathcal{B}$ onto $\mathcal{H}$ (cf. Siegel [18]). We now put

$$(1.19) \qquad \eta_{\mathcal{B}}(z) = 1_m - \bar{z}z, \qquad \eta_{\mathcal{H}}(z) = i(\bar{z} - z) \qquad (z \in \mathcal{S}).$$

These are primarily considered on $\mathcal{B}$ and $\mathcal{H}$, respectively. We have then

$$(1.20) \qquad \nu(\alpha)\eta_{\mathcal{B}}(z) = \overline{{}^t\mu_\alpha(z)}\,\eta_{\mathcal{B}}(\alpha z)\mu_\alpha(z) \qquad (\alpha \in G'),$$

$$(1.21) \qquad \nu(\alpha)\eta_{\mathcal{H}}(z) = \overline{{}^t\mu_\alpha(z)}\,\eta_{\mathcal{H}}(\alpha z)\mu_\alpha(z) \qquad (\alpha \in G_{\mathbf{R}}),$$

$$(1.22) \qquad \nu(\alpha)\eta_{\mathcal{B}}(z) = \overline{{}^t\mu_\alpha(z)}\,\eta_{\mathcal{H}}(\alpha z)\mu_\alpha(z) \qquad (\alpha \in G_{\mathbf{R}}\beta_0).$$

These formulas are valid as long as αz is meaningful. To prove (1.20), we observe that if $\alpha \in G'$, then

$$(1.23) \qquad \alpha\begin{pmatrix} 1_m & z \\ \bar{z} & 1_m \end{pmatrix} = \begin{pmatrix} 1_m & \alpha z \\ \overline{\alpha z} & 1_m \end{pmatrix}\begin{pmatrix} \overline{\mu_\alpha(z)} & 0 \\ 0 & \mu_\alpha(z) \end{pmatrix},$$

$$(1.24) \qquad \begin{pmatrix} 1_m & \bar{z} \\ z & 1_m \end{pmatrix} J \begin{pmatrix} 1_m & z \\ \bar{z} & 1_m \end{pmatrix} = \begin{pmatrix} 0 & -\eta_{\mathcal{B}}(z) \\ \overline{\eta_{\mathcal{B}}(z)} & 0 \end{pmatrix}.$$

Call X the product of (1.23) and form ${}^tXJX$. Then we obtain (1.20) by virtue of (1.24); (1.21) follows immediately from (1.9). To prove (1.22), observe that an element γ of $G_{\mathbf{R}} \beta_0$ has the form

$$(1.25) \qquad \gamma = \begin{pmatrix} i\bar{p} & p \\ i\bar{q} & q \end{pmatrix} \qquad (p, q \in M_m(\mathbf{C})).$$

If $\gamma z = w$ with $z \in \mathcal{B}$, then

$$(1.26) \qquad \gamma \begin{pmatrix} 1_m & z \\ \bar{z} & 1_m \end{pmatrix} = \begin{pmatrix} \bar{w} & w \\ 1_m & 1_m \end{pmatrix} \begin{pmatrix} i\,\overline{\mu_\gamma(z)} & 0 \\ 0 & \mu_\gamma(z) \end{pmatrix}.$$

Forming $'YJY$ with (1.26) as Y, we find (1.22).

2. Differential operators on $\mathcal{B}$ and $\mathcal{K}$. Let us now consider a rational representation

$$(2.1) \qquad \rho : GL_m(\mathbf{C}) \to GL(V)$$

with a complex vector space V of finite dimension. We shall later treat automorphic forms with $\rho(\mu_\alpha(z))$ as the factor of automorphy. For the moment, we consider an arbitrary V-valued meromorphic or C^∞-function f on $\mathcal{B}$ or on $\mathcal{K}$, or even on any domain contained in $\mathcal{S}$, and define a similar function $f|_\rho \alpha$ for $\alpha \in G_{\mathbf{C}}$ by

$$(2.2) \qquad (f|_\rho \alpha)(z) = \rho(\mu_\alpha(z))^{-1} f(\alpha(z)).$$

We shall be principally interested in the following three cases:

$$(2.3) \qquad f \text{ is defined on } \begin{cases} \mathcal{K}, \\ \mathcal{K}, \text{ and } \alpha \in \\ \mathcal{B}, \end{cases} \begin{cases} G_{\mathbf{R}}, \\ G_{\mathbf{R}} \beta_0, \qquad \nu(\alpha) > 0. \\ G', \end{cases}$$

We now view $\mathcal{S}$ as the global complex tangent space of $\mathcal{B}$ or of $\mathcal{K}$, identifying $(u_{ij}) \in \mathcal{S}$ with $\sum_{i,j=1}^m u_{ij}\partial_{ij} = \sum_{i<j} u_{ij}\partial/\partial z_{ij}$. If f is a function with values in V, we define Df to be a function with values in $\mathrm{Hom}(\mathcal{S}, V)$ (that is, the space of all C-linear maps of $\mathcal{S}$ into V) given by

$$(2.4) \qquad (Df)(u) = \sum_{i,j=1}^m u_{ij}\partial_{ij}f = \sum_{i<j} u_{ij}\partial f/\partial z_{ij} \qquad (u \in \mathcal{S}).$$

For $0 < k \in \mathbf{Z}$, we mean by $D^k f$ the effect of k-th iteration of D on f. Thus $D^k f$ has values in

$$(2.5) \qquad \mathrm{Hom}(\mathcal{S}, \mathrm{Hom}(\mathcal{S}, \ldots, \mathrm{Hom}(\mathcal{S}, V))).$$

We can identify this space with the space of all C-multilinear maps

$$\mathcal{S} \times \cdots \times \mathcal{S} \ (k \text{ copies}) \to V,$$

which we denote by $L_k(V)$. Then $D^k f$ can be viewed as a function with values in $L_k(V)$; thus

$$D^k f(z)(u_1, \ldots, u_k) = \left(\sum_{i,j} u_{ij}^k \partial_{ij}\right) \cdots \left(\sum_{i,j} u_{ij}^1 \partial_{ij}\right) f(z)$$

for $u_p = (u_{ij}^p) \in \mathfrak{S}$. We define a representation $\rho \otimes \tau^k$ of $GL_m(\mathbf{C})$ on $L_k(V)$ by

$$(2.6) \qquad \{(\rho \otimes \tau^k)(x)h\}(u_1, \ldots, u_k) = \rho(x)h({}^t x u_1 x, \ldots, {}^t x u_k x)$$

$$(x \in GL_m(\mathbf{C}), h \in L_k(V), u_p \in \mathfrak{S}).$$

We write simply τ for τ^1 and also for $1 \otimes \tau^1$. Then (1.10) implies

$$(2.7) \qquad \nu(\alpha)(Df) \circ \alpha = \tau(\mu_\alpha)D(f \circ \alpha) \qquad (\alpha \in G_{\mathbf{C}}).$$

Let us now define an operator D_ρ by

$$(2.8) \qquad (D_\rho f)(z) = \rho(\eta(z))^{-1} D(\rho(\eta(z))f(z))$$

for a V-valued function f, where $\eta = \eta_{\mathfrak{B}}$ or $\eta = \eta_{\mathcal{K}}$ according as f is defined on $\mathfrak{B}$ or on $\mathcal{K}$; $D_\rho f$ has values in $\mathrm{Hom}(\mathfrak{S}, V) = L_1(V)$.

LEMMA 2.1. $D_\rho(f|_\rho \alpha) = (D_\rho f)|_{\rho \otimes \tau} \alpha$ if $\nu(\alpha) = 1$, *in all three cases described in* (2.3), *where* $\rho \otimes \tau$ *is defined by* (2.6) *with* $k = 1$.

Proof. We prove only the case where $\alpha \in G_{\mathbf{R}} \beta_0$; other cases can be proved in the same way. By (1.22), we have

$$D_\rho(f|_\rho \alpha) = \rho(\eta_{\mathfrak{B}}(z))^{-1} D\big(\rho\big(\eta_{\mathfrak{B}}(z)\mu_\alpha(z)^{-1}\big)f(\alpha z)\big)$$

$$= \rho\big(\eta_{\mathfrak{B}}(z)^{-1}\big)D\big(\rho\big(\overline{{}^t\mu_\alpha(z)}\,\eta_{\mathcal{K}}(\alpha z)\big)f(\alpha z)\big)$$

$$= \rho\big(\eta_{\mathfrak{B}}(z)^{-1} \cdot \overline{{}^t\mu_\alpha(z)}\big)D((\rho(\eta_{\mathcal{K}})f) \circ \alpha)$$

$$= \Big[\rho\big(\mu_\alpha(z)^{-1}\eta_{\mathcal{K}}(\alpha z)^{-1}\big) \otimes \tau(\mu_\alpha(z))^{-1}\Big]D(\rho(\eta_{\mathcal{K}})f) \circ \alpha$$

by virtue of (2.7). The last function can be written $(D_\rho f)|_{\rho \otimes \tau} \alpha$.

Now, for $0 < k \in \mathbf{Z}$, we define an operator $D_\rho^{(k)}$ by

$$(2.9) \qquad D_\rho^{(k)} = D_{\rho \otimes \tau^{k-1}} \cdots D_{\rho \otimes \tau} D_\rho.$$

This sends a V-valued function to an $L_k(V)$-valued function. From the above lemma, we obtain immediately

$$(2.10) \quad D_\rho^{(k)}(f|_\rho \alpha) = \big(D_\rho^{(k)}f\big)|_{\rho \otimes \tau^k} \alpha \text{ if } \nu(\alpha) = 1, \text{ in all three cases of } (2.3).$$

As a representation space of $GL_m(\mathbf{C})$, $L_k(V)$ can be decomposed into various subspaces, from which we obtain various kinds of differential operators as follows. Let $\omega : GL_m(\mathbf{C}) \to GL(W)$ be a rational representation with a complex vector space W; suppose there is a $\mathbf{C}$-linear map φ of $L_k(V)$ into W such that $\varphi \circ (\rho \otimes \tau^k)(x) = \omega(x) \circ \varphi$ for every $x \in GL_m(\mathbf{C})$. If we put $D_\rho^{(k,\varphi)}f = \varphi D_\rho^{(k)}f$, then we obtain

$$(2.11) \qquad D_\rho^{(k,\varphi)}(f|_\rho \alpha) = \left(D_\rho^{(k,\varphi)}f\right)\big|_\omega \alpha \quad \text{if} \quad \nu(\alpha) = 1.$$

We employ this principle to obtain a differential operator which sends V-valued functions to V-valued functions. We first put, for every $h \in L_r(V)$,

$$(2.12) \qquad h^s(u_1, \ldots, u_r) = (r!)^{-1} \sum_\pi h(u_{\pi(1)}, \ldots, u_{\pi(r)}) \qquad (u_i \in \mathfrak{S}),$$

where π runs over all permutations of $\{1, \ldots, r\}$. Next, let $\partial_1', \ldots, \partial_N'$ with $N = m(m+1)/2$ be renamings of N operators $\partial_{11}, \partial_{12}, \ldots, \partial_{mm}$, and put

$$\Delta^k = \sum_{(i)} c_{(i)} \partial_{i_1}' \cdots \partial_{i_r}'$$

with $r = mk$ and $c_{(i)} \in \mathbf{Z}$, where Δ is defined by (1.12). We then define a map Φ_k of $L_{km}(V)$ into V for $0 < k \in \mathbf{Z}$ by

$$(2.13) \qquad \Phi_k(h) = \sum_{(i)} c_{(i)} h^s(\partial_{i_1}', \ldots, \partial_{i_r}') \qquad (h \in L_{km}(V)).$$

For example, $\Phi_1(h)$ can be given by

$$\Phi_1(h) = (m!)^{-1} \sum_{\sigma, \pi} \mathrm{sgn}(\sigma) h(\partial_{\pi(1)\sigma\pi(1)}, \ldots, \partial_{\pi(m)\sigma\pi(m)}),$$

where σ and π run independently over all permutations of $\{1, \ldots, m\}$. If $x = (x_{ij}) \in \mathfrak{S}$, we have ${}^t x \partial_{ij} x = \sum_{p,q} x_{ip} x_{jq} \partial_{pq}$. Therefore, by a straightforward calculation, we can easily verify that

$$(2.14) \quad \Phi_k\big((\rho \otimes \tau^{km})(x)h\big) = \det(x)^{2k} \rho(x) \Phi_k(h) \qquad (x \in GL_m, h \in L_{km}(V)).$$

Thus we view V also as the representation space of $(\det)^{2k}\rho$. We now put

$$(2.15) \qquad \Delta_\rho^{(k)} = \Phi_k D_\rho^{(km)} \qquad (0 < k \in \mathbf{Z}).$$

Notice that $\Delta_\rho^{(k)}f$ is V-valued for a V-valued f. From (2.11), we obtain

$$(2.16) \qquad \Delta_\rho^{(k)}(f|_\rho \alpha) = \left(\Delta_\rho^{(k)}f\right)\big|_\omega \alpha \text{ if } \nu(\alpha) = 1 \text{ } \textit{in all three cases of} \text{ (2.3)},$$

$$\textit{where } \omega(x) = \det(x)^{2k}\rho(x).$$

Our definition of Φ_k implies that

$$(2.17) \qquad \Phi_k D^{km} = \Delta^k$$

with operator Δ of (1.12). In Section 7, we shall give a formula for $\Delta_\rho^{(k)}$ in terms of Δ^k.

LEMMA 2.2. *Let $A_r f = \sum_{i,j} \partial_{ij}(g_{rij} f)$ for $r = 1, \ldots, n$ be differential operators acting on functions f defined in a neighborhood of the origin 0 of $\mathcal{B}$ with functions $g_{rij}(z)$, each of which has an expansion of the type*

$$(2.18) \qquad g(z) = c + \sum_\lambda M_\lambda(z) h_\lambda(\bar{z})$$

at $z = 0$, where c is a constant, $M_\lambda(z)$ is a monomial in z_{ij}, and $h_\lambda(\bar{z})$ is a holomorphic function in $\bar{z}$ such that $h_\lambda(0) = 0$. Then, for every C^∞-function f in a neighborhood of 0, we have

$$(A_n \cdots A_1 f)(0) = \left\{ \prod_{r=1}^{n} \left(\sum_{i,j} g_{rij}(0) \partial_{ij} \right) f \right\}(0).$$

Once stated, this is obvious and no proof is necessary.

Observe that every entry of $\rho(\eta_\mathcal{B}(z))$ is a function of type (2.18) for every rational representation ρ of GL_m. Therefore our lemma applied to $D_\rho^{(km)}$ shows that

$$(2.19) \qquad (\Delta_\rho^{(k)} f)(0) = (\Phi_k D^{km} f)(0) = (\Delta^k f)(0) \text{ if } f \text{ is defined on } \mathcal{B}.$$

LEMMA 2.3. *Put $\lambda_{q,r,\alpha}(z) = \det[\eta_{\mathcal{K}}(z)^{-r} \overline{\mu_\alpha(z)}^r \mu_\alpha(z)^{-r-q}]$ for $z \in \mathcal{K}$, $q \in \mathbf{Z}$, $r \in \mathbf{Z}$, and $\alpha \in G_\mathbf{R}$ with $\nu(\alpha) = 1$. Then*

$$(2.20) \qquad \lambda_{q,r,\alpha} \mid_\rho \beta = \lambda_{q,r,\alpha\beta} \text{ for every } \beta \in G_\mathbf{R} \text{ with } \nu(\beta) = 1;$$

$$(2.21) \qquad \Delta_\rho^{(k)}(\lambda_{q,r,\alpha}) = i^{km} \epsilon_k(-r-q) \lambda_{q,r+k,\alpha},$$

where $\rho(x) = \det(x)^{q+2r}$ and ϵ_k is defined as in (1.15).

Proof. The first relation can easily be derived from (1.21). Similarly, if $\gamma \in G_\mathbf{R} \beta_0$ and $\nu(\gamma) = 1$, we obtain from (1.22) a relation

$$(2.22) \qquad \lambda_{q,r,\alpha} \mid_\rho \gamma = \det\left[\eta_\mathcal{B}^{-r} \bar{\mu}_{\alpha\gamma}^r \mu_{\alpha\gamma}^{-r-q} \right].$$

Given $z \in \mathcal{K}$, we can take γ so that $\gamma(0) = z$. By (2.16), we have

$$\Delta_\rho^{(k)}(\lambda_{q,r,\alpha}) \circ \gamma = j_\gamma^{q+2r+2k} \Delta_\rho^{(k)}(\lambda_{q,r,\alpha} \mid_\rho \gamma).$$

Evaluate this at 0. From (2.22), (2.19), and Lemma 2.2, we obtain

$$(2.23) \qquad \Delta_\rho^{(k)}(\lambda_{q,r,a})(z) = j_\gamma(0)^{q+2r+2k} \, \overline{j_{\alpha\gamma}(0)}^{\,\prime} \Delta^k(j_{\alpha\gamma}^{-r-q})(0).$$

As observed in (1.25), we can put

$$\alpha\gamma = \begin{pmatrix} i\bar{a} & a \\ i\bar{b} & b \end{pmatrix}$$

with $a, b \in M_m(\mathbf{C})$. By (1.15), the right-hand side of (2.23) is

$$\epsilon_k(-r-q) j_\gamma(0)^{q+2r+2k} \det\left(\bar{b}^r(i\bar{b})^k b^{-r-q-k}\right).$$

Now $\eta_{\mathcal{K}}(z) = \overline{{}^t\mu_\gamma(0)}^{-1}\mu_\gamma(0)^{-1}$ by (1.22), and $b = \mu_{\alpha\gamma}(0) = \mu_\alpha(z)\mu_\gamma(0)$. Combining these, we obtain the desired relation.

Another proof of (2.22) will be given in Section 7. If $m = 1$, $\epsilon_k(s) \neq 0$ for $s < 0$. If $m > 1$, however, it can happen that $\epsilon_k(s) = 0$ even when $s < 0$.

So far we have treated only holomorphic differentiation. It is natural to consider also antiholomorphic differentiation $\bar{D}$ defined by

$$(2.24) \qquad (\bar{D}f)(u) = \sum_{i<j} u_{ij}\partial f/\partial \bar{z}_{ij} \qquad (u \in \mathfrak{S}).$$

We understand that $\bar{D}f$ is $L_1(V)$-valued when f is V-valued. Define a representation $\rho \otimes \psi$ of $GL_m(\mathbf{C})$ on $L_1(V)$ by

$$(2.25) \qquad \left[(\rho \otimes \psi)(x)h\right](u) = \rho(x)h(x^{-1}u \cdot {}^tx^{-1})$$

$$(x \in GL_m(\mathbf{C}), h \in L_1(V), u \in \mathfrak{S}),$$

and also an operator E by

$$(2.26) \qquad (Ef)(u) = (\bar{D}f)(\eta u \cdot {}^t\eta) \qquad (u \in \mathfrak{S}),$$

where $\eta = \eta_{\mathfrak{B}}$ or $\eta = \eta_{\mathcal{K}}$ according as f is defined on $\mathfrak{B}$ or on $\mathcal{K}$; Ef is again $L_1(V)$-valued. We can easily verify that

$$(2.27) \qquad E(f|_\rho\alpha) = (Ef)|_{\rho\otimes\psi}\alpha \text{ if } \nu(\alpha) = 1, \textit{ in all three cases of } (2.3).$$

If $m = 1$, we have $Ef = 4y^2\partial f/\partial\bar{z}$ on $\mathcal{K}$, where $y = (z - \bar{z})/(2i)$. Moreover, we can show that

$$(2.28) \qquad \langle f, y^{2-k}(\partial/\partial z)(y^{k-2}g)\rangle = \langle -y^2\partial f/\partial\bar{z}, g\rangle$$

for C^∞-automorphic forms f and g of weight k and $k-2$, respectively, where $\langle,\rangle$ is the Petersson inner product. This relation can be generalized in several ways in the case $m > 1$. We shall not go into details here, since it will not be needed in the present paper. The author hopes to treat it on a future occasion.

3. Arithmeticity at CM-points.

We now consider automorphic forms on the product $\mathcal{H}^r$ of r copies of $\mathcal{H}$ with respect to arithmetic discontinuous groups. Let F be a totally real algebraic number field of degree n, and let $\sigma_1, \ldots, \sigma_n$ be all the injections of F into $\mathbf{R}$. Further let B be a quaternion algebra over F which is unramified at $\sigma_1, \ldots, \sigma_r$ and ramified at $\sigma_{r+1}, \ldots, \sigma_n$; we assume $0 < r \leqslant n$. Define a $\mathbf{Q}$-rational algebraic group $\mathfrak{G}$ such that

$$(3.1) \qquad \mathfrak{G}_\mathbf{Q} = \left\{ \alpha \in GL_m(B) \,|\, {}^\iota\alpha^\iota\alpha = \nu(\alpha)1_m \quad \text{with} \quad \nu(\alpha) \in F \right\},$$

where ι denotes the main involution of B. Then there is an isomorphism

$$(3.2) \qquad \mathfrak{G}_\mathbf{R} \cong G_\mathbf{R}^r \times G^{*n-r}$$

with a group G^* defined by

$$(3.3) \qquad G^* = \left\{ \alpha \in GL_m(\mathbf{H}) \,|\, {}^\iota\alpha^\iota\alpha = \nu(\alpha)1_m \quad \text{with} \quad \nu(\alpha) \in \mathbf{R} \right\},$$

where $\mathbf{H}$ is the division ring of Hamilton quaternions. We can fix an isomorphism of (3.2) in such a way that if $\alpha \in \mathfrak{G}_\mathbf{Q}$ and $(\alpha_1, \ldots, \alpha_n)$ is the corresponding element of $G_\mathbf{R}^r \times G^{*n-r}$, then $\alpha_1, \ldots, \alpha_r$ belong to $M_{2m}(\overline{\mathbf{Q}})$. (See [7, §4.4], [14, §4].) Then we put, for $z = (z_1, \ldots, z_r) \in \mathcal{H}^r$,

$$(3.4) \qquad \mu_i(\alpha, z) = \mu(\alpha_i, z_i) \qquad (i = 1, \ldots, r).$$

Further we consider a $\mathbf{Q}$-linear embedding Ω_i of $M_m(B)$ into $M_{2m}(\overline{\mathbf{Q}})$ such that $\Omega_i(a1_m) = a^{\sigma_i}1_{2m}$ for $i = r+1, \ldots, n$ (which induces an embedding of the i-th factor G^* of (3.2) into $GL_{2m}(\mathbf{C})$).

Our automorphic forms are defined relative to a $\overline{\mathbf{Q}}$-rational representation

$$(3.5) \qquad \zeta : GL_m^r \times GL_{2m}^{n-r} \to GL_d$$

with a positive integer d. For $\alpha \in G_\mathbf{Q}$ with totally positive $\nu(\alpha)$, $z \in \mathcal{H}^r$, and a $\mathbf{C}^d$-valued function f on $\mathcal{H}^r$, we put

$$(3.6) \qquad (f|_\zeta\alpha)(z) = \Lambda_\zeta(\alpha, z)^{-1}f(\alpha(z)) \qquad (z \in \mathcal{H}^r),$$

$$(3.7) \qquad \Lambda_\zeta(\alpha, z) = \zeta(\mu_1(\alpha, z), \ldots, \mu_r(\alpha, z), \Omega_{r+1}(\alpha), \ldots, \Omega_n(\alpha)).$$

We denote by $\mathcal{C}_\zeta$ the set of all $\mathbf{C}^d$-valued meromorphic functions f on $\mathcal{H}^r$ such that $f|_\zeta\gamma = f$ for all γ in a congruence subgroup of $\mathfrak{G}_\mathbf{Q}$ (and which are meromorphic at cusps when $n = 1$ and $B = M_2(\mathbf{Q})$). To each CM-point w of $\mathcal{H}^r$ relative to $\mathfrak{G}$, we can attach a certain element $P_\zeta(w)$ of $GL_d(\mathbf{C})$; we then call an element f of $\mathcal{C}_\zeta$ *arithmetic* if $P_\zeta(w)^{-1}f(w)$ has components in $\overline{\mathbf{Q}}$ for every

CM-point w where f is holomorphic. For the precise definition of $P_\zeta(w)$, the reader is referred to [14, p. 319, p. 325], and also to the next section in the case $B = M_2(\mathbf{Q})$.

Now, for each $i \leqslant r$, we define D_i to be the operator D of (2.4) with respect to the variables on the i-th factor of $\mathcal{K}^r$. Then we put

$$(3.8) \quad D_{i,\zeta}f = \zeta(1, \ldots, 1, \eta_{\mathcal{K}}(z_i), 1, \ldots, 1)^{-1} D_i \{\zeta(1, \ldots, 1, \eta_{\mathcal{K}}(z_i), 1, \ldots, 1)f\}$$
$$(z = (z_1, \ldots, z_r) \in \mathcal{K}^r),$$

where we put $\eta_{\mathcal{K}}(z_i)$ into the i-th position. If we consider f as a function of z_i, this is the same as D_ρ of (2.8) with $\rho(x) = \zeta(1, \ldots, 1, x, 1, \ldots, 1)$. We naturally consider an iterated operator

$$(3.9) \qquad\qquad D_{p,\omega} \cdots D_{k,\theta} D_{j,\eta} D_{i,\zeta},$$

where $\eta = \zeta \otimes \tau_i$, $\theta = \eta \otimes \tau_j$, etc., and

$$(3.10) \quad \tau_i(a_1, \ldots, a_n) = \tau(a_i) \qquad \text{for} \quad (a_1, \ldots, a_n) \in GL_m^r \times GL_{2m}^{n-r}$$

with τ of (2.6). Thus if $\mathcal{S}_i$ denote the global tangent space of the i-th factor of $\mathcal{K}^r$, the representation spaces of η, θ, etc. are $\mathrm{Hom}(\mathcal{S}_i, \mathbf{C}^d), \mathrm{Hom}(\mathcal{S}_j, \mathrm{Hom}(\mathcal{S}_i, \mathbf{C}^d))$, etc. Let e_i be the number of factors $D_{i,\cdot}$ with index i in (3.9). Putting $e = (e_1, \ldots, e_r)$, we denote operator (3.9) by D_ζ^e. Then, from Lemma 2.1 or (2.10), we obtain

$$(3.11) \qquad D_\zeta^e(f|_\zeta \alpha) = (D_\zeta^e f)|_\psi \alpha \qquad \text{for} \quad \alpha \in \mathfrak{G}_\mathbf{R}, \nu(\alpha) = 1,$$

where $\psi = \zeta \otimes \tau_i \otimes \tau_j \otimes \cdots \otimes \tau_p$ which we may also write in the form

$$(3.12) \qquad\qquad \psi = \zeta \otimes \tau_1^{e_1} \otimes \cdots \otimes \tau_r^{e_r}.$$

The representation space of ψ, in which $D_\zeta^e f$ takes values, can be identified with the set of all multilinear maps

$$(3.13) \qquad \underbrace{\mathcal{S}_1 \times \cdots \times \mathcal{S}_1}_{e_1} \times \cdots \times \underbrace{\mathcal{S}_r \times \cdots \times \mathcal{S}_r}_{e_r} \to \mathbf{C}^d.$$

THEOREM 3.1. *Let f be an arithmetic element of $\mathcal{C}_\zeta$, and let D_ζ^e and ψ be defined as above. Then $\pi^{-|e|} P_\psi(w)^{-1}(D_\zeta^e f)(w)$ is $\overline{\mathbf{Q}}$-rational for every CM-point w on $\mathcal{K}^r$ relative to $\mathfrak{G}$ if f is holomorphic at w, where $|e| = e_1 + \cdots + e_r$.*

In other words, the nature of $\pi^{-|e|}(D_\zeta^e f)(w)$ is the same as that of the values of arithmetic elements of $\mathcal{C}_\psi$ at w. We do not give a detailed proof here, since it is almost a word-for-word translation of the proof of [14, Theorem 11.2], which concerns the case of orthogonal groups. We mention only that the following points are crucial:

(i) If f is an arithmetic element of $\mathcal{Q}_0$, where 0 indicates the trivial representation, then $D_{i,0}f$ is arithmetic. (See [15, Theorem 4], [13, Corollary 7.7, p. 587].)

(ii) The existence of an "arithmetic ζ-uniformizer," which is guaranteed by [14, Proposition 4.1] and [15, Theorem 3].

(iii) The problem can be reduced to the origin of $\mathcal{B}^r$ by means of a transformation obtained from an element of $(G_{\mathbf{R}}\,\beta_0)^r$. (We could have formulated everything on $\mathcal{B}^r$ from the beginning.)

(iv) The principle described in Lemma 2.2.

To state a result about operators of type $\Delta_\rho^{(k)}$, let $k = (k_1, \ldots, k_r)$ with $0 \leqslant k_i \in \mathbf{Z}$ and put

$$(3.14) \qquad \Delta_\zeta^k = \Phi_{1,k_1} \cdots \Phi_{r,k_r} D_\zeta^{mk},$$

where $mk = (mk_1, \ldots, mk_r)$ and Φ_{i,k_i} is operator Φ_{k_i} of (2.13) defined relative to $\mathsf{S}_i^{k_i}$. We see easily that Δ_ζ^k can be obtained also as an iteration of r operators of type (2.15). Therefore

$$(3.15) \qquad \Delta_\zeta^k(f|_\zeta\alpha) = (\Delta_\zeta^k f)|_\xi\alpha \qquad \text{for} \quad \alpha \in G_{\mathbf{R}}, \nu(\alpha) = 1$$

with

$$(3.16) \qquad \xi(a) = \det(a_1)^{2k_1} \cdots \det(a_r)^{2k_r}\zeta(a)$$

$$\text{for} \quad a = (a_1, \ldots, a_n) \in GL_m^r \times GL_{2m}^{n-r}.$$

THEOREM 3.2. *Let f be an arithmetic element of $\mathcal{Q}_\zeta$, and let Δ_ζ^k and ξ be defined as above. Then $\pi^{-m|k|}P_\xi(w)^{-1}(\Delta_\zeta^k f)(w)$ is $\overline{\mathbf{Q}}$-rational for every CM-point w on $\mathcal{K}^r$ relative to $\mathcal{G}$ if f is holomorphic at w, where $|k| = k_1 + \cdots + k_r$. In particular, suppose ζ is one-dimensional. Let g be an arithmetic element of $\mathcal{Q}_\xi$ which is holomorphic and nonvanishing at w. Then*

$$\pi^{-m|k|}g(w)^{-1}(\Delta_\zeta^k f)(w) \in \overline{\mathbf{Q}}.$$

Proof. Define ψ by (3.12) with $e = mk$. Let V_ψ and V_ξ be the representation spaces of ψ and ξ respectively, and let $\varphi = \Phi_{1,k_1} \cdots \Phi_{r,k_r}$. Then φ is a ($\overline{\mathbf{Q}}$-rational) linear map of V_ψ into V_ξ such that $\varphi \circ \psi(a) = \xi(a) \circ \varphi$ for $a \in GL_m^r \times GL_{2m}^{n-r}$. Therefore our definition of $P_\psi(w)$ in [13, (4.7)] shows that $\varphi \circ P_\psi(w) = P_\xi(w) \circ \varphi$. Since $\Delta_\zeta^k f = \varphi D_\zeta^{mk} f$, we obtain our first assertion immediately from Theorem 3.1. If ζ is one-dimensional, we can take $g(w) = P_\xi(w)$, and hence obtain the second assertion.

As mentioned in the introduction, these theorems are generalizations of [8, Main Theorem I] and [13, Theorem 7.6]. A result of the same nature in a different formulation was given also in [10, Theorem 4.1].

4. The consistency of two definitions of arithmeticity in the Siegel modular case. If $B = M_2(\mathbf{Q})$, $\mathfrak{G}$ can be identified with group G cf (1.1). In this case, we can also define the arithmeticity of automorphic forms by means of the $\overline{\mathbf{Q}}$-rationality of their Fourier coefficients as treated in [9] and [11]. That this coincides with the arithmeticity in terms of $P_\zeta(w)$ is implicit in [10] and [13]. For the reader's convenience, let us give a detailed proof of this fact here, recalling at the same time the definition of $P_\zeta(w)$ in this case, as we shall need it in the next section.

Let $Y = M_{n_1}(K_1) \oplus \cdots \oplus M_{n_t}(K_t)$ with CM-fields K_i and $0 < n_i \in \mathbf{Z}$, and let δ be a positive involution of Y. Suppose

$$(4.1) \qquad 2m = \sum_{i=1}^{t} n_i [K_i : \mathbf{Q}]$$

and there is a $\mathbf{Q}$-linear embedding h of Y into $M_{2m}(\mathbf{Q})$ such that

$$(4.2) \qquad h(y^\delta) = J \cdot {}^t h(y) J^{-1} \qquad \text{for all} \quad y \in Y.$$

Put

$$(4.3) \qquad Y^u = \{ x \in Y \,|\, x x^\delta = 1 \}.$$

Then $h(Y^u)$ is contained in $G_\mathbf{Q}$ and has a unique common fixed point w on $\mathcal{K}$; it is this type of point which we call a CM-point relative to G. Notice that w has algebraic coordinates. We have obviously

$$(4.4) \qquad h(a)\binom{w}{1_m} = \binom{w}{1_m} \mu(h(a), w) \qquad (a \in Y^u),$$

and therefore we can define an embedding ψ of Y into $M_m(\overline{\mathbf{Q}})$ such that $\psi(a) = \mu(h(a), w)$ for all $a \in Y^u$ and that

$$(4.5) \qquad h(x)\binom{w}{1_m} = \binom{w}{1_m} \psi(x) \qquad (x \in Y).$$

Let $\Lambda(w)$ be the lattice in $\mathbf{C}^m$ generated over $\mathbf{Z}$ by the columns of the "period matrix" $(w, 1_m)$. Take a projective model A of the abelian variety $\mathbf{C}^m/\Lambda(w)$; we hereafter identify A with $\mathbf{C}^m/\Lambda(w)$; we may also assume that A is $\overline{\mathbf{Q}}$-rational, as it has many complex multiplications. In fact, relation (4.5) shows that ${}^t\psi(x)$ defines an element of $\mathrm{End}(A) \otimes \mathbf{Q}$. From (1.21), we obtain

$$(4.6) \qquad \overline{{}^t\psi(x)} \,\eta_{\mathcal{K}}(w) = \eta_{\mathcal{K}}(w)\psi(x^\delta)$$

at least for $x \in Y^u$; this is actually true for all $x \in Y$, since Y is spanned by Y^u over $\mathbf{Q}$ as proved in [7, Proposition 1.6].

Let $K = K_1 \oplus \cdots \oplus K_t$; let J_K denote the set of all nontrivial homomorphisms of K into $\mathbf{C}$, and I_K the free $\mathbf{Z}$-module generated by the elements of J_K. Then we find m elements $\tau_1, \ldots, \tau_m$ of J_K and an element T of $GL_m(\overline{\mathbf{Q}})$ such that

$$(4.7) \qquad T\psi(x)T^{-1} = \mathrm{diag}[x^{\tau_1}, \ldots, x^{\tau_m}] \qquad (x \in K).$$

Moreover, we have $\sum_{i=1}^m \tau_i = \sum_{j=1}^t n_j \beta_j$ with a CM-type β_j of K_j.

Hereafter we speak of the $\mathbf{Q}$- or $\overline{\mathbf{Q}}$-rationality of modular forms exclusively in terms of Fourier coefficients. Given a $\mathbf{Q}$-rational representation $\rho : GL_m \to GL_d$ and a point z_0 of $\mathcal{K}$, we understand by a $\mathbf{Q}$-*rational ρ-uniformizer at z_0* an $M_d(\mathbf{C})$-valued meromorphic function S on $\mathcal{K}$ such that

(i) *every column of S belongs to $\mathcal{Q}_\rho$ and $\mathbf{Q}$-rational*;
(ii) *S is holomorphic at z_0 and $\det(S(z_0)) \neq 0$.*

If ϵ is the identity representation $GL_m \to GL_m$, the existence of a $\mathbf{Q}$-rational ϵ-uniformizer is proved in [10, Proposition 1.2]. If S is an ϵ-uniformizer at z_0, then obviously $\rho(S)$ is a ρ-uniformizer at z_0.

Now take a $\mathbf{Q}$-rational ϵ-uniformizer S at the above CM-point w. Let $u_1, \ldots, u_m$ be the coordinate functions on $\mathbf{C}^m$; put

$$(4.8) \qquad \begin{bmatrix} \omega_1 \\ \vdots \\ \omega_m \end{bmatrix} = \pi \cdot {}^t S(w) \begin{bmatrix} du_1 \\ \vdots \\ du_m \end{bmatrix}.$$

By [10, Theorem 1.5], $\omega_1, \ldots, \omega_m$ define $\overline{\mathbf{Q}}$-rational 1-forms on A. Take an element x_0 of K such that $x_0 x_0^\delta = 1$ and $K = \mathbf{Q}[x_0]$ (cf. [7, Proposition 1.5]). Put $S|_\epsilon \alpha = \mu_\alpha(z)^{-1} S(\alpha z)$ with $\alpha = h(x_0)^{-1}$ and $R = S^{-1} \cdot (S|_\epsilon \alpha)$. By [10, Proposition 1.4], each column of $S|_\epsilon \alpha$ is a $\overline{\mathbf{Q}}$-rational element of $\mathcal{Q}_\epsilon$, so that the entries of R are $\overline{\mathbf{Q}}$-rational modular functions with respect to congruence subgroups of $Sp(m, \mathbf{Q})$. Now we have $R(w) = S(w)^{-1} \psi(x_0) S(w)$. Since $R(w)$ is algebraic, we can find an element U of $GL_m(\overline{\mathbf{Q}})$ such that

$$(4.9) \qquad U^{-1} S(w)^{-1} \psi(x_0) S(w) U = \mathrm{diag}[x_0^{\tau_1}, \ldots, x_0^{\tau_m}].$$

From (4.7), we see that

$$(4.10) \qquad TS(w)U \text{ commutes with } \mathrm{diag}[x_0^{\tau_1}, \ldots, x_0^{\tau_m}].$$

Define 1-forms $\xi_1, \ldots, \xi_m$ on A by

$$(4.11) \qquad \begin{bmatrix} \xi_1 \\ \vdots \\ \xi_m \end{bmatrix} = {}^t U \begin{bmatrix} \omega_1 \\ \vdots \\ \omega_m \end{bmatrix} = \pi \cdot {}^t U \cdot {}^t S(w) \begin{bmatrix} du_1 \\ \vdots \\ du_m \end{bmatrix}.$$

Then (4.9) shows that the endomorphism of A represented by $^t\psi(x_0)$ sends ξ_i to $x_0^{\tau_i}\xi_i$. Since ξ_i is $\overline{\mathbf{Q}}$-rational, we have

$$(4.12) \qquad \int_c \xi_i \sim \pi \cdot p_K\left(\tau_i, \sum_{j=1}^{t} \beta_j\right)$$

for every 1-cycle c on A, where p_K is the symbol introduced in [14, (1.1)], and we write $a \sim b$ for two complex numbers a and b if $b \neq 0$ and $a/b \in \overline{\mathbf{Q}}$. This is explained in [10, p. 383] when K is a field; the general case can easily be reduced to such a special case; in fact, A is isogenous to a product $A_1^{n_1} \times \cdots \times A_t^{n_t}$ with abelian varieties A_j belonging to CM-types (K_j, β_j) respectively.

Now (4.11) shows that the period matrix of $\{\xi_1, \ldots, \xi_m\}$ is $\pi \cdot {}^tU \cdot {}^tS(w)(w, 1_m)$. Therefore, if we put

$$(4.13) \qquad \mathbf{P} = \mathrm{diag}[\, p_1, \ldots, p_m\,], \qquad p_i = p_K\left(\tau_i, \sum_{j=1}^{t} \beta_j\right),$$

then $\mathbf{P}^{-1} \cdot {}^tU \cdot {}^tS(w)$ must be algebraic, by virtue of (4.12). Now the symbol $P_\epsilon(w)$ of [14, (4.7)] is given by $P_\epsilon(w) = T^{-1}\mathbf{P}T$. By (4.10), we have

$$(4.14) \qquad P_\epsilon(w) = S(w)UPU^{-1}S(w)^{-1},$$

which shows that $P_\epsilon(w)^{-1}S(w)$ is algebraic.

Suppose now f is a $\overline{\mathbf{Q}}$-rational element of $\mathcal{Q}_\rho$ with a $\mathbf{Q}$-rational representation ρ of GL_m. Given a CM-point w, take S as above. Then the components of $\rho(S)^{-1}f$ are $\overline{\mathbf{Q}}$-rational modular functions, which take algebraic values at w. Therefore $\rho(P_\epsilon(w))^{-1}f(w)$ is algebraic. Since $P_\rho(w) = \rho(P_\epsilon(w))$, this shows that f is arithmetic in the sense of Section 3.

Conversely, suppose $f \in \mathcal{Q}_\rho$ and f is arithmetic in the sense of Section 3. Put $g = \rho(S)^{-1}f$ and observe that g takes algebraic values at every CM-point where f, S, and S^{-1} are holomorphic. Hence the components of g must be $\overline{\mathbf{Q}}$-rational modular functions. Since $f = \rho(S)g$, this shows that f is $\overline{\mathbf{Q}}$-rational.

We conclude this section by introducing a symbol $P_q(w)$ by

$$(4.15) \qquad P_q(w) = \det(\mathbf{P})^q = p_K\left(q\sum_{i=1}^{m}\tau_i, \sum_{j=1}^{t}\beta_j\right) \qquad (q \in \mathbf{Z}).$$

This means that we put $P_q(w) = P_\rho(w)$ if $\rho(x) = \det(x)^q$.

5. Singular values of Eisenstein series. We are going to apply our theorems to a nonholomorphic Eisenstein series of the form

$$(5.1) \qquad E_{q,\rho}(z) = \sum_{\alpha \in R} j_\alpha(z)^{-q}\rho\big(\mu_\alpha(z)^{-1}\overline{\mu_\alpha(z)}(z - \bar{z})^{-1}\big) \qquad (z \in \mathcal{H})$$

defined with a **Q**-rational polynomial representation

$$(5.2) \qquad\qquad \rho : GL_m \to GL_d$$

$(0 < d \in \mathbf{Z})$, and a complete set R of representatives of $\Gamma_\infty \backslash \Gamma$, where $\Gamma = Sp(m, \mathbf{Z})(= GL_{2m}(\mathbf{Z}) \cap G_\mathbf{Q})$ and Γ_∞ is the subgroup of Γ consisting of the elements of the form $\begin{pmatrix} a & b \\ 0 & d \end{pmatrix}$. From (1.21), we obtain

$$(5.3) \qquad \mu_\alpha(z)^{-1}\overline{\mu_\alpha(z)}(z - \bar{z})^{-1} = \mu_\alpha(z)^{-1}(\alpha z - \alpha \bar{z})^{-1} \cdot {}^t\mu_\alpha(z)^{-1}$$

$$(\alpha \in G_\mathbf{R}, \nu(\alpha) = 1).$$

Call this product s and let $y^{1/2}$ denote the positive definite real symmetric matrix whose square is $y = (z - \bar{z})/(2i)$. Then we see that ${}^t s = s$ and $4sy\bar{s}y = 1_m$, and hence

$$(5.4) \qquad\qquad y^{1/2}\mu_\alpha(z)^{-1}\overline{\mu_\alpha(z)}\, y^{-1/2} \text{ is a unitary matrix.}$$

It is well known that $\sum_{\alpha \in R} j_\alpha(z)^{-q}$ is absolutely and locally uniformly convergent on $\mathcal{H}$ if

$$(5.5) \qquad\qquad m + 1 < q \in 2\mathbf{Z}.$$

Therefore (5.4) implies that $E_{q,\rho}$ has the same convergence property under (5.5). From (5.3), we can easily derive that

$$(5.6) \qquad E_{q,\rho}(\gamma z) = j_\gamma(z)^q \rho(\mu_\gamma(z)) E_{q,\rho}(z) \rho({}^t\mu_\gamma(z)) \qquad \text{for every} \quad \gamma \in \Gamma.$$

In other words, $E_{q,\rho}|_\chi \gamma = E_{q,\rho}$ for $\gamma \in \Gamma$, where χ is a representation of GL_m into $GL(M_d(\mathbf{C}))$ defined by

$$(5.7) \qquad \chi(a)v = \det(a)^q \rho(a) v \rho({}^t a) \qquad (a \in GL_m(\mathbf{C}), v \in M_d(\mathbf{C})).$$

THEOREM 5.1.   *Suppose that $m + 1 < q \in 2\mathbf{Z}$ and ρ is obtained as a constituent of $x \mapsto x \otimes \cdots \otimes x$ with k copies of x (that is, the entries of $\rho(x)$ are homogeneous polynomials of degree k in the entries of x). Then, for every CM-point w of $\mathcal{H}$ relative to G, the entries of the matrix*

$$(5.8) \qquad\qquad \pi^{-k}P_q(w)^{-1}P_\rho(w)^{-1}E_{q,\rho}(w)P_{\rho'}(w)$$

are all algebraic provided $q > C_{m,k}$, where $\rho'(x) = \rho({}^t x^{-1}), P_\rho(w)$ is the symbol defined in Section 4, $P_q(w)$ is defined by (4.15), and $C_{m,k}$ is a constant depending only on m and k.

In other words, the nature of $\pi^{-k}E_{q,\rho}(w)$ is the same as that of the values of the arithmetic elements of $\mathcal{Q}_\chi$ at w.

Let us write $E_{q,r}$ for $E_{q,\rho}$ when $\rho(x) = \det(x)^r$ with $0 < r \in \mathbf{Z}$; we have thus

$$(5.9) \qquad E_{q,r}(z) = \det(z - \bar{z})^{-r} \sum_{\alpha \in R} \overline{j_\alpha(z)}^r j_\alpha(z)^{-r-q}.$$

Obviously Theorem 5.1 is applicable to $E_{q,r}$ with $k = rm$. In this case, however, we can drop the condition $q > C_{m,k}$ (so long as $m + 1 < q \in 2\mathbf{Z}$) as follows.

THEOREM 5.2. *For every CM-point w on $\mathcal{K}$, $\pi^{-rm} P_{-q-2r}(w) E_{q,r}(w)$ is algebraic, where $P_*(w)$ is defined by (4.15).*

Proof. As shown by Siegel [17], $E_{q,0}$ has $\mathbf{Q}$-rational Fourier coefficients. Now (2.21) implies that $\Delta_\sigma^{(r)} E_{q,0} = (-1)^{rm} \epsilon_r(-q) E_{q,r}$ if $\sigma(x) = \det(x)^q$. (This type of relation is due to Maass; see [4, p. 308].) Since $\epsilon_r(-q) \neq 0$ if $q > (m-1)/2$, we obtain our assertion from Theorem 3.2.

The idea of the proof of Theorem 5.1 is essentially the same. Namely we apply an operator of type (2.9) to $E_{q,0}$ and derive the desired result from Theorem 3.1. However, this procedure is not so simple. We start with a few elementary formulas:

$$(5.10) \qquad D\left\{ f(z^{-1}) \right\}(u) = -\left\{ (Df)(z^{-1}) \right\}(z^{-1} u z^{-1}) \qquad (u \in \mathfrak{S}),$$

$$(5.11) \qquad D\left\{ \det(z)^s \right\}(u) = s \cdot \det(z)^s \mathrm{tr}(z^{-1} u) \qquad (u \in \mathfrak{S}),$$

$$(5.12) \qquad D\left\{ \det(cz + d)^s \right\}(u) = s \cdot \det(cz + d)^s \mathrm{tr}\left((cz + d)^{-1} cu \right) \qquad (u \in \mathfrak{S}),$$

(5.13)

$$D\left\{ \mathrm{tr}\left((cz + d)^{-1} ch \right) \right\}(u) = -\mathrm{tr}\left((cz + d)^{-1} cu(cz + d)^{-1} ch \right) \qquad (u, h \in \mathfrak{S}),$$

where $cz + d = \mu_\gamma(z)$ for some $\gamma \in G_{\mathbf{C}}$, and h is a constant. Notice that $(cz + d)^{-1} c$ is symmetric. To prove (5.10), put $w = z^{-1}$. Then $0 = d(wz) = dw \cdot z + w \cdot dz$, so that $dw = -z^{-1} dz \cdot z^{-1}$. Put $g(z) = f(z^{-1})$ and $F = Df$ with a holomorphic f. Then $\mathrm{tr}(dz \cdot Dg) = dg = \mathrm{tr}(dw \cdot F(w)) = -\mathrm{tr}(z^{-1} dz \cdot z^{-1} F(z^{-1}))$, which proves (5.10). Next, to prove (5.12), let $x = (x_{ij})$ be a variable matrix of $M_m(\mathbf{C})$, and let $A(x) = \det(x) x^{-1} = (A_{ij}(x))$. Then $(\partial/\partial x_{ij})\det(x) = A_{ji}(x)$. Therefore

$$\partial_{ij} \det(cz + d)^s = s \cdot \det(cz + d)^{s-1} \sum_{p,q} A_{qp}(cz + d) \partial_{ij}(cz + d)_{pq}.$$

Since $2\partial_{ij}(cz + d)_{pq} = c_{pi}\delta_{qj} + c_{pj}\delta_{qi}$, we obtain (5.12), of which (5.11) is a special case. As for (5.13), it is sufficient to prove the case of invertible c. If c is invertible, we have $(cz + d)^{-1} c = (z + c^{-1}d)^{-1}$, and hence (5.13) can be obtained from (5.10) with $f(z) = \mathrm{tr}(zh)$.

Combining (5.12) with (5.13) and putting $\xi = (cz + d)^{-1}c$, we obtain

$$D^2\{\det(cz + d)^s\}(u, v) = \det(cz + d)^s\left[s^2\,\mathrm{tr}(\xi u)\mathrm{tr}(\xi v) - s \cdot \mathrm{tr}(\xi u \xi v)\right]$$

$$(u, v \in \mathfrak{S}).$$

More generally we can show, by induction, that

$$(5.14) \qquad D^k\{\det(cz + d)^s\}(u_1, \ldots, u_k)$$

$$= \det(cz + d)^s \sum_{\lambda=1}^{k} (-1)^{k-\lambda}s^\lambda \sum_{\epsilon} \varphi_\epsilon(\xi; u_1, \ldots, u_k)$$

$$(u_i \in \mathfrak{S}, \xi = (cz + d)^{-1}c).$$

Here ϵ runs over all permutations of $\{1, \ldots, k\}$ which can be expressed as the product of λ cycles (including trivial ones) without common numerals, and φ_ϵ is a function determined by ϵ by the rule illustrated by an example as follows:

$$(5.15) \qquad \varphi_\epsilon(\xi; u_1, \ldots, u_7) = \mathrm{tr}(\xi u_1 \xi u_2)\mathrm{tr}(\xi u_3)\mathrm{tr}(\xi u_4 \xi u_5 \xi u_6)\mathrm{tr}(\xi u_7)$$

if $k = 7$, $\lambda = 4$, and $\epsilon = (12)(3)(456)(7)$. Thus (5.14) consists of $k!$ such functions.

Let us now denote $D_\rho^{(k)}$ and $f|_\rho\alpha$ by $D_q^{(k)}$ and $f|_q\alpha$ when $\rho(x) = \det(x)^q$.

LEMMA 5.3.   *Let $\alpha \in G_\mathbf{R}$ with $\nu(\alpha) = 1$. Then*

$$D_q^{(k)}\big(j_\alpha(z)^{-q}\big)(u_1, \ldots, u_k) = j_\alpha(z)^{-q} \sum_{\lambda=1}^{k} q^\lambda \sum_{\epsilon} \varphi_\epsilon(\zeta; u_1, \ldots, u_k) \qquad (u_i \in \mathfrak{S}),$$

where $\zeta = \mu_\alpha(z)^{-1}\overline{\mu_\alpha(z)}(z - \bar{z})^{-1}$ and $\Sigma_\epsilon\varphi_\epsilon$ is the same as in (5.14).

Proof.   Take $\gamma \in G_\mathbf{R}\beta_0$ with $\nu(\gamma) = 1$. By (2.10), we have

$$(5.16) \qquad D_q^{(k)}(j_\alpha^{-q}) \circ \gamma = \sigma(\mu_\gamma)D_q^{(k)}(j_{\alpha\gamma}^{-q}),$$

where $\sigma(x) = \det(x)^q\tau^k(x)$ with τ^k of (2.6), that is,

$$(5.17) \qquad \sigma(x)F(u_1, \ldots, u_k) = \det(x)^q F({}^txu_1x, \ldots, {}^txu_kx)$$

$$(F \in L_k(\mathbf{C}), x \in GL_m(\mathbf{C}), u_i \in \mathfrak{S}).$$

Evaluate (5.16) at 0 and put $z = \gamma(0)$. By Lemma 2.2, we have

$$D_q^{(k)}\big(j_\alpha(z)^{-q}\big) = \sigma(\mu_\gamma(0))D^k(j_{\alpha\gamma}^{-q})(0).$$

Putting

$$\alpha\gamma = \begin{pmatrix} i\bar{a} & a \\ i\bar{b} & b \end{pmatrix}$$

with a and b in $M_m(\mathbf{C})$ as in the proof of Lemma 2.3, we obtain, from (5.14),

$$D^k(j_{\alpha\gamma}^{-q})(0)(u_1, \ldots, u_k) = \det(b)^{-q} \sum_{\lambda=1}^{k} (-1)^k q^\lambda \sum_\epsilon \varphi_\epsilon(\xi; u_1, \ldots, u_k)$$

with $\xi = ib^{-1}\bar{b}$. By (5.17), we have

$$\sigma(\mu_\gamma(0))D^k(j_{\alpha\gamma}^{-q})(0)(u_1, \ldots, u_k)$$

$$= j_\gamma(0)^q D^k(j_{\alpha\gamma}^{-q})(0)({}^t\mu_\gamma(0)u_1\mu_\gamma(0), \ldots, {}^t\mu_\gamma(0)u_k\mu_\gamma(0)).$$

As seen in the proof of Lemma 2.3, $b = \mu_\alpha(z)\mu_\gamma(0)$ and $i(z - \bar{z})^{-1} = \mu_\gamma(0) \cdot \overline{{}^t\mu_\gamma(0)}$, and hence if $\zeta = \mu_\gamma(z)^{-1}\overline{\mu_\gamma(z)}(z - \bar{z})^{-1}$, we have $-\zeta = \mu_\gamma(0)\xi \cdot {}^t\mu_\gamma(0)$. From expression (5.15), we obtain therefore

$$\varphi_\epsilon(\xi; {}^t\mu_\gamma(0)u_1\mu_\gamma(0), \ldots, {}^t\mu_\gamma(0)u_k\mu_\gamma(0)) = (-1)^k\varphi_\epsilon(\zeta; u_1, \ldots, u_k),$$

which proves our lemma.

To prove Theorem 5.1, it is sufficient to treat the case where $\rho(x) = x \otimes \cdots \otimes x$ with k copies of x. From (5.3), we see that $E_{q,\rho}$ takes values in $\mathcal{S} \otimes \cdots \otimes \mathcal{S}$. We identify this tensor product with $L_k(\mathbf{C})$ by the rule

$$(5.18) \quad (a_1 \otimes \cdots \otimes a_k)(u_1, \ldots, u_k) = \operatorname{tr}(a_1u_1) \cdots \operatorname{tr}(a_ku_k) \qquad (a_i, u_i \in \mathcal{S}).$$

Define a representation σ of GL_m on $L_k(\mathbf{C})$ by (5.17). Then, $E_{q,\rho}$, as a function with values in $L_k(\mathbf{C})$, satisfies $E_{q,\rho}\vert_\sigma \gamma = E_{q,\rho}$ for $\gamma \in \Gamma$. Therefore the algebraicity of (5.8) means that

$$(5.19) \qquad \pi^{-k}\det(\mathbf{P})^{-q}E_{q,\rho}(w)({}^t\mathbf{P}_w^{-1}u_1\mathbf{P}_w^{-1}, \ldots, {}^t\mathbf{P}_w^{-1}u_k\mathbf{P}_w^{-1}) \in \overline{\mathbf{Q}}$$

for every $u_1, \ldots, u_k$ of $\mathcal{S}$ with algebraic entries, where $\mathbf{P}_w = T^{-1}\mathbf{P}T$ with $\mathbf{P}$ of (4.13) and T of (4.7). On the other hand, applying Theorem 3.1 to $E_{q,0}$, we know that

$$(5.20) \qquad \pi^{-k}\det(\mathbf{P})^{-q}(D_q^{(k)}E_{q,0})(w)({}^t\mathbf{P}_w^{-1}u_1\mathbf{P}_w^{-1}, \ldots, {}^t\mathbf{P}_w^{-1}u_k\mathbf{P}_w^{-1}) \in \overline{\mathbf{Q}}$$

for every such algebraic u_i. Thus our task is to derive (5.19) from (5.20). If $k = 1$, this is immediate, since Lemma 5.3 shows that $qE_{q,\rho} = D_q^{(1)}E_{q,0}$. The matter is more complicated, however, if $k > 1$. To settle the problem, define an endomorphism A_ϵ of $L_k(\mathbf{C})$ for each permutation ϵ of $\{1, \ldots, k\}$ as follows. If, for example, $k = 7$ and $\epsilon = (12)(3)(456)(7)$ as in (5.15), we put

$$F_\epsilon(\alpha_1, \ldots, \alpha_7; u_1, \ldots, u_7) = \operatorname{tr}(\alpha_1u_1\alpha_2u_2)\operatorname{tr}(\alpha_3u_3)\operatorname{tr}(\alpha_4u_4\alpha_5u_5\alpha_6u_6)\operatorname{tr}(\alpha_7u_7)$$

$$(\alpha_i, u_i \in \mathcal{S}).$$

Then there is a $\mathbf{C}$-linear map

$$(5.21) \qquad A_\epsilon : \mathcal{S} \otimes \cdots \otimes \mathcal{S} \to L_7(\mathbf{C})$$

such that $A_\epsilon(\alpha_1 \otimes \cdots \otimes \alpha_7)(u_1, \ldots, u_7) = F_\epsilon(\alpha_1, \ldots, \alpha_7; u_1, \ldots, u_7)$. Identifying $S \otimes \cdots \otimes S$ with $L_7(\mathbf{C})$, we view A_ϵ as an endomorphism of $L_7(\mathbf{C})$. Then we have

$$(5.22) \qquad \varphi_\epsilon(\zeta; u_1, \ldots, u_k) = A_\epsilon(\zeta \otimes \cdots \otimes \zeta)(u_1, \ldots, u_k),$$

and therefore, by Lemma 5.3,

$$(5.23) \qquad D_q^{(k)}(j_\alpha^{-q}) = j_\alpha^{-q} B(\zeta \otimes \cdots \otimes \zeta)$$

with $B = \sum_{\lambda=1}^{k} q^\lambda \sum_\epsilon A_\epsilon$ and $\zeta = \mu_\alpha(z)^{-1} \overline{\mu_\alpha(z)}(z - \bar{z})^{-1}$. It follows that $D_q^{(k)}(E_{q,0}) = B(E_{q,\rho})$. Observe that A_1 is the identity map according to (5.18). Therefore B is invertible if $q > C_{m,k}$ with a sufficiently large constant $C_{m,k}$ depending only on m and k. Since B is $\mathbf{Q}$-rational and commutes with $\sigma(x)$ for $x \in GL_m(\mathbf{C})$, we obtain (5.19) by applying B^{-1} to (5.20). This completes the proof of Theorem 5.2.

Remark 5.4.   Let X denote the subspace of $L_k(\mathbf{C})$ consisting of all multilinear functions $F(u_1, \ldots, u_k)$ invariant under all permutations of $u_1, \ldots, u_k$. We observe that operator B maps X into itself, and also that $E_{q,\rho}$ is X-valued if $\rho(x) = x \otimes \cdots \otimes x$. Therefore we need only the invertibility of B on the subspace X for the validity of Theorem 5.1. As seen in the above proof, we can put $C_{m,1} = 0$. It can easily be verified that B is invertible on X for $q > 1/2$ if $k = 2$; hence we can put $C_{m,2} = 0$. To show that B is not invertible on X for some small values of q, apply Φ_1 of (2.13) to (5.23) with $k = m$. Then

$$j_\alpha^{-q} \Phi_1 B(\zeta \otimes \cdots \otimes \zeta) = \Delta_q^{(1)}(j_\alpha^{-q}) = (-1)^m \epsilon_1(-q) \bar{j}_\alpha j_\alpha^{-q-1} \det(z - \bar{z})^{-1}$$

by (2.21). Now $\epsilon_1(-q) = 0$ for $q = h/2$ with $h = 1, 2, \ldots, m - 1$. Observing that Φ_1 does not vanish identically on X, we find that B is not invertible on X for such values of q. Since we have imposed (5.5), this does not imply that condition $q > C_{m,k}$ in Theorem 5.1 is essential. In any case, it is an interesting problem to study the exact bound of the values of q for which B is invertible on X, or for which Theorem 5.1 is valid.

6.   An interpretation of the values of Eisenstein series at CM-points.   If $m = 1$, the values $E_{q,r}(w)$ are closely related to the values of L-functions of imaginary quadratic fields with certain Hecke characters. The same is true for the generalization of $E_{q,r}$ in the Hilbert modular case (see [8]). Therefore it is natural to ask whether such an interpretation exists in the present case with $m > 1$. The answer seems negative except in some special cases, if we consider only the known Hecke L-functions. However, we can express the value $E_{q,r}(w)$ as an infinite series obtained from a CM-type with no reference to $\mathfrak{K}$ nor to Γ, which will suggest a new type of zeta function attached to a CM-type. This is what we shall do in this section.

For some technical reasons, we consider K^n a vector space of n-dimensional *row* vectors with components in K, for every algebraic number field K (including $\mathbf{Q}$). Let X be a vector space over $\mathbf{Q}$ of dimension $2m$ and $\omega : X \times X \to \mathbf{Q}$ be a nondegenerate alternating form. Fix a **Z**-lattice Λ in X which is self-dual with respect to ω in the sense that

$$(6.1) \qquad \Lambda = \{x \in X \mid \omega(x, \Lambda) \subset \mathbf{Z}\}.$$

We call a submodule L of Λ *maximally ω-isotropic* if L has rank m and $\omega(x, y) = 0$ for all $x, y \in L$; also we call L *primitive* if Λ / L has no torsion.

LEMMA 6.1. *For every $\gamma \in \Gamma = \mathrm{Sp}(m, \mathbf{Z})$, let L_γ denote the submodule of $\mathbf{Z}^{2m}$ generated by the last m rows. Then $\gamma \mapsto L_\gamma$ gives a one-to-one correspondence between $\Gamma_\infty \backslash \Gamma$ and the set of all maximally J-isotropic primitive submodules of $\mathbf{Z}^{2m}$, where we put $J(x, y) = xJ \cdot {}^t y$ for $x, y \in \mathbf{Q}^{2m}$.*

We omit the proof which is completely elementary.

Now take CM-fields $K_1, \ldots, K_t$ and put

$$(6.2) \qquad X = K_1^{n_1} \oplus \cdots \oplus K_t^{n_t},$$

$$(6.3) \qquad Y = M_{n_1}(K_1) \oplus \cdots \oplus M_{n_t}(K_t).$$

We view X as a right Y-module by letting $M_{n_i}(K_i)$ act on $K_i^{n_i}$ by right multiplication. Take a positive involution δ of Y; naturally δ gives the complex conjugation on the center K_i of $M_{n_i}(K_i)$.

LEMMA 6.2. *Let $\omega : X \times X \to \mathbf{Q}$ be a $\mathbf{Q}$-bilinear form such that*

$$(6.4) \qquad \omega(x, y) = -\omega(y, x),$$

$$(6.5) \qquad \omega(xa, y) = \omega(x, ya^\delta) \qquad \textit{for every} \quad a \in Y.$$

Then there is an element $\zeta = (\zeta_1, \ldots, \zeta_t)$ of Y with $\zeta_i \in M_{n_i}(K_i)$ such that

$$(6.6) \qquad {}^t\bar\zeta_i = -\zeta_i, \quad x_i^\delta \zeta_i = \zeta_i \cdot {}^t\bar x_i \qquad \textit{for every} \quad x_i \in M_{n_i}(K_i),$$

$$(6.7) \qquad \begin{aligned} \omega(x, y) &= \sum_{i=1}^t \mathrm{Tr}_{K_i/\mathbf{Q}}(x_i \zeta_i \cdot {}^t\bar y_i) \\ \textit{for} \quad x &= (x_1, \ldots, x_t), \, y = (y_1, \ldots, y_t) \textit{ with } x_i, y_i \in M_{n_i}(K_i). \end{aligned}$$

Conversely, if we define ω by (6.7) with ζ satisfying (6.6), then ω satisfies (6.4) and (6.5).

We again omit the proof, as it is an easy exercise. Obviously ω is nondegenerate if and only if ζ is invertible.

Suppose $2m = \sum_{i=1}^t n_i[K_i : \mathbf{Q}]$ and take a $\mathbf{Q}$-linear embedding h of Y into $M_{2m}(\mathbf{Q})$ satisfying (4.2). Then there is a $\mathbf{Q}$-linear isomorphism θ of X onto $\mathbf{Q}^{2m}$ such that $\theta(xa) = \theta(x)h(a)$ for $a \in Y$. Put $\omega(x, y) = \theta(x)J \cdot {}^t\theta(y)$. Then (4.2) implies (6.5), so that we find ζ satisfying (6.6) and (6.7). Conversely, given an invertible $\zeta = (\zeta_1, \ldots, \zeta_t)$ of Y satisfying (6.6), define ω by (6.7). Since ω is a nondegenerate alternating form on X, we can find a $\mathbf{Q}$-linear isomorphism θ of X onto $\mathbf{Q}^{2m}$ such that $\omega(x, y) = \theta(x)J \cdot {}^t\theta(y)$. Then we can define a map h of Y into $M_{2m}(\mathbf{Q})$ by $\theta(xa) = \theta(x)h(a)$, which satisfies (4.2).

With such h and θ, we now consider the fixed point w of $h(Y^u)$ on $\mathcal{K}$ as in Section 4, and define ψ by (4.5). We view X as a subalgebra of Y by identifying K_i^n with the diagonal elements of $M_{n_i}(K_i)$ in a natural way. Then we can find m homomorphisms $\sigma_1, \ldots, \sigma_m$ of X into $\overline{\mathbf{Q}}$ and an element T of $GL_m(\overline{\mathbf{Q}})$ such that

$$(6.8) \qquad T\psi(x)T^{-1} = \operatorname{diag}[x^{\sigma_1}, \ldots, x^{\sigma_m}] \qquad (x \in X),$$

which is essentially the same as (4.7). In fact, decomposing ψ into irreducible representations of Y and restricting them to X, we find that $\sigma_1, \ldots, \sigma_m$ are determined by $\tau_1, \ldots, \tau_m$ of (4.7) in an obvious way, and vice versa.

Put $v = (v_1, \ldots, v_m) = \theta(1)\left(\begin{smallmatrix} w \\ 1_m \end{smallmatrix}\right)T^{-1}$. Then, for every $a \in X$, we have

$$\theta(a)\begin{pmatrix} w \\ 1_m \end{pmatrix}T^{-1} = \theta(1)h(a)\begin{pmatrix} w \\ 1_m \end{pmatrix}T^{-1} = (v_1 a^{\sigma_1}, \ldots, v_m a^{\sigma_m}).$$

Put $\Lambda = \theta^{-1}(\mathbf{Z}^{2m})$. Let L be a maximally ω-isotropic primitive submodule of Λ, and $\{a_1, \ldots, a_m\}$ a $\mathbf{Z}$-basis of L. By Lemma 6.1, there is an element γ of Γ whose last m rows are $\theta(a_1), \ldots, \theta(a_m)$. Putting $V = \operatorname{diag}[v_1, \ldots, v_m]$, we see that $\mu_\gamma(w) = (a_i^{\sigma_j})VT$. Now (4.6) implies that $\overline{T}(w - \overline{w})^{-1} \cdot {}^t T$ commutes with $\operatorname{diag}[x^{\tau_1}, \ldots, x^{\tau_m}]$ for all $x \in K$, and hence with $\mathbf{P}$ of (4.13). If ρ and ρ' are as in Theorem 5.1, then

$$P_\rho(w)^{-1}\rho\Big(\mu_\gamma(w)^{-1}\overline{\mu_\gamma(w)}(w - \overline{w})^{-1}\Big)P_{\rho'}(w)$$

$$= \rho\Big(T^{-1}\mathbf{P}^{-1}T T^{-1}V^{-1}(a_i^{\sigma_j})^{-1}\overline{(a_i^{\sigma_j})}\,\overline{V}\overline{T}(w - \overline{w})^{-1} \cdot {}^t T \mathbf{P}^{-1} \cdot {}^t T^{-1}\Big)$$

$$= \rho\Big(T^{-1}V^{-1}\mathbf{P}^{-1}(a_i^{\sigma_j})^{-1}\overline{(a_i^{\sigma_j})}\,\mathbf{P}^{-1}\overline{V}\overline{T}(w - \overline{w})^{-1}\Big).$$

We now define a quantity Z depending on X, $\{\sigma_i\}$, ω, Λ, q, and ρ by

$$(6.9) \qquad Z(X, \{\sigma_i\}, \omega, \Lambda, q, \rho) = \sum_L \det(a_i^{\sigma_j})^{-q}\rho\Big((a_i^{\sigma_j})^{-1}\overline{(a_i^{\sigma_j})}\Big),$$

where L runs over all maximally ω-isotropic primitive submodules of Λ; for each L, we choose a $\mathbf{Z}$-basis $\{a_1, \ldots, a_m\}$ of L, by which we form a matrix $(a_i^{\sigma_j})$ of size m. Observe that each term depends only on L and is independent of the choice of $\{a_i\}$. The above computation shows that

$$\rho(\mathbf{P})^{-1}Z\rho(\mathbf{P})^{-1} = \rho(VT)P_\rho(w)^{-1}E_{q,\rho}(w)P_{\rho'}(w)\rho\Big((w - \overline{w})\overline{T}^{-1}\overline{V}^{-1}\Big).$$

Since V, T, and $w - \overline{w}$ are algebraic, Theorem 5.1 implies that the entries of $\pi^{-k}\det(\mathbf{P})^{-q}\rho(\mathbf{P})^{-1}Z\rho(\mathbf{P})^{-1}$ are all algebraic. Before stating this fact as a theorem, let us examine the latitude we are allowed to have for the objects X, $\{\sigma_i\}$, ω, Λ.

First X is a direct sum of *CM*-fields as in (6.2); we put $2m = [X:\mathbf{Q}]$; $\sigma_1, \ldots, \sigma_m$ are m homomorphisms of X into $\overline{\mathbf{Q}}$; ω is defined by (6.7) with an invertible element $\zeta = (\zeta_1, \ldots, \zeta_t)$ of Y such that ${}^t\overline{\zeta}_i = -\zeta_i$ and that

$$(x_1, \ldots, x_t) \mapsto \left(\zeta_1^{-1} \cdot {}^t\overline{x}_1\zeta_1, \ldots, \zeta_t^{-1} \cdot {}^t\overline{x}_t\zeta_t\right)$$

defines a positive involution of Y; Λ is a $\mathbf{Z}$-lattice in X satisfying (6.1). Now we have to impose a condition on ζ to insure that $\{\sigma_i\}$ can be obtained as in (6.8). To state it, we take a *CM*-type $\sum_{j=1}^{g} \tau_{ij}$ of K_i for each i (which corresponds to β_i of Section 4), where $2g_i = [K_i:\mathbf{Q}]$. Write an element x of X in the form

$$x = (x_{11}, \ldots, x_{1n_1}, \ldots, x_{t1}, \ldots, x_{tn_t})$$

with $x_{ik} \in K_i$ for $1 \leqslant k \leqslant n_i$. Define σ_{ijk} to be the map $x \mapsto x_{ik}^{\tau_{ij}}$ for $1 \leqslant i \leqslant t$, $1 \leqslant j \leqslant g_i$, $1 \leqslant k \leqslant n_i$. Then $\{\sigma_{ijk}\}$ can be taken as $\{\sigma_1, \ldots, \sigma_m\}$ as a whole. Now the condition we need is

$$(6.10) \qquad -\sqrt{-1}\, \zeta_i^{\tau_{ij}} \text{ is positive definite for } 1 \leqslant i \leqslant t,\ 1 \leqslant j \leqslant g_i.$$

That this is necessary and sufficient for $\{\sigma_i\}$ to be obtained from the fixed point w as in (6.8) follows from the fact that ω must correspond to a Riemann form on an abelian variety. We are now ready to state our result as follows:

THEOREM 6.1. *Define Z by (6.9) with X, $\{\sigma_i\}$, ω, Λ as above and with a $\mathbf{Q}$-rational polynomial representation ρ of GL_m as in Theorem 5.1 and an integer q such that $\mathrm{Max}(C_{m,k}, m+1) < q \in 2\mathbf{Z}$, where $C_{m,k}$ is the constant of Theorem 5.1. Then $\pi^{-k} \det \cdot (\mathbf{P})^{-q}\rho(\mathbf{P})^{-1}Z\rho(\mathbf{P})^{-1}$ is algebraic, where $\mathbf{P}$ is the diagonal matrix of size m with $p_{K_i}(\tau_{ij}, \sum_{j=1}^{g}\tau_{ij})$ as the ν-th diagonal element if $\sigma_\nu = \sigma_{ijk}$ in the above described sense. In particular, if $\rho(x) = \det(x)^r$ with $0 \leqslant r \in \mathbf{Z}$, then $\pi^{-rm} \det \cdot (\mathbf{P})^{-q-2r}Z$ is algebraic for every q such that $m + 1 < q \in 2\mathbf{Z}$.*

If we define a series

$$\sum_L \left\{ \det \overline{(a_i^{q_j})} \big/ \det(a_i^{q_j}) \right\}^e \left| \det(a_i^{q_j}) \right|^{-2s} \qquad (s \in \mathbf{C})$$

with $e \in \mathbf{Z}$, we can consider Z (with $\rho(x) = \det(x)^r$) as its value at $s = q/2$ for $m + 1 < q \leqslant 2e$. This series seems to have no connection with known L-functions of number fields if $m > 2$.

7. A direct definition of $\Delta_\rho^{(k)}$.

To obtain operator $\Delta_\rho^{(k)}$ in Section 2, we had to iterate operators D_ρ and then take another operator Φ_k. Let us now show that $\Delta_\rho^{(k)}$ can be obtained more directly and more explicitly by considering only V-valued functions, and also that they are essentially the operators considered by Maass when the functions are scalar-valued.

Throughout this section we put $p = (m + 1)/2$. For $r \in \mathbf{R}$, $\alpha \in G_{\mathbf{C}}$, and a function f on a subset of $\mathfrak{S}$, define another function $f|_r\alpha$ by

$$(7.1) \qquad (f|_r\alpha)(z) = j_\alpha(z)^{-r} f(\alpha z).$$

If $r \in \mathbf{Z}$, this is a special case of (2.2); but if $r \notin \mathbf{Z}$, a branch of j_α^{-r} must be chosen, which we shall do whenever such necessity occurs.

LEMMA 7.1. $(\Delta^k f)|_{p+k}\alpha = \Delta^k(f|_{p-k}\alpha)$ for $\alpha \in G_\mathbf{C}$, $\nu(\alpha) = 1$, $0 \leqslant k \in \mathbf{Z}$, where we understand that $j_\alpha^{-p-k} = j_\alpha^{-2k} \cdot j_\alpha^{k-p}$.

Proof. This is proved in Maass [4] when $\alpha \in G_\mathbf{R}$ and $k = 1$. (Cf. also Resnikoff [5, Lemma 1].) It should be noted, however, that the case $k > 1$ cannot be obtained by iteration, even when $m = 1$. We first note that $G_\mathbf{C}$ is generated by J and the elements of the forms $\binom{a\ b}{0\ d}$. The formula is easy to verify for the last type of elements. In view of the purely algebraic nature of the formula, it is therefore sufficient to prove it for $\alpha = J$ or more generally for $\alpha \in G_\mathbf{R}$, $z \in \mathfrak{K}$, and for sufficiently many holomorphic functions f on $\mathfrak{K}$. For this purpose we recall the Cauchy formula

$$(7.2) \qquad f(z) = c \int_S \det(z - u)^{-p} f(u)\, du,$$

where $S = \mathfrak{S} \cap M_n(\mathbf{R})$ and $du = \prod_{i<j} du_{ij}$. This is valid with a suitable constant c for a holomorphic function f which is square integrable on $\mathfrak{K}$ (see Bochner [1, Theorem 2]). First assume that p is an integer. Applying Δ^k to (7.2) and substituting αz and αu for z and u, we obtain, from (1.13),

$$(\Delta^k f) \circ \alpha = cc' \int_S \det(\alpha z - \alpha u)^{-p-k} f(\alpha u)\, d(\alpha u)$$

with a constant c'. By (1.9) and (1.10), this is equal to

$$cc' j_\alpha(z)^{p+k} \int_S \det(z - u)^{-p-k}(f|_{p-k}\alpha)(u)\, du = j_\alpha(z)^{p+k}\Delta^k(f|_{p-k}\alpha),$$

again by means of (7.2), which proves the desired formula. Much care should be taken if $p \notin \mathbf{Z}$ (i.e., if m is even). Take a nontrivial two-sheet covering G^* of $\mathrm{Sp}(m, \mathbf{R})$, and let $h : G^* \to \mathrm{Sp}(m, \mathbf{R})$ be the covering map. Then we can define a factor of automorphy $j'(\beta, z)$ for $\beta \in G^*$ so that $j'(\beta, z)^2 = j(h(\beta), z)$. The action of β on $\mathfrak{K}$ is defined to be the same as that of $h(\beta)$. Now choosing a branch of $\det(z - u)^{1/2}$ (which is well defined), we have

$$\det(z - u)^{1/2} = t(\beta) j'(\beta, z) j'(\beta, u)\det(\beta z - \beta u)^{1/2} \qquad (\beta \in G^*)$$

with $t(\beta)^2 = 1$, where $j'(\beta, u)$ is the boundary value of $j'(\beta, z)$. We see easily that t defines a continuous character of G^*, which must be trivial, as G^* is connected. The above proof with $p \in \mathbf{Z}$ can now be repeated with $p \notin \mathbf{Z}$ and with β in place of α. This completes the proof.

PROPOSITION 7.2.   *Define* $\Delta_\rho^{(k)}$ *by* (2.15) *with a rational representation* ρ *as in* (2.1). *Then, for a V-valued function f defined on $\mathfrak{B}$ or on $\mathfrak{K}$, we have*

$$(\Delta_\rho^{(k)} f)(z) = \rho(\eta(z))^{-1}\delta(z)^{p-k}\Delta^k\big[\rho(\eta(z))\delta(z)^{k-p} f(z)\big],$$

where $p = (m + 1)/2$, $\delta(z) = \det(\eta(z))$, and $\eta = \eta_{\mathfrak{B}}$ or $\eta = \eta_{\mathfrak{K}}$ according as f is defined on $\mathfrak{B}$ or on $\mathfrak{K}$.

Proof. Denote by A the operator on the right-hand side of the desired equality. Then we have

$$(7.3) \qquad A(f|_\rho \alpha) = (Af)|_\omega \alpha \text{ if } \nu(\alpha) = 1 \text{ in all three cases of } (2.3),$$

where $\omega(x) = \det(x)^{2k}\rho(x)$. This can be derived from Lemma 7.1 in exactly the same fashion as in the proof of Lemma 2.1, by means of (1.20), (1.21), and (1.22). By Lemma 2.2, we have

$$(7.4) \qquad (Ag)(0) = (\Delta^k g)(0) \text{ for every function } g \text{ on } \mathfrak{B}.$$

Now, given a point z of $\mathfrak{K}$, take $\alpha \in G_{\mathbf{R}}\beta_0$ so that $z = \alpha(0)$ and $\nu(\alpha) = 1$. Suppose f is defined on $\mathfrak{K}$. Then (7.3) evaluated at 0, combined with (7.4), shows that

$$(Af)(z) = \omega(\mu_\alpha(0))A(f|_\rho \alpha)(0) = \omega(\mu_\alpha(0))\Delta^k(f|_\rho \alpha)(0).$$

By (2.16) and (2.19), we have the same result with $\Delta_\rho^{(k)}$ in place of A and hence $Af = \Delta_\rho^{(k)}f$ for functions f on $\mathfrak{K}$. The same reasoning applies to functions on $\mathfrak{B}$ if we take α from G' instead of $G_{\mathbf{R}}\beta_0$.

PROPOSITION 7.3. $\Delta_\omega^{(l)}\Delta_\rho^{(k)} = \Delta_\rho^{(l+k)}$ *if* $\omega(x) = \det(x)^{2k}\rho(x)$.

Proof. If Y denotes the operator on either side of the equality, we have $Y(f|_\rho \alpha) = (Yf)|_\psi \alpha$, where $\psi(x) = \det(x)^{2k+2l}\rho(x)$, and also, by Lemma 2.2, $(Yg)(0) = (\Delta^{k+l}g)(0)$ for every function g on $\mathfrak{B}$. Therefore the argument in the above proof proves our equality.

We now give the second proof of (2.21). Put $\delta = \det(\eta_{\mathfrak{K}})$ and substitute $\delta^{-r} = \lambda_{q,r,1}$ for f in (2.16) with $\rho(x) = \det(x)^{q+2r}$. Then

$$\Delta_\rho^{(k)}(\lambda_{q,r,\alpha}) = \Delta_\rho^{(k)}(\lambda_{q,r,1}|_\rho \alpha) = (\Delta_\rho^{(k)}\delta^{-r})|_\omega \alpha.$$

By Proposition 7.2 and (1.15), we have

$$\Delta_\rho^{(k)}\delta^{-r} = \delta^{p-k-q-2r}\Delta^k\delta^{q+k-p+r} = (-i)^{km}\epsilon_k(q + k - p + r)\delta^{-k-r}$$

$$= (-i)^{km}\epsilon_k(q + k - p + r)\lambda_{q,r+k,1},$$

and hence $\Delta_\rho^{(k)}(\lambda_{q,r,\alpha}) = (-i)^{km}\epsilon_k(q + k - p + r)\lambda_{q,r+k,\alpha}$, which proves (2.21), since one can easily verify that $\epsilon_k(-b) = (-1)^{km}\epsilon_k(b + k - p)$.

Some differential operators of the above type were introduced by Maass in [3] and [4]; they include the following two:

$$K_\alpha f = \alpha f 1_m + (z - \bar{z})Df,$$

$$M_\alpha f = \det(z - \bar{z})^{p-\alpha}\Delta\big(\det(z - \bar{z})^{\alpha+1-p}f\big) \qquad (\alpha \in \mathbf{R}).$$

(As for M_α, see [4, p. 317]; it was originally defined in [3] in a different way.) If we identify $L_1(\mathbf{C})$ with $\mathfrak{S}$ by (5.18) with $k = 1$, and if $\rho(x) = \det(x)^\alpha$, then $K_\alpha = (z - \bar{z})D_\rho$ and $M_\alpha = \det(z - \bar{z})\Delta_\rho^{(1)}$.

8. Differential operators on other symmetric domains. In addition to the space $\mathfrak{B}$ of Section 1 and the so-called exceptional ones, there are exactly three more classes of irreducible bounded symmetric domains, which are the quotients of the groups

$$G_{n,m}^{(1)} = SU(n, m)$$

$$= \left\{ \alpha \in SL_{n+m}(\mathbf{C}) \mid {}^t\bar{\alpha} I_{n,m} \alpha = I_{n,m} \right\}, \qquad I_{n,m} = \mathrm{diag}\left[1_n, -1_m \right],$$

$$G_m^{(2)} = \left\{ \alpha \in G_{m,m}^{(1)} \mid \alpha J_m = J_m \bar{\alpha} \right\},$$

$$G_m^{(3)} = \text{the identity component of}$$
$$\left\{ \alpha \in G_{m,2}^{(1)} \mid {}^t\alpha R \alpha = R \right\}, \qquad R = \mathrm{diag}\left[1_m, -\begin{pmatrix} 0 & 1 \\ 1 & 0 \end{pmatrix} \right].$$

The last group is isomorphic to $SO(m, 2)$; it is taken in the present form simply because our previous investigation [14] was made in that form. Let us now briefly describe the analogues of D_ρ and $\Delta_\rho^{(k)}$ as well as their properties in these three cases, hereafter referred to as Cases I, II, III, respectively.

Let $S_{n,m}^{(1)}$ denote the set of all $n \times m$ complex matrices, and let

$$S_m^{(2)} = \left\{ u \in S_{m,m}^{(1)} \mid {}^t u = -u \right\}, \qquad S_m^{(3)} = S_{m,1}^{(1)}.$$

Further define domains $B_{n,m}^{(1)}$ and $B_m^{(2)}$ by

$$B_{n,m}^{(1)} = \left\{ z \in S_{n,m}^{(1)} \mid 1_m - {}^t\bar{z}z > 0 \right\},$$

$$B_m^{(2)} = B_{m,m}^{(1)} \cap S_m^{(2)}.$$

We can let $G_{n,m}^{(1)}$ and $G_m^{(2)}$ act on $B_{n,m}^{(1)}$ and $B_m^{(2)}$ by the rule $\alpha(z) = (az + b)(cz + d)^{-1}$ for

$$\alpha = \begin{pmatrix} a & b \\ c & d \end{pmatrix} \in G_{n,m}^{(1)}$$

with $a \in M_n(\mathbf{C})$; moreover, two holomorphic factors of automorphy λ_α and μ_α can be defined by

$$\alpha \begin{bmatrix} 1_n & z \\ {}^t\bar{z} & 1_m \end{bmatrix} = \begin{bmatrix} 1_n & \alpha(z) \\ {}^t\alpha(z) & 1_m \end{bmatrix} \begin{bmatrix} \overline{\lambda_\alpha(z)} & 0 \\ 0 & \mu_\alpha(z) \end{bmatrix} \qquad (\alpha \in G_{n,m}^{(1)}, z \in B_{n,m}^{(1)}).$$

(See [6], [13, §3].) We have $\lambda_\alpha(z) = \mu_\alpha(z)$ if $\alpha \in G_m^{(2)}$ and $z \in B_m^{(2)}$. Put

$$\xi(z) = 1_n - \bar{z} \cdot {}^t z, \qquad \eta(z) = 1_m - {}^t\bar{z}z \qquad (z \in B_{n,m}^{(1)}).$$

As for $G_m^{(3)}$, we define $B_m^{(3)}$ to be the domain $\mathfrak{Z}_m$ of [14, (2.6)], and define also factors of automorphy λ_α, μ_α and functions ξ, η on $B_m^{(3)}$ by [14, (2.8), (2.12), (2.13), (2.21)]. (We note here a correction to [14]: $D_{-1}\eta$ on line 6 from bottom of page 321 should be $D_{-1}p$.) From now on we denote these domains and groups simply by $S^{(i)}$, $B^{(i)}$, and $G^{(i)}$, dropping the subscripts, if there is no fear of confusion. We have then

$$(8.1) \qquad \xi(z) = \overline{{}^t\lambda_\alpha(z)}\,\xi(\alpha(z))\lambda_\alpha(z) \qquad (z \in B^{(i)}, \alpha \in G^{(i)}, i = 1,3),$$

$$(8.2) \qquad \eta(z) = \overline{{}^t\mu_\alpha(z)}\,\eta(\alpha(z))\mu_\alpha(z) \qquad (z \in B^{(i)}, \alpha \in G^{(i)}, i = 1,2,3).$$

(See [13, (3.8)], [14, (2.14), (2.23)].) Now we view $S^{(i)}$ as the global complex tangent space of $B^{(i)}$. Given a rational representation

$$(8.3) \qquad \rho : \begin{cases} GL_n(\mathbf{C}) \times GL_m(\mathbf{C}) & \text{(Case I)} \\ GL_m(\mathbf{C}) & \\ GL_m(\mathbf{C}) \times GL_1(\mathbf{C}) \end{cases} \to GL(V) \qquad \begin{matrix} \text{(Case I)} \\ \text{(Case II)} \\ \text{(Case III)} \end{matrix}$$

with a finite-dimensional complex vector space V and a V-valued function f on $B^{(i)}$, we define $f|_\rho\alpha$ for $\alpha \in G^{(i)}$ and $D_\rho f$ by

$$f|_\rho\alpha = \begin{cases} \rho(\lambda_\alpha, \mu_\alpha)^{-1} f \circ \alpha & \text{(Cases I, III)} \\ \rho(\mu_\alpha)^{-1} f \circ \alpha & \text{(Case II)}, \end{cases}$$

$$D_\rho f = \begin{cases} \rho(\xi, \eta)^{-1} D(\rho(\xi,\eta)f) & \text{(Cases I, III)} \\ \rho(\eta)^{-1} D(\rho(\eta)f) & \text{(Case II)}, \end{cases}$$

where Df is a function on $B^{(i)}$ with values in $\mathrm{Hom}(S^{(i)}, V)$ defined by

$$(Df)(u) = \begin{cases} \sum_{i=1}^{n} \sum_{j=1}^{m} u_{ij}\partial f/\partial z_{ij} & (u \in S^{(1)}) \\ \sum_{i<j} u_{ij}\partial f/\partial z_{ij} & (u \in S^{(2)}) \\ \sum_{i=1}^{m} u_i\partial f/\partial z_i & (u \in S^{(3)}). \end{cases}$$

Then we have

$$(8.4) \qquad D_\rho(f|_\rho\alpha) = (D_\rho f)|_{\rho \otimes \tau}\alpha$$

with a representation

$$\rho \otimes \tau : \begin{cases} GL_n(\mathbf{C}) \times GL_m(\mathbf{C}) \to GL\big(\mathrm{Hom}(S^{(1)}, V)\big) & \text{(Case I)} \\ GL_m(\mathbf{C}) \to GL\big(\mathrm{Hom}(S^{(2)}, V)\big) & \text{(Case II)} \\ GL_m(\mathbf{C}) \times GL_1(\mathbf{C}) \to GL\big(\mathrm{Hom}(S^{(3)}, V)\big) & \text{(Case III)}, \end{cases}$$

defined by

$$[(\rho \otimes \tau)(x, y)h](u) = \rho(x, y)h({}^t x u y) \qquad \text{(Cases I, III)},$$

$$[(\rho \otimes \tau)(y)h](u) = \rho(y)h({}^t y u y) \qquad \text{(Case II)},$$

where $h \in \mathrm{Hom}(S^{(i)}, V)$, $u \in S^{(i)}$. Formula (8.4) can be verified in the same manner as in Lemma 2.1 by means of (8.1), (8.2). Iterated operators of type (2.9) or (2.11) can be defined in the present situation; then the analogues of Theorem 3.1 can be proved. (See [14, §11] for group $G^{(3)}$.) In Case I, we have $\det(\lambda_\alpha) = \det(\mu_\alpha)$ as proved in [13, (3.9)]. Hence the space of automorphic forms depends only on the restriction of ρ to

$$(8.5) \qquad \{(x, y) \in GL_n(\mathbf{C}) \times GL_m(\mathbf{C}) \,|\, \det(x) = \det(y)\}.$$

However, the arithmeticity of automorphic forms depends on ρ as a representation of $GL_n \times GL_m$, and cannot be defined by its restriction to (8.5). (See [12, p. 580, Remark], [14, p. 371].)

Now we can obtain analogues of $\Delta^{(k)}$ in a transparent way only for $G_{n,m}^{(1)}$ with $n = m$, $G_m^{(2)}$ with even m, and $G_m^{(3)}$ with any m. In fact, we first define Δ by

$$\Delta = \begin{cases} \det(\partial/\partial z_{ij}) & \text{(Case I, } m = n) \\ Pf(\partial/\partial z_{ij}) & \text{(Case II, } m = 2n) \\ \displaystyle\sum_{i=1}^{m} \partial^2/\partial z_i^2 & \text{(Case III)}, \end{cases}$$

where Pf is a polynomial function on $S_{2n}^{(2)}$ (called Pfaffian) such that $\det(u) = Pf(u)^2$; $Pf(\partial/\partial z_{ij})$ is obtained by substituting $\partial/\partial z_{ij}$ for z_{ij} with $i < j$ when $z = (z_{ij})$ is the variable matrix on $S_{2n}^{(2)}$. The analogues of Lemma 7.1 are:

$$(8.6) \qquad (\Delta^k f)|_{m+k}\alpha = \Delta^k(f|_{m-k}\alpha) \qquad (\alpha \in G_{m,m}^{(1)}),$$

$$(8.7) \qquad (\Delta^k f)|_{n+(k-1)/2}\alpha = \Delta^k(f|_{n-(k+1)/2}\alpha) \qquad (\alpha \in G_{2n}^{(2)}),$$

$$(8.8) \qquad (\Delta^k f)|_{k+(m/2)}\alpha = \Delta^k(f|_{(m/2)-k}\alpha) \qquad (\alpha \in G_m^{(3)}),$$

where subscript p stands for the representation ψ defined by

$$\psi(x, y) = \det(y)^p \qquad \text{(Case I)},$$

$$\psi(y) = \det(y)^p \qquad \text{(Case II)},$$

$$\psi(x, y) = y^p \qquad \text{(Case III)}.$$

If k is even in (8.7) or if m is odd in (8.8), we have to adjust factors of automorphy suitably as in Lemma 7.1; the best formulation is to state the formulas in terms of the two-sheet covering of $G_{2n}^{(2)}$ or $G_m^{(3)}$, as we did for $Sp(m, \mathbf{R})$ in the proof of Lemma 7.1.

Now, with ρ as in (8.3) and a V-valued function f on $B^{(i)}$, we put

$$\Delta_\rho^{(k)} f = \begin{cases} \rho(\xi,\eta)^{-1}\delta^{-q}\Delta^k(\rho(\xi,\eta)\delta^q f) & \text{(Cases I, III)}, \\ \rho(\eta)^{-1}\delta^{-q}\Delta^k(\rho(\eta)\delta^q f) & \text{(Case II)}, \end{cases}$$

where $\delta(z) = \det(\eta(z))$ and

$$q = \begin{cases} k - m & \text{(Case I)} \\ (k+1)/2 - n & \text{(Case II)} \\ k - (m/2) & \text{(Case III)}. \end{cases}$$

Notice that $\Delta_\rho^{(k)} f$ is V-valued. We have then

$$\Delta_\rho^{(k)}(f|_\rho \alpha) = (\Delta_\rho^{(k)} f)|_\sigma \alpha \qquad (\alpha \in G_{m,m}^{(1)}, \, G_{2n}^{(2)}, \text{ or } G_m^{(3)}),$$

where

$$\sigma(x, y) = \det(xy)^k \rho(x, y) \qquad \text{(Case I)},$$

$$\sigma(y) = \det(y)^k \rho(y) \qquad \text{(Case II)},$$

$$\sigma(x, y) = y^{2k} \rho(x, y) \qquad \text{(Case III)}.$$

In all three cases, there are analogues of (2.13), by which we can state formulas of type (2.15) for the present $\Delta_\rho^{(k)}$.

The domains $B_{m,m}^{(1)}$, $B_{2n}^{(2)}$, $B_m^{(3)}$ are holomorphically isomorphic, respectively, to tube domains

$$H_m^{(1)} = \left\{ z \in M_m(\mathbf{C}) \,|\, i('\bar{z} - z) > 0 \right\},$$

$$H_n^{(2)} = \left\{ z \in H_{2n}^{(1)} \,|\, {}^t z J_n = J_n z \right\},$$

$$H_m^{(3)} = \left\{ z \in \mathbf{C}^m \,|\, \mathrm{Im}(z_1) > \left[\sum_{j=2}^m \mathrm{Im}(z_j)^2 \right]^{1/2} \right\}.$$

In fact, $z \mapsto \beta_0(z) = i(z + 1)(1 - z)^{-1}$ maps $B_{m,m}^{(1)}$ onto $H_m^{(1)}$, and $z \mapsto \beta'(J_n z)$ maps $B_{2n}^{(2)}$ onto $H_n^{(2)}$, where β' is β_0 with $m = 2n$. We note here that some operators on $H_m^{(1)}$ similar to the Maass operators K_α and M_α were defined by Klingen [2]; operators of type Δ were considered by Resnikoff in [5], who attributed some of their properties to Selberg. Now it is easy to transfer the above operators D_ρ and $\Delta_\rho^{(k)}$ to those on these tube domains. Their explicit forms as well as their effect on the factors of automorphy similar to (2.21) or Lemma 5.3 can easily be obtained by the same methods as in the symplectic case, and therefore may be left to the reader.

REFERENCES

1. S. BOCHNER, *Group invariance of Cauchy's formula in several variables*, Ann. of Math. **45** (1944), 686–707.
2. H. KLINGEN, *Zur Theorie der hermitischen Modulfunktionen*, Math. Ann. **134** (1958), 355–384.
3. H. MAASS, *Die Differentialgleichungen in der Theorie der Siegelschen Modulfunktionen*, Math. Ann. **126** (1953), 44–68.
4. ————, *Siegel's modular forms and Dirichlet series*, Lecture Notes in Math. **216** (1971), Springer.
5. H. L. RESNIKOFF, *On a class of linear differential equations for automorphic forms in several complex variables*, Amer. J. Math. **95** (1973), 321–332.
6. G. SHIMURA, *On analytic families of polarized abelian varieties and automorphic functions*, Ann. of Math. **78** (1963), 149–192.
7. ————, *Algebraic number fields and symplectic discontinuous groups*, Ann. of Math. **86** (1967), 503–592.
8. ————, *On some arithmetic properties of modular forms of one and several variables*, Ann. of Math. **102** (1975), 491–515.
9. ————, *On the Fourier coefficients of modular forms of several variables*, Göttingen Nachr. Akad. Wiss. 1975, 261–268.
10. ————, *On the derivatives of theta functions and modular forms*, Duke Math. J. **44** (1977), 365–387.
11. ————, *On certain reciprocity-laws for theta functions and modular forms*, Acta math. **141** (1978), 35–71.
12. ————, *The arithmetic of automorphic forms with respect to a unitary group*, Ann. of Math. **107** (1978), 569–605.
13. ————, *Automorphic forms and the periods of abelian varieties*, J. Math. Soc. Japan, **31** (1979), 561–592.
14. ————, *The arithmetic of certain zeta functions and automorphic forms on orthogonal groups*, Ann. of Math. **111** (1980), 313–375.
15. ————, *On some problems of algebraicity*, Proc. of Int. Congress of Math. Helsinki 1978, 373–379.
16. C. L. SIEGEL, *Über die analytische Theorie der quadratischen Formen*, Ann. of Math. **36** (1935), 527–606 (= Abhandlungen I, 326–405).
17. ————, *Einführung in die Theorie der Modulfunktionen n-ten Grades*, Math. Ann. **116** (1939), 617–657 (= Abhandlungen II, 97–137).
18. ————, *Symplectic geometry*, Amer. J. of Math. **65** (1943), 1–86 (= Abhandlungen II, 274–359).

DEPARTMENT OF MATHEMATICS, PRINCETON UNIVERSITY, PRINCETON, NEW JERSEY 08544

Models of an abelian variety with complex multiplication over small fields

Journal of Number Theory, 15 (1982), 25-35

A few existence theorems are proved for the models of an abelian variety of *CM*-type with a given zeta function over a field which does not necessarily contain the reflex field.

INTRODUCTION

Let K be a *CM*-field of degree $2n$, Φ a *CM*-type of K, and (K', Φ') the reflex of (K, Φ). Our object of study is a structure (A, θ) formed by an abelian variety A of dimension n defined over an algebraic number field k and an injective homomorphism θ of K into $\mathrm{End}(A) \otimes \mathbf{Q}$ such that the representation of K via θ on the Lie algebra of A is equivalent to Φ. We then say that (A, θ) is of type (K, Φ). It is well known that if all elements of $\theta(K)$ are defined over k, then $K' \subset k$ and the zeta function of A over k is a product of the form

$$(0.1) \qquad \prod_{\tau \in J} L(s, \psi_\tau),$$

where J is the set of all embeddings of K into $\mathbf{C}$, and $L(s, \psi_\tau)$ is the L-function of k with a Hecke character ψ_τ. The characters ψ_τ are conjugates of each other in a certain sense and satisfy some simple conditions. Conversely, given an algebraic number field k and characters ψ_τ satisfying these conditions, it can be shown that there is a structure (A, θ) rational over k whose zeta function is (0.1). (See [2] for the proof.) Now let F be the maximal real subfield of K. Suppose A is defined over a field h, not containing K', over which the elements of $\theta(F)$ are rational. It was shown in

* Supported by NSF Grant MCS 7903631.

25

[2] that if we put $k = hK'$, then the elements of $\theta(K)$ are rational over k, and moreover the zeta function of A over h has the form

$$(0.2) \qquad \qquad \prod_{\tau \in J'} L(s, \psi_\tau),$$

where the ψ_τ are the characters of k as in (0.1), and J' is a subset of J consisting of n elements, which together with their complex conjugates form the whole J. We can naturally ask the following question. *Given a field h and characters ψ_τ of hK', can one find a model of (A, θ) such that A and the elements of $\theta(F)$ are rational over h, and that the zeta function of A over h has form (0.2)?* We gave in [2] a sufficient condition for h and ψ_τ, assuming that h has a real archimedean prime. The main purpose of the present paper is to give a more clear-cut necessary and sufficient condition for h and ψ_τ, removing the assumption on real archimedean primes. As a generalization, we shall study the same type of problem with F replaced by its subfield. Applying the results to $K = \mathbf{Q}(e^{2\pi i/l})$ with a prime l, we shall obtain an existence theorem of a model of (A, C) over its field of moduli for such K. A weaker result of this type was previously proved in [3].

Notation and terminology. By an *algebraic number field*, we always mean a finite algebraic extension of $\mathbf{Q}$ embedded in $\mathbf{C}$. If k is an algebraic number field, we denote by $\mathfrak{o}_k$, k_{ab}, $k_\mathbf{A}^\times$, $k_\infty^\times$, and $k_{\infty+}^\times$ the maximal order of k, the maximal abelian extension of k in $\mathbf{C}$, the idele group of k, the archimedean part of $k_\mathbf{A}^\times$, and the identity component of $k_\infty^\times$, respectively. If $x \in k_\mathbf{A}^\times$, x_∞ denotes the archimedean part of x, and $[x, k]$ the element of $\mathrm{Gal}(k_{ab}/k)$ which is the natural image of x; we put also $|x|_k = \prod_v |x_v|_v$, where v runs over all primes of k, $|\ |_v$ is the normalized valuation at v and x_v is the v-component of x. The completion of k at v is denoted by k_v. If k' is a subfield of k, we consider $k_\mathbf{A}'^\times$ a subgroup of $k_\mathbf{A}^\times$ in a natural manner. The complex conjugate of a complex number or a complex-valued function X is denoted by X^ρ. *The zeta function of an abelian variety* over an algebraic number field always refers to the one-dimensional part (see [2, (0.5)] for details).

1. Main Theorems

Throughout the paper, we let F denote a totally real algebraic number field of degree n, K a totally imaginary quadratic extension of F (embedded in $\mathbf{C}$), Φ a CM-type of K, and (K', Φ') the reflex of (K, Φ). Let (A, θ) be a structure of type (K, Φ) in the sense of the Introduction. Viewing Φ as a matrix representation of K acting on $\mathbf{C}^n$, we can find a lattice $\mathfrak{a}$ in K, a vector v_0 in $\mathbf{C}^n$ such that $\mathbf{C}^n/\Phi(\mathfrak{a})v_0$ is isomorphic to A and $\theta(b)$ is represented by $\Phi(b)$ for every $b \in K$. If we put $\omega(b) = \Phi(b) v_0$ for $b \in K$, then ω gives an isomorphism of $K/\mathfrak{a}$ onto the torsion subgroup of A. Suppose (A, θ) is rational over an algebraic number field k. Then $K' \subset k$ by [4, Sect. 8.5, Proposition 30] (cf. also [2, Proposition 2]). We can therefore define a homomorphism $g: k^\times \to K^\times$ by

$$(1.1) \qquad g(x) = \det \Phi'(N_{k/K'}(x)) \qquad (x \in k^\times).$$

This can be naturally extended to a continuous homomorphism of $k_{\mathbf{A}}^\times$ into $K_{\mathbf{A}}^\times$ which we denote by the same letter g. Let J denote the set of all injective homomorphisms of K into $\mathbf{C}$. Then we can identify $K_\infty^\times$ with the group

$$\{(y_\tau)_{\tau \in J} \in (\mathbf{C}^\times)^J \mid y_{\tau\rho} = y_\tau^\rho \text{ for all } \tau \in J\},$$

where ρ denotes the complex conjugation. Then, for every $x \in K_{\mathbf{A}}^\times$ and $\tau \in J$, we define its τ-*component* to be the τ-component of the projection of x to $K_\infty^\times$, and denote it by x_τ. Now the characters ψ_τ of (0.1) can be obtained by

$$(1.2) \qquad \psi_\tau(x) = (\alpha(x)/g(x))_\tau \qquad (x \in k_{\mathbf{A}}^\times, \tau \in J),$$

where α is a certain homomorphism of $k_{\mathbf{A}}^\times$ into $K^\times$. (For details, see [1, Sect. 7.8; 2, (1.9)].) We write ψ_ε simply ψ for the identity embedding ε of K into $\mathbf{C}$. We then say that (A, θ) *determines* ψ over k and call ψ *the (Hecke) character of* $k_{\mathbf{A}}^\times$ *determined by* (A, θ). This satisfies the following two conditions:

$$(1.3) \qquad \psi(x) = 1/g(x)_\varepsilon \qquad if \quad x \in k_\infty^\times,$$

$$(1.4) \qquad If\ x \in k_{\mathbf{A}}^\times\ and\ x_\infty = 1,\ then\ \psi(x) \in K^\times,\ \psi(x)\,\psi(x)^\rho = |x|_k^{-1},$$
$$and \quad \psi(x)\,\mathfrak{a} = g(x)\,\mathfrak{a},$$

where $\mathfrak{a}$ is the lattice in K determined by A as above. (See [2, (1.12), (1.13)].)

For our later purposes, it is necessary to consider a structure (A, C, θ) with a polarization C of A added to (A, θ). We always assume that

$$(1.5) \qquad \theta(K)\ is\ stable\ under\ the\ involution\ of\ \text{End}(A) \otimes \mathbf{Q}\ determined\ by\ C.$$

Such a C always exists for a given (A, θ). Moreover, if A is simple, every polarization C of A satisfies (1.5).

For a subfield X of K, we denote by θ_X the restriction of θ to X, and say that (A, θ_X) is *rational over* a field k if A and the elements of $\theta(X)$ are all rational over k. Let us now recall several basic facts by stating them as

PROPOSITION 1.   *Let h be a field of rationality for (A, θ_F), and F' the maximal real subfield of K'. Then the following assertions hold:*

(1.6) $$F' \subset h;$$

(1.7)   *hK' is the smallest field of rationality, containing h, for (A, θ);*

(1.8)   *every polarization of A satisfying (1.5) is rational over h.*

The last assertion is included in [2, Proposition 4], and the first and second ones in [2, Proposition 3]. (Here we note a correction to [2]: Read "S" for "K" on line 8 from bottom on page 515.)

Now suppose (A, θ_F) is rational over an algebraic number field h. If $K' \subset h$, (A, θ) is rational over h by (1.7), and hence the zeta function of A over h is given by (0.1). If $K' \not\subset h$, we have $[hK' : h] = 2$ by (1.6). In this case, the zeta function of A over h is given by (0.2) with ψ_τ of (1.2); ψ is the character of $(hK')_\mathbf{A}^\times$ determined by A. This result was proved in [2] as Theorem 7, though its formulation is somewhat different from the present statement. Notice that $L(s, \psi_\tau) = L(s, \psi_{\tau\rho})$ as will be shown below.

Thus our problem is as follows. Given h and a Hecke character ψ of $(hK')_\mathbf{A}^\times$, find a necessary and sufficient condition for h and ψ in order that there exist a structure (A, θ_F) rational over h such that (A, θ) determines ψ over hK'. We first give a necessary condition.

THEOREM 1.   *Let (A, θ) be a structure of type (K, Φ), and h an algebraic number field over which (A, θ_F) is rational. Suppose that $K' \not\subset h$ and put $k = hK'$. Let ψ be the Hecke character of $k_\mathbf{A}^\times$ determined by (A, θ), and χ the quadratic character of $h_\mathbf{A}^\times$ corresponding to the extension k of h. Then we have*

(1.9) $$\psi(z) = \chi(z) \, |z|_h^{-1} \qquad for \ z \in h_\mathbf{A}^\times.$$

To prove this, we need two lemmas.

LEMMA 1.   *Let W be an algebra of degree 2 over F with identity element isomorphic to a subalgebra of $\operatorname{End}(Y) \otimes \mathbf{Q}$ with an abelian variety Y of dimension n defined over a field of any characteristic. Then W is a totally imaginary field.*

Proof.   If W is a field, this follows from [2, Lemma 5]. If W is not a field, $W \cap \operatorname{End}(Y)$ contains an element c such that $0 < \dim(c(Y)) < n$. Then $\operatorname{End}(c(Y)) \otimes \mathbf{Q}$ contains an isomorphic image of F. By [4, p. 39, Proposition 2], n divides $2 \dim(c(Y))$, and hence $\dim(c(Y)) = n/2$. This is a contradiction, since a subfield of $\operatorname{End}(c(Y)) \otimes \mathbf{Q}$ of degree n must be totally imaginary by virtue of [2, Lemma 5].

LEMMA 2. *Let (S, Ω) be the reflex of (K', Φ'). Then $K = FS$, and $\theta(S)$ is the center of $\mathrm{End}(A) \otimes \mathbf{Q}$.*

Proof. Let (A_1, θ_1) be a structure of type (S, Ω). Then A_1 is simple and A is isogenous to the product of $[K: S]$ copies of A_1 as can be seen from the results of [4, Sects. 5, 6, 8]. Therefore we obtain our lemma.

Proof of Theorem 1. We first recall that the map $\alpha: k_\mathbf{A}^\times \to K^\times$ of (1.2) satisfies

$$(1.10) \qquad \alpha(x)\, \alpha(x)^\rho = |x_\infty/x|_k \qquad \text{for every } x \in k_\mathbf{A}^\times.$$

(See [1, Sect. 7.8; 2, Theorem 2].) Now the equality of (1.9) is obviously true if $z \in h^\times$. Therefore it is sufficient to prove the equality when z is a prime element of $h_\mathfrak{q}$ for a prime ideal $\mathfrak{q}$ of h, since the elements of this type together with $h^\times$ generate a dense subgroup of $h_\mathbf{A}^\times$. We can even exclude finitely many primes, and hence may assume that A has good reduction modulo $\mathfrak{q}$ and $\mathfrak{q}$ is unramified in k. Let $\mathfrak{p}$ be a prime ideal of k dividing $\mathfrak{q}$, and γ the generator of $\mathrm{Gal}(k/h)$. From (1.7) and Lemma 2, we see that $\theta(a)^\gamma = \theta(a^\rho)$ for every $a \in K$. Therefore, by [2, Proposition 1], we have $\alpha(x^\gamma) = \alpha(x)^\rho$ and $g(x^\gamma) = g(x)^\rho$, so that we have

$$(1.11) \qquad \psi(x^\gamma) = \psi(x)^\rho \qquad \textit{for all } x \in k_\mathbf{A}^\times.$$

(This was noted in [2, (4.6)].) Assuming $\mathfrak{p} \neq \mathfrak{p}^\gamma$, take a prime element π of $k_\mathfrak{p}$ and consider it an element of $k_\mathbf{A}^\times$; let $z = \pi\pi^\gamma$. Then z is a prime element of $h_\mathfrak{q}$, and $\psi(z) = \alpha(z) = \alpha(\pi)\,\alpha(\pi^\gamma) = \alpha(\pi)\,\alpha(\pi)^\rho = |z|_h^{-1}$ by (1.10), which proves the desired result. Next assume $\mathfrak{p} = \mathfrak{p}^\gamma$ and $N(\mathfrak{p}) = N(\mathfrak{q})^2$; let π be a prime element of $h_\mathfrak{q}$ and let $\beta = \alpha(\pi)$. Further let $(\tilde{A}, \tilde{\theta})$ be the structure obtained from (A, θ) by reduction modulo $\mathfrak{p}$ and φ the Frobenius endomorphism of $\tilde{A}$ of degree $N(\mathfrak{q})$. Then $\varphi^2 = \tilde{\theta}(\beta)$ as shown in the proof of [1, Theorem 7.42]. Now $\beta^\rho = \alpha(\pi)^\rho = \alpha(\pi^\gamma) = \alpha(\pi) = \beta$, and hence $\beta^2 = \beta\beta^\rho = \alpha(\pi)\,\alpha(\pi)^\rho = N(\mathfrak{p}) = N(\mathfrak{q})^2$ by (1.10). Thus $\beta = \pm N(\mathfrak{q})$. Let us identify K with $\tilde{\theta}(K)$. As shown in [1, p. 219] or in [2, pp. 529–530], φ does not belong to $\tilde{\theta}(F)$, and hence $[F[\varphi]: F] = 2$. By Lemma 1, $F[\varphi]$ is a totally imaginary field. Therefore $\beta = -N(\mathfrak{q})$, so that $\psi(\pi) = \alpha(\pi) = -N(\mathfrak{q}) = \chi(\pi)\,|\pi|_h^{-1}$, which completes the proof.

Remark 1. We have $\psi_\tau(x^\gamma) = \psi_\tau(x)^\rho$, and hence $L(s, \psi_\tau) = L(s, \psi_{\tau\rho})$. Also we can derive (1.11) formally from (1.3), (1.4), and (1.9) by taking $z = xx^\gamma$ in (1.9).

Now the converse of Theorem 1 is true in the sense that (1.9) combined with (1.3) and (1.4) gives a sufficient condition. More precisely, we have

THEOREM 2. *Let (A, θ) be a structure of type (K, Φ), C a polarization of*

A satisfying (1.5), *and h an algebraic number field which contains the field of moduli of* (A, C, θ_F), *but not* K'. *Put* $k = hK'$. *Further let* ψ *be a Hecke character of* $k_\mathbf{A}^\times$ *satisfying* (1.3), (1.4), *and* (1.9). *Then there exists a structure* (A', C', θ') *rational over* k *which is isomorphic to* (A, C, θ) *and determines* ψ, *and such that* (A', C', θ_F') *is rational over* h.

Notice that the field of moduli of (A, C, θ_F) contains F' and does not depend on the choice of C as shown by [2, Propositions 2 and 4].

Proof. By [2, Theorem 6], there is a structure P rational over k which determines ψ and is isomorphic to (A, C, θ). For simplicity we may assume that $P = (A, C. \theta)$. Let γ be the generator of $\mathrm{Gal}(k/h)$. Put $P^* = (A^\gamma, C^\gamma, \theta^*)$ with $\theta^*(a) = \theta(a^\rho)^\gamma$ for $a \in K$. As shown in Remark 1, we have (1.11). Therefore, by [2, Proposition 1], P^* determines ψ. Since h contains the field of moduli of (A, C, θ_F), there is an isomorphism λ of (A, C) onto (A^γ, C^γ) such that

$$(1.12) \qquad \lambda\theta(a) = \theta^*(a)\lambda \qquad \textit{for every } a \in F.$$

Let (S, Ω) be the reflex of (K', Φ'). By Lemma 2, $K = FS$ and $\theta(S)$ (resp. $\theta^*(S)$) is the center of $\mathrm{End}(A) \otimes \mathbf{Q}$ (resp. $\mathrm{End}(A^\gamma) \otimes \mathbf{Q}$). Therefore we can define an automorphism τ of S by $\lambda\theta(a) = \theta^*(a^\tau)\lambda$ for $a \in S$. Then we have $[K:S]\,\mathrm{Tr}(\Omega(a)) = \mathrm{Tr}(\Phi(a)) = \mathrm{Tr}(\Phi(a^\tau)) = [K:S]\,\mathrm{Tr}(\Omega(a^\tau))$ for all $a \in S$. Since (S, Ω) is primitive, τ must be the identity map of S. It follows that the equality of (1.12) is true for all $a \in K$. Thus λ is an isomorphism of P to P^*. By [2, Theorem 5], λ is rational over k. Observe that λ^γ is an isomorphism of P^* to P, and hence $\lambda^\gamma\lambda = \theta(\zeta)$ with a root of unity ζ in K. Then $\theta(\zeta) = \lambda^{-1}\lambda\lambda^\gamma\lambda = \lambda^{-1}\theta(\zeta)^\gamma\lambda = \theta(\zeta^\rho)$, and hence $\zeta = \pm 1$. Suppose $\zeta = -1$. Then $\lambda\lambda^\gamma = \theta^*(-1)$. Now take a prime ideal $\mathfrak{q}$ of h which remains prime in k and such that A has good reduction modulo the prime of k lying above $\mathfrak{q}$, say $\mathfrak{p}$. Let $\tilde{A}$, $\tilde{\lambda}$, and $\tilde{\theta}(a)$ denote the objects obtained from A, λ, and $\theta(a)$ by reduction modulo $\mathfrak{p}$. Indicate by the upper right q the action of $N(\mathfrak{q})$th power automorphism of the residue field modulo $\mathfrak{p}$. Let φ (resp. φ') be the Frobenius homomorphism of $\tilde{A}$ onto $\tilde{A}^q$ (resp. of $\tilde{A}^q$ onto $\tilde{A}$) of degree $N(\mathfrak{q})$. Then we see that $\tilde{A}^q$ and $\tilde{\lambda}^q$ are obtained from A^γ and λ^γ by reduction modulo $\mathfrak{p}$. Put $\mu = \varphi'\tilde{\lambda}$. Then $\mu \in \mathrm{End}(\tilde{A})$ and $\mu^2 = \varphi'\tilde{\lambda}\varphi'\tilde{\lambda} = \varphi'\tilde{\lambda}\tilde{\lambda}^q\varphi = -\varphi'\varphi$. Now $\varphi'\varphi$ is the Frobenius endomorphism of $\tilde{A}$ of degree $N(\mathfrak{p})$. Let π be a prime element of $h_\mathfrak{q}$. Condition (1.9) implies that $\alpha(\pi) = \psi(\pi) = -N(\mathfrak{q})$. Recall that the map ω of $K/\mathfrak{a}$ into A defined at the beginning of this section has the property

$$(1.13) \qquad \omega(u)^{[x, k]} = \omega(\alpha(x)\, g(x)^{-1}\, u) \qquad (x \in k_\mathbf{A}^\times, u \in K/\mathfrak{a}).$$

(See [1, Proposition 7.40; 2, Theorem 2].) If l is a rational prime not divisible by $\mathfrak{p}$ and if $u \in (K \otimes \mathbf{Q}_l)/(\mathfrak{a} \otimes \mathbf{Z}_l)$, then $\omega(u)^{[\pi, k]} = \omega(\alpha(\pi)u) =$

$-N(\mathfrak{q})\,\omega(u)$. Taking reduction modulo $\mathfrak{p}$, we find that $\varphi'\varphi = \tilde{\theta}(-N(\mathfrak{q}))$, so that $\mu^2 = \tilde{\theta}(N(\mathfrak{q}))$. On the other hand we have

$$\mu\tilde{\theta}(a) = \varphi'\tilde{\lambda}\tilde{\theta}(a) = \varphi'\tilde{\theta}(a^\rho)^q\,\tilde{\lambda} = \tilde{\theta}(a^\rho)\,\varphi'\tilde{\lambda} = \tilde{\theta}(a^\rho)\,\mu \qquad (a \in K).$$

Therefore $\mu \notin \tilde{\theta}(K)$. By Lemma 1, $F[\mu]$ must be a totally imaginary field, which is a contradiction since $\mu^2 = \tilde{\theta}(N(\mathfrak{q}))$. Therefore $\zeta = 1$, so that $\lambda^\gamma\lambda = id_A$. Applying Weil's criterion of descent, we find an abelian variety A' rational over h and an isomorphism ξ of A onto A' rational over k such that $\xi = \xi^\gamma \circ \lambda$. Let C' be the image of C under ξ, and let $\theta'(a) = \xi\theta(a)\xi^{-1}$ for $a \in K$. Then (A', C', θ') has all the required properties.

In the one-dimensional case, we can take Φ to be the natural injection of K into $\mathbf{C}$. Then we obtain

THEOREM 3. *Let K be an imaginary quadratic field (contained in $\mathbf{C}$), $\mathfrak{a}$ a lattice in K, j the invariant of an elliptic curve isomorphic to $\mathbf{C}/\mathfrak{a}$, and h an algebraic number field containing j but not K; put $k = hK$. Further let ψ be a Hecke character of $k_\mathbf{A}^\times$. Then there exists an elliptic curve which is rational over h, isomorphic to $\mathbf{C}/\mathfrak{a}$, and determines ψ over k, if and only if ψ satisfies the following three conditions:*

$$(1.14) \qquad \psi(x) = N_{k/K}(x)^{-1} \qquad \text{for } x \in k_\infty^\times;$$

$$(1.15) \quad \text{If } x \in k_\mathbf{A}^\times \text{ and } x_\infty = 1, \text{ then } \psi(x) \in K^\times \text{ and } \psi(x)\,\mathfrak{a} = N_{k/K}(x)\,\mathfrak{a};$$

$$(1.16) \qquad \psi(x) = \chi(x)\,|x|_h^{-1} \qquad \text{for } x \in h_\mathbf{A}^\times, \text{ where } \chi \text{ is the character}$$
$$\text{of } h_\mathbf{A}^\times \text{ corresponding to the extension } k \text{ of } h.$$

Notice that the equality $\psi(x)\,\psi(x)^\rho = |x|_k^{-1}$ in (1.4) is automatic in the present situation.

Remark 2. Let ψ^* be the ideal character associated with ψ, and $\mathfrak{c}$ the conductor of ψ. Then (1.3) and (1.9) can be written in terms of ψ^* as follows:

$$(1.17) \quad \psi^*(a\mathfrak{o}_k) = \det \Phi'(N_{k/K}(a)) \qquad \text{if } a \in k^\times \text{ and } a \equiv 1 \;(\text{mod } \mathfrak{c}),$$

$$(1.18) \quad \psi^*(\mathfrak{z}\mathfrak{o}_k) = \left(\frac{k/h}{\mathfrak{z}}\right) N(\mathfrak{z}) \qquad \text{for every ideal } \mathfrak{z} \text{ of } h \text{ prime to } \mathfrak{c}.$$

As for (1.4), let us assume for simplicity that $\mathfrak{a}$ is a fractional ideal of K. Then (1.4) is equivalent to

$$(1.19) \quad \text{For every ideal } \mathfrak{x} \text{ of } k \text{ prime to } \mathfrak{c}, \text{ we have } \psi^*(\mathfrak{x}) \in K^\times,$$

$$\psi^*(\mathfrak{x})\,\psi^*(\mathfrak{x})^\rho = N(\mathfrak{x}), \qquad \text{and} \qquad \psi^*(\mathfrak{x})\,\mathfrak{o}_K = N_{k/K}(\mathfrak{x})^{\Phi'},$$

where $\eta^{\Phi'}$ for an ideal η of K' denotes the ideal of K generated by $\det \Phi'(b)$ for all $b \in \eta$.

2. A Generalization of Theorem 2

Let us now generalize Theorem 2 by taking a smaller field in place of F. Before doing this, we give a theorem, whose essential content is included in [2, Theorem 12] and its proof.

Let (A, θ) be a structure of type (K, Φ), and k_0 an algebraic number field over which A is defined. We assume

$$(2.1) \qquad \theta(K) \text{ is stable under } \mathrm{Aut}(\mathbf{C}/k_0).$$

This is obviously so if A is simple. Let k be the smallest field of rationality for (A, θ) containing k_0. Under (2.1), k is a Galois extension of k_0 and we obtain an injective homomorphism π of $\mathrm{Gal}(k/k_0)$ into $\mathrm{Aut}(K)$ by

$$(2.2) \qquad \theta(a)^\sigma = \theta(a^{\pi(\sigma)}) \qquad (a \in K, \sigma \in \mathrm{Gal}(k/k_0)).$$

If A is simple, we have $k = k_0 K'$ by [4, Sect. 8.5, Proposition 30].

THEOREM 4. *The notation being as above, let T be a complete set of representatives of $\pi(\mathrm{Gal}(k/k_0))\backslash J$, and ψ the character of $k_{\mathbf{A}}^{\times}$ determined by A. Then the zeta function of A over k_0 is $\prod_{\tau \in T} L(s, \psi_\tau)$.*

Proof. By [2, Proposition 1], we have $\alpha(x^\sigma) = \alpha(x)^{\pi(\sigma)}$, $g(x^\sigma) = g(x)^{\pi(\sigma)}$, and $\psi(x^\sigma) = \psi_{\pi(\sigma)}(x)$ for $x \in k_{\mathbf{A}}^{\times}$ and $\sigma \in \mathrm{Gal}(k/k_0)$. Now we can repeat a part of the proof of [2, Theorem 12], starting from line 9 from bottom on page 529 of [2]. Then we find that $\zeta(s, A/k_0, D) = L(s, \psi_\tau)$, where D is the subfield of K such that

$$\mathrm{Gal}(K/D) = \pi(\mathrm{Gal}(k/k_0)).$$

Rewriting this result in terms of the ordinary zeta function, we obtain the desired fact in view of [1, Proposition 7.21].

If A is simple, we have

$$[K:D] = [k_0 K' : k_0] = [K' : k_0 \cap K'].$$

The last equality follows from the fact that K' is normal over $k_0 \cap K'$, which in turn follows from [2, (3.2)]. We also note that recently Yoshida [5] determined the zeta function of A over any algebraic number field which is a field of definition, with no condition on the rationality of $\theta(K)$.

Now, to generalize Theorem 2, we assume that K and K' have subfields D and D', respectively, satisfying the following conditions:

(2.3) $$D \subset F \quad and \quad D' \subset F';$$

(2.4) *K and K' are cyclic over D and D', respectively;*

(2.5) *$K = DS$, where S is the field generated over $\mathbf{Q}$ by $\mathrm{Tr}(\Phi'(a))$ for all $a \in K'$;*

(2.6) *there is an isomorphism $\sigma \mapsto [\sigma]$ of $\mathrm{Gal}(K'/D')$ onto $\mathrm{Gal}(K/D)$ such that $\mathrm{Tr}(\Phi(a))^\sigma = \mathrm{Tr}(\Phi(a^{[\sigma]}))$ for all $a \in K$.*

These conditions are obviously satisfied if $D = F$ and $D' = F'$.

THEOREM 5. *Let (A, θ) be a structure of type (K, Φ), C a polarization of A satisfying (1.5), k_0 an algebraic number field containing both D' and the field of moduli of (A, C, θ_D), and let $k = k_0 K'$ and $h = k_0 F'$. Further let ψ be a Hecke character of $k_\mathbf{A}^\times$. Suppose that K' and k_0 are linearly disjoint over D' and that ψ satisfies (1.3), (1.4), (1.9), and also*

(2.7) $$\psi(x)^{[\sigma]} = \psi(x^\sigma) \quad for\ x \in k_\mathbf{A}^\times,\ x_\infty = 1\ and\ \sigma \in \mathrm{Gal}(k/k_0),$$

where we identify $\mathrm{Gal}(k/k_0)$ with $\mathrm{Gal}(K'/D')$. Then there exists a structure (A', C', θ') which is rational over k, isomorphic to (A, C, θ) and determines ψ, and such that (A', C', θ'_D) is rational over k_0.

Proof. By Theorem 2, we may assume that (A, C, θ) is rational over k and determines ψ, and (A, C, θ_F) is rational over h. Let σ be a generator of $\mathrm{Gal}(k/k_0)$. Put $P = (A, C, \theta)$ and $P_1 = (A^\sigma, C^\sigma, \theta_1)$ with $\theta_1(a) = \theta(a^{[\sigma]^{-1}})^\sigma$ for $a \in K$. Since k_0 contains the field of moduli of (A, C, θ_D), there is an isomorphism μ of (A, C, θ_D) onto $(A^\sigma, C^\sigma, \theta_D^\sigma)$. We can then define an automorphism τ of S by $\mu\theta(a) = \theta_1(a^\tau)\mu$. For the same reason as in the proof of Theorem 2, τ is the identity map. By (2.5), we see that μ is an isomorphism of P onto P_1. By [2, Proposition 1] and (2.7), P_1 determines ψ, and hence by [2, Theorem 5], μ is rational over k. Put $[K:D] = 2m$, $\gamma = \sigma^m$, and $\lambda = \mu^{\sigma^{m-1}} \cdots \mu^\sigma \mu$. Then $\gamma = \rho$ on K' and $[\gamma] = \rho$ on K. As shown in the proof of Theorem 1, we have $\theta(a)^\gamma = \theta(a^\rho)$ for $a \in K$. Therefore we can easily verify that λ is an automorphism of P, and hence $\lambda = \theta(\zeta)$ with a root of unity ζ of K. Then $\mu^{\sigma^{2m-1}} \cdots \mu^\sigma \mu = \lambda^\gamma \lambda = \theta(\zeta^\rho \zeta) = id_A$. Applying Weil's criterion of descent to (A, C, θ_D) and μ, we obtain the desired structure.

Another type of result can be stated as follows.

THEOREM 6. *Remove assumption (2.3); suppose instead that $[K:D]$ is odd and D has no roots of unity other than ± 1. Let k_0, k, and ψ be as in Theorem 5. Suppose that K' and k_0 are linearly disjoint over D' and that ψ*

satisfies (1.3), (1.4), *and* (2.7), *but not necessarily* (1.9). *Then the conclusion of Theorem 5 holds.*

Proof. As in the proof of Theorem 2, we can start from a structure $P = (A, C, \theta)$ which is rational over k and determines ψ. Define σ, P_1, and μ as in the above proof; put $v = \mu^{\sigma^{t-1}} \cdots \mu^{\sigma}\mu$ with $t = [K:D]$. Then v is an automorphism of P, so that $v = \theta(\delta)$ with a root of unity δ of K. Moreover, we have $\theta(\delta)^{\sigma} = v^{\sigma} = \mu\theta(\delta)\mu^{-1} = \theta_1(\delta)$, and hence $\delta^{|\sigma|} = \delta$. Thus $\delta \in D$. Our assumption on D implies that $\delta = \pm 1$. Put $\eta = \mu\theta(\delta)$. Since t is odd, we have $\eta^{\sigma^{t-1}} \cdots \eta^{\sigma}\eta = id_A$. Therefore the desired structure can be obtained by Weil's criterion.

3. An Application to the Cyclotomic Case

Let us now consider the case where $K = \mathbf{Q}(\zeta)$, $\zeta = e^{2\pi i/l}$ with an odd prime l. We assume that (K, Φ) is primitive. Then $K' = K$. Let (A, θ) be of type (K, Φ), and let C be a polarization of A. We then assume that $\mathrm{End}(A) = \theta(\mathfrak{o}_k)$. This is so if and only if the lattice $\mathfrak{a}$ of (1.4) is a fractional ideal of K.

THEOREM 7. *The notation being as above, (A, C) has a model over its field of moduli.*

Proof. Let k_0 be the field of moduli of (A, C), and let $k = k_0 K$, $h = k_0 F$. By [2, Proposition 2], k is the field of moduli of (A, C, θ). Let $D = k_0 \cap K$. Then (2.4), (2.5), and (2.6) are satisfied with $D' = D$; the map $\sigma \mapsto [\sigma]$ of (2.6) is the identity map; (2.3) is satisfied if $[K:D]$ is even. Let $\sigma \in \mathrm{Aut}(\mathbf{C}/k_0)$. Then there is an isomorphism λ of (A, C) onto $(A, C)^{\sigma}$. We can define an automorphism τ of K by $\lambda\theta(a^{\tau}) = \theta(a)^{\sigma}\lambda$. Then $\mathrm{Tr}(\Phi(a^{\tau})) = \mathrm{Tr}(\Phi(a)^{\sigma})$. Since (K, Φ) is primitive, we have $\sigma = \tau$ on K, so that $\lambda\theta(a) = \theta(a)^{\sigma}\lambda$ for $a \in D$. This shows that k_0 is the field of moduli of (A, C, θ_D). Let $\mathfrak{l} = (1 - \zeta)\mathfrak{o}_K$ and let $\mathfrak{o}_{\mathfrak{l}}$ denote the $\mathfrak{l}$-completion of $\mathfrak{o}_K$. We are going to construct a Hecke ideal character ψ^* of k. The main theorem of complex multiplication (see [4, p. 128; 1, Corollary 5.16]) shows that if $\mathfrak{x}$ is an ideal in k, then $N_{k/K}(\mathfrak{x})^{\Phi'} = \beta\mathfrak{o}_k$ with an element β of K such that $\beta\beta^{\rho} = N(\mathfrak{x})$, where the upper right Φ' is defined as in (1.21). If $\mathfrak{x}$ is prime to l, then $N(\mathfrak{x}) \equiv 1 \pmod{l}$, so that $\beta \equiv \pm 1 \pmod{\mathfrak{l}\mathfrak{o}_{\mathfrak{l}}}$. Changing β for $\pm\zeta^m\beta$ with a suitable integer m, we may assume that $\beta \equiv 1 \pmod{\mathfrak{l}^2\mathfrak{o}_{\mathfrak{l}}}$. We then define $\psi^*(\mathfrak{x}) = \beta$. It can easily be verified that ψ^* is a Hecke character satisfying (1.17) and (1.19). (This is the same as the character we defined on [3, p. 78].) Moreover we have $\psi^*(\mathfrak{x}^{\sigma}) = \psi^*(\mathfrak{x})^{\sigma}$ for every $\sigma \in \mathrm{Gal}(k/k_0)$. Our aim is to show that there is a model (A, C) over k_0 such that (A, C, θ) determines ψ^* over k. If $K = D$, this is established by [2, Theorem 6]. If

$[K:D]$ is odd and $K \neq D$, the desired result follows from Theorem 6. Therefore assume that $[K:D]$ is even, that is, $D \subset F$. Then our assertion would follow from Theorem 5 if we could show that

$$(3.1) \qquad \psi^*(\mathfrak{p}\mathfrak{o}_k) = \left(\frac{k/h}{\mathfrak{p}}\right) N(\mathfrak{p})$$

for every prime ideal $\mathfrak{p}$ of h prime to l. Let γ be the generator of $\mathrm{Gal}(k/h)$. If $\mathfrak{p}\mathfrak{o}_k = \mathfrak{q}\mathfrak{q}^\gamma$ with a prime ideal $\mathfrak{q}$ of k, then $\psi^*(\mathfrak{p}\mathfrak{o}_k) = \psi^*(\mathfrak{q}) \psi^*(\mathfrak{q})^\rho = N(\mathfrak{q}) = N(\mathfrak{p})$, so that (3.1) is true for such $\mathfrak{p}$. Suppose $\mathfrak{p}$ remains prime in k. Let $\mathfrak{r} = \mathfrak{p} \cap F$ and $N_{h/F}(\mathfrak{p}) = \mathfrak{r}^t$. Since

$$\left(\frac{k/h}{\mathfrak{p}}\right) = \left(\frac{K/F}{\mathfrak{r}}\right)^t,$$

we have

$$\left(\frac{K/F}{\mathfrak{r}}\right) = \gamma$$

and t is odd. Therefore $N(\mathfrak{p}) = N(\mathfrak{r})^t \equiv -1 \pmod{l}$. On the other hand $N_{k/K}(\mathfrak{p}\mathfrak{o}_k)^{\Phi'} = N_{h/F}(\mathfrak{p})^{\Phi'} = N(\mathfrak{p})\,\mathfrak{o}_K$, and hence our definition of $\psi^*(\mathfrak{p}\mathfrak{o}_k)$ shows that $\psi^*(\mathfrak{p}\mathfrak{o}_k) = -N(\mathfrak{p})$, which proves (3.1) and completes the proof.

REFERENCES

1. G. SHIMURA, "Introduction to the Arithmetic Theory of Automorphic Functions," Publ. Math. Soc. Japan No. 11, Iwanami Shoten and Princeton Univ. Press, Princeton, N. J., 1971.
2. G. SHIMURA, On the zeta-function of an abelian variety with complex multiplication, *Ann. of Math.* **94** (1971), 504–533.
3. G. SHIMURA, On abelian varieties with complex multiplication, *Proc. London Math. Soc.* **34** (1977), 65–86.
4. G. SHIMURA AND Y. TANIYAMA, "Complex Multiplication of Abelian Varieties and Its Applications to Number Theory," Publ. Math. Soc. Japan No. 6, 1961.
5. H. YOSHIDA, Abelian varieties with complex multiplication and representations of the Weil groups, *Ann. of Math.* **114** (1981), 87–102.

Printed by the St. Catherine Press Ltd., Tempelhof 41, Bruges, Belgium

The periods of certain automorphic forms
of arithmetic type

Journal of the Faculty of Science, University of Tokyo, Sec. IA, 28 (1982), 605-632

To the memory of Takuro Shintani

The automorphic forms to be considered in this paper are the holomorphic ones on the upper half plane $\mathfrak{H}=\{z\in C\,|\,\mathrm{Im}\,(z)>0\}$ with respect to a congruence subgroup of the multiplicative group $B^\times$ of an indefinite quaternion algebra B over Q. As shown in our previous papers, we have the notion of rationality of such forms over the algebraic closure $\bar{Q}$ of Q. Let f be such a form of weight $2k$ with $0<k\in Z$ which is a $\bar{Q}$-rational eigenform of Hecke operators. Then we shall define two "fundamental periods" $u_+(f)$ and $u_-(f)$ which are nonzero complex numbers determined up to algebraic factors with the property that all the periods of f belong to $\bar{Q}u_+(f)+\bar{Q}u_-(f)$. It can be shown that $\pi\langle f,\,f\rangle/[u_+(f)\overline{u_-(f)}]$ is an algebraic number, where $\langle f,\,f\rangle$ is the normalized Petersson inner product (Theorem 4.4). Now it is well known that there is an elliptic cusp form g belonging to the same eigenvalues for Hecke operators as f. The main purpose of the present paper is to show that under certain assumptions on f, both $u_+(f)/u_+(g)$ and $u_-(f)/u_-(g)$ are algebraic numbers (Theorem 4.7). We obtain this result by first generalizing the integrals ·considered by Shintani in [16] and applying our results of [14] and [15] to the generalized integrals. It should be noted that the algebraicity of $\langle f,\,f\rangle/\langle g,\,g\rangle$ was proved in [15, II] in a more general case. Let us now give a summary of the contents.

We start with a somewhat more general setting with a totally real algebraic number field F of degree n as a basic field. We take a quaternion algebra B over F which is unramified at r archimedean primes of F with $0<r\leq n$. For a cusp form f of "even weight" with respect to a congruence subgroup Γ of $B^\times$, we consider an integral of the form

$$M(z,\,f)=\int_D \overline{\theta(z,\,w)}f(w)\,\mathrm{Im}\,(w)^{2k}d\mu(w)\qquad (D=\Gamma\backslash\mathfrak{H}^r)$$

with a certain theta function $\theta(z,\,w)$ on $\mathfrak{H}^n\times\mathfrak{H}^r$. We shall show that $M(z,\,f)$ is a Hilbert cusp form of half-integral weight and its Fourier coefficients are given by the "periods" of f (Theorems 2.2 and 3.1). Specializing this to the case $F=Q$, we shall show that

$$M(z,\,f\,|\,T(p))=M(z,\,f)\,|\,T(p^2)$$

for every odd prime p, where $T(p)$ and $T(p^2)$ are Hecke operators (Theorem 3.2).
These generalize the results of Shintani [16] which deal with the case $B = M_2(\boldsymbol{Q})$.
Our methods of proof are simpler than those of [16]. The quantities $u_+(f)$ and
$u_-(f)$ will be defined in Section 4 by means of the cohomology groups attached
to Γ. We shall show that the Fourier coefficients of $M(z, f)$ are $u_+(f)$ times
algebraic numbers if f is "primitive". This fact combined with several results
of arithmeticity proved in [15, II] and [14] concerning inner products will yield
the algebraicity of $u_+(f)/u_+(g)$ and $u_-(f)/u_-(g)$ under the assumption that
$L(z, f) \neq 0$.

1. Automorphic forms on $\mathfrak{H}^r$.

The symbols $\mathfrak{H}$, $\overline{\boldsymbol{Q}}$, F, and n will have the same meaning as in the introduction throughout the paper. We denote by $\tau_1, \cdots, \tau_n$ the embeddings of F into
$\boldsymbol{R}$, and by I_F the free $\boldsymbol{Z}$-module generated by $\tau_1, \cdots, \tau_n$; we put $CI_F = I_F \otimes_Z C$
$= \sum_{\nu=1}^{n} C\tau_\nu$. We embed F into $\boldsymbol{R}^n$ by the map $a \mapsto (a^{\tau_1}, \cdots, a^{\tau_n})$ and identify
$F \otimes_Q \boldsymbol{R}$ and $F \otimes_Q C$ with $\boldsymbol{R}^n$ and C^n through the map. If $p = \sum_{\nu=1}^{n} p_\nu \tau_\nu \in I_F$ and
$z = (z_1, \cdots, z_n) \in C^n$, we put $z^p = \prod_{\nu=1}^{n} z_\nu^{p_\nu}$; this is meaningful for $p \in CI_F$ if $0 < z_\nu$
$\in \boldsymbol{R}$ for all ν. We use the letter F also for the element $\sum_{\nu=1}^{n} \tau_\nu$ of I_F; thus z^F
$= z_1 \cdots z_n$; in particular, $a^F = N_{F/Q}(a)$ for $a \in F$. We put also

$$(1.1) \qquad\qquad e(x) = \exp(2\pi i x) \qquad \text{for} \quad x \in C,$$

$$(1.2) \qquad\qquad e_F(z) = e(\sum_{\nu=1}^{n} z_\nu) \qquad \text{for} \quad z = (z_1, \cdots, z_n) \in C^n.$$

We write $x \gg 0$ and also $0 \ll x$ for an element $x = (x_1, \cdots, x_n) \in \boldsymbol{R}^n$ if $x_\nu > 0$ for
all ν.

Let B be a quaternion algebra over F unramified at $\tau_1, \cdots, \tau_r$ and ramified
at $\tau_{r+1}, \cdots, \tau_n$; we assume $0 < r \leq n$; $M_2(F)$ is included as a special case. Put
$B_R = B \otimes_Q \boldsymbol{R}$. Then there is an isomorphism

$$(1.3) \qquad\qquad B_R \longrightarrow M_2(\boldsymbol{R})^r \times \boldsymbol{H}^{n-r},$$

which sends an element x of F to $(x^{\tau_1}, \cdots, x^{\tau_n})$, where $\boldsymbol{H}$ denotes the Hamilton
quaternions. For every $\alpha \in B_R$, we denote by α_ν its projection to the ν-th factor
of $M_2(\boldsymbol{R})^r \times \boldsymbol{H}^{n-r}$. Thus α_ν belongs to $M_2(\boldsymbol{R})$ or to $\boldsymbol{H}$ according as $\nu \leq r$ or $\nu > r$.
In particular, $x_\nu = x^{\tau_\nu}$ for $x \in F$.

For every integer $m \geq 0$, we obtain an $\boldsymbol{R}$-rational polynomial representation
$\sigma_m : \boldsymbol{H}^\times \to GL_{m+1}(C)$ by composing an injection of $\boldsymbol{H}^\times$ into $GL_2(C)$ with the representation of GL_2 by symmetric tensors of degree m; we understand that σ_0
is the trivial representation. Changing it for an equivalent representation, and
choosing a suitable isomorphism (1.3), we may assume:

(1.4) $\quad \alpha_\nu$ *has algebraic entries for every* $\alpha \in B$ *and every* $\nu \leq r$;

(1.5) $\sigma_m(\alpha_\nu)$ *has algebraic entries for every* $\alpha \in B^\times$, *every* $\nu > r$, *and every* $m \geqq 0$;

(1.6) $\sigma_m(x^\iota) = {}^\iota \overline{\sigma_m(x)}$ *for all* $x \in H^\times$.

Here and throughout the paper, ι denotes the main involution of any quaternion algebra $(B, M_2(R), H,$ etc.$)$. For an element α of B_R or $M_2(R)$ or H, we put $N(\alpha) = \alpha\alpha^\iota$ and $\mathrm{Tr}(\alpha) = \alpha + \alpha^\iota$. We let an element α of $B_R^\times$ act on $(C \cup \{\infty\})^r$ by the rule

$$\alpha(z_1, \cdots, z_r) = (\alpha_1 z_1, \cdots, \alpha_r z_r),$$

$$\alpha_\nu z_\nu = (a_\nu z_\nu + b_\nu)(c_\nu z_\nu + d_\nu)^{-1} \quad \text{for} \quad \alpha_\nu = \begin{pmatrix} a_\nu & b_\nu \\ c_\nu & d_\nu \end{pmatrix}.$$

We denote by $B_{R+}^\times$ the set of all α of B_R such that $N(\alpha) \gg 0$, and put $B_+^\times = B \cap B_{R+}^\times$.

Let $k = \sum_{\nu=1}^r k_\nu \tau_\nu \in I_F$ and $\kappa = \sum_{\nu=r+1}^n \kappa_\nu \tau_\nu \in I_F$ with $\kappa_\nu \geqq 0$. We define a factor of automorphy $J(\alpha, z)^k$ and a representation $\sigma_\kappa : B_R^\times \to GL_d(C)$ with $d = \prod_{\nu > r}(\kappa_\nu + 1)$ by

$$(1.7) \qquad J(\alpha, z)^k = \prod_{\nu=1}^r |N(\alpha_\nu)|^{-k\nu/2}(c_\nu z_\nu + d_\nu)^{k\nu} \quad \left(\alpha \in B_R^\times, \ z \in C^r, \ \alpha_\nu = \begin{pmatrix} * & * \\ c_\nu & d_\nu \end{pmatrix} \right),$$

$$(1.8) \qquad \sigma_\kappa(\alpha) = \sigma_{\kappa_{r+1}}(\alpha_{r+1}) \otimes \cdots \otimes \sigma_{\kappa_n}(\alpha_n).$$

For a C^d-valued function f on $\mathfrak{H}^r$ and an element α of $B_{R+}^\times$, we define a C^d-valued function $f\|_{k,\kappa}\alpha$ on $\mathfrak{H}^r$ by

$$(f\|_{k,\kappa}\alpha)(z) = \{\textstyle\prod_{\nu > r} N(\alpha_\nu)^{\kappa_\nu/2}\} \cdot J(\alpha, z)^{-k} \sigma_\kappa(\alpha)^{-1} f(\alpha z).$$

Let $\mathfrak{r}$ denote the maximal order of F and $\mathfrak{o}$ a maximal order of B. For every positive integer N, put

$$\Gamma_N = \{\gamma \in \mathfrak{o} \mid N(\gamma) = 1, \ \gamma - 1 \in N\mathfrak{o}\}.$$

By a congruence subgroup of B, we understand a subgroup Γ of $B_+^\times$ such that $\Gamma_N \subset \Gamma$ and $[\Gamma\mathfrak{r}^\times : \Gamma_N\mathfrak{r}^\times] < \infty$ for some N. We then denote by $\mathcal{M}_{k,\kappa}(\Gamma)$ the set of all C^d-valued holomorphic functions f on $\mathfrak{H}^r$ which are holomorphic at every cusp (if any), and which satisfy $f\|_{k,\kappa}\gamma = f$ for all $\gamma \in \Gamma$. (The cusp condition is necessary only when $B = M_2(Q)$.) The subspace of $\mathcal{M}_{k,\kappa}(\Gamma)$ consisting of the cusp forms (that is, the elements vanishing at all cusps) is denoted by $\mathcal{S}_{k,\kappa}(\Gamma)$; naturally $\mathcal{S}_{k,\kappa} = \mathcal{M}_{k,\kappa}$ if B is a division algebra. We write simply $\mathcal{M}_k$, $\mathcal{S}_k$, and $f\|_k\gamma$ for $\mathcal{M}_{k,0}$, $\mathcal{S}_{k,0}$ and $f\|_{k,0}\gamma$. We define a measure μ on $\mathfrak{H}^r$ by

$$(1.9) \qquad d\mu(z) = \prod_{\nu=1}^r y_\nu^{-2} dx_\nu dy_\nu \qquad (z_\nu = x_\nu + iy_\nu)$$

and the inner product $\langle f, g \rangle$ of two elements f and g of $\mathcal{S}_{k,\kappa}(\Gamma)$ by

$$(1.10) \qquad \langle f, g \rangle = \mu(D)^{-1} \int_D {}^\iota \overline{f(z)} g(z) \, \mathrm{Im}(z)^k d\mu(z) \qquad (D = \Gamma \backslash \mathfrak{H}^r).$$

If $B = M_2(F)$, we can introduce the notion of (Hilbert) modular forms of half-integral weight as follows. Put

$$\theta_F(z) = \sum_{x \in \mathfrak{r}} e(\sum_{\nu=1}^n x_\nu^2 z_\nu) \qquad (z = (z_1, \cdots, z_n) \in \mathfrak{H}^n),$$

(1.11)
$$\Gamma_\theta = \left\{ \begin{pmatrix} a & b \\ c & d \end{pmatrix} \in SL_2(\mathfrak{r}) \,\middle|\, c \in 4\mathfrak{b} \right\},$$

where $\mathfrak{b}$ denotes the different of F. Then we have, as shown in [13, p. 342],

(1.12)
$$\theta_F(\gamma z) = j(\gamma, z)\theta_F(z), \quad j(\gamma, z) = \zeta_\gamma (cz+d)^{F/2} \qquad (\gamma \in \Gamma_\theta)$$

with a root of unity ζ_γ, and moreover

$$j\left(\begin{pmatrix} a & b \\ c & d \end{pmatrix}, z\right) = \left(\frac{c}{d\mathfrak{r}}\right)(cz+d)^{F/2} \quad \text{if} \quad 0 \ll d \equiv 1 \qquad (\mathrm{mod}\ 4\mathfrak{b}),$$

where $(-)$ denotes the quadratic residue symbol in F. Now let $h = \sum_{\nu=1}^n h_\nu \tau_\nu \in (1/2)I_F$ with h_ν satisfying $2h_1 \equiv \cdots \equiv 2h_n \equiv 1$ (mod 2), and let Δ be a congruence subgroup of Γ_θ. Then we denote by $\mathcal{M}_h(\Delta)$ the set of all holomorphic functions f on $\mathfrak{H}^n$ holomorphic also at the cusps such that

$$f(\gamma z) = j(\gamma, z)(cz+d)^{h-(F/2)}f(z) \qquad \text{for every} \quad \gamma = \begin{pmatrix} * & * \\ c & d \end{pmatrix} \in \Delta.$$

The subspace of $\mathcal{M}_h(\Delta)$ consisting of the cusp forms is denoted by $\mathcal{S}_h(\Delta)$.

We now define a 3-dimensional subspace V of B over F by

(1.13)
$$V = \{\alpha \in B \mid \mathrm{Tr}(\alpha) = 0\}$$

and an F-valued F-bilinear symmetric form S on V by

(1.14)
$$S(\alpha, \beta) = \mathrm{Tr}(\alpha\beta') = -\alpha\beta - \beta\alpha \qquad (\alpha, \beta \in V).$$

We are going to apply our results of [13] and [15] to this form S. As in [13, §3], we denote by V_ν the completion of V at τ_ν, and by S_ν the C-bilinear extension of S to $V_\nu \otimes_R C$. The space $\mathfrak{Z}_m$ of [13, (2.6)] in the present case is isomorphic to $\mathfrak{H}$. We define a map $p: C \to M_2(C)$ by

(1.15)
$$p(w) = \begin{pmatrix} w & -w^2 \\ 1 & -w \end{pmatrix} \qquad (w \in C).$$

The restriction of p to $\mathfrak{H}$ is essentially a special case of [13, (2.7)]. If $\mathrm{Im}(w) \neq 0$, we have

(1.16)
$$\gamma p(w)\gamma' = (cw+d)^2 p(\gamma w) \qquad \text{for} \quad \gamma = \begin{pmatrix} * & * \\ c & d \end{pmatrix} \in GL_2(R).$$

We also put

(1.17)
$$[\alpha, w] = (w \quad 1)\varepsilon\alpha \begin{pmatrix} w \\ 1 \end{pmatrix}, \quad \varepsilon = \begin{pmatrix} 0 & 1 \\ -1 & 0 \end{pmatrix} \qquad (w \in C,\ \alpha \in V_\nu \otimes_R C).$$

Then $S_\nu(\alpha, p(w))=[\alpha, w]$ for $\nu \leq r$. For $\alpha \in V$, $w=(w_1, \cdots, w_r) \in C^r$, and $k \in I_F$ as above, we put

$$(1.18) \qquad [\alpha, w]^k = \prod_{\nu=1}^r [\alpha_\nu, w_\nu]^{k\nu}.$$

If $\mathrm{Im}(w_\nu) \neq 0$ for all ν, we have

$$(1.19) \qquad [\beta'\alpha\beta, w]^k = |N(\beta)|^k J(\beta, w)^{2k} [\alpha, \beta w]^k \qquad (\alpha \in V, \beta \in B^\times),$$

$$(1.20) \qquad [\beta\alpha\beta', \beta w]^k = |N(\beta)|^k J(\beta, w)^{-2k} [\alpha, w]^k \qquad (\alpha \in V, \beta \in B^\times).$$

For $\lambda = \sum_{\nu>r} \lambda_\nu \tau_\nu \in I_F$ with $\lambda_\nu \geq 0$, let $\mathscr{P}_\lambda$ denote the vector space over C of all polynomial functions on $V_{r+1} \times \cdots \times V_n$ which are S_ν-harmonic (in the sense of [13, p. 322]) and homogeneous of degree λ_ν on V_ν. Define the action of $B^\times$ on $\mathscr{P}_\lambda$ by

$$(1.21) \qquad t^\beta(\alpha) = t((\beta_\nu \alpha_\nu \beta'_\nu)_{\nu>r}) \qquad (\beta \in B^\times, \alpha \in \prod_{\nu>r} V_\nu, t \in \mathscr{P}_\lambda).$$

LEMMA 1.1. *The above representation of $B^\times$ on $\mathscr{P}_\lambda$ is equivalent to $\sigma_{2\lambda}$.*

PROOF. It is sufficient to prove the corresponding assertion for $GL_2(C)$ and its representation on the space $\mathscr{P}'_\lambda$ of homogeneous harmonic functions of degree λ on the 3-dimensional space $\{x \in M_2(C) | \mathrm{Tr}(x)=0\}$. To $p \in \mathscr{P}'_\lambda$, assign a function p^* on C^2 by $p^*(x) = p(\varepsilon^{-1} x \cdot {}^t x)$ for $x \in C^2$. It can easily be shown that $p \mapsto p^*$ gives an isomorphism of $\mathscr{P}'_\lambda$ onto the space of all homogeneous polynomial functions on C^2 of degree 2λ, which commutes with the action of $GL_2(C)$.

Thus we can find a C^d-valued function u on $\prod_{\nu>r} V_\nu$ whose components form $\bar{Q}$-rational basis of $\mathscr{P}_\lambda$ over C such that

$$(1.22) \qquad u(\beta\alpha\beta') = \sigma_{2\lambda}(\beta)u(\alpha) \qquad (\alpha \in V, \beta \in B^\times).$$

If $\lambda=0$, we understand that u is the constant 1.

2. A generalization of the Shintani integral.

In order to state our first main theorem, it is necessary to normalize Haar measures of certain subgroups of $SL_2(R)$. Put

$$a_t = \begin{pmatrix} t & 0 \\ 0 & t^{-1} \end{pmatrix}, \quad n_s = \begin{pmatrix} 1 & s \\ 0 & 1 \end{pmatrix}, \quad k_\theta = \begin{pmatrix} \cos\theta & \sin\theta \\ -\sin\theta & \cos\theta \end{pmatrix}$$

$$(0 \neq t \in R, \ s \in R, \ \theta \in R),$$

$$A = \{a_t | 0 < t \in R\}, \qquad A' = \{a_t | 0 \neq t \in R\},$$

$$N = \{n_s | s \in R\}, \qquad K = \{k_\theta | \theta \in R\}.$$

Let H be a subgroup of $SL_2(\boldsymbol{R})$ such that $\beta^{-1}H\beta$ for some $\beta\in SL_2(\boldsymbol{R})$ coincides with A, A', N, $N\{\pm1\}$, or K. Then we define an invariant measure dh on H so that: the whole measure of H is 1 if $\beta^{-1}H\beta=K$;

$$(2.1) \qquad \int_H \varphi(h)dh=\begin{cases} \displaystyle\int_0^\infty \varphi(\beta a_t\beta^{-1})t^{-1}dt & \text{if } \beta^{-1}H\beta=A, \\[2ex] \displaystyle\int_{-\infty}^\infty \varphi(\beta a_t\beta^{-1})|t^{-1}|dt & \text{if } \beta^{-1}H\beta=A'. \end{cases}$$

Notice that this is independent of the choice of β. If $\beta^{-1}H\beta=N$ or $N\{\pm1\}$, we fix an arbitrary invariant measure on H. In particular, if $H=N$, we take $d(n_s)$ $=ds$. Since $SL_2(\boldsymbol{R})=ANK$, we define an invariant measure dg on $SL_2(\boldsymbol{R})$ as usual by $d(ank)=dadndk$ ($a\in A$, $n\in N$, $k\in K$). This can also be characterized by

$$(2.2) \qquad \int_{SL_2(\boldsymbol{R})} \psi(g(i))dg=(1/2)\int_{\mathfrak{H}} \psi(z)d\mu(z)$$

for an integrable function ψ on $\mathfrak{H}$. We shall consider the product $SL_2(\boldsymbol{R})^r$ and its subgroups H such that $\beta^{-1}H\beta$ for some $\beta\in SL_2(\boldsymbol{R})^r$ is a product of copies of A, A', N, $N\{\pm1\}$, or K. We normalize Haar measures on $G=SL_2(\boldsymbol{R})^r$ and such H by taking the product of the above measures. If C is a discrete subgroup of H, then we normalize a measure $d(Ch)$ on $C\backslash H$ so that

$$\int_H \varphi(h)dh=\int_{C\backslash H} \{\textstyle\sum_{c\in C} \varphi(ch)\}\, d(Ch).$$

A measure on $H\backslash G$ can be normalized in a similar way. We shall simply write dg for $d(Hg)$ if there is no fear of confusion.

We now take a congruence subgroup Γ of $B^\times$ such that $N(\gamma)=1$ for all $\gamma\in\Gamma$, and let Γ act on V by $\alpha\mapsto\gamma\alpha\gamma^{-1}$ for $\alpha\in V$ and $\gamma\in\Gamma$. Put

$$(2.3) \qquad V^*=\{\alpha\in V \mid N(\alpha_\nu)<0 \quad\text{for all } \nu\leqq r\}.$$

This is stable under Γ. Identify B with a subset of $M_2(\boldsymbol{R})^r$ through the map $\alpha\mapsto(\alpha_1,\cdots,\alpha_r)$, and put, for each $\alpha\in V$, $\neq0$,

$$(2.4) \qquad H_\alpha=\{g\in SL_2(\boldsymbol{R})^r \mid g\alpha=\alpha g\}, \qquad \Gamma_\alpha=H_\alpha\cap\Gamma.$$

Given $f\in S_{2k,2\lambda}(\Gamma)$ and $0\neq\alpha\in V$, we put

$$(2.5) \qquad P(f,\alpha,\Gamma)=\int_{\Gamma_\alpha\backslash H_\alpha} [\alpha,\,hw]^k\cdot{}^t\overline{u(\alpha)}f(hw)d(\Gamma_\alpha h) \qquad\qquad (w\in\mathfrak{H}^r),$$

where u is the function of (1.22). Notice that the integrand is in fact Γ_α-invariant. Obviously H_α is a subgroup of $SL_2(\boldsymbol{R})^r$ of the above type, and therefore the measures of H_α and $\Gamma_\alpha\backslash H_\alpha$ can be normalized in the above described manner.

LEMMA 2.1. *Integral (2.5) is convergent and independent of w. Moreover, it is 0 unless $\alpha \in V^*$.*

PROOF. Postponing the proof of convergence, let us prove here only the independence from w and the vanishing. Call the right-hand side of (2.5) $\varphi(w)$. Then φ is holomorphic in w, and $\varphi(hw) = \varphi(w)$ for all $h \in H_\alpha$, and hence φ must be a constant. Now (1.19) shows that

$$\varphi = [\alpha, w]^k \int_{\Gamma_\alpha \backslash H_\alpha} J(h, w)^{-2k} \cdot {}^t\overline{u(\alpha)} f(hw) d(\Gamma_\alpha h).$$

Suppose $N(\alpha_\nu) > 0$ for some $\nu \leq r$. Since $\mathrm{Tr}(\alpha_\nu) = 0$, we have $\alpha_\nu(w_\nu) = w_\nu$ for some $w_\nu \in \mathfrak{H}$. With this choice of w_ν, we have $[\alpha, w]^k = 0$, and hence $\varphi = 0$. Next suppose $N(\alpha) = 0$. This happens only when $B = M_2(F)$. We can find an element δ of $SL_2(F)$ such that $\delta^{-1}\alpha\delta = \begin{pmatrix} 0 & p \\ 0 & 0 \end{pmatrix}$ with $0 \neq p \in F$. Put $H' = \delta^{-1}H_\alpha\delta$, $\Gamma' = \delta^{-1}\Gamma_\alpha\delta$, and $f' = f\|_{2k}\delta$. Then

$$(-p)^{-k}\varphi(\delta w) = \int_{\Gamma_\alpha \backslash H_\alpha} J(h\delta, w)^{-2k} f(h\delta w) d(\Gamma_\alpha h)$$

$$= \int_{\Gamma' \backslash H'} J(\delta h', w)^{-2k} f(\delta h' w) d(\Gamma' h')$$

$$= \int_{\Gamma' \backslash H'} f'(h'w) d(\Gamma' h').$$

Observe that $H' = (\{\pm 1\} N)^n$. Since f' is a cusp form, the last integral is 0. This completes the proof.

We note here a few easy relations:

$$(2.6) \qquad P(f, \beta\alpha\beta^{-1}, \Gamma) = P(f\|_{2k, 2\lambda}\beta, \alpha, \beta^{-1}\Gamma\beta) \qquad (\beta \in B_+^\times),$$

$$(2.7) \qquad P(f, \alpha, \Gamma') = [\Gamma_\alpha : \Gamma'_\alpha] P(f, \alpha, \Gamma) \quad \text{if} \quad [\Gamma : \Gamma'] < \infty,$$

$$(2.8) \qquad P(f, c\alpha, \Gamma) = c^{k+\lambda} P(f, \alpha, \Gamma) \qquad \text{if} \quad c \in (R^\times)^n.$$

A C-valued function η on V is called *locally constant* if there are two Z-lattices L and M in V such that $\eta(v) = 0$ for $v \notin L$ and $\eta(v) = \eta(v')$ for $v - v' \in M$; further we say that η is Γ-*invariant* if $\eta(v) = \eta(\gamma v \gamma^{-1})$ for all $\gamma \in \Gamma$.

THEOREM 2.2. *Let $k = \sum_{\nu=1}^{r} k_\nu \tau_\nu$ and $\lambda = \sum_{\nu=r+1}^{n} \lambda_\nu \tau_\nu$ be elements of I_F with $k_\nu > 0$ and $\lambda_\nu \geq 0$. Further let Γ be a congruence subgroup of $B^\times$ such that $N(\gamma) = 1$ for all $\gamma \in \Gamma$, f an element of $S_{2k, 2\lambda}(\Gamma)$, η a Γ-invariant locally constant function on V, and q an element of F such that $q^{\tau_\nu} > 0$ for $\nu \leq r$ and $q^{\tau_\nu} < 0$ for $\nu > r$. Let $R(\Gamma)$ be a complete set of representatives of V^* modulo the action $\alpha \mapsto \gamma\alpha\gamma^{-1}$ for all $\gamma \in \Gamma$. Then an infinite series*

$$(2.9) \qquad \sum_{\alpha \in R(\Gamma)} \eta(\alpha) |N(\alpha)|^{-\xi/2} P(f, \alpha, \Gamma) e_F(-qN(\alpha)z)$$

is convergent and defines a Hilbert cusp form of weight h (that is, an element of $S_h(\Delta)$ with a congruence subgroup Δ of $SL_2(F)$), where $\xi = \sum_{\nu=1}^{r} \tau_\nu$, and

$$(2.10) \qquad h = \sum_{\nu=1}^{n} h_\nu \tau_\nu, \qquad h_\nu = \begin{cases} k_\nu + (1/2) & \text{for } \nu \leq r, \\ \lambda_\nu + (3/2) & \text{for } \nu > r. \end{cases}$$

REMARK. That each term of (2.9) depends only on the class of α modulo Γ follows from (2.6). Series (2.9) is only superficially dependent on the choice of Γ. In fact, $\mu(\Gamma \backslash \mathfrak{H}^r)^{-1}$ times (2.9) depends only on f, η, and q, and is independent of Γ. This will follow from our proof of the theorem; it can easily be derived also from (2.7).

Our proof requires a theta function on $\mathfrak{H}^n \times \mathfrak{H}^r$ we introduced in [13] and [15], whose explicit form is

$$(2.11) \qquad \theta(z, w, t) = \theta(z, w, t; \eta, q)$$

$$= y^{\xi/2} \operatorname{Im}(w)^{-2k} \sum_{\alpha \in V} \overline{\eta(\alpha)} t(\alpha) [\alpha, \bar{w}]^k e(\sum_{\nu=1}^{n} q_\nu R_\nu[\alpha, z, w])$$

$$(z \in \mathfrak{H}^n, \ w \in \mathfrak{H}^r, \ t \in \mathcal{P}_\lambda),$$

$$(2.12) \qquad R_\nu[\alpha, z, w] = \begin{cases} N(\alpha_\nu)z_\nu + (i/2)y_\nu \operatorname{Im}(w_\nu)^{-2} |[\alpha_\nu, w_\nu]|^2 & (\nu \leq r), \\ N(\alpha_\nu)\bar{z}_\nu & (\nu > r). \end{cases}$$

Here $y_\nu = \operatorname{Im}(z_\nu)$; k, η, ξ, and q are the same as in the theorem. This is essentially the same as [15, I, (6.6)] specialized to the case $m = 1$. Notice that the present R_ν is $1/2$ times R_ν of [15, I, (6.7)] with $-\bar{z}$ in place of z. Now [15, I, (6.10)] in the present case becomes

$$(2.13) \qquad \theta(z, \beta w, t; \eta, q) = N(\beta)^k J(\beta, w)^{2k} \theta(z, w, t^\beta; \eta^\beta, N(\beta)^2 q) \qquad (\beta \in B_+).$$

where $\eta^\beta(\alpha) = \eta(\beta \alpha \beta^\iota)$. Our θ is also a special case of [13, (7.6)], and therefore, by [13, Proposition 7.1], it satisfies

$$(2.14) \qquad \theta(\gamma z, w, t) = \overline{j(\gamma, z)}(c\bar{z} + d)^{h - (F/2)} \theta(z, w, t)$$

$$\text{for every } \gamma = \begin{pmatrix} * & * \\ c & d \end{pmatrix} \in \Delta,$$

where Δ is a congruence subgroup of $SL_2(F)$, $j(\gamma, z)$ is defined by (1.12), and h is defined by (2.10) (cf. [15, I, (6.11)]). We now define $\Theta(z, w; \eta, q)$, or simply $\Theta(z, w)$, to be a column vector whose components are $\theta(z, w, t_i; \eta, q)$ for $i = 1, \cdots, d$, with components $t_1, \cdots, t_d$ of the column vector u of (1.22). From (2.13) we obtain

$$(2.15) \qquad \Theta(z, \beta w; \eta, q) = N(\beta)^k J(\beta, w)^{2k} \sigma_{2\lambda}(\beta) \Theta(z, w; \eta^\beta, N(\beta)^2 q) \qquad (\beta \in B_+).$$

Given Γ, f, and η as in the theorem, we consider an integral

$$(2.16) \qquad \int_D {}^t\overline{\Theta(z, w)} f(w) \operatorname{Im}(w)^{2k} d\mu(w) \qquad (D = \Gamma\backslash\mathfrak{H}^r).$$

This can be written $\mu(D)\langle\Theta(z, w), f(w)\rangle$. Since f is a cusp form, the convergence of (2.16) is straightforward. Now our theorem will follow from

PROPOSITION 2.3. *Series (2.9) is equal to $2^r q^{\xi/2}$ times (2.16).*

PROOF. Let R' be a complete set of representatives of $V - \{0\}$ modulo Γ. Since

$$V - \{0\} = \bigcup_{\alpha \in R'} \{\gamma^{-1}\alpha\gamma \mid \gamma \in \Gamma_\alpha\backslash\Gamma\},$$

integral (2.16) times $y^{-\xi/2}$ is equal (at least formally) to

$$\sum_{\alpha \in R'} \eta(\alpha) \sum_{\gamma \in \Gamma_\alpha\backslash\Gamma} \int_{\gamma D} [\alpha, w]^k \cdot {}^t\overline{u(\alpha)} f(w) e\left(-\sum_{\nu=1}^n q_\nu \overline{R_\nu[\alpha, z, w]}\right) d\mu(w)$$

$$= \sum_{\alpha \in R'} \eta(\alpha) \int_{D_\alpha} [\alpha, w]^k f_\alpha(w) e\left(-\sum_{\nu=1}^n q_\nu \overline{R_\nu[\alpha, z, w]}\right) d\mu(w),$$

where $D_\alpha = \Gamma_\alpha\backslash\mathfrak{H}^r$ and $f_\alpha(w) = {}^t\overline{u(\alpha)} f(w)$. This termwise integration will be justified later. For simplicity, put $G = SL_2(\mathbf{R})^r$ and $i = (i, \cdots, i)$ $(\in \mathfrak{H}^r)$. By (2.2), the integral over D_α can be written in the form

$$2^r \int_{\Gamma_\alpha\backslash G} [\alpha, gi]^k f_\alpha(gi) e\left(-\sum_{\nu=1}^n q_\nu \overline{R_\nu[\alpha, z, gi]}\right) dg$$

$$= 2^r \int_{H_\alpha\backslash G} e\left(-\sum_{\nu=1}^n q_\nu \overline{R_\nu[\alpha, z, gi]}\right) \left\{\int_{\Gamma_\alpha\backslash H_\alpha} [\alpha, hgi]^k f_\alpha(hgi) dh\right\} dg$$

$$= 2^r P(f, \alpha, \Gamma) \int_{H_\alpha\backslash G} e\left(-\sum_{\nu=1}^n q_\nu \overline{R_\nu[\alpha, z, gi]}\right) dg.$$

By Lemma 2.1, we may assume that $\alpha \in V^*$. The last integral over $H_\alpha\backslash G$ is equal to

$$(2.17) \qquad e\left(-\sum_{\nu \leq r} q_\nu N(\alpha_\nu)\bar{z}_\nu - \sum_{\nu > r} q_\nu N(\alpha_\nu) z_\nu\right)$$

$$\times \int_{H_\alpha\backslash G} \exp\left(-\pi\sum_{\nu=1}^r q_\nu y_\nu \operatorname{Im}(g_\nu i)^{-2} |[\alpha_\nu, g_\nu i]|^2\right) dg.$$

Take $\beta \in G$ so that $(\beta\alpha\beta^{-1})_\nu$ is diagonal for every $\nu \leq r$, and put $H = \beta H_\alpha \beta^{-1}$. Then $H = A'^r$. Changing g for $\beta^{-1}g$, we find that the last integral over $H_\alpha\backslash G$ is equal to

$$(2.17') \qquad \int_{H\backslash G} \exp\left(-4\pi\sum_{\nu=1}^r q_\nu |N(\alpha_\nu)| y_\nu \operatorname{Im}(g_\nu i)^{-2} |g_\nu i|^2\right) dg.$$

Observe that if ψ is an H-invariant function on $\mathfrak{H}^r$, then

$$2^r\int_{H\backslash G}\psi(gi)dg=\int_{R^r}\psi(i+u)du\,.$$

Applying this to (2.17′), we find that (2.17) is equal to

$$2^{-2r}|qN(\alpha)y|^{-\xi/2}e_F(-qN(\alpha)z)\,.$$

Thus (2.16) is equal to

$$2^{-r}q^{-\xi/2}\sum_{a\in R(\Gamma)}\eta(\alpha)|N(\alpha)|^{-\xi/2}P(f,\,\alpha,\,\Gamma)e_F(-qN(\alpha)z)\,,$$

and hence we obtain our proposition.

Now (2.14) shows that (2.16) as a function of z behaves like a Hilbert modular form of weight h under $\varDelta$. It remains to show that the transform of (2.16) under every element of $SL_2(F)$ has a Fourier expansion with constant term 0. Take an arbitrary $\alpha=\begin{pmatrix}*&*\\c&d\end{pmatrix}\in SL_2(F)$. By [13, Prop. 7.1], $(c\bar{z}+d)^{-h}\theta(\alpha z,\,w)$ is a finite linear combination of functions of the form $\theta(bz,\,w,\,t\,;\,\eta',\,q)$ with $0\ll b\in F$ and locally constant functions η' on V. As can be seen from [13, p. 340], this expression is independent of t. Therefore (2.16) transformed by α is a finite linear combination of integrals of the above type, and hence has a Fourier expansion with constant term 0. This completes the proof of Theorem 2.2.

Let us now justify the termwise integration in the above proof and prove the convergence of (2.5) and (2.9). We consider here only the case $B=M_2(F)$, since the case of division algebra can be treated in a similar and easier way. Write the variable w on $\mathfrak{H}^n$ as $w=u+iv$, and recall that $D\ (=\Gamma\backslash\mathfrak{H}^n)$ is contained in the union of a compact subset K of $\mathfrak{H}^n$ and $\bigcup_{\beta\in T}\beta(W)$, where T is a finite subset of $SL_2(F)$, and

$$W=\{u+iv\in\mathfrak{H}^n\,\big|\,|u_\nu|\leq a,\,v_\nu>1\ \text{ for all }\nu\}$$

with a positive constant a. Observe that w_ν/v_ν belongs to a compact subset of $\mathfrak{H}$ if $w\in W$. Fix one $\beta\in T$. Then we see that

$$\mathrm{Im}\,(R_\nu[\alpha,\,z,\,\beta w])=\mathrm{Im}\,(R_\nu[\beta^{-1}\alpha\beta,\,z,\,w])\geq v_\nu^{-2}y_\nu P_\nu[\alpha]\qquad\text{for all }\ w\in W$$

with a positive definite quadratic form P_ν independent of w. Similarly

$$|\eta(\alpha)[\alpha,\,\beta w]^k|=|\eta(\alpha)[\beta^{-1}\alpha\beta,\,w]^kJ(\beta,\,w)^{-2k}|\leq v^{2k}g(\alpha)|J(\beta,\,w)|^{-2k}$$

for all $w\in W$ with a polynomial function g on V. Thus $\theta(z,\,\beta w,\,t)$ has a majorant

$$y^{F/2}\sum_{a\in L}g(\alpha)\exp(-2\pi\textstyle\sum_{\nu=1}^n q_\nu y_\nu v_\nu^{-2}P_\nu[\alpha])|J(\beta,\,w)|^{2k}$$

on W, where L is a lattice of V. Multiply each term by $(f(w)v^{2k})\circ\beta$ and integrate over W. Since $|(f\|_{2k}\beta)(w)|\leq M\cdot\exp(-\lambda v^{F/n})$ on W with positive constants M and λ, each integral is majorized by

$$(2.18) \qquad My^{F/2}g(\alpha)\int_W \exp(-\lambda v^{F/n}-2\pi \sum_{\nu=1}^n q_\nu y_\nu v_\nu^{-2} P_\nu[\alpha])v^{2k-2F}du\,dv\,.$$

We now need an elementary

LEMMA 2.4. *Let* $0<a\in R$, $0<b\in R$, *and* $r\in R$. *Then for every* $q\in R$, >0, *there exists a constant* C *depending only on* a, q, r *such that*

$$\int_1^\infty \exp(-ax^{1/n}-bx^{-2})x^r dx \leqq Cb^{-q}\,.$$

PROOF. Put $p=r+2q+(1/2)$. Then the Schwarz inequality shows that the square of the integral in question is majorized by

$$\int_1^\infty \exp(-2ax^{1/n})x^{2p}dx\int_1^\infty \exp(-2bx^{-2})x^{-4q-1}dx\,.$$

Putting $x^{-2}=t$, we see that the last integral is smaller than

$$(1/2)\int_0^\infty e^{-2bt}t^{2q-1}dt=(1/2)\Gamma(2q)(2b)^{-2q}\,,$$

which proves our assertion.

This lemma shows that the integrals of (2.18) as well as their sum over all $\alpha\in L$ are convergent. The same can be shown for the integrals over K in a similar and simpler way. This justifies the termwise integration in the proof of Proposition 2.3, and at the same time proves the convergence of (2.9), provided $P(f, \alpha, \Gamma)$ is convergent. As for $P(f, \alpha, \Gamma)$, if $F[\alpha]$ is a field or $N(\alpha)=0$, we see easily that $\Gamma_\alpha\backslash H_\alpha$ is compact, so that there is no problem. If $N(\alpha)\neq 0$ and $F[\alpha]$ is not a field, then $N(\alpha)\ll 0$ and $F[\alpha]$ is isomorphic to $F\oplus F$. Before treating this case, let us first express $P(f, \alpha, \Gamma)$ for each $\alpha\in V^*$ as an integral of a holomorphic r-form over a cycle.

Given $\alpha\in V^*$, take $\beta\in SL_2(R)^r$ so that $(\beta\alpha\beta^{-1})_\nu=\begin{pmatrix} c_\nu & 0 \\ 0 & -c_\nu \end{pmatrix}$ for $\nu=1, \cdots, r$ with $0<c_\nu\in R$. Then $\beta H_\alpha\beta^{-1}=A^r$. Now $F[\alpha]$ is either a quadratic extension of F or $F\ominus F$. If $F[\alpha]$ is a field, it has exactly $2r$ real archimedean primes, and hence its unit group has rank $n+r-1$; therefore $\{\pm 1\}\Gamma_\alpha/\{\pm 1\}$ is isomorphic to Z^r. Define an isomorphism ω of $(R^\times)^r$ onto H_α by

$$(2.19) \qquad \omega(s)=\left(\beta_1^{-1}\begin{pmatrix} s_1 & 0 \\ 0 & s_1^{-1} \end{pmatrix}\beta_1, \cdots, \beta_r^{-1}\begin{pmatrix} s_r & 0 \\ 0 & s_r^{-1} \end{pmatrix}\beta_r\right)$$

for $s=(s_1, \cdots, s_r)\in(R^\times)^r$, and put $v(s)=\omega(s)v^0$ with a point v^0 of $\mathfrak{H}^r$. Let X be a fundamental domain of $(R^\times)^r/\omega^{-1}(\Gamma_\alpha)$. Then, with $f_\alpha={}^t\overline{u(\alpha)}f$, we have

$$(2.20) \qquad P(f, \alpha, \Gamma)=\int_{\Gamma_\alpha\backslash H_\alpha}[\alpha, hv^0]^k f_\alpha(hv^0)dh$$

$$=\int_X [\alpha,\, v(s)]^k f_\alpha(v(s))\,|s_1 \cdots s_r|^{-1}ds_1 \cdots ds_r\,.$$

We have $\beta_\nu v(s)_\nu = s_\nu^2 (\beta v^0)_\nu$, so that

$$(2.21) \qquad\qquad J(\beta,\, v(s))_\nu^{-2} dv(s)_\nu = 2s_\nu (\beta v^0)_\nu ds_\nu\,.$$

On the other hand, by (1.20),

$$[\alpha,\, v(s)]_\nu = J(\beta,\, v(s))_\nu^2 [\beta\alpha\beta^{-1},\, \beta v(s)]_\nu = -2c_\nu s_\nu^2 (\beta v^0)_\nu J(\beta,\, v(s))_\nu^2\,.$$

This combined with (2.21) yields

$$(2.22) \qquad\qquad [\alpha,\, v(s)]_\nu s_\nu^{-1} ds_\nu = -c_\nu dv(s)_\nu\,,$$

and hence

$$(2.23) \quad P(f,\, \alpha,\, \Gamma) = (-1)^r |N(\alpha)|^{\xi/2} \int_{v(X)} [\alpha,\, w]^{k-\xi} \cdot {}^t\overline{u(\alpha)}\, f(w)\,dw_1 \wedge \cdots \wedge dw_r$$

with a suitable orientation of $v(X)$.

Next let us assume that $F[\alpha] = F \oplus F$. This happens only when $B = M_2(F)$. Then we may assume that $\beta \in SL_2(F)$ and $\beta\alpha\beta^{-1} = \begin{pmatrix} c & 0 \\ 0 & -c \end{pmatrix}$ with $c \in F$. Observe that

$$\beta \Gamma_\alpha \beta^{-1} = \left\{ \begin{pmatrix} s & 0 \\ 0 & s^{-1} \end{pmatrix} \,\middle|\, s \in U \right\}$$

with a subgroup U of $\mathfrak{r}^\times$ of finite index. Defining again $\omega(s)$, $v(s)$, and X in the same fashion as above, we see that (2.23) holds also in the present case. Now the convergence of $P(f,\, \alpha,\, \Gamma)$ can be shown by decomposing X into two parts corresponding to $|s_1 \cdots s_n| \geqq 1$ and $|s_1 \cdots s_n| \leqq 1$. The integral over the first part is convergent, since f is rapidly decreasing at the cusp $i\infty$. The convergence of the other part can be seen by the transformation $w \mapsto \varepsilon(w)$ with $\varepsilon = \begin{pmatrix} 0 & 1 \\ -1 & 0 \end{pmatrix}$.

We now state special cases of some results of [15] concerning Θ with later applications in view.

PROPOSITION 2.5. *Let $\varDelta$ be a congruence subgroup of $SL_2(F)$ for which (2.14) holds. For $g \in \mathcal{M}_h(\varDelta)$, put*

$$f(w) = \int_\Phi g(z)\Theta(z,\, w\,;\, \eta,\, q)\,\mathrm{Im}\,(z)^h d\mu(z) \qquad (\Phi = \varDelta \backslash \mathfrak{H}^n)\,.$$

Then $f \in \mathcal{M}_{2k,2\lambda}(\Gamma)$ with a congruence subgroup Γ of $B^\times$.

PROOF. The function $\theta(-\bar{z},\, w,\, t\,;\, \eta,\, q)$ is a finite linear combination of functions of [15, I, (6.6)]. Therefore, changing z for $-\bar{z}$, we obtain our assertion immediately from [15, I, Theorem 6.2].

Let us now assume

(2.24)   *F has a subfield E such that $[E:Q]=r$ and the restrictions of $\tau_1, \cdots, \tau_r$ to E are all different.*

Let $\tau_1', \cdots, \tau_r'$ be the restrictions of $\tau_1, \cdots, \tau_r$ to E. Define a function $\Theta^*(z, w)$ on $\mathfrak{H}^r \times \mathfrak{H}^r$ by

$$\Theta^*(z, w) = y^\sigma \Theta(z^*, w; \eta, q),$$

$$z_\nu^* = \begin{cases} q^{-\tau_\nu} z_\nu & \text{if } \nu \leq r, \\ q^{-\tau_\nu} \bar{z}_\lambda & \text{if } \nu > r \text{ and } \tau_\nu = \tau_\lambda' \text{ on } E, \end{cases}$$

where σ is an element of $(1/2)I_E$ defined by

$$\sigma = \mathrm{Res}_{F/E}(\lambda) + (3/2)([F:E]-1)\sum_{\nu=1}^r \tau_\nu'.$$

Define also an element μ of $(1/2)I_E$ by

$$\mu = \sum_{\nu=1}^r (k_\nu+2)\tau_\nu' - \mathrm{Res}_{F/E}(\lambda) - (3/2)[F:E]\sum_{\nu=1}^r \tau_\nu'.$$

It can easily be seen that

$$\Theta^*(\gamma z, w) = \overline{j(\gamma, z)}(c\bar{z}+d)^{\mu-(E/2)}\Theta^*(z, w) \quad \text{for every } \gamma = \begin{pmatrix} * & * \\ c & d \end{pmatrix} \in \Delta'$$

with a congruence subgroup Δ' of $SL_2(E)$.

PROPOSITION 2.6.   *For $g \in \mathfrak{M}_\mu(\Delta')$ with such a Δ', put*

$$f(w) = \int_{\Phi'} g(z)\Theta^*(z, w)\,\mathrm{Im}\,(z)^\mu d\mu(z) \qquad (\Phi' = \Delta' \backslash \mathfrak{H}^r).$$

Then $f \in \mathfrak{M}_{2k,2\lambda}(\Gamma)$ with a congruence subgroup Γ of $B^\times$. Moreover, if g has algebraic Fourier coefficients, then f is arithmetic in the sense of [15, II, §2].

PROOF.   Our assertions follow immediately from [15, I, Theorems 6.3 and 6.4]. It should be noted that the $\bar{Q}$-rationality of f defined on [13, p. 329] is consistent with that of [15, II, §2] as can easily be shown.

We conclude this section by giving transformation formula (2.14) in a more explicit form, which will be needed in the next section. The symmetric form S being as in (1.14), choose a basis $\{\beta_1, \beta_2, \beta_3\}$ of V over F so that $qS(\beta_i, \beta_j) \in \mathfrak{r}$ for all i and j with an element q of F as in Theorem 2.2; put $L = \sum_{i=1}^3 \mathfrak{r}\beta_i$ and use the same letter S to denote the matrix $(S(\beta_i, \beta_j))$ of size 3. For $u, v \in V$, put

(2.25) $f(z, w; u, v)$

$$= y^{\xi/2}\,\mathrm{Im}\,(w)^{-2k}\sum_{\alpha-u \in L} e_F(qS(\alpha, v))t(\alpha)[\alpha, \bar{w}]^k e(\sum_{\nu=1}^n q_\nu R_\nu[\alpha, z, w]).$$

This is a special case of (2.11) and of [13, (7.6)]. Let $\gamma=\begin{pmatrix} a & b \\ c & d \end{pmatrix}\in SL_2(\mathfrak{r})$; suppose $bq\{S\}\equiv 0\ (2\mathfrak{b}^{-1})$, $cq^{-1}\{S^{-1}\}\equiv 0\ (2\mathfrak{b})$, $cq^{-1}S^{-1}\equiv 0\ (\mathfrak{b})$, and $0\ll d\equiv 1\ (4\mathfrak{b})$, where $\{S\}$ denotes the vector whose components are the diagonal elements of S. Then

$$(2.26) \qquad f(\gamma z, w; u, v)=\left(\frac{2c\cdot\det(qS)}{d\mathfrak{r}}\right)(c\bar{z}+d)^h e(X)f(z, w; u', v')$$

with $u'=au+cv$, $v'=bu+dv$, and $X=(1/2)\,\mathrm{Tr}_{F/Q}(qS(u, v)-qS(u', v'))$. This follows from [13, Lemma 7.2].

3. The commutativity with Hecke operators.

Let us now assume $F=Q$. We are going to specialize Theorem 2.2 to this case and then study the behavior of (2.9) under Hecke operators. For each prime p, let Z_p and Q_p be the p-completions of Z and Q. We fix a maximal order $\mathfrak{o}$ in B, and put $B_p=B\otimes_Q Q_p$, $V_p=V\otimes_Q Q_p$. For every lattice $\mathfrak{a}$ in B, we denote by $\mathfrak{a}_p$ its p-closure in B_p. Let e be the discriminant of B, that is, the product of all primes p ramified in B. If $p\nmid e$, we can find an isomorphism μ_p of B_p onto $M_2(Q_p)$ such that $\mu_p(\mathfrak{o}_p)=M_2(Z_p)$. Then

$$\mu_p(V_p)=\left\{\begin{pmatrix} a & b \\ c & -a \end{pmatrix}\in M_2(Q_p)\right\}.$$

Let us fix a positive integer m prime to e and define an "order of level m" to be the lattice $\mathfrak{o}'$ contained in $\mathfrak{o}$ such that

$$(3.1) \qquad \mathfrak{o}'_p=\begin{cases} \left\{\begin{pmatrix} a & b \\ c & d \end{pmatrix}\in M_2(Z_p)\,\middle|\,c\in mZ_p\right\} & \text{if }\ p\nmid e, \\[1em] \mathfrak{o}_p & \text{if }\ p\mid e. \end{cases}$$

We put then

$$(3.2) \qquad \Gamma'_m=\{\gamma\in\mathfrak{o}'\,|\,N(\gamma)=1\},$$

$$(3.3) \qquad \Gamma_m=\left\{\gamma\in\Gamma'_m\,\middle|\,\mu_p(\gamma)\equiv\begin{pmatrix} 1 & * \\ 0 & 1 \end{pmatrix}\ (\mathrm{mod}\ m\mathfrak{o}_p)\ \text{for all }\ p\mid m\right\}.$$

These groups will be denoted simply by Γ' and Γ when m is fixed. To define Hecke operators, let Y_m, or simply Y, be the set of all α in $B_+^{\times}\cap\mathfrak{o}'$ such that $\mu_q(\alpha)\equiv\begin{pmatrix} a & * \\ * & * \end{pmatrix}\ (\mathrm{mod}\ m\mathfrak{o}_q)$ for every prime factor q of m with an integer a prime to m. Given a Dirichlet character φ modulo m, we define a map $\varphi_Y:Y\to C^{\times}$ by $\varphi_Y(\alpha)=\varphi(a)^{-1}$ for such α and a. Then, for a positive integer h, we denote by $S_h(\Gamma', \varphi)$ the vector space of all cusp forms g on $\mathfrak{H}$ such that $g\|_h\gamma=\varphi_Y(\gamma)g$ for all $\gamma\in\Gamma'$. Now, for each prime number p, there is an element β of Y such that

$$\Gamma'\beta\Gamma' = \{\xi \in Y \mid N(\xi) = p\}.$$

Let $\Gamma'\beta\Gamma' = \cup_\lambda \Gamma'\beta_\lambda$ be a disjoint coset decomposition. Then a Hecke operator $T(p)$ acting on $S_h(\Gamma', \varphi)$ is defined by

$$(3.4) \qquad f \mid T(p) = p^{(h/2)-1} \sum_\lambda \varphi_Y(\beta_\lambda)^{-1} f \|_h \beta_\lambda \qquad (f \in S_h(\Gamma', \varphi)).$$

This is meaningful for all primes p including those dividing me. (For details, see [15, II, §1]. The present definition is different from that of [15] by a factor $p^{(h/2)-1}$.) If $B = M_2(Q)$, Γ'_m and Γ_m coincide with the groups

$$\Gamma_0(m) = \left\{ \begin{pmatrix} a & b \\ c & d \end{pmatrix} \in SL_2(Z) \,\Big|\, c \equiv 0 \pmod{m} \right\},$$

$$\Gamma_1(m) = \left\{ \begin{pmatrix} a & b \\ c & d \end{pmatrix} \in \Gamma_0(m) \,\Big|\, a \equiv d \equiv 1 \pmod{m} \right\}.$$

In this case, we write $S(m, \varphi, h)$ for $S_h(\Gamma'_m, \varphi)$. Then (3.4) coincides with the classical $T(p)$ defined by Hecke.

Let N be a positive integer divisible by 4, χ a character modulo N such that $\chi(-1) = 1$, and κ an odd positive integer. We then denote by $S(N, \chi, \kappa/2)$ the space of cusp forms f satisfying

$$f(\gamma z) = \chi(d) j(\gamma, z)^\kappa f(z) \qquad \text{for every} \quad \gamma = \begin{pmatrix} * & * \\ c & d \end{pmatrix} \in \Gamma_0(N),$$

where j is defined by (1.12).

To state a special case of Theorem 2.2, we consider θ of (2.11) with $q = 1$ and $t = 1$. Thus it has a simpler form

$$(3.5) \qquad \theta(z, w; \eta) = y^{1/2} v^{-2k} \sum_{\alpha \in V} \overline{\eta(\alpha)} [\alpha, \bar{w}]^k e(R[\alpha, z, w])$$

$$(z \in \mathfrak{H}, \ w \in \mathfrak{H}, \ y = \mathrm{Im}(z), \ v = \mathrm{Im}(w)).$$

We now take an explicitly defined η as follows. Let $\mathfrak{o}^*$ be the maximal order in B such that $\mathfrak{o}_p^* = \mathfrak{o}_p$ for all p prime to m and

$$\mathfrak{o}_p^* = \left\{ \alpha \in B_p \,\Big|\, \mu_p(\alpha) = \begin{pmatrix} a & b/m \\ cm & d \end{pmatrix} \text{ with } a, b, c, d \in Z_p \right\}$$

for every $p \mid m$. Given $\alpha \in \mathfrak{o}^*$, we can find an integer b such that

$$\mu_p(\alpha) \equiv \begin{pmatrix} * & b/m \\ * & * \end{pmatrix} \pmod{\mathfrak{o}'_p} \qquad \text{for every} \quad p \mid m.$$

For a Dirichlet character ψ modulo m, we then define a locally constant function η_ψ on V by $\eta_\psi(\alpha) = \psi(b)$ for such α contained in $\mathfrak{o}^* \cap V$ and $\eta_\psi(\alpha) = 0$ for $\alpha \notin \mathfrak{o}^* \cap V$. (We put $\psi(b) = 0$ for $(b, m) \neq 1$.) For simplicity, we fix ψ throughout the rest of this section and write simply η for η_ψ.

THEOREM 3.1. *Let m be a positive integer prime to the discriminant e of B, and ψ a character modulo m. For $f \in S_{2k}(\Gamma'_m, \psi^2)$ with a positive integer k, put*

$$(3.6) \qquad L(z, f) = (2\mu(\Gamma \backslash \mathfrak{H}))^{-1} \sum_{a \in R(\Gamma)} \eta(\alpha) |N(\alpha)|^{-1/2} P(f, \alpha, \Gamma) e(-N(\alpha)z)$$

with $\Gamma = \Gamma_m$ and with the above η, where $P(f, \alpha, \Gamma)$ and $R(\Gamma)$ are defined as in (2.5) and Theorem 2.2. Suppose $\psi(-1) = (-1)^k$. Then $L(z, f)$ as a function of z belongs to $S(N, \chi, (2k+1)/2)$, where N is $2me$ or $4me$ according as me is even or odd, and $\chi(d) = \left(\dfrac{-1}{d}\right)^k \psi(d)$.

PROOF. By Proposition 2.3, we have

$$(3.7) \qquad L(z, f) = \mu(D)^{-1} \int_D \overline{\theta(z, w\,;\,\eta)} f(w) \operatorname{Im}(w)^{2k} d\mu(w)$$

$$= \langle \theta(z, w\,;\,\eta), f \rangle \qquad (D = \Gamma \backslash \mathfrak{H}).$$

Let $L = \mathfrak{o}' \cap V$. Consider the matrix S relative to this L as defined at the end of §2. Observe that $\det(S) = 2m^2 e^2$, $\{S\} \equiv N\{S^{-1}\} \equiv 0 \pmod 2$, and $(N/2)S^{-1}$ is integral. Therefore, from (2.26), we obtain $\theta(\gamma z, w\,;\,\eta) = \overline{\chi(d)}\,\overline{j(\gamma, z)}^{2k+1} \theta(z, w\,;\,\eta)$ for every $\gamma = \begin{pmatrix} * & * \\ c & d \end{pmatrix} \in \Gamma_0(N)$ when $0 < d \equiv 1 \pmod 4$. Since $\Gamma_0(N)$ is generated by such γ and -1, the formula is true for all elements of $\Gamma_0(N)$. On the other hand, we can easily verify that $\eta(\delta \alpha \delta^{-1}) = \psi_Y(\delta)^{-2} \eta(\alpha)$ for every $\delta \in \Gamma'$, and hence, by (2.13), we have

$$(3.8) \qquad \theta(z, \delta w, \eta) = \psi_Y(\delta)^2 J(\delta, w)^{2k} \theta(z, w\,;\,\eta) \qquad \text{for every} \quad \delta \in \Gamma'.$$

Therefore we obtain our assertion from Theorem 2.2 and Proposition 2.3.

REMARK. Series $L(z, f)$ and θ can be defined for an arbitrary character ψ modulo m. It can easily be seen, however, that they are equal to 0 if $\psi(-1) \neq (-1)^k$.

For each prime number p, a Hecke operator $T(p^2)$ acting on $S(N, \chi, (2k+1)/2)$ was defined in [8]. If $g(z) = \sum_{n=1}^\infty a(n)e(nz) \in S(N, \chi, (2k+1)/2)$, we have $g|T(p^2) = \sum_{n=1}^\infty b(n)e(nz)$ with

$$(3.9) \qquad b(n) = a(p^2 n) + \chi'(p)\left(\frac{n}{p}\right) p^{k-1} a(n) + \chi(p^2) p^{2k-1} a(p^{-2}n),$$

where $\chi'(n) = \left(\dfrac{-1}{n}\right)^k \chi(n)$; we understand that $a(p^{-2}n) = 0$ if $n \notin p^2 \mathbf{Z}$ and $\chi'(n) = \chi(n) = 0$ if n is not prime to N (see [8, Theorem 1.7]).

THEOREM 3.2. *For every odd prime p and for every $f \in S_{2k}(\Gamma'_m, \psi^2)$, we have*

$$L(z, f)|T(p^2) = L(z, f|T(p)).$$

The same relation holds also for $p = 2$ if me is even.

PROOF. Define a locally constant function η' on V by

$$\eta'(\alpha)=\chi'(p)p^{k-1}\Big(\frac{-N(\alpha)}{p}\Big)\eta(\alpha)+p^{k}\chi(p^2)\eta(\alpha/p)+p^{k-1}\eta(p\alpha)$$

if $N(\alpha)\in Z$ and $\eta'(\alpha)=0$ if $N(\alpha)\notin Z$, where $\chi'(n)=\psi(n)$ if n is prime to $2me$ and $\chi'(n)=0$ otherwise. In view of (3.9) and (2.8), we see easily that $L(z,f)|T(p^2)$ is given by the right-hand side of (3.6) with η replaced by η'. Let us first assume p is prime to $2me$. Then we have

$$L(z,f|T(p))=\langle\theta(z,w\,;\eta),f|T(p)\rangle=\psi(p)^2\langle\theta(z,w\,;\eta)|T(p),f\rangle$$

(cf. [15, II, Lemma 1.3]). Taking $\{\beta_\lambda|0\le\lambda\le p\}$ as in (3.4), we have, by (2.13),

$$\theta(z,w\,;\eta)|T(p)=\sum_{\lambda,\alpha}\psi_\Gamma(\beta_\lambda)^{-2}p^{2k-1}y^{1/2}v^{-2k}\overline{\eta(\beta_\lambda\alpha\beta_\lambda^{-1})}[\alpha,\bar{w}]^k e(p^2R[\alpha,z,w]).$$

Substituting $p^{-1}\alpha$ for α, we find that $\theta(z,w\,;\eta)|T(p)=\theta(z,w\,;\eta^*)\cdot$ with

$$\eta^*(\alpha)=p^{k-1}\sum_{\lambda=0}^{p}\psi_\Gamma(\beta_\lambda)^2\eta(\beta_\lambda\alpha\beta_\lambda^{-1}).$$

Observe that $\eta'(\gamma\alpha\gamma^{-1})=\psi_\Gamma(\gamma)^{-2}\eta'(\alpha)$ and $\eta^*(\gamma\alpha\gamma^{-1})=\psi_\Gamma(\gamma)^{-2}\eta^*(\alpha)$ for $\gamma\in\Gamma'$, and hence $L(z,f)|T(p^2)$ and $L(z,f|T(p))$ are given by the right-hand side of (3.6) with η replaced by η' and $\psi(p)^2\eta^*$, respectively. We shall therefore prove the desired equality by showing

$$(3.10) \qquad\qquad \eta'(\alpha)=\psi(p)^2\eta^*(\alpha)$$

for all $\alpha\in V$ with $N(\alpha)<0$. For this purpose, we first observe that $\eta'(\alpha)\ne0$ or $\eta^*(\alpha)\ne0$ only if $N(\alpha)\in Z$ and $p\alpha\in\mathfrak{o}^*$.

LEMMA 3.3. *Let p be a prime not dividing $2me$, and let*

$$W=\{\beta\in V\cap p^{-1}\mathfrak{o}^*\,|\,0>N(\beta)\in Z\}.$$

Then for every $\beta\in W$, there exists an element γ of Γ such that $\alpha=\gamma\beta\gamma^{-1}$ belongs to the following three types:

(1) $\mu_p(\alpha)=\begin{pmatrix} a & pb \\ c/p & -a \end{pmatrix}$ *with $a,b,c\in Z_p$, $c\notin pZ_p$;*

(2) $\mu_p(\alpha)=\begin{pmatrix} a & b \\ c & -a \end{pmatrix}$ *with $a,b,c\in Z_p$, $c\notin pZ_p$;*

(3) $\alpha\in p\mathfrak{o}^*$.

PROOF. We first note a simple fact: for every $x\in Z_p$, there exist elements δ and ε of Γ such that

$$(3.11) \qquad \mu_p(\delta)\equiv\begin{pmatrix} 1 & 0 \\ x & 1 \end{pmatrix}, \qquad \mu_p(\varepsilon)\equiv\begin{pmatrix} 0 & -1 \\ 1 & 0 \end{pmatrix} \qquad (\mathrm{mod}\ p\mathfrak{o}_p).$$

This follows from the strong approximation theorem due to Eichler [1, Satz 5]. Now, given $\alpha \in W$, suppose $\alpha \notin \mathfrak{o}_p$. Put $N(\alpha) = -n$ and $p\mu_p(\alpha) = \begin{pmatrix} a & b \\ c & -a \end{pmatrix}$. Then $a^2 + bc = p^2 n$. Suppose $p \mid b$. Then $p \mid a$ but $p \nmid c$ since $\alpha \notin \mathfrak{o}_p$, so that $p^2 \mid b$. Thus α has form (1). If $p \nmid b$, choose $x \in \mathbf{Z}$ so that $a \equiv bx \pmod{p\mathbf{Z}_p}$ and take $\delta \in \Gamma$ as in (3.11). Then

$$p\mu_p(\delta\alpha\delta^{-1}) \equiv \begin{pmatrix} a-bx & b \\ * & * \end{pmatrix} \pmod{p\mathfrak{o}_p}.$$

Replacing α by $\delta\alpha\delta^{-1}$, we may assume that $p \mid a$. Then $p^2 \mid c$. Take ε as in (3.11). Then we find that $\varepsilon\alpha\varepsilon^{-1}$ has form (1). Next suppose $\alpha \in \mathfrak{o}_p$, $\in p\mathfrak{o}_p$. Put $\mu_p(\alpha) = \begin{pmatrix} a & b \\ c & -a \end{pmatrix}$. Take δ as in (3.11). Then $\mu_p(\delta\alpha\delta^{-1}) = \begin{pmatrix} * & * \\ c' & * \end{pmatrix}$ with $c' \equiv c + 2ax - bx^2 \pmod{p\mathbf{Z}_p}$. Since $\alpha \notin p\mathfrak{o}_p$ and $p > 2$, we can choose x so that $c' \in p\mathbf{Z}_p$. Then $\delta\alpha\delta^{-1}$ has form (2). Finally if $\alpha \in p^{-1}\mathfrak{o}^*$ and $\alpha \in p\mathfrak{o}_p$, then $\alpha \in p\mathfrak{o}^*$, which completes the proof.

We are going to prove (3.10) for the elements α of types (1), (2), (3) of the above lemma. First let α be of type (1). Then $\eta'(\alpha) = p^{k-1}\eta(p\alpha) = p^{k-1}\psi(p)\eta(\alpha)$. We can take $\{\beta_\lambda\}$ of (3.4) so that

$$(3.12) \qquad \mu_p(\beta_\lambda) \equiv \begin{cases} \begin{pmatrix} 1 & \lambda \\ 0 & p \end{pmatrix} & (0 \leq \lambda < p) \\ \begin{pmatrix} p & 0 \\ 0 & 1 \end{pmatrix} & (\lambda = p). \end{cases} \pmod{p^2 \mathfrak{o}'_p}$$

Then we find that $\beta_\lambda\alpha\beta_\lambda^{-1} \in \mathfrak{o}^*$ if and only if $\lambda = 0$. Therefore $\eta^*(\alpha) = p^{k-1}\psi_Y(\beta_0)^2 \cdot \eta(\beta_0\alpha\beta_0^{-1})$. For a prime factor q of m, let $\mu_q(\alpha) = \begin{pmatrix} a & b/m \\ cm & -a \end{pmatrix}$ and $\mu_q(\beta_0) = \begin{pmatrix} r & s \\ tm & u \end{pmatrix}$. Then

$$(3.13) \qquad \mu_q(\beta_0\alpha\beta_0^{-1}) \equiv \begin{pmatrix} * & r^2 b/(pm) \\ * & * \end{pmatrix} \pmod{\mathfrak{o}'_q},$$

and hence $\eta(\beta_0\alpha\beta_0^{-1}) = \psi(p)^{-1}\psi_Y(\beta_0)^{-2}\eta(\alpha)$. This shows that $\eta^*(\alpha) = p^{k-1}\psi(p)^{-1}\eta(\alpha)$. Next suppose α is of type (2); let $n = -N(\alpha)$. Then

$$\eta'(\alpha) = \psi(p)p^{k-1}\left(\frac{n}{p}\right)\eta(\alpha) + p^{k-1}\psi(p)\eta(\alpha).$$

Put $\mu_p(\alpha) = \begin{pmatrix} a & b \\ c & -a \end{pmatrix}$. Then $\beta_\lambda\alpha\beta_\lambda^{-1} \in \mathfrak{o}^*$ if and only if $b - 2a\lambda - c\lambda^2 \in p\mathbf{Z}_p$, $\lambda \neq p$. The number of such λ is $\left(\frac{n}{p}\right) + 1$. We have again $\eta(\beta_\lambda\alpha\beta_\lambda^{-1}) = \psi(p)^{-1}\psi_Y(\beta_\lambda)^{-2}\eta(\alpha)$, so that $\eta^*(\alpha)=p^{k-1}\left(1\div\left(\frac{n}{p}\right)\right)\psi(p)^{-1}\tau(\alpha)$, which proves (3.10). Finally suppose $\alpha\in p\mathfrak{o}^*$. Then $p^2|N(\alpha)$, and

$$\eta'(\alpha)=p^k\psi(p^2)\eta(\alpha/p)+p^{k-1}\tau(p\alpha)=(p+1)p^{k-1}\psi(p)\tau(\alpha).$$

In this case $\beta_\lambda\alpha\beta_\lambda^{-1}\in\mathfrak{o}^*$ for all λ, so that $\eta^*(\alpha)=p^{k-1}(p+1)\psi(p)^{-1}\tau(\alpha)$. Thus (3.10) is true for all $\alpha\in W$. This proves the equality of Theorem 3.2 for p not dividing $2me$.

Suppose $p|m$. Then $\eta'(\alpha)=p^{k-1}\eta(p\alpha)$ if $N(\alpha)\in Z$. Let δ be an element of Y such that $N(\delta)=p$ and $\mu_q(\delta)\equiv\begin{pmatrix}1 & 0\\ 0 & p\end{pmatrix}$ (mod $p^2m\mathfrak{o}_q$) for all prime factors q of m. Then we have $f|T(p)=p^{k-1}\sum_{\lambda=0}^{p-1}f\|\delta_\lambda$ with $\{\delta_\lambda\}$ such that $\Gamma\delta\Gamma=\bigcup_{j=0}^{p-1}\Gamma\delta_\lambda$. Let $\beta=\delta^\iota$ and $\Gamma\beta\Gamma=\bigcup_{\lambda=0}^{p-1}\Gamma\beta_\lambda$. Then we have

$$L(z, f|T(p))=\langle\theta(z, w ; \tau), f|T(p)\rangle$$

$$=p^{k-1}\sum_{\lambda=0}^{p-1}\langle\theta(z, w ; \eta)\|\beta_\lambda, f\rangle$$

by [7, Prop. 3.39], and hence, by (2.13),

$$L(z, f|T(p))=\langle\theta(z, w ; \eta^*), f\rangle$$

with $\eta^*(\alpha)=p^{k-1}\sum_{\lambda=0}^{p-1}\tau(\beta_\lambda\alpha\beta_\lambda^{-1})$. Now we can take β_λ so that $\mu_q(\beta_\lambda)\equiv\begin{pmatrix}p & 0\\ \lambda m & 1\end{pmatrix}$ (mod $p^2m\mathfrak{o}_q'$) for all prime factors q of m. For the same reason as in the case $p\nmid m$, it is sufficient to prove that $\eta^*(\alpha)=\eta'(\alpha)$ for α in a suitably chosen $R(\Gamma)$. We may naturally assume that $N(\alpha)\in Z$. First suppose $\alpha\in\mathfrak{o}^*$ and put $\mu_p(\alpha)=\begin{pmatrix}* & r/m\\ * & *\end{pmatrix}$. Then $\eta'(\alpha)=0$ and $\mu_p(\beta_\lambda\alpha\beta_\lambda^{-1})\equiv\begin{pmatrix}* & pr/m\\ * & *\end{pmatrix}$ (mod $p\mathfrak{o}_p$), so that $\eta^*(\alpha)=0$.

Next suppose $\alpha\in\mathfrak{o}^*$ and put $p\cdot\mu_p(\alpha)=\begin{pmatrix}a & b/m\\ cm & -a\end{pmatrix}$. If $b\in pZ_p$, we have again $\eta^*(\alpha)=\tau(\alpha)=0$ for the same reason. Therefore assume $b\notin pZ_p$. Take $x\in Z$ so that $a-bx\in pZ_p$, and take $\gamma\in\Gamma$ so that $\mu_p(\gamma)\equiv\begin{pmatrix}1 & 0\\ xm & 1\end{pmatrix}$ (mod $m\mathfrak{o}_p^*$). Then $p\cdot\mu_p(\gamma\alpha\gamma^{-1})=\begin{pmatrix}a' & b'/m\\ * & *\end{pmatrix}$ with $b'\notin pZ_p$, $a'\in pZ_p$. Replacing α by $\gamma\alpha\gamma^{-1}$, we may assume that $a\in pZ_p$. Since $a^2+bc\in p^2Z$, we see that $p^2|c$. Now a direct calculation shows that $\beta_\lambda\alpha\beta_\lambda^{-1}\in\mathfrak{o}^*$ if and only if $\lambda=0$, and

$$\mu_p(\beta_0\alpha\beta_0^{-1})\equiv\begin{pmatrix}a/p & b/m\\ cm/p^2 & -a/p\end{pmatrix} \quad (\text{mod } \mathfrak{o}_p').$$

A similar congruence can be found for each prime factor q of m. We then obtain $\eta^*(\alpha)=p^{k-1}\eta(p\alpha)=\eta'(\alpha)$.

Let us finally consider the case $p|e$. We have again $\eta'(\alpha)=p^{k-1}\eta(p\alpha)$. Take $\beta\in\mathfrak{o}'$ so that $N(\beta)=p$. Then $f|T(p)=p^{k-1}\psi_Y(\beta)^{-2}f\|\beta$, so that

$$L(z, f | T(p)) = p^{k-1} \psi_\Upsilon(\beta)^{-2} \langle \theta(z, w ; \eta), f \| \beta \rangle$$
$$= p^{k-1} \psi_\Upsilon(\beta)^{-2} \langle \theta(z, w ; \eta) \| \beta^{-1}, f \rangle = \langle \theta(z, w ; \eta^*), f \rangle$$

with $\eta^*(\alpha) = p^{k-1} \psi_\Upsilon(\beta)^{-2} \eta(\beta^{-1} \alpha \beta)$. A relation similar to (3.13) shows that $\eta(\beta^{-1} \alpha \beta) = \psi_\Upsilon(\beta)^2 \eta(p\alpha)$ and hence $\eta^* = \eta'$. This completes the proof of Theorem 3.2.

As mentioned in the introduction, Theorems 3.1 and 3.2 were given by Shintani in the case $B = M_2(Q)$ (with minor differences in formulation; see [16, Theorem 2]). His proof of the commutativity with Hecke operators relies on the theory of binary quadratic forms. Here we have presented a shorter and simpler proof which requires only local computations.

4. Main theorem on the periods.

Still with $F = Q$, for an arbitrary congruence subgroup Γ of $B^\times$, let $\mathcal{A}_0(\Gamma)$ denote the field of all Γ-invariant (meromorphic) automorphic functions on $\mathfrak{H}$ which take algebraic values at CM-points on $\mathfrak{H}$. Take an element g of $\mathcal{A}_0(\Gamma)$ other than the constants. We call an element f of $\mathcal{M}_h(\Gamma)$ arithmetic (or $\bar{Q}$-rational) if $\pi^h (dg/dz)^{-h} f^2$ belongs to $\mathcal{A}_0(\Gamma)$. (The arithmeticity can be defined in the general case with an arbitrary F. For details, see [12, §7], [13, §§4, 5], [15, II, §2].) It can be shown that $S_h(\Gamma'_m, \psi)$ is spanned by arithmetic elements (see [15, II, Lemma 2.3]); moreover Hecke operators send the set of arithmetic elements into itself.

We say that an element f of $S_h(\Gamma)$ is *primitive* if there exists m and φ with the following properties:

(4.1) $f \in S_h(\Gamma'_m, \varphi)$; f is an eigenform of all Hecke operators $T(p)$ (of level m);

(4.2)   *if l is a divisor of m smaller than m and g is an element of $S_h(\Gamma'_l, \varphi)$ which is an eigenform of $T(p)$ for almost all p, then the eigenvalues of $T(p)$ for g are different from those for f for infinitely many p.*

Then f is said to be of type (m, e, φ, h), where e is the discriminant of B defined in §3. Now the theory of Jacquet-Langlands in [2] (which sharpens that of Shimizu [5]) combined with the result of Miyake [3] shows the following facts.

(4.3)   *If f is a primitive form of type (m, e, φ, h), then there is a primitive form g of type $(me, 1, \varphi, h)$ with the same eigenvalues of Hecke operators Moreover, if f_1 is an element of $S_h(\Gamma'_m, \varphi)$ which is an eigenform of Hecke operators $T(p)$ for almost all p with the same eigenvalues as those for f, then f_1 is a constant multiple of f.*

As a consequence, every primitive form has a constant multiple which is arithmetic.

We shall now define the "fundamental periods" of a primitive form. This requires the cohomology group associated with $S_h(\Gamma)$. We first identify B_R with $M_2(R)$ and consider B a subring of $M_2(R)$. Then

$$V_R = V \otimes_Q R = \{x \in M_2(R) \mid \mathrm{Tr}\,(x)=0\}$$

for V defined by (1.13). Let $\begin{pmatrix} r & s \\ t & -r \end{pmatrix}$ be the variable element of V_R. For each integer $\kappa \geq 0$, we denote by $\mathscr{P}_R^\kappa$ the vector space over R of all R-valued homogeneous polynomial functions $\mathfrak{h}$ on V_R of degree κ such that $(\partial^2/\partial r^2 + 4\partial^2/\partial s\partial t)\mathfrak{h} = 0$. (This is similar to $\mathscr{P}_\lambda$ of Section 1, which was defined at the archimedean prime *ramified* in B.) We put then $\mathscr{P}_C^\kappa = \mathscr{P}_R^\kappa \otimes_R C$ and define a representation ρ_κ of $GL_2(R)$ on $\mathscr{P}_C^\kappa$ by

$$(4.4) \qquad [\rho_\kappa(\gamma)\mathfrak{h}](\xi) = \mathfrak{h}(\gamma'\xi\gamma) \qquad (\mathfrak{h} \in \mathscr{P}_C^\kappa,\ \gamma \in GL_2(R),\ \xi \in V_R).$$

For the same reason as in Lemma 1.1, ρ_κ is equivalent to the representation of $GL_2(R)$ by symmetric tensors of degree 2κ. Obviously $\mathscr{P}_R^\kappa$ is stable under ρ_κ. If Γ is a congruence subgroup of $B^\times$, we can let Γ act on $\mathscr{P}_R^\kappa$ through ρ_κ. Therefore we can define the (first) cohomology group

$$H(\Gamma,\ \mathscr{P}_R^\kappa) = Z(\Gamma,\ \mathscr{P}_R^\kappa)/B(\Gamma,\ \mathscr{P}_R^\kappa)$$

with $Z(\Gamma,\ \mathscr{P}_R^\kappa)$ and $B(\Gamma,\ \mathscr{P}_R^\kappa)$ given as follows (cf. [6], [7, Ch. 8]). An element $\mathfrak{x}$ of $Z(\Gamma,\ \mathscr{P}_R^\kappa)$, called a *cocycle*, is a $\mathscr{P}_R^\kappa$-valued function on Γ such that

$$(4.5) \qquad \mathfrak{x}(\gamma\delta) = \mathfrak{x}(\gamma) + \rho_\kappa(\gamma)\mathfrak{x}(\delta) \qquad for\ every\ \ \gamma,\ \delta \in \Gamma,$$

$$(4.6) \qquad \mathfrak{x}(\sigma) \in [1 - \rho_\kappa(\sigma)]\mathscr{P}_R^\kappa\ for\ every\ parabolic\ element\ \sigma\ of\ \Gamma.$$

$B(\Gamma,\ \mathscr{P}_R^\kappa)$ consists of all $\mathfrak{x}$, called *coboundaries*, for which there exists an element $\mathfrak{h}$ of $\mathscr{P}_R^\kappa$ such that

$$(4.7) \qquad \mathfrak{x}(\gamma) = [1 - \rho_\kappa(\gamma)]\mathfrak{h} \qquad for\ all\ \ \gamma \in \Gamma.$$

In the present setting, however, it is more convenient to view such a cocycle or a coboundary as an R-valued function $\mathfrak{x}(\gamma,\ \xi)$ with variables $\gamma \in \Gamma$ and $\xi \in V_R$. Then (4.5), (4.6), and (4.7) can be written as

$$(4.5') \qquad \mathfrak{x}(\gamma\delta,\ \xi) = \mathfrak{x}(\gamma,\ \xi) + \mathfrak{x}(\delta,\ \gamma^{-1}\xi\gamma) \qquad (\gamma,\ \delta \in \Gamma;\ \xi \in V_R);$$

$$(4.6') \qquad \mathfrak{x}(\sigma,\ \xi) = \mathfrak{z}_\sigma(\xi) - \mathfrak{z}_\sigma(\sigma^{-1}\xi\sigma)\ with\ \mathfrak{z}_\sigma \in \mathscr{P}_R^\kappa\ for\ each\ parabolic\ element\ \sigma\ of\ \Gamma;$$

$$(4.7') \qquad \mathfrak{x}(\gamma,\ \xi) = \mathfrak{h}(\xi) - \mathfrak{h}(\gamma^{-1}\xi\gamma) \qquad with\ \ \mathfrak{h} \in \mathscr{P}_R^\kappa.$$

We can similarly define the modules Z, B, H with $\mathscr{P}_C^\kappa$ instead of $\mathscr{P}_R^\kappa$; obviously $H(\Gamma,\ \mathscr{P}_C^\kappa) = H(\Gamma,\ \mathscr{P}_R^\kappa) \otimes_R C$. For an element $\mathfrak{a}$ of $Z(\Gamma,\ \mathscr{P}_C^\kappa)$ or $H(\Gamma,\ \mathscr{P}_C^\kappa)$, its complex conjugate, real part, and imaginary part can be defined in an obvious way; they are denoted by $\bar{a}$, $\mathrm{Re}(a)$, and $\mathrm{Im}(a)$. If $a \in Z(\Gamma, \mathscr{P}_C^k)$, we denote by $\mathrm{cl}(a)$ its cohomology class.

Let us now investigate certain integrals attached to the elements of $S_{2k}(\Gamma)$ and their periods. We first observe that for every $w \in \mathfrak{H}$, $[\xi, w]^k$ as a function of $\xi \in V_R$ belongs to $\mathscr{P}_C^k$. Given $f \in S_{2k}(\Gamma)$, $v \in \mathfrak{H}$, $\xi \in V$, and $\gamma \in \Gamma$, put

$$X(z, \xi, f, v) = \int_v^z [\xi, w]^{k-1} f(w) dw \qquad (z \in \mathfrak{H}),$$

$$\mathfrak{x}(\gamma, \xi, f, v) = X(\gamma v, \xi, f, v).$$

These as functions of ξ belong to $\mathscr{P}_C^{k-1}$. It can easily be seen that

$$X(\gamma z, \xi, f, v) = X(z, \gamma^{-1}\xi\gamma, f, v) + \mathfrak{x}(\gamma, \xi, f, v);$$

$\mathfrak{x}(\gamma, \xi, f, v)$ as a function of (γ, ξ) belongs to $Z(\Gamma, \mathscr{P}_C^{k-1})$; moreover, its cohomology class is independent of v. (For the treatment in a more general case, the reader is referred to [6] and [7, Chapter 8].) We put then $c[f] = \mathrm{cl}(\mathfrak{x}(\gamma, \xi, f, v))$.

PROPOSITION 4.1. *The map $f \mapsto \mathrm{Re}(c[f])$ gives an R-linear isomorphism of $S_{2k}(\Gamma)$ onto $H(\Gamma, \mathscr{P}_R^{k-1})$.*

This is a special case of [6, Théorème 1] and of [7, Theorem 8.4]. In fact, we can assign to each $p \in \mathscr{P}_C^{k-1}$ a homogeneous function p^* on R^2 by $p^*(x) = p(\varepsilon^{-1}x \cdot {}^t x)$ for $x \in R^2$ as in the proof of Lemma 1.1. Obviously $\mathrm{Re}(p^*) = (\mathrm{Re}(p))^*$. Therefore $f \mapsto \mathrm{Re}(c[f])$ is essentially the same as the maps of [6, Théorème 1] and [7, Theorem 8.4].

To simplify our notation, with fixed Γ and k, let us write $H(R)$ and $H(C)$ for $H(\Gamma, \mathscr{P}_R^{k-1})$ and $H(\Gamma, \mathscr{P}_C^{k-1})$, and further denote by $H(\bar{Q})$ the submodule of $H(C)$ consisting of all cohomology classes represented by the cocycles $\mathfrak{x}$ such that $\mathfrak{x}(\gamma, \xi) \in \bar{Q}$ for all $\gamma \in \Gamma$ and all $\xi \in V$.

PROPOSITION 4.2. *There exists a C-valued C-bilinear alternating form A on $H(C)$ with the following properties:*

(4.8) $i\mu(D)\langle f, g \rangle = A(\overline{c[f]}, c[g])$ *for $f, g \in S_{2k}(\Gamma)$, where $D = \Gamma \backslash H$;*

(4.9) $i\mu(D)\{\langle f, g \rangle - \langle g, f \rangle\} = 4A(\mathrm{Re}(c[f]), \mathrm{Re}(c[g]))$ *for $f, g \in S_{2k}(\Gamma)$;*

(4.10)   *A is R-valued on $H(R)$ and $\bar{Q}$-valued on $H(\bar{Q})$.*

PROOF. This is an easy consequence of [6]. In fact, by virtue of Proposition 4.1, we can define an R-valued alternating form A on $H(R)$ so that (4.9) holds. As shown in [6, §4], $A(\mathrm{cl}(a), \mathrm{cl}(b))$ can be explicitly expressed in terms of $a(\gamma)$ and $b(\gamma)$ with finitely many elements γ of Γ and some quantities determined at the cusps. If we extend A to $H(C)$ C-linearly, then the computation of [6, §4] shows that (4.8) holds. Assertion (4.10) can be proved by the same argument as in the proof of [6, Théorème 2].

Let us now consider the case where Γ is the group Γ_m defined by (3.3). For each prime number p, we can find an element β of Y_m such that $N(\beta)=p$ and $\mu_q(\beta)\equiv\begin{pmatrix}1&0\\0&p\end{pmatrix}$ (mod $m\mathfrak{o}_q$) for all prime factors q of m. Let $\Gamma\beta\Gamma=\cup_\lambda\Gamma\beta_\lambda$ be a disjoint coset decomposition. Then it can be shown that $f|T(p)=p^{k-1}\sum_\lambda f\|_{2k}\beta_\lambda$ for every $f\in S_{2k}(\Gamma)$. Now we can define the action of $\Gamma\beta\Gamma$ on $H(C)$ and on $H(R)$ in a purely algebraic way as in [7, §8.3]; we denote this action also by $T(p)$. Then we have $c[f|T(p)]=c[f]|T(p)$. This is proved in [7, Prop. 8.5] with $\mathrm{Re}(c)$ instead of c; the proof can easily be seen to be valid for c. We can also find an element δ of $\mathfrak{o}$ such that $N(\delta)=-1$ and $\mu_q(\delta)\equiv\begin{pmatrix}-1&0\\0&1\end{pmatrix}$ (mod $m\mathfrak{o}_q$) for all prime factors q of m. Then $\delta^2\in\Gamma$, $\delta\Gamma\delta^{-1}=\Gamma$, $\delta\Gamma'\delta^{-1}=\Gamma'$ and $\varphi_Y(\delta\alpha\delta^{-1})=\varphi_Y(\alpha)$ for all $\alpha\in Y$, where Γ' is defined by (3.2) and φ is a character modulo m. Now, for $f\in S_{2k}(\Gamma)$, we put

$$(4.11)\qquad (f|\delta)(w)=J(\delta,\,w)^{-2k}\overline{f(\delta\bar{w})}\qquad(w\in\mathfrak{H}).$$

Then $f|\delta\in S_{2k}(\Gamma)$; moreover $f|\delta\in S_{2k}(\Gamma',\,\bar{\varphi})$ if $f\in S_{2k}(\Gamma',\,\varphi)$; further $(f|\delta)|T(p)=(f|T(p))|\delta$ for every prime p. We can also define the action of δ on $Z(\Gamma,\,\mathscr{L}_C^{k-1})$ by

$$(4.12)\qquad (\mathfrak{x}|\delta)(\gamma,\,\xi)=(-1)^k\mathfrak{x}(\delta^{-1}\gamma\delta,\,\delta^{-1}\xi\delta),$$

which induces an action on $H(C)$ in a natural way. A direct calculation shows that

$$(4.13)\qquad c[f]|\delta=\overline{c[f|\delta]}.$$

Let f be a primitive form of type $(m,\,e,\,\varphi,\,2k)$ which is arithmetic, and let $f|T(p)=\lambda_p f$ with $\lambda_p\in C$ for each prime p. We have $(f|\delta)|T(p)=\bar{\lambda}_p(f|\delta)$. Now, as the nature of $\{\lambda_p\}_p$, the following two cases can occur.

Case I. $\lambda_p=\bar{\lambda}_p$ for all p. (This is so if $\lambda_p=\bar{\lambda}_p$ for almost all p.)

Case II. $\lambda_p\neq\bar{\lambda}_p$ for infinitely many p.

In Case I, we have $f|\delta=bf$ with a constant b. Then $b\bar{b}=1$. By [15, II, Lemma 4.2], $f|\delta$ is arithmetic, so that $b\in\bar{Q}$. Taking $a\in\bar{Q}$ so that $b=a/\bar{a}$ and replacing f by af, we may assume that $f|\delta=f$. Therefore we shall always assume $f|\delta=f$ in Case I in the following treatment. In Case II, f and $f|\delta$ are linearly independent over C. Put, in either case,

$$U_C=\{z\in H(C)\,\big|\,z|T(p)=\lambda_p z\text{ for all }p\},$$

$$U=U_C\cap H(\bar{Q}).$$

Obviously $U_C = U \otimes_{\bar{Q}} C$.

Lemma 4.3. *If f is a primitive form as above, $c[f]$ and $c[f]|\delta$ form a basis of U_C over C. Furthermore, U has a basis $\{\mathfrak{a}, \mathfrak{b}\}$ over $\bar{Q}$ such that $\mathfrak{a}|\delta = \mathfrak{a}$ and $\mathfrak{b}|\delta = -\mathfrak{b}$.*

Proof. In Case I, Proposition 4.1 shows that $\mathrm{Re}(c[f])$ and $\mathrm{Im}(c[f])$ form a basis of U_C over C, and hence $c[f]$ and $\overline{c[f]}$ $(=c[f]|\delta)$ form a basis of U_C over C. In Case II, we see again from Proposition 4.1 that $\mathrm{Re}(c[f])$, $\mathrm{Im}(c[f])$, $\mathrm{Re}(c[f|\delta])$, $\mathrm{Im}(c[f|\delta])$ are linearly independent over C, and hence $c[f]$ and $\overline{c[f|\delta]}$ $(=c[f]|\delta)$ form a basis of U_C over C. In either case, δ has eigenvalues ± 1 on U_C with multiplicity one, which proves the second assertion.

With a fixed arithmetic primitive element f as above, we take $\mathfrak{a}$ and $\mathfrak{b}$ as in Lemma 4.3. Since $c[f] \in U_C$, we can put

$$(4.14) \qquad c[f] = u_+(f)\mathfrak{a} + u_-(f)\mathfrak{b}$$

with complex numbers $u_+(f)$ and $u_-(f)$. We call these numbers *the fundamental periods of f*. They are determined by $\{\lambda_p\}_p$ up to algebraic factors. We have

$$(4.15) \qquad \overline{c[f|\delta]} = c[f]|\delta = u_+(f)\mathfrak{a} - u_-(f)\mathfrak{b},$$

so that $u_+(f)u_-(f) \neq 0$ in view of Lemma 4.3. In Case I, U is stable under the complex conjugation, and hence we can take $\mathfrak{a}$ and $\mathfrak{b}$ to be real; then $u_+(f)$ is real and $u_-(f)$ is pure imaginary.

Theorem 4.4. *If f is an arithmetic primitive element as above, $\pi \langle f, f \rangle / [u_+(f)\overline{u_-(f)}]$ is an algebraic number.*

Proof. Put $u = u_+(f)$ and $v = u_-(f)$ for simplicity. By (4.8), we have

$$i\mu(D)\langle f, f \rangle = A(\overline{u\mathfrak{a} + v\mathfrak{b}}, u\mathfrak{a} + v\mathfrak{b}),$$

$$i\mu(D)\langle f|\delta, f|\delta \rangle = A(u\mathfrak{a} - v\mathfrak{b}, \overline{u\mathfrak{a} - v\mathfrak{b}}).$$

Since $\langle f, f \rangle = \langle f|\delta, f|\delta \rangle$, we have

$$(4.16) \qquad i\mu(D)\langle f, f \rangle = \bar{u}vA(\bar{\mathfrak{a}}, \mathfrak{b}) + u\bar{v}A(\bar{\mathfrak{b}}, \mathfrak{a}).$$

In Case I, $\mathfrak{a}$, $\mathfrak{b}$, u, and iv are real, so that

$$(4.17) \qquad i\mu(D)\langle f, f \rangle = 2u\bar{v}A(\bar{\mathfrak{b}}, \mathfrak{a}).$$

In Case II, we have

$$(4.18) \qquad 0 = i\mu(D)\langle f|\delta, f \rangle = A(u\mathfrak{a} - v\mathfrak{b}, u\mathfrak{a} + v\mathfrak{b}) = 2uvA(\mathfrak{a}, \mathfrak{b}).$$

Now $\mathrm{Re}(c[f+f|\delta])=u\mathfrak{a}+\bar{u}\bar{\mathfrak{a}}$ and $\mathrm{Re}(c[f-f|\delta])=v\mathfrak{b}+\bar{v}\bar{\mathfrak{b}}$. Since $\langle f+f|\delta, f-f|\delta\rangle=0$, we obtain, from (4.9) and (4.18),

$$0=A(u\mathfrak{a}+\bar{u}\bar{\mathfrak{a}}, v\mathfrak{b}+\bar{v}\bar{\mathfrak{b}})=u\bar{v}A(\mathfrak{a}, \bar{\mathfrak{b}})+\bar{u}vA(\bar{\mathfrak{a}}, \mathfrak{b}).$$

This combined with (4.16) yields again (4.17), which proves our assertion, since $\mu(D)/\pi$ and $A(\mathfrak{a}, \bar{\mathfrak{b}})$ are algebraic.

PROPOSITION 4.5. *For an arithmetic primitive f of type $(m, e, \psi^2, 2k)$ with $e>1$, put $2\mu(\Gamma\backslash\mathfrak{H})L(z, f)=\sum_{n=1}^{\infty}a_ne(nz)$ with $\Gamma=\Gamma_m$ and with $L(z, f)$ of (3.6). Then $a_n/u_+(f)\in\bar{\boldsymbol{Q}}$ for all n.*

PROOF. Let τ denote the order of $\Gamma\cap\{\pm1\}$. Given $\alpha\in R(\Gamma)$ such that $n=-N(\alpha)$, let γ be an element of Γ_α which generates $\Gamma_\alpha\{\pm1\}/\{\pm1\}$. Define an isomorphism ω of $\boldsymbol{R}^\times$ onto H_α as in (2.19). In the present situation, we have

$$\beta\alpha\beta^{-1}=\begin{pmatrix} n^{1/2} & 0 \\ 0 & -n^{1/2} \end{pmatrix}, \qquad \omega(s)=\beta^{-1}\begin{pmatrix} s & 0 \\ 0 & s^{-1} \end{pmatrix}\beta \qquad (s\in\boldsymbol{R}^\times)$$

with $\beta\in SL_2(\boldsymbol{R})$. Changing γ for γ^{-1} if necessary, we may assume that $\gamma=\omega(t)$ with $t>0$. Then (2.20) together with (2.22) shows that

$$P(f, \alpha, \Gamma)=-(2n^{1/2}/\tau)\int_v^{\gamma v}[\alpha, w]^{k-1}f(w)dw,$$

with $v\in\mathfrak{H}$. Let $\mathfrak{x}$ be a cocycle in the class $c[f]$. Since $\gamma\alpha=\alpha\gamma$, we see that $\mathfrak{x}(\gamma, \alpha)$ depends only on $c[f]$, and

$$n^{-1/2}P(f, \alpha, \Gamma)=-(2/\tau)\mathfrak{x}(\gamma, \alpha).$$

Put $\alpha_1=\delta^{-1}\alpha\delta$, $\gamma_1=\delta^{-1}\gamma\delta$. Then $\eta(\alpha_1)=\psi(-1)\eta(\alpha)$ and

$$n^{-1/2}P(f, \alpha_1, \Gamma)=-(2/\tau)\mathfrak{x}(\gamma_1, \alpha_1)=-(2/\tau)(\mathfrak{x}|\delta)(\gamma, \alpha).$$

Since $\psi(-1)=(-1)^k$, we have

$$(4.19) \qquad \eta(\alpha)n^{-1/2}P(f, \alpha, \Gamma)+\eta(\alpha_1)n^{-1/2}P(f, \alpha_1, \Gamma)$$

$$=-(2/\tau)\eta(\alpha)[\mathfrak{x}(\gamma, \alpha)+(\mathfrak{x}|\delta)(\gamma, \alpha)].$$

If α and α_1 are not conjugate under Γ, this sum is a part of the coefficient a_n. Suppose $\varepsilon\alpha\varepsilon^{-1}=\alpha_1$ with $\varepsilon\in\Gamma$; then $\psi(-1)\eta(\alpha)=\eta(\alpha_1)=\eta(\alpha)$, so that $\eta(\alpha)=0$ or $\psi(-1)=1$; thus (4.19) is 0 or its half contributes to a_n. Now, with a cocycle $\mathfrak{a}_0$ in the class $\mathfrak{a}$ as in (4.14), we have

$$\mathfrak{x}(\gamma, \alpha)+(\mathfrak{x}|\delta)(\gamma, \alpha)=2u_+(f)\mathfrak{a}_0(\gamma, \alpha).$$

Since $\mathfrak{a}_0(\gamma, \alpha)$ is algebraic, we see that (4.19) is $u_+(f)$ times an algebraic number, which completes the proof.

For an elliptic modular form $g(z)=\sum_{n=1}^{\infty}b_n e(nz)$ and a Dirichlet character ξ, put

$$D(s, g, \xi)=\sum_{n=1}^{\infty}\xi(n)b_n n^{-s}.$$

Suppose that g is a primitive element of type $(N, 1, \varphi, k)$ with $b_1=1$. As shown in [10], there exist two constants $v_+(g)$ and $v_-(g)$ with the following properties: *for $t\in Z$, $0<t<k$, one has*

$$(4.20) \qquad \pi^{-t}D(t, g, \xi)\sim\begin{cases} v_+(g) & \text{if } \xi(-1)=(-1)^t, \\ v_-(g) & \text{if } \xi(-1)=(-1)^{t-1}, \end{cases}$$

$$(4.21) \qquad \pi\langle g, g\rangle\sim v_+(g)v_-(g).$$

Here and henceforth, we write $a\sim b$ for two complex numbers a and b if $b\neq 0$ and $a/b\in\overline{Q}$.

LEMMA 4.6. *If k is even, $v_+(g)\sim\overline{v_+(g)}$ and $v_-(g)\sim\overline{v_-(g)}$; if k is odd, $v_+(g)\sim\overline{v_-(g)}$.*

PROOF. Put $g'=\sum_{n=1}^{\infty}\bar{b}_n e(nz)$. Since $b_n=\varphi(n)\bar{b}_n$ for $(n, N)=1$, we have

$$(4.22) \qquad D(t, g, \xi)=a\cdot D(t, g', \xi\varphi)$$

for $0<t<k$ with $a\in\overline{Q}$ (depending on t). If $k>2$, the quantities of (4.22) are not 0 for $t=k-1$ by [9, Prop. 2] (cf. also [11, Prop. 4.16]). If $k=2$, we have to choose ξ so that (4.22) is not 0 for $t=1$, but this is guaranteed by [10, Theorem 2]. In any case, we have $v_+(g')\sim v_+(g)$ and $v_-(g')\sim v_-(g)$ if k is even, and $v_+(g')\sim v_-(g)$ if k is odd. As shown in [10, Theorem 1], $\overline{v_+(g)}\sim v_+(g')$ and $\overline{v_-(g)}\sim v_-(g')$, and hence we obtain our lemma.

We are now ready to state and prove our main theorem.

THEOREM 4.7. *Let ψ be a character modulo m such that $\psi(-1)=(-1)^k$, and f an arithmetic primitive form of type $(m, e, \psi^2, 2k)$ with $e>1$. Further let g be the primitive form of type $(me, 1, \psi^2, 2k)$ whose first Fourier coefficient is 1 and whose eigenvalues for Hecke operators are the same as those for f. Suppose $L(z, f)$ defined by (3.6) is not identically equal to 0. Then both $u_+(f)/v_+(g)$ and $u_-(f)/v_-(g)$ are algebraic numbers.*

PROOF. Put $h(z)=u_+(f)^{-1}\mu(\Gamma_m\backslash\mathfrak{H})L(f, z)$. By Proposition 4.5, h is $\overline{Q}$-rational, and by Theorems 3.1 and 3.2, $h\in S(N, \chi, (2k+1)/2)$ and $h|T(p^2)=\lambda_p h$ for all odd primes p. We have also

$$u_+(f)h(z)=\int_D \overline{\theta(z, w; \eta)}f(w)\operatorname{Im}(w)^{2k}d\mu(w) \qquad (D=\Gamma_m\backslash\mathfrak{H})$$

as in (3.7). Put

$$q(w)=\int_{D_1} h(z)\theta(z,\,w\,;\,\eta)\,\mathrm{Im}\,(z)^{k+(1/2)}\,d\mu(z) \qquad (D_1=\Gamma_1'(N)\backslash\mathfrak{H}).$$

Then

$$\mu(D_1)u_+(f)\langle h,\,h\rangle=\int_{D_1}\overline{h\,(z)}\int_{D}\overline{\theta(z,\,w\,;\,\eta)}\,f(w)\,\mathrm{Im}\,(w)^{2k}\,\mathrm{Im}\,(z)^{k+(1/2)}\,d\mu(w)d\mu(z)$$

$$=\mu(D)\langle q,\,f\rangle.$$

By Proposition 2.6, q is $\overline{Q}$-rational and by (3.8) belongs to $\mathcal{S}_{2k}(\Gamma_m)$. Since f is primitive, we have $q=bf+r$ with $b\in\overline{Q}$ and an element r of $\mathcal{S}_{2k}(\Gamma_m)$ orthogonal to f. Thus

$$\mu(D_1)u_+(f)\langle h,\,h\rangle=\mu(D)\bar{b}\langle f,\,f\rangle\sim u_+(f)\overline{u_-(f)}.$$

Now Theorem 1 of [14] asserts that $\pi\langle h,\,h\rangle\sim v_-(g)$, and hence $u_-(f)\sim\overline{v_-(g)}\sim v_-(g)$ by Lemma 4.6. On the other hand, $\langle f,\,f\rangle\sim\langle g,\,g\rangle$ by [15, II, Theorem 3.8], and hence $u_+(f)\overline{u_-(f)}\sim v_+(g)v_-(g)$ by Theorem 4.4 and (4.21), so that $u_+(f)\sim v_+(g)$. This completes the proof.

Let us now add a few concluding remarks. In the case $e=1$ (i.e. if $B=M_2(Q)$), we may put $f=g$. Then it can be shown that $u_+(f)\sim v_+(f)$ and $u_-(f)\sim v_-(f)$ more directly by the same technique as in the proof of [10, Theorem 3].

Recently, Ribet proved in [4] that the jacobian of $\Gamma_1'\backslash\mathfrak{H}$ is isogenous over Q to the primitive part of the jacobian of $\Gamma_0(e)\backslash\mathfrak{H}$. We can derive from this the conclusion of the above theorem when $k=1$ and $m=1$. Conversely, if we could prove the rationality of $u_+(f)/v_+(g)$ and $u_-(f)/v_-(g)$ over a specific number field, that would extend the result of Ribet to the case of arbitrary m.

In the above theorem, we assumed that m is prime to e and $L(z,\,f)\neq0$. It is naturally desirable to remove these conditions.

We treated in this section only the case $F=Q$. It seems that our methods can be generalized to a totally indefinite B over a field F of higher degree. The case of a partially definite B (i.e. the case with $r<n$), however, is not so transparent. The periods of the elements of $\mathcal{S}_{2k,2\lambda}(\Gamma)$ in this case seem to be related to the inner products of the nonholomorphic pullback of a Hilbert modular form of half-integral weight with holomorphic forms, which are similar to the inner products considered in [15, II, Theorems 3.6 and 3.7].

References

[1] Eichler, M., Allgemeine Kongruenzklasseneinteilungen der Ideale einfacher Algebren über algebraischen Zahlkörpern und ihre L-Reihen, J. Reine Angew. Math. **179** (1938), 227-251.

[2] Jacquet, H. and R. P. Langlands, Automorphic forms on $GL(2)$, Lecture Notes in Mathematics 114, Springer, 1970.

[3] Miyake, T., On automorphic forms on GL_2 and Hecke operators, Ann. of Math. **94** (1971), 174-189.

[4] Ribet, K., Sur les variétés abéliennes à multiplications réelles, C. R. Acad. Sci. Paris **291** (1980), Ser. A, 121-123.

[5] Shimizu, H., On zeta functions of quaternion algebras, Ann. of Math. **81** (1965), 166-193.

[6] Shimura, G., Sur les intégrales attachées aux formes automorphes, J. Math. Soc. Japan **11** (1959), 291-311.

[7] Shimura, G., Introduction to the arithmetic theory of automorphic functions, Iwanami Shoten and Princeton Univ. Press, 1971.

[8] Shimura, G., On modular forms of half integral weight, Ann. of Math. **97** (1973), 440-481.

[9] Shimura, G., The special values of the zeta functions associated with cusp forms, Comm. Pure Appl. Math. **29** (1976), 783-804.

[10] Shimura, G., On the periods of modular forms, Math. Ann. **229** (1977), 211-221.

[11] Shimura, G., The special values of the zeta functions associated with Hilbert modular forms, Duke Math. J. **45** (1978), 637-679.

[12] Shimura, G., Automorphic forms and the periods of abelian varieties, J. Math. Soc. Japan **31** (1979), 561-592.

[13] Shimura, G., The arithmetic of certain zeta functions and automorphic forms on orthogonal groups, Ann. of Math. **111** (1980), 313-375.

[14] Shimura, G., The critical values of certain zeta functions associated with modular forms of half-integral weight, J. Math. Soc. Japan. **33** (1981), 649-672.

[15] Shimura, G., On certain zeta functions attached to two Hilbert modular forms I, II, Ann. of Math. **114** (1981), 127-164, 569-607.

[16] Shintani, T., On construction of holomorphic cusp forms of half integral weight, Nagoya Math. J. **58** (1975), 83-126.

(Received June 2, 1981)

Department of Mathematics
Princeton University
Fine Hall
Princeton, N. J. 08540
U. S. A.

82c

Confluent hypergeometric functions
on tube domains

Mathematische Annalen, 260 (1982), 269-302

Our problem specialized to the case of the Siegel upper half space H_m of degree m concerns the Fourier expansion of a series

$$S(z;\alpha,\beta) = \sum_{a \in L} \det(z+a)^{-\alpha} \det(\bar{z}+a)^{-\beta}.$$

Here z is a variable on H_m, L is a lattice in the space V of all real symmetric matrices of size m, and $(\alpha, \beta) \in \mathbf{C}^2$. It can be shown that this is convergent if $\mathrm{Re}(\alpha+\beta) > m$, and also that if $\mathrm{Re}(\alpha) > m/2$ and $\mathrm{Re}(\beta) > m/2$, it has a Fourier expansion of the form

$$\mu(V/L)S(x+iy;\alpha,\beta)$$
$$= 2^m \pi^{m(m+1)/2} i^{m\beta-m\alpha} \Gamma_m(\alpha)^{-1} \Gamma_m(\beta)^{-1} \sum_{h \in L'} e^{2\pi i \,\mathrm{tr}(hx)} \eta(2y, \pi h; \alpha, \beta),$$

where $\mu(V/L)$ is the measure of V/L, and

$$L' = \{h \in V | \mathrm{tr}(hL) \subset \mathbf{Z}\},$$

$$\Gamma_m(\alpha) = \pi^{m(m-1)/4} \prod_{k=0}^{m-1} \Gamma(\alpha-(k/2)),$$

$$\eta(g,h;\alpha,\beta) = \int_Q e^{-\mathrm{tr}(gx)} \det(x+h)^{\alpha-\kappa} \det(x-h)^{\beta-\kappa} dx \quad (0 < g \in V)$$

with $\kappa = (m+1)/2$, $Q = \{x \in V | x \pm h > 0\}$; we write $x > 0$ for $x \in V$ when x is positive definite. If $m = 1$, η can be expressed in terms of a classical confluent hypergeometric (or Whittaker) function

$$\zeta(y;\alpha,\beta) = \int_0^\infty e^{-yt}(1+t)^{\alpha-1} t^{\beta-1} dt \quad (0 < y \in \mathbf{R}),$$

and therefore can be continued as a meromorphic function in (α, β) to the whole $\mathbf{C}^2$. Functions η in the higher-dimensional case were first introduced by Koecher in [4]. (See also Maass [5, Sect. 18]. In fact, the above Fourier expansion is given

---

*   Supported by NSF Grant MCS-8100744

388

in [4, p. 65] and [5, p. 304].) Their analytic continuation, however, has been an open question, though the case $m=2$ was settled by Kaufhold [3]; we shall actually employ some of his ideas in our investigation. The study of Whittaker functions in the framework of representation theory was initiated by Jacquet [2] and further developed by Schiffmann and Shahidi. However, there is little overlap between their investigations and ours, though the integrals in question are similar and related.

Now the purpose of the present paper is to study the series and functions of the same type as S and η on all four major families of tube domains. In particular, we shall prove that η and its analogues in the three other cases, multiplied by certain elementary factors, can be continued as holomorphic functions to the whole $\mathbf{C}^2$. To illustrate our results, let us take h in V with p positive and q negative eigenvalues, and assume $m=p+q$. We consider then

$$\omega(g,h;\alpha,\beta)=2^{-p\alpha-q\beta}\Gamma_p(\beta-(q/2))^{-1}\Gamma_q(\alpha-(p/2))^{-1}$$
$$\cdot\delta_+(hg)^{\kappa-\alpha-(q/4)}\delta_-(hg)^{\kappa-\beta-(p/4)}\det(g)^{\alpha+\beta-\kappa}\eta(g,h;\alpha,\beta),$$

where $\delta_+(x)$ is the product of all positive eigenvalues of x, and $\delta_-(x)=\delta_+(-x)$. Our main theorem (Theorem 4.2) specialized to this case asserts that:

(i) *ω can be continued as a holomorphic function in (α,β) to the whole $\mathbf{C}^2$*;

(ii) $$\omega(g,h;\alpha,\beta)=\omega(g,h;\kappa-\beta,\kappa-\alpha);$$

(iii) $$|\omega(g,h;\alpha,\beta)|\leq Ae^{-\tau(hg)}(1+\mu(hg)^{-B})$$

if (α,β) stays in a compact subset T of $\mathbf{C}^2$, where A and B are positive constants depending only on T, $\tau(x)$ is the sum of the absolute values of all eigenvalues of x, and $\mu(x)$ is the minimum absolute value of eigenvalues of x.

The proof of (i) and (ii) for definite h is relatively easy. In fact, if $h>0$, (ii) can be shown by a simple change of order of integration in a double integral, which is valid for all $m\geq1$ (Theorem 3.1). Thus no knowledge of the classical function ζ will be required in this paper. To complete the proof in the most general case, we employ two types of integral expressions of ω with definite and indefinite h in terms of two ω's with definite h of lower degree. Properties (i)–(iii) for indefinite h will follow from those for definite h, which, in turn, can be reduced to the case of lower degree. In particular, (iii) is crucial for the convergence of the integral expressions.

As a consequence of (i)–(iii), we shall show that $\Gamma_m(\alpha+\beta-\kappa)^{-1}S(z;\alpha,\beta)$ can be continued as a holomorphic function in (α,β) to the whole $\mathbf{C}^2$, provided L consists of matrices with algebraic entries (Theorem 6.1). Moreover, we shall determine the nature of

(*) $$\lim_{s\to0}S(z;\alpha+s,s)=\lim_{s\to0}\sum_{a\in L}\det(z+a)^{-\alpha}|\det(z+a)|^{-2s}$$

as a function of s (Theorems 6.4 and 6.5).

Although functions S and η are objects of no little interest on their own, a principal motive of our investigation is in the application to the Eisenstein series of arithmetic groups such as $\mathrm{Sp}(m,\mathbf{Z})$ acting on tube domains, which the author hopes to treat on a future occasion. For example, our function ω seems particularly useful for the analysis of the Eisenstein series in the limit process similar to (∗). Such a problem for Eisenstein series on $SO(2, m)$ is being successfully studied by Indik in his thesis [1] by means of ω.

A final technical remark may be added. In our exposition, one type of domain obtained from a quadratic form of signature $(1, m-1)$ often needs a separate discussion. A more uniform treatment could have been made with the systematic use of Jordan algebras, which would have allowed also the inclusion of exceptional domains. We think, however, that our formulation is easier to read, and better suited to future applications.

1. Definition of Domains and Functions

There are four types of tube domains to be dealt with, which we refer to as Cases I–IV. The first three of them are defined with respect to the basic division ring $\mathbf{K}$, which denotes the real number field $\mathbf{R}$, the complex number field $\mathbf{C}$, and the Hamilton quaternion algebra $\mathbf{H}$ in Cases I–III, respectively. We put then

$$(1.1) \qquad\qquad\qquad \iota = [\mathbf{K}:\mathbf{R}].$$

The formulas in this paper will be cross-referred to, for example, as (4.6.K) and (4.7.IV), which are valid in Cases I–III, and in Case IV, respectively. We denote by $\mathbf{K}_n^m$ and also by $\mathbf{K}(m, n)$ the set of all $m \times n$ matrices with entries in $\mathbf{K}$; we write $\mathbf{K}^m$ for $\mathbf{K}_1^m$. The identity element and the zero element of $\mathbf{K}_m^m$ are denoted by 1_m and 0_m, with subscript m dropped when there is no fear of confusion. If $X_1, \ldots, X_r$ are square matrices, $\mathrm{diag}[X_1, \ldots, X_r]$ denotes the matrix with $X_1, \ldots, X_r$ in the diagonal blocks and 0 in all other blocks. The transpose of a matrix X is denoted by $'X$. For $X \in \mathbf{K}_n^m$, we put $X^* = '\bar{X}$, where the bar denotes the quaternion conjugate if $\mathbf{K} = \mathbf{H}$, and the complex conjugate otherwise.

Now, with a fixed positive integer m, we define a vector space V_m over $\mathbf{R}$ by

$$(1.2.\mathrm{K}) \qquad\qquad V_m = \{x \in \mathbf{K}_m^m \mid x^* = x\},$$

$$(1.2.\mathrm{IV}) \qquad\qquad V_m = \mathbf{R}^m.$$

We assume $m \geqq 2$ in Case IV. In Case III, we identify $\mathbf{H}_m^m$ with

$$\left\{\begin{pmatrix} a & b \\ -\bar{b} & \bar{a} \end{pmatrix} \in \mathbf{C}_{2m}^{2m} \,\middle|\, a, b \in \mathbf{C}_m^m\right\}$$

and define $\det(x)$ and $\mathrm{tr}(x)$ for $x \in \mathbf{H}_m^m$ to be the trace and the determinant of the corresponding element of $\mathbf{C}_{2m}^{2m}$. Thus V_m in Case III can be identified with

$$(1.3.\mathrm{III}) \qquad\qquad \left\{\begin{bmatrix} a & b \\ -\bar{b} & \bar{a} \end{bmatrix} \in \mathbf{C}_{2m}^{2m} \,\middle|\, a^* = a,\, 'b = -b\right\}.$$

We define the complexification $V_{\mathbf{C}}$ of V_m by

$$(1.4.\mathrm{I}) \qquad\qquad V_{\mathbf{C}} = \{x \in \mathbf{C}_m^m \mid 'x = x\},$$

$$(1.4.\mathrm{II}) \qquad\qquad V_{\mathbf{C}} = \mathbf{C}_m^m,$$

$$(1.4.\mathrm{III}) \qquad\qquad V_{\mathbf{C}} = \{z \in \mathbf{C}_{2m}^{2m} \mid 'zJ = Jz\}, \qquad J = \begin{bmatrix} 0 & -1_m \\ 1_m & 0 \end{bmatrix},$$

$$(1.4.\mathrm{IV}) \qquad\qquad V_{\mathbf{C}} = \mathbf{C}^m.$$

Obviously $V_{\mathbf{C}} = V_m \otimes_{\mathbf{R}} \mathbf{C} = V_m + iV_m$ [with V_m identified with (1.3.III) in Case III].

To define a domain of positivity in Case IV, we fix an **R**-valued **R**-bilinear symmetric form $\sigma : V_m \times V_m \to \mathbf{R}$ of signature $(1, m-1)$. We put then

(1.5.K)$\qquad\qquad\qquad\qquad P_m = \{x \in V_m | x > 0\}$.

(1.5.IV)$\qquad\qquad\qquad\quad P_m = \{x \in V_m | \sigma(x, x) > 0, \sigma(x, \varepsilon) > 0\}$.

Here, in Case IV, we fix an element ε of V_m such that $\sigma(\varepsilon, \varepsilon) = 1$; in Cases I–III, we write $x > y$ for $x, y \in V_m$ when $u^*(x-y)u > 0$ for all $u \in \mathbf{K}^m$, $\neq 0$; in Case IV, we write $x > y$ if $x - y \in P_m$.

Our functions will be considered on the "upper half space" and the "right half space"

(1.6)$\qquad\qquad\qquad H_m = \{x + iy \in V_\mathbf{C} | x \in V_m, y \in P_m\}$,

(1.7)$\qquad\qquad\qquad H'_m = \{x + iy \in V_\mathbf{C} | x \in P_m, y \in V_m\}$.

We consider also a group G_m defined by

(1.8.K)$\qquad\quad G_m = \mathrm{GL}_m(\mathbf{K})$,

(1.8.IV)$\qquad\quad G_m = \{a \in \mathrm{GL}_m(\mathbf{R}) | \sigma(ax, ay) = v(a)\sigma(x, y) \text{ with } 0 < v(a) \in \mathbf{R}\}$,

whose elements a give automorphisms of V_m by the action

(1.9.K)$\qquad\qquad\qquad\qquad x \mapsto axa^*$,

$\qquad\qquad\qquad\qquad\qquad\qquad\qquad\quad (x \in V_m, a \in G_m)$.

(1.9.IV)$\qquad\qquad\qquad\qquad x \mapsto ax$,

Obviously P_m is stable under this action. When the integer m is fixed, which will be the case for the most part in the following treatment, the symbols V_m, P_m, H_m, H'_m, G_m, and $\kappa(m)$ [see (1.14) below] will be simply denoted by V, P, H, H', G, and κ, respectively.

We now define various functions, the simplest of which is δ on H' given by

(1.10.I, II)$\qquad\qquad\qquad \delta(z) = \det(z)$,

(1.10.III)$\qquad\qquad\qquad \delta(z) = \det(z)^{1/2}$,$\qquad (z \in H')$,

(1.10.IV)$\qquad\qquad\qquad \delta(z) = \sigma(z, z)$,

In Case III, we take δ to be positive when $z \in P$; actually $\delta(z)$ in Case III can be defined as the "Pfaffian" of Jz. Then $\delta(z)^s$ for $s \in \mathbf{C}$ and $z \in H'$ is defined in all cases by $\delta(z)^s = \exp(s \cdot \log \delta(z))$ with $\log \delta(z) \in \mathbf{R}$ for $z \in P$. We also define $\delta(z)^s$ and $\delta(\bar{z})^s$ for $z \in H$ by

(1.11)$\qquad\qquad\qquad \delta(z)^s = i^{\varrho s}\delta(-iz)^s, \quad \delta(\bar{z})^s = i^{-\varrho s}\delta(i\bar{z})^2 \quad (z \in H, s \in \mathbf{C})$,

where $i^x = e^{x\pi i/2}$ for $x \in \mathbf{C}$, and

(1.12.K)$\qquad\qquad\qquad\qquad\qquad\qquad \varrho = m$,

(1.12.IV)$\qquad\qquad\qquad\qquad\qquad\qquad \varrho = 2$.

Notice that $\delta(u) > \delta(v) > 0$ if $u > v > 0$ in all four cases and also that $\delta(\bar{z})^s = \delta(z^*)^s$ for $z \in H$ in Cases I–III.

We define Euclidean measures on $\mathbf{C}_n^m$ and $\mathbf{H}_n^m$ by viewing them as $\mathbf{R}^{2mn}$ and $\mathbf{R}^{4mn}$ in a standard way. We then define a Euclidean measure dx on V by viewing V as $\mathbf{R}^{m(m+1)/2}$, $\mathbf{R}^m \times \mathbf{C}^{m(m-1)/2}$, $\mathbf{R}^m \times \mathbf{H}^{m(m-1)/2}$, and $\mathbf{R}^m$ in Cases I–IV, respectively, in an obvious fashion. We have then

(1.13.K) $$\qquad\qquad d(axa^*)=\delta(aa^*)^\kappa dx$$
(1.13.IV) $$\qquad\qquad d(ax)=v(a)^\kappa dx \qquad (a\in G),$$

where

(1.14.K) $$\qquad\qquad \kappa=\kappa(m)=1+\iota(m-1)/2,$$

(1.14.IV) $$\qquad\qquad \kappa=\kappa(m)=m/2.$$

More explicitly, κ is $(m+1)/2$, m, and $2m-1$ in Cases I, II, and III, respectively. Notice that the dimension of V is $\varrho\kappa$.

In Case IV, we extend σ to a **C**-bilinear form on $V_\mathbf{C}$. We define also in Cases I–III a **C**-valued **C**-bilinear form σ on $V_\mathbf{C}$ by

(1.15.I, II) $$\qquad\qquad \sigma(x,y)=\operatorname{tr}(xy),$$

(1.15.III) $$\qquad\qquad \sigma(x,y)=(1/2)\operatorname{tr}(xy).$$

In these three cases, we shall also write $\sigma(x,y)=\sigma(xy)$, using the letter σ also for a **C**-linear function on $V_\mathbf{C}$. Notice that $\sigma(x,y)>0$ if $x,y\in P$. Now, in all four cases, we have a well known formula

(1.16) $$\qquad \int_P e^{-\sigma(z,x)}\delta(x)^{s-\kappa}dx=\Gamma_m(s)\delta(z)^{-s} \qquad [z\in H',\ \operatorname{Re}(s)>\kappa-1]$$

with

(1.17.K) $$\qquad \Gamma_m(s)=\pi^{\iota m(m-1)/4}\prod_{k=0}^{m-1}\Gamma(s-(\iota k/2)),$$

(1.17.IV) $$\qquad \Gamma_m(s)=|\sigma|^{-1/2}2^{2s-1}\pi^{\kappa-1}\Gamma(s)\Gamma(s-\kappa+1),$$

where $|\sigma|=\det(\sigma(b_i,b_j))$ with the standard basis $\{b_i\}$ of $V=\mathbf{R}^m$ (see Siegel [7, (110)], [8, (20)] and Maass [5, p. 323]). In Cases I–III, we have

(1.18.K) $$\qquad \Gamma_{p+q}(s)=\pi^{\iota pq/2}\Gamma_p(s)\Gamma_q(s-(\iota p/2)).$$

Given an L^1-function f on V, its Fourier transform f^σ relative to σ is defined by

(1.19) $$\qquad f^\sigma(x)=\int_V e^{-2\pi i\sigma(x,y)}f(y)dy \qquad (x\in V).$$

Then the Fourier inversion formula gives (under a suitable condition)

(1.20) $$\qquad \int_V e^{2\pi i\sigma(x,y)}f^\sigma(y)dy=|\sigma|^{-1}f(x) \qquad (x\in V).$$

Here and henceforth, we put

(1.21.K) $$\qquad\qquad |\sigma|=2^{m(\kappa-1)}.$$

Now, define, with any fixed $b\in P$, a function g on V by

$$g(u)=\begin{cases} e^{-\sigma(u,b)}\delta(u)^{s-\kappa} & (u\in P),\\ 0 & (u\notin P).\end{cases}$$

Then (1.16) shows that

$$(1.22) \qquad \int_V e^{-2\pi i\sigma(u,t)} g(u)\,du = \Gamma_m(s)\delta(b+2\pi it)^{-s} \qquad (t \in V).$$

This is valid if g is integrable, that is, if $\mathrm{Re}(s) > \kappa - 1$. It follows that g is L^2 if $\mathrm{Re}(s) > \kappa - (1/2)$. Therefore, its Fourier transform, and hence $\delta(b+2\pi it)^{-s}$, is L^2 if $\mathrm{Re}(s) > \kappa - (1/2)$. Thus $\delta(b+2\pi it)^{-s}$ is L^1 if $\mathrm{Re}(s) > 2\kappa - 1$. Since g is continuous if $\mathrm{Re}(s) > \kappa$, (1.20) yields

$$(1.23) \qquad |\sigma| \Gamma_m(s) \int_V e^{2\pi i\sigma(t,u)} \delta(b+2\pi it)^{-s}\,dt = \begin{cases} e^{-\sigma(u,b)}\delta(u)^{s-\kappa} & (u \in P), \\ 0 & (u \notin P), \end{cases}$$

if $b \in P$ and $\mathrm{Re}(s) > 2\kappa - 1$.

We now consider an infinite series

$$(1.24) \qquad S(z, L; \alpha, \beta) = \sum_{a \in L} \delta(z+a)^{-\alpha}\delta(\bar{z}+\bar{a})^{-\beta}$$

with $(\alpha, \beta) \in \mathbf{C}^2$, $z \in H$, and any lattice (i.e., a discrete subgroup of maximal rank) L in V. [In Cases I–III, we may use $\delta(z^* + a)$ instead of $\delta(\bar{z}+\bar{a})$.] Postponing the proof of convergence, let us first investigate its (formal) Fourier expansion. Put, for $g \in P$, $h \in V$, and $(\alpha, \beta) \in \mathbf{C}^2$,

$$(1.25) \qquad \xi(g, h; \alpha, \beta) = \int_V e^{-2\pi i\sigma(h,x)} \delta(x+ig)^{-\alpha}\delta(x-ig)^{-\beta}\,dx,$$

$$(1.26) \qquad \eta(g, h; \alpha, \beta) = \int_{Q(h)} e^{-\sigma(g,x)} \delta(x+h)^{\alpha-\kappa}\delta(x-h)^{\beta-\kappa}\,dx,$$

$$Q(h) = \{x \in V \mid x > h,\ x > -h\}.$$

Obviously we have

$$(1.27) \qquad \eta(g, -h; \alpha, \beta) = \eta(g, h; \beta, \alpha).$$

As to the convergence of these integrals, we first note

Lemma 1.1. *There is a constant $\lambda > 0$ depending only on P such that*

$$e^{-\lambda|\mathrm{Im}(s)|}|\delta(z)|^{\mathrm{Re}(s)} \leqq |\delta(z)^s| \leqq e^{\lambda|\mathrm{Im}(s)|}|\delta(z)|^{\mathrm{Re}(s)}$$

for all $z \in H'$. The same is true with H instead of H'.

Proof. This follows easily from the existence of λ such that $|\arg(\delta(z))| \leqq \lambda$ for all $z \in H'$, whose proof in Case I is as follows. Given $z \in H'$, there is an element a of G such that $aza^* = \mathrm{diag}[1+ib_1, \ldots, 1+ib_m]$ with $b_v \in \mathbf{R}$ (see Lemma 2.9 below). Then $\delta(z) = \det(a)^{-2} \prod_{v=1}^m (1+ib_v)$. Therefore we can take $\lambda = m\pi/2$ in this case. The other three cases can be treated in the same way.

This lemma implies that

$$(1.28) \qquad |\delta(x+ig)^{-\alpha}\delta(x-ig)^{-\beta}| \leqq C|\delta(x+ig)|^{-\mathrm{Re}(\alpha+\beta)} \qquad (x \in V,\ g \in P)$$

with a constant C independent of x and g. Therefore, the convergence of (1.25) for $\mathrm{Re}(\alpha+\beta) > 2\kappa - 1$ follows from the convergence of (1.23). Now, by (1.16) and (1.23), we have

$$\xi(g,h;\alpha,\beta)=i^{\varrho\beta-\varrho\alpha}\int_V e^{-2\pi i\sigma(h,x)}\delta(g-ix)^{-\alpha}\delta(g+ix)^{-\beta}dx$$

$$=i^{\varrho\beta-\varrho\alpha}\Gamma_m(\alpha)^{-1}\int_V\int_P e^{-2\pi i\sigma(h,x)-\sigma(u,g-ix)}\delta(u)^{\alpha-\kappa}\delta(g+ix)^{-\beta}dudx$$

$$=i^{\varrho\beta-\varrho\alpha}\Gamma_m(\alpha)^{-1}\int_P e^{-\sigma(u,g)}\delta(u)^{\alpha-\kappa}\int_V e^{i\sigma(x,u-2\pi h)}\delta(g+ix)^{-\beta}dxdu$$

$$=i^{\varrho\beta-\varrho\alpha}(2\pi)^{\varrho\kappa}|\sigma|^{-1}\Gamma_m(\alpha)^{-1}\Gamma_m(\beta)^{-1}$$

$$\cdot\int_{u>0,\,u>2\pi h} e^{\sigma(\pi h-u,\,2g)}\delta(u)^{\alpha-\kappa}\delta(u-2\pi h)^{\beta-\kappa}du,$$

since $\varrho\kappa=\dim(V)$. Putting $v=u-\pi h$, we obtain

$$(1.29)\qquad \xi(g,h;\alpha,\beta)=i^{\varrho\beta-\varrho\alpha}|\sigma|^{-1}(2\pi)^{\varrho\kappa}\Gamma_m(\alpha)^{-1}\Gamma_m(\beta)^{-1}\eta(2g,\pi h;\alpha,\beta).$$

Our computation shows that (1.26) is convergent and equality (1.29) holds at least for $\mathrm{Re}(\alpha)>\kappa-1$ and $\mathrm{Re}(\beta)>2\kappa-1$. [We shall later prove better results on the convergence (see Theorem 3.1 and Remark 4.3).] Now ξ is holomorphic in (α,β) if $\mathrm{Re}(\alpha+\beta)>2\kappa-1$. Therefore, if $\mathrm{Re}(\alpha+\beta)>2\kappa-1$ and η can be continued to a neighborhood of (α,β), then (1.29) holds there. For example, suppose $h=0$. By (1.16), we have

$$(1.30)\qquad\qquad \eta(g,0;\alpha,\beta)=\Gamma_m(\alpha+\beta-\kappa)\delta(g)^{\kappa-\alpha-\beta}$$

if $\mathrm{Re}(\alpha+\beta)>2\kappa-1$, and hence

$$(1.31)\quad \xi(g,0;\alpha,\beta)=\int_V\delta(x+ig)^{-\alpha}\delta(x-ig)^{-\beta}dx$$

$$=i^{\varrho\beta-\varrho\alpha}|\sigma|^{-1}(2\pi)^{\varrho\kappa}\Gamma_m(\alpha)^{-1}\Gamma_m(\beta)^{-1}\Gamma_m(\alpha+\beta-\kappa)\delta(2g)^{\kappa-\alpha-\beta}$$

if $\mathrm{Re}(\alpha+\beta)>2\kappa-1$.

Now the Poisson summation formula shows at least formally that

$$(1.32)\qquad\qquad \mu(V/L)S(z,L;\alpha,\beta)=\sum_{h\in L'} e^{2\pi i\sigma(h,x)}\xi(y,h;\alpha,\beta)$$

for $z=x+iy\in H$, where $L'=\{h\in V|\sigma(h,L)\subset\mathbf{Z}\}$ and $\mu(V/L)$ is the measure of V/L. It is this formula which gives the raison d'être to functions ξ and their companions η. As explained in the introduction, the main purpose of this paper is to study the analytic continuation of ξ and η in (α,β) and to prove certain asymptotic formulas for them. To prove the convergence of $S(z,L;\alpha,\beta)$, we need

Lemma 1.2. *Let D be a domain in $\mathbf{C}^n$. For every compact subset T of D, there exists a bounded open set U containing T whose closure is contained in D and a constant C such that $|f(w)|\leqq C\int_U|f(z)|dv(z)$ for every $w\in T$ and every holomorphic function f on D, where $dv(z)$ is the Euclidean measure on $\mathbf{C}^n$.*

This is well known and follows easily from the Cauchy integral formula.

Lemma 1.3. *Series $S(z,L;\alpha,\beta)$ is locally uniformly convergent on*

$$(1.33)\qquad\qquad H\times\{(\alpha,\beta)\in\mathbf{C}^2|\mathrm{Re}(\alpha+\beta)>2\kappa-1\}.$$

Proof. If $r > 2\kappa - 1$, (1.31) shows that

$$(1.34) \qquad \int_V |\delta(x+iy)|^{-r} dx = |\sigma|^{-1} (2\pi)^{\varrho\kappa} \Gamma_m(r/2)^{-2} \Gamma_m(r-\kappa) \delta(2y)^{\kappa-r} .$$

Given a compact subset T of H, take U as in Lemma 1.2 with H as D, and further take compact subsets F of V and K of P so that $U \subset F \times K$. Let N be the number of elements a of L such that $F \cap (F+a) \neq \emptyset$. If A is a finite subset of L and $r > 2\kappa - 1$, we have

$$\int_{F \times K} \sum_{a \in A} |\delta(x+iy+a)|^{-r} dx dy \leq N \int_{V \times K} |\delta(x+iy)|^{-r} dx dy .$$

By (1.34), if r stays in a compact set J, the last integral is smaller than a constant B depending only on K and J. By Lemma 1.2, we have $\sum_{a \in A} |\delta(z+a)|^{-r} \leq CNB$ for $z \in T$ with a constant C. This together with (1.28) proves the desired convergence.

This lemma shows that S is holomorphic in (α, β) and continuous on (1.33). Let us now prove

Lemma 1.4. $S(z, L; \alpha, \beta)$ *is real analytic on*

$$H \times \{(\alpha, \beta) \in \mathbf{C}^2 | \mathrm{Re}(\alpha) > \kappa - (1/2),\ \mathrm{Re}(\beta) > \kappa - (1/2)\} .$$

Proof. With complex variables z and w such that $z \in H$ and $\bar{w} \in H$, we consider

$$(1.35) \qquad \sum_{a \in L} \delta(z+a)^{-\alpha} \delta(w+\bar{a})^{-\beta} .$$

Since

$$\left\{ \sum_{a \in L} |\delta(z+a)^{-\alpha} \delta(w+\bar{a})^{-\beta}| \right\}^2 \leq \sum_{a \in L} |\delta(z+a)^{-2\alpha}| \sum_{a \in L} |\delta(w+\bar{a})^{-2\beta}| ,$$

(1.35) is locally uniformly convergent if $\mathrm{Re}(2\alpha) > 2\kappa - 1$ and $\mathrm{Re}(2\beta) > 2\kappa - 1$ by Lemma 1.3. Thus (1.35) is holomorphic in (α, β, z, w) for such (α, β), which proves our lemma.

By virtue of this lemma, we see that (1.32) is valid at least when $\mathrm{Re}(\alpha) > \kappa - (1/2)$ and $\mathrm{Re}(\beta) > \kappa - (1/2)$.

2. Some Elementary Lemmas

The first eight lemmas of this section concern only Cases I–III. We shall deal with Case IV towards the end of the section. Now put

$$(2.1.\mathrm{K}) \qquad\qquad U_m = \{u \in \mathrm{GL}_m(\mathbf{K}) | uu^* = 1_m\} .$$

It is well known that given $a \in V$, there is an element u of U_m such that $uau^* = \mathrm{diag}[\lambda_1, ..., \lambda_m]$ with $\lambda_v \in \mathbf{R}$. Naturally $\lambda_1, ..., \lambda_m$ are called *the eigenvalues of* a. If $a \neq 0$, we denote by $\mu(a)$ the minimum absolute value of nonzero eigenvalues of a. We have $a \in P$ if and only if $\lambda_v > 0$ for all v. We then define an element a^r of P for $r \in \mathbf{R}$ by $a^r = u^* \mathrm{diag}[\lambda_1^r, ..., \lambda_m^r] u$. This is characterized by the property that $ax = \lambda x$ with $0 < \lambda \in \mathbf{R}$ and $x \in \mathbf{K}^m$ implies $a^r x = \lambda^r x$. If $a, b \in P$ and $ab = ba$, then $a^r b^s = b^s a^r$ for $r, s \in \mathbf{R}$.

Lemma 2.1. *Let* $h = \begin{bmatrix} a & b \\ b^* & c \end{bmatrix} \in V_m$ *with* $a \in V_p$, $b \in \mathbf{K}_q^p$, $c \in V_q$, $p + q = m$. *Then*

$$h > 0 \Leftrightarrow \begin{Bmatrix} a > 0 \\ c > b^* a^{-1} b \end{Bmatrix} \Leftrightarrow \begin{Bmatrix} c > 0 \\ a > bc^{-1} b^* \end{Bmatrix},$$

$$\delta(h) = \delta(a)\delta(c - b^* a^{-1} b) = \delta(a - bc^{-1} b^*)\delta(c).$$

Proof. This follows from

$$\begin{bmatrix} 1 & -bc^{-1} \\ 0 & 1 \end{bmatrix} \begin{bmatrix} a & b \\ b^* & c \end{bmatrix} \begin{bmatrix} 1 & 0 \\ -c^{-1} b^* & 1 \end{bmatrix} = \begin{bmatrix} a - bc^{-1} b^* & 0 \\ 0 & c \end{bmatrix}$$

and a similar equality which produces $\operatorname{diag}[a, c - b^* a^{-1} b]$.

Lemma 2.2. *Put* $u = 1_p + sxx^*$, $v = 1_q + sx^* x$ *with* $s \in \mathbf{R}$ *and* $x \in \mathbf{K}_q^p$. *Then* $\delta(u) = \delta(v)$; $u > 0$ *if and only if* $v > 0$. *Moreover, if* $u > 0$, *we have* $u^r x = x v^r$ *for every* $r \in \mathbf{R}$.

Proof. The second assertion is obvious if $s \geq 0$. The case $s < 0$ can be seen from Lemma 2.1 with $a = 1_p$, $b = (-s)^{1/2} x$, $c = 1_q$ this proves at the same time the first assertion for $s < 0$, from which the case $s \geq 0$ follows since $\delta(u)$ and $\delta(v)$ are polynomials in s. Now observe that $ux = xv$. Suppose $vy = \lambda y$ with $0 < \lambda \in \mathbf{R}$ and $y \in \mathbf{K}^q$. Then $uxy = \lambda xy$, so that $u^r xy = \lambda^r xy = xv^r y$, and hence $u^r x = xv^r$.

Lemma 2.3. *Put* $\varphi(x) = (1_p + xx^*)^{-1/2} x$ *for* $x \in \mathbf{K}_q^p$. *Then* $\varphi(x) = x(1_q + x^* x)^{-1/2}$, *and* φ *gives a one-to-one map of* $\mathbf{K}_q^p$ *onto*

$$\{y \in \mathbf{K}_q^p \mid yy^* < 1_p\}.$$

The inverse map ψ *of* φ *is given by*

$$\psi(y) = (1_p - yy^*)^{-1/2} y = y(1_q - y^* y)^{-1/2}.$$

Moreover, if $y = \varphi(x)$, *we have*

$$1_p + xx^* = (1_p - yy^*)^{-1}, \qquad 1_q + x^* x = (1_q - y^* y)^{-1}.$$

Proof. Let $y = \varphi(x)$. By Lemma 2.2, $y = x(1_q + x^* x)^{-1/2}$, and hence

$$yy^* = x(1_q + x^* x)^{-1} x^* = (1_p + xx^*)^{-1} xx^* = 1_p - (1_p + xx^*)^{-1}.$$

Thus $1_p - yy^* = (1_p + xx^*)^{-1} > 0$ and $x = (1_p + xx^*)^{1/2} y = (1_p - yy^*)^{-1/2} y$. The remaining part of our lemma can be proved by the same technique.

Lemma 2.4. *Given* $u \in \mathbf{K}_q^p$ *with* $p \leq q$, *there exist two elements* $s \in U_p$ *and* $t \in U_q$ *such that* $sut = [\Lambda \quad 0]$, $\Lambda = \operatorname{diag}[\lambda_1, \ldots, \lambda_p]$ *with* $\lambda_v \in \mathbf{R}$.

Proof. Left multiplication by u defines a right $\mathbf{K}$-linear map of $\mathbf{K}^q$ into $\mathbf{K}^p$. Let $X = \operatorname{Ker}(u)$, $Y = \{y \in \mathbf{K}^q \mid y^* x = 0 \text{ for all } x \in X\}$, and $r = \dim_{\mathbf{K}} Y$. Then $r \leq p$. Thus we can find a basis $\{e_1, \ldots, e_q\}$ of $\mathbf{K}^q$ over $\mathbf{K}$ such that $e_i^* e_j = \delta_{ij}$ and $ue_{r+1} = \ldots = ue_q = 0$. In other words, there is an element a of U_q such that $ua = [w \quad 0]$ with $w \in \mathbf{K}_r^p$. Applying this result to w^*, we find an element b of U_p such that $bw = \begin{pmatrix} x \\ 0 \end{pmatrix}$ with $x \in \mathbf{K}_r^r$. Take $c \in U_r$ so that $c^* x^* xc = \operatorname{diag}[\mu_1, \ldots, \mu_r]$ with

$0 < \mu_i \in \mathbf{R}$. Put $f = xcd^{-1/2}$. Then $s = \mathrm{diag}[f^{-1}, 1_{p-r}]b$ and $t = a \cdot \mathrm{diag}[c, 1_{q-r}]$ have the required properties.

Lemma 2.5. *If $u \in \mathbf{K}_q^p$ and $p \leqq q$, the eigenvalues of $u^* u$ consist of the eigenvalues of uu^* and $q - p$ zeros.*

This follows easily from Lemma 2.4.

Lemma 2.6. *The jacobian of the map φ of Lemma 2.3 at $x \in \mathbf{K}_q^p$ is $\delta(1_p + xx^*)^{-\kappa(p+q)}$.*

Proof. We compute the jacobian of the inverse map ψ of φ. Put

$$(1 - \alpha)^{-1/2} = \sum_{n=0}^{\infty} c_n \alpha^n$$

for $\alpha \in \mathbf{R}$ with $c_n \in \mathbf{R}$. If $x = \psi(y) = (1 - yy^*)^{-1/2}y$, we have $x = \sum_{n=0}^{\infty} c_n(yy^*)^n y$. This is convergent for $yy^* < 1$. Then $dx = L_y(dy)$ with an $\mathbf{R}$-linear endomorphism L_y of $\mathbf{K}_q^p$ defined by

$$L_y(u) = u + c_1(uy^* y + yu^* y + yy^* u)$$
$$+ c_2(uy^* yy^* y + yu^* yy^* y + yy^* uy^* y + yy^* yu^* y + yy^* yy^* u) + \ldots.$$

Our problem is to compute $\det(L_y)$. Take $r \in U_p$ and $s \in U_q$ so that $rys = [\Lambda \quad 0]$, $\Lambda = \mathrm{diag}[\lambda_1, \ldots, \lambda_p]$ with $\lambda_\nu \in \mathbf{R}$ as in Lemma 2.4. (Here we assume $p \leqq q$.) Then $L_{rys}(rus) = rL_y(u)s$, so that $\det(L_y) = \det(L_{rys})$. Thus we may assume that $y = [\Lambda \quad 0]$. Then a straightforward calculation shows that $\det(L_y) = \prod_{i=1}^{p} (1 - \lambda_i^2)^{-t}$ with $t = 1 + (p + q - 1)\iota/2$, which proves our lemma.

Lemma 2.7. *If $x = (x_{ij}) \in P$, we have*

$$\delta(x) \leqq x_{11} \ldots x_{mm} \leqq \{\sigma(x)/m\}^m \quad and \quad \delta(1 + x) \leqq \{1 + m^{-1}\sigma(x)\}^m.$$

Proof. Put $x = \begin{bmatrix} x_{11} & y \\ y^* & z \end{bmatrix}$. By Lemma 2.1, $\delta(x) = (x_{11} - yz^{-1}y^*)\delta(z) \leqq x_{11}\delta(z)$. By induction on m, we have $\delta(x) \leqq x_{11} \ldots x_{mm}$. Now $(x_{11} \ldots x_{mm})^{1/m} \leqq \sum_i x_{ii}/m$. Applying this to $1 + x$, we obtain the last inequality.

Lemma 2.8. *For every compact subset J of $\mathbf{R}$, there exist two positive constants A and B depending only on P and J such that*

$$\int_{\mathbf{K}(n,m)} e^{-\sigma(xbx^*)}\delta(1_n + xx^*)^r dx \leqq A\delta(b)^{-\iota n/2}(1 + \mu(b)^{-B})$$

for $b \in P$ and $r \in J$, where dx is the Euclidean measure on $\mathbf{K}_m^n$.

Proof. We may assume that $b = \mathrm{diag}[b_1, \ldots, b_m]$. Then

$$(2.3.\mathrm{K}) \qquad \int_{\mathbf{K}(n,m)} e^{-\sigma(xbx^*)} dx = \{\pi^m/(b_1 \ldots b_m)\}^{\iota n/2}.$$

If $r \leq 0$, we have $\delta(1 + xx^*)^r \leq 1$, so that the integral in question is majorized by (2.3.K). Suppose $r > 0$. Applying $\partial^{|c|}/\partial b_1^{c_1} \dots \partial b_m^{c_m}$ to (2.3.K), we can evaluate

$$\int_{K(n,m)} e^{-\sigma(xbx^*)}(x_1^* x_1)^{c_1} \dots (x_m^* x_m)^{c_m} dx,$$

where x_i is the ith column of x. In this way, we see that

$$\int_{K(n,m)} e^{-\sigma(xbx^*)} \left(1 + \sum_{i=1}^m x_i^* x_i\right)^N dx = \delta(b)^{-tn/2} f(b_1^{-1}, \dots, b_m^{-1})$$

for $0 < N \in \mathbf{Z}$ with a polynomial function f depending only on n, N, and P. If $0 < r \leq N/m$, we have

$$\delta(1 + x^* x)^r \leq \delta(1 + x^* x)^{N/m} \leq \left(1 + \sum_{i=1}^m x_i^* x_i\right)^N$$

by Lemma 2.7, and hence we obtain our lemma.

Let us now introduce the notion of eigenvalues of an element of V in Case IV. Given $h \in V$, we define *the eigenvalues of h* to be the roots of a quadratic equation $\sigma(t\varepsilon - h, t\varepsilon - h) = 0$ in the variable t. In other words, if λ and λ' are the eigenvalues of h, we have $\sigma(h, h) = \lambda\lambda'$ and $2\sigma(h, \varepsilon) = \lambda + \lambda'$. Observe that $h \in P$ if and only if the eigenvalues of h are both positive, and also that $h = 0$ if and only if both eigenvalues of h are 0. When $h \neq 0$, we denote by $\mu(h)$ the smallest nonzero absolute value of eigenvalues of h. Also we put

$$U_m = \{a \in G_m | a\varepsilon = \varepsilon\}.$$

This is a compact group isomorphic to $O(m-1)$. Now we can find a basis $\{e_1, \dots, e_m\}$ of V over $\mathbf{R}$ such that

(2.4.IV) $$\varepsilon = e_1 + e_2,$$

(2.5.IV) $$\sigma\left(\sum_{k=1}^m x_k e_k, \sum_{k=1}^m y_k e_k\right) = (1/2)(x_1 y_2 - x_2 y_1) - \sum_{k>2} x_k y_k$$

for $x_k, y_k \in \mathbf{R}$. Fixing such a basis, we mean, by a *diagonal element* of V, an element of $\mathbf{R}e_1 + \mathbf{R}e_2$. In fact, $\lambda_1 e_1 + \lambda_2 e_2$ has λ_1 and λ_2 as its eigenvalues.

Let p, q, r be nonnegative integers such that $p + q + r = \varrho$. Then, in all four cases, we denote by $V(p, q, r)$ the subset of V_m consisting of the elements with p positive, q negative, and r zero eigenvalues. Obviously, $\text{diag}[1_p, -1_q, 0_r]$ belongs to $V(p, q, r)$. In Case IV, $\text{diag}[1_p, -1_q, 0_r]$ is defined to be $e_1 + e_2$ if $p = 2$, $e_1 - e_2$ if $p = q = 1$, $-e_1 - e_2$ if $q = 2$, e_1 if $p = r = 1$, $-e_1$ if $q = r = 1$, 0 if $r = 2$.

Lemma 2.9. *Let $g \in P_m$ and $h \in V(p, q, r)$. Then, in Cases I–III, the following assertions hold:*

(i) there exists an element a of G such that aga^* is diagonal and $aha^* = \text{diag}[1_p, -1_q, 0_r]$;

(ii) there exists an element b of G such that bgb^* is diagonal and

$$h = b^* \text{diag}[1_p, -1_q, 0_r]b;$$

(iii) there exists an element c of G such that $cgc^* = \varepsilon$ and chc^* is diagonal, where $\varepsilon = 1_m$.

Moreover, these assertions hold also in Case IV, if we replace axa^* and a^*xa by ax and $v(a)a^{-1}x$.

Proof. In Cases I–III, take $s\in U_m$ so that

$$sg^{-1/2}hg^{-1/2}s^* = \mathrm{diag}[x_1^2, \ldots, x_p^2, -y_1^2, \ldots, -y_q^2, 0_r]$$

with positive x_j and y_k. Then

$$a = \mathrm{diag}[x_1^{-1}, \ldots, x_p^{-1}, y_1^{-1}, \ldots, y_q^{-1}, 1_r]sg^{-1/2}, \qquad c = sg^{-1/2}$$

have the required properties of (i) and (iii). Applying (i) to g^{-1} and h, we find $f\in G$ such that $fg^{-1}f^*$ is diagonal and $fhf^* = \mathrm{diag}[1_p, -1_q, 0_r]$. Then $b = (f^*)^{-1}$ has the properties of (ii). In Case IV, we first observe that, given $x\in P$, there is an element a of G such that $ax=\varepsilon$. Applying this to g, we find an element d of G such that $v(d)=1$ and $dg=\beta\varepsilon$ with $\beta\in\mathbf{R}$. Let $W=\{x\in V\,|\,\sigma(\varepsilon,x)=0\}$, $\varepsilon'=e_1-e_2$, and $dh=\alpha\varepsilon+x$ with $\alpha\in\mathbf{R}$ and $x\in W$. Now σ is negative on W. Hence there exists an element v of $GL(W)$ which leaves σ invariant and such that $vx=\alpha'\varepsilon'$ with $\alpha'\in\mathbf{R}$. Let u be an element of G such that $u\varepsilon=\varepsilon$ and $u=v$ on W. Put $c=\beta^{-1}ud$. Then $cg=\varepsilon$ and ch is diagonal, which proves (iii). To prove (i), put $ch=\gamma_1e_1+\gamma_2e_2$ and suppose for example $\gamma_1<0$ and $\gamma_2>0$. Define $z\in G$ by $ze_1=-\gamma_1^{-1}e_2$, $ze_2=\gamma_2^{-1}e_1$, and $ze_k=(-\gamma_1\gamma_2)^{-1/2}e_k$ for $k>2$. Put $a=zc$. Then $a\in G$, ag is diagonal, and $ah=e_1-e_2$. Put $b=v(a)^{-1}a$. Then $h=v(b)b^{-1}(e_1-e_2)$. This proves (i) and (ii) when $\gamma_1<0$ and $\gamma_2>0$. The cases with other signatures of γ_i can be proved in a similar fashion.

3. Analytic Continuation of η in (α, β) when $h>0$

We first observe that

$$(3.1.\mathrm{K}) \qquad \eta(a^*ga, h; \alpha, \beta) = \delta(aa^*)^{\kappa-\alpha-\beta}\eta(g, aha^*; \alpha, \beta) \quad (a\in G),$$

$$(3.1.\mathrm{IV}) \qquad \eta(v(a)a^{-1}g, h; \alpha, \beta) = v(a)^{\kappa-\alpha-\beta}\eta(g, ah; \alpha, \beta) \quad (a\in G).$$

In this section, we consider η with $h>0$. Given $h\in P$ in Case IV, we can find an element a of G such that $ah=\varepsilon$. In Cases I–III, put $\varepsilon=1_m$. Then, for every $h\in P$, there is an element a of G such that $aha^*=\varepsilon$. Thus in all four cases, we can reduce the problem to the case $h=\varepsilon$, by virtue of (3.1). To treat this special case, it is convenient to introduce a function

$$(3.2) \qquad \zeta(z; \alpha, \beta) = \int_P e^{-\sigma(z,x)}\delta(x+\varepsilon)^{\alpha-\kappa}\delta(x)^{\beta-\kappa}dx \quad (z\in H', (\alpha,\beta)\in\mathbf{C}^2).$$

We see easily that

$$(3.3) \qquad \eta(g, \varepsilon; \alpha, \beta) = e^{-\sigma(g,\varepsilon)}2^{(\alpha+\beta-\kappa)\varrho}\zeta(2g; \alpha, \beta) \quad (g\in P),$$

$$(3.4.\mathrm{K}) \qquad \int_P e^{-\sigma(z,x)}\delta(x+h)^{\alpha-\kappa}\delta(x)^{\beta-\kappa}dx = \delta(h)^{\alpha+\beta-\kappa}\zeta(h^{1/2}zh^{1/2}; \alpha, \beta)$$

$$\text{if} \quad h\in P,$$

$$(3.4.\mathrm{IV}) \qquad \int_P e^{-\sigma(z,x)}\delta(x+a\varepsilon)^{\alpha-\kappa}\delta(x)^{\beta-\kappa}dx$$

$$= \nu(a)^{\alpha+\beta-\kappa}\zeta(\nu(a)a^{-1}z;\alpha,\beta) \quad \text{for}\quad a\in G,$$

$$(3.5.\mathrm{K}) \qquad \zeta(z;\alpha,\beta)=\zeta(aza^*;\alpha,\beta) \quad \text{for}\quad a\in U_m,$$

$$(3.5.\mathrm{IV}) \qquad \zeta(z;\alpha,\beta)=\zeta(az;\alpha,\beta) \quad \text{for}\quad a\in U_m.$$

Therefore, if $g\in P$, $\zeta(g;\alpha,\beta)$ depends only on the eigenvalues of g.

Theorem 3.1. *The integral of* (3.2) *is convergent for* $z\in H'$ *and* $\mathrm{Re}(\beta)>\kappa-1$, *and defines a holomorphic function in* (z,α,β). *Moreover, put*

$$(3.6) \qquad \omega(z;\alpha,\beta)=\Gamma_m(\beta)^{-1}\delta(z)^\beta\zeta(z;\alpha,\beta) \quad (z\in H',(\alpha,\beta)\in\mathbf{C}^2).$$

Then ω *can be continued as a holomorphic function to the whole* $H'\times\mathbf{C}^2$ *and satisfies*

$$(3.7) \qquad \omega(z;\kappa-\beta,\kappa-\alpha)=\omega(z;\alpha,\beta).$$

Furthermore, for every compact subset T *of* $\mathbf{C}^2$, *there exist two positive constants* A *and* B *depending only on* T, P, *and* σ *such that*

$$(3.8) \qquad |\omega(g;\alpha,\beta)|\leqq A(1+\mu(g)^{-B}) \quad \text{for}\quad g\in P \quad \text{and}\quad (\alpha,\beta)\in T.$$

Proof. For $z\in V_\mathbf{C}$, put

$$(3.9.\mathrm{K}) \qquad \Re(z)=(z+z^*)/2,$$

$$(3.9.\mathrm{IV}) \qquad \Re(z)=(z+\bar z)/2.$$

Let $g=\Re(z)$. Since $\sigma(g,x)=\mathrm{Re}(\sigma(z,x))$ for $x\in V$, we have

$$\int_P |e^{-\sigma(z,x)}\delta(x+\varepsilon)^{\alpha-\kappa}\delta(x)^{\beta-\kappa}|dx$$

$$\leqq \int_P e^{-\sigma(g,x)}|\delta(x)^{\beta-\kappa}|dx=\Gamma_m(\mathrm{Re}(\beta))\delta(g)^{-\mathrm{Re}(\beta)}$$

if $\mathrm{Re}(\beta)>\kappa-1$ and $\mathrm{Re}(\alpha)\leqq\kappa$. Hence ζ is holomorphic in (z,α,β) for such (α,β). To prove the analytic continuation, we employ a well known principle that if the integral of (3.2) is convergent locally uniformly in (z,α,β) in a domain, then the same is true for every partial complex derivative of the integrand. Define now a differential operator Δ^n of degree ϱn on $V_\mathbf{C}$ to be the nth power of an operator Δ given by

$$(3.10.\mathrm{I}) \qquad \Delta=\det(2^{-1}(1+\delta_{jk})\partial/\partial z_{jk}),$$

$$(3.10.\mathrm{II, III}) \qquad \Delta=\delta(\partial/\partial z_{jk}) \quad \text{for}\quad z=(z_{jk})\in V_\mathbf{C},$$

$$(3.10.\mathrm{IV}) \qquad \Delta=\sum_{j,k=1}^m t_{jk}\partial^2/\partial z_j\partial z_k,$$

where $(t_{jk})=(\sigma(b_j,b_k))^{-1}$ if $\{b_j\}$ is the standard basis of $\mathbf{R}^m$. Then we see easily that $\Delta^n e^{\sigma(u,z)}=\delta(u)^n e^{\sigma(u,z)}$, and hence

$$(3.11) \qquad (-1)^{\varrho n}\Delta^n\{e^{-\sigma(z,\varepsilon)}\zeta(z;\alpha,\beta)\}=e^{-\sigma(z,\varepsilon)}\zeta(z;\alpha+n,\beta).$$

This proves the convergence of (3.2) for all $\alpha \in \mathbf{C}$ and $\mathrm{Re}(\beta) > \kappa - 1$. To prove the functional equation, assume $\mathrm{Re}(\alpha) > \kappa - 1$ and $\mathrm{Re}(\beta) > \kappa - 1$. Then

$$\begin{aligned}
\Gamma_m(\beta)\zeta(g\,;\kappa-\beta,\alpha) &= \int_P e^{-\sigma(g,x)}\Gamma_m(\beta)\delta(x+\varepsilon)^{-\beta}\delta(x)^{\alpha-\kappa}dx \\
&= \int_P e^{-\sigma(g,x)}\int_P e^{-\sigma(u,x+\varepsilon)}\delta(u)^{\beta-\kappa}du\,\delta(x)^{\alpha-\kappa}dx \\
&= \int_P e^{-\sigma(u,\varepsilon)}\delta(u)^{\beta-\kappa}\int_P e^{-\sigma(g+u,x)}\delta(x)^{\alpha-\kappa}dx\,du \\
&= \Gamma_m(\alpha)\int_P e^{-\sigma(u,\varepsilon)}\delta(u+g)^{-\alpha}\delta(u)^{\beta-\kappa}du \\
&= \Gamma_m(\alpha)\delta(g)^{\beta-\alpha}\zeta(\dot g\,;\kappa-\alpha,\beta)
\end{aligned}$$

by (3.4). [In Case IV, we take $a \in G$ so that $a\varepsilon = g$ and observe that $v(a)a^{-1}$ has the same eigenvalues as g.] This proves (3.7) when $z \in P$, $\mathrm{Re}(\alpha) < 1$, and $\mathrm{Re}(\beta) > \kappa - 1$. At the same time, we can use (3.7) to continue $\omega(z\,;\alpha,\beta)$ analytically to the domain

$$H' \times \{(\alpha,\beta) \in \mathbf{C}^2 | \mathrm{Re}(\alpha) < 1 \text{ or } \mathrm{Re}(\beta) > \kappa - 1\}.$$

Now, from (3.11), we obtain

$$\begin{aligned}
(3.12) \qquad (-1)^{\varrho n}\Delta^n\{e^{-\sigma(z,\varepsilon)}\delta(z)^{-\beta}\omega(z\,;\alpha,\beta)\} \\
= e^{-\sigma(z,\varepsilon)}\delta(z)^{-\beta}\omega(z\,;\alpha+n,\beta),
\end{aligned}$$

which guarantees the analytic continuation of ω to $H' \times \mathbf{C}^2$.

To prove (3.8) in Cases I–III, write ζ_m and ω_m for ζ and ω to emphasize m; we have

$$(3.13) \qquad \zeta_1(g\,;\alpha,\beta) = \int_0^\infty e^{-gx}(x+1)^{\alpha-1}x^{\beta-1}dx \quad (0 < g \in \mathbf{R}).$$

This is a classical confluent hypergeometric function. Integration by parts shows that

$$\beta\zeta_1(g\,;\alpha,\beta) = g\zeta_1(g\,;\alpha,\beta+1) + (1-\alpha)\zeta_1(g\,;\alpha-1,\beta+1),$$

and hence

$$(3.14) \qquad \omega_1(g\,;\alpha,\beta) = \omega_1(g\,;\alpha,\beta+1) + (1-\alpha)g^{-1}\omega_1(g\,;\alpha-1,\beta+1).$$

(This gives another proof of the analytic continuation of ω_1.) If $\mathrm{Re}(\alpha-1) \leq n$ with $0 \leq n \in \mathbf{Z}$, we have

$$|(1+x)^{\alpha-1}| \leq (1+x)^n = \sum_{k=0}^n \binom{n}{k}x^k$$

for $x \geq 0$, and hence, if $\mathrm{Re}(\beta) = b > 0$, we obtain

$$\begin{aligned}
|\omega_1(g\,;\alpha,\beta)| &\leq |\Gamma(\beta)|^{-1}g^b \sum_{k=0}^n \binom{n}{k}\int_0^\infty e^{-gx}x^{k+b-1}dx \\
&= \sum_{k=0}^n \binom{n}{k}|\Gamma(\beta)|^{-1}\Gamma(k+b)g^{-k}.
\end{aligned}$$

This proves (3.8) for ω_1 when $\mathrm{Re}(\beta)>0$. By (3.14), we have

$$\omega_1(g;\alpha,\beta)=\sum_{k=0}^{n} f_{n,k}(\alpha)g^{-k}\omega_1(g,\alpha-k,\beta+n)$$

with polynomial functions $f_{n,k}$. Thus (3.8) for ω_1 in the general case can be derived from the special case $\mathrm{Re}(\beta)>0$. In order to prove (3.8) for ω_m, we proceed by induction on m. Since $\omega_m(g;\alpha,\beta)$ depends only on the eigenvalues of g, we may assume that g is diagonal. We first deal with Cases I–III. Write the variable on P in the form $\begin{bmatrix} x & z \\ z^* & y \end{bmatrix}$ with $z\in\mathbf{K}_q^p$, $p+q=m$. Put $g=\mathrm{diag}[a,b]$ with $a\in P_p$ and $b\in P_q$. Then

$$\zeta(g;\alpha,\beta)=\int_Q e^{-\sigma(ax)-\sigma(by)}\delta\begin{bmatrix} x+1 & z \\ z^* & y+1 \end{bmatrix}^{\alpha-\kappa}\delta\begin{bmatrix} x & z \\ z^* & y \end{bmatrix}^{\beta-\kappa} dxdydz,$$

$$Q=\{(x,y,z)|x>0,y>z^*x^{-1}z\}.$$

We make the change of variables $(x,y,z)\mapsto(u,v,w)$ by

$$x=u,\qquad z=(u^2+u)^{1/2}w,\qquad rvr=y-w^*(x+1)w,\qquad r=(1+w^*w)^{1/2}.$$

This gives a bijective map of Q onto

$$R=\{(u,v,w)|u>0,v>0,w\in\mathbf{K}_q^p\}=P_p\times P_q\times\mathbf{K}_q^p.$$

In fact

$$w^*(x+1)w=z^*(x^2+x)^{-1/2}(x+1)(x^2+x)^{-1/2}z=z^*x^{-1}z.$$

Thus $rvr=y-z^*x^{-1}z$, and hence $y>z^*x^{-1}z$ if and only if $v>0$. The bijectivity is obvious. Now we have $z^*(x+1)^{-1}z=w^*xw$, and hence

$$y+1-z^*(x+1)^{-1}z=y-w^*(x+1)w+1+w^*w=r(1+v)r.$$

Therefore, by Lemma 2.1, we have

$$\delta\begin{pmatrix} x+1 & z \\ z^* & y+1 \end{pmatrix}=\delta(x+1)\delta(y+1-z^*(x+1)^{-1}z)=\delta(1+u)\delta(1+v)\delta(1+ww^*),$$

$$\delta\begin{pmatrix} x & z \\ z^* & y \end{pmatrix}=\delta(x)\delta(y-z^*x^{-1}z)=\delta(u)\delta(v)\delta(1+w^*w).$$

A simple calculation shows that

$$\partial(x,y,z)/\partial(u,v,w)=\delta(u^2+u)^{\iota q/2}\delta(1+w^*w)^{\kappa(q)}.$$

Thus we obtain

$$\zeta_m(g;\alpha,\beta)=\int_R e^{-\sigma(au)-\sigma(brvr+bw^*(u+1)w)}\delta(u+1)^{\alpha-\kappa(p)}\delta(u)^{\beta-\kappa(p)}$$

$$\cdot\delta(v+1)^{\alpha-\kappa(m)}\delta(v)^{\beta-\kappa(m)}\delta(1+w^*w)^{\alpha+\beta-\kappa(m)-\iota p/2}dudvdw$$

$$=\int_{\mathbf{K}(p,q)} \zeta_p(a+wbw^*;\alpha,\beta)\zeta_q(rbr;\alpha-\iota p/2,\beta-\iota p/2)$$

$$\cdot e^{-\sigma(wbw^*)}\delta(1+w^*w)^{\alpha+\beta-\kappa(m)-\iota p/2}dw.$$

In view of (1.18.K), we have therefore

$$\omega_m(g;\alpha,\beta)=\pi^{-\imath pq/2}\int_{\mathbf{K}(p,q)}\omega_p(a+wbw^*;\alpha,\beta)\omega_q(rbr;\alpha-\imath p/2,\beta-\imath p/2)$$

$$\cdot e^{-\sigma(wbw^*)}\delta(a)^\beta\delta(b)^{\imath p/2}\delta(a+wbw^*)^{-\beta}\delta(1+w^*w)^{\alpha-\kappa(m)}dw.$$

Now $a+wbw^*\geqq a$ and $rbr\geqq b$ since $r\geqq1$. Therefore $\mu(a+wbw^*)\geqq\mu(a)\geqq\mu(g)$ and $\mu(rbr)\geqq\mu(b)\geqq\mu(g)$. Assuming that (3.8) is true for ω_p and ω_q, we have, for a compact subset T of $\mathbf{C}^2$,

$$|\omega_m(g;\alpha,\beta)|\leqq A(1+\mu(g)^{-B})\delta(b)^{\imath p/2}$$

$$\cdot\int_{\mathbf{K}(p,q)}e^{-\sigma(wbw^*)}|\delta(1+ss^*)^{-\beta}\delta(1+w^*w)^{\alpha-\kappa(m)}|dw$$

for $(\alpha,\beta)\in T$ with constants A and B depending only on p, q, and T, where $s=a^{-1/2}wb^{1/2}$. Suppose $\mathrm{Re}(\beta)\geqq0$. Then $|\delta(1+ss^*)^{-\beta}|\leqq1$, and hence (3.8) for ω_m follows from Lemma 2.8. Next suppose $\mathrm{Re}(\beta)\leqq0$. We may assume that $q=1$ and $b=\mu(g)$. Then $a\geqq b1_p$,

$$1+ss^*=1+ba^{-1/2}ww^*a^{-1/2}\leqq1+ww^*,$$

so that $|\delta(1+ss^*)^{-\beta}|\leqq|\delta(1+ww^*)^{-\beta}|$. Hence the desired result follows again from Lemma 2.8. This completes the proof of (3.8) in Cases I–III.

To prove (3.8) in Case IV, we may assume that σ has form (2.5.IV) with respect to the standard basis $\{e_i\}$ of $V=\mathbf{R}^m$; also we may take $g=be_1+ae_2$ with $0<b\leqq a$. Let $x,y,z_1,\ldots,z_n$ with $n=m-2$ be the coordinate functions on V with respect to $\{e_i\}$. Then

$$\zeta(g;\alpha,\beta)=\int_P e^{-(ax+by)/2}[(x+1)(y+1)-\|z\|^2]^{\alpha-m/2}(xy-\|z\|^2)^{\beta-m/2}dxdydz,$$

$$P=\{(x,y,z)\in\mathbf{R}^{2+n}|x>0,y>0,xy>\|z\|^2\},$$

where we write $\|z\|^2=\sum_{i=1}^n z_i^2$ for $z=(z_1,\ldots,z_n)\in\mathbf{R}^n$. We make a change of variables $(x,y,z)\mapsto(u,v,w)$ by

$$u=x,\qquad w_i=(x^2+x)^{-1/2}z_i,\qquad v=(1+\|w\|^2)^{-1}(y-(x+1)\|w\|^2).$$

Then $z_i=(u^2+u)^{1/2}w_i$, $y=(1+\|w\|^2)v+(1+u)\|w\|^2$, $\|w\|^2=(x^2+x)^{-1}\|z\|^2$, $xy-\|z\|^2=uv(1+\|w\|^2)$, $(x+1)(y+1)-\|z\|^2=(1+u)(1+v)(1+\|w\|^2)$. It can easily be verified that this maps P bijectively onto

$$R=\{(u,v,w)\in\mathbf{R}^{2+n}|u>0,v>0\},$$

and also that $\partial(x,y,z)/\partial(u,v,w)=(u^2+u)^{n/2}(1+\|w\|^2)$. Therefore, putting $W=\|w\|^2$, we have

$$\zeta(g;\alpha,\beta)=\int_R e^{-[au+b(1+W)v+b(u+1)W]/2}(u+1)^{\alpha-1}u^{\beta-1}$$

$$\cdot(v+1)^{\alpha-m/2}v^{\beta-m/2}(1+W)^{\alpha+\beta-m+1}dudvdw$$

$$= \int_{\mathbf{R}^n} \zeta_1((a+bW)/2;\alpha,\beta)\zeta_1((b+bW)/2;\alpha-n/2,\beta-n/2)$$

$$\cdot e^{-bW/2}(1+W)^{\alpha+\beta-n-1}dw,$$

where ζ_1 is function (3.13). Therefore

$$|\sigma|^{-1/2}\omega(g;\alpha,\beta)=2^{1-n/2}\pi^{1-m/2}b^{n/2}\int_{\mathbf{R}^n}e^{-bW/2}(1+a^{-1}bW)^{-\beta}(1+W)^{\alpha-m/2}$$

$$\cdot\omega_1((a+bW)/2;\alpha,\beta)\omega_1((b+bW)/2;\alpha-n/2,\beta-n/2)dw.$$

If $n=0$, we should simply put $W=0$ and ignore $\int dw$. Since we assumed $a\geqq b$, we have $|(1+a^{-1}bW)^{-\beta}|\leqq|(1+W)|^{|\mathrm{Re}(\beta)|}$, and hence (3.8) in Case IV follows again from Lemma 2.8 and (3.8) for ω_1 which was already established. This completes the proof of Theorem 3.1.

Remark. The integral expression of ω_m in terms of ω_p and ω_q gives another proof of analytic continuation as well as that of (3.7) by induction on m. This requires however an independent proof of (3.7) for ω_1. An easy proof of (3.7) for ω_1 is given in [6, Lemma 2].

Proposition 3.2. *The functions* $\delta(z)^{\varrho n}\omega(z;n+\kappa,\beta)$ *and* $\delta(z)^{\varrho n}\omega(z;\alpha,-n)$ *for non-negative integers* n *are polynomial functions in* z.

Proof. From (1.16) and (3.7), we obtain

$$(3.15)\qquad\qquad\omega(z;\kappa,\beta)=\omega(z;\alpha,0)=1,$$

and hence, by (3.12),

$$(3.16)\qquad\omega(z;n+\kappa,\beta)=(-1)^{\varrho n}e^{\sigma(z,\varepsilon)}\delta(z)^{\beta}\Delta^n\{e^{-\sigma(z,\varepsilon)}\delta(z)^{-\beta}\}.$$

Also (3.7) implies that $\omega(z;\alpha,-n)=\omega(z;n+\kappa,\kappa-\alpha)$. Therefore we obtain our assertion.

If $m=1$, an easy calculation shows that

$$(3.17)\qquad\omega_1(z;n+1,\beta)=\sum_{k=0}^{n}\binom{n}{k}\beta(\beta+1)\ldots(\beta+k-1)z^{-k},$$

$$(3.18)\qquad\omega_1(z;\alpha,-n)=\sum_{k=0}^{n}\binom{n}{k}(1-\alpha)(2-\alpha)\ldots(k-\alpha)z^{-k}\qquad(0<n\in\mathbf{Z}).$$

4. Functions η and ω in the General Case

In this section, we consider $\eta(g,h;\alpha,\beta)$ with h not necessarily contained in P. We first define *the eigenvalues of* h *relative to* g for $h\in V$ and $g\in P$. In Cases I–III, we define them to be the eigenvalues of $g^{1/2}hg^{1/2}$. In Case IV, they are defined to be the roots of a quadratic equation

$$t^2-2\sigma(g,h)t+\sigma(h,h)\sigma(g,g)=0$$

in the variable t. If $g = \varepsilon$, they are the eigenvalues of h. We then put

$$(4.1) \quad \begin{cases} \delta_+(hg) = \text{the product of all positive eigenvalues of } h \text{ relative to } g, \\ \delta_-(hg) = \delta_+((-h)g), \\ \quad \delta(hg) = \delta_+(hg)\delta_-(hg) \quad \text{(only in Case IV)}, \\ \tau_+(hg) = \text{the sum of all positive eigenvalues of } h \text{ relative to } g, \\ \tau_-(hg) = \tau_+((-h)g), \\ \quad \tau(hg) = \tau_+(hg) + \tau_-(hg), \\ \mu(hg) = \text{the smallest absolute value of nonzero eigenvalues of } h \\ \qquad\qquad \text{relative to } g \text{ if } h \neq 0; \; \mu(hg) = 1 \text{ if } h = 0. \end{cases}$$

These quantities are invariant under the maps

$$(4.2.\mathrm{K}) \qquad\qquad (g, h) \mapsto (a^{*-1} g a^{-1}, a h a^*),$$

$$(4.2.\mathrm{IV}) \qquad\qquad (g, h) \mapsto (v(a)^{-1} a g, a h)$$

for every $a \in G$. Put, in all four cases,

$$(4.3) \qquad\qquad \eta^*(g, h; \alpha, \beta) = \delta(g)^{\alpha + \beta - \kappa} \eta(g, h; \alpha, \beta).$$

Then we obtain, from (3.1), for every $a \in G$,

$$(4.4.\mathrm{K}) \qquad\qquad \eta^*(g, h; \alpha, \beta) = \eta^*(a^{*-1} g a^{-1}, a h a^*; \alpha, \beta),$$

$$(4.4.\mathrm{IV}) \qquad\qquad \eta^*(g, h; \alpha, \beta) = \eta^*(v(a)^{-1} a g, a h; \alpha, \beta).$$

Let us restrict ourselves for a moment to Cases I–III. Given g and h, we can find, by Lemma 2.9, $a \in G$ such that

$$(4.5) \qquad\qquad a^* g a = \mathrm{diag}[b, c], \quad h = a \cdot \mathrm{diag}[k, 0_r] a^*$$

with $b \in P_s$, $c \in P_r$, $k \in V_s$, $\delta(k) \neq 0$, $r + s = m$.

Proposition 4.1. *The symbols being the same as in (4.5), we have*

$$\delta_\pm(hg) = \delta_\pm(kb), \qquad \tau_\pm(hg) = \tau_\pm(kb),$$

$$\eta^*(g, h; \alpha, \beta) = \eta^*(\mathrm{diag}[b, c], \mathrm{diag}[k, 0_r]; \alpha, \beta)$$

$$= \pi^{\imath r s/2} \Gamma_r(\alpha + \beta - \kappa(m)) \eta^*(b, k; \alpha - \imath r/2, \beta - \imath r/2).$$

Proof. The first two equalities are obvious. The next one is a special case of (4.4.K). Now

$$\eta(\mathrm{diag}[b, c], \mathrm{diag}[k, 0_r]; \alpha, \beta)$$

$$= \int_Q e^{-\sigma(bu) - \sigma(cv)} \delta \begin{pmatrix} u + k & w \\ w^* & v \end{pmatrix}^{\alpha - \kappa} \delta \begin{pmatrix} u - k & w \\ w^* & v \end{pmatrix}^{\beta - \kappa} du\,dv\,dw,$$

$$Q = \{(u, v, w) \in V_s \times P_r \times \mathbf{K}_r^s \mid u \pm k > w v^{-1} w^* \}.$$

Put $y = u - w v^{-1} w^*$. Then $(u, v, w) \mapsto (y, v, w)$ sends Q bijectively onto

$$Q' = \{(y, v, w) \in V_s \times P_r \times \mathbf{K}_r^s \mid y \pm k > 0 \}.$$

Therefore the above integral becomes

$$\int_{Q'} e^{-\sigma(by)-\sigma(cv)-\sigma(bwv^{-1}w^*)}\delta(v)^{\alpha+\beta-2\kappa}\delta(y+k)^{\alpha-\kappa}\delta(y-k)^{\beta-\kappa}dydvdw.$$

It can easily be seen that

$$\int_{K(s,r)} e^{-\sigma(bwv^{-1}w^*)}dw = \pi^{\iota rs/2}\delta(b)^{-\iota r/2}\delta(v)^{\iota s/2}.$$

Substituting this into $\int_{Q'}$, we obtain the desired formula in a straightforward way.

We now define a function $\omega(g,h;\alpha,\beta)$ for $g\in P$, $h\in V$, and $(\alpha,\beta)\in \mathbf{C}^2$ in all four cases as follows. In Cases I–III, we put, for $h\in V(p,q,r)$ with $p+q+r=m$,

$$(4.6.\mathrm{K}) \qquad \omega(g,h;\alpha,\beta)=2^{-p\alpha-q\beta}\Gamma_p(\beta-\iota(m-p)/2)^{-1}\Gamma_q(\alpha-\iota(m-q)/2)^{-1}$$
$$\cdot\Gamma_r(\alpha+\beta-\kappa(m))^{-1}\delta_+(hg)^{\kappa-\alpha-\iota q/4}\delta_-(hg)^{\kappa-\beta-\iota p/4}\eta^*(g,h;\alpha,\beta).$$

We understand that Γ_0 is the constant function 1. In Case IV, we put $n=m-2$ and

$$(4.6.\mathrm{IV}) \qquad\qquad \omega(g,h;\alpha,\beta)=\eta^*(g,h;\alpha,\beta)$$

$$\begin{cases} 2^{-2\alpha}\Gamma_m(\beta)^{-1}\delta(hg)^{\kappa-\alpha} & \text{if } h\in V(2,0,0), \\ 2^{-2\beta}\Gamma_m(\alpha)^{-1}\delta(hg)^{\kappa-\beta} & \text{if } h\in V(0,2,0), \\ |\sigma|^{1/2}2^{-2\alpha-2\beta}\Gamma(\alpha-n/2)^{-1}\Gamma(\beta-n/2)^{-1}\delta_+(hg)^{1-\alpha+n/4}\delta_-(hg)^{1-\beta+n/4} \\ \quad \text{if } h\in V(1,1,0), \\ |\sigma|^{1/2}2^{-2\alpha-2\beta}\Gamma(\alpha+\beta-\kappa)^{-1}\Gamma(\beta-n/2)^{-1}\delta(hg)^{\kappa-\alpha} & \text{if } h\in V(1,0,1), \\ |\sigma|^{1/2}2^{-2\alpha-2\beta}\Gamma(\alpha+\beta-\kappa)^{-1}\Gamma(\alpha-n/2)^{-1}\delta(hg)^{\kappa-\beta} & \text{if } h\in V(0,1,1), \\ \Gamma_m(\alpha+\beta-\kappa)^{-1} & \text{if } h=0. \end{cases}$$

Then we have

$$(4.7.\mathrm{K}) \qquad \left.\begin{array}{l} \omega(a^{*-1}ga^{-1},aha^*;\alpha,\beta) \\ \omega(v(a)^{-1}ag,ah;\alpha,\beta) \end{array}\right\} = \omega(g,h;\alpha,\beta) \quad \text{for every } a\in G,$$
$$(4.7.\mathrm{IV})$$

$$(4.8) \qquad\qquad \omega(g,-h;\alpha,\beta)=\omega(g,h;\beta,\alpha),$$

$$(4.9) \qquad\qquad \omega(g,0;\alpha,\beta)=1,$$

$$(4.10) \qquad\qquad \omega(g,\varepsilon;\alpha,\beta)=2^{-\varrho\kappa}e^{-\sigma(g,\varepsilon)}\omega(2g;\alpha,\beta).$$

These follow immediately from (4.4), (1.27), (1.30), and (3.3), respectively. Also, in Cases I–III, we obtain, from Proposition 4.1,

$$(4.11.\mathrm{K}) \qquad\qquad \omega(\mathrm{diag}[b,c],\mathrm{diag}[k,0_r];\alpha,\beta)$$
$$= (\pi/2)^{\iota r(p+q)/2}\omega(b,k;\alpha-\iota r/2,\beta-\iota r/2)$$
$$\text{if } b\in P_{p+q}, \quad c\in P_r, \quad k\in V(p,q,0).$$

Now our main theorem can be stated as follows.

Theorem 4.2. *Function ω can be continued as a holomorphic function in (α, β) to the whole $\mathbf{C}^2$ and satisfies*

$$(4.12.\mathrm{K}) \qquad \omega(g, h; \alpha, \beta) = \omega(g, h; \kappa + (\imath r/2) - \beta, \kappa + (\imath r/2) - \alpha),$$

$$(4.12.\mathrm{IV}) \qquad \omega(g, h; \alpha, \beta) = \begin{cases} \omega(g, h; \kappa - \beta, \kappa - \alpha) & \text{if } r = 0 \text{ or } 2, \\ \omega(g, h; m - 1 - \beta, m - 1 - \alpha) & \text{if } r = 1, \end{cases}$$

where r is the number of zero eigenvalues of h. Moreover, for every compact subset T of $\mathbf{C}^2$, there exist two positive constants A and B depending only on T and σ such that

$$(4.13.\mathrm{K}) \qquad |\omega(g, h; \alpha, \beta)| \leq A e^{-\tau(hg)}(1 + \mu(hg)^{-B}),$$

$$(4.13.\mathrm{IV}) \qquad |\omega(g, h; \alpha, \beta)| \leq A e^{-\tau(hg)/2}(1 + \mu(hg)^{-B}),$$

for every $(g, h) \in P \times V$ and every $(\alpha, \beta) \in T.$

We first note that if $\pm h \in P$ or $h = 0$, our assertions follow from (4.7), (4.8), (4.9), (4.10), and Theorem 3.1. To prove the remaining cases, let us first treat Cases I–III. By virtue of (4.7), (4.11), and Lemma 2.9, we may assume that g is diagonal and $h = \mathrm{diag}[1_p, -1_q]$ with $m = p + q$, $pq \neq 0$. Put

$$(4.14) \qquad \varepsilon_{p,q} = \mathrm{diag}[1_p, -1_q], \quad \varepsilon_p = \mathrm{diag}[1_p, 0_q], \quad \varepsilon_q' = \mathrm{diag}[0_p, 1_q],$$

$$(4.15) \qquad U_{p,q} = \{s \in \mathrm{GL}_{p+q}(\mathbf{K}) | s\varepsilon_{p,q}s^* = \varepsilon_{p,q}\},$$

$$(4.16) \qquad \zeta_{pq}(g; \alpha, \beta) = e^{-\sigma(g)/2} \int_X e^{-\sigma(gx)} \delta(x + \varepsilon_p)^{\alpha - \kappa} \delta(x + \varepsilon_q')^{\beta - \kappa} dx,$$

$$X = \{x \in V | x + \varepsilon_p > 0, x + \varepsilon_q' > 0\},$$

$$(4.17) \qquad \omega_{pq}(g; \alpha, \beta) = \Gamma_q(\alpha - \imath p/2)^{-1} \Gamma_p(\beta - \imath q/2)^{-1}$$

$$\cdot \delta_+(\varepsilon_{p,q}g)^{\beta - \imath q/4} \delta_-(\varepsilon_{p,q}g)^{\alpha - \imath p/4} \zeta_{pq}(g; \alpha, \beta),$$

where $\kappa = \kappa(m)$. We see easily that

$$(4.18) \qquad \eta(g, \varepsilon_{p,q}; \alpha, \beta) = 2^{m(\alpha + \beta - \kappa)} \zeta_{pq}(2g; \alpha, \beta),$$

$$(4.19) \qquad \omega(g, \varepsilon_{p,q}; \alpha, \beta) = 2^{(\imath pq/2) - m\kappa} \omega_{pq}(2g; \alpha, \beta),$$

$$(4.20) \qquad \omega_{pq}(sgs^*; \alpha, \beta) = \omega_{pq}(g; \alpha, \beta) \quad \text{for every} \quad s \in U_{p,q}.$$

Therefore Lemma 2.9 together with (4.7.K) reduces our problems to the task of showing that ω_{pq} can be continued as a holomorphic function in (α, β) to the whole $\mathbf{C}^2$, and that

$$(4.21) \qquad \omega_{pq}(g; \kappa - \beta, \kappa - \alpha) = \omega_{pq}(g; \alpha, \beta),$$

$$(4.22) \qquad |e^{\tau(hg)/2} \omega_{pq}(g; \alpha, \beta)| \leq A(1 + \mu(hg)^{-B}) \quad (h = \varepsilon_{p,q})$$

for $(\alpha, \beta) \in T$ with A and B depending only on T. As mentioned above, we may assume that $g = \mathrm{diag}[a, b]$ with $a \in P_p$ and $b \in P_q$. Let $\begin{pmatrix} x & z \\ z^* & y \end{pmatrix}$ with $z \in \mathbf{K}_q^p$ be the variable on V. Then

$$e^{\sigma(g)/2} \zeta_{pq}(g; \alpha, \beta) = \int_X e^{-\sigma(ax) - \sigma(by)} \delta\begin{pmatrix} x + 1 & z \\ z^* & y \end{pmatrix}^{\alpha - \kappa} \delta\begin{pmatrix} x & z^* \\ z^* & y + 1 \end{pmatrix}^{\beta - \kappa} dx\,dy\,dz,$$

$$X = \{(x, y, z) | x > 0, y > 0, x + 1 > zy^{-1}z^*, y + 1 > z^*x^{-1}z\}$$

$$= \{(x, y, z) | x + 1 > 0, y + 1 > 0, x > z(y + 1)^{-1}z^*, y > z^*(x + 1)^{-1}z\}$$

by Lemma 2.1. Put $f=(x+1)^{-1/2}z(y+1)^{-1/2}$. Then

$$1-ff^*=(x+1)^{-1/2}\{x+1-z(y+1)^{-1}z^*\}(x+1)^{-1/2}>0.$$

Put $r=(1-f^*f)^{1/2}$ and $s=(1-ff^*)^{1/2}$. Now we make a change of variables $(x,y,z)\mapsto(u,v,w)$ by putting $u=x-ww^*$, $v=y-w^*w$, $w=s^{-1}f$. Notice that $w=fr^{-1}$ by Lemma 2.2. This maps X bijectively onto $Y=P_p\times P_q\times \mathbf{K}_q^p$. In fact, by Lemma 2.3, $(1+ww^*)^{-1}=1-ff^*$ and $f=(1+ww^*)^{-1/2}w$, so that

$$(4.23)\qquad (x+1)^{1/2}(1+ww^*)^{-1}(x+1)^{1/2}=x+1-z(y+1)^{-1}z^*>1;$$

hence $1+ww^*<x+1$, so that $u>0$. Similarly $v>0$. The inverse map is given by $x=u+ww^*$, $y=v+w^*w$, and

$$z=(u+1+ww^*)^{1/2}sw(v+1+w^*w)^{1/2}=(x+1)^{1/2}sw(y+1)^{1/2}$$

with $s=(1+ww^*)^{-1/2}$. If $u>0$ and $v>0$, obviously $x>0$, $y>0$, $x+1>1+ww^*$ and

$$1<(x+1)^{1/2}(1+ww^*)^{-1}(x+1)^{1/2}=x+1-z(y+1)^{-1}z^*,$$

and hence $x>z(y+1)^{-1}z^*$. Similarly $y>z^*(x+1)^{-1}z$. Now Lemma 2.6 together with a simple calculation shows that

$$\partial(x,y,z)/\partial(u,v,w)=\delta(1+x)^{iq/2}\delta(1+y)^{ip/2}\delta(1+w^*w)^{-\kappa(m)}.$$

By (4.23), we have

$$x-z(y+1)^{-1}z^*=(x+1)^{1/2}(1+ww^*)^{-1}(x+1)^{1/2}-1$$
$$=(x+1)^{1/2}(1+ww^*)^{-1}\{x+1-(1+ww^*)\}(x+1)^{-1/2}$$
$$=(x+1)^{1/2}(1+ww^*)^{-1}u(x+1)^{-1/2}.$$

Therefore

$$\delta\begin{pmatrix}x & z\\ z^* & y+1\end{pmatrix}=\delta(y+1)\delta(x-z(y+1)^{-1}z^*)$$
$$=\delta(v+1+w^*w)\delta(u)\delta(1+ww^*)^{-1},$$

and similarly

$$\delta\begin{pmatrix}x+1 & z\\ z^* & y\end{pmatrix}=\delta(u+1+ww^*)\delta(v)\delta(1+ww^*)^{-1}.$$

Thus putting $W=ww^*$ and $W'=w^*w$, we obtain

$$e^{\sigma(g)/2}\zeta_{pq}(g;\alpha,\beta)=\int_Y e^{-\sigma(au+aW)-\sigma(bv+bW')}\delta(u+1+W)^{\alpha-\kappa(p)}\delta(u)^{\beta-\kappa}$$
$$\cdot\delta(v+1+W')^{\beta-\kappa(q)}\delta(v)^{\alpha-\kappa}\delta(1+W)^{\kappa-\alpha-\beta}\,du\,dv\,dw.$$

Let ζ_m and ω_m be the functions considered in Sect. 3. By (3.4.K), the last integral is equal to

$$(4.24)\qquad \int_F e^{-\sigma(aW)-\sigma(bW')}\delta(1+W)^{\alpha+\beta-\kappa}$$
$$\cdot\zeta_p(ZaZ;\alpha,\beta-iq/2)\zeta_q(Z'bZ';\beta,\alpha-ip/2)\,dw,$$

where $F = \mathbf{K}_q^p$, $Z = (1_p + W)^{1/2}$, and $Z' = (1_q + W')^{1/2}$. Since $\delta_+(hg) = \delta(a)$, $\delta_-(hg) = \delta(b)$, and $\tau(hg) = \sigma(g)$ for $h = \varepsilon_{p,q}$, we have

$$(4.25) \qquad e^{\tau(hg)/2} \omega_{pq}(g; \alpha, \beta) = \delta(a)^{\iota q/4} \delta(b)^{\iota p/4}$$

$$\cdot \int_F e^{-\sigma(aW) - \sigma(bW')} \omega_p\left(ZaZ; \alpha, \beta - \frac{\iota q}{2}\right)$$

$$\cdot \omega_q\left(Z'bZ'; \beta, \alpha - \frac{\iota p}{2}\right) \delta(1 + W)^{(\iota/2) - 1} dw.$$

Applying (3.8) to ω_p and ω_q, we find that

$$|e^{\tau(hg)/2} \omega_{pq}(g; \alpha, \beta)|$$

$$\leq A(1 + \mu(hg)^{-B}) \delta(a)^{\iota q/4} \delta(b)^{\iota p/4} \int_F e^{-\sigma(aW) - \sigma(bW')} \delta(1 + W)^{(\iota/2) - 1} dw$$

if (α, β) stays in a compact set T, where A and B (>0) depend only on T. Since $\delta(1 + W)^{(\iota/2) - 1} \leq \delta(1 + W)$, the last integral over F is majorized by

$$\left\{ \int_F e^{-2\sigma(aW)} \delta(1 + W) dw \int_F e^{-2\sigma(bW')} \delta(1 + W) dw \right\}^{1/2},$$

which can be estimated by Lemma 2.8. This proves the analytic continuation of ω_{pq} as well as (4.22). Functional equation (4.21) can be obtained by applying (3.7) to ω_p and ω_q in (4.25). This completes the proof of Theorem 4.2 in Cases I–III.

To treat Case IV, we first observe that if $\sigma'(x, y) = \sigma(ax, ay)$ with $a \in \mathrm{GL}_m(\mathbf{R})$ and if ω' denotes the function defined with respect to σ', then

$$(4.26) \qquad \omega'(g, h; \alpha, \beta) = \omega(ag, ah; \alpha, \beta).$$

Therefore we may assume that σ has form (2.5.IV) with respect to the standard basis $\{e_i\}$ of $V = \mathbf{R}^m$. For the reasons already explained, it is sufficient to treat the cases $h = e_1 - e_2$ and $h = e_1$; at the same time we may put $g = be_1 + ae_2$ with positive a and b. Suppose $h = e_1 - e_2$. Then $\delta_+(hg) = a$, $\delta_-(hg) = b$, and $\tau(hg) = a + b$. Let $x, y, z_1, \ldots, z_n$ with $n = m - 2$ be the coordinate functions on V with respect to $\{e_i\}$. Then

$$\eta(g, h; \alpha, \beta)$$

$$= \int_Q e^{(-ax - by)/2} \{(x+1)(y-1) - \|z\|^2\}^{\alpha - \kappa} \{(x-1)(y+1) - \|z\|^2\}^{\beta - \kappa} dx\, dy\, dz,$$

$$Q = \{(x, y, z) \in \mathbf{R}^{2+n} | x > 1, \ y > 1, \ (x+1)(y-1) > \|z\|^2, \ (x-1)(y+1) > \|z\|^2\},$$

where $\|z\|^2 = \sum_{i=1}^{n} z_i^2$. Let us now make a change of variables

$$(4.27) \qquad (x, y, z) \mapsto (u, v, w): u = x - 1 - 2\|w\|^2, \qquad v = y - 1 - 2\|w\|^2,$$

$$w_i = (1 - \|r\|^2)^{-1/2} r_i, \qquad r_i = \{(x+1)(y+1)\}^{-1/2} z_i.$$

For simplicity, put $W = \|w\|^2$. Then $1 + W = (1 - \|r\|^2)^{-1}$,

$$(x+1)(y+1) - \|z\|^2 = (x+1)(y+1)(1 - \|r\|^2) = (x+1)(y+1)(1 + W)^{-1},$$

and hence

$$(*) \qquad (x+1)(y-1)-\|z\|^2=(x+1)\{(y+1)(1+W)^{-1}-2\}=(x+1)(1+W)^{-1}v.$$

Similarly

$$(**) \qquad (x-1)(y+1)-\|z\|^2=(y+1)(1+W)^{-1}u.$$

Thus $u>0$ and $v>0$, and hence (4.27) maps Q into

$$R=\{(u,v,w)\in\mathbf{R}^{2+n}|u>0,\ v>0\}.$$

This is bijective. In fact, we have

$$x=u+1+2W, \quad y=v+1+2W, \quad z_i=w_i\{(x+1)(y+1)/(1+W)\}^{1/2}.$$

If $u>0$ and $v>0$, we have $x>1$ and $y>1$. Therefore, in view of (*) and (**), we see that $(x,y,z)\in Q$. Now an easy calculation shows that

$$\partial(x,y,z)/\partial(u,v,w)=\{(x+1)(y+1)\}^{n/2}(1+W)^{-m/2}.$$

Therefore we obtain

$$\eta(g,h;\alpha,\beta)=\int_R e^{\{-a(u+1+2W)-b(v+1+2W)\}/2}(1+W)^{\kappa-\alpha-\beta}$$
$$\cdot(u+2+2W)^{\alpha-1}u^{\beta-\kappa}(v+2+2W)^{\beta-1}v^{\alpha-\kappa}dudvdw.$$

By (3.4.K), this is equal to

$$(4.28) \qquad 2^{2\alpha+2\beta-m}e^{-(a+b)/2}\int_{\mathbf{R}^n} e^{-(a+b)W}(1+W)^{\alpha+\beta-\kappa}$$
$$\cdot\zeta_1(a(1+W);\alpha,\beta-n/2)\zeta_1(b(1+W);\beta,\alpha-n/2)dw$$

with ζ_1 of Sect. 3. Therefore

$$\omega(g,h;\alpha,\beta)=2^{-m}e^{-(a+b)/2}(ab)^{n/4}\int_{\mathbf{R}^n} e^{-(a+b)W}(1+W)^{m/2-2}$$
$$\cdot\omega_1(a(1+W);\alpha,\beta-n/2)\omega_1(b(1+W);\beta,\alpha-n/2)dw.$$

By (3.8), if (α,β) belongs to a compact set T, the last integral is majorized by

$$A(1+\mu(hg)^{-B})\int_{\mathbf{R}^n} e^{-(a+b)W}(1+W)^{m/2-2}dw$$

with positive constants A and B depending only on T. (If $m=2$, we need to put $W=0$ and ignore $\int dw$.) Therefore we obtain our assertion in the same fashion as for ω_{pq} by virtue of Lemma 2.8.

Next suppose $h=e_1$. Again with $g=be_1+ae_2$, we have

$$\eta(g,h;\alpha,\beta)$$
$$=\int_Q e^{(-ax-by)/2}\{(x+1)y-\|z\|^2\}^{\alpha-\kappa}\{(x-1)y-\|z\|^2\}^{\beta-\kappa}dxdydz,$$

$$Q=\{(x,y,z)\in\mathbf{R}^{2+n}|x>1,\ y>0,\ (x-1)y>\|z\|^2\}.$$

Putting $2u = x - 1 - y^{-1}\|z\|^2$, we find that the integral is equal to

$$2^{\alpha+\beta+1-m} \int_R e^{-\{a(2u+1+t)+by\}/2}(u+1)^{\alpha-\kappa}u^{\beta-\kappa}y^{\alpha+\beta-m}\,du\,dy\,dz,$$

$$t = y^{-1}\|z\|^2, \qquad R = \{(u,y,z)\in\mathbf{R}^{2+n}|u>0,\ y>0\}.$$

Thus $\eta(g,h;\alpha,\beta)$ is equal to

$$\pi^{n/2}2^{2\alpha+2\beta-m}e^{-a/2}a^{-n/2}b^{\kappa-\alpha-\beta}\Gamma(\alpha+\beta-\kappa)\zeta_1(a;\alpha-n/2,\beta-n/2).$$

This is valid even when $m=2$. Hence we obtain

(4.29) $\qquad \omega(g,h;\alpha,\beta) = 2^{-1-m}\pi^{n/2}e^{-\tau(hg)/2}\omega_1(\mu(hg);\alpha-n/2,\beta-n/2).$

Our assertion for $h = e_1$ now follows from this relation. This completes the proof of Theorem 4.2.

Remark 4.3. In Sect. 1, we showed that the integral of (1.26) expressing η is convergent for $\mathrm{Re}(\alpha)>\kappa-1$ and $\mathrm{Re}(\beta)>2\kappa-1$. Examining carefully the proofs of Proposition 4.1 and Theorem 4.2, we can easily show that the integral with $h\in V(p,q,r)$ is convergent at least under the following conditions:

All four cases, $h>0$: $\mathrm{Re}(\beta)>\kappa-1$;
All four cases, $h<0$: $\mathrm{Re}(\alpha)>\kappa-1$;
All four cases, $h=0$: $\mathrm{Re}(\alpha+\beta)>2\kappa-1$;
Cases I–III: $\mathrm{Re}(\alpha)>\kappa-1$, $\mathrm{Re}(\beta)>\kappa-1$, $\mathrm{Re}(\alpha+\beta)>\kappa(m)+\kappa(r)-1$, where $\kappa(0) = -1$;
Case IV, $p=q=1$: $\mathrm{Re}(\alpha)>\kappa-1$, $\mathrm{Re}(\beta)>\kappa-1$;
Case IV, $p=r=1$: $\mathrm{Re}(\beta)>\kappa-1$, $\mathrm{Re}(\alpha+\beta)>\kappa$;
Case IV, $q=r=1$: $\mathrm{Re}(\alpha)>\kappa-1$, $\mathrm{Re}(\alpha+\beta)>\kappa$.

Remark 4.4. The factors $\delta_+(hg)$ and $\delta_-(hg)$ are introduced mainly for the purpose of stating inequality (4.13) in as sharp a form as possible. We can formulate the functional equation of η without them at least when $r=0$. In fact, put

(4.30.K) $\qquad \psi(g,h;\alpha,\beta) = 2^{-p\alpha-q\beta}\Gamma_p(\beta-\iota(m-p)/2)^{-1}\Gamma_q(\alpha-\iota(m-q)/2)^{-1}$

$$\cdot\Gamma_r(\alpha+\beta-\kappa)^{-1}\{\delta_+(hg)\delta_-(hg)\}^{(\kappa-\alpha-\beta)/2+\iota r/4}$$

$$\cdot\eta^*(g,h;\alpha,\beta),$$

(4.30.IV : $p=q=1$) $\qquad \psi(g,h;\alpha,\beta) = 2^{-2\alpha-2\beta}\Gamma(\alpha-n/2)^{-1}\Gamma(\beta-n/2)^{-1}$

$$\cdot\delta(hg)^{(\kappa-\alpha-\beta)/2}\eta^*(g,h;\alpha,\beta).$$

Observe that $\delta_+(hg)\delta_-(hg) = |\delta(h)\delta(g)|$ if $r=0$. Now we can easily verify that

(4.31) $\qquad \psi(g,h;\alpha,\beta) = \psi(g,h;\kappa+(\iota r/2)-\beta,\kappa+(\iota r/2)-\alpha),$

(4.32) $\qquad \psi(g,-h;\alpha,\beta) = \psi(g,h;\beta,\alpha),$

(4.33.K) $\qquad \psi(\mathrm{diag}[b,c],\mathrm{diag}[k,0_r];\alpha,\beta)$

$$= (\pi/2)^{\iota r(p+q)/2}\psi(b,k;\alpha-\iota r/2,\beta-\iota r/2)$$

$$\text{if}\quad b\in P_{p+q},\quad c\in P_r,\quad k\in V(p,q,0).$$

Coming back to functions $\xi(g,h;\alpha,\beta)$ in expansion (1.32), we obtain, from (1.29), (4.3), and (4.6),

$$(4.34.\text{K}) \quad \xi(g,h;\alpha,\beta)=|\sigma|^{-1}i^{m\beta-m\alpha}2^{\varphi}\pi^{\psi}\Gamma_r(\alpha+\beta-\kappa)\Gamma_{m-q}(\alpha)^{-1}\Gamma_{m-p}(\beta)^{-1}$$
$$\cdot\delta(g)^{\kappa-\alpha-\beta}\delta_+(hg)^{\alpha-\kappa+\iota q/4}$$
$$\cdot\delta_-(hg)^{\beta-\kappa+\iota p/4}\omega(2\pi g,h;\alpha,\beta)$$

if $h\in V(p,q,r)$, where

$$\varphi=(2p-m)\alpha+(2q-m)\beta+(m+r)\kappa+\iota pq/2,$$

$$\psi=p\alpha+q\beta+r+(\iota/2)\{r(r-1)-pq\}.$$

In Case IV, we have

$$(4.34.\text{IV}) \qquad \xi(g,h;\alpha,\beta)=|\sigma|^{-1/2}i^{2\beta-2\alpha}\delta(2g)^{\kappa-\alpha-\beta}\omega(2\pi g,h;\alpha,\beta)$$

$$\begin{cases}
2^{2\alpha+1}\pi^{2\alpha+1-\kappa}\Gamma(\alpha)^{-1}\Gamma(\alpha-n/2)^{-1}\delta_+(hg)^{\alpha-\kappa} & \text{if}\quad h>0,\\[4pt]
2^{2\beta+1}\pi^{2\beta+1-\kappa}\Gamma(\beta)^{-1}\Gamma(\beta-n/2)^{-1}\delta_-(hg)^{\beta-\kappa} & \text{if}\quad h<0,\\[4pt]
2^{\alpha+\beta+\kappa+1}\pi^{\alpha+\beta+1-\kappa}\Gamma(\alpha)^{-1}\Gamma(\beta)^{-1}\delta_+(hg)^{\alpha-1-n/4}\delta_-(hg)^{\beta-1-n/4}\\
\quad\text{if}\quad h\in V(1,1,0),\\[4pt]
2^{\alpha+2+\kappa}\pi^{\alpha+2-\kappa}\Gamma(\alpha+\beta-\kappa)\Gamma(\alpha)^{-1}\Gamma(\beta)^{-1}\Gamma(\alpha-n/2)^{-1}\delta_+(hg)^{\alpha-\kappa}\\
\quad\text{if}\quad h\in V(1,0,1),\\[4pt]
2^{\beta+2+\kappa}\pi^{\beta+2-\kappa}\Gamma(\alpha+\beta-\kappa)\Gamma(\alpha)^{-1}\Gamma(\beta)^{-1}\Gamma(\beta-n/2)^{-1}\delta_-(hg)^{\beta-\kappa}\\
\quad\text{if}\quad h\in V(0,1,1),\\[4pt]
|\sigma|^{1/2}2^{m+2-2\alpha-2\beta}\pi^2\Gamma_m(\alpha+\beta-\kappa)\Gamma(\alpha)^{-1}\Gamma(\beta)^{-1}\Gamma(\alpha-n/2)^{-1}\Gamma(\beta-n/2)^{-1}\\
\quad\text{if}\quad h=0.
\end{cases}$$

We conclude this section by noting

$$(4.35.\text{K}) \qquad \omega(g,h;\alpha,\iota r/2)=\omega(g,h;\kappa,\beta)=2^{-p\kappa}\pi^{\iota pr/2}e^{-\sigma(g,h)}$$
$$\text{if}\quad h\in V(p,0,r),$$

$$(4.35.\text{IV}) \quad \omega(g,h;\kappa,\beta)$$
$$=\begin{cases}
\omega(g,h;\alpha,0)=2^{-m}e^{-\sigma(g,h)} & \text{if}\quad h>0,\\[4pt]
\omega(g,h;\alpha,n/2)=2^{-1-m}\pi^{(m/2)-1}e^{-\sigma(g,h)} & \text{if}\quad h\in V(1,0,1).
\end{cases}$$

The case of $h\in P$ follows from (3.15), (4.7), and (4.10). If $h\in V(p,0,r)$ with $r>0$ in Cases I–III, the formula is reduced to that special case by (4.11.K). If $h\in V(1,0,1)$ in Case IV, it follows from (4.29) and (3.15).

5. The Real Analyticity of ω in g

The above proof of Theorem 4.2 does not tell the real analyticity of $\omega(g,h;\alpha,\beta)$ in g, which is actually true as will be shown in this section. We first consider only Cases I–III. For $x=(x_1,\ldots,x_m)\in\mathbf{C}^m$, we put $\|x\|=\left(\sum_{k=1}^m|x_k|^2\right)^{1/2}$, and for $a\in\mathbf{C}^m_m$, define $\|a\|$ by $\|a\| = \underset{\|x\|=1}{\mathrm{Max}} \|ax\|$. Obviously $\|a\|^2$ is the maximum eigenvalue of a^*a. We have also $|a_{jk}| \leq \|a\|$ if $a = (a_{jk})$, and $\|a\| \leq C_m \cdot \mathrm{Max}|a_{jk}|$ with a constant C_m depending only on m. In particular, $\|a\|$ for $a \in V_{\mathbf{C}}$ in Case III is meaningful, since $V_{\mathbf{C}} \subset C_{2m}^{2m}$.

Lemma 5.1. *For $g = \Re(z)$ (defined by (3.9)) with $z \in V_{\mathbf{C}}$, we have*
   (i) $\|g\| \leq \|z\|$,
   (ii) $\delta(g) \leq |\delta(z)|$ *if $g \in P$,*
   (iii) $\|z^{-1}\| \leq \mu(g)^{-1}$ *if $g \in P$,*
   (iv) $|(z^{-1})_{jk}| \leq \mu(g)^{-1}$ *if $g \in P$.*

Proof. It is sufficient to prove these assertions in Case II. Put $z = g + ih$. Take $p \in U_m$ so that $pgp^* = \mathrm{diag}[r_1, \ldots, r_m]$ with $|r_1| \geq \ldots \geq |r_m|$. Let e_1 be the first member of the standard basis of $\mathbf{C}^m$, and s the $(1, 1)$-entry of php^*. Then

$$\|pzp^*e_1\| = \|(r_1 + ip_1)e_1 + \ldots\| \geq |r_1| = \|g\|,$$

which proves (i). Suppose $g \in P$. By Lemma 2.9, we have

$$z = a^*\mathrm{diag}[r_1 + is_1, \ldots, r_m + is_m]a$$

with $a \in G$ and real r_k and s_k. Then

$$|\delta(z)| \geq \delta(a^*a) \prod_{k=1}^{m} r_k = \delta(g),$$

which proves (ii). Put $b = g^{1/2}$ and $c = b^{-1}hb^{-1}$. Then $z = b(1 + ic)b$. For every $x \in \mathbf{C}^m$,

$$\|(1 + ic)x\|^2 = x^*(1 - ic)(1 + ic)x = x^*(1 + c^2)x \geq \|x\|^2.$$

Putting $y = (1 + ic)x$, we have $\|(1 + ic)^{-1}y\|^2 \leq \|y\|^2$, and hence $\|(1 + ic)^{-1}\| \leq 1$. Therefore

$$\|z^{-1}\| = \|b^{-1}(1 + ic)^{-1}b^{-1}\| \leq \|b^{-1}\|^2 \leq \mu(g)^{-1}.$$

This proves (iii). As observed above, $|(z^{-1})_{jk}| \leq \|z^{-1}\|$. This together with (iii) proves (iv).

Lemma 5.2. *Let $z_1, \ldots, z_N$ be the Euclidean complex coordinate functions on $V_{\mathbf{C}}$. If $z \in H'$, $g = \Re(z)$, and $(\beta, \delta(g^{-1}z))$ stays in a compact subset T of $\mathbf{C}^2$ and $\mu(g) > r > 0$, then*

$$|\delta(z)^\beta \partial^n/\partial z_{k_1} \ldots \partial z_{k_n} \delta(z)^{-\beta}| \leq A\delta(g)^B$$

with positive constants A and B depending only on T, r, and n.

Proof. We have $\delta(z)^\beta \partial^n/\partial z_{k_1} \ldots \partial z_{k_n} \delta(z)^{-\beta} = \delta(z)^{-n} f(z)$ with a homogeneous polynomial function f. If $w = z^{-1}$, then $z_k = \delta(z)^2 g_k(w)$ with polynomial functions g_k. (The exponent 2 is necessary only in Case III.) Therefore $\delta(z)^{-n} f(z) = \delta(z)^M \varphi(w)$ with $0 < M \in \mathbf{Z}$ and a polynomial function φ. Hence, by Lemma 5.1, (iv),

$$|\delta(z)^{-n} f(z)| \leq C|\delta(g^{-1}z)|^M \delta(g)^M$$

if $\mu(g) > r$, with a constant C depending only on r, n, and β. This proves our lemma.

Lemma 5.3. *Let $f(z_1, \ldots, z_n)$ be a holomorphic function such that $|f| \leqq M$ for*

$$|z_1| < r_1, \ldots, |z_n| < r_n.$$

Then

$$|(\partial^a f/\partial z_1^{a_1} \ldots \partial z_n^{a_n})(0)| \leqq M a_1! \ldots a_n! r_1^{-a_1} \ldots r_n^{-a_n}.$$

This is well known.

Lemma 5.4. *For every $r > 0$, there exists a positive real number t_r depending only on r and m such that if $z, z' \in V_C$, $g = \Re(z)$, $g' = \Re(z')$, $g > 0$, $\mu(g) \geqq r$, and $|z_{jk} - z'_{jk}| \leqq t_r$, then $g' > 0$ and $\mu(g') \geqq r/2$.*

Proof. By Lemma 5.1, (i), we have $\|g - g'\| \leqq \|z - z'\| \leqq Ct$ if $|z_{jk} - z'_{jk}| \leqq t$, where C is a constant depending only on m. For $x \in \mathbf{C}^m$, we have then

$$\|g'x\| = \|gx - (g - g')x\| \geqq \|gx\| - \|(g - g')x\| \geqq \mu(g)\|x\| - Ct\|x\|.$$

Therefore $t_r = r/(2C)$ has the required property.

Lemma 5.5. *Let r and t_r be the same as in Lemma 5.4. Given $a > 0$, there exists a constant $b > 0$ depending only on r, a, and m such that if $z, z' \in V_C$, $g = \Re(z) > 0$, $g' = \Re(z')$, $\mu(g) \geqq r$, $|z_{jk} - z'_{jk}| \leqq t_r$, $|\delta(g^{-1}z)| \leqq a$, then $|\delta(g'^{-1}z')| \leqq b$.*

Proof. Suppose z, z', g, g' are given as above. We have again $\|g - g'\| \leqq \|z - z'\| \leqq Ct_r$, and $\|g'^{-1}\| \leqq 2/r$, $\|z'^{-1}\| \leqq 2/r$, $\|g^{-1}\| \leqq 1/r$, $\|z^{-1}\| \leqq 1/r$ by (iii) of Lemma 5.1 and Lemma 5.4. Since $z^{-1}z' = 1 + z^{-1}(z' - z)$, $z^{-1}z'$ belongs to a compact set. Similarly $g'^{-1}g$ belongs to a compact set. Thus $\delta(g'^{-1}g) \leqq A$ and $|\delta(z^{-1}z')| \leqq A$ with a constant A depending only on r and m. Therefore

$$|\delta(g'^{-1}z')| = |\delta(g'^{-1}gg^{-1}zz^{-1}z')| \leqq A^2 a,$$

which proves our lemma.

Lemma 5.6. *Given $a > 0$, $r > 0$, and a compact subset T of $\mathbf{C}^2$, there exist nonnegative constants A and B depending only on a, r, T, and m such that $|\omega(z; \alpha, \beta)| \leqq A\delta(g)^B$ if $z \in H'$, $g = \Re(z)$, $|\delta(g^{-1}z)| \leqq a$, $\mu(g) \geqq r$, and $(\alpha, \beta) \in T$.*

Proof. Suppose $\mathrm{Re}(\beta) \geqq \kappa$. Then the integral expression of ζ in (3.2) shows that

$$|\omega(z; \alpha, \beta)| \leqq |\Gamma(\beta)^{-1}\Gamma(\mathrm{Re}(\beta))\delta(g^{-1}z)^\beta \omega(g; \mathrm{Re}(\alpha), \mathrm{Re}(\beta))|,$$

and hence the desired inequality with $B = 0$ follows from (3.8). This result together with (3.7) proves the inequality (again with $B = 0$) for $\mathrm{Re}(\alpha) \leqq 0$. By (3.12), we have

$$(5.3) \qquad \omega(z; \alpha + n, \beta) = (-1)^{mn} e^{\sigma(z)} \delta(z)^\beta \Delta^n \{ e^{-\sigma(z)} \delta(z)^{-\beta} \omega(z; \alpha, \beta) \}.$$

Given $r > 0$ and $a > 0$, take t_r and b as in Lemmas 5.4 and 5.5. Then $|\omega(z; \alpha, \beta)| \leqq M$ if $|\delta(g^{-1}z)| \leqq b$, $\mu(g) \geqq r/2$, $(\alpha, \beta) \in T$, and $\mathrm{Re}(\alpha) \leqq 0$, where M is a constant depending on b, r, and T. By Lemmas 5.3–5.5, we have

$$|\partial^p/\partial z_{k_1} \ldots \partial z_{k_p} \omega(z; \alpha, \beta)| \leqq M' t_r^{-p}$$

if $\mu(g) \geqq r$, $|\delta(g^{-1}z)| \leqq a$, and $(\alpha, \beta) \in T$ with a constant M' depending only on a, r, p, and T. By Lemma 5.2,

$$|\delta(z)^\beta \partial^p / \partial z_{k_1} \ldots \partial z_{k_p} \delta(z)^{-\beta}| \leqq A\delta(g)^B$$

under the same conditions on z and (α, β) with A and B as in that lemma. Therefore we obtain our assertion for $\mathrm{Re}(\alpha) \leqq n$ (and hence for all α) from (5.3).

Lemma 5.7. *There is a holomorphic function $f(u, v; \alpha, \beta)$ in $(u, v, \alpha, \beta) \in H'_p \times H'_q \times \mathbf{C}^2$ such that*

$$f(a, b; \alpha, \beta) = \omega_{pq}(\mathrm{diag}[a, b]; \alpha, \beta) \quad for \quad a \in P_p, \quad b \in P_q, \quad (\alpha, \beta) \in \mathbf{C}^2.$$

Proof. Substituting u and v for a and b in (4.25), we put

$$f(u, v; \alpha, \beta) = \delta(u)^{iq/4} \delta(v)^{ip/4} e^{(-\sigma(u) - \sigma(v))/2}$$

$$\cdot \int_F e^{-\sigma(uW') - \sigma(vW'')} \omega_p\left(ZuZ; \alpha, \beta - \frac{iq}{2}\right) \omega_q\left(Z'vZ'; \beta, \alpha - \frac{ip}{2}\right) \delta(1 + W)^{(n/2)-1} dw.$$

By Lemma 5.6, if $(u, v; \alpha, \beta)$ stays in a compact subset of $H'_p \times H'_q \times \mathbf{C}^2$, the integral is convergent and defines a holomorphic function as desired.

Lemma 5.8. *Given $h \in V(p, q, r)$ and $g_0 \in P$ in Cases I–III, there exists a neighborhood U of g_0 in P and a real analytic map $a: U \to G$ such that*

$$h = a(g)^* \mathrm{diag}[1_p, -1_q, 0_r] a(g) \quad and \quad a(g) g a(g)^* = \mathrm{diag}[u, v, w]$$

with $u \in P_p$, $v \in P_q$, and $w \in P_r$ for every $g \in U$.

Proof. By Lemma 2.9, there exists $b \in G$ such that

$$h = b^* \mathrm{diag}[1_p, -1_q, 0_r] b \quad and \quad b g_0 b^* = \mathrm{diag}[u_0, v_0, w_0]$$

with $u_0 \in P_p$, $v_0 \in P_q$, and $w_0 \in P_r$. Now, for every $s \in \mathbf{K}_q^p$, put $\gamma(s) = \begin{bmatrix} t & s \\ s^* & t' \end{bmatrix}$ with

$$t = (1_p + ss^*)^{1/2} \quad and \quad t' = (1_q + s^*s)^{1/2}.$$

Then $\gamma(s) \in U_{p,q}$ by virtue of Lemma 2.2. It can easily be verified that the map

$$(u, v, s) \mapsto \gamma(s) \mathrm{diag}[u, v] \gamma(s)^*$$

of $P_p \times P_q \times \mathbf{K}_q^p$ into P_{p+q} has nonvanishing jacobian at $(u, v, w) = (u_0, v_0, 0)$, so that it has a real analytic local inverse. Put $bgb^* = \begin{pmatrix} x & z^* \\ z & y \end{pmatrix}$ with $x \in P_{p+q}$ for g in a small neighborhood of g_0. Then the relation $\gamma(s) \mathrm{diag}[u, v] \gamma(s)^* = x$ determines u, v, and s as real analytic functions in g. Now the required map a is given by

$$a(g) = \begin{pmatrix} \gamma(s)^{-1} & 0 \\ -zx^{-1} & 1_r \end{pmatrix} b.$$

Theorem 5.9. *Functions $\omega(g, h; \alpha, \beta)$ in all four cases are real analytic in $(g, \alpha, \beta) \in P \times \mathbf{C}^2$.*

Proof. If $\pm h \in P$ or $h=0$, our assertion follows from (4.10), Theorem 3.1, (4.8), and (4.9). Suppose, $h \in V(p,q,r)$ in Cases I–III. Given $g_0 \in P$, take U and $a(g)$ and put $a(g)ga(g)^* = \mathrm{diag}[u,v,w]$ as in Lemma 5.8. By (4.7) and (4.11), we have

$$\omega(g,h;\alpha,\beta) = (\pi/2)^{ir(p+q)/2}\omega(\mathrm{diag}[u,v],\varepsilon_{p,q};\alpha-ir/2,\beta-ir/2).$$

If $pq=0$, this proves our assertion. If $pq \neq 0$, (4.19) shows that

$$\omega(g,h;\alpha,\beta) = A\omega_{pq}(2\,\mathrm{diag}[u,v];\alpha-ir/2,\beta-ir/2)$$

with a constant A. Therefore Lemma 5.7 takes care of this case. It remains to consider Case IV. It is sufficient to treat $h=e_1-e_2$ and $h=e_1$. Suppose $h=e_1-e_2$. For $w \in \mathbf{R}^n$, $n=m-2$, put $r=(1+\|w\|^2)^{1/2}$ and

$$\gamma(w) = \begin{bmatrix} (r+1)/2 & (r-1)/2 & w^* \\ (r-1)/2 & (r+1)/2 & w^* \\ w/2 & w/2 & (1_n+ww^*)^{1/2} \end{bmatrix},$$

$$\xi(u,v,w) = \gamma(w)(ue_1+ve_2) = xe_1+ye_2+\sum_{k=1}^{n} z_k e_k.$$

Then $\gamma(w) \in G$, $\nu(\gamma(w))=1$, $x=u+(r-1)(u+v)/2$, $y=v+(r-1)(u+v)/2$, $z=(u+v)w/2$, $\gamma(w)(e_1-e_2)=e_1-e_2$, and $\partial c(x,y,z)/\partial(u,v,w) \neq 0$ at $(u,v,w)=(b,a,0)$ for $a>0$, $b>0$. Therefore, if $g_0 = be_1+ae_2 \in P$ and g belongs to a small neighborhood of g_0, then the equality $g=\gamma(w)(ue_1+ve_2)$ determines u, v, and w as real analytic functions in g. Observe that $\delta_+(hg)=\tau_+(hg)=v$ and $\delta_-(hg)=\tau_-(hg)=u$. As shown in the proof of Theorem 4.2,

$$\omega(g,h;\alpha,\beta) = 2^{-m}e^{-(u+v)/2}(uv)^{n/4}\int_{\mathbf{R}^n} e^{-(u+v)W}(1+W)^{m/2-2}$$

$$\cdot\omega_1(v(1+W);\alpha,\beta-n/2)\omega_1(u(1+W);\beta,\alpha-n/2)dw,$$

where $W=\|w\|^2$. Therefore our assertion in this case follows from Lemma 5.6 as in the proof of Lemma 5.7. Finally suppose $h=e_1$. Then

$$\mu(hg) = \delta_+(hg) = \tau(hg) = \tau_+(hg) = 2\sigma(g,h),$$

and ω is given by (4.29). Hence the real analyticity in (g,α,β) is obvious. This completes the proof.

Proposition 5.10. *Functions* $\delta_+(hg)$, $\delta_-(hg)$, $\tau_+(hg)$, $\tau_-(hg)$ *are real analytic in* $g \in P$.

Proof. Suppose $h \in V(p,q,r)$ in Cases I–III. With

$$a(g)ga(g)^* = \mathrm{diag}[u,v,w]$$

as in Lemma 5.8, we have $\delta_+(hg)=\delta(u)$, $\delta_-(hg)=\delta(v)$, $\tau_+(hg)=\sigma(u)$, $\tau_-(hg)=\sigma(v)$. Hence our assertion is obvious. In Case IV, our assertion is already included in the above proof of Theorem 5.9.

Finally we note that function ψ defined by (4.30) can be extended to a function $\psi(z,h;\alpha,\beta)$ of $(z,h;\alpha,\beta) \in H' \times V \times \mathbf{C}^2$ which is holomorphic in (z,α,β). This can be proved in the same fashion as in the analytic continuation of $\omega(z;\alpha,\beta)$ in the proof of Theorem 3.1.

6. Analytic Continuation of $S(z, L; \alpha, \beta)$

In Sect. 1, we defined a series

$$S(z, L; \alpha, \beta) = \sum_{a \in L} \delta(z + a)^{-\alpha} \delta(\bar{z} + \bar{a})^{-\beta} \quad (z \in H, (\alpha, \beta) \in \mathbf{C}^2)$$

with a lattice L in V, and showed that this has a Fourier expansion

$$(6.0) \qquad \mu(V/L) S(z, L; \alpha, \beta) = \sum_{h \in L'} e^{2\pi i \sigma(h, x)} \xi(y, h; \alpha, \beta) \quad (z = x + iy)$$

at least for $\mathrm{Re}(\alpha) > \kappa - 1/2$ and $\mathrm{Re}(\beta) > \kappa - 1/2$, where

$$L' = \{h \in V \mid \sigma(h, L) \subset \mathbf{Z}\}.$$

Now that we established the analytic continuation of ξ, we can investigate the analytic continuation of this series. We first introduce the notion of an *algebraic lattice L in $\mathbf{R}^n$*, which means a lattice whose elements have algebraic components. Since V has a natural **Q**-rational structure, we can speak of algebraic lattices in V. We are going to prove

Theorem 6.1. *Let L be an algebraic lattice in V; suppose σ is algebraic-valued on $L \times L$ in Case IV. Then*

$$\Gamma_m(\alpha + \beta - \kappa)^{-1} S(z, L; \alpha, \beta)$$

can be continued as a holomorphic function in (α, β) to the whole $\mathbf{C}^2$.

We first introduce a few symbols and prove some inequalities. Let $\lambda(h)$ [resp. $\lambda(hg)$] denote, for $0 \neq h \in V$ and $g \in P$, the maximum absolute value of the eigenvalues of h (resp. of h relative to g). We observe, in Cases I–III, that $\lambda(h) = \|h\|$ with $\|h\|$ defined as in Sect. 5. In Case IV, if we put $\|h\| = \left(\sum_{i=1}^{m} h_i^2\right)^{1/2}$ for $h = (h_i) \in \mathbf{R}^m = V$, then we have

$$(6.1.\mathrm{IV}) \qquad \|h\|/A \leqq \lambda(h) \leqq A\|h\|$$

with a constant $A > 1$ independent of h. In fact, by Lemma 2.9, $h = a(\lambda e_1 + \mu e_2)$ with $a \in U_m$ and $\lambda, \mu \in \mathbf{R}$. Since U_m is compact, we obtain (6.1.IV). In Cases I–III, we have $\lambda(aha^*) = \|aha^*\| \leqq \|a\|^2 \|h\|$, and hence

$$(6.2.\mathrm{K}) \qquad \|a^{-1}\|^{-2} \leqq \lambda(aha^*)/\lambda(h) \leqq \|a\|^2 \quad \text{for every} \quad a \in G.$$

Similarly in Case IV, we obtain, from (6.1.IV),

$$(6.2.\mathrm{IV}) \qquad \|a^{-1}\|^{-1}/B \leqq \lambda(ah)/\lambda(h) \leqq B\|a\| \quad \text{for every} \quad a \in G$$

with a constant $B > 1$ independent of h and a. Here and henceforth we always exclude the case $h = 0$ from our inequalities. We have also

$$(6.3.\mathrm{K}) \qquad \|a^{-1}\|^{-2} \leqq \mu(aha^*)/\mu(h) \leqq \|a\|^2 \quad \text{for every} \quad a \in G,$$

$$(6.3.\mathrm{IV}) \qquad \|a^{-1}\|^{-1}/C \leqq \mu(ah)/\mu(h) \leqq C\|a\| \quad \text{for every} \quad a \in G$$

with a constant $C>1$ independent of h and a. In fact, if h is invertible in Cases I–III, we obtain (6.3.K) from (6.2.K) by means of the relation $\mu(h)=\lambda(h^{-1})^{-1}$. As for noninvertible h, we may assume that $h=\mathrm{diag}[p,0]$ with an invertible $p\in V_n$, $n<m$. Write an element a of G in the form $a=\begin{pmatrix} x & y \\ u & v \end{pmatrix}$ with $x\in \mathbf{K}_n^n$. Then aha^* and ha^*a have the same eigenvalues. Let $x^*x+u^*u=b^2$ with $b\in P_n$. Since $ha^*a=\begin{pmatrix} pb^2 & * \\ 0 & 0 \end{pmatrix}$, we have $\mu(aha^*)=\mu(bpb)\leq\mu(p)\|b\|^2\leq\mu(h)\|a\|^2$, from which we obtain (6.3.K) in the general case. In Case IV, if $\delta(h)\neq 0$, we have $\lambda(h)\mu(h)=|\sigma(h,h)|$, so that

$$(6.4.\mathrm{IV}) \qquad \mu(ah)/\mu(h)=v(a)\lambda(h)/\lambda(ah) \quad \text{for every} \quad a\in G.$$

It can easily be verified that

$$(6.5.\mathrm{IV}) \qquad \|a\|/D\leq\|v(a)a^{-1}\|\leq D\|a\| \quad \text{for every} \quad a\in G$$

with a constant $D>1$ independent of a. Combining these we obtain (6.3.IV) when $\delta(h)\neq 0$. If $\delta(h)=0$, we have $\lambda(h)=\mu(h)$ and $\lambda(ah)=\mu(ah)$, and hence (6.3.IV) follows from (6.2.IV).

Now the above inequalities (6.2) and (6.3) can be given also in the forms

$$(6.6.\mathrm{K}) \qquad \mu(g)\leq\lambda(hg)/\lambda(h)\leq\|g\|,$$

$$(6.6.\mathrm{IV}) \qquad \mu(g)/E\leq\lambda(hg)/\lambda(h)\leq E\|g\|,$$

$$(6.7.\mathrm{K}) \qquad \mu(g)\leq\mu(hg)/\mu(h)\leq\|g\|,$$

$$(6.7.\mathrm{IV}) \qquad \mu(g)/F\leq\mu(hg)/\mu(h)\leq F\|g\|$$

for $0\neq h\in V$ and $g\in P$, with constants $E>1$ and $F>1$ independent of h and g. In fact, in Cases I–III, we obtain them by taking $a=g^{1/2}$ in (6.2.K) and (6.3.K). In Case IV, we first prove

$$(6.8.\mathrm{IV}) \qquad \|a\|\leq J\lambda(a\varepsilon), \qquad \|a^{-1}\|^{-1}\geq K\mu(a\varepsilon) \quad \text{for every} \quad a\in G$$

with positive constants J and K. Given $a\in G$, take $b\in U_m$ so that $ba\varepsilon=\lambda e_1+\mu e_2$ with $\lambda\geq\mu>0$; define $c\in G$ by $ce_1=\lambda e_1$, $ce_2=\mu e_2$, $ce_k=(\lambda\mu)^{1/2}e_k$ for $k>2$; put $d=c^{-1}ba$. Since $c\varepsilon=ba\varepsilon$, we see that $d\in U_m$. Hence

$$\|a\|=\|b^{-1}cd\|\leq J\|c\|\leq J\lambda=J\lambda(a\varepsilon)$$

with a constant J. Similarly

$$\|a^{-1}\|=\|d^{-1}c^{-1}b\|\leq J\|c^{-1}\|\leq J\mu^{-1}=J\mu(a\varepsilon)^{-1},$$

from which we obtain (6.8.IV). Now, if $g=a\varepsilon$ with $a\in G$ and $a'=v(a)a^{-1}$, we have $\mu(a'h)=\mu(hg)$ and $\lambda(a'h)=\lambda(hg)$. Therefore we obtain (6.7.IV) from (6.3.IV), (6.5.IV), and (6.8.IV).

Lemma 6.2. *Let F be a complex-valued polynomial function on $\mathbf{R}^n$ with algebraic coefficients, and L an algebraic lattice in $\mathbf{R}^n$. Then there exist two positive constants A and B depending only on F and L such that $|F(x)|\geq A\|x\|^{-B}$ for every $x\in L$ with $\|x\|\geq 1$, $F(x)\neq 0$.*

Proof. Changing the coordinate system, we may assume that $L = \mathbf{Z}^n$. Let M be the field generated over $\mathbf{Q}$ by the coefficients of F, and I the set of all isomorphic embeddings of M into $\mathbf{C}$. Denote by F^τ for $\tau \in I$ the polynomial function whose coefficients are the images of those of F under τ. Put $E(x) = \prod_{\tau \in I} F^\tau(x)$. Then there is a positive integer r such that rE has coefficients in $\mathbf{Z}$. Then $|rE(x)| \geq 1$ if $E(x) \neq 0$ and $x \in L$. Now $\left| \prod_{\tau \neq 1} F^\tau(x) \right| \leq \alpha \|x\|^\beta$ for $\|x\| \geq 1$ with positive constants α and β depending only on F. Thus, if $x \in L$, $\|x\| \geq 1$, and $F(x) \neq 0$, we have

$$|F(x)| = r^{-1} \left| rE(x) \Big/ \prod_{\tau \neq 1} F^\tau(x) \right| \geq \|x\|^{-\beta}/(r\alpha). \quad \text{Q.E.D.}$$

Lemma 6.3. *Let L be an algebraic lattice in V; suppose σ is algebraic-valued on $L \times L$ in Case IV. Then there exist two positive constants C and D such that $\mu(h) \geq C\|h\|^{-D}$ for $h \in L$ with $\|h\| \geq 1$.*

Proof. Lemma 6.2 is obviously true with any norm in $\mathbf{R}^n$ equivalent to the standard norm $\|x\|$ in $\mathbf{R}^n$. Therefore we can apply it to V with the norm $\|h\|$ defined in Sect. 5. It is also sufficient to prove the present lemma with $L \cap V(p, q, r)$ instead of L for fixed p, q, r. In Cases I–III, let $\sum_{k=0}^{m} t^k F_k(h) = \delta(t1_m - h)$ with an indeterminate t. Then $|F_r(h)| = \delta(h1_m)$. Now each eigenvalue of h has absolute value $\leq \|h\|$, so that

$$\mu(h) \geq |F_r(h)|/\|h\|^{m-r-1}.$$

Applying Lemma 6.2 to F_r, we obtain our assertion in Cases I–III. Now, in Case IV, $\lambda(h) \leq E\|h\|$ with a constant E by (6.1.IV). If $\delta(h) \neq 0$, we have

$$\mu(h) = |\delta(h)|/\lambda(h) \geq |\delta(h)|/(E\|h\|).$$

If $\delta(h) = 0$, we have $\mu(h) = |2\sigma(h, \varepsilon)|$. Therefore we obtain our assertion by applying Lemma 6.2 to δ and $\sigma(h, \varepsilon)$.

Proof of Theorem 6.1. If L is algebraic, its dual lattice L' is algebraic. Let $S(p, q, r)$ denote the partial sum

$$\sum_{h} e^{2\pi i \sigma(h, x)} \zeta(y, h; \alpha, \beta)$$

of (6.0) with h running over $L' \cap V(p, q, r)$. We are going to prove that this multiplied by a suitable gamma factor is uniformly convergent for (y, α, β) in $J \times T$ for any compact subsets J of P and T of $\mathbf{C}^2$. Since there are only finitely many terms with $\|h\| < 1$, it is sufficient to consider the terms with $\|h\| \geq 1$. By (6.1), (6.6), (6.7), and Lemma 6.3, we have

$$\tau(hy) \geq \lambda(hy) \geq D\mu(y)\|h\|,$$

$$\delta_+(hy) \geq \mu(hy)^p \geq A\mu(y)^p\|h\|^{-B},$$

$$\delta_+(hy) \leq \lambda(hy)^p \leq C\lambda(y)^p\|h\|^p$$

with positive constants A, B, C, and D independent of y and h. Similar inequalities hold also for $\delta_-(hy)$. Therefore, in Cases I, II, and III, if $(\alpha, \beta) \in T$, the part of

$$\Gamma_r(\alpha + \beta - \kappa)^{-1} \Gamma_{m-q}(\alpha) \Gamma_{m-p}(\beta) S(p, q, r)$$

with $\|h\| \geqq 1$ is majorized by

$$E(\lambda(y)^F + \mu(y)^{-F}) \sum_h e^{-M\mu(y)\|h\|} \|h\|^N$$

with some positive constants E, F, M, and N by virtue of (4.13) and (4.34.K). This is obviously convergent locally uniformly in y. Since $\Gamma_m^{-1}\Gamma_r$ is entire, we obtain our theorem in Cases I–III. Case IV can be treated in exactly the same fashion.

Let us now consider

$$(6.9) \qquad S(z, L; \alpha) = \sum_{a \in L} \delta(z + a)^{-\alpha} \quad (z \in H, \alpha \in \mathbf{C}).$$

By Lemma 1.3, if $\mathrm{Re}(\alpha) > 2\kappa - 1$, this is convergent and defines a holomorphic function in (z, α). Therefore expansion (6.0) is valid for such α and for an arbitrary lattice L which is not necessarily algebraic. Observing that all terms with $h \notin P$ vanish for $\beta = 0$ because of the gamma factors of such terms, we obtain, by virtue of (4.34) and (4.35),

$$(6.10) \qquad \mu(V/L)S(z, L; \alpha) = |\sigma|^{-1} i^{-\varrho\alpha}(2\pi)^{\varrho\alpha}\Gamma_m(\alpha)^{-1}$$
$$\cdot \sum_{0 < h \in L'} \delta(h)^{\alpha - \kappa}e^{2\pi i\sigma(h, z)} \quad (\mathrm{Re}(\alpha) > 2\kappa - 1).$$

This can be obtained also more directly from (1.23). It can easily be seen that $\Gamma_m(\alpha)S(z, L; \alpha)$ can be continued as a holomorphic function in (z, α) to $H' \times \mathbf{C}$ for an arbitrary lattice L which is not necessarily algebraic.

There is another way to deal with the series of type (6.9). Indeed, we put

$$(6.11) \qquad S^*(z, L; \alpha) = \lim_{s \to 0} S(z, L; \alpha + s, s).$$

This is obviously equal to $S(z, L; \alpha)$ if $\mathrm{Re}(\alpha) > 2\kappa - 1$, but is not necessarily so for smaller values of $\mathrm{Re}(\alpha)$. We have, however,

Theorem 6.4. *Suppose L is algebraic and σ is algebraic-valued on $L \times L$ in Case IV. Then $S^*(z, L; \alpha)$ coincides with $S(z, L; \alpha)$ at least for $\mathrm{Re}(\alpha) > \kappa$ except for the following values of α:*
Case I: $\alpha = (m/2) + 1$;
Case IV: m odd, $\alpha \in \mathbf{Z}$, $m/2 < \alpha \leqq m - 1$.

Proof. Suppose $h \in V(p, q, r)$ with $m > p$. In Cases II and III, the gamma factor appearing in $\xi(y, h; \alpha + s, s)$ is $\Gamma_r(\alpha + 2s - \kappa)\Gamma_{p+r}(\alpha + s)^{-1}\Gamma_{q+r}(s)^{-1}$ as shown by (4.30.K). This vanishes at $s = 0$ if $\mathrm{Re}(\alpha) > \kappa$, and hence $\mu(V/L)S^*$ coincides with (6.10). The same is true in Cases I and IV, except that we have to exclude the above mentioned values, for which some terms with $p < m$ can appear.

The critical case $\alpha = \kappa$ produces an interesting phenomenon:

Theorem 6.5. *Suppose L is algebraic and σ is algebraic-valued on $L \times L$ in Case IV. Suppose also m is even in Case IV. Then, excluding Case I, we have*

$$\mu(V/L)S^*(z, L; \kappa) = |\sigma|^{-1}i^{-\varrho\kappa}(2\pi)^{\varrho\kappa}\Gamma_m(\kappa)^{-1} \sum_h 2^{-r(h)}e^{2\pi i\sigma(h, z)},$$

where h runs over all the elements of L' with no negative eigenvalues and $r(h)$ denotes the number of zero eigenvalues of h.

We leave the details of the proof to the reader, as it is again a straightforward examination of $\xi(g, h; \kappa + s, s)$ when s tends to 0, with the use of (4.34) and (4.35).

We conclude this paper by mentioning a series

$$\sum_{a \in L} \varrho_1(z + a) \otimes \varrho_2(\bar{z} + \bar{a}) \delta(z + a)^{-\alpha} \delta(\bar{z} + \bar{a})^{-\beta}$$

as a generalization of $S(z, L; \alpha, \beta)$. Here ϱ_1 and ϱ_2 are rational representations of $GL_m(\mathbf{C})$ and $GL_{2m}(\mathbf{C})$ in Cases I, II, and in Case III, respectively; they are homogeneous polynomial functions on $V_{\mathbf{C}}$ in Case IV. This leads naturally to similar generalizations of functions ξ and η. Some of these can be obtained by applying certain differential operators to the original ξ, η, and S. A full treatment of these generalizations, however, may be left as an open problem to those interested in the subject.

References

1. Indik, R.: Thesis, Princeton University 1982
2. Jacquet, H.: Fonctions de Whittaker associées aux groupes de Chevalley. Bull. Soc. Math. France **95**, 243–309 (1967)
3. Kaufhold, G.: Dirichletsche Reihe mit Funktionalgleichung in der Theorie der Modulfunktion 2. Grades. Math. Ann. **137**, 454–476 (1959)
4. Koecher, M.: Über Thetareihen indefiniter quadratischer Formen. Math. Nachr. **9**, 51–85 (1953)
5. Maass, H.: Siegel's modular forms and Dirichlet series. Lecture Notes in Mathematics, Vol.216. Berlin, Heidelberg, New York: Springer 1971
6. Shimura, G.: On the holomorphy of certain Dirichlet series. Proc. London Math. Soc. **31**, 79–98 (1975)
7. Siegel, C.L.: Über die analytische Theorie der quadratischen Formen. Ann. Math. **36**, 527–606 (1935)
8. Siegel, C.L.: Über die Zetafunktionen indefiniter quadratischer Formen. Math. Z. **43**, 682–708 (1938)

Received November 19, 1981

Algebraic relations between critical values
of zeta functions and inner products

American Journal of Mathematics, 105 (1983), 253-285

Let us first describe our results in four most transparent cases which can be obtained as applications of a more general principle. For an algebraic number field E, let $J(E)$, or simply J_E, denote the set of all non-trivial homomorphisms of E into $\mathbf{C}$. Suppose E is totally real; let $\{\chi(\mathfrak{a})\}$ and $\{\chi'(\mathfrak{a})\}$ be two systems of eigenvalues of Hecke operators $T(\mathfrak{a})$ on the spaces of holomorphic cusp forms on $\mathfrak{H}^{J(E)}$ of weight m and m', respectively, with respect to congruence subgroups of $GL_2(E)$, where $\mathfrak{H}$ denotes the complex upper half plane and $\mathfrak{a}$ the integral ideals of E. One type of zeta function to be studied has the form

$$D(s, \chi, \chi') = \sum_{\mathfrak{a}} \chi(\mathfrak{a})\chi'(\mathfrak{a})N(\mathfrak{a})^{-s} \qquad (s \in \mathbf{C}).$$

In our previous paper [2], we showed that $D(s_0, \chi, \chi')$ for certain integers or half-integers s_0 is an algebraic number times $\pi^{|m|} \langle \mathbf{f}, \mathbf{f} \rangle$ if $m_\tau > m'_\tau$ for all $\tau \in J_E$ and if $m_\tau - m'_\tau \pmod 2$ is independent of τ, where m_τ denotes the τ-component of m, $\mathbf{f}$ is a normalized eigenform such that $\mathbf{f} | T(\mathfrak{a}) = \chi(\mathfrak{a})\mathbf{f}$ for all $\mathfrak{a}$, $\langle \ , \ \rangle$ denotes the inner product, and $|m| = \Sigma_\tau m_\tau$.

In the present paper, we consider $D(s, \chi, \chi')$ under the condition that $m_\tau > m'_\tau$ for $\tau \in \alpha$ and $m'_\tau > m_\tau$ for $\tau \in \beta$ with two disjoint nonempty subsets α and β of J_E whose union is J_E. The assumption that $m_\tau - m'_\tau$ $\pmod 2$ is independent of τ must be retained. Let B and B' be quaternion algebras over E which are unramified at $\tau \in \alpha$ and $\tau \in \beta$, and ramified at $\tau \in \beta$ and $\tau \in \alpha$, respectively. Suppose that there are nonzero automorphic eigenforms $\mathbf{g}$ on B and $\mathbf{g}'$ on B' such that $\mathbf{g} | T_B(\mathfrak{a}) = \chi(\mathfrak{a})\mathbf{g}$ and $\mathbf{g}' | T_{B'}(\mathfrak{a}) = \chi'(\mathfrak{a})\mathbf{g}'$, where $T_B(\mathfrak{a})$ and $T_{B'}(\mathfrak{a})$ are Hecke operators defined on B and on B'. We can take such $\mathbf{g}$ and $\mathbf{g}'$ to be $\bar{\mathbf{Q}}$-rational in the sense defined in [4]. Then our first main application asserts:

Manuscript received August 23, 1982.

253

I. $D(s_0, \chi, \chi')$ *for certain integers or half-integers s_0 is an algebraic number times $\pi^c \langle \mathbf{g}, \mathbf{g} \rangle \langle \mathbf{g}', \mathbf{g}' \rangle$, where $c = \Sigma_{\tau \in \alpha} m_\tau + \Sigma_{\tau \in \beta} m'_\tau$ (Theorem 5.3).*

Another result also concerns such "complementary" quaternion algebras B and B'. Suppose there is a nonzero automorphic eigenform $\mathbf{g}^*$ on B' such that $\mathbf{g}^* | T_{B'}(\mathfrak{a}) = \chi(\mathfrak{a})\mathbf{g}^*$ with the same χ as above. Assuming that $\mathbf{g}^*$ is $\overline{\mathbf{Q}}$-rational and taking a Hilbert cusp form $\mathbf{f}$ as above, we can show that:

II. *The quotient $\langle \mathbf{g}, \mathbf{g} \rangle \langle \mathbf{g}^*, \mathbf{g}^* \rangle / \langle \mathbf{f}, \mathbf{f} \rangle$ is an algebraic number (Theorem 5.4).*

As an easy consequence of this and other auxiliary facts, it can be shown that:

III. *The inner product $\langle \mathbf{g}, \mathbf{g} \rangle$ modulo algebraic factors depends only on χ and α, and is independent of the choice of B and $\mathbf{g}$, at least when $[E:\mathbf{Q}]$ or the number of elements in α (denoted by $|\alpha|$) is even and $m_\tau > 1$ for all $\tau \in J_E$: the same conclusion holds with no assumptions on $[E:\mathbf{Q}]$ and $|\alpha|$, if $m_\tau > 2$ for all $\tau \in \alpha$ (Theorem 5.6).*

The algebraicity of the quotient in II is reminiscent of the linearity of the period symbol $p_K(\varphi, \psi)$ introduced in [3] in the second variable ψ. In fact, we can show that:

IV. *If $\Sigma \chi(\mathfrak{a})N(\mathfrak{a})^{-s}$ is the L-series of a Hecke character of a CM-field K, and if $m_\tau > 1$ for all $\tau \in \alpha$, then $\pi^{|\alpha|} \langle \mathbf{g}, \mathbf{g} \rangle$ is an algebraic number times $p_K(\varphi, \psi)$ with suitable φ and ψ which depend only on χ and α (Theorem 5.8).*

These theorems will be derived as special cases of more general results of algebraicity concerning a series of the form

$$Z(s) = \sum_{(x,y)} \omega\left(\sum_{i=1}^{r} \mathrm{Tr}_{F_i/E}(\zeta_i x_i) + \sum_{i=r+1}^{t} \mathrm{Tr}_{K_i/E}(\zeta_i y_i y_i^\rho) \right)$$

$$\cdot \prod_{i=1}^{r} \chi_i(x_i \mathfrak{a}_i) x_i^{p_i - q_i s} \prod_{i=r+1}^{t} y_i^{\varphi_i} |y_i^{\psi_i}|^{-2s}.$$

Here $F_1, \ldots, F_r$ are totally real algebraic number fields and $K_{r+1}, \ldots, K_t$ are CM-fields, all containing E; χ_i for each i is a system of eigenvalues of Hecke operators on the space of automorphic forms on a quaternion algebra B_i over F_i; $\mathfrak{a}_i$ is a fractional ideal of F_i; $\zeta_i \in F_i$, where F_i for $i > r$

denotes the maximal real subfield of K_i; p_i and q_i (resp. φ_i and ψ_i) are
Z-linear combinations of the elements of $J(F_i)$ (resp. $J(K_i)$); y^ρ denotes
the complex conjugate of y; $\omega(a)$ denotes the Fourier coefficients of a
Hilbert modular form $\Omega(z) = \Sigma_{a \in E} \omega(a)\exp(2\pi i \Sigma_\tau a_\tau z_\tau)$ with respect to a
congruence subgroup of $GL_2(E)$; U is a subgroup of the group of units of
E of finite index, determined by Ω; $(x, y) = (x_1, \ldots, x_r, y_1, \ldots, y_t)$ runs
over a lattice in $\Pi_{i=1}^r F_i \times \Pi_{i=r+1}^t K_i$ under arbitrarily given congruence
conditions, modulo the action of U defined by

$$u(x_1, \ldots, x_r, y_{r+1}, \ldots, y_t) = (u^2 x_1, \ldots, u^2 x_r, uy_{r+1}, \ldots, uy_t) \qquad (u \in U);$$

in addition, $x_1, \ldots, x_r$ must be totally positive. Some natural conditions
are imposed on p_i, q_i, φ_i, ψ_i, and ζ_i. It will be shown that Z can be continued as a meromorphic function to the whole s-plane. Then our princi-
pal theorem will assert that:

*$Z(s_0)$ for certain integers s_0 is an algebraic number times $\pi^d \, \Pi_{i=1}^r$
$\langle \mathbf{g}_i, \mathbf{g}_i \rangle \, \Pi_{i=r+1}^t p_{K_i}(\varphi_i, 2\psi_i)$, where $\mathbf{g}_i$ is a $\overline{\mathbf{Q}}$-rational eigenform on B_i
belonging to χ_i, and d is a positive integer determined by φ_i, ψ_i, and the
weights of $\mathbf{g}_i$; moreover, the residue of Z at a certain point, divided by the
regulator of E, under some conditions on p_i, q_i, φ_i, ψ_i, and the weights of
$\mathbf{g}_i$, has the same property of algebraicity as $Z(s_0)$ (Theorem 4.4).*

This is a generalization of the main theorem of [4, Part II], which
deals with the case $t = r = 1$. We can actually generalize not only this
result on Z but also practically all the results of [4] in such a way that the
previous results become the special case $t = 1$. That is what we shall do in
the first four sections. Once the principle is presented, the process of generalization is straightforward and tedious. Therefore, without going into
details, we shall explain the procedure only to the extent that the reader
will be able to state and prove the generalized theorems easily. Accordingly, in some instances, we shall dispense with the explicit statements of
the theorems. The four results mentioned above will be proved in Section
5, in which we shall also make some conjectures about the general nature
of the constants of type $\langle \mathbf{g}, \mathbf{g} \rangle$ as well as the periods of $\mathbf{g}$.

1. Preliminaries.   The notation and terminology will be the same as
in [4]. We recall here that $\mathbf{e}(z) = \exp(2\pi i z)$ for $z \in \mathbf{C}$, $\mathfrak{H}$ denotes the com-
plex upper half plane $\{z \in \mathbf{C} \,|\, \mathrm{Im}(z) > 0\}$, $\overline{\mathbf{Q}}$ the algebraic closure of $\mathbf{Q}$ in
$\mathbf{C}$, and $\mathbf{A}$ the ring of adeles of $\mathbf{Q}$. For $a, b \in \mathbf{C}$, we write $a \sim b$ if $b \neq 0$ and

$a/b \in \overline{\mathbf{Q}}$. We shall often denote by x^ρ the complex conjugate of $x \in \overline{\mathbf{Q}}$. The formulas, theorems, etc. of [4] are referred to, for example, as Theorem I.4.3, (II.3.6), etc., which mean Theorem 4.3 of Part I, formula (3.6) of Part II of [4], etc.

Let K be the direct sum of algebraic number fields $K_1, \ldots, K_t$. We denote by $J(K)$ or J_K the set of all nontrivial homomorphisms of K into $\mathbf{C}$, and by $I(K)$ or I_K the free $\mathbf{Z}$-module generated by the elements of J_K. For $\sigma \in J_K$ and $x \in K$, we denote by x^σ the image of x under σ. Obviously J_K consists of $\pi_i\sigma_i$ with $\sigma_i \in J(K_i)$ for $i = 1, \ldots, t$, where π_i denotes the projection map of K onto K_i. We shall often identify $\pi_i\sigma_i$ with σ_i and view J_K as the disjoint union of $J(K_1), \ldots, J(K_t)$. Similarly I_K is the direct sum of $I(K_1), \ldots, I(K_t)$. We put $\mathbf{R}I_K = I_K \otimes_{\mathbf{Z}} \mathbf{R}$ and $\mathbf{C}I_K = I_K \otimes_{\mathbf{Z}} \mathbf{C}$. Every $\sigma \in J_K$ can be extended to a $\mathbf{C}$-linear homomorphism of $K \otimes_{\mathbf{Q}} \mathbf{C}$ into $\mathbf{C}$, which we denote by the same letter σ. If $m \in \mathbf{C}I_K$, we denote by m_σ or $m(\sigma)$ the σ-component of m for every $\sigma \in J_K$; thus $m = \Sigma_\sigma m_\sigma\sigma$. If $m \in I_K$, we put $x^m = \Pi_{\sigma \in J(K)} (x^\sigma)^{m_\sigma}$ for $x \in K \otimes_{\mathbf{Q}} \mathbf{C}$; x^m is well-defined for $m \in \mathbf{C}I_K$ if $0 < x^\sigma \in \mathbf{R}$ for all $\sigma \in J_K$. Also we put $|m| = \Sigma_\sigma m_\sigma$ and write $m \geq m'$ if $m_\sigma \geq m'_\sigma$ for all $\sigma \in J_K$. We shall often identify $K \otimes_{\mathbf{Q}} \mathbf{C}$ with $\mathbf{C}^{J(K)}$; thus if $x = (x_\sigma)_{\sigma \in J(K)} \in \mathbf{C}^{J(K)}$, we have $x^\sigma = x_\sigma$. We view every subset β of J_K also as an element $\Sigma_{\sigma \in \beta} \sigma$ of I_K; thus $x^\beta = \Pi_{\sigma \in \beta} x^\sigma$, and $|\beta|$ is the number of elements of β.

Throughout the paper, we denote by F the direct sum of totally real algebraic number fields $F_1, \ldots, F_t$; we put

$$(1.1) \qquad \mathbf{e}_F(z) = \mathbf{e}\left(\sum_\tau z_\tau\right) \quad \text{for} \quad z = (z_\tau)_{\tau \in J(F)} \in \mathbf{C}^{J(F)} = F \otimes_{\mathbf{Q}} \mathbf{C}.$$

In this case, we can also identify $F \otimes_{\mathbf{Q}} \mathbf{R}$ with $\mathbf{R}^{J(F)}$. For $b = (b_\tau) \in \mathbf{R}^{J(F)}$, we write $b \gg 0$ if $b_\tau > 0$ for all $\tau \in J_F$. Now the basic principle of our generalization of the results of [4] is to define everything taking this F in place of the ground field in [4] which was also denoted by F.

Let $I^0(F_i)$ denote the submodule of $\mathbf{R}I(F_i)$ consisting of all $r = \Sigma_\tau r_\tau\tau$ with $2r_\tau \in \mathbf{Z}$ and $\tau \in J(F_i)$ such that $2r_\tau \pmod 2$ is independent of τ. In Section I.1, we defined, for every $r \in I^0(F_i)$ and every congruence subgroup Γ_i of $SL_2(F_i)$, the space $\mathcal{C}_r(\Gamma_i)$ of automorphic C^∞-functions on $\mathfrak{H}^{J(F_i)}$ of weight r with respect to Γ_i and also its subspace $\mathfrak{M}_r(\Gamma_i)$ of all holomorphic Hilbert modular forms of weight r. We denote them also by $\mathcal{C}(r, \Gamma_i, F_i)$ and $\mathfrak{M}(r, \Gamma_i, F_i)$, respectively, and by $\mathcal{C}(r, F_i)$ and $\mathfrak{M}(r, F_i)$ their unions for all congruence subgroups Γ_i of $SL_2(F_i)$. Let $I^0(F)$ be the set of all $k = (k_1, \ldots, k_t) \in \mathbf{R}I_F$ with $k_i \in I^0(F_i)$, and let $\Gamma = \Pi_{i=1}^t \Gamma_i$. Then we denote by $\mathfrak{M}(k, \Gamma, F)$ the space of holomorphic functions f on $\mathfrak{H}^{J(F)} = \Pi_{i=1}^t \mathfrak{H}^{J(F_i)}$ such that f as a function on $\mathfrak{H}^{J(F_i)}$ belongs to $\mathfrak{M}(k_i, \Gamma_i, F_i)$. Then $\mathfrak{M}(k, F)$ denotes the union of $\mathfrak{M}(k, \Gamma, F)$ for all possible Γ. We define $\mathcal{C}(k, \Gamma, F)$ and $\mathcal{C}(k, F)$ in a similar way.

Let $f_i \in \mathcal{C}(k_i, \Gamma_i, F_i)$ and $f(z) = f_1(z_1) \cdots f_t(z_t)$ for $z = (z_1, \ldots, z_t) \in \mathfrak{H}^{J(F)}$ with $z_i \in \mathfrak{H}^{J(F_i)}$. Then $f \in \mathcal{C}(k, \Gamma, F)$. We write in this situation $f = f_1 \otimes \cdots \otimes f_t$. That every element of $\mathfrak{M}(k, \Gamma, F)$ is a finite sum of such functions with $f_i \in \mathfrak{M}(k_i, \Gamma_i, F_i)$ follows from a purely algebraic principle which can be stated as follows.

LEMMA 1.1. *Let A be a finite set of indices. For each $\alpha \in A$, let X_α be a set, U_α a finite-dimensional vector space over a field Φ, and S_α a finite-dimensional vector space over Φ consisting of some U_α-valued functions on X_α. Put $X = \Pi_{\alpha \in A} X_\alpha$ and $U = \otimes_{\alpha \in A} U_\alpha$. Further, for each $\beta \in A$, put $Y_\beta = \Pi_{\alpha \neq \beta} X_\alpha$ and $V_\beta = \otimes_{\alpha \neq \beta} U_\alpha$. Let f be a U-valued function on X with the property that, for every $\beta \in A$ and every $y \in Y_\beta$, $f(x, y)$ as a function of $x \in X_\beta$ is a finite sum $\Sigma_i u_i \otimes g_i(x)$ with $u_i \in V_\beta$ and $g_i \in S_\beta$. Then $f((x_\alpha)_{\alpha \in A})$ is a finite sum of functions of the form $\otimes_{\alpha \in A} h_\alpha(x_\alpha)$ with $h_\alpha \in S_\alpha$. Moreover, $(h_\alpha)_{\alpha \in A} \mapsto \otimes_{\alpha \in A} h_\alpha(x_\alpha)$ gives an isomorphism of $\otimes_{\alpha \in A} S_\alpha$ onto the space of all such functions f.*

Proof. The last assertion follows immediately from the first one, whose proof proceeds by induction on the cardinality of A. There is no problem if A consists of a single element. Suppose A has at least two elements; pick an element α of A; let $K = \{k\}$ and $H = \{h\}$ be bases of U_α and S_α over Φ, respectively. Since

$$\{0\} = \bigcap_{x \in X_\alpha} \{(c_h) \in \Phi^H \mid \sum_h c_h h(x) = 0\},$$

we can find a finite subset P of X_α such that

$$\{0\} = \bigcap_{p \in P} \{(c_h) \in \Phi^H \mid \sum_h c_h h(p) = 0\}.$$

Put $h(x) = \Sigma_{k \in K} h_k(x) k$ with Φ-valued functions h_k on X_α. Then

$$(c_h)_{h \in H} \mapsto \left(\sum_{h \in H} h_k(p) c_h\right)_{k \in K, p \in P}$$

gives an injective map of Φ^H into $\Phi^{K \times P}$. Therefore we can find a matrix (d_{kp}^h) of type $(H, K \times P)$ with entries in Φ such that $\Sigma_{k \in K, p \in P} d_{kp}^i h_k(p) = \delta_{ih}$. Put $B = A - \{\alpha\}$, $Z_\beta = \Pi_{\beta \neq \gamma \in B} X_\gamma$, and $W_\beta = \otimes_{\beta \neq \gamma \in B} U_\gamma$ for $\beta \in B$. Further put $f(p, y) = \Sigma_{k \in K} k \otimes g_{kp}(y)$ with V_α-valued functions g_{kp} on Y_α. Observe that for each $\beta \in B$ and every $z_0 \in Z_\beta$, $g_{kp}(z_0, u)$ as a function of $u \in X_\beta$ is a finite sum $\Sigma_j a_j \otimes b_j(u)$ with $a_j \in W_\beta$ and $b_j \in S_\beta$. Therefore applying the induction assumption to g_{kp}, we find that g_{kp} belongs to $\otimes_{\beta \neq \alpha} S_\beta$. Now our assumption on f implies that $f(x, y) = \Sigma_{h \in H} h(x) \otimes w_h(y)$ with V_α-valued functions w_h on Y_α. Then $g_{kp}(y) = \Sigma_{h \in H} h_k(p) w_h(y)$, so that $w_h(y) = \Sigma_{k, p} d_{kp}^h g_{kp}(y)$. Hence $w_h \in \otimes_{\beta \neq \alpha} S_\beta$. This completes the proof.

Applying this lemma to our modular forms, we find that $\mathfrak{M}(k, \Gamma, F)$ is naturally isomorphic to the tensor product of $\mathfrak{M}(k_i, \Gamma_i, F_i)$ for $i = 1, \ldots, t$ over $\mathbf{C}$, and $\mathfrak{M}(k, F)$ is isomorphic to the tensor product of $\mathfrak{M}(k_i, F_i)$ for $i = 1, \ldots, t$. The subset of $\mathfrak{M}(k_i, F_i)$ consisting of all cusp forms is denoted by $\mathcal{S}(k_i, F_i)$; then $\mathcal{S}(k, F)$ denotes the subset of $\mathfrak{M}(k, F)$ corresponding to the tensor product of $\mathcal{S}(k_i, F_i)$ for $i = 1, \ldots, t$.

Every element f of $\mathfrak{M}(k, F)$ has its Fourier expansion

$$f(z) = \sum_{b \in F} \xi(b) \mathbf{e}_F(bz) \qquad (z \in \mathfrak{H}^{J(F)})$$

with $\xi(b) \in \mathbf{C}$, where b runs over a lattice in F. If in particular $f = f_1 \otimes \cdots \otimes f_t$, this is simply the product of the Fourier expansions of $f_1, \ldots, f_t$. For $f, g \in \mathcal{C}(k, F)$, we define its inner product $\langle f, g \rangle$ by

$$\langle f, g \rangle = \mu(D)^{-1} \int_D \overline{f(z)} g(z) \mathrm{Im}(z)^k d\mu(z) \qquad (D = \Gamma \backslash \mathfrak{H}^{J(F)}),$$

$$d\mu(z) = \prod_{\tau \in J(F)} y_\tau^{-2} dx_\tau dy_\tau \qquad (z_\tau = x_\tau + iy_\tau),$$

whenever the integral is convergent; here we take Γ so that both f and g belong to $\mathcal{C}(k, \Gamma, F)$. If $f = f_1 \otimes \cdots \otimes f_t$ and $g = g_1 \otimes \cdots \otimes g_t$ with f_i and g_i in $\mathcal{C}(k_i, F_i)$, we have obviously

$$\langle f, g \rangle = \langle f_1, g_1 \rangle \cdots \langle f_t, g_t \rangle.$$

Let us now assume that $F_1, \ldots, F_t$ have a common subfield E, and consider F an E-algebra. The reduced trace and the reduced norm of F to

E are denoted by $\mathrm{Tr}_{F/E}$ and $N_{F/E}$. For each $\tau \in J_F$, $\mathrm{Res}_{F/E}(\tau)$ denotes its restriction to E. We extend this to a **C**-linear map $\mathrm{Res}_{F/E}: CI_F \to CI_E$. Further we fix a subset ϵ of J_F with the property

$$(1.3) \qquad \mathrm{Res}_{F/E} \text{ gives a one-to-one map of } \epsilon \text{ onto } J_E; \qquad \epsilon \cap \pi_i J(F_i) \neq \emptyset$$

$$\text{for} \quad i = 1, \ldots, t.$$

Here and throughout the paper, π_i denotes the projection map of F onto F_i. Naturally ϵ has exactly $[E:\mathbf{Q}]$ elements. We fix, in addition, an element ζ of F satisfying

$$(1.4) \qquad \zeta^\tau > 0 \quad \text{for} \quad \tau \in \epsilon \quad \text{and} \quad \zeta^\tau < 0 \quad \text{for} \quad \tau \in J_F - \epsilon.$$

For a C^∞-function g on $\mathfrak{H}^{J(F)}$, we define its *pullback relative to ζ* to be a function g^ζ on $\mathfrak{H}^{J(E)}$ defined by

$$(1.5) \qquad\qquad\qquad g^\zeta(z) = g(z*),$$

$$(1.6) \qquad\qquad (z*)_\tau = \begin{cases} \zeta^\tau z_\sigma & \text{if} \quad \tau \in \epsilon \quad \text{and} \quad \sigma = \mathrm{Res}_{F/E}(\tau), \\ \zeta^\tau \bar{z}_\sigma & \text{if} \quad \tau \notin \epsilon \quad \text{and} \quad \sigma = \mathrm{Res}_{F/E}(\tau). \end{cases}$$

Suppose $g \in \mathcal{C}(m, F)$ and let $k = \mathrm{Res}_{F/E}(\Sigma_{\tau \in \epsilon} m_\tau \tau)$, $\mu = \mathrm{Res}_{F/E}(\Sigma_{\tau \notin \epsilon} m_\tau \tau)$, and $h(z) = \mathrm{Im}(z)^\mu g^\zeta(z)$ for $z \in \mathfrak{H}^{J(E)}$. Then we observe, as a generalization of Lemma I.3.3., that $h \in \mathcal{C}(k - \mu, E)$. The consistency of the factors of automorphy can be proved by taking g to be the product of theta functions as in the proof of Lemma I.3.3. If $g = g_1 \otimes \cdots \otimes g_t$ with $g_i \in \mathcal{C}(m_i, F_i)$, we see easily that h is rapidly decreasing in the sense of Section I.2 if at least one of the g_i is rapidly decreasing and if all others are slowly increasing. This is so, for example, if $g_i \in \mathfrak{M}(m_i, F_i)$ for every i, and at least one of the g_i is a cusp form.

Now let f and Ω be elements of $\mathfrak{M}(m, F)$ and $\mathfrak{M}(l, E)$ with $m \in I_F^0$, $l \in I_E^0$ and with Fourier expansions

$$(1.7) \qquad\qquad f(w) = \sum_{b \in F} \xi(b) \mathbf{e}_F(bw) \qquad (w \in \mathfrak{H}^{J(F)}),$$

$$(1.8) \qquad\qquad \Omega(z) = \sum_{a \in E} \omega(a) \mathbf{e}_E(az) \qquad (z \in \mathfrak{H}^{J(E)}).$$

Let U_E be the group of all totally positive units of E and U a subgroup of U_E of finite index. We say that f is U-*invariant* if $\xi(va) = v^{m/2}\xi(a)$ for every $a \in F$ and every $v \in U$ (see [4, I, p. 134]). Assuming both f and Ω to be U-invariant, we form an infinite series

$$(1.9) \qquad \Lambda(s) = [U] \sum_{0 \ll b \in F/U} \xi(b)\omega_p(\mathrm{Tr}_{F/E}(\zeta b))b^{d-s\epsilon} \qquad (s \in \mathbf{C})$$

with some fixed $d \in I_F$ and $0 \le p \in I_E$. Here $\omega_p(a) = a^p\omega(a)$ and

$$(1.10) \qquad\qquad\qquad [U] = [U_E:U]^{-1};$$

the sum is extended over all cosets bU with $0 \ll b \in F$. This is a generalization of (I.3.2). We assume

$$(1.11) \qquad l + 2p + \mathrm{Res}_{F/E}(m + 2d) = \alpha J_E \quad \text{with} \quad \alpha \in 2^{-1}\mathbf{Z},$$

$$(1.12) \qquad\qquad\qquad d_\tau \ge 0 \quad \text{for every} \quad \tau \notin \epsilon.$$

Condition (1.11) makes each term of (1.9) dependent only on bU. Now Proposition I.3.1, Theorems I.3.2 and I.3.4 can be generalized to the present case in an obvious way with no essential change of proofs. The conditions concerning $m_i + d_i$ in Theorems I.3.2 and I.3.4 should be rephrased as follows:

$$2(m_\tau + d_\tau) > \alpha \quad \text{for at least one} \quad \tau \in \epsilon \quad (\text{Theorem I.3.2});$$

$$2(m_\tau + d_\tau) = \alpha \quad \text{for every} \quad \tau \in \epsilon \quad (\text{Theorem I.3.4}).$$

The elements μ, k, κ, v, and $\Pi_{i=1}^{r}\,\Gamma(m_i)$ of Theorem I.3.4 in the generalized formulation should read: $\mu = \mathrm{Res}_{F/E}(\Sigma_{\tau \notin \epsilon}\, m_\tau\tau)$, $k = \mathrm{Res}_{F/E}(\Sigma_{\tau \in \epsilon}\, m_\tau\tau)$, $\kappa = k - \mu$, $v = \Sigma_{\tau \notin \epsilon}\, v_\tau\tau$, and $\Pi_{\tau \in \epsilon}\,\Gamma(m_\tau)$. The condition that f is a *cusp form* should now read: *f is a finite sum of elements of the form* $f_1 \otimes \cdots \otimes f_t$ *where at least one of the f_i is a cusp form.*

In the next sections, we shall treat similar generalizations of other theorems of [4], but shall not necessarily state explicitly how each symbol or condition should be interpreted or rephrased in the generalized situation, since such modification will be obvious from the context as the above case exemplifies.

2. The case of Hecke characters of *CM*-fields.   Let V be a finite-dimensional vector space over $\mathbf{Q}$. We call a C-valued function η on V *locally constant* if there exist two $\mathbf{Z}$-lattices L and M of V such that $\eta(x) = 0$ for $x \notin L$ and $\eta(x)$ depends only on x modulo M. If V is a vector space over the field E of Section 1, we can find a subgroup U of U_E of finite index such that $\eta(ux) = \eta(x)$ for all $u \in U$; we then call η *U-invariant*.

Let K be the direct sum of CM-fields $K_1, \ldots, K_t$ such that $[K_i : F_i] = 2$ with F_i of Section 1. (A CM-field is a totally imaginary quadratic extension of a totally real algebraic number field.) We denote by ρ the involution of K whose restriction to K_i is the complex conjugation for every i. For $m = \Sigma_\sigma \, m_\sigma \sigma \in I_K$, we put $\rho m = m\rho = \Sigma_\sigma \, m_\sigma \sigma\rho$. We define C-linear maps $\mathrm{Res}_{K/F} : CI_K \to CI_F$ and $\mathrm{Res}_{K/E} : CI_K \to CI_E$ in the same fashion as for $\mathrm{Res}_{F/E}$.

Given ϵ as in (1.3), let us fix a subset ψ of J_K such that $\mathrm{Res}_{K/F}$ gives a one-to-one map of ψ onto ϵ. We then define a series

$$(2.1) \qquad D(s) = [U] \sum_{xU} \eta(x)\omega_p(\mathrm{Tr}_{F/E}(\zeta xx^\rho))x^\Phi |x^\psi|^{-2s} \qquad (s \in \mathbf{C}).$$

Here η is a locally constant function on K; ω_p is defined with $\Omega \in \mathfrak{M}(l, E)$ as in (1.9); we assume $l \in I_E$; ζ is an element of F satisfying (1.4); U is a subgroup of U_E of finite index such that η and Ω are U-invariant; the sum is taken over all xU with invertible x in K; Φ is an element of I_K such that

$$(2.2) \qquad \Phi = \sum_\sigma c(\sigma)\sigma \quad with \quad c(\sigma) \geq 0 \;\; for \;\; \sigma \notin \psi \cup \psi\rho.$$

To make each term of (2.1) dependent only on xU, we assume

$$(2.3) \qquad \mathrm{Res}_{K/E}(\Phi) + l + 2p = \beta J_E \quad with \quad \beta \in \mathbf{Z}.$$

Further, changing σ for $\rho\sigma$ if necessary, we assume

$$(2.4) \qquad c(\sigma) \geq c(\rho\sigma) \quad for\ all \quad \sigma \in \psi.$$

Then, as to the convergence and the analytic continuation of D, Proposition I.4.1 and Theorem I.4.2 can be generalized to the present case with obvious modifications. In fact, (2.1) is a special case of (1.9) for the same reason as explained in [4, Part I, p. 144]. (See Corrections to [4] at the end of the present paper.) Notice that the form f in the present case is a cusp form if $\mathrm{Res}_{K/K_i}(\Phi) \notin (1 + \rho)I(K_i)$ for all i, and also that $\mathrm{Im}(z)^\mu f^\varsigma$ is rapidly decreasing if $\Phi \notin (1 + \rho)I_K$.

As for the critical values and the residue of D, Theorems I.4.3, I.4.6, and I.4.7 hold in the generalized situation. The symbol p_K should be understood in the sense of [3, (1.1)]; the symbol E as an element of I_E should now be J_E. No change is necessary for the proof of the equivalence of Theorems I.4.6 and I.4.7. In order to prove the generalized version of Theorem I.4.3, we define an E-valued symmetric form S on K by

$$(2.5) \qquad S(x, y) = -\mathrm{Tr}_{K/E}(\varsigma xy^\rho) \qquad (x, y \in K),$$

and repeat the proof of Theorem I.4.3 in Section I.8, which in turn is a modification of the argument of [3, pp. 364–365]. Similarly, the proof of Theorem I.4.7 in Section I.9 can be repeated with obvious changes. That K is not necessarily a field produces no difficulty, since [3, Theorem 11.2] is applicable to any CM-point obtained from K of the general type.

3. Automorphic forms on the product of orthogonal groups.

Let V_i be a vector space over F_i of dimension $m_i + 2$ with a nonnegative integer m_i, and let $S_i(x, y)$ be an F_i-valued F_i-bilinear symmetric form on V_i. We denote by S the collection $\{S_1, \ldots, S_t\}$. Each $\tau \in J_F$ can be written $\tau = \pi_i\sigma$ with $\sigma \in J(F_i)$ for some i. We then denote by V_τ and S_τ the completion of V_i at σ and $\mathbf{C}$-bilinear extension of S_i to $V_\tau \otimes_\mathbf{R} \mathbf{C}$; we put $m_\tau = m_i$. We assume that S_τ on V_τ has signature $(m_\tau, 2)$ for $\tau \in \epsilon$ and S_τ is positive definite for $\tau \notin \epsilon$, where ϵ is a fixed subset of J_F such that $\epsilon \cap J(F_i) \neq \emptyset$ for every i; for the moment we do not assume the existence of E as in (1.3). We include the case $m_i = 0$ in our treatment, and assume $m_i > 0$ for $i = 1, \ldots, r$ and $m_i = 0$ for $i = r + 1, \ldots, t$; here $0 < r \le t$. Thus, for $i > r$, we may assume, without losing generality, that $V_i = K_i$ with a CM-field K_i such that $[K_i:F_i] = 2$ and

$$(3.1) \qquad S_i(x, y) = \mathrm{Tr}_{K_i/F_i}(\delta_i xy^\rho) \qquad (x, y \in K_i)$$

with an element δ_i of F_i such that $\delta_i^\tau < 0$ or $\delta_i^\tau > 0$ for $\tau \in J(F_i)$ according as $\tau \in \epsilon$ or $\tau \notin \epsilon$. We put $K = \Pi_{i>r} K_i$ and

$$(3.2) \quad \alpha = \bigcup_{i=1}^{r} J(F_i) \cap \epsilon, \quad \beta = \bigcup_{i>r} J(F_i) \cap \epsilon, \quad \alpha' = J(F_1 \times \cdots \times F_r) - \alpha.$$

For $x, y \in \Pi_{\tau \in J(F)} (V_\tau \otimes_{\mathbf{R}} \mathbf{C})$, we denote by $S(x, y)$ the element $(S_\tau(x_\tau, y_\tau))$ of $\mathbf{C}^{J(F)}$. We put $V = \Pi_{i=1}^t V_i$ and embed V into $\Pi_\tau V_\tau$ in an obvious way. Thus, for x, y in V, $S(x, y)$ is the element $(S_i(x_i, y_i))_{i=1}^t$ of F, viewed also as an element of $\mathbf{C}^{J(F)}$. We define groups $G(S_i)$, $G_+(S_i)$, and $G(S_\tau)$ as in [3], and put $G(S) = \Pi_{i=1}^r G(S_i)$, $G_+(S) = \Pi_{i=1}^r G_+(S_i)$; $G(S)$ is embedded into $\Pi_\tau G(S_\tau)$ in a natural way. If $\tau \in \alpha$, we can also define a domain $\mathcal{3}_{m_\tau}$ embedded in $\mathbf{C}^{m_\tau}$ relative to S_τ as in [3]; we denote it by $\mathcal{3}_\tau$ and put $\mathcal{3} = \Pi_{\tau \in \alpha} \mathcal{3}_\tau$. Then $G_+(S)$ acts on $\mathcal{3}$ in a natural way. Also a map A_τ of $V_\tau \otimes_{\mathbf{R}} \mathbf{C}$ onto $\mathbf{C}^{m_\tau+2}$ is defined as in [3, Proposition 2.1] relative to S_τ.

Take a $\bar{\mathbf{Q}}$-rational representation

$$\sigma : (GL_1)^\alpha \times \prod_{\tau \in \alpha} GL(m_\tau) \times \prod_{\tau \in \alpha'} G(S_\tau) \to GL(W, \mathbf{C})$$

with a finite-dimensional complex vector space W with a $\bar{\mathbf{Q}}$-rational structure. We then define the spaces $\mathfrak{A}_\sigma$ (resp. $\mathfrak{M}_\sigma$) of meromorphic (resp. holomorphic) automorphic forms on $\mathcal{3}$ with respect to congruence subgroups of $G_+(S)$ with a factor of automorphy $\Lambda_\sigma(\alpha, z)$ as in [3, Section 3].

To define a CM-point on $\mathcal{3}$, we take a $\mathbf{Q}$-algebra $Y = Y_1 \times \cdots \times Y_r$ with an F_i-algebra Y_i for each $i \leq r$ which is a direct sum of CM-fields and copies of F_i and such that $[Y_i:F_i] = m_i + 2$. Let h be an embedding of Y into $\text{End}(\Pi_{i=1}^r V_i)$, which consists of F_i-linear embeddings h_i of Y_i into $\text{End}(V_i, F_i)$ for $i = 1, \ldots, r$. Assume that (Y_i, h_i) satisfies [3, (5.2), (5.4)]. Then defining Y^u by [3, (5.3)], we see that $h(Y^u) \subset G_+(S)$ and $h(Y^u)$ has a unique common fixed point on $\mathcal{3}$, which in fact consists of the fixed points of $h_i(Y_i^u)$ for $i = 1, \ldots, r$. We call this point a CM-*point of* $G(S)$ *on* $\mathcal{3}$. Then arithmetic (or $\bar{\mathbf{Q}}$-rational) elements of $\mathfrak{A}_\sigma$ can be defined by the same idea as in [3, Section 5]. Suppose in particular $\sigma = \sigma_1 \otimes \cdots \otimes \sigma_r$ with representations

$$\sigma_i : (GL_1)^{\epsilon_i} \times GL(m_i)^{\epsilon_i} \times \prod_{\tau \in J(F_i) - \epsilon_i} G(S_\tau) \to GL(W_i, \mathbf{C}),$$

where $\epsilon_i = \epsilon \cap J(F_i)$; of course $W = W_1 \otimes \cdots \otimes W_r$. Then Lemma 1.1 shows that $\mathfrak{M}_\sigma$ can be identified with $\otimes_{i=1}^r \mathfrak{M}_{\sigma_i}$. Let $\mathfrak{M}_\sigma(\bar{\mathbf{Q}})$ denote the set of all arithmetic elements of $\mathfrak{M}_\sigma$. Then we have

LEMMA 3.1. (i) $\mathfrak{M}_\sigma(\bar{\mathbf{Q}}) \otimes_{\bar{\mathbf{Q}}} \mathbf{C} = \mathfrak{M}_\sigma$;
(ii) $\mathfrak{M}_\sigma(\bar{\mathbf{Q}}) = \otimes_{i=1}^r \mathfrak{M}_{\sigma_i}(\bar{\mathbf{Q}})$.

Proof.   We first prove

$$(3.3) \qquad\qquad \mathfrak{M}_{\sigma_i}(\overline{Q}) \otimes_{\overline{Q}} C = \mathfrak{M}_{\sigma_i}.$$

For this we need a fact corresponding to Lemma II.2.2 for σ_i, which can be proved in exactly the same fashion. (Notice that the existence of $\overline{Q}$-rational σ_i-uniformizer is established in [3, Theorem 6.5]. See also Corrections to [4] at the end of this paper.) Then (3.3) can be proved by the same argument as in the proof of Theorem II.2.1, invoking [3, Proposition 6.6]. Now, to prove (i) and (ii), let $f_1, \ldots, f_s$ be elements of $\mathfrak{M}_{\sigma}(\overline{Q})$ linearly independent over $\overline{Q}$. Let $T = \otimes_{i=1}^r T_i$ with $\overline{Q}$-rational σ_i-uniformizers T_i. Then the components of $T^{-1}f_1, \ldots, T^{-1}f_s$ are $\overline{Q}$-rational automorphic functions, so that $T^{-1}f_1, \ldots, T^{-1}f_s$ are linearly independent over C. Thus $f_1, \ldots, f_s$ are linearly independent over C. Now obviously the right-hand side of (ii) is contained in the left-hand side. Combining this fact with (3.3) and the linear independence which we have just proved, we obtain both (i) and (ii).

For $\kappa = \Sigma_{\tau\in\alpha'} \kappa_\tau \tau \in I_F$ with $\kappa_\tau \geq 0$, we denote by $\mathcal{P}_\kappa(S)$ the vector space of C-valued polynomial functions on $\Pi_{\tau\in\alpha'} V_\tau$ which are homogeneous of degree κ_τ and S_τ-harmonic on V_τ in the sense of [3, p. 322]. We now assume the existence of E satisfying (1.3) with the present ϵ, and define an infinite series f by

$$(3.4) \quad f(w,s) = f(w,s;q,\xi,\Omega) = [U] \sum_{0 \neq x \in V/U} \xi(x)\omega_h(\mathrm{Tr}_{F/E}(-S[x])q(x)$$

$$\cdot S[x]^j x^\varphi S(x, A^{-1}p(w))^{-k} |S(x, A^{-1}p(w))^\alpha x^\psi|^{-2s}.$$

Here $w \in \mathfrak{Z}, s \in C, q \in \mathcal{P}_\kappa(S)$; ξ is a locally constant function on V; Ω and ω_h are the same as in (1.8) and (1.9); U is a subgroup of U_E of finite index such that both ξ and Ω are U-invariant; $S[x] = S(x,x), 0 \leq j \in I(F_1 \times \cdots \times F_r)$, $k = \Sigma_{\tau\in\alpha} k_\tau \tau \in I_F, \varphi \in I_K$; ψ is a subset of J_K such that $\mathrm{Res}_{K/F'}(\psi) = \beta$, where $F' = \Pi_{i>r} F_i$; we write x^φ and x^ψ for y^φ and y^ψ when y is the projection of an element x of V to K; $S(x, A^{-1}p(w))$ is the element of C^α whose τ-component is $S_\tau(x_\tau, A_\tau^{-1}p(w_\tau))$; p is the map of [3, (2.7)]; the sum is extended over all Ux with $0 \neq x \in V$. We assume

$$(3.5) \qquad k_\tau > 0 \quad \textit{for all} \quad \tau \in \alpha \quad \textit{and} \quad \varphi_\tau \geq 0 \quad \textit{for} \quad \tau \notin \psi \cup \psi\rho,$$

$(3.6)\quad \mathrm{Res}_{F/E}(k - \kappa - 2j) - \mathrm{Res}_{K/E}(\varphi) - l - 2h = a_f J_E\quad \text{with}\quad a_f \in 2^{-1}\mathbf{Z},$

where l is the weight of Ω; (3.6) makes each term of (3.4) dependent only on Ux. Series (3.4) is a generalization of (I.5.2). Notice that

$$\mathrm{Tr}_{F/E}(S[x]) = \sum_{i \leq r} \mathrm{Tr}_{F_i/E}(S_i(x_i)) + \sum_{i > r} \mathrm{Tr}_{K_i/E}(\delta_i x_i x_i^\rho)$$

for $x = (x_1, \ldots, x_t) \in V$ with $x_i \in V_i$. It can easily be seen that $S(x, A^{-1}p(w))^\alpha x^\psi \neq 0$ if $0 \neq x \in V$ and $\omega(\mathrm{Tr}_{F/E}(-S[x])) \neq 0$ (cf. [3, (8.8)]). Now the results of Section I.5 can be generalized in a natural way. First we observe, as a generalization of Proposition I.5.1, that (3.4) is absolutely convergent if

$$(3.7)\qquad\qquad \mathrm{Re}(2s) > \sum_{i=1}^{t} (m_i + 2)[F_i:E] - a_f + 2\theta,$$

where θ is the number as in Proposition I.3.1. Next, Theorem I.5.2 has the following generalization.

THEOREM 3.2. *Put* $b_f = a_f + 2 - \Sigma_{i=1}^{t} (m_i + 2)[F_i:E]/2$, *and suppose, in addition to* (3.5) *and* (3.6), *that*

$$(3.8)\qquad\qquad 2l_\gamma \equiv \sum_{i=1}^{r} m_i[F_i:E] \ (\mathrm{mod}\ 2)\quad \textit{for all}\quad \gamma \in J_E.$$

Then $f(w, s)$ *can be continued to the whole plane as a meromorphic function of* s. *Moreover, suppose*

$$(3.9)\qquad\qquad \varphi_\tau \geq \varphi_{\tau\rho}\quad \textit{for every}\quad \tau \in \psi\quad (\textit{if}\ \ t > r),$$

$$(3.10)\quad 2\varphi_\tau > -b_f\quad \textit{for at least one}\quad \tau \in \psi\ \ \textit{if}\ \ t > r;\qquad b_f > 0\ \ \textit{if}\ \ t = r.$$

Then f *is holomorphic for* $\mathrm{Re}(s) \geq (1 - b_f)/2$; *if* $b_f > 0$, $f(w, 0)$ *is holomorphic in* w *provided that*

$$(3.11)\qquad\qquad E \neq \mathbf{Q}\quad \textit{or}\quad b_f \neq 2.$$

It should be noted that (3.8) holds if and only if $b_f \in \mathbf{Z}$.

To state generalizations of Theorems I.5.3 and I.5.4, take a $\overline{\mathbf{Q}}$-rational basis $\{q_1, \ldots, q_d\}$ of $\mathcal{P}_\kappa(S)$ over $\mathbf{C}$, and define a representation

$$\sigma' : \prod_{\tau \in \alpha'} G(S_\tau) \to GL_d(\mathbf{C})$$

by

$$(q_1 \circ \gamma, \ldots, q_d \circ \gamma) = (q_1, \ldots, q_d) \cdot {}^t\sigma'(\gamma),$$

and also a representation $\sigma : (GL_1)^\alpha \times \Pi_{\tau \in \alpha} GL(m_\tau) \times \Pi_{\tau \in \alpha'} G(S_\tau) \to GL_d(\mathbf{C})$ by

$$\sigma(c_1, c_2, c_3) = c_1^k \sigma'(c_3) \quad \text{for} \quad c_1 \in (GL_1)^\alpha, c_2 \in \prod_{\tau \in \alpha} GL(m_\tau), c_3 \in \prod_{\tau \in \alpha'} G(S_\tau).$$

Assuming (3.10) and that $b_f > 0$, define a $\mathbf{C}^d$-valued function $\mathfrak{f}$ on $\mathfrak{Z}$ by

$$(3.12) \qquad\qquad \mathfrak{f}(w) = (f(w, 0; q_i, \xi, \Omega))_{i=1}^d$$

with fixed ξ and Ω. Then $\mathfrak{f} \in \mathfrak{M}_\sigma$. Now we have

THEOREM 3.3. *Under conditions* (3.5), (3.6), (3.8), (3.9), *and* (3.11), *suppose further that* $b_f > 0$ *and also that*

$$(3.13) \qquad\qquad \varphi_\tau \geq 0 \quad \text{for all} \quad \tau \in \psi \quad (if\ t > r),$$

$$(3.14) \qquad\qquad \xi\ is\ \overline{\mathbf{Q}}\text{-valued and} \quad \omega(a) \in \overline{\mathbf{Q}} \quad \text{for all} \quad a \in E.$$

Then $\pi^{-|k|-e} p_K(-\varphi, 2\psi)\mathfrak{f}$ *is arithmetic, where* $e = \Sigma_{\tau \in \psi}(\varphi_\tau - \varphi_{\tau\rho})$.

THEOREM 3.4. *Under conditions* (3.5), (3.6), (3.8), *and* (3.9), *suppose further that* $b_f = 0$, $h = 0$, $j_\tau = 0$ *for all* $\tau \in \alpha$, *and* $\varphi_\tau = 0$ *for all* $\tau \in \psi$. *Then* $f(w, s)$ *is holomorphic in* s *for* $\mathrm{Re}(s) > 1$ *and has at most a simple pole at* $s = 1$. *Put, with* η *of* [3, (2.13)],

$$\mathfrak{g}(w) = \eta(w)^\alpha \lim_{s \to 1} (s - 1)(f(w, s; q_i, \xi, \Omega))_{i=1}^d.$$

Then $\mathfrak{g}$ *is holomorphic in* w *and defines an element of* $\mathfrak{M}_\sigma$. *If further* $\varphi_{\tau\rho} < 0$ *for all* $\tau \in \psi$ *and* (3.14) *is satisfied, then* $R_E^{-1} \pi^{-|k|-e} p_K(-\varphi, 2\psi)\mathfrak{g}$ *is arithmetic, where* R_E *is the regulator of* E *and* $e = -\Sigma_{\tau \in \psi} \varphi_{\tau\rho}$.

We understand that $p_K(-\varphi, 2\psi) = 1$ if $t = r$. (This applies also to the theorems in the next section.)

The proof of the above theorems requires a theta function of the following form:

$$(3.15) \quad \theta(z, w) = \prod_{\tau \in \alpha} \mathrm{Im}(z_\tau)^{m_\tau/2} \eta(w)^{-k}$$

$$\cdot \sum_{u \in V} \xi(u) q(u) S[u]^h u^\chi \overline{S(u, A^{-1} p(w))^k} \mathbf{e}_F(aR[u; z, w]).$$

Here $z \in \mathfrak{H}^{J(F)}$, $w \in \mathfrak{Z}$, $q \in \mathcal{P}_\kappa(S)$, $a \in F$; $a^\tau > 0$ if and only if $\tau \in \epsilon$; $\eta(w) = (\eta(w_\tau))_{\tau \in \alpha}$ with $\eta(w_\tau)$ defined by [3, (2.13)]; ξ is a locally constant function on V; $0 \leq k = \Sigma_{\tau \in \alpha} k_\tau, \tau \in I_F, 0 \leq h = \Sigma_{\tau \notin \alpha} h_\tau, \tau \in I_F, 0 \leq \chi \in I_K$;

$$R_\tau[u; z, w] = \begin{cases} -x_\tau S_\tau[u] + iy_\tau P_\tau[u; w] & (\tau \in \alpha), \\ -z_\tau S_\tau[u] & (\tau \notin \alpha) \end{cases}$$

with $P_\tau[u; w]$ defined by (I.6.2a) and $z_\tau = x_\tau + iy_\tau$.

This is a generalization of (I.6.6). The pullback of (3.15) relative to a^{-1} gives a generalization of (I.6.12). By using them (and choosing suitable h and χ depending on j and φ of (3.4)), we can prove the above theorems as well as the generalizations of Theorems I.6.2, I.6.3, and I.6.4 in the same fashion as in Sections I.6, I.7, and I.8. (Since (3.15) is a finite linear combination of the products of previous θ's defined relative to (S_i, V_i, F_i) for $i = 1, \ldots, t$, Theorem I.6.2 implies immediately its generalization.) In order to prove Theorem 3.3, we take a CM-point w_0 on $\mathfrak{Z}$ and evaluate $f(w, s)$ at $w = w_0$. Then $f(w_0, 0)$ can be viewed as a critical value of a series of type (2.1), and therefore the desired arithmeticity of $f(w, 0)$ follows from the generalization of Theorem I.4.3, which is already established. A technique of the same type applies to the arithmetic part of Theorem 3.4.

4. The case of eigenvalues occurring on quaternion algebras. Our final generalization concerns quaternion algebras $B_1, \ldots, B_r$ over $F_1, \ldots, F_r$, respectively. Here $0 < r \leq t$. We also consider CM-fields K_i such that $[K_i : F_i] = 2$ for $i = r + 1, \ldots, t$ as in Section 3. Take a subset ϵ of J_F such that $\epsilon \cap J(F_i) \neq \emptyset$ for all $i \leq t$, and define α, α', and β by (3.2). We then assume that B_i is unramified or ramified at $\tau \in J(F_i)$ according as

$\tau \in \alpha$ or $\tau \in \alpha'$. Put $B = B_1 \times \cdots \times B_r$ and $B_{\mathbf{A}} = B \otimes_{\mathbf{Q}} \mathbf{A}$. Denote by ι the $\mathbf{A}$-linear involution of $B_{\mathbf{A}}$ whose restriction to B_i is its main involution for every i, and put $N(x) = xx^\iota$, $\mathrm{Tr}(x) = x + x^\iota$ for $x \in B_{\mathbf{A}}$. We put $G_{\mathbf{A}} = B_{\mathbf{A}}^\times$, $G_{\mathbf{Q}} = B^\times$, and let G_∞, $G_{\infty+}$, and $G_{\mathbf{A}+}$ denote the infinite part of $G_{\mathbf{A}}$, its identity component, and the subgroup of $G_{\mathbf{A}}$ consisting of all x whose infinite parts belong to $G_{\infty+}$. Now various objects of Section II.1 can be generalized to the present situation. For $k = \Sigma_{\tau \in \alpha} k(\tau) \tau \in I_F$ and $\kappa = \Sigma_{\tau \in \alpha'} \kappa(\tau) \tau \in I_F$ with $\kappa(\tau) \geq 0$, we can define a factor of automorphy $j(x, z)^k$ for $x \in G_{\mathbf{A}}$ and $z \in \mathfrak{H}^\alpha$ and also a representation $\sigma_\kappa : G_{\mathbf{A}} \to GL_d(\mathbf{C})$ with $d = \Pi_{\tau \in \alpha'} (\kappa(\tau) + 1)$ by suitably modifying (II.1.5) and (II.1.4); then $f \|_{k,\kappa} x$ can be defined for $x \in G_{\mathbf{A}+}$ and a $\mathbf{C}^d$-valued function f on $\mathfrak{H}^\alpha$. Let $\Gamma = \Pi_{i=1}^r \Gamma_i$ with congruence subgroups Γ_i of B_i. We define $\mathfrak{M}_{k,\kappa}(\Gamma)$ (resp. $\mathfrak{A}_{k,\kappa}(\Gamma)$) to be the set of all $\mathbf{C}^d$-valued holomorphic (resp. meromorphic) functions f on $\mathfrak{H}^\alpha$ which are holomorphic (resp. meromorphic) at the cusps and which satisfy $f \|_{k,\kappa} \gamma = f$ for all $\gamma \in \Gamma$. The union of these spaces for all such Γ is denoted by $\mathfrak{M}_{k,\kappa}$ (resp. $\mathfrak{A}_{k,\kappa}$). By Lemma 1.1, $\mathfrak{M}_{k,\kappa}(\Gamma)$ is naturally isomorphic to $\otimes_{i=1}^r \mathfrak{M}_{k_i, \kappa_i}(\Gamma_i)$ if $k = (k_1, \ldots, k_r)$ and $\kappa = (\kappa_1, \ldots, \kappa_r)$ with k_i and κ_i in $I(F_i)$. The spaces of cusp forms $\mathcal{S}_{k,\kappa}(\Gamma)$ and $\mathcal{S}_{k,\kappa}$ can be defined in a similar way. The inner product $\langle f, g \rangle$ of elements f and g of $\mathcal{S}_{k,\kappa}$ is defined as in (II.1.9b).

Let $\mathfrak{c} = \mathfrak{c}_1 \times \cdots \times \mathfrak{c}_r \subset F_1 \times \cdots \times F_r$ with integral ideals $\mathfrak{c}_i$ of F_i, and let μ be a character of $(\Pi_{i \leq r} F_i)_{\mathbf{A}}^\times$ which is the "product" of Hecke characters μ_i of $(F_i)_{\mathbf{A}}^\times$ for $i = 1, \ldots, r$. Suppose $\mathfrak{M}_{k_i, \kappa_i}(\mathfrak{c}_i, \mu_i)$ of Section II.1 is meaningful for each i. We then define $\mathfrak{M}_{k,\kappa}(\mathfrak{c}, \mu)$ to be the space of $\mathbf{C}^d$-valued functions on $G_{\mathbf{A}}$ which is naturally isomorphic to $\otimes_{i=1}^r \mathfrak{M}_{k_i, \kappa_i}(\mathfrak{c}_i, \mu_i)$ in the sense of Lemma 1.1; $\mathcal{S}_{k,\kappa}(\mathfrak{c}, \mu)$ is defined similarly. The inner product $\langle \mathbf{f}, \mathbf{g} \rangle$ of two elements $\mathbf{f}$ and $\mathbf{g}$ of $\mathcal{S}_{k,\kappa}(\mathfrak{c}, \mu)$ can be defined in such a way that $\langle \mathbf{f}, \mathbf{g} \rangle = \Pi_{i=1}^r \langle \mathbf{f}_i, \mathbf{g}_i \rangle$ if $\mathbf{f} = \mathbf{f}_1 \otimes \cdots \otimes \mathbf{f}_r$ and $\mathbf{g} = \mathbf{g}_1 \otimes \cdots \otimes \mathbf{g}_r$ with $\mathbf{f}_i$ and $\mathbf{g}_i$ in $\mathcal{S}_{k_i, \kappa_i}(\mathfrak{c}_i, \mu_i)$. To emphasize that the spaces are defined relative to B, we write $\mathcal{S}_{k,\kappa}(\mathfrak{c}, \mu, B)$ for $\mathcal{S}_{k,\kappa}(\mathfrak{c}, \mu)$, and denote by $\mathcal{S}_{k,\kappa}(B)$ the union of such spaces for all possible $\mathfrak{c}$ and μ.

As for Hecke operators, given $\mathfrak{a} = \mathfrak{a}_1 \times \cdots \times \mathfrak{a}_r \subset F_1 \times \cdots \times F_r$ with fractional ideals $\mathfrak{a}_i$ of F_i, we let $T_\mathfrak{c}(\mathfrak{a})$ denote the $\mathbf{C}$-linear endomorphism of $\mathfrak{M}_{k,\kappa}(\mathfrak{c}, \mu)$ which is the tensor product of the $T_{\mathfrak{c}_i}(\mathfrak{a}_i)$ (defined in Section II.1) for $i = 1, \ldots, r$. Such an $\mathfrak{a}$ is called a *fractional ideal of* $F_1 \times \cdots \times F_r$. Naturally $T_\mathfrak{c}(\mathfrak{a}) = 0$ unless all $\mathfrak{a}_i$ are integral. There is no difficulty in defining the arithmeticity (or $\overline{\mathbf{Q}}$-rationality) of the elements of $\mathfrak{M}_{k,\kappa}$. We define it in terms of the CM-points on $\mathfrak{H}^\alpha$, and find, by the same reasoning as in the proof of Lemma 3.1, that the set of arithmetic elements of $\mathfrak{M}_{k,\kappa}$ is $\otimes_{i=1}^{r} \mathfrak{M}_{k_i,\kappa_i}(\bar{\mathbf{Q}})$.

Let $\{\mathbf{g}_p\}_{p\in P}$ be a $\bar{\mathbf{Q}}$-rational basis of $S_{k,\kappa}(\mathfrak{c},\,\mu)$. Define a complex matrix $(\chi_{pq}(\mathfrak{a}))$ of size P by

$$(4.1) \qquad \mathbf{g}_p\,|\,T_{\mathfrak{c}}(\mathfrak{a}) = \sum_{q\in P} \chi_{pq}(\mathfrak{a})\mathbf{g}_q \qquad (p\in P).$$

We now assume the existence of E as in (1.3), and put

$$(4.2) \qquad \Lambda_{pq}(s) = [U] \sum_{(x,y)\in L/U} \eta(x,\,y)\chi_{pq}(x\mathfrak{a})\omega_h(\mathrm{Tr}_{F/E}((x,\,yy^\rho)\zeta))$$

$$\cdot x^c y^\varphi \,|\, x^\alpha y^{2\psi} \,|^{-s}.$$

Here $s\in\mathbf{C}$, $L = F_1\times\cdots\times F_r\times K_{r+1}\times\cdots\times K_t$; η is a locally constant function on L; $x\in F_1\times\cdots\times F_r$, $y\in K = K_{r+1}\times\cdots\times K_t$; $\mathfrak{a}$ is a fractional ideal of $F_1\times\cdots\times F_r$; ω_h is defined with $0\le h\in I_E$ and $\Omega\in\mathfrak{M}(l,\,E)$ as in (1.9) and (3.4); ζ is an element of F satisfying (1.4); $c\in 2^{-1}I(F_1\times\cdots\times F_r)$, $\varphi\in I_K$, and ψ is a subset of J_K such that $\mathrm{Res}_{K/F'}(\psi) = \beta$, where $F' = F_{r+1}\times\cdots\times F_t$; we let an element u of U_E act on L by $u(x,\,y) = (u^2 x,\,uy)$; U is a subgroup of U_E of finite index such that η and Ω are U-invariant; the sum is extended over all $U(x,\,y)$ with $x\gg 0$; we understand that

$$\mathrm{Tr}_{F/E}((x,\,yy^\rho)\zeta) = \sum_{i\le r} \mathrm{Tr}_{F_i/E}(\zeta_i x_i) + \sum_{i>r} \mathrm{Tr}_{K_i/E}(\zeta_i y_i y_i^\rho),$$

where subscript i indicates the i-component. To make each term of (4.2) dependent only on $U(x,\,y)$, we assume

$$(4.3) \qquad l + 2h + 2\mathrm{Res}_{F/E}(c) + \mathrm{Res}_{K/E}(\varphi) = a_\Lambda J_E \quad \text{with} \quad a_\Lambda\in\mathbf{Z}.$$

We assume also

$$(4.4\mathrm{a}) \qquad \varphi_\sigma \ge 0 \quad \text{for} \quad \sigma\notin\psi\cup\psi\rho \quad \text{and} \quad \varphi_\sigma \ge \varphi_{\sigma\rho} \quad \text{for} \quad \sigma\in\psi;$$

$$(4.4\mathrm{b}) \qquad 2c_\tau \equiv k_\tau \pmod 2 \quad \text{for every} \quad \tau\in\alpha;$$

$$(4.4\mathrm{c}) \qquad \kappa_\tau \le 2c_\tau \equiv \kappa_\tau \pmod 2 \quad \text{for every} \quad \tau\in\alpha'.$$

Series Λ_{pq} is a special case of (1.9) for the same reason as explained in [4, I, p. 144, II, pp. 581–582]; condition (1.12) follows from (4.4a, b, c). Notice that (2.1) would have been a special case of (4.2), if we had allowed r to be 0. From the generalization of Theorem I.3.2, we can easily derive

PROPOSITION 4.1. *Under (4.3) and (4.4a, b, c), Λ_{pq} is convergent for sufficiently large* Re(s) *and can be continued to a meromorphic function on the whole s-plane, which is holomorphic for* Re$(s) > b_\Lambda/2$ *and has at most a simple pole at $s = b_\Lambda/2$, where*

$$(4.5) \qquad b_\Lambda = a_\Lambda + 2 \sum_{i \le r} [F_i : E] + \sum_{i > r} [F_i : E].$$

Moreover, Λ_{pq} is holomorphic for Re$(s) \ge (b_\Lambda - 1)/2$ *unless* $2c_\tau + 2 + k_\tau \le b_\Lambda$ *for all $\tau \in \alpha$ and $2\varphi_\sigma + 2 \le b_\Lambda$ for all $\sigma \in \psi$.*

To state our main results on arithmeticity, we naturally need a condition

$$(4.6) \qquad \Omega \quad and \quad \{g_p\} \quad are \ \overline{\mathbf{Q}}\text{-}rational \ and \quad \eta \quad is \ \overline{\mathbf{Q}}\text{-}valued.$$

THEOREM 4.2. *Let $\Lambda(s)$ and Ξ be the matrices of size P with $\Lambda_{pq}(s)$ and $\langle g_p, g_q \rangle$ as their (p, q)-entries, respectively. Then, under (4.3), (4.4a, b, c), and (4.6), the entries of the matrix $\pi^{-|k|-e} p_K(-\varphi, 2\psi) \cdot {}^t\Xi^{-1}\Lambda(s_0)$ are all algebraic for every integer s_0 satisfying*

$$(4.7a) \qquad (b_\Lambda - 1)/2 \le s_0 \le (k_\tau/2) + c_\tau \quad for \ every \quad \tau \in \alpha,$$

$$(4.7b) \qquad s_0 \le \varphi_\sigma \quad for \ every \quad \sigma \in \psi,$$

$$(4.7c) \qquad 2s_0 \ne b_\Lambda \quad if \quad E = \mathbf{Q},$$

where b_Λ is defined by (4.5) and $e = \Sigma_{\sigma \in \psi}(\varphi_\sigma - \varphi_{\sigma\rho})$.

THEOREM 4.3. *Suppose, besides (4.3), (4.4a, b, c), and (4.6), that*

$$(4.8a) \qquad 2c_\tau + k_\tau + 2 = b_\Lambda \quad for \ every \quad \tau \in \alpha,$$

$$(4.8b) \qquad \varphi_{\sigma\rho} < \varphi_\sigma = (b_\Lambda/2) - 1 \quad for \ every \quad \sigma \in \psi.$$

Define Λ, Ξ, and e as in Theorem 4.2. Then the residue of $R_E^{-1}\pi^{-|k|-e}\cdot$
$p_K(-\varphi, 2\psi)\cdot{}^t\Xi^{-1}\Lambda$ at $s = b_\Lambda/2$ is an algebraic matrix, where R_E is the
regulator of E.

To state the generalization of Theorem II.3.7, we consider a primitive
system of eigenvalues χ_i of Hecke operators occurring in $\mathcal{S}_{k_i,\kappa_i}(B_i)$ for each
i, and take nonzero $\overline{\mathbf{Q}}$-rational elements $\mathbf{g}_i$ in the space such that $\mathbf{g}_i \mid T(\mathfrak{a}_i) =$
$\chi_i(\mathfrak{a}_i)\mathbf{g}_i$ for all ideals $\mathfrak{a}_i$ of F_i. Put

$$(4.9) \qquad Z(s) = [U] \sum_{(x,y)\in L/U} \eta(x, y)\omega_h(\mathrm{Tr}_{F/E}((x, yy^\rho)\zeta))$$

$$\cdot \prod_{i=1}^{r} \chi_i(x_i\mathfrak{a}_i)x^c y^\varphi |x^\alpha y^{2\psi}|^{-s}$$

with fractional ideals $\mathfrak{a}_i$ of F_i, where $x = (x_1, \ldots, x_r)$ with $x_i \in F_i$ and
other symbols are the same as in (4.2). Then we obtain

THEOREM 4.4. *The notation being as above, suppose that Ω is
$\overline{\mathbf{Q}}$-rational, η is $\overline{\mathbf{Q}}$-valued, and (4.3), (4.4a, b, c) are satisfied. Then, for
every integer s_0 satisfying (4.7a, b, c), Z is finite at s_0 and $Z(s_0) \sim$
$\pi^{|k|+e}p_K(\varphi, 2\psi) \prod_{i=1}^{r} \langle \mathbf{g}_i, \mathbf{g}_i \rangle$. Furthermore, if (4.8a, b) are satisfied, the
residue of Z at $s = b_\Lambda/2$ is an algebraic number times $R_E\pi^{|k|+e}p_K(\varphi, 2\psi)\cdot$
$\prod_{i=1}^{r} \langle \mathbf{g}_i, \mathbf{g}_i \rangle$.*

These theorems can be proved in exactly the same fashion as in
Sections II.3 and II.6 with obvious rephrasing. The proofs require a suitable generalization of series $H(z, w, s)$ of (II.4.11), which can be obtained
as a special case of (3.4).

5. Applications and conjectures.   In this section, we consider qua-
ternion algebras over E. Let $\{\chi(\mathfrak{a})\}$ and $\{\chi'(\mathfrak{a})\}$ be systems of eigenvalues
of Hecke operators on the spaces of Hilbert cusp forms $\mathcal{S}_{m,0}(\mathfrak{c}, \mu, M_2(E))$
and $\mathcal{S}_{m',0}(\mathfrak{c}', \mu', M_2(E))$, respectively, where $m, m' \in I_E$, and μ, μ' are
Hecke characters of $E_\mathbf{A}^\times$ defined modulo ideals $\mathfrak{c}$, $\mathfrak{c}'$ of E. We consider a
Dirichlet series

$$(5.1) \qquad D(s, \chi, \chi') = \sum_{\mathfrak{a}} \chi(\mathfrak{a})\chi'(\mathfrak{a})N(\mathfrak{a})^{-s} \qquad (s \in \mathbf{C}),$$

where $\mathfrak{a}$ runs over all integral ideals of E. To eliminate the numerator part
of the Euler product of D, we take an L-function of E of the form

$$(5.2) \qquad L_e(s, \mu\mu') = \sum_{(\mathfrak{a},e)=1} (\mu\mu')^*(\mathfrak{a})N(\mathfrak{a})^{-s}$$

with any common multiple e of $\mathfrak{c}$ and $\mathfrak{c}'$, where $(\mu\mu')^*$ is the ideal character attached to $\mu\mu'$, and put

$$(5.3) \qquad \mathfrak{D}_e(s, \chi, \chi') = L_e(2s - 2, \mu\mu')D(s, \chi, \chi').$$

PROPOSITION 5.1. *Suppose $m_\tau - m_\tau'$ (mod 2) for $\tau \in J_E$ is independent of τ. Then $D(s, \chi, \chi')$ is convergent for sufficiently large $\mathrm{Re}(s)$, and can be continued as a meromorphic function to the whole s-plane. Moreover,*

$$\prod_{\tau \in J(E)} \Gamma(s - 2 + (m_\tau + m_\tau')/2)\Gamma(s - 1 + |m_\tau - m_\tau'|/2)\mathfrak{D}_e(s, \chi, \chi')$$

is holomorphic except for possible simple poles at $s = 2$ and $s = 1$. The poles may occur only when $m = m'$. Furthermore, if $m = m'$, the residue of $D(s, \chi, \bar\chi)$ at $s = 2$ is a nonzero algebraic number times $R_E\pi^{|m|}\langle \mathbf{f}, \mathbf{f}\rangle$, where $\mathbf{f}$ is a $\bar{\mathbf{Q}}$-rational primitive element of $\mathbb{S}_{m,0}(M_2(E))$ such that $\mathbf{f}|T(\mathfrak{a}) = \chi(\mathfrak{a})\mathbf{f}$ for all $\mathfrak{a}$, and R_E is the regulator of E.

This was proved in [2, Proposition 4.13] when $m \geq m'$. The same proof applies to the general case with suitable modifications. We observe that $D(s, \chi, \chi')$ is holomorphic for $\mathrm{Re}(s) > 2$, and even for $\mathrm{Re}(s) \geq 3/2$ if $m \neq m'$. This fact follows also from Theorem I.3.2.

PROPOSITION 5.2. *For an arbitrary integral ideal $\mathfrak{f}$ of E, put*

$$D^\mathfrak{f}(s, \chi, \chi') = \sum_{(\mathfrak{a},\mathfrak{f})=1} \chi(\mathfrak{a})\chi'(\mathfrak{a})N(\mathfrak{a})^{-s}.$$

Then $D^\mathfrak{f}(s, \chi, \chi') \neq 0$ for $\mathrm{Re}(s) \geq 2$.

Proof. Let $r = 2 + it$ with $t \in \mathbf{R}$, $e = \mathfrak{f} \cap \mathfrak{c} \cap \mathfrak{c}'$, and let

$$C(s, \chi, \chi') = L_e(2s - 2, \mu\mu')D^\mathfrak{f}(s, \chi, \chi'),$$

$$A(s) = C(s + r, \chi, \chi')C(s + \bar{r}, \bar\chi, \bar\chi')$$

$$\cdot C(s + 2, \chi, \bar\chi)C(s + 2, \chi', \bar\chi').$$

Then C has an Euler product expression consisting of the Euler $\mathfrak{p}$-factors of $\mathfrak{D}_e$ for $(\mathfrak{p}, \mathfrak{f}) = 1$. Hence we see easily that

$$\log A(s) = \sum_{n=1}^{\infty} n^{-1} \sum_{\mathfrak{p}} a(n, \mathfrak{p}) N(\mathfrak{p})^{-s}$$

with $a(n, \mathfrak{p}) \geq 0$, where $\mathfrak{p}$ runs over all prime ideals of E prime to $\mathfrak{f}$. Let $\mathrm{Re}(s) = \sigma_0$ be the line of convergence of the series expressing $\log A(s)$. The gamma factors of $\mathfrak{D}_e$ given in Proposition 5.1 produce many zeros of $A(s)$, so that σ_0 is a finite value. Since $a(n, \mathfrak{p}) \geq 0$, $\log A(s)$ is not holomorphic at σ_0. If $\sigma_0 > 0$, we have $A(\sigma) = \exp(\log A(\sigma)) \geq 1$ for $\sigma > \sigma_0$. Since A is holomorphic at σ_0, we have $A(\sigma_0) \geq 1$, and hence $\log A$ is holomorphic at σ_0, which is a contradiction. Therefore $\sigma_0 \leq 0$, and $A(s) = \exp(\log A(s)) \neq 0$ for $\mathrm{Re}(s) > 0$. Suppose $D(r, \chi, \chi') = 0$. Then A is entire. Repeating the same argument as above at σ_0, we obtain a contradiction, which completes the proof.

Our main interest is in the arithmetic nature of the values of D at some critical points and also of the residue of D. In [2], we studied such values under the conditions

$$(5.4) \qquad\qquad m_\tau - m_\tau' \ (\mathrm{mod}\ 2) \quad \textit{is independent of } \tau,$$

$$(5.5) \qquad\qquad m_\tau > m_\tau' \quad \textit{for all} \quad \tau \in J_E.$$

In fact, *suppose χ is primitive in the sense of Section* II.3 (*cf.* [2, p. 652]); *let s_0 be an integer or a half-integer such that*

$$(5.6) \qquad\qquad 1 < s_0 \leq 1 + (m_\tau - m_\tau')/2 \quad \textit{for all} \quad \tau \in J_E,$$

$$(5.7) \qquad\qquad 2s_0 \equiv m_\tau - m_\tau' \ (\mathrm{mod}\ 2) \quad \textit{for all} \quad \tau \in J_E.$$

Then $D(s_0, \chi, \chi') \sim \pi^{|m|} \langle \mathbf{f}, \mathbf{f} \rangle$ with $\mathbf{f}$ of Proposition 5.1. (See [2, Theorems 4.1 and 4.2].) This result follows also from Theorem II.3.7, (i).

We are going to study $D(s_0, \chi, \chi')$ under a condition different from (5.5). The result is the first main application of our generalization, which can be stated as

THEOREM 5.3.   *Let $J_E = \alpha \cup \beta$ with nonempty disjoint subsets α and β; let B (resp. B') be a quaternion algebra over E unramified at every*

$\tau \in \alpha$ (resp. $\tau \in \beta$) and ramified at every $\tau \in \beta$ (resp. $\tau \in \alpha$). Suppose the above χ and χ' are primitive and occur in $\mathcal{S}_{h,\lambda}(B)$ and $\mathcal{S}_{h',\lambda'}(B')$, respectively. Let $\mathbf{g}$ and $\mathbf{g}'$ be $\overline{\mathbf{Q}}$-rational nonzero elements in these spaces such that $\mathbf{g} \mid T(\mathfrak{a}) = \chi(\mathfrak{a})\mathbf{g}$ and $\mathbf{g}' \mid T'(\mathfrak{a}) = \chi'(\mathfrak{a})\mathbf{g}'$ for all ideals $\mathfrak{a}$ of E, where $T(\mathfrak{a})$ and $T'(\mathfrak{a})$ are Hecke operators. Assume further, in addition to (5.4), a condition

$$(5.8) \qquad m_\tau > m_\tau' \quad for \quad \tau \in \alpha \quad and \quad m_\tau < m_\tau' \quad for \quad \tau \in \beta.$$

Then $D(s_0, \chi, \chi') \sim \pi^{|h+h'|} \langle \mathbf{g}, \mathbf{g} \rangle \langle \mathbf{g}', \mathbf{g}' \rangle$ for every integer or half-integer s_0 satisfying (5.7) and

$$(5.9) \qquad 1 < s_0 \le 1 + (|m_\tau - m_\tau'|/2) \quad for\ all \quad \tau \in J_E.$$

Proof. Notice that $m = h + \lambda + 2\beta$ and $m' = h' + \lambda' + 2\alpha$ (see (II.3.3)). We are going to apply Theorem 4.4 to $B \times B'$ with $L = F = E \times E$. Let p and p' be the projection maps of F onto the first and the second factors E. Let d be either 0 or 1 according as the differences $m_\tau - m_\tau'$ are even or odd. We consider Z of (4.9) by taking $\epsilon = p\alpha + p'\beta$, $k = ph + p'h'$, $\kappa = p\lambda + p'\lambda'$, $2c = \kappa - p\lambda' - p'\lambda + d\epsilon$, $\Omega = 1$, $\mathfrak{a} = \eta \times \eta$ with a fractional ideal η of E, and $\zeta = (\zeta_0, -\zeta_0)$ with an element ζ_0 of E such that $\zeta_0^\tau > 0$ if and only if $\tau \in \alpha$. Then we find that conditions (4.3), (4.4a, b, c) are satisfied. Moreover, $\mathrm{Tr}_{F/E}(\zeta b) = 0$ for $b = (a_1, a_2) \in F = E \times E$ if and only if $a_1 = a_2$. Therefore we obtain a series of the form

$$Z_\eta(s) = \sum_{0 \ll a \in E/U} \chi(a\eta)\chi'(a\eta) N_{E/\mathbf{Q}}(a)^{(d/2)-s}$$

with $U = U_E$ as a special case of (4.9). Let Y be a complete set of representatives of the ideal classes of E modulo the product of all archimedean primes. Then $D(s, \chi, \chi') = \sum_{\eta \in Y} N(\eta)^{-s} Z_\eta(s + (d/2))$. Hence our assertion follows immediately from Theorem 4.4.

With χ occurring in $\mathcal{S}_{h,\lambda}(B)$, let $W(\chi, B)$ denote the space of all $\mathbf{g}$ in $\mathcal{S}_{h,\lambda}(B)$ such that $\mathbf{g} \mid T(\mathfrak{p}) = \chi(\mathfrak{p})\mathbf{g}$ for all except a finite number of prime ideals $\mathfrak{p}$ of E, and $W(\chi, B, \overline{\mathbf{Q}})$ the subset of $W(\chi, B)$ consisting of all $\overline{\mathbf{Q}}$-rational elements in it. We have of course $W(\chi, B) = W(\chi, B, \overline{\mathbf{Q}}) \otimes_{\overline{\mathbf{Q}}} \mathbf{C}$.

THEOREM 5.4.   *Let χ, B, χ', B' be the same as in Theorem 5.3 without condition (5.8), and let $\mathbf{f}$ be a form as in Proposition 5.1. Further let $\mathbf{g}$ and $\mathbf{h}$ (resp. $\mathbf{g}'$ and $\mathbf{h}'$) be elements of $W(\chi, B, \bar{\mathbf{Q}})$ (resp. $W(\chi', B', \bar{\mathbf{Q}})$). Suppose $\chi = \chi'$ or $\bar{\chi} = \chi'$. Then $\langle \mathbf{g}, \mathbf{h}\rangle\langle \mathbf{g}', \mathbf{h}'\rangle \sim \langle \mathbf{f}, \mathbf{f}\rangle$; in particular, $\langle \mathbf{g}, \mathbf{g}\rangle\langle \mathbf{g}', \mathbf{g}'\rangle \sim \langle \mathbf{f}, \mathbf{f}\rangle$.*

Proof.   The notation being the same as in the above proof, suppose $\bar{\chi} = \chi'$. Then (4.8) is satisfied, since $m = m'$. Take an integral ideal $\mathfrak{c}$ of E so that $\mathbf{g}$, $\mathbf{h} \in \mathcal{S}_{h,\lambda}(\mathfrak{c}, \mu, B)$ and $\mathbf{g}'$, $\mathbf{h}' \in \mathcal{S}_{h',\lambda'}(\mathfrak{c}, \mu', B')$. Let $\mathcal{U} = W(\chi, B) \cap \mathcal{S}_{h,\lambda}(\mathfrak{c}, \mu, B)$, and let $\mathcal{V}$ be the orthogonal complement of $\mathcal{U}$ in $\mathcal{S}_{h,\lambda}(\mathfrak{c}, \mu, B)$; define $\mathcal{U}'$ and $\mathcal{V}'$ similarly with χ' and B' in place of χ and B. Notice that $\mathbf{j} \mid T(\mathfrak{a}) = \chi(\mathfrak{a})\mathbf{j}$ for every $\mathbf{j} \in \mathcal{U}$ and every $\mathfrak{a}$ prime to $\mathfrak{c}$. As observed in the proof of Theorems II.3.7 and II.3.8, the spaces $\mathcal{U}$, $\mathcal{V}$, $\mathcal{U}'$, $\mathcal{V}'$ have $\bar{\mathbf{Q}}$-rational bases (over $\mathbf{C}$), say $\mathfrak{A}$, $\mathfrak{B}$, $\mathfrak{A}'$, $\mathfrak{B}'$. Define $\chi_{pq}(\mathfrak{a})$ with respect to $\{\mathbf{h}_p\} = \mathfrak{A} \cup \mathfrak{B}$ by $\mathbf{h}_p \mid T_{\mathfrak{c}}(\mathfrak{a}) = \Sigma_q \chi_{pq}(\mathfrak{a})\mathbf{h}_q$ and define similarly $\chi'_{p'q'}(\mathfrak{a})$ with respect to $\mathfrak{A}' \cup \mathfrak{B}'$. Let Y be the same as in the proof of Theorem 5.3. Let Θ (resp. Θ') be the matrix of size $\mathfrak{A}$ (resp. $\mathfrak{A}'$) whose entries are $\langle \mathbf{g}, \mathbf{h}\rangle$ with $\mathbf{g}$ and $\mathbf{h}$ in $\mathfrak{A}$ (resp. $\mathfrak{A}'$). For $\mathfrak{y} \in Y$, take a locally constant function η on E so that $\eta(b) = 1$ if $b\mathfrak{y}$ is an integral ideal prime to $\mathfrak{c}$ and $\eta(b) = 0$ otherwise. Put

$$Z_{\mathfrak{y}}^{*}(s) = [U] \sum_{0 \ll b\in E/U} \eta(b)\chi(by)\chi'(b\mathfrak{y})N(b)^{-s}.$$

Then Theorem 4.3 implies that the residue of $Z_{\mathfrak{y}}^{*}(s)$ at $s = 2$ times $R_E^{-1}\pi^{-|h+h'|}.{}^{t}(\Theta \otimes \Theta')^{-1}$ is an algebraic matrix. Now $D^{\mathfrak{c}}(s, \chi, \chi') = \Sigma_{\mathfrak{y}\in Y} N(\mathfrak{y})^{-s}Z_{\mathfrak{y}}^{*}$. The residue of $D^{\mathfrak{c}}$ at $s = 2$ is a nonzero algebraic number times $R_E\pi^{|m|}\langle \mathbf{f}, \mathbf{f}\rangle$ for the same reason as explained in the proof of Theorem II.3.8. This proves the case $\bar{\chi} = \chi'$, since $m = h + h'$. Now we can easily define a map similar to $J_{\mathfrak{c}}$ of [2, (2.46)] which sends $W(\chi, B, \bar{\mathbf{Q}})$ to $W(\bar{\chi}, B, \bar{\mathbf{Q}})$ and which leaves the inner product invariant. Therefore the assertion for $\chi = \chi'$ follows from the other case.

Let us now rephrase Conjecture II.3.9 as

CONJECTURE 5.5.   *Let B_1 and B_2 be quaternion algebras over E which are unramified exactly at the same archimedean primes of E. Suppose χ occurs in $\mathcal{S}_{h,\lambda}(B_1)$ and also in $\mathcal{S}_{h,\lambda}(B_2)$ with some h, $\lambda \in I_E$. Let $\mathbf{g}$, $\mathbf{h} \in W(\chi, B_1, \bar{\mathbf{Q}})$ and $0 \neq \mathbf{j} \in W(\chi, B_2, \bar{\mathbf{Q}})$. Then $\langle \mathbf{g}, \mathbf{h}\rangle \sim \langle \mathbf{j}, \mathbf{j}\rangle$.*

This is supported by

THEOREM 5.6. *Let α be the set of all archimedean primes of E where B_1 and B_2 are unramified. Then the above conjecture is true at least in the following three cases:*

$$(5.11a) \qquad\qquad\qquad\qquad \alpha = J_E;$$

$$(5.11b) \qquad h_\tau \geq 2 \quad \text{for all} \quad \tau \in \alpha; \qquad [E:\mathbf{Q}] \quad \text{or} \quad |\alpha| \quad \text{is even};$$

$$(5.11c) \qquad\qquad\qquad\qquad h_\tau \geq 3 \quad \text{for all} \quad \tau \in \alpha.$$

Proof. That Conjecture 5.5 is true if $\alpha = J_E$ has been shown in Theorem II.3.8. To treat case (5.11b), let the notation be the same as in the above conjecture. Our problem is to find a quaternion algebra B' which is complementary to B_1 in the sense of Theorem 5.3, such that χ occurs in $\mathcal{S}_{h',\lambda'}(B')$. Since $h_\tau = \lambda'_\tau + 2$ for $\tau \in \alpha$, we have to assume $h_\tau \geq 2$. Suppose $[E:\mathbf{Q}]$ is even. Then we can take B' so that B' is ramified exactly at the same finite primes of E as B_1. The result of Shimizu and Jacquet-Langlands (see [1]) tells that χ occurs in $\mathcal{S}_{h',\lambda'}(B')$. Taking a nonzero element $\mathbf{g}'$ of $W(\chi, B', \bar{\mathbf{Q}})$, we have $\langle \mathbf{g}, \mathbf{h} \rangle \sim \langle \mathbf{f}, \mathbf{f} \rangle / \langle \mathbf{g}', \mathbf{g}' \rangle \sim \langle \mathbf{j}, \mathbf{j} \rangle$ by Theorem 5.4. If $|\alpha|$ is even, we can find a quaternion algebra B' over E which is ramified at every τ in α and unramified at all other finite and infinite primes. Then χ occurs in $\mathcal{S}_{h',\lambda'}(B')$, and the same conclusion holds. This proves our assertion under (5.11b).

Before showing that (5.11c) is another sufficient condition, we consider one more type of application which involves a CM-field K such that $[K:E] = 2$. Let Ξ be a (primitive or an imprimitive) Hecke character of the ideal group of K defined modulo an integral ideal $\mathfrak{h}$ such that

$$(5.13) \qquad \Xi(a\mathfrak{r}_K) = a^\xi/|a^\xi| \quad \text{if} \quad a \in K \quad \text{and} \quad a \equiv 1 \,(\mathrm{mod}^\times \mathfrak{h}),$$

where $\mathfrak{r}_K$ is the maximal order of K and $\xi \in I_K$. We assume

$$(5.14) \qquad \xi = \sum_{\sigma \in \Phi} \xi_\sigma \sigma \quad \text{with} \quad \xi_\sigma \geq 0 \quad \text{and with a CM-type } \Phi \text{ of } K.$$

With χ occurring in $\mathcal{S}_{m,0}(M_2(E))$ as before, we define a series $D(s, \chi, \Xi)$ by

$$(5.15) \qquad D(s, \chi, \Xi) = \sum_{\mathfrak{b}} \Xi(\mathfrak{b})\chi(\mathfrak{b}\mathfrak{b}^\rho)N(\mathfrak{b})^{-s},$$

where $\mathfrak{b}$ runs over all integral ideals of K prime to $\mathfrak{h}$. If Ξ is primitive, there is a primitive system χ' of eigenvalues of Hecke operators occurring in $\mathcal{S}_{m',0}(M_2(E))$ such that

$$(5.16) \qquad \sum_{\mathfrak{b}} \Xi(\mathfrak{b}) N(\mathfrak{b})^{-s} = \sum_{\mathfrak{a}} \chi'(\mathfrak{a}) N(\mathfrak{a})^{-(1/2)-s},$$

$$(5.17) \qquad m' = \mathrm{Res}_{K/E}(\xi + \Phi)$$

(see [2, Section 5]). In this situation, we say that χ' *is of type* (K, ξ). Notice that χ' is also of type $(K, \xi\rho)$ and that

$$(5.18) \qquad D(s, \chi, \Xi) = D(s + (1/2), \chi, \chi').$$

THEOREM 5.7. *Let χ, m, α, β, B, h, λ, and $\mathbf{g}$ be the same as in Theorem 5.3, and let K, Ξ, Φ, and ξ be the same as above. (Ξ is not necessarily primitive.) Define m' by (5.17). Then, under (5.4) and (5.8), we have*

$$D(s_0 - (1/2), \chi, \Xi) \sim \pi^{|h|+e} p_K(\xi, 2\psi)\langle \mathbf{g}, \mathbf{g}\rangle$$

for every integer or half-integer s_0 satisfying (5.7) and (5.9), where ψ is the subset of Φ such that $\mathrm{Res}_{K/E}(\psi) = \beta$, and $e = \Sigma_{\sigma\in\psi} \xi_\sigma$.

Proof. We apply Theorem 4.4 to $B \times K$ with $F = E \times E$. Defining p, p', and ζ as in the proof of Theorem 5.3, we consider Z of (4.9) with $\Omega = 1$, $\epsilon = p\alpha + p'\beta$, $k = h$, $\kappa = \lambda$, $\varphi = \Sigma_{\sigma\notin\psi} \xi_\sigma\sigma - \Sigma_{\sigma\in\psi} \xi_\sigma\sigma\rho$, $2c = p(dJ_E - \mathrm{Res}_{K/E}(\varphi))$, where $d = \mathrm{Max}\{\kappa_\tau - \xi_\sigma \mid \mathrm{Res}_{K/E}(\sigma) = \tau, \sigma \in \psi\}$. Then we find a series of the form

$$(5.19) \qquad Z'_{\mathfrak{r}}(s) = \sum_{b V} \chi(bb^\rho \mathfrak{r}\mathfrak{r}^\rho) b^\xi |b^\xi|^{-1} N_{K/\mathbf{Q}}(b)^{(d/2)-s}$$

as a special case of (4.9), where $\mathfrak{r}$ is a fractional ideal of K, prime to $\mathfrak{h}$, and b runs over $\{b \in \mathfrak{r}^{-1} \mid 0 \neq b \equiv 1 \ (\mathrm{mod}^\times \mathfrak{h})\}$ modulo

$$V = \{u \in \mathfrak{r}_K^\times \mid u \equiv 1 \ (\mathrm{mod}\ \mathfrak{h})\}.$$

Let X be a complete set of representatives of the ideal classes of K modulo $\mathfrak{h}$. We have then

$$(5.20) \qquad D(s, \chi, \Xi) = \sum_{\mathfrak{r}\in X} \Xi(\mathfrak{r}) N(\mathfrak{r})^{-s} Z'_{\mathfrak{r}}(s + (d/2)).$$

Therefore our assertion follows from Theorem 4.4.

THEOREM 5.8. *Let χ, m, α, β, B, h, λ, and $\mathbf{g}$ be the same as in Theorem 5.3. Suppose that χ is of type (K, ξ) with a CM-field K and $\xi \in I_K$ satisfying (5.14). Suppose further that $m_\tau > 1$ for all $\tau \in \alpha$. Then $\pi^{|\alpha|}\langle \mathbf{g}, \mathbf{h} \rangle \sim p_K(\xi, 2\eta)$ for every $\mathbf{g}$ and $\mathbf{h}$ in $W(\chi, B, \overline{\mathbf{Q}})$, where η is the subset of Φ such that $\mathrm{Res}_{K/E}(\eta) = \alpha$.*

Proof. The notation and the assumption being the same as in Theorem 5.7 and its proof, suppose $\overline{\chi} = \chi'$. Then $m = m'$ with m' of (5.16) and $d = -1$, so that (4.8) is satisfied with $b_\Lambda = 2$. Define Θ as in the proof of Theorem 5.4, taking $\mathfrak{c}$ to be a multiple of $\mathfrak{h}$. Let $Z_\mathfrak{r}''$ denote function (5.19) defined with $\mathfrak{c}$ in place of $\mathfrak{h}$. Then Theorem 4.4 implies that $R_E^{-1}\pi^{|\beta|-|m|}p_K(-\xi, 2\psi)\Theta^{-1}$ multiplied by the residue of $Z_\mathfrak{r}''$ at $s = 1$ is an algebraic matrix. Therefore the reasoning at the end of the proof of Theorem 5.4 shows that

$$\pi^{|m|-|\beta|}p_K(\xi, 2\psi)\langle \mathbf{g}, \mathbf{h} \rangle \sim \pi^{|m|}\langle \mathbf{f}, \mathbf{f} \rangle$$

for $\mathbf{g}$, $\mathbf{h} \in W(\chi, B, \overline{\mathbf{Q}})$ with $\mathbf{f}$ of Proposition 5.1. By [2, Theorem 5.5], $\pi^n \langle \mathbf{f}, \mathbf{f} \rangle \sim p_K(\xi, 2\Phi)$, where $n = [E:\mathbf{Q}]$. Our assertion now follows from the linearity of symbol p_K in the second variable.

Remark 5.9. Theorem II.3.8 together with [2, Theorem 5.5] implies that our assertion is true even when $\alpha = J_E$. Thus the above theorem verifies Conjecture II.3.10 in a stronger form. In fact, the existence of subfield E in that conjecture is unnecessary. (The present E corresponds to F there.)

Let us now complete the proof of Theorem 5.6 under assumption (5.11c). We consider the situation of Theorem 5.7 and its proof with an arbitrarily chosen K. Suppose $h_\tau \geq 3$ for all $\tau \in \alpha$. Take ξ of (5.13) with $\xi_\sigma = h_\tau - 3$ for $\sigma \in \Phi - \psi$ and $\xi_\sigma = \kappa_\tau + 3$ for $\sigma \in \psi$, where $\tau = \mathrm{Res}_{K/E}(\sigma)$. We can then find a primitive Hecke character Ξ of K of type (5.13). Define Θ as in the proof of Theorem 5.4, taking $\mathfrak{c}$ to be a multiple of the conductor of Ξ; let $Z_\mathfrak{r}''$ be the function in (5.19) defined with $\mathfrak{c}$ in place of $\mathfrak{h}$. Then Theorem 4.2 shows that $\pi^{-|h|-e}p_K(-\xi, 2\psi)Z_\mathfrak{r}''(0)\Theta^{-1}$ is an algebraic matrix. Take χ' as in (5.16). Then (5.18) together with (5.20) shows that the same conclusion holds with $D^\mathfrak{c}(2, \chi, \chi')$ in place of $Z_\mathfrak{r}''(0)$. By Proposition 5.2, this means that

$$\pi^{|h|+e}p_K(\xi, 2\psi)\langle \mathbf{g}, \mathbf{h} \rangle \sim D(2, \chi, \chi')$$

for $\mathbf{g}, \mathbf{h} \in W(\chi, B, \overline{\mathbf{Q}})$. Since $D(2, \chi, \chi')$ depends only on χ and χ', this completes the proof of Theorem 5.6.

Our results tempt us to make a conjecture which is more comprehensive than Conjecture 5.5:

CONJECTURE 5.10. *Given a totally real algebraic number field E, a subset α of J_E, and a primitive system χ of eigenvalues of Hecke operators on $S_{m,0}(M_2(E))$ with $m_\tau > 1$ for all $\tau \in J_E$, there exists a nonzero complex constant $Q(\chi, \alpha)$, which is determined modulo $\overline{\mathbf{Q}}^\times$ and depends only on χ and α, with the following properties:*

(Q1) $Q(\chi, \alpha) \sim 1$ *if* $\alpha = \emptyset$.

(Q2) $Q(\chi, \alpha)Q(\chi, \beta) \sim Q(\chi, \alpha \cup \beta)Q(\chi, \alpha \cap \beta)$ *for* $\alpha, \beta \subset J_E$.

(Q3) *Suppose χ occurs in $S_{h,\lambda}(B)$ with a quaternion algebra B over E and suitable h, λ, and suppose B is unramified at every $\tau \in \alpha$ and ramified at every $\tau \in J_E - \alpha$. Then $\langle \mathbf{g}, \mathbf{h} \rangle \sim Q(\chi, \alpha)$ for every $\mathbf{g}$ and $\mathbf{h}$ in $W(\chi, B, \overline{\mathbf{Q}})$.*

(Q4) *A collection of properties of the same nature as described in Theorem 4.4. (For details, see below.)*

(Q5) *If K is a CM-field, $[K:E] = 2$, ξ, Φ, and η are as in Theorem 5.8, and χ is of type (K, ξ), then $\pi^{|\alpha|}Q(\chi, \alpha) \sim p_K(\xi, 2\eta)$.*

The precise meaning of (Q4) is as follows. Let L, F_i, K_i, ϵ, η, ζ, c, φ, α, ψ, and ω_h be the same as in (4.2); let χ_i be a system of eigenvalues occurring in $S_{m_i,0}(M_2(F_i))$ for each $i \leq r$. Put $\epsilon = \Sigma_{i=1}^{t} \pi_i\alpha_i$ with $\alpha_i \subset J(F_i)$. Define Z by (4.9). Assume, in addition to (4.3) and (4.4a), the conditions

$$(5.21) \quad 2c(\pi_i\tau) \geq m_i(\tau) - 2 \quad \textit{for every} \quad \tau \in J(F_i) - \alpha_i, \quad i = 1, \ldots, r.$$

$$(5.22) \quad 2c(\pi_i\tau) \equiv m_i(\tau) \,(\mathrm{mod}\, 2) \quad \textit{for every} \quad \tau \in J(F_i), \quad i = 1, \ldots, r.$$

Then the precise meaning of (Q4) is that the assertions of Theorem 4.4 concerning $Z(s_0)$ and the residue of Z hold with $\pi^{d+e}\,\Pi_{i=1}^{r}\,Q(\chi_i, \alpha_i)$ in place of $\pi^{|k|+e}\,\Pi_{i=1}^{r}\,\langle \mathbf{g}_i, \mathbf{g}_i \rangle$, where $d = \Sigma_{i=1}^{r}\,\Sigma_{\tau\in\alpha_i}\,m_i(\tau)$.

Once (Q4) is established, we can derive from it a result which extends Theorem 5.3 to the case where χ and χ' do not necessarily occur in $S_{h,\lambda}(B)$ and $S_{h',\lambda'}(B')$. The above array of properties (Q1–5) is redundant in the sense that (Q5) can be derived from (Q4) by the same argument as in the proof of Theorem 5.8. Also, (Q4) in the general case can be derived from its special case $r = t$ combined with (Q5).

It will be interesting to know to what extent the above conjecture can be generalized to the case with $m_\tau = 1$ for some $\tau \in J_E$. In fact, Theorem 4.3 includes the case of such m; furthermore, as shown in Theorem II.3.8, Conjecture 5.5 is true even for such m at least when B_1 and B_2 are totally indefinite.

For a Hecke character ψ of $E_{\mathbf{A}}^\times$ of finite order and χ as in Conjecture 5.10, we put

$$(5.23) \qquad L(s, \chi, \psi) = \sum_\mathfrak{a} \psi^*(\mathfrak{a})\chi(\mathfrak{a})N(\mathfrak{a})^{-s},$$

where $\mathfrak{a}$ ranges over all integral ideals of E and ψ^* is the ideal character attached to ψ. In [2, Theorem 4.3], we showed that the critical values of $L(s, \chi, \psi)$ can be described in terms of $2^{J(E)}$ quantities. This can be reformulated in a weaker form as

THEOREM 5.11. *Let χ and m be the same as in Conjecture 5.10, and let $m_0 = \mathrm{Max}\{m_\tau \mid \tau \in J_E\}$. Suppose $m_\tau > 2$ and $m_\tau \equiv m_0 \pmod 2$ for every $\tau \in J_E$. Then, for every $\gamma \subset J_E$, there exists a nonzero complex constant $U(\chi, \gamma)$, which is determined modulo $\overline{\mathbf{Q}}^\times$ and depends only on χ and γ, with the following properties.*

(i) $L(\mu + 1 - (m_0/2), \chi, \psi) \sim \pi^{n\mu}U(\chi, \gamma)$ for every integer μ such that

$$(m_0 - m_\tau)/2 < \mu < (m_0 + m_\tau)/2 \quad \text{for all} \quad \tau \in J_E$$

and for every Hecke character ψ of $E_{\mathbf{A}}^\times$ such that $\psi(x) = \mathrm{sgn}(x^{\gamma + \mu J(E)})$ for $x \in E_\infty^\times$, where $n = [E:\mathbf{Q}]$.

(ii) If $\gamma \cup \delta = J_E$, $\gamma \cap \delta = \emptyset$ and if $\mathbf{f}$ is a form as in Proposition 5.1, then $U(\chi, \gamma)U(\chi, \delta) \sim \pi^t\langle \mathbf{f}, \mathbf{f}\rangle$, where $t = n(1 - m_0) + |m|$.

(iii) $U(\chi, \gamma) \sim \pi^v p_K(\xi, \Phi)$ for every $\gamma \subset J_E$, if χ is of type (K, ξ), where Φ is a CM-type as in (5.14) and $v = (|m| - nm_0)/2$.

The last property is a weaker form of [2, Theorem 5.4] (which actually holds whenever $m_\tau > 1$ for all $\tau \in J_E$).

The second property implies the decomposition of $\pi^t Q(\chi, J_E)$ into two quantities. This suggests another conjecture about the decomposition of $Q(\chi, \alpha)$ into some quantities which are the periods of elements of $\mathcal{S}_{h,\lambda}(B)$. To state it in a simplified form, let us write $a \approx b$ for $a, b \in \mathbf{C}$ if $a/b \in \bigcup_{i\in\mathbf{Z}} \pi^i\overline{\mathbf{Q}}$.

CONJECTURE 5.12. *Given χ, α as in Conjecture 5.10 and given a subset γ of α, there exists a nonzero complex constant $P(\chi, \alpha, \gamma)$, which is determined modulo $\cup_{i \in \mathbf{Z}} \pi^i \bar{\mathbf{Q}}^\times$ and depends only on χ, α, and γ, with the following properties.*

(P1) $P(\chi, J_E, \gamma) \approx U(\chi, \gamma)$.

(P2) *The "periods" of the form* **g** *as in* (Q3) *are* $\bar{\mathbf{Q}}$-*rational linear combinations of* $P(\chi, \alpha, \gamma)$, *multiplied by suitable powers of* π, *for all* $\gamma \subset \alpha$.

(P3) $Q(\chi, \alpha) \approx P(\chi, \alpha, \gamma)P(\chi, \alpha, \delta)$ if $\gamma \cup \delta = \alpha$ and $\gamma \cap \delta = \emptyset$.

(P4) $P(\chi, \alpha \cup \beta, \gamma \cup \delta) \approx P(\chi, \alpha, \gamma)P(\chi, \beta, \delta)$ if $\alpha \cap \beta = \gamma \cap \delta = \emptyset$.

(P5) $P(\chi, \alpha, \gamma) \approx p_K(\xi, \eta)$ for all $\gamma \subset \alpha$ in the setting of (Q5).

Constants $P(\chi, \alpha, \gamma)$ are closely connected with Hilbert modular forms of half-integral weight, as shown in [6] in the elliptic modular case. We gave in [5] some evidence for the definability of $P(\chi, \alpha, \gamma)$ when $E = \mathbf{Q}$. Generalizing the methods of [5] and [6], and applying Theorem 4.4 to

$$\{x \in B \mid \mathrm{tr}(x) = 0\} \times \{x \in B' \mid \mathrm{tr}(x) = 0\},$$

we can at least prove a result of type (P4) similar to Theorem 5.4, although it may be too optimistic to expect (P4) in such a form. As to (P5), we note that $p_K(\xi, \eta)$ can be given also by the periods relative to the reflex of (K, ξ), by means of [3, Theorem 1.2].

It should also be mentioned that there are a few more types of generalizations of the results of [4] along the line of ideas developed in the present paper. One concerns the product of several unitary groups over CM-fields treated in Section I.10. We can even consider the product of such groups and several orthogonal groups of Section 3, and define automorphic forms of arithmetic type on it. Lastly, we can make conjectures which are generalized versions of those in Section II.7; they can even be proved if the basic field E is $\mathbf{Q}$. We leave the details of all these to the reader, since they can be carried out in the same manner as explained in the previous sections.

6. A supplement to [2]. So far we have been considering only the $\bar{\mathbf{Q}}$-rationality, whereas the rationality over specific number fields was studied in [2], [6], and other earlier papers of the author. Such improvements of the results in the previous sections in the most general case seem rather difficult. We insert here, as a relatively easy example, a refinement of Theorem 5.11, (iii), which may be viewed as a variation of [2, Theorem 5.4]. We first have to define the transform χ^σ of a primitive system χ of eigenvalues as in Conjecture 5.10 for $\sigma \in \mathrm{Gal}(\overline{\mathbf{Q}}/\mathbf{Q})$. Suppose χ occurs in $\mathcal{S}_{m,0}(M_2(E))$; let $m_0 = \mathrm{Max}\{m_\tau \mid \tau \in J_E\}$. We assume

$$(6.1) \qquad\qquad m_\tau \equiv m_0 \,(\mathrm{mod}\ 2) \quad \textit{for every} \quad \tau \in J_E.$$

Let an element σ of $\mathrm{Gal}(\overline{\mathbf{Q}}/\mathbf{Q})$ act on I_E on the right in a natural way. Then there is a primitive system of eigenvalues χ^σ occurring in $\mathcal{S}_{m\sigma,0}(M_2(E))$ such that

$$(6.2) \qquad\qquad N(\mathfrak{a})^{m_0/2}\chi^\sigma(\mathfrak{a}) = \{N(\mathfrak{a})^{m_0/2}\chi(\mathfrak{a})\}^\sigma$$

for every integral ideal $\mathfrak{a}$ of E. (This is shown in [2, Proposition 2.6]. Notice that k and $C(\mathfrak{a}, \mathbf{f})$ there correspond to m and $N(\mathfrak{a})^{(m_0/2)-1}\chi(\mathfrak{a})$ here.)

Now let K be a CM-field such that $[K:E] = 2$, and Ξ a Hecke ideal character of K defined modulo an integral ideal $\mathfrak{h}$ satisfying (5.13) with $\xi \in I_K$ and a CM-type Φ as in (5.14). Define the L-function of Ξ as usual by $L(s, \Xi) = \Sigma_\mathfrak{b}\, \Xi(\mathfrak{b})N(\mathfrak{b})^{-s}$, where $\mathfrak{b}$ ranges over all integral ideals of K prime to $\mathfrak{h}$. We put $\xi_0 = \mathrm{Max}\{\xi_\tau \mid \tau \in \Phi\}$ and assume that

$$(6.3) \qquad\qquad 0 < \xi_\tau \equiv \xi_0 \,(\mathrm{mod}\ 2) \quad \textit{for every} \quad \tau \in \Phi.$$

Suppose Ξ is primitive. Then we can find a primitive system χ of eigenvalues of type (K, ξ) occurring in $\mathcal{S}_{m,0}(M_2(E))$ such that $L(s, \Xi) = L(s + (1/2), \chi, 1)$ and $m = \mathrm{Res}_{K/E}(\xi + \Phi)$ (cf. [2, p. 677]). Now we have

THEOREM 6.1. *Define Hecke ideal characters ω of K and η of E, both defined modulo $\mathfrak{h}\mathfrak{h}^\rho$, by $\omega(\mathfrak{b}) = \Xi(\mathfrak{b})\Xi(\mathfrak{b}^\rho)^\rho$ and $\eta(\mathfrak{a}) = \Xi(\mathfrak{a}\mathfrak{r}_K)$, where $\mathfrak{r}_K$ denotes the maximal order of K. Put, for $\mu \in \mathbf{Z}$,*

$$B(\mu, \Xi) = i^{n\xi_0}\pi^{n(1-2\mu+\xi_0)}|D_K/D_E|^{1/2}L(\mu - (\xi_0/2), \Xi)^2/[\tau(\eta)L(1, \omega)],$$

where D_E and D_K are the discriminants of E and K, and $\tau(\eta)$ denotes the Gauss sum of η defined as in [2, (3.9)]. Then $B(\mu, \Xi)$ is algebraic and belongs to the field generated over $\mathbf{Q}$ by the values $N(\mathfrak{a})^{m_0/2}\chi(\mathfrak{a})$ for all $\mathfrak{a}$, if

$$(\xi_0 - \xi_\tau)/2 < \mu \le (\xi_0 + \xi_\tau)/2 \quad \textit{for all} \quad \tau \in \Phi.$$

Moreover, we have

$$(6.4) \qquad B(\mu, \Xi)^\sigma = B(\mu, \Xi^\sigma) \quad \text{for every} \quad \sigma \in \mathrm{Gal}(\bar{\mathbf{Q}}/\mathbf{Q})$$

and for every such integer μ,

where Ξ^σ *is defined by* $\Xi^\sigma(\mathfrak{b}) N(\mathfrak{b})^{\xi_0/2} = \{\Xi(\mathfrak{b}) N(\mathfrak{b})^{\xi_0/2}\}^\sigma$.

As an immediate consequence, we obtain

COROLLARY 6.2. *If* ξ_0 *is odd and* $L(1/2, \Xi) = 0$, *then* $L(1/2, \Xi^\sigma) = 0$ *for every* $\sigma \in \mathrm{Gal}(\bar{\mathbf{Q}}/\mathbf{Q})$.

To prove the theorem, we put $\beta(\mathfrak{a}) = \left(\frac{K/E}{\mathfrak{a}}\right)$ and

$$(6.5) \qquad A(\mu, \chi, \psi) = (2\pi i)^{-\mu n} \tau(\psi)^{-1} L(\mu + (1 - \xi_0)/2, \chi, \psi),$$

$$(6.6) \qquad E(\mathbf{f}) = (2\pi i)^{-\xi_0 n} \pi^{n + |\xi|} \tau(\beta\eta) \langle \mathbf{f}, \mathbf{f} \rangle,$$

where $\mathbf{f}$ is the normalized primitive eigenform in $\mathcal{S}_{m,0}(M_2(E))$ with eigenvalues $\chi(\mathfrak{a})$. Then [2, (4.16)] gives assertions similar to our theorem concerning $A(\mu, \chi, 1)A(\mu, \chi, \beta)/E(\mathbf{f})$, since β is totally ramified. Now we see easily that

$$(6.7) \qquad L(s, \chi, \beta) = L(s, \chi, 1) \prod_{\mathfrak{p} | \mathfrak{d}} [1 - \Xi(\mathfrak{p}) N(\mathfrak{p})^{(1/2) - s}],$$

where $\mathfrak{d}$ is the different of K relative to E. By [2, (5.13) and the last line on p. 677], we have

$$(6.8) \qquad \langle \mathbf{f}, \mathbf{f} \rangle = R \pi^{-2n - |\xi|} |D_E/D_K|^{1/2} L(1, \omega)$$

with a rational number R. (The character ω in the proof of [2, Theorem 5.5] has $\mathfrak{d}\mathfrak{h}\mathfrak{h}^\rho$ as its defining ideal, but this change of defining ideal multiplies $L(1, \omega)$ only by a nonzero rational number.) From these relations and our definition of $B(\mu, \Xi)$, we obtain

$$(6.9) \qquad B(\mu, \Xi) = R' \prod_{\mathfrak{p} | \mathfrak{d}} [1 - \Xi(\mathfrak{p}) N(\mathfrak{p})^{(\xi_0/2) - \mu}]^{-1}$$

$$\cdot [\tau(\beta)\tau(\eta\beta)/\tau(\eta)] A(\mu, \chi, 1) A(\mu, \bar{\chi}, \beta)/E(\mathbf{f})$$

with a rational number R', if $\mu \neq \xi_0/2$. Since Ξ^σ corresponds to χ^σ, we obtain (6.4) from [2, (4.16) and Lemma 4.12] provided $\mu \neq \xi_0/2$.

Suppose $\mu = \xi_0/2$. Since the factor $1 - \Xi(\mathfrak{p})$ may become 0, we have to take a different route in this case. Since $\xi_0(=2\mu)$ is even, (6.3) implies that $\xi_\tau \geq 2$ for all $\tau \in \Phi$, and hence $m_\alpha \geq 3$ for all $\alpha \in J_E$. Put $\lambda = 1 + \mu$. Then

$$(m_0 - m_\alpha)/2 < \mu < \lambda < (m_0 + m_\alpha)/2 \quad \text{for all} \quad \alpha \in J_E.$$

Therefore [2, Theorem 4.3] is applicable to both $A(\lambda, \chi, \psi)$ and $A(\mu, \chi, \psi)$. We have $\lambda + (1 - \xi_0)/2 = 3/2$, $L(3/2, \chi, 1) = L(1, \Xi) \neq 0$, and $L(3/2, \chi, \beta) = S(\Xi)L(1, \Xi)$ with $S(\Xi) = \Pi_{\mathfrak{p}|\mathfrak{b}}[1 - \Xi(\mathfrak{p})N(\mathfrak{p})^{-1}]$. Define symbols $u(r, \mathbf{f})$ for $r \in (\mathbf{Z}/2\mathbf{Z})^{J(E)}(=I_E/2I_E)$ as in [2, Theorem 4.3]. Let r be 0 or J_E according as μ is even or odd, and let $q = J_E - r$. By (6.5), (6.6), and (6.8), we have

$$B(\mu, \Xi) = R'' \frac{A(\mu, \chi, 1)^2}{u(r, \mathbf{f})^2} \cdot \frac{u(r, \mathbf{f})^2}{E(\mathbf{f})} \cdot \frac{\tau(\eta\beta)}{\tau(\eta)}$$

with a rational constant R''. Now [2, Theorem 4.3, (I)] implies that we can take $A(\lambda, \chi^\sigma, 1)$ and $A(\lambda, \chi^\sigma, \beta)$ as $u(q, \mathbf{f}^\sigma)$ and $u(r, \mathbf{f}^\sigma)$. With this choice of $u(q, \mathbf{f})$ and $u(r, \mathbf{f})$, we have

$$u(r, \mathbf{f}) = A(\lambda, \chi, \beta) = \tau(\beta)^{-1}S(\Xi)A(\lambda, \chi, 1) = \tau(\beta)^{-1}S(\Xi)u(q, \mathbf{f}),$$

and hence

$$(6.10) \qquad B(\mu, \Xi) = R''S(\Xi)\frac{A(\mu, \chi, 1)^2}{u(r, \mathbf{f})^2} \cdot \frac{u(q, \mathbf{f})u(r, \mathbf{f})}{E(\mathbf{f})} \cdot \frac{\tau(\eta\beta)}{\tau(\beta)\tau(\eta)}.$$

The action of σ on the right-hand side replaces $\mathbf{f}$, χ, and Ξ by $\mathbf{f}^\sigma$, χ^σ, and Ξ^σ by virtue of [2, Theorem 4.3 and Lemma 4.12]. Thus we obtain (6.4) for $\mu = \xi_0/2$.

Suppose $\chi^\sigma = \chi$ for $\sigma \in \mathrm{Gal}(\bar{\mathbf{Q}}/\mathbf{Q})$. Then we see easily that the right-hand sides of (6.9) and (6.10) are invariant under σ (though Ξ^σ may or may not be equal to Ξ). This proves the first assertion of our theorem.

PRINCETON UNIVERSITY

REFERENCES

[1] H. Jacquet and R. P. Langlands, Automorphic forms on $GL(2)$, *Lecture Notes in Mathematics* **114**, Springer, 1970.

[2] G. Shimura, The special values of the zeta functions associated with Hilbert modular forms, *Duke Math. J.* **45** (1978), 637-679.

[3] ———, The arithmetic of certain zeta functions and automorphic forms on orthogonal groups, *Ann. of Math.* **111** (1980), 313-375.

[4] ———, On certain zeta functions attached to two Hilbert modular forms I, II, *Ann. of Math.* **114** (1981), 127-164, 569-607.

[5] ———, The periods of certain automorphic forms of arithmetic type, *J. Fac. of Science, Univ. of Tokyo, Sec. IA, Math.* **28** (1982), 605-632.

[6] ———, The critical values of certain zeta functions associated with modular forms of half-integral weight, *J. Math. Soc. Japan*, **33** (1981), 649-672.

Corrections to [4]

Part I

Page 137, line 16: "$g(z)$" should be "$g(\alpha(z))$".

Page 144, line 4 from bottom: "$i \geq 0$" should be "$i > r$".

Page 145, line 2: "$c(\gamma)$" should be "$c(\gamma)\gamma$".

Part II

Page 579, line 13: The symbols involving Γ should be "$\Gamma_M z_0$".

Page 579, line 7 from bottom: "$\mathfrak{M}_{se}$ with some $s \in \mathbf{Z}$" should be "$\mathfrak{M}_s$ with some $s \in I_F$".

Page 579, line 6 from bottom: "$\mathfrak{M}_{(l-s)e}$" should be "$\mathfrak{M}_{le-s}$"; "$> s$" should be deleted.

Page 583, line 12 from bottom: "$\mathbf{Q}$" should be "$\bar{\mathbf{Q}}$".

Page 584, line 14: "$M_2(F)$" should be "$M_2(E)$".

Page 601, line 9 from bottom: Superscript "$-2s$" should be "$-2sE$".

Page 605, line 1: "(3.10)" should be "(3.13)".

On Eisenstein series

Duke Mathematical Journal, 50 (1983), 417-476

The groups on which our Eisenstein series will be defined are symplectic and unitary ones, which include as a special case a group G defined by

$$G = \left\{ \alpha \in \mathrm{SL}_{2m}(K) : {}^t\bar{\alpha}\eta\alpha = \eta \right\}, \qquad \eta = \begin{bmatrix} 0 & -1_m \\ 1_m & 0 \end{bmatrix}$$

with an imaginary quadratic field K and a positive integer m. Let H be the domain consisting of all complex square matrices z of size m such that $i({}^t\bar{z} - z)$ is positive definite. Writing a typical element α of G in the form $\alpha = \left(\begin{smallmatrix} a & b \\ c & d \end{smallmatrix}\right)$ with matrices a, b, c, d of size m, we define the action of α on H by $\alpha(z) = (az + b) \cdot (cz + d)^{-1}$ for $z \in H$, and a factor of automorphy j by $j(\alpha, z) = \det(cz + d)$. Let P be the subgroup of G consisting of the elements α for which $c = 0$. Then we consider Eisenstein series of the following two types:

$$E(z, s; k, \Gamma) = \sum_{\alpha \in (P \cap \Gamma)\backslash\Gamma} j(\alpha, z)^{-k}|j(\alpha, z)|^{-s},$$

$$E(z, s; k, \psi, \mathbf{b}) = \sum_{\alpha \in (P \cap \Gamma_{\mathbf{b}})\backslash\Gamma_{\mathbf{b}}} \psi(\det(d))j(\alpha, z)^{-k}|j(\alpha, z)|^{-s}.$$

Here $z \in H$, $s \in \mathbf{C}$, $k \in \mathbf{Z}$, Γ is a congruence subgroup of G, $\mathbf{b}$ is an ideal of $\mathbf{Z}$, ψ is a Dirichlet character modulo $\mathbf{b}$, and $\Gamma_{\mathbf{b}}$ is the congruence subgroup of G consisting of all α whose entries are all integral and such that $c \equiv 0 \pmod{\mathbf{b}}$. As will be shown in the text, $\det(d) \in \mathbf{Q}$ for every $\alpha \in G$. We naturally assume that $\psi(-1) = (-1)^k$. These series are convergent for $\mathrm{Re}(s) + k > 2m$. Moreover, by virtue of the result of Langlands [15], they can be continued as meromorphic functions in s to the whole complex plane. To state our problems, let us denote by $E(z, s)$ any one of the series. Then we ask the following questions.

(A) *Is $E(z, s)$ holomorphic in s at $s = 0$?*

(B) *If so, is $E(z, 0)$ holomorphic in z?*

(C) *If this is so too, does $E(z, 0)$ have cyclotomic Fourier coefficients?*

If $m = 1$, group G in this specialized setting coincides with $\mathrm{SL}_2(\mathbf{Q})$. The above questions in this case were answered by Hecke in the well known paper [9]. It should also be noted that if $k > 2m$, E is convergent at $s = 0$, and hence the answers to (A) and (B) are trivially true.

Received October 21, 1982. Supported by NSF Grant MCS-8100744.

Now our main results specialized to the group G of the present type are as follows:

(I) *The answer to* (A) *is affirmative if* $k \geqslant m$.

(II) *The answers to* (B) *and* (C) *are affirmative if* $k > m + 1$ *or* $k = m$.

In the case $k = m + 1$, the answers are more complicated and in a sense more interesting:

(III) *If* $k = m + 1$, *the answers to* (B) *and* (C) *are affirmative for* $E(z, s; k, \psi, \mathbf{b})$ *except when* $\psi = \theta^{m-1}$, *where* θ *is the quadratic Dirichlet character associated with* K.

(IV) *Define a differential operator* Δ *on* H *by* $\Delta = \det(\partial / \partial z_{jk})$ *with respect to the variable matrix* (z_{jk}) *on* H. *Still with* $k = m + 1$, *there exist, for each* $E(z, s)$, *two holomorphic functions* p_1 *and* p_2 *on* H *such that* $\Delta p_2 = 0$ *and*

$$E(z, 0) = \Delta \big\{ p_1(z) + p_2(z) \log \big[\det(z - {}^t\bar{z}) \big] \big\}.$$

Moreover, $\pi^m p_2$ *is an automorphic form of weight* $m - 1$ *with cyclotomic Fourier coefficients, and* Δp_1 *has a Fourier expansion with cyclotomic coefficients.*

There is one more type of result worth mentioning which concerns a similar but somewhat different problem:

(V) *If* $k = m - 1$, $E(z, s)$ *has at most a simple pole at* $s = 2$, *and its residue is* $\pi^{-m} \det(z - {}^t\bar{z})^{-1}$ *times a holomorphic automorphic form* f *of weight* $m - 1$ *with cyclotomic Fourier coefficients such that* $\Delta f = 0$.

Actually we consider the series of the same types on the unitary group over an arbitrary CM-field K as well as on the symplectic group over an arbitrary totally real number field F. In the unitary case, we give complete answers to the above questions and also prove the generalization of (V) (Theorems 7.1, 7.2, and 7.3). The nonholomorphic series of (IV) does not occur if $[K : \mathbf{Q}] > 2$. The Eisenstein series in the symplectic case can be defined in a similar and obvious way with $\mathrm{Sp}(m, F)$ and (the product of copies of) the Siegel upper half-space in place of G and H. Then we shall show that the answers to (A, B, C) are all affirmative if $k > m > 1$. The convergence at $s = 0$ in this case holds if $k > m + 1$. Thus our results in the symplectic case are not as good as those in the unitary case. The reason for this will be explained later.

It should be noted that all previous investigations on the question of type (C) concern the convergent case, except for those on $\mathrm{SL}_2(F)$ and those by the author and Indik on the orthogonal and unitary groups mentioned below. Even our result on $E(z, s; k, \Gamma)$ on $\mathrm{Sp}(m, F)$ in the convergent case does not seem to have previously been obtained for an arbitrary congruence subgroup Γ. (For successful and unsuccessful previous attempts, see Remark 7.5.)

Let us now explain our methods by giving a summary of contents. In Sections 1 and 2, we introduce the adele group $G_\mathbf{A}$, define series of type $E(z, s)$ and a certain series as a function on $G_\mathbf{A}$, and discuss their mutual relationship. We then choose a special type of series denoted by E^* and show that other series can be expressed as linear combinations of the transforms of several series of type E^*. In Section 3, we consider the Fourier expansion of E^* and express each Fourier coefficient as an integral on a certain adele space in a standard way. The integral can be decomposed into the archimedean part and the nonarchimedean part. The former turns out to be the product of confluent hypergeometric functions on tube domains which we studied in our paper [26] with this application in view. The latter is an Euler product whose factors are infinite series, which Siegel considered in the symplectic case, and which most researchers call Whittaker integrals, a rather inappropriate designation. It is one of our key ideas to reduce the problems to the Fourier expansion of E^* which is far simpler than those of other series.

In the next three sections, we determine the Euler **p**-factor explicitly as a rational function in $N(\mathbf{p})^{-s}$ for almost all primes **p** of the basic field. The easier part of our main theorems will be proved in Section 8 by means of a careful analysis of each Fourier coefficient at $s = 0$ or at $s = 2$. Here our results of [26] play crucial roles. Since there is no complete information on the bad Euler factors, we must at least gain some control of their poles, which can actually be found in the unitary case (Proposition 8.2). In the symplectic case, however, our result in this respect is incomplete, and only a certain conjecture on the poles of bad Euler factors can be made. This is why our theorem for $\mathrm{Sp}(m, F)$ is restricted to the case $k > m$. In Section 10, we show that the results for smaller weights in the symplectic case analogous to (I–V) follow from the conjecture.

The proof of (IV) is more difficult. Its analytic part and the statement concerning $\pi^m p_2$ will be proved in Section 9, where we shall also introduce a certain family $\mathscr{L}$ of nonholomorphic automorphic forms including $E(z, 0)$ of weight $m + 1$. That the Fourier coefficients of Δp_1 are cyclotomic will be proved in Section 11 as a consequence of a theorem concerning $\mathscr{L}$, which may be of independent interest (Theorem 11.1).

Although we treat in this paper only two types of groups, we believe that our methods are general enough to be applicable to other groups acting on tube domains. In fact, the case of $\mathrm{SO}(m, 2)$ has been studied by Indik in his thesis [10] by the same method. The papers [23], [24] may also be mentioned as investigations of the same topic including the cocompact case by different methods in a different framework.

Notation. For an associative ring R with identity element, we denote by $R^{\times}$ the group of all invertible elements of R and by R_n^m the module of all $m \times n$ matrices with entries in R. If X is a matrix, ^tX, $\det(X)$, and $\mathrm{tr}(X)$ stand for its transpose, determinant, and trace. We put $M_m(R) = R_m^m$, $\mathrm{GL}_m(R) = M_m(R)^{\times}$, and $\mathrm{SL}_m(R) = \{X \in M_m(R) : \det(X) = 1\}$ (when R is commutative). The identity and zero elements of $M_m(R)$ are denoted by 1_m and 0_m, respectively (when m needs to be stressed). For complex hermitian matrices X and Y, we write $X > Y$ to indicate that $X - Y$ is positive definite. If $X > 0$, we denote by $X^{1/2}$ its positive definite square root. If $X_1, \ldots, X_r$ are square matrices, $\mathrm{diag}[X_1, \ldots, X_r]$ denotes the matrix with $X_1, \ldots, X_r$ in the diagonal blocks and 0 in all other blocks. The adelization of an algebraic group Y over $\mathbf{Q}$ is denoted by $Y_{\mathbf{A}}$ or $Y(\mathbf{A})$. The disjoint union of sets $Z_1, \ldots, Z_s$ is denoted by $\coprod_{i=1}^{s} Z_i$.

1. Preliminaries on the adele groups. Throughout the paper, we denote by F a totally real algebraic number field of finite degree, and by K a totally imaginary quadratic extension of F. The nontrivial automorphism of K over F is denoted by ρ. We consider two types of algebraic groups G, which we refer to as Case SP and Case SU and which are defined by

$$(1.0) \qquad G = \begin{cases} \{\alpha \in \mathrm{GL}_{2m}(F) : {}^t\alpha\eta\alpha = \eta\} & \text{(Case SP)}, \\ \{\alpha \in \mathrm{SL}_{2m}(K) : \alpha^*\eta\alpha = \eta\} & \text{(Case SU)}, \end{cases}$$

where m is a fixed positive integer and $\alpha^* = {}^t\alpha^\rho$ (even for nonsquare α), and

$$(1.1) \qquad \eta = \begin{pmatrix} 0 & -1_m \\ 1_m & 0 \end{pmatrix}.$$

To make our exposition uniform, we shall often use α^* instead of ${}^t\alpha$ for $\alpha \in F_n^m$.

We let ∞ and $\mathfrak{f}$ denote the sets of archimedean primes and nonarchimedean primes of F, respectively. An element of ∞ determines an embedding of F into $\mathbf{R}$. We fix once for all $[F:\mathbf{Q}]$ embeddings of K into $\mathbf{C}$ which are extensions of the elements of ∞, and let ∞ denote also this (particular) set of embeddings of K into $\mathbf{C}$, which may be viewed as the set of archimedean primes of K. For $v \in \infty \cup \mathfrak{f}$, we denote by F_v the v-completion of F and put $K_v = K \otimes_F F_v$. If v splits in K, K_v is isomorphic to $F_v \times F_v$. For $v \in \infty$, F_v and K_v can be identified with $\mathbf{R}$ and $\mathbf{C}$ by the embeddings just mentioned. For each $v \in \infty \cup \mathfrak{f}$, we can define G_v to be the group of (1.0) with F_v and K_v in place of F and K, respectively, and with ρ naturally extended to K_v. We consider G an algebraic group over $\mathbf{Q}$ and define its adelization $G_\mathbf{A} = G(\mathbf{A})$; we put $G_\infty = \prod_{v \in \infty} G_v$ and $G_\mathfrak{f} = G_\mathbf{A} \cap \prod_{v \in \mathfrak{f}} G_v$. We use the letter G to denote the group of all its $\mathbf{Q}$-rational points embedded in $G_\mathbf{A}$. Thus G stands for the standard symbol $G_\mathbf{Q} = G(\mathbf{Q})$ that will not be used in this paper. This convention applies also to the groups S, P, Q, etc. defined below. If $x = \begin{pmatrix} a & b \\ c & d \end{pmatrix} \in G_\mathbf{A}$ with $a,b,c,d \in M_m(F_\mathbf{A})$ or $M_m(K_\mathbf{A})$, we often write $a = a_x$, $b = b_x$, $c = c_x$, $d = d_x$; we also define $x_\mathfrak{f}$ and x_∞ by $x = x_\mathfrak{f}x_\infty$, $x_\mathfrak{f} \in G_\mathfrak{f}$, $x_\infty \in G_\infty$.

Define a vector space S over F by

$$(1.2a) \qquad S = \begin{cases} \{\alpha \in M_m(F) : {}^t\alpha = \alpha\} & \text{(Case SP)}, \\ \{\alpha \in M_m(K) : \alpha^* = \alpha\} & \text{(Case SU)}, \end{cases}$$

and put $S_v = S \otimes_F F_v$. If $v \in \infty$, S_v is the space of real symmetric matrices (resp. complex hermitian matrices) in Case SP (resp. Case SU). Define a symmetric domain H by

$$(1.2b) \qquad H = \begin{cases} \{z \in M_m(\mathbf{C}) : {}^tz = z, \ \mathrm{Im}(z) > 0\} & \text{(Case SP)}, \\ \{z \in M_m(\mathbf{C}) : i(z^* - z) > 0\} & \text{(Case SU)}, \end{cases}$$

where naturally $z^* = {}^t\bar{z}$. This can also be given as

$$H = \{x + iy \in S_v \otimes_{\mathbf{R}} \mathbf{C} : x, y \in S_v, y > 0\}.$$

For $\alpha \in G_v$ and $z \in H$ we can define the image $\alpha(z)$ of z under α as well as holomorphic factors of automorphy $\kappa(\alpha, z)$ and $\mu(\alpha, z)$ by the rule

$$(1.3) \qquad \alpha\begin{pmatrix} z^* & z \\ 1_m & 1_m \end{pmatrix} = \begin{pmatrix} \alpha(z)^* & \alpha(z) \\ 1_m & 1_m \end{pmatrix}\begin{bmatrix} \overline{\kappa(\alpha, z)} & 0 \\ 0 & \mu(\alpha, z) \end{bmatrix}.$$

(See [21] for example.) We have $\kappa(\alpha, z) = \bar{c}_\alpha \cdot {}^t z + \bar{d}_\alpha$, $\mu(\alpha, z) = c_\alpha z + d_\alpha$, and $\alpha(z) = (a_\alpha z + b_\alpha)(c_\alpha z + d_\alpha)^{-1}$.

LEMMA 1.1.   *If $\begin{pmatrix} a & b \\ c & d \end{pmatrix} \in G$ in Case SU, then $\det(a), \det(b), \det(c), \det(d)$ are all contained in F.*

Proof.   It is sufficient to prove that $\det(d) \in F$. Suppose $\det(d) \neq 0$. Since

$$\begin{pmatrix} 1 & -bd^{-1} \\ 0 & 1 \end{pmatrix}\begin{pmatrix} a & b \\ c & d \end{pmatrix} = \begin{pmatrix} a - bd^{-1}c & 0 \\ c & d \end{pmatrix}$$

and $dc^* = cd^*$, we have $\det(d)^{-1} = \det(a - bd^{-1}c) = \det(a - bc^*d^{-*})$. On the other hand, $ad^* - bc^* = 1$, so that $\det(d^*)^{-1} = \det(a - bc^*d^{-*})$. Hence $\det(d) = \det(d^*) = \det(d)^\rho$ as desired. In [21, (1.19)], we showed that $\det(\kappa(\alpha, z)) = \det(\mu(\alpha, z))$ (since $\det(\alpha) = 1$), which gives another proof.

We consider a copy of H for each $v \in \infty$, and denote by H^∞ their product, and define the action of G_∞ on H^∞ componentwise. (Thus H^∞ is the set of $(z_v)_{v \in \infty}$ with $z_v \in H$.) Furthermore, for $x \in G_{\mathbf{A}}$, we define the action of x on H^∞ to be the same as that of x_∞. We denote by $\mathbf{g}$ and $\mathbf{h}$ the maximal orders of F and K, respectively, and by $\mathbf{g}_v$ and $\mathbf{h}_v$ their v-closures in F_v and K_v if $v \in \mathbf{f}$. Then a maximal compact subgroup C_v of G_v can be defined by

$$(1.4) \qquad C_v = \begin{cases} \{\alpha \in G_v : \alpha(\mathbf{i}) = \mathbf{i}\} & (v \in \infty), \\ G_v \cap \mathrm{GL}_{2m}(\mathbf{g}_v) & (v \in \mathbf{f}, \text{ Case SP}), \\ G_v \cap \mathrm{GL}_{2m}(\mathbf{h}_v) & (v \in \mathbf{f}, \text{ Case SU}), \end{cases}$$

where $\mathbf{i} = (i1_m, \ldots, i1_m) \in H^\infty$. Then we put $C = \prod_{v \in \mathbf{f} \cup \infty} C_v$.

We define (algebraic) subgroups P, Q, R of G by

$$(1.5) \qquad \begin{aligned} P &= \{\alpha \in G : c_\alpha = 0\}, \qquad Q = \{\alpha \in P : b_\alpha = 0\}, \\ R &= \{\alpha \in P : a_\alpha = d_\alpha = 1\}. \end{aligned}$$

Then $P = QR = RQ$. Observe that $R = \{\begin{pmatrix} 1 & b \\ 0 & 1 \end{pmatrix} : b \in S\}$. It is well known that

$$(1.6) \qquad G_{\mathbf{A}} = P_{\mathbf{A}}C.$$

LEMMA 1.2. $P\eta P = \{x \in G : \det(c_x) \neq 0\}$.

Proof. This follows immediately from a simple relation

$$\begin{bmatrix} a & b \\ c & d \end{bmatrix} = \begin{bmatrix} 1 & ac^{-1} \\ 0 & 1 \end{bmatrix}\begin{bmatrix} 0 & -1 \\ 1 & 0 \end{bmatrix}\begin{bmatrix} c & d \\ 0 & c^{-*} \end{bmatrix}.$$

To study the "cusps" of G, we consider the idele group $F_{\mathbf{A}}^{\times}$ of F and its subgroup U defined by

$$(1.7) \qquad U = F_{\infty}^{\times}\prod_{v\in\mathbf{f}}\mathbf{g}_v^{\times}, \qquad F_{\infty}^{\times} = \prod_{v\in\infty}F_v^{\times}.$$

Then $[F_{\mathbf{A}}^{\times} : F^{\times}U]$ is the class number of F, which we denote by h (only in this and the next sections). In view of Lemma 1.1, we can define a homomorphism $\lambda : P_{\mathbf{A}} \to F_{\mathbf{A}}^{\times}$ by $\lambda(y) = \det(d_y)$ for $y \in P_{\mathbf{A}}$.

LEMMA 1.3. *Let $t_1, \ldots, t_h$ be elements of $P_{\mathbf{A}}$ such that $F_{\mathbf{A}}^{\times} = \coprod_{i=1}^{h}F^{\times}U\lambda(t_i)$, and let Y be an open subgroup of $P_{\mathbf{A}}$ containing P_{∞} such that $U = \lambda(Y)$. Then $P_{\mathbf{A}} = \coprod_{i=1}^{h}Pt_iY$.*

Proof. Put $t_i = q_ir_i$ with $q_i \in Q_{\mathbf{A}}$ and $r_i \in R_{\mathbf{A}}$. Let $Q^0 = \{y \in Q : \lambda(y) = 1\}$. Then Q^0 is isomorphic to $\mathrm{SL}_m(F)$ or $\mathrm{SL}_m(K)$. By the strong approximation theorem, we have $Q_{\mathbf{A}}^0 = Q^0(Q_{\mathbf{A}}^0 \cap Y)$. Since $\lambda(Q) = F^{\times}$ and $\lambda(Q_{\mathbf{A}}) = F_{\mathbf{A}}^{\times} = \coprod_{i=1}^{h}F^{\times}\lambda(q_i)U$, we have $Q_{\mathbf{A}} \subset \bigcup_{i=1}^{h}Qq_iYR_{\mathbf{A}}$; hence $P_{\mathbf{A}} = Q_{\mathbf{A}}R_{\mathbf{A}} \subset \bigcup_i Qq_iYR_{\mathbf{A}} = \bigcup_i Qq_ir_iR_{\mathbf{A}}Y = \bigcup_i Qt_iR_{\mathbf{A}}Y = \bigcup_i QR_{\mathbf{A}}t_iY$. Observe that $R_{\mathbf{A}} = RW$ with any open subgroup W of $R_{\mathbf{A}}$. Take $W = \bigcap_{i=1}^{h}(t_iYt_i^{-1}) \cap R_{\mathbf{A}}$. Then $P_{\mathbf{A}} = \bigcup_i QR_{\mathbf{A}}t_iY = \bigcup_i QRWt_iY = \bigcup_i Pt_iY$. Since $\lambda(Pt_iY) = F^{\times}\lambda(t_i)U$, the union must be a disjoint one.

For every $s \in F_{\mathbf{A}}^{\times}$, let $s\mathbf{g}$ denote the fractional ideal of F such that $(s\mathbf{g})_v = s_v\mathbf{g}_v$ for all $v \in \mathbf{f}$. In view of (1.6), we assign an ideal $\mathrm{il}(x)$ to each $x \in G_{\mathbf{A}}$ by

$$(1.8) \qquad \mathrm{il}(yw) = \lambda(y)\mathbf{g} \qquad \text{for } y \in P_{\mathbf{A}} \text{ and } w \in C.$$

This is well defined. For $x \in G_{\mathbf{A}}$ and $v \in \mathbf{f}$, we see easily that $(c_x)_vM_m(\mathbf{g}_v) + (d_x)_vM_m(\mathbf{g}_v)$ (resp. $(c_x)_vM_m(\mathbf{h}_v) + (d_x)_vM_m(\mathbf{h}_v)$) is a right ideal of $M_m(\mathbf{g}_v)$ (resp. $M_m(\mathbf{h}_v)$) whose (reduced) norm is $\mathrm{il}(x)_v$ (resp. $\mathrm{il}(x)_v\mathbf{h}_v$).

LEMMA 1.4. *Assigning the ideal class of $\mathrm{il}(x)$ to x, we obtain a one-to-one correspondence between $P\backslash G_{\mathbf{A}}/CG_{\infty}$ and the ideal classes of F. Moreover, if $t_1, \ldots, t_h$ are the same as in Lemma 1.3, we have $G_{\mathbf{A}} = \coprod_{i=1}^{h}Pt_iCG_{\infty}$.*

Proof. Let $Y = P_{\mathbf{A}} \cap CG_{\infty}$. Then $\lambda(Y) = U$, so that $P_{\mathbf{A}} = \coprod_{i=1}^{h}Pt_iY$ by Lemma 1.3. Therefore $G_{\mathbf{A}} = P_{\mathbf{A}}CG_{\infty} = \bigcup_{i=1}^{h}Pt_iCG_{\infty}$. Observe that if $x \in Pt_iCG_{\infty}$, $\mathrm{il}(x)$ and $\mathrm{il}(t_i)$ belong to the same ideal class. Hence we obtain our lemma.

In order to deal with congruence subgroups of G, we fix an integral ideal $\mathbf{b}$ of F, and define an open subgroup D of C of "level $\mathbf{b}$" by $D = \prod_v D_v$ with

$$(1.9) \qquad D_v = \begin{cases} C_v & (v \in \infty), \\ \{x \in C_v : c_x \in \mathbf{b}M_m(\mathbf{g}_v)\} & (v \in \mathbf{f}, \text{ Case SP}), \\ \{x \in C_v : c_x \in \mathbf{b}M_m(\mathbf{h}_v)\} & (v \in \mathbf{f}, \text{ Case SU}). \end{cases}$$

LEMMA 1.5. *Every fractional ideal of F can be given as* il(α) *with* $\alpha \in G \cap P_{\mathbf{A}}D$.

Proof. Given an ideal $\mathbf{a}$, we can find an element t of $P_{\mathbf{A}}$ so that il$(t) = \mathbf{a}$. By the strong approximation theorem, we have $t \in \alpha D G_{\infty}$ with $\alpha \in G$. Then il$(\alpha) =$ il(t) and $\alpha \in G \cap P_{\mathbf{A}}D$.

LEMMA 1.6. *There exist h elements $\beta_1, \ldots, \beta_h$ of G such that $P_{\mathbf{A}}D$* $= \coprod_{i=1}^{h} P\beta_i D G_{\infty}$. *Moreover, if we put $\Gamma_0(\mathbf{b}) = G \cap D G_{\infty}$, then $G \cap P_{\mathbf{A}}D$* $= \coprod_{i=1}^{h} P\beta_i \Gamma_0(\mathbf{b})$ *with any choice of such β_i.*

Proof. By Lemma 1.5, we can find $\beta_1, \ldots, \beta_h$ in $G \cap P_{\mathbf{A}}D$ such that il$(\beta_1), \ldots,$ il(β_h) represent the ideal classes of F. Let $Y = P_{\mathbf{A}} \cap D G_{\infty}$. Then $\lambda(Y) = U$, so that $P_{\mathbf{A}} = \coprod_{i=1}^{h} P t_i Y$ with t_i as in Lemma 1.3. Then $P_{\mathbf{A}}D$ $= \bigcup_{i=1}^{h} P t_i D G_{\infty}$. This is a disjoint union by virtue of Lemma 1.4. Since each coset $P x D G_{\infty}$ is determined by the ideal class of il(x), we see that the cosets $P\beta_i D G_{\infty}$ coincide with the $P t_i D G_{\infty}$ as a whole. This proves the first assertion. Then $G \cap P_{\mathbf{A}}D = \coprod_{i=1}^{h} (G \cap P\beta_i D G_{\infty}) = \coprod_{i=1}^{h} P\beta_i \Gamma_0(\mathbf{b})$ as desired.

2. Definition of Eisenstein series.

To consider automorphic forms on H^{∞} and on $G_{\mathbf{A}}$, we put, for $x \in G_{\mathbf{A}}$, $z = (z_v)_{v \in \infty} \in H^{\infty}$, $k \in \mathbf{Z}$, and $s \in \mathbf{C}$,

$$(2.0a) \qquad j(x,z) = \prod_{v \in \infty} \det(\mu(x_v, z_v)),$$

$$(2.0b) \qquad J(x,z) = J_{k,s}(x,z) = j(x,z)^k |j(x,z)|^s.$$

If $v \in \infty$ and $x_v \in C_v$, we see that $\mu(x_v, i1_m)$ is a unitary matrix, and hence

$$(2.0c) \qquad |j(x,\mathbf{i})| = 1 \quad \text{if} \quad x \in C.$$

Our Eisenstein series will be a special case of a function f_0 on $G_{\mathbf{A}}$ such that

$$(2.1) \qquad f_0(\alpha x w) = f_0(x) J(w,\mathbf{i})^{-1} \quad \text{for} \quad \alpha \in G \quad \text{and} \quad w \in C'$$

with an open subgroup C' of C. Given such f_0 and C', let $\Gamma' = G \cap C' G_{\infty}$, and define a function f on H^{∞} by

$$(2.2) \qquad f(x(\mathbf{i})) = f_0(x) J(x,\mathbf{i}) \quad \text{for} \quad x \in C' G_{\infty}.$$

This is well defined and satisfies

$$(2.3) \qquad f(\gamma z) = f(z) J(\gamma, z) \quad \text{for} \quad \gamma \in \Gamma' \quad \text{and} \quad z \in H^{\infty}.$$

Conversely, given a function f satisfying (2.3), we can define a function f_0 on $G_{\mathbf{A}}$ satisfying (2.1) and (2.2) by

$$(2.4) \qquad f_0(\alpha x) = f(x(\mathbf{i})) J(x,\mathbf{i})^{-1} \quad \text{for} \quad \alpha \in G \quad \text{and} \quad x \in C' G_{\infty}.$$

(Notice that $G_{\mathbf{A}} = G C' G_{\infty}$ by virtue of the strong approximation theorem.) For a

C-valued function ξ on H^∞ and $\alpha \in G$, we define a function $\xi\|_{k,s}\alpha$ (written also simply $\xi\|\alpha$) on H^∞ by

$$(2.5) \qquad (\xi\|_{k,s}\alpha)(z) = J_{k,s}(\alpha,z)^{-1}\xi(\alpha(z)) \qquad (z \in H^\infty).$$

We can easily verify that

$$(2.6) \qquad \begin{array}{l} \textit{If } g = f\|\alpha^{-1} \textit{ with } \alpha \in G \textit{ and with } f \textit{ satisfying } (2.3), \textit{ then} \\ g_0(x) = f_0(x\alpha_f), \textit{ and vice versa.} \end{array}$$

For an integral ideal $\mathbf{b}$ of F, we define subgroups $\Gamma_0(\mathbf{b})$, $\Gamma_u(\mathbf{b})$, and $\Gamma(\mathbf{b})$ of G by

$$(2.7a) \quad \Gamma_0(\mathbf{b}) = \left\{ \alpha \in G \cap CG_\infty : c_\alpha \equiv 0 \ (\mathrm{mod}\, \mathbf{b}) \right\} \ (= G \cap DG_\infty \text{ of Lemma 1.6}),$$

$$(2.7b) \quad \Gamma_u(\mathbf{b}) = \left\{ \alpha \in \Gamma_0(\mathbf{b}) : \det(d_\alpha) \equiv e \ (\mathrm{mod}\, \mathbf{b}) \text{ for some } e \in \mathbf{g}^\times \right\},$$

$$(2.7c) \quad \Gamma(\mathbf{b}) = \left\{ \alpha \in \Gamma_0(\mathbf{b}) : \alpha \equiv 1_{2m} \ (\mathrm{mod}\, \mathbf{b}) \right\}.$$

By a *congruence subgroup of G*, we mean a subgroup Γ of G which contains $\Gamma(\mathbf{b})$ for some $\mathbf{b}$ as a subgroup of finite index. If $\alpha \in P \cap \Gamma$ with such a Γ, then $\lambda(\alpha) \in \mathbf{g}^\times$, so that $j(\alpha,z) = N_{F/\mathbf{Q}}(\lambda(\alpha)) = \pm 1$.

Now our Eisenstein series (as a function on H^∞) is defined by

$$(2.8) \qquad E(z,s;k,\Gamma) = \sum_{\alpha \in (P \cap \Gamma)\backslash \Gamma} J(\alpha,z)^{-1}$$

with a congruence subgroup Γ of G, under the condition

$$(2.9) \qquad j(\alpha,z)^k = 1 \qquad \text{for all} \quad \alpha \in P \cap \Gamma.$$

For the moment, let us assume the convergence of the series and treat it only formally. It can easily be seen that this is a function of type (2.3). Observe that $\Gamma(\mathbf{b})$ satisfies (2.9) if and only if

$$(2.10) \qquad N_{F/\mathbf{Q}}(e)^k = 1 \textit{ for every } e \in \mathbf{g}^\times \textit{ such that } e \equiv 1 \ (\mathrm{mod}\, \mathbf{b}).$$

Assuming (2.10), let $\alpha \in \Gamma_u(\mathbf{b})$ and $\det(d_\alpha) \equiv e$ $(\mathrm{mod}\, \mathbf{b})$ with $e \in \mathbf{g}^\times$. Put $\xi(\alpha) = N_{F/\mathbf{Q}}(e)^k$. This does not depend on the choice of e.

PROPOSITION 2.1.
   (i) $\Gamma_u(\mathbf{b}) = (P \cap \Gamma_u(\mathbf{b}))\Gamma(\mathbf{b})$.
   (ii) $E(z,s;k,\Gamma(\mathbf{b})) = \sum_{\alpha \in (\Gamma_u(\mathbf{b}) \cap P)\backslash \Gamma_u(\mathbf{b})} \xi(\alpha)J(\alpha,z)^{-1}$ *if (2.10) is satisfied.*
   (iii) *If* $[\Gamma : \Gamma'] < \infty$ *and* Γ *satisfies* (2.9), *then*

$$[P \cap \Gamma : P \cap \Gamma']E(z,s;k,\Gamma) = \sum_{\tau \in \Gamma'\backslash\Gamma} E(z,s;k,\Gamma')\|\tau.$$

Proof. Let $\alpha \in \Gamma_u(\mathbf{b})$. Then $\det(d_\alpha) \equiv e$ $(\mathrm{mod}\, \mathbf{b})$ with $e \in \mathbf{g}^\times$. By the strong approximation theorem on SL_m, there exists an element q of $\mathrm{GL}_m(\mathbf{g})$ or of

$\mathrm{GL}_m(\mathbf{h})$ such that $q \equiv d_\alpha \ (\mathrm{mod} \ \mathbf{b})$ and $\det(q) = e$. Put

$$\beta = \begin{bmatrix} q^* & -d_\alpha^* b_\alpha q^{-1} \\ 0 & q^{-1} \end{bmatrix}.$$

Then we see easily that $\beta \in P \cap \Gamma_u(\mathbf{b})$ and $\beta\alpha \in \Gamma(\mathbf{b})$. This proves (i). Then (ii) follows immediately from (i). To prove (iii), take $T \subset \Gamma$ so that $\Gamma = \coprod_{\tau \in T} \Gamma'\tau$. Then

$$[P \cap \Gamma : P \cap \Gamma'] E(z,s;k,\Gamma) = \sum_{\alpha \in (P \cap \Gamma')\backslash \Gamma} J(\alpha,z)^{-1}$$

$$= \sum_{\tau \in T} \sum_{\beta \in (P \cap \Gamma')\backslash \Gamma'} J(\beta\tau,z)^{-1}$$

$$= \sum_{\tau \in T} E(z,s;k,\Gamma')\|\tau, \quad \text{Q.E.D.}$$

To consider Eisenstein series on $G_{\mathbf{A}}$, define a function ϵ on $G_{\mathbf{A}}$ by

$$(2.11) \quad \epsilon(yw) = |\lambda(y)| = \prod_{v \in \mathbf{f} \cup \infty} |\lambda(y)_v|_v \quad \text{for} \quad y \in P_{\mathbf{A}} \quad \text{and} \quad w \in C,$$

where $| \ |_v$ denotes the normalized valuation of F_v. It can easily be seen that this is well defined, and

$$(2.12) \quad \epsilon(x) = \epsilon(x_{\mathbf{f}})|j(x,\mathbf{i})|, \quad \epsilon(x_{\mathbf{f}}) = N(\mathrm{il}(x))^{-1} \quad (x \in G_{\mathbf{A}}).$$

Now we take a *Hecke character* ψ of F, by which we mean a character ψ of $F_{\mathbf{A}}^\times$ such that $\psi(F^\times) = 1$ and $|\psi| = 1$. With a fixed integral ideal $\mathbf{b}$ and $k \in \mathbf{Z}$ as before, we assume:

$(2.13a)$ *the finite part of the conductor of ψ divides* $\mathbf{b}$;

$$(2.13b) \quad \psi(a) = \prod_{v \in \infty} \mathrm{sgn}(a_v)^k \quad \text{for} \quad a \in F_\infty^\times.$$

Then ψ is of finite order. Such a ψ exists if and only if (2.10) is satisfied. We denote by ψ^* the ideal-character attached to ψ.

Let D be the subgroup of C defined in Section 1. Then we define a "modified" factor of automorphy J_ψ by

$$(2.14a) \quad J_\psi(w,z) = J(w,z) \prod_{v \mid \mathbf{b}} \psi_v(\det(d_w)) \quad (w \in DG_\infty, z \in H^\infty),$$

where ψ_v denotes the v-component of ψ in the usual sense. Observe that the map $w \mapsto J_\psi(w,\mathbf{i})$ is a character of D. We now define a function φ on $G_{\mathbf{A}}$ by

$$(2.14b) \quad \varphi(x) = \begin{cases} \psi(\lambda(y))^{-1} J_\psi(w,\mathbf{i})^{-1} & \text{if } x = yw \text{ with } y \in P_{\mathbf{A}} \text{ and } w \in D, \\ 0 & \text{if } x \notin P_{\mathbf{A}}D. \end{cases}$$

This is well defined, and

$$\varphi(x) = \varphi(x_{\mathfrak{f}})j(x,\mathbf{i})^{-k}|j(x,\mathbf{i})|^{k} \qquad (x \in G_{\mathbf{A}}).$$

Furthermore we put

$$(2.15) \qquad \epsilon_{\varphi}(x,s) = \epsilon(x,s) = \varphi(x)\epsilon(x)^{-k-s} \qquad (x \in G_{\mathbf{A}}, s \in \mathbf{C}).$$

It can easily be verified that

$$(2.16a) \qquad \epsilon(x,s) = \epsilon(x_{\mathfrak{f}},s)J(x,\mathbf{i})^{-1} \qquad (x \in G_{\mathbf{A}}),$$

$$(2.16b) \qquad \epsilon(xw,s) = \epsilon(x,s)J_{\psi}(w,\mathbf{i})^{-1} \qquad (x \in G_{\mathbf{A}}, w \in D),$$

$$(2.16c) \qquad \epsilon(\beta x) = \epsilon(x), \qquad \varphi(\beta x) = \varphi(x), \qquad \epsilon(\beta x,s) = \epsilon(x,s)$$

$$(\beta \in P, x \in G_{\mathbf{A}}).$$

LEMMA 2.2. *Let* $\alpha \in G \cap P_{\mathbf{A}}D$, $\mathbf{a} = \mathrm{il}(\alpha)$, *and* $x \in G_{\infty}$. *If* $\mathbf{b} \neq \mathbf{g}$, *then* $\det(d_{\alpha}) \neq 0$ *and* $\det(d_{\alpha})\mathbf{a}^{-1}$ *is prime to* $\mathbf{b}$. *Moreover we have*

$$\epsilon(\alpha x,s) = \begin{cases} N(\mathbf{a})^{s+k}\psi^{*}(\mathbf{a}^{-1})J(\alpha x,\mathbf{i})^{-1} & \text{if } \mathbf{b} = \mathbf{g}, \\ N(\mathbf{a})^{s+k}\mathrm{sgn}\big[\,N_{F/\mathbf{Q}}(\det(d_{\alpha}))^{k}\big]\psi^{*}(\det(d_{\alpha})\mathbf{a}^{-1})J(\alpha x,\mathbf{i})^{-1} & \text{if } \mathbf{b} \neq \mathbf{g}. \end{cases}$$

Proof. We prove here only the case $\mathbf{b} \neq \mathbf{g}$; the case $\mathbf{b} = \mathbf{g}$ is similar and simpler. Let $\alpha_{\mathfrak{f}} = yw$ with $y \in P_{\mathbf{A}} \cap G_{\mathfrak{f}}$ and $w \in D$. Then $\det(d_{w})_{v} \neq 0$ for $v|\mathbf{b}$, and hence $\det(d_{\alpha}) \neq 0$, $\det(d_{\alpha})\mathbf{a}^{-1} = \det(d_{\alpha})\lambda(y)^{-1}\mathbf{g} = \det(d_{w})\mathbf{g}$. This is prime to $\mathbf{b}$. Now we have

$$\psi^{*}\big(\det(d_{\alpha})\mathbf{a}^{-1}\big)\mathrm{sgn}\big[\,N_{F/\mathbf{Q}}(\det(d_{\alpha}))^{k}\big]$$

$$= \prod_{v \in \mathfrak{f},\, v \nmid \mathbf{b}} \psi_{v}\big(\det(d_{\alpha})\lambda(y)^{-1}\big) \prod_{v \in \infty} \psi_{v}(\det(d_{\alpha}))$$

$$= \prod_{v|\mathbf{b}} \psi_{v}\big(\det(d_{\alpha})^{-1}\big) \prod_{v \nmid \mathbf{b}} \psi_{v}(\lambda(y))^{-1}$$

$$= \psi(\lambda(y))^{-1} \prod_{v|\mathbf{b}} \psi_{v}(\det(d_{w}))^{-1}$$

$$= \varphi(\alpha_{\mathfrak{f}}) = \epsilon(\alpha_{\mathfrak{f}},s)|\lambda(y)|^{s+k} = \epsilon(\alpha_{\mathfrak{f}},s)N(\mathbf{a})^{-k-s},$$

and hence we obtain the desired formula from (2.16a).

We now define our Eisenstein series on $G_{\mathbf{A}}$ by

$$(2.17a) \quad E_{0}(x) = E_{0}(x,s;k,\psi,\mathbf{b}) = \sum_{\alpha \in P\backslash G} \epsilon(\alpha x,s) \qquad (x \in G_{\mathbf{A}}, s \in \mathbf{C}).$$

We see easily that

$$(2.17b) \qquad E_{0}(\alpha xw) = E_{0}(x)J_{\psi}(w,\mathbf{i})^{-1} \qquad (\alpha \in G, x \in G_{\mathbf{A}}, w \in D).$$

It can easily be verified that this series is a special case of the series of Arthur [1, p. 926], [2]. Therefore E_0 converges for sufficiently large $\mathrm{Re}(s)$ and can be continued as a meromorphic function in s to the whole $\mathbf{C}$ by virtue of the result of Langlands [15] (and by its reformulation by Arthur).

If $\alpha \in G$ and $x \in DG_\infty$, we see that $\epsilon(\alpha x, s) \neq 0$ only when $\alpha \in G \cap P_{\mathbf{A}} D$. Take $\beta_i \in G \cap P_{\mathbf{A}} D$ as in Lemma 1.6, and put $\Gamma_i^P = P \cap \beta_i \Gamma_0(\mathbf{b}) \beta_i^{-1}$. By Lemma 1.6, $P \backslash (G \cap P_{\mathbf{A}} D)$ is equivalent to $\coprod_{i=1}^h \Gamma_i^P \backslash \beta_i \Gamma_0(\mathbf{b})$. Hence

$$E_0(x) = \sum_{i=1}^h \sum_{\alpha \in T_i} \epsilon(\alpha x, s) \qquad \text{if} \quad x \in DG_\infty,$$

where $T_i = \Gamma_i^P \backslash \beta_i \Gamma_0(\mathbf{b})$. Now E_0 corresponds to a function E on H^∞ by relations (2.2) and (2.4). Denote it by $E(z, s; k, \psi, \mathbf{b})$. Putting $\mathbf{a}_i = \mathrm{il}(\beta_i)$, we obtain, from Lemma 2.2,

$$(2.18) \quad E(z, s; k, \psi, \mathbf{b})$$

$$= \begin{cases} \displaystyle\sum_{i=1}^h N(\mathbf{a}_i)^{k+s} \sum_{\alpha \in T_i} \mathrm{sgn}\left[N_{F/\mathbf{Q}}(\det(d_\alpha))^k \right] \psi^*(\det(d_\alpha)\mathbf{a}_i^{-1}) J(\alpha, z)^{-1} \\ \qquad\qquad\qquad\qquad\qquad\qquad\qquad\qquad\qquad\qquad \text{if} \quad \mathbf{b} \neq \mathbf{g}, \\ \displaystyle\sum_{i=1}^h N(\mathbf{a}_i)^{k+s} \psi^*(\mathbf{a}_i^{-1}) \sum_{\alpha \in T_i} J(\alpha, z)^{-1} \qquad \text{if} \quad \mathbf{b} = \mathbf{g}. \end{cases}$$

To study the analytic behavior of E, it is more convenient to consider a series E_0^* defined by

$$(2.19) \qquad\qquad E_0^*(x) = E_0^*(x, s) = E_0\big(x \eta_{\mathfrak{f}}^{-1}, s; k, \psi, \mathbf{b}\big),$$

where η is defined by (1.1). Let $E^*(z, s; k, \psi, \mathbf{b})$ or simply $E^*(z, s)$ be the function on H^∞ corresponding to E_0^*. By (2.6), we have

$$(2.20) \qquad E^*(z, s) = E^*(z, s; k, \psi, \mathbf{b}) = E(\eta z, s; k, \psi, \mathbf{b}) J(\eta, z)^{-1}.$$

LEMMA 2.3. *Suppose* $\mathbf{b} \neq \mathbf{g}$; *define* R *by* (1.5). *Then*

$$E_0^*(x, s) = \sum_{\alpha \in \eta R} \epsilon\big(\alpha x \eta_{\mathfrak{f}}^{-1}, s\big) \qquad \textit{if} \quad x \in P_{\mathbf{A}} G_\infty.$$

Proof. Suppose $x \in P_{\mathbf{A}} G_\infty$ and $\epsilon(\alpha x \eta_{\mathfrak{f}}^{-1}, s) \neq 0$ with $\alpha \in G$. Then $\alpha x \in P_{\mathbf{A}} D \eta_{\mathfrak{f}}$. Since a_y and d_y for $y \in D_v$ are invertible if $v \mid \mathbf{b}$, we see easily that $\det(c_\alpha) \neq 0$, and hence $\alpha \in P \eta P$ by Lemma 1.2. We have $P \eta P = P \eta Q R = P Q \eta R = P \eta R$. Since $P \cap \eta R \eta^{-1} = \{1\}$, ηR gives a complete set of representatives of $P \backslash P \eta R$. This proves our lemma.

PROPOSITION 2.4. (i) *Let Γ be a congruence subgroup of G satisfying* (2.9) *and containing* $\Gamma(\mathbf{b})$. *Then*

$$r \cdot E(z,s;k,\Gamma) = \sum_{\tau \in \Gamma(\mathbf{b})\backslash\Gamma} \sum_{\psi \in X} E(z,s;k,\psi,\mathbf{b})\|\tau$$

with $r = \nu[P \cap \Gamma : P \cap \Gamma(\mathbf{b})]$, *where* X *is the set of all characters* ψ *satisfying* (2.13a, b) *and* ν *is the number of such characters.*

(ii) *If* $\mathbf{b}$ *satisfies* (2.10) *and* $\mathbf{c}$ *is a multiple of* $\mathbf{b}$, *then*

$$E(z,s;k,\psi,\mathbf{b}) = \sum_{\xi \in \Gamma_0(\mathbf{c})\backslash\Gamma_0(\mathbf{b})} J_\psi(\xi,z)^{-1} E(\xi z,s;k,\psi,\mathbf{c}).$$

Proof. If $\Gamma(\mathbf{b}) \subset \Gamma$ and Γ satisfies (2.9), then (2.10) is satisfied, and hence X is not empty. To prove (i), let us assume $\mathbf{b} \neq \mathbf{g}$; the case $\mathbf{b} = \mathbf{g}$ is similar and simpler. Take $\beta_1 = 1$ in the above discussion, and consider the sum of the expressions of (2.18) for all $\psi \in X$. Observe that

$$\sum_{\psi \in X} \psi^*\big(\det(d_\alpha)\mathbf{a}_i^{-1}\big) \neq 0 \qquad \text{only when} \quad i = 1 \quad \text{and} \quad \det(d_\alpha)\mathbf{g} = t\mathbf{g}$$

with $t \in F^\times$ such that $t - 1 \in \mathbf{b}_v$ for all $v \,|\, \mathbf{b}$, in which case the sum is $\nu \cdot \mathrm{sgn}[N_{F/\mathbf{Q}}(t)^k]$. Putting $\det(d_\alpha) = te$, we see that $e \in \mathbf{g}^\times$ and $\det(d_\alpha) \equiv e \pmod{\mathbf{b}}$, and hence $\alpha \in \Gamma_u(\mathbf{b})$, and

$$\mathrm{sgn}\big[N_{F/\mathbf{Q}}(\det(d_\alpha))^k \big] \sum_{\psi \in X} \psi^*(\det(d_\alpha)\mathbf{g}) = \nu \cdot N_{F/\mathbf{Q}}(e)^k.$$

Therefore $\nu^{-1}\sum_{\psi \in X} E(z,s;k,\psi,\mathbf{b})$ is exactly the right-hand side of (ii) of Proposition 2.1. Taking $\Gamma(\mathbf{b})$ as Γ' of Proposition 2.1, (iii), we obtain our first assertion. To prove (ii), let D', $\epsilon'(x,s)$, and $E_0'(x)$ denote the objects corresponding to D, $\epsilon(x,s)$, and $E_0(x)$ defined with $\mathbf{c}$ in place of $\mathbf{b}$ and with the same ψ. Take a finite subset $\{r\}$ of $G_\mathbf{f}$ so that $D = \coprod_{r \in \{r\}} D'r$. Then $P_\mathbf{A}D = \coprod_{r \in \{r\}} P_\mathbf{A}D'r$ and $\epsilon(x,s) = \sum_{r \in \{r\}} \epsilon'(xr^{-1},s)J_\psi(r,\mathbf{i})^{-1}$. Therefore $E_0(x) = \sum_{r \in \{r\}} E_0'(xr^{-1})J_\psi(r,\mathbf{i})^{-1}$. By strong approximation in G, we have $D \subset D'G_\infty G$, so that $DG_\infty = D'G_\infty\Gamma_0(\mathbf{b})$. Therefore we can take $\{\xi_\mathbf{f} : \xi \in \Gamma_0(\mathbf{c})\backslash\Gamma_0(\mathbf{b})\}$ as $\{r\}$. Then we obtain the equality of (ii) by virtue of (2.6).

Remark 2.5. We can consider a series of a more general type than (2.17a) as follows. Put

$$(2.21a) \qquad E_{t0}(x) = \sum_{\alpha \in P\backslash G} \epsilon_t(\alpha x),$$

$$(2.21b) \qquad \epsilon_t(x) = t(x)\epsilon(x)^{-k-s}j(x,\mathbf{i})^{-k}|j(x,\mathbf{i})|^k \qquad (x \in G_\mathbf{A})$$

with a **C**-valued function t on $G_\mathbf{A}$ such that

$$(2.22) \quad t(\xi xw) = t(x)\mathrm{sgn}\big[N_{F/\mathbf{Q}}(\lambda(\xi)) \big]^k \qquad (\xi \in P, x \in G_\mathbf{A}, w \in C'G_\infty),$$

where C' is an open subgroup of C. Then E_{t0} satisfies (2.1), and includes (2.17a) as a special case. However, this can be expressed as a linear combination of the transforms of several series of type (2.17a) (in the sense described below). To see this, it is sufficient to treat the case where there is a coset $Y = P\beta C'G_\infty$ with $\beta \in G_A$ such that

$$t(x) = \begin{cases} \mathrm{sgn}\left[N_{F/\mathbf{Q}}(\lambda(\xi))\right]^k & \text{if} \quad x \in \xi\beta C'G_\infty \quad \text{with} \quad \xi \in P, \\ 0 & \text{if} \quad x \notin Y. \end{cases}$$

By strong approximation, we may assume that $\beta \in G$. Let $x \in G_\infty$. Then $\epsilon_t(\alpha x) \neq 0$ implies that $\alpha x \in Y$, and hence $\alpha \in G \cap P\beta C'G_\infty = P\beta\Gamma$, where $\Gamma = G \cap C'G_\infty$. Put $T = (P \cap \beta\Gamma\beta^{-1})\backslash\beta\Gamma\beta^{-1}$. Then, by (2.12), we have

$$E_{t0}(x)J(x,\mathbf{i}) = \sum_{\alpha \in T} \epsilon_t(\alpha\beta x)J(x,\mathbf{i})$$

$$= t(\beta)N(\mathrm{il}(\beta))^{k+s} \sum_{\alpha \in T} J(\alpha\beta, z)^{-1}$$

$$= t(\beta)N(\mathrm{il}(\beta))^{k+s}E(z,s;k,\beta\Gamma\beta^{-1})\|\beta.$$

By virtue of Proposition 2.4, this can be expressed as a linear combination of transforms of series of type (2.17a) as expected.

3. The Fourier expansion of E^*. For a rational prime p, let $\mathbf{Q}_p$ denote the field of p-adic numbers. We put $\mathbf{e}(s) = e^{2\pi i s}$ for $s \in \mathbf{C}$, and define characters $\mathbf{e}_p$ and $\mathbf{e}_v$ of the additive groups $\mathbf{Q}_p$ and F_v for $v \in \mathbf{f} \cup \infty$ by

(3.1a) $\quad \mathbf{e}_p(t) = \mathbf{e}(\text{the fractional part of } -t) \qquad \text{for} \quad t \in \mathbf{Q}_p$,

(3.1b) $\quad e_v(t) = \begin{cases} \mathbf{e}_p\left(\mathrm{Tr}_{F_v/\mathbf{Q}_p}(t)\right) & \text{for} \quad t \in F_v \qquad \text{if} \quad v | p, \\ \mathbf{e}(t) & \text{for} \quad t \in F_v = \mathbf{R} \quad \text{if} \quad v \in \infty. \end{cases}$

Observe that $\mathrm{tr}(xy) \in F$ for every x and y in the set S of (1.2a). Therefore we can define a complex number $\chi(xy)$ for $x, y \in S_A$ by

(3.2a) $\qquad\qquad \chi(xy) = \prod_v \chi_v(x_v y_v) \qquad (x, y \in S_A),$

(3.2b) $\qquad\qquad \chi_v(x_v y_v) = \mathbf{e}_v(\mathrm{tr}(x_v y_v)) \qquad (x_v, y_v \in S_v).$

Now we fix a Haar measure μ on S_A so that $\mu(S_A/S) = 1$. This is self-dual with respect to χ. If g is a function on S_A/S, then under a suitable condition on g, we have

(3.3a) $\qquad\qquad g(x) = \sum_{h \in S} a(h)\chi(hx),$

(3.3b) $\qquad\qquad a(h) = \int_{S(A)/S} g(x)\chi(-hx)\,d\mu(x).$

We apply this to E_0^*. To simplify our notation, put

$$(3.4) \qquad \tau(x) = \begin{bmatrix} 1 & x \\ 0 & 1 \end{bmatrix} \qquad (x \in S_\mathbf{A}).$$

Consider $E_0^*(\tau(x)w, s)$ a function of $x \in S_\mathbf{A}$ and $w \in G_\infty$. Since this is invariant under $x \mapsto x + u$ with $u \in S$, we have, by (3.3),

$$(3.5a) \qquad E_0^*(\tau(x)w, s) = \sum_{h \in S} b(h, w, s)\chi(hx) \qquad (x \in S_\mathbf{A}, w \in G_\infty),$$

when s belongs to the half-plane of convergence, where

$$(3.5b) \qquad b(h, w, s) = \int_{S(\mathbf{A})/S} E_0^*(\tau(x)w, s)\chi(-hx)\, d\mu(x) \qquad (w \in G_\infty).$$

Define a lattice L in S and also L_v in S_v for $v \in \mathbf{f}$ by

$$(3.6a) \qquad L = \begin{cases} S \cap M_m(\mathbf{g}) \\ S \cap M_m(\mathbf{h}) \end{cases} \qquad L_v = \begin{cases} S_v \cap M_m(\mathbf{g}_v) & \text{(Case SP)}, \\ S_v \cap M_m(\mathbf{h}_v) & \text{(Case SU)}. \end{cases}$$

With $\mathbf{b}$ and D as in (1.9), observe that $\tau(-u)^* = \eta^{-1}\tau(u)\eta \in D_v$ for $u \in \mathbf{b}L_v$. Therefore, if $u \in \mathbf{b}L_v$ and $w \in G_\infty$, we have

$$E_0^*(\tau(x + u)w) = E_0^*(\tau(x)w\tau(u)) = E_0\big(\tau(x)w\tau(u)\eta_\mathbf{f}^{-1}\big)$$

$$= E_0\big(\tau(x)w\eta_\mathbf{f}^{-1}\tau(-u)^*\big) = E_0^*(\tau(x)w)$$

by (2.17b), and hence

$$b(h, w, s) = \int_{S(\mathbf{A})/S} E_0^*(\tau(x + u)w)\chi(-h(x + u))\, d\mu(x)$$

$$= \chi(-hu)b(h, w, s) \qquad \text{if} \quad u \in \mathbf{b}L_v.$$

Thus $b(h, w, s) \neq 0$ only when $h \in \mathbf{b}^{-1}L'$ with the lattice L' dual to L, which is defined by

$$(3.6b) \qquad L' = \big\{ y \in S : \mathrm{tr}(yL) \subset \mathbf{d}^{-1} \big\},$$

where $\mathbf{d}$ is the different of F over $\mathbf{Q}$. Putting $M = \mathbf{b}^{-1}L'$, we have

$$(3.7) \qquad E_0^*(\tau(x)w, s) = \sum_{h \in M} b(h, w, s)\chi(hx) \qquad (x \in S_\mathbf{A}, w \in G_\infty).$$

To express this on H^∞, for $z = (z_v)_{v \in \infty} \in H^\infty$ with $z_v = x_v + iy_v$, $x_v \in S_v$, $0 < y_v \in S_v$, put

$$(3.8) \qquad w_v = \mathrm{diag}\big[y_v^{1/2}, y_v^{-1/2} \big].$$

(See *Notation* for $y^{1/2}$.) Then with $x = (x_v)_{v \in \infty}$, we have

$$(3.9a) \qquad E^*(z, s) = E_0^*(\tau(x)w, s)J(\tau(x)w, \mathbf{i})$$

$$= \sum_{h \in M} a(h, y, s)\mathbf{e}\left(\sum_{v \in \infty} \mathrm{tr}(h_v x_v)\right),$$

$$(3.9b) \quad a(h, y, s) = b\left(h, \left(\mathrm{diag}[\, y_v^{1/2}, y_v^{-1/2}]\right)_{v \in \infty}, s\right) \prod_{v \in \infty} \det(y_v)^{-(k+s)/2}.$$

Suppose now $\mathbf{b} \neq \mathbf{g}$. Since $\tau(x)w \in P_{\mathbf{A}}G_\infty$, we obtain, from Lemma 2.3,

$$b(h, w, s) = \int_{S(\mathbf{A})/S} \sum_{\alpha \in \eta R} \epsilon\left(\alpha\tau(x)w\eta_{\mathfrak{f}}^{-1}, s\right)\chi(-hx)\, d\mu(x)$$

$$= \int_{S(\mathbf{A})/S} \sum_{t \in S} \epsilon\left(\eta\tau(x + t)w\eta_{\mathfrak{f}}^{-1}, s\right)\chi(-hx)\, d\mu(x)$$

$$= \int_{S(\mathbf{A})} \epsilon\left(\eta\tau(x)\eta_{\mathfrak{f}}^{-1}w, s\right)\chi(-hx)\, d\mu(x).$$

With w as in (3.8) and $x \in S_{\mathbf{A}}$,

$$(3.10) \qquad \left(\eta\tau(x)\eta_{\mathfrak{f}}^{-1}w\right)_v = \begin{cases} \begin{bmatrix} 0 & -y_v^{-1/2} \\ y_v^{1/2} & x_v y_v^{-1/2} \end{bmatrix} & (v \in \infty), \\ \tau(-x_v)^* & (v \in \mathfrak{f}). \end{cases}$$

Our definition of ϵ implies a product expression

$$\epsilon\left(\eta\tau(x)\eta_{\mathfrak{f}}^{-1}w, s\right) = \prod_{v \in \infty} \delta_v(x_v, y_v) \prod_{v \in \mathfrak{f}} \delta_v(x_v)$$

with a certain function δ_v for each $v \in \mathfrak{f}$ and

$$(3.11) \quad \delta_v(x_v, y_v) = \det(y_v)^{(k+s)/2}\det(x_v + iy_v)^{-k}|\det(x_v + iy_v)|^{-s} \quad (v \in \infty).$$

Take a measure μ_v on S_v for each $v \in \mathfrak{f} \cup \infty$ so that $\mu_v(L_v) = 1$ for every $v \in \mathfrak{f}$. Then

$$(3.12) \qquad \mu = c_\mu \prod_{v \in \mathfrak{f} \cup \infty} \mu_v$$

with a constant c_μ (depending on the choice of μ_v for $v \in \infty$), and we obtain

$$(3.13a) \quad c_\mu^{-1}a(h, y, s) = \prod_{v \in \infty} a_v(h, y_v, s) \prod_{v \in \mathfrak{f}} a_v(h, s),$$

$$(3.13b) \quad a_v(h, y_v, s) = \det(y_v)^{-(k+s)/2}\int_{S_v} \delta_v(x, y_v)\chi_v(-hx)\, d\mu_v(x) \qquad (v \in \infty),$$

$$(3.13c) \qquad a_v(h, s) = \int_{S_v} \delta_v(x)\chi_v(-hx)\, d\mu_v(x) \qquad (v \in \mathfrak{f}).$$

To obtain an explicit description of δ_v for $v \in \mathbf{f}$, we first introduce a function ν on S_v by

$$(3.14) \qquad \nu(x) = \epsilon(\tau(x)^*) \qquad (x \in S_v).$$

In other words, given $x \in S_v$, take a decomposition $\tau(x)^* = yw$ with $y \in P_v$ and $w \in C_v$. Then $1 = d_y d_w$, $x = d_w^{-1} c_w$, and

$$(3.15) \qquad \nu(x) = |\det(d_w)^{-1}|_v .$$

Notice that $\nu(x) = \nu(-x)$ and $\nu(x + u) = \nu(x)$ for $u \in L_v$.

LEMMA 3.1.

$$\delta_v(x) = \begin{cases} \psi_v(\nu_0(x))\nu(x)^{-k-s} & \text{if} \quad v \nmid \mathbf{b}, \\ 1 & \text{if} \quad v \mid \mathbf{b} \quad \text{and} \quad x \in \mathbf{b}L_v, \\ 0 & \text{if} \quad v \mid \mathbf{b} \quad \text{and} \quad x \notin \mathbf{b}L_v, \end{cases}$$

where $\nu_0(x)$ is the ideal of $\mathbf{g}_v$ such that $\nu(x) = [\mathbf{g}_v ; \nu_0(x)]$ and $\psi_v(a\mathbf{g}_v)$ is defined to be $\psi_v(a)$ for $a \in F_v^\times$ when $v \nmid \mathbf{b}$.

Proof. By (2.14b) and (3.10), $\delta_v(x) = 0$ for a given $x \in S_v$ unless $\tau(-x)^* \in P_v D_v$. Let $\tau(-x)^* = yw$ with $y \in P_v$ and $w \in D_v$. Then, from (2.15), we obtain

$$\delta_v(x) = \psi(\lambda(y))^{-1}|\lambda(y)|^{-k-s} \cdot \begin{cases} 1 & \text{if} \quad v \nmid \mathbf{b}, \\ \psi(\det(d_w))^{-1} & \text{if} \quad v \mid \mathbf{b} \end{cases}$$

If $v \nmid \mathbf{b}$, we have $|\lambda(y)| = \nu(x)$, which proves the first case. Suppose $v \mid \mathbf{b}$. Then $d_y^{-1} = d_w \in \mathrm{GL}_m(\mathbf{g}_v)$ or $\in \mathrm{GL}_m(\mathbf{h}_v)$; hence $x = -d_w^{-1} c_w \in \mathbf{b}L_v$, and $\delta_v(x) = 1$. This shows that $\tau(-x)^* \notin P_v D_v$ if $x \notin \mathbf{b}L_v$. Conversely, if $x \in \mathbf{b}L_v$ and $\tau(-x)^* = yw$ with $y \in P_v$ and $w \in C_v$, we have $c_w = -d_w x \equiv 0 \pmod{\mathbf{b}_v}$, so that $w \in D_v$. This completes the proof.

Combining Lemma 3.1 with (3.13c), we obtain

$$(3.16) \qquad a_v(h,s) = \begin{cases} [\mathbf{g}_v : \mathbf{b}_v]^{m(m+1)/2} & (h \in \dot{M}, v \mid \mathbf{b}, \text{Case SP}), \\ [\mathbf{g}_v : \mathbf{b}_v]^{m^2} & (h \in M, v \mid \mathbf{b}, \text{Case SU}). \end{cases}$$

If $v \in \infty$, (3.11) implies that

$$(3.17) \qquad a_v(h, y_v, s) = \xi(y_v, h_v; k + (s/2), s/2)$$

with a function ξ defined by

$$(3.18) \qquad \xi(g, h; \alpha, \beta) = \int_V \mathbf{e}(-\mathrm{tr}(hx))\det(x + ig)^{-\alpha}\det(x - ig)^{-\beta} dx$$

$$(\alpha, \beta \in \mathbf{C}, V = S_v, dx = d\mu_v(x), 0 < g \in V, h \in V).$$

This is exactly the integral studied in [26], provided the measure μ_v is defined as in [26, §1], that is,

$$(3.19) \qquad d\mu_v((x_{jk})) = \begin{cases} \left| \bigwedge_{j \leqslant k} dx_{jk} \right| & \text{(Case SP)}, \\ 2^{m(1-m)/2} \left| \bigwedge_{j=1}^{m} dx_{jj} \bigwedge_{j < k} dx_{jk}\, d\bar{x}_{jk} \right| & \text{(Case SU)}. \end{cases}$$

As proved in [26], ξ multiplied by a suitable gamma factor can be continued as a holomorphic function in (α, β) to the whole $\mathbf{C}^2$. With this choice of μ_v for $v \in \infty$, we can easily verify that the constant c_μ of (3.12) is given by

$$(3.20) \qquad c_\mu = \begin{cases} N(\mathbf{d})^{-m(m+1)/4} & \text{(Case SP)}, \\ N(\mathbf{d})^{-m/2} N(2^{-1}\mathbf{d}')^{-m(m-1)/4} & \text{(Case SU)}, \end{cases}$$

where $\mathbf{d}$ and $\mathbf{d}'$ are the differents of F and K over $\mathbf{Q}$, respectively.

Suppose $h \in M$, $v \in \mathbf{f}$, and $v \nmid \mathbf{b}$. By (3.13c) and Lemma 3.1, we have

$$a_v(h,s) = \int_{S_v} \psi_v(\nu_0(x))\nu(x)^{-k-s} \chi_v(-hx)\, d\mu_v(x).$$

Since the integrand is invariant under $x \mapsto x + u$ with $u \in L_v$, we can write

$$(3.21) \qquad a_v(h,s) = \sum_{y \in S_v/L_v} \psi_v(\nu_0(y))\nu(y)^{-k-s} \chi_v(-hy).$$

Let π_v be a prime element of F_v and let $q_v = |\pi_v|_v^{-1}$. We can now define a formal power-series $\alpha(v,h,t)$ in an indeterminate t for every $h \in S \cap L_v'$ and every $v \in \mathbf{f}$ by

$$(3.22) \qquad \alpha(v,h,q_v^{-s}) = \sum_{y \in S_v/L_v} \chi_v(-hy)\nu(y)^{-s}.$$

We call this a *Siegel series* or a *Siegel function*, because it was introduced by Siegel in his investigations of quadratic forms and Eisenstein series. This appellation seems more appropriate than the "Whittaker integral" which is prevalent. Observe that the leading coefficient of $\alpha(v,h,t)$ is 1 and that

$$(3.23) \qquad a_v(h,s) = \alpha\!\left(v,h,\psi(\pi_v)q_v^{-k-s}\right) \qquad (v \in \mathbf{f},\ v \nmid \mathbf{b}).$$

Thus our task is to determine $\alpha(v,h,t)$ as a function of t as explicitly as possible. We first prove that it is a rational function of t. For this purpose, we note

$$(3.24) \qquad \xi(g,h;\alpha,\beta) \neq 0 \qquad \text{if} \quad \kappa - 1 < \alpha \in \mathbf{R} \quad \text{and} \quad \kappa - 1 < \beta \in \mathbf{R},$$

where $\kappa = (m+1)/2$ in Case SP and $\kappa = m$ in Case SU. This follows from a more general result:

LEMMA 3.2. *Let $\omega(g, h; \alpha, \beta)$ and κ be defined as in [26, (4.6.K, IV), (1.14.K, IV)]. Then $\omega(g, h; \alpha, \beta) > 0$ if $\kappa - 1 < \alpha \in \mathbf{R}$ and $\kappa - 1 < \beta \in \mathbf{R}$.*

Proof. If $h > 0$, this follows immediately from [26, (4.7.K, IV), (4.20), (3.6), (3.2)]. The general case can be reduced to the case $h > 0$ by virtue of [26, (4.8), (4.9), (4.25), and the equality following (4.28)].

Now $\xi(g, h; \alpha, \beta)$ is $\omega(g, h; \alpha, \beta)$ times a certain gamma factor as shown by [26, (4.34.K)]. Checking the nonvanishing of this factor, we obtain (3.24).

PROPOSITION 3.3. *For every $v \in \mathbf{f}$ and $h \in S \cap L_v'$, $\alpha(v, h, t)$ is a rational function of t with coefficients in $\mathbf{Q}$.*

Proof. To avoid confusion, let us write the given prime v_0, keeping v for "generic ones". Given h and v_0, we take an integral ideal $\mathbf{b} \neq \mathbf{g}$ such that $v_0 \nmid \mathbf{b}$. Consider E^* with this choice of $\mathbf{b}$ and with trivial ψ and $k = 0$. As already mentioned, the result of Langlands [15] guarantees that $E^*(z, s)$ can be continued as a meromorphic function in s to the whole s-plane; more specifically, if s_0 is a pole, it does not involve z, and $(s - s_0)^N E^*(z, s)$ is holomorphic at $s = s_0$ and C^∞ in (z, s) with a suitable positive integer N (see [15, Theorem 7.1, p. 167]). Therefore $(s - s_0)^N a(h, y, s)$, as a Fourier coefficient of $(s - s_0)^N E^*(z, s)$, is holomorphic at s_0. We can thus conclude that $a(h, y, s)$ can be continued as a meromorphic function to the whole s-plane. Now (3.17) and (3.24), together with the analytic continuation of ξ proved in [26], show that $a_v(h, y, s)$ for $v \in \infty$ is a nonvanishing meromorphic function in the whole s-plane. Therefore

$$(3.25) \qquad \prod_{v \in \mathbf{f}} a_v(h, s)$$

must be meromorphic in the whole s-plane. This infinite product is a Dirichlet series of the form $\sum_{\mathbf{a}} c(\mathbf{a}) N(\mathbf{a})^{-s}$ with $c(\mathbf{a}) \in \mathbf{C}$, $c(\mathbf{g}) \neq 0$, where $\mathbf{a}$ ranges over all integral ideals of F. This is absolutely convergent for sufficiently large $\mathrm{Re}(s)$. (This fact is elementary. We shall later give an explicit form of its "majorant"

$$\prod_{v \in \mathbf{f}} \left\{ \sum_{y \in S_v / L_v} \nu(y)^{-s} \right\},$$

which is convergent for $\mathrm{Re}(s) > m + 1$ in Case SP and for $\mathrm{Re}(s) > 2m$ in Case SU; see Proposition 5.2 below.) Since the series tends to $c(\mathbf{g})$ when $\mathrm{Re}(s) \to \infty$, (3.25) is a nonvanishing function. Call this $a_{\mathbf{f}}(h, s, \mathbf{b})$. Let $\mathbf{b}'$ be the product of $\mathbf{b}$ and v_0. Now change $\mathbf{b}$ for $\mathbf{b}'$ and consider E^* with $\mathbf{b}'$ in place of $\mathbf{b}$. Then the factor $a_{v_0}(h, s)$ of $a_{\mathbf{f}}(h, s, \mathbf{b})$ is replaced by a constant of type (3.16). Thus $a_{v_0}(h, s)$ can be obtained as a constant times

$$a_{\mathbf{f}}(h, s, \mathbf{b}) / a_{\mathbf{f}}(h, s, \mathbf{b}').$$

Therefore $a_{v_0}(h, s)$ is meromorphic in the whole s-plane. We can easily conclude from this fact that $\alpha(v, h, t)$ is meromorphic on the whole t-plane. Now (3.22) shows that $'\alpha(v, h, t)$ is a power-series in t with coefficients in $\mathbf{Z}$, since the sum of $\chi_v(-hy)$ for all $y \in S_v/L_v$ such that $v(y) = q^n$ for a fixed n is invariant under every automorphism of the maximal cyclotomic field. A well known theorem of Borel [3] asserts that a power-series in t with coefficients in $\mathbf{Z}$ is a rational function if it is meromorphic in a disc of radius > 1. Applying this to $\alpha(v, h, t)$, we obtain the desired result.

4. The Siegel series $\alpha(v, h, t)$ for nonsingular h. Throughout Sections 4, 5, and 6, we fix a finite prime v of F and deal with the local objects defined relative to v. Therefore we shall sometimes drop subscript v, particularly when a simpler notation is desirable. For example, π denotes a prime element of $\mathbf{g}_v$ and $q = |\pi|_v^{-1}$; we put also $\mathbf{p} = \pi\mathbf{g}_v$ and use $\mathbf{d}$ (instead of $\mathbf{d}_v$) for the different of F_v over $\mathbf{Q}_p$ when $v \mid p$. The normalized valuation $| \ |_v$ of F_v is simply denoted by $| \ |$. In Case SU, we extend this to a function on K_v by putting

$$(4.1) \qquad |a| = |N_{K/F}(a)|_v^{1/2} \qquad \text{for} \quad a \in K_v.$$

This is the normalized valuation of K_v only when v is ramified in K.

LEMMA 4.1. *Let $c \in M_m(K_v)$, $d \in \mathrm{GL}_m(K_v)$, $y = d^{-1}c$, and $tyu = \mathrm{diag}[b_1, \ldots, b_m]$ with $b_i \in K_v$ and $t, u \in \mathrm{GL}_m(\mathbf{h}_v)$. Suppose that (c, d) is primitive in the sense that it is the lower half of an element of $\mathrm{GL}_{2m}(\mathbf{h}_v)$. Suppose further that $b_i \in F_v$ for all i when K_v is not a field. Then $\det(d)\mathbf{h}_v$ is the product of $b_i^{-1}\mathbf{h}_v$ for all b_i not contained in $\mathbf{h}_v$. The same assertion holds with $\mathbf{g}_v$ and F_v in place of $\mathbf{h}_v$ and K_v.*

Proof. Put $\Lambda = (\mathbf{h}_v)_1^m$. Then (c, d) is primitive if and only if $c\Lambda + d\Lambda = \Lambda$. Let $\{e_1, \ldots, e_m\}$ be the standard basis of Λ. Assume that $b_i \in \mathbf{h}_v$ if and only if $i > s$ $(0 \leqslant s \leqslant m)$. Then $td^{-1}\Lambda = td^{-1}c\Lambda + t\Lambda = tyu\Lambda + \Lambda = \sum_{i=1}^m (\mathbf{h}_v b_i e_i + \mathbf{h}_v e_i) = \sum_{i \leqslant s} \mathbf{h}_v b_i e_i + \sum_{i > s} \mathbf{h}_v e_i$, and hence $\det(d^{-1})\mathbf{h}_v = b_1 \ldots b_s \mathbf{h}_v$, Q.E.D.

Given $y \in M_m(K_v)$, take $t, u \in \mathrm{GL}_m(\mathbf{h}_v)$ so that $tyu = \mathrm{diag}[b_1, \ldots, b_m]$ with $b_i \in K_v$. Suppose either K_v is a field or $b_i \in F_v$ for all i. We then put

$$(4.2) \qquad v(y) = \prod_{|b_i| > 1} |b_i|.$$

In Case SP, we do the same within $M_m(F_v)$. The above lemma shows that this is consistent with the function v which was previously defined only on S_v by (3.15).

Our immediate aim is to determine an explicit form of $\alpha(v, h, t)$. In this section, we treat the case $\det(h) \neq 0$ under the assumption

$$(4.3) \qquad v \nmid 2 \text{ in Case SP, and } v \text{ is unramified in } K \text{ in Case SU}.$$

Then we can define $\left(\dfrac{c}{\mathbf{p}}\right)$ for $c \in \mathbf{g}_v^\times$ and $\left(\dfrac{K/F}{\mathbf{p}}\right)$ in the usual way. These symbols can be extended to characters of the ideal groups of F. In particular, we put, for an ideal $\mathbf{a}$ of F,

$$(4.4) \quad \theta(\mathbf{a}) = \begin{cases} \left(\dfrac{-1}{\mathbf{a}} \right) & \text{in Case SP for } \mathbf{a} \text{ prime to } 2, \\[2ex] \left(\dfrac{K/F}{\mathbf{a}} \right) & \text{in Case SU for } \mathbf{a} \text{ prime to the discriminant of } K \text{ over } F. \end{cases}$$

To determine $\alpha(h, \iota)$, we employ certain Gauss sums, following Siegel's idea. Let $a \in F_v^{\times}$; suppose $a\mathbf{d} = \mathbf{p}^{-\lambda}$ with $0 \leqslant \lambda \in \mathbf{Z}$. We then define a Gauss sum $G(a)$ by

$$(4.5) \qquad\qquad G(a) = \begin{cases} \displaystyle\sum_{x \in \mathbf{g}_v/\mathbf{p}^{\lambda}} \mathbf{e}_v(ax^2) & \text{(Case SP)}, \\[3ex] \displaystyle\sum_{y \in \mathbf{h}_v/\mathbf{p}^{\lambda}\mathbf{h}_v} \mathbf{e}_v(ayy^{\rho}) & \text{(Case SU)}. \end{cases}$$

LEMMA 4.2.

$$G(ca) = \left(\frac{c}{\mathbf{p}} \right)^{\lambda} G(a) \qquad \text{if} \quad c \in \mathbf{g}_v^{\times} \qquad (\text{Case SP}),$$

$$G(a)^2 = q^{\lambda}\theta(\mathbf{p}^{\lambda}) \qquad\qquad\qquad (\text{Case SP}),$$

$$G(a) = q^{\lambda}\theta(\mathbf{p}^{\lambda}) \qquad\qquad\qquad (\text{Case SU}).$$

Proof. The first two relations are well known. In fact, the first relation is a special case of Hecke [8, p. 223, Satz 155]. As shown in (171) on the same page, one has

$$G(a) = \sum_{x \bmod \mathbf{p}} \left(\frac{x}{\mathbf{p}} \right) \mathbf{e}_v(ax) \qquad \text{if} \quad \lambda = 1.$$

Then the standard argument shows that $|G(a)|^2 = q$ and $\overline{G(a)} = \theta(\mathbf{p})G(a)$, and hence $G(a)^2 = \theta(\mathbf{p})q$. The case $\lambda > 1$ can be reduced to the case $\lambda = 1$ as proved in [8, p. 222, Hilfssatz b]. In Case SU, the same technique shows that $G(a) = q^2 G(\pi^2 a)$ if $\lambda \geqslant 2$. Therefore $G(a) = q^{\lambda}$ if λ is even, and $G(a) = q^{\lambda-1}G(\pi^{\lambda-1}a)$ if λ is odd. Thus it is sufficient to consider the case $\lambda = 1$. Now, given $x \in \mathbf{g}_v$, we see easily that

$$\#\left\{ y \in \mathbf{h}_v/\mathbf{p}\mathbf{h}_v \, ; \, yy^{\rho} \equiv x \,(\mathrm{mod}\,\mathbf{p}) \right\} = \begin{cases} q - \theta(\mathbf{p}) & \text{if} \quad x \notin \mathbf{p}, \\ q + \theta(\mathbf{p})(q - 1) & \text{if} \quad x \in \mathbf{p}. \end{cases}$$

Hence

$$G(a) = q + \theta(\mathbf{p})(q - 1) + (q - \theta(\mathbf{p})) \sum_{0 \neq x \in \mathbf{g}/\mathbf{p}} \mathbf{e}_v(ax) = \theta(\mathbf{p})q,$$

which completes the proof.

LEMMA 4.3. *Let $y \in S_v$. Under (4.3), there is an element u of $\mathrm{GL}_m(\mathbf{g}_v)$ in Case SP (resp. $\mathrm{GL}_m(\mathbf{h}_v)$ in Case SU) such that $u^* y u$ is diagonal.*

This is well known.

For $0 \leqslant n \in \mathbf{Z}$ and $0 < k \in \mathbf{Z}$, let $X(n, m \times k)$ denote $(\mathbf{g}_v/\mathbf{p}^n)_k^m$ in Case SP and $(\mathbf{h}_v/\mathbf{p}^n \mathbf{h}_v)_k^m$ in Case SU. Then, for $y \in \mathbf{p}^{-n} \mathbf{d}^{-1} L_v$, we put

$$(4.6) \qquad G_n(y) = \sum_{x \in X(n, m \times 1)} \mathbf{e}_v(x^* y x).$$

We have obviously

$$(4.7) \qquad G_n(y)^k = \sum_{x \in X(n, m \times k)} \mathbf{e}_v(\mathrm{tr}(x^* y x)) \qquad (0 < k \in \mathbf{Z}).$$

LEMMA 4.4. *Let δ be an element of F_v such that $\mathbf{d} = \delta \mathbf{g}_v$ and let $y \in \mathbf{p}^{-n} \mathbf{d}^{-1} L_v$. Then*

$$q^{2mn} \theta(\nu_0(\delta y)) \nu(\delta y)^{-1} = \begin{cases} G_n(y)^2 & (\textit{Case SP}), \\ G_n(y) & (\textit{Case SU}), \end{cases}$$

where $\nu_0(z)$ is the ideal of $\mathbf{g}_v$ such that $\nu(z) = [\mathbf{g}_v : \nu_0(z)]$. Moreover, in Case SP, put $G_n(y) = \omega(y) q^{mn} \nu(\delta y)^{-1/2}$. Then $\omega(y)$ depends only on y modulo $\mathbf{d}^{-1} L_v$ and is independent of n; further $\omega(y)^2 = \theta(\nu_0(\delta y))$ and $\omega(cy) = \left(\dfrac{c}{\nu_0(\delta y)}\right) \omega(y)$ for every $c \in \mathbf{g}_v^{\times}$.

Proof. We prove our assertions only in Case SP; Case SU can be treated in a similar and simpler way. By Lemma 4.3, we may assume that $y = \mathrm{diag}[a_1, \ldots, a_m]$ with $a_i \in F_v$. Let $a_i \mathbf{d} = \mathbf{p}^{-\lambda}$ if $a_i \notin \mathbf{d}^{-1}$; put $\lambda_i = 0$ if $a_i \in \mathbf{d}^{-1}$. By (4.2), we have $\nu(\delta y) = q^\mu$ with $\mu = \lambda_1 + \cdots + \lambda_m$, and $G_n(y) = G_n(a_1) \cdots G_n(a_m) = q^{nm-\mu} G(a_1) \ldots G(a_m)$, where we put $G(a) = 1$ if $a \in \mathbf{d}^{-1}$. Therefore we obtain our assertion from Lemma 4.2.

Suppressing the letter v, we write $\alpha(v, h, t)$ simply $\alpha(h, t)$. In Case SP, we generalize this by defining a power-series $\alpha_\lambda(h, t)$ in t for $h \in S \cap L_v'$ and $\lambda \in \mathbf{Z}$ by

$$(4.8a) \qquad \alpha_\lambda(h, q^{-s}) = \sum_{y \in S_v/L_v} \omega(y/\delta)^\lambda \chi_v(-hy) \nu(y)^{-s}$$

with a fixed element δ as in Lemma 4.4. We take $\delta = 1$ whenever $\mathbf{d} = \mathbf{g}_v$. Further we define $\alpha_\lambda^n(h, t)$ by

$$(4.8b) \qquad \alpha_\lambda^n(h, q^{-s}) = \sum_{y \in Y(n, m)} \omega(y/\delta)^\lambda \chi_v(-hy) \nu(y)^{-s},$$

where $Y(n, m) = \mathbf{p}^{-n} L_v / L_v$. Obviously $\alpha_\lambda^n(h, t)$ is a polynomial in t. We see easily from Lemma 4.4 that

$$(4.9) \qquad \alpha_{\lambda+2\mu}(h,t) = \alpha_\lambda\big(h,\theta(\mathbf{p})^\mu t\big), \qquad \alpha^n_{\lambda+2\mu}(h,t) = \alpha^n_\lambda\big(h,\theta(\mathbf{p})^\mu t\big).$$

Therefore the determination of α_λ can be reduced to the cases $\lambda = 0$ and $\lambda = 1$. However, we shall consider α_λ for an arbitrary λ, since such will occur in a natural way. In Case SU, this generalization is unnecessary. To keep our notation uniform, we use symbols α_λ and α^n_λ also in Case SU, but understand that λ is always 0 and $\omega(y/\delta)^\lambda = 1$.

We are going to relate $\alpha_\lambda(h,t)$ with the number of representations of a quadratic (or a hermitian) form by another. Write L_v also $L_v^{(m)}$ to emphasize the size of matrices. For $l \in L_v^{(m)}$ and $j \in L_v^{(k)}$, let $A(j,l,n)$ denote the number of elements x of $X(n, m \times k)$ such that $xjx^* \equiv l \pmod{\mathbf{p}^n}$. Now fix an element h of $\delta^{-1}L_v$ and a positive integer k. In Case SP, we have, by Lemma 4.4 and (4.7),

$$(4.10a) \quad \alpha^n_k(h, q^{-k/2}) = q^{-mnk} \sum_{y \in Y(n,m)} \chi_v(-hy) G_n(y/\delta)^k$$

$$= q^{-mnk} \sum_{y \in Y(n,m)} \sum_{x \in X(n, m \times k)} \mathbf{e}_v\big(\delta^{-1}\mathrm{tr}\big[(xx^* - \delta h)y\big]\big)$$

$$= q^{nm(m+1)/2 - nmk} A(1_k, \delta h, n).$$

In Case SU, we have similarly

$$(4.10b) \qquad\qquad \alpha^n_0\big(h, \theta(\mathbf{p})^k q^{-k}\big) = q^{nm(m-2k)} A(1_k, \delta h, n).$$

Therefore, if series $\alpha_\lambda(h,t)$ is convergent at $t = q^{-k}$, we have

$$(4.11) \quad \alpha_\lambda\big(h, \theta(\mathbf{p})^k q^{-k-\lambda/2}\big) = \lim_{n\to\infty} \alpha^n_\lambda\big(h, \theta(\mathbf{p})^k q^{-k-\lambda/2}\big)$$

$$= \lim_{n\to\infty} \begin{cases} q^{nm(m+1)/2 - nm(2k+\lambda)} A(1_{2k+\lambda}, \delta h, n) & \\ & \text{(Case SP),} \\ q^{nm(m-2k)} A(1_k, \delta h, n) & \text{(Case SU).} \end{cases}$$

The limit of this type in Case SP is exactly the density of representation numbers of a quadratic form by another form studied by Siegel; the case of hermitian forms was treated by Braun in [4]. Here we translate their results in [27, vol. I, pp. 503–4] and [4] in terms of α^n_λ. (Though the results of [4] are stated only in the case $[K:\mathbf{Q}] = 2$, they can easily be generalized to the case of higher degree.)

LEMMA 4.5. *Suppose $h \in \delta^{-1}L_v$ and $\det(h) \neq 0$; suppose also $2k + \lambda \geqslant m$ in Case SP and $k \geqslant m$ in Case SU. Then there exists an integer γ depending only on $\det(\delta h)$ such that*

$$\alpha_\lambda^n\big(h,\theta(\mathbf{p})^k q^{-k-\lambda/2}\big) = \alpha_\lambda^\gamma\big(h,\theta(\mathbf{p})^k q^{-k-\lambda/2}\big) \qquad if \quad n \geqslant \gamma.$$

Moreover, suppose $\det(\delta h)$ *is a* v-*unit. Then we can take* $\gamma = 1$, *and* $\alpha_\lambda^1(h, \theta^k q^{-k-\lambda/2})$ *can be given as follows:*

$$\big(1 - \theta^k q^{-k}\big)\big(1 + \xi\theta^{k-m/2}q^{m/2-k}\big) \prod_{i=1}^{m/2-1} \big(1 - q^{2i-2k}\big) \qquad (Case\ SP,\ m\ even,\ \lambda = 0),$$

$$\big(1 - \theta^k q^{-k}\big) \prod_{i=1}^{(m-1)/2} \big(1 - q^{2i-2k}\big) \qquad (Case\ SP,\ m\ odd,\ \lambda = 0),$$

$$\big(1 + \xi\theta^{k-(m-1)/2}q^{(m-1)/2-k}\big) \prod_{i=1}^{(m-1)/2} \big(1 - q^{2i-2k-2}\big) \qquad (Case\ SP,\ m\ odd,\ \lambda = 1),$$

$$\prod_{i=1}^{m/2} \big(1 - q^{2i-2k-2}\big) \qquad (Case\ SP,\ m\ even,\ \lambda = 1),$$

$$\prod_{i=0}^{m-1} \big(1 - \theta^{k-i}q^{i-k}\big) \qquad (Case\ SU).$$

Here $\theta = \theta(\mathbf{p})$ *and* $\xi = \left(\dfrac{\det(\delta h)}{\mathbf{p}}\right)$.

Now $\alpha_\lambda(h,t)$ is a power series in t convergent in a neighborhood of 0. The above lemma shows that it coincides with a polynomial $\alpha_\lambda^\gamma(h,t)$ for $t = \theta^k q^{-k-\lambda/2}$ with sufficiently large k's. Therefore $\alpha_\lambda(h,t) = \alpha_\lambda^\gamma(h,t)$. Thus we obtain

PROPOSITION 4.6. *Suppose* $h \in \delta^{-1}L_v$ *and* $\det(h) \neq 0$. *Then, under* (4.3), $\alpha_\lambda(h,t)$ *is a polynomial in* t *with coefficients in* $\mathbf{Q}$. *Moreover, suppose* $\det(\delta h)$ *is a* v-*unit. Then* $\alpha_\lambda(h,t)$ *is equal to:*

$$(1 - t)(1 + \xi\theta^{m/2}q^{m/2}t) \prod_{i=1}^{m/2-1} (1 - q^{2i}t^2) \qquad (Case\ SP,\ m\ even,\ \lambda = 0),$$

$$(1 - t) \prod_{i=1}^{(m-1)/2} (1 - q^{2i}t^2) \qquad (Case\ SP,\ m\ odd,\ \lambda = 0),$$

$$(1 + \xi\theta^{(m-1)/2}q^{m/2}t) \prod_{i=1}^{(m-1)/2} (1 - q^{2i-1}t^2) \qquad (Case\ SP,\ m\ odd,\ \lambda = 1),$$

$$\prod_{i=1}^{m/2} (1 - q^{2i-1}t^2) \qquad (Case\ SP,\ m\ even,\ \lambda = 1)$$

$$\prod_{i=0}^{m-1} (1 - \theta^i q^i t) \qquad (Case\ SU).$$

Here θ *and* ξ *are the same as in Lemma* 4.5.

PROPOSITION 4.7. *Suppose $h \in \delta^{-1} L_v$ and $\det(h) \neq 0$. Then, under (4.3), $\alpha_\lambda(h, t)$ as a polynomial is divisible by*

$$\prod_{i=0}^{[(m-1)/2]} (1 - \theta^i q^i t) \qquad (\textit{Case SP}, \lambda = 0),$$

$$\prod_{i=1}^{[m/2]} (1 - q^{2i-1} t^2) \qquad (\textit{Case SP}. \lambda = 1),$$

$$\prod_{i=0}^{m-1} (1 - \theta^i q^i t) \qquad (\textit{Case SU}).$$

Proof. Formula (4.11) is true for an arbitrary positive integer k; it is also true for $k = 0$ in Case SP if $\lambda = 1$. Suppose $2k + \lambda < m$ in Case SP and $k < m$ in Case SU. Then, for sufficiently large n, $A(1_{2k+\lambda}, \delta h, n) = 0$ in Case SP and $A(1_k, \delta h, n) = 0$ in Case SU. Hence $\alpha_\lambda(h, \theta^k q^{-k-\lambda/2}) = 0$ for such k. Now, since $h \neq 0$, we have

$$\alpha_0^n(h, 1) = \sum_{y \in Y(n,m)} \chi_v(-hy) = 0$$

for sufficiently large n. Therefore we obtain our assertion in Case SU and in Case SP, $\lambda = 0$. If $\lambda = 1$ in Case SP, we find that $\alpha_1(h, t)$ is divisible by $\prod_{i=1}^{[m/2]}(1 - \theta^{i-1} q^{i-1/2} t)$. Take $c \in \mathbf{g}_v^\times$ so that $\left(\dfrac{c}{\mathbf{p}} \right) = -1$. Substituting cy for y in (4.8), we find, by Lemma 4.4, that (for an arbitrary h)

$$(4.12) \qquad\qquad \alpha_\lambda(h, t) = \alpha_\lambda\big(ch, (-1)^\lambda t\big).$$

Thus $\alpha_1(h, t)$ is divisible by $\prod_{i=1}^{[m/2]}(1 + \theta^{i-1} q^{i-1/2} t)$. This completes the proof.

5. Computation of $\alpha_\lambda(0, t)$. The determination of $\alpha_\lambda(h, t)$ for singular h is more difficult than that for nonsingular h. We treat the case $h = 0$ in this section and the general case in the next section. If $h = 0$, our series has the form

$$(5.0) \qquad\qquad \alpha_\lambda(0, q^{-s}) = \sum_{y \in S_v / L_1} \omega(y/\delta)^\lambda \nu(y)^{-s}.$$

with δ as in Lemma 4.4. We always take $\delta = 1$ if v is unramified over $\mathbf{Q}$. Putting $h = 0$ in (4.12), we find that

$$(5.1) \qquad\qquad \alpha_1(0, t) \textit{ is a power-series in } t^2.$$

Now we have

PROPOSITION 5.1.   *$\alpha_\lambda(0, t)$ can be given as follows*:

$$\frac{1-t}{1-q^m t} \prod_{i=1}^{[m/2]} \frac{1-q^{2i}t^2}{1-q^{2m+1-2i}t^2} \qquad (Case\ SP,\ \lambda = 0),$$

$$\prod_{i=0}^{[(m-1)/2]} \frac{1-q^{2i+1}t^2}{1-q^{2m-2i}t^2} \qquad (Case\ SP,\ \lambda = 1,\ v \nmid 2),$$

$$\prod_{i=0}^{m-1} \frac{1-\theta^i q^i t}{1-\theta^{m+i-1}q^{m+i}t} \qquad (Case\ SU,\ v\ unramified\ in\ K),$$

$$\prod_{i=0}^{[(m-1)/2]} \frac{1-q^{2i}t}{1-q^{2m-2i-1}t} \qquad (Case\ SU,\ v\ ramified\ in\ K).$$

Here $\theta = \theta(\mathbf{p})$ and $[x]$ is the largest integer $\leqslant x$.

The formulas in Case SU when K_v is a field are due to Casselman, who informed the author how to derive them as an application of the methods of [5]. Similar formulas with minimal parabolic subgroups instead of the present P were obtained by Langlands [14] for Chevalley groups and by Lai [13] for groups of more general types. Their methods are probably applicable to our α_0.

Here we shall derive the formula for α_0 in Case SP and for v splitting in K in Case SU by elementary methods, and also that for α_1 in Case SP from α_0 in Case SU. We shall also indicate the possibility of obtaining the formulas for nonsplitting v in Case SU by elementary methods. Before engaging in this task, we state an immediate consequence of the above expressions:

PROPOSITION 5.2.   *The series $\sum_{y \in S_v/L_v} \nu(y)^{-s}$ is convergent for $\mathrm{Re}(s) > m$ in Case SP and for $\mathrm{Re}(s) > 2m - 1$ in Case SU. Moreover*

$$\prod_{v \in \mathbf{f}} \left\{ \sum_{y \in S_v/L_v} \nu(y)^{-s} \right\}$$

is convergent for $\mathrm{Re}(s) > m + 1$ in Case SP and for $\mathrm{Re}(s) > 2m$ in Case SU.

Define symbols B, T, U by

$$(5.2) \quad B = \begin{cases} M_m(\mathbf{g}_v), \\ M_m(\mathbf{h}_v), \end{cases} \qquad T = \begin{cases} \mathrm{GL}_m(F_v), \\ \mathrm{GL}_m(K_v), \end{cases} \qquad U = \begin{cases} \mathrm{GL}_m(\mathbf{g}_v), & (Case\ SP) \\ \mathrm{GL}_m(\mathbf{h}_v), & (Case\ SU). \end{cases}$$

We fix a Haar measure dx of T so that $\int_U dx = 1$.

LEMMA 5.3.   *Let φ be the characteristic function of B. Suppose that K_v is a field in Case SU. Define $|a|$ for $a \in K_v$ by (4.1). Then, for every $y \in M_m(F_v)$ or $\in M_m(K_v)$, we have*

$$\int_{T \cap B} \varphi(yx)|\det(x)|^s\, dx = \gamma(s)^{-1}\nu(y)^{-s}$$

with γ given by

$$\gamma(s) = \begin{cases} \displaystyle\prod_{i=0}^{m-1} (1 - q^{i-s}) & (\textit{Case SP}), \\[2em] \displaystyle\prod_{i=0}^{m-1} (1 - q^{2i-s}) & (\textit{Case SU, } v \textit{ unramified in } K), \\[2em] \displaystyle\prod_{i=0}^{m-1} (1 - q^{i-s/2}) & (\textit{Case SU, } v \textit{ ramified in } K). \end{cases}$$

The integral is convergent for $\mathrm{Re}(s) > m - 1$ *in Case SP and* $\mathrm{Re}(s) > 2m - 2$ *in Case SU.*

Proof. Obviously the integral in question depends only on UyU. Therefore we may assume that $y = \mathrm{diag}[a_1, \ldots, a_m]$. Assume that $|a_i| \leqslant 1$ if and only if $i > t$ ($0 \leqslant t \leqslant m$). Let $z = \mathrm{diag}[a_1, \ldots, a_t, 1_{m-t}]$. For $x \in B \cap T$, we have $yx \in B$ if and only if $zx \in B$. Therefore our integral is equal to

$$|\det(z)|^{-s} \int_{T \cap B} |\det(x)|^s \, dx = |\det(z)|^{-s} \sum_{\alpha \in (B \cap T)/U} |\det(\alpha)|^s.$$

The last sum is a well known local zeta function which is equal to $\prod_{i=0}^{m-1} (1 - q^{i-s})^{-1}$ in Case SP; the result in Case SU is similar. A care is necessary because of our choice of $|a|$ for $a \in K_v$. In any case, we have $|\det(z)| = \nu(y)$ by (4.2); hence our lemma follows.

Let us now compute $\alpha(0, t)$ in Case SU when v splits in K. In this case $M_m(K_v)$ can be identified with $M_m(F_v) \oplus M_m(F_v)$. Moreover, if $x = (a, b) \in M_m(K_v)$ with a, b in $M_m(F_v)$, we have $x^* = ({}^t b, {}^t a)$. Now $a \mapsto (a, {}^t a)$ maps $M_m(F_v)$ onto S_v and $M_m(\mathfrak{g}_v)$ onto L_v; moreover $\nu((a, {}^t a)) = \nu(a)$. Therefore, (5.0) can be written as $\sum_{a \in A/B} \nu(a)^{-s}$, where $A = M_m(F_v)$ and $B = M_m(\mathfrak{g}_v)$. Defining a Haar measure da on A so that $\int_B da = 1$, we have, by Lemma 5.3 (in Case SP),

$$\alpha(0, q^{-s}) = \int_A \nu(a)^{-s} \, da = \gamma(s) \int_A \int_{T \cap B} \varphi(ax) |\det(x)|^s \, dx \, da$$

$$= \gamma(s) \int_{T \cap B} |\det(x)|^s \int_A \varphi(ax) \, da \, dx$$

with symbols γ, φ, T, B in Case SP. Now $\int_A \varphi(ax) \, da = |\det(x)|^{-m}$, and hence

$$\alpha(0, q^{-s}) = \gamma(s) \int_{T \cap B} |\det(x)|^{s-m} \, dx = \gamma(s)/\gamma(s - m),$$

which proves the formula of Proposition 5.1 in Case SU, $\theta = 1$.

To treat α_0 in Case SP, put, for integers $c_1, \ldots, c_m$ such that $0 \leqslant c_1 \leqslant \cdots \leqslant c_m$,

$$\delta(c_1, \ldots, c_m) = [U : U \cap \alpha U \alpha^{-1}] \qquad \text{for} \quad \alpha = \mathrm{diag}[\pi^{c_1}, \ldots, \pi^{c_m}],$$

where π is a prime element of F_v. Also, for $0 < m \in \mathbf{Z}$, put

$$(5.3b) \qquad b(m) = b(m, q) = \prod_{i=1}^{m} (1 - q^{-i}).$$

Suppose $0 \leqslant e_1 < \cdots < e_\lambda$, $e_i \in \mathbf{Z}$, $r_1 + \cdots + r_\lambda = m$, $0 < r_i \in \mathbf{Z}$. Then

$$(5.4) \qquad \delta(\overbrace{e_1, \ldots, e_1}^{r_1}, \overbrace{e_2, \ldots, e_2}^{r_2}, \ldots, \overbrace{e_\lambda, \ldots, e_\lambda}^{r_\lambda})$$

$$= b(m)\{b(r_1) \ldots b(r_\lambda)\}^{-1} \prod_{i<j} q^{(e_j - e_i)r_i r_j}.$$

To prove this, let $\alpha = \mathrm{diag}[\pi^{e_1}, \ldots, \pi^{e_\lambda}]$ with π^{e_i} repeated r_i times. Write an element ξ of $M_m(\mathbf{g}_v)$ in the form (ξ_{ij}) with $r_i \times r_j$-matrix ξ_{ij} in the (i, j)-block. If $\xi \in U \cap \alpha U \alpha^{-1}$, ξ_{ij} must be divisible by $\pi^{e_i - e_j}$ if $i > j$. Let X be the image of $U \cap \alpha U \alpha^{-1}$ under the natural map of U onto $\mathrm{GL}_m(\mathbf{g}_v / \pi^{e_\lambda} \mathbf{g}_v)$. Observe that ξ_{ii} $\mathrm{mod}\, \pi^{e_\lambda}$ is invertible and that $b(n)q^{n^2 e}$ is the order of $\mathrm{GL}_n(\mathbf{g}_v / \pi^e \mathbf{g}_v)$. Therefore the order of X is

$$\prod_{i=1}^{\lambda} q^{r_i^2 e_\lambda} b(r_i) \prod_{i<j} q^{e_\lambda r_i r_j} \prod_{i>j} q^{(e_\lambda - e_i + e_j)r_i r_j} = q^{m^2 e_\lambda} \prod_{i=1}^{\lambda} b(r_i) \prod_{i>j} q^{(e_j - e_i)r_i r_j}.$$

Dividing the order of $\mathrm{GL}_m(\mathbf{g}_v / \pi^{e_\lambda} \mathbf{g}_v)$ by the last product, we obtain (5.4).

For integers r and n such that $0 \leqslant r \leqslant n$, put

$$(5.5) \qquad c_r^n = c_r^n(q) = \prod_{i=0}^{r-1} (q^n - q^i)/(q^r - q^i).$$

Notice that

$$(5.6) \qquad c_r^n = c_{n-r}^n = q^{r(n-r)}b(n)/\left[b(r)b(n-r)\right].$$

To simplify our notation, let us hereafter write, until the end of this section, S, L, and dy for S_v, L_v, and $d\mu_v(y)$. Then

$$(5.7) \qquad \alpha_0(0, q^{-s}) = \int_S \nu(y)^{-s} \, dy = \gamma(s) \int_S \int_{T \cap B} \varphi(yx)|\mathrm{det}(x)|^s \, dx \, dy$$

$$= \gamma(s) \int_{T \cap B} |\mathrm{det}(x)|^s \int_S \varphi(yx) \, dy \, dx.$$

Suppose $axb = \mathrm{diag}[\pi^{e_1}, \ldots, \pi^{e_m}]$ with $0 \leqslant e_1 \leqslant \cdots \leqslant e_m$ and $a, b \in U$. Observe that $\int_S \varphi(yx) \, dy = \int_S \varphi(yaxb) \, dy$. Put $y = (y_{ij})$ with $y_{ij} \in F_v$. Then $yaxb \in B$ if and only if $y_{ij} \in \pi^{-e_i} \mathbf{g}_v$ for $i \leqslant j$. Therefore

$$\int_S \varphi(yaxb) \, dy = q^{\sigma(x)} \qquad \text{with} \quad \sigma(x) = \sum_{i=1}^{m} (m + 1 - i)e_i .$$

Combining this with (5.7), we obtain

$$(5.8) \qquad \gamma(s)^{-1}\alpha_0(0, q^{-s}) = \sum_{\xi \in (T \cap B)/U} q^{\sigma(\xi)}|\det(\xi)|^s.$$

If $\alpha = \mathrm{diag}[\pi^{e_1}, \ldots, \pi^{e_m}]$, the number of cosets ξU contained in $U\alpha U$ is $\delta(e_1, \ldots, e_m)$. Thus

$$(5.9) \qquad \gamma(s)^{-1}\alpha_0(0, q^{-s}) = \sum_{0 \leqslant e_1 \leqslant \cdots \leqslant e_m} \delta(e_1, \ldots, e_m)q^{\sigma(e,s)}$$

with $\sigma(e,s) = \sum_{i=1}^m (m + 1 - i - s)e_i$. Put $t = q^{-s}$, $x_1 = q^m t$, $x_2 = q^{m-1}t, \ldots,$ $x_m = qt$; let $A_m(t)$ denote the function of (5.9). Then

$$(5.10) \qquad A_m(t) = \sum_{0 \leqslant e_1 \leqslant \cdots \leqslant e_m} x_1^{e_1} \cdots x_m^{e_m}\delta(e_1, \ldots, e_m)$$

$$= \sum_{e_1 = 0}^{\infty} (x_1 \cdots x_m)^{e_1} \sum_{0 \leqslant a_2 \leqslant \cdots \leqslant a_m} x_2^{a_2} \cdots x_m^{a_m}\delta(0, a_2, \ldots, a_m)$$

$$= (1 - x_1 \cdots x_m)^{-1} \sum_{r=1}^{m} H_r^m(t),$$

where $H_m^m = 1$ and

$$H_r^m(t) = \sum_{0 < a_{r+1} \leqslant \cdots \leqslant a_m} x_{r+1}^{a_{r+1}} \cdots x_m^{a_m}\delta(\overbrace{0, \ldots, 0}^{r}, a_{r+1}, \ldots, a_m)$$

for $0 < r < m$. By (5.4) and (5.6), we see easily that

$$(5.11) \qquad \delta(0, \ldots, 0, a_{r+1}, \ldots, a_m) = \delta(a_{r+1}, \ldots, a_m)c_r^m q^{r(r-m)}q^{r\Sigma a_i},$$

and hence $H_r^m(t)$ is equal to

$$c_r^m q^{r(r-m)} \sum_{0 < a_{r+1} \leqslant \cdots \leqslant a_m} (q^r x_{r+1})^{a_{r+1}} \cdots (q^r x_m)^{a_m}\delta(a_{r+1}, \ldots, a_m)$$

$$= c_r^m x_{r+1} \cdots x_m \sum_{0 < b_{r+1} \leqslant \cdots \leqslant b_m} (q^r x_{r+1})^{b_{r+1}} \cdots (q^r x_m)^{b_m}\delta(b_{r+1}, \ldots, b_m)$$

$$= c_r^m x_{r+1} \cdots x_m A_{m-r}(q^r t).$$

Substituting this into (5.10) and putting $k = m - r$, we obtain

$$(5.12) \qquad (1 - q^{m(m+1)/2}t^m)A_m(t) = \sum_{k=0}^{m-1} c_k^m q^{k(k+1)/2}t^k A_k(q^{m-k}t)$$

with $A_0(t) = 1$. Shifting $q^{m(m+1)/2}t^m A_m(t)$ to the right-hand side, we can express the formula in the form

$$(5.13) \qquad A_m(t) = \sum_{k=0}^{m} c_k^m q^{k(k+1)/2} t^k A_k(q^{m-k}t).$$

LEMMA 5.4. $A_m(t) = \prod_{i=1}^{m}(1 + q^i t)(1 - q^{m+i}t^2)^{-1}$.

Proof. Since A_m is uniquely determined by the recurrence formula (5.12) with $A_0 = 1$, it is sufficient to show that the A_m as stated in our lemma satisfy (5.13). For this purpose, we invoke a formula

$$(5.14) \quad \prod_{i=0}^{m-1}(x - q^i y) = \sum_{k=0}^{m}(-1)^k q^{k(k-1)/2} c_k^m \prod_{i=0}^{k-1}(y - q^i) \prod_{i=k}^{m-1}(x - q^i),$$

due to Goldman and Rota [7, Proposition 1], where x, y, and q are indeterminates. Put $x = -q^{-1}t^{-1}$ and $y = q^{-m-1}t^{-2}$ in (5.14) and multiply both sides by $(-1)^m t^{2m} q^{m(m+3)/2}\prod_{i=1}^{m}(1 - q^{m+i}t^2)^{-1}$. Then we obtain the equality which shows that the A_m as given in our lemma satisfy (5.13).

Combining this result with (5.9), we obtain

$$\alpha_0(0,t) = \gamma(s)A_m(t) = \prod_{i=1}^{m}(1 - q^{i-1}t)(1 + q^i t)(1 - q^{m+i}t^2)^{-1},$$

which proves the first case of Proposition 5.1, since the last product can easily be reduced to the product as given in the proposition.

Next let us consider Case SU, assuming that K_v is a field. Observe that (5.7) is valid in this case; namely

$$\alpha(0, q^{-s}) = \gamma(s)\int_{T \cap B}|\det(x)|^s \int_S \varphi(yx)\,dy\,dx.$$

Let $x \in \mathrm{GL}_m(K_v)$; suppose $axb = \mathrm{diag}[\pi_0^{e_1}, \ldots, \pi_0^{e_m}]$ with $a, b \in U$ and $0 \leqslant e_1 \leqslant \cdots \leqslant e_m$, where π_0 is a prime element of K_v. Put $y = (y_{ij})$ with $y_{ii} \in F_v$ and $y_{ij} \in K_v$ for $i \neq j$. Then $yaxb \in B$ if and only if $y_{ij} \in \pi_0^{-e_i}\mathfrak{h}_v$ for $i \leqslant j$. Therefore

$$\int_S \varphi(yx)\,dy = \int_S \varphi(yaxb)\,dy = q^{\sigma(x)},$$

$$\sigma(x) = \begin{cases} \displaystyle\sum_{i=1}^{m}(2m + 1 - 2i)e_i & \text{if } v \text{ is unramified in } K, \\[2ex] \displaystyle\sum_{i=1}^{m}\{(m - i)e_i + [e_i/2]\} & \text{if } v \text{ is ramified in } K. \end{cases}$$

Then (5.8) and (5.9) are valid in the present case with

$$\sigma(e,s) = \begin{cases} \displaystyle\sum_{i=1}^{m} (2m + 1 - 2i - s)e_i & \text{if } v \text{ is unramified in } K, \\ \displaystyle\sum_{i=1}^{m} \{(m - i - s/2)e_i + [e_i/2]\} & \text{if } v \text{ is ramified in } K, \end{cases}$$

and with $\delta(e_1, \ldots, e_m)$ defined by (5.3a) with π_0 instead of π. If v is ramified in K, (5.4) is valid with no change; if v is unramified, q and $b(n)$ must be replaced by q^2 and $b(n, q^2)$.

Let us now assume that v is unramified in K, and put $t = q^{-s}$ and $x_k = q^{2m+1-2k}t$ for $k = 1, \ldots, m$. Call the right-hand side of (5.9) in this case $D_m(t)$. Then

$$D_m(t) = \sum_{0 \le e_1 \le \cdots \le e_m} x_1^{e_1} \cdots x_m^{e_m} \delta_2(e_1, \ldots, e_m),$$

where δ_2 indicates δ defined with q^2 in place of q. Putting

$$J_r^m(t) = \sum_{0 < a_{r+1} \le \cdots \le a_m} x_{r+1}^{a_{r+1}} \cdots x_m^{a_m} \delta_2(\overbrace{0, \ldots, 0}^{r}, a_{r+1}, \ldots, a_m),$$

$(J_m^m = 1)$, we have $D_m(t) = (1 - x_1 \cdots x_m)^{-1}\sum_{r=1}^{m} J_r^m(t)$ in exactly the same fashion as in Case SP. By (5.11) with q^2 in place of q, we have

$$\delta_2(0, \ldots, 0, a_{r+1}, \ldots, a_m) = \delta_2(a_{r+1}, \ldots, a_m)c_r^m(q^2)q^{2r(r-m)}q^{2r\sum_i a_i},$$

so that

$$J_r^m = c_r^m(q^2)q^{2r(r-m)} \sum_{0 < a_{r+1} \le \cdots \le a_m} (q^{2r}x_{r+1})^{a_{r+1}} \cdots (q^{2r}x_m)^{a_m}\delta_2(a_{r+1}, \ldots, a_m)$$

$$= c_r^m(q^2)x_{r+1} \cdots x_m D_{m-r}(q^{2r}t).$$

Therefore, putting $k = m - r$, we obtain

$$(1 - q^{m^2}t^m)D_m(t) = \sum_{k=0}^{m-1} c_k^m(q^2)q^{k^2}t^k D_k(q^{2(m-k)}t).$$

This together with $D_0 = 1$ determines the D_m completely. Shifting $q^{m^2}t^m D_m$ to the right-hand side, we have

$$(5.15) \qquad D_m(t) = \sum_{k=0}^{m} c_k^m(q^2)q^{k^2}t^k D_k(q^{2(m-k)}t).$$

As already mentioned, Casselman's unpublished computation gives the third formula of Proposition 5.1 with $\theta = -1$. Since $\alpha(0, q^{-s}) = \gamma(s)D_m(q^{-s})$, we obtain

$$(5.16) \qquad D_m(t) = \prod_{i=0}^{m-1} \left(1 - (-q)^i t\right)\left(1 - q^{2i}t\right)^{-1}\left(1 + (-q)^{m+i}t\right)^{-1}.$$

This proves that $D_m(t)$ defined by (5.16) satisfies a recurrence formula (5.15), which may be of an independent interest. Conversely, a direct proof of this fact determines $\alpha(0, q^{-s})$; the author has no such proof, however.

Let us now assume that v is ramified in K. Let $\gamma(s)$ be defined as in the last case of Lemma 5.3. Then

$$(5.17) \qquad \gamma(2s)^{-1}\alpha(0, q^{-2s}) = \sum_{0 \leqslant e_1 \leqslant \cdots \leqslant e_m} \delta(e_1, \ldots, e_m) q^{\tau(e,s)}$$

with $\tau(e,s) = \sum_{i=1}^{m}\{(m - i - s)e_i + [e_i/2]\}$. Again the unpublished result of Casselman in this case establishes $\alpha(0, q^{-2s})$ as stated in Proposition 5.1, and consequently shows that

$$(5.18) \qquad \sum_{0 \leqslant e_1 \leqslant \cdots \leqslant e_m} \delta(e_1, \ldots, e_m) q^{\tau(e,s)}$$

$$= \prod_{i=0}^{m-1} \left(1 - q^{i-s}\right)^{-1} \prod_{i=0}^{[(m-1)/2]} \left(1 - q^{2i-2s}\right)\left(1 - q^{2m-2i-1-2s}\right)^{-1}.$$

We can actually prove a system of two recurrence formulas similar to (5.15) satisfied by the series of (5.18) and another power-series. For the same reason as above, a direct solution of these formulas will give another proof of the last case of Proposition 5.1.

Finally let us consider $\alpha_1(0, t)$ in Case SP, assuming that v does not divide 2. For simplicity, put $y[z] = {}^t\!zyz$ for $y \in S$ and $z \in (F_v)_1^m$; put also $Z = (\mathbf{g}_v)_1^m$. Define a measure dz on Z so that $\int_Z dz = 1$. If $y \in \mathbf{p}^{-n}L$, we have, by Lemma 4.4,

$$\omega(y/\delta)\nu(y)^{-1/2} = q^{-mn} \sum_{u \in X(n, m \times 1)} \mathbf{e}_v(y[u]/\delta)$$

$$= \int_Z \mathbf{e}_v(y[z]/\delta)\, dz.$$

This combined with Lemma 5.3 gives

$$\omega(y/\delta)\nu(y)^{-s-1/2} = \gamma(s) \int_{T \cap B} \varphi(yx)|\det(x)|^s\, dx \int_Z \mathbf{e}_v(y[z]/\delta)\, dz.$$

Therefore we obtain

$$(5.19) \quad \alpha_1(0, q^{-s-1/2}) = \sum_{y \in S/L} \omega(y/\delta)\nu(y)^{-s-1/2}$$

$$= \gamma(s) \int_{T \cap B} |\det(x)|^s \int_Z \int_S \varphi(yx)\mathbf{e}_v(y[z]/\delta)\, dy\, dz\, dx.$$

The integral over $Z \times S$ depends only on UxU; hence we may assume that x is diagonal. Let $M(x) = S \cap Bx^{-1}$ and $x = \mathrm{diag}[\pi^{e_1}, \ldots, \pi^{e_m}]$ with $0 \leqslant e_1 \leqslant \cdots \leqslant e_m$. Then

$$\int_S \varphi(yx)\mathbf{e}_v(y[z]/\delta)\,dy = \int_{M(x)} \mathbf{e}_v(y[z]/\delta)\,dy,$$

$$M(x) = \{(y_{ij}) \in S : y_{ij} \in \mathbf{p}^{-e_i} \text{ for } i \leqslant j\}.$$

Now $\mathbf{e}_v(y[z]/\delta) = \mathbf{e}_v(\sum_{i=1}^m \delta^{-1}y_{ii}z_i^2 + 2\sum_{i<j}\delta^{-1}y_{ij}z_iz_j)$, and hence

$$\int_{M(x)} \mathbf{e}_v(y[z]/\delta)\,dy = \prod_{i=1}^m \int_{\mathbf{p}(i)} \mathbf{e}_v(y_{ii}z_i^2/\delta)\,dy_{ii} \prod_{i<j} \int_{\mathbf{p}(i)} \mathbf{e}_v(2y_{ij}z_iz_j/\delta)\,dy_{ij},$$

where $\mathbf{p}(i) = \mathbf{p}^{-e_i}$. This is nonvanishing if and only if $z_i^2 \in \mathbf{p}^{e_i}$, i.e., $z_i \in \mathbf{p}^{[(e_i+1)/2]}$, for all i, in which case the value is $q^{\sigma(e)}$ with $\sigma(e) = \sum_{i=1}^m (m+1-i)e_i$. Thus

$$\int_Z \int_S \varphi(yx)\mathbf{e}_v(y[z]/\delta)\,dy\,dz = q^{\sigma'(e)}$$

with $\sigma'(e) = \sigma(e) - \sum_{i=1}^m [(e_i+1)/2]$. Substituting this into (5.19), we obtain

$$\alpha_1(0, q^{-s-1/2}) = \gamma(s) \sum_{0 \leqslant e_1 \leqslant \cdots \leqslant e_m} \delta(e_1, \ldots, e_m)q^{\tau(e,s)}$$

with $\tau(e,s) = \sum_{i=1}^m \{(m+1-i-s)e_i - [(e_i+1)/2]\}$. Observe that this $\tau(e,s)$ is the same as that of (5.17). Therefore $\alpha_1(0, q^{-s-1/2})$ is $\gamma(s)$ times the series of (5.18), which establishes the second case of Proposition 5.1.

6. $\alpha_\lambda(h, t)$ in the general case.   We start with an easy

LEMMA 6.1.   *Let $w = \left(\begin{smallmatrix} a & b \\ b* & d \end{smallmatrix}\right) \in S_v$ with matrices a of size n and d of size $m - n$. Suppose that w and a are invertible. Then $w^{-1} = \left(\begin{smallmatrix} * & * \\ * & e \end{smallmatrix}\right)$ with an invertible e of size $m - n$ and $\det(a) = \det(w)\det(e)$.*

Proof.   Put $f = d - b*a^{-1}b$. Then

$$\begin{pmatrix} a & b \\ b* & d \end{pmatrix} = \begin{pmatrix} a & 0 \\ b* & f \end{pmatrix}\begin{pmatrix} 1 & a^{-1}b \\ 0 & 1 \end{pmatrix}.$$

Taking the inverse of this relation, we obtain our assertions.

Our problem is to determine $\alpha_\lambda(h, t)$ when $\det(h) = 0$. We fix an element h of $\delta^{-1}L_v$ (with δ such that $\mathbf{d}_v = \delta g_v$), and assume condition (4.3). In view of Lemma 4.3, we may assume that

$$(6.1) \qquad\qquad h = \mathrm{diag}[j, 0_{m-r}], \qquad r = \mathrm{rank}(h)$$

with $j \in \delta^{-1}L_v^{(r)}$, $\det(j) \neq 0$.

PROPOSITION 6.2. *If* $\det(\delta j)$ *is a v-unit, we have, under* (4.3),

$$\alpha_0(h,t) = (1 - t) \prod_{i=1}^{[m/2]} (1 - q^{2i}t^2)$$

$$\cdot \begin{cases} (1 - \xi\theta^{r/2}q^{m-r/2}t)^{-1} \displaystyle\prod_{i=1}^{[(m-r)/2]} (1 - q^{2m-r+1-2i}t^2)^{-1} & (\textit{Case SP, } r \textit{ even}), \\[2em] \displaystyle\prod_{i=0}^{[(m-r-1)/2]} (1 - q^{2m-r-2i}t^2)^{-1} & (\textit{Case SP, } r \textit{ odd}), \end{cases}$$

$$\alpha_1(h,t) = \prod_{i=0}^{[(m-1)/2]} (1 - q^{2i+1}t^2)$$

$$\cdot \begin{cases} \displaystyle\prod_{i=0}^{[(m-r-1)/2]} (1 - q^{2m-r-2i}t^2)^{-1} & (\textit{Case SP, } r \textit{ even}), \\[2em] (1 - \xi\theta^{(r-1)/2}q^{m-r/2}t)^{-1} \displaystyle\prod_{i=1}^{[(m-r)/2]} (1 - q^{2m-r+1-2i}t^2)^{-1} & (\textit{Case SP, } r \textit{ odd}), \end{cases}$$

$$\alpha(h,t) = \prod_{i=0}^{m-1} (1 - \theta^i q^i t) \prod_{i=0}^{m-r-1} (1 - \theta^{m+i-1}q^{m+i}t)^{-1} \qquad (\textit{Case SU}).$$

Here $\theta = \theta(\mathbf{p})$ *and* $\xi = \left(\dfrac{\det(\delta j)}{\mathbf{p}} \right)$.

Proof. These formulas are true even when $r = m$ or $r = 0$ as can be seen from Propositions 4.6 and 5.1; we understand that $\xi = 1$ if $h = 0$. Therefore we may assume that $0 < r < m$ in our proof. Our idea is to find a relation between $A(1_k, \delta h, n)$ and $A(1_k, \delta j, n)$. Fix an integer $k > r$, and put $\sigma = k - r$ and $\tau = m - r$. We first treat Case SU. Suppose $xx^* \equiv \delta h \pmod{\mathbf{p}^n}$ with $x \in (\mathbf{h}_v)_k^m$. Put $x = \binom{u}{v}$ with $u \in (\mathbf{h}_v)_k^r$ and $v \in (\mathbf{h}_v)_k^\tau$. Then $uu^* \equiv \delta j$, $uv^* \equiv 0$, $vv^* \equiv 0$ $\pmod{\mathbf{p}^n}$. The number of such $u \pmod{\mathbf{p}^n}$ is $A(1_k, \delta j, n)$. Let us now count the number of v's with a fixed u. Given u, take $a \in \mathrm{SL}_r(\mathbf{h}_v)$ and $b \in \mathrm{SL}_k(\mathbf{h}_v)$ so that $aub = (c\ 0)$ with $c = \mathrm{diag}[c_1, \ldots, c_r]$. Put $d = b^{-1}b^{-*}$, $e = a\delta ja^*$, and $d = \binom{f\ *}{*\ *}$ with f of size r. Then $aubdb^*u^*a^* \equiv e \pmod{\mathbf{p}^n}$, and hence $cfc^* \equiv e \pmod{\mathbf{p}^n}$. Therefore

(6.2) $$N_{K/F}(\det(c))\det(f) \equiv \det(e) = \det(\delta j) \pmod{\mathbf{p}^n}.$$

Let $d^{-1} = \binom{*\ *}{*\ z}$ with z of size σ. By Lemma 6.1, we have $\det(z) = \det(f)$ since $\det(d) = 1$. Put $vb^{-*} = w$. Then $(c\ 0)w^* \equiv 0$, $wd^{-1}w^* \equiv 0 \pmod{\mathbf{p}^n}$ if and only if $uv^* \equiv 0$, $vv^* \equiv 0 \pmod{\mathbf{p}^n}$. Since $c \in \mathrm{GL}_r(\mathbf{h}_v)$ by virtue of (6.2), $(c\ 0)w^* \equiv 0$ implies that $w \equiv (0\ y) \pmod{\mathbf{p}^n}$ with $y \in (\mathbf{h}_v)_\sigma^\tau$. Then $wd^{-1}w^* \equiv 0$ if and only if $yzy^* \equiv 0 \pmod{\mathbf{p}^n}$. Thus the number of v's with a fixed u is equal to $A(z, 0_\tau, n)$.

Now we have

$$q^{n\tau^2}A(z,0_\tau,n) = \sum_{x \in X(n,\sigma \times \tau)} \sum_{y \in Y(n,\tau)} \mathbf{e}_v(\mathrm{tr}(x^*zxy)/\delta),$$

where X and Y are the same as in Section 4. To compute the right-hand side, take $s \in \mathrm{GL}_\sigma(\mathbf{h}_v)$ so that $s^*zs = \mathrm{diag}[\zeta_1,\ldots,\zeta_\sigma]$ with $\zeta_i \in \mathbf{g}_v$. Since $\det(z) = \det(f)$, relation (6.2) implies that the ζ_i are all v-units. Therefore, by Lemma 4.4, we obtain

$$\sum_{x \in X(n,\sigma \times \tau)} \mathbf{e}_v(\mathrm{tr}(x^*zxy)/\delta) = \prod_{i=1}^{\sigma} G_n(\zeta_i y/\delta)$$

$$= q^{2n\sigma\tau}\theta(\nu_0(y))^\sigma \nu(y)^{-\sigma},$$

and hence

$$q^{n\tau(\tau-2\sigma)}A(z,0_\tau,n) = \sum_{y \in Y(n,\tau)} \theta(\nu_0(y))^\sigma \nu(y)^{-\sigma}.$$

This value depends only on σ and τ. Therefore

$$\frac{q^{nm(m-2k)}A(1_k,\delta h,n)}{q^{nr(r-2k)}A(1_k,\delta j,n)} = \sum_{y \in Y(n,\tau)} \theta(\nu_0(y))^\sigma \nu(y)^{-\sigma}.$$

Taking the limit when n tends to ∞, we obtain, from (4.11),

$$\alpha(h,\theta^k q^{-k})/\alpha(j,\theta^k q^{-k}) = \alpha(0_\tau,\theta^{k-r}q^{r-k}), \qquad \theta = \theta(\mathbf{p}).$$

Since this holds for sufficiently large k, we obtain

(6.3) $$\alpha(h,t) = \alpha(j,t)\alpha(0_\tau,\theta^r q^r t),$$

which together with Propositions 4.6 and 5.1 proves the desired formula in Case SU.

In Case SP, we have to modify our argument at a few points. First of all, we have

(6.4) $$\det(c)^2\det(f) \equiv \det(\delta j) \qquad (\mathrm{mod}\ \mathbf{p}^n)$$

instead of (6.2). Defining z and ζ_i in a similar way, we obtain

$$q^{n\tau(\tau+1)/2}A(z,0_\tau,n) = \sum_{x \in X(n,\sigma \times \tau)} \sum_{y \in Y(n,\tau)} \mathbf{e}_v(\mathrm{tr}({}^txzxy)/\delta)$$

$$= \sum_{y \in Y(n,\tau)} \prod_{i=1}^{\sigma} G_n(\zeta_i y/\delta)$$

$$= \sum_{y \in Y(n,\tau)} q^{n\sigma\tau}\nu(y)^{-\sigma/2} \prod_{i=1}^{\sigma} \omega(\zeta_i y/\delta)$$

$$= \sum_{y \in Y(n,\tau)} q^{n\sigma\tau}\left(\frac{\det(z)}{\nu_0(y)}\right)\omega(y/\delta)^\sigma \nu(y)^{-\sigma/2}$$

by Lemma 4.4. Since $\det(z)$ in the last sum can be replaced by $\det(\delta j)$, the sum depends only on j. Therefore, substituting $2l + \lambda$ for k and making n large, we obtain, from (4.11),

$$\frac{\alpha_\lambda(h, \theta' q^{-l-\lambda/2})}{\alpha_\lambda(j, \theta' q^{-l-\lambda/2})} = \sum_\gamma \left(\frac{\det(\delta j)}{\nu_0(y)} \right) \omega(y/\delta)^{2l+\lambda-r} \nu(y)^{(r-\lambda)/2-l}$$

$$= \alpha_{\lambda-r}(0_\tau, \xi\theta' q^{(r-\lambda)/2-l})$$

for sufficiently large integer l, where $\xi = \left(\dfrac{\det(\delta j)}{\mathbf{p}} \right)$. Therefore we obtain

$$(6.5) \qquad \alpha_\lambda(h, t) = \alpha_\lambda(j, t)\alpha_{\lambda-r}(0_{m-r}, \xi q^{r/2}t),$$

which is valid for every $\lambda \in \mathbf{Z}$. Specializing λ to 0 and 1, we obtain, in view of (4.9) and (5.1),

$$(6.6a) \qquad \frac{\alpha_0(h, t)}{\alpha_0(j, t)} = \begin{cases} \alpha_0(0_\tau, \xi\theta^{r/2}q^{r/2}t) & \text{if } r \text{ is even,} \\ \alpha_1(0_\tau, q^{r/2}t) & \text{if } r \text{ is odd,} \end{cases}$$

$$(6.6b) \qquad \frac{\alpha_1(h, t)}{\alpha_1(j, t)} = \begin{cases} \alpha_1(0_\tau, q^{r/2}t) & \text{if } r \text{ is even,} \\ \alpha_0(0_\tau, \xi\theta^{(r-1)/2}q^{r/2}t) & \text{if } r \text{ is odd,} \end{cases}$$

where $\tau = m - r$. Applying Propositions 4.6 and 5.1 to $\alpha_\lambda(j, t)$ and $\alpha_\lambda(0_\tau, t)$, we complete the proof.

As to the nature of $\alpha_\lambda(h, t)$ in the general case, we are tempted to make

CONJECTURE 6.3. *Under* (4.3), $\alpha_\lambda(h, t)$ *with* $\lambda = 0, 1$ *can be expressed as a rational function in* t *whose denominator is*:

$$(1 - \xi\theta^{[r/2]}q^{m-r/2}t) \prod_{i=1}^{[(m-r)/2]} (1 - q^{2m-r+1-2i}t^2) \qquad (\textit{Case SP}, \lambda + r \textit{ even}),$$

$$\prod_{i=0}^{[(m-r-1)/2]} (1 - q^{2m-r-2i}t^2) \qquad (\textit{Case SP}, \lambda + r \textit{ odd}),$$

$$\prod_{i=0}^{m-r-1} (1 - \theta^{m+i-1}q^{m+i}t) \qquad (\textit{Case SU}),$$

where ξ *is the same as in Proposition 6.2 if* $\det(\delta j)$ *is a v-unit and* $\xi = 0$ *otherwise.*

In fact, we shall later show in Lemma 8.2 a piece of evidence which supports this conjecture at least in Case SU and in Case SP with $\lambda = 0$, r odd. In general, by Lemma 5.3, we have

$$\alpha_0(h, q^{-s}) = \gamma(s) \int_{T \cap B} |\det(x)|^s \int_S \chi_v(-hy)\varphi(yx)\,dy\,dx$$

$$= \gamma(s) \sum_{x \in (B \cap T)/U} |\det(x)|^s \int_S \chi_v(-hy)\varphi(yx)\,dy.$$

In the simplest case where $m = 2$ and $r = 1$, we can compute the last sum in a straightforward way and verify the above conjecture both in Case SP and Case SU if $m = 2$, $r = 1$, and $\lambda = 0$. It should be noted that $\alpha_0(h, t)$ was explicitly determined by Kaufhold in [11] when $G = \mathrm{Sp}(2, \mathbf{Q})$.

7. Main theorems. To state our main results, it is necessary to introduce the notion of rationality of automorphic forms over a number field. For $k \in \mathbf{Z}$, $\alpha \in G$, and a C-valued function f on H^∞, we define a function $f\|_k \alpha$ on H^∞ by

$$(7.1) \qquad (f\|_k \alpha)(z) = j(\alpha, z)^{-k} f(\alpha(z)) \qquad (z \in H^\infty).$$

This is the special case of (2.5) with $s = 0$. For a congruence subgroup Γ of G, let $\mathscr{M}_k(\Gamma)$ denote the set of all holomorphic functions f on H^∞ which satisfy $f\|_k \gamma = f$ for all $\gamma \in \Gamma$ and which satisfy the well known cusp condition at each cusp when $m = 1$ and $F = \mathbf{Q}$. Further let $\mathscr{M}_k$ denote the union of $\mathscr{M}_k(\Gamma)$ for all congruence subgroups Γ of G. If $f \in \mathscr{M}_k$, it has a Fourier expansion of the form

$$(7.2) \qquad f(z) = \sum_{h \in \Lambda} c(h) \mathbf{e}\!\left(\sum_{v \in \infty} \mathrm{tr}(h_v z_v) \right) \qquad (z = (z_v)_{v \in \infty} \in H^\infty)$$

with a lattice Λ in S. Given a subfield Φ of $\mathbf{C}$, we denote by $\mathscr{M}_k(\Phi)$ (resp. $\mathscr{M}_k(\Gamma, \Phi)$) the set of all $f \in \mathscr{M}_k$ (resp. $\mathscr{M}_k(\Gamma)$) for which $c(h) \in \Phi$ for all h.

We now consider series $E(z, s; k, \Gamma)$ and $E(z, s; k, \psi, \mathbf{b})$ defined by (2.8) and (2.18). Here Γ is an arbitrary congruence subgroup of G satisfying (2.9), $k \in \mathbf{Z}$, $\mathbf{b}$ is an integral ideal of F, and ψ is a Hecke character of F satisfying (2.13a, b). As explained in the introduction, we ask the following questions, in which $E(z, s)$ denotes any of our series.

(7.3a)   *Is $E(z, s)$ finite at $s = 0$?*
(7.3b)   *If so, is $E(z, 0)$ holomorphic in z?*
(7.3c)   *If this is also the case, does $E(z, 0)$ belong to $\mathscr{M}_k(\mathbf{Q}_{\mathrm{ab}})$?*

Here $\mathbf{Q}_{\mathrm{ab}}$ denotes the maximal abelian extension of $\mathbf{Q}$ in $\mathbf{C}$. Our first main result can be stated as follows.

THEOREM 7.1. *Suppose $k \geqslant m + 1$ in Case SP and $k \geqslant m$ in Case SU. Then the answer to (7.3a) is affirmative for both $E(z, s; k, \Gamma)$ and $E(z, s; k, \psi, \mathbf{b})$; the answers to (7.3b, c) are affirmative except in the following two cases:*
    (i) *$F = \mathbf{Q}$, $m = 1$, and $k = 2$ in Case SP;*
    (ii) *$F = \mathbf{Q}$, $k = m + 1$ in Case SU.*
Furthermore, the answers to (7.3b, c) are affirmative for $E(z, s; k, \psi, \mathbf{b})$ even in these exceptional cases if $\psi \neq 1$ in Case SP and $\psi \neq \theta^{m-1}$ in Case SU, where θ is the character defined by (4.4).

We note that our series are convergent for $\mathrm{Re}(k + s) > m + 1$ in Case SP and $\mathrm{Re}(k + s) > 2m$ in Case SU. Therefore the answers to (7.3a, b) are obviously affirmative if $k > m + 1$ in Case SP and $k > 2m$ in Case SU. If $m = 1$, group G

in both Cases SP and SU coincides with $\mathrm{SL}_2(\mathbf{Q})$. Therefore the first exceptional case of the above theorem is included in the second case.

The functions $E(z, 0)$ in these exceptional cases have interesting properties. In fact we have

THEOREM 7.2. *Suppose* $F = \mathbf{Q}$; *let* $E(z, s)$ *denote any of the functions* $E(z, s; m + 1, \Gamma)$ *and* $E(z, s; m + 1, \psi, \mathbf{b})$ *in Case SU. Define a differential operator* Δ *on* H *by*

$$(7.4) \qquad\qquad \Delta = \det(\partial / \partial z_{jk})$$

with respect to the variable matrix (z_{jk}) *on* H. *Then the following assertions hold.*

(i) *There exist two holomorphic functions* p_1 *and* p_2 *on* H *such that* $\Delta p_2 = 0$ *and that*

$$E(z, 0) = \Delta\big\{ p_1(z) + p_2(z)\log\big[\det(z - z^*)\big]\big\}.$$

(ii) Δp_1 *and* p_2 *are uniquely determined for* $E(z, 0)$ *by the properties of* (i).

(iii) $\pi^m p_2 \in \mathscr{M}_{m-1}(\mathbf{Q}_{\mathrm{ab}})$, *and* $(\Delta p_1)\|_{m+1}\gamma = \Delta p_1 + \Delta[p_2\log\{ j(\gamma, z)\}]$ *for every* $\gamma \in G$ *such that* $E(z, 0)\|_{m+1}\gamma = E(z, 0)$.

(iv) $\Delta p_1(z) = \sum_h a_h \mathbf{e}(\mathrm{tr}(hz))$ *with* $a_h \in \mathbf{Q}_{\mathrm{ab}}$, *where* h *ranges over nonnegative elements of a lattice in* S.

If $\psi \neq \theta^{m-1}$, Theorem 7.1 asserts that $E(z, 0; m + 1, \psi, \mathbf{b})$ belongs to $\mathscr{M}_{m+1}(\mathbf{Q}_{\mathrm{ab}})$, and hence $p_2 = 0$ for this $E(z, 0)$. Theorem 7.2 adds new information that $E(z, 0; m + 1, \psi, \mathbf{b}) = \Delta p_1$ with a holomorphic function p_1 on H. If $\psi = \theta^{m-1}$, we have in fact a nontrivial p_2, so that $E(z, 0)$ is nonholomorphic. If $m = 1$, p_2 is a constant, and hence our assertions imply that

$$(7.5) \qquad\qquad E(z, 0) = b\pi^{-1}(z - \bar{z})^{-1} + \sum_{h \geqslant 0} a_h \mathbf{e}(hz)$$

with b and a_n in $\mathbf{Q}_{\mathrm{ab}}$. This is a classical result due to Hecke; see Remark 7.5 below.

The restriction $k \geqslant m$ in Case SU (and perhaps $k \geqslant (m + 1)/2$ in Case SP) seems natural for the purpose of obtaining positive answers to our questions. However, the weight $m - 1$ produces another interesting phenomenon:

THEOREM 7.3. *If* $k = m - 1$ *in Case SU, our Eisenstein series have the following properties at* $s = 2$.

(i) *If* $\psi \neq \theta^{m-1}$, $E(z, s; m - 1, \psi, \mathbf{b})$ *is finite at* $s = 2$.

(ii) $E(z, s; m - 1, \Gamma)$ *and* $E(z, s; m - 1, \theta^{m-1}, \mathbf{b})$ *have at most a simple pole at* $s = 2$. *The residue at* $s = 2$ *of each one of these series is* $\pi^{-mn}R_F\prod_{v \in \infty} \det(z_v - z_v^*)^{-1}$ *times an element* $f(z)$ *of* $\mathscr{M}_{m-1}(\mathbf{Q}_{\mathrm{ab}})$ *of form* (7.2) *with* $c(h) \neq 0$ *only when* $\det(h) = 0$, *where* $n = [F : \mathbf{Q}]$ *and* R_F *is the regulator of* F.

Theorems 7.1 and 7.3 will be proved in Section 8, and Theorem 7.2 in Sections 9 and 11.

Remark 7.4. In our theorems, only the rationality over $\mathbf{Q}_{ab}$ is stated. However, we can actually determine the action of $\mathrm{Gal}(\mathbf{Q}_{ab}/\mathbf{Q})$ on $E(z,0)$ as follows. We defined in [17], [18], and [22] the action of a certain adele group on $\mathcal{M}_k(\mathbf{Q}_{ab})$ when $G = \mathrm{SL}_2(F)$ or $G = \mathrm{Sp}(m,\mathbf{Q})$. This can easily be generalized to our group in both cases. In particular we can show how the action of η is related with that of $\mathrm{Gal}(\mathbf{Q}_{ab}/\mathbf{Q})$. Now the action of $\mathrm{Gal}(\mathbf{Q}_{ab}/\mathbf{Q})$ on $E^*(z,0)$ can be determined in a straightforward way. Then, applying η to E^*, we obtain the action of $\mathrm{Gal}(\mathbf{Q}_{ab}/\mathbf{Q})$ on $E(z,0)$. This was in fact done in the case $G = \mathrm{Sp}(m,\mathbf{Q})$ by Sturm (see [28, p. 340]).

Remark 7.5. If $m = 1$, our group G in both cases coincides with $\mathrm{SL}_2(F)$. In the case $G = \mathrm{SL}_2(\mathbf{Q})$, Hecke obtained in [9] the results stated in Theorem 7.1; this was generalized by Kloosterman [12] to the case $G = \mathrm{SL}_2(F)$ and was complemented by Klingen. Hecke observed also that (7.5) holds if $k = 2$. The behavior of $E(z,s;0,\psi,\mathbf{b})$ at $s = 2$ when $G = \mathrm{SL}_2(\mathbf{Q})$ is also classical. Notice that f of Theorem 7.3, (ii) is a constant if $m = 1$.

Eisenstein series of type

$$(7.6) \qquad \sum_{\alpha \in (P \cap \Gamma)\backslash \Gamma} j(\alpha,z)^{-k}$$

were considered in [27, II, No. 32] by Siegel when $\Gamma = \mathrm{Sp}(m,\mathbf{Z})$ and the series is convergent, that is, when $k > m + 1$. He showed that the Fourier coefficients are rational by finding their relation with the special values of the zeta and L-functions of $\mathbf{Q}$. In fact, our treatment of $\alpha_0(h,q^{-s})$ in Section 4 is an adaptation of his method. Generalizations of this result were attempted by several authors; Siegel himself considered (7.6) for some other congruence subgroups of $\mathrm{Sp}(m,\mathbf{Q})$. As one of the recent investigations on this topic, the following article may be mentioned:

M. Karel, Eisenstein series and fields of definition, Compositio Mathematica 37 (1978), 121–169.

The author claims in this paper that for a rather large class of groups, the series of type (7.6) under a certain condition on k, if convergent, has Fourier coefficients in $\mathbf{Q}_{ab}$. This statement may very well be true, but unfortunately, his proof fails (at least) at one crucial point. He expresses each Fourier coefficient as a finite sum of integrals, and says that each integral is essentially the product of local integrals. The last fact is given as (2) on p. 143 of this paper, which is actually false. In fact, the formula is based on Theorem 3.2.4 on p. 142, which is false even for $G = \mathrm{SL}_2(\mathbf{Q})$. It should also be mentioned that the statement "$P(\mathbf{Q}_p)\Delta = \Delta$ for almost all p" on p. 142, line 18 is false. Therefore his proof does not apply even to $\mathrm{SL}_2(\mathbf{Q})$.

In [23], [24], we obtained some results similar to the above theorems for certain series on the group of type $\mathrm{SO}(m,2)^r \times \mathrm{SO}(m+2)^s$ or $\mathrm{SU}(m,1)^r \times \mathrm{SU}(m+1)^s$. They are somewhat different from (and in a sense more general than) the series considered in the present paper. Their Fourier expansions in the case of $\mathrm{SO}(m,2)$ have been obtained by Indik [10].

Our proof of the rationality of $E(z,0)$ requires one basic fact:

PROPOSITION 7.6.   *If $f \in \mathcal{M}_k(\mathbf{Q}_{ab})$ and $\alpha \in G$, then $f\|_k\alpha \in \mathcal{M}_k(\mathbf{Q}_{ab})$.*

This fact was proved in [17], [18], [22] for $G = SL_2(F)$ and $G = Sp(m, \mathbf{Q})$. (See also [21, Theorems 4.4 and 6.1] in Case SU with $[K : \mathbf{Q}] = 2$.) These results can easily be extended to the general case by the same methods. In fact, the case $G = Sp(m, F)$ has been treated by Garrett [6]. One remark should be added, however. In the proof of [17, Theorem 6], we employed the $\mathbf{Q}$-rationality of Eisenstein series to obtain a meromorphic modular form $g \neq 0$ such that

(7.7a) $g = p/q$ with $p \in \mathcal{M}_k(\Gamma, \mathbf{Q})$ and $q \in \mathcal{M}_l(\Gamma, \mathbf{Q})$, $k > l$, where Γ is a congruence subgroup of $Sp(m, \mathbf{Q})$.

(7.7b)   *The divisor of g (written $\mathrm{div}(g)$) defined on a $\mathbf{Q}$-rational canonical model of $\Gamma\backslash H$ is $\mathbf{Q}$-rational.*

Such a g can be constructed also from theta series, instead of Eisenstein series. Since we need Proposition 7.6 for our proof of the $\mathbf{Q}_{ab}$-rationality of Eisenstein series, let us now briefly explain such a construction, in order to make the proof of Proposition 7.6 "Eisenstein series-free". For simplicity, we consider here only the case $G = Sp(m, \mathbf{Q})$. (The general case can be proved by embedding G into $Sp(n, \mathbf{Q})$ with sufficiently large n, and pulling back theta series of $Sp(n, \mathbf{Q})$ to G.) We first take theta functions

$$\theta(u,z;r,s) = \sum_{x-r\in\mathbf{Z}^n} \mathbf{e}\left(\frac{1}{2} \cdot {}^t x z x + {}^t x(u + s)\right)$$

with $u \in \mathbf{C}^m$, $z \in H$, $r \in \mathbf{Q}^m$, $s \in \mathbf{Q}^m$. Fix a positive integer $\lambda \geqslant 3$ and let $R = \lambda^{-1}\mathbf{Z}^m/\mathbf{Z}^m$. Then $u \mapsto \Theta(u,z) = (\theta(u,z;r,0))_{r\in R}$ gives a projective embedding of $\mathbf{C}^m/(z\ \lambda 1_m)\mathbf{Z}^{2m}$; let A_z denote the image variety. Then the torsion points on A_z have the forms

$$\Theta(zq + q',z) = (\theta(zq + q',z;r,0))_{r\in R}$$

with $q, q' \in \mathbf{Q}^m$. As shown in [20, p. 369], we have

(7.7c) $$\frac{\theta(zq + q',z;r',0)}{\theta(zq + q',z;r,0)} = \frac{\theta(0,z;r' + q,q')}{\theta(0,z;r + q,q')}.$$

Observe that the right-hand side is a modular function in z. Since such torsion points are dense in A_z, A_z is defined over the field generated by the quotients of form (7.7c). Now the field of moduli of A_z for generic z is the function field of $\Gamma\backslash H$ for some Γ. Therefore we can find N algebraically independent functions of form (7.7c), where $N = m(m + 1)/2$. Multiplying these by the square of the product of the denominators, we obtain algebraically independent functions $f_1/f_0, \ldots, f_N/f_0$ with $f_k \in \mathcal{M}_\mu(\Phi)$ for some $\mu > 0$ and $\Phi \subset \mathbf{Q}_{ab}$, $[\Phi : \mathbf{Q}] < \infty$. Let $h_k = f_k/f_0$ and

$$l = (\pi i)^{-N}\partial(h_1, \ldots, h_N)/\partial(z_{11},z_{12}, \ldots, z_{mm}).$$

Then $f_0^{2N}l \in \mathcal{M}_\nu(\Phi)$ with $\nu = 2N\mu + m + 1$. Moreover, the divisor of l is the same as the divisor of $dh_1 \wedge \cdots \wedge dh_N$, which is Φ-rational on a (standard) **Q**-rational canonical model, by virtue of [17, Prop. 5], or rather by its analogue (see [17, §6] and [18]). Let $\sigma \in \mathrm{Gal}(\Phi/\mathbf{Q})$. Denote by the upper right σ the action of σ on the Fourier coefficients, and extend this to the quotients of modular forms. Then $f_k^\sigma \in \mathcal{M}_\mu(\Phi)$ since $\theta(0, z; r, s)^\sigma = \theta(0, z; r, s')$ with some s'. Observe that $l^\sigma = (\pi i)^{-N} \partial(h_1^\sigma, \ldots, h_N^\sigma)/\partial(z_{11}, \ldots, z_{mm})$. In view of the remark in [17, p. 503, line 10 from bottom], we see that $\mathrm{div}(l)^\sigma = \mathrm{div}(dh_1^\sigma \wedge \cdots \wedge dh_N^\sigma) = \mathrm{div}(l^\sigma)$. Put $g = \prod_{\sigma \in \mathfrak{g}} l^\sigma$ with $\mathfrak{g} = \mathrm{Gal}(\Phi/\mathbf{Q})$. Then g has the required properties (7.7a, b). This shows that no Eisenstein series is necessary for the proof of Proposition 7.6.

To prove Theorem 7.1, we consider E^* of (2.20), which is the transform of $E(z, s; k, \psi, \mathbf{b})$ by η. By virtue of Propositions 2.4 and 7.6, it is sufficient to show that the answers to questions (7.3a, b, c) for $E^*(z, s)$ are affirmative in the said cases. We may even replace $\mathbf{b}$ by its suitable multiple, for the reason explained by (ii) of Proposition 2.4. However, we start with k, $\mathbf{b}$, and ψ only under (2.13a, b) and another condition $\mathbf{b} \neq \mathbf{g}$ which we need in order to make Lemma 2.3 and (3.13a, b, c) valid. Then we consider the Fourier expansion (3.9a) of E^*. To proceed further, put, for each $h \in \mathbf{b}^{-1}L' = M$,

$$(7.8a) \qquad a_{\mathfrak{f}}(h, s) = \prod_{v \in \mathfrak{f}} a_v(h, s) \qquad\qquad (s \in \mathbf{C}),$$

$$(7.8b) \qquad a_\infty(h, y, s) = \prod_{v \in \infty} \xi(y_v, h_v; k + (s/2), s/2) \qquad (0 < y_v \in S_v),$$

$$(7.8c) \qquad a(h, y, s) = c_\mu a_\infty(h, y, s) a_{\mathfrak{f}}(h, s)$$

with a_v of (3.13c) and (3.23), ξ of (3.18), and c_μ of (3.20). Let us also recall the Fourier expansion (3.9a) of E^*:

$$(7.9) \quad E^*(x + iy, s) = \sum_{h \in M} a(h, y, s) \mathbf{e}\left(\sum_{v \in \infty} \mathrm{tr}(h_v x_v) \right) \qquad (x + iy \in H^\infty).$$

The analytic nature of functions ξ was studied in [26]. Put

$$(7.10a) \qquad \Gamma_m(s) = \begin{cases} \pi^{m(m-1)/4} \prod_{\nu=0}^{m-1} \Gamma(s - (\nu/2)) & \text{(Case SP)}, \\ \pi^{m(m-1)/2} \prod_{\nu=0}^{m-1} \Gamma(s - \nu) & \text{(Case SU)}, \end{cases}$$

$$(7.10b) \qquad \kappa = \kappa(m) = \begin{cases} (m+1)/2 & \text{(Case SP)}, \\ m & \text{(Case SU)}. \end{cases}$$

We put $\Gamma_0 = 1$. Let $S_v^+ = \{ y \in S_v \mid y > 0 \}$ for $v \in \infty$. In [26], we obtained a function $\omega(g, h; \alpha, \beta)$ defined for $(g, h; \alpha, \beta) \in S_v^+ \times S_v \times \mathbf{C}^2$ which is holomorphic in $(\alpha, \beta) \in \mathbf{C}^2$ and satisfies

$$(7.11) \quad \xi(g,h;\alpha,\beta) = i^{m\beta - m\alpha} 2^\tau \pi^\theta \Gamma_\iota(\alpha + \beta - \kappa) \Gamma_{m-q}(\alpha)^{-1} \Gamma_{m-p}(\beta)^{-1}$$

$$\cdot \det(g)^{\kappa - \alpha - \beta} \delta_+ (hg)^{\alpha - \kappa + \iota q/4} \delta_- (hg)^{\beta - \kappa + \iota p/4}$$

$$\cdot \, \omega(2\pi g, h; \alpha, \beta).$$

(See [26, (4.34.K)]].) Here p (resp. q) is the number of positive (resp. negative) eigenvalues of h and $\iota = m - p - q$; $\delta_+ (x)$ is the product of all positive eigenvalues of x and $\delta_- (x) = \delta_+ (-x)$; $\iota = 1$ in Case SP and $\iota = 2$ in Case SU;

$$\tau = (2p - m)\alpha + (2q - m)\beta + m + \iota\kappa + (\iota pq/2),$$

$$\theta = p\alpha + q\beta + \iota + (\iota/2)\{\iota(\iota - 1) - pq\}.$$

In particular, we have

$$(7.12) \quad \xi(g,h;\alpha,0) = 2^{(1-\kappa)m} i^{-m\alpha} (2\pi)^{m\alpha} \Gamma_m(\alpha)^{-1} \det(h)^{\alpha - \kappa} e(i\operatorname{tr}(gh)),$$

$$(7.13) \quad \xi(g,0;\alpha,\beta) = i^{m\beta - m\alpha} 2^{m(\kappa + 1 - \alpha - \beta)} \pi^{m\kappa} \Gamma_m(\alpha + \beta - \kappa) \Gamma_m(\alpha)^{-1} \Gamma_m(\beta)^{-1}$$

$$\cdot \det(g)^{\kappa - \alpha - \beta},$$

$$(7.14) \quad \lim_{s \to 0} \xi(g,h;\kappa + (s/2), s/2) = 2^\sigma i^{-m\kappa} \pi^{m\kappa} \Gamma_m(\kappa)^{-1} e(i\operatorname{tr}(gh))$$

$$\text{if } q = 0 \text{ with } \sigma = [(m + p)/2] \text{ in Case SP}, \sigma = p \text{ in Case SU}.$$

These follow from [26, (1.31), (4.35.K)] and (7.11).

To express $a_f(h,s)$ in an explicit form, we have to consider the L-function of a Hecke (idele) character φ of F. For an integral ideal $\mathbf{a}$ of F, we put

$$(7.15) \quad L_\mathbf{a}(s,\varphi) = \prod_{v \nmid \mathbf{a}} (1 - \varphi(\pi_v) q_v^{-s})^{-1} = \sum_{(\mathbf{x},\mathbf{a}) = 1} \varphi^*(\mathbf{x}) N(\mathbf{x})^{-s},$$

where π_v is a prime element of F_v, $q_v = |\pi_v|_v^{-1}$, φ^* is the ideal character attached to φ, and $\mathbf{x}$ ranges over all integral ideals of F prime to $\mathbf{a}$. We shall often use the same symbol φ for φ^*. We write $L(s,\varphi)$ for $L_\mathbf{a}(s,\varphi)$ if $\mathbf{a} = \mathbf{g}$. Obviously we have

$$(7.16) \quad L_\mathbf{a}(s,\varphi) = L(s,\varphi) \prod_{\mathbf{p} \mid \mathbf{a}} (1 - \varphi^*(\mathbf{p}) N(\mathbf{p})^{-s}).$$

We are going to express $a_f(h,s)$ as a product of several L-functions of type (7.15). For this purpose, we have to introduce, in Case SP, a quadratic character ξ_h for each $h \in S$, $\neq 0$, of even rank. Given such an h, take $w \in \mathrm{GL}_m(F)$ so that

$'whw = \text{diag}[l, 0_{m-r}]$ with $l \in GL_r(F)$, $0 < r \leqslant m$. Then we put

$$(7.17) \qquad \xi_h(\mathbf{x}) = \left(\frac{F\big(\det(l)^{1/2}\big)/F}{\mathbf{x}} \right)$$

for every ideal $\mathbf{x}$ of F prime to the conductor of $F(\det(l)^{1/2})$ over F.

Given $\mathbf{b}$ and $h \in \mathbf{b}^{-1}L'$, we define an integral ideal $\mathbf{n}$ depending on $\mathbf{b}$ and h as follows:

$$(7.18) \qquad \mathbf{n} = \mathbf{n}(\mathbf{b}, h) = \begin{cases} \mathbf{n}_0\mathbf{n}_1 & \text{if } \quad h \neq 0, \\ \mathbf{g} & \text{if } \quad h = 0. \end{cases}$$

Here $\mathbf{n}_0$ is the product of all $v \in \mathbf{f}$, not dividing $\mathbf{b}$, which divide 2 in Case SP and which are ramified in K in Case SU; $\mathbf{n}_1$ is the product of all $v \in \mathbf{f}$ which do not divide $\mathbf{bn}_0$ and for which $\delta_v h$ has an elementary divisor other than 0 and v-units, where δ_v is an element of F_v such that $\delta_v \mathbf{g}_v = \mathbf{d}_v$. Observe that $\xi_h(\mathbf{x})$ is well-defined for $\mathbf{x}$ prime to $\mathbf{bn}$.

Given a character ψ modulo $\mathbf{b}$, we put

$$(7.19) \qquad D(h, \psi, s) = \prod_{v \nmid \mathbf{bn}} \alpha(v, h, \psi(\pi_v)q_v^{-s})$$

with α of (3.22). By (3.16) and (3.23), we have, for every $h \in \mathbf{b}^{-1}L'$,

$$(7.20) \qquad a_{\mathfrak{f}}(h, s) = N(\mathbf{b})^{m\kappa} D(h, \psi, k + s) \prod_{v \mid \mathbf{n}} \alpha\big(v, h, \psi(\pi_v)q_v^{-k-s}\big).$$

Now, from Propositions 4.6, 5.1, and 6.2, we obtain

THEOREM 7.7. *$D(h, \psi, s)$ is given as follows.*
I: $h = 0$.

$$\frac{L_{\mathbf{b}}(s - m, \psi)}{L_{\mathbf{b}}(s, \psi)} \prod_{i=1}^{[m/2]} \frac{L_{\mathbf{b}}(2s - 2m - 1 + 2i, \psi^2)}{L_{\mathbf{b}}(2s - 2i, \psi^2)} \qquad (Case\ SP),$$

$$\prod_{i=0}^{m-1} \frac{L_{\mathbf{b}}(s - m - i, \psi\theta^{m+i-1})}{L_{\mathbf{b}}(s - i, \psi\theta^i)} \qquad (Case\ SU).$$

Here θ^j for even j means the trivial character of conductor 1 so that the Euler factors for v ramified in K may be included.

II: $0 < r = \text{rank}(h) \leqslant m$.

$$\frac{L_{\mathbf{bn}}\big(s - m + (r/2), \psi\xi_h\theta^{r/2}\big)\prod_{i=1}^{[(m-r)/2]} L_{\mathbf{bn}}(2s - 2m + r - 1 + 2i, \psi^2)}{L_{\mathbf{bn}}(s,\psi)\prod_{i=1}^{[m/2]} L_{\mathbf{bn}}(2s - 2i, \psi^2)}$$

$$(\textit{Case SP}, r \textit{ even}),$$

$$\frac{\prod_{i=0}^{[(m-r-1)/2]} L_{\mathbf{bn}}(2s - 2m + r + 2i, \psi^2)}{L_{\mathbf{bn}}(s,\psi)\prod_{i=1}^{[m/2]} L_{\mathbf{bn}}(2s - 2i, \psi^2)} \qquad (\textit{Case SP}, r \textit{ odd}),$$

$$\frac{\prod_{i=0}^{m-r-1} L_{\mathbf{bn}}(s - m - i, \psi\theta^{m+i-1})}{\prod_{i=0}^{m-1} L_{\mathbf{bn}}(s - i, \psi\theta^i)} \qquad (\textit{Case SU}).$$

Here the product $\prod_{i=a}^{b}$ means 1 if $a > b$.

8. Proofs of Theorems 7.1 and 7.3. In the rest of the paper, we put $n = [F : \mathbf{Q}]$. We first recall a basic fact on the values of L-functions at integers. Let φ, $\mathbf{a}$, and $L_{\mathbf{a}}$ be the same as in (7.15); suppose $\varphi(x) = \prod_{v \in \infty} \text{sgn}(x_v)^\lambda$ for $x \in F_\infty^\times$ with $\lambda = 0$ or 1. Then it is well known (due to Klingen) that:

(8.0)   *The numbers $\pi^{-n(2l-\lambda)}L_{\mathbf{a}}(2l - \lambda, \varphi)$ and $L_{\mathbf{a}}(\lambda + 1 - 2l, \varphi)$ belong to $\mathbf{Q}_{\mathbf{ab}}$*

   for every positive integer l.

Now, as explained in the proof of Proposition 3.3, for every $s_0 \in \mathbf{C}$, there is a positive integer ν such that $(s - s_0)^\nu E^*(z,s)$ is holomorphic at s_0 for all $z \in H^\infty$. It follows that $(s - s_0)^\nu a(h, y, s)$ as a Fourier coefficient of $(s - s_0)^\nu E^*(z,s)$ is holomorphic at s_0. Therefore we can conclude that $E^*(z,s)$ is holomorphic at s_0 if and only if $a(h, y, s)$ is holomorphic at $s = s_0$ for every $h \in M$. Thus Theorem 7.1 can be obtained by carefully examining the nature of $a(h, y, s)$ at $s = 0$ for each given ψ and k. To illustrate this procedure, let us first treat Case SP.

Suppose $k \geqslant m + 1$. By Proposition 5.2, the last finite product $\prod_{v \mid \mathbf{n}}$ of (7.20) is finite at $s = 0$. Therefore (7.11) shows that $a(h, y, s)$ is finite or vanishes at $s = 0$ according as

(8.1) $$D(h, \psi, k + s) \prod_{v \in \infty} \left\{ \Gamma_{m-r_v}(k - \kappa + s)\Gamma_{m-q_v}\left(k + \frac{s}{2}\right)^{-1}\Gamma_{m-p_v}\left(\frac{s}{2}\right)^{-1} \right\}$$

is finite or vanishes at $s = 0$, where p_v and q_v are the numbers of positive and negative eigenvalues of h_v, and $r_v = p_v + q_v$. Since $k \geqslant m + 1$, the factors Γ_{m-r} and Γ_{m-q}^{-1} are finite and nonvanishing at $s = 0$; $\Gamma_{m-p}(s/2)^{-1}$ is nonvanishing only when $m = p$. Now the expressions for D given in Theorem 7.6 show that

$D(h,\psi,k + s)$ is finite at $s = 0$ except when

(8.2) $h = 0, \quad \psi = 1, \quad \text{and} \quad k = m + 1.$

In this exceptional case, it has a simple pole at $s = 0$; k must be even because of (2.13b). However, this pole is cancelled by the pole of $\Gamma_m(s/2)$ at $s = 0$; moreover we see that (8.1) under (8.2) vanishes if $F \neq \mathbf{Q}$ or $m > 1$. Therefore $E^*(z,s)$ is always finite at $s = 0$; the Fourier coefficient is nonvanishing only when h is totally positive definite or when $F = \mathbf{Q}$, $m = 1$, $\psi = 1$, and $k = 2$. Let us now exclude the last case. If h is totally positive definite, (7.12) shows that
$$a_\infty(h, y, 0) = \{2^{(1-\kappa)m}(-2\pi i)^{mk}\Gamma_m(k)^{-1}\}^n N_{F/\mathbf{Q}}(\det(h))^{k-\kappa}\mathbf{e}(i\sum_{v \in \infty} \text{tr}(h_v y_v)).$$
Substituting this into (7.9), we see that $E^*(z,0)$ has an expansion of form (7.2), and hence is holomorphic in z. On the other hand (8.0), together with (2.13b), shows that if h is totally positive definite, $D(h,\psi,k)$ is a number of $\mathbf{Q}_{\text{ab}}$ times $\pi^{(\lambda^2 - km)n}$ or $\pi^{(\lambda(\lambda+1)-km)n}$ according as m is even or odd, where $\lambda = [m/2]$. By Proposition 4.6 (or 3.3), the last product $\prod_{v|\mathfrak{n}}$ of (7.20) at $s = 0$ is a number of $\mathbf{Q}_{\text{ab}}$. Combining all pieces of information, we see that $E^*(z,0) \in \mathcal{M}_k(\mathbf{Q}_{\text{ab}})$. This proves Theorem 7.1 in Case SP.

The same argument applies to Case SU if $k \geq 2m$ except when $F = \mathbf{Q}$, $m = 1$, $\psi = 1$, and $k = 2$. In order to obtain our results for smaller values of k, we have to make an estimate of $\prod_{v|\mathfrak{n}}$ sharper than that of Proposition 5.2. For this purpose, we first prove

LEMMA 8.1.   *Let $0 < r \leq m$; suppose r is odd in Case SP. Then there exists a finite subset Z of $\mathfrak{f}$ with the following property: for every finite subset Y of $\mathfrak{f}$ disjoint with Z and for $j_v \in L_v$ of rank r for each $v \in Y$, there exists an element $h = \text{diag}[h', O_{m-r}]$ of L such that:*
   (i) $u_v h u_v^* = e_v j_v$ *with $u_v \in \text{GL}_m(\mathfrak{g}_v)$ resp. $\text{GL}_m(\mathfrak{h}_v)$ in Case SP resp. SU and $e_v \in \mathfrak{g}_v^\times$ for each $v \in Y$ (e_v is unnecessary in Case SU);*
   (ii) $h_v' < 0$ *for every $v \in \infty$;*
   (iii) $\det(h')$ *is a v-unit for every $v \notin Y \cup Z$.*

Proof.   We first treat Case SP. Let $\mathfrak{C}$ be a complete set of representatives of the ideal classes of F modulo ∞ consisting of integral ideals. Take Z to be the set consisting of the prime factors of 2 and the members of $\mathfrak{C}$. Suppose Y and j_v for $v \in Y$ are given as above. By Lemma 4.3, we may assume that j_v is diagonal. Therefore it is sufficient to treat the case $r = m$. Now we can find an integral ideal $\mathbf{a}$ such that $\mathbf{a}_v = \det(j_v)\mathfrak{g}_v$ for $v \in Y$ and $\mathbf{a}_v = \mathfrak{g}_v$ for $v \notin Y$. Then $\mathbf{a} = \mathbf{b}\mathbf{c}^{-1}$ for some $\mathbf{c} \in \mathfrak{C}$ with a totally negative element b of F. Then $b \in \mathfrak{g}$. If $m = 1$, we can take b to be the desired h. Suppose $m \geq 3$. If $v \in Y$, we have $b = e_v \det(j_v)$ with $e_v \in \mathfrak{g}_v^\times$, and hence $be_v^{m-1} = \det(e_v j_v)$. Since m ($= r$) is odd, b represents the discriminant of the quadratic form given by $e_v j_v$. For every nondegenerate quadratic form $f = \sum_{i=1}^m a_i x_i^2$ with $a_i \in F_v$, define its Hasse v-invariant $H_v(f)$ by

$$H_v(f) = \prod_{i < k}\left(\frac{a_i, a_k}{v}\right)$$

with Hilbert's symbol $\left(\dfrac{\ }{v}\right)$. Fix any prime w in Z and define $f^v \in S_v$ for each $v \in \infty \cup \mathbf{f}$ by $f^v = e_v j_v$ for $v \in Y$, and $f^v = \mathrm{diag}[b, -1_{m-1}]$ for $v \notin Y \cup \{w\}$. Then we choose $f^w \in L_w$ so that $H_w(f^w) = \prod_{v \neq w} H_v(f^v)$ and $b^{-1}\det(f^w)$ is a square in F_w. This is possible since $m \geqslant 3$. By a classical existence theorem due to Hasse, we can find an element f of S such that $b^{-1}\det(f)$ is a square in F and $f = {}^t x_v f^v x_v$ with $x_v \in \mathrm{GL}_m(F_v)$ for every $v \in \mathbf{f} \cup \infty$. We may assume that $f \in L$. Then there is a $\mathbf{g}$-lattice Λ in F^m such that

$$\Lambda_v = \begin{cases} x_v \mathbf{g}_v^m & \text{if } v \in Y \cup \{v \in \mathbf{f} : v \notin Z, v \mid \det(f)\}, \\ \mathbf{g}_v^m & \text{otherwise.} \end{cases}$$

By a well-known fact on lattices in F^m, we have

$$s\Lambda = \{(a_i) \in \mathbf{g}^m : a_m \in \mathbf{c}'\}$$

with $s \in \mathrm{GL}_m(F)$ and $\mathbf{c}' \in \mathfrak{C}$. Put $h = {}^t sfs$. If $v \in Y$ we have $h = {}^t s \, {}^t x_v e_v j_v x_v s$ and $sx_v \mathbf{g}_v^m = \mathbf{g}_v^m$. Suppose $v \notin Y \cup Z$. Then b is a v-unit. If $v \mid \det(f)$, then we have $sx_v \mathbf{g}_v^m = \mathbf{g}_v^m$ and

$$h = {}^t s \cdot {}^t x_v \mathrm{diag}[b, -1_{m-1}]x_v s,$$

so that $h \in L_v$ and $\det(h)$ is a v-unit. If $v \nmid \det(f)$, we have $h = {}^t sfs$ and $s\mathbf{g}_v^m = \mathbf{g}_v^m$ so that $h \in L_v$ and $\det(h)$ is a v-unit. Finally suppose $v \in Z$. Then $s\mathbf{g}_v^m \subset \mathbf{g}_v^m$, so that $h \in L_v$. Thus h has the required properties.

Case SU is somewhat simpler. Let $\mathfrak{C}'$ be a complete set of representatives of the ideal classes of K consisting of integral ideals. Take Z to be the set consisting of the prime factors of the members of $\mathfrak{C}$ and of $N_{K/F}(\mathbf{c})$ for $\mathbf{c} \in \mathfrak{C}'$, and the primes of F ramified in K. We may again assume $m = r$. Given Y and j_v, take a totally negative element b of $\mathbf{g}$ so that $b^{-1}\det(j_v)$ is a v-unit for $v \in Y$, and b is a v-unit for $v \notin Y \cup Z$. Put $f = \mathrm{diag}[b, -1_{m-1}]$. Then $f = x_v^* j_v x_v$ with $x_v \in \mathrm{GL}_m(K_v)$ for every $v \in Y$. Take an $\mathbf{h}$-lattice Λ in K^m and an element s of $\mathrm{GL}_m(K)$ in exactly the same fashion as in Case SP with $\mathbf{h}$ in place of $\mathbf{g}$ and with $x_v = 1$ for $v \notin Y$. Then $h = s^* fs$ has the required properties.

We note here that the above lemma is false for even r in Case SP, even when the condition at archimedean primes is dropped.

LEMMA 8.2. *Let $r \in \mathbf{Z}$, $k \in \mathbf{Z}$, $0 < r < m$, $k \geqslant \kappa(m)$; suppose r is odd in Case SP. Given such m, r, k, an integral ideal $\mathbf{b}$, and a character ψ satisfying (2.13a, b), there exists a finite subset $P(m, r, k, \mathbf{b}, \psi)$ of $\mathbf{f}$ including all prime factors of $\mathbf{b}$ such that*

$$(8.3) \quad a_v(h, s) \cdot \begin{cases} \displaystyle\prod_{i=0}^{[(m-r-1)/2]} \left(1 - \psi(\pi_v)^2 q_v^{2m-r-2i-2k-2s}\right) & (\text{Case SP}) \\[2em] \displaystyle\prod_{i=0}^{m-r-1} \left(1 - \psi(\pi_v)\theta(\pi_v)^{m+i-1} q_v^{m+i-k-s}\right) & (\text{Case SU}) \end{cases}$$

is finite at $s = 0$ for every $v \notin P(m, r, k, \mathbf{b}, \psi)$ and every $h \in \mathbf{b}^{-1}L'$ of rank r.

Proof. Let us prove here only Case SU, as Case SP can be treated in the same way. Suppose no such $P(m, r, k, \mathbf{b}, \psi)$ exists. Then there exists an infinite subset W of $\mathbf{f}$ and $j^v \in \mathbf{b}^{-1} L'$ for each $v \in W$ such that

$$(8.4) \qquad a_v(j_v, s) \prod_{i=0}^{m-r-1} \left(1 - \psi(\pi_v)\theta(\pi_v)^{m+i-1} q_v^{m+i-k-s} \right)$$

has a pole at $s = 0$. Take a finite subset Z of $\mathbf{f}$ as in Lemma 8.1. We may assume that Z contains all the primes of F ramified in K or ramified over $\mathbf{Q}$. Let $\mathbf{b}'$ be the least common multiple of $\mathbf{b}$ and the members of Z. Take any finite subset Y of W including no prime factors of $\mathbf{b}'$. By Lemma 8.1, we can find an element $h = \mathrm{diag}[-l, O_{m-r}]$ of L with totally positive definite l of size r, such that $c_v^* h c_v = j^v$ with $c_v \in \mathrm{GL}_m(\mathbf{h}_v)$ for every $v \in Y$, and that $\det(l)$ is a v-unit for $v \notin Y \cup Z$. By [26, (4.7.K), (4.8), (4.10), (4.11.K)], if $v \in \infty$ and $0 < b = b^* \in M_r(\mathbf{C})$, we have

$$\omega(\mathrm{diag}[\, b, 1_{m-r}], h_v; \alpha, \beta) = (\pi/2)^{(m-r)r} \omega(b, l_v; \beta - m + r, \alpha - m + r)$$

$$= (\pi/2)^{(m-r)r} \omega(l_v^{1/2} b l_v^{1/2}, 1_r; \beta - m + r, \alpha - m + r)$$

$$= 2^{-rm} \pi^{(m-r)r} e^{-\mathrm{tr}(bl_v)} \omega(2 l_v^{1/2} b l_v^{1/2}; \beta - m + r, \alpha - m + r).$$

This is positive for $\alpha - m + r > r - 1$ and $\beta \in \mathbf{R}$ by virtue of [26, (3.2), (3.6), and Theorem 3.1]. Therefore (7.8b) and (7.11) show that

$$\left\{ \Gamma_{m-r}(s + k - m)^{-1} \Gamma_{m-r}(k + (s/2)) \Gamma_m(s/2) \right\}^n a_\infty(h, y, s)$$

is finite and nonvanishing at $s = 0$ (if $k \geqslant m$), and hence $a_\infty(h, y, s)$ has a zero of order N at $s = 0$, with a positive integer N depending only on m, r, k, and n. Now consider E^* with $\mathbf{b}'$ in place of $\mathbf{b}$ and examine its Fourier coefficient $a_\mathbf{f}(h, s)$, By (7.20) and Theorem 7.7, we have

$$(8.5) \quad N(\mathbf{b}')^{-m\kappa} a_\mathbf{f}(h, s)$$

$$= \prod_{i=0}^{m-r-1} L_{\mathbf{b}'}(s + k - m - i, \psi\theta^{m+i-1})$$

$$\cdot \prod_{i=0}^{m-1} L_{\mathbf{b}'\mathbf{n}'}(s + k - i, \psi\theta^i)^{-1}$$

$$\cdot \prod_{v \mid \mathbf{n}'} \left\{ a_v(h, s) \prod_{i=0}^{m-r-1} \left(1 - \psi(\pi_v)\theta(\pi_v)^{m+i-1} q_v^{m+i-k-s} \right) \right\},$$

where $\mathbf{n}' = \mathbf{n}(\mathbf{b}', h)$. The second product $\prod_{i=0}^{m-1}$ is finite and nonvanishing at $s = 0$ if $k \geqslant m$. (Observe that $k - i = 1$ is incompatible with $\psi\theta^i = 1$.) Let ν ($\geqslant -1$) be the order of zero of the first product $\prod_{i=0}^{m-r-1}$ at $s = 0$. Then ν depends only on m, r, k, ψ, and $\mathbf{b}'$. Now consider a prime $v \nmid \mathbf{b}'$. If $v \notin Y$, $\det(l)$ is a v-unit, so that

$v \nmid \mathbf{n}'$. If $v \in Y$, we have $a_v(h,s) = a_v(j^v,s)$ and hence

$$(8.6) \qquad a_v(h,s) \prod_{i=0}^{m-r-1} \left(1 - \psi(\pi_v)\theta(\pi_v)^{m+i-1} q_v^{m+i-k-s}\right)$$

has a pole at $s = 0$. This means that $v \mid \mathbf{n}'$, since if $v \nmid \mathbf{b}'\mathbf{n}'$, our definition of $\mathbf{n}'$ together with Proposition 6.2 implies that (8.6) is finite for all $s \in \mathbf{C}$. This shows that $\mathbf{n}'$ is exactly the product of all primes in Y, and hence the product $\prod_{v \mid \mathbf{n}'}$ of (8.5) has a pole of order at least λ, where λ is the number of elements in Y. It follows that $a(h, y, s)$ has a pole of order at least $\lambda - N - \nu$ at $s = 0$. Now without changing $\mathbf{b}'$, N, and ν, we can make λ arbitrarily large, since W is an infinite set. This is a contradiction, since the order of the pole of $a(h, y, s)$ at $s = 0$ is not greater than that of $E^*(z,s)$ at $s = 0$. This completes the proof in Case SU. Case SP can be proved in exactly the same fashion.

Let us now prove Theorem 7.1 in Case SU. Given k ($\geqslant m$), ψ, and $\mathbf{b}$, we take $P(m,r,k,\mathbf{b},\psi)$ as in Lemma 8.2 for each positive $r > m$. Let $\mathbf{b}'$ be the least common multiple of $\mathbf{b}$ and the members of $P(m,r,k,\mathbf{b},\psi)$ for all such r. We then consider $E^*(z,s)$ with $\mathbf{b}'$ in place of $\mathbf{b}$. For the reason already explained, it is sufficient to show that $E^*(z,0)$ has the desired properties. Let $h \in \mathbf{b}'^{-1}L'$ and $r = \mathrm{rank}(h)$.

I: $h = 0$. We have

$$a_\infty(0, y, s)$$

$$= \left\{ i^{-km} 2^{m(m+1-k-s)} \pi^{mk} \Gamma_m(s + k - m) \Gamma_m(k + (s/2))^{-1} \Gamma_m(s/2)^{-1} \right\}^n$$

$$\cdot \prod_{v \in \infty} \det(y_v)^{m-k-s},$$

$$a_{\mathbf{f}}(0,s) = N(\mathbf{b}')^{mk} \prod_{i=0}^{m-1} L_{\mathbf{b}'}(s + k - m - i, \psi\theta^{m+i-1}) / L_{\mathbf{b}'}(s + k - i, \psi\theta^i).$$

Therefore we can easily verify that

$$(8.7) \qquad a(0, y, 0) = \begin{cases} \text{a number of } \mathbf{Q}_{\mathrm{ab}} & \text{if } k = m, \\ \text{a number of } \mathbf{Q}_{\mathrm{ab}} \text{ times } \pi^{-m}\det(y)^{-1} \\ \qquad \text{if} \quad F = \mathbf{Q}, \quad k = m + 1, \quad \text{and} \quad \psi = \theta^{m-1}, \\ 0 & \text{otherwise.} \end{cases}$$

II: $\det(h) \neq 0$. In this case, the analysis is similar to Case SP; $a(h, y, s)$ is nonvanishing only if h is totally positive definite, in which case its value at $s = 0$ is $\mathbf{e}(i\sum_{v \in \infty} \mathrm{tr}(h_v y_v))$ times a number of $\mathbf{Q}_{\mathrm{ab}}$.

III: $0 < r < m$. We have again (8.5). By our choice of $\mathbf{b}'$, the product $\prod_{v \mid \mathbf{n}'}$ of (8.5) is finite at $s = 0$, and its value belongs to $\mathbf{Q}_{\mathrm{ab}}$. Therefore, examining carefully the behavior of $D(h, \psi, s + k)$ and $a_\infty(h, y, s)$ at $s = 0$, we find that $a(h, y, s)$ is nonvanishing at $s = 0$ only in the following two cases (A) and (B).

(A) $k = m$ and h_v has no negative eigenvalues for every $v \in \infty$. In this case, we see from (7.12) that $a(h, y, 0)$ is $\mathbf{e}(i\sum_{v \in \infty} \operatorname{tr}(h_v y_v))$ times a number of $\mathbf{Q}_{ab}$.

(B) $k = m + 1$, $F = \mathbf{Q}$, $\psi = \theta^{m-1}$, and h has no negative eigenvalues. In this case, we obtain, from (7.11) and Theorem 7.7,

$$(8.8) \qquad a(h, y, 0) = \gamma_h \pi^{(p+1)(p-m)} \det(y)^{-1} \delta_+ (hy) \omega(2\pi y, h; m + 1, 0)$$

with $\gamma_h \in \mathbf{Q}_{ab}$, where $p = \operatorname{rank}(h)$.

Concluding all, we see that $E^*(z, 0)$ is holomorphic in z and belongs to $\mathcal{M}_k(\mathbf{Q}_{ab})$ except in Case (B). This together with Propositions 2.4 and 7.6 completes the proof of Theorem 7.1.

Proof of Theorem 7.3.   Let $\mathcal{M}_k^0$ be the set of all elements f of $\mathcal{M}_k$ of the form

$$(8.9) \qquad f(z) = \sum_{h \in S} c(h) \mathbf{e}\left(\sum_{v \in \infty} \operatorname{tr}(h_v z_v) \right)$$

with $c(h) \neq 0$ only when $\det(h) = 0$. Define differential operators Δ_∞ on H^∞ by

$$(8.10) \qquad \Delta_\infty = \prod_{v \in \infty} \Delta_v , \qquad \Delta_v = \det(\partial/\partial w_{jk}^v)$$

with respect to the variable (w_{jk}^v) on the v-th factor H of H^∞. If f is given by (8.9), we have

$$\Delta_\infty f(z) = (2\pi i)^{mn} \sum_{h \in S} c(h) N_{F/\mathbf{Q}}(\det(h)) \mathbf{e}\left(\sum_{v \in \infty} \operatorname{tr}(h_v z_v) \right).$$

Therefore $f \in \mathcal{M}_k^0$ if and only if $\Delta_\infty f = 0$ (and, in fact, if and only if $\Delta_v f = 0$ for any $v \in \infty$). Now, by [25, (8.6)], we have

$$(8.11) \qquad (\Delta_\infty \varphi)\|_{m+1}\alpha = \Delta_\infty(\varphi\|_{m-1}\alpha) \qquad (\alpha \in G)$$

for C^∞-functions φ on H^∞. (The formula of [25, (8.6)] concerns the operator on the bounded domain $\{z \in \mathbf{C}_m^m : z^*z < 1\}$, but it can easily be verified that it is valid also on the present domain H^∞.) We see from (8.11) that $f\|_{m-1}\alpha \in \mathcal{M}_{m-1}^0$ for every $\alpha \in G$ and $f \in \mathcal{M}_{m-1}^0$. Now define a function Y on H^∞ by

$$(8.12) \qquad Y(z) = \prod_{v \in \infty} \det(z_v - z_v^*).$$

Then we have

$$(8.13) \qquad Y(\alpha(z)) = |j(\alpha, z)|^{-2} Y(z) \qquad \text{for every} \quad \alpha \in G.$$

Therefore, in view of Propositions 2.4 and 7.6, it is sufficient to examine the behavior of $E^*(z, s; m - 1, \psi, \mathbf{b})$ at $s = 2$ with a suitable choice of $\mathbf{b}$. In fact, we replace **b** by **b**$'$ as in the proof of Theorem 7.1 in Case SU. To stress the weight k, let $a(h, y, s; k)$, $a_\infty(h, y, s; k)$, and $a_{\mathfrak{f}}(h, s; k)$ denote the Fourier coefficients of $E^*(z, s; k, \psi, \mathbf{b}')$ and their factors defined by (7.8a, b, c). Then we see easily that

$$a_{\mathfrak{f}}(h, s; m - 1) = a_{\mathfrak{f}}(h, s - 2; m + 1),$$

so that the behavior of $a_{\mathfrak{f}}(h, s; m - 1)$ at $s = 2$ is the same as that of $a_{\mathfrak{f}}(h, s; m + 1)$ at $s = 0$, which we examined in the proof of Theorem 7.1. Therefore checking the gamma factors of $a_\infty(h, y, s; m - 1)$, we find that $a(h, y, s; m - 1)$ has a pole at $s = 2$ only when $\psi = \theta^{m-1}$, h is totally nonnegative and $\det(h) = 0$. If such an h has rank p, we see from (7.11) and [26, (4.35.K)] that

$$a_\infty(h, y, 2; m - 1)$$

$$= \left\{ i^{m(1-m)} 2^{p+1-m} \pi^{m^2} \Gamma_m(m)^{-1} \right\}^n \prod_{v \in \infty} \det(y_v)^{-1} \mathbf{e}\left(i \sum_{v \in \infty} \mathrm{tr}(h_v y_v) \right).$$

The pole of $a(h, y, s; m - 1)$ at $s = 2$ is caused by the factor $L_{\mathbf{b}'\mathfrak{n}}(s - 1, \theta^0)$, whose residue at $s = 2$ is $D_F^{-1/2} R_F$ times a rational number, where D_F is the discriminant of F. Therefore our assertions of Theorem 7.3 can be obtained from these observations combined with (8.0) in the same fashion as in the proof of Theorem 7.1.

9. A class of nonholomorphic automorphic forms. The proof of Theorem 7.2 requires a few algebraic formulas concerning operator Δ, which we are going to state in two lemmas below. First, for $x \in \mathrm{GL}_m(\mathbf{C})$ and $0 \leqslant r \leqslant m$, let $\rho_r(x)$ denote the action of x on $\bigwedge^r \mathbf{C}^m$; we consider ρ_r a matrix representation defined with respect to the basis $\{e_{i_1} \wedge \cdots \wedge e_{i_r} : i_1 < \cdots < i_r\}$ of $\bigwedge^r \mathbf{C}^m$, where $\{e_k\}$ is the standard basis of $\mathbf{C}^m$. Thus $\rho_r(x)$ is a matrix of size $m! / [r! (m - r)!]$ composed of the subdeterminants of x of degree r; we have $\rho_0(x) = 1$, $\rho_1(x) = x$, $\rho_m(x) = \det(x)$, and $\rho_r({}^tx) = {}^t\rho_r(x)$. Define another representation ρ_r^* of GL_m by $\rho_r^*(x) = \det(x) \cdot {}^t\rho_{m-r}(x)^{-1}$; in other words, $\rho_{m-r}^*(x)$ is the matrix representing the action of x on $\bigwedge^{m-r} \mathbf{C}^m$ with respect to the basis dual to the above basis of $\bigwedge^r \mathbf{C}^m$. Since these are polynomial representations, $\rho_r(x)$ and $\rho_r^*(x)$ are meaningful for $x \in M_m(\mathbf{C})$.

Let $\partial/\partial z$ denote the matrix (∂_{jk}) of size m, where

$$(9.1) \qquad \partial_{jk} = \begin{cases} 2^{-1}(1 + \delta_{jk}) \partial/\partial z_{jk} & \text{(Case SP)}, \\ \partial/\partial z_{jk} & \text{(Case SU)} \end{cases}$$

with respect to the variable matrix $z = (z_{jk})$ on H. Then the differential operators $\rho_r(\partial/\partial z)$ and $\rho_r^*(\partial/\partial z)$ are meaningful; these assign to a $\mathbf{C}$-valued C^∞-function on H an $M_t(\mathbf{C})$-valued function on H, where $t = m! / [r! (m - r)!]$. In particular, we put

$$(9.2) \qquad \Delta = \rho_m(\partial/\partial z) = \rho_m^*(\partial/\partial z).$$

LEMMA 9.1. (i) *If f and g are $\mathbf{C}$-valued functions on H, then*

$$\Delta(fg) = \sum_{r=0}^{m} \mathrm{tr}\left[\,{}^{t}\rho_r(\partial/\partial z)f \cdot \rho^*_{m-r}(\partial/\partial z)g\right].$$

(ii)
$$\rho_r(\partial/\partial z)\det(z)^{\alpha} = c_r(\alpha)\det(z)^{\alpha-1}\rho^*_{m-r}(z),$$

$$\rho_r^*(\partial/\partial z)\det(z)^{\alpha} = c_r(\alpha)\det(z)^{\alpha-1}\rho_{m-r}(z)$$

for $\alpha \in \mathbf{C}$, where

$$c_r(\alpha) = \begin{cases} 1 & \text{if } r = 0, \\ \displaystyle\prod_{k=0}^{r-1}(\alpha + (\iota k/2)) & \text{if } r > 0, \end{cases}$$

$\iota = 1$ *in Case SP and* $\iota = 2$ *in Case SU.*

(iii)
$$\rho_r(\partial/\partial z)\log\{\det(z)\} = (\iota/2)^{r-1}(r-1)!\det(z)^{-1}\rho^*_{m-r}(z),$$

$$\rho_r^*(\partial/\partial z)\log\{\det(z)\} = (\iota/2)^{r-1}(r-1)!\det(z)^{-1}\rho_{m-r}(z)$$

if $r > 0$.

Proof. The first formula can be verified in a straightforward way. To prove (ii) in Case SP, let $f(z) = \rho_r(\partial/\partial z)\det(z)^{\alpha}$. Changing the variable z for ${}^{t}aza$ with $a \in \mathrm{GL}_m(\mathbf{C})$, we see that

$$\rho_r(a)f({}^{t}aza) \cdot {}^{t}\rho_r(a) = \det(a)^{2\alpha}f(z);$$

in particular, $f({}^{t}aa) = \det(a)^{2\alpha}\rho_r(a)^{-1}f(1)\rho_r({}^{t}a)^{-1}$. Hence $f(1)$ commutes with $\rho_r(a)$ if ${}^{t}aa = 1$. It follows that $f(1) = c_r(\alpha)1$, with a constant $c_r(\alpha)$ depending on r and α. Thus $f(z) = c_r(\alpha)\det(z)^{\alpha} \cdot \rho_r(z)^{-1}$. Let P denote the set of all positive definite real symmetric (resp. complex hermitian) matrices of size m in Case SP (resp. Case SU). To determine $c_r(\alpha)$, we need an integral formula over P. We first put

$$D(z; s_1, \ldots, s_m) = \prod_{r=1}^{m} \det_r(z)^{s_r} \qquad (s_r \in \mathbf{C}, iz \in H),$$

where $\det_r(z)$ is the determinant of the upper left r^2 entries of z; we put $\det_r(z)^s = \exp(s \cdot \log\{\det_r(z)\})$ with $\log\{\det(z)\} \in \mathbf{R}$ when $z \in P$. Now we have

$$(9.3) \qquad \int_P e^{-\mathrm{tr}(zx)}D(x; s_1, \ldots, s_{m-1}, s_m - \kappa)\,dx$$

$$= D(z^{-1}; s_1, \ldots, s_m)\pi^{\iota m(m-1)/4}\prod_{\nu=0}^{m-1}\Gamma\left(\sum_{r=\nu+1}^{m} s_r - (\iota\nu/2)\right)$$

$$\left(iz \in H, \ \sum_{r=\nu+1}^{m} \mathrm{Re}(s_r) > \iota\nu/2 \text{ for } 0 \leqslant \nu \leqslant m-1\right).$$

Here κ is defined by (7.10b); dx is the Euclidean measure on P as defined in [26, p. 272]. This is well known and can be obtained by reducing it to the case $z = 1_m$ and making the change of variables $x = u^*u$ which transforms the integral into that over the space of upper triangular matrices u. If $s_1 = \cdots = s_{m-1} = 0$, it reduces to

$$\int_P e^{-\mathrm{tr}(zx)}\det(x)^{s-\kappa}\,dx = \Gamma_m(s)\det(z)^{-s} \qquad (iz \in H).$$

Applying $\det_r(\partial/\partial z)$ to this, we find

$$\int_P e^{-\mathrm{tr}(zx)}\det_r(x)\det(x)^{s-\kappa}\,dx = (-1)^r\Gamma_m(s)\det_r(\partial/\partial z)\det(z)^{-s}.$$

The integral on the left-hand side is a special case of (9.3) with $s_r = 1$, $s_m = s$, and $s_k = 0$ for all other k's. Substituting these values of s_k into the right-hand side of (9.3), we obtain

$$\det_r(\partial/\partial z)\det(z)^{-s} = \det_r(z^{-1})\det(z)^{-s}\prod_{\nu=0}^{r-1}(s - (\nu/2)).$$

Putting $\alpha = -s$, we obtain $c_r(\alpha)$ as stated in (ii), since $\det_r(\partial/\partial z)$ is the first entry of $\rho_r(\partial/\partial z)$. The formula for ρ_r^* and also Case SU can be obtained in the same way. Now we have

$$\det(z)^{\alpha} = 1 + \alpha \cdot \log\{\det(z)\} + \sum_{k=2}^{\infty}\alpha^k(\log\{\det(z)\})^k/k!.$$

Therefore, if $r > 0$, we have

$$\rho_r(\partial/\partial z)\log\{\det(z)\} = \lim_{\alpha\to 0}\alpha^{-1}\rho_r(\partial/\partial z)\det(z)^{\alpha}$$

and hence (iii) can be derived from (ii).

LEMMA 9.2. *Define polynomial functions* $\lambda_r(X)$ *of* $X \in M_m(\mathbf{C})$ *by*

$$\det(t1_m - X) = \sum_{r=0}^{m}(-1)^r\lambda_r(X)t^{m-r}$$

with an indeterminate t. *Then*

(i) $\lambda_r(X) = \mathrm{tr}[\rho_r(X)]$,

(ii) $\Delta(e^{\mathrm{tr}(uz)}\det(z)^{\alpha}) = e^{\mathrm{tr}(uz)}\det(z)^{\alpha-1}\sum_{r=0}^{m}c_{m-r}(\alpha)\lambda_r(uz)$,

(iii) $\Delta(e^{\mathrm{tr}(uz)}\log\{\det(z)\}) = e^{\mathrm{tr}(uz)}\det(u)\log\{\det(z)\} + e^{\mathrm{tr}(uz)}\det(z)^{-1}$
$\sum_{r=1}^{m}(\iota/2)^{r-1}(r-1)!\lambda_{m-r}(uz)$ $(\alpha \in \mathbf{C},\ u \in M_m(\mathbf{C}))$.

(iv) $\omega(2\pi y, h; \kappa + 1, 0) = 2^{-p\kappa}\pi^{\iota p(m-p)/2}\mathbf{e}(i \cdot \mathrm{tr}(hy))\delta_+(4\pi hy)^{-1}\sum_{\nu=0}^{p}(-1)^{\nu}$
$c_{\nu}(\iota(m-p)/2)\lambda_{p-\nu}(4\pi hy)$, *if* h *is nonnegative and has rank* p.

Proof. The first formula is obvious. The second and third ones can be obtained by taking $\det(z)^{\alpha}$ and $e^{\mathrm{tr}(uz)}$ as f and g of (i) of Lemma 9.1, and employing (ii) and (iii) of the same lemma. For a nonnegative h of rank p, we can find, by [26, Lemma 2.9], a matrix b such that $bhb^* = \mathrm{diag}[1_p, 0_{m-p}]$ and $y = b^*\mathrm{diag}[g, g']b$ with g of size p. By [26, (4.7.K), (4.11.K), (4.10), (3.7)], we have

$$\omega(2\pi y, h; \kappa + 1, 0) = (\pi/2)^{\wp(m-p)/2}\omega(2\pi g, 1_p; \kappa(p) + 1, \iota(p - m)/2)$$

$$= 2^{-p\kappa}\pi^{\wp(m-p)/2}\mathbf{e}(i \cdot \mathrm{tr}(g))\omega_p(4\pi g; \kappa(p) + 1, \iota(p - m)/2).$$

where ω_p is the function of [26, (3.6)] defined for the matrices of size p. By [26, (3.16)], we have

$$\omega_p(w; \kappa(p) + 1, -\beta) = (-1)^p e^{\mathrm{tr}(w)}\det(w)^{-\beta}\Delta_w\!\left[e^{-\mathrm{tr}(w)}\det(w)^\beta\right] \qquad (\beta \in \mathbf{C}),$$

where w is the variable on H (of size p), and Δ_w is the operator defined by (7.4) with w instead of z. Applying (ii) to the right-hand side, we obtain

$$(9.4) \qquad \omega_p(w; \kappa(p) + 1, -\beta) = \det(w)^{-1}\sum_{\nu=0}^{p}(-1)^\nu c_\nu(\beta)\lambda_{p-\nu}(w).$$

Since $bhyb^{-1} = \mathrm{diag}[g, 0_{m-p}]$, we have

$$\det(t1_m - 4\pi hy) = t^{m-p}\det(t1_p - 4\pi g),$$

and hence $\lambda_{p-\nu}(4\pi g) = \lambda_{p-\nu}(4\pi hy)$. Therefore, putting $w = 4\pi g$ and $\beta = \iota(m - p)/2$ in (9.4), we obtain (iv).

We insert here an elementary

LEMMA 9.3. *Let* $u_j = \sum_{k=1}^{n}(b_{jk}z_k + c_{jk}\bar{z}_k)$ *for* $1 \leqslant j \leqslant n$ *with complex variables* $z_1, \ldots, z_n$ *and complex constants* b_{jk} *and* c_{jk} *such that* $\det(c_{jk}) \neq 0$. *Let* A *be a finite set of polynomial functions in* n *complex variables, linearly independent over* **C**. *Then the* $\alpha(u_1, \ldots, u_n)$ *for* $\alpha \in A$ *as functions of* $z_1, \ldots, z_n$ *are linearly independent over the ring of holomorphic functions on a domain in* $\mathbf{C}^n$.

The problem can be reduced to the case where $(c_{jk}) = 1_n$ and A consists of monomials. Then the assertion can easily be proved by induction on n.

We now assume that $F = \mathbf{Q}$ and also that m is odd in Case SP. We then consider a function f on H of the form

$$(9.5) \qquad f(z) = f_1(z) + \Delta\!\left[f_0(z)\log\{\det(z - z^*)\}\right]$$

with holomorphic functions f_0 and f_1 on H such that $\Delta f_0 = 0$. Let $\mathscr{L}_0$ denote the set of all such f. By Lemma 9.1, we have

$$(9.6) \quad \det(z - z^*)f(z) = \det(z - z^*)f_1(z) + \sum_{r=0}^{m-1}(\iota/2)^{m-r-1}(m - r - 1)!$$

$$\times \mathrm{tr}\!\left[\rho_r(z - z^*) \cdot {}^t\!\rho_r(\partial/\partial z)f_0\right].$$

We see from Lemma 9.3 that f_0 and f_1 are unique for f. Hereafter f_0 and f_1 for any $f \in \mathcal{L}_0$ always mean the functions uniquely determined by (9.5). Now we denote by $\mathcal{L}_0^*$ the subset of $\mathcal{L}_0$ consisting of f such that $f_1 = \Delta g$ with a holomorphic function g on H, and by $\mathcal{L}(\Gamma)$, for a congruence subgroup Γ of G, the set of all $f \in \mathcal{L}_0$ such that $f\|_{\kappa+1}\gamma = f$ for all $\gamma \in \Gamma$; we put $\mathcal{L}^*(\Gamma) = \mathcal{L}_0^* \cap \mathcal{L}(\Gamma)$. Further we denote by $\mathcal{L}$ (resp. $\mathcal{L}^*$) the union of $\mathcal{L}(\Gamma)$ (resp. $\mathcal{L}^*(\Gamma)$) for all congruence subgroups Γ of G.

PROPOSITION 9.4. *Suppose $F = \mathbf{Q}$ and $m > 1$; suppose m is odd in Case SP. Then the following assertions hold.*

(i) *If $f \in \mathcal{L}_0$ and $\alpha \in G$, then $f\|_{\kappa+1}\alpha \in \mathcal{L}_0$, and*

$$(f\|_{\kappa+1}\alpha)_0 = f_0\|_{\kappa-1}\alpha,$$

$$(f\|_{\kappa+1}\alpha)_1 = f_1\|_{\kappa+1}\alpha - \Delta\big[(f_0\|_{\kappa-1}\alpha)\log\{j(\alpha,z)\}\big].$$

(ii) *If $f \in \mathcal{L}_0^*$ and $\alpha \in G$, then $f\|_{\kappa+1}\alpha \in \mathcal{L}_0^*$.*

(iii) *An element f of $\mathcal{L}_0$ belongs to $\mathcal{L}(\Gamma)$ if and only if $f_0 \in \mathcal{M}_{\kappa-1}(\Gamma)$ and $f_1\|_{\kappa+1}\gamma = f_1 + \Delta[f_0\log\{j(\gamma,z)\}]$ for every $\gamma \in \Gamma$.*

(iv) *If $f \in \mathcal{L}(\Gamma)$, then f_1 has a Fourier expansion*

$$f_1(z) = \sum_h c_h \mathbf{e}(\mathrm{tr}(hz))$$

with $c_h \in \mathbf{C}$, where h ranges over all nonnegative elements of a lattice in S.

We assume $m > 1$ in order to dispense with the cusp condition.

Proof. We first recall a relation

(9.7) $$(\Delta\varphi)\|_{\kappa+1}\alpha = \Delta(\varphi\|_{\kappa-1}\alpha) \qquad (\alpha \in G).$$

In Case SU, this is (8.11); in Case SP, it is given in [25, Lemma 7.1]. For $\alpha \in G$, we have $\det(\alpha(z) - \alpha(z)^*) = |j(\alpha,z)|^{-2}\det(z - z^*)$, and hence

$$f\|_{\kappa+1}\alpha = f_1\|_{\kappa+1}\alpha + \Delta\big[(f_0\|_{\kappa-1}\alpha)\log\{\det(z - z^*)\}\big]$$

$$- \Delta\big[(f_0\|_{\kappa-1}\alpha)\log\{j(\alpha,z)\}\big] - \Delta\big[(f_0\|_{\kappa-1}\alpha)\log\{j(\alpha,\bar{z})\}\big].$$

The last term vanishes since $\Delta(f_0\|_{\kappa-1}\alpha) = (\Delta f_0)\|_{\kappa+1}\alpha = 0$. In view of the uniqueness of f_0 and f_1, we obtain assertion (i). If $f_1 = \Delta g$, we have $f_1\|_{\kappa+1}\alpha = \Delta(g\|_{\kappa-1}\alpha)$, and hence we obtain (ii). Assertion (iii) follows immediately from (i). Finally if $f \in \mathcal{L}(\Gamma)$, we see from (iii) that $f_1\|_{\kappa+1}\gamma = f_1$ for every $\gamma \in P \cap \Gamma$. Therefore a well-known reasoning guarantees the Fourier expansion of f_1 as stated in (iv).

Now the analytic part of Theorem 7.2 is included in

PROPOSITION 9.5. *Let $E(z,s)$ be as in Theorem 7.2. Then $E(z,0)$ belongs to $\mathcal{L}^*$.*

Proof. As already remarked, Theorem 7.2 is known for $m = 1$; so we may assume $m > 1$. In the proof of Theorem 7.1, we have shown, if $F = \mathbf{Q}$, that

$$(9.8a) \qquad E^*(x + iy, 0; m + 1, \psi, \mathbf{b}') = \sum_{h > 0} \gamma_h' \mathbf{e}(\mathrm{tr}(hz)) + \gamma_0' \pi^{-m} \det(y)^{-1}$$

$$+ \sum_{0 < \mathrm{rank}(h) < m} a(h, y, 0) \mathbf{e}(\mathrm{tr}(hx))$$

with $\gamma_h' \in \mathbf{Q}_{\mathrm{ab}}$. If $\psi \neq \theta^{m-1}$, only the terms with $h > 0$ appear; if $\psi = \theta^{m-1}$, $a(h, y, 0)$ is given by (8.8), and h ranges over all nonnegative elements $\neq 0$ of a lattice in S. Denote this function simply by $E_\psi^*(z)$. Combining (8.8), (9.8a), and (iv) of Lemma 9.2, we obtain

$$(9.8b) \qquad E_\psi^*(z) = \sum_{h > 0} a_h \mathbf{e}(\mathrm{tr}(hz)) + \pi^{-m} \det(z - z^*)^{-1}$$

$$\sum_{\det(h) = 0} a_h \mathbf{e}(\mathrm{tr}(hz)) \sum_{\nu = 0}^{p} (-1)^\nu c_\nu (m - p) \lambda_{p - \nu}(4\pi h y)$$

with $a_h \in \mathbf{Q}_{\mathrm{ab}}$, where $p = \mathrm{rank}(h)$, and h ranges over all nonnegative elements of a lattice in S, including 0; $a_h = 0$ for $\det(h) = 0$ if $\psi \neq \theta^{m-1}$. Define two functions u and v on H by

$$(9.9a) \qquad u(z) = (2\pi i)^{-m} \sum_{h > 0} a_h \det(h)^{-1} \mathbf{e}(\mathrm{tr}(hz)),$$

$$(9.9b) \qquad v(z) = \pi^{-m} \sum_{p = 0}^{m - 1} \left\{ (-1)^p / (m - p - 1)! \right\} \sum_{\mathrm{rank}(h) = p} a_h \mathbf{e}(\mathrm{tr}(hz)).$$

By (iii) of Lemma 9.2, we have

$$E_\psi^*(z) = \Delta \left[u(z) + v(z) \log\{ \det(z - z^*) \} \right]$$

and $\Delta v = 0$. Thus $E_\psi^* \in \mathcal{L}^*$. By (iii) of Proposition 9.4 and (9.9b), we see that $\pi^m v \in \mathcal{M}_{m-1}(\mathbf{Q}_{\mathrm{ab}})$. By Proposition 2.4, $E(z, 0)$ is a finite $\mathbf{Q}_{\mathrm{ab}}$-linear combination of $E_\psi^* \|_{m+1} \alpha$ with various ψ and $\alpha \in G$. Therefore our proposition follows from Proposition 9.4. Moreover, if p_2 is the function of (i) of Theorem 7.2, p_2 is a finite $\mathbf{Q}_{\mathrm{ab}}$-linear combination of the $v\|_{m-1}\alpha$, by virtue of (i) of Proposition 9.4. Therefore $\pi^m p_2 \in \mathcal{M}_{m-1}(\mathbf{Q}_{\mathrm{ab}})$ by Proposition 7.6.

Thus the remaining point of Theorem 7.2 is the fact that $a_h \in \mathbf{Q}_{\mathrm{ab}}$ in (iv). This will be proved in Section 11.

We conclude this section by noting a connection between $E(z, s; \kappa - 1)$ and $E(z, s; \kappa + 1)$; we do not assume $F = \mathbf{Q}$ here.

PROPOSITION 9.6. *Define Y by (8.12) in both Cases SP and SU; put also $\Delta_\infty = \prod_{v \in \infty} \Delta_v$, where Δ_v denotes Δ on the v-th factor H of H^∞. Then*

$$\Delta_\infty \left[Y(z)^{s/2} E(z, s; \kappa - 1) \right] = c_m (s/2)^n Y(z)^{(s/2) - 1} E(z, s - 2; \kappa + 1),$$

where $E(z, s; k)$ denotes $E(z, s; k, \psi, \mathbf{b})$ or $E(z, s; k, \Gamma)$.

Proof. From (ii) of Lemma 9.1, we obtain $\Delta_\infty Y^{s/2} = c_m(s/2)^n Y^{(s/2)-1}$. Taking $Y^{s/2}$ to be φ of (8.11) or (9.7), we find that

$$\Delta_\infty\left(Y(z)^{s/2}j(\alpha,z)^{1-\kappa}|j(\alpha,z)|^{-s}\right)$$

$$= c_m(s/2)^n Y(z)^{(s/2)-1}j(\alpha,z)^{-1-\kappa}|j(\alpha,z)|^{2-s} \qquad (\alpha \in G).$$

Our assertion is an immediate consequence of this equality.

10. Conjectures in Case SP. Theorem 7.1 in Case SP concerns only the weight $k \geqslant m + 1$. The obstacle in extending our results beyond the bound $m + 1$ is the lack of information on the last product $\prod_{\nu|\mathfrak{n}}$ of (7.20) for h of even rank. For h of odd rank, we have some control of the product by means of Lemma 8.2. As to the nature of these "bad factors", we already stated plausible forms in Conjecture 6.3. Assuming this conjecture to be true, let us now show that the results parallel to those in Case SU should hold in Case SP for $k \geqslant (m + 1)/2$ and $k = (m - 1)/2$. Since the case $m = 1$ is well known (or covered by our results in Case SU), we assume $m > 1$ throughout this section. The precise statements are given in the following three propositions.

PROPOSITION 10.1. *Suppose Conjecture 6.3 is true in Case SP for $\lambda = 0$ and r even; suppose also $m > 1$. Let $E(z,s)$ denote any one of $E(z,s;k,\Gamma)$ and $E(z,s;k,\psi,\mathbf{b})$. Then the following assertions hold in Case SP.*
  (i) *$E(z,s)$ is finite at $s = 0$ if $k \geqslant (m + 1)/2$.*
  (ii) *$E(z,0)$ belongs to $\mathcal{M}_k(\mathbf{Q}_{\mathrm{ab}})$ if $k \geqslant (m + 4)/2$ or $k = (m + 1)/2$.*
  (iii) *$E(z,0;k,\psi,\mathbf{b})$ for $k = (m + 3)/2$ and $k = (m + 2)/2$ belongs to $\mathcal{M}_k(\mathbf{Q}_{\mathrm{ab}})$ if $F \neq \mathbf{Q}$ or $\psi^2 \neq 1$.*

PROPOSITION 10.2. *Under the same assumptions as in Proposition 10.1, suppose m is odd, $k = (m + 3)/2$, and $F = \mathbf{Q}$. Then all assertions of Theorem 7.2 hold in Case SP with weights $(m + 3)/2$ and $(m - 1)/2$ in place of $m + 1$ and $m - 1$. (No change is necessary for the power π^m in assertion (iii).)*

PROPOSITION 10.3. *Under the same assumptions as in Proposition 10.1, suppose m is odd and $k = (m - 1)/2$. Then our Eisenstein series have the following properties at $s = 2$.*
  (i) *If $\psi^2 \neq 1$, $E(z,s;(m - 1)/2,\psi,\mathbf{b})$ is finite at $s = 2$.*
  (ii) *$E(z,s;(m - 1)/2,\Gamma)$ and $E(z,s;(m - 1)/2,\psi,\mathbf{b})$ with $\psi^2 = 1$ have at most a simple pole at $s = 2$. The residue at $s = 2$ of each one of these series is $\pi^{-mn}R_F\prod_{v\in\infty}\det(z_v - \bar{z}_v)^{-1}$ times an element f of $\mathcal{M}_{(m-1)/2}(\mathbf{Q}_{\mathrm{ab}})$ of form (7.2) with $c(h) \neq 0$ only when $\det(h) = 0$.*

The proof of these propositions can be given in the same fashion as in Case SU, though the analysis of $a(h, y, s)$ is somewhat subtler in the present case. Here we examine as a typical example the behavior of the Fourier coefficient $a(h, y, s)$ of E^* at $s = 0$ when m is odd, $k = (m + 1)/2$, and $0 < r = \mathrm{rank}(h) \equiv 0 \pmod{2}$.

In this case, by Theorem 7.7 and (7.11), we have

$$a_{\mathbf{f}}(h,s) = N(\mathbf{b})^{mk} \prod_{v \mid \mathbf{n}} \left\{ \alpha\big(v,h,\psi(\pi_v)q_v^{-k-s}\big) \prod_{i=1}^{m-r-1} \big(1 - \psi(\pi_v)^2 q_v^{m-r-2i-2s}\big) \right\}$$

$$\cdot L_{\mathbf{bn}}(k+s,\psi)^{-1} \prod_{i=1}^{(m-1)/2} L_{\mathbf{bn}}(2s+m+1-2i,\psi^2)^{-1}$$

$$\cdot L_{\mathbf{bn}}\big(s+(r+1-m)/2,\psi\xi_h\theta^{r/2}\big) \prod_{i=1}^{(m-r-1)/2} L_{\mathbf{b}}(2s+r-m+2i,\psi^2),$$

$$a_\infty(h,y,s) = g(y,s) \prod_{v \in \infty} \Gamma_{m-r}(s)/\Gamma_{m-p_v}(s/2),$$

where $g(y,s)$ is a function finite at $s = 0$ and p_v is the number of positive eigenvalues of h at v. Now Conjecture 6.3 implies that $\prod_{v \mid \mathbf{n}}$ is finite at $s = 0$. The second line of product of $a_{\mathbf{f}}$ at $s = 0$ is $\pi^{-\beta n}$ times a number of $\mathbf{Q}_{\mathrm{ab}}$, where $\beta = (m+1)^2/4$. The last product over $1 \leqslant i \leqslant (m-r-1)/2$ for $a_{\mathbf{f}}$ is finite at $s = 0$, since $r - m + 2i \leqslant -1$ for such i. Moreover, since $r - m + 2i \equiv 1$ (mod 2), the value of the product at $s = 0$ belongs to $\mathbf{Q}_{\mathrm{ab}}$. Thus our question is reduced to the behavior of

$$(10.1) \qquad L_{\mathbf{bn}}\big(s+(r+1-m)/2,\psi\xi_h\theta^{r/2}\big) \prod_{v \in \infty} \Gamma_{m-r}(s)/\Gamma_{m-p_v}(s/2)$$

at $s = 0$. Let $t_v = r - p_v$. Then

$$\big(\psi\xi_h\theta^{r/2}\big)(x) = \prod_{v \in \infty} \mathrm{sgn}(x_v)^{t_v+(m+1+r)/2} \qquad (x \in F_\infty^\times).$$

Let ϵ_v be 0 or 1 according as $t_v + (m+1+r)/2$ is even or odd. Then

$$(10.2) \qquad\qquad L_{\mathbf{bn}}\big(s,\psi\xi_h\theta^{r/2}\big) \prod_{v \in \infty} \Gamma((s+\epsilon_v)/2)$$

is finite at $s \leqslant 0$ (even if $\psi\xi_h\theta^{r/2} = 1$ and $s = 0$, since $\mathbf{b} \neq \mathbf{g}$.) Take $s = (r+1-m)/2$. Then $s + \epsilon_v \equiv t_v + 1$ (mod 2). Suppose $t_v \geqslant 2$ for some v. Then $\Gamma_{m-r}(s)/\Gamma_{m-p_v}(s/2) = 0$ at $s = 0$, and hence (10.1) vanishes at $s = 0$. Therefore $a(h,y,s) \neq 0$ at $s = 0$ only if $t_v \leqslant 1$ for all v. Suppose $t_v = 1$ for some v. If $m = r + 1$, we see that $\epsilon_v = 0$, so that the finiteness of (10.2) at $s = 0$ implies the vanishing of (10.1) at $s = 0$. If $m \geqslant r + 3$, we have $0 \geqslant \epsilon_v + (r+1-m)/2 \equiv t_v + 1 \equiv 0$ (mod 2), and hence the same conclusion holds. Thus $a(h,y,s) \neq 0$ at $s = 0$ only when $t_v = 0$ for all v, that is, h is totally nonnegative. For such an h, we can show, in view of (7.14), that $a(h,y,0) = c_h e(i\sum_{v \in \infty} \mathrm{tr}(h_v y_v))$ with $c_h \in \mathbf{Q}_{\mathrm{ab}}$.

If $m = 2$, we don't need Conjecture 6.3, since only possible ranks of h are 0, 1, and 2. Therefore we have

THEOREM 10.4. *If $m = 2$ in Case SP, $E(z,s)$ is finite at $s = 0$ for $k \geqslant 2$. Moreover, $E(z,0;2,\psi,\mathbf{b})$ belongs to $\mathcal{M}_2(\mathbf{Q}_{\mathrm{ab}})$ if $F \neq \mathbf{Q}$ or $\psi^2 \neq 1$.*

The case of weight $k > 2$ is covered by Theorem 7.1.

11. $\mathbf{Q}_{ab}$-rational elements of $\mathscr{L}$. Throughout this section, we assume $F = \mathbf{Q}$ and $m > 1$. In Section 9, we introduced a certain set $\mathscr{L}$ of functions on H. Recall that every element f of $\mathscr{L}$ has form (9.5) with holomorphic functions f_0 and f_1 on H and that f_1 has a Fourier expansion

$$(11.1) \qquad f_1(z) = \sum_h c_h \mathbf{e}(\mathrm{tr}(hz)).$$

Given a subfield $\mathscr{F}$ of $\mathbf{C}$, we denote by $\mathscr{L}(\mathscr{F})$ the set of all $f \in \mathscr{L}$ such that $\pi^m f_0 \in \mathscr{M}_{\kappa-1}(\mathscr{F})$ and c_h of (11.1) belongs to $\mathscr{F}$ for every h. We are going to prove

THEOREM 11.1. *If $f \in \mathscr{L}(\mathbf{Q}_{ab})$ and $\alpha \in G$, then $f\|_{\kappa+1}\alpha \in \mathscr{L}(\mathbf{Q}_{ab})$.*

Assertion (iv) of Theorem 7.2 (as well as its analogue in Case SP under Conjecture 6.3) is an immediate consequence of this fact. Indeed, let E_ψ^* be defined as in the proof of Proposition 9.5. Then we know that $E_\psi^* \in \mathscr{L}(\mathbf{Q}_{ab})$. If $E(z,s)$ is as in Theorem 7.2, $E(z,0)$ is a finite $\mathbf{Q}_{ab}$-linear combination of $E_\psi^*\|_{m+1}\alpha$ with several ψ and $\alpha \in G$, and hence $E(z,0) \in \mathscr{L}(\mathbf{Q}_{ab})$ by the above theorem, which implies (iv) of Theorem 7.2.

As a preliminary step to the proof of Theorem 11.1, we consider, for $0 \leqslant r \in \mathbf{Z}$, the set $\mathscr{A}_r$ of all functions f on H_1 of the form

$$(11.2) \qquad f(z) = \sum_{s=0}^{r} (\pi y)^{-s} \sum_a c_s(a)\mathbf{e}(az) \qquad (z \in H_1, y = \mathrm{Im}(z))$$

with $c_s(a) \in \mathbf{C}$, where a ranges over nonnegative elements of a lattice in $\mathbf{Q}$, and H_1 denotes the standard upper half complex plane. Further, for $0 < k \in \mathbf{Z}$ and a congruence subgroup Γ of $\mathrm{SL}_2(\mathbf{Q})$, we denote by $\mathscr{B}_k(\Gamma)$ the set of all functions f on H_1 such that

$$(11.3a) \qquad f\|_k\gamma = f \qquad \qquad \textit{for every} \quad \gamma \in \Gamma,$$

$$(11.3b) \qquad f\|_k\alpha \in \mathscr{A}_{[(k-1)/2]} \qquad \textit{for every} \quad \alpha \in \mathrm{SL}_2(\mathbf{Q}).$$

We denote also by $\mathscr{B}_k$ the union of $\mathscr{B}_k(\Gamma)$ for all congruence subgroups Γ of $\mathrm{SL}_2(\mathbf{Q})$, and by $\mathscr{B}_k(\mathscr{F})$, for a subfield $\mathscr{F}$ of $\mathbf{C}$, the set of all $f \in \mathscr{B}_k$ for which $c_s(a)$ of (11.2) belongs to $\mathscr{F}$ for every s and a. Define a differential operator δ_t^r for $t \in \mathbf{Z}$ and $0 \leqslant r \in \mathbf{Z}$ by

$$(11.4) \qquad \delta_t^r = \delta_{t+2r-2} \cdots \delta_{t+2}\delta_t, \qquad \delta_t = (2\pi i)^{-1}\big[t(z-\bar{z})^{-1} + \partial/\partial z\big];$$

we understand that δ_t^0 is the identity operator. We now need the vector space of elliptic modular forms $\mathscr{M}_t(\mathscr{F})$ on H_1 with respect to $\mathrm{SL}_2(\mathbf{Q})$. To emphasize that this is defined on H_1, let us write it $\mathscr{M}_t^1(\mathscr{F})$, retaining $\mathscr{M}_t(\mathscr{F})$ for the space relative to H and G of degree m.

LEMMA 11.2. *For every subfield $\mathscr{F}$ of $\mathbf{C}$, $\mathscr{B}_k(\mathscr{F})$ consists of all functions of the form $\sum_{s=0}^r \delta_{k-2s}^s h_s$ with $h_s \in \mathscr{M}_{k-2s}^1(\mathscr{F})$, where $r = [(k-1)/2]$.*

This was shown, in a somewhat different formulation, in [19, Lemma 7], when $\mathscr{F} = \mathbf{C}$. Since the h_s are unique for a given function, the assertion for an arbitrary $\mathscr{F}$ can easily be derived.

LEMMA 11.3. *Let M be an imaginary quadratic field embedded in $\mathbf{C}$, and let $w \in M \cap H_1$. If $f \in \mathscr{B}_k(\mathbf{Q}_{ab})$, we have $(f/g)(w) \in M_{ab}$ for every $g \in \mathscr{M}_k^1(\mathbf{Q}_{ab})$ such that $g(w) \neq 0$, where M_{ab} denotes the maximal abelian extension of M. Conversely, if an element f of $\mathscr{B}_k$ has this property for all possible M, w, and g, then $f \in \mathscr{B}_k(\mathbf{Q}_{ab})$.*

Proof. The first assertion follows immediately from Lemma 11.2 and [17, Main Theorem II]. To prove the converse, let $f = \sum_{s=0}^r \delta_{k-2s}^s h_s$ with $h_s \in \mathscr{M}_{k-2s}^1(\mathbf{C})$. Take any imaginary quadratic field M; let σ be an automorphism of $\mathbf{C}$ over M_{ab}. We can let σ act on $\mathscr{M}_k^1(\mathbf{C})$ by putting $(\sum c_a \mathbf{e}(az))^\sigma = \sum c_a^\sigma \mathbf{e}(az)$. Suppose f has the property described in our lemma. Then, for every $g \in \mathscr{M}_k^1(\mathbf{Q}_{ab})$ and $w \in M \cap H_1$ such that $g(w) \neq 0$, we have $(f/g)(w) = (f/g)(w)^\sigma = \sum_{s=0}^r (g^{-1}\delta_{k-2s}^s h_s)(w)^\sigma = \sum_{s=0}^r (g^{-1}\delta_{k-2s}^s h_s^\sigma)(w)$ by [17, Main Theorem III]. Since such points w are dense on H_1, we have $f = \sum_{s=0}^r \delta_{k-2s}^s h_s^\sigma$, and hence $h_s = h_s^\sigma$. Thus $h_s \in \mathscr{M}_{k-2s}^1(M_{ab})$. Since $\mathscr{M}_k^1$ has a $\mathbf{Q}$-rational basis, we can find a finite Galois extension L of $\mathbf{Q}$ such that $h_s \in \mathscr{M}_{k-2s}^1(L)$ for all s. Take another imaginary quadratic field M' whose discriminant is prime to the discriminant of L. For the same reason as above, we have $h_s \in \mathscr{M}_{k-2s}^1(M'_{ab})$. Now, by [16, 1.3], we have $L \cap M'_{ab} = \mathbf{Q}_{ab}$. Hence $h_s \in \mathscr{M}_{k-2s}^1(M'_{ab})$. Q.E.D.

As our next step, we consider an embedding $\epsilon_p : H_1 \to H$ defined by $\epsilon_p(w) = wp$ for $w \in H_1$, where p is any fixed positive definite element of S. For $\alpha = \begin{pmatrix} a & b \\ c & d \end{pmatrix} \in \mathrm{SL}_2(\mathbf{Q})$, we put

$$\alpha^p = \begin{pmatrix} a1_m & bp \\ cp^{-1} & d1_m \end{pmatrix}.$$

Then $\alpha \mapsto \alpha^p$ defines an injection of $\mathrm{SL}_2(\mathbf{Q})$ into G and $\epsilon_p \circ \alpha = \alpha^p \circ \epsilon_p$; moreover $(f \circ \epsilon_p)\|_{km}\alpha = (f\|_k \alpha) \circ \epsilon_p$.

Now let f be an element of $\mathscr{L}$. By (9.6), we have, for $w \in H_1$,

$$f(\epsilon_p(w)) = f_1(wp) + \det(p)^{-1} \sum_{r=0}^{m-1} (\iota/2)^{m-r-1}(m-r-1)!(w-\bar{w})^{r-m}$$

$$\cdot \mathrm{tr}\left[\rho_r(p) \cdot {}^t\rho_r(\partial/\partial z)f_0\right](wp).$$

To simplify our notation, let us hereafter put $\nu = \kappa + 1$.

LEMMA 11.4. (i) *If $f \in \mathscr{L}$, then $f \circ \epsilon_p \in \mathscr{B}_{\nu m}$.*
(ii) *If $f \in \mathscr{L}(\mathbf{Q}_{ab})$, then $f \circ \epsilon_p \in \mathscr{B}_{\nu m}(\mathbf{Q}_{ab})$.*

(iii) *If $f \in \mathscr{L}(\mathbf{Q}_{ab})$, then $f\|_\nu\gamma \in \mathscr{L}(\mathbf{Q}_{ab})$ for every $\gamma \in P$.*

(iv) *If $f \in \mathscr{L}$ and $(f\|_\nu\beta) \circ \epsilon_p \in \mathscr{B}_{\nu m}(\mathbf{Q}_{ab})$ for a fixed p and every $\beta \in R$, then $f \in \mathscr{L}(\mathbf{Q}_{ab})$.*

Here P and R are subgroups of G defined in (1.5).

Proof. Since $2m < \nu m$, the first three assertions can be verified in a straightforward way. To prove (iv), let $\beta = \tau(b) \in R$ with $b \in S$ (see (3.4)), and let

$$f_0(z) = \sum_{h \in \Lambda} c_h \mathbf{e}(\mathrm{tr}(hz)), \qquad f_1(z) = \sum_{h \in \Lambda} d_h \mathbf{e}(\mathrm{tr}(hz)) \qquad (z \in H),$$

where Λ is a lattice in S. Then, for $w \in H_1$, we have

$$(f\|_\nu\beta)(\epsilon_p(w)) = \sum_h d_h \mathbf{e}(\mathrm{tr}(hb))\mathbf{e}(\mathrm{tr}(hp)w) + \sum_{r=0}^{m-1} (\iota/2)^{m-r-1}(m-r-1)!$$

$$\cdot (2\pi i)^r (w - \overline{w})^{r-m}\det(p)^{-1}\sum_h \lambda_r(ph)c_h\mathbf{e}(\mathrm{tr}(hb))\mathbf{e}(\mathrm{tr}(hp)w)$$

with λ_r of Lemma 9.2. Given a nonnegative $q \in \mathbf{Q}$, there exist only finitely many h in Λ such that $\mathrm{tr}(hp) = q$. If $(f\|_\nu\beta) \circ \epsilon_p \in \mathscr{B}_{\nu m}(\mathbf{Q}_{ab})$, we have

$$\sum_{\mathrm{tr}(hp)=q,\, h \in \Lambda} d_h \mathbf{e}(\mathrm{tr}(hb)) \in \mathbf{Q}_{ab}$$

for every q. This holds for an arbitrary $b \in S$, and hence $d_h \in \mathbf{Q}_{ab}$. Similarly we see that $\pi^m c_h \in \mathbf{Q}_{ab}$, and so $f \in \mathscr{L}(\mathbf{Q}_{ab})$.

Proof of Theorem 11.1. Let $f \in \mathscr{L}(\mathbf{Q}_{ab})$. Since G is generated by η of (1.1) and P, it is sufficient to show, in view of (iii) of Lemma 11.4, that $f\|_\nu\eta \in \mathscr{L}(\mathbf{Q}_{ab})$. Let M, w, and g be as in Lemma 11.3 with $k = \nu m$. Let $\beta = \tau(b) \in R$ with $b \in S$; let $\zeta = (w - \overline{w})/2$ and $a = w - \zeta$. We are going to apply (iv) of Lemma 11.4 to f with $p = 1_m$. Put $c = a1_m + b$. Then

$$(f\|_\nu\eta\beta)(\epsilon_1(w)) = \det(\zeta 1_m + c)^{-\nu} f\big(-(\zeta 1_m + c)^{-1}\big).$$

Put $t = (c^2 - \zeta^2 1_m)^{-1}$. Then $-(\zeta 1_m + c)^{-1} = \zeta t - tc$. Put $\gamma = \tau(-tc)$. Then $f(-(\zeta 1_m + c)^{-1}) = (f\|_\nu\gamma)(\epsilon_t(\zeta))$. By Lemma 11.4, $(f\|_\nu\gamma) \circ \epsilon_t \in \mathscr{B}_{\nu m}(\mathbf{Q}_{ab})$, and hence, by Lemma 11.3, $(f\|_\nu\gamma)(\epsilon_t(\zeta))/h(\zeta) \in M_{ab}$ for every $h \in \mathscr{M}_{\nu m}^1(\mathbf{Q}_{ab})$ such that $h(\zeta) \neq 0$. Since $h(\zeta)/g(w) \in M_{ab}$, we see that $(f\|_\nu\eta\beta)(\epsilon_1(w))/g(w) \in M_{ab}$. By Lemma 11.3, this implies that $(f\|_\nu\eta\beta) \circ \epsilon_1 \in \mathscr{B}_{\nu m}(\mathbf{Q}_{ab})$. Therefore, by (iv) of Lemma 11.4, $f\|_\nu\eta \in \mathscr{L}(\mathbf{Q}_{ab})$, which completes the proof.

The above proof involves the values of an element of $\mathscr{L}(\mathbf{Q}_{ab})$ at CM-points on H of a special type. It is natural to ask the nature of the value at an arbitrary CM-point. The author hopes to treat this subject in a subsequent paper.

REFERENCES

1. J. ARTHUR, *A trace formula for reductive groups* I, Duke Math. J. **45** (1978), 911–952.
2. ——, *Eisenstein series and the trace formula*, Proc. Symp. Pure Math. **33** (1979), Part 1, 253–274.
3. E. BOREL, *Sur une application d'un théorème de M. Hadamard*, Bull. Sci. Math. Sér. 2, **18** (1894), 22–25.
4. H. BRAUN, *Zur Theorie der Hermitischen Formen*, Abh. Math. Sem. Hamburg **14** (1941), 61–150.
5. W. CASSELMAN, *A new non-unitary argument for p-adic representations*, J. Fac. of Science, Univ. of Tokyo, Sec. IA, Math. **28** (1982), 907–928.
6. P. GARRETT, *Arithmetic properties of Fourier–Jacobi expansions of automorphic forms in several variables*, Amer. J. of Math. **103** (1981), 1103–1134.
7. J. GOLDMAN AND G.-C. ROTA, *On the foundations of combinatorial theory* IV, Studies in Appl. Math. **49** (1970), 239–258.
8. E. HECKE, *Vorlesungen über die Theorie der algebraischen Zahlen*, 1923, Chelsea, 1948.
9. ——, *Theorie der Eisensteinschen Reihen höherer Stufe und ihre Anwendung auf Funktionentheorie und Arithmetik*, Abh. Math. Sem. Hamburg **5** (1927), 199–224.
10. R. INDIK, *Fourier coefficients of nonholomorphic Eisenstein series on a tube domain associated to an orthogonal group*, Thesis, Princeton Univ. 1982.
11. G. KAUFHOLD, *Dirichletsche Reihe mit Funktionalgleichung in der Theorie der Modulfunktion 2. Grades*, Math. Ann. **137** (1959), 454–476.
12. H. D. KLOOSTERMAN, *Theorie der Eisensteinschen Reihen von mehreren Veränderlichen*, Abh. Math. Sem. Hamburg **6** (1928), 163–188.
13. K. F. LAI, *Tamagawa number of reductive algebraic groups*, Compositio Mathematica **41** (1980), 153–188.
14. R. P. LANGLANDS, *Euler Products*, Yale University Press, 1971.
15. ——, *On the functional equations satisfied by Eisenstein series*, Lecture notes in Math. 544, Springer, 1976.
16. G. SHIMURA, *Construction of class fields and zeta functions of algebraic curves*, Ann. of Math. **85** (1967), 58–159.
17. ——, *On some arithmetic properties of modular forms of one and several variables*, Ann. of Math. **102** (1975), 491–515.
18. ——, *On the Fourier coefficients of modular forms of several variables*, Göttingen Nachr. Akad. Wiss. 1975, 261–268.
19. ——, *The special values of the zeta functions associated with cusp forms*, Comm. pure appl. Math. **29** (1976), 783–804.
20. ——, *On the derivatives of theta functions and modular forms*, Duke Math. J. **44** (1977), 365–387.
21. ——, *The arithmetic of automorphic forms with respect to a unitary group*, Ann. of Math. **107** (1978), 569–605.
22. ——, *On certain reciprocity-laws for theta functions and modular forms*, Acta Math. **141** (1978), 35–71.
23. ——, *The arithmetic of certain zeta functions and automorphic forms on orthogonal groups*, Ann. of Math. **111** (1980), 313–375.
24. ——, *On certain zeta functions attached to two Hilbert modular forms*: I, II, Ann. of Math. **114** (1981), 127–164, 569–607.
25. ——, *Arithmetic of differential operators on symmetric domains*, Duke Math. J. **48** (1981), 813–843.
26. ——, *Confluent hypergeometric functions on tube domains*, Math. Ann. **260** (1982), 269–302.
27. C. L. SIEGEL, *Gesammelte Abhandlungen*, I–IV, Springer.
28. J. STURM, *The critical values of zeta functions associated to the symplectic group*, Duke Math. J. **48** (1981), 327–350.

DEPARTMENT OF MATHEMATICS, PRINCETON UNIVERSITY, PRINCETON, NEW JERSEY 08544

Differential operators and the singular values of Eisenstein series

Duke Mathematical Journal, 51 (1984), 261-329

There are three main themes in this paper:

I. The values of holomorphic and nonholomorphic Eisenstein series of symplectic and unitary groups at CM-points;

II. The critical values of certain zeta functions;

III. Differential operators of arithmetic type on symmetric domains and their adjoint operators.

To explain our problems, let us first define the group, the domain, and the series in less general forms than what is actually treated in the text:

$$G = \left\{ \alpha \in SL_{2m}(K) \,|\, {}'\bar{\alpha}\iota_m\alpha = \iota_m \right\}, \qquad \iota_m = \begin{bmatrix} 0 & -1_m \\ 1_m & 0 \end{bmatrix},$$

$$H = \left\{ z \in M_m(\mathbf{C}) \,|\, i({}'\bar{z} - z) \text{ is positive definite} \right\},$$

where K is an imaginary quadratic field. For $z \in H$ and $\alpha = \begin{pmatrix} a & b \\ c & d \end{pmatrix} \in G$ with a, b, c, d of size m, we put

$$\alpha(z) = (az + b)(cz + d)^{-1}, \qquad \eta(z) = i({}'\bar{z} - z), \qquad \delta(z) = \det(\eta(z)),$$

$$\kappa_\alpha(z) = \bar{c} \cdot {}'z + \bar{d}, \qquad \mu_\alpha(z) = cz + d, \qquad j_\alpha(z) = \det(cz + d),$$

and define two types of Eisenstein series as follows:

$$\mathbf{E}(z, s; k, r) = \delta(z)^{s-r} \sum_{\alpha \in (\Gamma \cap P) \backslash \Gamma} j_\alpha(z)^{-k-r} \overline{j_\alpha(z)}^{\,r} |j_\alpha(z)|^{-2s},$$

$$\mathbf{E}(z, s; k, \sigma) = \delta(z)^s \sum_{\alpha \in (\Gamma \cap P) \backslash \Gamma} j_\alpha(z)^{-k} |j_\alpha(z)|^{-2s} \sigma\big(\mu_\alpha(z)^{-1} \overline{\kappa_\alpha(z)} \eta(z)^{-1} \big).$$

Here $s \in \mathbf{C}$, $k \in \mathbf{Z}$, $0 \leqq r \in \mathbf{Z}$, Γ is a congruence subgroup of G, P is a parabolic subgroup of G consisting of all α for which $c = 0$, and σ is a $\overline{\mathbf{Q}}$-rational polynomial representation $GL_m(\mathbf{C}) \to GL_n(\mathbf{C})$. The former series can be obtained as a special case of the latter by taking $\sigma(x) = \det(x)^r$. For simplicity, let us write $\mathbf{E}(z, s)$ for any of these series.

Received June 25, 1983. Supported by NSF Grant MCS-8100744.

If $m = 1$, G coincides with $SL_2(\mathbf{Q})$. In this case, the value $\mathbf{E}(z_0, 0)$ for an imaginary quadratic point z_0 on the upper half plane is closely connected with a critical value of an L-function of $\mathbf{Q}(z_0)$ with an algebraic-valued Hecke character. In fact, the algebraicity-theorem on the values of the L-functions of the same type of an arbitrary CM-field can be obtained by means of such a connection. Therefore it is natural to look for its higher-dimensional analogue, which is the main motive of the present investigation. More explicitly, we ask the following two questions:

(A) *What is the character of the value* $\mathbf{E}(z_0, 0)$ *for a CM-point* z_0 *on H?*

(B) *Can one construct zeta-functions, similar to the above-mentioned L-functions, whose values can be expressed by means of such* $\mathbf{E}(z_0, 0)$*?*

We shall answer these questions for the Eisenstein series defined on the unitary group of the above type over an arbitrary CM-field as well as on the symplectic group over an arbitrary totally real algebraic number field.

If $r = 0$ and $\mathbf{E}(z, s)$ is convergent at $s = 0$ (which is so when $k > 2m$), then $\mathbf{E}(z, 0)$ is a holomorphic modular form of weight k. If $r > 0$, $\mathbf{E}(z, s; k, r)$ can be obtained by applying a certain differential operator $\Delta_k^{(r)}$ to $\mathbf{E}(z, s; k, 0)$; thus intervene differential operators in our problems. The basic ideas of how such operators can be used in the arithmetical questions were laid out in our previous papers [5], [10], and [11]. By means of the methods developed there, we can readily conclude that $\mathbf{E}(z_0, 0; k, r)$, for $r \geqq 0$, $k > 2m$, and for a CM-point z_0, is an algebraic number times $\pi^{mr} g(z_0)$, once we know that $\mathbf{E}(z, 0; k, 0)$ has algebraic Fourier coefficients. Here g is any holomorphic modular form of weight $k + 2r$, with respect to a congruence subgroup of G, which has algebraic Fourier coefficients and such that $g(z_0) \neq 0$. Now, in our recent paper [14], we investigated the properties of $\mathbf{E}(z, 0)$ both in the convergent and divergent cases for the above two types of groups. In particular, we showed that $\mathbf{E}(z, 0; k, 0)$, for the present G, for example, is a holomorphic modular form with cyclotomic Fourier coefficients if $k > m + 1$ or $k = m$. Therefore the above statement about $\mathbf{E}(z_0, 0; k, r)$ is valid in those cases.

However, this leaves two questions unanswered:

(C) *What is the nature of* $\mathbf{E}(z_0, 0; k, \sigma)$ *for an arbitrary* σ*?*

(D) *What happens in the exceptional case* $k = m + 1$*?*

Our answer to (C) in [11] for $Sp(m, \mathbf{Q})$ was not clear-cut, since k had to be assumed sufficiently large compared with σ. It is one of our main purposes of this paper to settle this point in a more satisfactory way. In fact, we introduce a differential operator Δ_k^σ which is a certain twist of $\sigma(\partial/\partial z_{ij})$ and such that

$$\Delta_k^\sigma \mathbf{E}(z, s; k, 0) = b(s) \cdot {}'\mathbf{E}(z, s; k, \sigma)$$

with a scalar $b(s)$. This enables us to state an algebraicity-theorem on $\mathbf{E}(z_0, 0; k, \sigma)$ in the unitary case for every σ and every $k \geqq m$, including $k = m + 1$ for the reason explained below (Theorem 7.1). In the symplectic case, however, we still have to assume that every irreducible constituent of σ is irreducible on $O(m, \mathbf{R})$.

In the exceptional case $k = m + 1$, $\mathbf{E}(z, 0; k, 0)$ is not necessarily holomorphic in z; it actually belongs to a certain family $\mathfrak{L}$ of nonholomorphic functions f of the form

$$f(z) = f_1(z) + \det(\partial/\partial z_{ij})\left[\, f_0(z)\log\{\det(z - {}^t\bar{z})\}\,\right]$$

with two holomorphic functions f_0 and f_1 on H, as proved in [14]. Now we shall show, as another main result of this paper, that:

The value of $\Delta_{m+1}^{(r)} f$ at a CM-point z_0 is an algebraic number times $\pi^{mr} g(z_0)$ with g of the above type, of weight $m + 1 + 2r$, provided that both f_1 and $\pi^m f_0$ have algebraic Fourier coefficients. A similar algebraicity about $(\Delta_{m+1}^\sigma f)(z_0)$ holds for every σ as above (Theorem 3.4).

If $r = 0$, this amounts to the algebraicity of $f(z_0)/g(z_0)$; even in this special case, the result is far from trivial, because of the nonholomorphic nature of f. Anyway, as a consequence of this fact, the case $k = m + 1$ can be included in the above result concerning $\mathbf{E}(z_0, 0)$.

As an answer to question (B), we define a certain infinite series Z with respect to a direct sum of *CM*-fields containing K, and prove an algebraicity-theorem on its critical values (Theorem 8.1). Since the definition of Z in the most general case is too involved to be presented here, we mention only one simplest case:

$$(1) \qquad W(s) = \sum_{x \in U \backslash X} c(x\beta \cdot {}^t\bar{x})\left[\, \overline{\det(x)}\,/|\det(x)|\right]^e |\det(x)|^{-2s}.$$

Here $0 < e \in \mathbf{Z}$, $X = GL_m(K) \cap T$ with a coset T of $M_m(K)$ modulo a lattice, and U is a sufficiently small subgroup of finite index of the group $\{u \in GL_m(K) \mid uT = T\}$; the $c(b)$ are the Fourier coefficients of a "theta series of a hermitian form" defined by

$$(2) \qquad \sum_b c(b)\mathbf{e}(\mathrm{tr}(bz)) = \sum_{x \in T'} \mathbf{e}(\mathrm{tr}(x\alpha \cdot {}^t\bar{x}z)) \qquad (z \in H, \mathbf{e}(w) = e^{2\pi i w})$$

with a coset T' of $M_m(K)$ modulo a lattice; β of (1) and α of (2) are positive definite hermitian elements of $M_m(K)$. Our result (Corollary 8.2) specialized to this case asserts:

If $\det(\alpha\beta) \in N_{K/\mathbf{Q}}(K^\times)$, $k \in \mathbf{Z}$, $m \leqq k \leqq e$, and $k \equiv e \pmod 2$, then $W(k/2)$ is an algebraic number times $\pi^b p(\tau)$, where $b = me - m(m-1)/2$, p is an elliptic *modular form of weight $2me$ with algebraic Fourier coefficients, and τ is an element of K with positive imaginary part.*

The series Z is defined as the sum extended over a lattice modulo a "group of units" of the same type as the above U. In order to find a relation between this and the series $\mathbf{E}$ which is defined as a sum over $(P \cap \Gamma)\backslash\Gamma$, we have to study a series of the form

$$\sum_{x \in U\backslash X} \left[\det(x)/|\det(x)| \right]^e |\det(x)|^{-2s}.$$

In fact, we shall show that a series of a more general type defined on a matrix algebra over an arbitrary algebraic number field is a finite linear combination of products of classical L-functions of the field (Proposition 9.2), which may be of independent interest.

Comparison of our result on the critical values of Z with those on the values of zeta functions in our previous papers [12] and [13] suggests the existence of more general theorems which include both as special cases. In the present paper, however, we have restricted our treatment to the series which are directly connected to the Eisenstein series of the above types.

Though our main theorems on Eisenstein series concern the symplectic groups and the unitary groups of maximally isotropic hermitian forms, we include also in our exposition the discussion of $SU(l,m)$ with arbitrary l and m. This group modulo a maximal compact subgroup produces a domain $B_{l,m}$ consisting of all complex $l \times m$ matrices $z = (z_{ij})$ such that $1_m - {}^t\bar{z}z$ is positive definite. If $l = m$, $\sigma(\partial/\partial z_{ij})$ is again meaningful for any representation σ as above. To treat the case $l \neq m$, we consider $\sigma(\partial/\partial z_{ij})$ for a polynomial map σ which sends $l \times m$ matrices to $n' \times n$ matrices and satisfies $\sigma(axb) = \sigma'(a)\sigma(x)\sigma''(b)$ with representations $\sigma' : GL_l \to GL_{n'}$ and $\sigma'' : GL_m \to GL_n$. Then Δ_k^σ can be defined on $B_{l,m}$ as a certain twist of $\sigma(\partial/\partial z_{ij})$. The arithmeticity of $\Delta_k^\sigma f(z_0)$ for a CM-point z_0 and $B_{l,m}$ can also be proved in this case.

Our final topic, treated in Section 11, is a purely geometric or analytic question of finding the adjoint operator of Δ_k^σ. We shall show that it is given by a twist of $\sigma(\partial/\partial\bar{z}_{ij})$ combined with a trace operator (Theorems 11.4, 11.5). This generalizes the fact that $-\mathrm{Im}(z)^2\partial/\partial\bar{z}$ is the adjoint of $k(z - \bar{z})^{-1} + \partial/\partial z$ with respect to the Petersson inner product on the upper half plane. It should be noted that the existence of the (formal) adjoint of an arbitrary differential operator on a differentiable manifold is well known. Our emphasis here, however, is in its explicit description for the operator Δ_k^σ which is introduced for specific arithmetical purposes. Composing Δ_k^σ with its adjoint for various σ, we obtain a family of self-adjoint operators which includes the Laplace–Beltrami operator as a special case (Corollary 11.8). We shall then show that $\delta(z)^s$ for $s \in \mathbf{C}$ is an eigenfunction of these operators at least when $\sigma(x) = x$ or $\sigma(x) = \det(x)^r$, by giving each eigenvalue as a polynomial in s (Proposition 11.12).

Contents

1. Differential operators 266
2. The operator E_σ and a further study of Δ_ρ^σ 274
3. Arithmeticity at CM-points 278
4. Lemmas on tensor-valued functions 285
5. Proof of Theorem 3.4 in Case SU 288
6. Proof of Theorem 3.4 in Case SP 293
7. The singular values of Eisenstein series 299
8. Critical values of certain zeta functions 302
9. Certain zeta functions on GL_m 309
10. Proof of Theorem 8.1 313
11. The adjoint operators of Δ_ρ^σ and E_σ 316
12. Appendix: Polynomial maps σ 326

Notation and terminology. Our general notation, for the most part, is the same as in the previous paper [14] except that particular letters such as η and ϵ of [14] will no longer have the same meaning in the present paper. Here we recall the definition of a few symbols and introduce some new ones. For $z \in \mathbf{C}$, we put $\mathbf{e}(z) = e^{2\pi i z}$. For an associative ring R with identity element and positive integers m and n, we denote by $R^\times$ and R_n^m the group of all invertible elements of R and the module of all $m \times n$ matrices with entries in R, and put $R^m = R_1^m$; R_m^m is occasionally denoted by $M_m(R)$. For a complex matrix A, we put $A^* = {}^t\bar{A}$. The bar denotes as usual the complex conjugate. One exception is the symbol $\overline{\mathbf{Q}}$, which is the algebraic closure of the rational number field $\mathbf{Q}$ in $\mathbf{C}$. Another kind of exception is ρ_r^* which is defined in Section 2 and which is different from ${}^t\bar\rho_r$. We write $a > b$ for two hermitian matrices a and b if $a - b$ is positive definite. For a complex vector space V, $\mathrm{End}(V)$ denotes the ring of all its C-linear endomorphisms, and $GL(V) = \mathrm{End}(V)^\times$. By a representation $\{\rho, V\}$ of a group X, we mean a complex vector space V and a homomorphism ρ of X into $GL(V)$. The letter ρ is often employed as a representation of a group and also as the complex conjugation of a CM-field. The distinction will be clear from the context. By a *locally constant function* on a finite-dimensional vector space W over $\mathbf{Q}$, we understand a C-valued function f on W for which there exist two lattices L and M of W such that $f(x) = 0$ for $x \notin L$ and $f(x)$ depends only on x modulo M. If Y is a semi-simple algebra over a field F, $\mathrm{Tr}_{Y/F}$ and $N_{Y/F}$ denote the reduced trace and norm of Y to F. In contrast to this, the trace of a matrix α is denoted by $\mathrm{tr}(\alpha)$. Given two sets S and φ, we denote by S^φ "the product of φ copies of S" as usual, which consists of all elements $(x_v)_{v \in \varphi}$ with $x_v \in S$. For $x = (x_v)_{v \in \varphi}$ and $y = (y_v)_{v \in \varphi}$ in $\mathbf{R}^\varphi$, we write $x \geqq y$ and also $y \leqq x$ if $x_v \geqq y_v$ for all $v \in \varphi$. Whenever we consider differentiation of a function on a (complex) manifold, we assume it to be C^∞ or the product of a C^∞-function and a meromorphic function. The symbol $\amalg$ denotes a disjoint union of sets.

1. Differential operators. We treat in this paper symplectic and unitary groups which are defined in three types referred to as Cases SP, SU, and SUB, respectively. In this section we discuss them over $\mathbf{R}$. Their explicit forms are:

$$\mathfrak{G} = \mathfrak{G}_m = \begin{cases} \{\alpha \in SL_{2m}(\mathbf{R}) \mid {}^t\alpha \iota_m \alpha = \iota_m\} & \text{(Case SP)}, \\ \{\alpha \in SL_{2m}(\mathbf{C}) \mid \alpha^* \iota_m \alpha = \iota_m\} & \text{(Case SU)}, \end{cases}$$

$$\mathfrak{G}' = \mathfrak{G}'_{l,m} = \{\alpha \in SL_{l+m}(\mathbf{C}) \mid \alpha^* \iota'_{l,m} \alpha = \iota'_{l,m}\} \quad \text{(Case SUB)}.$$

Here l and m are nonnegative integers (naturally $m > 0$ for $\mathfrak{G}_m$ and $l + m > 0$ for $\mathfrak{G}'_{l,m}$), and

$$(1.1) \qquad \iota_m = \begin{bmatrix} 0 & -1_m \\ 1_m & 0 \end{bmatrix}, \qquad \iota'_{l,m} = \begin{bmatrix} 1_l & 0 \\ 0 & -1_m \end{bmatrix}.$$

We shall later consider algebraic groups whose localizations become $\mathfrak{G}$ or $\mathfrak{G}'$, but in this section we deal only with the Lie groups.

Define domains H and B by

$$H = H_m = \begin{cases} \{z \in \mathbf{C}_m^m \mid {}^tz = z, \ \mathrm{Im}(z) > 0\} & \text{(Case SP)}, \\ \{z \in \mathbf{C}_m^m \mid i(z^* - z) > 0\} & \text{(Case SU)}, \end{cases}$$

$$B = B_{l,m} = \{z \in \mathbf{C}_m^l \mid z^*z < 1_m\} \quad \text{(Case SUB)},$$

assuming $lm > 0$ in the last case. For $\alpha \in \mathfrak{G}$ (resp. $\alpha \in \mathfrak{G}'$) and $z \in H$ (resp. $z \in B$), we can define $\alpha(z)$ and factors of automorphy $\kappa(\alpha, z) = \kappa_\alpha(z)$ and $\mu(\alpha, z) = \mu_\alpha(z)$ by

$$(1.2a) \quad \alpha \begin{bmatrix} z^* & z \\ 1_m & 1_m \end{bmatrix} = \begin{bmatrix} \alpha(z)^* & \alpha(z) \\ 1_m & 1_m \end{bmatrix} \begin{bmatrix} \overline{\kappa_\alpha(z)} & 0 \\ 0 & \mu_\alpha(z) \end{bmatrix} \quad \text{(Cases SP, SU)},$$

$$(1.2b) \quad \alpha \begin{bmatrix} 1_l & z \\ z^* & 1_m \end{bmatrix} = \begin{bmatrix} 1_l & \alpha(z) \\ \alpha(z)^* & 1_m \end{bmatrix} \begin{bmatrix} \overline{\kappa_\alpha(z)} & 0 \\ 0 & \mu_\alpha(z) \end{bmatrix} \quad \text{(Case SUB)};$$

κ_α and μ_α have size m except for κ_α in Case SUB which has size l. (Cf. [8], [9, §3].) If either $l = 0$ or $m = 0$ in Case SUB, we understand that $B_{l,m}$ consists of a single element 1_{l+m}, $\mathfrak{G}'_{l,m}$ acts trivially on $B_{l,m}$, and

$$(1.2c) \qquad \kappa_\alpha(z) = \kappa(\alpha, z) = \bar{\alpha} \qquad \text{if} \quad m = 0,$$

$$(1.2d) \qquad \mu_\alpha(z) = \mu(\alpha, z) = \alpha \qquad \text{if} \quad l = 0.$$

The domain $B_{l,m}$ with $lm = 0$ will be needed in Section 3. In the rest of this section, however, we assume $lm > 0$ in Case SUB. Notice that $\kappa_\alpha = \mu_\alpha$ in Case

SP. As shown in [8, (1.19)] and [9, (3.9)], we have

$$(1.3) \qquad \det \kappa_\alpha(z) = \det \mu_\alpha(z) \qquad \text{(Cases SU, SUB).}$$

We define a scalar-valued factor of automorphy j_α and functions η, ξ_B, η_B, and δ on H or on B as follows:

$$(1.4) \qquad j_\alpha(z) = j(\alpha, z) = \det \mu_\alpha(z) \qquad \text{(all three cases),}$$

$$(1.5\text{a}) \qquad \eta(z) = i(z^* - z) \qquad (z \in H),$$

$$(1.5\text{b}) \qquad \xi_B(z) = 1_l - \bar{z} \cdot {}^t z, \qquad \eta_B(z) = 1_m - {}^t \bar{z} z \qquad (z \in B),$$

$$(1.6) \qquad \delta(z) = \begin{cases} \det \eta(z) & (z \in H, \text{ Cases SP and SU}), \\ \det \eta_B(z) = \det \xi_B(z) & (z \in B, \text{ Case SUB}). \end{cases}$$

As shown in [9, (3.7), (3.8)] and [8, (1.15), (1.16), (1.17)], we have

$$(1.7\text{a}) \qquad dz \circ \alpha = {}^t\kappa_\alpha(z)^{-1} dz \mu_\alpha(z)^{-1} \qquad \text{(all three cases),}$$

$$(1.7\text{b}) \quad \eta(z) = \mu_\alpha(z)^* \eta(\alpha(z)) \mu_\alpha(z) = {}^t\kappa_\alpha(z) \eta(\alpha(z)) \overline{\kappa_\alpha(z)} \qquad (\alpha \in \mathfrak{G}, z \in H),$$

$$(1.7\text{c}) \quad \xi_B(z) = \kappa_\alpha(z)^* \xi_B(\alpha(z)) \kappa_\alpha(z), \qquad \eta_B(z) = \mu_\alpha(z)^* \eta_B(\alpha(z)) \mu_\alpha(z)$$

$$(\alpha \in \mathfrak{G}', z \in B).$$

Taking the determinant, we obtain

$$(1.7\text{d}) \qquad \delta(\alpha(z)) = |j_\alpha(z)|^{-2} \delta(z) \qquad \text{(all three cases).}$$

Notice that the group $\mathfrak{G}$ and the domain H in Case SP are contained in those in Case SU, and the formulas in Case SP are special cases of those in Case SU.

Let us now briefly recall the definition of differential operators on H and B introduced in [11]. We first define a group $\mathscr{g}$ by

$$(1.8) \qquad \mathscr{g} = \begin{cases} GL_m(\mathbf{C}) & \text{(Case SP),} \\ GL_m(\mathbf{C}) \times GL_m(\mathbf{C}) & \text{(Case SU),} \\ GL_l(\mathbf{C}) \times GL_m(\mathbf{C}) & \text{(Case SUB),} \end{cases}$$

and take a rational representation

$$(1.9) \qquad \rho : \mathscr{g} \to GL(V)$$

with a finite-dimensional complex vector space V. For a V-valued function f on H (resp. B) and $\alpha \in \mathfrak{G}$ (resp. $\mathfrak{G}'$), we define a V-valued function $f\|_\rho \alpha$ on H (resp.

B) by

$$(1.10) \qquad f\|_\rho \alpha = \begin{cases} \rho(\mu_\alpha)^{-1}(f \circ \alpha) & \text{(Case SP)}, \\ \rho(\kappa_\alpha, \mu_\alpha)^{-1}(f \circ \alpha) & \text{(Cases SU, SUB)}. \end{cases}$$

If $\rho(x, y) = \det(x)^r \det(y)^{-r} \sigma(x, y)$ with $r \in \mathbf{Z}$ in Case SU or in Case SUB, we have $f\|_\rho \alpha = f\|_\sigma \alpha$ by virtue of (1.3). Put

$$(1.11) \qquad T = \begin{cases} \{ u \in \mathbf{C}_m^m \mid {}^t u = u \} & \text{(Case SP)}, \\ \mathbf{C}_m^m & \text{(Case SU)}, \\ \mathbf{C}_m^l & \text{(Case SUB)}, \end{cases}$$

and view T as a global complex tangent space of H or B by identifying $u = (u_{ij}) \in T$ with $\sum_{i=1}^l \sum_{j=1}^m u_{ij} \partial_{ij}$, where $l = m$ in Cases SP and SU, and

$$(1.12) \qquad \partial_{ij} = \begin{cases} 2^{-1}(1 + \delta_{ij})\partial/\partial z_{ij}(= \partial_{ji}) & \text{(Case SP)}, \\ \partial/\partial z_{ij} & \text{(Cases SU, SUB)}, \end{cases}$$

with respect to the variable matrix $z = (z_{ij})$ on H or B. In Case SP, ∂_{ij} as an element of T is equal to $(e_{ij} + e_{ji})/2$ with the standard matrix units e_{ij}. For a V-valued function f on H or B, we define Df to the function on the same space with values in $\mathrm{Hom}(T, V)$ such that

$$(1.13) \qquad (Df)(u) = \sum_{i=1}^l \sum_{j=1}^m u_{ij}\partial_{ij}f \qquad (u = (u_{ij}) \in T),$$

where $l = m$ in Cases SP and SU, and $\mathrm{Hom}(T, V)$ denotes the vector space of all C-linear maps of T into V. The kth iterate D^k of D can naturally be defined so that $D^k f$ takes values in

$$\mathrm{Hom}(T, \mathrm{Hom}(T, \ldots, \mathrm{Hom}(T, V) \cdots)).$$

This vector space can be identified with the space of all multilinear maps of T^k into V, which we denote by $\mathrm{Ml}_k(T, V)$. We define representations $\{\rho \otimes \tau^k, \mathrm{Ml}_k(T, V)\}$ and $\{\rho \otimes \pi^k, \mathrm{Ml}_k(T, V)\}$ of $\mathscr{G}$ by

$$(1.14a) \quad \{(\rho \otimes \tau^k)(x, y)h\}(u_1, \ldots, u_k) = \rho(x, y)h({}^t x u_1 y, \ldots, {}^t x u_k y),$$

$$(1.14b) \qquad \{(\rho \otimes \pi^k)(x, y)h\}(u_1, \ldots, u_k)$$

$$= \rho(x, y)h(x^{-1}u_1 \cdot {}^t y^{-1}, \ldots, x^{-1}u_k \cdot {}^t y^{-1})$$

for $h \in \mathrm{Ml}_k(T, V)$ and $u_1, \ldots, u_k \in T$. Here and until the end of Section 2, we adopt the following notational conventions: (x, y) means y and ${}^t x u y = {}^t y u y$ (or rather $x = y$) in Case SP; $l = m$, $\xi(z) = {}^t\eta(z)$ in Cases SP and SU; $\xi = \xi_B$ and $\eta = \eta_B$ in Case SUB; further we put

$$(1.15) \qquad \kappa = \begin{cases} (m+1)/2 & \text{(Case SP)}, \\ m & \text{(Case SU)}, \\ (l+m)/2 & \text{(Case SUB)}. \end{cases}$$

Coming back to (1.14a, b), we write τ and π for τ^1 and π^1; τ^k and π^k stand for $\rho \otimes \tau^k$ and $\rho \otimes \pi^k$ with trivial ρ and $V = \mathbf{C}$; naturally τ^0 and π^0 denote the trivial representation.

Given a V-valued function f as above, we define a function $D_\rho f$ with values in $\mathrm{Hom}(T, V)$ by

$$(1.16) \qquad D_\rho f = \rho(\xi, \eta)^{-1} D(\rho(\xi, \eta) f),$$

where $\rho(\xi, \eta) = \rho(\eta)$ in Case SP by our convention. Then we have

$$(1.17) \qquad D_\rho(f\|_\rho \alpha) = (D_\rho f)\|_{\rho \otimes \tau} \alpha \qquad (\alpha \in \mathfrak{G} \text{ or } \alpha \in \mathfrak{G}').$$

In Case SP, this is [11, Lemma 2.1]; the same type of proof applies to Cases SU and SUB. Next, an iterated operator $D_\rho^{(k)}$ for $0 \leqq k \in \mathbf{Z}$ acting on such functions f can be defined by

$$(1.18) \qquad D_\rho^{(k)} = D_{\rho \otimes \tau^{k-1}} \cdots D_{\rho \otimes \tau} D_\rho ;$$

naturally $D_\rho^{(0)}$ denotes the identity operator. Thus $D_\rho^{(k)} f$ takes values in $\mathrm{Ml}_k(T, V)$. From (1.17), we obtain

$$(1.19) \qquad D_\rho^{(k)}(f\|_\rho \alpha) = \left(D_\rho^{(k)} f\right)\|_{\rho \otimes \tau^k} \alpha \qquad (\alpha \in \mathfrak{G} \text{ or } \alpha \in \mathfrak{G}').$$

If $\{\sigma, W\}$ is another representation of $\mathscr{G}$ and g is a W-valued function on H, then

$$(1.20a) \qquad D_{\rho \otimes \sigma}(f \otimes g) = D_\rho(f) \otimes g + f \otimes D_\sigma(g).$$

More generally, we can easily verify that

$$(1.20b) \qquad D_{\rho \otimes \sigma}^{(k)}(f \otimes g)(u_1, \ldots, u_k)$$

$$= \sum_{a+b=k} D_\rho^{(a)}(f)(u_{i_1}, \ldots, u_{i_a}) \otimes D_\sigma^{(b)}(g)(u_{j_1}, \ldots, u_{j_b})$$

for $u_1, \ldots, u_k \in T$, where the sum is extended over all disjoint decompositions $\{1, \ldots, k\} = \{i_1, \ldots, i_a\} \cup \{j_1, \ldots, j_b\}$, $i_1 < \cdots < i_a, j_1 < \cdots < j_b$.

Before proceeding further, we make some algebraic preliminaries. For $p \in \mathrm{Ml}_r(T, \mathbf{C})$ and $h \in \mathrm{Ml}_r(T, V)$, we define a kind of their contraction $\{p, h\}$ to be the element of V given by

$$(1.21) \qquad \{p, h\} = \sum_{i,j,\ldots,s,t} p(\partial_{ij}, \ldots, \partial_{st}) h(\partial_{ij}, \ldots, \partial_{st}),$$

where $(i, \ldots, s; j, \ldots, t)$ ranges over $\{1, \ldots, l\}^r \times \{1, \ldots, m\}^r$. It can easily be seen that

$$(1.22) \qquad \{\pi'(x, y)p, (\rho \otimes \tau')(x, y)h\} = \{\tau'(x, y)p, (\rho \otimes \pi')(x, y)h\}$$

$$= \rho(x, y)\{p, h\}.$$

We call an element h of $\mathrm{Ml}_r(T, V)$ *symmetric* if

$$(1.23) \qquad h(u_1, \ldots, u_r) = h(u_{\alpha(1)}, \ldots, u_{\alpha(r)}) \qquad (u_i \in T)$$

for every permutation α of $1, \ldots, r$. Given an arbitrary h of $\mathrm{Ml}_r(T, V)$, we define its symmetrization h_{sym} by

$$(1.24) \qquad h_{sym}(u_1, \ldots, u_r) = r!^{-1} \sum_\alpha h(u_{\alpha(1)}, \ldots, u_{\alpha(r)}),$$

where α ranges over all permutations of $1, \ldots, r$.

The representation $\rho \otimes \tau'$ may be decomposed into several representations. Suppose a representation $\{\omega, W\}$ of $\mathcal{g}$ is contained in $\rho \otimes \tau'$ in the sense that $\omega(x, y) \circ \varphi = \varphi \circ (\rho \otimes \tau')(x, y)$ with a linear map φ of $\mathrm{Ml}_r(T, V)$ onto W. Then we obtain a differential operator $\varphi D_\rho^{(r)}$ which satisfies $\varphi D_\rho^{(r)}(f\|_\rho \alpha) = (\varphi D_\rho^{(r)}f)\|_\omega \alpha$. To obtain such ω and φ, we fix a system $\{\sigma, \sigma', \sigma''\}$ of a polynomial map and two polynomial representations

$$(1.25a) \qquad\qquad\qquad \sigma : \mathbf{C}_m^l \to \mathbf{C}_n^{n'},$$

$$(1.25b) \qquad \sigma' : GL_l(\mathbf{C}) \to GL_{n'}(\mathbf{C}), \qquad \sigma'' : GL_m(\mathbf{C}) \to GL_n(\mathbf{C}),$$

such that $\sigma(xay) = \sigma'(x)\sigma(a)\sigma''(y)$ and that σ is homogeneous of degree r in the entries of the variable matrix in $\mathbf{C}_m^l$. We assume $n = n'$ and $\sigma = \sigma' = \sigma''$ if $l = m$. All possible types of $\{\sigma, \sigma', \sigma''\}$ will be determined in Section 12.

Given a homogeneous polynomial function p on T of degree r, we can find a unique symmetric element p' of $\mathrm{Ml}_r(T, \mathbf{C})$ such that $p(x) = p'(x, \ldots, x)$. Identifying p with p', we may consider p a symmetric element of $\mathrm{Ml}_r(T, \mathbf{C})$. We now define a linear map

$$(1.26a) \qquad\qquad\qquad \Phi_\sigma : \mathrm{Ml}_r(T, V) \to \mathrm{Hom}(\mathbf{C}_n^{n'}, V)$$

by

$$(1.26b) \quad (\Phi_\sigma h)(w) = \sum_{a=1}^{n'} \sum_{b=1}^{n} w_{ab}\{\sigma_{ab}, h\} \qquad (w \in \mathbf{C}_n^{n'}, h \in \mathrm{Ml}_r(T, V)),$$

where σ_{ab} is the (a, b)-entry of σ viewed as a symmetric element of $Ml_r(T, \mathbf{C})$. Define also representations $\rho \otimes_1 \sigma$ and $\rho \otimes_2 \sigma$ of $\mathscr{g}$ on $\mathrm{Hom}(\mathbf{C}_n^{n'}, V)$ by

$$(1.27a) \qquad \{(\rho \otimes_1 \sigma)(x, y)q\}(w) = \rho(x, y)q\big({}^t\sigma'(x)w \cdot {}^t\sigma''({}^ty)\big),$$

$$(1.27b) \qquad \{(\rho \otimes_2 \sigma)(x, y)q\}(w) = \rho(x, y)q\big({}^t\sigma'({}^tx^{-1})w \cdot {}^t\sigma''(y^{-1})\big)$$

for $q \in \mathrm{Hom}(\mathbf{C}_n^{n'}, V)$ and $w \in \mathbf{C}_n^{n'}$. Formula (1.22) together with straightforward calculation shows that

$$(1.28a) \qquad \Phi_\sigma \circ (\rho \otimes \tau')(x, y) = (\rho \otimes_1 \sigma)(x, y) \circ \Phi_\sigma,$$

$$(1.28b) \qquad \Phi_\sigma \circ (\rho \otimes \pi')(x, y) = (\rho \otimes_2 \sigma)(x, y) \circ \Phi_\sigma.$$

Let D denote the $l \times m$-matrix whose (i, j)-entry is ∂_{ij}. With σ as above, we can define a matrix differential operator $\sigma(D)$ in an obvious formal way. In particular, we put, for $0 \leqq p \in \mathbf{Z}$,

$$(1.29) \qquad \Delta^p = \det(D)^p,$$

assuming $l = m$ in Case SUB. Our definition of Φ_σ implies that

$$(1.30) \qquad (\Phi_\sigma D'f)(w) = \sum_{a=1}^{n'} \sum_{b=1}^{n} w_{ab}\sigma(D)_{ab} f \qquad (w \in \mathbf{C}_n^{n'})$$

for V-valued functions f on H or on B. Now we put

$$(1.31) \qquad \Delta_\rho^\sigma = \Phi_\sigma D_\rho^{(r)}.$$

This maps V-valued functions to $\mathrm{Hom}(\mathbf{C}_n^{n'}, V)$-valued ones. We have $\Delta_\rho^\sigma = D_\rho$ if σ is the identity map of $\mathbf{C}_m^l$ onto itself. From (1.19) and (1.28a), we obtain immediately

$$(1.32) \qquad \Delta_\rho^\sigma(f\|_\rho \alpha) = (\Delta_\rho^\sigma f)\|_{\rho \otimes_1 \sigma}\alpha \qquad (\alpha \in \mathfrak{G} \text{ or } \mathfrak{G}').$$

In particular, let us put $\Delta_\rho^{(p)} = \Delta_\rho^\sigma$ when $\sigma(x) = \det(x)^p$, excluding Case SUB with $l \neq m$. Identifying $\mathrm{Hom}(\mathbf{C}, V)$ with V, we have

$$(\rho \otimes_1 \sigma)(x, y) = \det(xy)^p \rho(x, y) \quad \text{and} \quad (\rho \otimes_2 \sigma)(x, y) = \det(xy)^{-p}\rho(x, y).$$

In this case, Φ_σ coincides with the map Φ_p introduced in [11, (2.13)]. If $p = 1$, we have

$$(1.33) \qquad \Phi_1 h = m!^{-1}\sum_{\alpha, \beta} \mathrm{sgn}(\beta)h\big(\partial_{\alpha(1)\beta\alpha(1)}, \ldots, \partial_{\alpha(m)\beta\alpha(m)}\big),$$

where α and β run independently over all permutations of $1, \ldots, m$.

In [11], we employed isomorphisms of H onto a bounded domain in Case SP. The same idea works in Case SU. In fact, put

$$(1.34) \qquad \mathfrak{E} = \left\{ \alpha \in SL_{2m}(\mathbf{C}) \mid \alpha^* \iota_m \alpha = - i \iota'_{m,m} \right\}$$

with ι_m and $\iota'_{m,m}$ of (1.1). We can let every $\alpha = \left(\begin{smallmatrix} a & b \\ c & d \end{smallmatrix}\right)$ of $\mathfrak{E}$ act on $B_{m,m}$ by $\alpha(z) = (az + b)(cz + d)^{-1}$ for $z \in B_{m,m}$. This maps $B_{m,m}$ onto H bijectively. Further we define "factors of automorphy" ν_α and μ_α by

$$(1.35) \qquad \alpha \begin{bmatrix} 1_m & z \\ z^* & 1 \end{bmatrix} = \begin{bmatrix} \alpha(z)^* & \alpha(z) \\ 1_m & 1_m \end{bmatrix} \begin{bmatrix} i \cdot \overline{\nu_\alpha(z)} & 0 \\ 0 & \mu_\alpha(z) \end{bmatrix} \qquad (\alpha \in \mathfrak{E}, z \in B_{m,m}).$$

Then we can easily verify that

$$(1.36a) \qquad d(\alpha z) = {}^t\nu_\alpha(z)^{-1} dz\, \mu_\alpha(z)^{-1},$$

$$(1.36b) \quad \xi_B(z) = \nu_\alpha(z)^* \cdot {}^t\eta(\alpha(z))\nu_\alpha(z), \qquad \eta_B(z) = \mu_\alpha(z)^*\eta(\alpha(z))\mu_\alpha(z),$$

$$(1.36c) \qquad \det(\nu_\alpha(z)) = \det(\mu_\alpha(z)) \qquad (\alpha \in \mathfrak{E}, z \in B_{m,m}).$$

Let V and ρ be the same as in (1.9). For a V-valued function f on H and $\alpha \in \mathfrak{E}$, we define $f\|_\rho \alpha$ as a function on $B_{m,m}$ by

$$(1.37) \qquad f\|_\rho \alpha = \rho(\nu_\alpha, \mu_\alpha)^{-1}(f \circ \alpha).$$

Then formulas (1.17), (1.19), and (1.32) hold for $\alpha \in \mathfrak{E}$. We note here

$$(1.38) \qquad \Delta_\rho^{(p+q)}f = \Delta_\lambda^{(q)}\Delta_\rho^{(p)}f \qquad (\lambda(x, y) = \det(xy)^p \rho(x, y)),$$

$$(1.39) \qquad \Delta_\rho^{(p)}f = \rho(\xi, \eta)^{-1} \delta^{\kappa - p}\Delta^p\left[\delta^{p - \kappa}\rho(\xi, \eta)f\right]$$

with κ of (1.15). In Case SP, these are already given in [11, Propositions 7.2 and 7.3]. In all three cases, we have

$$(1.40) \qquad \Delta^p(f\|_{\kappa - p}\alpha) = (\Delta^p f)\|_{\kappa + p}\alpha \qquad (\alpha \in \mathfrak{G} \text{ or } \mathfrak{G}' \text{ or } \mathfrak{E}),$$

where $f\|_\kappa \alpha = \det(\mu_\alpha)^{-k}f \circ \alpha$. This is [11, Lemma 7.1 and (8.6)] and can be proved by the same method as in [11, Lemma 7.1]. Then our formulas can be derived in the same manner as [11, Propositions 7.2 and 7.3] (cf. Proof of Proposition 2.1 below).

We insert here two lemmas on the explicit action of differential operators.

LEMMA 1.1. *Exclude Case SUB with $l \neq m$. Suppose σ of (1.25a) is irreducible and has the highest weight $\lambda(x) = \prod_{i=1}^{m} \det_i(x)^{a_i}$ in the sense that $\sigma(x)e = \lambda(x)e$ for all upper triangular matrices x in $GL_m(\mathbf{C})$ with $0 \leq a_i \in \mathbf{Z}$ and with a nonzero vector $e \in \mathbf{C}^n$, where $\det_i(x)$ is the determinant of the first $i \times i$ entries of x. Suppose in addition, in Case SP, that the restriction of σ to the orthogonal group*

$\{x \in GL_m(\mathbf{R}) \mid {}^txx = 1\}$ *is irreducible. Then*

$$(1.41) \qquad \sigma(D)\det(z)^s = \beta_\lambda(s)\det(z)^s\sigma({}^tz^{-1}) \qquad (s \in \mathbf{C};\ z \in H \text{ or } B)$$

with a constant $\beta_\lambda(s)$ *given by*

$$\beta_\lambda(s) = \prod_{\nu=0}^{m-1} \prod_{j=0}^{b_\nu - 1} (s - j + (d\nu/2)),$$

$$b_\nu = \sum_{i=\nu+1}^{m} a_i, \qquad d = \begin{cases} 1 & (\textit{Case SP}), \\ 2 & (\textit{Cases SU, SUB}). \end{cases}$$

Proof. This can be proved by the same technique as for [14, Lemma 9.1], which concerns the case where σ is the representation of GL_m on $\bigwedge^r\mathbf{C}^m$. As shown there, we see easily that (1.41) holds with a constant $\beta_\lambda(s)$. To determine $\beta_\lambda(s)$, we observe that ${}^te'\sigma(x)e = \lambda(x)$ for all $x \in GL_m$ with a suitable choice of $e' \in \mathbf{C}^n$, $\neq 0$ (see [1, 5-04], for example). In other words, we may assume, changing the coordinate system of $\mathbf{C}^n$, that the (i, j)-entry $\sigma_{ij}(x)$ of $\sigma(x)$ is exactly $\lambda(x)$ for some i and j. Applying $\sigma_{ij}(D)$ $(=\lambda(D))$ to the integral formula [14, (9.3)], we find that

$$\lambda(D)\det(z)^{-s} = (-1)^r\lambda(z^{-1})\det(z)^{-s} \prod_{\nu=0}^{m-1} \Gamma(s + b_\nu - (d\nu/2))/\Gamma(s - (d\nu/2)),$$

where r is the degree of σ. Changing s for $-s$, we obtain our assertion.

It should be noted that (1.41) in Case SP has a counter-example when σ is reducible on $O(m, \mathbf{R})$.

LEMMA 1.2. *Let* ρ *be a representation of* $\mathcal{g}$ *defined by* $\rho(x, y) = \det(y)^k$ *with* $k \in \mathbf{Z}$, *and let* σ *be the same as in Lemma 1.1. Then, for* $\alpha \in \mathfrak{G}$, *one has*

$$\Delta_\rho^\sigma(\delta^s j_\alpha^{-k}|j_\alpha|^{-2s}) = i^r\beta_\lambda(-k-s)\delta^s j_\alpha^{-k}|j_\alpha|^{-2s}\sigma({}^t(\mu_\alpha^{-1}\bar{\kappa}_\alpha\eta^{-1})),$$

where r *is the degree of* σ *and* β_λ *is given in Lemma 1.1.*

Here we identify an element a of $\mathrm{Hom}(\mathbf{C}_n^r, \mathbf{C})$ with an element b of $\mathbf{C}_n^r$ by the relation $a(u) = \sum u_{jk}b_{jk}$ for $u \in \mathbf{C}_n^r$.

Proof. Our idea is the same as in the proof of [11, Lemma 2.3]. We first prove

$$(1.42) \qquad \Delta_\rho^\sigma(\delta^s) = \beta_\lambda(-k-s)\delta^s\sigma(i \cdot {}^t\eta^{-1}) \qquad \text{on } H.$$

Given $w \in H$, take $\gamma \in \mathfrak{E}$ so that $w = \gamma(0)$, where 0 is the origin of B. To distinguish δ on H and δ on B, write them δ_H and δ_B, respectively. For functions g on B, we have

$$(1.43) \qquad (\Delta_\rho^\sigma g)(0) = (\Phi_\sigma D_\rho^{(r)}g)(0) = (\Phi_\sigma D^r g)(0) = \sigma(D)g(0)$$

by virtue of the principle of [11, Lemma 2.2] and (1.30). Now (1.32) shows

$$(\Delta_\rho^\sigma \delta_H^s)\|_\rho\gamma = \sigma(\nu_\gamma)\big[\Delta_\rho^\sigma(\delta_H^s\|_\rho\gamma)\big]\sigma('\mu_\gamma).$$

Put $j_\gamma = \det(\mu_\gamma)$. Taking $g = \delta_H^s\|_\rho\gamma$ in (1.43) and evaluating it at 0, we obtain

$$j_\gamma^{-k}(0)(\Delta_\rho^\sigma \delta_H^s)(w) = \sigma(\nu_\gamma(0))\big[\sigma(D)(\delta_H^s\|_\rho\gamma)(0)\big]\sigma('\mu_\gamma(0)).$$

We have $\delta_H^s\|_\rho\gamma = \delta_B^s j_\gamma^{-k}|j_\gamma|^{-2s}$ by (1.36b), and by [11, Lemma 2.2],

$$\sigma(D)\big[\delta_B^s j_\gamma^{-k}|j_\gamma|^{-2s}\big](0) = \sigma(D)\big(j_\gamma^{-k-s}\bar{j}_\gamma^{-s}\big) = \bar{j}_\gamma^{-s}\sigma(D)\big[\det(cz+d)^{-k-s}\big]_{z=0},$$

where (c,d) is the lower half of γ. By Lemma 1.1, we have

$$(1.44)\qquad \sigma(D)\det(cz+d)^s = \beta_\lambda(s)\det(cz+d)^s\sigma('c\cdot{}'(cz+d)^{-1}).$$

Since $\gamma(0) = w$ and $\xi_B(0) = \eta_B(0) = 1$, we see from (1.35) and (1.36b) that $c = i\nu_\gamma(0)$, $d = \mu_\gamma(0)$, $\delta_H(w)^s = |j_\gamma(0)|^{-2s}$, and $\eta(w)^{-1} = \overline{\nu_\gamma(0)}\cdot{}'\nu_\gamma(0)$. Combining all these, we obtain (1.42). Taking f to be δ_H^s in (1.32), we can derive the desired formula from (1.42).

2. The operator E_σ and a further study of Δ_ρ^σ.

Let us first put

$$(2.1)\qquad \bar\partial_{ij} = \begin{cases} 2^{-1}(1+\delta_{ij})\partial/\partial\bar z_{ij} & \text{(Case SP)}, \\ \partial/\partial\bar z_{ij} & \text{(Cases SU, SUB)}, \end{cases}$$

and denote by $\overline D$ the matrix with $\bar\partial_{ij}$ as its (i,j)-entry. Then $\sigma(\overline D)$ is defined in an obvious way for σ as in (1.25a). For simplicity, we let X denote our space H or B. Given a complex vector space V and a V-valued function f on X, we let $\overline D^k f$ denote the function on X with values in $\mathrm{Ml}_k(T, V)$, similar to $D^k f$, given by

$$\big(\overline D^k f(z)\big)(u_1, \ldots, u_k) = \prod_{p=1}^{k}\bigg(\sum_{i=1}^{l}\sum_{j=1}^{m} u_{ij}^p \bar\partial_{ij}\bigg) f(z) \qquad (u_p = (u_{ij}^p) \in T).$$

Now we define an operator E_1 by

$$(2.2)\qquad (E_1 f)(u) = (\overline D f)(\xi u\cdot{}'\eta) \qquad (u \in T).$$

This sends V-valued functions on X to $\mathrm{Hom}(T, V)$-valued ones. It can easily be seen from (1.7a, b, c) and (1.36a, b) that

$$(2.3)\qquad E_1(f\|_\rho\alpha) = (E_1 f)\|_{\rho\otimes\pi}\alpha \qquad (\alpha \in \mathfrak{G}, \mathfrak{G}', \text{ or } \mathfrak{E})$$

with $\rho\otimes\pi$ of (1.14b). The iterated operator E_1^r is well defined, and sends V-valued functions to $\mathrm{Ml}_r(T, V)$-valued ones. Given σ of degree r as in (1.25a, b), we put

$$（2.4）\qquad E_\sigma = \Phi_\sigma E_1^r .$$

This sends V-valued functions to $\mathrm{Hom}(\mathbf{C}_n^{n'}, V)$-valued ones. We have $E_\sigma = E_1$ if σ is the identity map of $\mathbf{C}_m^l$ onto itself. From (2.3) and (1.28b), we obtain immediately

$$（2.5a）\qquad E_1^r(f\|_\rho \alpha) = (E_1^r f)\|_{\rho\otimes\pi'} \alpha,$$

$$（2.5b）\qquad E_\sigma(f\|_\rho \alpha) = (E_\sigma f)\|_{\rho\otimes_2\sigma} \alpha \qquad (\alpha \in \mathfrak{G}, \mathfrak{G}', \text{ or } \mathfrak{E}).$$

Hereafter in this section, we exclude Case SUB with $l \neq m$. We consider the space $\bigwedge^r \mathbf{C}^m$ of r-vectors of $0 \leqq r \leqq m$ and identify $\bigwedge^r \mathbf{C}^m$ with $\mathbf{C}^{A(r)}$, where $A(r)$ is the set of all sequences $a = \{a_1 < \cdots < a_r\}$ of r numerals among $1, \ldots, m$. For such an a, we put $e_a = e_{a_1} \wedge \cdots \wedge e_{a_r}$, where $\{e_k\}$ is the standard basis of $\mathbf{C}^m$; we shall write A for $A(r)$ when r is fixed. We define a polynomial representation

$$（2.6a）\qquad \rho_r : GL_m(\mathbf{C}) \to GL(\textstyle\bigwedge^r \mathbf{C}^m) = GL(\mathbf{C}^A) = GL_A(\mathbf{C})$$

by taking $\rho_r(x)$ to be the matrix representing the action of x on $\bigwedge^r \mathbf{C}^m$ with respect to $\{e_a\}_{a\in A}$. We have $\rho_0(x) = 1$, $\rho_1(x) = x$, $\rho_m(x) = \det(x)$, and $\rho_r(^tx) = {}^t\rho_r(x)$. We define another polynomial representation ρ_r^* of GL_m by

$$（2.6b）\qquad \rho_r^*(x) = \det(x) \cdot {}^t\rho_{m-r}(x)^{-1}(\in GL_{A(r)}(\mathbf{C})),$$

which represents the action of x on $\bigwedge^{m-r}\mathbf{C}^m$ with respect to the basis dual to $\{e_a\}_{a\in A(r)}$.

In Case SP, let T_r denote the subspace of $\mathrm{End}(\bigwedge^r \mathbf{C}^m)$ $(= \mathbf{C}_A^A)$ generated by $\rho_r(x)$ for all symmetric $x \in \mathbf{C}_m^m$; in Cases SU and SUB, we simply put $T_r = \mathrm{End}(\bigwedge^r \mathbf{C}^m)$. In particular we have $T_1 = T$ and identify T_m with $\mathbf{C}$. For $h \in \mathrm{Hom}(T_r, V)$ with a complex vector space V, we define its components $h_{ab} \in V$ for $a, b \in A$ by

$$（2.7）\qquad h(u) = \sum_{a,b\in A} u_{ab} h_{ab} \qquad (u = (u_{ab}) \in T_r).$$

In Case SP, we have to impose the condition that $(\varphi(h_{ab}))_{a,b\in A} \in T_r$ for every $\varphi \in \mathrm{Hom}(V, \mathbf{C})$, under which h_{ab} is uniquely determined. Similarly, for a function g on X with values in $\mathrm{Hom}(T_r, V)$, its components g_{ab} are defined as V-valued functions on X. Every element h of $\mathrm{Hom}(T_r, V)$ can be viewed as an element of $\mathrm{Hom}(\mathbf{C}_A^A, V)$ by defining $h(u)$ for $u \in \mathbf{C}_A^A$ by (2.7).

We are going to give an explicit expression for Δ_ρ^σ similar to (1.39) when $\sigma = \rho_r$. First we observe that if $\sigma = \rho_r$, Φ_σ maps $\mathrm{Ml}_r(T, V)$ into $\mathrm{Hom}(T_r, V)$. Next, for a V-valued function f on X, we define $\rho_r(D)f$ and $\rho_r(\overline{D})f$ to be the functions on X

with values in $\mathrm{Hom}(T_r, V)$ by

$$(2.8) \quad \left[\rho_r(D)f\right]_{ab} = \rho_r(D)_{ab}f, \quad \left[\rho_r(\overline{D})f\right]_{ab} = \rho_r(\overline{D})_{ab}f \qquad (a, b \in A).$$

That $(\rho_r(D)_{ab}f)_{a,b\in A}$ has values in T_r in Case SP if $V = \mathbf{C}$ must be and can easily be verified.

PROPOSITION 2.1. *Let $\{\rho, V\}$ be a rational representation of $\mathscr{g}$, and let $\sigma = \rho_r$ with $0 \leqq r \leqq m$. Then*

$$(2.9a) \qquad\qquad \Delta_\rho^\sigma f = \delta^\lambda \rho(\xi, \eta)^{-1} \rho_r(D)\left[\delta^{-\lambda}\rho(\xi, \eta)f\right],$$

$$(2.9b) \qquad (E_a f)(u) = \delta^\lambda\left[\rho_r(\overline{D})(\delta^{-\lambda}f)\right]\left(\rho_r(\xi)u\rho_r('\eta)\right) \qquad (u \in T_r),$$

$$(2.9c) \qquad\qquad \lambda = \begin{cases} (r-1)/2 & (\textit{Case SP}), \\ r-1 & (\textit{Cases SU, SUB}). \end{cases}$$

Notice that $\rho_r(D)$ and Δ_ρ^σ coincide with D and D_ρ if $r = 1$, and with Δ and $\Delta_\rho^{(1)}$ if $r = m$. Thus our result is trivial if $r = 1$; if $r = m$, (2.9a) is (1.39) with $p = 1$. To prove our proposition, we need two lemmas.

LEMMA 2.2. *Let $\sigma: GL_m \to GL_k$ be a polynomial representation. Then, in Cases SP and SU, each of $\sigma(\kappa_\alpha^{-1}\overline{\mu}_\alpha)$, $\sigma(\overline{\mu}_\alpha\kappa_\alpha^{-1})$, $\sigma(\mu_\alpha^{-1}\overline{\kappa}_\alpha)$, $\sigma(\overline{\kappa}_\alpha\mu_\alpha^{-1})$, for every $\alpha \in \mathscr{G}$, can be expressed as a polynomial function in the entries of $z - z^*$, whose coefficients are $M_k(\mathbf{C})$-valued holomorphic functions on H, and whose constant term is 1_k.*

Proof. It is sufficient to prove the case where $\sigma(x) = \bigotimes^p x$ with $0 < p \in \mathbf{Z}$. Put $\theta = z - z^*$. If (c, d) is the lower half of α, we have $\kappa_\alpha - \overline{\mu}_\alpha = \overline{c} \cdot {}'\theta$, so that $\kappa_\alpha^{-1}\overline{\mu}_\alpha = 1 - \kappa_\alpha^{-1}\overline{c} \cdot {}'\theta$, and hence $\sigma(\kappa_\alpha^{-1}\overline{\mu}_\alpha) = \bigotimes^p(1 - \kappa_\alpha^{-1}\overline{c} \cdot {}'\theta)$. This proves our assertion for $\sigma(\kappa_\alpha^{-1}\overline{\mu}_\alpha)$. The other three cases can be treated in a similar way.

LEMMA 2.3. *Let f_i, for $i = 1, 2$, be $\mathbf{C}$-valued functions on H of the form $f_i = \det(z - z^*)^{-1}\sum_{k=r}^m \mathrm{tr}[{}'\rho_k(z - z^*)g_{ik}]$ with holomorphic T_k-valued functions g_{ik} on H, where r is an integer such that $0 < r < m$. Suppose $f_1 \circ \alpha = j_\alpha^s f_2$ for an element α of $\mathscr{G}$ with $s \in \mathbf{C}$. Then $g_{1r} \circ \alpha = j_\alpha^{s-2}\rho_r(\kappa_\alpha)g_{2r} \cdot {}'\rho_r(\mu_\alpha)$.*

Proof. For simplicity, let us put $\theta = z - z^*$, $\epsilon = \det(z - z^*)$, and write κ, μ, and j for κ_α, μ_α, and j_α. Then, by (1.7b),

$$\epsilon f_2 = j^{1-s}\overline{j}(\epsilon f_1) \circ \alpha = j^{1-s}\overline{j}\sum_{k=r}^m \mathrm{tr}\left[{}'\rho_k(\mu^{-1}) \cdot {}'\rho_k(\theta)\rho_k(\overline{\mu})^{-1}(g_{1k} \circ \alpha)\right].$$

Since

$$\overline{j}\rho_k(\overline{\mu})^{-1} = {}'\rho_{m-k}^*(\overline{\mu}) = {}'\rho_{m-k}^*(\kappa^{-1}\overline{\mu}) \cdot {}'\rho_{m-k}^*(\kappa) = j \cdot {}'\rho_{m-k}^*(\kappa^{-1}\overline{\mu})\rho_k(\kappa^{-1}),$$

we have

$$\epsilon f_2 = j^{2-s} \sum_{k=r}^{m} \mathrm{tr}\Big[{}^t\rho_k(\theta) \cdot {}^t\rho_{m-k}^*(\kappa^{-1}\bar{\mu})\rho_k(\kappa^{-1})(g_{1k} \circ \alpha) \cdot {}^t\rho_k(\mu^{-1})\Big].$$

Applying Lemma 2.2 to $\rho_{m-k}^*(\kappa^{-1}\bar{\mu})$, we see that the right-hand side is a polynomial function in the entries of θ whose coefficients are holomorphic functions on H. Moreover, the lowest terms have degree r, and are given by

$$(2.10) \qquad j^{2-s}\mathrm{tr}\Big[{}^t\rho_r(\theta)\rho_r(\kappa^{-1})(g_{1r} \circ \alpha) \cdot {}^t\rho_r(\mu^{-1})\Big].$$

On the other hand, $\epsilon f_2 = \sum_{k=r}^{m} \mathrm{tr}[{}^t\rho_k(\theta)g_{2k}]$, and hence the terms of degree r are given by $\mathrm{tr}[{}^t\rho_r(\theta)g_{2r}]$, which must coincide with (2.10) because of the uniqueness of such a polynomial expression (see [14, Lemma 9.3]). (That g_{ik} is T_k-valued is essential in Case SP.) Therefore we obtain our assertion.

Proof of Proposition 2.1. We first prove, for C-valued f on X,

$$(2.11) \quad \big[\rho_r(D)(f\|_\lambda\alpha)\big](u) = \big\{\big[\rho_r(D)f\big]\|_\lambda\alpha\big\}\big(\rho_r({}^t\kappa_\alpha^{-1})u\rho_r(\mu_\alpha^{-1})\big) \qquad (u \in T_r),$$

where $f\|_s\alpha = j_\alpha^{-s}(f \circ \alpha)$. Since this is a purely algebraic formula, it is sufficient to prove it for holomorphic functions f and $\alpha \in \mathfrak{G}$. Now put $\Delta_s f = \delta^{\kappa-1-s}\Delta(\delta^{s+1-\kappa}f)$ for $s \in \mathbf{C}$. Then

$$(2.12) \qquad \Delta_s(f\|_s\alpha) = (\Delta_s f)\|_{s+2}\alpha.$$

This follows easily from (1.40). By [14, Lemma 9.1], we have

$$\Delta_s f = \delta^{\kappa-1-s} \sum_{k=0}^{m} \mathrm{tr}\Big[\rho_{m-k}^*(D)\delta^{s+1-\kappa} \cdot {}^t\rho_k(D)f\Big]$$

$$= \det(z - z^*)^{-1} \sum_{k=0}^{m} c_{m-k}(s + 1 - \kappa)\mathrm{tr}\Big[\rho_k(z - z^*) \cdot {}^t\rho_k(D)f\Big]$$

with c_k as given in that lemma. Fix r and take s to be λ of (2.9c); observe that $c_{m-r}(\lambda + 1 - \kappa) \neq 0$ and $c_{m-k}(\lambda + 1 - \kappa) = 0$ for $k < r$. Therefore

$$\Delta_\lambda f = \det(z - z^*)^{-1} \sum_{k=r}^{m} c_{m-k}(\lambda + 1 - \kappa)\mathrm{tr}\Big[\rho_k(z - z^*) \cdot {}^t\rho_k(D)f\Big].$$

Take $f\|_\lambda\alpha$ in place of f. Then we obtain (2.11) by virtue of Lemma 2.3. Now let R and S denote the right-hand sides of (2.9a, b). Then we have, with $\sigma = \rho_r$,

$$(2.13) \quad R(f\|_\rho\alpha) = (Rf)\|_{\rho\otimes_1\sigma}\alpha, \quad S(f\|_\rho\alpha) = (Sf)\|_{\rho\otimes_2\sigma}\alpha \quad (\alpha \in \mathfrak{G}, \mathfrak{G}', \text{ or } \mathfrak{E}).$$

These can be derived from (2.11) and its "complex conjugate" by straightforward calculations similar to the proof of [11, Lemma 2.1]. Our aim is to show that

$Rf = \Delta_\rho^\sigma f$ and $Sf = E_\sigma f$. Put $P = \Delta_\rho^\sigma$ and $\omega = \rho \otimes_\iota \sigma$. If g is a function on B, we have, by virtue of the principle of [11, Lemma 2.2], $(Rg)(0) = [\delta(D)g](0)$. For the same reason, (1.30) shows that $(Pg)(0) = [\sigma(D)g](0)$. Given $z \in H$, take $\alpha \in \mathfrak{G}$ so that $\alpha(0) = z$. By (2.13), we have $(Rf)(z) = \omega(\nu_\alpha(0), \mu_\alpha(0))R(f\|_\rho \alpha)(0)$; similarly, by (1.32), $(Pf)(z) = \omega(\nu_\alpha(0), \mu_\alpha(0))P(f\|_\rho \alpha)(0)$. Since $(Rg)(0) = (Pg)(0)$ for any g, we have $Pf = Rf$. The same idea is applicable to functions on B. Equality $S = E_\sigma$ can be proved in a similar way by means of the anti-holomorphic version of [11, Lemma 2.2].

We end this section by noting that

$$(2.14) \qquad E_\zeta f = \delta^{\kappa + p}\bar{\Delta}^p[\delta^{p-\kappa}f] \qquad \text{if} \quad \zeta(x) = \det(x)^p, \quad 0 \leqq p \in \mathbf{Z},$$

where $\bar{\Delta}^p = \det(\bar{D})^p$. This can be proved in the same manner as in the above proof, by first showing that the operator on the right-hand side of (2.14) has the same property (2.5b) as E_ζ, which follows easily from (1.40). Similarly, as a generalization of (1.38), we can prove

$$(2.15) \qquad \Delta_\rho^\epsilon = \Delta_{\rho \otimes_\iota \sigma}^{(p)}\Delta_\rho^\sigma = \Delta_\lambda^\sigma \Delta_\rho^{(p)} \qquad \text{if} \quad \lambda(x, y) = \det(xy)^p \rho(x, y) \quad \text{and}$$
$$\epsilon(x) = \det(x)^p \sigma(x) \qquad \text{with} \quad 0 \leqq p \in \mathbf{Z},$$

$$(2.16) \qquad E_\epsilon = E_\sigma E_\zeta \qquad \text{if} \quad \epsilon(x) = \zeta(x)\sigma(x) \quad \text{and}$$
$$\zeta(x) = \det(x)^p \qquad \text{with} \quad 0 \leqq p \in \mathbf{Z}.$$

3. Arithmeticity at CM-points.

Now we define an algebraic group G by

$$(3.0) \qquad G = \begin{cases} \{\alpha \in SL_{2m}(F) \,|\, {}^t\alpha \iota_m \alpha = \iota_m\} & \text{(Case SP)}, \\ \{\alpha \in SL_{2m}(K) \,|\, \alpha^* \iota_m \alpha = \iota_m\} & \text{(Case SU)}, \\ \{\alpha \in SL_n(K) \,|\, \alpha \iota \alpha^* = \iota\} & \text{(Case SUB)}. \end{cases}$$

Here and throughout the rest of the paper, we denote by F a totally real algebraic number field of finite degree, by K a totally imaginary quadratic extension of F, and by ρ the nontrivial automorphism of K over F; we put $z^* = {}^t z^\rho$ for matrices z with entries in K. In Case SUB, ι is an arbitrary element of $GL_n(K)$ such that $\iota^* = -\iota$. In Cases SU and SUB, we fix a CM-type φ of K, which is, by definition, a set of $[F : \mathbf{Q}]$ embeddings of K into $\mathbf{C}$, whose restrictions to F are all different. To make our exposition uniform, we use, in Case SP, the letter φ to denote the set of all embeddings of F into $\mathbf{R}$. (Thus φ is a new symbol for ∞ of [14]; also ι_m replaces η of [14].) For each $v \in \varphi$, we denote by a_v the image of $a \in K_m^n$ (or $a \in F_m^n$) under v. In Case SUB, let (p_v, q_v) be the signature of $-i\iota_v$. Take $Q_v \in GL_n(\overline{\mathbf{Q}})$ so that

$$(3.1) \qquad -i\iota_v = \overline{Q}_v \cdot \mathrm{diag}[\, 1_{p_v}, -1_{q_v}] \cdot {}^tQ_v .$$

Then the map

$$(3.2) \qquad \alpha \mapsto \left(Q_v^{-1}\overline{\alpha}_v Q_v \right)_{v \in \varphi}$$

gives an injection of G into $\prod_{v \in \varphi} \mathfrak{G}'_{p_v,q_v}$. Put $\mathscr{B} = \prod_{v \in \varphi} B_{p_v,q_v}$ and $\mathscr{g}_v = GL_{p_v}(\mathbf{C}) \times GL_{q_v}(\mathbf{C})$ (GL_0 must be disregarded). Then we can let G act on $\mathscr{B}$ through (3.2). We take our group G in Case SUB and its action on $\mathscr{B}$ in these forms simply in order to conform to the formulation of our previous papers. (For example, ι corresponds to T of [6].) In Cases SP and SU, we consider $\mathfrak{G}^\varphi$ and H^φ which are the products of φ copies of G and H (see Notation), and let $\mathfrak{G}^\varphi$ act on H^φ componentwise, and embed G into $\mathfrak{G}^\varphi$ by $\alpha \mapsto (\alpha_v)_{v \in \varphi}$, so that G acts on H^φ.

Take a finite-dimensional complex vector space V and a rational representation

$$(3.3) \qquad \zeta : \left\{ \begin{array}{l} \mathscr{g}^\varphi \\[4pt] \displaystyle\prod_{v \in \varphi} \mathscr{g}_v \end{array} \right\} \to GL(V) \qquad \begin{array}{l} \text{(Cases SP, SU)}, \\[4pt] \text{(Case SUB)}. \end{array}$$

Given $\alpha \in G$, $z \in H^\varphi$ or $z \in \mathscr{B}$, and a V-valued function f on H^φ or on $\mathscr{B}$, we define a V-valued function $f\|_\zeta\alpha$ by

$$(3.4\text{a}) \qquad (f\|_\zeta\alpha)(z) = \Lambda_\zeta(\alpha,z)^{-1} f(\alpha(z)),$$

$$(3.4\text{b}) \qquad \Lambda_\zeta(\alpha,z) = \begin{cases} \zeta\big((\mu_v(\alpha,z))_{v\in\varphi}\big) & \text{(Case SP)}, \\[6pt] \zeta\big((\kappa_v(\alpha,z),\, \mu_v(\alpha,z))_{v\in\varphi}\big) & \text{(Cases SU, SUB)}, \end{cases}$$

$$\kappa_v(\alpha,z) = \kappa(\alpha_v, z_v), \quad \mu_v(\alpha,z) = \mu(\alpha_v, z_v) \qquad \text{(Cases SP, SU)},$$

$$\kappa_v(\alpha,z) = \kappa\big(Q_v^{-1}\overline{\alpha}_v Q_v, z_v\big), \quad \mu_v(\alpha,z) = \mu\big(Q_v^{-1}\overline{\alpha}_v Q_v, z_v\big) \qquad \text{(Case SUB)}.$$

In Case SUB, κ or μ must be disregarded according as $p_v = 0$ or $q_v = 0$.

For a congruence subgroup Γ of G, we denote by $\mathfrak{A}_\zeta(\Gamma)$ (resp. $\mathfrak{M}_\zeta(\Gamma)$) the set of all meromorphic (resp. holomorphic) V-valued functions f on H^φ or on $\mathscr{B}$ which satisfy $f\|_\zeta\gamma = f$ for all $\gamma \in \Gamma$, and which are meromorphic (resp. holomorphic) at all cusps when G is isomorphic to $SL_2(\mathbf{Q})$; further we denote by $\mathfrak{A}_\zeta$ (resp. $\mathfrak{M}_\zeta$) the union of $\mathfrak{A}_\zeta(\Gamma)$ (resp. $\mathfrak{M}_\zeta(\Gamma)$) for all congruence subgroups Γ of G.

We now recall briefly the notion of CM-points and also that of arithmetic elements of $\mathfrak{A}_\zeta$ (cf. [10, pp. 319, 325, 371], [11, §§3, 4]).

For $Y = Y_1 \oplus \cdots \oplus Y_t$ with algebraic number fields Y_i of finite degree, we denote by $J_Y = J(Y)$ the set of all nontrivial homomorphisms of Y into $\mathbf{C}$, and by $I_Y = I(Y)$ the module of all formal sums $c = \sum_{\tau \in J(Y)} c_\tau \tau$ with $c_\tau \in \mathbf{Z}$. We consider such a c a character of $Y^\times$ by putting

$$(3.5) \qquad x^c = \prod_{\tau \in J(Y)} (x^\tau)^{c_\tau} \qquad (x \in Y^\times).$$

Naturally J_Y (resp. I_Y) can be identified with the union of $J(Y_i)$ (resp. the direct sum of $I(Y_i)$) for $1 \leq i \leq t$.

Suppose that the Y_i are all CM-fields. Then we denote by ρ the automorphism of Y whose restriction to each Y_i is the complex conjugation. An element ω of I_Y is called a CM-type of Y if $\omega + \rho\omega = \sum_{\tau \in J(Y)} \tau$. Such an ω can be viewed also as a subset of $J(Y)$. In [10, §1], we introduced a "period symbol" $p_Y(\alpha, \beta)$, which is a complex number determined modulo $\overline{\mathbf{Q}}^\times$ by $(\alpha, \beta) \in I_Y \times I_Y$. This is basic to our definition of arithmeticity.

To define CM-points of G in a uniform way, let us make the following convention: K denotes F in Case SP; $n = 2m$ and ι denotes ι_m in Cases SP and SU; in Case SUB, ι is the element employed in the definition of G. Now we consider a K-algebra

$$(3.6) \qquad Z = M_{n_1}(Y_1) \oplus \cdots \oplus M_{n_t}(Y_t)$$

with CM-fields Y_i containing K. Suppose that there exists a positive involution δ of Z and a K-linear injective homomorphism

$$(3.7) \qquad h : Z \to M_n(K)$$

such that $h(x^\delta) = \iota h(x)^* \iota^{-1}$ for all $x \in Z$; suppose also $n = \sum_{i=1}^t n_i [Y_i : K]$. Put

$$(3.8) \qquad Z[\delta] = \{ x \in Z \mid xx^\delta = 1 \}.$$

Then $h(Z[\delta])$ is contained in G and has a unique common fixed point w on H^φ or on $\mathscr{B}$, which we call a CM-point of G on H^φ or on $\mathscr{B}$.

Put $Y = Y_1 \oplus \cdots \oplus Y_t$. Now we can define a CM-type ψ_i of Y_i for each i so that

$$\prod_{i=1}^t a^{n_i \psi_i} = \prod_{v \in \varphi} \begin{cases} \det\left[\mu_v(h(a), w) \right] & \text{(Case SP),} \\ \det\left[\kappa_v(h(a), w) \right] \det\left[\mu_v(h(a), w) \right] & \text{(Cases SU, SUB)} \end{cases}$$

for all $a \in Y[\rho]$. Put $\psi = \psi_1 + \cdots + \psi_t$. Then ψ is a CM-type of Y.

To define the arithmetic elements of $\mathfrak{A}_\zeta$, we hereafter assume that V of (3.3) has a fixed $\overline{\mathbf{Q}}$-rational structure and ζ is $\overline{\mathbf{Q}}$-rational with respect to the standard $\overline{\mathbf{Q}}$-rational structure of $\mathscr{J}^\varphi$ or $\prod_{v \in \varphi} \mathscr{J}_v$; for simplicity we identify V with $\mathbf{C}^d$ for some d so that $\overline{\mathbf{Q}}^d$ represents the $\overline{\mathbf{Q}}$-rational elements of V.

With a CM-point w as above, we observe that $a \mapsto \Lambda_\zeta(h(a), w)$ is a $\overline{\mathbf{Q}}$-rational representation of $Y[\rho]$. Hence we have

$$\Lambda_\zeta(h(a), w) = U \cdot \mathrm{diag}[a^{\sigma_1}, \ldots, a^{\sigma_d}] U^{-1} \qquad \text{for all} \quad a \in Y[\rho]$$

with $U \in GL_d(\overline{\mathbf{Q}})$ and $\sigma_1, \ldots, \sigma_d \in I_Y$. We then put

$$(3.9\text{a}) \qquad P_\zeta(w) = U \cdot \mathrm{diag}\big[\, p_Y(\sigma_1, \psi), \ldots, p_Y(\sigma_d, \psi)\,\big] U^{-1}.$$

Our definition of Λ_ζ shows that

$$(3.9\text{b}) \qquad P_\zeta(w) = \zeta(\mathscr{P}_w)$$

with an element $\mathscr{P}_w$ of $\mathscr{g}^\varphi$ or $\prod_{v \in \varphi} \mathscr{g}_v$, which is independent of ζ (see [10, p. 319]).

Now we call an element f of $\mathfrak{A}_\zeta$ *arithmetic*, or $\overline{\mathbf{Q}}$-*rational*, if $P_\zeta(w)^{-1}f(w) \in \overline{\mathbf{Q}}^d$ for every CM-point w where f is holomorphic. We denote by $\mathfrak{A}_\zeta(\overline{\mathbf{Q}})$, $\mathfrak{M}_\zeta(\overline{\mathbf{Q}})$, $\mathfrak{A}_\zeta(\Gamma, \overline{\mathbf{Q}})$, and $\mathfrak{M}_\zeta(\Gamma, \overline{\mathbf{Q}})$ the sets of $\overline{\mathbf{Q}}$-rational elements of $\mathfrak{A}_\zeta$, $\mathfrak{M}_\zeta$, $\mathfrak{A}_\zeta(\Gamma)$, and $\mathfrak{M}_\zeta(\Gamma)$, respectively.

In Cases SP and SU, every element f of $\mathfrak{M}_\zeta$ has a Fourier expansion of the form

$$(3.10) \qquad f(z) = \sum_{h \in L} c(h)\mathbf{e}\!\left(\mathrm{tr}\!\left(\sum_{v \in \varphi} h_v z_v\right)\right) \qquad (z = (z_v)_{v \in \varphi} \in H^\varphi)$$

with $c(h) \in V\,(= \mathbf{C}^d)$, where L is a lattice in the space of all symmetric elements of F_m^m (resp. hermitian elements of K_m^m) in Case SP (resp. Case SU), and $\mathbf{e}(w) = e^{2\pi i w}$. Then f is $\overline{\mathbf{Q}}$-rational if and only if $c(h) \in \overline{\mathbf{Q}}^d$ for every $h \in L$; an element g of $\mathfrak{A}_\zeta$ is $\overline{\mathbf{Q}}$-rational if and only if $g = f_0^{-1}f_1$ with $0 \neq f_0 \in \mathfrak{M}_\sigma(\overline{\mathbf{Q}})$ and $f_1 \in \mathfrak{M}_{\sigma\zeta}(\overline{\mathbf{Q}})$, where σ is a scalar-valued $\overline{\mathbf{Q}}$-rational representation. This consistency of two definitions of $\overline{\mathbf{Q}}$-rationality is proved in [11, §4] for $G = Sp(m, \mathbf{Q})$; the same type of argument applies to other cases.

In Case SU, suppose

$$(3.11) \qquad \xi\big((x_v, y_v)_{v \in \varphi}\big) = \prod_{v \in \varphi} \det\big(x_v y_v^{-1}\big)^{k_v} \zeta\big((x_v, y_v)_{v \in \varphi}\big)$$

with $k \in \mathbf{Z}^\varphi$. Then $\mathfrak{A}_\xi = \mathfrak{A}_\zeta$ and $\mathfrak{M}_\xi = \mathfrak{M}_\zeta$ in view of (1.3), and we have even

PROPOSITION 3.1. *In Case SU, one has* $\mathfrak{A}_\xi(\overline{\mathbf{Q}}) = \mathfrak{A}_\zeta(\overline{\mathbf{Q}})$ *under* (3.11). *The same conclusion holds in Case SUB if* $p_v = q_v$ *for all* $v \in \varphi$.

The proof will be given at the end of this section.

We now take a system $\sigma = \{\sigma_v\}_{v \in \varphi}$ consisting of maps σ_v of type (1.25a, b) given for each $v \in \varphi$, and assume that each σ_v is $\overline{\mathbf{Q}}$-rational. For $0 \leqq k \in \mathbf{Z}$ and $v \in \varphi$, we let D_v, $D_{v,\zeta}^{(k)}$, and $\Delta_{v,\zeta}^\sigma$ denote the operators D, $D_{\zeta_v}^{(k)}$, and $\Delta_{\zeta_v}^{\sigma_v}$ of (1.13), (1.18), and (1.31) on the v-factor of H^φ or $\mathscr{B}$, where ζ_v is the restriction of ζ to the v-factor of $\mathscr{g}^\varphi$ or $\prod_{v \in \varphi} \mathscr{g}_v$. Let $0 \leqq e \in \mathbf{Z}^\varphi$. Then we define an operator $D_\zeta^{(e)}$ (resp. Δ_ζ^σ) to be, roughly speaking, the product of $D_{v,\zeta}^{(e_v)}$ (resp. $\Delta_{v,\zeta}^\sigma$) for all $v \in \varphi$. To explain this more precisely, suppose for simplicity that φ consists of two elements v and w. Now $D_{v,\zeta}^{(e_v)}f$ takes values in $\mathrm{Ml}_{e_v}(T_v, V)$, which is the representation space of $\zeta \otimes \tau_v^{e_v}$, where T_v is the tangent space of the v-factor of

H^φ or $\mathscr{B}$, and $\tau_v^k(x) = \tau^k(x_v)$ for $x \in \mathscr{g}$ or $x \in \prod_{v \in \varphi} \mathscr{g}_v$ with τ^k of (1.14a). Put $\psi = \zeta \otimes \tau_v^{e_v}$. Then $D_{w,\psi}^{(e_w)} D_{v,\zeta}^{(e_v)} f$ is meaningful, and takes values in $\mathrm{Ml}_{e_w}(T_w, \mathrm{Ml}_{e_v}(T_v, V))$, which we can identify with the vector space of all multilinear maps of $T_w^{e_w} \times T_v^{e_v}$ into V. Then we put $D_\zeta^{(e)} = D_{w,\psi}^{(e_w)} D_{v,\zeta}^{(e_v)}$. (Since the restriction of ψ to the w-factor of $\mathscr{g}^\varphi$ is the direct sum of several copies of ζ_w, $D_{w,\psi}^{(e_w)}$ may also be written $D_{w,\zeta}^{(e_w)}$ by "abuse of notation".) We define $D_\zeta^{(e)}$ and Δ_ζ^σ in the general case in a similar way. Thus $D_\zeta^{(e)} f$ (resp. $\Delta_\zeta^\sigma f$) takes values in $\mathrm{Ml}_e(T, V)$ (resp. $\mathrm{Ml}(\prod_{v \in \varphi} \mathbf{C}_{n_v}^{n'}, V)$), by which we mean the vector space of all multilinear maps of $\prod_{v \in \varphi} T_v^{e_v}$ (resp. $\prod_{v \in \varphi} \mathbf{C}_{n_v}^{n'_v}$) into V, where $\mathbf{C}_{n_v}^{n'_v}$ is the image space of σ_v. Group $\mathscr{g}^\varphi$ or $\prod_{v \in \varphi} \mathscr{g}_v$ acts on these spaces via the representation $\omega = \zeta \otimes (\bigotimes_{v \in \varphi} \tau_v^{e_v})$ and a representation χ defined by

$$(3.12) \qquad [\chi(x, y)q]((w_v)_{v \in \varphi}) = \zeta(x, y)q\left(\left({}'\sigma_v'(x_v)w_v \cdot {}'\sigma_v''({}'y_v)\right)_{v \in \varphi}\right)$$

$$\left(q \in \mathrm{Ml}\left(\prod_{v \in \varphi} \mathbf{C}_{n_v}^{n'}, V\right), (w_v)_{v \in \varphi} \in \prod_{v \in \varphi} \mathbf{C}_{n_v}^{n'}\right).$$

From (1.19) and (1.32), we obtain

$$(3.13a) \qquad D_\zeta^{(e)}(f\|_\zeta \alpha) = \left(D_\zeta^{(e)} f\right)\|_\omega \alpha \qquad (\alpha \in G),$$

$$(3.13b) \qquad \Delta_\zeta^\sigma(f\|_\zeta \alpha) = (\Delta_\zeta^\sigma f)\|_\chi \alpha \qquad (\alpha \in G).$$

Now the basic theorem of the arithmeticity of $D_\zeta^{(e)} f$ and $\Delta_\zeta^\sigma f$ can be given as follows:

THEOREM 3.2. *Let $f \in \mathfrak{A}_\zeta(\overline{\mathbf{Q}})$ with a $\overline{\mathbf{Q}}$-rational representation ζ and let w be a CM-point where f is holomorphic. Let $\sigma = \{\sigma_v\}_{v \in \varphi}$ be a system of $\overline{\mathbf{Q}}$-rational maps of type (1.25a, b), and let r_v be the degree of σ_v. Then $\pi^{-|e|} P_\omega(w)^{-1}(D_\zeta^{(e)} f)(w)$ and $\pi^{-|r|} P_\chi(w)^{-1}(\Delta_\zeta^\sigma f)(w)$ are $\overline{\mathbf{Q}}$-rational, where $\omega = \zeta \otimes (\bigotimes_{v \in \varphi} \tau_v^{e_v})$, χ is defined by (3.12), $|e| = \sum_{v \in \varphi} e_v$, and $|r| = \sum_{v \in \varphi} r_v$.*

In other words, the nature of the values of $\pi^{-|e|} D_\zeta^{(e)} f$ and $\pi^{-|r|} \Delta_\zeta^\sigma f$ at CM-points is the same as those of the elements of $\mathfrak{A}_\omega(\overline{\mathbf{Q}})$ and $\mathfrak{A}_\chi(\overline{\mathbf{Q}})$. Therefore a special case of the theorem can be stated as

COROLLARY 3.3. *Suppose that ζ and σ_v for all $v \in \varphi$ are one-dimensional. Let f and w be as in Theorem 3.2, and g an element of $\mathfrak{A}_\chi(\overline{\mathbf{Q}})$ such that $g(w) \neq 0$. Then $\pi^{-|r|} \Delta_\zeta^\sigma f(w)/g(w)$ is an algebraic number.*

The assertion of Theorem 3.2 concerning $D_\zeta^{(e)} f$ can be proved in exactly the same fashion as in [10, Theorem 11.2], which concerns the orthogonal case. The key points in the proof in the symplectic case are also explained in [11, p. 825, (i–iv)]. The assertion for $\Delta_\zeta^\sigma f$ can be derived from (1.31) by the argument of the proof of [11, Theorem 3.2].

The above theorem concerns *analytic* f. To state a theorem on nonanalytic f, let us confine ourselves to the case $F = \mathbf{Q}$, and write $\mathfrak{A}_k, \mathfrak{M}_k, f\|_k \alpha, D_k^{(e)}$, and Δ_k^σ

for $\mathfrak{A}_\zeta$, $\mathfrak{M}_\zeta$, $f\|_\zeta\alpha$, $D_\zeta^{(e)}$, and Δ_ζ^σ and when $\zeta(x) = \det(x)^\kappa$ in Case SP and $\zeta(x, y) = \det(y)^k$ in Cases SU and SUB with $k \in \mathbf{Z}$. We now exclude Case SUB, define κ by (1.15), and assume that m is odd in Case SP. Then we consider the family $\mathfrak{L}$ of functions introduced in [14, §9], which consists of all functions f on H satisfying the following two conditions:

(3.16a) $f\|_{\kappa+1}\gamma = f$ *for all γ in a congruence subgroup of G,*

(3.16b) $f(z) = f_1(z) + \Delta[f_0(z)\log\{\det(z - z^*)\}]$ *with holomorphic functions f_0 and f_1 on H such that $\Delta f_0 = 0$,*
where Δ is defined by (1.29) with $p = 1$.

To avoid the cusp condition, let us always assume that $m > 1$, though the case $m = 1$ can be included by imposing a suitable cusp condition (cf. Section 6). By [14, Proposition 9.4], we know that $f_0 \in \mathfrak{M}_{\kappa-1}$ and f_1 has a Fourier expansion of type (3.10), though f_1 is not necessarily an automorphic form. We denote by $\mathfrak{L}(\overline{\mathbf{Q}})$ the subset of $\mathfrak{L}$ consisting of all f for which $\pi^m f_0 \in \mathfrak{M}_{\kappa-1}(\overline{\mathbf{Q}})$ and all Fourier coefficients of f_1 belong to $\overline{\mathbf{Q}}$ (see [14, §11]). Now we have

THEOREM 3.4.   *Suppose $F = \mathbf{Q}$. Let $\sigma: GL_m \to GL_n$ be a $\overline{\mathbf{Q}}$-rational homogeneous polynomial representation of degree r; let $f \in \mathfrak{L}(\overline{\mathbf{Q}})$, $0 \leqq e \in \mathbf{Z}$, and let w be a CM-point of H. Then*

$$\pi^{-e}P_\omega(w)^{-1}(D_{\kappa+1}^{(e)}f)(w) \quad and \quad \pi^{-r}P_\chi(w)^{-1}(\Delta_{\kappa+1}^\sigma f)(w)$$

are $\overline{\mathbf{Q}}$-rational, where $\omega(x, y) = \det(y)^{\kappa+1}\tau^e(x, y)$ and $[\chi(x, y)q](w) = \det(y)^{\kappa+1}q({}'\sigma(x)w \cdot {}'\sigma({}'y))$ for $q \in \mathrm{Hom}(\mathbf{C}_n^n, V)$ and $w \in \mathbf{C}_n^n$.

Notice that if $e = 0$, our theorem implies the algebraicity of $f(w)/g(w)$ for every $g \in \mathfrak{A}_{\kappa+1}(\overline{\mathbf{Q}})$ such that $g(w) \neq 0$. The result is far from trivial even in that special case.

The proof would be simpler if we could find a holomorphic $h \in \mathfrak{M}_k(\overline{\mathbf{Q}})$ such that $f = \Delta_k^{(p)}h$ for some k and p. Indeed, if that were the case, we would have $\Delta_{\kappa+1}^\sigma f = \Delta_k^\varepsilon h$ by (2.15), and so our assertion for $\Delta_{\kappa+1}^\sigma f$ would follow from Theorem 3.2. In general, however, such an h does not exist. Suppose, for example, $m = 2$ in Case SU. Since f has weight $m + 1$, we have $k + 2p = m + 1 = 3$, so that $k = p = 1$. By (1.39), $f = \Delta_1^{(1)}h = \Delta h$, and hence f must be holomorphic, which is false in general. Thus we have to find a more complicated expression for f to which Theorem 3.2 is applicable. That will be our task in the next two sections, which deal only with Case SU. We prove Case SP in Section 6 by a different method.

Remark 3.5.   Every CM-point can be obtained from $\{Z, h\}$ with commutative Z. In other words, *changing Z and h suitably but without changing w, we may assume that $Z = Y$ and $[Y : K] = n$ in the definition of CM-point.* To show this, consider the objects Z, Y_i, δ, and h from which we obtain a CM-point w. Let Z act on K_n^1 by right multiplication through h. Since $n = \sum_{i=1}^t n_i[Y_i : K]$ and h is injective, we see that h is equivalent to the reduced representation of Y, and hence there is a

K-linear isomorphism

$$l : (Y_1)^1_{n_1} \oplus \cdots \oplus (Y_t)^1_{n_t} \to K^1_n \qquad (n = 2m \text{ in Cases SP and SU})$$

such that $l(xa) = l(x)h(a)$ for $a \in Z$. Then we can find an element ζ_i of $GL_{n_i}(Y_i)$ for each i such that $\zeta_i^* = -\zeta_i$, $u^\delta = \zeta_i u^* \zeta_i^{-1}$ for all $u \in M_{n_i}(Y_i)$, and that

$$(3.17) \qquad l(x)\mathit{l}(y)^* = \sum_{i=1}^{t} \mathrm{Tr}_{Y_i/K}(x_i \zeta_i y_i^*)$$

for $x = (x_1, \ldots, x_t)$, $y = (y_1, \ldots, y_t)$ with $x_i, y_i \in (Y_i)^1_{n_i}$ (cf. [11, Lemma 6.2]; $M_{n_i}(K_i)$ of [11, p. 834, (6.7)] should be $K_i^{n_i}$). Take $g_i \in GL_{n_i}(Y_i)$ so that $g_i \zeta_i g_i^*$ is diagonal. Put $g = (g_1, \ldots, g_t)$, $\zeta_i' = g_i \zeta_i g_i^*$, and define δ', l', and h' by $u^{\delta'} = \zeta_i' u^* \zeta_i'^{-1}$ for $u \in M_{n_i}(Y_i)$, $l'(x) = l(xg)$, and $h'(a) = h(g^{-1}ag)$. Then $h'(Z[\delta']) = h(Z[\delta])$. This means that we may assume, without changing Z and w, that the ζ_i are all diagonal. Consider now $Y' = Y_1^{n_1} \oplus \cdots \oplus Y_t^{n_t}$ as a subalgebra of Z by embedding $Y_i^{n_i}$ diagonally into $M_{n_i}(Y_i)$. Then the restriction of δ to Y' is ρ. Now we see easily that w is a unique common fixed point of $h'(Y'[\rho])$. Therefore, taking Y', h', and ρ in place of Z, h, and δ, we may assume that $Z = Y$ and $[Y : K] = n$. (This does not change $P_\zeta(w)$.)

Writing again $Y = Y_1 \oplus \cdots \oplus Y_t$ (for this new Y), but assuming $[Y : K] = n$, we have a map $l : Y \to K_n^1$ such that $l(xa) = l(x)h(a)$ for $a \in Y$ and that

$$(3.18) \qquad l(x)\mathit{l}(y)^* = \mathrm{Tr}_{Y/K}(\zeta x y^\rho) \qquad (x, y \in Y)$$

with an element ζ of Y such that $\zeta^\rho = -\zeta$, where $\mathrm{Tr}_{Y/K}$ denotes the reduced trace. Moreover, it can easily be shown (see Lemma 8.3 below) that

(3.19) $\mathrm{Im}(\zeta^\tau) > 0$ *for every τ belonging to the CM-type ψ of Y determined at w.*

Proof of Proposition 3.1. Let Y, Y_i, h, w, and ψ be as in the definition of $P_\zeta(w)$. We are going to prove that $P_\xi(w) = P_\zeta(w)$ under (3.11). By Remark 3.5, we may assume that $[Y : K] = 2m$. Consider the maps $\mathrm{Res}_{Y_i/K} : I_{Y_i} \to I_K$ and $\mathrm{Inf}_{Y_i/K} : I_K \to I_{Y_i}$ defined in [10, p. 317]. We can define similar maps $\mathrm{Res}_{Y/K} : I_Y \to I_K$ and $\mathrm{Inf}_{Y/K} : I_K \to I_Y$ by putting $\mathrm{Res}_{Y/K}(\sum_i \alpha_i) = \sum_i \mathrm{Res}_{Y_i/K}(\alpha_i)$ for $\alpha_i \in I_{Y_i}$ and $\mathrm{Inf}_{Y/K}(\gamma) = \sum_i \mathrm{Inf}_{Y_i/K}(\gamma)$ for $\gamma \in I_K$. Then we have

$$(3.20a) \qquad p_Y(\alpha, \mathrm{Inf}_{Y/K}(\gamma)) = p_K(\mathrm{Res}_{Y/K}(\alpha), \gamma) \qquad (\alpha \in I_Y, \gamma \in I_K),$$

$$(3.20b) \qquad p_Y(\mathrm{Inf}_{Y/K}(\gamma), \alpha) = p_K(\gamma, \mathrm{Res}_{Y/K}(\alpha)) \qquad (\alpha \in I_Y, \gamma \in I_K).$$

These follow from [10, Theorem 1.1, (3), (4)] in a straightforward way. Now, for each $v \in \varphi$, we can define elements α_v and β_v of I_Y so that

$$a^{\alpha_v} = \det[\kappa_v(h(a), w)], \qquad a^{\beta_v} = \det[\mu_v(h(a), w)] \qquad (a \in Y[\rho]).$$

Then we see that $\psi = \sum_{v \in \varphi}(\alpha_v + \beta_v)$, and also from (1.2a) that $\alpha_v \rho + \beta_v = \mathrm{Inf}_{Y/K}(v)$. Therefore $\mathrm{Res}_{Y/K}(\psi) = m(\varphi + \varphi\rho)$ and

$$p_Y(\alpha_v \rho + \beta_v, \psi) = p_Y(\mathrm{Inf}_{Y/K}(v), \psi) = p_K(v, \mathrm{Res}_{Y/K}(\psi)) = p_K(v, m(\varphi + \varphi\rho)) = 1$$

by [10, Theorem 1.1], and hence $p_Y(\alpha_v, \psi) = p_Y(\beta_v, \psi)$. Our definition of $P_\zeta(w)$, together with (3.11), shows that

$$(3.21) \qquad P_\xi(w) = p_Y\left(\sum_{v \in \varphi} k_v(\alpha_v - \beta_v), \psi\right) P_\zeta(w) = P_\zeta(w),$$

from which we obtain our assertion. The same proof applies to Case SUB if $p_v = q_v$ for all $v \in \varphi$.

4. Lemmas on tensor-valued functions. Let $\bigwedge^r \mathbf{C}^m$, ρ_r, and ρ_r^* be as in Section 2. In this and the next sections, we treat only Case SU. Thus $T = \mathbf{C}_m^m$ and $T_r = \mathrm{End}(\bigwedge^r \mathbf{C}^m)$. We define a representation $\{\sigma_r, T_r\}$ of $\mathscr{g}$ by

$$(4.1) \qquad \sigma_r(x, y)h = \rho_r(x)h\rho_r(^t y) \qquad (h \in T_r).$$

We identify $\bigwedge^r \mathbf{C}^m$ with its dual by the pairing

$$(4.2) \qquad \langle u_1 \wedge \cdots \wedge u_r, v_1 \wedge \cdots \wedge v_r \rangle = \det(^t u_i v_j) \qquad (u_i, v_j \in \mathbf{C}^m).$$

Notice that $\langle \rho_r(^t x)a, b \rangle = \langle a, \rho_r(x)b \rangle$ for $a, b \in \bigwedge^r \mathbf{C}^m$ and $x \in GL_m(\mathbf{C})$. For $h \in T_r$ and $v_1, \ldots, v_r, w_1, \ldots, w_r \in \mathbf{C}^m$, there is a unique element h' of $\mathrm{Ml}_r(T, \mathbf{C})$ such that

$$(4.3) \qquad h'(v_1 \cdot {}^t w_1, \ldots, v_r \cdot {}^t w_r) = \langle v_1 \wedge \cdots \wedge v_r, h(w_1 \wedge \cdots \wedge w_r) \rangle.$$

Then $h \mapsto h'$ defines an injection of T_r into $\mathrm{Ml}_r(T, \mathbf{C})$. Moreover, this maps $\sigma_r(x, y)h$ onto $\tau^r(x, y)h'$. Hereafter we embed T_r into $\mathrm{Ml}_r(T, \mathbf{C})$ by this map and identify h with h'. We note here a simple fact that an element h of T_r considered an element of $\mathrm{Ml}_r(T, \mathbf{C})$ is symmetric in the sense of (1.23). This follows immediately from (4.3).

Our first lemma concerns the map Φ_1 of $\mathrm{Ml}_m(T, \mathbf{C})$ into $\mathbf{C}$ defined by (1.33). For brevity, we put $\Phi = \Phi_1$. Let e_{ij} be the (i, j)-matrix unit of $\mathbf{C}_m^m$. As a tangent vector, this coincides with ∂_{ij}. Thus we can put $e_{\alpha(i)\beta\alpha(i)}$ in place of $\partial_{\alpha(i)\beta\alpha(i)}$ in (1.33).

LEMMA 4.1. *Given $x \in GL_m(\mathbf{C})$, $A \in \mathrm{Ml}_m(T, \mathbf{C})$ and $0 \leqq s \leqq m$, define $A_{s,x} \in \mathrm{Ml}_m(T, \mathbf{C})$ by*

$$A_{s,x}(u_1, \ldots, u_m) = A(u_1 x, \ldots, u_s x, u_{s+1}, \ldots, u_m) \qquad (u_i \in T).$$

Then there is an element B of T_{m-s}, uniquely determined by A and s, such that

$\Phi(A_{s,x}) = \det(x)\mathrm{tr}(\rho_{m-s}({}^t x^{-1})B)$ *for every* $x \in GL_m(\mathbf{C})$. *If further A has the form*

$$(4.5) \qquad A(u_1, \ldots, u_m) = \mathrm{tr}(u_1) \cdots \mathrm{tr}(u_x)C(u_{s+1}, \ldots, u_m)$$

with $C \in T_{m-s}$, then $B = \xi C$ with $\xi = s!(m-s)!^2/m!$.

Proof. As remarked above, we have, by (1.33),

$$m!\Phi(A_{s,x}) = \sum_{i_1, \ldots, i_s} \sum_{\alpha,\beta} \mathrm{sgn}(\beta)x_{\beta\alpha(1)i_1} \cdots x_{\beta\alpha(s)i_s}$$

$$\cdot A(e_{\alpha(1)i_1}, \ldots, e_{\alpha(s)i_s}, e_{\alpha(s+1)\beta\alpha(s+1)}, \ldots, e_{\alpha(m)\beta\alpha(m)}).$$

Fix α, and for each $K = \{k_1, \ldots, k_s\} \subset (1, \ldots, m\}$, consider all β such that $\beta\alpha(\{1, \ldots, s\}) = K$. Then the last sum is a linear combination of

$$(4.6) \qquad \sum_{\sigma \in S(K)} \mathrm{sgn}(\sigma)x_{\sigma(k_1)i_1} \cdots x_{\sigma(k_s)i_s}$$

with coefficients independent of x, where $S(K)$ is the group of all permutations of K. Since the sum of (4.6) is ± 1 times an entry of $\rho_s^*(x) = \det(x)\rho_{m-s}({}^t x^{-1})$, we obtain the first assertion. To prove the second one, suppose A has form (4.5), and put $r = m - s$. Then

$$m!\Phi(A_{s,x}) = \sum_{\alpha,\beta} \mathrm{sgn}(\beta)x_{\beta\alpha(1)\alpha(1)} \cdots x_{\beta\alpha(s)\alpha(s)}C(e_{\alpha(s+1)\beta\alpha(s+1)}, \ldots, e_{\alpha(m)\beta\alpha(m)}).$$

For each $I = \{i_1, \ldots, i_s\} \subset \{1, \ldots, m\}$, consider all α such that $\alpha(\{1, \ldots, s\})) = I$, and decompose the sum according to I. Denoting the complement of I by $I' = \{i'_1, \ldots, i'_r\}$, we obtain

$$m!\Phi(A_{s,x}) = \sum_{\beta,I} \sum_{\sigma \in S(I),\tau \in S(I')} \mathrm{sgn}(\beta)x_{\beta\sigma(i_1)\sigma(i_1)} \cdots x_{\beta\sigma(i_s)\sigma(i_s)}$$

$$\cdot C(e_{\tau(i'_1)\beta\tau(i'_1)}, \ldots, e_{\tau(i'_r)\beta\tau(i'_r)}).$$

Here the i_a and i'_b are arranged so that $i_1 < \cdots < i_s$ and $i'_1 < \cdots < i'_r$. Since C is symmetric, we have

$$m!\Phi(A_{s,x}) = s!r!\sum_{\beta,I} \mathrm{sgn}(\beta)x_{\beta(i_1)i_1} \cdots x_{\beta(i_s)i_s}$$

$$\cdot C(e_{i'_1\beta(i'_1)}, \ldots, e_{i'_r\beta(i'_r)}).$$

Consider another decomposition $\{1, \ldots, m\} = K \cup K'$ with $K = \{k_1, \ldots, k_s\}$, $K' = \{k'_1, \ldots, k'_r\}$, $k_1 < \cdots < k_s, k'_1 < \cdots < k'_r$. For each I and K, define a permutation π_{IK} by $\pi_{IK}(i_a) = k_a$ and $\pi_{IK}(i'_b) = k'_b$; consider all β such that

$\beta(I) = K$. Then we find that

$$m!\Phi(A_{s,x}) = s!r!\sum_{I,K}\mathrm{sgn}(\pi_{IK})\sum_{\gamma\in S(K),\delta\in S(K')}\mathrm{sgn}(\gamma)\mathrm{sgn}(\delta)$$

$$\cdot x_{\gamma(k_1)i_1}\cdots x_{\gamma(k_s)i_s}C\big(e_{i_1'\delta(k_1')},\ldots,e_{i_r'\delta(k_r')}\big)$$

$$= s!r!^2\sum_{I,K}\mathrm{sgn}(\pi_{IK})\det_{KI}(x)c_{I'K'},$$

where $\det_{KI}(x)$ is the (K, I)-entry of $\rho_s(x)$, and $c_{I'K'}$ is the (I', K')-entry of the matrix representing C as an element of T_r. Observing that $\mathrm{sgn}(\pi_{IK})\det_{KI}(x)$ is the (K', I')-entry of $\det(x)\rho_r({}^tx^{-1})$, we obtain

$$m!\Phi(A_{s,x}) = s!r!^2\det(x)\mathrm{tr}\big[\rho_r({}^tx^{-1})C\big],$$

which completes the proof.

Remark 4.2. In the above lemma, if A depends holomorphically on a complex parameter, say t, then B depends on t holomorphically as well. This is because $A \mapsto B$ is a linear map independent of x.

We now consider an element of $\mathrm{Ml}_n(T, \mathbf{C})$ of a special type. We first fix an element f of $\mathrm{Ml}_r(T, \mathbf{C})$ with $r \leqq n$, and take a disjoint decomposition

$$(4.7) \qquad \{1,\ldots,n\} = \alpha_1 \cup \cdots \cup \alpha_p \cup \beta_1 \cup \cdots \cup \beta_r,$$

with nonempty subsets α_i and β_j. For each α_i or β_j, we choose and fix an ordering of its members, and put

$$g_i = \mathrm{tr}(u_{k_1}\cdots u_{k_a}) \qquad \text{if} \quad \alpha_i = \{k_1,\ldots,k_a\},$$

$$w_j = u_{l_1}\cdots u_{l_b} \qquad \text{if} \quad \beta_j = \{l_1,\ldots,l_b\}$$

for $u_1,\ldots,u_n \in T$. Then we define an element A of $\mathrm{Ml}_n(T, \mathbf{C})$ by

$$A(u_1,\ldots,u_n) = g_1\cdots g_p f(w_1,\ldots,w_r).$$

To avoid the factor $i\ (=\sqrt{-1}\,)$ in $\eta(z) = i\,(z^* - z)$, we put

$$(4.8) \qquad \theta(z) = z - z^*, \qquad \epsilon(z) = \det(z - z^*) \qquad (z \in H),$$

and use θ and ϵ instead of η and δ. This change is not absolutely necessary, but will prove convenient later. With $\{k_1,\ldots,k_a\}$ and $\{l_1,\ldots,l_b\}$ as above, we define an element A^0 of $\mathrm{Ml}_n(T, \mathbf{C})$ by

$$(4.9) \qquad A^0(u_1,\ldots,u_n) = \prod \mathrm{tr}\big(u_{k_1}\theta^{-1}\cdots u_{k_a}\theta^{-1}\big)$$

$$\cdot f\big(\ldots,u_{l_1}\theta^{-1}u_{l_2}\theta^{-1}\cdots\theta^{-1}u_{l_b},\ldots\big).$$

More precisely, this is obtained from A by replacing u_h by $u_h \theta^{-1}$ except when h is the last numeral of β_j for some j. Let us now assume $n = m$, and put $s = m - r$. Observing that θ^{-1} occurs exactly s times on the right-hand side of (4.9), we find, by virtue of Lemma 4.1, that $\Phi A^0 = \epsilon^{-1}\mathrm{tr}(\rho_r({}^t\theta)B)$ with an element B of T_r. This B is determined by α_i, β_j, and f. If f is a holomorphic function on H with values in $\mathrm{Ml}_m(T, \mathbf{C})$, then we obtain B as a homolorphic function on H with values in T_r. In this setting, we have

LEMMA 4.3. *Let* $\{\rho, \mathrm{Ml}_r(T, \mathbf{C})\}$ *and* $\{\sigma, T_r\}$ *be representations of* g *defined by* $\rho(x, y) = \det(y)^\nu \tau^r(x, y)$ *and* $\sigma(x, y) = \det(y)^\nu \sigma_r(x, y)$ *with an integer* ν. *If* B *corresponds to* f *by the above process, then, for every* $\gamma \in G$, $B\|_\sigma \gamma$ *corresponds to* $f\|_\rho \gamma$.

Proof. Define A^0 as above. Fix γ; write simply j, κ, and μ for j_γ, κ_γ and μ_γ, and put $\zeta = \mu^{-1}\bar\kappa$, $f' = f\|_\rho \gamma$. By (1.7b), we have

$$j^{-\nu}(A^0 \circ \gamma)({}^t\kappa^{-1}u_1\mu^{-1}, \ldots, {}^t\kappa^{-1}u_m\mu^{-1})$$

$$= \prod \mathrm{tr}\big(u_{k_1}\zeta\theta^{-1}u_{k_2}\zeta\theta^{-1}\cdots u_{k_a}\zeta\theta^{-1}\big)f'\big(\ldots, u_{l_1}\zeta\theta^{-1}\cdots\zeta\theta^{-1}u_{l_b}, \ldots\big).$$

This is obtained from (4.9) by replacing θ and f by $\theta\zeta^{-1}$ and f'. Therefore, applying Φ, we obtain

$$j^{-\nu-2}\Phi(A^0 \circ \gamma) = \mathrm{tr}\big(\rho_s^*(\zeta\theta^{-1})B'\big) = \mathrm{tr}\big(\rho_s^*(\mu^{-1}\bar\kappa\theta^{-1})B'\big),$$

where B' is the function corresponding to f'. On the other hand,

$$\Phi(A^0 \circ \gamma) = \mathrm{tr}\big(\rho_s^*(\bar\kappa\theta^{-1}\cdot{}^t\kappa)(B \circ \gamma)\big),$$

and hence

$$B \circ \gamma = j^{\nu+2}\rho_s^*({}^t\kappa^{-1})B'\rho_s^*(\mu^{-1}) = j^\nu\rho_r(\kappa)B'\cdot{}^t\rho_r(\mu),$$

which proves our lemma.

5. Proof of Theorem 3.4 in Case SU. We need a few elementary formulas:

$$(5.1)\qquad D\big[f(z^{-1})\big](u) = -\big[(Df)(z^{-1})\big](z^{-1}uz^{-1}),$$

$$(5.2)\qquad D\big[\det(z)^s\big](u) = s \cdot \det(z)^s\mathrm{tr}(z^{-1}u),$$

$$(5.3)\qquad D(j_\alpha^s)(u) = sj_\alpha^s\,\mathrm{tr}(\mu_\alpha^{-1}cu),$$

$$(5.4)\qquad D\big[\mathrm{tr}(\mu_\alpha^{-1}cv)\big](u) = -\mathrm{tr}(\mu_\alpha^{-1}cu\mu_\alpha^{-1}cv).$$

Here f is a function on H, $z \in H$, $u, v \in T$, $\mu_\alpha = cz + d$ with $\alpha \in \mathfrak{G}$, and $s \in \mathbf{C}$.

These are valid in both Cases SP and SU. In fact, we proved them in Case SP in [11, p. 830]. The same type of proof applies to Case SU.

Let us now find an expression of $D_\rho^{(s)}f$ in the following situation: f is holomorphic and has values in $V = \mathrm{Ml}_r(T, \mathbf{C})$; $\rho(x, y) = \det(y)^\nu \tau^r(x, y)$ with an integer ν; $V = \mathbf{C}$ if $r = 0$. Naturally $D_\rho^{(s)}f$ has values in $\mathrm{Ml}_s(T, V)$. Now $\mathrm{Ml}_s(T, V)$ can be identified with $\mathrm{Ml}_{r+s}(T, \mathbf{C})$ if we put

$$h(u_1, \ldots, u_s, v_1, \ldots, v_r) = h(u_1, \ldots, u_s)(v_1, \ldots, v_r)$$

for u_p, $v_q \in T$ and $h \in \mathrm{Ml}_s(T, V)$. This is compatible with the action of $\mathscr{g}$; in fact, $(\rho \otimes \tau^s)(x, y) = \det(y)^\nu \tau^{r+s}(x, y)$.

With f as above, $f(v_1, \ldots, v_r)$ for fixed $v_1, \ldots, v_r$ in T is a $\mathbf{C}$-valued function on H. Here and in the following computation, we always suppress the variable z on H. Our definition (1.16) of D_ρ shows that

$$(D_\rho f)(u, v_1, \ldots, v_r) = \rho({}^t\theta, \theta)^{-1} D\big[\, \epsilon^\nu f(\theta v_1 \theta, \ldots, \theta v_r \theta)\,\big](u) \qquad (u \in T)$$

with θ and ϵ of (4.8). We see easily that $(D\theta)(u) = u$, and more generally

$$D(\theta v_1 \theta v_2 \cdots v_k \theta)(u) = u v_1 \theta v_2 \cdots v_k \theta$$
$$+ \theta v_1 u v_2 \cdots v_k \theta + \cdots + \theta v_1 \theta v_2 \cdots v_k u.$$

Since $(D\epsilon^\nu)(u) = \nu \epsilon^\nu \mathrm{tr}(\theta^{-1}u)$ by (5.2), we have

$$(D_\rho f)(u, v_1, \ldots, v_r) = (Df)(u, v_1, \ldots, v_r) + \nu \cdot \mathrm{tr}(\theta^{-1}u) f(v_1, \ldots, v_r)$$
$$+ \sum_{i=1}^{r} f(\ldots, v_{i-1}, u\theta^{-1}v_i + v_i\theta^{-1}u, v_{i+1}, \ldots).$$

Repeating this procedure s times and putting $v_i = u_{s+i}$, we obtain

$$(D_\rho^{(s)}f)(u_1, \ldots, u_s, u_{s+1}, \ldots, u_{s+r}) = \sum_{p=0}^{s} g_p,$$

where g_p is a finite sum of functions of the form

$$A\big(u_{\sigma(1)}\theta^{-1}, \ldots, u_{\sigma(p)}\theta^{-1}, u_{\sigma(p+1)}, \ldots, u_{\sigma(r+s)}\big)$$

with a permutation σ of $\{1, \ldots, r + s\}$ and a holomorphic function A on H with values in $\mathrm{Ml}_{r+s}(T, \mathbf{C})$. In particular, $g_s = \sum_{i=0}^{s} \nu^i h_i$ with functions h_i *independent of ν*. The explicit form of each h_i is unnecessary, except that we need to know:

(5.5a) $$h_s = \mathrm{tr}(u_1\theta^{-1}) \cdots \mathrm{tr}(u_s\theta^{-1}) f(u_{s+1}, \ldots, u_{s+r});$$

(5.5b) *each h_i is a finite linear combination of functions of type (4.9) in which θ^{-1} appears exactly s times.*

These facts can easily be verified by induction on s. We now assume $r + s = m$ and apply Φ $(= \Phi_1)$. By (1.28a) and (1.19), we have

$$\Phi\left[(\rho \otimes \tau^s)(x, y)q\right] = \det(x)\det(y)^{\nu+1}\Phi(q) \qquad ((x, y) \in \mathcal{G}, \ q \in \mathrm{Ml}_m(T, \mathbf{C})),$$

$$\Phi D_\rho^{(s)}(f\|_\rho \alpha) = \left(\Phi D_\rho^{(s)}f\right)\|_\varphi \alpha \qquad (\alpha \in \mathfrak{G})$$

with $\varphi(x, y) = \det(x)\det(y)^{\nu+1}$. By Lemma 4.1, we have

$$(5.6) \qquad \Phi D_\rho^{(s)}f = \epsilon^{-1} \sum_{k=r}^{m} \mathrm{tr}\left({}^t\rho_k(\theta)b_k\right)$$

with a function b_k on H with values in T_k for each k. By Remark 4.2, b_k is holomorphic. Thus we have proved

LEMMA 5.1. *Let f be a holomorphic map: $H \to \mathrm{Ml}_r(T, \mathbf{C})$ with $r \leq m$; let $\rho(x, y) = \det(y)^\nu \tau^r(x, y)$ and $s = m - r$. Then we have (5.6) with holomorphic maps $b_k : H \to T_k$ for $r \leq k \leq m$.*

Continuing the above analysis of $\Phi D_\rho^{(s)}f$, we find that

$$\Phi g_s = \sum_{i=0}^{s} \nu^i \Phi h_i = \epsilon^{-1}\mathrm{tr}\left({}^t\rho_r(\theta)b_r\right).$$

Applying Lemma 4.1 to h_i, we have $\Phi h_i = \epsilon^{-1}\mathrm{tr}({}^t\rho_r(\theta)c_i)$ with a holomorphic map $c_i : H \to T_r$. Then $b_r = \sum_{i=0}^{s} \nu^i c_i$. Now b_r and c_i depend linearly on f. We can even define linear maps β_r and γ_i of $\mathrm{Ml}_r(T, \mathbf{C})$ into T_r so that $b_r = \beta_r f$ and $c_i = \gamma_i f$. This can be seen from Lemma 4.1, since h_i is a finite sum of functions of form (4.9).

Let us now assume that f takes values in T_r. Then $c_s = \gamma_s f = \xi f$ with $\xi = s!r!^2/m!$ by Lemma 4.1. Therefore

$$b_r = \beta_r f = \left(\nu^s \xi + \sum_{i=0}^{s-1} \nu^i \gamma_i\right)f$$

for every holomorphic map $f : H \to T_r$. Moreover, the γ_i are independent of ν, and $\gamma_i(f\|_\sigma \alpha) = (\gamma_i f)\|_\sigma \alpha$ for every $\alpha \in \mathfrak{G}$ by Lemma 4.3; here σ is defined as in Lemma 4.3, and in fact coincides with the restriction of ρ to T_r. Since $\xi \neq 0$, β_r gives, for sufficiently large ν, an invertible map of T_r onto itself. Call its inverse α_r for a fixed ν. Then

$$(5.7) \qquad \alpha_r(f\|_\sigma \alpha) = (\alpha_r f)\|_\sigma \alpha \qquad (\alpha \in \mathfrak{G}).$$

We now assume that f has a Fourier expansion of the form

$$(5.8) \qquad f(z) = \sum_{h \in L} c(h)\mathbf{e}(\mathrm{tr}(hz)) \qquad (z \in H)$$

with $\overline{\mathbf{Q}}$-rational $c(h)$, where L is a lattice in the space of all hermitian elements of K_m^m. Then every partial derivative of f of order k has a Fourier expansion whose coefficients are π^k times $\overline{\mathbf{Q}}$-rational vectors. Observe that g_p is linear in the partial derivatives of f of order $s - p$, and also that the map $A \mapsto B$ is $\mathbf{Q}$-rational with respect to the obvious $\mathbf{Q}$-structures of $\mathrm{Ml}_m(T, \mathbf{C})$ and T_{m-s}. Therefore we see that $\pi^{r-k}b_k$ has a Fourier expansion with $\overline{\mathbf{Q}}$-rational coefficients. Observe also that β_r and γ_i are $\mathbf{Q}$-rational. Therefore α_r is $\mathbf{Q}$-rational as well. We now prove

PROPOSITION 5.2.   *Let q be a function on H of the form*

$$q = \epsilon^{-1} \sum_{k=0}^{m} \mathrm{tr}({}^t\rho_k(\theta)q_k)$$

with holomorphic maps $q_k : H \to T_k$. Suppose that $q\|_\lambda\gamma = q$ for every γ in a congruence subgroup Γ of G with an integer λ. If λ is greater than a constant λ_0 depending only on m, then q can be expressed in the form

$$q = f_m + \sum_{k=0}^{m-1} \Phi D_{\zeta_k}^{(m-k)}f_k$$

with $f_k \in \mathfrak{M}_{\zeta_k}(\Gamma)$ and a representation $\{\zeta_k, T_k\}$ of g defined by $\zeta_k(x, y) = \det(y)^{\lambda-2}\sigma_k(x, y)$. Moreover, suppose $\pi^{m-k}q_k$ for every k has a Fourier expansion of form (5.8) with $\overline{\mathbf{Q}}$-rational coefficients. Then we can take f_k from $\pi^{k-m}\mathfrak{M}_{\zeta_k}(\overline{\mathbf{Q}})$.

Proof.   We choose λ_0 so that the above maps β_r are invertible for $0 \leqq r < m$ if $\nu > \lambda_0 - 2$. Suppose $\lambda > \lambda_0$ and let $\nu = \lambda - 2$. Suppose also $q_0 = 0, \ldots, q_{r-1} = 0$, that is, q has the form

$$(5.9) \qquad q = \epsilon^{-1} \sum_{k=r}^{m} \mathrm{tr}({}^t\rho_k(\theta)q_k).$$

We are going to prove that

$$(5.10) \qquad q = f_m + \sum_{k=r}^{m-1} \Phi D_{\zeta_k}^{(m-k)}f_k$$

with $f_k \in \mathfrak{M}_{\zeta_k}(\Gamma)$. If $r = m$, this is trivial. Suppose (5.9) holds with $r < m$. Then $q_r \in \mathfrak{M}_{\zeta_r}(\Gamma)$ by Lemma 2.3. Let $f_r = \alpha_r q_r$ with the map $\alpha_r \in GL(T_r)$ which is the inverse of β_r, as above. By (5.7), we have $f_r \in \mathfrak{M}_{\zeta_r}(\Gamma)$. By Lemma 5.1, we have

$$(5.11) \qquad \Phi D_{\zeta_r}^{(m-r)}f_r = \epsilon^{-1} \sum_{k=r}^{m} \mathrm{tr}\left[{}^t\rho_r(\theta)b_k\right]$$

with holomorphic b_k. Moreover, $b_r = \beta_r f_r = q_r$. Therefore $q - \Phi D_{\zeta_r}^{(m-r)}f_r = \epsilon^{-1}\sum_{k>r} \mathrm{tr}[{}^t\rho_k(\theta)(q_k - b_k)]$. Applying induction to this, we obtain expression

(5.10). The case $r = 0$ is our first assertion. Suppose $\pi^{m-k}q_k$ has $\overline{\mathbf{Q}}$-rational Fourier coefficients in (5.9). If $r = m$, this means that $q = q_m$ is $\overline{\mathbf{Q}}$-rational. Suppose $r < m$. Then $\pi^{m-r}q_r \in \mathfrak{M}_{\zeta_r}(\overline{\mathbf{Q}})$, so that $\pi^{m-r}f_r \in \mathfrak{M}_{\zeta_r}(\overline{\mathbf{Q}})$. As observed above, we have (5.11) with b_k whose Fourier coefficients are π^{k-m} times $\overline{\mathbf{Q}}$-rational vectors. Therefore $q_k - b_k$ has the same property for every k. By induction on r, we obtain (5.10) with $\pi^{m-k}f_k$ in $\mathfrak{M}_{\zeta_k}(\overline{\mathbf{Q}})$, which proves the second assertion.

Let us now consider $D_\rho^{(b)}f$ with a V-valued function f on H or on $B_{m,m}$ when $V = \mathrm{Ml}_m(T, \mathbf{C})$ and $\rho(x, y) = \det(y)^\nu \tau^m(x, y)$ with $\nu \in \mathbf{Z}$. Then $D_\rho^{(b)}f$ has values in $\mathrm{Ml}_b(T, V)$. Define a map $\psi : \mathrm{Ml}_b(T, V) \to \mathrm{Ml}_b(T, \mathbf{C})$ by

$$(5.12) \quad (\psi h)(u_1, \ldots, u_b) = \Phi_1(h(u_1, \ldots, u_b)) \qquad (u_i \in T, h \in \mathrm{Ml}_b(T, V)).$$

Then $\psi D_\rho^{(b)}f$ has values in $\mathrm{Ml}_b(T, \mathbf{C})$. We can easily verify that

$$(5.13) \qquad \psi D_\rho^{(b)}(f\|_\rho \alpha) = \left(\psi D_\rho^{(b)}f\right)\|_\sigma \alpha \qquad (\alpha \in \mathfrak{G} \cup \mathfrak{E} \text{ or } \alpha \in \mathfrak{G}')$$

with $\sigma(x, y) = \det(y)^{\nu+2}\tau^b(x, y)$. On the other hand, $D_{\nu+2}^{(b)}\Phi f$ is well-defined, and has values in the same space; moreover, we have

$$(5.14) \qquad\qquad D_{\nu+2}^{(b)}\Phi(f\|_\rho \alpha) = \left(D_{\nu+2}^{(b)}\Phi f\right)\|_\sigma \alpha.$$

LEMMA 5.3. *In the above setting, we have* $D_{\nu+2}^{(b)}\Phi f = \psi D_\rho^{(b)}f$.

This can be proved in the same manner as in the last half of the proof of Proposition 2.1 by observing that $D^b \Phi = \psi D^b$.

We are now ready to prove Theorem 3.4 in Case SU. Let $f \in \mathfrak{L}(\overline{\mathbf{Q}})$, $0 \leq e \in \mathbf{Z}$, and let w be a CM-point of H. We choose an integer λ so that $\mathfrak{M}_{\lambda-m-1}(\overline{\mathbf{Q}}) \neq \{0\}$ and λ is greater than the constant λ_0 of Proposition 5.2. Take $p \in \mathfrak{M}_{\lambda-m-1}(\overline{\mathbf{Q}})$ so that $p(w) \neq 0$. As shown in [11, (9.6)], f is a function of the form

$$f = \epsilon^{-1} \sum_{k=0}^{m} \pi^{k-m}\mathrm{tr}\left['\rho_k(\theta)l_k\right]$$

with holomorphic maps $l_k : H \to T_k$ with $\overline{\mathbf{Q}}$-rational Fourier expansions. Applying Proposition 5.2 to pf, we have

$$pf = f_m + \sum_{k=0}^{m-1} \pi^{k-m}\Phi D_{\zeta_k}^{(m-k)}f_k$$

with $f_k \in \mathfrak{M}_{\zeta_k}(\overline{\mathbf{Q}})$. Put $s_k = \Phi D_{\zeta_k}^{(m-k)}f_k$. Then

$$(5.15) \qquad D_{m+1}^{(e)}f = D_{m+1}^{(e)}\left(p^{-1}f_m\right) + \sum_{k=0}^{m-1} \pi^{k-m}D_{m+1}^{(e)}\left(p^{-1}s_k\right).$$

Since $p^{-1}f_m \in \mathfrak{A}_{m+1}(\overline{\mathbf{Q}})$, Theorem 3.2 shows that $\pi^{-e}P_\omega(w)^{-1}D_{m+1}^{(e)}(p^{-1}f_m)(w)$ is $\overline{\mathbf{Q}}$-rational, where $\omega(x, y) = \det(y)^{m+1}\tau^e(x, y)$. Our task is to show that

$\pi^{k-m-e} P_\omega(w)^{-1} D_{m+1}^{(e)}(p^{-1}s_k)(w)$ is $\overline{\mathbf{Q}}$-rational. As noted in (3.9b), $P_\omega(w) = \omega(\not\!\! h, g)$ with an element $(\not\!\! h, g)$ of g determined by w. Therefore the $\overline{\mathbf{Q}}$-rationality of $\pi^{k-m-e} P_\omega(w)^{-1} D_{m+1}^{(e)}(p^{-1}s_k)(w)$ is equivalent to the $\overline{\mathbf{Q}}$-rationality of

$$(5.16) \quad \pi^{k-m-e} \det(g)^{-m-1} D_{m+1}^{(e)}(p^{-1}s_k)(w)({}^{t}\!\not\!\! h^{-1} u_1 g^{-1}, \ldots, {}^{t}\!\not\!\! h^{-1} u_e g^{-1})$$

for every $\overline{\mathbf{Q}}$-rational $u_1, \ldots, u_e \in T$. By (1.20b), $D_{m+1}^{(e)}(p^{-1}s_k)(w)(u_1, \ldots, u_e)$ is the sum of the quantities

$$(D_\nu^{(a)} p^{-1})(w)(u_{i_1}, \ldots, u_{i_a})(D_\lambda^{(b)} s_k)(w)(u_{j_1}, \ldots, u_{j_b}),$$

where $\nu = m + 1 - \lambda$, $a + b = e$, and $\{i_1, \ldots, i_a, j_1, \ldots, j_b\} = \{1, \ldots, e\}$. By Theorem 3.2, we know that

$$(5.17) \quad \pi^{-a} \det(g)^{-\nu}(D_\nu^{(a)} p^{-1})(w)({}^{t}\!\not\!\! h^{-1} u_{i_1} g^{-1}, \ldots, {}^{t}\!\not\!\! h^{-1} u_{i_a} g^{-1}) \in \overline{\mathbf{Q}}.$$

As for $D_\lambda^{(b)} s_k$, Lemma 5.3 shows that $D_\lambda^{(b)} s_k = D_\lambda^{(b)} \Phi D_{\xi_k}^{(m-k)} f_k = \psi D_{\xi_k}^{(b+m-k)} f_k$. By Theorem 3.2, we have

$$\pi^{k-m-b} \det(g)^{2-\lambda}(D_{\xi_k}^{(b+m-k)} f_k)(w)({}^{t}\!\not\!\! h^{-1} u_{j_1} g^{-1}, \ldots, {}^{t}\!\not\!\! h^{-1} u_{j_b} g^{-1})$$

$$\cdot({}^{t}\!\not\!\! h^{-1} v_1 g^{-1}, \ldots, {}^{t}\!\not\!\! h^{-1} v_m g^{-1}) \in \overline{\mathbf{Q}}$$

for $\overline{\mathbf{Q}}$-rational elements u_j and v_i of T. From (5.12) and (1.28a) we see that

$$\pi^{k-m-b} \det(g)^{-\lambda}(\psi D_{\xi_k}^{(b+m-k)} f_k)(w)({}^{t}\!\not\!\! h^{-1} u_{j_1} g^{-1}, \ldots, {}^{t}\!\not\!\! h^{-1} u_{j_b} g^{-1}) \in \overline{\mathbf{Q}}.$$

Combining this with (5.17), we find the $\overline{\mathbf{Q}}$-rationality of (5.16), which proves the first half of Theorem 3.4. Since $\Delta_{m+1}^\sigma = \Phi_\sigma D_{m+1}^{(r)}$, the second half follows immediately from the first one, by the same argument as in the proof of [11, Theorem 3.2].

Our methods of proof in Case SU will probably be applicable to Case SP if they are suitably modified, though the modification seems nontrivial and rather involved.

6. Proof of Theorem 3.4 in Case SP. This section concerns exclusively Case SP. Our idea is to consider a certain embedding of H_1^m into H_m and reduce the problem to the property of certain nonholomorphic functions at CM-points of H_1^m. This is in effect a refinement of [14, §11]. For simplicity, we use H for the space H_m of "degree m".

Let $X = X_1 \oplus \cdots \oplus X_t$ with totally real algebraic number fields $X_1, \ldots, X_t$. Define J_X and I_X as in Section 3. For $p = \sum_{\tau \in J(X)} p_\tau \tau \in I_X$ and $z = (z_\tau)_{\tau \in J(X)} \in \mathbf{C}^{J(X)}$, we put

$$(6.1) \qquad\qquad z^p = \prod_\tau z_\tau^{p_\tau}$$

whenever $z_\tau^{p_\tau}$ is meaningful. We write $p \geqq 0$ if $p_\tau \geqq 0$ for all τ. We can then define, for $p_i \in I(X_i)$, the space of holomorphic modular forms $\mathfrak{M}_{p_i}(\Gamma_i)$ of weight p_i on $H_1^{J(X_i)}$ with respect to a congruence subgroup Γ_i of $SL_2(X_i)$ as in Section 3. For $\Gamma = \Gamma_1 \times \cdots \times \Gamma_t$ and $p = p_1 + \cdots + p_t \in I(X)$ with $p_i \in I(X_i)$, we denote by $\mathfrak{M}_p^X(\Gamma)$ the set of all finite sums of functions f of the form

$$(6.2) \qquad f(z_1, \ldots, z_t) = f_1(z_1) \cdots f_t(z_t) \qquad \left(z_i \in H_1^{J(X_i)}\right)$$

with $f_i \in \mathfrak{M}_{p_i}(\Gamma_i)$. It can easily be shown that this coincides with the set of all holomorphic functions f on $H_1^{J(X)}$ such that f as a function of z_i belongs to $\mathfrak{M}_{p_i}(\Gamma_i)$ for every i (see [13, Lemma 1.1]). We denote by $\mathfrak{M}_p^X$ the union of $\mathfrak{M}_p^X(\Gamma)$ for all such Γ. An element f of $\mathfrak{M}_p^X$ has a Fourier expansion of the form

$$(6.3) \qquad f(z) = \sum_a \xi(a) \mathbf{e}\left(\sum_{\tau \in J(X)} a^\tau z_\tau \right) \qquad \left(z = (z_\tau)_{\tau \in J(X)} \in H_1^{J(X)}\right)$$

with $\xi(a) \in \mathbf{C}$, where a ranges over nonnegative elements of a lattice in X; we call a *nonnegative* if $a^\tau \geqq 0$ for all $\tau \in J_X$. For a subfield $\mathfrak{F}$ of $\mathbf{C}$, we denote by $\mathfrak{M}_p^X(\mathfrak{F})$ the set of all $f \in \mathfrak{M}_p^X$ such that $\xi(a)$ of (6.3) belongs to $\mathfrak{F}$ for every a.

We now define a differential operator δ_s^r on H_1 for $0 \leqq r \in \mathbf{Z}$ and $s \in \mathbf{Z}$ by

$$(6.4) \qquad \delta_s^r f = (2\pi i)^{-r} y^{1-r-s} (\partial/\partial z)^r (y^{r+s-1} f) \qquad (y = \mathrm{Im}(z))$$

for functions f on H_1. This is exactly $(2\pi i)^{-r} \Delta_s^{(r)}$ in the case $m = 1$, and coincides with the operator of [14, (11.4)], in view of (1.39). Further, for $p \in I_X$ and $0 \leqq q \in I_X$, we define an operator δ_p^q on $H_1^{J(X)}$ by

$$(6.5) \qquad \delta_p^q = \prod_{\tau \in J(X)} \delta_{p_\tau}^{q_\tau},$$

where each factor is the operator of (6.4) with $r = q_\tau$, and $s = p_\tau$ on the τth factor of $H_1^{J(X)}$.

For $0 \leqq r \in I_X$, we denote by A_r the set of all functions f on $H_1^{J(X)}$ of the form

$$(6.6) \qquad f(z) = \sum_{0 \leqslant s \leqslant r} (\pi y)^{-s} \sum_a c_s(a) \mathbf{e}\left(\sum_{\tau \in J(X)} a^\tau z_\tau \right)$$

with $c_s(a) \in \mathbf{C}$, $s \in J(X)$, $\pi y = (\pi y_\tau)_{\tau \in J(X)}$, $y_\tau = \mathrm{Im}(z_\tau)$, and a ranges over nonnegative elements of a lattice in X. We then denote by $\mathfrak{N}_p(\Gamma)$, for $0 \leqslant p \in I_X$, the set of all functions f on $H_1^{J(X)}$ such that

$$(6.7a) \qquad f\|_p \gamma = f \quad \textit{for every} \quad \gamma \in \Gamma,$$

$$(6.7b) \quad f\|_p \alpha \in A_r \quad \textit{for every} \quad \alpha \in SL_2(X) \quad \textit{with some} \quad r \in I_X, \; \geqq 0,$$

where $f\|_p \alpha = \prod_{\tau \in J(X)} j(\alpha^\tau, z_\tau)^{-p_\tau} f(\alpha(z))$. Here r may depend on α, but we can choose an r which depends only on f. In fact, if $f \in A_t$, we see easily that

$f\|_p\alpha = \sum_{0 < s < t}(\pi y)^{-s}h_s(z)$ with holomorphic h_s for every $\alpha \in SL_2(X)$; hence (6.7b) implies that $f\|_p\alpha \in A_t$. Now denote by $\mathfrak{N}_p^*(\Gamma)$ the subset of $\mathfrak{N}_p(\Gamma)$ consisting of all f for which r of (6.7b) can be taken so that

$$(6.7c) \qquad r \leq p' = \sum_{\tau \in J(X)} p'_\tau\tau, \qquad p'_\tau = \mathrm{Max}(0, [(p_\tau - 1)/2]).$$

We then denote by $\mathfrak{N}_p$ the union of $\mathfrak{N}_p(\Gamma)$ for all Γ, and by $\mathfrak{N}_p(\mathfrak{F})$, for a subfield $\mathfrak{F}$ of $\mathbf{C}$, the set of all $f \in \mathfrak{N}_p$ for which $c_s(a)$ of (6.6) belongs to $\mathfrak{F}$ for every s and a; we define $\mathfrak{N}_p^*$ and $\mathfrak{N}_p^*(\mathfrak{F})$ similarly. Then we have

LEMMA 6.1. *If $\mathfrak{F}$ contains all the conjugates of X_i over $\mathbf{Q}$ for all i, $\mathfrak{N}_p^*(\mathfrak{F})$ consists of all sums $\sum_{0 < r < p'}\delta_{p-2r}^r h_r$ with $h_r \in \mathfrak{M}_{p-2r}^X(\mathfrak{F})$.*

This can easily be proved by modifying the proof of [6, Lemma 7].

Let $Y = Y_1 \oplus \cdots \oplus Y_t$ with a totally imaginary quadratic extension Y_i of X_i for each i. Take a *CM*-type ψ of Y and take also $\xi \in Y$ so that $\mathrm{Im}(\xi^\sigma) > 0$ for every $\sigma \in \psi$. We identify $\mathbf{C}^\psi$ with $\mathbf{C}^{J(X)}$ through the bijection of ψ onto J_X. Then $(\xi^\sigma)_{\sigma \in \psi}$ defines an element of $H_1^{J(X)}$. By a *CM*-point of $H_1^{J(X)}$, we understand a point (ξ^σ) of this type obtained from arbitrary Y, ψ, and ζ of the above type. Then we have

LEMMA 6.2. *Let $f \in \mathfrak{N}_p(\overline{\mathbf{Q}})$, $g \in \mathfrak{M}_p^X(\overline{\mathbf{Q}})$, and let w be a CM-point of $H_1^{J(X)}$ such that $g(w) \neq 0$. Then $(f/g)(w) \in \overline{\mathbf{Q}}$.*

Proof. Suppose $f \in A_r$. Take $q \in I_X$, ≥ 0, and $f_1 \in \mathfrak{M}_q^X(\overline{\mathbf{Q}})$ so that $p_\tau + q_\tau > 2r_\tau$ for all $\tau \in J_X$ and $f_1(w) \neq 0$. Then $f_1 f \in \mathfrak{N}_{p+q}^*(\overline{\mathbf{Q}})$ and hence Lemma 6.1 is applicable to $f_1 f$. Therefore if $t = 1$, our assertion follows immediately from [5, Main Theorem 1]. The general case can be reduced to this special case by expressing each h_r of Lemma 6.1 as a linear combination of functions of type (6.2).

Assuming that X_i contains F for every i and that $[X : F] = m$, let us define embeddings of $H_1^{J(X)}$ into H^φ and $SL_2(X)$ into G as follows. Take a basis $\{b_1, \ldots, b_m\}$ of X over F. For each $v \in \varphi = J_F$, let $\tau_{v1}, \ldots, \tau_{vm}$ be the elements of J_X (in any fixed order) which coincide with v on F. Let B_v be the matrix of size m whose (i, j)-entry is $b_j^{\tau_{vi}}$. For every $z = (z_\tau) \in \mathbf{C}^{J(X)}$ (resp. $a \in X$), let $\Phi_v(z)$ (resp. $\Phi_v(a)$) denote the diagonal matrix of size m whose ith diagonal entry is $z_{\tau_{vi}}$ (resp. $a^{\tau_{vi}}$). Then we define a map ϵ_B of $H_1^{J(X)}$ into H^φ by

$$(6.8) \qquad \epsilon_B(z) = \left({}^tB_v\Phi_v(z)B_v\right)_{v \in \varphi} \qquad (z \in H_1^{J(X)}).$$

It can easily be shown that for $\alpha = \begin{pmatrix} a & b \\ c & d \end{pmatrix} \in SL_2(X)$, there is a unique element α^B of G such that

$$(6.9) \quad (\alpha^B)_v = \begin{bmatrix} {}^tB_v\Phi_v(a) \cdot {}^tB_v^{-1} & {}^tB_v\Phi_v(b)B_v \\ B_v^{-1}\Phi_v(c) \cdot {}^tB_v^{-1} & B_v^{-1}\Phi_v(d)B_v \end{bmatrix} \qquad \text{for every} \quad v \in J_F.$$

Then $\alpha \mapsto \alpha^B$ is an injective homomorphism; moreover $\epsilon_B \circ \alpha = \alpha^B \circ \epsilon_B$. Let

$\{\lambda, V\}$ be a representation of $\mathcal{g}^{\varphi}$ and f a V-valued function on H^{φ}. Put $f^B = \lambda((B_v)_{v \in \varphi}) f \circ \epsilon_B$. Then

$$(6.10) \quad (f\|_{\lambda}\alpha^B)^B(z) = \lambda\left(\left(\Phi_v((cz + d)_v^{-1})\right)_{v \in \varphi}\right) f^B(\alpha z) \qquad (z \in H_1^{J(X)})$$

for $\alpha = \left(\begin{smallmatrix} * & * \\ c & d \end{smallmatrix}\right) \in SL_2(X)$.

Now let w be a CM-point of H^{φ}. By Remark 3.5, w can be obtained as the fixed point of $h(Y[\rho])$ with $Y = Y_1 \oplus \cdots \oplus Y_t$ such that $[Y : F] = 2m$. Consider an F-linear map $l: Y \to F_{2m}^1$, an element ζ of Y, and a CM-type ψ of Y as in (3.18) and (3.19). (Recall that $K = F$ and $n = 2m$ in Case SP.) We now take X_i to be the maximal real subfield of Y_i. With $\{b_i\}$ as above, take another basis $\{c_i\}$ of X over F so that $\mathrm{Tr}_{X/F}(b_i c_j) = \delta_{ij}$. Since $\{b_i, (2\zeta)^{-1}c_i\}$ is a basis of Y over F, we can define a map $l_0: Y \to F_{2m}^1$ by

$$(6.11) \qquad l_0\left(\sum_{i=1}^{m}\left(s_i b_i + s_{m+i}(2\zeta)^{-1}c_i\right)\right) = (s_1, \ldots, s_{2m}) \qquad (s_i \in F).$$

Then we see easily that

$$(6.12) \qquad l_0(x)\iota_m \cdot {}^t l_0(y) = \mathrm{Tr}_{Y/F}(\zeta x y^{\rho}) \qquad (x, y \in Y).$$

Define $\alpha \in GL_{2m}(F)$ so that $l(x)\alpha = l_0(x)$ for all $x \in Y$. From (3.18) and (6.12), we see that $\alpha \in G$. Put $h_0(a) = \alpha^{-1}h(a)\alpha$. Then $l_0(xa) = l_0(x)h_0(a)$. Each τ_{vi} can be uniquely extended to an element of ψ, which we denote by τ_{vi} as well. For $y \in Y$, let y_v denote the column vector $(y^{\tau_{vi}})_{i=1}^{m}$ of $\mathbf{C}^m$. Let Ω_v be the $m \times 2m$ matrix defined by $\Omega_v = (B_v, \Phi_v(2\zeta)^{-1}C_v)$ with $C_v = (c_j^{\tau_{vi}})$. Then $C_v = {}^t B_v^{-1}$ and $y_v = \Omega_v \cdot {}^t l_0(y)_v$. Putting $\Phi_v(a) = \mathrm{diag}[a^{\tau_{v1}}, \ldots, a^{\tau_{vm}}]$ for $a \in Y$, we have $\Phi_v(a)\,\Omega_v \cdot {}^t l_0(y)_v = \Phi_v(a)y_v = (ay)_v = \Omega_v \cdot {}^t l_0(ay)_v = \Omega_v \cdot {}^t h_0(a)_v \cdot {}^t l_0(y)_v$, so that $h_0(a)_v \cdot {}^t\Omega_v = {}^t\Omega_v \Phi_v(a)$. If $aa^{\rho} = 1$, this shows that ${}^t B_v \Phi_v(2\zeta)B_v$ is the fixed point of $h_0(a)_v$ on H. Let ζ_{ψ} denote the point $(\zeta^{\sigma})_{\sigma \in \psi}$ of $\mathbf{C}^{\psi} = \mathbf{C}^{J(X)}$. Then $\epsilon_B(2\zeta_{\psi}) = ({}^t B_v \Phi_v(2\zeta)B_v)_{v \in \varphi}$, and this is the fixed point of $h_0(Y[\rho])$, and hence $w = \alpha(\epsilon_B(2\zeta_{\psi}))$.

Now we specialize this to the case $F = \mathbf{Q}$. Let f, e, and w be as in Theorem 3.4. The above argument shows that $w = \alpha(\epsilon_B(2\zeta_{\psi}))$ with suitably chosen X, Y, ψ, B, α, and ζ. Let $\chi(x) = \det(x)^{\kappa+1}\tau^e(x)$ for $x \in \mathcal{g}$, and let $w' = \epsilon_B(2\zeta_{\psi})$. Our definition of $P_{\chi}(w)$ shows that

$$P_{\chi}(w) = P_{\chi}(\alpha(w')) = \chi(\mu_{\alpha}(w'))P_{\chi}(w')\chi(\mu_{\alpha}(w'))^{-1}.$$

Therefore we have

$$P_{\chi}(w)^{-1}(D_{\kappa+1}^{(e)}f)(w) = \chi(\mu_{\alpha}(w'))P_{\chi}(w')^{-1}(D_{\kappa+1}^{(e)}(f\|_{\kappa+1}\alpha))(w').$$

Recall also that $f\|_{\kappa+1}\alpha \in \mathfrak{L}(\overline{\mathbf{Q}})$ by [14, Theorem 11.1]. Therefore, replacing f by

$f\|_{\kappa+1}\alpha$, we can reduce the problem to the algebraicity of $P_\chi(w')^{-1}D_{\kappa+1}^{(e)}f(w')$, since $\chi(\mu_\alpha(w'))$ is algebraic.

Since $F = \mathbf{Q}$, we write simply B, τ_i, and Φ for B_v, τ_{vi}, and Φ_v. Let S be the diagonal matrix of size m whose ith entry is $p_Y(\tau_i, \psi)$ with the symbol p_Y of [10, (1.1)]. Observe that

$$h_0(a)\begin{bmatrix} w' \\ 1 \end{bmatrix} = \begin{bmatrix} w' \\ 1 \end{bmatrix} B^{-1}\Phi(a)B \qquad (a \in Y).$$

Therefore $P_\chi(w') = \chi(B^{-1}SB)$, and hence, for $L \in \mathrm{Ml}_e(T, \mathbf{C})$ and $u_i \in T$, we have

$$(6.13) \qquad \left(\chi(B)P_\chi(w')^{-1}L\right)(u_1, \ldots, u_e)$$

$$= \det(S)^{-\kappa-1}(\chi(B)L)(S^{-1}u_1S^{-1}, \ldots, S^{-1}u_eS^{-1}).$$

Put $f' = \chi(B)(D_{\kappa+1}^{(e)}f) \circ \epsilon_B$. For $\gamma = \begin{pmatrix} a & b \\ c & d \end{pmatrix} \in SL_2(X)$, put $f_\gamma = f\|_{\kappa+1}\gamma^B$. By (6.10), we have

$$(6.14) \qquad \chi(\Phi(cz + d))^{-1}f'(\gamma z) = \chi(B)\left(D_{\kappa+1}^{(e)}f_\gamma\right) \circ \epsilon_B \qquad (z \in H_1^{J(X)}).$$

We insert here a simple lemma. Put $\theta(z) = z - \bar{z}$ for $z \in H$. Take any polynomial representation $\{\rho, V\}$ of $\mathscr{g}$ and a map $g : H \to V$ which is a finite sum of the form

$$(6.15) \qquad g(z) = \sum_\lambda \pi^{-|\lambda|}\lambda(\theta^{-1})g_\lambda(z) \qquad (z \in H),$$

where $\lambda(z)$ is a monomial of the entries of z with coefficient 1, $|\lambda|$ is the degree of λ, and g_λ is a holomorphic map of H into V which has a $\overline{\mathbf{Q}}$-rational Fourier expansion of type (3.10) with $F = \mathbf{Q}$.

LEMMA 6.3. *In the above setting, $\pi^{-e}D_\rho^{(e)}g$ has the same type of expression as* (6.15).

Proof. It is sufficient to prove the case $e = 1$. We have $(D_\rho g)(u) = (Dg)(u) + \rho(\theta^{-1})D[\rho(\theta)](u)g$ for $u \in T$. By (5.1), we have $(D\lambda(\theta^{-1}))(u) = -(D\lambda)(\theta^{-1})(\theta^{-1}u\theta^{-1})$. Since $\pi^{-1}Dg_\lambda$ is $\overline{\mathbf{Q}}$-rational, we see that $\pi^{-1}Dg$ has the required properties. To discuss the other term, we may assume that $\rho(x) = x \otimes \cdots \otimes x$ with k copies of x for a positive integer k. Then

$$\rho(\theta^{-1})D\big[\rho(\theta)\big](u) = \sum_{i=1}^{k} 1 \otimes \cdots \otimes \theta^{-1}u \otimes \cdots \otimes 1 \qquad (\theta^{-1}u \text{ in the } i\text{th factor}),$$

and hence we obtain our assertion.

Now, since $f_\gamma \in \mathfrak{L}(\overline{\mathbf{Q}})$, we have, by [14, (9.6) and the definition on p. 473],

$$(6.16) \qquad f_\gamma(z) = \sum_{k=0}^{m} \pi^{-k} \mathrm{tr}\big(\rho_k(\theta^{-1}) g_k\big)$$

with a holomorphic map $g_k : H \to \mathrm{End}(\bigwedge^k \mathbf{C}^m)$ which has a $\overline{\mathbf{Q}}$-rational Fourier expansion. Applying Lemma 6.3 to f_γ, we see that $\pi^{-e} D_{\kappa+1}^{(e)} f_\gamma$ has an expression of type (6.15). Then we can easily verify that $\pi^{-e}[\chi(B)(D_{\kappa+1}^{(e)} f_\gamma) \circ \epsilon_B]$ $(u_1, \ldots, u_e)$, for $\overline{\mathbf{Q}}$-rational $u_1, \ldots, u_e$, belongs to A_r for some $r \in I_X$ with $\overline{\mathbf{Q}}$-rational coefficients $c_s(a)$ in (6.6). Let $\{\partial\}$ denote the basis of T consisting of the elements ∂_{ij} of (1.12). Put $\xi = \Phi(cz + d)^{-1}$. Now, for $v_1, \ldots, v_e \in \{\partial\}$, we have, by (6.14),

$$(6.17) \quad \det(\xi)^{\kappa+1} f'(\gamma z)({}'\xi v_1 \xi, \ldots, {}'\xi v_e \xi)$$

$$= \Big[\chi(B)\big(D_{\kappa+1}^{(e)} f_\gamma\big) \circ \epsilon_B\Big](v_1, \ldots, v_e) \in A_r \qquad (z \in H_1^{J(X)})$$

for some $r \in I_X$. This holds for every $\gamma \in SL_2(X)$. If γ belongs to a sufficiently small congruence subgroup Γ, then $f_\gamma = f$, so that (6.17) coincides with $f'(v_1, \ldots, v_e)$. In other words, $f'(v_1, \ldots, v_e)\|_p \gamma = f'(v_1, \ldots, v_e)$ for every $\gamma \in \Gamma$ with some $p \in I_X$; p is given $p = (\kappa + 1)\sum_{\sigma \in J(X)} \sigma + \sum_{i=1}^e q_i$, where q_i is an element of I_X such that $\Phi(x)v_i\Phi(x) = x^{q_i} v_i$ for $x \in X$. By (6.17), we see that $\pi^{-e} f'(v_1, \ldots, v_e) \in \mathfrak{R}_p(\overline{\mathbf{Q}})$. Let ω be the element of $\mathbf{Z}^\psi$ whose "restriction to X" is p in an obvious sense. Then

$$(6.18) \qquad \det(S)^{-\kappa-1} L_1\big(S^{-1} v_1 S^{-1}, \ldots, S^{-1} v_e S^{-1}\big)$$

$$= p_Y(\omega, \psi)^{-1} L_1(v_1, \ldots, v_e)$$

for every $L_1 \in \mathrm{Ml}_e(T, \mathbf{C})$. By Lemma 6.2, we have

$$p_Y(\omega, \psi)^{-1} \pi^{-e} f'(2\zeta_\psi)(v_1, \ldots, v_e) \in \overline{\mathbf{Q}},$$

since $g(w)$ of the lemma is an algebraic number times $p_Y(\omega, \psi)$. (This is practically the definition of $p_Y(\omega, \psi)$.) Taking $\pi^{-e} f'(2\zeta_\psi)$ as L_1 of (6.18) and $L = D_{\kappa+1}^{(e)} f(w')$ in (6.13), we find that

$$\pi^{-e}\Big[\chi(B) P_X(w')^{-1} D_{\kappa+1}^{(e)} f(w')\Big](v_1, \ldots, v_e) \in \overline{\mathbf{Q}}$$

for every $v_1, \ldots, v_e \in \{\partial\}$. Since $\chi(B)$ is algebraic, this completes the proof of the first part of Theorem 3.4, from which follows the second one, since $\Delta_{\kappa+1}^\sigma = \Phi_\sigma D_{\kappa+1}^{(r)}$ (cf. the proof of [11, Theorem 3.2]).

Remark 6.4. In the above proof, we showed that $\pi^{-e} \chi(B)(D_{\kappa+1}^{(e)} f) \circ \epsilon_B$ belongs to $\mathfrak{R}_p(\overline{\mathbf{Q}})$. We can actually prove that it belongs to $\mathfrak{R}_p^*(\overline{\mathbf{Q}})$ if $m > 1$, by examining carefully how θ^{-1} occurs in $D_{\kappa+1}^{(e)} f$.

Remark 6.5.   In Case SU, we can similarly define an embedding of H_1^m into H_m. However, not every CM-point of H_m can be obtained by transforming the image of a CM-point of H_1^m with an element of G. For this reason, our method of proof in this section does not apply to Case SU.

7. The singular values of Eisenstein series.

For $c \in \mathbf{C}^\varphi$ and $r \in \mathbf{C}^\varphi$ with φ as in Section 3, we put

$$(7.1\mathrm{a}) \qquad c^r = \prod_{v \in \varphi} c_v^{r_v}$$

whenever the factors on the right-hand side are meaningful; we assume, for the most part, $c_v \neq 0$ and $r_v \in \mathbf{Z}$, or $0 < c_v \in \mathbf{R}$. In particular, we embed $\mathbf{C}$ into $\mathbf{C}^\varphi$ (and also $\mathbf{Z}$ into $\mathbf{Z}^\varphi$) diagonally, and define c^s for $s \in \mathbf{C}$ by

$$(7.1\mathrm{b}) \qquad c^s = \prod_{v \in \varphi} c_v^s .$$

For $\alpha \in G$ and $z \in H^\varphi$, we denote by $j_\alpha(z)$, $\mu_\alpha(z)$, and $\delta(z)$ the elements of $\mathbf{C}^\varphi$ whose v-components are $j(\alpha_v, z_v)$, $\mu(\alpha_v, z_v)$, and $\delta(z_v)$, respectively. Then $j_\alpha(z)^t$, $|j_\alpha(z)|^s$, $\delta(z)^t$, and $\delta(z)^s$ for $t \in \mathbf{Z}^\varphi$ and $s \in \mathbf{C}$ are all meaningful as special cases of (7.1a, b). We write $f\|_t \alpha$, Δ_t^σ, and $D_t^{(e)}$ for $f\|_\zeta \alpha$, Δ_ζ^σ, and $D_\zeta^{(e)}$ if $\zeta(y) = \det(y)^t$ in Case SP and $\zeta(x, y) = \det(y)^t$ in Case SU, where $\det(y) = (\det(y_v))_{v \in \varphi}$. In [14], we considered two types of Eisenstein series $E(z, s; k, \Gamma)$ and $E(z, s; k, \psi, \mathbf{b})$; the former is given by

$$E(z, s; k, \Gamma) = \sum_{\alpha \in (\Gamma \cup P)\backslash\Gamma} j_\alpha(z)^{-k} |j_\alpha(z)|^{-s}.$$

We recall that $z \in H^\varphi$, $s \in \mathbf{C}$, $k \in \mathbf{Z}$, Γ is an arbitrary congruence subgroup of G, and P consists of all the elements $\left(\begin{smallmatrix} a & b \\ c & d \end{smallmatrix}\right)$ of G for which $c = 0$. For the definition of $E(z, s; k, \psi, \mathbf{b})$, the reader is referred to [14, (2.18)]. To treat them uniformly, we denote each one of the series by $E(z, s; k, \Omega)$ with a symbol Ω which stands for either Γ or $(\psi, \mathbf{b})$. Then

$$(7.2) \qquad E(z, s; k, \Omega) = \sum_{\alpha \in A} b_\alpha c_\alpha^s j_\alpha(z)^{-k} |j_\alpha(z)|^{-s}$$

with a subset A of G, $b_\alpha \in \overline{\mathbf{Q}}$, and $0 < c_\alpha \in \mathbf{Q}$; A, b_α, c_α are determined by Ω. We now define a new series $\mathbf{E}$ for every $t \in \mathbf{Z}^\varphi$, $\geqq 0$, by

$$(7.3) \qquad \mathbf{E}(z, s; k, t, \Omega) = \delta(z)^{s-t} \sum_{\alpha \in A} b_\alpha c_\alpha^{2s} \overline{j_\alpha(z)}^t j_\alpha(z)^{-k-t} |j_\alpha(z)|^{-2s}.$$

Notice that the last exponent is $-2s$ instead of $-s$. This (as well as another series (7.6) below) can be continued as a meromorphic function in s to the whole plane by virtue of the result of Langlands [4]. We see easily that

$$(7.4) \qquad \mathbf{E}(z,s;k,\iota,\Omega)\|_{k+2\iota}\gamma = \mathbf{E}(z,s;k,\iota,\Omega) \qquad \textit{for all} \quad \gamma \in \Gamma_\Omega$$

with a congruence subgroup Γ_Ω depending on Ω. Series (7.3) can be generalized in the following way. Take a $\overline{\mathbf{Q}}$-rational polynomial representation

$$(7.5) \qquad \sigma : (GL_m)^\varphi \to GL_d \qquad (0 < d \in \mathbf{Z}).$$

With Ω as above, we define a $\mathbf{C}_d^d$-valued function $\mathbf{E}(z,s;k,\sigma,\Omega)$ by

$$(7.6) \quad \mathbf{E}(z,s;k,\sigma,\Omega) = \delta(z)^s \sum_{\alpha \in A} b_\alpha c_\alpha^{2s} j_\alpha(z)^{-k} |j_\alpha(z)|^{-2s} \sigma\big(\mu_\alpha(z)^{-1}\overline{\kappa_\alpha(z)}\,\eta(z)^{-1}\big).$$

This becomes (7.3) if $\sigma(x) = \det(x)^\iota$. We see that

$$\mathbf{E}(z,s;k,\sigma,\Omega)\|_\rho\gamma = \mathbf{E}(z,s;k,\sigma,\Omega) \qquad \text{for all} \quad \gamma \in \Gamma_\Omega$$

with a representation $\rho : \mathscr{g}^\varphi \to GL(\mathbf{C}_d^d)$ defined by

$$(7.7a) \qquad \rho(y)Z = \det(y)^k\sigma(y)Z\sigma({}^t y) \qquad \text{(Case SP)},$$

$$(7.7b) \qquad \rho(x,y)Z = \det(y)^k\sigma(y)Z\sigma({}^t x) \qquad \text{(Case SU)}$$

for $Z \in \mathbf{C}_d^d$. To state our theorem, we need two representations σ_1 and σ_2 of $\mathscr{g}^\varphi$ given by $\sigma_1(x,y) = {}^t\sigma({}^t x)$ and $\sigma_2(x,y) = \sigma(y)$ in Case SU; $\sigma_1(x) = {}^t\sigma({}^t x)$ and $\sigma_2(x) = \sigma(x)$ in Case SP. We make also the following assumption:

$$(7.8) \qquad \begin{aligned} &k \geqq \kappa \ (see \ (1.15)); \ if \ k = (m+2)/2 \ in \ Case \ SP, \\ &we \ assume \ F \neq \mathbf{Q} \ or \ \Omega = (\psi,\mathbf{b}) \ with \ \psi^2 \neq 1. \end{aligned}$$

THEOREM 7.1. *Let $0 \leqq r \in \mathbf{Z}^\varphi$ and $\sigma(x) = \bigotimes_{v \in \varphi} \sigma_v(x_v)$ with a $\overline{\mathbf{Q}}$-rational homogeneous polynomial representation σ_v of GL_m of degree r_v for each v. Suppose, in Case SP, that σ_v is irreducible on $O(m,\mathbf{R})$ for every $v \in \varphi$. Then, under (7.8), $\mathbf{E}(z,s;k,\sigma,\Omega)$ is finite at $s = 0$, and moreover, for every CM-point w on H^φ, the entries of the matrix*

$$\pi^{-|r|}g(w)^{-1}P_{\sigma_2}(w)^{-1}\mathbf{E}(w,0;k,\sigma,\Omega) \cdot {}^t P_{\sigma_1}(w)^{-1}$$

are all algebraic, where $|r| = \sum_{v \in \varphi} r_v$, $P_{\sigma_i}(w)$ is an element of $GL_d(\mathbf{C})$ explained in Section 3, and g is any element of $\mathfrak{A}_k(\overline{\mathbf{Q}})$ such that $g(w) \neq 0$.

A special case can be stated as

COROLLARY 7.2. *Let w be a CM-point of H^φ, and let $0 \leq t \in \mathbf{Z}^\varphi$, $h \in \mathfrak{A}_{k+2t}(\overline{\mathbf{Q}})$, $h(w) \neq 0$. Then, under (7.8), $\mathbf{E}(z, s; k, t, \Omega)$ is finite at $s = 0$, and moreover, $\pi^{-m|t|}h(w)^{-1}\mathbf{E}(w, 0; k, t, \Omega)$ is an algebraic number, where $|t| = \sum_{v \in \varphi} t_v$.*

Our theorem generalizes [11, Theorems 5.1, 5.2] which concern the case where $\Omega = Sp(m, \mathbf{Z})$ and $k > m + 1$; the result in the case $m = 1$ was obtained in [5, Theorem 1]. There is one more result of the character which was not considered in [11]:

THEOREM 7.3. *Suppose $k = \kappa - 1$. Let w be a CM-point of H^φ, and g an element of $\mathfrak{A}_{\kappa-1}(\overline{\mathbf{Q}})$ such that $g(w) \neq 0$. Then, under the same assumption on σ as in Theorem 7.1, $\mathbf{E}(w, s; \kappa - 1, \sigma, \Omega)$ has at most a simple pole at $s = 1$, and its residue has the form*

$$\pi^{|r| - m\nu}g(w)R_F P_{\sigma_2}(w)A \cdot {}^t P_{\sigma_1}(w)$$

with $A \in \overline{\mathbf{Q}}_d^d$, where R_F is the regulator of F and $\nu = [F : \mathbf{Q}]$.

To prove our theorems, we need two basic facts:

(7.9a)   *Under (7.8), $\mathbf{E}(z, s; k, 0, \Omega)$ is finite at $s = 0$, and $\mathbf{E}(z, 0; k, 0, \Omega)$ belongs to $\mathfrak{M}_k(\overline{\mathbf{Q}})$ or $\mathfrak{L}(\overline{\mathbf{Q}})$ as a function on H^φ.*

(7.9b)   *$\mathbf{E}(z, s; \kappa - 1, 0, \Omega)$ has at most a simple pole at $s = 1$, and has residue $\pi^{-m\nu}R_F h(z)$ with some $h \in \mathfrak{M}_{\kappa-1}(\overline{\mathbf{Q}})$.*

In Case SU, these were proved in [14]. In Case SP, we proved (7.9a) for $k > m$; for lower weights, we showed that the results hold under a certain conjecture. (See [14, Conjecture 6.3, Propositions 10.1, 10.2, 10.3]. The conjecture needs a correction: ξ should always be $((F'/F)/\mathbf{p})$ with $F' = F(\det(\delta j)^{1/2})$. This correction has no effect on the propositions.) Recently, Kitaoka [3] has obtained a result which validates the (corrected) conjecture, and therefore (7.9a, b) are true also in Case SP. Now, to prove Theorem 7.1, we may assume that each σ_v is irreducible. We then consider Δ_k^ϵ with $\epsilon = \{\epsilon_v\}_{v \in \varphi}$, $\epsilon_v(x) = {}^t\sigma_v({}^t x)$. As an immediate consequence of Lemma 1.2, we obtain

(7.10) $$\Delta_k^\epsilon \mathbf{E}(z, s; k, 0, \Omega) = \mathbf{b}(s) \cdot {}^t\mathbf{E}(z, s; k, \sigma, \Omega),$$

identifying $\mathrm{Hom}(\mathbf{C}_d^d, \mathbf{C})$ with $\mathbf{C}_d^d$, with a constant $\mathbf{b}(s)$ which is the product of several constants of the type $i^r\beta_\lambda(-k - s)$ of Lemma 1.2. Observing that $\mathbf{b}(0) \neq 0$ if $k \geq \kappa$ and $\mathbf{b}(1) \neq 0$ if $k = \kappa - 1$, we obtain Theorems 7.1 and 7.3 directly from Theorems 3.2, 3.4, and (7.9a, b).

Remark 7.4.   In Case SP, we needed the irreducibility of σ_v on $O(m, \mathbf{R})$. For σ of a more general type, we have at least:

(7.11)   *There is a constant $C_{m,r}$ depending only on m and r such that the conclusions of Theorems 7.1 and 7.3 are true if $k \geqq C_{m,r}$.*

This can be shown in the same manner as the proof of [11, Theorem 5.1]. For the reason explained in [11, Remark 5.4], we can take $C_{m,r} = 1$ if $r_v \leqq 2$ for all v, or somewhat more strongly, we can state:

(7.12)   *The conclusions of Theorems 7.1 and 7.3 in Case SP are true if, for every $v \in \varphi$, either $r_v \leqq 2$ or σ_v is irreducible on $O(m, \mathbf{R})$.*

Remark 7.5. $\mathbf{E}(w, s; \kappa - 1, \sigma, \Omega)$ is often finite at $s = 1$, and hence the assertions of Theorem 7.3 become trivial in such cases. For example, we have

(7.13) $\mathbf{E}(z, s; \kappa - 1, t, \Omega)$ *is finite at* $s = 1$ *if* $0 \neq t \geqq 0.$

In fact, if g is the residue of $\mathbf{E}(z, s; \kappa - 1, 0, \Omega)$ at $s = 1$, we have $\Delta_v g = 0$ for every $v \in \varphi$ by [14, Theorem 7.3 and Proposition 10.3]. Therefore $\Delta_{\kappa-1}^{(t)} g = 0$ if $t \neq 0$ by (1.38) and (1.39). This together with (7.10) proves (7.13).

8. Critical values of certain zeta functions.   We can re-formulate Theorem 7.1 by interpreting $\mathbf{E}(w, 0; k, \sigma, \Omega)$ as values of certain zeta functions attached to *CM*-fields. This was done in [5] when $m = 1$, and also in [11] for $G = Sp(m, \mathbf{Q})$. The interpretation can be generalized to our groups in both Cases SP and SU. In Case SU, however, it is more natural to take a formulation which is somewhat different from that of [11] and which we are going to treat in this section.

Given a *CM*-field K and its *CM*-type φ as before, we denote by **h** the maximal order of K. Let α be a totally positive definite hermitian element of $GL_p(K)$ with $0 \leqq p \leqq m$. Take a polynomial representation

(8.1a) $S : (GL_m)^{\varphi} \to GL_d$

and a polynomial function

(8.1b) $\epsilon : (\mathbf{C}_p^m)^{\varphi} \to \mathbf{C}^d$

such that $\epsilon(ax) = S(a)\epsilon(x)$ for every $a \in (GL_m)^{\varphi}$ and $x \in (\mathbf{C}_p^m)^{\varphi}$. With a C-valued locally constant function μ on K_p^m (see Notation and terminology), define a $\mathbf{C}^d$-valued "theta series" f_ϵ (depending on α, ϵ, and μ) by

(8.2a) $f_\epsilon(z) = \displaystyle\sum_{x \in K_p^m} \mu(x)\epsilon(x)\mathbf{e}\!\left(\mathrm{tr}\!\left(\sum_{v \in \varphi} x_v \alpha_v x_v^* z_v \right) \right)$ $(z \in H^{\varphi}).$

It can be shown that $f_\epsilon \in \mathfrak{M}_\zeta$ with $\zeta(x, y) = \det(y)^r S(y)$. For the formulation of our theorem, however, we need f_ϵ only as a formal series. If in particular S and ϵ are trivial, we obtain a scalar-valued series

$$(8.2\mathrm{b}) \qquad f_1(z) = \sum_{x \in K_p^m} \mu(x)\mathbf{e}\left(\mathrm{tr}\left(\sum_{v \in q} x_v \alpha_v x_v^* z_v \right) \right).$$

If $p = 0$, we understand that our series means the constant 1, and put $f_0 = 1$. We define the Fourier coefficients of these functions by putting

$$(8.2\mathrm{c}) \qquad f_*(z) = \sum_h c_*(h)\mathbf{e}\left(\mathrm{tr}\left(\sum_{v \in \varphi} h_v z_v \right) \right) \qquad (h = h^* \in K_m^m),$$

where $* = \epsilon$, 1, or 0. Obviously we have

$$(8.3\mathrm{a}) \qquad c_\epsilon(h) = \sum_{xax^* = h} \mu(x)\epsilon(x) \in \mathbf{C}^d,$$

$$(8.3\mathrm{b}) \qquad c_1(h) = \sum_{xax^* = h} \mu(x),$$

$$(8.3\mathrm{c}) \qquad c_0(h) = \begin{cases} 1 & \text{if} \quad h = 0, \\ 0 & \text{if} \quad h \neq 0. \end{cases}$$

Notice that $c_\epsilon(\gamma h \gamma^*) = S(\gamma)c_\epsilon(h)$ and $c_1(\gamma h \gamma^*) = c_1(h)$ for every γ in a congruence subgroup of $GL_m(\mathbf{h})$.

Now we consider an algebra $Y = Y_1 \oplus \cdots \oplus Y_t$ with CM-fields Y_i containing K such that $[Y : K] = m + q$ with $0 \leq q \leq m$. Put $X = X_1 \oplus \cdots \oplus X_t$ with the maximal real subfield X_i of Y_i. For each $v \in \varphi$, we take $m + q$ different elements $\tau_{v1}, \ldots, \tau_{vm}, \rho\sigma_{v1}, \ldots, \rho\sigma_{vq}$ of J_Y (see Section 3) which coincide with v on K, and also choose an element ξ of $X^\times$ such that

$$(8.4) \qquad \xi^{\tau_{vi}} > 0 \text{ for every } v \text{ and } i, \text{ and } \xi^\sigma < 0 \text{ for every } \sigma \text{ of}$$
$$J_X \text{ different from the restrictions of the } \tau_{vi} \text{ to } X.$$

For $a = (a_i)_{i=1}^m \in Y^m$, we put

$$(8.5\mathrm{a}) \qquad a\{\tau\} = (a\{\tau\}_v)_{v \in \varphi}, \qquad a\{\tau\}_v = (a_i^{\tau_{vj}})_{i,j=1}^m \in \mathbf{C}_m^m,$$

$$(8.5\mathrm{b}) \qquad a\{\sigma\} = (a\{\sigma\}_v)_{v \in \varphi}, \qquad a\{\sigma\}_v = (a_i^{\sigma_{vj}})_{\substack{1 \leq i \leq m \\ 1 \leq j \leq q}} \in \mathbf{C}_q^m,$$

$$(8.5\mathrm{c}) \qquad a[\tau] = (a[\tau]_v)_v, \qquad a[\tau]_v = \det(a\{\tau\}_v),$$

$$(8.6) \qquad R = \left\{ (a_i)_{i=1}^m \in Y^m \,\middle|\, \left[\sum_{i=1}^m Ka_i : K \right] = m \right\}.$$

Given Y, take a series f_* as above with $p = m - q$; we then define three types of series Z_0, Z_1, and Z_2 by

$$(8.7a) \qquad Z_0(s) = [\Delta_1 : \Delta]^{-1} \sum_{a \in \Delta \backslash R} \lambda(a)(\overline{a[\tau]}/|a[\tau]|)^e |a[\tau]|^{-2s}$$

$$\cdot c_0\big(\mathrm{Tr}_{Y/K}(\xi a_i a_j^\rho)_{i,j=1}^m\big) S\big(a\{\tau\}^{-1}\overline{a\{\sigma\}}\big)$$

(defined when $q = m$),

$$(8.7b) \qquad Z_1(s) = [\Delta_1 : \Delta]^{-1} \sum_{a \in \Delta \backslash R} \lambda(a)(\overline{a[\tau]}/|a[\tau]|)^e |a[\tau]|^{-2s}$$

$$\cdot c_1\big(\mathrm{Tr}_{Y/K}(\xi a_i a_j^\rho)_{i,j=1}^m\big) S\big(a\{\tau\}^{-1}\big)\zeta\big(\overline{a\{\sigma\}}\big)$$

(defined when $0 < q < m$),

$$(8.7c) \qquad Z_2(s) = [\Delta_1 : \Delta]^{-1} \sum_{a \in \Delta \backslash R} \lambda(a)(\overline{a[\tau]}/|a[\tau]|)^e |a[\tau]|^{-2s}$$

$$\cdot S\big(a\{\tau\}^{-1}\big) c_\epsilon\big(\mathrm{Tr}_{Y/K}(\xi a_i a_j^\rho)_{i,j=1}^m\big)$$

(defined when $p > 0$).

Here $s \in \mathbf{C}$, $0 \leqq e \in \mathbf{Z}^\varphi$, and λ is a $\mathbf{C}$-valued locally constant function on Y^m; x^e and y^{-2s} should be understood in the sense of (7.1a, b); $c_*(h)$ is the Fourier coefficient given by (8.3a, b, c); ζ is a polynomial map $(\mathbf{C}_q^m)^\varphi \to \mathbf{C}^d$ such that $\zeta(xy) = S(x)\zeta(y)$ for $x \in (GL_m)^\varphi$ and $y \in (\mathbf{C}_q^m)^\varphi$; $\Delta_1 = \underline{GL_m(\mathbf{h})}$ and Δ is a congruence subgroup of Δ_1 such that $\lambda(\delta a) = \lambda(a)$, $\overline{(\det(\delta)/|\det(\delta)|)}^e = 1$, $c_\epsilon(\delta h \delta^*) = S(\delta) c_\epsilon(h)$, and $c_1(\delta h \delta^*) = c_1(h)$ for all $\delta \in \Delta$. The existence of such a Δ can easily be verified. Obviously each term of $Z_i(s)$ depends only on Δa, and the factor $[\Delta_1 : \Delta]^{-1}$ makes each sum independent of the choice of Δ. To make each term meaningful, we need a simple fact which will be proved later:

$$(8.8) \qquad a[\tau]_v \neq 0 \qquad \textit{if } \big(\mathrm{Tr}_{Y/K}(\xi a_i a_j^\rho)_v\big)_{i,j=1}^m \textit{ is nonnegative.}$$

We shall show that each series Z_i is essentially a finite linear combination of the products of certain L-functions of K and $\mathbf{E}(z, s; k, \sigma, \Omega)$ with z specialized to a CM-point. This will prove the meromorphic continuation of Z_i to the whole plane. To state our theorem, we need the notion of the discriminant of X over F, denoted by $d(X/F)$, which is the class of $\det(\mathrm{Tr}_{X/F}(u_i u_j))$ modulo $\{a^2 \mid a$ where $\{u_i\}$ is a basis of X over F. Then we assume:

$$(8.9) \qquad (-1)^{m-p} \det(\alpha) N_{X/F}(\xi) d(X/F) \in N_{K/F}(K^\times)$$

(we understand that $\det(\alpha) = 1$ if $p = 0$);

$$(8.10) \qquad S, \epsilon, \zeta, \mu, \textit{ and } \lambda \textit{ are all } \overline{\mathbf{Q}}\textit{-rational.}$$

THEOREM 8.1. *Let k be an integer such that*

$$(8.11) \qquad m \leqq k \leqq e_v, \qquad k \equiv e_v \pmod 2 \quad \text{for every} \quad v \in \varphi.$$

Suppose $S(x) = \bigotimes_{v \in \varphi} S_v(x_v)$ with homogeneous S_v of degree t_v. Then, under (8.9) and (8.10), Z_0, Z_1, and Z_2 are finite at $k/2$, and their values can be written in the forms

$$Z_0(k/2) = \pi^b \hbar S(\mathcal{g}_\tau) A_0 S(\mathcal{g}_\sigma) \qquad \text{with} \quad A_0 \in \overline{\mathbf{Q}}_d^d,$$

$$Z_1(k/2) = \pi^b \hbar S(\mathcal{g}_\tau) A_1 \zeta \begin{pmatrix} \mathcal{g}_\sigma \\ 0 \end{pmatrix} \qquad \text{with} \quad A_1 \in \overline{\mathbf{Q}}_d^d,$$

$$Z_2(k/2) = \pi^b \hbar \varkappa S(\mathcal{g}_\tau) A_2 \qquad \text{with} \quad A_2 \in \overline{\mathbf{Q}}^d,$$

where

$$b = \sum_{v \in \varphi} (me_v + t_v) - [F:\mathbf{Q}]m(m-1)/2,$$

$$\hbar = p_Y\left(\sum_{v \in \varphi} \sum_{i=1}^{m} e_v \tau_{vi}, 2 \sum_{v \in \varphi} \sum_{i=1}^{m} \tau_{vi} \right),$$

$$\mathcal{g}_\tau = \left(\text{diag}[\, p_Y(\tau_{v1}, \psi), \ldots, p_Y(\tau_{vm}, \psi) \,] \right)_{v \in \varphi} (\in (GL_m)^\varphi),$$

$$\mathcal{g}_\sigma = \left(\text{diag}[\, p_Y(\sigma_{v1}, \psi), \ldots, p_Y(\sigma_{vq}, \psi) \,] \right)_{v \in \varphi} (\in (GL_q)^\varphi),$$

$$\varkappa = p_K\left(\sum_{v \in \varphi} t_v v, \varphi \right),$$

$$\psi = \sum_{v \in \varphi} \left(\sum_{i=1}^{m} \tau_{vi} + \sum_{i=1}^{q} \sigma_{vi} \right).$$

In particular, if S, ϵ, and ζ are trivial, the values of our series at $k/2$ are algebraic numbers times $\pi^b \hbar$. Notice that an integer k of type (8.11) exists if and only if $m \leqq e_v$ for all $v \in \varphi$ and e_v modulo 2 is independent of v.

Proof of (8.8). For a fixed v, put $B = \overline{a\{\sigma\}}_v$, $C = a\{\tau\}_v$ for $(a_i) \in R$, and $X = \text{diag}[\xi^{\tau_{v1}}, \ldots, \xi^{\tau_{vm}}]$, $X' = \text{diag}[-\xi^{\sigma_{v1}}, \ldots, -\xi^{\sigma_{vq}}]$. Then $\text{rank}[C, B] = m$, and hence (8.4) implies that

$$0 < (C, B)\text{diag}[X, X'](C, B)^* = CXC^* + BX'B^*.$$

On the other hand, we observe that

$$(\text{Tr}_{Y/K}(\xi a_i a_j^\rho)_v) = CXC^* - BX'B^*.$$

If this is nonnegative, we have $CXC^* > 0$, so that $\det(C) \neq 0$, which proves (8.8).

Postponing the proof of the theorem to Section 10, let us now show that the assertions on Z_1 and Z_2 follow from that on Z_0. Suppose $p > 0$. Series Z_2 is determined by $\{Y, \lambda, \tau_{vi}, \xi, \mu, \epsilon, \alpha\}$. Without losing generality, we may assume that α is diagonal, say $\alpha = \text{diag}[\alpha_1, \ldots, \alpha_p]$. Define Z_0 with a collection of data $\{Y', \lambda', \tau_{vi}, \sigma_{vi}, \xi'\}$ given as follows: $Y' = Y \oplus K^p$, $\lambda'(a, b) = \lambda(a)\mu(b)$ for $a \in Y^m$ and $b \in K^m_p$, $\xi' = (\xi, -\alpha_1, \ldots, -\alpha_p)$; we view τ_{vi} as elements of $J_{Y'}$; $\sigma_{v1}, \ldots, \sigma_{vm}$ are elements of $J_{Y'}$ which coincide with $v\rho$ on K and such that $\sum_{i=1}^m \sum_{v \in \varphi} (\tau_{vi} + \sigma_{vi})$ is a CM-type of Y'. We take σ_{vq+i} for $i = 1, \ldots, p$ to be $v\rho$ combined with the projection of Y' to the ith factor K of K^p. Now

$$Z_0 = [\Delta_1 : \Delta]^{-1} \sum \lambda'(x)(\overline{x[\tau]}/|x[\tau]|)^e |x[\tau]|^{-2s} S(x\{\tau\}^{-1}\overline{x\{\sigma\}}),$$

the sum being extended over all $x = (x_i) \in Y'^m$, modulo Δ, such that

$$(8.12) \quad \text{Tr}_{Y'/K}(\xi' x_i x_j^\rho) = 0 \quad \text{for all } i \text{ and } j, \text{ and } \quad \left[\sum_{i=1}^m K x_i : K \right] = m.$$

Put $x_i = (a_i, b_{i1}, \ldots, b_{ip})$ with $a_i \in Y$ and $b_{ij} \in K$. Then $x\{\tau\} = a\{\tau\}$, and $\overline{x\{\sigma\}} = (\overline{a\{\sigma\}}, b)$ with $b = (b_{ij}) \in K^m_p$. Put $u = \epsilon \begin{bmatrix} 0 \\ 1_p \end{bmatrix}$. Then we observe that $S([y, z])u = \epsilon(z)$ for $y \in \mathbf{C}^m_q$ and $z \in \mathbf{C}^m_p$. Therefore we have

$$Z_0 u = [\Delta_1 : \Delta]^{-1} \sum_{a, b} \lambda(a)\mu(b)(\overline{a[\tau]}/|a[\tau]|)^e$$

$$\cdot |a[\tau]|^{-2s} S(a\{\tau\}^{-1})\epsilon(b).$$

Now (8.12) holds if and only if $(\text{Tr}_{Y/K}(\xi a_i a_j^\rho)) = b\alpha b^*$ and $a \in R$. Therefore (8.3a) shows that $Z_0 u = Z_2$. Observing that $S(g_o)u = \varkappa u$, we can derive the assertion on $Z_2(k/2)$ from that on $Z_0(k/2)$. Similarly the problem about Z_1 can be reduced to that about Z_0.

The above Z_2 includes as a special case a series of the following type:

$$(8.13) \qquad W(s) = [\Delta_1 : \Delta]^{-1} \sum_{x \in \Delta \backslash GL_m(K)} \lambda(x)(\overline{\det(x)}/|\det(x)|)^e$$

$$\cdot |\det(x)|^{-2s} S(x^{-1}) c_\epsilon(x\beta x^*).$$

Here λ is a locally constant function on K^m_m; c_ϵ is the same as in (8.3a); we take $p = m$; β is a totally positive definite hermitian element of $GL_m(K)$. In fact, assuming $\beta = \text{diag}[\beta_1, \ldots, \beta_m]$ without losing generality, we take $Y = K^m$, $\xi = (\beta_1, \ldots, \beta_m)$, and τ_{vi} to be the embedding v composed with the projection map of Y onto the ith factor K. Then Z_2 becomes W. Therefore we obtain

COROLLARY 8.2. *Suppose* $\det(\alpha\beta) \in N_{K/F}(K^\times)$. *If k is an integer satisfying* (8.11), *then, under* (8.10), *W is finite at $k/2$, and $W(k/2)$ is* $\pi^b p_K(\sum_{v \in \varphi} 2(me_v + t_v)v, \varphi)$ *times an algebraic vector, where b is the same as in Theorem 8.1.*

If $m = p = 1$, $Y = K$, and if $S(x) = \epsilon(x) = x^t$ with $0 \leqq t \in \mathbf{Z}^{\varphi}$, then Z_2 can be given as

$$[\Delta_1 : \Delta]^{-1} \sum_{a \in \Delta \backslash K^\times} \lambda(a) c_\epsilon(\xi a a^\rho) a^{-t} (\bar{a}/|a|)^\epsilon N_{K/\mathbf{Q}}(a)^{-s}$$

with $\Delta_1 = \mathbf{h}^\times$ and $\xi \in F^\times$. In our previous papers [7, §5], [10], [12, I], and [13], we investigated a series of this type with an arbitrary Hilbert modular form f which is not necessarily of type (8.2a), and proved theorems of algebraicity for its critical values. The most general results when Y is a field are given in [12, I, Theorems 4.3 and 4.6]; these can be generalized to the case of a direct sum of CM-fields as shown in [13, §2]. It is therefore natural to expect some theorems which are m-dimensional versions of these results and which include Theorem 8.1 as a special case. Although plausible forms of such theorems can actually be stated, we content ourselves with a mere allusion to them here, leaving the details to a subsequent paper.

In the rest of this section, we study, as a preliminary step, the algebraic nature of the factors of automorphy $\kappa(\gamma, z)$ and $\mu(\gamma, z)$ when z is specialized to a CM-point. To treat all cases SP, SU, and SUB uniformly, we make the same notational conventions as in Section 3: $K = F$ in Case SP; $n = 2m$, $\iota = \iota_m$, $p_v = q_v = m$ in Cases SP and SU.

For $x \in K_n^r$ $(0 < r \in \mathbf{Z})$, $v \in \varphi$, and $z \in H^\varphi$ or $\mathcal{B}$, we define matrices $T_v(z) \in GL_n(\mathbf{C})$, $\kappa_v[x, z] \in \mathbf{C}_{p_v}^r$ and $\mu_v[x, z] \in \mathbf{C}_{q_v}^r$ by

$$(8.14) \qquad T_v = T_v(z) = \begin{cases} \begin{bmatrix} z_v^* & z_v \\ 1_m & 1_m \end{bmatrix} & \text{(Cases SP, SU),} \\[2em] Q_v \begin{bmatrix} 1_p & z_v \\ z_v^* & 1_q \end{bmatrix} & \text{(Case SUB),} \end{cases}$$

$$(8.15) \qquad \left(\overline{\kappa_v[x, z]}, \mu_v[x, z] \right) = \begin{cases} x_v T_v(z) & \text{(Cases SP, SU),} \\ \bar{x}_v T_v(z) & \text{(Case SUB).} \end{cases}$$

Here Q_v is the matrix as in (3.1); we write p and q for p_v and q_v, which applies to all formulas in the rest of this section. These κ_v and μ_v are holomorphic in z and satisfy

$$(8.16\text{a}) \qquad \kappa_v[x\alpha, z] = \kappa_v[x, \alpha(z)] \kappa_v(\alpha, z) \qquad (\alpha \in G),$$

$$(8.16\text{b}) \qquad \mu_v[x\alpha, z] = \mu_v[x, \alpha(z)] \mu_v(\alpha, z) \qquad (\alpha \in G).$$

Observe that (in view of (3.1) in Case SUB)

$$(8.17\text{a}) \quad i\iota_m = T_v(z)\operatorname{diag}[\eta(z_v)^{-1}, -\eta(z_v)^{-1}] T_v(z)^* \qquad \text{(Cases SP, SU),}$$

$$(8.17\text{b}) \quad i\iota_v = \overline{T_v(z)} \operatorname{diag}[-\xi_B(z_v)^{-1}, {}^t\eta_B(z_v)^{-1}] \cdot {}^t T_v(z) \qquad \text{(Case SUB).}$$

Therefore, for $x, y \in K_n^r$, we have

$$(8.18) \quad i(x\iota y^*)_v = \begin{cases} \overline{\kappa_v[x,z]}\,\eta(z_v)^{-1}\cdot {}^t\kappa_v[y,z] \\ \quad -\mu_v[x,z]\eta(z_v)^{-1}\mu_v[y,z]^* \qquad \text{(Cases SP, SU)}, \\ -\kappa_v[x,z]\xi_B(z_v)^{-1}\kappa_v[y,z]^* \\ \quad +\overline{\mu_v[x,z]}\cdot {}^t\eta_B(z_v)^{-1}\cdot {}^t\mu_v[y,z] \qquad \text{(Case SUB)}. \end{cases}$$

Let $Y = Y_1 \oplus \cdots \oplus Y_t$ with CM-fields Y_i containing K, such that $[Y:K] = n$. Consider the maps $h: Y \to K_n^n$ and $l: Y \to K_n^1$ such that $h(x^\rho) = \iota h(x)^*\iota^{-1}$, $l(xa) = l(x)h(a)$ as in Section 3; let w be the fixed point of $h(Y[\rho])$; take $\zeta \in Y$ so that $\zeta^\rho = -\zeta$ and $l(a)\iota l(b)^* = \mathrm{Tr}_{Y/K}(\zeta ab^\rho)$ as in (3.18). We can find $\sigma_{v1}, \ldots, \sigma_{vp}, \tau_{v1}, \ldots, \tau_{vq} \in J_Y$, $U \in GL_p(\overline{\mathbf{Q}})$, and $V \in GL_q(\overline{\mathbf{Q}})$ such that

$$(8.19a) \qquad U_v\kappa_v(h(a),w)U_v^{-1} = \mathrm{diag}[a^{\sigma_{v1}}, \ldots, a^{\sigma_{vp}}],$$

$$(8.19b) \qquad V_v\mu_v(h(a),w)V_v^{-1} = \mathrm{diag}[a^{\tau_{v1}}, \ldots, a^{\tau_{vp}}]$$

for $a \in Y[\rho]$. Notice that

$$(8.20) \qquad \{\rho\sigma_{v1}, \ldots, \rho\sigma_{vp}, \tau_{v1}, \ldots, \tau_{vq}\}$$

$$= \begin{cases} \{\alpha \in J_Y \mid \alpha = v \text{ on } K\} & \text{(Cases SP, SU)}, \\ \{\alpha \in J_Y \mid \alpha = v\rho \text{ on } K\} & \text{(Case SUB)}. \end{cases}$$

We need a map $l_r: Y^r \to K_n^r$ defined by

$$(8.21) \qquad l_r\begin{bmatrix} a_1 \\ \vdots \\ a_r \end{bmatrix} = \begin{bmatrix} l(a_1) \\ \vdots \\ l(a_r) \end{bmatrix} \qquad (a_i \in Y).$$

LEMMA 8.3. *There exist diagonal matrices $C_v \in GL_p(\overline{\mathbf{Q}})$ and $D_v \in GL_q(\overline{\mathbf{Q}})$ such that*

$$\kappa_v\big[l_r({}^t(a_1, \ldots, a_r)), w\big] = (a_i^{\sigma_{vj}})C_v U_v,$$

$$\mu_v\big[l_r({}^t(a_1, \ldots, a_r)), w\big] = (a_i^{\tau_{vj}})D_v V_v,$$

$$-i\cdot\mathrm{diag}[\zeta^{\sigma_{v1}}, \ldots, \zeta^{\sigma_{vp}}] = \begin{cases} C_v U_v \cdot {}^t\eta(w_v)^{-1}U_v^* C_v^* & \text{(Cases SP, SU)}, \\ C_v U_v \xi_B(w_v)^{-1}U_v^* C_v^* & \text{(Case SUB)}, \end{cases}$$

$$-i\cdot\mathrm{diag}[\zeta^{\tau_{v1}}, \ldots, \zeta^{\tau_{vq}}] = \begin{cases} D_v V_v \eta(w_v)^{-1}V_v^* D_v^* & \text{(Cases SP, SU)}, \\ D_v V_v \eta_B(w_v)^{-1}V_v^* D_v^* & \text{(Case SUB)}, \end{cases}$$

Proof. Put $\kappa_v[l(1), w]U_v^{-1} = (c_1, \ldots, c_p)$. By (8.16a), we have for $a \in Y[\rho]$ and $r = 1$,

$$(8.22) \qquad \kappa_v\big[l(a), w\big] = \kappa_v\big[l(1), w\big]\kappa_v(h(a), w) = (c_j a^{\sigma_{vj}})U_v.$$

By linearity, this holds for every $a \in Y$. Hence we obtain the first formula with $C_v = \mathrm{diag}[c_1, \ldots, c_p]$. The second one can be proved in the same manner. We prove the remaining formulas in Case SUB; the other two cases can be treated in a similar way. For $a, b \in Y$, we have, by (8.18) and (8.22),

$$-i \cdot \mathrm{Tr}_{Y/K}(\zeta ab^\rho)_v$$

$$= -i \cdot l(a)_v \iota_v l(b)^*_v$$

$$= (a^{\sigma_{v1}}, \ldots, a^{\sigma_{vp}})C_v U_v \xi_B(w_v)^{-1} U_v^* C_v^* \cdot {}^t(b^{\rho\sigma_{v1}}, \ldots, b^{\rho\sigma_{vp}})$$

$$\quad - (a^{\rho\tau_{v1}}, \ldots, a^{\rho\tau_{vp}})\overline{D}_v \overline{V}_v \cdot {}^t\eta_B(w_v)^{-1} \cdot {}^tV_v \cdot {}^tD_v \cdot {}^t(b^{\tau_{v1}}, \ldots, b^{\tau_{vq}}).$$

On the other hand, $\mathrm{Tr}_{Y/K}(\zeta ab^\rho)_v = \sum_{i=1}^p (\zeta ab^\rho)^{\sigma_{vi}} + \sum_{j=1}^q (\zeta ab^\rho)^{\tau_{vj}}$. Comparing this with the above result, we obtain the desired equalities.

This lemma together with (8.20) implies

$$(8.23a) \qquad \{\alpha \in J_Y \mid \alpha = v \text{ on } K, \mathrm{Im}(\zeta^\alpha) > 0\}$$

$$= \begin{cases} \{\tau_{v1}, \ldots, \tau_{vm}\} & \text{(Cases SP, SU)}, \\ \{\sigma_{v1}, \ldots, \sigma_{vp}\} & \text{(Case SUB)}, \end{cases}$$

$$(8.23b) \qquad \{\alpha \in J_Y \mid \alpha = v\rho \text{ on } K, \mathrm{Im}(\zeta^\alpha) > 0\}$$

$$= \begin{cases} \{\sigma_{v1}, \ldots, \sigma_{vm}\} & \text{(Cases SP, SU)}, \\ \{\tau_{v1}, \ldots, \tau_{vq}\} & \text{(Case SUB)}. \end{cases}$$

9. Certain zeta functions on GL_m. The proof of Theorem 8.1 requires the study of a Dirichlet series

$$(9.1) \qquad \sum_{x \in G/\Delta} \xi(x)\big(\overline{\det(x)}/|\det(x)|\big)^e |N_{K/Q}(\det(x))|^{-s},$$

where $G = GL_m(K)$, ξ is a C-valued locally constant function on K_m^m, and Δ is a sufficiently small congruence subgroup of $GL_m(K)$. We are going to show that (9.1) is a finite linear combination of products of Hecke L-functions. This is well known when $m = 1$, but the author has been unable to find any reference to such a result for an arbitrary ξ in the case $m > 1$. We need also the unitary group $U(m, m)$ in addition to $SU(m, m)$. Therefore we start our discussion in this section by introducing two more types of algebraic groups, which we denote again by G and refer to as Case U and Case GL, respectively as follows:

$$(9.2) \qquad G = \begin{cases} \{\alpha \in GL_{2m}(K) \mid \alpha^* \iota_m \alpha = \iota_m\} & \text{(Case U)}, \\ GL_m(K) & \text{(Case GL)}. \end{cases}$$

Here K is an arbitrary algebraic number field of finite degree in Case GL, and a CM-field in Case U. We let φ (resp. $\mathbf{f}$) denote the set of all archimedian (resp. nonarchimedean) primes of K. We consider parabolic subgroups P of G given by

$$(9.3a) \qquad P = \left\{ \begin{pmatrix} a & b \\ 0 & d \end{pmatrix} \in G \,\middle|\, a, b, d \in K_m^m \right\} \qquad \text{(Case U)},$$

$$(9.3b) \qquad P = \left\{ \begin{bmatrix} x_{11} & \cdots & x_{1r} \\ \cdots & \cdots & \cdots \\ x_{r1} & \cdots & x_{rr} \end{bmatrix} \in G \,\middle|\, \begin{array}{l} x_{ij} \text{ has size } n_i \times n_j \\ x_{ij} = 0 \text{ for } i > j \end{array} \right\} \qquad \text{(Case GL)},$$

where $m = n_1 + \cdots + n_r$ is an arbitrary decomposition of m into r positive integers.

LEMMA 9.1. *Let* Γ *be a subgroup of finite index of* $G \cap GL_n(\mathbf{h})$, *where* $\mathbf{h}$ *is the maximal order of* K, *and* $n = m$ *in Case GL and* $n = 2m$ *in Case U. Then* $P \backslash G / \Gamma$ *is finite.*

This is well known. In fact, if $\Gamma = G \cap GL_n(\mathbf{h})$, we can show that $P \backslash G / \Gamma$ is in one-to-one correspondence with the product of $r - 1$ copies of the ideal class group of K in Case GL, and with the set of all ideal classes of K containing ideals of F in Case U. The method of proof is the same as in [14, Lemma 1.6]. However, such a precise result is unnecessary in the present paper.

In the rest of this section, we confine ourselves to Case GL, and study the series of (9.1). Our result can actually be obtained for a somewhat more general type of series

$$(9.4) \qquad [\Delta_1 : \Delta]^{-1} \sum_{x \in G/\Delta} \xi(x) f(x) |N_{K/\mathbf{Q}}(\det(x))|^{-s} \qquad (\Delta_1 = GL_m(\mathbf{h})),$$

with a $\mathbf{C}$-valued locally constant function ξ on K_m^m and a $\mathbf{C}$-valued function f on G such that

$$(9.5a) \quad f(ax) = \chi(a_{\mathbf{f}}) f(x) \qquad \textit{for every } a \in P^0, \textit{ where } a_{\mathbf{f}} \textit{ is the finite part of } a \\ \textit{in the adelization } P_{\mathbf{A}}^0 \textit{ of } P^0;$$

$$(9.5b) \quad f(x\gamma) = f(x) \qquad \textit{for every } \gamma \textit{ in a congrugence subgroup of } G.$$

Here P^0 is the minimal parabolic subgroup of G defined by (9.3b) with $r = m$, and χ is a character of $P_{\mathbf{A}}^0$ defined by $\chi(a) = \prod_{i=1}^m \chi_i(a_{ii})$ with m Hecke characters χ_i of the idele group $K_{\mathbf{A}}^{\times}$ of K. In (9.4), Δ is an arbitrary congruence subgroup of Δ_1 such that $\xi(x\gamma) = \xi(x)$ and $f(x\gamma) = f(x)$ for every $\gamma \in \Delta$. The factor $[\Delta_1 : \Delta]^{-1}$ makes the value of (9.4) independent of the choice of Δ.

Define the L-function of χ_i as usual by

$$L(s, \chi_i) = \sum_{\mathbf{a}} \chi_i^*(\mathbf{a}) N(\mathbf{a})^{-s},$$

where χ_i^* is the ideal-character attached to χ_i, and $\mathbf{a}$ ranges over all integral ideals prime to the conductor of χ_i. We call χ_i *algebraic-valued* if $\chi_i^*(\mathbf{a})$ is algebraic for every $\mathbf{a}$. Our aim is to prove

PROPOSITION 9.2. *The series of* (9.4) *can be expressed as a finite sum*

$$\sum_k a_k b_k^s \prod_{i=0}^{m-1} L(s - i, \chi_{m-i}\psi_{ki})$$

with $a_k \in \mathbf{C}$, $0 < b_k \in \mathbf{Q}$ *and Hecke characters* ψ_{ki} *of* $K_{\mathbf{A}}^{\times}$ *of finite order. Moreover, if* ξ *and* f *are algebraic-valued, then algebraic numbers can be chosen as the* a_k.

Notice that all the χ_i are algebraic-valued if f is nonvanishing and algebraic-valued.

We prove our proposition by induction on m. We first put

$$(9.6) \qquad \Delta_{\mathbf{c}} = \left\{ \gamma \in GL_m(\mathbf{h}) \,\middle|\, \gamma \equiv 1_m \,(\mathrm{mod}\,\mathbf{c}) \right\}$$

for every integral ideal $\mathbf{c}$ of K. Suppose $m = 1$. Then it is sufficient to treat the case where ξ is the characteristic function of a set X of the form

$$X = \left\{ x \in \mathbf{h} \,\middle|\, 0 \neq x \equiv y \,(\mathrm{mod}\,\mathbf{c}) \right\}$$

with some $\mathbf{c}$ and $y \in \mathbf{h}$. Changing $\mathbf{c}$ for its suitable multiple if necessary, we may assume that $f(x\gamma) = f(x)$ for every $\gamma \in \Delta_{\mathbf{c}}$ and that the finite part of the conductor of χ divides $\mathbf{c}$. Then (9.4) becomes

$$(9.7) \qquad f(1) \sum_{x \in X/\Delta} \chi(x_{\mathbf{f}}) N(x\mathbf{h})^{-s} \qquad (\Delta = \Delta_{\mathbf{c}}).$$

We may assume that $y \neq 0$. Put $\mathbf{a} = \mathbf{c} + y\mathbf{h}$, $\mathbf{b} = \mathbf{a}^{-1}\mathbf{c}$, and

$$W = \left\{ w \in y^{-1}\mathbf{a} \,\middle|\, w \neq 0,\, w - 1 \in \mathbf{b}_v \text{ for every } v \mid \mathbf{b},\, v \in \mathbf{f} \right\}.$$

It can easily be seen that $X = yW$, and hence (9.7) is $f(1)$ times

$$\chi(y_{\mathbf{f}}) N(y\mathbf{h})^{-s} \sum_{w \in W/\Delta} \chi(w_{\mathbf{f}}) N(w\mathbf{h})^{-s}.$$

This is 0 unless $\chi(\delta_{\mathbf{f}}) = 1$ for every $\delta \in \Delta_{\mathbf{b}}$. Therefore, assuming this to be so, let Θ be the set of all Hecke characters θ of $K_{\mathbf{A}}^{\times}$ such that $\theta_{\infty} = \chi_{\infty}$ and the finite part of the conductor θ divides $\mathbf{b}$, where θ_{∞} denotes the archimedean part of θ. Then Θ is nonempty. Observe that $w \mapsto wy\mathbf{a}^{-1}$ gives a bijection of $W/\Delta_{\mathbf{b}}$ onto the set of all integral ideals in the ray-class of $y\mathbf{a}^{-1}$ modulo $\mathbf{b}$. Therefore choosing any $\theta_1 \in \Theta$, we have

$$[\Theta : 1]^{-1} \sum_{\theta \in \Theta} \theta^*(y^{-1}\mathbf{a}) \sum_{x + \mathbf{b} = \mathbf{h}} \theta^*(x) N(x)^{-s}$$

$$= \sum_{w \in W/\Delta_\mathbf{b}} \theta_1^*(wy\mathbf{a}^{-1}) N(wy\mathbf{a}^{-1})^{-s}$$

$$= \theta_1^*(y\mathbf{a}^{-1}) N(y^{-1}\mathbf{a})^s \sum_{w \in W/\Delta_\mathbf{b}} \theta_1(w_t) N(w\mathbf{h})^{-s}$$

$$= \theta_1^*(y\mathbf{a}^{-1}) N(y\mathbf{a}^{-1})^{-s} [\Delta_\mathbf{b} : \Delta]^{-1} \sum_{w \in W/\Delta} \chi(w_t) N(w\mathbf{h})^{-s}.$$

This proves our proposition in the case $m = 1$, since

$$\sum_{x + \mathbf{b} = \mathbf{h}} \theta^*(x) N(x)^{-s} = L(s, \theta) \prod_{\mathbf{p} | \mathbf{b}} \left[1 - \theta^*(\mathbf{p}) N(\mathbf{p})^{-s} \right].$$

Assuming now $m > 1$, we consider the parabolic subgroup P defined by (9.3b) with $r = 2$, $n_1 = m - 1$, and $n_2 = 1$. Let

$$\Delta = \{ \gamma \in GL_m(\mathbf{h}) \,|\, f(x\gamma) = f(x), \xi(x\gamma) = \xi(x) \}.$$

By Lemma 9.1, we have $G = \coprod_{\alpha \in A} P\alpha\Delta$ with a finite subset A of G. Put $\Delta_\alpha = \alpha\Delta\alpha^{-1}$ for each $\alpha \in A$; let T_α be a complete set of representatives of $P/(P \cap \Delta_\alpha)$. Then G/Δ is represented by $\coprod_{\alpha \in A} T_\alpha\alpha$. Therefore (9.4) becomes

$$[\Delta_1 : \Delta]^{-1} \sum_{\alpha \in A} \sum_{x \in T_\alpha} \xi(x\alpha) f(x\alpha) N(\det(x\alpha)\mathbf{h})^{-s}.$$

Observe that $\xi(\left(\begin{smallmatrix} a & b \\ 0 & d \end{smallmatrix}\right)\alpha)$ is a locally constant function of $(a, b, d) \in K_{m-1}^{m-1} \times K^{m-1} \times K$. Thus our problem is reduced to the study of

$$(9.8) \qquad \sum_{x \in T_\alpha} \eta(x) f(x\alpha) N(\det(x)\mathbf{h})^{-s}$$

with a locally constant function η of (a, b, d); T_α may be replaced by $P/(P \cap \Delta')$ with any congruence subgroup Δ' of Δ_α. Without losing generality and choosing a suitable integral ideal $\mathbf{c}$, we may assume that $\Delta_\mathbf{c} \subset \Delta_\alpha$, η is the characteristic function of a set

$$X = \left\{ \begin{pmatrix} a & b \\ 0 & d \end{pmatrix} \in \mathbf{h}_m^m \,\bigg|\, a \equiv a_0, b \equiv b_0, c \equiv c_0 \,(\mathrm{mod}\,\mathbf{c}) \right\},$$

and the finite part of the conductor of χ_i divides $\mathbf{c}$. Define a function g on $GL_{m-1}(K)$ by

$$g(a) = \chi_m(d_t)^{-1} f\left(\begin{pmatrix} a & b \\ 0 & d \end{pmatrix}\alpha \right).$$

It can easily be seen that this is well-defined, independent of b and d, and satisfies (9.5a, b). In particular, $g(a\delta) = g(a)$ for every δ in

$$\Delta' = \{\delta \in GL_{m-1}(\mathbf{h}) \mid \delta \equiv 1_{m-1} \ (\mathrm{mod}\,\mathbf{c})\}.$$

Now $X/(P \cap \Delta_c)$ is given by

$$\left\{\begin{pmatrix} a & b \\ 0 & d \end{pmatrix} \middle| a \in E, b \in (b_0 + \mathbf{c}^{m-1})/a\mathbf{c}^{m-1}, d \in D\right\},$$

where $E = \{a \in GL_{m-1}(K) \cap \mathbf{h}_{m-1}^{m-1} \mid a \equiv a_0 \ (\mathrm{mod}\,\mathbf{c})\}/\Delta'$, and

$$D = \{d \in \mathbf{h} \mid 0 \neq d \equiv d_0 \ (\mathrm{mod}\,\mathbf{c})\}/\{w \in \mathbf{h}^\times \mid w - 1 \in \mathbf{c}\}.$$

Therefore (9.8) is equal to

$$\sum_{a \in E} \sum_{d \in D} \left[\mathbf{c}^{m-1} : a\mathbf{c}^{m-1}\right] g(a)\chi_m(d_\mathfrak{t})N(\det(a)\,d\mathbf{h})^{-s}$$

$$= \sum_{d \in D} \chi_m(d_\mathfrak{t})N(d\mathbf{h})^{-s} \sum_{a \in E} g(a)N(\det(a)\mathbf{h})^{1-s}.$$

The sum $\sum_{d \in D}$ is exactly a sum of type (9.7), and $\sum_{a \in E}$ is a sum in the $(m-1)$-dimensional case. Hence our induction completes the proof.

10. Proof of Theorem 8.1. We now consider our groups in Cases U and SU. To distinguish one from the other, we denote the group in Case U by $\tilde{G}$ and that in Case SU by G $(\subset \tilde{G})$. We let $\tilde{P}$ denote the parabolic subgroup of $\tilde{G}$ defined by (9.3a), and put

$$(10.1) \qquad R = \left\{\begin{bmatrix} 1 & b \\ 0 & 1 \end{bmatrix} \in \tilde{P} \middle| b \in K_m^m\right\},$$

$$(10.2) \qquad W = \{x \in K_{2m}^m \mid x\iota_m x^* = 0, \ \mathrm{rank}(x) = m\}.$$

For each $\alpha \in G$, let $x(\alpha)$ denote its last m rows. Then x gives a bijection of $R \backslash \tilde{G}$ onto W (see [8, Proposition 5.1]). For $x \in W$, $z \in H^\varphi$ and $r \in \mathbf{Z}^\varphi$, we put

$$(10.3) \qquad j[x,z]^r = \prod_{v \in \varphi} j_v[x,z]^{r_v}, \qquad j_v[x,z] = \det(\mu_v[x,z]).$$

Notice that $j[x(\alpha),z]^r = j_\alpha(z)^r$ for $\alpha \in G$, and $j_v[x,z] \neq 0$ for every x and z. We now consider a series

$$(10.4) \qquad T(z,s;\lambda) = \delta(z)^{s-r} \sum_{x \in \Delta \backslash W} \lambda(x)\overline{j[x,z]}^r j[x,z]^{-k-r}$$

$$\cdot |j[x,z]|^{-2s} S\big(\mu[x,z]^{-1}\overline{\kappa[x,z]}\,\eta(z)^{-1}\big),$$

where $s \in \mathbf{C}$, $k \in \mathbf{Z}$, $r \in \mathbf{Z}^{\varphi}$, λ is a locally constant **C**-valued function on K_{2m}^{m}, S is the same as in (8.1a), and Δ is a congruence subgroup of $GL_m(\mathbf{h})$ such that $\lambda(\epsilon x) = \lambda(x)$ and $\overline{\det(\epsilon)}^{r}\det(\epsilon)^{-k-r} = 1$ for all $\epsilon \in \Delta$. We are going to show that this can be reduced essentially to series of type (7.6) multiplied by series of type (9.1).

To prove this, put

$$(10.5) \qquad \Gamma = \{\gamma \in G \cap SL_{2m}(\mathbf{h}) \mid \lambda(x\gamma) = \lambda(x)\}.$$

Since $\tilde{G} = \tilde{P}G$, we have $\tilde{G} = \coprod_{\beta \in B} \tilde{P}\beta\Gamma$ with a finite subset B of G, by Lemma 9.1 or [14, Lemma 1.6]. For each β, Let $\Gamma_\beta = \beta\Gamma\beta^{-1}$, and let S_β be a complete set of representatives of $(\tilde{P} \cap \Gamma_\beta)\backslash\Gamma_\beta$. Then $\tilde{G} = \coprod_{\beta \in B} \tilde{P}S_\beta\beta$, and hence $W = x(\tilde{G}) = \bigcup_\beta x(\tilde{P}S_\beta\beta) = \bigcup_\beta GL_m(K)x(S_\beta)\beta$. Moreover, we see easily that the map $(d, \alpha, \beta) \mapsto dx(\alpha)\beta$ gives a bijection of $\coprod_{\beta \in B}(GL_m(K) \times S_\beta \times \{\beta\})$ onto W. Let μ_β be the function on K_m^m defined by $\mu_\beta(d) = \lambda(dx(\beta))$. We see easily that μ_β is locally constant, $\mu_\beta(\delta d) = \mu_\beta(d)$ for every $\delta \in \Delta$, and $\lambda(dx(\alpha)\beta) = \mu_\beta(d)$ for every $\alpha \in \Gamma_\beta$ by (10.5). Let D be a complete set of representatives of $\Delta\backslash GL_m(K)$. Then the elements $dx(\alpha)\beta$ for $d \in D$, $\alpha \in S_\beta$, and $\beta \in B$ represent $\Delta\backslash W$ without repetition. Therefore we obtain

$$(10.6) \qquad T(z, s; \lambda) = \sum_{\beta \in B} \sum_{d \in D} \mu_\beta(d)\overline{\det(d)}^{r}\det(d)^{-k-r}$$

$$\cdot |\det(d)|^{-2s}\mathbf{E}(z, s; k, \Lambda, \Omega)\|\beta,$$

where $\Lambda(x) = \det(x)^{r}S(x)$, which is the desired formula for T. Before proving Theorem 8.1, we state two lemmas.

LEMMA 10.1. *Let χ be a Hecke character of $K_{\mathbf{A}}^{\times}$ such that $\chi(a_v) = (\bar{a}_v/|a_v|)^{e_v}$ with $0 < e_v \in \mathbf{Z}$ for $a_v \in K_v^{\times}$, $v \in \varphi$. Let q be an integer such that $-e_v < q \leqq e_v$ and $q \equiv e_v \pmod{2}$ for all $v \in \varphi$. Then $L(q/2, \chi)$ is an algebraic number times $\pi^c p_K(\sum_{v \in \varphi} e_v v, \varphi)$, where $c = ([F : \mathbf{Q}]q + \sum_{v \in \varphi} e_v)/2$.*

This was proved in [5, Theorem 2] and [7, Theorem 5.4].

LEMMA 10.2. *Let $M(s) = \sum_{d \in \Delta\backslash U} \mu(d)\overline{\det(d)}^{r}\det(d)^{-k-r}|\det(d)|^{-2s}$, where $k \in \mathbf{Z}$, $0 \leqq r \in \mathbf{Z}^{\varphi}$, $U = GL_m(K)$, μ is a $\overline{\mathbf{Q}}$-valued locally constant function on K_m^m, and*

$$\Delta = \left\{\delta \in GL_m(\mathbf{h}) \mid \mu(\delta d) = \mu(d),\ \overline{\det(\delta)}^{r}\det(\delta)^{-k-r} = 1\right\}.$$

Suppose $k \geqq m - r_v$ for all $v \in \varphi$. Then $M(0)$ is an algebraic number times $\pi^b p_K(\sum_{v \in \varphi}(k + 2r_v)v, m\varphi)$, where

$$b = [F : \mathbf{Q}]\{mk - m(m - 1)/2\} + m\sum_{v \in \varphi} r_v.$$

Proof. Observe that $M(s - k/2)$ is a series of type (9.4) with Hecke characters χ_i, each of which is of the same type as χ of Lemma 10.1 with $e_v = k + 2r_v$. Therefore we obtain our assertion immediately from Proposition 9.2 and Lemma 10.1.

We are now ready to prove Theorem 8.1. As already explained, Z_0 is our only problem. Let Y, τ_{vi}, σ_{vi}, ξ, and λ be as in (8.7a). Then the sum is extended over $a \in \Delta \setminus Y^m$ under (8.12) with ξ and a in place of ξ' and x. Take $\theta \in K$ so that $\theta^\rho = -\theta$ and $\mathrm{Im}(\theta_v) > 0$ for all $v \in \varphi$, and put $\zeta = \theta\xi$. Then

$$(10.7) \qquad (x, y) \mapsto \mathrm{Tr}_{Y/K}(\zeta xy^\rho)$$

defines a K-valued anti-hermitian form on Y. By (8.4), this has signature (m, m) at every archimedean prime. Therefore (8.9) implies that (10.7) is equivalent to ι_m; namely, there exists a K-linear bijection l of Y onto K^1_{2m} such that $l(x)\iota_m l(y)^* = \mathrm{Tr}_{Y/K}(\zeta xy^\rho)$. Define $h: Y \to K^{2m}_2$ by $l(xa) = l(x)h(a)$. Then $\iota h(a^\rho)$ $= h(a)^*\iota$, and hence $h(Y[\rho])$ has a fixed point w on H^φ. We observe that the σ_{vi} and τ_{vi} in the definition of Z_0 are exactly those determined by (8.23a, b) at w, and that the map l_m of (8.21) gives a bijection of the set of elements of Y^m of type (8.12) onto W of (10.1).

Given e and k as in Theorem 8.1, define $r \in \mathbf{Z}^\varphi$ by $r_v = (e_v - k)/2$, and consider $T(z, s; \lambda')$ at $z = w$ with $\lambda'(l_m(a)) = \lambda(a)$. By Lemma 8.3, we have $\kappa_v[l_m(a), w] = a\{\sigma\}C_v U_v$, $\mu_v[l_m(a), w] = a\{\tau\} \cdot D_v V_v$ and $\eta(w)^{-1} = \overline{U}_v^{-1}\overline{C}_v^{-1}E_v \cdot {}^t C_v^{-1} \cdot {}^t U_v^{-1}$ with C_v, D_v, U_v, V_v as given there and a diagonal element E_v of $GL_m(\overline{\mathbf{Q}})$. Therefore

$$(10.8) \qquad Z_0(s + k/2) = c|c'|^s S(DV)T(w, s; \lambda')S({}^t U \cdot {}^t CE^{-1})$$

with some constants c and c' of $\overline{\mathbf{Q}}$. This combined with (10.6) implies the meromorphic continuation of Z_0.

To obtain our result on $Z_0(k/2)$ in Theorem 8.1, we first observe that the assertions of Thoerem 7.1 hold not only for $\mathbf{E}$ itself, but also for $\mathbf{E}\|\beta$ with every $\beta \in G$. Therefore, putting $s = 0$ and combining Theorem 7.1 with Lemma 10.2, we see that

$$Z_0(k/2) = \pi^b p_K\left(\sum_{v \in \varphi} e_v v, m\varphi\right) g(w)$$

$$\cdot S(DV)P_2(w)A \cdot {}^t P_1(w)S({}^t U \cdot {}^t CE^{-1})$$

with $A \in \overline{\mathbf{Q}}^d_d$, g and P_i $(= P_{\sigma_i})$ as in Theorem 7.1, and with b as in Theorem 8.1. In the present situation, we have

$$P_1(w) = {}^t\Lambda({}^t U g_\sigma \cdot {}^t U^{-1}) = \det(g_\sigma)^r \cdot {}^t S({}^t U g_\sigma \cdot {}^t U^{-1}),$$

$$P_2(w) = \Lambda(V^{-1}g_\tau V) = \det(g_\tau)^r S(V^{-1}g_\tau V).$$

Since g_σ, g_τ, C, D, and E are all diagonal, we have

$$(10.9) \qquad Z_0(k/2) = \pi^b \not{p} S(g_\tau) A' S(g_\sigma)$$

with $\not{p} = p_K(\sum_{v \in \varphi} e_v v, m\varphi)\det(g_\sigma g_\tau)' g(w)$ and $A' \in \overline{\mathbf{Q}}_d^d$. By Proposition 3.1 or (3.21), we have

$$(10.10) \qquad \det(g_\sigma g_\tau)' g(w) = p_Y\left(\sum_{v \in \varphi} \sum_{i=1}^m e_v \tau_{vi}, \sum_{v \in \varphi} \sum_{i=1}^m (\sigma_{vi} + \tau_{vi}) \right).$$

By (8.20), we have

$$\sum_{v,i} (\sigma_{vi} + \tau_{vi}) = \sum_{v,i} (\tau_{vi} - \tau_{vi}\rho) + \mathrm{Inf}_{Y/K}(\varphi)\rho,$$

and hence (10.10) is equal to

$$p_Y\left(\sum_{v,i} e_v \tau_{vi}, \sum_{v,i} (\tau_{vi} - \tau_{vi}\rho) \right) p_K\left(m \sum_v e_v v, \varphi\rho \right)$$

by (3.20a). Therefore $\not{p}$ is exactly the quantity as given in Theorem 8.1, by virtue of [10, Theorem 1.1, (2)]. This completes the proof.

11. The adjoint operators of Δ_ρ^σ and E_σ. We make the same notational conventions as in Sections 1 and 2. In particular, we let X denote the spaces H or B. Case SUB is included in our treatment, but we assume $l = m$ in Case SUB whenever $\Delta_\rho^{(p)}$ or ρ_r with $r > 1$ is dealt with. Let y_{ij} and $(y^{-1})_{ij}$ denote the matrix entries of matrices y and y^{-1}. We observe that X in all three cases is a Kähler manifold with

$$(11.1a) \quad ds^2 = \mathrm{tr}(\bar{\xi}^{-1} dz\, \eta^{-1} dz^*) = \sum_{j,p=1}^l \sum_{k,q=1}^m (\xi^{-1})_{jp}(\eta^{-1})_{kq}\, dz_{jk}\, d\bar{z}_{pq},$$

$$(11.1b) \qquad \Omega = (i/2) \sum_{j,p=1}^l \sum_{k,q=1}^m (\xi^{-1})_{jp}(\eta^{-1})_{kq}\, dz_{jk} \wedge d\bar{z}_{pq}$$

as its invariant metric and fundamental form. (By our conventions, $l = m$ and $\xi = {}^t\eta$ in Cases SP and SU, $\xi = \xi_B$ and $\eta = \eta_B$ in Case SUB.) This fact implicitly influences our discussion, though our proofs do not require this knowledge. We need however the invariant volume element ω on X defined by

$$(11.2) \qquad \omega = \delta(z)^{-2\kappa}(i/2)^N \bigwedge_{p=1}^N (dz_p \wedge d\bar{z}_p),$$

where the z_p are rearrangements of z_{ij} for the variable matrix z on X, κ is defined by (1.15), and N is the complex dimension of X. We then define a differential form ζ_{pq} of codegree 1 to be a constant times the exterior product of dz_{ij} and $d\bar{z}_{ij}$ for all (i, j) excluding dz_{pq}, the constant being determined by

$$(11.3a) \qquad dz_{pq} \wedge \zeta_{pq} = \delta(z)^{2\kappa}\omega \qquad (1 \leqslant p \leqslant l, 1 \leqslant q \leqslant m).$$

Then for the matrix form (ζ_{pq}), we have

$$(11.3b) \qquad (\zeta_{pq}) \circ \alpha = |j_\alpha|^{-4\kappa}\kappa_\alpha(z)(\zeta_{pq}) \cdot {}^t\mu_\alpha(z) \qquad (\alpha \in \mathfrak{G} \text{ or } \mathfrak{G}').$$

For $h \in \mathrm{Hom}(T, V)$ with a complex vector space V, we define its components $h_{jk} \in V$ by

$$(11.4) \qquad h(u) = \sum_{j=1}^{l} \sum_{k=1}^{m} u_{jk}h_{jk} \qquad (u = (u_{jk}) \in T).$$

LEMMA 11.1. *For a C^∞-function φ on X with values in $\mathrm{Hom}(T, \mathbf{C})$, define differential forms $\varphi\, dz$ and φ_* by*

$$\varphi\, dz = \sum_{i=1}^{l} \sum_{j=1}^{m} \varphi_{ij}\, dz_{ij}, \qquad \varphi_* = \delta^{-2\kappa} \sum_{i=1}^{l} \sum_{j=1}^{m} \varphi_{ij}\zeta_{ij}.$$

Then $(\varphi\|_\tau\alpha)\, dz = (\varphi\, dz) \circ \alpha$, $(\varphi\|_\pi\alpha)_* = (\varphi_*) \circ \alpha$ for $\alpha \in \mathfrak{G}$ or $\alpha \in \mathfrak{G}'$ with τ and π of (1.14a, b), and $d(\varphi_*) = \delta^{2\kappa}\sum_{i=1}^{l}\sum_{j=1}^{m} \partial_{ij}(\delta^{-2\kappa}\varphi_{ij})\omega$.

These equalities follow easily from (1.7a), (11.2), and (11.3b).

LEMMA 11.2. $\sum_{i=1}^{l}\sum_{j=1}^{m} \partial_{ij}(\delta^{-2\kappa}\xi_{pi}\eta_{qj}) = 0$ *for every p and q.*

Proof. For a variable square matrix x, we have $(\partial/\partial x_{hk})\det(x)^s = s \cdot \det(x)^s(x^{-1})_{kh}$ by (5.2). Therefore, in Case SUB, we have

$$\partial_{ij}(\delta^s) = -s\delta^s \sum_{h=1}^{l} \bar{z}_{hj}(\xi^{-1})_{ih} = -s\delta^s \sum_{k=1}^{m} \bar{z}_{ik}(\eta^{-1})_{jk},$$

and hence

$$\delta^{l+m} \sum_{i,j} (\partial_{ij}\delta^{-l-m})\xi_{pi}\eta_{qj}$$

$$= l \sum_{i,j,h} \bar{z}_{hj}(\xi^{-1})_{ih}\xi_{pi}\eta_{qj} + m \sum_{i,j,k} \bar{z}_{ik}(\eta^{-1})_{jk}\xi_{pi}\eta_{qj}$$

$$= l \sum_{j} \bar{z}_{pj}\eta_{qj} + m \sum_{i} \bar{z}_{iq}\xi_{pi} = -\sum_{i,j} \partial_{ij}(\xi_{pi}\eta_{qj}),$$

which proves our equality in Case SUB. The other two cases can be proved in a similar way. Alternatively, our formula can be derived from the fact that the power Ω^{N-1} of Ω of (11.1b) is closed.

Let us now consider the system $\{\sigma, \sigma', \sigma''\}$ of (1.25a, b). Changing σ, σ', and σ'' by suitable linear transformations, we may assume, and hereafter do so, that

$$(11.5) \quad \sigma'('x) = {}'\sigma'(x), \qquad \sigma''('y) = {}'\sigma''(y); \qquad \textit{both } \sigma' \textit{ and } \sigma'' \textit{ are } \mathbf{R}\textit{-rational.}$$

We fix a rational representation $\{\rho, V\}$ of $\mathscr{g}$ as in (1.9). For $h \in \operatorname{Hom}(\mathbf{C}_n^{n'}, V)$ with $\mathbf{C}_n^{n'}$ of (1.25a), we define its components $h_{ab} \in V$ by $h(w) = \sum_{a=1}^{n'} \sum_{b=1}^{n} w_{ab} h_{ab}$. (This is consistent with (11.4) if σ is the identity map, and with (2.7) if $\sigma = \rho_r$.) We then define a linear map

$$(11.6) \qquad\qquad \theta_\sigma : \operatorname{Hom}\!\big(\mathbf{C}_n^{n'}, \operatorname{Hom}(\mathbf{C}_n^{n'}, V)\big) \to V$$

by $\theta_\sigma p = \sum_{a=1}^{n'} \sum_{b=1}^{n} (p_{ab})_{ab}$. Let us write θ_1 for θ_σ when σ is the identity map of $\mathbf{C}_m^l$ onto itself. Then

$$(11.7) \qquad\qquad \theta_1 p = \sum_{i=1}^{l} \sum_{j=1}^{m} p(\partial_{ij}, \partial_{ij}) \qquad \text{if} \quad p \in \operatorname{Ml}_2(T, V).$$

We can easily verify that

$$(11.8) \qquad \theta_\sigma\big[((\rho \otimes_1 \sigma) \otimes_2 \sigma)(x, y)p\big] = \theta_\sigma\big[((\rho \otimes_2 \sigma) \otimes_1 \sigma)(x, y)p\big]$$

$$= \rho(x, y)\theta_\sigma p \qquad ((x, y) \in \mathscr{g}),$$

where $\otimes_i$ is defined by (1.27a, b). Also, for $\alpha \in \mathfrak{G}$, $\mathfrak{G}'$, or $\mathfrak{E}$, we have

$$(11.9a) \qquad \theta_\sigma \Delta_{\rho \otimes_2 \sigma}^\sigma (f \|_{\rho \otimes_2 \sigma} \alpha) = (\theta_\sigma \Delta_{\rho \otimes_2 \sigma}^\sigma f)\|_\rho \alpha,$$

$$(11.9b) \qquad (\theta_\sigma E_\sigma)(f \|_{\rho \otimes_1 \sigma} \alpha) = (\theta_\sigma E_\sigma f)\|_\rho \alpha,$$

$$(11.9c) \qquad (\theta_\sigma g)\|_\rho \alpha = \theta_\sigma(g\|_{\rho \otimes_1 \sigma \otimes_2 \sigma} \alpha) = \theta_\sigma(g\|_{\rho \otimes_2 \sigma \otimes_1 \sigma} \alpha).$$

These follow immediately from (1.32), (2.5b), and (11.8).

LEMMA 11.3. *If g is a $\operatorname{Hom}(T, V)$-valued function on X with $\{\rho, V\}$ as in (1.9), we have*

$$\theta_1 D_{\rho \otimes_\pi} g = \delta^{2\kappa} \rho(\xi, \eta)^{-1} \sum_{i=1}^{l} \sum_{j=1}^{m} \partial_{ij}\big[\delta^{-2\kappa} \rho(\xi, \eta) g_{ij}\big].$$

By our conventions, $\rho(\xi, \eta)$ means $\rho(\eta)$ in Case SP.

Proof. Put $h_{ij} = \sum_{r,s} (\xi^{-1})_{ri} (\eta^{-1})_{sj} g_{rs}$. By Lemma 11.2, we have

$$\sum_{i,j} \partial_{ij}\big[\rho(\xi, \eta)\delta^{-2\kappa} g_{ij}\big] = \sum_{i,j} \partial_{ij}\bigg[\rho(\xi, \eta)\delta^{-2\kappa} \sum_{p,q} \xi_{pi} \eta_{qj} h_{pq}\bigg]$$

$$= \sum_{i,j,p,q} \delta^{-2\kappa} \xi_{pi} \eta_{qj} \partial_{ij}\big[\rho(\xi, \eta) h_{pq}\big].$$

On the other hand, for $u, v \in T$, we have

$$\rho(\xi,\eta)(D_{\rho\otimes\pi}g)(u, \xi^{-1}v \cdot {}'\eta^{-1}) = \left[(\rho\otimes\pi)(\xi,\eta)(D_{\rho\otimes\pi}g)\right](u,v)$$

$$= \sum_{i,j} u_{ij}\partial_{ij}\left[\rho(\xi,\eta)\sum_{r,s}(\xi^{-1}v\cdot{}'\eta^{-1})_{rs}\, g_{rs}\right]$$

$$= \sum_{i,j,p,q} u_{ij}v_{pq}\partial_{ij}\left[\rho(\xi,\eta)h_{pq}\right],$$

and hence $\rho(\xi,\eta)\theta_1 D_{\rho\otimes\pi}g = \sum_{i,j,p,q}\xi_{pi}\eta_{qj}\partial_{ij}[\rho(\xi,\eta)h_{pq}]$, from which we obtain our equality.

Given a rational representation $\{\rho, V\}$ of $\mathscr{g}$, we can always find an inner product $\langle x, y\rangle_V$ on V (C-antilinear in x) such that

$$(11.10) \qquad \langle\rho(w)x, y\rangle_V = \langle x, \rho(w^*)y\rangle_V \qquad (w \in \mathscr{g}; \ x, y \in V),$$

where $(a, b)^* = (a^*, b^*)$ in Cases SU and SUB. This can easily be seen by decomposing ρ into irreducible constituents. We hereafter fix such $\langle\ ,\ \rangle_V$ and define $\langle\ ,\ \rangle_W$ on $W = \mathrm{Hom}(\mathbf{C}_n^{n'}, V)$ by

$$(11.11) \qquad \langle h, h'\rangle_W = \sum_{a=1}^{n'}\sum_{b=1}^{n} \langle h_{ab}, h'_{ab}\rangle_V \qquad (h, h' \in W).$$

This satisfies (11.10) with respect to both $\rho\otimes_1\sigma$ and $\rho\otimes_2\sigma$.

Let us now fix a discrete subgroup Γ of $\mathfrak{G}$ or $\mathfrak{G}'$ such that $\Gamma\backslash X$ is compact. We then denote by $C_\rho(\Gamma)$ the set of all V-valued C^∞-functions f on X such that $f\|_\rho\gamma = f$ for all $\gamma \in \Gamma$. For $f, g \in C_\rho(\Gamma)$, we define their inner product by

$$(11.12) \qquad \langle f, g\rangle_\Gamma = \int_{\Gamma\backslash X}\langle f(z), \rho(\xi(z),\eta(z))\, g(z)\rangle_V\omega.$$

We omit the subscripts V and Γ if there is no fear of confusion.

THEOREM 11.4. *Let $\{\rho, V\}$ be a representation of $\mathscr{g}$ with inner product as in (11.10), and Γ be as above. Then we have*

$$(11.13a) \quad \langle f, \theta_1 D_{\rho\otimes\pi}g\rangle_\Gamma = -\langle E_1 f, g\rangle_\Gamma \qquad for \ \ f \in C_\rho(\Gamma), \qquad g \in C_{\rho\otimes\pi}(\Gamma),$$

$$(11.13b) \qquad \langle f, D_\rho g\rangle_\Gamma = -\langle\theta_1 E_1 f, g\rangle_\Gamma \qquad for \ \ f \in C_{\rho\otimes\tau}(\Gamma), \quad g \in C_\rho(\Gamma).$$

where $\rho\otimes\tau$ and $\rho\otimes\pi$ are defined by (1.14a, b).

Proof. Put $W = \mathrm{Hom}(T, V)$. For $f \in C_\rho(\Gamma)$ and $g \in C_{\rho\otimes\pi}(\Gamma)$, define a function φ with values in $\mathrm{Hom}(T, \mathbf{C})$ by $\varphi(u) = \langle f, \rho(\xi,\eta)g(u)\rangle_V$ for $u \in T$ and define φ_* as in Lemma 11.1. We see easily that $\varphi \in C_\pi(\Gamma)$, and hence, by Lemma 11.1, φ_* is Γ-invariant and

$$d\varphi_* = \delta^{2\kappa}\sum_{i,j}\partial_{ij}\left(\delta^{-2\kappa}\langle f, \rho(\xi,\eta)g_{ij}\rangle\right)\omega.$$

This can be written $(A + B)\omega$ with

$$A = \sum_{i,j} \langle \bar{\partial}_{ij} f, \rho(\xi, \eta) g_{ij} \rangle,$$

$$B = \langle f, \delta^{2\kappa} \sum_{i,j} \partial_{ij} [\rho(\xi, \eta) \delta^{-2\kappa} g_{ij}] \rangle.$$

Now we have

$$\langle E_1 f, (\rho \otimes \pi)(\xi, \eta) g \rangle = \sum_{i,j} \langle (E_1 f)_{ij}, \rho(\xi, \eta) [\pi(\xi, \eta) g]_{ij} \rangle$$

$$= \sum_{i,j} \left\langle \sum_{p,q} \xi_{pi} \eta_{qj} \bar{\partial}_{pq} f, \rho(\xi, \eta) \sum_{r,s} (\xi^{-1})_{ri} (\eta^{-1})_{sj} g_{rs} \right\rangle = A.$$

On the other hand, Lemma 11.3 shows that $B = \langle f, \rho(\xi, \eta) \theta_1 D_{\rho \otimes \pi} g \rangle$. Since $\Gamma \backslash X$ is compact, $\int_{\Gamma \backslash X} d\psi = 0$ for every Γ-invariant C^∞-form ψ on X of codegree 1. Applying this to $d\varphi_*$, we obtain (11.13a). The proof of (11.13b) is similar. In fact, for $f \in C_{\rho \otimes \tau}(\Gamma)$ and $g \in C_\rho(\Gamma)$, define a function φ with values in $\mathrm{Hom}(T, \mathbf{C})$ by $\varphi(u) = \langle f({}^t\xi \bar{u} \eta), \rho(\xi, \eta) g \rangle_V$ for $u \in T$. Again $\varphi \in C_\pi(\Gamma)$ and φ_* is Γ-invariant. We have

$$\varphi_{ij} = \sum_{p,q} \xi_{pi} \eta_{qj} \langle f_{pq}, \rho(\xi, \eta) g \rangle,$$

and hence, by Lemmas 11.1 and 11.2, $d\varphi_* = (A + B)\omega$ with

$$A = \left\langle \sum_{i,j,p,q} \xi_{ip} \eta_{jq} \bar{\partial}_{ij} f_{pq}, \rho(\xi, \eta) g \right\rangle,$$

$$B = \sum_{i,j,p,q} \xi_{pi} \eta_{qj} \langle f_{pq}, \partial_{ij} [\rho(\xi, \eta) g] \rangle.$$

We see easily that $\theta_1 E_1 f = \sum_{i,j,p,q} \xi_{ip} \eta_{jq} \bar{\partial}_{ij} f_{pq}$ and $B = \langle f, (\rho \otimes \tau)(\xi, \eta) D_\rho g \rangle$, from which (11.13b) follows.

Let us now determine the adjoint operators of Δ_ρ^σ and E_σ, still in all three cases, generalizing the above result:

THEOREM 11.5. *Let Γ and $\{\rho, V\}$ be as in Theorem 11.4, and let σ be a homogeneous polynomial map of $\mathbf{C}_m^l$ into $\mathbf{C}_n^{n'}$ of degree r as in (1.25a, b). Then*

$$(11.14a) \quad \langle f, \theta_\sigma \Delta_{\rho \otimes_2 \sigma}^\sigma g \rangle_\Gamma = (-1)^r \langle E_\sigma f, g \rangle_\Gamma \quad \textit{for } f \in C_\rho(\Gamma) \textit{ and } g \in C_{\rho \otimes_2 \sigma}(\Gamma),$$

$$(11.14b) \quad \langle f, \Delta_\rho^\sigma g \rangle_\Gamma = (-1)^r \langle \theta_\sigma E_\sigma f, g \rangle_\Gamma \quad \textit{for } f \in C_{\rho \otimes_1 \sigma}(\Gamma) \textit{ and } g \in C_\rho(\Gamma).$$

The proof requires a few preliminary algebraic observations. Since σ is a homogeneous map, there is a unique symmetric multilinear $\mathbf{R}$-rational map S: $(\mathbf{C}_m^l)^r \to \mathbf{C}_n^{n'}$ such that $S(u, \dots, u) = \sigma(u)$ for $u \in \mathbf{C}_m^l$. Then $S(xu_1 y, \dots, xu_r y)$

$= \sigma'(x)S(u_1, \ldots, u_r)\sigma''(y)$ for $(x, y) \in \mathscr{g}$. Also, we can take the components S_{ab} of S as the σ_{ab} in the definition (1.26b) of Φ_σ.

Put $M = \mathrm{Ml}_r(T, V)$ and define an inner product on M by

$$\langle x, y \rangle_M = \sum_{i,j,\ldots,s,t} \langle x(\partial_{ij}, \ldots, \partial_{st}), y(\partial_{ij}, \ldots, \partial_{st}) \rangle_V ,$$

where the indices run over $\{1, \ldots, l\}^r \times \{1, \ldots, m\}^r$. This satisfies (11.10) with respect to $\rho \otimes \tau'$ and $\rho \otimes \pi'$.

LEMMA 11.6. *Let* $W = \mathrm{Hom}(\mathbf{C}_n^{n'}, V)$ *and* $M = \mathrm{Ml}_r(T, V)$. *Define a linear map* $\chi : W \to M$ *by* $\chi h = h \circ S$ *for* $h \in W$. *Then*

(11.15)
$$\chi \circ (\rho \otimes {}_1\sigma)(x, y) = (\rho \otimes \tau')(x, y) \circ \chi,$$

$$\chi \circ (\rho \otimes {}_2\sigma)(x, y) = (\rho \otimes \pi')(x, y) \circ \chi \qquad ((x, y) \in \mathscr{g}),$$

(11.16)
$$\langle \chi h, y \rangle_M = \langle h, \Phi_\sigma y \rangle_W \qquad (h \in W, y \in M)$$

with Φ_σ *of* (1.26a).

Proof. The first two equalities are straightforward. As for (11.16), we have

$$\langle \chi h, y \rangle_M = \sum_{i,j,\ldots,s,t} \langle h(S(\partial_{ij}, \ldots, \partial_{st})), y(\partial_{ij}, \ldots, \partial_{st}) \rangle_V$$

$$= \sum_{i,j,\ldots,s,t} \left\langle \sum_{a,b} S_{ab}(\partial_{ij}, \ldots, \partial_{st})h_{ab}, y(\partial_{ij}, \ldots, \partial_{st}) \right\rangle_V$$

$$= \sum_{a,b} \langle h_{ab}, \{S_{ab}, y\} \rangle_V = \langle h, \Phi_\sigma y \rangle_W$$

by (1.21) and (1.26b).

Proof of Theorem 11.5. For simplicity, put $\theta = \theta_1$ and $E = E_1$. Let h be a function on X with values in $\mathrm{Ml}_r(T, V)$ which is symmetric in the sense of (1.23). Then $\theta E h$ can be defined naturally as a function with values in $\mathrm{Ml}_{r-1}(T, V)$, whose explicit form is

$$\theta E h(u_1, \ldots, u_{r-1}) = \sum_{i,j,p,q} \xi_{pi}\eta_{qj}\bar{\partial}_{pq}h(u_1, \ldots, u_{r-1}, \partial_{ij}).$$

Applying this successively, we obtain a function $(\theta E)^r h$ with values in V, given by

$$(\theta E)^r h = \sum_{i_1 j_1, \ldots, i_r j_r} \prod_{\alpha = 1}^{r} \left(\sum_{p_\alpha, q_\alpha} \xi_{p_\alpha i_\alpha}\eta_{q_\alpha j_\alpha}\bar{\partial}_{p_\alpha q_\alpha} \right) h(\partial_{i_1 j_1}, \ldots, \partial_{i_r j_r}).$$

If h is a function on B, the principle of [8, Lemma 2.2] shows that

$$(\theta E)^r h(O_B) = \sum_{i,j,\ldots,s,t} \bar{\partial}_{ij} \cdots \bar{\partial}_{st} h(\partial_{ij}, \ldots, \partial_{st})(O_B),$$

where O_B is the origin of B. Take $h = \chi f$ with a function f on B with values in $\mathrm{Hom}(C_n^{n'}, V)$. Then we find

$$(\theta E)'\chi f(O_B) = \sum_{a,b} \sum_{i,j,\ldots,s,t} S_{ab}(\partial_{ij}, \ldots, \partial_{st})\bar{\partial}_{ij} \cdots \bar{\partial}_{st} f_{ab}(O_B).$$

On the other hand, $\theta_\sigma E_\sigma f(O_B) = \theta_\sigma \Phi_\sigma \bar{E}_1^r f(O_B) = \theta_\sigma \Phi_\sigma \bar{D}' f(O_B)$ by the principle of [8, Lemma 2.2], and hence

$$\theta_\sigma E_\sigma f(O_B) = \sum_{a,b} (\Phi_\sigma \bar{D}' f_{ab})_{ab}(O_B) = \sum_{a,b} \left\{ S_{ab}, \bar{D}' f_{ab}(O_B) \right\}$$

$$= \sum_{a,b} \sum_{i,j,\ldots,s,t} S_{ab}(\partial_{ij}, \ldots, \partial_{st})\bar{\partial}_{ij} \cdots \bar{\partial}_{st} f_{ab}(O_B)$$

by (1.21) and (1.26b). Hence we obtain

$$(11.17) \qquad (\theta E)'\chi f = \theta_\sigma E_\sigma f \qquad \text{at } O_B.$$

Observe that $(\theta E)'\chi$ has the same property (11.9b) as $\theta_\sigma E_\sigma$. Therefore the idea in the last half of the proof of Proposition 2.1 together with (11.17) shows that

$$(11.18) \qquad (\theta E)'\chi f = \theta_\sigma E_\sigma f \qquad \text{for functions } f \text{ on } X.$$

Now let f and g be as in (11.14b). Successive application of (11.13b) shows that $(-1)^r \langle (\theta E)'\chi f, g \rangle = \langle \chi f, D_\rho^{(r)} g \rangle$. Combining this with (11.18), (11.15), and (11.16), we obtain

$$(-1)^r \langle \theta_\sigma E_\sigma f, g \rangle = \langle \chi f, D_\rho^{(r)} g \rangle = \langle f, \Phi_\sigma D_\rho^{(r)} g \rangle = \langle f, \Delta_\rho^\sigma g \rangle,$$

which establishes (11.14b). The other formula (11.14a) can be derived from (11.13a) in a similar way by observing that

$$(11.19) \qquad \theta_\sigma \Delta_{\rho \otimes_2 \sigma}^\sigma = (\theta D_{\rho \otimes \pi})'\chi,$$

which can be shown by the same technique as for (11.18).

CORROLARY 11.7. *Let the notation be the same as in Theorem* 11.5. *If f is a holomorphic element of $C_\rho(\Gamma)$ and $\zeta = \rho \otimes_2 \sigma$, then $\langle f, \theta_\sigma \Delta_\zeta^\sigma g \rangle_\Gamma = 0$ for every $g \in C_\zeta(\Gamma)$. Similarly, if f is a holomorphic element of $C_{\rho \otimes_1 \sigma}(\Gamma)$, then $\langle f, \Delta_\rho^\sigma g \rangle_\Gamma = 0$ for every $g \in C_\rho(\Gamma)$.*

These follow immediately from Theorem 11.5, (2.2), and (2.4).

COROLLARY 11.8. *The notation being as in Theorem* 11.5, *let L denote the following four types of operators: $(-1)^r \theta_\sigma \Delta_{\rho \otimes_2 \sigma}^\sigma E_\sigma$ and $(-1)^r \theta_\sigma E_\sigma \Delta_\rho^\sigma$ acting on V-valued functions on X; $(-1)^r \Delta_\rho^\sigma \theta_\sigma E_\sigma$ and $(-1)^r E_\sigma \theta_\sigma \Delta_{\rho \otimes_2 \sigma}^\sigma$ acting on $\mathrm{Hom}(C_n^{n'}, V)$-valued functions on X. Then the following assertions hold:*

(i) $(Lf)\|_\zeta \alpha = L(f\|_\zeta \alpha)$ *for* $\alpha \in \mathfrak{G}$, $\mathfrak{G}'$ *or* $\mathfrak{E}$, *where* $\zeta = \rho$ *for the first two operators, and* $\zeta = \rho \otimes_1 \sigma$ *and* $\zeta = \rho \otimes_2 \sigma$ *for the last two, respectively.*

(ii) *L is self-adjoint in the sense that* $\langle Lf, g\rangle_\Gamma = \langle f, Lg\rangle_\Gamma$ *for* $f, g \in C_\zeta(\Gamma)$.

(iii) *L is nonnegative in the sense that* $\langle Lf, f\rangle_\Gamma \geqq 0$ *for* $f \in C_\zeta(\Gamma)$.

These follow directly from Theorem 11.5. For example, if $L = (-1)^r E_\sigma \Delta_\rho^\sigma$, then by (11.14b), $\langle Lf, f\rangle = \langle \Delta_\rho^\sigma f, \Delta_\rho^\sigma f\rangle \geqq 0$.

Remark 11.9. If $l = m$ and $\sigma(x) = \det(x)^p \rho_r(x)$ with $0 \leqq p \in \mathbf{Z}$ and $0 \leqq r \leqq m$, then we may take $\mathrm{Hom}(T_r, V)$ instead of $\mathrm{Hom}(\mathbf{C}_n^{n'}, V)$, since $\mathrm{Hom}(T_r, V)$ is embedded in $\mathrm{Hom}(\mathbf{C}_A^A, V)$ as noted in Section 2. (The above r becomes $mp + r$ for this σ.) If in particular $\sigma(x) = \det(x)^p$, all functions involved are V-valued, and θ_σ is trivial.

Remark 11.10. We assumed in the above treatment that $\Gamma \backslash X$ is compact. All our results can be extended to the case where $\Gamma \backslash X$ has finite measure. Obviously the whole argument is valid as long as $\int_{\Gamma \backslash X} d\varphi_*$ in the proof of Theorem 11.4 vanishes, which is true if we assume a suitable growth condition on f, g, and their derivatives at cusps. Another method is to imitate the argument of [12, I, Lemma 2.1], introducing a complex parameter s. This is better suited for certain arithmetical questions. As we hope to treat them on a future occasion, we will not go into details here, dealing only with the co-compact case, which explains at least how things are contrived and tells what the adjoint operators should be in the most general case.

We note here the explicit forms of L in four simplest cases:

$$(11.20a) \qquad -\theta_1 E_1 D_\rho f = -\sum_{i,p=1}^{l} \sum_{j,q=1}^{m} \xi_{pi} \eta_{qj} \cdot \bar{\partial}_{pq}\{\rho(\xi,\eta)^{-1}\partial_{ij}[\rho(\xi,\eta)f]\},$$

$$(11.20b) \qquad -\theta_1 D_{\rho \otimes \pi} E_1 f = -\sum_{i,p=1}^{l} \sum_{j,q=1}^{m} \xi_{pi}\eta_{qj} \cdot \rho(\xi,\eta)^{-1}\partial_{ij}[\rho(\xi,\eta)\bar{\partial}_{pq}f],$$

$$(11.21a) \qquad (-1)^{mp} E_\sigma \Delta_\rho^\sigma f = (-1)^{mp}\delta^{\kappa+p}\bar{\Delta}^p\{\rho(\xi,\eta)^{-1}\Delta^p[\delta^{p-\kappa}\rho(\xi,\eta)f]\}$$

$$\text{if} \quad \sigma(x) = \det(x)^p,$$

$$(11.21b) \qquad (-1)^{mp}\Delta_\lambda^\sigma E_\sigma f = (-1)^{mp}\delta^{\kappa+p}\rho(\xi,\eta)^{-1}\Delta^p[\rho(\xi,\eta)\bar{\Delta}^p(\delta^{p-\kappa}f)]$$

$$\text{if} \quad \lambda(x,y) = \det(xy)^{-p}\rho(x,y) \quad \text{and} \quad \sigma(x) = \det(x)^p.$$

The first two formulas were merely the combinations of our definitions; the latter two are obtained from (1.39) and (2.14).

PROPOSITION 11.11. *The difference* $\theta_1 E_1 D_\rho - \theta_1 D_{\rho \otimes \pi} E_1$ *acts as multiplication by a constant element of* $\mathrm{End}(V)$. *In particular, if ρ is irreducible, this constant* *element is a scalar, and moreover every holomorphic V-valued function on X is an eigenfunction of both $\theta_1 E_1 D_\rho$ and $\theta_1 D_{\rho \otimes \pi} E_1$.*

Proof. We have $\theta_1 E_1 D_\rho f - \theta_1 D_{\rho \otimes \pi} E_1 f = A(z)f$ with

$$A(z) = \sum_{i,p=1}^{l} \sum_{j,q=1}^{m} \xi_{pi} \eta_{qj} \bar{\partial}_{pq} \left[\rho(\xi, \eta)^{-1} \partial_{ij} \rho(\xi, \eta) \right].$$

By (i) of Corollary 11.8, $A(\alpha(z))\rho(\kappa_\alpha, \mu_\alpha) = \rho(\kappa_\alpha, \mu_\alpha)A(z)$ for every $\alpha \in \mathfrak{G}$. Taking α in a maximal compact subgroup, and decomposing ρ into irreducible constituents, we obtain the first two assertions. If f is holomorphic, we have $\theta_1 D_{\rho \otimes \pi} E_1 f = 0$, and hence the last assertion follows from the second one.

PROPOSITION 11.12. *Suppose $\rho(x, y) = \det(y)^k$ with $k \in \mathbf{Z}$ in Cases SP and SU. Then δ^s with $s \in \mathbf{C}$ is an eigenfunction of the operators (11.20a, b) and (11.21a, b) with eigenvalues*

$$m(\kappa - s)(k + s),$$

$$m(\kappa - k - s)s,$$

$$\prod_{a=0}^{p-1} \prod_{b=0}^{m-1} (\kappa - s + a - (db/2))(k + s + a - (db/2)),$$

$$\prod_{a=0}^{p-1} \prod_{b=0}^{m-1} (\kappa - k - s + a - (db/2))(s + a - (db/2)),$$

respectively, where $d = 1$, $\kappa = (m + 1)/2$ in Case SP, and $d = 2$, $\kappa = m$ in Case SU. Moreover, $\delta^{(\kappa-k)/2}\log(\delta)$ is an eigenfunction of all these operators whose eigenvalues are the same as those of $\delta^{(\kappa-k)/2}$.

Proof. The assertion concerning δ^s can be obtained by direct calculations, employing (5.2), (5.4), and Lemma 1.1. For each operator L, we have $L(\delta^s) = \lambda(s)\delta^s$ with a polynomial λ. Applying $\partial/\partial s$ to this relation, we find that $\delta^s\log(\delta)$ is an eigenfunction when $\lambda'(s) = 0$. Let $m\alpha_i(s)$ with $i = 1, 2$ denote the first two eigenvalues of the above list. Then $\alpha_2(s) = \alpha_1(s) - \kappa k$, and the last two eigenvalues can be written in the form

$$\prod_{a=0}^{p-1} \prod_{b=0}^{m-1} \left[\alpha_i(s) + \left(\kappa - (-1)^i k \right)(a - (db/2)) + (a - (db/2))^2 \right].$$

We have $\alpha_i'((\kappa - k)/2) = 0$, and hence $\lambda'((\kappa - k)/2) = 0$ in all cases, which completes the proof.

It follows from this result and (i) of Corollary 11.8 that $\mathbf{E}(z, s; k, 0, \Omega)$ of (7.3) (with $t = 0$) is an eigenfunction of the above operators with the same eigenvalues as δ^s.

In Case SUB, δ^s is not necessarily an eigenfunction even when $l = m$. Indeed, we have, for $\rho(x, y) = \det(y)^k$,

$$-\theta_1 E_1 D_\rho \delta^s = \delta^s (k + s)\big[(m - s)l + s \cdot \mathrm{tr}(\xi)\big],$$

and hence δ^s is an eigenfunction of $-\theta_1 E_1 D_\rho$ if and only if $s = 0$ or $s = -k$.

In order to obtain eigenfunctions for $SU(l, m)$ with $l > m$, we take the group and the domain in the following forms:

$$(11.22a) \quad G_0 = \big\{ \alpha \in SL_{l+m}(\mathbf{C}) \mid \alpha^* h \alpha = h \big\}, \qquad h = \mathrm{diag}\big[1_{l-m}, i1_m\big],$$

$$(11.22b) \qquad\qquad X_0 = \big\{ (w, z) \in \mathbf{C}_m^{l-m} \times \mathbf{C}_m^m \mid \eta(w, z) > 0 \big\},$$

$$\eta(w, z) = i(z^* - z) - w^* w.$$

Then G_0 and X_0 are isomorphic to $SU(l, m)$ and $B_{l,m}$. By a direct calculation, we can show that $\det(\eta)^s$ for $s \in \mathbf{C}$ is an eigenfunction of the operators $-\theta_1 E_1 D_\rho$ and $-\theta_1 D_{\rho \otimes \pi} E_1$ on X_0 with eigenvalues $m(l - s)(k + s)$ and $ms(l - k - s)$, respectively, if $\rho(x, y) = \det(y)^k$. (The operators on X_0 can be defined by means of ξ and η of $[8, (1.5), (1.6)]$. In fact, we have to take $'\xi$ instead of ξ there.)

That the last two eigenvalues of Proposition 11.12 are polynomials of $\alpha_i(s)$ might suggest some polynomial relations between each operator of (11.21a, b) and that of (11.20a). This is not the case, however. In fact, we have

PROPOSITION 11.13. *Let* $L_r = (-1)^r \theta_\sigma E_\sigma \Delta_\rho^\sigma$ *for* $\sigma = \rho_r$, $0 < r \leqq m$, *and* $\rho(x, y) = \det(y)^k$ *with a fixed* $k \in \mathbf{Z}$. *Then there is no polynomial relation among* $L_1, \ldots, L_m$.

Proof. It is known that all differential operators M on X such that $M(f\|_\rho \alpha) = (Mf)\|_\rho \alpha$ for every $\alpha \in G$ form a commutative algebra. (See Helgason $[2, \text{Ch. X, Theorems } 2.9, 6.15]$ for the case $k = 0$. Harish-Chandra kindly showed the author how this can be generalized to the case of one-dimensional ρ.) Now consider L_r a polynomial in ∂_{ij} and $\bar{\partial}_{ij}$. Then it has degree $2r$, and its "leading term" is

$$(-1)^r \rho_r(\eta^2)_{aa} \bar{\partial}_{11} \cdots \bar{\partial}_{rr} \partial_{11} \cdots \partial_{rr},$$

where $a = \{1 < \cdots < r\} \in A(r)$ in the notation of Section 2. Take any polynomial P in m indeterminates. Looking at the highest terms in ∂_{11}, next in ∂_{22}, and so on, we see in a purely formal way that $P(L_1, \ldots, L_m) = 0$ only when P is identically equal to 0.

One may naturally ask the following question:

Is the algebra of operators M in the above proof generated by $L_1, \ldots, L_m$?

To conclude this section, a few final remarks are in order. First of all, it can easily be seen that 4 times the operators of (11.20a, b) with trivial ρ coincide with the Laplace–Beltrami operator with respect to the invariant metric (11.1a) in all three cases. Furthermore, one can interpret Theorem 11.4 for trivial ρ in the framework of Kählerian geometry. In fact, on any Kähler manifold, we have well-known formulas

$$(11.23a) \qquad\qquad (d'\alpha, \beta) = (\alpha, - *d''*\beta),$$

$$(11.23b) \qquad\qquad (d''\alpha, \beta) = (\alpha, - *d'*\beta),$$

which are valid for a p-form α and a $(p+1)$-form β. If ρ is trivial, (11.13a, b) are reformulations of (11.23a, b) with $p = 0$. To see this, we first identify $C_\tau(\Gamma)$ and $C_\pi(\Gamma)$ with the sets of 1-forms and vectors fields on $\Gamma\backslash X$ of bidegree $(1,0)$ by assigning to $\varphi \in C_\tau(\Gamma)$ and $\psi \in C_\pi(\Gamma)$ a form $\sum \varphi_{ij}\, dz_{ij}$ and a vector field $\sum \psi_{ij}\partial_{ij}$. Write $*d''*$ of (11.23a) when $p = 0$ explicitly in terms of the Kähler metric. Then we find (11.13b) with Dg and θEf corresponding to $d'\alpha$ and $*d''*\beta$. As for (11.13a), we have to rewrite (11.23b) in terms of vector fields instead of 1-forms, employing the contraction with the Kähler metric tensor. Then we find (11.13a) with $\theta D_\pi g$ and Ef corresponding to $*d'*\beta$ and $d''\alpha$; $\theta D_\pi g$ turns out to be the divergence of g.

In this paper, we restricted ourselves to the unitary and symplectic groups. As our theorems and these remarks indicate, one should be able to write the operators of our type and their formulas, with no new difficulties, on any hermitian symmetric space of noncompact type, and possibly in a more general case.

12. Appendix: Polynomial maps σ.

12. Appendix: Polynomial maps σ.   Let us now determine all possible systems $\{\sigma, \sigma', \sigma''\}$ of polynomial maps of (1.25a, b). We say that $\{\sigma, \sigma', \sigma''\}$ and $\{\tau, \tau', \tau''\}$ are *equivalent* if $\sigma(u) = a\tau(u)b^{-1}$, $\sigma'(x) = a\tau'(x)a^{-1}$, and $\sigma''(y) = b\tau''(y)b^{-1}$ for some $a \in GL_{n'}(\mathbf{C})$ and $b \in GL_n(\mathbf{C})$. We call $\{\sigma, \sigma', \sigma''\}$ *irreducible* if both σ' and σ'' are irrediducble and σ is not identically equal to 0. To classify all such systems, we assume $l \geqq m$ and put $p = l - m$. (The case $l < m$ can be treated by switching the right and the left.) Let us first present a method of constructing $\{\sigma, \sigma', \sigma''\}$, starting from a polynomial representation $\sigma' : GL_l \to GL_{n'}$. Suppose a representation $\sigma'' : GL_m \to GL_n$ is contained in the restriction of σ' to

$$(12.1) \qquad\qquad \{\operatorname{diag}[\, y, 1_p]\, | y \in GL_m\}.$$

Then we may put, changing the coordinate systems,

$$(12.2) \qquad \sigma'(\operatorname{diag}[\, y, 1_p]) = \operatorname{diag}[\, \sigma''(y), \tau(y)] \qquad (y \in GL_m)$$

with a representation τ (which is void if $n = n'$). Define $\sigma : \mathbf{C}_m^l \to \mathbf{C}_n^{n'}$ by

$$(12.3) \qquad\qquad \sigma(u) = \sigma'(u, 0_{l,p}) \begin{bmatrix} 1_n \\ 0_{q,n} \end{bmatrix} \qquad (u \in \mathbf{C}_m^l),$$

where $q = n' - n$ and $0_{q,n}$ is the zero element of $\mathbf{C}_n^q$. Then it can readily be checked that $\sigma(xuy) = \sigma'(x)\sigma(u)\sigma''(y)$; if σ' is homogeneous of degree r, then so is σ unless $\sigma = 0$. The vanishing of σ happens if, for example, $\sigma'(x) = \det(x)$ and $l > m$. Before studying the nonvanishing, let us first prove an easier fact:

PROPOSITION 12.1. *Every system* $\{\sigma, \sigma', \sigma''\}$ *can be decomposed, up to equivalence, into irreducible ones and zero maps (in the sense described in the proof). Moreover, every irreducible system is equivalent to a system obtained by* (12.2) *and* (12.3).

Proof. Given $\{\sigma, \sigma', \sigma''\}$, decompose σ' and σ'' into irreducible constituents, say

$$\sigma' = \mathrm{diag}[\sigma'_1, \ldots, \sigma'_r], \qquad \sigma'' = \mathrm{diag}[\sigma''_1, \ldots, \sigma''_s].$$

Decompose also $\sigma(a)$ into rs blocks $\sigma_{ij}(a)$ for $1 \leq i \leq r$ and $1 \leq j \leq s$ according to these decompositions. Then we have $\sigma_{ij}(xay) = \sigma'_i(x)\sigma_{ij}(a)\sigma''_j(y)$. Thus our problem can be reduced to the case where σ' and σ'' are irreducible. Assuming σ' and σ'' to be irreducible, put

$$U = \sigma\begin{bmatrix} 1_m \\ 0_{p,m} \end{bmatrix}.$$

Then $U\sigma''(y) = \sigma'(\mathrm{diag}[y, 1_p])U$. Therefore either $U = 0$ or $\mathrm{rank}(U) = n$. If $U = 0$, we have $\sigma(a) = \sigma'(a, 0_{l,p})U = 0$ for all $a \in \mathbf{C}'_m$. If $\mathrm{rank}(U) = n$, we may assume, changing the coordinate systems, that $'U = (1_n, 0_{n,q})$ with $q = n' - n$. Then we find that

$$\sigma'\begin{bmatrix} y & 0 \\ 0 & 1_p \end{bmatrix} = \begin{bmatrix} \sigma''(y) & * \\ 0 & * \end{bmatrix} \qquad (y \in GL_m).$$

Therefore σ' can be transformed to a representation satisfying (12.2). Then we obtain (12.3).

As this proof shows, if $\{\sigma, \sigma', \sigma''\}$ is irreducible and $l \geq m$, then $n' \geq n$. To determine all irreducible systems, we need the notion of the highest weight of an irreducible representation σ'' of GL_m as in Lemma 1.1. (Of course it can be defined in terms of diagonal matrices instead of triangular matrices.) The highest weight λ of σ'' can naturally be considered the highest weight of a suitable irreducible representation of GL_l. With this convention, we have

PROPOSITION 12.2. *If a system* $\{\sigma, \sigma', \sigma''\}$ *is irreducible, then* σ' *and* σ'' *have the same highest weight. Conversely, given irreducible representations* σ' *of* GL_l *and* σ'' *of* GL_m *with the same highest weight, there is a system* $\{\sigma, \sigma', \sigma''\}$ *with nonvanishing* σ*, which is unique up to equivalence.*

Proof. Continuing the above proof, we have

$$\sigma'\begin{bmatrix} y & w \\ 0 & z \end{bmatrix}\begin{bmatrix} 1_n \\ 0 \end{bmatrix} = \sigma\begin{bmatrix} y \\ 0 \end{bmatrix} = \begin{bmatrix} \sigma''(y) \\ 0 \end{bmatrix}$$

for $y \in \mathbf{C}_m^m$, $w \in \mathbf{C}_p^m$, and $z \in \mathbf{C}_p^p$. Suppose $\sigma''(h)e = \lambda(h)e$ for all upper triangular $h \in GL_m$ with $0 \neq e \in \mathbf{C}^n$ and with λ as in Lemma 1.1. Then we see that

$$\sigma' \begin{bmatrix} h & w \\ 0 & z \end{bmatrix} \begin{bmatrix} e \\ 0 \end{bmatrix} = \lambda(h) \begin{bmatrix} e \\ 0 \end{bmatrix}.$$

Therefore σ' has the same highest weight as σ''. Conversely, suppose σ' and σ'' have the same highest weight λ. Then σ'' occurs in the restriction of σ' to the group of (12.1) with multiplicity 1. Now we have $\sigma'(x)f = \lambda(x)f$ with a suitable $f \in \mathbf{C}^{n'} \neq 0$ for all diagonal matrices $x \in GL_l$. Then there is a unique subspace of $\mathbf{C}^{n'}$ which contains f and is irreducible under the group of (12.1). Changing the coordinate system, we may assume that $f = {}'(1, 0, \ldots, 0)$ and present σ' and σ'' satisfy (12.2). Define σ by (12.3). Since $\sigma'(x)f = \lambda(x)f$ holds even for diagonal x with $x_i = 0$ for $i > m$, we see that $\sigma \neq 0$. The uniqueness follows easily from the fact that the multiplicity of σ'' in the restriction of σ' is 1.

Thus the equivalence classes of irreducible systems $\{\sigma, \sigma', \sigma''\}$ are in one-to-one correspondence with the equivalence classes of all irreducible polynomial representations of GL_m if $l \geqq m$. In particular, if $l = m$, we may put $\sigma = \sigma' = \sigma''$, which is why we assumed so in Section 1. A typical example of irreducible system can be obtained by taking σ' and σ'' to be the representations of GL_l on $\bigwedge^r \mathbf{C}^l$ and of GL_m on $\bigwedge^r \mathbf{C}^m$ with $r \leqq m$ ($\leqq l$). Then $\sigma(u)$ consists of all subdeterminants of u of degree r.

In conclusion, two remarks may be added. We first note that formula (1.44) is valid in Case SUB if $\{\sigma, \sigma', \sigma''\}$ is irreducible and σ'' has the highest weight λ. We can similarly prove analogues of Lemma 1.2 in Case SUB on $B_{l,m}$ and on the domain X_0 of (11.22b) by the same technique. Thus Case SUB (or rather $SU(l, m)$ with $l \neq m$) can be handled more or less in the same fashion as in Cases SP and SU.

Another remark concerns the operator $\varphi D_\rho^{(r)}$ defined in Section 1 with a "homomorphism" φ of $\{\rho \otimes \tau', \mathrm{Ml}_r(T, V)\}$ into a representation $\{\omega, W\}$ of $\mathcal{g}$ (see the paragraph containing (1.25a, b)). It can be shown in Cases SU and SUB that $\varphi D_\rho^{(r)}$ essentially coincides with Δ_ρ^σ of (1.31) for some σ if ω is the tensor product of ρ with an irreducible constituent of τ' and if φ induces the identity map on V. In this sense $D_\rho^{(r)}$ is a "direct sum" of operators Δ_ρ^σ with several σ's, and therefore little is lost in Cases SU and SUB by considering only Δ_ρ^σ instead of $\varphi D_\rho^{(r)}$ for general φ. In order to justify such a statement in Case SP, however, a further generalization, or reformulation, of Δ_ρ^σ is necessary.

REFERENCES

1. R. GODEMENT, *Où l'on généralise une intégrale* . . . , Exposé No. 5, Séminaires H. Cartan, Fonctions automorphes, 1957/58.
2. S. HELGASON, *Differential Geometry and Symmetric Spaces*, Academic Press, 1962.
3. Y. KITAOKA, *Dirichlet series in the theory of Siegel modular forms*, to appear.
4. R. P. LANGLANDS, *On the functional equations satisfied by Eisenstein series*, Lecture Notes in Math. 544, Springer, 1976.

5. G. SHIMURA, *On some arithmetic properties of modular forms of one and several variables*, Ann. of Math. **102** (1975), 491–515.

6. ———, *The special values of the zeta functions associated with cusp forms*, Comm. Pure Appl. Math. **29** (1976), 783–804.

7. ———, *The special values of the zeta functions associated with Hilbert modular forms*, Duke Math. J. **45** (1978), 637–679.

8. ———, *The arithmetic of automorphic forms with respect to a unitary group*, Ann. of Math. **197** (1978), 569–605.

9. ———, *Automorphic forms and the periods of abelian varieties*, J. Math. Soc. Japan **31** (1979), 561–592.

10. ———, *The arithmetic of certain zeta functions and automorphic forms on orthogonal groups*, Ann. of Math. **111** (1980), 313–375.

11. ———, *Arithmetic of differential operators on symmetric domains*, Duke Math. J. **48** (1981), 813–843.

12. ———, *On certain zeta functions attached to two Hilbert modular forms* I, II, Ann. of Math. **114** (1981), 127–164, 569–607.

13. ———, *Algebraic relations between critical values of zeta functions and inner products*, Amer. J. of Math. **104** (1983), 253–285.

14. ———, *On Eisenstein series*, Duke Math. J. **50** (1983), 417–476.

DEPARTMENT OF MATHEMATICS, PRINCETON UNIVERSITY, PRINCETON, NEW JERSEY 08544

On differential operators attached to certain representations of classical groups

Inventiones mathematicae, 77 (1984), 463-488

Our starting point is a classical formula

$$(1) \qquad \det(D)\det(z)^s = s(s+\tfrac{1}{2})\ldots\left(s+\frac{m-1}{2}\right)\det(z)^{s-1}$$

which was first proved by Gårding [1], where $s \in \mathbb{C}$, $z=(z_{ij})$ is a variable matrix on the space T of all complex symmetric matrices of size m, and D is the matrix whose entries are the derivations $(1/2)(1+\delta_{ij})\partial/\partial z_{ij}$. A similar and simpler formula for the matrices without symmetricity is known since Cayley's time. Now our first question can be asked, rather naively, as follows:

Can one compute $h(D)\det(z)^s$ in a reasonable way for an arbitrary homogeneous polynomial function h on T?

One of the main purposes of this paper is to answer the same type of question for the tangent space of every classical bounded symmetric domain. To describe our result, let us again take the above T, for example. We let $GL_m(\mathbb{C})$ act on T by $z \mapsto az \cdot {}^t a$ for $a \in GL_m(\mathbb{C})$ and $z \in T$. Then $GL_m(\mathbb{C})$ acts naturally on the space $S_r(T)$ of all homogeneous polynomial functions on T of degree r. We consider an irreducible representation $\sigma: GL_m(\mathbb{C}) \to GL_n(\mathbb{C})$ which occurs in $S_r(T)$. For such a σ we can find a nonvanishing homogeneous polynomial map ζ of T into $\mathbb{C}^n$ such that

$$\zeta(ax \cdot {}^t a) = \sigma(a)\zeta(x) \qquad (a \in GL_m(\mathbb{C}),\ x \in T).$$

One can show that:

(i) *the multiplicity of σ in $S_r(T)$ is one; consequently, ζ is unique for σ up to a constant factor;*

(ii) *σ occurs in $S_r(T)$ if and only if its highest weight $\{r_1, \ldots, r_m\}$ consists of even nonnegative integers r_k such that $r_1 + \ldots + r_m = 2r$.*

Now $\zeta(D)$ is meaningful as an operator which assigns a $\mathbb{C}^n$-valued function to a $\mathbb{C}$-valued function on T. Then our generalization of (1) can be stated as

Research supported by NSF Grant MCS-8100744

$$(2) \qquad \zeta(D)\det(z)^s = \beta_\sigma(s)\det(z)^s\,\zeta(z^{-1}),$$

$$\beta_\sigma(s) = \prod_{k=1}^{m} \prod_{i=1}^{r_k/2}\left(s-i+\frac{k+1}{2}\right),$$

where $\{r_k\}$ is the highest weight of σ. We may view this as an answer to the above question, since $S_r(T)$ is spanned by the components of ζ for all such σ.

Though this by itself is a natural topic of investigation, our motive lies also in its application to the arithmeticity problem on automorphic forms, in particular on Eisenstein series. In fact, to each such σ and another representation ρ of $GL_m(\mathbb{C})$, we can attach a certain nonholomorphic differential operator Δ_ρ^σ, which is a certain twist of $\zeta(D)$ and which sends an automorphic form of weight σ to a form of weight $\rho \otimes \sigma$, by virtue of the general principles developed in our previous papers [7] and [8]. As its important property, Δ_ρ^σ preserves the arithmeticity at CM-points. Another characteristic is that it maps a scalar-valued holomorphic Eisenstein series to a nonholomorphic vector-valued one. It is essential to know exactly when the latter is nonvanishing. In the previous paper [11], we were able to solve this problem for the unitary groups, but gave only a partial answer for the symplectic groups. By means of the explicit form of $\beta_\sigma(s)$, we can now settle this point, which is the main application.

In the terminology of an arbitrary hermitian symmetric space $\mathscr{Z}$ of non-compact type, our preliminary problem of determining all σ may be stated as follows. Let $\mathscr{Z} = \mathscr{G}/\mathscr{K}$ with a semi-simple group $\mathscr{G}$ and a maximal compact subgroup $\mathscr{K}$. Then $\mathscr{K}$ acts on the holomorphic tangent space T of $\mathscr{Z}$ at the origin. Let $S_r(T)$ denote the r-th symmetric product of T, on which $\mathscr{K}$ naturally acts. Then we have to find all irreducible constituents of this representation of $\mathscr{K}$. This was already done by Hua [4] for the classical domains, and by Schmid [6] in the general case including the exceptional domains. They determined the highest weights of all such constituents, and showed that each constituent has multiplicity one. However, they did not indicate how to construct irreducible subspaces. Johnson [5] settled this point by explicitly exhibiting the generators of the highest weight vectors.

In the present paper, we take $\mathscr{G}$ to be a classical group belonging to four major types, excluding the exceptional ones. We shall present in Sect. 2 an easy description of the highest weights for each type (an example being the above (ii)), and specify each irreducible subspace of $S_r(T)$ by constructing explicitly an eigenvector of highest weight, which amounts to the results of the above authors restricted to the classical case. Without relying on their methods, we shall give an elementary proof. As a by-product, we shall show that the irreducible representations of $GL_m(\mathbb{C})$ in (ii) are exactly those whose restrictions to $O(m,\mathbb{C})$ contain trivial representations (Proposition 1.3). In Sect. 4, we shall prove generalizations of (1) and (2). Since functions like $\det(z)$ exist only on tube domains, we shall prove formulas of type (2) only on tube domains. We can, however, generalize (2) to the form

$$\zeta(D)\det(cz+d)^s = \beta_\sigma(s)\det(cz+d)^s\,\zeta({}^tc\cdot{}^t(cz+d)^{-1}),$$

which has analogues in all classical domains, including non-tube types (Theorem 4.3). We introduce the operator Δ_p^σ and also its anti-holomorphic counterpart E_σ in Sect. 5. The final section concerns the arithmeticity of Eisenstein series at CM-points. In these two sections, we confine ourselves to the symplectic groups in order to minimize the length of the paper, though all other cases can be treated in a parallel way.

Notation. Besides the standard symbols $\mathbb{Z}, \mathbb{Q}, \mathbb{R}$, and $\mathbb{C}$, we let $\mathbb{H}$ denote the division ring of Hamilton quaternions, and $\overline{\mathbb{Q}}$ the algebraic closure of $\mathbb{Q}$ in $\mathbb{C}$. If $x \in \mathbb{C}$ or $x \in \mathbb{H}$, $\bar{x}$ denotes the complex conjugate or the quaternion conjugate of x. For an associative ring R with identity element and positive integers m and n, we denote by R_n^m the module of all $m \times n$ matrices with entries in R. We often put $R^m = R_1^m$ and $M_n(R) = R_n^n$ particularly when we view R_n^n as a ring. The identity element and the group of invertible elements of $M_n(R)$ are denoted by 1_n and $GL_n(R)$; then $SL_n(R) = \{a \in GL_n(R) | \det(a) = 1\}$ when R is commutative. For $x \in \mathbb{C}_n^m$ and $1 \le i \le \mathrm{Min}(m, n)$, $\det_i(x)$ denotes the determinant of the upper left i^2 entries of x. For $x \in \mathbb{C}_n^m$ or $x \in \mathbb{H}_n^m$, we put $x^* = {}^t\bar{x}$, where the upper left t indicates the transpose as usual. If $x = x^*$, $y = y^*$, and $x - y$ is positive definite, we write $x > y$. For square matrices $x_1, \ldots, x_k$, $\mathrm{diag}[x_1, \ldots, x_k]$ denotes the matrix with $x_1, \ldots, x_k$ in the diagonal blocks and 0 in all other blocks. By a representation $\{\sigma, W\}$ of a group G, we mean a pair formed by a complex finite-dimensional vector space W and a homomorphism σ of G into the group $GL(W)$ of all $\mathbb{C}$-linear automorphisms of W. When G is a complex Lie group, its representation is always assumed to be complex analytic. The identity component of a Lie group G is denoted by G^0.

Given a field F and $Q = {}^tQ \in GL_n(F)$, we put

$$O(Q, F) = \{\alpha \in GL_n(F) | {}^t\alpha\, Q\, \alpha = Q\},$$
$$SO(Q, F) = O(Q, F) \cap SL_n(F),$$
$$O(n, F) = O(1_n, F), \quad SO(n, F) = SO(1_n, F), \quad SO(n) = SO(n, \mathbb{R}).$$

We define also other classical groups as usual:

$$U(n, m) = \{\alpha \in GL_{n+m}(\mathbb{C}) | \alpha^* I_{n,m} \alpha = I_{n,m}\}, \quad I_{n,m} = \mathrm{diag}[1_n, -1_m],$$
$$SU(n, m) = U(n, m) \cap SL_{n+m}(\mathbb{C}), \quad SO(n, m) = SO(I_{n,m}, \mathbb{R}),$$
$$U(n) = U(n, 0), \quad SU(n) = SU(n, 0),$$
$$Sp(n, F) = \{\alpha \in GL_{2n}(F) | {}^t\alpha\, J_n\, \alpha = J_n\},$$
$$J_n = \begin{bmatrix} 0 & -1_n \\ 1_n & 0 \end{bmatrix}.$$

We shall abbreviate $GL_n(\mathbb{C})$ to GL_n, as this is the group of the highest frequency in this paper.

1. Preliminaries

Every irreducible hermitian symmetric space $\mathscr{X} = \mathscr{G}/\mathscr{K}$ of non-compact and nonexceptional type belongs to the four types A, B, C, and D listed below.

Instead of working with $\mathscr{K}$, we take a complex Lie group G which is either the complexification $\mathscr{K}_{\mathbb{C}}$ of $\mathscr{K}$, or its product with GL_1. Then G acts on the space T of holomorphic tangent vectors of $\mathscr{Z}$ at the origin. We denote by $S_r(T)$, or simply by S_r, the vector space of all $\mathbb{C}$-valued homogeneous polynomial functions on T of degree r. Then G acts naturally on S_r. We denote by $\tau_r(g)$ the action of an element g of G on S_r. Our first problem is to decompose S_r into G-irreducible subspaces, which are naturally $\mathscr{K}$-irreducible. We fix one particular element ε of T and consider its stabilizer

$$(1.1) \qquad\qquad H = \{a \in G \mid a\varepsilon = \varepsilon\}.$$

It is one of our main ideas to relate each irreducible constituent of S_r with a trivial representation of H occurring in S_r. Now these objects $G, T, \tau_r, \varepsilon$, and H are explicitly given, according to the standard classification, as follows. For the reader's convenience, we include also $\mathscr{G}$ and $\mathscr{K}$, though they are unnecessary for our treatment.

Type A: $\mathscr{G} = SU(n, m)$, $G = GL_n \times GL_m$, $T = \mathbb{C}_m^n$, G acts on T by $(a, b)x = ax \cdot {}^t b$ for $a \in GL_n$, $b \in GL_m$, and $x \in T$; $[\tau_r(a, b)h](x) = h({}^t a x b)$ for $h \in S_r$;

$$(1.2) \qquad\qquad \varepsilon = \begin{bmatrix} 1_m \\ 0_* \end{bmatrix} \quad \text{if } n \geq m,$$

where 0_* is the zero element of $\mathbb{C}_m^{n-m}$. Then

$$H = \left\{ \left(\begin{pmatrix} a & b \\ 0 & d \end{pmatrix}, {}^t a^{-1} \right) \middle| a \in GL_m, \ d \in GL_{n-m}, \ b \in \mathbb{C}_{n-m}^m \right\}.$$

If $n = m$, the matrices 0_*, b, and d must be disregarded. If $n < m$, we take $\varepsilon = (1_n, 0)$ and define H in a similar way. We have

$$\mathscr{K} = \{(a, b) \in U(n) \times U(m) \mid \det(a)\det(b) = 1\},$$

$$\mathscr{K}_{\mathbb{C}} = \{(a, b) \in GL_n \times GL_m \mid \det(a)\det(b) = 1\}.$$

Type B: $\mathscr{G} = SO(n, 2)^0$, $\mathscr{K} = SO(n) \times SO(2)$, $G = \mathscr{K}_{\mathbb{C}} = SO(n, \mathbb{C}) \times GL_1$, $T = \mathbb{C}^n$. We let G act on T by $(a, b)x = abx$ for $a \in SO(n, \mathbb{C})$, $b \in GL_1$, and $x \in T$; $[\tau_r(a, b)h](x) = h({}^t abx)$ for $h \in S_r$; $\varepsilon = {}^t(1, 0, \ldots, 0)$,

$$H = \{(\mathrm{diag}[b, c], b) \mid c \in O(n-1, \mathbb{C}), \ b = \det(c)\}.$$

We assume $n > 1$, since the case $n = 1$ is trivial.

Type C: $\mathscr{G} = Sp(m, \mathbb{R})$, $\mathscr{K} \cong U(m)$, $G = GL_m \cong \mathscr{K}_{\mathbb{C}}$, $T = \{x \in \mathbb{C}_m^m \mid {}^t x = x\}$; $a \in G$ acts on T by $x \mapsto ax \cdot {}^t a$; $[\tau_r(a)h](x) = h({}^t a x a)$ for $h \in S_r$; $\varepsilon = 1_m$, $H = O(m, \mathbb{C})$.

Type D: $\mathscr{G} = \{a \in SU(n, n) \mid aJ_n = J_n \bar{a}\}$ $(n > 1)$, $\mathscr{K} \cong U(n)$, $G = GL_n \cong \mathscr{K}_{\mathbb{C}}$, $T = \{x \in \mathbb{C}_n^n \mid {}^t x = -x\}$; $a \in G$ acts on T by $x \mapsto ax \cdot {}^t a$; $[\tau_r(a)h](x) = h({}^t a x a)$ for $h \in S_r$;

$$\varepsilon = \begin{cases} J_m & \text{if } n = 2m, \\ \mathrm{diag}[J_m, 0] & \text{if } n = 2m + 1 \ (\text{see } \textit{Notation}), \end{cases}$$

$$H = \begin{cases} Sp(m,\mathbb{C}) & \text{if } n=2m, \\ \left\{ \begin{pmatrix} a & b \\ 0 & d \end{pmatrix} \in GL_n \,\middle|\, a\in Sp(m,\mathbb{C}),\ b\in\mathbb{C}^{2m},\ d\in GL_1 \right\} & \text{if } n=2m+1. \end{cases}$$

In the first case, $\mathscr{K}_{\mathbb{C}}$ is different from G. However, a subspace of S_r irreducible under $\mathscr{K}$ if and only if it is so under G. Therefore we lose nothing by taking G instead of $\mathscr{K}_{\mathbb{C}}$.

Proposition 1.1. *Let* $\sigma: G \to GL_d$ *be an irreducible representation. Then* σ *occurs in* τ_r *if and only if there is a nontrivial polynomial map* $\zeta: T \to \mathbb{C}^d$, *homogeneous of degree* r, *such that*

$$(1.2) \qquad\qquad \zeta(a\,x\cdot{}'b) = \sigma(a,b)\zeta(x) \qquad \text{(Type A or B)},$$

$$(1.3) \qquad\qquad \zeta(a\,x\cdot{}'a) = \sigma(a)\zeta(x) \qquad \text{(Type C or D)}$$

for $x\in T$ *and* $(a,b)\in G$ *or* $a\in G$.

Proof. For example, consider Type A. Given σ, put $\rho(a,b)={}'\sigma({}'a,{}'b)$ for $(a,b)\in G$. Then ρ is equivalent to σ. If ρ occurs in τ_r, we can find linearly independent elements $h_1,\ldots,h_d$ of S_r such that $h_j({}'a\,x\,b)=\sum_{i=1}^{d}\rho_{ij}(a,b)h_i(x)$. Putting $\zeta={}'(h_1,\ldots,h_d)$, we obtain $\zeta(a\,x\cdot{}'b)=\sigma(a,b)\zeta(x)$. Conversely, the components of such a map ζ span an irreducible subspace of S_r. Other types can be similarly treated.

Proposition 1.2. *Let* $\{\sigma, W\}$ *be an irreducible polynomial representation of* G. *Then the following assertions hold:*

(i) *If* σ *occurs in* τ_r *for some* r, *then*

$$(1.4) \qquad\qquad \{v\in W \,|\, \sigma(H)v=v\} \neq \{0\}.$$

(ii) *If* σ *occurs in* τ_r, *its multiplicity is* 1, *and the subspace of* W *defined by* (1.4) *is one-dimensional.*

(iii) *If* σ *occurs in* τ_r, *every continuous map* ζ *of* T *into* W *satisfying* (1.2) *or* (1.3) *is a polynomial map; moreover it is unique up to a constant factor.*

(iv) *Let* $\mathscr{I}_r=\{f\in S_r\,|\,\tau_r(H)f=f\}$ *and* $\mu_r=\dim\mathscr{I}_r$. *Then* τ_r *has exactly* μ_r *inequivalent irreducible constituents.*

For Types A, C, and D, there is no ambiguity about the terminology "polynomial representation". As for Type B, it is known that every holomorphic representation of $SO(n,\mathbb{C})$ can be extended to a polynomial map of $\mathbb{C}_n^n$. Therefore, by a *polynomial representation* of $G=SO(n,\mathbb{C})\times GL_1$, we mean one which induces a map $x\mapsto x^k$ for $x\in GL_1$ with $0\leq k\in\mathbb{Z}$. Obviously every irreducible constituent of τ_r is a polynomial representation.

Proof. Let ζ and σ be as in Proposition 1.1. Then $\zeta(\varepsilon)=\sigma(H)\zeta(\varepsilon)$. Since the G-orbit of ε is dense in T, we have $\zeta(\varepsilon)\neq 0$, which proves (i). To prove (iii), suppose ζ' is a continuous map satisfying (1.2) or (1.3). Then $\zeta'(a\,\varepsilon)=\sigma(a)\zeta'(\varepsilon)$ for every $a\in G$, and $\zeta'(\varepsilon)$ belongs to the subspace of (1.4). If we assume (ii), $\zeta'(\varepsilon)$ is unique for σ up to a constant factor. Since $G\varepsilon$ is dense in T, ζ' must be a constant times the polynomial map ζ which is established by Proposition 1.1.

The proof of (ii) and (iv) will be completed when we establish all irreducible constituents of τ_r in the next section. Our program is as follows. Let μ_r be defined as in (iv), and let θ be a maximal set of mutually inequivalent irreducible constituents of τ_r. For each $\sigma \in \theta$, let d_σ be the dimension of the space of (1.4), and e_σ the multiplicity of σ in τ_r. Then $\mu_r \geq \sum_{\sigma \in \theta} d_\sigma e_\sigma$. We shall show in the next section that θ has at least μ_r members. Then, by (i), $\sum_{\sigma \in \theta} d_\sigma e_\sigma \geq \mu_r$, and hence $d_\sigma = e_\sigma = 1$, which will complete the proof of (ii) and (iv).

The converse of (i) of Proposition 1.2 is essentially true in the sense of the following proposition, which will play no role in our later treatment, though it may be of independent interest.

Proposition 1.3. *Let $\{\sigma, W\}$ be an irreducible polynomial representation of G; suppose $\sigma(H)v = v$ for some $v \in W$, $\neq 0$. Then σ occurs in τ_r for some r, if G is of Type A, C, or D. For G of Type B, there is a nonnegative integer s such that a representation σ' defined by $\sigma'(a, b) = b^{2s}\sigma(a, b)$ occurs in τ_r for some r.*

Proof. Take $\mathbb{C}^d$ as W. Given such a v, we are going to construct a map ζ as in Proposition 1.1. For this purpose, take a countable subfield k of $\mathbb{C}$ over which σ and v are rational. Take a generic point y of G over k. Then $y\varepsilon$ is a generic point of T over k. Let α be an isomorphism of $k(y)$ into $\mathbb{C}$ which is the identity map on $k(y\varepsilon)$. Then $y^\alpha \varepsilon = y\varepsilon$. Hence $y^{-1}y^\alpha \in H$, and so $[\sigma(y)v]^\alpha = \sigma(y^\alpha)v = \sigma(y)v$. This shows that $\sigma(y)v$ is rational over $k(y\varepsilon)$. Since $y\varepsilon$ is a generic point of T over k, we can define a k-rational map ζ of T into $\mathbb{C}^d$ by $\zeta(y\varepsilon) = \sigma(y)v$. To prove that ζ is a polynomial map, let $\xi(x)$ be a component of the vector $\zeta(x)$. This is a rational function on T, and hence $\xi(x) = p(x)/q(x)$ with two polynomials p and q. Let E denote the vector spaces $\mathbb{C}_n^n \times \mathbb{C}_m^m$, $\mathbb{C}_n^n \times \mathbb{C}$, $\mathbb{C}_m^m$, and $\mathbb{C}_n^n$ for Types A, B, C, and D, respectively. Since $\xi(y\varepsilon)$ is a component of $\sigma(y)v$ and σ is a polynomial map, there is a polynomial function f on E rational over k such that $\xi(y\varepsilon) = f(y)$. Now, for Types A, C, and D, the action of G on T can be extended in an obvious way to that of E on T. Since $p(y\varepsilon) = f(y)q(y\varepsilon)$, $E\varepsilon = T$, and y is generic on E over k, we see that p vanishes at every zero of q. This shows that ξ, and hence ζ, is a polynomial map. Specializing y to an arbitrary point of T, we obtain $\zeta(a\varepsilon) = \sigma(a)v$ for every $a \in E$. Taking aa' in place of a with another $a' \in E$, and writing x for $a'\varepsilon$, we obtain $\zeta(ax) = \sigma(a)\zeta(x)$ for every $x \in T$. Taking a to be a scalar, we find that ζ is homogeneous. By Proposition 1.1, this proves our assertion for Types A, C, and D.

For G of Type B, we have to be more careful. Let $\beta(x) = \sum_{i=1}^{n} x_i^2$ for $x \in \mathbb{C}^n (= T)$. Suppose $q(b) = 0$ and $\beta(b) \neq 0$ for some $b \in T$. We can find an element a of G so that $a\varepsilon = b$, and so we obtain $p(b) = 0$. This implies that $\beta^s \xi$ is a polynomial function on T for some positive integer s. Take s so large that $\beta^s \zeta$ is a polynomial map; put $\zeta_1 = \beta^s \zeta$ and define a representation σ_1 of G by $\sigma_1(b, c) = b^{2s}\sigma(b, c)$ for $b \in SO(n, \mathbb{C})$ and $c \in GL_1$. Then we have $\zeta_1(bcx) = \sigma_1(b, c)\zeta_1(x)$, which together with Proposition 1.1 completes the proof.

2. Classification and construction of irreducible subspaces of S_r

We first recall some elementary facts on the irreducible representations of GL_n. The equivalent classes of irreducible polynomial representations of GL_n are in one-to-one correspondence with the ordered sets of integers $\{r_1,\ldots,r_n\}$ such that $r_1 \geqq \ldots \geqq r_n \geqq 0$. If σ corresponds to $\{r_1,\ldots,r_n\}$, we have

$$\operatorname{tr}(\sigma(\operatorname{diag}[a_1,\ldots,a_n])) = a_1^{r_1}\ldots a_n^{r_n} + \sum_{s<r} \lambda_s a_1^{s_1}\ldots a_n^{s_n}$$

with $\lambda_s \in \mathbf{Z}$, where $<$ in the last sum is the lexicographic ordering. We call σ an *irreducible representation of highest weight* $\{r_1,\ldots,r_n\}$.

Let R_n denote the subgroup of GL_n consisting of all upper triangular matrices. Given an irreducible representation $\{\sigma,W\}$ of GL_n, there is a common eigenvector p of $\sigma(R_n)$ in W, unique up to constant factors. If $\{r_1,\ldots,r_n\}$ is the highest weight of σ, we have

$$(2.1) \qquad \sigma(a)\,p = \prod_{i=1}^{n} \det_i(a)^{e_i}\,p \qquad \text{for every } a \in R_n,$$

where $e_i = r_i - r_{i+1}$ for $i<n$, and $e_n = r_n$. (See e.g. Godement [3]; see *Notation* for $\det_i(a)$.) Obviously p is an eigenvector of highest weight.

Lemma 2.1. *Let* $\{\sigma,V\}$ *be a polynomial representation of* GL_n, *and let* p *be a nonzero element of* V *satisfying* (2.1) *with nonnegative integers* e_i. *Then there is a unique subspace of* V *which is irreducible under* GL_n *and contains* p.

Proof. Decompose $\{\sigma,V\}$ into the direct sum of irreducible representations $\{\sigma_i,V_i\}$ $(i=1,\ldots,t)$ so that $V=V_1\oplus\ldots\oplus V_t$ and $\sigma_i=\sigma|_{V_i}$. Let $p=p_1+\ldots+p_t$ with $p_i\in V_i$. Suppose $p_k \neq 0$ if and only if $k\leqq s$. Observe that p_k and σ_k satisfy (2.1). Hence the $\{\sigma_k,V_k\}$ for $k\leqq s$ are mutually equivalent. We can find a GL_n-isomorphism f_k of V_1 to V_k such that $f_k p_1 = p_k$. Put $W=\left\{\sum_{k=1}^{s} f_k v \,\middle|\, v\in V_1\right\}$. Then W has the desired property. The uniqueness is obvious.

Type D. We start with this case which is somewhat more complicated than the other cases. We have $G=GL_n$ and $T=\{x\in\mathbb{C}_n^n \mid {}^t x = -x\}$. For ${}^t x = -x\in\mathbb{C}_{2i}^{2i}$, we can define its Pfaffian to be a polynomial $\operatorname{Pf}(x)$ of degree i in the entries of x such that $\operatorname{Pf}(x)^2 = \det(x)$ and $\operatorname{Pf}({}^t a\,x\,a) = \det(a)\operatorname{Pf}(x)$ for every $a\in GL_{2i}$. (We may normalize this by assuming $\operatorname{Pf}(J_i)=1$, but this is unnecessary.) To state our result, we put $n=2m$ or $n=2m+1$ according as n is even or odd.

Theorem 2.D. *An irreducible polynomial representation* σ *of* GL_n *of highest weight* $\{r_1,\ldots,r_n\}$ *occurs in* $\{\tau_r,S_r\}$ *of Type D if and only if* $2r=\sum_{k=1}^{n} r_k$ *and* $r_{2i-1} = r_{2i}$ *for* $1\leqq i\leqq m$, *and in addition,* $r_n=0$ *if* n *is odd. Such a* σ *has multiplicity* 1 *in* τ_r, *and the corresponding subspace of* S_r *has the function* p *defined by*

$$(2.2) \qquad p(x) = \prod_{i=1}^{m} \operatorname{Pf}_i(x)^{e_i} \qquad (x\in T)$$

as an eigenvector of highest weight, where $e_i = r_{2i} - r_{2i+2}$ $(r_{2m+2}=0)$ *and* $\operatorname{Pf}_i(x)$ *denotes the Pfaffian of the upper left* $(2i)^2$ *entries of* x.

We first consider the case of even n. Put

$$\operatorname{Pf}(\lambda J_m - x) = \sum_{i=0}^{m} g_i(x)\,\lambda^{m-i} \qquad (x\in T)$$

with an indeterminate λ. We see easily that $g_i({}^t a x a) = g_i(x)$ for every $a \in Sp(m, \mathbb{C})$ and $g_i \in S_i(T)$.

Lemma 2.2. *If $n = 2m$, the space $\mathcal{I}_r$ of Proposition 1.2, (iv) is spanned by all the monomials $g_1^{e_1} \ldots g_m^{e_m}$ such that $\sum_{i=1}^{m} i e_i = r$; moreover $g_1, \ldots, g_m$ are algebraically independent over $\mathbb{C}$. Consequently,*

(2.3) *μ_r is the number of all elements $(e_1, \ldots, e_m) \in \mathbb{Z}^m$ such that*

$$\sum_{i=1}^{m} i e_i = r \quad \text{and} \quad e_i \geq 0 \quad \text{for all } i.$$

Proof. Let $\mathbb{H} = \mathbb{C} + \mathbb{C}\mathbf{j}$ with an element $\mathbf{j}$ such that $\mathbf{j}^2 = -1$ and $\mathbf{j}c = \bar{c}\mathbf{j}$ for $c \in \mathbb{C}$. For $z = u + v\mathbf{j} \in M_m(\mathbb{H})$ with $u, v \in M_m(\mathbb{C})$, put

(2.4)
$$\varphi(z) = \begin{bmatrix} u & -v \\ \bar{v} & \bar{u} \end{bmatrix}.$$

Then φ is an injective homomorphism of $M_m(\mathbb{H})$ into $M_{2m}(\mathbb{C})$, and

(2.5)
$$\varphi(z^*) = \varphi(z)^* = J_m \cdot {}^t\varphi(z) J_m^{-1} \quad \text{(see Notation).}$$

Given $z = z^* = u + v\mathbf{j}$, we can find an element q of $GL_m(\mathbb{H})$ such that $q^*q = 1$ and $z = q^* \operatorname{diag}[b_1, \ldots, b_m] q$ with $b_i \in \mathbb{R}$. Put $a = \varphi(q)$ and $b = \operatorname{diag}[b_1, \ldots, b_m]$. From (2.5) we see that $a \in Sp(m, \mathbb{C})$ and

(2.6)
$$\begin{bmatrix} -\bar{v} & -\bar{u} \\ u & -v \end{bmatrix} = J_m \varphi(z) = {}^t a \begin{bmatrix} 0 & -b \\ b & 0 \end{bmatrix} a.$$

Let k be a countable subfield of $\mathbb{C}$. Then we can find $u = u^* \in M_m(\mathbb{C})$ and $v = -{}^t v \in M_m(\mathbb{C})$ so that the left-hand side of (2.6) is a generic point of T over k. This means that there is a generic point x of T over k of the form $x = {}^t a y a$ with $a \in Sp(m, \mathbb{C})$ and $y = \begin{pmatrix} 0 & -b \\ b & 0 \end{pmatrix}$. Let f be an element of $\mathcal{I}_r$ rational over k. Then $f(y)$ must be a symmetric polynomial in $b_1, \ldots, b_m$ with coefficients in k. Since the $g_i(y)$ are exactly ± 1 times the elementary symmetric functions of $b_1, \ldots, b_m$, we have $f(y) = h(g_1(y), \ldots, g_m(y))$ with a k-rational polynomial h. Then $f(x) = f({}^t a y a) = f(y) = h(g_1(x), \ldots, g_m(x))$. Since x is generic over k, this shows that $f = h(g_1, \ldots, g_m)$. The algebraic independence of $g_1, \ldots, g_m$ is obvious, because (b_i) can be specialized to an arbitrary element of $\mathbb{C}^m$. Thus we obtain our lemma.

To prove our theorem for $n = 2m$, take (e_i) so that $e_i \geq 0$ and $\sum_{i=1}^{m} i e_i = r$; define p by (2.2) and put

(2.7)
$$\chi(a) = \prod_{i=1}^{m} \det_{2i}(a)^{e_i} \quad (a \in G = GL_n).$$

Then $p \in S_r$, and $p({}^t a x a) = \chi(a) p(x)$ for every $a \in R_n$. By Lemma 2.1, S_r has an irreducible subspace W with p as an eigenvector of highest weight $\{r_1, \ldots, r_{2m}\}$

with $r_{2i-1}=r_{2i}=\sum_{k\geq i} e_k$. Thus each such (e_i) produces an irreducible constituent of τ_r. By (2.3), this shows that τ_r has at least μ_r inequivalent irreducible constituents. This together with the observation in the proof of Proposition 1.2 completes the proof for even n.

Next suppose $n=2m+1$. Write the variable x on T in the form

$$x=\begin{bmatrix} v & w \\ -{}^t w & 0 \end{bmatrix} \quad \text{with} \quad v=-{}^t v\in\mathbb{C}^{2m}_{2m} \text{ and } w\in\mathbb{C}^{2m}.$$

Then an element f of S_r, as a polynomial in v and w, belongs to $\mathcal{I}_r$ if and only if it is invariant under

$$(v,w)\mapsto ({}^t b\,v\,b,\,{}^t b\,v\,c+d\cdot{}^t b\,w)$$

for every $b\in Sp(m,\mathbb{C})$, $c\in\mathbb{C}^{2m}$, and $d\in GL_1$. We see easily that f is a polynomial only in v, invariant under $Sp(m,\mathbb{C})$. Therefore μ_r is again given by (2.3). Defining p and χ by (2.2) and (2.7) in the present case, we can complete the proof of Theorem 2.D for odd n in exactly the same manner as for even n.

Type A. We have $G=GL_n\times GL_m$ and $T=\mathbb{C}^n_m$. Every irreducible representation σ of G is equivalent to $\sigma'\otimes\sigma''$ with irreducible representations σ' of GL_n and σ'' of GL_m.

Theorem 2.A. *Let σ' and σ'' be irreducible representations of GL_n and GL_m, respectively. Then $\sigma'\otimes\sigma''$ occurs in τ_r if and only if σ' and σ'' have the highest weights*

$$\{r_1,\ldots,r_m,\,0,\ldots,0\} \text{ and } \{r_1,\ldots,r_m\} \quad \text{when } n\geq m,$$
$$\{r_1,\ldots,r_n\} \text{ and } \{r_1,\ldots,r_n,\,0,\ldots,0\} \quad \text{when } m\geq n,$$

with the same r_i such that $r_1+\ldots+r_\nu=r$ and $r_\nu\geq 0$, where $\nu=\mathrm{Min}(n,m)$. Such a $\sigma'\otimes\sigma''$ has multiplicity one in τ_r, and the corresponding irreducible subspace of S_r contains an element p defined by

$$p(x)=\prod_{i=1}^{\nu} \det_i(x)^{e_i} \quad (x\in\mathbb{C}^n_m=T)$$

as an eigenvector of highest weight with respect to both σ' and σ'', where $e_i=r_i-r_{i+1}$ for $i<\nu$ and $e_\nu=r_\nu$.

Proof. It is sufficient to prove the case $n\geq m$. Let $x=\binom{y}{z}\in\mathbb{C}^n_m$ with $y\in\mathbb{C}^m_m$ and $z\in\mathbb{C}^{n-m}_m$. Then $\mathcal{I}_r$ is the set of all $f\in S_r$, as polynomials in y and z, invariant under $(y,z)\mapsto(\alpha y\alpha^{-1},\,(\beta y+\delta z)\alpha^{-1})$ for all $\alpha\in GL_m$, $\beta\in\mathbb{C}^{n-m}_m$, and $\delta\in GL_{n-m}$. Therefore f does not involve z. Thus $\mathcal{I}_r$ consists of all homogeneous polynomials in y of degree r invariant under $y\mapsto\alpha y\alpha^{-1}$ for all $\alpha\in GL_m$. Put

$$(2.8) \qquad \det(\lambda 1_m-y)=\sum_{i=0}^{m} g_i(y)\lambda^{m-i} \quad (y\in\mathbb{C}^m_m)$$

with an indeterminate λ. Then $\mathcal{I}_r$ consists of all polynomials of $g_1,\ldots,g_m$, homogeneous in y of degree r. (This is well known and can be proved along the same line of ideas as in the proof of Lemma 2.2.) Therefore (2.3) holds again in the present case. Now, given $\{r_i\}$ such that $r_1 \geq \ldots \geq r_m \geq 0$ and $r = \sum_{i=1}^{m} r_i$, put $e_i = r_i - r_{i+1}$ for $i < m$ and $e_m = r_m$. Define p as in our theorem and put

$$(2.9) \qquad \chi_e(a) = \prod_{i=1}^{m} \det_i(a)^{e_i} \qquad (a \in GL_n \text{ or } GL_m).$$

Then we have $\tau_r(a,b) p = \chi_e(a) \chi_e(b) p$ for $(a,b) \in R_n \times R_m$. Let A, B, and C be the subalgebras of $\mathrm{End}(S_r)$ generated by $\tau_r(GL_n \times 1_m)$, $\tau_r(1_n \times GL_m)$, and $\tau_r(GL_n \times GL_m)$, respectively. By Lemma 2.1, Ap and Bp are subspaces of S_r, irreducible under $GL_n \times 1_m$ and $1_n \times GL_m$, which have p as an eigenvector of the highest weights $\{r_1, \ldots, r_m, 0, \ldots, 0\}$ and $\{r_1, \ldots, r_m\}$, respectively. We see easily that $ap \otimes bp \mapsto abp$ for $a \in A$ and $b \in B$ defines a linear map of $Ap \otimes Bp$ onto Cp, and hence Cp is irreducible under G. Therefore we can complete our proof in the same fashion as for Type D.

Remark 2.3. If $\sigma' : GL_n \to GL_h$ and $\sigma'' : GL_m \to GL_k$ are irreducible representations, then we can take $\mathbb{C}_k^h$ to be the representation space of $\sigma = \sigma' \otimes \sigma''$ by putting $\sigma(a,b) x = \sigma'(a) x \sigma''({}^t b)$. Then the map ζ of Proposition 1.1. associated with σ is a homogeneous polynomial map of T into $\mathbb{C}_k^h$ such that $\zeta(axb) = \sigma'(a) \zeta(x) \sigma''(b)$ for $a \in GL_n$, $b \in GL_m$, and $x \in T$. This is exactly the formulation chosen in [11] in dealing with $SU(n,m)$. In this sense, the essential content of Theorem 2.A is already proved in [11, Proposition 12.2].

Type B. We have $G = SO(n, \mathbb{C}) \times GL_1$, $T = \mathbb{C}^n$, and $[\tau_r(a,b) f](x) = f(b a^{-1} x)$ for $f \in S_r$; we assume $n > 1$.

Lemma 2.4. *For Type B, we have $\mu_r = r + 1$ if $n = 2$, and $\mu_r = [r/2] + 1$ if $n > 2$.*

Proof. If $n = 2$, we have $\mathscr{I}_r = S_r$, and hence $\mu_r = r + 1$. Suppose $n > 2$; write an element f of $\mathscr{I}_r$ in the form $f(x) = \sum_{k=0}^{r} x_1^k g_k(x_2, \ldots, x_n)$. Then g_k is a homogeneous polynomial of degree $r - k$ invariant under $SO(n-1, \mathbb{C})$, and hence it must be a power of $\sum_{i=2}^{n} x_i^2$ times a constant. Thus $f(x) = \sum_{e=0}^{[r/2]} c_e x_1^{r-2e} \left(\sum_{i=2}^{n} x_i^2 \right)^e$ with constants c_e. Since such an f belongs to $\mathscr{I}_r$, we obtain our assertion for $n > 2$.

To determine the irreducible constituents of τ_r, it is sufficient to consider the restriction of τ_r to $SO(n, \mathbb{C})$. In view of our later applications, we take, instead of $SO(n, \mathbb{C})$, the group $SO(Q, \mathbb{C})$ with an arbitrary symmetric element Q of GL_n. We let it act on S_r by the rule $[\tau_r(a) h](x) = h(a^{-1} x)$ for $a \in SO(Q, \mathbb{C})$ and $h \in S_r$.

Theorem 2.B. *An irreducible representation of $SO(Q, \mathbb{C})$ occurs in τ_r if and only if its highest weight (in the sense explained in the proof) has the form*

$$\{r - 2e, 0, \ldots, 0\} \qquad \text{with } e \in \mathbb{Z}, \ 0 \leq e \leq [r/2] \qquad \text{if } n > 2,$$

$$\{r - 2e\} \qquad \text{with } e \in \mathbb{Z}, \ 0 \leq e \leq r \qquad \text{if } n = 2.$$

Its multiplicity is one, and the corresponding subspace of S_r, when $n>2$, is spanned by the functions h_u of the form

$$(2.10) \qquad h_u(x) = ({}^t x Q x)^e ({}^t u Q x)^{r-2e} \qquad (x \in T)$$

for all $u \in T$ such that ${}^t u Q u = 0$. If $n=2$, the corresponding subspace of S_r is one-dimensional, and spanned by a function h_u of form (2.10) with a suitable $u \in T$ such that ${}^t u Q u = 0$.

Proof. Put $n=2m$ or $n=2m+1$ according as n is even or odd. We take Q in the following form:

$$Q = Q_n = \begin{cases} \begin{bmatrix} 0 & 1_m \\ 1_m & 0 \end{bmatrix} & \text{if } n=2m, \\[2ex] \mathrm{diag}[Q_{2m}, 1] & \text{if } n=2m+1. \end{cases}$$

All diagonal matrices in $SO(Q, \mathbb{C})$ form a Cartan subgroup, which we denote by C; thus

$$C = \{ \mathrm{diag}[a_1, \ldots, a_m, a_1^{-1}, \ldots, a_m^{-1}, 1] \,|\, a_k \in GL_1 \},$$

where the last 1 should be ignored if $n=2m$. A "weight" of C is an m-tuple $\{e_1, \ldots, e_m\} \in \mathbb{Z}^m$ which corresponds to, or rather, is identified with, a character of C:

$$\mathrm{diag}[a_1, \ldots, a_m, a_1^{-1}, \ldots, a_m^{-1}, 1] \mapsto a_1^{e_1} \ldots a_m^{e_m}.$$

To prove our theorem for $n>2$, define an element p of S_r by $p(x) = ({}^t x Q x)^e ({}^t \varepsilon_1 Q x)^f$ for $x \in T$, where $f = r - 2e$ and $\varepsilon_1 = {}^t(1, 0, \ldots, 0)$. Then p is an eigenvector of C of weight $\{f, 0, \ldots, 0\}$. Let W_f be the subspace of S_r spanned by $\tau_r(g) p$ for all $g \in SO(Q, \mathbb{C})$. Then W_f consists of all functions of the form $h = \sum_u c_u h_u$ with $c_u \in \mathbb{C}$ and h_u of (2.10). Suppose such an h is an eigenvector of C. Then

$$(2.11) \qquad \sum_u c_u ({}^t u \, dQ x)^f = a_1^{s_1} \ldots a_m^{s_m} \sum_u c_u ({}^t u Q x)^f$$

for every $d = \mathrm{diag}[a_1, \ldots, a_m, a_1^{-1}, \ldots, a_m^{-1}, 1]$, with $s_i \in \mathbb{Z}$. Expressing the left-hand side of (2.11) as a linear combination of monomials in $a_1, \ldots, a_m$ (negative exponents allowed), we see that $\{s_1, \ldots, s_m\} \leq \{f, 0, \ldots, 0\}$. Therefore W_f contains an irreducible subspace of highest weight $\{f, 0, \ldots, 0\}$, whose multiplicity is exactly the multiplicity of weight $\{f, 0, \ldots, 0\}$ in W_f. Thus S_r contains at least μ_r $(= [r/2] + 1)$ inequivalent irreducible constituents. For the reason explained in the proof of Proposition 1.2, these are exactly the irreducible constituents of S_r, each with multiplicity 1. Consequently, the weight $\{f, 0, \ldots, 0\}$ has multiplicity 1 in W_f, and p is its eigenvector; hence W_f, being spanned by the transforms of p, must be irreducible. This completes the proof for $n>2$. If $n=2$, we have $SO(Q, \mathbb{C}) = C$. Let $h_e(x) = x_1^e x_2^{r-e} (= (x_1 x_2)^e x_2^{r-2e})$. Then S_r is the direct sum of the one-dimensional C-subspaces $\mathbb{C} h_e$ for $0 \leq e \leq r$, which completes the proof.

Remark 2.5. The subspace spanned by the functions of (2.10) coincides with the space of all spherical harmonics of degree $r - 2e$, up to the factor $({}^t x Q x)^e$. In this sense, the above theorem is classical. We gave a proof, partly because it can easily be given by the same ideas as in other cases.

Remark 2.6. If we take $O(Q, \mathbb{C})$ instead of $SO(Q, \mathbb{C})$, it can easily be seen that every $O(Q, \mathbb{C})$-irreducible subspace of S_r is spanned by the functions of (2.10) even in the case $n = 2$.

Type C. We have $G = GL_m$ and $T = \{x \in \mathbb{C}_m^m | {}^t x = x\}$.

Theorem 2.C. *An irreducible representation σ of GL_m of highest weight $\{r_1, \ldots, r_m\}$ occurs in τ_r of Type C if and only if all r_i are even, $r_m \geqq 0$, and $r_1 + \ldots + r_m = 2r$. Such a σ has multiplicity 1 in τ_r, and the corresponding irreducible subspace of S_r has the function p defined by*

$$(2.12) \qquad p(x) = \prod_{i=1}^{m} \det_i(x)^{e_i} \qquad (x \in T)$$

as an eigenvector of highest weight, where $e_i = (r_i - r_{i+1})/2$ for $i < m$ and $e_m = r_m/2$.

Proof. Define g_i by (2.8) and p by (2.12). Then $\mathscr{I}_r$ is spanned by the monomials of the g_i, and hence we have again (2.3). Now we have $p({}^t a x a) = \chi_{2e}(a) p(x)$ for all $a \in R_m$ with χ_* of (2.9). By Lemma 2.1, S_r has an irreducible subspace W containing p. The remaining part of the proof is completely parallel to other cases.

We conclude this section by making a simple remark that if a group G and a G-space T are given as the products $\prod_\alpha G_\alpha$ and $\prod_\alpha T_\alpha$ with $\{G_\alpha, T_\alpha\}$ belonging to our list, then the decomposition of $S_r(T)$ can easily be reduced to that of $S_q(T_\alpha)$ with $q \leqq r$.

3. Γ-integrals involving the polynomial maps ζ

We now consider the cases in which $\mathscr{G}/\mathscr{K}$ is isomorphic to a tube domain. This is so for Types B and C; we have to assume that $n = m$ for Type A, and n is even for Type D. In each case, we define a domain P of positivity in a real vector space U and a tube domain $\mathscr{H}$ as follows:

$$\text{Type A:} \quad U = \{x \in \mathbb{C}_m^m | x^* = x\}, \quad P = \{x \in U | x > 0\},$$
$$\mathscr{H} = \{z \in \mathbb{C}_m^m | i(z^* - z) > 0\}.$$

$$\text{Type B:} \quad U = \mathbb{R}^n, \quad n > 1, \quad P = \{x \in U | Q(x, x) > 0, Q(x, \varepsilon) > 0\},$$
$$\mathscr{H} = \{z \in \mathbb{C}^n | i(\bar{z} - z) \in P\}.$$

Here $Q(x, y) = {}^t x Q y$ with a symmetric element Q of $GL_n(\mathbb{R})$ of signature $(1, n - 1)$, and ε is a fixed element of U such that $Q(\varepsilon, \varepsilon) = 1$.

$$\text{Type C:} \quad U = \{x \in \mathbb{R}_m^m | {}^t x = x\}, \quad P = \{x \in U | x > 0\},$$
$$\mathscr{H} = \{z \in \mathbb{C}_m^m | {}^t z = z, i(\bar{z} - z) > 0\}.$$

$$\text{Type D:} \quad U = \{x \in \mathbb{C}_{2m}^{2m} | {}^t x = -x, (Kx)^* = Kx\},$$

$$K = K_m = \mathrm{diag}\,[J_1, \ldots, J_1] \text{ with } m \text{ copies of } J_1 = \begin{pmatrix} 0 & -1 \\ 1 & 0 \end{pmatrix},$$

$$P = \{x \in U \,|\, Kx < 0\},$$

$$\mathscr{H} = \{z \in \mathbb{C}_{2m}^{2m} \,|\, {}^t z = -z,\ i(Kz - (Kz)^*) > 0\}.$$

In the notation of Theorem 2.D, this is the case $n = 2m$.

In all cases, we can identify $U \otimes_{\mathbb{R}} \mathbb{C}$ with T and view $\mathscr{H}$ as a subset of T given by $\mathscr{H} = U + iP$. We define a $\mathbb{C}$-bilinear form Q on T by

$$Q(x, y) = \begin{cases} \mathrm{tr}(xy) & \text{(Type A or C)} \\ {}^t x Q y & \text{(Type B)}, \\ \mathrm{tr}(K x K y)/2 & \text{(Type D)}. \end{cases}$$

It is also convenient to consider, instead of the "upper half" space $\mathscr{H}$, the "right half" space $\mathscr{H}'$ given by

$$\mathscr{H}' = -i\mathscr{H} = P + iU \qquad (\subset T = U \otimes_{\mathbb{R}} \mathbb{C}).$$

We define a function δ on T by

$$\delta(z) = \begin{cases} \det(z) & \text{(Type A or C)}, \\ Q(z, z) & \text{(Type B)}, \\ \mathrm{Pf}(z) & \text{(Type D)} \end{cases}$$

for $z \in T$, where we normalize Pf so that $\mathrm{Pf}(K) = 1$. In all cases, δ is positive on P. Then $\delta(z)^s$ for $s \in \mathbb{C}$ and $z \in \mathscr{H}'$ can be defined by $\delta(z)^s = \exp(s \cdot \log \delta(z))$ with $\log \delta(z) \in \mathbb{R}$ for $z \in P$. As to Type D, we observe that U corresponds to the set of all quaternion hermitian elements of $\mathbb{H}_m^m$ through the map $u \mapsto K\psi(u)$ for $u \in \mathbb{H}_m^m$, where ψ is defined, employing $\mathbf{j}$ in the proof of Theorem 2.D, by

$$(3.1) \qquad \psi((p_{\alpha\beta} + q_{\alpha\beta}\mathbf{j})_{\alpha, \beta = 1, \ldots, m})$$

$$= \begin{bmatrix} x_{11} \cdots x_{1m} \\ \cdots \cdots \cdots \\ x_{m1} \cdots x_{mm} \end{bmatrix}, \qquad x_{\alpha\beta} = \begin{bmatrix} p_{\alpha\beta} & -q_{\alpha\beta} \\ \bar{q}_{\alpha\beta} & \bar{p}_{\alpha\beta} \end{bmatrix},$$

for $(p_{\alpha\beta})$, $(q_{\alpha\beta}) \in \mathbb{C}_m^m$. Thus U, P, and $\mathscr{H}$ for Type D are K times the symbols V, P, and H in Case III of [9, §1]. For Type A, C, or D, the function $\delta(z)^s$ can be generalized to a function $\delta(z; s)$ of $(z, s) \in \mathscr{H}' \times \mathbb{C}^m$ defined by

$$(3.2) \qquad \delta(z; s_1, \ldots, s_m) = \prod_{i=1}^{m} \delta_i(z)^{s_i},$$

where $\delta_i(z) = \det_i(z)$ for Type A or C, and $\delta_i(z) = \mathrm{Pf}_i(z)$ for Type D (see Theorem 2.D).

Define a Euclidean measure dx on U in a standard way (see [9, pp. 272–3]). Put

$$(3.3) \qquad \kappa = \begin{cases} 1 + \iota(m-1)/2 & \text{(Type A, C, or D),} \\ n/2 & \text{(Type B),} \end{cases}$$

$$(3.4) \qquad \iota = \begin{cases} 2 & \text{(Type A),} \\ 1 & \text{(Type C),} \\ 4 & \text{(Type D).} \end{cases}$$

Then we have the following integral formulas:

$$(3.5) \qquad \int_P e^{-Q(z,x)} \delta(x; s_1, \ldots, s_m) \delta(x)^{-\kappa} dx$$
$$= \Gamma_P(s_1, \ldots, s_m) \delta(z^\rho; s_1, \ldots, s_m)$$

$$\text{(Type A, C, or D; } z \in \mathscr{H}', \operatorname{Re}(\sum_{i>k} s_i) > \iota k/2 \text{ for } 0 \leqq k < m),$$

$$(3.6) \qquad \int_P e^{-Q(z,x)} \delta(x)^{s-\kappa} dx = \Gamma_P(s) \delta(z^\rho)^s = \Gamma_P(s) \delta(z)^{-s}$$

$$\text{(Type B; } z \in \mathscr{H}', \operatorname{Re}(s) > \kappa - 1).$$

Here z^ρ and Γ_P are defined by

$$(3.7) \qquad z^\rho = \begin{cases} z^{-1} & \text{(Type A or C),} \\ \delta(z)^{-1} z & \text{(Type B),} \\ K z^{-1} K & \text{(Type D),} \end{cases}$$

$$\Gamma_P(s_1, \ldots, s_m) = \pi^{\iota m(m-1)/4} \prod_{k=0}^{m-1} \Gamma(\sum_{i>k} s_i - (\iota k/2)) \qquad \text{(Type A, C, or D),}$$

$$\Gamma_P(s) = |\det(Q)|^{-1/2} \pi^{\kappa-1} 2^{2s-1} \Gamma(s) \Gamma(s - \kappa + 1) \qquad \text{(Type B).}$$

These formulas are well known (see [2] for example). For Types A and C (resp. Type D), the integral can be computed by reducing it to the case $z = 1_m$ (resp. $z = K$) and making the change of variables $x = u^* u$ (resp. $x = K \cdot \psi(u^* u)$) which transforms the integral into that over the space of upper triangular matrices u. For Type B, see Siegel [12, (20)]. In each case, $z \mapsto z^\rho$ maps $\mathscr{H}'$ onto itself.

We now take an irreducible representation $\{\sigma, \mathbb{C}^d\}$ of G and a homogeneous polynomial map ζ of T into $\mathbb{C}^d$ as in Proposition 1.1. By (iii) of Proposition 1.2, ζ is uniquely determined by σ up to a constant factor. Let us briefly recall the description of the highest weight of σ, including the cases of non-tube domains, determined by Theorems 2.A, 2.B, 2.C, and 2.D.

Type A: σ is the tensor product of two irreducible representations of highest weights $\{r_1, \ldots, r_v, 0, \ldots, 0\}$ and $\{r_1, \ldots, r_v\}$; $r_1 + \ldots + r_v = r$, $v = \operatorname{Min}(n, m)$; $n = m = v$ for tube domains.

Type B: We take G to be $SO(Q, \mathbb{C}) \times GL_1$ with the present Q and consider ζ satisfying (1.2) for $(a, b) \in SO(Q, \mathbb{C}) \times GL_1$ and $x \in \mathbb{C}^n$; σ is equivalent to the representation σ_e on the space of functions of (2.10); $0 \leqq e \leqq [r/2]$ if $n > 2$ and $0 \leqq e \leqq r$ if $n = 2$.

Type C: σ has the highest weight $\{r_1,\ldots,r_m\}$ with even r_i for all i; $r_1+\ldots+r_m=2r$.

Type D: σ has the highest weight $\{r_1,\ldots,r_n\}$, $r_{2i-1}=r_{2i}$ for $1\leq i\leq[n/2]$ and $r_1+\ldots+r_n=2r$; $r_n=0$ if n is odd; $n=2m$ for tube domains.

Proposition 3.1. *Suppose that $n=m$ for Type A and $n=2m$ for Type D. Let σ, ζ, r, $\{r_i\}$, and e be as above; let $z\in\mathscr{H}'$. Then*

$$\int_P e^{-Q(z,x)}\zeta(x)\delta(x)^{s-\kappa}\,dx=\gamma_\sigma(s)\delta(z)^{-s}\zeta(z^\rho)$$

for $\mathrm{Re}(s)>\kappa-1$, *where z^ρ is defined by (3.7) and γ_σ is given by*

$$\gamma_\sigma(s)=\begin{cases}\pi^{m(m-1)/2}\displaystyle\prod_{k=1}^{m}\Gamma(s+r_k-k+1) & \text{(Type A),}\\[2ex]|\det(Q)|^{-1/2}\pi^{\kappa-1}2^{2s+r-1}\Gamma(s+r-e)\Gamma(s+e-\kappa+1) & \text{(Type B),}\\[2ex]\pi^{m(m-1)/4}\displaystyle\prod_{k=1}^{m}\Gamma(s+(r_k-k+1)/2) & \text{(Type C),}\\[2ex]\pi^{m(m-1)}\displaystyle\prod_{k=1}^{m}\Gamma(s+r_{2k}-2k+2) & \text{(Type D).}\end{cases}$$

Proof. As will be seen in the proof of Theorem 4.1 below, the integral in question can be obtained by applying a holomorphic differential operator on $\mathscr{H}'$ to the integral

$$\int_P e^{-Q(z,x)}\delta(x)^{s-\kappa}\,dx$$

which is convergent for $\mathrm{Re}(s)>\kappa-1$. Therefore a well known principle guarantees the convergence of our integral. Let us now prove the formula for Type D. Define a function φ_s on $\mathscr{H}'$ by

$$\delta(z)^s\varphi_s(z)=\int_P e^{-Q(z^\rho,x)}\zeta(x)\delta(x)^{s-\kappa}\,dx\qquad(z\in\mathscr{H}').$$

Then φ_s is holomorphic on $\mathscr{H}'$. Put

$$G_0=\{\alpha\in GL_{2m}(\mathbb{C})\,|\,\bar{\alpha}K=K\alpha\}.$$

We see that ψ of (3.1) maps $GL_m(\mathbb{H})$ onto G_0, and for every $\alpha\in G_0$, $x\mapsto\alpha x\cdot{}^t\alpha$ maps U, P, and $\mathscr{H}'$ onto themselves. Changing z and x for $\alpha z\cdot{}^t\alpha$ and $\alpha x\cdot{}^t\alpha$, we find that $\varphi_s(\alpha z\cdot{}^t\alpha)=\sigma(\alpha)\varphi_s(z)$ for every $\alpha\in G_0$. In particular, we have $\varphi_s(K)=\sigma(\alpha)\varphi_s(K)$ for every α in the group

(3.8) $\{\alpha\in G_0\,|\,\alpha K\cdot{}^t\alpha=K\}.$

Now $\zeta(K)$ has the same property. Since the group of (3.8) is the compact form of $\{\alpha\in GL_{2m}(\mathbb{C})\,|\,\alpha K\cdot{}^t\alpha=K\}$, (ii) of Proposition 1.2 implies that $\varphi_s(K)=\gamma(s)\zeta(K)$ with a constant $\gamma(s)$. Now every element of P can be written as $\alpha K\cdot{}^t\alpha$ with $\alpha\in G_0$. Then $\varphi_s(\alpha K\cdot{}^t\alpha)=\sigma(\alpha)\varphi_s(K)=\gamma(s)\sigma(\alpha)\zeta(K)=\gamma(s)\zeta(\alpha K\cdot{}^t\alpha)$, and so φ_s coincides with $\gamma(s)\zeta$ on P. Since both are holomorphic, $\varphi_s=\gamma(s)\zeta$ on the whole

$\mathcal{H}'$. To compute $\gamma(s)$, observe that γ depends only on the highest weight of σ. Therefore we may assume that the first component ζ_1 of ζ is given by $\zeta_1(x) = \prod_{i=1}^{m} \mathrm{Pf}_i(x)^{e_i}$ as described in Theorem 2.D (cf. also the proof of Proposition 1.1). Then, by (3.5), we have

$$\gamma(s)\,\delta(z)^s\,\zeta_1(z) = \int_P e^{-Q(z^\rho, x)}\,\delta(x; e_1, \ldots, e_{m-1}, s+\dot{e}_m)\,\delta(x)^{-\kappa}\,dx$$

$$= \Gamma_P(e_1, \ldots, e_{m-1}, s+e_m)\,\zeta_1(z)\,\delta(z)^s,$$

and hence $\gamma(s) = \Gamma_P(e_1, \ldots, e_{m-1}, s+e_m)$, which proves the desired formula for Type D. Types A and C can be treated in a similar way.

As for Type B, changing x and z for axb and azb with $0 < b \in \mathbb{R}$ and with $a \in SO(Q, \mathbb{R})^0$, we find that

$$\int_P e^{-Q(z, x)}\,\zeta(x)\,\delta(x)^{s-\kappa}\,dx = \gamma(s)\,\delta(z)^{-s}\,\zeta(z^\rho)$$

with a constant $\gamma(s)$, by the same type of argument as above. To compute $\gamma(s)$, assume $n > 2$ and take the first component ζ_1 of the vector ζ to be $\zeta_1(x) = Q(x, x)^e\, Q(u, x)^f$ with $u \in \mathbb{C}^n$ such that $Q(u, u) = 0$ and with nonnegative integers e and f such that $2e + f = r$. If $f = 0$, we have $\gamma(s) = \Gamma_P(s+e)$ by (3.6). The case $f > 0$ can be obtained by applying $\left(\sum_{i=1}^{n} u_i \partial/\partial z_i \right)^f$ to the equality in this special case. If $n = 2$, the reasoning has to be changed slightly, since $r - 2e$ may be negative, but the final formula is the same.

4. The differential operator $\zeta(D)$

We take a variable matrix $z = (z_{ij})$ or a variable vector $z = (z_i)$ on T, including non-tube cases, and define the complex coordinate functions on T in a standard way as follows:

$$
\begin{array}{lll}
\text{Type A:} & z_{ij} & (1 \leq i \leq n,\ 1 \leq j \leq m); \\
\text{Type B:} & z_i & (1 \leq i \leq n); \\
\text{Type C:} & z_{ij} & (1 \leq i \leq j \leq m); \\
\text{Type D:} & z_{ij} & (1 \leq i < j \leq n).
\end{array}
$$

Then we define (global) holomorphic tangent vectors ∂_{ij} or ∂_i on T and a symbol D which is a matrix or a vector of the same shape as z, whose components are ∂_{ij} or ∂_i, as follows:

$$
\begin{array}{lll}
\text{Type A:} & D = (\partial_{ij}), & \partial_{ij} = \partial/\partial z_{ij}; \\
\text{Type B:} & D = (\partial_i), & \partial_i = \partial/\partial z_i; \\
\text{Type C:} & D = (\partial_{ij}), & \partial_{ij} = \partial_{ji} = (1/2)(1+\delta_{ij})\partial/\partial z_{ij} \quad \text{for } i \leq j;
\end{array}
$$

$$\text{Type D:} \quad D=(\hat{\partial}_{ij}), \quad \hat{\partial}_{ij}=\begin{cases} \partial/\partial z_{ij} & \text{if } i<j, \\ 0 & \text{if } i=j, \\ -\partial/\partial z_{ji} & \text{if } i>j. \end{cases}$$

In each case, D may be viewed as the operator which assigns, to a complex-valued C^{∞}-function f on an open subset of T, a T-valued function $Df=(\partial_{ij}f)$ or $Df=(\hat{\partial}_i f)$. Given $h\in S_r(T)$, we can define $h(D)$ to be a homogeneous differential operator of degree r in an obvious way by substituting ∂_{ij} or $\hat{\partial}_i$ for z_{ij} or z_i. For Type B, we need to consider $h(Q^{-1}D)$ which is defined to be $h\left(\left(\sum_{j=1}^{n} q'_{ij}\partial_j\right)_{i=1}^{n}\right)$, where $(q'_{ij})=Q^{-1}$.

We now restrict ourselves to the cases in which the tube domain can be defined. Namely, we assume $n=m$ for Type A, and $n=2m$ for Type D.

Theorem 4.1. *Let the assumptions and the symbols σ, ζ, r, $\{r_h\}$, and e be the same as in Proposition 3.1; let $s\in\mathbb{C}$. Then, for any fixed branch of $\delta(z)^s$ in an open subset of T on which δ^s is meaningful, one has*

$$\zeta(D)\,\delta(z)^s=\beta_\sigma(s)\,\delta(z)^s\,\zeta('z^{-1}) \qquad \text{(Type A, C, or D)},$$
$$\zeta(Q^{-1}D)\,\delta(z)^s=\beta_\sigma(s)\,\delta(z)^s\,\zeta(2\delta(z)^{-1}z) \quad \text{(Type B)},$$

with a scalar function $\beta_\sigma(s)$ given by

$$\beta_\sigma(s)=\begin{cases} \displaystyle\prod_{h=1}^{m}\prod_{i=1}^{r_h}(s-i+h) & \text{(Type A)}, \\[2ex] \displaystyle\prod_{i=1}^{r-e}(s-i+1)\prod_{j=1}^{e}\left(s-j+\frac{n}{2}\right) & \text{(Type B)}, \\[2ex] \displaystyle\prod_{h=1}^{m}\prod_{i=1}^{r_h/2}\left(s-i+\frac{h+1}{2}\right) & \text{(Type C)}, \\[2ex] \displaystyle\prod_{h=1}^{m}\prod_{i=1}^{r_{2h}}(s-i+2h-1) & \text{(Type D)}. \end{cases}$$

Proof. Let us prove this only for Type D; other types can be treated in a similar way. Putting $\gamma_1(s)=\Gamma_P(0,\ldots,0,s)$, we have

$$\gamma_1(s)\,\delta(z)^{-s}=\int_P e^{-Q(z,x)}\,\delta(x)^{s-\kappa}\,dx \qquad (z\in\mathcal{H}', \ \mathrm{Re}(s)>\kappa-1).$$

Applying $\zeta(D)$ to this and substituting $-KxK$ for K, we find that

$$\gamma_1(s)\,\zeta(D)\,\delta(z)^{-s}=\int_P e^{-Q(z,x)}\,\zeta(KxK)\,\delta(x)^{s-\kappa}\,dx$$
$$=(-1)^r\int_P e^{-Q(-KzK,x)}\,\zeta(x)\,\delta(x)^{s-\kappa}\,dx.$$

By Proposition 3.1, this is equal to $(-1)^r\gamma_\sigma(s)\,\delta(z)^{-s}\,\zeta('z^{-1})$, and hence

$$\zeta(D)\,\delta(z)^{-s}=(-1)^r\,\gamma_1(s)^{-1}\,\gamma_\sigma(s)\,\delta(z)^{-s}\,\zeta('z^{-1}).$$

Employing the explicit expression of $\gamma_\sigma(s)$ in Proposition 3.1, we obtain

$$\gamma_\sigma(s)/\gamma_1(s) = \prod_{k=1}^{m} \Gamma(s + r_{2k} - 2k + 2)/\Gamma(s - 2k + 2).$$

Changing s for $-s$, we obtain the desired formula for sufficiently small $\operatorname{Re}(s)$, and then by analytic continuation, for all s. So far z is in $\mathcal{H}'$; however, analytic continuation validates the formula in any open subset of T on which δ^s is meaningful.

Remark 4.2. In the above theorem, the formula for Type B is valid for ζ and σ defined with respect to $SO(Q, \mathbb{C})$ with an *arbitrary* symmetric element Q of $GL_n(\mathbb{C})$; we put of course $\delta(z) = {}^t z Q z$. In fact, the general case can easily be reduced to the case of real Q of signature $(1, n-1)$. Notice also that $\zeta(Q^{-1}x)$ as a function of x satisfies (1.2) for $(a, b) \in SO(Q^{-1}, \mathbb{C}) \times GL_1$.

Let us now show that Theorem 4.1 can be extended to the form which includes the results even for non-tube types. To do this, we take a certain function which is essentially the standard scalar factor of automorphy such as $\det(cz + d)$ for $\begin{pmatrix} a & b \\ c & d \end{pmatrix} \in Sp(m, \mathbb{R})$ and $z \in \mathcal{H}$ for Type C. Instead of speaking of such a factor, however, we consider some functions on $T \times \mathcal{G}_\mathbb{C}$ more formally and generally. To be precise, we define $\mathcal{G}_\mathbb{C}$ and a set L as follows:

Type A: $\mathcal{G}_\mathbb{C} = GL_{n+m}$,

$$L = \{(c, d) \in \mathbb{C}_n^m \times \mathbb{C}_m^m \mid \operatorname{rank}(c, d) = m\};$$

Type B: $\mathcal{G}_\mathbb{C} = SO\left(\operatorname{diag}\left[Q^{-1}, -\begin{pmatrix} 0 & 1 \\ 1 & 0 \end{pmatrix}\right], \mathbb{C}\right)$ with an arbitrary $Q = {}^t Q \in GL_n(\mathbb{C})$,

$$L = \{(c, d, e) \in \mathbb{C}_n^1 \times \mathbb{C} \times \mathbb{C} \mid cQ \cdot {}^t c = 2de, \, (c, d, e) \neq 0\};$$

Type C: $\mathcal{G}_\mathbb{C} = Sp(m, \mathbb{C})$,

$$L = \{(c, d) \in \mathbb{C}_m^m \times \mathbb{C}_m^m \mid c \cdot {}^t d = d \cdot {}^t c, \, \operatorname{rank}(c, d) = m\};$$

Type D: $\mathcal{G}_\mathbb{C} = SO(E, \mathbb{C})$, $E = \begin{bmatrix} 0 & 1_n \\ 1_n & 0 \end{bmatrix}$,

$$\tilde{L} = \{(c, d) \in \mathbb{C}_n^n \times \mathbb{C}_n^n \mid c \cdot {}^t d = -d \cdot {}^t c, \, \operatorname{rank}(c, d) = n\},$$

L = the connected component of $\tilde{L}$ containing $(0, 1_n)$.

Let $k = m$, 1, m, or n for Type A, B, C, or D, respectively. Assigning to every element of $\mathcal{G}_\mathbb{C}$ its last k rows, we obtain a bijective map of $\mathcal{S} \backslash \mathcal{G}_\mathbb{C}$ onto L with the complex Lie subgroup $\mathcal{S}$ of $\mathcal{G}_\mathbb{C}$ corresponding to $(0, 1_k)$ by this map. Therefore L is a connected complex manifold. If we take $O(E, \mathbb{C})$ instead of $SO(E, \mathbb{C})$ for Type D, we obtain the set $\tilde{L}$, which has two connected components.

We now consider $\mathbb{C}$-valued functions

(4.2a) $\det(cz+d)$ (Type A, C, or D),

(4.2b) $cz+(d/2)\cdot {}^tzQ^{-1}z+e$ (Type B),

with $z\in T$ and $(c,d)\in L$ or $(c,d,e)\in L$. If we take a suitable real form $\mathscr{G}$ of $\mathscr{G}_{\mathbb{C}}$ and z in a domain such as $\mathscr{H}$, then the functions are factors of automorphy. (For Type A, for example, we can take $\mathscr{G}=SU(n,m)$ and the domain to be $\{z\in\mathbb{C}^n_m\mid z^*z<1_m\}$. For other types, see [7, §2] and [8, §8].) For our present purpose, however, it is sufficient to consider (4.2a, b) as functions on $T\times L$ without $\mathscr{G}$. We take L instead of $\tilde{L}$ for Type D because $\det(cz+d)=0$ if $z\in T$, $(c,d)\in\tilde{L}$, and $(c,d)\notin L$, which can easily be verified.

Theorem 4.3. *Let $\zeta: T\to\mathbb{C}^d$ be a polynomial map of Proposition 1.1 attached to an irreducible representation σ of G (with $G=SO(Q,\mathbb{C})\times GL_1$ for Type B). Let $\{r_i\}$ and e be determined as in the paragraphs before Proposition 3.1. Then, for any fixed (c,d) or (c,d,e) of L, and in any open subset of T on which the function of (4.2a, b) is nonvanishing, we have*

$$\zeta(D)\det(cz+d)^s=\beta_\sigma(s)\det(cz+d)^s\,\zeta({}^tc\cdot{}^t(cz+d)^{-1}) \qquad \text{(Type A or C)},$$

$$\zeta(D)(cz+(d/2)\cdot{}^tzQ^{-1}z+e)^s$$
$$=\beta_\sigma(s)(cz+(d/2)\cdot{}^tzQ^{-1}z+e)^s\,\zeta(D[\log(cz+(d/2)\cdot{}^tzQ^{-1}z+e)]) \qquad \text{(Type B)},$$

$$\zeta(D)\det(cz+d)^s=\beta_\sigma(2s)\det(cz+d)^s\,\zeta({}^tc\cdot{}^t(cz+d)^{-1}) \qquad \text{(Type D)},$$

where β_σ is the polynomial given in Theorem 4.1 with the following conventions: for Type A, $\prod_{h=1}^{m}$ should be $\prod_{h=1}^{v}$ with $v=\mathrm{Min}(n,m)$; for Type D, $m=[n/2]$.

Proof. We first prove this for Type D. If $n=2m$ and $\det(c)\neq 0$, we have, by Theorem 4.1,

$$\zeta(D)\det(cz+d)^s=\det(c)^s\,\zeta(D)\,\delta(z+c^{-1}d)^{2s}$$
$$=\beta_\sigma(2s)\det(c)^s\,\delta(z+c^{-1}d)^{2s}\,\zeta({}^t(z+c^{-1}d)^{-1})$$
$$=\beta_\sigma(2s)\det(cz+d)^s\,\zeta({}^tc\cdot{}^t(cz+d)^{-1}).$$

Now we may view both sides of this equality as holomorphic functions on the complex manifold

$$\{(c,d,z)\in L\times T\mid\det(cz+d)\neq 0\}.$$

Then, by analytic continuation, we can remove the condition $\det(c)\neq 0$. The same type of reasoning applies to Type C. Next suppose $n=2m+1$; let W be the subspace of S_r of highest weight $\{r_1,\ldots,r_n\}$ as in Theorem 2.D. Write the variable matrix z in T in the form

$$z=\begin{bmatrix} u & v \\ -{}^tv & 0 \end{bmatrix} \qquad (u\in T',\ v\in\mathbb{C}^{2m}),$$

where $T' = \{u \in \mathbb{C}^{2m}_{2m} \mid {}^t u = -u\}$. Let W' be the subspace of W consisting of all functions which do not involve v. Let $f \in W'$ and $\alpha = \begin{pmatrix} a & e \\ 0 & b \end{pmatrix} \in R_n$ with $a \in R_{2m}$, $e \in \mathbb{C}^{2m}$ and $b \in \mathbb{C}$; write $f(z) = f'(u)$ with $f' \in S_r(T')$. Then we see that $f({}^t \alpha z \alpha) = f'({}^t a u a)$. Therefore if f' is an eigenvector of R_{2m}, then f is an eigenvector of R_n, which must be a constant times the function p of Theorem 2.D. Since $p \in W'$, this shows that W' is irreducible under GL_{2m} and has the highest weight $\{r_1, \ldots, r_{2m}\}$ with the same r_i. Put

$$W_i = \{f \in W \mid \tau_r(\mathrm{diag}[1_{2m}, b]) f = b^i f\}.$$

Then $W_0 = W'$ and $W = \overset{r}{\underset{i=0}{\bigoplus}} W_i$. Therefore we may assume, after changing the coordinate system, that

$$\sigma(\mathrm{diag}[a, b]) = \mathrm{diag}[\sigma_0(a), b\sigma_1(a), \ldots, b^r \sigma_r(a)] \qquad (a \in GL_m, \ b \in GL_1)$$

with representations σ_i of GL_{2m}; in particular, σ_0 is an irreducible representation of highest weight $\{r_1, \ldots, r_{2m}\}$. (This conclusion can be derived also from the well known „Verzweigungssatz".) Putting $b = 0$, we find that

$$\zeta \begin{pmatrix} u & 0 \\ 0 & 0 \end{pmatrix} = \begin{bmatrix} \zeta_0(u) \\ 0 \end{bmatrix} \qquad (u \in T')$$

with a function ζ_0 on T' such that $\zeta_0(au \cdot {}^t a) = \sigma_0(a)\zeta_0(u)$ for $a \in GL_{2m}$. Now fix s and put $h(e, z) = \zeta(D)\det(ez + 1_n)^s$ for $e \in T$ and $z \in T$. Put $w = {}^t aza$ with $a \in GL_n$ and let D_w denote the symbol D defined with respect to w in place of z. Then $D = aD_w \cdot {}^t a$, and hence

$$h(ae \cdot {}^t a, z) = \zeta(D)\det(ae \cdot {}^t az + 1)^s = \zeta(aD_w \cdot {}^t a)\det(ew + 1)^s$$
$$= \sigma(a)\zeta(D_w)\det(ew + 1)^s = \sigma(a)h(e, w) = \sigma(a)h(e, {}^t aza).$$

Let $\varepsilon = \mathrm{diag}[J_m, 0]$. Writing simply J for J_m, we see that $\det(\varepsilon z + 1) = \det(Ju + 1)$, and hence

$$h(\varepsilon, z) = \zeta(D)\det(Ju + 1)^s = \begin{bmatrix} \zeta_0(D_u) \\ 0 \end{bmatrix}\det(Ju + 1)^s$$

$$= \beta_\sigma(2s)\det(Ju + 1)^s \begin{bmatrix} \zeta_0({}^t J \cdot {}^t (Ju + 1)^{-1}) \\ 0 \end{bmatrix}$$

$$= \beta_\sigma(2s)\det(Ju + 1)^s \zeta(x) \quad \text{with} \quad x = \mathrm{diag}[{}^t J \cdot {}^t (Ju + 1)^{-1}, 0],$$

by virtue of our result for $n = 2m$. Now, for $a \in GL_n$, put ${}^t aza = \begin{pmatrix} U & * \\ * & 0 \end{pmatrix}$. Then we have

$$h(a\varepsilon \cdot {}^t a, z) = \sigma(a)h(\varepsilon, {}^t aza) = \beta_\sigma(2s)\det(JU + 1)^s \zeta(aX \cdot {}^t a)$$

with $X = \mathrm{diag}[{}^t J \cdot {}^t (JU + 1)^{-1}, 0]$. By a direct calculation, we can verify that $aX \cdot {}^t a = {}^t(a\varepsilon \cdot {}^t a) \cdot {}^t(a\varepsilon \cdot {}^t az + 1)^{-1}$ and $\det(JU + 1) = \det(a\varepsilon \cdot {}^t az + 1)$, and so, putting $a\varepsilon \cdot {}^t a = e$, we obtain

$$(4.4) \qquad \zeta(D)\det(ez + 1)^s = \beta_\sigma(2s)\det(ez + 1)^s \zeta({}^t e \cdot {}^t(ez + 1)^{-1}).$$

Given $(c,d) \in L$, if $\det(d) \neq 0$ and $\mathrm{rank}(c) = 2m$, the desired formula follows immediately from (4.4). The general case is settled by analytic continuation.

Next we consider Type A. If $n = m$, our formula follows from Theorem 4.1 together with analytic continuation. Suppose $n > m$. By the same reasoning similar to that for Type D, we may assume that

$$\zeta \begin{bmatrix} u \\ 0 \end{bmatrix} = \begin{bmatrix} \zeta_0(u) \\ 0 \end{bmatrix} \quad (u \in \mathbb{C}^m_m),$$

$$\sigma(\mathrm{diag}[a, b 1_{n-m}], c)$$
$$= \mathrm{diag}[\sigma_0(a,c), b\sigma_1(a,c), \ldots, b^r \sigma_r(a,c)] \quad (a, c \in GL_m, b \in GL_1),$$
$$\zeta_0(au \cdot {}^t c) = \sigma_0(a,c)\,\zeta_0(u),$$

with representations σ_i of $GL_m \times GL_m$; in particular, σ_0 is an irreducible representation of "the same highest weight" as σ in an obvious sense. Put $h(e,z) = \zeta(D)\det(ez + 1_m)^s$ for $e \in \mathbb{C}^m_n$ and $z \in \mathbb{C}^n_m = T$. We can easily verify that

$$h(be \cdot {}^t a, z) = \sigma(a,b)\,h(e, {}^t azb) \quad (a \in GL_n, b \in GL_m).$$

Put $\eta = (1_m, 0)$ and ${}^t z = ({}^t u, {}^t v)$ with $u \in \mathbb{C}^m_n$ and $v \in \mathbb{C}^{n-m}_m$. Then

$$h(\eta, z) = \zeta(D)\det(u+1)^s = \beta_\sigma(s)\det(u+1)^s \zeta \begin{bmatrix} {}^t(u+1)^{-1} \\ 0 \end{bmatrix}$$

by Theorem 4.1. After this we can complete the proof in the case $n > m$ in exactly the same fashion as for Type D. The case $n < m$ can be proved by a similar computation.

As for Type B, recall that $\delta(z) = {}^t z Q z$. If $(c, d, e) \in L$ and $d \neq 0$, we have

$$cz + (d/2)\cdot {}^t z Q^{-1} z + e = (d/2)\delta(Q^{-1}z + d^{-1}\cdot {}^t c).$$

Therefore the desired formula is obtained directly from Theorem 4.1 and Remark 4.2, together with analytic continuation.

5. Differential operators of arithmetic type

In our previous papers [7], [8], and [11], we introduced certain differential operators which send automorphic forms of a given weight to those of another weight, and which keep a certain arithmeticity at CM-points. Let us now complement this theory by introducing a particular type of operator Δ^σ_ρ by reformulating what was done in [11]. To make our exposition simpler, we treat here only Type C. The principles explained in [8, §8], [7, §11], and [11, §§1, 2, 11] will tell how to define and prove corresponding objects and formulas in other cases. In fact, so long as Type A is concerned, the result of [11] is complete and there is nothing new to be added.

Thus we take $\mathscr{G} = Sp(m, \mathbb{R})$, $T = \{z \in \mathbb{C}^m_m \mid {}^t z = z\}$, and

$$\mathscr{H} = \{z \in T \mid i(\bar{z} - z) > 0\}.$$

For $z \in \mathcal{H}$ and $\alpha = \begin{pmatrix} a & b \\ c & d \end{pmatrix} \in \mathcal{G}$ with a, b, c, d of size m, we put

$$\eta(z) = i(\bar{z} - z), \qquad\qquad \alpha(z) = (az + b)(cz + d)^{-1}$$

$$\mu(\alpha, z) = \mu_\alpha(z) = cz + d, \quad j(\alpha, z) = j_\alpha(z) = \det(cz + d).$$

Given a rational representation $\{\rho, V\}$ of GL_m, $\alpha \in \mathcal{G}$, and a V-valued function f on $\mathcal{H}$, we define another such function $f\|_\rho \alpha$ by

$$(f\|_\rho \alpha)(z) = \rho(\mu_\alpha(z))^{-1} f(\alpha(z)) \qquad (z \in \mathcal{H}).$$

If f is C^∞, we can define a function $Df: \mathcal{H} \to \mathrm{Hom}(T, V)$ by

$$(Df)(u) = \sum_{i,j=1}^{m} u_{ij} \partial_{ij} f \qquad (u = (u_{ij}) \in T)$$

with ∂_{ij} of Sect. 4. (If we view the $\partial_{ij} f$ as the components of Df, the present Df is consistent with that of Sect. 4.) For $0 < r \in \mathbb{Z}$, $D^r f$ is defined inductively by $D^r f = D(D^{r-1} f)$ and takes values in

$$\mathrm{Hom}(T, \mathrm{Hom}(T, \ldots, \mathrm{Hom}(T, V) \ldots)).$$

We identify this vector space with the space of all multilinear maps of T^r into V, denote it by $\mathrm{Ml}_r(T, V)$, and define a representation $\{\rho \otimes \tau^r, \mathrm{Ml}_r(T, V)\}$ of GL_m by

$$[(\rho \otimes \tau^r)(a)h](u_1, \ldots, u_r) = \rho(a) h({}^t a u_1 a, \ldots, {}^t a u_r a) \quad (a \in GL_m)$$

for $h \in \mathrm{Ml}_r(T, V)$ and $u_i \in T$. We write $\rho \otimes \tau$ for $\rho \otimes \tau^1$. Given a V-valued function f as above, we define a function $D_\rho f$ on $\mathcal{H}$ with values in $\mathrm{Hom}(T, V)$ by

$$(5.1) \qquad\qquad D_\rho f = \rho(\eta)^{-1} D(\rho(\eta) f),$$

and a function $D_\rho^{(r)} f: \mathcal{H} \to \mathrm{Ml}_r(T, V)$ inductively by

$$(5.2) \qquad\qquad D_\rho^{(r)} f = D_{\rho \otimes \tau}^{(r-1)}(D_\rho f), \qquad D_\rho^{(1)} = D_\rho.$$

Then we have, as proved in [8, Lemma 2.1, (2.10)],

$$(5.3) \qquad\qquad D_\rho^{(r)}(f\|_\rho \alpha) = (D_\rho^{(r)} f)\|_{\rho \otimes \tau^r} \alpha \qquad (\alpha \in \mathcal{G}).$$

Call an element h of $\mathrm{Ml}_r(T, V)$ *symmetric* if $h(u_1, \ldots, u_r)$ is invariant under all permutations of $u_1, \ldots, u_r$. Obviously $D^r f$ is symmetric. Somewhat nontrivially, we can show that $D_\rho^{(r)} f$ is symmetric. In fact, as shown in [8] and [11], we can define $D_\rho^{(r)}$ also on the domain $\mathcal{B} = \{z \in T \,|\, \bar{z}z < 1_m\}$ in a consistent way; moreover $(D_\rho^{(r)} f)(0) = (D^r f)(0)$ by [8, Lemma 2.2]. Then (5.3), or rather [8, (2.10)], proves the desired symmetricity of $D_\rho^{(r)} f$ at every point of $\mathcal{H}$ and $\mathcal{B}$.

Thus $D_\rho^{(r)} f$ takes its values in the subspace of $\mathrm{Ml}_r(T, V)$ consisting of all symmetric elements, which is essentially $S_r(T) \otimes V$. Therefore, decomposing $S_r(T)$ into GL_m-irreducible subspaces, we can decompose $D_\rho^{(r)}$ into several operators. To do this in an explicit way, take an irreducible polynomial representation $\{\sigma, W\}$ of GL_m that occurs in S_r and fix a nontrivial map ζ: $T \to W$ of degree r such that

$$(5.4) \qquad \zeta(ax \cdot {}^t a) = \sigma(a)\zeta(x) \qquad (a \in GL_m, \ x \in T).$$

There is a unique symmetric element ζ' of $Ml_r(T, W)$ such that $\zeta'(u, \ldots, u) = \zeta(u)$. For simplicity, let us write ζ also for ζ'. Then we define a linear map Φ_ζ: $Ml_r(T, V) \to V \otimes W$ by

$$\Phi_\zeta h = \sum_{i, j, \ldots, s, t} h(\varepsilon_{ij}, \ldots, \varepsilon_{st}) \otimes \zeta(\varepsilon_{ij}, \ldots, \varepsilon_{st})$$

where $\varepsilon_{ij} = (e_{ij} + e_{ji})/2 \ (\in T)$ with the standard matrix units e_{ij}, and $(i, j, \ldots, s, t)$ ranges over $\{1, \ldots, m\}^{2r}$. We see easily that

$$(5.5) \qquad \Phi_\zeta \circ (\rho \otimes \tau^r)(a) = (\rho \otimes \sigma)(a) \circ \Phi_\zeta \qquad (a \in GL_m).$$

Given a V-valued function f, define $\Delta_\rho^\sigma f$: $\mathscr{H} \to V \otimes W$ by

$$(5.6) \qquad \Delta_\rho^\sigma f = \Phi_\zeta D_\rho^{(r)} f.$$

From (5.3) and (5.5), we obtain

$$(5.7) \qquad \Delta_\rho^\sigma(f \|_\rho \alpha) = (\Delta_\rho^\sigma f) \|_{\rho \otimes \sigma} \alpha \qquad (\alpha \in \mathscr{G}).$$

We can define Δ_ρ^σ with $\mathrm{Hom}(W, V)$ instead of $V \otimes W$ as we did in [11, §1]. Two definitions are of course equivalent, but each has advantages and disadvantages.

Proposition 5.1. *Let ζ be a map of (5.4) of degree r with an irreducible representation σ of GL_m of highest weight $\{r_1, \ldots, r_m\}$ (cf. Theorem 2.C); let $\rho(x) = \det(x)^k$ with $k \in \mathbb{Z}$ and $s \in \mathbb{C}$. Then*

$$\Delta_\rho^\sigma(\det(\eta)^s j_\alpha^{-k} |j_\alpha|^{-2s})$$
$$= i^r \beta_\sigma(-k-s) \det(\eta)^s j_\alpha^{-k} |j_\alpha|^{-2s} \zeta(\mu_\alpha^{-1} \bar{\mu}_\alpha \eta^{-1}) \qquad (i = \sqrt{-1})$$

with β_σ of Theorem 4.1.
    Notice that $\mu_\alpha^{-1} \bar{\mu}_\alpha \eta^{-1}$ is symmetric. In fact, we have

$$(5.8) \qquad \mu_\alpha(z)^{-1} \overline{\mu_\alpha(z)} \eta(z)^{-1} = \mu_\alpha(z)^{-1} \eta(\alpha(z))^{-1} \cdot {}^t \mu_\alpha(z)^{-1} \in T \qquad (\alpha \in \mathscr{G}).$$

The proof of the proposition can be given in exactly the same fashion as in [11, Lemma 1.2] and [8, Lemma 2.3], employing Theorem 4.3.
    Taking anti-holomorphic derivations $\bar{\partial}_{ij} = 2^{-1}(1 + \delta_{ij})\partial/\partial \bar{z}_{ij}$ in place of ∂_{ij}, we define $\bar{D}f$: $\mathscr{H} \to \mathrm{Hom}(T, V)$ and $\bar{D}^r f$: $\mathscr{H} \to Ml_r(T, V)$ in a natural way. Then we define $E_1 f$: $\mathscr{H} \to \mathrm{Hom}(T, V)$ by

$$(5.9) \qquad (E_1 f)(u) = (\bar{D}f)(\eta u \eta) \qquad (u \in T).$$

Then $E_1^r f$ takes values in $Ml_r(T, V)$. Further we define $E_\sigma f$: $\mathscr{H} \to V \otimes W$ by $E_\sigma f = \Phi_\zeta E_1^r f$. Then, from [11, (2.5a)], we can derive

$$(5.10) \qquad E_\sigma(f \|_\rho \alpha) = (E_\sigma f) \|_{\rho \otimes \sigma_*} \alpha \qquad (\alpha \in \mathscr{G}),$$

where $\sigma_*(a)=\sigma({}^t a^{-1})$. Changing σ and ζ for equivalent objects, we may assume that $W=\mathbb{C}^N$, $\sigma({}^t a)={}^t\sigma(a)$, and both σ and ζ are $\mathbb{R}$-rational. Every element h of $V\otimes W\otimes W$ can be written in the form $h=\sum_{i,j=1}^{N} h_{ij}\otimes e_i\otimes e_j$ with $h_{ij}\in V$, where $\{e_1,\ldots,e_N\}$ is the standard basis of $\mathbb{C}^N$. Then we define a contraction operator $\theta_\sigma\colon V\otimes W\otimes W\to V$ by $\theta_\sigma h=\sum_{i=1}^{N} h_{ii}$. We can now generalize the results of [11, §11] to the present case. Namely, Δ_ρ^σ and $(-1)^r \theta_\sigma E_\sigma$ (resp. $\theta_\sigma \Delta_{\rho\otimes\sigma_*}^\sigma$ and $(-1)^r E_\sigma$) *are adjoint to each other in the sense described in* [11, Theorem 11.5]; *the composites of these pairs of operators are nonnegative self-adjoint operators similar to the Laplace-Beltrami operator* (see [11, Corollary 11.8]). The same type of results hold also for Types A, B, and D. The precise formulation and its proof are straightforward generalizations or modifications of [11, §11], and therefore may be left to the reader.

6. Applications to Eisenstein series

We now consider $SP(m,F)$ with a totally real algebraic number field F of finite degree. Let φ denote the set of all embeddings of F into $\mathbb{R}$. We identify $F\otimes_\mathbb{Q}\mathbb{R}$ with $\mathbb{R}^\varphi$ through the map $a\otimes b\mapsto(a^\tau b)_{\tau\in\varphi}$ for $a\in F$ and $b\in\mathbb{R}$; similarly we embed $Sp(m,F)$ into $Sp(m,\mathbb{R})^\varphi$, and let it act on $\mathscr{H}^\varphi$ componentwise. For $\alpha\in Sp(m,F)$ and $z=(z_\tau)_{\tau\in\varphi}\in\mathscr{H}^\varphi$, we denote by $\eta(z),\mu_\alpha(z)$, and $j_\alpha(z)$ the elements $(\eta(z_\tau))_{\tau\in\varphi}$, $(\mu(\alpha^\tau,z_\tau))_{\tau\in\varphi}$ and $(j(\alpha^\tau,z_\tau))_{\tau\in\varphi}$ of T^φ, GL_m^φ, and GL_1^φ.

Given a representation $\{\rho,V\}$ of GL_m^φ and a V-valued function f on $\mathscr{H}^\varphi$, we put $f\|_\rho\alpha=\rho(\mu_\alpha(z))^{-1}f(\alpha(z))$ for $\alpha\in Sp(m,F)$. Now take a polynomial representation

$$(6.1) \qquad\qquad \sigma\colon GL_m^\varphi\to GL_n$$

and a polynomial map

$$(6.2) \qquad\qquad \zeta\colon T^\varphi\to\mathbb{C}^n$$

such that $\zeta(ax\cdot{}^t a)=\sigma(a)\zeta(x)$, where $ax\cdot{}^t a=(a_\tau x_\tau\cdot{}^t a_\tau)_{\tau\in\varphi}$. We assume that σ is irreducible and both σ and ζ are $\overline{\mathbb{Q}}$-rational. More explicitly, for each $\tau\in\varphi$, we pick $\overline{\mathbb{Q}}$-rational σ_τ and ζ_τ as described in Theorem 2.C and Proposition 1.1, and put $\sigma(a)=\bigotimes_{\tau\in\varphi}\sigma_\tau(a_\tau)$ and $\zeta(x)=\bigotimes_{\tau\in\varphi}\zeta_\tau(x_\tau)$. Then Δ_ρ^σ can be defined as the "product" of $\Delta_{\rho_\tau}^{\sigma_\tau}$ for all $\tau\in\varphi$, where ρ_τ is the restriction of ρ to the τ-factor of GL_m^φ; see [11, §3] for a precise definition (cf. [7, §11], [8, §3]). Then the assertions of Theorems 3.2 and 3.4 of [11] can be modified so as to apply to the present Δ_ρ^σ (cf. also the proof of [8, Theorem 3.2]). Namely, if f belongs to the set $\mathscr{A}_\rho(\overline{\mathbb{Q}})$ or $\mathscr{L}$ defined in [11, §3], then $\Delta_\rho^\sigma f$ has the expected property of arithmeticity at every CM-point of $\mathscr{H}^\varphi$.

Before proceeding further, we make a notational convention: for $c=(c_\tau)_{\tau\in\varphi}\in\mathbb{C}^\varphi$ and $s\in\mathbb{C}$, we put

$$(6.3) \qquad\qquad c^s=\prod_{\tau\in\varphi} c_\tau^s,$$

whenever the product is meaningful. Now, taking an arbitrary congruence subgroup Γ of $Sp(m, F)$, we define a series $\mathbf{E}(z, s; k, \Gamma)$ by

$$\mathbf{E}(z, s; k, \Gamma) = \det(\eta(z))^s \sum_{\alpha \in (P \cap \Gamma) \backslash \Gamma} j_\alpha(z)^{-k} |j_\alpha(z)|^{-2s},$$

where $z \in \mathscr{H}^\varphi$, $k \in \mathbb{Z}$, $s \in \mathbb{C}$, and P is the parabolic subgroup of $Sp(m, F)$ consisting of all the elements of the form $\begin{pmatrix} a & b \\ 0 & d \end{pmatrix}$ with d of size m; $\det(\eta)^s$, j_α^{-k}, and $|j_\alpha|^{-2s}$ should be understood in the sense of (6.3). We assume that $j_\alpha^k = 1$ for all $\alpha \in P \cap \Gamma$.

We can view this series as a special case of a $\mathbb{C}^n$-valued series

$$\begin{aligned} \mathbf{E}(z, s; k, \zeta, \Gamma) \\ = \det(\eta(z))^s \sum_{\alpha \in (P \cap \Gamma) \backslash \Gamma} j_\alpha(z)^{-k} |j_\alpha(z)|^{-2s} \zeta(\mu_\alpha(z)^{-1} \overline{\mu_\alpha(z)} \eta(z)^{-1}) \end{aligned}$$

defined with ζ of (6.2). By (5.8), we see that

$$\mathbf{E}(z, s; k, \zeta, \Gamma) = \sum_{\alpha \in (P \cap \Gamma) \backslash \Gamma} [\det(\eta)^s \zeta(\eta^{-1})] \|_\rho \alpha,$$

where $\rho(x) = \det(x)^k \sigma(x)$ for $x \in GL_m^\varphi$. It follows that $\mathbf{E}$, as a function of z (when convergent), is an automorphic form of weight ρ with respect to Γ. Another type of series can be defined by

$$\mathbf{E}(z, s; k, \psi, \mathbf{b}) = \det(\eta(z))^s E(z, 2s; k, \psi, \mathbf{b})$$

with E of [10, (2.18)]. Here $\mathbf{b}$ is an integral ideal of F and ψ is a Hecke character of F satisfying the conditions [10, (2.13 a, b)]. This series can be written in the form

$$\mathbf{E}(z, s; k, \psi, \mathbf{b}) = \det(\eta(z))^s \sum_{\alpha \in A} b_\alpha c_\alpha^{2s} j_\alpha(z)^{-k} |j_\alpha(z)|^{-2s}$$

with a certain subset A of $Sp(m, F)$, $b_\alpha \in \overline{\mathbb{Q}}$, and $0 < c_\alpha \in \mathbb{Q}$. Then we put

$$\begin{aligned} \mathbf{E}(z, s; k, \zeta, \psi, \mathbf{b}) \\ = \det(\eta(z))^s \sum_{\alpha \in A} b_\alpha c_\alpha^{2s} j_\alpha(z)^{-k} |j_\alpha(z)|^{-2s} \zeta(\mu_\alpha(z)^{-1} \overline{\mu_\alpha(z)} \eta(z)^{-1}). \end{aligned}$$

In [8] and [11], we studied the nature of the values of these series when $s = 0$ and z is specialized to a CM-point. We gave in [11] a complete answer to the same problem for the Eisenstein series on the maximally isotropic unitary groups over an arbitrary CM-field, but had to impose a certain condition on ζ in the symplectic case. In fact, in [11], we treated only those ζ which are the restrictions to T of the representations of GL_m whose restrictions to $O(m, \mathbb{C})$ are irreducible. By virtue of the results of the preceding sections, we can now give a complete answer as follows:

Theorem 6.1. *Let σ and ζ be as above, and let $\rho(x) = \det(x)^k \cdot \sigma(x)$. Let $\mathbf{E}(z, s)$ denote $\mathbf{E}(z, s; k, \zeta, \Gamma)$ or $\mathbf{E}(z, s; k, \zeta, \psi, \mathbf{b})$. Further let w be a CM-point of $Sp(m, F)$*

on $\mathscr{H}^{\varphi}$. Suppose $k > (m+2)/2$ or $k = (m+1)/2$. Then $\mathbf{E}(z,s)$ is finite at $s=0$, and the components of the vector $\pi^{-d} \mathbf{P}_{\rho}(w)^{-1} \mathbf{E}(w,0)$ are all algebraic, where d is the total degree of the map ζ. The same assertions hold for $\mathbf{E}(z,s;k,\zeta,\psi,\mathbf{b})$ with $k = (m+2)/2$ if $F \neq \mathbb{Q}$ or $\psi^2 \neq 1$.

Here $\mathbf{P}_{\rho}(w)$ is an element of $GL_n(\mathbb{C})$ determined at each CM-point w of $\mathscr{H}^{\varphi}$ (see [7, (4.7)], [11, (3.9 b)]).

Proof. Let $\xi(x) = \det(x)^k$ for $x \in GL_m^{\varphi}$. By Proposition 5.1, we have

$$\Delta_{\xi}^{\sigma} \mathbf{E}(z,s;k,\Gamma) = b(s) \mathbf{E}(z,s;k,\zeta,\Gamma)$$

with a polynomial function b, which is a product of several factors $i^r \beta_{\sigma}(-k-s)$ of that proposition. From the exact form of β_{σ} given in Theorem 4.1, we see that $b(0) \neq 0$ if $k \geq (m+1)/2$. Therefore we obtain our theorem for the same reason as in [11, §7] (cf. also [8, §5]).

Theorem 6.2. *The notation being the same as in Theorem 6.1, suppose $k = (m-1)/2$. Then $\mathbf{E}(w,s)$ has at most a simple pole at $s=1$, and its residue has the form $\pi^e R_F \mathbf{P}_{\rho}(w) A$ with $A \in \overline{\mathbb{Q}}^n$, where R_F is the regulator of F, and $e = d - m[F:\mathbb{Q}]$.*

The proof is again the same as in [11, Theorem 7.3]; see also [11, Remark 7.5].

References

1. Gårding, L.: Extension of a formula by Cayley to symmetric determinants. Proc. Edinburgh Math. Soc. Ser. 2, **8**, 73-75 (1947)
2. Gindikin, S.G.: Analysis in homogeneous domains (in Russian). Uspekhi Mat. Nauk **19**, (118), 3-92 (1964)
3. Godement, R.: Ou l'on généralise une intégrale .... Exposé No. 5, Séminaires H. Cartan, Fonctions automorphes, 1957/58
4. Hua, L.K.: Harmonic analysis of functions of several complex variables in the classical domains, Translations of Mathematical Monographs, vol. 6, Amer. Math. Soc., 1963
5. Johnson, K.D.: On a ring of invariant polynomials on a hermitian symmetric spaces. J. of Algebra **67**, 72-81 (1980)
6. Schmid, W.: Die Randwerte holomorpher Funktionen auf hermitesch symmetrischen Räumen. Invent. math. **9**, 61-80 (1969).
7. Shimura, G.: The arithmetic of certain zeta functions and automorphic forms on orthogonal groups. Ann. of Math. **111**, 313-375 (1980)
8. Shimura, G.: Arithmetic of differential operators on symmetric domains. Duke Math. J. **48**, 813-843 (1981)
9. Shimura, G.: Confluent hypergeometric functions on tube domains. Math. Ann. **260**, 269-302 (1982)
10. Shimura, G.: On Eisenstein series. Duke Math. J. **50**, 417-476 (1983)
11. Shimura, G.: Differential operators and the singular values of Eisenstein series. Duke Math. J. **51**, 261-329 (1984)
12. Siegel, C.L.: Über die Zetafunktionen indefiniter quadratischer Formen. Math. Z. **43**, 682-708 (1938)

Oblatum 9-XI-1983

On Eisenstein series of half-integral weight

Duke Mathematical Journal, 52 (1985), 281-314

The definition of an Eisenstein series of $\mathrm{Sp}(m, \mathbf{Q})$ of half-integral weight is as follows. We first consider a theta series

$$\theta(z) = \sum_{x \in \mathbf{Z}^m} \exp(\pi i \cdot {}^t x z x),$$

where z is the standard variable in the space H_m of complex symmetric matrices with positive definite imaginary part. For $\gamma = \left(\begin{smallmatrix} a & b \\ c & d \end{smallmatrix}\right) \in \mathrm{Sp}(m, \mathbf{Q})$ with a, b, c, d of size m, we write $a = a_\gamma$, $b = b_\gamma$, $c = c_\gamma$, $d = d_\gamma$, and define subgroups P and $\Gamma_0(N)$ of $\mathrm{Sp}(m, \mathbf{Q})$ by

$$P = \{\gamma \in \mathrm{Sp}(m, \mathbf{Q}) \mid c_\gamma = 0\}.$$

$$\Gamma_0(N) = \{\gamma \in \mathrm{Sp}(m, \mathbf{Z}) \mid b_\gamma \equiv 0 \ (\mathrm{mod}\, 2),\ c_\gamma \equiv 0 \ (\mathrm{mod}\, N/2)\},$$

where N is a positive integer divisible by 4. We can show that

$$\theta(\gamma(z)) = h_\gamma(z)\theta(z) \qquad \text{for every} \quad \gamma \in \Gamma_0(4)$$

with a factor of automorphy h_γ such that $h_\gamma(z)^4 = \det(c_\gamma z + d_\gamma)^2$. Taking an odd integer k and a Dirichlet character ψ modulo N such that $\psi(-1) = 1$, we consider a series

$$E(z, s; k/2, \psi, N) = \sum_\gamma \psi(\det(d_\gamma)) h_\gamma(z)^{-k} \det(\mathrm{Im}(\gamma(z)))^s,$$

where $z \in H_m$, $s \in \mathbf{C}$, and γ runs over $[P \cap \Gamma_0(N)]\backslash\Gamma_0(N)$. We define also another type of series with respect to an arbitrary congruence subgroup Γ of $\Gamma_0(4)$ by

$$E(z, s; k/2, \Gamma) = \sum_{\gamma \in (P \cap \Gamma)\backslash\Gamma} h_\gamma(z)^{-k} \det(\mathrm{Im}(\gamma(z)))^s.$$

In the present paper, we investigate the series of these types for $\mathrm{Sp}(m, F)$ with an arbitrary totally real algebraic number field F. Our main theorems in Section 2 will describe the behavior of E at two critical points $s = 0$ and $s = (m + 1 - k)/2$ when $k > 0$. For simplicity, let us state here only the results in the easiest cases:

Let $E(z, s)$ denote any one of the series of the above two types.

Received August 20, 1984. Research supported by NSF Grant MCS 81-00744.

(I) *If $k = m + 1$ or $k > m + 3$, then $E(z,s)$ is finite at $s = 0$ and $E(z,0)$ is a holomorphic modular form of weight $k/2$ with cyclotomic Fourier coefficients.*

(II) *If $0 < k \leqslant m$, $E(z,s)$ has at most a simple pole at $s = (m + 1 - k)/2$, and the residue is of the form $Ag(z)$ with a constant*

$$A = \pi^{-\lambda} \prod_{j=1}^{[(m-k)/2]} \zeta(2j + 1)$$

and a holomorphic modular form g of weight $k/2$ with cyclotomic Fourier coefficients, where $\lambda = (m + 1 - k)[(m + 1)/2]$ and ζ is the Riemann zeta function.

It should be noted that there is no previous investigation of this nature except for the case of $SL_2(F)$ with $[F:Q] \leqslant 2$ (see references cited at the end of Section 6).

Problems of the same type have been studied in the case of integral weight in our previous paper [15]. Therefore the present work may be regarded as its natural continuation. The naturalness may be emphasized, because there is a perfect parallelism between the case of integral weight and that of half-integral weight. For example, the weight $(m + 3)/2$ produces some interesting phenomena. It is an integer only when m is odd, and so appears only in the theory of half-integral weight if m is even. For this and other reasons, it may be said that the theory is complete only with the inclusion of the series of half-integral weight. In [15], only the integral weights $\geqslant (m - 1)/2$ have been treated. We can actually prove a result similar to the above (II) for all positive integral weights $\leqslant m/2$ by the same technique. The precise statement for this will be given as Theorem 2.7.

Our methods are basically the same as in [15], except that we have to deal with the difficulties arising from the seeming ambiguity of the factor h_γ. Besides, it is indispensable to define h_γ not only for γ in a congruence subgroup but also for a more general type of elements of $Sp(m, F)$. We do this by employing the metaplectic group M_A of $Sp(m, F)$ of Weil. We then define the adelic version of $E(z, s; k/2, \psi, N)$ and its suitable transform as functions on M_A and show that each Fourier coefficient of the latter has an Euler product. The archimedean Euler factors are confluent hypergeometric functions of [14]. Each nonarchimedean Euler factor is an infinite series involving Gauss sums of quadratic forms, which we already introduced in [15] with this application in view. We determined there the explicit form of the series as a rational function for almost all primes. Thanks to a result of Paul Feit [2], we can determine the series even for "bad primes" up to polynomial factors. Once this "explicit" Euler-product expression is established, our results can easily be derived by the same technique as in [15]. As an appendix, we shall add a short exposition of theta series of an indefinite quadratic form.

The present paper has been written not only as a natural continuation of [15], but also with one particular application in mind. We mention here only that it concerns generalizations of the results of [13] and [16], leaving the explanation of the problems to Remark 6.6.

Notation and terminology. For an associative ring R with identity element, we denote by $R^\times$ the group of all its invertible elements, and by R_n^m the module of all $m \times n$-matrices with entries in R. We occasionally write $M_n(R)$ for R_n^n, particularly when it is viewed as a ring; we put also $R_1^m = R^m$. The zero and identity elements of $M_n(R)$ are denoted by 0_n and 1_n. For complex hermitian matrices x and y, we write $x > y$ if $x - y$ is positive definite. If $x > 0$, $x^{1/2}$ denotes its positive definite square root. For square matrices $x_1, \ldots, x_r$, we denote by $\mathrm{diag}[x_1, \ldots, x_r]$ the matrix with $x_1, \ldots, x_r$ in the diagonal blocks and 0 in all other blocks. The adelization of an algebraic group W over $\mathbf{Q}$ is denoted by $W_\mathbf{A}$ or $W(\mathbf{A})$. In particular, if F is an algebraic number field, $F_\mathbf{A}$ denotes the ring of its adeles, and $F_\mathbf{A}^\times$ the idele group of F. The disjoint union of sets $Z_1, \ldots, Z_s$ is denoted by $\coprod_{i=1}^s Z_i$. By a *locally constant function* on a finite-dimensional vector space V over $\mathbf{Q}$, we understand a function $f: V \to \mathbf{C}$ for which there exist two $\mathbf{Z}$-lattices L and M in V such that $f(x) = 0$ for $x \notin L$ and $f(x)$ depends only on x modulo M. Besides the standard symbols $\mathbf{Z}$, $\mathbf{Q}$, $\mathbf{R}$, and $\mathbf{C}$, we let $\mathbf{T}$ denote the group of all complex numbers of absolute value 1, $\mathbf{Q}_{ab}$ the maximal abelian extension of $\mathbf{Q}$ in $\mathbf{C}$, and $\mathbf{Q}_p$ the p-completion of $\mathbf{Q}$ for each rational prime p. We put $\mathbf{e}(z) = e^{2\pi i z}$ for $z \in \mathbf{C}$. For a rational number r, $[r]$ denotes the largest integer $\leqslant r$.

1. Factors of automorphy of half-integral weight. For any commutative ring A with identity element and a positive integer m, we define the symplectic group $\mathrm{Sp}(m, A)$ as usual by

$$(1.1a) \quad \mathrm{Sp}(m, A) = \left\{ \alpha \in \mathrm{GL}_{2m}(A) \,\middle|\, {}^t\alpha \iota \alpha = \iota \right\}, \qquad \iota = \iota_m = \begin{bmatrix} 0 & -1_m \\ 1_m & 0 \end{bmatrix},$$

and define a domain H by

$$(1.1b) \qquad\qquad H = H_m = \left\{ z \in \mathbf{C}_m^m \,\middle|\, {}^t z = z,\ \mathrm{Im}(z) > 0 \right\}.$$

If $\alpha = \begin{pmatrix} a & b \\ c & d \end{pmatrix} \in \mathrm{Sp}(m, A)$ with a, b, c, d of size m, we shall often write $a = a_\alpha$, $b = b_\alpha$, $c = c_\alpha$, and $d = d_\alpha$. For $\alpha \in \mathrm{Sp}(m, \mathbf{R})$ and $z \in H$ (and also for $-z \in H$), we put

$$\alpha(z) = (a_\alpha z + b_\alpha)(c_\alpha z + d_\alpha)^{-1}, \qquad \mu(\alpha, z) = c_\alpha z + d_\alpha,$$

$$(1.2)$$

$$j(\alpha, z) = \det(c_\alpha z + d_\alpha).$$

Throughout the paper, we denote by F a totally real algebraic number field of finite degree, and by $\mathbf{a}$ and $\mathbf{f}$ the sets of all archimedean primes and nonarchimedean primes of F, respectively. We view each element v of $\mathbf{a}$ an embedding of F into $\mathbf{R}$. The v-completion of F for each $v \in \mathbf{a} \cup \mathbf{f}$ is denoted by F_v; the archimedean and nonarchimedean factors of $F_\mathbf{A}$ are denoted by $F_\mathbf{a}$ and $F_\mathbf{f}$. For any set S, we denote by $S^\mathbf{a}$ the product of $\mathbf{a}$ copies of S, that is, the set of all indexed elements $(x_v)_{v \in \mathbf{a}}$ with $x_v \in S$. We identify $(F_n^m) \otimes_{\mathbf{Q}} \mathbf{C}$ with $(\mathbf{C}_n^m)^{\mathbf{a}}$ via the map $\alpha \otimes c \mapsto (\alpha^v c)_{v \in \mathbf{a}}$, where α^v denotes the image of α under v. Each $v \in \mathbf{a}$ defines the projection of $(\mathbf{C}_n^m)^{\mathbf{a}}$ to the v-th factor $\mathbf{C}_n^m$. For this reason, we write also $\alpha^v = \alpha_v$ for $\alpha \in F_n^m$. Now we put

$$(1.3a) \qquad G = \mathrm{Sp}(m, F), \qquad G_v = \mathrm{Sp}(m, F_v) \qquad (v \in \mathbf{a} \cup \mathbf{f}),$$

$$(1.3b) \qquad G_{\mathbf{a}} = \prod_{v \in \mathbf{a}} G_v = \mathrm{Sp}(m, \mathbf{R})^{\mathbf{a}},$$

and let G and $G_{\mathbf{a}}$ act on $H^{\mathbf{a}}$ componentwise. Then, for $z \in H^{\mathbf{a}}$ and $\alpha \in G$ or $\alpha \in G_{\mathbf{a}}$, we put

$$(1.4a) \qquad \mu_v(\alpha, z) = \mu(\alpha_v, z_v), \qquad j_v(\alpha, z) = j(\alpha_v, z_v),$$

$$(1.4b) \qquad j(\alpha, z) = \prod_{v \in \mathbf{a}} j_v(\alpha, z),$$

$$(1.4c) \qquad Y(z) = \prod_{v \in \mathbf{a}} \det(\mathrm{Im}(z_v)).$$

We recall a well known relation

$$(1.4d) \qquad Y(\alpha(z)) = Y(z)|j(\alpha, z)|^{-2} \qquad (\alpha \in G \text{ or } \in G_{\mathbf{a}}).$$

To deal with factors of automorphy of half-integral weight, we consider a group $\mathscr{G}$ which consists of all couples (α, h) formed by $\alpha \in G$ and a holomorphic function h on $H^{\mathbf{a}}$ such that $h(z)^2 = tj(\alpha, z)$ with a root of unity t, the group law being defined by

$$(1.5) \qquad (\alpha, h)(\alpha', h') = (\alpha\alpha', h(\alpha'(z))h'(z)).$$

Let $X = F_m^1$; let $\mathscr{L}(X)$ denote the set of all locally constant functions on X. We now introduce a theta function

$$(1.6a) \qquad \theta(z, u; l) = \sum_{x \in X} l(x)\mathbf{e}_{\mathbf{a}}((1/2)xz \cdot {}^t x + xu),$$

where $z \in H^{\mathbf{a}}$, $u \in (\mathbf{C}^m)^{\mathbf{a}}$, $l \in \mathscr{L}(X)$, and

$$(1.6b) \qquad \mathbf{e}_{\mathbf{a}}(x) = \mathbf{e}\left(\sum_{v \in \mathbf{a}} x_v\right) \qquad \text{for} \quad x \in \mathbf{C}^{\mathbf{a}};$$

$xz \cdot {}^t x$ and xu denote the elements of $\mathbf{C}^{\mathbf{a}}$ whose v-components are $x_v z_v \cdot {}^t x_v$ and $x_v u_v$; this type of convention applies to all similar expressions in this paper. We let $\mathrm{Sp}(m, \mathbf{R})$ act on $H \times \mathbf{C}^m$ by

$$(1.7) \quad \alpha(z, u) = \left(\alpha(z), {}^t\mu(\alpha, z)^{-1}u\right) \qquad \text{for} \quad \alpha \in \mathrm{Sp}(m, \mathbf{R}), \ z \in H, \ u \in \mathbf{C}^m.$$

Then G and $G_{\mathfrak{a}}$ act on $(H \times \mathbf{C}^m)^{\mathfrak{a}}$. It is convenient to introduce functions

$$(1.8a) \qquad g(z,u) = \mathbf{e}\left({}^t u(z - \bar{z})^{-1}u/2\right) \qquad (z \in H, u \in \mathbf{C}^m),$$

$$(1.8b) \qquad g_{\mathfrak{a}}(z,u) = \mathbf{e}_{\mathfrak{a}}\left({}^t u(z - \bar{z})^{-1}u/2\right) \qquad (z \in H^{\mathfrak{a}}, u \in (\mathbf{C}^m)^{\mathfrak{a}}),$$

and define a modified theta function θ^* by

$$(1.9) \qquad \theta^*(z,u;l) = g_{\mathfrak{a}}(z,u)\theta(z,u;l).$$

We see easily that

$$(1.10) \qquad g_{\mathfrak{a}}(\alpha(z,u)) = \zeta_\alpha(z,u)^{-1} g_{\mathfrak{a}}(z,u) \qquad (\alpha \in G \text{ or } \in G_{\mathfrak{a}}),$$

where ζ_α is given by

$$(1.11) \qquad \zeta_\alpha(z,u) = \mathbf{e}_{\mathfrak{a}}\left((1/2) \cdot {}^t u(c_\alpha z + d_\alpha)^{-1}c_\alpha u\right).$$

This shows that ζ_α is a factor of automorphy of G on $(H \times \mathbf{C}^m)^{\mathfrak{a}}$.

PROPOSITION 1.1. *There is an action of $\mathscr{G}$ on $\mathscr{L}(X)$, written $l \mapsto l^\beta$ for $l \in \mathscr{L}(X)$ and $\beta = (\alpha, h) \in \mathscr{G}$, such that*

$$(1.12a) \qquad \theta(\alpha(z,u);l) = h(z)\zeta_\alpha(z,u)\theta(z,u;l^\beta),$$

$$(1.12b) \qquad \theta^*(\alpha(z,u);l) = h(z)\theta^*(z,u;l^\beta).$$

Moreover, for every $l \in \mathscr{L}(X)$, there is a congruence subgroup Γ of G and an isomorphism $\gamma \mapsto \{\gamma\} = (\gamma, h_\gamma)$ of Γ onto a subgroup $\tilde{\Gamma}$ of $\mathscr{G}$ such that $h_\gamma^2 = j(\gamma, z)$ and $l^{\{\gamma\}} = l$ for every $\gamma \in \Gamma$.

If $F = \mathbf{Q}$, this, as well as the following proposition, follows easily from the transformation formula of classical theta functions (of which a short proof can be found in [11]). The general case can be proved by embedding $\mathrm{Sp}(m, F)$ into $\mathrm{Sp}(m[F:\mathbf{Q}], \mathbf{Q})$. Alternatively, one can prove the result by using the metaplectic group of Weil [20], which we shall do in Section 3. Notice that (1.12a) and (1.12b) are equivalent by virtue of (1.10).

Let $\mathfrak{g}$ denote the maximal order of F, and $\mathfrak{d}$ the different of F over $\mathbf{Q}$. Define a congruence subgroup Γ' of G by

$$(1.13) \qquad \Gamma' = \left\{ \alpha \in G \mid a_\alpha \in \mathfrak{g}_m^m, \, b_\alpha \in (2\mathfrak{d}^{-1})_m^m, \, c_\alpha \in (2\mathfrak{d})_m^m, \, d_\alpha \in \mathfrak{g}_m^m \right\}.$$

PROPOSITION 1.2. *Put $L = \mathfrak{g}_m^1$ and $L_* = (\mathfrak{d}^{-1})_m^1$. Let l_0 be the characteristic function of L. Then there is an isomorphism $\gamma \mapsto [\gamma] = (\gamma, h(\gamma, z))$ of Γ' onto a subgroup of $\mathscr{G}$ such that $l_0^{[\gamma]} = l_0$ for every $\gamma \in \Gamma'$, that is*

$$(1.14) \qquad \theta(\gamma(z,u);l_0) = h(\gamma,z)\zeta_\gamma(z,u)\theta(z,u;l_0) \qquad \text{for every} \quad \gamma \in \Gamma'.$$

Moreover, one has

$$h(\gamma, z)^2 = \operatorname{sgn}(N_{F/\mathbf{Q}}(\det(d_\gamma))) \left[\frac{F(\sqrt{-1})/F}{\det(d_\gamma)\mathfrak{g}} \right] j(\gamma, z),$$

$$\lim_{\rho \to 0} h(\gamma, \rho\mathbf{i}) = \sum_{x \in R} \mathbf{e}_\mathbf{a}(-xd_\gamma^{-1}c_\gamma \cdot {}^t x/2),$$

where $\mathbf{i} = (i1_m, \ldots, i1_m) \in H^\mathbf{a}$ *and* $R = L_* / L_* d_\gamma$.

The proof will be given in Section 3. The last two formulas are unnecessary for our main purposes, but included since they show at least some aspects of the nature of $h(\gamma, z)$.

By a *congruence subgroup of* $\mathcal{G}$, we understand a subgroup Δ of $\mathcal{G}$ satisfying the following two conditions:

(1.15a)   *The projection of* $\mathcal{G}$ *to* G *gives an isomorphism of* Δ *onto a congruence subgroup* Γ *of* G.

(1.15b)   *The inverse of this isomorphism coincides with the map* $\gamma \mapsto [\gamma]$ *of Proposition 1.2 on a congruence subgroup of* $\Gamma' \cap \Gamma$.

For $\alpha = (\alpha_0, h) \in \mathcal{G}$, let us write $h = h_\alpha$, and define the action of α on $H^\mathbf{a}$ to be the same as that of α_0. For a $\mathbf{C}$-valued function f on $H^\mathbf{a}$ and an integer k, we define a function $f\|_{k/2}\alpha$ on $H^\mathbf{a}$ by

(1.16) $(f\|_{k/2}\alpha)(z) = h_\alpha(z)^{-k}f(\alpha(z))$ $(z \in H^\mathbf{a})$.

We denote by $\mathcal{M}_{k/2}(\Delta)$ the vector space of all holomorphic functions f on $H^\mathbf{a}$ which satisfy $f\|_{k/2}\alpha = f$ for all $\alpha \in \Delta$, and also the cusp condition if $G = \mathrm{SL}_2(\mathbf{Q})$. Put

(1.17) $S = \{\sigma \in F_m^m \mid {}^t\sigma = \sigma\}$.

Then every element f of $\mathcal{M}_{k/2}(\Delta)$ has a Fourier expansion of the form

(1.18) $f(z) = \sum_{\sigma \in \Lambda} c(\sigma)\mathbf{e}_\mathbf{a}(\operatorname{tr}(\sigma z))$ $(z \in H^\mathbf{a})$,

where Λ is a lattice in S. Given a subfield K of $\mathbf{C}$, we denote by $\mathcal{M}_{k/2}(\Delta, K)$ the set of all f in $\mathcal{M}_{k/2}(\Delta)$ such that $c(\sigma) \in K$ for all σ, and by $\mathcal{M}_{k/2}(K)$ the union of $\mathcal{M}_{k/2}(\Delta, K)$ for all congruence subgroups Δ of $\mathcal{G}$.

PROPOSITION 1.3.   (i) *If* Δ *is a congruence subgroup of* $\mathcal{G}$, *then* $\beta\Delta\beta^{-1}$ *is a congruence subgroup of* $\mathcal{G}$ *for every* $\beta \in \mathcal{G}$.

(ii) *Let* $\mathbf{Q}_{ab}$ *denote the maximal abelian extension of* $\mathbf{Q}$. *Then* $\mathcal{M}_{k/2}(\mathbf{Q}_{ab})$ *is stable under* $f \mapsto f\|_{k/2}\alpha$ *for every* $\alpha \in \mathcal{G}$.

(iii) *The subgroup* $\tilde{\Gamma}$ *of* $\mathcal{G}$ *in Proposition 1.1 is a congruence subgroup of* $\mathcal{G}$.

Proof. If $m > 1$, (1.15b) follows from (1.15a), because of the congruence subgroup property of G (see Bass–Milnor–Serre [1, Cor. 12.5]). Then (i) and (iii) are immediate. The proof of (iii) for $m = 1$ will be given in the proof of Proposition 7.2. Let Δ and β be as in (i) in the case $m = 1$; further let l_0, Γ', and $[\gamma]$ be as in Proposition 1.2. Put $\Delta' = \{[\gamma] \mid \gamma \in \Gamma'\}$. Then $\beta^{-1}\Delta'\beta$ may be viewed as the group $\tilde{\Gamma}$ of Proposition 1.1 with l_0^β as l. Therefore, if we assume (iii), $\beta^{-1}\Delta'\beta$ is a congruence subgroup. Now $\Delta \cap \Delta'$ is a congruence subgroup of $\mathcal{G}$ by (1.15b). Let Γ'' and α be the projections of $\Delta \cap \Delta'$ and β. Then $\beta^{-1}(\Delta \cap \Delta')\beta$, being the inverse image of $\alpha^{-1}\Gamma''\alpha$ in $\beta^{-1}\Delta'\beta$, is a congruence subgroup, and hence $\beta^{-1}\Delta\beta$, which contains this group, must be a congruence subgroup. Now (ii) was proved in [11, Prop. 1.5] when $F = \mathbf{Q}$. The general case can be proved in the same way.

2. Main theorems on Eisenstein series. We denote by P the parabolic subgroup of G consisting of all α with $c_\alpha = 0$, and by $\mathcal{P}$ the subgroup of $\mathcal{G}$ consisting of the elements lying above those of P. Let Δ be a congruence subgroup of $\mathcal{G}$ such that

$$(2.1) \qquad h_\alpha = 1 \quad \text{for every} \quad \alpha \in \mathcal{P} \cap \Delta.$$

For each odd integer k, we define an Eisenstein series $E(z, s; k/2, \Delta)$ of weight $k/2$ by

$$(2.2) \qquad E(z, s; k/2, \Delta) = \sum_{\alpha \in (\mathcal{P} \cap \Delta)\backslash\Delta} Y^s \|_{k/2}\alpha$$

$$= \sum_{\alpha \in (\mathcal{P} \cap \Delta)\backslash\Delta} Y(z)^s |h_\alpha(z)|^{-4s} h_\alpha(z)^{-k},$$

where $s \in \mathbf{C}$, $z \in H^{\mathbf{a}}$, and Y is defined by (1.4c). This series, as well as another series (2.11) below, can be continued to the whole s-plane as a meromorphic function, by virtue of Langlands' result [4]. Our main problems are the properties of $E(z, s; k/2, \Delta)$ at $s = 0$ and at other natural values of s, in particular whether $E(z, 0; k/2, \Delta)$ is holomorphic in z, and whether it has cyclotomic Fourier coefficients. We investigate these problems by the analysis of the Fourier coefficients of E as functions in s and $\text{Im}(z)$. For this purpose, we introduce another type of series (2.11) below and express (2.2) as a linear combination of the transforms of several series of type (2.11). The Fourier coefficients of a certain transform of the series (2.11) have Euler products and therefore are more manageable than those of (2.2), which is one of the reasons why we consider (2.11), though (2.11) is a natural series on its own, and should not be viewed merely as an auxiliary object.

In order to introduce this new series, we consider the adelization $G_{\mathbf{A}}$ of G and also the associated metaplectic group. Postponing the introduction of the latter to the next section, we take only $G_{\mathbf{A}}$ here. We denote by $G_{\mathbf{f}}$ the nonarchimedean factor of $G_{\mathbf{A}}$ and view $G_{\mathbf{a}}$ as the archimedean factor of $G_{\mathbf{A}}$. For $x \in G_{\mathbf{A}}$, $x_{\mathbf{f}}$ and $x_{\mathbf{a}}$

denote its projections to G_f and G_a. We view G as a subgroup of G_A as usual. (Thus G is the same as G_Q in the standard notation, which will not be used in this paper. The same convention applies to all other algebraic groups.)

Let us consider again $X = F_m^1$, $L = \mathfrak{g}_m^1$, and $L_* = (\mathfrak{d}^{-1})_m^1$. We then define a compact subgroup C of G_A by

$$(2.3a) \qquad C = \prod_{v \in \mathfrak{f} \cup \mathfrak{a}} C_v,$$

$$(2.3b) \qquad C_v = \begin{cases} \{\alpha \in G_v \mid \alpha(i1_m) = i1_m\} & \text{if } v \in \mathfrak{a}, \\ \{\alpha \in G_v \mid (L \times L_*)_v \alpha = (L \times L_*)_v\} & \text{if } v \in \mathfrak{f}. \end{cases}$$

This is different from the group C of [15], but still we have $G_A = P_A C$ with the present C. Define a map $\lambda: G_A \to F_A$ by

$$(2.4) \qquad \lambda(\alpha) = \det(d_\alpha) \qquad (\alpha \in G_A)$$

and put

$$(2.5) \qquad \Omega_v = \{\alpha \in G_v \mid \det(c_\alpha) \neq 0\}, \qquad \Omega_A = \{\alpha \in G_A \mid \det(c_\alpha) \in F_A^\times\}.$$

Then $\Omega_v = P_v \iota P_v$, $\Omega_A = P_A \iota P_A$ with ι of (1.1a), and

$$\Omega_v \iota = \{\alpha \in G_v \mid \det(d_\alpha) \neq 0\}, \qquad \Omega_A \iota = \{\alpha \in G_A \mid \det(d_\alpha) \in F_A^\times\}.$$

To $\alpha \in G_A$, we can assign a fractional ideal $\mathrm{il}(\alpha)$ of F and a positive real number $\epsilon(\alpha)$ by

$$(2.6) \qquad \mathrm{il}(yw) = \lambda(y)\mathfrak{g}, \qquad \epsilon(yw) = |\lambda(y)|_A \qquad \text{for } y \in P_A \text{ and } w \in C,$$

where $x\mathfrak{g}$ for $x \in F_A^\times$ denotes the ideal such that $(x\mathfrak{g})_v = x_v \mathfrak{g}_v$ for every $v \in \mathfrak{f}$. Then $\epsilon(\alpha) = \epsilon(\alpha_f)\epsilon(\alpha_a)$, $\epsilon(\alpha_f) = N(\mathrm{il}(\alpha))^{-1}$, $\epsilon(\alpha_a) = Y(\alpha_a(\mathbf{i}))^{-1/2} = |j(\alpha_a, \mathbf{i})|$, where $\mathbf{i} = (i1_m, \ldots, i1_m) \in H^a$.

We now fix an integral ideal $\mathfrak{c}$ of F divisible by 4, and define a subgroup $\Gamma_0(\mathfrak{c})$ of G and a subgroup D of C by $D = \prod_v D_v$ with $D_v = C_v$ for $v \in \mathfrak{a}$ and

$$(2.7) \qquad D_v = \left\{\alpha \in C_v \mid b_\alpha \in (2\mathfrak{d}_v^{-1})_m^m, c_\alpha \in (2^{-1}\mathfrak{c}_v \mathfrak{d}_v)_m^m\right\} \qquad \text{if } v \in \mathfrak{f},$$

$$(2.8) \qquad \Gamma_0(\mathfrak{c}) = G \cap DG_a.$$

LEMMA 2.1. (i) *The map* $x \mapsto \mathrm{il}(x)$ *gives a bijection of* $P \backslash G_A / CG_a$ *onto the ideal class group of* F.

(ii) *Every fractional ideal of F can be given as* $\mathrm{il}(\alpha)$ *for some* $\alpha \in G \cap P_A D$.

(iii) *Let* $\{\beta\}$ *be a finite subset of* $G \cap P_A D$ *such that* $\{\mathrm{il}(\beta)\}$ *is a complete set of representatives of the ideal classes of F. (Such a set exists by virtue of* (ii).) *Then* $P_A D = \coprod_\beta P\beta DG_a$ *and* $G \cap P_A D = \coprod_\beta P\beta\Gamma_0(\mathfrak{c})$.

These can be proved in the same way as in [15, Lemmas 1.4, 1.5, and 1.6], only with trivial modifications.

In the next section (Proposition 3.2), we shall define a map $\alpha \mapsto (\alpha, h(\alpha, z))$ of $G \cap P_{\mathbf{A}} D$ into $\mathscr{G}$, independent of $\mathfrak{c}$, such that:

$$(2.9a) \qquad h(\alpha\beta\gamma, z) = h(\alpha, z) h(\beta, \gamma(z)) h(\gamma, z)$$

$$if \quad \alpha \in P, \quad \beta \in G \cap P_{\mathbf{A}} D, \quad and \quad \gamma \in \Gamma_0(\mathfrak{c});$$

$$(2.9b) \qquad h(\alpha, z) = |N_{F/\mathbf{Q}}(\det(d_\alpha))|^{1/2} \qquad if \quad \alpha \in P;$$

$$(2.9c) \qquad h(\gamma, z) \ coincides \ with \ that \ of \ (1.14) \qquad if \quad \gamma \in \Gamma_0(\mathfrak{c}).$$

Take a Hecke character $\psi : F_{\mathbf{A}}^\times / F^\times \to \mathbf{T}$ such that

$$(2.10) \qquad \psi(F_{\mathbf{a}}^\times) = 1 \quad and \ the \ conductor \ of \ \psi \ divides \ \mathfrak{c}.$$

We denote by the same letter ψ the ideal-character attached to ψ. Take a subset $\{\beta\}$ of $G \cap P_{\mathbf{A}} D$ as in Lemma 2.1, (iii), and also a complete set of representatives T_β of $(P \cap \beta\Gamma\beta^{-1})\backslash\beta\Gamma$ for each β, where $\Gamma = \Gamma_0(\mathfrak{c})$; put $\mathfrak{a}_\beta = \mathrm{il}(\beta)$. We then consider a series

$$(2.11) \qquad E(z, s; k/2, \psi, \mathfrak{c})$$

$$= \sum_\beta N(\mathfrak{a}_\beta)^{2s+k/2} \sum_{\alpha \in T_\beta} \psi\big(\det(d_\alpha)\mathfrak{a}_\beta^{-1}\big) Y(\alpha(z))^s h(\alpha, z)^{-k}.$$

This is independent of the choice of $\{\beta\}$.

LEMMA 2.2. *Given an integral ideal $\mathfrak{c}$ divisible by* 4, *let $\Delta(\mathfrak{c})$ be the subgroup of G defined by*

$$\Delta(\mathfrak{c}) = \left\{ (\alpha, h(\alpha, z)) \mid \alpha \in \Gamma_0(\mathfrak{c}), \, a_\alpha - 1 \in \mathfrak{c}_m^m, \, b_\alpha \in (2\mathfrak{c}\mathfrak{d}^{-1})_m^m \right\}$$

with $h(\alpha, z)$ of (2.9a, b, c). Then, for every congruence subgroup Δ of $\mathscr{G}$, we can find $\mathfrak{c}$ such that $\Delta(\mathfrak{c}) \subset \Delta$. Moreover, if $\Delta(\mathfrak{c}) \subset \Delta$, we have

$$r \cdot E(z, s; k/2, \Delta) = \sum_{\tau \in \Delta(\mathfrak{c})\backslash\Delta} \sum_\psi E(z, s; k/2, \psi, \mathfrak{c})\|_{k/2}\tau,$$

where ψ runs over all characters satisfying (2.10), and r is the number of such characters times $[\mathscr{P} \cap \Delta : \mathscr{P} \cap \Delta(\mathfrak{c})]$.

Proof. The existence of $\Delta(\mathfrak{c})$ follows immediately from (1.15b) and (2.9c). Then our formula can be proved in the same fashion as in [15, Propositions 2.1 and 2.4].

Now our first main theorem can be stated as follows:

THEOREM 2.3. *Let $E(z,s)$ denote any of the series defined by (2.2) or (2.11). Then the following assertions hold.*

(1) If $k = m + 1$ or $k > m + 3$, then $E(z,s)$ is finite at $s = 0$ and $E(z,0)$ belongs to $\mathcal{M}_{k/2}(\mathbf{Q}_{ab})$.

(2) Suppose $k = m + 2$ or $k = m + 3$. Then $E(z,s)$ is finite at $s = 0$. Moreover, if $F \neq \mathbf{Q}$, $E(z,0)$ belongs to $\mathcal{M}_{k/2}(\mathbf{Q}_{ab})$; if $F = \mathbf{Q}$ and $\psi^2 \neq 1$, $E(z,0;k/2,\psi,c)$ belongs to $\mathcal{M}_{k/2}(\mathbf{Q}_{ab})$.

(3) Suppose $0 < k \leqslant m$; put $s_k = (m + 1 - k)/2$. If $\psi^2 \neq 1$, $E(z,s;k/2,\psi,c)$ is finite at $s = s_k$.

(4) With k and s_k as in (3), $E(z,s)$ has at most a simple pole at $s = s_k$, and the residue is of the form $Ag(z)$ with a constant

$$A = \pi^{-n\lambda}R_F \prod_{j=1}^{[(m-k)/2]} \zeta_F(2j + 1)$$

and an element g of $\mathcal{M}_{k/2}(\mathbf{Q}_{ab})$, where R_F is the regulator of F, ζ_F is the zeta function of F, $n = [F:\mathbf{Q}]$, and $\lambda = (m + 1 - k)[(m + 1)/2]$. Moreover, $g(z) = \sum_{\sigma \in S} c(\sigma)\mathbf{e}_\mathbf{a}(\mathrm{tr}(\sigma z))$ with $c(\sigma) = 0$ for $\mathrm{rank}(\sigma) > k$. If E is of type (2.11) and $\psi^2 = 1$, then $g \neq 0$.

Our series is convergent if and only if $\mathrm{Re}(2s) + k/2 > m + 1$. Thus our assertions concern divergent series if $k \leqslant 2m + 1$. There is another type of result which may be viewed as an extension of (3) and (4) to higher weights:

THEOREM 2.4. *Suppose $k \geqslant m + 1$; let $s_k = (m + 1 - k)/2$. Then the product*

$$E(z,s;k/2,\psi,c) \prod_{i=0}^{[(m-1)2]} L_c(4s + k - 1 - 2i,\psi^2)$$

is finite at $s = s_k$, and its value at s_k is π^{en} times a nonzero element of $\mathcal{M}_{k/2}(\mathbf{Q}_{ab})$, where L_c is as in [15, (7.16)], $n = [F:\mathbf{Q}]$, $e = m(m + 2)/4$ if m is even, and $e = (m + 1)^2/4$ if m is odd.

If $k = m + 1$, this is essentially equivalent to (1) of Theorem 2.3. In order to treat one case excluded in (2) of Theorem 2.3, let us now assume that $k = m + 3$ and $F = \mathbf{Q}$. We then consider the set $\mathcal{L}$ consisting of all the functions f on H satisfying the following two conditions:

(2.13a) $f\|_{(m+3)/2}\gamma = f$ *for all γ in a congruence subgroup of $\mathcal{G}$;*

(2.13b) $f(z) = f_1(z) + \Delta[f_0(z)\log\{\det(z - \bar{z})\}]$ *with holomorphic functions f_0 and f_1 on H such that $\Delta f_0 = 0$, where $\Delta = \det(2^{-1}(1 + \delta_{ij})\partial/\partial z_{ij})$.*

This is similar to the family of functions introduced in [15, §9]. We observe that [15, Proposition 9.4] is valid even in the case of half-integral weight. In particular, if f, f_0, and f_1 are as in (2.13a, b), then $f_0 \in \mathcal{M}_{(m-1)/2}(\mathbf{C})$, and f_1 has a Fourier expansion of type (1.18). We denote by $\mathcal{L}(\mathbf{Q}_{ab})$ the set of all f in $\mathcal{L}$ for which

$\pi^m f_0 \in \mathscr{M}_{(m-1)/2}(\mathbf{Q}_{ab})$ and the Fourier coefficients of f_1 belong to $\mathbf{Q}_{ab}$. Then, as an analogue of [15, Theorem 7.2 and Proposition 10.2] (which is now unconditionally valid for the reason explained below), we can prove:

THEOREM 2.5. *Suppose* $F = \mathbf{Q}$ *and* $k = m + 3$ (*and hence* m *is even*). *Then* $E(z,s)$ *is finite at* $s = 0$, *and* $E(z,0) \in \mathscr{L}(\mathbf{Q}_{ab})$.

PROPOSITION 2.6. *If* $f \in \mathscr{L}(\mathbf{Q}_{ab})$ *and* $\alpha \in \mathscr{G}$, *then* $f\|_{(m+3)/2}\alpha \in \mathscr{L}(\mathbf{Q}_{ab})$.

The last proposition can be proved in exactly the same fashion as in [15, §11], which deals with the case of integral weight. The proofs of the above theorems will be given in Section 5.

The assertions similar to Theorem 2.3, (1) was proved in [15] for the series of integral weight $> m$. We showed also in [15, Propositions 10.1, 10.2, 10.3] that the results for integral weight $\geqslant (m + 1)/2$ or $= (m - 1)/2$ depended on a certain conjecture. This conjecture has been proved by Kitaoka [3] and Feit [2], and therefore the assertions of those propositions are now unconditionally true. The counterparts of (3), (4) of Theorem 2.3 and Theorem 2.4 were not given in [15]. We can actually prove them by the same methods. To present their precise statements, define $E(z,s; k, \psi, \mathbf{b})$ as in [15, (2.18)] in both Cases SP and SU, where $k \in \mathbf{Z}$, $\mathbf{b}$ is an integral ideal of F, and ψ is a Hecke character satisfying [15, (2.13a, b)]. Put

$$\mathbf{E}(z,s) = E(z, 2s; k, \psi, \mathbf{b}) \prod_{v \in \infty} \det(z_v - z_v^*)^s,$$

$$D(z,s) = \mathbf{E}(z,s) \cdot \begin{cases} L_{\mathbf{b}}(2s + k, \psi) \displaystyle\prod_{i=1}^{[m/2]} L_{\mathbf{b}}(4s + 2k - 2i, \psi^2) & \text{(Case SP)}, \\[2em] \displaystyle\prod_{i=0}^{m-1} L_{\mathbf{b}}(2s + k - 1, \psi\theta^i) & \text{(Case SU)}, \end{cases}$$

where θ is the quadratic character of F corresponding to K/F. (The symbol ∞ was employed in [15] instead of $\mathbf{a}$.) Then we have

THEOREM 2.7. *Let* $s_k = \kappa - k$ *with* $\kappa = (m + 1)/2$ *in Case* SP *and* $\kappa = m$ *in Case* SU.

(1) *Suppose* $0 \leqslant k < \kappa$. *Then* $\mathbf{E}(z,s)$ *has at most a simple pole at* s_k. *The pole occurs only when* $\psi^2 = 1$ ($\psi = 1$ *when* $k = 0$) *in Case* SP *and* $\psi = \theta^k$ *in Case* SU. *The residue is of the form* $Ag(z)$ *with an element* g *of* $\mathscr{M}_k(\mathbf{Q}_{ab})$ *and a constant*

$$A = \pi^{-nv} R_F \cdot \begin{cases} L(m + 1 - k, \psi)^{-1} \displaystyle\prod_{j=1}^{[m/2-k]} \zeta_F(2j + 1) & (\text{\textit{Case}} \ \mathrm{SP}), \\[2em] \displaystyle\prod_{j=2}^{m-k} L(j, \theta^{j-1}) & (\text{\textit{Case}} \ \mathrm{SU}), \end{cases}$$

where n, R_F, and ζ_F are the same as in (4) of Theorem 2.3, and

$$
\nu = \begin{cases}
m\left(\dfrac{m}{2} - k\right) & (\text{Case SP, } m \text{ even}), \\[2mm]
(m - 1)\left(\dfrac{m}{2} - k\right) - 1 & (\text{Case SP, } m \text{ odd}), \\[2mm]
m(m - k) & (\text{Case SU}).
\end{cases}
$$

Moreover, $g(z) = \sum_{\sigma \in S} c(\sigma)\mathbf{e_a}(\mathrm{tr}(\sigma z))$ with $c(\sigma) = 0$ for $\mathrm{rank}(\sigma) > 2k$ in Case SP and for $\mathrm{rank}(\sigma) > k$ in Case SU.

(2). Suppose $k \geqslant \kappa$; suppose also that $\mathbf{b} \neq \mathbf{g}$ or $k > 2\kappa$. Then $D(z,s)$ is finite at $s = s_k$ and $D(z,s_k)$ is π^{en} times an element of $\mathcal{M}_k(\mathbf{Q}_{ab})$, where e is the same as in Theorem 2.4 in Case SP, and $e = m(m + 1)/2$ in Case SU.

One can show, beyond these statements, that $g \neq 0$ or $D(z,s_k) \neq 0$ under certain conditions, by examining the nonvanishing of some Fourier coefficients. For example, g of (1) is not 0 if $\psi^2 = 1$, $\mathbf{b} \neq \mathbf{g}$ and $L_\mathbf{b}(1 - k, \psi) \neq 0$ in Case SP, or if $\psi = \theta^k$, $\mathbf{b} \neq \mathbf{g}$, and $L_\mathbf{b}(0, \theta) \neq 0$ in Case SU.

3. A factor of automorphy on the metaplectic group.

We define a character $\mathbf{e}_v$ of F_v for $v \in \mathbf{a} \cup \mathbf{f}$ and a character $\mathbf{e_A}$ of $F_\mathbf{A}$ by $\mathbf{e_A}(x) = \prod_v \mathbf{e}_v(x_v)$, $\mathbf{e}_v(x) = \mathbf{e}(x)$ if $v \in \mathbf{a}$, and

$$
\mathbf{e}_v(x) = \mathbf{e}\!\left(-\mathrm{Tr}_{F_v/\mathbf{Q}_p}(\text{the fractional part of } x)\right)
$$

if v divides a rational prime p. With $X = F_m^1$ and $L = \mathfrak{g}_m^1$ as before, we normalize, for $v \in \mathbf{f}$, a Haar measure on X_v so that L_v has measure $N(\mathfrak{d}_v)^{-m/2}$. This is self-dual with respect to $(x, y) \mapsto \mathbf{e}_v(x \cdot {}^t y)$. Under right multiplication, G leaves the alternating form $(x, y) \mapsto x\iota \cdot {}^t y$ on $X \times X$ invariant. We can therefore define the metaplectic group $\mathrm{Mp}(X_v)$ and $\mathrm{Mp}(X_\mathbf{A})$ of Weil [20], which we denote for simplicity by M_v and $M_\mathbf{A}$, respectively. We recall that $M_\mathbf{A}$ is a group of unitary operators on $L^2(X_\mathbf{A})$, which is an extension of $G_\mathbf{A}$ with kernel $\mathbf{T}$; the same is true with the subscript v in place of $\mathbf{A}$. We let pr denote the projection maps of $M_\mathbf{A}$ onto $G_\mathbf{A}$ and of M_v onto G_v. There are splitting homomorphisms

$$
\text{(3.1a)} \qquad\qquad r : G \to M_\mathbf{A},
$$

$$
\text{(3.1b)} \qquad\qquad r_P : P_\mathbf{A} \to M_\mathbf{A},
$$

and also a map

$$
\text{(3.1c)} \qquad\qquad r_\Omega : \Omega_\mathbf{A} \to M_\mathbf{A}.
$$

The latter two are given by

$$
\text{(3.2)} \qquad \left[r_P\!\begin{pmatrix} a & b \\ 0 & d \end{pmatrix}\! f \right]\!(x) = |\det(a)|_\mathbf{A}^{1/2}\,\mathbf{e_A}(xa \cdot {}^t b \cdot {}^t x/2)\,f(xa),
$$

$$
\text{(3.3)} \quad \left[r_\Omega(\alpha)f \right]\!(x) = |\det(c)|_\mathbf{A}^{1/2} \int_{X_\mathbf{A}} f(xa + yc)\,\mathbf{e_A}(q_\alpha(x, y))\,dy, \qquad \alpha = \begin{pmatrix} a & b \\ c & d \end{pmatrix},
$$

for $f \in L^2(X_\mathbf{A})$, where q_α is the quadratic form defined for $\alpha \in G_\mathbf{A}$ by

$$(3.4) \quad q_\alpha(w) = 2^{-1}w(\alpha B \cdot {}^t\alpha - B) \cdot {}^tw, \qquad B = \begin{bmatrix} 0 & 1_m \\ 0 & 0 \end{bmatrix} \quad (w \in X_\mathbf{A} \times X_\mathbf{A}).$$

These maps are consistent in the sense that $r_\Omega(\alpha\beta\gamma) = r_P(\alpha)r_\Omega(\beta)r_P(\gamma)$ for $\alpha, \gamma \in P_\mathbf{A}$ and $\beta \in \Omega_\mathbf{A}$; moreover, $r = r_P$ on P and $r = r_\Omega$ on $P\iota P$. There are also similar maps of P_v and Ω_v into M_v given by the formulas with the subscript v in place of $\mathbf{A}$. We shall hereafter identify G with its image in $M_\mathbf{A}$ under the map r of (3.1a).

To simplify our notation, we define the action of an element α of $G_\mathbf{A}$ or of an element τ of $M_\mathbf{A}$ on $H^\mathbf{a}$ and $(H \times \mathbf{C}^m)^\mathbf{a}$ to be the same as that of $\alpha_\mathbf{a}$ or of $\mathrm{pr}(\tau)_\mathbf{a}$. The same convention applies also to the action of an element of M_v on H and $H \times \mathbf{C}^m$ when $v \in \mathbf{a}$. Similarly, we put, for $\tau \in M_\mathbf{A}$,

$$\lambda(\tau) = \lambda(\mathrm{pr}(\tau)), \qquad \epsilon(\tau) = \epsilon(\mathrm{pr}(\tau)), \qquad \mathrm{il}(\tau) = \mathrm{il}(\mathrm{pr}(\tau)),$$

(3.5)

$$j(\tau, z) = j(\mathrm{pr}(\tau)_\mathbf{a}, z).$$

To describe M_v for $v \in \mathbf{a}$ in terms of a factor of automorphy, we define a function φ on $X_v \times H \times \mathbf{C}^m$ by

$$(3.6) \qquad \varphi(x; z, u) = \mathbf{e}(2^{-1}xz \cdot {}^tx + xu) \qquad (x \in X_v, z \in H, u \in \mathbf{C}^m).$$

PROPOSITION 3.1. *For every $\sigma \in M_v$ with $v \in \mathbf{a}$, there is a holomorphic function h_σ on H which is determined by*

(i) $\sigma\varphi(x; z, u) = h_\sigma(z)^{-1}\zeta_\alpha(z, u)^{-1}\varphi(x; \alpha(z, u))$ *if* $\alpha = \mathrm{pr}(\sigma)$, *where ζ_α is the function of (1.11). This h_σ satisfies*

(ii) $h_{\sigma\tau}(z) = h_\sigma(\tau z)h_\tau(z)$,

(iii) $h_\sigma(z)^2 = t \cdot j(\sigma, z)$ *with* $t \in \mathbf{T}$,

(iv) $h_\sigma(z)^4 = (-1)^m j(\alpha, z)^2$ *if* $\sigma = r_\Omega(\alpha)$.

Proof. The existence of h_σ satisfying (i), (iii), and (iv) can easily be seen from (3.2) and (3.3) by an explicit computation if $\sigma \in r_P(P_v)$ or $\sigma = r_\Omega(\iota)$. In particular, $h_\sigma(z) = \det(-iz)^{1/2}$ if $\sigma = r_\Omega(\iota)$. Since M_v is generated by such elements and $\mathbf{T}$, we obtain h_σ for all $\sigma \in M_v$. Then (ii) follows from the fact that ζ_α is a factor of automorphy.

Thus M_v is isomorphic to the group formed by all couples (α, h) with $\alpha \in G_v$ and a holomorphic function h on H such that $h(z)^2 = tj(\alpha, z)$ with $t \in \mathbf{T}$, the law of composition being (1.5).

Define a subgroup C' of C by $C' = \prod_v C'_v$ with

$$(3.7a) \qquad C'_v = C_v \qquad\qquad\qquad\qquad\qquad\qquad \text{if} \quad v \in \mathbf{a},$$

$$(3.7b) \qquad C'_v = \left\{ \alpha \in C_v \mid b_\alpha \in (2\mathfrak{d}_v^{-1})_m^m, c_\alpha \in (2\mathfrak{d}_v)_m^m \right\} \qquad \text{if} \quad v \in \mathbf{f}.$$

This is the group D of (2.7) with $\mathfrak{c} = 4\mathfrak{g}$. We fix once for all an element δ of $F_\mathbf{A}^\times$

such that

(3.8a) $\qquad \delta_v \mathfrak{g}_v = \delta_v \quad$ if $\quad v \in \mathbf{f} \quad$ and $\quad \delta_v = 1 \quad$ if $\quad v \in \mathbf{a},$

and define an element η of $G_{\mathbf{A}}$ by

$$(3.8b) \qquad \eta_v = \begin{cases} 1_{2m} & \text{if} \quad v \in \mathbf{a}, \\ \begin{bmatrix} 0 & -\delta_v^{-1} 1_m \\ \delta_v 1_m & 0 \end{bmatrix} & \text{if} \quad v \in \mathbf{f}. \end{cases}$$

Observe that $\eta \in C$ and $\eta C' \eta^{-1} = C'$. Put $C'' = C' \cup C'\eta$. We now define a function $\Phi(x; z, u)$ on $X_{\mathbf{A}} \times (H \times \mathbf{C}^m)^{\mathbf{a}}$ by

$$(3.9a) \qquad \Phi_{z,u}(x) = \Phi(x; z, u) = \Phi_{\mathbf{f}}(x_{\mathbf{f}}) \Phi_{\mathbf{a}}(x_{\mathbf{a}}),$$

$$(3.9b) \qquad \Phi_{\mathbf{f}}(x_{\mathbf{f}}) = \prod_{v \in \mathbf{f}} \varphi_v(x_v),$$

$$(3.9c) \qquad \Phi_{\mathbf{a}}(x_{\mathbf{a}}) = \Phi_{\mathbf{a}}(x_{\mathbf{a}}; z, u) = \prod_{v \in \mathbf{a}} \varphi(x_v; z_v, u_v)$$

for $x \in X_{\mathbf{A}}$, $z \in H^{\mathbf{a}}$, $u \in (\mathbf{C}^m)^{\mathbf{a}}$, where φ_v for $v \in \mathbf{f}$ is the characteristic function of L_v, and φ is defined by (3.6).

PROPOSITION 3.2. *For every* $\xi \in M_{\mathbf{A}}$ *such that* $\mathrm{pr}(\xi) \in P_{\mathbf{A}} C''$, *there is a nonvanishing holomorphic function* $h(\xi, z)$ *on* $H^{\mathbf{a}}$ *determined by*

$$(3.10a) \qquad (\xi \Phi_{z,0})(0) = N(\mathrm{il}(\xi))^{1/2} h(\xi, z)^{-1}.$$

Moreover, h has the following properties:

$$(3.10b) \qquad h(\xi, z)^2 = t \cdot j(\xi, z) \qquad \text{with} \quad t \in \mathbf{T},$$

$$(3.10c) \qquad h(\beta \xi \tau, z) = h(\beta, z) h(\xi, \tau z) h(\tau, z)$$

$$\text{if} \quad \mathrm{pr}(\beta) \in P_{\mathbf{A}}, \quad \mathrm{pr}(\xi) \in P_{\mathbf{A}} C'', \quad \text{and} \quad \mathrm{pr}(\tau) \in C'' G_{\mathbf{a}},$$

$$(3.10d) \qquad h(\gamma, z)^4 = j(\gamma, z)^2 \qquad \text{if} \quad \gamma \in G \cap P_{\mathbf{A}} C'.$$

Proof. Since C_v'' for $v \in \mathbf{f}$ is contained in the group $B_0(X_v, L_v)$ of [20, n° 19, n° 36], there is a splitting homomorphism r_v of C_v'' into M_v that coincides with those given by the local versions of (3.2) and (3.3) on $C_v'' \cap P_v$ and on $C_v'' \cap \Omega_v$. By means of these formulas, we see easily that $r_v(C_v'')$ leaves φ_v invariant. Combining this fact with Proposition 3.1, we find, for $\tau \in M_{\mathbf{A}}$ with $\mathrm{pr}(\tau) \in C'' G_{\mathbf{a}}$, a function $h(\tau, z)$ such that

$$(3.11a) \quad (\tau \Phi_{z,u})(x) = h(\tau, z)^{-1} \zeta_\alpha(z, u)^{-1} \Phi(x; \tau(z, u)), \qquad (\alpha = \mathrm{pr}(\tau)_{\mathbf{a}}),$$

$$(3.11b) \qquad h(\tau, z)^2 = t \cdot j(\tau, z) \qquad \text{with some} \quad t \in \mathbf{T}.$$

Let $\beta = t \cdot r_P(\gamma)$ with $t \in \mathbf{T}$ and $\gamma \in P_{\mathbf{a}}$. If $\xi \in M_{\mathbf{A}}$ and $\mathrm{pr}(\xi) \in P_{\mathbf{A}}C''$, we have, by (3.2),

$$(3.12) \qquad (\beta \xi_T \Phi_{z,0})(0) = t \cdot N(\mathrm{il}(\gamma))^{1/2} |\lambda(\gamma)_{\mathbf{a}}|^{-1/2} h(\tau, z)^{-1} (\xi \Phi_{\tau(z),0})(0).$$

Taking $\xi = 1$, we obtain (3.10a, b) (with $\beta\tau$ as ξ there). In particular, we have

$$(3.13) \qquad h(t \cdot r_P(\gamma), z) = t^{-1} |\lambda(\gamma)_{\mathbf{a}}|^{1/2} \qquad \text{for} \quad t \in \mathbf{T} \quad \text{and} \quad \gamma \in P_{\mathbf{A}}.$$

Then (3.10c) can be verified by means of (3.12). To prove (3.10d), we first observe that

$$(3.14a) \quad h(\xi, z)^4 = (-1)^{mn} j(\xi, z)^2 \qquad \text{if} \quad \xi \in r_\Omega(\Omega_{\mathbf{A}} \cap C'' G_{\mathbf{a}}) \quad (n = [F : \mathbf{Q}]),$$

which follows immediately from (iv) of Proposition 3.1. With η as in (3.8b), we can find, by strong approximation, an element η_0 of G such that $\eta_0 \eta \in C' G_{\mathbf{a}}$. Then we see that $\eta_0 \in P\iota P \cap C'' G_{\mathbf{a}}$, and hence

$$(3.14b) \qquad\qquad h(\eta_0, z)^4 = (-1)^{mn} j(\eta_0, z)^2.$$

Let $\gamma \in G \cap P_{\mathbf{A}}C'$ and $\zeta = \gamma\eta_0^{-1}$. Then $\zeta \in P\iota P \cap P_{\mathbf{A}}C''$. Hence $\zeta = \pi\omega$ with $\pi \in P_{\mathbf{f}}$ and $\omega \in C'' G_{\mathbf{a}}$. Then $\omega_{\mathbf{a}} = \zeta_{\mathbf{a}}$ and $\gamma = r_P(\pi)r_\Omega(\omega)\eta_0$, and hence, by (3.10c) and (3.13), we have $h(\gamma, z) = h(r_\Omega(\omega), \eta_0(z)) \cdot h(\eta_0, z)$. This together with (3.14a, b) proves (3.10d). (A refinement of this result will be given as Lemma 3.5.)

Notice that $h(\tau, z)$ for $\tau \in \mathrm{pr}^{-1}(C'' G_{\mathbf{a}})$ is a factor of automorphy. Restricting τ to the elements of $G \cap P_{\mathbf{A}}D$, we observe that $h(\tau, z)$ satisfies (2.9a, b). As for (2.9c), it follows from the proof of Propositions 1.1 and 1.2 at the end of this section.

Put

$$(3.15a) \qquad\qquad S = \{\sigma \in F_m^m \mid {}^t\sigma = \sigma\}, \qquad \Lambda = S \cap \mathfrak{g}_m^m.$$

We define the Gauss sum of an element σ of S_v or $S_{\mathbf{A}}$ by

$$(3.15b) \qquad \gamma_v(\sigma) = N(\mathfrak{d}_v)^{m/2} \int_{L_v} \mathbf{e}_v(x\sigma \cdot {}^tx/2)\, dx \qquad (\sigma \in S_v,\, v \in \mathfrak{f}),$$

$$(3.15c) \qquad\qquad \gamma(\sigma) = \prod_{v \in \mathfrak{f}} \gamma_v(\sigma_v) \qquad (\sigma \in S_{\mathbf{A}}).$$

We define also a function $\nu(\sigma)$ and an ideal $\nu_0(\sigma)$ of $\mathfrak{g}$ by

$$(3.16) \quad \nu(\sigma) = \epsilon\left(\begin{pmatrix} 1_m & 0 \\ \sigma & 1_m \end{pmatrix}_{\mathfrak{f}}\right), \qquad \nu_0(\sigma) = \mathrm{il}\left(\begin{pmatrix} 1_m & 0 \\ \sigma & 1_m \end{pmatrix}\right)^{-1} \qquad (\sigma \in S_{\mathbf{A}}).$$

Then $\nu(\sigma) = N(\nu_0(\sigma))$, and $\nu(\sigma_v)$ depends only on σ_v modulo $\mathfrak{d}_v \Lambda_v$. We define an ideal character θ of the ideal group of F by

$$(3.17) \qquad\qquad \theta(\mathfrak{a}) = \left(\frac{F(\sqrt{-1})/F}{\mathfrak{a}} \right).$$

LEMMA 3.3. *Suppose* $v \in \mathfrak{f}$ *and* $v \nmid 2$. *Put*

$$(3.18) \qquad\qquad \omega_v(\sigma) = v(\delta_v^2 \sigma)^{1/2} \gamma_v(\sigma) \qquad (\sigma \in S_v).$$

Then $\omega_v(\sigma)$ *depends only on* σ *modulo* $\delta_v^{-1}\Lambda_v$, *and moreover,* $\omega_v(\sigma)^2 = \theta(v_0(\delta_v^2\sigma))$
and $\omega_v(c\sigma) = \omega_v(\sigma)\left(\dfrac{c}{v_0(\delta_v^2\sigma)} \right)$ *for every* $c \in \mathfrak{g}_v^\times$, *where the last factor is the*
quadratic residue symbol.

Proof. Let v' denote the symbol v of [15, (3.14)]. Then the present $v(\sigma)$ is
equal to $v'(\delta_v^{-1}\sigma)$. Hence we obtain our assertions immediately from [15, Lemma
4.4].

LEMMA 3.4. *Suppose* $\xi \in M_\mathbf{A}$ *and* $\mathrm{pr}(\xi) \in P_\mathbf{A}C''$. *If* ξ *or* $\xi \cdot r_\Omega(\iota)^{-1}$ *is contained*
in $r_\Omega(\Omega_\mathbf{A} \cap P_\mathbf{A}C'')$, $h(\xi,z)$ *is completely determined by* (3.10b) *and the following*
formulas:

$$\lim_{\rho \to \infty} h(\xi,\rho\mathbf{i})/|h(\xi,\rho\mathbf{i})| = \gamma/|\gamma| \qquad with \quad \gamma = \gamma(-c_\alpha^{-1}d_\alpha)$$

$$if \quad \xi = r_\Omega(\alpha) \quad with \quad \alpha \in \Omega_\mathbf{A} \cap P_\mathbf{A}C'',$$

$$\lim_{\rho \to 0} h(\xi,\rho\mathbf{i})/|h(\xi,\rho\mathbf{i})| = \gamma/|\gamma| \qquad with \quad \gamma = \gamma(\delta^{-2}d_\alpha^{-1}c_\alpha)$$

$$if \quad \xi = r_\Omega(\alpha\iota^{-1})r_\Omega(\iota) \quad with \quad \alpha \in \Omega_\mathbf{A}\iota \cap P_\mathbf{A}C''.$$

These formulas can easily be verified by means of (3.3). We omit the detailed
proof, as they are unnecessary for the proof of our main theorems.

LEMMA 3.5. *Suppose* $\alpha \in G \cap P_\mathbf{A}C'$. *Then* $\lambda(\alpha) \neq 0$, $v_0(d_\alpha^{-1}c_\alpha) = \lambda(\alpha)\mathrm{il}(\alpha)^{-1}$,
$\lambda(a)\mathrm{il}(\alpha)^{-1}$ *is prime to* 2, *and*

$$h(\alpha,z)^2 = \mathrm{sgn}(N_{F/\mathbf{Q}}(\lambda(\alpha)))\theta(\lambda(\alpha)\mathrm{il}(\alpha)^{-1})j(\alpha,z).$$

This follows easily from the last formula of Lemma 3.4 and Lemma 3.3 (cf.
also Lemma 4.1 below). Notice that the formula of Proposition 1.2 concerning
$h(\gamma,z)^2$ is a special case.

Proof of Propositions 1.1 *and* 1.2. Given $l \in \mathscr{L}(X)$, take a lattice L' in X so
that l induces a function on X/L'. We then define a function Ψ on $X_\mathbf{A}$ by

$\Psi(x) = \Psi_{\mathfrak{f}}^{l}(x_{\mathfrak{f}})\Phi_{\mathfrak{a}}(x_{\mathfrak{a}}; z, u)$ with $\Phi_{\mathfrak{a}}$ of (3.9c) and with Ψ^{l} such that $\Psi_{\mathfrak{f}}^{l}(y) = l(y')$ if $y' \in X$ and $y' - y_{v} \in L'_{v}$ for all $v \in \mathfrak{f}$. Then we see that

$$\theta(z, u; l) = \sum_{\xi \in X} \Psi(\xi).$$

We now apply [20, Théorème 6] to this sum to obtain

$$(3.19) \qquad \sum_{\xi \in X} \Psi(\xi) = \sum_{\xi \in X} (\alpha\Psi)(\xi)$$

for every $\alpha \in G$. In view of Proposition 3.1, we see that

$$(3.20) \qquad (\alpha\Psi)(x) = \Psi_{\mathfrak{f}}^{p}(x_{\mathfrak{f}})j(\alpha, z)^{-1/2}\zeta_{\alpha}(z, u)^{-1}\Phi_{\mathfrak{a}}(x_{\mathfrak{a}}; \alpha(z, u))$$

with a suitable $p \in \mathscr{L}(X)$ and ζ_{α} of (1.11). Therefore (3.19) implies that

$$\theta(z, u; l) = j(\alpha, z)^{-1/2}\zeta_{\alpha}(z, u)^{-1}\theta(\alpha(z, u); p).$$

The first assertion of Proposition 1.1 is a reformulation of this fact. Take in particular l to be the characteristic function l_{0} of L so that $\Psi = \Phi_{z,u}$. Then (3.11a) shows that (1.14) holds for every $\gamma \in G \cap C''G_{\mathfrak{a}}$. The last two formulas of Proposition 1.2 follow easily from Lemmas 3.3 and 3.4. To prove the second part of Proposition 1.1, it is sufficient to treat the case where l has the form $l(x) = l_{0}(xg + y)$ with $g \in \mathrm{GL}_{m}(F)$ and $y \in X$. For $w = (p, p') \in X_{\mathbf{A}} \times X_{\mathbf{A}}$, define $U(w) \in \mathrm{Aut}(L^{2}(X_{\mathbf{A}}))$ by $[U(w)f](x) = \mathbf{e}_{\mathbf{A}}(x \cdot {}^{t}p')f(x + p)$. By [20, p. 157, (15)], we have

$$(3.21) \qquad \beta^{-1}U(w)\beta = \mathbf{e}_{\mathbf{A}}(q_{\beta}(w))U(w\beta) \qquad \text{for every} \quad \beta \in G$$

with q_{β} of (3.4). Writing $\Psi_{z,u}$ for Ψ with the present l, and putting $w = (y_{\mathfrak{f}}, 0)$ and $\alpha = \mathrm{diag}[g, {}^{t}g^{-1}]$, we have $\Psi_{\alpha(z,u)} = \alpha U(w)\Phi_{z,u}$. If $\beta^{-1} = \left(\begin{smallmatrix} a & b \\ c & d \end{smallmatrix}\right) \in \Gamma'$, we have, by (3.11a) and (3.21),

$$\beta U(w)\Phi_{z,u}(x) = h(\beta, z)^{-1}\zeta_{\beta}(z, u)^{-1}\mathbf{e}_{\mathfrak{a}}(-2^{-1}ya \cdot {}^{t}b \cdot {}^{t}y - yb \cdot {}^{t}x)$$

$$\cdot \Phi_{\beta(z,u)}(x + y_{\mathfrak{f}}a).$$

Therefore if β belongs to a sufficiently small congruence subgroup Γ_{y} of Γ' depending only on y, we have

$$\beta U(w)\Phi_{z,u} = h(\beta, z)^{-1}\zeta_{\beta}(z, u)^{-1}U(w)\Phi_{\beta(z,u)}.$$

If $\gamma \in \alpha\Gamma_{y}\alpha^{-1}$ and $\beta = \alpha^{-1}\gamma\alpha$, then $\gamma\Psi_{\alpha(z,u)} = h(\beta, z)^{-1}\zeta_{\beta}(z, u)^{-1} \cdot \Psi_{\gamma\alpha(z,u)}$. Since $\zeta_{\gamma}(\alpha(z, u)) = \zeta_{\beta}(z, u)$, we have $\gamma\Psi_{z,u} = h(\alpha^{-1}\gamma\alpha, z)^{-1}\Psi_{\gamma(z,u)}$ which together with (3.19) proves the last assertion of Proposition 1.1.

4. Eisenstein series on $M_{\mathbf{A}}$. We fix an odd integer k as before, an element ρ

of $\mathbf{Z}^{\mathbf{a}}$, and an element μ of $\mathbf{R}^{\mathbf{a}}$, and put

$$(4.0) \qquad J(\xi,z) = h(\xi,z)^k j_{\mathbf{a}}(\xi,z)^\rho |j_{\mathbf{a}}(\xi,z)|^{i\mu-\rho}$$

for every $\xi \in M_{\mathbf{A}}$ such that $\mathrm{pr}(\xi) \in P_{\mathbf{A}}C''$. ($\mu_v$ of (1.4) will not appear in this section.) Here h is defined by Proposition 3.2, $j_{\mathbf{a}}(\xi,z)$ is an element of $\mathbf{C}^{\mathbf{a}}$ whose v-component is $j_v(\mathrm{pr}(\xi)_{\mathbf{a}},z)$, and $x^\rho = \prod_{v\in\mathbf{a}} x_v^{\rho_v}$ for $x \in \mathbf{C}^{\mathbf{a}}$; similarly x^τ is defined for $\tau \in \mathbf{C}^{\mathbf{a}}$ if $0 < x_v \in \mathbf{R}$ for all v. Now an automorphic form with respect to the factor of automorphy J, as a function on $M_{\mathbf{A}}$, is a function $g_0 : M_{\mathbf{A}} \to \mathbf{C}$ such that

$$(4.1) \qquad g_0(\alpha\xi u) = g_0(\xi) J(u,\mathbf{i})^{-1}$$

$$\text{for} \quad \alpha \in G, \quad \xi \in M_{\mathbf{A}}, \quad u \in M_{\mathbf{A}}, \quad \mathrm{pr}(u) \in B$$

with an open subgroup B of C'', where $\mathbf{i} = (i1_m, \ldots, i1_m) \in H^{\mathbf{a}}$. (Recall that G is considered a subgroup of $M_{\mathbf{A}}$.) Given such a g_0, we can define a function g on $H^{\mathbf{a}}$ by

$$(4.2) \qquad g(\xi(\mathbf{i})) = g_0(\xi) J(\xi,\mathbf{i}) \qquad \text{for} \quad \xi \in M_{\mathbf{A}}, \quad \mathrm{pr}(\xi) \in BG_{\mathbf{a}}.$$

Put $\Gamma = G \cap BG_{\mathbf{a}}$. Then

$$(4.3a) \qquad g(\alpha z) = g(z) J(\alpha,z) \qquad \text{for} \quad \alpha \in \Gamma \quad \text{and} \quad z \in H^{\mathbf{a}},$$

$$(4.3b) \qquad g_0(\alpha\xi) = g(\xi(\mathbf{i})) J(\xi,\mathbf{i})^{-1} \qquad \text{for} \quad \alpha \in G \quad \text{and} \quad \mathrm{pr}(\xi) \in BG_{\mathbf{a}}.$$

Conversely, if g is a function on $H^{\mathbf{a}}$ satisfying (4.3a), we can define a function g_0 on $M_{\mathbf{A}}$ by (4.3b), which satisfies (4.1) and (4.2).

To define an Eisenstein series on $M_{\mathbf{A}}$ as such a g_0, we fix an integral ideal $\mathfrak{c}$ of F divisible by 4, consider the subgroup D of $G_{\mathbf{A}}$ as in (2.7), and take a Hecke character $\psi : F_{\mathbf{A}}^\times / F^\times \to \mathbf{T}$ such that:

$$(4.4a) \qquad \textit{the finite part of the conductor of } \psi \textit{ divides } \mathfrak{c};$$

$$(4.4b) \qquad \psi(x) = |x|^{i\mu}(x/|x|)^\rho \qquad \textit{for} \quad x \in F_{\mathbf{a}}^\times.$$

(The existence of such a ψ imposes a condition on $\{\rho, \mu\}$.) The ideal-character attached to ψ will be denoted by the same letter ψ. We define a function f on $M_{\mathbf{A}}$ by

$$(4.5) \qquad f(\xi) = \begin{cases} 0 & \text{if} \quad \mathrm{pr}(\xi) \notin P_{\mathbf{A}}D, \\ \psi_{\mathfrak{f}}(\lambda(p)^{-1})\psi_{\mathfrak{c}}(\lambda(w)^{-1})|J(\xi,\mathbf{i})|J(\xi,\mathbf{i})^{-1} & \\ & \text{if} \quad \mathrm{pr}(\xi) = pw \quad \text{with} \quad p \in P_{\mathbf{A}} \quad \text{and} \quad w \in D, \end{cases}$$

where $\psi_{\mathfrak{f}} = \prod_{v\in\mathfrak{f}}\psi_v$, $\psi_{\mathfrak{c}} = \prod_{v|\mathfrak{c}}\psi_v$, and λ is defined by (2.4). From (3.10c) and (3.13) we see that

$$(4.6) \quad f(\beta\xi u) = f(\xi)\psi_{\mathfrak{c}}(\lambda(u)^{-1})J(u,\mathbf{i})^{-1} \qquad \text{if} \quad \beta \in P \quad \text{and} \quad \mathrm{pr}(u) \in D.$$

We now define an Eisenstein series $E_{\mathbf{A}}(\xi, s)$ on $M_{\mathbf{A}}$ by

$$(4.7\text{a}) \qquad E_{\mathbf{A}}(\xi, s) = E_A(\xi, s; k/2, \rho, \psi, c)$$

$$= \sum_{\alpha \in P \backslash G} f(\alpha\xi)\epsilon(\alpha\xi)^{-2s-k/2} \qquad (\xi \in M_{\mathbf{A}},\ s \in \mathbf{C}),$$

where ϵ is defined by (2.6) and (3.5). It can easily be seen that $E_{\mathbf{A}}(\xi, s)$ is a function of type (4.1), and therefore we can define a function $E(z, s)$ on $H^{\mathbf{a}} \times \mathbf{C}$ by

$$(4.7\text{b}) \qquad E(z, s) = E(z, s; k/2, \rho, \psi, c) = E_{\mathbf{A}}(\xi, s)J(\xi, \mathbf{i})$$

$$\text{for} \quad z = \xi(\mathbf{i}) \quad \text{with} \quad \xi \in M_{\mathbf{A}}, \quad \mathrm{pr}(\xi) \in G_{\mathbf{a}}.$$

Put $\Gamma_0(c) = G \cap DG_{\mathbf{a}}$. Then we can easily verify that

$$E(\gamma(z), s) = \psi_c(\lambda(\gamma))J(\gamma, z)E(z, s) \qquad \text{for every} \quad \gamma \in \Gamma_0(c).$$

LEMMA 4.1. *Let* $\alpha \in G$, $\mathfrak{a} = \mathrm{il}(\alpha)$, *and* $\xi \in M_{\mathbf{A}}$; *suppose* $\mathrm{pr}(\xi) \in G_{\mathbf{a}}$ *and* $\mathrm{pr}(\alpha\xi) \in P_{\mathbf{A}}D$. *Then* $\alpha \in G \cap P_{\mathbf{A}}D$, $\lambda(\alpha) \neq 0$, $\lambda(\alpha)\mathfrak{a}^{-1}$ *is prime to* c, *and*

$$f(\alpha\xi)\epsilon(\alpha\xi)^{-2s-k/2} = N(\mathfrak{a})^{2s+k/2}\psi_{\mathbf{a}}(\lambda(\alpha))\psi\big(\lambda(\alpha)\mathfrak{a}^{-1}\big)|j(\alpha\xi, \mathbf{i})|^{-2s}J(\alpha\xi, \mathbf{i})^{-1}.$$

This can be proved in the same manner as in [15, Lemma 2.2].

Now take $\{\beta\}$, $\mathfrak{a}_\beta$, and T_β as in Lemma 2.1, (iii) and (2.11). Then the above lemma shows that

$$(4.7\text{c}) \quad E(z, s) = \sum_{\beta} N(\mathfrak{a}_\beta)^{2s+k/2} \sum_{\alpha \in T_\beta} \psi_{\mathbf{a}}(\lambda(\alpha))\psi\big(\lambda(\alpha)\mathfrak{a}_\beta^{-1}\big) \cdot Y(\alpha(z))^s J(\alpha, z)^{-1}.$$

This coincides with (2.11) if $\rho = \mu = 0$. The series with nontrivial ρ and μ is introduced with a future application in view. The reader who is interested only in the main theorems of Section 2 may simplify the treatment by always assuming $\rho = \mu = 0$.

With η defined as in (3.8b), there is a unique element $\tilde{\eta}$ of $M_{\mathbf{A}}$ such that $\mathrm{pr}(\tilde{\eta}) = \eta$ and $\tilde{\eta}\Psi = \Psi_{\mathbf{a}}\prod_{v \in \mathfrak{f}} r_\Omega(\eta_v)\Psi_v$ for every Ψ in $\mathscr{S}(X_{\mathbf{A}})$ of the form $\Psi(x) = \Psi_{\mathbf{a}}(x_{\mathbf{a}})\prod_{v \in \mathfrak{f}} \Psi_v(x_v)$. Then we have

$$(4.8) \qquad\qquad h(\tilde{\eta}, z) = J(\tilde{\eta}, z) = 1.$$

Now define a function $E'_{\mathbf{A}}$ on $M_{\mathbf{A}}$ by

$$(4.9\text{a}) \qquad E'_{\mathbf{A}}(\xi) = E'_{\mathbf{A}}(\xi, s) = E_{\mathbf{A}}(\xi\tilde{\eta}, s) \qquad (\xi \in M_{\mathbf{A}}).$$

This satisfies (4.1) with a suitable B, and hence we can define a function E' on $H^{\mathbf{a}} \times \mathbf{C}$ by

$$(4.9\text{b}) \quad E'(z) = E'(z, s) = E'_{\mathbf{A}}(\xi)J(\xi, \mathbf{i})^{\hat{}} \quad \text{for} \quad z = \xi(\mathbf{i}), \quad \mathrm{pr}(\xi) \in BG_{\mathbf{a}}.$$

Take B so that both $E_\mathbf{A}$ and $E_\mathbf{A}'$ satisfy (4.1). By the strong approximation theorem, we can find an element η_0 of G (and fix it once for all) such that $\eta_0\eta \in BG_\mathbf{a}$. Then $\eta_0 \in G \cap C''G_\mathbf{a}$, and hence $h(\eta_0,z)$ can be defined. We have then

$$(4.10) \qquad E'(z,s) = E(\eta_0(z),s)J(\eta_0,z)^{-1}.$$

To show this, take $\xi \in M_\mathbf{A}$ so that $z = \xi(\mathbf{i})$ and $\mathrm{pr}(\xi) \in G_\mathbf{a}$. Since $\xi\tilde\eta = \tilde\eta\xi$ and $\mathrm{pr}(\eta_0\tilde\eta\xi) \in BG_\mathbf{a}$, we have $E'(z) = E_\mathbf{A}(\tilde\eta\xi)J(\xi,\mathbf{i}) = E_\mathbf{A}(\eta_0\tilde\eta\xi)J(\xi,\mathbf{i}) = E(\eta_0(z))J(\eta_0\tilde\eta\xi,\mathbf{i})^{-1}J(\xi,\mathbf{i})$, which together with (4.8) proves (4.10).

We now consider the Fourier expansions of $E_\mathbf{A}'$ and E'. With S as in (3.15a), we put

$$(4.11a) \qquad \chi(u) = \mathbf{e}_\mathbf{A}(\mathrm{tr}(u)) \qquad (u \in S_\mathbf{A}),$$

$$(4.11b) \qquad \chi_v(u) = \mathbf{e}_v(\mathrm{tr}(u)) \qquad (u \in S_v),$$

$$(4.12) \qquad t(u) = \begin{pmatrix} 1 & u \\ 0 & 1 \end{pmatrix}, \qquad \tau(u) = r_P(t(u)) \qquad (u \in S_\mathbf{A}),$$

and define a Haar measure du on $S_\mathbf{A}$ so that $S_\mathbf{A}/S$ has measure 1.

Fix an element ξ of $M_\mathbf{A}$ such that $\mathrm{pr}(\xi) \in G_\mathbf{a}$. Then $E_\mathbf{A}'(\tau(u)\xi)$, being invariant under $u \mapsto u + a$ with $a \in S$, has a Fourier expansion

$$(4.13) \qquad E_\mathbf{A}'(\tau(u)\xi,s) = \sum_{\sigma \in S} b(\sigma,\xi,s)\chi(\sigma u) \qquad (u \in S_\mathbf{A}),$$

when $\mathrm{Re}(s)$ is sufficiently large, where

$$b(\sigma,\xi,s) = \int_{S_\mathbf{A}/S} E_\mathbf{A}'(\tau(u)\xi,s)\chi(-\sigma u)\,du \qquad (\sigma \in S).$$

Now we can show, by the same type of argument as in [15, p. 430], that $b(\sigma,\xi,s) \neq 0$ only when $\sigma \in 2\mathfrak{c}^{-1}\Lambda'$, where

$$(4.14) \qquad \Lambda' = \{\sigma \in S \,|\, \mathrm{tr}(\sigma\Lambda) \subset \mathfrak{g}\}.$$

If we put $R = t(S)$, we have, for the same reason as in [15, Lemma 2.3],

$$E_\mathbf{A}'(\tau(u)\xi) = \sum_{\alpha \in \iota R} f(\alpha\tau(u)\xi\tilde\eta)\epsilon(\alpha\tau(u)\xi\tilde\eta)^{-2s-k/2}$$

$$= \sum_{a \in S} f(\iota\tau(u+a)\xi\tilde\eta)\epsilon(\iota\tau(u+a)\xi\tilde\eta)^{-2s-k/2},$$

and hence

$$(4.15) \qquad b(\sigma,\xi,s) = \int_{S_\mathbf{A}} f(\iota\tau(u)\xi\tilde\eta)\epsilon(\iota\tau(u)\xi\tilde\eta)^{-2s-k/2}\chi(-\sigma u)\,du.$$

In order to express this in terms of $E'(z)$, given $z = (z_v)_{v \in \mathbf{a}} \in H^{\mathbf{a}}$ with $z_v = x_v + iy_v$, we define $w \in G_{\mathbf{A}}$ and $u^0 \in S_{\mathbf{A}}$ by

$$w_v = \begin{cases} 1_{2m} & (v \in \mathbf{f}), \\ \mathrm{diag}\big[\, y_v^{1/2},\, y_v^{-1/2} \,\big] & (v \in \mathbf{a}), \end{cases}$$

$$u_v^0 = \begin{cases} 0 & (v \in \mathbf{f}), \\ x_v & (v \in \mathbf{a}). \end{cases}$$

Then we have

$$(4.16) \qquad E'(z,s) = J\big(\tau(u^0)r_P(w), \mathbf{i}\big) E_A'\big(\tau(u^0)r_P(w)\big)$$

$$= \det(y)^{-p/4} E_A'\big(\tau(u^0)r_P(w)\big) \qquad \text{(by (3.13))}$$

$$= \sum_{\sigma \in \Lambda''} c(\sigma, y, s)\mathbf{e}_{\mathbf{a}}(\mathrm{tr}(\sigma x)),$$

where $\Lambda'' = 2\mathfrak{c}^{-1}\Lambda'$, $\det(y) = (\det(y_v))_{v \in \mathbf{a}}$, $p = (k + 2i\mu_v)_{v \in \mathbf{a}}$, and

$$(4.17) \qquad c(\sigma, y, s) = \det(y)^{-p/4} b(\sigma, r_P(w), s).$$

Our immediate aim is to compute $b(\sigma, r_P(w), s)$ by means of (4.15). We first observe that

$$(4.18) \qquad (\iota t(u)w\eta)_v = \begin{cases} \begin{bmatrix} -\delta_v 1_m & 0 \\ \delta_v u_v & -\delta_v^{-1} 1_m \end{bmatrix} & (v \in \mathbf{f}), \\[2.5em] \begin{bmatrix} 0 & -y_v^{-1/2} \\ y_v^{1/2} & u_v y_v^{-1/2} \end{bmatrix} & (v \in \mathbf{a}). \end{cases}$$

LEMMA 4.2. $f(\iota\tau(u)r_P(w)\tilde{\eta}) \neq 0$ *if and only if* $u_v \in (2^{-1}\mathfrak{c}_v\mathfrak{d}_v^{-1})_m^m$ *for every* v *dividing* $\mathfrak{c}$, *in which case we have*

$$f(\iota\tau(u)r_P(w)\tilde{\eta})\epsilon(\iota\tau(u)r_P(w)\tilde{\eta})^{-2s-k/2} = A \prod_{v \in \mathbf{a}} g_v(u_v, y_v) \prod_{v \in \mathbf{f},\, v \nmid \mathfrak{c}} g_v(u_v),$$

with A *and* g_v *defined by*

$$A = (-1)^{m|\rho|}\psi(\delta^m)N(\mathfrak{d})^{-m(2s+k/2)},$$

$$g_v(u_v) = \omega_v(u_v)^k \psi\big(\nu_0(\delta_v^2 u_v)\big)\nu\big(\delta_v^2 u_v\big)^{-2s-k/2} \qquad (v \in \mathbf{f},\, v \nmid \mathfrak{c}),$$

$$g_v(u_v, y_v) = \mathbf{e}(mk/8)\det_v(y)^{(\alpha+\beta)/2}\det_v(u + iy)^{-\alpha}\det_v(u - iy)^{-\beta} \qquad (v \in \mathbf{a}),$$

where $|\rho| = \sum_{v \in \mathbf{a}} \rho_v$, δ is given by (3.8a) $\det_v(x)^\alpha = \det(x_v)^{\alpha_v}$, $\alpha_v = s + (k + \rho_v + i\mu_v)/2$, $\beta_v = s - (\rho_v - i\mu_v)/2$, and the branches of $\det(z)^{-\alpha}$ and $\det(\bar{z})^{-\beta}$ are determined as in [14, (1.10), (1.11)].

Proof. Let $C^* = \beta^{-1}C\beta$ with $\beta \in \mathrm{GL}_{2m}(F_\mathbf{A})$ such that $\beta_\mathbf{a} = 1$ and $\beta_\mathbf{f} = \mathrm{diag}[1_m, 21_m]$. Then $G_\mathbf{A} = P_\mathbf{A}C^*$. Hence, given $u \in S_\mathbf{A}$, we can put $\iota t(u)\eta = px$ with $p = \begin{pmatrix} * & * \\ 0 & e \end{pmatrix} \in P_\mathbf{A}$ and $x = \begin{pmatrix} a & b \\ c & d \end{pmatrix} \in C^*$. Then $(ed)_\mathbf{f} = (-\delta^{-1})_\mathbf{f}$ and $(d^{-1}c)_\mathbf{f} = (-\delta^2 u)_\mathbf{f}$. Suppose $f(\iota\tau(u)r_P(w)\tilde{\eta}) \neq 0$. Then $\iota t(u)\eta \in P_\mathbf{A}D$, so that we can take x in D. Then $d_v \in \mathrm{GL}_m(\mathfrak{g}_v)$ for $v|\mathfrak{c}$, and hence $u_v = (-\delta^{-2}d^{-1}c)_v \in (2^{-1}\mathfrak{c}_v\mathfrak{d}_v^{-1})_m^m$. Conversely, if $u_v \in (2^{-1}\mathfrak{c}_v\mathfrak{d}_v^{-1})_m^m$ for $v|\mathfrak{c}$, we have $c_v = (-\delta^2 du)_v \in (2^{-1}\mathfrak{c}_v\mathfrak{d}_v)_m^m$ for such v's, so that $x \in D$. This proves our first assertion. With x in D, we have $\begin{bmatrix} 1 & 0 \\ -\delta_v^2 u_v & 1 \end{bmatrix} = \begin{bmatrix} * & * \\ 0 & -\delta_v e_v \end{bmatrix} x_v$. Hence

$$(4.19) \qquad \epsilon\big((\iota t(u)\eta)_v\big) = |\det(e_v)| = |\delta_v|^{-m}\nu(\delta_v^2 u_v) \qquad (v \in \mathbf{f}),$$

$$(4.20) \qquad \nu_0(\delta^2 u) = \det(\delta e)^{-1}\mathfrak{g} = \det(d)\mathfrak{g}.$$

This is prime to $\mathfrak{c}$. Further we have

$$(4.21) \qquad \psi_\mathbf{f}(\lambda(p))\psi_\mathfrak{c}(\lambda(x)) = (-1)^{m|\rho|}\psi(\delta^{-m}) \prod_{v \nmid \mathfrak{c}} \psi\big(\nu_0(\delta_v^2 u_v)\big)^{-1}.$$

From (4.18), we obtain

$$(4.22) \qquad \epsilon\big((\iota t(u)w\eta)_v\big) = |\det(iy_v^{1/2} + u_v y_v^{-1/2})| \qquad (v \in \mathbf{a}).$$

Next, we observe, in view of (4.8), that

$$h(\iota\tau(u)r_P(w)\tilde{\eta}, \mathbf{i}) = h(\iota\tau(u)r_P(w), \mathbf{i}).$$

Since $\iota\tau(u)r_P(w) = r_\Omega(\iota t(u)w)$, we find, employing (3.3), that

$$\big[\iota\tau(u)r_P(w)\Phi_{\mathbf{l},0}\big](0) = \prod_{v \in \mathbf{f}} \int \varphi_v(x)\mathbf{e}_v(xu_v \cdot {}^t x/2)\,dx$$

$$\cdot \prod_{v \in \mathbf{a}} \det(y_v)^{1/4} \int \varphi(xy_v^{1/2}; i1_m, 0)\mathbf{e}(xu_v \cdot {}^t x/2)\,dx$$

$$= N(\mathfrak{d})^{-m/2}\gamma(u) \prod_{v \in \mathbf{a}} \det(y_v)^{1/4}\det(y_v - iu_v)^{-1/2}.$$

Combining this with (4.19), (3.10a), and (3.17), we obtain

$$h(\iota\tau(u)r_P(w)\tilde{\eta}, \mathbf{i}) = \prod_{v \nmid \mathfrak{c}} \omega_v(u_v)^{-1} \prod_{v \in \mathbf{a}} \det(y_v)^{-1/4}\det(y_v - iu_v)^{1/2},$$

since $\gamma_v(u_v) = \nu(\delta_v^2 u_v) = 1$ for $v|\mathfrak{c}$. Then, our definition of f together with (4.21)

shows that

$$f(\iota\tau(u)r_P(w)\tilde{\eta}) = i^{m|\rho|}\psi(\delta^m)\prod_{v\nmid c}\omega_v(u_v)^k\psi\big(\nu_0(\delta_v^2 u_v)\big)$$

$$\cdot \prod_{v\in\mathbf{a}}\det_v(y)^{i\mu/2}|\det_v(u+iy)|^{\rho-i\mu+k/2}\det_v(y-iu)^{-\rho-k/2}.$$

This combined with (4.19) and (4.22) yields the expected formula.

5. The explicit forms of the Fourier coefficients of E'. Take a measure du_v on S_v so that $\int_{\Lambda_v}du_v = 1$ for $v\in\mathbf{f}$ and $\int_{S_v/N}du_v = 1$ for $v\in\mathbf{a}$, where $N = S_v\cap\mathbf{Z}_m^m$. Then the measure du on $S_\mathbf{A}$ of Section 4 coincides with $N(\mathfrak{d})^{-m(m+1)/4}\prod_v du_v$. Therefore, substituting the formula of Lemma 4.2 into (4.15) and (4.17), we obtain, for $\sigma\in 2c^{-1}\Lambda'$,

$$(5.1)\qquad c(\sigma,y,s) = (-1)^{m|\rho|}\psi(\delta^m)N(\mathfrak{d})^{-mq}\prod_{v\in\mathbf{f}}c_v(\sigma,s)\prod_{v\in\mathbf{a}}c_v(\sigma,y,s),$$

where $q = 2s + (k/2) - (m+1)/4$, and

$$c_v(\sigma,s) = \begin{cases} N\big((2c^{-1})_v\big)^{m(m+1)/2} & \text{if } v\,|\,c, \\[2mm] \displaystyle\sum_{u\in S_v/\delta_v^{-1}\Lambda_v}\omega_v(u)^k\chi_v(-\sigma u)\psi\big(\nu_0(\delta_v^2 u)\big)\nu(\delta_v^2 u)^{-2s-k/2} & \text{if } v\nmid c, \end{cases}$$

$$c_v(\sigma,y,s) = \mathbf{e}(mk/8)\det(y_v)^s\int_{S_v}\chi_v(-\sigma_v u)\det(u+iy_v)^{-\alpha_v}\det(u-iy_v)^{-\beta_v}\,du$$

$$\text{if } v\in\mathbf{a}\quad\text{(with }\alpha_v\text{ and }\beta_v\text{ as in Lemma 4.2).}$$

Let $\xi(g,h;\alpha,\beta)$ be the confluent hypergeometric function introduced in [14, (1.25)]. Then we obtain

$$c_v(\sigma,y,s) = \mathbf{e}(mk/8)\det(y_v)^s\xi(y_v,\sigma_v;\alpha_v,\beta_v).$$

To determine c_v with $v\nmid c$, we fix one v not dividing 2, and define a power-series $A_v^k(\sigma,t)$ of an indeterminate t by

$$(5.2)\qquad A_v^k(\sigma,q_v^{-s}) = \sum_u\omega_v(u)^k\chi_v(-\sigma u)\nu(\delta_v^2 u)^{-s},$$

where u runs over $S_v/\delta_v^{-1}\Lambda_v$, and $q_v = |\pi_v|^{-1}$ with a prime element π_v of $\mathfrak{g}_v$. This is meaningful for $\sigma\in\Lambda_v$, $v\nmid 2$. For $v\nmid c$, we have

$$(5.3)\qquad c_v(\sigma,s) = A_v^k\big(\sigma,\psi_v(\pi_v)q_v^{-2s-k/2}\big).$$

The series A_v^k is essentially the same as the series α_λ introduced in [15, (4.8a)]. Let us write ω' and ν' for the symbols ω and ν of [15]. Then $\omega_v(u) = \omega'(u/2)$ and $\nu(u) = \nu'(\delta^{-1}u)$, and hence we obtain

$$(5.4) \qquad A_v^k(\sigma, t) = \alpha_k(2\sigma/\delta_v, t) \qquad (\sigma \in \Lambda_v, v \nmid 2).$$

Moreover, by Lemma 3.3, we have

$$(5.5) \qquad A_v^k(\sigma, t) = A_v^1\big(\sigma, \theta(\pi_v \mathfrak{g})^{(k-1)/2}t\big).$$

Let $r = \mathrm{rank}(\sigma)$. If $r > 0$ (and $v \nmid 2$), we can find an element ζ of $\mathrm{GL}_m(\mathfrak{g}_v)$ such that

$$(5.6a) \qquad {}^t\zeta\sigma\zeta = \mathrm{diag}\big[\sigma', 0_{m-r}\big]$$

with ${}^t\sigma' = \sigma' \in M_r(\mathfrak{g}_v)$. Let $K_v = F_v(\det(2\sigma')^{1/2})$. Then we put

$$(5.6b) \qquad \xi_\sigma = \begin{cases} 1 & \text{if} \quad K_v = F_v, \\ 0 & \text{if} \quad v \text{ is ramified in } K_v, \\ -1 & \text{if} \quad [K_v : F_v] = 2 \quad \text{and } v \text{ is unramified in } K_v. \end{cases}$$

PROPOSITION 5.1. *Suppose $\sigma \in \Lambda_v$ and $v \nmid 2$; let $r = \mathrm{rank}(\sigma)$. Define θ by (3.17) and write simply q for q_v. Then*

$$A_v^k(\sigma, t) = g_v\big(\sigma, \theta^{(k-1)/2}q^{-1/2}t\big) \prod_{i=0}^{[(m-1)/2]} (1 - q^{2i+1}t^2)$$

$$\cdot \begin{cases} \displaystyle\prod_{i=0}^{[(m-r-1)/2]} (1 - q^{2m-r-2i}t^2)^{-1} & (r \text{ even}), \\[4mm] (1 - \xi_\sigma\theta^b q^{m-r/2}t)^{-1} \displaystyle\prod_{i=1}^{[(m-r)/2]} (1 - q^{2m-r+1-2i}t^2)^{-1} & (r \text{ odd}), \end{cases}$$

where $b = (r + k - 2)/2$, $\theta = \theta(\pi_v\mathfrak{g})$, and $g_v(\sigma, t)$ is a polynomial in t with rational coefficients depending on σ and v; we understand that $\prod_{i=\alpha}^{\beta} = 1$ if $\beta < \alpha$. Moreover, if either $\sigma = 0$ or σ' of (5.6a) belongs to $\mathrm{GL}_r(\mathfrak{g}_v)$, then $g_v(\sigma, t) = 1$.

Proof. The formula has been obtained in [15, Propositions 4.6, 5.1 and 6.2] when $\sigma = 0$ or $\det(\sigma')$ is a v-unit. The general case with nonunit $\det(\sigma')$ is due to Paul Feit [2]. Here we reproduce his proof. For a positive integer n and symmetric matrices Y and Z of size h and m with entries in $\mathfrak{g}_v$, let $A(Y, Z, n)$ denote the number of elements $x \in (\mathfrak{g}_v/\mathfrak{p}^n)_m^h$ such that ${}^t x Y x \equiv Z \pmod{\mathfrak{p}^n}$, where $\mathfrak{p}$ denotes the maximal ideal of $\mathfrak{g}_v$. We first prove, for any $\sigma \in \Lambda_v$,

$$(5.7) \qquad A(1_h, \mathrm{diag}[1, \sigma], n) = A(1_h, 1, n)A(1_{h-1}, \sigma, n).$$

Suppose ${}^t x x \equiv \mathrm{diag}[1, \sigma] \pmod{\mathfrak{p}^n}$ and let $x = (u, y)$ with $u \in (\mathfrak{g}_v)_1^h$ and $y \in (\mathfrak{g}_v)_m^h$.

Then we have

$$(5.8) \qquad 'uu \equiv 1, \qquad 'uy \equiv 0, \qquad 'yy \equiv \sigma \qquad (\mathrm{mod}\,\mathfrak{p}^n).$$

The number of u $(\mathrm{mod}\,\mathfrak{p}^n)$ satisfying the first congruence is $A(1_h, 1, n)$. Let us now count the number of y's $(\mathrm{mod}\,\mathfrak{p}^n)$ for a fixed u. Take $a \in \mathrm{GL}_h(\mathfrak{g}_v)$ so that $'ua = (b, 0)$ with $b \in \mathfrak{g}_v$. Let c be the first upper left entry of $a^{-1} \cdot {}'a^{-1}$, and let $w = a^{-1}y$. Then (5.8) holds if and only if $b^2c \equiv 1$, $(b, 0)w \equiv 0$, $'w \cdot {}'aaw \equiv \sigma$ $(\mathrm{mod}\,\mathfrak{p}^n)$. Therefore both b and c are v-units, and $'w \equiv (0, z)$ with $z \in (\mathfrak{g}_v)_{h-1}^m$. Let d be the lower right submatrix of $'aa$ of size $h - 1$. Then the number of y's is equal to the number of z's such that $zd \cdot {}'z \equiv \sigma$ $(\mathrm{mod}\,\mathfrak{p}^n)$, that is, $A(d, \sigma, n)$. Now $\det(d) = c \cdot \det(a)^2$ by [15, Lemma 6.1]. Since $b^2c \equiv 1$ $(\mathrm{mod}\,\mathfrak{p}^n)$, this means that $\det(d)$ is a square in $\mathfrak{g}_v$, and hence $d = {}'ee$ with $e \in \mathrm{GL}_{h-1}(\mathfrak{g}_v)$. Thus $A(d, \sigma, n) = A(1_{h-1}, \sigma, n)$, which proves (5.7). Using now the notation of [15, §4], we have, by [15, (4.11)] and (5.7),

$$\alpha_1(\sigma/\delta, \theta^h q^{-h-1/2}) = \lim_{n\to\infty} q^{nm(m+1)/2 - nm(2h+1)} A\left(1_{2h+1}, \sigma, n\right)$$

$$= \lim_{n\to\infty} \frac{q^{n(m+1)(m+2)/2 - n(m+1)(2h+2)} A\left(1_{2h+2}, \tau, n\right)}{q^{n - n(2h+2)} A\left(1_{2h+2}, 1, n\right)}$$

$$= \alpha_0(\tau/\delta, \theta^{h+1} q^{-h-1})/\alpha_0(1/\delta, \theta^{h+1} q^{-h-1}),$$

where $\tau = \mathrm{diag}[1, \sigma]$. This is true for all positive integers h. Since α_0 and α_1 are power-series in t convergent in a neighborhood of 0, we thus obtain

$$(5.9) \qquad \alpha_1(\sigma/\delta, t) = \alpha_0(\tau/\delta, \theta q^{-1/2} t)/\alpha_0(1/\delta, \theta q^{-1/2} t).$$

Now $\alpha_0(1/\delta, t) = 1 - t$ by [15, Proposition 4.6]. As for $\alpha_0(\tau/\delta, t)$, a recent result of Kitaoka [3] and Feit [2] shows that

$$\alpha_0(\sigma/\delta, t) = \beta(t)(1 - t) \prod_{i=1}^{[m/2]} (1 - q^{2i} t^2)$$

$$\cdot \begin{cases} \left(1 - \xi_\sigma \theta^{r/2} q^{m-r/2} t\right)^{-1} \displaystyle\prod_{i=1}^{[(m-r)/2]} \left(1 - q^{2m-r+1-2i} t^2\right)^{-1} & (r \text{ even}), \\[2em] \displaystyle\prod_{i=0}^{[(m-r-1)/2]} \left(1 - q^{2m-r-2i} t^2\right)^{-1} & (r \text{ odd}) \end{cases}$$

with a polynomial β with rational coefficients. Substituting this formula, with τ, $m + 1$, and $r + 1$ in place of σ, m, and r, into (5.9), and replacing σ by 2σ, we obtain the expected result in view of (5.4) and (5.5).

In order to express $\prod_{v \nmid \mathfrak{c}} c_v(\sigma, s)$ as a product of L-functions, we define an ideal $\mathfrak{e}$ depending on σ and $\mathfrak{c}$ as follows: $\mathfrak{e} = \mathfrak{g}$ if $\sigma = 0$; $\mathfrak{e}$ is the product of all the primes v such that $v \nmid c$ and $L_v / L_v \sigma$ has nontrivial torsion elements. Further if $r = \text{rank}(\sigma)$ is odd, we take $\alpha \in \text{GL}_m(F)$ and $\sigma' \in \text{GL}_r(F)$ so that $'\alpha\sigma\alpha = \text{diag}[\sigma', 0_{m-r}]$ and put $K = F(\det(2\sigma')^{1/2})$; we then define an ideal-character ξ_σ of F by putting

$$(5.10) \qquad \xi_\sigma(\mathfrak{p}) = \begin{cases} 1 & \text{if} \quad K = F \quad \text{or } \mathfrak{p} \text{ splits in } K, \\ -1 & \text{if} \quad K \neq F \quad \text{and } \mathfrak{p} \text{ remains prime in } K, \\ 0 & \text{if} \quad \mathfrak{p} \text{ is ramified in } K, \end{cases}$$

for every prime ideal $\mathfrak{p}$ not dividing 2. This is consistent with (5.6b). Combining (5.1), (5.3), (5.5), and Proposition 5.1, we obtain

$$(5.11) \qquad c(\sigma, y, s) = B(s) c_{\mathbf{a}}(\sigma, y, s) c_{\mathfrak{f}}(\sigma, s)$$

with B, $c_{\mathbf{a}}$, and $c_{\mathfrak{f}}$ given by

$$(5.12a) \qquad B(s) = (-1)^{m|\rho|} \psi(\delta^m) N(2\mathfrak{c}^{-1})^{m(m+1)/2} N(\mathfrak{d})^{-mq},$$

$$q = 2s + (k/2) - (m+1)/4,$$

$$(5.12b) \qquad c_{\mathbf{a}}(\sigma, y, s) = e(mk[F : \mathbf{Q}]/8) \prod_{v \in \mathbf{a}} \det(y_v)^s \xi(y_v, \sigma_v; \alpha_v, \beta_v),$$

$$(5.12c) \qquad c_{\mathfrak{f}}(\sigma, s) = D(\sigma, 2s + k/2) \prod_{v \mid e} g_v(\sigma, \psi(\pi_v) q_v^{-2s-(k+1)/2}),$$

where $g_v(\sigma, t)$ is a polynomial in t as in Proposition 5.1, and $D(\sigma, s)$ is a product of L-functions determined as follows:

$$D(\sigma, s) = \frac{\prod_{i=1}^{[(m-r-1)/2]} L_c(2s - 2m + r + 2i, \psi^2)}{\prod_{i=0}^{[(m-1)/2]} L_c(2s - 1 - 2i, \psi^2)} \qquad \text{if } r \text{ is even,}$$

$$D(\sigma, s) = L_c(s + (r/2) - m, \psi\xi_\sigma\theta^b)$$

$$\cdot \frac{\prod_{i=0}^{[(m-r)/2]} L_c(2s - 2m + r - 1 + 2i, \psi^2)}{\prod_{i=0}^{[(m-1)/2]} L_c(2s - 1 - 2i, \psi^2)} \qquad \text{if } r \text{ is odd,}$$

where $r = \text{rank}(\sigma)$, $b = (r + k - 2)/2$, and $L_c(s, \chi)$ denotes the Euler product obtained from the standard L-function $L(s, \chi)$ by eliminating the Euler v-factors for all v dividing $\mathfrak{c}$ (see [15, (7.15), (7.16)]).

From this point, the proofs of Theorems 2.3, 2.4, 2.5, and 2.7 proceed in exactly the same fashion as in [15]. We do not give details here, since the argument is completely parallel to that of [15, §§8, 10]. We content ourselves with noting a few essential points:

(i) The problems concerning $E(z, s)$ of type (2.2) or (2.11) can be reduced to those about $E'(z, s)$ of (4.10), by virtue of Proposition 1.3, (ii), Lemma 2.2, and

Proposition 2.6 (cf. also [15, Propositions 2.1 and 2.4]). Thus it is sufficient to analyze the behavior of each Fourier coefficient $c(\sigma, y, s)$ of $E'(z,s)$ at $s = 0$ or $s = s_k$.

(ii) The behavior of $c_a(\sigma, y, s)$, or rather, that of $\xi(y_v, \sigma_v; s + k/2, s)$ at $s = 0$ or $s = s_k$ can be examined by employing the formulas [14, (4.34.K), (4.35.K)]. It should be noted that the function ω in these formulas is always finite.

(iii) As for $c_f(\sigma, s)$, nothing beyond the well known algebraicity and vanishing of L-functions at integer points is needed. It should be noted that Conjecture 6.3 of [15] has been proved by Feit in [2].

(iv) The property $c(\sigma) = 0$ for $\mathrm{rank}(\sigma) > k$ in (4) of Theorem 2.3 can be preserved by the transformation with the elements of $\mathscr{G}$. This follows from Proposition 2.1 and (1.32) of our previous paper in this journal, vol. 51 (1984), pp. 261–329. The same can be said for a similar property in (1) of Theorem 2.7.

6. The one-dimensional case. We assume $m = 1$ throughout this section. In this case, the Fourier expansion of E' can be written in the form

$$(6.1) \qquad E'(z, s; k/2, \rho, \psi, \mathfrak{c}) = \sum_{\sigma \in \mathfrak{c}'} c(\sigma, y, s) \mathbf{e}_a(\sigma x),$$

$$c(\sigma, y, s) = (-1)^{|\rho|}\psi(\delta) N(\mathfrak{b})^{(1-k)/2 - 2s} N(2\mathfrak{c}^{-1}) \mathbf{e}(k[F : \mathbf{Q}]/8)$$

$$\cdot c_f(\sigma, s) \prod_{v \in \mathbf{a}} y_v^s \xi(y_v, \sigma_v; \alpha_v, \beta_v),$$

where $\mathfrak{c}' = 2\mathfrak{c}^{-1}$, $|\rho|$, α_v and β_v are given in Lemma 4.2, and c_f is given by (5.12c). In particular, we have

$$(6.2) \qquad c_f(0, s) = L_{\mathfrak{c}}(4s + k - 2, \psi^2)/L_{\mathfrak{c}}(4s + k - 1, \psi^2).$$

As for $\sigma \neq 0$, we can completely determine the "extra factors" for $v \mid e$ in (5.12c) as follows:

THEOREM 6.1. *Put* $\lambda = (k - 1)/2$. *For* $0 \neq \sigma \in 2\mathfrak{c}^{-1}$, *define an ideal character* ω_σ *of* F *by*

$$\omega_\sigma(\mathfrak{p}) = \psi(\mathfrak{p})\left(\frac{-1}{\mathfrak{p}}\right)^\lambda \left(\frac{F(\sqrt{2\sigma})/F}{\mathfrak{p}}\right)$$

for every prime ideal $\mathfrak{p}$ *prime to* $\mathfrak{c}$. (*We understand that* $\omega_\sigma(\mathfrak{p}) = 0$ *if* $\mathfrak{p}$ *is ramified in* $F(\sqrt{2\sigma})$.) *Then*

$$c_f(\sigma, s) = \beta(\sigma, s) L_{\mathfrak{c}}(2s + \lambda, \omega_\sigma)/L_{\mathfrak{c}}(4s + 2\lambda, \psi^2)$$

with a finite sum β *given by*

$$\beta(\sigma, s) = \sum_{\mathfrak{a}, \mathfrak{b}} \mu(\mathfrak{a})\omega_\sigma(\mathfrak{a})\psi(\mathfrak{b}^2) N(\mathfrak{a})^{-\lambda - 2s} N(\mathfrak{b})^{2 - k - 4s},$$

where μ denotes the Moebius function, and $(\mathfrak{a}, \mathfrak{b})$ runs over all ordered pairs of integral ideals of F prime to $\mathfrak{c}$ such that $\mathfrak{a}^2\mathfrak{b}^2 \supset \mathfrak{o}\mathfrak{c}$.

If $F = \mathbf{Q}$, this result is essentially the same as [10, Proposition 1]. To prove our theorem, we fix a nonarchimedean prime v not dividing 2, and write symbols such as ω_v, χ_v, δ_v, and S_v without the subscript v. Coming back to the general case $m \geqslant 1$ temporarily, we have

$$A^1(\sigma, q^{-s}) = \int_S \omega(y/\delta)\chi(-\sigma y/\delta)\nu(\delta y)^{-s}\, dy.$$

By [15, Lemma 5.3], if $B = (\mathfrak{g}_v)^m_m$, $T = \mathrm{GL}_m(F_v)$, $U = \mathrm{GL}_m(\mathfrak{g}_v)$, and φ is the characteristic function of B, then

$$\nu(\delta y)^{-s} = \gamma(s)\int_{T \cap B} \varphi(yx)|\det(x)|^s\, dx,$$

where the measure on T is normalized so that $\int_U dx = 1$ and $\gamma(s) = \prod_{i=0}^{m-1}(1 - q^{i-s})$. Now with $L = (\mathfrak{g}_v)^1_m$ and a measure $d'z$ on L such that $\int_L d'z = 1$, we have

$$\omega(y/\delta) = \nu(\delta y)^{1/2}\int_L \mathbf{e}_v(zy \cdot {}^tz/(2\delta))\, d'z.$$

Combining these, we obtain

$$(6.3) \qquad A^1(\sigma, q^{-s-1/2}) = \gamma(s)\int_{T \cap B}|\det(x)|^s\int_L\int_S\varphi(yx)\chi(-\sigma y/\delta)$$

$$\cdot \mathbf{e}_v(zy \cdot {}^tz/(2\delta))\, dy\, d'z\, dx.$$

We now specialize this to the case $m = 1$ to find

$$A^1(\sigma, q^{-s-1/2}) = \gamma(s)\sum_{n=0}^{\infty} q^{-ns}\int_L\int_{\mathfrak{p}^{-n}} \mathbf{e}_v\big((z^2 - 2\sigma)y/(2\delta)\big)\, dy\, d'z.$$

Let c_n denote the measure of the set $\{z \in \mathfrak{g}_v \mid z^2 - 2\sigma \in \mathfrak{p}^n\}$ with respect to $d'z$. Then $A^1(\sigma, q^{-s-1/2}) = (1 - q^{-s})\sum_{n=0}^{\infty} c_n q^{n(1-s)}$. Let $2\sigma = \pi^{2t+1}u$ or $2\sigma = \pi^{2t}u$ with a prime element π and a v-unit u. By computing c_n in an elementary way, we find that

$$A^1(\sigma, q^{-s-1/2}) = \frac{1 - q^{-2s}}{1 - \xi q^{-s}}\left\{ \frac{1 - q^{(t+1)(1-2s)}}{1 - q^{1-2s}} - \xi q^{-s} \cdot \frac{1 - q^{t(1-2s)}}{1 - q^{1-2s}} \right\},$$

where $\xi = 0$ if $2\sigma = \pi^{2t+1}u$ and $\xi = \left(\dfrac{u}{\mathfrak{p}}\right)$ if $2\sigma = \pi^{2t}u$. Substituting this into (5.5) and (5.12c), we obtain our theorem.

COROLLARY 6.2. *Define $\Gamma_v(s)$ for each $v \in \mathbf{a}$ by*

$$\Gamma_v(s) = \Gamma\left(s + \frac{\tau_v}{2}\right) \cdot \begin{cases} \Gamma(s + (k + \rho_v)/2) & \text{if } k + 1 \geqslant -2\rho_v, \\ \Gamma(s - \rho_v/2) & \text{if } k + 1 < -2\rho_v, \end{cases}$$

where τ_v is the smallest integer such that $\tau_v \geqslant (k-1)/2$ and

$$\tau_v \equiv \begin{cases} \rho_v & (\mathrm{mod}\,2) & \text{if } k+1 \geqslant -2\rho_v, \\ \rho_v + 1 & (\mathrm{mod}\,2) & \text{if } k+1 < -2\rho_v. \end{cases}$$

Suppose $\sum_{v \in \mathbf{a}} \mu_v = 0$. Then

$$\prod_{v \in \mathbf{a}} \Gamma_v(s + i\mu_v/2) L_{\mathfrak{c}}(4s + k - 1, \psi^2) E(z,s; k/2, \rho, \psi, \mathfrak{c})$$

is holomorphic on the whole s-plane except for a possible simple pole at $s = (3-k)/4$. The pole occurs if and only if the following two conditions are satisfied:

(i) $\psi^2 = 1$;

(ii) $\rho_v + (k-1)/2$ is either an even nonnegative integer or an odd negative integer.

This is a generalization of [10, Proposition 3] which concerns the case $F = \mathbf{Q}$. The proof can be given in a straightforward way by showing that $c(\sigma, y, s)$ multiplied by $L_{\mathfrak{c}}(4s + k - 1, \psi^2)$ and an appropriate gamma-factor has the desired property of holomorphy. We do not give details here, as it completely parallels that in the case $F = \mathbf{Q}$ given in [10, pp. 91–92]. Let us now add a few more explicit formulas:

PROPOSITION 6.3. *Suppose $k \geqslant 3$ and $\rho = \mu = 0$; let $s_k = 1 - k/2$. Then $L_{\mathfrak{c}}(4s + k - 1, \psi^2) E'(z, s; k/2, \psi, \mathfrak{c})$ is finite at $s = s_k$ and its value at s_k is π^n times a nonzero element of $\mathcal{M}_{k/2}(\mathbf{Q}_{ab})$. More explicitly, the value is given by*

$$C \cdot \left\{ L_{\mathfrak{c}}(2 - k, \psi^2) + \sum_{\sigma} \beta(\sigma, s_k) L_{\mathfrak{c}}((3-k)/2, \omega_\sigma) \mathbf{e}_{\mathbf{a}}(\sigma z) \right\},$$

where $C = \psi(\delta) 2^{n(k+2)/2} \pi^n N(\mathfrak{c}^{-1}) D_F^{(k-3)/2}$, $n = [F : \mathbf{Q}]$, D_F denotes the discriminant of F, and σ runs over all totally positive elements of $2\mathfrak{c}^{-1}$.

Notice that if $k = 3$ and $\psi = 1$, we obtain the class number of $F(\sqrt{-2\sigma})$ as a factor of the Fourier coefficient.

PROPOSITION 6.4. *Suppose $k = 1$. If $\psi^2 \neq 1$, $E'(z, s; 1/2, \psi, \mathfrak{c})$ is finite at $s = 1/2$. If $\psi^2 = 1$, $E'(z, s; 1/2, \psi, \mathfrak{c})$ has a simple pole at $s = 1/2$, and the residue is $\pi^{-n} R_F$ times an element of $\mathcal{M}_{1/2}(\mathbf{Q}_{ab})$. More explicitly, suppose $\psi(\mathfrak{p}) = \left(\dfrac{t}{\mathfrak{p}}\right)$ with a totally positive element t of F; let $\mathfrak{c}$ be an integral ideal of F divisible by 4 and by the conductor of ψ. Put*

$$f(z) = \sum_{b} \mathbf{e}_{\mathbf{a}}(tb^2 z/2) \qquad (z \in H_1^{\mathbf{a}}),$$

where b runs over all the elements of F such that $tb^2 \in 4\mathfrak{c}^{-1}$. Then the residue of

$E'(z, s; 1/2, \psi, \mathfrak{c})$ *at* $s = 1/2$ *is* $A \cdot f(z)$ *with*

$$A = 2^{(3n-4)/2} \pi^n D_F^{-1} N(\mathfrak{c})^{-1} \psi(\delta) \prod_{\mathfrak{p} | \mathfrak{c}} \left(1 + N(\mathfrak{p})^{-1}\right)^{-1}$$

$$\cdot \zeta_F(2)^{-1} \cdot \text{Residue}_{s=1} \zeta_F(s),$$

where ζ_F *denotes the zeta function of* F.

These two propositions follow from Theorem 6.1 in a straightforward way by means of [14, (4.34.K), (4.35.K)].

Remark 6.5. The above proposition concerns only the case $k = 1$. The residue in the case $k > 1$ can be obtained from this special case as follows. For $r \in \mathbf{R}$, $0 < p \in \mathbf{Z}$, and a function f on $H_1^{\mathbf{a}}$, put

$$\delta_r^{(p)} f = \prod_{v \in \mathbf{a}} y_v^{1-p-r} (\partial/\partial z_v)^p y_v^{p+r-1} f.$$

Then, in view of (3.10d), we have

(6.4) $$\delta_{h/2}^{(p)} E(z, s; h/2, \psi, \mathfrak{c}) = \beta(s)^n E(z, s - p; 2p + h/2, \psi, \mathfrak{c})$$

with $\beta(s) = (2i)^{-p} \prod_{j=0}^{p-1} (s + j + h/2)$ and $n = [F:\mathbf{Q}]$. The same formula holds with E' in place of E, though an extra factor $(-1)^{pn}$ is necessary because of (3.14b). Take $h = 1$ and $k = 4p + 1$. Let $A \cdot f(z)$ be as in Proposition 6.4. Then we find that the residue of $E'(z, s; k/2, \psi, \mathfrak{c})$ at $(3 - k)/4$ is $(-2i)^{pn}(p!)^{-n} A \delta_{1/2}^{(p)} f$. It should be noted that one can generalize (6.4) to the case $m > 1$ by defining $\delta_r^{(p)}$ suitably.

Remark 6.6. In our previous papers [12], [13], and [16], we investigated some problems of algebraicity of the critical values of a certain Dirichlet series $Z(s, \Omega, \chi_1, \chi_2, \ldots)$ formed by a Hilbert modular form Ω of integral or half-integral weight and several systems $\chi_1, \chi_2, \ldots$ of eigenvalues of Hecke operators. The theorems in these papers were obtained under a certain parity condition on the weight of Ω. As explained in [13, II, §7], one can remove the condition, once one has a satisfactory theory of Eisenstein series of half-integral weight on $\mathrm{SL}_2(F)$. Such is provided by our results in this section, and therefore all the statements C.7.1–5 of [13, II, §7] are now valid. Moreover, they can be generalized to a more general case by the same ideas as in [16]. We shall not go into details here, as we hope to give a full treatment of this subject, together with some applications, in a future paper.

We conclude this section by mentioning Petersson [8], Maass [5], [6], Siegel [18], Sturm [19], and Pei [7] (besides the author [10]) as previous investigations including some explicit forms of the Fourier coefficients of Eisenstein series of half-integral weight on $\mathrm{SL}_2(F)$ in the csae $[F:\mathbf{Q}] \leq 2$.

7. Appendix: Theta functions of indefinite quadratic forms.

The purpose of this section is to prove the transformation formula of a certain theta series of an indefinite quadratic form first introduced by Siegel. The formula, which may be called well known at least in principle, is treated here partly because we think it worthwhile to present a short proof in the framework of Section 1, and partly because the formula in this form, not found in the existing literature, will be needed in our future investigation. Though the series is closely related to Eisenstein series, we shall not touch on that aspect in this paper.

First we add two more notational conventions as follows: given $S = {}^tS \in \mathbf{C}_q^q$ and $A \in \mathbf{C}_m^q$, we put $S[A] = {}^tASA$; also for $X = (x_{hi}) \in \mathbf{C}_q^q$ and $Y \in \mathbf{C}_m^m$, we define $X \otimes Y \in \mathbf{C}_{qm}^{qm}$ by

$$X \otimes Y = \begin{bmatrix} x_{11}Y & \cdots & x_{1q}Y \\ \cdots & \cdots & \cdots \\ x_{q1}Y & \cdots & x_{qq}Y \end{bmatrix}.$$

LEMMA 7.1. *Let S and P be symmetric elements of $\mathrm{GL}_q(\mathbf{R})$. Suppose that $P > 0$ and S has signature (r,s). Then $PS^{-1}P = S$ if and only if there exists an element A of $\mathrm{GL}_q(\mathbf{R})$ such that $P = {}^tAA$ and $S = {}^tA \cdot \mathrm{diag}[1_r, -1_s]A$.*

This is an easy exercise.

Fix two positive integers m and q; fix also S, P, r, s, and A as in the above lemma. Notice that $P = S$ or $P = -S$ according as $s = 0$ or $r = 0$. For $z \in H_m$, $-w \in H_m$, and $\alpha \in \mathrm{Sp}(m, \mathbf{R})$, put

$$\psi(z,w) = \psi(z,w; S, P) = (1/2)\{S \otimes (z + w) + P \otimes (z - w)\},$$

$$\alpha_S = \begin{bmatrix} 1_q \otimes a_\alpha & S \otimes b_\alpha \\ S^{-1} \otimes c_\alpha & 1_q \otimes d_\alpha \end{bmatrix}.$$

Then $\psi(z,w) \in H_{mq}$, and $\alpha \mapsto \alpha_S$ gives a homomorphism of $\mathrm{Sp}(m, \mathbf{R})$ into $\mathrm{Sp}(mq, \mathbf{R})$. Put $\mathfrak{A} = A \otimes 1_m$. Then we can easily verify that

$$\alpha_S(\psi(z,w)) = \psi(\alpha(z), \alpha(w)),$$

$$\psi(z,w) = {}^t\mathfrak{A} \cdot \mathrm{diag}[1_r \otimes z, -1_s \otimes w]\mathfrak{A},$$

$$\mu(\alpha_S, \psi(z,w)) = \mathfrak{A}^{-1}\mathrm{diag}[1_r \otimes \mu(\alpha, z), 1_s \otimes \mu(\alpha, w)]\mathfrak{A},$$

$$j(\alpha_S, \psi(z,w)) = j(\alpha, z)^r j(\alpha, w)^s,$$

where μ and j are defined by (1.4a, b). Specializing the situation to the case $w = \bar{z}$, we find that $\alpha_S(\psi(z, \bar{z})) = \psi(\alpha(z), \overline{\alpha(z)})$, and

$$(7.1) \qquad j(\alpha_S, \psi(z, \bar{z})) = j(\alpha, z)^r \overline{j(\alpha, z)}^s.$$

We now consider $\mathrm{Sp}(m, F)$ instead of $\mathrm{Sp}(m, \mathbf{R})$, and take a symmetric element S of $\mathrm{GL}_q(F)$. For $v \in \mathbf{a}$, let (r_v, s_v) be the signature of S_v; thus $q = r_v + s_v$. Take

$A = (A_v)_{v \in \mathbf{a}} \in \mathrm{GL}_q(\mathbf{R})^{\mathbf{a}}$ so that $S_v = {}^t A_v \mathrm{diag}[1_{r_v}, -1_{s_v}]A_v$ and put $P_v = {}^t A_v A_v$. We then put, for $z \in H_m^{\mathbf{a}}$,

$$(7.2) \qquad \psi(z) = \big(\psi(z_v, \bar{z}_v; S_v, P_v) \big)_{v \in \mathbf{a}}.$$

Then ψ maps $H_m^{\mathbf{a}}$ into $H_{mq}^{\mathbf{a}}$; the map $\alpha \mapsto \alpha_S$ now defines an injection of $\mathrm{Sp}(m, F)$ into $\mathrm{Sp}(mq, F)$, and $\psi(\alpha(z)) = \alpha_S(\psi(z))$. We now define a function g, which may be called a theta series of S, by

$$(7.3) \quad g(z, u; \lambda) = \mathbf{e}_{\mathbf{a}}\big((1/2)\mathrm{tr}\{ (z - \bar{z})^{-1}[u] \} \big)$$

$$\cdot \sum_{x \in F_m^q} \lambda(x)\mathbf{e}_{\mathbf{a}}\big(\mathrm{tr}(xuA)$$

$$+ (1/4)\mathrm{tr}\{ S[x](z + \bar{z}) + P[x](z - \bar{z}) \} \big)$$

for $z \in H_m^{\mathbf{a}}$, $u \in (\mathbf{C}_q^m)^{\mathbf{a}}$, and $\lambda \in \mathscr{L}(F_m^q)$, where $\mathscr{L}(F_m^q)$ denotes the space of all locally constant functions on F_m^q. This is a pullback of θ^* of (1.9) defined with mq in place of m. In fact, we can easily verify that

$$(7.4) \qquad g(z, u; \lambda) = \theta^*\big(\psi(z), u^0; \lambda^0 \big),$$

where u^0 ($\in (\mathbf{C}^{mq})^{\mathbf{a}}$) and λ^0 ($\in \mathscr{L}(F^{mq})$) are determined by

$$u_v^0 = ({}^t A_v \otimes 1_m) \begin{bmatrix} u_{v1} \\ \vdots \\ u_{vq} \end{bmatrix} \quad \text{for} \quad u_v = (u_{v1}, \ldots, u_{vq}) \quad \text{with} \quad u_{vi} \in \mathbf{C}^m,$$

$$\lambda^0(x_1, \ldots, x_q) = \lambda \begin{bmatrix} x_1 \\ \vdots \\ x_q \end{bmatrix} \quad \text{for} \quad x_1, \ldots, x_q \in F_m^1.$$

Let us write $\mathscr{G}_m$ for the group $\mathscr{G}$ of Section 1, emphasizing the degree m. For a function $f(z, u)$ on $(H_m \times \mathbf{C}_q^m)^{\mathbf{a}}$ and $\beta = (\alpha, h) \in \mathscr{G}_m$, define $f| \beta = f|_{r,s}\beta$ to be the function on the same space given by

$$(f|_{r,s}\beta)(z, u) = h(z)^{-q} \prod_{v \in \mathbf{a}} \{ j_v(\alpha, z)/|j_v(\alpha, z)| \}^{s_v} f(\alpha(z), [\alpha, z]u),$$

where $([\alpha, z]u)_v = w_v$ with

$$w_{vj} = \begin{cases} {}^t \mu_v(\alpha, z)^{-1} u_{vj} & (1 \leqslant j \leqslant r_v), \\ {}^t \overline{\mu_v(\alpha, z)}^{-1} u_{vj} & (r_v < j \leqslant q), \end{cases}$$

(the subscript j means the j-th column). This action of β is associative in the sense that $(f\,|\,\beta)\,|\,\beta' = f\,|\,(\beta\beta')$.

PROPOSITION 7.2. *There is an action of $\mathcal{G}_m$ on $\mathcal{L}(F_m^q)$, which we write $\lambda \mapsto \lambda^\beta$ for $\lambda \in \mathcal{L}(F_m^q)$ and $\beta \in \mathcal{G}_m$, and which depends only on S and is independent of P, such that*

$$(7.5) \qquad g(z,u;\lambda)|_{r,s}\beta = g(z,u;\lambda^\beta).$$

Moreover, for every $\lambda \in \mathcal{L}(F_m^q)$, there is a congruence subgroup Δ of $\mathcal{G}_m$ in the sense of (1.15a, b) *such that $\lambda^\delta = \lambda$ for all $\delta \in \Delta$.*

Proof. Observe that if $(\alpha,h) \in \mathcal{G}_m$ and $h(z)^2 = t \cdot j(\alpha,z)$, then

$$h(z)^{2q} \prod_{v \in \mathbf{a}} \{ j_v(\alpha,z)/|j_v(\alpha,z)| \}^{-2s_v} = t^q j(\alpha_S,\psi(z)).$$

Therefore, in view of (7.1), we can define an element (α_S,k) of $\mathcal{G}_{mq}$ so that

$$k(\psi(z)) = h(z)^q \prod_{v \in \mathbf{a}} \{ j_v(\alpha,z)/|j_v(\alpha,z)| \}^{-s_v}.$$

It can easily be seen that $(\alpha,h) \mapsto (\alpha_S,k)$ is a homomorphism of $\mathcal{G}_m$ into $\mathcal{G}_{mq}$ which depends only on S and is independent of P, since all possible P for a fixed S form a connected set. Therefore our first assertion follows from Proposition 1.1 and (7.4). At the same time, we obtain the second assertion if $m > 1$, since (1.5.1b) follows from (1.5.1a) if $m > 1$, as noted in the proof of Proposition 1.3. Suppose $m = 1$; let $\gamma = \left(\begin{smallmatrix} a & b \\ c & d \end{smallmatrix}\right) \in \mathrm{SL}_2(F)$. If γ belongs to a sufficiently small congruence subgroup and d is totally positive, then we have (7.5) with $\beta = (\gamma, h_\gamma)$ with

$$(7.6) \qquad \lim_{z \to 0} h_\gamma(z) = |N_{F/\mathbf{Q}}(d)|^{1/2}\left(\frac{2c}{d\mathfrak{g}} \right),$$

as proved in [12, Lemma 7.2] in a somewhat stronger form. This property determines the map $\gamma \mapsto (\gamma, h_\gamma)$ on a small congruence subgroup for the reason explained in [12, Lemma 7.4]. If $q = 1$, $S = 1$, and $A_v = 1$ for all $v \in \mathbf{a}$, then g coincides with θ^* of (1.9), and therefore h_γ determined by (7.5) with $\lambda^\beta = \lambda$ coincides, on a small congruence subgroup, with that determined by (1.12b) or by (1.14). This proves the second assertion and also (iii) of Proposition 1.3 when $m = 1$.

Theta series of type (7.3) were considered by Siegel [17] without the vector variable u. If S is totally definite and $P = S$, then $g(z,0;\lambda)$ is a theta series $\sum_x \lambda(x)\mathbf{e}_{\mathbf{a}}((1/2)\mathrm{tr}(S[x]z))$. One advantage of introducing the variable u is that we can obtain theta series of more general types by taking the derivatives of g with respect to u. Their transformation formulas under $\mathcal{G}$ are merely the derivatives of (7.5), and therefore can easily be given explicitly.

REFERENCES

1. H. BASS, J. MILNOR AND J.-P. SERRE, *Solutions of the congruence subgroup problem for* SL_n *and* SP_{2n}, Publ. Math. I.H.E.S. **33** (1967), 59–137.
2. P. FEIT, *Finiteness and analyticity of Eisenstein series*, Thesis, Princeton University, 1985.
3. Y. KITAOKA, *Dirichlet series in the theory of Siegel modular forms*, Nagoya Math. J. **95** (1984), 73–84.
4. R. P. LANGLANDS, *On the functional equations satisfied by Eisenstein series*, Lecture notes in Math. 544, Springer, 1976.
5. H. MAASS, *Konstruktion ganzer Modulformen halbzahliger Dimension mit θ-Multiplikatoren in einer und zwei Variablen*, Abh. Math. Sem. Hamburg **12** (1937), 134–162.
6. ————, *Konstruktion ganzer Modulformen halbzahliger Dimension mit θ-Multiplikatoren in zwei Variablen*, Math. Z. **43** (1938), 709–738.
7. T.-Y. PEI, *Eisensten series of weight 3/2: I, II*, Trans. Amer. Math. Soc. **274** (1982), 573–606 and **283** (1984), 589–603.
8. H. PETERSSON, *Über die Entwicklungskoeffizienten der ganzen Modulformen und ihre Bedeutung für die Zahlentheorie*, Abh. Math. Sem. Hamburg **8** (1931), 215–242.
9. G. SHIMURA, *On modular forms of half-integral weight*, Ann. of Math. **97** (1973), 440–481.
10. ————, *On the holomorphy of certain Dirichlet series*, Proc. London Math. Soc. Ser. 3, **31** (1975), 79–98.
11. ————, *Theta functions with complex multiplication*, Duke Math. J. **43** (1976), 673–696.
12. ————, *The arithmetic of certain zeta functions and automorphic forms on orthogonal groups*, Ann. of Math. **111** (1980), 313–375.
13. ————, *On certain zeta functions attached to two Hilbert modular forms: I, II*, Ann. of Math. **114** (1981), 127–164, 569–607.
14. ————, *Confluent hypergeometric functions on tube domains*, Math. Ann. **260** (1982), 269–302.
15. ————, *On Eisenstein series*, Duke Math. J. **50** (1983), 417–476.
16. ————, *Algebraic relations between critical values of zeta functions and inner products*, Amer. J. Math. **104** (1983), 253–285.
17. C. L. SIEGEL, *On the theory of indefinite quadratic forms*, Ann. of Math. **45** (1944), 577–622.
18. ————, *Die Fuktionalgleichungen einiger Dirichletscher Reihen*, Math. Z. **63** (1956), 363–373.
19. J. STURM, *Special values of zeta functions and Eisenstein series of half integral weight*, Amer. J. Math. **102** (1980), 219–240.
20. A. WEIL, *Sur certains groupes d'opérateurs unitaires*, Acta math. **111** (1964), 143–211.

Corrections to [15]

Page 432, (3.16): The exponents $m(m + 1)/2$ and m^2 should be $-m(m + 1)/2$ and $-m^2$, respectively.

Page 449, line 5: $(m/1)$ should be $(m - 1)$.

Page 451, Conjecture 6.3: The number ξ should be defined as follows: $\xi = 0$ if $\mathrm{ord}_v(\det(\delta j))$ is odd; $\xi = \left(\dfrac{u}{\mathbf{p}}\right)$ if $\det(\delta j) = u\pi^{2n}$ with a v-unit u, $n \in \mathbf{Z}$, and a prime element π.

Page 457, (7.12): Insert "if $h > 0$" at the end of the line.

Page 457, (7.20); page 462, (8.5); page 463, line 22; page 472, line 2: The exponents $m\kappa$ and $-m\kappa$ should be $-m\kappa$ and $m\kappa$, respectively.

Page 470, line 10: There should be a center dot before the first $\sum$ indicating that the line is a factor of the second term of the right-hand side of (9.8b).

Page 474, line 23: The last M' should be $\mathbf{Q}$.

DEPARTMENT OF MATHEMATICS, PRINCETON UNIVERSITY, PRINCETON, NEW JERSEY 08544

On the Eisenstein series
of Hilbert modular groups

Revista Matemática Iberoamericana, 1 (1985), 1-42

Introduction

Throughout the paper, we let F denote a totally real algebraic number field of degree n, and $\mathbf{a}$ the set of all archimedean primes of F. Given a set X, we denote by $X^{\mathbf{a}}$ the product of $\mathbf{a}$ copies of X, that is, the set of all indexed elements $(x_v)_{v \in \mathbf{a}}$ with $x_v \in X$. If $y \in X^{\mathbf{a}}$, y_v will detnote its v-component. Putting $H = \{z \in \mathbf{C} \,|\, \mathrm{Im}(z) > 0\}$, we let $SL_2(F)$ act on $H^{\mathbf{a}}$ through the injection of $SL_2(F)$ into $SL_2(\mathbf{R})^{\mathbf{a}}$. For $\sigma \in \mathbf{Z}^{\mathbf{a}}$ and $v \in \mathbf{a}$, we define a differential operator L_v^{σ} on $H^{\mathbf{a}}$ by

$$L_v^{\sigma} = -4y_v^{2 - \sigma_v}(\partial/\partial z_v)y_v^{\sigma_v}(\partial/\partial \bar{z}_v),$$

where z_v is the variable on the v-factor of $H^{\mathbf{a}}$ and $y_v = \mathrm{Im}(z_v)$.

Given a congruence subgroup Γ of $SL_2(F)$ and $\lambda \in \mathbf{C}^{\mathbf{a}}$, we denote by $\mathcal{Q}(\sigma, \lambda, \Gamma)$ the set of all C^{∞}-functions f on $H^{\mathbf{a}}$ such that

   i) $f(\gamma(z)) = \prod_{v \in \mathbf{a}}(c_v z_v + d_v)^{\sigma_v} f(z)$ for every $\gamma = \left(\begin{smallmatrix} a & b \\ c & d \end{smallmatrix}\right) \in \Gamma$,

   ii) $L_v^{\sigma} f = \lambda_v f$ for every $v \in \mathbf{a}$,

   iii) f is slowly increasing at every cusp.

Further we let $\mathcal{S}(\sigma, \lambda, \Gamma)$ denote the set of all cusp forms, defined as usual, belonging to $\mathcal{Q}(\sigma, \lambda, \Gamma)$, and $\mathcal{N}(\sigma, \lambda, \Gamma)$ the orthogonal complement of $\mathcal{S}(\sigma, \lambda, \Gamma)$ in $\mathcal{Q}(\sigma, \lambda, \Gamma)$.

Now the main purpose of this paper is to show that $\mathcal{N}(\sigma, \lambda, \Gamma)$ can be spanned, in most cases, by certain Eisenstein series, which are functions $E(z, s; \rho)$ of the variable z on $H^{\mathbf{a}}$, a complex parameter s, and another

1

discrete parameter $\rho \in \mathbf{C}^{\mathbf{a}}$. Namely, given σ, λ, and Γ, we can choose s_0 and ρ so that a suitable finite set of $E(z, s_0; \rho)$ spans $\mathfrak{N}(\sigma, \lambda, \Gamma)$ (Theorem 7.3). If $F = \mathbf{Q}$, the parameter ρ indicates nothing but the weight σ, but if $F \neq \mathbf{Q}$, ρ involves a variable in $\mathbf{R}^{\mathbf{a}}$ which parametrizes the archimedean factors of Hecke (Grössen-) characters of F. There are two cases in which Eisenstein series by themselves cannot generate $\mathfrak{N}(\sigma, \lambda, \Gamma)$. In fact, if $4\lambda_v = (1 - \sigma_v)^2$ for every $v \in \mathbf{a}$, we need $\partial E/\partial s$ (Theorem 7.8); in the other case, we need the residues of the $E(z, s)$ (Theorem 7.9). These theorems are valid also for eigenforms of half-integral weight, which can be defined by making suitable modifications in the above definition. Our results are not complete in the sense that we have to exclude the case of «multiple» λ, which occurs only when $F \neq \mathbf{Q}$, $(1 - \sigma_v)^2 \leqslant 4\lambda_v \in \mathbf{R}$ for all $v \in \mathbf{a}$, and $(1 - \sigma_v)^2 < 4\lambda_v$ for at least one v. We believe, however, that our technique is applicable even to multiple λ, and therefore no serious difficulties are expected in the task of extending our results to the most general case.

As an application, we shall show that every holomorphic Hilbet modular form is a sum of a holomorphic cusp form and a holomorphic Eisenstein series. This holds for all integral and half-integral weights $\geqslant \frac{1}{2}$ (Theorems 8.3, 8.4, and formula (8.3)). The explicit Fourier expansions of certain Eisenstein series obtained in our previous papers [12] and [13] play an essential role in the proof of this result as well as in that of the theorems on $\mathfrak{N}(\sigma, \lambda, \Gamma)$.

Another application concerns an interpretation of the zeros of L-functions of F in the critical strip. To explain the idea, let us assume $F = \mathbf{Q}$ for simplicity. Given $\xi = \begin{pmatrix} * & * \\ c & d \end{pmatrix} \in SL_2(\mathbf{Q})$ and $f \in \mathfrak{N}(\sigma, \lambda, \Gamma)$ with $\lambda \in \mathbf{C}$, $\sigma \in \mathbf{Z}$ and $\Gamma \subset SL_2(\mathbf{Z})$, we can speak of a Fourier expansion of f at the cusp $\xi(\infty)$, which has the form

$$(cz + d)^{-\sigma} f(\xi(z)) = a_\xi y^{s_0 - \sigma/2} + a_\xi' y^{1 - s_0 - \sigma/2} + \sum_{n \in \mathbf{Z}} b_\xi(n, y) e^{2\pi i n x/N}$$

with $0 < N \in \mathbf{Z}$, constants a_ξ and a_ξ', and a complex number s_0 such that $\lambda = (s_0 - \sigma/2)(1 - s_0 - \sigma/2)$. Now we call f a *cyclopean form of exponent* $1 - s_0 - \sigma/2$ if $0 < \mathrm{Re}(s_0) < \frac{1}{2}$ and $a_\xi = 0$ for every $\xi \in SL_2(\mathbf{Q})$. Then we shall show that a nonzero cyclopean form exists if and only if there exists a Dirichlet character ψ such that $L(2s_0, \psi) = 0$ and $\psi(-1) = (-1)^\sigma$. The same type of assertion can be made also for $F \neq \mathbf{Q}$ (Theorem 9.1). This result is tautological if $\Gamma = SL_2(\mathbf{Z})$, in the sense that it follows immediately from the well-known Fourier expansion of the Eisenstein series of $SL_2(\mathbf{Z})$. The assertion in the general case, however, is nontrivial, even when $F = \mathbf{Q}$. In fact, the L-functions involve Euler-products and Dirichlet (or Hecke) characters ψ while our definition of cyclopean forms does not require any such multiplicative structure at least on the surface, which is why we think that the fact deserves a statement as we present here.

Let us conclude the introduction by mentioning the previous investigations. The eigenforms were first studied by Maass in [3] and [4] for the congruence subgroups of $SL_2(\mathbf{Z})$. In particular, he proved a certain bilinear relation of the coefficients of the constant terms of eigenforms and showed that $\mathfrak{N}(\sigma, \lambda, \Gamma)$ can be spanned by Eisenstein series when $\sigma = 0$, $\lambda \geqslant \frac{1}{4}$ and $\Gamma \subset SL_2(\mathbf{Z})$. In [7], Roelcke generalized these to the eigenforms of an arbitrary weight with respect to an arbitrary Fuchsian group. The present paper owes much to their ideas in those papers; in fact, one of the key points in our treatment is a generalization of their bilinear relations.

In the holomorphic case, the fact that an elliptic modular form of integral weight is the sum of a cusp form and an Eisenstein series was proved by Hecke [1]. This was extended by Kloosterman [2] to the Hilbert modular forms of weight $\geqslant 2$. The case of weight 1 was proved recently by Shimizu [8]. As for the forms of half-integral weight, Petersson [6] obtained a corresponding result for weight $\geqslant \frac{5}{2}$ when $F = \mathbf{Q}$. Recently the case of weight $\frac{3}{2}$ with $F = \mathbf{Q}$ was settled by Pei [5].

1. Congruence subgroups and factors of automorphy

The symbols F, n, $\mathbf{a}$, $X^{\mathbf{a}}$, and H we used in the introduction will have the same meaning throughout the paper. In addition, we let $\mathbf{f}$ denote the set of all nonarchimedean primes of F, $\mathfrak{g}$ the maximal order of F, $\mathfrak{g}^{\times}$ the group of all units of F, and $\mathfrak{b}$ the different of F. Each element of $\mathbf{a}$ will be viewed as an injection of F into $\mathbf{R}$. Then $F \otimes_{\mathbf{Q}} \mathbf{R}$ and $F \otimes_{\mathbf{Q}} \mathbf{C}$ can be identified naturally with $\mathbf{R}^{\mathbf{a}}$ and $\mathbf{C}^{\mathbf{a}}$, respectively, through the map $a \otimes b \mapsto (a_v b)_{v \in \mathbf{a}}$ for $a \in F$ and $b \in \mathbf{R}$ (or $\mathbf{C}$), where a_v denotes the image of a under v. We write $a \geqslant 0$ for $a \in \mathbf{R}^{\mathbf{a}}$ if $a_v > 0$ for all v. For two elements c and x of $\mathbf{C}^{\mathbf{a}}$, we put

$$(1.1) \qquad c^x = \prod_{v \in \mathbf{a}} c_v^{x_v}$$

whenever each factor is well-defined (according to the context). We denote by u the identity element of the ring $\mathbf{C}^{\mathbf{a}}$. We have then

$$(1.2) \qquad c^{su} = \prod_{v \in \mathbf{a}} c_v^{s} \quad \text{for} \quad s \in \mathbf{C}.$$

Given an associative ring R with identity element, we denote by $R^{\times}$ the group of all invertible elements of R, and by $M_2(R)$ the ring of all 2×2-matrices with entries in R, and put $SL_2(R) = \{\xi \in M_2(R) | \det(\xi) = 1\}$ when R is commutative. For $\xi = \begin{pmatrix} a & b \\ c & d \end{pmatrix} \in M_2(R)$, we write $a = a_{\xi}$, $b = b_{\xi}$, $c = c_{\xi}$, and $d = d_{\xi}$. For $\alpha \in SL_2(\mathbf{R})$ and $z \in \mathbf{C}$, we put

$$(1.3) \qquad \alpha(z) = (a_{\alpha} z + b_{\alpha})/(c_{\alpha} z + d_{\alpha}), \quad j(\alpha, z) = c_{\alpha} z + d_{\alpha}.$$

Further, for $\alpha = (\alpha_v)_{v \in \mathbf{a}} \in SL_2(\mathbf{R})^{\mathbf{a}}$ and $z = (z_v)_{v \in \mathbf{a}} \in \mathbf{C}^{\mathbf{a}}$, we put

$$(1.4a) \qquad \alpha(z) = (\alpha_v(z_v))_{v \in \mathbf{a}}, \quad j_v(\alpha, z) = j(\alpha_v, z_v),$$

$$(1.4b) \qquad j_\alpha(z) = j(\alpha, z) = (j_v(\alpha, z))_{v \in \mathbf{a}} \quad (\in \mathbf{C}^{\mathbf{a}}).$$

With u as in (1.2), we have $x^u = N_{F/\mathbf{Q}}(x)$ for $x \in F$ and also

$$j_\alpha(z)^u = \prod_{v \in \mathbf{a}} j_v(\alpha, z).$$

We define our basic group G and its parabolic subgroup P by

$$(1.5) \qquad G = SL_2(F), \qquad P = \{\alpha \in G \mid c_\alpha = 0\}.$$

We identify $M_2(F) \otimes_{\mathbf{Q}} \mathbf{R}$ with $M_2(\mathbf{R})^{\mathbf{a}}$ and embed $M_2(F)$ and G into $M_2(\mathbf{R})^{\mathbf{a}}$ and $SL_2(\mathbf{R})^{\mathbf{a}}$; then we let G act on $H^{\mathbf{a}}$ (or even on $\mathbf{C}^{\mathbf{a}}$) through this embedding.

Given an integral ideal $\mathfrak{z}$ and fractional ideals $\mathfrak{r}$ and $\mathfrak{y}$ in F such that $\mathfrak{r}\mathfrak{y}$ is integral, we define a subring $\mathfrak{o}[\mathfrak{r}, \mathfrak{y}]$ of $M_2(F)$ and subgroups $\Gamma[\mathfrak{r}, \mathfrak{y}]$ and $\Gamma[\mathfrak{z}]$ of G by

$$(1.6) \qquad \mathfrak{o}[\mathfrak{r}, \mathfrak{y}] = \{\alpha \in M_2(F) \mid a_\alpha \in \mathfrak{g}, d_\alpha \in \mathfrak{g}, b_\alpha \in \mathfrak{r}, c_\alpha \in \mathfrak{y}\},$$

$$(1.7a) \qquad \Gamma[\mathfrak{r}, \mathfrak{y}] = \mathfrak{o}[\mathfrak{r}, \mathfrak{y}] \cap G,$$

$$(1.7b) \qquad \Gamma[\mathfrak{z}] = \{\alpha \in G \mid a_\alpha \equiv d_\alpha \equiv 1, b_\alpha \equiv c_\alpha \equiv 0 \quad (\mathrm{mod}\,\mathfrak{z})\}.$$

A subgroup Γ of G is called a *congruence subgroup* of G if it contains $\Gamma[\mathfrak{z}]$ as a subgroup of finite index for some $\mathfrak{z}$.

We are going to consider automorphic forms of integral and half-integral weights with respect to congruence subgroups of G. A *weight* will be an element σ of $(1/2)\mathbf{Z}^{\mathbf{a}}$ such that $2\sigma_v (\mathrm{mod}\,2)$ is independent of v. Our treatment will be divided into two cases according to the parity: Case I for $\sigma \in \mathbf{Z}^{\mathbf{a}}$ (*integral weight*) and Case II for $\sigma \notin \mathbf{Z}^{\mathbf{a}}$ (*half-integral weight*). We consider the group $\mathcal{G}_\sigma$ consisting of all couples (α, l) formed by $\alpha \in G$ and a holomorphic function l on $H^{\mathbf{a}}$ such that $l(z)^2 = t j_\alpha(z)^{2\sigma}$ with a root of unity t, the group-law being defined by

$$(1.8) \qquad (\alpha, l)(\alpha', l') = (\alpha\alpha', l(\alpha'(z))l'(z)).$$

In Case I, $\mathcal{G}_\sigma$ is obviously isomorphic to the direct product of G and the group of all roots of unity. For $\xi = (\alpha, l) \in \mathcal{G}_\sigma$, we write $\alpha = \mathrm{pr}(\xi)$, $l = l_\xi$, $a_\xi = a_\alpha$, $b_\xi = b_\alpha$, $c_\xi = c_\alpha$, $d_\xi = d_\alpha$, and put $\xi(z) = \alpha(z)$ for $z \in H^{\mathbf{a}}$. The group $\mathcal{G}_\sigma$ is introduced for the purpose of dealing with Case II. We consider it even in Case I, simply in order to make our exposition uniform.

Let F_v denote the v-completion of F for each $v \in \mathbf{a} \cup \mathbf{f}$. If $\mathfrak{r}$ is a fractional ideal in F and $v \in \mathbf{f}$, we denote by $\mathfrak{r}_v$ its closure in F_v. We put $G_v = SL_2(F_v)$ and define the adelization $G_{\mathbf{A}}$ and $P_{\mathbf{A}}$ of G and P as usual. We denote by $G_{\mathbf{a}}$ and $G_{\mathbf{f}}$ the archimedean and nonarchimedean factors of $G_{\mathbf{A}}$; we identify G with its diagonal embedding into $G_{\mathbf{A}}$, and $SL_2(\mathbf{R})^{\mathbf{a}}$ with $G_{\mathbf{a}}$. For $\mathfrak{r}$ and $\mathfrak{y}$ as in (1.6), we put

$$(1.9a) \qquad D[\mathfrak{r}, \mathfrak{y}] = \prod_{v \in \mathbf{f} \cup \mathbf{a}} D_v[\mathfrak{r}, \mathfrak{y}],$$

$$(1.9b) \qquad D_v[\mathfrak{r}, \mathfrak{y}] = \begin{cases} SO(2) = \{x \in G_v \mid {}^t x x = 1\} & (v \in \mathbf{a}), \\ \mathfrak{o}[\mathfrak{r}, \mathfrak{y}]_v \cap G_v & (v \in \mathbf{f}), \end{cases}$$

where $\mathfrak{o}[\mathfrak{r}, \mathfrak{y}]_v$ is the closure of $\mathfrak{o}[\mathfrak{r}, \mathfrak{y}]$ in $M_2(F_v)$. We observe that $\Gamma[\mathfrak{r}, \mathfrak{y}] = G \cap D[\mathfrak{r}, \mathfrak{y}]G_{\mathbf{a}}$. There is another important subset

$$(1.10) \qquad W = G \cap P_{\mathbf{A}} \cdot D[2\mathfrak{b}^{-1}, 2\mathfrak{b}]$$

of $G_{\mathbf{A}}$. Obviously $P \cdot W \cdot \Gamma[2\mathfrak{b}^{-1}, 2\mathfrak{b}] = W$. In [13, Proposition 3.2], we assigned, to each $\beta \in W$, a holomorphic function h_β on $H^{\mathbf{a}}$ that satisfies the following conditions:

$$(1.11a) \quad h_\beta(z)^4 = j_\beta(z)^{2u} \qquad \text{(and hence $(\beta, h_\beta) \in \mathcal{G}_{u/2}$)};$$

$$(1.11b) \quad h_{\alpha\beta\gamma}(z) = h_\alpha(z)h_\beta(\gamma(z))h_\gamma(z) \text{ if } \alpha \in P, \ \beta \in W, \text{ and } \gamma \in \Gamma[2\mathfrak{b}^{-1}, 2\mathfrak{b}];$$

$$(1.11c) \quad h_\alpha(z) = |d_\alpha|^{u/2} \quad \text{if} \quad \alpha \in P;$$

$$(1.11d) \quad h_\gamma^2 / j_\gamma^u = (d_\gamma / |d_\gamma|)^u \left(\frac{F(\sqrt{-1})/F}{d_\gamma \mathfrak{g}} \right) \quad \text{if} \quad \gamma \in \Gamma[2\mathfrak{b}^{-1}, 2\mathfrak{b}].$$

As for the last two properties, see [13, Proposition 1.2 and (3.13)]. We then define, in Case II, a map $\Lambda_\sigma^k \colon W \to \mathcal{G}_\sigma$ for each odd integer k by

$$(1.12) \qquad \Lambda_\sigma^k(\beta) = (\beta, h_\beta^k j_\beta^{\mathfrak{g} - (k/2)u}) \qquad (\beta \in W).$$

Then $(1.11b)$ implies that

$$(1.13) \qquad \Lambda_\sigma^k(\alpha\beta\gamma) = \Lambda_\sigma^k(\alpha)\Lambda_\sigma^k(\beta)\Lambda_\sigma^k(\gamma) \quad \text{for} \quad \alpha, \beta, \gamma \quad \text{as in (1.11b)}.$$

In Case I, we define an injection $\Lambda_\sigma^0 \colon G \to \mathcal{G}_\sigma$ by

$$(1.14) \qquad \Lambda_\sigma^0(\beta) = (\beta, j_\beta^{\mathfrak{g}}) \qquad (\beta \in G).$$

Given an integral ideal $\mathfrak{c}$, we put

$$(1.15) \qquad \Delta_\sigma^k[\mathfrak{c}] = \begin{cases} \Lambda_\sigma^0(\Gamma[\mathfrak{c}]) & \text{(Case I)}, \\ \Lambda_\sigma^k(\{\beta \in \Gamma[2\mathfrak{c}\mathfrak{b}^{-1}, 2^{-1}\mathfrak{c}\mathfrak{b}] \mid a_\beta - 1 \in \mathfrak{c}\}) & \text{(Case II)}, \end{cases}$$

assuming that $\mathfrak{c} \subset 4\mathfrak{g}$ in Case II. Here and henceforward, we understand that $k = 0$ in Case I. There is a congruence subgroup Γ of G such that

$$(1.16) \qquad \Lambda_\sigma^k(\gamma) = \Lambda_\sigma^1(\gamma) \quad \text{for every} \quad \gamma \in \Gamma.$$

This follows from $(1.11d)$ and [10, Lemma 7.4].

Now, by a *congruence subgroup of* $\mathcal{G}_\sigma$, we understand a subgroup Δ of $\mathcal{G}_\sigma$ satisfying the following two conditions:

(1.17*a*) pr *gives an isomorphism of* Δ *onto a subgroup of* G;

(1.17*b*) Δ *contains* $\Delta_\sigma^k[\mathfrak{a}]$ *as a subgroup of finite index for some* $\mathfrak{a}$ *and* k, *where* k *should be* 0 *in Case I.*

If Δ is a congruence subgroup of $\mathcal{G}_\sigma$, then so is $\xi\Delta\xi^{-1}$ for every $\xi \in \mathcal{G}_\sigma$. This is trivial in Case I, and is proved in [13, Proposition 1.3] in Case II.

2. Automorphic eigenforms

For a function $f: H^\mathbf{a} \to \mathbf{C}$ and $\alpha \in \mathcal{G}_\sigma$, we define $f \| \alpha: H^\mathbf{a} \to \mathbf{C}$ by

$$(2.1) \qquad (f \| \alpha)(z) = l_\alpha(z)^{-1}f(\alpha(z)).$$

From now on, we always put $y_v = \mathrm{Im}(z_v)$, $y = (y_v)_{v\in\mathbf{a}}$, and view y as an $\mathbf{R}^\mathbf{a}$-valued function on $H^\mathbf{a}$. Then we have

$$(2.2) \qquad y^p \| \alpha = l_\alpha^{-1}|j_\alpha|^{-2p}y^p \qquad (p \in \mathbf{R}^\mathbf{a}, \alpha \in \mathcal{G}_\sigma).$$

For $v \in \mathbf{a}$ and $\sigma \in \mathbf{R}^\mathbf{a}$, we define differential operators ϵ_v, δ_v^σ, and L_v^σ acting on C^∞-functions f on $H^\mathbf{a}$ by

$$(2.3a) \qquad \epsilon_v f = -y_v^2 \cdot \partial f/\partial\bar{z}_v,$$

$$(2.3b) \qquad \delta_v^\sigma f = y_v^{-\sigma_v} \cdot \partial(y_v^{\sigma_v}f)/\partial z_v,$$

$$(2.3c) \qquad L_v^\sigma f = 4\delta_v^{\sigma'}\epsilon_v, \qquad \sigma_v' = \sigma_v - 2.$$

We have then

$$(2.4) \qquad \begin{aligned} L_v^\sigma &= -4y_v^2\partial^2/\partial z_v\partial\bar{z}_v + 2i\sigma_v y_v\partial/\partial\bar{z}_v \\ &= -\sigma_v + 4\epsilon_v\delta_v^\sigma. \end{aligned}$$

It can easily be seen, for every $\xi \in \mathcal{G}_\sigma$, that

$$(2.5a) \qquad \delta_v^\sigma(f \| \xi) = (\delta_v^\sigma f) \| \xi^* \quad \text{with} \quad \xi^* = (\mathrm{pr}(\xi), j_\xi^{2v}l_\xi),$$

$$(2.5b) \qquad \epsilon_v(f \| \xi) = (\epsilon_v f) \| \xi_* \quad \text{with} \quad \xi_* = (\mathrm{pr}(\xi), j_\xi^{-2v}l_\xi),$$

$$(2.5c) \qquad L_v^\sigma(f \| \xi) = (L_v^\sigma f) \| \xi.$$

Furthermore, if $p \in \mathbf{R}^\mathbf{a}$, we have

$$(2.6) \qquad L_v^\sigma y^p = p_v(1 - \sigma_v - p_v)y^p.$$

Let Δ be a congruence subgroup of $\mathcal{G}_\sigma$. By an *automorphic eigenform* with respect to Δ, we understand a real-analytic function f on $H^\mathbf{a}$ satisfying the following three conditions:

(2.7a) $f \| \alpha = f$ *for every* $\alpha \in \Delta$;

(2.7b) $L_v^\sigma f = \lambda_v f$ *with* $\lambda_v \in \mathbf{C}$ *for every* $v \in \mathbf{a}$;

(2.7c) *for every* $\xi \in \mathcal{G}_\sigma$, *there exist positive numbers A, B. and c (depending on f and ξ) such that* $y^{\sigma/2} |(f \| \xi)(x + iy)| \leqslant Ay^{cu}$ *if* $y^u > B$.

We denote by $\mathcal{C}(\sigma, \lambda, \Delta)$ the set of all such f, and by $\mathcal{C}(\sigma, \lambda)$ the union of $\mathcal{C}(\sigma, \lambda, \Delta)$ for all congruence subgroups Δ of $\mathcal{G}_\sigma$. Condition (2.7c) concerns, in essence, only finitely many elements ξ of $\mathcal{G}_\sigma$. In fact, put

$$(2.8) \qquad \mathcal{P}_\sigma = \{\xi \in \mathcal{G}_\sigma \mid \mathrm{pr}(\xi) \in P\}.$$

Then $\Delta \backslash \mathcal{G}_\sigma / \mathcal{P}_\sigma$ is a finite set as will be seen in Section 3. The inequality of (2.7c) is true for all $\xi \in \mathcal{G}_\sigma$ if it is true for the members of a complete set of representatives of $\Delta \backslash \mathcal{G}_\sigma / \mathcal{P}_\sigma$.

Given two continuous functions f and g satisfying (2.7a), we define their inner product $\langle f, g \rangle$ by

$$(2.9) \qquad \langle f, g \rangle = \mu(\Phi)^{-1} \int_\Phi \bar{f} g y^\sigma \, d\mu(z) \qquad (\Phi = \Delta \backslash H^\mathbf{a}),$$

where

$$\mu(\Phi) = \int_\Phi d\mu(z) \quad \text{and} \quad d\mu(z) = y^{-2u} \prod_{v \in \mathbf{a}} dx_v \, dy_v.$$

This does not depend on the choice of Δ. We see easily that

$$(2.10) \qquad \langle f, g \rangle = \langle f \| \alpha, g \| \alpha \rangle \quad \text{for every} \quad \alpha \in \mathcal{G}_\sigma.$$

To study the Fourier expansion of an eigenform, let us put

$$(2.11a) \qquad \mathbf{e}(w) = \mathbf{e}^{2\pi i w} \qquad \text{for} \quad w \in \mathbf{C},$$

$$(2.11b) \qquad \mathbf{e_a}(z) = \mathbf{e}\left(\sum_{v \in \mathbf{a}} z_v\right) \quad \text{for} \quad z \in \mathbf{C^a}.$$

If $f \in \mathcal{C}(\sigma, \lambda, \Delta)$, f has a Fourier expansion of the form

$$(2.12) \qquad f(x + iy) = \sum_{h \in \mathfrak{m}} b(h, y) \mathbf{e_a}(hx)$$

with a lattice $\mathfrak{m}$ in F, where $hx = (h_v x_v)_{v \in \mathbf{a}}$ (i.e., the product in the algebra $\mathbf{C^a}$). We can find a subgroup U of $\mathfrak{g}^\times$ of finite index such that

$$\Lambda_\sigma^k(\mathrm{diag}[a, a^{-1}]) \in \Delta \quad \text{for all} \quad a \in U.$$

Then

$$(2.13a) \qquad f(a^2z) = |a|^{-(k/2)u}a^{-\sigma+(k/2)u}f(z) \qquad \text{for every} \quad a \in U,$$

$$(2.13b) \qquad b(a^2h, y) = |a|^{(k/2)u}a^{\sigma-(k/2)u}b(h, a^2y) \qquad \text{for every} \quad a \in U.$$

Now (2.7b) implies that $b(h, y)$ as a function in y_v satisfies

$$(2.14) \qquad (y_v^2\partial^2/\partial y_v^2 + \sigma_v y_v\partial/\partial y_v - 4\pi^2 h_v^2 y_v^2 + 2\pi h_v\sigma_v y_v + \lambda_v)b = 0.$$

If $h_v \neq 0$, the solutions are given by Whittaker functions. To present them in a normalized form, we introduce a function $V(g; \alpha, \beta)$, defined for $0 < g \in \mathbf{R}$ and $(\alpha, \beta) \in \mathbf{C}^2$, which has an expression

$$(2.15) \qquad V(g; \alpha, \beta) = e^{-g/2}g^\beta\Gamma(\beta)^{-1}\int_0^\infty e^{-tg}(1 + t)^{\alpha-1}t^{\beta-1}\,dt$$

for $\mathrm{Re}(\beta) > 0$. This can be continued as a holomorphic function in (α, β) (and real-analytic in (g, α, β)) to the whole $\mathbf{C}^2$, and satisfies

$$(2.16) \qquad V(g; 1 - \beta, 1 - \alpha) = V(g; \alpha, \beta),$$

$$(2.17) \qquad \lim_{g\to\infty} e^{g/2}V(g; \alpha, \beta) = 1.$$

These facts are well known. For the reader's convenience, we give in Section 10 a self-contained treatment of Whittaker functions of this type including the proofs of these and other properties of V.

Now, given $\sigma \in \mathbf{C}^\mathbf{a}$ and $\lambda \in \mathbf{C}^\mathbf{a}$, we take α_v and β_v so that

$$(2.18) \qquad \sigma_v = \alpha_v - \beta_v, \qquad \lambda_v = \beta_v(1 - \alpha_v),$$

and define a function W_v for $t \in \mathbf{R}^\times$ by

$$(2.19) \qquad W_v(t; \sigma, \lambda) = \begin{cases} V(4\pi t; \alpha_v, \beta_v) & \text{if} \quad t > 0, \\ |4\pi t|^{-\sigma_v}V(-4\pi t; \beta_v, \alpha_v) & \text{if} \quad t < 0. \end{cases}$$

If (α_v, β_v) is a solution of (2.18), then the other solution is $(1 - \beta_v, 1 - \alpha_v)$ (which may be equal to (α_v, β_v)). In view of (2.16), W_v is well-defined. Now it can be verified that $W_v(h_v y_v; \sigma, \lambda)$ as a function of y_v satisfies (2.14); it is $O(y_v^c)$ with $c \in \mathbf{R}$ when $y_v \to \infty$, as can be seen from (2.17); moreover, such a solution of (2.14) is unique up to constant factors (see Proposition 10.1).

We now put, for $t \in (\mathbf{R}^\times)^\mathbf{a}$, $\sigma \in \mathbf{R}^\mathbf{a}$, and $\lambda \in \mathbf{C}^\mathbf{a}$,

$$(2.20) \qquad W(t; \sigma, \lambda) = \prod_{v\in\mathbf{a}} W_v(t_v; \sigma, \lambda).$$

Then conditions (2.7b, c) imply that $b(h, y)$ is a constant multiple of $W(hy; \sigma, \lambda)$, and hence

$$(2.21) \qquad f(x + iy) = b_0(y) + \sum_{0 \neq h \in \mathfrak{m}} b_h W(hy; \sigma, \lambda) \mathbf{e_a}(hx)$$

with a function $b_0(y)$ and $b_h \in \mathbf{C}$. The nature of b_0 will be examined in the next section. We call b_0 *the constant term of f,* and understand, by a *cusp form,* an element f of $\mathcal{A}(\sigma, \lambda)$ such that the constant term of $f \parallel \xi$ is 0 for every $\xi \in \mathcal{G}_\sigma$. We denote by $\mathcal{S}(\sigma, \lambda)$ the set of all such forms and put $\mathcal{S}(\sigma, \lambda, \Delta) = = \mathcal{A}(\sigma, \lambda, \Delta) \cap \mathcal{S}(\sigma, \lambda)$.

Proposition 2.1. *Let f and b_h be as in (2.21). Then:*

(1) *There exist constants $p > 0$ and $q \geqslant 0$ such that $|b_h| \leqslant p|h|^{qu + \sigma/2}$ for all $h \in \mathfrak{m}, \neq 0$.*

(2) *There exist positive constants A, B, and C such that*

$$y^{\sigma/2} \sum_{0 \neq h \in \mathfrak{m}} |b_h W(hy; \sigma, \lambda)| \leqslant A \cdot \exp(-By^{u/n}) \quad if \quad y^u \geqslant C.$$

(3) *$\langle f, g \rangle$ is meaningful if either f or g is a cusp form.*
(4) *If f is a cusp form, we can take $q = 0$ in (1).*

Proposition 2.2. *Put $\Delta^* = \{\xi^* \mid \xi \in \Delta\}$, $\Delta_* = \{\xi_* \mid \xi \in \Delta\}$ with ξ^* and ξ_* of (2.5a, b). Then*

$$\epsilon_v \mathcal{A}(\sigma, \lambda, \Delta) \subset \mathcal{A}(\sigma - 2v, \lambda - (\sigma_v - 2)v, \Delta_*),$$

$$\delta_v^\sigma \mathcal{A}(\sigma, \lambda, \Delta) \subset \mathcal{A}(\sigma + 2v, \lambda + \sigma_v v, \Delta^*),$$

where v is viewed as the element of $\mathbf{C}^\mathbf{a}$ of which the v-component is 1 and all other components are 0.

Proposition 2.3. *$\mathcal{A}(\sigma, \lambda, \Delta)$ is finite-dimensional over $\mathbf{C}$.*

These propositions will be proved in Section 11.
Define subsets $\mathfrak{N}(\sigma, \lambda)$ and $\mathfrak{N}(\sigma, \lambda, \Delta)$ of $\mathcal{A}(\sigma, \lambda)$ by

$$(2.22a) \qquad \mathfrak{N}(\sigma, \lambda) = \{g \in \mathcal{A}(\sigma, \lambda) \mid \langle f, g \rangle = 0 \quad \text{for all} \quad f \in \mathcal{S}(\sigma, \lambda)\},$$

$$(2.22b) \qquad \mathfrak{N}(\sigma, \lambda, \Delta) = \{g \in \mathcal{A}(\sigma, \lambda, \Delta) \mid \langle f, g \rangle = 0 \quad \text{for all} \quad f \in \mathcal{S}(\sigma, \lambda, \Delta)\}.$$

Then we see easily that

$$(2.23a) \qquad \mathcal{A}(\sigma, \lambda, \Delta) = \mathcal{S}(\sigma, \lambda, \Delta) \oplus \mathfrak{N}(\sigma, \lambda, \Delta),$$

$$(2.23b) \qquad \mathcal{A}(\sigma, \lambda) = \mathcal{S}(\sigma, \lambda) \oplus \mathfrak{N}(\sigma, \lambda),$$

$$(2.24) \qquad \mathfrak{N}(\sigma, \lambda, \Delta) = \mathfrak{N}(\sigma, \lambda) \cap \mathcal{A}(\sigma, \lambda, \Delta).$$

The inclusion $\mathfrak{N}(\sigma, \lambda, \Delta) \subset \mathfrak{N}(\sigma, \lambda)$ is not completely trivial. To see this, let $g \in \mathfrak{N}(\sigma, \lambda, \Delta)$; take any normal subgroup Δ' of Δ. Then $g = f + h$ with $f \in \mathfrak{S}(\sigma, \lambda, \Delta')$ and $h \in \mathfrak{N}(\sigma, \lambda, \Delta')$. For $\gamma \in \Delta$, we have $g = f \| \gamma + h \| \gamma$. Observing that $f \| \gamma \in \mathfrak{S}(\sigma, \lambda, \Delta')$ and $h \| \gamma \in \mathfrak{N}(\sigma, \lambda, \Delta')$. We obtain $f \| \gamma = f$ and $h \| \gamma = h$ and hence $f \in \mathfrak{S}(\sigma, \lambda, \Delta)$ and $h \in \mathfrak{N}(\sigma, \lambda, \Delta)$. Therefore $g = h \in \mathfrak{N}(\sigma, \lambda, \Delta')$, which shows that $g \in \mathfrak{N}(\sigma, \lambda)$.

Proposition 2.4. *For each* $v \in \mathbf{a}$, *we have*

$$(2.25) \qquad \langle \epsilon_v f, g \rangle = \langle f, \delta_v^\sigma g \rangle \quad \text{for} \quad f \in \mathfrak{A}(\sigma + 2v, \lambda), \qquad g \in \mathfrak{A}(\sigma, \lambda'),$$

if either f *or* g *is a cusp form*;

$$(2.26) \qquad \langle L_v f, g \rangle = \langle f, L_v g \rangle \quad \text{for} \quad f \in \mathfrak{S}(\sigma, \lambda), \qquad g \in \mathfrak{A}(\sigma, \lambda').$$

These formulas are actually true for C^∞-functions f and g satisfying only (2.7a, c), under a suitable condition on the convergence, as will be shown in Section 6; (2.26) follows from (2.25), since

$$\langle L_v f, g \rangle = 4\langle \delta_v^{\sigma - 2v} \epsilon_v f, g \rangle = 4\langle \epsilon_v f, \epsilon_v g \rangle = \langle f, L_v g \rangle.$$

(Formula (2.25) was proved also in [14, Lemma 2.3].) This shows also that $\langle f, L_v f \rangle = 4\langle \epsilon_v f, \epsilon_v f \rangle \geqslant 0$. Therefore $\mathfrak{S}(\sigma, \lambda) \neq \{0\}$ only if $0 \leqslant \lambda_v \in \mathbf{R}$ for every $v \in \mathbf{a}$.

Proposition 2.5. *Every holomorphic function on* $H^{\mathbf{a}}$ *satisfying* (2.7a, c) *belongs to* $\mathfrak{A}(\sigma, 0, \Delta)$. *Moreover, every element of* $\mathfrak{S}(\sigma, 0)$ *is holomorphic on* $H^{\mathbf{a}}$.

PROOF. A function f on $H^{\mathbf{a}}$ is holomorphic if and only if $\epsilon_v f = 0$ for every $v \in \mathbf{a}$. Thus the first assertion is obvious. If $f \in \mathfrak{S}(\sigma, 0)$, we have $4\langle \epsilon_v f, \epsilon_v f \rangle = \langle f, L_v f \rangle = 0$, so that $\epsilon_v f = 0$, which proves the second assertion.

3. The constant term of an eigenform

Let U be a subgroup of $\mathfrak{g}^\times$ of finite index. We call an element τ of $\mathbf{R}^{\mathbf{a}}$ U-*admissible* if

$$(3.1) \qquad\qquad |x|^{i\tau} = 1 \quad \text{for all} \quad x \in U \quad \text{and} \quad \sum_{v \in \mathbf{a}} \tau_v = 0,$$

and denote by T_U the set of all U-admissible τ. We call τ *admissible* if it is U-admissible for some U. We can easily prove

$$(3.2) \qquad \{p \in \mathbf{C}^{\mathbf{a}} \mid |x|^p = 1 \quad \text{for all} \quad x \in U\} = iT_U \oplus \mathbf{C}u.$$

Now let $b(y)$ be the constant term of an element of $\mathcal{Q}(\sigma, \lambda)$. Putting $h = 0$ in (2.14), we have

$$(3.3) \qquad (y_v^2 \partial^2 / \partial y_v^2 + \sigma_v y_v \partial / \partial y_v + \lambda_v) b = 0.$$

A pair of independent solutions of this equation can be given as follows:

(3.4a) y_v^p and y_v^q with the roots p and q of $X^2 - (1 - \sigma_v) X + \lambda_v$ if $4\lambda_v \neq (1 - \sigma_v)^2$,

(3.4b) y_v^q and $y_v^q \log y_v$ with $q = (1 - \sigma_v)/2$ if $4\lambda_v = (1 - \sigma_v)^2$.

Therefore $b(y)$ is a linear combination of the products of these functions for all $v \in \mathbf{a}$. However, not every product can appear. In fact, in view of (2.13b), we can find a subgroup U of $\mathfrak{g}^\times$ of finite index whose elements are all totally positive and such that

$$(3.5) \qquad b(a^2 y) = a^{-\sigma} b(y) \quad \text{for every} \quad a \in U.$$

This imposes a nontrivial condition on the combination of the solutions of (3.4a, b). To be precise, we have:

Proposition 3.1. *The constant term $b(y)$ of an element of $\mathcal{Q}(\sigma, \lambda)$ has one of the following forms*:

(i) *If $4\lambda_v = (1 - \sigma_v)^2$ for all $v \in \mathbf{a}$, then $b(y) = a_1 y^q + a_2 y^q \log y^u$ with $a_i \in \mathbf{C}$ and $q = (u - \sigma)/2$.*

(ii) *If $4\lambda_v \neq (1 - \sigma_v)^2$ for some $v \in \mathbf{a}$, then $b(y) = \sum_p a_p y^p$ with $a_p \in \mathbf{C}$ and $p \in \mathbf{C}^{\mathbf{a}}$. Each p must satisfy the following two conditions*:

$$(3.6a) \qquad \lambda_v = p_v(1 - \sigma_v - p_v) \quad \text{for all} \quad v \in \mathbf{a};$$

$$(3.6b) \qquad p = su - (\sigma - i\tau)/2 \text{ with } s \in \mathbf{C} \text{ and } \tau \in T_U.$$

Proof. Put $\mathbf{b} = \{v \in \mathbf{a} \mid 4\lambda_v = (1 - \sigma_v)^2\}$ and $(\log y)^{\mathbf{d}} = \prod_{v \in \mathbf{d}} \log y_v$ for $\mathbf{d} \subset \mathbf{b}$. Then $b(y) = \sum_{\mathbf{d} \subset \mathbf{b}} \sum_p A_{p, \mathbf{d}} y^p (\log y)^{\mathbf{d}}$ with constants $A_{p, \mathbf{d}}$ and p_v as in (3.6a). Take a maximal subset $\mathbf{d}$ of $\mathbf{b}$ such that $A_{p, \mathbf{d}} \neq 0$ for some p. Fix such a p. Then (3.5) implies that $a^{\sigma + 2p} = 1$ for $a \in U$; hence $\sigma + 2p = i\tau + 2su$ with $s \in \mathbf{C}$ and $\tau \in T_U$ by (3.2). Thus p must be as in (3.6b). Suppose $\mathbf{d} \neq \emptyset$ and let $\mathbf{d} = \mathbf{e} \cup \{w\}$ with an arbitrarily fixed w. Then (3.5) implies that $\sum A_{p, \mathbf{x}} \log a_v^2 = 0$ for every $a \in U$, where the sum is taken over all $(v, \mathbf{x})$ such that $\{v\} \cup \mathbf{e} = \mathbf{x} \subset \mathbf{b}$. Since $A_{p, \mathbf{d}} \neq 0$, this can happen only when $\mathbf{e} = \emptyset$ and $\mathbf{b} = \mathbf{a}$. Then $p = (u - \sigma)/2$, $b(y) = Ay^p + \sum_{v \in \mathbf{a}} A_v y^p \log y_v$, and $\sum_v A_v \log a_v^2 = 0$ for all $a \in U$, and hence we obtain (i). If $A_{p,\mathbf{d}} = 0$ for all $\mathbf{d} \neq \varnothing$, then we obtain (ii).

Thus, given σ and λ, $b(y)$ belongs to a two-dimensional space if $4\lambda_v = (1 - \sigma_v)^2$ for every $v \in \mathbf{a}$, and to a 2^n-dimensional space otherwise. The latter space can actually be reduced to a 2-dimensional space in most cases. In fact, take p, τ, and s as in (3.6a, b). Let $q = u - \sigma - p$. Then $q = (1 - s)u - (\sigma + i\tau)/2$, and hence q, $-\tau$, and $1 - s$ satisfy (3.6a, b). Since $4\lambda_v \neq (1 - \sigma_v)^2$ for some v, we have $p \neq q$, and therefore y^p and y^q form a two-dimensional vector space. Now our question is whether y^r, with r different from p and q, can occur. Suppose it can, and let $r = tu - (\sigma - ix)/2$ with $t \in \mathbf{C}$ and $x \in T_U$. Then, for each v, r_v must coincide with p_v or q_v. Decompose $\mathbf{a}$ into the disjoint union of three subsets $\mathbf{b}$, $\mathbf{c}$, and $\mathbf{d}$ so that $r_v = p_v = q_v$ for $v \in \mathbf{b}$, $r_v = p_v \neq q_v$ for $v \in \mathbf{c}$, and $r_v = q_v \neq p_v$ for $v \in \mathbf{d}$. Then $\mathbf{c} \neq \varnothing$ and $\mathbf{d} \neq \varnothing$; $t + ix_v/2 = s + i\tau_v/2$ for $v \in \mathbf{b} \cup \mathbf{c}$ and $t + ix_v/2 = 1 - s - i\tau_v/2$ for $v \in \mathbf{b} \cup \mathbf{d}$. Hence $\mathrm{Re}(s) = \mathrm{Re}(t) = \mathrm{Re}(1 - s)$, so that $\mathrm{Re}(s) = 1/2$. Therefore $\mathrm{Re}(p) = (u - \sigma)/2$. Observing that $(1 - \sigma_v)^2 \leqslant 4\lambda_v \in \mathbf{R}$ if and only if $\mathrm{Re}(p_v) = (1 - \sigma_v)/2$, we obtain

Proposition 3.2. *The constant term $b(y)$ of an element of $\mathcal{Q}(\sigma, \lambda)$, for fixed σ and λ, belongs to a two-dimensional vector space unless the following condition is satisfied:*

(3.7) $F \neq \mathbf{Q}$, $(1 - \sigma_v)^2 \leqslant 4\lambda_v \in \mathbf{R}$ *for all $v \in \mathbf{a}$, and $(1 - \sigma_v)^2 < 4\lambda_v$ for at least one v.*

If this is satisfied and if p is as in (3.6a, b), then $\mathrm{Re}(p) = (u - \sigma)/2$.

We call λ *critical* if $4\lambda_v = (1 - \sigma_v)^2$ for all $v \in \mathbf{a}$; otherwise we call λ *noncritical*. We call λ *simple* if either λ is critical or λ is noncritical and there are only two p's satisfying (3.6a, b). In the latter case, if p is one, the other is $u - \sigma - p$. We call λ *multiple* if it is not simple. Any p as in (3.6a, b) is called an *exponent attached to* λ. In Remark 5.5 below, we shall give an example of multiple λ.

Hereafter we fix a weight σ and write simply $\mathcal{G}$ and $\mathcal{P}$ for $\mathcal{G}_\sigma$ and $\mathcal{P}_\sigma$, where $\mathcal{P}_\sigma$ is defined by (2.8). Given an admissible τ, we put, throughout the rest of the paper,

(3.8) $$\rho = (\sigma - i\tau)/2 \quad (\in \mathbf{C}^{\mathbf{a}}).$$

Then, for λ and p of (3.6a, b), we have

(3.9a) $$\lambda_v = (s - \rho_v)(1 - s - \bar{\rho}_v),$$

(3.9b) $$p = su - \rho, \qquad u - \sigma - p = (1 - s)u - \bar{\rho}.$$

Let Δ be a congruence subgroup of $\mathcal{G}$ and let $\Gamma = \mathrm{pr}(\Delta)$. Then *pr* gives a bijective map of $\mathcal{P}\backslash\mathcal{G}/\Delta$ onto $P\backslash G/\Gamma$, which is a finite set corresponding bijectively to $\Gamma\backslash(F\cup\{\infty\})$ via the map $\alpha \mapsto \alpha^{-1}(\infty)$. Therefore we call a coset $\mathcal{P}\xi\Delta$ with $\xi \in \mathcal{G}$ a *cusp-class* of Δ. Given a cusp-class $\mathcal{P}\xi\Delta$, we call it *ρ-regular* if

$$(3.10a) \qquad y^{-\rho}\|\gamma = y^{-\rho} \quad \text{for every} \quad \gamma \in \mathcal{P}\cap\xi\Delta\xi^{-1}.$$

or equivalently,

$$(3.10b) \qquad l_\gamma^{-1}|d_\gamma|^{\sigma-i\tau} = 1 \quad \text{for every} \quad \gamma \in \mathcal{P}\cap\xi\Delta\xi^{-1}.$$

It can easily be seen that $(3.10a)$ is in fact a condition on $\mathcal{P}\xi\Delta$, independent of the choice of a representative ξ. The meaning of this condition is explained by:

Proposition 3.3.   *Let ρ, λ, and p be as above; let $f \in \mathcal{A}(\sigma, \lambda, \Delta)$ and $\xi \in \mathcal{G}$. Then y^p or $y^p \log y^u$ can appear nontrivially in the constant term of $f\|\xi^{-1}$ only if $\mathcal{P}\xi\Delta$ is ρ-regular.*

This follows immediately from our definition.

If all cusp-classes of Δ are ρ-regular, then the same is true for every congruence subgroup of Δ. Such a Δ indeed exists because of

Proposition 3.4.   *Given $\rho = (\sigma - i\tau)/2$ as above, there exists an integral ideal $\mathfrak{a}$ such that all cusp-classes of $\Delta_\sigma^k[\mathfrak{a}]$ are ρ-regular.*

PROOF.   Take an integral ideal $\mathfrak{b}$ so that $x \gg 0$ and $x^{i\tau} = 1$ if $x \in \mathfrak{g}^\times$ and $x - 1 \in \mathfrak{b}$. In Case II, choose $\mathfrak{b}$ so that $\mathfrak{b} \subset 4\mathfrak{g}$. Write simply $\Delta[\mathfrak{a}]$ for $\Delta_\sigma^k[\mathfrak{a}]$. In either case, we have $\mathcal{G} = \mathcal{P}Z\Delta[\mathfrak{b}]$ with a finite subset Z of $\mathcal{G}$. We can find an integral ideal $\mathfrak{a} \subset \mathfrak{b}$ such that $\Delta[\mathfrak{a}] \subset \bigcap_{\zeta\in Z}\zeta^{-1}\Delta[\mathfrak{b}]\zeta$. Obviously $\mathcal{G} = \mathcal{P}X\Delta[\mathfrak{a}]$ with a suitable subset X of $Z\Delta[\mathfrak{b}]$. If $\xi \in \zeta\Delta[\mathfrak{b}]$ with $\zeta \in Z$, then $\xi\Delta[\mathfrak{a}]\xi^{-1} = \zeta\Delta[\mathfrak{a}]\zeta^{-1} \subset \Delta[\mathfrak{b}]$. Let $\gamma \in \mathcal{P}\cap\Delta[\mathfrak{b}]$. Then, in Case II, we have $l_\gamma = |d_\gamma|^{ku/2}d_\gamma^{\sigma-(k/2)u}$ by $(1.11c)$ and (1.12). Hence our choice of $\mathfrak{b}$ implies $(3.10b)$ for $\Delta = \Delta[\mathfrak{a}]$. The same can be verified in Case I in a similar way.

4. Eisenstein series

Given a congruence subgroup Δ of $\mathcal{G}$, we define its Eisenstein series by

$$(4.1) \qquad E(z, s) = E(z, s; \rho, \Delta)$$

$$= \sum_{\alpha\in(\mathcal{P}\cap\Delta)\backslash\Delta} y^{su-\rho}\|\alpha.$$

Here $s \in \mathbf{C}$ and $\rho = (\sigma - i\tau)/2$ with an admissible τ. To make the sum meaningful, we have to assume that $y^{-\rho}\|\gamma = y^{-\rho}$ for every $\gamma \in \mathcal{P} \cap \Delta$, that is, $\mathcal{P}\Delta$ is ρ-regular. The series is covergent for $\mathrm{Re}(s) > 1$, and can be continued as a meromorphic function in s to the whole s-plane. (See Theorem 4.2 below for a precise statement.) Assuming this result, we have obviously $E(z, s) \| \gamma = {} = E(z, s)$ for every $\gamma \in \Delta$, and moreover, by (2.6) and (2.5c),

$$(4.2) \qquad L_v^\sigma E(z, s) = \lambda_v E(z, s) \quad \text{with} \quad \lambda_v = (s - \rho_v)(1 - s - \bar{\rho}_v)$$

for every $v \in \mathbf{a}$. Therefore, if $E(z, s)$ is finite at s, it satisfies (2.7a, b) as a function in z, and in fact belongs to $\mathcal{C}(\sigma, \lambda, \Delta)$ as (2.7c) can be shown in our later discussion.

From our definition, we can easily derive a relation

$$(4.3) \qquad [\mathcal{P} \cap \Delta : \mathcal{P} \cap \Delta'] E(z, s; \rho, \Delta) = \sum_{\gamma \in \Delta' \backslash \Delta} E(z, s; \rho, \Delta') \| \gamma$$

for every congruence subgroup $\Delta' \subset \Delta$. For each ρ-regular cusp-class $\mathcal{P}\xi\Delta$ (see (3.10a, b)), we put

$$(4.4) \qquad E(z, s; \rho, \xi, \Delta) = E(z, s; \rho, \xi\Delta\xi^{-1}) \| \xi.$$

Then we see easily that

$$(4.5) \quad E(z, s; \rho, \xi, \Delta) \| \gamma = E(z, s; \rho, \xi, \Delta) \quad \text{for every} \quad \gamma \in \Delta,$$

$$(4.6) \quad E(z, s; \rho, \alpha\xi\gamma, \Delta) = l_\alpha^{-1} |d_\alpha|^{2\rho - 2su} E(z, s; \rho, \xi, \Delta) \quad \text{if} \quad \alpha \in \mathcal{P} \quad \text{and} \quad \gamma \in \Delta.$$

Thus, ignoring elementary factors, we associate with Δ exactly as many Eisenstein series as its ρ-regular cusp-classes.

We now introduce another type of Eisenstein series, which is attached to an integral ideal $\mathfrak{c}$ in F and a Hecke character $\psi: F_\mathbf{A}^\times/F^\times \to \mathbf{C}^\times$. We assume

$$(4.7a) \qquad\qquad 4\mathfrak{g} \supset \mathfrak{c} \text{ in Case II};$$

$$(4.7b) \qquad\qquad |\psi| = 1;$$

$(4.7c)$ the finite part of the conductor of ψ divides $\mathfrak{c}$;

$$(4.7d) \qquad\qquad \psi(x) = |x|^{i\tau}(x/|x|)^{\sigma'} \quad \text{for} \quad x \in F_\mathbf{a}^\times;$$

where

$$(4.8) \qquad\qquad \sigma' = \begin{cases} \sigma & \text{(Case I)}, \\ \sigma - (k/2)u & \text{(Case II)}, \end{cases}$$

$F_\mathbf{A}^\times$ denotes the idele group of F, and $F_\mathbf{a}^\times$ its archimedean factor. We fix $\mathfrak{c}$, assume $\mathfrak{c} \subset 4\mathfrak{g}$ in Case II, and put

$$(4.9a) \qquad D = \begin{cases} D[\mathfrak{g}, \mathfrak{c}] & \text{(Case I)}, \\ D[2\mathfrak{b}^{-1}, 2^{-1}\mathfrak{c}\mathfrak{b}] & \text{(Case II)}, \end{cases}$$

$$(4.9b) \qquad \Gamma_0(\mathfrak{c}) = G \cap DG_{\mathbf{a}}.$$

Writing simply Γ for $\Gamma_0(\mathfrak{c})$, take a complete set of representatives B of $P\backslash(G \cap P_{\mathbf{A}}D)/\Gamma$. Take, for each $\beta \in B$, a complete set of representatives R_β of $(P \cap \beta\Gamma\beta^{-1})\backslash\beta\Gamma$. Then we put

$$(4.10) \qquad E_k(z, s; \rho, \psi, \mathfrak{c}) = \sum_{\beta \in B} N(\mathfrak{a}_\beta)^{2s} \sum_{\alpha \in R_\beta} \psi(d_\alpha \mathfrak{a}_\beta^{-1})\psi_{\mathbf{a}}(d_\alpha) y^{su - \rho} \| \Lambda_\sigma^k(\alpha),$$

where $\mathfrak{a}_\beta = c_\beta\mathfrak{g} + d_\beta\mathfrak{g}$ in Case I and $\mathfrak{a}_\beta = 2c_\beta\mathfrak{b}^{-1} + d_\beta\mathfrak{g}$ in Case II, and $\psi_{\mathbf{a}}$ is the archimedean part of ψ; we use the same letter ψ for the ideal character attached to ψ; we understand that $\psi(d_\alpha \mathfrak{a}_\beta^{-1})\psi_{\mathbf{a}}(d_\alpha) = \psi(\mathfrak{a}_\beta^{-1})$ if $\mathfrak{c} = \mathfrak{g}$. The right-hand side of (4.10) is convergent for $\mathrm{Re}(s) > 1$ and satisfies (4.2). In Case I, we have $k = 0$, and so we write simply E for E_k. As for the relation between the series of type (4.10) and that of (4.1), see (4.24) and Proposition 5.3 below.

Define the L-function $L(s, \psi)$ of ψ as usual and put

$$(4.11) \qquad L_{\mathfrak{c}}(s, \psi) = L(s, \psi) \prod_{\mathfrak{p} | \mathfrak{c}} [1 - \psi(\mathfrak{p})N(\mathfrak{p})^{-s}],$$

where $\mathfrak{p}$ denotes a prime ideal in F.

Theorem 4.1. *The series $E_k(z, s; \rho, \psi, \mathfrak{c})$ can be continued as a meromorphic function to the whole s-plane. More precisely, put*

$$D(z, s) = \begin{cases} \displaystyle\prod_{v \in \mathbf{a}} \Gamma(s + (|\sigma_v| + i\tau_v)/2)L_{\mathfrak{c}}(2s, \psi)E(z, s; \rho, \psi, \mathfrak{c}) & \text{(Case I)}, \\ \displaystyle\prod_{v \in \mathbf{a}} \Gamma_v(s + i\tau_v/2)L_{\mathfrak{c}}(4s - 1, \psi^2)E_k(z, s; \rho, \psi, \mathfrak{c}) & \text{(Case II)}, \end{cases}$$

where Γ_v in Case II is defined by

$$\Gamma_v(s) = \Gamma(s + (\theta_v/2) - (1/4)) \cdot \begin{cases} \Gamma(s + (\sigma_v/2)) & \text{if } 2\sigma_v \geqslant -1, \\ \Gamma(s - (\sigma_v/2)) & \text{if } 2\sigma_v < -1, \end{cases}$$

with the smallest nonnegative integer θ_v that is congruent modulo 2 to $\sigma_v - 1/2$ or $\sigma_v + 1/2$ according as $2\sigma_v \geqslant -1$ or $2\sigma_v < -1$. Then there is a real analytic function on $H^{\mathbf{a}} \times \mathbf{C}$ that is holomorphic in s and that coincides with $s(s - 1)D(z, s)$ in Case I and with $(s - 3/4)D(z, s)$ in Case II for $\mathrm{Re}(s) > 1$. (Thus we are able to speak of possible simple poles at $s = 0, 1,$ or $3/4$.) In Case I, the pole at $s = 0$ occurs if and only if $\mathfrak{c} = \mathfrak{g}$, $\psi = 1$, and $\sigma = \tau = 0$; the pole at $s = 1$ occurs if and only if $\psi = 1$ and $\sigma = \tau = 0$. In Case II, the pole at $s = 3/4$ occurs if and only if $\psi^2 = 1$ and, for every $v \in \mathbf{a}$, $\sigma_v - 1/2$ is either an even nonnegative integer or an odd negative integer.

The result in Case II is merely a paraphrase of [13, Corollary 6.2]. In fact, the symbols k, ρ, μ, and τ there correspond to k, σ', τ, and θ here. If we denote by $E^*(z, s)$ the function $E(z, s; k/2, \rho, \psi, \mathfrak{c})$ there, then, comparison of (4.10) with [13, (4.7c)] shows that

$$(4.12) \qquad E_k(z, s; \rho, \psi, \mathfrak{c}) = y^{(k/4)u - \rho}E^*(z, s - k/4),$$

and hence our assertion follows immediately from [13, Corollary 6.2].

The result in Case I can be obtained by modifying the formulation of [12]. To be more specific, take $m = 1$ in [12]; using the same notation, we define a function f on $G_\mathbf{A}$ by

$$(4.13) \quad f(x) = \begin{cases} 0 & \text{if } x \notin P_\mathbf{A}D, \\ \psi_\mathbf{f}(d_p)^{-1}\psi_\mathfrak{c}(d_w)^{-1}J(x, \mathbf{i})^{-1} & \text{if } x = pw \text{ with } p \in P_\mathbf{A} \text{ and } w \in D, \end{cases}$$

where

$$\psi_\mathbf{f}(a) = \prod_{v \in \mathbf{f}} \psi(a_v), \qquad \psi_\mathfrak{c}(a) = \prod_{v \mid \mathfrak{c}} \psi(a_v),$$

and

$$J(x, \mathbf{i}) = \prod_{v \in \mathbf{a}} j(x_v, i)^{\sigma_v}|j(x_v, i)|^{i\tau_v - \sigma_v}.$$

We then define a series $E_\mathbf{A}$ on $G_\mathbf{A}$ by

$$(4.14) \qquad E_\mathbf{A}(x, s) = \sum_{\alpha \in P \backslash G} f(\alpha x)\epsilon(\alpha x)^{-2s} \qquad (x \in G_\mathbf{A}, s \in \mathbf{C})$$

with ϵ of [12, (2.11)]. We can easily verify that

$$(4.15) \qquad E(z, s; \rho, \psi, \mathfrak{c}) = y^{-\rho}E_\mathbf{A}(x, s)J(x, \mathbf{i})$$

if $x \in G_\mathbf{a}$ and $x_v(i) = z_v$ for $v \in \mathbf{a}$. Put

$$(4.16) \qquad E'(z, s) = E(z, s; \rho, \psi, \mathfrak{c}) \| \Lambda_\sigma^0(\eta), \qquad \eta = \begin{pmatrix} 0 & -1 \\ 1 & 0 \end{pmatrix}.$$

This has a Fourier expansion of the form

$$(4.17) \qquad E'(z, s) = \sum_{h \in \mathfrak{b}} a(h, y, s)\mathbf{e}_\mathbf{a}(hx)$$

with $\mathfrak{b} = (\mathfrak{c}\mathfrak{d})^{-1}$. Applying the methods of [12] to E' with obvious modifications, we find that if $\mathfrak{c} \neq \mathfrak{g}$,

$$(4.18a) \quad a(h, y, s) = N(\mathfrak{b})^{-1/2}N(\mathfrak{c})^{-1}a_\mathbf{f}(h, s)y^{su - \rho} \prod_{v \in \mathbf{a}} \xi(y_v, h_v; s + \bar{\rho}_v, s - \rho_v),$$

$$(4.18b) \quad a_\mathbf{f}(h, s) = \prod_{v \in \mathbf{f}, v \nmid \mathfrak{c}} \alpha(v, h, \psi(\pi_v)q_v^{-2s}),$$

where ξ, α, π_v, and q_v are defined by [12, (3.18), (3.22), and (3.23)] (in the one-dimensional case). In particular

$$(4.19) \qquad \xi(g, h; \alpha, \beta) = \int_{\mathbf{R}} \mathbf{e}(-hx)(x + ig)^{-\alpha}(x - ig)^{-\beta}\, dx.$$

As for $a_{\mathfrak{f}}$, we have

$$(4.20) \qquad L_{\mathfrak{c}}(2s, \psi)a_{\mathfrak{f}}(h, s) = \begin{cases} L_{\mathfrak{c}}(2s - 1, \psi) & \text{if } h = 0, \\ \sum \psi(\mathfrak{a})N(\mathfrak{a})^{1-2s} & \text{if } h \neq 0, \end{cases}$$

where $\mathfrak{a}$ runs over all integral ideals in F prime to $\mathfrak{c}$ and dividing $h\mathfrak{c}\mathfrak{d}$. This result for $h = 0$ is already given in [12, Theorem 7.7, I]. If $h \neq 0$, the Euler v-factor for $v \nmid h\mathfrak{c}\mathfrak{d}$ is determined by [12, Proposition 4.6]. The «bad factors» can be determined by the methods of [13, §6]. In fact, the present case is easier than [13, §6]. Thus after an easy calculation, we obtain a final result as given in (4.20). As for ξ of (4.19), we have

$$(4.21a) \quad \prod_{v \in \mathbf{a}} \xi(y_v, 0; \alpha_v, \beta_v) =$$

$$= i^{-\{\sigma\}}(2\pi)^n (2y)^{u - \alpha - \beta} \prod_{v \in \mathbf{a}} \Gamma(\alpha_v + \beta_v - 1)\Gamma(\alpha_v)^{-1}\Gamma(\beta_v)^{-1},$$

$$(4.21b) \quad y^\beta \prod_{v \in \mathbf{a}} \xi(y_v, h_v; \alpha_v, \beta_v) =$$

$$= (-2i)^{\{\sigma\}}\pi^{\{\alpha\}}|h|^{\alpha - u} W(hy; \sigma, \lambda) \prod_{v \in \mathbf{a}} \Gamma(\gamma_v)^{-1}.$$

where σ and λ are determined by (2.18), $\{\alpha\} = \sum_{v \in \mathbf{a}} \alpha_v$, and $\gamma_v = \alpha_v$ or β_v according as $h_v > 0$ or $h_v < 0$ (see [11, (1.31), (4.34K)]). Therefore we obtain our assertion on D in Case I by examining the local behavior of each Fourier coefficient of E', provided that $\mathfrak{c} \neq \mathfrak{g}$. To treat the case $\mathfrak{c} = \mathfrak{g}$, we first observe that if $\mathfrak{c} \supset \mathfrak{e}$, we have (in both cases $\mathfrak{c} \neq \mathfrak{g}$ and $\mathfrak{c} = \mathfrak{g}$)

$$(4.22) \qquad E(z, s; \rho, \psi, \mathfrak{c}) = \sum_{\gamma \in \Gamma_0(\mathfrak{e}) \backslash \Gamma_0(\mathfrak{c})} \psi_{\mathfrak{c}}(d_\gamma)^{-1} j_\gamma^{-\sigma} E(\gamma(z), s; \rho, \psi, \mathfrak{e}),$$

which can be proved in the same manner as in [12, Proposition 2.4, (ii)]. Suppose $\mathfrak{c} = \mathfrak{g}$. Take an arbitrary $\mathfrak{e} \neq \mathfrak{g}$. Then our result on $E(z, s; \rho, \psi, \mathfrak{e})$ shows that the poles can occur only at $s = 0$ and $s = 1$; the pole at $s = 0$ is produced by the difference of $L(s, \psi)$ from $L_{\mathfrak{e}}(s, \psi)$. To see that these poles do occur when $\psi = 1$ and $\rho = 0$, we apply the method of [12] to $E(z, s)$ (instead of $E'(z, s)$) to find that

$$E(z, s; \rho, \psi, \mathfrak{c}) = y^{su - \rho} + \sum_{h \in \mathfrak{b}^{-1}} b(h, y, s)\mathbf{e}_{\mathbf{a}}(hx),$$

with Fourier coefficients b which are similar to but somewhat more complicated than the above $a(h, y, s)$. It is easy, however, to see that $b(h, y, s) = {} = a(h, y, s)$ if $\mathfrak{c} = \mathfrak{g}$, and hence the poles at $s = 0$ and $s = 1$ occur if $\psi = 1$ and $\rho = 0$. This completes the proof of Theorem 4.1.

Now observe that $E(z, s; \rho, \Delta_\sigma^k[\mathfrak{c}])$ is meaningful if and only if

$$(4.23) \qquad |x|^{i\tau}(x/|x|)^{\sigma'} = 1 \quad \text{for every} \quad x \in \mathfrak{g}^\times \quad \text{such that} \quad x \equiv 1 \quad (\mathrm{mod}\,\mathfrak{c}).$$

This holds if and only if a Hecke character ψ satisfying (4.7b, c, d) exists. Assuming (4.23), let Ψ be the set of all such characters ψ with fixed $\mathfrak{c}$ and ρ, and $|\Psi|$ the number of elements in Ψ. Then we have

$$(4.24) \qquad |\Psi| E(z, s; \rho, \Delta_\sigma^k[\mathfrak{c}]) = \sum_{\psi \in \Psi} E_k(z, s; \rho, \psi, \mathfrak{c})$$

for the same reason as in [12, Proposition 2.4].

Theorem 4.2. *$E(z, s; \rho, \Delta)$ can be continued as a meromorphic function in s to the whole plane in the sense that there exist a nonzero holomorphic function $A(s)$ and a real analytic function $B(z, s)$ on $H^\mathbf{a} \times \mathbf{C}$, holomorphic in s such that $A(s)E(z, s; \rho, \Delta) = B(z, s)$ for $Re(s) > 1$. Moreover, $E(z, s; \rho, \Delta)$ is holomorphic in s except at the following points:*

(1) *$0 \leqslant Re(s) < \frac{1}{2}$ in Case I and $\frac{1}{4} \leqslant Re(s) < \frac{1}{2}$ in Case II;*
(2) *a possible simple pole at $s = 1$ in Case I, which occurs only if $\rho = 0$;*
(3) *a possible simple pole at $s = 3/4$ in Case II, which occurs only if $\tau = 0$ and σ is given as at the end of Theorem 4.1;*
(4) *possible poles at the roots of a polynomial $T_\rho(s)$ given by*

$$T_\rho(s) = \prod_{v \in \mathbf{a}} \Gamma(s + (|\sigma_v| + i\tau_v)/2)/\Gamma(s + (\delta_v + i\tau_v)/2) \qquad (\textit{Case I}),$$

$$T_\rho(s) = \prod_{v \in \mathbf{a}} \Gamma_v(s + i\tau_v/2)/\left[\Gamma\left(s + (i\tau_v/2) - \tfrac{1}{4}\right)\Gamma\left(s + (i\tau_v/2) + \tfrac{1}{4}\right)\right] \qquad (\textit{Case II}),$$

where $\delta_v = 0$ or 1 according as σ_v is even or odd, and Γ_v is as in Theorem 4.1.

PROOF. In view of (4.3), it is sufficient to prove our theorem when $\Delta = \Delta_\sigma^k[\mathfrak{c}]$. Let D_ψ denote the function D defined in Theorem 4.1, and put

$$R_\psi(s) = \begin{cases} \displaystyle\prod_{v \in \mathbf{a}} \Gamma(s + (\delta_v + i\tau_v)/2)L_\mathfrak{c}(2s, \psi) & (\textit{Case I}), \\[2ex] \displaystyle\prod_{v \in \mathbf{a}} \Gamma\left(s + (i\tau_v/2) - \tfrac{1}{4}\right)\Gamma\left(s + (i\tau_v/2) + \tfrac{1}{4}\right)L_\mathfrak{c}(4s - 1, \psi^2) & (\textit{Case II}). \end{cases}$$

Then $R_\psi(s) \neq 0$ except at the points of (1). By (4.24), we have

$$(4.25) \qquad |\Psi| E(z, s; \rho, \Delta_\sigma^k[\mathfrak{c}]) = \sum_{\psi \in \Psi} D_\psi(z, s)/[T_\rho(s)R_\psi(s)].$$

Observing that $T_\rho(s)$ is indeed a polynomial in s, we obtain our assertions from Theorem 4.1.

The polynomial $T_\rho(s)$ has no zero when $\mathrm{Re}(s) \geqslant \frac{1}{2}$. Therefore $E(z, s; \rho, \Delta)$ *is holomorphic in s if $\mathrm{Re}(s) \geqslant \frac{1}{2}$ except for a possible simple pole described in (2) or (3) of the above theorem.* The pole at $s = 1$ does occur if $\rho = 0$. In fact we have

Proposition 4.3. *For a congruence subgroup Γ of G, let $r(\Gamma)$ be the residue of $E(z, s; 0, \Lambda_0^0(\Gamma))$ at $s = 1$. Then $r(\Gamma)$ is a positive number with the following properties*:

(i) $r(\Gamma)/r(\Gamma') = [\Gamma:\Gamma']/[P\cap\Gamma:P\cap\Gamma']$ *if* $\Gamma' \subset \Gamma$;

(ii) $r(SL_2(\mathfrak{g})) = 2^{n-2}\pi^n D_F^{-1}\zeta_F(2)^{-1}R_F$, *where D_F is the discriminant of F, ζ_F is the zeta function of F, and R_F is the regulator of F*;

(iii) $r(\Gamma)\mu(\Gamma\backslash H^{\mathfrak{a}}) = \mu_K(P\cap\Gamma)\backslash K)$,   *where* $K = \{z \in H^{\mathfrak{a}} \,|\, \mathrm{Im}(z)^u = 1\}$, *furnished with a certain invariant measure μ_K (see the proof below).*

PROOF. Assertion (ii) follows immediately from (4.24) and the explicit Fourier expansion given in the proof of Theorem 4.1. This together with (4.3) proves that $r(\Gamma)$ is a positive number satisfying (i). As for (iii), we give here only a sketch of the proof. Put $A = \{y \in \mathbf{R}^{\mathfrak{a}} \,|\, y \gg 0\}$ and $B = \{y \in A \,|\, y^u = 1\}$. Then every $y \in A$ can be written uniquely $y = t^{1/n}y'$ with $0 < t \in \mathbf{R}$ and $y' \in B$. Let $d^\times y = y^{-u}\,dy$ with the Euclidean measure dy on $\mathbf{R}^{\mathfrak{a}}$. Then $d^\times y = t^{-1}\,dt\,dy'$ with a Haar measure dy' on B. Since $K = \mathbf{R}^{\mathfrak{a}} \times B$, we can determine a measure μ_K on K by $d\mu_K(x + iy) = dx\,dy'$. Now take $\Gamma = SL_2(\mathfrak{g})$ and put $U = \{a^2 \,|\, a \in \mathfrak{g}^\times\}$. By a well known principle, we have

$$(4.26) \qquad \int_{A/U}\varphi(y^u)d^\times y = 2^{n-1}R_F\int_0^\infty\varphi(t)t^{-1}\,dt$$

for a continuous function φ on A. In particular, this implies

$$(4.27) \qquad \mu_K((P\cap\Gamma)\backslash K) = 2^{n-1}D_F^{1/2}R_F.$$

Take $\varphi(t) = e^{-t}t^s$. Then

$$2^{n-1}D_F^{1/2}R_F\Gamma(s) = \int_{(P\cap\Gamma)\backslash H^{\mathfrak{a}}}\exp(-y^u)y^{(s+1)u}\,d\mu(z) = \int_{\Gamma\backslash H^{\mathfrak{a}}}M(z, s)\,d\mu(z),$$

where

$$M(z, s) = \sum_{\alpha \in (P\cap\Gamma)\backslash\Gamma}\exp(-\mathrm{Im}(\alpha(z))^u)\,\mathrm{Im}(\alpha(z))^{(s+1)u}.$$

Since $1 - t \leqslant e^{-t} \leqslant 1$, we have, for $1 < s \in \mathbf{R}$,

$$E(z, s+1) - E(z, s+2) \leqslant M(z, s) \leqslant E(z, s+1),$$

where $E(z, s) = E(z, s; 0, \Lambda_0^0(\Gamma))$. Then we see that $\lim_{s \to 0} sM(z, s) = r(\Gamma)$, and hence $r(\Gamma)\mu(\Gamma \backslash H^{\mathbf{a}}) = 2^{n-1} D_F^{1/2} R_F$, which proves (iii) when $\Gamma = SL_2(\mathfrak{g})$. The general case can be proved in a similar way; alternatively, it follows from the special case by virtue of (i).

Combining (ii), (iii), and (4.27), we find that

$$(4.28) \qquad \mu(SL_2(\mathfrak{g}) \backslash H^{\mathbf{a}}) = 2\pi^{-n} D_F^{3/2} \zeta_F(2),$$

which is classical.

Proposition 4.4. *Let Q be a finite set of functions of the form $E(z, s) \| \alpha$ with $\alpha \in \mathcal{G}$ and E of type (4.1) or (4.10), and let $g(z, s) = \sum_{q \in Q} f_q(s) q(z, s)$ with meromorphic functions f_q on $\mathbf{C}$. Then, for every $s_0 \in \mathbf{C}$, there exists an integer m and a neighborhood V of s_0 such that $(s - s_0)^m g(z, s)$ is a real analytic function on $H^{\mathbf{a}} \times V$ that is holomorphic in s. If in particular, g is finite at $s = s_0$, then $g(z, s_0)$ is an element of $\mathcal{A}(\sigma, \lambda)$ with $\lambda_v = (s_0 - \rho_v)(1 - s_0 - \bar{\rho}_v)$.*

This will be proved in Section 11.

5. The constant terms of Eisenstein series

Lemma 5.1. *Let $\mathfrak{r}$ be a lattice in F and $\mathfrak{y}$ its dual lattice defined by*

$$\mathfrak{y} = \{ b \in F \mid \mathrm{Tr}_{F/\mathbf{Q}}(b\mathfrak{r}) \subset \mathbf{Z} \}.$$

Further let

$$S(z, \mathfrak{r}; \alpha, \beta) = \sum_{a \in \mathfrak{r}} (z + a)^{-\alpha}(\bar{z} + a)^{-\beta} \qquad (z \in H^{\mathbf{a}}; \alpha, \beta \in \mathbf{C}^{\mathbf{a}}).$$

Then this is convergent and real analytic (at least) on

$$H^{\mathbf{a}} \times \{(\alpha, \beta) \in \mathbf{C}^{\mathbf{a}} \times \mathbf{C}^{\mathbf{a}} \mid \mathrm{Re}(\alpha_v) > \tfrac{1}{2}, \mathrm{Re}(\beta_v) > \tfrac{1}{2} \quad \text{for every} \quad v \in \mathbf{a}\},$$

and has a Fourier expansion

$$\mu(\mathbf{R}^{\mathbf{a}}/\mathfrak{r}) S(x + iy, \mathfrak{r}; \alpha, \beta) = \sum_{h \in \mathfrak{y}} \mathbf{e}_{\mathbf{a}}(hx) \xi(y, h; \alpha, \beta),$$

$$\xi(y, h; \alpha, \beta) = \prod_{v \in \mathbf{a}} \xi(y_v, h_v; \alpha_v, \beta_v)$$

with ξ of (4.19).

This can be proved in the same fashion as in [11, (1.32), Lemma 1.4] (see the last sentence of [11, §1]).

Proposition 5.2. *Let Y be a complete set of representatives of ρ-regular cusp-classes of Δ in the sense that every ρ-regular cusp-class is given as $\mathcal{P}\xi\Delta$ with exactly one $\xi \in Y$. Let E_ξ, for $\xi \in Y$, denote $E(z, s; \rho, \xi, \Delta)$ with fixed ρ and Δ. Then, for $(\xi, \eta) \in Y \times Y$, we have*

$$E_\xi \| \eta^{-1} = \delta_{\xi\eta} y^{su - \rho} + f_{\xi\eta}(s) y^{u - su - \bar\rho} + \sum_{0 \neq h \in \mathfrak{y}} g_{\xi\eta}(h, s, y) \mathbf{e_a}(hx),$$

where $f_{\xi\eta}$ and $g_{\xi\eta}$ are meromorphic in s, $\delta_{\xi\eta}$ is Kronecker's delta, and $\mathfrak{y}$ is a lattice in F.

PROOF. The point of our assertion is merely in the shape of the constant term. Fix one ξ and put $\Delta_\xi = \xi\Delta\xi^{-1}$. Put $r(a) = \Lambda_\sigma^k \begin{pmatrix} 1 & a \\ 0 & 1 \end{pmatrix}$ for $a \in F$. Then $r(F) \cap \Delta_\xi = r(\mathfrak{r})$ with a lattice $\mathfrak{r}$ in F. Take a subset Φ of Δ so that $1 \notin \Phi$ and $\{1\} \cup \Phi$ is a complete set of representatives of $(\mathcal{P} \cap \Delta_\xi) \backslash \Delta_\xi / r(\mathfrak{r})$. Then 1 and the elements $\varphi r(a)$ with $\varphi \in \Phi$ and $a \in \mathfrak{r}$ represent $(\mathcal{P} \cap \Delta_\xi) \backslash \Delta_\xi$ without overlap. Therefore

$$E_\xi \| \xi^{-1} = y^{su - \rho} + \sum_{\varphi \in \Phi} \sum_{a \in \mathfrak{r}} y^{su - \rho} \| \varphi r(a).$$

Fix one $\varphi \in \Phi$ and put $c = c_\varphi$, $d = d_\varphi$. Then $c \neq 0$, and

$$\sum_{a \in \mathfrak{r}} y^{su - \rho} \| \varphi r(a) = t y^{su - \rho} c^{-\alpha - \beta} \sum_{a \in \mathfrak{r}} (z + c^{-1}d + a)^{-\alpha} (\bar z + c^{-1}d + a)^{-\beta}$$

with $\alpha = su + \bar\rho$, $\beta = su - \rho$, and a constant t such that $|t| = 1$. By Lemma 5.1, this has an expansion of the form

$$t y^{su - \rho} c^{-\alpha - \beta} \mu(\mathbf{R^a}/\mathfrak{r})^{-1} \sum_{h \in \mathfrak{y}} \mathbf{e_a}(h(x + c^{-1}d)) \xi(y, h; \alpha, \beta).$$

By (4.21a), the term $h = 0$ produces a function of the form $t c^{-\alpha - \beta} p(s) y^{u - su - \bar\rho}$ with a meromorphic fuction p independent of φ. Taking the sum over all $\varphi \in \Phi$, we obtain the Fourier expansion of $E_\xi \| \xi^{-1}$ in the form stated in our proposition. The meromorphic continuation of $E_\xi \| \xi^{-1}$ implies that of $f_{\xi\xi}$ and $g_{\xi\xi}$ to the whole s-plane.

Next let $\xi \neq \eta \in Y$. Let Z be a complete set of representatives of $(\mathcal{P} \cap \Delta_\xi) \backslash \xi\Delta\eta^{-1} / [r(F) \cap \Delta_\eta]$. Then the elements $\zeta r(a)$ with $\zeta \in Z$ and $r(a) \in r(F) \cap \Delta_\eta$ represent $(\mathcal{P} \cap \Delta_\xi) \backslash \xi\Delta\eta^{-1}$ without overlap. Therefore the same argument as above establishes the Fourier expansion of $E_\xi \| \eta^{-1}$; the only difference is that $y^{su - \rho}$ doesn't appear this time.

Proposition 5.3. *Let $\Gamma = \Gamma_0(\mathfrak{c})$, $\Gamma' = \{\gamma \in \Gamma \mid d_\gamma - 1 \in \mathfrak{c}\}$, $\Delta = \Lambda_\sigma^k(\Gamma)$, and $\Delta' = \Lambda_\sigma^k(\Gamma')$. Assume (4.23). Let D, B, and $\mathfrak{a}_\beta$ be as in (4.9a) and (4.10). Then $\mathcal{P}\alpha\Delta'$ is ρ-regular for every $\alpha \in \Lambda_\sigma^k(G \cap P_\mathbf{A}D)$. Moreover, if T_β is a complete* *set of representatives of* $(P \cap \beta \Gamma \beta^{-1}) \backslash \beta \Gamma / \Gamma'$, *then*

$$E_k(z, s; \rho, \psi, c) = \sum_{\beta \in B} N(\mathfrak{a}_\beta)^{2s} \sum_{\xi \in T_\beta} \psi(d_\xi \mathfrak{a}_\beta^{-1}) \psi_\mathbf{a}(d_\xi) \cdot E(z, s; \rho, \Lambda_\sigma^k(\xi), \Delta').$$

PROOF. Put $\tilde{\alpha} = \Lambda_\sigma^k(\alpha)$ for $\alpha \in G \cap P_\mathbf{A} D$. Observe that

$$\mathcal{P} \cap \tilde{\alpha} \Delta' \tilde{\alpha}^{-1} = \Lambda_\sigma^k(P \cap \alpha \Gamma' \alpha^{-1}).$$

Then the first assertion can easily be verified. Let S_ξ be a complete set of representatives of $(P \cap \xi \Gamma' \xi^{-1}) \backslash \xi \Gamma'$. Then the S_ξ for all $\xi \in T_\beta$ form a disjoint union, which gives a complete set of representatives of $(P \cap \beta \Gamma \beta^{-1}) \backslash \beta \Gamma$. Taking this union as R_β of (4.10), we obtain our formula.

Proposition 5.4. *Let E_ψ denote the function of (4.10), and let $\zeta \in \mathcal{G}$. Then the constant term of $E_\psi \| \zeta^{-1}$ contains $y^{su-\rho}$ nontrivially if and only if* $\mathrm{pr}(\zeta) \in$ $\in G \cap P_\mathbf{A} D$ *with D of (4.9a). Moreover, the term involving $y^{su-\rho}$ has the form $ab^s y^{su-\rho}$ with $a \in \mathbf{C}$ and $0 < b \in \mathbf{R}$.*

PROOF. Let $\alpha = \mathrm{pr}(\zeta)$. By Propositions 3.3, 5.2, and 5.3, $y^{su-\rho}$ appears nontrivially in $E_\psi \| \zeta^{-1}$ only if $\alpha \in P \beta \Gamma$ for some $\beta \in B$, that is, only if $\alpha \in G \cap P_\mathbf{A} D$. Conversely, if $\alpha \in P \beta \Gamma$ with $\beta \in B$, such a β is unique, and $\alpha \in P \xi \Gamma'$ with a unique $\xi \in T_\beta$. Therefore, Propositions 5.2 and 5.3 show that $y^{su-\rho}$ appears nontrivially in $E_\psi \| \zeta^{-1}$ in the form as claimed.

Remark 5.5. To show that the exceptional case of Proposition 3.2 can happen, take $[F : \mathbf{Q}] = 2$ and set $\mathbf{a} = \{v, w\}$. Take θ so that $T_U = \mathbf{Z}\theta$. Then $\theta_v = -\theta_w$. Let $p = su - (\sigma - im\,\theta)/2$ and $r = tu - (\sigma - in\,\theta)/2$ with $s = (1 + in\,\theta_v)/2$, $t = (1 + im\,\theta_v)/2$, and $m, n \in \mathbf{Z}$. Suppose $|m| \neq |n|$. Then y^p, $y^{u-\sigma-p}$, y^r, $y^{u-\sigma-r}$ belong to the same set of eigenvalues $\{\lambda_v, \lambda_w\}$, where $4\lambda_v = (1 - \sigma_v)^2 +$ $+ (m + n)^2 \theta_v^2$ and $4\lambda_w = (1 - \sigma_w)^2 + (m - n)^2 \theta_v^2$. If we put $\rho = (\sigma - im\,\theta)/2$ and $\rho' = (\sigma - in\,\theta)/2$, then y^p, $y^{u-\sigma-p}$, y^r, and $y^{u-\sigma-r}$ can appear nontrivially in $E(z, s; \rho, \Delta)$, $E(z, 1 - s; \bar{\rho}, \Delta)$, $E(z, s; \rho', \Delta)$, and $E(z, 1 - t; \bar{\rho}', \Delta)$, respectively, for a sufficiently small Δ.

6. Bilinear relations

Let Δ be a congruence subgroup of $\mathcal{G}$ and let $\Gamma = \mathrm{pr}(\Delta)$. Take a minimal finite subset X of $\mathcal{G}$ so that $\mathcal{G} = \bigcup_{\xi \in X} \mathcal{P} \xi \Delta$. For each $\xi \in X$, let $Q_\xi = P \cap \mathrm{pr}(\xi \Delta \xi^{-1})$. We consider a group

$$(6.1) \qquad \Theta = \left\{ \begin{bmatrix} a & b \\ 0 & a^{-1} \end{bmatrix} \,\middle|\, a \in U_1, b \in \mathfrak{m} \right\}$$

with a fractional ideal $\mathfrak{m}$ of F and a subgroup U_1 of $\mathfrak{g}^{\times}$ of finite index. We choose U_1 and $\mathfrak{m}$ so that $\Lambda_\sigma^k(\Theta) \subset \xi\Delta\xi^{-1}$ for all $\xi \in X$ and $a \geqslant 0$ for every $a \in U_1$.

Let $f \in \mathfrak{A}(\sigma, \lambda, \Delta)$ and $g \in \mathfrak{A}(\sigma, \lambda', \Delta)$. Assuming that both λ and λ' are noncritical, put, for each $\xi \in X$,

$$(6.2a) \qquad f \| \xi^{-1} = \sum_p a_{p,\xi} y^p + \ldots,$$

$$(6.2b) \qquad g \| \xi^{-1} = \sum_q b_{q,\xi} y^q + \ldots$$

with constants $a_{p,\xi}$, $b_{q,\xi}$ and $p \in \mathbf{C}^{\mathbf{a}}$ as given in Proposition 3.1, where . . . indicates the nonconstant terms of the Fourier expansions. If both λ and λ' are critical, we put

$$(6.3a) \qquad f \| \xi^{-1} = a_\xi y^p + a'_\xi y^p \log y^u + \ldots,$$

$$(6.3b) \qquad g \| \xi^{-1} = b_\xi y^p + b'_\xi y^p \log y^u + \ldots$$

with $p = (u - \sigma)/2$.

Theorem 6.1. *Suppose* $\lambda' = \bar{\lambda}$ *and* λ *is noncritical and simple. Fix one exponent* p *attached to* λ *and put* $q = u - \sigma - p$. *Then*

$$\sum_{\xi \in X} v_\xi(\bar{a}_{p,\xi} b_{\bar{q},\xi} - \bar{a}_{q,\xi} b_{\bar{p},\xi}) = 0,$$

where $v_\xi = [Q_\xi\{\pm 1\} : \Theta\{\pm 1\}]^{-1}$. *If* $\lambda = \lambda'$ *and* λ *is critical, one has*

$$\sum_{\xi \in X} v_\xi(\bar{a}_\xi b'_\xi - \bar{a}'_\xi b_\xi) = 0.$$

Proof. For $0 < r \in \mathbf{R}$, put

$$T_r = \{z \in H^{\mathbf{a}} \mid y^u > r\}, \qquad M_r = \{z \in H^{\mathbf{a}} \mid y^u = r\}.$$

We can find an r such that the sets $\xi^{-1}(Q_\xi \backslash T_r)$ can be embedded into $\Gamma \backslash H^{\mathbf{a}}$ without overlap. For each ξ, take a positive number $r(\xi) > r$. Also take a union J of small neighborhoods of elliptic fixed points on $\Gamma \backslash H^{\mathbf{a}}$. Let K be the complement of $\bigcup_{\xi \in X} \xi^{-1}(Q_\xi \backslash T_{r(\xi)}) \cup J$ in $\Gamma \backslash H^{\mathbf{a}}$. Then K is a compact manifold with boundary, and

$$\partial K = \sum_{\xi \in X} \xi^{-1}(Q_\xi \backslash M_{r(\xi)}) - \partial J.$$

Let φ be a Γ-invariant C^∞-form on $H^{\mathbf{a}}$ of codegree 1. Then

$$(6.4) \qquad \int_K d\varphi = \int_{\partial K} \varphi = \sum_{\xi \in X} v_\xi \int_{B_\xi} \varphi \cdot \xi^{-1} - \int_{\partial J} \varphi,$$

where $B_\xi = \Theta_\xi \backslash M_{r(\xi)}$ (with a natural orientation). We fix one $v \in \mathbf{a}$ and put

$$\omega = y^{-2u} \prod_{v \in \mathbf{a}} dx_v \wedge dy_v,$$

$$\zeta_v = (i/2)y^{-2u} d\bar{z}_v \wedge \prod_{w \neq v} dx_w \wedge dy_w,$$

and $\varphi = \bar{f}hy^\sigma \zeta_v$ with two C^∞-functions f and h on $H^{\mathbf{a}}$ satisfying (2.7a) with Δ and Δ_* (of Proposition 2.2), respectively. Then it is easy to see that

$$d\varphi = \bar{f}\delta_v^\tau hy^\sigma \omega - \overline{\epsilon_v f} \cdot hy^\tau \omega \qquad (\tau = \sigma - 2v).$$

Applying (6.4) to this and taking the limit when $r \to \infty$, we find

(6.5) $$\langle f, \delta_v^\tau h \rangle = \langle \epsilon_v f, h \rangle$$

provided that these inner products are meaningful, and that f or h is rapidly decreasing in the sense that the inequality of (2.7c) holds for *every* $c \in \mathbf{R}$. This proves (2.25). Now take $h = \epsilon_v g$ with g of weight σ. Then

$$d\varphi = \tfrac{1}{4}\bar{f}L_v^\sigma g \cdot y^\sigma \cdot \omega - \overline{\epsilon_v f} \cdot \epsilon_v g \cdot y^\tau \omega.$$

Putting similarly $\varphi' = \bar{g}\epsilon_v f \cdot y^\sigma \zeta_v$, we find that

$$d\varphi - \overline{d\varphi'} = \tfrac{1}{4}(\bar{f}L_v^\sigma g - \overline{L_v^\sigma f} \cdot g)y^\sigma \omega.$$

Applying (6.4) to this form, we obtain

$$\tfrac{1}{4}\int_K (\bar{f}L_v^\sigma g - \overline{L_v^\sigma f} \cdot g)y^\sigma \omega = \sum_{\xi \in X} v_\xi \int_{B_\xi} (\varphi - \bar{\varphi}') \circ \xi^{-1} - \int_{\partial J} (\varphi - \bar{\varphi}').$$

We now assume that f and g are eigenfunctions with expansions as in (6.2a, b). Then

$$\varphi \circ \xi^{-1} = -\tfrac{i}{2}\sum_{p,q} q_v \bar{a}_{p,\xi} b_{q,\xi} y^{\bar{p}+q+v+\sigma}\zeta_v + \dots,$$

$$\bar{\varphi}' \circ \xi^{-1} = \tfrac{i}{2}\sum_{p,q} \bar{p}_v \bar{a}_{p,\xi} b_{q,\xi} y^{\bar{p}+q+v+\sigma}\bar{\zeta}_v + \dots.$$

Here the unwritten terms contain some contributions to the «constant terms» of the Fourier expansions, but they tend to zero in our later limit process. (This can easily be shown by virtue of (2) of Proposition 2.1.) Put $U = \{a^2 \mid a \in U_1\}$. Then $\Theta \backslash M_r$ may be viewed as the product of $\mathbf{R}^{\mathbf{a}}/\mathfrak{m}$ and $\{y \in \mathbf{R}^{\mathbf{a}} \mid y \gg 0, y^u = r\}/U$. Then we can easily prove

Lemma 6.2. *Let* $t = su + i\tau \in \mathbf{C}^{\mathbf{a}}$ *with* $s \in \mathbf{C}$ *and* τ *in the set* T_U *of* (3.1). *Then*

$$\int_{\Theta \backslash M_r} y^{t+v}\zeta_v = (-i/2)\mu(\mathbf{R}^{\mathbf{a}}/\mathfrak{m})R_U r^{s-1},$$

667

where R_U is the regulator of U defined by $R_U = R_F[\mathfrak{g}^\times : U\{\pm\}]$ with the regulator R_F of F.

Applying this to the first terms of $(\varphi - \bar{\varphi}') \circ \xi^{-1}$, we find that

$$(\lambda'_v - \bar{\lambda}_v)\int_K \bar{f}gy^\sigma\omega + 4\int_{\partial_J}(\varphi - \bar{\varphi}')$$

$$= \mu(\mathbf{R}^\mathbf{a}/\mathrm{m})R_U \sum_{\xi\in X} \nu_\xi \sum_{p,q}(\bar{p}_v - q_v)\bar{a}_{p,\xi}b_{q,\xi}r_\xi^{e(p,q)} + \ldots$$

where $e(p,q) = \sum_{w\in\mathbf{a}}(\bar{p}_w + q_w + \sigma_w - 1)/[F:\mathbf{Q}]$. Suppose that $\lambda' = \bar{\lambda}$ and λ is simple. Then, with one exponent p fixed, the sum $\sum_{p,q}$ can be written as

$$(2\bar{p}_v - 1 + \sigma_v)(\bar{a}_{p,\xi}b_{u-\sigma-\bar{p},\xi} - \bar{a}_{u-\sigma-p,\xi}b_{\bar{p},\xi}).$$

Since λ is not critical, $2\bar{p}_v - 1 + \sigma_v \neq 0$ for at least one v. Therefore, taking the limit when $J \to \varnothing$ and $r_\xi \to \infty$, we obtain the first assertion of Theorem 6.1. The second one can be proved in a similar way.

If λ is not simple, $e(p,q)$ can be a pure imaginary number which is not necessarily equal to 0. Therefore we obtain certain linear relations even for multiple λ, whose nature is somewhat different from that for simple λ.

7. Construction of $\mathfrak{N}(\sigma, \lambda, \Delta)$ by Eisenstein series

We are going to show that the space $\mathfrak{N}(\sigma, \lambda, \Delta)$ of (2.23a) is generated by the series of type (4.1), their derivatives, and their residues. Given σ and λ, we are interested in the case where $\mathcal{A}(\sigma, \lambda, \Delta) \neq \mathcal{S}(\sigma, \lambda, \Delta)$, that is, the case in which nontrivial constant terms appear. Then λ must be given as in Proposition 3.1. We assume throughout that λ is simple. Then

$$(7.1) \qquad \lambda_v = (s_0 - \rho_v)(1 - s_0 - \bar{\rho}_v), \qquad \rho = (\sigma - i\tau)/2$$

with $s_0 \in \mathbf{C}$ and an admissible $\tau \in \mathbf{R}^\mathbf{a}$. Notice that (s_0, ρ) may be changed for $(1 - s_0, \bar{\rho})$ without changing (σ, λ). Notice also that (7.1) includes critical λ as a special case. In fact λ is critical if and only if $s_0 = \frac{1}{2}$ and $\tau = 0$; then $\rho = \frac{\sigma}{2}$. This is so if and only if $s_0u - \rho = u - s_0u - \bar{\rho}$.

In this section, we fix a complete set of representatives X of $\mathcal{P}\backslash\mathcal{G}/\Delta$, and also a subset Y of X that represents all ρ-regular cusp-classes of Δ; we then denote by $\varkappa$ the number of elements of Y. Further we let $\mathcal{E}[\rho, \Delta]$ denote the complex vector space spanned by the functions $E(z, s; \rho, \xi, \Delta)$ for all $\xi \in Y$. For a complex number s_0, we denote by $\mathcal{E}[s_0, \rho, \Delta]$ the subspace of $\mathcal{E}[\rho, \Delta]$ consisting of all functions $g(z, s)$ that are finite at s_0, and by $\mathcal{E}(s_0, \rho, \Delta)$ the vector space consisting of $g(z, s_0)$ for all $g \in \mathcal{E}[s_0, \rho, \Delta]$. Similarly we denote by $\mathcal{E}^*[s_0, \rho, \Delta]$ the set of elements of $\mathcal{E}[\rho, \Delta]$ that have at most a simple pole at s_0 and by $\mathcal{E}^*(s_0, \rho, \Delta)$ the residues at s_0 of all elements of $\mathcal{E}^*[s_0, \rho, \Delta]$.

Proposition 7.1. *Both $\mathcal{E}(s_0, \rho, \Delta)$ and $\mathcal{E}^*(s_0, \rho, \Delta)$ are contained in $\mathfrak{N}(\sigma, \lambda, \Delta)$ with λ of* (7.1).

PROOF. The spaces in question are contained in $\mathcal{Q}(\sigma, \lambda, \Delta)$ by virtue of Proposition 4.4. To prove that they are orthogonal to cusp forms, take a congruence subgroup $\Delta' \subset \Delta$ so that

$$\mathcal{P} \cap \Delta' = \left\{ \Lambda^k_\sigma\!\left(\begin{pmatrix} a & b \\ 0 & a^{-1} \end{pmatrix}\right) \,\middle|\, a \in U_1, b \in \mathfrak{b} \right\}$$

with an ideal $\mathfrak{b}$ and a subgroup U_1 of $\mathfrak{g}^\times$ of finite index consisting of totally positive units. Then $(\mathcal{P} \cap \Delta')\backslash H^\mathbf{a}$ can be represented by $B \times A$, with $B = \mathbf{R}^\mathbf{a}/\mathfrak{b}$ and $A = \{ y \in \mathbf{R}^\mathbf{a} \,|\, y \gg 0 \}/U$ where $U = \{ a^2 \,|\, a \in U_1 \}$. We now consider an integral

$$\int_A \int_B \overline{f(x + iy)}\, dx \cdot y^{(s-2)u + \bar\rho}\, dy$$

for $f \in \mathcal{S}(\sigma, \lambda', \Delta)$ with any fixed λ'. Since the constant term of f is 0, this is obviously 0. If $\mathrm{Re}(s)$ is sufficiently large and $\Phi = \Delta'\backslash H^\mathbf{a}$, the integral can be transformed to

$$\int_\Phi \left\{ \sum_{\gamma \in (\mathcal{P} \cap \Delta')\backslash \Delta'} (y^{su + \bar\rho}\bar{f}) \circ \gamma \right\} d\mu(z) = \mu(\Phi)\langle f, E(z, s; \rho, \Delta')\rangle.$$

Therefore $\langle f, E(z, s; \rho, \Delta')\rangle = 0$ for sufficiently large $\mathrm{Re}(s)$. The same holds with Δ instead of Δ', by virtue of (4.3). Then the desired orthogonality can easily be shown by analytic continuation.

Proposition 7.2. (i) dim $\mathcal{E}[\rho, \Delta] = \varkappa$.

(ii) *The map* $g(z, s) \rightarrow g(z, s_0)$ *gives an isomorphism of* $\mathcal{E}[s_0, \rho, \Delta]$ *onto* $\mathcal{E}(s_0, \rho, \Delta)$ *provided that* λ *of* (7.1) *is noncritical.*

PROOF. Assertion (i) follows immediately from Proposition 5.2. Let $g = \sum_{\xi \in Y} a_\xi E(z, s; \rho, \xi, \Delta) \in \mathcal{E}[s_0, \rho, \Delta]$. Then we have

$$(7.2) \qquad g(z, s_0)\|\eta^{-1} = a_\eta y^{s_0 u - \rho} + \left(\sum_{\xi \in Y} a_\xi f_{\xi\eta}\right)(s_0)y^{u - s_0 u - \bar\rho} + \dots$$

for every $\eta \in Y$. If λ is noncritical, we have $s_0 u - \rho \neq u - s_0 u - \bar\rho$, and therefore, if $g(z, s_0) = 0$, we have $a_\eta = 0$ for all $\eta \in Y$, so that $g = 0$. This proves (ii).

Theorem 7.3. *With* λ, s_0, *and* ρ *as in* (7.1), *suppose* λ *is noncritical and simple; suppose also that* $\mathcal{E}[\rho, \Delta] = \mathcal{E}[s_0, \rho, \Delta]$ *and* $\mathcal{E}[\bar\rho, \Delta] = \mathcal{E}[\bar{s}_0, \bar\rho, \Delta]$. *Then* $\mathfrak{N}(\sigma, \lambda, \Delta) = \mathcal{E}(s_0, \rho, \Delta)$, *and* $\mathcal{Q}(\sigma, \lambda, \Delta)$ *is the direct sum of* $\mathcal{S}(\sigma, \lambda, \Delta)$ *and* $\mathcal{E}(s_0, \rho, \Delta)$.

669

PROOF. Put $p = s_0 u - \rho$ and $q = u - \sigma - p$. Let Y' be the set of all $\xi \in X$ such that $\mathcal{P}\xi\Delta$ is $\bar{\rho}$-regular, and $\varkappa'$ the number of elements of Y'. Given $f \in \mathcal{Q}(\sigma, \lambda, \Delta)$ and $g \in \mathcal{Q}(\sigma, \bar{\lambda}, \Delta)$, we consider expansions

$$(7.3a) \qquad\qquad f\|\xi^{-1} = a_\xi y^p + a'_\xi y^q + \ldots$$

$$(7.3b) \qquad\qquad g\|\xi^{-1} = b_\xi y^{\bar{p}} + b'_\xi y^{\bar{q}} + \ldots$$

for each $\xi \in X$. By Proposition 3.3 and Theorem 6.1, we have

$$(7.4) \qquad\qquad \sum_{\xi \in Y} v_\xi \bar{a}_\xi b'_\xi - \sum_{\xi \in Y'} v_\xi \bar{a}'_\xi b_\xi = 0.$$

Moreover, the map

$$(7.5) \qquad\qquad f \mapsto ((a_\xi)_{\xi \in Y}, (a'_\xi)_{\xi \in Y'})$$

gives an injection of $\mathcal{Q}(\sigma, \lambda, \Delta)/\mathcal{S}(\sigma, \lambda, \Delta)$ into $\mathbf{C}^\mu$ with $\mu = \varkappa + \varkappa'$; a similar statement holds with g and $\bar{\lambda}$ instead of f and λ. By Proposition 7.2 and our assumption, $\mathcal{E}(s_0, \rho, \Delta)$ is $\varkappa$-dimensional, and $\mathcal{E}(\bar{s}_0, \bar{\rho}, \Delta)$ is $\varkappa'$-dimensional. Each g in the latter space produces a linear relation of type (7.4), and hence the image of the map of (7.5) is at most $\varkappa$-dimensional. This combined with (2.23a) completes the proof.

Remark 7.4. (1) If $\mathrm{Re}(s_0) \geq \frac{1}{2}$, then, by Theorem 4.2, $\mathcal{E}[\rho, \Delta] = \mathcal{E}[s_0, \rho, \Delta]$ except when $s_0 = 1$ and $\rho = 0$ in Case I and $s_0 = \frac{3}{4}$, $\tau = 0$, and σ is as in Theorem 4.1 in Case II. If $\mathcal{E}[\rho, \Delta] = \mathcal{E}[s_0, \rho, \Delta]$ and $\mathrm{Re}(s_0) \geq \frac{1}{2}$, then we have automatically $\mathcal{E}[\bar{\rho}, \Delta] = \mathcal{E}[\bar{s}_0, \bar{\rho}, \Delta]$, since $\bar{s}_0 = s_0$ and $\bar{\rho} = \rho$ in those exceptional cases.

(2) The pair (σ, λ) corresponds to (s_0, ρ) and $(1 - s_0, \bar{\rho})$. Therefore, changing (s_0, ρ) for $(1 - s_0, \bar{\rho})$ if necessary, we can take s_0 such that $\mathrm{Re}(s_0) \geq \frac{1}{2}$ without changing λ.

Proposition 7.5. *The number of ρ-regular cusp-classes of Δ is equal to the number of $\bar{\rho}$-regular cusp-classes of Δ.*

PROOF. Given ρ and Δ, we can find s_0 so that $\mathcal{E}[\rho, \Delta] = \mathcal{E}[s_0, \rho, \Delta] = \mathcal{E}[1 - \bar{s}_0, \rho, \Delta]$, $\mathcal{E}[\bar{\rho}, \Delta] = \mathcal{E}[\bar{s}_0, \bar{\rho}, \Delta] = \mathcal{E}[1 - s_0, \bar{\rho}, \Delta]$, and λ of (7.1) is noncritical and simple (cf. Proposition 3.2). Then we have $\mathfrak{N}(\sigma, \lambda, \Delta) = \mathcal{E}(s_0, \rho, \Delta) = \mathcal{E}(1 - s_0, \bar{\rho}, \Delta)$, which proves our proposition.

Proposition 7.6 *Suppose $\mathfrak{N}(\sigma, \lambda, \Delta) = \mathcal{E}(s_0, \rho, \Delta)$, and λ is noncritical and simple. For $f \in \mathcal{Q}(\sigma, \lambda, \Delta)$ and $\xi \in Y$, put*

$$f\|\xi^{-1} = a_\xi y^p + a'_\xi y^q + \ldots$$

with $p = s_0 u - \rho$ and $q = u - \sigma - p$. If $a_\xi = 0$ for all $\xi \in Y$, then f is a cusp form.

PROOF. Let $f = g(z, s_0) + h$ with $g \in \mathcal{E}[s_0, \rho, \Delta]$ and $h \in \mathcal{S}(\sigma, \lambda, \Delta)$. Writing g as in the proof of Proposition 7.2, we see that the assumption $a_\xi = 0$ implies that $g = 0$.

Theorem 7.7. *Suppose every ρ-regular cusp-class of Δ is also $\bar\rho$-regular. Define a $\mathbf{C}^Y$-valued meromorphic function $\mathbf{E}_\Delta(z, s, \rho)$ by*

$$(7.6) \qquad \mathbf{E}_\Delta(z, s, \rho) = (E(z, s; \rho, \xi, \Delta))_{\xi \in Y}.$$

Then there exists an $\mathrm{End}(\mathbf{C}^Y)$-valued meromorphic function $\Phi_\Delta(s, \rho)$ on $\mathbf{C}$ such that

$$(7.7a) \qquad \mathbf{E}_\Delta(z, s, \rho) = \Phi_\Delta(s, \rho)\mathbf{E}_\Delta(z, 1 - s, \bar\rho),$$

$$(7.7b) \qquad \Phi_\Delta(1 - s, \bar\rho)\Phi_\Delta(s, \rho) = 1.$$

Moreover, there is a diagonal matrix A, depending only on Δ and Y, whose diagonal entries are positive integers such that

$$(7.7c) \qquad \overline{\Phi_\Delta(s, \rho)A} \cdot {}^t\Phi_\Delta(1 - \bar s, \rho) = A.$$

PROOF. Put $\lambda_v(s) = (s - \rho_v)(1 - s - \bar\rho_v)$, $p = su - \rho$, and $q = u - \sigma - p$. Suppressing the symbols z and Δ for simplicity, we have

$$(7.8a) \qquad E(s, \rho, \xi)\|\eta^{-1} = \delta_{\xi\eta}y^p + f_{\xi\eta}(s)y^q + \dots,$$

$$(7.8b) \qquad E(1 - s, \bar\rho, \xi)\|\eta^{-1} = \delta_{\xi\eta}y^q + g_{\xi\eta}(1 - s)y^p + \dots \qquad (\xi, \eta \in Y)$$

with meromorphic $f_{\xi\eta}$ and $g_{\xi\eta}$. Then, for every $\zeta \in Y$, we have

$$(7.9) \qquad \left\{ E(1 - s, \bar\rho, \xi) - \sum_{\eta \in Y} g_{\xi\eta}(1 - s)E(s, \rho, \eta) \right\} \Big\| \zeta^{-1} =$$

$$= 0y^p + \left\{ \delta_{\xi\zeta} - \sum_{\eta \in Y} g_{\xi\eta}(1 - s)f_{\eta\zeta}(s) \right\}y^q + \dots.$$

Now we can find a nonempty open subset W of $\mathbf{C}$ such that $\mathcal{E}[\rho, \Delta] = \mathcal{E}[s, \rho, \Delta] = \mathcal{E}[1 - \bar s, \rho, \Delta]$, $\mathcal{E}[\bar\rho, \Delta] = \mathcal{E}[\bar s, \bar\rho, \Delta] = \mathcal{E}[1 - s, \bar\rho, \Delta]$, and that $\lambda(s)$ is noncritical and simple for every $s \in W$. (As to simple λ, see Proposition 3.2.) Now the left-hand side of (7.9) without $\| \zeta^{-1}$ belongs to $\mathfrak{N}(\sigma, \lambda(s), \Delta)$ for $s \in W$. By Proposition 7.6, we have

$$E(1 - s, \bar\rho, \xi) = \sum_{\eta \in Y} g_{\xi\eta}(1 - s)E(s, \rho, \eta),$$

$$\delta_{\xi\zeta} = \sum_{\eta} g_{\xi\eta}(1 - s)f_{\eta\zeta}(s).$$

Writing $\Phi(s, \rho)$ for the matrix $(f_{\xi\eta}(s))$, we obtain (7.7a, b). Now

$$E(1 - \bar{s}, \rho, \xi) \,\|\, \eta^{-1} = \delta_{\xi\eta} y^{\bar{q}} + f_{\xi\eta}(1 - \bar{s})y^{\bar{p}} + \ldots,$$

and $E(1 - \bar{s}, \rho, \xi)$ belongs to $\mathfrak{a}(\sigma, \bar{\lambda}, \Delta)$ for $s \in W$. By (7.4), we have

$$\sum_{\eta \in Y} \nu_\eta [\delta_{\xi\eta}\delta_{\zeta\eta} - \overline{f_{\xi\eta}(s)}f_{\zeta\eta}(1 - \bar{s})] = 0.$$

Denoting by D the diagonal matrix whose diagonal elements are ν_η, we obtain $D = \overline{\Phi(s, \rho)}D \cdot {}^t\Phi(1 - \bar{s}, \rho)$, which proves (7.7c).

By Remark 7.4, (1), $E(z, s; \rho, \xi, \Delta)$ is finite at $s = \frac{1}{2}$ and hence $\Phi_\Delta(s, \rho)$ is finite at $s = \frac{1}{2}$. Moreover, we have $\Phi_\Delta\left(\frac{1}{2}, \rho\right)^2 = 1$ if $\tau = 0$.

Theorem 7.8. *Suppose λ is critical $\left(\text{and hence } \rho = \frac{\sigma}{2}\right)$. Let $\mathcal{E}'\left(\frac{1}{2}, \rho, \Delta\right)$ denote the space spanned by $(\partial g/\partial s)\left(z, \frac{1}{2}\right)$ for all $g \in \mathcal{E}[\rho, \Delta]$, and $\mathcal{E}'_0\left(\frac{1}{2}, \rho, \Delta\right)$ the subspace of $\mathcal{E}'\left(\frac{1}{2}, \rho, \Delta\right)$ consisting of $(\partial g/\partial s)\left(z, \frac{1}{2}\right)$ for all such g satisfying $g\left(z, \frac{1}{2}\right) = 0$. Further let ν_+ (resp. ν_-) the multiplicity of 1 (resp. -1) in the eigenvalues of $\Phi\left(\frac{1}{2}, \rho\right)$. Then $\varkappa = \nu_+ + \nu_-$, $\dim \mathcal{E}\left(\frac{1}{2}, \rho, \Delta\right) = \nu_+$, $\dim \mathcal{E}'_0\left(\frac{1}{2}, \rho, \Delta\right) = \nu_-$, and*

$$\mathcal{E}'\left(\tfrac{1}{2}, \rho, \Delta\right) \subset \mathfrak{N}(\sigma, \lambda, \Delta) = \mathcal{E}\left(\tfrac{1}{2}, \rho, \Delta\right) \oplus \mathcal{E}'_0\left(\tfrac{1}{2}, \rho, \Delta\right).$$

Moreover, $\mathcal{E}\left(\frac{1}{2}, \rho, \Delta\right)$ consists of the elements of $\mathfrak{N}(\sigma, \lambda, \Delta)$ that do not involve $y^{(u - \sigma)/2} \log y^u$.

PROOF. For simplicity, let us suppress the symbols ρ and Δ occasionally. That an element of $\mathcal{E}'\left(\frac{1}{2}\right)$ satisfies (2.7a, b) can be verified immediately. That it satisfies (2.7c) is shown in the proof of Proposition 4.4 in Section 11, and hence $\mathcal{E}'\left(\frac{1}{2}\right) \subset \mathfrak{a}(\sigma, \lambda)$. The orthogonality with cusp forms can also be seen, because the integral expressing $\langle f, g(z, s) \rangle$ is uniformly convergent in a neighborhood of $s = \frac{1}{2}$ for every fixed cusp form f. Thus $\mathcal{E}'\left(\frac{1}{2}\right) \subset \mathfrak{N}(\sigma, \lambda)$. Put $p = (u - \sigma)/2$. From (7.8a) we obtain

$$E\left(\tfrac{1}{2}, \rho, \xi\right) \,\|\, \eta^{-1} = \left[\delta_{\xi\eta} + f_{\xi\eta}\left(\tfrac{1}{2}\right)\right]y^p + \ldots,$$

$$(\partial E/\partial s)\left(\tfrac{1}{2}, \rho, \xi\right) \,\|\, \eta^{-1} = \left[\delta_{\xi\eta} - f_{\xi\eta}\left(\tfrac{1}{2}\right)\right]y^p \log y^u + (df_{\xi\eta}/ds)\left(\tfrac{1}{2}\right)y^p + \ldots.$$

For $g(z, s) = \sum_\xi c_\xi E(s, \rho, \xi)$ with $c = (c_\xi)_{\xi \in Y} \in \mathbf{C}^Y$, we have $g\left(z, \frac{1}{2}\right) = 0$ if and only if ${}^t\Phi\left(\frac{1}{2}\right)c = -c$. Hence $\dim \mathcal{E}\left(\frac{1}{2}\right) = \nu_+$. If ${}^t\Phi\left(\frac{1}{2}\right)c = -c$, we have $(\partial g/\partial s)\left(z, \frac{1}{2}\right) \,\|\, \eta^{-1} = 2c_\eta y^p \log y^u + \ldots$, which shows that $\dim \mathcal{E}'_0\left(\frac{1}{2}\right) = \nu_-$. Since no element of $\mathcal{E}\left(\frac{1}{2}\right)$ involves $y^p \log y^u$, we see that $\mathcal{E}\left(\frac{1}{2}\right)$ and $\mathcal{E}'_0\left(\frac{1}{2}\right)$ form a direct sum of dimension $\varkappa$. Now Theorem 6.1 shows that $\mathfrak{N}(\sigma, \lambda, \Delta)$ has dimension $\leqslant \varkappa$. Therefore we obtain all the remaining assertions.

Theorem 7.9. *With* λ, s_0, *and* ρ *as in* (7.1), *suppose that* λ *is real, noncritical, and simple. Suppose* $\mathcal{E}[\rho, \Delta] = \mathcal{E}^*[s_0, \rho, \Delta]$ *and a cusp-class of* Δ *is* ρ-*regular if and only if it is* $\bar{\rho}$-*regular. Then* $\mathfrak{N}(\sigma, \lambda, \Delta)$ *has dimension* $\varkappa$, *and is the direct sum of* $\mathcal{E}(s_0, \rho, \Delta)$ *and* $\mathcal{E}^*(s_0, \rho, \Delta)$.

PROOF. Define $R: \mathcal{E}[\rho, \Delta] \to \mathcal{E}^*(s_0, \rho, \Delta)$ by $R(g) = \mathrm{Res}_{s_0} g(z, s)$. Then $\mathcal{E}[s_0, \rho, \Delta] = \mathrm{Ker}(R)$, so that, by Proposition 7.2,

$$\dim \mathcal{E}(s_0, \rho, \Delta) + \dim \mathcal{E}^*(s_0, \rho, \Delta) = \varkappa.$$

Let $h \in \mathcal{E}(s_0, \rho, \Delta) \cap \mathcal{E}^*(s_0, \rho, \Delta)$. Then $h(z) = r(z, s_0) = R(g)$ with $r \in \mathcal{E}[s_0, \rho, \Delta]$ and $g \in \mathcal{E}[\rho, \Delta]$. Put

$$(7.10) \qquad r = \sum_{\xi \in Y} a_\xi E(z, s; \rho, \xi, \Delta), \qquad g = \sum_{\xi \in Y} b_\xi E(z, s; \rho, \xi, \Delta)$$

with $a_\xi, b_\xi \in \mathbf{C}$. Then, for $\eta \in Y$, we have

$$h \parallel \eta^{-1} = a_\eta y^p + \left(\sum_\xi a_\xi f_{\xi\eta} \right)(s_0) y^q + \ldots$$

$$= 0 y^p + \left(\sum_\xi b_\xi \mathrm{Res}_{s_0} f_{\xi\eta} \right) y^q + \ldots,$$

where $p = s_0 u - \rho$ and $q = u - \sigma - p$. Hence $a_\eta = 0$ for all η, so that $h = 0$. Thus $\mathcal{E}(s_0, \rho, \Delta)$ and $\mathcal{E}^*(s_0, \rho, \Delta)$ form a direct sum of dimension $\varkappa$. Consider again the map of (7.5) of $\mathcal{Q}(\sigma, \lambda, \Delta)/\mathcal{S}(\sigma, \lambda, \Delta)$ into $\mathbf{C}^{2\varkappa}$. Now the relation of Theorem 6.1 shows that the image of the map has dimension at most $\varkappa$. This completes the proof.

Remark 7.10. Given σ and λ, we can take s_0 and τ so that $\mathrm{Re}(s_0) \geqslant \frac{1}{2}$ and (7.1) is satisfied. By Theorem 4.2, we have $\mathcal{E}[\rho, \Delta] = \mathcal{E}^*[s_0, \rho, \Delta]$ if $\mathrm{Re}(s_0) \geqslant \frac{1}{2}$; moreover, the pole occurs only when $s_0 = 1$ or $= \frac{3}{4}$, and $\rho = \bar{\rho}$. Theorem 7.9 is applicable to such cases.

Remark 7.11. If $\rho = 0$ and $s_0 = 1$, we see that $\mathcal{E}^*(1, 0, \Delta)$ consists of the constants, as shown in Proposition 4.3. Therefore we obtain

$$(7.11) \qquad \dim \mathcal{E}(s_0, \rho, \Delta) = \varkappa - 1 \quad \text{if} \quad \rho = 0 \quad \text{and} \quad s_0 = 1.$$

Combining Theorems 7.3, 7.8, 7.9 and Remarks 7.4, 7.10, we obtain

Theorem 7.12. *If* λ *is simple,* $\mathfrak{N}(\sigma, \lambda, \Delta)$ *has dimension* $\varkappa$.

In this section, we treated $\mathfrak{N}(\sigma, \lambda, \Delta)$ only for simple λ. If λ is multiple, $\mathfrak{N}(\sigma, \lambda, \Delta)$ is probably generated by Eisenstein series with several different (s_0, ρ), as Remark 5.5 suggests. The proof of this fact does not seem very difficult, though the author has no complete result.

8. Applications to holomorphic forms

Let $\mathcal{H}(\sigma, \Delta)$ denote the set of all holomorphic functions on $H^{\mathbf{a}}$ satisfying (2.7*a, c*), and $\mathcal{H}(\sigma)$ the union of $\mathcal{H}(\sigma, \Delta)$ for all congruence sugroups Δ of $\mathcal{G}_\sigma$. (It is well known that (2.7*c*) follows from (2.7*a*) and the holomorphy if $F \neq \mathbf{Q}$.) If $f \in \mathcal{H}(\sigma)$, it has an expansion

$$(8.1) \qquad f(z) = b_0 + \sum_{0 \prec h \in \mathfrak{m}} b_h \mathbf{e_a}(hz)$$

with a lattice $\mathfrak{m}$ in F and complex coefficients b_0 and b_h. Given a subfield K of $\mathbf{C}$, we denote by $\mathcal{H}(\sigma, K)$ and $\mathcal{H}(\sigma, \Delta, K)$ the subsets of $\mathcal{H}(\sigma)$ and $\mathcal{H}(\sigma, \Delta)$ consisting of all f such that the coefficients b_0 and b_h belong to K. We shall be especially interested in the case where K is the maximal abelian extension of $\mathbf{Q}$ which we denote by $\mathbf{Q}_{ab}$.

Proposition 8.1.

   (i) $\mathcal{S}(\sigma, 0, \Delta) \subset \mathcal{H}(\sigma, \Delta) \subset \mathcal{Q}(\sigma, 0, \Delta)$;

   (ii) $\mathcal{S}(\sigma, 0, \Delta) = \mathcal{H}(\sigma, \Delta)$ *if* $\sigma \notin \mathbf{Q}u$.

PROOF. Assertion (i) is a restatement of Proposition 2.5. If $f \in \mathcal{H}(\sigma, \Delta)$, $\xi \in \mathcal{G}$, and c_0 is the constant term of $f \,\|\, \xi$, then (2.13*b*) shows that $c_0 = a^\sigma c_0$ for every a in a subgroup of $\mathfrak{g}^\times$ of finite index. Therefore $c_0 = 0$ if $\sigma \notin \mathbf{Q}u$, which proves (ii).

In order to study the holomorphic elements of $\mathfrak{N}(\sigma, 0, \Delta)$, put

$$(8.2) \qquad \mathfrak{N}\mathcal{H}(\sigma, \Delta) = \mathcal{H}(\sigma, \Delta) \cap \mathfrak{N}(\sigma, 0, \Delta).$$

From (2.23*a*) and the above (i), we obtain

$$(8.3) \qquad \mathcal{H}(\sigma, \Delta) = \mathcal{S}(\sigma, 0, \Delta) \oplus \mathfrak{N}\mathcal{H}(\sigma, \Delta).$$

The main purpose of this section is to show that $\mathfrak{N}\mathcal{H}(\sigma, \Delta)$ can be obtained from Eisenstein series. By (ii) of the above proposition, the problem concerns only the case $\sigma \in \mathbf{Q}u$.

Proposition 8.2 *Let* $E(z, s)$ *denote any series of type* (4.1), (4.4), *or* (4.10) *with* $2\rho = \sigma = tu$, $0 < t \in \left(\frac{1}{2}\right)\mathbf{Z}$. *Suppose* $k = 2t$ *in Case II. Then the following assertions hold*:

   (i). *E is finite at* $s = t/2$.

   (ii) *If* $t > 2$ *or* $t = 1$, *$E(z, t/2)$ belongs to* $\mathcal{H}(tu, \mathbf{Q}_{ab})$.

   (iii) *Suppose* $t = 2$ *or* $t = 3/2$; *suppose also* $F \neq \mathbf{Q}$. *Then* $E(z, t/2)$ *belongs to* $\mathcal{H}(tu, \mathbf{Q}_{ab})$.

(iv) *Suppose $F = \mathbf{Q}$ and $t > 1$. Then $E_k(z, t/2; tu/2, \psi, c)$ belongs to $\mathcal{H}(tu, \mathbf{Q}_{ab})$ except in the following two cases; (A) $t = 2$ and $\psi = 1$; (B) $t = 3/2$ and $\psi^2 = 1$.*

(v) *Suppose $t = 1/2$. Then $E(z, s)$ has at most a simple pole at $s = 3/4$ and the residue is $\pi^{-n}R_F$ times an element of $\mathcal{H}(tu, \mathbf{Q}_{ab})$, where R_F is the regulator of F.*

PROOF. The assertions in Case II are included in [13, Theorem 2.3]. In Case I, the results are essentially due to Hecke [1] when $F = \mathbf{Q}$, and to Kloosterman [2] and Klingen in the case $[F: \mathbf{Q}] > 1$, though our formulation is different from theirs. In the present formulation, the assertions in Case I are included in [12, Theorem 7.1] as special cases.

Theorem 8.3. *Let $2\rho = \sigma = tu$ with $0 < t \in \left(\frac{1}{2}\right)\mathbf{Z}$. If $t > \frac{1}{2}$, one has*

$$(8.4) \qquad \mathfrak{N}\mathcal{H}(\sigma, \Delta) = \mathcal{E}\left(\tfrac{t}{2}, \rho, \Delta\right) \cap \mathcal{H}(\sigma, \Delta).$$

Moreover

$$(8.5) \qquad \mathfrak{N}\mathcal{H}(\sigma, \Delta) = \mathcal{E}\left(\tfrac{t}{2}, \rho, \Delta\right)$$

except in the following three cases: (i) $t = \frac{1}{2}$; (ii) $t = \frac{3}{2}$ and $F = \mathbf{Q}$; (iii) $t = 2$ and $F = \mathbf{Q}$.

PROOF. The last assertion follows from (8.4) and Proposition 8.2. Now Proposition 3.2 shows that λ is simple if $\lambda = 0$. Moreover, λ is critical if and only if $t = 1$. Therefore, putting $s_0 = \frac{t}{2}$ with $t > 1$ in Theorem 7.3, we obtain

$$(8.6) \qquad \mathfrak{N}(\sigma, 0, \Delta) = \mathcal{E}\left(\tfrac{t}{2}, \rho, \Delta\right) \quad \text{if} \quad t > 1,$$

which proves (8.4). If $t = 1$, the last part of Theorem 7.8 proves (8.4).
    As for the case $t = \frac{1}{2}$, we have

Theorem 8.4. $\mathfrak{N}\mathcal{H}\left(\frac{u}{2}, \Delta\right) = \mathcal{E}^*\left(\frac{3}{4}, \frac{u}{4}, \Delta\right)$.

PROOF. By Proposition 8.2, (v), $\mathcal{E}^*\left(\frac{3}{4}, \frac{u}{4}, \Delta\right) \subset \mathcal{H}\left(\frac{u}{2}, \Delta\right)$. In view of Theorem 7.9, it is sufficient to prove that 0 is the only holomorphic element of $\mathcal{E}\left(\frac{3}{4}, \frac{u}{4}, \Delta\right)$. To see this, let $r \in \mathcal{E}\left[\frac{3}{4}, \frac{u}{4}, \Delta\right]$ and express r as in (7.10). Then we see that

$$r\left(z, \tfrac{3}{4}\right) \big\| \, \eta^{-1} = a_\eta y^{u/2} + c_\eta + \ldots$$

with $c_\eta \in \mathbf{C}$ for every $\eta \in Y$. If $r\left(z, \frac{3}{4}\right)$ is holomorphic, we have $a_\eta = 0$ for every η, so that $r = 0$, which proves the desired fact.

Remark 8.5. The result of [13, Proposition 6.4] together with (4.3), (4.12), and (4.24) shows that the elements of $\mathcal{E}^*\left(\frac{3}{4}, \frac{u}{4}, \Delta\right)$ are theta series.

As to the previous investigations on $\mathfrak{N}\mathfrak{K}(\sigma, \Delta)$, the reader is referred to the papers mentioned in the introduction.

9. Cyclopean forms

We call an element f of $\mathcal{C}(\sigma, \lambda)$ a *cyclopean form* (or simply a *cyclops*) *of exponent q*, if the following conditions (9.1a, b, c) are satisfied:

(9.1a) $f \in \mathfrak{N}(\sigma, \lambda)$;

(9.1b) *for every $\xi \in \mathcal{G}$, the constant term of $f \| \xi$ is of the form $c_\xi y^q$ with $c_\xi \in \mathbf{C}$; that is, it has no term of the form by^p with p other than q;*

$$(9.1c) \quad (1 - \sigma_v)/2 < \mathrm{Re}(q_v) < \begin{cases} (2 - \sigma_v)/2 \\ (3 - 2\sigma_v)/4 \end{cases} \quad \textit{for every} \quad v \in \mathbf{a} \qquad \begin{matrix} (\textit{Case I}), \\ (\textit{Case II}). \end{matrix}$$

By Proposition 3.2, (9.1c) implies that λ is noncritical and simple. By (3.6b), we can put $q = (1 - s_0)u - \bar{\rho}$ and $\rho = (\sigma - i\tau)/2$ with $s_0 \in \mathbf{C}$ and an admissible τ. Then (9.1c) is equivalent to

$$(9.2) \qquad\qquad \tfrac{1}{2} > \mathrm{Re}(s_0) > \begin{cases} 0 & (\text{Case I}), \\ \frac{1}{4} & (\text{Case II}). \end{cases}$$

We also note that

$$(9.3) \qquad \lambda_v = q_v(1 - \sigma_v - q_v) = (s_0 - \rho_v)(1 - s_0 - \bar{\rho}_v).$$

Put $p = s_0 u - \rho$. If $f \in \mathcal{C}(\sigma, \lambda)$, we have, for $\xi \in \mathcal{G}$,

$$f \| \xi = b_\xi y^p + c_\xi y^q + \dots.$$

Thus (9.1b) means that $b_\xi = 0$ for *every* $\xi \in \mathcal{G}$.

Theorem 9.1. *Let $\rho = (\sigma - i\tau)/2$ and $q = (1 - s_0)u - \bar{\rho}$ with $s_0 \in \mathbf{C}$ and an admissible τ. In Case II, let k be an arbitrarily fixed odd integer. If there exists a nonzero cyclopean form of $\mathcal{C}(\sigma, \lambda)$ of exponent q, then there exists a Hecke character ψ of F such that*

(9.4a) $L(2s_0, \psi) = 0$ (*Case I*),

(9.4b) $L(4s_0 - 1, \psi^2) = 0$ (*Case II*),

(9.5) $\psi(x) = |x|^{\pm i\tau}(x/|x|)^{\sigma'}$ *for* $x \in F_\mathbf{a}^\times$, *where* $\sigma' = \sigma$ *in Case I and* $\sigma' = \sigma - ku/2$ *in Case II.*

Conversely, suppose there exist a Hecke character ψ of F and a complex number s_0 satisfing (9.2), (9.4a or b), and (9.5). Then there exists a nonzero cyclops of $\mathcal{Q}(\sigma, \lambda)$ of exponent q. More explicitly,

$$[L(2s, \psi)E(z, s; \rho, \psi, \mathfrak{c})]_{s = s_0} \qquad (Case\ I),$$

$$[L(4s - 1, \psi^2)E_k(z, s; \rho, \psi, \mathfrak{c})]_{s = s_0} \qquad (Case\ II)$$

are cyclopes, for every multiple $\mathfrak{c}$ of the conductor of ψ that is divisible by 4 in Case II.

PROOF. We prove this only in Case II; Case I can be treated in a similar way. Suppose $L(4s_0 - 1, \psi^2) \neq 0$ for every ψ of type (9.5). Then $\bar{s}_0$ has the same property. Let f be a cyclops of exponent q belonging to $\mathfrak{N}(\sigma, \lambda, \Delta)$. Theorem 4.1 together with (4.3) and (4.24) shows that $\mathcal{E}[\rho, \Delta] = \mathcal{E}[s_0, \rho, \Delta]$ and $\mathcal{E}[\bar{\rho}, \Delta] = \mathcal{E}[\bar{s}_0, \bar{\rho}, \Delta]$. By Theorem 7.3, we have $f(z) = h(z, s_0)$, $h = {}= \sum_{\xi \in Y} a_\xi E_k(z, s; \rho, \xi, \Delta)$ with $a_\xi \in \mathbf{C}$. Putting $p = s_0 u - \rho$ and employing the notation of Proposition 5.2, we have

$$f \| \eta^{-1} = a_\eta y^p + \left(\sum_{\xi \in Y} a_\xi f_{\xi\eta} \right) (s_0) y^q + \dots$$

for $\eta \in Y$. Hence $a_\eta = 0$ for all $\eta \in Y$, so that $f = 0$, a contradiction.

Conversely, suppose $L(4s_0 - 1, \psi^2) = 0$ for s_0 and ψ satisfying (9.2) and (9.5). Take any common multiple $\mathfrak{c}$ of 4 and the conductor of ψ, and put

$$g(z, s) = L_{\mathfrak{c}}(4s - 1, \psi^2)E_k(z, s; \rho, \psi, \mathfrak{c}).$$

By Theorem 4.1, g is finite at s_0. Hence $g(z, s_0)$ belongs to $\mathfrak{N}(\sigma, \lambda)$ by Propositions 7.1 and 5.3. Now, for every $\zeta \in \mathcal{G}$, we have, by Proposition 5.4.

$$g(z, s) \| \zeta = ac^s L_{\mathfrak{c}}(4s - 1, \psi^2) y^{su - \rho} + \dots$$

with $a \in \mathbf{C}$ and $0 < c \in \mathbf{R}$. Therefore $g(z, s_0)$ satisfies (9.1b). To show that $g(z, s_0) \neq 0$, we consider an element η_0 of G as in [13, (4.10)]. Then the Fourier coefficients of $g \| \Lambda_\sigma^k(\eta_0)$ has been determined in [13, §6]. In particular, its constant term at s_0 is a nonzero constant times $L_{\mathfrak{c}}(4s_0 - 2, \psi^2) y^q$. Since $-1 < 4s_0 - 2 < 0$, this term is nonvanishing. This completes the proof, since $L_{\mathfrak{c}}/L$ is nonvanishing for this value.

Proposition 9.2. *Let s_0 be a complex number satisfying (9.2). Define Φ_Δ as in Theorem 7.7 for each Δ such that a cusp-class of Δ is ρ-regular if and only if it is $\bar{\rho}$-regular. Then a Hecke character ψ of F satisfying (9.4a or b) and (9.5) exists if and only if $\det \Phi_\Delta(s, \rho)$ has a pole at s_0 for some Δ. Moreover, the maximum number of linearly independent cyclopes in $\mathfrak{N}(\sigma, \lambda, \Delta)$ with λ of (9.3) is $\varkappa - \mathrm{rank}\ \Phi_\Delta(1 - s_0, \bar{\rho})$.*

PROOF. By Theorem 7.3 and Remark 7.4, (1), we have $\mathfrak{N}(\sigma, \lambda, \Delta) = \mathcal{E}(1 - s_0, \bar{\rho}, \Delta)$. Given (a row vector) $c \in \mathbf{C}^Y$, we have

$$\sum_{\xi} c_\xi E(1 - s_0, \bar{\rho}, \xi) \,\big\|\, \eta^{-1} = c_\eta y^q + \sum_{\xi} c_\xi g_{\xi\eta}(1 - s_0) y^p + \dots$$

with the same notation as in (7.8b). This gives a nontrivial cyclops of exponent q if and only if $c \neq 0$ and $c\Phi_\Delta(1 - s_0, \bar{\rho}) = 0$, which proves the last assertion. The first assertion follows from this fact, Theorem 9.1, (7.7b), and Proposition 3.4.

10. Appendix I: Whittaker functions

For $y > 0$ and $(\alpha, \beta) \in \mathbf{C}^2$, we put

$$(10.1) \qquad \tau(y, \alpha, \beta) = \int_0^\infty e^{-yt}(1 + t)^{\alpha - 1} t^{\beta - 1}\, dt.$$

This is convergent if $\mathrm{Re}(\beta) > 0$. We have obviously

$$(10.2) \qquad \left(\frac{\partial}{\partial y}\right)\tau(y, \alpha, \beta) = -\tau(y, \alpha, \beta + 1).$$

Since $(1 + t)^\alpha = (1 + t)^{\alpha - 1}(1 + t)$, we obtain

$$(10.3) \qquad \tau(y, \alpha + 1, \beta) = \tau(y, \alpha, \beta) + \tau(y, \alpha, \beta + 1).$$

Integration by parts shows

$$(10.4) \qquad \beta\tau(y, \alpha + 1, \beta) = y\tau(y, \alpha + 1, \beta + 1) - \alpha\tau(y, \alpha, \beta + 1).$$

From these formulas, we obtain easily

$$(10.5) \qquad \left\{ y\left(\frac{\partial}{\partial y}\right)^2 + (\alpha + \beta - y)\cdot\frac{\partial}{\partial y} - \beta \right\}\tau(y, \alpha, \beta) = 0.$$

Let us now put

$$(10.6) \qquad V(y, \alpha, \beta) = e^{-y/2}y^\beta\Gamma(\beta)^{-1}\tau(y, \alpha, \beta).$$

From (10.3), we obtain

$$(10.7) \qquad V(y, \alpha + 1, \beta) = V(y, \alpha + 1, \beta + 1) - \alpha y^{-1}V(y, \alpha, \beta + 1).$$

This shows that V can be continued as a holomorphic function in (α, β) to the whole $\mathbf{C}^2$. Now we have

$$y^\beta\tau(y, \alpha, \beta) = \int_0^\infty e^{-t}(1 + y^{-1}t)^{\alpha - 1}t^{\beta - 1}\, dt.$$

Therefore we see, at least for $\mathrm{Re}(\beta) > 0$, that

$$(10.8) \qquad \lim_{y \to \infty} e^{y/2} V(y, \alpha, \beta) = 1.$$

Since this is consistent with (10.7), we can easily verify that (10.8) holds uniformly for (α, β) in any compact subset of $\mathbf{C}^2$.

We now consider a differential equation

$$(10.9) \qquad y^2 f''(y) + \sigma y f'(y) + (\lambda + A\sigma y - A^2 y^2) f(y) = 0$$

with $A \in \mathbf{R}^\times$, $(\sigma, \lambda) \in \mathbf{C}^2$, and $0 < y \in \mathbf{R}$.

Proposition 10.1. *Let α and β be complex numbers such that $\alpha - \beta = \sigma$ and $\beta(1 - \alpha) = \lambda$. For fixed α, β, and A, define a function f_A by*

$$f_A(y) = \begin{cases} V(2Ay, \alpha, \beta) & \text{if} \quad A > 0, \\ |2Ay|^{-\sigma} V(-2Ay, \beta, \alpha) & \text{if} \quad A < 0. \end{cases}$$

Then f_A is a solution of (10.9). Moreover, if f is a solution of (10.9) and $f(y) = O(y^B)$ with $B \in \mathbf{R}$ when $y \to \infty$, then f is a constant multiple of f_A.

PROOF. That f_A is a solution of (10.9) follows from (10.5) in a straightforward way. Let f be a solution of (10.9) such that $f(y) = O(y^B)$. Then

$$(y^\sigma f')' = y^\sigma (f'' + \sigma y^{-1} f') = y^\sigma (A^2 - A\sigma y^{-1} - \lambda y^{-2}) f = O(y^C)$$

with $C \in \mathbf{R}$ when $y \to \infty$. It follows that $y^\sigma f'$, as well as f', is $O(y^D)$ with $D \in \mathbf{R}$. Now put $h = f_A f' - f_A' f$. Then $h' = f_A f'' - f_A'' f = -\sigma y^{-1} h$, and hence $h = a y^{-\sigma}$ with a constant a. Since both f_A and f_A' are $O(e^{-|A|y/2})$ as can easily be seen from (10.8) and (10.2), we see that $a = 0$. Therefore f is a constant multiple of f_A.

In Proposition 10.1, we can change (α, β) for $(1 - \beta, 1 - \alpha)$ without changing σ and λ. Therefore $V(2Ay, 1 - \beta, 1 - \alpha)$ for $A > 0$ is a solution of (10.9), and hence must be a constant multiple of f_A. In view of (10.8), we thus obtain

$$(10.10) \qquad V(y, 1 - \beta, 1 - \alpha) = V(y, \alpha, \beta).$$

We note also that, given a compact subset K of $\mathbf{C}^2$, there exist two positive constants B and C depending only on K such that

$$(10.11) \quad |V(y, \alpha, \beta)| \leqslant C e^{-y/2}(1 + y^{-B}) \qquad \text{for } y > 0 \text{ and} \qquad (\alpha, \beta) \in K.$$

This can be proved in an elementary way by means of (10.1) and (10.8); for details, see [11, pp. 282-283].

With σ, λ, A, and f_A as in Proposition 10.1, define a function φ_A on H by

$$(10.12) \qquad \varphi_A(x + iy, \sigma, \lambda) = e^{iAx} f_A(y).$$

Further define operators ϵ and δ^σ on H by $\epsilon f = -y^2 \partial f / \partial \bar{z}$ and $\delta^\sigma f = y^{-\sigma} \partial(y^\sigma f) / \partial z$. Then we can easily verify, employing (10.2), (10.3), and (10.4), that

$$(10.13a) \qquad \epsilon \varphi_A(z, \sigma, \lambda) = \begin{cases} \lambda(4Ai)^{-1} \varphi_A(z, \sigma - 2, \lambda + 2 - \sigma) & \text{if } A > 0, \\ (4Ai)^{-1} \varphi_A(z, \sigma - 2, \lambda + 2 - \sigma) & \text{if } A < 0, \end{cases}$$

$$(10.13b) \qquad \delta^\sigma \varphi_A(z, \sigma, \lambda) = \begin{cases} iA \varphi_A(z, \sigma + 2, \lambda + \sigma) & \text{if } A > 0, \\ (\lambda + \sigma) iA \varphi_A(z, \sigma + 2, \lambda + \sigma) & \text{if } A < 0. \end{cases}$$

11. Appendix II: Proofs of Propositions 2.1, 2.2, 2.3, and 4.4

Throughout this section, we put $U_F = \{a \in \mathfrak{g}^\times \mid a \gg 0\}$, $\mu(y) = \mathrm{Min}\{y_v \mid v \in \mathfrak{a}\}$ for $y \in \mathbf{R}^{\mathfrak{a}}$, $|z| = (|z_v|)_{v \in \mathfrak{a}}$ and $\{z\} = \sum_{v \in \mathfrak{a}} z_v$ for $z \in \mathbf{C}^{\mathfrak{a}}$. For example, we have $\mathbf{e}_{\mathfrak{a}}(i|h|y) = \exp(-2\pi\{|h|y\})$ for $h \in F$ and $0 \ll y \in \mathbf{R}^{\mathfrak{a}}$.

Lemma 11.1 *Let $\mathfrak{a}$ be a fractional ideal of F, and β an element of $\mathbf{R}^{\mathfrak{a}}$. Then there exist positive constants A, B, and C such that*

$$\sum_{0 \neq h \in \mathfrak{a}} |h|^\beta \mathbf{e}_{\mathfrak{a}}(i|h|y) \leqslant A(1 + \mu(y)^{-B}) \exp(-Cy^{u/n}).$$

for $0 \ll y \in \mathbf{R}^{\mathfrak{a}}$.

Proof. Let $\|h\| = \{h^2\}^{1/2}$. If $c \geqslant 0$, then $|h_v|^c \leqslant \|h\|^c$, and

$$|h_v|^{-c} = \left| h^{-u} \prod_{w \neq v} h_w \right|^c \leqslant N(\mathfrak{a})^{-c} \|h\|^{(n-1)c} \quad \text{for} \quad 0 \neq h \in \mathfrak{a}.$$

Therefore $|h|^\beta \leqslant A\|h\|^b$ for $0 \neq h \in \mathfrak{a}$ with positive constants A and b. Now $\{|h|y\} \geqslant n|hy|^{u/n} \geqslant nN(\mathfrak{a})^{1/n} y^{u/n}$ for such h. Put $C = \pi n N(\mathfrak{a})^{1/n}$. Then $2\pi\{|h|y\} \geqslant \pi\{|h|y\} + Cy^{u/n} \geqslant \pi\mu(y)\|h\| + Cy^{u/n}$. Therefore we have $\sum_h |h|^\beta \exp(-2\pi\{|h|y\}) \leqslant A \cdot \exp(-Cy^{u/n}) \sum_h \|h\|^b \exp(-\pi\mu(y)\|h\|)$. Since there are only finitely many h's in $\mathfrak{a}$ such that $\|h\| < 1$, we may assume, changing A for a larger constant, that b is a positive integer. For $0 < m \in \mathbf{Z}$, let p_m be the number of elements h of $\mathfrak{a}$ such that $m - 1 < \|h\| \leqslant m$. Then $p_m \leqslant Dm^{n-1}$ with a constant D, and the last sum $\sum_h$ is majorized by $D \sum_{m=1}^{\infty} m^{b+n-1} e^{t-mt}$ with $t = \pi\mu(y)$. This is $\leqslant E(1 + t^{-b-n})$ with a constant E, which completes the proof.

Lemma 11.2. *Let Δ be a congruence subgroup of $\mathfrak{G}$, and f a continuous function on $H^{\mathfrak{a}}$ satisfying (2.7a, c). Then there exist two positive constants A and B such that*

$$(11.1) \qquad |y^{\sigma/2}f(x + iy)| \leqslant A(y^{Bu} + y^{-Bu}) \quad \text{for all} \quad x + iy \in H^{\mathfrak{a}}.$$

PROOF. With a compact fundamental domain M of $\mathbf{R}^{\mathfrak{a}}/\mathfrak{g}$ and $0 < c \in \mathbf{R}$, put

$$(11.2) \qquad T_c = \{x + iy \mid x \in M, \mu(y) > c\}.$$

Then we can take a finite subset X of $\mathfrak{G}$ so that $H^{\mathfrak{a}} = \bigcup_{\beta \in \Delta, \xi \in X} \beta\xi(T_c)$. By (2.7c), we can find two positive constants A and B such that $|y^{\sigma/2}(f \| \xi)(x + iy)| \leqslant Ay^{Bu}$ if $\mu(y) > c$ and $\xi \in X$. Given $z = x + iy \in H^{\mathfrak{a}}$, take $\beta \in \Delta$ and $\xi \in X$ so that $z = \beta\xi(z')$ with $z' = x' + iy' \in T_c$. Let $\mathrm{pr}(\xi^{-1}) = \begin{pmatrix} * & * \\ p & q \end{pmatrix}$ and $\mathrm{pr}(\beta\xi)^{-1} = \begin{pmatrix} * & * \\ r & s \end{pmatrix}$. To prove our lemma, we may assume that $\mathrm{pr}(\Delta) \subset SL_2(\mathfrak{g})$. Then $r, s \in p\mathfrak{g} + q\mathfrak{g}$. Let D be the smallest of $N(p\mathfrak{g} + q\mathfrak{g})$ for all $\xi \in X$. If $r \neq 0$, we have $|r^u| \geqslant D$. Now $y'^u = y^u|rz + s|^{-2u} \leqslant D^{-2}y^{-u}$, and hence $|y^{\sigma/2}f(z)| = |y'^{\sigma/2}(f \| \beta\xi)(z')| \leqslant Ay'^{Bu} \leqslant A(D^2y^u)^{-B}$ if $r \neq 0$. When $r = 0$, we have $y'^u = s^{-2u}y^u \leqslant D^{-2}y^u$, so that $|y^{\sigma/2}f(z)| \leqslant A(D^{-2}y^u)^B$. This proves our lemma.

PROOF OF PROPOSITION 2.1. Given $f \in \mathcal{Q}(\sigma, \lambda, \Delta)$, define b_h as in (2.21). By Lemma 11.2, we see easily that

$$(11.2') \qquad |y^{\sigma/2}b_h W(hy; \sigma, \lambda)| \leqslant A'(y^{Bu} + y^{-Bu})$$

with positive constants A' and B independent of y and h. let U be a subgroup of U_F of finite index such that $\Lambda_\sigma^k(\mathrm{diag}[a, a^{-1}]) \subset \Delta$ for every $a \in U$. Now we can find two positive constants c_1 and c_2 with the following property: given $0 \ll y \in \mathbf{R}^{\mathfrak{a}}$, there exists an element a of U such that $c_1 y^{u/n} \leqslant a_v^2 y_v \leqslant c_2 y^{u/n}$ for every $v \in \mathfrak{a}$. Hereafter c_m for $m = 3, 4, \ldots$ will denote constants independent of h and y. Given $0 \neq h \in F$, take $a \in U$ so that $c_1|h|^{u/n} \leqslant |a_v h_v| \leqslant \leqslant c_2|h|^{u/n}$. By (10.8), we can find a constant $d > 1$ so that $|V(g; \alpha_v, \beta_v)| \geqslant \geqslant 2^{-1}e^{-g/2}$ and $|V(g; \beta_v, \alpha_v)| \geqslant 2^{-1}e^{-g/2}$ if $g \geqslant d$. Put $t = c_1^{-1}d|h|^{-u/n}$. Then $ta_v|h_v| \geqslant d$, so that

$$(11.3) \qquad |(tah)^{\sigma/2}W(tah; \sigma, \lambda)| \geqslant c_3|tah|^{\sigma'/2}e^{-2\pi t\{|ah|\}},$$

where $\sigma_v' = \mathrm{sgn}(h_v)\sigma_v$. Taking ta to be y in (11.2'), we find that

$$|(ta)^{\sigma/2}b_h W(tah; \sigma, \lambda)| \leqslant A'(t^{nB} + t^{-nB}),$$

which together with (11.3) shows that

$$|h^{-\sigma/2}b_h| \leqslant c_4(t^{nB} + t^{-nB})|tah|^{-\sigma'/2}e^{2\pi t\{|ah|\}}.$$

Since $t = c_1^{-1}d|h|^{-u/n}$ and $|a_v h_v| \leq c_2|h|^{u/n}$, we have $|(tah)_v| \leq dc_2/c_1$, and hence $|h^{-\sigma/2}b_h| \leq c_5|h|^{Bu}$. This proves (1) of Proposition 2.1. Next, we see from (10.11) that

$$(11.4) \qquad |W(hy; \sigma, \lambda)| \leq c_6 \sum_{s \in S} |hy|^s \mathbf{e_a}(ihy)$$

with a finite subset S of $\mathbf{R^a}$. Hence, by Lemma 11.1, we obtain

$$(11.5) \qquad y^{\sigma/2} \sum_{h \neq 0} |b_h W(hy; \sigma, \lambda)| \leq c_7 \sum_{s \in S} y^s (1 + \mu(y)^{-B}) \exp(-Cy^{u/n})$$

with constants B and C independen of s. Now the left-hand side is invariant under $y \to a^2 y$ with $a \in U$. Given y, take $a \in U$ so that $c_1 y^{u/n} \leq (a^2 y)_v \leq c_2 y^{u/n}$ for every $v \in \mathbf{a}$. Then $\mu(a^2 y) \geq c_1 y^{u/n}$ and $(a^2 y)^s \leq c_8(y^{Du} + y^{-Du})$ with D independent of s. Hence, substituting $a^2 y$ for y in (11.5), we obtain

$$(11.6) \qquad y^{\sigma/2} \sum_{h \neq 0} |b_h W(hy; \sigma, \lambda)| \leq c_9(y^{Eu} + y^{-Eu}) \exp(-Cy^{u/n})$$

with a constant E, which proves (2) of Proposition 2.1. Assertion (3) is now an easy consequence of (2) and (2.7c). To prove (4), take $f \in \mathcal{S}(\sigma, \lambda, \Delta)$ and take X as in the proof of Lemma 11.2. Applying (2) to $f \| \xi$ for each $\xi \in X$, we see that $y^{\sigma/2}f$ is bounded on the whole $H^\mathbf{a}$. Therefore we can take $B = 0$ in (11.1) and also in (11.2'). Repeating the proof of (1) with $B = 0$, we can conclude that $h^{-\sigma/2}b_h$ is bounded. This completes the proof.

Lemma 11.3. *Let f be a C^∞-function of form (2.21) satisfying (2.7a, b). Suppose $|b_h| \leq p|h|^{qu + \sigma/2}$ for $0 \neq h \in \mathfrak{m}$ with positive constants p and q. Then f satisfies (2.7c).*

Proof. Applying the above proof of (2) to $f - b_0(y)$, we obtain, from (11.6) that

$$y^{\sigma/2}|f - b_0(y)| \leq A(y^{Eu} + y^{-Eu}) \exp(-Cy^{u/n}).$$

Since $b_0(y)$ is a linear combination of the functions of Proposition 3.1, we have $y^{\sigma/2}|f(z)| \leq A'(y^{Ju} + y^{-Ju})$ on $H^\mathbf{a}$ with constants A' and J. Then (2.7c) can easily be verified.

Proof of Proposition 2.2. Let $g \in \mathcal{Q}(\sigma, \lambda, \Delta)$. It is straightforward to see that $\epsilon_v g$ and $\delta_v^\sigma g$ satisfy (2.7a, b) with modified σ, λ, and Δ as stated in the proposition. To verify (2.7c), take $\xi \in \mathcal{G}_\sigma$. Then $(\epsilon_v g) \| \xi_* = \epsilon_v(g \| \xi)$ by (2.5b). Since $g \| \xi \in \mathcal{Q}(\sigma, \lambda)$, it has an expansion of type (2.21) with b_h as in (1) of Proposition 2.1. By (10.13a), we see that

$$(\epsilon_v g) \| \xi_* = \epsilon_v b_0 + \sum_{0 \neq h} c_h W(hy; \sigma - 2v, \lambda + (2 - \sigma_v)v)$$

with c_h satisfying (1) of Proposition 2.1 with $\sigma - 2v$ instead of σ. Therefore, by Lemma 11.3, $\epsilon_v g$ satisfies (2.7c). The assertion for $\delta^\sigma_v g$ can be proved in a similar way.

PROOF OF PROPOSITION 2.3. We first note that given two positive integers a and p, and a positive real number $r < 1$, one has

$$(11.7) \qquad \sum_{m=p}^{\infty} m^a x^m \leq C(a, r) p^a x^p \quad \text{for} \quad 0 \leqslant x \leqslant r$$

with a constant $C(a, r)$ independent of p and x. In fact,

$$\sum_{m=p}^{\infty} m^a x^{m-p} = \sum_{n=0}^{\infty} (n+p)^a x^n \leqslant \sum_{i=0}^{\infty} \binom{a}{i} p^{a-i} \sum_{n=0}^{\infty} n^i r^n.$$

Now take X as in the proof of Lemma 11.2. For $f \in \mathcal{S}(\sigma, \lambda, \Delta)$ and $\xi \in X$, put $M_f = \mathrm{Max}|y^{\sigma/2} f|$ and

$$f \| \xi = \sum_h b_{h, \xi} W(hy; \sigma, \lambda) \mathbf{e_a}(hx).$$

Since $|y^{\sigma/2}(f \| \xi)| \leqslant M_f$, we have $|b_{h, \xi}| \leqslant A M_f |h|^{\sigma/2}$ with a constant A independent of f, as can be seen from the proof of (1) and (4) of Proposition 2.1. Fix an integer $p > 1$ and suppose $b_{h, \xi} = 0$ for all $\xi \in X$ and all h such that $\|h\| < p$. Then, by (11.4), we have

$$|y^{\sigma/2} f \| \xi| < B M_f \sum_{s \in S} |hy|^{\sigma/2 + s} \mathbf{e_a}(i|h|y)$$

with a constant B independent of f. The same reasoning as in the proof of Lemma 11.1 shows that, for any fixed $q > 0$, we have

$$|y^{\sigma/2} f \| \xi| \leqslant C M_f \sum_{m=p}^{\infty} m^a e^{-\pi m \mu(y)} \quad \text{if} \quad \mu(y) > q$$

with a constant C and a positive integer a independent of f. By (11.7), we have $|y^{\sigma/2} f \| \xi| \leqslant D_q M_f p^a e^{-\pi p \mu(y)}$ for $\mu(y) > q$ with a constant D_q independent of f and p. Take q to be c of (11.2). For every $z \in H^\mathbf{a}$, take $\beta \in \Delta$ and $\xi \in X$ so that $z = \beta \xi(z')$ with $z' = x' + iy' \in T_c$. Then

$$|y^{\sigma/2} f(z)| = |y^{\sigma/2} f(\beta \xi(z'))| = |y'^{\sigma/2}(f \| \beta \xi)(z')|,$$

and hence $M_f \leqslant D_q M_f p^a e^{-\pi p q}$. If p is sufficiently large, we obtain $M_f = 0$. This shows that $f = 0$ if $b_{h, \xi} = 0$ for $\|h\| < p$ and for all $\xi \in X$. Thus $\mathcal{S}(\sigma, \lambda, \Delta)$ is finite-dimensional. Now the constant term of an element of $\mathcal{C}(\sigma, \lambda)$ belongs to a 2^n-dimensional space as shown in Section 3, and hence $\mathcal{C}(\sigma, \lambda, \Delta)/\mathcal{S}(\sigma, \lambda, \Delta)$ is finite-dimensional. This completes the proof.

Proof of Proposition 4.4. By (4.3) and (4.24), we may restrict E to the functions of type (4.10). Then our first assertion follows immediately from Theorem 4.1. Suppose g is finite at s_0. By (4.4) and analytic continuation, we see that $L_v^\sigma g(z, s_0) = \lambda_v g(z, s_0)$. In order to verify (2.7c) for $g(z, s_0)$, we consider a function D' which is obtained from D of Theorem 4.1 by replacing E by E', where E' is defined by (4.16) in Case I and by [13, (4.10)] in Case II. Then we take a Fourier expansion

$$(11.8) \qquad l(s)D'(z, s) = a_0(s, y) + \sum_{h \neq 0} a_h(s)W(hy; \sigma, \lambda)\mathbf{e_a}(hx),$$

where $l(s)$ is the polynomial 1, $s(s - 1)$, or $s - \frac{3}{4}$ that cancels the pole(s) of D'. Then for every compact subset K of $\mathbf{C}$, we have $|a_h(s)| \leqslant A|h|^{\sigma/2 + Bu}$ for $s \in K$ with positive constants A and B depending only on D' and K. This follows from the explicit form of $a_h(s)$ given by (4.20) in Case I and by [13, Theorem 6.1] in Case II. Then Lemma 11.3 shows that $l(s)D'(z, s)$ satisfies (2.7c). Put $q(z, s) = l(s)D'(z, s)\|\xi$ with any $\xi \in \mathcal{G}$. Then $q(z, s)$ belongs to $\mathcal{C}(\sigma, \lambda)$ and satisfies (2.7c) uniformly on K. By a well-known principle, the same type of estimate holds for $\partial^m q/\partial s^m$ for every m. Now we consider a finite linear combination $\sum f_q(s)q(z, s)$ with meromorphic functions f_q on $\mathbf{C}$. We observe that if it is finite at s_0, it satisfies (2.7c). This completes the proof.

References

[1] Hecke, E.   Theorie der Eisensteinschen Reihen höherer Stufe und ihre Anwendung auf Funktionentheorie und Arithmetik, *Abh. Math. Sem. Hamburg,* 5(1927), 199-224 (Werke, No. 24).

[2] Kloosterman, H. D. Theorie der Eisensteinschen Reihen von mehreren Veränderlichen, *Abh. Math. Sem. Hamburg,* **6** (1928), 163-188.

[3] Maass, H.   Über eine neue Art von nichtanalytischen automorphen Funktionen und die Bestimmung Dirichletscher Reihen durch Funktionalgleichungen, *Math. Ann.,* **121** (1949), 141-183.

[4] Maass, H. Die Differentialgleichungen in der Theorie der elliptischen Modulfunktionen, *Math. Ann.,* **125** (1953), 235-263.

[5] Pei, T-y.   Eisenstein series of weight 3/2: I, II, *Trans. Amer. Math. Soc.,* **274** (1982), 573-606, **283** (1984), 589-603.

[6] Petersson, H.   Über die Entwicklungskoeffizienten der ganzen Modulformen und ihre Bedeutung für die Zahlentheorie, *Abh. math. Sem. Hamburg,* **8** (1931), 215-242.

[7] Roelcke, W.   Das Eigenwertproblem der automorphen Formen in der hyperbolischen Ebene: I, II, *Math. Ann.,* **167** (1966), 292-337, **168** (1967), 261-324.

[8] Shimizu, H.   A remark on the Hilbert modular forms of weight 1, *Math. Ann.,* **265** (1983), 457-472.

[9] Shimura, G.   The special values of the zeta functions associated with Hilbert modular forms, *Duke Math. J.,* **45** (1978), 637-679.

[10] Shimura, G.   The arithmetic of certain zeta functions and automorphic forms on orthogonal groups, *Ann. of Math.,* **111** (1980), 313-375.
[11] Shimura, G.   Confluent hypergeometric functions on tube domains, *Math. Ann.,* **260** (1982), 269-302.
[12] Shimura, G.   On Eisenstein series, *Duke Math. J.,* **50** (1983), 417-476.
[13] Shimura, G.   On Eisenstein series of half-integral weight, *Duke Math. J.,* **52** (1985), 281-314.
[14] Shimura, G.   On certain zeta functions attached to two Hilbert modular forms: I. The case of Hecke characters, *Ann. of Math.,* **114** (1981), 127-164.

Goro Shimura
Princeton University
Fine Hall, Washington Road
Princeton, New Jersey 08544
U.S.A.

On a class of nearly holomorphic automorphic forms

Annals of Mathematics, 123 (1986), 347-406

Introduction

The class of automorphic forms to be investigated has the following function as its prototype:

$$f(z) = \left[\sum (cz + d)^{-k} |cz + d|^{-2s} \right]_{s=0}.$$

Here (c, d) runs over nonzero elements of $\mathbf{Z}^2$ under some congruence conditions, z is the variable on the complex upper half plane, and k is a positive integer. It is well known that if $k = 2$,

$$f(z) = a(z - \bar{z})^{-1} + \sum_{m=0}^{\infty} b_m e^{2\pi i m z / N}$$

with $a, b_m \in \mathbf{C}$ and $0 < N \in \mathbf{Z}$. The point that attracts our attention is its nonholomorphic nature. It is noteworthy that f is holomorphic if $k \neq 2$, and also such nonholomorphy doesn't appear in the Hilbert modular case. Therefore one may naturally ask whether this is an isolated oddity. In our previous papers [10] and [13], we showed that it was not so. In fact, a similar Eisenstein series on $\mathrm{Sp}(n, \mathbf{Q})$ or on $\mathrm{SU}(n, n)$ of a particular weight can have the form

$$\sum_{r=0}^{n} \mathrm{tr}\left[\rho_r (z - {}^t\bar{z})^{-1} g_r(z) \right],$$

where ρ_r is the representation of $\mathrm{GL}_n(\mathbf{C})$ on $\wedge^r \mathbf{C}^n$, z is the matrix variable on the symmetric (tube) domain in question, and g_r is a holomorphic function with values in $\mathrm{End}(\wedge^r \mathbf{C}^n)$.

The purpose of the present paper is first to define a class of functions in a more general framework in which the functions of the above type are comprehended, and secondly, as an application, to show that certain automorphic forms which are expressed as infinite series similar to Eisenstein series belong to

Supported by NSF Grant DMS-8401291.

that class. Our definition relies on a simple differential calculus. Given a complex manifold $\mathscr{D}$ of dimension n, we take n functions $r_1, \ldots, r_n$ and define n vector fields $\partial/\partial r_1, \ldots, \partial/\partial r_n$ on $\mathscr{D}$ so that $\partial/\partial \bar{z}_i = \sum_{k=1}^{n} \partial r_k/\partial \bar{z}_i \cdot \partial/\partial r_k$, where the z_i are local complex coordinates. Then our class is defined to consist of the functions annihilated by all monomials of the $\partial/\partial r_i$ of a fixed degree. If $\mathscr{D}$ is Kählerian, there is a natural (local) choice of the r_i; if in particular $\mathscr{D}$ is a bounded symmetric domain, these vector fields have a certain covariance property under the group action, and in fact are equivalent to the operator E introduced in [11] and [12], a crucial point when we want to study the local behavior of the effect of the $\partial/\partial r_i$ on explicitly given functions.

Although our calculus is introduced mainly for the purpose of dealing with the inevitable nonholomorphy as in the above examples, it should also be noticed that the above definition includes the characterization of holomorphic functions as those annihilated by the $\partial/\partial r_i$. This fact, if obvious, has a nontrivial consequence, particularly when there is a group action, as will be seen in our application, to which we now turn.

To avoid excessive details, let us define our series in one of the simplest cases. Let K be an imaginary quadratic field embedded in $\mathbf{C}$ and H a hermitian element of $\mathrm{GL}_q(K)$ of signature (l, m) with $l \geq m > 0$. Take $Q \in \mathrm{GL}_q(\mathbf{C})$ so that $H = \bar{Q} \cdot \mathrm{diag}[1_l, -1_m] \cdot {}^t Q$. Let K_q^m denote the set of all $m \times q$-matrices with entries in K, and $\mathscr{B}$ the space of all complex $l \times m$-matrices w such that $1_l - \bar{w} \cdot {}^t w$ is positive definite. For $x \in K_q^m$ and $w \in \mathscr{B}$, put $j[x, w] = \det[\bar{x}Q\begin{pmatrix} w \\ 1 \end{pmatrix}]$. Take a coset C of K_q^m modulo a lattice and put

$$C' = \left\{ x \in C \mid \mathrm{rank}(x) = m,\ xH \cdot {}^t\bar{x} = 0 \right\},$$

$$U = \left\{ a \in \mathrm{SL}_m(K) \mid aC = C \right\}.$$

Then our series is defined by

$$(*) \qquad f(w, s) = \det(1 - \bar{w} \cdot {}^t w)^s \sum_{x \in U \backslash C'} j[x, w]^{-k} |j[x, w]|^{-2s},$$

where $w \in \mathscr{B}$, $s \in \mathbf{C}$, and $0 < k \in \mathbf{Z}$. We shall show, as a preliminary result, that the series is convergent for sufficiently large $\mathrm{Re}(s)$, and can be continued as a meromorphic function of s to the whole plane. Now our main theorems specialized to this case assert:

(1) $f(w, s)$ *is finite at $s = 0$ if $k \geq l$.*

(2) $f(w, 0)$ *is holomorphic in w if $k \geq l$ except when $k = l + 1$.*

(3) *If $k = l + 1$, $f(w, 0)$ is a polynomial in the components of $(\bar{w} \cdot {}^t w - 1)^{-1}\bar{w}$ of degree $\leq m$ with holomorphic functions on $\mathscr{B}$ as coefficients.*

(4) *If $l - m \leq k < l$, $f(w, s)$ has at most a simple pole at $s = l - k$, and the residue is holomorphic in w except when $l = m + 1 > 2$ and $k = 1$.*

We shall actually investigate the problems of the same nature for a more general family of series defined with respect to a hermitian form over a CM-field and also with respect to a symmetric form over a totally real number field. These series were introduced in our previous papers [6], [7], and [9], and the results corresponding to assertions (1), (2), and (4) were proved in the case of $SU(l, 1)$, and also in the orthogonal case under a certain parity condition. However, the result of type (3) concerning the exceptional weight was missing. In the present paper, we extend the results to $SU(l, m)$ with $m \geq 1$, remove the parity condition, and above all, settle the case of exceptional weight by means of the above calculus with $\partial / \partial r_i$.

To explain our methods of proof, let us take again the series f of $(*)$. In this case, the r_i are the components of $(\overline{w} \cdot {}^t w - 1)^{-1}\overline{w}$. We prove (2) and (3) by showing that $f(w, 0)$ is annihilated by $\partial^\nu / \partial r_{i_1} \cdots \partial r_{i_\nu}$ with $\nu = 1$ and $\nu = m + 1$, respectively. Without further details, we list here several points worthy of mention.

(i) The derivatives of f are computed by means of our results in [12] concerning certain differential operators.

(ii) Each derivative of f in the general situation, as well as f itself, can be expressed as an inner product involving a theta function θ, an Eisenstein series $\mathscr{E}$, and an automorphic form denoted A (resp. Ω) in the unitary (resp. orthogonal) case. In the unitary case, θ belongs to a certain nonscalar factor of automorphy which is not so simple. For this reason, we prove in Section 7 rather general transformation formulas of theta functions of an indefinite hermitian form.

(iii) Assertions (1), (2), (3), and (4) (in particular, the annihilation of $f(w, 0)$ by $\partial^\nu / \partial r_{i_1} \cdots \partial r_{i_\nu}$) are reduced essentially, but not in a straightforward way, to the corresponding facts on $\mathscr{E}$, which have been proved in [10], [13], and [14]. Especially, to remove the parity condition in the orthogonal case, we employ the Eisenstein series of half-integral weight investigated in [13].

(iv) The present proof of holomorphy as in (2) is different from, and in fact simpler than, that in [7].

(v) There is an important question beyond holomorphy or near holomorphy, that is, the arithmeticity of $f(w, 0)$ and of the residue of f. In [7], we showed that if $m = 1$ and $k \geq l$, then $f(w, 0)$ is arithmetic up to a constant factor, except when $k = l + 1$. We can now prove that even the nonholomorphic $f(w, 0)$ with $k = l + 1$ (and $m = 1$) is arithmetic in the sense that its value at every CM-point is of the same nature as the value of an arithmetic holomorphic form. The same is true in the orthogonal case, too. In order to keep the paper a reasonable length, however, we only state the results of such arithmeticity, leaving the details of proof to a subsequent paper. For further comments on possible generalizations or reformulation, the reader is referred to the remark at the end of Section 4.

(vi) The form A (resp. Ω) is defined on a group of type $\mathrm{SU}(m, m)$ (resp. $\mathrm{SL}(2)$) over a field E which is possibly different from the field F of definition of the unitary or orthogonal group in question. The series of $(*)$ concerns the case in which $F = E$ and $A = 1$. The involvement of A of a general type and two different fields E and F in the unitary case is parallel to that in the orthogonal case, which, in turn, is a technical necessity in the proof of arithmeticity of $f(w, 0)$ even of the simplest type and also in the proof of related facts on the critical values of certain zeta functions, as explained in [6] and [7]. This point is reflected by a fact that a nontrivial A with $E \neq F$ appears naturally when a series of type $(*)$ is specialized to a CM-point. However, the reader may simplify the whole treatment by considering only the series of type $(*)$, if he is interested only in understanding the methods of the present paper in the most transparent way.

In concluding the introduction, the author wishes to thank K. Doi and the referee for eliminating some slips of the pen in the manuscript.

Notation

For an associative ring A with identity element, we denote by A_n^m the module of all $m \times n$-matrices with entries in A, and by $A^\times$ the group of all invertible elements of A. We put $A^m = A_1^m$, $M_n(A) = A_n^n$, $\mathrm{GL}_n(A) = M_n(A)^\times$, and $\mathrm{SL}_n(A) = \{ x \in M_n(A) | \det(A) = 1 \}$ (when A is commutative). The identity element of $M_n(A)$ is denoted by 1_n, when n needs to be stressed. The complex conjugate of a complex matrix z is denoted by $\bar{z}$. We then put $z^* = {}^t\bar{z}$. However, $\overline{\mathbf{Q}}$ denotes the algebraic closure of $\mathbf{Q}$ embedded in $\mathbf{C}$. For complex hermitian matrices X and Y, we write $X > Y$ (and also $Y < X$) if $X - Y$ is positive definite. By a *representation* $\{\rho, X\}$ of a group G, we understand a couple formed by a finite-dimensional complex vector space X and a homomorphism ρ of G into the group $\mathrm{GL}(X)$ of $\mathbf{C}$-linear automorphisms of X; $\mathrm{End}(X)$ then denotes the ring of all $\mathbf{C}$-linear endomorphisms of X. We call $\{\rho, X\}$ *K-rational* for a subfield K of $\mathbf{C}$ when G and X have K-rational structures and ρ is K-rational. If Y is a finite-dimensional vector space over $\mathbf{Q}$, we denote by $\mathscr{L}(Y)$ the module of all $\mathbf{C}$-valued functions f on Y for which there exist two lattices M and N in Y such that $f(x) = 0$ for $x \notin M$ and $f(x)$ depends only on x modulo N. We call such an f a *locally constant function* on Y. If U and V are C^∞ manifolds we denote by $C^\infty(U, V)$ the set of all C^∞ maps of U into V, and put $C^\infty(U) = C^\infty(U, \mathbf{C})$.

Not every symbol will have a consistent meaning. For example, the letter E stands for the operator introduced in Section 1, and also for a totally real algebraic number field in Section 4. The distinction will be clear from the context.

1. Groups, domains, and differential operators

As explained in the introduction, we consider orthogonal and unitary groups acting on bounded symmetric domains. In the unitary case, we need also a group acting on a tube domain, which is of an auxiliary nature. In this section, we take the groups as Lie groups, and introduce algebraic ones over number fields in later sections. Let G denote any group belonging to the three types of groups $G^{(1)}$, $G^{(2)}$, and $G^{(3)}$, which are referred to as Case UT (unitary-tube), Case UB (unitary-bounded), and Case O (orthogonal), and defined as follows:

$$G^{(1)} = \left\{ \alpha \in \mathrm{SL}_{2n}(\mathbf{C}) | \alpha^* J_n \alpha = J_n \right\} \qquad \text{(Case UT)},$$

$$G^{(2)} = \left\{ \alpha \in \mathrm{SL}_{l+m}(\mathbf{C}) | \alpha^* I_{l,m} \alpha = I_{l,m} \right\} \qquad \text{(Case UB)},$$

$$G^{(3)} = \text{the identity component of } \tilde{G},$$

$$(1.1\mathrm{a}) \qquad \tilde{G} = \left\{ \alpha \in \mathrm{SL}_{n+2}(\mathbf{C}) |\, {}^t\alpha R \alpha = R,\ \alpha^* Q \alpha = Q \right\} \qquad \text{(Case O)},$$

$$(1.1\mathrm{b}) \qquad J_n = \begin{bmatrix} 0 & -1_n \\ 1_n & 0 \end{bmatrix}, \qquad I_{l,m} = \mathrm{diag}[1_l, -1_m],$$

$$(1.1\mathrm{c}) \qquad R = \mathrm{diag}\left[1_n, \begin{pmatrix} 0 & -1 \\ -1 & 0 \end{pmatrix}\right], \qquad Q = I_{n,2}.$$

We also consider domains

$$\mathscr{H}_n = \left\{ z \in \mathbf{C}_n^n | i(z^* - z) > 0 \right\} \qquad \text{(Case UT)},$$

$$\mathscr{B}_{l,m} = \left\{ z \in \mathbf{C}_m^l | 1_m > z^* z \right\} \qquad \text{(Case UB)},$$

$$\mathscr{Z}_n = \left\{ z \in \mathbf{C}^n | z^* z < 1 + (1/4)|{}^t z z|^2 < 2 \right\} \qquad \text{(Case O)}.$$

When the distinction is unnecessary, we denote any one of these domains simply by $\mathscr{D}$, and suppress the subscripts l, m, and n, whenever they are clear from the context. To each point $z \in \mathscr{D}$, we assign a matrix $B(z)$ of the same size as the elements of G as follows:

$$(1.2\mathrm{a}) \qquad B(z) = \begin{cases} \begin{bmatrix} z^* & z \\ 1_n & 1_n \end{bmatrix} & (z \in \mathscr{H}), \\[2ex] \begin{bmatrix} 1_l & z \\ z^* & 1_m \end{bmatrix} & (z \in \mathscr{B}), \\[2ex] [q(z) \quad p'(z) \quad p(z)] & (z \in \mathscr{Z}), \end{cases}$$

$$(1.2b) \qquad {}^tp(z) = ({}^tz, W, 1), \qquad {}^tp'(z) = ({}^t\bar{z}, 1, \overline{W}), \qquad W = {}^tzz/2,$$

$$(1.2c) \qquad q(z) = \begin{bmatrix} 1_n - \eta(z)^{-1}z \cdot {}^t(\overline{W}z - \bar{z}) \\ {}^tz - \eta(z)^{-1}W \cdot {}^t(\overline{W}z - \bar{z}) \\ -\eta(z)^{-1} \cdot {}^t(\overline{W}z - \bar{z}) \end{bmatrix} \left(\in \mathbf{C}_n^{n+2} \right)$$

with $\eta(z)$ of (1.6a) below. (See [6, p. 321]; a correction must be made there: $D_{-1}\eta$ on line 6 from bottom should be $D_{-1}p$.) For $\alpha \in G$ and $z \in \mathcal{D}$, we can define a point $\alpha(z)$ of $\mathcal{D}$, written also αz, and two holomorphic factors of automorphy $\lambda(\alpha, z)$ and $\mu(\alpha, z)$ by the relation

$$(1.3) \qquad \alpha B(z) = \begin{cases} B(\alpha z)\mathrm{diag}\left[\overline{\lambda(\alpha, z)}, \mu(\alpha, z)\right] & \text{(Cases UT, UB)}, \\ B(\alpha z)\mathrm{diag}\left[\lambda(\alpha, z), \overline{\mu(\alpha, z)}, \mu(\alpha, z)\right] & \text{(Case O)}. \end{cases}$$

Here λ and μ take values in $\mathrm{GL}_l(\mathbf{C})$ and $\mathrm{GL}_m(\mathbf{C})$, respectively. (In this and the next two sections, we make conventions that $l = n$ and $m = 1$ in Case O, and $l = m = n$ in Case UT.) For the proof of the fact that αz, λ, and μ are well-defined by (1.3), the reader is referred to [4], [5, §3], and [6].

So far we assumed implicitly $lm \neq 0$ in Case UB. If $lm = 0$, we understand that $\mathcal{B}_{l,m}$ consists of a single point, say 0, and G acts on it trivially. Putting $B(0) = 1_{l+m}$, we have

$$(1.4) \qquad \lambda(\alpha, z) = \bar{\alpha} \quad \text{if } m = 0; \qquad \mu(\alpha, z) = \alpha \quad \text{if } l = 0.$$

These will become necessary in Section 5. In the rest of this section, however, we always assume $lm \neq 0$ in Case UB.

We define a scalar factor of automorphy $j_\alpha(z)$, two positive definite hermitian matrices $\xi(z)$ and $\eta(z)$ of size l and m respectively, and a positive number $\delta(z)$ by

$$(1.5) \qquad j_\alpha(z) = j(\alpha, z) = \det[\mu(\alpha, z)] \qquad (\alpha \in G),$$

$$(1.6a) \qquad \eta(z) = \begin{cases} i(z^* - z) & (z \in \mathcal{H}), \\ 1_m - z^*z & (z \in \mathcal{B}_{l,m}), \\ 1 + |W|^2 - z^*z & (z \in \mathcal{Z}, W = {}^tzz/2), \end{cases}$$

$$(1.6b) \qquad \xi(z) = \begin{cases} {}^t\eta(z) = i(\bar{z} - {}^tz) & (z \in \mathcal{H}), \\ 1_l - \bar{z} \cdot {}^tz & (z \in \mathcal{B}_{l,m}), \\ 1_n - \bar{z} \cdot {}^tz + \eta(z)^{-1}(W\bar{z} - z) \cdot {}^t(\overline{W}z - \bar{z}) & \\ \qquad\qquad\qquad\qquad\qquad\qquad (z \in \mathcal{Z}, W = {}^tzz/2), \end{cases}$$

$$(1.6c) \qquad \delta(z) = \det(\eta(z)).$$

As shown in [4, p. 574], [5, p. 572], and [6, pp. 321–322], we have

$$(1.7) \qquad dz \circ \alpha = {}^t\lambda(\alpha, z)^{-1} \, dz \, \mu(\alpha, z)^{-1} \qquad (\alpha \in G),$$

$$(1.8a) \quad \lambda(\alpha, z)^* \xi(\alpha z)\lambda(\alpha, z) = \xi(z), \qquad \mu(\alpha, z)^* \eta(\alpha z)\mu(\alpha, z) = \eta(z),$$

$$(1.8b) \qquad \delta(\alpha z) = |j_\alpha(z)|^{-2}\delta(z),$$

$$(1.9) \qquad \det[\lambda(\alpha, z)] = j_\alpha(z) \qquad \text{(Cases UT, UB)}.$$

Defining $\mathrm{SO}(n, \mathbf{C})$ as usual by $\mathrm{SO}(n, \mathbf{C}) = \{ X \in \mathrm{SL}_n(\mathbf{C}) | {}^t XX = 1_n \}$, we find

$$(1.10) \qquad \lambda(\alpha, z) \in \mathrm{SO}(n, \mathbf{C}), \qquad \xi(z) \in \mathrm{SO}(n, \mathbf{C}) \qquad \text{(Case O)}.$$

The first inclusion is mentioned in [6, (2.22), (2.24)]. This together with (1.8a) proves the second inclusion, since $\xi(0) = 1_n$.

Given a nonnegative integer e and finite-dimensional complex vector spaces X and Y, we denote by $\mathscr{S}_e(Y, X)$ the vector space of all homogeneous polynomial maps of Y into X of degree e, and by $\mathscr{M\ell}_e(Y, X)$ the vector space of all multilinear maps of $Y \times \cdots \times Y$ (e copies) into X. Given $\zeta \in \mathscr{S}_e(Y, X)$, we can find an element ζ' of $\mathscr{M\ell}_e(Y, X)$, symmetric in the sense of [11, (1.23)], such that $\zeta(u) = \zeta'(u, \ldots, u)$. We shall often identify ζ with ζ', and view $\mathscr{S}_e(Y, X)$ as a subspace of $\mathscr{M\ell}_e(Y, X)$. We understand that $\mathscr{S}_0(Y, X) = \mathscr{M\ell}_0(Y, X) = X$, and put $\mathscr{S}_e(Y) = \mathscr{S}_e(Y, \mathbf{C})$. Notice that $\mathscr{S}_1(Y, X) = \mathscr{M\ell}_1(Y, X) = \mathrm{Hom}(Y, X)$.

Our automorphic forms will be defined with respect to a rational representation of a group

$$(1.11) \qquad \Re = \begin{cases} \mathrm{GL}_n(\mathbf{C}) \times \mathrm{GL}_n(\mathbf{C}) & \text{(Case UT)}, \\ \mathrm{GL}_l(\mathbf{C}) \times \mathrm{GL}_m(\mathbf{C}) & \text{(Case UB)}, \\ \mathrm{SO}(n, \mathbf{C}) \times \mathrm{GL}_1(\mathbf{C}) & \text{(Case O)}. \end{cases}$$

Throughout this section, we put $T = \mathbf{C}_m^l$. The group $\Re$ acts on T by matrix multiplication, which is essentially the action of the maximal compact subgroup of G on the tangent space of $\mathscr{D}$. We specify the action in two different ways. Given a representation $\{ \rho, X \}$ of $\Re$, we define two representations $\rho \otimes \tau_e$ and $\rho \otimes \pi_e$ of $\Re$ on $\mathscr{M\ell}_e(T, X)$ by

$$(1.12a) \quad [(\rho \otimes \tau_e)(a, b)h](u_1, \ldots, u_e) = \rho(a, b)h({}^t a u_1 b, \ldots, {}^t a u_e b),$$

(1.12b)

$$[(\rho \otimes \pi_e)(a, b)h](u_1, \ldots, u_e) = \rho(a, b)h(a^{-1} u_1 \cdot {}^t b^{-1}, \ldots, a^{-1} u_e \cdot {}^t b^{-1})$$

for $(a, b) \in \Re$, $h \in \mathscr{M\ell}_e(T, X)$, and $u_i \in T$; we use the same symbols $\rho \otimes \tau_e$ and $\rho \otimes \pi_e$ for their restrictions to $\mathscr{S}_e(T, X)$. If $X = \mathbf{C}$ and ρ is the trivial representation, we write them simply τ_e and π_e; thus

(1.13) $\quad [\tau_e(a, b)h](u) = h({}^t aub), \qquad [\pi_e(a, b)h](u) = h(a^{-1}u \cdot {}^t b^{-1})$

for $h \in \mathscr{S}_e(T)$ and $u \in T$.

For $\varphi \in \mathscr{M}\ell_e(T, \mathbf{C})$ and $h \in \mathscr{M}\ell_e(T, X)$, we define an element $\{\varphi, h\}$ of X by

(1.14) $\qquad\qquad \{\varphi, h\} = \sum \varphi(c_1, \ldots, c_e)h(c_1, \ldots, c_e),$

where $c_1, \ldots, c_e$ run independently over the standard basis of $T = \mathbf{C}_m^l$. We see easily that

(1.15) $\quad \{\pi_e(a, b)\varphi, (\rho \otimes \tau_e)(a, b)h\} = \{\tau_e(a, b)\varphi, (\rho \otimes \pi_e)(a, b)h\}$
$$= \rho(a, b)\{\varphi, h\} \qquad ((a, b) \in \mathfrak{R}).$$

For $\alpha \in G$ and $f \in C^\infty(\mathscr{D}, X)$, we define $f\|_\rho \alpha \in C^\infty(\mathscr{D}, X)$ by

(1.16) $\qquad\qquad (f\|_\rho\alpha)(z) = \rho(\lambda(\alpha, z), \mu(\alpha, z))^{-1}f(\alpha z).$

We define also $\mathrm{Hom}(T, X)$-valued functions Df, $\overline{D}f$, $D_\rho f$, Cf, and Ef by

(1.17a) $\quad (Df)(u) = \sum_{i=1}^{l} \sum_{j=1}^{m} u_{ij}\, \partial f/\partial z_{ij}, \qquad (\overline{D}f)(u) = \sum_{i=1}^{l} \sum_{j=1}^{m} u_{ij}\, \partial f/\partial \bar{z}_{ij},$

(1.17b) $\qquad\qquad (D_\rho f)(u) = \rho(\xi, \eta)^{-1}D[\rho(\xi, \eta)f](u),$

(1.17c) $\qquad (Cf)(u) = (Df)({}^t\xi u\eta), \qquad (Ef)(u) = (\overline{D}f)(\xi u \cdot {}^t\eta),$

where $u = (u_{ij}) \in T$. Then iterated operators $D_\rho^{(e)}$, C^e, and E^e for $0 \leqq e \in \mathbf{Z}$ can be defined by

(1.18a) $\qquad D_\rho^{(e+1)} = D_{\rho \otimes \tau_e}D_\rho^{(e)}, \qquad D_\rho^{(1)} = D_\rho, \qquad D_\rho^{(0)}f = f,$

(1.18b) $\qquad\qquad C^{e+1} = CC^e, \qquad C^1 = C, \qquad E^{e+1} = EE^e,$
$$E^1 = E, \qquad C^0f = E^0f = f.$$

Formally $D_\rho^{(e)}f$, C^ef, and E^ef are defined to be $\mathscr{M}\ell_e(T, X)$-valued. As will be shown in Proposition 2.5 below, their values are contained in $\mathscr{S}_e(T, X)$. These operators except C^e have been introduced in our previous papers. In particular, as shown in [6, §11], [8], and [11], (and in fact, as can easily be derived from (1.7)), we have

(1.19a) $\qquad\qquad D_\rho^{(e)}(f\|_\rho\alpha) = \left(D_\rho^{(e)}f\right)\|_{\rho \otimes \tau_e}\alpha,$

(1.19b) $\qquad\qquad E^e(f\|_\rho\alpha) = (E^ef)\|_{\rho \otimes \pi_e}\alpha \qquad\qquad (\alpha \in G).$

The significance of this newly introduced operator C^e is explained by

(1.20) $\qquad\qquad D_\rho^{(e)}f = (\rho \otimes \tau_e)(\xi, \eta)^{-1}C^e[\rho(\xi, \eta)f].$

If $e = 1$, this follows immediately from our definition. The general case can be proved by induction on e.

Now we decompose $D_\rho^{(e)}$ and E^e into several operators according to the decomposition of $\mathscr{S}_e(T)$ into $\mathfrak{R}$-stable subspaces. To this end, first, for two representations $\{\rho, X\}$ and $\{\sigma, Y\}$ of $\mathfrak{R}$, we define a representation $\{\rho \otimes_0 \sigma, \mathrm{Hom}(Y, X)\}$ of $\mathfrak{R}$ by

$$(1.21) \qquad [(\rho \otimes_0 \sigma)(a, b)q](y) = \rho(a, b)q\big[\sigma({}^t a, {}^t b)y\big]$$

$$((a, b) \in \mathfrak{R}, \qquad q \in \mathrm{Hom}(Y, X), \qquad y \in Y).$$

Next, we take a $\mathfrak{R}$-stable subspace Z of $\mathscr{S}_e(T)$, and define a map $\Phi_Z\colon \mathscr{S}_e(T, X) \to \mathrm{Hom}(Z, X)$ by

$$(1.22) \qquad (\Phi_Z h)\zeta = \{\zeta, h\} \quad \text{for } h \in \mathscr{S}_e(T, X) \quad \text{and} \quad \zeta \in Z,$$

with the symbol $\{\ ,\ \}$ of (1.14). Let τ_Z and π_Z denote the restrictions of τ_e and π_e to Z, respectively. Then we see easily that

$$(1.23a) \qquad \Phi_Z \circ (\rho \otimes \tau_e)(g) = (\rho \otimes_0 \tau_Z)(g) \circ \Phi_Z,$$

$$(1.23b) \qquad \Phi_Z \circ (\rho \otimes \pi_e)(g) = (\rho \otimes_0 \pi_Z)(g) \circ \Phi_Z \qquad (g \in \mathfrak{R}).$$

Moreover, if $\mathscr{S}_e(T)$ is a direct sum of subspaces $\{Z\}$, then it can easily be seen that $h \mapsto (\Phi_Z h)_Z$ gives an isomorphism of $\mathscr{S}_e(T, X)$ onto the direct sum of $\mathrm{Hom}(Z, X)$ for $Z \in \{Z\}$. We can now define $\mathrm{Hom}(Z, X)$-valued functions $D_\rho^Z f$ and $E_Z f$ by

$$(1.24) \qquad D_\rho^Z f = \Phi_Z D_\rho^{(e)} f, \qquad E_Z f = \Phi_Z E^e f.$$

From (1.19a, b) and (1.23a, b), we obtain immediately

$$(1.25) \qquad D_\rho^Z\big(f \|_\rho \alpha\big) = \big(D_\rho^Z f\big)\|_{\rho \otimes_0 \tau_Z} \alpha,$$

$$E_Z\big(f \|_\rho \alpha\big) = (E_Z f)\|_{\rho \otimes_0 \pi_Z} \alpha \qquad (\alpha \in G).$$

These operators D_ρ^Z and E_Z are essentially the same as those introduced in [11] and [12]. We have presented them here in the form most suitable for our later applications.

2. Interpretations of C^e and D^e

We start with a rather general situation. Let $X_1, \ldots, X_n$ be mutually commutative C^∞ vector fields on an N-dimensional C^∞ manifold U, and $r_1, \ldots, r_n$ be elements of $C^\infty(U)$ such that the $n \times n$-matrix $(X_j r_k)$ is everywhere invertible. We assume $n \leq N$; we take, however, the X_j and the r_k to be complex-valued.

LEMMA 2.0. *Define vector fields $Y_1, \ldots, Y_n$ by $Y_i = \sum_{j=1}^n b_{ij} X_j$, where the b_{ij} are determined by $\sum_{j=1}^n b_{ij} X_j r_k = \delta_{ik}$. Then $Y_1, \ldots, Y_n$ are mutually commutative.*

Proof. The essential content of this lemma is contained in the well-known integrability theorem. We give here an easy direct proof for the reader's convenience. Put $a_{jk} = X_j r_k$. Then $X_h a_{jk} = X_j a_{hk}$. We have

$$Y_p Y_q = \sum_{i,j} b_{pi} b_{qj} X_i X_j + \sum_j \left(\sum_i b_{pi} X_i b_{qj} \right) X_j.$$

Therefore our task is to show that $\sum_i b_{pi} X_i b_{qj} = \sum_i b_{qi} X_i b_{pj}$ for every p, q, and j, which is equivalent to

$$(*) \qquad X_k b_{qj} = \sum_{i,p} a_{kp} b_{qi} X_i b_{pj}$$

for every k, q, and j. Applying X_i to $ab = 1_n$, we obtain

$$(**) \qquad \sum_p b_{pj} X_i a_{kp} = -\sum_p a_{kp} X_i b_{pj}.$$

Hence the right-hand side of $(*)$ is equal to

$$-\sum_{i,p} b_{pj} b_{qi} X_i a_{kp} = -\sum_{i,p} b_{pj} b_{qi} X_k a_{ip}.$$

By $(**)$, this is equal to $\sum_{i,p} b_{qi} a_{ip} X_k b_{pj} = X_k b_{qj}$ as desired.

We now assume that U is an n-dimensional complex manifold, and take n elements $r_1, \ldots, r_n$ of $C^\infty(U)$ such that $(\partial r_k / \partial \bar{z}_j)_{j,k=1}^n$ is everywhere invertible, where $z_1, \ldots, z_n$ are local complex coordinates. Define vector fields $\partial / \partial r_i$ and $\partial / \partial \bar{r}_i$ for $i = 1, \ldots, n$ by

$$(2.1) \qquad \partial / \partial r_i = \sum_{j=1}^n b_{ij}\, \partial / \partial \bar{z}_j, \qquad \partial / \partial \bar{r}_i = \sum_{j=1}^n \bar{b}_{ij}\, \partial / \partial z_j,$$

where the b_{ij} are determined by $\sum_{j=1}^n b_{ij}\, \partial r_k / \partial \bar{z}_j = \delta_{ik}$. It can easily be seen that these do not depend on the choice of local coordinates. (The reason for this somewhat awkward notation is merely that we shall primarily manipulate $\partial / \partial r_i$, but not much $\partial / \partial \bar{r}_i$.) Then each set of n vector fields $\partial / \partial r_1, \ldots, \partial / \partial r_n$ or $\partial / \partial \bar{r}_1, \ldots, \partial / \partial \bar{r}_n$ are mutually commutative; however, $\partial / \partial r_i$ and $\partial / \partial \bar{r}_j$ do not necessarily commute, as can easily be seen by a counterexample, even when $n = 1$. We put $\partial^k / \partial r_{i_1} \ldots \partial r_{i_k} = (\partial / \partial r_{i_1}) \ldots (\partial / \partial r_{i_k})$.

LEMMA 2.1. (i) *If $g(z, w)$ is a holomorphic function in $(z, w) \in U \times V$ with an open subset V of $\mathbf{C}^n$ and $h(z, w) = \partial^k g / \partial w_{i_1} \ldots \partial w_{i_k}$, then $h(z, r(z)) = (\partial^k / \partial r_{i_1} \ldots \partial r_{i_k}) g(z, r(z))$ whenever $g(z, r(z))$ is meaningful.*

Similarly, if g is antiholomorphic in z and holomorphic in w, then $h(z, \overline{r(z)}) = (\partial^k/\partial\bar{r}_{i_1} \ldots \partial\bar{r}_{i_k})g(z, \overline{r(z)})$.

(ii) *A* C^∞ *function f on U is a polynomial in* $r_1, \ldots, r_n$ *of degree* $< k$ *with holomorphic functions as coefficients if and only if all the derivatives of the form* $\partial^k f/\partial r_{i_1} \ldots \partial r_{i_k}$ *are 0.*

Proof. Assertion (i) can be verified in a straightforward way. We prove the "if"-part of (ii) by induction on k. The case $k = 1$ is obvious. Assuming $\partial^k f/\partial r_{i_1} \ldots \partial r_{i_k} = 0$ with $k > 1$, we have, by the induction assumption, a polynomial $Q_i(X)$ of degree $< k - 1$ whose coefficients are holomorphic functions on U such that $\partial f/\partial r_i = Q_i(r)$, where $X = (X_1, \ldots, X_n)$ denotes indeterminates. Let $Q_{ij} = \partial Q_i/\partial X_j$. By (i), we have $Q_{ij}(r) = \partial^2 f/\partial r_i \partial r_j = Q_{ji}(r)$, and hence $Q_{ij} = Q_{ji}$. Therefore we can find a polynomial $P(X)$ whose coefficients are holomorphic functions such that $\partial P/\partial X_i = Q_i$ for every i. Then $(\partial/\partial r_i)[f - P(r)] = 0$, and hence $f - P(r)$ is holomorphic. This completes our induction. The converse part follows immediately from (i).

Suppose for example U is a Kähler manifold whose fundamental form is $(i/2)d''d'\varphi$ with a real-valued function φ; put $r_p = \partial\varphi/\partial z_p$. Then $d''d'\varphi = \Sigma\, \partial r_p/\partial\bar{z}_q\, d\bar{z}_q \wedge dz_p$, and therefore the above lemma is applicable to the r_p. (Strictly speaking, only the module spanned by the r_p (resp. the $\partial/\partial r_p$) over the ring of holomorphic functions is globally well-defined; the set of functions annihilated by the $\partial^k/\partial r_{i_1} \ldots \partial r_{i_k}$ is globally well-defined as well.) This is in fact what we are going to do on our domain $\mathscr{D}$. Before presenting r on $\mathscr{D}$ explicitly, we note a simple fact, though it will not be needed in our later treatment:

(2.2)   *The Laplace-Beltrami operator with respect to the metric associated with* $(i/2)d''d'\varphi$ *is* $4\Sigma_{p=1}^n(\partial/\partial r_p)(\partial/\partial z_p)$.

LEMMA 2.2. *Define a T-valued function r on* $\mathscr{D}$ *by*

$$(2.3) \qquad r(z) = \begin{cases} {}^t(z - z^*)^{-1} & (\text{Case UT}), \\ -\,\xi(z)^{-1}\bar{z} = -\,\bar{z} \cdot {}^t\eta(z)^{-1} & (\text{Case UB}), \\ \eta(z)^{-1}(({}^t\bar{z}\bar{z}/2)z - \bar{z}) & (\text{Case O}). \end{cases}$$

Then we have, for $u \in T$,

$$(2.4a) \qquad \xi^{-1}(D\xi)(u) = r \cdot {}^t u, \qquad \eta^{-1}(D\eta)(u) = {}^t ru \quad (\text{Cases UT, UB}),$$

$$(2.4b) \qquad \xi^{-1}(D\xi)(u) = r \cdot {}^t u - u \cdot {}^t r, \qquad \eta^{-1}(D\eta)(u) = {}^t ru \quad (\text{Case O}),$$

$$(2.5) \qquad (Dr)(u) = \begin{cases} -\,r \cdot {}^t ur & (\text{Cases UT, UB}), \\ ({}^t rr/2)u - r \cdot {}^t ru & (\text{Case O}). \end{cases}$$

Furthermore, in all three cases, we have

$$(2.6) \qquad (\overline{D}r)(v) = - \xi^{-1}v \cdot {}^t\eta^{-1},$$

$$(2.7) \qquad (D\delta^s)(u) = s\delta^s \mathrm{tr}({}^tru) \qquad\qquad (s \in \mathbf{C}),$$

$$(2.8) \qquad (D \log \delta)(u) = \mathrm{tr}({}^tru),$$

$$(2.9) \qquad (\overline{D}D \log \delta)(u, v) = - \mathrm{tr}({}^tu\xi^{-1}v \cdot {}^t\eta^{-1}) \qquad (u, v \in T).$$

Proof. The formulas concerning $(\xi^{-1}D\xi)u$ and $(\eta^{-1}D\eta)u$ follow immediately from (1.6a, b), except $(\xi^{-1}D\xi)u$ in Case O. In this case, a direct calculation using (1.6b) shows that

$$(\eta D\xi)u = \left[\overline{W}(W\bar{z} - z) - \eta\bar{z}\right] \cdot {}^tu - \xi u \cdot {}^t(\overline{W}z - \bar{z})$$

with $W = {}^tzz/2$. Thus we can write $(\xi^{-1}D\xi)u = x \cdot {}^tu - u \cdot {}^tr$ with $x \in T$. Since ${}^t\xi\xi = 1$, $(\xi^{-1}D\xi)u$ must be alternating, and hence $x = r$. Formula (2.8) can be obtained by applying $\partial/\partial s$ to formula (2.7), which is the simplest case of [12, Theorem 4.3]. The last formula follows immediately from (2.8) and (2.6). Now (2.5) and (2.6) can be verified in a straightforward way, except that we need (2.10) below in Case O.

LEMMA 2.3. *With r as in (2.3), we have, in Case O,*

$$(2.10) \qquad {}^trr = \eta^{-1} \cdot {}^t\bar{z}\bar{z},$$

$$(2.11) \qquad \eta^{-1} = 1 + ({}^trr/2)({}^tzz/2) - {}^tzr,$$

$$(2.12) \qquad \eta(z)^{-1}\bar{z} = ({}^trr/2)z - r.$$

Proof. The first formula can be derived directly from (2.3) and (1.6a). Combining (2.10) with (2.3), we have $r = ({}^trr/2)z - \eta(z)^{-1}\bar{z}$, which proves (2.12). Now, from (1.6a), we obtain $1 = \eta^{-1} + \eta^{-1}\overline{W}W - {}^tz\eta^{-1}\bar{z}$, where $W = {}^tzz/2$. By (2.10) and (2.12), we obtain $1 = \eta^{-1} + ({}^trr/2)W - {}^tz[({}^trr/2) - r]$, from which (2.11) follows immediately.

Formula (2.9) can be written in the form

$$(- i/2)d'd''\log \delta(z) = (i/2) \sum_{j, p=1}^{l} \sum_{k, q=1}^{m} (\xi^{-1})_{jp}(\eta^{-1})_{kq} \, dz_{jk} \wedge d\bar{z}_{pq}.$$

Our space $\mathscr{D}$ is a Kähler manifold with this as its fundamental form. This fact is not absolutely necessary, but may supply a certain conceptual background for our calculation.

Now the operators C and E introduced in Section 1 are essentially $(\partial/\partial\bar{r}_{ij})$ and $(\partial/\partial r_{ij})$ as shown by the following proposition, which is a key to the proof of a central result of this paper.

PROPOSITION 2.4. *Define $r(z)$ by (2.3); define also $\partial/\partial r_{ij}$ and $\partial/\partial \bar{r}_{ij}$ as in (2.1). Then the following assertions hold for $f \in C^\infty(D)$:*

    (i) $(Cf)(u) = - \sum_{i=1}^{l}\sum_{j=1}^{m} u_{ij} \partial f/\partial \bar{r}_{ij}$ *for* $u \in T$;

    (ii) $(Ef)(u) = - \sum_{i=1}^{l}\sum_{j=1}^{m} u_{ij} \partial f/\partial r_{ij}$ *for* $u \in T$;

    (iii) $E^k f = 0$ *if and only if f is a polynomial in the r_{ij} of degree $< k$ whose coefficients are holomorphic functions.*

Proof. These follow immediately from (1.17c), (2.6), and Lemma 2.1.

PROPOSITION 2.5. *Let $\{\rho, X\}$ be a rational representation of $\Re$, and Z a $\Re$-stable subspace of $\mathscr{S}_e(T)$; let $f \in C^\infty(D, X)$. Then $D_\rho^{(e)}f$, $C^e f$, and $E^e f$ take values in $\mathscr{S}_e(T, X)$. Moreover, for every $\zeta \in Z$, we have*

    (i) $(D_\rho^Z f)\zeta = (-1)^e \rho(\xi, \eta)^{-1} \zeta'(\partial/\partial \bar{r})[\rho(\xi, \eta)f]$, *where* $\zeta'(u) = \zeta(\xi^{-1}u \cdot {}^t\eta^{-1})$ *for* $u \in T$;

    (ii) $(E_Z f)\zeta = (-1)^e \zeta(\partial/\partial r)f$.

Here $\zeta(\partial/\partial r)$ and $\zeta'(\partial/\partial \bar{r})$ are the operators obtained by substituting $\partial/\partial r_{ij}$ and $\partial/\partial \bar{r}_{ij}$ for u_{ij} in $\zeta(u)$ and $\zeta'(u)$, respectively.

Proof. That the values of $C^e f$ (resp. $E^e f$) are symmetric follows from the commutativity of the $\partial/\partial \bar{r}_{ij}$ (resp. the $\partial/\partial r_{ij}$) and Proposition 2.4. Then (1.20) proves the same for $D_\rho^{(e)}f$. To see (ii), let ε_{ij} denote the standard basis of $T = \mathbf{C}_m^l$, and let $\partial/\partial r = \sum_{i,j} \varepsilon_{ij} \partial/\partial r_{ij}$. Then, by (1.24), (1.22), (1.14), and Proposition 2.4, (ii), we have

$$(E_Z f)\zeta = \{\zeta, E^e f\} = \sum_{i,j,k,l,\ldots} \zeta(\varepsilon_{ij}, \varepsilon_{kl}, \ldots)(E^e f)(\varepsilon_{ij}, \varepsilon_{kl}, \ldots)$$

$$= (-1)^e \sum_{i,j,k,l,\ldots} \zeta(\varepsilon_{ij}, \varepsilon_{kl}, \ldots)(\partial/\partial r_{ij})(\partial/\partial r_{kl}) \ldots f$$

$$= (-1)^e \zeta(\partial/\partial r, \partial/\partial r, \ldots)f,$$

which proves (ii). Similarly, we have $\{\zeta, C^e g\} = (-1)^e \zeta(\partial/\partial \bar{r})g$, and therefore, by (1.20) and (1.15),

$$\left(D_\rho^Z f\right)\zeta = \left\{\zeta, (\rho \otimes \tau_e)(\xi, \eta)^{-1} C^e[\rho(\xi, \eta)f]\right\}$$

$$= \rho(\xi, \eta)^{-1}\{\zeta', C^e[\rho(\xi, \eta)f]\}$$

$$= (-1)^e \rho(\xi, \eta)^{-1} \zeta'(\partial/\partial \bar{r})[\rho(\xi, \eta)f].$$

3. A further study of the functions annihilated by E^k

In the easiest case in which $\mathscr{D} = \mathscr{H}_1$, assertion (iii) of Proposition 2.4 says that $(y^2 \partial/\partial \bar{z})^k f = 0$ if and only if $f(z) = \sum_{\nu=0}^{k-1} y^{-\nu} g_\nu(z)$ with holomorphic functions g_ν, where $y = \mathrm{Im}(z)$. We are going to study such functions on $\mathscr{D}$ in connection with $D_\rho^{(e)}$. Throughout this section, k denotes a fixed nonnegative integer which is positive whenever the context demands, and $\{\rho, X\}$ a rational representation of $\mathfrak{R}$.

LEMMA 3.1. *Let* $\varphi_a(u) = \mathrm{tr}({}^t a u)^k$ *for* $a, u \in T$. *Then* $\{\varphi_a, h\} = h(a)$ *for every* $h \in \mathscr{S}_k(T, X)$.

This follows immediately from (1.14).

Before proceeding further, let us introduce a notational convention. If g is a holomorphic function on $\mathscr{D}$ with values in $\mathscr{S}_k(T, X)$, we express it in the form $g(z, v)$ with $z \in \mathscr{D}$ and $v \in T$, the value $g(z, v)$ being in X. Then g is a homogeneous polynomial in v of degree k, and is holomorphic in z. Conversely, any such function on $\mathscr{D} \times T$ may be viewed as a holomorphic function on $\mathscr{D}$ with values in $\mathscr{S}_k(T, X)$. Now let f be an X-valued function such that $E^{k+1}f = 0$. By Proposition 2.4, we have

$$(3.1) \qquad f(z) = \sum_{j=0}^{k} g_j(z, r(z))$$

with $\mathscr{S}_j(T, X)$-valued holomorphic functions g_j on $\mathscr{D}$, using the notation just introduced. By virtue of (2.8), we may view $(D \log \delta)^j$, for $0 \leq j \in \mathbf{Z}$, as an $\mathscr{S}_j(T)$-valued function on $\mathscr{D}$ such that

$$(3.2) \qquad (D \log \delta)^j(u) = \mathrm{tr}({}^t r(z) u)^j \quad \text{for } u \in T.$$

Then, by Lemma 3.1, (3.1) can be written also

$$(3.3) \qquad f(z) = \sum_{j=0}^{k} \left\{ (D \log \delta)^j, g_j \right\},$$

where we understand that the term for $j = 0$ is g_0.

LEMMA 3.2. *If* $g\colon \mathscr{D} \to \mathscr{S}_k(T, X)$ *is holomorphic, then, for every subspace* Z *of* $\mathscr{S}_j(T)$, $j \leq k$ *and for every* $\zeta \in Z$, *we have*

$$\left[E_Z \{ (D \log \delta)^k, g \} \right] \zeta = (-1)^j [k!/(k-j)!] \left\{ \zeta \cdot (D \log \delta)^{k-j}, g \right\}.$$

Further, if f *has form* (3.3) *with holomorphic* $g_j\colon \mathscr{D} \to \mathscr{S}_j(T, X)$, *then* $E^k f = (-1)^k k! g_k$.

Proof. By Proposition 2.4, (ii), we see easily that

$$\left[E^j \mathrm{tr}({}^t r u)^k \right](v_1, \ldots, v_j) = (-1)^j [k!/(k-j)!] \mathrm{tr}({}^t r u)^{k-j} \mathrm{tr}({}^t v_1 u) \ldots \mathrm{tr}({}^t v_j u)$$

for $u, v_1, \ldots, v_j \in T$, and hence, by Lemma 3.1,

$$(3.4) \qquad \left[E_Z (D \log \delta)^k \right] \zeta = (-1)^j [k!/(k-j)!] (D \log \delta)^{k-j} \zeta,$$

which proves the first assertion. In particular, if $k = j$, we have $[E^k \mathrm{tr}('ru)^k](v)$ $= (-1)^k k! \mathrm{tr}('vu)^k$, and hence

$$(3.5) \qquad \left[E^k \left\{ (D \log \delta)^k, g \right\} \right](v) = (-1)^k k! g(v)$$

by Lemma 3.1. Therefore we obtain the second assertion by applying E^k to (3.3).

LEMMA 3.3. *Let* $\{ \sigma, Y \}$ *be a rational representation of* $\Re$; *let k and ν be nonnegative integers. If* $g \in C^\infty(\mathscr{D}, Y)$ *and* $E^\nu g = 0$, *then* $E^{k+\nu} D_\sigma^{(k)} g = 0$. *In particular,* $E^{k+1} D_\sigma^{(k)} g = 0$ *if g is holomorphic.*

Proof. Let $d\sigma$ be the homomorphism of the Lie algebra of $\Re$ into $\mathrm{End}(Y)$ corresponding to σ. In general, if $\varphi \colon \mathbf{R} \to \Re$ is a smooth map, then

$$(3.6) \qquad (d/dt)\sigma(\varphi(t)) = \sigma(\varphi(t)) \, d\sigma\left[\varphi(t)^{-1} \, d\varphi/dt\right].$$

In fact, putting $\exp[\psi(t)] = \varphi(t_0)^{-1}\varphi(t)$ in a neighborhood of t_0, we have $\sigma(\varphi(t_0)^{-1}\varphi(t)) = \exp(d\sigma[\psi(t)])$, and hence $\sigma(\varphi(t_0))^{-1}(d/dt)\sigma(\varphi(t)) = d\sigma[d\psi/dt]$ at $t = t_0$, which proves (3.6). Applying this to $\sigma(\xi, \eta)$, we find that

$$\sigma(\xi, \eta)^{-1} D\sigma(\xi, \eta)(u) = d\sigma\left(\xi^{-1}(D\xi)u, \eta^{-1}(D\eta)u\right) \quad \text{for } u \in T.$$

By Lemma 2.2, we have

$$(3.7) \quad \sigma(\xi, \eta)^{-1} D\sigma(\xi, \eta)(u) = \begin{cases} d\sigma(r \cdot {}^t u, {}^t ru) & \text{(Cases UT, UB),} \\ d\sigma(r \cdot {}^t u - u \cdot {}^t r, {}^t ru) & \text{(Case O).} \end{cases}$$

Denote the right-hand side in each case by $\omega_\sigma(r, u)$, which is an element of $\mathrm{End}(Y)$ bilinear in (r, u). From (1.17b) we obtain

$$(3.8) \qquad (D_\sigma g)(u) = (Dg)(u) + \omega_\sigma(r, u)g.$$

In view of (2.5), we observe that if g is a polynomial in r of degree $< \nu$ with holomorphic functions as coefficients, then Dg is a polynomial in r of degree $\leqq \nu$. Therefore (3.8) shows that $D_\sigma g$ has degree $\leqq \nu$ in r. By induction, we see that $D_\sigma^{(k)} g$ has degree $< \nu + k$. Combining this with Proposition 2.4, (iii), we obtain our assertions.

Applying (3.1) to $D_\sigma^{(k)} g$ with $\{ \sigma, Y \}$ as above and a holomorphic Y-valued function g, we find

$$(3.9) \qquad D_\sigma^{(k)} g = \sum_{j=0}^{k} h_j(z, r(z)),$$

with $h_j \colon \mathscr{D} \times T \to \mathscr{S}_k(T, Y)$ which is holomorphic on $\mathscr{D}$ and of degree j on T.

(We are taking $\{\sigma \otimes \tau_k, \mathscr{S}_k(T, Y)\}$ as $\{\rho, X\}$ there.) Now we have

(3.10) $$h_k(z, v)(u) = P_k(v, u)g(z) \qquad (u, v \in T)$$

with $P_k(v, u) \in \mathrm{End}(Y)$, which depends only on σ and k, and which is homogeneous of degree k in u and also in v. This can easily be shown by induction on k, employing (3.8); in fact, we have

(3.11)
$$P_{k+1}(r, u) = \left[DP_k(r, u) \right](u) + \omega_\sigma(r, u)P_k(r, u) + \left[\omega_{\tau_k}(r, u)p_k(r) \right](u),$$

where $p_k(r)$ is an element of $\mathscr{S}_k(T, \mathrm{End}(Y))$ defined by $p_k(r)(u) = P_k(r, u)$. By Lemma 3.2, we have $(-1)^k k! h_k = E^k D_\sigma^{(k)} g$. This combined with (1.19a, b) (or (3.11) combined with (2.5) and (3.7)) shows that

(3.12) $P_k(av \cdot {}^t b, u) = \sigma(a, b)P_k(v, {}^t aub)\sigma(a, b)^{-1}$ for every $(a, b) \in \mathfrak{R}$.

Given a representation $\{\rho, X\}$ of $\mathfrak{R}$, we define a "contraction"

(3.13a) $$\theta_X: \mathscr{M}\ell_k(T, \mathscr{M}\ell_k(T, X)) \to X$$

as follows. Write an element φ of $\mathscr{M}\ell_k(T, \mathscr{M}\ell_k(T, X))$ as

$$\varphi(u_1, \ldots, u_k; v_1, \ldots, v_k)$$

with two sets of k variables $u_1, \ldots, u_k$ and $v_1, \ldots, v_k$ in T. Then we put

(3.13b) $$\theta_X \varphi = \sum \varphi(c_1, \ldots, c_k; c_1, \ldots, c_k),$$

where $c_1, \ldots, c_k$ run independently over the standard basis of T. We see easily that

(3.14) $\theta_X \circ (\rho \otimes \tau_k \otimes \pi_k)(a, b) = \theta_X \circ (\rho \otimes \pi_k \otimes \tau_k)(a, b)$

$$= \rho(a, b) \circ \theta_X \quad \text{for every } (a, b) \in \mathfrak{R}.$$

Now we view $\mathscr{S}_k(T, \mathscr{S}_k(T, X))$ as a subspace of $\mathscr{M}\ell_k(T, \mathscr{M}\ell_k(T, X))$. Thus if g is an $\mathscr{S}_k(T, X)$-valued function, then $\theta_X D_{\rho \otimes \pi_k}^{(k)} g$ is meaningful as an X-valued function.

PROPOSITION 3.4. *Let* $\rho(a, b) = \det(b)^\nu \rho_0(a, b)$ *for* $(a, b) \in \mathfrak{R}$ *with* $\nu \in \mathbf{Z}$ *and a representation* $\{\rho_0, X\}$ *of* $\mathfrak{R}$*; let* $f \in C^\infty(\mathscr{D}, X)$*. Suppose* $E^{k+1}f = 0$ *and* $f\|_\rho \gamma = f$ *for every* γ *in a subgroup* Γ *of* G*. If* ν *is larger than an integer* $N(\rho_0, k)$ *depending only on* ρ_0 *and* k*, then*

$$f = \sum_{e=0}^{k} \theta_X D_{\rho \otimes \pi_e}^{(e)} g_e$$

with $\mathscr{S}_e(T, X)$*-valued holomorphic functions* g_e *on* $\mathscr{D}$ *such that* $g_e\|_{\rho \otimes \pi_e} \gamma = g_e$ *for every* $\gamma \in \Gamma$*.*

Proof. We prove this by induction on k. The case $k = 0$ is trivial. Put $\sigma_0 = \rho \otimes \pi_k$, $Y = \mathscr{S}_k(T, X)$, and $\sigma(a, b) = \det(b)^\nu \sigma_0(a, b)$. We have $d\sigma(x, y) = \nu \mathrm{tr}(y) 1_Y + d\sigma_0(x, y)$ and hence (3.7) shows that $\omega_\sigma(r, u) = \nu \mathrm{tr}({}^t r u) 1_Y + \omega_{\sigma_0}(r, u)$, where 1_Y is the identity element of $\mathrm{End}(Y)$. Now we observe that the polynomial $P_k(v, u)$ of (3.10) for the present σ can be written in the form

$$P_k(v, u) = \sum_{i=0}^{k} \nu^i q_i(v, u)$$

with polynomials q_i depending only on ρ and k, and moreover, $q_k(v, u) = \mathrm{tr}({}^t v u)^k 1_Y$. This can be verified by induction on k by virtue of (3.11). Identify an element $\varphi(u)$ of $\mathscr{S}_k(T, X)$ with an element of $\mathscr{M\ell}_k(T, X)$ and write its values $\varphi(u_1, \ldots, u_k)$ for $u_1, \ldots, u_k \in T$. Similarly, when v is fixed, write $P_k(v; u_1, \ldots, u_k)$ and $q_i(v; u_1, \ldots, u_k)$ for the elements of $\mathscr{M\ell}_k(T, \mathrm{End}(Y))$ identified with $P_k(v, u)$ and $q_i(v, u)$. Define $A, B_i \in \mathrm{End}(Y)$ by

$$(A\varphi)(v) = \theta_X\{[P_k(v, *)\varphi](*)\} = \sum_c [P_k(v; c_1, \ldots, c_k)\varphi](c_1, \ldots, c_k),$$

$$(B_i\varphi)(v) = \sum_c [q_i(v; c_1, \ldots, c_k)\varphi](c_1, \ldots, c_k)$$

for $\varphi \in Y$ with $c_1, \ldots, c_k$ as in (3.13b). By Lemma 3.1, we have $B_k = 1_Y$, and hence $A = \nu^k 1_Y + \sum_{i=0}^{k-1} \nu^i B_i$. Now with g and h_j as in (3.9), we have $\theta_X D_\sigma^{(k)} g = \sum_{j=0}^{k} \theta_X h_j$ and $\theta_X h_k = (Ag)(r)$. Observe that A is invertible for ν greater than an integer, say N_0. Suppose f is a given function of our proposition. Then, as observed before, $f = \sum_{j=0}^{k} s_j(r)$ with $\mathscr{S}_j(T, X)$-valued holomorphic functions s_j on $\mathscr{D}$. By Lemma 3.2, we have $E^k f = (-1)^k k! s_k$, and hence $s_k\|_\sigma \gamma = s_k$ for every $\gamma \in \Gamma$ by (1.19b). Define $N(\rho_0, k) = \mathrm{Max}(N_0, N(\rho_0, k-1))$. Suppose $\nu > N(\rho_0, k)$ and put $g = A^{-1} s_k$. Then g is holomorphic. Moreover, (3.12) and (3.14) show that A commutes with every element of $\sigma(\Re)$, and hence $g\|_\sigma \gamma = g$ for every $\gamma \in \Gamma$. Observe that $f - \theta_X D_\sigma^{(k)} g$ is a polynomial in r of degree $< k$. Applying our induction to this polynomial, we complete our proof.

Let us now specialize the above result to the case of degree 1, in which the result can be stated in a more clear-cut way.

LEMMA 3.5. *Let $\rho(a, b) = \det(b)^\nu$ with $\nu \in \mathbf{Z}$, and let g be a holomorphic $\mathscr{S}_1(T)$-valued function on $\mathscr{D}$. Then*

$$\theta_C D_{\rho \otimes \pi} g = \theta_C D g + (\nu - N)\{D \log \delta, g\},$$

where $N = 2n$, $l + m$, and n in Cases UT, UB, *and* O, *respectively.*

Proof. We see easily that if $h \in \mathscr{S}_1(T)$, then

$$[d(\rho \otimes \pi)(a, b)h](u) = \nu \mathrm{tr}(b)h(u) - h(au) - h(u \cdot {}^t b),$$

and therefore, by (3.7) and (3.8), we have, for $u, v \in T$,

$$(D_{\rho \otimes \pi} g)(u, v) = (Dg)(u, v) + \nu \operatorname{tr}({}^{t}rv)g(u) - g(u \cdot {}^{t}vr)$$

$$- \begin{cases} g(r \cdot {}^{t}vu) & \text{(Cases UT, UB),} \\ g((r \cdot {}^{t}v - v \cdot {}^{t}r)u) & \text{(Case O).} \end{cases}$$

Therefore we obtain our result by a direct calculation.

PROPOSITION 3.6. *Let $f \in C^{\infty}(\mathscr{D})$; suppose $E^2 f = 0$ and $f\|_{\rho}\gamma = f$ for every γ in a subgroup Γ of G with $\rho(a, b) = \det(b)^{\nu}$. If ν is different from the integer N of Lemma 3.5, then $f = \theta_{\mathbf{C}} D_{\rho \otimes \pi} p + q$ with holomorphic functions p and q with values in $\mathscr{S}_1(T)$ and $\mathbf{C}$ such that $q\|_{\rho}\gamma = q$ and $p\|_{\rho \otimes \pi}\gamma = p$ for every $\gamma \in \Gamma$.*

Proof. Let $p = (N - \nu)^{-1}Ef$. By Lemmas 3.2 and 3.5, we have $Ef = E\theta_{\mathbf{C}}D_{\rho \otimes \pi}p$, which proves our assertion.

Expression (3.1) or (3.3) simply tells that f is a polynomial in r with holomorphic functions as coefficients. Some of the functions we shall later produce in our main theorems involve only polynomials in r of rather restricted types. The rest of this section is devoted to the investigation of such types. We first treat Cases UB and UT assuming $l \geq m$ in Case UB. Our convention that $l = m = n$ in Case UT is still in force.

Let R_m be the subgroup of $\mathrm{GL}_m(\mathbf{C})$ consisting of all upper triangular matrices. Now each τ_e-irreducible subspace Z of $\mathscr{S}_e(T)$ is spanned by the functions of the form $\zeta(u) = \zeta_0(aub)$ with $a \in \mathrm{GL}_l(\mathbf{C})$ and $b \in \mathrm{GL}_m(\mathbf{C})$ with a special function ζ_0 defined by

$$(3.15) \qquad\qquad \zeta_0(u) = \prod_{i=1}^{m} \det{}_i(u)^{c_i} \qquad\qquad (u \in T)$$

with $0 \leq c_i \in \mathbf{Z}$, where $\det_i(u)$ is the determinant of the upper left i^2 entries of u. The "highest weight" of Z is then $(s_1, \ldots, s_m)$, $s_i = c_i + \cdots + c_m$. Moreover, ζ_0 is characterized, up to constant factors, by the property that it is an eigenvector of $\tau_e(R_l, R_m)$. (For these facts, see [12, Theorem 2.A], for example.)

LEMMA 3.7. (i) *Let Z and Z' be different irreducible subspaces of $\mathscr{S}_e(T)$. Then $\{Z, Z'\} = 0$, where $\{\ ,\ \}$ is defined by (1.14).*

(ii) *Let Z and W be irreducible subspaces of $\mathscr{S}_e(T)$ and $\mathscr{S}_{e+k}(T)$ of highest weight $(s_1, \ldots, s_m)$ and $(p_1, \ldots, p_m)$, respectively. Suppose $(s_1, \ldots, s_m) > (p_1, \ldots, p_m)$ in the lexicographic order. Then $\{hZ, W\} = 0$ for every $h \in \mathscr{S}_k(T)$.*

Proof. Define ζ_0 by (3.15) and define similarly ζ_0' for Z'. By (1.15), we have

$$(3.16) \qquad\qquad \{\tau_e(a, b)g, h\} = \{g, \tau_e({}^{t}a, {}^{t}b)h\}$$

for g, $h \in \mathscr{S}_r(T)$, $a \in \mathrm{GL}_l(\mathbf{C})$, and $b \in \mathrm{GL}_m(\mathbf{C})$. Taking a and b to be diagonal, we see immediately that $\{\zeta_0, \zeta_0'\} = 0$. Obviously $\{\zeta_0, \tau_e(R_l, R_m)\zeta_0'\} = 0$, and by (3.16), $\{\zeta_0, \tau_e({}^t R_l, {}^t R_m)\zeta_0'\} = 0$. Since ${}^t R_l R_l \times {}^t R_m R_m$ is dense in $\mathrm{GL}_l \times \mathrm{GL}_m$, we see that $\{\zeta_0, Z'\} = 0$, and hence $\{Z, Z'\} = 0$ again by (3.16). To prove (ii), take a vector ξ of W of the highest weight. Let $\{h_i\}$ be a basis of $\mathscr{S}_k(T)$ consisting of eigenvectors of the diagonal elements of $\mathrm{GL}_l \times \mathrm{GL}_m$. Suppose $(p_1, \ldots, p_m) < (s_1, \ldots, s_m)$. Then $\{h_i\zeta_0, \xi\} = 0$, and hence $\{h\zeta_0, \xi\} = 0$ for every $h \in \mathscr{S}_k(T)$. Again by (3.16), we see that $\{h \cdot \tau_e({}^t R_l, {}^t R_m)\zeta_0, \xi\} = 0$, so that $\{\mathscr{S}_k(T)Z, \xi\} = 0$, from which we obtain (ii).

Let Z be a τ_e-irreducible subspace of $\mathscr{S}_e(T)$ of highest weight $(s_1, \ldots, s_m)$. We call Z *elementary* or *nonelementary* according as $s_1 \leq 1$ or $s_1 > 1$. If $e \leq m$, $\mathscr{S}_e(T)$ contains a unique elementary subspace.

LEMMA 3.8. *For a holomorphic map* $g\colon \mathscr{D} \to \mathscr{S}_k(T, X)$, $k \leq m$, *the following four conditions are equivalent:*

 (i) $\Phi_W g = 0$ *for every nonelementary subspace W of $\mathscr{S}_k(T)$;*

 (ii) $E_W\{(D \log \delta)^k, g\} = 0$ *for every nonelementary $W \subset \mathscr{S}_k(T)$;*

 (iii) $E_Z\{(D \log \delta)^k, g\} = 0$ *for every nonelementary $Z \subset \mathscr{S}_j(T)$ and every* $j \leq k$;

 (iv) *If* $\{x_1, \ldots, x_N\}$ *is a basis of X over* $\mathbf{C}$, *then* $g = \sum_{i=1}^N g_i x_i$ *with holomorphic maps g_i of $\mathscr{D}$ into the elementary subspace of $\mathscr{S}_k(T)$.*

Proof. Assume (ii). By Lemma 3.2, $\{\zeta, g\} = 0$ for every $\zeta \in W$ and every nonelementary $W \subset \mathscr{S}_k(T)$; hence (i) follows. If Z_k is the elementary subspace of $\mathscr{S}_k(T)$, then by Lemma 3.7, Z_k consists of all $h \in \mathscr{S}_k(T)$ such that $\{W, h\} = 0$ for all nonelementary W in $\mathscr{S}_k(T)$. Therefore (i) and (iv) are equivalent. Assume (iv). By Lemma 3.7, $\{\mathscr{S}_{k-j}(T)Z, g\} = 0$ for every nonelementary $Z \subset \mathscr{S}_j(T)$, $j \leq k$. Therefore, by Lemma 3.2, $E_Z\{(D \log \delta)^k, g\} = 0$ for every such Z, which is (iii). Obviously (iii) implies (ii).

PROPOSITION 3.9. *Let* $f \in C^\infty(\mathscr{D}, X)$ *in Cases* UT *and* UB; *suppose* $E_Z f = 0$ *for every nonelementary irreducible subspace Z of $\mathscr{S}_j(T)$, $j \leq m + 1$. Then*

$$(3.17) \qquad f(z) = \sum_{k=0}^m \left\{(D \log \delta)^k, g_k\right\}$$

with a holomorphic map $g_k\colon \mathscr{D} \to \mathscr{S}_k(T, X)$ for each $k \leq m$ satisfying the conditions of Lemma 3.8.

Proof. Every irreducible subspace of $\mathscr{S}_{m+1}(T)$ is nonelementary. Thus $E^{m+1}f = 0$, and hence f can be given as $f = \sum_{k=0}^m \{(D \log \delta)^k, g_k\}$ with holomorphic maps $g_k\colon \mathscr{D} \to \mathscr{S}_k(T, X)$. If Z is a nonelementary subspace of $\mathscr{S}_j(T)$ and $j \leq m$, then $0 = E_Z f = \sum_{k=0}^m E_Z\{(D \log \delta)^k, g_k\}$. Since the k-th term of the last sum is 0 for $k < j$ and is of degree $k - j$ in r if $k \geq j$, each term must vanish, which proves our proposition.

We now turn to Case O and study the elements of $C^\infty(\mathscr{Z}, X)$ annihilated by EE_{Z_1} and E_Z, where X is a fixed complex vector space, and

$$(3.18a) \qquad\qquad Z_1 = \mathbf{C}\zeta_1, \qquad \zeta_1(u) = {}^t\! uu,$$

(3.18b) Z is the subspace of $\mathscr{S}_2(T)$ spanned by $({}^t\! cu)^2$ for all $c \in T$ such that ${}^t\! cc = 0$.

These symbols Z, Z_1, and ζ_1 will have the same meaning throughout the rest of this section. If $n = 1$, the statements concerning E_Z should be ignored, since $Z = \{0\}$.

LEMMA 3.10. *If $h \in \mathscr{S}_2(T, X)$ and $h(y) = 0$ for every $y \in T$ such that ${}^t\! yy = 0$, then $h = \zeta_1 h_0$ with a constant $h_0 \in X$.*

This is an easy exercise.

LEMMA 3.11. *If $f \in C^\infty(\mathscr{Z}, X)$, $E_Z f = 0$, and $(E_{Z_1} f)\zeta_1 = \varphi$, then $nE^2 f = \zeta_1 \varphi$.*

Proof. Put $g = E^2 f$ and $\zeta_c(u) = ({}^t\! cu)^2$ for $c \in T$, ${}^t\! cc = 0$. Then $0 = (E_Z f)\zeta_c = (\Phi_Z g)\zeta_c = \{\zeta_c, g\} = g(c)$ by Lemma 3.1. Hence, by Lemma 3.10, $g = \zeta_1 h$ with an X-valued function h. Then $(E_{Z_1} f)\zeta_1 = \{\zeta_1, g\} = nh$, which proves our lemma.

PROPOSITION 3.12. *The following three conditions on $f \in C^\infty(\mathscr{Z}, X)$ are equivalent:*

    (i) $E_Z f = 0$ and $EE_{Z_1} f = 0$;

    (ii) $E^2 f = 2\zeta_1 a$ with a holomorphic $a: \mathscr{Z} \to X$;

    (iii) $f = {}^t\! rr \cdot a + \{D \log \eta, b\} + c$ *with r of (2.3) and holomorphic functions a, b, and c on $\mathscr{Z}$ with values in X, $\mathscr{S}_1(T, X)$, and X, respectively.*

Proof. That (i) follows from (ii) can be verified immediately. Assuming (i), put $2na = (E_{Z_1} f)\zeta_1$. Then $Ea = 0$, so that a is holomorphic. Therefore (ii) follows from Lemma 3.11. Suppose $E^2 f$ has the form of (ii). Then $E^3 f = 0$, and hence, by Proposition 2.4, f has form (3.3) with $k = 2$. By Lemma 3.2, we have $E^2 f = 2g_2$. By our assumption, this is equal to $2\zeta_1 a$. Then, by Lemma 3.1, $\{(D \log \eta)^2, g_2\} = \zeta_1(r)a$, which proves that (ii) $\Rightarrow$ (iii). Conversely, assume (iii). Then $E^2 f = E^2({}^t\! rr \cdot a)$. Now ${}^t\! rr \cdot a = \{(D \log \eta)^2, \zeta_1 a\}$, and hence, by Lemma 3.2, we have $E^2({}^t\! rr \cdot a) = 2\zeta_1 a$, which is (ii).

Remark 3.13. In this paper, we treat only orthogonal and unitary groups. However, the results of Sections 1, 2, and 3 can be proved also for other types of groups acting on hermitian symmetric domains in a parallel way. We note here the objects G, T, $\mathscr{D}$, ξ, η, δ, and r for the groups of Types C and D.

Type C: $G = \mathrm{Sp}(n, \mathbf{R})$, $T = \{z \in \mathbf{C}_n^n | {}^t z = z\}$, $\mathscr{D} = T \cap \mathscr{H}_n$, $\xi(z) = \eta(z) = i(\bar{z} - z)$, $\delta = \det(\xi)$, $r(z) = (z - \bar{z})^{-1}$.

Type D: $G = \{\alpha \in \mathrm{SL}_{2n}(\mathbf{C}) | \alpha^* I_{n,n} \alpha = I_{n,n}, \ \alpha J_n = J_n \bar{\alpha}\}$, $T = \{z \in \mathbf{C}_n^n | {}^t z = -z\}$, $\mathscr{D} = T \cap \mathscr{B}_{n,n}$, $\xi(z) = \eta(z) = 1_n + \bar{z}z$, $\delta = \det(\xi)$, $r(z) = \xi(z)^{-1}\bar{z}$.

In both cases, formulas (2.4a), (2.6), (2.7), (2.8), and (2.9) hold with the modification that $(\overline{D}r)(v) = \xi^{-1}v \cdot {}^t\xi^{-1}$ and $(\overline{D}D \log \delta)(u, v) = \mathrm{tr}({}^t u \xi^{-1} v \cdot {}^t \xi^{-1})$ for Type D; $(Dr)(u) = -r \cdot {}^t ur$ for both types.

Remark 3.14. As can naturally be expected, the operators D_ρ and E correspond to the action of the elements of $\mathfrak{p}_+$ and $\mathfrak{p}_-$ (in the trade jargon) if we transfer the functions on $\mathscr{D}$ to those on G. This gives another proof of the first assertion of Proposition 2.5.

4. Main theorems in the orthogonal case

Given a direct sum K of algebraic number fields $K_1, \ldots, K_t$, we denote by $J(K)$ or J_K the set of all nontrivial homomorphisms of K into $\mathbf{C}$. We shall often identify J_K with the disjoint union of $J(K_1), \ldots, J(K_t)$. If α is a subset of J_K (often J_K itself) and W is a set, we denote by W^α "the product of α copies of W," which is the set of all indexed elements $(w_\sigma)_{\sigma \in \alpha}$ with w_σ in W. Now every $\sigma \in J_K$ can be extended to a C-linear homomorphism of $K \otimes_\mathbf{Q} \mathbf{C}$ into $\mathbf{C}$. For $x \in K \otimes_\mathbf{Q} \mathbf{C}$, we denote by x^σ or x_σ the image of x under this homomorphism, and identify $K \otimes_\mathbf{Q} \mathbf{C}$ with $\mathbf{C}^{J(K)}$ through the map $x \mapsto (x^\sigma)$. We put also $I_K = \mathbf{Z}^{J(K)}$, $\mathbf{R}I_K = \mathbf{R}^{J(K)}$, and $\mathbf{C}I_K = \mathbf{C}^{J(K)}$. For $x, c \in K \otimes_\mathbf{Q} \mathbf{C}$ $(= \mathbf{C}I_K)$, we put $x^c = \prod_{\sigma \in J(K)}(x_\sigma)^{c_\sigma}$ whenever each factor $(x_\sigma)^{c_\sigma}$ is well-defined. We shall often identify a subset α of J_K with the element $\sum_{\sigma \in \alpha}\sigma$ of I_K; thus $x^\alpha = \prod_{\sigma \in \alpha}x^\sigma$. In particular, we denote by ι_K or $\iota(K)$ the element of I_K corresponding to J_K; thus $x^{\iota(K)} = N_{K/\mathbf{Q}}(x)$ for $x \in K$. For $c, c' \in \mathbf{R}I_K$, we write $c \geqq c'$ if $c_\sigma \geqq c'_\sigma$ for all $\sigma \in J_K$. When $K_1, \ldots, K_t$ have a common subfield L, we denote by $\mathrm{Res}_{K/L}$ the linear map of $\mathbf{C}I_K$ into $\mathbf{C}I_L$ which sends each element of $J(K_i)$ to its restriction to L.

Throughout this section, we denote by F the direct sum of totally real algebraic number fields $F_1, \ldots, F_t$, and put

(4.1a) $\qquad \mathbf{e}_F(z) = \mathbf{e}\left(\sum_{\sigma \in J(F)}z_\sigma\right) \quad$ for $z = (z_\sigma) \in \mathbf{C}^{J(F)} = F \otimes_\mathbf{Q} \mathbf{C}$,

(4.1b) $\qquad\qquad\qquad \mathbf{e}(u) = e^{2\pi i u} \quad$ for $u \in \mathbf{C}$.

We can identify $F \otimes_Q \mathbf{R}$ with $\mathbf{R}^{J(F)}$ in an obvious way. For $x \in \mathbf{R}^{J(F)}$ we write $x \gg 0$ if $x_\sigma > 0$ for all $\sigma \in J_F$.

We now fix a common subfield E of $F_1, \ldots, F_t$ and assume that J_F has a subset ε with the following two properties:

$(4.2a)$ $\qquad\qquad\qquad$ $\mathrm{Res}_{F/E}$ *gives a bijection of ε onto J_E;*

$(4.2b)$ $\qquad\qquad\qquad$ $\varepsilon \cap J(F_i) \neq \varnothing$ *for every i.*

We let ε' denote the complement of ε in J_F, i.e., $\varepsilon' = \iota_F - \varepsilon$.

We take a vector space V_i of dimension $n_i + 2$ over F_i for each i with $n_i \geq 0$, and an F_i-bilinear symmetric form $S_i \colon V_i \times V_i \to F_i$. We put $V = \prod_{i=1}^t V_i$, $S = \{S_1, \ldots, S_t\}$ and

$$G(S) = \prod_{i=1}^t G(S_i),$$

$$G(S_i) = \{\alpha \in \mathrm{SL}(V_i) \mid S_i(\alpha x, \alpha x) = S_i(x, x)\}.$$

For each $\tau \in J_F$, take i so that $\tau \in J(F_i)$ and denote by V_τ the τ-completion of V_i and by S_τ the C-bilinear extension of S_i to $V_\tau \otimes_R C$; we put then $n_\tau = n_i$. We can identify $V \otimes_Q C$ with $\prod_{\tau \in J(F)} V_\tau \otimes_R C$. We always assume the following condition:

(4.3) $\quad$ S_τ *has signature $(n_\tau, 2)$ on V_τ for $\tau \in \varepsilon$, and $(n_\tau + 2, 0)$ for $\tau \in \varepsilon'$.*

We put $S_\tau[y] = S_\tau(y, y)$, and denote by $S[x]$, for $x \in V \otimes_Q C$, the element $(S_\tau[x_\tau])_\tau$ of $C^{J(F)}$. We include the case $n_i = 0$, and arrange the V_i so that $n_i > 0$ if and only if $i \leq r$; here $0 < r \leq t$. We put then

$(4.4a)$ $\quad$ $\delta = \varepsilon \cap J(F_1 \times \cdots \times F_r),$ $\qquad$ $\varepsilon - \delta = \varepsilon \cap J(F_{r+1} \times \cdots \times F_t),$

$(4.4b)$ $\quad$ $\delta' = \varepsilon' \cap J(F_1 \times \cdots \times F_r),$ $\qquad$ $\varepsilon' - \delta' = \varepsilon' \cap J(F_{r+1} \times \cdots \times F_t).$

(We shall no longer use δ of $(1.6c)$ in Case O; instead we use η.) For each $\tau \in \delta$ we take a C-linear map A_τ of $V_\tau \otimes_R C$ onto $C^{n_\tau + 2}$ such that $A_\tau(V) \subset \overline{\mathbf{Q}}^{n_\tau + 2}$ and that (cf. [6, Proposition 2.1])

(4.5) $\qquad$ $S_\tau(x, y) = {}^t(A_\tau x) R A_\tau y,$ $\qquad$ $S_\tau(\bar{x}, y) = (A_\tau x)^* Q A_\tau y$

with R and Q of $(1.1c)$, $n = n_\tau$. We then define $\mathscr{Z}_\tau$ to be the space $\mathscr{Z}_n$ of Section 1, and denote by $\mathscr{Z}$ the product of these $\mathscr{Z}_\tau$ for all $\tau \in \delta$. In this section, we use w for the variable on $\mathscr{Z}$, and z for that on $\mathscr{H}_1^{J(E)}$.

If $\alpha \in G(S)$, $A_\tau \alpha_\tau A_\tau^{-1}$ belongs to the group $\tilde{G}$ of $(1.1a)$. We denote by $G_+(S)$ the set of all α such that $A_\tau \alpha_\tau A_\tau^{-1} \in G$ $(=$ the identity component of $\tilde{G})$ for all $\tau \in \delta$. Then we put, for $\alpha \in G_+(S)$ and $w \in \mathscr{Z}$, $\eta_\tau(w) = \eta(w_\tau)$, $\lambda_\tau(\alpha, w) = \lambda(A_\tau \alpha_\tau A_\tau^{-1}, w_\tau)$, $\mu_\tau(\alpha, w) = \mu(A_\tau \alpha_\tau A_\tau^{-1}, w_\tau)$.

We need also certain "factors of automorphy" which are defined for $\tau \in \delta$, $x \in \Pi V_\tau$, and $w \in \mathscr{Z}$, by

$$(4.6a) \qquad \lambda_\tau[x, w] = {}^t(A_\tau x_\tau)Rq(w_\tau),$$

$$(4.6b) \qquad \mu_\tau[x, w] = S_\tau(x_\tau, A_\tau^{-1}p(w_\tau)) = {}^t(A_\tau x_\tau)Rp(w_\tau),$$

where p and q are the matrices of (1.2b, c); $\lambda_\tau[x, w] \in \mathbf{C}^1_{n_\tau}$ and $\mu_\tau[x, w] \in \mathbf{C}$; μ_τ is holomorphic in w. We see easily that

$$(4.7)$$
$$\lambda_\tau[\alpha^{-1}x, w] = \lambda_\tau[x, \alpha w]\lambda_\tau(\alpha, w), \qquad \mu_\tau[\alpha^{-1}x, w] = \mu_\tau[x, \alpha w]\mu_\tau(\alpha, w)$$

for every $\alpha \in G_+(S)$.

Given $\kappa \in \mathbf{Z}^{\delta'}$, ≥ 0, we define $\mathscr{P}_\kappa(S)$ to be the vector space of C-valued polynomial functions q on $\Pi_{\tau \in \delta'}V_\tau$ which are homogeneous of degree κ_τ and S_τ-harmonic on V_τ in the sense of [6, p. 322].

If $n_i = 0$, we may assume, without losing generality, that V_i is a CM-field such that $[V_i: F_i] = 2$ and

$$(4.8a) \qquad S_i(x, y) = \mathrm{Tr}_{V_i/F_i}(\beta_i x y^\rho)$$

with an element $\beta_i \in F_i$, where ρ denotes complex conjugation. Condition (4.3) implies that, for $\tau \in J(F_i)$, we have $\beta_i^\tau < 0$ if $\tau \in \varepsilon$ and $\beta_i^\tau > 0$ if $\tau \in \varepsilon'$. We put

$$(4.8b) \qquad K = V_{r+1} \times \cdots \times V_t$$

and fix a subset ψ of J_K such that

$$(4.8c) \qquad \mathrm{Res}_{V_i/F_i}(\psi \cap J(V_i)) = \varepsilon \cap J(F_i) \quad \text{for every } i > r.$$

Define a submodule I_E^0 of $\mathbf{R}I_E$ by $I_E^0 = I_E + (1/2)\mathbf{Z}\iota_E$. For each $l \in I_E^0$, we denote by $\mathscr{M}(l, E)$ the set of all holomorphic modular forms on $\mathscr{H}_1^{J(E)}$ of weight l with respect to congruence subgroups of $\mathrm{SL}_2(E)$; see [7, I, p. 131] for a precise definition.

In order to define a series $f(w, s)$ that is our main object of study, we first take an element

$$(4.9) \qquad \Omega(z) = \sum_a \omega(a)e_E(az) \qquad \left(z \in \mathscr{H}_1^{J(E)}\right)$$

of $\mathscr{M}(l, E)$ with $0 \leq l \in I_E^0$. (We include the case where $l = 0$ and Ω is a constant.) Let U_E denote the group of all totally positive units of E. There is a subgroup U' of U_E of finite index such that $\omega(ua) = u^{l/2}\omega(a)$ for every $u \in U'$ and every $a \in E$. We then say that Ω is U'-invariant, and put $[U'] =$

$[U_E: U']^{-1}$. We now define a series $f(w, s)$ by

$$(4.10) \quad f(w, s) = [U] \sum_{0 \neq v \in V/U} c(v)\omega\big(\mathrm{Tr}_{F/E}(-S[v])\big)S[v]^{j}q(v)v^{\varphi}$$

$$\cdot \mu[v, w]^{-k}\eta(w)^{s\delta}|\mu[v, w]^{\delta}v^{\psi}|^{-2s}.$$

Here $w \in \mathscr{Z}$, $s \in \mathbf{C}$, $c \in \mathscr{L}(V)$ (see *Notation*), $q \in \mathscr{P}_{\kappa}(S)$, $0 \leq \kappa \in \mathbf{Z}^{\delta'}$, $0 \leq j \in I_F$, $\varphi \in I_K$, $0 \leq k \in \mathbf{Z}^{\delta}$; we understand that v^{φ} and v^{ψ} mean x^{φ} and x^{ψ}, when x is the projection of v onto K; U is a subgroup of U_E of finite index such that Ω is U-invariant and $c(uv) = c(v)$ for every $u \in U$ and every $v \in V$; the sum is taken over all Uv with $v \neq 0$. We assume:

$$(4.11) \quad k_{\tau} > 0 \quad \text{for every } \tau \in \delta \quad \text{and} \quad \varphi_{\sigma} \geq 0 \quad \text{for } \sigma \notin \psi \cup \psi\rho;$$

$$(4.12) \quad \mathrm{Res}_{F/E}(k - \kappa - 2j) - \mathrm{Res}_{K/E}(\varphi) - l = a_f l_E \quad \text{with } a_f \in 2^{-1}\mathbf{Z};$$

$$(4.13) \quad \varphi_{\sigma} \geq \varphi_{\sigma\rho} \quad \text{for every } \sigma \in \psi,$$

where ρ denotes complex conjugation. The last condition is inessential, since we can change σ for $\sigma\rho$ without changing $|v^{\psi}|$. Under (4.12), each term of (4.10) depends only on Uv. Moreover, $\mu[v, w]^{\delta}v^{\psi} \neq 0$ if $0 \neq v \in V$ and $\omega(\mathrm{Tr}_{F/E}(-S[v])) \neq 0$ for the same reason as explained in [6, (8.8)], and hence $f(w, s)$ is meaningful at least formally. As shown in [6, Proposition 8.2] (cf. also [9, p. 265]), it converges for sufficiently large $\mathrm{Re}(s)$. Now we have

THEOREM 4.1. *The series $f(w, s)$ can be continued as a meromorphic function to the whole s-plane; more precisely, there exist a nonzero entire function $A(s)$ and a real analytic function $B(w, s)$ on $\mathscr{Z} \times \mathbf{C}$ that is holomorphic in s such that $A(s)f(w, s) = B(w, s)$ for sufficiently large $\mathrm{Re}(s)$. Further put*

$$(4.14) \qquad b_f = a_f + 2 - \sum_{i=1}^{t}(n_i + 2)[F_i: E]/2 \qquad (\in (1/2)\mathbf{Z}).$$

Suppose $b_f \geq 1$. Then $f(w, s)$ is finite at $s = 0$. Moreover, $f(w, 0)$ is holomorphic in w except in the following cases:

$$(4.15) \quad b_f = 2 \text{ or } 3/2, \ \delta \text{ consists of a single element, and } \varphi_{\sigma} < 0 \text{ for every }$$
$$\sigma \in \psi \ (\text{if } t > r).$$

This was given in [9, Theorem 3.2] when $b_f \in \mathbf{Z}$. The above theorem includes also half-integral b_f. The main new feature of the present paper, however, is in the occurrence of the functions studied in Sections 2 and 3, as stated in the following theorems, in which we still assume (4.11), (4.12), and (4.13). To state them, we first take a subset ψ' of J_K so that

$$(4.16) \quad \iota_K = \psi + \psi\rho + \psi' + \psi'\rho \quad \text{and} \quad \varphi_{\sigma} \geq \rho_{\sigma\rho} \quad \text{for every } \sigma \in \psi'.$$

For each $\tau \in \delta$, we let N_τ denote the coefficient of $\mathrm{Res}_{F/E}(\tau)$ in

$$\mathrm{Res}_{F/E}(j) + \mathrm{Res}_{K/E}\left(\sum_{\sigma \in \psi} \varphi_{\sigma\rho}\sigma \right).$$

THEOREM 4.2. *In case (4.15), suppose $b_f = 2$; let τ be the element of J_F that constitutes δ. Suppose further*

(4.17) $\qquad\qquad l_\tau \neq 1/2 \quad \text{if } E = F = \mathbf{Q} \quad \text{and} \quad n_\tau = 1.$

Then $f(w, 0)$ is a polynomial of degree at most $2N_\tau + 2$ in the components of $r(w_\tau)$ with holomorphic functions on $\mathscr{Z} = \mathscr{Z}_\tau$ as coefficients, where r is defined by (2.3). In particular, if $N_\tau = 0$ (still under (4.17)), $f(w, 0)$ is a function on $\mathscr{Z}_\tau$ of the type of Proposition 3.12.

THEOREM 4.3. *Suppose that $b_f = 0$ or $1/2$. Then $f(w, s)$ has at most a simple pole at $s = 1 - b_f$. Suppose further that*

(4.18) $\quad l_\tau \neq 1/2$ *for every $\tau \in \delta$ if $b_f = 0$, $E = F$, and $n_1 = 1$.*

Then the residue of f at $s = 1 - b_f$ is a polynomial in the components of the $r(w_\tau)$ for $\tau \in \delta$ with holomorphic functions on $\mathscr{Z}$ as coefficients; it has degree at most $2N_\tau$ in the components of $r(w_\tau)$. In particular, the residue is holomorphic on $\mathscr{Z}$ if $N_\tau = 0$ for all $\tau \in \delta$.

It should be noted that $N_\tau \neq 0$ in the exceptional case eliminated by (4.18), by virtue of (4.12).

That $f(w, 0)$ can indeed be nonholomorphic for $b_f = 2$ was shown by Indik [2] when $F = E$, $j = 0$ and Ω is a constant.

The holomorphy of the residue for $b_f = 0$ was given in [7, I, Theorem 5.4] and [9, Theorem 3.4]. In the latter, the condition on φ is misstated; it should be given as in the above Theorem 4.3. We add here a complementary result:

PROPOSITION 4.4. *If $t > r$, the following assertions hold:*

(i) If $b_f \geq 1$, the function $f(w, 0)$ vanishes unless $\varphi_{\sigma\rho} < 0$ for every $\sigma \in \psi$;

(ii) If $b_f = 0$, $f(w, s)$ is finite at $s = 1$ unless $\varphi_{\sigma\rho} \leq 0$ for every $\sigma \in \psi$.

This and all the above theorems will be proved in Section 10.

So far we haven't been viewing $f(w, s)$ as an *automorphic form* on $\mathscr{Z}$. Let us now touch on that aspect. Let $\{\rho, X\}$ be a $\overline{\mathbf{Q}}$-rational representation of $\prod_{\tau \in \delta}[\mathrm{SO}(n_\tau, \mathbf{C}) \times \mathbf{C}^\times] \times \prod_{\tau \in \delta'} G(S_\tau)$. Given $h \in C^\infty(\mathscr{Z}, X)$ and $\gamma \in G_+(S)$, we define $h\|_\rho\gamma \in C^\infty(\mathscr{Z}, X)$ by

(4.19) $\qquad (h\|_\rho\gamma)(w) = \rho(\lambda(\gamma, w), \mu(\gamma, w), \gamma')^{-1}h(\gamma w) \qquad (w \in \mathscr{Z}),$

where γ' is the projection of γ on $\prod_{\tau \in \delta'} G(S_\tau)$. We then define the set of all holomorphic automorphic forms $\mathcal{M}_\rho$ on $\mathcal{Z}$ of weight ρ with respect to congruence subgroups of $G(S)$, and also its subset $\mathcal{M}_\rho(\overline{\mathbf{Q}})$ consisting of all arithmetic elements, as in [6, Section 5] and [9, p. 263]. Here we recall only the following fact: for each CM-point w of $\mathcal{Z}$, there is an element $\mathfrak{P}_\rho(w)$ of $\mathrm{GL}(X)$ such that an element g of $\mathcal{M}_\rho$ is arithmetic if and only if $\mathfrak{P}_\rho(w)^{-1}g(w)$ is $\overline{\mathbf{Q}}$-rational for every CM-point w of $\mathcal{Z}$. More generally, given an element h of $C^\infty(\mathcal{Z}, X)$ such that $h\|_\rho\gamma = h$ for every γ in a congruence subgroup of $G(S)$, let us call h *arithmetic* if $\mathfrak{P}_\rho(w)^{-1}h(w)$ is $\overline{\mathbf{Q}}$-rational for every CM-point w of $\mathcal{Z}$.

Now coming back to $f(w, s)$, take $X = \mathcal{S}_1(\mathcal{P}_\kappa(S))$, and define ρ by $[\rho(a, b, c)\varphi]q = b^k\varphi(q \circ c)$, $(q \circ c)(x) = q(cx)$ for $a \in \prod_{\tau \in \delta}\mathrm{SO}(n_\tau, \mathbf{C})$, $b \in (\mathbf{C}^\times)^\delta$, $c \in \prod_{\tau \in \delta'}G(S_\tau)$, $\varphi \in \mathcal{S}_1(\mathcal{P}_\kappa(S))$, $q \in \mathcal{P}_\kappa(S)$, and $x \in \prod_{\tau \in \delta'}V_\tau$. This ρ is $\overline{\mathbf{Q}}$-rational, since $\mathcal{P}_\kappa(S)$ has a $\overline{\mathbf{Q}}$-rational structure. We can view $f(w, s)$ as a function with values in $\mathcal{S}_1(\mathcal{P}_\kappa(S))$. It can easily be verified, by means of (4.7), that $f(w, s)\|_\rho\gamma = f(w, s)$ for every γ in a congruence subgroup of $G(S)$. Now our main theorems of arithmeticity of f can be stated as follows:

THEOREM 4.5. *In addition to (4.11), (4.12) and (4.13), suppose that $b_f \geqq 1$ and also that*

(4.20) *c is $\overline{\mathbf{Q}}$-valued and $\omega(a) \in \overline{\mathbf{Q}}$ for all $a \in E$;*

(4.21) *$\varphi_\sigma \geqq 0$ for every $\sigma \in \psi$ if $t > r$;*

(4.22) *$b_f \neq 3/2$ if $E = \mathbf{Q}$;*

(4.23) *$l_\tau \neq 1/2$ if $b_f = 2$, $F = E = \mathbf{Q}$, and $n_\tau = 1$.*

Then $\pi^{-\nu}p_K(-\varphi, 2\psi)f(w, 0)$ is arithmetic, where $\nu = \sum_{\tau \in \delta}k_\tau + \sum_{\sigma \in \psi}(\varphi_\sigma - \varphi_{\sigma\rho})$, and p_K is the symbol introduced in [6, §1].

THEOREM 4.6. *Suppose $b_f = 0$ or $1/2$. Let $h(w)$ be the residue of $f(w, s)$ at $s = 1 - b_f$ obtained in Theorem 4.3 under (4.18). Suppose that (4.20) is satisfied and also that*

(4.24) *$\varphi_\sigma \geqq 0 > \varphi_{\sigma\rho}$ for every $\sigma \in \psi$ if $b_f = 0$ and $t > r$;*

(4.25) *$\varphi_\sigma \geqq 0$ for every $\sigma \in \psi$ if $b_f = 1/2$ and $t > r$.*

Then $R_E^{-1}\pi^{-\nu}p_K(-\varphi, 2\psi)h$ is arithmetic, where R_E is the regulator of E and

$$\nu = \sum_{\tau \in \delta}k_\tau + \sum_{\sigma \in \psi}(\varphi_\sigma - \varphi_{\sigma\rho}) - \begin{cases} 0 & \text{if } b_f = 0, \\ [E:\mathbf{Q}] & \text{if } b_f = 1/2. \end{cases}$$

The proofs of these two theorems will be given in a subsequent paper. The results of arithmeticity of this type were given in [6], [7], and [9] when $b_f \in \mathbf{Z}$ and the function $f(w, 0)$ or h is holomorphic. The new points here are that the cases of nonholomorphic functions and of half-integral b_f are included. Previously, the case with $b_f = 2$ and $E = \mathbf{Q}$ was always excluded. Now we can include this case under (4.23). We still have two exceptional cases (4.22) and (4.23), of which the nature is not quite clear.

In the above theorems, we have considered the value or the residue of f only at the points $s = 0$, 1, and $1/2$. This is not the best formulation. In fact, it is more natural to treat also the values of $f(w, s)$ at all integer points s such that $-b_f < s \leq 0$ (in so far as $b_f \in \mathbf{Z}$). However, in order to obtain satisfactory results, we have to choose c and Ω in (4.10) more carefully and multiply f by a certain L-function. Then the value at such an s is a polynomial in $r(w_\tau)$ as in Theorems 4.2 and 4.3. This is another raison d'être for the nonholomorphic functions of our type. If $E = \mathbf{Q}$, this is relatively simple, but in the general case, the adelization becomes inevitable. Although a full exposition of the theory in the adelized framework may not be so complicated, it will certainly obscure the essential ideas. Therefore, in the present paper, we content ourselves with merely mentioning this possibility.

5. Main theorems in the unitary case

Let F be a totally real algebraic number field, and K a totally imaginary quadratic extension of F. In this section, we consider a unitary group over K. Although a more general case involving several unitary groups over different CM-fields can be treated in the same manner as in the orthogonal case, we take here only a single unitary group for simplicity. In this section, complex conjugation will be indicated by a superscript ρ, though the bar will also be used (see *Notation*); we put $x^* = {}^t x^\rho$ for a matrix x.

Taking a positive integer q and a hermitian element H of $\mathrm{GL}_q(K)$, we define a group $G(H)$ by

$$(5.1) \qquad G(H) = \left\{ \alpha \in \mathrm{SL}_q(K) \,|\, \alpha H \alpha^* = H \right\}.$$

We fix a CM-type φ of K, identify $K_n^m \otimes_{\mathbf{Q}} \mathbf{R}$ with $(\mathbf{C}_n^m)^\varphi$ through the map $a \otimes b \mapsto (a^\tau b)_{\tau \in \varphi}$, and write a_τ for a^τ occasionally. Let (l_τ, m_τ) be the signature of H^τ for each $\tau \in \varphi$; take an element Q_τ of $\mathrm{GL}_q(\overline{\mathbf{Q}})$ so that

$$(5.2) \qquad \overline{Q}_\tau \cdot \mathrm{diag}\!\left[1_{l_\tau}, -1_{m_\tau}\right] \cdot {}^t Q_\tau = H^\tau.$$

Let $G_{l, m}$ denote the group $G^{(2)}$ of Section 1. Then the map $\alpha \mapsto (Q_\tau^{-1} \alpha^{\tau\rho} Q_\tau)_{\tau \in \varphi}$ embeds $G(H)$ into $\prod_{\tau \in \varphi} G_{l_\tau, m_\tau}$. Putting $\mathscr{B} = \prod_{\tau \in \varphi} \mathscr{B}_{l_\tau, m_\tau}$, we can let $G(H)$ act on $\mathscr{B}$. For $\alpha \in G(H)$ and $w \in \mathscr{B}$, we put

$$(5.3\text{a}) \quad \lambda_\tau(\alpha, w) = \lambda\!\left(Q_\tau^{-1} \alpha^{\tau\rho} Q_\tau, w_\tau\right), \qquad \mu_\tau(\alpha, w) = \mu\!\left(Q_\tau^{-1} \alpha^{\tau\rho} Q_\tau, w_\tau\right),$$

$$(5.3\text{b}) \qquad\qquad j_\tau(\alpha, w) = \det(\mu_\tau(\alpha, w)).$$

The case $l_\tau m_\tau = 0$ is included in (5.3a); see (1.4).

Let $\{\sigma, X\}$ be a $\overline{\mathbf{Q}}$-rational representation of $\prod_{\tau \in \varphi}[\mathrm{GL}(l_\tau) \times \mathrm{GL}(m_\tau)]$. For an X-valued function f on $\mathscr{B}$ and $\alpha \in G(H)$, we define another such function $f\|_\sigma\alpha$ by

$$(5.4) \qquad (f\|_\sigma\alpha)(w) = \sigma\big((\lambda_\tau(\alpha, w), \mu_\tau(\alpha, w))_{\tau \in \varphi}\big)^{-1} f(\alpha w) \qquad (w \in \mathscr{B}),$$

and denote by $\mathscr{M}_\sigma$ the set of all holomorphic X-valued functions on $\mathscr{B}$ which satisfy $f\|_\sigma\gamma = f$ for every γ in a congruence subgroup of $G(H)$ and which are holomorphic at the cusps when $G(H)$ is isomorphic to $\mathrm{SL}_2(\mathbf{Q})$. We can then define the arithmeticity of the elements of $\mathscr{M}_\sigma$ as in [6, §14] and [11, §3].

For $x \in K_q^n$ with a fixed positive integer n and $w \in \mathscr{B}$, we define two matrices $\lambda_\tau[x, w] \in \mathbf{C}_{l_\tau}^n$ and $\mu_\tau[x, w] \in \mathbf{C}_{m_\tau}^n$ by

$$(5.5) \qquad x^{\tau\rho} Q_\tau B(w_\tau) = \big(\overline{\lambda_\tau[x, w]}, \mu_\tau[x, w]\big),$$

where B is defined by (1.2a). Then we have

$$(5.6)$$
$$\lambda_\tau[x\alpha, w] = \lambda_\tau[x, \alpha w]\lambda_\tau(\alpha, w), \qquad \mu_\tau[x\alpha, w] = \mu_\tau[x, \alpha w]\mu_\tau(\alpha, w)$$

for every $\alpha \in G(H)$. (In (5.5) and (5.6), λ_τ or μ_τ should be ignored if $l_\tau = 0$ or $m_\tau = 0$.)

We are going to define a series on $\mathscr{B}$ similar to (4.10). We first define a group $G_{n, K}$ by

$$(5.7) \qquad G_{n, K} = \{\alpha \in \mathrm{SL}_{2n}(K) | \alpha^* J_n \alpha = J_n\}$$

with J_n of (1.1b). Let G_n denote the group $G^{(1)}$ of Section 1. Then $G_{n, K}$ can be embedded into G_n^φ and naturally acts on $\mathscr{H}_n^\varphi$. We put then, for $\alpha \in G_{n, K}$ and $z \in \mathscr{H}_n^\varphi$,

$$(5.8) \qquad \lambda_\tau(\alpha, z) = \lambda(\alpha_\tau, z_\tau), \qquad \mu_\tau(\alpha, z) = \mu(\alpha_\tau, z_\tau).$$

Given a $\overline{\mathbf{Q}}$-rational representation (χ, Y) of $\mathrm{GL}(n)^\varphi \times \mathrm{GL}(n)^\varphi$, we can define the space $\mathscr{M}_\chi$ of holomorphic Y-valued automorphic forms on $\mathscr{H}_n^\varphi$ with respect to congruence subgroups of $G_{n, K}$ in the same fashion as for $G(H)$. We emphasize the basic field K by writing $\mathscr{M}_\chi$ also $\mathscr{M}_{\chi, K}$.

We now take a subfield E of F, a subset ε of J_F, and a subset ψ of φ; we then assume:

$(5.9a)$ $\mathrm{Res}_{K/F}(\psi) = \varepsilon$ and $\mathrm{Res}_{F/E}(\varepsilon) = \iota_E;$

$(5.9b)$ $l_\tau = l \geqq m = m_\tau$ for $\tau \in \psi$ and $m_\tau = 0$ for $\tau \in \varphi - \psi$ with two positive integers l and m such that $q = l + m$.

$(5.9c)$ If $m > 1$, $K = FL$ with a quadratic extension L of E such that $\mathrm{Res}_{K/L}(\varphi) = [K : L]\mathrm{Res}_{K/L}(\psi).$

We fix such an L; (5.9a) implies that $\mathrm{Res}_{K/L}(\psi)$ is a CM-type of L. For this reason, we shall often view ψ as a CM-type of L.

To define our series, let us first assume $m > 1$. We take polynomial representations $\{R, X\}$ and $\{R', X'\}$ of $\mathrm{GL}(m)^{\varphi}$ and $\prod_{\tau \in \varphi} \mathrm{GL}(l_\tau)$, respectively, and also a polynomial map

$$S: \prod_{\tau \in \varphi} \mathbf{C}_m^{l_\tau} \to \mathrm{Hom}(X, X')$$

such that $S(axb) = R'(a)S(x)R(b)$ for $a \in \prod_{\tau \in \varphi} \mathrm{GL}(l_\tau)$, $x \in \prod_{\tau \in \varphi} \mathbf{C}_m^{l_\tau}$ and $b \in \mathrm{GL}(m)^{\varphi}$. We define a representation $\{R_0, X\}$ of $\mathrm{GL}(m)^{\psi}$ by $R_0(b) = R(b')$, where $b'_\tau = b_\sigma$ when $\sigma = \tau$ on L, and a representation $\{\chi, X\}$ of $\mathrm{GL}(m)^{\psi} \times \mathrm{GL}(m)^{\psi}$ by $\chi(a, b) = \det(b)^p R_0({}^t b^{-1})$ with $p \in \mathbf{Z}^{\psi}$. Take an element A of $\mathscr{M}_{X, L}$, which is an X-valued function on $\mathscr{H}_m^{\psi}$ with an expansion

$$(5.10a) \qquad A(z) = \sum a(h)\mathbf{e}\!\left(\sum_{\tau \in \psi} \mathrm{tr}(h_\tau z_\tau) \right) \qquad (z \in \mathscr{H}_m^{\psi}),$$

where $a(h) \in X$, and h runs over hermitian elements of $M_m(L)$. (Since R_0 is a polynomial representation, we see easily that $A = 0$ unless $p \geq 0$; thus we may assume $p \geq 0$ with no loss of generality. We see also that if $p = 0$, R_0 must be trivial, and hence A is a constant.) Let $\mathfrak{o}_L$ be the maximal order of L. We can find a subgroup U_1 of $\mathrm{GL}_m(\mathfrak{o}_L)$ of finite index such that $\det(U_1) \subset E$ and $A\|_\chi \mathrm{diag}[u^*, u^{-1}] = A$ for every $u \in U_1$. Then

$$(5.10b) \qquad a(uhu^*) = \det(u)^p R_0({}^t u^{-1})a(h) \quad \text{for every } u \in U_1.$$

We now define an X'-valued series f by

$$(5.11) \quad f(w, s) = [U] \sum_{v \in U \backslash \mathscr{V}} b(v)(vHv^*)^j \delta(w)^{s\psi} |\det \mu[v, w]|^{-2s\psi}$$

$$\cdot \det(\mu[v, w])^{-k} S({}^t\lambda[v, w]) a\!\left(-\mathrm{Tr}_{K/L}(vHv^*) \right).$$

Here $w \in \mathscr{B}$, $s \in \mathbf{C}$, $\mathscr{V} = \{x \in K_q^m \mid \mathrm{rank}(x) = m\}$, U is a subgroup of $\mathrm{GL}_m(\mathfrak{o}_L)$ of finite index, $[U] = [\mathrm{GL}_m(\mathfrak{o}_L): U]^{-1}$, $b \in \mathscr{L}(K_q^m)$ (see *Notation*), $0 \leq j \in \mathbf{Z}^{\varphi}$, $0 \leq k \in \mathbf{Z}^{\psi}$, δ is defined by (1.6c), and $\mathrm{Tr}_{K/L}(h) = (\mathrm{Tr}_{K/L}(h_{ij}))_{i, j=1}^m$ for $h = (h_{ij}) \in M_m(K)$. The factor $(vHv^*)^j$ should be taken into account only if $m = 1$, in which case the series is defined under the following conventions: $X = \mathbf{C}$, $R(b) = b^r$ with $r \in \mathbf{Z}^{\varphi}$, $A \in \mathscr{M}(p - \mathrm{Res}_{K/E}(r), E)$ with $p \in I_E$ (see Section 4), S is X'-valued and satisfies $S(axb) = b^r R'(a)S(x)$, $\mathrm{Tr}_{K/L}(vHv^*)$ should be replaced by $\mathrm{Tr}_{F/E}(vHv^*)$, U is a subgroup of $\mathfrak{o}_E^{\times}$ of finite index, and $[U] = [\mathfrak{o}_E^{\times} : U]^{-1}$.

We note here a simple fact which will be proved in Section 9:

$$(5.12) \quad \prod_{\tau \in \psi} \det \mu_\tau[v, w] \neq 0 \ \text{ if } \ \mathrm{rank}(v) = m \ \text{ and } \ -\mathrm{Tr}_{K/L}(vHv^*) \ \text{ is to-}$$
tally nonnegative.

Since $a(h) = 0$ unless h is totally nonnegative, the sum of (5.11) is taken over all v such that $-\operatorname{Tr}_{K/L}(vHv^*)$ is totally nonnegative. Therefore (5.12) shows that each term is meaningful. Now to make each term dependent only on Uv, we take U so that $\det(u)$ is totally positive and $b(uv) = b(v)$ for every $u \in U$, and (5.10b) holds with U as U_1. Then the substitution of uv for v multiplies each term with a certain power of $\det(u)$ which becomes 1 if we assume

$$(5.13) \qquad \operatorname{Res}_{K/E}(k - p - 2j) = c_f \iota_E$$

with an integer c_f, where we put $j = 0$ when $m > 1$. It is conjecturable that our series is convergent for sufficiently large $\operatorname{Re}(s)$. In Section 9, we shall at least prove:

PROPOSITION 5.1. *Under* (5.13), *the series of* (5.11) *is absolutely and locally uniformly convergent on* $\mathscr{B} \times \{s \in \mathbf{C} | \operatorname{Re}(s) > \gamma\}$ *for a sufficiently large* γ, *provided that* $m = 1$ *or A is a constant, or A is a cusp form.*

By virtue of (1.8b) and (5.6), we see easily that f is an automorphic form in the sense that

$$(5.14) \qquad f(\alpha w, s) = j_\alpha(w)^k R'\big({}^t\lambda(\alpha, w)^{-1}\big) f(w, s)$$

for every $\alpha \in G(H)$ such that $b(v\alpha) = b(v)$. Now our first main result in the unitary case is as follows:

THEOREM 5.2. *In addition to* (5.13), *suppose that* $k_\tau > 0$ *for every* $\tau \in \psi$, *and the series f of* (5.11) *is locally uniformly convergent on* $\mathscr{B} \times \{s \in \mathbf{C} | \operatorname{Re}(s) > \gamma\}$ *for some* $\gamma \in \mathbf{R}$. *Put* $\kappa = c_f + 2m - (l + m)[F : E]$. *Then the following assertions hold:*

(1) $f(w, s)$ can be continued as a meromorphic function to the whole s-plane in the same sense as in Theorem 4.1.

(2) $f(w, s)$ is finite at $s = 0$ if $\kappa \geq m$.

(3) $f(w, 0)$ is holomorphic in w if $\kappa \geq m$ except when $E = \mathbf{Q}$ and $\kappa = m + 1$.

(4) If $E = \mathbf{Q}$, $\kappa = m + 1$, and $j = 0$, then $f(w, 0)$ is a polynomial function of the components of $(1 - \overline{w} \cdot {}^t w)^{-1} \overline{w}$ of degree $\leq m$ with holomorphic functions on $\mathscr{B}$ as coefficients; more precisely, it is a function of the type described in Proposition 3.9.

(5) If $0 \leq \kappa < m$, $f(w, s)$ has at most a simple pole at $s = m - \kappa$. Moreover, if $j = 0$, its residue is holomorphic in w, except in the following two cases: (i) $\kappa = 0$, $F = E = \mathbf{Q}$, $l = m > 1$, $k = 1$, and $p = 1$; (ii) $\kappa = 0$, $F = E = \mathbf{Q}$, $l = m + 1 > 2$, $k = 1$, and $p = 0$.

It is an open question whether the last two cases of (5) produce counterexamples. If $m = 1$, we have the following additional results:

THEOREM 5.3. *The notation and the assumption being the same as in Theorem 5.2, suppose $m = 1$. For $\tau \in \varepsilon$, let N_τ be the coefficient of $\mathrm{Res}_{F/E}(\tau)$ in $\mathrm{Res}_{F/E}(j)$. Then the following assertions hold:*

(6) *If $\kappa = 0$, the residue of $f(w, s)$ at $s = 1$ is a polynomial function of the components of $r(w_\tau) = (\overline{w}_\tau \cdot {}^t w_\tau - 1)^{-1} \overline{w}_\tau$ for $\tau \in \delta$ with holomorphic functions on $\mathscr{B}$ as coefficients; it has degree at most N_τ in the components of $r(w_\tau)$.*

(7) *If $E = \mathbf{Q}$ and $\kappa = 2$, then $f(w, 0)$ is a function of the type of (6), with $N_\tau + 1$ instead of N_τ.*

To state the arithmeticity of f, we restrict ourselves to the case $m = 1$, though the same type of result is probably true in general. We define the arithmeticity of C^∞ automorphic forms on $\mathscr{B}$ in the same manner as in the orthogonal case.

THEOREM 5.4. *Suppose $m = 1$; let $C = \pi^e p_K(k + r, \varphi)$ with $e = \sum_{\tau \in \psi} k_\tau$. Then $C^{-1} f(w, 0)$ is arithmetic in cases (3) of Theorem 5.2 and (7) of Theorem 5.3. Moreover, let $h(w)$ be the residue of $f(w, s)$ at $s = 1$ in case (6) of Theorem 5.3. Then $R_E^{-1} C^{-1} h(w)$ is arithmetic, where R_E is the regulator of E.*

Theorems 5.2 and 5.3 will be proved in Section 8, and Theorem 5.4 in a subsequent paper. The remark made at the end of Section 4 applies also to the present case.

A special case of (5.11) is worthy of particular mention. Take $X = \mathbf{C}$, $R_0(a) = \det(a)^p$, $A = 1$, and $j = 0$. Then our series takes the form

$$(5.15) \qquad f(w, s) = [U] \sum_{v \in U \backslash \mathscr{V}'} b(v) S({}^t \lambda [v, w]) \delta(w)^{s\psi}$$

$$\cdot \det(\mu[v, w])^{-k} |\det(\mu[v, w])|^{-2s\psi},$$

where $\mathscr{V}' = \{ v \in K_q^m | \mathrm{rank}(v) = m, \mathrm{Tr}_{K/L}(vHv^*) = 0 \}$. This may be viewed as a kind of Eisenstein series. Suppose in particular $F = E$ and $p = 0$. Then S is a constant, and so f and $\mathscr{V}'$ are essentially the same as f and C' of the introduction, when $E = \mathbf{Q}$. (Thus assertions (1), (2), (3), and (4) of the introduction are included in Theorem 5.2.) Furthermore, suppose $l = m$ and $\mathscr{V}' \neq \varnothing$ with E not necessarily equal to $\mathbf{Q}$. Then H is equivalent to $I_{m, m}$, and hence $G(H)$ and $\mathscr{B}$ are isomorphic to $G_{m, K}$ and $\mathscr{H}_m^\varphi$. Therefore a suitable isomorphism of $\mathscr{B}$ to $\mathscr{H}_m^\varphi$ sends f to a series defined on $\mathscr{H}_m^\varphi$ of the form

$$[U] \sum_{(c, d) \in U \backslash \mathscr{X}} \beta(c, d) \delta(z)^{s\varphi} \det(cz + d)^{-k} |\det(cz + d)|^{-2s\varphi},$$

where $z \in \mathscr{H}_m^\varphi$, $\mathscr{X} = \{(c, d) \in K_m^m \times K_m^m | \mathrm{rank}(c, d) = m, \ cd^* = dc^*\}$, and $\beta \in \mathscr{L}(K_m^m \times K_m^m)$. This is a special case of [11, (10.4)]. As proved in [11, (10.6)], it is a finite linear combination of several Eisenstein series of $G_{m, K}$. Therefore the assertions of the above theorems in this special case follow immediately from [10, Theorems 7.1, 7.2, 7.3] and [13, Theorem 2.7]. Thus, in order to prove Theorems 5.2 and 5.3, we may assume that $F \neq E$ or $l > m$ or $p \neq 0$.

6. The effect of E_Z on factors of automorphy

The principal idea of the proof of our main theorems is as follows. We apply E_Z of (1.24) to the series $f(w, s)$ and prove that $E_Z f = 0$ at $s = 0$. If, for example, $Ef = 0$ at $s = 0$, it shows that $f(w, 0)$ is holomorphic in w. In the case of critical weight, we will have to show that $E^N f = 0$ for some $N > 1$. For these reasons, it is necessary to study $E_Z f$, or rather the effect of E_Z on each term of f.

We fix one $\tau \in \psi$ in Case UB and $\tau \in \delta$ in Case O, and put, for $s \in \mathbf{C}$,

$$(6.1) \quad \kappa[x, w, s] = h(w)[\det \eta(w)]^s \cdot \begin{cases} |\det \mu_\tau[x, w]|^{-2s} & (\text{Cases UB, O}), \\ |\det \mu(x, w)|^{-2s} & (\text{Case UT}), \end{cases}$$

with an arbitrary holomorphic function h on $\mathscr{D}$. Here, in Case UT, x is an element of G; in Case UB, x is a vector variable in $\mathbf{C}_q^m$; in Case O, $x \in V_r$; we assume that $\det \mu_\tau[x, w] \neq 0$. Since τ is fixed, we suppress it in the rest of this section. Thus the setting is similar to that of the first three sections; we use w instead of z for the variable on $\mathscr{D}$, however.

Let Z be an irreducible subspace of $\mathscr{S}_e(T)$. We recall a formula

$$(6.2) \quad \zeta(D)\det(aw + b)^s = \beta_Z(s)\det(aw + b)^s \zeta\left({}^t a \cdot {}^t(aw + b)^{-1}\right)$$

in Cases UT and UB, proved in [12, Theorem 4.3]. Here $\zeta \in Z$, $\zeta(D)$ is the operator $\zeta(\partial/\partial w_{ij})$, $a \in \mathbf{C}_l^m$, $b \in \mathbf{C}_m^m$; we assume that $\det(aw + b) \neq 0$; in Case UT, we understand that $l = m = n$; β_Z is a polynomial determined by the highest weight of Z. Its explicit form is given in [12, Theorem 4.1] (see (8.2) below). Now we have

LEMMA 6.1. *Let Z be an irreducible subspace of $\mathscr{S}_e(T)$, and let $\zeta \in Z$.*
  (i) *Case UT:*

$$(E_Z \kappa[x, w, s])\zeta = (-i)^e \beta_Z(-s)$$

$$\cdot \kappa[x, w, s]\zeta\left({}^t\lambda(x, w) \cdot \overline{{}^t\mu(x, w)}^{-1}\eta(w)\right).$$

  (ii) *Case UB:*

$$(E_Z \kappa[x, w, s])\zeta = \beta_Z(-s)$$

$$\cdot \kappa[x, w, s]\zeta\left({}^t\lambda[x, w] \cdot \overline{{}^t\mu[x, w]}^{-1}\eta(w)\right).$$

(iii) *Case* O: *If* $Z = \mathscr{S}_1(T)$ *or* Z *is spanned by the functions of the form* $\zeta(u) = ({}^t au)^e$ *with vectors* $a \in \mathbf{C}^n$ *such that* ${}^t aa = 0$, *then for every* $\zeta \in Z$, *we have*

$$(E_Z \kappa[x, w, s])\zeta = (-1)^e s(s+1)\ldots(s+e-1)$$

$$\cdot \kappa[x, w, s]\zeta\Big({}^t\lambda[x, w]\overline{\mu[x, w]}^{-1}\eta(w)\Big).$$

(iv) *Case* O: *If* $Z_1 = \mathbf{C}\zeta_1$ *and* $\zeta_1(u) = {}^t uu$, *then*

$$\big(E_{Z_1}\kappa[x, w, s]\big)\zeta_1 = s(2s + 2 - n)\kappa[x, w, s+1]\mu[x, w]^2$$

$$+ s(s+1)S[x]\kappa[x, w, s+2]\mu[x, w]^2.$$

Proof. By (6.2) and Lemma 2.1, (i), and Proposition 2.5, (ii), we have

$$(6.3) \quad \big[E_Z\det(ar + b)^s\big]\zeta = (-1)^e \beta_Z(s)\det(ar + b)^s\zeta\big({}^t a \cdot {}^t(ar + b)^{-1}\big),$$

where a and b are holomorphic maps of $\mathscr{D}$ into $\mathbf{C}_l^m$ and $\mathbf{C}_m^m$, respectively. To prove our formulas, it is sufficient to compute the effect of E_Z on $\delta^s\det\overline{\mu[x, w]}^{-s}$, since E_Z is antiholomorphic and κ differs from the last function by a holomorphic factor. In Case UB, we can put $\lambda[x, w] = \bar{c} + \bar{d}\cdot{}^t w$ and $\mu[x, w] = cw + d$ with $c \in \mathbf{C}_l^m$ and $d \in \mathbf{C}_m^m$. Observing that ${}^t\eta^{-1} = 1 - {}^t wr$, we have

$$\delta(w)^s\det\overline{\mu[x, w]}^{-s} = \det\big[(\overline{cw} + \bar{d})\cdot{}^t\eta^{-1}\big]^{-s} = \det\big[(-\bar{c} - \bar{d}\cdot{}^t w)r + \bar{d}\big]^{-s}.$$

Applying (6.3) to this, we find that

$$E_Z\Big(\delta^s\det\overline{\mu[x, w]}^{-s}\Big)\zeta = \beta_Z(-s)\delta^s\det\overline{\mu[x, w]}^{-s}$$

$$\cdot \zeta\Big({}^t\lambda[x, w]\cdot\overline{{}^t\mu[x, w]}^{-1}\eta(w)\Big),$$

which proves Case UB. A similar proof applies to Case UT. In Case O, the formula similar to (6.2) proved in [12, Theorem 4.3] is not sufficient for our present purpose. In fact, we need

LEMMA 6.2. *For* $w \in T$ (*in Case* O) *and* $x = ({}^t\alpha, \beta, \gamma) \in \mathbf{C}_{n+2}^1$ *with* $\alpha \in T$, $\beta, \gamma \in \mathbf{C}$, *put* $M[x, w] = {}^t\alpha w + \beta({}^t ww/2) + \gamma$ *and* $D = {}^t(\partial/\partial w_1, \ldots, \partial/\partial w_n)$.
(1) *If* $\zeta \in Z$ *with* Z *as in Lemma* 6.1, (iii), *then*

$$\zeta(D)M[x, w]^s = s(s-1)\ldots(s-e+1)M[x, w]^{s-e}\zeta(\alpha + \beta w).$$

(2) *If* $\zeta(u) = {}^t uu$, *then*

$$\zeta(D)M[x, w]^s = s(s + (n/2) - 1)M[x, w]^{s-2}\zeta(\alpha + \beta w)$$

$$- (ns/2)({}^t\alpha\alpha - 2\beta\gamma)M[x, w]^{s-2}.$$

These formulas can be verified in a straightforward way; we only need to observe that

$$(6.4) \qquad {}^t(\alpha + \beta w)(\alpha + \beta w) = 2\beta M[x, w] + {}^t\alpha\alpha - 2\beta\gamma.$$

Now, in order to prove (iii) and (iv) of Lemma 6.1, take A_τ as in (4.5) and put ${}^t(A_\tau x) = ({}^ta, b, c)$ with $a \in T$ and $b, c \in \mathbf{C}$. From (4.5), we see that $Q\bar{A}_\tau A_\tau^{-1} = R$, and hence $\bar{a} = a$ and $\bar{b} = c$. By (4.6a, b), (1.2b, c), and (2.3), we have

$${}^t\lambda[x, w] = a - r \cdot {}^twa + br - cw + c({}^tww/2)r,$$

$$\mu[x, w] = {}^taw - c({}^tww/2) - b.$$

Therefore, by virtue of Lemma 2.3, we have

$$\eta(w)^{-1}\overline{\mu[x, w]} = {}^t(cw - a)r + \left({}^taw - c({}^tww/2) - b\right)({}^trr/2) - c.$$

We view this as $M[X, r]$ with $X = ({}^t(cw - a), {}^taw - c({}^tww/2) - b, -c)$. Applying Proposition 2.5, (ii), to $E_Z M[X, r]^{-s}$, and employing again (6.4), we obtain (iii) and (iv) of Lemma 6.1 from (1) and (2) of Lemma 6.2.

LEMMA 6.3. *In Case* O, *let* $\zeta(u) = ({}^tau)^e$ *with* $a \in \mathbf{C}^n$ *and* $\zeta'(u) = ({}^tuu)^d\zeta(u)$; *suppose* $\zeta = 1$ *or* $e = 1$ *or* ${}^taa = 0$. *Then, for* $Z = \mathscr{S}_{2d+e}(T)$, $2d + e > 0$, *we have*

$$(E_Z\kappa[x, w, s])\zeta' = \zeta\left({}^t\lambda[x, w] \cdot \overline{\mu[x, w]}^{-1}\eta(w)\right)\mu[x, w]^{2d}$$

$$\cdot \sum_{i=0}^{d} [\Gamma(s + d + e + i)/\Gamma(s)]$$

$$\cdot q_{e, d, i}(s)S[x]^i|\eta(w)\mu[x, w]^{-2}|^{d+i}\kappa[x, w, s]$$

with polynomials $q_{e, d, i}$ *depending only on* $e, d,$ *and* i.

Proof. Let $\zeta_1(u) = {}^t uu$. By Proposition 2.5, (ii),

$$(E_Z\kappa)\zeta' = (-1)^e \cdot \zeta(\partial/\partial r)\zeta_1(\partial/\partial r)^d\kappa.$$

Therefore we obtain our formula immediately from Lemma 6.1, (iii) and (iv).

In this section, we gave formulas concerning E_Z. The same technique is applicable to D_ρ^Z. For example, if $\rho(a, b) = \det(b)^k$ in Case UT and Z is an irreducible subspace of $\mathscr{S}_e(T)$, then, for every $\zeta \in Z$ and every $\alpha \in G$,

$$\left[D_\rho^Z\left(\delta^s j_\alpha^{-k}|j_\alpha|^{-2s}\right)\right]\zeta = i^e\beta_Z(-k - s)\delta^s j_\alpha^{-k}|j_\alpha|^{-2s}\zeta\left(\xi^{-1} \cdot {}^t\bar{\lambda}_\alpha \cdot {}^t\mu_\alpha^{-1}\right).$$

This was proved in [11, Lemma 1.2] by reducing the problem to the origin of $\mathscr{B}_{n,n}$. A more direct proof can now be given by employing Proposition 2.5, (i) in the same fashion as in the proof of Lemma 6.1, (i).

7. Theta series of hermitian forms

The proofs of our main theorems in the unitary case require transformation formulas of theta series of indefinite hermitian forms. That such series are automorphic forms is well known, but the formulas we need have never been stated before. Therefore we present in this section an exposition of the subject. Our method is to obtain the series as the pullbacks of some theta series on a symplectic group. We first make some notational conventions. For $X = (x_{hi}) \in \mathbf{C}_n^m$ and $Y \in \mathbf{C}_q^p$, we define $X \otimes Y \in \mathbf{C}_{nq}^{mp}$ by

$$X \otimes Y = \begin{bmatrix} x_{11}Y & \cdots & x_{1n}Y \\ \cdots & \cdots & \cdots \\ x_{m1}Y & \cdots & x_{mn}Y \end{bmatrix};$$

if $X^* = X$ and $n = p$, we put $X\{Y\} = Y^*XY$; further if $^tX = X$ and $n = p$, we put $X[Y] = {}^tYXY$.

Let F, K, and φ be as in Section 5. We define $\mathrm{Sp}(n, F)$ and $\mathfrak{H}_n$ by

$$\mathrm{Sp}(n, F) = G_{n,K} \cap M_{2n}(F), \qquad \mathfrak{H}_n = \big\{ z \in \mathscr{H}_n \,\big|\, {}^tz = z \big\}.$$

The action of $\mathrm{Sp}(n, F)$ on $\mathfrak{H}_n^\varphi$ is the restriction of that of $G_{n,K}$ on $\mathscr{H}_n^\varphi$; the factors of automorphy $\lambda_\tau(\alpha, z)$ and $\mu_\tau(\alpha, z)$ can be defined for $\alpha \in \mathrm{Sp}(n, F)$ and $z \in \mathfrak{H}_n$ as special cases of (1.3) and (5.8). Notice that $\lambda_\tau(\alpha, z) = \mu_\tau(\alpha, z)$ if $\alpha \in \mathrm{Sp}(n, F)$ and $z \in \mathfrak{H}_n^\varphi$.

We now define embeddings

$$\omega: G_{n,K} \to \mathrm{Sp}(2n, F), \qquad \Omega: \mathscr{H}_n^\varphi \to \mathfrak{H}_{2n}^\varphi,$$

by first taking an element ζ of $K^\times$ such that $\zeta^\rho = -\zeta$ and putting

$$(7.1a) \qquad \Omega(z)_\tau = T_\tau \mathrm{diag}\big[{}^tz_\tau, z_\tau\big] T_\tau^{-1} U_\tau \qquad (z \in \mathfrak{H}_n^\varphi, \tau \in \varphi),$$

$$(7.1b) \qquad \omega \begin{bmatrix} a & b \\ c & d \end{bmatrix} = \begin{bmatrix} \omega_0(a) & \omega_0(b)U \\ U^{-1}\omega_0(c) & U^{-1}\omega_0(d)U \end{bmatrix} \quad (a, b, c, d \in K_n^n),$$

$$(7.2) \qquad \omega_0(a) = T\,\mathrm{diag}\big[a^\rho, a\big]T^{-1} \qquad\qquad (a \in K_n^n),$$

$$(7.3) \qquad T = \begin{bmatrix} 1_n & 1_n \\ \zeta 1_n & -\zeta 1_n \end{bmatrix}, \qquad U = \mathrm{diag}\big[1_n, \zeta\zeta^\rho 1_n\big].$$

We have then

$$\omega(\gamma)(\Omega(z)) = \Omega(\gamma z), \qquad \omega\big(J_n^{-1}\alpha^*J_n\big) = J_{2n}^{-1} \cdot {}^t\omega(\alpha)J_{2n},$$

$$\mu_\tau(\omega(\gamma), \Omega(z)) = U_\tau^{-1}T_\tau \mathrm{diag}\big[\lambda_\tau(\gamma, z), \mu_\tau(\gamma, z)\big] T_\tau^{-1}U_\tau$$

$$(\gamma \in G_{n,K}, z \in \mathscr{H}_n^\varphi),$$

$$(7.4) \qquad j_\tau(\omega(\gamma), \Omega(z)) = j_\tau(\gamma, z)^2.$$

Define a congruence subgroup Γ of $\mathrm{Sp}(m, F)$ by

$$(7.5) \quad \Gamma = \left\{ \gamma \in \mathrm{Sp}(m, F) \mid a_\gamma \in \mathfrak{o}_m^m, b_\gamma \in (2\mathfrak{d}^{-1})_m^m, c_\gamma \in (2\mathfrak{d})_m^m, d_\gamma \in \mathfrak{o}_m^m \right\},$$

where $a_\gamma, b_\gamma, c_\gamma, d_\gamma$ are the standard a-, b-, c-, d-blocks of γ, $\mathfrak{o}$ denotes the maximal order of F, and $\mathfrak{d}$ the different of F. In [13, Propositions 1.2 and 3.2], we defined a factor of automorphy $h(\gamma, z)$ of weight $1/2$ for $\gamma \in \Gamma$ with the following properties:

$$(7.6a) \qquad h(\gamma, z)^2 = \mathrm{sgn}\big(N_{F/\mathbf{Q}}(\det(d_\gamma)) \big) \left(\frac{F(\sqrt{-1})/F}{\det(d_\gamma)\mathfrak{o}} \right) j(\gamma, z),$$

$$(7.6b) \qquad\qquad \lim_{r \to 0} h(\gamma, ri) = \sum_{x \in R} \mathbf{e}_F\big(-{}^t x d_\gamma^{-1} c_\gamma x / 2 \big),$$

where $i = (i1_m, \ldots, i1_m) \in \mathfrak{H}_m^\varphi$ and $R = L/{}^t d_\gamma L$, $L = (\mathfrak{d}^{-1})_1^m$; $\mathbf{e}_F$ is defined by (4.1a). Now (7.4) together with (7.6a) shows that

$$(7.7) \qquad\qquad h(\omega(\gamma), \Omega(z)) = t_\gamma j_\gamma(z)^\varphi \quad \text{with } |t_\gamma| = 1,$$

where $j_\gamma(z)^\varphi = \prod_{\tau \in \varphi} j_\tau(\gamma, z)$. This notation is a special case of

$$(7.8) \qquad\qquad j_\gamma(z)^\kappa = \prod_{\tau \in \varphi} j_\tau(\gamma, z)^{\kappa_\tau} \overline{j_\tau(\gamma, z)}^{\kappa_{\tau\rho}},$$

defined for $\kappa = \sum_{\tau \in \varphi} [\kappa_\tau \tau + \kappa_{\tau\rho} \tau\rho] \in I_K$, which is in accordance with our notation introduced in Section 4; as before, ρ denotes complex conjugation.

PROPOSITION 7.1. *Let Γ and h be as in (7.5) and (7.6a, b) with $m = 2n$. Then there is a congruence subgroup Δ of $G_{n, K}$ such that $\omega(\Delta) \subset \Gamma$ and $h(\omega(\gamma), \Omega(z)) = j_\gamma(z)^\varphi$ for every $\gamma \in \Delta$.*

To prove this, we need a few facts on hermitian Gauss sums. Let $\mathfrak{o}_K$ and $\mathfrak{d}_K$ denote the maximal order and the different of K. Given an integral ideal $\mathfrak{a}$ of F and an element c of $F^\times$ such that $c \mathfrak{a} \mathfrak{d} + \mathfrak{a} = \mathfrak{o}$, put

$$H(\mathfrak{a}, c) = \sum_{x \in A} \mathbf{e}_F(cxx^\rho) \qquad\qquad (A = \mathfrak{o}_K/\mathfrak{a}\mathfrak{o}_K).$$

LEMMA 7.2. *If $\mathfrak{a}$ is prime to the different of K over F, then $H(\mathfrak{a}, c) = \left(\dfrac{K/F}{\mathfrak{a}} \right) N(\mathfrak{a})$ for every such c.*

Proof. This is similar to the results of Hecke [1, pp. 220–223] on the quadratic Gauss sums, and can be proved by the same technique. Namely, we can reduce the problem to the case where $\mathfrak{a}$ is a power of a prime ideal. In this special case, our result is proved in [10, Lemma 4.2].

LEMMA 7.3. *Let* $\gamma \in G_{n,K} \cap SL_{2n}(\mathfrak{o}_K)$. *If* $\det(d_\gamma)$ *is prime to* $\mathfrak{d}_K$, *then*

$$\sum_{x \in X} e_F\big(x^* b_\gamma d_\gamma^{-1} x\big) = \left(\frac{K/F}{\det(d_\gamma)\mathfrak{o}}\right)\big|N_{F/\mathbf{Q}}(\det(d_\gamma))\big|, \quad X = (\mathfrak{o}_K)^n/d_\gamma(\mathfrak{o}_K)^n.$$

Proof. Fixing γ, let us drop the subscript γ. Put $f = \det(d)$ and $h = fbd^{-1}$. Then $h \in (\mathfrak{o}_K)_n^n$. Since f is prime to $\mathfrak{d}_K$, we can find an element g of $(\mathfrak{o}_K)_n^n$ such that $\det(g)$ is prime to f and g^*hg is congruent to a diagonal matrix modulo $f\mathfrak{o}_K$. We see easily that

$$\big[d(\mathfrak{o}_K)^n\colon f(\mathfrak{o}_K)^n\big] \sum_{x \in X} e_F(x^* bd^{-1}x) = \sum_{y \in Y} e_F(y^* hy/f)$$

with $Y = (\mathfrak{o}_K)^n/f(\mathfrak{o}_K)^n$. Let $r_1, \ldots, r_n$ be the diagonal entries of g^*hg. Then the last sum is equal to

$$(*) \qquad \qquad \prod_{i=1}^{n} \sum_{w \in W} e_F(r_i ww^\rho/f), \qquad \qquad W = \mathfrak{o}_K/f\mathfrak{o}_K.$$

Put $r_i f^{-1}\mathfrak{d} = \mathfrak{a}_i^{-1}\mathfrak{b}_i$ with integral ideals $\mathfrak{a}_i$ and $\mathfrak{b}_i$ of F such that $\mathfrak{a}_i + \mathfrak{b}_i = \mathfrak{o}$. By Lemma 7.2, $(*)$ is equal to

$$(**) \qquad \qquad \prod_{i=1}^{n} [\mathfrak{a}_i\mathfrak{o}_K\colon f\mathfrak{o}_K]\left(\frac{K/F}{\mathfrak{a}_i}\right)N(\mathfrak{a}_i).$$

Put $L = (\mathfrak{o}_K)_n^1$. Since $Lb + Ld = L$, we have $Lh + Lf = Lfd^{-1}$. If f is divisible by a prime ideal $\mathfrak{p}$ of F, we have $L_\mathfrak{p}g^* = L_\mathfrak{p}g = L_\mathfrak{p}$ and $L_\mathfrak{p}fd^{-1}g = L_\mathfrak{p}g^*hg + L_\mathfrak{p}f = L_\mathfrak{p}r + L_\mathfrak{p}f$, where $r = \text{diag}[r_1, \ldots, r_n]$. This shows that L/Lfd^{-1} is isomorphic to $\prod_{i=1}^{n}\mathfrak{o}_K/(f\mathfrak{o}_K + r_i\mathfrak{o}_K)$. Since f is prime to $\mathfrak{d}$, we see that $\mathfrak{a}_i = f(f\mathfrak{o} + r_i\mathfrak{o})^{-1}$. Hence $\mathfrak{a}_1 \ldots \mathfrak{a}_n = f^n \det(fd^{-1})^{-1}\mathfrak{o} = f\mathfrak{o}$, which completes the proof.

Proof of Proposition 7.1. Let $\gamma \in G_{n,K} \cap SL_{2n}(\mathfrak{o}_K)$ and $\omega(\gamma) = \begin{pmatrix} A & B \\ C & D \end{pmatrix}$; suppose $\omega(\gamma) \in \Gamma$ and $2^{-1}b \in M_n(\mathfrak{o}_K)$. By (7.6b), we have

$$\lim_{z \to 0} h(\omega(\gamma), \Omega(z)) = \sum_{w \in W} e_F(-{}^t wD^{-1}Cw/2),$$

where $W = (\mathfrak{d}^{-1})^{2n}/{}^t D(\mathfrak{d}^{-1})^{2n}$. Putting $w = C^{-1}x$, we can transform the last sum to the form $\sum_{x \in X} e_F({}^t xBD^{-1}x/2)$ with $X = \mathfrak{o}^{2n}/D\mathfrak{o}^{2n}$. Define an F-linear isomorphism $p\colon F^{2n} \to K^n$ by $p\begin{pmatrix} u \\ v \end{pmatrix} = u + \zeta v$ for $u, v \in F^n$ with ζ of (7.3). Suppressing again the subscript γ, we have ${}^t xBD^{-1}x = p(x)^* bd^{-1}p(x)$, $p(\mathfrak{o}^{2n}) = \mathfrak{a}^n$, and $p(D\mathfrak{o}^{2n}) = d\mathfrak{a}^n$, where $\mathfrak{a} = \mathfrak{o} + \zeta\mathfrak{o}$. Hence

$$(7.9) \qquad \lim_{z \to 0} h(\omega(\gamma), \Omega(z)) = \sum_{y \in Y} e_F(y^* bd^{-1}y/2), \qquad Y = \mathfrak{a}^n/d\mathfrak{a}^n.$$

Let $\mathfrak{q}$ be the product of all the prime ideals $\mathfrak{p}$ of F such that $\mathfrak{a}_\mathfrak{p} \neq (\mathfrak{o}_K)_\mathfrak{p}$. Suppose $\det(d)$ is prime to $\mathfrak{q}\mathfrak{d}_K$. Then (7.9) holds with $\mathfrak{o}_K$ instead of $\mathfrak{a}$. Since $2^{-1}b \in M_n(\mathfrak{o}_K)$, Lemma 7.3 shows that (7.9) is equal to $\chi(\det(d))N_{F/\mathbf{Q}}(\det(d))$, where $\chi(s) = ((K/F)/s\mathfrak{o})\mathrm{sgn}(N_{F/\mathbf{Q}}(s))$ for s in F prime to $\mathfrak{d}_K$. Therefore t_γ of (7.7) is equal to $\chi(\det(d_\gamma))$, and hence $t_\gamma = 1$ if γ belongs to a sufficiently small congruence subgroup. This completes the proof.

We insert here a simple fact whose proof is an easy exercise:

LEMMA 7.4. *Let H and P be hermitian elements of $\mathrm{GL}_q(\mathbf{C})$. Suppose that $P > 0$ and H has signature (l, m). Then $PH^{-1}P = H$ if and only if $P = A^*A$ and $H = A^*I_{l,m}A$ with $A \in \mathrm{GL}_q(\mathbf{C})$.*

Let us now consider the groups $G(H)$ of (5.1) and $G_{n,K}$ of (5.7). Take $P \in \mathrm{GL}_q(\mathbf{C})^\varphi$ such that $P_\tau^* = P_\tau$ and $P_\tau H_\tau^{-1} P_\tau = H_\tau$. (Notice that $P_\tau = H_\tau$ if $m_\tau = 0$ and $P_\tau = -H_\tau$ if $l_\tau = 0$.) For $z \in \mathscr{H}_n^\varphi$ and $\alpha = \begin{pmatrix} a & b \\ c & d \end{pmatrix} \in G_{n,K}$, put

$$(7.10\mathrm{a}) \qquad \psi(z) = (\psi_\tau(z))_{\tau \in \varphi},$$

$$(7.10\mathrm{b}) \qquad \psi_\tau(z) = (1/2)\{ H_\tau \otimes (z_\tau + z_\tau^*) + P_\tau \otimes (z_\tau - z_\tau^*)\},$$

$$(7.10\mathrm{c}) \qquad \alpha_H = \begin{bmatrix} 1_q \otimes a & H \otimes b \\ H^{-1} \otimes c & 1_q \otimes d \end{bmatrix}.$$

Then $\psi(z) \in \mathscr{H}_{nq}^\varphi$, and $\alpha \mapsto \alpha_H$ gives an injection of $G_{n,K}$ into $G_{nq,K}$. Take $A \in \mathrm{GL}_q(\mathbf{C})^\varphi$ so that $P_\tau = A_\tau^* A_\tau$ and $H_\tau = A_\tau^* I_{l_\tau, m_\tau} A_\tau$, and put $\mathscr{A}_\tau = A_\tau \otimes 1_n$. It can easily be verified that

$$\alpha_H(\psi(z)) = \psi(\alpha(z)), \qquad \eta(\psi_\tau(z)) = P_\tau \otimes \eta(z_\tau),$$

$$\psi_\tau(z) = \mathscr{A}_\tau^* \mathrm{diag}\big[1_{l_\tau} \otimes z_\tau, \, -1_{m_\tau} \otimes z_\tau^*\big]\mathscr{A}_\tau,$$

$$\lambda_\tau(\alpha_H, \psi(z)) = \overline{\mathscr{A}}_\tau^{-1}\mathrm{diag}\big[1_{l_\tau} \otimes \lambda_\tau(\alpha, z), 1_{m_\tau} \otimes \overline{\mu_\tau(\alpha, z)}\big]\overline{\mathscr{A}}_\tau,$$

$$\mu_\tau(\alpha_H, \psi(z)) = \mathscr{A}_\tau^{-1}\mathrm{diag}\big[1_{l_\tau} \otimes \mu_\tau(\alpha, z), 1_{m_\tau} \otimes \overline{\lambda_\tau(\alpha, z)}\big]\mathscr{A}_\tau,$$

$$(7.11) \qquad j(\alpha_H, \psi(z))^\varphi = j_\alpha(z)^{l+m\rho}.$$

We define a series

$$(7.12) \quad f(u, u', z; b) = \mathbf{e}_\varphi^q\big(2 \cdot {}^t u(z - z^*)^{-1}u'\big) \sum_{x \in K_n^q} b(x)\mathbf{e}_\varphi^q(\bar{x}u\overline{A} + xu'A)$$

$$\cdot \mathbf{e}_\varphi^n((1/4)[H\{x\}(z + z^*) + P\{x\}(z - z^*)]),$$

where $u, u' \in (\mathbf{C}_q^n)^\varphi$, $z \in \mathscr{H}_n^\varphi$, $b \in \mathscr{L}(K_n^q)$, and

$$(7.13a) \qquad e_\varphi^n(X) = e_\varphi(\operatorname{tr}(X)) \qquad \text{for } X \in (\mathbf{C}_n^n)^\varphi,$$

$$(7.13b) \qquad e_\varphi(a) = \exp\!\left(2\pi i \sum_{\tau \in \varphi} a_\tau\right) \quad \text{for } a \in \mathbf{C}^\varphi.$$

For every function $g(u, u', z)$ on $(\mathbf{C}_q^n)^\varphi \times (\mathbf{C}_q^n)^\varphi \times \mathscr{H}_n^\varphi$ and $\alpha \in G_{n,K}$, we define $g|_{l,m}\alpha$ to be the function on the same space given by

$$(7.14) \qquad (g|_{l,m}\alpha)(u, u', z) = j_\alpha(z)^{-l-m\rho} g([\alpha, z](u, u'), \alpha(z)),$$

where $[\alpha, z](u, u') = (v, v') \in (\mathbf{C}_q^n)^\varphi \times (\mathbf{C}_q^n)^\varphi$ with

$$v_{\tau i} = \begin{cases} {}^t\mu_\tau(\alpha, z)^{-1} u_{\tau i} & (1 \leq i \leq l_\tau), \\[2mm] \overline{{}^t\lambda_\tau(\alpha, z)}^{\,-1} u_{\tau i} & (l_\tau < i \leq q), \end{cases}$$

$$v'_{\tau i} = \begin{cases} {}^t\lambda_\tau(\alpha, z)^{-1} u'_{\tau i} & (1 \leq i \leq l_\tau), \\[2mm] \overline{{}^t\mu_\tau(\alpha, z)}^{\,-1} u'_{\tau i} & (l_\tau < i \leq q), \end{cases}$$

the subscript i indicating the i-th column of the matrix in question.

PROPOSITION 7.5. *There is an action of $G_{n,K}$ on $\mathscr{L}(K_n^q)$, which is independent of P and A, and written $b \to b_\alpha$ for $b \in \mathscr{L}(K_n^q)$ and $\alpha \in G_{n,K}$, such that*

$$f(u, u', z; b)|_{l,m}\alpha = f(u, u', z; b_\alpha).$$

Moreover, for each b, the set $\{\alpha \in G_{n,K} | b_\alpha = b\}$ contains a congruence subgroup of $G_{n,K}$.

Proof. We combine the above maps ψ and α_H with the maps Ω and ω of (7.1a, b); for the latter maps, we take nq in place of n; thus

$$\mathscr{H}_n^\varphi \xrightarrow{\psi} \mathscr{H}_{nq}^\varphi \xrightarrow{\Omega} \mathfrak{H}_{2nq}^\varphi,$$

$$G_{n,K} \to G_{nq,K} \to \operatorname{Sp}(2nq, F),$$

$$\alpha \mapsto \alpha_H \mapsto \omega(\alpha_H).$$

Define $\Lambda : F^{2nq} \to K_n^q$ by

$$\Lambda\!\begin{pmatrix} h_1 \\ \vdots \\ h_{2q} \end{pmatrix} = \left({}^t h_i - \zeta \cdot {}^t h_{q+i}\right)_{i=1,\ldots,q} \qquad (h_i \in F^n)$$

with ζ of (7.3). Then we see easily that

$$2\Omega(\psi(z))[h] = \operatorname{tr}(\overline{H}\{\Lambda(h)\}(z + z^*) + \overline{P}\{\Lambda(h)\}(z - z^*)).$$

Next define $M: (\mathbf{C}_q^n \times \mathbf{C}_q^n)^\varphi \to (\mathbf{C}^{2nq})^\varphi$ by $M(u, u') = (w_\tau)_{\tau \in \varphi}$ with

$$w_\tau = T_\tau \begin{bmatrix} p_1 \\ \vdots \\ p_{2q} \end{bmatrix}, \qquad (p_1 \;\; \cdots \;\; p_{2q}) = (uA, u'\overline{A})_\tau \quad \text{with } p_i \in \mathbf{C}^n,$$

$$T = \begin{bmatrix} 1_{nq} & 1_{nq} \\ \zeta 1_{nq} & -\zeta 1_{nq} \end{bmatrix}.$$

If $\alpha \in G_{n,K}$, $z \in \mathscr{H}_n^\varphi$, $\beta = \omega(\alpha_H)$ and $Z_1 = \Omega(\psi(z))$, then

$${}^t h M(u, u') = \operatorname{tr}\left(\overline{\Lambda(h)} uA + \Lambda(h) u'\overline{A}\right) \quad \text{for } h \in F^{2nq},$$

$$(Z_1 - \overline{Z}_1)^{-1}[M(u, u')] = 4 \operatorname{tr}\left({}^t u(z - z^*)^{-1} u'\right),$$

$${}^t\mu(\beta, Z_1)^{-1} M(u, u') = M([\alpha, z](u, u')).$$

Let us now consider a theta function

$$\theta^*(U, Z; \sigma) = \mathbf{e}_\varphi\left((Z - \overline{Z})^{-1}[U]/2\right) \sum_{x \in F^{2nq}} \sigma(x) \mathbf{e}_\varphi\left(2^{-1} Z[x] + {}^t x U\right),$$

where $U \in (\mathbf{C}^{2nq})^\varphi$, $Z \in (\mathfrak{H}_{2nq})^\varphi$, and $\sigma \in \mathscr{L}(F^{2nq})$. This is the series introduced in [13, (1.9)]. Let $g(u, u', z; b)$ denote the function defined by taking $\overline{H}, \overline{P}, \overline{A}$ in place of H, P, A in (7.12). Then we see that

$$g(u, u', z; b) = \theta^*(M(u, u'), \Omega(\psi(z)); b \circ \Lambda).$$

Obviously it is sufficient to prove our assertion for g instead of f. By (7.4) and (7.11), we have

$$j_\alpha(z)^{l+m\rho} = j(\alpha_H, \psi(z))^\varphi = \pm j(\beta, \Omega(\psi(z)))^{\varphi/2}.$$

Let $k(Z)$ denote the square root of $j(\beta, Z)^\varphi$ such that $k(\Omega(\psi(z))) = j(\alpha_H, \psi(z))^\varphi$. Obviously this is independent of P and A, and $\alpha \mapsto (\omega(\alpha_H), k)$ gives a homomorphism of $G_{n,K}$ into the group $\mathscr{G}$ introduced in [13, §1]. Therefore, by [13, Proposition 1.1],

$$g(u, u', z; b)\big|_{l,m}\alpha = k(Z_1)^{-1}\theta^*\left({}^t\mu(\beta, Z_1)^{-1} M(u, u'), \beta(Z_1); b \circ \Lambda\right)$$

$$= \theta^*\left(M(u, u'), Z_1; (b \circ \Lambda)^{(\beta, k)}\right) = g(u, u', z; b')$$

with a suitable $b' \in \mathscr{L}(K_n^q)$. By Proposition 7.1, we have $k(Z) = h(\beta, Z)$ if α_H belongs to the group Δ there. Hence, by [13, Proposition 1.3, (iii)], we have $b' = b$ if α belongs to a sufficiently small congruence subgroup. This completes the proof.

We are going to parametrize the above P by the points on the domain $\mathscr{B} = \prod_{\tau \in \varphi} \mathscr{B}_{l_\tau, m_\tau}$ with a fixed H. For $w \in \mathscr{B}$, put

$$\Xi_\tau(w) = \mathrm{diag}\big[\xi(w_\tau), {}^t\eta(w_\tau)\big], \qquad A_\tau = \Xi_\tau(w)^{-1/2} \cdot {}^tB(w_\tau) \cdot {}^tQ_\tau,$$

$$P_\tau(w) = A_\tau^* A_\tau = \overline{Q_\tau} \overline{B(w_\tau)} \Xi_\tau(w)^{-1} \cdot {}^tB(w_\tau) \cdot {}^tQ_\tau.$$

with B of (1.2a) and Q_τ of (5.2). Then we have $H_\tau = A_\tau^* I_{l_\tau, m_\tau} A_\tau$, and

$$(7.15\mathrm{a}) \qquad (P_\tau(w) + H_\tau)\{x^*\} = 2\xi(w_\tau)^{-1}\{\lambda_\tau[x, w]^*\},$$

$$(7.15\mathrm{b}) \qquad (P_\tau(w) - H_\tau)\{x^*\} = 2 \cdot {}^t\eta(w_\tau)^{-1}\{{}^t\mu_\tau[x, w]\} \qquad (x \in K_q^n)$$

with λ_τ and μ_τ of (5.5). We then define a series Θ by

$$(7.16) \quad \Theta(u, u'; z, w; c)$$

$$= e_\varphi^q\big(2 \cdot {}^tu(z - z^*)^{-1}u'\Xi(w)\big) \sum_{x \in V} b(x)$$

$$\cdot e_\varphi^q\big({}^tu\big(\lambda[x, w], \overline{\mu[x, w]}\big) + {}^tu'\big(\overline{\lambda[x, w]}, \mu[x, w]\big)\big)$$

$$\cdot e_\varphi^n\big((1/4)\big[H\{x^*\}(z + z^*) + P(w)\{x^*\}(z - z^*)\big]\big).$$

Here $u, u' \in (C_q^n)^\varphi$, $z \in \mathscr{H}_n^\varphi$, $w \in \mathscr{B}$, $V = K_q^n$, and $b \in \mathscr{L}(V)$. This can be obtained by substituting $u\Xi^{1/2}$, $u'\Xi^{1/2}$, and x^* for u, u', and x in (7.12). We note that

$$(7.17) \qquad H\{x^*\}(z + z^*) + P(w)\{x^*\}(z - z^*)$$

$$= 2H\{x^*\}z + 2 \cdot {}^t\eta(w)^{-1}\{{}^t\mu[x, w]\}(z - z^*)$$

with the understanding that $\eta(w_\tau)$ and μ_τ should be ignored if $m_\tau = 0$. As an immediate consequence of Proposition 7.5, we obtain

PROPOSITION 7.6. *Let $V = K_q^n$. There is an action of $G_{n, K}$ on $\mathscr{L}(V)$, which is independent of w and written $b \mapsto b^\alpha$ for $b \in \mathscr{L}(V)$ and $\alpha \in G_{n, K}$, such that*

$$(7.18) \qquad \Theta(u, u'; z, w; b)|_{l, m}\alpha = \Theta(u, u'; z, w; b^\alpha),$$

where $|_{l, m}\alpha$ is defined by (7.14). Moreover, for each b, the set $\{\alpha \in G_{n, K} | b^\alpha = b\}$ contains a congruence subgroup of $G_{n, K}$.

From (1.3), (1.8a), and (5.3a), we can easily derive

$$(7.19\mathrm{a}) \qquad P_\tau(\beta w) = \beta_\tau P_\tau(w)\beta_\tau^* \qquad (\beta \in G(H)),$$

and hence

$$(7.19\mathrm{b}) \quad \Theta\big(u \cdot \mathrm{diag}\big[{}^t\lambda(\beta, w), {}^t\overline{\mu(\beta, w)}\big],$$

$$u' \cdot \mathrm{diag}\big[{}^t\overline{\lambda(\beta, w)}, {}^t\mu(\beta, w)\big]; z, \beta w; {}^\beta b\big) = \Theta(u, u'; z, w; b),$$

for every $\beta \in G(H)$, where ${}^\beta b(x) = b(x\beta)$.

We now take a polynomial representation $\{R, X\}$ of $GL_n(\mathbf{C})^\varphi \times GL_n(\mathbf{C})^\varphi$ and a polynomial map

$$(7.20) \qquad S:\ \prod_{\tau \in \varphi} \mathbf{C}_{l_\tau}^n \times \prod_{\tau \in \varphi} \mathbf{C}_{m_\tau}^n \to X$$

such that $S(ax, by) = R(a, b)S(x, y)$ for $a, b \in GL_n(\mathbf{C})$, $x \in \prod_\tau \mathbf{C}_{l_\tau}^n$, and $y \in \prod_\tau \mathbf{C}_{m_\tau}^n$. Then we define four types of series Θ_{00}, Θ_{01}, Θ_{10}, and Θ_{11} with values in X by

$$\Theta_{ij}(z, w; b) = \sum_{x \in V} b(x) S_{ij}(x, w)$$

$$\cdot\, e_\varphi^n\big((1/2)zH\{x^*\} + (1/2)(z - z^*) \cdot {}^t\eta(w)^{-1}\{{}^t\mu[x, w]\}\big)$$

for $z \in \mathscr{H}_n^\varphi$, $w \in \mathscr{B}$, and $b \in \mathscr{L}(V)$, where

$$S_{00}(x, w) = S(\lambda[x, w], \mu[x, w]), \qquad S_{01}(x, w) = S\big(\lambda[x, w], \overline{\mu[x, w]}\big),$$

$$S_{10}(x, w) = S\big(\overline{\lambda[x, w]}, \mu[x, w]\big), \qquad S_{11}(x, w) = S\big(\overline{\lambda[x, w]}, \overline{\mu[x, w]}\big).$$

PROPOSITION 7.7. *For every* $\alpha \in G_{n, K}$, *we have*

$$(7.21) \qquad \Theta_{ij}(\alpha z, w; b) = j_\alpha(z)^{l + m\rho} R_{ij}(\alpha, z) \Theta_{ij}(z, w; b^\alpha),$$

where b^α *is the same as in Proposition 7.6, and*

$$R_{00}(\alpha, z) = R\big(\mu(\alpha, z), \overline{\mu(\alpha, z)}\big), \qquad R_{01}(\alpha, z) = R\big(\mu(\alpha, z), \overline{\lambda(\alpha, z)}\big),$$

$$R_{10}(\alpha, z) = R\big(\lambda(\alpha, z), \overline{\mu(\alpha, z)}\big), \qquad R_{11}(\alpha, z) = R\big(\lambda(\alpha, z), \overline{\lambda(\alpha, z)}\big).$$

Proof. Take the last m_τ columns of u and the first l_τ columns of u' to be 0 for all $\tau \in \varphi$. Then the first factor e_φ^q of Θ involving Ξ becomes trivial. Taking the derivatives of (7.18) with respect to the remaining variable components of (u, u'), we obtain our assertion concerning Θ_{00} when S has the form $S(r, s) = \bigotimes_{\tau \in \varphi}(r_\tau \otimes \cdots \otimes r_\tau \otimes s_\tau \otimes \cdots \otimes s_\tau)$. The other three types of Θ_{ij} can be treated in a similar way.

To show that the case with S of a more general type can be reduced to this special case, we consider a group $\mathfrak{G} = \prod_{i=1}^k GL(V_i)$ with finite-dimensional complex vector spaces V_i and its representation $\{R, W\}$ given by $R(a) = \bigotimes_{i=1}^k R_i(a_i)$ and $W = \bigotimes_{i=1}^k W_i$, where $\{R_i, W_i\}$ is an irreducible polynomial representation of $GL(V_i)$ of degree p_i. We let $\mathfrak{G}$ act on $\prod_{i=1}^k V_i^{m_i}$ diagonally. Suppose $S: \prod_{i=1}^k V_i^{m_i} \to W$ is a polynomial map such that $S(ax) = R(a)S(x)$. (In the setting of (7.20), $V_i = \mathbf{C}^n$ and $(m_i)_{i=1}^k = (l_\tau, m_\tau)_{\tau \in \varphi}$. For obvious reasons, it is sufficient to consider R of the type $\bigotimes_i R_i$.) For $t = (t_1, \ldots, t_k)$ with scalars t_i, we have $S(tx) = t^p S(x)$, and hence S is homogeneous of degree p_i on $V_i^{m_i}$. Write an element $x \in \prod V_i^{m_i}$ in the form $x = (x_{ij})$ with $x_{ij} \in V_i$ for $j = 1, \ldots, m_i$.

Then $S(x)$ can be decomposed uniquely into the sum $S(x) = \sum_e S_e(x)$, where $e = (e_{ij})$, $p_i = \sum_{j=1}^{m_i} e_{ij}$, and S_e is homogeneous of degree e_{ij} in the components of x_{ij}. Now there is a unique multilinear map Q_e of $\Pi_{i,j} V_i^{e_{ij}}$ into W such that

$$S_e(x) = Q_e(x_{11}, \ldots, x_{11}, \ldots, x_{ij}, \ldots, x_{ij}, \ldots) \qquad \text{(with } e_{ij} \text{ copies of } x_{ij})$$

and that Q_e is symmetric in each set of e_{ij} vectors on $V_i^{e_{ij}}$. The uniqueness of S_e and Q_e implies that $Q_e(ax) = R(a)Q_e(x)$ for the diagonal action of $a \in \mathfrak{G}$. We can identify Q_e with the corresponding linear map of $\otimes_{i=1}^{k} \otimes_{j=1}^{m_i} (V_i \otimes \cdots \otimes V_i)$ (with e_{ij} copies of V_i) into W. Applying this result to our S of Proposition 7.7, we find that

$$S(x, y) = \sum_{e,e'} T_{e,e'}\left(\bigotimes_{\tau \in \varphi} \left\{ \bigotimes_{j=1}^{l_\tau} (x_{\tau j} \otimes \cdots \otimes x_{\tau j}) \right\} \otimes \left\{ \bigotimes_{j=1}^{m_\tau} (y_{\tau j} \otimes \cdots \otimes y_{\tau j}) \right\} \right),$$

where $x_{\tau j}$ (resp. $y_{\tau j}$) is the j-th column of x_τ (resp. y_τ), and repeated $e_{\tau j}$ times (resp. $e'_{\tau j}$ times), and $T_{e,e'}$ is a linear map which transfers the natural action of $(a, b) \in \mathrm{GL}_n^\varphi \times \mathrm{GL}_n^\varphi$ on the tensor product to $R(a, b)$. Therefore, applying $T_{e,e'}$ to the series obtained at the beginning by taking the derivatives of Θ, and adding over all (e, e'), we obtain the result in the general case.

Remark 7.8. The above proposition can be extended further to the series with coefficients somewhat more general than S_{ij}. To illustrate this, assume for simplicity that φ consists of two elements σ and τ. Then we may replace S_{ij} with

$$T(x, w) = S\big(\lambda_\sigma[x, w], \overline{\lambda_\tau[x, w]}, \overline{\mu_\sigma[x, w]}, \mu_\tau[x, w]\big),$$

for example. Then the new series satisfies (7.21) with R_{ij} replaced by

$$Q(\alpha, z) = R\big(\mu_\sigma(\alpha, z), \lambda_\tau(\alpha, z), \overline{\lambda_\sigma(\alpha, z)}, \overline{\mu_\tau(\alpha, z)}\big).$$

In other words, this series is of type (01) at σ and of type (10) at τ. In general, we can assign any of the four types to each place of φ. Then the corresponding series satisfies (7.21) with R_{ij} modified accordingly.

Remark 7.9. If $m_\tau = 0$ for all $\tau \in \varphi$, (that is, if H is totally positive), the series does not involve w, and Θ_{ij} is a holomorphic automorphic form with respect to a congruence subgroup of $G_{n, K}$.

Remark 7.10. In order to obtain the transformation formula of Θ_{ij} under $G(H)$ in a clear-cut form, we have to decompose S according to the action of $\Pi_{\tau \in \varphi}[\mathrm{GL}(l_\tau) \times \mathrm{GL}(m_\tau)]$ on the right of $\Pi_{\tau \in \varphi}(\mathbf{C}_{l_\tau}^n \times \mathbf{C}_{m_\tau}^n)$. More precisely, we take the above S so that

$$S(ax \cdot {}^t a', by \cdot {}^t b') = \tilde{R}(a, b, a', b')S(x, y)$$

for $(a, b, a', b') \in \prod_{\tau \in \varphi}[\mathrm{GL}(n) \times \mathrm{GL}(n) \times \mathrm{GL}(l_\tau) \times \mathrm{GL}(m_\tau)]$ with a polynomial representation $\tilde{R}$. (Thus $\tilde{R}(a, b, 1, 1)$ is the above R.) Put $R'(a', b') = \tilde{R}(1, 1, a', b')$. Then we can easily verify that

$$(7.22) \qquad \Theta_{ij}(z, \beta w; {}^\beta b) = R'_{ij}(\beta, w)\Theta_{ij}(z, w; b),$$

where ${}^\beta b(x) = b(x\beta)$ and

$$R'_{00}(\beta, w) = R'\big({}^t\lambda(\beta, w), {}^t\mu(\beta, w)\big)^{-1},$$

$$R'_{01}(\beta, w) = R'\big({}^t\lambda(\beta, w), {}^t\overline{\mu(\beta, w)}\big)^{-1},$$

$$R'_{10}(\beta, w) = R'\big({}^t\overline{\lambda(\beta, w)}, {}^t\mu(\beta, w)\big)^{-1},$$

$$R'_{11}(\beta, w) = R'\big({}^t\overline{\lambda(\beta, w)}, {}^t\overline{\mu(\beta, w)}\big)^{-1}.$$

8. Proofs in the unitary case

As mentioned in Section 6, we need to examine $E_Z f$ for the series of (5.11). We focus our attention on one $\tau \in \psi$, put $T = \mathbf{C}_m^l$ as in Section 1, and take an irreducible subspace Z of $\mathscr{S}_e(T)$ with $0 \leq e \in \mathbf{Z}$. We then define E_Z with respect to the variable w_τ on the τ-factor of $\mathscr{B}$. We are going to study the analytic continuation of f as well as that of $E_Z f$. Given $\zeta \in \mathscr{S}_e(T)$, let $f(w, s; \zeta)$ denote the series obtained from (5.11) by multiplying each term with $\zeta\big({}^t\lambda_\tau[v, w] \cdot {}^t\mu_\tau[v, w]^{-1}\eta(w_\tau)\big)$. The series is convergent for sufficiently large $\mathrm{Re}(s)$, as will be shown in Section 9. If $e = 0$, we take ζ to be the constant 1, and view the original f as a special case of $f(w, s; \zeta)$. Now Lemma 6.1, (ii) shows that

$$(8.1) \qquad [E_Z f(w, s)]\zeta = \beta_Z(-s)f(w, s; \zeta) \quad \text{for every } \zeta \in Z.$$

For the moment, this holds only for sufficiently large $\mathrm{Re}(s)$, but eventually for all s, by virtue of analytic continuation.

As mentioned in Section 3, Z is determined by its highest weight, and spanned by the transforms of the function ζ_0 of (3.15). Moreover, if $\{t_1, \ldots, t_m\}$ is the highest weight of Z, then, as shown in [12, Theorem 4.1], we have

$$(8.2) \qquad \beta_Z(s) = \prod_{i=1}^{m} \prod_{j=1}^{t_i} (s + i - j).$$

Notice that this is independent of l. (We have assumed $l \geq m$ in Section 5.)

In order to obtain $f(w, s; \zeta)$ as an integral involving a certain theta series, we first take a polynomial representation P and a polynomial map Q defined as follows:

$$P: \mathrm{GL}_m(\mathbf{C}) \times \mathrm{GL}_m(\mathbf{C}) \to \mathrm{GL}(\mathscr{S}_e(\mathbf{C}_m^m)),$$

$$Q: \mathbf{C}_l^m \times \mathbf{C}_m^m \to \mathrm{Hom}(\mathscr{S}_e(T), \mathscr{S}_e(\mathbf{C}_m^m)),$$

$$[P(a, b)\varphi](u) = \varphi({}^t aub) \quad \text{for } \varphi \in \mathscr{S}_e(\mathbf{C}_m^m), \qquad u \in \mathbf{C}_m^m,$$

$$[Q(x, y)\zeta](u) = \zeta({}^t xuy) \quad \text{for } \zeta \in \mathscr{S}_e(T), \qquad x \in \mathbf{C}_l^m, \qquad u, y \in \mathbf{C}_m^m.$$

Then $Q(ax, by) = P(a, b)Q(x, y)$. Hereafter we put $T' = \mathbf{C}_m^m$.

LEMMA 8.1. *Suppose $l \geq m$; let Z be an irreducible subspace of $\mathscr{S}_e(T)$. Then the image of Z under $Q(x, y)$ belongs to the irreducible subspace of $\mathscr{S}_e(T')$ of the same highest weight as Z.*

Proof. Define ζ_0 by (3.15). Similarly put $\xi_0(u) = \prod_{i=1}^m \det_i(u)^{c_i}$ for $u \in T'$. If $\zeta(v) = \zeta_0(avb)$ with $a \in \mathrm{GL}_l(\mathbf{C})$ and $b \in \mathrm{GL}_m(\mathbf{C})$, then, for $x \in \mathbf{C}_l^m$ and $y \in \mathbf{C}_m^m$, we have $[Q(x, y)\zeta](u) = \zeta_0(a \cdot {}^t xuyb) = \xi_0(wuyb)$, where w is the first m rows of $a \cdot {}^t x$. Thus the image of Z under $Q(x, y)$ belongs to the space spanned by the transforms of ξ_0 under $\mathrm{GL}_m \times \mathrm{GL}_m$. This proves our lemma.

To prove Theorem 5.2, we first assume $j = 0$. Take $\{R, X\}$, $\{R', X'\}$, and S as in Section 5, and put $Y = \mathrm{Hom}(\mathscr{S}_e(T), \mathscr{S}_e(T'))$. We now define a series $\Theta(z, w)$ with values in $\mathrm{Hom}(X, X') \otimes Y$ by

$$\Theta(z, w) = \sum_{v \in V} b(v)\delta(w)^{-k} \det \overline{\mu[v, w]}^{k - e\tau} \mathbf{e}_\varphi^m(R[v; w, z])$$

$$\cdot S({}^t\lambda[v, w]) \otimes Q(\lambda_\tau[v, w], \mathrm{ad}\overline{\mu_\tau[v, w]}).$$

Here $z \in \mathscr{H}_m^\varphi$, $w \in \mathscr{B}$, $V = K_q^m$; b and k are the same as in (5.11); $0 \leq e \leq k_\tau$; $\mathbf{ad}$ is defined by ${}^t M \cdot \mathbf{ad}(M) = \det(M)$;

$$R_\sigma[v; w, z] = (1/2)(vHv^*)_\sigma z_\sigma$$

$$+ \begin{cases} (1/2)(z - z^*)_\sigma \cdot {}^t\eta(w_\sigma)^{-1}\{{}^t\mu_\sigma[v, w]\} & \text{if } \sigma \in \psi, \\ 0 & \text{if } \sigma \notin \psi. \end{cases}$$

This is a series of type Θ_{01} of Proposition 7.7 with $n = m$. Therefore we have, by (7.21),

$$(8.3) \quad \Theta(\beta z, w) = j_\beta(z)^\nu \Big\{ R^r({}^t\mu(\beta, z)) \otimes P\Big(\mu_\tau(\beta, z), \overline{{}^t\lambda_\tau(\beta, z)}^{-1}\Big) \Big\} \Theta(z, w)$$

with $\nu = m(\psi\rho - \psi) + (l + m)\varphi + k\rho$ for every β in a congruence subgroup of $G_{m, K}$, where $R^r(a)$ means the right action of $R(a)$ on $\mathrm{Hom}(X, X')$. For $B \in \mathrm{Hom}(X, X')$, $C \in Y$, and $\zeta \in \mathscr{S}_e(T)$, define $(B \otimes C)\zeta$ to be an element of $\mathscr{S}_e(T', \mathrm{Hom}(X, X'))$ given by

$$[(B \otimes C)\zeta](u) = (C\zeta)(u)B \quad \text{for } u \in T'.$$

Then, for $u \in T'$, we have

$$[\Theta(z, w)\zeta](u) = \sum_{v \in V} b(v)\delta(w)^{-k} \det \overline{\mu[v, w]}^{k - e\tau} \mathbf{e}_\varphi^m(R[v; w, z])$$

$$\cdot \zeta({}^t\lambda_\tau[v, w]u \cdot \mathbf{ad}\overline{\mu_\tau[v, w]})S({}^t\lambda[v, w]).$$

Taking the embedding $z \mapsto z'$ of $\mathcal{H}_m^\psi$ into $\mathcal{H}_m^\varphi$ defined by $(z')_\sigma = z_\alpha$ for $\sigma \in \varphi$ and $\alpha \in \psi$ when $\sigma = \alpha$ on E, put $\theta(z, w; \zeta) = \Theta(z', w)\zeta$ for $\zeta \in \mathscr{S}_e(T)$ and $z \in \mathcal{H}_m^\psi$. Then, for $u \in T'$,

$$\theta(z, w; \zeta)(u) = \sum_{v \in V} b(v)\delta(w)^{-k} \det \overline{\mu[v, w]}^{k - e\tau}$$

$$\cdot \mathbf{e}_\psi^m\big((1/2)\mathrm{Tr}_{K/L}(vHv^*)z + (1/2)(z - z^*) \cdot {}^t\eta(w)^{-1}\{{}^t\mu[v, w]\}\big)$$

$$\cdot \zeta({}^t\lambda_\tau[v, w]u \cdot \mathbf{ad}\overline{\mu_\tau[v, w]})S({}^t\lambda[v, w]),$$

where $\mathrm{Tr}_{F/E}$ should replace $\mathrm{Tr}_{K/L}$ if $m = 1$. From (8.3) and (5.9c), we obtain easily

$$(8.4) \quad \theta(\gamma z, w; \zeta) = j_\gamma(z)^{\nu'}$$

$$\cdot \left\{ R_0^r({}^t\mu(\gamma, z)) \otimes P\big(\mu_\tau(\gamma, z), {}^t\overline{\lambda_\tau(\gamma, z)}^{-1}\big)\right\}\theta(z, w; \zeta)$$

for every γ in a suitable torsion-free congruence subgroup Γ of $G_{m, L}$, where $\nu' = k\rho + m(\psi\rho - \psi) + (l + m)[F : E]\psi$. ($G_{m, L}$ stands for $\mathrm{SL}_2(E)$ if $m = 1$.) Put

$$W = \{h \in \mathbf{C}_m^m | h^* = h\}, \qquad W_+ = \{h \in W | h > 0\},$$

$$W_L = \begin{cases} \{h \in L_m^m | h^* = h\} & \text{if } m > 1, \\ E & \text{if } m = 1. \end{cases}$$

Given A as in (5.10a), we put $A'(z) = A(z/2)$ and choose Γ so that $A'\|_\chi\Gamma = A'$. Take a subgroup U of $\mathrm{GL}_m(\mathfrak{o}_L)$ (of $\mathfrak{o}_E^\times$ if $m = 1$) of finite index and a lattice Λ in W_L so that $\det(u)$ is a totally positive unit of E for every $u \in U$ and that

$$(8.5) \qquad \left\{\begin{bmatrix} a^* & a^*b \\ 0 & a^{-1} \end{bmatrix}\Big| a \in U, b \in \Lambda\right\}$$

is a subgroup of Γ. We then consider an integral

$$I(y, w; \zeta)(u) = \int_M \theta(x + iy, w; \zeta)(u)A'(x + iy) \, dx \qquad (u \in T')$$

with values in X', where $M = W^\psi/\Lambda$, x and y are variables on W^ψ and on $(W_+)^\psi$ (so that $x + iy = z \in \mathcal{H}_m^\psi$), and dx is a fixed Euclidean measure on W^ψ.

(If $m > 1$, ψ is viewed as a CM-type of L; if $m = 1$, W^ψ is identified with $\mathbf{R}^{J(E)}$.) We see easily that

$$I(y, w; \zeta)(1_m) = \mathrm{vol}(M) \sum_{v \in V} b(v)\delta(w)^{-k}\det\overline{\mu[v, w]}^{k - e\tau}$$

$$\cdot \zeta({}^t\lambda_\tau[v, w]\mathrm{ad}\overline{\mu_\tau[v, w]})\mathbf{e}_\psi^m\big(iy \cdot {}^t\eta(w)^{-1}\{{}^t\mu[v, w]\}\big)$$

$$\cdot S({}^t\lambda_\tau[v, w])a\big(-\mathrm{Tr}_{K/L}(vHv^*)\big).$$

Next, take an integral

$$(8.6) \qquad \int_N I(y, w; \zeta)(1_m)\det(y)^{s\psi - m\psi + k}\, dy,$$

where N is a fundamental domain of W_+^ψ modulo the action of U defined by $y \mapsto a^*ya$ for $a \in U$. Now we recall a well known formula

$$\int_{W_+} \exp(-\mathrm{tr}(qh))\det(h)^{s-m}\, dh = \Gamma_m(s)\det(q)^{-s},$$

$$\Gamma_m(s) = \pi^{m(m-1)/2}\prod_{j=0}^{m-1}\Gamma(s - j),$$

valid for $q \in W_+$ and $\mathrm{Re}(s) > m - 1$. Write

$$(8.7) \qquad I(y, w; \zeta)(1_m)\det(y)^{s\psi - m\psi + k} = \sum_{v \in V} c(v, y)$$

and observe that $c(av, y) = c(v, a^*ya)$ for $a \in U$. Therefore we see that (8.6) is equal to

$$\mathrm{vol}(M)[U]^{-1}(2\pi)^{-r}\prod_{\sigma \in \psi}\Gamma_m(s + k_\sigma)f(w, s; \zeta)$$

with $r = m\sum_{\sigma \in \psi}(s + k_\sigma)$, under the condition that

(8.8) $I(y, w; \zeta)(1_m)$ *consists only of the terms with* $\det\overline{\mu[v, w]}^\psi \neq 0$.

So far our calculation is valid for sufficiently large $\mathrm{Re}(s)$. To obtain analytic continuation, we first note that the invariant volume element of $\mathcal{H}_m^\psi$ is given by $\omega = \det(y)^{-2m\psi}\, dx\, dy$. Now let $\mathcal{P}$ be the parabolic subgroup of $G_{m, L}$ consisting of all the elements α of $G_{m, L}$ such that $c_\alpha = 0$. Define an element $B(z, w; \zeta)$ of $\mathcal{S}_e(T', X')$ by

$$(8.9) \qquad B(z, w; \zeta)(u) = \det(y)^{s\psi + m\psi + k}\theta(z, w; \zeta)(u)A'(z) \qquad (u \in T').$$

Denote by $\{U \times \Lambda\}$ the group of (8.5). Then (8.6) is equal to

$$[\mathcal{P} \cap \Gamma: \{U \times \Lambda\}]\int_\Psi B(z, w; \zeta)(1_m)\omega,$$

where $\Psi = (\mathscr{P} \cap \Gamma) \backslash \mathscr{H}_m^\psi$. Taking a basis $\{x_i\}_{i \in I}$ of X' over $\mathbf{C}$, define the components k_i of $k \in \mathscr{S}_e(T', X')$ by $k(u) = \sum_{i \in I} k_i(u)x_i$; then $k_i \in \mathscr{S}_e(T')$. By (8.4) and (5.13), we see that

$$B_i(\gamma z, w; \zeta) = |j_\gamma|^{-2s\psi} j_\gamma^{-\kappa\psi} P\left(\mu_\tau(\gamma, z), \overline{{}^t\lambda_\tau(\gamma, z)}^{-1}\right) B_i(z, w; \zeta)$$

for every $\gamma \in \Gamma$ and $i \in I$ with κ of Theorem 5.2. Put $\Phi = \Gamma \backslash \mathscr{H}_m^\psi$, $R = (\mathscr{P} \cap \Gamma) \backslash \Gamma$, and define an Eisenstein series $\mathscr{E}_P$ with values in $\mathrm{Hom}(\mathscr{S}_e(T'), \mathbf{C})$ by

$$(8.10) \quad \mathscr{E}_P(z, s)h = \sum_{\gamma \in R} \delta(z)^{s\psi} |j_\gamma(z)|^{-2s\psi} j_\gamma(z)^{-\kappa\psi} h\left({}^t\mu_\tau(\gamma, z) \cdot {}^t\overline{\lambda_\tau(\gamma, z)}^{-1}\right)$$

for $h \in \mathscr{S}_e(T')$. Since Ψ is equivalent to $\bigcup_{\gamma \in R} \gamma \Phi$, we find that

$$(8.11) \qquad \int_\Psi B_i(z, w; \zeta)(1_m)\omega = \int_\Phi \mathscr{E}_P(z, s)B_i(z, w; \zeta)\omega \qquad (i \in I),$$

under the condition that B is rapidly decreasing, which will be examined afterward. Define another simpler Eisenstein series $\mathscr{E}$ by

$$(8.12) \qquad \mathscr{E}(z, s) = \sum_{\gamma \in R} \delta(z)^{s\psi} |j_\gamma(z)|^{-2s\psi} j_\gamma(z)^{-\kappa\psi}.$$

The series of (8.10) and (8.12) can be defined for an arbitrary integer κ, and have been investigated in [10] and [13]. For $h \in \mathscr{S}_e(T')$, define $h_z' \in \mathscr{S}_e(T')$ by $h_z'(u) = h({}^t u \cdot {}^t\eta(z_\tau)^{-1})$. If h belongs to an irreducible subspace W of $\mathscr{S}_e(T')$, then h_z' belongs to W, and by (1.8a) and Lemma 6.1, (i), we have

$$(8.13) \qquad [E_W \mathscr{E}(z, s)] h_z' = (-i)^e \beta_W(-s) \mathscr{E}_P(z, s)h,$$

where E_W is considered on the τ-factor.

Now fix an irreducible subspace Z of $\mathscr{S}_e(T)$; let Z' be the irreducible subspace of $\mathscr{S}_e(T')$ of the same highest weight as Z. Define $B'(z, w; \zeta) \in \mathscr{S}_e(T', X')$ by

$$B'(z, w; \zeta)(u) = B(z, w; \zeta)\left({}^t u \cdot {}^t\eta(z_\tau)^{-1}\right) \qquad (u \in T').$$

Our definition of θ and B together with Lemma 8.1 shows that if $\zeta \in Z$, then $B_i(z, w; \zeta)$ and $B_i'(z, w; \zeta)$ belong to Z'. We have $\beta_Z = \beta_{Z'}$ by (8.2), and therefore (8.13) implies that

$$(8.14) \quad [E_{Z'}\mathscr{E}(z, s)] B_i'(z, w; \zeta) = (-i)^e \beta_Z(-s) \mathscr{E}_P(z, s)B_i(z, w; \zeta).$$

Combining (8.1), (8.7), (8.11), and (8.14) together, we obtain

$$(8.15) \quad (2\pi)^{-r} \prod_{\sigma \in \psi} \Gamma_m(s + k_\sigma)[E_Z f(w, s)]\zeta$$

$$= C \sum_{i \in I} x_i \int_\Phi [E_{Z'}\mathscr{E}(z, s)] B_i'(z, w; \zeta)\omega$$

with a nonzero constant C. Now we need

LEMMA 8.2. (i) $\mathscr{E}(z, s)$ *can be continued as a meromorphic function in s to the whole plane.*

(ii) $\mathscr{E}(z, s)$ *is finite at* $s = 0$ *if* $\kappa \geq m$;

(iii) $\mathscr{E}(z, 0)$ *is holomorphic in z if* $\kappa \geq m$ *except when* $E = \mathbf{Q}$ *and* $\kappa = m + 1$.

(iv) *If* $E = \mathbf{Q}$ *and* $\kappa = m + 1$, *then* $E_{Z'}\mathscr{E}(z, 0) = 0$ *for every nonelementary* Z' *in the sense of Section* 3.

(v) *If* $0 \leq \kappa < m$, $\mathscr{E}(z, s)$ *has at most a simple pole at* $s = m - \kappa$, *and its residue is holomorphic in z.*

The first assertion is a special case of Langlands' result in [3]. The remaining ones were proved in [10] and [13]. In particular, (v) is included in [13, Theorem 2.7]. As for (iv), we proved in [10, Theorem 7.2 and (9.6)] that if $E = \mathbf{Q}$ and $\kappa = m + 1$, $E(z, 0)$ is a polynomial of degree $\leq m$ in $r(z)$ of (2.3) with holomorphic functions as coefficients. Therefore we have $E^{m+1}\mathscr{E}(z, 0) = 0$ by Proposition 2.4. Moreover, [10, (9.6)] shows that $\mathscr{E}(z, 0)$ is of type (3.17). Hence Proposition 3.9 together with (iii) of Lemma 3.8 implies (iv).

It should also be noted that if $(s - s_0)^n \mathscr{E}(z, s)$ is finite in a neighborhood of s_0 with $n \in \mathbf{Z}$, then the product, as well as $(s - s_0)^n E_{Z'}\mathscr{E}$, is slowly increasing at every cusp. This is included in the results of [3]. Now (8.13) shows that $\mathscr{E}_P$ is essentially the same as $E_{Z'}\mathscr{E}$ up to a certain factor. Our definition of Θ together with (7.17) shows that Θ is rapidly decreasing at each cusp on $\mathscr{H}_m^\psi$ if all its transforms under $G_{m, K}$ consist only of the terms with $\mathrm{rank}(v) = m$. By Proposition 7.7, every transform of Θ is of the same type as Θ. Therefore this condition is satisfied if $k_\tau > e$ or $E \neq \mathbf{Q}$, since we have assumed $k_\sigma > 0$ for every $\sigma \in \psi$. (Then (8.8) follows from (5.12).) Thus, if $e = 0$ and $\zeta = 1$, we see from (8.9) that B is also rapidly decreasing, and hence (8.15) shows that $f(w, s)$ has a meromorphic continuation to the whole s-plane. This proves (1) of Theorem 5.2, and also (2), in view of Lemma 8.2, (ii). Similarly, if $k_\tau > e$ or $E \neq \mathbf{Q}$, $f(w, s; \zeta)$ has a meromorphic continuation to the whole s-plane, and (8.15) is meaningful.

To complete the proof of Theorem 5.2, we may assume

$$(8.16) \qquad F \neq E \quad or \quad l > m \quad or \quad p \neq 0$$

for the reason explained at the end of Section 5. Now (5.13) shows that

$$(8.17) \qquad k_\sigma \geq p_\sigma + (l + m)[F : E] + \kappa - 2m \quad for\ every\ \sigma \in \psi.$$

Suppose $E \neq \mathbf{Q}$ and $\kappa \geq 0$. For a fixed $\tau \in \psi$, consider E_Z as above with respect to the variable w_τ with $Z = \mathscr{S}_1(T)$. By (8.15) and Lemma 8.2, (iv) and (v), we see that $E_Z f(w, 0) = 0$ if $\kappa \geq m$ and $(s - m + \kappa)E_Z f(w, s)$ vanishes at $s = m - \kappa$ if $0 \leq \kappa \leq m$. This proves (3) and (5) of Theorem 5.2 when $E \neq \mathbf{Q}$.

Suppose $E = \mathbf{Q}$. Then ψ consists of a single element τ, and so (8.16) combined with (8.17) implies that $k_\tau \geq \kappa + 1$. Therefore the above reasoning about $E_Z f$ is applicable even in this case and establishes (3) and (5) of Theorem 5.2 at least when $\kappa > 0$. If $\kappa = 0$ and $k_\tau = 1$, the factor $\det \overline{\mu[v, w]}^{k-\tau}$ disappears, and so it is not clear that B is in fact rapidly decreasing. If $m = 1$, however, the factor $\zeta({}^t\lambda_\tau \ldots)$ kills the term for $v = 0$, and therefore the conclusion holds. From (8.16) and (8.17), we see easily that $k_\tau = 1$, $\kappa = 0$, and $m > 1$ only in the two exceptional cases given in (5) of Theorem 5.2.

It remains to prove (4) of Theorem 5.2. In this case, we have $k_\tau \geqq m + 2$. Therefore B is rapidly decreasing if $e \leq m + 1$. Take a nonelementary irreducible subspace Z of $\mathscr{S}_e(T)$ with $e \leq m + 1$ in the sense of Section 3, and take also an irreducible subspace Z' of $\mathscr{S}_e(T')$ of the same highest weight as Z. By Lemma 8.2, (iv), the right-hand side of (8.15) vanishes at $s = 0$, so that $E_Z f(w, 0) = 0$, which together with Proposition 3.9 proves (4) of Theorem 5.2.

So far we assumed $j = 0$. We now consider the case of nontrivial j with $m = 1$. As observed in [7, I, Remark 5.5], a^j for $a \in F$ is a finite $\mathbf{Q}$-linear combination of $\mathrm{Tr}_{F/E}(a)^t a^u$ with $0 \leq t \in I_E$ and $0 \leq u \in \mathbf{Z}^{J(F)-\varepsilon}$. Then we obtain $f(w, s; \zeta)$ as a finite $\mathbf{Q}$-linear combination of several integrals of form (8.6) in which Θ and A' are replaced by $\mathbf{d}^u\Theta$ and $\mathbf{d}^t A'$ with $\mathbf{d}^u$ as in (10.7) below. By [7, I, (1.16a)], these functions can be expressed as linear combinations of $D_\rho^{(a)}\Theta$ and $D_\sigma^{(b)}A'$, and therefore the integrals can be reduced to those over Φ. We shall not go into details of this procedure here, since the same technique will be employed in the orthogonal case in Section 10, and the unitary case can be handled in a parallel way.

9. Convergence of $f(w, s)$ in the unitary case

Let m and n be two integers such that $0 < m \leq n$; let L, $\mathfrak{o}_L$, E, and ψ be as in Section 5. We may view ψ as a CM-type of L. Let $\mathfrak{X} = \{ x \in L^n_m |$ $\mathrm{rank}(x) = m \}$. For each $\tau \in \psi$, we take an element $P_\tau = P_\tau^* > 0$ of $\mathrm{GL}_n(\mathbf{C})$ and also a continuous function $D(x)$ on $(\mathbf{C}^n_m)^\psi$ such that $D(xy) = D(x)|\det(y)|^e$ for every $y \in \mathrm{GL}_m(\mathbf{C})^\psi$ with $0 \leq e \in \mathbf{Z}^\psi$. Then we consider a series

$$\sum_{x \in \mathfrak{X}/U} \lambda(x)D(x)(\det(x^*Px))^{-s\psi - e/2}$$

with $s \in \mathbf{R}$, $\lambda \in \mathscr{L}(L^n_m)$ and a subgroup U of $\mathrm{GL}_m(\mathfrak{o}_L)$ of finite index such that $\lambda(xu) = \lambda(x)$ for every $u \in U$ and $\det(U) \subset E$.

LEMMA 9.1. *The above series is convergent if $s > n$.*

Proof. Take a positive integer g and an element $S = S^*$ of $\mathrm{GL}_n(L) \cap M_n(\mathfrak{o}_L)$ so that $0 < S_\tau < gP_\tau$ for every $\tau \in \psi$. Then it is sufficient to prove the convergence with P_τ replaced by S_τ. It is also sufficient to treat the case where λ is the characteristic function of a lattice Λ in L^n_m. We may even reduce the problem to the case $\Lambda = (\mathfrak{o}_L)^n_m$. We first treat the case $D = 1$. Then the series becomes

$$(9.1) \qquad Z(s) = \sum_{xU} N(\det(x^*Sx))^{-s} \qquad (x \in \mathfrak{X} \cap \Lambda),$$

where $N = N_{E/\mathbf{Q}}$. The result in this case is perhaps well known; for the reader's convenience, we sketch a proof here. Let $\Theta = (\mathfrak{o}_L)^m_m$. For a totally positive element $h = h^* \in \mathrm{GL}_m(L) \cap \Theta$ let $\nu(S, h)$ denote the number of $x \in \mathfrak{X} \cap \Lambda$ such that $x^*Sx = h$, and $r(h)$ the order of the group $\{ u \in U | u^*hu = h \}$. Then we see easily that

$$Z(s) = \sum_h r(h)^{-1}\nu(S, h)N(\det(h))^{-s},$$

where h is taken modulo the equivalence under U. Let $W = \{ w \in \mathbf{C}^m_m | w^* = w > 0 \}$. Take a Siegel domain $\mathfrak{S} = \mathfrak{S}(M, C)$, which consists of all the elements $q = (q_\tau)_{\tau \in \psi} \in W^\psi$ with the following properties: for each τ, $q_\tau = t^*_\tau d_\tau t_\tau$ with $t_\tau \in C$ and $d_\tau = \mathrm{diag}[d_{\tau 1}, \ldots, d_{\tau m}]$, $d_{\tau i}/d_{\sigma i} \leqq M$, $0 < d_{\tau i} \leqq M d_{\tau i+1}$, where C is a compact set of unipotent matrices, and $0 < M \in \mathbf{R}$. The basic theorem of reduction theory tells that if M and C are suitably chosen, there exists a finite subset B of $\mathrm{GL}_m(L) \cap \Theta$ such that $W^\psi = \bigcup_{y \in B'U} y^*\mathfrak{S}y$, where $B' = \{ b^{-1} | b \in B \}$. Now let $h \in \Theta \cap W^\psi$; take $b \in B$ and $u \in U$ so that $k = b^*u^*hub \in \mathfrak{S}$. Since $\nu(S, h) \leqq \nu(S, k)$, we see that

$$(9.2) \qquad Z(s) \leqq \sum_{b \in B} N(\det(b^*b))^{-s}\sum_k \nu(S, k)N(\det(k))^{-s},$$

where k is taken from $\mathfrak{S} \cap \Theta$ modulo the equivalence under $\bigcap_{b \in B} b^{-1}Ub$. The definition of $\mathfrak{S}$ shows that if $k \in \mathfrak{S} \cap \Theta$, then

$$(9.3) \qquad N(\det(k)) \leqq N(k_{11} \ldots k_{mm}) \leqq c_1 N(\det(k))$$

with a constant c_1 independent of k. Moreover, we see easily that $\prod_i \nu(S, p_i)$ for totally positive $p_1, \ldots, p_m$ in $\mathfrak{o}_E$ is equal to the sum of $\nu(S, k)$ for all $k \in \Theta$ such that $k_{ii} = p_i$. Therefore the last sum of (9.2) is majorized by a constant times

$$c_1^s \sum_{p_1, \ldots, p_m} \nu(S, p_1) \ldots \nu(S, p_m)N(p_1 \ldots p_m)^{-s},$$

where each p_i is taken modulo a subgroup of $\mathfrak{o}_E$ of finite index. This is essentially the m-th power of $\sum_x N(x^*Sx)^{-s}$ with $0 \neq x \in (\mathfrak{o}_L)^n/\mathfrak{o}_E^\times$. We may view x^*Sx as a quadratic form on E^{2n}, and hence a well known result shows that the last series is convergent for $s > n$ (see [6, Lemma 8.1], for example). This proves the case $D = 1$.

To treat the case with nontrivial D, first observe that if $x_{\tau i}$ denotes the i-th column of $x_\tau \in \mathbf{C}_m^n$, then

$$D(t_{\tau 1} x_{\tau 1}, \ldots, t_{\tau m} x_{\tau m}) = D(x) \prod_{\tau \in \psi} |t_{\tau 1} \ldots t_{\tau m}|^{e_\tau} \quad \text{for } t_{\tau i} \in \mathbf{R}.$$

Let $\|x\| = (x^* x)^{1/2}$ for $x \in \mathbf{C}^n$. Then

$$(9.4) \qquad |D(x)| \leq \alpha \prod_{\tau \in \psi} \left(\|x_{\tau 1}\| \cdots \|x_{\tau m}\| \right)^{e_\tau}$$

with a constant α independent of x. Given $x \in \mathfrak{X} \cap \Lambda$, take $u \in U$ and $b \in B$ so that $S\{xub\} \in \mathfrak{S}$. Let $xub = y$ and $h = S\{y\}$. Since $h_{ii} = S\{y_i\}$, we see that there is a constant f_τ depending only on S such that $\|y_{\tau i}\|^2 \leq f_\tau h_{ii}^\tau$. Therefore, by (9.3) and (9.4), $|D(y)| \leq \beta \det(h)^{e/2}$ with a constant β, and hence

$$|D(x)| = |D(y) \det(bu)^{-e}| \leq \beta \det(y^* S y)^{e/2} |\det(bu)|^{-e}$$

$$= \beta \det(x^* S x)^{e/2}.$$

This reduces the problem to the case of trivial D and our proof is thereby completed.

Let us now consider $f(w, s)$ of (5.11), and first prove (5.12). For a fixed $\tau \in \psi$, let $\theta = \theta_\tau$ be the set of all $\alpha \in \varphi$ such that $\alpha = \tau$ on E. Then, by (5.9c), we have

$$(9.5) \qquad -\operatorname{Tr}_{K/L}(vHv^*)^\tau = -(vHv^*)^\tau - \sum_{\tau \neq \alpha \in \theta} (vHv^*)^\alpha.$$

Since $H^\alpha > 0$ for $\tau \neq \alpha \in \theta$, $-(vHv^*)^\tau$ is nonnegative if $-\operatorname{Tr}_{K/L}(vHv^*)$ is totally nonnegative. Therefore (7.15b) shows that ${}^t\eta(w_\tau)^{-1}\{{}^t\mu_\tau[v, w]\}$ is positive definite, which proves (5.12).

In order to prove the convergence of (5.11), let us first assume $m > 1$. For each $\sigma \in \varphi$, let K_σ denote the σ-completion of K, and L_σ the closure of L in K_σ; let $V^{(r)} = K_r^q$, $V_\sigma^{(r)} = V^{(r)} \otimes_K K_\sigma$, and $Y_\sigma^{(r)} = V^{(r)} \otimes_L L_\sigma$. Now fix $\tau \in \psi$. Then $Y_\tau^{(r)}$ is isomorphic to $\prod_{\alpha \in \theta_\tau} V_\alpha^{(r)}$. For a fixed $w \in \mathscr{B}$, we define a hermitian form T_τ on $Y_\tau^{(1)}$ by

$$(9.6) \qquad T_\tau\{x\} = P_\tau(w_\tau)\{x_\tau\} + \sum_{\tau \neq \alpha \in \theta} H_\alpha\{x_\alpha\}$$

for $x = (x_\alpha) \in \prod_{\alpha \in \theta} V_\alpha^{(1)} = Y_\tau^{(1)}$ with P_τ of (7.15a, b). Obviously T_τ is positive definite. Observe that (9.6) is meaningful also for $x = (x_\alpha) \in \prod V_\alpha^{(m)} = Y_\tau^{(m)}$. Now suppose $-\sum_{\alpha \in \theta}(vHv^*)_\alpha$ is nonnegative for a fixed $v^* \in Y_\tau^{(m)}$. Then we have

$$(9.7) \qquad vT_\tau v^* \leqq vT_\tau v^* - \sum_{\alpha \in \theta} (vHv^*)_\alpha$$

$$\leqq (P_\tau(w_\tau) - H_\tau)\{v_\tau^*\} = 2 \cdot {}^t\eta(w_\tau)^{-1}\{{}^t\mu_\tau[v, w]\}$$

by (7.15b).

Let us now assume that the X-valued form A of (5.10a) is a cusp form. We may assume that $X = \mathbf{C}^n$, $X' = \mathbf{C}^{n'}$, S takes values in $\mathbf{C}_n^{n'}$, $R({}^tb) = {}^tR(b)$, and R is $\mathbf{R}$-rational. Put $\chi(b) = \det(b)^p R_0({}^tb^{-1})$ and $\|a\| = (a^*a)^{1/2}$ for $a \in X$ or $a \in X'$. A standard argument shows that $\|\chi(\eta(z)^{1/2})A(z)\|$ is bounded for all $z \in \mathscr{H}_m^\psi$. Consequently $\|\chi(h^{-1/2})a(h)\| \leqq C_1$ with a constant C_1 for all the Fourier coefficients $a(h)$ of A. We are going to prove the convergence of ${}^txf(w, s)$ for every fixed $x \in X'$. We focus our attention on one term of (5.11) with $v \in \mathscr{V}$, and put $h = - \operatorname{Tr}_{K/L}(vHv^*)$. By (9.7), we have $h_\tau \leqq 2 \cdot {}^t\eta(w_\tau)^{-1}\{{}^t\mu_\tau[v, w]\}$. If $A \neq 0$, χ must be a polynomial representation, and so

$$\left|{}^txS({}^t\lambda[v, w])a(h)\right|^2 = \left|{}^txS({}^t\lambda[v, w])\chi(h^{1/2})\chi(h^{-1/2})a(h)\right|^2$$

$$\leqq C_1^2 \left\|{}^t\big({}^txS({}^t\lambda[v, w])\chi(h^{1/2})\big)\right\|^2$$

$$= C_1^2 \cdot {}^txS({}^t\lambda[v, w])\chi(h)S({}^t\lambda[v, w])^*\bar{x}$$

$$\leqq C_1^2 \cdot {}^txS({}^t\lambda[v, w])$$

$$\cdot \chi\big(2 \cdot {}^t\eta(w)^{-1}\{{}^t\mu[v, w]\}\big)S({}^t\lambda[v, w])^*\bar{x}.$$

Call the last product $D(v)$. This is well defined for ${}^tv \in \prod_{\tau \in \psi} V_\tau^{(m)} \cong (\mathbf{C}_m^{qr})^\psi$, where $r = [K : L]$, and $D(yv) = D(v)|\det(y)|^{2p}$ for $y \in \operatorname{GL}_m(\mathbf{C})^\psi$. By (9.7), we have

$$\left|2^m\delta(w)^{-1}\det \mu[v, w]\right|^{2}\Big]^{-s\psi} \leqq \det(vTv^*)^{-s\psi}$$

for $s > 0$. Therefore we have, for $x \in X'$,

$$(9.8) \quad |\delta(w)^{k/2} \cdot {}^txf(w, s)| \leqq C_2 \sum_v |b(v)D(v)^{1/2}| \det(vTv^*)^{-s\psi - k/2}$$

with a constant C_2. In view of (5.13), this is a series of type (9.1) with $e = p$, and hence it is convergent for $s > q[K : L] - (c_f/2)$ by Lemma 9.1. This proves Proposition 5.1 when A is a cusp form.

Next assume $A = 1$. Then our series has form (5.15). Again by (9.7), we have (9.8) with $D(v) = \|S({}^t\lambda[v, w])\|^2$ and the same conclusion holds.

Finally assume $m = 1$ with $j \geq 0$. Decompose our series into two parts Σ_1 and Σ_2, where Σ_1 consists of the terms with $\operatorname{Tr}_{K/L}(vHv^*) = 0$. Since $m = 1$, A is either a cusp form, or $A \in \mathscr{M}(\nu\iota_E, E)$ with $0 \leqq \nu \in \mathbf{Z}$. In the latter case, we have $a(h) = O(N_{E/\mathbf{Q}}(h)^\nu)$ for $h \neq 0$. Therefore the above proof in the case

400 GORO SHIMURA

$j = 0$ is applicable to both Σ_1 and Σ_2 with obvious modifications. This completes the proof of Proposition 5.1.

Let us finally consider $f(w, s; \zeta)$ defined in Section 8. This is obtained by multiplying each term of (5.11) with

$$\overline{\det \mu_\tau[v, w]}^{-e} \zeta\big({}^t\lambda_\tau[v, w] \cdot \overline{\mathrm{ad}\mu_\tau[v, w]}\,\eta(w_\tau)\big).$$

The change of v for vy with $y \in \mathrm{GL}_m(\mathbf{C})^\psi$ multiplies the latter factor by $\det(y)^{e\tau}$, and therefore, replacing the above D by its product with this ζ-factor and taking $k - e\tau$ instead of k, we obtain the convergence of $f(w, s; \zeta)$.

10. Proofs in the orthogonal case

Though the essential ideas are the same as in the unitary case, there are some differences. We first replace the series f of (4.10) by some series of a similar but slightly different type. Given $\varphi, \psi \in I_K$ as in (4.10), (4.11), and (4.13), take ψ' as in (4.16), and put

$$(10.1a) \qquad \chi = \sum_{\sigma \in \psi} (\varphi_\sigma - \varphi_{\sigma\rho})\sigma, \qquad \chi' = \sum_{\sigma \in \psi'} (\varphi_\sigma - \varphi_{\sigma\rho})\sigma,$$

$$(10.1b) \quad \nu = \sum_{\sigma \in \psi} \varphi_{\sigma\rho}\sigma, \qquad \nu' = \sum_{\sigma \in \psi'} \varphi_{\sigma\rho}\sigma, \qquad \xi = \chi + \chi' + \nu(1 + \rho),$$

where ρ denotes complex conjugation. Then $\varphi = \xi + \nu'(1 + \rho)$, so that $y^\varphi = y^\xi(yy^\rho)^{\nu'}$ for $y \in K$. Observe that $\chi \geq 0$, $\chi' \geq 0$, and $\nu' \geq 0$. Given $0 \leq j \in I_F$, we can easily show that a^j for $a \in F$ is a finite $\mathbf{Q}$-linear combination of the products $\mathrm{Tr}_{F/E}(a)^p a^t$ with $0 \leq p \in I_E$ and $0 \leq t \in \mathbf{Z}^{e'}$. Put $h = t + \mathrm{Res}_{K/F}(\nu')$. Then $0 \leq h \in \mathbf{Z}^{e'}$, and $S[x]^t x^\varphi$ for $x \in V$ coincides with $S[x]^h x^\xi$ up to an algebraic factor independent of x. Thus $S[x]^j x^\varphi$ is a finite $\overline{\mathbf{Q}}$-linear combination of the products $\mathrm{Tr}_{F/E}(S[x])^p S[x]^h x^\xi$ with p and h as above, and ξ as in (10.1b). Let $f'(w, s)$ be the series obtained by replacing $S[v]^j v^\varphi$ in (4.10) by $\mathrm{Tr}_{F/E}(S[v])^p S[v]^h v^\xi$. Then it is sufficient to prove the theorems of Section 4 for f' instead of f. Observe that

$$(10.2) \quad \mathrm{Res}_{F/E}(2j) + \mathrm{Res}_{K/E}(\varphi) = 2p + \mathrm{Res}_{F/E}(2h) + \mathrm{Res}_{K/E}(\xi),$$

and hence (4.12) implies that

$$(10.3) \qquad \mathrm{Res}_{F/E}(k - \kappa - 2h) - \mathrm{Res}_{K/E}(\xi) - 2p - l = a_{f} l_E.$$

We fix one τ in δ and consider E_Z with respect to the variable w_τ on $\mathcal{Z}_\tau$ and $Z \subset \mathcal{S}_e(T)$, where $T = \mathbf{C}^{n_\tau}$. Let $f(w, s; \zeta)$ and $f'(w, s; \zeta)$ denote the series obtained from $f(w, s)$ and $f'(w, s)$ by multiplying each term of (4.10) with $\zeta({}^t\lambda_\tau[v, w]\overline{\mu_\tau[v, w]}^{-1}\eta(w_\tau))$. We take Z of the type of Lemma 6.1, (iii). The lemma shows that

$$(10.4) \qquad\qquad [E_Z f'(w, s)]\zeta = \gamma_Z(s) f'(w, s; \zeta) \qquad\qquad (\zeta \in Z)$$

with a polynomial $\gamma_Z(s)$; in particular, $\gamma_Z(s) = -s$ if $Z = \mathscr{S}_1(T)$. Our task is to examine when the right-hand side vanishes.

To $w \in \mathscr{Z}$ and $\sigma \in \delta$, we assign a positive definite symmetric $\mathbf{R}$-bilinear form $P_\sigma(x, y; w)$ on V_σ as in [6, §2] and [7, I, §6], and put $P_\sigma[x; w] = P_\sigma(x, x; w)$.

LEMMA 10.1. *Fix* $\sigma \in \delta$ *and put* $n = n_\sigma$, $\mathbf{r} = \mathbf{r}(w_\sigma) = A_\sigma^{-1} p(w_\sigma)$, *and* $\mathbf{s} = \mathbf{s}(w_\sigma) = \sum_{i=1}^{n} a_i \partial \mathbf{r}/\partial w_{\sigma i}$ *with a fixed* $a \in \mathbf{C}^n$. *Further put* $\mathbf{t} = \mathbf{t}(w_\sigma) = \mathbf{s} + \eta(w_\sigma)^{-1} S_\sigma(\bar{\mathbf{r}}, \mathbf{s})\mathbf{r}$ *and* $\zeta(u) = ({}^t a u)^e$ *for* $u \in \mathbf{C}^n$ *with* $0 \leq e \in \mathbf{Z}$. *Then*
  (1) $S_\sigma[\mathbf{r}] = S_\sigma(\mathbf{r}, \mathbf{s}) = S_\sigma(\mathbf{r}, \mathbf{t}) = S_\sigma(\bar{\mathbf{r}}, \mathbf{t}) = 0$;
  (2) $P_0(x, \mathbf{r}; w) = -S_\sigma(x, \mathbf{r})$, $P_\sigma(x, \mathbf{t}; w) = S_\sigma(x, \mathbf{t})$;
  (3) $\mathbf{t} = A_\sigma^{-1} q(w_\sigma) a$ *with* q *of* (1.2c);
  (4) $\zeta({}^t\lambda_\sigma[x, w]) = S_\sigma(x, \mathbf{t})^e$;
  (5) $S_\sigma[\mathbf{s}] = S_\sigma[\mathbf{t}] = S_\sigma(\mathbf{s}, \mathbf{t}) = 0$ *if* ${}^t a a = 0$.

Proof. That $S_\sigma[\mathbf{r}] = 0$ is immediate from ${}^t p R p = 0$. As shown in [7, I, (6.2b), (6.3)], we have, suppressing w,

$$(10.5) \quad P_\sigma(x, y) = S_\sigma(x, y) + 2\eta(w_\sigma)^{-1}\{S_\sigma(x, \mathbf{r})S_\sigma(y, \bar{\mathbf{r}}) + S_\sigma(x, \bar{\mathbf{r}})S_\sigma(y, \mathbf{r})\}$$

and $\eta(w_\sigma) = -S_\sigma(\mathbf{r}, \bar{\mathbf{r}})$. Then we obtain the first half of (2) and $S_\sigma(\bar{\mathbf{r}}, \mathbf{t}) = 0$. Taking the derivatives of $S_\sigma[\mathbf{r}] = 0$, we obtain $S_\sigma(\mathbf{r}, \mathbf{s}) = 0$, and hence $S_\sigma(\mathbf{r}, \mathbf{t}) = 0$. Substituting $\mathbf{t}$ for y in (10.5), we obtain $P_\sigma(x, \mathbf{t}) = S_\sigma(x, \mathbf{t})$. Let q_i denote the i-th column of $q(w_\sigma)$. The definition of q in [6, p. 321] says that $q_i = \partial p/\partial w_{\sigma i} - \eta(w_\sigma)^{-1}\partial\eta(w_\sigma)/\partial w_{\sigma i} \cdot p(w)$. (There is a misprint in [6, p. 321, line 6 from bottom]: $D_{-1}\eta$ should be $D_{-1}p$.) Since $\eta = -S_\sigma(\bar{\mathbf{r}}, \mathbf{r})$, we have $\partial\eta/\partial w_{\sigma i} = -S_\sigma(\bar{\mathbf{r}}, \partial\mathbf{r}/\partial w_{\sigma i})$, so that $\mathbf{t} = A_\sigma^{-1}q(w_\sigma)a$. Then (4) follows from (4.6a). Now $\sum_i a_i \partial\mathbf{s}/\partial w_{\sigma i} = A_\sigma^{-1}\sum_{i, j} \partial^2 p(w_\sigma)/\partial w_{\sigma i}\partial w_{\sigma j} = 0$ if ${}^t a a = 0$, since ${}^t p(w) = ({}^t w, {}^t w w/2, 1)$. Hence $0 = \sum_i a_i(\partial/\partial w_{\sigma i})S_\sigma(\mathbf{r}, \mathbf{s}) = S_\sigma[\mathbf{s}] + S_\sigma(\mathbf{r}, \sum_i a_i \partial\mathbf{s}/\partial w_{\sigma i}) = S_\sigma[\mathbf{s}]$. Then we obtain $S_\sigma(\mathbf{s}, \mathbf{t}) = S_\sigma[\mathbf{t}] = 0$ from (1).

The above definition of P_σ applies only to $\sigma \in \delta$. If $\sigma \in \varepsilon - \delta$ (and hence $n_\sigma = 0$), we put $P_\sigma[u; w] = -S_\sigma[u]$ for $u \in V_\sigma$. We have then, in view of (10.5) and (4.8a),

$$(10.6a) \qquad P_\sigma[u; w] = S_\sigma[u] + \begin{cases} 4\eta(w_\sigma)^{-1}|\mu_\sigma[u, w]|^2 & (\sigma \in \delta), \\ 4|\beta^\sigma|u u^\rho & (\sigma \in \varepsilon - \delta), \end{cases}$$

where $\beta = (\beta_{r+1}, \ldots, \beta_t) \in F_{r+1} \times \cdots \times F_t$ with β_i of (4.8a).

We now define a theta series Θ_h by

$$\Theta_h(z, w) = \eta(w)^{e\tau - k}\sum_{v \in V} c(v)q(v)\overline{\mu[v, w]}^{k - e\tau}S[v]^h v^{x + x'}$$

$$\cdot \zeta({}^t\lambda_\tau[v, w])e(R[v; z, w]),$$

where $z \in \mathcal{H}_1^{J(F)}$, $w \in \mathcal{Z}$, and $R[v; z, w]$ is defined by

$$(10.6b) \qquad R_\sigma[v; z, w] = \begin{cases} x_\sigma S_\sigma[v] + iy_\sigma P_\sigma[v; w] & (\sigma \in \varepsilon), \\ z_\sigma S_\sigma[v] & (\sigma \in \varepsilon') \end{cases}$$

for $z_\sigma = x_\sigma + iy_\sigma$; c, q, and k are the same as in (4.10); h, χ, χ', τ, e, and ζ are the symbols introduced before Lemma 10.1; we take $0 \leq e \leq k_\tau$. For $g \in C^\infty(\mathcal{H}_1^J)$ and $0 \leq a \in \mathbf{Z}^J$ with $J = J_E$ or J_F, define $\mathbf{d}^a g$ by

$$(10.7) \qquad \mathbf{d}^a g = \prod_{\sigma \in J} \left[(2\pi i)^{-1} (\partial/\partial z_\sigma) \right]^{a_\sigma} g.$$

Since $h \in \mathbf{Z}^{\varepsilon'}$, we see that $\Theta_h = \mathbf{d}^h \Theta_0$. Now, if $F = F_1$, Θ_0 is a finite linear combination of the series of [6, (7.6)]. In fact, the factor $v^{x+x'} \cdot q \cdot \bar{\mu}^{k-e\tau} \zeta('\lambda)$ corresponds to χ of [6, (7.4)] by virtue of Lemma 10.1. In the general case where $F = F_1 \times \cdots \times F_t$, Θ_0 is a finite linear combination of the products $\theta_1 \ldots \theta_t$ with θ_i of the type of [6, (7.6)] defined on $\mathcal{H}_1^{J(F_i)}$. Therefore, by [6, Proposition 7.1], Θ_0 satisfies

$$\Theta_0(\gamma z, w) = j_\gamma(z)^A \overline{j_\gamma(z)}^B \Theta_0(z, w) \quad \text{for every } \gamma \in \Delta$$

with a congruence subgroup Δ of $\mathrm{SL}_2(F)$, where

$$(10.8a) \qquad A = (1/2) \sum_{i=1}^t (n_i + 2)\iota(F_i) - \varepsilon + e\tau + \kappa + \mathrm{Res}_{K/F}(\chi'),$$

$$(10.8b) \qquad B = \varepsilon + k - e\tau + \mathrm{Res}_{K/F}(\chi).$$

Take a sufficiently small congruence subgroup Γ of $\mathrm{SL}_2(E)$ and put $\Gamma_\infty = \left\{ \left(\begin{smallmatrix} a & b \\ 0 & d \end{smallmatrix} \right) \in \Gamma \right\}$; define an embedding $z \mapsto z^\circ$ of $\mathcal{H}_1^{J(E)}$ into $\mathcal{H}_1^{J(F)}$ by $(z^\circ)_\sigma = z_\alpha$ for $\sigma \in J_F$ if $\alpha = \mathrm{Res}_{F/E}(\sigma)$. Then we consider an integral

$$(10.9a) \qquad \int_L \mathbf{d}^P \Omega(z) \Theta_h(z^\circ, w) y^{s\varepsilon + \varepsilon + k - m} \{ dz \},$$

where $L = \Gamma_\infty \backslash \mathcal{H}_1^{J(E)}$, $m = \mathrm{Res}_{K/E}(\nu)$, $\{ dz \} = y^{-2\varepsilon} dx\,dy$; we identify $\mathbf{Z}^\varepsilon$ with $\mathbf{Z}^{J(E)}$. For the same reason as in Section 8 (cf. also [7, I, §2]), (10.9a) is equal to

$$(10.9b) \quad C_1 \prod_{\sigma \in \varepsilon - \delta} |\beta^\sigma|^{-s} (8\pi)^{-[E:\mathbf{Q}]s} \cdot \prod_{\sigma \in \delta} \Gamma(s + k_\sigma) \prod_{\sigma \in \psi} \Gamma(s - \varphi_{\sigma\rho}) f'(w, s; \zeta)$$

with a nonzero constant C_1. Let D_b^a denote the operator on $\mathcal{H}_1^J$ defined for $0 \leq a \in \mathbf{Z}^J$ and $b \in \mathbf{C}^J$ (with $J = J_E$ or J_F) by

$$(10.10a) \qquad D_b^a = \prod_{\sigma \in J} D_\sigma(b_\sigma + 2a_\sigma - 2) \ldots D_\sigma(b_\sigma + 2) D_\sigma(b_\sigma),$$

$$(10.10b) \qquad D_\sigma(b_\sigma) = (2\pi i)^{-1} \left[b_\sigma(z_\sigma - \bar{z}_\sigma)^{-1} + \partial/\partial z_\sigma \right].$$

By [7, I, (1.16a)], $\mathbf{d}^h\Theta_0 = \Sigma_{0 \le a \le h} c_a y^{a-h} D_A^a \Theta_0$ with constants c_a; similarly $\mathbf{d}^p\Omega = \Sigma_{0 \le q \le p} c_q' y^{q-p} D_l^q\Omega$. Observe that B belongs to $\mathbf{Z}^\varepsilon$ and therefore is disjoint with ε'; hence multiplication with y^B commutes with $\mathbf{d}^h$ and D_A^a. Thus (10.9a) is equal to a finite linear combination of integrals

$$(10.11) \qquad \int_L D_l^q\Omega(z) D_A^a\Theta(z^\circ, w) y^{\Xi + s\varepsilon}\{dz\},$$

with $\Xi = q - p + a^* - h^* + \varepsilon + k - m$, where $x^* = \mathrm{Res}_{F/E}(x)$ for $x \in I_F$. Let $g(z)$ be the integrand of (10.11). For a suitably chosen Γ, we see, in view of (10.3), that

$$g(\gamma z) = g(z) |j_\gamma(z)|^{-2s\iota} j_\gamma(z)^{-b\iota} \left(\overline{j_\gamma(z)}/j_\gamma(z)\right)^\beta$$

for every $\gamma \in \Gamma$, where $\iota = \iota_E$, b is the number b_f of (4.14), and

$$(10.12) \qquad \beta = p - q + h^* - a^* - e\tau + \mathrm{Res}_{K/E}\left(\sum_{\sigma \in \psi} \varphi_\sigma \sigma\right) \quad (\in I_E).$$

If $b \notin \mathbf{Z}$, $j_\gamma(z)^{-b\iota}$ should be defined by

$$(10.13) \qquad j_\gamma(z)^{-b\iota} = j_\gamma(z)^{\{(1/2)-b\}\iota} h_\gamma(z)^{-1}$$

with a factor of automorphy h_γ of weight $1/2$ of [13, Proposition 1.2], which is the same as that of [6, (7.28)]. Therefore, the well known principle, which was employed in Section 8, transforms (10.11) to an integral

$$(10.14a) \qquad \int_M D_l^q\Omega(z) D_A^a\Theta_0(z^\circ, w) \mathscr{E}_b^\beta(z, s) y^B\{dz\},$$

where $M = \Gamma \backslash \mathscr{H}_1^{J(E)}$ and

$$(10.14b) \qquad \mathscr{E}_b^\beta(z, s) = \sum_{\gamma \in R} y^{s\iota - \beta} |j_\gamma(z)|^{-2s\iota} \left(|j_\gamma(z)|/j_\gamma(z)\right)^{2\beta} j_\gamma(z)^{-b\iota}$$

$$(R = \Gamma_\infty \backslash \Gamma).$$

This is meaningful for every $b \in (1/2)\mathbf{Z}$ and $\beta \in (1/2)I_E$.

LEMMA 10.2. (i) $\mathscr{E}_b^\beta$ *can be continued as a meromorphic function in s to the whole plane.*

(ii) *If $b \in \mathbf{Z}$, $\mathscr{E}_b^\beta$ is finite for $\mathrm{Re}(s) \ge (1 - b)/2$ except for a possible simple pole at $s = 1 - (b/2)$, which occurs only when $b\iota + 2\beta = 0$.*

(iii) *If $b \notin \mathbf{Z}$, $\mathscr{E}_b^\beta$ is finite for $\mathrm{Re}(s) \ge (1 - b)/2$ except for a possible simple pole at $s = (3 - 2b)/4$, which occurs only when, for every $\sigma \in I_E$, $b - (1/2) + 2\beta_\sigma$ is an even nonnegative integer or an odd negative integer.*

This was given in [7, I, Proposition 1.1] for $b \in \mathbf{Z}$. The case of half-integral weight was treated in [13] and [14]. In particular, [14, Theorem 4.2] provides general results covering both integral and half-integral b; in fact, $\mathscr{E}_b^\beta(z, s)$ coincides with $E(z, s + (b/2); (bu/2) + \beta, \Gamma)$ of that paper.

This lemma combined with the expression of (10.14a) establishes the analytic continuation of $f'(w, s; \zeta)$, and also tells about the behavior of f' at a specific point s. To justify this method, however, we need to know that $D_A^a \Theta_0(z^\circ, w)$, as a function of z, is rapidly decreasing at every cusp. This is so if $k_\sigma > 0$ for all $\sigma \in \delta$ and $e \le k_\tau$. In fact, under this assumption, the term of Θ_0 for $v = 0$ vanishes. Now Θ_0 is a finite linear combination of products $\theta_1 \dots \theta_t$ with theta functions θ_i of $G(S_i)$. The vanishing of the constant term implies that at least one of the θ_i is rapidly decreasing, which guarantees that $D_A^a \Theta_0(z^\circ, w)$ is rapidly decreasing.

Suppose now $0 < b_f \in \mathbf{Z}$ and take $e = 0$. Then (10.12) shows that $\beta \geqq \operatorname{Res}_{K/E}(\Sigma_{\sigma \in \psi} \varphi_\sigma \sigma) \in \mathbf{Z}^{\varepsilon - \delta}$ so that $b\iota \ne -2\beta$. Hence, by Lemma 10.2, $f'(w, s)$ is finite at $s = 0$. Next take $e = 1$. We have then

$$b\iota + 2\beta = b\iota - 2\tau + 2\operatorname{Res}_{K/E}\left(\sum_{\sigma \in \psi} \varphi_\sigma \sigma\right) + 2\alpha$$

with $0 \leqq \alpha \in I_E$, and hence $b\iota = -2\beta$ can happen only under (4.15) (with $b = 2$). Therefore, excluding this special case, we know that $f'(w, s; \zeta)$ is finite at $s = 0$, so that $E_Z f'(w, 0) = 0$ with $Z = \mathscr{S}_1(T)$ by virtue of (10.4). This proves the holomorphy of $f(w, 0)$ for $0 < b \in \mathbf{Z}$ except in case (4.15).

If $3/2 < b \notin \mathbf{Z}$, we obtain the same conclusion in a similar way. Suppose $b = 3/2$. Then $\mathscr{E}_b^\beta$ (and hence $f'(w, s; \zeta)$) may have a simple pole at $s = 0$. This cannot happen if $e = 0$, since $\beta_\tau \geq 0$. Thus $f'(w, s)$ is finite at $s = 0$. Similarly, if $e = 1$, we see that at least one component of β is nonnegative except in case (4.15). Therefore $f'(w, s; \zeta)$ is finite at $s = 0$, which implies the holomorphy of $f'(w, 0)$. This completes the proof of Theorem 4.1.

To investigate further the cases $b_f \le 2$, it is necessary to consider $E_Z f$ of higher order as in Lemma 6.3. Thus, let $f(j', k', c\tau; \zeta)$ denote the series obtained from $f(w, s; \zeta)$ by changing j and k for j' and k', and multiplying each term with $\eta(w_\tau)^c |\mu_\tau[v, w]|^{-2c}$, where c is a nonnegative integer. Then Lemma 6.3 shows that

$$(10.15) \quad \left[E_Z f(j, k, 0; 1)\right]\zeta' = s \sum_{i=0}^{d} q_{e, d, i}^*(s) f(j + i\tau, k - 2d\tau, (d + i)\tau; \zeta)$$

with polynomials $q_{e, d, i}^*$, where the symbols ζ, ζ', d, and e are the same as in the lemma. Now the series $f(j + i\tau, \dots)$ can be obtained as a sum of integrals of type (10.9a) with the following changes: at the beginning, take $j + i\tau$ in place of j; take $k - 2d\tau - e\tau$ instead of $k - e\tau$ in the definition of θ_h; take $k + (i - d)\tau - m$ in place of $k - m$ in (10.9a). In this new setting, (10.2) and (10.3) should be replaced by

$$(10.16) \quad \mathrm{Res}_{F/E}(2j) + \mathrm{Res}_{K/E}(\varphi) + 2i\tau = 2p + \mathrm{Res}_{F/E}(2h) + \mathrm{Res}_{K/E}(\xi),$$

$$(10.17) \quad \mathrm{Res}_{F/E}(k - \kappa - 2h) - \mathrm{Res}_{K/E}(\xi) - 2p - l = a_f \iota_E - 2i\tau.$$

Then $f(j + i\tau, \dots)$ is a finite linear combination of several series f', each of which, multiplied by a gamma-factor, can be expressed as an integral of type (10.14a), in which $b = b_f$ as before, but β is replaced by

$$(10.18) \quad \beta' = p - q + h^* - a^* + \mathrm{Res}_{K/E}\!\left(\sum_{\sigma \in \psi} \varphi_\sigma \sigma \right) - (e + d + i)\tau.$$

Before proceeding further, we observe that (4.12) implies that

$$(10.19) \qquad\qquad k_\tau \geqq b_f - 2 + M + l_\tau + 2N_\tau,$$

where $M = \sum_{i=1}^t [F_i : E](n_i + 2)/2$ and N_τ is the coefficient of $\mathrm{Res}_{F/E}(\tau)$ in $\mathrm{Res}_{F/E}(j) + \mathrm{Res}_{K/E}(\sum_{\sigma \in \psi} \varphi_{\sigma\rho} \sigma)$. From (10.16) and (10.18), we see that $\beta'_\tau \leqq N_\tau - d - e$.

To prove Theorem 4.3, suppose $b_f = 0$ or $1/2$; put $s_f = 1 - b_f$. By Lemma 10.2 and the expression of (10.14a), we see that $f(w, s)$ has at most a simple pole at s_f. Coming back to (10.15), take $2d + e = 2N_\tau + 1$. Then $e > 0$ and $\beta'_\tau < 0$, and therefore $\mathscr{E}_b^{\beta'}$ (needed for $f(j + i\tau, k - 2d\tau, (d + i)\tau; \zeta)$) is finite at s_f. Thus the right-hand side of (10.15) is finite at s_f. This implies that $E^t f(w, s)$ is finite at s_f with $t = 2N_\tau + 1$, which is the second assertion of Theorem 4.3. In particular, if $N_\tau = 0$ for all $\tau \in \delta$, we obtain the last assertion. In this argument, however, we have to take $k_\tau \geq 2d + e = 2N_\tau + 1$. This is so if

$$(10.20) \qquad\qquad b_f + l_\tau + M \geqq 3,$$

by virtue of (10.19). This holds if $F \neq E$. Therefore let us assume $F = E$. Since $a_f - l_\tau \in \mathbf{Z}$, (10.20) fails only in the following three cases: (i) $b_f = 0$, $l = 0$, and $n_1 = 2$; (ii) $b_f = 0$, $l_\tau = 1/2$, and $n_1 = 1$; (iii) $b_f = 1/2$, $l = 0$, and $n_1 = 1$. (It should be noted that $l = 0$ if $l_\tau = 0$ for at least one τ.) If $l = 0$ and $F = E$, the sum of (4.10) is extended over all $v \in V$ such that $S[v] = 0$. Therefore $f = 0$ if $j \neq 0$. Thus we may assume $j = 0$. Then β of (10.12) is $-e\tau$, and hence we can conclude that Ef is finite at s_f. This completes the proof of Theorem 4.3, since the second case is eliminated by assumption (4.18).

Theorem 4.2 can be proved in a similar way. In fact, suppose $b_f = 2$ and $\delta = \tau$; take $2d + e = 2N_\tau + 3$. Then $e > 0$ and $b_f + 2\beta'_\tau < 0$, and hence $\mathscr{E}_b^{\beta'}$ (for $f(j + i\tau, \dots)$) is finite at $s = 0$. Thus (10.15) vanishes at $s = 0$. Our argument is justified if $2N_\tau + 3 \leq k_\tau$. By (10.19), this is so if $l_\tau + M \geq 3$. Since $\delta = \tau$, this fails only if $F = E = \mathbf{Q}$. Under (4.17), the only exceptional case is when $l = 0$ and $n_\tau = 2$. Excluding this case which will be treated at the end, we can conclude that $E^t f(w, 0) = 0$ with $t = 2N_\tau + 3$. Assume in particular $N_\tau = 0$.

Then $b_f + 2\beta'_\tau < 0$ if $(d, e) = (0, 2)$ or $(d, e) = (1, 1)$. Therefore the above reasoning shows that $f(w, 0)$ satisfies condition (i) of Proposition 3.12. Suppose finally $F = E = \mathbf{Q}$, $l = 0$, and $n_\tau = 2$. Then $f = 0$ if $j \neq 0$ for the same reason as above. By Lemma 6.1, (iii) and (iv), we have

$$[E_Z f(0, k, 0; 1)]\zeta' = \begin{cases} 2s^2 f(0, k - 2\tau, \tau; 1) & \text{if } (d, e) = (1, 0), \\ s(s + 1)f(0, k, 0; \zeta) & \text{if } (d, e) = (0, 2). \end{cases}$$

We have $b + 2\beta_\tau < 0$ for $f(0, k, 0; \zeta)$; as for the case $(d, e) = (1, 0)$, the factor s^2 produces a zero of order 2 at $s = 0$. Therefore $E^2 f = 0$ at $s = 0$. This completes the proof of Theorem 4.2.

It remains to prove Proposition 4.4. Our proof shows that the function of (10.9b) has at most a simple pole at $s = 1$ if $b_f = 0$. Therefore if $\varphi_{\sigma\rho} > 0$ for some $\sigma \in \psi$, then $f'(w, s)$ is finite at $s = 1$, which is (ii) of Proposition 4.4. Assertion (i) can be proved in a similar way.

Princeton University, Princeton, New Jersey

References

[1] E. Hecke, *Vorlesungen über die Theorie der algebraischen Zahlen*, 1923, Chelsea, 1948.

[2] R. Indik, Fourier coefficients of nonholomorphic Eisenstein series on a tube domain associated to an orthogonal group, Thesis, Princeton Univ., 1982.

[3] R. P. Langlands, *On the Functional Equations Satisfied by Eisenstein Series*, Lecture notes in Math. 544, Springer, 1976.

[4] G. Shimura, The arithmetic of automorphic forms with respect to a unitary group, Ann. of Math. **107** (1978), 569–605.

[5] ———, Automorphic forms and the periods of abelian varieties, J. Math. Soc. Japan **31** (1979), 561–592.

[6] ———, The arithmetic of certain zeta functions and automorphic forms on orthogonal groups, Ann. of Math. **111** (1980), 313–375.

[7] ———, On certain zeta functions attached to two Hilbert modular forms I, II, Ann. of Math. **114** (1981), 127–164, 569–607.

[8] ———, Arithmetic of differential operators on symmetric domains, Duke Math. J. **48** (1981), 813–843.

[9] ———, Algebraic relations between critical values of zeta functions and inner products, Amer. J. of Math. **104** (1983), 253–285.

[10] ———, On Eisenstein series, Duke Math. J. **50** (1983), 417–476.

[11] ———, Differential operators and the singular values of Eisenstein series, Duke Math. J. **51** (1984), 261–329.

[12] ———, On differential operators attached to certain representations of classical groups, Invent. Math. **77** (1984), 463–488.

[13] ———, On Eisenstein series of half-integral weight, Duke Math. J. **52** (1985), 281–314.

[14] ———, On the Eisenstein series of Hilbert modular groups, Revista Mat. Iberoamer **1**, No. 3 (1985), 1–42.

(Received May 28, 1985)

87a

Nearly holomorphic functions on hermitian symmetric spaces

Mathematische Annalen, 278 (1987), 1-28

To Friedrich Hirzebruch

The notion of nearly holomorphic function on a complex Kähler manifold V is defined as follows: Taking a Kähler form $\Omega = i\partial\bar\partial\varphi$ on V with a real-valued function φ in a coordinate neighborhood U, we call a function on U *nearly holomorphic* if it is a polynomial in $\partial\varphi/\partial z_1, \ldots, \partial\varphi/\partial z_n$ with holomorphic coefficients, where $\{z_1, \ldots, z_n\}$ is a set of complex coordinate functions on U. This class of functions was introduced in our previous paper [16] for the purpose of gaining a clearer comprehension of certain automorphic forms that are nonholomorphic but arithmetically behave very much like holomorphic automorphic forms. Such forms appear naturally as Eisenstein series and also as certain infinite series similar to them.

Our emphasis in [16] was laid on the *fact* that such series are nearly holomorphic and also on the *method* of proof. In the present paper, we lean more towards generalities; yet explicit examples such as Eisenstein series are considered at some places. Though nearly holomorphic functions can be defined on any complex Kähler manifold, hermitian symmetric spaces seem their most natural habitat. Therefore, for the most part, we restrict our present investigation to the functions on such spaces.

Now a summary of this paper can be given as follows. After reviewing the definition of near holomorphy and making some elementary observations, we shall determine the set of all functions which are nearly holomorphic everywhere on a complex projective space (Theorem 1.2), and more generally on a hermitian symmetric space of compact and nonexceptional type (Theorems 2.1 and 2.3). To illustrate our result, take for example the space to be the Grassmannian manifold V consisting of all $N \times m$ complex matrices w of rank m modulo $Gl_m(\mathbb{C})$. Then we can show that *every nearly holomorphic function on V is a polynomial in the entries of* $w({}^t\bar{w}w)^{-1} \cdot {}^t\bar{w}$, *and vice versa*.

In Sect. 3, we consider automorphic forms on a hermitian symmetric space of noncompact type. Our main interest is in the arithmeticity of holomorphic and nearly holomorphic forms, especially in connection with the differential operators introduced in our previous papers. Any such operator, written $D_\varrho^{(e)}$ or D_ϱ^Z, sends a form of weight ϱ to that of weight $\varrho \otimes \sigma$ with a certain σ; moreover it keeps near holomorphy but not holomorphy. Now let $\mathscr{W}$ be the set of all *CM*-points of our space $\mathscr{D}$ with respect to the algebraic group in question. We say that a C^∞ automorphic form f is *arithmetic* at a point w of $\mathscr{W}$ if $f(w)$ has the same property of algebraicity as the values of arithmetic holomorphic forms of the same weight at w. (For details, the reader is referred to the text and [8, 10, 12].) Then we prove, as one of the main results of the present paper, the following theorem:

Let $\mathscr{V}$ be a subset of $\mathscr{W}$ that is dense in $\mathscr{D}$, and let f be a nearly holomorphic automorphic form on $\mathscr{D}$. If f is arithmetic at every point of $\mathscr{V}$, then $D_\varrho^{(e)}f$, as well as f itself, is arithmetic at every point of $\mathscr{W}$ (Theorem 3.7).

This result will play a key role in the proof of the arithmeticity of certain automorphic forms of orthogonal and unitary groups which are given as infinite series similar to Eisenstein series. In fact, we proved in [16] the holomorphy and near holomorphy of such series, and stated a few theorems concerning their arithmeticity without proof. We hope to present the detailed proof in a subsequent paper.

Section 4 concerns the near holomorphy of certain Eisenstein series on a tube domain attached to the product of copies of $Sp\,(m, \mathbb{R})$ or $SU\,(m, m)$. To exemplify our results, let us take the case in which the series is defined with respect to the group

$$\Gamma = \left\{ \begin{pmatrix} a & b \\ c & d \end{pmatrix} \in Sp\,(m, \mathfrak{o}) \mid c \equiv 0 \,(\mathrm{mod}\,\mathfrak{c}) \right\},$$

where $\mathfrak{o}$ is the maximal order of a totally real algebraic number field F, and $\mathfrak{c}$ is an integral ideal of F. Let J be the set of all archimedean primes of F, and $\mathscr{H}_m$ the set of all complex symmetric matrices of size m with positive definite imaginary part; further let Γ_∞ be the subgroup of Γ consisting of the elements with vanishing c-block. Assuming, for simplicity, the class number of F to be 1, we consider a series

$$E(z,s) = \sum_\alpha \Psi_\infty (\det(d_\alpha)) \, \Psi^*(\det(d_\alpha)\mathfrak{o}) \prod_{\tau \in J} \det\,(\mathrm{Im}\,(\alpha z)_\tau)^{s - q_\tau/2} \det\,(c_\alpha^\tau z_\tau + d_\alpha^\tau)^{-q_\tau}.$$

Here $z = (z_\tau)_{\tau \in J} \in \mathscr{H}_m^J$, $s \in \mathbb{C}$, $q_\tau \in \mathbb{Z}$, α runs over $\Gamma_\infty \backslash \Gamma$, Ψ is a Hecke idele character of F defined modulo $\mathfrak{c}J$ such that $\Psi_\infty (a) = \prod_{\tau \in J} (|a_\tau|/a_\tau)^{q_\tau}$, Ψ^* is the ideal character attached to Ψ, and (c_α, d_α) is the lower half of α. Further put

$$D(z,s) = E(z,s)\,L_\mathfrak{c}(2s, \Psi) \prod_{i=1}^{[m/2]} L_\mathfrak{c}(4s - 2i, \Psi^2),$$

where $L_\mathfrak{c}$ is the L-function without the Euler-factors for the primes dividing $\mathfrak{c}$. Then, assuming again for simplicity that m is odd and $\mathfrak{c} \neq (1)$, we can prove:

Let μ be an integer such that $m - q_\tau < \mu \leq q_\tau$ and $\mu \equiv q_\tau \,(\mathrm{mod}\,2)$ for every $\tau \in J$. Then $\pi^{-\beta} D(z, \mu/2)$ is nearly holomorphic and arithmetic, where

$$\beta = m \sum_{\tau \in J} (q_\tau + \mu)/2 - [F: \mathbb{Q}]\,(m^2 - 1)/4.$$

In the text, we shall prove a more comprehensive theorem which covers even m, the field F of class number > 1, the series of half-integral weight, and the case of unitary groups (Theorems 4.2 and 4.3).

In Sect. 5, we consider automorphic forms on $\mathcal{H}_1^J$ with respect to a quaternion algebra over F [including $M_2(F)$]. We shall prove that if such a form is nearly holomorphic, it can be expressed as a linear combination of functions of the form $D_b^a g$, where D_b^a is the above operator $D_\varrho^{(e)}$ specialized to this case, and g is either a holomorphic form or the nonholomorphic Eisenstein series of weight 2 for $SL_2(\mathbb{Z})$; further such an expression is consistent with arithmeticity (Theorems 5.2 and 5.3).

The final section is a brief discussion of the theta functions (in the sense of Jacobi-Riemann) that are nearly holomorphic. We shall show that the results of Sect. 5 have analogues for such theta functions, which may be viewed as reformulations of some theorems of [6].

Notation

Given an associative ring A with identity element, we denote by A_n^m the module of all $m \times n$-matrices with entries in A. We put $A^m = A_1^m$ and also $M_n(A) = A_n^n$ when the set is viewed as a ring, and denote by 1_n its identity element. For two sets X and J, we denote by X^J the set of all indexed elements $(x_\tau)_{\tau \in J}$ with $x_\tau \in X$. We write $a \leqq b$ for a, $b \in \mathbb{R}^J$ if $a_\tau \leqq b_\tau$ for every $\tau \in J$ and put $|a| = \sum_{\tau \in J} a_\tau$. Given two C^∞ manifolds U and V, we denote by $C^\infty(U, V)$ the set of all C^∞ maps of U into V, and put $C^\infty(U) = C^\infty(U, \mathbb{C})$. By a *representation* $\{\varrho, X\}$ of a group G, we understand a couple formed by a finite-dimensional complex vector space X and a homomorphism ϱ of G into the group $GL(X)$ of $\mathbb{C}$-linear automorphisms of X. For a complex matrix z, we denote its complex conjugate by $\bar{z}$ and its transpose by ${}^t z$; we put $z^* = {}^t \bar{z}$. We write $h > k$ for two complex hermitian matrices h and k of the same size if $h - k$ is positive definite. For $z \in \mathbb{C}$, we put $\mathbf{e}(z) = \exp(2\pi i z)$.

1. Nearly Holomorphic Functions on Kähler Manifolds

Let V be an n-dimensional complex manifold, and U a coordinate neighborhood on V with complex coordinate functions $z_1, \ldots, z_n$. Suppose n elements $r_1, \ldots, r_n$ of $C^\infty(U)$ satisfy the condition that the $n \times n$-matrix $(\partial r_q / \partial \bar{z}_p)$ is invertible everywhere on U. Then we can define n vector fields $\partial/\partial r_1, \ldots, \partial/\partial r_n$ by

$$\partial/\partial \bar{z}_p = \sum_{q=1}^n \partial r_q / \partial \bar{z}_p \cdot \partial/\partial r_q \qquad (p = 1, \ldots, n). \tag{1.1}$$

Lemma 1.1. (i) *The vector fields $\partial/\partial r_1, \ldots, \partial/\partial r_n$ are mutually commutative.*

(ii) *A C^∞ function f on U is a polynomial in $r_1, \ldots, r_n$ of degree $< k$ with holomorphic functions on U as coefficients if and only if $(\partial/\partial r_{v_1}) \ldots (\partial/\partial r_{v_k}) f = 0$ for all $(v_1, \ldots, v_k) \in \{1, \ldots, n\}^k$.*

These are restatements of [16, Lemmas 2.0 and 2.1].

Suppose now that V is a Kähler manifold with a fundamental 2-form Ω, which is given in the form $\Omega = i \sum_{p,q=1}^n h_{pq} dz_q \wedge d\bar{z}_p$ on U. We can then define n vector fields $X_1, \ldots, X_n$ on U by the relations

$$\partial/\partial \bar{z}_p = \sum_{q=1}^n h_{pq} X_q \qquad (p = 1, \ldots, n). \tag{1.2}$$

If D is the interior of a compact polydisc contained in U, then the easiest case of the Dolbeault lemma shows the existence of n functions $r_1, \ldots, r_n$ in $C^\infty(D)$ such that $\partial r_q / \partial \bar{z}_p = h_{pq}$. (It is also a well known fact, already found in Kähler [2], that locally $\Omega = i \partial \bar{\partial} \varphi$ with a real valued function φ. Then we can take $r_q = \partial \varphi / \partial z_q$. Cf. also Weil [17, p. 72, Corollaire 2].) Then $X_q = \partial / \partial r_q$ on D. Therefore the above lemma shows that the X_p are mutually comutative.

Now the X_p may depend on the choice of $z_1, \ldots, z_n$. However, if $w_1, \ldots, w_n$ are coordinate functions on U' and $Y_1, \ldots, Y_n$ are defined relative to the w_p in the same way, we see that $X_p = \sum_q \partial z_p / \partial w_q \cdot Y_q$ on $U \cap U'$. Therefore if we denote by $\mathscr{N}^{k-1}(U)$ the set of all elements of $C^\infty(U)$ annihilated by $X_{v_1} \ldots X_{v_k}$ for all $(v_1, \ldots, v_k) \in \{1, \ldots, n\}^k$, this is well-defined independently of the choice of the z_p. Now, for every open subset W of V, let $\mathscr{N}^k(W)$ be the set of all elements f of $C^\infty(W)$ such that the restriction of f to any coordinate neighborhood U belongs to $\mathscr{N}^k(U)$, and let $\mathscr{N}(W) = \bigcup_{k=0}^{\infty} \mathscr{N}^k(W)$. We then call an element of $\mathscr{N}(W)$ a *nearly holomorphic function* on W relative to Ω. In this way we obtain a presheaf $\mathscr{N}^k$ of nearly holomorphic functions of degree $\leq k$ over V (relative to Ω). If D and $r_1, \ldots, r_n$ are given as above, then, by Lemma 1.1, $\mathscr{N}^k(D)$ consists of all the polynomials of $r_1, \ldots, r_n$ of degree $\leq k$ with holomorphic functions on D as coefficients. Obviously $\mathscr{N}^0(W)$ is the set of all holomorphic functions on W.

Let us now take V to be the complex projective space P^n of dimension n, and examine the nature of $\mathscr{N}(P^n)$. Let $(w_0, \ldots, w_n)$ denote homogeneous coordinates on P^n, and let U_α denote, for $\alpha = 0, \ldots, n$, the affine subspace of P^n defined by $w_\alpha \neq 0$. Put $\varphi_\alpha = \log \sum_{p=0}^{n} |w_p / w_\alpha|^2$. Then we can define a fundamental form Ω on P^n so that $\Omega = i \partial \bar{\partial} \varphi_\alpha$ on U_α. Now we put

$$z_{q\alpha} = w_q / w_\alpha, \quad r_{q\alpha} = \bar{z}_{q\alpha} \bigg/ \sum_{j=0}^{n} |z_{j\alpha}|^2 \qquad (\alpha, q \in \{0, \ldots, n\}), \tag{1.3}$$

and view $z_{q\alpha}$ and $r_{q\alpha}$ as C^∞ functions on U_α. We see easily that

$$r_{\alpha\alpha} = 1 - \sum_{q \neq \alpha} z_{q\alpha} r_{q\alpha}. \tag{1.4}$$

Fix α and take the $z_{q\alpha}$ for $q \neq \alpha$ as the coordinate functions on U_α. Then $r_{q\alpha} = \partial \varphi_\alpha / \partial z_{q\alpha}$ for $q \neq \alpha$, and hence the $r_{q\alpha}$ for $q \neq \alpha$ generate $\mathscr{N}(U_\alpha)$ over $\mathscr{N}^0(U_\alpha)$. Even $r_{\alpha\alpha}$ belongs to $\mathscr{N}^1(U_\alpha)$, be virtue of (1.4).

Now it is noteworthy that the $r_{q\alpha}$ can be extended to the elements of $\mathscr{N}^1(P^n)$. In fact, we have

$$r_{q\alpha} = z_{\alpha\beta} r_{q\beta} \qquad \text{on } U_\alpha \cap U_\beta, \tag{1.5}$$

and hence we can extend each $r_{q\alpha}$ to an element of $\mathscr{N}^1(P^n)$ that vanishes outside U_α. This is a marked contrast to the fact that $\mathscr{N}^0(P^n)$ consists only of the constants.

To determine the structure of $\mathscr{N}^k(P^n)$, let us denote by $\mathbb{N}$ the set of all nonnegative integers, and define symbols w^s, z_α^s, and r_α^s for $s = (s_0, \ldots, s_n) \in \mathbb{N}^{n+1}$ by $w^s = \prod_{q=0}^{n} w_q^{s_q}$, $z_\alpha^s = \prod_{q=0}^{n} (z_{q\alpha})^{s_q}$, and $r_\alpha^s = \prod_{q=0}^{n} (r_{q\alpha})^{s_q}$. We put $|s| = s_0 + \ldots + s_n$.

Theorem 1.2. *For two elements s and t of $\mathbb{N}^{n+1}$ such that $|s| = |t| = k$, there exists a unique element y_{st} of $\mathcal{N}^k(P^n)$ such that $y_{st} = z_\alpha^s r_\alpha^t$ on U_α for every α. Moreover, the y_{st} for all such s and t form a basis of $\mathcal{N}^k(P^n)$ over $\mathbb{C}$.*

Proof. In fact, we shall prove in the next section a more general theorem of which the present one is a special case. For the reader's convenience, however, we give here a direct proof. We see from (1.3) that $\left(\sum_{j=0}^{n} |w_j|^2 \right)^{-k} w^s \overline{w}^t = z_\alpha^s r_\alpha^t$ for every α, which proves the first assertion. Obviously the y_{st} are linearly independent over $\mathbb{C}$. Now let $g \in \mathcal{N}^k(P^n)$. Then the restriction of g to U_α is a polynomial in the $r_{q\alpha}$ with coefficients in $\mathcal{N}^0(U_\alpha)$. In view of (1.4), we can express g in the form $g = \sum_t f_t r_\alpha^t$ with $f_t \in \mathcal{N}^0(U_\alpha)$, where t runs over the elements of $\mathbb{N}^{n+1}$ such that $|t| = k$. Notice that the f_t are unique for g. Now (1.5) shows that $g = \sum_t f_t z_{\alpha\beta}^k r_\beta^t$ on $U_\alpha \cap U_\beta$ for $\beta \neq \alpha$. Hence $f_t z_{\alpha\beta}^k$, for each t, must be extended to an element of $\mathcal{N}^0(U_\beta)$. Therefore f_t can be extended to a meromorphic function on P^n. Being holomorphic on U_α, it is a polynomial in the $z_{q\alpha}$. Take a homogeneous polynomial $h(w_0, \ldots, w_n)$ so that $f_t = h(z_{0\alpha}, \ldots, z_{n\alpha})$. We may assume that h is not divisible by w_α. Let m be the degree of h. Then $f_t z_{\alpha\beta}^k = w_\alpha^{k-m} w_\beta^{-k} h(w_0, \ldots, w_n)$. Since this is holomorphic on U_β, we have $m \leq k$. Putting $w_\alpha^{k-m} h(w_0, \ldots, w_n) = \sum_s c_{st} w^s$ with $c_{st} \in \mathbb{C}$, we obtain the desired expression of g.

It seems that there are relatively few types of compact Kähler manifolds with nontrivial nearly holomorphic functions. It is more natural to investigate such functions on a domain V embedded in a finite-dimensional complex vector space T. In fact, suppose a Kähler form Ω is given globally on V by $\Omega = i\partial\bar{\partial}\varphi$ with a real-valued function φ on V. For $f \in C^\infty(V)$ and $c \in \mathbb{C}$, we can define an element $\delta_c f$ of $C^\infty(V, \mathrm{Hom}(T, \mathbb{C}))$ by

$$(\delta_c f)(u) = e^{-c\varphi} D_u(e^{c\varphi} f) \qquad (u \in T), \tag{1.6}$$

where D_u is the holomorphic derivation on T such that $D_u h = h(u)$ for $h \in \mathrm{Hom}(T, \mathbb{C})$. Taking coordinate functions $z_1, \ldots, z_n$ on T, we have

$$(\delta_c f)(u) = cf \sum_v u_v \partial\varphi/\partial z_v + \sum_v u_v \partial f/\partial z_v, \tag{1.7}$$

where $u_v = z_v(u)$. This is nearly holomorphic if f is holomorphic. Successive application of operators of this type produces nearly holomorphic functions, provided the higher derivatives of φ are nearly holomorphic, which is often the case as we shall see in later sections.

2. Nearly Holomorphic Functions on Compact Hermitian Symmetric Spaces

Every irreducible hermitian symmetric space of compact and nonexceptional type can be given as $V = G/K$ with compact groups G and K given below. V has a structure of projective variety with a G-invariant Kähler form Ω. In this section we shall determine the set of all nearly holomorphic functions on V relative to Ω.

There are four types of V, which we call Types A, B, C, and D according to the standard classification of G. Instead of working with G/K, we present V as a quotient space $V = L/GL_m(\mathbb{C})$ with an affine subset L of $\mathbb{C}_m^N$. There is an algebraic subgroup $G_{\mathbb{C}}$ of $GL_N(\mathbb{C})$ acting on the left of L. Then G is given as $G = G_{\mathbb{C}} \cap U(N)$, where $U(N) = \{x \in GL_N(\mathbb{C}) \mid x^* x = 1\}$, and K as the isotropy subgroup of G at any point of V. Now the explicit forms of these objects are given as follows:

Type A. $G_{\mathbb{C}} = GL_N(\mathbb{C})$, $G = U(N)$, $N = n + m$, $n \geq m > 0$, $K \cong U(n) \times U(m)$,

$$L = \{w \in \mathbb{C}_m^N \mid \operatorname{rank}(w) = m\}.$$

Type B. $G_{\mathbb{C}} = \{x \in SL_N(\mathbb{C}) \mid {}^t x R x = R\}$, $G = G_{\mathbb{C}} \cap U(N)$, $N = n + 2 > 2$,

$$R = \begin{bmatrix} 1_n & 0 & 0 \\ 0 & 0 & -1 \\ 0 & -1 & 0 \end{bmatrix},$$

$$L = \{w \in \mathbb{C}^N \mid {}^t w R w = 0, \; w \neq 0\}.$$

In this case we take $m = 1$, so that $V = L/\mathbb{C}^{\times}$. It can easily be seen that $G \cong SO(N, \mathbb{R})$ and $K \cong SO(n) \times SO(2)$.

Type C. $G_{\mathbb{C}} = \{x \in GL_{2m}(\mathbb{C}) \mid {}^t x J x = J\}$, $G = G_{\mathbb{C}} \cap U(2m)$, $N = 2m$, $K \cong U(m)$,

$$L = \{w \in \mathbb{C}_m^{2m} \mid {}^t w J w = 0, \; \operatorname{rank}(w) = m\},$$

$$J = \begin{pmatrix} 0 & -1_m \\ 1_m & 0 \end{pmatrix}.$$

Type D. $G_{\mathbb{C}} = \{x \in SL_{2m}(\mathbb{C}) \mid {}^t x J' x = J'\}$, $G = G_{\mathbb{C}} \cap U(2m)$, $N = 2m > 2$, $K \cong U(m)$,

$$J' = \begin{pmatrix} 0 & 1_m \\ 1_m & 0 \end{pmatrix},$$

$L =$ the connected component of $\tilde{L}$ containing ${}^t(0, 1_m)$,

$$\tilde{L} = \{w \in \mathbb{C}^{2m} \mid {}^t w J' w = 0, \; \operatorname{rank}(w) = m\}.$$

It can easily be seen that G is isomorphic to $SO(2m, \mathbb{R})$.

In each case, the isotopy subgroup of $G_{\mathbb{C}}$ at a point of V is a parabolic subgroup $P_{\mathbb{C}}$ with the property that $G_{\mathbb{C}} = GP_{\mathbb{C}}$. Then V is isomorphic to G/K with $K = G \cap P_{\mathbb{C}}$. This is an easy way to see that G acts transitively on V and also that the above list exhausts all the irreducible hermitian symmetric spaces of compact and nonexceptional type. It should also be noted that the set L is practically the same as that given in [13, p. 480].

In order to define the complex structure of V in an explicit form, we first take a complex vector space T as follows:

$$T = \mathbb{C}_m^n \qquad\qquad \text{(Type A or B)}, \qquad\qquad (2.1\text{a})$$

$$T = \{u \in \mathbb{C}_m^m \mid {}^t u = u\} \qquad \text{(Type C)}, \qquad\qquad (2.1\text{b})$$

$$T = \{u \in \mathbb{C}_m^m \mid {}^t u = -u\} \qquad \text{(Type D)}. \qquad\qquad (2.1\text{c})$$

We then define a map $\lambda: T \to L$ by

$$\lambda(u) = \begin{pmatrix} u \\ 1_m \end{pmatrix} \qquad \text{(Type A, C, or D)}, \qquad (2.2\text{a})$$

$$\lambda(u) = \begin{pmatrix} u \\ {}^t u u/2 \\ 1 \end{pmatrix} \qquad \text{(Type B)} . \qquad (2.2\text{b})$$

For every $\alpha \in G$ and $w \in L$, let w_α denote the square matrix consisting of the last m rows of αw, and let $U_\alpha = M_\alpha / GL_m(\mathbb{C})$ with

$$M_\alpha = \{ w \in L \mid \det(w_\alpha) \neq 0 \} . \qquad (2.3)$$

The U_α for $\alpha \in G$ define a covering of V; moreover, $\alpha U_\alpha = U_1$ and $M_1 = \lambda(T) GL_m(\mathbb{C})$. Therefore we can define a complex coordinate map $z_\alpha: U_\alpha \to T$ by

$$z_\alpha = z_1 \circ \alpha, \qquad z_1(\lambda(u) \bmod GL_m(\mathbb{C})) = u \qquad \text{for } u \in T. \qquad (2.4)$$

Then the (U_α, z_α) for $\alpha \in G$ define the complex structure of V.

To smooth our exposition, we identify a function on a subset of V with the corresponding function on a subset of L invariant under $GL_m(\mathbb{C})$. Using this convention, we define an L-valued function Z_α on U_α by

$$Z_\alpha(w) = w w_\alpha^{-1} . \qquad (2.5)$$

Then $\alpha Z_\alpha(w) = \lambda(z_\alpha(w))$ for $w \in M_\alpha$. We define also a holomorphic map $u_{\alpha\beta}: U_\beta \to \mathbb{C}_m^m$ by $u_{\alpha\beta}(w) = w_\alpha w_\beta^{-1}$ for $w \in M_\beta$. Then $Z_\alpha u_{\alpha\beta} = Z_\beta$ on $U_\alpha \cap U_\beta$.

To define a Kähler form on V, take $\varphi_\alpha \in C^\infty(U_\alpha)$ given by

$$\varphi_\alpha = \log [\det(Z_\alpha^* Z_\alpha)] . \qquad (2.6)$$

We see easily that

$$\varphi_\alpha(w) = \log [\det(w^* w)] - \log [\det(w_\alpha^* w_\alpha)] , \qquad (2.7\text{a})$$

$$\varphi_\beta - \varphi_\alpha = \log [\det(u_{\alpha\beta}^* u_{\alpha\beta})] , \qquad (2.7\text{b})$$

so that there is a unique 2-form Ω on V such that $\Omega = i \partial \bar\partial \varphi_\alpha$ on U_α. This is obviously G-invariant, and positive as will be shown in (2.10) below.

Let us now define matrix-valued functions ξ, η, and r on T by

$$\xi(u) = 1_n + \bar u \cdot {}^t u, \quad \eta(u) = 1_m + u^* u \qquad \text{(Type A, C, or D)}, \qquad (2.8\text{a})$$

$$\eta(u) = 1 + u^* u + |{}^t u u/2|^2 \qquad \text{(Type B)}, \qquad (2.8\text{b})$$

$$\xi(u) = 1_n + \bar u \cdot {}^t u - \eta(u)^{-1}(x\bar u + u) \cdot {}^t(\bar x u + \bar u) \qquad (x = {}^t u u/2, \text{ Type B}). \qquad (2.8\text{c})$$

$$r(u) = \xi(u)^{-1} \bar u = \bar u \cdot {}^t \eta(u)^{-1} \qquad \text{(Type A, C, or D)}, \qquad (2.9\text{a})$$

$$r(u) = \eta(u)^{-1}(\bar u + \bar x u) \qquad (x = {}^t u u/2, \text{ Type B}). \qquad (2.9\text{b})$$

Here and henceforth, we understand that $n = m$ for Types C and D. Observe that r is T-valued. It can easily be verified that the components of $r(z_1)$ are the partial derivatives $\partial\varphi_1/\partial z_{jk}$ with the components z_{jk} of z_1 up to nonzero constant factors, and moreover that

$$\Omega = i \sum_{p,q=1}^{n} \sum_{j,k=1}^{m} (\xi^{-1}\circ z_1)_{pq}\, (\eta^{-1}\circ z_1)_{jk}\, dz_{pj} \wedge d\bar{z}_{qk} \tag{2.10}$$

on U_1. The verification of these facts can be done in a more or less straightforward way, and is in fact parallel to the similar facts in the non-compact case studied in [16, Sects. 2, 3].

Let us now define $r_\alpha: U_\alpha \to T$ and $R_\alpha: U_\alpha \to L$ by

$$r_\alpha = r(z_\alpha), \qquad R_\alpha(w) = \bar{w}\cdot{}^t(w^*w)^{-1}\cdot{}^t w_\alpha \qquad (w\in M_\alpha). \tag{2.11}$$

Then we see that $R_\alpha = \bar{Z}_\alpha\cdot{}^t(Z_\alpha^* Z_\alpha)^{-1}$ and $R_\alpha = R_\beta\cdot{}^t u_{\alpha\beta}$ on $U_\alpha\cap U_\beta$. Define $\varepsilon: T\to G_{\mathbb{C}}$ by

$$\varepsilon(u) = \begin{pmatrix} 1_n & 0 \\ {}^t u & 1_m \end{pmatrix} \qquad \text{(Type A, C, or D)}, \tag{2.12a}$$

$$\varepsilon(u) = \begin{bmatrix} 1_n & u & 0 \\ 0 & 1 & 0 \\ {}^t u & {}^t uu/2 & 1 \end{bmatrix} \qquad \text{(Type B)}. \tag{2.12b}$$

Then we can easily verify that $\varepsilon(u)^{-1} = \varepsilon(-u)$ and

$$\varepsilon(z_\alpha)\,\bar{\alpha}\,R_\alpha = \lambda(r_\alpha). \tag{2.13}$$

Since the entries of r_α are the holomorphic derivatives of φ_α, $\mathcal{N}(U_\alpha)$ consists of all the polynomials in the entries of r_α with coefficients in $\mathcal{N}^0(U_\alpha)$. In particular, the entries of R_α are nearly holomorphic. Now observe that ${}^t Z_\alpha R_\alpha = 1_m$ and

$$Z_\alpha(w)\cdot{}^t R_\alpha(w) = w(w^*w)^{-1}w^* \qquad \text{for } w\in M_\alpha. \tag{2.14}$$

The right-hand side of the last equality is independent of α, and therefore its entries belong to $\mathcal{N}(V)$.

In order to determine the set $\mathcal{N}(V)$ explicitly, we fix a nonnegative integer k, and denote by $\mathcal{Y}_k$ the set of all $\mathbb{C}$-valued functions g on L such that $g(wb) = \det(b)^k g(w)$ for every $b\in GL_m(\mathbb{C})$ and that $g(w)$ can be expressed as rational expressions in the entries of w defined everywhere on L. As will be seen from Lemma 2.2 below, $\mathcal{Y}_k$ is finite-dimensional over $\mathbb{C}$. Define $Q_k: L\to\mathbb{C}^l$ by $Q_k = {}^t(q_1,\ldots,q_l)$ with an arbitrarily fixed basis $\{q_1,\ldots,q_l\}$ of $\mathcal{Y}_k$ over $\mathbb{C}$. Then we obtain a representation $\varrho_k: G_{\mathbb{C}}\to GL_l(\mathbb{C})$ by $Q_k(aw) = \varrho_k(a)Q_k(w)$ for $a\in G_{\mathbb{C}}$. Let us now consider a function g on L of the form

$$g(w) = \det(w^*w)^{-k}\cdot{}^t Q_k(\bar{w})\, A\, Q_k(w) \tag{2.15}$$

with any $A\in\mathbb{C}_l^l$. Obviously this defines a function on V. Moreover we have

$$g = {}^t Q_k(R_\alpha)\, A\, Q_k(Z_\alpha) \qquad \text{on } U_\alpha, \tag{2.16}$$

so that $g\in\mathcal{N}(V)$. Now we have, as the principal result of this section,

Theorem 2.1. *For V belonging to the above four types, $\mathcal{N}(V)$ consists of all the functions of form (2.15) with $A\in\mathbb{C}_l^l$ and $0\le k\in\mathbb{Z}$.*

We need some preliminaries. For a meromorphic function f on V, let $\mathrm{div}(f)$ denote its divisor, and let $\mathrm{div}(f) = \mathrm{div}_0(f) - \mathrm{div}_\infty(f)$, where the subscripts 0 and ∞ indicate the zero and the pole. Let $\mathscr{S}(T)$ denote the set of all polynomial functions on T, and $\mathscr{S}_k(T)$ the set of all homogeneous elements of $\mathscr{S}(T)$ of degree k. Now observe that $\det(w_\alpha) = 0$ defines a divisor X_α on V. (More precisely, $\det(u_{\alpha\beta}) = 0$ defines the part of X_α lying in U_β. We see easily that X_α is irreducible for Types A and C; for Type B, it is irreducible if and only if $n > 2$; for Type D, $(1/2)X_\alpha$ is irreducible.) Put

$$E_k = \{p \in \mathscr{S}(T) \mid \mathrm{div}(p \circ z_1) \geqq -kX_1\}, \tag{2.17}$$

where we view $p \circ z_1$ as a meromorphic function on V. Then E_k is a finite-dimensional vector space over $\mathbb{C}$.

Lemma 2.2. (i) $\mathscr{S}(T) = \bigcup_{k=0}^{\infty} E_k$;

(ii) $E_k = \{q \circ \lambda \mid q \in \mathscr{Y}_k\}$;

(iii) E_k consists of all the $p \in \mathscr{S}(T)$ such that $\det(u_{1\alpha})^k p(z_1)$ is finite on U_α for every $\alpha \in G$.

Proof. Since $V = X_\alpha \cup U_\alpha$ set-theoretically, we have $\mathrm{div}(f) \geqq -cX_\alpha$ for some $c > 0$ if and only if f is finite on U_α, that is, $f = p(z_\alpha)$ with $p \in \mathscr{S}(T)$. Therefore (i) is obvious. To prove (ii), given $p \in E_k$, define a rational function f on L by $f(w) = \det(w_1)^k p(z_1)$. Then $f(wb) = \det(b)^k f(w)$ for $b \in GL_m(\mathbb{C})$, and $f(Z_\alpha) = f(Z_1) \det(u_{1\alpha})^k = p(z_1) \det(u_{1\alpha})^k$. Since $\mathrm{div}(\det(u_{1\alpha})) = X_1 - X_\alpha$, we have $\mathrm{div}(f(Z_\alpha)) \geqq -kX_\alpha$, so that f is defined on M_α. Therefore f is defined everywhere on L, and so $f \in \mathscr{Y}_k$. Obviously $f \circ \lambda = p$. Thus $E_k \subset \mathscr{Y}_k \circ \lambda$. Next let $q \in \mathscr{Y}_k$. Then $\det(u_{1\alpha})^k q(\lambda(z_1)) = \det(u_{1\alpha})^k q(Z_1) = q(Z_1 u_{1\alpha}) = q(Z_\alpha)$. This is finite on U_α. In particular $q(\lambda(z_1))$, being finite on U_1, is a polynomial function of z_1. Hence $q \circ \lambda$ has the property of p of (iii). Therefore our only remaining task is to show that such a p belongs to E_k. Take α so that X_α and X_1 have no common component. The condition on p in (iii) implies that $\mathrm{div}(\det(u_{1\alpha})^k p(z_1)) \geqq -cX_\alpha$ for some $c > 0$. Then $\mathrm{div}(p(z_1)) \geqq (k - c) X_\alpha - kX_1$. Since $\mathrm{div}_\infty(p(z_1))$ cannot involve X_α, we have $\mathrm{div}(p(z_1)) \geqq -kX_1$, so that $p \in E_k$. This completes the proof.

To prove Theorem 2.1, take $f \in \mathscr{N}(V)$. Then f is a finite sum $\sum g_\nu h_\nu(r_1)$ on U_1 with $g_\nu \in \mathscr{N}^0(U_1)$ and $h_\nu \in \mathscr{S}(T)$. By Lemma 2.2, (i), we can find a positive integer k so that $h_\nu \in E_k$ for every ν. By Lemma 2.2, (ii), we have $h_\nu = H_\nu \circ \lambda$ with $H_\nu \in \mathscr{Y}_k$. Expressing each H_ν as a linear combination of the q_μ, we can write $f = {}^t P Q_k(\lambda(r_1))$ on U_1 with a holomorphic map $P: U_1 \to \mathbb{C}^l$. Let $S = {}^t \varrho_k(\varepsilon(z_1)) P$ with ε of (2.12a, b). By (2.13), we have $f = {}^t S Q_k(R_1)$ on U_1. For every $\alpha \in G$, we have $R_1 = R_\alpha \cdot {}^t u_{1\alpha}$, so that

$$f = {}^t S Q_k(R_\alpha) \det(u_{1\alpha})^k = \det(u_{1\alpha})^k \cdot {}^t S \varrho_k(\varepsilon(z_\alpha)\bar\alpha)^{-1} Q_k(\lambda(r_\alpha))$$

on $U_1 \cap U_\alpha$. Now the $q_\mu(\lambda(r_\alpha))$ are linearly independent over $\mathscr{N}^0(U_\alpha)$. Therefore, that $f \in \mathscr{N}(U_\alpha)$ implies that the components of $\det(u_{1\alpha})^k S$ can be extended to holomorphic functions on U_α. Consequently the components of S can be extended to meromorphic functions on V. Since S is holomorphic on U_1, we can put $'S = (s_1(z_1), \ldots, s_l(z_1))$ with $s_\nu \in \mathscr{S}(T)$. By Lemma 2.2, (iii), $s_\nu \in E_k$. Determine $A = (a_{\mu\nu}) \in \mathbb{C}_l^l$ by $s_\nu = \sum a_{\mu\nu} q_\mu \circ \lambda$. Then

$$f = \sum a_{\mu\nu} q_\mu (\lambda(z_1)) q_\nu(R_1) = {}'Q_k(R_1) A Q_k(Z_1) = \det(w^*w)^{-k} \cdot {}'Q_k(\bar{w}) A Q_k(w)$$

as expected.

Excluding Type D, the result can be given in a different way:

Theorem 2.3. *For V of Type A, B, or C, the following assertions hold:*

    (i) *Every element of $\mathscr{Y}_k$ can be expressed as a polynomial in the entries of $w \in L$;*

    (ii) *$\mathscr{N}(V)$ consists of all polynomial functions in the entries of $w(w^*w)^{-1}w^*$.*

Proof. Let us first treat Type A. A rational function of $\mathbb{C}_m^N$ finite on M_α can be written in the form $p(w)/\det(w_\alpha)^k$ with a polynomial p. Assertion (i) follows easily from this fact. Now, as shown by (2.14), every polynomial function in the entries of $w(w^*w)^{-1}w^*$ belongs to $\mathscr{N}(V)$ for all four types. To prove the converse for Type A, let g be the element of $\mathscr{Y}_k$ defined by $g(w) = \det(w_1)^k$. Then $\mathscr{Y}_k$ is spanned by $g(aw)$ with $a \in G_\mathbb{C}$. This follows easily from [13, Theorem 2.A] which describes the decomposition of the space of polynomial functions on $\mathbb{C}_m^N$ into irreducible subspaces. Let $\varDelta$ be the set of the determinants of the submatrices of w of degree m, and $\varDelta_k$ the set of all monomials of the members of $\varDelta$ of degree k. Then $g(aw)$ is a linear combination of the elements of $\varDelta_k$, and hence we can take the components q_ν of Q_k from $\varDelta_k$. If d and d' belong to $\varDelta$, then we see that $\det(w^*w)^{-1} d(w) d'(\bar{w})$ is a polynomial function in the entries of $w(w^*w)^{-1}w^*$. Therefore we obtain (ii) for Type A from Theorem 2.1.

The proof for Types B and C are more involved. Take Type C for example. We let an element a of $GL_m(\mathbb{C})$ act on $\mathscr{S}(T)$ by $(ah)(u) = h('aua)$ for $h \in \mathscr{S}(T)$ and $u \in T$. We decompose E_k into irreducible subspaces under this action and find that each irreducible subspace is spanned by the functions of the form $h('aua)$ with $a \in GL_m(\mathbb{C})$ and a special function

$$h(u) = \prod_{i=1}^{m} \det_i(u)^{e_i} \qquad (u \in T) \tag{2.18}$$

such that $0 \leq e_i \in \mathbb{Z}$ and $\sum_{i=1}^{m} e_i \leq k$, where $\det_i(u)$ denotes the determinant of the upper left i^2 entries of u. Such a decomposition of E_k can be done for all four types by means of "the eigenfunctions of highest weight" given in [13, Theorems 2.A, B, C, and D]. As a consequence of this fact, it can be shown that an element of $\mathscr{Y}_k$ for Type C is the restriction of an element of $\mathscr{Y}_k$ for Type A with $n = m$. Assertions (i) and (ii) for Type C then can be derived easily from those for Type A. Type B can be treated in a similar way.

The above result generalizes Theorem 1.2, since $P^n(\mathbb{C})$ is V of Type A with $m = 1$. Whether the assertions of Theorem 2.3 are true for Type D is an open question. We conclude this section by noting a curious fact of an algebraic nature:

Proposition 2.4. *Let z be a variable matrix on T, and let $\mathbb{C}(z)$ (resp. $\mathbb{C}(z, \bar{z})$) denote the field of all rational expressions in the components of z (resp. z and $\bar{z}$). Then there is an automorphism σ of $\mathbb{C}(z, \bar{z})$ over $\mathbb{C}(z)$ of order 2 such that $\bar{z}^\sigma = -r(z)$, $r(z)^\sigma = -\bar{z}$, $\xi(z)^\sigma = \xi(z)^{-1}$, and $\eta(z)^\sigma = \eta(z)^{-1}$, where r, ξ, and η are defined by (2.8a, b, c) and (2.9a, b).*

The proof is merely computational, and therefore may be left to the reader. Notice that $'\xi\xi = 1$ for Type B. It should also be noted that such an automorphism exists even in the case of hermitian symmetric spaces of noncompact type (see [16, Lemmas 2.2 and 2.3, Remark 3.13]).

We add a final remark that the holomorphic derivatives of r are quadratic polynomials in the components of r. Consequently, the operator of type (1.6) maps $\mathcal{N}(U_1)$ into itself.

3. The Arithmeticity of Nearly Holomorphic Automorphic Forms

As shown in our previous paper [16], nearly holomorphic functions appear naturally on hermitian symmetric spaces of noncompact type. If $\mathcal{D}$ is such a space given as a domain in a complex vector space T, we obtain a fundamental form of $\mathcal{D}$ in the form $i\partial\bar{\partial}\varphi$ with a globally defined real-valued function φ, and so we can define elements $r_1, \ldots, r_n$ of $\mathcal{N}^1(\mathcal{D})$ by $r_p = \partial\varphi/\partial z_p$ with the global complex coordinate functions $z_1, \ldots, z_n$ on T. Then $\mathcal{N}(\mathcal{D})$ consists of all polynomials in the r_p with holomorphic coefficients. The explicit forms of the r_p for the domain $\mathcal{D}$ given as a certain matrix-space have been determined for four classical types of domains in [16, Lemma 2.2 and Remark 3.13]. In this section, we shall investigate the arithmeticity of automorphic forms belonging to $\mathcal{N}(\mathcal{D})$, especially in connection with differential operators studied in [12, 13, 16]. We naturally employ the same notation as in [16, Sects. 1 ∼ 3]. (See also the *Notation* of the present paper.)

Our setting consists of a domain $\mathcal{D}$ and groups G and $\mathfrak{R}$ given as

$$\mathcal{D} = \prod_{\tau \in J} \mathcal{D}_\tau, \qquad G = \mathfrak{R}' \times \prod_{\tau \in J} G_\tau, \qquad \mathfrak{R} = \mathfrak{R}' \times \prod_{\tau \in J} \mathfrak{R}_\tau, \qquad (3.1)$$

where J is a finite set, $\mathcal{D}_\tau$ for each τ is one of the domains $\mathcal{H}_n$, $\mathcal{B}_{l.m}$, and $\mathcal{L}_n$ of [16, Sect. 1], G_τ is one of the groups $G^{(1)}$, $G^{(2)}$, and $G^{(3)}$ acting on $\mathcal{D}_\tau$, $\mathfrak{R}_\tau$ is the group defined by [16, (1.11)] corresponding to G_τ, and $\mathfrak{R}'$ is an arbitrary compact group. (We can also include G and $\mathcal{D}$ of Types C and D in [16, Remark 3.13]. The only necessary care is the proper definition of $\mathfrak{R}$, its action on T, π_e, τ_e, and the symbol $\{\ ,\ \}$ of [16, (1.14)], which can easily be given.) We let G act on $\mathcal{D}$ with trivial action of $\mathfrak{R}'$ on $\mathcal{D}$, and write an element of $\mathfrak{R}$ in the form (a, b, c) with $a \in \mathfrak{R}'$, $(b, c) = (b_\tau, c_\tau)_{\tau \in J}$, $(b_\tau, c_\tau) \in \mathfrak{R}_\tau$. We put $T = \prod_{\tau \in J} T_\tau$ and $T^e = \prod_{\tau \in J} T_\tau^{e_\tau}$ for $0 \leq e \in \mathbb{Z}^J$, where T_τ for each τ is the complex vector space $T = \mathbb{C}_m^l$ defined for $\mathcal{D}_\tau$ as in [16, Sect. 1]. We let $\mathfrak{R}$ act on T componentwise with trivial action of $\mathfrak{R}'$. Given a finite-dimensional complex vector space X and e as above, we denote by $\mathcal{M}\ell_e(T, X)$ the vector space of all $\mathbb{C}$-multilinear maps of T^e (that is, $\mathbb{C}$-linear on each single factor T_τ) into X, and by $\mathcal{S}_e(T, X)$ the vector space of all polynomial maps of T into X which are homogeneous of degree e_τ on T_τ; we then view $S_e(T, X)$ as a subspace of $\mathcal{M}\ell_e(T, X)$ consisting of all the elements that are symmetric on $T_\tau^{e_\tau}$ for every $\tau \in J$. Given a representation $\{\varrho, X\}$ of $\mathfrak{R}$, we can define two representations $\{\varrho \otimes \tau_e, \mathcal{M}\ell_e T, X)\}$ and $\{\varrho \otimes \pi_e, \mathcal{M}\ell_e(T, X)\}$ of $\mathfrak{R}$ by generalizing [16, (1.12a, b)] in an obvious way. Similarly, for $\varphi \in \mathcal{M}\ell_e(T, \mathbb{C})$ and $h \in \mathcal{M}\ell_e(T, X)$, we can define $\{\varphi, h\}$ by generalizing [16, (1.14)]. Then [16, (1.15)] holds in the generalized situation. We insert here

Lemma 3.1. *Define* $\varphi_a \in \mathscr{S}_e(T, X)$ *by* $\varphi_a(u) = \prod_{\tau \in J} \mathrm{tr}\,({}^t a_\tau u_\tau)^{e_\tau}$ *for* a, $u \in T$. *Then* $\{\varphi_a, h\} = h(a)$ *for every* $h \in \mathscr{S}_e(T, X)$.

This is a generalization of [16, Lemma 3.1] and can easily be verified.

With e and $\{\varrho, X\}$ as above and $f \in C^\infty(\mathscr{D}, X)$, we define $E^e f$, $C^e f$, and $D_\varrho^{(e)} f$ as elements of $C^\infty(\mathscr{D}, \mathscr{S}_e(T, X))$ by generalizing [16, (1.17 b, c), (1.18 a, b)] in a natural way; more precisely, $E^e = \prod_{\tau \in J} E_\tau^{e_\tau}$ and $C^e = \prod_{\tau \in J} C_\tau^{e_\tau}$, where E_τ and C_τ are the operators E and C on $\mathscr{D}_\tau$ defined by [16, (1.17 c)];

$$D_\varrho^{(e)} f = (\varrho \otimes \tau_e)(\Xi)^{-1} C^e [\varrho(\Xi) f], \qquad \Xi = (1, \xi_\tau, \eta_\tau) \in \mathfrak{R}, \tag{3.2}$$

where ξ_τ and η_τ are hermitian matrices of [16, (1.6 a, b)]. Then $D_\varrho^Z f$ and $E_Z f$ can be defined, as elements of $C^\infty(\mathscr{D}, \mathrm{Hom}(Z, X))$, for every $\mathfrak{R}$-stable subspace Z of $\mathscr{S}_e(T)$ by [16, (1.22), (1.24)], every symbol being understood in the generalized sense.

Lemma 3.2. *Let* $\{\varrho, X\}$ *and* $\{\sigma, Y\}$ *be representations of* $\mathfrak{R}$, *and let* $f \in C^\infty(\mathscr{D}, X)$ *and* $g \in C^\infty(\mathscr{D}, Y)$. *Then, for* $u \in T^e$, *one has*

$$D_{\varrho \otimes \sigma}^{(e)}(f \otimes g)(u) = \sum_{\alpha, \beta} (D_\varrho^{(a)} f)(u_\alpha) \otimes (D_\sigma^{(b)} g)(u_\beta),$$

where the sum is extended over all $2^{|e|}$ decompositions of T^e into two factors T_α and T_β isomorphic to T^a and T^b (thus $e = a + b$), and u_α and u_β are the projections of u to T_α and T_β, respectively.

This can be proved easily by first showing the corresponding formula for C^e, and then by applying (3.2) to it. See [12, (1.20 b)] for a special case.

Given $\alpha \in G$ and $z \in \mathscr{D}$, we denote by $\Lambda(\alpha, z)$ the element $(\alpha', \lambda(\alpha_\tau, z_\tau), \mu(\alpha_\tau, z_\tau))$ of $\mathfrak{R}$, where λ and μ are defined by [16, (1.3)], and α' is the $\mathfrak{R}'$-component of α. For $f \in C^\infty(\mathscr{D}, X)$, we define an element $f \|_\varrho \alpha$ of $C^\infty(\mathscr{D}, X)$ by

$$f \|_\varrho \alpha = \varrho(\Lambda(\alpha, z))^{-1} f(\alpha z) \qquad (z \in \mathscr{D}). \tag{3.3}$$

Then [16, (1.19 a, b) and (1.25)] hold in the present generalized setting.

Let us now denote by $\mathscr{N}^e(\mathscr{D}, X)$, or simply by $\mathscr{N}^e[X]$, the set of all the functions $f \in C^\infty(\mathscr{D}, X)$ such that $E_\tau^{1+e_\tau} f = 0$ for every $\tau \in J$. By [16, Proposition 2.4, (ii)], the union of $\mathscr{N}^e[X]$ for all $e \in \mathbb{Z}^J$, ≥ 0, is the set of all X-valued nearly holomorphic functions on $\mathscr{D}$. Let $r(z)$ be the T-valued function on $\mathscr{D}$ defined by $r(z) = (r_\tau(z_\tau))_{\tau \in J}$, where r_τ is the T_τ-valued function on $\mathscr{D}_\tau$ defined by [16, (2.3)]; further let $(D \log \delta)^e$ denote the element of $C^\infty(\mathscr{D}, \mathscr{S}_e(T, X))$ defined by

$$(D \log \delta)^e(u) = \prod_{\tau \in J} \mathrm{tr}(r_\tau(z_\tau) u_\tau)^{e_\tau} \qquad (z \in \mathscr{D}, u \in T).$$

This is a generalization of [16, (3.2)]. Then $\mathscr{N}^e[X]$ consists of all the functions f of the form

$$f(z) = \sum_{0 \leq j \leq e} \{(D \log \delta)^j, g_j\} = \sum_{0 \leq j \leq e} g_j(z, r(z)) \tag{3.4}$$

with holomorphic maps $g_j : \mathscr{D} \to \mathscr{S}_j(T, X)$, for the same reason as in [16, p. 360].

Let $\{\sigma, Y\}$ be a representation of $\mathfrak{R}$. Then, for $0 \leq k \in \mathbb{Z}^J$, we have

$$g \in \mathscr{N}^e[Y] \Rightarrow D_\sigma^{(k)} g \in \mathscr{N}^{k+e}[\mathscr{S}_k(T, Y)]. \tag{3.5}$$

This follows immediately from [16, Lemma 3.3].

To state a key proposition, we need a contraction operator θ_X: $\mathcal{M}\ell_k(T, \mathcal{M}\ell_k(T, X)) \to X$ defined by

$$\theta_X \varphi = \sum \varphi(\dots, c_{\tau 1}, \dots, c_{\tau k_\tau}, \dots; \dots, c_{\tau 1}, \dots, c_{\tau k_\tau}, \dots), \qquad (3.6)$$

where $c_{\tau 1}, \dots, c_{\tau k_\tau}$ run independently over the standard basis of T_τ for each $\tau \in J$; this is a generalization of [16, (3.13a, b)].

Proposition 3.3. *Let* $\{\varrho, X\}$ *and* $\{\varrho_0, X\}$ *be representations of* $\mathfrak{K}$ *such that* $\varrho(a, b, c) = \det(c)^v \varrho_0(a, b, c)$ *for* $(a, b, c) \in \mathfrak{K}$ *with* $v \in \mathbb{Z}^J$. *Let* f *be an element of* $\mathcal{N}^k[X]$ *such that* $f\|_\varrho \gamma = f$ *for every* γ *in a subgroup* Γ *of* G. *If* v_τ, *for every* $\tau \in J$, *is larger than an integer* $N(\varrho_0, k)$ *that depends only on* ϱ_0 *and* k, *then*

$$f = \sum_{0 \le e \le k} \theta_X D^{(e)}_{\varrho \otimes \pi_e} g_e \qquad (3.7)$$

with holomorphic maps $g_e \colon \mathcal{D} \to \mathcal{S}_e(T, X)$ *such that* $g_e\|_{\varrho \otimes \pi_e} \gamma = g_e$ *for every* $\gamma \in \Gamma$.

This is a generalization of [16, Proposition 3.4] and can be proved by induction on the lexicographic order of k with the same technique.

Given two representations $\{\varrho, X\}$ and $\{\sigma, Y\}$ of $\mathfrak{K}$, let φ be a $\mathbb{C}$-linear map of Y into X such that $\varphi\sigma(\alpha) = \varrho(\alpha)\varphi$ for every $\alpha \in \mathfrak{K}$. Then we can define, for $0 \le e \in \mathbb{Z}^J$, a map $\varphi^e \colon \mathcal{M}\ell_e(T, Y) \to \mathcal{M}\ell_e(T, X)$ by $\varphi^e(h) = \varphi \circ h$ for $h \in \mathcal{M}\ell_e(T, Y)$. Then we see easily that

$$\varphi^e \circ (\sigma \otimes \tau_e)(\alpha) = (\varrho \otimes \tau_e)(\alpha) \circ \varphi^e \qquad \text{for every } \alpha \in \mathfrak{K}.$$
$$\varphi^e \circ (\sigma \otimes \pi_e)(\alpha) = (\varrho \otimes \pi_e)(\alpha) \circ \varphi^e$$

Lemma 3.4. *Let* $\{\varrho, X\}$, $\{\sigma, Y\}$, e, *and* φ *be as above. Then* $\varphi^e C^e g = C^e(\varphi \circ g)$, $\varphi^e E^e g = E^e(\varphi \circ g)$, *and* $\varphi^e D^{(e)}_\sigma g = D^{(e)}_\varrho(\varphi \circ g)$ *for every* $g \in C^\infty(\mathcal{D}, Y)$.

Proof. The first two equalities follow immediately from [16, Proposition 2.4]. As for the last one, we have, by (3.2),

$$\varphi^e D^{(e)}_\sigma g = \varphi^e(\sigma \otimes \tau_e)(\Xi)^{-1} C^e[\sigma(\Xi)g] = (\varrho \otimes \tau_e)(\Xi)^{-1} \varphi^e C^e[\sigma(\Xi)g]$$
$$= (\varrho \otimes \tau_e)(\Xi)^{-1} C^e[\varphi\sigma(\Xi)g] = (\varrho \otimes \tau_e)(\Xi)^{-1} C^e[\varrho(\Xi)\varphi g]$$
$$= D^{(e)}_\varrho(\varphi \circ g).$$

To study the arithmeticity, we assume that our group G is the product of several groups for which the notion of *CM*-points and the arithmeticity of automorphic forms are well-defined. We recall here a few essential facts, which, together with (3.8) and Lemma 3.5 below, may be viewed as "the axioms of arithmeticity". They are in fact true for the groups that appear in later sections, as shown in our previous papers.

First of all, G must be the group of real points of an algebraic group defined over $\mathbb{Q}$, so that the group $G_\mathbb{Q}$ of $\mathbb{Q}$-rational points is meaningful; thus we can speak of congruence subgroups of $G_\mathbb{Q}$. Next, $\mathfrak{K}$ has a $\overline{\mathbb{Q}}$-rational structure, so that a $\overline{\mathbb{Q}}$-rational representation $\{\varrho, X\}$ of $\mathfrak{K}$ is meaningful. We let $\mathscr{C}_\varrho$ denote the set of all $f \in C^\infty(\mathcal{D}, X)$ such that $f\|_\varrho \gamma = f$ for every γ in a congruence subgroup of $G_\mathbb{Q}$. There is a dense subset $\mathscr{W}$ of $\mathcal{D}$, whose points are called *CM-points*, and which is stable under $G_\mathbb{Q}$. There is a map $\mathfrak{p} \colon \mathscr{W} \to \mathfrak{K}$ such that $\mathfrak{p}(\alpha w)^{-1} \Lambda(\alpha, w) \mathfrak{p}(w)$ is $\overline{\mathbb{Q}}$-rational for every $w \in \mathscr{W}$ and every $\alpha \in G_\mathbb{Q}$. We put $\mathfrak{P}_\varrho(w) = \varrho(\mathfrak{p}(w))$. We say that an element f of $\mathscr{C}_\varrho$ (or more generally a function f that is defined on a neighborhood of a *CM*-

point w) is *arithmetic at w* if $\mathfrak{P}_\varrho(w)^{-1}f(w)$ is $\overline{\mathbb{Q}}$-rational; we call f *arithmetic* if f is arithmetic at every $w \in \mathscr{W}$. We denote by $\mathscr{M}_\varrho$ the set of all $f \in \mathscr{C}_\varrho$ that are holomorphic on $\mathscr{D}$ as well as at cusps, and by $\mathscr{M}_\varrho(\overline{\mathbb{Q}})$ the set of arithmetic elements of $\mathscr{M}_\varrho$. We assume that

$$\mathscr{M}_\varrho = \mathscr{M}_\varrho(\overline{\mathbb{Q}}) \otimes_{\overline{\mathbb{Q}}} \mathbb{C} \qquad \text{for every } \overline{\mathbb{Q}}\text{-rational } \varrho. \tag{3.8}$$

We denote by $\mathscr{A}_\varrho$ the set of all quotients of the form f/g with $f \in \mathscr{M}_\sigma$ and $0 \neq g \in \mathscr{M}_\omega$ where $\sigma = \omega\varrho$ and $\omega(a, b, c) = \det(c)^v$ with an arbitrary $v \in \mathbb{Z}^J$, and by $\mathscr{A}_\varrho(\overline{\mathbb{Q}})$ the set of all such f/g with $f \in \mathscr{M}_\sigma(\overline{\mathbb{Q}})$ and $g \in \mathscr{M}_\omega(\overline{\mathbb{Q}})$. (For such ω and v, we shall write $D_v^{(e)}$, $\mathscr{A}_v$, and $\mathscr{M}_v$ for $D_\omega^{(e)}$, $\mathscr{A}_\omega$, and $\mathscr{M}_\omega$.)

Lemma 3.5. *If $h \in \mathscr{A}_\varrho(\overline{\mathbb{Q}})$, then $\pi^{-|e|}D_\varrho^Z h$ is arithmetic at every CM-point where h is finite, and for every $\mathfrak{R}$-stable $\overline{\mathbb{Q}}$-rational subspace Z of $\mathscr{S}_e(T)$.*

This was proved in [8, Theorem 11.2] when the group is orthogonal and $Z = \mathscr{S}_e(T)$. The same type of proof is applicable to other groups, at least to the symplectic and unitary ones (see [10, p. 825; 12, p. 282]). However, in all these three cases, one can give a simpler proof, which is completely parallel to the proof of [4, Main Theorem I], by means of Lemma 3.2. The assertion about D_ϱ^Z follows from that about $D_\varrho^{(e)}$ by virtue of the following

Lemma 3.6. *The notation being as in Lemma 3.4, one has $\varphi C_\sigma \subset C_\varrho$, $\varphi \mathscr{M}_\sigma \subset M_\varrho$, and $\varphi \mathscr{A}_\sigma \subset \mathscr{A}_\varrho$. Suppose moreover that ϱ, σ, and φ are $\overline{\mathbb{Q}}$-rational. If $f \in \mathscr{C}_\sigma$ (or $f \in \mathscr{A}_\sigma$) is arithmetic at w, then φf is arithmetic at w.*

This is an immediate consequence of our definition.

We denote by $\mathscr{N}_\varrho^k$ the set of all elements f of $\mathscr{N}^k[X] \cap \mathscr{C}_\varrho$ for which the g_j of (3.4) satisfy the cusp condition. (If $G_\mathbb{Q}$ has no factor isogenous to $SL_2(\mathbb{Q})$, the cusp condition is automatically satisfied, so that $\mathscr{N}_\varrho^k = \mathscr{N}^k[X] \cap \mathscr{C}_\varrho$.) Further we denote by $\mathscr{N}_\varrho^k(\overline{\mathbb{Q}})$ the set of all arithmetic elements of $\mathscr{N}_\varrho^k$. Obviously both $\mathscr{N}_\varrho^k$ and $\mathscr{N}_\varrho^k(\overline{\mathbb{Q}})$ are stable under $\|_\varrho\alpha$ for every $\alpha \in G_\mathbb{Q}$.

Now the principal result of this section is the following theorem, which is essential in the proof of arithmeticity of certain (nonholomorphic) series $f(w, 0)$ as given in [16, Theorems 4.5, 4.6, and 5.4].

Theorem 3.7. *Let $\mathscr{V}$ be a dense subset of $\mathscr{D}$ contained in $\mathscr{W}$, and $\{\varrho, X\}$ a $\overline{\mathbb{Q}}$-rational representation of $\mathfrak{R}$. Suppose an element f of $\mathscr{N}_\varrho^k$, $0 \leq k \in \mathbb{Z}^J$, is arithmetic at every point of $\mathscr{V}$. Then $\pi^{-|e|}D_\varrho^Z f$, as well as f itself, is arithmetic, where Z is the same as in Lemma 3.5.*

Proof. Fix k; let $\mathscr{N}_\varrho'$ be the set of all f in $\mathscr{N}_\varrho^k$ that is arithmetic at every point of $\mathscr{V}$. Let $\sigma(a, b, c) = \det(c)^v \cdot \varrho(a, b, c)$ for $(a, b, c) \in \mathfrak{R}$ with $0 \leq v \in \mathbb{Z}^J$. Let $\mathscr{N}_\sigma^*$ be the set of all $f \in \mathscr{N}_\varrho^k$ of the form

$$f = \sum_{0 \leq s \leq k} \pi^{-|s|} \theta_X D_{\sigma \otimes \pi_s}^{(s)} g_s \tag{3.9}$$

with $g_s \in \mathscr{M}_{\sigma \otimes \pi_s}(\overline{\mathbb{Q}})$. Now take any $f \in \mathscr{N}_\sigma^k$. If v_τ is sufficiently large for every τ, f can be written in form (3.9), by virtue of Proposition 3.3. Fix such a v. In view of (3.8), we see that $\mathscr{N}_\sigma^k$ is spanned by $\mathscr{N}_\sigma^*$ over $\mathbb{C}$. By Lemmas 3.5 and 3.6, $\pi^{-|s|}\theta_X D_{\sigma \otimes \pi_s}^{(s)} g_s$ is arithmetic for every $g_s \in \mathscr{M}_{\sigma \otimes \pi_s}(\overline{\mathbb{Q}})$, and hence

$\mathcal{N}_\sigma^* \subset \mathcal{N}_\sigma^k(\overline{\mathbb{Q}}) \subset \mathcal{N}_\sigma'$. Let us now prove that $\mathcal{N}_\sigma' \subset \mathcal{N}_\sigma^*$. Take a basis B of $\mathbb{C}$ over $\overline{\mathbb{Q}}$ including 1; let $f \in \mathcal{N}_\sigma'$. Then $f = \sum_{c \in B} cg_c$ with $g_c \in \mathcal{N}_\sigma^*$. For every $w \in \mathcal{V}$, we have $\mathfrak{P}_\sigma(w)^{-1} f(w) = \sum_c c \mathfrak{P}_\sigma(w)^{-1} g_c(w)$. Since $\mathfrak{P}_\sigma(w)^{-1} f(w)$ and $\mathfrak{P}_\sigma(w)^{-1} g_c(w)$ are algebraic, we have $g_c(w) = 0$ for $c \neq 1$, and hence $f(w) = g_1(w)$ for every $w \in \mathcal{V}$. Since $\mathcal{V}$ is dense, we have $f = g_1$, so that $\mathcal{N}_\sigma' \subset \mathcal{N}_\sigma^*$. To prove the arithmeticity of an element h of $\mathcal{N}_\varrho'$, take any $w_0 \in \mathcal{W}$. Take $v \in \mathbb{Z}^J$ and $q \in \mathcal{M}_v(\overline{\mathbb{Q}})$ so that $q(w_0) \neq 0$, and that the above result is applicable to this v. Obviously $qh \in \mathcal{N}_\sigma' = \mathcal{N}_\sigma^*$, and so qh is arithmetic. We can take $q(w_0)\mathfrak{P}_\varrho(w_0)$ as $\mathfrak{P}_\sigma(w_0)$. Then $\mathfrak{P}_\varrho(w_0)^{-1} h(w_0) = \mathfrak{P}_\sigma(w_0)^{-1}(qh)(w_0)$, which is $\overline{\mathbb{Q}}$-rational. Therefore h is arithmetic at w_0 as desired.

In order to prove that $\pi^{-|e|} D_\varrho^Z h$ is arithmetic, it is sufficient, in view of Lemma 3.6, to show that $\pi^{-|e|} D_\varrho^{(e)} h$ is arithmetic. Take w_0 and q again; write $qh = \sum_{0 \leq s \leq k} \pi^{-|s|} \theta_X D_{\sigma \otimes \pi}^{(s)} g_s$ with $g_s \in \mathcal{M}_{\sigma \otimes \pi}(\overline{\mathbb{Q}})$. Put $p_s = \pi^{-|s|} \theta_X D_{\sigma \otimes \pi}^{(s)} g_s$. Then $D_\varrho^{(e)} h = \sum_s D_\varrho^{(e)}(q^{-1} p_s)$. For a fixed s, we have, by Lemma 3.2,

$$D_\varrho^{(e)}(q^{-1} p_s)(u) = \sum_{\alpha, \beta} (D_{-v}^{(a)} q^{-1})(u_\alpha)(D_\sigma^{(b)} p_s)(u_\beta), \tag{3.10a}$$

where the notation should be understood in the sense of the lemma. Put $\mathfrak{p}(w_0) = (\mathfrak{p}', \mathfrak{p}_1, \mathfrak{p}_2)$ with $\mathfrak{p}' \in \mathfrak{R}'$ and $(\mathfrak{p}_1, \mathfrak{p}_2) \in \prod_\tau \mathfrak{R}_\tau$. For $\psi \in \mathcal{M\ell}_a(T, X)$ and $u \in T^a$, we have

$$[\mathfrak{P}_{\varrho \otimes \tau_e}(w_0)^{-1} \psi](u) = [(\varrho \otimes \tau_e)(\mathfrak{p}(w_0))^{-1} \psi](u)$$
$$= \varrho(\mathfrak{p}(w_0))^{-1} \psi({}^t\mathfrak{p}_1^{-1} u \mathfrak{p}_2^{-1}). \tag{3.10b}$$

Now $\pi^{-|a|} D_{-v}^{(a)} q^{-1}$ is arithmetic by Lemma 3.5. On the other hand, by Lemma 3.4, $D_\sigma^{(b)} p_s = \pi^{-|s|} D_\sigma^{(b)} \theta_X D_\omega^{(s)} g_s = \pi^{-|s|} \theta_X^b D_\omega^{(s+b)} g_s$ with $\omega = \sigma \otimes \pi_s$. Hence, by Lemmas 3.5 and 3.6, $\pi^{-|b|} D_\sigma^{(b)} p_s$ is arithmetic. By (3.10a, b), we have

$$\mathfrak{P}_{\varrho \otimes \tau_e}(w_0)^{-1}[\pi^{-|e|} D_\varrho^{(e)}(q^{-1} p_s)](u)$$
$$= \sum \mathfrak{P}_{-v}(w_0)^{-1}[\pi^{-|a|} D_{-v}^{(a)} q^{-1}]({}^t\mathfrak{p}_1^{-1} u_\alpha \mathfrak{p}_2^{-1})$$
$$\cdot \mathfrak{P}_\sigma(w_0)^{-1}[\pi^{-|b|} D_\sigma^{(b)} p_s]({}^t\mathfrak{p}_1^{-1} u_\beta \mathfrak{p}_2^{-1}),$$

which is algebraic for every $\overline{\mathbb{Q}}$-rational $u \in T^a$. This completes the proof.

Proposition 3.8. *Let $\{\varrho, X\}$ and k be as above, and let Z and e be as in Lemma 3.5, and $\omega = \varrho \otimes_0 \tau_Z$ (see [16, (1.21)]). Then*

$$\mathcal{N}_\varrho^k = \mathcal{N}_\varrho^k(\overline{\mathbb{Q}}) \otimes_{\overline{\mathbb{Q}}} \mathbb{C}, \tag{3.11}$$

$$\pi^{-|e|} D_\varrho^Z \mathcal{N}_\varrho^k(\overline{\mathbb{Q}}) \subset \mathcal{N}_\omega^{k+e}(\overline{\mathbb{Q}}). \tag{3.12}$$

Proof. Let $f_1, \ldots, f_m$ be elements of $\mathcal{N}_\varrho^k(\overline{\mathbb{Q}})$ linearly independent over $\overline{\mathbb{Q}}$. Suppose $\sum_{i=1}^m a_i f_i = 0$ with $a_i \in \mathbb{C}$. Let $\{b_1, \ldots, b_m\}$ be a basis of $\sum_{i=1}^m \overline{\mathbb{Q}} a_i$ over $\overline{\mathbb{Q}}$. Then $a_i = \sum_j c_{ij} b_j$ with $c_{ij} \in \overline{\mathbb{Q}}$, and $\sum_j b_j \sum_i c_{ij} f_i = 0$. The reasoning employed in the proof of Theorem 3.7 shows that $\sum_i c_{ij} f_i = 0$ for every j, and hence $c_{ij} = 0$. Thus the f_i are linearly independent over $\mathbb{C}$. Therefore, to prove (3.11), it is sufficient to show that $\mathcal{N}_\varrho^k$ is spanned by $\mathcal{N}_\varrho^k(\overline{\mathbb{Q}})$ over $\mathbb{C}$. Using the same notation as in the proof of Theorem 3.7, we see that (3.11) is true for $\mathcal{N}_\sigma^k$, since $\mathcal{N}_\sigma^* = \mathcal{N}_\sigma^k(\overline{\mathbb{Q}})$. Let $f \in \mathcal{N}_\varrho^k$. If $0 \neq u \in \mathcal{M}_v(\overline{\mathbb{Q}})$, we have $uf = \sum_{c \in B} cv_c$ with $v_c \in \mathcal{N}_\sigma^k(\overline{\mathbb{Q}})$, and $f = \sum_{c \in B} cv_c/u$.

Let $u'f = \sum_{c \in B} cv_c'$ with another nonzero element u' of $\mathcal{M}_v(\overline{\mathbb{Q}})$ and $v_c' \in \mathcal{N}_\sigma^k(\overline{\mathbb{Q}})$. Then we see that $v_c/u = v_c'/u'$ for every c. Now for every point (or cusp) w of $\mathcal{D}$, we can take u so that $u(w) \neq 0$. Therefore we can conclude that the quotients v_c/u belong to $\mathcal{N}_\varrho^k$. Moreover, they are arithmetic, as both u and v_c are arithmetic. This proves (3.11). As for (3.12), the inclusion without $\overline{\mathbb{Q}}$-rationality follows immediately from [16, (1.25) and Lemma 3.3]. Theorem 3.7 implies that the elements on the left-hand side of (3.12) are arithmetic, and hence we obtain (3.12).

We conclude this section by characterizing the elements of $\mathcal{N}_\varrho^k(\overline{\mathbb{Q}})$ in terms of Fourier coefficients. For this we restrict our treatment to the two cases, referred to as Case SP and Case SU, in which $G_{\mathbb{Q}}$ is given by

$$G_{\mathbb{Q}} = \{\alpha \in SL_{2m}(F) \mid {}^t\alpha \iota \alpha = \iota\} \qquad \text{(Case SP)}, \tag{3.13a}$$

$$G_{\mathbb{Q}} = \{\alpha \in SL_{2m}(K) \mid {}^t\bar{\alpha} \iota \alpha = \iota\} \qquad \text{(Case SU)}, \tag{3.13b}$$

$$\iota = \begin{pmatrix} 0 & -1_m \\ 1_m & 0 \end{pmatrix}.$$

Here m is a positive integer, F is a totally real algebraic number field, and K is a totally imaginary quadratic extension of F. We put

$$T = \{z \in \mathbb{C}_m^m \mid {}^tz = z\} \qquad \text{(Case SP)}, \tag{3.14a}$$

$$T = \mathbb{C}_m^m \qquad \text{(Case SU)}, \tag{3.14b}$$

$$\mathcal{H} = \{z \in T \mid i(z^* - z) > 0\}. \tag{3.15}$$

Then our domain $\mathcal{D}$ becomes $\mathcal{H}^J$, where J is the set of all archimedean primes of F in Case SP, and J is a CM-type of K in Case SU; the action of $G_{\mathbb{Q}}$ on $\mathcal{H}^J$ is the standard componentwise fractional linear transformation (see [12]); $\mathfrak{R} = GL_m(\mathbb{C})^J$ in Case SP, and $\mathfrak{R} = GL_m(\mathbb{C})^J \times GL_m(\mathbb{C})^J$ in Case SU. We define r to be the T^J-valued function on $\mathcal{H}^J$ defined by $r(z) = ({}^t(z_\tau - z_\tau^*)^{-1})_{\tau \in J}$.

We now consider $\mathcal{N}^k[X] = \mathcal{N}^k(\mathcal{H}^J, X)$ with $0 \leq k \in \mathbb{Z}^J$ and a $\overline{\mathbb{Q}}$-rational representation $\{\varrho, X\}$ of $\mathfrak{R}$. Every element f of $\mathcal{N}^k[X]$, being a polynomial in the components of r with holomorphic coefficients invariant under a group of translations, can be written in the form

$$f(z) = \sum_h p_h(r(z))\, \mathbf{e}\left(\sum_{\tau \in J} \operatorname{tr}(h_\tau z_\tau)\right) \qquad (z \in \mathcal{H}^J), \tag{3.16}$$

where $\mathbf{e}(x) = e^{2\pi i x}$, h runs over the symmetric elements of F_m^m in Case SP and over the hermitian elements of K_m^m in Case SU, and p_h is a polynomial map of T^J into X which is of degree $\leq k_\tau$ on T_τ. Now we have

Proposition 3.9. $\mathcal{N}_\varrho^k(\overline{\mathbb{Q}})$ *consists of all the elements f of $\mathcal{N}_\varrho^k$ of form (3.16) such that $p_h(\pi x)$ is $\overline{\mathbb{Q}}$-rational as a polynomial in $x \in T^J$ for every h.*

Proof. If $k = 0$, this amounts to the characterization of the elements of $\mathcal{M}_\varrho(\overline{\mathbb{Q}})$ by the algebraicity of Fourier coefficients, which was proved in [10, Sect. 4] for $Sp(m, \mathbb{Q})$. The same type of proof applies to the general case. Thus our question concerns nontrivial k. Let $\mathcal{F}(X)$ be the set of all elements f of $C^\infty(\mathcal{H}^J, X)$ of form (3.16) such that $p_h(\pi x)$ is $\overline{\mathbb{Q}}$-rational in x for every h. We see that $D_\varrho^{(e)}$ maps $\mathcal{F}(X)$ into $\mathcal{F}(\mathcal{S}_e(T, X))$. This can easily be seen from the proof of [16, Lemma 3.3], in particular [16, (3.8)]. Hence the functions of form (3.9) with arithmetic g_s are contained in $\mathcal{F}(X)$. Let $f \in \mathcal{N}_\varrho^k(\overline{\mathbb{Q}})$ and let

$$t(z) = \sum_{x \in \Lambda} \mathbf{e}\left(\sum_{\tau \in J} \operatorname{tr}(x_\tau^* z_\tau x_\tau) \right) \quad (z \in \mathcal{H}^J),$$

where Λ is a lattice in F^m or K^m. The inclusion $\mathcal{N}_\sigma^k(\overline{\mathbb{Q}}) \subset \mathcal{N}_\sigma^*$ as shown in the proof of Theorem 3.7 implies that $t^q f$ for a sufficiently large even integer q has form (3.9) with arithmetic g_s. Hence $t^q f \in \mathcal{F}(X)$, from which we can easily derive that $f \in \mathcal{F}(X)$. Conversely, let $f \in \mathcal{N}_\varrho^k \cap \mathcal{F}(X)$. By (3.11), f is a finite sum $\sum c_i f_i$ with $c_i \in \mathbb{C}$ and $f_i \in \mathcal{N}_\varrho^k(\overline{\mathbb{Q}})$. We may assume that the c_i are linearly independent over $\overline{\mathbb{Q}}$ and $c_1 = 1$. Let p_h and p_{ih} be the polynomial coefficients of f and f_i in the expressions of form (3.16). Then $p_h(\pi x) = \sum c_i p_{ih}(\pi x)$ for $x \in T^J$. The $\overline{\mathbb{Q}}$-rationality of $p_h(\pi x)$ and $p_{ih}(\pi x)$ implies that $p_h = p_{1h}$, so that $f = f_1 \in \mathcal{N}_\varrho^k(\overline{\mathbb{Q}})$, which completes the proof.

4. Nearly Holomorphic Eisenstein Series

We take our group and domain to be of types (3.13a, b) and (3.15). Our aim is to show that certain Eisenstein series defined as functions of $(z, s) \in \mathcal{H}^J \times \mathbb{C}$ provide nearly holomorphic forms of arithmetic type for certain integer values of s. This may be viewed as an analogue of the algebraicity of the critical values of certain zeta functions.

Hereafter we denote $G_{\mathbb{Q}}$ of (3.13a, b) simply by G. (We no longer need the group G introduced at the beginning of Sect. 3.) For $\gamma = \begin{pmatrix} a & b \\ c & d \end{pmatrix} \in G$ with a, b, c, and d of size m, we write $a = a_\gamma$, $b = b_\gamma$, $c = c_\gamma$, and $d = d_\gamma$. For $z = (z_\tau)_{\tau \in J} \in \mathcal{H}^J$, we put

$$j_\gamma(z) = (\det(c_\gamma^\tau z_\tau + d_\gamma^\tau))_{\tau \in J}, \qquad \delta(z) = (\det[i(z_\tau^* - z_\tau)])_{\tau \in J}. \tag{4.1}$$

These are elements of $\mathbb{C}^J$. (Recall that J is the archimedean primes of F or a *CM*-type of K.) We identify F and K with subsets of $\mathbb{C}^J$ through the map $x \mapsto (x^\tau)_{\tau \in J}$. For $c \in \mathbb{C}^J$ and $p \in \mathbb{C}^J$, we put

$$c^p = \prod_{\tau \in J} c_\tau^{p_\tau} \tag{4.2}$$

whenever each factor on the right-hand side is well-defined. We denote by u the identity element of the ring $\mathbb{C}^J$, and thus write $s^{su} = \prod_{\tau \in J} c_\tau^s$ for $s \in \mathbb{C}$.

Given a congruence subgroup Γ of G and $q \in \mathbb{Z}^J$, we define an Eisenstein series E on $\mathcal{H}^J$ by

$$E(z, s; q, \Gamma) = \sum_{\alpha \in (P \cap \Gamma) \backslash \Gamma} \delta(\alpha(z))^{su - (q/2)} j_\alpha(z)^{-q}. \tag{4.3}$$

Here $z \in \mathscr{H}^J$, $s \in \mathbb{C}$, and $P = \{\gamma \in G \mid c_\gamma = 0\}$. To make the sum meaningful, we assume

$$[|\det(d_\alpha)|/\det(d_\alpha)]^q = 1 \qquad \text{for every} \quad \alpha \in P \cap \Gamma. \tag{4.4}$$

We introduce another type of series defined relative to a congruence subgroup of type

$$\Gamma_0(\mathfrak{c}) = \{\alpha \in SL_{2m}(\mathfrak{o}) \mid c_\alpha \in \mathfrak{c}_m^m\}, \tag{4.5}$$

where $\mathfrak{c}$ is an integral ideal of F, and $\mathfrak{o}$ is the maximal order of F (resp. K) in Case SP (resp. SU). We consider the adelization $G_{\mathbf{A}}$ and $P_{\mathbf{A}}$ of G and P, and denote by $C_\mathfrak{c}$ the closure of $\Gamma_0(\mathfrak{c})G_\infty$ in $G_{\mathbf{A}}$. Taking a Hecke character ψ of $F_{\mathbf{A}}^\times$, we put

$$E(z, s; q, \psi, \mathfrak{c}) = \sum_{\beta \in B} N(\mathfrak{a}_\beta)^{2s} \tag{4.6}$$

$$\cdot \sum_{\alpha \in R_\beta} \psi_\infty(\det(d_\alpha))\psi^*(\det(d_\alpha)\mathfrak{a}_\beta^{-1})\, \delta(\alpha(z))^{su - (q/2)} j_\alpha(z)^{-q}.$$

Here B is a complete set of representatives of $P \setminus (G \cap P_{\mathbf{A}} C_\mathfrak{c})/\Gamma_0(\mathfrak{c})$, R_β is a complete set of representatives of $(P \cap \beta\Gamma\beta^{-1}) \setminus \beta\Gamma$ with $\Gamma = \Gamma_0(\mathfrak{c})$, $\mathfrak{a}_\beta = \mathrm{il}(\beta)$ with the symbol il of $[11, (1.8)]$, ψ_∞ is the archimedean factor of ψ, and ψ^* is the ideal-character attached to ψ. This time we assume, instead of (4.4), the conditions

$$\textit{the finite part of the conductor of } \psi \textit{ divides } \mathfrak{c}; \tag{4.7a}$$

$$\psi_\infty(x) = \prod_{\tau \in J} (|x_\tau|/x_\tau)^{q_\tau}. \tag{4.7b}$$

If $\mathfrak{c} = (1)$, we understand that $\psi_\infty(\det(d_\alpha))\, \psi^*(\det(d_\alpha)\mathfrak{a}_\beta^{-1})$ means $\psi^*(\mathfrak{a}_\beta^{-1})$. The series of (4.6) is well-defined independently of the choice of B and R_β (see $[11,$ p. 427]).

We define the L-function $L(s, \psi)$ of ψ as usual, and put

$$L_\mathfrak{c}(s, \psi) = L(s, \psi) \prod_{\mathfrak{p}|\mathfrak{c}}(1 - \psi^*(\mathfrak{p})\, N(\mathfrak{p})^{-s}), \tag{4.8}$$

where $\mathfrak{p}$ runs over the prime ideals of F dividing $\mathfrak{c}$. Then we modify the series E by multiplying with certain L-functions as follows:

$$D(z, s; q, \psi, \mathfrak{c}) = E(z, s; q, \psi, \mathfrak{c})$$

$$\cdot \begin{cases} L_\mathfrak{c}(2s, \psi) \displaystyle\prod_{i=1}^{[m/2]} L_\mathfrak{c}(4s - 2i, \psi^2) & \text{(Case SP)}, \\[2ex] \displaystyle\prod_{i=0}^{m-1} L_\mathfrak{c}(2s - i, \psi\theta^i) & \text{(Case SU)}. \end{cases} \tag{4.9}$$

Here θ is the Hecke character of $F_{\mathbf{A}}^\times$ corresponding to the quadratic extension K/F.

Recall that $\mathfrak{K} = GL_m(\mathbb{C})^J$ in Case SP and $\mathfrak{K} = GL_m(\mathbb{C})^J \times GL_m(\mathbb{C})^J$ in Case SU. Let ϱ be a representation of $\mathfrak{K}$ defined by $\varrho(a) = \det(a)^q$ in Case SP and $\varrho(a, b) = \det(b)^q$ in Case SU. Then we see that the functions of (4.3), (4.6), and (4.9) for a fixed $s_0 \in \mathbb{C}$ belong to $\mathscr{C}_\varrho$ as functions on $\mathscr{H}^J$, provided that they are finite at s_0. Now, for $0 \leq t \in \mathbb{Z}^J$, let $\mathscr{N}_q^t(\overline{\mathbb{Q}})$ denote the set $\mathscr{N}_\varrho^t(\overline{\mathbb{Q}})$ defined in Section 3, and $\mathbb{Q}_{ab}$ the maximal abelian extension of $\mathbb{Q}$ contained in $\mathbb{C}$. Then we define $\mathcal{N}_q^1(\mathbf{Q}_{ab})$ to be the set of all elements f of $\mathcal{N}_q^1(\overline{\mathbf{Q}})$ of form (3.16) such that $p_h(\pi x)$ as a polynomial in x is $\mathbf{Q}_{ab}$-rational for every h. We put $\mathcal{M}_q(\overline{\mathbf{Q}}) = \mathcal{N}_q^0(\overline{\mathbf{Q}})$ and $\mathcal{M}_q(\mathbf{Q}_{ab}) = \mathcal{N}_q^0(\mathbf{Q}_{ab})$; this is consistent with the notation of Sect. 3.

Proposition 4.1. *Let $\kappa = (m+1)/2$ in Case SP and $\kappa = m$ in Case SU; let $\mu \in \mathbb{Z}$.*

(i) *If $\mu \geqq \kappa$, $E(z, \mu/2; \mu u, \Gamma)$ belongs to $\mathcal{M}_{\mu u}(\mathbf{Q}_{ab})$ except when $F = \mathbf{Q}$ and $\kappa + (1/2) \leqq \mu \leqq \kappa + 1$.*

(ii) *If $F = \mathbf{Q}$ and $\mu = \kappa + 1$, $E(z, \mu/2; \mu, \Gamma)$ belongs to $\mathcal{N}_\mu^m(\mathbf{Q}_{ab})$.*

(iii) *If $\mu \geqq \kappa$, $E(z, \mu/2; \mu u, \psi, \mathfrak{c})$ belongs to $\mathcal{M}_{\mu u}(\mathbf{Q}_{ab})$ except in the following four cases:*

$\qquad$ (A) $\quad$ *Case SP:* $\quad \mu = (m+2)/2$, $F = \mathbf{Q}$, and $\psi^2 = 1$;

$\qquad$ (B) $\quad$ *Case SP:* $\quad m = 1$, $\mu = 2$, $F = \mathbf{Q}$, and $\psi = 1$;

$\qquad$ (C) $\quad$ *Case SP:* $\quad m > 1$, $\mu = (m+3)/2$, $F = \mathbf{Q}$, and $\psi^2 = 1$;

$\qquad$ (D) $\quad$ *Case SU:* $\quad \mu = m+1$, $F = \mathbf{Q}$, and $\psi = \theta^{m+1}$.

(iv) *In Cases (B), (C), and (D), $E(z, \mu/2; \mu u, \psi, \mathfrak{c})$ belongs to $\mathcal{N}_\mu^m(\mathbf{Q}_{ab})$.*

(v) *Suppose $\mu \leqq \kappa$; let $k = 2\kappa - \mu$. Then $D(z, \mu/2; \ ku, \ \psi, \mathfrak{c})$ belongs to $\pi^{en} \cdot \mathcal{M}_{ku}(\mathbf{Q}_{ab})$ except in the following three cases (E), (F), and (G), where $n = [F : \mathbf{Q}]$ and*

$$e = \begin{cases} m(m+2)/4 & \text{(Case SP, m even),} \\ \kappa(m+1)/2 & \text{(all other cases):} \end{cases}$$

$\qquad$ (E) $\quad$ *Case SP:* $\quad \mu = 0$, $\mathfrak{c} = (1)$, and $\psi = 1$;

$\qquad$ (F) $\quad$ *Case SP:* $\quad 0 < \mu \leqq m/2$, $\mathfrak{c} = (1)$, and $\psi^2 = 1$;

$\qquad$ (G) $\quad$ *Case SU:* $\quad 0 \leqq \mu < m$, $\mathfrak{c} = (1)$, and $\psi = \theta^\mu$.

(vi) *If $m = 1$ and $\mathfrak{c} = (1)$, $D(z, 0; 2u, \psi, \mathfrak{c})$ belongs to $\pi^n \mathcal{M}_{2u}(\mathbf{Q}_{ab})$ except when $F = \mathbf{Q}$, in which case it belongs to $\pi \mathcal{N}_2^1(\mathbf{Q}_{ab})$.*

Almost all of these assertions are restatements of the results of the previous papers; see [11, Theorems 7.1, 7.2; 14, Theorem 2.7], and the explanations in [14, p. 291]. The only new points are assertions (v) and (vi), which are somewhat different from [14, Theorem 2.7, (ii)]. If $\mathfrak{c} \neq (1)$, the assertion of (v) is the same as what is given there. A special care is necessary when $\mathfrak{c} = (1)$, since the computation of the Fourier coefficients in [11] was made under the assumption $\mathfrak{c} \neq (1)$. Suppose now $\mathfrak{c} = (1)$; take an arbitrary prime ideal $\mathfrak{p}$ of F. By [11, Proposition 2.4, (ii)], we have

$$E(z, s; q, \psi, (1)) = \sum_{\xi \in X} E(z, s; q, \psi, \mathfrak{p}) \|_q \xi$$

with a finite subset X of G. Hence

$$L(s) D(z, s; q, \psi, (1)) = \sum_{\xi \in X} D(z, s; q, \psi, \mathfrak{p}) \|_q \xi$$

with

$$L(s) = \begin{cases} (1 - \psi^*(\mathfrak{p}) N(\mathfrak{p})^{-2s}) \prod_{i=1}^{[m/2]} (1 - \psi^*(\mathfrak{p})^2 N(\mathfrak{p})^{2i-4s}) & \text{(Case SP),} \\ \prod_{i=0}^{m-1} (1 - (\psi\theta^i)^*(\mathfrak{p}) N(\mathfrak{p})^{i-2s}) & \text{(Case SU).} \end{cases}$$

Observe that we can find $\mathfrak{p}$ so that $L(\mu/2) \neq 0$ if we exclude Cases (E), (F), and (G). Therefore we obtain the desired conclusion of (v) for $\mathfrak{c} = (1)$ from that for $\mathfrak{c} = \mathfrak{p}$. Suppose $m = 1$ and $\mathfrak{c} = (1)$. In [15, pp. 16–18]], we gave the Fourier expansion of $E(z, s; q, \psi, (1))$. Taking $s = 0$ and $q = 2u$, we find that

$$D(z, 0; 2u, \psi, (1)) = L(0, \psi)\,\mathrm{Im}(z)^{-u} + \pi^u \sum_h c_h\, \mathbf{e}\!\left(\sum_{\tau \in J} h^\tau z_\tau\right)$$

with $c_h \in \mathbb{Q}_{\mathrm{ab}}$. Hence we obtain (vi).

Theorem 4.2. *Let $q \in \mathbb{Z}^J$. With κ as in Proposition 4.1, suppose $q_\tau \geqq \kappa$ for every $\tau \in J$ and $q_\tau \bmod 2$ is independent of τ.*

(i) Let μ be an integer such that $\kappa \leqq \mu \leqq q_\tau$ and $\mu \equiv q_\tau \,(\mathrm{mod}\ 2)$ for every $\tau \in J$. Then $E(z, \mu/2; q, \Gamma)$ belongs to $\pi^\alpha \mathcal{N}_q^\tau(\overline{\mathbb{Q}})$ except when $F = \mathbb{Q}$ and $\mu = (m+2)/2$ in Case SP, where $\alpha = (m/2)\sum_{\tau \in J}(q_\tau - \mu)$ and

$$t = \begin{cases} m(q - \mu + 2)/2 & \text{if}\ \ \mu = \kappa + 1\ \ \text{and}\ \ F = \mathbb{Q}, \\ m(q - \mu u)/2 & \text{otherwise}. \end{cases}$$

(ii) Let μ be an integer such that $2\kappa - q_\tau \leqq \mu \leqq q_\tau$ for every $\tau \in J$ and that

$$\mu \equiv q_\tau + 1\,(\mathrm{mod}\ 2) \quad \text{if}\ \ \mu < \kappa \notin \mathbb{Z},$$
$$\mu \equiv q_\tau\,(\mathrm{mod}\ 2) \quad \text{otherwise}.$$

Suppose that μ, F, ψ, and $\mathfrak{c}$ do not fall into Cases (A), (E), (F), and (G) of Proposition 4.1. Then $D(z, \mu/2; q, \psi, \mathfrak{c})$ belongs to $\pi^\beta \mathcal{N}_q^\tau(\overline{\mathbb{Q}})$, where

$$r = \begin{cases} m(q - \mu + 2)/2 & \text{in Cases } (B),\ (C),\ \text{and } (D), \\ m(q - |\mu - \kappa|u - \kappa u)/2 & \text{otherwise}, \end{cases}$$

and $\beta = (m/2)\sum_{\tau \in J}(q_\tau + \mu) - [F : \mathbb{Q}]e$ with

$$e = \begin{cases} m(m+2)/4 - \mu & (\text{Case SP},\ m\ \text{even},\ \mu \geqq \kappa), \\ m^2/4 & (\text{Case SP},\ m\ \text{even},\ \mu < \kappa), \\ (m^2 - 1)/4 & (\text{Case SP},\ m\ \text{odd}), \\ m(m-1)/2 & (\text{Case SU}). \end{cases}$$

(iii) Suppose $m = 1$, $\mu = 0$, and $2u \leqq q \in 2\mathbb{Z}^J$. Then $D(z, 0; q, \psi, \mathfrak{c})$ belongs to $\pi^\beta \mathcal{N}_q^\tau(\overline{\mathbb{Q}})$ with $r = (q - 2u)/2$ and $\beta = \sum_{\tau \in J} q_\tau/2$ except when $F = \mathbb{Q}$ and $\mathfrak{c} = (1)$. If $F = \mathbb{Q}$ and $\mathfrak{c} = (1)$, it belongs to $\pi^{q/2} \mathcal{N}_q^{q/2}(\overline{\mathbb{Q}})$.

Proof. Define a differential operator Δ_q^p on $\mathcal{H}^J$ with $p, q \in \mathbb{Z}^J$, $p \geqq 0$, acting on $f \in C^\infty(\mathcal{H}^J)$ by

$$\Delta_q^p f = \delta(z)^{\kappa u - p - q} \prod_{\tau \in J}(\Delta_\tau)^{p_\tau} [\delta(z)^{p + q - \kappa u} f], \tag{4.10a}$$

$$\Delta_\tau = \det((\partial_{ij}^\tau)_{i,j=1}^m), \tag{4.10b}$$

$$\partial_{ij}^\tau = \begin{cases} (1/2)(1 + \delta_{ij})\,\partial/\partial z_{\tau ij} & (\text{Case SP}), \\ \partial/\partial z_{\tau ij} & (\text{Case SU}), \end{cases} \tag{4.10c}$$

where $(z_{\tau ij})_{i,j=1}^{m}$ is the variable matrix on the τ-factor of $\mathscr{H}^J$. Then we have

$$\Delta_q^p \mathcal{N}_q^t(\overline{\mathbb{Q}}) \subset \pi^{m|p|} \mathcal{N}_{q+2p}^{t+mp}(\overline{\mathbb{Q}}), \tag{4.11}$$

where $|p| = \sum_{\tau \in J} p_\tau$. This is a special case of (3.12), since Δ_q^p is a special case of D_ϱ^z (see [12, pp. 271–272]). Now let $E(z, s; q)$ denote a series of type (4.3) or (4.6). For $0 \leqq p \in \mathbb{Z}^J$, we have

$$\Delta_q^p E(z, s; q) = c_q^p(s) \, i^\alpha E(z, s; q+2p), \tag{4.12}$$

where $\alpha = m|p|$ and

$$c_q^p(s) = \begin{cases} \displaystyle\prod_{\tau \in J} \prod_{a=1}^{m} \prod_{b=1}^{p_\tau} [-s - (q_\tau/2) - b + (a+1)/2] & \text{(Case SP)}, \\[2em] \displaystyle\prod_{\tau \in J} \prod_{a=1}^{m} \prod_{b=1}^{p_\tau} [-s - (q_\tau/2) - b + a] & \text{(Case SU)}. \end{cases}$$

This follows immediately from [12, Lemma 1.2] (cf. also [16, (6.5)]). Given μ as in (i), put $p = (q - \mu u)/2$. Then $0 \leqq p \in \mathbb{Z}^J$, and (4.12) with $q = \mu u$ yields

$$\Delta_{\mu u}^p E(z, \mu/2; \mu u) = c_{\mu u}^p(\mu/2) \, i^\alpha E(z, \mu/2; q). \tag{4.13a}$$

Observing that $c_{\mu u}^p(\mu/2) \neq 0$, we obtain our assertion of (i) from Proposition 4.1, (i), (ii), and (4.11).

Next let μ be given as in (ii). If $\mu \geqq \kappa$, the above proof is applicable also to this case, except that we have to multiply the functions by the product of certain values of L-functions, and hence a power of π appears as stated. If $\mu < \kappa$, we put $k = 2\kappa - \mu$, $p = (q - ku)/2$ and $s_k = \kappa - k$. Then $k > \kappa$, $0 \leqq p \in \mathbb{Z}^J$, and

$$\Delta_{ku}^p D(z, \mu/2; ku, \psi, \mathfrak{c}) = i^{m|p|} c_{ku}^p(\mu/2) D(z, \mu/2; q, \psi, \mathfrak{c}). \tag{4.13b}$$

Observing that $c_{ku}^p(\mu/2) \neq 0$, we obtain assertion (ii) from (v) of Proposition 4.1 and (4.11). Assertion (iii) follows from (vi) of Proposition 4.1 in a similar manner.

Assertion (iii) means that if $m = 1$, we have results even in Cases (E) and (G). It is conjecturable that similar results hold in Cases (E), (F), and (G) even for $m > 1$. We mention here Paul Feit [1] as a work which contains an investigation of the same type as the above theorem from a different angle.

In Case SP, we can consider also the series of half-integral weight. Namely, for an odd integer l, an element q of $\mathbb{Z}^J$, and a Hecke character ψ of $F_\mathbb{A}^\times$ as above, we put

$$E(z, s; l/2, q, \psi, \mathfrak{c}) = \sum_\beta N(\mathfrak{a}_\beta)^{2s} \sum_{\alpha \in T_\beta} \psi_\infty(\det(d_\alpha)) \tag{4.14}$$
$$\cdot \, \psi^*(\det(d_\alpha) \, \mathfrak{a}_\beta^{-1}) \, \delta(\alpha(z))^{su - (q/2) - (l/4)} j_\alpha(z)^{-q} h(\alpha, z)^{-l}.$$

Here $h(\alpha, z)$ is the factor of automorphy of weight $1/2$ defined in [14, (2.9a, b, c), Propositions 1.2 and 3.1]; $\{\beta\}$, $\mathfrak{a}_\beta$, and T_β are the same as in [14, (2.11) and (4.7c)]. We assume, in addition to (4.7a, b), that $\mathfrak{c}$ is divisible by 4.

Let ω denote the Hecke character of $F_\mathbb{A}^\times$ corresponding to $F((-1)^{1/2})/F$. Then, for the elements α of T_β appearing in (4.14), we have

$$h(\alpha, z)^2 j_\alpha(z)^{-u} = \omega_\infty(\det(d_\alpha)) \omega^*(\det(d_\alpha) \, \mathfrak{a}_\beta^{-1}). \tag{4.15}$$

as shown in [14, Lemma 3.5]. Therefore we have

$$E(z, s; l/2, q, \psi, \mathfrak{c}) = E(z, s; l'/2, q', \psi\omega^{(l-l')/2}, \mathfrak{c}) \tag{4.16}$$

if $q + (l/2)u = q' + (l'/2)u$. Now put

$$D(z, s; l/2, q, \psi, \mathfrak{c}) = E(z, s; l/2, q, \psi, \mathfrak{c}) \prod_{i=0}^{[(m-1)/2]} L_{\mathfrak{c}}(4s - 1 - 2i, \psi^2). \tag{4.17}$$

Then (4.16) holds also with D instead of E.

For an element p of $\mathbf{Q}^J$ such that $p - (u/2) \in \mathbf{Z}^J$, let C_p denote the set of elements f of $C^\infty(\mathscr{H}^J)$ such that

$$f(\gamma(z)) = h(\gamma, z) j_\gamma(z)^{p - (u/2)} f(z) \tag{4.18}$$

for every γ in a congruence subgroup of $Sp(m, F)$ contained in the group of [14, (1.13)]. Then, for $0 \leqq t \in \mathbf{Z}^J$, we define $\mathcal{N}_p^t$ (resp. $\mathcal{N}_p^t(\overline{\mathbf{Q}})$) to be the set of elements f of C_p of form (3.16) satisfying the cusp condition and such that $p_h(\pi x)$ is a polynomial (resp. $\overline{\mathbf{Q}}$-rational polynomial) in $x \in T^J$ which has degree $\leqq t_\tau$ on the τ-factor of T^J.

Theorem 4.3. *Let $\kappa = (m+1)/2$ and $r = q + (l/2)u$ with $q \in \mathbf{Z}^J$ and an odd integer l. Suppose that $r_\tau \geqq \kappa$ for every $\tau \in J$ and $q_\tau \bmod 2$ is independent of τ. Let λ be an element of $(1/2)\mathbf{Z}$ such that $2\kappa - r_\tau \leqq \lambda \leqq r_\tau$ for every $\tau \in J$ and that*

$$\lambda \equiv \begin{cases} r_\tau & (\bmod\, 2) & \text{if } \lambda \geqq \kappa, \\ r_\tau + m & (\bmod\, 2) & \text{if } \lambda < \kappa. \end{cases}$$

Then $D(z, \lambda/2; l/2, q, \psi, \mathfrak{c})$ belongs to $\pi^\gamma \mathcal{N}_r^t(\overline{\mathbf{Q}})$ except when $\lambda = (m+2)/2$, $F = \mathbf{Q}$, and $\psi^2 = 1$, where

$$t = \begin{cases} m(r - \lambda + 2)/2 & \text{if } 2\lambda = m + 3, \ F = \mathbf{Q}, \ \text{and} \ \psi^2 = 1, \\ (m/2)(r - |\lambda - \kappa|u - \kappa u) & \text{otherwise}, \end{cases}$$

and $\gamma = (m/2) \sum_{\tau \in J} (r_\tau + \lambda) - [F : \mathbf{Q}] e$ with

$$e = \begin{cases} (m+1)^2/4 - \lambda & (m \text{ odd}, \ \lambda \geqq \kappa), \\ (m^2 - 1)/4 & (m \text{ odd}, \ \lambda < \kappa), \\ m^2/4 & (m \text{ even}). \end{cases}$$

This can be derived from [14, Theorems 2.3–2.5] in the same fashion as in Theorem 4.2. A result similar to (i) of Theorem 4.2 can be obtained also in the present case; its precise statement may be left to the reader, as it is essentially contained in the above theorem in view of [14, Lemma 2.2].

5. Nearly Holomorphic Forms on the Upper Half Planes

In this section, we consider automorphic forms on the product of copies of the upper half plane $\mathscr{H} = \{z \in \mathbf{C} \mid \mathrm{Im}(z) > 0\}$ relative to a quaternion algebra B over a totally real algebraic number field F. We denote by J the set of all archimedean primes of F, and by α (res. α') the set of primes in J where B is unramified (resp.

ramified); we assume $\alpha \neq \emptyset$ throughout. Now $B \otimes_{\mathbb{Q}} \mathbb{R}$ can be identified with $M_2(\mathbb{R})^{\alpha} \times \mathbb{H}^{\alpha'}$, where $\mathbb{H}$ denotes the Hamilton quaternions. For each $\tau \in J$ and $x \in B$, we denote by x_τ the corresponding element in $M_2(\mathbb{R})$ or $\mathbb{H}$. Put

$$G = \{x \in B \mid xx^\iota = 1\}, \tag{5.0}$$

where ι denotes the main involution of B. Then G can be embedded into $SL_2(\mathbb{R})^{\alpha} \times Sp(1)^{\alpha'}$, where $Sp(1)$ denotes the group of all quaternions in $\mathbb{H}$ of norm 1. Let $\{\sigma, X\}$ be a continuous irreducible representation of $Sp(1)^{\alpha'}$, which we view also as a representation of G, via the map $G \to Sp(1)^{\alpha'}$. (We understand that this is the trivial representation of the identity group if $\alpha' = \emptyset$.) For $f \in C^{\infty}(H^{\alpha}, X)$, $k \in \mathbb{Z}^{\alpha}$, and $\xi \in G$, we define $f\|_{k,\sigma}\xi \in C^{\infty}(\mathcal{H}^{\alpha}, X)$ by

$$(f\|_{k,\sigma}\xi)(z) = \sigma(\xi)^{-1} \prod_{\tau \in \alpha}(c_\tau z_\tau + d_\tau)^{-k_\tau} f(\xi z) \qquad (z \in \mathcal{H}^{\alpha}), \tag{5.1}$$

where $\xi_\tau = \begin{pmatrix} * & * \\ c_\tau & d_\tau \end{pmatrix}$. This is a special case of (3.3). Therefore we can define the space $\mathcal{M}_{k,\sigma}$ of holomorphic automorphic forms on $\mathcal{H}^{\alpha}$ with respect to congruence subgroups of G in a natural way. If σ is trivial, we write simply $\mathcal{M}_k$ for $\mathcal{M}_{k,\sigma}$. To define arithmeticity, we assume that $\{\sigma, X\}$ as a representation of G is $\overline{\mathbb{Q}}$-rational. Then $\mathcal{M}_{k,\sigma}(\overline{\mathbb{Q}})$ is meaningful (see [9, II, Sect. 2] for details).

Lemma 5.1. $\mathcal{M}_{k,\sigma} \neq \{0\}$ *only if* $k_\tau > 0$ *for every* $\tau \in \alpha$, *or if* $k = 0$ *and* σ *is trivial.*

Proof. If $B = M_2(F)$, this was given in [7, Proposition 1.1]. Suppose B is a division algebra, $k_\tau \leqq 0$ for some τ, and $\mathcal{M}_{k,\sigma} \neq \{0\}$. By [3, Theorem 3.1], the integers k_τ for all $\tau \in \alpha$ must be the same, and so $k_\tau \leqq 0$ for every $\tau \in \alpha$. Then, by [3, Theorem 5.1], $\mathcal{M}_{k,\sigma} = \{0\}$ unless $k = 0$ and σ is trivial.

Given $k \in \mathbb{Z}^{\alpha}$ and $\{\sigma, X\}$ as above, we consider an X-valued function f on $\mathcal{H}^{\alpha}$ of the form

$$f(z) = \sum_{0 \leqq a \leqq A} y^{-a} f_a(z) \tag{5.2}$$

with $y = (\mathrm{Im}(z_\tau))_{\tau \in \alpha}$, $a \in \mathbb{Z}^{\alpha}$, $0 \leq A \in \mathbb{Z}^{\alpha}$, and X-valued holomorphic functions f_a on $\mathcal{H}^{\alpha}$. If $\xi \in G$ and c_τ, d_τ are as in (5.1), then

$$\mathrm{Im}(\xi(z))_\tau^{-1} = y_\tau^{-1}(cz+d)_\tau^2 - 2ic_\tau(cz+d)_\tau. \tag{5.3}$$

Hence $f\|_{k,\sigma}\xi = \sum_{0 \leqq a \leqq A} y^{-a} f_{\xi a}(z)$ with X-valued holomorphic functions $f_{\xi a}$. We denote by $\mathcal{N}_{k,\sigma}$ the set of all f of form (5.2) satisfying the following conditions:

$$f\|_{k,\sigma}\gamma = f \text{ for every } \gamma \text{ in a congruence subgroup of } G; \tag{5.4a}$$

$$\text{If } B = M_2(\mathbb{Q}), \text{ then for every } \xi \text{ and } a, f_{\xi a}(z) = \sum_{0 \leqq q \in \mathbb{Q}} c_q \mathbf{e}(qz)$$
$$\text{with constants } c_q \text{ depending on } \xi \text{ and } a. \tag{5.4b}$$

We write $\mathcal{N}_k$ for $\mathcal{N}_{k,\sigma}$ if σ is trivial, and denote by $\mathcal{N}_{k,\sigma}(\overline{\mathbb{Q}})$ the set of all arithmetic elements of $\mathcal{N}_{k,\sigma}$, which is defined as in Section 3. If $B = M_2(F)$, G is the group of (3.13a) with $m = 1$, and therefore the characterization of arithmeticity by means of Fourier coefficients as in Proposition 3.9 is valid.

Now the main purpose of this section is to prove a theorem which is of the same nature as Proposition 3.3, but more definitive. The requisite differential operators in the present case have simpler forms:

$$D_b^a = \prod_{\tau \in \alpha} \prod_{i=1}^{a_\tau} D_\tau(b_\tau + 2i - 2),$$ (5.5a)

$$D_\tau(c) = (2\pi i)^{-1}[c(z_\tau - \bar{z}_\tau)^{-1} + \partial/\partial z_\tau].$$ (5.5b)

Here $0 \leq a \in \mathbb{Z}^\alpha$, $b \in \mathbb{C}^\alpha$, and $c \in \mathbb{C}$.

Theorem 5.2. *For $f \in \mathcal{N}_{k,\sigma}$, the following assertions hold.*

(1) *$f \neq 0$ only when $k_\tau > 0$ for every $\tau \in \alpha$ or when $k = 0$ and σ is trivial.*

(2) *In (5.2), $f_a \neq 0$ only when $2a \leq k$; moreover, if $k \neq 0$ and $B \neq M_2(\mathbb{Q})$, then $f_a \neq 0$ only when $2a_\tau < k_\tau$ for every $\tau \in \alpha$.*

(3) *$\mathcal{N}_0$ consists only of the constants.*

(4) *f can be uniquely written in the form*

$$f(z) = \sum_{0 \leq p \leq k/2} D_{k-2p}^p g_p + \begin{cases} cD_2^{(k/2)-1} E_2 & \text{if } B = M_2(\mathbb{Q}) \text{ and } k \in 2\mathbb{Z}, \\ 0 & \text{otherwise}, \end{cases}$$

where $g_p \in \mathcal{M}_{k-2p,\sigma}$, $c \in \mathbb{C}$, and

$$E_2(z) = (4\pi y)^{-1} - (12)^{-1} + 2\sum_{n=1}^{\infty} \left(\sum_{0 < d|n} d \right) \mathbf{e}(nz).$$

Proof. With f_a as in (5.2) and Γ denoting a congruence subgroup as in (5.4a), let t be a maximal element of $\mathbb{Z}^\alpha$ with respect to the partial order "$\leq$" such that $f_t \neq 0$. Then we see easily that $f_t\|_{k-2t,\sigma}\xi = f_{\xi t}$, and in particular $f_t\|_{k-2t,\sigma}\gamma = f_t$ for every $\gamma \in \Gamma$. Therefore $f_t \in \mathcal{M}_{k-2t,\sigma}$. By Lemma 5.1, we see that either $k = 2t$ and σ is trivial, or $k_\tau > 2t_\tau$ for every $\tau \in \alpha$. This proves the first part of (2). Suppose $k = 2t$, σ is trivial, and $k \neq 0$. Then f_t is a nonzero constant. We are going to show that this cannot happen if $B \neq M_2(\mathbb{Q})$. Define, for $a \in Z^\alpha$ and $\tau \in \alpha$, operators ε_τ and L_τ^a by

$$\varepsilon_\tau = 2iy_\tau^2 \partial/\partial\bar{z}_\tau, \qquad L_\tau^a = -4\pi D_\tau(a_\tau - 2)\varepsilon_\tau.$$ (5.6)

Then L_τ^a coincides with the operator of [15, (2.3c)] with σ and v there corresponding to a and τ here. Also, using the notation $\partial/\partial r_p$ of (1.1), we may put $\varepsilon_\tau = \partial/\partial(y_\tau^{-1})$.

Since $k \neq 0$, we have $t_\tau > 0$ for some τ. Take any such τ and put $g = \left(\varepsilon_\tau^{t_\tau-1} \prod_{v \neq \tau} \varepsilon_v^{t_v} \right) f$. Then we see that $\varepsilon_\tau g = cf_t$ with a nonzero constant c, and $g = cy_\tau^{-1} f_t + h$ with a holomorphic function h. Observe that $g\|_{2\tau}\gamma = g$ for every $\gamma \in \Gamma$. Suppose B is a division algebra. Then we have

$$\langle \varepsilon_\tau u, v \rangle = \langle u, -4\pi D_\tau(a_\tau)v \rangle$$ (5.7)

for $u, v \in C^\infty(\mathcal{H}^\alpha)$ such that $u\|_{a+2\tau}\gamma = u$ and $v\|_a\gamma = v$ for every $\gamma \in \Gamma$. (This is well known; see [9, I, (2.18); 12, (11.13b)], for example.) Applying this to f_t and g with $a = 0$, we obtain

$$\langle cf_t, f_t \rangle = \langle \varepsilon_\tau g, f_t \rangle = \langle g, 2i\partial f_t/\partial z_\tau \rangle = 0,$$

since f_t is a constant, and so $f_t = 0$, a contradiction. Thus $k_v > 2t_v$ for every $v \in \alpha$ if B is a division algebra. Next suppose $B = M_2(F)$ and $F \neq \mathbf{Q}$. Since $g - cy_\tau^{-1}f_t$ is holomorphic, we see easily that $L_v^{2\tau}g = 0$ for *every* $v \in \alpha$. Therefore g is an automorphic eigenform belonging to $\mathscr{A}(2\tau, 0, \Gamma)$ of [15, Sect. 2]. Let $p = u - \tau$ with $u = \sum_{\tau \in J} \tau$ and $s_0 = 1$. Then $p = s_0 u - (2\tau)/2$. Thus the present s_0, 2τ, and p correspond to the symbols s_0, $\sigma(=2\varrho)$, and p of [15, Sect. 7] attached to the set of eigenvalues $\lambda = 0$. Observe that λ is noncritical and simple in the sense of [15, Sect. 3]. Therefore, by [15, Theorem 7.3 and Remark 7.4], we have $\mathscr{N}(2\tau, 0, \Gamma) = \mathscr{E}(s_0, \tau, \Gamma)$. Now, for every $\xi \in SL_2(F)$, $g\|_{2\tau}\xi = c'y_\tau^{-1} + h'$ with a constant c' and a holomorphic function h', and hence y^p doesn't appear in $g\|_{2\tau}\xi$, since $F \neq \mathbf{Q}$. Therefore we can apply [15, Proposition 7.6] to g to find that g is a cusp form. Then [15, Proposition 2.5] shows that g is holomorphic, and hence $f_t = 0$, a contradiction. This shows that $k_v > 2t_v$ for every $v \in \alpha$ if $k \neq 0$ and $B \neq M_2(\mathbf{Q})$. This proves (1) and (2), since (1) is true even when $B = M_2(\mathbf{Q})$. As a consequence of (2), we have $\mathscr{N}_0 = \mathscr{M}_0$, which proves (3).

Finally, as for (4), it is trivial if $k = 0$; so assume $k_v > 0$ for every $v \in \alpha$ and $B \neq M_2(\mathbf{Q})$. Take t as before. Since $k_v > 2t_v$ for every $v \in \alpha$, we see that $D_{k-2t}^t f_t = \sum_{0 \leq a \leq t} y^{-a}q_a$ with holomorphic functions q_a and $q_t = bf_t$ with $0 \neq b \in \mathbf{C}$. Then $f - b^{-1}D_{k-2t}^t f_t$ belongs to $\mathscr{N}_{k,\sigma}$ and has no terms of degree $\geq t$. Repeating the same procedure, we obtain the expression of f as in (4). Suppose now $B = M_2(\mathbf{Q})$ and $k \neq 0$; take t as before. Then $k \geq 2t$. If $k > 2t$, the same reasoning yields the desired result. Suppose $k = 2t$. Then f_t is a constant. Recall that E_2 of (4) is a well known Eisenstein series of weight 2, and so belongs to $\mathscr{N}_2$. Observe that $D_2^{t-1}E_2 = \sum_{a=0}^{t} y^{-a}p_a$ with holomorphic functions p_a finite at ∞, and that p_t is a nonzero constant. Therefore $f - (f_t/p_t)D_2^{t-1}E_2$ has degree $< t$, and hence the problem can be reduced to the case $k > 2t$. Our procedure of finding the expression for f in (4) shows clearly its uniqueness. This completes the proof.

A special case of the above theorem is worth noting:

Corollary 5.3. *Suppose* $k_\tau \leq 2$ *for every* $\tau \in \alpha$. *Then* $\mathscr{N}_{k,\sigma} = \mathscr{M}_{k,\sigma}$ *except when* $B = M_2(\mathbf{Q})$ *and* $k = 2$, *in which case* $\mathscr{N}_k = M_k \oplus \mathbf{C}E_2$ *with* E_2 *of Theorem 5.2.*

This follows immediately from (2) and (4) of Theorem 5.2.

Theorem 5.4. *Every element* f *of* $\mathscr{N}_{k,\sigma}(\overline{\mathbf{Q}})$ *can be written as in (4) of Theorem 5.2 with* $g_p \in \mathscr{M}_{k-2p,\sigma}(\overline{\mathbf{Q}})$ *and* $c \in \overline{\mathbf{Q}}$.

Proof. As shown in the proof of [4, Theorem 1], E_2 is arithmetic. Therefore, in view of (3.12), we obtain our assertion by the same type of reasoning as in the first part of the proof of Theorem 3.7.

If $B = M_2(F)$, the above theorems can be extended to the forms of half-integral weight as follows. Let k be an element of $\mathbf{Q}^J$ such that $k - (u/2) \in \mathbf{Z}^J$. In the paragraph preceding Theorem 4.3, we defined the sets $\mathscr{N}_k^t$ and $\mathscr{N}_k^t(\overline{\mathbf{Q}})$ for $0 \leq t \in \mathbf{Z}^J$. We denote by $\mathscr{N}_k$ (resp. $\mathscr{N}_k(\overline{\mathbf{Q}})$) the union of the sets $\mathscr{N}_k^t$ (resp. $\mathscr{N}_k^t(\overline{\mathbf{Q}})$) for all such t, and put $\mathscr{M}_k = \mathscr{N}_k^0$.

Theorem 5.5. *If $B = M_2(F)$, the assertions of Lemma 5.1, Theorems 5.2, 5.4, and Corollary 5.3 are all valid for $\mathcal{M}_k$ and $\mathcal{N}_k$ with $k \in \mathbb{Q}^J$ such that $k - (u/2) \in \mathbb{Z}^J$.*

Proof. Obviously $f^2 \in \mathcal{M}_{2k}$ if $f \in \mathcal{M}_k$, and therefore Lemma 5.1 implies the corresponding statement for half-integral k. The rest can be proved by checking the validity of the proof of Theorems 5.2 and 5.4 for the present k.

6. Nearly Holomorphic Theta Functions

Let V be a finite-dimensional complex vector space, and L a lattice in V such that V/L is an abelian variety. We fix on V a nondegenerate Riemann form H relative to L, by which we mean a positive definite hermitian form $H(u, v)$, $\mathbb{C}$-linear in v, such that $\mathrm{Im}\,(H(L, L)) \subset \mathbb{Z}$. An L-invariant Kähler form Ω on V can be defined by $\Omega = (i/2) \sum_{p,q} h_{pq} dz_q \wedge d\bar{z}_p$, where the constants h_{pq} are determined by $H(z, w) = \sum_{p,q} h_{pq} \bar{z}_p w_q$ with respect to an affine coordinate system $\{z_p\}$ on V. Then $\mathcal{N}^k(V)$ for $0 \leqq k \in \mathbb{Z}$ in this case is merely the set of all polynomial functions in the $\bar{z}_p$ of degree $\leqq k$ with holomorphic coefficients, and therefore is not an object of much interest unless our attention is focused on the functions of a specific type. We consider here nearly holomorphic theta functions, which are natural analogues of nearly holomorphic automorphic forms studied in the previous sections.

Let $E(u, v) = \mathrm{Im}\,[H(u, v)]$, and let Ψ be a map of L into the group of all complex numbers of absolute value 1 such that

$$\Psi(l + m) = \Psi(l)\,\Psi(m)\,\mathbf{e}\,(E(l, m)/2). \tag{6.1}$$

Then we denote by $C(H, \Psi, L)$ the set of all $f \in C^\infty(V)$ such that

$$f(z + l) = f(z)\,\Psi(l)\,\mathbf{e}\,((2i)^{-1} H(l, z + (l/2))) \tag{6.2}$$

for every $z \in V$ and $l \in L$, and put

$$\mathcal{N}^k(H, \Psi, L) = \mathcal{N}^k(V) \cap C(H, \Psi, L) \qquad (0 \leqq k \in \mathbb{Z}). \tag{6.3}$$

Let $\mathfrak{D}$ denote the complex Lie algebra of V, which consists of all the derivations $\sum a_p \partial/\partial z_p$ with $a_p \in \mathbb{C}$ and the z_p as above. To each $D \in \mathfrak{D}$, we can associate an element u of V by $Dh = h(u)$ for every $\mathbb{C}$-linear map h of V into $\mathbb{C}$. We put $u = [D]$, and thus $Dh = h([D])$. We define, for $D \in \mathfrak{D}$, an operator D^H acting on the elements f of $C^\infty(V)$ by

$$(D^H f)(z) = (Df)(z) - \pi H(z, [D])\,f(z) \qquad (z \in V). \tag{6.4}$$

This is the same as the operator of [6, (2.3)] and actually a special case of (1.7). The following facts can easily be verified:

$$D^H \text{ maps } C(H, \Psi, L) \text{ into itself.} \tag{6.5}$$

$$D^H \text{ maps } \mathcal{N}^k(H, \Psi, L) \text{ into } \mathcal{N}^{k+1}(H, \Psi, L). \tag{6.6}$$

Proposition 6.1. *$\mathcal{N}^k(H, \Psi, L)$ is spanned by the functions of the form $D_1^H \ldots D_r^H g$ with $g \in \mathcal{N}^0(H, \Psi, L)$ and $D_1, \ldots, D_r \in \mathfrak{D}, 0 \leqq r \leqq k$.*

Proof. Given $f \in \mathcal{N}^k(H, \Psi, L)$, express it as a polynomial in the $\bar{z}_p$, and observe that the coefficient of one of the terms of the highest total degree belongs to $\mathcal{N}^0(H, \Psi, L)$. Then our assertion can be proved by the same type of argument as in the proof of Theorem 5.2, (4).

Let us now assume that V/L has many complex multiplications in the sense of [5, Sect. 2]. Let M be the field generated over $\mathbb{Q}$ by $\operatorname{tr}(\alpha)$ for all $\alpha \in \operatorname{End}(V/L)$, where $\operatorname{tr}(\alpha)$ is the trace of α as a $\mathbb{C}$-linear endomorphism of V. Then M is a *CM*-field. (M coincides with K' of [5, p. 684].) Let M_{ab} denote the maximal abelian extension of M contained in $\mathbb{C}$. For every $f \in C^\infty(V)$, define $f_H \in C^\infty(V)$ by

$$f_H(z) = \mathbf{e}((i/4)\, H(z, z))\, f(z) \qquad (z \in V). \tag{6.7}$$

Assuming that $\Psi(l)$ is a root of unity for every $l \in L$, we call an element f of $\mathcal{N}^k(H, \Psi, L)$ *arithmetic* if $f_H(u) \in M_{ab}$ for every $u \in \mathbb{Q}L$, and call also an element D of $\mathfrak{D}$ *arithmetic* if the field of M_{ab}-rational meromorphic functions on V/L is stable under D. (For a more detailed explanation, the reader is referred to [6, p. 375].) We then denote by $\mathcal{N}_a^k(H, \Psi, L)$ (resp. $\mathfrak{D}_a$) the set of all arithmetic elements of $\mathcal{N}^k(H, \Psi, L)$ (resp. $\mathfrak{D}$). Now we have a result similar to Theorem 5.4:

Theorem 6.2. (i) $\mathcal{N}_a^k(H, \Psi, L)$ *spans* $\mathcal{N}^k(H, \Psi, L)$ *over* $\mathbb{C}$.

(ii) $\mathfrak{D}_a$ *spans* $\mathfrak{D}$ *over* $\mathbb{C}$.

(iii) *If* $D \in \mathfrak{D}_a$, D^H *maps* $\mathcal{N}_a^k(H, \Psi, L)$ *into* $\mathcal{N}_a^{k+1}(H, \Psi, L)$.

(iv) $\mathcal{N}_a^k(H, \Psi, L)$ *is spanned over* M_{ab} *by the functions of the form* $D_1^H \ldots D_r^H g$ *with* $g \in \mathcal{N}_a^0(H, \Psi, L)$ *and* $D_1, \ldots, D_r \in \mathfrak{D}_a$, $0 \leqq r \leqq k$.

Proof. Assertion (ii) is obvious. Assertion (i) for $k = 0$ follows from [5, Propositions 1.2 and 2.5]. To prove (iv), let $\mathcal{N}_*$ denote the set spanned over M_{ab} by all the functions $D_1^H \ldots D_r^H g$ as in (iv). By [6, Theorem 2.2], $\mathcal{N}_* \subset \mathcal{N}_a^k(H, \Psi, L)$. To prove the opposite inclusion, take $f \in \mathcal{N}_a^k(H, \Psi, L)$ and also a basis B of $\mathbb{C}$ over M_{ab}. By (i) for $k = 0$, (ii), and Proposition 6.1, we have $f = \sum_{c \in B} c k_c$ with finitely many $k_c \in \mathcal{N}_*$. Then $f_H(u) = \sum c(k_c)_H(u)$ for every $u \in \mathbb{Q}L$. Since both $f_H(u)$ and $(k_c)_H(u)$ belong to M_{ab}, we have $f_H(u) = (k_1)_H(u)$, so that $f = k_1$. This proves (iv). Then (iii) follows from (iv); (i) for general k can be obtained from its special case $k = 0$, (ii), (iv), and Proposition 6.1.

References

1. Feit, P.: Poles and residues of Eisenstein series for symplectic and unitary groups. Mem. of Am. Math. Soc. Vol. 61, No. 346, 1986
2. Kähler, E.: Über eine bemerkenswerte Hermitesche Metrik. Abh. Math. Semin. Univ. Hamb. **9**, 173–186 (1933)
3. Matsushima, Y., Shimura, G.: On the cohomology groups attached to certain vector valued differential forms on the product of the upper half planes. Ann. Math. **78**, 417–449 (1963)
4. Shimura, G.: On some arithmetic properties of modular forms of one and several variables. Ann Math. **102**, 491–515 (1975)
5. Shimura, G.: Theta functions with complex multiplication. Duke Math. J. **43**, 673–696 (1976)
6. Shimura, G.: On the derivatives of theta functions and modular forms. Duke Math. J. **44**, 365–387 (1977)

7. Shimura, G.: The special values of the zeta functions associated with Hilbert modular forms. Duke Math. J. **45**, 637–679 (1978)
8. Shimura, G.: The arithmetic of certain zeta functions and automorphic forms on orthogonal groups. Ann. Math. **111**, 313–375 (1980)
9. Shimura, G.: On certain zeta functions attached to two Hilbert modular forms. I, II. Ann. Math. **114**, 127–164, 569–607 (1981)
10. Shimura, G.: Arithmetic of differential operators on symmetric domains. Duke Math. J. **48**, 813–843 (1981)
11. Shimura, G.: On Eisenstein series. Duke Math. J. **50**, 417–476 (1983)
12. Shimura, G.: Differential operators and the singular values of Eisenstein series. Duke Math. J. **51**, 261–329 (1984)
13. Shimura, G.: On differential operators attached to certain representations of classical groups. Invent. Math. **77**, 463–488 (1984)
14. Shimura, G.: On Eisenstein series of half-integral weight. Duke Math. J. **52**, 281–314 (1985)
15. Shimura, G.: On the Eisenstein series of Hilbert modular groups. Rev. Mat. Ibero-Am. **1**, N° 3, 1–42 (1985)
16. Shimura, G.: On a class of nearly holomorphic automorphic forms. Ann. Math. **123**, 347–406 (1986)
17. Weil, A.: Introduction à l'étude des variétés kählériennes. Act. Sci. et Ind. 1267. Paris: Hermann 1958

Received February 24, 1986

87b

On Hilbert modular forms of
half-integral weight

Duke Mathematical Journal, 55 (1987), 765-838

There are two main themes in this paper: (i) the relation between the Fourier coefficients of a Hilbert modular form of half-integral weight and those of a form of integral weight; (ii) the arithmeticity of critical values of a zeta function attached to two forms of half-integral weight. Our results will generalize those in the elliptic modular case obtained in our previous papers. To describe them more explicitly, let F, $\mathfrak{o}$, $\mathfrak{d}$, $\mathbf{a}$, and $\mathbf{f}$ denote throughout the paper a totally real algebraic number field of finite degree, the maximal order of F, the different of F over $\mathbf{Q}$, the set of archimedean primes of F, and the set of nonarchimedean primes of F, respectively. We put $G = \mathrm{SL}_2(F)$ and let G act on $\mathscr{H}^{\mathbf{a}}$ as usual, where

$$\mathscr{H} = \{ z \in \mathbf{C} | \mathrm{Im}(z) > 0 \}.$$

For two fractional ideals $\mathfrak{x}$ and $\mathfrak{y}$ of F such that $\mathfrak{x}\mathfrak{y} \subset \mathfrak{o}$, we put

$$\Gamma[\mathfrak{x}, \mathfrak{y}] = \{ \gamma \in G | a_\gamma \in \mathfrak{o}, b_\gamma \in \mathfrak{x}, c_\gamma \in \mathfrak{y}, d_\gamma \in \mathfrak{o} \},$$

where a_γ, b_γ, c_γ, and d_γ are the entries of γ in the standard order. By a *half-integral weight* we mean an element k of $(1/2)\mathbf{Z}^{\mathbf{a}}$ such that $2k_v$ is odd for all $v \in \mathbf{a}$; naturally an *integral weight* is an element of $\mathbf{Z}^{\mathbf{a}}$. For $\gamma \in G$, $z \in \mathscr{H}^{\mathbf{a}}$, and a weight k, we define a factor of automorphy J_k by

$$J_k(\gamma, z) = \begin{cases} \displaystyle\prod_{v \in \mathbf{a}} (c_v z_v + d_v)^{k_v} & (k \in \mathbf{Z}^{\mathbf{a}}), \\ h(\gamma, z) \displaystyle\prod_{v \in \mathbf{a}} (c_v z_v + d_v)^{k_v - (1/2)} & (k \notin \mathbf{Z}^{\mathbf{a}}), \end{cases}$$

where $(c, d) = (c_\gamma, d_\gamma)$, and $h(\gamma, z)$ is a factor of weight $1/2$ introduced in [S8]. It should be noted that $h(\gamma, z)$ is defined only for γ in a certain subset of G, but at least for $\gamma \in \Gamma[2\mathfrak{d}^{-1}, 2\mathfrak{d}]$. Then we denote by $\mathscr{M}_k$ the set of all holomorphic modular forms on $\mathscr{H}^{\mathbf{a}}$ of weight k with respect to congruence subgroups of G, defined as usual relative to J_k.

Let us now assume k to be half-integral and put $m_v = k_v - (1/2)$ for $v \in \mathbf{a}$. In parallel to the elliptic modular case, we choose a "level" which is an integral

Received November 14, 1986. Research supported by NSF Grant DMS-8401291.

ideal c in F divisible by 4, and a Hecke idele character ψ of F, and then denote by $\mathcal{M}_k(c, \psi)$ the subset of $\mathcal{M}_k$ consisting of the functions f such that

$$(*) \qquad f(\gamma z) = \psi_c(a_\gamma) J_k(\gamma, z) f(z) \quad \text{for every } \gamma \in \Gamma[2\mathfrak{d}^{-1}, 2^{-1} c \mathfrak{d}],$$

under the assumptions that ψ is unramified outside $c \cup \mathbf{a}$ and that $\psi_c(-1) = (-1)^{\{m\}}$, where $\psi_c = \prod_{v|c} \psi_v$ and $\{m\} = \sum_{v \in \mathbf{a}} m_v$. We can define for every prime ideal $\mathfrak{p}$ in F a Hecke operator $T_\mathfrak{p}$ acting on $\mathcal{M}_k(c, \psi)$, which corresponds to a constant multiple of $T(p^2)$ of [S1] when $\mathfrak{p} = p\mathbf{Z}$.

Given $f \in \mathcal{M}_k(c, \psi)$ and a fractional ideal $\mathfrak{a}$ in F, we take the Fourier expansion of f at a cusp corresponding to $\mathfrak{a}$, which has the form

$$\sum_\xi \lambda_f(\xi, \mathfrak{a}) \exp\left(\pi i \sum_{v \in \mathbf{a}} \xi_v z_v \right),$$

where ξ runs over $\mathfrak{a}^{-2}$. The coefficients $\lambda_f(\xi, \mathfrak{a})$ have the property

$$\lambda_f(\xi b^2, \mathfrak{a}) = b^m \psi_\mathbf{a}(b) \lambda_f(\xi, b\mathfrak{a}) \quad \text{for} \quad 0 \neq b \in F.$$

That we consider $\lambda_f(\xi, \mathfrak{a})$ with two parameters ξ and $\mathfrak{a}$ is one of the essential points of our method. Now our first set of main results concerns a certain correspondence $\mathcal{M}_k \to \mathcal{M}_{2m}$, and can be stated in the following (I–III):

(I) *Let $\tau \mathfrak{o} = \mathfrak{q}^2 \mathfrak{r}$ with a totally positive element τ of $\mathfrak{o}$, an integral ideal $\mathfrak{q}$, and a squarefree integral ideal $\mathfrak{r}$. Suppose $0 \neq f \in \mathcal{M}_k(c, \psi)$ and $f|T_\mathfrak{p} = \omega_\mathfrak{p} f$ with $\omega_\mathfrak{p} \in \mathbf{C}$ for every $\mathfrak{p}$. Then we have a formal equality*

$$\sum_\mathfrak{a} \lambda_f(\tau, \mathfrak{q}^{-1}\mathfrak{a}) M(\mathfrak{a}) \sum_\mathfrak{a} (\psi' \varepsilon_\tau^*)(\mathfrak{a}) N(\mathfrak{a})^{-1} M(\mathfrak{a})$$

$$= \lambda_f(\tau, \mathfrak{q}^{-1}) \prod_\mathfrak{p} \left[1 - \omega_\mathfrak{p} M(\mathfrak{p}) + \psi'(\mathfrak{p})^2 N(\mathfrak{p})^{-1} M(\mathfrak{p}^2) \right]^{-1},$$

where $M(\mathfrak{a})$ is a formal symbol such that $M(\mathfrak{a}\mathfrak{b}) = M(\mathfrak{a})M(\mathfrak{b})$, $\mathfrak{a}$ (resp. $\mathfrak{p}$) runs over all the integral (resp. prime) ideals of F, ε_τ^* is the Hecke ideal character of F corresponding to $F(\tau^{1/2})/F$, and ψ' is the ideal character modulo c attached to ψ (Theorem 5.5).

(II) *Suppose that f is a cusp form and $k_v \geq 3/2$ for every $v \in \mathbf{a}$. Then there exists a holomorphic modular form $\mathbf{g}_\tau$ of weight $2m$ and of level $2^{-1}c$ on $\mathrm{GL}_2(F_\mathbf{A})$ with character ψ^2 in the sense of [S3] such that*

$$\sum_\mathfrak{a} c(\mathfrak{a}, \mathbf{g}_\tau) M(\mathfrak{a}) = \sum_\mathfrak{a} \lambda_f(\tau, \mathfrak{q}^{-1}\mathfrak{a}) M(\mathfrak{a}) \sum_\mathfrak{a} (\psi' \varepsilon_\tau^*)(\mathfrak{a}) N(\mathfrak{a})^{-1} M(\mathfrak{a}),$$

where $c(\mathfrak{a}, \mathbf{g}_\tau)$ is the Fourier coefficient of $\mathbf{g}_\tau$ as defined in [S3] and $\sum_\mathfrak{a}$ is the same *as in* (I). *This holds without the assumption that* $f|T_{\mathfrak{p}} = \omega_{\mathfrak{p}} f$. *If* $f|T_{\mathfrak{p}} = \omega_{\mathfrak{p}} f$ *for every* $\mathfrak{p}$, *then* $\mathbf{g}_\tau \neq 0$ *for some* τ, *and* $\mathbf{g}_\tau$ *is an eigenform with eigenvalues* $\omega_{\mathfrak{p}}$ (*Theorem* 6.1, I *and* II).

(III) *The form* $\mathbf{g}_\tau$ *is a cusp form if* $k_v \neq 3/2$ *for some* v, *or if* $k_v = 3/2$ *for all* $v \in \mathbf{a}$ *and* f *is orthogonal to every holomorphic theta series of weight* k (*Theorem* 6.1, III).

A few remarks are in order at this point as to our formulation. The form f as above corresponds to an automorphic form $f_\mathbf{A}$ defined as a function on the metaplectic covering $M_\mathbf{A}$ of $\mathrm{SL}_2(F_\mathbf{A})$. Then $\lambda_f(\xi, \mathfrak{a})$ can be given as the Fourier coefficients of $f_\mathbf{A}$ in a more natural way. Now a modular form of weight $2m$ as a function on $\mathrm{GL}_2(F_\mathbf{A})$ corresponds to a set of κ elements $g_1, \ldots, g_\kappa$ of $\mathcal{M}_{2m}$, where κ is the class number of F modulo $\mathbf{a}$. Thus the above result may suggest, on the surface, the correspondence $f_\mathbf{A} \mapsto \mathbf{g}_\tau$, or $f \mapsto \{g_1, \ldots, g_\kappa\}$, but this is not a proper explanation. In fact, the Hecke character ψ satisfying (*) is not necessarily unique for f, and so the $\lambda_f(\xi, \mathfrak{a})$ depend not only on f, but also on ψ. Therefore our correspondence is more accurately given as $\{f, \psi\} \mapsto \{g_1, \ldots, g_\kappa\}$.

The above results require the condition that $k_v \geqq 3/2$ for every $v \in \mathbf{a}$. When this is not so, we have:

(IV) *Suppose* $k_w = 1/2$ *for some* $w \in \mathbf{a}$, *and* $\mathcal{M}_k \neq \{0\}$. *Then* $k_v = 1/2$ *or* $3/2$ *for every* $v \in \mathbf{a}$, *and* $\mathcal{M}_k$ *consists of theta series* (*Theorem* 6.4).

Our second theme concerns the arithmeticity of a Dirichlet series

$$(**) \qquad D(s; f, g) = \sum \lambda_f(\xi, \mathfrak{a}) \overline{\lambda_g(\xi, \mathfrak{a})} \, N(\xi \mathfrak{a}^2)^{-s} \prod_{v \in \mathbf{a}} \xi_v^{(1 - k_v - l_v)/2}.$$

Here f is a cusp form in $\mathcal{M}_k(\mathfrak{c}, \psi)$, $g \in \mathcal{M}_l(\mathfrak{c}', \varphi)$ with half-integral l, $\mathfrak{a}$ runs over a complete set of representatives of the ideal classes of F, and ξ over all totally positive elements of $\mathfrak{a}^{-2}$ modulo the squares of the units. For simplicity, let us assume here that both ψ and φ are of finite order, $(\psi/\varphi)_\mathbf{a}(x) = \mathrm{sgn}(x)^{k-l}$, and $k_v > 3/2$ for every $v \in \mathbf{a}$, though a more general case will be treated in the text. We take the product

$$\mathcal{D}(s; f, g) = D(s; f, g) \sum_{\mathfrak{m}} (\psi/\varphi)(\mathfrak{m}) N(\mathfrak{m})^{-1 - 2s},$$

where $\mathfrak{m}$ runs over all the integral ideals prime to $\mathfrak{cc}'$, and prove that $\mathcal{D}(s; f, g)$ multiplied by suitable gamma factors has a holomorphic continuation to the whole s-plane (Theorem 10.4). Now our main results of arithmeticity are given as follows:

(V) *Suppose that* $\lambda_f(\xi, \mathfrak{o})$ *and* $\lambda_g(\xi, \mathfrak{o})$ *are algebraic and* $f|T_{\mathfrak{p}} = \omega_{\mathfrak{p}} f$ *for almost all primes* $\mathfrak{p}$. *Let* $\mathbf{h}$ *be a primitive form of type* $(2^{-1}\mathfrak{c}, \psi^2)$ *in the sense of* (II) *belonging to the eigenvalues* $\omega_{\mathfrak{p}}$. (*Such a primitive form always exists.*) *Let* t *be an integer such that*

$$l_v - k_v < t + 1 \leqq k_v - l_v \quad and \quad t + 1 \equiv k_v - l_v \pmod{2} \ for \ every \ v \in \mathbf{a};$$

let $\mathfrak{g}(\varphi)$ *be the Gauss sum of* φ *and* $d = [F : \mathbf{Q}]$. *Then* $(\pi i)^{-td} \mathfrak{g}(\varphi) \mathcal{D}(t/2; f, g)$ *is an algebraic number times a "period" of* **h** *which is determined only by* **h** *and* $\psi_{\mathbf{a}}$. *Moreover, if we denote by* $\alpha(t; f, g)$ *this algebraic number, then*

$$\alpha(t; f, g)^{\sigma} = \alpha(t; f^{\sigma}, g^{\rho\sigma\rho})$$

for every $\sigma \in \mathrm{Gal}(\overline{\mathbf{Q}}/\mathbf{Q})$, *where* ρ *is complex conjugation (Theorem* 10.6).

(VI) *The notation and the assumptions being as in* (V), *suppose that* $k = l$ *and* g *is a cusp form. Then the inner product* $\langle g, f \rangle$ *modified by a well-defined constant factor is an algebraic number times a "period" of* **h** *as in* (V). *Moreover, this algebraic number has the same property under* $\mathrm{Gal}(\overline{\mathbf{Q}}/\mathbf{Q})$ *as* α *of* (V) *(Theorem* 10.5).

The results of types (I) and (II) were first proved in [S1] in the elliptic modular case. The correspondence as integral transforms by means of theta functions was given by Shintani in the opposite direction $\mathcal{M}_{2m} \to \mathcal{M}_k$ and then in the direction as in [S1] by Niwa, still in the elliptic modular case. We adapt some of Niwa's ideas in [N] to the present setting. The generalization to the case of an arbitrary global field was successively and successfully investigated by Gelbart, Piatetski–Shapiro, Flicker, and Waldspurger in terms of representations of GL_2 and metaplectic coverings of GL_2 or SL_2 (see [GP1], [F], and [Wa]; a survey of the mutual relationship of these can be found in [GP2]).

The present investigation is restricted to the holomorphic case, and so the mere existence of the correspondence $\mathcal{M}_k \to \mathcal{M}_{2m}$ is not the point of particular interest. Our emphasis is laid on the explicit description of the correspondence in terms of the Fourier coefficients of individual modular forms. As exemplified by a recent result of Waldspurger, the coefficients of forms of half-integral weight in the holomorphic case provide an arithmetical object of great interest, and it is our hope that the present formulation will bring about a clearer understanding of this subject. Besides, (I) and (II) may be viewed also as preliminaries to our second theme on arithmeticity, on which we now make a few comments.

The special cases of (V) and (VI) when $F = \mathbf{Q}$ were obtained in [S6]. As explained in that paper, there are four types of series similar to ($* *$) classified according to the weights of the two forms involved. The case in which both are of integral weight was treated in [S3]. The present paper deals with the case of two forms of half-integral weight. The two remaining cases concerning a form of integral weight and another of half-integral weight are being investigated by J. Im.

As for the results of types (III) and (IV) for the forms of weights $\leq 3/2$, they were conjectured in the elliptic modular case in [S1]. The assertion for weight $1/2$ in that special case was first proved by Serre and Stark in [SS]. Then the corresponding fact in terms of representations was given by Gelbart and Piatetski–Shapiro for cusp forms in a more general case, and by Flicker in the most general setting [F]. The last article contains an assertion of type (III) about the forms of weight $3/2$ in terms of representations; the elliptic modular case of the latter half of (III) was also proved by Sturm in [St]. Because of the difference in formulation, it is not clear whether (IV) and the latter half of (III) are easy consequences of the results of [F], though this may well be the case. Anyway, our proof is relatively short and relies on our results in [S9].

To conclude the introduction, we add that our investigation has been motivated not only by the intrinsic interest of the subject, but also by one particular application to the arithmeticity problem on the periods of certain automorphic forms, which we hope to treat in a subsequent paper.

1. Notation and preliminaries on metaplectic groups. In addition to the symbols F, $\mathfrak{o}$, $\mathfrak{d}$, $\mathbf{a}$, $\mathbf{f}$, and $\mathscr{H}$ defined in the introduction, we let D_F, $F_\mathbf{A}$, and $F_\mathbf{A}^\times$ denote throughout the paper the discriminant, the ring of adeles, and the group of ideles of F. For $v \in \mathbf{a} \cup \mathbf{f}$, we denote by F_v the v-completion of F. If $v \in \mathbf{a}$, we view v as an embedding of F into $\mathbf{R}$ (and thus $F_v = \mathbf{R}$) and denote by x_v the image of $x \in F$ under v. For $c \in F_\mathbf{A}^\times$ and a fractional ideal $\mathfrak{m}$ in F, we denote by $c\mathfrak{m}$ the fractional ideal such that $(c\mathfrak{m})_v = c_v \mathfrak{m}_v$ for every $v \in \mathbf{f}$, where $\mathfrak{m}_v$ denotes the closure of $\mathfrak{m}$ in F_v. For any set X, we denote by $X^\mathbf{a}$ the set of all indexed elements $(x_v)_{v \in \mathbf{a}}$ with $x_v \in X$. For example, $\mathbf{R}^\mathbf{a}$ can be identified with the archimedean factor of $F_\mathbf{A}$, which we write also $F_\mathbf{a}$. Naturally $F_\mathbf{f}$ denotes the nonarchimedean factor of $F_\mathbf{A}$.

By a *Hecke character* of F we understand a character of $F_\mathbf{A}^\times$ which is trivial on $F^\times$ and has values in

$$(1.1) \qquad \mathbf{T} = \{ z \in \mathbf{C} \mid |z| = 1 \}.$$

For such a character ψ we denote by ψ^* the ideal character attached to ψ, and make a convention that the *conductor* of ψ always refers to its finite part. We then put $\psi^*(\mathfrak{m}) = 0$ for every $\mathfrak{m}$ not prime to the conductor.

For two elements c and n of $\mathbf{C}^\mathbf{a}$, we put

$$(1.2a) \qquad c^n = \prod_{v \in \mathbf{a}} c_v^{n_v}$$

whenever each factor is well-defined (according to the context), and

$$(1.2b) \qquad \{n\} = \sum_{v \in \mathbf{a}} n_v.$$

The symbol $\{n\}$ will appear only as an exponent, and so there will be no fear of confusion. The identity element of the ring $\mathbf{C}^\mathbf{a}$, particularly when it appears as an exponent, is denoted by u. Thus, for example, $c^{su} = \prod_{v \in \mathbf{a}} c_v^s$ and $(su)^n = s^{\{n\}}$ for $s \in \mathbf{C}$. For x, $y \in \mathbf{R}^\mathbf{a}$, we write $x \geq y$ and also $y \leq x$ if $x_v \geq y_v$ for all $v \in \mathbf{a}$; we write $x \gg 0$ and also $0 \ll x$ for $x \in \mathbf{R}^\mathbf{a}$ or $x \in F$ if $x_v > 0$ for all $v \in \mathbf{a}$.

Now we put

$$G = \mathrm{SL}_2(F), \qquad \tilde{G} = \mathrm{GL}_2(F),$$

$$G_v = \mathrm{SL}_2(F_v), \qquad \tilde{G}_v = \mathrm{GL}_2(F_v) \qquad (v \in \mathbf{a} \cup \mathbf{f}).$$

We consider the adelization $G_\mathbf{A}$ of G and denote by $G_\mathbf{a}$ and $G_\mathbf{f}$ its archimedean and nonarchimedean factors; $\tilde{G}_\mathbf{A}$, $\tilde{G}_\mathbf{a}$, and $\tilde{G}_\mathbf{f}$ are defined similarly. For x in $\tilde{G}_\mathbf{A}$ or $F_\mathbf{A}$, $x_\mathbf{a}$ and $x_\mathbf{f}$ denote its archimedean and nonarchimedean parts. We put then

$$\mathrm{GL}_2^+(\mathbf{R}) = \left\{ x \in \mathrm{GL}_2(\mathbf{R}) \mid \det(x) > 0 \right\},$$

$$\tilde{G}_{\mathbf{a}+} = \prod_{v \in \mathbf{a}} \mathrm{GL}_2^+(F_v), \qquad \tilde{G}_{\mathbf{A}+} = \tilde{G}_{\mathbf{a}+}\tilde{G}_\mathbf{f}, \qquad \tilde{G}_+ = \tilde{G} \cap \tilde{G}_{\mathbf{A}+}.$$

Our theory requires also the metaplectic groups $\mathrm{Mp}(F_\mathbf{A})$ and $\mathrm{Mp}(F_v)$ of Weil [We] with respect to the alternating form $(x, y) \mapsto x\iota \cdot {}^t y$ on F^2, where

$$(1.3) \qquad\qquad \iota = \begin{pmatrix} 0 & -1 \\ 1 & 0 \end{pmatrix}.$$

We put $M_\mathbf{A} = \mathrm{Mp}(F_\mathbf{A})$ and $M_v = \mathrm{Mp}(F_v)$ for simplicity. We recall that $M_\mathbf{A}$ is a group of unitary operators on $L^2(F_\mathbf{A})$, which is an extension of $G_\mathbf{A}$ with kernel $\mathbf{T}$; the same is true with the subscript v in place of $\mathbf{A}$. The projection map of $M_\mathbf{A}$ to $G_\mathbf{A}$ and that of M_v to G_v are denoted by pr.

Given a 2×2 matrix $\alpha = \begin{pmatrix} a & b \\ c & d \end{pmatrix}$, we write $a = a_\alpha$, $b = b_\alpha$, $c = c_\alpha$, and $d = d_\alpha$ whenever there is no fear of confusion. For $\alpha \in \mathrm{GL}_2^+(\mathbf{R})$ and $z \in \mathscr{H}$, we define $\alpha(z)$ and a factor of automorphy j by

$$(1.4)$$

$$\alpha(z) = \alpha z = (a_\alpha z + b_\alpha)/(c_\alpha z + d_\alpha), \qquad j(\alpha, z) = \det(\alpha)^{-1/2}(c_\alpha z + d_\alpha).$$

For $x \in \tilde{G}_{\mathbf{A}+}$, $\xi \in M_\mathbf{A}$, and $z \in \mathscr{H}^\mathbf{a}$, we put

$$(1.5a) \quad j(x, z) = (j_v(x, z))_{v \in \mathbf{a}}, \qquad j(\xi, z) = (j_v(\xi, z))_{v \in \mathbf{a}} \quad (\in \mathbf{C}^\mathbf{a}),$$

$$(1.5b) \qquad\qquad j_v(x, z) = j(x_v, z_v), \qquad j_v(\xi, z) = j(\mathrm{pr}(\xi)_v, z_v),$$

and let $x_\mathbf{a}$ act on $\mathscr{H}^\mathbf{a}$ componentwise; we then define the action of x (resp. ξ) on $\mathscr{H}^\mathbf{a}$ to be the same as that of $x_\mathbf{a}$ (resp. $\mathrm{pr}(\xi)_\mathbf{a}$). Further, we put $a_\xi = a_y$, $b_\xi = b_y$, $c_\xi = c_y$, and $d_\xi = d_y$ when $y = \mathrm{pr}(\xi)$.

For two fractional ideals $\mathfrak{x}$ and $\mathfrak{y}$ in F such that $\mathfrak{x}\mathfrak{y} \subset \mathfrak{o}$, we put

$$\tilde{D}[\mathfrak{x}, \mathfrak{y}] = \mathrm{SO}(2)^\mathbf{a} \cdot \prod_{v \in \mathbf{f}} \tilde{D}_v[\mathfrak{x}, \mathfrak{y}] \quad (\subset \tilde{G}_\mathbf{A}),$$

$$\tilde{D}_v[\mathfrak{x}, \mathfrak{y}] = \left\{ x \in \tilde{G}_v \mid a_x \in \mathfrak{o}_v, \, b_x \in \mathfrak{x}_v, \, c_x \in \mathfrak{y}_v, \, d_x \in \mathfrak{o}_v, \, |\det(x)|_v = 1 \right\},$$

$$D[\mathfrak{x}, \mathfrak{y}] = G_\mathbf{A} \cap \tilde{D}[\mathfrak{x}, \mathfrak{y}], \qquad D_v[\mathfrak{x}, \mathfrak{y}] = G_v \cap \tilde{D}_v[\mathfrak{x}, \mathfrak{y}],$$

$$\tilde{\Gamma}[\mathfrak{x}, \mathfrak{y}] = \tilde{G} \cap (\tilde{G}_{\mathbf{a}+}\tilde{D}[\mathfrak{x}, \mathfrak{y}]), \qquad \Gamma[\mathfrak{x}, \mathfrak{y}] = G \cap \tilde{\Gamma}[\mathfrak{x}, \mathfrak{y}].$$

We define a map $e: \mathbf{C} \to \mathbf{C}$ and characters $\mathbf{e}_\mathbf{A}$ and $\mathbf{e}_v$ of $F_\mathbf{A}$ by $\mathbf{e}(z) = e^{2\pi i z}$ for $z \in \mathbf{C}$, $\mathbf{e}_\mathbf{A}(x) = \prod_{v \in \mathbf{a} \cup \mathbf{f}} \mathbf{e}_v(x)$ for $x \in F_\mathbf{A}$, $\mathbf{e}_v(x) = \mathbf{e}(x_v)$ for $v \in \mathbf{a}$, and $\mathbf{e}_v(x) = \mathbf{e}(-y)$ for $v \in \mathbf{f}$, where $y \in \bigcap_{q \neq p}(\mathbf{Z}_q \cap \mathbf{Q})$, $y - \mathrm{Tr}_{F_v/\mathbf{Q}_p}(x_v) \in \mathbf{Z}_p$, $v | p$. We then put $\mathbf{e}_\mathbf{a}(x) = \mathbf{e}_\mathbf{A}(x_\mathbf{a})$ and $\mathbf{e}_\mathbf{f}(x) = \mathbf{e}_\mathbf{A}(x_\mathbf{f})$; we employ the symbol $\mathbf{e}_\mathbf{a}$ also for the function on $\mathbf{C}^\mathbf{a}$ defined by $\mathbf{e}_\mathbf{a}(z) = \prod_{v \in \mathbf{a}} \mathbf{e}(z_v)$.

In order to recall some basic properties of the action of $M_\mathbf{A}$ on $L^2(F_\mathbf{A})$, we first put

$$(1.6a) \qquad P_\mathbf{A} = \{w \in G_\mathbf{A} | c_w = 0\}, \qquad P_v = \{w \in G_v | c_w = 0\},$$

$$(1.6b) \qquad \Omega_\mathbf{A} = \{w \in G_\mathbf{A} | c_w \in F_\mathbf{A}^\times\}, \qquad \Omega_v = \{w \in G_v | c_w \neq 0\},$$

$$(1.6c) \qquad P = P_\mathbf{A} \cap G, \qquad \Omega = \Omega_\mathbf{A} \cap G = P\iota P.$$

To each $x \in G_\mathbf{A}$ or $\xi \in M_\mathbf{A}$, we can assign a fractional ideal $\mathfrak{a}_x$ or $\mathfrak{a}_\xi$ by

$$(1.7) \quad (\mathfrak{a}_x)_v = (c_x)_v \mathfrak{d}_v^{-1} + (d_x)_v \mathfrak{o}_v \quad \text{for every } v \in \mathbf{f}; \qquad \mathfrak{a}_\xi = \mathfrak{a}_{\mathrm{pr}(\xi)}.$$

If $x = py$ with $p \in P_\mathbf{A}$ and $y \in D[\mathfrak{d}^{-1}, \mathfrak{d}]$, then $\mathfrak{a}_x = d_p \mathfrak{o}$. (This ideal $\mathfrak{a}_x$ is the same as $\mathrm{il}(x)$ of [S8].)

Now there are three liftings:

$$(1.8) \qquad r: G \to M_\mathbf{A}, \qquad r_P: P_\mathbf{A} \to M_\mathbf{A}, \qquad r_\Omega: \Omega_\mathbf{A} \to M_\mathbf{A}.$$

We always view G as a subgroup of $M_\mathbf{A}$ through the map r. The action of $r_P(w)$ or $r_\Omega(w)$ on $f \in L^2(F_\mathbf{A})$ is given by

$$(1.9) \qquad [r_P(w)f](x) = |a_w|_\mathbf{A}^{1/2} e_\mathbf{A}(a_w b_w x^2/2) f(a_w x) \qquad (w \in P_\mathbf{A}),$$

$(1.10a)$

$$[r_\Omega(w)f](x) = |c_w|_\mathbf{A}^{1/2} \int_{F_\mathbf{A}} f(a_w x + c_w y) e_\mathbf{A}(q_w(x, y)) \, dy \qquad (w \in \Omega_\mathbf{A}),$$

$(1.10b)$

$$q_w(x, y) = (1/2)(a_w b_w x^2 + 2b_w c_w xy + c_w d_w y^2) \qquad (w \in G_\mathbf{A}; \; x, y \in F_\mathbf{A}).$$

Here $dy = \prod_v dy_v$, dy_v for $v \in \mathbf{a}$ is the standard measure on $\mathbf{R}$, and if $v \in \mathbf{f}$, dy_v is such that the measure of $\mathfrak{o}_v$ is $N(\mathfrak{d}_v)^{-1/2}$. The maps of (1.8) are consistent in the sense that $r = r_P$ on P, $r = r_\Omega$ on Ω, and

$$(1.11) \qquad r_\Omega(\alpha\beta\gamma) = r_P(\alpha)r_\Omega(\beta)r_P(\gamma) \quad \text{for } \alpha, \gamma \in P_\mathbf{A} \text{ and } \beta \in \Omega_\mathbf{A}.$$

There are also similar maps of P_v and Ω_v into M_v satisfying (1.9), $(1.10a, b)$, and (1.11) with the subscript v in place of $\mathbf{A}$.

Suppose $f \in L^2(F_\mathbf{A})$ is given in the form $f(x) = \prod_v f_v(x_v)$. In general, there is no canonical decomposition of the action of an element ξ of $M_\mathbf{A}$ into the product of its action on the local factors. However, if $\xi \in r_P(P_{\mathbf{A}})$ or $\xi \in r_\Omega(\Omega_{\mathbf{A}})$, then formulas (1.9), (1.10a), and their local versions guarantee such a decomposition. We state this fact symbolically as follows:

$$(1.12) \qquad \left[r_*(w) \prod_v f_v \right](x) = \prod_v [r_*(w_v)f_v](x_v) \qquad (w \in P_{\mathbf{A}} \text{ or } \Omega_{\mathbf{A}}).$$

Another simple fact should be noted:

$$(1.13) \qquad \textit{If } \xi, \eta \in M_{\mathbf{A}}, \mathrm{pr}(\xi) \in G_{\mathbf{a}}, \textit{ and } \mathrm{pr}(\eta) \in G_{\mathfrak{f}}, \textit{ then } \xi\eta = \eta\xi.$$

2. Factors of automorphy and Hermite polynomials.   Several types of theta functions will appear in our theory. One type involves the Hermite polynomials $H_n(x)$ defined by

$$(2.1) \quad H_n(x) = (-1)^n \exp(x^2/2)(d/dx)^n \exp(-x^2/2) \qquad (0 \leqq n \in \mathbf{Z}).$$

We list here some formulas concerning H_n:

$$(2.2) \qquad\qquad\qquad H_0(x) = 1, \qquad H_1(x) = x,$$

$$(2.3) \qquad\qquad\qquad H_n(-x) = (-1)^n H_n(x),$$

$$(2.4) \qquad\qquad\qquad H_{n+1}(x) = xH_n(x) - H_n'(x),$$

$$(2.5) \qquad\qquad\qquad (x + iy)^n = \sum_{k=0}^{n} \binom{n}{k} i^k H_k(y) H_{n-k}(x),$$

$$(2.6) \qquad \int_{\mathbf{R}} \mathrm{e}(ix^2/2) H_n(\sqrt{4\pi}\,x) \mathrm{e}(-xy)\, dx = (-i)^n \mathrm{e}(iy^2/2) H_n(\sqrt{4\pi}\,y),$$

$$(2.7) \qquad\qquad \sqrt{2} \int_{\mathbf{R}} H_n(\sqrt{4\pi}\,t) \mathrm{e}(it^2) \mathrm{e}(tx)\, dt = (i\sqrt{\pi}\,x)^n \mathrm{e}(ix^2/4),$$

$$(2.8) \quad (\sqrt{2\pi}\,i)^n \int_{\mathbf{R}} \exp(-\pi t^2 y^{-1}|cz + u|^2)(c\bar{z} + u)^n \mathrm{e}(-uv)\, du$$

$$= (\sqrt{y}/t)^{n+1} H_n(\sqrt{2\pi y}\,(ct + vt^{-1})) \mathrm{e}(cxv + (iy/2)(c^2 t^2 + v^2 t^{-2}))$$

$$(y = \mathrm{Im}(z) > 0, c \in \mathbf{R}, v \in \mathbf{R}, 0 < t \in \mathbf{R}),$$

$$(2.9) \quad \int_0^\infty H_n(\sqrt{4\pi y}\,a) \mathrm{e}(ia^2 y) y^{(s/2)-1}\, dy$$

$$= (s - 1)(s - 2) \cdots (s - n)\Gamma((s - n)/2) 2^{-n/2}(2\pi)^{-s/2} a^{-s}$$

$$(\mathrm{Re}(s) > n, a > 0).$$

These are well known except, possibly, (2.5), (2.8), and (2.9); (2.5) follows from $z^n e^{-z\bar{z}/2} = (-\partial/\partial x - i\,\partial/\partial y)^n e^{-z\bar{z}/2}$. If $n = 0$, (2.8) is merely the self-reciprocity of $\mathbf{e}(iu^2/2)$ combined with change of variables; the general case can be obtained by successive applications of $2\pi i c\bar{z} - \partial/\partial v$ in view of (2.4). Formula (2.9) is obtained by applying d^n/da^n to the case $n = 0$.

We now fix one v in $\mathbf{a}$ and examine the effect of an element of $M_v = \mathrm{Mp}(\mathbf{R})$ on the functions

$$(2.10) \quad \varphi_n(t, z) = y^{-n/2} H_n\big(\sqrt{4\pi y}\, t\big)\mathbf{e}(t^2 z/2) \qquad (t \in \mathbf{R},\ y = \mathrm{Im}(z) > 0).$$

For a fixed z, this belongs to $L^2(\mathbf{R})$ as a function of t, and so the action of $\sigma \in \mathrm{Mp}(\mathbf{R})$ on it is meaningful. To study this action, we need a unitary operator $U(u, u'; \xi)$ on $L^2(\mathbf{R})$ defined for $(u, u') \in \mathbf{R}^2$ and $\xi \in \mathbf{T}$ by

$$[U(u, u'; \xi)f](x) = \xi\mathbf{e}(u'x)f(x + u) \qquad (x \in \mathbf{R})$$

(see Weil [We, p. 149]). Now Bargman [B] showed that if we define a function Bf on $\mathbf{C}$ for $f \in L^2(\mathbf{R})$ by

$$(Bf)(z) = \exp(\pi z^2/2)\int_{\mathbf{R}} \exp(-\pi x^2)\mathbf{e}(xz)f(x)\, dx \qquad (z \in \mathbf{C}),$$

then B gives a unitary isomorphism of $L^2(\mathbf{R})$ onto the space $\mathfrak{H}(\mathbf{C})$ consisting of all holomorphic functions g on $\mathbf{C}$ such that

$$\int_{\mathbf{C}} |g(z)|^2 \exp(-\pi|z|^2)|dz \wedge d\bar{z}| < \infty.$$

LEMMA 2.1. *If $\tau \in \mathrm{Mp}(\mathbf{R})$ and*

$$\mathrm{pr}(\tau) = \begin{bmatrix} \cos\theta & \sin\theta \\ -\sin\theta & \cos\theta \end{bmatrix}$$

with $\theta \in \mathbf{R}$, then $\tau\varphi_n(t, i) = \varepsilon_\tau e^{in\theta}\varphi_n(t, i)$ with an element ε_τ of $\mathbf{T}$ independent of n.

Proof. For $\zeta = e^{i\theta}$, define unitary operators $C(\zeta)$ on $L^2(\mathbf{R})$ and $C'(\zeta)$ on $\mathfrak{H}(\mathbf{C})$ by $C(\zeta) = B^{-1}C'(\zeta)B$ and $[C'(\zeta)g](z) = g(\zeta z)$ for $g \in \mathfrak{H}(\mathbf{C})$. Then a direct (and somewhat tedious) calculation shows that

$$C(\zeta)^{-1}U(u, u'; \xi)C(\zeta) = U\big((u, u')\rho, \xi\mathbf{e}(q_\rho(u, u'))\big)$$

with $\rho = \mathrm{pr}(\tau)$ and q_ρ of (1.10b) (cf. [B, §3] and Igusa [I, p. 36]). This means that $C(\zeta) \in \mathrm{Mp}(\mathbf{R})$ and $\mathrm{pr}(C(\zeta)) = \rho$ (see [We, p. 157, p. 183]). Therefore $\tau = \varepsilon C(\zeta)$ with $\varepsilon \in \mathbf{T}$. Let $\eta_n(t) = \varphi_n(t, i)$. Then $(B\eta_n)(z) = 2^{-1/2}(i\pi^{1/2}z)^n$, since the equality holds for real z by virtue of (2.7). Obviously $C'(\zeta)$ maps z^n to $\zeta^n z^n$, and hence $\tau\eta_n = \varepsilon C(\zeta)\eta_n = \varepsilon\zeta^n\eta_n$, which proves our lemma.

LEMMA 2.2. *For every* $\sigma \in \mathrm{Mp}(\mathbf{R})$ *and* $0 \leq n \in \mathbf{Z}$, *there exists a function* $p_n(\sigma, z)$ *holomorphic in* $z \in \mathscr{H}$ *with the following properties:*
- (i) $(\sigma\varphi_n)(t, z) = p_n(\sigma, z)^{-1}\varphi_n(t, \sigma z)$,
- (ii) $p_n(\sigma, z) = p_0(\sigma, z)j(\sigma, z)^n$,
- (iii) $p_0(r_p(w), z) = |d_w|^{1/2}$, $p_0(r_\Omega(\iota), z) = (-iz)^{1/2}$,
- (iv) $p_0(\sigma, z)^2 = \zeta_\sigma j(\sigma, z)$ *with* $\zeta_\sigma \in \mathbf{T}$.

Proof. This was proved in [S8, Proposition 3.1] for $n = 0$. In fact, $\varphi_0(t, z)$ coincides with $\varphi(t; z, 0)$ there, and so the function h_σ there gives $p_0(\sigma, z)$. As for $n > 0$, if $\sigma = r_p(w)$ with $w \in P_v$, then (1.9) shows (i) with $p_n(\sigma, z) = |d_w|^{1/2}d_w^n$. Since $\mathrm{Mp}(\mathbf{R})$ is generated by $r_p(P_v)$, $\mathbf{T}$, and $r_\Omega(\iota)$, it is sufficient to establish $p_n(r_\Omega(\iota), z)$ for $n > 0$. For this purpose, let $z = \alpha(i)$ and $\iota(z) = \beta(i)$ with $\alpha \in P_v$ and $\beta \in P_v$. Put $r_\Omega(\iota)r_p(\alpha) = r_p(\beta)\tau$. Then

$$\mathrm{pr}(\tau) = \begin{bmatrix} \cos\theta & \sin\theta \\ -\sin\theta & \cos\theta \end{bmatrix}$$

with $\theta \in \mathbf{R}$. Putting $s_w = |d_w|^{1/2}$ for $w \in P_v$, we have, by Lemma 2.1,

$$r_\Omega(\iota)\varphi_n(t, z) = r_\Omega(\iota)\left[s_\alpha d_\alpha^n r_p(\alpha)\varphi_n(t, i)\right]$$

$$= s_\alpha d_\alpha^n r_p(\beta)\tau\varphi_n(t, i) = \left(\varepsilon_\tau s_\alpha/s_\beta\right)\left(e^{i\theta}d_\alpha/d_\beta\right)^n\varphi_n(t, \iota z).$$

Observe that $e^{i\theta}d_\alpha/d_\beta = z^{-1} = j(\iota, z)^{-1}$. Thus we obtain

$$r_\Omega(\iota)\varphi_n(t, z) = \left(\varepsilon_\tau s_\alpha/s_\beta\right)j(\iota, z)^{-n}\varphi_n(t, \iota z).$$

Putting $n = 0$, we obtain $\varepsilon_\tau s_\alpha/s_\beta = p_0(r_\Omega(\iota), z)^{-1}$, which establishes $p_n(r_\Omega(\iota), z)$ as desired.

We now consider functions on $L^2(F_\mathbf{A})$ and prove a global version of the above lemma. Define a subgroup C'' and an element ε of $G_\mathbf{A}$ by

$$(2.11) \qquad C'' = D[2\mathfrak{d}^{-1}, 2\mathfrak{d}] \cup D[2\mathfrak{d}^{-1}, 2\mathfrak{d}]\varepsilon,$$

$$(2.12) \qquad \varepsilon_v = \begin{cases} 1 & \text{if } v \in \mathbf{a}, \\ \begin{bmatrix} 0 & -\delta_v^{-1} \\ \delta_v & 0 \end{bmatrix} & \text{if } v \in \mathbf{f}, \end{cases}$$

Here and henceforth we fix an element δ of $F_\mathbf{f}^\times$ such that $\mathfrak{d}_v = \delta_v \mathfrak{o}_v$.

Let $\mathscr{S}(F_\mathbf{f})$ denote the Schwartz–Bruhat space of $F_\mathbf{f}$. Through the natural injection of F into $F_\mathbf{f}$, we can "restrict" an element of $\mathscr{S}(F_\mathbf{f})$ to F. We then easily see that this restriction gives an isomorphism of $\mathscr{S}(F_{\mathfrak{f}})$ onto the space $\mathscr{L}(F)$ of locally constant functions on F as defined in [S8, p. 283]. For every $\eta \in \mathscr{S}(F_{\mathfrak{f}})$, $0 \leq n \in \mathbf{Z}^{\mathbf{a}}$, and $z \in \mathscr{H}^{\mathbf{a}}$, put

$$(2.13) \qquad \eta_n(t, z) = \eta(t_{\mathfrak{f}}) \prod_{v \in \mathbf{a}} \varphi_{n_v}(t_v, z_v) \qquad (t \in F_{\mathbf{A}}).$$

Then we can let an element of $M_{\mathbf{A}}$ act on η_n.

PROPOSITION 2.3. *Let* $\omega = \prod_{v \in \mathfrak{f}} \omega_v \in \mathscr{S}(F_{\mathfrak{f}})$ *with the characteristic function of* $\mathfrak{o}_v$ *as* ω_v. *For every* $\tau \in \mathrm{pr}^{-1}(P_{\mathbf{A}}C'')$ *there is a nonvanishing holomorphic function* $h(\tau, z)$ *on* $\mathscr{H}^{\mathbf{a}}$ *with the following properties:*

(i) $[\tau\omega_0(t, z)]_{t=0} = N(\mathfrak{a}_\tau)^{1/2} h(\tau, z)^{-1}$, *where* $\mathfrak{a}_\tau$ *is defined by* (1.7);

(ii) $h(\tau, z)^2 = \zeta_\tau j(\tau, z)$ *with* $\zeta_\tau \in \mathbf{T}$;

(iii) $h(\sigma\tau\rho, z) = h(\sigma, z) h(\tau, \rho z) h(\rho, z)$ *if* $\mathrm{pr}(\sigma) \in P_{\mathbf{A}}$, $\mathrm{pr}(\tau) \in P_{\mathbf{A}}C''$, *and* $\mathrm{pr}(\rho) \in C''G_{\mathbf{a}}$;

(iv) $h(\zeta r_P(w), z) = \zeta^{-1} |(d_w)_{\mathbf{a}}|^{u/2}$ *for* $\zeta \in \mathbf{T}$ *and* $w \in P_{\mathbf{A}}$.

This was given in [S8, Proposition 3.2 and (3.13)].

PROPOSITION 2.4. *For* $\tau \in \mathrm{pr}^{-1}(P_{\mathbf{A}}C'')$ *and* $\eta \in \mathscr{S}(F_{\mathfrak{f}})$, *there is a unique element* $^\tau\eta$ *of* $\mathscr{S}(F_{\mathfrak{f}})$ *such that*

$$\tau\eta_n(t, z) = h(\tau, z)^{-1} j(\tau, z)^{-n} (^\tau\eta)_n(t, \tau z)$$

for every $n \in \mathbf{Z}^{\mathbf{a}}$, ≥ 0, *where* j^{-n} *is understood in the sense of* (1.2a).

Proof. Let $\sigma = \mathrm{pr}(\tau)$ and $S = \mathbf{a} \cup \{v \in \mathfrak{f} \,|\, \sigma_v \notin C''\}$. Take $\alpha_v \in M_v$ so that $\mathrm{pr}(\alpha_v) = \sigma_v$ for $v \in S$. Let

$$(2.14) \qquad r_v \colon B_0(F_v, \mathfrak{o}_v) \to M_v$$

denote the lift defined for $v \in \mathfrak{f}$ in [We, n°19, n°36]. Observing that $C_v'' \subset B_0(F_v, \mathfrak{o}_v)$, we can find an element β of $M_{\mathbf{A}}$ such that

$$\beta\left(\prod_v f_v\right) = \prod_{v \in S} \alpha_v f_v \prod_{v \notin S} r_v(\sigma_v) f_v$$

(see [We, n°38]). Then $\tau = \zeta\beta$ with $\zeta \in \mathbf{T}$, and hence $\tau\eta_n = \zeta\beta\eta_n = \zeta\eta'\prod_{v \in \mathbf{a}} \alpha_v \varphi_{n_v}$ with an element η' of $\mathscr{S}(F_{\mathfrak{f}})$ which is independent of n. By Lemma 2.2, we have

$$\tau\eta_n(t, z) = j(\tau, z)^{-n} \prod_{v \in \mathbf{a}} p_0(\alpha_v, z_v)^{-1} (\zeta\eta')_n(t, \tau z).$$

By Proposition 2.3(ii), and Lemma 2.2(iv), we have $h(\tau, z) = \zeta' \prod_{v \in \mathbf{a}} p_0(\alpha_v, z_v)$ with $\zeta' \in \mathbf{T}$. Then we obtain the desired formula with $^\tau\eta = \zeta\zeta'\eta'$.

It should be noted that the action of $\mathrm{pr}^{-1}(P_{\mathbf{A}}C'')$ on $\mathscr{S}(F_{\mathbf{f}})$ given by $(\tau, \eta) \mapsto {}^{\tau}\eta$ is only partially associative in the sense that

$$(2.15) \qquad {}^{(\sigma\tau\rho)}\eta = {}^{\sigma}\left({}^{\tau}({}^{\rho}\eta)\right) \qquad \textit{for } \sigma, \tau, \textit{ and } \rho \textit{ as in Proposition 2.3(iii).}$$

From the above proof we can easily derive:

$$(2.16\mathrm{a}) \qquad {}^{\tau}\eta = \eta \quad \textit{if } \mathrm{pr}(\tau) \in G_{\mathbf{a}}, \textit{ in particular, if } \tau \in \mathbf{T};$$

$$(2.16\mathrm{b}) \qquad \textit{If } \mathrm{pr}(\tau) = \alpha\beta \textit{ with } \alpha \in P_{\mathbf{A}} \textit{ and } \beta \in C'', \textit{ then}$$

$$ {}^{\tau}\left(\prod_{v \in \mathbf{f}} \eta_v\right) = \prod_{v \in \mathbf{f}} r_P(\alpha_v) r_v(\beta_v) \eta_v;$$

$$(2.16\mathrm{c}) \quad {}^{\tau}\left(\prod_{v \in \mathbf{f}} \eta_v\right) = \prod_{v \in \mathbf{f}} \left(r_{\Omega}(\sigma_v)\eta_v\right) \quad \textit{if } \tau = r_{\Omega}(\sigma) \textit{ with } \sigma \in \Omega \cap P_{\mathbf{A}}C'';$$

$$(2.16\mathrm{d})$$

$$\left\{\tau \in \mathrm{pr}^{-1}(P_{\mathbf{A}}C'') \,|\, {}^{\tau}\eta = \eta\right\} \textit{ is an open subgroup of } M_{\mathbf{A}} \textit{ for every } \eta \in \mathscr{S}(F_{\mathbf{f}}).$$

For $\lambda \in F_{\mathbf{A}}$ and $\mu \in F_v$, we define their Gauss sums by

$$(2.17\mathrm{a}) \qquad \gamma(\lambda) = \prod_{v \in \mathbf{f}} \gamma_v(\lambda_v) \qquad (\lambda \in F_{\mathbf{A}}),$$

$$(2.17\mathrm{b}) \qquad \gamma_v(\mu) = N(\mathfrak{d}_v)^{1/2} \int_{\mathfrak{o}_v} \mathbf{e}_v(\mu x^2/2)\, dx \qquad (\mu \in F_v, v \in \mathbf{f}).$$

Then we have, as shown in [S8, Lemmas 3.4 and 3.5],

$$(2.18\mathrm{a}) \qquad \lim_{\rho \to \infty} h(\xi, \rho\mathrm{i})/|h(\xi, \rho\mathrm{i})| = \gamma/|\gamma| \quad \text{with } \gamma = \gamma\left(-c_{\xi}^{-1} d_{\xi}\right)$$

$$\textit{if } \xi \in r_{\Omega}(\Omega_{\mathbf{A}} \cap P_{\mathbf{A}}C''),$$

$$(2.18\mathrm{b}) \qquad \lim_{\rho \to 0} h(\xi, \rho\mathrm{i})/|h(\xi, \rho\mathrm{i})| = \gamma/|\gamma| \quad \text{with } \gamma = \gamma\left(\delta^{-2} d_{\xi}^{-1} c_{\xi}\right)$$

$$\textit{if } \xi \in r_{\Omega}(\Omega_{\mathbf{A}}) r_{\Omega}(\iota) \cap \mathrm{pr}^{-1}(P_{\mathbf{A}}C''),$$

$$(2.19\mathrm{b}) \qquad h(\alpha, z)^2 = \mathrm{sgn}\left(N_{F/\mathbf{Q}}(d_{\alpha})\right) \theta^*\left(d_{\alpha}\mathfrak{a}_{\alpha}^{-1}\right) j(\alpha, z)^u$$

$$\textit{if } \alpha \in G \cap P_{\mathbf{A}} D[2\mathfrak{d}^{-1}, 2\mathfrak{d}].$$

Here $\delta = (\delta_v)_{v \in \mathbf{f}}$ with δ_v as in (2.12), θ is the Hecke character of F corresponding to $F((-1)^{1/2})/F$, and $\mathbf{i}$ is the origin of $\mathscr{H}^{\mathbf{a}}$ given by

$$(2.20) \qquad \mathbf{i} = (i, \ldots, i) = iu \in \mathscr{H}^{\mathbf{a}}.$$

LEMMA 2.5. *Let* $\sigma = r_P(w)$ *with* $w = \operatorname{diag}[q, q^{-1}]$, $q \in F_{\mathbf{f}}^{\times}$, *and let* $U = \operatorname{pr}^{-1}(D[2\mathfrak{d}^{-1}, 2\mathfrak{d}]G_{\mathbf{a}})$. *If* $x, y \in U$ *and* $x\sigma = \sigma y$, *then* $h(x, z) = h(y, z)$.

Proof. Write $x = x_1 x_2$ with $\operatorname{pr}(x)_{\mathbf{a}} = \operatorname{pr}(x_2)$. Then $\operatorname{pr}(x_1) \in G_{\mathbf{f}}$. In view of Proposition 2.3(iii), it is sufficient to prove $h(x_1, z) = h(yx_2^{-1}, z)$. Now, by (1.13), $\sigma y x_2^{-1} = x_1 \sigma$. This means that our problem can be reduced to the case where both x and y belong to $\operatorname{pr}^{-1}(G_{\mathbf{f}})$. Thus our task is to show that $h(x, \mathbf{i}) = h(y, \mathbf{i})$ for such x and y. Since $x \in U$, we have $x(\varphi_{\mathbf{a}} \prod_{v \in \mathbf{f}} \varphi_v) = t\varphi_{\mathbf{a}} \prod_{v \in \mathbf{f}} r_v(x_v)\varphi_v$ with r_v of (2.14), suitable $x_v \in D_v[2\mathfrak{d}^{-1}, 2\mathfrak{d}]$, and $t \in \mathbf{T}$. Then

$$y\left(\varphi_{\mathbf{a}} \prod_{v \in \mathbf{f}} \varphi_v\right) = t\varphi_{\mathbf{a}} \prod_{v \in \mathbf{f}} r_P(w)^{-1} r_v(x_v) r_P(w)\varphi_v.$$

We apply these formulas to the function ω_0 of Proposition 2.3. If $|q_v| = 1$, we have $r_P(w)\omega_v = r_v(x_v)\omega_v = \omega_v$. Hence we may assume that $x_v = 1$ whenever $|q_v| = 1$. Now suppose $|q_v| \neq 1$. Then $(c_y)_v = (q^2 c_x)_v \in (2q^2\mathfrak{d})_v$ and $(c_x)_v \in (2q^{-2}\mathfrak{d})_v$; hence $|d_x|_v = |d_y|_v = 1$. Thus $d_x \in F_{\mathbf{A}}^{\times}$, so that $\operatorname{pr}(x) \in \Omega_{\mathbf{A}}\iota$. Therefore we can put $x = t'r_\Omega(\alpha\iota^{-1})r_\Omega(\iota)$ with $t' \in \mathbf{T}$ and $\alpha \in \Omega_{\mathbf{A}}\iota$, $\alpha_{\mathbf{a}} = 1$. Let $\beta = w^{-1}\alpha w$. Then $\beta \in \Omega_{\mathbf{A}}\iota$. By (1.11) we see that $y = t'r_\Omega(w^{-1}\alpha\iota^{-1})r_\Omega(\iota w) = t'r_\Omega(\beta\iota^{-1}w)r_\Omega(w^{-1}\iota) = t'r_\Omega(\beta\iota^{-1})r_\Omega(\iota)$. By (2.18b) we have $h(x, \mathbf{i})/h(y, \mathbf{i}) = |\gamma'/\gamma|\gamma/\gamma'$ with $\gamma = \gamma(\delta^{-2}d_\alpha^{-1}c_\alpha)$ and $\gamma' = \gamma(\delta^{-2}d_\beta^{-1}c_\beta)$. Now $d_\alpha = d_\beta$ and $c_\beta = q^2 c_\alpha$. If $|q_v| = 1$, γ and γ' have the same v-factor. Suppose $|q_v| \neq 1$; then $|d_\alpha|_v = 1$. Since both $(c_\alpha)_v$ and $(c_\beta)_v$ belong to $2\mathfrak{d}_v$, we see that $\gamma_v = \gamma'_v = 1$ from (2.17b). This completes the proof.

3. Automorphic forms and their Fourier expansions. We are going to consider holomorphic and nonholomorphic automorphic forms of integral or half-integral weight. An *integral weight* is an element of $\mathbf{Z}^{\mathbf{a}}$; a *half-integral weight* is an element of $(1/2)\mathbf{Z}^{\mathbf{a}}$ of the form $(u/2) + m$ with $m \in \mathbf{Z}^{\mathbf{a}}$. Given a weight k, we define, for $\tau \in \tilde{G}_{\mathbf{A}+}$ or $M_{\mathbf{A}}$ and $z \in \mathscr{H}^{\mathbf{a}}$, a factor of automorphy $J_k(\tau, z)$ by

$$(3.1a) \qquad J_k(\tau, z) = j(\tau, z)^k \quad \text{if } k \in \mathbf{Z}^{\mathbf{a}} \text{ and } \tau \in \tilde{G}_{\mathbf{A}+},$$

(3.1b)

$$J_k(\tau, z) = h(\tau, z)j(\tau, z)^m \quad \text{if } k = (u/2) + m, \ m \in \mathbf{Z}^{\mathbf{a}}, \text{ and } \tau \in \operatorname{pr}^{-1}(P_{\mathbf{A}}C''),$$

using the notation of (1.2a), (1.5a, b), and h of Proposition 2.3. It should be remembered that if $k \notin \mathbf{Z}^{\mathbf{a}}$, J_k is only a partial factor of automorphy in the same sense as in Proposition 2.3(iii). Given a function f on $\mathscr{H}^{\mathbf{a}}$, we define a function $f\|_k\tau$ on $\mathscr{H}^{\mathbf{a}}$ by

$$(3.2) \qquad (f\|_k\tau)(z) = J_k(\tau, z)^{-1}f(\tau(z)) \qquad (z \in \mathscr{H}^{\mathbf{a}}).$$

For two functions f and g on $\mathcal{H}^{\mathbf{a}}$ satisfying $f\|_k\gamma = f$ and $g\|_k\gamma = g$ for every γ in a congruence subgroup Γ of G, we define their inner product $\langle f, g\rangle$ by

$$(3.3) \qquad \langle f, g\rangle = \mu(\Phi)^{-1} \int_\Phi \overline{f(z)}\, g(z)\, y^k\, d_H z \qquad (\Phi = \Gamma\backslash\mathcal{H}^{\mathbf{a}})$$

whenever the integral is convergent, where $y = \mathrm{Im}(z)$, $d_H(x + iy) = \prod_{v\in\mathbf{a}} y_v^{-2}\, dx_v\, dy_v$, and $\mu(\Phi) = \int_\Phi d_H z$. This inner product is independent of the choice of Γ, and

$$(3.4) \qquad \langle f\|_k\tau, g\|_k\tau\rangle = \langle f, g\rangle \qquad \left(\tau \in \tilde{G}_{\mathbf{A}+} \text{ or } \mathrm{pr}^{-1}(P_\mathbf{A}C'')\right).$$

Given a congruence subgroup Γ of G, we denote by $\mathscr{M}_k(\Gamma)$ the vector space of all holomorphic functions f on $\mathcal{H}^{\mathbf{a}}$ which satisfy $f\|_k\gamma = f$ for every $\gamma \in \Gamma$ and also the cusp condition when $F = \mathbf{Q}$; we assume $\Gamma \subset C''G_{\mathbf{a}}$ if k is half-integral. We then denote by $\mathscr{M}_k$ the union of $\mathscr{M}_k(\Gamma)$ for all such Γ, and by $\mathscr{S}_k$ the subset of $\mathscr{M}_k$ consisting of the cusp forms; we put $\mathscr{S}_k(\Gamma) = \mathscr{S}_k \cap \mathscr{M}_k(\Gamma)$.

In the rest of this section we treat only the forms of half-integral weight. For simplicity, we fix $k = (u/2) + m$ with $m \in \mathbf{Z}^{\mathbf{a}}$ and write J and $f\|\tau$ for J_k and $f\|_k\tau$. Now an automorphic form as a function on $M_\mathbf{A}$ is a map $p: M_\mathbf{A} \to \mathbf{C}$ such that

$$(3.5) \quad p(\alpha x w) = p(x)J(w, \mathbf{i})^{-1} \quad \textit{for every } \alpha \in G,\ x \in M_\mathbf{A},\ \textit{and } w \in \mathrm{pr}^{-1}(B),$$

where $\mathbf{i}$ is defined by (2.20), and B is an open compact subgroup of C''. Given such a p, there exists a unique function f on $\mathcal{H}^{\mathbf{a}}$ such that

$$(3.6) \qquad p(\alpha x) = (f\|x)(\mathbf{i}) \quad \textit{for every } \alpha \in G \textit{ and } x \in \mathrm{pr}^{-1}(BG_{\mathbf{a}}),$$

$$(3.7) \qquad f\|\gamma = f \quad \textit{for every } \gamma \in G \cap BG_{\mathbf{a}}.$$

Conversely, given any f satisfying (3.7), we can define a function p on $M_\mathbf{A}$ by (3.6); then it satisfies (3.5). We denote by $f_\mathbf{A}$ this function p on $M_\mathbf{A}$ corresponding to f in this fashion.

We now refine (3.5), (3.6), and (3.7) by taking a level $\mathfrak{c}$ and a character ψ into consideration. Here $\mathfrak{c}$ is an integral ideal divisible by 4, and ψ is a Hecke character of F of finite or infinite order such that

$$(3.8a) \qquad\qquad \textit{the conductor of } \psi \textit{ divides } \mathfrak{c};$$

$$(3.8b) \qquad\qquad \psi_{\mathbf{a}}(-1) = (-1)^{\{m\}},$$

where $\psi_{\mathbf{a}} = \prod_{v\in\mathbf{a}}\psi_v$. We define $\psi_{\mathbf{f}}$ in a similar way and put $\psi_{\mathfrak{c}} = \prod_{v|\mathfrak{c}}\psi_v$. We put also

$$(3.9) \qquad D_{\mathfrak{c}} = D[2\mathfrak{d}^{-1}, 2^{-1}\mathfrak{c}\mathfrak{d}], \qquad \Gamma_{\mathfrak{c}} = \Gamma[2\mathfrak{d}^{-1}, 2^{-1}\mathfrak{c}\mathfrak{d}].$$

We then consider a function f on $\mathscr{H}^{\mathbf{a}}$ such that

$$(3.10) \qquad f\|\gamma = \psi_{\mathfrak{c}}(a_\gamma)f \quad \text{for every } \gamma \in \Gamma_{\mathfrak{c}}.$$

This implies (3.7) with $B = \{w \in D_{\mathfrak{c}} | \psi_{\mathfrak{c}}(a_w) = 1\}$, and so $f_{\mathbf{A}}$ is meaningful. It satisfies

$$(3.11) \qquad f_{\mathbf{A}}(\alpha x w) = f_{\mathbf{A}}(x)\psi_{\mathfrak{c}}(a_w)^{-1}J(w,\mathbf{i})^{-1}$$

$$\text{for every } \alpha \in G,\ x \in M_{\mathbf{A}},\ \text{and } w \in \mathrm{pr}^{-1}(D_{\mathfrak{c}}),$$

$$(3.12) \quad f_{\mathbf{A}}(\alpha x) = \psi_{\mathfrak{c}}(a_x)^{-1}(f\|x)(\mathbf{i}) \quad \text{if } \alpha \in G \text{ and } x \in \mathrm{pr}^{-1}(D_{\mathfrak{c}}G_{\mathbf{a}}).$$

We denote by $\mathscr{M}_k(\mathfrak{c},\psi)$ the set of all f in $\mathscr{M}_k$ satisfying (3.10), and put $\mathscr{S}_k(\mathfrak{c},\psi) = \mathscr{S}_k \cap \mathscr{M}_k(\mathfrak{c},\psi)$.

Condition (3.10) concerns only $\psi_{\mathfrak{c}}$, and therefore $\mathscr{M}_k(\mathfrak{c},\psi) = \mathscr{M}_k(\mathfrak{c},\psi')$ if $\psi'_{\mathfrak{c}} = \psi_{\mathfrak{c}}$ and ψ' satisfies (3.8a, b). Thus, for a given f, we have a certain freedom of choosing ψ. It should be emphasized that our theory depends not only on f but also *on the choice of* ψ. In other words, our main results will concern the pair (f,ψ) which is compatible in the above sense.

Let us now study the Fourier expansions of f and $f_{\mathbf{A}}$, assuming that f is C^∞, but not necessarily holomorphic for the moment. Given an f satisfying (3.10), put

$$g(t,s) = f_{\mathbf{A}}\!\left(r_P\begin{bmatrix} t & s \\ 0 & t^{-1} \end{bmatrix}\right) \qquad (t \in F_{\mathbf{A}}^\times,\ s \in F_{\mathbf{A}}).$$

From (3.6) and (3.10), we see that g is continuous on $F_{\mathbf{A}}^\times \times F_{\mathbf{A}}$ and C^∞ in $(t,s)_{\mathbf{a}}$. Now (3.11) implies that

$$g(t,s) = g(at, as + bt^{-1}) \qquad (a \in F^\times, b \in F),$$

$$g(t,s) = \psi_{\mathfrak{c}}(c)g(tc, td + sc^{-1}) \qquad \left(c \in \prod_{v \in \mathfrak{f}} \mathfrak{o}_v^\times,\ d \in \prod_{v \in \mathfrak{f}} (2\mathfrak{d}^{-1})_v\right).$$

Therefore we easily see that

$$g(t,s) = \sum_{\xi \in F} \kappa(\xi,t)\mathbf{e}_{\mathbf{A}}(ts\xi/2)$$

with complex coefficients $\kappa(\xi,t)$ such that $\psi_{\mathfrak{c}}(c)\kappa(\xi,tc) = \kappa(\xi,t)$ for c as above, $\kappa(\xi,at) = \kappa(a^2\xi,t)$ for $a \in F^\times$, and $\kappa(\xi,t) = 0$ unless $\xi \in t^{-2}\mathfrak{o}$; thus $\psi_{\mathfrak{c}}(t)\kappa(\xi,t)$ depends only on $t\mathfrak{o}$ and $t_{\mathbf{a}}$. Putting $\kappa'(\xi,t\mathfrak{o},t_{\mathbf{a}}) = \psi_{\mathfrak{c}}(t)\kappa(\xi,t)$, we obtain

$$(3.13) \qquad f_{\mathbf{A}}\left(r_P\begin{bmatrix} t & s \\ 0 & t^{-1} \end{bmatrix}\right) = \sum_{\xi \in t^{-2}\mathfrak{o}} \psi_{\mathfrak{c}}(t)^{-1}\kappa'(\xi, t\mathfrak{o}, t_{\mathbf{a}})\mathbf{e}_{\mathbf{A}}(ts\xi/2).$$

PROPOSITION 3.1. *Suppose* $f \in \mathcal{M}_k(\mathfrak{c}, \psi)$. *Then there is a complex number* $\lambda(\xi, \mathfrak{m}; f, \psi)$ *determined for* $\xi \in F$ *and a fractional ideal* $\mathfrak{m}$ *in F such that*

$$\psi_{\mathbf{f}}(t)t_{\mathbf{a}}^{-m}|t|_{\mathbf{A}}^{-1/2}f_{\mathbf{A}}\left(r_P\begin{bmatrix} t & s \\ 0 & t^{-1} \end{bmatrix}\right) = \sum_{\xi \in F} \lambda(\xi, t\mathfrak{o}; f, \psi)\mathbf{e}_{\mathbf{a}}(it^2\xi/2)\mathbf{e}_{\mathbf{A}}(ts\xi/2).$$

Moreover $\lambda(\xi, \mathfrak{m}; f, \psi)$ *has the following properties:*

(3.14a) $\lambda(\xi, \mathfrak{m}; f, \psi) \neq 0$ *only when* $\xi \in \mathfrak{m}^{-2}$ *and* ξ *is totally nonnegative;*

(3.14b) $\lambda(\xi b^2, \mathfrak{m}; f, \psi) = b^m\psi_{\mathbf{a}}(b)\lambda(\xi, b\mathfrak{m}; f, \psi)$ *for every* $b \in F^\times$.

Furthermore, if $\beta \in G \cap P_{\mathbf{A}}D_{\mathfrak{c}}$, *then* $d_\beta \neq 0$, $d_\beta \mathfrak{a}_\beta^{-1}$ *is prime to* $\mathfrak{c}$; *moreover, if* $\beta \in \mathrm{diag}[t, t^{-1}]D_{\mathfrak{c}}G_{\mathbf{a}}$ *with* $t \in F_{\mathbf{A}}^\times$, *then*

$$(3.14c) \qquad \psi_{\mathbf{a}}(d_\beta)\psi^*\left(d_\beta \mathfrak{a}_\beta^{-1}\right)J(\beta, \beta^{-1}z)f(\beta^{-1}z)$$

$$= N(\mathfrak{a}_\beta)^{1/2}\sum_{\xi \in F} \lambda\left(\xi, \mathfrak{a}_\beta^{-1}; f, \psi\right)\mathbf{e}_{\mathbf{a}}(\xi z/2).$$

This means that for a fixed $\mathfrak{m}$, $\lambda(\xi, \mathfrak{m}; f, \psi)$ appears as a coefficient of the Fourier expansion of f at a cusp corresponding to $\mathfrak{m}$. In particular, we have

$$(3.15) \qquad f(z) = \sum_{\xi \in F} \lambda(\xi, \mathfrak{o}; f, \psi)\mathbf{e}_{\mathbf{a}}(\xi z/2).$$

The Fourier coefficient $\lambda(\xi, \mathfrak{m}; f, \psi)$ depends on the choice of ψ. However, this being understood, we shall often drop ψ and write it simply $\lambda(\xi, \mathfrak{m}; f)$ or $\lambda_f(\xi, \mathfrak{m})$. (See Remark 6.3 below.)

Proof. Let $w = \begin{pmatrix} t & s \\ 0 & t^{-1} \end{pmatrix} \in P_{\mathbf{A}}$, $z = w(i)$, $\sigma = r_P(w_{\mathbf{f}})$, and $\tau = r_P(w_{\mathbf{a}})$. By strong approximation, we have $w_{\mathbf{f}} \in \beta D_{\mathfrak{c}}G_{\mathbf{a}}$ with $\beta \in G$. Put $\beta^{-1}\sigma = \rho$ and $q = pr(\rho\tau)$. Then $q \in D_{\mathfrak{c}}G_{\mathbf{a}}$, $q(i) = \rho(z) = \beta^{-1}z$, and hence, by (3.12),

$$(3.16) \qquad f_{\mathbf{A}}(r_P(w)) = f_{\mathbf{A}}(\beta\rho\tau) = f(\beta^{-1}z)J(\rho\tau, i)^{-1}\psi_{\mathfrak{c}}(a_q)^{-1}.$$

By Proposition 2.3(iii) and (iv), we have $1 = J(\sigma, z) = J(\beta\rho, z) = J(\beta, \rho z)J(\rho, z)$, and $J(\rho\tau, i) = J(\rho, z)J(\tau, i) = J(\beta, \beta^{-1}z)^{-1}|t_{\mathbf{a}}|^{-u/2}t_{\mathbf{a}}^{-m}$. Since

$\beta q = w$, we have $a_q = t d_\beta$, and so $d_\beta \mathfrak{a}_\beta^{-1} = d_\beta t \mathfrak{o} = a_q \mathfrak{o}$, which is prime to $\mathfrak{c}$. Therefore $\psi_\mathfrak{f}(t)\psi_\mathfrak{c}(a_q)^{-1} = \psi_\mathfrak{f}(t)\psi_\mathfrak{c}(td_\beta)^{-1} = \psi_\mathfrak{a}(d_\beta)\psi^*(d_\beta\mathfrak{a}_\beta^{-1})$. Thus (3.16) can be written in the form

$$\psi_\mathfrak{f}(t)|t_\mathfrak{a}|^{-u/2}t_\mathfrak{a}^{-m}f_\mathcal{A}(r_P(w)) = \psi_\mathfrak{a}(d_\beta)\psi^*(d_\beta\mathfrak{a}_\beta^{-1})J(\beta,\beta^{-1}z)f(\beta^{-1}z).$$

The right-hand side, being an element of $\mathcal{M}_k$, has an expansion of the form $\Sigma_{\xi \in F}\omega_\xi e_\mathfrak{a}(\xi z/2)$ with $\omega_\xi \in \mathbf{C}$. Comparing this with (3.13), we find that

$$(\psi_\mathfrak{f}/\psi_\mathfrak{c})(t)|t_\mathfrak{a}|^{-u/2}t_\mathfrak{a}^{-m}\kappa'(\xi, t\mathfrak{o}, t_\mathfrak{a})e_\mathfrak{a}(-\mathfrak{i}t^2\xi/2) = \omega_\xi e_\mathfrak{f}(-ts\xi/2).$$

The left-hand side depends only on ξ, $t_\mathfrak{a}$, and $t\mathfrak{o}$, and the right-hand side only on ξ, $t_\mathfrak{f}$, and $s_\mathfrak{f}$, since β depends only on $w_\mathfrak{f}$. Therefore this number depends only on ξ and $t\mathfrak{o}$. Putting $\omega_\xi e_\mathfrak{f}(-ts\xi/2) = N(t\mathfrak{o})^{-1/2}\lambda_f(\xi, t\mathfrak{o})$, we obtain our first assertion, as well as the last one, which is a mere reformulation of the reasoning above in the case $s_\mathfrak{f} = 0$; (3.14a, b) follow from the corresponding facts on $\kappa(\xi, t)$.

PROPOSITION 3.2. *Let $\mathfrak{q}$ be a fractional ideal of F and let $0 \ll \tau \in \mathfrak{q}^2$. Then for every $f \in \mathcal{M}_k(\mathfrak{c}, \psi)$, there exists an element $g \in \mathcal{M}_k(\tau\mathfrak{q}^{-2}\mathfrak{c}, \psi\varepsilon_\tau)$ such that*

$$\lambda(\xi, \mathfrak{m}; g, \psi\varepsilon_\tau) = \lambda(\xi/\tau, \mathfrak{q}\mathfrak{m}; f, \psi),$$

where ε_τ is the Hecke character of F corresponding to $F(\tau^{1/2})/F$. In particular, if $\mathfrak{q} = \mathfrak{o}$, g is given by $g(z) = f(\tau z)$.

We first prove

LEMMA 3.3. *Let $0 \ll \tau \in \mathfrak{o}$, $\rho = \mathrm{diag}[\tau, 1]$, $\alpha \in G \cap P_\mathbf{A}D_{\tau\mathfrak{c}}$, and $\beta = \rho\alpha\rho^{-1}$. Then $d_\beta = d_\alpha \neq 0$, $\mathfrak{a}_\beta = \mathfrak{a}_\alpha$, $d_\beta\mathfrak{a}_\beta^{-1}$ is prime to $\tau\mathfrak{c}$, and $J(\beta, \tau z) = \varepsilon_\tau^*(d_\beta\mathfrak{a}_\beta^{-1})J(\alpha, z)$.*

Proof. Let $\alpha \in xD_{\tau\mathfrak{c}}$ with $x \in P_\mathbf{A}$, $y = \rho x\rho^{-1}$, $c = c_\beta$ and $d = d_\beta$. Then we easily see that $\beta \in yD_\mathfrak{c}$, $d_\alpha = d \neq 0$, and hence $\beta \in P_\mathbf{A}D_\mathfrak{c} \cap \Omega\iota$ and $\mathfrak{a}_\beta = d_y\mathfrak{o} = d_x\mathfrak{o} = \mathfrak{a}_\alpha$. By Proposition 3.1, $d_\alpha\mathfrak{a}_\alpha^{-1}$ $(= d\mathfrak{a}_\beta^{-1})$ is prime to $\tau\mathfrak{c}$. Since $j(\beta, \tau z) = j(\alpha, z)$, we have, by (2.18b),

$$(3.17) \qquad J(\beta, \tau z)/J(\alpha, z) = |\gamma'/\gamma|\gamma/\gamma'$$

with $\gamma = \gamma(\delta^{-2}d^{-1}c)$ and $\gamma' = \gamma(\delta^{-2}d^{-1}\tau c)$. If $v|\tau c$, we easily see that $d^{-1}c \in 2\mathfrak{d}_v$, and hence (2.17b) shows that $\gamma_v(\delta^{-2}d^{-1}c) = \gamma_v(\delta^{-2}d^{-1}\tau c) = 1$. Suppose $v \nmid \tau c$. Then

$$\gamma_v(\delta^{-2}d^{-1}\tau c)/\gamma_v(\delta^{-2}d^{-1}c) = \left(\frac{\tau}{\nu_0(d^{-1}c)_v}\right)$$

by [S8, Lemma 3.3], where $\nu_0(d^{-1}c)_v^{-1} = d^{-1}c\mathfrak{d}_v^{-1} + \mathfrak{o}_v = (d^{-1}\mathfrak{a}_\beta)_v$. Therefore (3.17) is equal to $\varepsilon_\tau^*(d\mathfrak{a}_\beta^{-1})$, which completes the proof.

To prove Proposition 3.2, we may assume that $\mathfrak{q} \subset \mathfrak{o}$ by changing τ and $\mathfrak{q}$ for $b^2\tau$ and $b\mathfrak{q}$ with a suitable $b \in F$. Let $p(z) = f(\tau z)$ and $\varphi = \psi\varepsilon_\tau$. With α and β as in Lemma 3.3, put

$$f_\beta(z) = \psi_{\mathbf{a}}(\dot{d}_\beta)\psi^*\left(d_\beta \mathfrak{a}_\beta^{-1}\right)J(\beta, \beta^{-1}z)f(\beta^{-1}z),$$

$$p_\alpha(z) = \varphi_{\mathbf{a}}(d_\alpha)\varphi^*\left(d_\alpha \mathfrak{a}_\alpha^{-1}\right)J(\alpha, \alpha^{-1}z)p(\alpha^{-1}z).$$

Lemma 3.3 implies immediately that $p_\alpha(z) = f_\beta(\tau z)$. In particular, if $\alpha \in \Gamma_{\tau\mathfrak{c}}$, then $\beta \in \Gamma_\mathfrak{c}$, so that $f_\beta = f$, and hence $p_\alpha = p$. This shows that $p \in \mathcal{M}_k(\tau\mathfrak{c}, \varphi)$. Moreover by Proposition 3.1 the equality $p_\alpha(z) = f_\beta(\tau z)$ shows that $\lambda_f(\xi/\tau, \mathfrak{m}) = \lambda_p(\xi, \mathfrak{m})$, since every fractional ideal can be given as $\mathfrak{a}_\beta$ with β as above. If $\mathfrak{q} = \mathfrak{o}$, we can take p as g. If $\mathfrak{q} \neq \mathfrak{o}$, we have to modify p in a certain way to obtain the desired g. For this purpose, take $q \in F_f^\times$ so that $\mathfrak{q} = q\mathfrak{o}$ and put $\sigma = r_P(\mathrm{diag}[q, q^{-1}])$ and $g_{\mathbf{A}}(x) = p_{\mathbf{A}}(x\sigma)$. Then $g_{\mathbf{A}}(\gamma x) = g_{\mathbf{A}}(x)$ for every $\gamma \in G$. Let $w \in \mathrm{pr}^{-1}(D[2\mathfrak{q}^2\mathfrak{d}^{-1}, 2^{-1}\tau\mathfrak{c}\mathfrak{q}^{-2}\mathfrak{d}])$ and $y = \sigma^{-1}w\sigma$. Then $\mathrm{pr}(y) \in D_{\tau\mathfrak{c}}$ and $g_{\mathbf{A}}(xw) = p_{\mathbf{A}}(x\sigma y) = \varphi_{\tau\mathfrak{c}}(a_y)^{-1}J(y, \mathbf{i})^{-1}p_{\mathbf{A}}(x\sigma)$, so that

$$(3.18) \qquad g_{\mathbf{A}}(xw) = \varphi_{\tau\mathfrak{c}}(a_w)^{-1}J(w, \mathbf{i})^{-1}g_{\mathbf{A}}(x),$$

since $J(y, \mathbf{i}) = J(w, \mathbf{i})$ by Lemma 2.5. Now we have

$$(3.19) \quad g_{\mathbf{A}}\left(r_P\begin{bmatrix} t & s \\ 0 & t^{-1} \end{bmatrix}\right) = p_{\mathbf{A}}\left(r_P\begin{bmatrix} tq & sq^{-1} \\ 0 & t^{-1}q^{-1} \end{bmatrix}\right)$$

$$= \varphi_{\mathfrak{f}}(tq)^{-1}|tq|_{\mathbf{A}}^{1/2}t_{\mathbf{a}}^m \sum_{\xi \in F} \lambda_p(\xi, t\mathfrak{q})\mathbf{e}_{\mathbf{a}}(it^2\xi/2)\mathbf{e}_{\mathbf{A}}(ts\xi/2).$$

Relation (3.18) means that $g_{\mathbf{A}}$ corresponds to a function g on $\mathscr{H}^{\mathbf{a}}$ by the principle of (3.6). Taking $t_{\mathfrak{f}} = 1$ and $s_{\mathfrak{f}} = 0$ in (3.19), we see that

$$g(z) = \varphi(\mathfrak{q})^{-1}N(\mathfrak{q})^{-1/2}\sum_\xi \lambda_f(\xi/\tau, \mathfrak{q})\mathbf{e}_{\mathbf{a}}(\xi z/2).$$

Thus g is holomorphic. Since $\lambda_f(\xi/\tau, \mathfrak{q}) \neq 0$ only when $\xi \in \tau\mathfrak{q}^{-2}$, we have $g(z + b) = g(z)$ for $b \in 2\mathfrak{d}^{-1}$. Moreover, (3.18) shows that $g\|_k\gamma = \varphi_{\tau\mathfrak{c}}(a_\gamma)g$ for every $\gamma \in \Gamma[2\mathfrak{q}^2\mathfrak{d}^{-1}, 2^{-1}\tau\mathfrak{c}\mathfrak{q}^{-2}\mathfrak{d}]$. Hence, by Lemma 3.4 below, we see that $g \in \mathcal{M}_k(\tau\mathfrak{c}\mathfrak{q}^{-2}, \varphi)$. The above expansion of $g_{\mathbf{A}}$ shows that $\lambda_g(\xi, \mathfrak{a}) = \varphi(\mathfrak{q})^{-1}N(\mathfrak{q})^{-1/2}\lambda_f(\xi/\tau, \mathfrak{a}\mathfrak{q})$. Therefore, modifying g by a constant factor, we obtain g as stated in our proposition.

LEMMA 3.4. *If $\mathfrak{x}' \subset \mathfrak{x}$ and $\mathfrak{y}' \subset \mathfrak{y}$, the group $\Gamma[\mathfrak{x}, \mathfrak{y}]$ is generated by $\Gamma[\mathfrak{x}', \mathfrak{y}']$ and the elements of the forms $\begin{pmatrix} 1 & b \\ 0 & 1 \end{pmatrix}$ and $\begin{pmatrix} 1 & 0 \\ c & 1 \end{pmatrix}$ with $b \in \mathfrak{x}$ and $c \in \mathfrak{y}$.*

This is completely elementary, and so the proof may be left to the reader.

4. Theta functions. We now introduce some theta series which are not necessarily holomorphic, and study their Fourier expansions. For $\eta \in \mathscr{S}(F_{\mathfrak{f}})$ as in Section 2 and $0 \leq n \in \mathbf{Z}^{\mathbf{a}}$, define η_n by (2.13) and put

$$(4.1) \qquad \theta_{n,\mathbf{A}}(\sigma, \eta) = \sum_{\xi \in F} (\sigma \eta_n)(\xi, \mathbf{i}) \qquad (\sigma \in M_{\mathbf{A}}),$$

viewing $\eta_n(t, \mathbf{i})$ with $t \in F_{\mathbf{A}}$ as an element of $L^2(F_{\mathbf{A}})$. By [We, Théorème 6], we have $\theta_{n,\mathbf{A}}(\alpha\sigma, \eta) = \theta_{n,\mathbf{A}}(\sigma, \eta)$ for $\alpha \in G$. Moreover, by Proposition 2.4, we have, putting $l = (u/2) + n$,

$$(4.2) \quad \theta_{n,\mathbf{A}}(\sigma w, \eta) = J_l(w, \mathbf{i})^{-1} \theta_{n,\mathbf{A}}(\sigma, {}^w\eta) \quad \textit{if } \mathrm{pr}(w) \in P_{\mathbf{A}} C'' \textit{ and } w(\mathbf{i}) = \mathbf{i}.$$

By (2.16d), we see that $\theta_{n,\mathbf{A}}(\sigma, \eta)$ is a function of type (3.5) and therefore corresponds to a function $\theta_n(z, \eta)$ of $z \in \mathscr{H}^{\mathbf{a}}$ by the relation

$$(4.3) \qquad \theta_n(w(\mathbf{i}), \eta) = \theta_{n,\mathbf{A}}(w, \eta) J_l(w, \mathbf{i}) \quad \textit{for } {}^w\eta = \eta, \mathrm{pr}(w) \in C''G_{\mathbf{a}}.$$

For $t \in F_{\mathbf{A}}^{\times}$ and $s \in F_{\mathbf{A}}$, we have, by (1.9),

$$\left(r_P \begin{bmatrix} t & s \\ 0 & 1/t \end{bmatrix} \eta_n \right)(x, \mathbf{i}) = |t|_{\mathbf{A}}^{1/2} \eta_n(tx, \mathbf{i}) e_{\mathbf{A}}(tsx^2/2) \qquad (x \in F_{\mathbf{A}}).$$

Hence we obtain

$$(4.4) \qquad \theta_{n,\mathbf{A}}\left(r_P \begin{bmatrix} t & s \\ 0 & 1/t \end{bmatrix}, \eta \right)$$

$$= |t|_{\mathbf{A}}^{1/2} \sum_{\xi \in F} \eta((t\xi)_{\mathfrak{f}}) H_n(\sqrt{4\pi}\, t_{\mathbf{a}}\xi_{\mathbf{a}}) e_{\mathbf{a}}(it^2\xi^2/2) e_{\mathbf{A}}(ts\xi^2/2),$$

$$(4.5) \qquad \theta_n(z, \eta) = y^{-n/2} \sum_{\xi \in F} \eta(\xi) H_n(\sqrt{4\pi y}\, \xi) e_{\mathbf{a}}(\xi^2 z/2),$$

where $H_n(x) = \prod_{v \in \mathbf{a}} H_{n_v}(x_v)$ for $x \in \mathbf{R}^{\mathbf{a}}$. Notice that if $n \leq u$, then θ_n is holomorphic, and (4.4) and (4.5) fit in the framework of Proposition 3.1.

LEMMA 4.1. $\theta_n(\alpha z, {}^{\alpha}\eta) = \theta_n(z, \eta) J_l(\alpha, z)$ if $\alpha \in G \cap P_{\mathbf{A}} C''$.

Proof. Since $\theta_n(z, \eta) = \sum_{\xi \in F} \eta_n(\xi, z)$, we have by [We, Théorème 6] $\theta_n(z, \eta) = \sum_{\xi \in F}(\alpha\eta_n)(\xi, z)$, which together with Proposition 2.4 proves our lemma.

Let us now take a Hecke character χ of F of conductor $\mathfrak{f}$ and of finite or infinite order such that

$$(4.6) \qquad\qquad \chi_{\mathbf{a}}(-1) = (-1)^{\{n\}}.$$

Let ω_v denote the characteristic function of $\mathfrak{o}_v$. Define $\eta \in \mathscr{S}(F_\mathfrak{f})$ by $\eta(x) = \Pi_{v \in \mathfrak{f}}\eta_v(x_v)$ with

$$(4.7a) \qquad\qquad \eta_v = \omega_v \quad \text{if } v \nmid \mathfrak{f},$$

$$(4.7b) \qquad\qquad \eta_v(t) = \begin{cases} \chi_v(t)^{-1} & \text{if } v|\mathfrak{f} \text{ and } |t|_v = 1, \\ 0 & \text{if } v|\mathfrak{f} \text{ and } |t|_v \neq 1. \end{cases}$$

LEMMA 4.2. *For ι of* (1.3) *and η as in* (4.7a, b), *one has*

$$'\eta(x) = D_F^{-1/2}N(\mathfrak{f})^{-1}\mathfrak{g}(\chi)\chi(e)^{-1}\bar{\eta}(ex) \qquad (x \in F_\mathfrak{f}),$$

where e is an element of $F_\mathfrak{f}^\times$ such that $e\mathfrak{o} = \mathfrak{d}\mathfrak{f}$, and $\mathfrak{g}(\chi)$ is the Gauss sum of χ defined by

$$(4.8) \quad \mathfrak{g}(\chi) = \begin{cases} \chi^*(\mathfrak{d}) & \text{if } \mathfrak{f} = \mathfrak{o}, \\ \Sigma_{b \in R}\chi_\mathbf{a}(b)\chi^*(b\mathfrak{d}\mathfrak{f})\mathbf{e}_\mathbf{a}(b) & \text{if } \mathfrak{f} \neq \mathfrak{o} \quad (R = \mathfrak{f}^{-1}\mathfrak{d}^{-1}/\mathfrak{d}^{-1}). \end{cases}$$

Proof. By (2.16c) we have $'\eta = \Pi_v r_\Omega(\iota)\eta_v$; by (1.10a), $r_\Omega(\iota)\eta_v$ is the Fourier transform of η_v. Then the result can be obtained by a direct calculation which is well known in a standard proof of the functional equation of $L(s, \chi)$.

Now put

$$(4.9) \qquad \theta_{n,\chi}(z) = (4\pi y)^{-n/2} \sum_{\xi \in \mathfrak{o}} \chi_\mathbf{a}(\xi)\chi^*(\xi\mathfrak{o})H_n(\sqrt{4\pi y}\,\xi)\mathbf{e}_\mathbf{a}(\xi^2 z/2),$$

where $z \in \mathscr{H}^\mathbf{a}$ and $y = \text{Im}(z)$. This is $(4\pi)^{-n/2}\theta_n(z, \eta)$ with η of (4.7a, b). Here, as well as in Lemma 4.4, (4.10), and (4.11) below, we understand that $\chi_\mathbf{a}(\xi)\chi^*(\xi\mathfrak{m})$ for $\xi = 0$ with any fractional ideal $\mathfrak{m}$ means $\chi^*(\mathfrak{m})$ or 0 according as $\mathfrak{f} = \mathfrak{o}$ or $\mathfrak{f} \neq \mathfrak{o}$. Observe that (4.9) is nonvanishing if and only if (4.6) is satisfied.

LEMMA 4.3. *Let $l = (u/2) + n$; then $\theta_{n,\chi}\|_l\gamma = \chi_\mathfrak{c}(a_\gamma)\theta_{n,\chi}$ for every $\gamma \in \Gamma_\mathfrak{c}$, where $\mathfrak{c} = 4\mathfrak{f}^2$.*

Proof. For $x = r_P(\tau)$, $\tau = \text{diag}[s, s^{-1}]$ with $s \in \Pi_{v \in \mathfrak{f}}\mathfrak{o}_v^\times$, we have $^x\eta = \Pi_v r_P(\tau_v)\eta_v = \chi(s)^{-1}\eta$ by (2.16b) and (1.9). Since $^w\eta = \eta$ for w in an open subgroup of $M_\mathbf{A}$, we have $^w\eta = \chi(a_w)^{-1}\eta$ for $w \in \text{pr}^{-1}(D[\mathfrak{a}, \mathfrak{a}])$ for some integral ideal $\mathfrak{a}$ contained in $2\mathfrak{d}$, and hence, by Lemma 4.1, the desired formula is true for $\gamma \in \Gamma[\mathfrak{a}, \mathfrak{a}]$. On the other hand, (4.9) shows that $\theta_{n,\chi}(z + b) = \theta_{n,\chi}(z)$ for $b \in 2\mathfrak{d}^{-1}$. From Lemmas 4.1 and 4.2, we see that

$$(4.10) \quad (\theta_{n,\chi}\|_l\iota)(z) = (-1)^{\{n\}}D_F^{-1/2}N(\mathfrak{f})^{-1}\mathfrak{g}(\chi)$$

$$\cdot (4\pi y)^{-n/2}\sum_\xi \bar{\chi}_\mathbf{a}(\xi)\bar{\chi}^*(\xi\mathfrak{d}\mathfrak{f})H_n(\sqrt{4\pi y}\,\xi)\mathbf{e}_\mathbf{a}(\xi^2 z/2),$$

where ξ runs over $\mathfrak{d}^{-1}\mathfrak{f}^{-1}$. This is invariant under $z \mapsto z + c$ for all $c \in 2\mathfrak{d}\mathfrak{f}^2$, which, together with Proposition 2.3(iii), implies that the desired formula for $\theta_{n,\chi}$ is true for $\gamma = \begin{pmatrix} 1 & 0 \\ c & 1 \end{pmatrix}$ with c in $2\mathfrak{d}\mathfrak{f}^2$. Therefore our lemma follows from Lemma 3.4.

LEMMA 4.4. *If $\beta \in G \cap \mathrm{diag}[t, t^{-1}]D_cG_a$ with $c = 4\mathfrak{f}^2$ and $t \in F_A^\times$, then*

$$\chi_a(d_\beta)\chi^*\!\left(d_\beta\mathfrak{a}_\beta^{-1}\right)J_l(\beta, \beta^{-1}z)\theta_{n,\chi}(\beta^{-1}z)$$

$$= N(\mathfrak{a}_\beta)^{1/2}(4\pi y)^{-n/2}\sum_{\xi \in \mathfrak{a}_\beta}\chi_a(\xi)\chi^*\!\left(\xi\mathfrak{a}_\beta^{-1}\right)H_n\!\left(\sqrt{4\pi y}\,\xi\right)\mathbf{e}_a(\xi^2 z/2).$$

This follows from (4.4) by the same argument as in the proof of Proposition 3.1.

If $n \leq u$, we have $H_n(x) = x^n$ for $x \in \mathbf{R}^a$, so that

$$(4.11) \qquad \theta_{n,\chi}(z) = \sum_{\xi \in \mathfrak{o}}\chi_a(\xi)\chi^*(\xi\mathfrak{o})\xi^n\mathbf{e}_a(\xi^2 z/2).$$

This is holomorphic. By Lemma 4.4 the Fourier coefficients of $\theta_{n,\chi}$ with $n \leq u$ in the sense of Proposition 3.1 are given by

$$(4.12) \quad \lambda(\zeta, \mathfrak{m}; \theta_{n,\chi}, \chi) = \begin{cases} 2\chi_a(\xi)\chi^*(\xi\mathfrak{m})\xi^n & \text{if } 0 \neq \zeta = \xi^2 \in \mathfrak{m}^{-2}, \\ \chi^*(\mathfrak{m}) & \text{if } \zeta = 0, \ \mathfrak{f} = \mathfrak{o}, \text{ and } n = 0, \\ 0 & \text{otherwise.} \end{cases}$$

If $F = \mathbf{Q}$, the series of (4.11) is essentially the same as that of [S1, Proposition 2.2]. Lemma 4.3 together with (4.10) generalizes that result. It should be noted however that if $F \neq \mathbf{Q}$, *the same character χ can correspond to several holomorphic series of different weights*, since n is not unique under (4.6).

5. Hecke operators. In this section we fix a half-integral weight $k = (u/2) + m$, a multiple c of 4, and a Hecke character ψ of F satisfying (3.8a, b); for simplicity, we put $D = D_c$, $\Gamma = \Gamma_c$, and $U = \mathrm{pr}^{-1}(DG_a)$. Then $\Gamma = G \cap U$. Moreover we have:

$(5.1a)$ *For every $y \in M_A$, G contains an element α such that $yU = \alpha U$;*

$(5.1b)$ $UyU = Uy\Gamma = \Gamma yU$ *for every $y \in M_A$;*

$(5.1c)$ $\Gamma\alpha\Gamma = G \cap U\alpha U$ *for every $\alpha \in G$;*

$(5.1d)$ $\Gamma\alpha\Gamma = \coprod_{\beta \in A}\Gamma\beta \Rightarrow U\alpha U = \coprod_{\beta \in A}U\beta.$

Here we employ $\amalg$ to indicate a disjoint union. These are easy consequences of strong approximation (cf. [S5, II, Lemmas 1.1 and 1.2]).

Let us now define subsets Δ and Ξ of $M_{\mathbf{A}}$ by

$$(5.2a) \qquad\qquad \Xi = U\Delta U,$$

$$(5.2b) \qquad\qquad \Delta = \left\{ r_P\!\left(\mathrm{diag}[\zeta^{-1}, \zeta]\right) \mid \zeta \in F_{\mathfrak{f}}^{\times},\ \zeta \mathfrak{o} \subset \mathfrak{o} \right\}.$$

To define Hecke operators, let us again employ the simplified notation J and $f\|\tau$, suppressing the subscript k. For $x = w_1 \eta w_2$ with $w_1, w_2 \in U$ and $\eta \in \Delta$, we put

$$(5.3) \qquad\qquad J_{\Xi}(x, z) = J(w_1 w_2, z) \qquad (z \in \mathscr{H}^{\mathbf{a}}).$$

This is well-defined. In fact, if $x = y_1 \theta y_2$ with $y_1, y_2 \in U$ and $\theta \in \Delta$, then $\theta = \tau\eta$, $\tau = r_P(\mathrm{diag}[\varepsilon^{-1}, \varepsilon])$ with $\varepsilon \in \prod_{v \in \mathfrak{f}} \mathfrak{o}_v^{\times}$. Then $w_1^{-1} y_1 \tau\eta = \eta w_2 y_2^{-1}$. Lemma 2.5 shows that $J(w_1^{-1} y_1 \tau, z) = J(w_2 y_2^{-1}, z)$. Now $J(\tau, z) = 1$ by Proposition 2.3(iv). Since $J(\alpha, z)$ is a factor of automorphy for $\alpha \in U$, we see that $J(w_1 w_2, z) = J(y_1 y_2, z)$, and hence (5.3) is well-defined.

From (5.3), we see immediately

$$(5.4) \quad J_{\Xi}(y_1 x y_2, z) = J(y_1, x y_2(z)) J_{\Xi}(x, y_2(z)) J(y_2, z) \quad \text{if } y_1, y_2 \in U.$$

Somewhat more nontrivially, we have:

LEMMA 5.1. *Let* $\sigma = r_P(\tau)$, $\tau = \mathrm{diag}[\zeta^{-1}, \zeta]$ *with* $\zeta \in F_{\mathfrak{f}}^{\times}$ *such that* $\zeta\mathfrak{o} + \mathfrak{c} = \mathfrak{o}$. *Then* $U\sigma^{-1}U = U\sigma U$ *and* $J_{\Xi}(\theta^{-1}, z) = J_{\Xi}(\theta, \theta^{-1}z)^{-1}$ *for every* $\theta \in U\sigma U$. *In particular,* $J_{\Xi}(\sigma^{-1}, z) = 1$.

Proof. Since $\zeta^2 \mathfrak{o} + \mathfrak{c} = \mathfrak{o}$, we have $e + t = 1$ with $e \in \zeta^2 \mathfrak{o}$ and $0 \neq t \in \mathfrak{c}$. Let $d = \zeta^{-2} e\ (\in F_{\mathbf{A}})$. Define $\alpha, \beta \in \Omega_{\mathbf{A}}$ by

$$\alpha_{\mathfrak{f}} = \begin{bmatrix} d & 2/\delta \\ -t\delta/2 & \zeta^2 \end{bmatrix}_{\mathfrak{f}}, \qquad \beta_{\mathfrak{f}} = \begin{bmatrix} \zeta^2 d & 2/\delta \\ -t\delta/2 & 1 \end{bmatrix}_{\mathfrak{f}}, \qquad \alpha_{\mathbf{a}} = \beta_{\mathbf{a}} = \iota$$

with δ of (2.12). Put $w = r_{\Omega}(\alpha)$ and $w' = r_{\Omega}(\beta)$. Then $w\sigma^{-1} = r_{\Omega}(\alpha\tau^{-1}) = r_{\Omega}(\tau\beta) = \sigma r_{\Omega}(\beta)$, and so $\sigma^{-1} = w^{-1}\sigma w'$; hence $U\sigma^{-1}U = U\sigma U$ and $J_{\Xi}(\sigma^{-1}, z) = J(w^{-1}w', z)$. To prove our last assertion, it is sufficient to show that $J(w, z) = J(w', z)$. By (2.18a) we have

$$J(w, z)/J(w', z) = h(w, z)/h(w', z) = |\gamma'/\gamma|\gamma/\gamma'$$

with $\gamma = \gamma(2\zeta^2 t^{-1}\delta^{-1})$ and $\gamma' = \gamma(2t^{-1}\delta^{-1})$. If $|\zeta_v| < 1$, then $|t_v| = 1$, so that $\gamma_v = \gamma_v' = 1$ by (2.17b); if $|\zeta_v| = 1$, we obviously have $\gamma_v = \gamma_v'$ by (2.17b), which proves the last assertion. Now let $\theta = q\sigma r$ with q and r in U. Then $\theta^{-1} =$

$r^{-1}w^{-1}\sigma w'q^{-1}$, and so $J_{\Xi}(\theta^{-1}, z) = J(r^{-1}w^{-1}w'q^{-1}, z) = J(r^{-1}q^{-1}, z)$, since $J(w^{-1}w', z) = 1$. On the other hand, $J_{\Xi}(\theta, \theta^{-1}z) = J(qr, (qr)^{-1}z) = J(r^{-1}q^{-1}, z)^{-1}$, which completes the proof.

We now define the Hecke operator $T_{\mathfrak{x}}$ on $\mathscr{M}_k(\mathfrak{c}, \psi)$ for each *squarefree* integral ideal $\mathfrak{x}$ in F as follows: Take an element ξ of $F_{\mathfrak{f}}^{\times}$ so that $\xi\mathfrak{o} = \mathfrak{x}$; let $\sigma = r_P(\mathrm{diag}[\xi^{-1}, \xi])$ and $G \cap U\sigma U = \amalg_{\alpha \in A}\Gamma\alpha$. Then we define, for $f \in \mathscr{M}_k(\mathfrak{c}, \psi)$, a function $f|T_{\mathfrak{x}}$ on $\mathscr{H}^{\mathbf{a}}$ by

(5.5)

$$(f|T_{\mathfrak{x}})(z) = N(\mathfrak{x})^{-3/2}(\psi/\psi_{\mathfrak{c}})(\xi) \sum_{\alpha \in A} \psi_{\mathfrak{c}}(a_{\alpha})^{-1}J_{\Xi}(\alpha, z)^{-1}f(\alpha z) \qquad (z \in \mathscr{H}^{\mathbf{a}}).$$

It can easily be verified that $f|T_{\mathfrak{x}}$ is independent of the choice of ξ and A, and belongs to $\mathscr{M}_k(\mathfrak{c}, \psi)$; moreover, if $U\sigma U = \amalg_{w \in W}Uw$ and $\mathrm{pr}(w) \in G_{\mathfrak{f}}$ for every $w \in W$, then

(5.6) $\quad (f|T_{\mathfrak{x}})_A(x) = N(\mathfrak{x})^{-3/2}(\psi/\psi_{\mathfrak{c}})(\xi) \sum_{w \in W} \psi_{\mathfrak{c}}(a_w)^{-1}J_{\Xi}(w, \mathbf{i})^{-1}f_A(xw^{-1})$

$$(x \in M_{\mathbf{A}}).$$

Since $T_{\mathfrak{x}}$ depends on ψ and $\mathfrak{c}$, perhaps $T_{\mathfrak{x}, \psi, \mathfrak{c}}$ is a better notation, but this will not be used, for ψ and $\mathfrak{c}$ will always be clear from the context.

PROPOSITION 5.2. *The operators $T_{\mathfrak{x}}$ generate a commutative ring. Moreover, $T_{\mathfrak{x}\mathfrak{y}} = T_{\mathfrak{x}}T_{\mathfrak{y}}$ if $\mathfrak{x}$ and $\mathfrak{y}$ are relatively prime.*

Proof. Let $\mathfrak{x}$, ξ, and σ be as above. For an integral ideal $\mathfrak{y}$ prime to $\mathfrak{x}$, let $\tau = r_P(\mathrm{diag}[\eta^{-1}, \eta])$ with $\eta \in F_{\mathfrak{f}}^{\times}$, $\eta\mathfrak{o} = \mathfrak{y}$, and let $\rho = \sigma\tau$. Then $D\,\mathrm{pr}(\rho)D = D\,\mathrm{pr}(\sigma)D\,\mathrm{pr}(\tau)D$, so that $U\rho U = U\sigma U\tau U$. Let $\alpha = x\sigma x'$ and $\beta = y\tau y'$ with x, x', y, and y' in U. Then $\sigma x'y\tau \in \sigma U\tau \subset U\rho U$, so that $\sigma x'y\tau = w\rho w'$ with w and w' in U. Put $q = \tau w'\tau^{-1}$. Obviously $\mathrm{pr}(q)_v \in D_v$ if $|\eta_v| = 1$; also, $q = \sigma^{-1}w^{-1}\sigma x'y$, and hence $\mathrm{pr}(q)_v \in D_v$ if $|\xi_v| = 1$. Since $\mathfrak{x}$ is prime to $\mathfrak{y}$, we see that $\mathrm{pr}(q)_v \in D_v$ for all $v \in \mathbf{f}$, so that $q \in U$. Now $\alpha\beta = xw\rho w'y'$, and so $J_{\Xi}(\alpha\beta, z) = J(xww'y', z)$, $J_{\Xi}(\alpha, z) = J(xx', z)$, and $J_{\Xi}(\beta, z) = J(yy', z)$. We are going to show

(5.7) $$J_{\Xi}(\alpha\beta, z) = J_{\Xi}(\alpha, \beta z)J_{\Xi}(\beta, z),$$

which is equivalent to $J(ww', z) = J(x'y, z)$. By Lemma 2.5 we have $J(w', z) = J(q, z)$. Since $\sigma x'y = w\sigma q$, we have $J(x'y, z) = J_{\Xi}(\sigma x'y, z) = J(wq, z) = J(ww', z)$, which proves (5.7). Now let $U\sigma U = \amalg_{\alpha \in A}U\alpha$ and $U\tau U = \amalg_{\beta \in B}U\beta$. Then $U\rho U = U\sigma U\tau U = \amalg_{\alpha \in A, \beta \in B}U\alpha\beta$, since the corresponding equality holds on $G_{\mathbf{A}}$. Then we see from (5.6) and (5.7) that $T_{\mathfrak{x}\mathfrak{y}} = T_{\mathfrak{x}}T_{\mathfrak{y}}$. Similarly $T_{\mathfrak{x}\mathfrak{y}} = T_{\mathfrak{y}}T_{\mathfrak{x}}$, and therefore we obtain our assertions.

PROPOSITION 5.3. *If* $f \in \mathcal{M}_k(\mathfrak{c}, \psi)$, $g \in \mathcal{S}_k(\mathfrak{c}, \psi)$, *and* $\mathfrak{x}$ *is prime to* $\mathfrak{c}$, *then*
$\langle f, g|T_{\mathfrak{x}} \rangle = \psi^*(\mathfrak{x})^2 \langle f|T_{\mathfrak{x}}, g \rangle$.

This follows immediately from our definition (5.5) and the equality $J_{\Xi}(\theta^{-1}, z) = J_{\Xi}(\theta, \theta^{-1}z)^{-1}$ in Lemma 5.1 by the standard argument.

PROPOSITION 5.4. *For* $f \in \mathcal{M}_k(\mathfrak{c}, \psi)$ *and a prime ideal* $\mathfrak{p}$, *we have*

$$\lambda(\xi, \mathfrak{m}; f|T_{\mathfrak{p}}) = \lambda_f(\xi, \mathfrak{m}\mathfrak{p})$$

$$+ \begin{cases} \psi^*(\mathfrak{p})N(\mathfrak{p})^{-1}\left(\dfrac{\xi c^2}{\mathfrak{p}}\right)\lambda_f(\xi, \mathfrak{m}) \\ \qquad + \psi^*(\mathfrak{p})^2 N(\mathfrak{p})^{-1}\lambda_f(\xi, \mathfrak{m}\mathfrak{p}^{-1}) & \text{if } \mathfrak{p} \not\supset \mathfrak{c}, \\ 0 & \text{if } \mathfrak{p} \supset \mathfrak{c}, \end{cases}$$

provided $\xi \in \mathfrak{m}^{-2}$, *where* c *is an element of* $F_{\mathfrak{p}}$ *such that* $\mathfrak{m}_{\mathfrak{p}} = c\mathfrak{o}_{\mathfrak{p}}$, *and* $\left(\dfrac{d}{\mathfrak{p}}\right)$ *is the quadratic residue symbol, that is, the number of solutions of* $x^2 \equiv d \pmod{\mathfrak{p}}$ *minus* 1.

Proof. This is a generalization of [S1, Theorem 1.7] which deals with the case $F = \mathbf{Q}$; the present proof is completely parallel to that special case. We first assume that $\mathfrak{p}$ is prime to $\mathfrak{c}$. Take $\pi \in F_{\mathfrak{p}}^{\times}$ (embedded in $F_{\mathbf{A}}^{\times}$) so that $\pi\mathfrak{o} = \mathfrak{p}$ and put

$$\alpha_b = \begin{pmatrix} 1/\pi & 0 \\ 0 & \pi \end{pmatrix}\begin{pmatrix} 1 & 2b/\delta \\ 0 & 1 \end{pmatrix}_{\mathfrak{f}}, \qquad (b \in \mathfrak{o}/\mathfrak{p}^2),$$

$$\beta_h = \begin{pmatrix} 1 & 2h/(\pi\delta) \\ 0 & 1 \end{pmatrix}_{\mathfrak{f}} = \begin{pmatrix} 1 & 0 \\ e\pi\delta/2 & 1 \end{pmatrix}_{\mathfrak{f}}\alpha_0\begin{pmatrix} \pi & 2h/\delta \\ -e\delta/2 & q \end{pmatrix}_{\mathfrak{f}}, \qquad \left(h \in (\mathfrak{o}/\mathfrak{p})^{\times}\right),$$

$$\alpha' = \begin{pmatrix} \pi & 0 \\ 0 & 1/\pi \end{pmatrix}.$$

Here e and q are chosen and fixed for each h so that $\pi q + he = 1$, $0 \neq e \in \mathfrak{c}$, $q \in F_{\mathbf{A}}^{\times}$, and $\pi q \in \mathfrak{p}$. The images of these under r_P form a complete set of representatives of $U \backslash Ur_P(\alpha_0)U$. By (5.3), Proposition 2.3(iv), and Lemma 5.1, we have $J_{\Xi}(r_P(\alpha_b), z) = J_{\Xi}(r_P(\alpha'), z) = 1$. As for β_h, put

$$\sigma = \begin{bmatrix} 1 & 0 \\ -e\pi\delta/2 & 1 \end{bmatrix}_{\mathfrak{f}}, \qquad \tau = \begin{bmatrix} \pi & 2h/\delta \\ -e\delta/2 & q \end{bmatrix}_{\mathfrak{f}}.$$

By (1.11), $r_{\Omega}(\sigma)r_P(\beta_h) = r_{\Omega}(\sigma\beta_h) = r_{\Omega}(\alpha_0\tau) = r_P(\alpha_0)r_{\Omega}(\tau)$, and so

$$J_{\Xi}(r_P(\beta_h), z) = J\left(r_{\Omega}(\sigma)^{-1}r_{\Omega}(\tau), z\right)$$

$$= h\left(r_{\Omega}(\tau), z\right)/h\left(r_{\Omega}(\sigma), z\right) = |\gamma/\gamma'|\gamma'/\gamma$$

by (2.18a), where $\gamma = \gamma(2(e\pi\delta)^{-1})$ and $\gamma' = \gamma(2q(e\delta)^{-1})$. Since $q(e\delta)^{-1} - (e\pi\delta)^{-1} = -h(\pi\delta)^{-1}$, we have $\mathbf{e}_v(xq(e\delta)^{-1}) = \mathbf{e}_v(x(e\pi\delta)^{-1})$ for $x \in \mathfrak{o}_v$ if $v \neq \mathfrak{p}$, and hence $\gamma_v = \gamma'_v$ for $v \neq \mathfrak{p}$. On the other hand, $|e|_\mathfrak{p} = 1$, so that $\gamma'_\mathfrak{p} = 1$, and therefore

$$(5.8a) \qquad J_\Xi(r_P(\beta_h), z) = |\gamma_\mathfrak{p}|/\gamma_\mathfrak{p}.$$

Since $(\pi e)^{-1} - \pi^{-1}h = e^{-1}q$ and this is $\mathfrak{p}$-integral, we have, by (2.17b),

$$\gamma_\mathfrak{p} = N(\mathfrak{d}_\mathfrak{p})^{1/2} \int_{\mathfrak{o}_\mathfrak{p}} \mathbf{e}_\mathfrak{p}(\delta^{-1}\pi^{-1}hx^2)\, dx,$$

which is essentially a Gauss sum, so that

$$(5.8b) \qquad \gamma_\mathfrak{p}/|\gamma_\mathfrak{p}| = N(\mathfrak{p})^{-1/2} \sum_{x \in \mathfrak{o}/\mathfrak{p}} \mathbf{e}_\mathfrak{p}(\delta^{-1}\pi^{-1}hx^2).$$

Now if $g = f|T_\mathfrak{p}$, (5.6) shows that

$$N(\mathfrak{p})^{3/2}\psi^*(\mathfrak{p})^{-1}g_\mathbf{A}(x) = \sum_b f_\mathbf{A}\big(xr_P(\alpha_b)^{-1}\big)$$

$$+ \sum_h f_\mathbf{A}\big(xr_P(\beta_h)^{-1}\big)J_\Xi(r_P(\beta_h), z)^{-1} + f_\mathbf{A}\big(xr_P(\alpha')^{-1}\big).$$

If $x = r_P\!\begin{bmatrix} t & s \\ 0 & t^{-1} \end{bmatrix}$, we see from Proposition 3.1 that

$$\psi_\mathfrak{f}(t)t_\mathbf{a}^{-m}|t|_\mathbf{A}^{-1/2}\sum_b f_\mathbf{A}\big(xr_P(\alpha_b)^{-1}\big)$$

$$= \psi_\mathfrak{f}(\pi)^{-1}N(\mathfrak{p})^{-1/2} \sum_{\xi \in F} \lambda_f(\xi, t\mathfrak{p}) \sum_b \mathbf{e}_\mathfrak{f}(-\xi bt^2/\delta)\mathbf{e}_\mathbf{a}(it^2\xi/2)\mathbf{e}_\mathbf{A}(ts\xi/2)$$

$$= \psi^*(\mathfrak{p})^{-1}N(\mathfrak{p})^{3/2} \sum_\xi \lambda_f(\xi, t\mathfrak{p})\mathbf{e}_\mathbf{a}\big(it^2\xi/2\big)\mathbf{e}_\mathbf{A}(ts\xi/2) \qquad (\xi \in t^{-2}\mathfrak{o}).$$

Next, as for Σ_h, we have, by (5.8a, b),

$$\psi_\mathfrak{f}(t)t_\mathbf{a}^{-m}|t|_\mathbf{A}^{-1/2}\sum_h f_\mathbf{A}\big(xr_P(\beta_h)^{-1}\big)J_\Xi(r_P(\beta_h), z)^{-1}$$

$$= N(\mathfrak{p})^{-1/2} \sum_{\xi \in F} \lambda_f(\xi, t\mathfrak{o})S_\xi\mathbf{e}_\mathbf{a}\big(it^2\xi/2\big)\mathbf{e}_\mathbf{A}(ts\xi/2)$$

with $S_\xi = \Sigma_{h,\,x}\mathbf{e}_\mathfrak{p}(h(x^2 - \xi t^2)/(\pi\delta))$, where h runs over $(\mathfrak{o}/\mathfrak{p})^\times$ and x over $\mathfrak{o}/\mathfrak{p}$.

We see easily that $S_\xi = N(\mathfrak{p})\left(\dfrac{\xi t^2}{\mathfrak{p}}\right)$. Finally

$$\psi_{\mathfrak{f}}(t)t_{\mathfrak{a}}^{-m}|t|_{\mathbf{A}}^{-1/2}f_{\mathbf{A}}\left(xr_P(\alpha')^{-1}\right)$$

$$= \psi_{\mathfrak{f}}(\pi)N(\mathfrak{p})^{1/2}\sum_{\xi}\lambda_f(\xi, t\mathfrak{p}^{-1})\mathbf{e}_{\mathfrak{a}}(it^2\xi/2)\mathbf{e}_{\mathbf{A}}(ts\xi/2).$$

Summing up all these, we obtain the formula for $\lambda_g(\xi, \mathfrak{m})$ when $\mathfrak{p}$ is prime to $\mathfrak{c}$. If $\mathfrak{p}$ divides $\mathfrak{c}$, we can repeat the above computation only with the sum involving the α_b, and obtain the desired result.

In order to attach several formal Dirichlet series to f, it is convenient to introduce a system of formal multiplicative symbols $M(\mathfrak{m})$ for the fractional ideals $\mathfrak{m}$ of F as follows: the $M(\mathfrak{p})$ for the prime ideals $\mathfrak{p}$ are independent indeterminates; $M(\mathfrak{m}\mathfrak{n}) = M(\mathfrak{m})M(\mathfrak{n})$ and $M(\mathfrak{o}) = 1$. Then we can speak of the commutative ring of formal Dirichlet series $\sum_{\mathfrak{m} \subset \mathfrak{o}}c_{\mathfrak{m}}M(\mathfrak{m})$ with $c_{\mathfrak{m}} \in \mathbf{C}$.

THEOREM 5.5.   *Let* $0 \neq f \in \mathscr{M}_k(\mathfrak{c}, \psi), 0 \ll \tau \in \mathfrak{o}$, *and* $\tau\mathfrak{o} = \mathfrak{q}^2\mathfrak{r}$ *with integral ideals* $\mathfrak{q}$ *and* $\mathfrak{r}$. *Suppose there is an integral ideal* $\mathfrak{t}$ *such that* $\mathfrak{r}$ *has no square factor prime to* $\mathfrak{t}$ *and suppose* $f|T_{\mathfrak{p}} = \omega_{\mathfrak{p}}f$ *for every prime ideal* $\mathfrak{p}$ *not dividing* $\mathfrak{t}$. *Then*

$$\sum_{\mathfrak{m}+\mathfrak{t}=\mathfrak{o}} \lambda_f(\tau, \mathfrak{q}^{-1}\mathfrak{m})M(\mathfrak{m})$$

$$= \lambda_f(\tau, \mathfrak{q}^{-1})\prod_{\mathfrak{p}\nmid\mathfrak{t}}\left[1 - N(\mathfrak{p})^{-1}(\psi'\varepsilon_\tau^*)(\mathfrak{p})M(\mathfrak{p})\right]$$

$$\cdot\left[1 - \omega_{\mathfrak{p}}M(\mathfrak{p}) + \psi'(\mathfrak{p})^2 N(\mathfrak{p})^{-1}M(\mathfrak{p}^2)\right]^{-1},$$

where $\mathfrak{m}$ *runs over all the integral ideals in F prime to* $\mathfrak{t}$, $\mathfrak{p}$ *over all the prime ideals in F prime to* $\mathfrak{t}$, ε_τ *is defined as in Proposition 3.2, and* $\psi'(\mathfrak{p}) = \psi^*(\mathfrak{p})$ *or* $= 0$ *according as* $\mathfrak{p} \nmid \mathfrak{c}$ *or* $\mathfrak{p}|\mathfrak{c}$.

Proof.   Write simply λ for λ_f. Fix a prime ideal $\mathfrak{p}$ not dividing $\mathfrak{t}$, and an integral ideal $\mathfrak{a}$ prime to $\mathfrak{p}$. By Proposition 5.4 and (3.14a), we have

$$(*)\quad \omega_{\mathfrak{p}}\lambda(\tau, \mathfrak{q}^{-1}\mathfrak{a}) = \lambda(\tau, \mathfrak{q}^{-1}\mathfrak{p}\mathfrak{a}) + (\psi'\varepsilon_\tau^*)(\mathfrak{p})N(\mathfrak{p})^{-1}\lambda(\tau, \mathfrak{q}^{-1}\mathfrak{a}),$$

$$(**)\quad \omega_{\mathfrak{p}}\lambda(\tau, \mathfrak{q}^{-1}\mathfrak{p}^n\mathfrak{a}) = \lambda(\tau, \mathfrak{q}^{-1}\mathfrak{p}^{n+1}\mathfrak{a}) + \psi'(\mathfrak{p})^2 N(\mathfrak{p})^{-1}\lambda(\tau, \mathfrak{q}^{-1}\mathfrak{p}^{n-1}\mathfrak{a})$$

for $0 < n \in \mathbf{Z}$. Let $H_{\mathfrak{a}} = H_{\mathfrak{a}}(x) = \sum_{n=0}^{\infty}\lambda(\tau, \mathfrak{q}^{-1}\mathfrak{p}^n\mathfrak{a})x^n$ with an indeterminate x. Adding x times $(*)$ and x^{n+1} times $(**)$ for all $n > 0$, we obtain

$$\omega_{\mathfrak{p}}xH_{\mathfrak{a}} = H_{\mathfrak{a}} - \lambda(\tau, \mathfrak{q}^{-1}\mathfrak{a}) + \psi'(\mathfrak{p})^2 N(\mathfrak{p})^{-1}x^2 H_{\mathfrak{a}}$$

$$+ (\psi'\varepsilon_\tau^*)(\mathfrak{p})N(\mathfrak{p})^{-1}\lambda(\tau, \mathfrak{q}^{-1}\mathfrak{a})x,$$

and hence $H_a(x) = \lambda(\tau, q^{-1}a)E_{\mathfrak{p}}(x)$ with

$$E_{\mathfrak{p}}(x) = \left[1 - (\psi'\varepsilon_{\tau}^*)(\mathfrak{p})N(\mathfrak{p})^{-1}x\right]\left[1 - \omega_{\mathfrak{p}}x + \psi'(\mathfrak{p})^2 N(\mathfrak{p})^{-1}x^2\right]^{-1}.$$

Therefore

$$\sum_{\mathfrak{m}} \lambda(\tau, q^{-1}\mathfrak{m})M(\mathfrak{m}) = \sum_{\mathfrak{p}\nmid a} H_a(M(\mathfrak{p}))M(a)$$

$$= E_{\mathfrak{p}}(M(\mathfrak{p})) \sum_{\mathfrak{p}\nmid a} \lambda(\tau, q^{-1}a)M(a).$$

Repeating this procedure successively with all $\mathfrak{p}$ prime to $\mathfrak{t}$, we obtain our theorem.

Suppose, in the setting of Theorem 5.5, that $F = \mathbf{Q}$, $f = \sum_{n=0}^{\infty}\lambda(n)e(nz/2)$, $\tau = q^2 r$, and $\mathfrak{t} = t\mathbf{Z}$ with positive integers q, r, and t. Then $\lambda_f(\tau, q^{-1}n\mathbf{Z}) = (q/n)^m \lambda(rn^2)$ by (3.14b) and (3.15), so that

$$(5.9) \qquad \sum_{(n,\,t)=1} \lambda_f(\tau, q^{-1}n\mathbf{Z})n^{-s} = q^m \sum_{n} \lambda(rn^2)n^{-m-s}.$$

Thus the content of the above theorem, when $F = \mathbf{Q}$, is essentially the same as that of our previous paper [S1, Corollary 1.8 and Theorem 1.9].

6. Main theorems on the correspondence. Let us first recall the theory of Hecke operators on the forms of integral weight, following the formulation of [S4, pp. 352–353], which is somewhat (but not much) more general than that of [S3]. Let $\mathfrak{b}$ be an integral ideal in F, and φ a Hecke character of F such that

$(6.1a)$ *the conductor of φ divides* $\mathfrak{b}$,

$(6.1b)$ $\varphi_{\mathbf{a}}(x) = \mathrm{sgn}(x_{\mathbf{a}})^n|x_{\mathbf{a}}|^{2i\mu},$

where $0 \leq n \in \mathbf{Z}^{\mathbf{a}}$ and $\mu \in \mathbf{R}^{\mathbf{a}}$; we assume $\{\mu\} = 0$. We then consider a function $\mathbf{g}: \tilde{G}_{\mathbf{A}} \to \mathbf{C}$ such that

$(6.2a)$ $\mathbf{g}(\alpha xw) = \varphi_{\mathfrak{b}}(d_w)j(w, \mathbf{i})^{-n}\mathbf{g}(x)$ *for $\alpha \in \tilde{G}$, and $w \in \tilde{D}[\mathfrak{b}^{-1}, \mathfrak{b}\mathfrak{b}]$*;

$(6.2b)$ *for every $x \in \tilde{G}_{\mathfrak{f}}$, there is an element f_x of $\mathcal{M}_n$ such that*

$$\mathbf{g}(xy) = \det(y)^{i\mu}(f_x\|_n y)(\mathbf{i}) \quad \textit{for } y \in \tilde{G}_{\mathbf{a}+};$$

$(6.2c)$ $\mathbf{g}(sx) = \varphi(s)\mathbf{g}(x)$ *for $s \in F_{\mathbf{A}}^{\times}$ and $x \in \tilde{G}_{\mathbf{A}}$.*

We denote by $\mathscr{M}_n(\mathfrak{b}, \varphi)$ the set of all such functions g. We also put

$$(6.3) \qquad W_\mathfrak{b} = \tilde{D}[\mathfrak{d}^{-1}, \mathfrak{b}\mathfrak{d}]\tilde{G}_{\mathbf{a}+}.$$

Taking a disjoint coset decomposition of $\tilde{G}_\mathbf{A}$ of the form

$$(6.4a) \qquad \tilde{G}_\mathbf{A} = \coprod_{\lambda=1}^{\kappa} \tilde{G}x_\lambda W_\mathfrak{b}, \qquad x_\lambda = \mathrm{diag}[1, t_\lambda]$$

with t_λ in $F_\mathbf{f}^\times$, we put

$$(6.4b) \qquad \tilde{\Gamma}_\lambda = \tilde{\Gamma}\big[t_\lambda^{-1}\mathfrak{d}^{-1}, t_\lambda\mathfrak{b}\mathfrak{d}\big] = \tilde{G} \cap x_\lambda W_\mathfrak{b} x_\lambda^{-1} \qquad (1 \le \lambda \le \kappa).$$

Let $\mathscr{M}_n(\tilde{\Gamma}_\lambda, \varphi_\mathfrak{b}, \mu)$ denote the set of all f in $\mathscr{M}_n$ satisfying

$$(6.5) \qquad f\|_n\gamma = \varphi_\mathfrak{b}(a_\gamma)\det(\gamma)^{i\mu}f \quad \text{for every } \gamma \in \tilde{\Gamma}_\lambda.$$

We put then $\mathscr{M}_n(\varphi_\mathfrak{b}, \mu) = \prod_{\lambda=1}^{\kappa}\mathscr{M}_n(\tilde{\Gamma}_\lambda, \varphi_\mathfrak{b}, \mu)$. Given $(g_1, \ldots, g_\kappa) \in \mathscr{M}_n(\varphi_\mathfrak{b}, \mu)$, we can define a function g on $\tilde{G}_\mathbf{A}$ by

$$(6.6) \quad \mathbf{g}(\alpha x_\lambda^* w) = \varphi_\mathfrak{b}(d_w)\det(w_\mathbf{a})^{i\mu}(g_\lambda\|_n w_\mathbf{a})(\mathbf{i}) \quad \text{for } \alpha \in \tilde{G} \text{ and } w \in W_\mathfrak{b},$$

where $x_\lambda^* = \mathrm{diag}[t_\lambda^{-1}, 1]$. Then g satisfies (6.2a, b); moreover, every g satisfying (6.2a, b) can be obtained from such $(g_1, \ldots, g_\kappa)$. Identifying g with $(g_1, \ldots, g_\kappa)$, we may view $\mathscr{M}_n(\mathfrak{b}, \varphi)$ as a subset of $\mathscr{M}_n(\varphi_\mathfrak{b}, \mu)$. We denote by $\mathscr{S}_n(\mathfrak{b}, \varphi)$ and $\mathscr{S}_n(\varphi_\mathfrak{b}, \mu)$ the subsets of these consisting of all those g for which the g_λ are cusp forms.

Now each g_λ has a Fourier expansion

$$(6.7a) \qquad g_\lambda(z) = \sum_{\xi\in F} c_\lambda(\xi)\mathbf{e}_\mathbf{a}(\xi z) \qquad (z \in \mathscr{H}^\mathbf{a}).$$

Then for every fractional ideal $\mathfrak{m}$ we put

$$(6.7b) \qquad c(\mathfrak{m}, \mathbf{g}) = \begin{cases} c_\lambda(\xi)\xi^{-(n/2)-i\mu} & \text{if } \mathfrak{m} = \xi t_\lambda^{-1}\mathfrak{o} \subset \mathfrak{o}, \\ 0 & \text{if } \mathfrak{m} \not\subset \mathfrak{o}. \end{cases}$$

The coefficients $c(\mathfrak{m}, \mathbf{g})$ can be obtained from g as a function on $\tilde{G}_\mathbf{A}$ without g_λ; hence they are independent of the choice of t_λ (see [S4, (9.28)]).

For each integral ideal $\mathfrak{n}$ we can define the Hecke operator $\mathfrak{T}(\mathfrak{n})$ acting on $\mathscr{M}_n(\varphi_\mathfrak{b}, \mu)$ with the property that

$$c(\mathfrak{m}, \mathbf{g}|\mathfrak{T}(\mathfrak{n})) = \sum_\mathfrak{a}\varphi^*(\mathfrak{a})N(\mathfrak{a}^{-1})c(\mathfrak{a}^{-2}\mathfrak{m}\mathfrak{n}, \mathbf{g}),$$

where $\mathfrak{a}$ runs over all the integral ideals prime to $\mathfrak{c}$ and dividing $\mathfrak{m} + \mathfrak{n}$. (For details, see [S4, p. 353] and [S3, §2]. We take the present $\mathfrak{T}(\mathfrak{n})$ to be $N(\mathfrak{n})^{-1}$ times those of [S3, (2.20)] and [S4, (9.29)].) Then the following three conditions on $\mathbf{g} \in \mathcal{M}_n(\varphi_{\mathfrak{b}}, \mu)$ are equivalent:

(6.8a)

$\mathbf{g} \in M_n(\mathfrak{b}, \varphi)$ *and* $\mathbf{g}|\mathfrak{T}(\mathfrak{n}) = \omega(\mathfrak{n})\mathbf{g}$ *with* $\omega(\mathfrak{n}) \in \mathbf{C}$ *for every integral ideal* $\mathfrak{n}$;

(6.8b)

$\mathbf{g} \in \mathcal{M}_n(\mathfrak{b}, \varphi)$ *and* $\mathbf{g}|\mathfrak{T}(\mathfrak{p}) = \omega(\mathfrak{p})\mathbf{g}$ *with* $\omega(\mathfrak{p}) \in \mathbf{C}$ *for every prime ideal* $\mathfrak{p}$;

(6.8c)

$$\sum_{\mathfrak{m} \subset \mathfrak{o}} c(\mathfrak{m}, \mathbf{g}) M(\mathfrak{m}) = c(\mathfrak{o}, \mathbf{g}) \prod_{\mathfrak{p}} \left[1 - \omega(\mathfrak{p}) M(\mathfrak{p}) + \varphi'(\mathfrak{p}) N(\mathfrak{p})^{-1} M(\mathfrak{p}^2) \right]^{-1}$$

where $\varphi'(\mathfrak{p}) = \varphi^*(\mathfrak{p})$ *or* $= 0$ *according as* $\mathfrak{p} \not\supset \mathfrak{b}$ *or* $\mathfrak{p} \supset \mathfrak{b}$.

We call $\mathbf{g}$ a *normalized eigenform* if these conditions are satisfied and $c(\mathfrak{o}, \mathbf{g}) = 1$. Now the first set of our main results can be stated as the following three theorems (6.1, 6.2, and 6.4):

THEOREM 6.1. *Let* $0 \neq f \in \mathscr{S}_k(\mathfrak{c}, \psi)$ *with a half-integral weight* $k = (u/2) + m \geq 3u/2$, *an integral ideal* $\mathfrak{c}$ *divisible by* 4, *and a Hecke character* ψ *of* F *as in* (3.8a, b). *Let* τ *be an arbitrary totally positive element of* $\mathfrak{o}$, *and let* $\tau\mathfrak{o} = \mathfrak{q}^2\mathfrak{r}$ *with an integral ideal* $\mathfrak{q}$ *and a squarefree integral ideal* $\mathfrak{r}$. *Put* $\mathfrak{b} = 2^{-1}\mathfrak{c}$ *and suppose* $\psi(x) = |x|^{i\mu}$ *for* $0 \ll x \in F_{\mathbf{a}}^{\times}$ *with* $\mu \in \mathbf{R}^{\mathbf{a}}$, $\{\mu\} = 0$. *Then the following three assertions hold*:

(I) *For every fractional ideal* $\mathfrak{y}$ *of* F, *there exists an element* $g(z) = \Sigma_{\xi \in F} c(\xi) \mathbf{e}_{\mathbf{a}}(\xi z)$ *of* $\mathcal{M}_{2m}(\tilde{\Gamma}[\mathfrak{y}^{-1}\mathfrak{d}^{-1}, \mathfrak{b}\mathfrak{y}\mathfrak{d}], \psi_{\mathfrak{b}}^2, \mu)$ *such that*

$$\sum_{0 \ll \xi \in \mathfrak{y}/\mathfrak{o}_+^{\times}} c(\xi) \xi^{-m-i\mu} M(\xi\mathfrak{y}^{-1})$$

$$= \sum_{\mathfrak{m}} \lambda_f(\tau, \mathfrak{q}^{-1}\mathfrak{m}) M(\mathfrak{m}) \sum_{\mathfrak{n}\mathfrak{m}\mathfrak{y} \sim 1} (\psi\varepsilon_\tau)^*(\mathfrak{n}) N(\mathfrak{n})^{-1} M(\mathfrak{n}),$$

where $\mathfrak{o}_+^{\times} = \{a \in \mathfrak{o}^{\times} | a \gg 0\}$, ε_τ *is the Hecke character of* F *corresponding to* $F(\tau^{1/2})/F$, $\mathfrak{m}$ *runs over all the integral ideals of* F, *and* $\mathfrak{n}$ *over all the integral ideals of* F *which are prime to* $\mathfrak{r}\mathfrak{c}$ *and equivalent to* $(\mathfrak{y}\mathfrak{m})^{-1}$ *modulo* $\mathbf{a}$.

(II) *Let* $\mathbf{g}_\tau = (g_1, \ldots, g_\kappa)$ *with* g_λ *as the form* g *of* (I) *for* $\mathfrak{y} = t_\lambda \mathfrak{o}$. *Then* $\mathbf{g}_\tau \in \mathcal{M}_{2m}(\mathfrak{b}, \psi^2)$, *and*

$$\sum_{\mathfrak{m}} c(\mathfrak{m}, \mathbf{g}_\tau) M(\mathfrak{m})$$

$$= \sum_{\mathfrak{m}} \lambda_f(\tau, \mathfrak{q}^{-1}\mathfrak{m}) M(\mathfrak{m}) \prod_{\mathfrak{p} \nmid \mathfrak{r}\mathfrak{c}} \left[1 - (\psi\varepsilon_\tau)^*(\mathfrak{p}) N(\mathfrak{p})^{-1} M(\mathfrak{p}) \right]^{-1},$$

where m *runs over all the integral ideals of F, and* $\mathfrak{p}$ *over all the prime ideals which do not divide* $\mathfrak{r}\mathfrak{c}$.

(III) *The form g of* (I) *is a cusp form, if* $m \neq u$ *or if* $m = u$ *and f is orthogonal to every theta series p of the form* $p(z) = \theta_u(az, \eta)$ *with* $0 \ll a \in F$ *and* $\theta_u(z, \eta)$ *of type* (4.5).

The proof will be given in Section 7. As we already said, our formulation requires a *compatible pair* (f, ψ). In general, f alone cannot determine the form g of integral weight except for the case $F = \mathbf{Q}$; the choice of a Hecke character ψ must be made.

THEOREM 6.2. *The notation being the same as in the above theorem, suppose* $f|T_{\mathfrak{p}} = \omega_{\mathfrak{p}}f$ *with* $\omega_{\mathfrak{p}} \in \mathbf{C}$ *for every prime ideal* $\mathfrak{p}$. *Then there exists a nonzero element* $\mathbf{h}$ *of* $\mathcal{M}_{2m}(\mathfrak{b}, \psi^2)$ *such that* $\mathbf{h}|\mathfrak{X}(\mathfrak{p}) = \omega_{\mathfrak{p}}\mathbf{h}$ *for every* $\mathfrak{p}$.

Proof. For $\mathbf{g}_\tau$ of Theorem 6.1(II), we have, by Theorem 5.5,

$$\sum_{\mathfrak{m}} c(\mathfrak{m}, \mathbf{g}_\tau) M(\mathfrak{m}) = \lambda_f(\tau, q^{-1}) \prod_{\mathfrak{p}} \left[1 - \omega_{\mathfrak{p}} M(\mathfrak{p}) + \psi'(\mathfrak{p})^2 N(\mathfrak{p})^{-1} M(\mathfrak{p}^2) \right]^{-1}.$$

Since $f \neq 0$, $\lambda_f(\tau, q^{-1}\mathfrak{m}) \neq 0$ for some τ and an integral $\mathfrak{m}$. Then $\mathbf{g}_\tau$ for such a τ gives the desired $\mathbf{h}$, for (6.8c) implies (6.8a, b).

Remark 6.3. If $0 \neq f \in \mathcal{M}_k(\mathfrak{c}, \psi)$, then $f \in \mathcal{M}_k(\mathfrak{c}, \psi')$ if and only if $\psi' = \eta\psi$ with a character η which is unramified at every $v \in \mathbf{f}$. From Proposition 3.1, we see that

$$(6.9) \qquad \lambda_f(\xi, \mathfrak{m}; \eta\psi) = \eta^*(\mathfrak{m})\lambda_f(\xi, \mathfrak{m}; \psi).$$

Therefore, if g corresponds to (f, ψ) in the sense of Theorem 6.1(I), then $\eta^*(\mathfrak{y}^{-1})g$ corresponds to $(f, \eta\psi)$. If $\mathbf{g} = (g_\lambda)$ corresponds to (f, ψ) in the sense of (II), then $\mathbf{g}' = (\eta(t_\lambda)^{-1}g_\lambda)$ corresponds to $(f, \eta\psi)$. As a function on $\tilde{G}_\mathbf{A}$, $\mathbf{g}'$ is given by $\mathbf{g}'(x) = \eta(\det(x))\mathbf{g}(x)$. Furthermore, the eigenvalue $\omega_{\mathfrak{p}}$ is replaced by $\eta^*(\mathfrak{p})\omega_{\mathfrak{p}}$.

The above theorems concern only the case in which $k \geq 3u/2$. For the weights which do not satisfy this condition, we have:

THEOREM 6.4. *Suppose* $k_v = 1/2$ *for some* $v \in \mathbf{a}$. *Then* $\mathcal{M}_k \neq \{0\}$ *only if* $k \leq 3u/2$, *that is, only if* $m \leq u$, *in which case* $\mathcal{M}_k$ *is spanned by the theta series of the form* $\theta_m(az, \eta)$ *with* $0 \ll a \in F$ *and* $\theta_m(z, \eta)$ *of type* (4.5).

This will be proved in Section 7bis.

Let us now show that the proof of Theorem 6.1 can be reduced to the case $\tau = 1$. Given f, τ, q, and $\mathfrak{r}$, Proposition 3.2 guarantees an element h of $\mathcal{S}_k(\mathfrak{r}\mathfrak{c}, \psi\varepsilon_\tau)$ such that $\lambda_h(\xi, \mathfrak{m}) = \lambda_f(\xi/\tau, q\mathfrak{m})$. By (3.15) and (3.14c) we have

$$(6.10) \qquad h(z) = \sum \lambda_f(\xi, q)\mathbf{e}_\mathbf{a}(\tau\xi z/2) = cJ(\beta, \beta^{-1}(\tau z))f(\beta^{-1}(\tau z))$$

with a constant c and an element β of G. This, together with Lemma 4.1, shows that h satisfies the orthogonality condition of (III) if f satisfies it (when $m = u$). Assuming (I) and (III) of Theorem 6.1 to be true for h in the case $\tau = 1$, we can find, for every fractional ideal $\mathfrak{x}$, an element $g' = \Sigma c'(\xi)e_\mathfrak{a}(\xi z)$ of $\mathcal{M}_{2m}(\tilde{\Gamma}[\mathfrak{x}^{-1}\mathfrak{d}^{-1}, \mathfrak{b}\mathfrak{r}\mathfrak{x}\mathfrak{d}], \psi^2_{\mathfrak{b}\mathfrak{r}}, \mu)$, which is a cusp form under the conditions of (III) and such that

$$\sum_{0 \ll \xi \in F/\mathfrak{o}^\times_+} c'(\xi)\xi^{-m-i\mu}M(\xi\mathfrak{x}^{-1})$$

$$= \sum_{m'} \lambda_h(1, m')M(m')\sum_{n}(\psi\varepsilon_r)^*(n)N(n)^{-1}M(n),$$

where m' runs over all the integral ideals, and n over all those which are prime to $\mathfrak{r}\mathfrak{c}$ and equivalent to $(\mathfrak{x}m')^{-1}$. By (3.14b) we have

$$\lambda_h(1, m') = \lambda_f(1/\tau, \mathfrak{q}m') = \lambda_f(\tau, \mathfrak{q}^{-1}\mathfrak{r}^{-1}m')\tau^{-m-i\mu}.$$

By (3.14a) this is not 0 only when $m'^2 \subset \mathfrak{r}$, that is, when $m' \subset \mathfrak{r}$, since $\mathfrak{r}$ is squarefree. Putting $\mathfrak{x} = \mathfrak{r}^{-1}\mathfrak{y}$ and $m' = m\mathfrak{r}$, we see that the Fourier coefficients of $\tau^{m+i\mu}g'$ have the property of Theorem 6.1(I). This shows that $c'(\xi) \neq 0$ only when $\xi \in \mathfrak{y}$, so that $g(z + b) = g(z)$ for $b \in \mathfrak{d}^{-1}\mathfrak{y}^{-1}$. Hence $g' \in \mathcal{M}_{2m}(\tilde{\Gamma}[\mathfrak{y}^{-1}\mathfrak{d}^{-1}, \mathfrak{b}\mathfrak{y}\mathfrak{d}], \psi^2_{\mathfrak{b}}, \mu)$ as desired.

As for (II), if $\mathbf{g}_r = (g_1, \ldots, g_\kappa)$ as given there, we see, by taking the sum of the equalities for g_λ in (I) for all λ, that the $c(m, \mathbf{g}_r)$ have the stated property. The nontrivial point of (II) is that $\mathbf{g}_r$ satisfies (6.2c), that is, $\mathbf{g}_r(sx) = \psi(s)^2\mathbf{g}_r(x)$ for $s \in F_\mathbf{A}^\times$. Suppose this is so for the above h when $\tau = 1$, and let $\mathbf{g}'$ be the corresponding function on $\tilde{G}_\mathbf{A}$ belonging to $\mathcal{M}_{2m}(\mathfrak{b}\mathfrak{r}, \psi^2)$. Take an element r of $F_\mathbf{f}^\times$ so that $r\mathfrak{o} = \mathfrak{r}$ and identify $\mathbf{g}'$ with an element (g_λ') of

$$\prod_{\lambda=1}^\kappa \mathcal{M}_{2m}(\tilde{\Gamma}[rt_\lambda^{-1}\mathfrak{d}^{-1}, r^{-1}t_\lambda\mathfrak{b}\mathfrak{r}\mathfrak{d}], \psi^2_{\mathfrak{b}\mathfrak{r}}, \mu)$$

by the principle of (6.6), but with respect to $\{r^{-1}t_\lambda\}$ instead of $\{t_\lambda\}$, that is,

$$(6.11) \qquad g'(\alpha \operatorname{diag}[rt_\lambda^{-1}, 1]w) = \psi^2_{\mathfrak{b}\mathfrak{r}}(d_w)\det(w_\mathbf{a})^{i\mu}(g_\lambda'\|w_\mathbf{a})(\mathbf{i})$$

for $\alpha \in \tilde{G}$ and $w \in W_{\mathfrak{b}\mathfrak{r}}$. Our assumption implies that $\mathbf{g}'(sx) = \psi(s)^2\mathbf{g}'(x)$ for $s \in F_\mathbf{A}^\times$. Since $\mathbf{g}'$ corresponds to h, g_λ' is the form g of (I) corresponding to $(h, \psi\varepsilon_r)$ with $r^{-1}t_\lambda\mathfrak{o}$ as $\mathfrak{y}$. Let $g_\lambda = \tau^{m+i\mu}g_\lambda'$. We have seen that g_λ belongs to $\mathcal{M}_{2m}(\tilde{\Gamma}_\lambda, \psi^2_{\mathfrak{b}}, \mu)$ and has the property of (I) with $\mathfrak{y} = t_\lambda\mathfrak{o}$. Define $\mathbf{g}$ by (6.6) with these g_λ and with ψ^2 as φ. Comparing (6.6) with (6.11), we easily see that $\mathbf{g}(x) = \tau^{m+i\mu}\mathbf{g}'(x \cdot \operatorname{diag}[r, 1])$, and hence $\mathbf{g}$ satisfies (6.2c) as expected.

The rest of this section is devoted to the discussion of certain Eisenstein series needed for the proof of Theorem 6.1. We first put

(6.12a) $$P_+ = \{\alpha \in P \mid d_\alpha \gg 0\}.$$

Then we can find a finite subset B of G such that

(6.12b) $$G \cap P_A D_{\mathfrak{c}} = \coprod_{\beta \in B} P_+ \beta \Gamma_{\mathfrak{c}}$$

with $D_{\mathfrak{c}}$ and $\Gamma_{\mathfrak{c}}$ of (3.9), $\mathfrak{c} \subset 4\mathfrak{o}$. This is similar to [S7, Lemma 1.6] and [S8, Lemma 2.1], and can be proved in the same way. We see that $\{\mathfrak{a}_\beta \mid \beta \in B\}$ is a complete set of representatives of the ideal classes of F modulo $\mathfrak{a}$. With ψ and μ as in Theorem 6.1, we put, for $(z, s) \in \mathscr{H}^{\mathbf{a}} \times \mathbf{C}$ and $n \in \mathbf{Z}^{\mathbf{a}}$,

(6.13a) $$E^+(z, s; n, \psi, \mathfrak{c}) = \sum_{\beta \in B} E_\beta^+(z, s; n, \psi, \mathfrak{c}),$$

(6.13b) $$E_\beta^+(z, s; n, \psi, \mathfrak{c}) = N(\mathfrak{a}_\beta)^{2s} \sum_{\alpha \in S_\beta} \psi_{\mathbf{a}}(d_\alpha)$$

$$\cdot \psi^*\!\left(d_\alpha \mathfrak{a}_\beta^{-1}\right) \mathrm{Im}(z)^{su+(i\mu-n)/2} \|_n \alpha,$$

where $S_\beta = (P_+ \cap \beta \Gamma_{\mathfrak{c}} \beta^{-1}) \backslash \beta \Gamma_{\mathfrak{c}}$. Observe that this is well-defined and E_β^+ depends only on $P_+ \beta \Gamma_{\mathfrak{c}}$. These series are similar to but somewhat different from those considered in [S7] and [S8]. We define another type of series $\dot{E}_{\mathfrak{x}}$, for a fractional ideal $\mathfrak{x}$ of F, by

(6.14) $$E_{\mathfrak{x}}(z, s; n, \psi, \mathfrak{c}) = N(\mathfrak{x})^{2s} \sum_{(c, d)} \psi_{\mathbf{a}}(d) \psi^*(d\mathfrak{x}^{-1})$$

$$\cdot (cz + d)^{-n} |cz + d|^{n-i\mu-2su} \mathrm{Im}(z)^{su+(i\mu-n)/2},$$

where (c, d) runs over $T[\mathfrak{x}]/\mathfrak{o}_+^\times$ with

(6.15) $$T[\mathfrak{x}] = \left\{(c, d) \in 2^{-1}\mathfrak{c}\,\mathfrak{d}\mathfrak{x} \times \mathfrak{x} \mid d\mathfrak{x}^{-1} + \mathfrak{c} = \mathfrak{o}\right\}.$$

We easily see that $E_{\mathfrak{x}}$ depends only on the ideal class of $\mathfrak{x}$ modulo $\mathbf{a}$, and that both E_β^+ and $E_{\mathfrak{x}}$ identically vanish unless the following condition is satisfied:

(6.16) $$\psi_{\mathbf{a}}(a) = \mathrm{sgn}(a)^n |a|^{i\mu} \quad \textit{for every } a \in \mathfrak{o}^\times.$$

We consider also two types of L-functions of ψ:

(6.17a) $$L_{\mathfrak{c}}(s, \psi) = \sum_{\mathfrak{m}} \psi^*(\mathfrak{m}) N(\mathfrak{m})^{-s}$$

(6.17b) $$L_{\mathfrak{c}}^+(s, \psi, \mathfrak{x}) = \sum_{\mathfrak{m} \sim \mathfrak{x}} \psi^*(\mathfrak{m}) N(\mathfrak{m})^{-s}.$$

Here $\mathfrak{m}$ runs over all the integral ideals prime to $\mathfrak{c}$, with the additional condition that $\mathfrak{m} = \alpha\mathfrak{x}$ with $0 \ll \alpha \in F$ in (6.17b). Now we have

LEMMA 6.5.   (i) $E_{\mathfrak{x}}(z, s; n, \psi, \mathfrak{c}) = \sum_{\beta \in B} L_{\mathfrak{c}}^{+}(2s, \psi, \mathfrak{x}^{-1}\mathfrak{a}_{\beta}) E_{\beta}^{+}(z, s; n, \psi, \mathfrak{c})$,

(ii) $L_{\mathfrak{c}}(2s, \psi) E^{+}(z, s; n, \psi, \mathfrak{c}) = \sum_{\mathfrak{x}} E_{\mathfrak{x}}(z, s; n, \psi, \mathfrak{c})$, where $\mathfrak{x}$ runs over a complete set of representatives of the ideal classes of F modulo $\mathbf{a}$.

Proof.   Put $l(\alpha) = (c_{\alpha}, d_{\alpha})$ for $\alpha \in G$. Then we have

$$(6.18) \quad T[\mathfrak{x}]/\mathfrak{o}_{+}^{\times} = \coprod_{\beta \in B} \left\{ tl(\alpha) \mid 0 \ll t \in \mathfrak{x}\mathfrak{a}_{\beta}^{-1}/\mathfrak{o}_{+}^{\times}, t\mathfrak{a}_{\beta}\mathfrak{x}^{-1} + \mathfrak{c} = \mathfrak{o}, \alpha \in S_{\beta} \right\}.$$

More precisely, there is no repetition among the vectors $tl(\alpha)$ on the right-hand side, and moreover they give a complete set of representatives of $T[\mathfrak{x}]/\mathfrak{o}_{+}^{\times}$. Our equality of (i) is an immediate consequence of this fact; (ii) is merely the sum over the $\mathfrak{x}$. Thus our task is to prove (6.18). That there is no repetition modulo $\mathfrak{o}_{+}^{\times}$ can easily be seen. Now let $\alpha \in S_{\beta}$ with $\beta \in B$; put $\alpha = \beta\gamma$ with $\gamma \in \Gamma_{\mathfrak{c}}$ and $\beta = pw$ with $p \in P_{\mathbf{A}}$ and $w \in D_{\mathfrak{c}}$. Then $l(\alpha) = d_p l(w\gamma)$ and $d_p\mathfrak{o} = \mathfrak{a}_{\beta}$. Hence, if $t \in \mathfrak{x}\mathfrak{a}_{\beta}^{-1}$ and $t\mathfrak{a}_{\beta}\mathfrak{x}^{-1}$ is prime to $\mathfrak{c}$, we have $td_p\mathfrak{o} \subset \mathfrak{x}$ and $tl(\alpha) = td_p(c_{w\gamma}, d_{w\gamma})$, so that $tl(\alpha) \in T[\mathfrak{x}]$. Conversely, let $(c, d) \in T[\mathfrak{x}]$. Then $(c, d) = l(\xi)$ with $\xi \in G$. We can put $\xi = qy$ with $q \in P_{\mathbf{A}}$ and $y \in D[2\mathfrak{d}^{-1}, 2^{-1}\mathfrak{d}]$. Since $\mathfrak{o} = 2c_y\mathfrak{d}^{-1} + d_y\mathfrak{o}$ and $d\mathfrak{x}^{-1}$ is prime to $\mathfrak{c}$, we have $d_q\mathfrak{x}^{-1} = 2c\mathfrak{x}^{-1}\mathfrak{d}^{-1} + d\mathfrak{x}^{-1} = c\mathfrak{d}^{-1}\mathfrak{x}^{-1} + d\mathfrak{x}^{-1} = \mathfrak{a}_{\xi}\mathfrak{x}^{-1} \subset \mathfrak{o}$, and hence $d_q\mathfrak{x}^{-1}$ is prime to $\mathfrak{c}$. Thus $c_y = d_q^{-1}c \in (d_q\mathfrak{x}^{-1})^{-1}c\mathfrak{x}^{-1} \subset (2^{-1}c\mathfrak{d})_v$, for every $v|\mathfrak{c}$, so that $y \in D_{\mathfrak{c}}$. Therefore $\xi \in P_{\mathbf{A}}D_{\mathfrak{c}}$, and hence $\xi \in P_{+}\beta\Gamma_{\mathfrak{c}}$ for some $\beta \in B$. Consequently we can put $(c, d) = tl(\alpha)$ with $0 \ll t \in F$ and $\alpha \in S_{\beta}$. By Proposition 3.1, $d_{\alpha}\mathfrak{a}_{\beta}^{-1}$ is prime to $\mathfrak{c}$, and hence $t\mathfrak{a}_{\beta}\mathfrak{x}^{-1}$ is prime to $\mathfrak{c}$. Moreover $t\mathfrak{a}_{\beta} = t\mathfrak{a}_{\alpha} = c\mathfrak{d}^{-1} + d\mathfrak{o} \subset \mathfrak{x}$, so that $t \in \mathfrak{x}\mathfrak{a}_{\beta}^{-1}$. This completes the proof of (6.18), as well as that of our lemma.

7. Proof of Theorem 6.1.   We first consider a theta function of a ternary quadratic form on the vector space

$$(7.0) \qquad\qquad V = \left\{ \alpha \in M_2(F) \mid \mathrm{tr}(\alpha) = 0 \right\},$$

where $M_2(R)$, for any associative ring R, denotes the ring of 2×2 matrices with entries in R. For $\alpha = \begin{pmatrix} a & b \\ c & -a \end{pmatrix} \in M_2(\mathbf{C})$ and $w \in \mathbf{C}$, we put, with ι of (1.3),

$$(7.1) \qquad\qquad [\alpha, w] = -(w \quad 1)\iota\alpha\begin{pmatrix} w \\ 1 \end{pmatrix} = cw^2 - 2aw - b.$$

For $\alpha \in V$ and $w \in \mathbf{C}^{\mathbf{a}}$ we put $[\alpha, w] = ([\alpha_v, w_v])_{v \in \mathbf{a}}$; then $[\alpha, w]^n$ for $0 \leq n \in \mathbf{Z}^{\mathbf{a}}$ is meaningful, and

$$(7.2) \qquad [\beta^{-1}\alpha\beta, w]^n = J_{2n}(\beta, w)[\alpha, \beta w]^n \quad \text{for every } \beta \in \tilde{G}_{+}.$$

Given $k = (u/2) + m$ with $0 \le m \in \mathbf{Z}^{\mathbf{a}}$, we define a theta function Θ on $\mathscr{H}^{\mathbf{a}} \times \mathscr{H}^{\mathbf{a}}$ by

$$(7.3a) \quad \Theta(z, w; \eta) = \mathrm{Im}(z)^{u/2}\mathrm{Im}(w)^{-2m} \sum_{\alpha \in V} \eta(\alpha)[\alpha, \bar{w}]^m \mathbf{e}_{\mathbf{a}}(2^{-1}R[\alpha, z, w]).$$

Here $z \in \mathscr{H}^{\mathbf{a}}$, $w \in \mathscr{H}^{\mathbf{a}}$, η is a locally constant function on V, that is, the restriction of an element of $\mathscr{S}(V_{\mathbf{f}})$ to V, and

$$(7.3b) \qquad R[\alpha, z, w] = (R_v[\alpha, z, w])_{v \in \mathbf{a}},$$

$$(7.3c) \qquad R_v[\alpha, z, w] = \det(\alpha_v)z_v + (i/2)\mathrm{Im}(z_v)\mathrm{Im}(w_v)^{-2}|[\alpha_v, w_v]|^2.$$

This is a special case of [S4, (7.6)] (and also of [S5, I, (6.6)] and [S10, p. 401]), and therefore, by [S4, Proposition 7.1] or by Proposition 11.8 below, we have

$$(7.4) \qquad\qquad \Theta(\gamma z, w; \eta) = \overline{J_k(\gamma, z)}\,\Theta(z, w; \eta) \quad \textit{for } \gamma \in \Gamma$$

with a congruence subgroup Γ of G depending on η. From (7.2) we obtain easily

$$(7.5) \qquad \Theta(z, \beta w; \eta)J_{-2m}(\beta, w) = \Theta(z, w; \eta^\beta) \quad \textit{for every } \beta \in \tilde{G}_+,$$

where $\eta^\beta(\alpha) = \eta(\beta\alpha\beta^{-1})$.

PROPOSITION 7.1. *Suppose $m_v > 0$ for every $v \in \mathbf{a}$. For $f \in \mathscr{M}_k(\Gamma)$ with Γ as in (7.4), put*

$$(7.6) \qquad g(w) = \int_\Phi f(z)\Theta(z, w; \eta)\mathrm{Im}(z)^k\,d_H z \qquad (\Phi = \Gamma \backslash \mathscr{H}^{\mathbf{a}}).$$

Then the integral is convergent, and g is holomorphic in w and belongs to $\mathscr{M}_{2m}$. Suppose moreover that f is a cusp form. Then g is a cusp form if $m \ne u$, or if $m = u$ and f is orthogonal to every theta series of type (4.5) with $n = u$.

This is a special case of [S5, I, Theorem 6.2] except for the last assertion. It follows also from [S10, Theorem 4.3], since the integral can easily be seen to be the residue given in that theorem, as will be explained in Section 7bis. The last assertion will be proved at the end of this section.

We are going to apply the above proposition to the given f of Theorem 6.1 with a special choice of η, which is as follows: Given $\mathfrak{c}$, ψ, and μ as in Theorem 6.1 and a fractional ideal $\mathfrak{x}$ in F prime to $\mathfrak{c}$, we define η by

$$(7.7) \qquad\qquad\qquad \eta\!\begin{pmatrix} a & b \\ c & -a \end{pmatrix} = \omega_1(a)\omega_2(b)\omega_3(c).$$

Here ω_1 resp. ω_3 is the characteristic function of $\mathfrak{o}$ resp. $\mathfrak{c}\mathfrak{d}\mathfrak{x}$, and

$$(7.8) \quad \omega_2(b) = \begin{cases} 0 & \text{if } b \notin (\mathfrak{c}\mathfrak{d}\mathfrak{x})^{-1}, \\ D_F^{-1/2} N(\mathfrak{c}\mathfrak{x})^{-1} \sum_t \overline{\psi_\mathfrak{c}(t)}\, \mathbf{e_a}(-bt) & \text{if } b \in (\mathfrak{c}\mathfrak{d}\mathfrak{x})^{-1}, \end{cases}$$

where t runs over $\mathfrak{x}/\mathfrak{c}\mathfrak{x}$ under the condition $|t|_v = 1$ for $v|\mathfrak{c}$. Then we see easily that

$$(7.9) \quad \eta(\beta\alpha\beta^{-1}) = \psi_\mathfrak{c}\big(a_\beta^2/\det(\beta)\big)\eta(\alpha) \quad \text{for every } \beta \in \tilde{\Gamma}\big[(2\mathfrak{d}\mathfrak{x})^{-1}, \mathfrak{c}\mathfrak{d}\mathfrak{x}\big].$$

Let us hereafter denote $\Theta(z, w; \eta)$ with this η simply by $\Theta(z, w)$. Then we have, by (7.5) and (7.9),

$$(7.10) \quad \Theta(z, \beta w) J_{-2m}(\beta, w) = \psi_\mathfrak{c}\big(a_\beta^2/\det(\beta)\big)\Theta(z, w)$$

$$\text{for } \beta \in \tilde{\Gamma}\big[(2\mathfrak{d}\mathfrak{x})^{-1}, \mathfrak{c}\mathfrak{d}\mathfrak{x}\big].$$

Now, given $f \in \mathcal{S}_k(\mathfrak{c}, \psi)$, define g by (7.6), in which we take Γ so that $\Gamma \subset \{\gamma \in \Gamma_\mathfrak{c}|a_\gamma - 1 \in \mathfrak{c}\}$. By (7.10) and Proposition 7.1, g belongs to $\mathcal{M}_{2m}(\tilde{\Gamma}[\mathfrak{y}^{-1}\mathfrak{d}^{-1}, \mathfrak{b}\mathfrak{y}\mathfrak{d}], \psi_\mathfrak{b}^2, \mu)$ with $\mathfrak{y} = 2\mathfrak{x}$ and $\mathfrak{b} = 2^{-1}\mathfrak{c}$. Put $g(w) = \Sigma_{\xi \in F}c(\xi)\mathbf{e_a}(\xi w)$. We prove Theorem 6.1(I) by showing that the $c(\xi)$, up to a constant factor, have the desired property. Put

$$(7.11) \quad K = \{y \in \mathbf{R^a}|y \gg 0\}/\mathfrak{o}_+^\times, \qquad \mathfrak{o}_+^\times = \{a \in \mathfrak{o}^\times|a \gg 0\}.$$

Take $\rho \in \mathbf{R^a}$ so that $\{\rho\} = 0$ and $|a|^{i\rho} = 1$ for every $a \in \mathfrak{o}^\times$, and observe that

$$(7.12) \quad \int_K g(ir) r^{m+su-u+i\rho+i\mu}\, dr = (2\pi)^{-\{m+su\}} \prod_{v \in \mathbf{a}} \Gamma(s + m_v + i\rho_v + i\mu_v)$$

$$\cdot \sum_{0 \ll \xi \in F/\mathfrak{o}_+^\times} c(\xi)\xi^{-m-i\rho-i\mu}N(\xi\mathfrak{o})^{-s}$$

for sufficiently large $\mathrm{Re}(s)$. Here we have to assume that $c(0) = 0$. However, if we could somehow show that the integral is convergent for such s, which is in fact the case for the present g, as will be shown later, then that would imply $c(0) = 0$. Anyway, the last series is essentially the series of Theorem 6.1(I). We have of course, formally for the moment,

$$(7.13) \quad \int_K g(ir) r^s\, dr = \int_\Phi f(z)\mathrm{Im}(z)^k \int_K \Theta(z, ir) r^s\, dr\, d_H z.$$

Our immediate task is to examine the nature of $\Theta(z, ir)$. From (7.1) and (7.3c) we obtain, dropping the subscript v for simplicity,

$$[\alpha, ir]^m = (-2ai - br^{-1} - cr)^m r^m,$$

$$R\left[\begin{pmatrix} a & b \\ c & -a \end{pmatrix}, z, ir\right] = -a^2\bar{z} - bcx + (iy/2)(c^2r^2 + b^2r^{-2}),$$

and therefore, by virtue of (2.5),

$$y^{(m-u)/2}(\sqrt{\pi}\,r)^m\overline{\Theta(z, ir)}$$

$$= (-1)^{\{m\}} \sum_{0\leqslant n\leqslant m} \binom{m}{n} i^{\{m-n\}} y^{(m-n)/2}\theta_{m-n}(z)A_n(z, r),$$

where we put

$$\binom{m}{n} = \prod_{v\in\mathbf{a}}\binom{m_v}{n_v},$$

$$\theta_l(z) = y^{-l/2}\sum_{a\in\mathfrak{o}} H_l(\sqrt{4\pi y}\,a)\mathbf{e}_\mathbf{a}(a^2z/2),$$

$$A_n(z, r) = \sum_{b\in F,\, c\in F} \overline{\omega_2(b)}\,\omega_3(c)H_n(\sqrt{\pi y}\,(cr + br^{-1}))$$

$$\cdot\mathbf{e}_\mathbf{a}((bc/2)x + (iy/4)(c^2r^2 + b^2r^{-2})).$$

Observe that θ_n is $\theta_n(z, \omega_1)$ of (4.5), which is identically equal to 0 if $\{n\}$ is odd. Now we have

LEMMA 7.2.

$$i^{-\{n\}}(2\pi)^{-\{n\}/2}(2r^2/y)^{-(n+u)/2}A_n(z, r)$$

$$= \sum_{(c, d)} \psi_\mathfrak{c}(d)(c\bar{z} + d)^n\mathbf{e}_\mathbf{a}(ir^2|cz + d|^2/y),$$

where (c, d) runs over the set $T[\mathfrak{x}]$ of (6.15).

Proof. If $\hat{h}$ is the Fourier transform of a function h on $\mathbf{R}^\mathbf{a}$, then under a suitable convergence condition, we have, by the Poisson summation formula,

$$\sum_d \psi_\mathfrak{c}(d)h(d) = \sum_b \overline{\omega_2(b)}\,\hat{h}(b),$$

where d runs over the elements of $\mathfrak{x}$ prime to $\mathfrak{c}$, and b over $(\mathfrak{c}\mathfrak{d}\mathfrak{x})^{-1}$. Applying this to $h(u) = ((c\bar{z}/2) + u)^n e_{\mathbf{a}}(ir^2|(cz/2) + u|^2/y)$ whose Fourier transform is given by (2.8), we obtain our formula.

Let $C_n(z, r)$ denote the function of Lemma 7.2. Then

$$(7.14) \qquad g(ir) = i^{\{m\}} \sum_{0 \leqslant n \leqslant m} \binom{m}{n} 2^{\{n+u/2\}} \pi^{\{n-m\}/2}$$

$$\cdot \int_\Phi f(z) \overline{\theta_{m-n}(z) \, C_n(z, r)} \, r^{u-m+n} y^{k-n} \, d_H z.$$

Though $\theta_l = 0$ for odd $\{l\}$, we keep even those vanishing terms for some technical reason. With K of (7.11), we have, for sufficiently large $\mathrm{Re}(s)$,

$$(7.15) \qquad \int_K \overline{C_n(z, r)} r^{n+su+i\rho+i\mu} \, dr$$

$$= (2\pi)^{-\{n+su+u\}/2} 2^{-\{u\}}$$

$$\cdot \prod_{v \in \mathbf{a}} \Gamma((s + 1 + n_v + i\rho_v + i\mu_v)/2)(\psi\chi)^*(\mathfrak{x})$$

$$\cdot N(\mathfrak{x})^{-s-1} E_{\mathfrak{x}}(z, (s + 1)/2; -n, \psi\chi, \mathfrak{c})$$

with $E_{\mathfrak{x}}$ of (6.14), where χ is any Hecke character of F such that

$$(7.16) \qquad \chi(x) = |x_{\mathbf{a}}|^{i\rho} \quad \text{for } x \in F_{\mathbf{a}}^\times \prod_{v \in \mathbf{f}} \mathfrak{o}_v^\times.$$

As remarked in Section 6, $E_{\mathfrak{x}}$ may vanish for some n, but we include even such vanishing terms in our calculation. We observe that the integral of (7.15) is absolutely convergent for sufficiently large $\mathrm{Re}(s)$ for every n, and

$$\int_K |C_n(z, r)| r^{n+\sigma u} \, dr \leqq (2\pi)^{-\{n+\sigma u+u\}/2} 2^{-\{u\}} y^{n/2}$$

$$\cdot \prod_{v \in \mathbf{a}} \Gamma((\sigma + 1 + n_v)/2) N(\mathfrak{x})^{-\sigma-1}$$

$$\cdot E_{\mathfrak{x}}(z, (\sigma + 1)/2; 0, \chi_0, \mathfrak{c})$$

for sufficiently large real σ, where χ_0 is the identity character. Thus we have, at least formally,

$$(7.17) \quad \int_K g(ir)r^{m+su-u+i\rho+i\mu}\,dr$$

$$= i^{\{m\}} \sum_{0\leqslant n\leqslant m} \binom{m}{n} 2^{\{n-su-2u\}/2}\pi^{-\{m+su+u\}/2}$$

$$\cdot \prod_{v\in\mathfrak{a}} \Gamma((s+1+n_v+i\rho_v+i\mu_v)/2)(\psi\chi)^*(\mathfrak{x})N(\mathfrak{x})^{-s-1}$$

$$\cdot \int_\Phi f(z)\overline{\theta_{m-n}(z)}\,E_{\mathfrak{x}}(z,(s+1)/2;-n,\psi\chi,\mathfrak{c})y^{k-n}\,d_H z.$$

Since f is a cusp form, the last integral is convergent for sufficiently large $\mathrm{Re}(s)$. Replacing every integrand by its absolute value, we see that $\int_K|g(ir)|r^{m+\sigma u-u}\,dr$ is convergent for sufficiently large σ, so that we can justify our formal calculation, and at the same time we have proved that $c(0)=0$.

By Lemma 6.5(i), the last integral over Φ is equal to

$$\sum_{\beta\in B} L_{\mathfrak{c}}^+\left(s+1,\psi\chi,\mathfrak{x}^{-1}\mathfrak{a}_\beta\right)$$

$$\cdot \int_\Phi f(z)\overline{\theta_{m-n}(z)}\,E_\beta^+(z,(s+1)/2;-n,\psi\chi,\mathfrak{c})y^{k-n}\,d_H z.$$

We choose each β so that $\beta\in\mathrm{diag}[t^{-1},t]D_{\mathfrak{c}}G_{\mathfrak{a}}$ with $t\in F_A^\times$ and $t\mathfrak{o}\ (=\mathfrak{a}_\beta)$ is prime to $\mathfrak{c}$. Fix n and for each $\beta\in B$ put

$$f_\beta(z) = \psi_{\mathfrak{c}}(d_\beta)^{-1}J_k(\beta,\beta^{-1}z)f(\beta^{-1}z),$$

$$\theta_{l,\beta}(z) = J_{l+u/2}(\beta,\beta^{-1}z)\theta_l(\beta^{-1}z),$$

$$E_\beta'(z,s) = \psi_{\mathfrak{c}}(d_\beta)E_\beta^+(\beta^{-1}z,s;-n,\psi\chi,\mathfrak{c})j(\beta^{-1},z)^n.$$

Then we easily see that

$(7.18a)$

$$f_\beta(z)\overline{\theta_{m-n,\beta}(z)} = \psi_{\mathfrak{c}}(d_\beta)^{-1}j(\beta^{-1},z)^{-n}|j(\beta^{-1},z)|^{2n-2k}(f\overline{\theta}_{m-n})(\beta^{-1}z),$$

$(7.18b)$

$$\left(f_\beta\overline{\theta}_{m-n,\beta}\right)(\gamma z) = \psi_{\mathfrak{c}}(a_\gamma)j(\gamma,z)^n|j(\gamma,z)|^{2k-2n}\left(f_\beta\overline{\theta}_{m-n,\beta}\right)(z)$$

$$\textit{for every }\gamma\in\beta\Gamma_{\mathfrak{c}}\beta^{-1},$$

and moreover the last integral over Φ is equal to

$$\int_{\beta\Phi} f_\beta(z)\overline{\theta_{m-n,\beta}(z)}\, E'_\beta(z,(s+1)/2)\, y^{k-n}\, d_H z.$$

Our choice of β shows that

(7.19a) $$\beta\Gamma_c\beta^{-1} = \Gamma\big[2\mathfrak{d}^{-1}\mathfrak{a}_\beta^{-2},\, 2^{-1}\mathfrak{c}\mathfrak{d}\mathfrak{a}_\beta^{2}\big],$$

(7.19b) $$P_+ \cap \beta\Gamma_c\beta^{-1} = \left\{ \begin{pmatrix} a & b \\ 0 & 1/a \end{pmatrix} \Big| a \in \mathfrak{o}_+^\times,\, b \in 2\mathfrak{d}^{-1}\mathfrak{a}_\beta^{-2}\right\}.$$

Putting $q = k + (1/2)(su + u - n + i\rho + i\mu)$ and $T_\beta = S_\beta\beta^{-1}$, we obtain, from (7.18b),

$$\sum_{\gamma\in T_\beta} \big(f_\beta\bar{\theta}_{m-n,\beta}y^q\big)\circ\gamma$$

$$= (\psi\chi)^*(\mathfrak{a}_\beta)N(\mathfrak{a}_\beta)^{-s-1}f_\beta\bar{\theta}_{m-n,\beta}y^{k-n}E'_\beta(z,(s+1)/2),$$

and hence

$$(\psi\chi)^*(\mathfrak{a}_\beta)N(\mathfrak{a}_\beta)^{-s-1}\int_\Phi f\bar{\theta}_{m-n}E^+_\beta(z,(s+1)/2;-n,\psi\chi,\mathfrak{c})y^{k-n}d_H z$$

$$= \int_{\beta\Phi}\sum_{\gamma\in T_\beta}\big(f_\beta\bar{\theta}_{m-n,\beta}y^q\big)\circ\gamma\, d_H z$$

$$= 2[\Gamma_c:\{\pm1\}\Gamma]\int_\Psi f_\beta\bar{\theta}_{m-n,\beta}y^q\, d_H z,$$

where $\Psi = (P_+ \cap \beta\Gamma_c\beta^{-1})\backslash\mathscr{H}^\mathbf{a}$. By Proposition 3.1 and Lemma 4.4, we have

$$f_\beta(z) = \psi^*(\mathfrak{a}_\beta)N(\mathfrak{a}_\beta)^{1/2}\sum_{\xi\in F}\lambda_f\big(\xi,\mathfrak{a}_\beta^{-1}\big)\mathbf{e_a}(\xi z/2),$$

$$\theta_{m-n,\beta}(z) = N(\mathfrak{a}_\beta)^{1/2}y^{(n-m)/2}\sum_{\xi\in\mathfrak{a}_\beta}H_{m-n}\big(\sqrt{4\pi y}\,\xi\big)\mathbf{e_a}(\xi^2 z/2).$$

In view of (7.19b), we may take $\Psi = [\mathbf{R}^\mathbf{a}/2\mathfrak{d}^{-1}\mathfrak{a}_\beta^{-2}]\times K'$ with $K' = \{y\in\mathbf{R}^\mathbf{a}\,|\,y\gg 0\}/\{a^2\,|\,a\in\mathfrak{o}_+^\times\}$, so that

$$\int_\Psi f_\beta\bar{\theta}_{m-n,\beta}y^q\, d_H z = \psi^*(\mathfrak{a}_\beta)N(\mathfrak{a}_\beta)D_F^{1/2}N\big(2\mathfrak{d}^{-1}\mathfrak{a}_\beta^{-2}\big)$$

$$\cdot\int_{K'}\sum_{0\neq\xi\in\mathfrak{a}_\beta}\lambda_f\big(\xi^2,\mathfrak{a}_\beta^{-1}\big)H_{m-n}\big(\sqrt{4\pi y}\,\xi\big)$$

$$\cdot\mathbf{e_a}(i\xi^2 y)y^{q-2u+(n-m)/2}\, dy.$$

Write the last integrand in the form $\Sigma_\xi\Lambda(\xi,y)$. By (3.14b) we see that $\Lambda(a\xi,y) =$

$\Lambda(\xi, a^2 y)$ for every $a \in \mathfrak{o}_+^\times$, and therefore, by (2.9), the last integral over K' is equal to

$$\gamma(s, m, n, \rho + \mu) \sum_{0 \neq \xi \in \mathfrak{a}_\beta/\mathfrak{o}_+^\times}{}' \mathrm{sgn}(\xi)^{m-n} \lambda_f\!\left(\xi^2, \mathfrak{a}_\beta^{-1}\right) |\xi|^{-su-m-i\rho-i\mu},$$

$$\gamma(s, m, n, \rho + \mu) = 2^{\{n-m\}/2} (2\pi)^{-\{su+m\}/2}$$

$$\cdot \prod_{v \in \mathbf{a}} \Gamma\!\left((s + n_v + i\rho_v + i\mu_v)/2\right) \prod_{j=1}^{m_v - n_v} (s + m_v - j + i\rho_v + i\mu_v).$$

Take $h \in \mathbf{Z}^\mathbf{a}$ so that $\psi_\mathbf{a}(x) = \mathrm{sgn}(x_\mathbf{a})^h |x_\mathbf{a}|^{i\mu}$. Observing that $\chi^*(\xi\mathfrak{o}) = |\xi|^{-i\rho}$, we obtain, by (3.14b),

$$\chi^*(\mathfrak{a}_\beta)^{-1} N(\mathfrak{a}_\beta)^s \mathrm{sgn}(\xi)^{m-n} \lambda_f\!\left(\xi^2, \mathfrak{a}_\beta^{-1}\right) |\xi|^{-su-m-i\rho-i\mu}$$

$$= \mathrm{sgn}(\xi)^{h-n} \lambda_f\!\left(1, \xi\mathfrak{a}_\beta^{-1}\right) \chi^*\!\left(\xi\mathfrak{a}_\beta^{-1}\right) N\!\left(\xi\mathfrak{a}_\beta^{-1}\right)^{-s}.$$

Therefore, combining the whole calculation together, we obtain

$$\int_K g(ir) r^{m+su-u+i\rho+i\mu}\, dr$$

$$= A 2^\alpha \pi^{\alpha'} \prod_{v \in \mathbf{a}} \Gamma(s + m_v + i\rho_v + i\mu_v)$$

$$\cdot N(\mathfrak{r})^{-s-1} (\psi\chi)^*(\mathfrak{r}) \sum_{\beta \in B} L_c^+\!\left(s + 1, \psi\chi, \mathfrak{r}^{-1}\mathfrak{a}_\beta\right)$$

$$\sum_{0 \neq \xi \in \mathfrak{a}_\beta/\mathfrak{o}_+^\times} \lambda_f\!\left(1, \xi\mathfrak{a}_\beta^{-1}\right) \chi^*\!\left(\xi\mathfrak{a}_\beta^{-1}\right) N\!\left(\xi\mathfrak{a}_\beta^{-1}\right)^{-s} \sum_{0 \leqslant n \leqslant m} \binom{m}{n} \mathrm{sgn}(\xi)^{h-n},$$

where $\alpha = \{u - 2su - m\}$, $\alpha' = -\{su + m\}$, and $A = i^{\{m\}} 2[\Gamma_c : \{\pm 1\}\Gamma] \cdot D_F^{-1/2}$. The last sum over n is equal to $(1 + \mathrm{sgn}(\xi))^m \mathrm{sgn}(\xi)^{h-m}$, which is $2^{\{m\}}$ or 0 according as $\xi \gg 0$ or otherwise. Therefore the last two lines of the above product can be written in the form

$$2^{\{m\}} (\psi\chi)^*(\mathfrak{r}) N(\mathfrak{r})^{-s-1} \sum_{\mathfrak{m}} \lambda_f(1, \mathfrak{m}) \chi^*(\mathfrak{m}) N(\mathfrak{m})^{-s}$$

$$\cdot \sum_{\mathfrak{n}} \psi^*(\mathfrak{n}) N(\mathfrak{n})^{-1} \chi^*(\mathfrak{n}) N(\mathfrak{n})^{-s},$$

where $\mathfrak{m}$ and $\mathfrak{n}$ are as in Theorem 6.1(I), with $\mathfrak{r} = \mathfrak{o}$ and $\mathfrak{y} = 2\mathfrak{x}$. Comparing this with (7.12) and putting $M_{\chi,s}(\mathfrak{m}) = \chi^*(\mathfrak{m})N(\mathfrak{m})^{-s}$ and $A' = A2^{(m+u)}\psi^*(\mathfrak{x})$ $N(\mathfrak{x})^{-1}$, we find that the equality of Theorem 6.1(I), for $\tau = 1$ holds with $A'\lambda_f$ and $M_{\chi,s}$ in place of λ_f and M. This proves assertion (I) itself, since $M(\mathfrak{m}) \mapsto M_{\chi,s}(\mathfrak{m})$ is "faithful" in an obvious sense. Strictly speaking, we have proved only the case $\mathfrak{y} = 2\mathfrak{x}$ with $\mathfrak{x}$ prime to $\mathfrak{c}$, but clearly this is sufficient.

To prove (II) of Theorem 6.1, we may again assume $\tau = 1$. Let $\mathfrak{p}$ be a prime ideal in F prime to $\mathfrak{c}$ and π a prime element of $F_\mathfrak{p}$ embedded in $F_\mathbf{A}^\times$. For $\mathbf{g} = (g_1,\ldots,g_\kappa) \in \mathscr{M}_n(\varphi_\mathfrak{b}, \mu)$ as in Section 6, let $\mathbf{g}'(x) = \mathbf{g}(\pi x)$ for $x \in \tilde{G}_\mathbf{A}$ and let $\mathbf{g}' = (g_1',\ldots,g_\kappa')$. As explained in [S3, p. 648] and [S4, (9.24)], the g_ν' can be obtained from the g_λ by

$$(7.20) \qquad g_\nu' = \varphi_\mathfrak{b}(a_{\beta_\lambda})^{-1}\det(\beta_\lambda)^{-i\mu}g_\lambda\|_n\beta_\lambda$$

with an element β_λ of $\tilde{G}_+ \cap \pi x_\lambda W_\mathfrak{b} x_\nu^{-1}$, $W_\mathfrak{b}$ and x_λ being as in (6.3) and (6.4a). The element $\mathbf{g}$ belongs to $\mathscr{M}_n(\mathfrak{b}, \varphi)$ if it satisfies (6.2c), that is, if $g_\lambda' = \varphi(\pi)g_\lambda$ for every λ (cf. [S3, Proposition 2.1]). Thus our problem can be formulated as follows: Let $\mathfrak{x}$ and $\mathfrak{x}'$ be two fractional ideals prime to $\mathfrak{c}$; let $2\mathfrak{x} = t\mathfrak{o}$ and $2\mathfrak{x}' = t'\mathfrak{o}$ with t and t' in $F_\mathfrak{f}^\times$; let g and g' be the forms of weight $2m$ obtained in (I) with $2\mathfrak{x}$ and $2\mathfrak{x}'$ as $\mathfrak{y}$ there. Then our task is to show that

$$(7.21) \qquad g\|_{2m}\beta = \psi_\mathfrak{c}(a_\beta)^2\det(\beta)^{i\mu}\psi^*(\mathfrak{p})^2g'$$

for an element β of $\tilde{G}_+$ such that

$$(7.22) \qquad \beta = \pi \cdot \operatorname{diag}[1, t]w \cdot \operatorname{diag}[1, t']^{-1}$$

with $w \in W_\mathfrak{b}$, $\mathfrak{b} = 2^{-1}\mathfrak{c}$. To prove this, let η and η' be the functions defined by (7.7) for $\mathfrak{x}$ and $\mathfrak{x}'$. We have obtained g as the integral of (7.6) times $A_0\psi^*(\mathfrak{x})^{-1}N(\mathfrak{x})$, where A_0 is a nonzero constant independent of $\mathfrak{x}$. Therefore, in view of (7.5), it is sufficient to show that

$$(7.23) \qquad \eta(\beta\alpha\beta^{-1}) = \psi_\mathfrak{c}(a_\beta)^2\det(\beta)^{i\mu}\psi^*(\mathfrak{p}^2\mathfrak{x}\mathfrak{x}'^{-1})N(\mathfrak{x}^{-1}\mathfrak{x}')\eta'(\alpha).$$

From (7.22) we obtain $\det(\beta)\mathfrak{o} = \mathfrak{p}^2\mathfrak{x}\mathfrak{x}'^{-1}$, so that $\det(\beta)^{i\mu}\psi^*(\mathfrak{p}^2\mathfrak{x}\mathfrak{x}'^{-1}) = \psi_\mathfrak{c}(\det(\beta))^{-1}$. Therefore we obtain (7.23) from (7.7) and (7.8) in an elementary way. Indeed, $\beta \in D_v[\mathfrak{b}^{-1}, \mathfrak{b}\mathfrak{d}]$ for every $v|\mathfrak{c}$, and hence, localizing the problem, we see that (7.23) is of the same nature as (7.9).

It remains to prove (III) of Theorem 6.1. Our assumption of (I) implies $m \geqq u$. Since g is of weight $2m$, it is a cusp form if $m \notin \mathbf{Z}u$. Hence we may assume that $m = pu$ with $0 < p \in \mathbf{Z}$. Our task is to show that the constant term of $g\|_{2m}\alpha$ is 0 for every $\alpha \in G$. By (7.5) we have

$$(7.24) \qquad g\|_{2m}\alpha = \int_\Phi f(z)\Theta(z, w; \eta^\alpha)\operatorname{Im}(z)^k\, d_H z.$$

Now η^α is a linear combination of functions of type (7.7) with some locally constant functions on F in place of the ω_i. For the same reason as in (7.14), we have

$$(g\|_{2m}\alpha)(ir) = \sum_{j=1}^{N} \sum_{0 \leqslant n \leqslant m} \int_{\Phi'} f(z)\overline{\theta_{j,m-n}(z)\,C_{j,n}(z,r)}\, r^{u-pu+n}y^{k-n}\,d_H z,$$

where $\theta_{j,l}$ is of type (4.5), and

$$C_{j,n}(z,r) = \sum_{(c,d)} \xi_{j,n}(c,d)(c\bar{z}+d)^n \mathbf{e}_\mathbf{a}\!\left(ir^2|cz+d|^2/y\right)$$

with a locally constant function $\xi_{j,n}$ on F^2; $\Phi' = \Gamma'\backslash\mathscr{H}^\mathbf{a}$ with a sufficiently small subgroup Γ' of Γ. If $n = 0$, $C_{j,0}$ may have a nonvanishing term $\xi_{j,0}(0,0)$. (This was not so for $C_0(z,r)$ in the above proof.) Put

$$A = \sum_{j=1}^{N} \overline{\xi_{j,0}(0,0)} \int_{\Phi'} f\bar{\theta}_{j,m}\, y^k\, d_H z.$$

Then, for the same reason as before, we see that

$$(7.25) \qquad \int_L |(g\|_{2m}\alpha)(ir) - Ar^{u-pu}|r^{\sigma u}\,dr$$

is convergent for sufficiently large σ, where L is $\{0 \ll y \in \mathbf{R}^\mathbf{a}\}/\mathfrak{u}$ with a suitable subgroup $\mathfrak{u}$ of $\mathfrak{o}_+^\times$. If $p > 1$, this implies that $g\|_{2m}\alpha$ has no constant term. If $p = 1$ and $\langle \theta_{j,m}, f \rangle = 0$ for every j, then $A = 0$, and hence we obtain the same conclusion. This proves (III), and the last assertion of Proposition 7.1 as well, since our argument is valid for an arbitrary η.

7bis. Proof of Theorem 6.4. We first define a Laplace–Beltrami operator L_v for each $v \in \mathbf{a}$ by

$$(7.26) \qquad L_v = -4\,\mathrm{Im}(w_v)^2 \partial^2/\partial w_v\,\partial\bar{w}_v,$$

where w_v is the v-th variable on $\mathscr{H}^\mathbf{a}$.

PROPOSITION 7.3. *For $f \in \mathscr{S}_k(\Gamma)$ with $k = (u/2) + m$, $0 \leq m \in \mathbf{Z}^\mathbf{a}$, and Γ as in (7.4), define g by (7.6). Then $\partial g/\partial\bar{w}_v = 0$ if $m_v > 0$ and $L_v g = 0$ if $m_v = 0$.*

Postponing the proof, we fix k and m, and assume throughout the rest of this section (except when the contrary is stated) that $m_v = 0$ (that is, $k_v = 1/2$) for some v. To prove Theorem 6.4, we first note that $\mathscr{M}_k = \mathscr{S}_k$ unless $k = u/2$, and $\mathscr{M}_{u/2} = \mathscr{S}_{u/2} \oplus \mathscr{E}_{u/2}$, where $\mathscr{E}_{u/2}$ is spanned by $\theta_0(az,\eta)$ with $0 \ll a \in F$ and θ_0

of type (4.5). This follows from the results of [S9, §8] and [S8, Proposition 6.4], as will be explained in Section 9, particularly in the proof of Theorem 9.1. Therefore, in order to prove Theorem 6.4, it is sufficient to show that an element f of $\mathscr{S}_k$ must vanish under either of the following two conditions:

(7.27a) $$m_t > 1 \quad \textit{for some } t \in \mathbf{a};$$

(7.27b)

$\quad$ *$m \leqq u$ and f is orthogonal to every function p of the form $p(z) = \theta_m(az, \eta)$*

$\quad\quad$ *with $0 \ll a \in F$ and $\theta_m(z, \eta)$ of type (4.5).*

Let f and g be as in Proposition 7.3. Then $L_v g = 0$ for every v and $g\|_{2m}\gamma = g$ for every γ in a congruence subgroup Γ^* of G by virtue of (7.5). It can easily be seen that g is slowly increasing at every cusp, and hence g is an automorphic eigenform in the sense of [S9, §2]. Therefore, by [S9, (2.21) and Proposition 3.2], we have, for every $\alpha \in G$,

$$(g\|_{2m}\alpha)(w) = cr^p + c'r^{u-2m-p} + \sum_{0 \neq \xi \in F} c_\xi W(\xi r)\mathbf{e}_\mathbf{a}(\xi \operatorname{Re}(w)),$$

where c, c', and c_ξ are complex constants, $r = \operatorname{Im}(w)$, W is a certain Whittaker function, and $p \in \mathbf{C}^\mathbf{a}$. From conditions (3.6a, b) of [S9] on p, we easily see that $p_v = 0$ or $1 - 2m_v$ and $p = su - m$ with $s \in \mathbf{Z}$. Since $m_v = 0$ for some v, s must be 0 or 1, so that $p = -m$ or $p = u - m$. This contradicts the condition that $p_v = 0$ or $1 - 2m_v$ for every v unless $m \leqq u$. Therefore a nontrivial constant term can appear only if $m \leqq u$, in which case the constant term is of the form $cr^{-m} + c'r^{u-m}$. Under (7.27b), we see, by the same argument as at the end of Section 7, that

$$\int_L |(g\|_{2m}\alpha)(ir)| r^{\sigma u + m} \, dr,$$

with L as in (7.25), is convergent for sufficiently large σ. This implies that $c = c' = 0$, and hence g is a cusp form. Then by [S9, Proposition 2.5] g must be holomorphic, and consequently $g = 0$, since $\mathscr{S}_n \neq \{0\}$ only if $n_v > 0$ for every v (cf. [S3, Proposition 1.1]). In the other case (7.27a), we have seen that g is a cusp form, and hence $g = 0$ for the same reason.

Now, take an arbitrary $f_0 \in \mathscr{S}_k$ under (7.27a) or (7.27b). We can find a totally positive element q of F and a multiple $\mathfrak{e}$ of 4 such that $f_0(qz) \in \mathscr{S}_k(\Gamma')$, where $\Gamma' = \{\gamma \in \Gamma_{\mathfrak{e}} | a_\gamma - 1 \in \mathfrak{e}\}$. Put $f_q = f_0(qz)$. Then $f_q = \sum_\varphi f_\varphi$, where φ runs over all the characters of $(\mathfrak{o}/\mathfrak{e})^\times$ such that $\varphi(-1) = (-1)^{\{m\}}$ and

$$f_\varphi = [\Gamma_\mathfrak{e} : \Gamma']^{-1} \sum_{\gamma \in R} \varphi(d_\gamma) f_q\|_k\gamma \quad (R = \Gamma_\mathfrak{e}/\Gamma').$$

We see that f_φ satisfies (7.27b) if $m \leq u$, and belongs to $\mathscr{S}_k(\mathfrak{e}, \chi)$ with a Hecke character χ such that $\chi_e = \varphi$ on $(\mathfrak{o}/\mathfrak{e})^\times$. Thus our task is to prove that $f_\varphi = 0$. Take τ, $\mathfrak{q}$, and $\mathfrak{r}$ as in Theorem 6.1, and take $f \in \mathscr{S}_k(\mathfrak{r}\mathfrak{e}, \chi\varepsilon_\tau)$ so that $\lambda_f(\xi, m) = \lambda(\xi/\tau, \mathfrak{q}\mathfrak{m}; f_\varphi)$, as guaranteed by Proposition 3.2. For the same reason as in (6.10), f satisfies (7.27b) if $m \leq u$. Define g by (7.6) with this f and η of (7.7); we take of course $\mathfrak{c} = \mathfrak{r}\mathfrak{e}$ and $\psi = \chi\varepsilon_\tau$. We have $g = 0$ as shown above. The computation of Section 7 is valid in this case and shows that $\lambda_f(1, m) = 0$ for every m, and hence $\lambda(\tau, \mathfrak{o}; f_\varphi) = 0$. Since this holds for every totally positive τ in $\mathfrak{o}$, we obtain $f_\varphi = 0$ as expected.

Thus it only remains to prove Proposition 7.3. For this purpose, put, for α in V of (7.0) and $s \in \mathbf{C}$,

$$(7.28) \qquad \kappa[\alpha, w, s] = [\alpha, w]^{-m} \mathrm{Im}(w)^{2su} |[\alpha, w]|^{-2su},$$

where $[\alpha, w]$ is defined by (7.1). Keeping the letter v for a generic element of $\mathbf{a}$, let t be an element of $\mathbf{a}$ such that $m_t = 0$. Then a direct calculation shows that

$$(7.29) \qquad L_t \kappa[\alpha, w, s] = 2s(1 - 2s)\kappa[\alpha, w, s]$$

$$- 16s^2 \kappa[\alpha, w, s] \det(\alpha_t) \mathrm{Im}(w)^{2t} |[\alpha, w]|^{-2t}.$$

Given an element $f(z) = \sum_\xi \lambda(\xi) \mathbf{e}_\mathbf{a}(\xi z/2) \in \mathscr{S}_k(\Gamma)$ and $\eta \in \mathscr{S}(V_\mathbf{f})$, put

$$P(w, s) = \sum_{0 \neq \alpha \in V/U} \eta(\alpha)\lambda(-\det(\alpha))\kappa[\alpha, w, s],$$

$$Q(w, s) = \sum_{0 \neq \alpha \in V/U} \eta(\alpha)\lambda(-\det(\alpha))\det(\alpha_t)\mathrm{Im}(w)^{2t}|[\alpha, w]|^{-2t}\kappa[\alpha, w, s],$$

where U is a subgroup of $\mathfrak{o}_+^\times$ of finite index such that $\eta(a\alpha) = \eta(\alpha)$ and $\mathrm{diag}[a, a^{-1}] \in \Gamma$ for every $a \in U$.

LEMMA 7.4. *Both P and Q converge for sufficiently large $\mathrm{Re}(s)$ and can be continued as meromorphic functions to the whole s-plane with at most a simple pole at $s = 1$. Moreover, the residues of P and Q at $s = 1$ are $Ag(w)$ and $-(A/8)g(w)$, respectively, where A is a nonzero constant and g is the function of (7.6) defined for the present f and η.*

Proof. Here the meromorphic continuation should be understood in the sense of [S10, Theorem 4.1]. In fact, P is a special case of $f(w, s)$ of that theorem, and Q is of the same type as far as the convergence is concerned, and therefore the convergence is a special case of [S4, Proposition 8.2]. Put $\Psi = (P \cap \Gamma) \backslash \mathscr{H}^\mathbf{a}$. For the same reason as in Section 7, we have

$$\int_\Psi f(z)\Theta(z, w; \eta) y^{k+su} d_H z = A_0 P(w, s)(\pi/2)^{\{-m-su\}} \prod_{v \in \mathbf{a}} \Gamma(s + m_v)$$

with a constant A_0. For $n \in \mathbf{Z}^{\mathbf{a}}$, put

$$\mathscr{E}_{\Gamma}^{n}(z, s) = \sum_{\gamma \in (P \cap \Gamma) \backslash \Gamma} \mathrm{Im}(z)^{su-n} \|_{2n}\gamma.$$

Then we have

$$\int_{\Psi} f(z) \Theta(z, w; \eta) y^{k+su} d_H z = \int_{\Phi} f(z) \Theta(z, w; \eta) \mathscr{E}_{\Gamma}^{0}(z, s) y^k d_H z.$$

Now $\mathscr{E}_{\Gamma}^{0}$ has a meromorphic continuation to the whole s-plane and has a simple pole at $s = 1$ with a positive constant, say R, as its residue. (For the proof, see, for example, [S9, Theorem 4.2 and Proposition 4.3].) Our assertion about P follows immediately from these facts. The residue of P at $s = 1$ is $Ag(w)$ with

$$A = RA_0^{-1}(\pi/2)^{\{m+u\}} \prod_{v \in \mathbf{a}} \Gamma(m_v + 1)^{-1}.$$

Therefore the assertion $\partial g / \partial \overline{w}_v = 0$ when $m_v > 0$ in Proposition 7.3, as well as the holomorphy of g in Proposition 7.1, follows from [S10, Theorem 4.3], or rather its proof, which is valid if f is a cusp form and $m_v > 0$, or if $f \in \mathscr{M}_k$ and $m_v > 0$ for every $v \in \mathbf{a}$.

Our assertion on Q can be proved in a similar but more complicated way. We introduce differential operators

$$\delta_v(c) = (2\pi i)^{-1}\left[c(z_v - \bar{z}_v)^{-1} + \partial / \partial z_v\right],$$

$$\bar{\delta}_v(c) = (-2\pi i)^{-1}\left[c(\bar{z}_v - z_v)^{-1} + \partial / \partial \bar{z}_v\right],$$

for $v \in \mathbf{a}$ and $c \in \mathbf{R}$, and put

$$\Theta_t(z, w; \eta) = \prod_{v \neq t} \bar{\delta}_v(-1/2) \Theta(z, w; \eta).$$

Then we easily see that

$$\Theta_t(z, w; \eta) = 8^{\{t-u\}} \mathrm{Im}(z)^{u/2} \mathrm{Im}(w)^{2t-2u-2m}$$

$$\cdot \sum_{\alpha \in V} \eta(\alpha)[\alpha, \overline{w}]^m |[\alpha, w]|^{2u-2t} \mathbf{e_a}(2^{-1}R[\alpha, z, w]).$$

Since $(\pi i)^{-1}\partial f / \partial z_t = \sum_{\xi} \xi_t \lambda(\xi) \mathbf{e_a}(\xi z/2)$, we have

$$(7.30) \quad \int_{\Psi} (\pi i)^{-1} \partial f / \partial z_t \Theta_t(z, w; \eta) y^{k+su+u} d_H z$$

$$= -A_0 8^{\{t-u\}} Q(w, s)(\pi/2)^{\{-m-su-u\}} \prod_{v \in \mathbf{a}} \Gamma(s + 1 + m_v).$$

We have $\bar{\delta}_v(-1/2) = \bar{\delta}_v(k_v) + (m_v + 1)(4\pi y_v)^{-1}$, and hence

$$\Theta_t(z, w; \eta) = \prod_{v \neq t} \left[\bar{\delta}_v(k_v) + (m_v + 1)(4\pi y_v)^{-1} \right] \Theta(z, w; \eta)$$

$$= \sum_{0 \leq a \leq u - t} M_a y^{a + t - u} \overline{D}_a \Theta(z, w; \eta),$$

where M_a is a constant and $\overline{D}_a = \prod_{v \leq a} \bar{\delta}_v(k_v)$. Similarly, we have $(\pi i)^{-1} \partial f / \partial z_t$ $= 2\delta_t(1/2)f + (4\pi y_t)^{-1}f$, and therefore the integral of (7.30) is equal to

$$\sum_a 2M_a \int_\Psi \delta_t(1/2)f \cdot \overline{D}_a \Theta \cdot y^{k + t + a + su} d_H z$$

$$+ \sum_a (4\pi)^{-1} M_a \int_\Psi f \cdot \overline{D}_a \Theta \cdot y^{k + a + su} d_H z.$$

The a-th integral of the first resp. second sum is equal to

$$\int_\Phi \delta_t(1/2)f \cdot \overline{D}_a \Theta \cdot \mathscr{E}_\Gamma^{a-t} y^{k+2a} d_H z \quad \text{resp.} \quad \int_\Phi f \overline{D}_a \Theta \cdot \mathscr{E}_\Gamma^{a} y^{k+2a} d_H z.$$

This (or (7.31) below) establishes the meromorphic continuation of $Q(w, s)$. By [S9, Theorem 4.2], $\mathscr{E}_\Gamma^n$ is holomorphic at $s = 1$ if $n \neq 0$, and hence Q has at most a simple pole at $s = 1$, and the residue is contributed only by the term involving $\mathscr{E}_\Gamma^0$. Observing that $M_0 = (4\pi)^{\{t-u\}} \prod_{v \in \mathbf{a}} (m_v + 1)$, we find after an easy calculation that the residue is $(-A/8)g(w)$, which proves Lemma 7.4.

Now (7.29) implies that

$$(7.31) \qquad L_t P(w, s) = 2s(1 - 2s)P(w, s) - 16s^2 Q(w, s).$$

Taking the residue at $s = 1$, we see from Lemma 7.4 that $L_t g = 0$, which completes the proof of Proposition 7.3.

A few remarks may be in order at this point, as our method may look somewhat roundabout. In fact, instead of applying L_t to the series P, one could have examined $L_t \Theta$. However, our technique is applicable to the study of the analytic nature of the series $f(w, s)$ introduced in [S10] in a more general case, in which $L_t \Theta$ is not of any help. Therefore we have presented the above proof partly in order to show the efficacy of this method. It should be noted also that $P(w, s)$, or more generally $f(w, s)$, is not an eigenfunction of invariant differential operators for generic s, which attests to the subtlety of the problem.

8. The rationality of modular forms. Throughout the rest of the paper, we denote by $\overline{\mathbf{Q}}$ and $\mathbf{Q}_{ab}$ the algebraic closure of $\mathbf{Q}$ and the maximal abelian extension of $\mathbf{Q}$, both embedded in $\mathbf{C}$. We let $\mathrm{Aut}(\mathbf{C})$ denote the group of all ring-automorphisms of $\mathbf{C}$, and for $\sigma \in \mathrm{Aut}(\mathbf{C})$ and $x \in \mathbf{C}$ we denote by x^σ the image of x under σ. We also define the right action of σ on $\mathbf{Q}^{\mathbf{a}}$ as follows: for $a = (a_v)_v \in \mathbf{Q}^{\mathbf{a}}$, we put $a\sigma = (b_v)_v$ with $b_{v\sigma} = a_v$, where $v\sigma$ is the embedding of F into $\mathbf{R}$ given by $a \mapsto (a_v)^\sigma$. Then for $c \in \mathbf{C}^{\mathbf{a}}$ and $a \in \mathbf{Z}^{\mathbf{a}}$, we have $(c^a)^\sigma = c^{a\sigma}$, where c^a is defined by (1.2a).

We consider a function f on $\mathscr{H}^{\mathbf{a}}$ of the form

$$(8.1) \qquad f(z) = \sum_{\xi \in F} \sum_{0 \leqslant a \leqslant A} c(a, \xi)(\pi y)^{-a} \mathbf{e}_{\mathbf{a}}(\xi z) \qquad (z \in \mathscr{H}^{\mathbf{a}}),$$

where $A \in \mathbf{Z}^{\mathbf{a}}$, $c(a, \xi) \in \mathbf{C}$, $y = \mathrm{Im}(z)$, and ξ runs over the elements of F which are either totally positive or equal to 0. We then define f^σ, for $\sigma \in \mathrm{Aut}(\mathbf{C})$, to be the formal series

$$(8.2) \qquad f^\sigma(z) = \sum_{\xi \in F} \sum_{0 \leqslant a \leqslant A} c(a, \xi)^\sigma (\pi y)^{-a\sigma} \mathbf{e}_{\mathbf{a}}(\xi z).$$

Given a weight $k \geq 0$ as in Section 3, we denote by $\mathscr{N}_k$ the set of all functions f of form (8.1) such that $f\|_k\gamma = f$ for every γ in some congruence subgroup of G (contained in $C''G_{\mathbf{a}}$ if k is half-integral) and also that if $F = \mathbf{Q}$, $f(\alpha z)(c_\alpha z + d_\alpha)^{-k}$ can be written in form (8.1) for every $\alpha \in \mathrm{SL}_2(\mathbf{Z})$ (cf. [S11, (5.4a, b)]). For a subfield L of $\mathbf{C}$, we denote by $\mathscr{N}_k(L)$ the set of all elements f of $\mathscr{N}_k$ such that $c(a, \xi)$ of (8.1) belongs to L for every a and ξ. Further we put $\mathscr{M}_k(L) = \mathscr{M}_k \cap \mathscr{N}_k(L)$ and $\mathscr{S}_k(L) = \mathscr{S}_k \cap \mathscr{N}_k(L)$. We call the elements of $\mathscr{N}_k(L)$ L-rational.

Given two fractional ideals $\mathfrak{x}$ and $\mathfrak{y}$ of F such that $\mathfrak{x}\mathfrak{y}$ is integral and a character φ of $(\mathfrak{o}/\mathfrak{x}\mathfrak{y})^\times$, we denote by $\mathscr{M}_k(\mathfrak{x}, \mathfrak{y}; \varphi)$ resp. $\mathscr{N}_k(\mathfrak{x}, \mathfrak{y}; \varphi)$ the set of all f in $\mathscr{M}_k$ resp. $\mathscr{N}_k$ such that $f\|_k\gamma = \varphi(a_\gamma)f$ for every $\gamma \in \Gamma[\mathfrak{x}, \mathfrak{y}]$; we assume $\mathfrak{x} \subset 2\mathfrak{d}^{-1}$ and $\mathfrak{y} \subset 2\mathfrak{d}$ when k is half-integral. We put $\mathscr{S}_k(\mathfrak{x}, \mathfrak{y}; \varphi) = \mathscr{S}_k \cap \mathscr{M}_k(\mathfrak{x}, \mathfrak{y}; \varphi)$.

PROPOSITION 8.1. *Let k be a weight, and let $\sigma \in \mathrm{Aut}(\mathbf{C})$. Let $\mathscr{X}$ denote any of the three symbols $\mathscr{M}$, $\mathscr{N}$, and $\mathscr{S}$. Then the following assertions hold:*

(1) If $f \in \mathscr{X}_k$, f^σ defines a function on $\mathscr{H}^{\mathbf{a}}$ belonging to $\mathscr{X}_{k\sigma}$;

(2) $\mathscr{X}_k(\mathfrak{x}, \mathfrak{y}; \varphi)^\sigma = \mathscr{X}_{k\sigma}(\mathfrak{x}, \mathfrak{y}; \varphi^\sigma)$;

(3) $\mathscr{X}_k = \mathscr{X}_k(\Phi) \otimes_\Phi \mathbf{C}$ where Φ is the subfield of $\overline{\mathbf{Q}}$ corresponding to $\{\tau \in \mathrm{Gal}(\overline{\mathbf{Q}}/\mathbf{Q})|k\tau = k\}$;

(4) $\mathscr{X}_k(\overline{\mathbf{Q}})$ is stable under $f \mapsto (c_\alpha z + d_\alpha)^{-k}f(\alpha z)$ for every $\alpha \in G$;

(5) $\mathscr{X}_k(\mathfrak{x}, \mathfrak{y}; \varphi)$ is spanned by its $\overline{\mathbf{Q}}$-rational elements.

Proof. These assertions were proved in [S3, Theorem 1.5, Propositions 1.6, 1.7, and 1.8] for $\mathscr{X} = \mathscr{M}$ and $\mathscr{X} = \mathscr{S}$ when k is integral. Since $\mathscr{X}_k$ is spanned by the $\mathscr{X}_k(\mathfrak{x}, \mathfrak{y}; \varphi)$ for all possible $\mathfrak{x}$, $\mathfrak{y}$, and φ, (1) follows from (2). Also, (5) follows from (2) and (3); cf. [S6, Lemma 3]. Suppose now k is half-integral. Let

$n = (u/2) + k$, $q = n\sigma$, and

$$(8.3) \quad \theta(z) = \sum_{\xi \in \mathfrak{o}} \mathbf{e_a}(\xi^2 z/2), \qquad \chi(d) = \mathrm{sgn}\big(N_{F/\mathbf{Q}}(d)\big)\left(\frac{F(\sqrt{-1})/F}{d\mathfrak{o}}\right).$$

By Lemma 4.3, $\theta \in \mathcal{M}_{u/2}(2\mathfrak{d}^{-1}, 2\mathfrak{d}; 1)$, and hence $\theta f \in \mathcal{M}_n(\mathfrak{x}, \mathfrak{y}; \chi\varphi)$ by (2.19) if $f \in \mathcal{M}_k(\mathfrak{x}, \mathfrak{y}; \varphi)$. By (1) with integral k, we have $(\theta f)^\sigma \in \mathcal{M}_q(\mathfrak{x}, \mathfrak{y}; \chi\varphi^\sigma)$. Put $\theta^{-1}(\theta f)^\sigma = g$. Then g is meromorphic and $g^2 = (f^2)^\sigma$, and hence g is holomorphic on $\mathscr{H}^{\mathbf{a}}$ (and even at the cusps when $F = \mathbf{Q}$). Obviously $g = f^\sigma$, and hence we obtain (2) for $\mathscr{X} = \mathcal{M}$. If $f \in \mathscr{S}_k$, then $f^2 \in \mathscr{S}_{2k}$ and hence $g^2 \in \mathscr{S}_{2k\sigma}$, so that g must be a cusp form. To prove (3) for $\mathcal{M}_k$ with half-integral k, take a positive integer N divisible by $2\mathfrak{d}$ and take the model V_N over $\mathbf{Q}$ of $\Gamma_N \backslash \mathscr{H}^{\mathbf{a}}$ considered in the proof of [S3, Proposition 1.7], where

$$(8.4) \qquad\qquad \Gamma_N = \big\{ \gamma \in \Gamma[N\mathfrak{o}, N\mathfrak{o}] \,|\, a_\gamma - 1 \in N\mathfrak{o} \big\}.$$

As mentioned there, we can find a positive integer t and a nonzero element p of $\mathcal{M}_{tu}(\Gamma_N, \mathbf{Q})$ such that $\mathrm{div}(p)$ on V_N is $\mathbf{Q}$-rational. Then θ^{2t}/p is a $\mathbf{Q}$-rational function on V_N, so that $\mathrm{div}(\theta)$ on V_N is $\mathbf{Q}$-rational. The proof there shows that $\mathcal{M}_n(\Gamma_N, \Phi)$, with Φ as in (3), contains an element h such that $\mathrm{div}(h)$ is $\overline{\mathbf{Q}}$-rational. Then $\mathrm{div}(h/\theta)$ is $\overline{\mathbf{Q}}$-rational. Using the isomorphism between the space of modular forms and the linear system of $\mathrm{div}(h/\theta)$, we find that $\mathcal{M}_k$ is spanned by its $\overline{\mathbf{Q}}$-rational elements. Therefore, to prove (3) for $\mathscr{X} = \mathcal{M}$, it is sufficient to show that $\mathcal{M}_k(\overline{\mathbf{Q}})$ is spanned by $\mathcal{M}_k(\Phi)$ over $\overline{\mathbf{Q}}$, since the linear disjointness is obvious. Let $f \in \mathcal{M}_k(\overline{\mathbf{Q}})$ and $g = \theta f$; let K_f resp. K_g denote the field generated by the Fourier coefficients of f resp. g. We easily see that $K_f = K_g$. Moreover, [S3, Proposition 1.3] shows that K_g is finitely generated over $\mathbf{Q}$. Therefore we can find a finite Galois extension L of Φ containing K_f. Let $\mathscr{g} = \mathrm{Gal}(L/\Phi)$. Then $\sum_{\sigma \in \mathscr{g}} b^\sigma f^\sigma \in \mathcal{M}_k(\Phi)$ for every $b \in L$, so that f is an L-linear combination of some elements of $\mathcal{M}_k(\Phi)$. This proves (3) for $\mathscr{X} = \mathcal{M}$. Assertion (4) for $\mathcal{M}_k$ and $\mathscr{S}_k$ with $k \in \mathbf{Z}^{\mathbf{a}}$ can be derived from the corresponding facts for weight $2k$ by taking the squares of the forms; cf. [S8, Proposition 1.3]. To prove (3) for $\mathscr{S}_k$, take a basis B of $\mathbf{C}$ over $\overline{\mathbf{Q}}$. Given $f \in \mathscr{S}_k$, we have $f = \sum_{b \in B} b g_b$ with $g_b \in \mathcal{M}_k(\overline{\mathbf{Q}})$ by (3) with $\mathscr{X} = \mathcal{M}$. Then for every $\alpha \in G$, we have $(c_\alpha z + d_\alpha)^{-k} f(\alpha z) = \sum_{b \in B} b(c_\alpha z + d_\alpha)^{-k} g_b(\alpha z)$. Comparing the constant terms, we see from (4) that the g_b vanish at every cusp, so that $g_b \in \mathscr{S}_k(\overline{\mathbf{Q}})$. This proves (3) for $\mathscr{S}_k$ with $\overline{\mathbf{Q}}$ instead of Φ. The assertion with Φ can then be proved in the same manner as for $\mathcal{M}_k$.

In order to prove our assertions for $\mathscr{X} = \mathcal{N}$, we first define differential operators $\delta_v(c)$ and δ_k^a on $\mathscr{H}^{\mathbf{a}}$ for $v \in \mathbf{a}$, $c \in \mathbf{R}$, and $0 \leq a \in \mathbf{Z}^{\mathbf{a}}$ by

$$(8.5a) \qquad\qquad \delta_v(c) = (2\pi i)^{-1}\big[c(z_v - \bar{z}_v)^{-1} + \partial/\partial z_v\big],$$

$$(8.5b) \qquad\qquad \delta_k^a = \prod_{v \in \mathbf{a}} \prod_{j=1}^{a_v} \delta_v(k_v + 2j - 2).$$

Then one immediately sees that $\delta_k^a \mathcal{N}_k \subset \mathcal{N}_{k+2a}$ and

$$(8.6) \qquad (\delta_k^a f)^\sigma = \delta_{k\sigma}^{a\sigma} f^\sigma \quad \text{for } f \in \mathcal{M}_k \text{ and } \sigma \in \mathrm{Aut}(\mathbf{C}).$$

In [S11, Theorems 5.2 and 5.5], we proved

> **LEMMA 8.2.** *Every f in $\mathcal{N}_k$ can be uniquely written in the form*

$$f = \sum_{0 \leqslant p \leqslant k/2} \delta_{k-2p}^p g_p + \begin{cases} c\delta_2^{(k/2)-1} E_2 & \text{if } F = \mathbf{Q} \text{ and } k \in 2\mathbf{Z}, \\ 0 & \text{otherwise}, \end{cases}$$

where $g_p \in \mathcal{M}_{k-2p}$, $c \in \mathbf{C}$, and

$$E_2(z) = (4\pi y)^{-1} - 12^{-1} + 2 \sum_{n=1}^\infty \left(\sum_{0 < d \mid n} d \right) e(nz).$$

This together with (8.6) proves (1), (2), (4), and (5) for $\mathcal{N}$; (3) for $\mathcal{N}$ can be proved by the same argument as for $\mathcal{M}$.

> **LEMMA 8.3.** *Suppose k is half-integral. Let Γ' be a congruence subgroup of G which is a normal subgroup of Γ_c for some fixed c, and let $\sigma \in \mathrm{Aut}(\mathbf{C})$. Then there is an automorphism of Γ_c/Γ', written $\gamma \mapsto \gamma_\sigma \pmod{\Gamma'}$, which doesn't change $a_\gamma \pmod{c}$ and such that $(f\|_k\gamma)^\sigma = f^\sigma\|_{k\sigma}\gamma_\sigma$ for every $f \in \mathcal{M}_k(\Gamma')$ and every $\gamma \in \Gamma_c$.*

Proof. For an integral ideal $\mathfrak{m}$ divisible by 4, put

$$(8.7a) \quad X_\mathfrak{m} = G_\mathbf{a}\left\{ x \in D[2\mathfrak{d}^{-1}\mathfrak{m}, 2^{-1}\mathfrak{d}\mathfrak{m}] \mid (a_x)_v - 1 \in \mathfrak{m}_v \text{ for every } v \in \mathbf{f} \right\},$$

$$(8.7b) \qquad\qquad\qquad \Gamma'_\mathfrak{m} = G \cap X_\mathfrak{m}.$$

The given Γ' always contains a group of type $\Gamma'_\mathfrak{m}$, and therefore we may assume $\Gamma' = \Gamma'_\mathfrak{m}$ for some $\mathfrak{m}$. In view of Proposition 8.1(5), it is sufficient to prove the desired property for $\overline{\mathbf{Q}}$-rational f. Let $s = \mathrm{diag}[1, t^{-1}]$ with an element t of $\prod_p \mathbf{Z}_p^\times$ whose action on $\mathbf{Q}_{ab}$ is the same as σ. Given $\gamma \in \Gamma_c$, we can find $\gamma_\sigma \in G$ such that $s^{-1}\gamma s \in X_\mathfrak{m}\gamma_\sigma$, by virtue of strong approximation. Then $\gamma \mapsto \gamma_\sigma \pmod{\Gamma'}$ defines an automorphism of Γ_c/Γ'. Let f be a $\overline{\mathbf{Q}}$-rational element of $\mathcal{M}_k(\Gamma')$, and let $m = k - (u/2)$. In [S3, Theorem 1.5] we defined the action of a certain group G on meromorphic modular forms of integral weight. Employing the notation and the properties of the action stated there, we see that $g^{(x,1)} = g$ for every $x \in X_\mathfrak{m}$ and g in the set $\mathcal{A}_m(\Gamma')$ of [S3, (1.7b)], since this holds for $x \in \Gamma'$ and $X_\mathfrak{m}$ is the closure of $\Gamma'G_\mathbf{a}$. By Proposition 8.1(2), we have $f^\sigma \in \mathcal{M}_{k\sigma}(\Gamma')$. Hence $\theta^{-1}f$ resp. $(\theta^{-1}f)^\sigma$, with θ of (8.3), belongs to $\mathcal{A}_m(\Gamma')$ resp. $\mathcal{A}_{m\sigma}(\Gamma')$, so that they are invariant under $(x,1)$ with $x \in X_\mathfrak{m}$. By [S3, Theorem 1.5], $(\theta^{-1}f\|_m\gamma)^\sigma = (\theta^{-1}f)^{(\gamma s,\, \sigma)} = (\theta^{-1}f)^\sigma\|_{m\sigma}\gamma_\sigma$. Since $\theta^\sigma = \theta = \theta\|_{u/2}\gamma = \theta\|_{u/2}\gamma_\sigma$, we obtain the desired property of γ_σ.

Obviously a similar statement holds for integral k, but that will not be needed in this paper (cf. [S6, Lemma 1]).

LEMMA 8.4.   *Let* $\beta \in G \cap ZY$, *where*

$$Z = G_{\mathbf{a}}\{w \in D[\mathfrak{x}, \mathfrak{y}] \| (a_w)_v - 1 \in (\mathfrak{x}\mathfrak{y})_v \text{ for every } v \in \mathbf{f}\},$$

$$Y = \left\{ \begin{pmatrix} 0 & -b \\ 1/b & 0 \end{pmatrix} \Big| b \in F_{\mathbf{A}}^{\times} \right\}.$$

Then, if $f \in \mathcal{M}_k(\mathfrak{x}, \mathfrak{y}; \varphi)$, $\sigma \in \mathrm{Aut}(\mathbf{C})$, *and* k *is integral, we have*

(8.8) $$(f \|_k \beta)^{\sigma} = \varphi(r)^{\sigma} f^{\sigma} \|_{k\sigma} \beta$$

with an integer r *prime to* $\mathfrak{x}\mathfrak{y}$ *such that* $\mathbf{e}(1/q)^{\sigma} = \mathbf{e}(r/q)$, *where* $0 < q \in \mathbf{Z}$, $q\mathbf{Z} = \mathfrak{x}\mathfrak{y} \cap \mathbf{Z}$. *If* k *is half-integral,* (8.8) *holds for* $\beta \in Z\varepsilon \cap G$ *with* ε *of* (2.12).

Proof.   By Proposition 8.1(3), we may assume $f \in \mathcal{M}_k(\overline{\mathbf{Q}})$. Take t and s as in the above proof and put $w = \mathrm{diag}[t^{-1}, t]$. Suppose $\beta \in Zy$ with $y \in Y$. Observe that $f^{(x,1)} = f$ for every $x \in Z$. Hence $(f\|_k\beta)^{\sigma} = f^{(\beta s, \sigma)} = f^{(ys, \sigma)} = f^{(wsy, \sigma)}$, since $ys = wsy$. By strong approximation, $w \in Z\gamma$ with $\gamma \in G$. Then $\gamma \in \Gamma[\mathfrak{x}, \mathfrak{y}]$ and $a_{\gamma} \equiv r \pmod{\mathfrak{x}\mathfrak{y}}$. Therefore $f^{(wsy, \sigma)} = f^{(\gamma sy, \sigma)} = (f\|_k\gamma)^{(sy,1)} = (\varphi(r)f)^{(sy, \sigma)} = \varphi(r)^{\sigma}(f^{\sigma})^{(y,1)} = \varphi(r)^{\sigma}f^{\sigma}\|_{k\sigma}\beta$, since $f^{\sigma} \in \mathcal{M}_{k\sigma}(x, y; \varphi_{*}^{\sigma})$, which proves the case of integral weight. Next, we have $\theta\|_{u/2}\gamma = \theta$ for θ of (8.3) and for every $\gamma \in G \cap C''G_{\mathbf{a}}$ by Lemma 4.1 and (2.16b), and in particular for $\gamma \in Z\varepsilon \cap G$. Therefore (8.8) for half-integral weight can be obtained from the case of integral weight by considering $\theta^{-1}f$ or θf as in the proof of Lemma 8.3 or Proposition 8.1.

LEMMA 8.5.   *Given* $\alpha \in G$, $f \in \mathcal{M}_k$, *and* $\sigma \in \mathrm{Aut}(\mathbf{C})$, *there exists an element* $\beta \in G$ *such that*

$$\left[j(\alpha, z)^{-k} f(\alpha z) \right]^{\sigma} = j(\beta, z)^{-k\sigma} f^{\sigma}(\beta z),$$

where, for half-integral k, *the branch of* $j(\beta, z)^{-k\sigma}$ *must be chosen suitably according to the choice of the branch of* $j(\alpha, z)^{-k}$.

Proof.   The half-integral case can be reduced to the integral case by considering f^2. Thus assume k integral. By Proposition 8.1, $f = \sum_{b \in B} b g_b$ with a finite subset B of $\mathbf{C}$ and $g_b \in \mathcal{M}_k(\overline{\mathbf{Q}})$. As in the proof of Lemma 8.3, we can find an open subgroup U of $G_{\mathbf{A}}$ such that $g_b^{(x,1)} = g_b$ and $g_b^{\sigma} = g_b^{\sigma(x,1)}$ for every $x \in U$ and every $b \in B$. With s in that proof, we take $\beta \in G$ and $w \in U$ so that $s^{-1}\alpha s = w\beta$. By [S3, Theorem 1.5(iv)] we have $(g_b\|\alpha)^{\sigma} = g_b^{(\alpha s, \sigma)} = g_b^{(sw\beta, \sigma)} = g_b^{\sigma}\|\beta$, and hence $(f\|\alpha)^{\sigma} = f^{\sigma}\|\beta$ as desired.

PROPOSITION 8.6.   *Let $\sigma \in \mathrm{Aut}(\mathbf{C})$ and $f \in \mathcal{M}_k(\mathfrak{c}, \psi)$ with $k = (u/2) + m$, $\mathfrak{c}$, and ψ as in* (3.8a, b). *Suppose ψ is of finite order. Then $f^\sigma \in \mathcal{M}_{k\sigma}(\mathfrak{c}, \psi^\sigma)$, $\lambda(\xi, \mathfrak{m}; f, \psi)^\sigma = \lambda(\xi, \mathfrak{m}; f^\sigma, \psi^\sigma)$, and $(f | T_\mathfrak{p})^\sigma = f^\sigma | T_\mathfrak{p}$ for every prime ideal $\mathfrak{p}$.*

Proof.   The first assertion is a special case of Proposition 8.1(2). To prove the second one, take t and s as in the proof of Lemma 8.3. Given an ideal $\mathfrak{m}$ prime to $\mathfrak{c}$, take $e \in F_\mathfrak{f}^\times$ and $\alpha \in G$ so that $\mathfrak{m} = e\mathfrak{o}$ and $\alpha \in \mathrm{diag}[e, e^{-1}]D_\mathfrak{c}G_\mathbf{a}$. Further take $\beta \in G$ and w in the group $X_\mathfrak{c}$ of (8.7a) (with $\mathfrak{c}$ as $\mathfrak{m}$ there) so that $s\alpha s^{-1}w = \beta$, which is guaranteed by strong approximation. Put $g = \theta^{-1}f$. By [S3, Theorem 1.5] we have $g^\sigma \|_{m\sigma} \alpha^{-1} = g^{(s\alpha^{-1}, \sigma)} = g^{(w\beta^{-1}s, \sigma)} = (g\|_m \beta^{-1})^\sigma$. Observe that $\mathfrak{a}_\alpha = \mathfrak{a}_\beta = \mathfrak{m}^{-1}$ and $\psi_\mathfrak{c}(d_\alpha) = \psi_\mathfrak{c}(d_\beta)$. By Proposition 3.1 and Lemma 4.4, if $p = f^\sigma$, we have

$$\psi^*(\mathfrak{m})N(\mathfrak{m})^{1/2}J_k(\beta, \beta^{-1}z)f(\beta^{-1}z) = \psi_\mathfrak{c}(d_\beta)\sum_\xi \lambda_f(\xi, \mathfrak{m})\mathbf{e}_\mathbf{a}(\xi z/2),$$

$$\psi^*(\mathfrak{m})^\sigma N(\mathfrak{m})^{1/2}J_{k\sigma}(\alpha, \alpha^{-1}z)f^\sigma(\alpha^{-1}z) = \psi_\mathfrak{c}(d_\alpha)^\sigma \sum_\xi \lambda_p(\xi, \mathfrak{m})\mathbf{e}_\mathbf{a}(\xi z/2),$$

$$h(\alpha, \alpha^{-1}z)\theta(\alpha^{-1}z) = h(\beta, \beta^{-1}z)\theta(\beta^{-1}z) = N(\mathfrak{m})^{-1/2}\sum_\zeta \mathbf{e}_\mathbf{a}(\zeta^2 z/2),$$

where ζ runs over $\mathfrak{m}^{-1}$. Therefore

$$\psi^*(\mathfrak{m})^\sigma g^\sigma \|_{m\sigma}\alpha^{-1} = \psi_\mathfrak{c}(d_\alpha)^\sigma \left[\sum_\zeta \mathbf{e}_\mathbf{a}(\zeta^2 z/2)\right]^{-1}\sum_\xi \lambda_p(\xi, \mathfrak{m})\mathbf{e}_\mathbf{a}(\xi z/2),$$

$$\psi^*(\mathfrak{m})g\|_m\beta^{-1} = \psi_\mathfrak{c}(d_\beta)\left[\sum_\zeta \mathbf{e}_\mathbf{a}(\zeta^2 z/2)\right]^{-1}\sum_\xi \lambda_f(\xi, \mathfrak{m})\mathbf{e}_\mathbf{a}(\xi z/2).$$

Since σ sends the latter to the former, we obtain $\lambda_f(\xi, \mathfrak{m})^\sigma = \lambda_p(\xi, \mathfrak{m})$ at least for $\mathfrak{m}$ prime to $\mathfrak{c}$. This combined with (3.14b) proves the formula for $\mathfrak{m}$ not necessarily prime to $\mathfrak{c}$. The last assertion then follows immediately from Proposition 5.4.

Let $f \in \mathcal{M}_k(\mathfrak{c}, \psi)$. By (3.15), $\lambda(\xi, \mathfrak{o}; f, \psi) \in \overline{\mathbf{Q}}$ for every ξ if f is $\overline{\mathbf{Q}}$-rational. If further ψ is of finite order, then (3.14c) together with (4) of Proposition 8.1 implies that $\lambda(\xi, \mathfrak{m}; f, \psi) \in \overline{\mathbf{Q}}$ for every ξ and every $\mathfrak{m}$. This is not necessarily so, however, for ψ of infinite order. Even ψ^σ may not be a Hecke character for such a ψ and an arbitrary σ.

If ψ is of infinite order, we can formulate some properties of algebraicity of the elements of $\mathcal{M}_k(\mathfrak{c}, \psi)$ at least in the following way. Suppose $\psi_\mathbf{a}(x) = \mathrm{sgn}(x_\mathbf{a})^r |x_\mathbf{a}|^{i\mu}$ with $r \in \mathbf{Z}^\mathbf{a}$ and $\mu \in \mathbf{R}^\mathbf{a}$, $\{\mu\} = 0$. Given a fractional ideal $\mathfrak{m}$ of F, take a positive integer p so that $\mathfrak{m}^p = \alpha\mathfrak{o}$ with $0 \ll \alpha \in F$. Then we define a symbol $|\mathfrak{m}|^{i\mu}$ to be the coset of $\mathbf{C}^\times$ modulo the group of all roots of unity represented by $|\alpha|^{i\mu/p}$. If $\mathfrak{m}$ is prime to the conductor of ψ, $|\mathfrak{m}|^{i\mu}$ is represented by $\psi^*(\mathfrak{m})^{-1}$. By abuse of notation, we use $|\mathfrak{m}|^{i\mu}$ to denote any complex number in this coset.

LEMMA 8.7. *Let* $0 \neq f \in \mathcal{M}_k(\mathfrak{c}, \psi) \cap \mathcal{M}_k(\overline{\mathbf{Q}})$. *Then with* μ *as above,* $|\mathfrak{m}|^{i\mu}\lambda(\xi, \mathfrak{m}; f, \psi) \in \overline{\mathbf{Q}}$ *for every* ξ *and every* $\mathfrak{m}$. *Moreover, if* $f|T_\mathfrak{p} = \omega_\mathfrak{p} f$ *for a prime ideal* $\mathfrak{p}$, *then* $|\mathfrak{p}|^{i\mu}\omega_\mathfrak{p} \in \overline{\mathbf{Q}}$, *and* $|\mathfrak{p}|^{i\mu}T_\mathfrak{p}$ *maps the* $\overline{\mathbf{Q}}$-*rational elements of* $\mathcal{M}_k(\mathfrak{c}, \psi)$ *into themselves.*

This follows immediately from (3.14c), (5.5) (or Proposition 5.4), and Proposition 8.1(4).

LEMMA 8.8. *Suppose* $k = (u/2) + m$ *with* $u \leq m \in \mathbf{Z}^\mathbf{a}$. *Let* $0 \neq f \in \mathcal{S}_k(\mathfrak{c}, \psi)$; *suppose* $f|T_\mathfrak{p} = \omega_\mathfrak{p} f$ *with* $\omega_\mathfrak{p} \in \mathbf{C}$ *for every* $\mathfrak{p}$ *in a subset* S *of* $\mathfrak{f}$ *containing almost all primes. Then there exists a nonzero element* $\mathbf{g}$ *of* $\mathcal{M}_{2m}(2^{-1}\mathfrak{c}, \psi^2)$ *which is a common eigenfunction of* $\mathfrak{X}(\mathfrak{p})$ *for all primes* $\mathfrak{p}$, *and such that* $\mathbf{g}|\mathfrak{X}(\mathfrak{p}) = \omega_\mathfrak{p}\mathbf{g}$ *for every* $\mathfrak{p} \in S$.

Let V be the set of all $g \in \mathcal{S}_k(\mathfrak{c}, \psi)$ such that $g|T_\mathfrak{p} = \omega_\mathfrak{p} g$ for every $\mathfrak{p} \in S$. The $T_\mathfrak{p}$ for all primes $\mathfrak{p}$ restricted to V form a commutative ring, and therefore have a common eigenfunction, say h. Applying Theorem 6.2 to h, we find a form $\mathbf{g}$ with the expected properties.

In the setting of Lemma 8.8, suppose that either $m \neq u$, or $m = u$ and f is orthogonal to every theta series of type (4.5) with $n = u$. By Theorem 6.1(III), $\mathbf{g}$ is a cusp form. Then there exists an element $\mathbf{h}$ of $\mathcal{S}_{2m}(2^{-1}\mathfrak{c}, \psi^2)$ which is *primitive* in the sense of [S3, p. 652] (the notion is based on Miyake's result in [M]) and such that $\mathbf{h}|\mathfrak{X}(\mathfrak{p}) = \omega_\mathfrak{p}\mathbf{h}$ at least for every $\mathfrak{p}$ prime to $\mathfrak{c}$. (The exact "level" of $\mathbf{h}$ may or may not be equal to $2^{-1}\mathfrak{c}$.) In this way we can associate a primitive form of weight $2m$ to f. This observation leads us to the following definition.

We start from k, m, $\mathfrak{c}$, and ψ as in (3.8a, b) and take a primitive form $\mathbf{h}$ in $\mathcal{S}_{2m}(2^{-1}\mathfrak{c}, \psi^2)$; let $\mathbf{h}|\mathfrak{X}(\mathfrak{p}) = \omega_\mathfrak{p}\mathbf{h}$ for every prime $\mathfrak{p}$ not dividing $\mathfrak{c}$. We then denote by $\mathcal{S}_k(\mathfrak{c}, \psi, \mathbf{h})$ or $\mathcal{S}_k(\mathfrak{c}, \psi, \{\omega_\mathfrak{p}\})$ the set of all f in $\mathcal{S}_k(\mathfrak{c}, \psi)$ such that $f|T_\mathfrak{p} = \omega_\mathfrak{p} f$ for *almost all* $\mathfrak{p}$.

If in particular ψ is of finite order, we can define $\mathbf{h}^\sigma$ for $\sigma \in \mathrm{Gal}(\overline{\mathbf{Q}}/\mathbf{Q})$ by $\mathbf{h}^\sigma = (h_1^\sigma, \ldots, h_\kappa^\sigma)$ when $\mathbf{h} = (h_1, \ldots, h_\kappa)$. Then $\mathbf{h}^\sigma$ is a primitive form belonging to $\mathcal{S}_{2m\sigma}(2^{-1}\mathfrak{c}, \psi^{2\sigma})$, and $\mathbf{h}^\sigma|\mathfrak{X}(\mathfrak{p}) = \omega_\mathfrak{p}^\sigma\mathbf{h}^\sigma$ as proved in [S3, Proposition 2.6]. We denote by $\mathbf{Q}(\mathbf{h})$ the subfield of $\overline{\mathbf{Q}}$ such that $\mathbf{h}^\sigma = \mathbf{h}$ if and only if $\sigma = \mathrm{id}$. on it. (See [S3, Proposition 2.8] for some basic properties of $\mathbf{Q}(\mathbf{h})$.)

PROPOSITION 8.9. *The notation being as above, the following assertions hold:*
(1) $f|T_\mathfrak{p} = \omega_\mathfrak{p} f$ *for every* $f \in \mathcal{S}_k(\mathfrak{c}, \psi, \{\omega_\mathfrak{p}\})$ *and every* $\mathfrak{p}$ *prime to* $\mathfrak{c}$.
(2) $\mathcal{S}_k(\mathfrak{c}, \psi, \mathbf{h})$ *is spanned by its* $\overline{\mathbf{Q}}$-*rational elements.*
(3) *If* ψ *is of finite order,* $\mathcal{S}_k(\mathfrak{c}, \psi, \mathbf{h})^\sigma = \mathcal{S}_{k\sigma}(\mathfrak{c}, \psi^\sigma, \mathbf{h}^\sigma)$ *for every* $\sigma \in \mathrm{Aut}(\mathbf{C})$.
(4) *Let* $\mathcal{T}_k(\mathfrak{c}, \psi, \mathbf{h})$ *be the orthogonal complement of* $\mathcal{S}_k(\mathfrak{c}, \psi, \mathbf{h})$ *in* $\mathcal{S}_k$. *Then* $\mathcal{T}_k(\mathfrak{c}, \psi, \mathbf{h})$ *is spanned by its* $\overline{\mathbf{Q}}$-*rational elements. Moreover,* $\mathcal{T}_k(\mathfrak{c}, \psi, \mathbf{h})^\sigma = \mathcal{T}_{k\sigma}(\mathfrak{c}, \psi^\sigma, \mathbf{h}^\sigma)$ *for every* $\sigma \in \mathrm{Aut}(\mathbf{C})$ *if* ψ *is of finite order.*

(5) *If $m \neq u$, $\mathscr{S}_k(\mathfrak{c}, \psi)$ is the direct sum of subspaces $\mathscr{S}_k(\mathfrak{c}, \psi, \mathbf{h})$ for finitely many $\mathbf{h}$'s.*

Proof. Naturally we assume that $\mathscr{S}_k(\mathfrak{c}, \psi, \mathbf{h}) \neq \{0\}$. By Propositions 5.2 and 5.3, we see that $\mathscr{S}_k(\mathfrak{c}, \psi)$ has an orthogonal basis $\{q_1, \ldots, q_t\}$ consisting of common eigenfunctions of the $T_\mathfrak{p}$ for all $\mathfrak{p}$ prime to $\mathfrak{c}$. Let $q_a | T_\mathfrak{p} = \omega_\mathfrak{p}^a q_a$. Changing the order if necessary, we may assume that $q_a \in \mathscr{S}_k(\mathfrak{c}, \psi, \mathbf{h})$ if and only if $a \leq r$. Then it can easily be seen that $\mathscr{S}_k(\mathfrak{c}, \psi, \mathbf{h})$ is spanned by $q_1, \ldots, q_r$. By Lemma 8.8, each q_a with $a \leq r$ corresponds to a form $\mathbf{g}_a \in \mathscr{M}_{2m}(2^{-1}\mathfrak{c}, \psi^2)$ such that $\mathbf{g}_a | \mathfrak{T}(\mathfrak{p}) = \omega_\mathfrak{p}^a \mathbf{g}_a$ for every $\mathfrak{p}$ prime to $\mathfrak{c}$. Since $\omega_\mathfrak{p}^a = \omega_\mathfrak{p}$ for almost all $\mathfrak{p}$ and $\mathbf{h}$ is primitive, we see that $\omega_\mathfrak{p}^a = \omega_\mathfrak{p}$ for every $\mathfrak{p}$ prime to $\mathfrak{c}$. This proves (1). Now $\mathscr{S}_k(\mathfrak{c}, \psi, \mathbf{h})$ consists of the elements f of $\mathscr{S}_k(\mathfrak{c}, \psi)$ such that $|\mathfrak{p}|^{i\mu} f | T_\mathfrak{p} = |\mathfrak{p}|^{i\mu} \omega_\mathfrak{p} f$. Therefore we obtain (2) from Lemma 8.7. Assertion (3) follows from Propositions 8.1 and 8.6. Let $W_k(\mathfrak{c}, \psi)$ be the orthogonal complement of $\mathscr{S}_k(\mathfrak{c}, \psi)$ in $\mathscr{S}_k$. Then $\mathscr{T}_k(\mathfrak{c}, \psi, \mathbf{h}) = W_k(\mathfrak{c}, \psi) + \sum_{a > r} \mathbf{C} q_a$. The space

$$\left\{ f \in \mathscr{S}_k(\mathfrak{c}, \psi) \mid |\mathfrak{p}|^{i\mu} f | T_\mathfrak{p} = |\mathfrak{p}|^{i\mu} \omega_\mathfrak{p}^a f \text{ for every } \mathfrak{p} \text{ prime to } \mathfrak{c} \right\}$$

contains q_a and is spanned by its $\overline{\mathbf{Q}}$-rational elements. If ψ is of finite order, we have $q_a^\sigma | T_\mathfrak{p} = (\omega_\mathfrak{p}^a)^\sigma q_a^\sigma$, so that q_a^σ is orthogonal to q_b^σ if $a > r \geq b$. Therefore, to prove (4), it is sufficient to show that $W_k(\mathfrak{c}, \psi)$ is spanned by $\overline{\mathbf{Q}}$-rational elements and moreover $W_k(\mathfrak{c}, \psi)^\sigma = W_{k\sigma}(\mathfrak{c}, \psi^\sigma)$ if ψ is of finite order. For this purpose take $\Gamma'_\mathfrak{m}$ of (8.7b) with a multiple $\mathfrak{m}$ of $\mathfrak{c}$. Then $\mathscr{S}_k$ is a union of $\mathscr{S}_k(\Gamma'_\mathfrak{m})$ for all such $\mathfrak{m}$. Fixing one $\mathfrak{m}$, put, for $f \in \mathscr{S}_k(\Gamma'_\mathfrak{m})$,

$$P(f) = P_k(f, \psi) = [\Gamma_\mathfrak{c} : \Gamma'_\mathfrak{m}]^{-1} \sum_{\gamma \in R} \psi_\mathfrak{c}(d_\gamma) f \|_k \gamma \qquad (R = \Gamma'_\mathfrak{m} \backslash \Gamma_\mathfrak{c}).$$

Then $P(f) \in \mathscr{S}_k(\mathfrak{c}, \psi)$ and $\langle g, f \rangle = \langle g, P(f) \rangle$ for $g \in \mathscr{S}_k(\mathfrak{c}, \psi)$ and $f \in \mathscr{S}_k(\Gamma'_\mathfrak{m})$. Therefore $f \in W_k(\mathfrak{c}, \psi)$ if and only if $P(f) = 0$. In view of Proposition 8.1(4), this shows that $W_k(\mathfrak{c}, \psi) \cap \mathscr{S}_k(\Gamma'_\mathfrak{m})$, and hence $W_k(\mathfrak{c}, \psi)$ itself, is spanned by $\overline{\mathbf{Q}}$-rational elements. If ψ is of finite order and $\sigma \in \mathrm{Aut}(\mathbf{C})$, then $\mathscr{S}_k(\Gamma'_\mathfrak{m})^\sigma = \mathscr{S}_{k\sigma}(\Gamma'_\mathfrak{m})$ by Proposition 8.1(2), and $P_k(f, \psi)^\sigma = P_{k\sigma}(f^\sigma, \psi^\sigma)$ by virtue of Lemma 8.3, and hence $W_k(\mathfrak{c}, \psi)^\sigma = W_{k\sigma}(\mathfrak{c}, \psi^\sigma)$ as desired. Finally if $m \neq u$, Theorem 6.1(III) combined with Lemma 8.8 shows that each one of the above q_a corresponds to some primitive form belonging to $\mathscr{S}_{2m}(2^{-1}\mathfrak{c}, \psi^2)$. This proves (5). (If $m = u$, we have to take the orthogonal complement of the theta series in $\mathscr{S}_k(\mathfrak{c}, \psi)$.)

9. Some auxiliary facts on Eisenstein series.

For an integral or a half-integral weight k, put

$$(9.1) \qquad\qquad \mathscr{E}_k = \{ f \in \mathscr{M}_k \mid \langle f, g \rangle = 0 \text{ for every } g \in \mathscr{S}_k \}.$$

Then $\mathscr{M}_k = \mathscr{S}_k \oplus \mathscr{E}_k$. Recall that $\mathscr{E}_k = \{0\}$ if $k \notin \mathbf{Z}u/2$ (see [S9, Proposition 8.1]).

THEOREM 9.1. *For $0 < k \in \mathbf{Z}u/2$, $\mathscr{E}_k$ is spanned by its $\mathbf{Q}$-rational elements.*

Proof. Let $k = tu$ with $t \in (1/2)\mathbf{Z}$. Exclude the following three cases: (i) $t = 1/2$; (ii) $t = 2$ and $F = \mathbf{Q}$; (iii) $t = 3/2$ and $F = \mathbf{Q}$. Let us first treat the case where $t = m + (1/2)$ with $0 < m \in \mathbf{Z}$. Take the function $E'(z, s; 1/2, mu, \psi, \mathfrak{c})$ of [S8, (6.1)] with a Hecke character ψ of F such that $\psi_{\mathbf{a}}(x) = \mathrm{sgn}(x_{\mathbf{a}})^{mu}$ and a common multiple $\mathfrak{c}$ of 4 and the conductor of ψ. From its Fourier expansion given in [S8, (6.1) and Theorem 6.1], we see that it is finite at $s = m/2$ and

$$(9.2) \qquad y^{-mu/2}E'(z, m/2; 1/2, mu, \psi, \mathfrak{c}) = A(\psi)\sum_{\xi} b_{\xi}(\psi)\mathbf{e}_{\mathbf{a}}(\xi z),$$

where $\xi = 0$ or $0 \ll \xi \in 2\mathfrak{c}^{-1}$, $A(\psi)$ is a nonzero constant depending only on m, $\mathfrak{c}$, and ψ, and

$$b_{\xi}(\psi) = N_{F/\mathbf{Q}}(\xi)^{t-1}L_{\mathfrak{c}(\xi)}(m, \varepsilon_{2\xi}\psi)\beta_{\xi}(\psi) \quad \text{if } \xi \gg 0.$$

Here $\mathfrak{c}(\xi)$ is $\mathfrak{c}$ times the conductor of $\varepsilon_{2\xi}$, ε_* is as in Proposition 3.2, and $\beta_{\xi}(\psi)$ is an algebraic number given as $\beta(\xi, m/2)$ with β of [S8, Theorem 6.1]. Observe that $\beta_{\xi}(\psi)^{\sigma} = \beta_{\xi}(\psi^{\sigma})$ for every $\sigma \in \mathrm{Gal}(\overline{\mathbf{Q}}/\mathbf{Q})$. Put $P_{\mathfrak{c}}(m, \varphi) = D_F^{1/2}(2\pi i)^{-md}\mathfrak{g}(\varphi)^{-1}L_{\mathfrak{c}}(m, \varphi)$ with $\mathfrak{g}(*)$ of (4.8) for $0 < m \in \mathbf{Z}$ and a Hecke character φ of F such that $\varphi_{\mathbf{a}}(x) = \mathrm{sgn}(x_{\mathbf{a}})^{mu}$. Here and throughout the rest of the paper, we put

$$(9.3) \qquad\qquad\qquad d = [F : \mathbf{Q}].$$

We recall that $P_{\mathfrak{c}}(m, \varphi)^{\sigma} = P_{\mathfrak{c}}(m, \varphi^{\sigma})$ for every $\sigma \in \mathrm{Gal}(\overline{\mathbf{Q}}/\mathbf{Q})$ (see [S3, Proposition 3.1]). Now we need two lemmas.

LEMMA 9.2. *For any number of Hecke characters $\varphi_1, \ldots, \varphi_r$ of F of finite order and for $\sigma \in \mathrm{Gal}(\overline{\mathbf{Q}}/\mathbf{Q})$, one has*

$$\left[\mathfrak{g}(\varphi_1) \cdots \mathfrak{g}(\varphi_r)/\mathfrak{g}(\varphi_1 \cdots \varphi_r)\right]^{\sigma} = \mathfrak{g}(\varphi_1^{\sigma}) \cdots \mathfrak{g}(\varphi_r^{\sigma})/\mathfrak{g}(\varphi_1^{\sigma} \cdots \varphi_r^{\sigma}).$$

This was given in [S3, Lemma 4.12].

LEMMA 9.3. *If ε is the Hecke character of F corresponding to $F(\sqrt{c})/F$ with $0 \neq c \in F$, then $\mathfrak{g}(\varepsilon) = i^n N(\mathfrak{f})^{1/2}$, where n is the number of $v \in \mathbf{a}$ such that $c_v < 0$ and $\mathfrak{f}$ is the conductor of ε. Moreover, $c^{-1}\mathfrak{f}$ is the square of a fractional ideal in F.*

Proof. The first assertion is due to Hecke, and in fact follows immediately from the functional equation of $L(s, \varepsilon)$. Let $K = F(\sqrt{c})$ and let $\mathfrak{b}$ be the different of K relative to F. The conductor-discriminant theorem tells that $\mathfrak{f} = N_{K/F}(\mathfrak{b})$. Now $\mathfrak{b} = \sqrt{c}\,\mathfrak{a}\,\mathfrak{o}_K$ with a fractional ideal $\mathfrak{a}$ in F, and hence $\mathfrak{f} = c\mathfrak{a}^2$ as expected.

Coming back to the proof of Theorem 9.1, let $C(z, \psi)$ denote the function of (9.2) times $A(\psi)^{-1}D_F^{1/2}(2\pi i)^{-md}2^{d/2}\mathfrak{g}(\psi)^{-1}$. Then

$$C(z, \psi) = \sum_{\xi} c_{\xi}(\psi)\mathbf{e}_{\mathbf{a}}(\xi z),$$

$$c_{\xi}(\psi) = \left[\mathfrak{g}(\varepsilon_{2\xi})\mathfrak{g}(\psi)/\mathfrak{g}(\psi\varepsilon_{2\xi})\right]^{-1}2^{d/2}N_{F/\mathbf{Q}}(\xi)^{t-1}\mathfrak{g}(\varepsilon_{2\xi})P_{\mathfrak{c}(\xi)}(m, \psi\varepsilon_{2\xi})\beta_{\xi}(\psi)$$

if $\xi \gg 0$. The above two lemmas show that $c_{\xi}(\psi)^{\sigma} = c_{\xi}(\psi^{\sigma})$, and hence

$$(9.4) \qquad\qquad C(z, \psi)^{\sigma} = C(z, \psi^{\sigma}) \quad \text{for every } \sigma \in \mathrm{Gal}(\overline{\mathbf{Q}}/\mathbf{Q}).$$

By [S9, (4.24) and (8.5)], we see that $\mathscr{E}_k$ is spanned by the functions $j(\alpha, z)^{-k}C(\alpha z, \psi)$ for all $\alpha \in G$. Let $\mathscr{E}_k^0$ denote the $\mathbf{Q}_{ab}$-linear span of all such functions. Our result (9.4) together with [S8, Proposition 1.3] and Lemma 8.5 shows that $\mathscr{E}_k^0$ is contained in $\mathscr{M}_k(\mathbf{Q}_{ab})$ and stable under $\mathrm{Gal}(\mathbf{Q}_{ab}/\mathbf{Q})$. With K_f defined as in the proof of Proposition 8.1 for $f \in \mathscr{E}_k^0$ and with $\mathscr{g} = \mathrm{Gal}(K_f/\mathbf{Q})$, we observe that $\Sigma_{\sigma \in \mathscr{g}} c^{\sigma}f^{\sigma} \in \mathscr{E}_k^0 \cap \mathscr{M}_k(\mathbf{Q})$ for every $c \in K_f$. Thus f is a $\mathbf{Q}_{ab}$-linear combination of elements of $\mathscr{E}_k^0 \cap \mathscr{M}_k(\mathbf{Q})$, which completes the proof for half-integral k, excluding cases (i) and (iii). In case (iii), it was shown by Pei [P] that $\mathscr{E}_{3/2}$ is spanned by $j(\alpha, z)^{-3/2}f(\alpha z)$ for all $\alpha \in G$ and all f belonging to a certain explicitly given subset of $\mathscr{E}_{3/2}$. It can easily be verified that this set consists of $\mathbf{Q}_{ab}$-rational forms and stable under $\mathrm{Gal}(\mathbf{Q}_{ab}/\mathbf{Q})$. Therefore we obtain the desired conclusion in the same fashion as in the nonexceptional case. Suppose now $t = 1/2$. In this case, [S9, Proposition 8.2 and Theorem 8.4] say that $\mathscr{E}_k$ is spanned by the transforms under G of the residues of $E'(z, s; 1/2, 0, \psi, \mathfrak{c})$ at $s = 1/2$. Each residue is a constant times $\Sigma_{\xi \in \mathfrak{m}}\mathbf{e}_{\mathbf{a}}(\tau\xi^2 z/2)$ with $\tau \gg 0$ and a fractional ideal $\mathfrak{m}$, as proved in [S8, Proposition 6.4]. Therefore, the desired fact can be derived in the same way.

The case of integral k can be proved by the same technique. That $\mathscr{E}_k$ is spanned by the Eisenstein series is guaranteed by [S9, (8.5)]; the $\mathbf{Q}_{ab}$-rationality of the Eisenstein series is well known (see [S9, Proposition 8.2], for example). Therefore the above reasoning applies to this case. Care must be taken in case (ii), in which the Eisenstein series belong to $\mathscr{N}_2$ and are not necessarily holomorphic. However, subtracting a suitable constant multiple of E_2 of Lemma 8.2 from each series, we again obtain the expected result.

PROPOSITION 9.4. *For each integral or half-integral weight k, there is a* C-*linear map p_k of $\mathscr{N}_k$ into $\mathscr{S}_k$ with the following properties:*
  (i) $\langle f, h \rangle = \langle p_k(f), h \rangle$ *for every $f \in \mathscr{N}_k$ and $h \in \mathscr{S}_k$;*
  (ii) $p_k(f)^{\sigma} = p_{k\sigma}(f^{\sigma})$ *for every $\sigma \in \mathrm{Aut}(\mathbf{C})$.*

Proof. For $f \in \mathscr{N}_k$, take g_p as in Lemma 8.2. By [S5, I, (2.19)] we have $\langle f, h \rangle = \langle g_0, h \rangle$ for every $h \in \mathscr{S}_k$. (If $F = \mathbf{Q}$ and $k = 2$, E_2 may appear, but $\langle E_2, h \rangle = 0$, since E_2 is an Eisenstein series.) Write $g_0 = q + r$ with $q \in \mathscr{S}_k$ and $r \in \mathscr{E}_k$. Since q is uniquely determined by f, we can define a C-linear map

p_k of $\mathcal{N}_k$ into $\mathcal{S}_k$ by $p_k(f) = q$. Then (i) follows immediately from our construction, and (ii) from (8.6) and the fact that $q^\sigma \in \mathcal{S}_k$, and $r^\sigma \in \mathscr{E}_k$ by virtue of Proposition 8.1 and Theorem 9.1.

Our later treatment requires a refinement of the above proposition, which is as follows: Let $\mathcal{S}_k(\mathfrak{c}, \psi, \mathbf{h})$ and $\mathcal{T}_k(\mathfrak{c}, \psi, \mathbf{h})$ be as in Proposition 8.9. Then $\mathcal{S}_k = \mathcal{S}_k(\mathfrak{c}, \psi, \mathbf{h}) \oplus \mathcal{T}_k(\mathfrak{c}, \psi, \mathbf{h})$, and hence we can define the projection map q_k of $\mathcal{S}_k$ into $\mathcal{S}_k(\mathfrak{c}, \psi, \mathbf{h})$ relative to this decomposition. Then $q_k \circ p_k$ is a C-linear map with the following properties:

$$(9.5) \qquad \langle q_k(p_k(f)), g \rangle = \langle f, g \rangle \quad \text{for } f \in \mathcal{N}_k \text{ and } g \in \mathcal{S}_k(\mathfrak{c}, \psi, \mathbf{h}),$$

$$(9.6) \qquad q_k(p_k(f)) \text{ is } \overline{\mathbf{Q}}\text{-rational if } f \text{ is } \overline{\mathbf{Q}}\text{-rational},$$

$$(9.7)$$

$$q_k(p_k(f))^\sigma = q_{k\sigma}(p_{k\sigma}(f^\sigma)) \quad \text{for every } \sigma \in \text{Aut}(\mathbf{C}) \text{ if } \psi \text{ is of finite order},$$

$$\text{where } q_{k\sigma} \text{ denotes the projection map of } \mathcal{S}_{k\sigma} \text{ into } \mathcal{S}_{k\sigma}(\mathfrak{c}, \psi^\sigma, h^\sigma).$$

The first formula is obvious; the second and third ones follow from Proposition 8.9(2), (3), and (4).

The rest of this section concerns some results of rationality and near holomorphy of Eisenstein series of integral weight. Though the same type of results hold also for half-integral weight, we state here only what we need in the next section. Given an integral ideal $\mathfrak{c}$ in F divisible by 4, a Hecke character χ of F, $q \in \mathbf{Z}^\mathbf{a}$, and $\tau \in \mathbf{R}^\mathbf{a}$ such that $\{\tau\} = 0$, we put

$$(9.8) \qquad C(z, s; q, \chi, \mathfrak{c}) = L_\mathfrak{c}(2s, \chi) E(z, s; q, \chi, \mathfrak{c}),$$

$$(9.9) \qquad E(z, s; q, \chi, \mathfrak{c}) = \sum_{\beta \in B} E_\beta(z, s; q, \chi, \mathfrak{c}),$$

$$(9.10) \qquad E_\beta(z, s; q, \chi, \mathfrak{c})$$

$$= N(\mathfrak{a}_\beta)^{2s} \sum_{\alpha \in R_\beta} \chi_\mathbf{a}(d_\alpha) \chi^*\left(d_\alpha \mathfrak{a}_\beta^{-1}\right) y^{su + (i\tau - q)/2} \|_q \alpha.$$

Here $z \in \mathcal{H}^\mathbf{a}$, $s \in \mathbf{C}$, $B = P \backslash [G \cap P_\mathbf{A} D_\mathfrak{c}]/\Gamma_\mathfrak{c}$, and $R_\beta = (P \cap \beta \Gamma_\mathfrak{c} \beta^{-1}) \backslash \beta \Gamma_\mathfrak{c}$; we assume that

$$(9.11a) \qquad\qquad \text{the conductor of } \chi \text{ divides } \mathfrak{c};$$

$$(9.11b) \qquad\qquad \chi_\mathbf{a}(x) = \text{sgn}(x_\mathbf{a})^q |x_\mathbf{a}|^{i\tau}.$$

The difference of E from E^+ of (6.13a, b) is that we take P here instead of P_+ there, and that we are assuming (9.11b).

PROPOSITION 9.5. *There is a real analytic function of $(z, s) \in \mathcal{H}^{\mathbf{a}} \times \mathbf{C}$ which coincides with*

$$(s - 1)C(z, s; q, \chi, \mathfrak{c}) \prod_{v \in \mathbf{a}} \Gamma(s + (|q_v| + i\tau_v)/2)$$

for $\mathrm{Re}(s) > 1$. The factor $s - 1$ is necessary only when $\chi = 1$ and $q = \tau = 0$, in which case C indeed has a simple pole at $s = 1$; the residue is $2^{2d-2}\pi^d D_F^{-2} N(\mathfrak{c})^{-1}\kappa_0 R_F \prod_{\mathfrak{p}|\mathfrak{c}}[1 - N(\mathfrak{p})^{-1}]$, where κ_0 is the class number of F and R_F is the regulator of F.

PROPOSITION 9.6. *Suppose $\tau = 0$; put*

$$C'(z, s; q, \chi, \mathfrak{c}) = z^{-q}C(-1/z, s; q, \chi, \mathfrak{c}).$$

(I) *If $1 < \mu \in \mathbf{Z}$, we have*

$$C'(z, \mu/2; \mu u, \chi, \mathfrak{c}) = D_F^{-3/2}N(2\mathfrak{c}^{-1})(-2\pi i)^{\mu d}\Gamma(\mu)^{-d}$$

$$\cdot \left\{ A(8\pi y)^{-1} + \sum_h \mathbf{e}_{\mathbf{a}}(hz) \sum_{\mathfrak{x}} \chi^*(\mathfrak{x}) N(h\mathfrak{x}^{-1})^{\mu - 1} \right\},$$

where h runs over all totally positive elements of $2\mathfrak{c}^{-1}\mathfrak{d}^{-2}$, $\mathfrak{x}$ over all the integral ideals prime to $\mathfrak{c}$ and dividing $h\mathfrak{d}^2$, and

$$A = \begin{cases} \prod_{p|\mathfrak{c}}(1 - p^{-1}) & \text{if } F = \mathbf{Q}, \mu = 2, \text{ and } \chi = 1, \\ 0 & \text{otherwise.} \end{cases}$$

(II) *If $0 < \lambda \in \mathbf{Z}$, we have*

$$C'(z, (2 - \lambda)/2; \lambda u, \chi, \mathfrak{c}) = D_F^{-3/2}N(2\mathfrak{c}^{-1})(-2i)^{\lambda d}\pi^d$$

$$\cdot \left\{ 2^{-d}L_{\mathfrak{c}}(1 - \lambda, \chi) \right.$$

$$\left. + \sum_h \mathbf{e}_{\mathbf{a}}(hz) \sum_{\mathfrak{x}} \chi^*(\mathfrak{x}) N(\mathfrak{x})^{\lambda - 1} \right\},$$

where h and $\mathfrak{x}$ are the same as in (I).

These two propositions are essentially proved in [S7] and [S9]. It should be noted however that the present $\mathfrak{a}_\beta$ and $\Gamma_{\mathfrak{c}}$ are different from those of [S9, Case I]. If they were indeed the same, then Proposition 9.5 would be completely included in [S9, Theorem 4.1] and Proposition 9.6 would follow immediately from [S9, (4.17), (4.18a), (4.20)] and [S7, (7.11), (7.12), (7.13), (7.14)]. Since the present functions are defined in a somewhat different way, some minor modifications are necessary, but still we obtain the stated results in the same manner.

PROPOSITION 9.7. *Suppose* $\tau = 0$ *and* $q \geq 0$. *Given an integer* t *such that* $-q_v < t \leq q_v$ *and* $t \equiv q_v \pmod{2}$ *for every* $v \in \mathbf{a}$, *put*

$$C^*(z, t, q, \chi, \mathfrak{c})$$

$$= (2\pi)^{-\{tu+q\}/2} i^{td} \mathfrak{g}(\chi)^{-1} D_F^{1/2} C(z, t/2; q, \chi, \mathfrak{c}).$$

Then this belongs to $\mathcal{N}_q(\overline{\mathbf{Q}})$, *and for every* $\sigma \in \mathrm{Gal}(\overline{\mathbf{Q}}/\mathbf{Q})$, *we have*

$$(9.12) \qquad C^*(z, t, q, \chi, \mathfrak{c})^\sigma = C^*(z, t, q\sigma, \chi^\sigma, \mathfrak{c}).$$

Proof. Suppose $t > 1$, and put $p = (q - tu)/2$. As shown in [S11, (4.13a)], we have

$$\delta_{tu}^p E(z, t/2; tu, \chi, \mathfrak{c}) = b(p, t)(2\pi)^{-\{p\}} E(z, t/2; q, \chi, \mathfrak{c})$$

with a nonzero rational number $b(p, t)$ whose explicit form is given there, and hence

$$\delta_{tu}^p C^*(z, t, tu, \chi, \mathfrak{c}) = b(p, t) C^*(z, t, q, \chi, \mathfrak{c}).$$

Combining Lemma 8.4 with (I) of Proposition 9.6, we see that

$$C^*(z, t, tu, \chi, \mathfrak{c})^\sigma = C^*(z, t, tu, \chi^\sigma, \mathfrak{c}).$$

Observing that $b(p, t) = b(p\sigma, t)$, we obtain (9.12) from (8.6). Next, suppose $t \leq 1$ and put $\lambda = 2 - t$ and $p = (q - \lambda u)/2$. Then by [S11, (4.13b)],

$$\delta_{\lambda u}^p C^*(z, t, \lambda u, \chi, \mathfrak{c}) = b'(p, t) C^*(z, t, q, \chi, \mathfrak{c})$$

with a nonzero rational number $b'(p, t)$ similar to $b(p, t)$, and therefore we obtain our assertion from (II) of Proposition 9.6 for the same reason as in the previous case.

10. Main theorems on algebraicity. Given $\mathbf{g} = (g_1, \ldots, g_\kappa) \in \mathscr{S}_n(\mathfrak{b}, \varphi)$ with n, $\mathfrak{b}$, and φ as in (6.1a, b) and a Hecke character χ of F, we put

$$(10.1a) \qquad D(s, \mathbf{g}) = \sum_{\mathfrak{m}} c(\mathfrak{m}, \mathbf{g}) N(\mathfrak{m})^{-s},$$

$$(10.1b) \qquad D(s, \mathbf{g}, \chi) = \sum_{\mathfrak{m}} c(\mathfrak{m}, \mathbf{g}) \chi^*(\mathfrak{m}) N(\mathfrak{m})^{-s},$$

where $\mathfrak{m}$ runs over all the integral ideals of F. Let us first recall some of our previous results in [S3] concerning the special values of (10.1a, b), which are necessary for the formulation of our present question. We put

$$(10.2) \qquad A(t, \mathbf{g}, \chi) = (2\pi i)^{-td} \mathfrak{g}(\chi)^{-1} D(t, \mathbf{g}, \chi) \qquad (t \in (1/2)\mathbf{Z}),$$

$$(10.3a) \qquad E(\mathbf{g}) = (2\pi i)^d \pi^{\{n\}} \mathfrak{g}(\varphi) \langle \mathbf{g}, \mathbf{g} \rangle,$$

$$(10.3b) \qquad \langle \mathbf{g}, \mathbf{g} \rangle = \sum_{\lambda=1}^{\kappa} \langle g_\lambda, g_\lambda \rangle,$$

where $d = [F : \mathbf{Q}]$, $\mathfrak{g}(*)$ is the Gauss sum of (4.8), and $\langle \ , \ \rangle$ is the inner product of (3.3). We are interested in the nature of $A(t, \mathbf{g}, \chi)$ for $t \in (1/2)\mathbf{Z}$ satisfying

$$(10.4) \qquad 2t \equiv n_v \pmod 2 \quad and \quad -n_v < 2t < n_v \ for\ every\ v \in \mathbf{a}.$$

Such a t exists if and only if

$$(10.5) \qquad n \geqq 2u \quad and \quad n_v \bmod 2\ is\ independent\ of\ v.$$

No additional assumption is necessary if $F = \mathbf{Q}$ or $n_v > 2$ for every $v \in \mathbf{a}$, but if $F \neq \mathbf{Q}$ and $n_v = 2$ for some v, we have to assume a condition:

$$(10.6) \qquad$$ *For every $r \in \mathbf{Z}^{\mathbf{a}}/2\mathbf{Z}^{\mathbf{a}}$ and every integral ideal $\mathfrak{m}$, there exists a Hecke character χ such that $\chi_\mathbf{a}(x) = \mathrm{sgn}(x_\mathbf{a})^r |x_\mathbf{a}|^{-i\mu}$ with μ of (6.1b), $\mathfrak{m}$ divides the conductor of χ, and $D(0, \mathbf{g}, \chi) \neq 0$.*

This is in fact true for $F = \mathbf{Q}$ as proved in [S2, Theorem 2], and probably always so even when $F \neq \mathbf{Q}$.

THEOREM 10.1. *Let $\mathbf{g}$ be a normalized eigenform in $\mathscr{S}_n(\mathfrak{b}, \varphi)$. Suppose that (10.6) is satisfied if $F \neq \mathbf{Q}$ and $n_v = 2$ for some $v \in \mathbf{a}$. Then there exists, for each $r \in \mathbf{Z}^{\mathbf{a}}/2\mathbf{Z}^{\mathbf{a}}$, a complex number $U(r, \mathbf{g})$ with the following properties:*

(I) *Let t be an element of $(1/2)\mathbf{Z}$ satisfying (10.4) and let $t' = t$ or $t' = t + (1/2)$ according as $t \in \mathbf{Z}$ or $t \notin \mathbf{Z}$. If $\chi_\mathbf{a}(x) = \mathrm{sgn}(x_\mathbf{a})^{t'u+r}|x_\mathbf{a}|^{-i\mu}$ with μ of (6.1b), then $\pi^{-td} D(t, \mathbf{g}, \chi)/U(r, \mathbf{g}) \in \overline{\mathbf{Q}}$.*

(II) *If $q, r \in \mathbf{Z}^{\mathbf{a}}/2\mathbf{Z}^{\mathbf{a}}$ and $q + r - u \in 2\mathbf{Z}^{\mathbf{a}}$, then $U(q, \mathbf{g})U(r, \mathbf{g})/E(\mathbf{g}) \in \overline{\mathbf{Q}}$.*

Proof. If $\mu = 0$, this is a weaker version of [S3, Theorem 4.3]. The general case can also be reduced to that theorem as follows: Given χ as in (I), we have, by [S4, Proposition 9.7], $D(s, \mathbf{g}, \chi) = D(s, \mathbf{h})$ with some $\mathbf{h}$ in $\mathscr{S}_n(\mathfrak{a}, \varphi\chi^2)$, where $\mathfrak{a}$ is a certain multiple of $\mathfrak{b}$. Observe that $\varphi\chi^2$ is of finite order. Let $\mathbf{k}$ be the primitive form associated with $\mathbf{h}$. Let $n_0 = \mathrm{Max}\{n_v | v \in \mathbf{a}\}$ and $a = t + (n_0/2)$. Then $D(s - (n_0/2), \mathbf{k})$ coincides with the series of [S3, (2.25)], and hence, by [S3, Theorem 4.3], $D(t, \mathbf{g}, \chi)$ is an algebraic number times $\pi^{ad}u(au, \mathbf{k})$, where

$u(*, \mathbf{k})$ is the invariant given in that theorem. Here we have to assume that $F = \mathbf{Q}$ or $n \geq 3u$. If $p \in (1/2)\mathbf{Z}$ and χ' is a character such that $\chi'_\mathbf{a}(x) = \mathrm{sgn}(x_\mathbf{a})^{p'u+r}|x_\mathbf{a}|^{-i\mu}$ ($p' = p$ or $p + (1/2)$) and $\mathbf{k}'$ is defined for χ' in a similar way, then we see that $D(s, \mathbf{k}, \chi'/\chi)$ differs from $D(s, \mathbf{k}')$ only by finitely many Euler factors which are finite and nonvanishing at least for $\mathrm{Re}(s) > 0$. The definition of $u(*, \mathbf{k})$ in [S3, (4.34)] shows that $u(au, \mathbf{k})$ coincides with $u(pu + (n_0/2)u, \mathbf{k}')$ modulo $\overline{\mathbf{Q}}^\times$. Therefore, putting $U(r, \mathbf{g}) = \pi^{n_0 d/2} u(au, \mathbf{k})$, we obtain the desired U with the property of (I) in a consistent way. Then (II) follows from [S3, Theorem 4.3(II)], since it can easily be seen from [S3, Proposition 4.13] that $\langle \mathbf{g}, \mathbf{g} \rangle / \langle \mathbf{k}, \mathbf{k} \rangle$ is algebraic. If $F \neq \mathbf{Q}$ and $n_v = 2$ for some v, we can define $u(r, \mathbf{k})$ under (10.6) in view of [S3, (4.16)], and therefore the desired U can be obtained also in this case.

A stronger version of the above theorem can be stated in a special case:

THEOREM 10.2. *Suppose* $\mu = 0$, $n \in 2\mathbf{Z}^\mathbf{a}$, *and* $\mathbf{g}$ *is a primitive form; suppose also that* (10.6) *is satisfied if* $F \neq \mathbf{Q}$ *and* $n_v = 2$ *for some* v. *Then there exists, for each* $r \in \mathbf{Z}^\mathbf{a}/2\mathbf{Z}^\mathbf{a}$ *and* $\mathbf{g}^\sigma$ *with* $\sigma \in \mathrm{Gal}(\overline{\mathbf{Q}}/\mathbf{Q})$, *a complex number* $V(r, \mathbf{g}^\sigma)$ *with the following properties:*

(I) *If* $\chi_\mathbf{a}(x) = \mathrm{sgn}(x_\mathbf{a})^{r+tu}$ *with an integer* t *such that* $|t| < n_v/2$ *for every* $v \in \mathbf{a}$, *then* $A(t, \mathbf{g}, \chi)/V(r, \mathbf{g}) \in \overline{\mathbf{Q}}$, *and moreover*

$$[A(t, \mathbf{g}, \chi)/V(r, \mathbf{g})]^\sigma = A(t, \mathbf{g}^\sigma, \chi^\sigma)/V(r, \mathbf{g}^\sigma)$$

for every $\sigma \in \mathrm{Gal}(\overline{\mathbf{Q}}/\mathbf{Q})$.

(II) *If* $q, r \in \mathbf{Z}^\mathbf{a}/2\mathbf{Z}^\mathbf{a}$ *and* $q + r - u \in 2\mathbf{Z}^\mathbf{a}$, *then* $V(q, \mathbf{g})V(r, \mathbf{g})/E(\mathbf{g}) \in \overline{\mathbf{Q}}$, *and moreover*

$$[V(q, \mathbf{g})V(r, \mathbf{g})/E(\mathbf{g})]^\sigma = V(q, \mathbf{g}^\sigma)V(r, \mathbf{g}^\sigma)/E(\mathbf{g}^\sigma)$$

for every $\sigma \in \mathrm{Gal}(\overline{\mathbf{Q}}/\mathbf{Q})$.

For the same reason as in the above proof, this follows immediately from [S3, Theorem 4.3] if we put $V(r, \mathbf{g}^\sigma) = (2\pi i)^{n_0 d/2} u(r + (n_0/2)u, \mathbf{g}^\sigma)$. In that theorem, the case in which $n_v = 2$ for some v was excluded, but (10.6) together with [S3, (4.16)] makes the definition of $u(*, *)$ in [S3, (4.34)] valid also in this exceptional case.

We note here a related fact:

LEMMA 10.3. *If* $\mathbf{g}$ *is a normalized eigenform, then* $D(s, \mathbf{g}, \chi) \neq 0$ *for* $\mathrm{Re}(s) \geq 1/2$.

If $\mu = 0$ and χ is of finite order, this follows immediately from [S3, Propositions 4.5 and 4.16]. We see easily that the proof of [S3, Proposition 4.16] is valid even for $D(s, \mathbf{g}, \chi)$ without those assumptions.

We now turn to our main problems for the forms of half-integral weight. Let $f \in \mathscr{S}_k(\mathfrak{c}, \psi)$ and $g \in \mathscr{M}_l(\mathfrak{c}', \varphi)$ with $k = (u/2) + m$ and $l = (u/2) + n$. Here

m and n are elements of $\mathbf{Z}^{\mathbf{a}}$; ψ and φ are Hecke characters of F such that $\psi_{\mathbf{a}}(-1) = (-1)^{\{m\}}$ and $\varphi_{\mathbf{a}}(-1) = (-1)^{\{n\}}$. We take μ and ν of $\mathbf{R}^{\mathbf{a}}$ so that $\psi_{\mathbf{a}}(y) = |y|^{i\mu}$ and $\varphi_{\mathbf{a}}(y) = |y|^{i\nu}$ for $0 \ll y \in \mathbf{R}^{\mathbf{a}}$ and $\{\mu\} = \{\nu\} = 0$. We then assume that

$$(10.7) \qquad (\varphi/\psi)_{\mathbf{a}}(x) = \mathrm{sgn}(x_{\mathbf{a}})^{n-m}|x_{\mathbf{a}}|^{i\nu - i\mu}.$$

In Proposition 3.1, we defined the Fourier coefficients $\lambda(\xi, \mathfrak{m}; f, \psi)$ for every $(\xi, \mathfrak{m})$ such that $0 \ll \xi \in \mathfrak{m}^{-2}$. Let us call two such $(\xi, \mathfrak{m})$ and $(\xi', \mathfrak{m}')$ *equivalent* if $\xi' = \alpha^2 \xi$ and $\mathfrak{m} = \alpha \mathfrak{m}'$ for some $\alpha \in F$. We then define a Dirichlet series $D(s; f, g)$ by

$$(10.8) \qquad D(s; f, g) = D(s; f, \psi; g, \varphi)$$

$$= \sum_{(\xi, \mathfrak{m})} \lambda(\xi, \mathfrak{m}; f, \psi)\overline{\lambda(\xi, \mathfrak{m}; g, \varphi)}$$

$$\cdot \xi^{-(m+n+i\mu-i\nu)/2} N\left(\xi \mathfrak{m}^2\right)^{-s},$$

where $(\xi, \mathfrak{m})$ runs over a complete set of representatives of the equivalence classes in the above sense. This is well-defined in view of (3.14b) and (10.7), and convergent for sufficiently large $\mathrm{Re}(s)$. Taking a common multiple $\mathfrak{e}$ of $\mathfrak{c}$ and $\mathfrak{c}'$, we put

$$(10.9) \qquad \mathscr{D}_{\mathfrak{e}}(s; f, g) = L_{\mathfrak{e}}(2s + 1, \psi/\varphi)D(s; f, g),$$

with $L_{\mathfrak{e}}$ of (6.17a). To simplify our notation, we put also

$$(10.10a) \quad q = (1/2)(m + n + i\mu - i\nu),$$

$$(10.10b) \quad q' = (q'_v)_{v \in \mathbf{a}}, \qquad q'_v = (1/2)(1 + |m_v - n_v| + i\mu_v - i\nu_v).$$

THEOREM 10.4. *The product*

$$(2s - 1)\mathscr{D}_{\mathfrak{e}}(s; f, g) \prod_{v \in \mathbf{a}} \Gamma(s + q_v)\Gamma(s + q'_v)$$

can be continued as a holomorphic function to the whole plane. The factor $2s - 1$ is necessary only when $m = n$ and $\psi = \varphi$, in which case $D(s; f, g)$ has at most a simple pole at $s = 1/2$ with residue $A2^{d/2}\pi^{\{m\}}R_F\langle g, f\rangle$, where A is a positive rational number and R_F is the regulator of F.

Now our main results of algebraicity can be stated in the following two theorems:

THEOREM 10.5. *Let $k = (u/2) + m$ with $0 \le m \in \mathbf{Z}^{\mathbf{a}}$; let ψ and $\mathfrak{c}$ be as in (3.8a, b); suppose $\psi_{\mathbf{a}}(x) = \mathrm{sgn}(x_{\mathbf{a}})^r|x_{\mathbf{a}}|^{i\mu}$ with $r \in \mathbf{Z}^{\mathbf{a}}/2\mathbf{Z}^{\mathbf{a}}$ and $\mu \in \mathbf{R}^{\mathbf{a}}$, $\{\mu\} = 0$.*

Let $\mathbf{h}$ *be a primitive form belonging to* $\mathcal{S}_{2m}(2^{-1}\mathfrak{c}, \psi^2)$. *In case* $F \neq \mathbf{Q}$ *and* $m_v = 1$
for some $v \in \mathbf{a}$, *assume that* (10.6) *is satisfied with* $\mathbf{h}$ *as* $\mathbf{g}$. *For* $f \in \mathcal{S}_k(\mathfrak{c}, \psi, \mathbf{h})$ *and*
$p \in \mathcal{S}_k$, *put*

$$(10.11) \qquad\qquad I(p, f) = 2^{d/2} i^d \pi^{d+\{m\}} \mathfrak{g}(\psi)\langle p, f\rangle.$$

Then the following assertions hold:
   (I) $\pi^{d+\{m\}}\langle p, f\rangle / U(u - r, \mathbf{h}) \in \overline{\mathbf{Q}}$ *if both* p *and* f *are* $\overline{\mathbf{Q}}$-*rational*.
   (II) *If* ψ *is of finite order, then for every* $\sigma \in \mathrm{Aut}(\mathbf{C})$ *one has*

$$[I(p, f)/V(u - r, \mathbf{h})]^\sigma = I(p^{\rho\sigma\rho}, f^\sigma)/V(u - r, h^\sigma),$$

where ρ *is complex conjugation*.

THEOREM 10.6.   *Let* $f \in \mathcal{S}_k(\mathfrak{c}, \psi, \mathbf{h})$ *and* $g \in \mathcal{M}_l(\mathfrak{c}', \varphi)$ *with half-integral*
weights $k = (u/2) + m$ *and* $l = (u/2) + n$. *Suppose that* $(\varphi/\psi)_\mathbf{a}(x) =$
$\mathrm{sgn}(x_\mathbf{a})^{n-m}$; *if* $F \neq \mathbf{Q}$ *and* $m_v = 1$ *for some* $v \in \mathbf{a}$, *suppose in addition that* (10.6)
is satisfied with $\mathbf{h}$ *as* $\mathbf{g}$. *Let* r *be as in Theorem* 10.5 *and let* $\mathfrak{e} = \mathfrak{c} \cap \mathfrak{c}'$. *Then the*
following assertions hold:
   (I) *If* f *and* g *are* $\overline{\mathbf{Q}}$-*rational, then* $\pi^{-td}\mathscr{D}_\mathfrak{e}(t/2; f, g)/U(u - r, \mathbf{h}) \in \overline{\mathbf{Q}}$ *for*
every integer t *such that*

(10.12)

$$l_v - k_v < t + 1 \leq k_v - l_v \quad \text{and} \quad t + 1 \equiv k_v - l_v \pmod 2 \; \text{for every } v \in \mathbf{a}.$$

   (II) *Suppose both* ψ *and* φ *are of finite order. For every* t *satisfying* (10.12), *put*

$$J(t; f, g) = (\pi i)^{-td} \mathfrak{g}(\varphi)\mathscr{D}_\mathfrak{e}(t/2; f, g).$$

Then for every $\sigma \in \mathrm{Aut}(\mathbf{C})$ *one has*

$$[J(t; f, g)/V(u - r, \mathbf{h})]^\sigma = J(t; f^\sigma, g^{\rho\sigma\rho})/V(u - r, \mathbf{h}^\sigma).$$

If $F = \mathbf{Q}$, these theorems are essentially the same as Theorems 1 and 2 of [S6]
except for a minor difference of formulation.

The rest of this section is devoted to the proof of the above three theorems.
The notation being the same as in Theorem 10.4, put $\chi = \varphi/\psi$ and $\Phi = \Gamma_e \backslash \mathscr{H}^\mathbf{a}$,
and then consider an integral

$$\int_\Phi \overline{f(z)g(z)\, E(z, \bar{s} + (1/2); m - n, \chi, \mathfrak{e})}\, y^k d_H z$$

with E of (9.9). For each β in the set B of (9.9), put

$$f_\beta(z) = \psi_\mathbf{a}(d_\beta)\psi^*\!\left(d_\beta \mathfrak{a}_\beta^{-1}\right) J_k(\beta, \beta^{-1}z) f(\beta^{-1}z),$$

$$g_\beta(z) = \varphi_\mathbf{a}(d_\beta)\varphi^*\!\left(d_\beta \mathfrak{a}_\beta^{-1}\right) J_l(\beta, \beta^{-1}z) g(\beta^{-1}z).$$

The above integral is the sum $\Sigma_{\beta \in B} I_\beta$ with

$$I_\beta = \int_\Phi f(z)\overline{g(z)\, E_\beta(z, \bar{s} + (1/2);\, m - n, \chi, e)}\, y^k\, d_H z.$$

The same reasoning as in the proof of Theorem 6.1 following formula (7.18b) shows that

$$I_\beta = N(\mathfrak{a}_\beta)^{2s+1} \int_\Psi f_\beta \bar{g}_\beta y^{su+u+q}\, d_H z,$$

where $\Psi = (P \cap \beta \Gamma_e \beta^{-1}) \backslash \mathscr{H}^{\mathbf{a}}$ and q is as in (10.10a). By the same type of calculation as in Section 7, we find that

$$I_\beta = 2^d D_F^{-1/2}(2\pi)^{-sd - \{q\}} \prod_{v \in \mathbf{a}} \Gamma(s + q_v)$$

$$\cdot \sum \lambda_f\big(\xi, \mathfrak{a}_\beta^{-1}\big)\overline{\lambda_g\big(\xi, \mathfrak{a}_\beta^{-1}\big)}\, \xi^{-q} N\big(\xi \mathfrak{a}_\beta^{-2}\big)^{-s},$$

where ξ runs over all totally positive elements of $\mathfrak{a}_\beta^2$ modulo $\{a^2 | a \in \mathfrak{o}^\times\}$. Taking the sum over all β in B and multiplying by $L_e(2s + 1, \psi/\varphi)$, we obtain

$$(10.13) \qquad 2^d(2\pi)^{-sd - \{q\}} D_F^{-1/2} \mathscr{D}_e(s; f, g) \prod_{v \in \mathbf{a}} \Gamma(s + q_v)$$

$$= \int_\Phi f(z)\overline{g(z)\, C(z, \bar{s} + (1/2);\, m - n, \chi, e)}\, y^k\, d_H z$$

with C of (9.8). Therefore Theorem 10.4 follows from Proposition 9.5, since f is a cusp form and C is slowly increasing at every cusp.

In order to prove Theorem 10.5, we take $f \in \mathscr{S}_k(\mathfrak{c}, \psi, \mathbf{h})$ and $n \in \mathbf{Z}^{\mathbf{a}}$ so that $0 \leq n \leq u$ and $m - n \in 2\mathbf{Z}^{\mathbf{a}}$. Then we choose a Hecke character φ_1 of conductor $\mathfrak{f}$ so that

$$(10.14) \qquad (\varphi_1 \psi)_{\mathbf{a}} = 1 \ \textit{and every prime factor of } \mathfrak{c} \textit{ divides } \mathfrak{f}.$$

The existence of such a φ_1 will be proved later. Put again $l = (u/2) + n$, and take any $\tau \in \mathfrak{o}$, $\gg 0$. Then we can find integral ideals $\mathfrak{q}$ and $\mathfrak{r}$ such that $\tau \mathfrak{o} = \mathfrak{q}^2 \mathfrak{r}$, $\mathfrak{r}$ has no square factors prime to $\mathfrak{f}$, and $\mathfrak{q}$ is prime to $\mathfrak{f}$. Combining Lemma 4.3, (4.12), and Proposition 3.2, we find an element g of $\mathscr{M}_l(4\mathfrak{r}\mathfrak{f}^2, \bar{\varphi}_1 \varepsilon_\tau)$ such that

$$(10.15) \quad \overline{\lambda_g(\xi, m)} = \begin{cases} \varphi_{1\mathbf{a}}(\zeta)\varphi_1^*(\zeta\mathfrak{q}\mathfrak{m})\zeta^n & \text{if } 0 \neq \xi = \tau\zeta^2 \in \mathfrak{r}\mathfrak{m}^{-2}, \\ 0 & \text{otherwise.} \end{cases}$$

For this choice of g, we have

$$D(s; f, g) = \tau^{(-m-n)/2} N(\mathfrak{r})^{-s} \sum_{\mathfrak{m}} \varphi_1^*(\mathfrak{m}) \lambda_f(\tau, \mathfrak{q}^{-1}\mathfrak{m}) N(\mathfrak{m})^{-2s},$$

where $\mathfrak{m}$ runs over all the integral ideals. Let $\mathfrak{e} = \mathfrak{c} \cap 4\mathfrak{r}\mathfrak{f}^2$. By Theorem 5.5 with $\mathfrak{f}$ as $\mathfrak{t}$, we have

$$\tau^{(m+n)/2} N(\mathfrak{r})^s \mathcal{D}_{\mathfrak{e}}(s; f, g) = \lambda_f(\tau, \mathfrak{q}^{-1}) D(2s, \mathbf{h}, \varphi_1),$$

and therefore (10.13) shows that

(10.16)

$$\lambda_f(\tau, \mathfrak{q}^{-1}) D(2s, \mathbf{h}, \varphi_1) 2^d (2\pi)^{-sd-\{m+n\}/2} D_F^{-1/2} \prod_{v \in \mathbf{a}} \Gamma(s + (m_v + n_v)/2)$$

$$= \tau^{(m+n)/2} N(\mathfrak{r})^s \int_\Phi \overline{f\bar{g}C(z, \bar{s} + (1/2); m - n, \chi, \mathfrak{e})}\, y^k \, d_H z,$$

where $\chi = \bar{\psi}\bar{\varphi}_1 \varepsilon_\mathfrak{r}$. Evaluate (10.16) at $s = 1/2$ and observe that

$$C(z, 1; m - n, \chi, \mathfrak{e}) = D_F^{-1/2}(2\pi)^{d+\{m-n\}/2}(-1)^d \mathfrak{g}(\chi) C^*(z, \chi)$$

with an element C^* of $\mathcal{N}_{m-n}(\overline{\mathbf{Q}})$ by Proposition 9.7. Here we have to assume that $m \geqq 2u$; we suppress symbols t, $m - n$, and $\mathfrak{e}$ for simplicity. By (10.2) we have

(10.17) $\lambda_f(\tau, \mathfrak{q}^{-1}) i^d \mathfrak{g}(\varphi_1) A(1, \mathbf{h}, \varphi_1)$

$$= B(m, n, \mathfrak{e}) N(2\mathfrak{r})^{1/2} \tau^{(m+n)/2} \pi^{d+\{m\}} \overline{\mathfrak{g}(\chi)} \langle gC^*, f \rangle$$

with a nonzero rational number $B(m, n, \mathfrak{e})$ depending only on m, n, and $\mathfrak{e}$. Observe that $gC^* \in \mathcal{N}_k$ and put $b = q_k(p_k(gC^*))$ with p_k and q_k of (9.5). By (9.5) we have $\langle gC^*, f \rangle = \langle b, f \rangle$. Since b depends on τ (for a fixed φ_1), write it b_τ. We have $A(1, \mathbf{h}, \varphi_1) \neq 0$ by Lemma 10.3, and hence (10.17) shows that

(10.18) $$\lambda_f(\tau, \mathfrak{q}^{-1}) = 0 \quad \Leftrightarrow \quad \langle b_\tau, f \rangle = 0.$$

Suppose now $f \neq 0$. Then $\lambda_f(\tau, \mathfrak{o}) \neq 0$ for some τ, and Theorem 5.5 (with $\mathfrak{f}$ as $\mathfrak{t}$) implies that $\lambda_f(\tau, \mathfrak{q}^{-1}) \neq 0$. From (10.18) we see that $\mathcal{S}_k(\mathfrak{c}, \psi, \mathbf{h})$ is spanned by the b_τ for all $\tau \in \mathfrak{o}$, $\gg 0$. Suppose further that f is $\overline{\mathbf{Q}}$-rational. Then $|q|^{-i\mu}\lambda_f(\tau, \mathfrak{q}^{-1}) \in \overline{\mathbf{Q}}$ and $|q|^{i\mu}g$ is $\overline{\mathbf{Q}}$-rational, and so $|q|^{i\mu}b_\tau$ is $\overline{\mathbf{Q}}$-rational by (9.6). Writing it b'_τ, we obtain, from (10.17),

(10.19) $$\pi^{2d+\{m\}}\langle b'_\tau, f \rangle / D(1, \mathbf{h}, \varphi_1) \in \overline{\mathbf{Q}}.$$

Now every $\overline{\mathbf{Q}}$-rational element p of $\mathscr{S}_k(\mathfrak{c}, \psi, \mathbf{h})$ is a finite linear combination $\sum a_\tau b_\tau'$ with $a_\tau \in \overline{\mathbf{Q}}$. Therefore assertion (I) of Theorem 10.5 (when $m \geq 2u$) follows immediately from (10.19) and (I) of Theorem 10.1.

To prove (II) of Theorem 10.5, we employ Lemma 9.3 to transform (10.17) into the form

$$(10.20) \qquad I(b_\tau, f) = B_\tau(m, n, \mathfrak{e}) \tau^{(-m-n)/2} \lambda\left(\tau, \mathfrak{q}^{-1}; f, \psi\right)$$

$$\cdot \left[\mathfrak{g}(\psi) \mathfrak{g}(\varphi_1) \mathfrak{g}(\varepsilon_\tau) / \mathfrak{g}(\psi \varphi_1 \varepsilon_\tau)\right] A(1, \mathbf{h}, \varphi_1)$$

with a nonzero rational number $B_\tau(m, n, \mathfrak{e})$. Suppose ψ is of finite order. Then b_τ is $\overline{\mathbf{Q}}$-rational. Let $\sigma \in \mathrm{Aut}(\mathbf{C})$. Then $f^\sigma \in \mathscr{S}_{k\sigma}(\mathfrak{c}, \psi^\sigma, \mathbf{h}^\sigma)$, $g^\sigma \in \mathscr{M}_{l\sigma}(4\mathfrak{r}\mathfrak{f}^2, \varphi_1^{\sigma\rho}\varepsilon_\tau)$, and g^σ corresponds to φ_1^σ in the same way as g does to φ_1. Moreover, by (9.7) and (9.12), the change of f and g for f^σ and g^σ sends b_τ to b_τ^σ. Therefore (10.20) still holds after b_τ, f, m, n, ψ, φ_1, and $\mathbf{h}$ are replaced by their images under σ. It is easy to see that $B_\tau(m, n, \mathfrak{e}) = B_\tau(m\sigma, n\sigma, \mathfrak{e})$. Applying σ to (10.20), we obtain, from Proposition 8.6, Lemma 9.2, and (II) of Theorem 10.2,

$$(10.21) \qquad \left[I(b_\tau, f)/V(r - u, \mathbf{h})\right]^\sigma = I(b_\tau^\sigma, f^\sigma)/V(r - u, \mathbf{h}^\sigma).$$

Now let $p \in \mathscr{S}_k$. Then $p = \sum a_\tau b_\tau + c$ with $a_\tau \in \mathbf{C}$ and $c \in \mathscr{T}_k(\mathfrak{c}, \psi, \mathbf{h})$. Since $I(p, f)$ is anti-$\mathbf{C}$-linear in p, we obtain (II) of Theorem 10.5 from (10.21) and Proposition 8.9(4).

So far we have assumed $m \geq 2u$. Suppose $m_v = 1$ for some v. This time choose n, φ_1, and $\mathfrak{f}$ so that $0 \leq n \leq u$, $m - n - u \in 2\mathbf{Z}^\mathbf{a}$, $(\varphi_1 \psi)_\mathbf{a}(x) = \mathrm{sgn}(x_\mathbf{a})^u$, and every prime factor of $\mathfrak{c}$ divides $\mathfrak{f}$. Evaluate (10.16) at $s = 0$. Then, instead of (10.17), we have

$$(10.22) \qquad \lambda_f\left(\tau, \mathfrak{q}^{-1}\right) i^d \mathfrak{g}(\varphi_1) A(0, \mathbf{h}, \varphi_1)$$

$$= B(m, n, \mathfrak{e}) 2^{d/2} \tau^{(m+n)/2} \pi^{d + \{m\}} \overline{\mathfrak{g}(\chi)} \langle gC^*, f \rangle$$

with $B(m, n, \mathfrak{e}) \in \mathbf{Q}^\times$ and $C^* \in \mathscr{N}_{m-n}(\overline{\mathbf{Q}})$. Therefore, under assumption (10.6) on $\mathbf{h}$, we can repeat the same reasoning as before. We only have to observe that $\tau^{(m+n)/2} = \tau^{(m+n-u)/2} \tau^{u/2}$, $m + n - u \in 2\mathbf{Z}^\mathbf{a}$, and $\mathfrak{g}(\varepsilon_\tau) \tau^{u/2} \in \mathbf{Q}$ by Lemma 9.3.

It remains to prove the existence of φ_1. We first observe that the composite of F and a suitable cyclotomic extension of $\mathbf{Q}$ yields a totally real cyclic extension of F whose conductor is divisible by any given prime power. Consequently, we can find a Hecke character ω of F such that $\omega_\mathbf{a} = 1$ and $\mathfrak{c}^2$ divides the conductor of ω. Let $\varphi_1 = \omega \psi^{-1}$. Then $(\varphi_1 \psi)_\mathbf{a} = 1$ and $\mathfrak{c}^2$ divides the conductor of φ_1.

To prove Theorem 10.6, let t be an integer satisfying (10.12). We evaluate (10.13) at $s = t/2$ to find that

$$(10.23) \quad (\pi i)^{-td} \mathscr{D}_\mathfrak{e}(t/2; f, g) = a(m, n, \mathfrak{e}) \mathfrak{g}(\psi/\varphi) 2^{d/2} i^d \pi^{d + \{m\}} \langle gC^*, f \rangle,$$

where $a(m, n, e)$ is a rational number depending only on m, n, and e, and C^* is the function $C^*(z, t + 1, m - n, \varphi/\psi, e)$ of Proposition 9.7. Let $p = p_k(gC^*)$ with p_k of Proposition 9.4. Then $\langle gC^*, f \rangle = \langle p, f \rangle$ and (10.23) can be written in the form

$$J(t; f, g) = a(m, n, e)\mathfrak{g}(\psi/\varphi)\mathfrak{g}(\varphi)\mathfrak{g}(\psi)^{-1}I(p, f).$$

Therefore we obtain Theorem 10.6 from Theorem 10.5 and Lemma 9.2.

11. Transformation formulas of theta series. As preliminaries to our subsequent investigation, we prove some transformation formulas of a theta series attached to an indefinite quadratic form. Though the automorphy property of such a series is well known in principle, little has been proved about the precise form of the factor of automorphy with respect to a naturally defined discrete subgroup which is not too small, particularly when the basic field is not $\mathbf{Q}$, much less about the transformation formula under a group element of a more general type. We are going to present such formulas which are at least general enough for our purposes, and which may be viewed as improvements on our previous results in [S4, §7] and [S8, §7].

In this section we put $G = \mathrm{Sp}(m, F)$ and $X = F_m^1$ (row vectors) and consider the global and local metaplectic groups $M_\mathbf{A}$ and M_v ($v \in \mathbf{a} \cup \mathbf{f}$) of G, using the notation of [S8] with a few changes. In particular, we put

$$(11.1) \qquad \mathscr{H}_m = \left\{ z \in \mathbf{C}_m^m \mid {}^t z = z,\ \mathrm{Im}(z) > 0 \right\},$$

$\mathbf{i} = (i1_m, \dots, i1_m) \in \mathscr{H}_m^\mathbf{a}$, and define the action of an element of $G_\mathbf{A}$ or $M_\mathbf{A}$ on $\mathscr{H}_m^\mathbf{a}$ to be the same as that of its projection to $G_\mathbf{a}$. The groups $G_\mathbf{a}$, $G_\mathbf{f}$, G_v for $v \in \mathbf{a} \cup \mathbf{f}$ and other symbols are obvious generalizations of those of Section 1; for details, see [S8]. Among such generalizations, the following ones should be mentioned specifically: for two fractional ideals $\mathfrak{x}$ and $\mathfrak{y}$ such that $\mathfrak{x}\mathfrak{y}$ is integral, we put

$$(11.2a) \qquad D[\mathfrak{x}, \mathfrak{y}] = \left\{ w \in G_\mathbf{A} \mid w(\mathbf{i}) = \mathbf{i},\ w_v \in D_v[\mathfrak{x}, \mathfrak{y}] \text{ for all } v \in \mathbf{f} \right\},$$

$$(11.2b)$$

$$D_v[\mathfrak{x}, \mathfrak{y}] = \left\{ w \in G_v \mid a_w \in (\mathfrak{o}_v)_m^m,\ b_w \in (\mathfrak{x}_v)_m^m,\ c_w \in (\mathfrak{x}_v)_m^m,\ d_w \in (\mathfrak{o}_v)_m^m \right\},$$

$$(11.2c) \qquad \Gamma[\mathfrak{x}, \mathfrak{y}] = G \cap (D[\mathfrak{x}, \mathfrak{y}]G_\mathbf{a}).$$

Let $\mathscr{S}(X_\mathbf{f})$ denote the Schwartz–Bruhat space of $X_\mathbf{f}$, which we identify with the space of locally constant functions on X in the sense of [S8]. For every

$l \in \mathscr{S}(X_{\mathfrak{f}})$, $z \in \mathscr{H}_{m}^{\mathbf{a}}$, and $u \in (\mathbf{C}^{m})^{\mathbf{a}}$, we put

(11.3)

$$l_{z,u}(x) = l(x_{\mathfrak{f}})\mathbf{e}_{\mathbf{a}}\!\left(xu + 2^{-1}xz \cdot {}^{t}x + 2^{-1} \cdot {}^{t}u(z - \bar{z})^{-1}u\right) \qquad (x \in X_{\mathbf{A}}),$$

(11.4)
$$\theta^{*}(z, u; l) = \sum_{\xi \in X} l_{z,u}(\xi).$$

In [S8, Proposition 3.2] we defined a factor of automorphy $h(\tau, z)$ of weight $1/2$ for every τ in $M_{\mathbf{A}}$ such that $\mathrm{pr}(\tau) \in P_{\mathbf{A}}C''$, where C'' is the subgroup of $G_{\mathbf{A}}$ defined in [S8, p. 294]. For such a τ and $l \in \mathscr{S}(X_{\mathfrak{f}})$ we can define a unique element ${}^{\tau}l$ of $\mathscr{S}(X_{\mathfrak{f}})$ by the formula

(11.5)
$$\tau(l_{z,u}) = h(\tau, z)^{-1}({}^{\tau}l)_{\tau(z,u)},$$

where $\tau(z, u) = (\tau(z), {}^{t}(c_{\tau}z + d_{\tau})^{-1}u)$. This is similar to Proposition 2.4 and can be proved in exactly the same fashion. The action $(\tau, l) \mapsto {}^{\tau}l$ has properties (2.15) and (2.16a, b, c, d). Since ${}^{\tau}l$ is uniquely determined by l and $\mathrm{pr}(\tau)$, we denote it also by ${}^{\sigma}l$ when $\sigma = \mathrm{pr}(\tau)$.

LEMMA 11.1. $\theta^{*}(z, u; l) = h(\alpha, z)^{-1}\theta^{*}(\alpha(z, u); {}^{\alpha}l)$ if $\alpha \in G \cap P_{\mathbf{A}}C''$.

Proof. By [We, Théorème 6] we have $\theta^{*}(z, u; l) = \sum_{\xi \in X}\alpha(l_{z,u})(\xi)$ for every $\alpha \in G$, which together with (11.5) proves our formula.

PROPOSITION 11.2. *Given* $l \in \mathscr{S}(X_{\mathfrak{f}})$, *put*

$$U_{l} = \left\{\sigma \in D[2\mathfrak{d}^{-1}, 2\mathfrak{d}]G_{\mathbf{a}} \,\big|\, {}^{\sigma}l = l\right\}.$$

If $\beta \in G \cap \mathrm{diag}[\,p, {}^{t}p^{-1}]U_{l}$ *with* $p \in GL_{m}(F_{\mathfrak{f}})$, *then*

$$h(\beta, \beta^{-1}z)\theta^{*}(\beta^{-1}(z, u); l) = \theta^{*}(z, u; {}^{\beta}l),$$

and ${}^{\beta}l(x) = |\det(p)|^{1/2}l(xp)$ *for all* $x \in X_{\mathfrak{f}}$.

Proof. Substituting $\beta^{-1}(z, u)$ for (z, u) of Lemma 11.1 with $\beta = \alpha$, we obtain the first formula. Let $w = \mathrm{diag}[\,p, {}^{t}p^{-1}]$. Then ${}^{\beta}l = {}^{w}l$ and ${}^{w}l(x) = |\det(p)|^{1/2}l(xp)$ by [S8, (3.2)], which proves the second one.

We now introduce a theta series attached to a symmetric matrix S and its majorant P as follows:

(11.6) $\quad g(z, u; \eta) = \mathbf{e}_{\mathbf{a}}\!\left(\mathrm{tr}({}^{t}u(z - \bar{z})^{-1}u)/2\right)$

$$\cdot \sum_{\xi \in F_{m}^{q}} \eta(\xi)\mathbf{e}_{\mathbf{a}}\!\left(\mathrm{tr}(\xi uA) + 2^{-1}\mathrm{tr}({}^{t}\xi S\xi x + {}^{t}\xi P\xi \cdot iy)\right).$$

Here $'S = S \in \mathrm{GL}_q(F)$, $P = {}'AA$, A is an element of $\mathrm{GL}_q(\mathbf{R})^{\mathbf{a}}$ such that $S_v = {}'A_v \mathrm{diag}[1_{r_v}, -1_{s_v}]A_v$ (naturally $q = r_v + s_v$), $z = x + iy \in \mathcal{H}_m^{\mathbf{a}}$, $u \in (\mathbf{C}_q^m)^{\mathbf{a}}$, and $\eta \in \mathcal{S}((F_m^q)_{\mathbf{f}})$. The series is the same as that of [S8, (7.3)]. If $q = 1$ and $P = S = A = 1_m$, this coincides with $\theta^*(z, u; \eta)$, and therefore the following treatment includes θ^* as a special case.

Define embeddings ψ of $\mathcal{H}_m^{\mathbf{a}}$ into $\mathcal{H}_{mq}^{\mathbf{a}}$ and $\alpha \mapsto \alpha_S$ of G into $\mathrm{Sp}(mq, F)$ by

$$(11.7) \qquad \psi(z) = (S_v \otimes x_v + P_v \otimes iy_v)_{v \in \mathbf{a}} \qquad (z = x + iy \in \mathcal{H}_m^{\mathbf{a}}),$$

$$(11.8) \qquad \alpha_S = \begin{bmatrix} 1_q \otimes a_\alpha & S \otimes b_\alpha \\ S^{-1} \otimes c_\alpha & 1_q \otimes d_\alpha \end{bmatrix} \qquad (\alpha \in G),$$

and define a factor of automorphy $J_S(\alpha, z)$ by

$$(11.9) \qquad J_S(\alpha, z) = h(\alpha, z)^q \prod_{v \in \mathbf{a}} |\det(c_v z_v + d_v)|^{s_v} \det(c_v z_v + d_v)^{-s_v}$$

$$\left(c = c_\alpha, d = d_\alpha, \alpha \in G \cap P_A C'', z \in \mathcal{H}_m^{\mathbf{a}} \right).$$

PROPOSITION 11.3. *Suppose $S \in \mathfrak{o}_q^q$ (only here); let χ be the Hecke character of F corresponding to $F(\det(S)^{1/2})/F$. Then for $\alpha \in G \cap P_A D[2\mathfrak{d}^{-1}, 2\det(S)\mathfrak{d}]$, we have*

$$h(\alpha_S, \psi(z)) = \chi_{\mathbf{a}}(\det(d_\alpha))\chi^*(\det(d_\alpha)\mathfrak{a}_\alpha^{-1})J_S(\alpha, z),$$

where $\mathfrak{a}_\alpha = \mathrm{il}(\alpha)$ with il of [S8, (12.6)].

Proof. By [S8, (7.1)], we have

$$(11.10) \qquad h(\alpha_S, \psi(z)) = tJ_S(\alpha, z)$$

with $t \in \mathbf{T}$. For α in our proposition, we observe that $\alpha_S \in P_A D[2\mathfrak{d}^{-1}, 2\mathfrak{d}]$. Therefore, by [S8, Lemma 3.4] we have

$$\lim_{z \to 0} h(\alpha, z)/|h(\alpha, z)| = \gamma/|\gamma|, \qquad \lim_{Z \to 0} h(\alpha_S, Z)/|h(\alpha_S, Z)| = \gamma'/|\gamma'|$$

with $\gamma = \gamma(\delta^{-2}d_\alpha^{-1}c_\alpha)$ and $\gamma' = \gamma(\delta^{-2}S^{-1} \otimes d_\alpha^{-1}c_\alpha)$. Let $\alpha \in P_A \tau$ with $\tau \in D[2\mathfrak{d}^{-1}, 2\det(S)\mathfrak{d}]$. For simplicity, write c and d for c_τ and d_τ; obviously $d_\alpha^{-1}c_\alpha = d^{-1}c$. Now (11.10) implies that $\gamma'/|\gamma'| = t\gamma^q/|\gamma^q|\chi_{\mathbf{a}}(\det(d_\alpha))$. If $v \nmid 2\det(S)$, then S^{-1} is v-integral and locally equivalent to $\mathrm{diag}[\sigma_1, \ldots, \sigma_q]$ with $\sigma_i \in \mathfrak{o}_v^\times$. By [S8, Lemma 3.3] we then see that

$$\gamma_v(\delta^{-2}S^{-1} \otimes d^{-1}c) = \prod_{i=1}^q \gamma_v(\delta^{-2}\sigma_i d^{-1}c) = \gamma_v(\delta^{-2}d^{-1}c)^q \left(\frac{\det(S)}{\nu_0(d^{-1}c)_v} \right).$$

Observe that $\nu_0(d^{-1}c) = \det(d)^{-1}\mathfrak{o} = \det(d_\alpha)^{-1}\mathfrak{a}_\alpha$. If $v|2\det(S)$, then $|\det(d)_v|$ $= 1$, and hence both $\delta^{-1}d^{-1}c$ and $\delta^{-1}S^{-1} \otimes d^{-1}c$ are v-integral. Therefore, by [S8, (3.15b)], we have $\gamma_v = \gamma'_v = 1$. Combining all these, we obtain our proposition.

For $\alpha \in G$, we define the action of α on $(z, u) \in (\mathscr{H}_m \times \mathbf{C}^m)^{\mathbf{a}}$ by

$$\alpha(z, u) = (\alpha z, w), \qquad w_{vj} = \begin{cases} {}^t(c_v z_v + d_v)^{-1} u_{vj} & (1 \leqq j \leqq r_v) \\ {}^t(c_v \bar{z}_v + d_v)^{-1} u_{vj} & (r_v < j \leqq q), \end{cases}$$

where $c = c_\alpha$, $d = d_\alpha$, $v \in \mathbf{a}$ and the subscript j means the j-th column.

PROPOSITION 11.4. *For $\beta \in G \cap P_A C''$ and $\eta \in \mathscr{S}((F_m^q)_{\mathbf{f}})$, there is a unique element ${}^\beta\eta$ of $\mathscr{S}((F_m^q)_{\mathbf{f}})$ such that*

$$J_S(\beta, z)^{-1} g(\beta(z, u); {}^\beta\eta) = g(z, u; \eta).$$

Moreover, ${}^\beta\eta$ is independent of A and P.

This is merely a reformulation of [S8, Proposition 7.2]. Substituting $\beta^{-1}(z, u)$ for (z, u), we can rewrite the formula in the form

$$(11.11) \qquad J_S(\beta, \beta^{-1}z) g(\beta^{-1}(z, u); \eta) = g(z, u; {}^\beta\eta).$$

Now ${}^\beta\eta$ can be determined at least to the following extent.

PROPOSITION 11.5. *Let $\mathfrak{c}$ be the conductor of the character χ of Proposition 11.3. Given $\eta \in \mathscr{S}((F_m^q)_{\mathbf{f}})$, there is an open subgroup U of $D[2\mathfrak{d}^{-1}, 2\mathfrak{c}\mathfrak{d}]G_{\mathbf{a}}$ with the property that if $\beta \in G \cap \mathrm{diag}[p, {}^t p^{-1}]U$ with $p \in \mathrm{GL}_m(F_{\mathbf{f}})$, then*

$${}^\beta\eta(x) = \chi_{\mathbf{a}}(\det(d_\beta))\chi^*(\det(d_\beta)\mathfrak{a}_\beta^{-1})N(\mathfrak{a}_\beta)^{q/2}\eta(xp) \qquad (x \in (F_m^q)_{\mathbf{f}}).$$

Proof. Let $S' = {}^t BSB$, with $B \in \mathrm{GL}_q(F)$, $A' = AB$, and $P' = {}^t A'A'$. If g' is the function g defined with S', A', and P' instead of S, A, and P, we see that $g(z, u; \eta) = g'(z, u; \eta')$ with $\eta'(x) = \eta(Bx)$. Hence our assertion is true for g' if it is true for g. For this reason, we may assume that $S \in \mathfrak{o}_q^q$. Now, as shown in [S8, (7.4)], there is an isomorphism $u \mapsto u^0$ of $(\mathbf{C}_q^m)^{\mathbf{a}}$ onto $(\mathbf{C}^{mq})^{\mathbf{a}}$ and also an isomorphism $\eta \mapsto \eta^0$ of $\mathscr{S}((F_m^q)_{\mathbf{f}})$ onto $\mathscr{S}(F_{\mathbf{f}}^{mq})$ such that

$$(11.12) \qquad g(z, u; \eta) = \theta^*(\psi(z), u^0; \eta^0),$$

where θ^* is the function of (11.4) defined with respect to $G' = \mathrm{Sp}(mq, F)$ instead of $\mathrm{Sp}(m, F)$. Take U_l as in Proposition 11.2 with $l = \eta^0$. Now the map $\alpha \mapsto \alpha_S$ can be extended to a map of G_A into G'_A. Let U be the intersection of the inverse image of U_l with $D[2\mathfrak{d}^{-1}, 2\det(S)\mathfrak{d}]G_\mathbf{a}$. Let $\beta \in G \cap \mathrm{diag}[\,p, {}'p^{-1}]U$ with $p \in \mathrm{GL}_m(F_\mathfrak{f})$ and let $\gamma = \beta_S$. By Proposition 11.2 we have

$$(11.13) \quad h\big(\beta_S, \beta_S^{-1}\psi(z)\big)g\big(\beta^{-1}(z, u); \eta\big)$$

$$= h\big(\gamma, \gamma^{-1}\psi(z)\big)\theta*\big(\gamma^{-1}(\psi(z), u^0); l\big) = \theta*\big(\psi(z), u^0; {}^\gamma l\big),$$

and ${}^\gamma l(x) = N(\det(p)\mathfrak{o})^{-q/2}l(x(1_q \otimes p))$. Let η' be the element of $\mathscr{S}((F_m^q)_\mathfrak{f})$ such that $(\eta')^0 = {}^\gamma l$. Then $\eta'(x) = N(\mathfrak{a}_\beta)^{q/2}\eta(xp)$, and the last function of (11.13) is equal to $g(z, u; \eta')$, and hence our assertion follows immediately from Proposition 11.3.

LEMMA 11.6. *Let*

$$\iota = \begin{bmatrix} 0 & -1_m \\ 1_m & 0 \end{bmatrix} \quad and \quad \eta \in \mathscr{S}((F_m^q)_\mathfrak{f}).$$

Then

$$'\eta(x) = i^{-p}\big|N_{F/\mathbf{Q}}\det(S)\big|^{-m/2} \int_Y \eta(y)\mathbf{e}_\mathfrak{f}\big(-\mathrm{tr}({}'xSy)\big)\, dy,$$

where $p = m\sum_{v \in \mathbf{a}}s_v$, $Y = (F_m^q)_\mathfrak{f}$, *and dy is determined so that the measure of $\mathfrak{o}_v$ is* $N(\mathfrak{d}_v)^{-1/2}$ *for each $v \in \mathfrak{f}$.*

Proof. Let $\gamma = \iota_S$. By [S8, Lemma 3.4] we easily see that $h(\iota, z) = \det(-iz)^{u/2}$ and $h(\gamma, Z) = |\det(S)_\mathbf{a}|^{-m/2}\det(-iZ)^{u/2}$, where the branch of each square root is taken so that it is positive when the variable matrix is pure imaginary. Taking z to be pure imaginary in (11.10) with $\alpha = \iota$, we find that $h(\iota_S, \psi(z)) = i^p J_S(\iota, z)$. If $l = \eta^0$ as in the above proof, we see from (11.13) that $('\eta)^0 = i^{-p} \cdot {}^\gamma l$. (We don't need the integrality of S here.) By [S8, (3.3)] and (2.16c) we have

$$'^\gamma l(x) = \big|\det(c_\gamma)_\mathfrak{f}\big|^{-1/2} \int_W l(y)\mathbf{e}_\mathfrak{f}\big(xb_\gamma \cdot {}'y\big)\, dy \qquad \big(W = (F_{mq}^1)_\mathfrak{f}\big),$$

which proves our formula for $'\eta$.

PROPOSITION 11.7. *Let χ and $\mathfrak{c}$ be as in Proposition 11.5. Given $\eta \in \mathscr{S}((F_m^q)_\mathfrak{f})$, let M be an $\mathfrak{o}$-lattice in F_m^q such that $\eta(x + u) = \eta(x)$ for every $u \in M$. Further let $\mathfrak{x}$, $\mathfrak{y}$, and $\mathfrak{z}$ be fractional ideals of F with the following properties:*
   (i) *${}'xSx$ has entries in $\mathfrak{x}$ for every $x \in F_m^q$ such that $\eta(x) \neq 0$;*
   (ii) *${}'xSx$ has entries in $\mathfrak{y}$ for every $x \in F_m^q$ such that $\mathrm{tr}({}'xSy) \in \mathfrak{d}^{-1}$ for every $y \in M$;*
   (iii) *if $a \in \prod_{v \in \mathfrak{f}}\mathrm{GL}_m(\mathfrak{o}_v)$ and $a_v - 1 \in (\mathfrak{z}_v)_m^m$ for every $v \in \mathfrak{f}$, then $\eta(xa) = \eta(x)$.*

Then $^\gamma\eta(x) = \chi_c(\det(d_\gamma))\eta(x(a_\gamma)_\delta)$ for every $\gamma \in \Gamma[2\mathfrak{d}^{-1}\mathfrak{a}, 2^{-1}\mathfrak{d}\mathfrak{a}^{-1}\mathfrak{b}]$, where $\mathfrak{a} = \mathfrak{x}^{-1} \cap \mathfrak{o}$, $\mathfrak{b} = \mathfrak{c} \cap \mathfrak{z} \cap 4\mathfrak{a} \cap 4\mathfrak{d}^{-2}\mathfrak{a}\mathfrak{y}^{-1}$, and $(a_\gamma)_\delta$ is the projection of a_γ to $\prod_{v|\delta}\mathrm{GL}_m(F_v)$.

Proof. Let $P_1 = \{\alpha \in G | a_\alpha = 1_m, c_\alpha = 0\}$ and let $\alpha \in P_1$. Since $h(\alpha, z) = 1$, we easily see that $^\alpha\eta(x) = \eta(x)\mathbf{e}_\mathbf{f}(\mathrm{tr}(^t xSxb_\alpha)/2)$. Therefore $^\alpha\eta = \eta$ if b_α has entries in $2\mathfrak{d}^{-1}\mathfrak{x}^{-1}$. Put $\iota^{-1}\alpha\iota = \beta$ and $\eta' =\ ^\iota \eta$. Lemma 11.6 shows that $\eta'(x) = \mathbf{e}_\mathbf{f}(\mathrm{tr}(^t xSy))\eta'(x)$ for every $y \in M$, and hence $\eta'(x) \neq 0$ only if $^t xSx$ has entries in $\mathfrak{y}$. Therefore $^\alpha\eta' = \eta'$ if b_α has entries in $2\mathfrak{d}^{-1}\mathfrak{y}^{-1}$, and hence $^\beta\eta = \eta$ if β is of the form $\beta = \begin{pmatrix} 1 & 0 \\ c & 1 \end{pmatrix}$ and c has entries in $2\mathfrak{d}^{-1}\mathfrak{y}^{-1} \cap 2\mathfrak{d}$. Proposition 11.5 shows that the expected formula for $^\gamma\eta$ is true for $\gamma \in \Gamma[\mathfrak{e}, \mathfrak{e}]$ with a suitable multiple $\mathfrak{e}$ of $2\mathfrak{d}^{-1}\mathfrak{a} \cap 2^{-1}\mathfrak{d}\mathfrak{a}^{-1}\mathfrak{b}$. Now $\Gamma[2\mathfrak{d}^{-1}\mathfrak{a}, 2^{-1}\mathfrak{d}\mathfrak{a}^{-1}\mathfrak{b}]$ can be generated by its intersection with $P_1 \cup \iota P_1 \iota^{-1}$ and $\Gamma[\mathfrak{e}, \mathfrak{e}]$. This fact is a generalization of Lemma 3.4 and can be proved by combining strong approximation and the corresponding fact on $D[*, *]$ which can be proved in an elementary way. Our formula, being true for these generators, must be true for the whole group in our assertion.

Notice that $^\beta(^\gamma\eta) =\ ^{\beta\gamma}\eta$ is true by virtue of [S8, (3.10c)] if β and γ belong to $\Gamma[2\mathfrak{d}^{-1}, 2\mathfrak{d}]$, but not necessarily so in general, which is why we have to take $\mathfrak{a} \subset \mathfrak{o}$ and $\mathfrak{b} \subset 4\mathfrak{a}$.

Let us now derive the transformation formula of a series of the form

$$(11.14) \qquad f(z, \eta) = \sum_{\xi \in F^q} \eta(\xi)\sigma(\xi)\mathbf{e}_\mathbf{a}\left(2^{-1}(^t\xi S\xi x +\ ^t\xi P\xi \cdot iy)\right)$$

$$(z = x + iy \in \mathscr{H}^\mathbf{a}),$$

assuming $m = 1$ for simplicity; the case $m > 1$ can be treated in a similar way. The notation is the same as before except for a new function σ on $(\mathbf{R}^q)^\mathbf{a}$ $(= (F^q)_\mathbf{a})$, which has the form

$$(11.15) \qquad \sigma(w) = (^t\rho Sw)^\lambda (^t\rho' Sw)^{\lambda'} \qquad (w \in (\mathbf{R}^q)^\mathbf{a})$$

with $0 \leq \lambda \in \mathbf{Z}^\mathbf{a}$, $0 \leq \lambda' \in \mathbf{Z}^\mathbf{a}$, $\rho \in (\mathbf{C}^q)^\mathbf{a}$, $\rho' \in (\mathbf{C}^q)^\mathbf{a}$. We assume

$$(11.16a) \qquad S_v\rho_v = P_v\rho_v, \qquad S_v\rho'_v = -P_v\rho'_v \qquad (v \in \mathbf{a}),$$

$$(11.16b) \qquad {}^t\rho_v S_v\rho_v = 0 \ \text{ if } \lambda_v > 1; \qquad {}^t\rho'_v S_v\rho'_v = 0 \ \text{ if } \lambda'_v > 1.$$

PROPOSITION 11.8. *For $\beta \in G \cap P_\mathbf{A}C''$, put*

$$(11.17) \qquad K(\beta, z) = h(\beta, z)^q j(\beta, z)^{\lambda - s}\overline{j(\beta, z)}^{\lambda'}|j(\beta, z)|^s.$$

Then for every $\eta \in \mathscr{S}(F_{\mathbf{i}}^q)$ and every such β, one has

$$K(\beta, z)^{-1} f(\beta z, {}^\beta\eta) = f(z, \eta),$$

with ${}^\beta\eta$ of Proposition 11.4.

Proof. Put $A\rho = \tau$ and $A\rho' = \tau'$. Then (11.16a) implies that $\tau_{vj} = 0$ for $j > r_v$ and $\tau'_{vj} = 0$ for $j \leq r_v$. Hence we easily see that

$$(-1)^{\{\lambda'\}}(2\pi i)^{\{\lambda+\lambda'\}} f(z, \eta)$$

$$= \left[\prod_{v \in \mathbf{a}} \left(\sum_{j=1}^{q} \tau_{vj}\, \partial/\partial u_{vj} \right)^{\lambda_v} \left(\sum_{j=1}^{q} \tau'_{vj}\, \partial/\partial u_{vj} \right)^{\lambda'_v} g(z, u; \eta) \right]_{u=0}.$$

Therefore our formula follows immediately from Proposition 11.4.

Remark 11.9.   If q is even, we can formulate somewhat better results in the following way. We first put, instead of (11.9) and (11.17),

$$(11.18) \qquad J^S(\alpha, z) = \prod_{v \in \mathbf{a}} \det(c_v z_v + d_v)^{(q/2)-s_v} |\det(c_v z_v + d_v)|^{s_v}$$

$$(c = c_\alpha,\, d = d_\alpha,\, \alpha \in G,\, z \in \mathscr{H}_m^{\mathbf{a}}),$$

$$(11.19)\quad K'(\alpha, z) = j(\alpha, z)^{(q/2)+\lambda-s} \overline{j(\alpha, z)}^{\lambda'} |j(\alpha, z)|^s \qquad (\alpha \in \mathrm{SL}_2(F)).$$

By [S8, Proposition 7.2], we can define η^α *for every* $\alpha \in G$ by

$$(11.20) \qquad\qquad J^S(\alpha, z)^{-1} g(\alpha(z, u); \eta) = g(z, u; \eta^\alpha).$$

Differentiating g as in the proof of Proposition 11.8, we obtain

$$(11.21) \qquad K'(\alpha, z)^{-1} f(\alpha z, \eta) = f(z, \eta^\alpha) \qquad (\alpha \in \mathrm{SL}_2(F)).$$

Now, in view of [S8, Lemma 3.5], the assertion of Proposition 11.5 can be changed as follows:

If $\beta \in G \cap U \cdot \mathrm{diag}[p, {}^t p^{-1}]$ *with* $p \in \mathrm{GL}_m(F_{\mathbf{i}})$, *then*

$$(11.22) \quad \eta^\beta(x) = \omega_{\mathbf{a}}\big(\det(d_\beta)\big)\,\omega^*\big(\det(d_\beta)\mathfrak{a}_\beta^{-1}\big) N(\mathfrak{a}_\beta)^{-q/2} \eta(xp^{-1}),$$

where ω is the Hecke character of F corresponding to $F((-1)^{q/4}\det(S)^{1/2})/F$.

Similarly, the assertion of Proposition 11.7, when q is even, can be changed to

$$(11.23) \qquad\qquad \eta^\gamma(x) = \omega_{\mathfrak{c}}\big(\det(d_\gamma)\big)\,\eta\big(x(a_\gamma)_\delta^{-1}\big)$$

for every $\gamma \in \Gamma[2\mathfrak{d}^{-1}\mathfrak{x}^{-1}, 2^{-1}\mathfrak{d}\mathfrak{x}\mathfrak{b}]$, $\mathfrak{b} = \mathfrak{z} \cap \mathfrak{c} \cap 4\mathfrak{d}^{-2}\mathfrak{x}^{-1}\mathfrak{y}^{-1}$, *where* $\mathfrak{c}$ *is the conductor of* ω.

REFERENCES

[B]   V. Bargmann, *On a Hilbert space of analytic functions and an associated integral transform: Part I*, Comm. Pure Appl. Math. **14** (1961), 187–214.

[F]   Y. Flicker, *Automorphic forms on covering groups of GL(2)*, Invent. Math. **57** (1980), 119–182.

[GP1]   S. Gelbart and I. Piatetski-Shapiro, "On Shimura's correspondence for modular forms of half-integral weight," *Proc. Int. Coll. Auto. Forms, Rep. Theory and Arith.* (1979), Tata Institute, Springer, 1981.

[GP2]   ______, *Some remarks on metaplectic cusp forms and the correspondences of Shimura and Waldspurger*, Israel J. of Math. **44** (1983), 97–126.

[I]   J. Igusa, *Theta Functions*, Springer, 1972.

[M]   T. Miyake, *On automorphic forms on GL_2 and Hecke operators*, Ann. of Math. **94** (1971), 174–189.

[N]   S. Niwa, *Modular forms of half integral weight and the integral of certain theta-functions*, Nagoya Math. J. **56** (1974), 147–161.

[P]   T-Y. Pei, *Eisenstein series of weight 3/2: I, II*, Trans. Amer. Math. Soc. **274** (1982), 573–606, **283** (1984), 589–603.

[SS]   J.-P. Serre and H. Stark, "Modular forms of weight 1/2," *Mod. f. of One Var. VI*, Lecture Notes in Math. **627** (1977), 27–67.

[S1]   G. Shimura, *On modular forms of half-integral weight*, Ann. of Math. **97** (1973), 440–481.

[S2]   ______, *On the periods of modular forms*, Math. Ann. **229** (1977), 211–221.

[S3]   ______, *The special values of the zeta functions associated with Hilbert modular forms*, Duke Math. J. **45** (1978), 637–679.

[S4]   ______, *The arithmetic of certain zeta functions and automorphic forms on orthogonal groups*, Ann. of Math. **111** (1980), 313–375.

[S5]   ______, *On certain zeta functions attached to two Hilbert modular forms: I, II*, Ann. of Math. **114** (1981), 127–164, 569–607.

[S6]   ______, *The critical values of certain zeta functions associated with modular forms of half-integral weight*, J. Math. Soc. Japan **33** (1981), 649–672.

[S7]   ______, *On Eisenstein series*, Duke Math. J. **50** (1983), 417–476.

[S8]   ______, *On Eisenstein series of half-integral weight*, Duke Math. J. **52** (1985), 281–314.

[S9]   ______, *On the Eisenstein series of Hilbert modular groups*, Revista Mat. Iberoamer. **1**, 3 (1985), 1–42.

[S10]   ______, *On a class of nearly holomorphic automorphic forms*, Ann. of Math. **123** (1986), 347–406.

[S11]   ______, *Nearly holomorphic functions on hermitian symmetric spaces*, to appear in Math. Ann.

[St]   J. Sturm, *Theta series of weight 3/2*, J. of Number Theory **14** (1982), 353–361.

[Wa]   J.-L. Waldspurger, *Correspondance de Shimura*, J. Math. pure et appl. **59** (1980), 1–133.

[We]   A. Weil, *Sur certains groupes d'opérateurs unitaires*, Acta math. **111** (1964), 143–211.

DEPARTMENT OF MATHEMATICS, PRINCETON UNIVERSITY, PRINCETON, NEW JERSEY 08544

CORRECTIONS TO [S8]

Page 291, line 10 from bottom (formula in Case SU): $2s + k - 1$ should be $2s + k - i$.

Page 294, line 6 from bottom: $C_v'' \cap \Omega_v$ should be $(P_v \cap C_v'')\eta_v$.

Page 295, (3.14a): This formula is not always true, but at least valid for $\xi \in G \cap P_A C' \eta$. Therefore (3.10d) and (3.14b) must be proved differently. First note that (3.10d) follows immediately from Lemma 3.5. Also Lemma 3.4 shows that $h(\iota, z) = \prod_{v \in \mathbf{a}} \det(-iz_v)^{1/2}$. Take η_0 in $G \cap C' G_{\mathbf{a}} \eta^{-1}$; let $\gamma = \iota \eta_0^{-1}$. Then $\gamma \in \iota \eta C' G_{\mathbf{a}} \subset P_A C'$, and hence, by (3.10c), $h(\iota, z) = h(\gamma, \eta_0 z) h(\eta_0, z)$, which together with (3.10d) proves (3.14b). If $\xi \in G \cap P_A C' \eta$, then $\xi = \alpha \eta_0$ with $\alpha \in G \cap P_A C'$, since $P_A C' \eta = P_A C' \eta_0$. Then (3.14a) for this ξ follows from (3.10c, d) and (3.14b).

Page 306, line 12 from bottom: $\prod_{i=1}^{[(m-r-1)/2]}$ should be $\prod_{i=0}^{[(m-r-1)/2]}$.

Page 306, line 10 from bottom: $\prod_{i=0}^{[(m-r)/2]}$ should be $\prod_{i=1}^{[(m-r)/2]}$.

Corrections to [S9]

Page 22, line 19: The last equality should be $\theta_v = -\theta_w$.

Page 24, Lemma 6.2: The equality is true only when $\tau = 0$. The integral vanishes if $\tau \neq 0$.

Page 25, line 6: Insert "$\Sigma_{p,q}$ is taken over all those (p, q) such that $\bar{p} + q + \sigma \in Cu$, and" before $e(p, q)$.

Page 25, lines 12, 13, and 14 should be deleted.

Page 35, line 5 from bottom: (10.3) should be (10.4).

On the critical values of certain Dirichlet series and the periods of automorphic forms

Inventiones mathematicae, 94 (1988), 245-305

Introduction

This paper consists of several types of results forming a certain logical chain, which eventually leads to some conjectures. Each result concerns the arithmeticity of the critical values or the residue of a certain Dirichlet series, or that of automorphic forms similar to Eisenstein series; the conjectures concern the same objects, as well as the periods of automorphic forms, in particular, of abelian integrals on an arithmetic curve. Though each case has its raison-d'être, it seems advantageous, in the introduction, to concentrate on the zeta functions of the following form:

$$(1) \qquad D(s) = \sum \lambda(b)\mu(b)b^{\alpha}N_{E/\mathbb{Q}}(b)^{-s}.$$

Here E is a totally real algebraic number field of degree v, b runs over all the totally positive elements of E modulo a group of units, α is an element of $\mathbb{Q}^J$, where J denotes the set of archimedean primes of E, and λ resp. μ represents the Fourier coefficients of a holomorphic Hilbert modular form f resp. g of integral or half-integral weight with respect to a congruence subgroup of $SL_2(E)$. An integral weight is an element of $\mathbb{Z}^J$, and a half-integral weight is an element of $\mathbb{Q}^J$ each of whose components is half an odd integer. The arithmetic nature of the critical values of D depends on whether the weights of f and g are integral or half-integral, and also on the mutual relationship between them. Denoting by k and l the weights of f and g respectively, we can classify the weight combinations into the following seven cases in which δ and δ' are nonempty subsets of J whose union is J:

(IA) $\qquad\qquad k \in \mathbb{Z}^J,\ l \in \mathbb{Z}^J;\quad k_v > l_v \quad$ for all $v \in J$.

(IB) $\qquad\qquad k \in \mathbb{Z}^J,\ l \in \mathbb{Z}^J;\quad k_v > l_v \quad$ for $\quad v \in \delta \quad$ and $\quad k_v < l_v$ for $v \in \delta'$.

(IIA) $\qquad\qquad k \in \mathbb{Z}^J,\ l \notin \mathbb{Z}^J;\quad k_v > l_v \quad$ for all $v \in J$.

(IIIA) $\qquad\qquad k \notin \mathbb{Z}^J,\ l \in \mathbb{Z}^J;\quad k_v > l_v \quad$ for all $v \in J$.

Research supported by NSF Grant DMS-8401291

$$\text{(IIB)} = \text{(IIIB)} \qquad k \in \mathbb{Z}^J,\ l \notin \mathbb{Z}^J; \qquad k_v > l_v \quad \text{for} \quad v \in \delta \quad \text{and} \quad k_v < l_v$$
$$\text{for} \quad v \in \delta'.$$

$$\text{(IVA)} \qquad k \notin \mathbb{Z}^J,\ l \notin \mathbb{Z}^J; \qquad k_v > l_v \quad \text{for all} \quad v \in J.$$

$$\text{(IVB)} \qquad k \notin \mathbb{Z}^J,\ l \notin \mathbb{Z}^J; \qquad k_v > l_v \quad \text{for} \quad v \in \delta \quad \text{and} \quad k_v < l_v$$
$$\text{for} \quad v \in \delta'.$$

For some natural reasons, we exclude the cases in which $k_v = l_v$ for some v. Now Cases (IA) and (IB) were treated in [78] and [83a], and Case (IVA) in [87b]; Cases (IIA) and (IIIA) are being investigated by J. Im in his thesis [I]. Thus the main interests of the present paper, as far as the series of type (1) is concerned, are in Cases (IIB) and (IVB).

To explain the results in these cases, however, it is necessary to recall the previous results in the other cases. Also, we need to introduce, in addition to (1), two series of the forms

$$\text{(2a)} \qquad D(s, \chi, \varphi) = \sum \chi(\mathfrak{a}) \varphi(\mathfrak{a}) N(\mathfrak{a})^{-s-1},$$

$$\text{(2b)} \qquad D(s, \chi, \chi') = \sum \chi(\mathfrak{a}) \chi'(\mathfrak{a}) N(\mathfrak{a})^{-s},$$

where $\mathfrak{a}$ runs over all the integral ideals in E, χ and χ' are primitive systems of eigenvalues of Hecke operators on the spaces of Hilbert modular forms of integral weight k and l respectively, and φ is a Hecke character of E of finite order. Then we can associate a complex number, or rather a coset of $\mathbb{C}^\times/\bar{\mathbb{Q}}^\times$, $V(\chi, r)$, to χ and each $r \in \mathbb{Z}^J/2\mathbb{Z}^J$, with which we can state the following results ([78], [87b]) in which we write $a \sim b$ if a/b is algebraic:

(i) *Let t be an integer such that $|t| < k_v \equiv t \pmod 2$ for every $v \in J$ and let $t' = t/2$ or $t' = (t+1)/2$ according as t is even or odd. Suppose $\varphi_v(-1) = (-1)^{r_v + t'}$ for every $v \in J$. Then $D(t/2, \chi, \varphi) \sim \pi^{tv/2} V(\chi, r)$.*

(ii) *Suppose $k_v > l_v$ for all $v \in J$. Then for every integer t such that*

$$\text{(3)} \qquad 2 < t \le |k_v - l_v| + 2 \quad \text{and} \quad t \equiv k_v - l_v \pmod 2 \quad \text{for every} \quad v \in J,$$

we have $D(t/2, \chi, \chi') \sim \pi^a \langle \mathbf{f}, \mathbf{f} \rangle$, where $a = \sum_{v \in J} k_v$, $\mathbf{f}$ is a $\bar{\mathbb{Q}}$-rational eigenform belonging to χ, and $\langle\ ,\ \rangle$ is the Petersson inner product.

(iii) *$V(\chi, r) V(\chi, s) \sim \pi^{v+a} \langle \mathbf{f}, \mathbf{f} \rangle$ if $r_v + s_v \equiv 1 \pmod 2$ for every $v \in J$.*

Result (ii) pertains to Case (IA). The corresponding statement in Case (IB) requires a pair of quaternion algebras B and B' over E which are complementary of type (δ, δ') in the sense that δ resp. δ' is exactly the set of archimedean primes unramified in B resp. B'. Then the result in Case (IB) is as follows [83a]:

(iv) *Suppose k and l are as in (IB); suppose also that there exist $\bar{\mathbb{Q}}$-rational eigenforms $\mathbf{h}$ and $\mathbf{h}'$ on the spaces $\mathscr{S}(B)$ and $\mathscr{S}(B')$ of automorphic forms on B and B', belonging to χ and χ', respectively. Then for every integer t satisfying (3), we have $D(t/2, \chi, \chi') \sim \pi^b \langle \mathbf{h}, \mathbf{h} \rangle \langle \mathbf{h}', \mathbf{h}' \rangle$, where $b = \sum_{v \in \delta} k_v + \sum_{v \in \delta'} l_v$.*

As a supplementary result, obtained also in [83a], we should mention:

(v) *If $\chi' = \chi$ or $\chi' = \bar{\chi}$, then $\langle \mathbf{h}, \mathbf{h} \rangle \langle \mathbf{h}', \mathbf{h}' \rangle \sim \langle \mathbf{f}, \mathbf{f} \rangle$ with $\mathbf{f}$ as in (ii).*

One more result in Case (IVA) needs to be recalled (see [87b] or Theorem 8.3 of the present paper).

(vi) *Suppose the $\lambda(b)$ and $\mu(b)$ in (1) are all algebraic; suppose moreover that both f and g are of half-integral weight, and f corresponds to a form of integral weight belonging to χ. Take $\alpha = (\iota - k - l)/2$ in (1), where ι is the identity element of the ring $\mathbb{Q}^J$. Then $D(t/2) \sim \pi^{-\nu} V(\chi, s)$ with some $s \in \mathbb{Z}^J/2\mathbb{Z}^J$ for every integer t such that*

$$(4) \qquad 0 \leq t < |k_v - l_v| \quad and \quad t \equiv k_v + l_v \pmod 2 \quad for\ every \quad v \in J.$$

Now comparison of (iii) with (v) indicates a possibility that the constant $V(\chi, r)$ may be decomposed into two factors. We notice also that passing from (ii) to (vi), the quantity $\langle \mathbf{f}, \mathbf{f} \rangle$ is reduced to $V(\chi, s)$, which is a factor of $\langle \mathbf{f}, \mathbf{f} \rangle$ as shown in (iii). These observations lead us to the following crude conjecture, in which we assume $\bar{\chi} = \chi$ for simplicity:

(X) *If r and s are as in (iii) and $\mathbf{f}, \mathbf{h}, \mathbf{h}', \chi, \chi'$ are as in (ii) and (iv), then there exist constants $P(\chi, \delta, r)$ such that $V(\chi, r) \approx P(\chi, \delta, r) P(\chi, \delta', r)$, $\langle \mathbf{h}, \mathbf{h} \rangle \approx P(\chi, \delta, r) P(\chi, \delta, s)$, and $\langle \mathbf{h}', \mathbf{h}' \rangle \approx P(\chi', \delta', r) P(\chi', \delta', s)$, where we write $a \approx b$ if $a/b \sim \pi^n$ for some $n \in \mathbb{Z}$.*

Indeed, the values $D(t/2, \chi, \varphi)$ can be obtained as ν-dimensional integrals, and therefore (i) shows that $V(\chi, r)$ is a kind of period of $\mathbf{f}$ on a ν-cycle. On the other hand, $\langle \mathbf{f}, \mathbf{f} \rangle$ and $\langle \mathbf{h}, \mathbf{h} \rangle$ may be viewed respectively as periods on a 2ν-cycle and a $2d$-cycle, where d is the number of elements in δ. Thus the desired factors $P(\chi, \delta, r)$ are would-be periods on d-cycles, and the $P(\chi', \delta', r)$ are those on $(\nu - d)$-cycles, which provides a geometric interpretation of (X).

If (X) were really true, comparison of (vi) with (iv) would suggest another conjecture:

(Y) *For the series D of (1) in Case (IVB) and for every integer t satisfying (4), one has $D(t/2) \approx P(\chi, \delta, s) P(\chi', \delta', s')$ for some s and s'.*

After these considerations, we can now come to our new result in Case (IVB). We assume that f and g in (1) are obtained as inner products

$$(5) \qquad f(z) = \langle \theta(z, w), \mathbf{h}(w) \rangle, \quad g(z) = \langle \theta'(z, w), \mathbf{h}'(w) \rangle$$

with $\mathbf{h}$ and $\mathbf{h}'$ as in (iv), where θ and θ' are certain ($\bar{\mathbb{Q}}$-rational) theta functions. Then, as one of the main results of this paper, we obtain:

(vii) *For the series D of (1) in Case (IVB) with $\alpha = (\iota - k - l)/2$ and with such f and g, and for every integer t satisfying (4), one has $D(t/2) \sim \pi^e \langle \mathbf{h}, \mathbf{h} \rangle \langle \mathbf{h}', \mathbf{h}' \rangle$, where $e = \sum_{v \in \delta} k_v + \sum_{v \in \delta'} l_v - \nu$ (Theorem 9.1).*

A similar result holds also in Case (IIB) (Theorem 9.2). It may appear that (vii) contradicts (Y), but this point can be explained in the following way: the Hilbert modular forms f and f' given in (5) are not necessarily $\bar{\mathbb{Q}}$-rational. Indeed, we can add one more conjecture:

(Z) *If f and g are as in (5), then $f/P(\chi, \delta, r)$ and $g/P(\chi', \delta', r')$ are $\bar{\mathbb{Q}}$-rational for some r and r'.*

This together with (X) makes the new result (vii) compatible with (Y).

Since (X) and (Y) are self-contained, the involvement of theta functions in (vii) and (Z) may look merely technical, but there is at least one significant aspect of the inner products as in (5), which were investigated in [82]. As shown in that paper, if f and $\mathbf{h}$ are as in (5), then the Fourier coefficients of f can be viewed as periods of $\mathbf{h}$, which are d-dimensional. This combined with (Z) explains the nature of $P(\chi, \delta, r)$ as a period of $\mathbf{h}$.

In Sect. 9, after proving the above-mentioned results in Cases (IIB) and (IVB), we shall present refined versions of (X, Y, Z) along with other conjectures, and show that the assertion of (Z) concerning f, when extended to the case $\delta = J$, is true (Theorem 9.4). In Sect. 10, we shall treat the periods of $\mathbf{h}$ when $d = 1$ and prove the same assertion of (Z) in such a case (Theorem 10.4). The last part of Sect. 10 will be devoted to the discussion of some consequences of our theorems in support of the conjectures. In particular, under a certain assumption of nonvanishing, we shall prove a formula (10.18) that is the first equivalence given in (X) in the case $v = 2$.

One of the most noteworthy consequences of (X) will be treated in Sect. 11. The refined form of (X) predicts that with a suitable choice of (r, s), one can have $P(\chi, \delta', r) = P(\chi, \delta', s)$ even when $P(\chi, \delta, r)$ and $P(\chi, \delta, s)$ are different, so that the following nontrivial proportionality should hold:

$$\text{(P)} \qquad\qquad P(\chi, \delta, r)/P(\chi, \delta, s) \approx V(\chi, r)/V(\chi, s).$$

Suppose now that $d = 1$ and $\mathbf{h}$ in (iv) is of weight 2. Then we can associate with $\mathbf{h}$ an abelian variety A that is a factor of the jacobian of the upper half plane modulo a congruence subgroup of $B^\times$. Now $P(\chi, \delta, r)$ and $P(\chi, \delta, s)$ are periods of $\mathbf{h}$ as remarked above, and so A is determined by several quantities of the form $P(\chi, \delta, r)/P(\chi, \delta, s)$. On the other hand, $V(\chi, r)$ is a period of a Hilbert modular form. Therefore (P) proposes the possibility of obtaining A from the periods of Hilbert modular forms. Moreover, (i) combined with (P) implies

$$\text{(Q)} \qquad\qquad P(\chi, \delta, r)/P(\chi, \delta, s) \approx D(0, \chi, \varphi)/D(0, \chi, \varphi').$$

for some φ and φ', thus conjecturally establishing a connection of A with the values of the zeta function pertaining to A. We shall explain these ideas in clearer terms by presenting (P) and (Q) in precise forms.

The last three sections being thus summarized, we need to do the same for the first eight ones. Sections 1 and 2 are generalizations of the results of [80] and [81], in which several theorems of arithmeticity were proved under a certain parity condition and also with an exception. We shall remove the condition and also include the exceptional case. In particular, we shall prove some theorems that were stated in [86] without proof. In these sections we shall merely outline the proof, since the ideas are the same as in those papers except that we need some results of [85a, b] and [87a]. Sections 3 and 4 concern generalizations of the first four sections of [83a] in the same sense. Here, however, our proof will be somewhat more detailed, since that in [83a] was sketchy. One of the new points in these generalizations is the perpetual presence of Hilbert modular forms of half-integral weight. Section 5 consists of some applications. One of them is a result (Theorem 5.2) in Cases (IA, IIA), which is stated for the purpose of comparison with the later results in other cases.

The topic of Sect. 6 is of a different nature. We shall consider the inner products of (5), and shall show that the map $\mathbf{h} \mapsto f = \langle \theta, \mathbf{h} \rangle$ commutes with Hecke operators. Then in Sect. 7, we shall reformulate the results of Sect. 3 by incorporating those of Sect. 6. After recalling the properties of $V(\chi, r)$ and the results in Case (IA) as in (i), (ii), (iii) above, we shall prove in Sect. 8 a result (Theorem 8.3) that is somewhat more comprehensive than (iv).

A conjecture including (X) was already stated in our previous paper [83a, p. 281], in which we also indicated its close connection with forms of half-integral weight, as well as some ideas of obtaining a result in this direction. The present paper is intended for the fulfilment of the "promise" made there. It should be noted that (P) and (Q) are implicit in, or rather, immediate consequences of [83a, p. 281, (P1), (P3), (P4)].

Thus the reader with a certain predilection may view the whole paper as a long winding road to a locked gate through which one may glimpse at a new territory. As a final remark, it may be mentioned that though (P) and (Q) are conjectures, we shall indicate in Sect. 11 a possible approach to them, that is, a possible key to the lock, which seems to have a fair chance of fitting.

0. Notation

Given a finite-dimensional vector space V over $\mathbb{Q}$, we denote by $\mathscr{L}(V)$ the set of all $\mathbb{C}$-valued functions f on V for which there exist two lattices L and M in V such that $f(x) = 0$ for $x \notin L$ and $f(x+y) = f(x)$ for every $y \in M$. We call such an f a *locally constant function on V*.

For $K = \prod_{i=1}^{t} K_i$ with algebraic number fields K_i of finite degree, we denote by $J(K)$ or J_K the set of all nontrivial homomorphisms of K into $\mathbb{C}$. J_K is naturally identified with the disjoint union of the $J(K_i)$. For $\alpha \subset J_K$ and a set X, we denote by X^α the set of all indexed elements $(x_\tau)_{\tau \in \alpha}$ with x_τ in X. We extend each $\tau \in J_K$ to a $\mathbb{C}$-linear homomorphism of $K \otimes_{\mathbb{Q}} \mathbb{C}$ into $\mathbb{C}$ and denote, for $x \in K \otimes_{\mathbb{Q}} \mathbb{C}$, by x^τ or x_τ the image of x under this homomorphism. Then $K \otimes_{\mathbb{Q}} \mathbb{C}$ can be identified with $\mathbb{C}^{J(K)}$ through the map $x \mapsto (x_\tau)$. We put $I_K = I(K) = \mathbb{Z}^{J(K)}$, $\mathbb{R}I_K = \mathbb{R}^{J(K)}$, and $\mathbb{C}I_K = \mathbb{C}^{J(K)}$; further we put $x^c = \prod_{\tau \in J(K)} (x_\tau)^{c_\tau}$ for x, $c \in K \otimes_{\mathbb{Q}} \mathbb{C} = \mathbb{C}^{J(K)}$, whenever $(x_\tau)^{c_\tau}$ is well-defined. A subset α of J_K will often be identified with the element $\sum_{\tau \in \alpha} \tau$ of I_K, so that $x^\alpha = \prod_{\tau \in \alpha} x^\tau$. In particular, we put $\iota_K = \iota(K) = \sum_{\tau \in J(K)} \tau \ (\in I_K)$; thus $x^{\iota(K)} = N_{K/\mathbb{Q}}(x)$. We write $c \geqq c'$ for $c, c' \in \mathbb{R}I_K$ if $c_\tau \geqq c'_\tau$ for every $\tau \in J_K$, and write $c \gg 0$ if $c_\tau > 0$ for every $\tau \in J_K$. We also put $\|c\| = \sum_{\tau \in J(K)} c_\tau$. (This notation will be employed only when c appears as an exponent.) When $F = \prod_{i=1}^{s} F_i$ is a similar product and if there is a fixed embedding of F into K, then we denote by $\mathrm{Res}_{K/F}$ or simply by $\mathrm{Res}_{/F}$ the $\mathbb{C}$-linear map of $\mathbb{C}I_K$ into $\mathbb{C}I_F$ which sends each element of J_K onto its composition with the embedding.

By a *CM-field*, we understand a totally imaginary quadratic extension of a totally real algebraic number field of finite degree. If the K_i are CM-fields, we define the period symbol $p_K(\xi, \eta)$ for $(\xi, \eta) \in I_K \times I_K$ as in [80, p. 319], and denote by ϱ the element of I_K which induces complex conjugation on each K_i.

Suppose now the K_i are all totally real. We put then

$$(0.0\text{a}) \qquad \mathbf{e}_K(z) = \mathbf{e}\left(\sum_{\tau \in J(K)} z_\tau\right) \quad \text{for} \quad z = (z_\tau) \in \mathbb{C}^{J(K)} = K \otimes_{\mathbb{Q}} \mathbb{C},$$

$$(0.0\text{b}) \qquad \mathbf{e}(v) = e^{2\pi i v} \quad \text{for} \quad v \in \mathbb{C}.$$

For $0 \leq h \in I_K$, we define differential operators $\mathbf{d}^h$ and $\bar{\mathbf{d}}^h$ on $\mathbb{C}^{J(K)}$ by

$$(0.1) \qquad \mathbf{d}^h = \prod_{\tau \in J(K)} [(2\pi i)^{-1} \partial/\partial z_\tau]^{h_\tau}, \quad \bar{\mathbf{d}}^h f = \overline{(\mathbf{d}^h \bar{f})}.$$

When K itself is a totally real field, we let R_K denote the regulator of K, and U_K the group of all totally positive units of K; we then put

$$(0.2) \qquad [U] = [U_K : U]^{-1}$$

for every subgroup U of U_K of finite index. We denote by $\mathcal{M}(m, K)$ or $\mathcal{M}_m(K)$ the set of all Hilbert modular forms of weight m with respect to congruence subgroups of $SL_2(K)$ defined as holomorphic functions on $H^{J(K)}$, where H denotes the upper half complex plane:

$$(0.3) \qquad H = \{z \in \mathbb{C} \mid \text{Im}(z) > 0\}.$$

A *weight* is either an *integral weight* which is an element of I_K, or a *half-integral weight* which is an element m of $(1/2)I_K$ such that $2m_\tau$ is odd for every $\tau \in J_K$. (For the precise definition of $\mathcal{M}_m(K)$, see [81, I, § 1] or [85a].) The set of all integral and half-integral weights with respect to K is denoted by $I^0(K)$, and the set of all cusp forms in $\mathcal{M}_m(K)$ by $\mathcal{S}_m(K)$. For two elements f and g of $\mathcal{S}_m(K)$, we define their inner product $\langle f, g \rangle$ by

$$(0.4) \qquad \langle f, g \rangle = \text{vol}(D)^{-1} \int_D \bar{f}(z) g(z) \, \text{Im}(z)^m d^* z \quad (D = \Gamma \backslash H^{J(K)})$$

with a sufficiently small congruence subgroup Γ which makes the integral meaningful, where

$$(0.5) \qquad d^* z = \prod_{\tau \in J(K)} y_\tau^{-2} dx_\tau dy_\tau \quad (z_\tau = x_\tau + iy_\tau).$$

We employ $\langle f, g \rangle$ even for C^∞ automorphic forms f and g whenever the integral is convergent.

The Hecke operators $\mathfrak{T}(\mathfrak{a})$ acting on quaternionic automorphic forms (including Hilbert modular forms of integral weight) are the same as those in [78, p. 648], [80, (9.24)], and [81, II, p. 576] (see (6.22) below). It should be noted that $\mathfrak{T}(\mathfrak{a})$ is $N(\mathfrak{a})$ times that of [87b]. In the elliptic modular case, $n^{(k/2)-1} T(n\mathbb{Z})$ coincides with the one defined by Hecke.

The algebraic closure and the maximal abelian extension of $\mathbb{Q}$, both in $\mathbb{C}$, are denoted by $\bar{\mathbb{Q}}$ and $\mathbb{Q}_{\text{ab}}$. We write $a \sim b$ for $a, b \in \mathbb{C}$ if $b \neq 0$ and $a/b \in \bar{\mathbb{Q}}$.

The adele ring and the idele group of an algebraic number field F are denoted by $F_{\mathbf{A}}$ and $F_{\mathbf{A}}^\times$, respectively. By a *Hecke character* of F, we understand a continuous homomorphism of $F_{\mathbf{A}}^\times$ into $\{z \in \mathbb{C} \mid |z| = 1\}$ which is trivial on $F^\times$. By the *conductor* of such a character, we always mean the finite part, ignoring the infinite part.

For a 2×2-matrix x, we often denote by a_x, b_x, c_x, and d_x the entries of x in the standard order, whenever there is no fear of confusion.

1. The critical values of a Dirichlet series attached to a Hilbert modular form and several CM-fields

Let $K = \prod\limits_{i=1}^{t} K_i$ and $F = \prod\limits_{i=1}^{t} F_i$ with CM-fields K_i and their maximum real subfields F_i. We assume that the F_i have a common subfield E, and fix a subset ε of J_F and a subset ψ of J_K such that

$$(1.1) \qquad \mathrm{Res}_{K_i/F_i}(\psi \cap J(K_i)) = \varepsilon \cap J(F_i) \neq \emptyset \quad \textit{for every } i \textit{ and } \quad \mathrm{Res}_{F/E}(\varepsilon) = \iota_E.$$

We put $\varepsilon' = \iota_F - \varepsilon$. Our object of study in this section is an infinite series

$$(1.2) \qquad D(s) = [U] \sum_{xU} c(x)\omega_q(\mathrm{Tr}_{F/E}(\zeta x x^\varrho)) x^\varphi |x^\psi|^{-2s}.$$

Here $s \in \mathbb{C}$, U is a subgroup of U_E of finite index, the sum is taken over all $x \in K^\times/U$, $c \in \mathscr{L}(K)$, $\omega_q(a) = a^q \omega(a)$ with $0 \leq q \in I_E$ and with the Fourier coefficients $\omega(a)$ of an element

$$(1.3) \qquad \Omega(z) = \sum_{a \in E} \omega(a) \mathbf{e}_E(az) \quad (z \in H^{J(E)})$$

of $\mathscr{M}(l, E)$, $0 \leq l \in I^0(E)$, $\zeta \in F^\times$, $\varphi \in I_K$; ϱ denotes the automorphism of K whose restriction to each K_i is complex conjugation.

For various technical reasons, we assume:

$$(1.4\mathrm{a}) \qquad \zeta^\tau > 0 \quad \textit{for} \quad \tau \in \varepsilon \quad \textit{and} \quad \zeta^\tau < 0 \quad \textit{for} \quad \tau \in \varepsilon';$$

$$(1.4\mathrm{b}) \qquad \varphi_\sigma > \varphi_{\sigma\varrho} \quad \textit{for every} \quad \sigma \in \psi; \quad \varphi_\sigma \geq 0 \quad \textit{for} \quad \sigma \in J_K - \psi - \psi\varrho;$$

$$(1.5\mathrm{a}) \qquad c(ux) = c(x) \quad \textit{and} \quad \omega(ux) = u^{l/2}\omega(x) \quad \textit{for every} \quad u \in U;$$

$$(1.5\mathrm{b}) \qquad \mathrm{Res}_{K/E}(\varphi) + l + 2q = \beta\iota_E \quad \textit{with} \quad \beta \in (1/2)\mathbb{Z}.$$

Observe that $\beta \notin \mathbb{Z}$ if and only if $l \notin I_E$, that is, if and only if l is half-integral.

Given c and Ω, we can always find a subgroup U of U_E of finite index satisfying condition (1.5a), which makes each term of D dependent only on xU. Because of the factor $[U]$, our series is independent of the choice of U. We allow Ω to be a constant, in which case $l = 0$ and the sum is extended over all xU such that $\mathrm{Tr}_{F/E}(\zeta x x^\varrho) = 0$, $x \neq 0$.

To smooth our exposition, we fix, once for all, a CM-type Ψ of K containing ψ such that $\varphi_\sigma \geq \varphi_{\sigma\varrho}$ for every $\sigma \in \Psi$, and put $\psi' = \Psi - \psi$.

Proposition 1.1. *The series D is convergent for sufficiently large $\mathrm{Re}(s)$, and can be continued as a meromorphic function of s to the whole plane. It is holomorphic for $\mathrm{Re}(s) \geq (\beta + [F:E] - 1)/2$ except for a possible simple pole at $(\beta + [F:E])/2$ if $\beta \in \mathbb{Z}$, and at $(2\beta + 2[F:E] - 1)/4$ if $\beta \notin \mathbb{Z}$. The pole at $(\beta + [F:E])/2$ occurs only when*

$$(1.6) \qquad \mathrm{Res}_{K/E}\left(\sum_{\sigma \in \psi} \varphi_\sigma \sigma\right) \leq (1/2)(\beta + [F:E] - 2)\iota_E$$

$$\leq q + \mathrm{Res}_{K/E}\left(\sum_{\sigma \in \psi} \varphi_\sigma \sigma + \sum_{\sigma \in \psi'} \varphi_{\sigma\varrho}\sigma\right).$$

As our proof will show, this proposition holds with the condition

(1.7) $\qquad\qquad\qquad \varphi_\sigma \geqq \varphi_{\sigma\varrho} \quad$ *for every* $\quad \sigma \in \psi$;

$\qquad\qquad\qquad\quad \varphi_\sigma \geqq 0 \quad$ *for* $\quad \sigma \in J_K - \psi - \psi\varrho$; $\ \varphi \notin (1+\varrho)I_K$;

instead of (1.4b). See also Theorem 4*.1 for a generalization.

Our main results of this section will concern the values and the residue of D at certain points. Naturally we assume

(1.8) $\quad c(x)$ *and* $\omega(a)$ *are algebraic numbers for every* $x \in K$ *and* $a \in E$.

We also employ the following symbols throughout this section:

(1.9) $\qquad\qquad\qquad m = [E : \mathbb{Q}], \quad e = \sum_{\sigma \in \psi} (\varphi_\sigma - \varphi_{\sigma\varrho}) .$

Theorem 1.2. *Under* (1.4a, b) *and* (1.5a, b), *let* μ *be an integer such that*

(1.10a) $\qquad\qquad (\beta + [F : E] - 1)/2 \leqq \mu \leqq \varphi_\sigma \quad$ *for every* $\quad \sigma \in \psi$,

(1.10b) $\qquad\qquad 2\mu \neq \beta + [F : \mathbb{Q}] - (1/2) \quad$ *if* $\quad E = \mathbb{Q}$.

Then D *is finite at* μ, *and* $D(\mu) \sim \pi^e p_K(\varphi, 2\psi)$.

Theorem 1.3. *Let the assumptions be the same as above.*

(i) *Suppose* $\beta \in \mathbb{Z}$ *and* $2\varphi_\sigma = \beta + [F : E] - 2$ *for every* $\sigma \in \psi$; *suppose further*

(1.11) $\qquad\qquad\qquad l \neq 0 \quad$ *when* $\quad [F : E] = 2$.

Then the residue of D *at* $(\beta + [F : E])/2$ *is an algebraic number times* $\pi^e R_E p_K(\varphi, 2\psi)$.

(ii) *Suppose* $2\beta + 2[F : E] \equiv 3 \pmod 4$, *and* $2\varphi_\sigma \geqq \beta + [F : E] - (3/2)$ *for every* $\sigma \in \psi$. *Then the residue of* D *at* $(2\beta + 2[F : E] - 1)/4$ *is an algebraic number times* $\pi^{e-m} R_E p_K(\varphi, 2\psi)$.

The proof of these theorems is inseparably connected with that of arithmeticity of the series f introduced in [86, §4]. To illustrate the mutual dependence of our theorems, let us symbolize them as follows:

$$
\begin{aligned}
(D) \ \ &= \text{Theorem 1.2,} \\
(DR) \ \ &= \text{Theorem 1.3 } (R \text{ means "residue"}), \\
(D_0) \ \ &= \text{Theorem 1.2 with } F = E, \\
(DR_0) &= \text{Theorem 1.3 with } F = E, \\
(f) \ \ \ \ \ &= [86, \text{Theorem 4.5}], \\
(fR) \ \ &= [86, \text{Theorem 4.6}], \\
(f_0) \ \ &= [86, \text{Theorem 4.5}] \text{ with } F = E, \\
(fR_0) &= [86, \text{Theorem 4.6}] \text{ with } F = E.
\end{aligned}
$$

Our proof will be given in the following way:

$$(D_0) \Rightarrow (f_0) \Rightarrow (D) \Rightarrow (f),$$

$$(DR_0) \Rightarrow (fR_0) \Rightarrow (DR) \Rightarrow (fR).$$

The implication $(D) \Rightarrow (f)$ includes $(D_0) \Rightarrow (f_0)$ as a special case; the same is true for (DR) and (fR). Therefore our task is to prove (D_0) and (DR_0), and then show that

$(f_0) \Rightarrow (D) \Rightarrow (f)$ and $(fR_0) \Rightarrow (DR) \Rightarrow (fR)$. This idea is the same as in [80], [81], and [83a]. In fact, (D) and (DR) with $\beta \in \mathbb{Z}$ and the corresponding cases of (f) and (fR) were proved in those papers. However we always had to exclude the special case of (D) in which

$$(1.12) \qquad\qquad E = \mathbb{Q} \neq F, \quad \beta + [F:E] = 2\mu,$$

and also the corresponding special case of f, in which the series $f(w, 0)$ becomes nonholomorphic. Thus the new points of the present paper on D and f are the inclusion of the case of half-integral β and also the special case (1.12). In order to deal with half-integral β, we employ the Eisenstein series of $SL_2(E)$ of half-integral weight, investigated in [85a, b]. As for (1.12), it corresponds to nonholomorphic (or rather, nearly holomorphic) f; even (fR) involves nearly holomorphic functions. To overcome the difficulty of dealing with such nonholomorphy, we proved in [87a] a theorem of arithmeticity of nearly holomorphic automorphic forms, which can be effectively applied to the special cases of (f) and (fR) involving such forms.

In this section, we shall first prove Proposition 1.1 and then (D_0) and (DR_0). We add two technical remarks here:

(I) In order to prove Proposition 1.1, (D), and (DR), we may assume $q=0$ for the same reason as explained in [81, I, Remark 4.4]. The consideration of ω_q instead of ω is made for the purpose of smoothing the proof of $(D) \Rightarrow (f)$ and that of $(DR) \Rightarrow (fR)$.

(II) Condition (1.11) may be replaced by

$$(1.13) \qquad\qquad l_\tau > 2 - [F:E] \quad \text{for every} \quad \tau \in J_E.$$

In fact, (1.13) is trivially true if $[F:E] > 2$; it is equivalent to (1.11) if $[F:E] = 2$. Suppose $F = E$. By the above remark, we may assume $q = 0$. Then $\varphi_\sigma + \varphi_{\sigma\varrho} + l_\tau = \beta$ for every $\sigma \in \psi$, where $\tau = \operatorname{Res}_{K/E}(\sigma)$. If φ_σ is as in (i) of Theorem 1.3, we have $\beta = 2\varphi_\sigma + 1$, and hence $l_\tau = \varphi_\sigma - \varphi_{\sigma\varrho} + 1 > 1$ which implies (1.13).

Proof of Proposition 1.1. Let us assume (1.7) instead of (1.4b). The convergence of D was proved in [80, Proposition 9.1] and [83a, §2], and the meromorphic continuation when $\beta \in \mathbb{Z}$ in [80, Theorem 9.2], [81, I, Theorem 4.2], and [83a, §2]. The case of half-integral β can be proved by the same idea by employing the Eisenstein series of half-integral weight of [85a, b]. We give here a uniform (if somewhat sketchy) proof applicable to both cases. Put

$$\xi = \sum_{\sigma \in \Psi} (\varphi_\sigma - \varphi_{\sigma\varrho})\sigma, \quad p = \sum_{\sigma \in \psi} \varphi_{\sigma\varrho}\sigma, \quad h' = \sum_{\sigma \in \psi'} \varphi_{\sigma\varrho}\sigma,$$

$$h = \operatorname{Res}_{K/F}(h'), \quad v = \iota_F + \operatorname{Res}_{K/F}(\xi),$$

$$f(z) = \sum_{b \in K} c(b) b^\xi \mathbf{e}_F(bb^\varrho z) \quad (z \in H^{J(F)}).$$

Then $\varphi = \xi + (p + h')(1 + \varrho)$ and $f \in \mathcal{M}(v, F)$. Thus $f(z)$ is a linear combination of products $f_1(z_1) \ldots f_t(z_t)$ with Hilbert modular forms $f_i(z_i)$ on $H^{J(F_i)}$, where $z = (z_i)$, $z_i \in H^{J(F_i)}$. The condition $\varphi \notin (1 + \varrho)I_K$ of (1.7) implies that at least one of the f_i is a cusp form. Define g^ζ as in [83a, (1.5), (1.6)]. As shown in [81, I, (3.7)], we have

$$(\mathbf{d}^h f)^\zeta(z) = \sum_{b \in K} c(b) b^{\xi + h'(1+\varrho)}$$

$$\cdot \mathbf{e}_E(x \operatorname{Tr}_{F/E}(\zeta b b^\varrho) + iy[2\zeta b b^\varrho - \operatorname{Tr}_{F/E}(\zeta b b^\varrho)]) \quad (z = x + iy \in H^{J(E)}).$$

Put $\Omega_\varrho(z) = \sum_a \overline{\omega(a)} \mathbf{e}_E(az)$. Take a lattice $\mathfrak{a}$ in E so that both Ω_ϱ and $(\mathbf{d}^h f)^\zeta$ are invariant under $z \mapsto z + a$ for every $a \in \mathfrak{a}$ and $U\mathfrak{a} \subset \mathfrak{a}$. Put $L = \mathbb{R}^{J(E)}/\mathfrak{a}$. Then we see that

$$\operatorname{vol}(L)^{-1} \int_L \overline{\Omega_\varrho(z)} (\mathbf{d}^h f)^\zeta(z) dx$$

$$= \sum_{b \in K} c(b) \omega(\operatorname{Tr}_{F/E}(\zeta b b^\varrho)) b^{\xi + h'(1+\varrho)} \mathbf{e}_E(2iy\zeta b b^\varrho).$$

Express the right-hand side times $y^{(s+1)\iota - d}$ with $\iota = \iota_E$ and $d = \operatorname{Res}_{K/E}(p)$ in the form $\sum_{b \in K} t(b, y)$, and observe that $t(ub, y) = t(b, u^2 y)$ for every $u \in U$. Therefore, putting $M = \Gamma'_\infty \backslash H^{J(E)}$ with

$$\Gamma'_\infty = \left\{ \begin{pmatrix} u & a \\ 0 & u^{-1} \end{pmatrix} \middle| u \in U, a \in \mathfrak{a} \right\},$$

we find that

$$(1.14) \qquad [U] \int_M \overline{\Omega_\varrho(z)} (\mathbf{d}^h f)^\zeta(z) y^{(s+1)\iota - d} d*z$$

$$= \operatorname{vol}(\mathbb{R}^{J(E)}/\mathfrak{a}) D(s) \prod_{\sigma \in \psi} (4\pi\zeta^\sigma)^{\varphi_{\sigma\varrho} - s} \Gamma(s - \varphi_{\sigma\varrho}).$$

By [81, I, (1.16a)], we have $\mathbf{d}^h f = \sum_{0 \le k \le h} a_k (\pi y)^{k-h} D_v^k f$ with rational constants a_k, where D_*^* is defined in [86, p. 402]. Therefore, the integral of (1.14) is a linear combination of the integrals

$$(1.15) \qquad \int_M \bar{\Omega}_\varrho (D_v^k f)^\zeta y^{(s+1)\iota + u} d*z$$

with $u = \operatorname{Res}_{F/E}(k - h) - d$. Let $\Phi = \Gamma \backslash H^{J(E)}$ with a sufficiently small congruence subgroup Γ of $SL_2(E)$. Then a well known technique transforms (1.15) to a positive rational number times

$$\int_\Phi \bar{\Omega}_\varrho (D_v^k f)^\zeta \mathscr{E}_\alpha^A(z, s+1) y^{A+u} d*z,$$

where $\mathscr{E}_\alpha^A$ is defined by [86, (10.14b)], $\alpha = -\beta - [F:E]$, and

$$A = (\beta + [F:E] - 1)\iota - \operatorname{Res}_{K/E}\left(\sum_{\sigma \in \psi} \varphi_\sigma \sigma \right) - \operatorname{Res}_{F/E}(h - k).$$

Therefore our proposition follows from [86, Lemma 10.2].

Proof of (D_0) *and* (DR_0). Suppose $F = E$ and $q = 0$. Then K is a CM-field, ψ is a CM-type of K, and D has the form

$$D(s) = [U] \sum_{xU} c(x)\omega(\zeta x x^\varrho) x^\varphi N_{K/\mathbb{Q}}(x)^{-s}.$$

If Ω is a constant, this is trivially equal to 0. For this reason, we may assume $l \neq 0$. Now take $F = E$ in the proof of Proposition 1.1. Then $\varepsilon = \iota$, $\Psi = \psi$, and $h = 0$. Put $g(z) = f(\zeta z)$. Since this is $(\mathbf{d}^h f)^\zeta$, (1.14) takes the form

$$(1.16) \qquad [U] \int_M \overline{\Omega_\varrho(z)} g(z) y^{(s+1)\iota - d d^*} z$$

$$= \mathrm{vol}(\mathbb{R}^{J(E)}/\mathfrak{a}) D(s) \prod_{\sigma \in \psi} (4\pi\zeta^\sigma)^{\varphi_{\sigma\varrho} - s} \Gamma(s - \varphi_{\sigma\varrho}),$$

which is equal to a positive rational number times

$$(1.17) \qquad \int_\Phi \bar{\Omega}_\varrho \cdot g \cdot \mathscr{E}_\alpha^A(z, s+1) y^l d^* z,$$

where $A = \beta\iota - \mathrm{Res}_{K/E}\left(\sum_{\sigma \in \psi} \varphi_\sigma \sigma\right) = l + \mathrm{Res}_{K/E}\left(\sum_{\sigma \in \psi} \varphi_{\sigma\varrho} \sigma\right)$ and $\alpha = -\beta - 1$. Given an integer μ, let $\lambda = 2\mu + 1 - \beta$ and $r = (\beta - \mu)\iota - A$. Observe that $y^l \mathscr{E}_\alpha^A(z, \mu+1) = y^v \mathscr{E}_\lambda^r(z, 0)$. Hence, putting $v = \sum_{\sigma \in \psi} \varphi_{\sigma\varrho} - m\mu$, we find that

$$(1.18) \qquad \pi^v \prod_{\sigma \in \psi} \Gamma(\mu - \varphi_{\sigma\varrho}) D(\mu) \sim \int_\Phi \bar{\Omega}_\varrho \overline{\mathscr{E}_\lambda^r(z, 0)}\, g y^v d^* z$$

$$= \mathrm{vol}(\Phi)\langle \Omega_\varrho \mathscr{E}_\lambda^r(z, 0), g \rangle.$$

Observe that $r \in I_E$ and

$$(1.19) \qquad D_\lambda^r \mathscr{E}_\lambda^0(z, s) = (-4\pi)^{-\|r\|} \prod_{\tau \in J(E)} \frac{\Gamma(s + \lambda + r_\tau)}{\Gamma(s + \lambda)} \mathscr{E}_\lambda^r(z, s)$$

if $r \geq 0$. Now we need a few lemmas.

Lemma 1.4. (i) *If* $1 \leq \lambda \in (1/2)\mathbb{Z}$, $\mathscr{E}_\lambda^0(z, s)$ *is finite at* $s = 0$. *Moreover,* $\mathscr{E}_\lambda^0(z, 0)$ *is a* $\mathbb{Q}_{\mathrm{ab}}$-*rational holomorphic modular form except in the following two cases:* (A) $\lambda = 3/2$ *and* $E = \mathbb{Q}$; (B) $\lambda = 2$ *and* $E = \mathbb{Q}$.

(ii) *If* $\lambda = 2$ *and* $E = \mathbb{Q}$, *we have* $\mathscr{E}_\lambda^0(z, 0) = b/(\pi y) + \sum_{a \in \mathbb{Q}} c(a)\mathbf{e}(az)$ *with* b *and* $c(a)$ *in* $\mathbb{Q}_{\mathrm{ab}}$.

(iii) *If* $\lambda = 1/2$, $\mathscr{E}_\lambda^0(z, s)$ *has at most a simple pole at* $s = 1/2$, *and the residue is* $\pi^{-m} R_E$ *times a* $\mathbb{Q}_{\mathrm{ab}}$-*rational holomorphic modular form.*

(iv) *If* $\alpha\iota + 2A = 0$, $\mathscr{E}_\alpha^A(z, s)$ *has at most a simple pole at* $s = 1 - (\alpha/2)$; *its residue is* $\pi^{-m} R_E D_E^{1/2}$ *times a positive rational number.*

These are well known when $\lambda \in \mathbb{Z}$ or $\alpha \in \mathbb{Z}$. The cases of half-integral λ and α have been proved in [85a, Theorem 2.3] (cf. also [85b, Propositions 4.3 and 8.2]).

Lemma 1.5. *Let* $g(z) = \sum_{b \in K} \eta(b) b^\xi \mathbf{e}_E(\zeta b b^\varrho z)$ *with* $\eta \in \mathscr{L}(K)$, $\xi = \sum_{\sigma \in \psi} \xi_\sigma \sigma$, *and a totally positive* $\zeta \in E$. *Suppose* η *is* $\bar{\mathbb{Q}}$-*valued,* $[K : E] = 2$, ψ *is a CM-type of* K, *and* $\xi_\sigma > 0$ *for every* $\sigma \in \psi$. *Then* $\langle h, g \rangle \sim \pi^{-m} p_K(\xi, 2\psi)$ *for every* $\bar{\mathbb{Q}}$-*rational holomorphic* $h \in \mathscr{M}(\mathrm{Res}_{K/E}(\xi) + \iota, E)$.

This is a restatement of [80, Theorem 9.8].

Lemma 1.6. *If Γ is a congruence subgroup of $SL_2(E)$, then $\mathrm{vol}(\Gamma \backslash H^{J(E)})$ is a rational number times π^m.*

This is well known (see [85b, Proposition 4.3], for example).

Lemma 1.7. *If f is a cusp form belonging to $\mathcal{M}(k, E)$ and $h \in \mathcal{M}(k - 2s, E)$ with $s \neq 0$, then $\langle D_{k-2s}^s h, f \rangle = 0$.*

This follows immediately from [81, I, Lemma 2.3].

To continue our proof, take μ as in (1.10a). Then $\lambda \geq 1$ and $r \geq 0$. Assume that $E \neq \mathbb{Q}$ or $\lambda \notin \{3/2, 2\}$. By Lemma 1.4, $\mathscr{E}_\lambda^0(z, 0)$ is $\bar{\mathbb{Q}}$-rational. Therefore (1.20) shows that $\pi^{-\|r\|} \Omega_\varrho \mathscr{E}_\lambda^r(z, 0)$ is a polynomial in $\mathrm{Im}(z)$ with holomorphic coefficients and that it is arithmetic. Hence we have, by [87a, Theorems 5.4 and 5.5],

$$\pi^{-\|r\|} \Omega_\varrho \mathscr{E}_\lambda^r(z, 0) = \sum_{0 \leq s \leq r} D_{v-2s}^s h_s,$$

where h_s is a $\bar{\mathbb{Q}}$-rational element of $\mathcal{M}(v - 2s, E)$. Now our assumption on μ together with (1.5b) implies that $\mu - \varphi_{\sigma\varrho} \geq l_\tau + \varphi_\sigma - \mu \geq l_\tau > 0$ for every $\sigma \in \psi$, where $\tau = \mathrm{Res}_{K/E}(\sigma)$. Hence we obtain

$$D(\mu) \sim \pi^{\|r\| - v + m} \left\langle \sum_{0 \leq s \leq r} D_{v-2s}^s h_s, g \right\rangle \quad \text{by (1.18) and Lemma 1.6}$$

$$\sim \pi^{\|r\| - v + m} \langle h_0, g \rangle \qquad \text{by Lemma 1.7}$$

$$\sim \pi^e p_K(\varphi, 2\psi) \qquad \text{by Lemma 1.5,}$$

since $p_K(\varphi, 2\psi) = p_K(\xi, 2\psi)$ by [80, Theorem 1.1, (2)].

If $E = \mathbb{Q}$ and $\lambda = 2$, $\mathscr{E}_\lambda^0(z, 0)$ is of the type described in (ii) of Lemma 1.4. Since Ω_ϱ is of weight $l > 0$, our argument is still applicable to this case with $0 \leq s \leq r + 1$ instead of $0 \leq s \leq r$; we then obtain the same conclusion. This completes the proof of Theorem 1.2 when $F = E$.

To prove Theorem 1.3 with $F = E$, suppose β, φ, and l are given as in (i). Then $\varphi_\sigma = (\beta - 1)/2$ and $\varphi_{\sigma\varrho} = (\beta + 1)/2 - l_\tau$ for every $\sigma \in \psi$, $\alpha = -2A$, and $v = l$. By Lemma 1.6 and (iv) of Lemma 1.4, the function of (1.17) has at most a simple pole at $(\beta + 1)/2$, and the residue is an algebraic number times $R_E \langle \Omega_\varrho, g \rangle$. Therefore we obtain (i) of Theorem 1.3 when $F = E$.

Next take β and φ as in (ii); put $r = (2\beta + 1)(\iota/4) - A$. Then $0 \leq r \in I_E$ and

$$y^l \mathscr{E}_\alpha^A(z, s + 1) = y^v \overline{\mathscr{E}_{1/2}^r(z, \bar{s} + (1 - 2\beta)/4)} .$$

Applying (1.19) to the right-hand side and putting $S = s + (1 - 2\beta)/4$, we see that (1.17) is equal to

$$(-4\pi)^{\|r\|} \prod_{\tau \in J(E)} \frac{\Gamma(S + 1/2)}{\Gamma(S + 1/2 + r_\tau)} \, \mathrm{vol}(\Phi) \langle \Omega_\varrho D_{1/2}^r \mathscr{E}_{1/2}^0(z, \bar{S}), g \rangle .$$

By (iii) of Lemma 1.4, the last inner product has at most a simple pole at $S = 1/2$, and its residue is of the form $\pi^{-m} R_E \langle \Omega_\varrho D_{1/2}^r h, g \rangle$ with a $\bar{\mathbb{Q}}$-rational holomorphic form h of weight $1/2$. Therefore we obtain (ii) of Theorem 1.3 when $F = E$ by the same technique as in the above proof of Theorem 1.2.

2. Proof of Theorems 1.1, 1.2, and the arithmeticity theorems of [86]

Proof of $(f_0) \Rightarrow (D)$. This can be done by practically repeating the argument of [80, §12] and [81, I, pp. 156–157] (see also [83a, p. 262]). We only have to make sure that the method works well for half-integral β and also in the special case (1.12). To be explicit, let the symbols be the same as in Theorem 1.2; put

$$(2.1a) \qquad \Xi = \sum_{\sigma \in \psi} (\varphi_\sigma - \mu)(\sigma + \sigma\varrho) + \sum_{\sigma \in \psi'} (\varphi_\sigma \sigma + \varphi_{\sigma\varrho} \sigma\varrho),$$

$$(2.1b) \qquad \Delta = \sum_{\sigma \in \psi} (\varphi_\sigma - \varphi_{\sigma\varrho})\sigma, \quad d = \mathrm{Res}_{K/E}(\Delta), \quad e = \mathrm{Res}_{K/E}(\Xi).$$

Then $\Xi \geqq 0$, $\Delta \geqq 0$, and $\varphi - \mu\psi(1+\varrho) = \Xi - \Delta\varrho$. Therefore

$$(2.2) \qquad D(s+\mu) = [U] \sum_{xU} c(x)\omega(\mathrm{Tr}_{F/E}(\zeta x x^\varrho))x^{\Xi - \Delta\varrho}|x^\phi|^{-2s}.$$

We are assuming $q=0$ here for the reason explained in Sect. 1.

Put $[K:E] = n+2$ and $S(x,y) = -\mathrm{Tr}_{K/E}(\zeta x y^\varrho)$ for $x, y \in K$. Since assertions (D_0) and (DR_0) in which $n=0$ have been proved, we may assume $n \geqq 2$ in the proof of (D) and (DR). Viewing K as a vector space over E, we define $G(S)$ as in [86, §4] with $F = E$ and $V = K$. Let $h : K \to \mathrm{End}_E(V)$ be the regular representation of K over E, and let $K^u = \{x \in K | x x^\varrho = 1\}$. Then $h(K^u)$ has a fixed point w_0 on the domain $\mathscr{Z}$. Let V_σ denote, for each $\sigma \in \psi$, the completion of V over E with respect to $\mathrm{Res}_{K/E}(\sigma)$. Then $V \otimes_{\mathbb{Q}} \mathbb{R}$ can be identified with $\prod_{\sigma \in \psi} V_\sigma$ in view of (1.1). Define a polynomial function g on $\prod_{\sigma \in \psi} V_\sigma$ so that $g(x) = x^\Xi$ for $x \in K = V$. Repeating the argument of [80, pp. 364–365] or [81, I, p. 157], we find that $D(\mu)$, or rather the value of (2.2) at $s = 0$, is a finite $\bar{\mathbb{Q}}$-linear combination of the quantities of the form $[D_k^{(e)} f(w_0, 0)](u)$, where u is a $\bar{\mathbb{Q}}$-rational element of $T^e = \prod_\tau T_\tau^{e_\tau}$ ($T_\tau =$ the tangent space of $\mathscr{Z}_\tau$), $D_k^{(e)}$ is the operator of Sect. 2, and

$$(2.3) \quad f(w, s) = [U] \sum_{0 \neq v \in V/U} c(v)\omega(-S[v])S[v]^p \mu[v, w]^{-k} \eta(w)^{s\iota} |\mu[v, w]|^{-2s\iota}$$

as in [86, (4.10)] with $k = d - e + 2p$, $0 \leqq p \leqq e/2$, $\iota = \iota_E$; $u = (u_\alpha, \dots, u_\zeta)$ as explained in [80, Lemma 12.1].

Define the constants a_f and b_f by [86, (4.12) and (4.14)]. Then $b_f = 2\mu + 2 - \beta - [F:E]$. Thus, from (1.10a,b), we obtain $b_f \geqq 1$ and $b_f \neq 3/2$ if $E = \mathbb{Q}$. Therefore $\pi^{-\|k\|} f(w, 0)$ is arithmetic by (f_0), so that the conclusion of (D) follows in exactly the same fashion as in [80, p. 365]. In the exceptional case (1.12), $f(w, 0)$ is nearly holomorphic; also $2\beta \equiv 2b_f \pmod 2$. In any case, the arithmeticity is guaranteed by (f_0) and [87a, Theorem 3.7].

Proof of $(fR_0) \Rightarrow (DR)$. This can be given by modifying the above proof. (In [81, I, §9], a different, more complicated method was adopted when $\beta \in \mathbb{Z}$. Here we choose a simpler way.) We again define Ξ, Δ, d, and e by (2.1a,b) with $\mu = (\beta + [F:E] - 2)/2$, under the assumptions of Theorem 1.3, (i). Then $\Xi \geqq 0$, $\Delta \geqq 0$, and (2.2) holds again.

Put

$$L_{k,s}^{(e)}(x,u) = D_k^{(e)}[\eta(w)^{s\iota}\mu[x,u]^{-k-s\iota}]_{w=w_0}(u),$$

where $0 \leq e$, $k \in I_E$, $s \in \mathbb{C}$, $x \in V$, $u \in T^e$, and $w \in \mathscr{Z}$. Then $L_{k,1}^{(e)}$ is essentially the same as $L_{k+\iota,0}^{(e)}$ for the reason explained in [80, Lemma 11.1]. Therefore, [80, Lemma 12.1] implies the following fact: *if $d_\tau - e_\tau > (n-4)/2$ for every $\tau \in J_E$, then $x^{\Xi - \varrho\Delta}$ is a finite $\overline{\mathbb{Q}}$-linear combination of $S[x]^p L_{d-e+2p,1}^{(e-2p)}(x,u)$ with $0 \leq p \leq e/2$ and $u = (u_\alpha, \ldots, u_\zeta) \in T^e$ as given in that lemma.*

Now, with d and e given by (2.1a,b), we have $d - e = l + (n-2)(\iota/2) \geq \mathrm{Max}(1,(n-2)/2)\iota$ because of (1.11). Therefore we find that the residue of $D(s+\mu)$ at $s=1$ is a finite $\overline{\mathbb{Q}}$-linear combination of the residue of $[D_k^{(e-2p)}f(w_0,s)](u)$ at $s=1$, where f is defined by (2.3) with $k = d-e+2p$. Since $k \geq \iota$ and $b_f = 0$, the residue of $\pi^{-\|k\|}R_E^{-1}f(w,s)$ at $s=1$ is arithmetic by (f_0), and hence [87a, Theorem 3.7] implies the arithmeticity of the residue of $\pi^{\|2p-e-k\|}R_E^{-1}D_k^{(e-2p)}f(w_0,s)$. From this we can derive the desired conclusion of (DR) when $\beta \in \mathbb{Z}$.

The case of half-integral β can be proved in the same way. Let β, φ, and l be given as in Theorem 1.3, (ii). Define Ξ, Δ, d, and e by (2.3) with $\mu = (\beta + [F:E])/2 - 3/4$. Then $\Xi \geq 0$. Observe that [80, Lemma 12.1] is valid even with half-integral d_i. Therefore, taking $L_{*,1/2}^*$ instead of $L_{*,1}^*$, we obtain the assertion of Theorem 1.3, (ii) from (f_0) and [87a, Theorem 3.7] in the same fashion as for integral β.

Proof of $(D) \Rightarrow (f)$. The principal idea is as follows. Define the series f as in [86, (4.10)]. We then take a *CM*-point w_0 of $\mathscr{Z}$ of a special type and observe that $f(w_0,0)$ is given as $D(\mu)$ with a suitable D of type (1.2). Then (D) implies the arithmeticity of $cf(w,0)$ at w_0 with a certain constant c as specified in (f). This argument is valid for $\alpha(w_0)$ in place of w_0 for every $\alpha \in G_+(S)$. Therefore [87a, Theorem 3.7] establishes the arithmeticity of $cf(w,0)$, since $f(w,0)$ belongs to $\mathscr{N}^{(k)}$ for some k as proved in [86, Theorems 4.1 and 4.2]. The same process was taken in the proof given in [80, §10] and [81, I, §8], though only holomorphic functions and the series D for which K is a field were considered there. In the present situation, the group $G(S)$ is a product $\prod_{i=1}^{t} G(S_i)$. Therefore we choose w_0 in the following way. Let the notation be the same as in [86, §4]. For each i, take an F_i-algebra Y_i and an F_i-linear embedding h_i of Y_i into $\mathrm{End}(V_i,F_i)$. As shown in [80, Lemma 10.1], we can take Y_i so that $Y_i = L_i \oplus Y_{i0}$ with a *CM*-field L_i and $Y_{i0} \cong F_i^{n_i}$, $[L_i:F_i]=2$ and that $h_i(Y_i^u)$ has a fixed point w_i on $\mathscr{Z}_i$. Then we put $w_0 = (w_1, \ldots, w_t)$. Modifying the procedure of [80, §10] in a more or less straightforward (if somewhat lengthy) way, we can show the desired fact about $f(w_0,0)$ mentioned above.

Proof of $(DR) \Rightarrow (fR)$. This is essentially the same as above, except that we take the residues of D and f instead of their values. We have to verify that the various conditions in Theorem 1.3, (i) and (ii) are satisfied, which can easily be done by virtue of the assumptions made in [86, Theorem 4.6].

Proof of [86, Theorem 5.4]. This concerns the arithmeticity in the *unitary* case. The idea of the proof is the same as in the orthogonal case. More specifically, we take a direct sum L of *CM*-fields containing K such that $[L:K]=l+1$, and take a K-linear embedding h of L into $M_{l+1}(K)$ such that $h(x^\varrho)H = Hh(x)^*$ for every $x \in L$. Then

$h(\{x \in L | xx^\varrho = 1\})$ has a unique common fixed point w_0 on $\mathscr{B}$. Let $f(w, s)$ be given as in [86, (5.11)] with $m = 1$. Then we observe that $f(w_0, s)$ is a finite $\bar{\mathbb{Q}}$-linear combination of series D of type (1.2) defined with L as K there. Define β for these D by (1.5b), and κ as in [86, Theorem 5.2]. Then $\kappa = 2 - \beta - [L:E]/2$. If $\kappa \geq 1$, we can apply Theorem 1.2 to $D(\mu)$ with $\mu = 0$ to obtain the desired arithmeticity of $f(w, 0)$ at w_0 up to the constant C given in [86, Theorem 5.4].

That the residue of $f(w, s)$ at $s = 1$, when $\kappa = 0$, has a similar property can be proved in the same way by means of Theorem 1.3, (i). There is one nontrivial point: the condition that $l \neq 0$ when $[F:E] = 2$ stated there is not necessarily satisfied in the present case. However, we can show, in this exceptional case, that $\varphi = \psi$ and $f(w_0, s)$ is a finite $\bar{\mathbb{Q}}$-linear combination of some products of the form

$$\mathscr{E}_0^0(z, s) \sum_{c \in K/U} \lambda(c)(c^\varrho/c)^k |c^\varphi|^{-2s},$$

where $\mathscr{E}_*^*$ is defined by [86, (10.14b)] and $\lambda \in \mathscr{L}(K)$. Applying Theorem 1.2 to the last sum with $\mu = 1$, we obtain the desired property of the residue of $f(w_0, s)$ at $s = 1$.

Let us conclude this section by noting the plausibility of more general results than [86, Theorem 4.5] which concerns the arithmeticity of $f(w, s)$ at $s = 0$. Namely, it is conjecturable that $f(w, s_0)$ is arithmetic up to a constant factor for the integer points s_0 in a certain critical strip. Of course such a property of arithmeticity must be preceded by the near holomorphy of $f(w, s_0)$, as remarked in [86, p. 373]. In fact, A. Bluher has shown in her thesis [B] such near holomorphy. It is our hope that the methods employed in this section will be effective also for the proof of the expected arithmeticity of $f(w, s_0)$.

We insert here some corrections to [86].

Page 393, (8.6), (8.7): ζ should be ζ_w.

Page 393, line 8: Insert "$\zeta_w(u) = \zeta(u\eta(w_\tau))$ and" after "where".

Page 393, (8.9): $\theta(z, w; \zeta)$ should be $\theta(z, w; \zeta_w)$.

Page 401, Lemma 10.1, (2): P_0 should be P_σ.

Page 401, the last line: $\mathbf{e}$ should be $\mathbf{e}_F$.

Page 403, line 4 from bottom: I_E should be J_E.

Page 406: The proof of Theorem 4.2 in the case $F = E = \mathbb{Q}$, $l = 0$, and $n_\tau = 2$ given on this page is erroneous. It should be as follows. In this case, f is the sum over the elements v of V such that $S[v] = 0$. Hence we may assume that $j = 0$ and $k_\tau = 2$; S is a quadratic form over $\mathbb{Q}$ of signature $(2, 2)$ that represents 0. If S is totally isotropic, we may take $V = M_2(\mathbb{Q})$ and $S[x] = \det(x)$ for $x \in V$. Otherwise we may take

$$(2.4) \qquad V = \left\{ \begin{bmatrix} a & b \\ c & -a^\sigma \end{bmatrix} \middle| b, c \in \mathbb{Q}, \, a \in K \right\},$$

where K is a real quadratic field and σ is the generator of $\mathrm{Gal}(K/\mathbb{Q})$; $S[x] = q \cdot \det(x)$ for $x \in V$ with $0 < q \in \mathbb{Q}$. In either case, put $V_0 = \{\alpha \in V | \mathrm{rank}(\alpha) = 1\}$, and

$$[\alpha; z, w] = (-1, z)\alpha \binom{w}{1} \qquad (\alpha \in V, \, (z, w) \in H^2).$$

Thus our series f is essentially the same as

$$f(z, w; s) = \mathrm{Im}(z)^s \mathrm{Im}(w)^s \sum_{\alpha \in V_0} p(\alpha)[\alpha; z, w]^{-2} |[\alpha; z, w]|^{-2s}$$

with $p \in \mathcal{L}(V)$. Let $v = \begin{bmatrix} 0 & 1 \\ 0 & 0 \end{bmatrix}$ and let R be a complete set of representatives of $P \backslash SL_2(K)$, where $P = \{\alpha \in SL_2(K) \mid c_\alpha = 0\}$. It can easily be seen that $(c, \gamma) \mapsto c\gamma^{-1} v\gamma^\sigma$ gives a bijective map of $\mathbb{Q}^\times \times R$ onto V_0 of (2.4). Take a congruence subgroup Γ of $SL_2(K)$ so that $p(\gamma^{-1} \alpha \gamma^\sigma) = p(\alpha)$ for every $\gamma \in \Gamma$. Let B be a complete set of representatives of $P \backslash SL_2(K) / \Gamma$ and let $R_\beta = (\beta \Gamma \beta^{-1} \cap P) \backslash \beta \Gamma$ for each $\beta \in B$. With $R = \bigcup_{\beta \in B} R_\beta$, we obtain

$$f(z, w; s) = \operatorname{Im}(z)^s \operatorname{Im}(w)^s \sum_{\beta \in B} \sum_{c \in \mathbb{Q}^\times} p(c\beta^{-1} v\beta^\sigma) |c|^{-2-2s}$$
$$\cdot \sum_{\gamma \in R_\beta} j(\gamma; z, w)^{-2} |j(\gamma; z, w)|^{-2s}.$$

The last sum over R_β is an Eisenstein series of $SL_2(K)$. Such a series at $s = 0$ is well known to be holomorphic in (z, w). (See [83b], for example.) This settles the case in which V has form (2.4). If $V = M_2(\mathbb{Q})$, we obtain, by a similar argument, the product of two Eisenstein series of $SL_2(\mathbb{Q})$ in place of a single series on $SL_2(K)$. Again a well known classical result shows that such a product at $s = 0$ is a linear combination of 1, $\operatorname{Im}(z)^{-1}$, $\operatorname{Im}(w)^{-1}$, and $\operatorname{Im}(z)^{-1} \operatorname{Im}(w)^{-1}$ with holomorphic functions as coefficients. This completes the proof.

3. The critical values of Dirichlet series attached to several systems of eigenvalues of Hecke operators and forms of half-integral weight

The series in the title are defined with respect to automorphic forms on several quaternion algebras. Let us first define them and introduce necessary symbols. Let B be a quaternion algebra over a totally real algebraic number field K; let α resp. β be the set of elements of J_K where B is unramified resp. ramified. Put $B_{\mathbb{R}} = B \otimes_{\mathbb{Q}} \mathbb{R}$. Then there is an $\mathbb{R}$-linear isomorphism

$$(3.1) \qquad B_{\mathbb{R}} \to M_2(\mathbb{R})^\alpha \times \mathbb{H}^\beta,$$

that coincides on K with the injection: $K \to K \otimes_{\mathbb{Q}} \mathbb{R} = \mathbb{R}^{J(K)}$ (see Sect. 0). Here $\mathbb{H}$ denotes the Hamilton quaternions, and $M_n(A)$ the ring of all matrices of size n with entries in a ring A. For each integer $m \geq 0$, there is an $\mathbb{R}$-rational irreducible polynomial representation $\sigma_m : \mathbb{H}^\times \to GL_{m+1}(\mathbb{C})$ of degree m, unique up to equivalence. Changing it for an equivalent representation and choosing a suitable isomorphism in (3.1), we may assume:

$$(3.2a) \qquad x_\tau \in M_2(\bar{\mathbb{Q}}) \text{ for every } \tau \in \alpha \text{ and every } x \in B;$$

$$(3.2b) \qquad \sigma_m(x_\tau) \in M_{m+1}(\bar{\mathbb{Q}}) \text{ for every } \tau \in \beta, \text{ every } x \in B \text{ and every } m \geq 0;$$

$$(3.2c) \qquad \sigma_m(x^*) = {}^t\sigma_m(x) \text{ for every } x \in \mathbb{H}^\times.$$

Here x_τ is the projection of an element x of $B_{\mathbb{R}}$ to the τ-factor via the chosen isomorphism of (3.1); $x \mapsto x^*$ denotes the main involution in any quaternion algebra $B, M_2(\mathbb{R})$, or $\mathbb{H}$, and also the $\mathbb{R}$-linear extension to $B_{\mathbb{R}}$ of the main involution of B.

We put $N(x) = xx^*$ and $\mathrm{Tr}(x) = x + x^*$ for x in $B_{\mathbb{R}}$, $M_2(\mathbb{R})$, or $\mathbb{H}$, and also

$$(3.3) \qquad\qquad B_+ = B \cap B_{\mathbb{R}+}, \quad B_{\mathbb{R}+} = \{ x \in B_{\mathbb{R}} \mid N(x) \geqslant 0 \}.$$

For $\xi \in B_{\mathbb{R}}$ and $z \in \mathbb{C}^\alpha$, we define $\xi(z)$ and $j(\xi, z)$ as elements of $\mathbb{C}^\alpha$ by

$$(3.4a) \qquad\qquad \xi z = \xi(z) = ((a_\tau z_\tau + b_\tau)(c_\tau z_\tau + d_\tau)^{-1})_{\tau \in \alpha},$$

$$(3.4b) \qquad\qquad j(\xi, z) = (|\det(\xi_\tau)|^{-1/2}(c_\tau z_\tau + d_\tau))_{\tau \in \alpha},$$

where $a_\tau, b_\tau, c_\tau, d_\tau$ are the entries of ξ_τ in the standard order. For $0 \leq \kappa \in \mathbb{Z}^\beta$, we define a representation $\sigma_\kappa : B_{\mathbb{R}}^\times \to GL_d(\mathbb{C})$ with $d = \prod_{\tau \in \beta}(\kappa_\tau + 1)$ by

$$(3.5) \qquad\qquad \sigma_\kappa(\xi) = \bigotimes_{\tau \in \beta} \sigma_{\kappa_\tau}(\xi_\tau).$$

For $\xi \in B_{\mathbb{R}+}$, $k \in \mathbb{Z}^\alpha$, such a κ, and a map $f : H^\alpha \to \mathbb{C}^d$, we define a map $f\|_{k,\kappa}\xi : H^\alpha \to \mathbb{C}^d$ by

$$(3.6) \qquad\qquad (f\|_{k,\kappa}\xi)(z) = j(\xi, z)^{-k} N(\xi)^{\kappa/2} \sigma_\kappa(\xi)^{-1} f(\xi z).$$

Given a congruence subgroup Γ of B_+, we denote by $\mathcal{M}_{k,\kappa}(\Gamma)$ the set of all holomorphic maps $f : H^\alpha \to \mathbb{C}^d$ which satisfy $f\|_{k,\kappa}\gamma = f$ for every $\gamma \in \Gamma$ and also the cusp condition when $B = M_2(\mathbb{Q})$; further we put

$$\mathcal{S}_{k,\kappa}(\Gamma) = \begin{cases} \mathcal{M}_{k,\kappa}(\Gamma) & \text{if } B \text{ is a division algebra,} \\ \text{the set of all cusp forms in } \mathcal{M}_{k,\kappa}(\Gamma) & \text{if } B = M_2(K). \end{cases}$$

The union of $\mathcal{M}_{k,\kappa}(\Gamma)$ resp. $\mathcal{S}_{k,\kappa}(\Gamma)$ for all congruence subgroups Γ of B_+ is denoted by $\mathcal{M}_{k,\kappa}$ resp. $\mathcal{S}_{k,\kappa}$. If $\beta = \emptyset$, we write these simply $\mathcal{M}_k$ and $\mathcal{S}_k$. For two elements f and g of $\mathcal{S}_{k,\kappa}(\Gamma)$, we put

$$\langle f, g \rangle = \mathrm{vol}(D)^{-1} \int_D {}^t\overline{f(z)} g(z) \,\mathrm{Im}(z)^k d^* z \quad (D = \Gamma \backslash H^\alpha).$$

It is well known that $\mathrm{vol}(D) \sim \pi^{\|\alpha\|}$ (see Shimizu [S, p. 192]). The symbol $\langle f, g \rangle$ will be used more generally for two continuous functions f and g with the same behavior as the elements of $\mathcal{M}_{k,\kappa}$ whenever the integral is convergent, and even for the functions defined on the product of several spaces of type H^α under the rule

$$\langle f_1 \otimes \ldots \otimes f_t, g_1 \otimes \ldots \otimes g_t \rangle = \prod_{i=1}^t \langle f_i, g_i \rangle \quad (\text{cf. [83a, pp. 257–258]}). \text{ We write also}$$

$\mathcal{S}_{k,\kappa}(B)$ for the above $\mathcal{S}_{k,\kappa}$ when B needs to be specified. So much for the notation in general.

To define our series, we take totally real algebraic number fields $F_1, \ldots, F_t$ of finite degree, and for each i, a vector space V_i over F_i of dimension $m_i + 2$ such that $0 \leq m_i \leq 2$; we put then

$$(3.7a) \qquad\qquad F = F^{(0)} \times F^{(1)} \times F^{(2)}, \quad F^{(\nu)} = \prod_{m_i = \nu} F_i,$$

$$(3.7b) \qquad\qquad V = V^{(0)} \times V^{(1)} \times V^{(2)}, \quad V^{(\nu)} = \prod_{m_i = \nu} V_i.$$

Each V_i is assumed to have an additional structure according to m_i as follows:

Case $m_i=0$: V_i is a totally imaginary quadratic extension of F_i.
Case $m_i=1$: $V_i=\{x\in B_i|x^*=-x\}$ with a quaternion algebra B_i over F_i.
Case $m_i=2$: V_i is a quaternion algebra over F_i.

We now take subsets ε, ε', ε_i, ε_i', δ_v, δ_v', and δ of J_F, viewed also as elements of I_F, satisfying the following conditions:

$$(3.8a) \qquad \varepsilon_i=\varepsilon\cap J(F_i), \quad \varepsilon_i'=\varepsilon'\cap J(F_i), \quad \varepsilon_i+\varepsilon_i'=\iota(F_i).$$

(3.8b) *If $m_i=1$ or 2, then ε_i is the set of all τ in $J(F_i)$ unramified in B_i or V_i.*

$$(3.8c) \qquad \delta_v=J(F^{(v)})\cap\varepsilon, \quad \delta_v'=J(F^{(v)})\cap\varepsilon', \quad \delta=\delta_1+\delta_2.$$

We fix a common subfield E of the F_i and also two subsets ψ and ψ' of $J(V^{(0)})$, viewed also as elements of $I(V^{(0)}\times F^{(1)}\times F^{(2)})$; we then assume:

$$(3.9a) \qquad \mathrm{Res}_{F/E}(\varepsilon)=\iota_E;$$

$$(3.9b) \qquad \varepsilon_i\neq\emptyset \quad \textit{for every} \quad i;$$

$$(3.9c) \qquad \mathrm{Res}_{/F}(\psi)=\delta_0, \quad \mathrm{Res}_{/F}(\psi')=\delta_0'.$$

We shall often consider the products or sums of the objects with indices i in the range $1\leq i\leq t$ under the condition $m_i=v$, such as $\prod\limits_{m_i=v}$, $\sum\limits_{m_i=v}$, and $\bigotimes\limits_{m_i=v}$. For simplicity, we shall denote them by $\prod\limits_{(v)}$, $\sum\limits_{(v)}$, and $\bigotimes\limits_{(v)}$, respectively. We put also $e^{(v)}(*)=e_X(*)$ and $H^{(v)}=H^{J(X)}$ when $X=F^{(v)}$ (see (0.0a) for e_X).

We allow each factor $V^{(v)}$ and also the product of two of them to become trivial; we assume however that $V^{(1)}\times V^{(2)}$ is nontrivial.

For each factor V_i of $V^{(2)}$, we consider $\mathscr{S}_{k_i,\kappa_i}(V_i)$, and for each factor of $V^{(1)}$, we consider $\mathscr{S}_{2k_i,2\kappa_i}(B_i)$, where k_i, $\kappa_i\in I(F_i)$. We put

$$(3.10) \qquad k=k^{(1)}+k^{(2)}, \quad k^{(v)}=\sum_{(v)}^{} k_i;$$

$$\kappa=\kappa^{(1)}+\kappa^{(2)}, \quad \kappa^{(v)}=\sum_{(v)}^{} \kappa_i.$$

Our aim is to state two theorems of arithmeticity of the special values of a certain series $Z(s)$, which involves, roughly speaking, the forms in these spaces. More precisely, we take a system of eigenvalues χ_i occurring in $\mathscr{S}_{k_i,\kappa_i}(V_i)$ which is primitive in the sense of [81, II, p. 584], and also an element r of $\bigotimes\limits_{(1)} \mathscr{S}_{2k_i,2\kappa_i}(B_i)$; we then associate with r an element $I(z,r)$ of $\bigotimes\limits_{(1)} \mathscr{S}_{t_i}(F_i)$ with

$$(3.11) \qquad t_i=k_i+\kappa_i+(1/2)\varepsilon_i+(3/2)\varepsilon_i',$$

and take its Fourier expansion

$$(3.12) \qquad I(z,r)=\sum_{a\in F^{(1)}} \mu_r(a)e^{(1)}(az) \quad (z\in H^{(1)}).$$

(The map $r \mapsto I(z, r)$, which will be explained below, depends on the choice of a locally constant function on $V^{(1)}$. For the moment, however, we suppress it with the understanding that the locally constant function is fixed.) Further we take an element

$$(3.13) \qquad \Omega(z) = \sum_{a \in E} \omega(a) \mathbf{e}_E(az) \quad (z \in H^{J(E)})$$

of $\mathcal{M}_l(E)$ with $l \in I^0(E)$. Then our series Z^r is given by

$$(3.14) \qquad Z^r(s) = [U] \sum_{(x, y) \in K/U} c(x'', y) \omega(\mathrm{Tr}_{F/E}(\zeta(x, yy^\varrho))) x^\varDelta y^\varphi$$

$$\cdot |x^\delta y^{2\psi}|^{-s} \mu_r(x') \prod^{(2)} \chi_i(x_i \mathfrak{a}_i) .$$

Here $K = F^{(1)} \times F^{(2)} \times V^{(0)}$, $c \in \mathcal{L}(F^{(2)} \times V^{(0)})$, $x = (x', x'') \in F^{(1)} \times F^{(2)}$ with $x' \in F^{(1)}$ and $x'' \in F^{(2)}$, $y \in V^{(0)}$ (and thus $(x, yy^\varrho) \in F$), $\varDelta \in (1/2) I(F^{(1)} \times F^{(2)})$, $\varphi \in I(V^{(0)})$, $\zeta \in F^\times$, $\mathfrak{a}_i$ is a fractional ideal in F_i, and U is a subgroup of U_E of finite index; we let U act on K by $u(x, y) = (u^2 x, uy)$ for $u \in U$; the sum is extended over all $(x, y) U$ with $x \gg 0$ and $y \neq 0$. To make each term dependent only on $(x, y) U$, we have to assume

$$(3.15) \qquad l + \mathrm{Res}_{/E}(\varphi) + \mathrm{Res}_{/E}(2\varDelta + k^{(1)} + \kappa^{(1)} - \delta_1) = a_Z 1_E \text{ with } a_Z \in (1/2)\mathbb{Z} ,$$

$$(3.16) \qquad c(u^2 x'', uy) = c(x'', y) \text{ and } \omega(u^2 a) = u^l \omega(a) \text{ for every } u \in U .$$

The last condition is inessential, since it is satisfied with a sufficiently small subgroup U of U_E of finite index, and the factor $[U]$ in (3.14) makes Z^r independent of the choice of U. We also assume:

$$(3.17) \qquad \zeta^\tau > 0 \text{ for } \tau \in \varepsilon \quad and \quad \zeta^\tau < 0 \text{ for } \tau \in \varepsilon' ;$$

$$(3.18a) \qquad \varphi_\sigma \geqq \varphi_{\sigma\varrho} \text{ for } \sigma \in \psi \cup \psi' \quad and \quad \varphi_\sigma \geqq 0 \text{ for every } \sigma \in \psi' \cup \psi'\varrho ;$$

$$(3.18b) \qquad 2\varDelta_\tau \equiv 1 \pmod 2 \text{ for every } \tau \in \delta_1 \quad and \quad 0 \leq \varDelta_\tau \in \mathbb{Z} \text{ for every } \tau \in \delta_1' ;$$

$$(3.18c) \qquad 2\varDelta_\tau \equiv k_\tau \pmod 2 \text{ for every } \tau \in \delta_2 \quad and$$

$$\kappa_\tau \leq 2\varDelta_\tau \equiv \kappa_\tau \pmod 2 \text{ for every } \tau \in \delta_2' ;$$

(3.19) $c(x'', y) \in \bar{\mathbb{Q}}$ and $\omega(a) \in \bar{\mathbb{Q}}$ for every (x'', y) and every $a \in E$; r is $\bar{\mathbb{Q}}$-rational in the sense of $[81, \text{II}, \S 2]$; the locally constant function on $V^{(1)}$ involved in $I(z, r)$ is $\bar{\mathbb{Q}}$-rational in the sense explained below.

Naturally the factor $\mu_r(x')$ or $\prod^{(2)} \chi_i(x_i \mathfrak{a}_i)$ means 1 when $V^{(1)}$ or $V^{(2)}$ is trivial; similarly y should be ignored when $V^{(0)}$ is trivial. The series Z^r should be written simply Z if $V^{(1)}$ is trivial.

As a preliminary result, we have:

Proposition 3.1. Z^r *can be continued as a meromorphic function to the whole plane. Put*

$$(3.20a) \qquad b_Z = a_Z + \sum_{i=1}^t [F_i : E](m_i + 2)/2 ,$$

$$(3.20\mathrm{b}) \qquad N_\tau = \begin{cases} 2\varphi_\sigma + 2 & \text{if} \quad \sigma \in \psi \quad \text{and} \quad \mathrm{Res}_{/F}(\sigma) = \tau \in \delta_0, \\ 2k_\tau + 2\varDelta_\tau + 1 & \text{if} \quad \tau \in \delta_1, \\ k_\tau + 2\varDelta_\tau + 2 & \text{if} \quad \tau \in \delta_2. \end{cases}$$

Then Z^r is holomorphic for $\mathrm{Re}(s) \geqq (b_Z - 1)/2$ except for a possible simple pole at $s = b_Z/2$ or $s = (2b_Z - 1)/4$ according as $b_Z \in \mathbb{Z}$ or $b_Z \notin \mathbb{Z}$. The pole does not occur if $N_\tau > b_Z$ for some $\tau \in \varepsilon$ in the case $b_Z \in \mathbb{Z}$, or if $b_Z - (1/2) < N_\tau \equiv b_Z - (1/2) \pmod 2$ for some $\tau \in \varepsilon$ in the case $b_Z \notin \mathbb{Z}$.

This will be proved in Sect. 4*.

To state our theorems, we fix a congruence subgroup Γ_i of B_i, and take a $\overline{\mathbb{Q}}$-rational basis R of $\overset{(1)}{\otimes} \mathscr{S}_{2k_i, 2\kappa_i}(\Gamma_i)$. We are interested in the value of Z^r at an integer s_0 such that

$$(3.21\mathrm{a}) \qquad (b_Z - 1)/2 \leqq s_0 \leqq (N_\tau - 2)/2 \quad \text{for every} \quad \tau \in \varepsilon;$$

$$(3.21\mathrm{b}) \qquad 2s_0 \neq b_Z - (1/2) \quad \text{if} \quad E = \mathbb{Q};$$

$$(3.21\mathrm{c}) \qquad l \neq (1/2)\iota_E \quad \text{if} \quad 2s_0 = b_Z, \quad F = E = \mathbb{Q}, \quad \text{and} \quad m_1 = 1.$$

Now the main theorems of this section can be stated as follows:

Theorem 3.2. *Let $\mathbf{g}_i$ be a $\overline{\mathbb{Q}}$-rational nonzero eigenform belonging to χ_i. Let $(a_{rr'})_{r, r' \in R}$ be the inverse of the matrix $(\langle r, r' \rangle)_{r, r' \in R}$. Then, under (3.17), (3.18a, b, c) and (3.19),*

$$\sum_{r \in R} a_{rr'} Z^r(s_0) \sim \pi^d p_W(\varphi, 2\psi) \overset{(2)}{\prod} \langle \mathbf{g}_i, \mathbf{g}_i \rangle$$

for every $r' \in R$ and every integer s_0 satisfying (3.21a, b, c), where $d = \|k\| - \|\delta_1\| + \sum_{\sigma \in \psi} (\varphi_\sigma - \varphi_{\sigma\varrho})$ and $W = V^{(0)}$, and the left-hand side means $Z(s_0)$ if $V^{(1)}$ is trivial.

If $V^{(0)}$ or $V^{(2)}$ is trivial, the factor $p_W(\varphi, 2\psi)$ or $\overset{(2)}{\prod} \langle \mathbf{g}_i, \mathbf{g}_i \rangle$ means 1. Hereafter we use, for simplicity, the period symbol p_W always in the above sense, that is, with $W = V^{(0)}$.

Theorem 3.3. *In addition to (3.17), (3.18a, b, c), and (3.19), assume:*
(Ra) Case $b_Z \in \mathbb{Z}$: $N_\tau = b_Z$ for every $\tau \in \varepsilon$ and $\varphi_{\sigma\varrho} < \varphi_\sigma$ for every $\sigma \in \psi$;

$$l_\tau \neq 1/2 \quad \text{for every} \quad \tau \in J_E \quad \text{if} \quad E = F \quad \text{and} \quad m_1 = 1;$$

(Rb) Case $b_Z \notin \mathbb{Z}$: $N_\tau \geqq b_Z + (1/2)$ for every $\tau \in \varepsilon$.

Then Z^r has at most a simple pole at $s = b_Z/2$ or $s = (2b_Z - 1)/4$ according as $b_Z \in \mathbb{Z}$ or $b_Z \notin \mathbb{Z}$, and the residue of $\sum_{r \in R} a_{rr'} Z^r(s)$ (the residue of $Z(s)$ when $V^{(1)}$ is trivial) at this point is an algebraic number times $R_E \pi^{d'} p_W(\varphi, 2\psi) \overset{(2)}{\prod} \langle \mathbf{g}_i, \mathbf{g}_i \rangle$, where

$$d' = \|k\| + \sum_{\sigma \in \psi} (\varphi_\sigma - \varphi_{\sigma\varrho}) - \begin{cases} \|\delta_1\| & \text{if} \quad b_Z \in \mathbb{Z}, \\ [E : \mathbb{Q}] & \text{if} \quad b_Z \notin \mathbb{Z}. \end{cases}$$

These results were given in [83a, Theorem 4.4] when l is integral and $V^{(1)}$ is trivial under the condition

$$(3.22) \qquad\qquad 2s_0 \neq b_Z \quad or \quad E \neq \mathbf{Q}.$$

The present theorems include the case of half-integral l, involve the Fourier coefficients μ_r, and remove (3.22). Conditions (3.21b,c) concern the case of half-integral l, and are perhaps removable under certain assumptions on c and r.

To prove our theorems, we reformulate them as follows. First we take an integral ideal c_i in F_i and a Hecke character λ_i of F_i, so that $\mathbf{g}_i$ of Theorem 3.2 belongs to the space $\mathscr{S}_{k_i, \kappa_i}(c_i, \lambda_i)$ of forms of type (c_i, λ_i) on $B_{\mathbf{A}}^{\times}$ defined in [81, II, Sect. 1]. We then take a $\bar{\mathbf{Q}}$-rational basis P of $\overset{(2)}{\otimes} \mathscr{S}_{k_i, \kappa_i}(c_i, \lambda_i)$. On the last space, we can define a Hecke operator $T(\mathfrak{a}) = \overset{(2)}{\otimes} T(\mathfrak{a}_i)$ for each set $\mathfrak{a} = \{\mathfrak{a}_i\}$ of integral ideals $\mathfrak{a}_i$ in F_i as in (3.14), with $T(\mathfrak{a}_i)$ defined as in [81, II, p. 576]. Now we put

$$(3.23) \qquad\qquad p|T(\mathfrak{a}) = \sum_{q \in P} \chi_{pq}(\mathfrak{a})q \quad (p \in P)$$

with $\chi_{pq}(\mathfrak{a}) \in \mathbf{C}$, and define a series Z_{pq}^r by (3.14) with $\chi_{pq}(x''\mathfrak{a})$ in place of $\overset{(2)}{\prod} \chi_i(x_i \mathfrak{a}_i)$. (If $V^{(2)}$ is trivial, nothing new is introduced here.)

Proposition 3.4. *Suppose $V^{(2)}$ is nontrivial. Then Z_{pq}^r has the same analytic properties as those of Z^r stated in Proposition 3.1. Moreover, let $(b_{pq})_{p,q \in P}$ be the inverse of the matrix $(\langle p, q \rangle)_{p,q \in P}$. Then, under the same assumptions as in Theorem 3.2, and with the same d, we have*

$$\sum_{r \in R, p \in P} a_{rr'} b_{pp'} Z_{pq}^r(s_0) \sim \pi^d p_W(\varphi, 2\psi)$$

for every $r' \in R$ and every $(p', q) \in P \times P$. The assertions of Theorem 3.3 correspond similarly to the assertions about the residue of $\sum_{r \in R, p \in P} a_{rr'} b_{pp'} Z^r$.

Let us first show that Proposition 3.1 as well as Theorems 3.2 and 3.3 follow from Proposition 3.4, when $V^{(2)}$ is nontrivial. Let X_i be the set of all $\bar{\mathbf{Q}}$-rational elements of $\mathscr{S}_{k_i, \kappa_i}(c_i, \lambda_i)$, and Y_i the set of all elements of X_i belonging to the eigenvalues $\chi_i(*)$; let $X = \overset{(2)}{\otimes} X_i$ and $Y = \overset{(2)}{\otimes} Y_i$. Since χ_i is primitive, we see that X is the direct sum of Y and its orthogonal complement. (The key point here is that this is so *within the set of $\bar{\mathbf{Q}}$-rational elements*; cf. the proof of [81, I, Theorem 3.7].) Therefore we see that Proposition 3.4 implies the corresponding assertions with a basis Q of Y over $\bar{\mathbf{Q}}$ in place of P. Put $S = \pi^{-d} p_W(-\varphi, 2\psi) \sum_{r \in R} a_{rr'} Z^r(s_0)$. If $S = 0$, there is no problem; so assume $S \neq 0$. Now the matrix

$$\left(\pi^{-d} p_W(-\varphi, 2\psi) \sum_{r \in R} a_{rr'} Z_{pq}^r(s_0) \right)_{p, q \in Q}$$

is S times the identity matrix, and hence $S^{-1}(\langle p, q \rangle)_{p, q \in Q}$ is algebraic, and in particular $S^{-1}\langle p, p \rangle$ is algebraic for every $p \in Q$. Taking p to be $\overset{(2)}{\otimes} \mathbf{g}_i$, we obtain

Theorem 3.2. Theorem 3.3 can be derived in a similar way. As for Proposition 3.1, our assertion is obvious, since $Z^r = Z^r_{pp}$ with that choice of p. (If $V^{(2)}$ is trivial, Proposition 3.4 is unnecessary.)

We note here another type of reduction. Given s_0 as in Theorem 3.2, we observe that $Z^r(s + s_0)$ is again a series of type Z^r, in which Δ, φ, N_τ, and b_z are replaced by $\Delta - s_0 \delta$, $\varphi - s_0(1 + \varrho)\psi$, $N_\tau - 2s_0$, and $b_z - 2s_0$. Therefore, considering $Z^r(s + s_0)$ instead of $Z^r(s)$, we may assume that $s_0 = 0$.

The definition if $I(r, z)$, as well as our proof, requires certain theta functions. To define them, we first consider an F_i-bilinear symmetric form $S_i : V_i \times V_i \to F_i$ given by

$$(3.24) \qquad S_i(x, y) = \begin{cases} -\zeta_i(xy^\varrho + yx^\varrho) & \text{if } m_i = 0, \\ \operatorname{Tr}(xy^*) & \text{if } m_i > 0, \end{cases}$$

where ζ_i is the i-th component of the element ζ employed in (3.14). We put $S[x] = (S_i[x_i])^t_{i=1}$, $S_i[x_i] = S_i(x_i, x_i)$ for $x_i \in V_i$. Conditions (3.8b) and (3.17) imply that S_i is definite exactly at $\tau \in \varepsilon'_i$. Thus our set of objects $\{F_i, V_i, S_i, E\}$ fits in the framework of [86, §4] as a special case. We shall apply the theorems of [86] to this set.

Next, we put, for κ of (3.10),

$$(3.25a) \qquad \sigma^{(\nu)}_\kappa(\xi) = \overset{(\nu)}{\otimes} \sigma_{\kappa_i}(\xi_i) \left(\xi \in \overset{(1)}{\prod} B_i \ \text{ if } \nu = 1; \ \xi \in V^{(2)} \ \text{ if } \nu = 2 \right).$$

As shown in [82, p. 609], we can find a nontrivial $\bar{\mathbb{Q}}$-rational map q' of $V^{(1)}$ into $\mathbb{C}^d$ with $d = \prod_{\tau \in \delta'_1} (2\kappa_\tau + 1)$ such that

$$(3.25b) \qquad q'(\beta\alpha\beta^*) = \sigma^{(1)}_{2\kappa}(\beta)q'(\alpha) \quad \left(\alpha \in V^{(1)}, \ \beta \in \overset{(1)}{\prod} B_i \right),$$

and that the components of q' span the space of harmonic polynomials of degree κ in the sense explained in [82, p. 609]; we understand that $q' = 1$ if $\kappa = 0$. We then put

$$(3.25c) \qquad q(u) = q'(u') \otimes \sigma^{(2)}_\kappa(u'') \quad \text{for} \quad u \in V,$$

where u' and u'' are the projections of u to $V^{(1)}$ and $V^{(2)}$. Next we put

$$(3.26a) \qquad [\alpha; w, w'] = \left((-1, w_\tau)\alpha_\tau \begin{bmatrix} w'_\tau \\ 1 \end{bmatrix} \right)_{\tau \in \delta} \quad (\in \mathbb{C}^\delta),$$

$$(3.26b) \qquad \eta(w, w') = (4\operatorname{Im}(w_\tau)\operatorname{Im}(w'_\tau))_{\tau \in \delta} (\in \mathbb{C}^\delta) \quad (\alpha \in V, (w, w') \in (\mathbb{C}^\delta)^2),$$

$$(3.27) \qquad \mathfrak{H} = \{(w, w') \in H^\delta \times H^\delta | w_\tau = w'_\tau \text{ for } \tau \in \delta_1\}.$$

Now our theta series is given by

$$(3.28) \qquad \Theta(z; w, w') = \operatorname{Im}(z)^{(\delta_1/2) + \delta_2} \eta(w, w')^{-k} \sum_{\alpha \in V} C(\alpha)q(\alpha)\alpha^\lambda$$

$$\cdot [\alpha; \bar{w}, \bar{w}']^k \mathbf{e}_F((1/2)R[\alpha; z, w, w']).$$

Here $z \in H^{J(F)}$, $(w, w') \in \mathfrak{H}$, $C \in \mathscr{L}(V)$, k is as in (3.10),

$$(3.29) \qquad \lambda = \sum_{\sigma \in \psi + \psi'} (\varphi_\sigma - \varphi_{\sigma\varrho})\sigma,$$

$$(3.30) \qquad R[\alpha; z, w, w'] = (R_\tau[\alpha; z, w, w'])_{\tau \in J(F)},$$

$$R_\tau[\alpha; z, w, w'] = \begin{cases} S[\alpha]_\tau z_\tau & \text{if } \tau \in \varepsilon', \\ S[\alpha]_\tau \bar{z}_\tau & \text{if } \tau \in \delta_0, \\ S[\alpha]_\tau z_\tau + 4 i y_\tau \eta(w, w')_\tau^{-1} |[\alpha; w, w']_\tau|^2 & \text{if } \tau \in \delta, \end{cases}$$

where $y_\tau = \mathrm{Im}(z_\tau)$. This is a special case of the series of [86, p. 401], up to the factor $\mathrm{Im}(z)^*$. We assume that C is $\bar{\mathbb{Q}}$-valued.

Let us now take C in the form $C(v) = \prod_{i=1}^{t} C_i(v_i)$ with $\bar{\mathbb{Q}}$-valued C_i in $\mathscr{L}(V_i)$. Then Θ is the product of t theta series defined on $H^{J(F_i)}$, $1 \le i \le t$. Instead of these series, however, we consider their transforms given as follows. With ζ as in (3.17), define a map $z \mapsto \zeta^{-1}\{z\}$ of $H^{J(F)}$ into itself by $\zeta^{-1}\{z\}_\tau = \zeta_\tau^{-1} z_\tau$ for $\tau \in \varepsilon$ and $\zeta^{-1}\{z\}_\tau = \zeta_\tau^{-1} \bar{z}_\tau$ for $\tau \in \varepsilon'$; take $b \in V^{(2)}$ so that $N(b)_\tau = -\zeta_\tau^{-1}$ for every $\tau \in J(F^{(2)})$. Then we have

$$(3.31) \qquad \Theta(\zeta^{-1}\{z\}; w, b(\bar{w}')) j(b, \bar{w}')^{-k^{(2)}} \sigma_\kappa^{(2)}(b)$$

$$= A_0 \left[\overset{(0)}{\prod} \theta_i(z_i) \right] \cdot \left[\overset{(1)}{\otimes} \theta_i(z_i, w_i) \right] \otimes \left[\overset{(2)}{\otimes} \theta_i(z_i; w_i, w'_i) \right]$$

with an algebraic number A_0 and certain theta functions θ_i, where we understand that $b(\bar{w}')_\tau = w_\tau$ for $\tau \in \varepsilon_1$. The explicit forms of these θ_i are as follows:

Case $m_i = 0$. Putting $C^{(0)}(v) = \overset{(0)}{\prod} C_i(v_i)$, we have

$$(3.32\mathrm{a}) \qquad \overset{(0)}{\prod} \theta_i(z_i) = \sum_{v \in V^{(0)}} C^{(0)}(v) v^\lambda \mathbf{e}^{(0)}(-v v^\varrho \bar{z}) \quad (z \in H^{(0)}).$$

Notice that $\bar{\theta}_i$ is holomorphic and belongs to $M_u(F_i)$ with $u = \iota(F_i) + \mathrm{Res}_{/F}(\lambda_i)$, where λ_i is the i-part of λ.

Case $m_i = 1$.

$$(3.32\mathrm{b}) \qquad \theta_i(z, w) = \mathrm{Im}(z)^{\varepsilon_i/2} \eta(w, w)^{-k_i} \sum_{v \in V_i} C_i(v) q_i(v) [v; \bar{w}, \bar{w}]^{k_i}$$

$$\cdot \mathbf{e}_{F_i}(\zeta_i^{-1} R'[v; z, w]) \quad (z \in H^{J(F_i)}, \ w \in H^{\varepsilon_i}),$$

$$R'[v; z, w]_\tau = \begin{cases} N(v)_\tau \bar{z}_\tau & \text{if } \tau \in \varepsilon_i', \\ N(v)_\tau z_\tau + (i/2) y_\tau \mathrm{Im}(w_\tau)^{-2} |[v; w, w]_\tau|^2 & \text{if } \tau \in \varepsilon_i. \end{cases}$$

This is essentially the same as the function Θ of [82, p. 612, line 4 from bottom].

Case $m_i = 2$.

$$(3.32\mathrm{c}) \qquad \theta_i(z; w, w') = \mathrm{Im}(z)^{\varepsilon_i} \eta(w, w')^{-k_i} \sum_{v \in V_i} C_i(v b^{-1}) [v; \bar{w}, w']^{k_i} \sigma_{\kappa_i}(v)$$

$$\cdot \mathbf{e}_{F_i}(-R'[v; z, w, w']) \quad (z \in H^{J(F_i)}, \ (w, w') \in H^{\varepsilon_i} \times H^{\varepsilon_i}),$$

$$R'[v; z, w, w']_\tau = \begin{cases} N(v)_\tau \bar{z}_\tau & \text{if } \tau \in \varepsilon_i', \\ N(v)_\tau z_\tau + 2 i y_\tau \eta(\bar{w}, w')_\tau^{-1} |[v; \bar{w}, w']_\tau|^2 & \text{if } \tau \in \varepsilon_i. \end{cases}$$

This is essentially the same as θ_1 of [81, II, (5.6)] up to the transformation $z \mapsto -\bar{z}$.

Now define $I(z, g)$ for $g \in \mathcal{S}_{2k_i, 2\kappa_i}(B_i)$ and $I(z, w', g)$ for $g \in \mathcal{S}_{k_i, \kappa_i}(V_i)$ by

$$(3.33a) \qquad I(z, g) = \langle \theta_i(z, w), g(w) \rangle \quad (m_i = 1, \ z \in H^{J(F_i)}),$$

$$(3.33b) \qquad I(z, w', g) = \langle \theta_i(z, w, w'), g(w) \rangle \quad (m_i = 2, \ z \in H^{J(F_i)}, \ w' \in H^{\varepsilon_i}).$$

By [82, (2.15), Theorem 2.2, and Proposition 2.3], we see that $I(z, g)$ is well-defined and belongs to $\mathcal{S}_{t_i}(F_i)$, where

$$(3.34a) \qquad t_i = k_i + (1/2)\varepsilon_i + \kappa_i + (3/2)\varepsilon_i'.$$

Similary, by [81, II, (5.7), Propositions 5.1 and 5.4], $I(z, w', g)$ is well-defined, and as a function of z belongs to $S_{t_i}(F_i)$, where

$$(3.34b) \qquad t_i = k_i + \kappa_i + 2\varepsilon_i'.$$

To simplify our notation, let us put

$$(3.35) \qquad \mathcal{S}^{(1)} = \overset{(1)}{\otimes} \mathcal{S}_{2k_i, 2\kappa_i}(B_i), \quad \mathcal{S}^{(2)} = \overset{(2)}{\otimes} \mathcal{S}_{k_i, \kappa_i}(V_i).$$

Define similarly $\mathcal{M}^{(1)}$ and $\mathcal{M}^{(2)}$ with $\mathcal{M}$ in place of $\mathcal{S}$. Putting $I\left(z, \overset{(1)}{\otimes} g_i\right)$

$= \overset{(1)}{\otimes} I(z_i, g_i)$, we can define $I(z, r)$ for $r \in \mathcal{S}^{(1)}$ as a function of $z \in H^{(1)}$ belonging to

$\overset{(1)}{\otimes} \mathcal{S}_{t_i}(F_i)$. Similarly we can define $I(z, w', g)$ for $z \in H^{(2)}$, $w' \in H^{\delta_2}$, and $g \in \mathcal{S}^{(2)}$,

which as a function of z belongs to $\overset{(2)}{\otimes} \mathcal{S}_{t_i}(F_i)$. These depend on the choice of the θ_i,

or rather on the choice of the C_i, but we suppress them with the understanding that they are fixed for the moment. Anyway it is this $I(z, r)$ that we employ in (3.12). The $\bar{\mathbb{Q}}$-rationality mentioned in (3.19) means that C_i is $\bar{\mathbb{Q}}$-valued.

4. Proof of Theorems 3.2, 3.3, and Proposition 3.4

For the reason explained in the previous section, we take $s_0 = 0$. This means that we assume conditions (3.21a,b,c) with $s_0 = 0$. We define an element j of I_F and elements ξ, μ, μ' of $I(V^{(0)})$ by

$$(4.1a) \qquad j = \Delta + k^{(1)} + (1/2)(k^{(2)} - \delta_1 - \kappa^{(2)}),$$

$$(4.1b) \qquad \xi = \lambda + \mu(1 + \varrho), \quad \mu = \sum_{\sigma \in \psi} \varphi_{\sigma\varrho} \sigma, \quad \mu' = \sum_{\sigma \in \psi'} \varphi_{\sigma\varrho} \sigma.$$

Conditions (3.18b,c) and (3.21a) with $s_0 = 0$ imply that $0 \le j \in I(F^{(1)} \times F^{(2)})$. Now one can easily find a finite sum expression

$$(4.2) \qquad a^j = \sum_{p,q} c_{pq} \operatorname{Tr}_{F/E}(\zeta a)^p a^q,$$

valid for every $a \in F$ with $0 \le p \in I_E$, $0 \le q \in \mathbb{Z}^{\varepsilon'}$, and constants $c_{pq} \in \bar{\mathbb{Q}}$. Put $h = q + \operatorname{Res}_{/F}(\mu')$ and write $\gamma_{ph} = c_{pq}$. Then $0 \le h \in \mathbb{Z}^{\varepsilon'}$ and we have, for $a = (x, yy^\varrho) \in F$

with $x \in F^{(1)} \times F^{(2)}$ and $y \in V^{(0)}$,

$$(4.3) \qquad x^{\Delta} y^{\varphi} = \sum_{p,h} \gamma_{ph} \mathrm{Tr}_{F/E}(\zeta(x, yy^{\varrho}))^p a^{\Delta - j + h} y^{\xi}.$$

For each pair of such (p, h), we consider an infinite series

$$(4.4) \qquad f_{ph}(w, w'; s) = [U] \sum_{0 \neq v \in V/U} C(v) \omega_p(\mathrm{Tr}_{F/E}(-S[v])) S[v]^h v^{\xi} q(v)$$
$$\cdot [v; w, w']^{-k} \eta(w, w')^{s\delta} |[v; w, w']^{\delta} v^{\psi}|^{-2s}.$$

Here $(w, w') \in H$ and $s \in \mathbb{C}$; $\omega_p(a) = a^p \omega(a)$ with ω of (3.13); $C, S, q, [v; w, w'], \eta, k, \delta$, and ψ are the same as in (3.14) and (3.28). This series is a special case of those considered in [86, §§4, 10], or more precisely, that of the series $f'(w, s)$ of [86, p. 400]. (The series of [86, (4.10)] is somewhat different, but essentially the same as (4.4) for the reason explained in [81, I, Remark 5.5] and [86, p. 400].)

It is easy to see that

$$\mathrm{Res}_{F/E}(2j) + \mathrm{Res}_{/E}(\varphi) = 2p + \mathrm{Res}_{F/E}(2h) + \mathrm{Res}_{/E}(\xi).$$

Combining this with (3.15), we obtain

$$\mathrm{Res}_{/E}(k - \kappa - 2h) - \mathrm{Res}_{/E}(\xi) - 2p - l = \mathrm{Res}_{/E}(k - \kappa - 2j) - \mathrm{Res}_{/E}(\varphi) - l = -a_Z l_E,$$

which shows that $-a_Z$ is the invariant a_f defined for f_{ph} in [86, (4.12), (10.3)]. As for the invariant b_f of [86, (4.14)], we have $b_f = 2 - b_Z$. Therefore (3.21a) with $s_0 = 0$ implies that $b_f \geq 1$ as well as [86, (4.21)]; (3.21b) and (3.21c) imply [86, (4.22) and (4.23)] respectively. Thus [86, Theorems 4.1, 4.2, and 4.5] are applicable to f_{ph}. Namely, $f_{ph}(w, w'; 0)$ is "nearly holomorphic"; it is nonholomorphic only when

$$(4.5) \qquad\qquad\qquad b_Z = 0 \quad \text{and} \quad E = \mathbb{Q}.$$

In all cases, $\pi^{-e} p_W(-\varphi, 2\psi) f_{ph}(w, w'; 0)$ is arithmetic, where $e = \|k\| + \sum_{\sigma \in \psi} (\varphi_\sigma - \varphi_{\sigma\varrho})$.

To prove our theorems, we may assume that $c(x'', y)$ of (3.14) can be factored as $c(x'', y) = c^{(0)}(y) c^{(2)}(x'')$, $c^{(0)}(y) = \prod^{(0)} c_i(y_i)$, $c^{(2)}(x'') = \prod^{(2)} c_i(x_i'')$ with $c_i \in \mathscr{L}(V_i)$ or $c_i \in \mathscr{L}(F_i)$, all $\overline{\mathbb{Q}}$-valued. Also Ω may be changed for $\sum \overline{\omega(a)} \mathbf{e}_E(az)$; or rather, we shall derive the desired results for the series Z^r and Z^r_{pq} defined with $\overline{\omega(a)}$ in place of $\omega(a)$.

Now we consider $C = \prod_{i=1}^{t} C_i$ of $\mathscr{L}(V)$ with $C_i \in \mathscr{L}(V_i)$, taking $C_i(y) = \bar{c}_i(y^\varrho)$ when $m_i = 0$. Given $r \in \mathscr{S}^{(1)}$ and $g \in \mathscr{S}^{(2)}$, we put

$$(4.6) \qquad M(z, w') = \prod^{(0)} \overline{\theta_i(z_i)} \cdot I(z^{(1)}, r) \otimes I(z^{(2)}, w', g)$$

for $w' \in H^{\delta_2}$ and $z \in H^{J(F)}$ with projection $z^{(v)}$ on $H^{(v)}$. By (3.32), we have

$$(4.7) \qquad\qquad \prod^{(0)} \overline{\theta_i(z_i)} = \sum_{v \in V^{(0)}} c^{(0)}(v) v^{\lambda} \mathbf{e}^{(0)}(vv^{\varrho} z).$$

Calling the function on the right-hand side of (3.31) $\Theta^*(z;w,w')$, we see that

$$(4.8) \qquad \langle \bar{\mathbf{d}}^h \Theta^*(z;w,w'), r \otimes g \rangle = A_1 \mathbf{d}^h M(z,w'),$$

where $\bar{\mathbf{d}}^h$ is defined by (0.1), and A_1 is a nonzero algebraic number. In the following pages, we denote by $A_2, A_3, \ldots$ nonzero algebraic constants which we don't have to specify; A_n is real and positive if it appears in the form A_n^s with $s \in \mathbf{C}$.

Define now two embeddings $z \mapsto z^0$ and $z \mapsto \zeta[z]$ of $H^{J(E)}$ into $H^{J(F)}$ as follows: if $z \in H^{J(E)}$, $\tau \in J_F$, and $\sigma = \mathrm{Res}_{F/E}(\tau)$, then

$$(4.9a) \qquad\qquad\qquad (z^0)_\tau = z_\sigma,$$

$$(4.9b) \qquad \zeta[z]_\tau \text{ is } \zeta_\tau z_\sigma \text{ or } \zeta_\tau \bar{z}_\sigma \text{ according as } \tau \in \varepsilon \text{ or } \tau \in \varepsilon'.$$

Obviously $\zeta^{-1}\{\zeta[z]\} = z^0$. For a congruence subgroup Γ of $SL_2(E)$, let Γ_∞ denote the isotropy subgroup of Γ at ∞. Putting $L = \Gamma_\infty \backslash H^{J(E)}$, we consider an integral

$$(4.10a) \qquad \int_L \mathbf{d}^p \Omega(z)(\bar{\mathbf{d}}^h \Theta^*)(\zeta[z];w,w') y^{\beta+s\imath+\imath} d^*z \quad (y = \mathrm{Im}(z)),$$

where $\imath = \imath_E$ and $\beta = \mathrm{Res}_{/E}(k - \delta_2 - (1/2)\delta_1 - \mu)$. This is a special case of [86, (10.9a)] combined with the transformation $w' \mapsto b(\bar{w}')$ considered in (3.31), up to constant factors. Therefore we see that the integral is equal to

$$(4.10b) \qquad A_2 A_3^s \pi^{-\gamma - s\|\varepsilon\|} \prod_{\tau \in \delta} \Gamma(s + k_\tau) \prod_{\sigma \in \psi} \Gamma(s - \varphi_{\sigma\varrho})$$

$$\cdot f_{ph}(w, b(\bar{w}'); s) j(b, \bar{w}')^{-k} \sigma_\kappa^{(2)}(b),$$

where $\gamma = \|k\| - \|\mu\|$. Call this function $\mathscr{A}(w,w';s)$. By (4.8), we have

$$(4.11) \qquad \langle r \otimes g, \mathscr{A}(w,w';s) \rangle = \bar{A}_1 \int_L \mathbf{d}^p \Omega(z) \cdot \overline{{}^t(\mathbf{d}^h M(\zeta[z],w')} y^{\beta+s\imath+\imath} d^*z.$$

Put $M(z,w') = \sum_{a \in F} \mu(a,w') \mathbf{e}_F(az)$. Applying [81, I, (3.7)] to $(\mathbf{d}^h M)(\zeta[z],w')$, we find that the right-hand side of (4.11) is equal to

$$A_4 A_5^s \prod_{\sigma \in \varepsilon} \Gamma(s + T_\sigma) \pi^{-\|T + s\varepsilon\|} \sum_{0 \neq a \in F/U} \omega_p(\mathrm{Tr}_{F/E}(\zeta a)) \cdot \overline{{}^t \mu(a,w')} a^{h - T - s\varepsilon},$$

where $T = k - \delta_2 - (1/2)\delta_1 - \mathrm{Res}_{/F}(\mu)$ and U is a sufficiently small subgroup of U_E of finite index, which we may assume to be the same as that of (3.14) or (4.4).

Now the arithmeticity of f_{ph} mentioned earlier combined with Lemmas 4.1 and 4.2 of [81, I] (or rather their generalizations to the functions on $\mathfrak{H}$) enables us to put

$$(4.12) \qquad \pi^{\gamma - e} p_W(-\varphi, 2\psi) \left\{ \prod_{\sigma \in \psi} \Gamma(s - \varphi_{\sigma\varrho})^{-1} \mathscr{A}(w,w';s) \right\}_{s=0}$$

$$= \sum \mathfrak{f}(w) \cdot {}^t \mathfrak{g}(w') \quad (w \in H^\delta, w' \in H^{(2)})$$

with finitely many $\bar{\mathbf{Q}}$-rational elements $\mathfrak{f}$ of $\mathscr{M}^{(1)} \times \mathscr{M}^{(2)}$ and $\mathfrak{g}$ of $\mathscr{M}^{(2)}$. Here we exclude the special case (4.5) which will be treated afterward. Taking the inner product with $r \otimes g$ and observing that $\prod_{\sigma \in \delta} \Gamma(T_\sigma) \sim \pi^{\|\delta_1\|/2}$, we find that

$$(4.13) \qquad A_6 \pi^{e-\|\delta\|} p_W(\varphi, 2\psi) \sum \langle \mathfrak{f}, r \otimes g \rangle \mathfrak{g}(w')$$

$$= \left[\sum_{0 \neq a \in F/U} \bar{\omega}_p(\mathrm{Tr}_{F/E}(\zeta a)) \mu(a, w') a^{h-T-s\varepsilon} \right]_{s=0}.$$

By [81, II, Proposition 5.1], we have

$$I(z, w', g) = A_7 \pi^{-\|\delta_2\|} \sum_{\alpha \in Y} \bar{C}^{(2)}(\alpha) N(\alpha)^n (g\|\alpha)(w') \mathbf{e}^{(2)}(N(\alpha)z),$$

where $n = (1/2)(k^{(2)} + \kappa^{(2)}) - \delta_2$, $C^{(2)} = \overset{(2)}{\prod} C_i$, and $Y = \overset{(2)}{\prod} Y_i$ with a complete set of

representatives Y_i of $\Gamma_i \backslash V_{i+}$. Here we take a sufficiently small congruence subgroup Γ_i of V_{i+}. Combining this with (3.12) and (4.7), we can write the inside of the brackets of (4.13) in the form

$$(4.14) \qquad A_8 \pi^{-\|\delta_2\|} \sum_{a, y, \alpha} \bar{\omega}_p(\mathrm{Tr}_{F/E}(\zeta a)) c^{(0)}(y) \mu_r(a^{(1)}) y^\xi$$

$$\cdot a^{h+\Delta-j-s\varepsilon} \bar{C}^{(2)}(\alpha)(g\|\alpha)(w'),$$

where $0 \neq a \in F/U$, y runs over $V^{(0)}$ under the condition $yy^\rho = a^{(0)}$, and α over the set $Y(a^{(2)})$ defined by

$$Y(x) = \left\{ \overset{(2)}{\underset{t_0}{\prod}} \Gamma_i \right\} \backslash \left\{ \alpha \in \overset{(2)}{\prod} V_{i+} | N(\alpha) = x \right\};$$

$a^{(v)}$ denotes the projection of $a F^{(v)}$.

Let $Z^{rg}(s, w')$ denote the series obtained from (3.14) by replacing ω and $c^{(2)}(x'') \overset{(2)}{\prod} \chi_i(x_i \mathfrak{a}_i)$ by $\bar{\omega}$ and $\sum_{\alpha \in Y(x'')} \bar{C}^{(2)}(\alpha)(g\|\alpha)(w')$; further let $W_{ph}^{rg}(s, w')$ denote

the series obtained from Z^{rg} by substituting $\bar{\omega}_p$ and $(x, yy^\rho)^{h+\Delta-j} y^\xi$ for $\bar{\omega}$ and $x^\Delta y^\varphi$. Then the sum of (4.14) is a positive rational number times W_{ph}^{rg}, and moreover, (4.3) shows that $Z^{rg} = \sum_{p, h} \gamma_{ph} W_{ph}^{rg}$. The above computation shows that

$$(4.15) \qquad Z^{rg}(0, w') = \pi^d p_W(\varphi, 2\psi) \sum \langle \mathfrak{f}, r \otimes g \rangle \mathfrak{g}(w')$$

with $d = e - \|\delta_1\|$ and finitely many $\bar{\mathbf{Q}}$-rational $\mathfrak{f}$ and $\mathfrak{g}$ (which may be different from those of (4.12)). With R as in Theorem 3.2, we can put $\mathfrak{f} = \sum_{r' \in R} r' \otimes g_{r'} + \mathfrak{f}'$ with $\bar{\mathbf{Q}}$-rational elements $g_{r'}$ of $\mathscr{S}^{(2)}$ and an element f' orthogonal to $r \otimes g$, and therefore

$$Z^{rg}(0, w') = \pi^d p_W(\varphi, 2\psi) \sum_{r' \in R} \sum_t \langle r', r \rangle \langle t, g \rangle q_{r't}(w')$$

with finitely many $\bar{\mathbf{Q}}$-rational elements t of $\mathscr{S}^{(2)}$ and $q_{r't}$ of $\mathscr{M}^{(2)}$. Hence if $a_{rr'}$ is defined as in Theorem 3.2, then

$$(4.16) \qquad \pi^{-d} p_W(-\varphi, 2\psi) \sum_{r \in R} a_{rr'} Z^{rg}(0, w') = \sum_t \langle t, g \rangle q_{r't}(w').$$

If $V^{(2)}$ is trivial, this, or rather (4.15), proves Theorem 3.2. In fact, in such a case, we can ignore g and $\mathfrak{g}$ in (4.15) to conclude that $Z^r(0) \sim \pi^d p_W(\varphi, 2\psi)$ as stated in the theorem.

Suppose $V^{(2)}$ is nontrivial. Then the technique employed in [81, II, the lower half of p. 600 and pp. 603–604] reduces the problem to the arithmeticity of

$$\pi^{-d} p_W(-\varphi, 2\psi) \sum_{q \in Q} b_{gg'} Z^{rg}(0, w') \text{ for every } g' \in Q, \text{ where } Q \text{ is a } \bar{\mathbb{Q}}\text{-rational basis of}$$

$\prod_{(2)} \mathscr{S}_{k_i, \kappa_i}(\Gamma_i, \varphi_i)$ and $(b_{gg'})$ is the inverse of $(\langle g, g' \rangle)_{g, g' \in Q}$. $(\mathscr{S}_{*, *}(\Gamma, \varphi)$ is the space of cusp forms defined in [81, II, p. 573].) Since we can take the t in (4.16) to be the members of Q and some elements orthogonal to them, the desired assertion follows immediately from (4.16).

The assertions of Theorem 3.3 or Proposition 3.4 concerning the residue can be proved in the same way. Considering $Z^r(s + s_0)$ instead of $Z^r(s)$ with $s_0 = (b_z - 2)/2$ or $(2b_z - 3)/4$ according as $b_z \in \mathbb{Z}$ or $b_z \notin \mathbb{Z}$, we may assume that $b_z = 2$ or $3/2$. Define again j by (4.1a). Conditions (3.18b,c) together with (Ra, b) of Theorem 3.3 imply that $0 \leq j \in I(F^{(1)} \times F^{(2)})$, and so we can again define f_{ph} by (4.4). This time we have $b_f = 0$ or $1/2$. Suppose for the moment $b_f = 0$. By [86, Theorem 4.3], f_{ph} has at most a simple pole at $s = 1$, and the residue is nearly holomorphic. Moreover, the residue times $R_E^{-1} \pi^{-e} p_W(-\varphi, 2\psi)$ is arithmetic, by virtue of [86, Theorem 4.6]. If the residue is holomorphic, we can repeat the above proof by taking the residue at $s = 1$ instead of the value at $s = 0$. The same reasoning applies to the case $b_z = 2/3$. Therefore it remains to settle the cases in which the residue or the value at $s = 0$ is nonholomorphic. (The latter can happen when (4.5) holds.)

For this purpose, we consider the symbols B, k, κ, α and d as in (3.1), (3.5), and (3.6), and define $\mathscr{N}_{k, \kappa}(B)$ to be the set of all nearly holomorphic functions on H^α with values in $\mathbb{C}^d$ and with the same behavior under the elements of $\mathscr{M}_{k, \kappa}(B)$. We can then speak of the set $\mathscr{N}_{k, \kappa}^*(B)$ of all the $\bar{\mathbb{Q}}$-rational elements of $\mathscr{N}_{k, \kappa}(B)$. (For the precise definition, see [87a, §5].) Put then $\mathscr{S} = \mathscr{S}^{(1)} \otimes \mathscr{S}^{(2)} \otimes \mathscr{S}^{(2)}$ and

$$\mathscr{N} = [\otimes^{(1)} \mathscr{N}_{2k_i, 2\kappa_i}(B_i)] \otimes [\otimes^{(2)} \mathscr{N}_{k_i, \kappa_i}(V_i)]^{\otimes 2},$$

$$\mathscr{N}^* = [\otimes^{(1)} \mathscr{N}_{2k_i, 2\kappa_i}^*(B_i)] \otimes [\otimes^{(2)} \mathscr{N}_{k_i, \kappa_i}^*(V_i)]^{\otimes 2}$$

where $A^{\otimes 2}$ means $A \otimes A$. (Thus an element of $\mathscr{N}$ is a function on $\mathfrak{H}$.) Then $\mathscr{N}^*$ is the set of all arithmetic elements of $\mathscr{N}$, and spans $\mathscr{N}$ over $\bar{\mathbb{Q}}$. (See [83a, Lemma 1.1] and [87a, (3.11)].)

Lemma 4.1. *Every element $\mathfrak{f}$ of $\mathscr{N}^*$ can be written in the form $\mathfrak{f} = \mathfrak{f}_0 + \mathfrak{f}_1$ with $\mathfrak{f}_0 \in \mathscr{S} \cap \mathscr{N}^*$ and an element $\mathfrak{f}_1$ orthogonal to $\mathscr{S}$.*

Proof. Obviously it is sufficient to prove the case of a single quaternion algebra, that is, the case in which $\mathscr{N} = \mathscr{N}_{k, \kappa}(B)$. In such a case, we have proved in [87a, Theorem 5.4] that $\mathfrak{f} = \mathfrak{g}_0 + \sum_{n=1}^{r} D_n \mathfrak{g}_n$ with a $\bar{\mathbb{Q}}$-rational holomorphic element $\mathfrak{g}_0$ of $\mathscr{N}$ and nearly holomorphic elements $\mathfrak{g}_n$ of lower weights, where each D_n is a nontrivial differential operator which increases the weight. Now $D_n \mathfrak{g}_n$ is orthogonal to $\mathscr{S}$, since $\langle D_n \mathfrak{g}_n, \mathfrak{h} \rangle = \langle \mathfrak{g}_n, E\mathfrak{h} \rangle$ with an operator E which annihilates holomorphic functions, as shown in [81, I, (2.18)] and [84, Theorems 11.4 and 11.5]. The element

$\mathfrak{g}_0$ belongs to $\mathscr{S}$ if B is a division algebra. If B is not a division algebra, we can express $\mathfrak{g}_0$ as a sum of an Eisenstein series and a $\bar{\mathbb{Q}}$-rational cusp form $\mathfrak{f}_0$, a fact that follows from [87b, Theorem 9.1]. This completes the proof.

In order to settle case (4.5), we first observe that (4.12) holds with $\bar{\mathbb{Q}}$-rational nearly holomorphic $\mathfrak{f}$ and $\mathfrak{g}$, by virtue of the arithmeticity of $f_{ph}(w, w'; 0)$ guaranteed by [86, Theorem 4.5]. Applying the above lemma to these $\mathfrak{f}$, we find that (4.16) is valid with $\bar{\mathbb{Q}}$-rational holomorphic elements t, which leads to the same conclusion as before. The problems about the residue can be handled in the same way.

4*. Proof of Proposition 3.1

Taking an element $f(z) = \sum\limits_{b \in F} \xi(b) \mathbf{e}_F(bz)$ of $\overset{t}{\underset{i=1}{\otimes}} \mathscr{S}_{n_i}(F_i)$ with the F_i as in Sect. 3, we define a series Λ by

$$(4^*.1) \qquad \Lambda(s) = [U] \sum_{0 \ll b \in F/U} \xi(b) \omega_p(\mathrm{Tr}_{F/E}(\zeta b)) b^{d-s\varepsilon},$$

where $s \in \mathbb{C}$, $\omega_p(a) = a^p \omega(a)$ with $0 \leq p \in I_E$ and ω of (3.13), $\zeta \in F$, and $d \in (1/2) I_F$. We assume, in addition to (3.9a,b) and (3.17),

$$(4^*.2a) \qquad\qquad d_\tau \geq 0 \quad \textit{for every} \quad \tau \in \varepsilon',$$

$$(4^*,2b) \qquad l + 2p + \mathrm{Res}_{F/E}(n+2d) = \alpha \iota_E \quad \textit{with} \quad \alpha \in 2^{-1}\mathbb{Z}.$$

For the same reason as in (3.14), we can find a subgroup U of U_E of finite index that makes the sum of (4*.1) meaningful. This type of series was introduced in [81, I, (3.2)] and [83a, (1.9)].

Theorem 4*.1. *The series Λ of (4*.1) is convergent for sufficiently large $\mathrm{Re}(s)$. Moreover Λ has a meromorphic continuation to the whole plane. It is holomorphic for $\mathrm{Re}(s) \geq (\alpha-1)/2$ except for a possible simple pole at $s = \alpha/2$ when $\alpha \in \mathbb{Z}$, at $s = (2\alpha-1)/4$ when $\alpha \notin \mathbb{Z}$. The pole at $\alpha/2$ does not occur if $2n_\tau + 2d_\tau > \alpha$ for some $\tau \in \varepsilon$; the pole at $(2\alpha-1)/4$ does not occur if $\alpha - (1/2) < 2n_\tau + 2d_\tau \equiv \alpha - (1/2) \pmod 2$ for some $\tau \in \varepsilon$.*

The assertions in the special case $F = F_1$ was proved in [81, I, pp. 140–142]; the proof can easily be adapted to the general case. In fact, the idea of the proof is the same as in the proof of Proposition 1.1. In [81] and also in [83a], α was always assumed to be an integer. The case of half-integral α can be handled by employing an Eisenstein series of half-integral weight whose properties are recalled in Lemma 1.4.

Now the series Z^r of (3.14) is essentially a special case of (4*.1). To see this, we denote by f_0 the function of (4.7), and put $f(z) = f_0(z^{(0)}) I(r, z^{(1)}) f_2(z^{(2)})$ with

$$f_2(z) = \sum_{a \in F^{(2)}} c^{(2)}(a) a^u \overset{(2)}{\prod} \chi_i(a_i \mathfrak{a}_i) \mathbf{e}^{(2)}(az) \quad (z \in H^{(2)}),$$

where $u = (1/2)(k^{(2)} + \kappa^{(2)}) - \delta_2$. Then $f \in \overset{t}{\underset{i=1}{\otimes}} \mathscr{S}_{n_i}(F_i)$ with

$$n = k + \kappa + \mathrm{Res}_{/F}(\lambda) + \delta_0 + \delta_0' + (1/2)\delta_1 + (3/2)\delta_1' + 2\delta_2'.$$

Given $x^4 y^\varphi$ as in (3.14), we see that if $b=(x, yy^\varrho)\in F$, then $x^4 y^\varphi = y^\lambda b^{d+u}$ with

$$d = \Delta + \operatorname{Res}_{/F}\left(\sum_{\sigma\in\psi+\psi'} \varphi_{\sigma\varrho}\sigma\right) + \delta_2 - (1/2)(k^{(2)}+\kappa^{(2)})\,.$$

Therefore, if $c(x'', y) = c^{(0)}(y)c^{(2)}(x'')$, then Z^r is a constant times the series Λ defined with these f, d, and $p=0$. Observe that (4*.2a) follows from (3.18a,b,c), (3.15) implies (4*.2b) with $\alpha=b_Z$, and N_τ of (3.20b) coincides with $2n_\tau + 2d_\tau$. Therefore Proposition 3.1 follows immediately from Theorem 4*.1.

5. Applications of easier types

First of all, the removal of condition (3.22) has the following proposition as its immediate consequence:

Proposition 5.1. *In assertion* (ii) *of* [81, II, Theorem 3.11], *the condition "$\mu \neq 4$ when $E=\mathbb{Q}$" is unnecessary.*

We can also eliminate [81, I, (4.8b); II, (3.12b)], but this point is already incorporated in Theorems 1.2 and 3.2.

Our main applications concern the special values or the residue of a series of the form

$$(5.1) \qquad D(s;f,g) = [U] \sum_{0\,\ll\,b\in E/U} \lambda(b)\mu(b)b^{-(k+l)/2} N_{E/\mathbb{Q}}(b)^{-s}$$

which is defined in the following way. We fix a basic totally real algebraic number field E, take $f\in\mathcal{M}_k(E)$ and $g\in\mathcal{M}_l(E)$ with integral or half-integral weights k and l, consider their Fourier expansions

$$f(z) = \sum_{b\in E} \lambda(b)\mathbf{e}_E(bz), \quad g(z) = \sum_{b\in E} \mu(b)\mathbf{e}_E(bz) \quad (z\in H^{J(E)})\,,$$

and observe that the sum of (5.1) is meaningful for a sufficiently small subgroup U of U_E of finite index. We always assume that either f or g is a cusp form. Obviously (5.1) is a special case of (4*.1), and therefore it has analytic properties stated in Theorem 4*.1.

Now the nature of the special values or the residue of D depends on whether the weights are integral or half-integral, and also on the difference between them. In fact we shall give our results on the special values according to the classification ((IA–IVB) of (k, l) explained in the introduction, where we employed symbols J, δ, and δ'. In the present setting we have of course $J=J_E=\delta\cup\delta'$.

Ir $k_v=l_v$ for some v, the shape of the gamma factors of D eliminates the possibilities of the integer points where D may have a reasonable property of algebraicity, except that we can investigate the case $f=g$ by taking the symmetric square Euler product instead of D (see Theorem 5.5 below). Case (*A) may be viewed as an extreme case of (*B), but the author thinks that separate formulation of cases (*A) will make the exposition easier to understand.

Now the results in Case (IA) were already given in [78], and those in Case (IB) in [83a] as immediate consequences of a special case of Theorem 3.2. In this section, we treat only (IIA), since the remaining cases require a detailed analyis of the correspondence between the forms of half-integral weight and those of integral weight occurring on quaternion algebras, which will be done in Sect. 6.

In Case (IIA), we formulate the results in terms of a series of the form

$$(5.2) \qquad W(s) = [U] \sum_{0 \,\ll\, x \in E/U} \eta(x)\omega(x)\chi(x\mathfrak{a})x^{(a\imath - l)/2} N_{E/\mathbb{Q}}(x)^{-s}$$

with symbols and assumptions as follows: $\Omega = \sum_{a \in E} \omega(a)\mathbf{e}_E(az) \in \mathcal{M}_l(E)$, $l \in I^0(E)$, $\eta \in \mathscr{L}(E)$, χ is a primitive system of eigenvalues on $\mathscr{S}_{k,0}(M_2(E))$, $k \in I_E$, $\mathfrak{a}$ is a fractional ideal of E, $a \in (1/2)\mathbb{Z}$, and $\imath = \imath_E$; we assume

$$(5.3) \qquad k_v + l_v \bmod 2 \quad for \quad v \in J_E \quad is \ independent \ of \quad v,$$

and take a so that $a \equiv k_v + l_v \pmod 2$; U is a subgroup of U_E of finite index such that $\eta(ux) = \eta(x)$ and $\omega(ux) = u^{l/2}\omega(x)$, and the sum is taken over all Ux with $0 \ll x \in E$.

The series W is a special case of (3.14) with $F = F^{(2)} = E$, $B = M_2(E)$, $\zeta = 1$, and $\Delta = (a\imath - l)/2$. Thus W has the analytic properties as in Proposition 3.1. As special cases of Theorems 3.2 and 3.3, we obtain

Theorem 5.2. *Suppose that η is $\bar{\mathbb{Q}}$-valued and Ω is $\bar{\mathbb{Q}}$-rational. Let $\mathbf{g}$ be a $\bar{\mathbb{Q}}$-rational nonzero eigenform in $\mathscr{S}_{k,0}(M_2(E))$ belonging to χ, and let s_0 be an integer such that*

$$a + 1 \leqq 2s_0 \leqq k_v - l_v + a \quad for \ every \quad v \in J_E,$$

$$2s_0 \neq a + (3/2) \quad if \quad E = \mathbb{Q} \quad and \quad l \notin I_E.$$

Then W is finite at s_0, and $W(s_0) \sim \pi^{\|k\|}\langle \mathbf{g}, \mathbf{g} \rangle$.

Theorem 5.3. *With η, Ω, and $\mathbf{g}$ as above, suppose either $k_v = l_v$ for all $v \in J_E$ or $k_v - l_v \geqq 1/2$ for all $v \in J_E$. Then W has at most a simple pole at $(a/2) + 1$ if $l \in I_E$, and at $(2a + 3)/4$ if $l \notin I_E$; the residue at either point is an algebraic number times $R_E \pi^{\|k\| - v}\langle \mathbf{g}, \mathbf{g} \rangle$, where $v = 0$ if $l \in I_E$ and $v = [E : \mathbb{Q}]$ if $l \notin I_E$.*

If $l \in I_E$, Theorem 5.2 is essentially included in [78, Theorem 4.1]. When $l \notin I_E$ and $E = \mathbb{Q}$, the result is also included in the work [St] of Sturm; see Remark 5.6 below. Now we can state a generalization of Lemma 1.5 as follows:

Corollary 5.4. *Let χ and $\mathbf{g}$ be as in Theorem 5.2, and let*

$$q(z) = \sum_{0 \,\ll\, b \in E} \eta(b)\chi(b\mathfrak{a})b^{(k/2) - \imath}\mathbf{e}_E(bz) \quad (z \in H^{J(E)})$$

with $\mathfrak{a}$ and a $\bar{\mathbb{Q}}$-valued η as above.. Then $q \in \mathscr{S}_k(E)$, and $\langle p, q \rangle \sim \langle \mathbf{g}, \mathbf{g} \rangle$ for every $\bar{\mathbb{Q}}$-rational element p of $\mathcal{M}_k(E)$.

Proof. The first assertion is included in [81, II, Theorem 3.1]. To prove the second one, let $p(z) = \sum_b h(b)\mathbf{e}_E(bz) \in \mathcal{M}_k(E)$ and

$$W(s) = [U] \sum_{0 \,\ll\, b \in E/U} \eta(b)\overline{h(b)}\chi(b\mathfrak{a})b^{-\imath - (k/2)} N_{E/\mathbb{Q}}(b)^{-s}.$$

This is a series of type (5.2) with $l=k$ and $a=-2$. As shown in [81, I, (2.15)], we have

$$(4\pi)^{-\|k\|-vs} \prod_{v \in J(E)} \Gamma(s+k_v) W(s) = C \int_\Phi \overline{p(z)} q(z) \mathscr{E}_0^0(z,s+1) y^k d^*z,$$

where $C \in \bar{\mathbb{Q}}^\times$, $\Phi = \Gamma \backslash H^{J(E)}$ with a sufficiently small congruence subgroup Γ of $SL_2(E)$, and $\mathscr{E}_0^0$ is defined by [86, (10.14b)]. Comparing the residues at $s=0$ of both sides, we obtain the desired result from Lemma 1.4, (iv) and Theorem 5.3.

To obtain another application, we take a Hecke character φ of E of finite order, denote by φ^* the ideal-character attached to φ, and put

$$(5.4) \qquad D(s) = \sum_{\mathfrak{x}} \varphi^*(\mathfrak{x}) \chi(\mathfrak{x}^2) N(\mathfrak{x})^{-s},$$

where $\mathfrak{x}$ runs over all the integral ideals of E, and χ is the same as in (5.2). We know that D has an Euler product of the form

$$(5.5) \qquad D(s) \prod_{\mathfrak{p} \nmid \mathfrak{b}} [1 - (\varphi\psi)^*(\mathfrak{p}^2) N(\mathfrak{p})^{2-2s}]^{-1} = \prod_{\mathfrak{p}} R_\mathfrak{p}(N(\mathfrak{p})^{-s})^{-1},$$

where ψ is the Hecke character of E to which χ belongs in the sense of [81, II, (1.21)], $\mathfrak{b}$ is the product of the conductor of φ and that of χ, and $R_\mathfrak{p}$ is a polynomial of degree ≤ 3, whose explicit form in the case $E=\mathbb{Q}$ is given in [75]. Take $r \in I_E$ so that $0 \leq r \leq \iota_E$ and

$$(5.6) \qquad \varphi(x) = \operatorname{sgn}(x)^r \quad \text{for} \quad x \in E_\infty^\times.$$

Theorem 5.5. *Let* χ, k, *and* $\mathbf{g}$ *be the same as in Theorem 5.2. Suppose that* $k_v - r_v$ (mod 2) *is independent of* v. *Then the following assertions hold:*

(i) D can be continued to the whole plane as a meromorphic function, and has at most a simple pole at $s=2$.

(ii) Let μ be an integer such that

$$1 < \mu \leq k_v - r_v \equiv \mu \pmod 2 \quad \text{for every} \quad v \in J_E,$$

$$\mu \neq 2 \quad \text{if} \quad E = \mathbb{Q}.$$

Then D is finite at μ and $D(\mu) \sim \pi^{\|k\|} \langle \mathbf{g}, \mathbf{g} \rangle$.

(iii) Suppose $k_v \geq r_v + 1$ for every $v \in J_E$. Then the residue of D at $s=2$ is an algebraic number times $R_E \pi^{\|k\|-v} \langle \mathbf{g}, \mathbf{g} \rangle$, where $v = [E:\mathbb{Q}]$.

Proof. Let A be a complete set of representatives of the ideal classes of E. Then

$$D(s) = \sum_{\mathfrak{a} \in A} N(\mathfrak{a})^s \sum_{0 \neq b \in \mathfrak{a}/V} \varphi^*(b\mathfrak{a}^{-1}) \chi(b^2 \mathfrak{a}^2) |N_{E/\mathbb{Q}}(b)|^{-s},$$

where V denotes the group of units in E. With a fixed $\mathfrak{a}$, put

$$\Omega(z) = \sum_{b \in \mathfrak{a}} \operatorname{sgn}(b)^r \varphi^*(b\mathfrak{a}^{-1}) b^r \mathbf{e}_E(b^2 z) \quad (z \in H^{J(E)}).$$

Observe that $\operatorname{sgn}(b)^r \varphi^*(b\mathfrak{a}^{-1})$ as a function of b, defined to be 0 outside $\mathfrak{a}$, belongs to $\mathscr{L}(E)$. Hence $\Omega \in \mathscr{M}(r + (1/2), E)$ by [80, Proposition 7.1]. Consider W of (5.2) with this Ω and with $\mathfrak{a}^{-2}$ in place of $\mathfrak{a}$, taking η to be the characteristic function of $\mathfrak{a}^2$.

Then

$$W((2s+2a-1)/4)=2^m \sum_{0 \neq b \in \mathfrak{a}/V} \varphi^*(b\mathfrak{a}^{-1})\chi(b^2\mathfrak{a}^{-2})|N_{E/\mathbb{Q}}(b)|^{-s}$$

with $m \in \mathbb{Z}$. Therefore we obtain our assertions immediately from Theorems 5.2 and 5.3.

Remark 5.6. We investigated in [75] some analytic properties of D when $E=\mathbb{Q}$ and showed in particular the holomorphy of the function of (5.5) times certain gamma factors. The algebraicity of $D(\mu)$ was given by Sturm [St] when $E=\mathbb{Q}$. The above assertion (ii) generalizes a portion (roughly 1/2) of Sturm's results. Thorough investigations of Cases (IIA) and (IIIA) are being carried out in the thesis of J. Im [I], which will give complete generalizations of [St] and the result of type (IIIA) in [81z]. In the present paper, we have contented ourselves with stating the results which are direct consequences of those in Sect. 3 with the hope that the reader may gain some insight by comparing the above theorems with the subsequent ones in other cases.

6. Forms of half-integral weight corresponding to forms on quaternion algebras

In this section, we fix a basic totally real algebraic number field E, and we let $\mathbf{a}$ and $\mathbf{f}$ denote the sets of archimedean and nonarchimedean primes of E, respectively, $\mathfrak{g}$ the maximal order of E, and $\mathfrak{d}_E$, or simply $\mathfrak{d}$, the different of E over $\mathbb{Q}$. (Thus $\mathbf{a}$ is a new symbol for J_E.) we put $G=SL_2(E)$ and define its parabolic subgroup P by $P=\{x \in G|c_x=0\}$. The adelization of G is denoted by $G_\mathbf{A}$, its archimedean and nonarchimedean factors by $G_\mathbf{a}$ and $G_\mathbf{f}$, respectively. This use of subscripts will apply to other algebraic groups. Given two fractional ideals $\mathfrak{y}$ and $\mathfrak{z}$ such that $\mathfrak{y}\mathfrak{z} \subset \mathfrak{g}$, we put

(6.0a)
$$D[\mathfrak{y}, \mathfrak{z}]=SO(2)^\mathbf{a} \prod_{v \in \mathbf{f}} D_v[\mathfrak{y}, \mathfrak{z}] \quad (\subset G_\mathbf{A}),$$

(6.0b)
$$D_v[\mathfrak{y}, \mathfrak{z}]=\{x \in G_v|a_x \in \mathfrak{g}_v, b_x \in \mathfrak{y}_v, c_x \in \mathfrak{z}_v, d_x \in \mathfrak{g}_v\},$$

(6.0c)
$$\Gamma[\mathfrak{y}, \mathfrak{z}]=G \cap D[\mathfrak{y}, \mathfrak{z}],$$

where the subscript v means the localization at v as usual.

In [85] and [87b, §2], we defined a factor of automorphy $h(\alpha, z)$ for $\alpha \in G \cap P_\mathbf{A} D[2\mathfrak{d}^{-1}, 2\mathfrak{d}]$. For such an α and a half-integral weight $n \in I^0(E)$, we define a factor of automorphy J_n by

(6.1)
$$J_n(\alpha, z)=j(\alpha, z)^m h(\alpha, z), \quad m=n-(1/2)\iota_E.$$

For a function f on $H^\mathbf{a}$, we put $(f\|_n\alpha)(z)=J_n(\alpha, z)^{-1}f(\alpha z)$.

Let $\mathfrak{c}$ be an integral ideal in E divisible by 4, and ψ a Hecke character of E (of finite or infinite order) such that

(6.2)
$$\psi_\mathbf{a}(-1)=(-1)^{\|m\|} \quad \text{and the conductor of } \psi \text{ divides } \mathfrak{c}.$$

Then we denote by $\mathscr{S}_n(\mathfrak{c}, \psi)$ the set of all $f \in \mathscr{S}_n(E)$ such that $f\|_n\gamma=\psi_\mathfrak{c}(a_\gamma)f$ for every $\gamma \in \Gamma[2\mathfrak{d}^{-1}, 2^{-1}\mathfrak{c}\mathfrak{d}]$. For such an f and each fractional ideal $\mathfrak{x}$ in E, we can consider the Fourier expansion of f at the cusp corresponding to $\mathfrak{x}$ as follows (see [87b, Proposition 3.1]):

$$(6.3) \qquad \psi_{\mathbf{a}}(d_\beta) \prod_{v\nmid\mathfrak{c}} \psi_v(d_\beta t) J_n(\beta, \beta^{-1}z) f(\beta^{-1}z)$$

$$= N(\mathfrak{x})^{-1/2} \sum_{0 \ll \xi \in E} \lambda(\xi, \mathfrak{x}; f, \psi) \mathbf{e}_E(\xi z/2).$$

Here $\beta \in G \cap \operatorname{diag}[t, t^{-1}] D[2\mathfrak{d}^{-1}, 2^{-1}\mathfrak{c}\mathfrak{d}] G_{\mathbf{a}}$ and t is an element of $E_{\mathfrak{f}}^\times$ such that $\mathfrak{x} = t\mathfrak{g}$. We often write simply $\lambda_f(\xi, \mathfrak{x})$ for $\lambda(\xi, \mathfrak{x}; f, \psi)$. If $\mathfrak{x} = \mathfrak{g}$, (6.3) is merely the expansion $f(z) = \sum \lambda_f(\xi, \mathfrak{g}) \mathbf{e}_E(\xi z/2)$.

In Sect. 3, we employed a form of half-integral weight as $I(z, g)$ with a form g on a quaternion algebra. The purpose of this section is a closer examination of the map $g \mapsto I(z, g)$; in particular, we shall show its commutativity with Hecke operators, which was first proved by Shintani when the algebra in question is $M_2(\mathbb{Q})$, and by the author in [82] more generally when $E = \mathbb{Q}$. Thus we fix a quaternion algebra B over E, and define $B_{\mathbf{A}}$, $B_{\mathbf{a}}$, $B_{\mathfrak{f}}$, $B_{\mathbf{A}}^\times$, etc. as usual, and assume that $B_{\mathbf{a}} (= B \otimes_{\mathbb{Q}} \mathbb{R})$ is isomorphic to $M_2(\mathbb{R})^\delta \times \mathbb{H}^{\delta'}$ with subsets δ and δ' of J_E. In other words, the setting is that of (3.1) with E, δ, and δ' instead of K, α, and β there; we assume $\delta \neq \emptyset$. We denote by $\mathfrak{d}_B$ the product of all the prime ideals of E ramified in B.

Given an integral ideal $\mathfrak{m}$ of E, we define an order $\mathfrak{o}_1$ of level $\mathfrak{m}$ in B as follows. Fix a maximal order $\mathfrak{o}$ in B and also an E_v-linear isomorphism μ_v, for each $v \in \mathfrak{f}$ prime to $\mathfrak{d}_B$, of B_v onto $M_2(E_v)$ so that $\mu_v(\mathfrak{o}_v) = M_2(\mathfrak{g}_v)$. Then we define $\mathfrak{o}_1$ to be a g-lattice in B such that

$$(6.4) \qquad \mathfrak{o}_{1v} = \begin{cases} \mu_v^{-1}(\{x \in M_2(\mathfrak{g}_v) \mid b_x \in \mathfrak{d}_v^{-1}, \ c_x \in \mathfrak{m}_v \mathfrak{d}_v\}) & \text{if } v \nmid \mathfrak{d}_B, \\ \mathfrak{g}_v + \mathfrak{m}_v \mathfrak{o}_v & \text{if } v \mid \mathfrak{d}_B, \end{cases}$$

and also a subgroup $W_{\mathfrak{m}}$, written also simply W, of $B_{\mathbf{A}}^\times$ by

$$W = W_{\mathfrak{m}} = \{x \in B_{\mathbf{A}}^\times \mid N(x_{\mathbf{a}}) \gg 0, \ x_v \in \mathfrak{o}_{1v}^\times \text{ for every } v \in \mathfrak{f}\}.$$

We take a coset decomposition

$$(6.5) \qquad B_{\mathbf{A}} = \bigcup_{\lambda=1}^{\omega} B^\times x_\lambda W = \bigcup_{\lambda=1}^{\omega} W x_\lambda B$$

with elements x_λ of $B_{\mathfrak{f}}^\times$ such that $x_{\lambda v} = 1$ for $v \mid \mathfrak{d}_B$ and $\mu_v(x_{\lambda v}) = \operatorname{diag}[1, t_{\lambda v}]$ for $v \nmid \mathfrak{d}_B$. Here $t_\lambda \in E_{\mathfrak{f}}^\times$ and $t_{\lambda v} = 1$ for $v \mid \mathfrak{d}_B$. Obviously $N(x_\lambda) = t_\lambda$.

Let Φ be a Hecke character of E such that

$$(6.6) \quad \textit{the conductor of } \Phi \textit{ is prime to } \mathfrak{d}_B \textit{ and divides } \mathfrak{m}; \ \Phi_{\mathbf{a}}(x) = \operatorname{sgn}(x_{\mathbf{a}})^{q+r} |x_{\mathbf{a}}|^{2i\mu},$$

where $q \in \mathbb{Z}^\delta$, $r \in \mathbb{Z}^{\delta'}$, and $\mu \in \mathbb{R}^{\mathbf{a}}$, $\|\mu\| = 0$. We then define $\mathscr{S}_{q,r}(\mathfrak{m}, \Phi; B)$, with B often suppressed, to be the set of all the vector-valued functions $\mathbf{h}$ on $B_{\mathbf{A}}^\times$ satisfying the following conditions:

$$(6.7\text{a}) \qquad \mathbf{h}(\beta x u) = \Phi_{\mathfrak{m}}(d_u) \mathbf{h}(x) \quad \textit{if} \quad \beta \in B^\times \quad \textit{and} \quad u \in W, u_{\mathbf{a}} = 1 \ ;$$

$$(6.7\text{b}) \qquad \mathbf{h}(sx) = \Phi(s) \mathbf{h}(x) \quad \textit{for} \quad s \in E_{\mathbf{A}}^\times \ ;$$

$$(6.7\text{c}) \qquad \textit{For every } x \in B_{\mathfrak{f}}^\times, \textit{ there is an element } g_x \textit{ of } \mathscr{S}_{q,r}(B) \textit{ such that}$$

$$\mathbf{h}(xy) = N(y)^{i\mu} (g_x \|_{q,r} y)(\mathbf{i}) \quad \textit{for every} \quad y \in B_{\mathbf{a}+} .$$

Here $\mathbf{i}$ is the origin $(i,\ldots,i)$ of H^δ and $\Phi_{\mathfrak{m}} = \prod_{v|\mathfrak{m}} \Phi_v$; $d_\mathfrak{u}$ is the element of $E_\mathfrak{f}^\times$ whose v-component is 1 or the d-entry of $\mu_v(u)$ according as $v\nmid\mathfrak{e}$ or $v|\mathfrak{e}$, where $\mathfrak{e}$ is the conductor of Φ. We define $a_\mathfrak{u}$ similarly with a-entries instead of d-entries. This notation will be employed always in the form $\Phi_{\mathfrak{m}}(d_\mathfrak{u})$ or $\Phi_{\mathfrak{m}}(a_\mathfrak{u})$, and so no confusion will arise. Condition (6.7c) implies that $\mathbf{h}$ takes its values in $\mathbb{C}^e$ with $e=\prod_{v\in\delta'} (r_v+1)$ (see a few paragraphs after (3.5)).

Put $\Gamma_\lambda = B^\times \cap x_\lambda W x_\lambda^{-1}$ and denote by $\mathscr{S}_{q,r}(\Gamma_\lambda, \Phi)$ the subset of $\mathscr{S}_{q,r}(B)$ consisting of the elements f such that

$$(6.8a) \qquad f\big\|_{q,r}\gamma = \Phi_{\mathfrak{m}}(a_\gamma) N(\gamma)^{i\mu} f \quad \text{for every} \quad \gamma\in\Gamma_\lambda.$$

Given $(f_1,\ldots,f_\omega)\in \prod_{\lambda=1}^{\omega} \mathscr{S}_{q,r}(\Gamma_\lambda, \Phi)$, define, in view of (6.5), a function $\mathbf{h}$ on $B_\mathbf{A}^\times$ by

$$(6.8b) \qquad \mathbf{h}(\alpha x_\lambda u) = \Phi(t_\lambda)\Phi_{\mathfrak{m}}(d_\mathfrak{u}) N(u_\mathfrak{a})^{i\mu} (f_\lambda\big\|_{q,r} u_\mathfrak{a})(\mathbf{i}).$$

Then it satisfies (6.7a,c). Conversely every $\mathbf{h}$ satisfying (6.7a,c) can be obtained in that way. We will often identify $\mathbf{h}$ with $(f_1,\ldots,f_\omega)$ and view $\mathscr{S}_{q,r}(\mathfrak{m}, \Phi)$ as a subset of $\prod_{\lambda=1}^{\omega} \mathscr{S}_{q,r}(\Gamma_\lambda, \Phi)$.

We are going to associate an element of $\mathscr{S}_n(F)$ with an element of $\mathscr{S}_{2k,2\kappa}(\mathfrak{m}, \Phi)$ with $k\in\mathbb{Z}^\delta$ and $\kappa\in\mathbb{Z}^{\delta'}$, where

$$(6.9) \qquad n = k + \kappa + (1/2)\delta + (3/2)\delta'.$$

This requires a theta function $\theta(z, w; \eta, \tau)$, with τ often suppressed, defined by

$$(6.10) \quad \theta(z, w; \eta, \tau) = \theta(z, w; \eta)$$
$$= \operatorname{Im}(z)^{\delta/2} \operatorname{Im}(w)^{-2k} \sum_{\alpha\in V} \eta(\alpha) q(\alpha) [\alpha, \bar{w}]^k e_E((\tau/2) R'[\alpha; z, w]).$$

Here $z\in H^\mathfrak{a}$, $w\in H^\delta$, $\eta\in\mathscr{L}(V)$, $V=\{\alpha\in B | \operatorname{Tr}(\alpha)=0\}$, q is the function q' of (3.25b) (with $V^{(1)}=V$, $\kappa^{(1)}=\kappa$, $F^{(1)}=E$),

$$(6.10a) \qquad [\alpha, w] = \left((-1, w_v)\alpha_v \begin{bmatrix} w_v \\ 1 \end{bmatrix}\right)_{v\in\delta} \quad (\in\mathbb{C}^\delta),$$

$$(6.10b) \qquad R'[\alpha; z, w]_v = \begin{cases} N(\alpha)_v \bar{z}_v & \text{if } v\in\delta', \\ N(\alpha)_v z_v + (iy_v/2)\operatorname{Im}(w_v)^{-2}|[\alpha_v, w_v]|^2 & \text{if } v\in\delta; \end{cases}$$

τ is an element of E such that $\tau_v > 0$ for $v\in\delta$ and $\tau_v < 0$ for $v\in\delta'$. Thus $\theta(z, w; \eta)$ is essentially the same as θ_i of (3.32b) (if B_i, F_i, and ε_i there are B, E, and δ here). We easily see that

$$(6.11) \qquad N(\gamma)^\kappa \sigma_{2\kappa}(\gamma)^{-1} j(\gamma, w)^{-2k}\theta(z, \gamma w; \eta) = \theta(z, w; \eta^\gamma) \quad \text{for } \gamma\in B_+,$$
$$\text{where } \eta^\gamma(\alpha) = \eta(\gamma\alpha\gamma^{-1}).$$

It can easily be seen that $\mathscr{L}(V)$ consists of the restrictions to V of the Schwartz-Bruhat functions on $V_\mathfrak{f}$. Therefore, identifying each element of $\mathscr{L}(V)$ with its natural extension to $V_\mathfrak{f}$, we view $\mathscr{L}(V)$ as the set of such functions on $V_\mathfrak{f}$.

Lemma 6.1. *For $\eta \in \mathscr{L}(V)$ and $\beta \in G \cap P_{\mathbf{A}} D [2\mathfrak{d}^{-1}, 2\mathfrak{d}]$, there is an element $^{\beta}\eta$ of $\mathscr{L}(V)$ such that $\theta(\beta z, w; {}^{\beta}\eta) = \overline{J_n(\beta, z)} \, (z, w; \eta)$ with n of* (6.9).

Proof. This is a special case of [87b, Proposition 11.8]. In fact, $y^{-\delta/2}\theta$ is a special case of $f(z, \eta)$ of [87b, (11.14)] with $S[\alpha] = \tau N(\alpha)$; the present symbols $q(\alpha) [\alpha, \bar{w}]^k$, 0, $k + \kappa$, and $2\delta + 3\delta'$ correspond to $\sigma(\xi)$, λ, λ', and s there. Notice also that $h(\beta, z)^4 = j(\beta, z)^2$ by [85a, (3.10d)].

Lemma 6.2. *Let φ be the Hecke character of E corresponding to the extension $E(\tau^{1/2})/E$, and $\mathfrak{f}$ the conductor of φ. Given $\eta \in \mathscr{L}(V)$, there exists an open subgroup U of $D[2\mathfrak{d}^{-1}, 2\mathfrak{f}\mathfrak{d}] G_{\mathbf{a}}$ with the property that if $\beta \in G \cap \operatorname{diag}[p, 1/p] U$ with $p \in E_{\mathfrak{f}}^{\times}$, then $^{\beta}\eta$ of Lemma 6.1 can be given by $^{\beta}\eta(x) = |p|^{3/2} \varphi(p) \eta(px)$.*

Proof. If S is a 3×3-matrix representing the quadratic form $\alpha \mapsto \tau N(\alpha)$, then we see that $\tau \det(S)$ is a square of an element of E. Hence we can derive our assertion immediately from [87b, Proposition 11.5], by taking U so that $(d_u)_v - 1 \in \mathfrak{f}_v$ for every $u \in U$ and every $v | \mathfrak{f}$.

Formula (6.11) makes $\langle \theta(z, w; \eta), h(w) \rangle$ meaningful for $h \in \mathscr{S}_{2k, 2\kappa}(B)$. This inner product as a function of z behaves like an element of $\mathscr{S}_n(E)$ by virtue of Lemma 6.2. In fact, we have

Proposition 6.3. *Let Γ be a congruence subgroup of $\{\alpha \in B | N(\alpha) = 1\}$ such that $\eta^{\gamma} = \eta$ for every $\gamma \in \Gamma$, and let $f(z) = \langle \theta(z, w; \eta), h(w) \rangle$ with $h \in \mathscr{S}_{2k, 2\kappa}(\Gamma)$. Then $f \in \mathscr{S}_n(E)$ with n of* (6.9). *Moreover f has a Fourier expansion of the form*

$$f(z) = \sum_{\alpha \in R(\Gamma)} \overline{\eta(\alpha)} |N(\alpha)|^{-\delta/2} Q(h, \alpha, \Gamma) \mathbf{e}_E(-\tau N(\alpha) z / 2),$$

where $R(\Gamma)$ is a complete set of representatives of the set

$$V^* = \{\alpha \in V | -\tau N(\alpha) \gg 0\}$$

modulo the transformations $\alpha \mapsto \gamma \alpha \gamma^{-1}$ for all $\gamma \in \Gamma$, and $Q(h, \alpha, \Gamma)$ is a certain period of h relative to α.

This is a restatement of [82, Theorem 2.2 and Proposition 2.3]. The period $Q(h, \alpha, \Gamma)$ is $(2\tau)^{-\delta/2} \operatorname{vol}(\Gamma \backslash H^{\delta})^{-1}$ times the number $P(h, \alpha, \Gamma)$ of [82, (2.5), (2.23)].

We now take η in a more specific form. Given k, κ, n, and $\mathfrak{m}$ as above, we take a Hecke character ψ_1 of E such that

(6.12) $\psi_{1\mathbf{a}}(-1) = (-1)^{\|k + \kappa\|}$; $\psi_1(x) = x^{i\mu}$ for $0 \ll x \in E_{\mathbf{a}}^{\times}$ with $\mu \in \mathbb{R}^{\mathbf{a}}$; $\|\mu\| = 0$; *the conductor of ψ_1 is prime to $\mathfrak{d}_B$ and divides $\mathfrak{m}$.*

Then we take $\eta \in \mathscr{L}(V)$ satisfying the following conditions:

(6.13a) $\tau N(\alpha) \in \mathfrak{g}$ *if* $\eta(\alpha) \neq 0$;

(6.13b) $\eta(sx) = \psi_1(s) \eta(x)$ *for every* $s \in \prod_{v \in \mathfrak{f}} \mathfrak{g}_v^{\times}$;

(6.13c) $\eta(wxw^{-1}) = \psi_{1\mathfrak{m}}(a_w^2 / N(w)) \eta(x)$ *for every* $w \in W_{\mathfrak{m}}$.

Such an η can be obtained, for example, in the following way. Take τ of (6.10) to be integral, and define φ and $\mathfrak{f}$ as in Lemma 6.2. Take an integral ideal $\mathfrak{b}$ prime to $\mathfrak{d}_B$ and divisible by the conductor of ψ_1. Further take $e \in E_{\mathfrak{f}}^{\times}$ so that $e\mathfrak{g} = \mathfrak{b}\mathfrak{d}$. (For a fractional ideal $\mathfrak{a}$ and $x \in E_{\mathbf{A}}^{\times}$, $x\mathfrak{a}$ denotes the fractional ideal such that $(x\mathfrak{a})_v = x_v \mathfrak{a}_v$ for every $v \in \mathfrak{f}$.) Define $\varepsilon_v \in B_v$ so that $\varepsilon_v = 1$ if $v | \mathfrak{d}_B$ and $\mu_v(\varepsilon_v) = \mathrm{diag}\,[1, e_v]$ if $v \nmid \mathfrak{d}_B$. We then define η by $\eta(x) = \prod_{v \in \mathfrak{f}} \eta_v(x_v)$ for $x \in V_{\mathfrak{f}}$ with η_v given as follows:

(6.14a) $\qquad v \nmid \mathfrak{b}$: η_v is the characteristic function of $V_v \cap \varepsilon_v \mathfrak{o}_v \varepsilon_v^{-1}$.

(6.14b) $\qquad v | \mathfrak{b}$: If $\alpha \in V_v \cap \varepsilon_v \mathfrak{o}_v \varepsilon_v^{-1}$, $\mu_v(\alpha) = \begin{pmatrix} a & b \\ c & -a \end{pmatrix}$, and $|e_v b| = 1$, then

$$\eta_v(\alpha) = \psi_{1v}(b); \quad \text{otherwise } \eta_v(\alpha) = 0.$$

We easily see that this η satisfies (6.13a, b, c) with $\mathfrak{m} = \mathfrak{b}$; we call it a *standard function of type* $(\mathfrak{b}, \psi_1)$.

Coming back to an arbitrary η satisfying (6.13a, b, c), define $\eta_\lambda \in \mathscr{L}(V)$ for $1 \le \lambda \le \omega$ by

(6.15) $$\eta_\lambda(y) = \psi_1(t_\lambda)^{-1} \eta(x_\lambda^{-1} y x_\lambda).$$

From (6.13c), we obtain

(6.16) $$\eta_\lambda(w y w^{-1}) = \psi_{1\mathfrak{m}}(a_w^2/N(w)) \eta_\lambda(y) \quad \text{for} \quad w \in x_\lambda W x_\lambda^{-1},$$

and hence by (6.11),

(6.17) $$N(\gamma)^\kappa \sigma_{2\kappa}(\gamma)^{-1} j(\gamma, w)^{-2k} \theta(z, \gamma w; \eta_\lambda) = \psi_{1\mathfrak{m}}(a_\gamma^2/N(\gamma)) \theta(z, w; \eta_\lambda)$$

$$\text{for every} \quad \gamma \in B^\times \cap x_\lambda W x_\lambda^{-1}.$$

Applying [87b, Proposition 11.7] to η_λ, we find that

(6.18) $$^\gamma\eta_\lambda(y) = (\psi_1 \varphi)_{\mathfrak{c}}(a_\gamma) \eta_\lambda(y) \quad \text{for every} \quad \gamma \in \Gamma[2\mathfrak{d}^{-1}, 2^{-1} \mathfrak{c}\mathfrak{d}]$$

with some integral ideal $\mathfrak{c}$. If $\tau \in \mathfrak{g}$ and η is a standard function of type $(\mathfrak{b}, \psi_1)$, then we can take the ideal $\mathfrak{c}$ determined by

(6.19) $$\mathfrak{c}_v = (4\mathfrak{g} \cap \tau \mathfrak{d}_B)_v \quad \text{if } v | \mathfrak{d}_B \quad \text{and} \quad \mathfrak{c}_v = \tau(4\mathfrak{g} \cap \mathfrak{b})_v \quad \text{if} \quad v \nmid \mathfrak{d}_B.$$

In any case, (6.18) together with Lemma 6.1 shows

(6.20) $$\theta(\gamma z, w; \eta_\lambda) = (\psi_1 \varphi)_{\mathfrak{c}}(a_\gamma)^{-1} \overline{J_n(\gamma, z)}\, \theta(z, w; \eta_\lambda)$$

$$\textit{for every} \quad \gamma \in \Gamma[2\mathfrak{d}^{-1}, 2^{-1} \mathfrak{c}\mathfrak{d}].$$

Given $\mathbf{h} = (h_1, \ldots, h_\omega) \in \mathscr{S}_{2k, 2\kappa}(\mathfrak{m}, \psi_1^2)$, we put

(6.21) $$I(z : \eta, \tau : \mathbf{h}) = \sum_{\lambda=1}^{\omega} \langle \theta(z, w; \eta_\lambda, \tau), h_\lambda(w) \rangle.$$

Lemma 6.4. *Let* $f(z) = I(z; \eta, \tau; \mathbf{h})$ *and* $\psi = \varphi\psi_1$ *with* φ *of Lemma 6.2. Then* $f \in \mathscr{S}_n(\mathfrak{c}, \psi)$ *with some* $\mathfrak{c}$. *Moreover, let* $\mathfrak{a} = t\mathfrak{g}$ *with* $t \in E_{\mathfrak{f}}^{\times}$, $\eta^t(x) = \psi(t)^{-1}|t|\eta(tx)$, *and* $\eta_\lambda^t(x) = \psi(t_\lambda)^{-1} \eta^t(x_\lambda^{-1} x x_\lambda)$. *Define the Fourier coefficients* $\lambda_f(\xi, \mathfrak{a})$ *of* f *as in* (6.3). *Then*

$$\sum_{\xi \in E} \lambda_f(\xi, \mathfrak{a}) \mathbf{e}_E(\xi\, z/2) = \sum_{\lambda=1}^{\omega} \langle \theta(z, w; \eta_\lambda^t), h_\lambda(w) \rangle.$$

Furthermore, with a sufficiently small Γ, we have

$$N(\mathfrak{a})|\xi/\tau|^{\delta/2}\lambda_f(\xi,\mathfrak{a})=(\varphi\psi)(t)\sum_{\lambda=1}^{\omega}\sum_{\alpha\in R(\Gamma,\xi)}\overline{\eta_\lambda(t\alpha)}\,Q(h_\lambda,\alpha,\Gamma),$$

where $R(\Gamma,\xi)=\{\alpha\in R(\Gamma)\,|\,N(\alpha)=-\xi/\tau\}$.

Proof. The first assertion is an immediate consequence of Proposition 6.3 and (6.20). Let U be the subgroup of $D[2\mathfrak{d}^{-1},2^{-1}\mathfrak{c}\mathfrak{d}]G_{\mathbf{a}}$ with the property of Lemma 6.2 for the η_λ; we may assume that $(a_w)_v-1\in\mathfrak{c}_v$ for every $v\in\mathbf{f}$ and every $w\in U$. Take $\beta\in G\cap\mathrm{diag}\,[t,1/t]\,U$. By (6.3) we have

$$\sum_\xi\lambda_f(\xi,\mathfrak{a})\mathbf{e}_E(\xi z/2)=N(\mathfrak{a})^{1/2}\psi(t)J_n(\beta,\beta^{-1}z)f(\beta^{-1}z)$$
$$=\sum_\lambda\langle N(\mathfrak{a})^{1/2}\psi(t)^{-1}\overline{J_n(\beta,\beta^{-1}z)}\theta(\beta^{-1}z,w;\eta_\lambda),h_\lambda\rangle.$$

By Lemmas 6.1 and 6.2, the last inner product can be written $\langle\theta(z,w;\eta_\lambda^t),h_\lambda\rangle$. Therefore the last assertion follows from Proposition 6.3.

Lemma 6.5. *Given an arbitrary $f\in\mathscr{S}_n(\mathfrak{c},\psi)$ and an ideal $\mathfrak{a}$ such that $\mathfrak{a}^{-1}$ is integral, put $f_\mathfrak{a}(z)=\sum_{\xi\in E}\lambda_f(\xi,\mathfrak{a})\mathbf{e}_E(\xi z/2)$. Then $f_\mathfrak{a}\in\mathscr{S}_n(\mathfrak{c}\mathfrak{a}^{-2},\psi)$ and $(f|T_v)_\mathfrak{a}=f_\mathfrak{a}|T_v$ for every v prime to $\mathfrak{a}$, where T_v is the Hecke operator of [87b, (5.5)]. Moreover, suppose $f=I(z;\eta,\tau;\mathbf{h})$ as in Lemma 6.4. Then η^t of Lemma 6.4 satisfies (6.13a,b,c) and $f_\mathfrak{a}=I(z;\eta^t,\tau;\mathbf{h})$.*

Proof. Proposition 3.2 of [87b] guarantees an element g of $\mathscr{S}_n(\mathfrak{c}\mathfrak{a}^{-2},\psi)$ such that $\lambda_g(\xi,\mathbf{x})=\lambda_f(\xi,\mathfrak{a}\mathbf{x})$. Then $g=f_\mathfrak{a}$. By [87b, Proposition 5.4], we can easily verify that $\lambda(\xi,\mathfrak{g};g|T_v)=\lambda(\xi,\mathfrak{a};f|T_v)$ if $v\nmid\mathfrak{a}^{-1}$. (Notice that the equality holds even if $\xi\notin\mathfrak{a}^{-2}$, in which case both sides vanish.) This proves that $(f|T_v)_\mathfrak{a}=f_\mathfrak{a}|T_v$ for such a v. That η^t satisfies (6.13a,b,c) is obvious; $\mathfrak{a}^{-1}$ has to be integral in order that η^t satisfies (6.13a). The last assertion follows immediately from Lemma 6.4.

Let us now recall the definition of the Hecke operator $\mathfrak{T}_v$ acting on $\mathscr{S}_{2k,2\kappa}(\mathfrak{m},\psi_1^2)$ for $v\in\mathbf{f}$. We put $\psi=\varphi\psi_1$: obviously $\psi^2=\psi_1^2$. We first take a prime element π of E_v and an element y_0 of $B_v^\times$, viewed as an element of $B_{\mathbf{A}}^\times$, such that $\pi=N(y_0)$ with an additional specification that $\mu_v(y_0)=\mathrm{diag}\,[1,\pi]$ if $v\nmid\mathfrak{d}_B$. If $\mathbf{h}=(h_1,\ldots,h_\omega)\in\mathscr{S}_{2k,2\kappa}(\mathfrak{m},\psi^2)$, then $\mathbf{h}|\mathfrak{T}_v=(h_1',\ldots,h_\omega')$ with

$$(6.22)\qquad\qquad h_v'=\sum_{\alpha\in A}\psi_\mathfrak{m}(a_\alpha)^{-2}N(\alpha)^{-i\mu}h_\lambda\|\alpha,$$

where A is a subset of B_+ such that $Wy_0W=\coprod_{\alpha\in A}Wx_\lambda^{-1}\alpha x_v$. (Here and henceforth, we use $\coprod$ for a disjoint union.) The index v is determined by v and λ under the condition that $x_\lambda y_0\in B^\times x_v W$. (See [80, p. 353], [81, II, p. 576].) Now this operator is meaningful also for any set $(h_1,\ldots,h_\omega)$ of C^∞ functions h_λ on H^δ satisfying (6.8a) with $\Phi=\psi^2$. In view of (6.17), we can therefore apply $\mathfrak{T}_v$ to $(\theta(z,w;\eta_\lambda))_{\lambda=1}^\omega$.

Lemma 6.6. $(\theta(z,w;\eta_\lambda))_{\lambda=1}^\omega|\mathfrak{T}_v=(\theta(z,w;\eta_\lambda'))_{\lambda=1}^\omega$ *with* $\eta_\lambda'(x)=\psi_1(t_\lambda)^{-1}\eta'(x_\lambda^{-1}xx_\lambda)$, $\eta'(x)=\sum_{y\in Y}\psi_\mathfrak{m}(a_y)^{-2}\psi_1(N(y))\eta(yxy^{-1})$, *where Y is a subset of $B_\mathbf{f}^\times$ such that* $Wy_0W=\coprod_{y\in Y}Wy$ *with y_0 as above.*

Proof. By (6.22) and (6.11), we have $(\theta(\dots\eta_\lambda))_\lambda|\mathfrak{T}_v=(\theta(\dots\eta'_\lambda))_\lambda$ with $\eta'_v(x)$ $=\sum\limits_{\alpha\in A} \psi_\mathfrak{m}(a_\alpha)^{-2}N(\alpha)^{-i\mu}\eta_\lambda(\alpha x\alpha^{-1})$ with A and v as in (6.22). Taking $Y=\{(x_\lambda^{-1}\alpha x_v)_\mathfrak{f}|\alpha\in A\}$, we obtain our assertion for this particular Y, which is enough, since η' is well-defined by the formula independently of the choice of Y.

Theorem 6.7. *Let* $\mathbf{h}\in\mathscr{S}_{2k,2\kappa}(\mathfrak{m},\psi^2)$, $\psi=\varphi\psi_1$ *with* ψ_1 *satisfying* (6.12), *and let* η *satisfy* (6.13a,b,c). *Define* $I(z;\eta,\tau;\mathbf{h})$ *as in* (6.21). *Then*

$$N(v)I(z;\eta,\tau;\mathbf{h})|T_v=I(z;\eta,\tau;\mathbf{h}|\mathfrak{T}_v)$$

for almost all primes v, *where* $N(v)=|\pi|^{-1}$ *with a prime element* π *of* E_v, *and* T_v *is the operator of* [87b, (5.5)] *defined relative to* ψ. *Suppose in particular* $\tau\in\mathfrak{g}$ *and* η *is a standard function of type* $(\mathfrak{b},\psi_1)$, *and determine* $\mathfrak{c}$ *by* (6.19). *Then our equality holds for every* v *prime to* τ *provided* T_v *is defined at level* $\mathfrak{c}$ *and* $\mathfrak{T}_v$ *at level* $2\mathfrak{b}\mathfrak{d}_B$.

Proof. We first prove the last assertion for such τ and η; v will always be assumed to be prime to τ. Put $f=I(z;\eta,\tau;\mathbf{h})$ and $g=I(z;\eta,\tau;\mathbf{h}|\mathfrak{T}_v)$. Let us first treat the case $v\nmid\mathfrak{c}$. By [81, II, Lemma 1.3], we have $g=\psi(\pi)^2\langle(\theta(\dots\eta_\lambda))|\mathfrak{T}_v,\mathbf{h}\rangle$. Hence, by Lemma 6.6 and Proposition 6.3, we have

$$|\xi/\tau|^{\delta/2}\lambda_g(\xi,\mathfrak{g})=\psi(\pi)^2\sum_{\lambda=1}^{\omega}\sum_{\alpha\in R(\xi,\Gamma)}\overline{\eta'_\lambda(\alpha)}\,q_\lambda(\alpha)$$

with η'_λ of Lemma 6.6 and $q_\lambda(\alpha)=Q(h_\lambda,\alpha,\Gamma)$. (We take a sufficiently small Γ.) On the other hand, if $\xi\in\mathfrak{g}$, we have, putting $\mathfrak{p}=\pi\mathfrak{g}$,

$$\lambda(\xi,\mathfrak{g};f|T_v)=\lambda_f(\xi,\mathfrak{p})+\psi(\pi)N(v)^{-1}\left(\frac{\xi}{\mathfrak{p}}\right)\lambda_f(\xi,\mathfrak{g})+\psi(\pi)^2N(v)^{-1}\lambda_f(\xi,\mathfrak{p}^{-1})$$

by virtue of [87b, Proposition 5.4]. By Lemma 6.4, we have

$$|\xi/\tau|^{\delta/2}\lambda_f(\xi,\mathfrak{g})=\sum_{\lambda,\alpha}\overline{\eta_\lambda(\alpha)}\,q_\lambda(\alpha),$$

$$|\xi/\tau|^{\delta/2}\lambda_f(\xi,\mathfrak{p})=\psi_1(\pi)N(v)^{-1}\sum_{\lambda,\alpha}\overline{\eta_\lambda(\pi\alpha)}\,q_\lambda(\alpha),$$

$$|\xi/\tau|^{\delta/2}\lambda_f(\xi,\mathfrak{p}^{-1})=\psi_1(\pi)^{-1}N(v)\sum_{\lambda,\alpha}\overline{\eta_\lambda(\pi^{-1}\alpha)}q_\lambda(\alpha),$$

where $\sum\limits_{\lambda,\alpha}=\sum\limits_{\lambda=1}^{\omega}\sum\limits_{\alpha\in R(\Gamma,\xi)}$. Observe that $\left(\dfrac{\xi}{\mathfrak{p}}\right)=\varphi(\pi)\left(\dfrac{-N(\alpha)}{\mathfrak{p}}\right)$ if $\tau N(\alpha)=-\xi$. Hence we can write, when $\xi\in\mathfrak{g}$,

$$(6.23)\qquad\qquad |\xi/\tau|^{\delta/2}\lambda(\xi,\mathfrak{g};f|T_v)=\sum_{\lambda,\alpha}\overline{\eta_\lambda^*(\alpha)}\,q_\lambda(\alpha)$$

with η_λ^* defined by

$$\eta_\lambda^*(\alpha)=\psi_1(\pi)^{-1}N(v)^{-1}\left\{\eta_\lambda(\pi\alpha)+\left(\frac{-N(\alpha)}{\mathfrak{p}}\right)\eta_\lambda(\alpha)+N(v)\eta_\lambda(\pi^{-1}\alpha)\right\}.$$

Therefore our task is to show that $\eta'_\lambda(\alpha) = N(v)\psi(\pi)^2 \eta^*_\lambda(\alpha)$, or equivalently

$$(6.24) \qquad \sum_{y \in Y} \eta(y\alpha y^{-1}) = \eta(\pi\alpha) + \left(\frac{-N(\alpha)}{\mathfrak{p}}\right)\eta(\alpha) + N(v)\eta(\pi^{-1}\alpha)$$

for every $\alpha \in V$ such that $\tau N(\alpha) \in \mathfrak{g}$, where Y is the set of Lemma 6.6 such that $N(y) = \pi$ for every $y \in Y$ and $Y \subset B_v^\times$. Obviously it is sufficient to show (6.24) for η_v and $\alpha \in V_v$, $N(\alpha) \in \mathfrak{g}_v$. Identifying B_v with $M_2(E_v)$ through μ_v, take Y to be the set of the $N(v) + 1$ elements

$$(6.25) \qquad y' = \begin{pmatrix} \pi & 0 \\ 0 & 1 \end{pmatrix}, \quad y_i = \begin{pmatrix} 1 & i/r \\ 0 & \pi \end{pmatrix} \quad (i \in \mathfrak{g}_v/\pi\mathfrak{g}_v),$$

where r is an element of E_v such that $r\mathfrak{g}_v = \mathfrak{d}_v$. Then the local version of (6.24), which we call (6.24v), can be proved in an elementary way. Without going into all the details, let us prove here only a typical case. We first observe that η_v is the characteristic function of $\mathfrak{o}_{1v} \cap V_v$ and (6.24v) is obviously true if $\pi\alpha \notin \mathfrak{o}_{1v}$. So let

$$\alpha = \begin{pmatrix} a & b/r \\ cr & -a \end{pmatrix} \in \pi^{-1}\mathfrak{o}_{1v}, \; N(\alpha) \in \mathfrak{g}_v; \text{ suppose } a \notin \mathfrak{g}_v. \text{ Since } a^2 + bc \in \mathfrak{g}_v, \text{ we see that } \pi a,$$

πb, and πc are v-units. Now the right-hand side of (6.24v) is 1. As for the left-hand side, we observe that $y'\alpha y'^{-1} \notin \mathfrak{o}_{1v}$, and

$$y_i \alpha y_i^{-1} = \begin{pmatrix} a + ci & (b - 2ai - ci^2)/(\pi r) \\ \pi cr & -a - ci \end{pmatrix}.$$

There is a unique $i \in \mathfrak{g}_v/\pi\mathfrak{g}_v$ such that $a + ci \in \mathfrak{g}_v$. For that i, we have $c(b - 2ai - ci^2) = -(a + ci)^2 - N(\alpha) \in \mathfrak{g}_v$, and hence $y_i \alpha y_i^{-1} \in \mathfrak{o}_{1v}$. Thus the left-hand side is also 1 as expected. This proves the case $a \notin \mathfrak{g}_v$. The other cases can be proved in the same fashion. (Cf. the proof in [82, pp. 621–623], which deals with the case $E = \mathbb{Q}$. In fact, the present case, being formulated locally, is simpler than that of [82] which is only partly local.)

Next suppose $v|2\mathfrak{b}$ and $v \nmid \mathfrak{d}_B$. Again by [87b, Proposition 5.4], we have (6.23) with $\eta^*_\lambda(\alpha) = \psi_1(\pi)^{-1} N(v)^{-1} \eta_\lambda(\pi\alpha)$. Let A be the set of (6.22) with fixed λ and v, and let $\alpha_0 \in A$. Since $\Gamma_\lambda \cap E$ and $\mathrm{vol}(\Gamma_\lambda \backslash H^\delta)$ are independent of λ, $\Gamma_\lambda \backslash \Gamma_\lambda \alpha_0 \Gamma_v$ and $\Gamma_\lambda \alpha_0 \Gamma_v / \Gamma_v$ have the same cardinality. Hence we can take A so that $\Gamma_\lambda \alpha_0 \Gamma_v = \coprod_{\alpha \in A} \Gamma_\lambda \alpha = \coprod_{\alpha \in A} \alpha \Gamma_v$. Let $\alpha \mapsto \alpha^*$ denote the main involution of B. Then $\Gamma_v \alpha_0^* \Gamma_\lambda = \coprod_{\alpha \in A} \Gamma_v \alpha^*$, and therefore if h'_v is defined by (6.22), then

$$\langle \theta(z, w; \eta_v), h'_v \rangle = \sum_{\alpha \in A} \psi_\mathfrak{b}(a_\alpha)^{-2} N(\alpha)^{-i\mu} \langle \theta(z, w; \eta_v) \| \alpha^*, h_\lambda \rangle.$$

This can be written $\langle \theta'', h_\lambda \rangle$ with

$$\theta'' = \sum_{\alpha \in A} \psi_\mathfrak{b}(a_\alpha)^2 N(\alpha)^{i\mu} j(\alpha^*, w)^{-2k} N(\alpha)^\kappa \sigma_{2\kappa}(\alpha^*)^{-1} \theta(z, \alpha^* w; \eta_v).$$

Putting $C = A^*$, we find, by (6.11), that $\theta'' = \theta(z, w; \eta''_\lambda)$ with

$$\eta''_\lambda(x) = \sum_{\beta \in C} \psi_\mathfrak{b}(d_\beta)^2 N(\beta)^{i\mu} \eta_v(\beta x \beta^{-1}).$$

Now the same type of reasoning as in Lemma 6.6 shows that

$$\eta''_\lambda(x)=\psi_1(t_\lambda)^{-1}\eta''(x_\lambda^{-1}xx_\lambda),$$

$$\eta''(x)=\sum_{y\in Z}\psi_\mathfrak{b}(d_y)^2\psi_1(N(y))^{-1}\eta(yxy^{-1}),$$

where Z *is any subset of* $B_\mathfrak{f}^\times$ *such that* $Wy_0^*W=\coprod_{y\in Z}Wy$; for instance $Z=\{(x_\lambda^{-1}\alpha x_v)_\mathfrak{f}^*|\alpha\in A\}$. Instead, we take

$$Z=\left\{\begin{pmatrix}\pi&0\\cs&1\end{pmatrix}\middle|c\in\mathfrak{g}_v/\pi\mathfrak{g}_v\right\}$$

with a fixed element s of E_v such that $s\mathfrak{g}_v=(\mathfrak{b}\mathfrak{d})_v$. Then our task is to show that

$$(6.26)\qquad\eta_v(\pi x)=\sum_{y\in Z}\eta_v(yxy^{-1})\quad\text{for}\quad x\in V_v,\ N(x)\in\mathfrak{g}_v,$$

which can be achieved in the same manner as in the case $v\nmid\mathfrak{c}$.

It remains to treat the case $v|\mathfrak{d}_B$. We again have (6.23) with $\eta_\lambda^*(x)=\psi_1(\pi)^{-1}N(v)^{-1}\eta_\lambda(\pi x)$. In this case, A of (6.22) consists of a single element α and $\Gamma_\lambda\alpha=\alpha\Gamma_v$, so that we can apply the reasoning in the last case with A consisting of a single element. Thus our task is to show that $\eta_v(\pi x)=\eta_v(yxy^{-1})$ for $x\in V_v$, $N(x)\in\mathfrak{g}_v$, with any element y of $\mathfrak{o}_v$ such that $N(y)=\pi$. But this is trivial, since if $N(x)\in\mathfrak{g}_v$, both x and yxy^{-1} must be contained in $\mathfrak{o}_v$, a well known fact on local division algebras, and hence both sides of the desired equality are equal to 1. This completes the proof of the special case in which η is a standard function.

As for η of a general type, the problem can again be reduced to (6.24) if $v\nmid\tau\mathfrak{m}\mathfrak{c}$. For $x\in V_\mathfrak{f}$, let x' denote its projection to $\prod_{v\ne u\in\mathfrak{f}}V_u$. If η can be written in the form $\eta(x)=\eta_v(x_v)\zeta(x')$ with the characteristic function η_v of $V_v\cap\mathfrak{o}_{1v}$ and some ζ, then our previous proof of (6.24v) is applicable. Since such a decomposition of η holds for almost all v, we obtain the first assertion of our theorem.

As a simple remark, we note here that if $\delta'=\varnothing$, we can take $\tau=1$, in which case $\mathfrak{c}$ of (6.19) is $4\mathfrak{g}\cap\mathfrak{b}\mathfrak{d}_B$, and the equality of our theorem holds for *all* primes v when η is a standard function.

Lemma 6.8. *Given η and η_λ as in* (6.14a, b) *and* (6.15), *there exists a function* $\Theta(z,x)$ *of* $z\in H^\mathbf{a}$ *and* $x\in B_\mathbb{A}^\times$ *such that* $\Theta(z,sx)=\psi(s)^2\Theta(z,x)$ *for every* $s\in E_\mathbb{A}^\times$ *and that*

$$\Theta(z,\beta x_\lambda u)=\psi(t_\lambda)^2\psi_\mathfrak{m}(d_u)^2N(u_\mathbf{a})^{\kappa+i\mu}j(u_\mathbf{a},\mathbf{i})^{-2k}$$

$$\cdot\sigma_{2\kappa}(u_\mathbf{a})^{-1}\Theta(z,u_\mathbf{a}(\mathbf{i});\eta_\lambda)$$

for $\beta\in B^\times$, $1\le\lambda\le\omega$, *and* $u\in W_\mathfrak{m}$.

Proof. Put, for $x\in B_\mathbb{A}^\times$ and $z\in H^\mathbf{a}$,

$$\Theta(z,x)=\text{Im}(z)^{\delta/2}\sum_{\alpha\in V}\eta((x^{-1}\alpha x)_\mathfrak{f})\psi_1(N(x))\,\text{sgn}(N(x_\mathbf{a}))^kj(x_\mathbf{a},-\mathbf{i})^{2k}$$

$$\cdot[\alpha,x_\mathbf{a}(-\mathbf{i})]^kN(x_\mathbf{a})^\kappa\sigma_{2\kappa}(x_\mathbf{a})^{-1}q(\alpha)\mathbf{e}_E((\tau/2)R'[\alpha;z,x_\mathbf{a}(\mathbf{i})]).$$

Then we easily see that this has the desired properties.

This lemma shows that $(\theta(z, w; \eta_\lambda))_{\lambda=1}^\omega$ corresponds to $\Theta(z, x)$ in exactly the same fashion as $(h_1, \ldots, h_\omega)$ does to the function $\mathbf{h}$ on $B_\mathbf{A}^\times$.

7. Reformulation of Theorems 3.2 and 3.3

Throughout the rest of the paper, we will consider automorphic forms on $B_\mathbf{A}^\times$ or Hilbert modular forms defined with respect to Hecke characters *of finite order*; namely, whenever we speak of $\mathscr{S}_{q,r}(\mathfrak{b}, \Phi; B)$ as in Sect. 6 or $\mathscr{S}_n(\mathfrak{c}, \psi)$, we assume that Φ or ψ is of finite order. Similarly, when we employ the symbol $I(z; \eta, \tau; \mathbf{h})$ of (6.21), we assume that ψ_1 of (6.12) is of finite order. We also fix a basic field E as in Sect. 6, use symbols $\mathbf{a}$ and $\mathbf{f}$, and put

$$(7.1) \qquad \qquad \iota = \iota_E, \quad v = [E : \mathbb{Q}].$$

Now given B, δ, δ', $q \in \mathbb{Z}^\delta$, and $r \in \mathbb{Z}^{\delta'}$ as in Sect. 6, put $t = q + r + 2\delta'$. It is well known that every system of eigenvalues occurring in $\mathscr{S}_{q,r}(B)$ always occurs in the space $\mathscr{S}_t(E)$ of Hilbert cusp forms, as proved by Eichler, Shimizu, Jacquet, and Langlands. We thus start with a primitive system of eigenvalues $\{\chi(\mathfrak{a})\}$ which occurs in $\mathscr{S}_t(E)$. (See [78, p. 652] and [81, II, p. 584]. We normalize the Hecke operators as in (6.22) so that $\sum \chi(\mathfrak{a}) N(\mathfrak{a})^{-s}$ satisfies a functional equation under the map $s \mapsto 2 - s$. This definition coincides with that of [81, II, p. 576], but differs from those of [78] and [87b] by certain powers of $N(\mathfrak{a})$.) If $\mathfrak{p}$ is the prime ideal at a prime $v \in \mathbf{f}$, let us write $\chi(v)$ for $\chi(\mathfrak{p})$. Then we denote by $\mathscr{W}_{q,r}^\iota(\chi, B)$ the set of all elements $\mathbf{h}$ in $\mathscr{S}_{q,r}(\mathfrak{b}, \Phi; B)$ with any $\mathfrak{b}$ such that $\mathbf{h}|T_v = \chi(v)\mathbf{h}$ for almost all v, and by $\mathscr{W}_{q,r}^\iota(\chi, B, \bar{\mathbb{Q}})$ the set of all such $\mathbf{h}$'s that are $\bar{\mathbb{Q}}$-rational. We will often write simply $\mathscr{W}$ for $\mathscr{W}_{q,r}^\iota$.

Lemma 7.1. *The notation χ, t, δ, δ', and B being as above, there exists a nonzero complex constant $Q(\chi, \delta)$ depending only on χ and δ such that $\langle \mathbf{h}, \mathbf{k} \rangle \sim Q(\chi, \delta)$ for every $\mathbf{h}$ and $\mathbf{k}$ in $\mathscr{W}_{q,r}^\iota(\chi, B, \bar{\mathbb{Q}})$, provided that χ and δ fall into one of the following four cases:*

    (i) $\delta = \iota$;
    (ii) *v or $\|\delta\|$ is even, and $t_v \geq 2$ for every $v \in \delta$;*
    (iii) *$t_v \geq 3$ for every $v \in \delta$;*
    (iv) *$t_v \geq 2$ for every $v \in \delta$, and $\sum\limits_{\mathfrak{a}} \chi(\mathfrak{a}) N(\mathfrak{a})^{-s}$ is the L-function of a Hecke character of a totally imaginary quadratic extension of E.*

This is merely a restatement of [83a, Theorems 5.6 and 5.8]. It is conjecturable that such a constant $Q(\chi, \delta)$ exists unconditionally. Our subsequent theorems will be stated under the assumption that $Q(\chi, \delta)$ exists for (χ, δ) in question. In fact, in order to formulate our results, we need only the existence of a constant $Q(\chi, B)$ which may depend on B and such that $\langle \mathbf{h}, \mathbf{k} \rangle \sim Q(\chi, B)$ for every $\mathbf{h}$ and $\mathbf{k}$ in $\mathscr{W}(\chi, B, \bar{\mathbb{Q}})$. However, since the existence of $Q(\chi, \delta)$ independent of B is highly plausible, as can be seen from the above lemma, we have chosen the formulation in terms of $Q(\chi, \delta)$ instead of $Q(\chi, B)$. It should be noted that if $Q(\chi, \delta)$ is well-defined, so is $Q(\bar{\chi}, \delta)$ and $Q(\bar{\chi}, \delta) \sim Q(\chi, \delta)$. An important property of Q obtained in [83a,

Theorem 5.4] should also be noted:

(7.2) $Q(\chi, \iota) \sim Q(\chi, \delta) Q(\chi, \iota - \delta)$ *for every* $\delta \in (\mathbb{Z}/2\mathbb{Z})^{\mathbf{a}}$, *where we understand that* $Q(\chi, 0) = 1$.

Our aim is to reformulate Theorems 3.2 and 3.3 by incorporating the results of Sect. 6. Using the same notation F_i, V_i, and m_i as in Sect. 3, we consider, for each i such that $m_i = 1$, a theta function $\theta(z_i, w_i; \eta_i, \tau_i)$ of (6.10) with $\eta_i \in \mathscr{L}(V_i)$ and $\tau_i \in F_i$ defined relative to B_i. Let φ_i be the Hecke character of F_i corresponding to $F_i(\tau_i^{1/2})/F_i$, and let $\mathbf{h}_i \in \mathscr{S}_{2k_i, 2\kappa_i}(\mathfrak{b}_i, \psi_i^2)$ with some $\mathfrak{b}_i$ and ψ_i. Assuming that $\varphi_i \psi_i$, η_i, τ_i, and $\mathfrak{b}_i$ satisfy (6.12) and (6.13 a, b, c), we put $f_i(z_i) = I(z_i; \eta_i, \tau_i; \mathbf{h}_i)$ with the symbol I of (6.21), and consider the Fourier expansion

$$(7.3)\qquad \overset{(1)}{\prod} f_i(z_i) = \sum_a \mu_{\mathbf{h}}(a) \mathbf{e}^{(1)}(az/2) \quad (z = (z_i) \in H^{(1)})$$

with the symbols $\overset{(1)}{\prod}$, $\mathbf{e}^{(1)}$, and $H^{(1)}$ as in Sect. 3. Suppose $\mathbf{h}_i \in \mathscr{W}(\chi_i', B_i)$ with some primitive χ_i' occurring in $\mathscr{S}_{t_i}(F_i)$, where $t_i = 2k_i + 2\kappa_i + 2\varepsilon_i'$; naturally χ_i' must be consistent with ψ_i^2. Now Theorems 3.2 and 3.3 can be reformulated as follows:

Theorem 7.2. *The notation being as above, define* $Z^{\mathbf{h}}$ *to be the series obtained from* Z^r *of* (3.14) *by replacing* μ_r *with* $\mu_{\mathbf{h}}$ *of* (7.3). *Then* $Z^{\mathbf{h}}$ *has the same properties of analyticity as those of* Z^r *stated in Proposition 3.1. Suppose moreover that the* $\mathbf{h}_i$ *are* $\bar{\mathbb{Q}}$-*rational and the* η_i *are* $\bar{\mathbb{Q}}$-*valued, and that the constant* $Q(\chi_i', \varepsilon_i)$ *as in Lemma 7.1 exists for each* χ_i'. *Then the following two assertions hold:*

(I) *Under the same assumptions as in Theorem 3.2, one has*

$$Z^{\mathbf{h}}(s_0) \sim \pi^d p_W(\varphi, 2\psi) \overset{(1)}{\prod} Q(\chi_i', \varepsilon_i) \overset{(2)}{\prod} \langle \mathbf{g}_i, \mathbf{g}_i \rangle$$

for every integer s_0 *satisfying* (3.21 a, b, c), *where* d *and* W *are the same as in Theorem 3.2.*

(II) *Under the assumption (Ra) or (Rb) of Theorem 3.3, the residue of* $Z^{\mathbf{h}}$ *at* $s = b_Z/2$ *or* $s = (2b_Z - 1)/4$, *according as* $b_Z \in \mathbb{Z}$ *or* $b_Z \notin \mathbb{Z}$, *is an algebraic number times*

$$R_E \pi^{d'} p_W(\varphi, 2\psi) \overset{(1)}{\prod} Q(\chi_i', \varepsilon_i) \overset{(2)}{\prod} \langle \mathbf{g}_i, \mathbf{g}_i \rangle,$$

where d' *is the same as in Theorem 3.3.*

Proof. Though these can perhaps be derived directly from Theorems 3.2 and 3.3, we will give a proof here modifying our argument in Sect. 4. In order not to obscure the ideas, let us assume for simplicity that $F^{(1)} = F_1$ and write simply η for η_1. We consider then the η_λ as in (6.15). Now θ_1 of (3.32b) with $C_1 = 2^{2k_1}\eta_\lambda$ and $\zeta_1 = 2/\tau_1$ is exactly $\theta(z, w; \eta_\lambda)$. Define f_{ph}^λ to be the series of (4.4) with this choice of C_1, and view it as a function of $w \in H^{\varepsilon_1}$, suppressing other variables. Then we see that $(f_{ph}^\lambda)_{\lambda=1}^\omega$ corresponds to a function $\mathbf{f}_{ph}$ on $B_{\mathbf{A}}^\times$ satisfying (6.7a, b) with ψ^2 in place of Φ. This is because (4.10b) is equal to (4.10a), and Θ of (4.10a) as a function of $w \in H^{\varepsilon_1}$ has the required property by virtue of Lemma 6.8. Therefore, if $b_Z \neq 0$ or

$E \neq \mathbb{Q}$, and if we write $\mathcal{A}_\lambda$ for the function of (4.10b) with $f_{ph} = f_{ph}^\lambda$, then there is a $\bar{\mathbb{Q}}$-rational element $\mathfrak{f} = (\mathfrak{f}_1, \ldots, \mathfrak{f}_\omega)$ of $\mathcal{M}_{2k,2\kappa}(\mathfrak{b}, \psi^2)$ such that (4.12) holds with $\mathcal{A}_\lambda$ and $\mathfrak{f}_\lambda$ in place of $\mathcal{A}$ and $\mathfrak{f}$ there. Then, taking the sum $\sum_\lambda \langle \mathfrak{f}_\lambda, h_\lambda \otimes g \rangle$ instead of $\langle \mathfrak{f}, r \otimes g \rangle$ in (4.13), we eventually find that the assertion of Theorem 3.2 is valid with a $\bar{\mathbb{Q}}$-rational basis of $\mathscr{S}_{2k,2\kappa}(\mathfrak{b}_1, \psi_1^2)$ in place of R there. As explained in the proof of [81, II, Theorem 3.8] and that of [83a, Theorem 5.4], we can find such a basis that is a union of a subset S of $\mathscr{W}(\chi_1', B_1, \bar{\mathbb{Q}}) \cap \mathscr{S}_{2k,2\kappa}(\mathfrak{b}_1, \psi_1^2)$ and a set S' orthogonal to S. Since $\langle \mathbf{h}, \mathbf{k} \rangle \sim Q(\chi_1', \varepsilon_1)$ for $\mathbf{h}$ and $\mathbf{k}$ in S, we obtain the desired conclusion about $Z^{\mathbf{h}}(s_0)$. Assertion (II) as well as the cases involving nonholomorphic functions can be settled in a similar way. In the general case in which $F^{(1)}$ has more factors, we only have to take the sum involving the indices λ_i corresponding to the factors F_i of $F^{(1)}$, and therefore there is no difficulty.

8. Some preliminary observations

Given a primitive system of eigenvalues χ occuring in $\mathscr{S}_h(E)$ and a Hecke character φ of E of finite order, we put

$$(8.0) \qquad D(s, \chi, \varphi) = \sum \varphi(\mathfrak{a}) \chi(\mathfrak{a}) N(\mathfrak{a})^{-s-1},$$

where $\mathfrak{a}$ runs over all the integral ideals of E. (The factor $N(\mathfrak{a})^{-1}$ makes (8.0) coincide with the series of [87b, (10.1b)]. The center of the critical strip of D is 0.) In [78] and [87b], we defined certain invariants by which we could state some theorems of algebraicity of the critical values of the above series. To recall them, let us assume $h = 2m$ with $m \in \mathbb{Z}^{\mathbf{a}}$, as we need the invariants only in this case in our treatment. If $E \neq \mathbb{Q}$ and $2m_v = 2$ for some $v \in \mathbf{a}$, we have to assume, in addition, that χ satisfies:

(8.1)   *For every $p \in \mathbb{Z}^{\mathbf{a}}/2\mathbb{Z}^{\mathbf{a}}$ and every integral ideal $\mathfrak{n}$, there exists a Hecke character η such that $\eta_{\mathbf{a}}(x) = \mathrm{sgn}(x_{\mathbf{a}})^p$, $\mathfrak{n}$ divides the conductor of η, and $D(0, \chi, \eta) \neq 0$.*

Under these conditions, we have a complex number $V(\chi, r)$, determined modulo $\bar{\mathbb{Q}}^\times$, for each $r \in \mathbb{Z}^{\mathbf{a}}/2\mathbb{Z}^{\mathbf{a}}$ with the following properties:

(8.2a)   *If $\varphi_{\mathbf{a}}(x) = \mathrm{sgn}(x_{\mathbf{a}})^{t+r}$ with an integer t such that $|t| < m_v$ for every $v \in \mathbf{a}$, then $D(t, \chi, \varphi) \sim \pi^{tv} V(\chi, r)$;*

(8.2b)   *If $\mathbf{g}$ is a $\bar{\mathbb{Q}}$-rational eigenform in $\mathscr{S}_{2m}(E)$ belonging to χ, then $Q(\chi, \iota) \sim \langle \mathbf{g}, \mathbf{g} \rangle \sim \pi^{-\|2m\| - v} V(\chi, q) V(\chi, \iota - q)$ for every $q \in \mathbb{Z}^{\mathbf{a}}/2\mathbb{Z}^{\mathbf{a}}$.*

In [87b], $V(\chi, r)$ was written $V(r, \mathbf{g})$; for their finer properties, the reader is referred to [87b, Theorem 10.2] and [78, Theorem 4.1, 4.2, and 4.3]; the latter paper includes the results in the case of weight $h \notin 2\mathbb{Z}^{\mathbf{a}}$. In (8.1), the condition that the conductor of η is divisible by $\mathfrak{n}$ is unnecessary for the existence of $V(\chi, r)$ satisfying (8.2a, b). It is necessary however for (8.3) below.

Lemma 8.1. *One can take a real number as $V(\chi, r)$. Moreover, one has $V(\bar{\chi}, r) \sim V(\chi, r)$.*

Proof. The functional equation of $D(s, \chi, \varphi)$ (see [78, (2.48), (2.50), and Proposition 4.5]) implies that $V(\bar{\chi}, r) \sim V(\chi, r)$. Obviously $\overline{V(\chi, r)} \sim V(\bar{\chi}, r)$. Hence if we put $b = \overline{V(\chi, r)}/V(\chi, r)$, then $b \in \bar{\mathbb{Q}}$ and $b\bar{b} = 1$. Take $c \in \bar{\mathbb{Q}}$ so that $b = \bar{c}/c$. Then $cV(\chi, r) \in \mathbb{R}$, which proves our lemma.

Let $n = m + (\iota/2)$ with m as above. Then n is a half-integral weight. With $\mathfrak{c}$ and ψ as in (6.2), suppose χ occurs in $\mathscr{S}_{2m}(2^{-1}\mathfrak{c}, \psi^2)$. We then denote by $\mathscr{S}_n(\mathfrak{c}, \psi, \chi)$ the subset of $\mathscr{S}_n(\mathfrak{c}, \psi)$ consisting of the elements f such that $f|T_v = N(v)^{-1}\chi(v)f$ for almost all primes v. As proved in [87b, Theorem 10.5], we have

(8.3) If $f \in \mathscr{S}_n(\mathfrak{c}, \psi, \chi)$, $p \in \mathscr{S}_n(E)$, $\psi_{\mathbf{a}}(x) = \mathrm{sgn}(x_{\mathbf{a}})^r$, and both f and p are $\bar{\mathbb{Q}}$-rational, then $\langle p, f \rangle \sim \pi^{-\|n\| - (v/2)} V(\chi, \iota - r)$.

Lemma 8.2. $\mathscr{S}_n(\mathfrak{c}, \psi, \chi)$ *has an orthogonal (but not necessarily orthonormal) basis over* $\mathbb{C}$ *consisting of* $\bar{\mathbb{Q}}$-*rational elements.*

Proof. In [87b, Proposition 8.9], we showed that $\mathscr{S}_n(\mathfrak{c}, \psi, \chi)$ is spanned by its $\bar{\mathbb{Q}}$-rational elements. If f and g are two such nonzero elements, then $\langle f, g \rangle \sim \langle f, f \rangle$ by (8.3). Therefore the standard construction of an orthogonal basis can be done within the set of $\bar{\mathbb{Q}}$-rational elements.

To state some results preliminary to our main applications, we consider $f \in \mathscr{S}_n(\mathfrak{c}, \psi)$ and $g \in \mathscr{M}_l(E)$. Here n is half-integral and l may be integral or half-integral. With the Fourier coefficients $\lambda(b, \mathfrak{x}; f, \psi)$ of f as in (6.3), where $\mathfrak{x}$ is a fixed fractional ideal of E, and also with the Fourier expansion $g(z) = \sum_{b \in E} \lambda'(b)\mathbf{e}_E(bz/2)$, we define a series

(8.4) $$D(s; f, \psi, \mathfrak{x}; g) = [U] \sum_{b \in E/U} \lambda(b, \mathfrak{x}; f, \psi)\lambda'(b)b^{(e\iota - n - l)/2} N_{E/\mathbb{Q}}(b)^{-s},$$

where U is a sufficiently small subgroup of U_E of finite index, and $e = 1/2$ or 1 according as $l \in I_E$ or $l \notin I_E$. This series is a special case of (4*.1). In fact, take $\Omega = g$, $F = E$, $p = 0$, $\zeta = 1$, and $d = (e\iota - n - l)/2$ in (4*.1). Then (4*.2b) holds with $\alpha = e$. Therefore D has a meromorphic continuation to the whole plane; it is holomorphic for $\mathrm{Re}(s) \geq (e - 1)/2$ except for a possible simple pole at $s = 0$ or $s = 1/2$ according as $l \in I_E$ or $l \notin I_E$. If $l \notin I_E$ and g is a cusp form, the pole occurs only when $l = n$.

Theorem 8.3. *Denote the series of* (8.4) *simply by* D. *Suppose that both* f *and* g *are* $\bar{\mathbb{Q}}$-*rational,* $f \in \mathscr{S}_n(\mathfrak{c}, \psi, \chi)$, *and* $\psi_{\mathbf{a}}(x) = \mathrm{sgn}(x_{\mathbf{a}})^r$ *with* $r \in \mathbb{Z}^{\mathbf{a}}/2\mathbb{Z}^{\mathbf{a}}$. *Then the following assertions hold:*

(I) $D(t/2) \sim \pi^{-v} V(\chi, \iota - r)$ *for every integer* t *such that*
(8.5) $0 \leq t < n_v - l_v + e - 1$ *and* $t \equiv n_v - l_v + e \pmod{2}$ *for every* $v \in \mathbf{a}$
except when $l \in I_E$, $E = \mathbb{Q}$, *and* $t = 0$.

(II) *If* $n = l$, D *has a possible simple pole at* $s = 1/2$, *and its residue is an algebraic number times* $\pi^{-v} R_E V(\chi, \iota - r)$.

(III) *If* $l \in I_E$, D *has at most a simple pole at* $s = 0$, *which occurs only when* $|n_v - l_v| - 1/2 \in 2\mathbb{Z}$ *for every* $v \in \mathbf{a}$. *If moreover* $n_v > l_v$ *for every* $v \in \mathbf{a}$, *the residue is an algebraic number times* $\pi^{-v} R_E V(\chi, \iota - r)$.

According to the classification in the introduction, this theorem concerns Cases (IIIA) and (IVA). Our assertions are essentially included in the results of [87b, § 10] in Case (IVA), and in those of Im [I] in Case (IIIA), but given here for the purpose of comparison with other cases as explained at the end of Sect. 5.

Proof. Since $\lambda_f(ba^2, x) = a^{n-(1/2)}\lambda_f(b, ax)$ for $0 \ll a \in E$, we may assume that x^{-1} is integral. Furthermore, by Lemma 6.5, we may, changing f for $\sum \lambda_f(b, x)\mathbf{e}_E(bz/2)$, assume that $x = \mathfrak{g}$. Put $g_\varrho(z) = \overline{g(-\bar{z})}$. Then $g_\varrho \in \mathcal{M}_l(E)$. Let $\Gamma = \{\gamma \in \Gamma[\mathfrak{a}, \mathfrak{a}] | a_\gamma - 1 \in \mathfrak{a}\}$ with an ideal $\mathfrak{a} \subset 4\mathfrak{b}$. Take $\mathfrak{a}$ so that $g_\varrho \in \mathcal{M}_l(\Gamma)$ and $f \in \mathcal{S}_n(\Gamma)$. Let $R = (P \cap \Gamma)\backslash\Gamma$ and

$$(8.6) \qquad E(z, s, q) = \sum_{\gamma \in R} \mathrm{Im}(z)^{s\mathfrak{l} - (q/2)} \|_q \gamma \qquad (z \in H^{\mathbf{a}})$$

for every $q \in I^0(E)$. The method explained in [81, I, (2.8), (2.9), (2.15)] and employed in Sect. 1 shows that

$$(8.7) \qquad D(s)(2\pi)^{\| -s\mathfrak{l} - (n+l-e\mathfrak{l})/2\|} \prod_{v \in \mathfrak{a}} \Gamma(s + (l_v + n_v - e)/2)$$

$$= A D_E^{1/2} \mathrm{vol}(\Gamma\backslash H^{\mathbf{a}}) \langle g_\varrho E(z, \bar{s} + 1 - (e/2), n - l), f \rangle$$

with $0 < A \in \mathbb{Q}$. Put $\mu = t + 2 - e$ and $q = n - l$ with t as in (8.5). By Lemma 1.4 and (1.19) (or by [87a, Theorems 4.2 and 4.3]), $\pi^{\|\mu\mathfrak{l} - q\|/2} E(z, \mu/2, q)$ is an arithmetic element of $\mathcal{N}_q(M_2(E))$, which multiplied with g_ϱ produces an arithmetic element of $\mathcal{N}_n(M_2(E))$. Now [87b, Proposition 9.4] guarantees a $\mathbb{C}$-linear map p_n of $\mathcal{N}_n$ into $\mathcal{S}_n$ which preserves arithmeticity and such that $\langle h, h' \rangle = \langle p_n(h), h' \rangle$ for every $h \in \mathcal{N}_n$ and every $h' \in \mathcal{S}_n$. Therefore our first assertion follows immediately from (8.3). As for (II), we recall that $E(z, s, 0)$ has a simple pole at $s = 1$ and its residue is a positive rational number times $D_E^{1/2}\pi^{-\nu}R_E$ (see [85b, Proposition 4.3] or Lemma 1.4). This together with (8.3) proves (II). Assertion (III) can be proved in the same fashion by virtue of (1.19), Lemma 1.4, (iii), and [86, Lemma 10.2, (iii)].

9. The critical values of $D(s, f, g)$ in Cases (IIB) and (IVB)

The title refers to the classification given in the introduction. Namely we are interested in $D(s, f, g)$ with two forms f and g of weight n and n' respectively, under the condition

$$(9.1) \qquad n_v > n'_v \text{ for every } v \in \delta \text{ and } n_v < n'_v \text{ for every } v \in \delta'$$

with a nontrivial decomposition $J_E = \delta \cup \delta'$. We first treat Case (IVB) by taking g of (8.4) in a more specific form. Thus, with $f \in \mathcal{S}_n(\mathfrak{c}, \psi)$ and $f' \in \mathcal{S}_{n'}(\mathfrak{c}', \psi')$, we put

$$(9.2) \qquad D(s; f, \psi, x; f', \psi', x')$$

$$= [U] \sum_{b \in E/U} \lambda(b, x; f, \psi)\lambda(b, x'; f', \psi')b^{(t-n-n')/2} N_{E/\mathbb{Q}}(b)^{-s}.$$

Here n and n' are half-integral; x and x' are fixed fractional ideals in E. We assume

$$(9.3) \qquad (\psi\psi')_{\mathbf{a}}(u) = \mathrm{sgn}(u)^{n+n'-t} \quad \text{for every} \quad u \in \mathfrak{g}^\times.$$

Then we can take $U = \{u^2 | u \in \mathfrak{g}^\times\}$ in (9.2). We consider the values of the series at $t/2$, where t is an integer satisfying

$$(9.4) \qquad 0 \leq t \leq |n_v - n'_v| - 1, \ t \equiv n_v + n'_v \ (\mathrm{mod}\, 2) \quad \text{for every} \quad v \in \mathbf{a}.$$

If $n_v > n'_v$ for all $v \in \mathbf{a}$, the answer is given by Theorem 8.3. Therefore we naturally assume (9.1). To obtain such a value as an application of Theorem 7.2, we consider two quaternion algebras B and B' over E such that

$$(9.5) \qquad \delta = \{v \in \mathbf{a} | B' \text{ is ramified at } v\}, \quad \delta' = \{v \in \mathbf{a} | B \text{ is ramified at } v\}$$

with δ and δ' as in (9.1). We call (B, B') a *complementary pair of quaternion algebras (over E) of type (δ, δ').*

Our next theorem concerns the case where $f(z) = I(z; \eta, \tau; \mathbf{h})$ and $f'(z) = I(z; \eta', \tau'; \mathbf{h}')$ with $\mathbf{h} \in \mathscr{W}^q_{2k, 2\kappa}(\chi, B, \overline{\mathbf{Q}})$ and $\mathbf{h}' \in \mathscr{W}^{q'}_{2k', 2\kappa'}(\chi', B', \overline{\mathbf{Q}})$ for such a pair. We naturally assume both δ and δ' to be nonempty; χ and χ' are primitive systems of eigenvalues occurring in $\mathscr{S}_q(E)$ and $\mathscr{S}_{q'}(E)$, respectively. Then $f \in \mathscr{S}_n(\mathfrak{c}, \psi, \chi)$ and $f' \in \mathscr{S}_{n'}(\mathfrak{c}', \psi', \chi')$ with

$$n = k + \kappa + (1/2)\delta + (3/2)\delta', \quad n' = k' + \kappa' + (3/2)\delta + (1/2)\delta',$$

and with $\mathfrak{c}, \psi, \mathfrak{c}', \psi'$ as in Theorem 6.7. We assume that η and η' are $\overline{\mathbf{Q}}$-valued. (This doesn't imply that f or f' is $\overline{\mathbf{Q}}$-rational.)

Theorem 9.1. *Suppose that $Q(\chi, \delta)$ and $Q(\chi', \delta')$ are well-defined. Then the following assertions hold for the series D of (9.2):*
  (I) *Under (9.1), we have, for every integer t satisfying (9.4),*

$$D(t/2) \sim \pi^{\|k + k'\| - v} Q(\chi, \delta) Q(\chi', \delta').$$

  (II) *If $n = n'$, D has at most a simple pole at $s = 1/2$. The pole occurs only when $\chi' = \bar{\chi}$, in which case the residue is an algebraic number times $\pi^{\|n\| - (3v/2)} R_E Q(\chi, \iota)$.*

Proof. In view of Lemma 6.5, we may assume that $\mathfrak{x} = \mathfrak{x}' = \mathfrak{g}$. To prove (I), we naturally assume the existence of at least one t satisfying (9.4). Then we see that

$$(9.6) \qquad k_u + \kappa'_u \equiv k'_v + \kappa_v \pmod{2} \text{ for every } u \in \delta \text{ and every } v \in \delta'.$$

Specialize the setting of Sect. 3 as follows: $F = F^{(1)} = E \times E$, $\varepsilon = p_1 \delta + p_2 \delta'$, $\varepsilon' = p_1 \delta' + p_2 \delta$, where p_1 and p_2 are the projection maps of F onto the frist and second factors, respectively, and

$$\Delta = (-1/2)[p_1(k + \kappa' + e\delta) + p_2(k' + \kappa + e\delta')],$$

where e is 0 or 1 according as both sides of (9.6) are odd or even. Then (3.18b) is satisfied. Take $\Omega = 1$ and $\zeta = (\tau, -\tau)$ in (3.14). For $(a, b) \in E \times E$, we have $\omega(Tr_{F/E}(\zeta(a, b))) = \omega(\tau(a - b))$, which is nonzero if and only if $a = b$, and hence the sum of (3.14) becomes

$$[U] \sum_{b \in E/U'} (b, b)^\Delta \mu_r(b, b) N(b)^{-s},$$

where $U' = \{u^2 | u \in U\}$. Thus, if we define $Z^{(\mathbf{h}, \mathbf{h}')}$ as in Sect. 7, then $Z^{(\mathbf{h}, \mathbf{h}')}(s + (1 - e)/2)$ coincides with $[U : U']D(s)$. Observing that $b_Z = 2 - e$, we obtain (I) immediately from Theorem 7.2, (I). To prove (II), we take the same series with $e = 0$. Since $n = n'$ this time, the sums in (9.6) are odd. As shown in the proof of Theorem 8.3, (II), D has a pole at $s = 1/2$ only if $\langle f_1, f_2 \rangle \neq 0$, where

$f_1 = \sum \overline{\lambda_f(b, \mathfrak{x})} \mathbf{e}_E(bz/2)$ and $f_2 = \sum \lambda_{f'}(b, \mathfrak{x}') \mathbf{e}_E(bz/2)$. By [87b, Prop. 5.3] and Lemma 6.5, such a nonvanishing occurs only if $\chi' = \bar{\chi}$. Therefore (II) follows from Theorem 7.2, (II) and (7.2).

Next we consider Case (IIB) by defining another type of series

$$(9.7) \qquad D(s; f, \psi, \mathfrak{x}; \chi', \mathfrak{y})$$

$$= [U] \sum_{b \in E/U} \lambda(b, \mathfrak{x}; f, \psi) \chi'(b\mathfrak{y}^{-1}) b^{-(n/2)-(3/4)\iota} N(b)^{-s},$$

where $f \in \mathscr{S}_n(\mathfrak{c}, \psi)$ and χ' is a primitive system of eigenvalues occurring in $\mathscr{S}_{n'}(E)$; $\mathfrak{x}$ and $\mathfrak{y}$ are fractional ideals in E. Thus n is half-integral and n' integral. If $n_v > n'_v$ for all v or if $n_v < n'_v$ for all v, Theorem 8.3 or Theorems 5.2 and 5.3 tell about the nature of the critical values and the residue of (9.7).

Theorem 9.2. *Let* $f = I(z; \eta, \tau; \mathbf{h})$ *with* $\bar{\mathbb{Q}}$-*rational* η *and* $\mathbf{h} \in \mathscr{W}^q_{2k, 2\kappa}(\chi, B, \bar{\mathbb{Q}})$. *Suppose* χ' *occurs in* $\mathscr{S}_{k', \kappa'}(B')$, (B, B') *is a complementary pair as in* (9.5), *and both* $Q(\chi, \delta)$ *and* $Q(\chi', \delta')$ *are well-defined; suppose moreover both* δ *and* δ' *are nonempty and* (9.1) *is satisfied. Then the following assertions hold for the series* D *of* (9.7):

(I) *We have* $D(t/2) \sim \pi^{\|k+k'-\delta\|} Q(\chi, \delta) Q(\chi', \delta')$ *for every integer* t *satisfying*

$$(9.8) \qquad 0 \leq t < |n_v - n'_v| - (1/2) \quad \text{and} \quad t \equiv |n_v - n'_v| + (1/2) \ (\mathrm{mod}\, 2) \quad \text{for every } v \in \mathbf{a}.$$

(II) *Suppose* $|n_v - n'_v| - (1/2) \in 2\mathbb{Z}$ *for every* $v \in \mathbf{a}$. *Then* D *has at most a simple pole at* $s = 0$, *and the residue is an algebraic number times* $\pi^{\|k+k'\|-v} R_E Q(\chi, \delta) Q(\chi', \delta')$.

Proof. This time we take $F = F^{(1)} \times F^{(2)}$, $F^{(1)} = F^{(2)} = E$, $\varepsilon = p_1 \delta + p_2 \delta'$, $\varepsilon' = p_1 \delta' + p_2 \delta$ with p_i as in the proof of Theorem 9.1, and

$$\Delta = (1/2)[p_1(-k - \kappa' + (1-c)\delta) + p_2(\kappa' - \kappa - c\delta')],$$

where c is 0 or 1 according as $|n_v - n'_v| - (1/2)$ is even or odd (for every v); take $\Omega = 1$ and $\zeta = (\tau, -\tau)$. Then $Z^{\mathbf{h}}(s + (3-c)/2)$ coincides with a positive integer times $D(s)$. Therefore we obtain our assertions immediately from Theorem 7.2 in the same fashion as in the proof of Theorem 9.1.

It may be said that the assertions of the above two theorems are not stated in the most desirable forms, because one will probably be able to state such results with no reference to the fact that f and f' are obtained as in (6.21) by means of some theta functions. Indeed, f or f' in Theorem 9.1 or 9.2 is not necessarily $\bar{\mathbb{Q}}$-rational. To clarify this point, let us now present possible better versions of the theorems among a set of conjectural statements. For convenience, we view every subset δ of J_E as an idempotent of the ring $\mathbb{Z}^{\mathbf{a}}$ and also as an element of the ring $(\mathbb{Z}/2\mathbb{Z})^{\mathbf{a}}$, so that for every $r \in \mathbb{Z}^{\mathbf{a}}$ or $r \in (\mathbb{Z}/2\mathbb{Z})^{\mathbf{a}}$, we have $r\delta = \sum_{v \in \delta} r_v v$. (If we identify an element r of $(\mathbb{Z}/2\mathbb{Z})^{\mathbf{a}}$ with the corresponding subset of $\mathbf{a}$, then $r\delta = r \cap \delta$. Likewise, the element t in the following conjecture may be viewed as a subset of δ.)

Conjecture 9.3. *Let* χ *be a primitive system of eigenvalues occurring in* $\mathscr{S}_{2m}(\mathfrak{b}, \psi^2; M_2(E))$ *with some* $\mathfrak{b}$, ψ, *and* $m \in \mathbb{Z}^{\mathbf{a}}$. *For each subset* δ *of* J_E *and each* $t \in (\mathbb{Z}/2\mathbb{Z})^{\delta}$, *there is a nonzero complex constant* $P(\chi, \delta, t)$ *determined modulo* $\bar{\mathbb{Q}}^{\times}$ *with the following properties* (C1–C9):

(C1) $P(\chi, 0, 0) = 1$, $P(\chi, \iota, t) = \pi^{-\|m\| - \nu} V(\chi, t)$.

(C2) $P(\bar{\chi}, \delta, t) \sim \overline{P(\chi, \delta, t)}$.

(C3) $V(\chi, r) \sim \pi^{\|m\| + \nu} P(\chi, \delta, r\delta) P(\bar{\chi}, \iota - \delta, r - r\delta)$ for every $r \in (\mathbb{Z}/2\mathbb{Z})^{\mathbf{a}}$ and every $\delta \subset J_E$.

(C4) $Q(\chi, \delta) \sim \pi^{\|\delta\|} P(\chi, \delta, t) P(\bar{\chi}, \delta, \delta - t)$ for every $t \in (\mathbb{Z}/2\mathbb{Z})^{\delta}$.

(C5) *Suppose* $f = I(z; \eta, \tau; \mathbf{h})$ *with* $\bar{\mathbb{Q}}$-*rational* η *and* $\mathbf{h} \in \mathscr{W}^{2m}_{2k, 2\kappa}(\chi, B, \bar{\mathbb{Q}})$, B *is unramified at* $v \in \delta$ *and ramified at* $v \in \iota - \delta$, *and* $\psi_{1\mathbf{a}}(x) = \mathrm{sgn}(x_{\mathbf{a}})^q$ *with* $q \in (\mathbb{Z}/2\mathbb{Z})^{\delta}$, *where* ψ_1 *is the Hecke character involved in* η *as in* (6.13b,c). *Then* $P(\chi, \delta, q\delta)^{-1} f$ *is* $\bar{\mathbb{Q}}$-*rational.*

(C6) *If* $f \in \mathscr{S}_n(\mathfrak{c}, \psi, \chi)$, $f' \in \mathscr{S}_{n'}(\mathfrak{c}', \psi', \chi')$, $\psi_{\mathbf{a}}(x) = \mathrm{sgn}(x_{\mathbf{a}})^r$, $\psi'_{\mathbf{a}}(x) = \mathrm{sgn}(x_{\mathbf{a}})^{r'}$, *and both* f *and* f' *are* $\bar{\mathbb{Q}}$-*rational, then under* (9.1), *we have, for every integer* t *satisfying* (9.4),

$$D(t/2; f, \psi, \mathfrak{x}; f', \psi', \mathfrak{x}') \sim \pi^{\|n\delta + n'\delta'\| - (\nu/2)} P(\bar{\chi}, \delta, \delta - r\delta) P(\bar{\chi}', \delta', \delta' - r'\delta').$$

(C7) *If* f, χ, ψ, *and* r *are as in* (C6) *and* χ' *is a primitive system of eigenvalues occurring in* $\mathscr{S}_{n'}(E)$, *then under* (9.1), *we have*

$$D(t/2; f, \psi, \mathfrak{x}; \chi', \mathfrak{y}) \sim \pi^{\|n\delta + n'\delta' - (\delta/2)\|} P(\bar{\chi}, \delta, \delta - r\delta) Q(\chi', \delta')$$

for every integer t *satisfying* (9.8).

(C8) *Let* f, χ, ψ, r, *and* χ' *be as in* (C7). *Suppose that* (9.1) *is satisfied and* $|n_v - n'_v| - (1/2) \in 2\mathbb{Z}$ *for every* $v \in \mathbf{a}$. *Then* $D(s; f, \psi, \mathfrak{x}; \chi', \mathfrak{y})$ *has a possible simple pole at* $s = 0$, *and the residue is an algebraic number times* $R_E \pi^{\|n\delta + n'\delta' - (\delta/2) - \delta'\|} P(\bar{\chi}, \delta, \delta - r\delta) Q(\chi', \delta')$.

(C9) *Suppose that* $\sum_{\mathfrak{a}} \chi(\mathfrak{a}) N(\mathfrak{a})^{-s - 1/2}$ *coincides with the L-function of a Hecke character* ω *of a totally imaginary quadratic extension* K *of* E *such that* $\omega(x) = |x^{\xi}|/x^{\xi}$ *for* $x \in K_{\mathbf{a}}$ *with* $0 \leq \xi \in \mathbb{Z}^{\Phi}$, *where* Φ *is a CM-type of* K. *Let* $\delta = \mathrm{Res}_{K/E}(\eta)$ *with* $\eta \subset \Phi$. *Then* $\pi^{\|\delta\|} P(\chi, \delta, t) \sim p_K(\xi, \eta)$ *for every* $t \in (\mathbb{Z}/2\mathbb{Z})^{\delta}$.

In these statements, δ can be either ι or 0, except in (C5), in which $\delta \neq 0$. We understand that $Q(\chi', 0) = 1$.

We now present some pieces of evidence for our conjecture. First of all, the plausibility of (C5) can be seen by comparing Theorem 9.1, (II) with Theorem 8.3, (II), or even from Theorem 7.2. Once (C4) and (C5) are assumed, (C6), (C7), and (C8) can be seen to be consistent with Theorems 9.1 and 9.2 when $\delta \neq 0$ and $\delta' \neq 0$, and with Theorem 8.3 when $\delta = \iota$; (C3) is consistent with (8.2b), (8.3), and Theorem 9.1, (II). In the next section we shall prove, in support of (C5), that if $\|\delta\| = 1$, $p^{-1} I(z; \eta, \tau; \mathbf{h})$ is $\bar{\mathbb{Q}}$-rational with a certain period p of $\mathbf{h}$. The validity of (C5) in the other extreme case $\delta = \iota$ can be shown as follows:

Theorem 9.4. *Let* $f(z) = I(z; \eta, \tau; \mathbf{h})$ *as in* (6.21) *with* $\mathbf{h} \in \mathscr{W}^{2k}_{2k, 0}(\chi, B, \bar{\mathbb{Q}})$. *Suppose that* B *is totally indefinite,* η *is* $\bar{\mathbb{Q}}$-*valued, and* $\psi_{1\mathbf{a}}(x) = \mathrm{sgn}(x_{\mathbf{a}})^r$ *with* $r \in (\mathbb{Z}/2\mathbb{Z})^{\mathbf{a}}$, *where* ψ_1 *is the Hecke character of* E *as in* (6.12) *and* (6.13c). *Then* $\pi^{\|k\| + \nu} V(\chi, r)^{-1} f$ *is* $\bar{\mathbb{Q}}$-*rational, that is, if* $\delta = \iota$, (C5) *is true with* $P(\chi, \iota, q) = \pi^{-\|k\| - \nu} V(\chi, q)$.

Proof. Theorem 6.7 shows that $f \in \mathscr{S}_n(\mathfrak{c}, \psi_1, \chi)$ with some $\mathfrak{c}$ and $n = k + (\iota/2)$. Take a $\bar{\mathbb{Q}}$-rational orthogonal basis $\{p\}$ of $\mathscr{S}_n(\mathfrak{c}, \psi_1, \chi)$ as guaranteed by Lemma 8.2. Then

$f = \sum_p c_p p$ with $c_p \in \mathbb{C}$. Take any p, put $p_\varrho(z) = \overline{p(-\bar{z})}$, and define $Z^{\mathbf{h}}$ as in Sect. 7 with $F = F^{(1)} = E$, $B_1 = B$, $\Omega(z) = p_\varrho(2z)$, $\zeta = 1$, and $\Delta = -k$. Then $Z^{\mathbf{h}}$ coincides with a positive integer times $D(s; f, \psi, \mathfrak{g}; p_\varrho)$. From (8.7) and Lemma 1.4, (iv), we see that its residue at $s = 1/2$ is a positive algebraic number times $R_E \pi^{\|k\|} \langle p, f \rangle$. On the other hand, by Theorem 7.2, (II), the residue is an algebraic number times $R_E \pi^{\|k\| - \nu} Q(\chi, \iota)$, and hence

$$c_p = \langle p, f \rangle / \langle p, p \rangle \sim \pi^{-\nu} Q(\chi, \iota) / \langle p, p \rangle \sim \pi^{-\|k\| - \nu} V(\chi, r)$$

by (8.2b) and (8.3). Since this holds for every p, we obtain our theorem.

A few more remarks may be added as to the nature of the constants $P(\chi, \delta, t)$. As Proposition 6.3 combined with [82, (2.23)] shows, the coefficients of $I(z; \eta, \tau; \mathbf{h})$ are "periods" of $\mathbf{h}$. Thus (C5) means that these periods are algebraic numbers times $P(\chi, \delta, q\delta)$. It is also conjecturable that all the periods of $\mathbf{h}$ (defined in a suitable way) have the same $\overline{\mathbb{Q}}$-linear span as that of the $P(\chi, \delta, t)$ for all $t \in (\mathbb{Z}/2\mathbb{Z})^\delta$, up to a certain power of π. In the next section we will show that this is in fact the case when $\|\delta\| = 1$ (see (10.6) and Theorem 10.4 below).

In [83a], we already mentioned the possibility of defining the constants P and their connection with Hilbert modular forms of half-integral weight. The above conjecture may be viewed as an improved version of the one made there. (See [83a, Conjecture 5.12] and the paragraph subsequent to it. We should have assumed $\chi = \bar{\chi}$ in that conjecture, which could be false if $\chi \neq \bar{\chi}$.) Further comments on, or consequences of, the above conjecture, (except for the one we are about to make) the reader is referred to the last part of Sect. 10 and also to Sect. 11.

We conclude this section by one more remark. The characterization of $P(\chi, \delta, q\delta)$ by (C5) is effective only if we can find η and τ so that $I(z; \eta, \tau; \mathbf{h}) \neq 0$, which poses a highly nontrivial question. Though we have no definite result, let us make some observations which lead to some interesting arithmetic and analytic questions. Let the notation be the same as in Theorem 6.7 and (C5). Then $f \in \mathscr{S}_n(\mathfrak{c}, \psi)$ with some integral ideal $\mathfrak{c}$ and n as in (6.9). As shown in [87b, p. 63, p. 65], we can find a Hecke character ε of E of finite order such that $(\varepsilon\psi)_{\mathbf{a}} = 1$ and every prime factor of $\mathfrak{c}$ divides the conductor of ε. Take any $\sigma \in \mathfrak{g}$, $\gg 0$ and also $m \in I_E$ so that $n - (\iota/2) - m \in 2I_E$ and $0 \leq m \leq \iota$. By [87b, (4.12), (10.15)], we have an equality of the form

$$(9.9) \qquad \lambda_f(\sigma, \mathfrak{q}^{-1}) D(2s, \chi, \varepsilon) A_1 B_1^s \pi^{-s\nu - \|\mu\|} \prod_{v \in \mathbf{a}} \Gamma(s + \mu_v)$$

$$= L_*(2s + 1, \psi\varepsilon\varphi') \int_\Phi f(z) \overline{g(z)} \, \overline{E(z, \bar{s} + (1/2))} \, y^n d^* z.$$

Here $\mathfrak{q}$ is a certain integral ideal, $\mu = (m + n - (\iota/2))/2$, g is a theta function belonging to $\mathscr{M}(m + (1/2), E)$, $\Phi = \Gamma \backslash H^{\mathbf{a}}$ with a sufficiently small congruence subgroup of $SL_2(E)$, φ' is the Hecke character corresponding to $E(\sigma^{1/2})$, L_* is the L-function some of whose Euler factors are dropped, and $E(z, s)$ is an Eisenstein series of [87b, (9.9)]; A_1 and B_1 (as well as A_2 and B_2 of (9.11) below) are positive real numbers that are algebraic. Obviously the integral of (9.9) can be written in the form $\sum\limits_{\lambda=1}^{\omega} \langle q_\lambda(w,\bar{s}), h_\lambda(w)\rangle$ with

(9.10) $$q_\lambda(w,s) = \int_\Phi g(z)\theta(z,w;\eta_\lambda,\tau)E(z,s+(1/2))y^n d*z .$$

The technique explained in [87b, p. 39, p. 63] transforms the last integral into

$$\sum_{\beta\in\mathscr{B}} N(\mathfrak{a}_\beta)^{2s+2} \int_{\Psi_\beta} g_\beta(z)\theta_\beta(z,w)y^{st+\iota+\mu}d*z ,$$

where $\mathscr{B}$ is a certain finite subset of G, g_β and θ_β are transforms of g and θ under β^{-1}, and $\Psi_\beta = [P\cap\beta\Gamma\beta^{-1}]\backslash H^{\mathfrak{a}}$. Employing the Fourier expansions of g_β and θ_β, we can express the last integral over Ψ_β as an infinite series. Combining the computation with (9.9), we obtain

(9.11) $$\lambda_f(\sigma,\mathfrak{q}^{-1})D(2s,\chi,\varepsilon)A_2 B_2^s \pi^{\|\delta\|/2} \prod_{v\in\delta} \Gamma(s+\mu_v+(1/2))^{-1}\Gamma(s+\mu_v)$$

$$= L_*(2s+1,\psi\varepsilon\varphi') \sum_{\lambda=1}^{\omega} \langle Q_\lambda(w,s), h_\lambda(w)\rangle,$$

$$Q_\lambda(w,s) = \sum_{\mathfrak{x}} N(\mathfrak{x})^{-2s} \sum_{\xi,\alpha} c_\mathfrak{x}^\lambda(\xi,\alpha)\xi^m q(\alpha)N(\alpha)^{-s\delta'-X}$$

$$\cdot [\alpha,w]^{-k}(|[\alpha,w]|/\mathrm{Im}(w))^{k-\delta-m\delta-2s\delta},$$

where $\mathfrak{x}$ runs over a complete set of representatives of the ideal classes of E modulo $\mathfrak{a}$, (ξ,α) over $(E^\times \times V)/\mathfrak{g}^\times$ under the condition $\sigma\xi^2 = -\tau N(\alpha)$, $X = \sum\limits_{v\in\delta'} (\kappa_v + m_v + 1)/2$, and $c_\mathfrak{x}^\lambda$ is an element of $\mathscr{L}(E\times V)$ that depends on λ, $\mathfrak{x}$, ε, q, and η. We can show that $L_*(2s+1,\psi\varepsilon\varphi')Q_\lambda(w,s)$ has a meromorphic continuation to the whole s-plane which is holomorphic except at $s=1/2$. The analytic behavior of Q_λ at specific values of s can be studied by the methods of [86]. In fact, if $\delta=\iota$, it is a special case of the series $f(w,s)$ of [86]. If $\delta\neq\iota$, however, Q_λ is quite different from f.

For simplicity, assume that $m=0$, $k=k_0\delta$ with $k_0\in\mathbf{Z}$, and the serie for Q_λ is convergent at $s=(k_0-1)/2$. Then k_0 is even, and $\{Q_\lambda(w,(k_0-1)/2)\}_{\lambda=1}^{\omega}$ defines an element $\mathbf{q}$ of $\mathscr{M}_{2k,2\kappa}(B)$, and thus $\lambda_f(\sigma,\mathfrak{q}^{-1})D(k_0-1,\chi,\varepsilon)$ is an elementary nonvanishing factor times $L_*(k_0,\psi\varepsilon\varphi')\langle\mathbf{q},\mathbf{h}\rangle$. Therefore $f\neq0$ if $\langle\mathbf{q},\mathbf{h}\rangle\neq0$, which is so at least for some $\mathbf{h}$, since we easily see that $\mathbf{q}\neq0$ for some choices of k, κ, and η. Now the present computation together with (C5) implies that $\langle\mathbf{q},\mathbf{h}\rangle\sim\pi^\gamma P(\chi,\delta,q\delta)D(k_0-1,\chi,\varepsilon)$ with some γ. This suggests that in general there is no good constant c such that $c\mathbf{q}$ is arithmetic, unless $\delta=\iota$. Even so, Q_λ seems an interesting function worth investigating.

10. The periods of automorphic forms in the one-dimensional case

The periods to be investigated are those of the elements of $\mathscr{S}_{2k,2\kappa}(B)$ with $k\in\mathbf{Z}^\delta$ and $\kappa\in\mathbf{Z}^{\delta'}$, where B, δ, and δ' are the same as in Sect. 6, under the condition that $\|\delta\|=1$, which will be assumed throughout this section. Naturally δ may be viewed as an element of J_E. We have thus $k=k_0\delta$ with an integer k_0; for simplicity, we use the letter k to denote this integer k_0. Similarly we use H instead of H^δ. Let $V = \{x \in B \mid \mathrm{Tr}(x) = 0\}$ as before; let V_v for $v \in J_E$ be the completion of V at v, and let $V_{\delta'} = \prod_{v \in \delta'} V_v$. Denote by S_v the quadratic form $x \mapsto N(x)$ on V_v, and by $\mathscr{P}_\kappa$ the vector space over $\mathbb{R}$ consisting of the $\mathbb{R}$-valued polynomial functions on $V_{\delta'}$ which are S_v-harmonic in the sense of [80, p. 322] and homogeneous of degree κ_v on V_v. Define similarly $\mathscr{P}_{k-1}$ with $(k-1)\delta$ instead of κ, and put $\mathscr{P} = \mathscr{P}_{k-1} \otimes_{\mathbb{R}} \mathscr{P}_\kappa$. We view the elements of $\mathscr{P}$ as functions on V through the natural embedding of V into $V_\delta \times V_{\delta'}$. We take a column vector q whose components form an $\mathbb{R}$-basis of $\mathscr{P}_\kappa$ and are $\bar{\mathbb{Q}}$-valued on V. Then we may assume, as explained in [82, p. 609], that $q(\alpha\xi\alpha^*) = \sigma_{2\kappa}(\alpha)q(\xi)$ with the representation $\sigma_{2\kappa}$ with properties (3.2a,b,c). Since this is $\mathbb{R}$-rational, we have $\sigma_{2\kappa}(\alpha^*) = {}^t\sigma_{2\kappa}(\alpha)$. (We use q instead of q' of (3.25b); this is consistent with the notation of Sect. 6)

We can define a representation $\varrho : B^\times \to GL(\mathscr{P})$ by $[\varrho(\alpha)p](\xi) = p(\alpha^{-1}\xi\alpha)$ for $\alpha \in B^\times$, $\xi \in V$, and $p \in \mathscr{P}$. Given a congruence subgroup Γ of $B^\times$, we let Γ act on $\mathscr{P}$ through ϱ, and define the (first) cohomology group $H(\Gamma, \mathscr{P}) = Z(\Gamma, \mathscr{P})/B(\Gamma, \mathscr{P})$ with the set $Z(\Gamma, \mathscr{P})$ of cocycles and the set $B(\Gamma, \mathscr{P})$ of coboundaries as usual. A *cocycle* is a map $\mathfrak{x} : \Gamma \to \mathscr{P}$ such that $\mathfrak{x}(\alpha\beta) = \mathfrak{x}(\alpha) + \varrho(\alpha)\mathfrak{x}(\beta)$ for every $\alpha, \beta \in \Gamma$. (Here we exclude the case $B = M_2(\mathbb{Q})$. For the discussion in this case, see [71, Ch. 8] and [82, §4].) Since an element of $\mathscr{P}$ is a function on V, we can view $\mathfrak{x}$ as an $\mathbb{R}$-valued function on $\Gamma \times V$. Thus a cocycle is a map $\mathfrak{x} : \Gamma \times V \to \mathbb{R}$ which belongs to $\mathscr{P}$ as a function on V and such that $\mathfrak{x}(\alpha\beta, \xi) = \mathfrak{x}(\alpha, \xi) + \mathfrak{x}(\beta, \alpha^{-1}\xi\alpha)$ for $\alpha, \beta \in \Gamma$ and $\xi \in V$; a *coboundary* is such a map given by $\mathfrak{x}(\alpha, \xi) = \mathfrak{y}(\xi) - \mathfrak{y}(\alpha^{-1}\xi\alpha)$ for every $\alpha \in \Gamma$ and $\xi \in V$ with $\mathfrak{y} \in \mathscr{P}$. Hereafter we always deal with cocycles as functions on $\Gamma \times V$. We can similarly define $H(\Gamma, \mathscr{P}_{\mathbb{C}}) = Z(\Gamma, \mathscr{P}_{\mathbb{C}})/B(\Gamma, \mathscr{P}_{\mathbb{C}})$ with $\mathscr{P}_{\mathbb{C}} = \mathscr{P} \otimes_{\mathbb{R}} \mathbb{C}$ by considering $\mathbb{C}$-valued functions on $\Gamma \times V$. Obviously $H(\Gamma, \mathscr{P}_{\mathbb{C}}) = H(\Gamma, \mathscr{P}) \otimes_{\mathbb{R}} \mathbb{C}$. For an element $\mathfrak{a}$ of $Z(\Gamma, \mathscr{P}_{\mathbb{C}})$ or $H(\Gamma, \mathscr{P}_{\mathbb{C}})$, its complex conjugate, real part, and imaginary part, denoted by $\bar{\mathfrak{a}}$, $\mathrm{Re}(\mathfrak{a})$, and $\mathrm{Im}(\mathfrak{a})$, can be defined in an obvious way; we denote also by $\mathrm{cl}(\mathfrak{a})$ the cohomology class of $\mathfrak{a}$.

Let us now associate a cohomology class to an element f of $\mathscr{S}_{2k,2\kappa}(\Gamma)$. We first observe that $\xi \mapsto [\xi, w]^{k-1}$ with $[\xi, w]$ of (6.10a), for a fixed $w \in H$, defines an element of $\mathscr{P}_{k-1} \otimes_{\mathbb{R}} \mathbb{C}$. Put

$$X(\xi, f; z, v) = \int_v^z [\xi, w]^{k-1} \cdot {}^t q(w) f(w) dw \quad (v, z \in H, \ \xi \in V),$$

$$\mathfrak{x}(\gamma, \xi; f, v) = X(\xi, f; \gamma v, v) \quad (\gamma \in \Gamma).$$

These as functions of ξ belong to $\mathscr{P}_{\mathbb{C}}$, and satisfy, as can easily be verified,

$$X(\alpha^{-1}\xi\alpha, f\|\alpha; z, v) = X(\xi, f; \alpha z, \alpha v) \quad (\alpha \in B_+),$$

$$X(\xi, f; \gamma z, v) = X(\gamma^{-1}\xi\gamma, f; z, v) + \mathfrak{x}(\gamma, \xi; f, v) \quad (\gamma \in \Gamma).$$

Moreover, $\mathfrak{x}$ as a function of (γ, ξ) belongs to $Z(\Gamma, \mathscr{P}_{\mathbb{C}})$ and $\mathrm{cl}(\mathfrak{x})$ is independent of v. We denote by $c(f)$ this cohomology class.

Proposition 10.1. *The map $f \mapsto \mathrm{Re}(c(f))$ gives an $\mathbb{R}$-linear isomorphism of $\mathscr{S}_{2k,2\kappa}(\Gamma)$ onto $H(\Gamma, \mathscr{P})$. Moreover there exists a $\mathbb{C}$-valued $\mathbb{C}$-bilinear alternating form A on*

$H(\Gamma, \mathscr{P}_{\mathbb{C}})$ *with the following properties:*

(10.1a) $i \cdot \mathrm{vol}(\Gamma \backslash H)[\langle f, g \rangle - \langle g, f \rangle] = 4 A(\mathrm{Re}(c(f)), \mathrm{Re}(c(g)))$,

(10.1b) $i \cdot \mathrm{vol}(\Gamma \backslash H)\langle f, g \rangle = A(\overline{c(f)}, c(g))$,

(10.1c)   *A is $\mathbb{R}$-valued on $H(\Gamma, \mathscr{P})$ and $\bar{\mathbb{Q}}$-valued on $H(\bar{\mathbb{Q}})$,*

where $H(\bar{\mathbb{Q}})$ is the set of all cohomology classes in $H(\Gamma, \mathscr{P}_{\mathbb{C}})$ represented by $\bar{\mathbb{Q}}$-valued cocycles.

Proof. The first assertion is a special case of [71, Theorem 8.4]. Once we know that $f \mapsto \mathrm{Re}(c(f))$ is an isomorphism, we can define an $\mathbb{R}$-bilinear alternating form A on $H(\Gamma, \mathscr{P})$ so that (10.1a) holds. Then its $\mathbb{C}$-bilinear extension to $H(\Gamma, \mathscr{P}_{\mathbb{C}})$ satisfies (10.1b). As whown in [59, §4], $A(\mathrm{cl}(\mathfrak{a}), \mathrm{cl}(\mathfrak{b}))$ has an explicit expression in terms of $\mathfrak{a}(\gamma)$ and $\mathfrak{b}(\gamma)$ with some $\gamma \in \Gamma$ and some quantities determined at the cusps. Therefore (10.1c) can be proved by the same argument as in the proof of [59, Théorème 2].

We now consider $\mathscr{S}_{2k,2\kappa}(\mathfrak{b}, \Phi; B)$ with $\mathfrak{b}$ prime to $\mathfrak{d}_B$ and a Hecke character Φ such that $\Phi_{\mathfrak{a}} = 1$. Let e be the element of $B_{\mathbb{A}}^{\times}$ such that $e_{\delta} = \mathrm{diag}[-1, 1]$ and $e_v = 1$ for $\delta \neq v \in \mathfrak{a} \cup \mathfrak{f}$. For $\mathbf{h} \in \mathscr{S}_{2k,2\kappa}(\mathfrak{b}, \Phi)$, define a function $\mathbf{h}^e$ on $B_{\mathbb{A}}^{\times}$ by

(10.2) $\mathbf{h}^e(x) = \overline{\mathbf{h}(xe)}$ $(x \in B_{\mathbb{A}}^{\times})$.

To show that $\mathbf{h}^e \in \mathscr{S}_{2k,2\kappa}(\mathfrak{b}, \bar{\Phi})$, we first observe that $\mathbf{h}^e$ satisfies (6.7a,b) with $\bar{\Phi}$ in place of Φ. Let (h_λ) be the element of $\prod_\lambda \mathscr{S}_{2k,2\kappa}(\Gamma_\lambda, \Phi)$ determined by (6.8b). Let

$$W_{\mathfrak{b}}^1 = \{x \in W_{\mathfrak{b}} \mid a_v(x) \equiv 1 \ (\mathrm{mod}\ \mathfrak{b}_v) \ \text{for every}\ v | \mathfrak{b}\},$$

where $a_v(x)$ denotes the a-entry of $\mu_v(x)$ with μ_v of Sect. 6. Given an index λ, we have $x_\lambda e \in B^{\times} x_\mu W_{\mathfrak{b}}^1$ with a unique index μ. Then $x_\lambda w = \varepsilon x_\mu e$ with $w \in W_{\mathfrak{b}}^1$ and $\varepsilon \in B^{\times}$. Fixing such an ε, we have, for $u \in W_{\mathfrak{b}}$,

$$\mathbf{h}^e(x_\mu u) = \overline{\mathbf{h}(x_\mu ue)} = \mathbf{h}(\varepsilon^{-1} x_\lambda weue) = \bar{\Phi}(t_\lambda) \bar{\Phi}_{\mathfrak{b}}(d_w d_u) \overline{(h_\lambda \| weue)(\mathbf{i})}.$$

Since $w_{\mathfrak{a}} = (\varepsilon e)_{\mathfrak{a}}$ and $\Phi_{\mathfrak{b}}(d_w) = \Phi(t_\lambda^{-1} t_\mu)$, this shows that $\mathbf{h}^e$ belongs to $\mathscr{S}_{2k,2\kappa}(\mathfrak{b}, \bar{\Phi})$ and corresponds to (h'_μ) with

(10.3) $h'_\mu(z) = j(\varepsilon, z)^{-2k} N(\varepsilon)^\kappa \sigma_{2\kappa}(\varepsilon)^{-1} \overline{h_\lambda(\varepsilon \bar{z})}$.

From [81, II, (1.23)], we immediately see that $(\mathbf{h}|\mathfrak{T}_v)^e = \mathbf{h}^e|\mathfrak{T}_v$ for every $v \in \mathfrak{f}$. Obviously $(\mathbf{h}^e)^e = \mathbf{h}$.

Put $\Gamma_\lambda^1 = B^{\times} \cap x_\lambda W_{\mathfrak{b}}^1 x_\lambda^{-1}$. Then both h_λ and h'_λ belong to $\mathscr{S}_{2k,2\kappa}(\Gamma_\lambda^1)$. We then consider $\prod_{\lambda=1}^{\omega} H(\Gamma_\lambda^1, \mathscr{P}_{\mathbb{C}})$, define $c(h_\lambda)$ as an element of $H(\Gamma_\lambda^1, \mathscr{P}_{\mathbb{C}})$, and put $c(\mathbf{h}) = (c(h_\lambda))_{\lambda=1}^{\omega}$. With λ, μ, and ε as above, we see that $\varepsilon^{-1} x_\lambda W_{\mathfrak{b}}^1 x_\lambda^{-1} \varepsilon = x_\mu W_{\mathfrak{b}}^1 x_\mu^{-1}$, and hence $\Gamma_\lambda^1 \varepsilon = \varepsilon \Gamma_\mu^1$. Given $(\mathfrak{x}_\lambda) \in \prod_{\lambda=1}^{\omega} Z(\Gamma_\lambda^1, \mathscr{P}_{\mathbb{C}})$, we can define $(\mathfrak{x}'_\lambda)$ in the same product space by

(10.4) $\mathfrak{x}'_\mu(\gamma, \xi) = (-1)^k \mathfrak{x}_\lambda(\varepsilon \gamma \varepsilon^{-1}, \varepsilon \xi \varepsilon^{-1})$ $(\gamma \in \Gamma_\mu^1, \xi \in V)$.

Then $(\mathbf{x}_\lambda) \mapsto (\mathbf{x}'_\lambda)$ defines an automorphism of $\prod_{\lambda=1}^{\omega} H(\Gamma_\lambda^1, \mathscr{P}_{\mathbb{C}})$, which we denote by $c \mapsto c^e$. (Notice that $\mathbf{h} \mapsto \mathbf{h}^e$ is anti-$\mathbb{C}$-linear, but $c \mapsto c^e$ is $\mathbb{C}$-linear.) It can easily be seen from (10.3) that

$$(10.5) \qquad\qquad\qquad\qquad c(\mathbf{h})^e = \overline{c(\mathbf{h}^e)} \, .$$

Here, for $c = (c_\lambda) \in \prod_{\lambda} H(\Gamma_\lambda^1, \mathscr{P}_{\mathbb{C}})$, we put $\bar{c} = (\bar{c}_\lambda)$; similarly we put $\mathrm{Re}(c) = (\mathrm{Re}(c_\lambda))$ and $\mathrm{Im}(c) = (\mathrm{Im}(c_\lambda))$.

Let us now assume that $\mathbf{h}$ is a $\bar{\mathbb{Q}}$-rational primitive eigenform belonging to a system of eigenvalues χ, keeping the assumption that $\mathfrak{b}$ is prime to $\mathfrak{d}_B$. In addition, we now assume that $\mathfrak{b}$ is the exact level of $\mathbf{h}$ in the sense that χ does not occur in $\mathscr{S}_{2k, 2\kappa}(\mathfrak{b}', \Phi; B)$ with any proper divisor $\mathfrak{b}'$ of $\mathfrak{b}$. Obviously $\mathbf{h}^e$ is a $\bar{\mathbb{Q}}$-rational primitive form belonging to $\bar{\chi}$.

For each integral ideal $\mathfrak{a}$, we can define a Hecke operator $T_{\mathfrak{a}}$ on $\prod_{\lambda} H(\Gamma_\lambda^1, \mathscr{P}_{\mathbb{C}})$ purely algebraically so that $c(\mathbf{g}|\mathfrak{T}_{\mathfrak{a}}) = c(\mathbf{g})|T_{\mathfrak{a}}$ (see [62, §9] and [71, §8.3]). Put

$$U_{\mathbb{C}} = \left\{ x \in \prod_{\lambda} H(\Gamma_\lambda^1, \mathscr{P}_{\mathbb{C}}) \mid x|T_{\mathfrak{a}} = \chi(\mathfrak{a})x \text{ for every } \mathfrak{a} \right\},$$

$$U = \text{the set of all } \bar{\mathbb{Q}}\text{-rational classes in } U_{\mathbb{C}}.$$

Then $U_{\mathbb{C}} = U \otimes_{\mathbb{Q}} \mathbb{C}$. Now it can happen that $\chi = \bar{\chi}$. In this case, $\mathbf{h}^e = b\mathbf{h}$ with $b \in \bar{\mathbb{Q}}$. Then $b\bar{b} = 1$. Taking $a \in \bar{\mathbb{Q}}$ so that $b = a/\bar{a}$ and replacing $\mathbf{h}$ by $a\mathbf{h}$, we may assume $\mathbf{h}^e = \mathbf{h}$. Therefore we hereafter normalize $\mathbf{h}$ so that $\mathbf{h}^e = \mathbf{h}$ when $\bar{\chi} = \chi$.

Lemma 10.2. *With $\mathbf{h}$ as above, $c(\mathbf{h})$ and $c(\mathbf{h})^e$ form a basis of $U_{\mathbb{C}}$ over $\mathbb{C}$. Moreover U has a basis $\{x, y\}$ over $\bar{\mathbb{Q}}$ such that $x^e = x$ and $y^e = -y$.*

Proof. Our assumptions on $\mathbf{h}$ and b together with Proposition 10.1 show that $\mathbf{h}$ is uniquely determined by χ at level $\mathfrak{b}$ up to constant algebraic factors. If $\bar{\chi} = \chi$, Proposition 10.1 shows that $\mathrm{Re}(c(\mathbf{h}))$ and $\mathrm{Im}(c(\mathbf{h}))$ form a basis of $U_{\mathbb{C}}$ over $\mathbb{C}$, and hence $c(\mathbf{h})$ and $\overline{c(\mathbf{h})}$ form a basis of $U_{\mathbb{C}}$ over $\mathbb{C}$. Suppose $\bar{\chi} \neq \chi$. Then the same proposition shows that $\mathrm{Re}(c(\mathbf{h}))$, $\mathrm{Im}(c(\mathbf{h}))$, $\mathrm{Re}(c(\mathbf{h}^e))$ and $\mathrm{Im}(c(\mathbf{h}^e))$ are linearly independent over $\mathbb{C}$, and hence $c(\mathbf{h})$ and $c(\mathbf{h})^e$ form a basis of $U_{\mathbb{C}}$ over $\mathbb{C}$. In either case, $(c(\mathbf{h})^e)^e = \{\overline{c(\mathbf{h}^e)}\}^e = \overline{c(\mathbf{h}^e)^e} = c(\mathbf{h})$. Thus the map $c \mapsto c^e$ on $U_{\mathbb{C}}$ has ± 1 as its eigenvalues, which proves the second assertion.

We now define *the fundamental periods* $\mathfrak{p}_+(\mathbf{h})$ and $\mathfrak{p}_-(\mathbf{h})$ of $\mathbf{h}$ to be the complex numbers such that

$$(10.6) \qquad\qquad\qquad c(\mathbf{h}) = \mathfrak{p}_+(\mathbf{h})x + \mathfrak{p}_-(\mathbf{h})y$$

with x and y as in the above lemma. We have

$$(10.7) \qquad\qquad \overline{c(\mathbf{h}^e)} = c(\mathbf{h})^e = \mathfrak{p}_+(\mathbf{h})x - \mathfrak{p}_-(\mathbf{h})y \, ,$$

and hence the above lemma shows that $\mathfrak{p}_+(\mathbf{h})\mathfrak{p}_-(\mathbf{h}) \neq 0$. The cosets $\mathfrak{p}_+(\mathbf{h})\bar{\mathbb{Q}}^{\times}$ and $\mathfrak{p}_-(\mathbf{h})\bar{\mathbb{Q}}^{\times}$ are determined by χ and B independently of the choice of x, y, and $\mathbf{h}$. We denote these cosets by $\mathfrak{p}_+(\chi, B)$ and $\mathfrak{p}_-(\chi, B)$. We are going to show in Theorem 10.4 below that they have the property of $P(\chi, \delta, t)$ stated in (C5) of Sect. 9. Though the symbols $\mathfrak{p}_\pm(\chi, B)$ are more informative and therefore better notation than $\mathfrak{p}_\pm(\mathbf{h})$, we employ the latter symbols in the following treatment for simplicity. Observing that $\mathfrak{p}_+(\mathbf{h}^e) \sim \overline{\mathfrak{p}_+(\mathbf{h})}$ and $\mathfrak{p}_-(\mathbf{h}^e) \sim \overline{\mathfrak{p}_-(\mathbf{h})}$, we find

$$(10.8) \qquad \overline{\mathfrak{p}_+(\chi, B)} = \mathfrak{p}_+(\bar{\chi}, B), \quad \overline{\mathfrak{p}_-(\chi, B)} = \mathfrak{p}_-(\bar{\chi}, B).$$

If $\bar{\chi} = \chi$, U is stable under complex conjugation, and hence we can choose x and y so that $\bar{x} = x$ and $\bar{y} = y$. Then $\mathfrak{p}_+(\mathbf{h})$ is real and $\mathfrak{p}_-(\mathbf{h})$ pure imaginary.

Proposition 10.3. $\mathfrak{p}_+(\mathbf{h})\overline{\mathfrak{p}_-(\mathbf{h})} \sim \pi\langle \mathbf{h}, \mathbf{h}\rangle$.

Proof. For simplicity, put $p = \mathfrak{p}_+(\mathbf{h})$ and $q = \mathfrak{p}_-(\mathbf{h})$. Put $x = (x_\lambda)$ and $y = (y_\lambda)$ with x_λ, $y_\lambda \in H(\Gamma_\lambda^1, \mathscr{P}_\mathbf{C})$; let A_λ be the alternating form on $H(\Gamma_\lambda^1, \mathscr{P}_\mathbf{C})$ as in Proposition 10.1; further let $D_\lambda = \Gamma_\lambda^1 \backslash H$. Then, for $\mathbf{h} = (h_\lambda)$ and $\mathbf{h}^e = (h'_\lambda)$, we have

$$i \cdot \mathrm{vol}(D_\lambda)\langle h_\lambda, h_\lambda\rangle = A_\lambda\overline{(c(h_\lambda), c(h_\lambda))} = A_\lambda(\overline{px_\lambda + qy_\lambda}, px_\lambda + qy_\lambda),$$

$$i \cdot \mathrm{vol}(D_\lambda)\langle h'_\lambda, h'_\lambda\rangle = A_\lambda\overline{(c(h'_\lambda), c(h'_\lambda))} = A_\lambda(\overline{px_\lambda - qy_\lambda}, \overline{px_\lambda} - \overline{qy_\lambda}).$$

Since $\mathrm{vol}(D_\lambda)$ is independent of λ and $\langle \mathbf{h}, \mathbf{h}\rangle = \langle \mathbf{h}^e, \mathbf{h}^e\rangle$, we have

$$(10.9) \qquad i \cdot \mathrm{vol}(D_1)\langle \mathbf{h}, \mathbf{h}\rangle = \sum_\lambda \{\bar{p}q A_\lambda(\bar{x}_\lambda, y_\lambda) + p\bar{q} A_\lambda(\bar{y}_\lambda, x_\lambda)\}.$$

If $\bar{\chi} = \chi$, this can be written

$$(10.10) \qquad i \cdot \mathrm{vol}(D_1)\langle \mathbf{h}, \mathbf{h}\rangle = 2p\bar{q} \sum_\lambda A_\lambda(\bar{y}_\lambda, x_\lambda).$$

If $\bar{\chi} \ne \chi$, we have

$$(10.11) \qquad 0 = i \cdot \mathrm{vol}(D_1)\langle \mathbf{h}^e, \mathbf{h}\rangle = \sum_\lambda A_\lambda\overline{(c(h'_\lambda), c(h_\lambda))}$$

$$= \sum_\lambda A_\lambda(\overline{px_\lambda - qy_\lambda}, px_\lambda + qy_\lambda) = 2pq \sum_\lambda A_\lambda(x_\lambda, y_\lambda).$$

Now $\mathrm{Re}(c(\mathbf{h} + \mathbf{h}^e)) = px + \bar{p}\bar{x}$ and $\mathrm{Re}(c(\mathbf{h} - \mathbf{h}^e)) = qy + \bar{q}\bar{y}$. Since $\langle \mathbf{h} + \mathbf{h}^e, \mathbf{h} - \mathbf{h}^e\rangle = 0$, we obtain, from (10.1a) and (10.11),

$$0 = \sum_\lambda A_\lambda(\overline{px_\lambda + \bar{p}x_\lambda}, qy_\lambda + \overline{qy_\lambda})$$

$$= p\bar{q} \sum_\lambda A_\lambda(x_\lambda, \bar{y}_\lambda) + \bar{p}q \sum_\lambda A_\lambda(\bar{x}_\lambda, y_\lambda).$$

This combined with (10.9) yields (10.10) when $\bar{\chi} \ne \chi$, which completes the proof since $\mathrm{vol}(D_1) \sim \pi$ and $\sum_\lambda A_\lambda(\bar{y}_\lambda, x_\lambda) \sim 1$.

Theorem 10.4. *Let ψ_1 be a Hecke character of E of finite order whose conductor divides $\mathfrak{b}$ and such that $\psi_{1\mathbf{a}}(-1) = (-1)^{\|k + \kappa\|}$, and let $\mathbf{h}$ be a $\bar{\mathbf{Q}}$-rational primitive form in $\mathscr{S}_{2k, 2\kappa}(\mathfrak{b}, \psi_1^2)$ as above. Let $f(z) = I(z; \eta, \tau; \mathbf{h})$ with $\bar{\mathbf{Q}}$-valued η satisfying (6.13a,b,c) for this ψ_1 and a multiple $\mathfrak{m}$ of $\mathfrak{b}$. Then $(\pi/\mathfrak{p})f$ is $\bar{\mathbf{Q}}$-rational, where $\mathfrak{p} = \mathfrak{p}_+(\mathbf{h})$ or $\mathfrak{p}_-(\mathbf{h})$ according as $\psi_{1\delta}(-1) = (-1)^k$ or $(-1)^{k-1}$.*

Proof. Take a congruence subgroup Γ of $\{\alpha \in B \,|\, N(\alpha)=1\}$ so that $-1 \notin \Gamma \subset \Gamma_\lambda$ for every λ. By Lemma 6.5, we have

$$|\xi/\tau|^{\delta/2} \lambda_f(\xi, \mathfrak{g}) = \sum_{\alpha \in R(\Gamma, \xi)} \sum_{\lambda=1}^{\omega} \overline{\eta_\lambda(\alpha)} \, Q(h_\lambda, \alpha, \Gamma).$$

As explained earlier, $Q(h, \alpha, \Gamma) = (2\tau)^{-\delta/2} \operatorname{vol}(\Gamma \backslash H)^{-1} P(h, \alpha, \Gamma)$ with P of [82, (2.5)]. Let $\alpha \in R(\Gamma, \xi)$, $0 \ll \xi \in F$. Then $N(\alpha) = -\xi/\tau$ and as explained in [82, p. 615], $\Gamma \cap E[\alpha]$ is a free cyclic group. Let γ be one of its generators. Take $\beta \in SL_2(\mathbb{R})$ so that $\beta \alpha_\delta \beta^{-1} = \operatorname{diag}[a, -a]$ with $a > 0$. Then $a^2 = -(\xi/\tau)_\delta$. Changing γ for γ^{-1} if necessary, we may assume that $\beta \gamma_\delta \beta^{-1} = \pm\operatorname{diag}[t, t^{-1}]$ with $t > 1$. Then [82, (2.20) and (2.22)] show that

$$P(f, \alpha, \Gamma) = -2a \int_v^{\gamma v} [\alpha, w]^{k-1} \cdot {}^t q(\alpha) f(w) \, dw.$$

Let $\mathfrak{x}_\lambda$ be a cocycle representing $c(h_\lambda)$. Since $\gamma\alpha = \alpha\gamma$, we see that $\mathfrak{x}(\gamma, \alpha)$ for any cocycle $\mathfrak{x}$ depends only on $\operatorname{cl}(\mathfrak{x})$, γ, and α. Hence $P(f, \alpha, \Gamma) = -2|\xi/\tau|^{\delta/2} \mathfrak{x}_\lambda(\gamma, \alpha)$, and

$$(10.12) \qquad \operatorname{vol}(\Gamma \backslash H) \lambda_f(\xi, \mathfrak{g}) = -|2/\tau|^{\delta/2} \sum_{\alpha \in R(\Gamma, \xi)} \sum_{\lambda=1}^{\omega} \overline{\eta_\lambda(\alpha)} \, \mathfrak{x}_\lambda(\gamma, \alpha).$$

With μ and ε as in (10.3) and (10.4), take $w \in W_{\mathfrak{b}}^1$ so that $x_\lambda w = \varepsilon x_\mu e$. Put $\alpha_1 = \varepsilon \alpha \varepsilon^{-1}$ and $\gamma_1 = \varepsilon \gamma \varepsilon^{-1}$. If $\beta_1 = \operatorname{diag}[N(\varepsilon)_\delta, 1]\beta\varepsilon^{-1}$, then $\beta_1 \in SL_2(\mathbb{R})$, $\beta_1 \alpha_{1\delta} \beta_1^{-1} = \operatorname{diag}[a, -a]$, and $\beta_1 \gamma_{1\delta} \beta_1^{-1} = \pm\operatorname{diag}[t, t^{-1}]$. Hence

$$P(f, \alpha_1, \Gamma) = -2|\xi/\tau|^{\delta/2} \mathfrak{x}_\lambda(\gamma_1, \alpha_1).$$

By (6.28c), we have $\eta_\lambda(\alpha_1) = \psi_1(t_\lambda)^{-1} \eta(x_\lambda^{-1} \varepsilon \alpha \varepsilon^{-1} x_\lambda) = \psi_1(t_\lambda)^{-1} \eta(w x_\mu^{-1} \alpha x_\mu w^{-1}) = \psi_1(t_\lambda)^{-1} \psi_{1\mathfrak{b}}(N(w)^{-1}) \eta(x_\mu^{-1} \alpha x_\mu) = \psi_{1\delta}(-1) \eta_\mu(\alpha)$. This combined with (10.4) shows

$$\sum_\lambda \overline{\eta_\lambda(\alpha_1)} \, \mathfrak{x}_\lambda(\gamma_1, \alpha_1) = (-1)^k \psi_{1\delta}(-1) \sum_\mu \overline{\eta_\mu(\alpha)} \, \mathfrak{x}_\mu'(\gamma, \alpha),$$

where $(\mathfrak{x}_\mu')$ is the cocycles representing $c(\mathbf{h})^e$. Put $s = (-1)^k \psi_{1\delta}(-1)$. Then we have

$$(10.13) \qquad \sum_\lambda \left\{ \overline{\eta_\lambda(\alpha)} \, \mathfrak{x}_\lambda(\gamma, \alpha) + \overline{\eta_\lambda(\alpha_1)} \, \mathfrak{x}_\lambda(\gamma_1, \alpha_1) \right\}$$

$$= \sum_\lambda \overline{\eta_\lambda(\alpha)} \left\{ \mathfrak{x}_\lambda(\gamma, \alpha) + s\mathfrak{x}_\lambda'(\gamma, \alpha) \right\}.$$

The left-hand side times $1/2$ or 1 contributes to $\lambda_f(\xi, \mathfrak{g})$ in the sense of (10.12) accordings as α_1 and α are conjugae under Γ or not. Let $p = \mathfrak{p}_+(\mathbf{h})$ and $q = \mathfrak{p}_-(\mathbf{h})$; let $(\mathfrak{a}_\lambda)$ and $(\mathfrak{b}_\lambda)$ be the cocycles representing the classes x and y of Lemma 10.2. Then

$$\mathfrak{x}_\lambda(\gamma, \alpha) = p\mathfrak{a}_\lambda(\gamma, \alpha) + q\mathfrak{b}_\lambda(\gamma, \alpha), \qquad \mathfrak{x}_\lambda'(\gamma, \alpha) = p\mathfrak{a}_\lambda(\gamma, \alpha) - q\mathfrak{b}_\lambda(\gamma, \alpha).$$

Therefore (10.13) is equal to

$$2p \sum_\lambda \overline{\eta_\lambda(\alpha)} \, \mathfrak{a}_\lambda(\gamma, \alpha) \quad \text{or} \quad 2q \sum_\lambda \overline{\eta_\lambda(\alpha)} \, \mathfrak{b}_\lambda(\gamma, \alpha)$$

according as $s = 1$ or $s = -1$. Since $\eta_\lambda(\alpha)$, $\mathfrak{a}_\lambda(\gamma, \alpha)$, and $\mathfrak{b}_\lambda(\gamma, \alpha)$ are all algebraic, we obtain our theorem.

Let us now state some facts, as well as a few conjectures, which can be derived or inferred from the above theorem, with or without certain assumptions.

(I) It is conjecturable that the periods $\mathfrak{p}_\pm(\chi, B)$ depend only on χ and δ, and are independent of B. If $E = \mathbb{Q}$, this will be seen from (10.16) below. Therefore suppose $E \neq \mathbb{Q}$. Let (B, B') be a complementary pair of type (δ, δ'), and let $\mathbf{h}$ and $\mathbf{h}'$ be $\bar{\mathbb{Q}}$-rational eigenforms in $\mathscr{S}_{2k, 2\kappa}(B)$ and $\mathscr{S}_{2k', 2\kappa'}(B')$, respectively, both belonging to the same χ. Let $f = I(z; \eta, \tau; \mathbf{h}^e)$, $f' = I(z; \eta', \tau'; \mathbf{h}')$, and $f_\varrho(z) = \overline{f(-\bar{z})}$. Here we assume $\|\delta\| = 1$, and η and η' are $\bar{\mathbb{Q}}$-rational and of type (6.13a, b, c). Further we choose η and η' so that f_ϱ and f' belong to $\mathscr{S}_n(\mathfrak{c}, \psi)$ with the same ψ. The proof of Theorem 8.3 shows that the residue of (9.2) at $s = 1/2$, when $\mathfrak{x} = \mathfrak{x}' = \mathfrak{g}$, is a nonzero algebraic number times $R_E \pi^{\|n\| - (v/2)} <f_\varrho, f'>$. By Theorem 9.1, (II), we have

$$(10.14) \qquad \langle f_\varrho, f' \rangle \sim \pi^{-v} Q(\chi, \iota).$$

Suppose now that tf' is $\bar{\mathbb{Q}}$-rational with a nonzero complex number t and also that $\langle f_\varrho, f' \rangle \neq 0$. By Theorem 10.4, $\pi \mathfrak{p}_+(\bar{\chi}, B)^{-1} f$ or $\pi \mathfrak{p}_-(\bar{\chi}, B)^{-1} f$ is $\bar{\mathbb{Q}}$-rational according as $\psi_\delta(-1) = (-1)^k$ or $= (-1)^{k-1}$. This together with (8.3), (10.14), and (8.2b) shows that

$$(10.15) \qquad \mathfrak{p}_\pm(\bar{\chi}, B) \sim t^{-1} \pi^{1 + \|n\| - (v/2)} Q(\chi, \iota) / V(\chi, \iota - r)$$

$$\sim t^{-1} \pi^{1 - \|m\| - v} V(\chi, r),$$

where $\mathfrak{p}_\pm$ is $\mathfrak{p}_+$ or $\mathfrak{p}_-$ in the manner specified above, and r is determined by $\psi_\mathbf{a}(x) = \operatorname{sgn}(x_\mathbf{a})^r$. This supports the conjecture that $\mathfrak{p}_\pm(\chi, B)$ depends only on χ and δ.

(II) If $E = \mathbb{Q}$, we have $\psi_\delta(-1) = (-1)^k$, and hence $\mathfrak{p}$ is always $\mathfrak{p}_+(\chi, B)$ in Theorem 10.4. This result was proved in [82, Prop. 4.5]. Suppose that $I(z; \eta, \tau; \mathbf{h}) \neq 0$ for some η and τ. Then comparison of the last result with Theorem 9.4 (when $E = \mathbb{Q}$) shows the first part of (10.16) below, which, combined with (8.2b), Lemma 8.1, and Proposition 10.3, proves the second part:

$$(10.16) \qquad \pi^{\|k\|} \mathfrak{p}_+(\chi, B) \sim V(\chi, k), \qquad \pi^{\|k\|} \mathfrak{p}_-(\chi, B) \sim V(\chi, k-1) \quad \text{if} \quad E = \mathbb{Q}.$$

This was also given in [82, Theorem 4.7]. However, the present result is an improvement in the sense that there is some latitude for τ and η here, while they were fixed in [82].

(III) Coming back to the general case, we observe that if $\mathfrak{p}_\pm(\chi, B)$ were indeed independent of B as discussed in (I), then Theorem 10.4 would imply (C5) of Conjecture 9.3 when $\|\delta\| = 1$, with

$$(10.17) \qquad P(\chi, \delta, s\delta) = \begin{cases} \mathfrak{p}_+(\chi, B)/\pi & \text{if} \quad s \equiv k \pmod 2, \\ \mathfrak{p}_-(\chi, B)/\pi & \text{if} \quad s \equiv k-1 \pmod 2. \end{cases}$$

This together with Proposition 10.3 implies (C4) of Conjecture 9.3 when $\|\delta\| = 1$.

(IV) Suppose $[E : \mathbb{Q}] = 2$; let $J_E = \{\delta, \delta'\}$. Take B, B', f, f', and ψ etc. as in (I). Observe that $k = \kappa' + \delta$ and $k' = \kappa + \delta'$. Let φ and φ' be the Hecke characters of E corresponding to $E(\tau^{1/2})$ and $E(\tau'^{1/2})$, respectively. Naturally we take $\bar{\psi}\varphi$ and $\psi\varphi'$ to be the characters ψ_1 and ψ_1' involved in η and η' in the sense of (6.13b, c). By Theorem 10.4 we can take $\pi^{-1} \mathfrak{p}_\pm(\chi, B')$ to be t in the discussion of (I), and hence

(10.15) implies, under the assumption that $\langle f_\varrho, f' \rangle \neq 0$, that

$$(10.18) \qquad \pi^{-\|m\|} V(\chi, r) \sim \begin{cases} \mathfrak{p}_+(\bar{\chi}, B)\mathfrak{p}_+(\chi, B') & \text{if} \quad \psi_\delta(-1) = (-1)^k, \\ \mathfrak{p}_-(\bar{\chi}, B)\mathfrak{p}_-(\chi, B') & \text{if} \quad \psi_\delta(-1) = (-1)^{k-1}. \end{cases}$$

Observe that $r \equiv k + \kappa + \delta'$ or $\equiv k + \kappa + \delta \pmod{2\mathbb{Z}^\mathbf{a}}$ according as $\psi_\delta(-1) = (-1)^k$ or $= (-1)^{k-1}$. Therefore (10.18) combined with (10.17) proves (C3) of Conjecture 9.3 when $v = 2$ and η, τ, η', τ' can be chosen so that $\langle f_\varrho, f' \rangle \neq 0$.

There is one more comment which requires a somewhat lengthy treatment, and so we devote the following section to its full discussion.

11. The abelian variety associated with χ, its zeta function, and its periods

The constant $V(\chi, r)$ is essentially a critical value of a Dirichlet series of (8.0) associated with some Hilbert modular forms, and therefore, can be given as a linear combination of certain integrals of such forms of the Mellin transform type. Thus $V(\chi, r)$ can be viewed as a period of a form on a v-cycle lying on $\Gamma \backslash H^\mathbf{a}$. On the other hand, $\mathfrak{p}_\pm(\chi, B)$ are periods on 1-cycles. Therefore (10.18) expresses the decomposition of a 2-dimensional period into two one-dimensional periods. More generally, (C3) of Conjecture 9.3 is a decomposition of a v-dimensional period into a $\|\delta\|$-dimensional one and a $(v - \|\delta\|)$-dimensional one.

Now, there is an important consequence of this decomposition: *While in general each individual period $P(\chi, \delta, t)$ cannot be obtained from the periods $V(\chi, r)$, the quotient of two P's can be obtained as the quotient of two V's.* In fact, (C3) implies the following proportionality which is also conjectural:

$$(11.1) \qquad P(\chi, \delta, t)/P(\chi, \delta, t') \sim V(\chi, t+s)/V(\chi, t'+s) \quad \text{for every } t, t' \in (\mathbb{Z}/2\mathbb{Z})^\delta$$

$$\text{and} \quad s \in (\mathbb{Z}/2\mathbb{Z})^{1-\delta}.$$

The significance of this relation, when $\|\delta\| = 1$ and χ occurs in $\mathscr{S}_{2i}(E)$, is in the following two points:

(i) *It establishes algebraic relations between the periods of an abelian variety and the critical values of its zeta function.*

(ii) *It suggests a likelihood of an abelian variety being constructed from the periods $V(\chi, r)$ of Hilbert modular forms.*

To explain these ideas, let us first recall the results obtained in our previous paper [77] that concerns the elliptic modular case. Let $h(z) = \sum_{n=1}^\infty \chi(n)\mathbf{e}(nz)$ be a primitive element of $\mathscr{S}_2(\Gamma_0(N))$ with $0 < N \in \mathbb{Z}$, K the field generated by the $\chi(n)$ over $\mathbb{Q}$, and A the factor of $\mathrm{Jac}(\Gamma_0(N) \backslash H)$ attached to h as in [73, Theorem 1]. Then K is totally real and embeddable in $\mathrm{End}(A) \otimes \mathbb{Q}$, and $\dim(A) = [K : \mathbb{Q}]$. In general, if an abelian variety A and a number field K satisfy these conditions (with no reference to h), then A is isomorphic to a complex torus

$$(11.2) \qquad \mathbb{C}^{J(K)}/\{(a^\tau w_\tau + b^\tau)_{\tau \in J(K)} | (a, b) \in L\},$$

where L is a lattice in K^2 and $w = (w_\tau) \in H^{J(K)}$. The endomorphism of A corresponding to an element x of K is represented by $\mathrm{diag}[x^\tau]_{\tau \in J(K)}$ on $\mathbb{C}^{J(K)}$. If A is obtained from h as above, then A is isomorphic to

$$(11.3) \qquad \mathbb{C}^{J(K)}/P \quad \text{with} \quad P = \left\{ \left(\int_q^{\gamma q} h_\tau(z)\,dz \right)_{\tau \in J(K)} \middle| \tau \in \Gamma_0(N) \right\},$$

where q is any fixed point on H and $h_\tau(z) = \sum \chi(n)^\tau e(nz)$ (see [73, Proposition 3]). Now Theorem 3 and a part of Theorem 1 of [77] can be stated as follows:

Theorem 11.1. *There exist two nonzero complex numbers v_τ^0 and v_τ^1 for each $\tau \in J_K$ with the following two properties:*

(11.4)   *For every primitive Dirichlet character φ modulo r, put*

$$C(\chi, \varphi) = D(0, \chi, \varphi)/[2\pi i \sum_{n=1}^r \varphi(n)e(n/r)].$$

Let $\varphi(-1) = (-1)^\varepsilon$ with $\varepsilon = 0$ or 1. Then $C(\chi, \varphi)/v_1^\varepsilon$ belongs to the field generated by the values $\varphi(n)$ over K, and moreover $[C(\chi, \varphi)/v_1^\varepsilon]^\sigma = C(\chi^\tau, \varphi^\sigma)/v_\tau^\varepsilon$ for every $\sigma \in \mathrm{Gal}(\bar{\mathbb{Q}}/\mathbb{Q})$, where τ is the restriction of σ to K.

(11.5)   *The $\mathbb{Q}$-linear span of P of (11.3) coincides with*

$$\{(a^\tau v_\tau^1 + b^\tau v_\tau^0)_{\tau \in J(K)} \mid a, b \in K\},$$

and consequently the factor A of $\mathrm{Jac}(\Gamma_0(N) \backslash H)$ attached to h is isomorphic to the torus of (11.2) with $\pm v_\tau^0/v_\tau^1$ as w_τ for a suitable L.

It should be noted that A has a well-defined $\mathbb{Q}$-rational model whose zeta function (or rather its one-dimensional part) is $\prod_{\tau \in J(K)} (\sum \chi(n)^\tau n^{-s})$ and that given ε, there exists a Dirichlet character φ_ε such that $D(0, \chi, \varphi_\varepsilon) \neq 0$ and $\varphi_\varepsilon(-1) = (-1)^\varepsilon$ (see [73, Theorem 1] and [77, Theorem 2]). From (11.4) we obtain a relation

$$(11.6) \quad v_\tau^0/v_\tau^1 = b^\sigma C(\chi^\tau, \varphi_0^\sigma)/C(\chi^\tau, \varphi_1^\sigma) \quad \text{for every } \sigma \in \mathrm{Gal}(\bar{\mathbb{Q}}/\mathbb{Q}), \ \tau = \mathrm{Res}_{/K}(\sigma)$$

with $0 \neq b \in \bar{\mathbb{Q}}$ and such φ_ε. If we can take *real* characters as such φ_ε, then b can be taken from K. Now our principal question is:

To what extent can this result be generalized to the case of a basic field E of an arbitrary degree?

To attempt an answer, we naturally start with χ that occurs in $\mathscr{S}_{2t}(E)$ and assume that χ occurs also in $\mathscr{S}_{2\delta,0}(B)$ for some B with $\|\delta\| = 1$ as in Section 10. Suppose for simplicity $\bar{\chi} = \chi$; let K be the field generated by the values of χ. Then we can associate with χ a factor A of $\prod_{\lambda=1}^\omega \mathrm{Jac}(\Gamma_\lambda^1 \backslash H)$ such that: (i) K is embeddable in $\mathrm{End}(A) \otimes \mathbb{Q}$; (ii) $\dim(A) = [K : \mathbb{Q}]$; (iii) A is defined over E and its zeta function is $\prod_{\tau \in J(K)} D(s, \chi^\tau, \theta_0)$, where θ_0 denotes the trivial character. (See Hida [H, Theorems 4.4 and 4.12]. Several examples of Γ of level 1 such that $E = \mathbb{Q}(5^{1/2})$ and $\Gamma \backslash H$ itself is an elliptic curve are given in [67, p. 158]. For further examples, see [H, §8].) For simplicity, let us assume $\omega = 1$ and put $\Gamma = \Gamma_1^1$. Then $\mathbf{h}$ of Section 10 is represented by a $\bar{\mathbb{Q}}$-rational eigenform h in $\mathscr{S}_{2\delta,0}(\Gamma)$. Similarly we have a $\bar{\mathbb{Q}}$-rational eigenform h_τ in $\mathscr{S}_{2\delta,0}(\Gamma)$ with eigenvalues χ^τ for each $\tau \in J_K$. Since $(k, \kappa) = (2\delta, 0)$, we have $\mathscr{P} = \mathbb{R}$, and $H(\Gamma, \mathscr{P}_\mathbb{C})$ is the vector space of all homomorphisms of Γ into $\mathbb{C}$.

Let c_τ be the cohomology class attached to h_τ. In the present case we have

$$c_\tau(\gamma) = \int_q^{\gamma q} h_\tau(z)dz \qquad (\gamma \in \Gamma).$$

Define $c : \Gamma \to \mathbb{C}^{J(K)}$ by $c(\gamma) = (c_\tau(\gamma))_{\tau \in J(K)}$. By [73, Prop. 3], A is isomorphic to $\mathbb{C}^{J(K)}/c(\Gamma)$, which generalizes (11.3). As remarked above, we can find a point w of $H^{J(K)}$ such that A is isomorphic to the torus of (11.2) with a suitable L. This means that

$$c(\Gamma) \otimes_{\mathbb{Z}} \mathbb{Q} = \{(a^\tau u_\tau + b^\tau v_\tau)_{\tau \in J(K)} | a, b \in K\}$$

with some nonzero complex numbers u_τ and v_τ such that $w_\tau = u_\tau/v_\tau$. Now, from this, (10.6), and (10.7), we see that for each fixed τ, the $(\mathbb{R} \cap \bar{\mathbb{Q}})$-linear span of $\mathfrak{p}_\pm(\chi^\tau, B)$ coincides with that of u_τ and v_τ, and hence

(11.7) $\mathfrak{p}_-(\chi^\tau, B)/\mathfrak{p}_+(\chi^\tau, B) = \alpha_\tau(w_\tau)$

for each $\tau \in J_K$ with some $\alpha_\tau \in GL_2(\mathbb{R} \cap \bar{\mathbb{Q}})$. By (10.17), (11.1), and (8.2a), we have

(11.8) $\alpha_\tau(w_\tau) \sim V(\chi^\tau, s)/V(\chi^\tau, s+\delta)$

$$\sim D(0, \chi^\tau, \varphi)/D(0, \chi^\tau, \varphi')$$

for every $\tau \in J_K$, where φ and φ' are Hecke characters of E such that $\varphi_\mathbf{a}(x) = \mathrm{sgn}(x_\mathbf{a})^s$ and $\varphi'_\mathbf{a}(x) = \mathrm{sgn}(x_\mathbf{a})^{\delta+s}$, and s is an arbitrary element of $(\mathbb{Z}/2\mathbb{Z})^{1-\delta}$. This holds provided that (C3) and (C5) of Conjecture 9.3 are true, χ occurs in $\mathscr{S}_{2\delta,0}(B)$ for some B, and a certain inner product, as well as the values $D(0, ...)$, is nonvanishing. If $[E : \mathbb{Q}] = 2$, the existence of B and the last nonvanishing are the only necessary assumptions, since (C5) is true in the sense of Theorem 10.4 and (C3) is shown as in (10.18) under those assumptions.

Relation (11.8) is a generalization of a weaker form of Theorem 11.1, or rather, of (11.6). It should be emphasized however that if $E \neq \mathbb{Q}$, the critical values $D(0, \chi^\tau, \varphi)$ by themselves cannot represent the periods of h_τ (or those of A) that are in fact *their factors* as expressed by (C3). One significant aspect of (11.8) is that it does not involve B. This suggests, as we said earlier, the possibility of obtaining w_τ and consequently (the isogeny class of) A from the values $D(0, \chi^\tau, \varphi)$ or $V(\chi^\tau, r)$, with no reference to quaternion algebras of type B.

Finally let us indicate a possible approach to our conjectures. All the results in the present paper are stated over $\bar{\mathbb{Q}}$. If we could prove them over a well specified number field as we did in our previous papers [77], [78], and [87b], then we would obtain (11.8) in a stronger form which really includes (11.6) as a special case, and which establishes a direct connection of A (or the corresponding point w on $H^{J(K)}$) with the critical values $D(0, \chi^\tau, \varphi)$, or with the periods $V(\chi^\tau, r)$ of Hilbert modular forms. Another important point to be proved, besides the algebraicity in an exact form, is the nonvanishing of the forms $I(z; \eta, \tau; \mathbf{h})$ and some of their inner products. Though the above discussion as well as Section 10 is confined to the case $\|\delta\| = 1$, our ideas seem applicable to the general case. It will not be an easy task to materialize all these, but it does not seem unapproachably difficult. The author hopes that the present paper will arouse the interest of more researchers in this subject.

References

[B]   Bluher, A.: Thesis, Princeton University, 1988
[H]   Hida, H.: On abelian varieties with complex multiplication as factors of the Jacobians of Shimura curves. Am. J. Math. **103**, 727–776 (1981)
[I]   Im, J.: Thesis, Princeton University, 1988
[S]   Shimizu, H.: On zeta functions of quarternion algebras. Ann. Math. **81**, 166–193 (1965)
[59]   Shimura, G.: Sur les intégrales attachées aux formes automorphes. J. Math. Soc. Japan **11**, 291–311 (1959)
[62]   Shimura, G.: On Dirichlet series and abelian varieties attached to automorphic forms. Ann. Math. **76**, 237–294 (1962)
[67]   Shimura, G.: Construction of class fields and zeta functions of algebraic curves. Ann. Math. **85**, 58–159 (1967)
[71]   Shimura, G.: Introduction to the arithmetic theory of automorphic functions. (Publ. Math. Soc. of Japan, Vol. 11). Iwanami Shoten and Princeton Univ. Press, 1971
[73]   Shimura, G.: On the factors of the jacobian variety of a modular function field. J. Math. Soc. Japan **25**, 523–544 (1973)
[75]   Shimura, G.: On the holomorphy of certain Dirichlet series. Proc. London Math. Soc. (Ser. 3) **31**, 79–98 (1975)
[77]   Shimura, G.: On the periods of modular forms. Math. Ann. **229**, 211–221 (1977)
[78]   Shimura, G.: The special values of the zeta functions associated with Hilbert modular forms. Duke Math. J. **45**, 637–679 (1978)
[80]   Shimura, G.: The arithmetic of certain zeta functions and automorphic forms on orthogonal groups. Ann. Math. **111**, 313–375 (1980)
[81]   Shimura, G.: On certain zeta functions attached to two Hilbert modular forms I, II. Ann. Math. **114**, 127–164, 569–607 (1981)
[81z]   Shimura, G.: The critical values of certain zeta functions associated with modular forms of half-integral weight. J. Math. Soc. Japan **33**, 649–672 (1981)
[82]   Shimura, G.: The periods of certain automorphic forms of arithmetic type. J. Fac. Sci. Univ. Tokyo (Sec. IA) **28**, 605–632 (1982)
[83a]   Shimura, G.: Algebraic relations between critical values of zeta functions and inner products. Am. J. Math. **104**, 253–285 (1983)
[83b]   Shimura, G.: On Eisenstein series. Duke Math. J. **50**, 417–476 (1983)
[84]   Shimura, G.: Differential operators and the singular values of Eisenstein series. Duke Math. J. **51**, 261–329 (1984)
[85a]   Shimura, G.: On Eisenstein series of half-integral weight. Duke Math. J. **52**, 281–324 (1985)
[85b]   Shimura, G.: On the Eisenstein series of Hilbert modular groups. Rev. Mat. Ib. **1**, (No. 3) 1–42 (1985)
[86]   Shimura, G.: On a class of nearly holomorphic automorphic forms. Ann. Math. **123**, 347–406 (1986)
[87a]   Shimura, G.: Nearly holomorphic functions on hermitian symmetric spaces. Math. Ann. **278**, 1–28 (1987)
[87b]   Shimura, G.: On Hilbert modular forms of half-integral weight. Duke Math. J. **55**, 765–838 (1987)
[St]   Sturm, J.: Special values of zeta functions and Eisenstein series of half integral weight. Am. J. Math. **102**, 219–240 (1980); Addendum, ibid. pp. 781–783

Oblatum 24-XI-1987

Notes III

Numbers in brackets set in boldface such as [**79a**] mean items in the list of articles; those in roman such as [17] and [94b] are references originally cited in the article to which notes are given. Some of the corrections were already made in the original articles, but they are superseded by the corrections made in these notes.

78a. On certain reciprocity-laws for theta functions and modular forms

p. 42, line 3: For $\Gamma_S \cap G_{\mathbf{Q}} = S$ read $\Gamma_S = G_{\mathbf{Q}} \cap S$.

p. 70, line 7: For $0 < m \in \mathbf{Z}$ read $0 \neq m \in \mathbf{Z}$.

78b. The arithmetic of automorphic forms with respect to a unitary group

p. 577, line 5: Insert "$c(\overline{x})$" after "constant".

p. 585, last line: The symbols $\mathcal{A}_\rho(\Psi)$ and $\mathfrak{A}_\sigma^{(m)}(\Psi)$ should be interchanged.

p. 601, line 7 from the bottom: For "ν" read "v".

78c. The special values of the zeta functions associated with Hilbert modular forms

The typesetter of this article was incredibly incompetent, and the editor made unwarranted changes after my proof-reading. A partial list of corrections was given in [**81e**]. Here we present a corrected version with minor changes, adding the proof of some easy facts; such was given in my graduate course at Princeton University in the Fall Term, 1978–79.

A more systematic treatment of modular forms for symplectic and unitary groups can be found in [**00**]; in particular, generalizations of Theorem 1.5, Propositions 1.6, 1.7, and 1.8 can be obtained in such cases; see [**00**, Section 10]. The space $\mathcal{M}_k(\mathfrak{c}, \psi)$ can be defined in a more transparent way, including the case of ψ of infinite order; see [**87b**, Section 6] and [**91**, Section 2].

The convergence of $D(s, f)$ of (2.41a) requires an estimate of $a(\xi)$; meromorphic continuation of $R(s, f)$ of (2.41b) can be reduced to the Mellin transform of $f(iy)$. These were more or less well-known. Easy expositions of these in more general cases are given in [**00**, Sections A6, A7]. Similarly, a detailed treatment of the series of type (3.1), including the results of type (3.6) and Proposition 3.1, is given in [**00**, Section 18]. In the paragraph subsequent to (3.5) we excluded the case in which $n = 1$ and $\kappa = 2$, but that case is treated in [**00**, Section 18].

Changing our formulation, we can give Theorems 4.1, 4.2, and 4.3 in stronger forms; see [**91**, Theorems 7.2, 11.3, and 8.1]. In particular, the case $k^0 = 2$, excluded in Theorem 4.3, is included in [**91**, Theorem 8.1]. The case of Hilbert modular forms of half-integral weight is also treated in [**81a**] and [**87b**, Theorem 10.6]. Most comprehensive results, including those of [17], are given in [**91**, §§7, 9, and 10].

A more clear-cut form of Lemma 4.10 is given in [**87a**, Theorems 5.2 and 5.5]. Proposition 4.15 is true also for $k = \mathbf{1}$, since every form of that weight is the sum of a cusp form and an Eisenstein series as proved in [**85b**]; see also [**91**, Theorem 7.1].

Higher-dimensional generalizations of forms of type (5.2) and their automorphy properties are discussed in [**86**, §7], [**97a**, §A7], and [**00**, §A5]. A result similar to, but somewhat different from, Theorem 5.5 is given in [**80**, Theorem 9.8].

79a. Automorphic forms and the periods of abelian varieties

p. 579, line 9 from the bottom: For "du_j" read "du_j^ν".
p. 581, line 1: For "k_j" read "h_j".

It should be noted that the functions $P \circ \varepsilon$, R_ν, S_ν, and Y of (5.3) can be chosen so that they are holomorphic and invertible at any given point z_0 of $\mathfrak{D}$. Indeed, since $\{\alpha(z_0) | \alpha \in G_{\mathbf{Q}}\}$ is dense in $\mathfrak{D}$, we can find $\alpha \in G_{\mathbf{Q}}$ so that $P \circ \varepsilon$ and Y are holomorphic and invertible at $\alpha(z_0)$. Let $[C \ D]$ be the lower half of the element α^* of $Sp(n, \mathbf{Q})$ as in (4.8). Put $P_1(Z) = (CZ + D)^{-1} P\big(\alpha^*(Z)\big)$ and $Y_1 = Y \circ \alpha$. Then, in view of (4.12), we easily see that the equality on Page 578, line 5 is true with P_1 and Y_1 in place of P and Y. Then (5.3) with P_1 and Y_1 in place of P and Y gives the desired fact, since ω_2 is everywhere finite and invertible.

Also, there exists a holomorphic function f on $\mathfrak{D}$ such that the $f R_\nu$ and $f S_\nu$ are holomorphic everywhere and $f(z_0) \neq 0$. Indeed, by [**77c**, (1.20)] there exists a nonzero holomorphic function $\theta(Z)$ such that θP is holomorphic everywhere. Choose the above α so that $\theta\big(\varepsilon(\alpha(z_0))\big) \neq 0$ and $\det(\theta P)\big(\alpha(z_0)\big) \neq 0$. Define P_1 and Y_1 as above. Now every element of $\mathfrak{A}_0(\overline{\mathbf{Q}})$ holomorphic at z_0 is a quotient g/h with holomorphic functions g and h on $\mathfrak{D}$ such that $h(z_0) \neq 0$. This follows from the projective embedding of $\Gamma \backslash \mathfrak{D}$

by means of automorphic forms of a sufficiently large weight, as explained in the proof of [**75c**, Theorem 5]. Apply this fact to the entries of Y_1^{-1}. Take f to be the product of $\theta\big(\varepsilon(\alpha(z))\big)$ and the denominators h for all the entries of Y_1^{-1}. Again (5.3) with P_1 and Y_1 in place of P and Y gives the desired fact.

79b. On some problems of algebraicity

p. 377, line 4: For "$h(a)^\delta$" read "$h(a^\delta)$".

p. 377, line 5: For "$x \in Y$" read "$x \in Y^\times$".

p. 377, line 15 from the bottom: The diagonal matrix for $v > r$ should read

$$\mathrm{diag}\big[P_{v1}^{-1} \otimes 1_{q_1}, \ldots, P_{vt}^{-1} \otimes 1_{q_t}, P_{v1} \otimes 1_{q_1}, \ldots, P_{vt} \otimes 1_{q_t} \big].$$

Theorems 3 and 4, when $n = 1$, were proved in [**79a**]. Though the general case can be proved in a similar way, the proof is somewhat more involved. Therefore we prove here these two theorems for $n \geq 1$ in detail.

1. Let the symbols be as in §3; write G and G_+ for the groups $G_{\mathbf{Q}}$ and $G_{\mathbf{Q}+}$. Define $\mathfrak{H}_n$ and the action of G_+ on $\mathfrak{H}_n^r$ as described there. Take a totally imaginary quadratic extension K of F and put $B_K = B \otimes_F K$. In [**67d**, Proposition 6.2] we found an embedding of G into the group of similitudes of a hermitian form defined in $GL_n(B_K)$. For that purpose we took a positive involution ρ of B_K. We now make the following observation: ρ *can be changed for any positive involution of* B_K. Indeed, if σ is another positive involution of B_K, then by a well-known principle, $x^\rho = ax^\sigma a^{-1}$ for every $x \in B_K$ with an element $a \in B_K^\times$ such that $a^\sigma = a$. We easily see that in the proof of that proposition we can change v and ρ for va and σ, which gives the desired fact.

Take in particular K so that $B_K = M_2(K)$. Then $GL_n(B_K)$ can be identified with $GL_{2n}(K)$. Clearly the restriction of ρ to K is the generator of $\mathrm{Gal}(K/F)$; we denote complex conjugation in $\mathbf{C}$ also by ρ if there is no fear of confusion. Then we put $X^* = {}^t X^\rho$ for a matrix X with entries in K or in $\mathbf{C}$. We define various symbols as follows:

$$\mathfrak{S}_n = \big\{ \mathfrak{z} \in M_n(\mathbf{C}) \,\big|\, 1 - \mathfrak{z}^* \mathfrak{z} > 0 \big\}, \quad \mathfrak{D}_n = \big\{ z \in \mathfrak{S}_n \,\big|\, {}^t z = z \big\},$$

$$B_0(z) = \begin{bmatrix} \bar{z} & z \\ 1_n & 1_n \end{bmatrix} \ (z \in \mathfrak{H}_n), \quad B(\mathfrak{z}) = \begin{bmatrix} 1_n & \mathfrak{z} \\ \mathfrak{z}^* & 1_n \end{bmatrix} \ (\mathfrak{z} \in \mathfrak{S}_n),$$

$$U(n, n) = \big\{ \alpha \in GL_{2n}(\mathbf{C}) \,\big|\, \alpha I_{n,n} \alpha^* = I_{n,n} \big\}, \quad I_{n,n} = \mathrm{diag}[1_n, -1_n],$$

$$G(T) = \big\{ \alpha \in GL_{2n}(K) \,\big|\, \alpha T \alpha^* = v(\alpha) T \text{ with } v(\alpha) \in F \big\},$$

$$G_+(T) = \big\{ \alpha \in G(T) \,\big|\, v(\alpha) \text{ is totally positive} \big\}.$$

Here T is an element of $GL_{2n}(K)$ such that $T = -T^*$. For $\alpha \in \mathbf{R}^\times \cdot U(n, n)$ and $\mathfrak{z} \in \mathfrak{S}_n$ we can define $\lambda(\alpha, \mathfrak{z}), \mu(\alpha, \mathfrak{z}) \in GL_n(\mathbf{C})$ and $\alpha\mathfrak{z} \in \mathfrak{S}_n$ by the formula

$$\alpha B(\mathfrak{z}) = B(\alpha\mathfrak{z}) \mathrm{diag}[\overline{\lambda(\alpha, \mathfrak{z})}, \mu(\alpha, \mathfrak{z})].$$

We now fix a CM-type $\{\tau_v\}_{v=1}^g$ of K so that the restriction of τ_v to F is θ_v. We identify $K_{\mathbf{R}} = K \otimes_{\mathbf{Q}} \mathbf{R}$ with $\mathbf{C}^g$ through $x \mapsto (x^{\tau_v})_{v=1}^g$ for $x \in K$. In [**67d**, Proposition 6.6] we found an F-linear ring-injection i of $M_n(B)$ into $M_{2n}(K)$ that maps G into

$G(T)$ with a suitable T and also a holomorphic bijection j of $\mathfrak{H}_n^r$ onto $\mathfrak{D}_n^r$. Let us describe them explicitly. First we may assume that $'\xi^\rho$ of [**67d**, Proposition 6.2] is ξ^* for the reason explained at the beginning, so that we can take the group $G(T)$ there to be the present $G(T)$. By [**67d**, (6.3.8)], $-\sqrt{-1}T^{\tau_v}$ has signature (n, n) for $v \le r$ and $(2n, 0)$ for $v > r$. Also, every element of $G_+(T)$ acts on $\mathfrak{S}_n^r$. To be precise, we first identify $M_n(B_K)$ with $M_{2n}(K)$ as follows: let $\xi = (x_{ij}) \in M_n(B_K)$ with $x_{ij} = \begin{bmatrix} a_{ij} & b_{ij} \\ c_{ij} & d_{ij} \end{bmatrix} \in B_K = M_2(K)$; then we identify ξ with $\begin{bmatrix} A & B \\ C & D \end{bmatrix} \in M_{2n}(K)$, $A = (a_{ij})$, $B = (b_{ij})$, $C = (c_{ij})$, $D = (d_{ij})$. Then $\varphi_v^n(\xi)$ of [**67d**, §6.3] coincides with ξ^{τ_v}. Take $W_v \in GL_{2n}(\overline{\mathbf{Q}})$ so that $\sqrt{-1}W_v(T^{\tau_v})^{-1}W_v^*$ is $I_{n,n}$ or 1_{2n} according as $v \le r$ or $v > r$, as in [**67d**, (6.4.3)]. Put $\xi_v = {}^t W_v^{-1}\xi^{\rho\tau_v} \cdot {}^t W_v$. Then $\xi \mapsto (\xi_v)_{v=1}^g$ maps $G_+(T)$ into $\left[\mathbf{R}^\times \cdot U(n, n)\right]^r \times \left[\mathbf{R}^\times \cdot U(2n)\right]^{g-r}$, and the action of $G_+(T)$ on $\mathfrak{S}_n^r$ is defined with respect this map and the action of $\left[\mathbf{R}^\times \cdot U(n, n)\right]^r$ on $\mathfrak{S}_n^r$ defined above.

For $\alpha \in G_+(T)$ and $\mathfrak{z} \in \mathfrak{S}_n^r$ we put

$$\lambda_v(\alpha, \mathfrak{z}) = \lambda(\alpha_v, \mathfrak{z}_v), \quad \mu_v(\alpha, \mathfrak{z}) = \mu(\alpha_v, \mathfrak{z}_v) \text{ if } v \le r,$$

and $\lambda_v(\alpha, \mathfrak{z}) = (\alpha_v)^\rho$ if $v > r$; we do not define μ_v for $v > r$. (All these are essentially the same as what is done in [**79a**, §4], though W_v and $\mathfrak{S}_n^r$ here correspond to $'Q_v$ and $\mathfrak{D}$ of [**79a**, (4.2) and (4.4)].) Define α_v for $\alpha \in G$ and χ_v as in §3 of the present paper.

As for the maps $i : G \to G(T)$ and $j : \mathfrak{H}_n^r \to \mathfrak{D}_n^r$, the proof of [**67d**, Proposition 6.6] shows that they are given by (or satisfy) $i(\alpha)_v = A_v \alpha_v A_v^{-1}$ for $\alpha \in G$ and $j(z)_v = (p_v z_v + q_v)(r_v z_v + s_v)^{-1}$ with $A_v = \begin{bmatrix} p_v & q_v \\ r_v & s_v \end{bmatrix} = UZ_v$, where $U = \begin{bmatrix} 1_n & -i1_n \\ 1_n & i1_n \end{bmatrix}$ and $Z_v \in \mathbf{R}^\times \cdot Sp(n, \mathbf{R})$. Examining the proof and multiplying by a suitable constant, we may assume that $Z_v \in Sp(n, \mathbf{R}) \cap GL_{2n}(\overline{\mathbf{Q}})$. Also observe that $U B_0(z) = B(w)\mathrm{diag}[\overline{z} - i1_n, z + i1_n]$ if $w = (z - i1_n)(z + i1_n)^{-1}$.

For $z \in \mathfrak{H}_n^r$ put $\kappa_v(z) = r_v z_v + s_v$. Then for $\mathfrak{z} = j(z)$ we have $A_v B_0(z_v) = B(\mathfrak{z}_v)\mathrm{diag}\left[\overline{\kappa_v(z)}, \kappa_v(z)\right]$. Now we have

(1) $\lambda_v\left(i(\alpha), \mathfrak{z}\right) = \mu_v\left(i(\alpha), \mathfrak{z}\right) = \kappa_v\left(\alpha(z)\right)\chi_v(\alpha, z)\kappa_v(z)^{-1}$ for every $\alpha \in G_+$.

To prove this, put $\beta = i(\alpha)$ and suppress the subscript v. Then

$$B\left(j(\alpha(z))\right)\mathrm{diag}\left[\overline{\kappa\left(\alpha(z)\right)}\overline{\chi(\alpha, z)}, \kappa\left(\alpha(z)\right)\chi(\alpha, z)\right]$$
$$= A B_0\left(\alpha(z)\right)\mathrm{diag}\left[\overline{\chi(\alpha, z)}, \chi(\alpha, z)\right]$$
$$= A\alpha B_0(z) = \beta A B_0(z) = \beta B(\mathfrak{z})\mathrm{diag}\left[\overline{\kappa(z)}, \kappa(z)\right]$$
$$= B\left(\beta(\mathfrak{z})\right)\mathrm{diag}\left[\overline{\lambda(\beta, \mathfrak{z})}, \mu(\beta, \mathfrak{z})\right]\mathrm{diag}\left[\overline{\kappa(z)}, \kappa(z)\right],$$

from which we obtain (1). At the same time we obtain $\beta\left(j(z)\right) = j\left(\alpha(z)\right)$.

2. Let the symbols $Y, Y_i, L, L_i, \delta, h, \Psi_v,$ and E_v be as in §3 of the present article. Take K so that $K \otimes_F L_i$ is a field for every i in addition to the condition that $B_K = M_2(K)$. Put $Q_i = L_i \otimes_F K$ and $Q = \bigoplus_{i=1}^t Q_i$. By [**67d**, (4.7.1)], Y_i belongs to the same Brauer class as $B \otimes_F L_i$. Therefore $Y_i \otimes_F K$ is isomorphic to $M_{q_i}(Q_i)$.

Put $Y_K = Y \otimes_F K$ and define a ring-injection $p : Y_K \to M_{2n}(K)$ by $p(cy) = c \cdot i\left(h(y)\right)$ for $c \in K$ and $y \in Y$; similarly extend δ to Y_K by $(cy)^\delta = c^\rho y^\delta$. Now the map i is obtained by $i(\alpha) = i_0(\beta\alpha\beta^{-1})$ with an element β of G and the map i_0 of [**67d**, Proposition 6.2], whose proof shows that $i_0('\alpha^\iota) = Ti_0(\alpha)^*T^{-1}$. Therefore we can easily verify that $p(x^\delta) = Tp(x)^*T^{-1}$ for every $x \in Y_K$. Put $Y_K^u = \left\{y \in Y_K \big| yy^\delta = 1\right\}$ and $\mathfrak{w} = j(w)$

with the fixed point w of $h(Y[\delta])$ on $\mathfrak{H}_n^r$. Then we see that $\mathfrak{w}$ is the fixed point of $h(Y_K^\mu)$ on $\mathfrak{S}_n^r$.

Define two representations $\Psi_\nu^\pm$ of Q so that $\Psi_\nu^+(ca) = c^{\tau_\nu}\Psi_\nu(a)$ and $\Psi_\nu^-(ca) = c^{\rho\tau_\nu}\Psi_\nu(a)$ for $c \in K$ and $a \in L$. From [**79a**, (4.9)] we see that $\lambda_\nu(p(c), \mathfrak{z}) = c^{\tau_\nu}1_n$ for every ν and $\mu_\nu(p(c), \mathfrak{z}) = c^{\rho\tau_\nu}1_n$ for $\nu \leq r$. Put $C_\nu = E_\nu\kappa_\nu(w)^{-1}$. Then, from (1) we obtain, for $\nu \leq r$,

$$(2) \qquad C_\nu\lambda_\nu(p(b), \mathfrak{w})C_\nu^{-1} = \Psi_\nu^+(b) \text{ and } C_\nu\mu_\nu(p(b), \mathfrak{w})C_\nu^{-1} = \Psi_\nu^-(b)$$

for every $b \in Q$ such that $bb^\delta = 1$. If $\nu > r$, we can find an element κ_ν of $GL_{2n}(\overline{\mathbf{Q}})$ such that $(i(\alpha)_\nu)^\rho = \kappa_\nu\chi_\nu(\alpha)\kappa_\nu^{-1}$ for every $\alpha \in M_n(B)$. Then, putting $C_\nu = E_\nu\kappa_\nu^{-1}$, for $\nu > r$ we have

$$(3) \qquad C_\nu\lambda_\nu(p(b), \mathfrak{w})C_\nu^{-1} = C_\nu(p(b)_\nu)^\rho C_\nu^{-1} = \mathrm{diag}[\Psi_\nu^-(b)^\rho, \Psi_\nu^+(b)]$$

for every b as above. We are going to apply the results of [**79a**, §5] to the present situation; $\Psi_\nu^\pm$ here correspond to $\Psi_\nu^\pm$ there if $\nu \leq r$; however, if $\nu > r$, Ψ_ν^+ there is the right-hand side of (3). (In the latter half of [**79a**, §5] we assumed that the fixed point, written z_0 there, belongs to a somewhat restricted type. However, as already noted in the paragraph preceding [**79a**, (5.5a)], the formulas obtained there are applicable to the present case with obvious modifications, as explicitly described below.)

To prove Theorem 3, we use the symbol p_K for a CM-field K and its generalization p_L for L as above defined in [**80**, Theorem 1.1 and (1.1)]. In [**79a**, §5] we obtained meromorphic maps $R_\nu, S_\nu : \mathfrak{S}_n^r \to M_n(\mathbf{C})$ for $\nu \leq r$ and $R_\nu : \mathfrak{S}_n^r \to M_{2n}(\mathbf{C})$ for $\nu > r$, such that $R_\nu(\gamma(z)) = \lambda_\nu(\gamma, z)R_\nu(z)$ and $S_\nu(\gamma(z)) = \mu_\nu(\gamma, z)S_\nu(z)$ for every $\nu \leq g$ and every γ in a congruence subgroup of $G(T)$, where we ignore S_ν if $\nu > r$. We can choose these so that their values at any fixed point $\mathfrak{z}_0$ of $\mathfrak{S}_n^r$ are finite and invertible. (See notes to [**79a**].) Put $T_\nu(z) = \delta_\nu^{-1}\kappa_\nu(z)^{-1}R_\nu(j(z))$ and $U_\nu(z) = \delta_\nu\kappa_\nu(z)^{-1}S_\nu(j(z))$ for each ν and $z \in \mathfrak{H}_n^r$, where $\kappa_\nu(z) = \kappa_\nu$ if $\nu > r$, and $\delta_\nu = p_K(\tau_\nu, \sum_{\mu>r}\tau_\mu)$. Then T_ν and U_ν have property (ii) of Theorem 3, and property (i) as well, if R_ν and S_ν are suitably chosen. Notice that $\kappa_\nu(w)$ has algebraic entries, since the matrices A_ν and w_ν have algebraic entries.

Define embeddings $\psi_{\nu ij}$ and $\psi'_{\nu ij}$ of L_i into $\mathbf{C}$ by $\psi_{\nu ij} = \psi'_{\nu ij} = \sigma_{ij}^\nu$ on L_i and $\psi_{\nu ij} = \rho\psi'_{\nu ij} = \tau_\nu$ on K; put $\omega_i = \sum_{j=1}^{m_i}\left(\sum_{\nu=1}^g \psi_{\nu ij} + \sum_{\nu=1}^r \psi'_{\nu ij} + \sum_{\nu>r} \rho\psi'_{\nu ij}\right)$. Then ω_i is a CM-type of Q_i. By [**79a**, (5.9a,b)], if R_ν and S_ν are finite and invertible at $\mathfrak{w}$, then

$$(4a) \qquad \mathrm{diag}\left[\ldots, p_{Q_i}(\psi_{\nu ij}, \omega_i)1_{q_i}, \ldots\right]^{-1}C_\nu R_\nu(\mathfrak{w}) \in GL_n(\overline{\mathbf{Q}}) \quad \text{if } \nu \leq r,$$

$$(4b) \qquad \mathrm{diag}\left[\ldots, p_{Q_i}(\psi'_{\nu ij}, \omega_i)1_{q_i}, \ldots\right]^{-1}C_\nu S_\nu(\mathfrak{w}) \in GL_n(\overline{\mathbf{Q}}) \quad \text{if } \nu \leq r,$$

$$(5) \qquad \mathrm{diag}\left[\ldots, p_{Q_i}(\rho\psi'_{\nu ij}, \omega_i)1_{q_i}, \ldots, p_{Q_i}(\psi_{\nu ij}, \omega_i)1_{q_i}, \ldots\right]^{-1}C_\nu R_\nu(\mathfrak{w})$$

$$\in GL_{2n}(\overline{\mathbf{Q}}) \text{ if } \nu > r.$$

Let p_{ij}^ν and $P_\nu(w)$ be defined as in the paragraph preceding Theorem 3. By [**80**, Theorem 1.1 (2)] we can put $p_{ij}^\nu = p_{L_i}(\sigma_{ij}^\nu, \sum_{k=1}^{m_i}\sum_{\mu=1}^r \sigma_{ik}^\mu)$.

By [**80**, Theorem 1.1 (4)] we have, for every fixed (v, i, j),

$$p_{ij}^v = p_{Q_i}\left(\psi_{vij}, \mathrm{Inf}_{Q_i/L_i}\left(\sum_{k=1}^{m_i}\sum_{\mu=1}^{r}\sigma_{ik}^\mu\right)\right) = p_{Q_i}\left(\psi_{vij}, \sum_{k=1}^{m_i}\sum_{\mu=1}^{r}(\psi_{\mu ik} + \psi_{\mu ik}')\right),$$

$$\delta_v = p_{Q_i}\left(\psi_{vij}, \mathrm{Inf}_{Q_i/K}\left(\sum_{\mu>r}\tau_\mu\right)\right) = p_{Q_i}\left(\psi_{vij}, \sum_{k=1}^{m_i}\sum_{\mu>r}(\psi_{\mu ik} + \rho\psi_{\mu ik}')\right),$$

and hence $\delta_v p_{ij}^v = p_{Q_i}(\psi_{vij}, \omega_i)$; similarly $\delta_v^{-1} p_{ij}^v = p_{Q_i}(\psi_{vij}', \omega_i)$. Therefore the right-hand sides of (4a), (4b), and (5) are $P_v(w)^{-1} E_v T_v(w)$, $P_v(w)^{-1} E_v U_v(w)$ for $v \le r$, and $P_v(w)^{-1} E_v T_v(w)$ for $v > r$, respectively. Thus T_v for every v and also U_v for $v \le r$ have property (iii) of Theorem 3 at w.

At the beginning we fixed K so that $B_K = M_2(K)$ and $L_i \otimes_F K$ is a field for every i, and then chose R_v and S_v. Let D denote the set of data $\{K, R_v, S_v\}$ and let $\mathcal{W}(D)$ denote the set of fixed points w of the above type such that R_v and S_v are finite and invertible at $j(w)$. We have proved that T_v has property (iii) for $w \in \mathcal{W}(D)$. In the next section we prove property (iii) for *every* CM-point where T_v is finite.

3. If w is a fixed point of $h(Y[\delta])$ and $h_1 : T \to M_n(B)$ is defined by $h_1(a) = \alpha h(a)\alpha^{-1}$ with $\alpha \in G_+$, then $\alpha(w)$ is the fixed point of $h_1(Y[\delta])$. Let w' be the fixed point of $h'(Y'[\delta'])$ for some (Y', δ', h'). Then we can find $D' = \{K', R_v', S_v'\}$ such that $w' \in \mathcal{W}(D')$, and functions T_v' with property (iii) for the points of $\mathcal{W}(D')$, and such that $T_v'(w')$ is finite and invertible. Our task is to show that T_v has property (iii) at w'.

Now we can find a CM-field P such that $[P : F] = 2$, B_P is isomorphic to $M_2(P)$, and both $P \otimes_F K$ and $P \otimes_F K'$ are fields. Let $Y_0 = M_n(P)$. By [**67d**, (4.7.2), (4.7.3)], Y_0 has a positive involution δ_0, and there is an F-linear ring-injection h_0 of Y_0 into $M_n(B)$ such that $h_0(a^\delta) = {}^t h(a)^t$ for every $a \in Y_0$. Let w_0 be the fixed point of $h_0(Y_0[\delta_0])$. Let X be the set of points $z \in \mathfrak{H}_n^r$ such that the R_v, R_v', S_v, S_v' are all finite and invertible at $j(z)$. Then X is an open subset of $\mathfrak{H}_n^r$. We can find a holomorphic function f on $\mathfrak{S}_n^r$ such that the $f R_v$ and $f S_v$ are all holomorphic everywhere on $\mathfrak{S}_n^r$; moreover f can be chosen so that $f \circ j$ is nonzero at any given point of $\mathfrak{H}_n^r$. (See notes to [**79a**].) We choose f so that $\det(f R_v) \circ j$ and $\det(f S_v) \circ j$ are all nonzero functions on $\mathfrak{H}_n^r$. Let Z be the set of all $z \in \mathfrak{H}_n^r$ such that all those functions, as well as $f \circ j$, are nonzero. Then Z is an open dense subset of $\mathfrak{H}_n^r$. Define similarly Z' with R_v' and S_v' in place of R_v and S_v. Then Z' is open and dense in $\mathfrak{H}_n^r$, and so is $Z \cap Z'$. Since $Z \cap Z' \subset X$, X is dense in $\mathfrak{H}_n^r$.

Put $G^1 = \{\alpha \in G | v(\alpha) = 1\}$, $G_{\mathbf{R}}^1 = \{\alpha \in G_{\mathbf{R}}^1 | v(\alpha) = 1\}$, and $\mathfrak{X} = \{\alpha \in G_{\mathbf{R}}^1, |\alpha(w_0) \in X\}$. Then $\mathfrak{X}$ is open and dense in $G_{\mathbf{R}}^1$. Let $\mathcal{W}_0$ be the set of all points of the form $\alpha(w_0)$ contained in X with $\alpha \in G^1$. Since G^1 is dense in $G_{\mathbf{R}}^1$, we see that $\mathcal{W}_0$ is dense in $\mathfrak{H}_n^r$, and clearly contained in $\mathcal{W}(D) \cap \mathcal{W}(D')$. Put $g_v = (T_v')^{-1} T_v$. Then there is a congruence subgroup Γ of G_+ such that the entries of g_v are Γ-invariant meromorphic functions on $\mathfrak{H}_n^r$. Since T_v and T_v' have property (ii) at the points of $\mathcal{W}_0$, we see that g_v is finite and takes algebraic values at the points of $\mathcal{W}_0$. Since $\mathcal{W}_0$ is dense in $\mathfrak{H}_n^r$, the entries of g_v must belong to $\mathfrak{A}_0(\overline{\mathbf{Q}})$, as will be shown below. Now $T_v = T_v' g_v$. Suppose T_v is finite at w'; then both T_v' and g_v are finite at w'. Since the entries of g_v belong to $\mathfrak{A}_0(\overline{\mathbf{Q}})$, $g_v(w')$ is algebraic; also T_v' has property (iii) at w'. Consequently T_v has property (iii) at w' as expected. We can similarly show that U_v has property (iii) at every CM-point of $\mathfrak{H}_n^r$.

We need to prove that if p is a Γ-invariant meromorphic function on $\mathfrak{H}_n^r$ and $p(w) \in \overline{\mathbf{Q}}$ for $w \in \mathcal{W}_0$ whenever p is finite at w, then $p \in \mathfrak{A}_0(\overline{\mathbf{Q}})$. To show this, take a $\overline{\mathbf{Q}}$-rational canonical model V of $\Gamma \backslash \mathfrak{H}_n^r$ established in [**67d**] or [**70a**]; let φ be the natural Γ-invariant map $\mathfrak{H}_n^r \to V$. Then $p = q \circ \varphi$ with an element q of the function-field of V over $\mathbf{C}$ in the sense of algebraic geometry. Now every point of $\varphi(\mathcal{W}_0)$ is a $\overline{\mathbf{Q}}$-rational point of V, so that for every $\sigma \in \mathrm{Aut}(\mathbf{C}/\overline{\mathbf{Q}})$ and $x \in \varphi(\mathcal{W}_0)$ we have $q^\sigma(x) = q^\sigma(x^\sigma) = q(x)^\sigma = q(x)$, provided q is finite at x. Since all such points x form a dense subset of V, we have $q^\sigma = q$, that is, q is $\overline{\mathbf{Q}}$-rational. Thus $p \in \mathfrak{A}_0(\overline{\mathbf{Q}})$, Q.E.D.

4. *Proof of Theorem 4.* Let $\mathcal{A}_0(\overline{\mathbf{Q}})$ denote the field of all $\overline{\mathbf{Q}}$-rational automorphic functions on $\mathfrak{S}_n^r$ with respect to $G(T)$ in the sense of [**79a**, §5]. Let k be the complex dimension of $\mathfrak{H}_n^r$; then we can find k elements $\mathfrak{f}_1, \ldots, \mathfrak{f}_k$ of $\mathcal{A}_0(\overline{\mathbf{Q}})$ such that the $\mathfrak{f}_i \circ j$ are algebraically independent over $\overline{\mathbf{Q}}$. Put $g_i = \mathfrak{f}_i \circ j$. Then the $\partial/\partial g_i$ are well-defined derivations of $\mathfrak{A}_0(\overline{\mathbf{Q}})$, and $\Delta_\nu f = \sum_{i=1}^k (\partial f/\partial g_i)\Delta_\nu g_i$ for every $f \in \mathfrak{A}_0(\overline{\mathbf{Q}})$. Therefore, to prove Theorem 4, it is sufficient to treat the case where $f = \mathfrak{f} \circ j$ with $\mathfrak{f} \in \mathcal{A}_0(\overline{\mathbf{Q}})$. Define Δ_ν^* on $\mathfrak{S}_n^r$ by $\Delta_\nu^* \mathfrak{f} = \big(\partial \mathfrak{f}/\partial \mathfrak{z}_{ij}^\nu\big)_{i,j=1}^n$, where $\mathfrak{z}^\nu = (\mathfrak{z}_{ij}^\nu)$ is the matrix variable on the ν-th factor of $\mathfrak{S}_n^r$. For $z \in \mathfrak{H}_n$, put $w = U(z) = (z-i)(z+i)^{-1}$; then $dw = 2i(z+i)^{-1}dz(z+i)^{-1}$. Now $j(z)_\nu = U\big(Z_\nu(z_\nu)\big)$. Since $Z_\nu \in Sp(n, \mathbf{R})$, we have $dZ_\nu(z_\nu) = {}^t(c_\nu z_\nu + d_\nu)^{-1}dz_\nu(c_\nu z_\nu + d_\nu)^{-1}$, where $[c_\nu \ d_\nu]$ is the lower half of Z_ν. Thus, for $\mathfrak{z} = j(z)$ we have

$$
\begin{aligned}
d\mathfrak{z}_\nu &= 2i\big(Z_\nu(z_\nu) + i\big)^{-1} \cdot {}^t(c_\nu z_\nu + d_\nu)^{-1}dz_\nu(c_\nu z_\nu + d_\nu)^{-1}\big(Z_\nu(z_\nu) + i\big)^{-1} \\
&= 2i \cdot {}^t\kappa_\nu(z)^{-1}dz_\nu\kappa_\nu(z)^{-1}.
\end{aligned}
$$

Therefore, for $f = \mathfrak{f} \circ j$ we have

$$
\begin{aligned}
\sum_{\nu=1}^r \mathrm{tr}\big(\Delta_\nu f \cdot dz_\nu\big) = df &= \sum_{\nu=1}^r \mathrm{tr}\big(\Delta_\nu^* \mathfrak{f} \cdot d\mathfrak{z}_\nu\big) \circ j \\
&= 2i \sum_{\nu=1}^r \mathrm{tr}\big((\Delta_\nu^* \mathfrak{f}) \circ j \cdot {}^t\kappa_\nu(z)^{-1}dz_\nu\kappa_\nu(z)^{-1}\big),
\end{aligned}
$$

so that

$$
(6) \qquad \Delta_\nu f = i \cdot \kappa_\nu(z)^{-1}\big[\big(\Delta_\nu^* \mathfrak{f} + {}^t(\Delta_\nu^* \mathfrak{f})\big) \circ j\big] \cdot {}^t\kappa_\nu(z)^{-1}.
$$

By [**79a**, Theorem 6.1], $\Delta_\nu^* \mathfrak{f} = \pi R_\nu W_\nu \cdot {}^t S_\nu$ with $W_\nu \in M_n\big(A_0(\overline{\mathbf{Q}})\big)$. Let T_ν and U_ν be defined as in §2. Then the right-hand side of (6) is $\pi i(T_\nu V_\nu \cdot {}^t U_\nu + U_\nu \cdot {}^t V_\nu \cdot {}^t T_\nu)$ with $V_\nu(z) = W_\nu\big(j(z)\big)$. Since $\mathfrak{f} \circ j$ is meaningful, the entries of $(\Delta_\nu^* \mathfrak{f}) \circ j$ are meaningful as meromorphic functions on $\mathfrak{H}_n^r$; hence the same is true for the entries of V_ν. Put $X_\nu = T_\nu^{-1}U_\nu$. Then the entries of V_ν and X_ν belong to $\mathfrak{A}_0(\overline{\mathbf{Q}})$, since their values at the points in the set $\mathcal{W}_0$ are algebraic. Now

$$
\Delta_\nu f = \pi i T_\nu(V_\nu \cdot {}^t X_\nu + X_\nu \cdot {}^t V_\nu) \cdot {}^t T_\nu,
$$

which completes the proof of Theorem 4.

As for the functions R_ν, S_ν and the symbol p_K, an exposition more detailed than [**79a**] and [**80**] can be found in [**98**, §§32 and 33] and also in [**00**, §11], [**00**, Proposition 11.14], in particular.

80. The arithmetic of certain zeta functions and automorphic forms on orthogonal groups

p. 318, Theorem 1.2: We can add one more fact: $p_K(\mathrm{id}_K, \xi\beta) = p_L(\eta, \beta)$ *for every* $\beta \in J_L$. This can be proved in the same way. A better exposition for the contents of Section 1 was given in [**98**, Section 32]. In particular, Theorem 1.2 with the above addition is [**98**, Theorem 32.8].

p. 321, line 6 from the bottom: For "$D_{-1}\eta$" read "$D_{-1}p$".

p. 325, line 16 from the bottom: For "G_q" read "$\mathfrak{G}_{\mathbf{Q}+}$".

p. 325, line 2 from the bottom: For "T" read "U".

p. 327, line 14: For "$h(Y^u)$" read " the closure of $h(Y^u)$".

p. 330, line 4 from the bottom: For "$\mathfrak{F}$" read "F".

p. 339, (7.13): For "$g - r \in \mathbf{Z}^n$" read "$g - r \in \mathbf{Z}^{nq}$".

p. 339, (7.14): For "$\theta(u, z; r, s)$" read "$\theta(u, Z; r, s)$".

p. 339, last line: For "$({}^t r, {}^t s)\gamma$" read "$({}^t r, {}^t s)\sigma$".

p. 340, (7.19): For "r_{vj}" and "r'_{vj}" read "$(Sr)_{vj}$" and "$(Sr')_{vj}$".

p. 341, line 2: For "p. 70" read "p. 69".

p. 343, Proposition 7.6: For "$\mathcal{M}_k^r(\mathbf{Q})$" read "$\mathcal{M}_k^r(\overline{\mathbf{Q}})$".

p. 352, line 15 from the bottom: Insert "$\sum_{v=1}^n m_v = 0$ and" at the end of the line.

p. 353, line 7 from the bottom: For "(2.21)" read "(2.20)".

p. 358, line 16 from the bottom: Insert "a finite $\overline{\mathbf{Q}}$-linear combination of functions of the form" at the end of the line.

p. 358, line 15 from the bottom: Delete "$f(w, 0; t, \xi, \chi) = M_U d'$".

p. 361, line 9: Insert " by Lemma 11.1" at the end of the line.

p. 362, line 3 from the bottom: For "$\mathfrak{g}$" read "g".

p. 363, line 2: Insert "(x)" before "$(u_\alpha, \cdots$".

p. 363, line 17: Insert " independent of k" after "$e - 2j$".

p. 363, (12.8): The sum should read "$\sum_{0 \le j \le e/2}$".

p. 374, Reference [7]: This paper appeared in Nagoya Math. J. 76 (1979), 153–171.

81a. The critical values of certain zeta functions associated with modular forms of half-integral weight

p. 667, line 6 from the bottom: For "a(t)" read "$a(t)$".

81b. On certain zeta functions attached to two Hilbert modular forms: I. The case of Hecke characters

p. 137, line 16: For "$(cz + d)^{-1}g(z)$" read "$(cz + d)^{-1}g(\alpha(z))$".

p. 144, line 7; p. 145, line 7: For "p_k" read "p_K".

p. 144, line 4 from the bottom: For "$i \ge 0$" read "$i > r$".

p. 144, Theorem 4.3: Condition (4.8b) is unnecessary. Indeed, this theorem is included in [**88**, Theorem 1.2] as a special case. Notice that condition (1.10b) of [**88**,

Theorem 1.2] is always satisfied if the Hilbert modular form $\Omega(z)$ of [**88**, (1.3)]] is of integral weight.

p. 145, line 2: Insert "γ" after "$c(\gamma)$".

81c. On certain zeta functions attached to two Hilbert modular forms: II. The case of automorphic forms on a quaternion algebra

p. 576, last line: For "$N(\mathfrak{p})^{1-2s}$" read "$N(\mathfrak{p})^{-1-2s}$".

p. 579, line 13: The symbol involving Γ should read "$\Gamma_M z_0$".

p. 579, line 7 from the bottom: For "$\mathcal{M}_{se}$ with some $s \in \mathbf{Z}$" read "$\mathcal{M}_s$ with some $s \in I_F$".

p. 579, line 6 from the bottom: For "$\mathcal{M}_{(t-s)e}$" read "$\mathcal{M}_{te-s}$"; delete also "$> s$".

p. 582, Theorem 3.3: Condition (3.12b) is unnecessary. In fact, this result is contained in [**88**, Theorem 3.2] as a special case. In the setting of Theorem 3.3, the number b_Z is an integer, and hence [**88**, (3.21b)] is satisfied.

p. 583, line 12 from the bottom: For "$\mathbf{Q}$" read "$\overline{\mathbf{Q}}$".

p. 584, line 14: For "F" read "E".

p. 588, Theorem 3.11, (ii): The condition "$\mu \neq 4$ if $E = \mathbf{Q}$" is unnecessary; see [**88**, Proposition 5.1].

p. 592, line 6 from the bottom: The last "k" should read "k_i".

p. 601, line 9 from the bottom: For "$-2s$" read "$-2sE$".

p. 605, line 1: For "(3.10)" read "(3.13)".

p. 605, line 9; For "on" read "in".

p. 606, line 6: For "$L(S_0)$" read "$L(s_0)$".

81d. Arithmetic of differential operators on symmetric domains

p. 834, (6.7): For "$M_{n_i}(K_i)$" read "$K_i^{n_i}$".

82b. The periods of certain automorphic forms of arithmetic type

p. 612, line 7: For "$\mu(\Gamma\backslash\mathfrak{H}^r)^{-1}$" read "$[\Gamma \cap \{\pm 1\} : 1]\mu(\Gamma\backslash\mathfrak{H}^r)^{-1}$". This change is unnecesssary if we assume that $-1 \notin \Gamma$. Otherwise we have to insert the factor $[\Gamma \cap \{\pm 1\} : 1]$ or its inverse at various places on pp. 613–614. In particular, $2^r q^{\xi/2}$ on page 613, line 5 should be multiplied by $[\Gamma \cap \{\pm 1\} : 1]^{-1}$. .

p. 615, line 8 from the bottom: For "A^r" read "A''^r".

p. 620, (3.7): The left-hand side should be multiplied by $[\Gamma \cap \{\pm 1\} : 1]$.

p. 626, line 5: For "$\xi \in V$" read "$\xi \in V_{\mathbf{R}}$".

p. 626, line 9: The inverted f should be f.

82c. Confluent hypergeometric functions on tube domains

p. 272, line 14 from the bottom: The last comma should be a period.

p. 272, line 9 from the bottom: For "$\delta(i\bar{z})^2$" read "$\delta(i\bar{z})^s$".

p. 274, Lemma 1.1: The proof shows that $\delta(z) \neq 0$ for $z \in H \cup H'$.

p. 277, line 10: Insert a semicolon after "$c = 1_q$".

p. 278, line 1: Inseert "with $d = \mathrm{diag}[\mu_1, \ldots, \mu_r]$" after "$xcd^{-1/2}$".

p. 279, line 10: The eigenvalues of an element of V in Case IV defined here are real numbers. Indeed, given $h \in V$ we can put $h = c\varepsilon + f$ with $c \in \mathbf{R}$ and $f \in V$ such that $\sigma(\varepsilon, f) = 0$. Since σ has signature $(1, m - 1)$, we have $\sigma(f, f) < 0$. Now $\sigma(h, h) = c^2 + \sigma(f, f)$ and $\sigma(h, e) = c$. Then it is easy to see that the eigenvalues are $c \pm |\sigma(f, f)|^{1/2}$.

p. 279, (2.5.IV): For "$x_1 y_2 -$" read "$x_1 y_2 +$".

p. 288, line 3 from the bottom: For "$\begin{pmatrix} x & z^* \\ z^* & y+1 \end{pmatrix}$" read "$\begin{pmatrix} x & z \\ z^* & y+1 \end{pmatrix}$".

The product $\delta(y)^s e^{2\pi i \sigma(h,x)} \xi(y, h; \alpha, \beta)$ as a function of $z = x + iy$ for any fixed $h \in V$ and suitable s, α, β in $\mathbf{C}$ is an eigenfunction of all covariant differential operators on H_m of an appropriate type. For details, see Theorem 2 of notes to [**99c**].

83a. Algebraic relations between critical values of zeta functions and inner products

p. 263, line 2 from the bottom; p. 264, line 2: For "$\overline{Q}$" read "$\overline{\mathbf{Q}}$".

p. 265, line 4: For "$(x_i]$" read "$[x_i]$".

p. 268, line 12: For "B_i" read "$B_i^{\times}$".

p. 269, line 8 from the bottom: For "$\mathrm{Tr}_{K_i/E}$" read "$\mathrm{Tr}_{F_i/E}$".

p. 273, line 3 from the bottom: For "$N(\mathfrak{p})^{-s}$" read "$N(\mathfrak{p})^{-ns}$".

p. 275, line 15 from the bottom: For "$\chi(by)$" read "$\chi(b\eta)$".

p. 278, line 2: Delete "and $\mathbf{g}$".

No detailed proof for Theorems 4.2, 4.3, and 4.4 are given. Generalizations of these are proved in detail in [**88**].

83b. On Eisenstein series

p. 419, line 16: For "Proposition" read "Lemma".

p. 432 (3.16): For "$m(m + 1)/2$" and "m^2" read "$-m(m + 1)/2$" and "$-m^2$".

p. 435, line 15: This statement is wrong and unnecessary, and therefore should be deleted.

p. 436, Proof of Lemma 4.2: Actually the Gauss sum of Hecke is $\overline{G(a)}$. Still, his argument and result apply to the present case.

pp. 435–452: A self-contained treatment of the series α is given in [**97**, Sections 13, 14, and 15] and [**00**, Section A1]. In particular see [**97**, Theorem 13.6] and [**00**, Theorem

16.2]. Conjecture 6.3 requires some modification; the above theorems of [**97**] and [**00**] contain the proof of the modified conjecture in a better form.

p. 449, line 5: For "$m/1$" read "$m - 1$".

p. 457, (7.12): Insert "if $h > 0$" at the end of the line.

p. 458, (7.20): For "$m\kappa$" read "$-m\kappa$".

p. 461, line 8: For "$x_v \mathbf{g}_v^m$" read "$^t x_v \mathbf{g}_v^m$".

p. 461, line 11: For "$s\Lambda$" read "$^t s\Lambda$".

p. 461, lines 13, 14: For "sx_v" read "$^t s \cdot {}^t x_v$".

p. 462, line 3: Insert "$\mathrm{rank}(j^v) = r$ and" at the end of the line.

p. 462, (8.5): For "$-m\kappa$" read "$m\kappa$".

p. 463, line 22: For "$m\kappa$" read "$-m\kappa$".

p. 463, line 13: For "$r > m$" read "$r < m$".

p. 464, line 4: For "(7.12)" read "(7.14)".

p. 470, line 10: Put a center dot in front of the line.

p. 472, line 2: For "$m\kappa$" read "$-m\kappa$".

p. 474, line 23: For the last "M'_{ab}" read "$\mathbf{Q}_{\mathrm{ab}}$".

p. 474, line 7 from the bottom: For "α" read "α^p".

84a. Differential operators and the singular values of Eisenstein series

p. 311, line 2 from the bottom: Insert "of" after "conductor".

p. 312, line 4: For "$\theta_1^*(wy\mathbf{a}^{-1})$" read "$\theta_1^*(w\mathbf{h})$".

p. 312, lines 5, 6: Delete "$\theta_1^*(y\mathbf{a}^{-1})$".

p. 325, lines 14–17 from the bottom: The communication of Harish-Chandra consisted of [**90a**, Proposition 2.1] and the reference to Lepowsky's paper cited in that article. These concern the operators on G of an appropriate type, but not those on X, and therefore we have to show that the operators on X in question correspond to those on G bijectively. For the proof of this fact and the commutativity of the ring of operators, see Theorem D in notes to [**90a**].

p. 325, line 5 from the bottom: The algebra is indeed generated by the L_r. See notes to [**90a**], the last paragraph of the section concerning differential operators on G/K.

For better treatments of differential operators discussed in this article, see [**84b**, Section 5], [**90a**], [**94b**], and [**00**, Sections 12 and 13].

84b. On differential operators attached to certain representations of classical groups

p. 464, line 10: For "σ" read "ρ".

p. 485, Proposition 5.1: For the formulas of this type, see also [**86**, Section 6] and [**00**, Lemma 13.9]; the latter gives a more direct proof of Proposition 5.1.

85a. On Eisenstein series of half-integral weight

p. 291, line 13: The paper [3] is unreliable and therefore should be disregarded. In this paper a certain result is claimed for all primes p citing a result in a previous paper which assumes $p \neq 2$; the paper has other unacceptable aspects. However this has no effect on the validity of the results in the present article.

p. 291, line 10 from the bottom: For "$2s + k - 1$" read "$2s + k - i$".

p. 294, line 6 from the bottom: For "$C_v'' \cap \Omega_v$" read "$(P_v \cap C_v'')\eta_v$".

p. 295, line 1: For "$P_{\mathbf{a}}$" read "$P_{\mathbf{A}}$".

p. 295, (3.14a): This formula is not always true, but at least valid for $\xi \in G \cap P_{\mathbf{A}}C'\eta$. Therefore (3.10d) and (3.14b) must be proved differently. First note that (3.10d) follows immediately from Lemma 3.5. Also Lemma 3.4 shows that $h(\iota, z) = \prod_{v \in \mathbf{a}} \det(-iz_v)^{1/2}$. Take η_0 in $G \cap C'G_{\mathbf{a}}\eta^{-1}$; let $\gamma = \iota\eta_0^{-1}$. Then $\gamma \in \iota\eta C'G_{\mathbf{a}} \subset P_{\mathbf{A}}C'$, and hence, by (3.10c), $h(\iota, z) = h(\gamma, \eta_0 z)h(\eta_0, z)$, which together with (3.10d) proves (3.14b). If $\xi \in G \cap P_{\mathbf{A}}C'\eta$, then $\xi = \alpha\eta_0$ with $\alpha \in G \cap P_{\mathbf{A}}C'$, since $P_{\mathbf{A}}C'\eta = P_{\mathbf{A}}C'\eta_0$. Then (3.14a) for such a ξ follows from (3.10c, d) and (3.14b).

p. 302, line 9 from the bottom: Insert "(3.10c) and" before "(4.8)".

p. 303, line 6: For "$\mathbf{Z}_m^m$" read "$\mathfrak{g}_m^m$".

p. 306, line 12 from the bottom: For "$\prod_{i=1}^{[(m-r-1)/2]}$" read "$\prod_{i=0}^{[(m-r-1)/2]}$".

p. 306, line 10 from the bottom: For "$\prod_{i=0}^{[(m-r)/2]}$" read "$\prod_{i=1}^{[(m-r)/2]}$".

p. 310, line 3 from the bottom: For "csae" read "case".

p. 312, line 14: For "F^{mq}" read "F_{mq}^1".

85b. On the Eisenstein series of Hilbert modular groups

p. 19, line 2 from the bottom: For "$1 < s$" read "$0 < s$".

p. 22, line 19 : For "$\theta_{\%}$" read "θ_v".

p. 24, Lemma 6.2: The equality is true only when $\tau = 0$. The integral vanishes if $\tau \neq 0$.

p. 25, line 6: Insert "$\sum_{p,q}$ is taken over all the (p, q) such that $\overline{p} + q + \sigma \in C u$, and" after "where".

p. 25, lines 12–14: Delete these three lines.

p. 35, line 5 from the bottom: For "(10.3)" read "(10.4)".

86. On a class of nearly holomorphic automorphic forms

p. 363, line 7: For "ρ" read "ρ_0".

p. 363, Lemma 3.5: The assumption that g is holomorphic is unnecessary.

p. 383, line 8: For "$2^{-1}b$" read "$2^{-1}b_\gamma$".

p. 388, line 4: For "$GL_n(\mathbf{C})$" read "$GL_n(\mathbf{C})^\varphi$".

p. 393, (8.6), (8.7): For "ζ" read "ζ_w".

p. 393, line 8: Insert "$\zeta_w(u) = \zeta\big(u\eta(w_\tau)\big)$ and" after "where".

p. 393, (8.9): For "$\theta(z, w; \zeta)$" read "$\theta(z, w; \zeta_w)$".

p. 393, last line: Both $\mathcal{P} \cap \Gamma$ and $\{U \times \Lambda\}$ must be multiplied by the center of $G_{m,L}$. Instead, we can take a sufficiently small Γ such that $\mathcal{P} \cap \Gamma = \{U \times \Lambda\}$.

p. 401, line 9: For "P_0" read "P_σ".

p. 401, last line: For "$\mathbf{e}$" read "$\mathbf{e}_F$".

p. 403, line 4 from the bottom: For "I_E" read "J_E".

p. 406: The proof of Theorem 4.2 in the case $F = E = \mathbf{Q}, l = 0$, and $n_\tau = 2$ given on this page is erroneous. See the correct proof given in [**88**, p. 259].

More systematic and detailed expositions of the contents of Sections 1 through 3 are given in [**94c**] and [**00**, Sections 12 and 13].

87a. Nearly holomorphic functions on hermitian symmetric spaces

p. 4, line 6: For "comutative" read "commutative".

p. 6, line 13 from the bottom: For "$\mathbf{C}^{2m}$" read "$\mathbf{C}_m^{2m}$".

p. 6, line 11 from the bottom: For "isotopy" read "isotropy".

p. 14, line 18: For "M_ρ" read "$\mathcal{M}_\rho$".

p. 14, line 6 from the bottom: For "$\mathcal{N}_\rho^k$" read "$\mathcal{N}_\sigma^k$".

Section 3: A more detailed treatment of this part is given in [**00**, Section 14] for symplectic and unitary groups. In particular, Proposition 3.3 and Lemma 3.5 for such groups are proved as [**00**, Proposition 14.2 and Theorem 14.7] in detail; similarly [**00**, Theorem 14.9] includes Theorem 3.7 for such groups, and is proved essentially in the same way, but in more detail.

Equality (4.12) is obtained by termwise application of Δ_q^p to the series of (4.3). Since this is not completely trivial, we present here two proofs. The key idea of the first of these is the technique used in the proof of [**82c**, Lemma 1.4].

First proof. Let the notation be as in Section 4; put

$$(1) \qquad \varepsilon(z, w) = \left(\det[i(w_\tau^* - z_\tau)]\right)_{\tau \in J} \qquad (z, w \in \mathcal{H}^J).$$

By the principle of (4.2) we can define $\varepsilon(z, w)^p$ for every $p \in \mathbf{C}^J$, where we take $\det(-iu)^q = \exp\left(q \cdot \log[\det(-iu)]\right)$ for $u \in \mathcal{H}$ and $q \in \mathbf{C}$ with the branch of $\log[\det(-iu)]$ that is real for $u = i1_m$. We then put, for $z, w \in \mathcal{H}^J$ and $s \in \mathbf{C}$,

$$(2) \qquad S(z, w; s) = \sum_{\alpha \in A} \varepsilon\left(\alpha(z), \alpha(w)\right)^{su - q/2} j_\alpha(z)^{-q}, \quad A = (P \cap \Gamma)\backslash\Gamma.$$

This is well-defined and locally uniformly convergent for $(z, w, s) \in \mathcal{H}^J \times \mathcal{H}^J \times \{s \in \mathbf{C} | \operatorname{Re}(s) > \sigma_0\}$, where $\sigma_0 = m$ in Case SU and $\sigma_0 = (m + 1)/2$ in Case SP. We shall prove a result more general than this fact in notes to [**95a**]. Now each term of (2) is holomorphic in $(z, \overline{w}, s)$, and hence $S(z, w; s)$ is a holomorphic function of $(z, \overline{w}, s)$ for $(z, w) \in \mathcal{H}^J \times \mathcal{H}^J$ and $\operatorname{Re}(s) > \sigma_0$. By a well-known principle, application of $(\partial/\partial z)^a$ and $(\partial/\partial \overline{w})^b$ to (2) can be done termwise. Now $S(z, z; s) = E(z, s; q, \Gamma)$, and hence Δ_q^p can be applied to the latter function termwise. This justifies (4.12) for $\operatorname{Re}(s) > \sigma_0$. Since both sides of (4.12) are real analytic in (z, s), we have (4.12) for every $s \in \mathbf{C}$ where the functions are finite. See also notes to [**95a**].

Second proof. We first consider the following setting. Let M be an open subset of $\mathbf{R}^m$, and D a C^∞ differential operator on M. It is well-known that there exists a differential operator E on M such that

$$(3) \qquad \int_M \varphi(x)(D\psi)(x)dx = \int_M (E\varphi)(x)\psi(x)dx$$

for every C^∞ functions φ and ψ of compact support. Clearly (3) is valid under the weaker assumption that either φ or ψ has compact support.

Lemma A. (i) *Let μ be a measure on a set N in the standard sense of measure space; let $f(x, y)$ be a measurable function of $(x, y) \in M \times N$ which is C^∞ in x for every $y \in N$. Suppose that both $f(x, y)$ and $(Df)(x, y)$ are integrable on N locally uniformly with respect to $x \in M$ in the following sense: for every compact subset C of M there exists a constant A such that*

$$(4) \qquad \int_N |f(x, y)|d\mu(y) \le A \quad and \quad \int_N |(Df)(x, y)|d\mu(y) \le A \text{ for every } x \in C.$$

Put

$$p(x) = \int_N f(x, y)d\mu(y) \quad and \quad q(x) = \int_N (Df)(x, y)d\mu(y).$$

Then $Dp = q$, provided p is C^∞.
(ii) *Suppose that $Qf(x, y) = 0$ for every $y \in N$ with a real analytic elliptic differential operator Q on M, and $f(x, y)$ is integrable on N locally uniformly with respect to $x \in M$. Then p is real analytic and $Qp = 0$.*

PROOF.   To prove (i), it is sufficient to show that $\int_M \varphi q\, dx = \int_M \varphi(Dp)dx$ for every C^∞ function φ on M of compact support. Given such a φ, we have

$$\int_M \varphi q\, dx = \int_M \int_N \varphi(x)(Df)(x, y)d\mu(y)dx = \int_N \int_M \varphi(x)(Df)(x, y)dx\, d\mu(y)$$

$$= \int_N \int_M (E\varphi)(x)f(x, y)dx\, d\mu(y) = \int_M \int_N (E\varphi)(x)f(x, y)d\mu(y)dx$$

$$= \int_M (E\varphi)p\, dx = \int_M \varphi(Dp)dx$$

as desired. Here the second and fourth equalities follow from (4).

Next, to prove (ii), let R be the adjoint of Q in the sense of (3). Then the above argument with Q and R in place of D and E shows that $\int_M (R\varphi) \cdot p\, dx = 0$, which means that p as a distribution satisfies $Qp = 0$. A well-known principle guarantees the real analyticity of p. This completes the proof.

We can take N to be the set of natural numbers and μ to be the point measure. Then $f(x, y)$ can be written $\{f_n\}_{n=1}^\infty$ with C^∞ functions f_n on M; then $p(x) = \sum_{n=1}^\infty f_n(x)$ and $q(x) = \sum_{n=1}^\infty Df_n(x)$. In this case we can replace condition (4) by the local uniform (not necessarily absolute) convergence of $\sum_{n=1}^\infty f_n$ and $\sum_{n=1}^\infty Df_n$. Under either condition, we have $D \sum_{n=1}^\infty f_n = \sum_{n=1}^\infty Df_n$, provided $\sum_{n=1}^\infty f_n$ is C^∞. Similarly assertion (ii) is applicable to $\sum_{n=1}^\infty f_n$.

We apply this lemma to our function $E(z, \sigma; q, \Gamma)$ of (4.3). The series defining it is locally uniformly convergent for sufficiently large $\mathrm{Re}(s)$, and C^∞. Now, termwise application of Δ_q^p produces the series of the same type that defines the right-hand side of (4.12). Thus Lemma A is applicable, and therefore (4.12) can be justified, first for sufficiently large Re(s), and eventually for every s, as noted at the end of the first proof.

We can justify termwise application of operators of type D_ρ^Z or E_Z of [**86**] in a similar way. This is clear if we use S of (2). If we want to use Lemma A, then, the only point is to prove the convergence of the series $\sum_{\alpha \in A} D_\rho^Z(\delta^{su-q/2}\|_q\alpha)$. By [**00**, Lemma 13.9], application of D_ρ^Z multiplies each term with a polynomial of s independent of α and $\zeta(\xi^{-1}\lambda_\alpha^* \cdot {}^t\mu_\alpha^{-1})$. To show that the last quantity is "bounded," let $\mathfrak{G}$ be the localization of our algebraic group G at any fixed archimedean prime; thus $\mathfrak{G} = Sp(m, \mathbf{R})$ in Case SP. This acts on the space $\mathcal{H}$ of (3.15); we have $\lambda_\alpha(z) = \bar{c}_\alpha \cdot {}^t z = \bar{d}_\alpha$ and $\mu_\alpha(z) = c_\alpha z + d_\alpha$. Put $h(\alpha, z) = {}^t\overline{\lambda_\alpha(z)} \cdot {}^t\mu_\alpha(z)^{-1}$. Then, our boundedness means: for every compact subset C of $\mathcal{H}$, the entries of $h(\alpha, z)$ are bounded for $(\alpha, z) \in \mathfrak{G} \times C$. To prove this, put $\mathfrak{P} = \{\gamma \in \mathfrak{G} | c_\gamma = 0\}$ and $\mathfrak{K} = \{\gamma \in \mathfrak{G} | \gamma(i1_m) = i1_m\}$. Clearly h is a continuous function of $(\alpha, z) \in \mathfrak{G} \times \mathcal{H}$, and so its entries are bounded on $\mathfrak{K} \times C$. Now $h(\pi\alpha, z) = h(\alpha, z)$ for every $\pi \in \mathfrak{P}$. Since $\mathfrak{G} = \mathfrak{P}\mathfrak{K}$, we obtain the desired fact.

87b. On Hilbert modular forms of half-integral weight

p. 775, (2.14): For "B_0" read "Ps".

p. 776, (2.16c): For "$\sigma \in \Omega \cap P_A C''''$" read "$\sigma \in P_A \cdot \{\alpha \in C''| L_v(c_\alpha)_v = \delta_v L_v\}$".

p. 781, line 15: The colon on the right-hand side should be a semicolon.

p. 781, line 18: This line should be set in roman.

p. 792, last line: For "$\mathfrak{c}$" read "$\mathfrak{b}$".

p. 793, line 5: For "M_n" read "$\mathcal{M}_n$".

p. 795, lines 8, 12: For "m$'$" read "m$'$".

p. 820, line 13: For "h^σ" read "$\mathbf{h}^\sigma$".

p. 830, line 4 from the bottom: For "$c_w \in (\mathfrak{r}_v)_m^m$" read "$c_w \in (\mathfrak{h}_v)_m^m$".

p. 832, line 4: For "1_m" read "1".

p. 832, line 12 from the bottom: For "(12.6)" read "(2.6)".

p. 833, line 1: For "$\det(d)^{-1}\mathfrak{o}$" read "$\det(d)\mathfrak{o}$".

p. 833, line 5: For "$\mathbf{C}^m$" read "$\mathbf{C}_q^m$".

p. 833, line 5 from the bottom: For "$F_\mathfrak{r}^{mq}$" read "$(F_{mq}^1)_\mathfrak{r}$".

88. On the critical values of certain Dirichlet series and the periods of automorphic forms

p. 253, line 6: For "f" read "(f)".

p. 255, line 8 from the bottom: Delete "at most".

p. 263, line 3 from the bottom: Set "Proposition 3.1" in boldface.

p. 266, line 8: For "if" read "of".

p. 271, line 15: For "$a F^{(\nu)}$" read "a to $F^{(\nu)}$".

p. 274, line 8 from the bottom: For "Ir" read "If".

p. 279, (6.8b): Insert "$(\alpha \in B^\times, u \in W)$" at the end of the line.

p. 280, line 2: Insert "θ" before "$(z, w; \eta)$".

p. 280, line 21: Insert "$\left[\Gamma \cap \{\pm 1\} : 1\right]^{-1}$" before "$f(z)$". This is unnecessary if we take Γ so that $\Gamma \cap \{\pm 1\} = \{1\}$ at the beginning.

p. 281, lines 3, 4 from the bottom: For "ψ" read "ψ_1".

p. 283, line 11: For "$2\mathfrak{bd}_B$" read "$\mathfrak{b} \prod_{v \mid \mathfrak{d}_B} (2\mathfrak{g})_v$".

p. 296, line 9 from the bottom: For "$q(w)$" read "$q(\xi)$".

p. 298 line 16: For "$U \otimes_{\mathbf{Q}} \mathbf{C}$" read "$U \otimes_{\overline{\mathbf{Q}}} \mathbf{C}$".

p. 300, line 2: For "6.5" read "6.4".

p. 300, lines 12, 17: For "$P(f$" read "$P(h_\lambda$".

p. 300, line 18: For "(6.28c)" read "(6.13c)".

Vorwort des Verfassers

Die Theorie der elektromechanischen Schaltungen (Kontaktschaltungen) hat in den letzten Jahren eine wesentliche Ausweitung und Vertiefung erfahren, so daß ihre Anwendung in der technischen Praxis Erfolge verspricht. Da sie sich auf physikalische und mathematische Grundlagen stützt, gewährt sie einen gründlichen Einblick in die zwischen Form und Wirkung bestehenden Zusammenhänge, die häufig weder naheliegend noch anschaulich sind. Die Kenntnis der Gesetze, denen die elektrischen Schaltungen folgen, ist jedoch die erste Voraussetzung für den Entwurf von Bestformen mit einfachster Wirkungsweise und geringstem Aufwand an Bauteilen.

Das vorliegende Buch soll den Leser auf kürzestem Wege in diese Theorie einführen. Es wendet sich in erster Linie an die mit dem Entwurf elektrischer Schaltungen und Schaltgeräte betrauten Techniker, die eine vollständige Beherrschung ihres Fachgebietes anstreben; ferner soll es den Lehrern und vorgeschrittenen Studierenden höherer technischer Schulen Gelegenheit geben, den gegenwärtigen Stand dieses Zweiges der Elektrotechnik kennenzulernen. Gute Kenntnis der allgemeinen Elektrotechnik sowie der elementaren Mathematik durfte daher vorausgesetzt und der dargebotene Stoff auf das wesentliche beschränkt werden. Hingegen erschien es wichtig, durch eine größere Zahl von Anwendungsbeispielen die praktische Verwertung der dargelegten Theorie zu zeigen. Die Beispiele entstammen durchwegs der technischen Praxis und wurden den verschiedensten Sondergebieten der Elektrotechnik entnommen; sie beziehen sich auf den Bau von Hochspannungsanlagen, von Fernmelde- und Fernwirkeinrichtungen, die Technik der elektrischen Steuerungen sowie die Installationstechnik und die Konstruktion elektrischer Schaltgeräte.

Es sei dem Leser anheimgestellt, die Schaltaufgaben, die den behandelten Beispielen zugrunde liegen, noch vor dem Studium des Buches nach seinen bisherigen Gepflogenheiten und technischen Erfahrungen zu lösen; der nachträgliche Vergleich mit den Verfahren und Lösungen des Buches zeigt dann am besten den Gebrauchswert der Schaltungstheorie. Diese hat ihr Ziel erreicht, wenn auch für sie das Wort von BOLTZMANN gilt: „Es gibt nichts praktischeres als eine Theorie."

Wien, im Mai 1954

Otto Plechl

Vorwort des Bearbeiters

Als engstem Mitarbeiter Dr. Otto Plechls war es mir eine selbstverständliche Pflicht, das Manuskript dieses Buches aus seinen hinterlassenen Schriften zusammenzustellen, zu ergänzen und druckfertig zu machen. Ich hoffe, daß mir diese Arbeit gelungen ist, ohne durch die unvermeidbaren Eingriffe die Einheitlichkeit der Darstellung zu gefährden.

Es war Dozent Plechls Wunsch, die vorliegende Arbeit noch zu erweitern und auch die Theorie der Impulsschaltungen einzubeziehen. Bei Behandlung derartiger Probleme erhielt er statt einfacher Leitwertgleichungen Differentialgleichungen, deren Deutung ihn noch bis kurz vor seinem Tode beschäftigte, deren ausführliche Untersuchung ihm jedoch nicht mehr vergönnt war.

Möge deshalb die vorliegende Darstellung der mathematischen Möglichkeiten zur Ermittlung von Bestformen elektrischer Schaltungen nicht nur die praktische Anwendung der noch wenig bekannten neuen Methodik verbreiten, sondern auch die weitere Entwicklung der Theorie fördern.

Wien, im März 1956

Werner Rieder

Inhaltsverzeichnis

Seite

I. Die Grundlagen der Schaltungstechnik 1

   1. Die Aufgaben und Arbeitsverfahren der Schaltungstechnik 1
   2. Die Arten und Bestandteile der elektrischen Schaltungen 13
   3. Die Grundbegriffe der Schaltungstechnik 19
   4. Die technische und mathematische Darstellung der elektrischen Schaltungen 30

II. Die Kombinatorik der Schaltzustände (Strompfade) 44

   1. Die Leitwerte von Kontaktkombinationen 44
   2. Die Leitwerte von Stromkreisen 50
   3. Die Verknüpfung von Leitbedingungen 53
   4. Die Kürzung von Leitbedingungen 57
   5. Die Variation der Leitbedingungen 62
   6. Die Umkehrung von Leitbedingungen 70
   7. Die Freiheitsgrade von Bedingungskomplexen 74
   8. Die Kombinatorik der Leitbedingungen 84

III. Die Kombinatorik der Schaltungsformen 106

   1. Die Zusammenlegung von Leitungen und Kontakten 106
   2. Die natürlichen Schaltungsformen 113
   3. Die Schaltungsformen mit erzwungener Vermaschung 121
   4. Die natürliche Kontaktfolge 127
   5. Die mehrteiligen Schaltungsformen und die topologische Umkehrung 142
   6. Die topologische Äquivalenz 147
   7. Die Funktion bei vorgegebener Schaltungsform................... 154
   8. Die Kombinatorik der Spannungszustände 160

IV. Die Kombinatorik der elektromechanischen Schaltgeräte (Drehschalter) ... 168

   1. Die Bauteile und Bauformen der Drehschalter 168
   2. Die Drehschalter mit einfach benützten axialen Schaltbrücken (einreihige und mehrfach-einreihige Walzenschalter).................... 173
   3. Die Drehschalter mit mehrfach benützten axialen Schaltbrücken (mehrreihige Walzenschalter) 189
   4. Die Drehschalter mit tangentialen und radialen Schaltbrücken (Reihenschalter und Paketschalter) 201
   5. Die Drehschalter mit Brücken in gemischter Anordnung 212

Namen- und Sachverzeichnis...................................... 218

I. Die Grundlagen der Schaltungstechnik

1. Die Aufgaben und Arbeitsverfahren der Schaltungstechnik

Elektrische Schaltungen werden heute in vielen Zweigen der Technik benützt. Besonders wichtige Anwendungsgebiete finden sich im Schaltanlagenbau, in der Technik der elektrischen Steuerungen, im Signalwesen und in der Fernmeldetechnik. Für bestimmte, sich häufig wiederholende Aufgaben sind dort Schaltungen entworfen und erprobt worden, deren Kenntnis zum selbstverständlichen Rüstzeug des Fachmannes gehört. Die ständig fortschreitende Entwicklung der Technik bringt es jedoch mit sich, daß bekannte Schaltungen laufend verbessert und gelegentlich auch vollkommen neuartige Schaltungen entwickelt werden müssen. Hierbei zeigt sich nun, daß alle Hilfsmittel und Kenntnisse, die in der übrigen Technik zur Lösung elektrophysikalischer oder konstruktiver Aufgaben führen, grundsätzlich versagen, wenn es sich um die Ausmittlung elektrischer Schaltungen handelt. Die Ursache hierfür liegt darin, daß die Wirkungsweise einer Schaltung weder an den Ablauf irgendwelcher physikalischer Vorgänge noch an die Bemessung ihrer Bestandteile gebunden ist, sondern ausschließlich von der gewählten Anordnung, also von der topologischen Gestalt der Schaltung, bestimmt wird. Die Technik kennt aber bisher noch kein allgemein anwendbares Verfahren zur Ausmittlung von Anordnungen; deren Auffindung wird vielmehr als Erfindungstätigkeit gewertet, sofern sie über die einfache Zusammensetzung aus bekannten Teilen ohne gegenseitige Rückwirkung hinausgeht. Die Ausmittlung elektrischer Schaltungen ist aber eine so häufige Aufgabe des Technikers, daß die Anwendung systematischer Entwicklungsverfahren unumgänglich nötig ist. Die Technik der elektrischen Schaltungen erscheint daher als eigener Zweig der Elektrotechnik mit besonderer Aufgabenstellung, die auch besondere Hilfsmittel und Arbeitsverfahren verlangt.

Welche Aufgaben im einzelnen beim Entwurf einer neuen Schaltung auftreten, kann sehr verschieden sein; dies hängt in erster Linie vom Zweck der gesuchten Schaltung und häufig auch von zusätzlichen Beschränkungen konstruktiver Art ab. Im wesentlichen lassen sich drei Teilaufgaben, die bei der Ausmittlung einer Schaltung zu lösen sind, unterscheiden. Diese sind:

1. Die Ermittlung der erforderlichen *Schaltgeräte*.

2. Die Ermittlung der erforderlichen *Kontakte* und *Verbindungsleitungen*.

3. Die Ermittlung der erforderlichen *Schaltstücke* innerhalb der Schaltgeräte.

Je nach Art der Schaltaufgabe tritt oft eine dieser Teilaufgaben gegenüber den anderen zurück, weil man beispielsweise auf vorhandene Vorbilder zurückgreifen kann. Sehr häufig und besonders bei vollständig neuen Schaltungen treten jedoch alle drei genannten Teilprobleme nacheinander in Erscheinung. Daß dies schon bei verhältnismäßig recht einfachen Aufgaben der technischen Praxis der Fall ist, soll durch ein Beispiel näher erläutert werden; dieses soll auch klarlegen, welche grundsätzlichen Überlegungen bei der Ausmittlung einer neuen Schaltung jedesmal angestellt werden müssen.

Eine Ladestation für Sammlerbatterien umfaßt zwei Gleichrichtergruppen G_1 und G_2, die für je 30 A bemessen und untereinander gleich sind. Ferner sind drei Ladeanschlüsse L_1, L_2 und L_3 vorhanden. — An den Anschlüssen L_1 und L_2 können gleichzeitig zwei Sammler zu 30 A geladen werden. Der Anschluß L_3 dient dazu, unter Parallelschaltung der beiden Gleichrichter auch Sammler doppelter Stromstärke laden zu können. Gesucht ist die Schalteinrichtung zur Herstellung und Trennung aller erforderlichen Verbindungen zwischen den Gleichrichtern und den Ladeanschlüssen. Es muß möglich sein, jeden Anschluß spannungslos zu machen. Ferner muß eine Zusammenschaltung von Sammlern zwangläufig verhindert sein, da sonst bei ungleichem Ladezustand schädliche Ausgleichsströme zustande kommen. Die Parallelschaltung der beiden Gleichrichter bei Einschaltung des Anschlusses L_3 soll zwangläufig sein.

Die Schaltaufgabe macht keinerlei Vorschrift über Zahl und Art der Schaltgeräte, die zur Verwendung gelangen sollen. Diese Frage muß aber geklärt sein, bevor man versuchen kann, irgendeine Schaltung aufzuzeichnen. Wir müssen daher in erster Linie untersuchen, wie viele Schaltgeräte überhaupt notwendig sind, um die gewünschte Wirkung zu erzielen. Der erfahrene Techniker wird sofort erkennen, daß Schütze oder sonstige elektrisch gesteuerte Geräte nicht erforderlich sind und daß man die Aufgabe mit handbedienten Geräten, z. B. Drehschaltern üblicher Bauart, lösen kann. Es ist also nur mehr festzustellen, wieviele solcher Schalter nötig oder zweckmäßig sind. Hiezu untersuchen wir vorerst die verschiedenen Betriebszustände, die mit der Schalteinrichtung herstellbar sein müssen. Jedem Schaltzustand lassen sich dann gewisse Verbindungen zwischen Gleichrichtern und Ladeanschlüssen zuordnen, wie die nachstehende Tabelle zeigt.

Nr.	Zustand (Schalterstellung)	Verbindungen
0	Ausschaltung	—
1	Einschaltung von L_1	$G_1 - L_1$
2	Einschaltung von L_1 und L_2	$G_1 - L_1,\quad G_2 - L_2$
3	Einschaltung von L_2	$G_2 - L_2$
4	Einschaltung von L_3	$\left.\begin{array}{c} G_1 \\ G_2 \end{array}\right\} - L_3$

Die gesuchte Schaltung muß also mindestens fünf verschiedene Zustände einnehmen können. Diese lassen sich beispielsweise mit einem einzigen Umschalter der Reihe nach herstellen, woraus sich eine Schaltung nach Abb. 1 ergibt. Ein Nachteil dieser Lösung liegt darin, daß man vom *Ausschaltzustand* nicht unmittelbar auf jeden anderen Betriebszustand übergehen kann. Auch ist der erforderliche Schalter verhältnismäßig groß; man bevorzugt im allgemeinen Schaltungsformen mit möglichst einfachen und handelsüblichen Schaltgeräten. Die Schalter lassen sich in der Regel dadurch vereinfachen, daß man ihre Zahl erhöht und somit ihre Funktion auf mehrere Geräte aufteilt. Auch im vorliegenden Fall ist dies möglich.

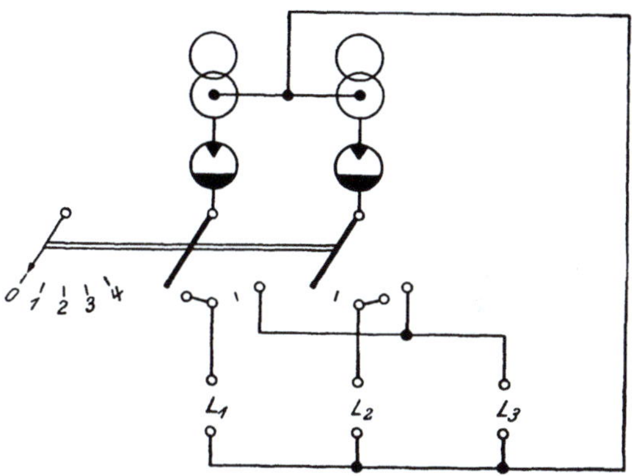

Abb. 1. Ladestelle mit einem Schalter

Eine Ausführung mit zwei Schaltern ergibt sich dadurch, daß die Umschaltung von 30 A auf 60 A einem eigenen Schalter zugeordnet wird. Dieser erhält zweckmäßigerweise noch eine dritte Stelle für die vollständige Ausschaltung. Zur Herstellung der übrigen Verbindungen, die sich nunmehr auf die Ladeanschlüsse L_1 und L_2 beschränken, dient ein zweiter Schalter. Diese Lösung ist in Abb. 2 dargestellt. Sie gestattet, die gewünschten Verbindungen in beliebiger Reihenfolge und unabhängig voneinander herzustellen. Dieser Vorteil wurde mit der vergrößerten Zahl der Schalterstellungen erkauft. Jeder der beiden Schalter hat nunmehr drei Stellungen, so daß insgesamt $3^2 = 9$ Schaltzustände einstellbar sind. Die Schaltaufgabe verlangt aber, wie die Lösung mit einem Schaltgerät zeigt, nur fünf verschiedene Zustände. Der Aufwand zur Lösung der gestellten Aufgabe ist also bei dieser Ausführung größer.

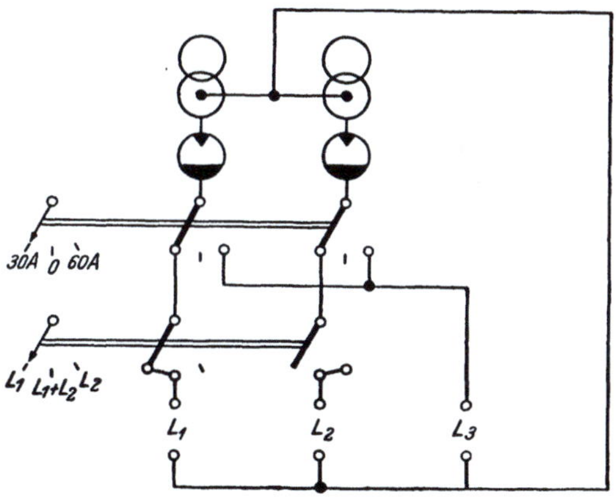

Abb. 2. Ladestelle mit zwei Schaltern

Eine weitere Möglichkeit besteht in der Ausführung mit drei Schaltern, von denen jeder einem der drei Ladeanschlüsse zugeordnet ist. Abb. 3 zeigt diese Schaltung, die sich durch besonders sinnfältige Handhabung auszeichnet. Auch bietet sie den weiteren Vorteil, daß sie nur Schalter mit je zwei Stellungen verlangt. Durch die weitere Erhöhung der Schalterzahl auf drei sind ferner nicht nur die Schalter selbst einfacher geworden,

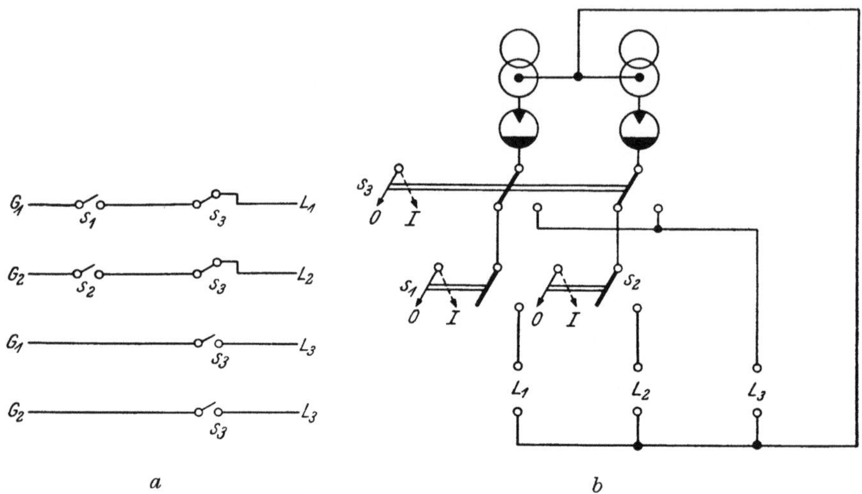

Abb. 3. Ladestelle mit drei Schaltern. *a* Strompfade, *b* Stromlaufplan

sondern es hat sich auch deren Ausnützungsgrad verbessert. Den fünf unbedingt nötigen Schaltzuständen stehen nur mehr $2^3 = 8$ herstellbare Zustände gegenüber. Während also die Freizügigkeit in der Bedienung bei zwei Schaltern zu vier unbenützten Stellungen führt, vermindert sich deren Zahl bei drei Schaltern auf drei.

Man könnte nun versuchen, die Schalter noch weiter zu vereinfachen, indem man beispielsweise den zweipoligen Umschalter für den Ladeanschluß L_3 in zwei einpolige zerlegt. Eine solche Lösung mit vier Schaltern enthält aber nicht mehr die geforderte Zwangläufigkeit in der Parallelschaltung der beiden Gleichrichter. Die Höchstzahl der Schaltgeräte ist also durch die Schaltaufgabe selbst mit drei festgelegt. Mit dieser Feststellung ist die erste Teilaufgabe zur Ausmittlung der Schaltung gelöst und die Zahl der Schaltgeräte kann innerhalb der beschriebenen Grenzen gewählt werden. Sofern nicht weitere Nebenbedingungen konstruktiver Art, z. B. Rücksicht auf geringen Platzbedarf, vorliegen, erscheint die Ausführung mit drei Schaltern am zweckmäßigsten. Es soll daher diese Schaltungsform weiter untersucht werden. Das vorliegende Beispiel ist so einfach, daß die verschiedenen Varianten der Schaltung ohne größere Mühe aufgezeichnet werden können. Im allgemeinen ist dies nicht der Fall; man muß dann die Zahl und Art der

Schaltgeräte, die der Aufgabe am besten entsprechen, bestimmen, ohne die zugehörige Form der Schaltung zu kennen. Wir wollen daher annehmen, daß wir auch mit unserem Beispiel so verfahren sind und vorläufig nur wissen, daß wir drei Schalter mit je zwei Stellungen benützen wollen.

Der nächste Schritt zur Ermittlung der gesuchten Schaltung besteht in der Festlegung der Kontakte, die in den einzelnen Strompfaden enthalten sein müssen. Aus der Angabe läßt sich entnehmen, daß insgesamt vier Strompfade nötig sind. Zwei davon führen zum Ladeanschluß L_3 und enthalten die Kontakte des zugehörigen Schalters s_3. Die beiden übrigen Strompfade gehören zu den Ladeanschlüssen L_1 und L_2 und enthalten je einen Kontakt des Schalters s_1 und s_2. Um die in der Aufgabe geforderte Sperrung gegen Ausgleichsströme zu erreichen, müssen diese Strompfade noch mit gegenläufigen Kontakten des Schalters s_3 ausgestattet werden. Abb. 3 *a* zeigt diesen Aufbau der Strompfade; die Reihenfolge der Kontakte ist noch nicht endgültig und kann sich im Verlauf der weiteren Entwicklung der Schaltung noch verändern.

Eine Darstellung, entsprechend Abb. 3 *a*, in der jeder Strompfad für sich eingetragen ist, entspricht einer vollständig unverketteten Schaltung. Es ist aber nahezu stets möglich, einzelne Schaltelemente für zwei oder mehrere Strompfade gemeinsam zu verwenden, das heißt, die Schaltung zu verketten.

Das Aufsuchen der zweckmäßigsten Verkettung bildet die zweite große Teilaufgabe bei der Ausmittlung der Schaltungen. Sie läuft letzten Endes darauf hinaus, die günstigste Reihenfolge der Kontakte innerhalb der Stromkreise festzustellen; durch die Kontaktfolge ist dann bereits eindeutig bestimmt, welche Leitungen und Kontakte sich zusammenlegen lassen. Beim vorliegenden Beispiel zeigt sich, daß man in den Strompfaden für die Ladeanschlüsse L_1 und L_2 die Kontakte des Schalters s_3 unmittelbar an die Gleichrichter G_1 und G_2 anzuschließen hat; sie lassen sich dann mit den Kontakten der beiden Strompfade für L_3 zu Umschaltkontakten vereinigen. Eine weitere Zusammenlegung ist bei diesem einfachen Beispiel nicht möglich. Man kann daher den Stromlaufplan der gesuchten Schaltung zeichnen, wie er in Abb. 3 *b* dargestellt ist. Die vorliegende **Schaltung, die als** einführendes Beispiel besonders einfach sein mußte, darf nicht den Eindruck erwecken, daß die Ermittlung der besten Kontaktfolge stets eine leichte Aufgabe von untergeordneter Bedeutung sei. Wer jemals mit Schaltungen größeren Umfanges zu tun hatte, wird aus Erfahrung wissen, daß es sich dabei um ein Teilproblem der Schaltungstechnik handelt, das keineswegs einfach ist und manchmal außerordentliche Schwierigkeiten bietet.

Durch den Stromlaufplan ist die Form der sogenannten äußeren Schaltung, das ist die Gesamtheit aller Kontaktverbindungen, festgelegt. Die dritte und letzte Teilaufgabe besteht nun in der Ermittlung der inneren Schaltung der Schaltgeräte. Hierbei handelt es sich darum, die Zahl und Anordnung der Schaltstücke, die jedes Schaltgerät haben soll, zu bestimmen. Auch hierbei bestehen in der Regel mehrere verschiedene

Möglichkeiten; dies ist auch dann der Fall, wenn man die konstruktiven Einschränkungen in Rechnung stellt, die mit Rücksicht auf die Technik des Schalterbaues vorhanden sind. In der Regel wird man trachten, mit

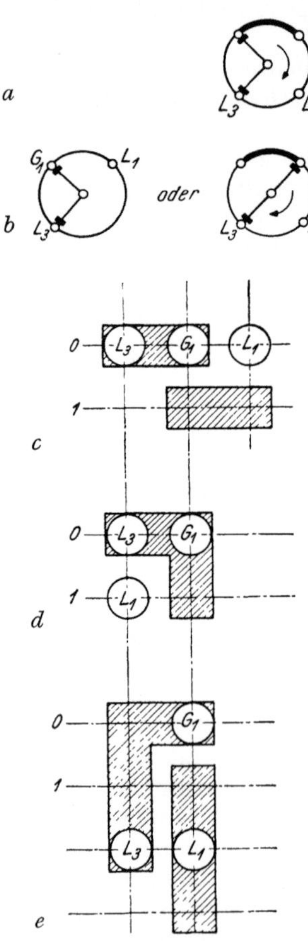

vorhandenen und üblichen Bauteilen das Auslangen zu finden. Es bleibt dann aber noch die Wahl, ob man beispielsweise Paketschalter, Walzenschalter oder Reihenschalter verwenden soll. Aber selbst nach Wahl der Bauart der Schalter sind noch verschiedene Anordnungen der festen und beweglichen Schaltstücke möglich. Abb. 4 zeigt beispielsweise verschiedene Formen eines einpoligen Umschalters, der als Drehschalter ausgebildet ist. Wählt man eine Ausführung als Paketschalter, so ergibt sich bei einem Schrittwinkel von 180° ein bewegliches Schaltstück in Form einer Winkelbrücke entsprechend Abb. 4 a. Legt man den Schrittwinkel mit 90° fest, so ergibt sich eine gewinkelte oder gestreckte Schaltbrücke nach Abb. 4 b. Walzenschalter bieten noch eine größere Zahl von Möglichkeiten, weil man außer dem Schrittwinkel noch die Zahl der Fingerreihen variieren kann. Die Abb. 4 c bis 4 e zeigen Abwicklungen des Schaltbelages von Walzenschaltern verschiedener Form, die durchwegs einpolige Umschalter sind. Welche Form man fallweise wählt, hängt von konstruktiven Bedingungen, wie Schrittwinkel des Antriebes, Zugänglichkeit der Kontaktfinger und dergleichen, ab. Sehr oft kann man auch auf vorhandene Konstruktionen zurückgreifen oder handelsübliche Geräte mit passendem Schaltbelag heranziehen. Erst nach Festlegung der Bauart und der inneren Schaltung der Schaltgeräte ist die gestellte Schaltaufgabe vollständig gelöst. Man kann den Wirkschaltplan, der nunmehr sämtliche Schaltstücke, Anschlußklemmen und Verbindungsleitungen enthält, aufzeichnen.

Abb. 4. Bauformen eines einpoligen Drehumschalters. *a* Paketschalter mit 180° Schrittwinkel, *b* Paketschalter mit 90° Schrittwinkel, *c* Einreihiger Walzenschalter mit 180° Schrittwinkel, *d* Zweireihiger Walzenschalter mit 180° Schrittwinkel, *e* Zweireihiger Walzenschalter mit 90° Schrittwinkel

Dem im vorstehenden behandelten Beispiel lag eine verhältnismäßig einfache Schaltaufgabe zugrunde. Trotzdem ergab sich eine überraschende Mannigfaltigkeit von Lösungen. Zu beachten ist, daß nicht nur die hier angeführten, sondern auch noch weitere, weniger zweckmäßige Schaltungsformen möglich wären, die stillschweigend übergangen wurden. Bei größeren und schwierigeren Schaltaufgaben ist die Zahl der möglichen Lösungen meist erheblich größer, als man fürs erste zu glauben geneigt ist. Hierauf gründet sich die Notwendigkeit einer systematischen Behandlung schaltungstechnischer Aufgaben, denn nur die Kenntnis aller Möglichkeiten erlaubt die Auswahl der zweckmäßigsten Ausführung.

Jede systematische Entwicklung einer Schaltung muß von der gestellten Aufgabe ausgehen. An Hand des erläuterten Beispieles sei nun klargelegt, welche Bestimmungsstücke eine Schaltaufgabe überhaupt enthalten muß. Da sie sich auf die fallweise Herstellung und Sperrung elektrischer Verbindungen bezieht, muß sie in erster Linie die zu verbindenden Elemente vollständig anführen. Aus der Zahl dieser Elemente läßt sich dann ohne weiteres ermitteln, wieviel Verbindungen überhaupt möglich sind. Die beschriebene Ladestation enthält zwei Gleichrichter und drei Ladeanschlüsse. Von diesen fünf Elementen kann höchstens jedes mit jedem anderen verbunden werden; dabei ist zu beachten, daß eine Verbindung von L_1 nach L_2 mit der umgekehrten Verbindung von L_2 nach L_1 gleichbedeutend ist. Die fünf Elemente der Angabe ermöglichen also nach der Formel $\frac{1}{2} \cdot 5 \, (5 - 1) = 10$ höchstens zehn Verbindungen (Abb. 5 a). Von diesen sollen fünf fallweise hergestellt werden, nämlich die vier Verbindungen zwischen den Gleichrichtern einerseits und den Ladeanschlüssen andererseits, sowie die Parallel-Schaltung der beiden Gleichrichter selbst (Abb. 5 b). Die zeichnerische Darstellung dieser Verbindungen zeigt einen geschlossenen Linienzug in Form des Dreiecks G_1, G_2, L_3. **Es ist nun leicht** einzusehen, daß man von einem solchen Dreieck nur zwei Seiten durch Schaltverbindungen herstellen muß, während sich die dritte jeweils von selbst ergibt. Verbindet man nämlich L_3 mit G_1 und G_2, so ist ohne weitere Maßnahme auch G_1 und G_2

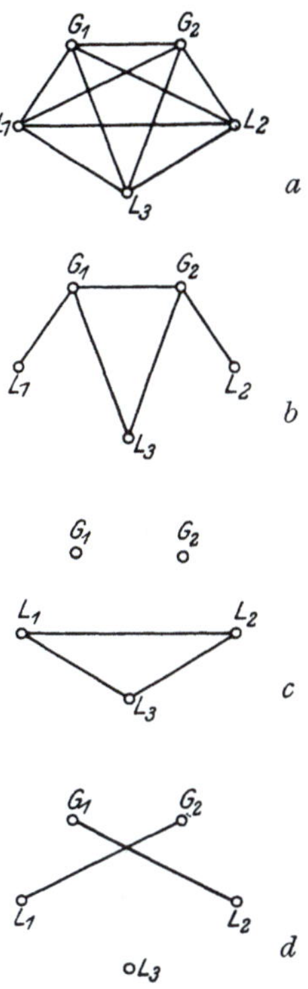

Abb. 5. Die bestimmenden Komplexe einer Schaltaufgabe. a Mögliche Verbindungen, b Verlangte Verbindungen, c Verbotene Verbindungen, d Zugelassene Verbindungen

mittelbar verbunden. Voraussetzung hierbei ist, daß die beiden unmittelbaren Verbindungen gleichzeitig bestehen, was beim vorliegenden Beispiel auch zutrifft. Es ist dann vollständig gleichgültig, welches Seitenpaar des Dreiecks in Form elektrischer Verbindungen hergestellt wird. Die Gleichzeitigkeit einzelner gewünschter Verbindungen hat also zur Folge, daß einzelne dieser Verbindungen eingespart werden können, weil sie sich als mittelbare Verbindungen von selbst ergeben. Dementsprechend zeigt die behandelte Schaltung nur vier Strompfade, obwohl fünf Verbindungen vorhanden sind.

Den gewünschten Verbindungen stehen jene gegenüber, die durch die Schaltaufgabe verboten werden. Es sind dies die drei Verbindungen zwischen den Ladeanschlüssen L_1, L_2 und L_3 (Abb. 5c). Von den zehn möglichen Verbindungen werden also durch die Angabe fünf verlangt und drei verboten. Somit verbleiben noch zwei Verbindungen, nämlich $G_1 L_2$ und $G_2 L_1$ (Abb. 5 d), über die keine Aussage vorliegt. Es handelt sich demnach um zugelassene Verbindungen, die in der Schaltung vorhanden sein dürfen, aber ebensogut auch fehlen können. In den als Beispiele angeführten Schaltungsformen sind diese Verbindungen nicht enthalten, weil sie keine Vorteile eingebracht hätten. Es gibt aber viele Schaltaufgaben, bei denen sich unter Verwendung zugelassener Verbindungen eine Lösung mit geringerem Aufwand ergibt.

Jede Schaltaufgabe bestimmt somit eine gewisse Zahl von möglichen Verbindungen; ferner unterteilt sie diese in verlangte, verbotene und zugelassene. Von den verlangten Verbindungen können je nach Lage und Gleichzeitigkeit einzelne ausgeschieden werden, weil sie sich als mittelbare Verbindungen von selbst ergeben. Besondere Beachtung verlangen die zugelassenen Verbindungen, weil sie in der Aufgabe gewöhnlich nicht ausdrücklich angeführt sind und sich häufig zur Vereinfachung der Schaltung heranziehen lassen. Auch diese Betrachtung zeigt, daß eine Schaltung in den seltensten Fällen durch die gestellte Aufgabe eindeutig bestimmt wird.

Man erkennt ferner, daß sich die Aufgabe der Schaltungstechnik keineswegs darin erschöpft, bestimmte fallweise wirksame Verbindungen mit einem möglichst geringen Aufwand an Kontakten und Leitungen herzustellen. Bevor man an diese Frage überhaupt herantreten kann, ist es vielmehr nötig, sich über die zweckmäßigste Wirkungsweise klar zu werden; man hat also vorerst zu ermitteln, welche Verbindungen in der Schaltung überhaupt vorkommen sollen und wie die Handhabung der Schaltung zur Herstellung dieser Verbindungen vor sich gehen soll. Erst nach Bestimmung der Wirkungsweise und der für sie notwendigen Schaltgeräte ist es möglich, die günstigste Form der Schaltung zu suchen und die erforderlichen Kontakte und Verbindungsleitungen zu ermitteln. Die große Mannigfaltigkeit der möglichen Lösungen, die sich bei den meisten Schaltaufgaben zeigt, hat also zwei Gründe. Zu einer bestimmten Aufgabe bestehen in der Regel mehrere Lösungen mit verschiedener Wirkungsweise; eine bereits eindeutig festgelegte Wirkungsweise läßt sich ferner noch durch Schaltungen verschiedener Form erreichen. Die Schaltaufgabe bestimmt die Grenzen, innerhalb welcher die Wirkungsweise der Schaltung variiert werden darf; zu jeder bestimmten Wirkungsweise besteht in der Regel eine größere Zahl wirkungsgleicher Schaltungsformen, von denen die zweckmäßigste zu suchen ist.

Wir können uns nunmehr der Frage zuwenden, welche systematischen Verfahren dem Techniker zur Lösung schaltungstechnischer Aufgaben zur Verfügung stehen[1].

Diese sind im wesentlichen:

1. Die Zusammensetzung aus bekannten Elementarschaltungen.

2. Die schrittweise Annäherung durch Probieren.

3. Die Entwicklung nach Schaltregeln.

4. Die Berechnung.

Welches dieser Verfahren fallweise den Vorzug verdient, richtet sich nach der Art der Schaltaufgabe. Im allgemeinen ist es kaum zweckmäßig, sich auf die Anwendung eines einzigen Verfahrens zu beschränken. Jedes Verfahren hat seine Vorzüge und Grenzen. Diese seien im nachstehenden kurz erläutert.

Handelt es sich um eine Schaltaufgabe, die sich aus einzelnen, bereits gelösten Teilaufgaben zusammensetzt, ist es sicherlich am einfachsten, die gesuchte Schaltung aus bekannten und erprobten Teilschaltungen aufzubauen. Kennt man Wirkungsweise und Eigenschaften der einzelnen Teilschaltungen hinreichend genau, so hat man auch nicht zu befürchten, daß bei der Zusammenfügung irgendwelche störende Strompfade und ungewollte Nebenwirkungen auftreten, wovon man sich durch nachträgliche Überprüfung zu überzeugen hat. Das Verfahren führt verhältnismäßig rasch zum Ziel, bietet aber selbstverständlich keine Gewähr dafür, daß die gefundene Lösung die beste und sparsamste ist. Es setzt ferner die genaue Kenntnis einer großen Zahl von Elementarschaltungen

---

[1] Es sei hier vorausgeschickt, daß die technische Praxis gegenwärtig noch weit davon entfernt ist, alle anfallenden Schaltaufgaben systematisch und exakt zu behandeln. Die Schaltungstechnik ist einer der jüngsten Zweige der Elektrotechnik. Der Bedarf an Schalteinrichtungen größeren Umfanges entstand erst durch die allgemeine Einführung der elektrischen Energie- und Nachrichtenübertragung. Es mußten vorerst die nötigen Maschinen und Geräte zur Erzeugung und Verwendung elektrischen Stromes vorhanden sein, bevor sich die Notwendigkeit ergab, den Stromfluß durch Schalteinrichtungen zu steuern. Im Zuge der Entwicklung mußte also die Frage der Energieumsetzung zeitlich an erster und diejenige der schaltungstechnischen Anordnungen an zweiter Stelle stehen. Daraus erklärt sich, daß Berechnung und Bau elektrischer Stromerzeuger und Stromverbraucher einen hohen Grad der Vollkommenheit erreicht haben, wogegen die Schaltungstechnik augenblicklich noch an ihrem Anfang steht. Die Grundformen der heute benützten elektrischen Schaltungen entstanden im allgemeinen nicht durch systematische Untersuchung der technischen Möglichkeiten, sondern vorwiegend auf Grund erfinderischer Begabung einer großen Zahl von Technikern. Der gegenwärtige Stand der Schaltungstechnik ist dadurch gekennzeichnet, daß in der Praxis wohl für jede anfallende Schaltungsaufgabe eine Lösung gefunden werden kann, oft aber nicht die zweckmäßigste und wirt-

voraus und eignet sich daher nur für sehr erfahrene Techniker, die ihr engeres Fachgebiet vollkommen beherrschen. Große Verbreitung hat dieses Verfahren im Fernmeldewesen sowie in der Technik der elektrischen Steuerungen für Werkzeugmaschinen gefunden.

Liegt eine Schaltaufgabe vor, die vollkommen neu ist oder sich von bekannten und bereits gelösten Aufgaben in wesentlichen Punkten unterscheidet, wird meistens das Verfahren der schrittweisen Annäherung durch Probieren benützt. Der Vorteil dieses Verfahrens liegt in erster Linie darin, daß es keinerlei Vorkenntnisse verlangt. Dafür ist es meistens sehr zeitraubend; es verleitet daher auch dazu, daß man sich mit der ersten Lösung, die halbwegs entspricht, zufrieden gibt. Zu Bestlösungen führt das Verfahren in den seltensten Fällen; technisch vertretbar erscheint es eigentlich nur als Vorstufe, die einen ersten Einblick in eine vorliegende Schaltaufgabe verschaffen soll.

Beide im vorstehenden beschriebenen Verfahren sind also wenig befriedigend. Bessere Ergebnisse kann ein Verfahren liefern, das eine Ausmittlung der Schaltungen nach allgemein gültigen Regeln, die auf Grund der Zusammenhänge zwischen Form und Wirkungsweise abgeleitet sind, gestattet. Dieser Weg wurde bereits zu Ende des vorigen Jahrhunderts von M. BODA[1], der eine Anleitung zum Entwurf von Eisenbahnsignalschaltungen zusammenstellte, beschritten. R. EDLER[2] veröffentlichte 1903 erstmalig Regeln zur Entwicklung von Schaltungen beliebiger Art. Späterhin wurden diese Anleitungen und Regeln teils von EDLER selbst[3], teils durch andere Autoren vervollständigt und zusammengefaßt. Zu beachten ist die Schaltlehre von R. LISCHKE[4], der als erster den kombinatorischen Charakter der Schaltungen klar erkannte. Die praktische

---

schaftlichste. Dieser Umstand bleibt meistens unbemerkt, weil sich ohne Kenntnis der wesentlichen Zusammenhänge, die zwischen Form und Wirkung einer Schaltung bestehen, überhaupt nicht beurteilen läßt, ob man mit einer ausgeführten Schaltung ihre Bestform erreicht hat oder nicht. Während sich beispielsweise der Gütegrad einer elektrischen Maschine durch Zahlenangaben, wie Wirkungsgrad, Leistungsfaktor, Füllfaktor, genau kennzeichnen und beurteilen läßt, verfügt die Schaltungstechnik derzeit über keinerlei Anhaltspunkte für die Bewertung schaltungstechnischer Anordnungen. Erst in jüngster Zeit beginnt sich die Erkenntnis Bahn zu brechen, daß eine systematische Behandlung schaltungstechnischer Aufgaben noch große Fortschritte bringen kann, aber besondere Arbeitsverfahren verlangt und keineswegs leichter und einfacher ist als etwa die Berechnung eines Stromerzeugers oder eines sonstigen zur Umsetzung elektrischer Energie bestimmten Gerätes.

[1] BODA, M.: Die Schaltungstheorie der Blockwerke. Wiesbaden: Kreidel. 1899.

[2] EDLER, R.: Z. Elektrotechn. 1903, Heft 31 und 32.

[3] EDLER, R.: Entwurf von Schaltungen und Schaltapparaten. Hannover—Leipzig: Dr. Max Jänecke. 1905; sowie EDLER, R.: Schalterbau, Band 2 (Schaltlehre). Leipzig: Dr. Max Jänecke. 1927.

[4] LISCHKE, R.: Schaltlehre. Leipzig: Hachmeister u. Thal. 1911.

Schwierigkeit liegt darin, Schaltregeln zu finden, die alle vorhandenen Gesetzmäßigkeiten und Möglichkeiten vollständig und genau erfassen. Beispielsweise zeigte sich, daß Regeln, die nach langjähriger Erfahrung aufgestellt und für allgemein gültig gehalten wurden, gelegentlich und zwar in wichtigen Fällen versagen. So findet sich in einzelnen Lehrbüchern eine Regel über die Zusammenlegung gleichartiger Kontakte, die in mehreren parallelen Strompfaden vorkommen; sie besagt, daß man zweckmäßigerweise zuerst jene Kontakte zusammenfaßt, die in den Strompfaden am häufigsten vertreten sind, weil man dadurch die größte Ersparnis erzielt. Tatsächlich hat diese Regel keine allgemeine Gültigkeit und versagt gerade dann, wenn sie sich nur schwer überprüfen läßt. Manche Erfahrungen der Praxis bewirken geradezu eine Irreführung des Benützers. Als Beispiel diene die oft anzutreffende Meinung, daß sich eine Symmetrie in der Schaltaufgabe auch in der Form der Schaltung widerspiegeln müsse. Diese Ansicht ist irrig, weil die betreffende Aufgabe meistens mehrere Lösungen hat, die zueinander symmetrisch, selbst aber unsymmetrisch sind. Mangelhafte Kenntnis der schaltungstechnischen Möglichkeiten hat auch zur Aufstellung unrichtiger Regeln von grundlegender Bedeutung geführt. In allen Veröffentlichungen, die sich mit der Vereinfachung der Schaltungsformen befassen, findet sich die Regel, daß Zahl und Aufbau der Strompfade bei vorgegebener Wirkungsweise nicht verändert werden dürfen; auch diese grundsätzliche Schaltregel ist falsch und gerade die Verminderung oder Vereinfachung der Strompfade durch Kürzung der Schaltbedingungen bringt unter Umständen große Ersparnisse. Irrtümer dieser Art sollen jedoch nicht dazu beitragen, das Verfahren zur Lösung von Schaltaufgaben nach allgemeinen Schaltregeln in Mißkredit zu bringen. Die Untersuchung der Gesetze, denen die elektrischen Schaltungen folgen, erlaubt es, Regeln von allgemeiner Gültigkeit oder genau bestimmtem Geltungsbereich mit mathematischer Exaktheit aufzustellen. Eine solche Regel ist beispielsweise die eingangs angeführte Anweisung, daß man bei einer Schaltung zuerst die Bestimmung der Wirkungsweise, dann die Zusammenlegung der Leitungen und Kontakte und erst zuletzt die Zusammenfassung der Schaltstücke in den Schaltgeräten vorzunehmen hat. Die Verwendung solcher verhältnismäßig einfacher Regeln ist für die praktische Schaltungstechnik von ausschlaggebender Bedeutung. Der Vorteil der Schaltregeln liegt hauptsächlich darin, daß sie zeigen, wie die einzelnen Teilaufgaben, die bei der Ausmittlung einer Schaltung auftreten, anzufassen sind; sie ersparen dem Techniker meist eine Anzahl grundsätzlicher und schwieriger Überlegungen und führen ihn auf kürzestem Wege zum Ziel. Es liegt aber im Wesen aller Regeln, daß sich diese nur auf Zusammenhänge allgemeiner Art beziehen können; der Eigenart einer bestimmten Schaltaufgabe können sie selbstverständlich nicht Rechnung tragen. Ihre Anwendung verlangt daher eine Ergänzung durch ein Verfahren, das eine Lösung von Schaltaufgaben auf primärem Wege gestattet. Ein solches Verfahren ist die rechnerische

Ermittlung von Schaltungen mit Hilfe von Leitwertgleichungen und Leitwertmatrizen[1].

Die Stromverteilung innerhalb der elektrischen Schaltungen geht nach den KIRCHHOFFschen Gesetzen vor sich. Die Kontakte erscheinen als Strecken veränderlicher Leitfähigkeit mit den Leitwerten 0 und ∞. Aus den physikalischen Gesetzen für die Reihen- und Parallelschaltung von Leitern ergeben sich durch Grenzübergang die Rechenregeln der algebraischen Logik, die auch den Namen Logistik führt. Daraus folgt die Möglichkeit, die Strompfade beliebiger Schaltungen in Form von Gleichungen anzuschreiben und exakt zu berechnen. Weiters läßt sich zeigen, daß sich die geometrische Gestalt jeder Schaltung in einen Streckenkomplex transformieren läßt, der nach den Regeln der kombinatorischen Topologie rechnerisch behandelt werden kann. Die Verwendung dieser mathematischen Hilfsmittel stützt sich demnach auf die physikalischen und geometrischen Gegebenheiten, die den Schaltungen zugrunde liegen, und ermöglicht eine exakte Bearbeitung aller schaltungstechnischen Probleme. Die Hilfsmittel der Kombinatorik liefern auch den erforderlichen Überblick über sämtliche konstruktive Möglichkeiten, die bei der Lösung einer Schaltaufgabe vorhanden sind. Damit trat die Schaltlehre in das Stadium der Exaktheit, das ihr in früherer Zeit noch abging.

Die Fortschritte, die während der letzten Jahre in der Theorie der elektrischen Schaltungen gemacht wurden, gestatten es nunmehr, die gewonnenen Erkenntnisse in der Praxis mit Erfolg zu verwerten. So weit dies bereits geschehen ist, zeigte sich, daß die exakte Bearbeitung der Schaltaufgaben zu wesentlich besseren und sparsameren Lösungen führte, als die früher vorwiegend üblichen Verfahren der Zusammensetzung aus bekannten Teilschaltungen und der schrittweisen Annäherung. Mit diesen Verfahren könnten zwar stets Schaltungen entwickelt werden, die ihren Zweck erfüllen, jedoch sind die so gefundenen Lösungen hinsichtlich des Aufwandes an Schaltelementen häufig weit von der möglichen Bestform entfernt. Erst die rechnerische Überprüfung hat gezeigt, daß sich oft auch scheinbar sehr vollkommene Schaltungen in überraschendem Ausmaß verbessern lassen. Von besonderem Interesse ist

---

[1] Schon frühzeitig wurde versucht, die Zusammenhänge zwischen Form und Wirkungsweise der elektrischen Schaltungen exakt zu erfassen und in mathematischer Form darzustellen. Den ersten Versuch hierzu unternahm R. LISCHKE mit seinem 1911 erschienenen Lehrbuch, in dem er die formale Ähnlichkeit zwischen der Zusammenlegung von Kontakten und der Kürzung algebraischer Brüche zur Grundlage eines besonderen Rechenverfahrens machte. Die von LISCHKE aufgestellten Rechenregeln entbehren jedoch einer exakten Grundlage und gelten nur unter bestimmten Voraussetzungen, so daß das Verfahren keine praktische Bedeutung erlangt hat. W. BADER entdeckte 1934, daß sich Wechselschaltungen durch eindimensionale Matrizen abbilden und rechnerisch ausmitteln lassen. Im Jahre 1938 konnten A. NAKASIMA und M. HANZAWA zeigen, daß für die Umformung verketteter Schaltungen die gleichen Gesetze gelten, wie für die Transformation algebraischer

die Feststellung, daß die jeweiligen Bestlösungen durchaus nicht sinnfällig sind. Bei einer größeren, auf höchste Einfachheit gebrachten Schaltung läßt sich oft nur mehr rechnerisch nachweisen, daß sie tatsächlich allen Bedingungen der gestellten Aufgabe entspricht. Dieser vorerst befremdliche Umstand ist durch den kombinatorischen Charakter der Schaltungen bedingt. Schon eine geringe Anzahl von Elementen führt zu einer Vielfalt von Kombinationen, die sich jedem Überblick entzieht, soferne man nicht besondere Hilfsmittel zur Darstellung in Anspruch nimmt. Beispielsweise sind bei einer Schaltung mit nur fünf Anschlußpunkten zehn unmittelbare und 150 mittelbare Verbindungen möglich. Die Ersparnis an Bauelementen folgt aber nicht nur aus der teilweisen Zusammenlegung von Verbindungen, sondern auch aus der Zusammenfassung textlich vorgegebener Funktionsbedingungen. Das Ziel der Schaltungstechnik sind Bauformen, die nur wenige Bauteile und daher auch wenige direkte Verbindungen enthalten; die Wirkungsweise wird hierbei ohne zusätzlichen Aufwand durch Benützung mittelbarer Funktionsbedingungen und Verbindungen erreicht. Solche Schaltungsformen sind zwar zweckmäßig, aber keinesfalls naheliegend; sinnfällig ist nur ihre Wirkungsweise, das heißt die Art der Betätigung zur Herstellung und Trennung der Verbindungen.

2. Die Arten und Bestandteile der elektrischen Schaltungen

Unter einer elektrischen *Schaltung* im weitesten Sinne versteht man eine Einrichtung, die eine Fortleitung von elektrischer Energie auf verschiedenen Wegen gestattet. Jene Stellen der Schaltung, deren Potentiale durch Anschluß an eine Stromquelle oder Stromsenke festgelegt sind, heißen ihre *Pole*. Die möglichen Wege der Strömung zwischen zwei Polen verschiedener Potentiale sind die *Strompfade* der Schaltung. Die zeitlich wechselnde Intensität oder Richtung des Stromes kann verursacht sein durch Veränderung der angelegten Potentiale; die Schaltung kann dann einen stets gleichbleibenden Leitwert aufweisen. Solche *feste Schaltungen* finden vorwiegend in der Meßtechnik Verwendung. Sind die Potentiale konstant, so dient die Schaltung selbst zur Steuerung des Stromverlaufes.

Gleichungen. Im gleichen Jahre erbrachte A. RITTER den Nachweis, daß es sich hierbei um die Rechenregeln der algebraischen Logik handelt. Schließlich gelang es 1943 dem Verfasser, die gemeinsame physikalische Grundlage für diese verschiedenartigen mathematischen Darstellungen zu finden und daraus allgemein gültige Schaltregeln und Lösungsverfahren abzuleiten. Vgl. LISCHKE, R.: Schaltlehre. Leipzig: Hachmeister u. Thal. 1911; BADER, W.: Wechselschaltungen. Arch. Elektrotechn. 1935, Heft 2; NAKASIMA, A., und M. HANZAWA: The Theory of Equivalent Transformation of Simple Partial Paths in the Relay Circuit. Nippon El. Comm. Engng. 1938; RITTER, A.: Beiträge zur Schaltlehre, Dissertation an der Technischen Hochschule Wien 1938; PLECHL, O.: Die Kombinatorik der Strompfade elektromechanischer Schaltungen, Dissertation an der Technischen Hochschule Wien 1943.

Die Leitfähigkeit der Strompfade ist dann variabel. Läßt sich der Leitwert stetig oder in Stufen verändern (regeln), so spricht man von einer *Regelschaltung*. Die Strompfade können aber auch Elemente enthalten, die ihren Leitwert nur sprunghaft zwischen einem sehr großen und einem verschwindend kleinen Betrag wechseln. Solche Schaltelemente werden als *Kontakte* bezeichnet. Für Schaltungen, bei denen der Stromfluß nur durch Kontakte beeinflußt wird, besteht gegenwärtig keine Artbezeichnung. Man könnte sie folgerichtig „*Kontaktschaltungen*" nennen. Da die Kontakte fast ausnahmslos auf mechanischem Wege betätigt werden, ist auch der Name „*elektromechanische Schaltungen*" zutreffend. Dieser bringt überdies zum Ausdruck, daß Schaltungen dieser Art nicht auf Elektrizität als strömendes Mittel beschränkt sind. Sie können vielmehr in sinngemäßer Abwandlung ebenso für mechanische Kraftübertragung beliebiger Art benützt werden. Die Analogie zwischen elektrischen und hydraulischen Schaltungen liegt auf der Hand. Die gleichen Gesetzmäßigkeiten gelten jedoch auch für kinematische Anordnungen, z. B. Klinkwerke, Gesperre usw., wenn man statt des elektrischen Leitwertes den Elastizitätsmodul und statt der Kontakte die Eingriffstellen der Übertragungsglieder in Rechnung stellt[1]. Es gibt auch kaum eine elektrische Kontaktschaltung, die ohne mechanische Bauteile auskommt. Ob für eine bestimmte Teilfunktion fallweise die elektrische oder mechanische Lösung gewählt wird, ist für die Wirkungsweise der Schaltung im wesentlichen belanglos. Als Beispiel diene die bekannte Schaltung für Gefahrmeldungen, die zur Überwachung elektrischer Maschinen und Schaltanlagen ständig benützt wird. Sie dient dazu, einen Störungszustand, der durch vorübergehendes Schließen eines Hilfsstromkreises erfaßt wird, anzuzeigen und diese Meldung bis zur Abstellung von Hand aus aufrechtzuerhalten. Die Aufgabe läßt sich sowohl auf rein elektrischem Wege mit Hilfe eines Halteschützes, als auch mit einem mechanischen Fallzeichenmelder lösen. Die beiden wirkungsgleichen Ausführungen sind in Abb. 6 dargestellt.

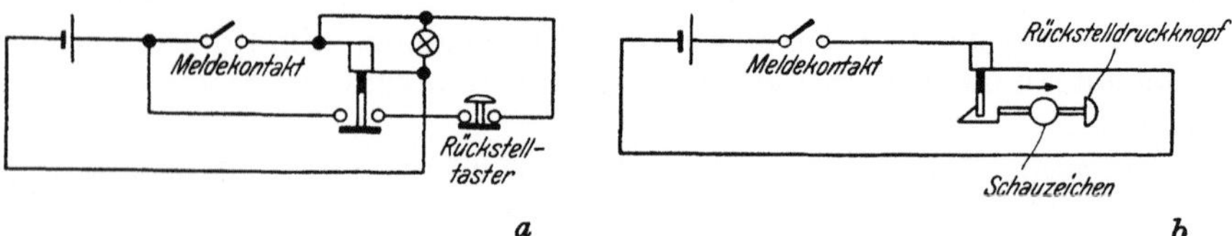

Abb. 6. Äquivalenz elektromechanischer Schaltungen. *a* Elektrische Selbsthaltung, *b* Mechanische Selbsthaltung

Das Anwendungsgebiet der schaltungstechnischen Kombinatorik bezieht sich im wesentlichen auf elektromechanische Schaltungen. Im Hinblick auf deren Wirkungsweise lassen sich zwei Hauptarten unterscheiden; ein Beispiel diene zur Erläuterung. Die Abb. 7*a* zeigt die bekannte

---

[1] FRANKE: Eine vergleichende Schalt- und Getriebelehre. München: Oldenbourg. 1930.

Wechselschaltung, mit der sich eine Lampe von zwei Stellen aus ein-
und ausschalten läßt. Der jeweilige Zustand der Lampe ist ausschließlich
von den Stellungen der beiden Schalter abhängig. Legt man einen der
beiden Schalter um, so läßt sich durch eine gegenläufige Betätigung desselben Schalters der ursprüngliche Zustand der Lampe wieder herstellen.
Diese Wirkungsweise erfährt auch keine Veränderung, wenn man die
Lampe nicht unmittelbar, sondern über ein Schütz oder einen Fernschalter steuert. Der Sprachgebrauch der technischen Praxis, der noch

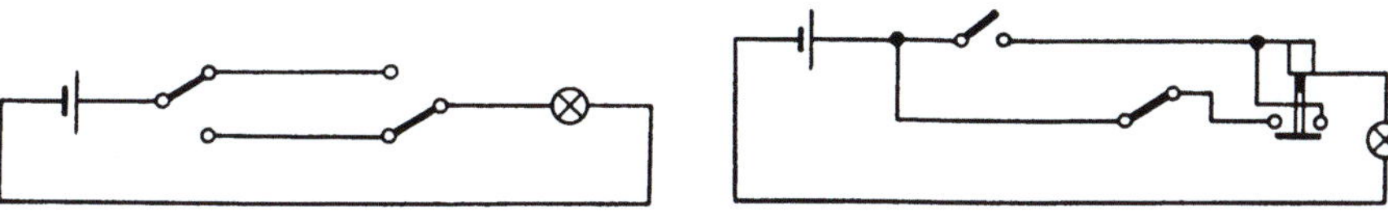

a   b

Abb. 7. Hauptarten elektromechanischer Schaltungen. *a* Dauerschaltung,
b Impulsschaltung

wenig systematisch ist, kennt gegenwärtig keine Sammelbezeichnung für
die große Gruppe von Schaltungen mit dieser grundlegenden Eigenschaft. Gänzlich anders verhält sich die in Abb. 7 *b* gezeigte Anordnung.
Sie besteht aus einem Schütz mit Haltekontakt, das durch zwei Schalter
gesteuert wird. Von diesen dient einer zur Einschaltung, der andere zur
Ausschaltung. Bringt man das Schütz durch Umlegen des ersten Schalters
in seine Arbeitsstellung, so verbleibt es in dieser Stellung, wenn der ursprüngliche Zustand der Steuergeräte wieder hergestellt ist. Es besteht
also kein eindeutiger Zusammenhang zwischen den Stellungen der Steuerschalter und dem Zustand des gesteuerten Gerätes. Diese Wirkungsweise
ist durch den Haltekontakt des Schützes bedingt; sie tritt immer auf,
wenn sich im steuernden Stromkreis ein Kontakt befindet, der vom gesteuerten Gerät selbst betätigt wird. Die Zustandsänderungen solcher
Schaltungen verlangen nur eine kurzzeitige Betätigung der Steuergeräte,
weil deren Funktion sofort durch die selbststeuernden Kontakte (Selbst-
halter und Selbstunterbrecher) übernommen wird. Daher werden diese
Schaltungen in der Praxis *Impulsschaltungen oder Anreizschaltungen* genannt. **Im Gegensatz hiezu** könnte man die früher beschriebene Haupt-
gruppe von Schaltungen als *Dauerschaltungen* bezeichnen.

In der technischen Praxis ist ferner eine Unterteilung in *Arbeitsstrom-
und Ruhestromschaltungen* üblich. Auch diese Benennungen beziehen sich
auf die Wirkungsweise der Schaltungen. Die Unterteilung gilt nur für
Schaltungen, die von Tastern oder Schützen gesteuert werden. Schaltgeräte dieser Art besitzen eine ausgeprägte *Arbeitsstellung*, in die sie durch
äußere mechanische oder elektrische Einwirkung gebracht werden; bei
Aufhören der Einwirkung kehren sie durch Federkraft oder Gewicht
selbsttätig in ihre Ruhestellung zurück. Ist die Schaltung so beschaffen,
daß sie in der Arbeitsstellung der Steuergeräte Strom führt, wird sie
Arbeitsstromschaltung genannt; im gegenteiligen Falle bezeichnet man

sie als Ruhestromschaltung. Dienen zur Steuerung Schalter, deren Stellungen untereinander gleichwertig sind, ist eine solche Unterscheidung nicht möglich.

Die technische Praxis bevorzugt Schaltungen, bei denen eine Seite der Verbraucher unmittelbar an einen geerdeten Pol der Stromquelle angeschlossen ist. Der Vorteil liegt darin, daß ein allfälliger Erdschluß innerhalb der Schaltung den betroffenen Verbraucher kurzschließt und dadurch gegen ungewollte Einschaltungen schützt. Abb. 8 a zeigt eine solche Schaltungsform; ihr Kennzeichen ist, daß alle steuernden Kontakte auf einer Seite der gesteuerten Geräte liegen, so daß man diese Anordnungen als *einseitige Schaltungen* bezeichnen kann. Manchmal besteht aber die Möglichkeit, die Zahl der Kontakte durch Zusammenlegung zu vermindern, wenn man die Kontakte zu beiden Seiten der Verbraucher anordnet. Abb. 8 b zeigt eine solche *zweiseitige Schaltung*.

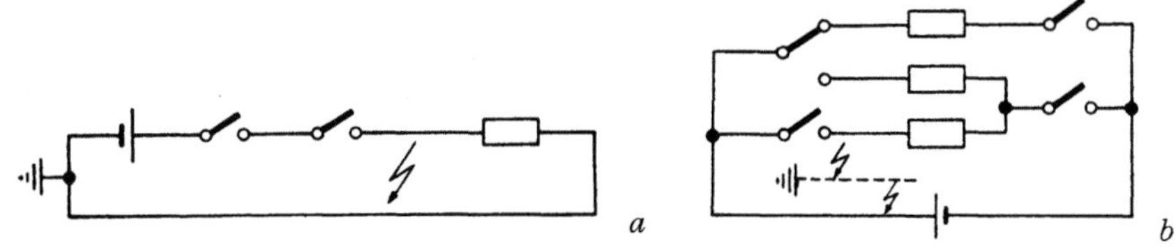

Abb. 8. *a* Einseitige Schaltung, *b* Zweiseitige Schaltung

Eine besondere Ausführung der zweiseitigen Formen sind die *mehrpoligen Schaltungen*. Sie bestehen ihrem Wesen nach aus zwei oder mehreren vollständig gleichen, einseitigen Schaltungen (eine für jeden Spannungspol) zwischen denen die Verbraucher liegen. Je nach Polzahl der Stromquelle entstehen auf diese Weise zweipolige und mehrpolige Schaltungen. Sie verhalten sich hinsichtlich Gefährdung durch Erdschlüsse genau so wie einseitige Schaltungen, verlangen aber keine Erdung der Stromquelle. Voraussetzung hiefür ist, daß die Auftrennung durch Kontakte an sämtlichen Spannungspolen durchgeführt wird, daß also die Schaltungen *allpolig* ausgeführt sind.

In den Steuergeräten mehrpoliger Schaltungen ist jeder Kontakt mehrmals vorhanden, und zwar für jeden Spannungspol einmal. Im technischen Sprachgebrauch wird daher diese Bezeichnung auch auf die Bauart der Schaltgeräte angewendet. Das Kennzeichen eines mehrpoligen Schalters sollte darin bestehen, daß er aus mehreren vollständig gleichen Kontaktgruppen, die voneinander elektrisch getrennt sind, besteht. Die mangelnde Systematik der Praxis hat allerdings dazu geführt, daß auch Kontakte, die in verschiedenen Schalterstellungen geschlossen, aber elektrisch getrennt sind, Schalterpole genannt werden. Häufig werden sogar die einzelnen Kontaktbahnen eines mehrstelligen Schalters als Schalterpole bezeichnet, selbst wenn sie mit anderen Bauteilen elektrisch zusammenhängen. Im Interesse der Klarheit und Verständlichkeit der technischen Ausdrücke sollten solche Ungenauigkeiten vermieden werden.

Auch für die verschiedenartigen *Bestandteile* elektromechanischer Schaltungen läßt sich eine systematische Unterteilung durchführen. Die Bezeichnungen sind nicht nur durch den allgemeinen Sprachgebrauch, sondern größtenteils auch durch Normung festgelegt worden. Es lassen sich zwei Hauptgruppen von Schaltelementen unterscheiden, und zwar solche mit veränderbarem und solche mit festem Leitwert.

Unveränderlichen Leitwert zeigen Leitungen, Stromerzeuger und Stromverbraucher. Unter einer *Leitung* verstehen wir jedes Gebilde, das einen unwesentlichen Strömungswiderstand aufweist und dessen Punkte daher jederzeit praktisch gleiches Potential haben. Es ist hierbei gleichgültig, welche Gestalt und wieviele Anschlußstellen die Leitung hat; eine Leitung mit nur zwei Anschlüssen wird *Leiter* genannt. Eine Leitung besteht im allgemeinen aus mehreren Leitern, die nach Belieben aneinandergereiht oder auch zu einem Stern verbunden sein können. Abb. 9 zeigt die beiden Elementarformen einer Leitung, und zwar Abb. 9 *a* das Leiterpolygon, Abb. 9 *b* den Leiterstern. Die Gestalt der Leitung hat auf deren Wirkungsweise keinen Einfluß. Ein Schaltelement, das die Form einer Leitung, aber einen wesentlichen und unveränderlichen Widerstand aufweist, wird als *Verbraucher* bezeichnet. Ist innerhalb der Leitung eine elektromotorische Kraft wirksam, so heißt sie *Stromquelle*.

Von den Schaltelementen mit veränderbarem Widerstand sind besonders jene von Bedeutung, deren Widerstand sprunghaft zwischen einem sehr großen Wert (Isolationswiderstand) und einem verschwindend kleinen Betrag (Übergangswiderstand) wechseln kann; sie führen den Namen *Kontakte*. Jeder Kontakt besteht aus wenigstens zwei *Schaltstücken*, von denen mindestens eines beweglich sein muß (Abb. 10 *a*). Häufig werden Kontakte aus drei Schaltstücken gebildet, **wovon zwei ortsfest sind und** das dritte beweglich ist (Abb. 10 *b*). Das bewegliche Schaltstück nennt man dann *Schaltbrücke*, die festen hingegen *Schaltfinger*. Kontakte, die zum selben Schaltgerät gehören, können teilweise gemeinsame Schaltstücke haben (Abb. 10 *c*).

Die Vereinigung eines oder mehrerer Kontakte mit einer gemeinschaftlichen *Antriebsvorrichtung*, z. B. einem Handgriff, nennt man *Schalter*. Ferner wurde durch Normung festgelegt, daß Geräte, die sich nur bei Einwirkung äußerer Kräfte bewegen, den Namen *Stellschalter* führen sollen (Drehschalter, Druckschalter, Zugschalter). Handbediente Geräte mit selbsttätiger Rückstellvorrichtung, beispielsweise mit Rückholung

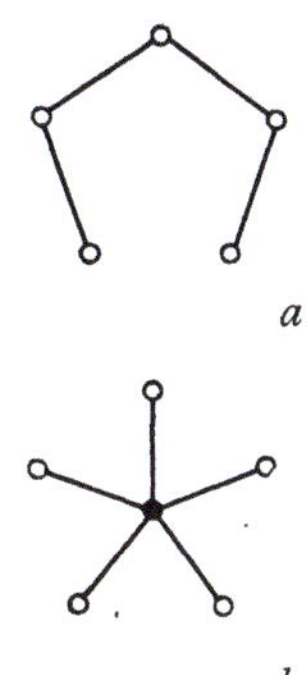

Abb. 9. Hauptformen von Leitungen. *a* Leiterpolygon, *b* Leiterstern

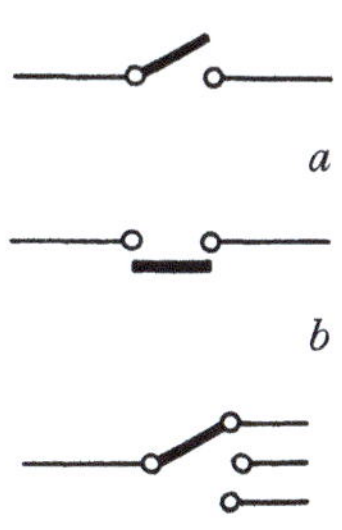

Abb. 10. Kontaktformen. *a* Mit zwei Schaltstükken, *b* Mit drei Schaltstücken, *c* Mit einem gemeinsamen Schaltstück

durch Feder oder Gewicht, werden *Taster* genannt (Drucktaster, Zug-
taster, Drehtaster)[1]. Erfolgt die Betätigung durch ein Übertragungs-
mittel, beispielsweise elektromagnetisch oder pneumatisch, so heißt das
dem Taster entsprechende Schaltgerät *Schütz*. Ein mittelbar ange-
triebener Schalter ist ein *Fernschalter*. Schalter, deren Kontakte minde-
stens teilweise gemeinschaftliche Schaltbrücken und gemeinsame Anschlußstellen aufweisen, heißen *Umschalter* (Abb. 11 *a*). Sind die Brücken gemein-

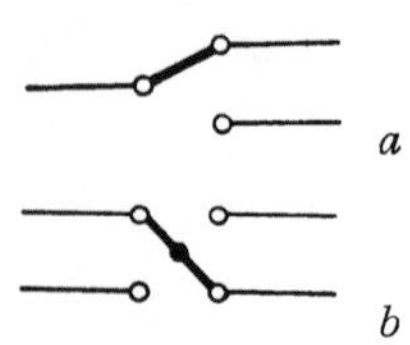

Abb. 11. Kontakt-
kombinationen in
einem Schalter.
a Umschalter,
b Kreuzschalter

sam, die Anschlußstellen aber verschieden, so wird
das Gerät als *Kreuzschalter* bezeichnet (Abb. 11 *b*).
Für Umschalter mit einer größeren Zahl von Stellungen ist auch die Bezeichnung *Wahlschalter oder
Wähler* üblich.

Die im vorstehenden angeführten Einteilungen und
Bezeichnungen beziehen sich auf die konstruktive
Ausführung der Schaltungen und Schaltgeräte. Darüber hinaus sind noch Unterscheidungen möglich und
üblich, die sich auf die fallweise verschiedene Verwendung der Schaltelemente beziehen. Geräte, die zur primären Betätigung einer Schaltung
dienen, führen den Namen *Befehlsschalter*. Sie sind gewöhnlich handbediente Geräte, doch ist dies nicht unbedingt notwendig; eine Schalt-
uhr ist beispielsweise ebenfalls ein Befehlsschalter. Viele Schaltungen, darunter alle Anreizschaltungen, enthalten außer den Steuergeräten
noch weitere Schaltgeräte, für die der Sammelname *Relais* gebraucht
wird. Diese lassen sich auf zwei völlig verschiedene Arten verwenden.
In Dauerschaltungen dienen sie dazu, die Befehlsschalter hinsichtlich
Stromstärke oder Kontaktzahl zu entlasten; auf die Wirkungsweise der Schaltung haben sie keinen Einfluß. In Anreizschaltungen
dient hingegen ein Teil ihrer Kontakte zur *Selbsthaltung* oder *Selbstunterbrechung*, so daß diesen Geräten vorübergehend eine eigene Steuerwirkung
zukommt; sie sind dann Geräte, mit deren Hilfe sich erst die Gesamtwirkung der Schaltung ergibt[2].

---

[1] Vgl. die Normen DIN **40 713** und **43 605**.

[2] Es muß beachtet werden, daß diese streng systematische Einteilung
im lebenden Sprachgebrauch der Praxis noch nicht klar hervortritt. Dies
gilt besonders für die Bezeichnungen. Innerhalb einzelner beschränkter
Sondergebiete haben sich sprachliche Gepflogenheiten herausgebildet, die
zu den in anderen Zweigen der Elektrotechnik üblichen in Widerspruch
stehen. Beispielsweise gebraucht man in der Fernmeldetechnik die Be-
nennung „Relais" für Schütze kleiner Stromstärke und verwendet somit
diesen Namen zur Kennzeichnung einer Bauform. Im Schaltanlagenbau
der Starkstromtechnik bezeichnet man hingegen als „Relais" ein Schaltgerät mit eingebautem Meßwerk; ein Gerät dieser Art heißt aber in der
Technik der elektrischen Antriebe „Wächter", womit das Gerät wesentlich
treffender gekennzeichnet ist. Es wird eines Tages unumgänglich notwendig
sein, diese Unstimmigkeiten durch eine das Gesamtgebiet der elektrischen

3. Die Grundbegriffe der Schaltungstechnik

Für die Beurteilung, Verbesserung oder Neuentwicklung einer Schaltung ist die Kenntnis der Zusammenhänge zwischen Form und Wirkungsweise der Schaltung erste Voraussetzung; über diese Grundbegriffe muß daher volle Klarheit bestehen. Wir legen uns daher die Frage vor, was man sich unter der Wirkungsweise oder Funktion einer Schaltung eigentlich vorzustellen hat. Es handelt sich dabei nicht darum, diesen Begriff durch eine Definition formal einzugrenzen; vielmehr ist eine physikalische Deutung, die als Grundlage zu einer mathematischen Behandlung dienen kann, erforderlich. Es ist nun das Kennzeichen aller elektromechanischen Schaltungen, daß sich die Leitfähigkeit ihrer Strompfade durch Betätigung von Schaltgeräten verändern läßt. Die *Wirkungsweise* oder *Funktion* einer Schaltung ist dann nichts anderes als die gegenseitige Beziehung zwischen dem jeweiligen Zustand der Schaltgeräte und den zugehörigen Leitwerten der Strompfade. Diese Definition führt den Begriff der Wirkungsweise auf physikalische Größen zurück, die sich durch Messung und Rechnung bestimmen lassen. Der Zustand eines Schaltgerätes, seine Schaltstellung, läßt sich nach Festlegung einer Nullage, bei der die Zählung beginnen soll, durch einen Längen- oder Winkelabstand ausdrücken; auch die elektrischen Leitwerte zwischen allen beliebigen Punkten der Schaltung können durch Messung erfaßt und in physikalischen Einheiten, beispielsweise in Ω^{-1}, angegeben werden. Es ist daher auch möglich, die Funktion einer Schaltung als mathematische Beziehung zwischen deren Variablen darzustellen. Dies gilt sowohl für Kontaktschaltungen als auch für Regelschaltungen.

Jeder einfache Weg, der zwei Punkte verschiedenen Potentials miteinander verbindet, ist ein *Strompfad* der Schaltung. Als einfache Wege bezeichnet man solche, die jede Leitung nur einmal berühren, also keine in sich geschlossenen Schleifen aufweisen. Es läßt sich nämlich mit allgemeiner Gültigkeit nachweisen, daß solche Schleifen für die Wirkungsweise der Schaltung, also für ihre Leitwertfunktion, gegenstandslos sind. Die Veränderung der Leitfähigkeit eines Strompfades ist durch den

---

Schaltungen umfassende Normung zu beseitigen. Bedauerlicherweise weichen auch die **österreichischen Normen (ÖVE-S 5)** hinsichtlich der Nomenklatur wesentlich von den entsprechenden deutschen Vorschriften (VDE 0660) ab. So nennt man in Österreich die Befehlsschalter Steuergeräte, obwohl dieses Wort nach VDE alle Geräte bezeichnet, welche betriebsmäßig häufig geschaltet werden. Der Begriff „Steuerschalter (VDE)" deckt sich weitgehend mit dem Begriff „Verbraucherschaltgeräte (ÖVE)". Vgl. RIEDER, W.: Die Uneinheitlichkeit der Benennungen und Definitionen elektrischer Schaltgeräte innerhalb des deutschen Sprachraumes. EuM 72 (1955), S. 162—165. Die im vorstehenden benützten Benennungen stellen einen Durchschnitt durch den gegenwärtigen technischen Sprachgebrauch dar und beruhen auf den sachlichen Unterschieden in der Bauart und schaltungstechnischen Verwendung der Geräte.

wechselnden Betriebszustand der Kontakte verursacht. Unter einem *Kontakt* haben wir demnach nicht einen bestimmten Bauteil zu verstehen, sondern jede Strecke beliebiger Art, deren Leitwert abwechselnd 0 und ∞ ist. Der jeweilige Zustand des Kontaktes wird durch die *Stellung* des Schaltgerätes, dem er angehört, bestimmt. Maßgebend für den jeweiligen Leitwert eines Strompfades sind daher die Stellungen sämtlicher Schalter, deren Kontakte im Strompfad vorkommen. Die Reihenfolge der Kontakte spielt hierbei keine Rolle. Um die Wirkungsweise eines Strompfades anzugeben, genügt demnach die Aufzählung aller Schalterstellungen, in denen Kontakte des betreffenden Strompfades geschlossen sind. Als Beispiel denken wir uns eine Schaltung mit vier Schaltgeräten a, b, c und d. Jedes dieser Geräte habe zwei Stellungen, 0 und 1. Dann können wir unter anderem einen Strompfad herstellen, der über alle vier Schalter führt und Kontakte enthält, die in den Stellungen a_0, b_1, c_0 und d_1 geschlossen sind. Bestimmend für den Leitwert des Strompfades sind dann diese vier angeführten Schalterstellungen. In den Veröffentlichungen über die Theorie der elektrischen Schaltungen wird die Gesamtheit der für einen Strompfad maßgebenden Schalterstellungen als *Stellungskombination* oder *Stellungskomplex* bezeichnet; auch der Ausdruck *Schaltreihe* ist üblich. Die letztere Bezeichnung soll zum Ausdruck bringen, daß die betreffenden Kontakte in Reihe geschaltet sind; es darf aber nicht übersehen werden, daß eine Schaltreihe nicht ein Komplex von Kontakten, sondern von Stellungen der Schaltgeräte ist.

Die Gesamtheit der Stellungen aller Geräte einer Schaltung, die jeweils gleichzeitig vorhanden sind, wird als *Schaltzustand* bezeichnet. Durch ihn ist der Stromfluß innerhalb der Schaltung und somit auch der Zustand der Stromverbraucher eindeutig bestimmt. Aus der Zahl der vorhandenen Schaltgeräte und ihrer Stellungen läßt sich stets ermitteln, wie viel Zustände die ganze Einrichtung einnehmen kann. Deren Anzahl ergibt sich, wenn man die Zahl der Stellungen sämtlicher Schaltgeräte miteinander multipliziert. Sind beispielsweise zwei Geräte mit je drei Stellungen und eines mit vier Stellungen vorhanden, so kann die Schaltung insgesamt nur $3 \cdot 3 \cdot 4 = 36$ verschiedene Schaltzustände haben. Für jeden Stromverbraucher der Schaltung läßt sich angeben, bei welchen Zuständen er Strom führt. Die Wirkungsweise einer Schaltung, die als Beziehung zwischen Schalterstellungen und Leitwerten definiert wurde, ist also nichts anderes, als die Zuordnung der Stromverbraucher zu den für sie wirksamen Schaltzuständen. Die physikalische Definition des Funktionsbegriffs ermöglicht also auch eine kombinatorische Deutung. Die Funktion erscheint hiebei als Summe von Stellungskombinationen und jede solche als Summe von Schaltzuständen (Stellungsvariationen).

Hat man beim Entwurf einer Schaltung die notwendigen Verbindungen sowie die Zahl der Schaltgeräte und ihrer Stellungen bereits festgelegt, dann kann man auch die Stellungskombinationen, die den Strompfaden zugrunde liegen, bestimmen. Es kommt nun häufig vor, daß das Zusammenwirken von zwei oder mehreren Stellungskombinationen eine Vereinfachung erlaubt; es ist dann möglich, in diesen Kombinationen einzelne Schalterstellungen als überflüssig zu streichen. Dieser Vorgang wird *Kürzung* genannt. Ein ganz einfaches Beispiel sei an Hand der Abb. 12 gezeigt. Die Schaltaufgabe sei so beschaffen, daß sich eine Teilschaltung aus drei Schaltgeräten a, b und c mit den Stellungskombinationen a_0, c_1 und a_1, b_1, c_1 ergibt. Der Schalter a sei hiebei ein unterbrechungslos wirkender Umschalter. Entwickelt man die Strompfade genau nach den angeführten Stellungskombinationen, so erhält man eine Ausführung nach Abb. 12 a. Es läßt sich aber ohne Veränderung der Wirkungsweise die Stellung a_1 in der zweiten Kombination streichen,

woraus eine wesentlich einfachere Schaltung gemäß Abb. 12 b folgt. Solche Vereinfachungen, die sich an Hand der zeichnerischen Darstellung nur schwer auffinden lassen, erscheinen bei der rechnerischen Behandlung als Kürzung der algebraischen Leitwertgleichungen; der Name *Kürzung* ist auch deswegen zutreffend und sinnfällig, weil die Strompfade einer gekürzten Schaltung weniger Kontakte in Reihe enthalten und somit kürzer wurden.

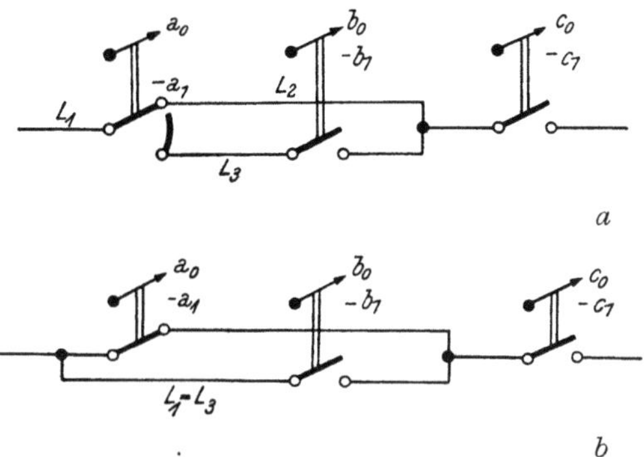

Abb. 12. Kürzung. *a* Ungekürzte Form, *b* Gekürzte Form

Eine zweite Möglichkeit, Schaltungen zu vereinfachen, besteht in der Zusammenlegung paralleler Leitungen und Kontakte; hierfür ist die Bezeichnung *Verkettung* üblich. Ein Beispiel hierfür zeigt die Abb. 13. Aus zwei Stellungskombinationen a_1, b_1 und a_1, c_1 ergibt sich unmittelbar eine Schaltungsform entsprechend Abb. 13 a; es ist aber jedem Elektro-

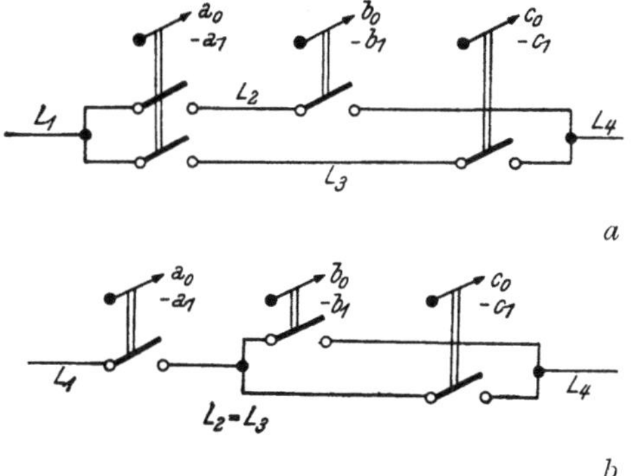

Abb. 13. Verkettung. *a* Unverkettete Form, *b* Verkettete Form

techniker geläufig, daß sich die Kontakte des Schalters a sowie die anschließenden Leitungen L_2 und L_3 vereinigen lassen, wie es Abb. 13 b zeigt. Wesentlich ist, daß sich bei dieser Art der Vereinfachung die Zahl der Kontakte in den einzelnen Strompfaden nicht verändert.

Während die Vereinfachung von Schaltungsformen durch Verkettung seit langem in Gebrauch steht, blieben die Möglichkeiten zur Kürzung

lange Zeit völlig unerkannt; erst die Einführung mathematischer Hilfsmittel setzte die Schaltungstechnik instand, auch Kürzungen systematisch anzuwenden. Die anfänglichen Versuche, zur Lösung von Schaltaufgaben eine algebraähnliche Symbolik zu entwickeln, führten sogar
zu einer Verwechslung der Kürzung mit der Verkettung. In verschiedenen
älteren Veröffentlichungen wurde überhaupt jede Vereinfachung einer
Schaltung als Kürzung bezeichnet. Es sollen daher die wesentlichen Unterschiede zwischen den beiden Arten der Vereinfachung von Schaltungsformen vergleichend betrachtet werden. Die Kürzung beruht auf der
fallweisen Erkenntnis, daß einzelne funktionelle Bedingungen, die aus
der Schaltaufgabe folgen, überflüssig sind, weil sie in anderen Teilbedingungen bereits versteckt enthalten sind. Die Kürzung führt daher
zur vollständigen Streichung einzelner Schalterstellungen in den Stellungskombinationen, so daß die zugehörigen Strompfade weniger Kontakte
enthalten. Aus der Kürzung ergibt sich auch eine Verminderung der verbindenden Leitungen; wie Abb. 12 erkennen läßt, ermöglicht die Kürzung
der Schalterstellung a_1 eine Vereinigung der beiden in Serie liegenden
Leitungen L_1 und L_3. Im Gegensatz hierzu bleiben bei der Verkettung
die Stellungskombinationen und demnach auch die Strompfade unverändert erhalten. Es kann auch niemals ein einzelner Kontakt, sondern
immer nur einer von zwei gleichartigen weggelassen werden; die Verkettung ist eine örtliche Vereinigung, aber kein Entfall eines Kontaktes.
Auch in der Zusammenlegung der Leitungen zeigt sich der Unterschied.
Die Leitungen L_1 und L_3, die bei der Kürzung nach Abb. 13 vereinigt
wurden, lagen ursprünglich in Serie, wogegen bei der Verkettung gemäß
Abb. 13 die Leitungen L_2 und L_3, die zueinander parallel waren, zusammengelegt werden konnten. Die schaltungstechnische Definition der Verkettung deckt sich übrigens vollständig mit dem energetischen Verkettungsbegriff, der in der Technik der mehrphasigen Wechselströme in
Verwendung steht. Auch ein verkettetes Wechselstromsystem ist dadurch gekennzeichnet, daß einzelne parallele Leiter und Kontakte, die
zu verschiedenen Phasen und somit Strompfaden gehören, zu je einem
Bauteil zusammengelegt sind.

Die vergleichende Untersuchung verketteter Schaltungen zeigt das
Bestehen von drei verschiedenen Grundformen, die sich in ihrem Verhalten wesentlich unterscheiden. Schaltungen, bei denen stets nur je
zwei parallele Strompfade durch Verkettung verbunden sind, wie es
Abb. 14 *a* zeigt, nennt man *einfach verkettet*. Kennzeichen dieser Bauart
ist, daß sich von jedem Kontakt der Schaltung unabhängig vom jeweiligen Schaltzustand angeben läßt, zu welchen anderen Kontakten er
parallel und zu welchen er in Serie liegt; ferner wird jeder Kontakt stets
in gleicher Richtung vom Strom durchflossen. Die rechnerische Behandlung solcher Schaltungsformen zeigt, daß sie sich durch Leitwertgleichungen so abbilden lassen, daß jeder Variablen der Gleichung ein
bestimmter Kontakt der Schaltung entspricht. Eine höhere Stufe der
Verkettung ist die *Vermaschung*. Unter bestimmten Voraussetzungen
lassen sich zwei verkettete Paare von Strompfaden nochmals verketten,

wie es Abb. 14 *b* zeigt. Die Anordnung ist dann dadurch gekennzeichnet, daß die beiden Strompfade, die durch einen gemeinsamen Kontakt führen, in diesem Kontakt entgegengesetzte Stromrichtung aufweisen. Während also die Kontakte einer einfach verketteten Schaltung untereinander voll-

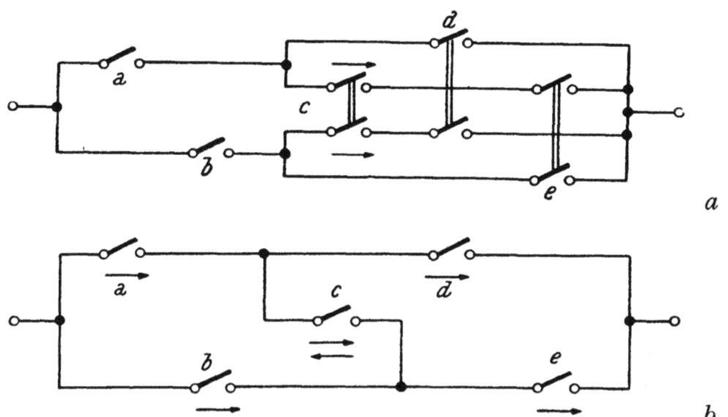

Abb. 14. Vermaschung. *a* Unvermaschte Form, *b* Vermaschte Form

ständig gleichwertig sind, ist dies bei einer vermaschten Schaltungsform nicht mehr der Fall. Von den fünf in Abb. 14 *b* dargestellten Kontakten sind die vier Kontakte *a*, *b*, *d* und *e* sogenannte *Längskontakte* mit stets gleichbleibender Stromrichtung; der *Querkontakt c*, der auch *Maschen-kontakt* oder *Brückenkontakt* genannt wird, zeigt jedoch Richtungswechsel des Stromes. Zu beachten ist, daß nur die Stromrichtung in den Kontakten, nicht aber diejenige in den Leitungen Bedeutung hat. Wie aus der Abb. 15 ersichtlich ist, kann auch in einer unvermaschten **Schaltung** die Strom-richtung innerhalb einzelner **Leiter** wechseln; ob dieser Richtungswechsel eintritt, hängt ausschließlich von der Gestalt der Leitung ab. In Abb. 15 *a* und 15 *b* sind zwei gleichwirkende Schaltungen dargestellt; ist die Leitung *L* gemäß Abb. 15 *a* sternförmig,

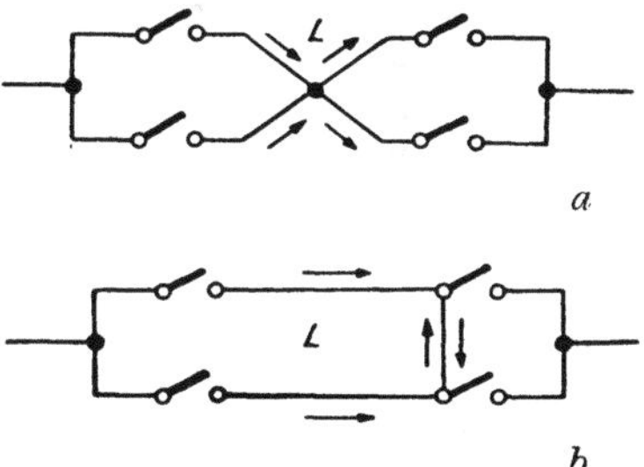

Abb. 15. Die Stromrichtung in Leitungen. *a* Ohne Wechsel, *b* Mit Wechsel

bleibt die Stromrichtung unverändert, wogegen bei Ausführung als Leiterpolygon nach Abb. 15 *b* ein Richtungswechsel eintritt. Der Strom-fluß in den Leitern hat also mit der Zusammenlegung von Leitungen und Kontakten nichts zu tun. Maßgeblich ist nur die Stromrichtung inner-halb der Kontakte; in der unvermaschten Schaltung bleibt sie stets gleich, in Maschenkontakten wechselt sie mit dem Schaltzustand. Die beiden Schaltungsformen unterscheiden sich auch im Zahlenverhältnis ihrer Bestandteile; bei der einfachen Verkettung entfällt mit jeder

Leitung auch ein Kontakt, wogegen bei der Vermaschung stets mehr Kontakte erspart werden als Leitungen. Auch in der rechnerischen Behandlung besteht ein wesentlicher Unterschied. Vermaschte Schaltungsformen lassen sich nicht mehr durch Leitwertgleichungen konform abbilden, sondern können nur tabellarisch, z. B. durch zweidimensionale Matrizen, so dargestellt werden, daß jedem Kontakt ein einziges Rechenzeichen entspricht.

In Schaltungen größeren Umfanges ist häufig noch eine dritte Stufe der Verkettung möglich, nämlich die *Kreuzung*, die in der teilweisen Zusammenlegung zweier bereits vermaschter Teilschaltungen besteht. Abb. 16 *d* zeigt die einfachste Form einer gekreuzten Schaltung; deren Kennzeichen liegt darin, daß sie sich im Gegensatz zu einfach verketteten oder einfach vermaschten Schaltungen nicht mehr ohne Kreuzung von Leitungen in einer Ebene abbilden läßt. Die Kreuzung ist also — im Gegensatz zur einfachen oder ebenen Vermaschung — eine räumliche Verkettung. Wie sich an Hand der Abb. 16 verfolgen läßt, setzt diese Art der Zusammenlegung mindestens acht Strompfade voraus; auch diese Schaltungsform zeigt selbstverständlich den Wechsel der Stromrichtung in einzelnen Kontakten.

Zusammenfassend läßt sich also feststellen, daß die Vereinfachung einer größeren Schaltung durch Zusammenlegen von Leitungen und Kontakten in drei Stufen vor sich geht, wobei sich die Merkmale und Eigenschaften der Schaltungsform sprunghaft ändern. Die Abb. 16 zeigt die verschiedenen Entwicklungsstufen einer Schaltung, bei der die zur Kreuzung erforderlichen Symmetriebedingungen erfüllt sind. Die Schaltung enthält insgesamt acht Schalter mit je zwei Stellungen; der Deutlichkeit halber sind die einzelnen Kontakte mit Nummern versehen, wobei gleiche Ziffern besagen, daß die betreffenden Kontakte zum selben Schalter gehören und in der gleichen Stellung geschlossen sind. Die einzelnen Strompfade sind in Abb. 16 *a* voneinander getrennt, also unverkettet, wiedergegeben. Würde man die Schaltung so ausführen, so hätte sie 32 Kontakte und 26 verbindende Leitungen. Die erste Stufe der Verkettung, die in Abb. 16 *b* gezeigt ist, ergibt sich durch schrittweise Zusammenfassung gleichartiger Kontakte von je zwei parallelen Strompfaden; diese einfach verkettete Form besteht aus 22 Kontakten und 16 Leitungen. Durch weitere Zusammenfassung von je zwei verketteten Strompfadpaaren folgt als nächste Entwicklungsstufe die vermaschte Form nach Abb. 16 *c*; sie enthält nur mehr 14 Kontakte und 10 Leitungen. Die beiden vermaschten Strompfadvierer lassen sich nun nochmals teilweise vereinigen, woraus die gekreuzte Form gemäß Abb. 16 *d* entsteht. Diese letzte Entwicklungsstufe der Schaltung verlangt nur 8 Kontakte und 6 Leitungen; jeder Schalter ist nur mehr mit einem Kontakt vertreten.

Die Schaltungen, die in der Praxis vorkommen, lassen sich nahezu immer verketten. Hingegen sind die Bedingungen zur Vermaschung und Kreuzung häufig nicht gegeben. Es ist dann wiederum oft möglich, die Voraussetzung hierfür künstlich zu schaffen; hierzu dient eine Maß-

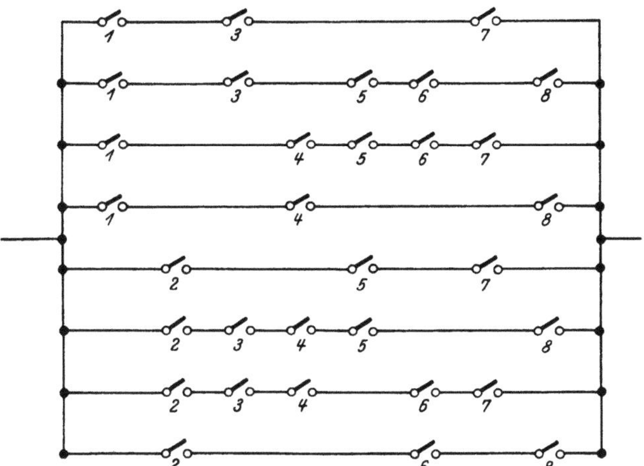

nahme, die als *Erweiterung* bezeichnet wird und einen zur Kürzung gegenteiligen Vorgang darstellt. Ein einfaches Beispiel diene zur Erläuterung. Gegeben sei eine einfach verkettete Schaltung gemäß Abb. 17 a; sie umfaßt drei Strompfade, die auf den Stellungskombinationen $a_1 c_1$, $b_1 c_1$ und $b_1 d_1$ beruhen. Diese Kombinationen bilden keine Grundlage für eine weitere Vereinfachung der Schaltungsform, weil zu einer Vermaschung mindestens vier Strompfade erforderlich sind. Es läßt sich aber ohne Veränderung der Wirkungsweise ein vierter Strompfad hinzufügen, dem eine Stellungskombination $a_1 a_0 d_1$ zugrunde liegt. Diese Kombination enthält zwei Stellungen des Schalters a, die nicht gleichzeitig

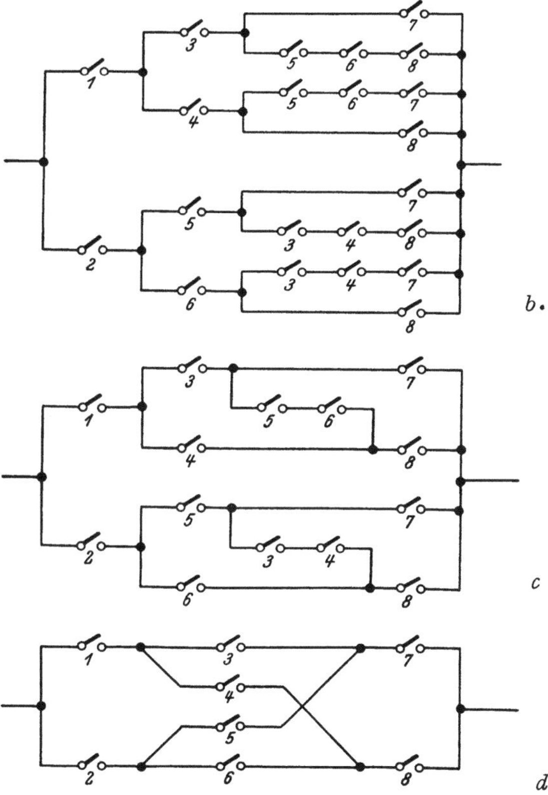

Abb. 16. Stufen der Verkettung. *a* Unverkettete Schaltung, *b* Einfach verkettete Form, *c* Vermaschte Form, *d* Gekreuzte Form

vorhanden sein können; ein Strompfad aus Kontakten, die zu diesen Schalterstellungen gehören, ist daher stets an einer Stelle unterbrochen und somit wirkungslos (Abb. 17 *b*). Die Erweiterung der Schaltung um einen solchen *scheinbaren Strompfad* schafft aber die Möglichkeit zu einer Vermaschung gemäß Abb. 17 *c*. Durch diesen Vorgang wurde zwar die Zahl der Kontakte und Leitungen nicht vermindert, weil der sogenannte Sperrkon-

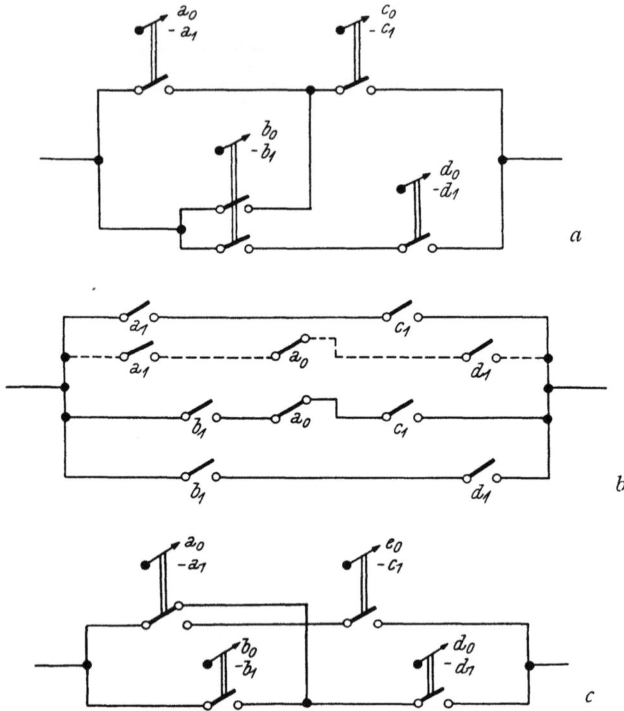

Abb. 17. Formale Vermaschung. *a* Verkettung von drei Strompfaden, *b* Erweiterte Schaltung, *c* Vermaschung der drei Strompfade

takt a_0 neu hinzukam; die beiden Parallelkontakte am Schalter *b* wurden aber durch Umschaltekontakte am Schalter *a* ersetzt, so daß sich die Zahl der Schaltbrücken innerhalb der Geräte um eines vermindert hat. Vereinfachungen dieser Art sind also durchaus zweckmäßig und lassen sich sehr häufig durchführen. Da einer der vier Strompfade niemals Strom führt, fließt bei dieser Schaltungsform im Querkontakt a_0 nur in einer Richtung Strom. Es handelt sich also nicht um eine *funktionelle*, sondern um eine rein *formale Vermaschung*. Möglichkeiten dieser Art bestehen selbstverständlich auch für Vereinfachungen von Schaltungsformen durch Kreuzung. Ferner gibt es Schaltungen, bei denen sich ohne zusätzliche Sperrkontakte wirkungslose Strompfade bilden und somit formal vermaschte und gekreuzte Formen mit kleinerer Kontaktzahl herstellen lassen; die in der Installationstechnik benützten Wechselschaltungen zeigen diese Eigenschaft.

Der als Erweiterung bezeichnete Vorgang zeigt übrigens einen weiteren wesentlichen Unterschied zwischen Kürzung und Verkettung. Letztere führt stets zu einer Verminderung von Kontakten und Leitungen. Bei Kürzungen ist dies häufig, aber durchaus nicht immer der Fall: wenn die Möglichkeit zu einer formalen Vermaschung oder zu einer formalen Kreuzung vorhanden ist, erhält man die einfachste Schaltung nicht durch möglichst weitgehende Kürzung; die teilweise ungekürzten oder nachträglich erweiterten Stellungskombinationen ergeben dann einen geringeren Aufwand an Kontakten und Leitungen.

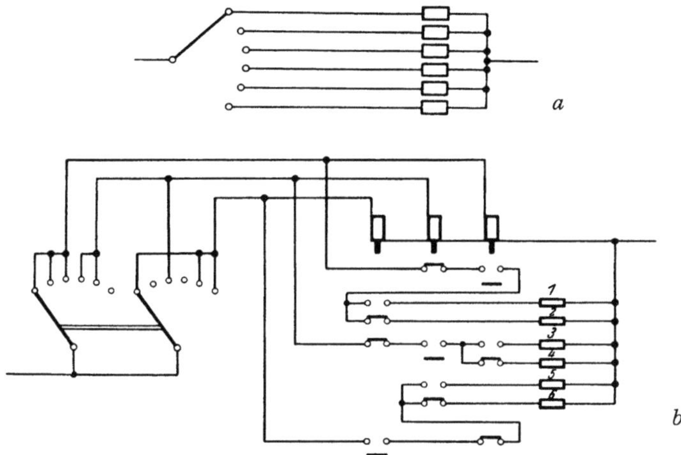

Abb. 18. Mittelbare Verkettung. *a* Unverkettete Umschaltung, *b* Verkettete Umschaltung

Bei allen diesen Arten der Verkettung werden stets Kontakte vereinigt, die zu gleichen Schaltern gehören und gleichzeitig geschlossen sind; es handelt sich also um vollständig gleichwertige oder *äquivalente Kontakte*. Das Wesen der Verkettung liegt aber nicht in der Vereinigung der Kontakte, sondern in der Zusammenlegung der Leitungen. Da sich auch äquivalente Kontakte nicht immer vereinigen lassen, bemühte man sich lange vergebens, Regeln für die Zusammenlegung von Kontakten zu finden, ließ aber dabei die anschließenden Leitungen außer acht. Erst eine genaue Untersuchung zeigte, daß die Verkettung der Leitungen der primäre Vorgang ist; *zwei Kontakte lassen sich dann und nur dann vereinigen, wenn die vier anschließenden Leitungen paarweise verbunden werden können.*

Es ist ferner möglich, Leitungen zu ersparen, wenn die Schaltung überhaupt keine äquivalenten Kontakte enthält. Ein Beispiel hierfür zeigt die Abb. 18. Eine größere Anzahl von Verbrauchern soll durch einen gemeinschaftlichen Umschalter abwechselnd eingeschaltet werden; die unverkettete Form der Schaltung ist in der Abb. 18 *a* dargestellt. Der Umschalter befindet sich dabei in größerer Entfernung von den Verbrauchern, so daß man die Zahl der Verbindungsleitungen möglichst klein halten will. Es besteht nun die Möglichkeit, unter Vermehrung der Kontakte Leitungen einzusparen; diese Aufgabe ist jedoch grundsätzlich nur unter Verwendung von Zwischenrelais lösbar. Abb. 18 *b* zeigt eine der möglichen Ausführungsformen, bei der nur drei Fernleitungen vorhanden sind; diese Leitungen steuern ebensoviele Zwischenschütze, über deren Kontakte die Verbraucher angespeist werden. Eine solche *mittelbare Verkettung* über Zwischenrelais senkt also den Aufwand an Leitungen unter gleichzeitiger Vermehrung der Kontakte. Schaltungsformen dieser Art finden sich vorwiegend in der Technik der Fernwirkanlagen, wo es auf äußerste Ersparnis an Fernleitungen ankommt. Die mittelbare Verkettung ist übrigens keine Zusammenlegung wie die unmittelbare, sondern

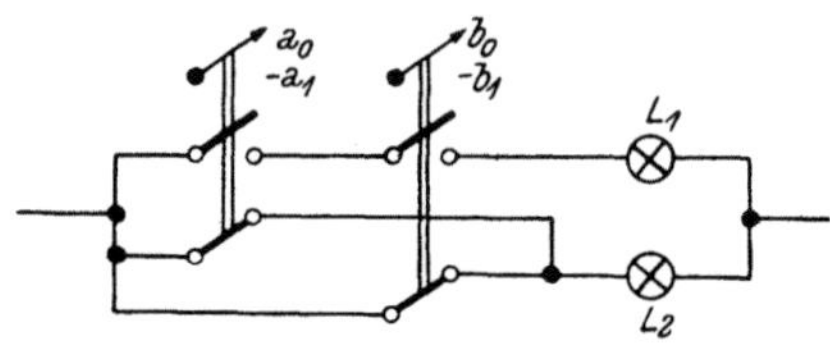

Abb. 19. Komplementäre Schaltungen

beruht auf der Bildung von Kombinationen aus den Spannungszuständen der Fernleitungen.

Außer der Kürzung, Erweiterung und Verkettung ist in der Schaltungstechnik noch eine vierte Operation von Bedeutung, nämlich die *Umkehr oder Inversion*. Als Beispiel hierfür diene die in Abb. 19 wiedergegebene Schaltung. Zwei an sich beliebige Geräte einer Anlage, etwa Ventile einer Dampfleitung, seien mit je einem Meldeschalter a und b versehen. Eine Meldelampe L_1 soll leuchten, wenn beide Ventile geöffnet sind; die zugehörigen Meldeschalter mögen sich hierbei in ihren Stellungen a_1 und b_1 befinden. Diese Wirkung erreicht man durch Serienschaltung je eines Kontaktes der beiden Meldeschalter mit der Lampe L_1. Es soll nun verhindert werden, daß bei einem Defekt der Lampe L_1 eine Fehlanzeige zustande kommt. Zu diesem Zweck ist eine zweite Lampe L_2 vorgesehen, die immer dann brennt, wenn die Lampe L_1 abgeschaltet ist; die beiden Lampen leuchten also abwechselnd, wobei sich jeder Defekt durch das Erlöschen beider Lampen bemerkbar macht. Die Teilschaltung für die Lampe L_2 hat die genau gegenteilige Funktion wie diejenige für die Lampe L_1; von den beiden Teilschaltungen führt eine immer nur dann Strom, wenn die andere stromlos ist. Die beiden Teilschaltungen sind also zueinander *komplementär*. Schaltungspaare dieser Art kommen in der Technik häufig vor; beispielsweise stehen sämtliche Ruhestromschaltungen zu den zugehörigen Arbeitsstromschaltungen im gleichen Verhältnis. Die zu einer vorgegebenen Schaltung komplementäre Schaltung wird als ihre *Kehrschaltung* bezeichnet. Von besonderer Wichtigkeit ist, daß sich nicht nur die Schaltungen selbst, sondern auch die zu ihnen führenden Bedingungen einer Schaltaufgabe nach den gleichen Regeln umkehren lassen. Von dieser Möglichkeit kann beim Entwurf von Schaltungen sehr häufig Gebrauch gemacht werden. Schaltaufgaben sind häufig so gestellt, daß sie teilweise Aussagen über den gewünschten Stromfluß und im übrigen Bedingungen für die Sperrung einzelner Verbindungen enthalten. Um solche verschiedenartige Angaben auf gleiche Basis zu bringen, werden vorerst die Sperrbedingungen nach den Regeln der schaltungstechnischen Inversion in Leitbedingungen umgewandelt; die so umgeformte Schaltaufgabe enthält dann nur Bedingungen gleicher Art, die sich durchwegs auf den Stromfluß beziehen und gemeinschaftlich verarbeitet werden können.

Die im vorstehenden erläuterten Begriffe und Benennungen beziehen sich auf die Zustände und Formen der Schaltungen. Für die Ausmittlung von Schaltungen ist aber auch der *Zustandswechsel* von Bedeutung. Viele Schaltaufgaben bestimmen die gewünschte Wirkungsweise dadurch, daß sie Beziehungen zwischen den Stellungsänderungen der Schaltgeräte und dem Zustandswechsel von Verbrauchern vorschreiben. Es kann beispielsweise gewünscht sein, daß ein Motor bei Drücken eines Tasters zu laufen beginnt und bei Betätigung eines zweiten Schaltgerätes zum Stillstand kommt. Diese Vorschrift ist eine ausschließliche Beziehung zwischen *Schaltvorgängen*. Um Ungenauigkeiten im Ausdruck zu vermeiden, wollen wir festhalten, daß sich ein Schaltvorgang stets auf die Bewegung eines einzigen Schaltgerätes von seiner jeweiligen Stellung in eine unmittelbar benachbarte beziehen soll; ein Schaltvorgang ist demnach die Ausführung eines einzelnen *Schaltschrittes* und die damit verbundene Veränderung des Stromflusses innerhalb der Schaltung. Betrachtet man die Stellungen der Schaltgeräte vor und nach einem Schaltvorgang, so zeigt sich, daß sich die beiden Schaltzustände nur hinsichtlich einer einzigen Schalterstellung voneinander unterscheiden; sie sind *benachbarte Schaltzustände*, die durch einen einzigen Schaltvorgang ineinander übergeführt werden können. Jedem Schaltzustand der Steuergeräte kann ein bestimmter Zustand der Verbraucher zugeordnet sein. Je nachdem, ob sich zwei benachbarte Schaltzustände hinsichtlich des Stromflusses in den Verbrauchern gleich oder ungleich verhalten, können wir wirksame und unwirksame Schaltschritte unterscheiden. Ferner kommt jedem Schaltschritt ein bestimmter Richtungssinn zu. Ein Schritt von x_n auf x_{n+1} soll positiv, der entgegengesetzte Schritt von x_{n+1} auf x_n negativ genannt werden. Zwischen zwei benachbarten Stellungen eines Schaltgerätes liegen stets zwei Schaltschritte entgegengesetzter Richtung. Alle Schaltschritte treten also paarig auf. Führt ein Schrittpaar zum ursprünglichen Zustand des Stromflusses zurück, nennen wir die beiden damit verbundenen Schaltvorgänge *umkehrbar*. Dauerschaltungen ermöglichen nur umkehrbare (reversible) Vorgänge. Schaltvorgänge, die nicht umkehrbar sind (irreversible Vorgänge), sind das Kennzeichen aller Anreizschaltungen.

Zwei oder mehrere Schaltvorgänge, die unmittelbar aufeinanderfolgen, bilden eine *Schaltfolge*. Innerhalb jeder Schaltfolge wechseln Vorgänge und Zustände miteinander ab; auf jeden Vorgang folgt ein Zustand, an jeden Zustand schließt sich ein Vorgang an. Die Schaltfolgen zeigen also einen ähnlichen Aufbau wie die Strompfade einer Schaltung. Auch in diesen findet sich ein ständiger Wechsel zwischen zwei Arten von Elementen, da im Zuge des Stromflusses stets ein Kontakt auf eine Leitung und eine Leitung auf einen Kontakt folgt. Eine weitere Übereinstimmung liegt im zyklischen Aufbau. Entsprechend dem KIRCHHOFFschen Gesetz muß jeder Strompfad zu einem *Stromkreis* geschlossen sein; anderenfalls könnte er niemals Strom führen. Eine einfache Überlegung zeigt, daß auch jede Schaltfolge in sich zu einem *Schaltzyklus* geschlossen sein muß. Bekanntlich läßt sich mit einer endlichen Anzahl von Schaltgeräten und Schalterstellungen nur eine ganz bestimmte und ebenfalls endliche Anzahl von Schaltzuständen herstellen. Betätigen wir nun die Schaltgeräte in einer an sich beliebigen Reihenfolge, so muß diese Schaltfolge, soferne sie nur lange genug ist, wiederum zu einem bereits früher bestandenen Zustand zurückführen. Die Schaltfolgen zeigen also grundsätzlich den gleichen Aufbau wie die Strompfade der elektrischen Schaltungen; Schaltaufgaben, die auf gewünschten Schaltvorgängen beruhen, lassen sich daher im wesentlichen mit den gleichen Hilfsmitteln und Verfahren lösen, wie solche, die sich auf Schaltzustände und Schaltverbindungen beziehen.

Wie aus dem Vorhergehenden ersichtlich ist, läßt sich die Wirkungsweise der elektrischen Schaltungen auf zwei verschiedene Arten beschreiben. Sie kann entweder gegeben sein durch die leitenden Verbindungen, die bei den einzelnen Schaltzuständen bestehen sollen, oder aber durch das gewünschte Spiel der Schaltvorgänge. Durch die Angabe sämtlicher Schaltzyklen sind auch sämtliche Schaltzustände der Steuergeräte eindeutig bestimmt. Hingegen machen die Schaltzyklen keine Aussage über allfällig notwendige Hilfsschaltgeräte. Bei Dauerschaltungen, die keine solchen zusätzlichen Geräte enthalten, ist es daher völlig gleichgültig, auf welche der beiden Arten die Wirkungsweise angegeben wird. Bei Anreizschaltungen ist dies jedoch nicht der Fall. Die Gesamtheit der möglichen Schaltzyklen bestimmt nur jenen Teil der Wirkungsweise, die sich auf die Steuergeräte bezieht, also die *Steuerwirkung* oder *Steuerfunktion*. Für die *Gesamtwirkung*, die auch das Verhalten der Hilfsgeräte umfaßt, bestehen dann in der Regel noch mehrere Freiheitsgrade.

4. Die technische und mathematische Darstellung der elektrischen Schaltungen

Die technische Praxis kennt als Hilfsmittel zur Darstellung von Schaltungen ausschließlich den *Schaltplan*. Dieser soll eine sinnfällige Abbildung der Schaltung sein und ist daher so aufgebaut, daß sich die Strompfade möglichst leicht verfolgen lassen. Die einzelnen Bestandteile der Schaltung, wie Leiter, Kontakte, Schaltgeräte und so weiter, werden durch *Schaltzeichen* dargestellt, die in Anlehnung an übliche Bauformen gewählt und zum größten Teil durch Normung festgelegt wurden[1]. Den verschiedenen Verwendungszwecken entsprechend, entwickelte die technische Praxis zwei verschiedene Ausführungsformen, die in Abb. 20

---

[1] Vgl. Normen DIN 40 710—719.

wiedergegeben sind. Der *Wirkschaltplan* (Abb. 20 *a*) ist eine vollständige
Wiedergabe der Schaltung, die sowohl die elektrischen als auch die
mechanischen Zusammenhänge erkennen läßt. Er enthält also nicht nur
die Schaltzeichen für die elektrisch wirksamen Bestandteile, sondern zeigt
auch die Zusammenfassung der Kontakte und Antriebe vermittels
isolierender Gestänge. Ein Wirkschaltplan ist daher ohne zusätzliche Bezeichnungen oder Erläuterungen verständlich. Er dient in erster Linie
dazu, eine bereits entwickelte oder ausgeführte Schaltung zeichnerisch

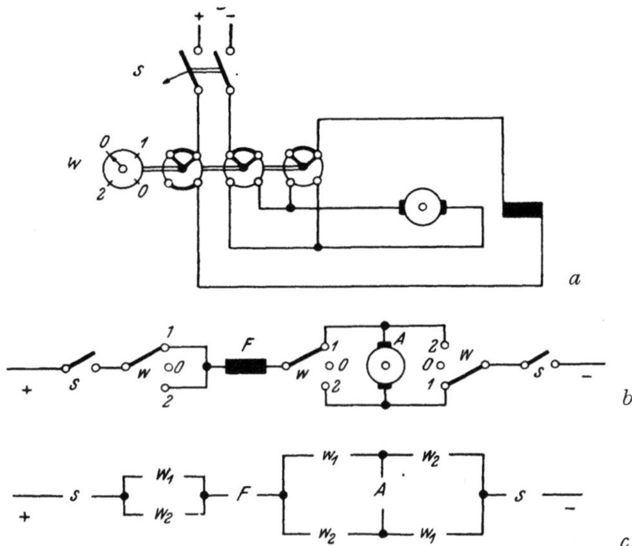

Abb. 20. Technische Darstellung von Schaltungsformen. *a* Beispiel eines Wirk-
schaltplanes, *b* Zugehöriger Stromlaufplan mit Schaltzeichen, *c* Der gleiche Stromlaufplan ohne Schaltzeichen

festzuhalten. Als Hilfsmittel zum Entwurf neuer Schaltungen ist diese
Darstellung wenig geeignet. Die örtliche Zusammenfassung der Kontakte zu Schaltgeräten führt meistens zu Kreuzungen der Verbindungsleitungen, die elektrisch nicht begründet sind; hinsichtlich des Stromverlaufes ist der Wirkschaltplan daher wenig übersichtlich und läßt allfällige
Möglichkeiten zur Vereinfachung kaum erkennen. Die technische Praxis
schuf daher eine zweite Ausführungsform von Schaltplänen, bei der auf
die Wiedergabe der mechanischen Zusammenhänge verzichtet wurde.
Diese Darstellung führt den Namen *Stromlaufplan* (Abb. 20 *b*). Der Entfall der mechanischen Verbindungen macht es möglich, die Strompfade
der Schaltung sinnfällig anzuordnen, so daß sie leicht verfolgt werden
können. Der Stromlaufplan eignet sich daher wesentlich besser zur Entwicklung und Umformung von Schaltungen. Zu beachten ist jedoch, daß
diese zeichnerische Darstellung noch einer Ergänzung durch Bezugszeichen bedarf, damit die Zusammengehörigkeit der Kontakte und somit

die Wirkungsweise der Schaltung erkennbar ist; ein Stromlaufplan ohne Beschriftung ist nicht eindeutig und daher unverständlich. Mit der Einführung dieser Darstellung wurde also der Weg der rein zeichnerischen Wiedergabe zum Teil verlassen und durch eine schriftliche Symbolik ersetzt; nur die elektrischen Verbindungen erscheinen als Plan, die mechanischen hingegen als ein System von Buchstaben oder Ziffern. Dieses muß für jeden Kontakt angeben, zu welchem Schaltgerät er gehört und in welcher Stellung dieses Gerätes er geschlossen ist. Der in Abb. 20 b gezeigte Stromlaufplan enthält kleine Buchstaben zur Kennzeichnung der Schalter und Ziffern als Stellungsbezeichnung. Der Schutzschalter des Motors ist mit s und der Wendeschalter mit w bezeichnet; dem Anker und der Feldwicklung des Motors sind die Großbuchstaben A und F zugeordnet. Die Bezeichnung der Verbraucher hätte bei diesem Beispiel unterbleiben können; sie ist nur dann erforderlich, wenn gleiche Schaltzeichen mehrmals vorkommen und verwechselt werden können.

Die in Abb. 20 b gezeigte Ausführung eines Stromlaufplanes entstand in der Praxis; sie ist jedoch nicht ganz folgerichtig. Wenn man zur Darstellung der mechanischen Zusammenhänge ein System von Bezugszeichen aus Ziffern und Buchstaben verwendet, sind zusätzliche Schaltzeichen für die Kontakte vollständig entbehrlich. Der Plan ist genau so eindeutig und verständlich, wenn er ausschließlich aus Bezugszeichen und verbindenden Leitungen besteht. Diese Ausführung des Stromlaufplanes, die auf EDLER und LISCHKE zurückgeht, ist in Abb. 20 c wiedergegeben. Die Abbildung zeigt auch, daß sie nicht nur eine Ersparnis von Zeichenarbeit mit sich bringt, sondern auch die Form der Schaltung noch schärfer hervortreten läßt.

Wirkschaltplan und Stromlaufplan sind rein technische Darstellungen. Um Schaltungsformen auf exaktem Wege untersuchen und entwickeln zu können, ist jedoch noch eine mathematische Darstellung erforderlich. Diese ergibt sich auf Grund physikalischer und geometrischer Überlegungen. Wie bereits erläutert wurde, haben wir unter einem Kontakt eine Strecke mit variablem Leitwert (Schaltstrecke) zu verstehen. Die Leitungen sind hingegen Gebilde, deren Punkte stets gleiches Potential aufweisen; eine Leitung ist also in physikalischer Deutung ein *Potentialpunkt*. Eine Darstellung, die diesen physikalischen Gegebenheiten Rechnung trägt, muß also die Kontakte als Strecken und die Leitungen als Punkte zeigen. Stromquellen und Verbraucher erscheinen ebenfalls als Strecken, weil sie aus vielen Punkten verschiedenen Potentials bestehen. Wir können also eine Schaltung mit n Leitungen als n-Eck abbilden, wobei die Eckpunkte den Leitungen und die verbindenden Strecken den Stromquellen, Kontakten und Verbrauchern zugeordnet sind. Die Mathematik nennt ein solches aus Punkten und Strecken bestehendes Gebilde einen *Streckenkomplex* (Abb. 21 c). Der Vergleich mit dem in Abb. 21 b gezeigten Stromlaufplan läßt erkennen, daß beide Arten der Darstellung durch eine echte Transformation ineinander übergeführt werden können, da jedem Bestandteil des Stromlaufplanes ein wohldefiniertes Element des Streckenkomplexes entspricht und umgekehrt.

Der Vorteil dieser Transformation liegt darin, daß sie den Weg zu einer mathematischen Behandlung aller Aufgaben, die sich auf Schaltungsformen beziehen, eröffnet. In Streckenkomplexen bestehen bestimmte gesetzmäßige Zusammenhänge zwischen Punkten und Strecken, deren Untersuchung Gegenstand eines Sonderzweiges der Geometrie ist. Die Mathematik bezeichnet jenen Teil der allgemeinen Geometrie, der sich ausschließlich mit der gegenseitigen Lage und Anordnung geometrischer Gebilde ohne Rücksicht auf deren Größe beschäftigt, als *kombinatorische Topologie*; die Wiedergabe durch einen Streckenkomplex ist eine topologische Darstellung der Schaltungsform. Hierdurch ergibt sich unmittel-

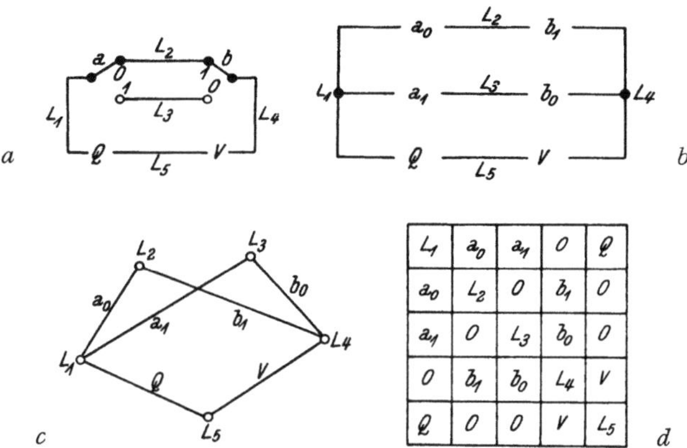

Abb. 21. Mathematische Darstellung von Schaltungsformen. *a* Wechselschaltung im Stromlaufplan mit Schaltzeichen für die Kontakte, *b* Derselbe Stromlaufplan ohne Schaltzeichen, *c* Darstellung als Streckenkomplex, *d* Darstellung als Matrix

bar die Möglichkeit zu einer rein rechnerischen Behandlung. Jeder Streckenkomplex mit n Punkten läßt sich auch als quadratische n-reihige *Matrix* anschreiben, wie Abb. 21 *d* zeigt. In der Hauptdiagonale der Matrix stehen die Rechenzeichen (Operanden) der Punkte, in den übrigen Feldern diejenigen der verbindenden Strecken. Die Matrix ist zur Hauptdiagonale symmetrisch, so daß alle Streckenoperanden doppelt erscheinen; dies bedeutet, daß jeder Strecke zwei Richtungen für den Stromfluß zukommen. Jeder Streckenoperand findet sich in jenem Feld der Matrix, in dem sich Zeile und Spalte der anschließenden Punkte kreuzen. Beispielsweise liegt die Schaltstrecke a_1 in Abb. 21 *c* zwischen den Leitungspunkten L_1 und L_3; in der Matrix (Abb. 21 *d*) steht daher in den beiden Kreuzungsfeldern von Zeile und Spalte L_1 mit Zeile und Spalte L_3 ebenfalls der Operand a_1. Die Matrix ist nichts anderes als ein Stromlaufplan, bei dem die verbindenden Linien durch eine tabellarische Zuordnung ersetzt sind. Im Gegensatz zu allen planartigen Darstellungen erlaubt eine Matrix die Durchführung mathematischer Operationen; Matrizen lassen sich bekanntlich durch Addition und Subtraktion zusammensetzen und zerlegen, sowie auch multiplizieren und potenzieren; sie gestatten also eine exakte rechnerische Behandlung der Formen elektrischer Schaltungen.

Abgesehen von den Wirkschaltplänen verlangen alle technischen und mathematischen Darstellungen elektrischer Schaltungen ein Bezeichnungssystem für die gegenseitige Zuordnung von Kontakten und Schalterstellungen. Es ist von großer Wichtigkeit, daß dieses System sinnfällig, eindeutig und universell ist; diese Eigenschaften kann es nur haben, wenn es den physikalischen und mathematischen Zusammenhängen Rechnung trägt. Auf dieser Grundlage wurde ein Bezeichnungssystem entwickelt, das jeder praktischen Erprobung standgehalten hat. Es geht davon aus, daß für einen Kontakt nur maßgeblich ist, in welcher Schalterstellung er geschlossen ist. Kontakte, die zum gleichen Schaltgerät und zur gleichen Stellung gehören, dürfen und müssen gleich bezeichnet werden, da sie vollständig gleichwertig sind und miteinander beliebig vertauscht werden können. Die Schaltgeräte sind stets so gebaut, daß auf jede ihrer Stellungen eine ganz bestimmte andere Stellung folgt und somit eine natürliche Reihenfolge besteht; es ist daher naheliegend, die Schalterstellungen durch aufeinanderfolgende Zahlen zu kennzeichnen. Es ist zweckmäßig, die Bezifferung der Schalterstellungen stets mit Null zu beginnen, auch wenn das betreffende Schaltgerät keine bevorzugte Nullstellung (z. B. Ruhestellung) besitzt. Zum Unterschied von dieser Stellungsbezeichnung erfolgt die Bezeichnung der Schaltgeräte mit Buchstaben. Es wurden hierfür kleine lateinische Buchstaben gewählt; sie nehmen weniger Platz ein und sind handlicher als Großbuchstaben. Letztere dienen zur Kennzeichnung von Leitungen, Stromverbrauchern und Stromquellen. Es bedeutet somit ein kleiner Buchstabe, daß es sich um ein Element mit variablem Widerstand handelt, während ein Großbuchstabe stets einen Bestandteil mit gleichbleibendem Leitwert ausdrückt. Die *Stellungsnummern* werden den Buchstaben der Schaltgeräte als Zeiger (Indizes) beigefügt. Zu beachten ist, daß einem solchen zusammengesetzten Rechenzeichen eine im Wesen der Kontaktschaltungen begründete Doppelbedeutung zukommt. Man hat beispielsweise unter a_1 sowohl die Stellung 1 des Schalters a, als auch sämtliche in dieser Stellung geschlossenen Kontakte zu verstehen. Bei Anreizschaltungen ist es ferner zweckmäßig, zwischen Steuergeräten und Hilfsgeräten zu unterscheiden; letztere werden dann durch kleine griechische Buchstaben dargestellt. Beispielsweise heißt α_6 die Stellung 6 und jeder in dieser Stellung geschlossene Kontakt eines elektromagnetischen Drehwählers α.

Beispiele dieses Bezeichnungssystems zeigt Abb. 22. Ein Umschalter a mit drei Stellungen, wie er in Abb. 22 a gezeichnet ist, enthalte drei Kontakte, nämlich die Strecken AB, AC und AD. Entsprechend den drei Schalterstellungen heißen diese Strecken a_0, a_1 und a_2. Abb. 22 b zeigt einen Ausschalter b mit zwei äquivalenten Kontakten; da sie in der Stellung b_1 geschlossen sind, werden beide b_1 genannt.

Festzuhalten ist, daß sich diese Kontaktbezeichnungen stets auf jene Schalterstellungen beziehen, in denen die betreffenden Kontakte geschlossen sind. Manchmal ist es notwendig oder zweckmäßig, anzugeben, daß ein Kontakt in einer bestimmten Stellung offen sein soll. Hierfür mußte ein eigenes Zeichen eingeführt werden; die Mathematik verwendet zur Kennzeichnung einer Umkehrung einen über das betreffende Symbol gesetzten Querstrich, der *Umkehrungszeichen* (Inversionsoperator) genannt wird. Es bedeutet somit $\overline{a}_1$ (sprich *a* eins quer) einen Kontakt, der in allen Stellungen des Schaltgerätes mit Ausnahme der Stellung 1 geschlossen ist. Streng genommen bezieht sich dieses zusammengesetzte Zeichen nicht auf einen Einzelkontakt, sondern auf eine Schar paralleler Schaltstrecken, die durch ein verlängertes Schaltstück (Schleifstück) gebildet wird. Ein Beispiel hiefür zeigt Abb. 22 c. Der Schalter *c* mit drei

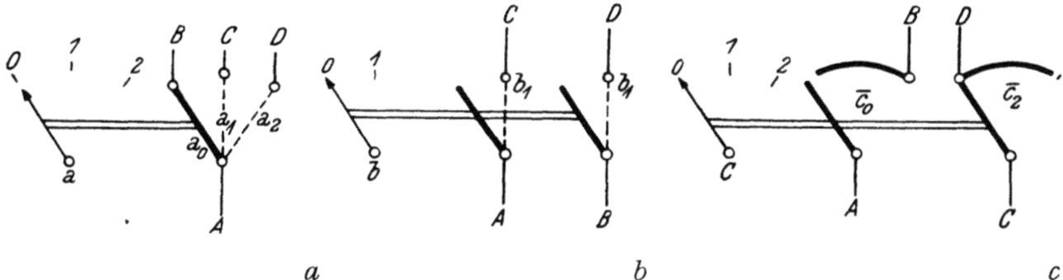

Abb. 22. Bezeichnung von Kontakten. *a* Verschiedene Kontakte eines Schalters, *b* Gleiche Kontakte eines Schalters, *c* Kontakte mit verlängerten Schaltstücken (Schleifkontakte)

Stellungen hält eine Verbindung AB nur in der Stellung O offen; die Strecke AB ist daher mit $\overline{c}_0$ bezeichnet. Eine zweite Verbindung CD ist nur in Stellung 2 unterbrochen und heißt daher $\overline{c}_2$. Verbindet man die anschließenden Leitungen AC, so wirkt das Gerät als unterbrechungsloser Umschalter. Das Umkehrungszeichen gestattet also auch, einander übergreifende Kontakte von solchen, die mit Unterbrechung umschalten, zu unterscheiden. Diese Kennzeichnung ist besonders für die Haltestrompfade von Anreizschaltungen wichtig. Bei Dauerschaltungen kommt es im allgemeinen nicht darauf an, wie sich die Kontakte zwischen zwei Betriebsstellungen verhalten. Für diesen Fall ist eine Vereinfachung zulässig; wenn es sich um Geräte mit nur zwei Stellungen handelt, kann man dann näherungsweise statt $\overline{a}_1$ auch a_0 und statt $\overline{a}_0$ einfach a_1 setzen.

Je nach Art der Schaltung läßt sich das beschriebene Bezeichnungssystem fallweise vereinfachen. Enthält eine Schaltung durchwegs Schaltgeräte, deren Kontakte in einer einzigen Stellung geschlossen sind, kann man auf das Anschreiben der Zeiger (Indizes) verzichten; der Kennbuchstabe eines Gerätes bedeutet dann gleichzeitig die Schließstellung der Kontakte. Die gegenteilige Stellung wird dann mit Hilfe des Umkehrungszeichens angegeben. Bei einem Schalter *a* mit nur zwei Stellungen läßt sich also statt a_0 und a_1 vereinfacht $\overline{a}$ und a setzen. Bei der Ausmittlung der inneren Verbindungen eines Schaltgerätes ist es überflüssig, den Kennbuchstaben des Gerätes anzuschreiben; es genügen dann zur Kontaktbezeichnung die Stellungsnummern allein.

Die Großbuchstaben, die den Bestandteilen mit festem Leitwert zugeordnet wurden, können ebenfalls mit Zeigern zur näheren Kennzeichnung versehen werden. Diese können sowohl Ziffern als auch Buchstaben sein. Beispielsweise ist es zweckmäßig, ungepolte Leitungen einheitlich mit L zu benennen und in beliebiger Reihenfolge zu numerieren; sinngemäß heißen sie dann L_1, L_2, L_3 usw. Die Spule S eines Relais a heißt dementsprechend S_a; seine Kontakte sind dann a_0 oder a_1.

In Berechnungen kommen selbstverständlich zu den angeführten Zeichen, die den Bestandteilen der Schaltungen entsprechen, noch weitere Rechenzeichen hinzu, die der allgemeinen Physik und Mathematik entnommen sind; sie haben die dort übliche Bedeutung. Beispielsweise werden elektrische Leitwerte mit Λ, allgemeine Zahlen mit n usw. bezeichnet; der griechische Großbuchstabe Σ dient wie üblich als Summenzeichen.

Alle im vorstehenden erläuterten Darstellungen und Bezeichnungen beschränken sich ausschließlich auf die Formen der elektrischen Schaltungen; die Wirkungsweise geht daraus nur mittelbar hervor; Schaltaufgaben sind aber in der überwiegenden Zahl der Fälle so beschaffen, daß sie sich ausschließlich auf die Funktion beziehen. Es war ein ganz wesentlicher Fortschritt der schaltungstechnischen Theorie, auch für die Wirkungsweise rechnerische und zeichnerische Darstellungen zu finden.

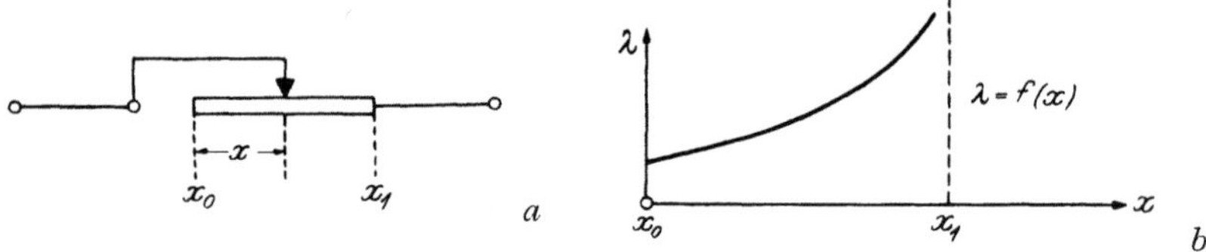

Abb. 23. Leitwerte einer Regelschaltung. *a* Schaltung, *b* Leitwertfunktion

Dadurch ergibt sich die Möglichkeit, in Worten vorgegebene Funktionsbedingungen in übersichtlicher Form festzuhalten und sodann in gegenseitige Beziehung zu bringen. Auch die nachträgliche Überprüfung bereits entwickelter Schaltungen wird dadurch sehr erleichtert. Für die rechnerische Behandlung von Schaltaufgaben ist ein mathematischer Ausdruck für Funktionsbedingungen und Gesamtwirkungen von ausschlaggebender Bedeutung. Es soll nunmehr gezeigt werden, welche Möglichkeiten hierfür bestehen.

Gemäß Definition verstehen wir unter der Funktion oder Wirkungsweise einer Schaltung den Zusammenhang zwischen den Stellungen der Schaltgeräte und dem Leitwert der Schaltung. Wie üblich soll der Leitwert, der dem reziproken Wert des elektrischen Widerstandes gleich ist, mit λ bezeichnet werden. Wir betrachten vorerst den Leitwert einer ganz einfachen Schaltung, die aus einem einzigen veränderlichen Widerstand besteht. Abb. 23 *a* zeigt einen Schiebewiderstand, bei dem der Weg des Schiebers mit x bezeichnet ist; die beiden Endstellungen des Schiebers sind x_0 und x_1. Der Leitwert λ läßt sich dann gemäß Abb. 23 *b* als Funktion von x darstellen. Diese Funktion, die sowohl als Leitwertkurve als auch als Leitwertgleichung abgebildet werden kann, bringt eindeutig zum Ausdruck, wie die Schaltung wirkt. Man ersieht also, daß die Wirkungsweise oder Funktion einer Schaltung auch als mathematische Funktion aufgefaßt werden darf. Dies gilt ganz allgemein für Schaltungen beliebiger Art. Sind mehrere veränderliche Widerstände vorhanden, so gelten für die Leitwertfunktionen das KIRCHHOFFsche und OHMsche Gesetz. Darnach sind Leitwerte paralleler Widerstände zu addieren. Liegen also beispielsweise zwei Schiebewiderstände x und y nebeneinander, so ergibt sich eine Leitwertgleichung folgender Form:

$$\lambda = f_1(x) + f_2(y).$$

Liegen die beiden Widerstände hintereinander, so gilt für den Leitwert der Schaltung in bekannter Weise:

$$\lambda = \frac{f_1(x) \cdot f_2(y)}{f_1(x) + f_2(y)} .$$

Da sich jede Schaltung in einzelne Strompfade auflösen und somit auf Reihen- und Parallelschaltungen zurückführen läßt, besteht die Möglichkeit, die Wirkungsweise jeder wie immer gearteten Schaltung als Leitwertgleichung anzuschreiben. Eine zeichnerische Darstellung durch Leitwertkurven ist jedoch nicht möglich, weil hierzu bei n Widerständen ein $n + 1$ dimensionales Koordinatensystem nötig wäre.

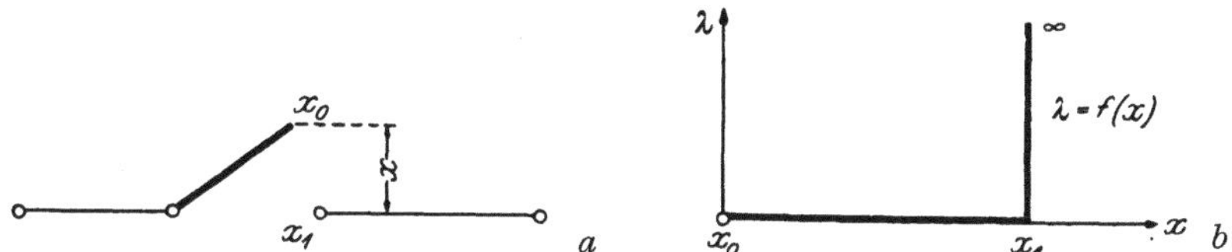

Abb. 24. Leitwerte einer Kontaktschaltung. a Schaltung, b Leitwertfunktion

Die gleiche Betrachtung läßt sich auch auf Kontaktschaltungen anwenden. Abb. 24 a zeigt den einfachsten Fall, der sich auf einen einzelnen Kontakt beschränkt. Der Weg des beweglichen Schaltstückes ist wiederum mit x, die Ausschaltstellung mit x_0 und die Einschaltstellung mit x_1 bezeichnet. Auch hier ist der Leitwert eine Funktion des Schaltweges x, deren Verlauf die Abb. 24 b wiedergibt. Der Unterschied gegenüber einer Regelschaltung nach Abb. 23 liegt nur darin, daß die Leitwertkurve *unstetig* verläuft. Die Gültigkeit der Formeln für Parallel- und Serienschaltung werden hierdurch jedoch nicht berührt.

Wie der Abb. 24 b zu entnehmen ist, zeichnet sich die Einschaltstellung x_1 des betrachteten Kontaktes dadurch aus, daß ihr und nur ihr der Leitwert ∞ zukommt. Im ganzen übrigen Bereich von x ist der Leitwert Null. Daher ist der Verlauf der Funktion $\lambda = f(x)$ überhaupt nicht von Interesse. Aus diesem Grunde ist für Kontaktschaltungen eine ganz wesentliche Vereinfachung möglich. Wir bezeichnen den Kontakt, der in der Stellung x_1 des Schalters x geschlossen ist, ebenfalls mit x_1 und verstehen unter diesem Rechenzeichen unmittelbar den variablen Leitwert des Kontaktes. Ist der Kontakt geschlossen, so gilt $x_1 = \infty$, im anderen Falle $x_1 = 0$. Diese Festsetzung gibt uns die Möglichkeit, Stromlaufpläne in einfachster Weise in Leitwertgleichungen zu transformieren. Ein Beispiel hierfür zeigt Abb. 25. Die Schaltung bestehe aus zwei Kontakten, a_1 und b_1, die gemäß Abb. 25 a parallelgeschaltet sind. Entsprechend der früher angeführten Formel für den Leitwert paralleler Widerstände lautet dann die darstellende Gleichung dieser Schaltung

$$\lambda = a_1 + b_1.$$

Eine Betrachtung der möglichen Schaltzustände zeigt die Richtigkeit dieser mathematischen Darstellung. In Abb. 25 b ist jener Zustand wiedergegeben, in dem beide Kontakte offen stehen. Es ist dann $a_1 = 0$ und $b_1 = 0$, woraus $\lambda = 0 + 0 = 0$ folgt. Die Rechnung zeigt also, daß die Schaltung bei diesem Zustand der Kontakte keinen Strom leitet. Abb. 25 c stellt jenen Zustand dar, der durch $a_1 = \infty$ und $b_1 = 0$ gekennzeichnet ist; der Leitwert der Schaltung ist dann $\lambda = \infty + 0 = \infty$. Die dritte Möglichkeit zeigt Abb. 25 d; es sind beide Kontakte geschlossen und der Leitwert der Schaltung $\lambda = a_1 + b_1 = \infty + \infty$ wird wiederum ∞.

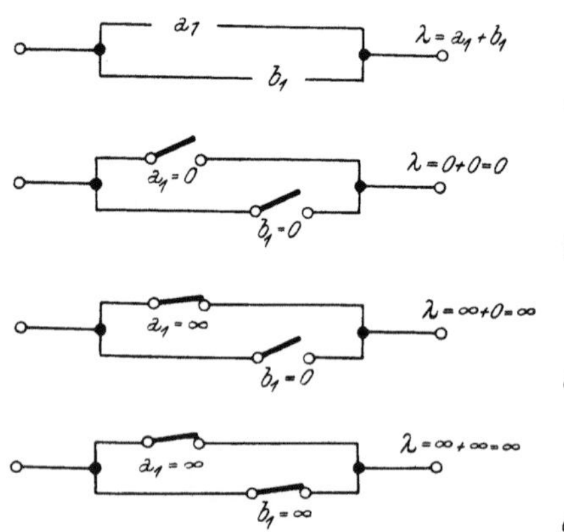

Abb. 25. Schaltzustand und Leitwert. *a* Parallelschaltung zweier Kontakte, *b* Beide Kontakte offen, *c* Ein Kontakt geschlossen, *d* Beide Kontakte geschlossen

Mit der Darstellung elektrischer Schaltungen durch Leitwertgleichungen ist die Möglichkeit zur rechnerischen Behandlung von Schaltaufgaben gegeben. Da diese Gleichungen auf den bekannten elektrischen und mathematischen Gesetzen beruhen, gelten für sie auch sämtliche Regeln der Algebra. Ein Unterschied ist nur insofern vorhanden, als für Kontakte ausschließlich zwei Zahlenwerte vorkommen, nämlich 0 und ∞. Diese Beschränkung des Zahlenbereiches hat aber ein weiteres Rechengesetz zur Folge, das in der allgemeinen Algebra nicht gilt und für die Schalttechnik von entscheidender und ausschlaggebender Bedeutung ist. Zur Ableitung dieses zusätzlichen Rechengesetzes betrachten wir den Fall der Parallelschaltung zweier gleicher Kontakte, das heißt, die Summe gleicher Leitwerte. Da jeder Kontakt nur zwei Leitwerte, 0 oder ∞, haben kann, ergeben sich für den Wert der Summe zwei Möglichkeiten, die in nachstehender Tabelle wiedergegeben sind.

x	$x + x$
0	0
∞	∞

Die Tabelle besagt, daß der Leitwert einer Summe aus zwei beliebigen Elementen x stets den gleichen Wert hat wie x selbst. Das gleiche gilt auch für die Multiplikation.

x	$x \cdot x$
0	0
∞	∞

Für die auf das Wertepaar 0 und ∞ eingeschränkte Algebra besteht also ein zusätzliches Grundgesetz, das sowohl für die Addition als auch für die Multiplikation gilt; es besagt, daß diese und alle daraus folgenden höheren Operationen, sofern sie sich auf gleiche Variable beziehen, keine Änderung des Funktionswertes bewirken. Es bestehen also die Identitäten:

$$x + x + x + \ldots = x,$$
$$x \cdot x \cdot x \cdot \ldots = x.$$

In der Mathematik wird dieses Gesetz *Äquivalenz-Gesetz* oder *Tautologie-Satz* bezeichnet. Die spezielle Algebra, die auf ein Paar einander ausschließender Werte beschränkt ist und diesem Gesetz folgt, ist die *algebraische Logik* oder *Logistik*. Dieser Zweig der Mathematik bildet sonach die Grundlage für die Berechnung elektrischer Kontaktschaltungen[1].

Das Äquivalenz-Gesetz hat weiters zur Folge, daß sämtliche additiven und multiplikativen Verknüpfungen der Werte 0 und ∞ ein eindeutiges Ergebnis haben, was in der allgemeinen Algebra nicht der Fall ist. Aus

---

[1] In der Schaltungstechnik findet die Logistik erstmalig eine technische Anwendung. In der theoretischen Logik wird sie zur Ableitung der Gesetze für Aussagenverknüpfungen benützt. Die Mathematik bedient sich der Logistik, um die elementaren Rechenoperationen auf die Axiome der allgemeinen Logik zurückzuführen. Aus diesem Grund wurden Operationen rein mathematischer Art in der Logistik bisher nicht benützt. Die moderne Wahrscheinlichkeitslehre führt zwar in Form der prozentuellen Wahrscheinlichkeit den Zahlbegriff, aber keinerlei mathematische Operationen ein. Erst die hier gezeigte technische Anwendung macht es möglich und erforderlich, unter Behaltung der logistischen Grundgesetze alle Arten von mathe-

$x + x = x$ folgt unmittelbar $n \cdot x = x$, wobei n eine beliebige Zahl sein kann. Für $n = \infty$ wird dann

$$\infty \cdot x = x.$$

Setzt man $x = 0$, so folgt

$$\infty \cdot 0 = 0.$$

Während also in der uneingeschränkten Mathematik das Produkt $\infty \cdot 0$ einen unbestimmten Wert hat, ist es in der Logistik eindeutig bestimmt und stets 0. Die Logistik beruht also auf den nachstehenden eindeutigen Verknüpfungen der beiden betrachteten Werte:

$$0 + 0 = 0 \qquad\qquad 0 \cdot 0 = 0$$
$$\infty + 0 = \infty \qquad\qquad \infty \cdot 0 = 0$$
$$\infty + \infty = \infty \qquad\qquad \infty \cdot \infty = \infty$$

Unbestimmte Resultate kennt die Logistik erst bei Umkehrung dieser Operationen, nämlich bei der Subtraktion und Division; so sind beispielsweise $0 : 0$ und $\infty - \infty$ entweder 0 oder ∞, also unbestimmt. Das Auftreten mehrerer Lösungen zu einer Schaltaufgabe findet seinen mathematischen Ausdruck oft darin, daß die rechnerische Entwicklung der Schaltung Operationen mit unbestimmten Resultaten enthält.

Leitwertgleichungen der beschriebenen Art zeigen die Funktionen der Schaltungen in Abhängigkeit von den Schaltstellungen der Geräte. Bei Anreizschaltungen kann man daher die Gleichungen in dieser Form nur ansetzen, wenn die Zahl der Geräte bereits bekannt ist. Sehr häufig besteht aber die Schaltaufgabe gerade darin, daß eine vorgeschriebene Wirkung mit möglichst wenigen Geräten erzielt werden soll. Um solche Funktionsbedingungen mathematisch ansetzen zu können, ist es nötig, die den Relais zugeordneten Variablen aus den Gleichungen auszuscheiden. Es handelt sich also darum, die Beziehung zwischen den Leitwerten und den Bewegungen der Schalter, also die Schaltvorgänge darzustellen.

Ein Beispiel hierfür zeigt Abb. 26. Es handle sich um eine an sich bekannte Wechselschaltung mit einer Stromquelle Q, einem Verbraucher V und drei Schaltern a, b, c, mit je zwei Stellungen, 0 und 1; Abb. 26 a zeigt den Stromlaufplan. Wir nehmen vorerst an, daß uns die Schaltung nicht bekannt sei, sondern erst aus den Funktionsbedingungen entwickelt werden soll. Die Schaltaufgabe lautet in Worten: der Verbraucher V soll von drei Stellen a, b, c

---

matischen Operationen auszuführen. Deswegen und auch mit Rücksicht auf die physikalisch-mathematische Vorbildung des Technikers wurden in der vorliegenden Abhandlung für die logistischen Begriffe und Operationen die üblichen mathematischen Formelzeichen und nicht die davon stark abweichenden Symbole der theoretischen Logik verwendet. Vgl. WHITEHEAD and RUSSEL: Principia Mathematica. Cambridge 1910; HILBERT und ACKERMANN: Grundzüge der theoretischen Logik. Berlin 1928; CARNAP, R.: Abriß der Logistik. Wien 1919; REICHENBERG, H.: Wahrscheinlichkeitslehre. Leiden 1935.

jederzeit aus- und einschaltbar sein. Daraus können wir vorerst nur entnehmen, daß mindestens drei Schaltgeräte a, b, c vorhanden sein müssen. Wir bezeichnen nun je eine Stellung dieser Geräte mit 0 und setzen willkürlich fest, daß bei einem Zustand a_0, b_0, c_0 der Verbraucher V ausgeschaltet, der Leitwert der Schaltung also 0 sein soll. Diesen Schaltzustand bilden wir nun, wie in Abb. 26 b dargestellt ist, durch einen Kreis ab, in den wir den Leitwert 0 eintragen. Da drei Schalter vorhanden sind, bestehen nur drei voneinander unabhängige Möglichkeiten, von diesem Zustand in einen benachbarten überzugehen. Es ist dem Bedienenden vollkommen freigestellt, ob er den Schalter a, b oder c bedienen will.

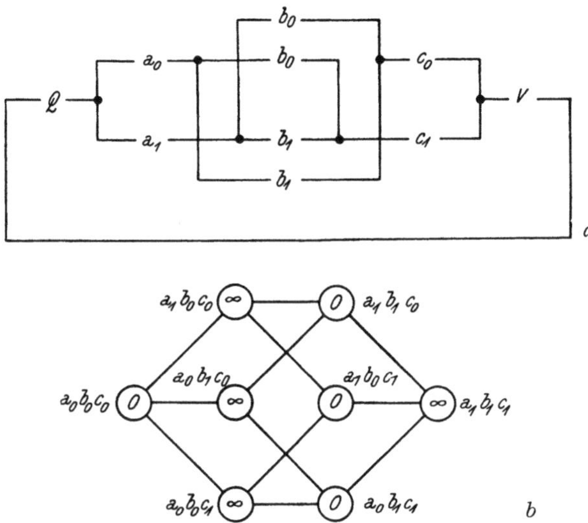

Abb. 26. Zeichnerische Darstellung der Wirkungsweise einer Schaltung. *a* Stromlaufplan einer Wechselschaltung mit drei zweistelligen Schaltern, *b* Schaltfolgeplan der Wechselschaltung

Wir können also an den Kreis, der den betrachteten Zustand abbildet, drei Linien anschließen, die zu je einem benachbarten Zustand führen. Die Nachbarzustände stellen wir wieder als Kreise dar; da jeder Schaltschritt einen Zustandwechsel des Verbrauchers bewirken soll, tragen wir in die drei neuen Kreise den Leitwert ∞ ein. Auch die zugehörigen Kombinationen der Schalterstellungen können wir anschreiben, da wir vom Zustand a_0, b_0, c_0 ausgegangen sind und mit jedem Schaltschritt nur einen Schalter verstellen, heißen die neuen Stellungskombinationen a_1, b_0, c_0, a_0, b_1, c_0 und a_0, b_0, c_1. Wie wir diese Stellungskombinationen den einzelnen Kreisen zuordnen, stellt uns die Aufgabe frei. Von den mit ∞ bezeichneten Kreisen müssen wiederum je drei Linien ausgehen, so daß wir unsere Figur schrittweise vervollständigen können. Da insgesamt drei Schalter mit je zwei Zuständen vorhanden sind, ergeben sich

$2 \cdot 2 \cdot 2 = 2^3 = 8$ Zustände; unsere Abbildung ist fertig, wenn alle acht Punkte und die zugehörigen Verbindungen gezeichnet sind. Die so entstandene Figur ist ein eindeutiges Abbild der Steuerwirkung, wie sie in der Aufgabe textlich beschrieben wurde. Sie zeigt die Aufeinanderfolge von leitenden und nichtleitenden Zuständen und die verbindenden Schaltschritte; es sind also sämtliche Schaltfolgen, die sich mit der Einrichtung durchlaufen lassen, erkennbar, so daß wir unsere Figur als *Schaltfolgeplan*[1] bezeichnen können. Betrachtet man die den Schaltzuständen entsprechenden Kreise als Punkte der Figur, so besteht diese nur aus Punkten und Strecken. Sie ist demnach ein Streckenkomplex im Sinne der kombinatorischen Topologie. Da von jedem Punkt gleich viele, und zwar drei Strecken ausgehen, ist der Plan ein sogenannter *regelmäßiger* oder *regulärer Komplex* dritter Ordnung. Die Regelmäßigkeit zeigt sich auch darin, daß sämtliche Schaltschritte, die zum gleichen Schalter gehören, als parallele Strecken erscheinen. Es läßt sich zeigen, daß die Schaltfolgepläne aller Dauerstromschaltungen, deren Steuergeräte keine mechanisch gesperrten Stellungen enthalten, durchwegs reguläre Komplexe sind. Bei Schaltern mit zwei Stellungen, bei denen die Drehrichtung keine Bedeutung hat, ist die Ordnungszahl stets gleich der Anzahl der Schaltgeräte. Haben die Steuergeräte mehr als zwei Stellungen, so besteht ein Unterschied, ob man auf die nächst höhere oder nächst tiefere Stellung übergeht. Es gibt dann für jedes Steuergerät zwei Arten der Betätigung, nämlich die Aufwärtsschaltung und die Abwärtsschaltung. Bei n Geräten gehören dann zu jedem Punkt des Komplexes $2n$ benachbarte Punkte; die Ordnungszahl des Komplexes ist also gleich der doppelten Anzahl der Schaltgeräte. Hat man auf solche Weise die textlich vorgegebene Schaltaufgabe in den Schaltfolgeplan übersetzt, so läßt sich aus diesem durch Anwendung der Rechengesetze, die für die kombinatorische Topologie und die Logistik gelten, die Form der Schaltung in sämtlichen Variationen ausmitteln.

Das hauptsächlichste Anwendungsgebiet für Schaltfolgepläne ist die Entwicklung von Anreizschaltungen. Diese sind durch irreversible Schaltvorgänge gekennzeichnet, was in den Schaltfolgeplänen deutlich zum Ausdruck kommt. Jede mechanische oder elektrische Haltevorrichtung eines Relais bewirkt eine Störung der Regelmäßigkeit des Streckenkomplexes, wie es Abb. 27 veranschaulicht. Der einfachste Fall ergibt sich bei einer Schaltung mit nur einer Stromquelle Q, einem Verbraucher V und zwei Steuerkontakten a_1 und b_0. Abb. 27 a zeigt die Schaltung ohne Relais, das heißt als Dauerstromschaltung. Im Zustand $a_0 b_0$ ist der Leitwert 0; durch Betätigung von a wird der Verbraucher eingeschaltet und bei darauffolgendem Umlegen des Schalters b wieder

---

[1] Der Schaltfolgeplan ist wesensverschieden von dem in der Fernmeldetechnik benützten *Schaltfolgediagramm*. Dieses ist kein Schaltplan, sondern ein Zeitwegdiagramm zur Darstellung der Schaltzeiten elektromagnetischer Relais. Zur Wiedergabe und Ausmittlung schaltungstechnischer Verknüpfungen ist dieses Diagramm weder gedacht, noch geeignet.

ausgeschaltet. Beide Vorgänge sind umkehrbar; der in Abb. 27 b gezeichnete zugehörige Schaltfolgeplan ist daher ein regulärer Komplex zweiter Ordnung mit vier Punkten. Jede Strecke des Komplexes kann in beiden Richtungen durchlaufen werden und ist somit ein Streckenpaar. Nun vergrößern wir die Schaltung um ein Relais, und zwar um das Schütz a, wie es in Abb. 27 c dargestellt ist. Die Schützenspule S_a liegt parallel zum Verbraucher V, der Selbsthaltekontakt a_1 parallel zum Steuerkontakt a_1. Geht man nun wiederum vom Zustand $a_0 b_0$ aus, so läßt sich auch in diesem Falle der Verbraucher mit dem Schalter a einschalten. Gleichzeitig spricht aber das Schütz a an und überbrückt den Kontakt a_1, so daß die Rückstellung des Schalters a wirkungslos ist; der erste Schaltvorgang ist also nicht umkehrbar. Im Schaltfolgeplan darf also die

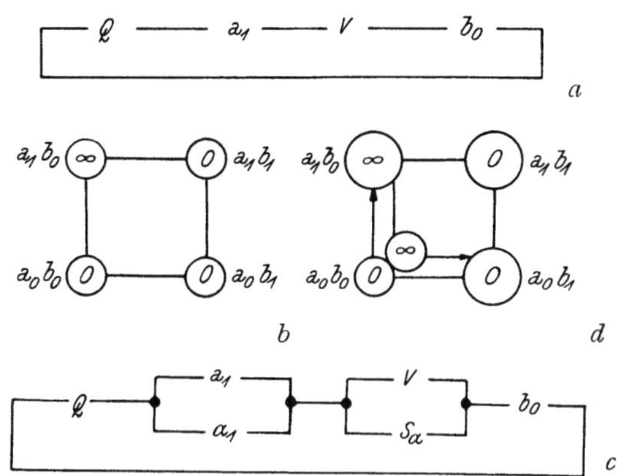

Abb. 27. Veränderung der Steuerfunktion durch ein Hilfsschaltgerät. a Dauerschaltung mit zwei Steuerkontakten, b Schaltfolgeplan der Dauerschaltung, c Anreizschaltung mit zwei Steuerkontakten, d Schaltfolgeplan der Anreizschaltung

zugehörige Strecke zwischen den Zuständen $a_0 b_0$ und $a_1 b_0$ nur in einer Richtung durchlaufen werden, was durch einen Richtungspfeil angedeutet ist. Befindet sich die Schaltung im Zustand $a_1 b_0$, in dem das Schütz a erregt ist, so ist jede weitere Betätigung des Schalters a ohne Wirkung auf den Leitwert der Schaltung. Alle Schaltschritte von a bewirken nur den Wechsel zwischen zwei Zuständen $a_1 b_0$ und $a_0 b_0$, die beide dem Leitwert ∞ der Schaltung entsprechen. Im Schaltfolgeplan findet sich daher zwischen diesen beiden Zuständen neben der gerichteten Strecke des ersten Schaltschrittes noch ein Streckenpaar, das ebenfalls dem Schalter a zukommt und einen umkehrbaren Schaltvorgang darstellt. Der Schaltzustand $a_0 b_0$ tritt hierbei zweimal auf und zwar einmal in Verbindung mit dem Leitwert 0 und das andere Mal mit dem Leitwert ∞. Die Schaltung zeigt also fünf Schaltzustände, weil der Zustand $a_0 b_0$ in zwei solche mit verschiedenen Leitwerten gespalten ist. Der in Abb. 28 b vorhandene Linienzug, der aus vier Streckenpaaren besteht und zu einem Zyklus geschlossen ist, wurde also durch Einfügung des Hilfsrelais gestört. Es sind wohl wiederum vier Streckenpaare vorhanden, doch bilden diese wegen der Aufspaltung des Punktes $a_0 b_0$ keinen Zyklus mehr; dieser schließt sich erst über zwei neue gerichtete Strecken. Es läßt sich nachweisen, daß jede solche Spaltung eines Punktes im Streckenkomplex, die stets mit dem Auftreten eines neuen Paares gerichteter Strecken verbunden ist, eine elektrische oder mechanische Halteeinrichtung in der Schaltung erforderlich macht. Diese Einrichtung kann mechanischer oder elektrischer Art sein, also durch Schütze oder Fernschalter verwirklicht werden. Aus den Schalterstellungen der Zustände, zwischen denen die gerichteten Strecken liegen, läßt sich fallweise ermitteln, welche Strompfade zum Ein- und Ausschalten der Relais erforderlich sind. Aus dem Schaltfolgeplan geht also unmittelbar hervor, wie viele Relais die Schaltung haben muß und wie diese Geräte gesteuert werden können.

Mit dem Schaltfolgeplan ist der Kreis der möglichen Darstellungen elektrischer Schaltungen geschlossen. Er beginnt bei der technischen Wiedergabe durch den Wirkschaltplan und Stromlaufplan. Letzterer verlangt bereits eine symbolische Bezeichnung für Schaltgeräte und Schaltstellungen und bildet somit den Übergang zur algebraischen Darstellung. Er läßt sich auch in ein geometrisches Gebilde ähnlich einem Streckenkomplex transformieren und gestattet damit die Anwendung der Gesetze der kombinatorischen Topologie. In grundsätzlich gleicher Weise wie Schaltungsformen lassen sich auch Schaltvorgänge mathematisch und zeichnerisch darstellen.

II. Die Kombinatorik der Schaltzustände (Strompfade)

1. Die Leitwerte von Kontaktkombinationen

Der Leitwert jeder Schaltung ist bestimmt durch Zahl und Aufbau der Strompfade. Diese sind zueinander parallel und enthalten Kontakte in Serie. Kennt man die resultierenden Leitwerte einer Parallel- und Serienschaltung, so kann man die Leitwertfunktion jeder gewünschten Kontaktverknüpfung ermitteln. Für beliebige Leitwerte x und y gelten bekanntlich die beiden physikalischen Relationen:

$$\Lambda_{\text{parallel}} = x + y,$$

$$\Lambda_{\text{Serie}} = \frac{x \cdot y}{x + y}.$$

Bei Kontakten kommen den Variablen nur die Alternativwerte 0 und ∞ zu. Unter Verwendung der logistischen Äquivalenzregeln $x + x = x$ und $x \cdot x = x$ kann nun in der Formel für die Serienschaltung das Produkt $x\,y$ um einen Faktor $(x + y)$ erweitert und dieser gegen den Nenner gekürzt werden:

$$\Lambda_{\text{Serie}} =$$

$$= \frac{x \cdot y}{x + y} = \frac{x \cdot y + x \cdot y}{x + y} = \frac{x \cdot x \cdot y + x \cdot y \cdot y}{x + y} = \frac{(x + y) \cdot x \cdot y}{x + y} = x\,y.$$

Wir erhalten demnach den Leitwert einer *Parallelschaltung* als Summe

$$\Lambda_{\text{parallel}} = x + y$$

und denjenigen einer *Serienschaltung* als Produkt

$$\Lambda_{\text{Serie}} = x \cdot y$$

der beiden Variablen.

Wie haben wir nun dieses Ergebnis inhaltlich zu deuten? Eine Parallelschaltung hat nur dann den Leitwert 0, wenn alle ihre Elemente diesen Zustand einnehmen. Das gleiche gilt aber auch für eine mathematische Summe. Die Serienschaltung verhält sich genau umgekehrt; ihr Leitwert ist 0, wenn ein einziges Element seine Leitfähigkeit verliert. Auch das mathematische Produkt wird Null, wenn ein einziger Faktor diesen Wert annimmt. Die Analogie zwischen den genannten mathematischen und schaltungstechnischen Operationen führt auch zu einer sinnfälligen Deutung der beiden Äquivalenzregeln als Kontaktverdoppelung ohne Änderung der Wirkungsweise (Abb. 28).

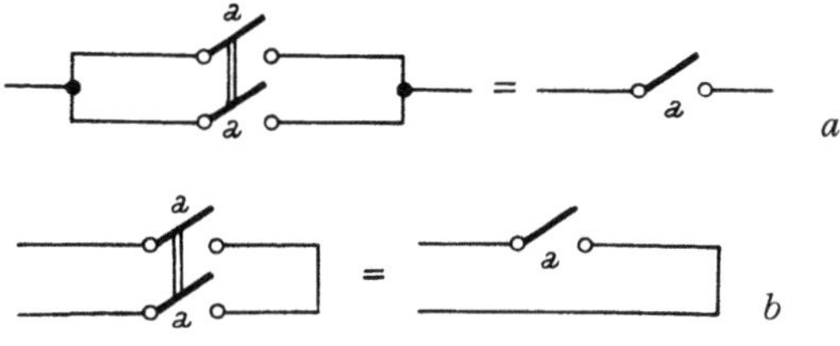

Abb. 28. Verknüpfungen gleicher Kontakte.
a Parallelschaltung: $a + a = a$, *b* Serienschaltung: $a \cdot a = a$

Für die praktische Anwendung der Logistik ist es wichtig, daß sie aus der allgemeinen Algebra durch Verjüngerung, das heißt Einschränkung auf bestimmte Werte, abgeleitet werden kann. Es behalten daher alle algebraischen Rechenregeln ihre Gültigkeit. So bestehen beispielsweise für Kontaktverknüpfungen ebenfalls die Gesetze der Kommutation (Vertauschung), Assoziation (Zusammenfassung) und Distribution (Verteilung) gemäß Abb. 29. Das Gesetz der Distri-

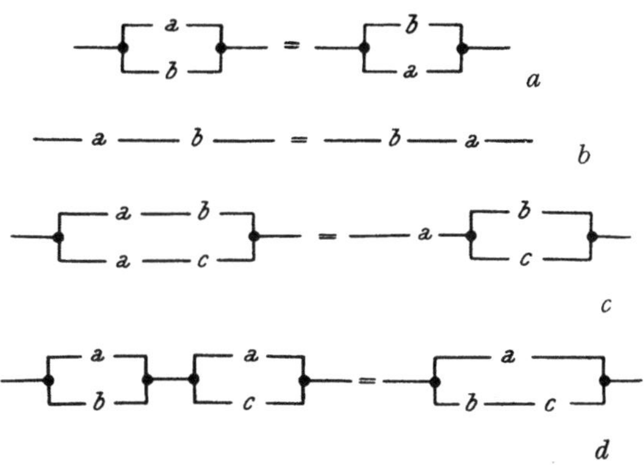

Abb. 29. Verknüpfungen verschiedener Kontakte.
a Parallelschaltung: $a + b = b + a$, *b* Serienschaltung: $a\,b = b\,a$, *c* Gemeinsamer Serienkontakt: $a\,b + a\,c = a(b + c)$, *d* Gemeinsamer Parallelkontakt: $(a + b)\,(a + c) = a + b\,c$

bution gilt in der Zahlenrechnung nur für die Summe von Produkten. Die Rechengesetze der Logistik erlauben es aber, diese Operation auch für den Fall des Produktes von Summen anzuwenden. Die in Abb. 29 *d* gezeigte Umformung ergibt sich aus folgender Rechenoperation, bei der nach dem Äquivalenzgesetz $a = a + a$ und $a = a \cdot \infty$ gesetzt wird:

$$(a + b) \cdot (a + c) = a + a \cdot b + a \cdot c + b \cdot c$$
$$= (a + a \cdot b) + (a + a \cdot c) + b \cdot c$$
$$= a \cdot (\infty + b) + a \cdot (\infty + c) + b \cdot c$$
$$= a \cdot \infty \qquad + a \cdot \infty \qquad + b \cdot c$$
$$= a + b \cdot c.$$

Die vollständige Analogie zwischen den Rechengesetzen der Logistik und den Verknüpfungsregeln für Kontakte erlaubt es, die Funktion jeder beliebigen Kontaktschaltung durch eine *Leitwertgleichung* darzustellen. Die Gültigkeit des distributiven Gesetzes (Abb. 29 c, d) hat ferner zur Folge, daß sich auch Formen von Schaltungen auf diese Weise wiedergeben lassen; dies gilt jedoch nur für unvermaschte Formen. Wenn jeder

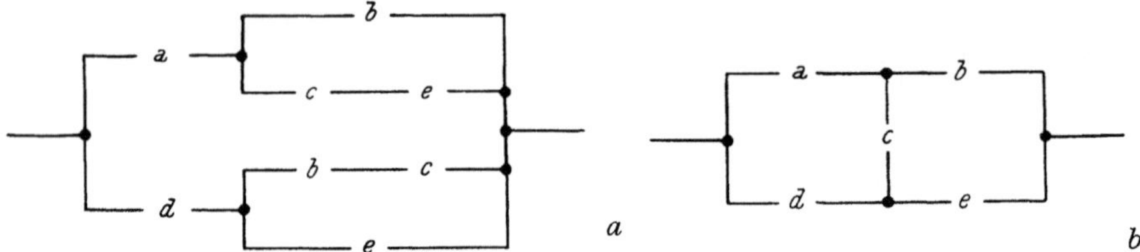

Abb. 30. Funktion und Formen einer vermaschten Schaltung. Leitwertgleichung:
$\lambda = a(b + c\,e) + d(b\,c + e)$. *a* Unvermaschte Form, *b* Vermaschte Form

Kontakt mit allen anderen eindeutig verknüpft ist, (entweder parallel oder in Serie), dann läßt sich die Leitwertgleichung stets so umformen, daß jeder in ihr enthaltene Operand einem bestimmten Kontakt der verketteten Schaltung entspricht; dies ist aber nicht möglich, wenn die Verknüpfung der Kontakte wechselt, wie es bei vermaschten Schaltungen der Fall ist. Die in Abb. 30 genannte Leitwertgleichung läßt sich nicht weiter vereinfachen; die ihr entsprechende Schaltung nach Abb. 30 *a* erlaubt hingegen eine Vermaschung gemäß Abb. 30 *b*, die nicht mehr durch eine Gleichung beschrieben werden kann.

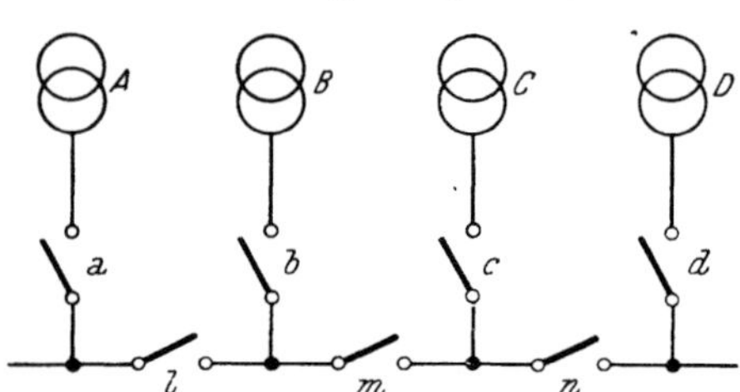

Abb. 31. Hochspannungsanlage mit vier Transformatoren

Da in der Praxis viele Schaltungen vorkommen, bei denen eine Vermaschung unmöglich ist oder keine Vorteile einbringt, reicht die Leitwertgleichung sehr oft aus, um eine gesuchte Schaltung aus den gestellten Bedingungen vollständig zu entwickeln. Hierzu ein Beispiel aus dem Schaltanlagenbau: Vier Umspanner A, B, C, D speisen über zugehörige Leistungsschalter a, b, c, d vier Abschnitte eines Sammelschienensystems, die durch drei Kuppelschalter l, m und n verbunden werden können. Abb. 31 zeigt den einpoligen Schaltplan der Hochspannungsanlage; die von den Sammelschienen abgehenden Leitungen der Anlage sind in diesem Plan nicht dargestellt, da sie für die Schaltaufgabe ohne Belang sind. Zur Begrenzung der Kurzschlußströme soll ein Parallelbetrieb von Umspannern vermieden und der Wärter auf diesen verbotenen Schaltzustand durch ein Warnzeichen W, z. B. einen Leuchtmelder, aufmerksam gemacht werden. Gesucht ist die Schaltung zwischen dem Warnzeichen W und einer Hilfsstromquelle Q. Als Schaltmittel sollen keine Relais, sondern nur Hilfskontakte an den Hochspannungsschaltern benützt werden; von den verschiedenen Schaltungsformen, mit denen sich die Aufgabe lösen läßt, ist jene auszuwählen, bei der die Hilfskontakte möglichst gleichmäßig auf die einzelnen Hochspannungsschalter verteilt sind.

Wir gehen davon aus, daß die Hauptkontakte und die mit ihnen gleichzeitig geschlossenen Hilfskontakte stets gleiche Leitwerte haben. Die möglichen Verbindungen beim Parallelbetrieb zwischen A, B, C, D können durch sechs Leitwertgleichungen beschrieben werden; diese lassen sich an Hand der vorgegebenen Schaltung (Abb. 31) sofort ansetzen:

$$\lambda(A, B) = a\,l\,b,$$
$$\lambda(A, C) = a\,l\,m\,c,$$
$$\lambda(A, D) = a\,l\,m\,n\,d,$$
$$\lambda(B, C) = b\,m\,c,$$
$$\lambda(B, D) = b\,m\,n\,d,$$
$$\lambda(C, D) = c\,n\,d.$$

Wenn eine dieser Verbindungen besteht, soll das Warnzeichen W mit der Stromquelle Q verbunden sein. Der Leitwert $\lambda(Q, W)$ der Meldeschaltung muß daher gleich der Summe aus den sechs vorstehenden Leitwertfunktionen sein; er wird nur 0, wenn alle Summanden 0 sind, das heißt, wenn keine verbotene Verbindung vorhanden ist.

$$\lambda(Q, W) = a\,l\,b + a\,l\,m\,c + a\,l\,m\,n\,d + b\,m\,c + b\,m\,n\,d + c\,n\,d.$$

In dieser Gleichung kommen die Operanden a, l in denselben Produkten vor; auch der Schaltplan zeigt, daß die Schalter a und l nur gemeinsam vom Strom durchflossen werden können, weil sich dazwischen keine Abzweigung befindet. **Das gleiche gilt auch für das Schalterpaar n, d.** Die Rechnung läßt sich deshalb übersichtlicher gestalten, da man für $a\,l$ einen neuen Operanden x und für $n\,d$ einen Operanden y einsetzen kann. Die Leitwertgleichung hat dann die handlichere Form:

$$\lambda(Q, W) = x\,b + x\,m\,c + x\,m\,y + b\,m\,c + b\,m\,y + c\,y.$$

Durch gruppenweises Herausheben gemeinsamer Faktoren läßt sich diese Gleichung auf verschiedene Weise umformen. Jeder Gestalt der Gleichung entspricht eine verkettete Schaltungsform. Alle diese Formen sind wirkungsgleich, da sich durch das Herausheben von Faktoren die Aussage der Leitwertgleichung nicht ändert. Die Aufgabe schreibt nun jene Schaltungsform vor, bei der die Kontakte möglichst gleichmäßig verteilt sind. Diese Lösung muß symmetrisch sein und wir finden sie durch Herausheben des Operanden m, der dem in der Mitte der vorgegebenen Schaltung liegenden Hochspannungsschalter zugeordnet ist.

$$\lambda(Q, W) = x\,b + m(x\,c + x\,y + b\,c + b\,y) + c\,y$$
$$= x\,b + m[x(c + y) + b(c + y)] + c\,y$$
$$= x\,b + m(c + y)\,(x + b) + c\,y.$$

Nun ordnen wir die Glieder so, daß die Reihenfolge der Operanden mit derjenigen der Hochspannungsschalter übereinstimmt; dadurch vermeiden wir überflüssige Schleifen in den Verbindungsleitungen zwischen den Kontakten.

$$\lambda(Q, W) = x\,b + (x + b)\,m(c + y) + c\,y.$$

Durch Rückeinsetzen für x und y ergibt sich die darstellende Gleichung der gesuchten Schaltungsform.

$$\lambda(Q, W) = a\,l\,b + (a\,l + b)\,m(c + n\,d) + c\,n\,d.$$

Diese Gleichung läßt sich ohne weiteres in einen Stromlaufplan übersetzen, da jedes Produkt eine Reihenschaltung und jedes Summenzeichen eine Parallelschaltung bedeutet. Abb. 32 zeigt die Lösung. Der Vollständigkeit halber sei erwähnt, daß man außer den verschiedenen ein-

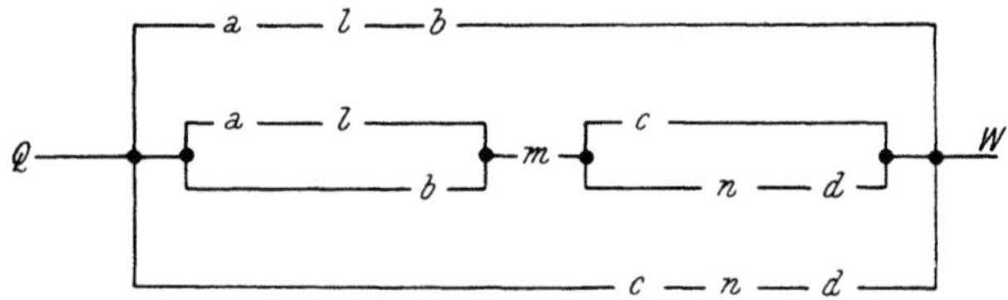

Abb. 32. Meldeschaltung für den Parallellauf von Transformatoren

fach verketteten Formen, die sich unmittelbar aus den Gleichungen ablesen lassen, noch weitere Lösungen durch Vermaschung finden kann. Hierdurch läßt sich die Zahl der Kontakte noch weiter vermindern; diese Formen verlangen aber an einzelnen Schaltern mehr als zwei Hilfskontakte und entsprechen daher nicht der gestellten Forderung nach einer möglichst gleichmäßigen Kontaktaufteilung.

Wie das vorstehende Beispiel zeigt, ist die Leitwertgleichung ein ebenso einfaches wie zweckmäßiges Hilfsmittel, das von einer textlich vorgegebenen Schaltaufgabe zu den Strompfaden der gesuchten Schaltung führt; sie erleichtert somit den ersten und wichtigsten Schritt, bei dem sich die Schaltung einer zeichnerischen Darstellung noch entzieht.

Zum vollen Verständnis der Leitwertgleichungen sei noch eine Betrachtung physikalischer Art beigefügt. Der aufmerksame Leser wird bemerkt haben, daß diese Gleichungen hinsichtlich ihrer physikalischen Dimensionen scheinbar fehlerhaft sind. Reihenschaltungen sind durch Produkte von Leitwerten dargestellt; die einzelnen Glieder der Summen zeigen eine ungleiche Zahl von Faktoren, sind also scheinbar dimensionell verschieden. In einer physikalischen Größengleichung dürfen aber Summanden verschiedener Dimension nicht auftreten; außerdem sind höhere Potenzen von Leitwerten oder Widerständen physikalisch sinnlos. Die Operanden, das heißt die Rechenzeichen für die Variablen, haben also sichtlich nicht die physikalische Bedeutung von Leitwerten, wie es in der vorstehenden Ableitung der Einfachheit halber angenommen wurde. Die Erklärung finden wir aus der Einschränkung auf das Wertepaar Null und Unendlich. Bekanntlich definiert die Mathematik diese Grenzwerte als Zahlen, die größer oder kleiner als jede beliebig gewählte Vergleichszahl sind. Die Gleichungen werden daher sofort dimensionsrichtig, wenn wir jeden Leitwert eines Kontaktes durch einen beliebigen Vergleichswert, der ebenfalls die Dimension eines Leitwertes hat, dividieren. Die Operanden sind also selbst keine Leitwerte, sondern Leitwertverhältnisse und somit reine Zahlen (Skalare). Die Logistik ist ihrem Wesen nach keine Größenrechnung, sondern eine Rechnung mit Verhältniszahlen, was durch das Äquivalenzgesetz zum Ausdruck kommt.

Abschließend seien die für alle Verknüpfungen von Kontakten gültigen Gesetze nochmals und zusammenhängend aufgezählt; sie bilden die grundlegenden *Schaltregeln für Kontaktkombinationen*. Diese lauten:

1. Äquivalenzregel. Eine Vervielfachung von Kontakten durch Parallel- oder Serienschaltung ist wirkungslos. Es gilt

$$x + x = x \qquad \text{und} \qquad x \cdot x = x.$$

2. Leitungsregel. Die Wirkung eines Kontaktes wird durch eine in Serie liegende Leitung nicht verändert. Es gilt:

$$x \cdot \infty = x.$$

3. Kurzschlußregel. Eine zu einem Kontakt parallele Leitung macht diesen wirkungslos. Es gilt

$$x + \infty = \infty.$$

4. Vertauschungsregel. In einer Kombination mit gleichartiger Verknüpfung (nur parallel oder nur Serie) sind sämtliche Kontakte untereinander vertauschbar. Es gilt

$$x + y = y + x \qquad \text{und} \qquad x\,y = y\,x.$$

5. Heraushebungsregel. Ist eine Anzahl gleichartiger Kontakte in einer beliebigen Kombination zu allen anderen Kontakten in Serie oder parallel, so kann sie durch einen einzigen Serien- oder Parallelkontakt ersetzt werden. Es gilt

$$x\,y + x\,z = x(y + z) \qquad \text{und} \qquad (x + y) \cdot (x + z) = x + y\,z.$$

Die meisten dieser Regeln sind anschaulich und beinahe selbstverständlich; nur die Heraushebung eines gemeinschaftlichen Parallelkontaktes ist nicht mehr ganz sinnfällig.

2. Die Leitwerte von Stromkreisen

Die Beschränkung der Rechnung auf die extremen Leitwerte 0 und ∞ erlaubt es, für Strompfade mit Kontakten und verbindenden Leitungen algebraische Leitwertfunktionen aufzustellen. Jede Funktion bezieht sich hierbei auf den Leitwert zwischen einer Stromquelle und einem Verbraucher. Wollen wir Schaltungen mit mehreren Verbrauchern behandeln, so müssen wir diese in die Strompfade einbeziehen und den Leitwert der Schaltung zwischen den Anschlußstellen der Stromquelle bestimmen. Nun haben aber Verbraucher keinen unendlichen, sondern einen festen endlichen Leitwert. Es soll nun gezeigt werden, daß das logistische Rechenverfahren auch für diesen Fall anwendbar bleibt.

Ein beliebiges Gebilde betrachten wir als leitend, solange sein Leitwert nicht Null ist. Für den Begriff der Verschiedenheit von Null wollen wir ein eigenes Rechenzeichen einführen, und zwar versehen wir hierzu das Zeichen 0 (Null) mit dem in der Logistik gebräuchlichen Querstrich als Umkehrungszeichen (Inversionsoperator). Der so entstandene Operand $\overline{0}$ bedeutet dann eine von Null verschiedene, sonst aber beliebige Größe des Leitwertes; $\overline{0}$ bezeichnet also weder eine Variable noch einen bestimmten Festwert, sondern einen unbestimmten Wert, der nicht 0 sein kann. Addiert oder multipliziert man einen solchen unbestimmten Wert mit einem anderen eben solchen, so ist das Ergebnis auch stets von Null verschieden. Es gelten daher auch für den Wert $\overline{0}$ die Äquivalenzregeln:

$$\overline{0} + \overline{0} = \overline{0}$$

und

$$\overline{0} \cdot \overline{0} = \overline{0}.$$

Wir können also mit dem endlichen Wertbereich $\overline{0}$ genau so rechnen wie mit ∞. Der Wert ∞ stellt einen speziellen Wert im Bereich der möglichen Werte von $\overline{0}$ dar; die für 0 und ∞ abgeleiteten Äquivalenzregeln sind daher nur Sonderfälle der Regeln für 0 und $\overline{0}$.

Aus Vorstehendem folgt, daß sich der Geltungsbereich des logistischen Rechenverfahrens nicht nur auf Elemente mit extremen Leitwerten, also Kontakte und Leitungen, beschränkt, sondern auch Stromquellen und Verbraucher umfaßt. Wir können also Leitwertfunktionen für vollständige Stromkreise aufstellen, und zwar ohne weitere Rechenregeln; die Verbraucher und Stromquellen dürfen genau so behandelt werden wie Kontakte.

Für einen Stromkreis, der aus einer Stromquelle Q, einer beliebigen Kontaktkombination a und einem Verbraucher A besteht, lautet dann die Leitwertfunktion

$$\lambda = Q\,a\,A \qquad \text{oder} \qquad \frac{\lambda}{Q\,A} = a.$$

Liegt eine Schaltung mit mehreren Verbrauchern vor, läßt sich für jeden davon eine Leitwertgleichung ansetzen; jede Gleichung beschreibt dann die Teilfunktion, die sich auf den zugehörigen Verbraucher bezieht. Im allgemeinen liegen die Verbraucher zueinander parallel, so daß wir diese Gleichungen addieren können. Da im allgemeinen eine gemeinsame Stromquelle vorhanden ist, hat dann die Gesamtfunktion die Gestalt

$$\frac{\lambda}{Q} = x\,A + y\,B + z\,C + \dots$$

Hierin können x, y, z beliebige Kombinationen von Schalterstellungen oder Kontakten sein.

Wenn nicht mehrere Stromquellen vorhanden sind, kann man sich das Anschreiben des Operanden Q im Nenner des Bruches ersparen. Die Gleichung bleibt trotzdem mathematisch richtig. Der Stromquelle kommt der Leitwert $\overline{0}$ zu und jeder mögliche Wert von λ, nämlich 0, $\overline{0}$ oder ∞, bleibt bei Division durch $\overline{0}$ unverändert erhalten. Das gleiche gilt auch für gemeinschaftliche Verbraucher; das Anschreiben eines Nenners vom Wert $\overline{0}$ ist also nur ein Vermerk, zwischen welchen Punkten der Schaltung der Leitwert gelten soll.

Die praktische Anwendung sei an einem Beispiel aus der Installationstechnik erläutert. Die Lichtanlage einer photographischen Dunkelkammer bestehe aus drei Lampen in den Farben rot, gelb und weiß. Die Lampen sollen durch zwei Schalter gesteuert werden, von denen sich einer bei der Türe und der andere am Arbeitsplatz befindet. Beim Einschalten ist eine bestimmte Reihenfolge einzuhalten, und zwar soll zuerst die rote Lampe, dann die gelbe und zuletzt die weiße Strom erhalten; für das Ausschalten gilt sinngemäß die gegenteilige Farbenfolge. Wir bezeichnen die beiden Schalter mit a und b, die rote Lampe mit R, die gelbe mit G und die weiße mit W. Mit jedem der beiden Schalter sollen sich vier Zustände der Beleuchtung herstellen lassen, nämlich dunkel und die drei Lichtfarben. Jeder Schalter muß daher vier Stellungen haben. Wir bezeichnen willkürlich die Stellungen der beiden Schalter, bei denen keine Lampe leuchtet, mit a_0 und b_0. Den eben beschriebenen Schaltzustand, dem der Leitwert 0 zukommt, können wir dann durch das Produkt $a_0 \cdot b_0 \cdot 0$ kennzeichnen. Bringen wir nun den Schalter b in seine Stellung b_1, so soll die rote Lampe eingeschaltet werden, was einem Zustand $a_0 b_1 R$ entspricht. Bei Weiterdrehen des Schalters b folgt dann $a_0 b_2 G$ und zuletzt $a_0 b_3 W$. Auf diesen Zustand folgt dann wiederum die Ausgangslage $a_0 b_0 0$. Wir können also die Gleichung einer Teilfunktion anschreiben, die sich auf die Wirkung des Schalters b bei unveränderter Stellung a_0 des Schalters a bezieht. Sie lautet:

$$\frac{\lambda_1}{Q} = a_0(b_0\,0 + b_1\,R + b_2\,G + b_3\,W).$$

In analoger Weise läßt sich eine zweite Gleichung ansetzen, die sich ebenfalls auf die Wirkungsweise des Schalters b bezieht, der aber die Stellung a_1 zugrunde liegt. Beim Übergang von der Ausgangslage $a_0 b_0$ auf die Stellungskombination $a_1 b_0$ soll ebenfalls die rote Lampe eingeschaltet werden, so daß sich ein Zustand $a_1\, b_0\, R$ ergibt. Schaltet man nun mit dem Schalter b weiter, so folgen entsprechend dem verlangten Schaltzyklus die weiteren Zustände $a_1\, b_1\, G$, dann $a_1\, b_2\, W$ und schließlich $a_1\, b_3\, 0$. Die Gleichung der zweiten Teilfunktion lautet also:

$$\frac{\lambda_2}{Q} = a_1\,(b_0\, R + b_1\, G + b_2\, W + b_3\, 0).$$

In Fortsetzung dieses Verfahrens erhalten wir dann noch zwei weitere analoge Gleichungen für die Schalterstellungen a_2 und a_3. Durch Addition aller vier Gleichungen ergibt sich die Gesamtfunktion der Schaltung:

$$\frac{\lambda}{Q} = a_0(b_0\, 0 + b_1\, R + b_2\, G + b_3\, W) +$$
$$+ a_1(b_0\, R + b_1\, G + b_2\, W + b_3\, 0) +$$
$$+ a_2(b_0\, G + b_1\, W + b_2\, 0 + b_3\, R) +$$
$$+ a_3(b_0\, W + b_1\, 0 + b_2\, R + b_3\, G).$$

Die Gleichung enthält auch jene Stellungskombinationen, bei denen keine Lampe leuchtet und zeigt daher das gesamte Schaltspiel der gesuchten Anordnung. Sie umfaßt sämtliche Stellungskombinationen der Schalter a und b, die überhaupt möglich sind, und bietet dadurch die vollständige Sicherheit, daß beim Ansatz kein Schaltzustand übersehen wurde. Durch die Multiplikation der unwirksamen Stellungskombinationen mit 0 kommt zum Ausdruck, daß diesen der Leitwert 0 der Schaltung entsprechen soll.

In der vorstehenden Gleichung kommt jeder der Operanden R, G, W mehrmals vor. Die zugehörigen Lampen sollen aber in der Schaltung nur je einmal vertreten sein. Daher haben wir die Gleichung so umzuformen, daß sich die Operanden R, G, W als gemeinsame Faktoren herausheben lassen. Bei dieser Operation können wir auch die mit 0 multiplizierten Glieder entfallen lassen. Die Gleichung erhält dadurch eine Gestalt, in der sie bereits die verkettete Form der gesuchten Schaltung beschreibt.

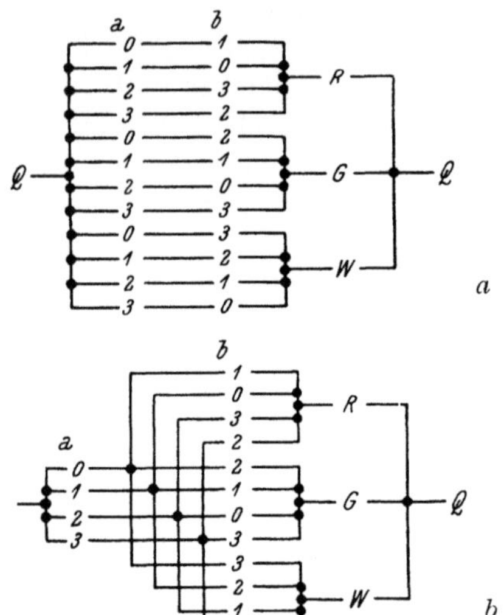

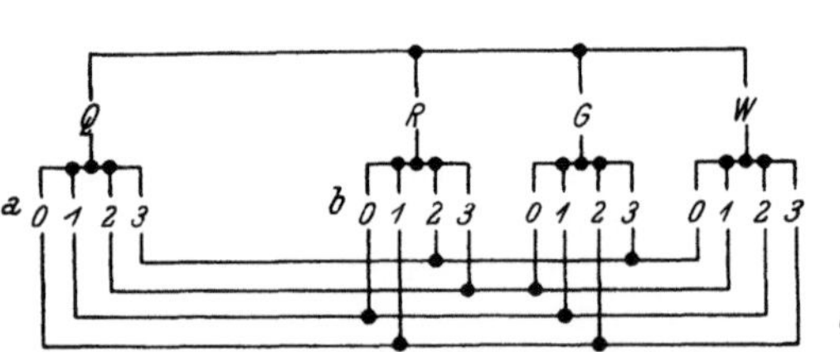

Abb. 33. Dunkelkammerschaltung. a Einfach verkettete Form, b Vermaschte und gekreuzte Form, c Zyklisch geordneter Stromlaufplan

$$\frac{\lambda}{Q} = (a_0\, b_1 + a_1\, b_0 + a_2\, b_3 + a_3\, b_2)\, R +$$

$$+ (a_0\, b_2 + a_1\, b_1 + a_2\, b_0 + a_3\, b_3)\, G +$$

$$+ (a_0\, b_3 + a_1\, b_2 + a_2\, b_1 + a_3\, b_0)\, W.$$

An Hand dieser Gleichung läßt sich die Schaltung in Form eines Stromlaufplanes aufzeichnen, der in Abb. **33** *a* wiedergegeben ist.

Wir haben nun noch zu untersuchen, ob die gefundene Schaltungsform mit einfacher Verkettung die Zweckmäßigste ist oder ob sich eine Ersparnis an Kontakten durch Vermaschung erreichen läßt. Aus der Leitwertgleichung gehen solche Möglichkeiten nicht hervor. Der Stromlaufplan zeigt aber, daß eine Zusammenlegung äquivalenter Kontakte des Schalters *a* naheliegend ist. Durch Vereinigung gleichnamiger Kontakte ergibt sich ohne jegliche Rechenoperation eine vermaschte und gekreuzte Schaltungsform gemäß Abb. **33** *b*. Da für die praktische Ausführung der Schaltung nur Drehschalter, vorzugsweise Paketschalter, in Frage kommen, empfiehlt es sich, den Schaltplan zyklisch zu ordnen, was die Herstellung der Leitungsanschlüsse wesentlich erleichtert. Abb. **33** *c* zeigt den so umgeformten Stromlaufplan; aus ihm geht hervor, daß der Schalter *a* ein einfacher Umschalter mit vier Stellungen ist, wogegen der Schalter *b* aus drei ebensolchen Umschaltern besteht.

Als **Schaltregel** können wir festhalten: Die schaltungstechnische Verknüpfung der Stromquellen und Stromverbraucher folgt den gleichen Gesetzen wie diejenige der Kontakte.

3. Die Verknüpfung von Leitbedingungen

Schalttechnische Aufgaben werden meistens so gestellt, daß sie eine bestimmte Wirkungsweise der gesuchten Schaltung festlegen. Die Beschreibung der gewünschten Funktion erfolgt hierbei fast immer durch Teilangaben, die wir als *Funktionsbedingungen* bezeichnen wollen; auf Grund dieser vorgegebenen Bedingungen haben wir dann die Leitwertfunktion aufzustellen, die von der Schaltung erfüllt werden soll. Hierzu ist es vorerst notwendig, die einzelnen Funktionsbedingungen als Leitwertgleichungen anzusetzen. Das ist aber nur möglich, wenn wir die Verknüpfungen der Leitwerte in den Gleichungen gleichzeitig als Verknüpfungen von Funktionsbedingungen deuten können. Das Rechnen mit Bedingungen (Prämissen) fällt nun in das Gebiet der reinen Logistik; wir haben also unseren Funktionsgleichungen neben der physikalischen Deutung als Leitwertfunktionen noch eine zweite, logistische Deutung zu unterlegen. Dies ist ohne weiteres möglich, da die Einschränkung auf die Leitwerte 0 und ∞ bereits zu den Rechenregeln der Logistik geführt hat und die Leitwertgleichungen auf diesen Regeln beruhen; die Erklärung hierfür liefert uns die nachstehende Betrachtung.

Die Verschiedenheit des Leitwertes von Null ist die Voraussetzung für jeden Stromfluß. Der Leitwert $\overline{0}$ und daher auch sein spezieller Wert ∞

sind also *Bedingungen* für die Leitfähigkeit einer Schaltung. Liegen uns irgendwelche Funktionsbedingungen vor, so haben wir sie nur so zu fassen, daß sie Bedingungen für die Stromleitung ausdrücken: dann können wir mit diesen Bedingungen nach den gleichen Regeln rechnen wie mit Leitwerten. Dabei dürfen wir die Variablen für den Leitwert von Kontakten ganz allgemein als *Bedingungsoperanden*[1] betrachten; es bedeutet dann beispielsweise x eine beliebige Bedingung, die erfüllt sein kann (Wert $\overline{0}$) oder nicht (Wert 0). Wir haben nun noch zu untersuchen, welche Deutung wir der Addition und Multiplikation in dieser Bedingungsrechnung zu geben haben. Dazu denken wir uns zwei Variable x und y, wobei die Bedingung x erfüllt, also $x = \overline{0}$ sein soll; für die Summe $x + y$ gibt es dann zwei Möglichkeiten entsprechend der folgenden Tabelle:

x	y	$x + y$
$\overline{0}$	0	$\overline{0} + 0 = \overline{0}$
$\overline{0}$	$\overline{0}$	$\overline{0} + \overline{0} = \overline{0}$

Das Resultat ist in beiden Fällen $\overline{0}$ und vom Wert von y unabhängig. Das gleiche ergibt sich, wenn wir x und y vertauschen. Bei der Addition ist also die Erfüllung von x oder y allein schon hinreichend, um das Resultat $\overline{0}$ zu erhalten. Wir erkennen, daß die Addition eine Verkettung von Bedingungen ist, von denen jede für sich allein zur Leitfähigkeit führt; die Bedingungen bleiben in ihrer Wirkung unabhängig voneinander und die Logik nennt daher die Addition eine äußere Verknüpfung (Disjunktion). Liegen also in einer Schaltaufgabe Bedingungen vor, die *unabhängig* voneinander zur Stromleitung führen sollen, so haben wir die zugehörigen Operanden zueinander zu *addieren*. Umgekehrt verhält sich die Multiplikation. Für sie lautet die Werttabelle:

x	y	$x \cdot y$
$\overline{0}$	0	$\overline{0} \cdot 0 = 0$
$\overline{0}$	$\overline{0}$	$\overline{0} \cdot \overline{0} = \overline{0}$

---

[1] Die den Stellungen der Schaltgeräte zugeordneten Variablen sind in physikalischer Hinsicht als Leitwertverhältnisse aufzufassen. Ihre Deutung als Funktionsbedingungen ist möglich, weil physikalische Bedingungen stets Verhältniswerte und demnach Skalare sind. Beispielsweise besteht die Bedingung zum Betreten eines Raumes darin, daß irgend ein Eingang vorhanden ist. Dabei wird aber stillschweigend vorausgesetzt, daß der Eingang größer ist als der eintretende Gegenstand; andernfalls wäre er für diesen kein Eingang. Mehrere Eingänge neben- oder nacheinander wirken so wie ein einziger, was im logistischen Äquivalenzgesetz zum Ausdruck kommt.

Es müssen also beide Variable den Wert $\bar{0}$ haben, damit das Ergebnis $\bar{0}$ wird; in der Logik wird daher die Multiplikation als innere Verknüpfung (Konjunktion) bezeichnet. Enthält eine Schaltaufgabe Bedingungen, die *gleichzeitig* erfüllt sein müssen, damit der Stromfluß eintritt, so haben wir demnach die Bedingungsoperanden miteinander zu *multiplizieren.*

Die eben gegebenen Regeln für Summen und Produkte von Leitbedingungen gelten selbstverständlich nicht für einzelne Variable, sondern auch für Verknüpfungen von solchen. Ihre praktische Anwendung soll an einem Beispiel gezeigt werden. Eine Anlage habe drei einphasige Verbraucher, z. B. Heizwicklungen, von denen je nach Bedarf ein oder zwei eingeschaltet sein und der dritte als Reserve verbleiben soll; durch die Schaltung soll erzwungen werden, daß stets mindestens ein beliebiger Verbraucher ausgeschaltet bleibt. Wir nennen die Verbraucher A, B, C und die zugehörigen Steuerschalter a, b, c. Die Ruhestellung jedes Schalters bezeichnen wir mit dem Index 0 und die Arbeitsstellung mit 1. Daß die Einschaltung der Verbraucher an die Stellung 1 der Schalter gebunden sein soll, können wir durch drei Bedingungsgleichungen ausdrücken:

$$\lambda_1 = a_1 A,$$
$$\lambda_2 = b_1 B,$$
$$\lambda_3 = c_1 C.$$

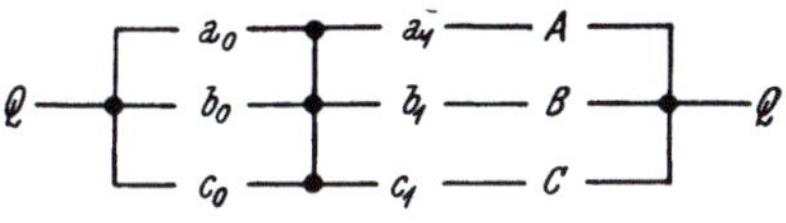

Abb. 34. Schaltung für drei Verbraucher mit erzwungener Reservehaltung

Ferner soll verhindert werden, daß alle drei Verbraucher gleichzeitig eingeschaltet werden. Diese Bedingung ist vorerst so zu fassen, daß sie sich auf die Stromleitung bezieht; sie hat also zu lauten: „Die Schaltung soll nur Strom führen, wenn sich mindestens ein Schalter in seiner Ruhestellung befindet." Diese Bedingung λ_4 kann von jedem Schalter allein erfüllt werden und erscheint daher als Summe der Ruhestellungen der Schalter:

$$\lambda_4 = a_0 + b_0 + c_0.$$

Die Bedingungen λ_1 bis λ_3 gelten unabhängig voneinander und sind demnach ebenfalls zu addieren; sie sind aber alle drei an die Bedingung λ_4 gebunden und ihre Summe ist daher mit dieser vierten Bedingung zu multiplizieren. Die Verknüpfung aller vier Bedingungen lautet also:

$$\lambda = \lambda_4(\lambda_1 + \lambda_2 + \lambda_3).$$

Durch Einsetzen ergibt sich die Leitwertfunktion der Schaltung:

$$\lambda = (a_0 + b_0 + c_0) \cdot (a_1 A + b_1 B + c_1 C).$$

In dieser Gleichung kommt jeder Operand nur einmal vor, so daß die Funktion λ zugleich eine Schaltungsform mit maximaler Verkettung abbildet. Abb. 34 zeigt die gefundene Schaltung.

Als weiteres Beispiel soll die vorbeschriebene Aufgabe in einer etwas schwierigeren Abwandlung behandelt werden. Es sei angenommen, daß die drei Verbraucher mehrphasig sind und über Schütze durch Taster gesteuert werden. Die Selbsthaltung der Schütze verhindert hierbei, daß eine versuchte Einschaltung des dritten Verbrauchers die beiden anderen ausschaltet. Die Steuerschaltung zwischen den Einschalttastern a, b, c und den zugehörigen Schützenspulen ist gefragt. Der Unterschied gegen die frühere Aufgabe liegt darin, daß die Taster ihre Stellung nicht beibehalten und daher nicht nur miteinander, sondern auch mit Hilfskontakten der Schütze verriegelt werden müssen. Man könnte nun jeden Taster eines Schützes über Ruhekontakte aller anderen Schütze anspeisen; eine solche Schaltung führt bei einer größeren Zahl von Verbrauchern zu einem unerträglichen Aufwand an Hilfskontakten und Leitungen. Die Bedingungsrechnung liefert hingegen sofort die einfachste Form. Wir gehen davon aus, daß ein Schütz α nur dann in seiner Ruhelage α_0 sein kann, wenn sich auch sein Befehlsschalter a in der Ruhestellung a_0 befindet. In diesem Zustand bilden beide Geräte vorübergehend eine Einheit, die einem handbedienten Befehlsschalter entspricht. Wir nennen diesen gedachten Schalter x und setzen x_0 statt der Kombination von a_0 und α_0 in die Rechnung ein. Da diese Substitution nur gilt, wenn die beiden Geräte gleichzeitig in der Ruhestellung sind, so erscheint x_0 als Bedingungsprodukt; es ist

$$a_0\,\alpha_0 = x_0, \qquad \text{bzw.} \qquad b_0\,\beta_0 = y_0 \qquad \text{und} \qquad c_0\,\gamma_0 = z_0.$$

In Analogie zum früheren Beispiel ergibt sich dann die Gesamtfunktion für die Verbindung der Schützenspulen $S_\alpha, S_\beta, S_\gamma$ mit der Stromquelle:

$$\lambda = (x_0 + y_0 + z_0) \cdot (a_1\,S_\alpha + b_1\,S_\beta + c_1\,S_\gamma)$$
$$= (a_0\,\alpha_0 + b_0\,\beta_0 + c_0\,\gamma_0) \cdot (a_1\,S_\alpha + b_1\,S_\beta + c_1\,S_\gamma).$$

Auch diese Gleichung enthält jeden Operanden nur einmal und stellt somit gleichzeitig eine Schaltungsform maximaler Verkettung dar. Abb. 35 zeigt die Schaltung im Stromlaufplan und Wirkschaltplan, wobei die für die Wirkungsweise belanglosen Hauptkontakte der Schütze nicht dargestellt sind. Der Vollständigkeit halber wurden im Wirkschaltplan auch die Haltestrompfade der Schütze mit den Tastern für die Ausschaltung eingetragen.

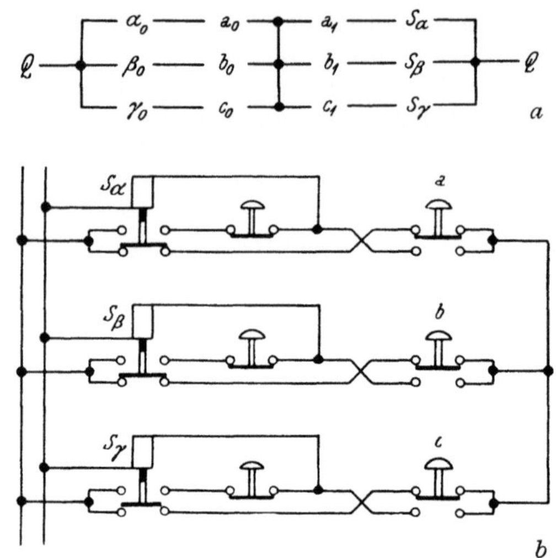

Abb. 35. Schaltung für drei Schütze mit erzwungener Reservehaltung. a Stromlaufplan der Befehlsstromkreise, b Wirkschaltplan der Gesamtschaltung

Das vorstehende Beispiel wird vielleicht den Einwand hervorrufen, daß die gestellte Aufgabe nur zum geringeren Teil durch die rechnerische Behandlung gelöst wurde und noch eine beachtliche Menge an Überlegung und schaltungstechnischer Erfahrung für den Ansatz der Rechnung erforderlich war. Es wird aber an späterer Stelle gezeigt werden, daß sich die Denkarbeit bei der Entwicklung von Schaltungen in weit größerem Umfang durch Rechenarbeit ersetzen läßt; daß dies hier nicht geschehen ist, liegt nur daran, daß sich die Rechnung auf die beiden elementarsten Operationen der Addition und Multiplikation beschränken sollte. Unter Benützung höherer Rechenoperationen lassen sich auch schwierigere Aufgaben mühelos mathematisch formulieren und rein schematisch lösen.

Die Tatsache, daß sich die Operanden der Rechnung gleichzeitig als Variable für die Leitwerte der Kontakte wie auch als Symbole für die Leitbedingungen deuten lassen, gestattet die Aufstellung einer sehr wichtigen Schaltregel. Diese gibt eine klare Anweisung für den ersten Ansatz der Schaltung auf Grund vorgeschriebener Funktionsbedingungen und lautet:

Ansatzregel. Funktionsbedingungen, die unabhängig voneinander den Stromfluß vorschreiben, führen zu einer Parallelschaltung von Kontakten oder Kontaktkombinationen: Bedingungen, deren gleichzeitige Erfüllung für den Stromfluß notwendig ist und die somit einschränkende Aussagen darstellen, führen zu einer Serienschaltung der Kontakte oder Kontaktkombinationen.

4. Die Kürzung von Leitbedingungen

Bei vielen Schaltungen enthalten die Schaltgeräte Kontakte, die in verschiedenen Stellungen geschlossen sind; Verknüpfungen solcher Kontakte miteinander sollen nun untersucht werden. Hierbei wollen wir vorerst Schalter mit je zwei Stellungen betrachten und dann das Ergebnis auf Geräte mit beliebiger Stellungszahl erweitern.

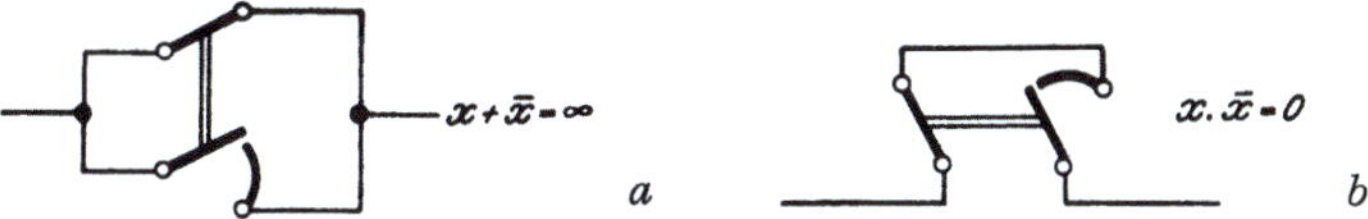

Abb. 36. Verknüpfung komplementärer Kontakte. *a* Parallelschaltung, *b* Serienschaltung

Die beiden Stellungen eines zweistelligen Schaltgerätes können wir unter Verwendung des Inversionsoperanden mit $\bar{x}$ und x bezeichnen; es sind dann alle Kontakte x geschlossen, wenn die Kontakte $\bar{x}$ offen sind und umgekehrt. Schalten wir nun zwei Kontakte x und $\bar{x}$ zueinander parallel, so ist diese Kombination auf alle Fälle leitend, weil stets einer der parallelen Kontakte geschlossen ist (Abb. **36** *a*). Es gilt

$$x + \overline{x} = \infty.$$

Zueinander inverse Leitwerte ergänzen sich bei der Addition zu einer dauernden Leitfähigkeit; sie sind also *komplementär*. Schalten wir die beiden Kontakte hintereinander (Abb. 36 *b*), so ist der Strompfad stets an einer Stelle offen und es folgt daraus

$$x \cdot \overline{x} = 0.$$

Beide Gesetze lassen sich auch in allgemein gültiger Weise mathematisch ableiten: Wir betrachten die Werttabelle

x	$\overline{x}$	$x + \overline{x}$
0	∞	$0 + \infty = \infty$
∞	0	$\infty + 0 = \infty$

und ersehen, daß für beide möglichen Werte von x die Summe $x + \overline{x}$ den Wert ∞ annimmt.

Die analoge Werttabelle für das Produkt $x \cdot \overline{x}$ läßt sich nur auswerten, wenn das Produkt $0 \cdot \infty$ einen bestimmten Wert darstellt. Infolge des logistischen Äquivalenzgesetzes ist dies tatsächlich der Fall.

Aus

$$x + x = x$$

oder

$$n \cdot x = x$$

folgt für $n = \infty$ unmittelbar

$$\infty \cdot x = x$$

und für $x = 0$ der bestimmte Wert

$$\infty \cdot 0 = 0.$$

Die Werttabelle für ein Produkt aus zwei zueinander inversen Operanden lautet also:

x	$\overline{x}$	$x \cdot \overline{x}$
0	∞	$0 \cdot \infty = 0$
∞	0	$\infty \cdot 0 = 0$

Die Rechnung verläuft im übrigen genau so, wenn man statt ∞ den allgemeinen Wert $\overline{0}$ einsetzt.

Die Verknüpfungen inverser Variabler führen also stets zu einem konstanten Wert 0 oder ∞ und ermöglichen daher *Kürzungen* in den Leitwertgleichungen. Diese Operationen laufen stets darauf hinaus, daß entweder ein Faktor von der Gestalt $(x + \overline{x})$ oder ein Summand von der Gestalt $x \cdot \overline{x}$ gebildet wird, durch den der Funktionswert unverändert bleibt.

Für die Addition von $x\,\overline{x} = 0$ ist dies sofort einzusehen; daß die Multiplikation mit $x + \overline{x} = \infty$ den Funktionswert nicht ändert, können wir an Hand der Werttabelle leicht feststellen.

λ	$\lambda \cdot \infty = \lambda$
0	$0 \cdot \infty = 0$
∞	$\infty \cdot \infty = \infty$

Durch die Kürzung von Leitwertgleichungen lassen sich häufig Vereinfachungen von Schaltungen finden, die wenig sinnfällig sind und erst bei der rechnerischen Behandlung aufscheinen. Ein Beispiel hierfür wurde in der Abb. 12 gezeigt; nunmehr soll der rechnerische Beweis für die Richtigkeit dieser Umformung geliefert werden.

Die ungekürzte Schaltung gemäß Abb. 12 a läßt sich durch eine Gleichung

$$\lambda = (a_0 + \overline{a}_0\, b_1)\, c_1$$

beschreiben. Wir multiplizieren den Operanden a_0 mit einem Faktor $(b_1 + \overline{b}_1)$, der immer ∞ ist und somit den Wert von a_0 nicht ändert. Die Rechnung verläuft dann folgendermaßen:

$$\begin{aligned}
\lambda &= (a_0 + \overline{a}_0\, b_1)\, c_1 = \\
&= [a_0(b_1 + \overline{b}_1) + \overline{a}_0\, b_1]\, c_1 = \\
&= (a_0\, b_1 + a_0\, \overline{b}_1 + \overline{a}_0\, b_1)\, c_1 = \\
&= (a_0\, b_1 + a_0\, \overline{b}_1 + a_0\, b_1 + \overline{a}_0\, b_1)\, c_1 = \\
&= [a_0(b_1 + \overline{b}_1) + b_1(a_0 + \overline{a}_0)]\, c_1 = \\
&= (a_0 \infty + b_1 \infty)\, c_1 = \\
&= (a_0 + b_1)\, c_1.
\end{aligned}$$

Das Resultat besagt, daß der Kontakt $\overline{a}_0$ ohne Änderung der Wirkungsweise gestrichen und die Schaltung nach Abb. 12 b ausgeführt werden darf. Das Beispiel zeigt uns eine allgemein gültige Regel; es ist stets

$$\begin{aligned}
x + \overline{x}\, y &= x(\overline{y} + y) + \overline{x}\, y = \\
&= x\,\overline{y} + x\, y + \overline{x}\, y = \\
&= x\,\overline{y} + x\, y + x\, y + \overline{x}\, y = \\
&= x(y + \overline{y}) + y(x + \overline{x}) = \\
&= x \cdot \infty + y \cdot \infty = \\
&= x + y.
\end{aligned}$$

Diese Umformung bildet das weniger sinnfällige Gegenstück zur trivialen Kürzung

$$x + x\,y = x(\infty + y) = x \cdot \infty = x.$$

Mit vorstehendem wurde der Nachweis geführt, daß Summen und Produkte von Variablen, die zueinander invers sind, selbst nicht mehr Variable sondern Festwerte darstellen. Da hierzu keinerlei Einschränkungen nötig waren, gilt dieses Gesetz nicht nur für Schaltgeräte mit zwei Stellungen, sondern auch für solche mit beliebig großer Stellungszahl. Wir betrachten einen Schalter a und greifen zwei Kontakte, die in verschiedenen Stellungen m und n geschlossen sind, heraus. Schalten wir die beiden Kontakte hintereinander, so ergibt sich ein Strompfad, der stets unterbrochen ist. Es gilt also allgemein für mehrstellige Schaltgeräte

$$a_m \cdot a_n = 0.$$

Ferner ist jede Schalterstellung a_x zum übrigen Stellungsbereich $\overline{a}_x$ invers. Wegen

$$a_x + \overline{a}_x = \infty$$

gilt daher für mehrstellige Schaltgeräte die Beziehung

$$a_1 + a_2 + a_3 \ldots + a_0 = \infty.$$

Ein praktisches Beispiel möge die Bedeutung dieser Beziehungen für die Schaltungstechnik zeigen.

Wir wollen eine Wählerschaltung, die in der Fernmeldetechnik und für Fernwirkanlagen benützt wird, rechnerisch ableiten. Es handle sich um einen Drehwähler üblicher Bauart; dieser ist ein vielstelliger Umschalter mit Schrittschaltwerk und elektromagnetischem Antrieb, der einen Ruhekontakt als Selbstunterbrecher trägt (Abb. 37 b). Die Aufgabe sei durch folgende Funktionsbedingungen umrissen:

1. Die Spule S_a des Wählerantriebes a soll über ein fernbetätigtes Impulsrelais i angespeist werden, so daß der Wähler w_0 nach einem das Relais i anregenden Impuls um einen Schritt weiterschaltet.

2. Zu Beginn jeder Impulsfolge wird ein Impuls längerer Dauer (Anlaufimpuls) gegeben und durch ein dem Impulsrelais i nachgeschaltetes Relais v mit Ansprechverzögerung erfaßt. Der Wähler soll seine Ruhelage w_0 nur nach einem solchen Anlaufimpuls verlassen; bei kürzeren Impulsen soll seine Spule S_α daher nicht erregt werden.

3. Befindet sich der Wähler bei Eintreffen eines Anlaufimpulses nicht in seiner Ruhestellung w_0, so soll die Antriebsspule S_a über den Unterbrechungskontakt a_0 solange angespeist werden, bis der Wähler die Stellung w_0 erreicht hat; dabei sei vorausgesetzt, daß der Anlaufimpuls länger dauert als ein Wählerumlauf.

Gesucht ist die Steuerschaltung für die Antriebsspule S_a des Wählers. Da alle drei Bedingungen gleichzeitig gelten sollen, müssen wir sie miteinander multiplizieren; diese Operation stellt sicher, daß keine einschränkende Vorschrift der Angabe übersehen wird und ungewollte Nebenwirkungen auftreten. Wir setzen daher

$$\frac{\lambda}{S_\alpha} = \lambda_1\,\lambda_2\,\lambda_3.$$

Die Bedingung 1 besagt nur, daß die Spule S bei Ansprechen des Impulsrelais i erregt werden soll; sie lautet daher als Gleichung

$$\lambda_1 = i_1.$$

Bedingung 2 verlangt die Erregung der Spule bei eingeschaltetem Relais v in der Wählerstellung w_0 oder symbolisch $v_1\,w_0$; was im Stellungsbereich $\overline{w}_0$ geschehen soll, bleibt freigestellt und wir haben diese Möglichkeit einer zusätzlichen Anspeisung durch Beifügung des Summanden $\overline{w}_0$ anzuschreiben. Es folgt daraus

$$\lambda_2 = v_1\,w_0 + \overline{w}_0.$$

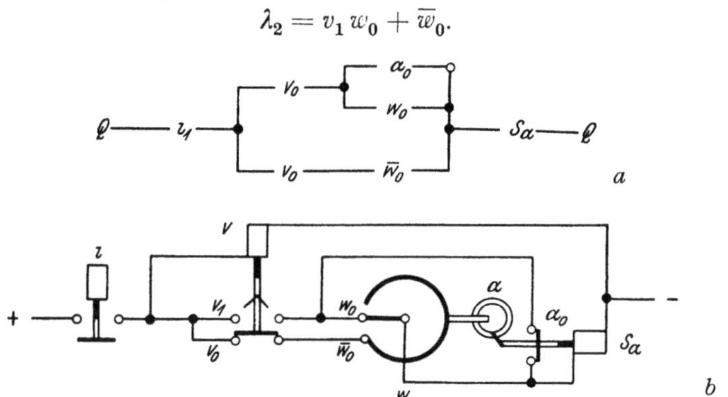

Abb. 37. Steuerung eines Drehwählers. a Stromlaufplan, b Wirkschaltplan

Diese Formel können wir sofort kürzen zu

$$\lambda_2 = v_1 + \overline{w}_0.$$

Die Bedingung 3 beschreibt eine Anspeisung der Spule S_a über den Ruhekontakt a_0 im Stellungsbereich $\overline{w}_0$ bei erregtem Relais v, was wir mit $a_0\,\overline{w}_0\,v_1$ ausdrücken können. Das Verhalten bei w_0 und v_0 bleibt hierbei unbeschränkt, so daß wir diese Operanden addieren müssen. Zu beachten ist, daß auch über $\overline{a}_0$ keine Angabe besteht und dieser Operand trotzdem nicht als freie Bedingung beigefügt werden darf, weil a kein selbständiger Steuerschalter, sondern der gesteuerte (unfreie) Antrieb ist. Wir erhalten demnach

$$\lambda_3 = a_0\,\overline{w}_0\,v_1 + w_0 + v_0.$$

Auch diese Gleichung kürzen wir sofort zu

$$\lambda_3 = a_0\,v_1 + w_0 + v_0.$$

Aus den oben formulierten Bedingungen erhalten wir nun die Funktion der Schaltung:

$$\frac{\lambda}{S_a} = i_1(v_1 + \overline{w}_0) \cdot (a_0 v_1 + w_0 + v_0) =$$
$$= i_1(a_0 v_1 + a_0 v_1 \overline{w}_0 + v_1 w_0 + w_0 \overline{w}_0 + v_1 v_0 + v_0 \overline{w}_0 =$$
$$= i_1[a_0 v_1(\infty + \overline{w}_0) + v_1 w_0 + 0 + 0 + v_0 \overline{w}_0] =$$
$$= i_1(a_0 v_1 + v_1 w_0 + v_0 \overline{w}_0) =$$
$$= i_1[v_1(a_0 + w_0) + v_0 \overline{w}_0].$$

Die gefundene Leitwertgleichung zeigt gleichzeitig die vollständig verkettete Form der Schaltung; eine Vermaschung kommt nicht in Frage, weil jeder Kontakt nur mehr einmal in der Schaltung vorhanden ist. Abb. 37 zeigt die Schaltung, von der wir nun auf Grund der Rechnung mit Bestimmtheit sagen können, daß sie die einfachste überhaupt mögliche Lösung der gestellten Aufgabe ist.

Die Möglichkeiten der Vereinfachung, die sich aus der Verknüpfung von zueinander inversen Variablen ergeben, lassen sich auch als Schaltregeln formulieren. Diese seien im nachstehenden angeführt:

Kürzungsregel für Serienschaltungen. Enthält ein Strompfad hintereinandergeschaltete Kontakte oder Kontaktkombinationen, die nicht gleichzeitig Strom führen, ist er wirkungslos; er kann nach Bedarf aus der Schaltung gestrichen oder dieser beigefügt werden. Es gilt:

$$x \cdot \overline{x} = 0.$$

Kürzungsregel für Parallelschaltungen. Zwei oder mehrere zueinander parallele Kontakte oder Strompfade, die stets abwechselnd Strom führen, können durch eine Leitung ersetzt werden. Es gilt

$$x + \overline{x} = \infty.$$

Kürzungsregeln für gemischte Schaltungen. Enthält ein Strompfad einen Kontakt oder eine Kontaktkombination, die zu einem parallelen Strompfad invers (komplementär) ist, so hat der Kontakt, bzw. die Kombination keine Wirkung und kann durch eine Leitung ersetzt werden. Es gilt

$$x + \overline{x}\, y = x + \infty\, y = x + y.$$

Enthält hingegen ein Strompfad sämtliche Kontakte eines parallelen Strompfades, so ist er wirkungslos und kann zur Gänze entfallen. Es gilt

$$x + x\, y = x.$$

5. Die Variation der Leitbedingungen

Wir haben gesehen, daß der Leitwert eines Strompfades ausschließlich davon abhängt, in welchen Schalterstellungen die Kontakte, die er enthält, geschlossen sind. Durch die Zahl der Schalter und ihrer Stellungen ist daher bestimmt, wieviel funktionsverschiedene Strompfade eine Schaltung höchstens haben kann. Dies wird sofort verständlich, wenn man eine Wählerschaltung der Fernmeldetechnik betrachtet, deren grundsätzlicher Aufbau in Abb. 38 dargestellt ist.

Hat der erste Schalter S_1 Stellungen, so spaltet er die anspeisende Leitung in S_1 parallele Zweige. Nach einem zweiten Schalter mit S_2 Stellungen sind bereits $S_1 \cdot S_2$ Zweige vorhanden. Jeder weitere Schalter vervielfacht die Zahl der möglichen Strompfade um einen Faktor, der seiner Stellungszahl gleich ist. Die Höchstzahl der verschiedenartigen Strompfade, die über alle Schalter führen, ist also gleich dem Produkt der Stellungszahlen der Schaltgeräte. Schreibt man sämtliche Strompfade, die sich auf diese Weise bilden lassen, untereinander in algebraischer Form an, so ergibt sich eine Tabelle, in der die Stellungsnummern der Schalter als Variationen geordnet sind und die deshalb *Variationstabelle* genannt werden soll. Nachstehend ein Beispiel für eine solche Tabelle, die sich auf drei Schalter mit je zwei Stellungen bezieht:

a	b	c		a	b	c
a_0	b_0	c_0		0	0	0
a_0	b_0	c_1		0	0	1
a_0	b_1	c_0		0	1	0
a_0	b_1	c_1	oder	0	1	1
a_1	b_0	c_0		1	0	0
a_1	b_0	c_1		1	0	1
a_1	b_1	c_0		1	1	0
a_1	b_1	c_1		1	1	1

Diese Tabelle entspricht einer Schaltung, die in Abb. 39 wiedergegeben ist. Die Tabelle zeigt, daß die Stellungsnummern 0 und 1 als Elemente einer Variation aufgefaßt werden können. Die Schaltgeräte a, b, c sind die Plätze, auf denen die Elemente erscheinen. In der Tabelle bilden somit die Spalten die Plätze, wogegen jede Zeile eine andere Stellungsvariation zeigt. Die Reihenfolge der Zeilen ist an sich belanglos. Die Tabelle wird jedoch übersichtlicher, wenn man die Variationen nach einer natürlichen Reihenfolge ordnet. Diese ist durch steigenden Zahlenwert der Stellungsnummern gekennzeichnet; betrachtet man die durch Stellungs- nummern dargestellten Variationen als dekadische Zahlen, so nimmt deren Wert von Zeile zu Zeile zu.

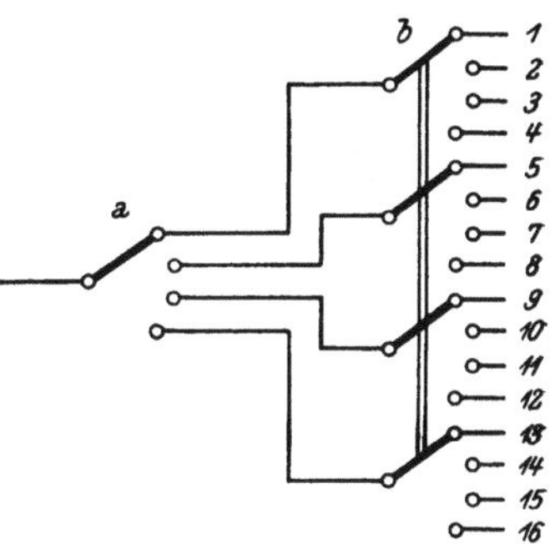

Abb. 38. Wählerschaltung

Bei den Wählerschaltungen der Fern- sprechtechnik, deren Aufbau der Abb. 38 entspricht, ist jede Stellungsvariation der Geräte einem anderen Ver- braucher zugeordnet. Im allgemeinen ist dies jedoch nicht der Fall. Als Beispiel betrachten wir eine Wechselschaltung mit drei zweistelligen Schaltern, wie sie in der Installationstechnik benützt wird. Die Leit- wertgleichung dieser Schaltung, die in Abb. 26 a gezeigt wurde, lautet:

$$\frac{\lambda}{Q\,V} = a_0\,b_0\,c_1 + a_0\,b_1\,c_0 + a_1\,b_0\,c_0 + a_1\,b_1\,c_1.$$

Vergleicht man diese Leitwertgleichung mit der Schaltung nach Abb. 39, so erkennt man, daß von den insgesamt acht möglichen Strompfaden genau die Hälfte benützt und an den gemeinschaftlichen Verbraucher angeschlossen ist. Dieses sehr kennzeichnende Merkmal läßt sich

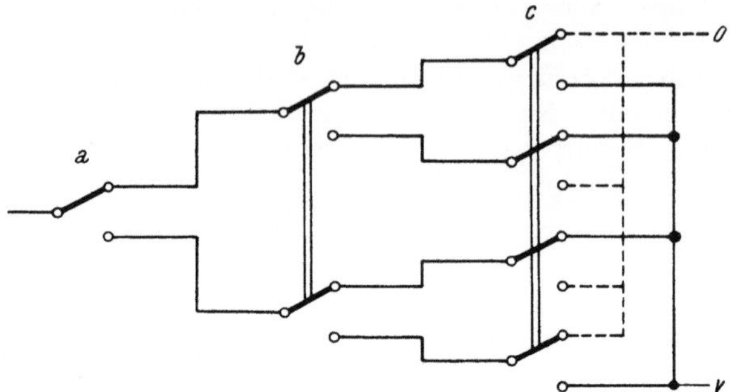

Abb. 39. Stellungsvariationen einer Wechselschaltung mit drei Schaltern

aus dem Stromlaufplan nach Abb. 26 a und auch aus der zugehörigen Leitwertgleichung nicht auf den ersten Blick erkennen; es tritt aber mit voller Deutlichkeit zutage, wenn man die nachstehend gezeigte Variationstabelle der Schaltung anschreibt.

a	b	c	λ
0	0	0	0
0	0	1	∞
0	1	0	∞
0	1	1	0
1	0	0	∞
1	0	1	0
1	1	0	0
1	1	1	∞

Die Tabelle sagt mehr aus als der Schaltplan und die Leitwertgleichung; sie zeigt nicht nur jene Strompfade, die tatsächlich vorhanden sind, sondern überhaupt alle, die sich mit den benützten Schaltgeräten herstellen lassen. Durch Beifügung des jeweiligen Leitwertes 0 oder ∞ scheidet sie die möglichen Strompfade in benützte und unbenützte. Sie beschreibt somit nicht nur die Wirkungsweise der Schaltung, sondern zeigt auch die Möglichkeiten zu deren Veränderung.

Das im vorstehenden beschriebene Beispiel betraf eine Schaltung, deren Strompfade durchwegs über Kontakte sämtlicher Schaltgeräte führen; daher konnte die Variationstabelle unmittelbar aus der Leitwertgleichung entwickelt werden. Bei den meisten Schaltungen ist diese Voraussetzung jedoch nicht gegeben. Es läßt sich aber jede beliebige Leitwertgleichung durch Erweiterung so umformen, daß sie eine Summe von Stellungsvariationen darstellt. Als Beispiel betrachten wir die einfache Gleichung

$$\lambda = a_0 + b_0,$$

die sich auf zwei Schalter mit je zwei Stellungen beziehen möge. Da wir die Zwischenstellung beim Überschalten vernachlässigen können, ist $a_0 = \overline{a}_1$ und $a_1 = \overline{a}_0$. Wir multiplizieren nun den Summanden a_0 mit einem Faktor $b_0 + b_1 = \infty$, der den Wert der Funktion nicht verändert; in analoger Weise erweitern wir das Glied b_0 der Gleichung um einen Faktor $a_0 + a_1$. Hierdurch erhalten wir:

$$\begin{aligned}
\lambda = a_0 + b_0 &= \\
&= a_0(b_0 + b_1) + b_0(a_0 + a_1) = \\
&= a_0\,b_0 + a_0\,b_1 + a_0\,b_0 + a_1\,b_0 = \\
&= a_0\,b_0 + a_1\,b_0 + a_0\,b_1.
\end{aligned}$$

Nach dieser Umformung enthält die Leitwertgleichung durchwegs Glieder, die sich auf beide Schalter beziehen und somit Stellungsvariationen, das heißt Schaltzustände beschreiben. Diese *vollständig erweiterte Form* (Vollform) der Gleichung gestattet nunmehr den Ansatz der Variationstabelle. Im vorliegenden Falle lautet diese:

a	b	λ
0	0	∞
0	1	∞
1	0	∞
1	1	0

Die Erweiterung durch algebraische Rechnung wurde im vorstehenden nur zum besseren Verständnis gezeigt. Die Variationstabelle läßt sich auch unmittelbar aus der gekürzten Form der Gleichung ansetzen. Um die Spalte λ auszufüllen, **hätten wir** in allen Zeilen, in denen a_0 oder b_0 vorkommt, das Zeichen ∞ eintragen können; die übrigbleibende letzte Zeile muß dann durch 0 ergänzt werden. In analoger Weise läßt sich die gekürzte Leitwertgleichung aus der Tabelle ermitteln. Faßt man die beiden ersten Zeilen durch Addition zusammen, so erscheint b in allen beiden Stellungen und kann gekürzt werden, so daß a_0 übrigbleibt; ebenso ergibt die erste mit der dritten Zeile zusammen b_0. Die Kürzung besteht also in einem schrittweisen Vergleich der Zeilen; dabei können solche Zeilen zusammengefaßt werden, die sich nur hinsichtlich einer einzigen Stellungsnummer voneinander unterscheiden. Es ist hierbei nicht nötig, jede Zeile mit jeder anderen zu vergleichen. Handelt es sich um Schaltgeräte mit n Stellungen, so liegt zwischen zusammengehörigen Zeilen stets ein Abstand, der gleich einer Potenz von n ist. Im vorstehenden Beispiel entspricht der Abstand zwischen der ersten und zweiten Zeile der nullten, zwischen der ersten und dritten Zeile der ersten Potenz von 2. Bei größeren Tabellen erleichtert diese Gesetzmäßigkeit der Tabelle ganz wesentlich die Umwandlung in eine maximal gekürzte Gleichung[1]. Die Variationstabelle ist daher ein bequemes Hilfsmittel zur Erweiterung und Kürzung von Leitwertfunktionen.

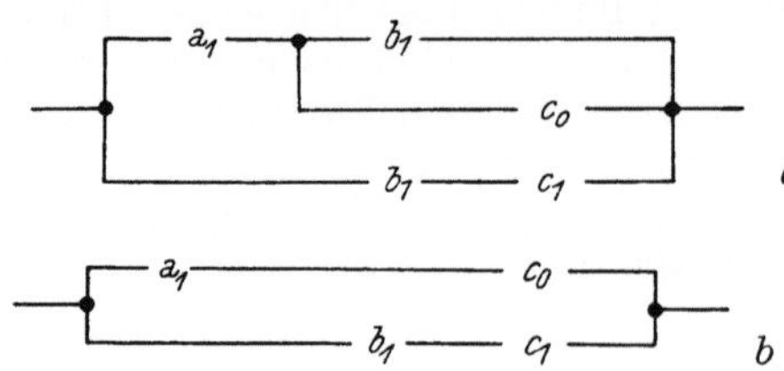

Abb. 40. Entfall eines Strompfades durch Kürzung. *a* Form mit drei Strompfaden, *b* Form mit zwei Strompfaden

Die Umwandlung einer Leitwertgleichung in ihre Vollform läßt häufig Kürzungsmöglichkeiten erkennen, die auf andere Weise nur schwer zu finden sind. Hierzu ein Beispiel: Gegeben sei eine Schaltung gemäß Abb. 40 *a*. Hierbei sei angenommen, daß die Reihenfolge der Schalter in den Strompfaden mit Rücksicht auf die Leitungsführung festgelegt sei und nicht verändert werden darf. Der Schaltung entspricht eine Leitwertgleichung

$$\lambda = a_1\,b_1 + a_1\,c_0 + b_1\,c_1 = a_1(b_1 + c_0) + b_1\,c_1.$$

Weder dem Stromlaufplan noch der beschreibenden Gleichung ist anzusehen, daß sich die Schaltung durch Kürzung vereinfachen läßt. Wir übertragen nun die Funktion in die Variationstabelle. Da es sich um drei Schalter mit je zwei Stellungen handelt, sind insgesamt acht Stellungsvariationen möglich, die sich ohne Schwierigkeit anschreiben lassen.

Nr.	a	b	c
1	0	0	0
2	0	0	1
3	0	1	0
4	0	1	1
5	1	0	0
6	1	0	1
7	1	1	0
8	1	1	1

Nunmehr können wir zu den einzelnen Summanden der Gleichung der Reihe nach die zugehörigen Zeilen der Tabelle aufsuchen und dort den

---

[1] Voraussetzung zur Kürzung sind komplementäre Stellungsvariationen. Diese unterscheiden sich voneinander nur hinsichtlich je einer Schalterstellung; sie entsprechen benachbarten Schaltzuständen, was sich an Hand des Schaltfolgeplanes leicht erkennen läßt. Da alle Schaltfolgen zyklisch sein müssen, kann zur Ermittlung komplementärer Variationen auch die Modulrechnung herangezogen werden. Auf dieses dem Techniker ganz ungewohnte Hilfsmittel kann jedoch verzichtet werden, weil die Variationstabellen für den praktischen Gebrauch völlig ausreichen. Es genüge ein Hinweis auf diese Möglichkeit, von der die theoretische Schaltlehre später zur Lösung besonderer Aufgaben Gebrauch machen könnte.

Leitwert ∞ eintragen. Wir beginnen mit dem ersten Glied $a_1 b_1$. Diese Stellungskombination findet sich in der Tabelle in der siebenten und achten Zeile, die wir beide mit dem Leitwert ∞ kennzeichnen.

Nr.	a	b	c	λ
1	0	0	0	.
2	0	0	1	.
3	0	1	0	.
4	0	1	1	.
5	1	0	0	.
6	1	0	1	.
7	1	1	0	∞
8	1	1	1	∞

Der nächsten Kombination $a_1 c_0$ entspricht die fünfte und siebente Zeile; letztere ist bereits ausgefüllt, so daß wir nur die fünfte Zeile mit ∞ zu ergänzen haben.

Nr.	a	b	c	λ
1	0	0	0	.
2	0	0	1	.
3	0	1	0	.
4	0	1	1	.
5	1	0	0	∞
6	1	0	1	.
7	1	1	0	∞
8	1	1	1	∞

Schließlich finden wir zum letzten Glied $b_1 c_1$ die vierte und achte Zeile, denen ebenfalls der Leitwert ∞ zukommt.

Nr.	a	b	c	λ
1	0	0	0	.
2	0	0	1	.
3	0	1	0	.
4	0	1	1	∞
5	1	0	0	∞
6	1	0	1	.
7	1	1	0	∞
8	1	1	1	∞

Da die Gleichung keine weiteren Summanden mehr enthält, bleiben in der Spalte λ der Tabelle noch vier Plätze frei, die wir mit 0 ausfüllen.

Nr.	a	b	c	λ
1	0	0	0	0
2	0	0	1	0
3	0	1	0	0
4	0	1	1	∞
5	1	0	0	∞
6	1	0	1	0
7	1	1	0	∞
8	1	1	1	∞

Damit ist die Variationstabelle fertig und zeigt, daß in unserer Schaltung vier Variationen benützt und die restlichen vier unbenützt sind. Nunmehr können wir darangehen, aus der Tabelle die maximal gekürzte Leitwertfunktion zu ermitteln. Die ersten drei Zeilen können wir überspringen, da sie den Leitwert 0 haben und somit nicht zur Leitwertfunktion gehören. Wir beginnen daher mit Zeile 4 und vergleichen diese mit den übrigen Zeilen, in denen ∞ steht. Da die Schalter je zwei Stellungen haben, müssen nicht alle diese Zeilen verglichen werden; es genügt jeweils die nächste, zweite und vierte. Wir vergleichen daher vorerst die Zeilen 4 und 5. Diese unterscheiden sich um mehr als eine Stellung und ermöglichen keine Kürzung. Das gleiche gilt auch für das Zeilenpaar 4 und 6. Hingegen ist das Zeilenpaar 4 und 8 nur in der Stellung von a verschieden, so daß wir zu $b_1 c_1$ kürzen können. Diese Kombination ist das erste Glied der gekürzten Gleichung. Wir können daher den Ansatz

$$\lambda = b_1 c_1 + \ldots$$

anschreiben und in der Tabelle die zugehörigen Zeilen 4 und 8 streichen. Nunmehr setzen wir den Zeilenvergleich fort und betrachten das Zeilenpaar 5 und 7. Wir sehen, daß sich $a_1 c_0$ herausheben und $b_0 + b_1$ zu ∞ kürzen läßt. Daher dürfen wir auch diese beiden Zeilen der Tabelle streichen und erhalten die vollständige gekürzte Funktion

$$\lambda = b_1 c_1 + a_1 c_0.$$

Aus der ursprünglich vorgegebenen Gleichung ist also der Summand $a_1 b_1$ vollständig verschwunden. Abb. 40 b zeigt die maximal gekürzte Form.

An vorstehendem Beispiel verdient ein Umstand besondere Beachtung: Die vorgegebene Funktion und die zugehörige Schaltung umfaßt drei Strompfade, während die gekürzte Form nur zwei solche aufweist. Die Kürzung geht hier so weit, daß der ganze Stromkreis $a_1 b_1$ als überflüssig erkannt und ausgeschieden wird. Es ist ein besonderer Vorteil der rechnerischen Behandlung, solche Möglichkeiten aufzuzeigen, die in der Praxis häufig übersehen werden, weil sie aus dem Schaltplan außerordentlich schwer erkennbar sind. Ferner vermittelt uns die vorstehende Rechnung eine grundlegende Erkenntnis: Die Wirkungsweise einer Schaltung ist *nicht* durch Zahl und Aufbau ihrer Strompfade bestimmt, sondern einzig und allein durch jene Stellungsvariationen der Schaltgeräte, die den Strompfaden zugrunde liegen. Durch Zusammenfassung von Variationen und Kürzung der komplementären Stellungen ergeben sich die Strompfade; aus der teilweisen Zusammenlegung der Strompfade folgt dann die verkettete Schaltung.

Der Zusammenhang zwischen der Variationstabelle und der gekürzten Leitwertgleichung führt zu einer sehr wichtigen Schaltregel. Diese lautet:

Funktionsregel. Die Wirkungsweise einer Schaltung ist ausschließlich durch die Stellungsvariationen der Schaltgeräte, bei denen sie Strom führt, bestimmt; die den Strompfaden zugrunde liegenden Stellungskombinationen sind gekürzte Summen dieser Stellungsvariationen. Um die für eine vorgegebene Wirkungsweise notwendigen Strompfade zu finden, hat man daher jene Stellungsvariationen, denen ein von 0 verschiedener Leitwert der Schaltung zugeordnet ist, zu addieren und die Summe vollständig zu kürzen.

Die feste Beziehung zwischen der Höchstzahl der Strompfade und der Anzahl der Schalter und deren Stellungen ermöglicht ferner die Lösung mancher Aufgaben, bei denen die Zahl der Schaltgeräte und ihrer Stellungen *nicht* gegeben ist. Dies sei durch ein Beispiel gezeigt:

Zur selbsttätigen Herstellung von Fernsprech- oder Fernwirkverbindungen ist es erforderlich, eine gemeinsame Fernleitung fallweise mit verschiedenen Ortsanschlüssen zu verbinden; hierzu benützt man Wahlschalter (Wähler), die entsprechend Abb. 38 in Reihe geschaltet sind. In Fernsprechanlagen verwendet man meist Geräte, bei denen die Zahl der Stellungen zehn oder ein Vielfaches davon beträgt, weil dann die Nummern der Strompfade zugleich die Rufnummern der Teilnehmer sind. Für Fernwirkanlagen besteht diese Beschränkung nicht und es können ohne weiteres Geräte mit anderer Stellungszahl gewählt werden, wenn daraus technische Vorteile erwachsen. Wir wollen nun die Forderung stellen, daß die Zahl der Schaltschritte aller Wähler, die bei der Herstellung einer Verbindung ausgeführt werden müssen, möglichst klein sein soll; damit erzielen wir die größte Arbeitsgeschwindigkeit der Schaltung und gleichzeitig die größte Lebensdauer der Geräte. Wir wollen nun untersuchen, bei welcher Stellungszahl der Wähler diese Forderung erfüllt ist.

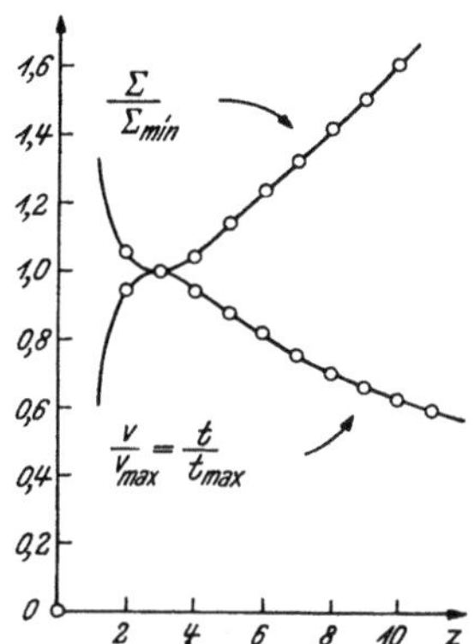

Abb. 41. Arbeitsgeschwindigkeit und Lebensdauer von Wählerschaltungen.

$z\ldots$ Zahl der Stellungen,
$\Sigma\ldots$ Zahl der Schritte,
$v\ldots$ Geschwindigkeit,
$t\ldots$ Lebensdauer

Nennen wir die Zahl der Wählerstellungen z und die Zahl der hintereinandergeschalteten Wähler n, so sind insgesamt z^n Stellungsvariationen **und daher ebensoviele Strompfade herstellbar;** wir bezeichnen die Zahl der Strompfade mit s und schreiben:

$$s = z^n \qquad \text{oder} \qquad n = \frac{\ln s}{\ln z} .$$

Zur Herstellung und nachfolgenden Trennung einer Verbindung macht jeder Wähler einen vollen Umlauf. Die Zahl der Schaltschritte je Verbindung ist daher z für einen Wähler und $\Sigma = n\,z$ für die ganze Einrichtung. Der Wert Σ soll nun bei vorgegebener Anzahl s der Strompfade ein Minimum werden. Wir setzen daher

$$\sum = n\,z = z\,\frac{\ln s}{\ln z} = \frac{z}{\ln z}\ln s .$$

Abb. 41 zeigt den Verlauf dieser Funktion. Durch Differentiation ermitteln wir die Stelle des Minimums:

$$\frac{d\Sigma}{dz} = \frac{\ln z - 1}{\ln^2 z} \cdot \ln s = 0$$

$$\ln z = 1 \quad \text{und daraus} \quad z = e.$$

Die der Zahl e nächstliegende Zahl ist 3; wir haben also die Wähler mit je drei Stellungen auszuführen. Ein Vergleich mit dekadischen Wählern zeigt eine Verbesserung im Verhältnis $\dfrac{10}{3} \cdot \dfrac{\ln 3}{\ln 10}$ oder rund 3 : 2. Die technische Praxis benützt allerdings nicht genau diese Ausführung, sondern begnügt sich mit einer Annäherung, die durch Schaltgeräte mit je zwei Stellungen gekennzeichnet ist (Dualschaltung). Wie die Abb. 41 zeigt, liegt bei dieser Schaltung die Arbeitsgeschwindigkeit und Lebensdauer der Geräte dem Höchstwert sehr nahe. Durch die Beschränkung auf nur zwei Stellungen je Wähler ergibt sich aber eine konstruktive Vereinfachung, weil man an Stelle von Drehwählern Schütze verwenden und somit die Schrittschaltwerke ersparen kann.

6. Die Umkehrung von Leitbedingungen

Eine in der Schaltungstechnik häufige Aufgabe besteht darin, daß zu einer vorgegebenen Schaltung die zugehörige Kehrschaltung zu suchen ist. Man versteht unter *Kehrschaltungen* oder *komplementären Schaltungen* ein Schaltungspaar, von dem die eine Anordnung immer dann Strom führt, wenn die andere stromlos ist und umgekehrt; haben die benützten Schaltgeräte eine bevorzugte Ruhestellung, so bezeichnet man eine der beiden komplementären Anordnungen als Ruhestromschaltung und die andere als Arbeitsstromschaltung. Die Aufgabe, eine Teilschaltung oder eine Verknüpfung von Schaltbedingungen umzukehren, tritt übrigens bei der Ausmittlung aller Schaltungen auf, wenn sich irgendwelche Funktionsbedingungen auf das Nichtfließen des Stromes beziehen. Um eine Leitwertfunktion anschreiben zu können, müssen alle Teilbedingungen so formuliert sein, daß sie Leitbedingungen für den Strom darstellen. Gegenteilige Bedingungen müssen also vorerst umgekehrt werden; diese *Umkehr* oder *Inversion* läßt sich auch rein gedanklich durchführen, doch ist das Rechenverfahren sicherer und bequemer.

Vorgegebene Schaltungen oder Schaltbedingungen lassen sich nun stets auf Summen und Produkte der einzelnen Elemente zurückführen; es genügt also, wenn wir das Verhalten dieser beiden Verknüpfungen bei der Inversion kennen. Wir betrachten also zwei beliebige Variable x, y und suchen zuerst die Inversion von $\lambda = x + y$. Zu diesem Zweck erweitern wir die Funktion zu einer vollständigen Variationstabelle:

$$x + y = \left| \begin{array}{l} \overline{x}\,\overline{y} \cdot 0 \\ \overline{x}\,y \cdot \infty \\ x\,\overline{y} \cdot \infty \\ x\,y \cdot \infty \end{array} \right.$$

Die komplementäre Schaltung soll immer dann Strom leiten, wenn die ursprüngliche nicht leitend ist und umgekehrt. Wir finden also die komplementäre Funktion, indem wir in der zugehörigen Variationstabelle die Zeichen 0 und ∞ miteinander vertauschen. Die Inversion von $x + y$ ergibt also:

$$\overline{x + y} = \begin{vmatrix} \overline{x}\,\overline{y} \cdot \infty \\ \overline{x}\,y \cdot 0 \\ x\,\overline{y} \cdot 0 \\ x\,y \cdot 0 \end{vmatrix} = \overline{x} \cdot \overline{y}.$$

Als Ergebnis zeigt sich, daß das Komplement einer Summe zweier Elemente gleich dem Produkt der komplementären Elemente ist. In Anwendung auf die Schaltungstechnik bedeutet dies, daß sich bei der Umkehrung der Schaltung nicht nur die Kontakte umkehren, sondern an Stelle der Parallelschaltung eine Serienschaltung tritt.

In Analogie hierzu erhalten wir als Komplement eines beliebigen Produktes $x\,y$:

$$\overline{x\,y} = \begin{vmatrix} \overline{x}\,\overline{y} \cdot 0 \\ \overline{x}\,y \cdot 0 \\ x\,\overline{y} \cdot 0 \\ x\,y \cdot \infty \end{vmatrix}$$

und hieraus durch Inversion

$$\overline{x\,y} = \begin{vmatrix} \overline{x}\,\overline{y} \cdot \infty \\ \overline{x}\,y \cdot \infty \\ x\,\overline{y} \cdot \infty \\ x\,y \cdot 0 \end{vmatrix} = \overline{x} + \overline{y}.$$

Das Komplement eines Produktes von Elementen ist also gleich der Summe der komplementären Elemente. Bei der Umkehrung einer Serienschaltung **sind daher die komplementären Kontakte** parallel zu schalten. Die beiden gefundenen Inversionsregeln besagen ferner, **daß die beiden** Operationen der Addition und der Multiplikation zueinander invers sind.

Die praktische Anwendung dieser Regeln soll an einem Beispiel gezeigt werden. Wir betrachten eine Hochspannungsanlage, die zwei Sammelschienensysteme und zwei Fernleitungsabzweige F_1 und F_2 enthält (Abb. 42 a). Diese Leitungen sollen betriebsmäßig über eines der beiden Schienensysteme durchgeschaltet werden; wenn dies nicht der Fall ist, soll ein Signal an eine entfernte Stelle, beispielsweise an eine Lastverteilstelle, über eine Meldeleitung M gegeben werden. Gesucht ist die aus Hilfskontakten der Hochspannungsschalter zu bildende Schaltung zwischen einer Hilfsquelle Q und der Meldeleitung M; sie soll immer Strom führen, wenn zwischen F_1 und F_2 keine Verbindung besteht.

Die gewünschte Beziehung zwischen der gegebenen Hochspannungs-schaltung $\dfrac{\lambda}{F_1 F_2}$ und der gesuchten Meldeschaltung $\dfrac{\lambda}{Q M}$ ist die Inversion:

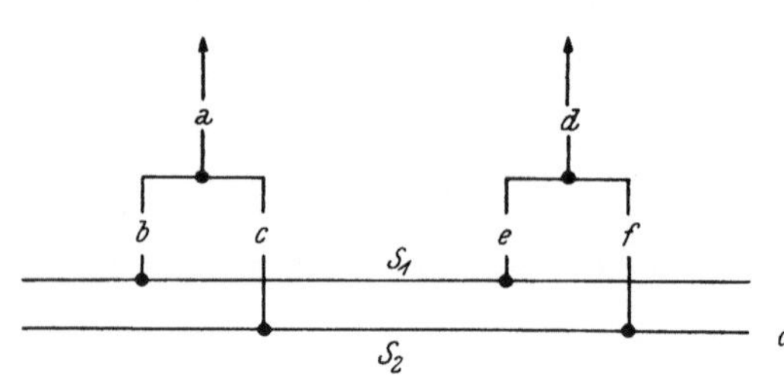

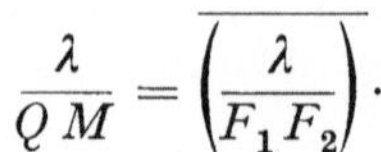

$$\frac{\lambda}{Q M} = \overline{\left(\frac{\lambda}{F_1 F_2}\right)}.$$

Aus dem einpoligen Schaltplan (Abb. 42 a), bzw. dem Stromlaufplan (Abb. 42 b) können wir die Leitwertgleichung durch Verfolgen der möglichen Strompfade leicht herauslesen und erhalten:

$$\frac{\lambda}{F_1 F_2} = a(b\,e + c\,f)\,d =$$
$$= a\,b\,d\,e + a\,c\,d\,f.$$

Auf Grund der Inversionsregeln folgt daraus

$$\frac{\lambda}{Q M} = \overline{\left(\frac{\lambda}{F_1 F_2}\right)} = (\overline{a} + \overline{b} + \overline{d} + \overline{e}) \cdot$$
$$\cdot (\overline{a} + \overline{c} + \overline{d} + \overline{f}) =^{1}$$
$$= \overline{a} + \overline{b}\,\overline{c} + \overline{c}\,\overline{e} + \overline{d} + \overline{b}\,\overline{f} + \overline{e}\,\overline{f} =$$
$$= \overline{a} + \overline{b}(\overline{c} + \overline{f}) + \overline{d} + \overline{e}(\overline{c} + \overline{f}) =$$
$$= \overline{a} + (\overline{b} + \overline{e})\,(\overline{c} + \overline{f}) + \overline{d}$$

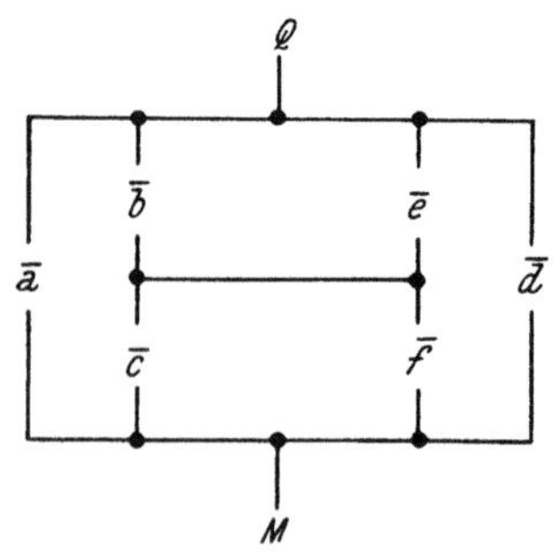

Abb. 42. Meldeschaltung für die Trennung zweier Hochspannungsleitungen. *a* Hochspannungsanlage, *b* Stromlaufplan, *c* Meldeschaltung

Der Stromlaufplan dieser Meldeschaltung ist in Abb. 42 c dargestellt.

Eine zweimalige Inversion führt zur ursprünglichen Funktion zurück. Diese Regel läßt sich zur Umformung von Schaltungen benützen. Gegeben sei eine Wechselschaltung in der üblichen Gestalt nach Abb. 43 a, ihre Funktion ist

$$\lambda = a_0\,b_1 + a_1\,b_0.$$

Die zweimalige Inversion zerlegt die Gleichung in ihre Wurzeln:

$$\overline{\lambda} = (a_1 + b_0) \cdot (a_0 + b_1) = a_1\,b_1 + a_0\,b_0,$$
$$\overline{\overline{\lambda}} = \lambda = (a_0 + b_0) \cdot (a_1 + b_1).$$

Abb. 43 b zeigt die zweite Form der Schaltung.

Die beschriebenen Gesetze für die Umkehrung gelten nicht nur für Schaltungen, sondern auch für die zu ihnen führenden Funktionsbe-

---

[1] Kürzungsregeln!

dingungen. Enthält eine Schaltaufgabe Bedingungen, die sich auf eine Sperrung des Stromflusses beziehen, so können diese genau so wie Leitbedingungen angeschrieben werden; durch Inversion findet man dann die wirklichen Leitbedingungen, die den gleichen Aussagewert haben wie die vorgegebenen Sperrbedingungen. Als Beispiel sei auf die Schaltung nach Abb. 34 in Abschnitt II, 3, verwiesen. Durch die Schaltung soll erzwungen werden, daß von drei beliebigen Verbrauchern stets einer ausgeschaltet bleibt. Bei der Lösung des Beispiels war diese Vorschrift nicht mathematisch, sondern durch Überlegung in eine Leitbedingung λ_4 verwandelt worden. Einfacher ist es, unmittelbar die Sperrbedingung anzuschreiben

$$\overline{\lambda_4} = a_1\, b_1\, c_1.$$

Unter Anwendung der Inversionsregel folgt hieraus ohne jegliche Denkarbeit die gleichwertige Leitwertbedingung

$$\lambda_4 = a_0 + b_0 + c_0,$$

die in der Rechnung verwendet wurde und zur Lösung der Aufgabe geführt hat.

Abschließend sei die Schaltregel für die Umkehrung von Schaltungen und Schaltbedingungen nochmals angeführt. Sie lautet:

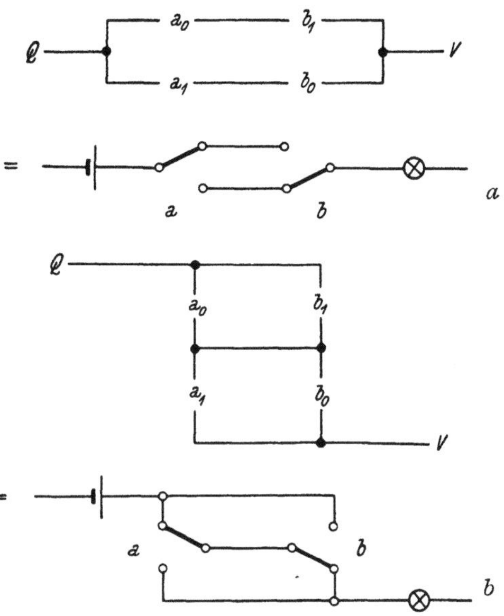

Abb. **43**. Die beiden Formen einer Wechselschaltung. *a* Form mit zwei parallelen Serienverbindungen, *b* Form mit zwei hintereinanderliegenden Parallelverbindungen

Umkehrregel. (Inversionsregel.) Bei der Umkehrung einer Schaltung treten an Stelle der **Kontakte** die komplementären Kontakte und an Stelle aller Parallelschaltungen Serienschaltungen **und umgekehrt**; bei der Umkehrung von Funktionsbedingungen ist der stromleitende **Zustand** mit dem nicht leitenden zu vertauschen, wobei voneinander unabhängige Bedingungen zu gleichzeitigen werden und umgekehrt. Es gilt:

$$\overline{x+y} = \overline{x}\cdot\overline{y} \qquad \text{und} \qquad \overline{x\,y} = \overline{x} + \overline{y}.$$

Zu beachten ist, daß es keine Schaltelemente gibt, die zu Stromquellen oder Verbrauchern komplementär sind; diese haben endliche Leitwerte und die Inversion von 0 führt stets zu Null.

7. Die Freiheitsgrade von Bedingungskomplexen

Es soll nun untersucht werden, ob es außer der logischen Summe und dem logischen Produkt noch weitere Verknüpfungen von Leitbedingungen gibt und welche Eigenschaften diese zeigen. Wie wir gesehen haben, gilt in der Rechnung mit Bedingungen das distributive Gesetz; dessen Umkehrung

$$x\,y + x\,z = x(y + z)$$

beruht auf der Division der Produkte $x\,y$ und $x\,z$ durch den gemeinsamen Faktor x. Die Operation der Division liefert nun eine weitere Art der Verknüpfung, nämlich den Quotienten zweier Leitbedingungen. Wir wollen diese Verknüpfung so definieren, daß wir unter dem Quotienten $q = x/y$ einen variablen Wert verstehen, der die Bedingung $q \cdot y = x$ erfüllt. Während nun der Quotient endlicher Größen eine eindeutige Funktion dieser Größen ist, bewirkt die Geltung des Äquivalenzgesetzes eine *Mehrdeutigkeit*. Es ist nämlich sowohl

$$x = x \cdot \overline{0} \qquad \text{und} \qquad \frac{x}{x} = \overline{0},$$

als auch

$$x = x \cdot x \qquad \text{und} \qquad \frac{x}{x} = x.$$

Der Quotient $q = x/x$ kann also $\overline{0}$ oder x sein; in beiden Fällen erfüllt er die Bedingung $q\,x = x$. Für $\overline{0}$ kann hierbei stets auch der spezielle Wert ∞ gesetzt werden, wenn x der Leitwert eines Kontaktes oder einer Kontaktkombination ist.

Eine analoge Betrachtung läßt sich auch für die Differenz zweier Leitbedingungen anstellen:

$$x = x + 0, \qquad x - x = 0,$$
$$x = x + x, \qquad x - x = x.$$

Auch die Subtraktion liefert also ein mehrdeutiges Ergebnis (unbestimmtes Resultat). Ein Beispiel möge die Anwendung dieser Rechenregeln erläutern.

Gegeben sei ein dreiphasiger Verbraucher U, V, W (z. B. ein Widerstandsofen), bei dem die einzelnen Phasen im Dreieck liegen und getrennt durch je einen Schalter u, v, w gesteuert werden (Abb. 44 a). Die Schaltung soll nun so abgeändert werden, daß der Verbraucher einen geerdeten Sternpunkt erhält; die Schaltung soll wirkungsgleich sein mit der Dreieckschaltung. Gefragt ist, welche Elemente in den Sternleitungen an den Stellen x, y, z (Abb. 44 b) vorhanden sein müssen.

Die Funktion der gegebenen Dreieckschaltung läßt sich durch drei Leitwertgleichungen beschreiben, die auch bei der gesuchten Schaltung erfüllt sein sollen. Hieraus ergeben sich drei Doppelgleichungen:

$$\frac{\lambda}{R\,S} = u\,U + v\,V\,w\,W = x\,y,$$

$$\frac{\lambda}{S\,T} = v\,V + w\,W\,u\,U = y\,z,$$

$$\frac{\lambda}{T\,R} = w\,W + u\,U\,v\,V = z\,x.$$

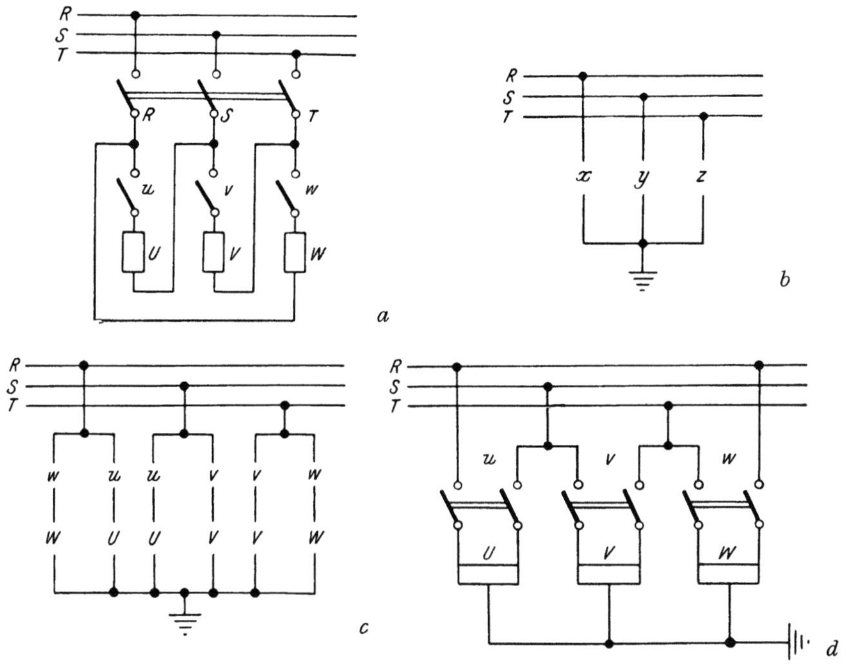

Abb. 44. Umformung einer Dreieckschaltung in eine wirkungsgleiche Sternschaltung. *a* Dreieckschaltung, *b* Kontaktstellen der Sternschaltung, *c* Sternschaltung, Stromlaufplan, **d Sternschaltung, Wirkschaltplan**

Wir substituieren $u\,U = a$, $v\,V = b$, $w\,W = c$ und erhalten:

$$a + b\,c = x\,y,$$
$$b + c\,a = y\,z,$$
$$c + a\,b = z\,x.$$

Die Ausrechnung von x ergibt

$$x = \frac{a + b\,c}{b + c\,a} \cdot \frac{c + a\,b}{x} = \frac{a\,b + b\,c + c\,a}{(b + c\,a)\,x}$$

und nach Multiplikation beider Seiten mit x:

$$x = \frac{a\,b + b\,c + c\,a}{b + c\,a}.$$

Bei der Durchführung der Division machen wir von der Mehrdeutigkeit der Differenz Gebrauch und benützen jeweils den Wert, der eine Division ohne Rest gestattet:

$$(a\,b + b\,c + c\,a) : (b + c\,a) = a + c.$$

$$\frac{(a\,b \qquad + c\,a)}{0 + b\,c + c\,a}$$

$$\frac{(b\,c + c\,a)}{0 + 0}$$

Das Resultat lautet daher:

$$x = c + a = w\,W + u\,U$$

und analog hierzu:

$$y = a + b = u\,U + v\,V,$$
$$z = b + c = v\,V + w\,W.$$

Abb. 44 c zeigt die gefundene Schaltung.

Subtraktion und Division sind zueinander inverse Operationen wie Addition und Multiplikation. Dies läßt sich leicht zeigen. Wir setzen:

$$\frac{x}{y} = q,$$
$$x = q\,y,$$
$$\overline{x} = \overline{q} + \overline{y},$$
$$\overline{q} = \overline{x} - \overline{y}.$$

Es ist also:

$$\overline{\left(\frac{x}{y}\right)} = \overline{x} - \overline{y}$$

und analog auch

$$\overline{x - y} = \frac{\overline{x}}{\overline{y}}.$$

Dieser Zusammenhang gibt uns die Möglichkeit, jede Division auf eine Subtraktion zurückzuführen und umgekehrt, was oft eine wesentliche Vereinfachung der Rechnung zur Folge hat. Das Verfahren beruht auf einer zweimaligen Inversion nach dem Schema

$$\frac{x}{y} = \overline{\overline{x - y}}.$$

Als Beispiel wollen wir uns die Aufgabe stellen, sämtliche Quotienten des Bruches $\dfrac{a\,b}{b}$ zu bestimmen. Hierzu schreiben wir nach obigem

$$\frac{a\,b}{b} = \overline{(\overline{a} + \overline{b}) - \overline{b}}$$

und führen zuerst die Subtraktion aus, wobei wir die einzelnen Glieder zur Vollform erweitern

$$\overline{a}\,\overline{b} + \overline{a}\,b + a\,\overline{b}$$
$$- \overline{a}\,\overline{b} \qquad\qquad - a\,\overline{b}$$
$$\begin{Bmatrix} 0 \\ \text{oder} \\ \overline{a}\,\overline{b} \end{Bmatrix} + \overline{a}\,b + \begin{Bmatrix} 0 \\ \text{oder} \\ a\,\overline{b} \end{Bmatrix}$$

Wir erhalten insgesamt vier Lösungen für die Differenz d:

Lösung 1: $d_1 = 0 + \overline{a}\,b + 0 = \overline{a}\,b$ (kleinste Differenz),

Lösung 2: $d_2 = 0 + \overline{a}\,b + a\,\overline{b} = \overline{a}\,b + a\,\overline{b}$,

Lösung 3: $d_3 = \overline{a}\,\overline{b} + \overline{a}\,b + 0 = \overline{a}$,

Lösung 4: $d_4 = \overline{a}\,\overline{b} + \overline{a}\,b + a\,\overline{b} = \overline{a} + \overline{b}$ (größte Differenz).

Durch Inversion der Resultate ergeben sich die gesuchten Quotienten:

Lösung 1: $q_1 = a + \overline{b} = \overline{a}\,\overline{b} + a\,\overline{b} + a\,b$ (größter Quotient),

Lösung 2: $q_2 = \overline{a}\,\overline{b} + a\,b$,

Lösung 3: $q_3 = a = a\,\overline{b} + a\,b$,

Lösung 4: $q_4 = a\,b$ (kleinster Quotient).

Die Resultate lassen sich entsprechend der Zahl ihrer Glieder in der Vollform nach der Größe ordnen; hierbei entspricht die *kleinste Differenz* dem *größten Quotienten* und umgekehrt. Die größte Differenz und der kleinste Quotient sind stets gleich dem Minuenden, bzw. Dividenden. Von Wichtigkeit ist jenes Resultat, das die stärkste Kürzung erlaubt und daher zur einfachsten Schaltung führt; in der vorstehenden Rechnung ist dies die Lösung 3 mit $d_3 = \overline{a}$, bzw. $q_3 = a$. Wir wollen diese speziellen Differenzen und Quotienten *kürzeste Resultate* nennen. Ist die gegebene Funktion hinsichtlich ihrer Elemente ganz oder teilweise symmetrisch, so treten mehrere voneinander verschiedene Resultate mit maximaler **Kürzung auf. Ein Beispiel hierfür ist**

$$d = (a + b) - (\overline{a}\,b + a\,\overline{b}) = \quad \overline{a}\,b + a\,b + a\,\overline{b}$$
$$- (\overline{a}\,b \qquad\qquad + a\,\overline{b})$$
$$\begin{Bmatrix} 0 \\ \text{oder} \\ \overline{a}\,b \end{Bmatrix} + a\,b + \begin{Bmatrix} 0 \\ \text{oder} \\ a\,\overline{b} \end{Bmatrix}$$

Lösung 1: $d_1 = a\,b$ kleinste Differenz,

$\left. \begin{array}{l} \text{Lösung 2:} \quad d_2 = a\,b + a\,\overline{b} = a \\ \text{Lösung 3:} \quad d_3 = a\,b + \overline{a}\,b = b \end{array} \right\}$ kürzeste Differenzen,

Lösung 4: $d_4 = a\,b + \overline{a}\,b + a\,\overline{b} = a + b$ größte Differenz.

Zu beachten ist, daß die verschiedenen Resultate der Subtraktion und Division verschiedene Funktionen sind und nicht etwa durch Kürzung oder Erweiterung ineinander übergeführt werden können. Alle Resultate erfüllen die gestellte mathematische Bedingung. Beispielsweise hat die Division

$$x = \frac{a\,b + b\,c + c\,a}{b + c\,a}$$

die zur früher gezeigten Sternschaltung (Abb. 44 c) führte, insgesamt acht Lösungen, und zwar

$$q_1 = a\,b + b\,c + c\,a \quad \text{(kleinster Quotient)},$$
$$q_2 = a\,b + b\,c + c\,a + \overline{a}\,\overline{b}\,\overline{c},$$
$$q_3 = a + b\,c,$$
$$q_4 = a\,b + c,$$
$$q_5 = b\,c + a + \overline{b}\,\overline{c},$$
$$q_6 = a\,b + c + \overline{a}\overline{b},$$
$$q_7 = c + a \quad \text{(kürzester Quotient)},$$
$$q_8 = c + a + \overline{b} \quad \text{(größter Quotient)}.$$

Der kleinste Quotient hat in seiner Vollform vier Glieder (Stellungsvariationen), der größte hingegen sieben. Alle erfüllen die vorgegebene funktionelle Bedingung, die also nicht eine bestimmte Wirkungsweise, sondern einen *Wirkungsbereich* beschreibt:

Auch für die mehrdeutigen Operationen der Logistik gelten die Gesetze der Kommutation, Assoziation und Distribution; die Gesamtheit der Resultate ist also von der Reihenfolge und Zusammenfassung der Elemente unabhängig. Wegen der aus dem Äquivalenzgesetz folgenden Unbestimmtheit ist es jedoch im allgemeinen zweckmäßig, die Rechnung aus der Vollform der Teilfunktionen zu entwickeln, weil nur dann sämtliche Lösungen aufscheinen; andernfalls erhält man nur Einzelresultate, die je nach Zusammenfassung der Glieder im Ansatz voneinander verschieden sein können. Hingegen läßt sich die Umwandlung in die Vollform ersparen, wenn die mehrdeutige Operation ohne Anwendung der Äquivalenzregeln ausführbar ist; man kann dann einfach nach den Regeln der Arithmetik subtrahieren und dividieren und erhält sofort kürzeste Resultate ohne Rücksicht auf die Zusammenfassung der Glieder. So ist beispielsweise

$$(a + b) - (a - b) = a + b - a + b = b + b = b,$$

genau so wie

$$(a + b - a) + b = b + b = b.$$

Da solche *arithmetische Resultate* stets aus vollständig gekürzten Teilfunktionen hervorgehen, müssen sie zur Gruppe der kürzesten Lösungen gehören; sie sind ferner singulär, weil die arithmetische Subtraktion und Division von Variablen eindeutig ist. Der Verzicht auf die Äquivalenzregeln bedeutet aber eine ganz beträchtliche Einschränkung, so daß die

Mehrzahl aller logistischen Operationen keine arithmetischen Resultate liefert.

Zu beachten ist weiter, daß der Subtraktionsoperator (Minuszeichen) ebenso wie die übrigen logistischen Operatoren ein reines Verknüpfungszeichen und kein Vorzeichen im Sinne der Zahlenrechnung darstellt. Eine negative Bedingung kann für sich allein betrachtet nicht existieren und wäre in logischer Hinsicht sinnlos. Mathematisch drückt sich dies dadurch aus, daß eine Gleichung

$$x = -y, \qquad \text{das heißt} \qquad x + y = 0$$

für den Fall $x = \overline{0}$ nicht erfüllbar ist; ein Strompfad x kann durch zu ihm parallele Kontakte y nicht ausgeschaltet werden. Da die Subtraktion zur Division invers ist, hat auch die Gleichung $x\,y = \overline{0}$ für $x = 0$ keine Lösung. Es können daher auch keine reziproken Bedingungen existieren. Die logistische Subtraktion und Division ist also nur dann möglich und sinnvoll, wenn sie ohne Rest aufgeht. Differenzen und Quotienten sind daher nur Verknüpfungen von Leitbedingungen und lassen sich nicht als Symbole für Kontaktverknüpfungen auffassen.

Hingegen gestatten es die Operationen der Subtraktion und Division, den logistischen Begriff der Inversion auf algebraische Verknüpfungen zurückzuführen.

Aus

$$x + \overline{x} = \overline{0} \qquad \text{oder} \qquad \overline{0} - x = \overline{x}$$

und

$$x + \overline{0} = \overline{0} \qquad \text{oder} \qquad \overline{0} - x = \overline{0}$$

folgt

$$\overline{x} = (\overline{0} - x)_{min}.$$

Ebenso ergibt sich aus

$$x \cdot \overline{x} = 0 \qquad \text{oder} \qquad \frac{0}{x} = \overline{x}$$

und

$$x \cdot 0 = 0 \qquad \text{oder} \qquad \frac{0}{x} = 0$$

analog

$$\overline{x} = \left(\frac{0}{x}\right)_{max}.$$

Die Inversion läßt sich also durch eine Subtraktion oder Division ersetzen; sie ist nicht nur eine logische, sondern auch eine mathematische Operation und das Inversionszeichen (Querstrich) ist nur eine abgekürzte Schreibweise für die Extremwerte der unbestimmten Verknüpfungen mit 0 und $\overline{0}$.

Zum tieferen Verständnis der schaltungstechnischen Zusammenhänge und vor allem zur systematischen Entwicklung mehrfach vermaschter

Schaltungsformen ist die Kenntnis der beschriebenen logistischen Grundoperationen unbedingt erforderlich. Besonders anschaulich lassen sich diese mit Hilfe der Variationstabellen darstellen. Diese Tabellen beschreiben die Funktion einer Schaltung dadurch, daß einem Teil ihrer Zeilen der Leitwert 0 und dem Rest der Wert ∞ zugeordnet ist. Sie teilen damit den Variationsbereich, der sich mit den vorhandenen Schaltgeräten herstellen läßt, in zwei Teile, die wir *Leitbereich* und *Sperrbereich* nennen wollen. Es soll nun gezeigt werden, daß sich die logistischen Grundoperationen aus der *Überdeckung* von Leitbedingungen, das heißt gleichnamiger Bereiche zweier Variationstabellen anschaulich erklären lassen. Wir betrachten zuerst die Addition.

Gegeben sei eine beliebige Summe $a + b$. Entwickeln wir die auf zwei Elemente bezogene Vollform des Summanden a, so erhalten wir:

$$a = a\,\overline{b} + a\,b,$$

analog ergibt sich hierzu:

$$b = \overline{a}\,b + a\,b.$$

Schreiben wir die beiden erweiterten Funktionen in tabellarischer Form an, so lautet die Addition:

$$
\begin{vmatrix} \overline{a}\,\overline{b}\ 0 \\ \overline{a}\,b\ 0 \\ a\,\overline{b}\ \infty \\ a\,b\ \infty \end{vmatrix}
+
\begin{vmatrix} \overline{a}\,\overline{b}\ 0 \\ \overline{a}\,b\ \infty \\ a\,\overline{b}\ 0 \\ a\,b\ \infty \end{vmatrix}
=
\begin{vmatrix} \overline{a}\,\overline{b}\ 0 \\ \overline{a}\,b\ \infty \\ a\,\overline{b}\ \infty \\ a\,b\ \infty \end{vmatrix}
$$

Das Wesen der Operation tritt klarer hervor, wenn wir von den Tabellen nur die letzte Spalte, die dem Leitwert entspricht, anschreiben:

$$
\begin{vmatrix} 0 \\ 0 \\ \infty \\ \infty \end{vmatrix}
+
\begin{vmatrix} 0 \\ \infty \\ 0 \\ \infty \end{vmatrix}
=
\begin{vmatrix} 0 \\ \infty \\ \infty \\ \infty \end{vmatrix}
$$

Man erkennt, daß die Summentabelle den Wert ∞ in allen Zeilen enthält, die in einem der beiden Summanden mit ∞ bezeichnet sind. Für Funktionen beliebiger Größe verläuft demnach die Addition nach folgendem Schema:

$$
\begin{vmatrix} 0 \\ \hline \infty \\ \infty \\ \infty \\ \infty \\ \hline 0 \\ 0 \end{vmatrix}
+
\begin{vmatrix} 0 \\ 0 \\ \hline \infty \\ \infty \\ \infty \\ \infty \\ \infty \end{vmatrix}
=
\begin{vmatrix} 0 \\ \hline \infty \\ \infty \\ \infty \\ \infty \\ \infty \\ \infty \end{vmatrix}
$$

Noch deutlicher wird das Bild, wenn wir Leit- und Sperrbereich durch schwarze und weiße Farbe unterscheiden.

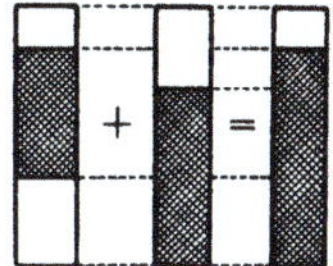

Der Leitbereich der Summenfunktion ist also jener Bereich, der von den Leitbereichen der Summanden mindestens einfach überdeckt wird.

Nunmehr untersuchen wir die Multiplikation an Hand eines Produktes $a\,b$ oder in tabellarischer Darstellung:

$$
\begin{vmatrix}
\bar{a}\,\bar{b}\ 0 \\
\bar{a}\,b\ 0 \\
a\,\bar{b}\ \infty \\
a\,b\ \infty
\end{vmatrix}
\cdot
\begin{vmatrix}
\bar{a}\,\bar{b}\ 0 \\
\bar{a}\,b\ \infty \\
a\,\bar{b}\ 0 \\
a\,b\ \infty
\end{vmatrix}
=
\begin{vmatrix}
\bar{a}\,\bar{b}\ 0 \\
\bar{a}\,b\ 0 \\
a\,\bar{b}\ 0 \\
a\,b\ \infty
\end{vmatrix}
$$

In jedem der beiden Faktoren sind die Zeilen Summanden eines Polynomes. Die Multiplikation ist daher so auszuführen, daß wir jede Zeile einer Tabelle mit jeder Zeile der anderen zu multiplizieren haben. Man erkennt sofort, daß die Multiplikation von Gliedern, die in verschiedenen Zeilen stehen, auf jeden Fall zum Wert Null führt. Beispielsweise ergibt die vorletzte Zeile des ersten Faktors mit der letzten des zweiten unabhängig vom jeweiligen Leitwert stets 0, weil $a\,\bar{b}\cdot a\,b$ wegen $\bar{b}\cdot b = 0$ selbst schon 0 wird. Es genügt also für die Multiplikation, wenn wir gleiche Zeilen miteinander multiplizieren. Wie bereits mehrfach erläutert, gilt hierbei für die Leitwerte:

$$0 \cdot 0 = 0,$$
$$0 \cdot \infty = 0,$$
$$\infty \cdot \infty = \infty.$$

In der Tabellenrechnung zeigt das Produkt daher nur in jenen Zeilen, die auch in allen Faktoren den Wert ∞ aufweisen, ∞. Im nachstehenden die bildliche Darstellung beliebiger **Multiplikationen.**

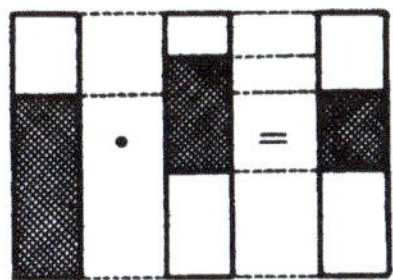

Der Leitbereich des Produktes ist also jener Bereich, der von den Leitbereichen *sämtlicher* Faktoren überdeckt wird.

An Hand der beiden eben erläuterten Operationen können wir nunmehr den mathematischen Charakter der Variationstabellen feststellen. Sowohl bei der Addition als auch bei der Multiplikation genügt es, Glieder gleicher Zeilen miteinander zu vergleichen. Zwei oder mehrere Tabellen können also zeilenweise addiert und multipliziert werden. Wie wir ferner gesehen haben, kommt es hierbei nur auf die Leitwerte und nicht auf die Variablen an, so daß wir diese in den Tabellen weglassen konnten. Die Rechenoperationen beschränkten sich also ausschließlich auf die Felder, in denen die Leitwerte eingetragen sind. Die Mathematik bezeichnet nun Tabellen, in denen gleichliegende Felder addiert und multipliziert werden dürfen, als *eindimensionale Matrizen*. Von dieser Bezeichnung soll jedoch in der Schaltungstechnik nicht Gebrauch gemacht werden, da diese zur Darstellung von Schaltungsformen auch zweidimensionale Matrizen, die anderen Rechenregeln folgen, kennt. Um Verwechslungen vorzubeugen, empfiehlt es sich daher, die eindimensionalen Matrizen einfach als Tabellen anzusprechen und den Namen *Matrix* für die zweidimensionalen Matrizen freizuhalten.

Nun wenden wir uns der Subtraktion zu und betrachten zuerst den Fall, daß die abzuziehende Funktion in der anderen vollständig enthalten sei. Die Tabellen zeigen dann folgendes Bild:

$$
\begin{vmatrix} 0 \\ \infty \\ \infty \\ \infty \end{vmatrix} - \begin{vmatrix} 0 \\ 0 \\ \infty \\ \infty \end{vmatrix} = \;?
$$

Für die Subtraktion von Leitwerten gilt:

$$0 - 0 = 0,$$

$$\infty - 0 = \infty.$$

Eine Differenz $0 - \infty$ ist, wie bereits früher erläutert, in der Logistik nicht existenzfähig. Die Subtraktion $\infty - \infty$ liefert hingegen ein unbestimmtes Ergebnis, für das nach Belieben 0 oder ∞ eingesetzt werden darf. Wir schreiben symbolisch:

$$\infty - \infty = \;?$$

Unter Verwendung dieser Regeln liefert die Tabellenrechnung

$$
\begin{vmatrix} 0 \\ \infty \\ \infty \\ \infty \end{vmatrix} - \begin{vmatrix} 0 \\ 0 \\ \infty \\ \infty \end{vmatrix} = \begin{vmatrix} 0 \\ \infty \\ ? \\ ? \end{vmatrix}
$$

Das Resultat dieser Operation hat also nicht zwei, sondern drei verschiedene Bereiche; außer dem Leitbereich und dem Sperrbereich ist noch ein *Wahlbereich* vorhanden. In unserem Beispiel besteht dieser aus zwei Feldern. In jedes davon darf willkürlich 0 oder ∞ gesetzt werden. Es bestehen somit $2 \cdot 2 = 4$ Möglichkeiten des Einsetzens und somit vier Lösungen.

$$\left|\begin{matrix} 0 \\ \infty \\ ? \\ ? \end{matrix}\right| = \left|\begin{matrix} 0 \\ \infty \\ 0 \\ 0 \end{matrix}\right| \text{ oder } \left|\begin{matrix} 0 \\ \infty \\ 0 \\ \infty \end{matrix}\right| \text{ oder } \left|\begin{matrix} 0 \\ \infty \\ \infty \\ 0 \end{matrix}\right| \text{ oder } \left|\begin{matrix} 0 \\ \infty \\ \infty \\ \infty \end{matrix}\right|$$

Die resultierende Tabelle hat also vier Freiheitsgrade. Die erste Lösung zeigt nur in einer Spalte den Wert ∞ und ist somit die kleinste Differenz. Die Lösung 2 läßt sich zu b kürzen und ist somit die kürzeste Differenz. Lösung 4 zeigt nur in der ersten Zeile den Wert 0 und ist somit die größte Differenz. Sie ist gleich dem Minuenden. Neben der kürzesten Differenz (Lösung 2) ist noch die kleinste Differenz (Lösung 1) von besonderer Bedeutung, wie das nachstehende Bild zeigt:

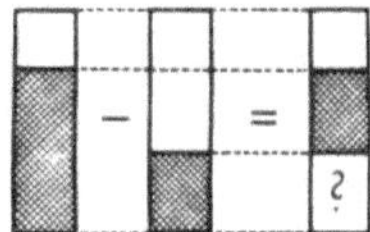

Der Bereich der resultierenden Funktion entspricht jenem Bereich des Minuenden und Subtrahenden, in dem diese verschiedene Leitwerte aufweisen. Die Bestimmung der kürzesten Differenz ist also gleichbedeutend mit einem Aussuchen jener Variationen, die in den beiden vorgegebenen Funktionen hinsichtlich des Leitwertes verschieden sind.

Bei der Subtraktion kann nun auch der Fall eintreten, daß eine der beiden Funktionen in der anderen nicht vollständig enthalten ist. Es treten dann Differenzen $0 - \infty$ auf, so daß sich eine echte Subtraktion nicht durchführen läßt. Wir können uns aber darauf beschränken, den absoluten Wert der Differenz zu bestimmen und dürfen dann jeweils $\infty - 0$ statt $0 - \infty$ setzen. Die so erhaltene *absolute Differenz* läßt sich aus Funktionen beliebiger Art ohne jegliche Einschränkung bilden; für sie gilt selbstverständlich

$$\lambda_1 - \lambda_2 = \lambda_2 - \lambda_1.$$

Auch absolute Differenzen können teilweise unbestimmt sein:

$$\left|\begin{matrix} 0 \\ \infty \\ \infty \\ \infty \end{matrix}\right| - \left|\begin{matrix} \infty \\ \infty \\ 0 \\ 0 \end{matrix}\right| = \left|\begin{matrix} \infty \\ ? \\ \infty \\ \infty \end{matrix}\right|$$

Auch für die absolute Differenz gilt, daß ihr Kleinstwert jenen Leitbereich umfaßt, in dem die Leitwerte des Minuenden und Subtrahenden verschieden sind.

Abschließend sei noch die Division betrachtet; sie verläuft vollständig analog zur Subtraktion. Da ein Wert $\infty : 0$ nicht existenzfähig ist, lassen sich echte Divisionen nur ausführen, wenn die Funktion des Dividenden in derjenigen des Divisors vollständig enthalten ist.

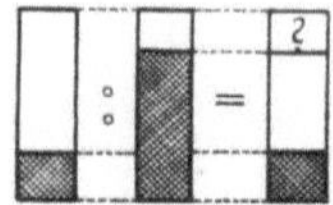

Auch hierbei treten unbestimmte Resultate auf; jede Zeile, in der die beiden vorgegebenen Funktionen den Leitwert 0 haben, vermehrt die Zahl der Freiheitsgrade des Resultates um den Faktor 2. Der kleinste Quotient ist gleich dem Dividenden. Die Bildung des größten Quotienten ist ein Aussuchen der Bereiche, in denen die beiden Funktionen gleiche Leitwerte haben. Überdecken sich die vorgegebenen Funktionen nicht vollständig, so läßt sich kein echter, sondern nur ein *absoluter Quotient* bilden. Wie bei der Kombinatorik der Schaltungsformen gezeigt werden wird, ist dieser Quotient bestimmend für die Querverbindungen aller vermaschten Schaltungen.

Das Ergebnis der vorstehenden Betrachtung sei nun als Schaltregel formuliert:

Unbestimmtheitsregel. Schaltaufgaben, die auf Unterschieden (Differenzen) oder Verhältnissen (Quotienten) von Funktionsbedingungen beruhen, sind mehrdeutig; sie beschreiben keine bestimmte Wirkungsweise, sondern einen Wirkungsbereich, innerhalb dessen die Funktion frei wählbar ist.

8. Die Kombinatorik der Leitbedingungen

Um die Wirkungsweise einer Schaltung, die bestimmte gestellte Bedingungen erfüllt, zu ermitteln, ist es notwendig, sämtliche Stellungsvariationen der zugehörigen Schaltgeräte in Betracht zu ziehen; nur dann besteht die Sicherheit, daß die Schaltung keine ungewollten Nebenwirkungen zeigt. Das mathematische Hilfsmittel hierzu ist die Tabelle der Stellungsvariationen. Wie in den vorhergehenden Abschnitten gezeigt wurde, lassen sich auch sämtliche Rechenoperationen zur Verknüpfung vorgegebener Bedingungen mit Hilfe von Variationstabellen wesentlich einfacher und übersichtlicher ausführen als durch eine algebraische Rechnung; dies gilt vor allem für die Operationen mit teilweise unbestimmten Resultaten. Auch die maximal gekürzte Funktion ist in vielen Fällen nur aus der vollständig erweiterten Form, also aus der Variationstabelle, zu entnehmen. Die Kombinatorik der Leitbedingungen stützt sich daher im wesentlichen auf die *Tabellenrechnung*. Nun haftet aber den Variationstabellen in der bisher gezeigten Form noch ein wesentlicher Nachteil an. Die Schreibweise, die für jede Stellungsvariation eine eigene Zeile benötigt, wird bei größeren Schaltungen unhandlich und unübersichtlich. Beispielsweise gehören zu einer Schaltung mit sechs zweistelligen Schaltern 64 Stellungsvariationen; jede Variation enthält 7 Operanden, nämlich sechs für die Schalterstellungen und einen für den Leitwert, so daß insgesamt 448 Operanden in 64 Zeilen anzuschreiben wären. Eine Tabelle solcher Größe gewährt keinen Überblick mehr und verlangt bei der Ausführung einer Rechenoperation eine unerträgliche Schreibarbeit. Es war daher unbedingt nötig, für die Variationstabellen eine einfachere und sparsamere Form zu finden; diese soll im nachstehenden gezeigt und erläutert werden.

Die Zeilen einer Variationstabelle müssen nicht in der natürlichen oder einer anderen bestimmten Reihenfolge angeschrieben werden. Beispielsweise ergibt sich bei der algebraischen Entwicklung einer Vollform eine Summe, bei der die einzelnen Glieder, die durch Pluszeichen verbunden sind, nebeneinander stehen; ist die Formel sehr lang, so kann sie sich auch über mehrere Zeilen erstrecken. Wir dürfen daher auch die zugehörige Variationstabelle so anschreiben, daß die einzelnen Variationen teilweise nebeneinander und teilweise untereinander stehen. Es ist nun zweckmäßig, alle jene Variationen, die einen gemeinsamen Faktor enthalten, zu einer Zeile zusammenzufassen. Eine Tabelle, die ursprünglich sechs Zeilen zu je einer Variation enthielt, erscheint nach dieser Umformung mit zwei Zeilen und drei Spalten:

$$
\begin{vmatrix} a_0\,b_0\ \ 0 \\ a_0\,b_1\ \infty \\ a_0\,b_2\ \ 0 \\ a_1\,b_0\ \infty \\ a_1\,b_1\ \ 0 \\ a_1\,b_2\ \ 0 \end{vmatrix}
=
\begin{array}{|c|c|c|} \hline a_0\,b_0\ 0 & a_0\,b_1\ \infty & a_0\,b_2\ 0 \\ \hline a_1\,b_0\ \infty & a_1\,b_1\ 0 & a_1\,b_2\ 0 \\ \hline \end{array}
$$

Nunmehr heben wir in jeder Zeile den gemeinsamen Faktor, im vorliegenden Falle a_0 und a_1, heraus und setzen diese Operanden als Zeilenbezeichnung an den linken Rand der Tabelle. Ebenso verfahren wir mit den gleichen Faktoren der drei Spalten, indem wir b_0, b_1, b_2 über die Tabelle setzen.

$$
\begin{array}{|c|} \hline a_0\,b_0\ 0 \\ \hline a_0\,b_1\ \infty \\ \hline a_0\,b_2\ 0 \\ \hline a_1\,b_0\ \infty \\ \hline a_1\,b_1\ 0 \\ \hline a_1\,b_2\ 0 \\ \hline \end{array}
= \left\{
\begin{array}{c|c|c|c} & b_0 & b_1 & b_2 \\ \hline a_0 & 0 & \infty & 0 \\ \hline a_1 & \infty & 0 & 0 \end{array}
\right.
$$

In den Feldern verbleiben somit nur die Operanden für den Leitwert, wodurch die Tabelle wesentlich übersichtlicher wird. Diese Darstellung läßt sich auch verwenden, wenn die Schaltung mehr als zwei Schaltgeräte aufweist. Die Bezeichnung einer Reihe (Zeile oder Spalte) besteht dann aus zwei oder mehreren Stellungsoperanden.

$$
\begin{array}{|c|c|}
\hline
a_0\,b_0\,c_0\ 0 & a_0\,b_0\,c_1\ 0 \\
\hline
a_0\,b_1\,c_0\ 0 & a_0\,b_1\,c_1\ \infty \\
\hline
a_1\,b_0\,c_0\ 0 & a_1\,b_0\,c_1\ \infty \\
\hline
a_1\,b_1\,c_0\ \infty & a_1\,b_1\,c_1\ 0 \\
\hline
\end{array}
=
\left\{
\begin{array}{c|c|c}
 & c_0 & c_1 \\
\hline
a_0\,b_0 & 0 & 0 \\
\hline
a_0\,b_1 & 0 & \infty \\
\hline
a_1\,b_0 & 0 & \infty \\
\hline
a_1\,b_1 & \infty & 0 \\
\end{array}
\right.
$$

Wie das vorstehende Beispiel, das drei Schaltgeräte mit je zwei Stellungen umfaßt, erkennen läßt, bildet die Gesamtheit der Zeilenbezeichnungen eine vollständige Variation der Stellungen von a und b; die Reihenbezeichnungen sind also Teilvariationen der Schaltung.

Die mehrreihige Schreibart der Variationstabellen bringt um so größere Ersparnisse, je umfangreicher die darzustellende Schaltung ist. Sind beispielsweise vier Schalter zu je drei Stellungen vorhanden, so ergeben sich $3^4 = 81$ Variationen. Bei einreihiger Darstellung hätte die zugehörige Tabelle also 81 Zeilen mit $4 \times 81 = 324$ Kontaktoperanden. Faßt man hingegen die vier Schalter zu 2 Paaren zusammen, so ergibt sich eine quadratische neunreihige Tabelle mit $2 \times 18 = 36$ Operanden zur Reihenbezeichnung. Statt 324 Kontaktoperanden sind also nur 36 anzuschreiben, so daß der Arbeitsaufwand auf ein Neuntel herabgesetzt ist.

Die Rechenregeln, die für Variationstabellen gelten, werden von der Art der Anschrift selbstverständlich nicht berührt. Wir dürfen mit mehrreihigen Tabellen genau so rechnen, wie mit einreihigen, das heißt gleichliegende Felder addieren, subtrahieren, multiplizieren und dividieren. Nachstehend ein Beispiel für eine Multiplikation:

$$
\begin{array}{c|c|c}
 & c_0 & c_1 \\
\hline
a_0\,b_0 & 0 & 0 \\
a_0\,b_1 & 0 & \infty \\
a_1\,b_0 & \infty & 0 \\
a_1\,b_1 & \infty & \infty \\
\end{array}
\cdot
\begin{array}{c|c}
c_0 & c_1 \\
\hline
0 & \infty \\
\infty & \infty \\
0 & 0 \\
0 & \infty \\
\end{array}
=
\begin{array}{c|c}
c_0 & c_1 \\
\hline
0 & 0 \\
0 & \infty \\
0 & 0 \\
0 & \infty \\
\end{array}
$$

$$(a_1\,c_0 + b_1\,c_1) \cdot (a_0\,b_1 + a_0\,c_1 + b_1\,c_1) = b_1\,c_1.$$

Vergleicht man die Rechenarbeit bei algebraischer und tabellarischer Durchführung der Multiplikation, so zeigt sich deutlich der praktische Wert der Tabellenrechnung. Diese liefert das Produkt in einfachster Weise dadurch, daß zuerst die Nullen der ersten Tabelle und dann diejenigen der zweiten Tabelle in das Resultat übertragen werden; die freibleibenden Stellen werden mit ∞ ergänzt. Die Rechenoperationen verlangen also praktisch keinerlei Denkarbeit.

Auch die Kürzung läßt sich bei mehrreihigen Tabellen genau so durchführen wie bei einreihigen. Daß Kürzungen, die sich innerhalb einer Reihe abspielen, in gleicher Weise verlaufen, ist selbstverständlich. Innerhalb einer Zeile oder Spalte kann man also komplementäre Variationen genau so zusammenlegen wie bei einreihigen Tabellen.

$$
\begin{array}{c|cc}
 & c_0 & c_1 \\
\hline
a_0\,b_0 & 0 & 0 \\
a_0\,b_1 & \infty & \infty \\
a_1\,b_0 & 0 & 0 \\
a_1\,b_1 & 0 & 0
\end{array} = a_0\,b_1
$$

oder auch

$$
\begin{array}{c|cc}
 & c_0 & c_1 \\
\hline
a_0\,b_0 & 0 & 0 \\
a_0\,b_1 & 0 & \infty \\
a_1\,b_0 & 0 & 0 \\
a_1\,b_1 & 0 & \infty
\end{array} = b_1\,c_1.
$$

Zu größeren Kürzungen, die mehrere Schaltgeräte betreffen, müssen Zeilen und Spalten gemeinschaftlich herangezogen werden. Ihrem Wesen nach ist jede Kürzung eine Zerlegung in zwei oder mehrere Summanden, von denen jeder einem gekürzten Strompfad entspricht. Wegen des Äquivalenzgesetzes $x = x + x$ darf daher jeder Operand der Summentabelle beliebig oft benützt, das heißt in mehrere oder alle Summanden übertragen werden. Beispielsweise ergibt sich:

$$
\begin{array}{c|cc}
 & c_0 & c_1 \\
\hline
a_0\,b_0 & 0 & 0 \\
a_0\,b_1 & \infty & \infty \\
a_1\,b_0 & 0 & 0 \\
a_1\,b_1 & 0 & \infty
\end{array} =
\begin{array}{c|cc}
 & c_0 & c_1 \\
\hline
 & 0 & 0 \\
 & \infty & \infty \\
 & 0 & 0 \\
 & 0 & 0
\end{array} +
\begin{array}{c|cc}
 & c_0 & c_1 \\
\hline
 & 0 & 0 \\
 & 0 & \infty \\
 & 0 & 0 \\
 & 0 & \infty
\end{array} =
$$

$$
= a_0\,b_1 + b_1\,c_1 = b_1(a_0 + c_1).
$$

Schon nach geringer Übung in der kombinatorischen Tabellenrechnung erkennt man mögliche Zusammenfassungen leicht, ohne die Zerlegung in Summanden durchführen zu müssen. Die Operation läßt sich dadurch anschaulicher machen, daß man den einzelnen gekürzten Kombinationen (Strompfaden) Ziffern zuordnet und diese in die Felder der Tabelle einträgt. In allen zum gleichen Strompfad zugehörigen Feldern der Tabelle erscheint dann die gleiche Ziffer. Dadurch lassen sich Rechenfehler auch bei großen Tabellen mit Sicherheit vermeiden:

$$
\begin{array}{c|c|c}
 & c_0 & c_1 \\
\hline
a_0\,b_0 & 0 & 0 \\
\hline
a_0\,b_1 & \overset{\infty}{1} & \overset{\infty}{12} \\
\hline
a_1\,b_0 & 0 & 0 \\
\hline
a_1\,b_1 & 0 & \overset{\infty}{2}
\end{array}
\quad
\begin{array}{c}
= a_0\,b_1 + b_1\,c_1 \\
 1 \qquad 2
\end{array}
$$

Es soll nun ein praktisches Beispiel für die tabellarische Kürzung einer größeren Funktion gegeben werden:

Eine Werkzeugmaschine habe fünf Supporte, die nebeneinander angeordnet und selbsttätig verstellbar sind. Um das Werkstück an jeder beliebigen Stelle bearbeiten zu können, haben benachbarte Supporte übergreifende Verstellbereiche. Damit eine gegenseitige Behinderung der benachbarten Supporte vermieden wird, dürfen diese nicht gleichzeitig in Betrieb sein. Um Fehlbedienungen auszuschließen, soll das Hauptschütz S der Maschine über Endschalter a, b, c, d, e der fünf Supporte verriegelt werden; es darf nur Spannung erhalten, wenn sich keine benachbarten Endschalter in der Arbeitsstellung befinden.

Wir bezeichnen die der Ruhelage der Supporte zugeordneten Stellungen der Endschalter mit 0 und den gegenteiligen Zustand mit 1. Dann lautet die Bedingung für die Sperrung der Anspeisung:

$$\overline{\left(\frac{\lambda}{Q\,S}\right)} = a_1\,b_1 + b_1\,c_1 + c_1\,d_1 + d_1\,e_1.$$

Nun übertragen wir diese Funktion in eine Variationstabelle, wobei wir zugleich die Inversion durchführen und 0 statt ∞ einsetzen; die Tabelle zeigt dann sofort die gesuchte Leitwertfunktion für die Anspeisung des Hauptschützes.

	d_0 e_0	d_0 e_1	d_1 e_0	d_1 e_1
$a_0\,b_0\,c_0$	∞	∞	∞	0
$a_0\,b_0\,c_1$	∞	∞	0	0
$a_0\,b_1\,c_0$	∞	∞	∞	0
$a_0\,b_1\,c_1$	0	0	0	0
$a_1\,b_0\,c_0$	∞	∞	∞	0
$a_1\,b_0\,c_1$	∞	∞	0	0
$a_1\,b_1\,c_0$	0	0	0	0
$a_1\,b_1\,c_1$	0	0	0	0

Nun fassen wir die Variationen, die teilweise komplementär sind und somit Kürzungen ermöglichen, additiv zu Kombinationen zusammen. Die Kombinationen versehen wir mit fortlaufenden Ziffern, die wir in die Tabelle eintragen. Zueinander komplementär sind bei der

$$\text{Kombination 1:}\quad e_0 \text{ und } e_1,$$
$$\text{Kombination 2:}\quad d_0 \text{ und } d_1,$$
$$\text{Kombination 3:}\quad b_0 \text{ und } b_1,$$
$$\text{Kombination 4:}\quad a_0 \text{ und } a_1.$$

	d_0 e_0	d_0 e_1	d_1 e_0	d_1 e_1
$a_0\,b_0\,c_0$	1234	13	24	0
$a_0\,b_0\,c_1$	1	1	0	0
$a_0\,b_1\,c_0$	23	3	2	0
$a_0\,b_1\,c_1$	0	0	0	0
$a_1\,b_0\,c_0$	14	1	4	0
$a_1\,b_0\,c_1$	1	1	0	0
$a_1\,b_1\,c_0$	0	0	0	0
$a_1\,b_1\,c_1$	0	0	0	0

Die durch gleiche Ziffern gekennzeichneten Variationen des Leitbereiches ergeben zusammen je eine gekürzte Kombination, die einem
Strompfad der Schaltung entspricht. Aus der bezifferten Tabelle läßt sich
daher sofort die Leitwertgleichung ablesen.

$$\frac{\lambda}{Q\,S} = \underset{1}{b_0\,d_0} + \underset{2}{a_0\,c_0\,e_0} + \underset{3}{a_0\,c_0\,d_0} + \underset{4}{b_0\,c_0\,e_0} =$$

$$= a_0\,c_0(d_0 + e_0) + b_0(c_0\,e_0 + d_0).$$

Wir müssen die Variationstabelle nur einmal anschreiben, um die
vorgegebenen Sperrbedingungen zu addieren, umzukehren und zu kürzen;
die Tabellenrechnung erlaubt die Durchführung aller drei Operationen in
einem Zug. Die gefundene Leitwertgleichung beschreibt die zugehörige
verkettete Schaltung. Sie zeigt auch, daß die maximale Kürzung im
vorliegenden Falle nicht zur günstigsten Verkettung führt. Die beiden
Klammerausdrücke unterscheiden sich nur um den Faktor c_0 beim
Operanden e_0. Da vor der ersten Klammer bereits c_0 steht, dürfen wir auch
innerhalb der Klammer $c_0\,e_0$ statt e_0 schreiben und erhalten dann gleiche
Klammerausdrücke.

$$\frac{\lambda}{Q\,S} = a_0\,c_0(c_0\,e_0 + d_0) + b_0(c_0\,e_0 + d_0) = (c_0\,e_0 + d_0) \cdot (a_0\,c_0 + b_0).$$

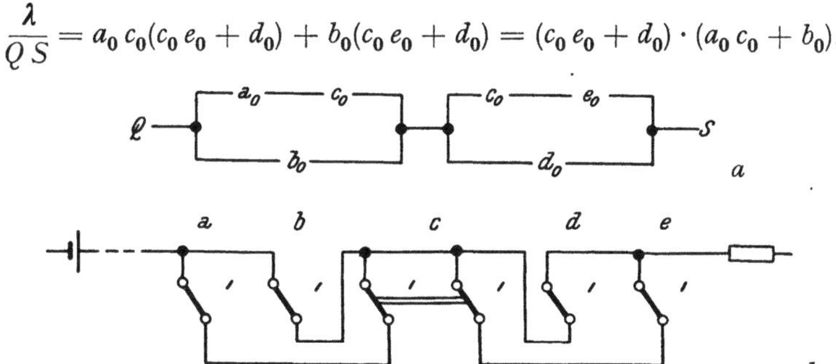

Abb. 45. Verriegelung benachbarter Supporte einer Werkzeugmaschine. *a* Stromlaufplan, *b* Wirkschaltplan

Die Gleichung erscheint nunmehr als Produkt zweier Summanden und entspricht somit einer zweiteiligen Schaltung; diese verlangt um zwei Kontakte weniger, als es bei vollständiger Kürzung der Fall gewesen wäre. Abb. 45 zeigt die Lösung.

Besonders zweckmäßig ist die Tabellenrechnung für Operationen mit teilweise unbestimmtem Resultat; sie zeigt dann in übersichtlicher Weise sämtliche möglichen Lösungen. Zu beachten ist, daß auch die Kürzung nicht immer eindeutig ist. Da sie ihrem Wesen nach eine Zerlegung in mehrere Summanden, also eine Subtraktion, darstellt, ist auch die maximale Kürzung nicht immer eindeutig. Ein Beispiel hierfür ist die im Abschnitt II, 4 (Abb. 37) gezeigte Steuerschaltung für einen elektromagnetischen Drehwähler. Die Leitwertgleichung dieser Schaltung lautet:

$$\lambda = i(v_0\,\overline{w}_0 + v_1\,w_0 + a_0\,v_1)\,S_a.$$

Wir betrachten nun das in der Klammer stehende Polynom und übertragen es in eine Variationstabelle:

	w_0	$\overline{w}_0$
$a_0\,v_0$	0	∞
$a_0\,v_1$	∞	∞
$a_1\,v_0$	0	∞
$a_1\,v_1$	∞	0

Diese Tabelle läßt sich in drei Summanden zerlegen, die den drei Gliedern des Polynoms obiger Gleichung entsprechen:

	w_0	$\overline{w}_0$
$a_0\,v_0$	0	∞
$a_0\,v_1$	0	0
$a_1\,v_0$	0	∞
$a_1\,v_1$	0	0

$+$

	w_0	$\overline{w}_0$
$a_0\,v_0$	0	0
$a_0\,v_1$	∞	0
$a_1\,v_0$	0	0
$a_1\,v_1$	∞	0

$+$

	w_0	$\overline{w}_0$
$a_0\,v_0$	0	0
$a_0\,v_1$	∞	∞
$a_1\,v_0$	0	0
$a_1\,v_1$	0	0

$=$

$$= \quad v_0\,\overline{w}_0 \quad + \quad v_1\,w_0 \quad + \quad a_0\,v_1.$$

Es gibt nun noch eine zweite Art der Zerlegung. Eindeutig sind nur die Zusammenfassungen der beiden ersten Glieder. Die Variation Nr. 4 in der zweiten Zeile und zweiten Spalte muß nicht mit der Variation Nr. 3 in der gleichen Zeile zusammengelegt werden; sie läßt sich auch mit der Variation Nr. 2 in der ersten Zeile und zweiten Spalte vereinigen.

	w_0	$\overline{w}_0$
$a_0\,v_0$	0	∞
$a_0\,v_1$	0	0
$a_1\,v_0$	0	∞
$a_1\,v_1$	0	0

$+$

	w_0	$\overline{w}_0$
$a_0\,v_0$	0	0
$a_0\,v_1$	∞	0
$a_1\,v_0$	0	0
$a_1\,v_1$	∞	0

$+$

	w_0	$\overline{w}_0$
$a_0\,v_0$	0	∞
$a_0\,v_1$	0	∞
$a_1\,v_0$	0	0
$a_1\,v_1$	0	0

$$= \quad v_0\,\overline{w}_0 \quad + \quad v_1\,w_0 \quad + \quad a_0\,\overline{w}_0.$$

Die aus den Funktionsbedingungen entwickelte Schaltung führt also zu zwei Wahlformen. Beide sind maximal gekürzt und haben somit die gleiche Zahl von Kontakten. Auch die Zahl der Strompfade ist gleich. Zwei davon sind beiden Formen gemeinsam, während der dritte Strompfad entweder nach einer Stellungskombination $a_0 v_1$ oder nach einer Kombination $a_0 \overline{w}_0$ ausgeführt werden darf. Abb. 46 zeigt die beiden Formen im Stromlaufplan. Da sie auf den gleichen Stellungsvariationen beruhen, ist auch ihre Wirkungsweise genau dieselbe. In der verketteten Schaltung besteht nur insofern ein Unterschied, als der Kontakt a_0 einmal zu w_0 und das andere Mal zu v_0 parallel ist. Da a_0 und w_0 Kontakte desselben Schaltgerätes sind, ist diese Form günstiger; sie verlangt einen Leiter weniger zur Verbindung des Drehwählers mit dem Relais i. Dieser Unter-

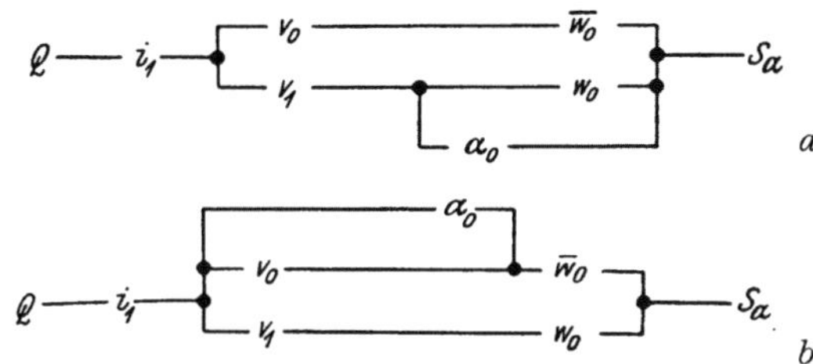

Abb. 46. Wahlformen mit verschiedener Kürzung.
a Form I, *b* Form II

schied in der Zahl der Leitungsverbindungen zeigt, daß die Kenntnis sämtlicher Möglichkeiten der Kürzung praktische Bedeutung hat. Aus dem Schaltplan sind solche Möglichkeiten nicht zu erkennen. Auch ein sehr geübter Praktiker der Schaltungstechnik wird beim Vergleich der beiden in Abb. 46 gezeigten Stromlaufpläne unbedingt eine Verschiedenheit der Wirkungsweise vermuten und sich vergebens bemühen, den vermeintlichen funktionellen Unterschied zu finden. Die Einsparung einer Verbindung zwischen Wähler und Impulsrelais durch andere Kürzung ist so wenig naheliegend, daß sie nur aus der kombinatorischen Theorie verständlich und nur aus einer mathematischen Darstellung erkennbar ist. Die Tabelle zeigt:

$$
\begin{array}{cc}
0 & \infty \uparrow \\
\infty \leftarrow & \infty \\
0 & \infty \\
\infty & 0
\end{array}
$$

Die Ausmittlung der Strompfade bei vorgegebener Funktion der Schaltung besteht also im wesentlichen darin, daß die Gesamtfunktion in Teilfunktionen, die je einem Strompfad entsprechen, zerlegt wird. Die Zusammenfassung von Stellungsvariationen zu Teilfunktionen ist völlig willkürlich; zweckmäßig ist es, die Teilfunktionen so zu wählen, daß die in ihnen enthaltenen Stellungsvariationen möglichst komplementär sind und dadurch Kürzungen zulassen. Besonders anschaulich zeigt sich dies, wenn man die Variationen nicht den Feldern einer Tabelle, sondern Eckpunkten eines Streckenkomplexes (Schaltfolgeplans) zuordnet. Der Vorteil dieser topologischen Darstellung liegt darin, daß benachbarte Stellungsvariationen, also solche, die sich voneinander nur durch einen Schritt eines einzigen Schalters unterscheiden, durch Linien miteinander verbunden sind; komplementäre Variationen erscheinen daher im Streckenkomplex als geschlossene Linienzüge:

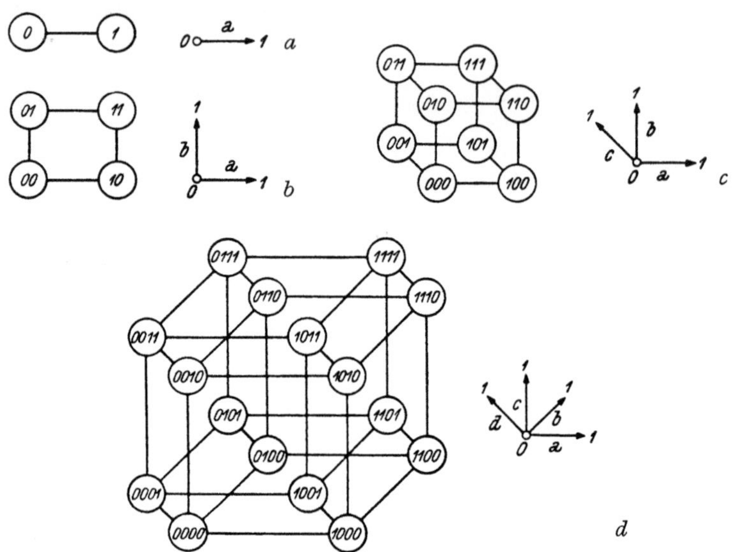

Abb. 47. Streckenkomplexe zu Variationen aus zwei Elementen 0 und 1. *a* Komplex erster Ordnung, *b* Komplex zweiter Ordnung, *c* Komplex dritter Ordnung, *d* Komplex vierter Ordnung

In Abb. 47 sind verschiedene Streckenkomplexe dargestellt, die sich auf Schaltungen mit zweistelligen Schaltgeräten beziehen. Ist nur ein Schalter vorhanden, so kann die Schaltung nur zwei Zustände haben und der darstellende Komplex besteht aus zwei Punkten und einer Verbindungslinie (Abb. 47 *a*). Sind zwei Schalter vorhanden, so enthält der zugehörige Komplex $2 \times 2 = 4$ Punkte und vier Verbindungen, entsprechend Abb. 47 *b*; an jeden Punkt schließen zwei Verbindungslinien an, so daß ein Komplex zweiter Ordnung vorliegt. Tritt ein weiterer zweistelliger Schalter hinzu, so verdoppelt sich die Zahl der Punkte und die Ordnungszahl erhöht sich auf 3 (Abb. 47 *c*). Bei weiterer Erhöhung der Schalterzahl verdoppelt sich jeweils die Zahl der Punkte bei steigender Ordnungszahl des Komplexes. Abb. 47 *d* zeigt einen Komplex vierter Ordnung, der den Stellungsvariationen einer Schaltung mit vier zweistelligen Schaltern entspricht.

Jeder Punkt dieser Komplexe gehört zu einer bestimmten Variation aus den Elementen (Stellungsnummern) 0 und 1. Wie die Abb. 47 *a* bis *d* zeigen, ist jeder Komplex höherer Ordnung aus mehreren Komplexen kleinerer Ordnungszahl zusammengesetzt. Beispielsweise ist der Komplex nach Abb. 47 *c* nichts anderes als die ebene Projektion eines Würfels, dessen Flächen Komplexe zweiter Ordnung gemäß Abb. 47 *b* sind. Die vier Seiten jeder Fläche sind wiederum Komplexe erster Ordnung nach Abb. 47 *a*. Eine einzelne Variation entspricht einem Komplex nullter Ordnung, das heißt einem Punkt. Die Komplexe, deren Ordnungszahl größer als drei ist, sind ebene Projektionen von räumlichen Gebilden höherer Dimension.

Führt ein Strompfad einer Schaltung über Kontakte sämtlicher Schaltgeräte, so entspricht er einer einzelnen Stellungsvariation; er erscheint demnach als Punkt des darstellenden Komplexes. Handelt es sich hingegen um

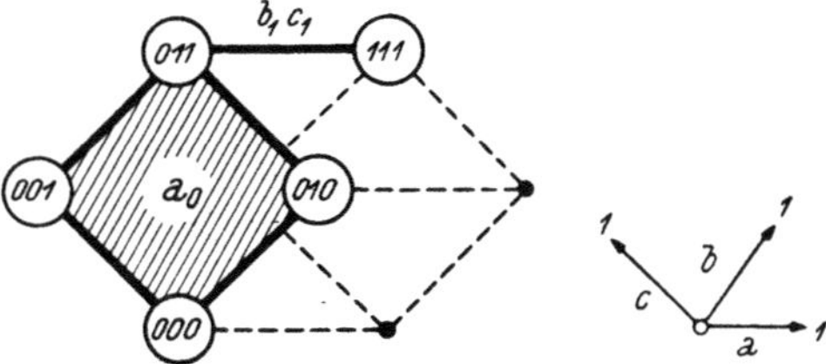

Abb. 48. Zerlegung in Teilkomplexe (Strompfade)

einen Strompfad, bei dem ein Schalter durch Kürzung entfallen ist, so umfaßt er zwei Kombinationen, die hinsichtlich dieses einen Schalters komplementär sind. Diese beiden Variationen bilden also einen Unterkomplex erster Ordnung nach Abb. 47 *a*, so daß der Strompfad im Gesamtkomplex als Verbindungslinie zweier benachbarter Punkte erscheint. Entfallen in einem Strompfad zwei oder mehrere Schalter durch Kürzung, so erscheint er als Unterkomplex, der sämtliche Variationen der gekürzten Schalter enthält. Abb. 48 zeigt eine Schaltung mit drei zweistelligen Schaltern *a, b, c* mit zwei Strompfaden entsprechend einer Leitwertgleichung

$$\lambda = a_0 + b_1\,c_1.$$

In der Darstellung durch einen Streckenkomplex erscheint der Strompfad a_0, bei dem die beiden Schalter *b, c* durch Kürzung entfallen sind, als Unterkomplex mit vier Punkten, der sämtliche vier Stellungsvariationen der gekürzten Schalter *b, c* enthält. Der andere Strompfad $b_1\,c_1$ erscheint als **Unterkomplex erster Ordnung**, der nur die beiden Stellungsvariationen des Schalters *a* umfaßt. Ein **Vergleich der vier Variationen** des Unterkomplexes a_0 zeigt, daß ihnen die in der Schaltung verbleibende Stellung a_0 gemeinsam ist, während die Stellungsnummern der beiden gekürzten Schalter *b* und *c* eine vollständige Variation bilden.

Wie Abb. 48 zeigt, läßt die topologische Darstellung der Schaltungsfunktion die Zerlegung in Teilfunktionen, die den einzelnen Strompfaden entsprechen, sehr deutlich erkennen. Sie zeigt daher auch, ob für die Kürzung mehrere Möglichkeiten vorhanden sind. In Abb. 49 ist die Leitwertfunktion der Wählerschaltung gemäß Abb. 46 wiedergegeben. Der Leitbereich umfaßt insgesamt fünf Variationen. Die zugehörigen Punkte des Komplexes liegen so, daß sie durch einen zusammenhängenden Linienzug verbunden sind. Der Zug enthält aber keinen vollständigen Komplex zweiter Ordnung (Viereck), so daß sich in keinem Strompfad zwei Schalter kürzen lassen. Somit bedeutet jede Verbindungslinie des Zuges einen Strompfad. Die Schaltung darf also höchstens vier Strompfade mit je zwei Schaltern erhalten. Es ist aber nicht notwendig, alle diese Strompfade auszuführen. Wie Abb. 49 b zeigt, läßt sich jede zweite

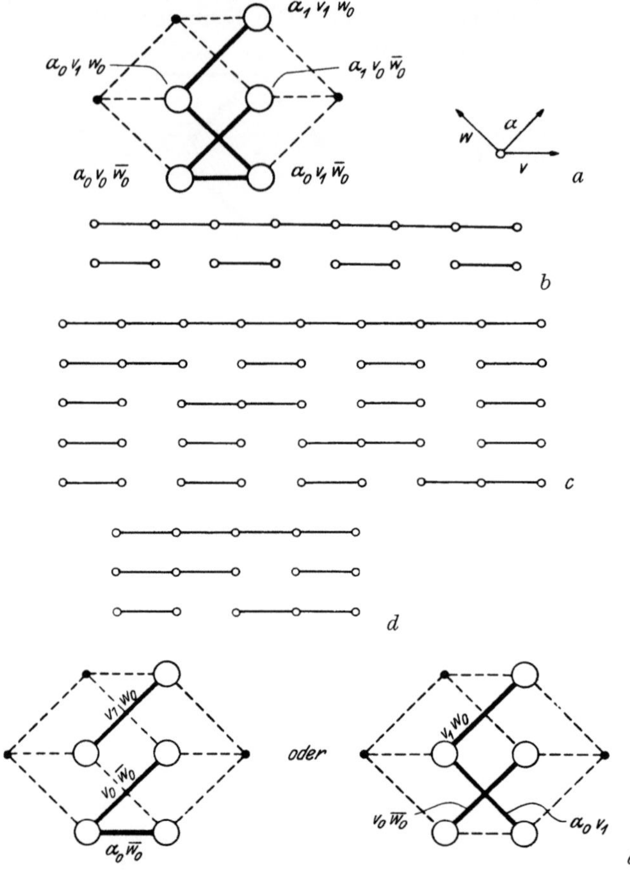

Abb. 49. Zerlegung in Teilkomplexe. *a* Mögliche Strompfade, *b* Zerlegung bei gerader Punktezahl, *c* Zerlegungen bei ungerader Punktezahl, *d* Zerlegungen bei fünf Punkten, *e* Notwendige Strompfade

Strecke des Zuges streichen, ohne daß irgendein Punkt des Komplexes dadurch entfällt. Es kommt dies daher, daß bei der Ausführung aller möglichen Strompfade nur die Endpunkte des Zuges einfach benützt sind, die übrigen hingegen in je zwei Strompfaden vorkommen. Hat der Zug eine gerade Anzahl von Punkten, so gibt es nur eine Art der Zerlegung mit einem Minimum von Strompfaden (Abb. 49 b). Ist die Anzahl der Punkte ungerade, so müssen mindestens an einer Stelle zwei Strecken aneinander anschließen, wobei ein Punkt für zwei Strecken benützt wird.

Wie Abb. 49 *c* zeigt, läßt sich dann der Zug auf mehrere Arten so zerlegen, daß die Zahl der Strecken ein Minimum bleibt. Bei *n* Punkten beträgt die Zahl der möglichen Lösungen $(1/2)\ (n-1)$. Da die betrachtete Schaltung einen Zug mit fünf Punkten umfaßt, ermöglicht sie zwei verschiedene Zerlegungen und somit zwei Wahlformen mit verschiedenen Strompfaden (Abb. 49 *d*); die Endstrecken des Zuges sind beiden Formen gemeinsam, die beiden mittleren Strecken können gegeneinander vertauscht werden. Abb. 49 *e* zeigt die beiden Lösungen.

Die Ausmittlung der notwendigen Strompfade einer Schaltung an Hand des zugehörigen Streckenkomplexes (Schaltfolgeplanes) hat den Vorteil größter Anschaulichkeit. Der Aufwand an Schreib- und Zeichenarbeit ist allerdings beträchtlich größer als bei der tabellarischen Ausrechnung. Für umfangreichere Schaltungen sind die Variationstabellen wesentlich handlicher und für den praktischen Gebrauch geeigneter. Bei den meisten Schaltungen der Praxis beruht übrigens die Vielfalt der Wahlformen nur zum geringen Teil auf der Unbestimmtheit der Kürzung. Die Mehrzahl der Freiheitsgrade folgt vielmehr daraus, daß die Wirkungsweise in der gestellten Schaltaufgabe nicht eindeutig umschrieben ist. Die Aufgabe legt meistens bestimmte Leit- und Sperrbereiche der Funktion fest, zwischen denen sich ein Wahlbereich aus zugelassenen Stellungsvariationen befindet. Zur Kennzeichnung der gewöhnlich großen Zahl von Variationen, die nach Be-

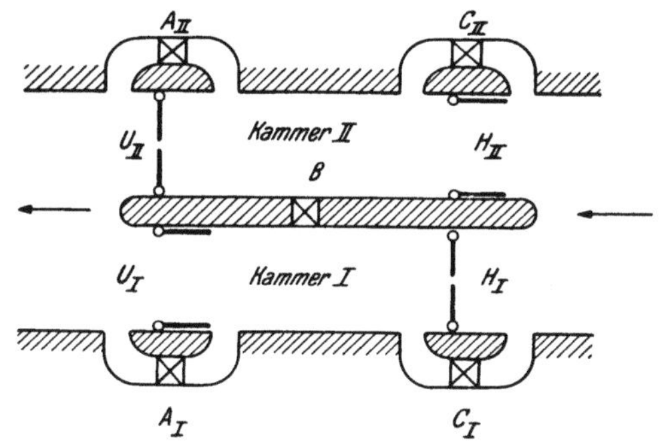
Abb. 50 *a*. Steuerung einer Zweikammerschleuse. Anordnung der Tore und Schieber

lieben dem Leitbereich oder Sperrbereich zugewiesen werden dürfen, sind die Variationstabellen besser geeignet als die topologische Darstellung. Als Beispiel sei eine größere Schaltung gezeigt, die mehrere Verbraucher sowie **mehrere Schaltgeräte mit** verschiedener Stellungszahl umfaßt.

Für eine Schiffsschleuse mit zwei Kammern *I* und *II*, deren Anordnung in Abb. 50 *a* wiedergegeben ist, soll die elektrische Schaltung zur Steuerung der Tore und Schieber entworfen werden. Gegen das Unterwasser sind die Kammern durch zwei Stemmtore U_I und U_{II}, gegen das Oberwasser mit zwei ebensolchen Toren H_I und H_{II} abzuschließen. Zur Füllung und Entleerung der Kammern dienen vier Schieber A_I, A_{II} und C_I, C_{II}. Um die Wasserverluste beim Schleusen zu vermindern, ist noch ein weiterer Schieber *B* in die Trennwände zwischen den Kammern eingebaut. Abb. 50 *a* zeigt den Zustand bei entleerter Kammer *I* und gefüllter Kammer *II*. Der Betrieb wird so geführt, daß die beiden Kammern im Gegentakt arbeiten. Vom abgebildeten Zustand ausgehend, schließen sich zuerst die beiden Tore U_I und H_{II} sowie die ihnen parallelgeschalteten Schieber A_I und C_{II}. Dann wird der Ausgleichschieber B geöffnet und es strömt solange Wasser von Kammer II in Kammer I, bis die beiden Wasserspiegel annähernd gleich sind. Dann wird der Schieber B wieder geschlossen und dafür das andere Schieberpaar C_I und A_{II} geöffnet; die Kammer I füllt sich dadurch mit Oberwasser, während sich die Kammer II in das Unterwasser entleert. Sobald diese Vorgänge beendet sind, öffnen sich die Tore U_{II} und H_I, womit das Arbeitsspiel beendet ist. Nach Aus- und Einfahrt der durchzuschleusenden Fahrzeuge wiederholt sich dann das beschriebene Spiel in umgekehrtem Sinn, bis der in Abb. 50 a gezeigte Ausgangszustand wieder erreicht ist.

Die Betätigung der Tore und Schieber kann hydraulisch oder elektromotorisch erfolgen; für die gesuchte Steuerschaltung ist die Ausführung belanglos, weil die Einleitung der Vorgänge in beiden Fällen durch die Einschaltung von Erregerspulen elektromagnetischer Ventile oder Schütze eingeleitet wird. Es ist ausreichend und auch zweckmäßig, jeden Schieber nur mit einem Steuerventil oder Steuerschütz auszurüsten; bei Erregung der Magnetspule öffnet sich der Schieber, bei Stromunterbrechung schließt er sich selbsttätig (*Sicherheitsschaltung*). Den Toren müssen hingegen für Schließen und Öffnen getrennte Ventile oder Schütze zugeordnet werden; ein selbsttätiges Schließen im Störungsfalle wäre gefahrvoll, weil es gerade bei Ein- oder Ausfahrt eines Schiffes eintreten könnte. Wir bezeichnen die beiden Endstellungen des Tores U_I mit U_{I0} (geschlossener Zustand) und U_{I1} (offener Zustand); diese Bezeichnung soll auch gleichzeitig für die zugehörigen Steuerspulen gelten, die die genannten Zustände herbeiführen. In analoger Weise benennen wir die Erregerspulen zur Steuerung der Schieber A_I bis C_{II}. Da das Arbeitsspiel nur fortgesetzt werden darf, wenn die Tore und Schieber den elektrischen Steuerbefehlen auch tatsächlich gefolgt sind, ist es noch nötig, ihre jeweiligen Stellungen durch angebaute Meldeschalter zu erfassen. Zur Kennzeichnung der Zugehörigkeit der Meldeschalter bezeichnen wir diese genau so wie die Tore und Schieber, jedoch mit kleinen Buchstaben; es bedeutet somit b den Meldeschalter des Ausgleichschiebers B, wobei b_0 den geschlossenen und b_1 den geöffneten Zustand des Schiebers angibt.

Die Bedienung der Schaltung soll in einfachster Weise durch einen einzigen Steuerschalter s mit zwei Stellungen s_0 und s_1 erfolgen. Bringt man den Schalter in seine Stellung s_0, so stellt sich der in Abb. 50 a gezeichnete Zustand her; durch Umlegen des Schalters in dessen Stellung s_1 wird selbsttätig der gegenteilige Endzustand herbeigeführt. Mit weiteren Handgriffen zur Betätigung von Toren oder Schiebern soll der Schleusenwärter nicht belastet werden, damit er sein Augenmerk vollständig dem Schiffsverkehr in der Schleuse widmen kann. Die Ein- und Ausschaltung der verschiedenen Steuerspulen wird also durch den Handschalter s nur vorbereitet und muß im übrigen in Abhängigkeit der Wasserstände in den beiden Kammern erfolgen. Beispielsweise darf der Schieber B nur so lange geöffnet sein, bis der Spiegelausgleich zwischen beiden Kammern stattgefunden hat; dann hat er sich selbsttätig zu schließen, wogegen eines der Schieberpaare A_I, C_{II}, bzw. A_{II} und C_I geöffnet werden muß. Da in einem natürlichen Flußlauf der Wasserspiegel sowohl im Oberwasser als auch im Unterwasser größeren Schwankungen unterworfen ist, kommt es hierbei nicht auf die absolute Spiegelhöhe, sondern auf die Spiegelunterschiede an. Man kann diese durch manometrische Messung mit Hilfe von Differenzdruckschaltern erfassen. Wir ordnen parallel zum Schieber A_I einen Differenzdruckschalter x_I an; dessen Stellung x_{I0} möge dem Zustand der Spiegelgleichheit und seine Stellung x_{I1} der Spiegelverschiedenheit zwischen Unterwasser und Kammer entsprechen. Einen gleichartigen Druckschalter z_I legen wir parallel zum Schieber C_I; auch er habe zwei Stellungen z_{I0} und z_{I1} für Gleichheit und Verschiedenheit der Wasserspiegel. Ein dritter Differenzdruckschalter y liegt parallel zum Schieber B zwischen den beiden Kammern. Dieser Druckschalter hat nicht zwei, sondern drei Zustände zu unterscheiden. Steht das Wasser in der Kammer I tiefer als in der Kammer II, möge dies einer Schalterstellung y_1 entsprechen. Einem tieferen Wasserstand in der Kammer II ordnen wir die Stellung y_2 des Druckschalters zu. Die Spiegelgleichheit der beiden Kammern sei durch die Schalterstellung y_0 ausgedrückt. Der Differenzdruckschalter y ist beiden Kammern gemeinsam. Die Kammer II hat daher noch zwei weitere solche Schalter x_{II} und z_{II} zu erhalten.

Schaltelemente		Unter-wasser	Aus-gleich	Ober-wasser
Steuerung der Tore	Spulen für Schließen	U_0	—	H_0
	Spulen für Öffnen	U_1	—	H_1
Endschalter an den Toren	Tor geschlossen	u_0	—	h_0
	Tor offen	u_1	—	h_1
Schiebersteuerung	Spulen	A	B	C
Endschalter an den Schiebern	Schieber geschlossen	a_0	b_0	c_0
	Schieber offen	a_1	b_1	c_1
Differenz-druckschalter	Kammer I tiefer	—	y_1	z_1
	Kammer I gleich	x_0	y_0	z_0
	Kammer I höher	x_1	y_2	—

Wie die vorstehenden Ausführungen über die Wirkungsweise und die Aufzählung der Bestandteile zeigen, handelt es sich um eine ziemlich umfangreiche Einrichtung. Die Steuerschaltung umfaßt insgesamt 15 Schaltgeräte, nämlich einen Drehschalter für die Handsteuerung, vier Meldeschalter für die Tore, fünf Meldeschalter für die Schieber und fünf Differenzdruckschalter, wovon einer drei Stellungen aufweist; ferner sind 13 Stromverbraucher vorhanden, und zwar acht Erregerspulen für die Betätigung der Tore und fünf solche für die Schieber. Die gewünschte Gegentaktsteuerung der beiden Kammern gestattet jedoch eine wesentliche Vereinfachung; sämtliche Betrachtungen können auf eine der beiden Kammern beschränkt bleiben, weil sich die Vorgänge in der zweiten in analoger Weise gegenläufig abspielten. Es genügt also, die Steuerschaltung für die Kammer I auszumitteln, wobei nur die mit I bezeichneten Schaltelemente in Erscheinung treten; es ist dann auch nicht nötig, das Kennzeichen I, das nur zur Unterscheidung gegenüber der Kammer II dient, anzuschreiben. Der Übersicht halber seien die gewählten Bezeichnungen für die Schaltelemente der Kammer I in einer Tabelle zusammengestellt.

Jede Höhe des Wasserspiegels in der Kammer hat eine bestimmte Stellungskombination der Differenzdruckschalter zur Folge. Auf Grund dieser Kombination erfolgt die Betätigung der Tore und Schieber. Wir können nun in einer zweiten Tabelle die Stellungskombinationen so der Reihe nach anschreiben, wie sie sich beim Füllen der Kammer I der Reihe nach einstellen.

Wasserstand im Vergleich zu			Stellungen		
Unterwasser	Kammer II	Oberwasser	x	y	z
gleich	tiefer	tiefer	0	1	1
höher	tiefer	tiefer	1	1	1
höher	gleich	tiefer	1	0	1
höher	höher	tiefer	1	2	1
höher	höher	gleich	1	2	0

Das Arbeitsspiel wird dadurch eingeleitet, daß man den Steuerschalter s in seine Stellung 1 bringt. Die Stellungen der drei Druckschalter, die hierbei vorhanden sind, entsprechen der ersten Zeile der vorstehenden Tabelle. Der Beginn der Füllung ist also durch eine Stellungskombination $s_1\,x_0\,y_1\,z_1$ gekennzeichnet. Das Auftreten dieser Stellungskombination soll bereits den ersten Schaltvorgang bewirken, das heißt, das untere Tor U durch Anspeisung der Steuerspule U_0 schließen. Elektrisch gesteuerte Verstellantriebe werden in zweckmäßiger Weise so ausgebildet, daß der Steuerstrom nach Erreichen der Endlage selbsttätig von einem Endschalter unterbrochen wird. Mit der Steuerspule U_0 liegt daher ein Endschalterkontakt u_0 ständig in Serie. Der erste Strompfad, der bei der Füllung der Kammer zu schließen ist, lautet daher $s_1\,x_0\,y_1\,z_1\,\overline{u}_0\,U_0$. Sofort nach Schließen des Tores U soll der Ausgleichschieber B geöffnet werden, damit Wasser von der Kammer II einströmen kann. Zu der vorangeführten Stellungskombination der Schaltgeräte gehört also noch ein zweiter Steuervorgang, nämlich die Anspeisung der Spule B. Der zugehörige Strompfad darf aber erst geschlossen werden, wenn das Tor U geschlossen, der Endschalter somit in seiner Stellung u_0 ist. Auch der Schieber A muß geschlossen, das heißt der zugehörige Endschalter in seiner Stellung a_0 sein. Die Formel für den zweiten Strompfad muß daher $s_1 x_0 z_1 u_0 a_0 B$ lauten.

Die Schieber A und C sollen geschlossen sein, ihre Steuerspulen dürfen nicht erregt werden. Auch das Tor H soll in seiner geschlossenen Stellung H_0 verbleiben; die zugehörige Steuerspule ist aber durch einen Endschalterkontakt $\overline{h}_0$ unterbrochen, so daß eine allfällige Anspeisung dieser Spule keine Wirkung hat. Außer den beiden notwendigen Strompfaden für U_0 und B ergibt sich daher noch ein zulässiger Strompfad $s_1 x_0 y_1 z_1 \overline{h}_0 H_0$.

Durch die Formeln für die drei genannten Strompfade ist der Zustand sämtlicher Verbraucher bei der ersten Stellungskombination der Schaltgeräte beschrieben. Wir können ihn als erste Zeile einer Tabelle, die uns die Vorgänge während eines Arbeitsspieles zeigen soll, eintragen.

	s	x	y	z	U_1	U_0	A	B	C	H_0	H_1
füllen	1	0	1	1	0	$\overline{u}_0 U_0$	0	$u_0 a_0 B$	0	$\overline{h}_0 H_0$?	0

Daß die Anspeisung der Spule H_0 freigestellt ist, wurde in der letzten Spalte der Tabelle durch Beifügung eines Fragezeichens vermerkt. Dieses Zeichen können wir als algebraischen Operanden auffassen, dem wir in der späteren Rechnung nach Bedarf den Wert 0 oder ∞ zuweisen dürfen.

Infolge der Erregung der Steuerspule B öffnet sich der Ausgleichschieber, so daß Wasser von der Kammer II in die Kammer I fließen kann. Dort beginnt der Wasserspiegel zu steigen, wodurch der Differenzdruckschalter x in seine Stellung x_1 gelangt. Es kommt somit eine neue Stellungskombination zustande, die wir in die zweite Zeile der begonnenen Tabelle eintragen. Eine Änderung in den Schieberstellungen soll hierbei nicht stattfinden, so daß wir den Zustand der Verbraucher unverändert von Zeile 1 übernehmen können; zu U_0 dürfen wir ein Fragezeichen setzen wie in der ersten Zeile zu H_0.

	s	x	y	z	U_1	U_0	A	B	C	H_0	H_1
füllen	1	0	1	1	0	$\overline{u}_0 U_0$	0	$u_0 a_0 B$	0	$\overline{h}_0 H_0$?	0
	1	1	1	1	0	$\overline{u}_0 U_0$?	0	$u_0 a_0 B$	0	$\overline{h}_0 H_0$?	0

Erst bei ungefährer Spiegelgleichheit der beiden Kammern soll wieder ein Schaltvorgang eintreten; Schieber B soll geschlossen und Schieber C geöffnet werden. Die Spiegelgleichheit bewirkt einen Zustandswechsel des Differenzdruckschalters y, der von seiner Stellung y_1 in die Stellung y_0 kommt.

Der neuen Stellungskombination $s_1\, x_1\, y_0\, z_1$ entspricht die dritte Zeile der Tabelle. Sie zeigt in der Spalte des Schiebers B den Operanden 0 entsprechend dem geschlossenen Zustand dieses Gerätes. Der Schieber C darf erst öffnen, wenn der Ausgleichschieber B seine Bewegung vollendet hat, da sonst die zu entleerende Kammer II mit dem Oberwasser in Verbindung käme. Die Steuerspule C ist daher mit einem Kontakt b_0 des Ausgleichschiebers B in Serie zu schalten; wir setzen daher in die Spalte C der Tabelle das Produkt $b_0\, C$. Spalte H bleibt unverändert, so daß wir auch die dritte Zeile anschreiben können.

	s	x	y	z	U_1	U_0	A	B	C	H_0	H_1
füllen	1	0	1	1	0	$\overline{u_0}\,U_0$	0	$u_0 a_0 B$	0	$\overline{h_0}\,H_0$?	0
	1	1	1	1	0	$\overline{u_0}\,U_0$?	0	$u_0 a_0 B$	0	$\overline{h_0}\,H_0$?	0
	1	1	0	1	0	$\overline{u_0}\,U_0$?	0	0	$b_0 C$	$\overline{h_0}\,H_0$?	0

Das weitere Ansteigen des Wasserspiegels in der Kammer I bringt den Differenzdruckschalter y in seine Stellung y_2, was noch keinen Schaltvorgang bewirken soll. Die vierte Zeile der Tabelle ist somit gleich der dritten mit Ausnahme der Stellung von y. Erst bei Spiegelgleichheit mit dem Oberwasser, das heißt bei Übergang von z_1 auf z_0, ist wieder ein Steuervorgang nötig. Es kann nunmehr das Tor H durch Erregung seiner Steuerspule H_1 geöffnet werden. Dieser Spule liegt ein Endschalterkontakt $\overline{h_1}$ in Serie, so daß wir in die Tabelle das Produkt $\overline{h_1}\, H_1$ zu setzen haben. Ob der Schieber C geschlossen wird oder offen bleibt, ist belanglos, so daß wir in der Spalte C das Produkt $b_0\, C$ mit einem Fragezeichen versehen.

	s	x	y	z	U_1	U_0	A	B	C	H_0	H_1
füllen	1	0	1	1	0	$\overline{u_0}\,U_0$	0	$u_0 a_0 B$	0	$\overline{h_0}\,H_0$?	0
	1	1	1	1	0	$\overline{u_0}\,U_0$?	0	$u_0 a_0 B$	0	$\overline{h_0}\,H_0$?	0
	1	1	0	1	0	$\overline{u_0}\,U_0$?	0	0	$b_0 C$	$\overline{h_0}\,H_0$?	0
	1	1	2	1	0	$\overline{u_0}\,U_0$?	0	0	$b_0 C$	$\overline{h_0}\,H_0$?	0
gefüllt	1	1	2	0	0	$\overline{u_0}\,U_0$?	0	0	$b_0 C$?	0	$\overline{h_1}\,H_1$

Die Tabelle zeigt nunmehr das Arbeitsspiel für die Füllung der Kammer I. In vollkommen analoger Weise können wir auch weitere fünf Zeilen für das gegensinnige Spiel zur Entleerung der Kammer ansetzen, womit sich der Schaltzyklus der Einrichtung schließt.

	s	x	y	z	U_1	U_0	A	B	C	H_0	H_1
füllen	1	0	1	1	0	$\overline{u}_0 U_0$	0	$u_0 a_0 B$	0	$\overline{h}_0 H_0$?	0
	1	1	1	1	0	$\overline{u}_0 U_0$?	0	$u_0 a_0 B$	0	$\overline{h}_0 H_0$?	0
	1	1	0	1	0	$\overline{u}_0 U_0$?	0	0	$b_0 C$	$\overline{h}_0 H_0$?	0
	1	1	2	1	0	$\overline{u}_0 U_0$?	0	0	$b_0 C$	$\overline{h}_0 H_0$?	0
gefüllt	1	1	2	0	0	$\overline{u}_0 U_0$?	0	0	$b_0 C$?	0	$\overline{h}_1 H_1$
entleeren	0	1	2	0	0	$\overline{u}_0 U_0$?	0	$h_0 c_0 B$	0	$\overline{h}_0 H_0$	0
	0	1	2	1	0	$\overline{u}_0 U_0$?	0	$h_0 c_0 B$	0	$\overline{h}_0 H_0$?	0
	0	1	0	1	0	$\overline{u}_0 U_0$?	$b_0 A$	0	0	$\overline{h}_0 H_0$?	0
	0	1	1	1	0	$\overline{u}_0 U_0$?	$b_0 A$	0	0	$\overline{h}_0 H_0$?	0
leer	0	0	1	1	$\overline{u}_1 U_1$	0	$b_0 A$?	0	0	$\overline{h}_0 H_0$?	0

Aus der Tabelle ersehen wir, daß jeder Stellungsvariation der Steuerschalter ein ganz bestimmter Zustand der anzuspeisenden Steuerspulen entspricht. Daraus folgt, daß die gesuchte Schaltung eine Dauerschaltung sein wird, das heißt keinerlei Hilfsgeräte mit Kontakten zur Selbsthaltung oder Selbststeuerung verlangt. Wäre im Zuge des Arbeitsspieles irgendeine Stellungsvariation zweimal, und zwar mit verschiedenem Zustand der Verbraucher vorgekommen, so hätte sich die Aufgabe nur durch eine Impulsschaltung lösen lassen.

Die angesetzte Tabelle ist eine unvollständige Variationstabelle; drei Schalter mit je zwei Stellungen und ein dreistelliger Schalter ermöglichen insgesamt $2 \cdot 2 \cdot 2 \cdot 3 = 24$ Stellungsvariationen. Die zur Beschreibung der Steuervorgänge angesetzte Tabelle umfaßt jedoch nur zehn Variationen. Die restlichen vierzehn Zustände der Steuergeräte können nicht auftreten; dies ist dadurch bedingt, daß das Wasser stets nur in einer Richtung fließen kann. Für die gesuchte Steuerschaltung folgt daraus, daß ihr Verhalten bei den restlichen Stellungsvariationen der Schalter völlig belanglos ist. Es besteht somit eine große Zahl von Freiheitsgraden für die Gesamtfunktion und wir können diese so wählen, daß die Schaltung möglichst einfach wird. Um all diese Möglichkeiten zu erkennen und zu berücksichtigen, ziehen wir aus der angeschriebenen Tabelle die vollständigen Variationstabellen für die einzelnen Verbraucher heraus. Wir beginnen mit der Teilfunktion für die Steuerspule U_1.

$$
\frac{\lambda}{Q\,\overline{u}_1\,U_1} =
\begin{cases}
\end{cases}
$$

	x_0 z_0	x_0 z_1	x_1 z_0	x_1 z_1
$s_0\,y_1$	?	∞	?	0
$s_0\,y_0$	?	?	?	0
$s_0\,y_2$	?	?	0	0
$s_1\,y_1$	?	0	?	0
$s_1\,y_0$	?	?	?	0
$s_1\,y_2$	?	?	0	0

Aus der Tabelle ist übrigens der Zyklus des Schaltspieles deutlich zu ersehen, womit wir sie auch auf ihre Vollständigkeit nachprüfen können.

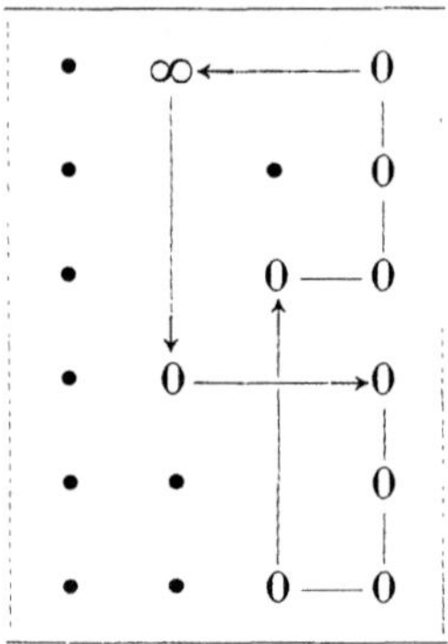

Nun setzen wir in die mit einem Fragezeichen versehenen Felder willkürlich 0 oder ∞ ein, und zwar so, daß sich die stärkste Kürzung vornehmen läßt. Aus Spalte 2 ist leicht zu erkennen, daß wir ihre obere Hälfte mit ∞ und die untere mit 0 ergänzen müssen, wodurch der Schalter y herausfällt. Die erste Spalte füllen wir vollkommen gleich aus und kürzen dadurch den Schalter z. Es ergibt sich also

$$
\frac{\lambda}{Q\,\overline{u}_1\,U_1} = \; \cdots \; = s_0\,x_0.
$$

	x_0 z_0	x_0 z_1	x_1 z_0	x_1 z_1
$s_0\,y_1$	∞	∞	0	0
$s_0\,y_0$	∞	∞	0	0
$s_0\,y_2$	∞	∞	0	0
$s_1\,y_1$	0	0	0	0
$s_1\,y_0$	0	0	0	0
$s_1\,y_2$	0	0	0	0

In vollkommen analoger Weise finden wir für die Steuerspule H_1 des oberen Tores

$$
\frac{\lambda}{Q\,\overline{h}_1\,H_1} = s_1\,z_0.
$$

Eine besonders weitgehende Kürzung gestatten die Teilfunktionen für die Spulen U_0 und H_0:

$$\frac{\lambda}{Q\,\overline{u}_0\,U_0} =$$

	x_0 z_0	x_0 z_1	x_1 z_0	x_1 z_1		x_0 z_0	x_0 z_1	x_1 z_0	x_1 z_1
$s_0\,y_1$	?	0	?	?		0	0	0	·0
$s_0\,y_0$	?	?	?	?		0	0	0	0
$s_0\,y_2$	?	?	?	?	=	0	0	0	0
$s_1\,y_1$	?	∞	?	?		∞	∞	∞	∞
$s_1\,y_0$	?	?	?	?		∞	∞	∞	∞
$s_1\,y_2$	?	?	?	?		∞	∞	∞	∞

$= s_1$

und analog hierzu:

$$\frac{\lambda}{Q\,\overline{h}_0\,H_0} = s_0.$$

In gleicher Weise ermitteln wir die Strompfade für die Steuerspulen A und C.

$$\frac{\lambda}{Q\,b_0\,A} =$$

	x_0 z_0	x_0 z_1	x_1 z_0	x_1 z_1		x_0 z_0	x_0 z_1	x_1 z_0	x_1 z_1
$s_0\,y_1$	?	?	?	∞		∞	∞	∞	∞
$s_0\,y_0$	?	?	?	∞		∞	∞	∞	∞
$s_0\,y_2$	?	?	0	0	=	0	0	0	0
$s_1\,y_1$	?	0	?	0		0	0	0	0
$s_1\,y_0$	?	?	?	0		0	0	0	0
$s_1\,y_2$	?	?	0	0		0	0	0	0

$=$

$$= s_0(y_1 + y_0) = s_0\,\overline{y}_2$$

und analog hiezu:

$$\frac{\lambda}{Q\,b_0\,C} = s_1\,\overline{y}_1.$$

Die Steuerspule für den Ausgleichschieber B liegt einmal in Reihe mit Endkontakten $u_0\,a_0$, das andere Mal mit Kontakten $h_0\,c_0$. Wir stellen ihre anspeisenden Strompfade durch zwei getrennte Variationstabellen dar:

$$\frac{\lambda}{Q\,B} = a_0\,u_0$$

	x_0 z_0	x_0 z_1	x_1 z_0	x_1 z_1		x_0 z_0	x_0 z_1	x_1 z_0	x_1 z_1
$s_0\,y_1$	?	0	?	0		?	0	?	0
$s_0\,y_0$	?	?	?	0		?	?	?	0
$s_0\,y_2$	?	?	0	0		?	?	∞	∞
$s_1\,y_1$	?	∞	?	∞	$+\ c_0\,h_0$	?	0	?	0
$s_1\,y_0$	?	?	?	0		?	?	?	0
$s_1\,y_2$	?	?	0	0		?	?	0	0

$=$

$$= a_0\,u_0 \begin{cases} s_0\,y_1 \\ s_0\,y_0 \\ s_0\,y_2 \\ s_1\,y_1 \\ s_1\,y_0 \\ s_1\,y_2 \end{cases} \begin{array}{|cccc|} 0 & 0 & 0 & 0 \\ 0 & 0 & 0 & 0 \\ 0 & 0 & 0 & 0 \\ \infty & \infty & \infty & \infty \\ 0 & 0 & 0 & 0 \\ 0 & 0 & 0 & 0 \end{array} + c_0\,h_0 \begin{array}{|cccc|} 0 & 0 & 0 & 0 \\ 0 & 0 & 0 & 0 \\ \infty & \infty & \infty & \infty \\ 0 & 0 & 0 & 0 \\ 0 & 0 & 0 & 0 \\ 0 & 0 & 0 & 0 \end{array} =$$

$$= a_0\,u_0\,s_1\,y_1 + c_0\,h_0\,s_0\,y_2.$$

Damit haben wir die maximal gekürzten Teilfunktionen für sämtliche Verbraucher der Schaltung ermittelt. Durch Addition erhalten wir die Leitwertgleichung der Steuerschaltung für die Kammer I.

$$\frac{\lambda}{\varrho} = s_0\,x_0\,\overline{u}_1\,U_1 + s_1\,z_0\,\overline{h}_1\,H_1 + s_1\,\overline{u}_0\,U_0 + s_0\,\overline{h}_0\,H_0 + s_0\,\overline{y}_2\,b_0\,A +$$

$$+ s_1\,\overline{y}_1\,b_0\,C + (s_1\,y_1\,a_0\,u_0 + s_0\,y_2\,c_0\,h_0)\,B.$$

In dieser Gleichung kommt der Differenzdruckschalter y mit vier verschiedenen Stellungen vor. Nun ist es aber mit Rücksicht auf die Arbeitsgenauigkeit dieses Schalters nicht zulässig, ihn mit vier Kontakten mechanisch zu belasten. Wir bilden ihn daher als einfachen Umschalter aus und steuern durch ihn zwei Zwischenschütze a und γ. Zu ihrer Betätigung bilden wir die Strompfade der Spulen S_a und S_γ:

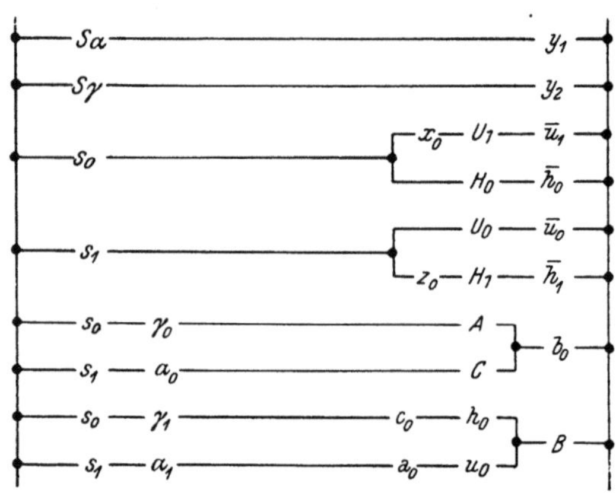

Abb. 50 b.   Steuerstrompfade für Kammer I

$$\frac{\lambda}{\varrho\,S_a} = y_1 \qquad \text{und}$$

$$\frac{\lambda}{\varrho\,S_\gamma} = y_2.$$

An Stelle der vier Kontakte des Differenzdruckschalters dürfen wir dann einsetzen:

$$y_1 = a_1, \qquad y_2 = \gamma_1,$$
$$\overline{y}_1 = a_0, \qquad \overline{y}_2 = \gamma_0.$$

Die Leitwertgleichung der Steuerschaltung erhält damit die Form:

$$\frac{\lambda}{\varrho} = y_1\,S_a + y_2\,S_\gamma + s_0\,x_0\,\overline{u}_1\,U_1 + s_1\,z_0\,\overline{h}_1\,H_1 + s_0\,\overline{h}_0\,H_0 + s_1\,\overline{u}_0\,U_0 +$$

$$+ s_0\,\gamma_0\,b_0\,A + s_1\,a_0\,b_0\,C + (s_0\,\gamma_1\,c_0\,h_0 + s_1\,a_1\,a_0\,u_0)\,B.$$

Die verkettete Schaltung, die der gefundenen Leitwertgleichung entspricht, ist in Abb. 50 *b* dargestellt. Wir haben nunmehr noch zu untersuchen, ob sich durch Vermaschung eine Vereinfachung erzielen läßt. Aus dem Stromlaufplan ist leicht zu entnehmen, daß die gleichwertigen Kontakte des Steuerschalters *s* in den Strompfaden für die Spulen *A*, *B*, *C* zusammengelegt werden können. Ferner läßt sich in den Strompfaden

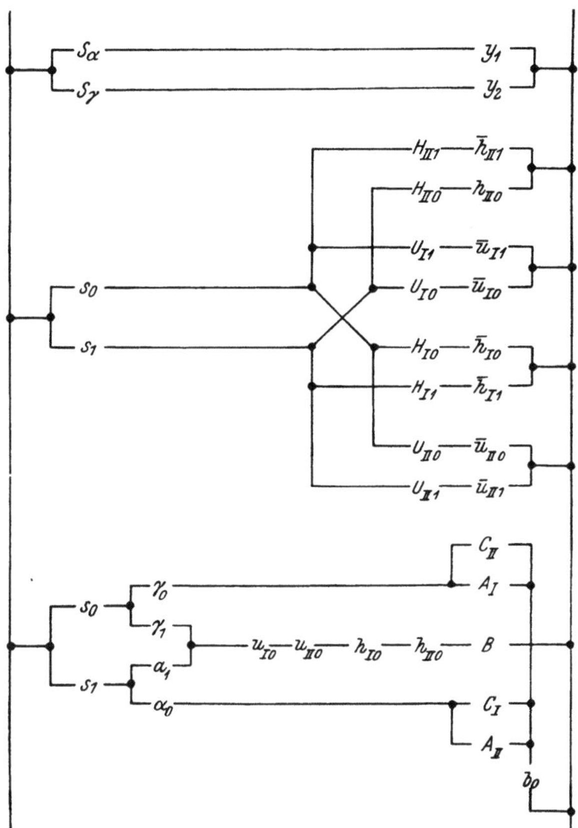

Abb. 50 *c*. Stromlaufplan der Steuerschaltung

für die Spule *B* noch eine Verbindungsleitung ersparen. Die Rechnung ergab, daß die Spule *B* beim Entleeren mit den Endschaltern *c*, *h* und beim Füllen mit den Endschaltern *a*, *u* verriegelt sein muß. Über diese Mindestforderung hinaus darf aber *B* auch mit allen vier Endschaltern gleichzeitig verriegelt sein, weil bei offenem Schieber *B* auch die Schieber *A* und *C* sowie die beiden Tore geschlossen sind. Wir dürfen also die Anspeisung der Spule *B* über eine gemeinschaftliche Leitung und alle vier

Endschalter in Reihe vornehmen. Die so gefundene Schaltung haben wir ferner noch durch Parallelschaltung der Strompfade für die Kammer *II* zu vervollständigen. Diese Operation ist so einfach, daß sie sich ohne jegliche Rechnung durchführen läßt. Abb. 50 *c* zeigt den Stromlaufplan der gesamten Steuerschaltung.

III. Die Kombinatorik der Schaltungsformen

1. Die Zusammenlegung von Leitungen und Kontakten

Hat man im Zuge der Entwicklung einer Schaltung die Wirkungsweise und die Gesamtheit der erforderlichen Strompfade, festgelegt, so kann man sich der Frage der günstigsten *Schaltungsform* zuwenden. Dem Kommutationsgesetz entsprechend, darf die Reihenfolge der Kontakte innerhalb der Stromkreise beliebig verändert werden; sie ist ohne Einfluß auf die Wirkungsweise der Strompfade. Alle größeren Schaltungen ermöglichen daher auch nach Festlegung der Wirkungsweise noch eine größere Anzahl verschiedener Schaltungsformen, die sich voneinander in der Kontaktfolge unterscheiden.

Größere Schaltungen enthalten in der Regel Strompfade, in denen teilweise *äquivalente Kontakte* vorkommen. Als äquivalent bezeichnen wir hierbei Kontakte, die dem gleichen Schalter zugehören und in gleichen Stellungen dieses Schalters geschlossen sind. Dann ist es häufig möglich, zwei oder mehrere Kontakte zusammenzulegen, das heißt einzelne Kontakte für mehr als einen Strompfad zu benützen. Schaltungsformen dieser Art nennt man *verkettete Formen*. Die Zusammenlegung von Kontakten ist jedoch an bestimmte Voraussetzungen gebunden, so daß sich äquivalente Kontakte nicht immer vereinigen lassen. Wir wollen nun vorerst untersuchen, wann die Zusammenlegung von Kontakten möglich ist und wann nicht.

Jeder Kontakt stellt eine fallweise Verbindung zwischen zwei Leitungen dar; zu zwei Kontakten gehören daher im allgemeinen vier Leitungen. Vereinigt man die beiden Kontakte zu einem einzigen, so legt man damit auch die anschließenden Leitungen paarweise zusammen. Aus der Anordnung:

$$L_1 - x - L_2,$$
$$L_3 - x - L_4$$

ergibt sich bei Verkettung:

$$(L_1 = L_3) - x - (L_2 = L_4).$$

Die Zusammenlegung ist also immer dann und nur dann statthaft, wenn eine solche paarige Vereinigung von Leitungen zulässig ist. Die Operation der Verkettung verläuft also nach folgendem Schema:

$$\left. \begin{array}{l} L_1 - x - L_2 \\ L_3 - x - L_4 \end{array} \right\} \rightarrow (L_1 = L_3) - \left[\begin{array}{c} x \\ x \end{array} \right] - (L_2 = L_4) \rightarrow (L_1 = L_3) - x - (L_2 = L_4).$$

Man erkennt, daß die Zusammenfassung von Kontakten kein primärer Vorgang, sondern nur die Folge einer Vereinigung von Leitungen ist.

Wir wollen daher untersuchen, welche Voraussetzungen erfüllt sein müssen, damit sich zwei Leitungen einer Schaltung zusammenlegen lassen.

Die Leitungen vollständig getrennter, das heißt unverketteter Strompfade sind einfache Leiter. Jeder Leiter hat nur zwei Enden und teilt die Kontakte des zugehörigen Strompfades in zwei Gruppen, wovon eine im Zuge des Stromflusses vor dem Leiter, die andere hingegen hinter ihm liegt. Somit können wir jedem Leiter einer unverketteten Schaltung zwei Leitwerte λ_1 und λ_2 zuordnen:

$$\lambda_1 \underline{\quad\quad L \quad\quad} \lambda_2$$

Durch den Leiter L werden die beiden Kontaktgruppen hintereinandergeschaltet, so daß sich ein Strompfad mit dem Leitwert $\lambda = \lambda_1 \lambda_2$ ergibt. Wir dehnen nun unsere Betrachtung auf zwei parallele Leiter L_1 und L_2 aus:

$$\lambda_1 \underline{\quad\quad L_1 \quad\quad} \lambda_2,$$

$$\lambda_3 \underline{\quad\quad L_2 \quad\quad} \lambda_4.$$

Da die beiden Strompfade zueinander parallel liegen, ergibt sich der Leitwert der unverketteten Anordnung zu

$$\lambda = \lambda_1 \lambda_2 + \lambda_3 \lambda_4.$$

Vereinigen wir L_1 und L_2 zu einer gemeinschaftlichen Leitung L, so verändern wir damit den Leitwert der Anordnung; diese erhält die Form

$$\begin{matrix} \lambda_1 \text{—} \\ \lambda_3 \text{—} \end{matrix} \underline{\quad L = L_1 = L_2 \quad} \begin{matrix} \text{—}\lambda_2 \\ \text{—}\lambda_4 \end{matrix}$$

und den Leitwert

$$\lambda = (\lambda_1 + \lambda_3) \cdot (\lambda_2 + \lambda_4) = \lambda_1 \lambda_2 + \lambda_1 \lambda_4 + \lambda_2 \lambda_3 + \lambda_3 \lambda_4.$$

Durch die Zusammenlegung hat sich somit der Leitwert um die beiden Glieder $\lambda_1 \lambda_4$ und $\lambda_2 \lambda_3$ vergrößert. Die Operation ist also nur dann zulässig, wenn in der Schaltung auch diese beiden Strompfade vorkommen. Wir können somit jedes mathematische Hilfsmittel, das uns die zulässigen Strompfade oder die den Strompfaden zugrunde liegenden Stellungsvariationen zeigt, heranziehen, um die Möglichkeiten der Verkettung zu erkennen. Besonders geeignet hierfür sind die Variationstabellen.

Vorerst betrachten wir zwei parallele Strompfade, deren Leitwerte wir in je zwei Faktoren zerlegen; die Leitwertgleichung der Anordnung hat dann die Form

$$\lambda = \lambda_1 \lambda_2 + \lambda_3 \lambda_4.$$

Innerhalb einer Variationstabelle der Schaltung erscheinen die beiden Produkte als Felder mit dem Leitwert ∞:

$$\lambda = \begin{cases} & \lambda_2 \quad \lambda_4 \\ \lambda_1 & \infty \\ \lambda_3 & \qquad \infty \end{cases}$$

Die Serienschaltung der Teilfunktionen λ_1 und λ_2 sowie λ_3 und λ_4 muß über zwei Leiter erfolgen, die wir L_1 und L_2 nennen. An Stelle der allgemeinen Leitwerte ∞ dürfen wir dann in die Tabelle die speziellen Operanden L_1 und L_2, denen ebenfalls der Leitwert ∞ zukommt, einsetzen:

$$\lambda = \begin{cases} & \lambda_2 \quad \lambda_4 \\ \lambda_1 & L_1 \\ \lambda_3 & \qquad L_2 \end{cases}$$

Nun nehmen wir an, daß in der Schaltung noch eine weitere Kombination der vier Teilfunktionen, und zwar $\lambda_1 \lambda_4$ vorkommen möge. Dann lautet die Leitwertgleichung

$$\lambda = \lambda_1 \lambda_2 + \lambda_3 \lambda_4 + \lambda_1 \lambda_4 = \lambda_1(\lambda_2 + \lambda_4) + \lambda_3 \lambda_4.$$

Die Formel umfaßt wiederum zwei Produkte entsprechend zwei Verbindungsleitungen L_1 und L_2 der Schaltung; die beiden Strompfade, denen die Teilfunktion λ_1 gemeinsam ist, verlaufen hierbei über die gemeinschaftliche Leitung L_1. Dies zeigt auch die Variationstabelle nach Einsetzen der Leitungsoperanden:

$$\lambda = \begin{cases} & \lambda_2 \quad \lambda_4 \\ \lambda_1 & \infty \quad \infty \\ \lambda_3 & \qquad \infty \end{cases} - \begin{array}{cc} \lambda_2 & \lambda_4 \\ L_1 & L_1 \\ & L_2 \end{array} \qquad \text{Schaltung nach Abb. 52 } b.$$

Die Leitwertgleichung erlaubt aber noch eine andere Art der Heraushebung:

$$\lambda = \lambda_1 \lambda_2 + (\lambda_1 + \lambda_3) \lambda_4.$$

In die Variationstabelle können wir daher auch folgendermaßen einsetzen:

$$\begin{array}{c|cc} & \lambda_2 & \lambda_4 \\ \hline \lambda_1 & L_1 & L_2 \\ \lambda_3 & & L_2 \end{array} \qquad \text{Schaltung nach Abb. 52 } a.$$

Für die Verkettung bestehen also zwei Möglichkeiten. Der Strompfad $\lambda_1 \lambda_4$ kann entweder zusammen mit $\lambda_1 \lambda_2$ über eine gemeinschaftliche Leitung L_1 oder zusammen mit $\lambda_3 \lambda_4$ über eine Leitung L_2 geführt werden. Die drei Strompfade lassen sich, da sie keine allen gemeinsame Teilfunktion aufweisen, nur paarweise gemäß Abb. 51 verketten; der hierbei übrig bleibende dritte Strompfad muß unverkettet bleiben, wie Abb. 52 zeigt. Die Verdoppelung von Kontakten ist hierbei unvermeidlich. Die beiden

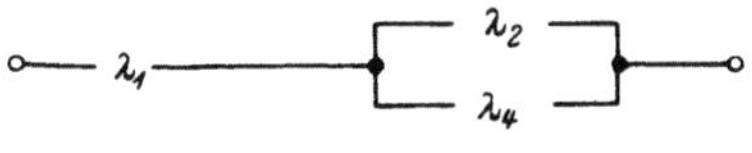

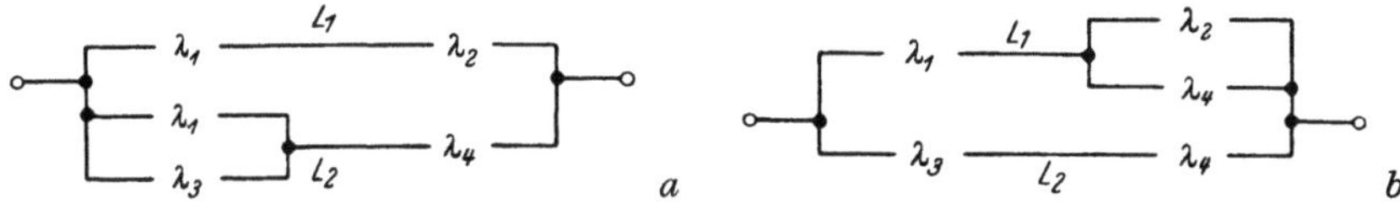

Abb. 51. Verkettung von zwei Strompfaden

möglichen Formen, die in den Abb. 52 *a* und *b* wiedergegeben sind, unterscheiden sich voneinander dadurch, daß von der Kontaktverdoppelung einmal die Kombination λ_1 und bei der anderen Form die Kombination λ_4 betroffen ist. Die beiden Formen sind zueinander symmetrisch.

Abb. 52. Paarige Verkettung von zwei Strompfaden. *a* Kontaktverdoppelung am Anfang, *b* Kontaktverdoppelung am Ende

Wir fügen nun der betrachteten Schaltung noch einen vierten Strompfad $\lambda_2 \lambda_3$ hinzu. Die Leitwertgleichung lautet dann:

$$\lambda = \lambda_1 \lambda_2 + \lambda_1 \lambda_4 + \lambda_2 \lambda_3 + \lambda_3 \lambda_4.$$

Die Gleichung läßt sich in ein Produkt mit zwei Polynomen verwandeln:

$$\lambda = (\lambda_1 + \lambda_3) \cdot (\lambda_2 + \lambda_4).$$

Die Kontaktkombinationen vor und nach den Leitern lassen sich demnach unmittelbar parallelschalten, so daß nur eine Verbindungsleitung notwendig ist. Obwohl also keine allen vier Strompfaden gemeinsame Kontaktkombination vorhanden und somit nur eine paarige Verkettung möglich ist, lassen sich trotzdem alle vier Strompfade über eine einzige

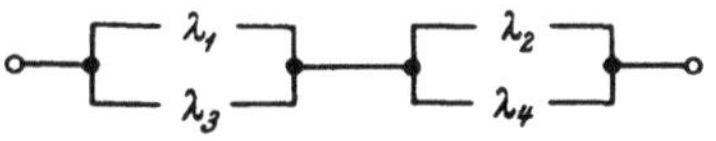

Abb. 53. Paarige Verkettung von vier Strompfaden

gemeinschaftliche Leitung entsprechend Abb. 53 führen. In der tabellarischen Darstellung zeigt sich diese Zusammenlegung folgendermaßen:

$$\lambda = \left\{ \begin{array}{c|c|c} & \lambda_2 & \lambda_4 \\ \hline \lambda_1 & \infty & \infty \\ \hline \lambda_3 & \infty & \infty \end{array} \right. = \left\{ \begin{array}{c|c|c} & \lambda_2 & \lambda_4 \\ \hline \lambda_1 & L_1 & L_1 \\ \hline \lambda_3 & L_1 & L_1 \end{array} \right. = (\lambda_1 + \lambda_3) \begin{array}{|c|} \hline (\lambda_2 + \lambda_4) \\ \hline L_1 \\ \hline \end{array}$$

Aus den vorstehend gezeigten Tabellen für zwei bis vier Strompfade läßt sich bereits das Gesetz für die Zuordnung von Strompfaden und

Leitungen erkennen. Strompfade, deren Stellungskombinationen in einer Reihe (Zeile oder Spalte) stehen, können über dieselbe Leitung geführt werden. Reihen, in denen gleichliegende Felder mit gleichen Operanden versehen sind, entsprechen Leitungen, die zusammengelegt werden können. Die Zusammenlegung läßt sich willkürlich nach Zeilen oder Spalten vornehmen.

Für die Anwendung dieser Rechenregel sei ein praktisches Beispiel angeführt. Die Umspanner zur Speisung großer Stromversorgungsnetze werden meistens so ausgeführt, daß sie eine Regelung der abgegebenen Spannung gestatten. Sie haben hierzu eigene Regelwicklungen mit einer großen Zahl von Anzapfungen, die über motorisch bewegte Stufenschalter mit dem Sternpunkt verbunden werden können. Die Steuerung der Motorantriebe erfolgt selbsttätig durch ein Spannungsrelais. Gewöhnlich arbeiten mehrere solche Umspanner parallel. Es muß dann

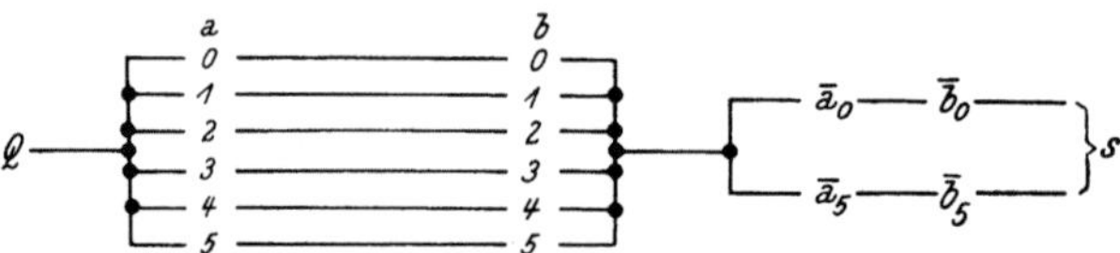

Abb. 54. *a*   Gleichlaufschaltung für Regelumspanner.   Schaltung für erzwungene
Lastgleichheit

durch eine Verriegelungsschaltung verhindert werden, daß bei Störung in einem Antrieb die übrigen Schalter verstellt werden; eine größere Verschiedenheit in den Schalterstellungen hätte Ausgleichströme, die das Ausmaß von Kurzschlüssen annehmen können, zur Folge. Die Verriegelung geschieht auf elektrischem Wege dadurch, daß die Kontakte des Spannungsrelais über Hilfskontakte an den Regelschaltern angespeist werden.

Abb. 54 zeigt eine ältere Ausführung dieser Verriegelung, die unter dem Namen „Gleichlaufschaltung" bekannt ist. Der Einfachheit halber sind in der Abbildung nur zwei Regelschalter *a* und *b* dargestellt, wobei die Zahl der Stellungen auf sechs beschränkt ist. Zwischen der Hilfsstromquelle Q und den Kontakten des Spannungsrelais s befinden sich außer den Hilfskontakten der Schalter *a* und *b* noch Endkontakte $\bar{a}_0$, $\bar{a}_5$ und $\bar{b}_0$, $\bar{b}_5$, die ein Überlaufen der Endstellungen verhindern. Bei der Ausführung gemäß Abb. 54 *a* sind die gleichnamigen Hilfskontakte der beiden Schalter durch je eine Leitung verbunden. Die Schaltung führt nur Strom, wenn die Regelschalter in gleichen Stellungen stehen. Bleibt infolge einer Störung einer der beiden Schalter um einen Schritt zurück, so wird der Steuerstromkreis unterbrochen und die Stellungsgleichheit muß durch einen Eingriff von Hand aus wieder hergestellt werden.

Die Ausführung nach Abb. 54 *a* verlangt so viele Verbindungsleitungen zwischen den beiden Umspannern, als Reglerstellungen vorhanden sind. Die Zahl der Stellungen liegt im allgemeinen zwischen 20 und 30, so daß der Aufwand an Leitungen recht groß ist. Ein weiterer Nachteil tritt in Erscheinung, wenn es sich um mehr als zwei Umspanner handelt. Wird einer davon vorübergehend ausgeschaltet, müssen seine Hilfskontakte überbrückt werden, damit die Steuerung der übrigen Regelschalter ungestört bleibt. Bei 20 bis 30 Stellungen ergibt sich somit ein fühlbarer Aufwand an Kontakten. Seit mehreren Jahren benützt man daher eine wesentlich zweckmäßigere Gleichlaufschaltung, die im folgenden rechnerisch abgeleitet werden soll.

Die Schaltung soll Strom führen, wenn sich die beiden Schalter in gleichen Stellungen befinden. Diese Bedingung läßt sich als Teilfunktion anschreiben:

$$\lambda_1 = a_0 \, b_0 + a_1 \, b_1 + \ldots + a_5 \, b_5 = \Sigma \, a_n \, b_n.$$

Bleibt eines der Geräte bei der Auf- oder Abwärtssteuerung um einen Schritt zurück, soll die Stromleitung unterbrochen und damit die selbsttätige Steuerung stillgesetzt werden. Diese Sperrbedingung können wir als weitere Teilfunktion festhalten.

$$\overline{\lambda}_2 = \Sigma \, a_n \cdot b_{n \pm 1}.$$

Wir übertragen nun die beiden Bedingungen in eine Variationstabelle; die restlichen Felder der Tabelle entsprechen Variationen mit frei wählbarem Leitwert und können mit einem Fragezeichen versehen werden.

	b_0	b_1	b_2	b_3	b_4	b_5
a_0	∞	0	?	?	?	?
a_1	0	∞	0	?	?	?
a_2	?	0	∞	0	?	?
a_3	?	?	0	∞	0	?
a_4	?	?	?	0	∞	0
a_5	?	?	?	?	0	∞

Die Felder des Wahlbereiches werden nun so ausgefüllt, daß sich möglichst viele Zeilen oder Spalten mit gleicher Feldbesetzung ergeben. Für benachbarte Reihen läßt sich die angestrebte Gleichheit nicht herstellen; hingegen ist es möglich, jede zweite Zeile oder Spalte gleich zu machen. Für die Funktion der gesuchten Schaltung lautet dann die Variationstabelle:

	b_0	b_1	b_2	b_3	b_4	b_5
a_0	∞	0	∞	0	∞	0
a_1	0	∞	0	∞	0	∞
a_2	∞	0	∞	0	∞	0
a_3	0	∞	0	∞	0	∞
a_4	∞	0	∞	0	∞	0
a_5	0	∞	0	∞	0	∞

Durch Einsetzen gleicher Leitungsoperanden in Zeilen mit gleicher Feldbesetzung erhält man die *Leitungstabelle* für die Verbindungen zwischen den beiden Schaltern a und b; gleiche Zeilen können dann zusammengelegt werden, so daß sich schließlich eine Tabelle mit vier Feldern und zwei Leitungsoperanden ergibt.

$$
\begin{array}{c|cccccc}
 & b_0 & b_1 & b_2 & b_3 & b_4 & b_5 \\
\hline
a_0 & L_1 & 0 & L_1 & 0 & L_1 & 0 \\
a_1 & 0 & L_2 & 0 & L_2 & 0 & L_2 \\
a_2 & L_1 & 0 & L_1 & 0 & L_1 & 0 \\
a_3 & 0 & L_2 & 0 & L_2 & 0 & L_2 \\
a_4 & L_1 & 0 & L_1 & 0 & L_1 & 0 \\
a_5 & 0 & L_2 & 0 & L_2 & 0 & L_2
\end{array}
=
$$

$$
= \left\{
\begin{array}{c|cc}
 & b_0 + b_2 + b_4 & b_1 + b_3 + b_5 \\
\hline
a_0 + a_2 + a_4 & L_1 & 0 \\
a_1 + a_3 + a_5 & 0 & L_2
\end{array}
\right.
$$

Da die Variationstabelle hinsichtlich der beiden Schalter symmetrisch ist, hätte eine Zusammenlegung nach Spalten statt nach Zeilen zur gleichen Leitungstabelle geführt. Diese besagt, daß zur Verbindung der beiden Schalter nur zwei Leitungen erforderlich sind; sie läßt auch erkennen, welche Kontakte an die beiden Enden jeder Leitung angeschlossen

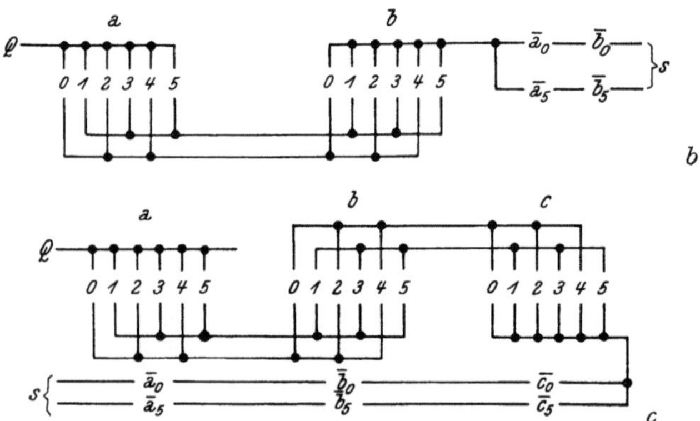

Abb. 54. b Schaltung für erzwungene Lastaufteilung, c Schaltung für mehrere Umspanner

werden müssen. Die gefundene Schaltung ist in Abb. 54 b dargestellt. Die Schaltung ist so einfach, daß sie sich ohne besondere Überlegung oder zusätzliche Rechnung für mehr als zwei Schalter erweitern läßt. Abb. 54 c zeigt die Ausführung für drei Regelumspanner. Zu beachten ist, daß durch die Zusammenlegung der Leitungen nicht nur der Aufwand herabgesetzt, sondern auch eine wesentliche funktionelle Verbesserung erzielt wurde; die Schaltung sichert den Gleichlauf nicht nur bei Stellungsgleichheit, das heißt bei gleichmäßiger Lastaufteilung auf die Umspanner; die Belastung kann vielmehr von Hand aus verschieden aufgeteilt werden und es ist nur notwendig, daß sich entweder alle Schalter in einer geradzahligen oder in einer ungeradzahligen Stellung befinden. Für die Führung des Betriebes besteht daher keine merkliche Beschränkung.

Abschließend sei noch das gefundene Gesetz für die Zusammenlegung von Leitungen als Schaltregel formuliert. Die kennzeichnende Gleichheit der Feldbesetzung in den Reihen der Tabellen kommt nur dann zustande, wenn jeder Kontakt oder jede Kontaktkombination, die einer Schar paralleler Leitungen vorgeschaltet ist, mit jeder nachgeschalteten Kontaktkombination zu einem Strompfad verbunden ist. Die Leitungsschar teilt den topologischen Bereich der Schaltung in zwei Plätze; auf dem Platz vor der Leitungsschar können alle dortselbst befindlichen Kontakte oder Kontaktkombinationen fallweise in Wirksamkeit treten und bilden somit die Elemente einer Variation; das gleiche gilt für den Platz hinter der Leitungsschar. Legt man alle Leitungen der Schar zusammen, so entsprechen die dadurch entstehenden Strompfade einem vollständigen Variationskomplex aus den Elementen der beiden Plätze. Da zwischen Kontakten, Verbrauchern und Stromquellen in schaltungstechnischer Hinsicht kein wesentlicher Unterschied besteht, gilt dieser Zusammenhang für Schaltelemente beliebiger Art. Die Schaltregel lautet daher:

Verkettungsregel. Parallele Leitungen einer Schaltung lassen sich immer dann und nur dann zu einer gemeinschaftlichen Leitung zusammenlegen, wenn jedes den Leitungen vorgeschaltete Schaltelement mit jedem nachgeschalteten zu einem Strompfad verbunden ist. Die Gesamtheit der Strompfade, die über eine gemeinsame Leitung geführt werden darf, muß somit einen vollständigen Variationskomplex aus den vorgeschalteten und nachgeschalteten Elementen bilden. Äquivalente Kontakte dürfen zu einem einzigen Kontakt vereinigt werden, wenn sowohl für die vorgeschalteten wie auch für die nachgeschalteten Leitungen die Bedingungen zur Zusammenlegung erfüllt sind.

2. Die natürlichen Schaltungsformen

Das im vorhergehenden abgeleitete Verkettungsgesetz gilt ohne jede Einschränkung; es bildet daher auch die Grundlage für alle Arten der Vermaschung. Das Kennzeichen vermaschter Schaltungen besteht darin, daß die Stromrichtung in einzelnen Kontakten (Maschenkontakten) mit dem Schaltzustand fallweise wechselt. Bei Kontakten, die unmittelbar an eine festgepolte Leitung anschließen, kann dieser Richtungswechsel nicht eintreten. Für eine Vermaschung besteht daher die Voraussetzung, daß die Schaltung Strompfade mit drei oder mehr Kontakten enthält; die Strompfade enthalten dann auch hintereinandergeschaltete Leitungen. Da jede Verkettung eine Zusammenlegung paralleler Leitungen ist, kommen auch für die Vermaschung nur Leitungen in Frage, die zueinander parallel sind. Um eine vermaschte Form zu finden, hat man daher die Leiter der einzelnen Strompfade zu einzelnen Scharen paralleler Leiter zusammenzufassen und für jede Schar zu untersuchen, ob die Bedingungen für eine Zusammenlegung gegeben sind. Hierbei kann man von jenen Leitern ausgehen, die an Kontakte mit fester Polung (Anfangs- oder Endkontakte der Schaltung) anschließen; diese bilden die erste Schar paralleler Leitungen. Alle Leiter, die mit dieser ersten Leitungsschar durch je einen Kontakt verbunden sind, gehören dann zu einer zweiten Schar. An diese schließt sich wiederum über einfache Kontakte die dritte Schar an und so fort. Jede dieser Leitungsscharen läßt sich durch eine Leitungstabelle abbilden und hinsichtlich der Verkettung untersuchen. Man kann also eine vermaschte Schaltungsform durch ein schrittweises Verfahren finden, indem man die einzelnen Leitungsscharen nacheinander verkettet und aus ihnen die Gesamtschaltung zusammensetzt. Im nachstehenden sei ein Beispiel dafür gezeigt:

Vier beliebige Verbraucher A, B, C und D sollen durch zugeordnete zweistellige Schalter a, b, c, d einschaltbar sein. Die Schaltung soll jedoch verhindern, daß jeweils mehr als ein Verbraucher Strom erhält; beispielsweise darf der Verbraucher A nur Strom führen, wenn der zugehörige Schalter a in seiner Stellung a_1, die übrigen Schalter jedoch in ihrer Nullstellung stehen. Hieraus folgt die Leitwertgleichung

$$\frac{\lambda}{Q} = a_1 b_0 c_0 d_0 A + a_0 b_1 c_0 d_0 B + a_0 b_0 c_1 d_0 C + a_0 b_0 c_0 d_1 D.$$

Um die Leitungstabellen der Schaltung zu finden, übertragen wir vorerst die Leitwertgleichung in eine Variationstabelle. Bekanntlich ist das Format dieser Tabellen frei wählbar; es ist unserem Belieben überlassen, welche Operanden wir zur Zeilenbezeichnung und welche wir als Spaltenbezeichnung verwenden. Nun suchen wir aber zuerst die Leitungstabelle für jene Leiter, die unmittelbar an einen Schalter mit fester Polung anschließen; wir nehmen an, daß dies der Schalter a sein soll und setzen die Variationstabelle so an, daß sie aus zwei Zeilen, die den beiden Stellungen des Schalters a entsprechen, besteht. Die Operanden der Verbraucher A, B, C, D dürfen wir vorerst in der Rechnung vernachlässigen; sie kommen immer gemeinsam mit den Stellungsoperanden a_1, b_1, c_1, d_1 vor und können nachträglich mit diesen Kontakten in Serie gelegt werden. Vorläufig schreiben wir daher die Variationstabelle so an, als ob nur ein Verbraucher vorhanden wäre und erhalten:

$\frac{\lambda}{Q} =$		$b_0 c_0 d_0$	$b_0 c_0 d_1$	$b_0 c_1 d_0$	$b_0 c_1 d_1$	$b_1 c_0 d_0$	$b_1 c_0 d_1$	$b_1 c_1 d_0$	$b_1 c_1 d_1$
	a_0	0	∞	∞	0	∞	0	0	0
	a_1	∞	0	0	0	0	0	0	0

Aus der Tabelle ist ersichtlich, daß die Funktion keinerlei Kürzung erlaubt; jede Variation stellt einen eigenen Strompfad dar. Die unmittelbar an die Kontakte des Schalters a anschließenden Leiter der vier Strompfade bilden zusammen die erste Schar paralleler Leitungen. Da die Schar zwischen den Schaltern a und b liegt und eine Summe von Leitern darstellt, wollen wir sie mit $\overset{b}{\underset{a}{\Sigma}} (L)$ bezeichnen; ihr tabellarisches Abbild erhalten wir aus der Variationstabelle, indem wir die zur Funktion gehörigen Felder jeder Zeile mit gleichen Leitungsoperanden versehen. Die Leitungstabelle der ersten Schar lautet daher:

$$\overset{b}{\underset{a}{\Sigma}} (L) =$$

	b_0	b_0	b_0	b_0	b_1	b_1	b_1	b_1
a_0	0	L_1	L_1	0	L_1	0	0	0
a_1	L_2	0	0	0	0	0	0	0

Der nächste Schalter in der Richtung des Stromflusses möge der Schalter b sein. An die erste Leitungsschar schließt dann eine zweite an, die zwischen den Schaltern b und c liegt. Die zugehörige Leitungstabelle finden wir aus der ersten, indem wir die oberste Spaltenbezeichnung als Zeilenbezeichnung verwenden. Die letzte Zeile muß nicht mehr angeschrieben werden, weil sie keinen Leitungsoperanden enthält.

$$\overset{c}{\underset{b}{\Sigma}} (L) =$$

		c_0	c_0	c_1	c_1
L_1	b_0	0	L_3	L_3	0
L_1	b_1	L_4	0	0	0
L_2	b_0	L_4	0	0	0

Auf genau gleiche Weise finden wir die Tabelle für die dritte Leitungsschar zwischen c und d.

$$\overset{d}{\underset{c}{\Sigma}} (L) =$$

		d_0	d_1
L_3	c_0	0	L_5
L_3	c_1	L_6	0
L_4	c_0	L_6	0

Jede der drei Tabellen läßt sich unmittelbar in einen Schaltplan übersetzen. Dieser zeigt die Leitungen der Schar und die anschließenden Kontakte. In Abb. 55 a bis c sind die drei gefundenen Leitungsscharen zeichnerisch dargestellt; durch Zusammensetzen der drei Teilpläne und Einfügen der Operanden A bis D hinter die zugehörigen Kontakte a_1 bis d_1 ergibt sich die vermaschte Form der Schaltung (Abb. 55 d).

Die Entwicklung der Schaltung führte deshalb zu einer vermaschten Form, weil die Strompfade Kontakte enthalten, die zu verschiedenen Stellungen des gleichen Schalters gehören und deshalb niemals zu gleicher Zeit geschlossen sein können. Diese Kontakte bewirken eine Sperrung aller jener Wege, die sich bei der Vermaschung ergeben, aber keine Strompfade der Schaltung sind. Ein solcher Weg ist beispielsweise $a_1 A b_0 B b_1 b_0 c_1 C d_0$; wegen der Serienschaltung der Kontakte b_0 und b_1 wird dieser Weg niemals als Strompfad wirksam. Die Schaltungsform ist also eine *Sperrform*. Da ferner jeder Strompfad nur jene Kontakte enthält, die er mit Rücksicht auf die Wirkungsweise mindestens haben muß, ist die Anordnung auch eine *natürliche Schaltungsform*. Wie das Beispiel zeigt, können auch natürliche Formen zugleich Sperrformen sein. Es sind sogar die meisten Schaltungen der technischen Praxis, die eine natürliche Vermaschung gestatten, reine Umschaltungen, die zu Sperrformen führen; andernfalls sind die Bedingungen für eine natürliche Vermaschung in den seltensten Fällen gegeben.

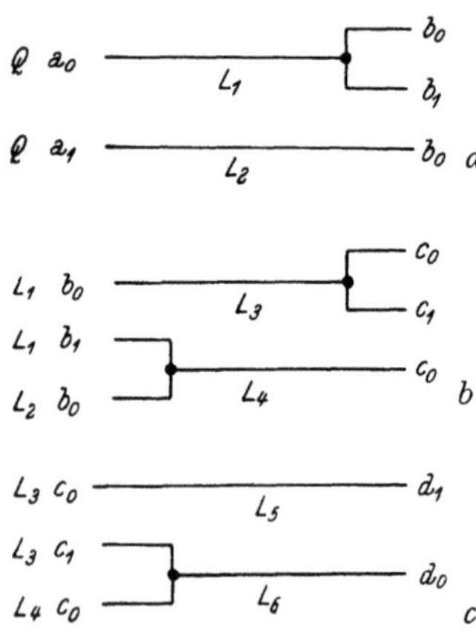

Das beschriebene Verfahren der Zusammensetzung aus Scharen paralleler Leitungen läßt sich auch dann zweckmäßig anwenden, wenn die behandelte Schaltung keine vermaschte, sondern nur eine einfach verkettete Form aufweist. Solche Schaltungen lassen sich zwar auch durch Leitwertgleichungen formgenau abbilden. Dies ist aber nur möglich, wenn die Funktion eindeutig vorgegeben ist. Häufig enthält die Aufgabe funktionelle Freiheitsgrade und die Wirkungsweise soll so gewählt werden, daß sich eine möglichst einfache Schaltungsform ergibt. Dann ist es

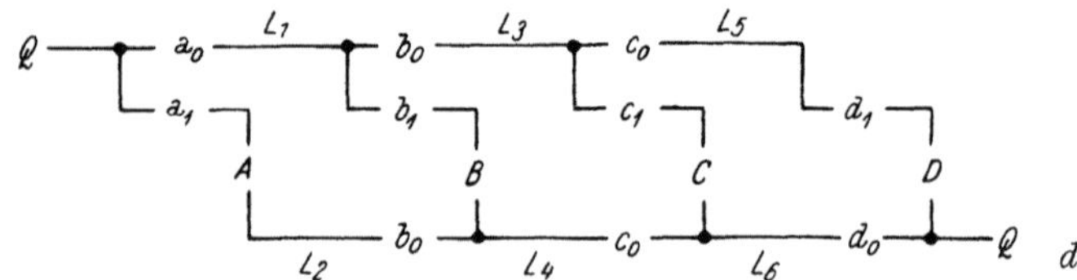

Abb. 55. Zusammensetzung einer Schaltung aus ihren Leitungsscharen. *a* Erste Schar, *b* Zweite Schar, *c* Dritte Schar, *d* Zusammengesetzte Schaltung

zweckmäßig, die Funktion nicht als Leitwertgleichung, sondern als Variationstabelle anzusetzen, weil diese durch teilweise Gleichheit ihrer Reihen sofort die Möglichkeiten zur Zusammenlegung von Leitungen erkennen läßt. Ein Beispiel aus der Installationstechnik möge dies näher erläutern.

Die gebräuchlichen Wechselschaltungen dienen dazu, eine Lampe oder Lampengruppe von zwei oder mehreren Stellen aus ein- und auszuschalten. Eine weitere bekannte Schaltung, die sogenannte Gruppen- oder Lusterschaltung, hat den Zweck, mehrere Lampengruppen eines Beleuchtungskörpers der Reihe nach einzeln und gleichzeitig in Betrieb zu setzen. Durch Kombination beider Aufgaben ergibt sich die Wirkungsweise der *Lusterwechselschaltung*; die Lampen des Beleuchtungskörpers sollen so gesteuert werden wie durch die einfache Lusterschaltung, aber von zwei oder mehreren Stellen aus. Im einfachsten Falle umfaßt die Schaltung zwei Lampengruppen A und B sowie zwei vierstellige Schalter a und b; die einfachste Form dieser Schaltung soll ermittelt werden.

Mit zwei Verbrauchern A und B lassen sich insgesamt vier Zustände herstellen, nämlich 0, A, $A + B$ und B. Mit jedem der beiden Schalter soll dieser Zyklus von Zuständen verwirklicht werden. Wir betrachten nun die Schaltung vorerst bei festgehaltener Stellung a_0 des Schalters a. Durch Betätigung des Schalters b gelangen wir vom Ausgangszustand $a_0 b_0 0$ zuerst zu $a_0 b_1 A$, dann zu $a_0 b_2 (A + B)$ und schließlich zu $a_0 b_3 B$. Wir könnten also die Wirkungsweise durch eine Leitwertgleichung folgender Form ausdrücken:

$$\frac{\lambda}{Q} = a_0 b_0 0 + a_0 b_1 A + a_0 b_2 (A + B) + \ldots \text{ usw.}$$

Zweckmäßiger ist die Abbildung durch eine Variationstabelle. Hierbei sind die Stellungsoperanden als Bezeichnungen der Zeilen und Spalten herausgehoben, wogegen die Operanden der Verbraucher in den Feldern verbleiben.

$$\frac{\lambda}{Q} = \begin{cases} & \end{cases}$$

	b_0	b_1	b_2	b_3
a_0	0	A	$A + B$	B
a_1	A	$A + B$	B	0
a_2	$A + B$	B	0	A
a_3	B	0	A	$A + B$

Variationstabellen dieser Art sind identisch mit den eindimensionalen Matrizen, mit deren Hilfe W. BADER[1] seine allgemeine Theorie der Wechselschaltungen entwickelte; diese Theorie findet ihre Erklärung in der Deutung der Matrizen als Leitwerttabellen. Untersuchen wir die einzelnen Zeilen oder Spalten auf ihre Gleichheit, so finden wir, daß sämtliche Reihen nicht nur insgesamt, sondern auch hinsichtlich der einzelnen Verbraucher A und B voneinander verschieden sind. Die tabellarisch dargestellte Funktion gestattet somit keinerlei Zusammenlegung. Man kann aber die Funktion dadurch abändern, daß man den Drehsinn des Schaltzyklus von Reihe zu Reihe ändert. Durch Trennung und Heraushebung der Operanden A und B ergibt sich dann eine Variationstabelle gewohnter Art mit paarweise gleichen Reihen.

---

[1] BADER, W.: Wechselschaltungen. Arch. Elektrotechn. 1935, Heft 2.

118   Die Kombinatorik der Schaltungsformen

$$\frac{\lambda}{Q} = \left\{ \begin{array}{c|cccc} & b_0 & b_1 & b_2 & b_3 \\ \hline a_0 & 0 & A & A+B & B \\ a_1 & A & 0 & B & A+B \\ a_2 & A+B & B & 0 & A \\ a_3 & B & A+B & A & 0 \end{array} \right\} = \left\{ \begin{array}{c|cccc} & b_0 & b_1 & b_2 & b_3 \\ \hline A\,a_0 & 0 & \infty & \infty & 0 \\ A\,a_1 & \infty & 0 & 0 & \infty \\ A\,a_2 & \infty & 0 & 0 & \infty \\ A\,a_3 & 0 & \infty & \infty & 0 \\ B\,a_0 & 0 & 0 & \infty & \infty \\ B\,a_1 & 0 & 0 & \infty & \infty \\ B\,a_2 & \infty & \infty & 0 & 0 \\ B\,a_3 & \infty & \infty & 0 & 0 \end{array} \right.$$

Durch Umwandlung der Variationstabelle in Leitungstabellen für die parallelen Leitungen findet man die Form der Schaltung zwangläufig ohne zusätzliche Überlegungen. Die erste Leitungsschar besteht aus zwei Verbindungsleitungen von den Verbrauchern zum Schalter a; sie ist so einfach, daß sich der Ansatz der zugehörigen Leitungstabelle erübrigt. Aus der Variationstabelle läßt sich sofort die Tabelle für die Leitungsschar zwischen den beiden Schaltern a, b entnehmen.

$$\overset{b}{\underset{a}{\Sigma}}(L) = \left\{ \begin{array}{c|cccc} & b_0 & b_1 & b_2 & b_3 \\ \hline a_0 + a_3 & 0 & L_1 & L_1 & 0 \\ a_1 + a_2 & L_2 & 0 & 0 & L_2 \\ a_0 + a_1 & 0 & 0 & L_3 & L_3 \\ a_2 + a_3 & L_4 & L_4 & 0 & 0 \end{array} \right.$$

Durch Übertragen dieser Tabelle in einen Stromlaufplan erhält man bereits die gesuchte Schaltung, die in Abb. 56 a wiedergegeben ist. Wie der Schaltplan zeigt, besteht jeder der beiden Schalter aus zwei vierstelligen Umschaltern. Bei Verwendung von Dosenschaltern oder Paketschaltern üblicher Bauart muß jedes dieser Geräte zwei Kontaktbahnen aufweisen, die den Verbrauchern A und B zugeordnet sind. Abb. 56 b zeigt den Wirkschaltplan dieser Ausführung. Auf gleiche Weise läßt sich die Schaltung auch für mehr als zwei Schalter oder Verbraucher ausmitteln; dann ergeben sich vermaschte und gekreuzte Schaltungsformen, die durchwegs natürliche Sperrformen sind.

Die tabellarische Darstellung auch für unvermaschte Schaltungen hat, wie das Beispiel zeigt, den Vorteil, daß zweckmäßige Abänderungen der Wirkungsweise sichtbar sind. Die Leitwertgleichung läßt solche Möglichkeiten in der Regel nicht erkennen. Auch geht aus ihr nicht hervor, ob sich die Schaltung durch Vermaschung vereinfachen läßt. Die Leitwerttabellen führen hingegen auf jeden Fall zur einfachsten Form, und zwar unabhängig davon, ob diese nur einfach verkettet oder vermascht ist.

Wie bereits eingangs erwähnt wurde, ergeben sich nur dann vermaschte Formen, wenn die Schaltung Strompfade mit drei oder mehr Schaltelementen enthält; bei nur zwei Elementen in Serie kann nur eine einfache Verkettung zustande kommen. Die Vermaschung ist keine Zusammenlegung, sondern eine Verbindung paralleler Leitungen über Kon-

takte. Durch die Auflösung der Schaltungsform in einzelne Scharen
paralleler Leitungen wird die Vermaschung auf die leichter lösbare Aufgabe der einfachen Verkettung zurückgeführt. Dies läßt sich als Schaltregel von allgemeiner Gültigkeit zum Ausdruck bringen:

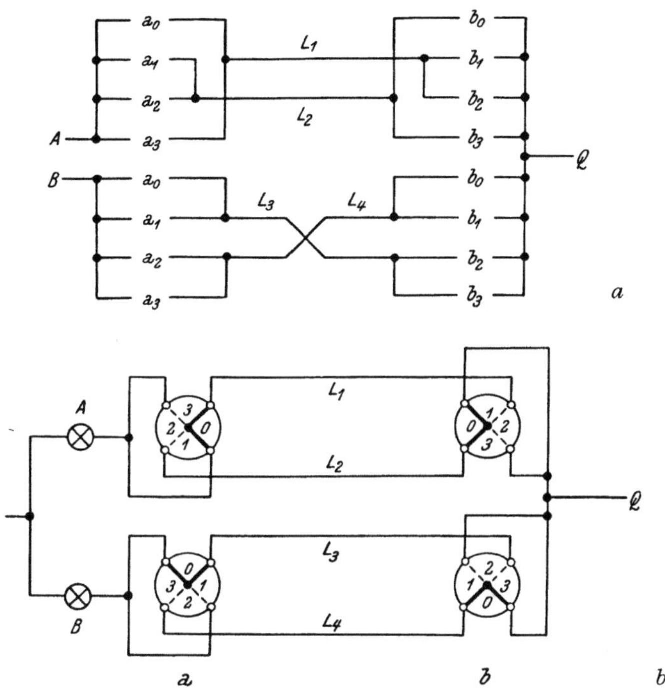

Abb. 56. Lusterwechselschaltung. *a* Stromlaufplan, *b* Wirkschaltplan

Vermaschungsregel. Eine Vermaschung ist nur bei Schaltungen möglich, die Strompfade mit mindestens drei Schaltelementen in Reihe enthalten. Jede vermaschte Schaltung mit zwei Spannungspolen besteht
aus einzelnen Scharen paralleler Leitungen, deren Leiter miteinander einfach verkettet sind. Die an einem gemeinsamen Spannungspol liegenden
Kontakte führen stets zu Leitungen, die eine zusammengehörige Leitungsschar bilden. Alle mit dieser Schar über einfache Kontakte verbundenen
Leitungen bilden eine zweite Schar paralleler Leitungen usw. Die Schaltung mit geringstem Gesamtaufwand an Leitungen und Kontakten ergibt
sich aus der Verbindung der einzelnen verketteten Leitungsscharen über
die gemeinsamen Kontakte.

Die vorstehende Regel gilt auch für Schaltungen, bei denen Kürzungen
möglich sind. Gekürzte Schaltungen, die eine Vermaschung gestatten,
kommen allerdings äußerst selten vor. Die technische Praxis zeigt fast nur
Umschaltungen sowie Formen mit erzwungener Vermaschung. Die Vereinigung von Kürzung und Vermaschung erlaubt jedoch, ein wichtiges

Gesetz der Schaltungstechnik abzuleiten. Wir betrachten den einfachsten Fall einer vermaschten Schaltung ohne Sperrkontakte; deren Leitwertgleichung lautet:

$$\lambda = a_1\,c_1\,e_1 + a_1\,d_1 + b_1\,c_1\,d_1 + b_1\,e_1.$$

Die Reihenfolge der Schalter in den Strompfaden möge der alphabetischen entsprechen. Wir stellen die Frage, wie viele Leitungen zwischen dem Schalter c und dem Schalter d notwendig sind. Dies läßt sich auch so ausdrücken, daß die Mindestzahl der Leitungen der zwischen c und d liegenden Schar gesucht werden soll. Hierzu verwandeln wir die Leitwertgleichung in eine Variationstabelle mit den Variationen von a, b, c als Zeilen und denjenigen von d, e als Spalten.

$$
\lambda = \;
$$

	$d_0\,e_0$	$d_0\,e_1$	$d_1\,e_0$	$d_1\,e_1$
$a_0\,b_0\,c_0$	0	0	0	0
$a_0\,b_0\,c_1$	0	0	0	0
$a_0\,b_1\,c_0$	0	∞	0	∞
$a_0\,b_1\,c_1$	0	∞	∞	∞
$a_1\,b_0\,c_1$	0	0	∞	∞
$a_1\,b_0\,c_1$	0	∞	∞	∞
$a_1\,b_1\,c_0$	0	∞	∞	∞
$a_1\,b_1\,c_1$	0	∞	∞	∞

Durch Kürzung entfallen sämtliche Zeilen mit den Operanden a_0 und b_0; in den übrigbleibenden Zeilen lassen sich die Variationen mit c_0 mit den in der gleichen Spalte befindlichen Variationen, die c_1 enthalten, zusammenfassen. Somit ergibt sich:

$$
\lambda = \;
$$

	$d_0\,e_0$	$d_0\,e_1$	$d_1\,e_0$	$d_1\,e_1$
$b_1\,c_0$	0	∞	0	∞
$b_1\,c_1$	0	∞	∞	∞
$a_1\,c_0$	0	0	∞	∞
$a_1\,c_1$	0	∞	∞	∞

$$
= \;
$$

	$d_0\,e_0$	$d_0\,e_1$	$d_1\,e_0$	$d_1\,e_1$
$b_1\,\infty$	0	∞	0	∞
$b_1\,c_1$	0	?	∞	?
$a_1\,\infty$	0	0	∞	∞
$a_1\,c_1$	0	∞	?	?

Aus diesem Ergebnis folgt eine Leitungstabelle mit paarweise gleichen Zeilen.

$$
\overset{d}{\underset{c}{\Sigma}}\,(L) = \;
$$

	$d_0\,e_0$	$d_0\,e_1$	$d_1\,e_0$	$d_1\,e_1$
$b_1\,\infty$	0	L_1	0	L_1
$b_1\,c_1$	0	0	L_2	L_2
$a_1\,\infty$	0	0	L_2	L_2
$a_1\,c_1$	0	L_1	0	L_1

$$
= \;
$$

	d_1	e_1
$b_1\,\infty$	0	L_1
$b_1\,c_1$	L_2	0
$a_1\,\infty$	L_2	0
$a_1\,c_1$	0	L_1

Die Tabelle besagt, daß die gesuchte Leitungsschar zwischen den Schaltern c und d aus nur zwei Leitungen besteht; diese sind über einen Querkontakt c miteinander verbunden. Das Beispiel lehrt uns, daß sich die Zahl der Leitungen an einer beliebig herausgegriffenen Stelle der Schaltung ohne Kenntnis der Gesamtschaltung ermitteln läßt. Man kann sich davon überzeugen, daß die Leitungstabelle unverändert bleibt, wenn die Reihenfolge der Schalter a und b oder d und e geändert wird. Für die betrachtete Leitungsschar zwischen c und d ist nur maßgeblich, welche Elemente ihr vor- und nachgeschaltet sind; im übrigen ist deren Reihenfolge gleichgültig. Diese Tatsache ist für die Schaltungen des Fernmeldewesens und der Fernwirktechnik von Bedeutung. Die Leitungsverbindung zwischen der Sende- und Empfangsstelle einer elektrischen Übertragungseinrichtung wird nur durch die beiden Teilfunktionen der Schaltgeräte an diesen Stellen bestimmt; die schaltungstechnische und konstruktive Ausbildung der Ausrüstung ist ohne Einfluß auf die Fernverbindung. Wir können als weitere Schaltregel festhalten:

Leitungsregel. Bei vorgegebener Wirkungsweise ist die Mindestzahl der Leitungen an einer bestimmten Stelle einer Schaltung nur davon abhängig, welche Schaltelemente der betrachteten Stelle vorgeschaltet und welche ihr nachgeschaltet sind. Man findet die erforderliche Leitungsschar durch Kürzung der Strompfade und einfache Verkettung der Leiter an dieser Stelle.

3. Die Schaltungsformen mit erzwungener Vermaschung

Schaltungen, bei denen Kürzungen möglich und auch vollständig durchgeführt sind, erlauben in der Regel keine Vermaschungen. Man kann jedoch die Voraussetzung hiezu künstlich schaffen, indem man auf die Kürzung ganz oder teilweise verzichtet. Ist die Reihenfolge der Schaltgeräte im Zuge des Stromflusses vorgegeben, so bedingt die vollständige Kürzung eine ganz bestimmte natürliche Schaltungsform, die je nach Zahl der Spannungspole in mehreren polsymmetrischen Varianten auftritt. Damit ist auch eine bestimmte Anzahl von Kontakten an den einzelnen Schaltgeräten und von Verbindungsleitungen zwischen den Geräten vorgegeben. Durch den Verzicht auf mögliche Kürzungen erreicht man zwei Vorteile. Erstens lassen sich die erforderlichen Kontakte anders verteilen, so daß gewisse Schaltgeräte bevorzugt und kleiner ausgeführt werden können; zweitens treten statt äquivalenter Kontakte, die getrennte Schaltbrücken erfordern, Umschaltekontakte mit gemeinschaftlichen Brücken auf, woraus sich ebenfalls eine Vereinfachung der Schaltgeräte ergibt. Schaltungsformen, deren Vermaschung auf· Kosten der Kürzung erzwungen ist, sind daher für die technische Praxis von großer Bedeutung. Ihr Kennzeichen liegt darin, daß die Strompfade außer jenen Kontakten, die für die Wirkungsweise nötig sind, noch Sperrkontakte enthalten, die bei vollständiger Kürzung entfallen würden und nur dazu dienen, die sonst unmögliche Vermaschung zu erzwingen.

Bei bekannter Reihenfolge der Schaltgeräte in den Strompfaden lassen sich solche künstliche Schaltungsformen auf gleiche Weise ermitteln, wie die natürlichen. Hierfür ein Beispiel:

Die Auslösevorrichtungen von Starkstromschaltern werden üblicherweise von zugehörigen Befehlsschaltern angespeist. In Fällen der Gefahr besteht meistens der Wunsch, mehrere oder alle Auslöser gleichzeitig durch ein gemeinschaftliches Schaltgerät, z. B. einen Notschalter, in Tätigkeit zu setzen. Für die Entwicklung der hierzu erforderlichen Schaltung genügt es, drei Auslöser A, B, C zu betrachten; sie sollen einzeln durch zugeordnete Schalter a, b, c und gemeinschaftlich durch einen Summenschalter s einschaltbar sein. Die beschriebene Wirkungsweise führt zu einer Leitwertfunktion

$$\lambda = (a_1 + s_1)\,A + (b_1 + s_1)\,B + (c_1 + s_1)\,C.$$

Die natürliche Form der Schaltung ist nur einfach verkettet und entspricht der angeführten Gleichung; sie ist in Abb. 57 a wiedergegeben. Der Schalter s muß hierbei soviele Kontakte haben, als Verbraucher vorhanden sind. Meist ist aber der Summenschalter kein Handschalter, sondern ein mit einem Meßwerk versehenes Relais. Mit Rücksicht auf die Meßgenauigkeit ist dann die mechanische Belastung durch mehrere Kontakte unerwünscht. Es stellt sich somit die Aufgabe, die Schaltung so umzuformen, daß das Schaltgerät s möglichst wenige Kontakte erhält.

Wir haben zuerst die Reihenfolge der Schaltelemente festzulegen; sie ist nicht ausdrücklich vorgegeben, aber naheliegend. Damit der Schalter s wenig Kontakte enthält, setzen wir ihn an das Ende der Schaltung; an den anderen Pol der Stromquelle legen wir die Verbraucher A, B, C, womit ungewollte Serienschaltungen von vornherein vermieden werden. In der Mitte der Schaltung liegen dann die Schalter a, b, c für die Einzelsteuerung. Nunmehr sind wir in der Lage, die Variationstabelle der Schaltung anzusetzen:

$$\frac{\lambda}{Q} =$$

	A	B	C
$s_0\,a_0\,b_0\,c_0$	0	0	0
$s_0\,a_0\,b_0\,c_1$	0	0	∞
$s_0\,a_0\,b_1\,c_0$	0	∞	0
$s_0\,a_0\,b_1\,c_1$	0	∞	∞
$s_0\,a_1\,b_0\,c_0$	∞	0	0
$s_0\,a_1\,b_0\,c_1$	∞	0	∞
$s_0\,a_1\,b_1\,c_0$	∞	∞	0
$s_0\,a_1\,b_1\,c_1$	∞	∞	∞
$s_1\,a_0\,b_0\,c_0$	∞	∞	∞
$s_1\,a_0\,b_0\,c_1$	∞	∞	∞
$s_1\,a_0\,b_1\,c_0$	∞	∞	∞
$s_1\,a_0\,b_1\,c_1$	∞	∞	∞
$s_1\,a_1\,b_0\,c_0$	∞	∞	∞
$s_1\,a_1\,b_0\,c_1$	∞	∞	∞
$s_1\,a_1\,b_1\,c_0$	∞	∞	∞
$s_1\,a_1\,b_1\,c_1$	∞	∞	∞

Die Tabelle läßt sich so kürzen, daß die Schalterstellung s_0 verschwindet:

$$\frac{\lambda}{Q} = \begin{cases}
& A & B & C \\
\infty\ a_0\ b_0\ c_0 & 0 & 0 & 0 \\
\infty\ a_0\ b_0\ c_1 & 0 & 0 & \infty \\
\infty\ a_0\ b_1\ c_0 & 0 & \infty & 0 \\
\infty\ a_0\ b_1\ c_1 & 0 & \infty & \infty \\
\infty\ a_1\ b_0\ c_0 & \infty & 0 & 0 \\
\infty\ a_1\ b_0\ c_1 & \infty & 0 & \infty \\
\infty\ a_1\ b_1\ c_0 & \infty & \infty & 0 \\
\infty\ a_1\ b_1\ c_1 & \infty & \infty & \infty \\
s_1\ a_0\ b_0\ c_0 & \infty & \infty & \infty \\
s_1\ a_0\ b_0\ c_1 & \infty & \infty & ? \\
s_1\ a_0\ b_1\ c_0 & \infty & ? & \infty \\
s_1\ a_0\ b_1\ c_1 & \infty & ? & ? \\
s_1\ a_1\ b_0\ c_0 & ? & \infty & \infty \\
s_1\ a_1\ b_0\ c_1 & ? & \infty & ? \\
s_1\ a_1\ b_1\ c_0 & ? & ? & \infty \\
s_1\ a_1\ b_1\ c_1 & ? & ? & ?
\end{cases}$$

Für die weitere Kürzung bestehen nunmehr zwei Möglichkeiten. Setzt man in die unbestimmten Felder durchwegs den Leitwert ∞ ein, so lassen sich auch die Schalterstellungen a_0, b_0 und c_0 kürzen; es ergibt sich dann die natürliche Schaltungsform gemäß Abb. 57 a. Man kann aber auch auf die maximale Kürzung verzichten, indem man den fraglichen Feldern den Leitwert 0 zuweist und dann gleiche Zeilen zusammenfaßt. Die ·Durchführung dieser Operation kann hier übergangen werden, da sie nichts Neues bietet; ihr Ergebnis lautet:

$$\frac{\lambda}{Q} = \begin{cases}
& & & & A & B & C \\
\infty & \infty & \infty & c_1 & 0 & 0 & \infty \\
\infty & \infty & b_1 & \infty & 0 & \infty & 0 \\
\infty & a_1 & \infty & \infty & \infty & 0 & 0 \\
s_1 & a_0 & \infty & \infty & \infty & 0 & 0 \\
s_1 & \infty & b_0 & \infty & 0 & \infty & 0 \\
s_1 & \infty & \infty & c_0 & 0 & 0 & \infty
\end{cases}$$

Daraus folgt unmittelbar die Leitungstabelle:

$$\overset{ABC}{\underset{sabc}{\Sigma}}(L) = \begin{cases}
& & & & A & B & C \\
\infty & \infty & \infty & c_1 & 0 & 0 & L_1 \\
\infty & \infty & b_1 & \infty & 0 & L_2 & 0 \\
\infty & a_1 & \infty & \infty & L_3 & 0 & 0 \\
s_1 & a_0 & \infty & \infty & L_3 & 0 & 0 \\
s_1 & \infty & b_0 & \infty & 0 & L_2 & 0 \\
s_1 & \infty & \infty & c_0 & 0 & 0 & L_1
\end{cases}$$

Die Tabelle entspricht einer Schaltung nach Abb. 57 *b*, deren Stromlaufplan in Abb. 57 *c* wiedergegeben ist.

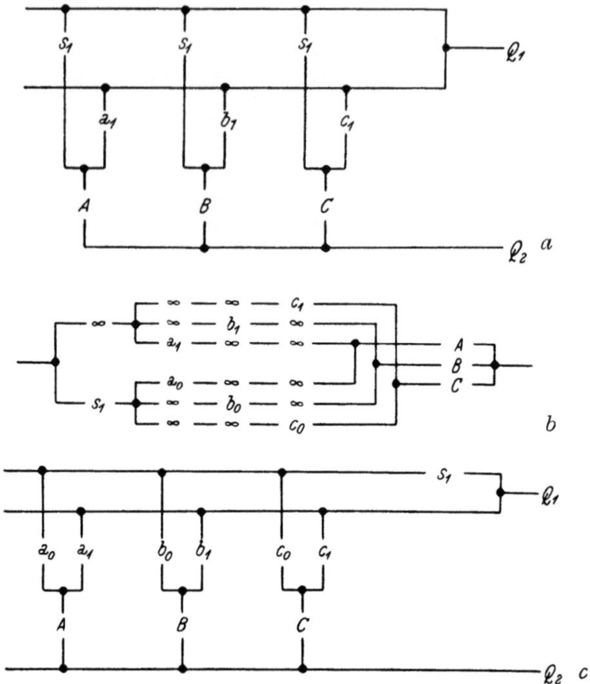

Abb. 57. Schaltung für Gefahrauslösung. *a* Natürliche Form, *b* Entwicklung der Sperrform, *c* Stromlaufplan der Sperrform

Die meisten Schaltungen, die in der technischen Praxis vorkommen, zeigen nur eine paarige Verkettung ihrer Strompfade; die Stellungskombinationen haben dann paarweise gemeinsame Faktoren. Die zugehörigen Leitwertgleichungen entsprechen also der Form:

$$\lambda = a\,b + b\,c + c\,d.$$

Es sei nun untersucht, welche Schaltungsformen zu dieser Funktion überhaupt bestehen können. Bei natürlicher Vermaschung erhält man zwei Formen, die hinsichtlich der Spannungspole symmetrisch und in Abb. 58 *a* wiedergegeben sind. Entwickelt man die Formen mit erzwungener Vermaschung, so findet man wiederum zwei symmetrische Formen entsprechend Abb. 58 *b*. Wie der Vergleich der vier Schaltungsformen zeigt, bleibt die Zahl der Kontakte stets dieselbe; auch die beiden Strompfade $a_1\,b_1$ und $c_1\,d_1$, die zueinander teilerfremd sind, erscheinen in allen vier Formen. Der dritte Strompfad $b_1\,c_1$, der mit den beiden anderen verkettet ist, führt zu einem fünften Kontakt, der einem beliebigen Schalter zugewiesen werden kann; in den natürlichen Formen erscheint er als Kontakt c_1 oder d_1, in den Sperrformen als Kontakt a_0 oder d_0.

Man ist also in der Lage, die unvermeidlichen Doppelkontakte an jenem Gerät der Schaltung anzubringen, wo sie am wenigsten stören. Bei größeren Schaltungen läßt sich damit auch die Gesamtzahl der Kontakte beeinflussen. Beispielsweise kann a ein Einzelschalter sein, wogegen die übrigen Elemente b, c, d Kombinationen von mehreren Geräten darstellen; dann wird man die Sperrform mit einem Kontakt a_0 wählen, weil die Verdoppelung der anderen Schaltelemente zu einer höheren Kontaktzahl führen würde.

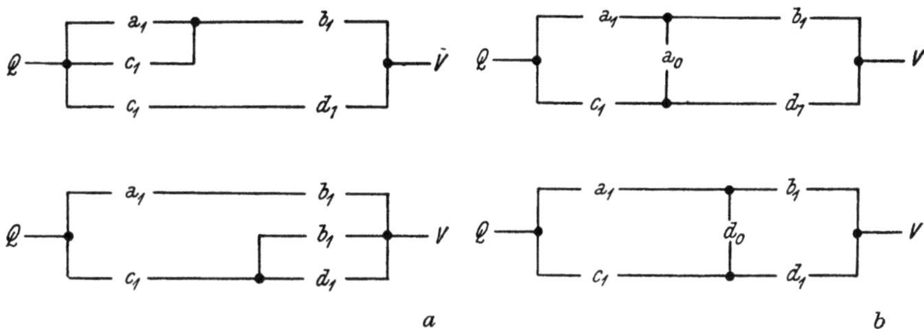

Abb. 58. Die vier Formen der paarigen Verkettung. a Die beiden natürlichen Formen, b Die beiden Sperrformen

Die Kenntnis der in Abb. 58 gezeigten Formen genügt sehr oft, um Schaltungen mit erzwungener Vermaschung ohne jede Rechnung zu finden. Man geht hierbei nicht von der ungekürzten Schaltung aus, sondern beginnt mit der natürlichen Form; diese ist gewöhnlich nur einfach verkettet und daher leicht zu finden. Die Vermaschung führt man in der Weise durch, daß man äquivalente Kontakte vorerst ohne Rücksicht auf die Strompfade zusammenlegt. An jeder Stelle einer solchen Zusammenfassung läßt man Platz für einen Sperrkontakt frei. Zu jedem der freien Plätze untersucht man dann die vor- und nachgeschalteten Kontaktkombinationen und bildet die möglichen Wege, die keinem Strompfad der Schaltung entsprechen. Diese Wege müssen gesperrt werden; hierzu kann man jeden Kontakt verwenden, der in den zu sperrenden Wegen vorkommt und für die **wirksamen** Strompfade zulässig ist. Das Kennzeichen für die Zulässigkeit besteht darin, daß der Sperrkontakt oder die Sperrkombination zu den vor- oder nachgeschalteten parallelen Kontakten komplementär sein muß. Dieses Verfahren hat den Vorteil großer Anschaulichkeit; es sei in nachstehendem Beispiel erläutert:

Gegeben sei eine Schaltung mit einfacher Verkettung, entsprechend Abb. 59 a. Die Ableitung dieser Schaltung erfolgte im Abschnitt II, 1. In der natürlichen Schaltungsform kommen die Kontakte a_1, l_1, n_1, d_1 doppelt vor, so daß ihre Zusammenfassung naheliegt. Führt man diese Operation im Stromlaufplan aus, so ergibt sich eine Form nach Abb. 59 b. Durch die Zusammenlegung entstehen zwei neue Knotenpunkte; in den abzweigenden Leitern ist je ein Platz für einen Sperrkontakt vorzusehen.

Ferner entsteht ein neuer Weg, der über einen Kontakt b_1, die Sperrstelle und nochmals einen Kontakt b_1 führt. Für die Sperrung dieses Weges besteht somit nur eine Möglichkeit, nämlich die Einführung eines zu b_1 komplementären Kontaktes b_0. In analoger Weise kommt ein Weg über zwei Kontakte c_1 und die zweite Sperrstelle zustande; dortselbst ist daher ein Sperrkontakt c_0 anzubringen. Abb. 59 c zeigt die gefundene Sperrform. Das zeichnerische Verfahren zur Auffindung solcher Schaltungsformen ist allerdings nur dann anwendbar, wenn die Stromkreise der behandelten Schaltung ausschließlich paarig verkettet sind; bei weniger einfachen Verhältnissen läßt es die schaltungstechnischen Möglichkeiten nicht erkennen.

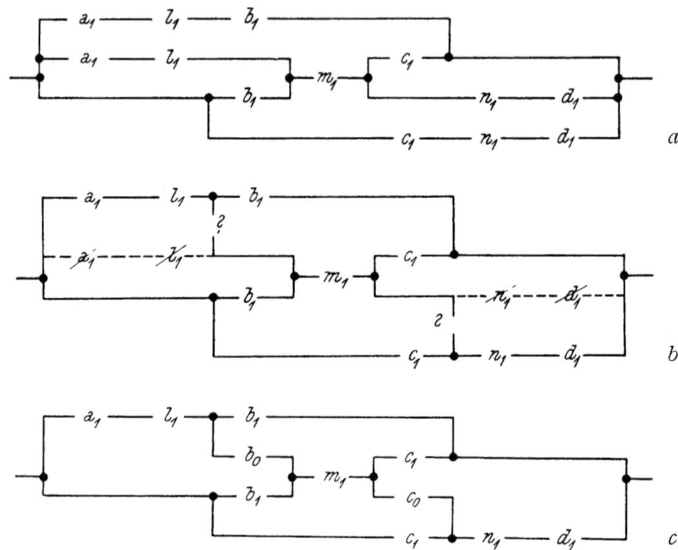

Abb. 59. Zeichnerische Entwicklung von Sperrformen. *a* Natürliche Formen, *b* Zusammenlegung äquivalenter Kontakte, *c* Sperrform

Ferner ist zu beachten, daß die Schaltungsformen, die sich durch Verkettung und Vermaschung ergeben, im wesentlichen durch die Reihenfolge der Schaltgeräte in den Strompfaden bedingt sind. Bei Änderung der Reihenfolge ändern sich auch die Voraussetzungen zur Zusammenlegung von Leitungen und Kontakten. Da die Schaltaufgaben der Praxis meistens keine bestimmten Vorschriften über die Schalterfolge machen, liegt das Problem, die zweckmäßigste Schaltungsform zu finden, in erster Linie in der Bestimmung der richtigen Schalterfolge. Diese weitaus schwierigste Frage der schaltungstechnischen Topologie soll im folgenden Abschnitt gesondert behandelt werden.

Das Ergebnis der vorstehenden Betrachtungen sei abschließend als Schaltregel zusammengefaßt:

Sperrkontaktregel. In Schaltungen mit paarig verketteten Strompfaden, bei denen die Voraussetzungen zu einer natürlichen Vermaschung nicht vorliegen, läßt sich durch Sperrkontakte eine formale Vermaschung erzwingen. Als Sperrkontakte eignen sich solche Kontakte, die zu beliebigen Kontakten aller funktionsfremden Wege sowie zu allen der Sperrstelle vor- oder nachgeschalteten Kontaktkombinationen der Strompfade komplementär sind.

4. Die natürliche Kontaktfolge

Bei manchen Schaltaufgaben ist eine bestimmte Reihenfolge der Schaltgeräte in den Strompfaden vorgeschrieben. Dieser Fall ist immer dann gegeben, wenn die einzelnen Geräte voneinander weiter entfernt und über einen gemeinschaftlichen Leitungsstrang, z. B. ein Kabel, verbunden sind. Die Zahl der Leiter im Leitungsstrang wird dann am kleinsten, wenn die Schalterfolge in allen Strompfaden gleich ist und der räumlichen Anordnung der Geräte entspricht. Meistens sind jedoch die zu einer Schaltung gehörenden Geräte örtlich vereinigt, so daß die Leiterlängen keine wesentliche Rolle spielen. Die Kontaktfolge in den einzelnen Strompfaden kann dann frei gewählt werden. Zur Bestimmung der zweckmäßigsten Ausführung sind in diesem Falle zusätzliche Schaltbedingungen konstruktiver Art (Konstruktionsbedingungen) erforderlich. Beispielsweise kann verlangt werden, daß die Gesamtzahl der Kontakte ein Minimum sein soll. Ferner kann der Wunsch vorliegen, die Kontakte möglichst gleichmäßig auf die einzelnen Geräte aufzuteilen, um Schaltertypen mit vielen Kontakten zu vermeiden. Häufig besteht schließlich die Einschränkung, daß ein bestimmtes Schaltgerät, z. B. ein empfindliches Relais, mit möglichst wenigen oder nur einem Kontakt ausgestattet werden darf. Solchen verschiedenartigen Forderungen läßt sich durch Wahl einer ganz bestimmten Kontaktfolge in den Strompfaden Rechnung tragen. Es soll nun gezeigt werden, welche Möglichkeiten hierzu bestehen.

Der Einfluß der Kontaktfolge auf die Form der Schaltung rührt daher, daß die verschiedenen Stellen, an denen sich Kontakte befinden können, untereinander nicht gleichwertig sind. Jedem Elektrotechniker ist aus Erfahrung bekannt, daß sich gleichartige Kontakte gewöhnlich nur dann zusammenlegen lassen, wenn sie am Anfang oder Ende der Strompfade angeordnet und somit Randkontakte der Schaltung sind; die Randstellen sind also gegenüber anderen Stellen sichtlich bevorzugt. Hingegen müssen Kontakte, die als echte Maschenkontakte oder Sperrkontakte benützt werden, im Inneren der Schaltung angeordnet sein. Um dies einzusehen, betrachten wir eine ideale Schaltung, in der jede Schalterstellung nur durch einen einzigen Kontakt vertreten ist. Schließt man die Schaltung an eine Gleichstromquelle an, so tritt der Strom bei allen unmittelbar an den Pluspol anschließenden Kontakten in die Schaltung ein und verläßt diese bei allen negativ gepolten Kontakten. Ein Richtungswechsel des Stromes ist bei den Randkontakten unmöglich; alle Kontakte, die am gleichen Spannungspol liegen, sind dauernd zueinander parallel, wogegen ihre Verknüpfung mit den anderen Kontakten in Abhängigkeit vom Schaltzustand wechseln kann. Um eine Schar gleichgepolter Randkontakte zu finden, hat man daher aus der vorgegebenen Funktion festzustellen, welche Kontakte zueinander parallel und niemals in Serie vorkommen. Auf dieser Überlegung beruht ein von H. Piesch[1] angegebenes Verfahren, das im nachstehenden erläutert sei:

Vorgegeben sei eine Schaltung, die eine natürliche Vermaschung erlaubt; ihre Leitwertgleichung sei:

$$\lambda = a\,b + a\,d\,e + b\,c\,e + c\,d.$$

Wir verwandeln die Gleichung in eine Tabelle, deren Zeilen den Strompfaden und deren Spalten den wirksamen Schalterstellungen zugeordnet sind (*Stellungstabelle*).

a	b	∞	∞	∞
a	∞	∞	d	e
∞	b	c	∞	e
∞	∞	c	d	∞

Nun fassen wir jene Spalten zusammen, die bei Multiplikation gleichliegender Felder eine resultierende Spalte ergeben, die in jedem Feld einen Stellungsoperanden aufweist; diese Spalte nennt eine zum gleichen Spannungspol gehörige Schar von Kontakten. Bei zwei Spannungspolen der Schaltung haben wir zwei solche Randspalten zu bilden. Führen wir diese Operation mit der vorstehenden Tabelle durch, so ergibt sich:

a	∞	b
a	e	d
c	e	b
c	∞	d

Die Tabelle besagt, daß sich die Kontakte a, c an einem, die Kontakte b, d am anderen Ende der Strompfade und die Kontakte e in der Mitte der Schaltung befinden müssen; bei jeder anderen Reihenfolge ergibt sich ein höherer Kontaktbedarf.

In einfacheren Fällen reicht das beschriebene Verfahren allein aus, um die gesuchte Schaltung vollständig zu entwickeln. Als Beispiel diene die bekannte Schaltung zur Serien-Parallelschaltung zweier Verbraucher, die häufig zur Regelung von Heizkörpern verwendet wird. Durch einen gemeinsamen Umschalter s werden zwei Heizwiderstände A und B zuerst

---

[1] PIESCH, H.: Über die Vereinfachung von allgemeinen Schaltungen. Arch. Elektrotechn. 1939, Heft 11.

hintereinander, dann einzeln und schließlich parallel an die Stromquelle gelegt. Diese Wirkungsweise läßt sich durch folgende Leitwertgleichung ausdrücken:

$$\frac{\lambda}{Q} = s_1\,A\,B + s_2\,A + s_3\,B + s_4(A + B).$$

Aus ihr ergibt sich die Stellungstabelle:

s_1	∞	∞	∞	A	B
∞	s_2	∞	∞	A	∞
∞	∞	s_3	∞	∞	B
∞	∞	∞	s_4	A	∞
∞	∞	∞	s_4	∞	B

Man wäre versucht, die ersten vier Spalten dieser Tabelle zu einer Randspalte zusammenzulegen. Die beiden restlichen Spalten ergeben aber zusammen keine Randspalte für den zweiten Pol der Schaltung; es wäre nur eine teilweise Zusammenlegung möglich, so daß einer der beiden Heizwiderstände verdoppelt werden müßte. Es ist zweckmäßiger, eine Verdoppelung von Kontakten in Kauf zu nehmen und die beiden Spalten A und B zu Randspalten zu ergänzen. Die Tabelle erhält dann die Form:

A	s_1	B
A	∞	s_2
s_3	∞	B
A	∞	s_4
s_4	∞	B

Die so geordnete Stellungstabelle läßt sich unmittelbar in den Stromlaufplan der gesuchten Schaltung übertragen, wie Abb. 60 zeigt.

Das beschriebene Verfahren gibt nur Auskunft über die Randkontakte der Schaltung; führen die Strompfade über mehr als drei Schaltgeräte, so läßt sich die zweckmäßigste Reihenfolge der übrigen Kontakte auf

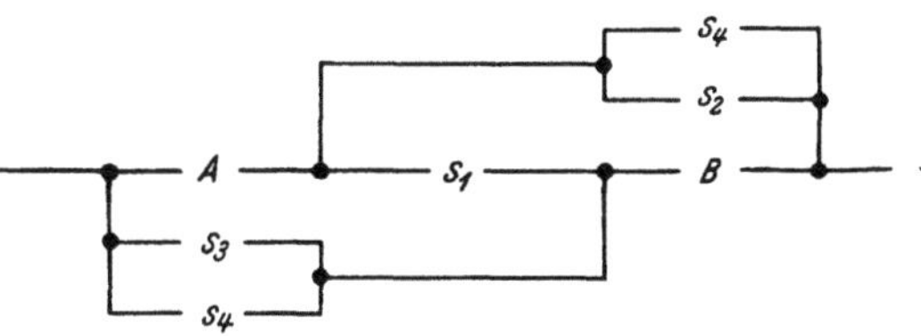

Abb. 60. Heizstufenschaltung

diesem Wege nicht bestimmen. Besondere Bedeutung kommt jedoch dem Umstand zu, daß es für die Randkontakte vollkommen gleichgültig ist, welche Form die Schaltung sonst hat. Da an den Anschlußstellen der Stromquelle kein Richtungswechsel eintreten kann, nehmen die Randkontakte an einer funktionellen oder formalen Vermaschung niemals teil; durch diese Maßnahme wird daher die Wahl der zweckmäßigsten Randkontakte in keiner Weise berührt. Als Beispiel diene die im Abschnitt III,3 gezeigte Auslöseschaltung (Abb. 57 *a* bis *c*). Die natürliche Form und die

Sperrform haben gleiche Randkontakte, wie eine Gegenüberstellung der zugehörigen Stellungstabellen zeigt.

a_1		A
s_1		A
b_1		B
s_1		B
c_1		C
s_1		C

a_1	∞	A
s_1	a_0	A
b_1	∞	B
s_1	b_0	B
c_1	∞	C
s_1	c_0	C

Bei der Festlegung der Randkontakte einer Schaltung können drei verschiedene Fälle in Erscheinung treten. Der einfachste Fall liegt vor, wenn sich genau zwei Randspalten der Stellungstabelle und somit zwei eindeutig bestimmte Scharen von Randkontakten bilden lassen. Sehr häufig zeigt sich, daß die Zahl der möglichen Randspalten größer ist. Dann führt die Schaltaufgabe zu mehreren Lösungen. Kommen bestimmte Kontakte stets miteinander in Serie vor, so kann an ihrer Stelle ein gemeinschaftlicher Operand in die Rechnung eingeführt werden. Die Reihenfolge der einzelnen zugehörigen Kontakte ist dann gleichgültig und die Schaltung auf eine einfachere zurückgeführt. Verbleiben auch nach der Substitution noch mehrere Möglichkeiten zur Bildung von Randspalten, so wird die jeweils günstigste Form von den Möglichkeiten zur Vermaschung bestimmt; die möglichen Lösungen müssen einzeln daraufhin untersucht werden. Ein Beispiel hierfür ist die in Abschnitt III, 2 (Abb. 55) gezeigte Schaltung. Bei freigestellter Kontaktfolge und vier Verbrauchern lassen sich fünf verschiedene Randspalten bilden, und zwar a_0, a_1 oder b_0, b_1 oder c_0, c_1 oder d_0, d_1 und schließlich A, B, C, D. Davon können zwei beliebige willkürlich herausgegriffen und der Schaltungsform zugrunde gelegt werden. Die Formen, bei denen die Verbraucher nicht am Rand liegen, sind gleichwertig. Verwendet man hingegen die Verbraucher als Randelemente, so zeigt die vermaschte Schaltung um vier Kontakte mehr; man wird diese Anordnung nur dann wählen, wenn eine einseitige Schaltung mit Anschluß der Verbraucher an einen geerdeten Spannungspol angestrebt wird. Der letzte und schwierigste Fall ist dann vorhanden, wenn sich aus den vorgegebenen Strompfaden nur eine oder überhaupt keine Schar natürlicher Randelemente finden läßt. Dann muß die Schaltung unter allen Umständen äquivalente Kontakte enthalten, das heißt es müssen einzelnen Schalterstellungen mehrere Kontakte zugeordnet werden. Die Entwicklung der günstigsten Form kann dann nicht von den Randkontakten ausgehen; man muß den genau gegenteiligen Weg einschlagen und vorerst untersuchen, welche Schaltelemente sich zur Bildung von Querverbindungen eignen und somit eine funktionelle oder wenigstens formale Vermaschung gestatten.

Ob sich für jeden Spannungspol eine Schar natürlicher Randkontakte, die sämtliche Strompfade umfaßt, finden läßt, hängt von der Funktion der betreffenden Schaltung ab. Es sei kurz erläutert, welche Voraussetzungen hierfür bestehen müssen. Die einfachste Schaltung mit zwei Spannungspolen, bei der die notwendige Bedingung nicht erfüllt ist, hat die Funktion

$$\lambda = a\,b + b\,c + c\,a.$$

Da von den drei Schaltgeräten a, b, c keines bevorzugt ist, bestehen drei Möglichkeiten für den Ansatz der zugehörigen Stellungstabelle:

a	b		b	c		c	a
c	b	oder	a	c	oder	b	a
c	a		a	b		b	c

In jedem der drei Fälle kommt ein Stellungsoperand in zwei verschiedenen Spalten vor; welcher Operand verdoppelt wird, ist der freien Wahl überlassen. Der Grund hierfür ist aus der Tabelle ersichtlich und liegt in der *zyklischen Verkettung* der einzelnen Glieder der Funktion. Wäre der Zyklus nicht geschlossen und beispielsweise statt des Strompfades $c\,a$ ein anderer Strompfad $c\,d$ vorhanden, so wären sämtliche Kontakte natürliche Randelemente. Dieser Umstand hat insoferne praktische Bedeutung, als sich zwei äquivalente Kontakte, die an verschiedenen Spannungspolen liegen, weder zusammenlegen noch durch ein Paar von Umschaltekontakten ersetzen lassen; Funktionen dieser Art haben demnach zur Folge, daß unter allen Umständen einzelnen Schalterstellungen mehrere gleichartige Kontakte zugeordnet werden müssen.

Das Kennzeichen der zyklischen Verkettung ist jedoch allein noch nicht maßgebend. Wir vergrößern die betrachtete Funktion auf sechs Strompfade:

$$\lambda = a\,b + b\,c + c\,d + d\,e + e\,f + f\,a.$$

Die zugehörige Stellungstabelle besteht dann aus zwei Spalten mit durchwegs verschiedenen Operanden.

a	b
c	b
c	d
e	d
e	f
a	f

Sämtliche Kontakte der Schaltung sind also natürliche Randelemente. Dies kommt daher, daß die Zahl der Glieder der Funktion gerade ist. Der Operand am Anfang der ersten Zeile erscheint deshalb in der letzten Zeile wiederum am Anfang, das heißt in derselben Spalte. Das Hindernis für die Bildung vollständiger Randspalten mit natürlicher eindeutiger Zuordnung der Stellungsoperanden liegt also in einer zyklischen Verkettung mit ungerader Gliederzahl. Dieses Kennzeichen ist auch dann vorhanden, wenn die Zahl der Strompfade gerade ist, aber in einzelnen davon zusätzliche Elemente vorkommen, die sich auch in den Randspalten finden.

a	∞	b
c	∞	b
c	a	d
e	∞	d
e	∞	f
a	∞	f

In den sechs Strompfaden, die der vorstehenden Tabelle entsprechen, ist ein dreigliedriger, das heißt ungeradzahliger Zyklus von Serienverbindungen enthalten, nämlich $a\,b$, $b\,c$, $c\,a$; es ist daher auch in diesem Falle unmöglich, jeder Schalterstellung einen eindeutig bestimmten, natürlichen Platz innerhalb der Schaltung zuzuweisen.

Das Ergebnis der vorstehenden Betrachtung sei nun als Schaltregel zusammengefaßt. Diese lautet:

Polungsregel. Schaltelemente, die durchwegs in verschiedenen Strompfaden und daher nicht in Serie miteinander liegen, bilden eine Schar natürlicher Randelemente, die dem gleichen Spannungspol zugehören. Enthält die Schaltung eine ungerade Anzahl zyklisch verketteter Serienverbindungen, so läßt sich nicht für jeden Spannungspol eine zugehörige Schar von Randelementen finden, die sämtliche Strompfade umfaßt, so daß äquivalente Randkontakte mit verschiedener Polung unvermeidlich sind.

Die meisten Schaltungen der technischen Praxis haben Strompfade, die über eine größere Zahl von Schaltgeräten führen. Nach Bestimmung der Randelemente, die Kontakte oder Verbraucher sein können, bleiben gewöhnlich noch viele Kontakte übrig, deren günstigste Reihenfolge zu bestimmen ist. Die im Inneren einer Schaltung liegenden Kontakte können entweder Längskontakte (Serienkontakte) oder Querkontakte (Maschenkontakte) sein. Letztere können in beiden Richtungen Strom führen und sind daher besser ausgenützt; je größer der Anteil an Maschenkontakten ist, desto geringer ist die Gesamtzahl der Kontakte bei gleicher Wirkungsweise. Um Schaltungsformen mit einem Minimum von Kontakten zu finden, hat man daher zuerst festzustellen, welchen Schalterstellungen der vorgegebenen Leitwertfunktion Maschenkontakte zugeordnet werden können. Das Verfahren zur Lösung dieser Aufgabe, die sich niemals auf trivialem Wege, sondern nur mit Hilfe der Kombinatorik bewältigen läßt, sei im nachstehenden erläutert.

Wir betrachten in einer Schaltung beliebiger Art zwei willkürlich herausgegriffene Leitungen L_1 und L_2. Für dieses Leitungspaar sind drei verschiedene Leitwerte von Interesse, zwei davon beziehen sich auf die Verbindung der beiden Leitungen mit der Stromquelle Q, der dritte betrifft die Verbindung der Leitungen miteinander. Nennen wir den Anschluß an die Stromquelle L_0, so sind die Verbindungen $L_0\,L_1$ und $L_0\,L_2$ reine Längsverbindungen ohne Richtungswechsel des Stromes; die Verbindung $L_1\,L_2$ ist hingegen eine Querverbindung (Maschenverbindung), in der sich die Stromrichtung ändern kann. Abb. 61 zeigt die gegenseitige

Lage der drei betrachteten Leitungen und ihrer Verbindungen. Um feststellen zu können, welche Kontakte in den Längs- und Querverbindungen auftreten, müssen wir vorerst die Leitwerte der drei Verbindungen kennen. Die Längsverbindungen $L_0 L_1$ und $L_0 L_2$ sind auch in einer unvermaschten Schaltung vorhanden; die zugehörigen Leitwerte λ_{01} und λ_{02} lassen sich daher aus der Leitwertfunktion der Gesamtschaltung ermitteln. Die Aufgabe reduziert sich also auf die Ausrechnung des Leitwertes λ_{12}, der zur Maschenverbindung $L_1 L_2$ gehört; dieser Leitwert ist bestimmend für die Maschenkontakte zwischen den beiden Leitungen. Es läßt sich nun zeigen, daß der Leitwert λ_{12} durch die beiden anderen Leitwerte λ_{01} und λ_{02} eindeutig festgelegt ist und aus ihnen berechnet werden kann.

Vorerst wollen wir uns darüber klar werden, bei welchen Stellungsvariationen der Schaltgeräte eine Verbindung der Leitungen L_1 und L_2 überhaupt zulässig ist, ohne die Funktion der Schaltung zu verändern. Offenbar dürfen die Leitungen miteinander verbunden sein, wenn sie

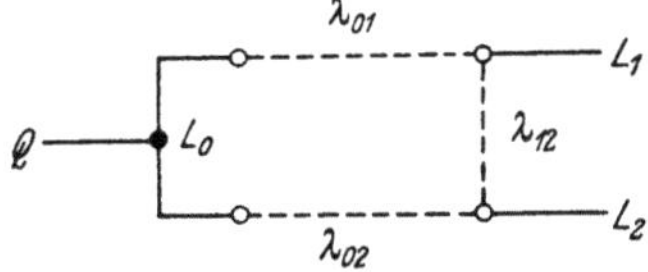

Abb. 61. Längs- und Querverbindungen von Leitungen

gleiches elektrisches Potential haben, das heißt wenn sie entweder beide spannungslos oder beide mit der Stromquelle verbunden sind. Diese Bedingung kennzeichnet den größten Funktionswert, den die Querverbindung λ_{12} einnehmen darf. Es gilt somit die Leitwertgleichung:

$$(\lambda_{12})_{max} = \overline{\lambda}_{01}\,\overline{\lambda}_{02} + \lambda_{01}\,\lambda_{02}.$$

In dieser Gleichung ist der erste Summand ein Produkt aus den komplementären Leitwerten der Längsverbindungen; der Wert dieses Produktes ist nur dann von Null verschieden, wenn beide Längsverbindungen zur Stromquelle unterbrochen sind. Der zweite Summand ist das Produkt der beiden Leitwerte selbst und tritt nur in Erscheinung, wenn beide Längsverbindungen gleichzeitig leiten. Die Gleichung für $(\lambda_{12})_{max}$ ist somit nichts anderes als die mathematisch formulierte Bedingung für die Potentialgleichheit der beiden Leitungen L_1 und L_2. Die Gleichung läßt sich noch nach den Regeln der Logistik vereinfachen, was eine beträchtliche Erleichterung für den praktischen Gebrauch mit sich bringt. Es ist nämlich:

$$(\lambda_{12})_{max} = \overline{\lambda}_{01}\,\overline{\lambda}_{02} + \lambda_{01}\,\lambda_{02} = \left[\frac{\lambda_{01}}{\lambda_{02}}\right]_{max}.$$

In Worten:

Der größte zulässige Leitwert einer Verbindung von zwei ungepolten Leitungen ist gleich dem größten Absolutwert des Quotienten aus den Leitwerten der Verbindungen der Leitungen mit der Stromquelle.

Hat man von einer gesuchten Schaltung eine Schar Randelemente bestimmt, so kann man die an diese Elemente anschließenden Leitungen daraufhin untersuchen, welche Querverbindungen zwischen ihnen zulässig und zweckmäßig sind. Der im vorstehenden abgeleitete Satz bietet die Möglichkeit hiezu. Der Quotient aus den beiden Leitwerten nennt alle

Stellungsvariationen, die den Querverbindungen zwischen den beiden Leitwerten zugrunde gelegt werden dürfen. Welche davon tatsächlich benützt werden, ist der freien Wahl überlassen. In der Regel handelt es sich darum, die Variationen so auszuwählen, daß sich für die Gesamtschaltung ein möglichst geringer Bedarf an Kontakten ergibt. Das mathematische Hilfsmittel hierzu besteht in der Kürzung der Leitwertquotienten. An Hand der Abb. 62 sei dies näher erläutert. Es sei angenommen, daß die Längsverbindung von L_0 nach L_1 aus nur zwei Stellungsvariationen v_1 und v_2 besteht; in der Verbindung $L_0 L_2$ sei ebenfalls die Variation v_1 sowie eine weitere Variation v_3 vorhanden. Bei einem der Variation v_1 entsprechenden Schaltzustand sind beide Leitungen mit der

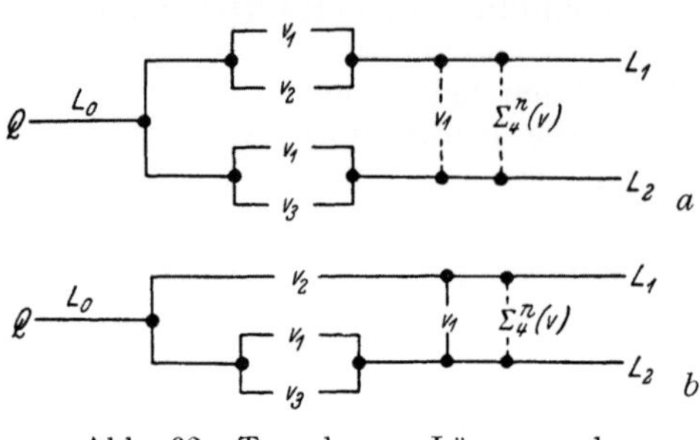

Abb. 62. Tausch von Längs- und Querverbindungen

Stromquelle Q verbunden und führen gemeinschaftlich Spannung. Sie dürfen daher während dieses Zustandes auch miteinander verbunden sein, so daß in der Querverbindung $L_1 L_2$ die Variation v_1 vorkommen darf; sie entspricht dem Produkt der Leitwerte aus den beiden Längsverbindungen, weil dieses nur die gemeinschaftlichen Variationen enthält. Ferner dürfen die beiden Leitungen verbunden sein, wenn sie spannungslos sind; dieser Fall ist bei jedem Schaltzustand gegeben, der von den drei Variationen v_1, v_2, v_3 verschieden ist. In der Querverbindung $L_1 L_2$ dürfen daher von allen n Variationen der Schaltung sämtliche Variationen von v_4 bis v_n vorkommen. Abb. 62 a zeigt diese Verteilung der Stellungsvariationen. Man erkennt, daß die gemeinschaftliche Variation v_1 an drei Stellen auftreten kann, aber nur in zweien davon auftreten muß. Die Wirkungsweise der Schaltung ändert sich nicht, wenn sie nach Abb. 62 b ausgeführt, das heißt die Variation v_1 nur in einer der Längsverbindungen vorhanden ist. Beim Schaltzustand v_1 erhält dann eine der beiden Leitungen ihre Spannung über die Querverbindung. Man wird daher zweckmäßigerweise die Variation v_1 an jene Stelle legen, wo sie sich mit den übrigen Variationen kürzen läßt. Am besten ist dies an einem Beispiel zu erkennen. Wir wählen hierzu eine Schaltung mit der Funktion:

$$\frac{\lambda}{Q\,V} = a\,b\,c + b\,c\,d + c\,d\,e + d\,e\,f.$$

Vorerst bestimmen wir die Randkontakte:

a	b	c	∞	∞	∞		a	b	c
∞	b	c	d	∞	∞		d	b	c
∞	∞	c	d	e	∞	$=$	d	e	c
∞	∞	∞	d	e	f		d	e	f

Die Tabelle zeigt zwei Paare von Randkontakten a, d und c, f; welches Paar an den Anfang und welches an das Ende der Schaltung gelegt wird, ist wegen der Symmetrie der Funktion belanglos. Wir legen willkürlich fest, daß die Kontakte a, d unmittelbar an die Spannungsquelle und die Kontakte c, f an den Verbraucher angeschlossen werden sollen. Die zur Stromquelle führende Leitung nennen wir L_0, den Anschluß an den Verbraucher L_1. Ferner können wir zwei weitere Leitungen L_2, L_3, die über die Endkontakte c, f mit L_1 verbunden sind, identifizieren. Für die Längsverbindungen dieser vier Leitungen gilt dann die Beziehung gemäß Abb. 63 a:

$$\frac{\lambda}{Q\,V} = \lambda_{01} = \lambda_{02}\,\lambda_{21} + \lambda_{03}\,\lambda_{31}.$$

Die Gleichung entspricht der ursprünglichen Leitwertgleichung nach Heraushebung der Endkontakte c und f:

$$\frac{\lambda}{Q\,V} = (a\,b + b\,d + d\,e)\,c + (d\,e)f.$$

Durch Vergleich der beiden Formeln erhalten wir:

$$\lambda_{02} = a\,b + b\,d + d\,e,$$
$$\lambda_{21} = c,$$
$$\lambda_{03} = d\,e,$$
$$\lambda_{31} = f.$$

Die möglichen Querverbindungen zwischen den beiden ungepolten Leitungen L_2 und L_3 finden wir durch Ausrechnung des Quotienten.

$$(\lambda_{23})_{max} = \left[\frac{\lambda_{02}}{\lambda_{03}}\right]_{max}.$$

Die Division läßt sich mit Hilfe der Variationstabelle leicht ausführen:

	$\frac{\bar d}{\bar e}$	$\frac{\bar d}{e}$	$\frac{d}{\bar e}$	$\frac{d}{e}$
$\bar a\ \bar b$	0	0	0	∞
$\bar a\ b$	0	0	∞	∞
$a\ \bar b$	0	0	0	∞
$a\ b$	∞	∞	∞	∞

$:$

	$\frac{\bar d}{\bar e}$	$\frac{\bar d}{e}$	$\frac{d}{\bar e}$	$\frac{d}{e}$
$\bar a\ \bar b$	0	0	0	∞
$\bar a\ b$	0	0	0	∞
$a\ \bar b$	0	0	0	∞
$a\ b$	0	0	0	∞

$=$

	$\frac{\bar d}{\bar e}$	$\frac{\bar d}{e}$	$\frac{d}{\bar e}$	$\frac{d}{e}$
$\bar a\ \bar b$	?	?	?	∞
$\bar a\ b$	?	?	0	∞
$a\ \bar b$	?	?	?	∞
$a\ b$	0	0	0	∞

Durch Kürzung der resultierenden Tabelle finden wir den größten zulässigen Leitwert für die Verbindung der beiden Leitungen:

$$(\lambda_{23})_{max} = \bar b + \bar a\,\bar d + \bar a\,e + d\,e.$$

Nun haben wir zu entscheiden, welchen Teil dieser Maximalfunktion wir tatsächlich für eine Querverbindung benützen wollen. Aus der tabellarischen Darstellung ersehen wir, daß das Glied $\bar a\,\bar d$ der Gleichung durch eine Zusammenziehung von Variationen entstanden ist, die durchwegs dem Wahlbereich der Quotientenfunktion angehören. Die Kombination

$\overline{a}\,\overline{d}$ führt somit zu einer Verbindung, die nur in spannungslosem Zustand leitend ist und somit niemals Strom führt; die Verbindung braucht daher nicht ausgeführt zu werden. Das letzte Glied der Gleichung ist eine Kombination $d\,e$; sie besteht durchwegs aus Variationen, die dem Leitbereich des Quotienten angehören. Auch diese Kombination ist als Querverbindung zwecklos; wie Abb. **62 b** erkennen läßt, müssen die gemeinschaftlichen Variationen der Längsverbindungen auf alle Fälle zweimal vorhanden sein und man gewinnt nichts dadurch, daß man sie ohne sonstige Zusammenfassung von einer Längsverbindung in eine Querverbindung überträgt. Ein Ersparnis tritt nur dann ein, wenn man in der Querverbindung Variationen des Leit- und Wahlbereiches zusammenfaßt, weil sich dann Kürzungen ergeben, die sonst unmöglich wären. Um zur Schaltungsform mit geringstem Kontaktaufwand zu gelangen, haben wir also aus der Quotientenfunktion jene Variationen herauszugreifen, die eine Kürzung des Leitbereiches mit dem Wahlbereich gestattet. Gehen wir nach dieser Regel vor, so erhalten wir den Leitwert der zweckmäßigsten Maschenverbindung

$$\lambda_{23} = \overline{b} + \overline{a}\,e.$$

Das erste Glied der Formel zeigt uns, daß in der gesuchten Schaltungsform ein Kontakt $\overline{b}$ zwischen den beiden Leitungen L_2 und L_3 anzuordnen ist. Das zweite Glied $\overline{a}\,e$ läßt hingegen noch verschiedene Deutungen zu. Es kann beispielsweise eine mittelbare Maschenverbindung, die aus einem Längs- und einem Querkontakt besteht, darstellen. Um entscheiden zu können, ob einer der beiden Kontakte mit einem Längskontakt der Schaltung identisch ist, benützen wir einen einfachen Kunstgriff. Hierzu denken wir uns den bereits bekannten Teil der Schaltung kurz geschlossen und bestimmen die Endkontakte der verbleibenden Restschaltung; deren Leitwertgleichung finden wir durch Addition der Leitwerte der Längsverbindungen zwischen L_0 und L_2, L_3:

$$\lambda_{02} + \lambda_{03} = a\,b + b\,d + d\,e.$$

Die zugehörigen Randkontakte ermitteln wir aus der Kontakttabelle.

a	b	∞	∞		a	b
∞	b	d	∞		d	b
∞	∞	d	e		d	e

Daraus ersehen wir, daß der Kontakt e ein Randkontakt der Restschaltung und somit ein Längskontakt der Gesamtschaltung ist. Die Kombination $\overline{a}\,e$ ist daher eine mittelbare Maschenverbindung. Wir können sie vorläufig vernachlässigen und die Längskontakte b, e an die Leitungen L_2, L_3 anfügen. Sie führen dann zu zwei Leitungen L_4, L_5. Die Leitwerte der zugehörigen Längsverbindungen finden wir durch Herausheben der Endkontakte aus der Leitwertformel der Restschaltung:

$$\lambda_{02} + \lambda_{03} = (a + d)\,b + (d)\,e$$
$$= \lambda_{04}\,\lambda_{42} + \lambda_{05}\,\lambda_{53}.$$

Daraus folgen die Leitwerte:

$$\lambda_{42} = b,$$
$$\lambda_{35} = e,$$
$$\lambda_{04} = a + d,$$
$$\lambda_{05} = d.$$

Wie man sieht, enthalten die Längsverbindungen λ_{04} und λ_{05} nur mehr die Anfangskontakte der Gesamtschaltung; wir haben daher nur noch die Querverbindung zwischen L_4 und L_5 auszurechnen:

$$\lambda_{45} = \frac{\lambda_{04}}{\lambda_{05}} = \left\{ \begin{array}{c|cc} & \overline{d} & d \\ \hline \overline{a} & 0 & \infty \\ a & \infty & \infty \end{array} : \begin{array}{c|cc} & \overline{d} & d \\ \hline & 0 & \infty \\ & 0 & \infty \end{array} = \begin{array}{c|cc} & \overline{d} & d \\ \hline & ? & \infty \\ & 0 & \infty \end{array} \right.$$

Von den beiden Variationen der ersten Zeile des Resultates gehört die eine zum Wahlbereich, die andere zum Wirkbereich; durch zusammenfassen erhalten wir die Querverbindung:

$$\lambda_{45} = \overline{a}.$$

Damit kennen wir sämtliche Verbindungen der gesuchten Schaltung; durch Aufzeichnen der sechs Leitungen und der dazwischenliegenden Verbindung erhalten wir den Stromlaufplan, der in Abb. 63 b wiedergegeben ist.

Die im vorstehenden gezeigte Rechnung läßt sich noch vereinfachen und dadurch auch anschaulicher gestalten. Bei der Ermittlung der ersten Maschenverbindung λ_{23} ergab sich durch Kürzung der resultierenden Variationstabelle

$$\lambda_{23} = \overline{b} + \overline{a}\,e.$$

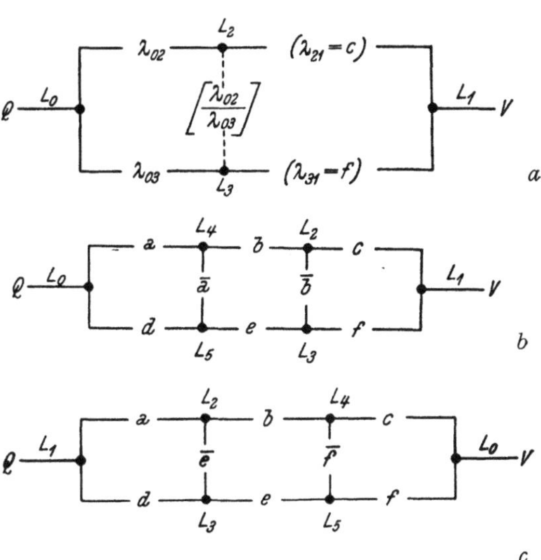

Abb. 63. Schrittweise Vermaschung. a Erste Stufe der Entwicklung, b Vollständige Form, c Symmetrische Zweitform

Diese Formel beruht auf einer maximalen Kürzung, wobei die Variationen $\overline{a}\,b\,\overline{d}\,e$ und $\overline{a}\,b\,d\,e$ je zweimal benützt sind. Diese Operation ist nach dem Äquivalenzgesetz zulässig, aber für die Ermittlung von Maschenverbindungen nicht zweckmäßig. Schaltungen mit Sperrkontakten enthalten auch Maschenverbindungen, die einer teilweise erweiterten Leitwertfunktion entsprechen. Wenn die Formeln für die Maschenverbindungen auch solche Möglichkeiten zeigen sollen, dürfen wir in diesem Falle vom Äquivalenzgesetz keinen Gebrauch machen. Wir haben also bei dieser Kürzung jede Variation des Wirkbereiches nur einmal zu verwenden und erhalten dann Gleichungen, die ein eindeutiges Abbild der mittelbaren und unmittelbaren Maschenverbindung sind. Gehen wir so vor, dann erhalten wir in unserem Beispiel

$$\lambda_{23} = \overline{b} + \overline{a}\,b\,e.$$

Wir vergleichen diese Formel mit dem in Abb. 63 b wiedergegebenen Stromlaufplan; er zeigt, daß die beiden Summanden nunmehr kontaktgetreue Darstellungen der beiden Maschen sind, die an die Leitungen L_2 und L_3 anschließen. Durch Ausrechnen eines einzigen Quotienten lassen sich also sämtliche Maschen auffinden. Welche Kontakte dieser Maschen Längs- und Querkontakte sind, erkennt man durch schrittweises Kurzschließen und durch Bestimmung der Randkontakte der Restschaltungen.

Die überwiegende Mehrzahl aller Schaltungen arbeitet mit nur einer Stromquelle und hat daher zwei Spannungspole. Für die Schaltungsform bedeutet dies in der Regel, daß zwei hinsichtlich der beiden Pole symmetrische Lösungen möglich sind. Auch im vorstehend gezeigten Beispiel besteht eine zweite Form mit kleinstem Kontaktaufwand, die in Abb. 63 c wiedergegeben ist. Die beiden Lösungen können auch miteinander identisch sein; dieser Fall tritt immer ein, wenn die vorgegebene Leitwertfunktion zyklisch ist. Hat die Schaltung mehr als zwei Spannungspole, so ist die Zahl der Lösungen entsprechend größer; geht man von einem bestimmten Spannungspol aus, so müssen die Bedingungen zur Bildung der Maschenverbindungen für sämtliche übrige Spannungspole gleichzeitig erfüllt sein. Man hat demnach die Leitwerte der Längsverbindungen und die zugehörigen Quotienten für jeden Pol getrennt zu ermitteln und das Produkt der Quotienten zu bilden; das einem Leitungspaar zugehörige Quotientenprodukt bildet dann die Grundlage für die Maschenverbindungen. Da Schaltungen mit mehr als zwei Spannungspolen in der Praxis sehr selten vorkommen, erübrigt es sich, auf dieses Verfahren genauer einzugehen.

Das Ergebnis der vorstehenden Untersuchung läßt sich in zwei Schaltregeln zusammenfassen. Diese lauten:

Kürzungsregel für Maschenverbindungen. Bei vorgegebener Wirkungsweise und unbekannter Kontaktfolge findet man die unmittelbaren und mittelbaren Maschenverbindungen mit geringstem Kontaktaufwand, die zwischen zwei beliebigen Leitungen der Schaltung möglich sind, durch unvollständige Kürzung des absoluten Quotienten aus den Leitwerten zwischen den beiden Leitungen und den Spannungspolen. Die Kürzung beschränkt sich hierbei auf die Zusammenfassung von teilweise komplementären Stellungsvariationen des Wirk- und Wahlbereiches.

Kontaktfolgeregel. Bei freigestellter Kontaktfolge läßt sich die Reihenfolge der Längs- und Querkontakte, die zum geringsten Gesamtaufwand an Kontakten führt, schrittweise ermitteln. Hierzu bestimmt man zuerst eine Schar gleichgepolter Randelemente sowie die Leitwerte zwischen den anschließenden Leitungen und den übrigen Spannungspolen. Aus diesen Leitwerten lassen sich die Maschenverbindungen zwischen den Leitungen errechnen. Sodann bildet man durch Addition der gleichen Leitwerte die Funktion der noch unbekannten Restschaltung, aus der sich wiederum Randelemente und anschließend Querverbindungen ermitteln lassen. Durch mehrmalige Wiederholung dieses Vorganges findet man alle erforderlichen Leitungen und die zwischen ihnen liegenden Schaltelemente.

Das in den beiden vorstehenden Regeln gekennzeichnete Verfahren bietet die Möglichkeit, auch schwierige Schaltaufgaben der technischen Praxis exakt zu lösen und Anordnungen zu finden, die sich durch kleinstmöglichen Aufwand an Schaltelementen auszeichnen. Als Beispiel diene eine Aufgabe aus dem Bau elektrischer Schaltanlagen. Es handle sich um ein Umspannwerk mit drei Transformatoren, die auf drei über Kuppelschalter verbundene Abschnitte einer Ringsammelschiene arbeiten. Durch eine über Meldekontakte der Transformatoren- und Kupplungsschalter führende Hilfsschaltung soll erfaßt werden, ob ein Parallellauf zwischen einzelnen Transformatoren besteht oder nicht und über welchen Kupplungsschalter der Ausgleichstromkreis führt. Hierzu ist jedem Kupplungsschalter ein Verbraucher in Gestalt einer Meldelampe oder eines Auslösemagneten zugeordnet; dieser soll mit einer Hilfsstromquelle verbunden sein, wenn der zugehörige Kupplungsschalter geschlossen ist und zwei eingeschaltete Transformatoren miteinander verbindet. Die Anordnung der beschriebenen Schaltanlage ist in Abb. 64 a als einpoliger Schaltplan dargestellt. Die gesuchte Hilfsschaltung soll mit einer möglichst kleinen Zahl von Hilfskontakten ausgeführt werden.

Wir bezeichnen die drei Transformatoren mit A, B, C und die zugehörigen Abzweigschalter mit a, b, c. Die zwischen den Abschnitten der Ringsammelschiene liegenden Kupplungsschalter seien x, y, z genannt; ihnen sind Meldelampen oder Auslöser X, Y, Z zugeordnet. Aus dem Schaltplan nach Abb. 64 a lassen sich die Bedingungen für den Parallellauf der Transformatoren sofort ablesen. Sie lauten:

$$\frac{\lambda}{A\,B} = a\,b\,x + a\,b\,y\,z,$$

$$\frac{\lambda}{B\,C} = b\,c\,y + b\,c\,z\,x,$$

$$\frac{\lambda}{C\,A} = c\,a\,z + c\,a\,x\,y.$$

Jeder Summand der drei Gleichungen, der den Faktor x enthält, beschreibt eine Verbindung zwischen zwei Transformatoren, die über den Kupplungsschalter x führt; in diesem Falle soll der zugehörige Verbraucher X mit der Hilfsstromquelle Q verbunden werden. Durch Addition dieser Summanden erhalten wir somit die Leitwertgleichung für die Anspeisung des Verbrauchers X. Auf analoge Weise ergeben sich die Teilfunktionen für Y und Z. Die Addition der drei Teilfunktionen ergibt die Leitwertgleichung der gesuchten Gesamtschaltung. Sie lautet:

$$\frac{\lambda}{Q} = a\,b\,x\,X + b\,c\,z\,x\,X + c\,a\,x\,y\,X +$$
$$+ b\,c\,y\,Y + c\,a\,x\,y\,Y + a\,b\,y\,z\,Y +$$
$$+ c\,a\,z\,Z + a\,b\,y\,z\,Z + b\,c\,z\,x\,Z.$$

Diese Funktion ist hinsichtlich der Schalterstellungen ungeradzahlig zyklisch; es läßt sich daher keine Schar natürlicher Randkontakte bilden. Nur die drei Verbraucher X, Y, Z liegen zueinander nicht in Serie und sind natürliche Randelemente. Wir gehen daher von ihnen aus und nennen die anschließenden Leitungen der gesuchten Schaltung L_1 bis L_3. Die Leitwerte zwischen diesen Leitungen und dem anderen Spannungspol sind identisch mit den drei Teilfunktionen der Verbraucher und lauten:

$$\frac{\lambda}{Q\,X} = a\,b\,x + b\,c\,z\,x + c\,a\,x\,y,$$

$$\frac{\lambda}{Q\,Y} = b\,c\,y + c\,a\,x\,y + a\,b\,y\,z,$$

$$\frac{\lambda}{Q\,Z} = c\,a\,z + a\,b\,y\,z + b\,c\,z\,x.$$

Nun haben wir zu untersuchen, ob Querverbindungen, die unmittelbar an die Verbraucher anschließen und zwischen den Leitungen L_1 bis L_3 liegen, möglich und zweckmäßig sind. Hierbei soll auf einen Umstand besonders aufmerksam gemacht werden. In der Leitwertgleichung der Schaltung kommen stets die Kombinationen $x\,X$, $y\,Y$, $z\,Z$ vor und es liegt nahe, diese Paare durch unmittelbare Serienschaltung zusammenzufassen. Überlegungen solcher Art können aber leicht zu Trugschlüssen führen. Darüber klärt uns der Schaltplan nach Abb. 64 *a* auf. Bei ausgeschaltetem Transformatorschalter b liegen die beiden Kuppelschalter x und y stets miteinander in Serie, so daß die zugehörigen Verbraucher X und Y den gleichen Schaltzustand haben müssen; man könnte sie also durch einen Maschenkontakt $\bar{b}$ unmittelbar miteinander verbinden. In analoger Weise wären Querverbindungen $\bar{a}$ und $\bar{c}$ möglich. Ob diese Maßnahme zweckmäßig ist, läßt sich jedoch nur in Kenntnis aller Kürzungsmöglichkeiten, das heißt auf Grund einer rechnerischen Untersuchung entscheiden. Wir bilden daher die Quotienten aus den Leitwerten der Längsverbindungen. Da die Durchführung dieser Operation bereits erläutert wurde, sei nachstehend nur ihr Ergebnis angeführt:

$$\lambda_{12} = x\,y(b\,z + c\,a\,x\,y),$$
$$\lambda_{23} = y\,z(c\,x + a\,b\,y\,z),$$
$$\lambda_{31} = z\,x(a\,y + b\,c\,z\,x).$$

Diese drei Leitwerte haben paarweise gemeinschaftliche Faktoren. Die Schalterstellung x kommt sowohl in der Verbindung von L_1 zu L_2

wie auch in der Verbindung von L_1 zu L_3 vor. Die beiden Querverbindungen zweigen daher erst nach einem gemeinsamen Kontakt x ab; damit wird aber dieser Kontakt zu einem Längskontakt und die errechneten Leitwerte beziehen sich auf mittelbare Maschenverbindungen. Das gleiche gilt selbstverständlich auch für die Schalterstellungen y und z. Ferner läßt sich aus der Variationstabelle des Quotienten, auf deren Anschrift verzichtet wurde, entnehmen, daß die früher erwähnten unmittelbaren Querverbindungen $\overline{a}, \overline{b}, \overline{c}$ durchwegs aus Variationen des Wahlbereiches bestehen und somit vollständig wirkungslos sind. Die gesuchte Schaltungsform hat daher keine Querkontakte zwischen den Leitungen L_1, L_2 und L_3. Diese sind vielmehr über die Längskontakte x, y, z an die nächste Leitungsschar L_4, L_5, L_6 angeschlossen. Für diese Leitungen bestimmen wir nun wiederum die Leitwerte der Längsverbindungen und deren Quotienten; wir erhalten:

$$\lambda_{45} = b\,z + c\,a\,x\,y,$$
$$\lambda_{56} = c\,x + a\,b\,y\,z,$$
$$\lambda_{64} = a\,y + b\,c\,z\,x.$$

Diese drei Leitwerte gehören, wie sich leicht erkennen läßt, zu unmittelbaren Querverbindungen; sie enthalten keine gemeinsamen Faktoren und auch keine Elemente, die natürliche Randelemente der verbleibenden Restschaltung sind. **Für letztere ergibt sich durch** Kürzung eine dreigliedrige zyklische Funktion:

$$\lambda_{04} + \lambda_{05} + \lambda_{06} = a\,b + b\,c + c\,d.$$

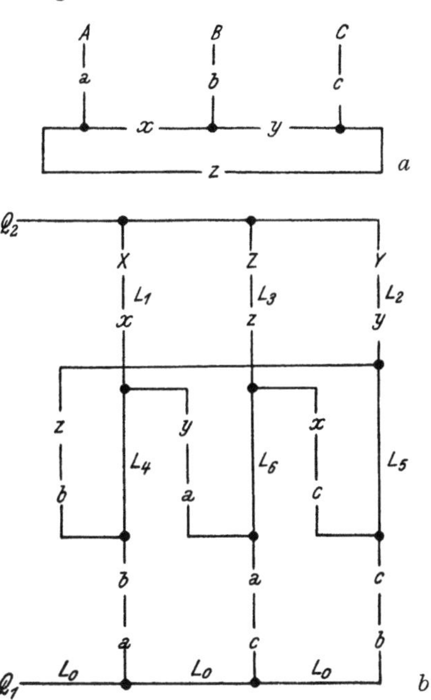

Abb. 64. Parallellaufverriegelung für drei Umspanner mit **Ringsammelschiene**. *a* Schaltplan der Starkstromanlage, *b* Verriegelungsschaltung

Diese Funktion gestattet keine Vereinfachung mehr; die restlichen Längsverbindungen sind daher:

$$\frac{\lambda}{L_0 L_4} = a\,b, \qquad \frac{\lambda}{L_0 L_5} = b\,c, \qquad \frac{\lambda}{L_0 L_6} = c\,a.$$

Die gefundenen Leitwertgleichungen für die Längs- und Querbindungen lassen sich nunmehr in den Stromlaufplan übertragen, der in Abb. 64 *b* dargestellt ist. Auch zu dieser Schaltungsform gibt es eine symmetrische Variante mit den Querverbindungen $a\,x$, $b\,y$, $c\,z$.

Die Schaltung ist in mehrfacher Hinsicht von Interesse. Sie ist ein Beispiel für den verhältnismäßig seltenen Fall einer natürlichen Vermaschung ohne Sperrkontakte. Vor allem aber zeigt sie sehr deutlich, daß die scheinbare Anschaulichkeit eines Schaltplanes zur Entwicklung höherer Schaltungsformen nicht ausreicht. Die zulässigen Maschenverbindungen $\overline{a}, \overline{b}, \overline{c}$, die man aus dem Stromlaufplan der Starkstromschaltung herleiten kann, erweisen sich als zwecklos; die günstigste Form geben hingegen zusammengesetzte Querverbindungen, die sich auf Grund des Stromlaufplanes nicht erkennen lassen. Vermaschte Schaltungen, die nicht sehr einfach sind, lassen sich daher nur mit den Hilfsmitteln der Kombinatorik entwickeln. Auch die in Abb. 64 b gezeigte Form konnte erst durch Rechnung gefunden werden; früher bekannte empirische Lösungen enthielten mindestens achtzehn Kontakte.

5. Die mehrteiligen Schaltungsformen und die topologische Umkehrung

Enthält eine Schaltung eine größere Anzahl von Strompfaden, so läßt sie sich sehr häufig in mehrere Teilschaltungen, die zueinander parallel und miteinander nicht verkettet sind, zerlegen. Entwickelt man zu einer solchen Schaltung die komplementäre Anordnung, so besteht diese aus mehreren hintereinander liegenden Teilschaltungen. Dieser Zusammenhang ermöglicht es, zusammengesetzte Schaltungsformen auf einfache Weise zu ermitteln.

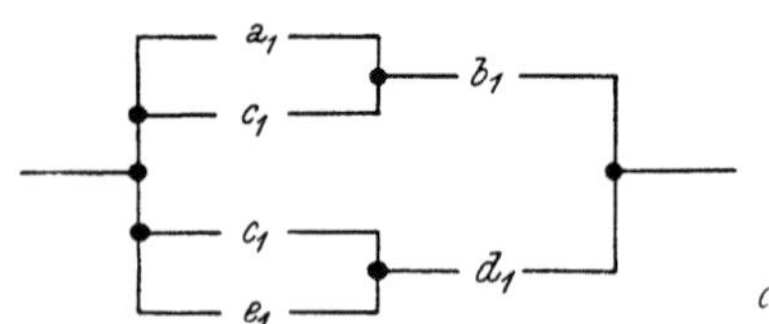

Als Beispiel betrachten wir die im Abschnitt II, 8 (Abb. 45) gezeigte Verriegelungsschaltung für eine Werkzeugmaschine. Für diese gilt die Sperrbedingung

$$a \qquad \overline{\lambda} = a_1 b_1 + b_1 c_1 + c_1 d_1 + d_1 e_1.$$

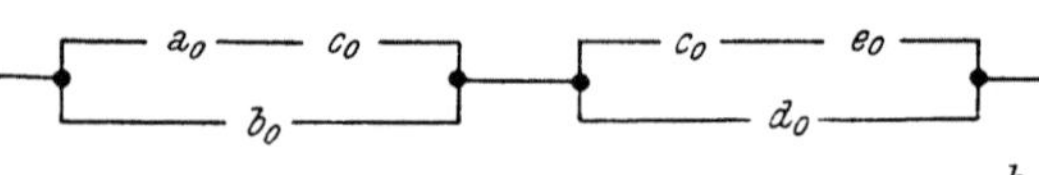

Wir entwickeln nun eine Hilfsschaltung, deren Leitbedingung dieser Sperrbedingung gleich ist. Die Hilfsschaltung ist dann zur gesuchten Schaltung komplementär.

b

Abb. 65. Entwicklung mehrteiliger Schaltungen. a Komplementäre Hilfsschaltung, b Zweiteilige Form

Aus der angesetzten Gleichung lassen sich zwei Scharen natürlicher Randelemente finden, und zwar a_1, c_1, e_1 sowie b_1, d_1. Wir beginnen mit dem Paar b_1, d_1 und heben diese Faktoren aus der Gleichung heraus.

$$\overline{\lambda} = (a_1 + c_1) b_1 + (c_1 + e_1) d_1.$$

Diese Gleichung entspricht der verketteten Form der Hilfsschaltung, die in Abb. 65 a wiedergegeben ist. Die Schaltung besteht aus zwei getrennten, zueinander parallelen Teilschaltungen.

$$\overline{\lambda}_1 = (a_1 + c_1) b_1,$$
$$\overline{\lambda}_2 = (c_1 + e_1) d_1.$$

Um die gesuchte Schaltung zu finden, haben wir jede dieser Teilschaltungen umzukehren und die gefundenen Kehrschaltungen in Serie zu legen. Es ergibt sich:

$$\lambda_1 = a_0\, c_0 + b_0,$$
$$\lambda_2 = c_0\, e_0 + d_0$$

und hieraus die zweiteilige Form der gesuchten Schaltung

$$\lambda = (a_0\, c_0 + b_0) \cdot (c_0\, e_0 + d_0).$$

Abb. 65 b zeigt diese verkettete Form.

Die Entwicklung aus einer komplementären Hilfsform läßt sich häufig ohne jegliche Rechnung durch ein zeichnerisches (topologisches) Verfahren ausführen. Bei der Umkehrung tritt an Stelle jeder Serienschaltung eine Parallelschaltung und umgekehrt. Ferner besteht bei jedem zusammengehörigen Paar von Kehrschaltungen eine Wechselbeziehung zwischen Leitungen und Maschen. Unter einer *Masche* verstehen wir jeden in sich geschlossenen Weg innerhalb der Schaltung, der jede in ihm enthaltene Leitung nur einmal berührt. Ist eine Masche so beschaffen, daß ein Teil ihrer Elemente selbst wiederum eine kleine Masche bildet, so sprechen wir von einer zusammengesetzten, andernfalls von einer einfachen Masche. Letztere bildet die vollständige Berandung einer isolierenden Fläche. Diese hat den Leitwert 0 und ist somit das Komplement zu einer Leitung, der ein Leitwert ∞ zukommt.

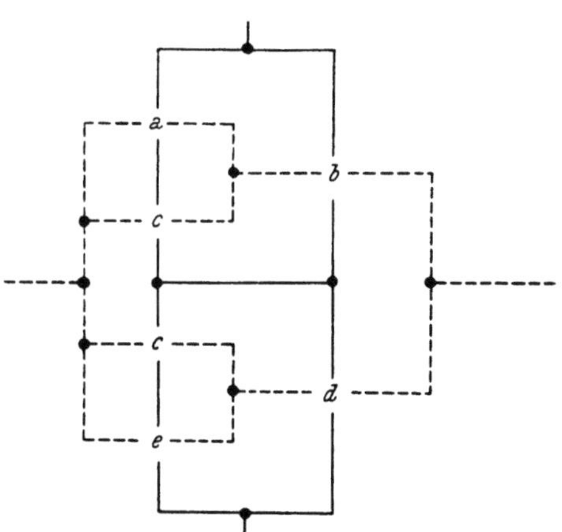

Abb. 66. Topologische Umkehrung

Kehren wir die Schaltung um, wird aus jeder einfachen Masche eine Leitung der Kehrschaltung oder umgekehrt. A. RITTER[1] hat nachgewiesen, daß dieses Gesetz für alle ebenen, das heißt ungekreuzten Schaltungsformen gilt. Beispielsweise lassen sich die beiden in Abb. 65 a und b dargestellten komplementären Formen so aufzeichnen, daß jede einfache Masche der einen Form eine Leitung der anderen Form umschließt. Abb. 66 zeigt diese Verflechtung der beiden komplementären Formen. Kontakte, die an eine bestimmte Leitung anschließen — beispielsweise a, b, c in Abb. 66 — sind in der Kehrschaltung die Bestandteile der entsprechenden einfachen Masche. Ebenso werden bei der Umkehrung die Kontakte einer einfachen Masche — zum Beispiel c, b, d, c — zu Begrenzungen der entsprechenden Leitung.

---

[1] RITTER, A.: Verfahren zur Entwicklung von Kehrschaltungen. Arch. Elektrotechn. **1941**.

Auf Grund dieser Regel lassen sich Kehrformen zu ungekreuzten Schaltungen rasch und mühelos auffinden, soferne sie nur eine Stromquelle und einen Verbraucher enthalten. Für mehrere Verbraucher ist das Verfahren nicht geeignet, weil die Umkehrung zu einer Serienschaltung führen würde. Trotz dieser Beschränkung läßt sich das Verfahren in der Praxis häufig anwenden und führt oft auf sehr einfache und elegante Weise zum Ziel. Hierfür ein Beispiel:

Zur elektrischen Befehlsübermittlung verwendet man die sogenannte Nachsteuerschaltung; auf ihr beruhen die Maschinentelegraphen, die auf Schiffen, in Kraftwerken usw. benützt werden. Zur Einstellung der Befehle, die übertragen werden sollen, dient ein Umschalter mit einer größeren Zahl von Stellungen. Der Empfänger besteht aus einem elektromagnetischen Relais, das mittels eines Schrittschaltwerkes oder Elektromotors einen Zeiger bewegt. Das Relais wird so lange erregt, bis die Zeigerstellung auf der Empfangsseite mit der Umschalterstellung auf der Gebenseite übereinstimmt.

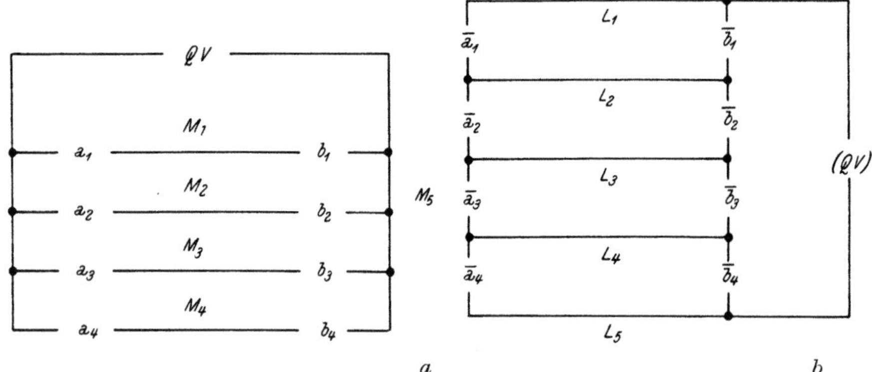

Abb. 67. Nachsteuerschaltung.   *a* Ruhestromschaltung,   *b* Arbeitsstromschaltung

Die Sperrbedingung der Schaltung läßt sich auf Grund der Angabe sofort ansetzen; die Verbindung zwischen der Stromquelle Q und dem Verbraucher V soll unterbrochen sein, wenn sich der Geber a und der Empfänger b in gleichen Schaltstellungen befinden. Hieraus folgt die Formel der Sperrbedingung:

$$\overline{\lambda} = a_1\,b_1 + a_2\,b_2 + a_3\,b_3 + \dots.$$

Auf Grund dieser Formel können wir eine zur gesuchten Schaltung komplementäre Hilfsschaltung aufzeichnen; diese ist in Abb. 67 *a* wiedergegeben. Die Schaltung wäre auch in dieser Form für die Befehlsübermittlung geeignet, doch stünde das elektromagnetische Relais dauernd unter Strom (Ruhestromschaltung). Man verwendet daher besser die zugehörige Arbeitsstromschaltung, die man durch Umkehrung findet. Die in Abb. 67 *a* gezeigte Schaltung zeigt fünf einfache Maschen M_1 bis M_5, da die Fläche außerhalb der Schaltung ebenfalls als Masche in Erscheinung tritt. Die gesuchte Schaltung hat daher fünf Leitungen. Da die gegenseitige Verknüpfung von Stromquelle und Verbraucher nicht umgekehrt, sondern eine Serienschaltung bleiben soll, betrachten wir die Kombination $Q\,V$ als ein einziges Schaltelement, das wie ein Kontakt behandelt werden kann. Es kommt in den Maschen M_1 und M_5 gemeinschaftlich vor und erscheint daher in der Kehrschaltung als Verbindung zwischen den Leitungen L_1 und L_5. Das Maschenpaar M_1, M_2 zeigt gemeinsame Kontakte a_1 und b_1; zwischen den zugehörigen Leitungen L_1, L_2 liegen daher zu diesen Elementen komplementäre Kontakte a_1 und b_1. Man erkennt, daß sich die Transformation der vorgegebenen Ruhestromschaltung in die gesuchte Arbeitsstromschaltung rein schematisch durchführen läßt: ihr Ergebnis ist die in Abb. 67 b gezeigte mehrteilige Schaltungsform, die aus einer Serienschaltung von vier einfachen Parallelschaltungen besteht.

Das Ergebnis läßt sich in einer Schaltregel zusammenfassen; diese lautet:

Kehrformregel. Ungekreuzte Schaltungsformen lassen sich durch topologische Transformation umkehren; hierbei wird aus jeder einfachen Masche eine Leitung und umgekehrt. Besteht die vorgegebene Schaltung aus mehreren parallelen Teilen, die miteinander nicht verkettet sind, so ist die Kehrschaltung eine Serienschaltung aus mehreren Teilschaltungen.

Mehrteilige Schaltungen sind häufig auch dann zweckmäßig, wenn mehrere Verbraucher vorhanden sind. Die maximale Vermaschung führt nämlich oft zu Formen, bei denen die Verbraucher in der Mitte der Schaltung liegen (zweiseitige Schaltungsformen). Mit Rücksicht auf die Gefahr, daß ein Kontakt durch Doppelerdschluß überbrückt und ein Verbraucher fälschlich eingeschaltet wird, bevorzugt man jedoch bei wichtigeren Schaltungen die einseitigen Formen; sämtliche Verbraucher werden hierbei mit einem Ende an einen geerdeten Pol der Spannungsquelle gelegt, so daß jeder Erdschluß einer Leitung den betreffenden Verbraucher kurzschließt. Die Umwandlung einer zweiseitigen Schaltungsform in eine **einseitige läßt sich häufig durch Heraushebung einer gemeinsamen Teil**schaltung bewerkstelligen. Ein Beispiel hierfür zeigt Abb. 68. Den geringsten Kontaktaufwand hat die in Abb. 68 a wiedergegebene zweiseitige Form. Wie Abb. 68 c zeigt, läßt sich eine für sämtliche Verbraucher gemeinsame Teilschaltung herausheben. Deren Leitwertfunktion findet man in einfachster Weise dadurch, daß man die Verbraucher gleich ∞ setzt und ihre Teilfunktionen addiert.

$$\lambda_1 = a_1\,b_0\,c_0\,A\,,$$
$$\lambda_2 = a_0\,b_1\,c_0\,B$$
$$\lambda_3 = a_0\,b_0\,c_1\,C$$
$$\overline{\lambda_0 = a_1\,b_0\,c_0 + a_0\,b_1\,c_0 + a_0\,b_0\,c_1.}$$

Hierbei bedeutet λ_0 die Teilfunktion, die allen drei Verbrauchern gemeinsam ist. Zwischen der herausgehobenen Teilschaltung und den Verbrauchern liegen weitere Teilschaltungen, die sich auf einfache Weise berechnen lassen. Bezeichnen wir die Restschaltung für den Verbraucher A mit x, so gilt gemäß Abb. 68 b

$$\lambda_1 = \lambda_0 \cdot x \qquad \text{oder} \qquad x = \frac{\lambda_1}{\lambda_0}.$$

Der Quotient enthält stets einen Wahlbereich, dessen Variationen so benützt werden dürfen, daß eine maximale Kürzung möglich ist. Im vorliegenden Falle erhält man:

	c_0	c_1		c_0	c_1		c_0	c_1		c_0	c_1	
$a_0 b_0$	0	0		0	∞		?	0		0	0	
$a_0 b_1$	0	0	:	∞	0	=	0	?	=	0	0	$= a_1$
$a_1 b_0$	∞	0		∞	0		∞	?		∞	∞	
$a_1 b_1$	0	0		0	0		?	?		∞	∞	

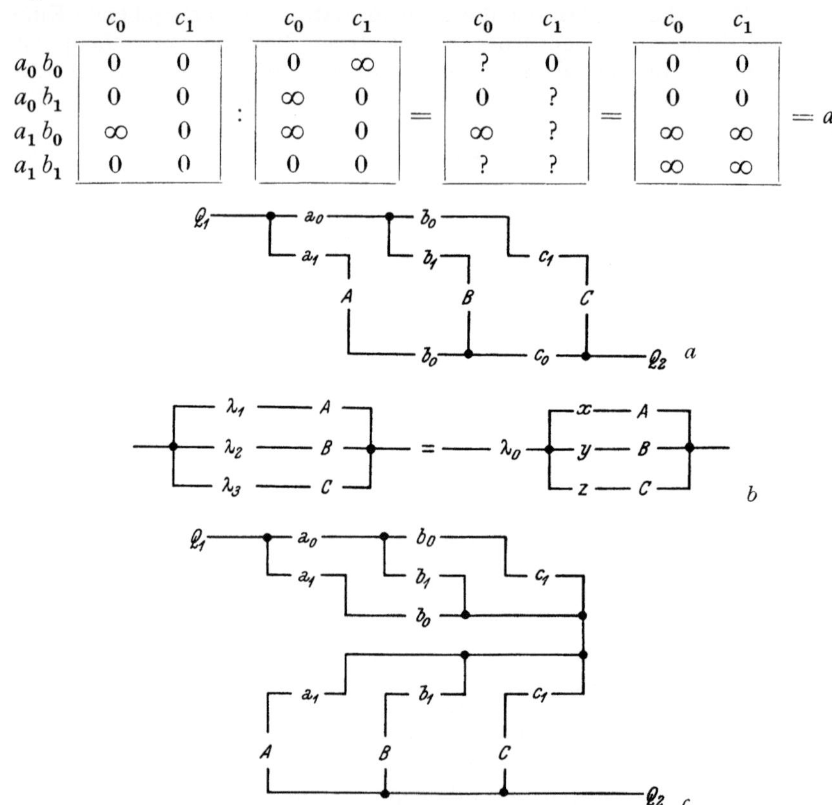

Abb. 68. Heraushebung einer gemeinsamen Teilschaltung. a Einteilige zweiseitige Schaltungsform, b Schema der Heraushebung, c Mehrteilige einseitige Schaltungsform

Auf analoge Weise erhält man für den Verbraucher B eine Restschaltung $y = b_1$ und für den Verbraucher C ebenso $z = c_1$. Das Ergebnis entspricht somit der in Abb. 68 c dargestellten einseitigen Schaltungsform.

Für die Entwicklung mehrteiliger Schaltungen mit einseitig geerdeten Verbrauchern gilt demnach folgende Regel:

Zerlegungsregel. Aus einer Funktion, die mehrere Verbraucher umfaßt, läßt sich eine gemeinschaftliche Teilfunktion herausheben; sie ist gleich der Summe aus den Teilfunktionen der Verbraucher. Zwischen der herausgehobenen Teilschaltung und den einzelnen Verbrauchern liegen Restschaltungen, deren Funktionen den Quotienten aus den Verbraucherfunktionen und der herausgehobenen gemeinsamen Funktion gleich sind.

6. Die topologische Äquivalenz

In den vorhergehenden Abschnitten wurden Aufgaben behandelt, die sich auf Schaltungen mit nur einer Stromquelle, das heißt mit zwei Spannungspolen beschränken. Viele Schaltaufgaben der Praxis betreffen jedoch Verbindungen zwischen mehreren vorgegebenen Punkten, die als Pole in Erscheinung treten können. Solche Aufgaben kommen vor allem in der Starkstromtechnik vor. Beispielsweise kann jede Leitung einer Schaltanlage, die auf ein mehrfach angespeistes Verteilnetz arbeitet, Rückstrom führen; je nach dem Betriebszustand des Netzes wirkt sie dann für die Schaltanlage als Stromquelle oder Stromverbraucher. Auch im Bau elektrischer Steuergeräte treten solche Schaltaufgaben auf. Im allgemeinen werden diese Geräte, beispielsweise Walzenschalter, so gebaut, daß sie für mehrere ähnliche Verwendungszwecke brauchbar sind. Hierbei ist die Polung der Anschlußstücke nicht eindeutig bestimmt; bekannt sind nur die Verbindungen, die zwischen diesen Punkten hergestellt werden sollen. Die Lösung solcher Schaltaufgaben verlangt die Kenntnis der topologischen Gesetze, nach denen sich Schaltungen so umformen und vereinfachen lassen, daß ihre Wirkungsweise bei beliebiger Polung ihrer Leitungen unverändert bleibt. Verschiedene Schaltungsformen, deren Wirkungsweise unabhängig von der jeweiligen Polung gleich ist, bezeichnen wir als *äquivalente Formen.*

Die Wirkungsweise einer Schaltung mit wechselnder Polung läßt sich nicht mehr durch eine bestimmte Leitwertfunktion darstellen. Ein elektrischer Leitwert bezieht sich stets auf zwei Punkte, zwischen denen er gemessen werden kann, und gilt daher nur für ein bestimmtes Polpaar. **Zu einer Schaltung mit mehreren Punkten, die Pole sein können, gehört daher nicht ein bestimmter Leitwert, sondern eine Leitwertschar.** Die Transformation einer Schaltungsform mit unbestimmter Polung verlangt daher eine mathematische Darstellung des Zusammenhanges zwischen der Verknüpfung der Schaltelemente und der zugehörigen Leitwertschar. Im nachstehenden soll diese Darstellung auf Grund der topologischen Gesetze entwickelt werden.

Die Stromquellen, Verbraucher und Kontakte einer Schaltung sind als Strecken mit bestimmtem Leitwert, der fest oder veränderlich ist, aufzufassen. Die verbindenden Leitungen haben im Vergleich hierzu einen geringen Widerstand, der vernachlässigt werden kann; sämtliche Punkte einer Leitung zeigen stets praktisch gleiches Potential und bilden in energetischer Hinsicht je einen Potentialpunkt. Hierauf gründet sich die Darstellung von Schaltungsformen durch Streckenkomplexe, in denen die Leitungen als Punkte und die übrigen Schaltelemente als Strecken abgebildet sind (Abb. 21 c). Diese Darstellung zeigt auch, daß das Zahlenverhältnis zwischen Leitungen und Widerstandsverbindungen an bestimmte Grenzen gebunden ist. Zwischen n Punkten können höchstens $1/2\, n\,(n-1)$, das heißt $\binom{n}{2}$ verbindende Strecken vorhanden sein, wenn die Strecken in beiden Richtungen gleiche Wirkung haben. Da es auch Schaltungen gibt, bei denen dies nicht zutrifft, muß bei einer allgemeinen Darstellung jede Richtung einer Widerstandsverbindung gesondert betrachtet werden; ungerichtete Verbindungen bilden dann ein zusammengehöriges Streckenpaar. Die Gesamtzahl der Schaltelemente, die eine Schaltung mit n Leitungen überhaupt haben kann, ist daher gleich der Summe aus den Leitungen und den gerichteten Widerstandsverbindungen; sie beträgt:

$$\Sigma = n + n(n-1) = n^2.$$

Es ist daher möglich, jede beliebige Schaltung mit n Leitungen durch eine *quadratische Matrix* mit n^2 Feldern abzubilden. Man kann sich eine solche Matrix in einfacher Weise aus einer Tabelle der zwischen den Leitungen liegenden Widerstandsverbindungen entstanden denken:

	L_1	L_2	L_3
L_1	∞	x	y
L_2	x	∞	z
L_3	y	z	∞

Der Verbindung jeder Leitung mit sich selbst kommt der Leitwert ∞ zu; die Tabelle enthält daher in sämtlichen Feldern der Hauptdiagonale den Operanden ∞. Den gleichen Leitwert haben aber die Leitungen selbst, so daß man ihre Operanden in die Hauptdiagonale setzen darf; die Tabelle benötigt demnach keine Randbezeichnungen für die Zeilen und Spalten und läßt sich als echte zweidimensionale Matrix anschreiben.

$$\left\| \begin{matrix} L_1 & x & y \\ x & L_2 & z \\ y & z & L_3 \end{matrix} \right\|$$

Sind die Widerstandsverbindungen, wie es meistens der Fall ist, in beiden Richtungen gleichwertig, so ist die Matrix hinsichtlich der Hauptdiagonale symmetrisch; es genügt dann eine abgekürzte Schreibweise, die nur eine Hälfte der Matrix umfaßt:

$$\left\| \begin{matrix} L_1 & x & y \\ & L_2 & z \\ & & L_3 \end{matrix} \right\|$$

Jedes Feld einer solchen Matrix entspricht einer Schaltverbindung; die Hauptdiagonale enthält die widerstandslosen Verbindungen, in den übrigen Feldern finden sich die Widerstandsverbindungen der Schaltung.

Die Matrix soll daher als *Verbindungsmatrix* bezeichnet werden. Sie läßt sich aus dem Stromlaufplan der Schaltung unmittelbar durch Transformation ansetzen, da jedem Element des Schaltplanes ein Operand der Matrix entspricht oder umgekehrt.

Die Verbindungsmatrix ist eine Darstellung der *unmittelbaren Verbindungen* der einzelnen Leitungen der Schaltung. Für die Leitwerte, die zwischen den einzelnen Leitungen bestehen, sind aber auch die mittelbaren Verbindungen von Bedeutung. Unter einer *mittelbaren Verbindung* zwischen einem Leitungspaar verstehen wir hierbei solche Verbindungen, die über eine oder mehrere andere Leitungen führen. Die Regeln der Matrixrechnung gestatten es, zu einer vorgegebenen Schaltungsform die Leitwerte zwischen den einzelnen Leitungen auf einfache und sichere Weise zu ermitteln. Als Beispiel betrachten wir drei Leitungen L_1, L_2, L_3 und zwei Kontakte x, y. Es sei angenommen, daß L_1 über x mit L_2 und L_2 über y mit L_3 verbunden sei. Die Verbindungsmatrix dieser einfachen Schaltung lautet:

$$\left\| \begin{array}{ccc} L_1 & x & 0 \\ x & L_2 & y \\ 0 & y & L_3 \end{array} \right\|$$

Unabhängig von der allfälligen *Polung* der drei Leitungen enthält die Anordnung noch eine mittelbare Verbindung zwischen L_1 und L_3; diese führt über die Leitung L_2 und lautet $x\,y$. Tragen wir auch diese Verbindung in die Matrix ein, so erhalten wir:

$$\left\| \begin{array}{ccc} L_1 & x & x\,y \\ x & L_2 & y \\ x\,y & y & L_3 \end{array} \right\|$$

Durch diese Darstellung wird die vorgegebene Schaltung nicht mehr konform abgebildet, denn die Matrix enthält mehr Operanden, als Schaltelemente vorhanden sind. Dafür zeigt sie aber sämtliche Leitwerte und somit auch sämtliche Funktionen, die sich mit der Schaltung bei entsprechender **Polung der Leitungen herstellen lassen. Diese Matrix soll** daher als *Funktionsmatrix* bezeichnet werden.

Die Umwandlung der Verbindungsmatrix in die zugehörige Funktionsmatrix läßt sich durch eine einfache Rechenoperation ausführen. Ein Beispiel diene zur Erläuterung: Gegeben sei eine Schaltung entsprechend der nachstehend angeführten Verbindungsmatrix:

$$\begin{array}{c|cccc} & L_1 & L_2 & L_3 & L_4 \\ \hline L_1 & \infty & a & b & c \\ L_2 & a & \infty & d & e \\ L_3 & b & d & \infty & f \\ L_4 & c & e & f & \infty \end{array}$$

Um das Wesen der Operation klar hervortreten zu lassen, sind in die Felder der Hauptdiagonale nicht die Operanden der Leitungen, sondern die zugehörigen Leitwerte ∞ eingesetzt. Nun betrachten wir das Leitungspaar L_1, L_4. Die Verbindungsmatrix zeigt nur die unmittelbare Verbindung c. Wegen $L_1 a L_2$ und $L_2 e L_4$ besteht aber auch eine mittelbare Verbindung $a e$, die von L_1 über L_2 zu L_4 führt. Außerdem ist noch eine zweite mittelbare Verbindung $b f$ über L_3 vorhanden. Der Leitwert zwischen L_1 und L_4 ist die Summe aller mittelbaren und unmittelbaren Verbindungen und hat daher den Leitwert $c + a e + b f$ oder genauer $c + a e + b f + c$. Die ersten Faktoren dieser vier Produkte sind die Operanden der Zeile L_1, die zweiten Faktoren entsprechen der Spalte L_4. Der Leitwert zwischen L_1 und L_4 ergab sich also dadurch, daß die Zeile L_1 und die Spalte L_4 feldweise multipliziert und die Produkte addiert wurden; das ist aber die Regel, nach der zweidimensionale Matrizen quadriert werden. Wegen der Symmetrie von Zeilen und Spalten läßt sich diese Operation auch so durchführen, daß man die Zeile L_1 mit der Zeile L_4 statt mit der Spalte L_4 multipliziert; zusammengehörige Felder stehen dann übereinander, wodurch sich die Rechnung übersichtlich gestaltet.

$$\left\| \begin{array}{cccc} \infty & a & b & c \\ c & e & f & \infty \end{array} \right\|$$
$$\infty\, c + a\, e + b\, f + c\, \infty.$$

Durch die einmalige Multiplikation der Verbindungsmatrix mit sich selbst erhält man eine Matrix, in der auch Verbindungen von drei Leitungen erscheinen. Um sämtliche Leitungen zu berücksichtigen, müssen wir die Operation mehrmals wiederholen, das heißt bei n Leitungen die $(n-1)$-te Potenz der Verbindungsmatrix ausrechnen. Hierfür ein Beispiel:

$$\left\| \begin{array}{ccccc} L_1 & a & b & 0 & 0 \\ a & L_2 & 0 & c & 0 \\ b & 0 & L_3 & 0 & d \\ 0 & c & 0 & L_4 & e \\ 0 & 0 & d & e & L_5 \end{array} \right\|^4 = \left\| \begin{array}{ccccc} L_1 & a & b & ac & bd \\ a & L_2 & ab & c & ce \\ b & ab & L_3 & de & d \\ ac & c & de & L_4 & e \\ bd & ce & d & e & L_5 \end{array} \right\|^2 =$$

$$= \left\| \begin{array}{ccccc} L_1 & a+bcde & acde+b & ac+bde & ace+bd \\ a+bcde & L_2 & ab+cde & abde+c & abd+ce \\ acde+b & ab+cde & L_3 & abc+de & abce+d \\ ac+bde & abde+c & abc+de & L_4 & abcd+e \\ ace+bd & abd+ce & abce+d & abcd+e & L_5 \end{array} \right\|$$

Die Verbindungsmatrix ist das Abbild einer topologischen Anordnung, die Funktionsmatrix hingegen die Darstellung einer logistischen Verknüpfung. Letztere unterliegt deshalb dem Äquivalenzgesetz und verändert sich bei weiterem Potenzieren nicht mehr. Ferner beschreibt sie nicht eine einzige Schaltungsform, sondern gilt für alle Schaltungen die unabhängig von ihrer Polung gleiche Wirkung haben; zu allen äquivalenten Schaltungsformen gehört die gleiche Funktionsmatrix. Dieser Zusammenhang läßt sich zur Berechnung von Schaltungen mit mehr als zwei Spannungspolen benützen. Als Beispiel diene die in Abschnitt I, 1 (Abb. 1 bis 3) gezeigte Ladestelle mit zwei Gleichrichtern G_1, G_2, und drei Ladeanschlüssen L_1, L_2, L_3. Zu suchen sind drei gewünschte Polverbindungen x, y, z entsprechend den drei Teilbedingungen:

$$\lambda_1 = G_1\, x\, L_1,$$
$$\lambda_2 = G_2\, y\, L_2,$$
$$\lambda_3 = (G_1 + G_2)\, z\, L_3.$$

Diese noch unbekannten Verbindungen lassen sich als Matrix anschreiben.

$$\left\|\begin{array}{ccccc} G_1 & 0 & x & 0 & z \\ & G_2 & 0 & y & z \\ & & L_1 & 0 & 0 \\ & & & L_2 & 0 \\ & & & & L_3 \end{array}\right\|$$

Die zugehörige Funktionsmatrix finden wir durch Ausrechnen der vierten Potenz, also durch zweimaliges Quadrieren.

$$\left\|\begin{array}{ccccc} \infty & 0 & x & 0 & z \\ 0 & \infty & 0 & y & z \\ x & 0 & \infty & 0 & 0 \\ 0 & y & 0 & \infty & 0 \\ z & z & 0 & 0 & \infty \end{array}\right\|^4 = \left\|\begin{array}{ccccc} \infty & z & x & 0 & z \\ z & \infty & 0 & y & z \\ x & 0 & \infty & 0 & xz \\ 0 & y & 0 & \infty & yz \\ z & z & xz & yz & \infty \end{array}\right\|^2 =$$

$$= \left\|\begin{array}{ccccc} \infty & z & x & yz & z \\ z & \infty & zx & y & z \\ x & zx & \infty & xyz & xz \\ yz & y & xyz & \infty & yz \\ z & z & xz & yz & \infty \end{array}\right\|$$

Wir stellen nun diese Matrix einer ebensolchen gegenüber, in der wir die verbotenen Polverbindungen mit 0 bezeichnen.

$$\left\|\begin{array}{ccccc} \infty & z & x & yz & z \\ & \infty & xz & y & z \\ & & \infty & xyz & xz \\ & & & \infty & yz \\ & & & & \infty \end{array}\right\| = \left\|\begin{array}{ccccc} G_1 & ? & ? & 0 & ? \\ & G_2 & 0 & ? & ? \\ & & L_1 & 0 & 0 \\ & & & L_2 & 0 \\ & & & & L_3 \end{array}\right\|$$

Hieraus folgt:

$$x\,z = 0,$$
$$y\,z = 0.$$

Die triviale Lösung $z = 0$ scheidet aus, da z eine wirksame Schaltverbindung sein soll. Für die drei Unbekannten sind vielmehr drei Variable einzusetzen, und zwar so, daß die beiden Gleichungen $x\,z = 0$ und

$y\,z = 0$ erfüllt sind. Jede Substitution, die diesen beiden Bedingungsgleichungen Genüge leistet, ist eine Lösung der gestellten Aufgabe. Beispielsweise erhält man unter Verwendung eines einzigen Schaltgerätes:

$$x = a_1 + a_2,$$
$$y = a_2 + a_3,$$
$$z = a_4$$

$$x\,y = (a_1 + a_2)\,(a_2 + a_3) = a_2$$
$$y\,z = (a_2 + a_3)\,a_4 = 0$$
$$z\,x = (a_1 + a_2)\,a_4 = 0$$

Sieht man zwei Schaltgeräte vor, so ergibt sich:

$$x = (a_1 + a_2)\,b_0$$
$$y = (a_2 + a_3)\,b_0$$
$$z = b_1$$

$$x\,y = a_2\,b_0$$
$$y\,z = (a_2 + a_3)\,b_0\,b_1 = 0$$
$$z\,x = (a_1 + a_2)\,b_0\,b_1 = 0$$

Schließlich läßt sich unter Benützung von drei Schaltgeräten folgendermaßen einsetzen:

$$x = a_1\,c_0$$
$$y = b_1\,c_0$$
$$z = c_1$$

$$x\,y = a_1\,b_1\,c_0$$
$$y\,z = b_1\,c_0\,c_1 = 0$$
$$z\,x = a_1\,c_0\,c_1 = 0$$

Hat man sich für eine bestimmte Lösung entschieden, so findet man die Form der Schaltung durch Eintragen der substituierten Variablen in die zuerst angesetzte Verbindungsmatrix. Beispielsweise folgt aus der Lösung mit drei Schaltgeräten

$$\left\|\begin{array}{ccccc} G_1 & 0 & a_1\,c_0 & 0 & c_1 \\ & G_2 & 0 & b_1\,c_0 & c_1 \\ & & L_1 & 0 & 0 \\ & & & L_2 & 0 \\ & & & & L_3 \end{array}\right\|$$

Diese Verbindungsmatrix läßt sich unmittelbar in den Stromlaufplan (Abb. 3 b) transformieren.

Alle Schaltungen, deren Verbindungsmatrizen beim wiederholten Potenzieren zur gleichen Funktionsmatrix führen, sind äquivalent, das heißt unabhängig von ihrer jeweiligen Polung, wirkungsgleich. Dieser Zusammenhang spielt eine bedeutende Rolle beim Entwurf elektrischer

Steuergeräte. Hierfür ein Beispiel: Abb. 69 a zeigt den Stromlaufplan und den abgewickelten Schaltbelag eines einfachen Walzenschalters. Die vier Schaltfinger sind mit A, B, C, D und die Stellungen mit s_0, s_1, s_2 bezeichnet. Führt man die Walze genau nach dem in Abb. 69 a wiedergegebenen Stromlaufplan aus, so ist eine isolierte Verbindung zwischen den Schaltstücken der Stellung s_2 unvermeidlich. Es sei nun an Hand der Funktionsmatrix festgestellt, wie viele und welche Bauformen bei gleicher Wirkungsweise noch möglich sind. Die Funktionsmatrix läßt sich sofort anschreiben; sie lautet:

$$\begin{Vmatrix} A & s_1 & s_1 & s_1\,s_2 \\ & B & s_1 & s_2 \\ & & C & s_1\,s_2 \\ & & & D \end{Vmatrix}$$

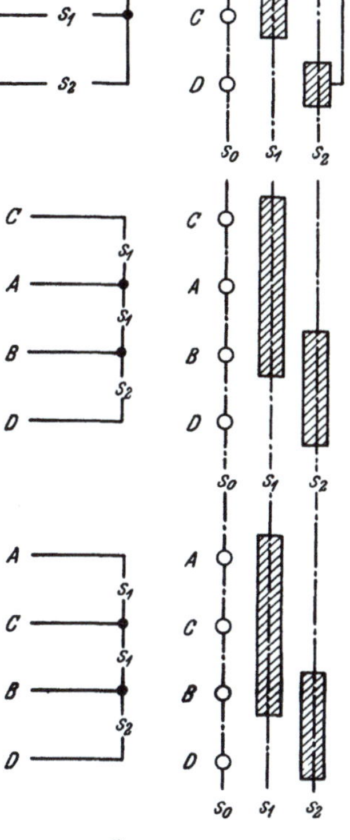

Abb. 69. Äquivalente Formen eines Walzenschalters

Sie zeigt, daß in der Stellung s_1 eine zyklische Verbindung der Finger A, B, C vorhanden ist, nämlich:

$$A\,s_1\,B + B\,s_1\,C + C\,s_1\,A.$$

Würde man für jede dieser drei Verbindungen ein eigenes Schaltstück vorsehen, so ergäbe sich eine *Ringverbindung*, bei der jeder Schaltfinger von zwei Seiten her Strom erhielte. Da eine einfache Anspeisung jedes Fingers genügt, kann die Ringverbindung an einer beliebigen Stelle aufgetrennt, das heißt eines der drei Schaltstücke ausgelassen werden. Zur vorliegenden Funktionsmatrix bestehen daher drei erzeugende Verbindungsmatrizen, die drei äquivalenten Formen des Walzenschalters entsprechen.

$$\begin{Vmatrix} A & s_1 & 0 & 0 \\ & B & s_1 & s_2 \\ & & C & 0 \\ & & & D \end{Vmatrix} = \begin{Vmatrix} A & s_1 & s_1 & 0 \\ & B & 0 & s_2 \\ & & C & 0 \\ & & & D \end{Vmatrix} = \begin{Vmatrix} A & 0 & s_1 & 0 \\ & B & s_1 & s_2 \\ & & C & 0 \\ & & & D \end{Vmatrix}$$

Die erste der drei Matrizen entspricht der Schaltungsform nach Abb. 69 a. Die zweite Matrix enthält hingegen drei Verbindungen, die hinsichtlich der Schaltfinger paarig verkettet sind:

$$C\,s_1\,A + A\,s_1\,B + B\,s_2\,D.$$

Die Matrix läßt sich nun durch Vertauschung ihrer Reihen so umformen, daß die *Fingerfolge* mit derjenigen der vorstehenden *Schaltkette* übereinstimmt; sie entspricht dann einer Walzenform mit durchwegs blanken Verbindungen:

$$\left\| \begin{array}{ccc} C & s_1 & 0 & 0 \\ A & s_1 & 0 \\ & B & s_2 \\ & & D \end{array} \right\|$$

Abb. 69 *b* zeigt Stromlaufplan und Schaltbelag dieser Lösung.

Die dritte Verbindungsmatrix enthält ebenfalls paarig verkettete Verbindungen:

$$A\,s_1\,C + C\,s_1\,B + B\,s_2\,D.$$

Die Fingerfolge der Kette führt zur Verbindungsmatrix:

$$\left\| \begin{array}{ccc} A & s_1 & 0 & 0 \\ C & s_1 & 0 \\ & B & s_2 \\ & & D \end{array} \right\|$$

und entspricht der dritten äquivalenten Walzenform nach Abb. 69 *c*.

Für die Äquivalenz von Schaltungsformen gilt somit die folgende Schaltregel:

Äquivalenzregel. Enthält eine Schaltung mittelbare oder unmittelbare Verbindungen aus Kontakten, die in der gleichen Schaltstellung geschlossen sind (äquivalente Kontakte) und einen geschlossenen Ring bilden, so sind so viele äquivalente Schaltungsformen möglich, als Verbindungen im Ring enthalten sind. Man erhält die Formen durch Auslassung je einer Verbindung des Ringes. Enthält die Schaltung mehrere solcher Ringe, so ist die Zahl der äquivalenten Schaltungsformen gleich dem Produkt aus der Zahl der Verbindungen der einzelnen Ringe. Alle äquivalenten Schaltungsformen sind ohne Rücksicht auf die jeweilige Polung wirkungsgleich; sie enthalten die gleiche Anzahl von Schaltelementen (Leitungen und Kontakten), unterscheiden sich aber durch deren gegenseitige Anordnung.

7. Die Funktion bei vorgegebener Schaltungsform

Eine in der Praxis häufig vorkommende Aufgabe besteht in der Bestimmung der Wirkungsweise einer vorgegebenen Schaltung. Bei der Entwicklung größerer Schaltungen ist es zweckmäßig, die gefundene Lösung auf ihre Funktion nachzuprüfen. Hierbei hat man in erster Linie festzustellen, welche Strompfade bei vorgegebener Polung der Schaltung zustande kommen. Diese Ermittlung ist umso schwieriger, je stärker vermascht die Schaltung ist, das heißt, je verschlungener die Strompfade verlaufen. In einfachen Fällen lassen sich die Strompfade an Hand des

Stromlaufplanes feststellen. Bei größeren Schaltungen gewährt dieser Vorgang aber keinen Schutz dagegen, daß ein Strompfad übersehen wird. Volle Sicherheit bietet nur ein rechnerisches Verfahren, das im nachstehenden erläutert sei:

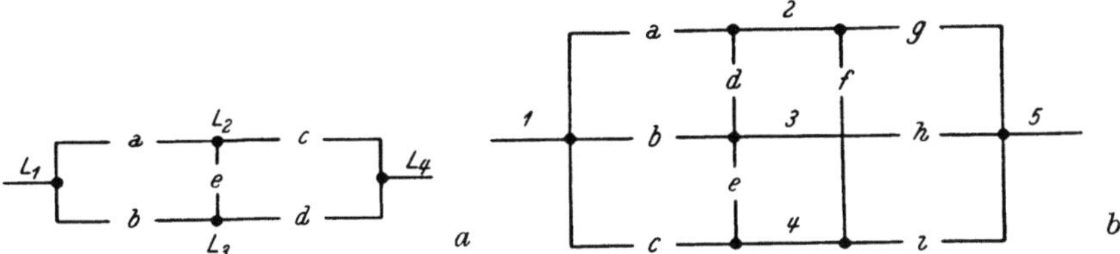

Abb. 70. Die Strompfade vermaschter Schaltungen. *a* Schaltung mit vier Strompfaden, *b* Schaltung mit 15 Strompfaden

Wie bereits gezeigt wurde, läßt sich die Form jeder beliebigen Schaltung durch eine Verbindungsmatrix abbilden. Diese Matrix läßt sich auch als Determinante auffassen, wobei die Operanden als Koeffizienten eines Systems von Leitwertgleichungen zu deuten sind. Als Beispiel betrachten wir eine vermaschte Schaltung nach Abb. 70 *a*. Nach Eintragung von Bezeichnungen für Leitungen und Kontakte läßt sich die zugehörige Verbindungsmatrix anschreiben:

$$
\left\|\begin{array}{cccc}
L_1 & a & b & 0 \\
 & L_2 & e & c \\
 & & L_3 & d \\
 & & & L_4
\end{array}\right\|
=
\left\|\begin{array}{cccc}
\infty & a & b & 0 \\
a & \infty & e & c \\
b & e & \infty & d \\
0 & c & d & \infty
\end{array}\right\|
$$

Diese Matrix läßt sich in das nachstehend angeführte Gleichungssystem auflösen:

$$
\frac{\lambda}{L_1} = \infty\, L_1 + a\, L_2 + b\, L_3 + 0\, L_4,
$$

$$
\frac{\lambda}{L_2} = a\, L_1 + \infty\, L_2 + e\, L_3 + c\, L_4,
$$

$$
\frac{\lambda}{L_3} = b\, L_1 + e\, L_2 + \infty\, L_3 + d\, L_4,
$$

$$
\frac{\lambda}{L_4} = 0\, L_1 + c\, L_2 + d\, L_3 + \infty\, L_4.
$$

Zu beachten ist, daß dieser Darstellung durch Gleichungen eine bestimmte Voraussetzung über die *Polung* zugrunde liegt. Da nur die unmittelbaren Verbindungen in Erscheinung treten, muß angenommen werden, daß der Reihe nach je eine Leitung als Stromquelle und alle anderen als Verbraucheranschlüsse betrachtet sind. In physikalischer Hinsicht liegt also kein echtes Gleichungssystem vor, weil für jede Gleichung eine andere Polung gilt, so daß die Gleichungen nur alternativ und nicht gleichzeitig gültig sein können. Auf diesen Umstand muß bei der Anwendung der Rechenregeln geachtet werden. Andere Einschränkungen bestehen nicht. Daher ist es auch statthaft, die Kontaktoperanden als Koeffizienten des Systems und die Verbindungsmatrix als zugehörige *Determinante* aufzufassen. Es dürfen aber weder die Gleichungen, noch die Reihen der Determinante addiert werden, wie es in der Determinantenrechnung sonst üblich ist. Ein weiterer Unterschied besteht hinsichtlich des Wertes der Determinante:

Eine Schaltung erhält ihre Funktion erst durch den Anschluß an eine Stromquelle, also durch ihre Polung. Ändert man die Polung, so wird auch die Funktion eine andere. Daraus folgt aber, daß der darstellenden Determinante nicht nur ein einziger Wert zukommt, sondern so viele, als verschiedene Polungen möglich sind. Wollen wir einen bestimmten Wert herausgreifen, so haben wir die zugehörige Polung anzugeben; wir müssen also in der Determinante vermerken, zwischen welchen Leitungen (Reihen) der Leitwert ausgerechnet werden soll. Beispielsweise schreiben wir

$$\lambda = \begin{vmatrix} L_1 & a & b & 0 \\ & L_2 & e & c \\ & & L_3 & d \\ & & & L_4 \end{vmatrix}_{L_1}^{L_4}$$

und kennzeichnen so die Polung der Schaltung. Die Ausrechnung des nunmehr eindeutig festgelegten Wertes der Determinante läßt sich schrittweise durchführen. Wir beginnen bei einer gepolten Leitung, also beispielsweise bei L_1. Von dort gehen so viele parallele Strompfade aus, als Verbindungen von L_1 zu anderen Leitungen vorhanden sind. Wir betrachten nun den Strompfad, der mit dem Kontakt a beginnt: sein Leitwert ist bestimmt durch den von L_1 nach L_2 führenden Kontakt a und den Leitwert einer in Serie liegenden Restschaltung zwischen L_2 und L_4. In der Rechnung erscheint dieser Strompfad als Produkt von a mit einer *Unterdeterminante*, in der die Reihe L_1 entfallen ist und dafür die Reihe L_2 als Pol auftritt. Da wir die gleiche Betrachtung für alle von L_1 ausgehenden Strompfade anstellen können, dürfen wir die Determinante mit einer Summe gleichsetzen, deren Glieder aus den Produkten der Operanden einer gepolten Reihe mit den zugehörigen Unterdeterminanten bestehen. Im vorliegenden Fall haben wir demnach zu setzen:

$$\begin{vmatrix} L_1 & a & b & 0 \\ a & L_2 & e & c \\ b & e & L_3 & d \\ 0 & c & d & L_4 \end{vmatrix}_{L_1}^{L_4} = a \cdot \begin{vmatrix} L_2 & e & c \\ e & L_3 & d \\ c & d & L_4 \end{vmatrix}_{L_2}^{L_4} + b \cdot \begin{vmatrix} L_2 & e & c \\ e & L_3 & d \\ c & d & L_4 \end{vmatrix}_{L_3}^{L_4}$$

Für die Bildung der Unterdeterminanten gilt also eine von der sonstigen Determinantenrechnung etwas abweichende Regel. Es entfallen nicht Zeile und Spalte des betreffenden Feldes, sondern nur die beiden gepolten symmetrischen Reihen; die beiden ungepolten Reihen, die zum betreffenden Feldpaar gehören, bleiben erhalten und bestimmen die

Polung der Unterdeterminante. Wir können nun dieses Verfahren so lange fortsetzen, bis wir auf Unterdeterminanten kommen, die nur mehr je einen Kontaktoperanden enthalten:

$$\lambda = a \cdot \begin{vmatrix} L_2 & e & c \\ e & L_3 & d \\ c & d & L_4 \end{vmatrix}^{L_4}_{L_2} + b \cdot \begin{vmatrix} L_2 & e & c \\ e & L_3 & d \\ c & d & L_4 \end{vmatrix}^{L_4}_{L_3} =$$

$$= a\,e \cdot \begin{vmatrix} L_3 & d \\ d & L_4 \end{vmatrix}^{L_4}_{L_3} + a\,c \cdot \begin{vmatrix} L_3 & d \\ d & L_4 \end{vmatrix}^{L_4}_{L_4} + b\,e \cdot \begin{vmatrix} L_2 & c \\ c & L_4 \end{vmatrix}^{L_4}_{L_2} + b\,d \cdot \begin{vmatrix} L_2 & c \\ c & L_4 \end{vmatrix}^{L_4}_{L_4}$$

Nun ist aber der Leitwert von L_4 nach L_4 gleich ∞ und wir erhalten die gesuchte Funktion:

$$\lambda = a\,e\,d + a\,c + b\,e\,c + b\,d.$$

Das Verfahren läßt sich noch merklich vereinfachen. Da man aus der vorgegebenen Determinante auch alle Unterdeterminanten herauslesen kann, erübrigt es sich, diese neuerlich anzuschreiben. Es genügt eine symbolische Schreibweise, die nur angibt, welche Reihen die betreffende Unterdeterminante noch enthält; auch genügt es, von den Leitungsoperanden nur die Indizes anzuschreiben. Wir bezeichnen also abgekürzt:

$$\begin{vmatrix} L_2 & e & c \\ e & L_3 & d \\ c & d & L_4 \end{vmatrix}^{L_4}_{L_2} = 2 \mid 2, 3, 4 \mid 4.$$

Für den Überblick über die Rechnung ist es ferner vorteilhaft, in dieser Kurzbezeichnung die jeweils zuletzt entfallene Reihe zu vermerken und beispielsweise $2 \mid 1, 2, 3, 4 \mid 4$ statt $2 \mid 2, 3, 4 \mid 4$ zu schreiben. In dieser Form ist das Verfahren hinreichend handlich, um auch große und verwickelte Schaltungen damit auswerten zu können. Ein Beispiel soll dies darlegen:

Wir wollen die Leitwertfunktion der in Abb. 70 *b* dargestellten Schaltung ermitteln. Hierzu tragen wir in den Schaltplan vorerst Bezeichnungen für die Leitungen ein und übertragen dann deren Verbindungen in die Determinante.

$$= \begin{vmatrix} 1 & a & b & c & 0 \\ & 2 & d & f & g \\ & & 3 & e & h \\ & & & 4 & i \\ & & & & 5 \end{vmatrix}^{5}_{1}$$

Ihre Ausrechnung von 1 nach 5 ergibt:

$$= a \cdot 2 \mid 1\,2\,3\,4\,5 \mid 5 + b \cdot 3 \mid 1\,2\,3\,4\,5 \mid 5 + c \cdot 4 \mid 1\,2\,3\,4\,5 \mid 5 + 0 =$$
$$= a\,d \cdot 3 \mid 2\,3\,4\,5 \mid 5 + a\,f \cdot 4 \mid 2\,3\,4\,5 \mid 5 + a\,g +$$
$$+ b\,d \cdot 2 \mid 2\,3\,4\,5 \mid 5 + b\,e \cdot 4 \mid 2\,3\,4\,5 \mid 5 + b\,h +$$
$$+ c\,f \cdot 2 \mid 2\,3\,4\,5 \mid 5 + c\,e \cdot 3 \mid 2\,3\,4\,5 \mid 5 + c\,i =$$

$$= a\,d\,e \cdot 4\,|\,3\,4\,5\,|\,5 + a\,d\,h + a\,f\,e \cdot 3\,|\,3\,4\,5\,|\,5 + a\,f\,i + a\,g +$$
$$+\, b\,d\,f \cdot 4\,|\,2\,4\,5\,|\,5 + b\,d\,g + b\,e\,f \cdot 2\,|\,2\,4\,5\,|\,5 + b\,c\,i + b\,h =$$
$$= a\,d\,e\,i + a\,d\,h + a\,f\,e\,h + a\,f\,i + a\,g +$$
$$+\, b\,d\,f\,i + b\,d\,g + b\,e\,f\,g + b\,e\,i + b\,h +$$
$$+\, c\,d\,f\,h + c\,f\,g + c\,d\,e\,g + c\,e\,h + c\,i.$$

Wie man sieht, verursacht das Verfahren keinerlei Denkarbeit oder Gedächtnisbelastung. Man setzt jeweils jenen Operanden vor die Unterdeterminante, der sich im Kreuzungsfeld der gestrichenen mit der neu gepolten Reihe befindet. Beispielsweise heißt das erste Glied $a \cdot 2\,|\,1\,2\,3\,4\,5\,|\,5$, weil man in Zeile 1 und Spalte 2 den Operanden a findet. Der wesentliche Vorzug liegt aber darin, daß die Rechnung mit voller Sicherheit sämtliche Strompfade der Schaltung zeigt. Bei der Überprüfung größerer vermaschter Schaltungen ist die Rechnung auch wesentlich einfacher und rascher als ein Aufsuchen der möglichen Wege aus

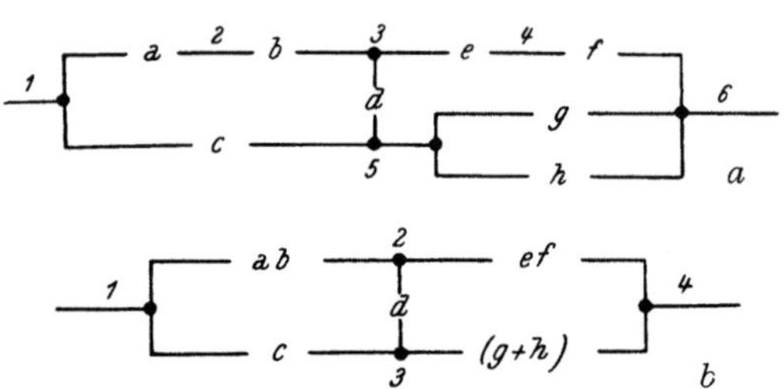

Abb. 71. Reduktion von Schaltungsformen. _a_ Teilweise vermaschte Schaltung, _b_ Reduzierte Form

dem Schaltplan. Dies kommt daher, daß die Zahl der Wege größer ist als diejenige der Strompfade. So gibt es im vorstehenden Beispiel unter anderem auch einen Weg $a\,d\,e\,f$, der aber nicht als Strompfad in Erscheinung tritt, weil sich diese Kombination gegen die ebenfalls vorhandene unmittelbare Verknüpfung $a\,g$ kürzt. Das Rechenverfahren berücksichtigt diese Kürzung bereits und führt daher rascher zum Ziel.

Die rechnerische Nachprüfung vorgegebener Schaltungen wird dadurch vereinfacht, daß die Verbindungsmatrix nicht sämtliche Leitungen, sondern nur diejenigen mit drei oder mehr anschließenden Kontaktstellen umfassen muß. Kontakte, die durch eine unverzweigte Leitung miteinander verbunden sind, können als einziges Schaltelement betrachtet und zusammengefaßt werden. Im zugehörigen Feld der Matrix erscheint dann ein Produkt aus zwei oder mehreren Operanden. In analoger Weise lassen sie zwei oder mehrere Kontakte, die unmittelbar parallel liegen, zusammenfassen, wobei in der Matrix eine Operandensumme erscheint. Abb. 71 gibt ein Beispiel hierfür. Der in Abb. 71 _a_ gezeigte Stromlaufplan betrifft eine Schaltung mit acht Kontakten _a_ bis _h_ und sechs verbindenden Leitungen. Die Schaltung ist über den Kontakt _d_ vermascht, bezüglich der übrigen Kontakte jedoch einfach verkettet. Durch Zusammenfassung aller Kontakte, die zueinander unmittelbar parallel oder in Serie liegen, ergibt sich eine wesentlich vereinfachte _reduzierte Schaltungsform_, die vollständig vermascht ist (Abb. 71 _b_); diese Form enthält nicht mehr sechs, sondern nur vier Leitungen. Die Reduktion der Form vereinfacht auch die darstellende Verbindungsmatrix und erleichtert die Ausrechnung als Determinante. Es ist:

$$
\begin{vmatrix}
1 & a & 0 & 0 & c & 0 \\
  & 2 & b & 0 & 0 & 0 \\
  &   & 3 & e & d & 0 \\
  &   &   & 4 & 0 & f \\
  &   &   &   & 5 & g+h \\
  &   &   &   &   & 6
\end{vmatrix}
=
\begin{vmatrix}
1 & a\,b & c & 0 \\
  & 2 & d & e\,f \\
  &   & 3 & g+h \\
  &   &   & 4
\end{vmatrix}
$$

Die Möglichkeit, eine nur teilweise vermaschte Schaltungsform durch Reduktion auf eine einfachere Form mit vollständiger Vermaschung zurückzuführen, läßt auch den Unterschied zwischen der einfachen Verkettung und der Vermaschung deutlich erkennen. Jede unvermaschte Schaltung läßt sich durch Reduktion so weit vereinfachen, daß sie nur mehr zwei Leitungen enthält. Die Darstellung als Streckenkomplex zeigt diese beiden Leitungen als Potentialpunkte und die ganze Kontaktverknüpfung als ein einziges Streckenpaar; unvermaschte Schaltungen erscheinen demnach als eindimensionale, vermaschte hingegen als zweidimensionale Verknüpfungen.

Die Bestimmung der Strompfade vorgegebener Schaltungen durch die Determinantenrechnung läßt auch deutlich erkennen, daß die Vermaschung eine außerordentlich wirksame Maßnahme zur Einsparung von Kontakten und Leitungen ist. Die Zahl der Leitungen ist bei jeder Schaltung geringer als diejenige der Kontakte. Bei vollständiger Vermaschung kann jede Leitung mit jeder anderen über eine Kontaktstelle verbunden sein; bei n Leitungen beträgt daher die Höchstzahl der möglichen Kontaktstellen

$$
\sum (k) = \frac{1}{2}\, n\,(n-1) = \binom{n}{2}.
$$

Die Zahl der Strompfade läßt sich aus der Zerlegung in Unterdeterminanten leicht ermitteln. Es ist stets nur ein Strompfad vorhanden, der eine einzige Kontaktstelle **aufweist, weil** die Determinante nur ein Feldpaar enthält, in dem sich die Reihe der beiden gepolten **Leitungen** kreuzen. Die Zahl der Strompfade mit zwei Kontaktstellen entspricht der Feldzahl einer Reihe abzüglich dem Leitungsfeld und dem Kreuzungsfeld der gepolten Reihen; bei einer n-reihigen Determinante, die einer Schaltung mit n Leitungen entspricht, ergeben sich somit $n-2$ Strompfade mit je zwei Kontaktstellen. Zu jedem der $n-2$ Felder läßt sich eine $(n-1)$-reihige Unterdeterminante bilden. Daraus folgt sofort die Zahl der Strompfade mit drei Kontaktstellen zu $(n-2)\,(n-3)$. Für i Kontakte in Reihe ist dann die Zahl der Strompfade

$$
\sum (i) = (n-2)\,(n-3)\,(n-4) \ldots (n-i) = \frac{(n-2)!}{(n-i-1)!}.
$$

Der längste Strompfad kann bei n Leitungen höchstens $n-1$ Kontaktstellen aufweisen, woraus sich die Zahl aller überhaupt möglichen Strompfade errechnen läßt. Sie ist:

$$\sum (s) = \frac{(n-2)!}{(n-2)!} + \frac{(n-2)!}{(n-3)!} + \cdots \frac{(n-2)!}{(n-n)!} =$$

$$= (n-2)! \left[1 + \frac{1}{1!} + \frac{1}{2!} + \frac{1}{3!} + \frac{1}{(n-2)!} \right].$$

Die Zahl der Strompfade einer vermaschten Schaltung kann also wesentlich höher sein, als man auf den ersten Blick zu glauben geneigt ist. Die nachstehende Tabelle zeigt die Höchstzahl der Strompfade in Abhängigkeit von der Anzahl der Leitungen.

Leitungen	2	3	4	5	6	7	8
Kontaktstellen	1	3	6	10	15	21	28
Strompfade							
mit einer Kontaktstelle	1	1	1	1	1	1	1
mit zwei Kontaktstellen		1	2	3	4	5	6
mit drei Kontaktstellen			2	6	12	20	30
mit vier Kontaktstellen				6	24	60	120
mit fünf Kontaktstellen					24	120	360
mit sechs Kontaktstellen						120	720
mit sieben Kontaktstellen							720
Summe	1	2	5	16	65	326	1957

Die Zahlen in der Tabelle gelten für vollständige Vermaschung, das heißt für Schaltungen, bei denen sämtliche möglichen Kontaktstellen benützt und mit unabhängigen Kontakten besetzt sind. Diese Voraussetzung ist selten zur Gänze erfüllt, doch lassen sich viele Schaltungen der Praxis weitgehend vermaschen, so daß man bei großen Schaltungen mit verhältnismäßig wenigen Kontakten und vor allem mit wenigen Leitungen auskommen kann.

8. Die Kombinatorik der Spannungszustände

Alle elektrischen Fernmelde- und Fernwirkeinrichtungen sind dadurch gekennzeichnet, daß sich zwischen ihren Teilen Verbindungsleitungen von großer Länge befinden. Es besteht daher das Bedürfnis, die Anzahl dieser Leitungen auf ein Mindestmaß einzuschränken. Die Umformung der Schaltung durch Verkettung und Vermaschung führt nun im allgemeinen nicht zum gewünschten Minimum. Hiervon kann man sich leicht durch eine einfache Betrachtung überzeugen. Sind mehrere Verbraucher durch einen gemeinsamen Umschalter wahlweise einzuschalten, so müssen zwischen diesem und den Verbrauchern so viele Leitungen verlegt werden, als Schalterstellungen und Verbraucher vorhanden sind. Eine Zusammenlegung dieser alternativ wirksamen Leitung ist auf dem Wege der Verkettung und Vermaschung nicht möglich. Mit Hilfe der Kombinationsrechnung läßt sich jedoch leicht nachweisen, daß man die gestellte Aufgabe auch mit einer wesentlich geringeren Zahl von Leitungen lösen kann. Da es sich um eine reine Umschaltung handelt, entspricht jede Schalterstellung einem Zustand der gesamten Einrichtung. Die Schar der Fernleitungen, durch die der jeweilige Zustand ferngemeldet werden soll, muß in der Lage sein, alle diese Zustände eindeutig abzubilden. Die Zahl der Spannungszustände, die innerhalb der Leitungsschar herstellbar ist, muß also der Zahl der Schaltzustände der gesamten Einrichtung gleich sein. Nun ist aber nach dem KIRCHHOFFschen Gesetz die Summe der Spannungen in einem beliebigen elektrischen System stets gleich Null, so daß bei n Leitungen $n - 1$ Spannungen frei wählbar sind. Jede Spannung zwischen zwei Leitern, die willkürlich angelegt oder beseitigt werden kann, vermehrt nun die Zahl der möglichen Spannungszustände um den Faktor 2. Eine Schar von n Leitungen kann daher, sofern man nicht weitere Kriterien, wie Stromrichtung oder Frequenz, benützt, insgesamt 2^{n-1} Spannungszustände alternativ einnehmen. Die Zahl der erforderlichen Fernleitungen ist also ganz wesentlich geringer als die Zahl der Verbraucher und der Stellungen des Umschalters. Es sei nun untersucht, durch welche Schaltungsformen und Schaltelemente sich die absolute Mindestzahl von Leitungen erreichen läßt.

Da es durch Verkettung und Vermaschung nicht möglich ist, die Zahl der Leitungen in so hohem Maße herabzusetzen, ist die Verwendung zusätzlicher Schaltelemente unvermeidlich. An Stelle des Umschalters mit nur einer Kontaktbahn muß ein solcher mit mehreren Bahnen treten, weil gleichzeitig mehrere Leitungen unter Spannung gesetzt werden müssen. Durch die Schaltelemente auf der Verbraucherseite soll erfaßt werden, zwischen welchen Leitungen jeweils Spannung liegt; es sind also spannungsabhängige Schaltgeräte mit je zwei Stellungen, das heißt Schütze, erforderlich. Da die Stellung dieser Schütze durch den jeweiligen Spannungszustand des zugehörigen Leitungspaares bestimmt ist, benötigen sie keine Selbsthaltung, sondern sind reine Zwischenschütze. Ihre Anzahl ist dadurch bestimmt, daß bei n Leitungen $n - 1$ voneinander unabhängige Spannungszustände bestehen können. Zu n Leitungen gehören also auch $n - 1$ Schütze; sie werden gewöhnlich so geschaltet, daß die Wicklungsanfänge ihrer Erregerspulen an je einer von $n - 1$ Leitungen liegen, während sämtliche Wicklungsenden miteinander und mit der n-ten Leitung verbunden sind. Abb. 72 zeigt diese Schaltung.

Die Strompfade der Verbraucher lassen sich aus Kontakten der Zwischenschütze bilden. Jedem Strompfad entspricht eine andere Stellungsvariation der $n - 1$ Schütze (Variationsschaltung). Der Aufbau

dieser sehr sinnfälligen und einfachen Kontaktschaltung entspricht der
Formel:

$$\lambda = a_0\,b_0\,c_0\,d_0\,V_1 + a_0\,b_0\,c_0\,d_1\,V_2 + a_0\,b_0\,c_1\,d_0\,V_3 + a_0\,b_0\,c_1\,d_1\,V_4 + \ldots,$$

wobei die Operanden a bis d den Zwischenschützen einer Schaltung mit
fünf Übertragungsleitungen entsprechen. Wie Abb. 72 b zeigt, läßt sich
die Schaltung so weit verketten, daß die Hälfte der Kontakte durch Zusammenlegung eingespart wird; es sind dann genau doppelt so viel
Kontakte vorhanden, als Verbraucher. Bei größeren Einrichtungen

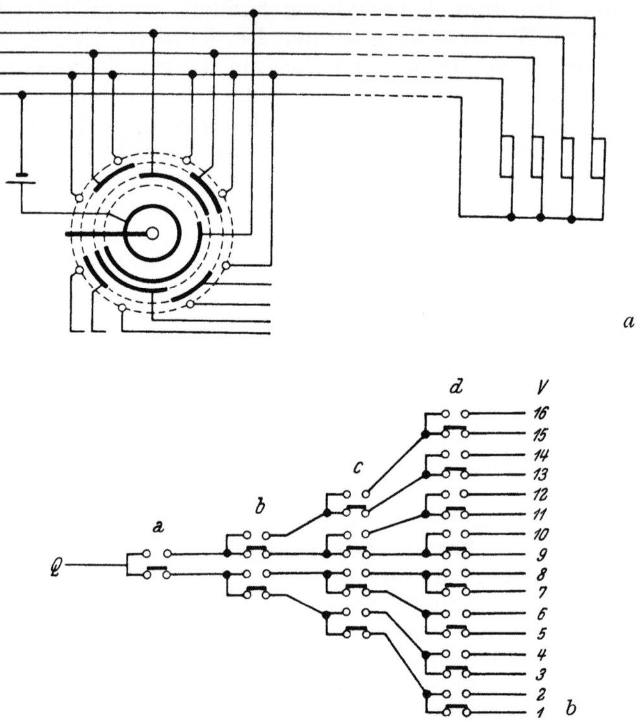

Abb. 72. Variationsschaltung. a Strompfade der Schützenspulen, b Strompfade
der Schützenkontakte

werden die Kontaktzahlen einzelner Schütze so hoch, daß sich diese nicht
mehr an einem Gerät unterbringen lassen. Man pflegt dann eine entsprechend große Zahl von Schützen parallel zu schalten, so daß sie wie
ein einzelnes Gerät wirken.

Der beschriebenen Schaltung haftet der Nachteil an, daß sie gegen
Störungen sehr empfindlich ist. Bei Erdschluß oder sonstigem Ausfall
einer Leitung kommt es nicht nur zum Versagen der Einrichtung, sondern
zur fehlerhaften Einschaltung eines anderen Strompfades. Dieser Mangel
läßt sich dadurch beheben, daß die Umschaltschütze durch Schützenpaare

ersetzt werden, die voneinander unabhängig und auf eine bestimmte Polarität oder Frequenz des Betriebsstromes abgestimmt sind. An Stelle des spannungslosen Zustandes der Leitung tritt dann eine Anspeisung mit Gleichstrom umgekehrter Polung oder mit Wechselstrom anderer Frequenz. Bezeichnet man die beiden Gruppen von Schützen, die sich durch verschiedene Abstimmung voneinander unterscheiden, mit a, b, c, d und $\alpha, \beta, \gamma, \delta$, so folgt die Schaltung der Verbraucher dem Schema:

$$\lambda = a\,b\,c\,d\,V_1 + a\,b\,c\,\delta\,V_2 + a\,b\,\gamma\,d\,V_3 + a\,b\,\gamma\,\delta\,V_4 + \ldots$$

Bei dieser Anordnung kommt nur dann ein Stromkreis zustande, wenn eine bestimmte und stets gleich bleibende Zahl von Schützen erregt ist; der Ausfall einer Leitung hat nur das Versagen der Einrichtung, aber keine Fehlwirkung zur Folge. Dieser Vorteil ist mit der Einführung einer zweiten Spannung verkehrter Polarität oder anderer Frequenz erkauft (Dualschaltung).

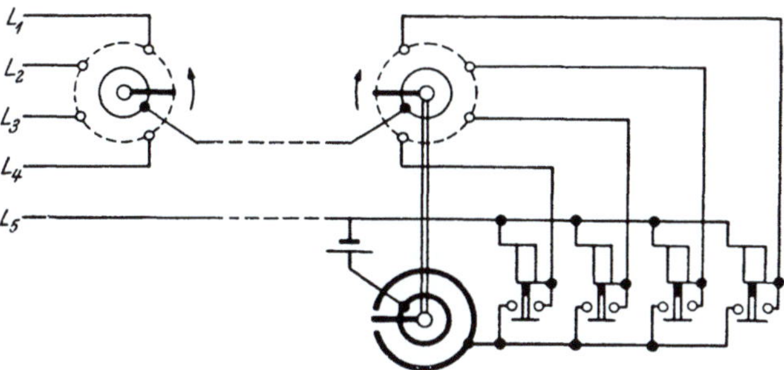

Abb. 73. Fernübertragung von Spannungszuständen mit synchronen Drehwählern

Bei größerer Zahl der Strompfade verlangt auch diese Schaltung noch ziemlich viele Übertragungsleitungen. Ist die Entfernung, die mit Hilfe der Fernmeldeeinrichtung überbrückt werden soll, sehr groß, so besteht das Bedürfnis, die Schar der Fernleitungen noch weiter zu beschränken und möglichst auf einen einzigen Stromkreis herabzusetzen. Dies ist nur da**durch möglich, daß man die Spannungszustände** der Leitungen nicht gleichzeitig und unmittelbar überträgt, sondern durch eine Wählereinrichtung der Reihe nach abtastet. Im Zug der Fernleitungen befindet sich ein Paar synchronlaufender Wähler, die durch eine einzige Leiterschleife miteinander verbunden sind, wie es Abb. 73 zeigt; die Verbindungsleitung zwischen den beiden Wählern tritt der Reihe nach an die Stelle der Leitungen der ursprünglichen Schaltung. Den Schützen wird hierbei nur je ein kurzzeitiger Stromstoß, der für das Ansprechen genügt, zugeführt. Da die Strompfade der Verbraucher über die Kontakte mehrerer Schütze führen, müssen diese jedoch gleichzeitig erregt sein und bedürfen daher einer Selbsthaltung. Nach Vollendung eines vollen Umlaufes der Wählereinrichtung kann die elektrische Haltung wieder unterbrochen werden. Hieraus folgt weiters, daß diese Schaltung eine zweite, getrennte Stromquelle am Ort der Verbraucher verlangt.

In vielen Fällen ist die Erhöhung des Aufwandes, die aus dem Betrieb mit zwei Trägerwellen verschiedener Frequenz folgt, nicht erwünscht. Die Zwischenschaltung der Wählereinrichtung macht es auch überflüssig, die Zahl der Leitungen auf das absolute Mindestmaß zu beschränken; diese treten nicht mehr als Fernleitungen in Erscheinung, so daß eine Erhöhung ihrer Zahl nur die Stellungszahl der Wähler und damit die Übertragungsdauer vergrößert. Unter geringfügiger Vermehrung der Leitungszahl läßt sich die Variationsschaltung so abändern, daß auch bei einer einzigen Trägerwelle keine Fehlwirkungen zustande kommen können. Voraussetzung ist hierbei wiederum, daß keine Ruhekontakte, das heißt keine Umschaltschütze verwendet werden. Die Schaltung enthält daher $n-1$ Schütze mit äquivalenten Kontakten, wobei jedem herzustellenden Strompfad eine andere Schützenkombination zugeordnet ist (Kombinationsschaltung). Die Zahl der gleichzeitig erregten Schütze muß hierbei stets gleich sein; anderenfalls wären Ruhekontakte zur Unterscheidung der Strompfade unvermeidlich. Der Aufbau der Schaltung entspricht sonach dem Schema:

$$\lambda = a\,b\,V_1 + a\,c\,V_2 + a\,d\,V_3 + b\,c\,V_4 + \ldots\,.$$

Die Schütze treten hierbei als *Elemente* der Kombinationen auf; die Zahl der gleichzeitig erregten Schütze nennt die *Klasse* der Kombinationen. Die vorstehende Formel beschreibt die Summe aller Kombinationen der zweiten Klasse aus vier Elementen a bis d. Bezeichnet man die Zahl der Elemente mit e und die Klassenzahl mit k, so lassen sich insgesamt $\binom{e}{k}$ verschiedene Kombinationen bilden. Hierbei ist:

$$\binom{e}{k} = \frac{e!}{k!\,(e-k)!} = \frac{e(e-1)\,(e-2)\,\ldots\,(e-k+1)}{1\cdot 2\cdot 3 \ldots k}\,.$$

Ein besonderer Vorteil dieser Kombinations-Schaltungen liegt darin, daß sich die Kontakte bei passender Wahl der Elementenzahl und Klassenzahl völlig gleichmäßig auf die einzelnen Schütze verteilen lassen. Bei e Elementen zerfällt dann die ganze Schaltung in e gleiche Zweige, so daß jedes Schütz den Anfangskontakt einer verketteten Teilschaltung trägt (Abb. 74). Voraussetzung hierfür ist, daß die Zahl der Kombinationen (Strompfade) durch die Anzahlen aller Kombinationen mit kleinerer Klasse teilbar ist; dies ist nur dann der Fall, wenn die Elementenzahl durch alle Zahlen von 1 bis k unteilbar ist. Als Beispiel diene eine Schaltung mit elf Schützen, von denen je vier gleichzeitig erregt werden. Die Zahl der Strompfade ist dann

$$\binom{11}{4} = \frac{11\cdot 10\cdot 9\cdot 8}{1\cdot 2\cdot 3\cdot 4} = 330.$$

Diese Zahl ist ohne Rest teilbar durch $\binom{11}{3} = 165$ sowie durch $\binom{11}{2} = 55$ und $\binom{11}{1} = 11$. Die ganze Schaltung besteht somit aus elf gleichen verketteten Teilschaltungen mit je einem Kontakt an erster Stelle (Anfangskontakt), $55/11 = 5$ Kontakten an zweiter Stelle, $165/11 = 15$ an dritter und $330/11 = 30$ Kontakten an letzter Stelle. Auf jedes Schaltelement entfallen daher $1 + 5 + 15 + 30 = 51$ Kombinationskontakte und ein Haltekontakt. Aus konstruktiven Gründen ist auch hier die Parallelschaltung mehrerer Schütze unvermeidlich.

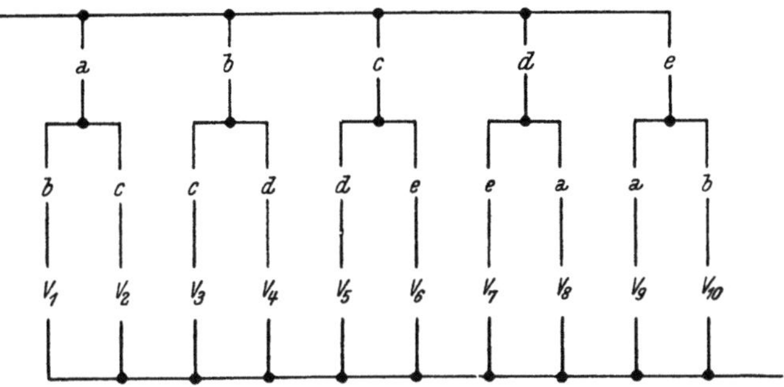

Abb. 74. Kombinationsschaltung mit gleichmäßiger Kontaktaufteilung

Die Zahl der Kombinationen, die sich aus vorgegebenen Elementen bilden lassen, steigt so lange mit der Klassenzahl, bis $k = e/2$ bei geradzahligem e und $k = \dfrac{e-1}{2}$ bei ungeradzahligem e ist. Abb. 75 zeigt diesen Zusammenhang. Die Zahl der Kontakte steigt ebenfalls mit der Klassenzahl, und zwar anfangs nur wenig und später sehr stark. In Abb. 75 sind die Kontaktzahlen für eine verkettete Schaltung mit gleichmäßiger Kontaktaufteilung eingetragen. Der Kontaktaufwand entspricht hierbei der Formel

$$\sum (k) = \binom{e}{0} + \binom{e}{1} + \binom{e}{2} + \ldots + \binom{e}{k}.$$

In der Regel ist nur die Zahl der Strompfade vorgegeben, so daß die Elementenzahl e und die Klassenzahl k frei gewählt werden dürfen. Man wird dann diese beiden Werte so festlegen, daß sowohl die Übertragungsdauer als auch der Kontaktaufwand möglichst klein bleiben. Wie aus Abb. 75 hervorgeht, ist die Zahl der Kontakte bei kleiner Klassenzahl k nicht wesentlich höher als die Zahl der Strompfade. Bei einem bestimmten Verhältnis von e/k ändert sich jedoch dieser Zusammenhang. Die Kurve S, auf der die Zahlenwerte für die Strompfade liegen, hat an einer bestimmten Stelle eine Wendetangente W. Links von dieser befindet sich ein Bereich der Kurve S, in dem die Zahl der Strompfade mit zunehmender Klassenanzahl rasch ansteigt. In diesem Bereich ist somit eine Erhöhung der Klassenzahl zweckmäßig und wirtschaftlich. Der rechts von der Tangente liegende Bereich zeigt ein gegenteiliges Verhalten der Kurve S, wogegen der Kontaktaufwand K nach wie vor stark zunimmt. Die zu diesem Bereich gehörenden Klassenzahlen ergeben somit keine wirtschaftlichen Schaltungen. Es erscheint daher zweckmäßig, das Verhältnis von k/e nur bis zu jenem Wert zu steigern, der dem gemeinsamen Punkt der Kurve S und der Wendetangente W entspricht. Das zugehörige Verhältnis k/e ergibt sich aus dem Ansatz:

$$\binom{e}{k} - \binom{e}{k-1} = \binom{e}{k+1} - \binom{e}{k}.$$

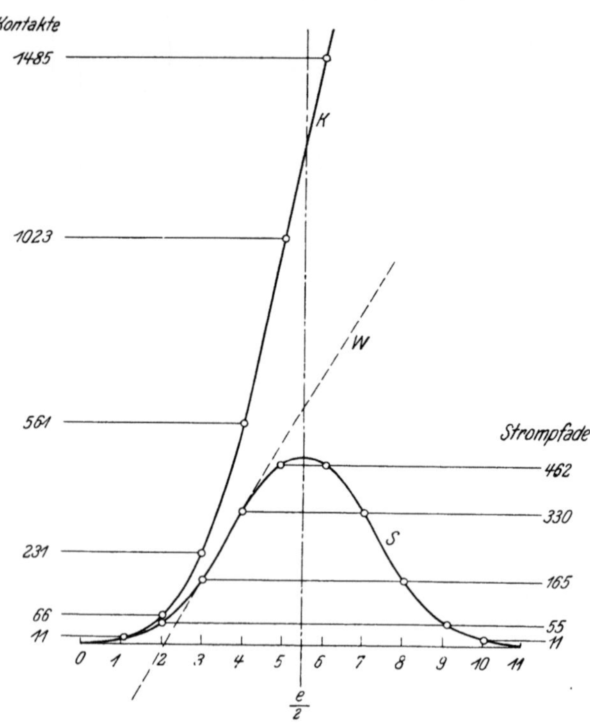

Abb. 75. Zahl der Strompfade und Kontakte von Kombinationsschaltungen

Die Ausrechnung liefert:

$$k = \frac{e^2 - e - 2}{2(2e-1)}.$$

Da die Zahl der Elemente stets wesentlich größer als 1 ist, ergibt sich mit sehr guter Näherung:

$$k = \frac{e}{4}.$$

Bei diesem Verhältnis von k/e entfallen auf jeden Strompfad 4/5 Kontakte. Dieser Aufwand ist wesentlich kleiner als derjenige der Variationsschaltungen, die für jeden Strompfad zwei Kontakte verlangen. Dafür

ist die Zahl der notwendigen Elemente und somit auch die Übertragungsdauer eine größere. Bezeichnen wir die Zahl der Elemente einer Variationsschaltung, die gleich viele Strompfade liefert, mit x, so gilt für den
Vergleich der beiden Schaltungsarten:

$$\binom{e}{k} = e^x.$$

Für $k = e/4$ ergibt sich hieraus auf Grund der Näherungsformel von
STIRLING:

$$x = \frac{2}{3}\, e.$$

Bei den Kombinationsschaltungen mit günstigstem Verhältnis k/e
ist also die Elementenzahl um 50% größer als bei gleichwertigen Variationsschaltungen. Der Übergang von zwei Trägerwellen auf eine bedingt
also keine Verdoppelung, sondern nur eine 50%ige Erhöhung der Übertragungsdauer der Wählereinrichtung. Der Kontaktaufwand ist hierbei
im gleichen Verhältnis kleiner, als die Elementenzahl größer ist.

Die Variationsschaltungen lassen sich auch so ausführen, daß an Stelle
von Schützen Drehwähler vorhanden sind. Damit vermindert sich die
Zahl der Erregerspulen, wodurch sich besonders bei großen Schaltungen
die Kosten und der Platzbedarf bedeutend verringern. Für Kombinationen aus e Elementen mit der Klasse k sind hierbei $k + 1$ Wähler mit
je $e + 1 - k$ Arbeitsstellungen erforderlich. Die Summe der Schaltschritte sämtlicher Wähler ist für alle Kombinationen gleich und beträgt
$e + 1$. Eine Schaltung mit $\binom{11}{3} = 165$ Strompfaden enthält somit
vier Wähler a, b, c, d mit je neun Stellungen; die Zahl der Schaltschritte
aller vier Wähler zusammen beträgt 12. Die Kontaktverbindungen verlaufen nach folgendem Schema:

$$\lambda = a_1\, b_1\, c_1\, d_9\, V_1 + a_1\, b_1\, c_2\, d_8\, V_2 + a_1\, b_1\, c_3\, d_7\, V_3 + \ldots + a_9\, b_1\, c_1\, d_1\, V_{165}.$$

Eine gleichmäßige Aufteilung der Kontakte auf die einzelnen Wählerstellungen läßt sich bei diesen Schaltungsformen nicht erreichen.

Die beschriebenen Schaltungen, die auf Variationen oder Kombinationen von Zwischenrelais (Schützen oder Drehwählern) beruhen, finden
vorwiegend in der Technik der Fernwirkanlagen, das heißt zur Fern-
überwachung und Fernsteuerung Verwendung. Auch bei der Konstruktion von Fernschreibern wurde davon Gebrauch gemacht. In den üblichen
Fernsprechanlagen mit selbsttätiger Gesprächsvermittlung werden diese
Schaltungen nicht verwendet; dort besteht kein Bedürfnis, die Zahl der
Wählerstellungen auf ein Mindestmaß zu verringern, weil die Vermitt-
lungsdauer gegenüber der Gesprächsdauer keine wesentliche Rolle spielt.

IV. Die Kombinatorik der elektromechanischen Schaltgeräte (Drehschalter)

1. Die Bauteile und Bauformen der Drehschalter

Eine sehr bedeutende Gruppe der elektrischen Schaltgeräte bilden die Drehschalter. Ausführungen, bei denen die beweglichen Schaltstücke eine Schubbewegung ausführen, beschränken sich im allgemeinen auf Geräte mit nur zwei Stellungen. Schaltgeräte mit mehreren Stellungen und Schubkontakten finden sich nur gelegentlich als Zellenschalter für Sammlerbatterien (Spindelschalter). In den weitaus meisten Fällen macht

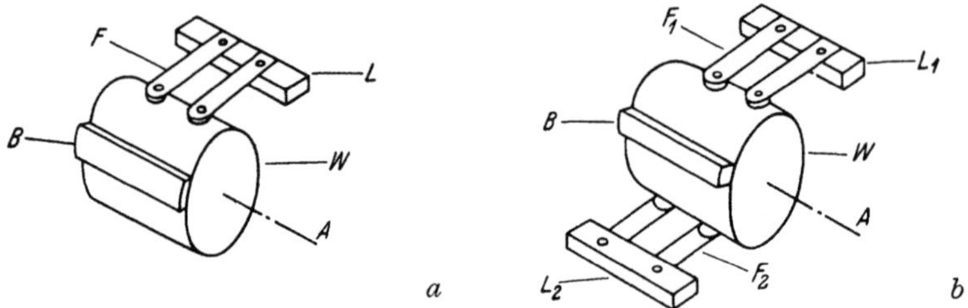

Abb. 76. Kontaktanordnung bei Walzenschaltern. *a* Einreihiger Walzenschalter, *b* Zweireihiger Walzenschalter

man von den Vorteilen der Drehbewegung, die zu wesentlich einfacheren Konstruktionen führt, Gebrauch. Die wichtigsten Bauformen der Drehschalter sind die *Walzenschalter, die Reihenschalter* und *die Paketschalter.* Das gemeinsame Kennzeichen aller dieser Bauformen sind Kontakte mit je zwei Unterbrechungsstellen in Serie. Jeder Kontakt wird hierbei durch drei Schaltstücke gebildet, indem zwei fest angeordnete *Schaltfinger* fallweise durch eine bewegliche *Schaltbrücke* miteinander verbunden werden. Die Unterschiede der einzelnen Bauformen liegen in der Anordnung der Brücken, die entweder parallel zur Drehachse oder in der Richtung des Umfanges oder radial angeordnet sein können. Es lassen sich auch an ein und demselben Schaltgerät Brücken in verschiedener Anordnung verwenden, wodurch gemischte Bauformen entstehen, die dem Konstrukteur viele Möglichkeiten zum Entwurf kleiner und wirtschaftlicher Schaltgeräte bieten.

Die axiale Anordnung der Schaltbrücken ist das Merkmal der *Walzenschalter.* Abb. 76 *a* zeigt das Wesentliche dieser Bauform. Die Schaltfinger *F* sind nebeneinander auf einer Fingerleiste *L* befestigt und berühren die Walze *W* längs einer Erzeugenden. Dazu parallel liegt die Schaltbrücke *B*, so daß sie die Kontaktfinger *F* bei Drehung der Walze fallweise verbindet oder trennt. Bei einer größeren Anzahl von Fingern läßt sich manchmal die Walzenfläche und die Zahl der Brücken dadurch vermindern, daß man die Finger in mehreren Reihen anordnet, wobei ein und dieselbe Brücke zur fallweisen Verbindung verschiedener Fingerpaare benützt wird. (*Mehrreihige Walzenschalter.*) Abb. 76 *b* zeigt ein Beispiel für einen *zweireihigen Walzenschalter*. Es sind zwei Fingerleisten L_1 und L_2 vorhanden, die je ein Fingerpaar F_1 und F_2 tragen; zur Verbindung der Finger eines Paares dient die Schaltbrücke B, die beiden Fingerpaaren gemeinsam ist. Die Zuordnung gemeinschaftlicher Schaltbrücken zu mehreren Fingerpaaren ist nicht immer möglich, sondern setzt eine bestimmte Beschaffenheit des Schaltprogrammes voraus. *Einreihige Walzenschalter* lassen sich hingegen für jedes beliebig geartete Schaltprogramm entwerfen, wobei für jede Anschlußleitung nur ein Schaltfinger erforderlich ist. Diese Eigenschaft, die keiner anderen Bauart von Drehschaltern zukommt, erklärt die häufige Verwendung einreihiger Walzenschalter.

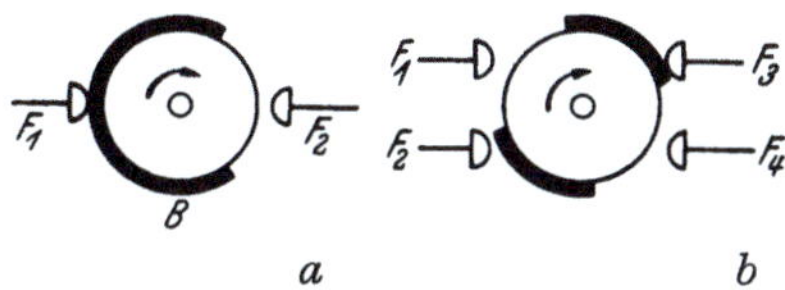

Abb. 77. Kontaktanordnung bei Reihenschaltern. *a* Kontaktbahn mit einer Brücke, *b* Kontaktbahn mit zwei Brücken

Das Kennzeichen der *Reihenschalter* sind Brücken in Form von Ringausschnitten (tangentiale Brücken). Abb. 77 *a* zeigt ein Beispiel hierfür. Die beiden zu einem Kontakt gehörigen Finger F_1 und F_2 und die verbindende Schaltbrücke B liegen in einer Ebene, die zur Drehachse senkrecht steht; der Strom fließt in der Richtung des Umfanges. Der Vorteil

Abb. 78. Kontaktanordnung bei Paketschaltern. *a* Mit Winkelbrücke, *b* Mit gestreckter Brücke, *c* Mit Mehrfachbrücke

dieser Bauart liegt darin, daß sich größere Schaltgeräte durch Aneinanderreihen von einzelnen scheibenförmigen Bauteilen in beliebiger Anordnung zusammensetzen lassen. Wie Abb. 77 *b* zeigt, kann eine *Kontaktbahn* auch zwei oder mehr Brücken enthalten. Die in Abb. 77 *b* wiedergegebene Anordnung läßt sich beispielsweise als zweipoliger Ausschalter verwenden. Bei Reihenschaltern ist es daher nicht notwendig, für verschiedene Spannungspole getrennte Kontaktbahnen vorzusehen; zwischen den Kontaktbahnen und den Polen der Schaltung besteht kein fester Zusammenhang; die in der Praxis anzutreffende Bezeichnung der Bahnen (Scheiben) als Schalterpole ist daher irreführend und nicht zweckmäßig.

Außer der axialen und tangentialen Anordnung der Schaltbrücken ist ferner noch eine Bauform möglich, bei der sich die Brücken in radialer Richtung erstrecken. Schaltgeräte dieser Art sind die *Paketschalter*. Jede Brücke besteht hierbei aus zwei *Schalthebeln*, die miteinander leitend verbunden sind. Abb. 78 zeigt ein Beispiel hierfür. Je nach dem Winkel, den die Hebelarme miteinander einschließen, ergeben sich *Winkelbrücken* nach Abb. 78 *a* oder *gestreckte Brücken* nach Abb. 78 *b*. Wie bei den

Reihenschaltern, liegen auch bei den Paketschaltern sämtliche Finger einer Kontaktbahn mit der zugehörigen Brücke in einer Ebene; auch diese Geräte ermöglichen die einfache Zusammensetzung aus gleichartigen Bauteilen. Der Vorteil der Paketschalter liegt darin, daß auch Finger, die nicht unmittelbar benachbart sind, durch blanke Brücken miteinander verbunden werden können; Reihenschalter benötigen in diesem Falle isolierte Brückenverbindungen. Da aber jede Brücke aus zwei Schalthebeln (Schaltarmen), die sich an der Achse treffen, gebildet wird, kann jede Kontaktbahn eines Paketschalters nur eine einzige Schaltbrücke enthalten. Anordnungen nach Abb. 77 *b*, die bei Reihenschaltern möglich sind, lassen sich mit Paketschaltern nicht herstellen. Hingegen können auch mit radialen Brücken gleichzeitig mehr als zwei Finger miteinander verbunden werden; es ergeben sich dann mehrarmige Brücken (Mehrfachbrücken) nach Abb. 78 *c*.

Abb. 79. Flachbahnschalter

Radiale Schaltbrücken finden sich noch bei einer weiteren Bauform, die als Flachbahnschalter bezeichnet wird. Hierbei liegen zwei oder mehrere Kontaktbahnen konzentrisch in einer Ebene. Die Zahl der Bahnen, die sich auf diese Weise unterbringen lassen, ist allerdings sehr beschränkt, so daß Flachbahnschalter nur gelegentlich für Sonderzwecke (Anlasser, Feldregler und dergleichen) benützt werden; sie haben meist nur zwei Kontaktbahnen, wobei die innere als Schleifring ausgebildet ist. Abb. 79 zeigt ein Beispiel dieser Bauform, die mit Rücksicht auf die Zugänglichkeit der Schaltstücke auf eine einzige *Schaltebene* beschränkt ist.

Unter den Schaltgeräten mit Brücken verschiedener Art sind an erster Stelle die Walzenschalter mit zusätzlichen Ringbrücken zu nennen. Sie enthalten axiale und tangentiale Schaltbrücken und sind demnach eine Vereinigung von Reihenschaltern und Walzenschaltern. Durch passende Zuordnung der Kontaktverbindungen des Schaltprogrammes zu den beiden Arten von Brücken lassen sich Geräte mit geringsten Abmessungen und kleinstem Aufwand an Fingern und Brücken entwerfen. Schaltgeräte dieser gemischten Bauart sind daher besonders für große Schaltprogramme sehr zweckmäßig.

Eine weitere und oft benützte Mischform entsteht durch Vereinigung der radialen und axialen Brückenanordnung. Bei Paketschaltern werden häufig benachbarte Kontaktbahnen entlang der Achse miteinander verbunden. Da jede Bahn nur eine Brücke enthalten kann, ist auch die Zahl der Axialverbindungen zwischen zwei Bahnen auf je eine beschränkt. Sie genügt in vielen Fällen, um den Paketschalter einfacher und kleiner zu gestalten.

Manchmal ist es zweckmäßig, das Schaltprogramm, das einem Schaltgerät zugrunde liegen soll, auf zwei oder mehrere getrennte Geräte aufzuteilen und diese miteinander mechanisch zu kuppeln. (*Mehrteilige Schaltgeräte.*) Beispielsweise pflegt man Stufenschalter für große Schaltleistungen aus einem *Lastschalter* mit zwei Stellungen und einem vielstelligen *Anzapfschalter* (*Wähler*) zusammenzusetzen. Abb. 80 zeigt die

Aufteilung der Gesamtschaltung auf die beiden Teilgeräte. Eine häufig benützte Bauart für mehrteilige Schaltgeräte findet sich in den *Nockenschaltern*. Bei diesen ist jedem Kontakt ein eigener Schalter mit nur zwei Stellungen zugeordnet. Die Betätigung der beweglichen Schaltstücke der Kontakte erfolgt abwechselnd durch eine mechanische Antriebs-

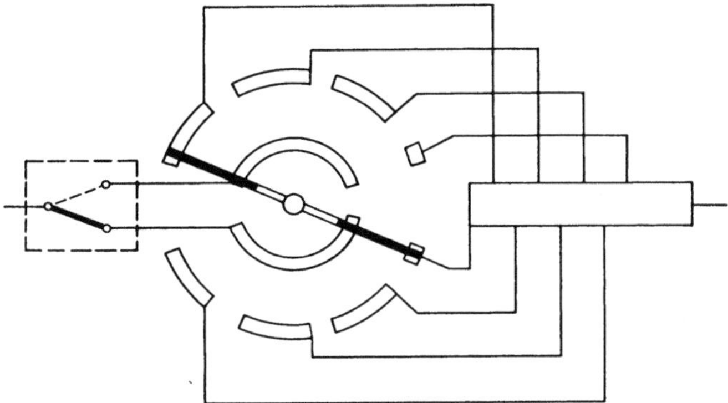

Abb. 80. In Lastschalter und Wähler geteilter Stufenschalter

vorrichtung (Nockenwalze). Das Schaltprogramm hat hierbei nur auf die Gestalt des mechanischen Teiles Einfluß, wogegen volle Freiheit für die Ausbildung der Kontakte besteht. Nockenschalter eignen sich daher besonders für hohe Stromstärken und Schaltleistungen. Wenn es das Schaltprogramm zuläßt, können für je zwei oder mehrere Kontakte gemeinsame Nockenscheiben benützt werden; es ergeben sich dann Nockenschalter mit zwei oder mehr Kontaktreihen (mehrreihige Nockenschalter). Abb. 81 zeigt eine Kontaktbahn eines Nockenschalters, die aus einer Nockenscheibe und zwei von ihr betätigten Kontakten besteht. Auch diese Bauform hat den Vorteil, daß sich größere Schaltgeräte durch Aneinanderreihung gleichartiger Einzelteile herstellen lassen.

Bei allen Arten von Drehschaltern liegen die Berührungsstellen von Fingern und Brücken auf der Oberfläche eines Zylinders. Schaltfinger, die in einer Reihe angeordnet sind, berühren den Zylinder entlang einer Erzeugenden. Gewöhnlich haben die Schalter eine ausgeprägte Nullstellung, in der die anspeisenden Schaltfinger außer Eingriff stehen. Die zugehörige Erzeugende des Zylindermantels wird Nullinie oder ebenfalls

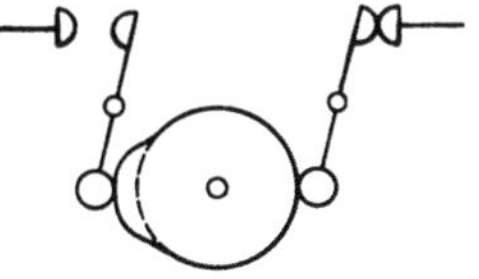

Abb. 81. Kontaktanordnung bei einem zweireihigen Nockenschalter

Nullstellung genannt. Zu beachten ist, daß zu jeder Fingerreihe eine andere Nullinie des Zylindermantels gehört. Bei einem Walzenschalter mit zwölf Stellungen und zwei gegenüberliegenden Fingerreihen ist beispielsweise die Nullstellung für eine Fingerreihe gleichzeitig Stellung 6 für die andere Reihe und umgekehrt. Die einzelnen Stellungen der Drehschalter sind gewöhnlich durch Rasten, die im Antrieb angebracht sind, markiert. Stellungen, in denen der Schalter betriebsmäßig nicht angehalten werden soll, bleiben gewöhnlich ungerastet. Der Begriff „*Stellung*" kennzeichnet dann nicht eine bestimmte Erzeugende des Zylindermantels,

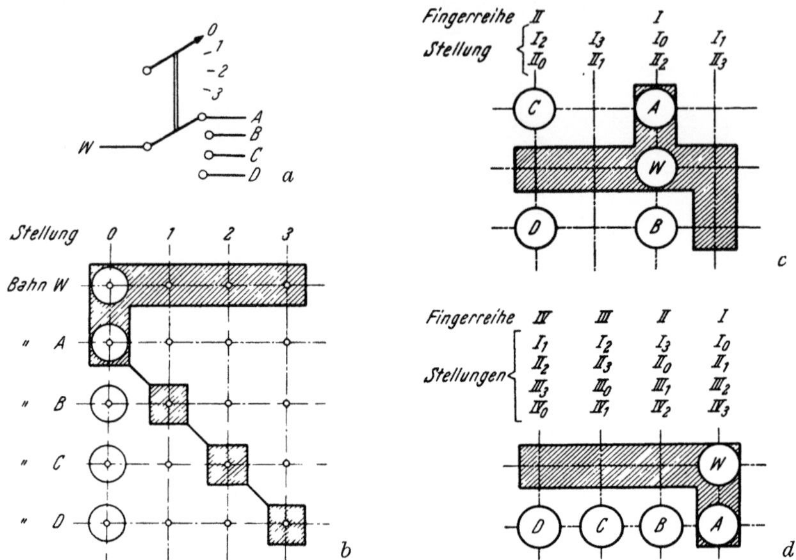

Abb. 82. Walzenumschalter für vier Stromkreise. *a* Schaltprogramm, *b* Bauform mit 1 Fingerreihe und 5 Schaltbahnen, Füllfaktor 11/20 = 55%, *c* Bauform mit 2 Fingerreihen und 3 Schaltbahnen, Füllfaktor 9/12 = 75%, *d* Bauform mit 4 Fingerreihen und 2 Schaltbahnen, Füllfaktor 8/8 = 100%

sondern den ganzen Bereich, innerhalb dessen die elektrischen Verbindungen unverändert bleiben. Es ist stets zulässig, diesen Bereich durch seine Mittellinie zu ersetzen. Der Abstand zwischen zwei benachbarten Stellungen wird *Schaltschritt* genannt. Der dazugehörige Drehwinkel des Schalters heißt *Schrittwinkel,* der auf den Zylindermantel bezogene Bogen *Schritteilung.* Die Gesamtheit der Schaltschritte, die mit dem Schalter ausgeführt werden können, führt die Bezeichnung *Schaltbereich,* bzw. *Schaltwinkel.* Die Bahnen der Schaltfinger auf der Zylinderfläche sind Kreise; sie werden *Schaltbahnen* oder *Kontaktbahnen* genannt. Der Abstand zwischen benachbarten Schaltbahnen heißt *Fingerteilung* oder *Bahnteilung.*

Zeichnet man die Abwicklung des Zylindermantels eines Drehschalters auf, so bilden die Stellungslinien und die Schaltbahnen ein System voneinander rechtwinkelig schneidenden Geraden (Abb. 82). Durch Eintragen der Schaltfinger und Schaltbrücken ergibt sich ein Bild für die innere Schaltung des Gerätes. Die Zahl der Stellungen ist gewöhnlich

durch das Schaltprogramm vorgeschrieben, wogegen die Zahl der Schaltbahnen vom Konstrukteur bestimmt werden kann. Um die Abmessungen
des Gerätes klein zu halten, wird man trachten, mit möglichst wenig
Schaltbahnen auszukommen. Die Grenze ist dann erreicht, wenn fast alle
Schnittpunkte von Stellungen und Bahnen mit Brücken oder Fingern besetzt sind, so daß der Raster keine freien Knotenpunkte aufweist. Das
Verhältnis aus der Zahl der belegten Knoten zur Gesamtzahl aller Knoten
des Rasters bildet ein anschauliches Maß für den Ausnützungsgrad (Füllfaktor) der Konstruktion. Die Abb. 82 *a* bis *d* zeigen die Abwicklung eines
Umschalters für vier Stromkreise in verschiedenen Bauformen mit ein
bis vier Fingerreihen und zugehörigen Füllfaktoren von 55 bis 100%.

2. Die Drehschalter mit einfach benützten axialen Schaltbrücken
(einreihige und mehrfach-einreihige Walzenschalter)

Schaltbrücken, die den Strom parallel zur Drehachse des Schalters
führen, sind kennzeichnend für alle Walzenschalter. Hinsichtlich der konstruktiven Ausbildung lassen sich zwei Grundformen der Schaltwalzen
unterscheiden, nämlich die *Rohr-*
walzen und die *Skelettwalzen.* Da sich
die Kontaktgabe bei einem Drehschalter stets auf der Mantelfläche
eines Zylinders abspielt, liegt es sehr
nahe, die beweglichen Schaltstücke
an einem rohrförmigen Isolierkörper
zu befestigen, wie es Abb. 83 *a* und *b*
zeigen. Diese Bauform ist durch
große Freizügigkeit in der Anordnung
der Brücken gekennzeichnet. Vor
allem lassen sich Verbindungen, die
einzelne Schaltbahnen überspringen
sollen und daher vom übrigen Schaltbelag isoliert sein müssen, in das
Innere der Rohrwalze verlegen. Die
Isolierbrücke besteht dann aus zwei
getrennten Schaltstücken, die durch
eine Lasche gemäß Abb. 83 *b* miteinander verbunden sind. Dem Vorteil
dieser Bauweise stehen jedoch auch
verschiedene Nachteile gegenüber,
die durch die schlechte Zugänglichkeit der Verschraubungen innerhalb
des Rohres und die verhältnismäßig
hohen Kosten der Herstellung ge-

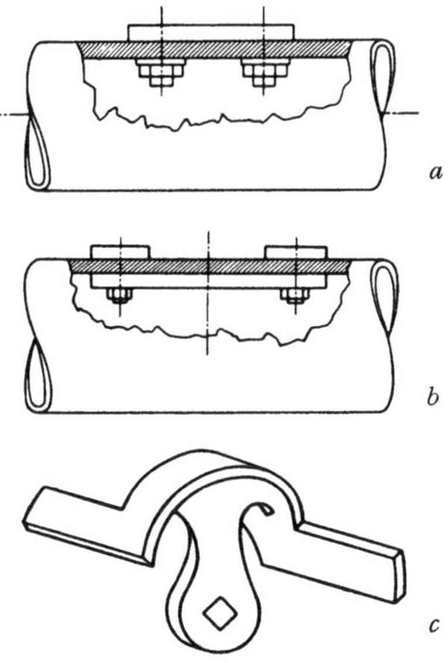

Abb. 83. Brückenformen von Walzenschaltern. *a* Blanke Brücke einer
Rohrwalze, *b* Isolierte Brücke einer
Rohrwalze, *c* Zu einem Schaltstück
vereinigte Brücke einer Skelettwalze

geben sind. Handelt es sich um Geräte, die in hohen Stückzahlen gefertigt
werden, so läßt sich diese Bauart dadurch verbessern, daß der Isolier-

körper mit Ausnehmungen versehen wird. Die Mantelfläche erhält hierdurch die Form eines Gitters, durch dessen Öffnungen das Innere der Rohrwalze zugänglich ist. Eine weitere Erleichterung für den Zusammenbau läßt sich durch Zusammensetzen des Rohres aus einzelnen ringförmigen Scheiben erzielen. Verzichtet man auf die Verwendung von Isolierbrücken, so empfiehlt sich die Bauart als Skelettwalze. Der Zylindermantel des Schalters besteht dann ausschließlich aus dem metallischen Schaltbelag, wobei die einzelnen Schaltstücke durch Arme auf einer isolierten Welle befestigt sind. Gewöhnlich ist es möglich, zwei oder mehrere benach-

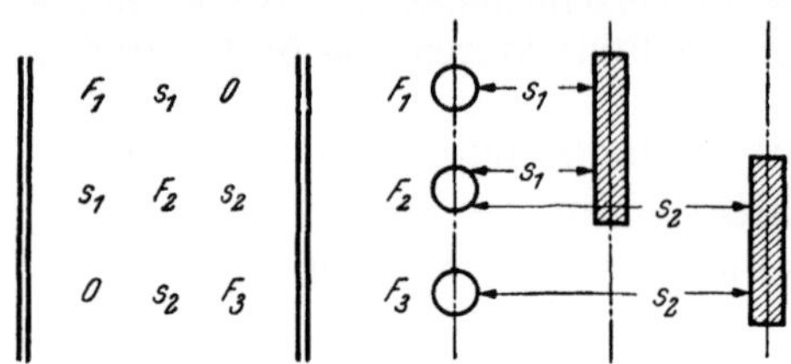

Abb. 84. Verbindungsmatrix und Schaltplan eines Walzenschalters

barte Brücken zu größeren Schaltstücken zu vereinigen und mit einem gemeinsamen Tragarm auszustatten; die Zahl der Bestandteile, aus denen die Walze besteht, läßt sich auf diese Weise stark vermindern. Abb. 83 c gibt ein Beispiel für ein Schaltstück einer Skelettwalze, das aus zwei axialen Brücken, die miteinander und mit einem gemeinschaftlichen Arm verbunden sind, besteht. Bei Skelettwalzen für kleinere Stromstärken und Spannungen sind die Schaltstücke gewöhnlich gestanzte und gebogene Blechteile; für größere Skelettwalzen werden gegossene Schaltstücke verwendet.

Sehr häufig werden die Walzenschalter einreihig ausgeführt, wobei sämtliche Schaltfinger auf einer einzigen Fingerleiste befestigt sind; die Berührungsstellen zwischen Fingern und Brücken liegen in diesem Falle auf einer Erzeugenden der Zylinderfläche. Eine einfache Überlegung läßt erkennen, daß sich mit einer Fingerreihe und axialen Schaltbrücken jedes beliebige Schaltprogramm erfüllen läßt und daß dabei für jeden Schaltzustand nur eine Schalterstellung und für jeden äußeren Anschluß nur ein Schaltfinger erforderlich ist. Wir denken uns, daß der Zylindermantel in jedem Knotenpunkt des Rasters, der aus den Stellungslinien und den Kontaktbahnen gebildet wird, mit einem isoliert befestigten Schaltstück belegt sei; durch Verbindung je zweier Schaltstücke, die zur selben Stellung der Walze gehören, können dann Brücken gebildet werden, die jede gewünschte Fingerverbindung herstellen. Sind die Schaltstücke benachbart, so dürfen sie blank verbunden werden, andernfalls sind Isolierbrücken erforderlich.

Wie jede andere Schaltung, läßt sich auch die Anordnung der Schaltfinger und Schaltbrücken innerhalb eines Schaltgerätes durch eine Matrix mathematisch abbilden. Abb. 84 zeigt eine Gegenüberstellung von Schaltplan und zugehöriger Verbindungsmatrix. Da sich die Matrix nur auf ein einziges Schaltgerät bezieht, läßt sich die Schreibweise etwas vereinfachen; es genügt, die Schalterstellungen durch Stellungsnummern zu kennzeichnen und beispielsweise statt s_2 einfach 2 zu setzen. In jene Plätze der Matrix, die keiner Verbindung entsprechen, darf dann nicht der

Operand 0 eingetragen werden, weil das Zeichen 0 nunmehr die Bedeutung von s_0 hat. Plätze, die fehlenden Verbindungen zugeordnet sind, müssen bei dieser Schreibweise freigelassen werden. Es bedeutet demnach:

$$\left\|\begin{array}{ccc} F_1 & s_1 & 0 \\ & F_2 & s_2 \\ & & F_3 \end{array}\right\| = \left\|\begin{array}{ccc} F_1 & 1 & \\ & F_2 & 2 \\ & & F_3 \end{array}\right\| .$$

Der Vorteil der mathematischen Darstellung liegt darin, daß diese die Wirkungsweise wesentlich klarer zum Ausdruck bringt als der Schaltplan. Die Zeichnung kann stets nur einen einzigen Zustand abbilden, wogegen die Matrix die Verbindungen in sämtlichen Stellungen des Gerätes gleichzeitig zeigt. Daher läßt sie auch alle Möglichkeiten zur konstruktiven Vereinfachung wesentlich besser erkennen als der Schaltplan und ist somit das gegebene Hilfsmittel zur Bestimmung der jeweils zweckmäßigsten Anordnung von Fingern und Brücken.

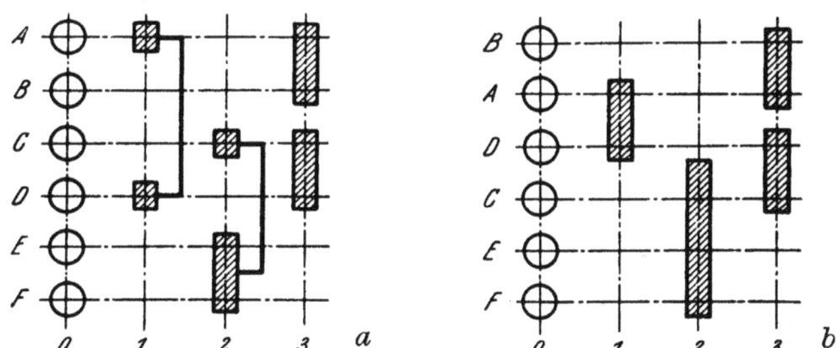

Abb. 85. Fingerfolge und Schaltbrücken eines Walzenschalters. *a* Mit isolierten Brücken, *b* Mit blanken Brücken

Um den Arbeitsaufwand zur Herstellung eines Walzenschalters gering zu halten, bemüht man sich in erster Linie, isolierte Verbindungen auf der Walze zu vermeiden. Ob sich dieses Ziel überhaupt erreichen läßt, hängt von der Beschaffenheit des Schaltprogramms ab; Voraussetzung ist ferner eine bestimmte Reihenfolge der Schaltfinger. Die Ermittlung der zweckmäßigsten Fingerfolge ist also die erste Aufgabe bei der Bestimmung der zweckmäßigsten Schaltung des Gerätes.

Der Aussagewert einer Verbindungsmatrix bleibt unverändert, wenn man die Reihenfolge ihrer Zeilen, bzw. Spalten verändert. Diese Operation entspricht im vorliegenden Fall einer Änderung der Fingerfolge und somit einer anderen Ausführung der dargestellten Schaltwalze. Es läßt sich nun aus der Verbindungsmatrix sofort erkennen, welche Verbindungen bei der gewählten Fingerfolge blank verlegt und welche isoliert sein müssen. Abb. 85 *a* zeigt den Schaltplan eines Walzenschalters mit sechs Fingern *A* bis *F*, wobei der Walzenbelag aus sechs blanken und zwei isolierten

Brücken besteht. Die Übertragung des Schaltplanes in eine Verbindungsmatrix ergibt:

$$\begin{Vmatrix} A & 3 & . & 1 & . & . \\ & B & . & . & . & . \\ & & C & 3 & 2 & . \\ & & & D & . & . \\ & & & & E & 2 \\ & & & & & F \end{Vmatrix}$$

In dieser Matrix erscheinen die blanken Verbindungen in einer Nebendiagonale, die unmittelbar an die Hauptdiagonale anschließt; alle an anderen Stellen der Matrix eingetragenen Verbindungen gehören zu Isolierbrücken. Will man nur blanke Brücken verwenden, hat man demnach die Reihen der Matrix so miteinander zu vertauschen, daß sämtliche Stellungsnummern in der ersten Nebendiagonale zu stehen kommen. Die Matrix erscheint dann in folgender Form:

$$\begin{Vmatrix} B & 3 & . & . & . & . \\ & A & 1 & . & . & . \\ & & D & 3 & . & . \\ & & & C & 2 & . \\ & & & & E & 2 \\ & & & & & F \end{Vmatrix}$$

Hierbei sind die Finger miteinander über je eine Schaltbrücke paarweise verkettet; die Schaltung zeigt eine zusammenhängende Folge (Kette) von Verbindungen nach dem Schema:

$$B\,3\,A\,1\,D\,3\,C\,2\,E\,2\,F.$$

Um die Kette aus einer vorgegebenen beliebigen Matrix zu finden, ist es zweckmäßig, diese durch Einsetzen der Stellungsnummern in die beiden symmetrischen Hälften zu vervollständigen.

$$\begin{Vmatrix} A & 3 & . & 1 & . & . \\ 3 & B & . & . & . & . \\ . & . & C & 3 & 2 & . \\ 1 & . & 3 & D & . & . \\ . & . & 2 & . & E & 2 \\ . & . & . & . & 2 & F \end{Vmatrix}$$

Jede Verbindungskette beginnt und endet bei einer Reihe (Zeile oder Spalte), in der sich nur eine einzige Stellungsnummer befindet. In den zur Hauptdiagonale gehörigen Feldern dieser Reihen finden sich die Operanden jener Schaltfinger, die *Randfinger* der Schaltwalze sein müssen. Von diesen ausgehend, läßt sich die Verbindungskette leicht ablesen. Im vorliegenden Beispiel gelangt man von *B* über 3 zu *A*, von dort über 1 zu *D*, dann über 3 zu *C* usw. Aus der Verbindungskette läßt sich sofort der

Schaltplan nach Abb. 85 *b* entwerfen; es ist nicht mehr nötig, die zugehörige Matrix mit geänderter Reihenfolge der Finger anzuschreiben.

Da die Verbindungsmatrix als tabellarische Darstellung eines Streckenkomplexes aufgefaßt werden kann, läßt sich die Fingerfolge eines Walzenschalters mit blanken Brücken auch durch ein zeichnerisches Verfahren ermitteln. Man transformiert den vorgegebenen Stromlaufplan oder die Verbindungsmatrix in einen Streckenkomplex, dessen Punkte den Schaltfingern und dessen Strecken den Kontaktverbindungen zugeordnet sind, wie es Abb. 86 zeigt. Isolierte Brücken lassen sich vermeiden, wenn die Kontaktverbindungen des Komplexes einen offenen Linienzug oder mehrere solche bilden. Die Reihenfolge der Punkte im Linienzug nennt die Fingerfolge des Walzenschalters. Auch in der dem Komplex entsprechenden Matrix tritt der gleiche Linienzug in Erscheinung.

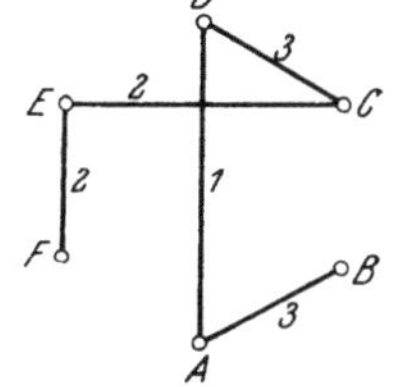

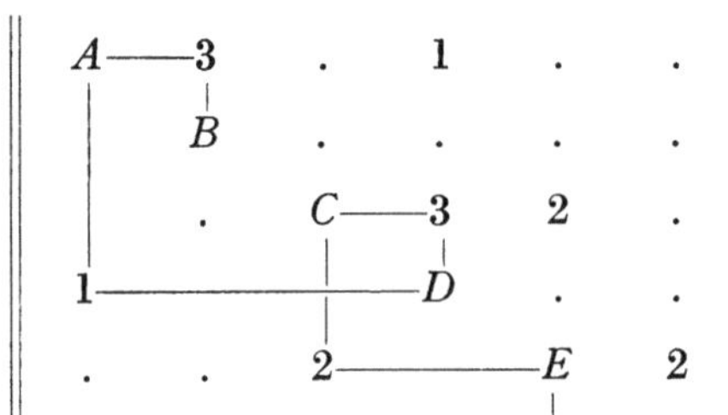

Abb. 86. Ermittlung der Fingerfolge für blanke Schaltbrücken

Man kann also nach Belieben die Matrix oder den Streckenkomplex als Hilfsmittel zur Ausmittlung der Fingerfolge verwenden. Als praktisches Beispiel hierfür diene die in Abschnitt III, 4 und Abb. 60 erläuterte Schaltung von zwei Heizkörpern *A* und *B*. Durch Serien-, Einzel- und Parallelschaltung lassen sich vier Heizstufen herstellen. Es soll nun ein einreihiger Walzenschalter mit einer Nullstellung und vier Arbeitsstellungen für dieses Schaltprogramm entworfen werden. Abb. 87 zeigt die Ausmittlung des Schalters auf zeichnerischem Wege. Der Stromlaufplan ist in Abb. 87 *a* wiedergegeben, die auf die gewünschte Netztrennung in der Nullage Rücksicht nimmt. In Abb. 87 *b* ist dieser Schaltplan in einen Streckenkomplex transformiert; die nicht zum Schalter gehörigen inneren Verbindungen der Heizkörper und der Stromquelle sind durch punktierte Strecken angedeutet. Der Komplex liefert durch einen zusammenhängenden Linienzug die natürliche Fingerfolge für eine Schaltwalze mit ausschließlich blanken Brücken. Abb. 87 *c* zeigt den gefundenen Schaltplan als Abwicklung des Walzenbelages. Die Ausmittlung läßt sich auch rein rechnerisch durchführen. Aus der Leitwertgleichung

$$\frac{\lambda}{Q} = (s_1\,A\,B + s_2\,A + s_3\,B + s_4\,A + s_4\,B)$$

findet man durch Einführung der Anfänge und Enden für Verbraucher und Stromquelle die erweiterte Gleichung

$$\lambda = Q_1(s_1\,A_1\,A_2\,s_1\,B_1\,B_2\,s_1 + s_2\,A_1\,A_2\,s_2 + s_3\,B_1\,B_2\,s_3 + s_4\,A_1\,A_2\,s_4 +$$
$$+ s_4\,B_1\,B_2\,s_4)\,Q_2.$$

Aus dieser Gleichung folgt unmittelbar die Verbindungsmatrix:

$$\left\|\begin{array}{ccccccc} Q_1 & 1,2,4 & . & 3,4 & . & . \\ & A_1 & . & . & . & . \\ & & A_2 & 1 & . & 2,4 \\ & & & B_1 & . & . \\ & & & & B_2 & 1,3,4 \\ & & & & & Q_2 \end{array}\right\|$$

und daraus die Verbindungskette:

$$A_1 - 1,2,4 - Q_1 - 3,4 - B_1 - 1 - A_2 - 2,4 - Q_2 - 1,3,4 - B_2.$$

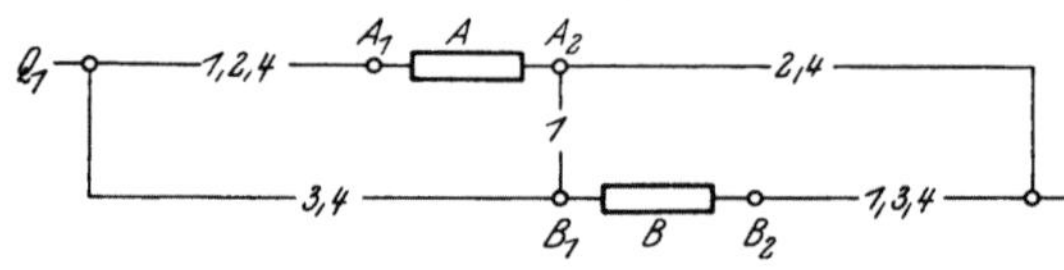

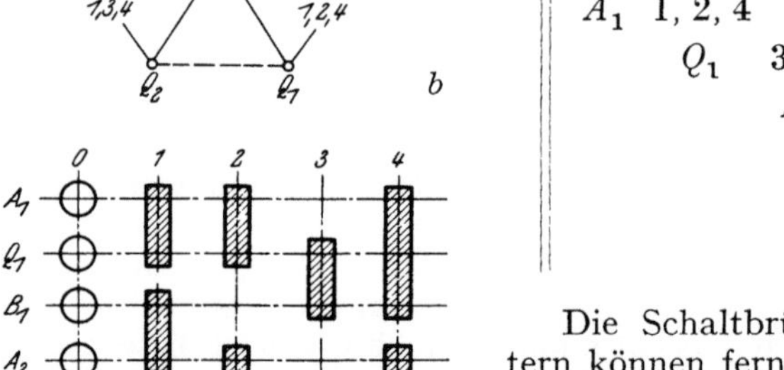

Abb. 87. Heizkörperumschalter.
a Stromlaufplan, b Streckenkomplex, c Abwicklung der Schaltwalze

Damit ist die Reihenfolge der Kontaktfinger bestimmt; man kann entweder den Schaltplan gemäß Abb. 87 c aufzeichnen oder an seiner Stelle die Verbindungsmatrix des gesuchten Schalters anschreiben. Die Matrix mit richtiger Fingerfolge lautet dann:

$$\left\|\begin{array}{ccccccc} A_1 & 1,2,4 & . & . & . & . \\ & Q_1 & 3,4 & . & . & . \\ & & B_1 & 1 & . & . \\ & & & A_2 & 2,4 & . \\ & & & & Q_2 & 1,3,4 \\ & & & & & B_2 \end{array}\right\|$$

Die Schaltbrücken von Walzenschaltern können ferner teilweise miteinander verbunden, das heißt zu gemeinschaftlichen Schaltstücken vereinigt werden. Hierdurch vermindert sich die Zahl der Schaltstücke und der Bauteile zu ihrer Befestigung. Ein weiterer Vorteil dieser konstruktiven Ausbildung liegt darin, daß der Stromfluß nicht bei jedem Schaltschritt an allen Kontaktstellen unterbrochen wird. Auch ist der Kraftbedarf für den Antrieb geringer, weil die Finger nicht bei jedem Schritt von sämtlichen Schaltbrücken angehoben werden müssen. Der elektrische und mechanische Verschleiß der Schaltfinger bleibt daher kleiner als bei getrennten Schaltbrücken. Als Nachteil muß hingegen in Kauf genommen werden, daß sich für die gemeinsamen Schaltstücke meist komplizierte Formen ergeben, deren Herstellung sich nur bei großer Stückzahl lohnt. Welche Schaltbrücken zusammengelegt werden dürfen, läßt sich aus der Verbindungsmatrix ersehen. Brücken, deren Stellungsnummern im gleichen Feld der Matrix stehen, dürfen stets elektrisch verbunden und zu einem Schaltstück vereinigt werden. Das gleiche gilt für Brücken, die in zwei benachbarten Feldern angeschrieben sind. Eine Zusammenfassung über mehr als zwei aneinander anschließende Felder ist nur dann zulässig, wenn in ihnen jede Stellungsnummer

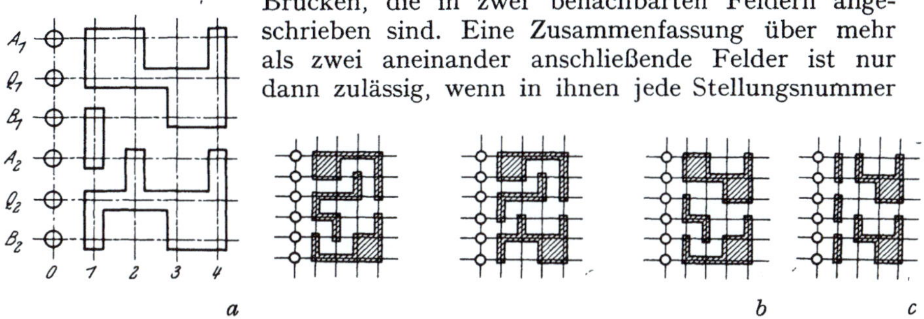

a   b   c

Abb. 88. Vereinigung axialer Schaltbrücken. *a* Bauform mit Schaltstücken geringster Zahl und Größe, *b* Äquivalente Bauformen, *c* Bauform mit gleichartigen Schaltstücken geringster Zahl und Größe

entweder nur einmal oder in benachbarten Feldern erscheint. Zerlegt man die Matrix unter Beachtung dieser Regel in mehrere Teilmatrizen, so entspricht jede von diesen einem Schaltstück des Walzenschalters. Die im vorstehenden wiedergegebene Verbindungsmatrix läßt sich beispielsweise in drei Teilmatrizen zerlegen:

$$\left\| \begin{matrix} A_1 & 1,2,4 & \\ & Q_1 & 3,4 \\ & B_1 & \end{matrix} \right\| + \left\| \begin{matrix} B_1 & 1 \\ & A_2 \end{matrix} \right\| + \left\| \begin{matrix} A_2 & 2,4 & . \\ & Q_2 & 1,3,4 \\ & B_2 \end{matrix} \right\|$$

Diese Zusammenfassung der Schaltbrücken führt zu einer **Form des** Schaltbelages der Walze gemäß Abb. 88 *a*. Bei der Zusammenfassung ist es jedoch keineswegs notwendig, die im gleichen Feld der Gesamtmatrix abgebildeten Brücken zu einem Schaltstück zu vereinigen. Es sind auch andere Formen des Schaltbelages möglich. Unzulässig wäre die Vereinigung der mit 1 bezeichneten Brücken, da sich diese Stellungsnummern nicht in benachbarten Feldern der Matrix befinden. Ebenso darf die Brücke 2 der Verbindung $A_1 Q_1$ nicht mit der gleichnamigen Brücke von $A_2 Q_2$ zusammengelegt werden; das gleiche gilt für die Brücken 3 der Verbindungen $Q_1 B_1$ und $Q_2 B_2$. Alle übrigen Arten der Zusammenfassung sind zulässig und bilden äquivalente Formen des Schaltbelages. Abb. 88 *b* zeigt drei weitere Möglichkeiten zur Vereinigung der Brücken. Diese Formen entsprechen den Matrixsummen:

$$
\left\|\begin{array}{ll} A_1 & 1,2,4 \quad . \\ & Q_1 \quad 4 \\ & B_1 \end{array}\right\| +
\left\|\begin{array}{lll} Q_1 & 3 & . \quad . \\ B_1 & 1 & . \\ & A_2 & 2 \\ & & Q_2 \end{array}\right\| +
\left\|\begin{array}{ll} A_2 & 4 \quad . \\ & Q_2 \quad 1,3,4 \\ & B_2 \end{array}\right\|
$$

oder

$$
\left\|\begin{array}{ll} A_1 & 1,2,4 \quad . \\ & Q_1 \quad 4 \\ & B_1 \end{array}\right\| +
\left\|\begin{array}{ll} Q_1 & 3 \quad . \\ & B_1 \quad 1 \\ & A_2 \end{array}\right\| +
\left\|\begin{array}{ll} A_2 & 2,4 \quad . \\ & Q_2 \quad 1,3,4 \\ & B_2 \end{array}\right\|
$$

oder

$$
\left\|\begin{array}{ll} A_1 & 1,2,4 \quad . \\ & Q_1 \quad 3,4 \\ & B_1 \end{array}\right\| +
\left\|\begin{array}{ll} B_1 & 1 \quad . \\ & A_2 \quad 2 \\ & Q_2 \end{array}\right\| +
\left\|\begin{array}{ll} A_2 & 4 \quad . \\ & Q_2 \quad 1,3,4 \\ & B_2 \end{array}\right\|
$$

Im Gegensatz zur Ausführung nach Abb. 87 a erstreckt sich bei diesen Formen das mittlere Schaltstück über drei oder vier Fingerteilungen; diese Bauformen sind also keine Lösungen mit kleinsten Abmessungen der Schaltstücke. Es kann aber trotzdem zweckmäßig sein, auf solche Möglichkeiten zurückzugreifen, wenn mit Rücksicht auf vorhandene Konstruktionen eine bestimmte Gestalt der Schaltstücke angestrebt wird. Besonders zu beachten ist die erste der drei Formen nach Abb. 88 b. Sie zeigt an ihrem mittleren Schaltstück, daß Brücken auch dann verbunden werden dürfen, wenn sie zu verschiedenen Spannungspolen der Schaltung gehören. Die Zusammenfassung von Brücken zu größeren Schaltstücken wird also von der Polung nicht beeinflußt. Das mittlere Schaltstück unterscheidet sich von den übrigen nur dadurch, daß es in verschiedenen Schalterstellungen verschiedenes Potential hat. Oft ist es zweckmäßig, nicht sämtliche Möglichkeiten zur Verbindung von Brücken zu benützen. Zur Vereinfachung der Herstellung des Schaltbelages empfiehlt es sich, die einzelnen Schaltstücke möglichst gleichartig zu gestalten.

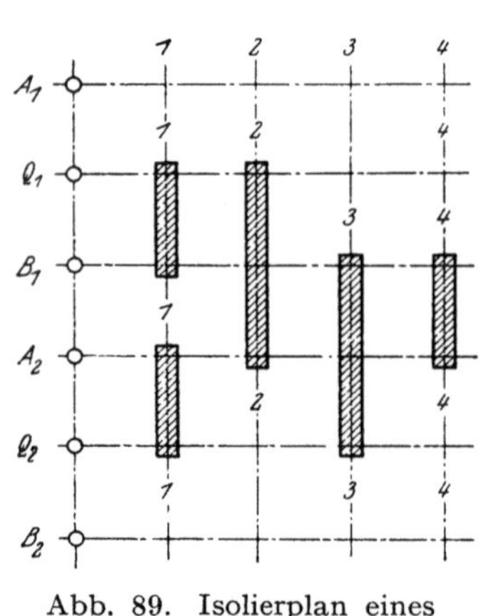

Abb. 89. Isolierplan eines Walzenschalters

Die in Abb. 87 b wiedergegebene Ausführung mit getrennten Brücken umfaßt neun Schaltstücke, wovon sich sieben über je zwei Fingerteilungen und zwei über je drei Teilungen erstrecken. Bei maximaler Vereinigung von Brücken ermäßigt sich die Zahl der Schaltstücke auf drei, die jedoch durchwegs verschiedene Gestalt haben. Führt man die Zusammenlegung gemäß Abb. 88 *a* nur hinsichtlich der Schalterstellungen 2 bis 4 durch, so kommt man ebenfalls mit nur zwei Arten von Schaltstücken aus, wobei sich deren Zahl von neun auf fünf ermäßigt. Abb. 88 *c* zeigt diese Bauform, die für die Herstellung sehr zweckmäßig sein kann.

Besonders anschaulich zeigen sich die verschiedenen Möglichkeiten, wenn man in den Raster des Zylindermantels die verbotenen Brückenverbindungen einträgt. Man erhält dann den in Abb. 89 dargestellten Isolierplan des Walzenschalters. *Der Plan muß durch so viele senkrecht zur Achse verlaufende Schnitte geteilt werden, daß jede verbotene Verbindung wenigstens einmal von einer Schnittlinie gekreuzt wird.* Bei dem in Abb. 89 wiedergegebenen Beispiel genügen zwei Schnitte, die entweder in den Bahnen Q_1, A_2 oder B_1, A_2 oder B_1, Q_2 verlaufen. In diesen Schaltbahnen dürfen keine verbindenden Ringsegmente angeordnet werden.

Ein Schaltbelag mit getrennten blanken Brücken läßt sich nur herstellen, wenn die aus Fingern und Schaltbrücken gebildeten Kontakte paarig verkettet sind und eine oder mehrere offene Ketten bilden. Ist eine Kette in sich geschlossen, sind isolierte Brückenverbindungen bei einreihiger **Anordnung der** Finger unvermeidlich. Als Beispiel diene ein Wendeschalter, dessen Stromlaufplan in Abb. 90 *a* wiedergegeben ist. Die Transformation des Schaltbelages in einen Streckenkomplex nach Abb. 90 *b* zeigt, daß die Kontaktverbindungen einen geschlossenen Ring bilden. Auch aus der Verbindungsmatrix ist diese Eigenschaft des Schaltprogrammes ersichtlich:

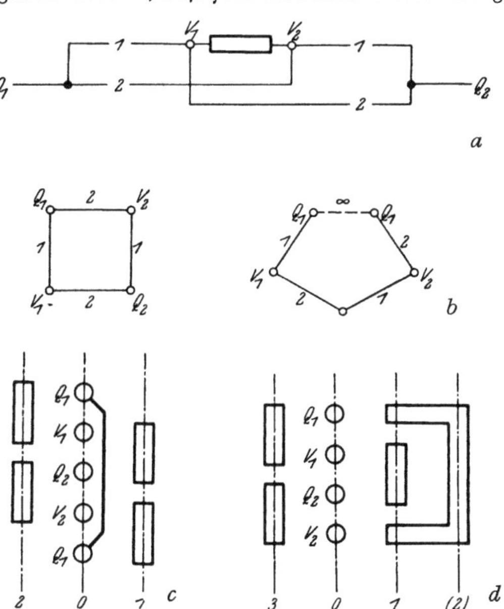

Abb. 90. Einreihiger Walzenumschalter mit drei Stellungen. *a* Stromlaufplan, *b* Streckenkomplexe mit geschlossener und offener Verbindungskette, *c* **Bauform mit zusätzlichem Schaltfinger**, *d* Bauform mit zusätzlicher Stellung (Sperrstellung)

Sollen isolierte Verbindungen auf der Schaltwalze unbedingt vermieden werden, muß ein Schaltprogramm dieser Art entsprechend abgeändert werden. Eine Möglichkeit hierzu besteht in der Vermehrung der Schaltfinger. Im vorliegenden Fall genügt die Verdoppelung irgend eines der vier Finger, um die geschlossene Kette in eine offene zu verwandeln, wie Abb. 90 *b* und *c* zeigt. Durch diese Maßnahme wird die isolierte Verbindung auf der Walze durch eine ebensolche äußere Verbindung zwischen

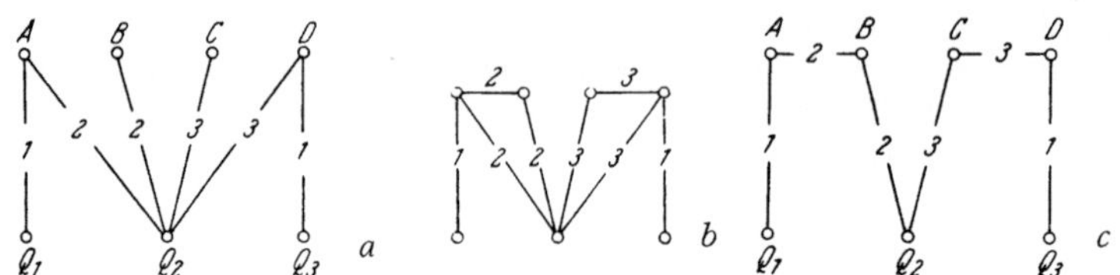

Abb. 91. Äquivalenz von Verbindungskomplexen verschiedener Ordnung. *a* Komplex vierter Ordnung, *b* Funktionskomplex, *c* Äquivalenter Komplex zweiter Ordnung

zwei äquivalenten Fingern ersetzt. Eine weitere Möglichkeit besteht in der Einführung einer zusätzlichen Schalterstellung, die auf mechanischem Wege gesperrt werden muß. Blanke Verbindungen, die dieser Stellung zugeordnet sind, kommen deshalb niemals als Schaltbrücken zur Wirkung und sind mit isolierten Verbindungen gleichwertig. Abb. 90 *d* zeigt diese Ausführung des Wendeschalters. Die Vergrößerung der Stellungszahl hat im allgemeinen keine wesentliche Erhöhung des Aufwandes zur Folge.

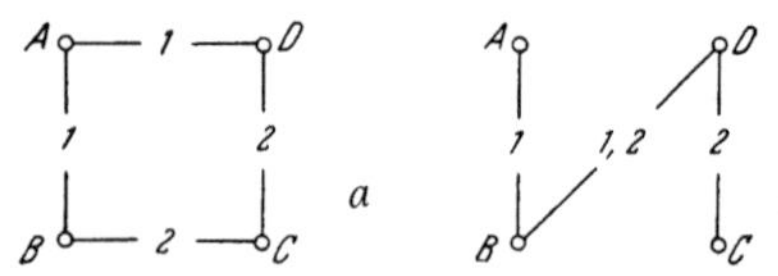

Abb. 92. Äquivalenz von geschlossenen und offenen Verbindungskomplexen. *a* Geschlossener Komplex, *b* Offener Komplex

Umfaßt das Schaltprogramm viele wirksame Stellungen, so fällt die Vermehrung um eine Totstellung nicht ins Gewicht: sind nur zwei oder drei Arbeitsstellungen vorgeschrieben, so ist die Einführung von Leerstellungen meist unvermeidlich, weil der Schrittwinkel mit Rücksicht auf die bequeme Betätigung nicht größer als 90° sein soll. Die Sperrung einzelner Stellungen hat selbstverständlich zur Folge, daß der *Schaltwinkel* des Gerätes kleiner als 360° sein muß; elektromagnetische oder motorische Antriebe müssen daher für beide Drehrichtungen ausgelegt werden.

Ist das Schaltprogramm so beschaffen, daß ein Finger mit mehr als zwei anderen verbunden werden muß, so lassen sich im allgemeinen isolierte Brücken nicht vermeiden. Die Darstellung als Streckenkomplex zeigt dann Punkte, in denen sich mehr als zwei Strecken treffen (Punkte dritter oder höherer Ordnung). Gehören diese Strecken teilweise zu gleichen Stellungen, so lassen sich manchmal äquivalente Formen finden, die nur Punkte zweiter Ordnung aufweisen. Ein Beispiel für eine solche Transformation zeigt Abb. 91. Auch Schaltprogramme mit geschlossener Verbindungskette lassen sich auf analoge Weise transformieren, wenn äquivalente Kontakte vorhanden sind und zu gleichen Schaltfingern gehören. Abb. 92 zeigt die Umformung eines geschlossenen Streckenkomplexes in einen gleichwertigen offenen. In der Verbindungsmatrix zeigt sich das Bestehen äquivalenter Formen dadurch an, daß eine oder mehrere Stellungsnummern öfter als einmal in der gleichen Reihe auftreten. Zur Transformation bildet man durch Einsetzen der mittelbaren Verbindungen die Funktionsmatrix; aus dieser sucht man die offene Verbindungskette heraus, die sämtliche Schaltfinger umfaßt und jede Verbindung mindestens einfach enthält. Der Transformation gemäß Abb. 92 entspricht also folgender Rechnungsgang:

$$
\begin{Vmatrix}
A & 1 & . & . \\
1 & B & 2 & 1 \\
. & 2 & C & 2 \\
. & 1 & 2 & D
\end{Vmatrix}
=
\begin{Vmatrix}
A & 1 & . & 1 \\
1 & B & 2 & 1,2 \\
. & 2 & C & 2 \\
1 & 1,2 & 2 & D
\end{Vmatrix}
=
\begin{Vmatrix}
A\!-\!1 & . & . \\
. & B & . & . \\
. & & C\!-\!2 \\
. & 1,2\!-\!\!-\!\!-\!D
\end{Vmatrix}
$$

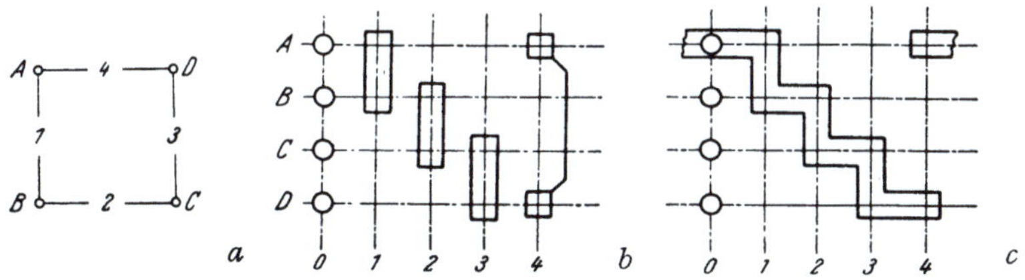

Abb. 93. Ersatz von isolierten Brücken durch zusammengesetzte Schaltstücke. *a* Schaltprogramm, *b* Schaltbelag mit isolierter Verbindung, *c* Blanker Schaltbelag

Enthält die geschlossene Verbindungskette keine äquivalenten Kontakte, so gestattet das Schaltprogramm keine Transformation. Bei Ausführung mit getrennten Brücken müssen dann einzelne Verbindungen isoliert werden. Vereinigt man hingegen die Brücken durch Tangentialverbindungen zu größeren Schaltstücken, so lassen sich manchmal **isolierte** Verbindungen durch bereits vorhandene Blankverbindungen ersetzen. Abb. **93 gibt ein Beispiel** hierfür. Das Schaltprogramm ist in Abb. 93 *a* als Streckenkomplex **wiedergegeben**; Abb. 93 *b* zeigt den Schaltbelag der Walze mit getrennten Brücken und einer isolierten Verbindung. Schließt man die Brücken durch Ringsegmente zu einem einzigen Schaltstück zusammen, so ersetzt dieses die isolierte Verbindung, wie aus Abb. 93 *c* hervorgeht. Von dieser Möglichkeit kann man auch Gebrauch machen, wenn das Schaltprogramm mehrere Verbindungen enthält, die vom selben Finger ausgehen und in verschiedenen Stellungen geschlossen sind. Das Schaltprogramm wird dann so abgeändert, daß je zwei dieser Verbindungen durch eine weitere zusätzliche Verbindung zu einem geschlossenen Ring vereinigt werden. Die zusätzliche Verbindung kann entweder nach Abb. 94 *a* in eine Sperrstellung verlegt oder gemäß Abb. 94 *b* aus zwei in verschiedenen Stellungen geschlossenen Verbindungen zusammengesetzt werden. In diesem zweiten Fall entsteht im Streckenkomplex ein zusätzlicher Punkt P, der einem unbenützten Fingerplatz der Schaltwalze entspricht. Es ist also dem Belieben überlassen, ob zum Ersatz für isolierte Brückenverbindung eine gesperrte Schaltstellung oder ein leerer Fingerplatz herangezogen wird. Abb. 94 c zeigt die beiden Ausführungen.

Abschließend sei nunmehr die allgemeine Regel für die Verbindung axialer Brücken durch tangentiale Ringsegmente abgeleitet. Hierzu betrachten wir ein Schaltprogramm entsprechend der Verbindungsmatrix:

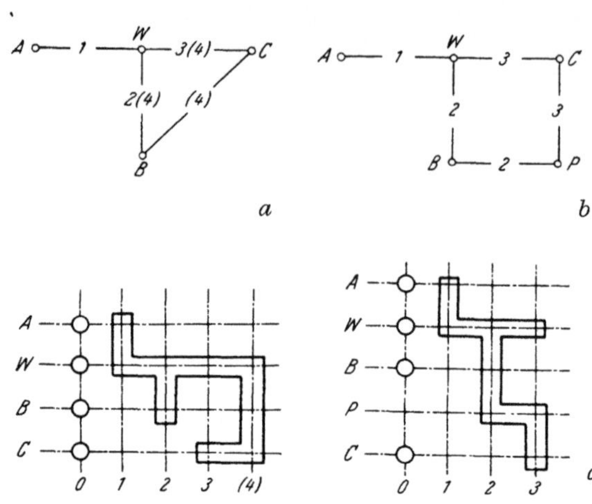

$$\left\| \begin{matrix} A & 1 & . & . \\ & B & 3 & . \\ & & C & 2 \\ & & & D \end{matrix} \right\|$$

Abb. 94. Einreihiger Walzenumschalter für drei Stromkreise mit Nullstellung. a Schaltprogramm mit gesperrter Schalterstellung, b Schaltprogramm mit leerem Fingerplatz, c Wahlformen des Walzenschalters

Stellt man die zusätzliche Forderung, daß in der Nullstellung der Walze kein Eingriff zwischen Fingern und Schaltbrücken vorhanden sein soll, so sind nur zwei Bauformen entsprechend Abb. 95 a möglich. Ringverbindungen, die alle drei axialen Brücken verbinden und nicht über die Nullstellung der Walze führen, wären nicht statthaft, weil hierbei eine unzulässige Verbindung der Finger B, C in der Stellung 2 zustande käme (Abb. 95 b). Die Vereinigung aller drei Brücken zu einem gemeinsamen Schaltstück ist nur dann möglich, wenn eines der beiden Ringsegmente durch die Nullstellung geht, wie es Abb. 95 c zeigt. Man erkennt hieraus die Regel, daß jedes Schaltstück an jeder Stelle der Walze nur eine Tangentialverbindung aufweisen darf. Diese Forderung ist nur dann erfüllt, wenn die Stellungsnummern der miteinander verbundenen Brücken eine Zahlenfolge bilden, die nur zunehmende oder abnehmende Zahlenwerte zeigt. Hat der Schalter n Stellungen, so kann hierbei der Nullstellung nach Bedarf die Stellungsnummer 0 oder n zugeordnet werden; eine beliebige andere Stellung x kann in gleicher Weise als Stellung $n + x$ aufgefaßt werden. Beispielsweise kommt bei einem vierstelligen Schalter der Nullstellung auch die Stellungsnummer 4 und der ersten Arbeitsstellung die Nummer 5 zu. Die vorgegebene Verbindungsmatrix läßt sich daher so umformen, daß die erste Nebendiagonale eine Ziffernfolge mit abnehmendem Zahlenwert enthält:

$$
\left\|\begin{array}{cccc} A & 1 & . & . \\ & B & 3 & . \\ & & C & 2 \\ & & & D \end{array}\right\|
=
\left\|\begin{array}{cccc} A & 5 & . & . \\ & B & 3 & . \\ & & C & 2 \\ & & & D \end{array}\right\|
$$

Gleiche Stellungsnummern können dann nur in benachbarten Feldern der Matrix auftreten. Ferner muß das Intervall zwischen den äußersten Stellungen kleiner als der Walzenumfang sein; die Differenz zwischen größter und kleinster Stellungsnummer muß daher kleiner als die Zahl der Schalterstellungen bleiben. Diese Regel umfaßt auch den Fall der zu

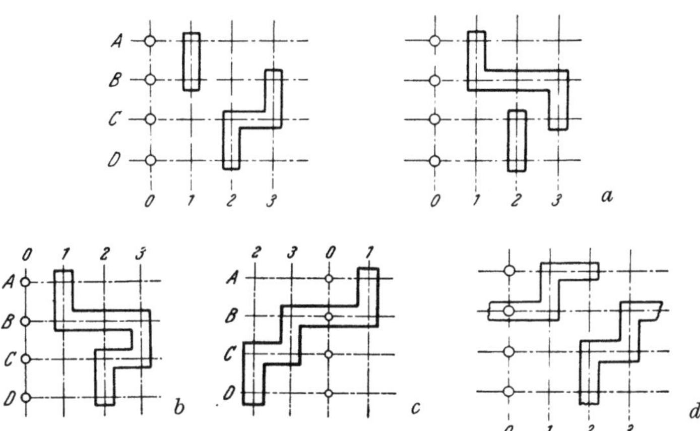

Abb. 95. Schaltstücke aus verbundenen Axialbrücken. *a* Zulässige Verbindungen ohne Fingereingriff in der Nullage, *b* Verbotene Verbindung, *c* Verbindung über die Nullinie, *d* Geschlossene Verbindungskette

einem Ring geschlossenen Verbindungskette. Auch hierbei lassen sich isolierte Schaltbrücken nur dann durch blanke Tangentialverbindungen ersetzen, wenn die beschreibende Matrix so umgeformt werden kann, daß die Zahlenfolge nur steigende oder fallende Werte aufweist. Hierfür ein Beispiel:

$$
\left\|\begin{array}{cccc} A & 1 & . & 2 \\ & B & 3 & . \\ & & C & 2 \\ & & & D \end{array}\right\|
=
\left\|\begin{array}{cccc} A & 5 & . & . \\ . & B & 3 & . \\ . & . & C & 2 \\ 2 & . & . & D \end{array}\right\|
$$

Die beiden mit der Nummer 2 bezeichneten Brücken schließen an den Schaltfinger D an und sind somit benachbart. In der Verbindungsmatrix gehört auch das Feld, in dem sich die Reihen A und D kreuzen, zur ersten Nebendiagonale, die nunmehr eine Nummernfolge mit absteigendem Zahlenwert 5, 3, 2, 2 enthält. Abb. 95 *d* zeigt diesen Schaltbelag, der aus einem einzigen Schaltstück besteht und der vorstehenden Verbindungsmatrix entspricht. Bilden die vom Schaltprogramm vorgeschriebenen Verbindungen einen geschlossenen Ring, so umfassen die tangentialen Brückenverbindungen den ganzen Umfang der Schaltwalze; es ist dann unvermeidlich, daß ein Ringsegment über die Nullstellung der Walze führt.

Zu beachten ist, daß die Ringsegmente zur Verbindung der Brücken von Walzenschaltern keine tangentialen Schaltbrücken, wie sie bei Reihenschaltern vorkommen, sind. Sie dienen nicht der unmittelbaren Verbindung von Schaltfingern, sondern stellen tangentiale Verbindungsstücke axialer Brücken dar; als solche folgen sie daher anderen topologischen Gesetzen als die tangentialen Schaltbrücken der Reihenschalter.

Abschließend seien die gefundenen Gesetze für Schalter mit einer Fingerreihe und axialen Schaltbrücken als Schaltregel zusammengefaßt:

Schaltregel für einreihige Walzenschalter. Ein Walzenschalter mit einer Fingerreihe läßt sich für jedes beliebige Schaltprogramm so bauen, daß für jede Anschlußleitung nur ein Schaltfinger und eine Kontaktbahn und für jeden Schaltzustand nur eine Schalterstellung vorhanden ist. Bilden die durch das Schaltprogramm vorgeschriebenen axialen Verbindungen durchwegs offene Ketten, so läßt sich stets eine Reihenfolge der Schaltfinger finden, bei der sämtliche Schaltbrücken blank verlegt werden können. Brücken, die zum gleichen Fingerpaar gehören, dürfen miteinander tangential zu gemeinschaftlichen Schaltstücken verbunden werden. Das gleiche gilt für benachbarte Brücken, wenn das dadurch entstehende Schaltstück an jeder Stelle der Walze nur eine Tangentialverbindung aufweist; dieser Fall ist immer gegeben, wenn die Stellungsnummer der verbundenen Brücken eine Zahlenreihe bilden, die nur zunehmende oder nur abnehmende Zahlenwerte aufweist und höchstens einen einmaligen Schaltzyklus umfaßt. Eine solche Vereinigung von Brücken durch blanke Tangentialverbindungen ist auch möglich, wenn die vorgeschriebenen axialen Verbindungen eine geschlossene Kette bilden; das von den verbundenen Brücken gebildete Schaltstück umfaßt dann den gesamten Umfang der Schaltwalze einschließlich der Nullstellung.

Die in der technischen Praxis benützten Walzenschalter haben meist eine große Anzahl von Schaltfingern; bei einreihiger Bauart folgt daraus eine große axiale Länge. Diese läßt sich dadurch vermindern, daß man die Finger in zwei oder mehreren Reihen anordnet. Unabhängig vom jeweiligen Schaltprogramm läßt sich jeder einreihige Walzenschalter dadurch in einen n reihigen verwandeln, daß jede Schaltstellung n-mal ausgeführt wird; die Länge der Schaltwalze beträgt dann nur ein n-tel bei n-mal so großem Durchmesser. Die Walzenfläche sowie auch die Zahl der Schaltbrücken bleibt hierbei unverändert. Ein Beispiel hierfür zeigt Abb. 96; in dieser ist der in Abb. 88 a dargestellte Walzenschalter in einen solchen mit zwei Fingerreihen umgeformt. Schaltbrücken zur Verbindung von Fingern, die zwei verschiedenen Reihen angehören, müssen hierbei im allgemeinen isoliert werden. Lassen sich diese Brücken an den

Rand der Walze verlegen, wie es im vorliegenden Beispiel der Fall ist,
kann die isolierte Brücke durch eine blanke Tangentialverbindung ersetzt werden.   Die Wiederholung der Schaltstellungen am Umfang hat
zur Folge, daß der gesamte Schaltwinkel des Gerätes nicht mehr 360°,
sondern bei n Fingerreihen nur ein n-tel davon betragen darf.   Der Schaltwinkel darf jede Stellung nur einmal enthalten, wogegen die übrigen
Stellungen mechanisch gesperrt werden müssen.   Auch der Schrittwinkel

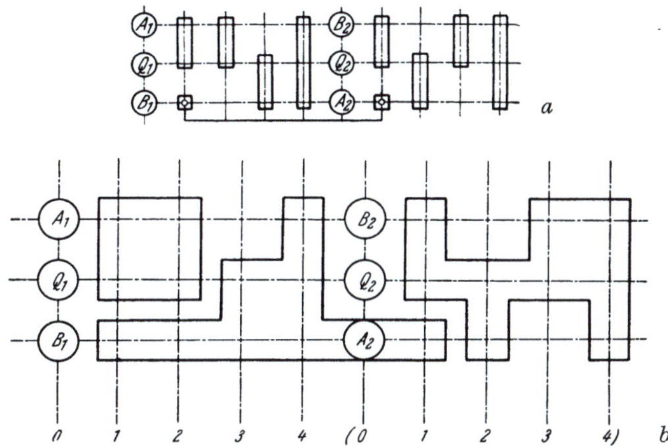

Abb. 96. Doppelt-einreihiger Walzenschalter. a Mit getrennten Brücken, b Mit drei
blanken Schaltstücken

verringert sich bei n Fingerreihen auf ein n-tel. Wie Abb. 96 zeigt, lassen
sich die Schaltbrücken solcher Geräte in genau gleicher Weise zu Schaltstücken verbinden, wie es bei einreihigen Walzenschaltern der Fall ist.
Da bei Schaltern dieser Art jede Brücke nur einfach benützt ist, besteht
im wesentlichen kein Unterschied gegenüber der einreihigen Bauart; es
handelt sich nicht um echte mehrreihige, sondern um einreihige Walzenschalter mit Teilung der Fingerreihe in zwei oder mehrere Reihen (Mehrfach-einreihige Walzenschalter).   Abgesehen von der geringeren axialen
Länge bringt diese Bauform im allgemeinen keine Vorteile.   Der Rauminhalt der Schaltwalzen ist sogar größer als derjenige der einreihigen Form,
weil der Walzenquerschnitt bei n Reihen auf das n^2-fache wächst, während
die Länge nur auf ein n-tel zurückgeht.   Eine Ersparnis an Walzenfläche
und Rauminhalt läßt sich nur dann erzielen, wenn die Walze in einzelnen
Stellungen brückenlos wird; diese Stellungen können dann ausgelassen
werden.   Ein Beispiel hierfür zeigt Abb. 97, die einen Umschalter für vier
Strompfade in einreihiger und zweireihiger Form wiedergibt.   Durch Einfügung zweier leerer Fingerplätze läßt sich der Schalter mit ausschließlich
blankem Schaltbelag entsprechend Abb. 97 a bauen.   Bei Übergang auf
zwei Fingerreihen bleiben die Stellungen 3 und 4 der ersten Walzenhälfte
unbenützt, wie aus Abb. 97 b hervorgeht.   Diese beiden Stellungen können

daher gestrichen werden, so daß der Schalter außer den vier wirksamen Stellungen nur zwei Sperrstellungen gemäß Abb. 97 c erhält. Ein Vergleich mit der einreihigen Baumform zeigt eine Verminderung der Länge im Verhältnis 4 : 7 und eine Vergrößerung des Walzenumfanges von 4 auf 6 Stellungen. Die Walzenflächen der beiden Formen verhalten sich

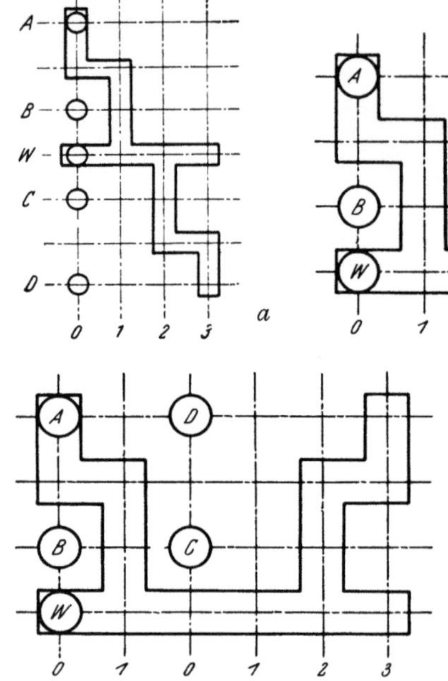

Abb. 97. Bauformen eines Walzenumschalters für vier Strompfade. a Einreihige Form, b Zweireihige Form mit Leerstellungen, c Zweireihige Form ohne Leerstellungen

daher wie 6 : 7 zu Gunsten der zweireihigen Form. Zusammenfassend können wir festhalten:

Schaltregel für mehrfach-einreihige Walzenschalter. Unabhängig vom jeweiligen Schaltprogramm läßt sich jeder einreihige Walzenschalter dadurch in einen mehrreihigen verwandeln, daß die Schaltfinger und die zugehörigen axialen Brücken auf mehrere Teile des Walzenumfanges aufgeteilt werden. Die Zahl der Stellungen, die bei der mehrreihigen Bauart vorhanden sein müssen, ist gleich der Summe der benützten Stellungen der einzelnen Teilprogramme. Der Schrittwinkel vermindert sich im Verhältnis den Stellungszahlen; die durch die Vergrößerung der Stellungszahl hinzugekommenen Schaltstellungen müssen mechanisch gesperrt werden. Die Vereinigung der axialen Brücken zu Schaltstücken läßt sich genau wie bei einreihigen Walzenschaltern durchführen. Schaltbrücken, die Finger verschiedener Reihen verbinden, müssen isoliert sein, soferne sie nicht an den Rand der Schaltwalze verlegt werden können.

3. Die Drehschalter mit mehrfach benützten axialen Schaltbrücken (mehrreihige Walzenschalter)

Der besondere Vorzug der mehrreihigen Bauart liegt jedoch darin, daß sich häufig einzelne Brücken zusammenlegen, das heißt für je zwei oder mehrere Fingerverbindungen verwenden lassen. Hierdurch vermindert sich nicht nur die Baulänge und Oberfläche der Schaltwalze, sondern auch die Anzahl der Brücken. Voraussetzung hierfür ist allerdings eine besondere Gestalt des Schaltprogrammes. Dieses muß so beschaffen sein, daß es sich ganz oder teilweise in einzelne Teilprogramme zerlegen läßt, deren Verbindungen gegeneinander um eine beliebige, aber stets gleich bleibende Zahl von Schaltschritten versetzt sind. In der Verbindungsmatrix zeigt sich diese Eigenschaft dadurch an, daß sich die Stellungsnummern der Brücken in die Felder einer beliebigen Nebendiagonale einordnen lassen. Eine Matrix, in der nur die Hauptdiagonale und die n-te Nebendiagonale benützt ist, läßt sich stets in eine Summe von n Teilmatrizen transformieren. Die Summanden bilden dann je eine Fingerreihe mit zugehörigen Brücken ab. Diese stehen miteinander nicht in Verbindung und können als getrennte einreihige Walzenschalter mit gemeinschaftlichem Antrieb aufgefaßt werden. Die Brücken dieser Teilschalter lassen sich zusammenlegen, wenn sich durch Permutation der Fingerfolge Teilmatrizen mit gleichen Intervallen der Stellungsnummern ergeben. Hierfür ein Beispiel:

$$
\begin{Vmatrix}
A & . & 1 & . & . & . \\
  & B & . & 4 & . & . \\
  &   & C & . & 3 & . \\
  &   &   & D & . & 2 \\
  &   &   &   & E & . \\
  &   &   &   &   & F
\end{Vmatrix} =
$$

$$
= \begin{Vmatrix}
A & 1 & . \\
  & C & 3 \\
  &   & E
\end{Vmatrix} +
\begin{Vmatrix}
B & 4 & . \\
  & D & 2 \\
  &   & F
\end{Vmatrix} =
$$

$$
= \begin{Vmatrix}
A & 1 & . \\
  & C & 3 \\
  &   & E
\end{Vmatrix} +
\begin{Vmatrix}
F & 2 & . \\
  & D & 4 \\
  &   & B
\end{Vmatrix}
$$

Die Stellungsnummern 1 und 3 der ersten Teilmatrix gehören zu zwei Brücken, die doppelt benützt sind und denen in der zweiten Teilmatrix die Nummern 2 und 4 zukommen. Abb. 98 a zeigt die der vorstehenden Rechnung entsprechende Ausführung des Schalters. Eine Vereinigung der beiden Brücken zu einem gemeinsamen Schaltstück ist in diesem Falle nicht möglich; das hierzu nötige Ringsegment würde eine unzulässige Verbindung der beiden Schaltfinger C und D bewirken. Man erkennt unmittelbar aus der zeichnerischen Darstellung, daß die tangentiale Erstreckung der Schaltstücke kleiner sein muß, als der Abstand der beiden

Fingerreihen. Im vorliegenden Falle beträgt dieser Abstand nur einen Schaltschritt, so daß überhaupt keine Tangentialverbindungen zur Vereinigung von Schaltbrücken möglich sind. Bei mehrreihigen Walzenschaltern lassen sich aus diesem Grunde Schaltprogramme mit geschlossener Verbindungskette nur durch isolierte Schaltbrücken verwirklichen. Der Ersatz durch Tangentialverbindungen würde Schaltstücke verlangen, die sich über den ganzen Umfang erstrecken, was jedoch bei mehr als einer Fingerreihe nicht mehr zulässig ist. Abb. 98 *b* zeigt ein Beispiel für einen zweireihigen Walzenschalter mit einer isolierten Brücke. Der Schaltbelag entspricht der nachstehenden Matrix:

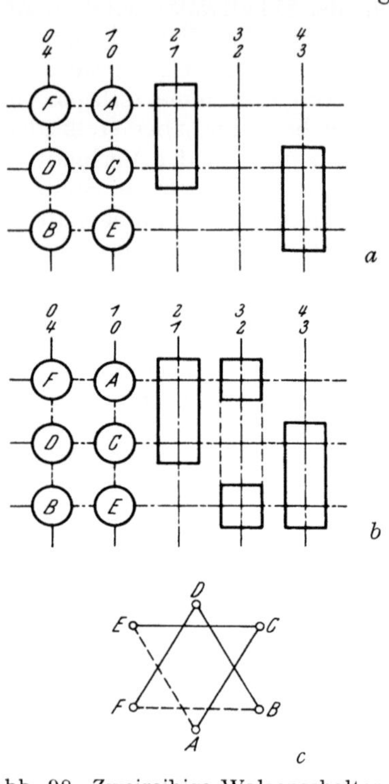

$$
\begin{Vmatrix}
A & . & 1 & . & 2 & . \\
  & B & . & 4 & . & 3 \\
  &   & C & . & 3 & . \\
  &   &   & D & . & 2 \\
  &   &   &   & E & . \\
  &   &   &   &   & F
\end{Vmatrix} =
$$

$$
= \begin{Vmatrix}
A & . & 1 & . & . & . \\
. & B & . & 4 & . & . \\
. & . & C & . & 3 & . \\
. & . & . & D & . & 2 \\
2 & . & . & . & E & . \\
3 & . & . & . & . & F
\end{Vmatrix} =
$$

$$
= \begin{Vmatrix} A & 1 & 2 \\ & C & 3 \\ & & E \end{Vmatrix} + \begin{Vmatrix} F & 2 & 3 \\ & D & 4 \\ & . & B \end{Vmatrix}
$$

Abb. 98. Zweireihige Walzenschalter mit gemeinsamen Schaltbrücken. *a* Für zwei offene Verbindungsketten, *b* Für zwei geschlossene Verbindungsketten, *c* Teilkomplexe des Schaltprogrammes

Stellt man das Schaltprogramm in Form eines Streckenkomplexes dar, so besteht dieser aus zwei getrennten Teilkomplexen, die den beiden Teilmatrizen der Rechnung entsprechen. Abb. 98 *c* zeigt diese Form des Komplexes; die punktierten Verbindungen gehören zu den isolierten Verbindungen der zweireihigen Schaltwalze.

Meistens ist das Schaltprogramm nicht in einer Form vorgegeben, die unmittelbar zu einem mehrreihigen Walzenschalter führt. Es ist dann vorerst zu untersuchen, ob sich das Programm in eine geeignete äquivalente Form transformieren läßt. Ein Hilfsmittel hierzu, das fast immer zum Ziele führt, ist die Vervielfachung einzelner Schaltfinger; entsteht hierbei eine Kontaktbahn, die ausschließlich äquivalente Finger enthält, so können diese zusammengelegt werden, wobei an die Stelle der äußeren Fingerverbindung ein tangentiales Schleifsegment (Schleifring) auf der Walze tritt. Als Beispiel diene ein zweireihiger Umschalter für vier Strompfade. Seine Verbindungsmatrix lautet:

$$
\left\|
\begin{array}{ccccc}
W & 0 & 1 & 2 & 3 \\
A & . & . & & . \\
B & & . & & . \\
C & & & . & . \\
D & & & &
\end{array}
\right\|
$$

Unter zweimaliger Benützung des Wurzelanschlusses W läßt sich die Matrix in eine Summe von zwei Teilmatrizen verwandeln:

$$
\left\|
\begin{array}{ccc}
W & 0 & 1 \\
A & & . \\
B & &
\end{array}
\right\|
+
\left\|
\begin{array}{ccc}
W & 2 & 3 \\
C & & . \\
D & &
\end{array}
\right\|
$$

Die beiden Teilmatrizen zeigen gleiche Intervalle der Stellungsnummern und entsprechen nach Herstellung der geeigneten Fingerfolge einem zweireihigen Walzenschalter.

$$
\left\|
\begin{array}{ccc}
A & 0 & . \\
W & 1 & \\
B & &
\end{array}
\right\|
+
\left\|
\begin{array}{ccc}
C & 2 & . \\
W & 3 & \\
D & &
\end{array}
\right\|
$$

Die mittlere Zeile der beiden Matrizen enthält ausschließlich den Wurzelanschluß W; der Schalter kann daher mit einem Finger W und einem zugehörigen Schleifring ausgeführt werden, wie aus Abb. 99 a hervorgeht. In welcher Fingerreihe der Schaltfinger W angebracht wird, ist hierbei belanglos; wählt man eine Reihe A, W, B entsprechend Abb. 99 a, so bleibt in der anderen Fingerreihe C, D ein freier Fingerplatz in der Kontaktbahn des Schaltfingers W übrig.

Die vorgegebene Verbindungsmatrix erlaubt übrigens noch eine zweite Art der Zerlegung, und zwar:

$$
\left\|
\begin{array}{ccc}
A & 0 & . \\
W & 2 & \\
C & &
\end{array}
\right\|
+
\left\|
\begin{array}{ccc}
B & 1 & . \\
W & 3 & \\
D & &
\end{array}
\right\|
$$

Abb. 99 b zeigt diese zweite Form des Schaltbelages, bei der die beiden Fingerreihen benachbart sind. Mit Rücksicht auf den Platzbedarf der Schaltfinger bevorzugt man im allgemeinen solche Bauformen, bei denen der Abstand der Fingerreihen mehrere Schaltschritte beträgt, wie es bei der Ausführung nach Abb. 99 a der Fall ist.

Die Zerlegung des Schaltprogrammes in mehrere gleichartige Teilprogramme läßt sich auch mit Hilfe des in einen Streckenkomplex verwandelten Schaltplanes durchführen, wie in Abb. 99 c wiedergegeben ist.

Die Möglichkeiten zur Verbindung von Schaltbrücken zu größeren Schaltstücken lassen sich jedoch an Hand der Matrix deutlicher erkennen, so daß das rechnerische Verfahren dem zeichnerischen überlegen ist.

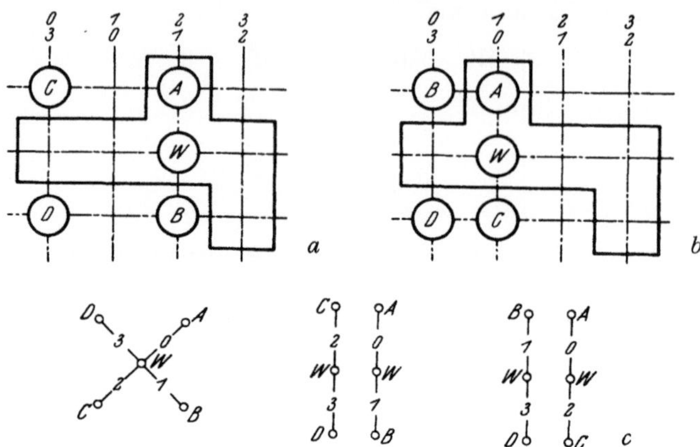

Abb. 99. Zweireihiger Walzenschalter für vier Strompfade. *a* Mit gegenüberliegenden Fingerreihen, *b* Mit benachbarten Fingerreihen, *c* Zerlegungen des Schaltprogrammes

Das besondere Anwendungsgebiet der mehrreihigen Walzenschalter sind die mehrpoligen und vielstelligen Umschalter. Als Beispiel diene ein Wendeschalter für einen Reihenschlußmotor, dessen Schaltprogramm aus dem Stromlaufplan gemäß Abb. 100 *a* zu entnehmen ist. Als Matrix angeschrieben, lautet das Programm:

$$\begin{Vmatrix} Q_1 & 1 & 0 & . \\ & A_1 & . & 0 \\ & & A_2 & 1 \\ & & & F_1 \end{Vmatrix}$$

entsprechend einem Streckenkomplex mit vier Punkten zweiter Ordnung nach Abb. 100 *b*. Wie bereits an Hand der Abb. 90 *a* bis *d* gezeigt wurde, läßt sich ein solches Schaltprogramm bei einreihiger Bauweise

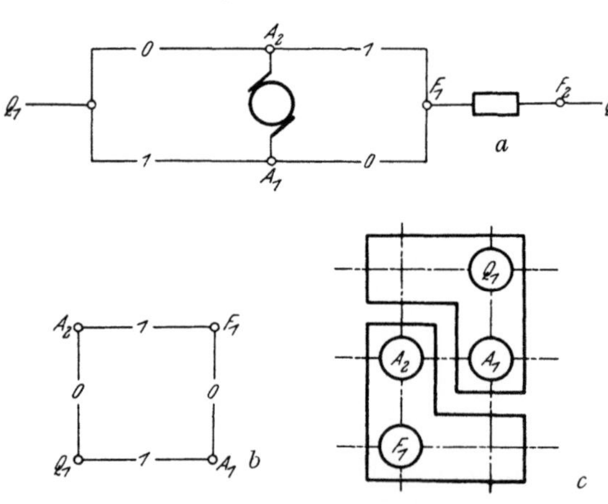

Abb. 100. Zweireihiger Wendeschalter. *a* Stromlaufplan, *b* Schaltprogramm, *c* Schaltbelag

des Walzenschalters nur mit einer isolierten Brücke oder mit zusätzlichen Schaltfingern oder Totstellungen verwirklichen. Die zweireihige Bauform ermöglicht jedoch einen durchwegs blanken Schaltbelag ohne

Vermehrung von Fingern oder Stellungen. Die Zerlegung der vorgegebenen Verbindungsmatrix in zwei Summanden ergibt:

$$\begin{Vmatrix} Q_1 & 1 & . \\ A_1 & 0 & \\ & F_1 & \end{Vmatrix} + \begin{Vmatrix} Q_1 & 0 & . \\ A_2 & 1 & \\ & F_1 & \end{Vmatrix}$$

Da bei einem zweistelligen Schalter die Nullstellung mit der Stellung 2 identisch ist, folgt daraus weiters:

$$\begin{Vmatrix} Q_1 & 1 & . \\ A_1 & 0 & \\ & F_1 & \end{Vmatrix} + \begin{Vmatrix} Q_1 & 2 & . \\ A_2 & 1 & \\ & F_1 & \end{Vmatrix}$$

Es ergibt sich somit eine Ausführung nach Abb. 100 c mit zwei blanken Schaltstücken, zwei Fingerreihen und drei Kontaktbahnen.

Der an Hand der Abb. 87 ausgemittelte Heizkörperumschalter soll als zweireihiger Walzenschalter ausgeführt werden. Zu diesem Zweck untersuchen wir, ob sich Brückenpaare finden lassen, deren Stellungsnummern durch Addition einer gleichbleibenden Zahl ineinander überführen lassen. Das Schaltprogramm wurde bereits abgeleitet und lautet:

$$A_1 - 1, 2, 4 - Q_1 - 3, 4 - B_1 - 1 - A_2 - 2, 4 - Q_2 - 1, 3, 4 - B_2.$$

Wir beginnen bei der Brücke $A_1 Q_1$ mit den Stellungsnummern 1, 2, 4 und addieren der Reihe nach die Zahlen 1 bis 4 als mögliche Anzahl der Schaltschritte zwischen den beiden Fingerreihen. Da der Schalter fünf Stellungen aufweist, ist seine Stellung 5 mit der Nullstellung identisch, so daß die Nummer 5 durch 0 ersetzt werden kann. Es ergibt sich dann die nachstehende Permutation der Stellungsnummern.

1	2	4
2	3	0
3	4	1
4	0	2
0	1	3

Der Vergleich dieser Zahlen mit den Stellungsnummern der übrigen Brücken zeigt, daß die zweite Permutation, die 3, 4, 1 lautet, zur Brücke $Q_2 B_2$ gehört. Wir können also zwei Fingerreihen im Abstand von zwei Schaltschritten anordnen, wobei sich für die Fingerpaare $A_1 Q_1$ und $Q_2 B_2$ die gleichen Brücken ergeben. Führen wir die gleiche Rechnung für die restlichen Brücken durch, so zeigt sich, daß weitere Zusammenlegungen von Brücken nicht mehr möglich sind. Der Schalter kann also nur teilweise zweireihig gebaut werden. Die angeführten Fingerpaare, die zum zweireihigen Teil gehören, müssen gemeinsame Schaltbahnen erhalten. In Abb. 101 a sind diese Bahnen strichpunktiert in das Schaltprogramm eingetragen. Die Figur läßt bereits die bauliche Anordnung des Schalters erkennen.

Die erste und letzte Schaltbahn gehört zum einreihigen Teil, die beiden mittleren Bahnen zum zweireihigen Teil des Gerätes. Die Verbindung $A_2 B_1$ muß als isolierte Brücke ausgebildet werden, wenn man nicht weitere Schaltbahnen und Finger verwenden will. Vergleicht man diese Ausführung nach Abb. 101 *b* mit der einreihigen Bauart gemäß Abb. 87 *c*, so zeigt sich eine Verminderung der Schaltbahnen von 6 auf 4 bei gleicher Anzahl der Schaltstellungen; Oberfläche und Volumen des Gerätes sind um ein Drittel kleiner geworden.

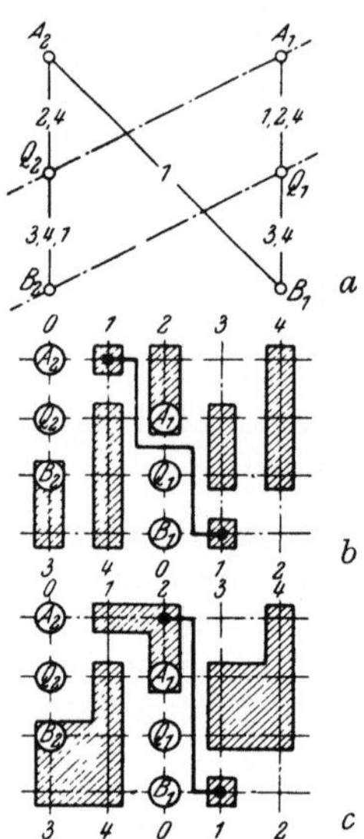

Abb. 101. Teilweise zweireihiger Schalter. *a* Schaltprogramm, *b* Anordnung mit getrennten Brücken, *c* Anordnung mit vereinigten Brücken

Auch bei mehrreihigen Walzenschaltern lassen sich die Schaltbrücken teilweise durch Ringverbindungen zu größeren Schaltstücken vereinigen. Hierfür gelten die gleichen Regeln wie für Axialbrücken einreihiger Schalter. Eine zusätzliche Beschränkung besteht nur insoferne, als die Ausdehnung eines Schaltstückes in der Richtung des Umfanges kleiner sein muß als der Abstand der Fingerreihen; anderenfalls käme durch das Schaltstück eine Verbindung von Fingern verschiedener Brücken zustande. Abb. 101 *c* gibt ein Beispiel für einen Walzenschalter mit verbundenen Brücken.

Besondere Bedeutung hat die mehrreihige Bauform bei Walzenanlassern. Diese sind unterbrechungslos wirkende Umschalter mit einem Schaltprogramm, wie es in Abb. 102 *a* wiedergegeben ist. Ein Programm dieser Art gestattet es, das ganze Gerät mehrreihig auszuführen und damit die Abmessungen ganz wesentlich zu verkleinern. Bei der Zerlegung des Schaltprogrammes in mehrere Teilprogramme muß der Wurzelpunkt des Umschalters mehrfach wiederholt werden, wie Abb. 102 *b* zeigt. Da jedoch sämtliche Wurzelpunkte der gleichen Schaltbahn angehören, genügt die Anordnung eines gemeinschaftlichen Wurzelfingers und einer Schleifbahn auf der Walze. Sehr anschaulich erscheint diese konstruktive Möglichkeit, wenn die Zerlegung des Schaltprogramms an Hand einer Matrix durchgeführt wird.

$$
\left\|\begin{array}{ccccccc}
W & 1,2 & 2,3 & 3,4 & 4,5 & 5,6 & 6 \\
A & 2 & . & . & . & . & . \\
 & B & 3 & . & . & . & . \\
 & & C & 4 & . & . & . \\
 & & & D & 5 & . & \\
 & & & & E & 6 & \\
 & & & & & F &
\end{array}\right\| =
$$

$$= \left\| \begin{array}{ccc} W & 1,2 & 2,3 \\ & A & 2 \\ & B & \end{array} \right\| + \left\| \begin{array}{ccc} W & 3,4 & 4,5 \\ & C & 4 \\ & D & \end{array} \right\| + \left\| \begin{array}{ccc} W & 5,6 & 6 \\ & E & 6 \\ & F & \end{array} \right\|$$

Damit die Bedingung für die Formgleichheit der drei Summanden erfüllt ist, muß in der letzten Teilmatrix im Feld der Verbindung WF statt der Stellungsnummer 6 die Nummerngruppe 6, 7 gesetzt werden; hieraus geht hervor, daß der Schalter bei mehrreihiger Ausführung eine zusätzliche Stellung haben muß. Diese achte Stellung ist nicht durch die geforderten Schaltzustände, sondern durch die Bauform bedingt.

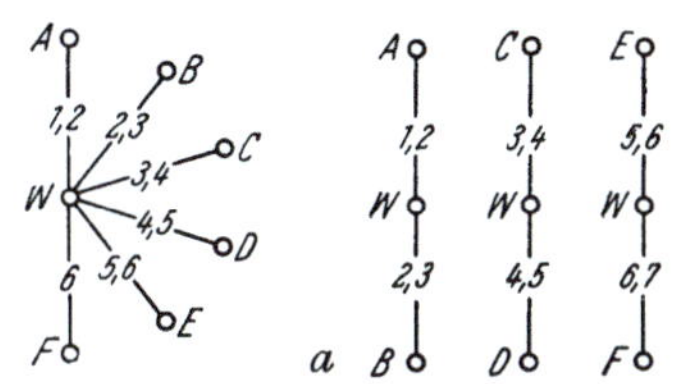

Die Ersparnis, die mit der mehrreihigen Ausführung verbunden ist, geht aus Abb. 102 hervor; Walzenfläche und Rauminhalt der einreihigen Bauform nach Abb. 102 c sind mehrfach größer als bei der mehrreihigen Form nach Abb. 102 d. Die Ersparnis ist um so größer, je mehr Fingerreihen vorhanden sind. Der Abstand und damit die Zahl der Fingerreihen ist durch die Fingerkonstruktion, und zwar durch die Länge der Finger in der Umfangsrichtung des Schalters bestimmt. In der Regel läßt

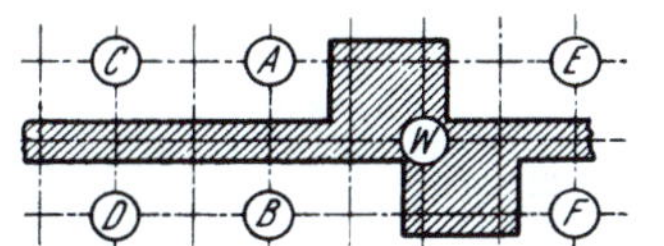

Abb. 102. Walzenanlasser für Gleichstrommaschinen. *a* Schaltprogramm, *b* Zerlegung des Programmes, *c* Einreihige Form (Füllfaktor 49%), *d* Mehrreihige Form (Füllfaktor 88%)

sich der Reihenabstand ohne Schwierigkeiten auf zwei bis drei Schrittteilungen **herabsetzen; bei vielstelligen** Anlassern ergibt sich hierdurch eine große Zahl von Fingerreihen und eine außerordentliche Ersparnis an Walzenfläche.

Als weiteres Beispiel sei die Ausmittlung einer Anlaßwalze eines Reihenschlußmotors für Drehrichtungsumkehr gezeigt (Abb. 103). Der Einfachheit halber sei angenommen, daß der Anlaßwiderstand nur vier Stufen aufweisen soll. Da die Umpolung des Ankers und die Schaltung des Widerstandes nicht miteinander zusammenhängen, ergeben sich für das Gerät zwei getrennte Teile, deren Determinanten vorerst gesondert angeschrieben und untersucht werden können. Gemeinsam ist die Gesamtzahl der Schalterstellungen. Diese ergibt sich aus den vier Stufen des Widerstandes; es sind je fünf Stellungen für die beiden Drehrichtungen sowie eine Ausschaltstellung, insgesamt also elf Stellungen notwendig.

13*

Wir bezeichnen die Ausschaltstellung sinngemäß mit Null und die Vorwärtsstellungen mit 1 bis 5; die Rückwärtsstellungen haben dann die Stellungsnummern $11 - (1 \text{ bis } 5) = 6 \text{ bis } 10$. Die Ankerumschaltung betrifft vier Pole, nämlich die zwei Ankerbürsten A_1, A_2 und die Anschlüsse F_1 und W_1 an Feld und Widerstand. Das Schaltprogramm läßt sich sofort als Matrix anschreiben.

$$\left\| \begin{array}{lll} A_1 & 1 \text{ bis } 5 & 6 \text{ bis } 10 \\ A_2 & 6 \text{ bis } 10 & 1 \text{ bis } 5 \\ F_1 & & \\ & W_1 & \end{array} \right\|$$

Die Schaltung läßt sich in zwei Teilmatrizen zerlegen, wenn man entweder die Fingerpaare F_1, W_1 oder die Paare A_1, A_2 doppelt anschreibt. Wir wählen von diesen beiden symmetrischen Lösungen willkürlich die erstere:

$$\left\| \begin{array}{lll} F_1 & 1 \text{ bis } 5 & . \\ A_1 & 6 \text{ bis } 10 & \\ & W_1 & \end{array} \right\| + \left\| \begin{array}{lll} F_1 & 6 \text{ bis } 10 & . \\ A_2 & 1 \text{ bis } 5 & \\ & W_1 & \end{array} \right\|$$

Die beiden Teilmatrizen stellen noch keine zweireihige Schaltwalze dar; zwischen den Verbindungen zu F_1 liegen $(6 - 1) = 5$ Schaltschritte, zwischen denen zu W_1 jedoch $(11 + 1 - 6) = 6$ Schritte. Die Gleichheit läßt sich herstellen, wenn wir das Gerät mit 12 Stellungen ausführen. Dann erhalten die Rückwärtsstellungen die Nummern 7 bis 11 und die Teilmatrizen lauten:

$$\left\| \begin{array}{lll} F_1 & 1 \text{ bis } 5 & \\ A_1 & 7 \text{ bis } 11 & \\ & W_1 & \end{array} \right\| + \left\| \begin{array}{lll} F_1 & 7 \text{ bis } 11 & \\ A_2 & 1 \text{ bis } 5 & \\ & W_1 & \end{array} \right\|$$

Jetzt beträgt die Stellungsdifferenz einheitlich sechs Schritte als Reihenabstand, womit die erste Teilaufgabe als gelöst erscheint. Die zweireihige Bauart dieses Walzenteiles erspart somit in axialer Richtung eine Fingerteilung. Dafür erscheinen in der Determinante zwei Finger mehr als bei einreihiger Bauart. Dies läßt sich vermeiden, wenn man die äußeren Verbindungen der aufdeckenden Fingerpaare F_1 bzw. W_1 durch Schleifringe auf der Walze ersetzt. Allerdings ist das Gerät dann kein reiner Walzenschalter mehr, weil Finger verschiedener Reihen miteinander verbunden werden; da jedoch keine schaltenden Reihenbrücken, sondern nur Schleifringe als Ersatz für Finger vorhanden sind, werden auch diese Bauformen als Walzenschalter bezeichnet. In der Determinante zeigt sich diese Ausführung folgendermaßen:

$$\left\| \begin{array}{llll} F_1 & 1 \text{ bis } 5 & & 1 \text{ bis } 11 \\ A_1 & 7 \text{ bis } 11 & & \\ & W_1 & & \\ & & (F_1) & 7 \text{ bis } 11 \\ & & & 1 \text{ bis } 11 \\ & & A_2 & 1 \text{ bis } 5 \\ & & (W_1) & \end{array} \right\|$$

Durch diese Maßnahme erhalten wir die verkürzte zweireihige Bauart
ohne Vermehrung der Finger. Der zugehörige Schaltplan ist aus Abb. 103 *b*
zu entnehmen.

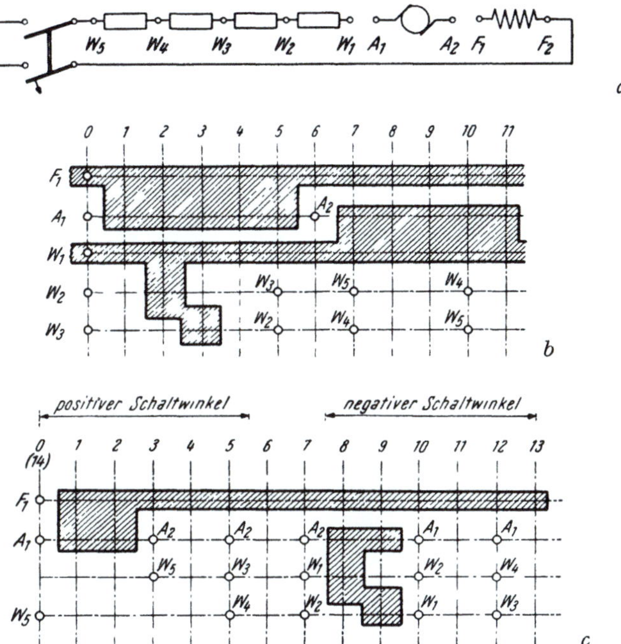

Abb. 103. Umkehranlasser für einen Reihenschlußmotor als Beispiel eines mehrreihigen Walzenschalters. *a* Bezeichnung der Anschlüsse, *b* Zweiteiliger Schalter
mit kleinster Baulänge jedes Schalterteiles, *c* Einteiliger Schalter kleinster Baulänge

Die Einschaltung des ganzen Anlaßwiderstandes ist bereits im Schaltspiel der **Ankerumpolung** enthalten, da die Ankerbürste in den Stellungen
$\pm\,1$ an das Ende W_1 des Gesamtwiderstandes gelegt wird. Der restliche
Teil des Schalters hat demnach nur den Widerstand schrittweise kurz-
zuschließen. Es ist also in Stellung $\pm\,2$ das Ende W_1 mit der Anzapfung
W_2 zu verbinden und so fort. Die Matrix lautet daher:

$$\left\|\begin{array}{ccccc} W_1 & 2,\,10 & 3,\,9 & 4,\,8 & 5,\,7 \\ & W_2 & \cdot & \cdot & \cdot \\ & & W_3 & \cdot & \cdot \\ & & & W_4 & \cdot \\ & & & & W_5 \end{array}\right\|$$

Durch wiederholtes Anschreiben des Fingers W_1 könnten wir acht Teilmatrizen mit je einer Verbindung bilden, was einem achtreihigen Walzenschalter entspräche. Die axiale Länge wäre dann mit zwei Fingerteilungen

ein absolutes Minimum. Der Reihenabstand betrüge dann aber nur einen Schaltschritt, was im allgemeinen zur Unterbringung der Finger nicht ausreicht. Wir wollen daher zwischen den Reihen zwei Schritte annehmen und das Gerät mit nur vier Fingerreihen versehen. Daraus folgt die Schaltung:

$$
\text{Reihe 0 (12)} \qquad \text{Reihe 10} \qquad \text{Reihe 7} \qquad \text{Reihe 5}
$$

$$
\begin{Vmatrix} W_1 & 2 & 3 \\ & W_2 & \cdot \\ & & W_3 \end{Vmatrix} + \begin{Vmatrix} (W_1) & 4 & 5 \\ & W_4 & \cdot \\ & & W_5 \end{Vmatrix} + \begin{Vmatrix} (W_1) & 7 & 8 \\ & W_5 & \cdot \\ & & W_4 \end{Vmatrix} + \begin{Vmatrix} (W_1) & 9 & 10 \\ & W_3 & \cdot \\ & & W_2 \end{Vmatrix}
$$

Die aufdeckenden Finger (W_1) lassen sich wieder durch einen Schleifring ersetzen, so daß sich insgesamt neun Finger ergeben. Gegenüber der einreihigen Bauart mit fünf Fingern wurde die axiale Länge um zwei Fingerteilungen vermindert.

Durch Zusammensetzen der beiden Schaltungsteile erhält man einen Mehrfachschalter, dessen Teile für sich minimale Baulänge besitzen. Dabei zeigt sich, daß zwei Schleifringe mit je einem Finger W_1 nebeneinander kommen. Man kann daher sowohl die Schleifringe wie auch die Finger zusammenlegen und erspart so einen Finger und eine Teilung in der Länge. Abb. 103 *b* zeigt den Schaltplan des mehrteiligen Walzenschalters.

Die Ausführung als zweiteilige Walze mit Teilen minimaler Länge führt jedoch nicht zum absoluten Minimum der Baulänge. Um diese zu erhalten, hat man aus der Gesamtschaltung einen einteiligen Walzenschalter zu entwickeln. Die vorhergehende Untersuchung der Teilschaltungen ist jedoch zweckmäßig, weil sie die erforderlichen Vermehrungen der Stellungen und Finger aufzeigt und dadurch die Zerlegung der Gesamtschaltung in Teilmatrizen erleichtert. Im vorliegenden Falle ergab sich bereits, daß wir der Matrix nicht 11 sondern 12 Schalterstellungen zugrunde legen müssen. Auch geben die Teilmatrizen der Teilschaltungen schon einen Hinweis auf die Zerlegung in Fingerreihen. Es ergibt sich daher ohne weiteres:

$$
\begin{Vmatrix}
F_1 & 1\text{–}5 & 7\text{–}11 & \cdot & \cdot & \cdot & \cdot & \cdot \\
& A_1 & \cdot & 11 & 10 & 9 & 8 & 7 \\
& & A_2 & 1 & 2 & 3 & 4 & 5 \\
& & & W_1 & \cdot & \cdot & \cdot & \cdot \\
& & & & W_2 & \cdot & \cdot & \cdot \\
& & & & & W_3 & \cdot & \cdot \\
& & & & & & W_4 & \cdot \\
& & & & & & & W_5
\end{Vmatrix} =
$$

$$
\text{Reihe 0 (12)} \qquad\qquad \text{Reihe 10} \qquad\qquad \text{Reihe 8}
$$

$$
= \begin{Vmatrix} F_1 & 1,2 & \cdot & \cdot \\ & A_1 & 7 & 8 \\ & & W_5 & \cdot \\ & & & W_4 \end{Vmatrix} + \begin{Vmatrix} F_1 & 3,4 & \cdot & \cdot \\ & A_1 & 9 & 10 \\ & & W_3 & \cdot \\ & & & W_2 \end{Vmatrix} + \begin{Vmatrix} F_1 & 5,6 & \cdot & \cdot \\ & A_1 & 11 & 0 \\ & & W_1 & \cdot \\ & & & 0 \end{Vmatrix} +
$$

$$
+\begin{Vmatrix}\text{Reihe 7}\\ F_1 \quad 6,7 \quad . \\ A_2 \quad 0 \quad 1 \\ \quad 0 \quad . \\ \quad\quad W_1\end{Vmatrix}
+\begin{Vmatrix}\text{Reihe 5}\\ F_1 \quad 8,9 \quad . \quad . \\ A_2 \quad 2 \quad 3 \\ \quad W_2 \quad . \\ \quad\quad W_3\end{Vmatrix}
+\begin{Vmatrix}\text{Reihe 3}\\ F_1 \quad 10,11 \quad . \quad . \\ A_2 \quad 4 \quad 5 \\ \quad W_4 \quad . \\ \quad\quad W_5\end{Vmatrix}
$$

Die so gefundene Schaltung enthält wieder Fingerreihen im Abstand eines Schaltschrittes und weiters noch ausgeschlossene Stellungen. Wir führen daher zwei weitere Stellungen ein und erhalten die Konstruktionsmatrizen:

$$
\begin{Vmatrix}\text{Reihe 0 (14)}\\ F_1 \quad 1,2 \quad . \quad . \\ A_1 \quad (8) \quad 9 \\ \quad 0 \quad . \\ \quad\quad W_5\end{Vmatrix}
+\begin{Vmatrix}\text{Reihe 12}\\ F_1 \quad 3,4 \quad . \quad . \\ A_1 \quad 10 \quad 11 \\ \quad W_4 \quad . \\ \quad\quad W_3\end{Vmatrix}
+\begin{Vmatrix}\text{Reihe 10}\\ F_1 \quad 5,6 \quad . \quad . \\ A_1 \quad 12 \quad 13 \\ \quad W_2 \quad . \\ \quad\quad W_1\end{Vmatrix}+
$$

$$
+\begin{Vmatrix}\text{Reihe 7}\\ F_1 \quad 8,9 \quad . \quad . \\ A_2 \quad 1 \quad 2 \\ \quad W_1 \quad . \\ \quad\quad W_2\end{Vmatrix}
+\begin{Vmatrix}\text{Reihe 5}\\ F_1 \quad 10,11 \quad . \quad . \\ A_2 \quad 3 \quad 4 \\ \quad W_3 \quad . \\ \quad\quad W_4\end{Vmatrix}
+\begin{Vmatrix}\text{Reihe 3}\\ F_1 \quad 12,13 \quad . \quad . \\ A_2 \quad 5 \quad (6) \\ \quad W_5 \quad . \\ \quad\quad 0\end{Vmatrix}
$$

Den zugehörigen Schaltplan zeigt Abb. 103 c. Es ist nun noch zu untersuchen, ob die aus konstruktiven Gründen vorgesehenen und funktionsfremden Stellungen mechanisch gesperrt werden müssen. Es sind dies die Stellungen (6) und (8). Sie kommen nur je einmal vor als Verbindungen F_1 (6) A_1 und F_1 (8) A_2, so daß sie keinen Stromkreis bilden. Die Schaltung erfordert also keine Begrenzung des Schaltwinkels. Mit Rücksicht auf die Bedienung wird man allerdings eine Sperre vorsehen, die ein Durchdrehen der Walze verhindert.

Abschließend noch ein Vergleich der möglichen Bauformen: Die einreihige Ausführung erfordert ein Minimum an Fingern bei maximaler Länge des Gerätes. Mit zunehmender Zahl der Fingerreihen wird die Walze kürzer und der Belag einfacher; hingegen nimmt die Zahl der Stellungen und der Finger zu. Als anderes Extrem erscheint die einteilige vielreihige Form. Die mehrteiligen Walzen liegen hinsichtlich Baulänge und Aufwand dazwischen. Für das behandelte Beispiel gibt die nachstehende Tabelle eine Zahlenübersicht über die Konstruktionsdaten. Die zweite Zeile bezieht sich auf eine Walze mit zweireihigem Wender und einreihigem Anlasser.

Wie die Tabelle zeigt, liegt das wirtschaftliche Optimum zwischen den Extremen. Die mehrteilige Bauform kommt dem Minimum der Baulänge und Oberfläche schon recht nahe, benötigt aber wesentlich weniger Finger. In der Praxis finden sich daher mehrreihige Schaltwalzen meistens in mehrteiliger Ausführung.

Bauform	Länge in Teilungen	Zahl der Stellungen	Walzenfläche in Teilungen	Zahl der Finger
einreihig	8	11	88	8
teilweise mehrreihig	7	12	88	8
mehrteilig mehrreihig	5	12	60	12
einteilig mehrreihig	4	14	52	17

Die Zweckmäßigkeit der Anordnung hinsichtlich ihres Aufwandes läßt sich durch eine Verhältniszahl in erster Näherung erfassen. Wir legen einen funktionsgleichen einreihigen Walzenschalter als Vergleichsbasis zugrunde und nennen das Verhältnis des Aufwandes ξ. Dann gilt

$$\text{für die Stellungszahl } z \qquad \xi_z = z/z_1,$$
$$\text{für die axiale Länge } a \qquad \xi_a = a/a_1,$$
$$\text{für die Zahl der Kontaktfinger } k \quad \xi_k = k/k_1 = k/a_1.$$

Die Wertungszahl für den Gesamtaufwand ist dann

$$\xi = \xi_z \cdot \xi_a \cdot \xi_k = \frac{z \cdot a \cdot k}{z_1 \cdot a_1{}^2}$$

Hierbei ist a_1 die Zahl der äußeren Anschlußleitungen und z_1 die Stellungszahl des Schaltprogrammes. Der Aufwandszahl liegt demnach keine bestimmte Konstruktion zugrunde, sondern die vorgegebene Schaltaufgabe; der einreihige Walzenschalter hat aber entsprechend seiner Anordnung immer $\xi = 1$.

Für die vier Bauformen des bekannten Beispiels ergibt sich dann:

$$\text{Einreihige Form } \xi_1 = 1.$$
$$\text{Gemischte Form } \xi_2 = 0,95.$$
$$\text{Mehrteilige Form } \xi_3 = 1,2.$$
$$\text{Einteilige Form } \xi_4 = 1,35.$$

Schaltregeln für mehrreihige Walzenschalter. Läßt sich das Schaltprogramm eines Walzenschalters in einzelne Teilprogramme zerlegen, deren Verbindungen gegeneinander um eine beliebige, aber stets gleichbleibende Zahl von Schaltschritten versetzt sind, dann können zur Herstellung analoger Verbindungen dieselben Schaltbrücken verwendet werden. Die entsprechenden Finger müssen in diesem Falle um die betreffende Zahl von Schaltschritten versetzt angeordnet werden. Ist diese Bedingung nicht erfüllt, dann kann ein Programm unter Umständen durch Vervielfachung einzelner Schaltfinger in eine geeignete äquivalente Form transformiert werden. Enthält eine Kontaktbahn nur äquivalente Finger, können diese bis auf einen durch einen Schleifring ersetzt werden. Zeigt nur ein Teil des Programmes die erforderliche Periodizität, kann der

Schalter teilweise mehrreihig ausgeführt werden. Soll das Gerät eine ausgeprägte Nullstellung erhalten, in der der Verbraucher vom Netz völlig getrennt wird, dann dürfen jene Stellen, die in der Nullstellung mit Fingern zusammen kommen, keine störenden Brücken aufweisen.

4. Die Drehschalter mit tangentialen und radialen Schaltbrücken (Reihenschalter und Paketschalter)

Reihenschalter, Hebelumschalter, Flachbahnschalter und Paketschalter haben als gemeinschaftliches Kennzeichen, daß Schaltfinger, die zur selben Schaltbahn gehören, miteinander durch Brücken fallweise verbunden werden. Schaltgeräte dieser Art haben daher stets zwei oder mehr Fingerreihen. Sollen zwei Schaltfinger, deren Abstand nur einen Schaltschritt beträgt, miteinander verbunden werden, so ist es belanglos, ob hierfür Ringbrücken oder Hebelbrücken verwendet

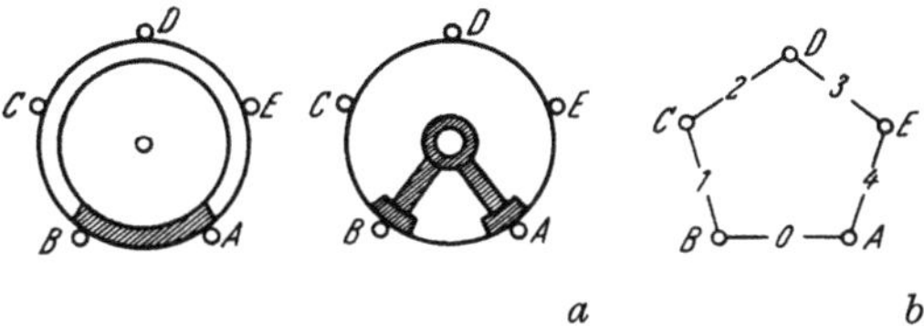

Abb. 104. Brücken zwischen benachbarten Fingern einer Schaltbahn. *a* Brückenformen, *b* Schaltprogramm

werden. Abb. 104 *a* zeigt diese beiden Brückenformen. Das Schaltprogramm entspricht einem zyklischen Komplex zweiter Ordnung nach Abb. 104 *b*; wird das Programm als Matrix angeschrieben, so enthält diese fortlaufende Stellungsnummern in den beiden ersten Nebendiagonalen.

$$\begin{Vmatrix} A & 0 & . & . & 4 \\ 0 & B & 1 & . & . \\ . & 1 & C & 2 & . \\ . & . & 2 & D & 3 \\ 4 & . & . & 3 & E \end{Vmatrix}$$

Sind zwei Finger, die miteinander verbunden werden sollen, um mehr als einen Schaltschritt voneinander entfernt, so müssen entweder isolierte Tangentialbrücken oder Hebelbrücken verwendet werden, wie aus Abb. 105 *a* ersichtlich ist. Das zu diesen Brückenformen gehörige Schaltprogramm ist in Abb. 105 *b* dargestellt; es entspricht ebenfalls einem zyklischen Komplex zweiter Ordnung, jedoch mit geänderter Reihenfolge der Punkte. Schreibt man das Schaltprogramm als Matrix an und wählt die Reihenfolge der Zeilen und Spalten so, daß sie der Fingerfolge der Schaltbahn entspricht, dann erscheinen die Stellungsnummern in jenen Nebendiagonalen, deren Abstand von der Hauptdiagonale gleich dem in Schaltschritten gemessenen Fingerabstand ist. Beispielsweise entspricht den Brückenformen nach Abb. 105 *a* eine Matrix, in der die zweite Nebendiagonale mit Stellungsnummern versehen ist.

$$\begin{Vmatrix} A & . & 0 & 3 & . \\ . & B & . & 1 & 4 \\ 0 & . & C & . & 2 \\ 3 & 1 & . & D & . \\ . & 4 & 2 & . & E \end{Vmatrix}$$

Hat der Schalter z_0 Stellungen, so ist eine Brücke, die z Schaltschritte umfaßt, identisch mit einer solchen, deren Öffnungswinkel $z_0 - z$ Schaltschritte beträgt. In der Matrix müssen daher stets zwei Nebendiagonalen mit Stellungsnummern besetzt sein, was auch aus dem symmetrischen Aufbau dieser Darstellung zwangläufig folgt.

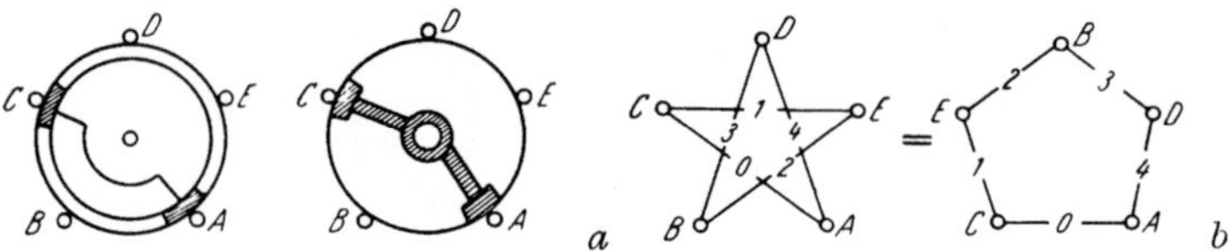

Abb. 105. Brücken zwischen beliebigen Fingern einer Schaltbahn. *a* Brückenformen, *b* Schaltprogramm

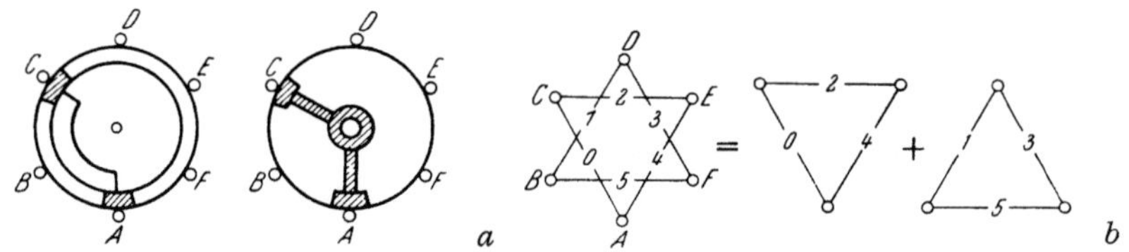

Abb. 106. Brücken zwischen Fingern von Schaltern mit gerader Stellungszahl. *a* Brückenformen, *b* Schaltprogramm

Ist die Zahl z_0 durch z ganzzahlig teilbar, so zerfällt das Schaltprogramm in mehrere zyklische Teilprogramme, die miteinander nicht zusammenhängen. Abb. 106 *b* zeigt als Beispiel einen zyklischen Komplex zweiter Ordnung, der aus sechs Punkten besteht und in zwei getrennte Komplexe zu je drei Punkten zerlegt werden kann. Auch die Matrixrechnung zeigt die Umformung solcher Schaltkomplexe mit gerader Punktezahl in eine Summe aus zwei Matrizen mit halber Anzahl von Zeilen und Spalten.

$$\begin{Vmatrix} A & . & 0 & . & 4 & . \\ & B & . & 1 & . & 5 \\ & & C & . & 2 & . \\ & & & D & . & 3 \\ & & & & E & . \\ & & & & & F \end{Vmatrix} =$$

$$= \begin{Vmatrix} A & 0 & 4 \\ & C & 2 \\ & & E \end{Vmatrix} + \begin{Vmatrix} B & 1 & 5 \\ & D & 3 \\ & & F \end{Vmatrix}$$

Ist z_0 gleich $2\,z$, so entsteht eine Brücke nach Abb. 107 a, die gegenüberliegende Schaltfinger verbindet. Wegen $z = z_0 - z$ fallen in der darstellenden Matrix die beiden mit Stellungsnummern besetzten Hauptdiagonalen zusammen, so daß jedem Feld zwei Stellungsnummern zukommen.

$$\left\|\;\begin{matrix} A & . & . & 0,\,3 & . & . \\ . & B & . & . & 1,\,4 & . \\ . & . & C & . & . & 2,\,5 \\ 0,\,3 & . & . & D & . & . \\ . & 1,\,4 & . & . & E & . \\ . & . & 2,\,5 & . & . & F \end{matrix}\;\right\|$$

Die Brücke stellt während eines Umlaufes jede Verbindung zweimal her. Der zyklische Komplex zweiter Ordnung zerfällt hierbei in z getrennte Komplexe erster Ordnung, wie es in Abb. 107 b gezeigt wird. Schalter mit gestreckten Brücken gestatten es, durch entsprechende äußere Verbindung von Schaltfingern jedes beliebige Schaltprogramm zu verwirklichen und werden daher in der technischen Praxis häufig verwendet, wobei die Stellungszahl z_0 doppelt so groß gewählt wird wie die Zahl der Schaltzustände des Schaltprogramms. Der Entwurf von Schaltgeräten nach diesem Verfahren verursacht zwar wenig Mühe, führt aber zu einem großen Aufwand an Schaltstellungen und Schaltfingern.

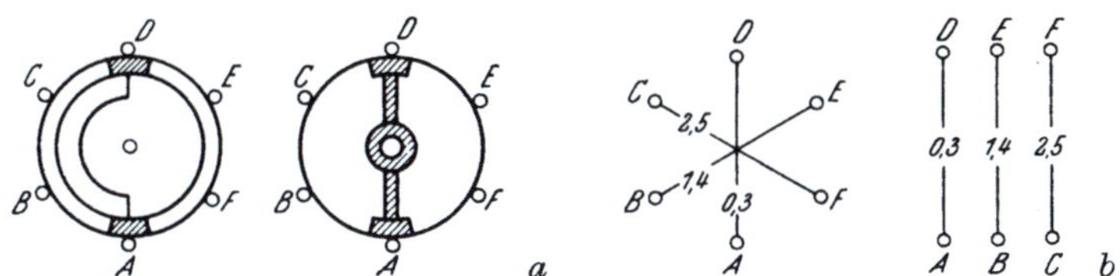

Abb. 107. Brücken zwischen gegenüberliegenden Fingern einer Schaltbahn.
a Brückenformen, b Schaltprogramm

Schaltgeräte mit axialen oder tangentialen Brücken, denen ein einfaches und übersichtliches Schaltprogramm zugrunde liegt, lassen sich mit Hilfe der Streckenkomplexe entwickeln. Als Beispiel soll der Aufbau eines Paketschalters zur fallweisen Verbindung zweier Verbraucher mit zwei Stromquellen abgeleitet werden. Der Schalter soll vier Stellungen haben. In der Nullage darf keine Verbindung zwischen den Stromquellen und den Verbrauchern vorhanden sein. In seiner Stellung 1 soll der Schalter den Netzanschluß $N\,1$ mit beiden Verbrauchern verbinden. In der Stellung 2 ist das Netz $N\,1$ mit dem Verbraucher $V\,1$, sowie das Netz $N\,2$ mit $V\,2$ zusammenzuschalten. In Stellung 3 sollen schließlich beide Verbraucher mit dem Netzanschluß $N\,2$ verbunden werden. Abb. 108 a zeigt dieses Schaltprogramm in Form eines Streckenkomplexes. Durch Eintragen der mittelbaren Verbindungen ergibt sich sofort der zugehörige Funktionskomplex, der in Abb. 108 b wiedergegeben ist. Dieser enthält zwei zyklische Teilkomplexe gemäß Abb. 108 c, die alle sechs notwendigen Verbindungen des Schaltprogrammes enthalten.

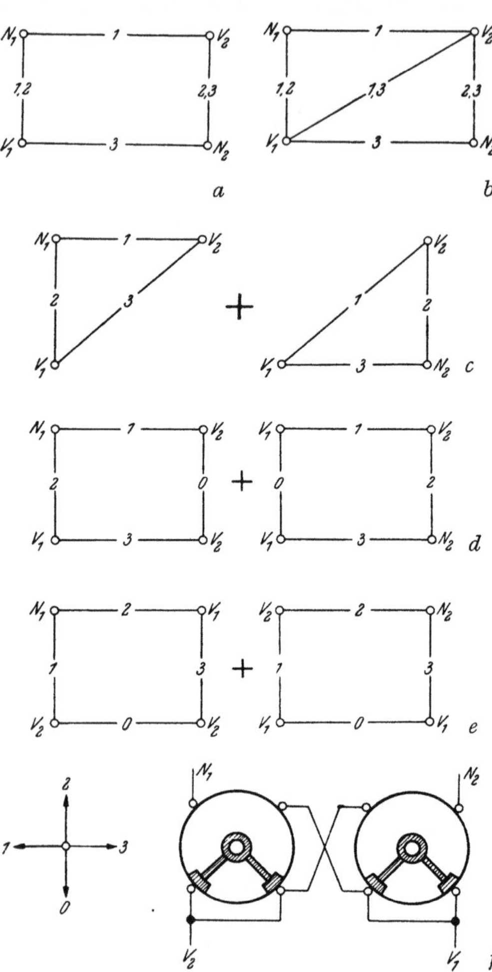

Jedes dieser zyklischen Teilprogramme entspricht einer Schaltbahn des gesuchten Paketschalters. Die Zyklen umfassen je drei Schaltschritte; da das Schaltprogramm vier Schaltzustände vorsieht, müssen sie um eine weitere Strecke, nämlich eine Verbindung in der Nullstellung, erweitert werden. Hierzu besteht eine einfache Möglichkeit, und zwar die Verdoppelung je eines Punktes der beiden Teilkomplexe; diese erhalten durch die Erweiterung eine Form, die in Abb. 108 d dargestellt ist. Um für die Schaltbrücken gleichen Drehsinn zu erhalten, ist noch eine Umformung gemäß Abb. 108 e notwendig; nach dieser zeigen die beiden Schaltzyklen der beiden Teilprogramme die Fingerfolge für die beiden Schaltbahnen des Paketschalters. Man kann nunmehr den Wirkschaltplan des Paketschalters, der in Abb. 108 f wiedergegeben ist, aufzeichnen.

Abb. 108. Netzumschalter für zwei Verbraucher. a Vorgegebenes Schaltprogramm, b Funktionskomplex, c Zerlegung des Schaltprogrammes in zwei unvollständige zyklische Teilprogramme, d Erweiterung des Programmes auf vier Zustände, e Schaltprogramm für zwei gleichsinnig drehende Schaltbrücken, f Schaltplan des Umschalters

Handelt es sich um eine Schaltaufgabe, die ein größeres und weniger einfaches Programm vorschreibt, ist es zweckmäßig, die Matrixrechnung zu Hilfe zu nehmen. Deren praktische Anwendung sei an einem Beispiel erläutert. Drehstrommotoren mit Kurzschlußanker werden gewöhnlich durch Sterndreieckschalter angelassen. Ein solcher Schalter soll als Paketschalter entworfen und für Anlauf des Motors in beiden

Drehrichtungen ausgebildet werden. Das Gerät soll fünf Stellungen aufweisen, und zwar:

> Stellung 0 . . . . . . Ausschaltung,
> Stellung 1 . . . . . Rechtslauf im Stern,
> Stellung 2 . . . . . Rechtslauf im Dreieck,
> Stellung —1 . . . . Linkslauf im Stern,
> Stellung —2 . . . . Linkslauf im Dreieck.

An Stelle der Stellungsnummern —1 und —2 können auch die Nummern 4 und 3 gesetzt werden, womit sich das Anschreiben der Minusreihen erübrigt. Der Schaltschritt zwischen den Stellungen 2 und 3 wird mechanisch gesperrt.

Wir bezeichnen die drei Anschlußpunkte des Drehstromnetzes mit R, S, T, die Anfänge der Motorwicklungen mit U, V, W und die Wicklungsenden mit X, Y, Z. Bilden wir das Schaltprogramm als Streckenkomplex ab, so entspricht jedem dieser Anschlüsse ein Punkt des Komplexes. Durch Eintragen der zu den einzelnen Schalterstellungen gehörigen Verbindungen zwischen den neun Punkten des Komplexes ergibt sich das Bild des Schaltprogrammes entsprechend Abb. 109 a.

Die Zerlegung in Teilprogramme, die den einzelnen Schaltbahnen zukommen, soll nun an Hand von Matrizen durchgeführt werden. Da der Schalter fünf Stellungen hat, läßt sich jede Bahn durch eine fünfreihige Matrix darstellen. In dieser erscheint jede Brücke als zyklische Folge von Stellungsnummern innerhalb einer Nebendiagonale. Man kann die Nummernfolge sofort anschreiben, wenn man sich für einen bestimmten Brückenwinkel entschieden hat. Der Brückenwinkel und damit die Nebendiagonale der Matrix, die der Brücke zugeordnet ist, läßt sich unmittelbar aus dem Streckenkomplex entnehmen. Wir beginnen beispielsweise mit dem Punkt R. Die anschließenden Strecken $R\,U$ und $R\,V$ sind durch die Nummernpaare 1, 2 und 3, 4 gekennzeichnet. Die beiden Paare sind um die Zahl 2 voneinander verschieden, so daß eine Brücke, die zwei Schaltschritte umgreift, zweckmäßig ist. Wir haben daher die **zweite Nebendiagonale** der zugehörigen Matrix mit einer zyklischen Nummernfolge auszufüllen.

$$
\left\|\begin{array}{ccccc}
\cdot & \cdot & 0 & 3 & \cdot \\
\cdot & \cdot & 1 & 4 & \\
\cdot & \cdot & 2 & & \\
\cdot & \cdot & & & \\
& \cdot & & &
\end{array}\right\|
$$

Die vorletzte Spalte der Matrix enthält das Ziffernpaar 1, 3; die zugehörigen Strecken des Komplexes nach Abb. 109 a treffen sich im Punkt R, so daß diese Reihe der Matrix in der Hauptdiagonale mit R zu bezeichnen ist. Die Endpunkte der Strecken sind die Punkte V und U, denen sodann die erste und zweite Zeile der Matrix entspricht. Auf diese Weise erhalten wir:

$$
\begin{array}{ccccc}
V & . & 0 & 3 & . \\
U & . & 1 & 4 & \\
  & . & . & 2 & \\
  & R & . & & \\
  & & . & &
\end{array}
$$

In der Zeile U zeigt diese Matrix die noch unbenützte Stellungsnummer 4; aus dem Streckenkomplex ist zu entnehmen, daß vom Punkt U eine Strecke 4 zu S führt. Wir haben daher in der Spalte 5 der Matrix S einzusetzen. Diese Spalte enthält weiters die freie Stellungsnummer 2. Im Streckenkomplex führt nun eine Strecke 2 von S nach V. Tragen wir auch diese Verbindungen in die Matrix ein, so erhalten wir einen vollständigen zyklischen Komplex als Abbild einer zur Gänze ausgenützten Schaltbahn.

$$
\begin{array}{ccccc}
V & . & 0 & 3 & . \\
U & . & 1 & 4 & \\
V & . & . & 2 & \\
R & . & & & \\
S & & & &
\end{array}
$$

Die mit der Matrix erfaßten Strecken können im Schaltprogramm gestrichen werden. Zwischen den vier Punkten R, U, S, V verbleiben somit noch vier von acht Strecken übrig, die sich in analoger Weise in eine zweite Schaltbahn einordnen lassen.

$$
\begin{array}{ccccc}
U & . & 0 & 3 & . \\
V & . & 1 & 4 & \\
U & . & . & 2 & \\
S & . & & & \\
R & & & &
\end{array}
$$

Die dritte Schaltbahn soll zur Verbindung der Punkte T, W, die in allen Stellungen mit Ausnahme der Nullage herzustellen ist, dienen. Diese Schaltbahn kann mit einer Winkelbrücke, die einen Schaltschritt umfaßt, ausgestattet werden, wie sich durch Einsetzen in eine Matrix mit Stellungsnummern in der ersten Nebendiagonale feststellen läßt.

Zweckmäßiger ist ein Brückenwinkel von zwei Schritten, weil dann gleichnamige Schaltfinger nebeneinander zu liegen kommen.

$$
\begin{array}{ccccc}
W & . & 0 & 3 & . \\
. & W & . & 1 & 4 \\
. & . & W & . & 2 \\
. & . & . & T & . \\
. & . & . & . & T
\end{array}
$$

Diese Ausbildung der Schaltbahn für die Finger T, W ist auch deshalb zweckmäßiger, weil dann alle drei Bahnen für den Netzanschluß Brücken gleicher Form aufweisen.

Die Schaltung für die Wicklungsenden verlangt zwei Sternverbindungen in den Stellungen 1, 4 und zwei Dreiecksverbindungen in den Stellungen 2, 3. Wie aus dem Streckenkomplex nach Abb. 109 a zu entnehmen ist, bilden die von einem Wicklungsende ausgehenden Verbindungen einen zusammenhängenden Streckenzug

$$Y-1-X-2-V-3-$$
$$-X-4-Y.$$

Dieses Schaltprogramm läßt sich durch eine Brücke, die einen Schaltschritt umfaßt, verwirklichen.

$$
\begin{Vmatrix}
Y & 1 & . & . & 0 \\
 & X & 2 & . & . \\
 & & V & 3 & . \\
 & & & X & 4 \\
 & & & & Y
\end{Vmatrix}
$$

Es ist aber zweckmäßiger, auch für diese Verbindungen einen Brückenwinkel von zwei Schritten zu benützen, damit sämtliche Schaltbahnen des Gerätes gleich werden. Unter Benützung der zweiten Nebendiagonale ergibt sich hierbei als Fingerfolge:

$$
\begin{Vmatrix}
V & . & 0 & 3 & . \\
 & Y & . & 1 & 4 \\
 & & V & . & 2 \\
 & & & X & . \\
 & & & & X
\end{Vmatrix}
$$

In analoger Weise erhält man für die Verbindung Y, Z, W:

$$
\begin{Vmatrix}
W & . & 0 & 3 & . \\
 & Z & . & 1 & 4 \\
 & & W & . & 2 \\
 & & & Y & . \\
 & & & & Y
\end{Vmatrix}
$$

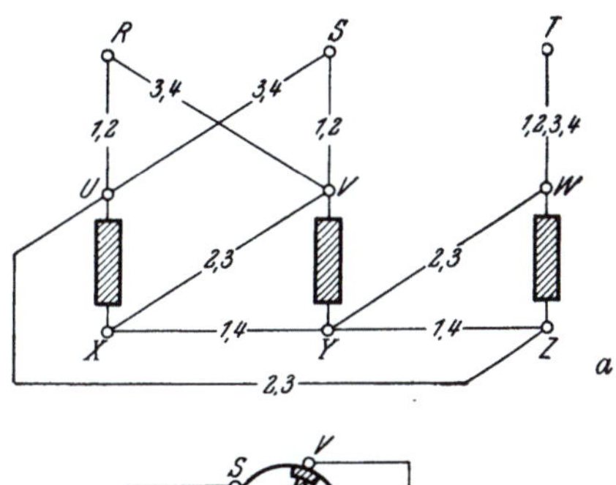

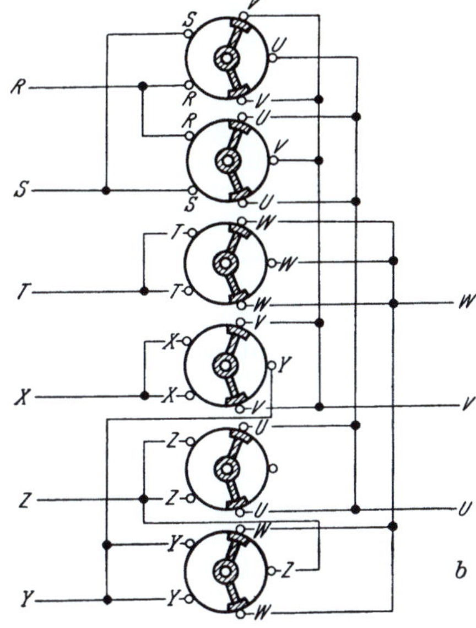

Abb. 109. Paketschalter als Sterndreieck-Anlasser für beide Drehrichtungen.
a Schaltprogramm, b Wirkschaltplan

In der dritten Phase sind nur die Verbindungen Z, U herzustellen; die Matrix zeigt, daß hierdurch ein Finger für den Anschluß X entfallen kann.

$$
\begin{Vmatrix}
U & . & 0 & 3 & . \\
 & . & . & 1 & 4 \\
 & & U & . & 2 \\
 & & & Z & . \\
 & & & & Z
\end{Vmatrix}
$$

Der vollständige Schalter besteht somit aus sechs Schaltbahnen und läßt sich durch eine Summe von sechs Teilmatrizen darstellen; die Reihenfolge der Bahnen wählt man zweckmäßigerweise so, daß gleichnamige Schaltfinger tunlichst in benachbarte Bahnen kommen. Die Matrixsumme lautet sodann:

$$
\begin{array}{ccccc}
V & . & 0 & 3 & . \\
  & U & . & 1 & 4 \\
  &   & V & . & 2 \\
  &   &   & R & . \\
  &   &   &   & S
\end{array}
+
\begin{array}{ccccc}
V & . & 0 & 3 & . \\
  & Y & . & 1 & 4 \\
  &   & V & . & 2 \\
  &   &   & X & . \\
  &   &   &   & X
\end{array}
+
\begin{array}{ccccc}
U & . & 0 & 3 & . \\
  & V & . & 1 & 4 \\
  &   & U & . & 2 \\
  &   &   & S & . \\
  &   &   &   & R
\end{array}
+
$$

$$
+
\begin{array}{ccccc}
W & . & 0 & 3 & . \\
  & Z & . & 1 & 4 \\
  &   & W & . & 2 \\
  &   &   & Y & . \\
  &   &   &   & Y
\end{array}
+
\begin{array}{ccccc}
W & . & 0 & 3 & . \\
  & W & . & 1 & 4 \\
  &   & W & . & 2 \\
  &   &   & T & . \\
  &   &   &   & T
\end{array}
+
\begin{array}{ccccc}
U & . & 0 & 3 & . \\
  & . & . & 1 & 4 \\
  &   & U & . & 2 \\
  &   &   & Z & . \\
  &   &   &   & Z
\end{array}
$$

Wie das vorstehende Beispiel zeigt, zwingt die Zerlegung des Schaltprogrammes in zyklische Teilprogramme dazu, für einzelne oder alle Anschlußpunkte des Schalters mehrere gleichnamige Schaltfinger zu benützen. Anordnungen ohne axiale Schaltbrücken verlangen in der Regel wesentlich mehr Finger als Geräte mit axialen oder gemischten Brücken. Die Vermehrung der Schaltfinger kann trotzdem wirtschaftlich sein, wenn die Finger sehr einfach und billig sind. Beispielsweise ist bei Paketschaltern üblicher Bauweise die Federung der Kontakte in die Schaltbrücken verlegt, so daß als Schaltfinger gestanzte Blechteile verwendet werden können.

Bei Reihenschaltern, die für größere Schaltleistungen zweckmäßiger sind als Paketschalter, müssen die Schaltfinger gefedert sein; man trachtet dann mit Rücksicht auf die Baukosten, die Zahl der Finger klein zu halten. Enthält das Schaltprogramm Verbindungen, die in mehreren Stellungen geschlossen sein sollen, dann läßt sich die Zahl der Finger durch Vermehrung der Brücken herabsetzen. Eine Vermehrung der Schaltbahnen muß damit nicht verbunden sein, sondern es können die Bahnen mit je zwei oder mehr Brücken ausgestattet werden. Um diese in der Schaltbahn unterzubringen, kann eine Vermehrung der Schalterstellungen erforderlich werden. Als Beispiel für diese konstruktiven Möglichkeiten sei der früher behandelte Sterndreieckschalter betrachtet. Er soll als Reihenschalter mit einer möglichst geringen Fingerzahl gebaut werden.

Wie der in Abb. 109 *a* wiedergegebene Schaltplan zeigt, sind zwischen den drei Anschlüssen U, R, V die Verbindungen

$$U - 1, 2 - R - 3, 4 - V$$

herzustellen. Trägt man diese Verbindungen in eine fünfreihige Matrix mit zyklischer Folge der Stellungsnummern ein, so lautet diese:

$$
\begin{array}{ccccc}
V & . & 0, 1 & 3, 4 & . \\
  & U & . & 1, 2 & 4, 0 \\
  &   & . & . & 2, 3 \\
  &   &   & R & . \\
  &   &   &   & .
\end{array}
$$

Die Matrix ist das Abbild einer Schaltbahn mit zwei Brücken, da sie sich in zwei Summanden, die je einer Brücke entsprechen, zerlegen läßt:

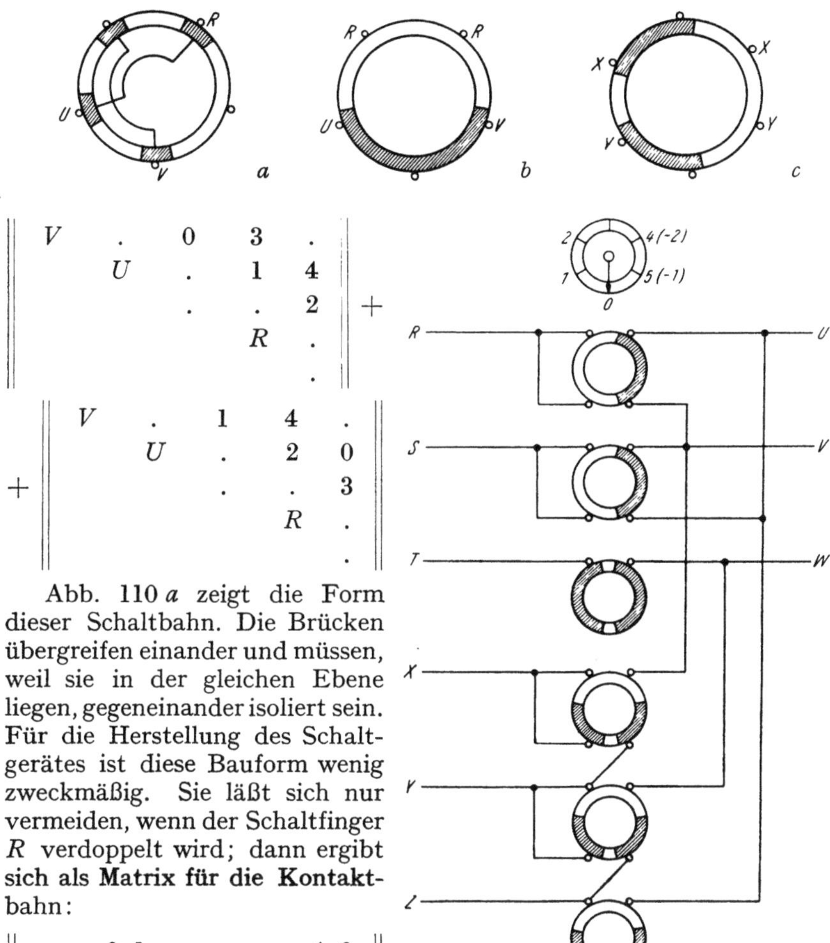

Abb. 110 a zeigt die Form dieser Schaltbahn. Die Brücken übergreifen einander und müssen, weil sie in der gleichen Ebene liegen, gegeneinander isoliert sein. Für die Herstellung des Schaltgerätes ist diese Bauform wenig zweckmäßig. Sie läßt sich nur vermeiden, wenn der Schaltfinger R verdoppelt wird; dann ergibt sich als **Matrix für die Kontakt**bahn:

$$\begin{Vmatrix} . & 0,1 & . & . & 4,0 \\ U & 1,2 & . & . & \\ & R & 2,3 & . & \\ & & R & 3,4 & \\ & & & V & \end{Vmatrix}$$

Um die Gewißheit zu haben, daß die zugehörige Schaltbahn keine Fehlverbindungen herbeiführt, erweitern wir die Verbindungsmatrix zur Funktionsmatrix und erhalten:

Abb. 110. Reihenschalter als Sterndreieck-Anlasser für beide Drehrichtungen.
a Schaltbahn mit zwei isolierten Brücken, b Schaltbahn mit zwei zusammenhängenden blanken Brücken, c Schaltbahn mit zwei getrennten blanken Brücken, d Wirkschaltplan

$$
\begin{array}{c|ccccc|}
 & \cdot & 0,1 & 1 & 4 & 4,0 \\
U & & 1,2 & 2 & & 0 \\
R & & & 2,3 & 3 & \\
R & & & & 3,4 & \\
 & & & & & V
\end{array}
$$

Wie daraus hervorgeht, kommt in der Stellung 0 des Schalters eine unbeabsichtigte Verbindung $U\,V$ zustande; da jedoch die Netzfinger R in dieser Stellung abgetrennt sind, ist diese zusätzliche Verbindung nicht störend und daher statthaft. Man kann also die Schaltbahn entsprechend dieser Matrix ausführen; sie hat die in Abb. 110 b gezeigte Form.

Weniger einfach ist die Entwicklung der Schaltbahnen für die Wicklungsenden. Wir betrachten die Verbindungsfolge

$$Y - 1, 4 - X - 2, 3 - V.$$

Zwischen den Stellungen 1 und 4 befindet sich die Nullage des Schalters, so daß diese Stellungen durch zwei Schaltschritte getrennt sind. Die Finger X, V sind hingegen in den Stellungen 2, 3, die unmittelbar aufeinander folgen, zu verbinden. Es ist also nicht möglich, für die Verbindungen $Y\,X$ und $X\,V$ die gleichen Brücken zu verwenden. Will man trotzdem mit einer einzigen Schaltbahn das Auslangen finden, so muß man eine zusätzliche, mechanisch gesperrte Schaltstellung einführen. Den Stellungen — 1 und — 2 des Schaltprogrammes entsprechen dann nicht die Stellungsnummern 3, 4, sondern 4, 5. Die Verbindungsfolge zwischen den Fingern Y, X, V lautet dann:

$$Y - 1, 5 - X - 2, 4 - V.$$

Zwischen den Stellungen der Nummernpaare 5, 1 und 2, 4 liegen dann je zwei Schaltschritte. Für die zugehörige Schaltbahn folgt daraus die Matrix:

$$
\begin{array}{c|ccccc|}
 & \cdot & 0,4 & \cdot & \cdot & \cdot & 5,3 \\
Y & & 1,5 & \cdot & \cdot & & \cdot \\
X & & & 2,0 & \cdot & & \cdot \\
 & & & \cdot & 3,1 & & \cdot \\
X & & & & & 4,2 & \\
 & & & & & & V
\end{array}
$$

Die Kontrolle durch Entwicklung der Funktionsmatrix ist hier überflüssig, weil in keiner Reihe zwei gleiche Stellungsnummern vorkommen und somit keine mittelbaren Verbindungen auftreten können. Die Schaltbahn darf entsprechend der Abb. 110 c ausgeführt werden.

Da sich für die Verbindung der Wicklungsenden Schaltbahnen mit sechs Stellungen ergaben, müssen auch die Bahnen für die Wicklungsanfänge eine zusätzliche Stellung erhalten. Aus dieser Änderung des Schaltprogramms ergibt sich die Verbindungsfolge

$$U - 1, 2 - R - 4, 5 - V$$

und die Matrix:

$$\left\|\begin{array}{cccccc} 0,1 & . & . & . & 5,0 \\ U & 1,2 & . & . & . \\ R & 2,3 & . & . \\ & . & 3,4 & . \\ & & R & 4,5 \\ & & & V \end{array}\right\|$$

Durch die Einfügung der sechsten Stellung hat sich die Form der Schaltbahn gegenüber der Abb. 110 *b* nur insofern geändert, als zwischen den beiden Schaltfingern R ein unbenützter Fingerplatz vorhanden ist. Abschließend ist noch die Verbindung

$$T - 1, 2, 4, 5 - W$$

zu untersuchen. Hierfür genügt eine Schaltbahn entsprechend der Matrix:

$$\left\|\begin{array}{cccccc} T & 1245 & . & . & . & . \\ W & 2350 & . & . & . \\ & . & 3401 & . & . \\ & . & 4512 & . \\ & . & 5023 \\ & & . \end{array}\right\|$$

Somit sind für sämtliche Verbindungen des Schaltprogramms die Schaltbahnen, Schaltbrücken und Fingerfolgen bestimmt. Durch Vereinigung dieser Einzelteile ergibt sich der Wirkschaltplan des Reihenschalters, der in Abb. 110 *d* dargestellt ist. Ein Vergleich mit dem Schaltplan des wirkungsgleichen Paketschalters nach Abb. 109 *b* zeigt eine Verminderung der Schaltfinger von 29 auf 20. Diese Ersparnis kommt dadurch zustande, daß sich bei Reihenschaltern zwei voneinander elektrisch getrennte Brücken in derselben Schaltbahn unterbringen lassen, was bei Paketschaltern nicht möglich ist.

Schaltregel für Reihen- und Paketschalter. Zur Konstruktion von Reihen- und Paketschaltern muß das Schaltprogramm in zyklische Teilprogramme zerlegt werden, die je einer Kontaktbahn zuzuordnen sind; dies ist häufig nur durch Vermehrung der Schaltfinger möglich. Sollen Finger benachbarter Reihen miteinander verbunden werden, kann dies mit Hilfe blanker Ringbrücken oder auch durch Axialbrücken erfolgen. Sind die zu verbindenden Reihen nicht benachbart, sind radiale Hebelbrücken oder isolierte Tangentialbrücken zu verwenden.

5. Die Drehschalter mit Brücken in gemischter Anordnung

Wie mit vorstehendem gezeigt wurde, eignet sich die Bauart als Walzenschalter besonders für Schaltaufgaben, bei denen die verlangten Verbindungen keinen geschlossenen Zyklus bilden; für Reihenschalter ist hingegen die Herstellung solcher Zyklen unbedingt Voraussetzung.

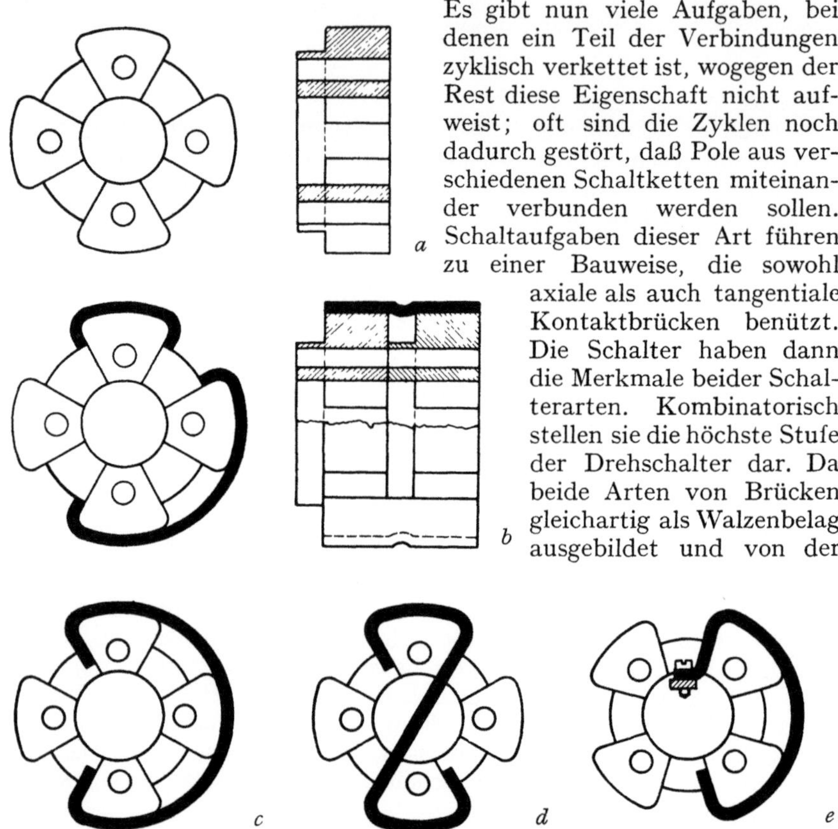

Es gibt nun viele Aufgaben, bei denen ein Teil der Verbindungen zyklisch verkettet ist, wogegen der Rest diese Eigenschaft nicht aufweist; oft sind die Zyklen noch dadurch gestört, daß Pole aus verschiedenen Schaltketten miteinander verbunden werden sollen. Schaltaufgaben dieser Art führen zu einer Bauweise, die sowohl axiale als auch tangentiale Kontaktbrücken benützt. Die Schalter haben dann die Merkmale beider Schalterarten. Kombinatorisch stellen sie die höchste Stufe der Drehschalter dar. Da beide Arten von Brücken gleichartig als Walzenbelag ausgebildet und von der

Abb. 111. Bauteile für Schaltwalzen mit beliebiger Anordnung der Schaltbrücken. *a* Isolierscheibe, *b* Befestigung axialer Schaltbrücken, *c* Befestigung tangentialer Schaltbrücken, *d* Befestigung radialer Schaltbrücken, *e* Verbindung tangentialer Schaltbrücken durch axiale Schienen

Welle isoliert sind, besteht keinerlei konstruktive Beschränkung hinsichtlich der Anordnung der Finger und Brücken.

Bei vorgegebenem Schaltprogramm eines Drehschalters erhält man also im allgemeinen die kleinste Bauform, wenn nicht ausschließlich radiale oder axiale oder tangentiale Brücken verwendet, sondern Brücken in verschiedener Anordnung nach Bedarf gleichzeitig benützt werden. Bei

Schaltern, die mit Rohrwalzen versehen sind, lassen sich ohne weiters axiale oder tangentiale Brücken mischen. Das Bestreben, die Isolierkörper der Schaltwalzen aus einzelnen, untereinander gleichen Bauteilen in gewünschter Länge zusammensetzen zu können, hat auch zu Bauformen geführt, bei denen Schaltbrücken in allen drei möglichen Anordnungen blank verlegt werden können. Abb. 111 gibt ein Beispiel für eine solche Konstruktion. Das Isolierrohr wird aus einzelnen Scheiben, die je einer Schaltbahn zugeordnet und durch Spannbolzen miteinander verbunden sind, zusammengesetzt. Wie Abb. 111 *a*, in der eine solche Isolierscheibe wiedergegeben ist, erkennen läßt, sind die Scheiben mit seitlich vorspringenden Butzen versehen. Diese dienen zur Befestigung der Schaltstücke und lassen zwischen den einzelnen Stellungen Öffnungen zur Durchführung radialer Brücken frei. Abb. 111 *b* zeigt die Befestigung axialer Schaltbrücken, die sich über zwei oder mehrere Isolierscheiben erstrecken. Die Schaltstücke umgreifen die Butzen der Isolierscheiben und werden von diesen ohne Verschraubung oder sonstige zusätzliche Befestigung in ihrer Lage gehalten. Abb. 111 *c* stellt eine Schaltbahn dar, die mit einem aus zwei tangentiellen Brücken zusammengesetzten Schaltstücke versehen ist. Das Schaltstück ist aus einem Metallstreifen hergestellt; dessen abgebogene Enden sind so weit nach innen geführt, daß das Schaltstück von den beiden benachbarten Scheiben gehalten und gegen seitliche Verschiebung gesichert wird. Radiale Brücken werden zur Gänze durch den Hohlraum zwischen den Butzen der Isolierscheiben hindurchgeführt, wie Abb. 111 *d* zeigt. Konstruktionen dieser Art gewähren größtmögliche Freiheit in der Anordnung der Brücken und erlauben daher den Entwurf raumsparender Drehschalter. Sie gestatten es auch, tangentiale oder radiale Brücken miteinander zu verbinden, und zwar auch dann, wenn diese Brücken nicht in benachbarten Schaltbahnen liegen; hierzu dienen blanke Schienen, die im Hohlraum der Isolierkörper untergebracht und mit den Brücken verschraubt werden, wie es in Abb. 111 *e* dargestellt ist.

Zur Ausmittlung eines Schaltbelages mit Brücken in gemischter Anordnung ist es oft zweckmäßig, zuerst aus dem Schaltprogramm jene Verbindungen auszusuchen, die sich für mehrfach benützte Axialbrücken eignen. Die übrigen Verbindungen werden sodann den tangentialen oder radialen Brücken zugeordnet. Ein Beispiel diene zur Erläuterung.

Der an Hand der Abb. 87 beschriebene Umschalter für einen Heizkörper mit zwei Wicklungen und vier Heizstufen soll als Walzenschalter mit gemischter Brückenanordnung und möglichst geringen Abmessungen entworfen werden. Hierzu legen wir dem Gerät ein Schaltprogramm zugrunde, das von demjenigen der Abb. 87 *a* etwas abweicht und in Abb. 112 *a* als Stromlaufplan wiedergegeben ist. Bei dieser Schaltung wird der Widerstand A abwechselnd in verschiedener Richtung vom Strom durchflossen. Das Ende A_2 der Wicklung A ist hierbei mit dem Anfang $B\,1$ der Wicklung B fest verbunden; der darstellende Streckenkomplex, der in Abb. 112 *b* wiedergegeben ist, umfaßt daher nur fünf Punkte. Für einen Schalter mit ausschließlich axialen Brücken ist diese Ausführung der Schaltung wenig geeignet, weil der Linienzug des Schaltprogrammes in sich geschlossen ist. Für eine Bauform, die auch tangentiale Brücken enthält, wirkt dieser Umstand nicht störend.

An Hand des Streckenkomplexes nach Abb. 112 *b* stellen wir nun fest, welche Verbindungen sich durch mehrfach benützte Axialbrücken bilden lassen. Der Komplex enthält zwei Strecken mit je einer Stellungsnummer, und zwar:

$$Q_2 - 2 - A_2 \quad \text{und} \quad A_1 - 4 - B_2.$$

Diesen beiden Verbindungen entspricht eine gemeinsame Axialbrücke, die mit zwei Fingerreihen im Abstand von zwei Schaltschritten zusammenarbeitet. Ferner sind zwei weitere Strecken mit je zwei Stellungsnummern, und zwar:

$$A_1 - 1, 2 - Q_1 \quad \text{und} \quad Q_1 - 3, 4 - A_2.$$

Diese Verbindungen lassen sich durch zwei gemeinschaftliche Brücken mit ebenfalls zwei Fingerreihen verwirklichen.

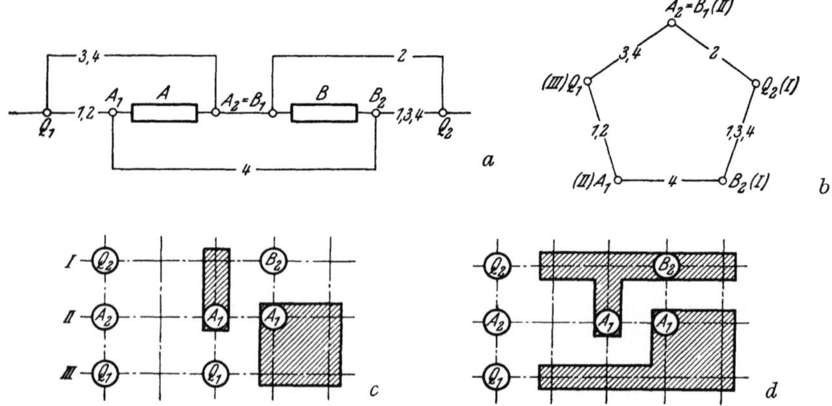

Abb. 112. Heizkörperumschalter mit gemischten Schaltbrücken. *a* Stromlaufplan, *b* Schaltprogramm als Streckenkomplex, *c* Axiale Brücken des Schaltbelages, *d* Vollständiger Schaltbelag

Die übrig bleibende fünfte Verbindung des Komplexes

$$Q_2 - 1, 3, 4 - B_2$$

läßt sich durch ein aus Tangentialbrücken zusammengestelltes Schaltstück herstellen. Daraus folgt, daß die Schaltfinger Q_2 und B_2 in der gleichen Schaltbahn liegen müssen. Dadurch ist aber die Zuordnung sämtlicher Finger zu den Schaltbahnen des Gerätes gegeben und kann in den Streckenkomplex nach Abb. 112 *b* eingetragen werden. Wir vermerken eine Schaltbahn *I* bei den Punkten B_2 und Q_2 des Komplexes. Den anschließenden Strecken mit den Stellungsnummern 2 und 4 entspricht eine gemeinschaftliche Brücke, so daß die Punkte A_1 und A_2

ebenfalls in ein und derselben Schaltbahn liegen müssen; wir vermerken bei diesen Punkten eine Bahn *II*. Der letzte Punkt Q_1 des Komplexes liegt dann in einer dritten Bahn *III*, die mit der Bahn *II* durch zwei axiale Brücken verbunden ist. Nach dieser Festsetzung ist es bereits möglich, die Abwicklung der Walzenfläche zu zeichnen und die Schalt-finger sowie die axialen Brücken einzu-tragen. Abb. 112 *c* zeigt diesen Teil des Schaltbelages.

Es verbleibt somit noch die Aus-mittlung der Tangentialbrücken zwischen den Punkten Q_2 und B_2, was sich an Hand einer Verbindungsmatrix leicht durchführen läßt. Da die Schaltfinger Q_2 und B_2 durch zwei Schaltschritte von-einander getrennt sind, haben wir die Stellungsnummern in die zweite Neben-diagonale einzusetzen. Die Matrix lautet daher:

$$\begin{Vmatrix} Q_2 & \cdot & 134 & 412 & \cdot \\ & \cdot & \cdot & 240 & 023 \\ & B_2 & \cdot & 301 & \\ & & \cdot & & \cdot \\ & & & \cdot & \end{Vmatrix}$$

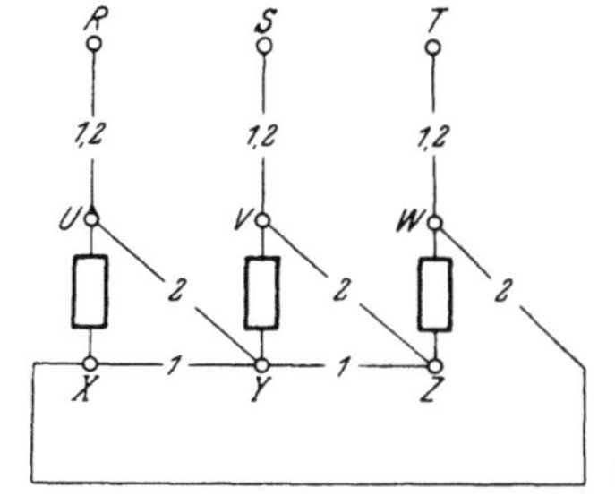

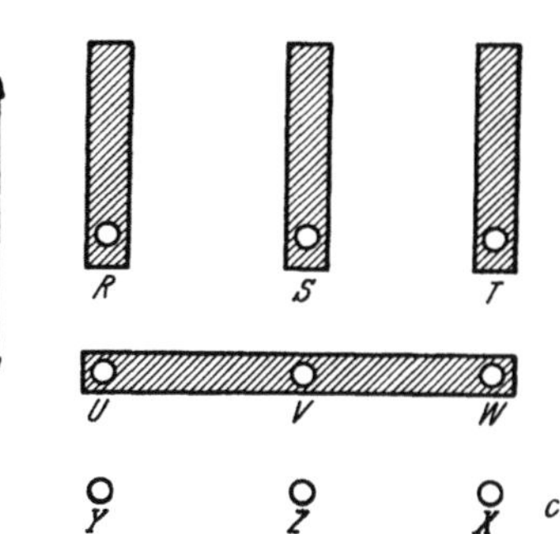

Abb. 113. Einfacher Sterndreieck-Schalter mit gemischten Schalt-brücken. *a* Schaltprogramm, *b* Extremform mit kleinstem Durch-messer, *c* Extremform mit kleinster Baulänge

In der Matrix kommt jede Stellungs-nummer dreimal vor, so daß die Bahn ein aus drei Tangentialbrücken zu-sammengesetztes Schaltstück erhält. Aus der Lage der Stellungsnummer 0 ist ersichtlich, welche Punkte des Walzen-rasters in der Nullage miteinander ver-bunden sind, so daß das Schaltstück eingezeichnet werden kann. Damit ergibt sich die **Abwicklung** des vollständigen Schaltbelages, der in Abb. 112 *d* dar-gestellt ist. Hierbei wurde einer der beiden Finger Q_1, die in der gleichen Bahn *III* liegen, eingespart und durch eine tangentiale Verlängerung des Schalt-stückes ersetzt. Ein Vergleich mit der Ausführung des Gerätes als einreihiger Walzenschalter nach Abb. 87 und 88 zeigt, daß sich die Zahl der Schaltbahnen von 6 auf 3 und diejenige der Schaltstücke von 3 auf 2 vermindert hat. Die Anzahl der Schaltfinger sowie der Schalterstellungen ist unverändert geblieben. Die Schaltwalze enthält einen ausschließlich blanken Belag und kann sowohl als einfache oder zusammengesetzte Rohrwalze sowie auch als Skelettwalze ausgebildet werden.

Zur kombinatorischen Darstellung und Ausmittlung gemischter Brückenanordnungen kann man auch aus einem vorgegebenen Schaltprogramm zuerst die Bahnen mit ihren Reihenverbindungen festlegen und dann erst diese durch die erforderlichen Axialbrücken verbinden oder ergänzen. Das Verfahren sei an Hand eines Beispieles erläutert. Als Schaltprogramm diene das eines einfachen Stern-Dreieck-Schalters (Abb. 113 a), wobei vorerst die Form mit kleinster Stellungszahl gesucht werden soll. Wir fassen zunächst Zyklen zusammen, so daß die Restverbindungen keine Kette mehr bilden; die Zyklen entsprechen dann den einzelnen Bahnen β und die Restfunktionen den Axialverbindungen α. Von den möglichen Kombinationen greifen wir vorerst willkürlich eine heraus, wobei wir uns wegen der Symmetrie der Schaltung auf eine Phase beschränken können:

$$\beta_1 = R\,1\,U + U\,2\,R + R\,0\,R,$$
$$\beta_2 = Z\,1\,Y + Y\,2\,U + U\,0\,Z.$$

Es ergibt sich also, daß das Teilprogramm durch zwei Bahnen eines reinen Reihenschalters erfüllt werden kann, was bereits an früherer Stelle gezeigt wurde. Dabei erscheint aber der Anschluß U in beiden Bahnen, was eine Verdoppelung des Fingers bedeutet. Dies wollen wir verhindern und streichen aus β_2 die Glieder mit U. Das Glied $Z\,1\,Y$ verbleibt dann als Rest der Schaltkette, während die beiden gestrichenen Glieder als Teilfunktion axialer Brücken auftreten:

$$\beta_1 = R\,1\,U + U\,2\,R + R\,0\,R,$$
$$\beta_2 = Y\,1\,Z + Z\,2\,.\, + .\,0\,Y,$$
$$\alpha_{12} = U\,2\,Y \qquad \text{oder} \qquad R\,2\,Y$$

setzen wir die Tangentialverbindungen der Bahnen in ihre Matrizen ein und schreiben wir zwischen die Matrizen benachbarter Bahnen, in welcher Stellung benachbarte Finger durch Axialbrücken verbunden werden, dann erhalten wir folgende Lösung:

$$
\left\|\begin{array}{ccc} R & 0 & 2 \\ & R & 1 \\ & & U \end{array}\right\|
\left\|\begin{array}{c} 1 \\ 2 \\ 0 \end{array}\right\|
\left\|\begin{array}{ccc} . & 0 & 2 \\ & Y & 1 \\ & & Z \end{array}\right\|
$$

Die Angabe gestattet nun noch eine andere Zusammenfassung

$$\beta_1 = U\,1\,R + R\,2\,Y + Y\,0\,U,$$
$$\beta_2 = 0,$$
$$\alpha_{12} = Z\,1\,Y + (R\,2\,U \quad \text{oder} \quad Y\,2\,U).$$

In diesem Fall enthält die Bahn β_2 keine Tangentialbrücke und die Schaltung (Abb. 113 b) lautet:

$$
\left\|\begin{array}{ccc} U & 1 & 0 \\ & R & 2 \\ & & Y \end{array}\right\|
\left\|\begin{array}{c} 1 \\ 2 \\ 0 \end{array}\right\|
\left\|\begin{array}{ccc} R & . & . \\ & U & . \\ & & Z \end{array}\right\|
$$

oder als weitere äquivalente Lösung

$$\left\|\begin{array}{ccc|c|ccc} U & 1 & 0 & 2 & R & \cdot & \cdot \\ & R & 2 & 0 & & \cdot & \cdot \\ & & Y & 1 & & & Z \end{array}\right\|$$

Alle drei Formen haben gleichen Aufwand; ihre Aufwandszahl ermittelt sich aus

$$z_1 = z = 3, \quad a_1 = 9, \quad a = 5, \quad k = 14 \qquad \text{zu}$$

$$\xi = \frac{3 \cdot 5 \cdot 14}{3 \cdot 9 \cdot 9} = 0{,}86.$$

Schließlich wollen wir noch die Extremform mit kleinster Baulänge aufsuchen. Hierzu haben wir die Schaltung durch Vermehrung der Stellungen in eine möglichst geringe Zahl von Zyklen so einzuordnen, daß diese alle Anschlüsse enthalten und die Restverbindungen den Bedingungen eines mehrreihigen Walzenschalters genügen. Durch Einsetzen in die Matrix erhalten wir für die erste Bahn

$$\beta_1 = \left\|\begin{array}{cccc} R & 1,2 & 2 & 0 & 0,1 \\ U & 2,3 & 3 & 1 \\ Y & 3,4 & 4 \\ & & 4,0 \\ & \cdot \end{array}\right\|$$

Die Matrizen β_2 und β_3 für die Phasen S und T lauten analog und es ergibt sich bereits die Gesamtschaltung

$$\left\|\begin{array}{ccccc} R & 1,2 & 2 & 0 & 0,1 \\ U & 2,3 & 3 & 1 \\ & Y & 3,4 & 4 \\ & & \cdot & 4,0 \\ & & \cdot \end{array}\right\| \begin{array}{c}(4)\\0\\1\\2\\3\end{array} \left\|\begin{array}{ccccc} S & 1,2 & 2 & 0 & 0,1 \\ V & 2,3 & 3 & 1 \\ & Z & 3,4 & 4 \\ & & \cdot & 4,0 \\ & & \cdot \end{array}\right\| \begin{array}{c}(4)\\0\\1\\2\\3\end{array} \left\|\begin{array}{ccccc} T & 1,2 & 2 & 0 & 0,1 \\ W & 2,3 & 3 & 1 \\ & X & 3,4 & 4 \\ & & \cdot & 4,0 \\ & & \cdot \end{array}\right\|$$

Wie die Matrix zeigt, bedarf die **Walze einer Begrenzung des Schaltwinkels, da sonst in der** Stellung (4) eine unzulässige Verbindung der drei Netzpole R, S, T zustande käme. Die Schaltung (Abb. 113 c) zeichnet sich dadurch aus, daß sie mit dem Minimum an Fingern auskommt; ihre Aufwandkennzahl ist daher sehr klein und beträgt

$$\xi = \frac{5 \cdot 3 \cdot 9}{3 \cdot 9 \cdot 9} = 0{,}55.$$

Abb. 113 b, c zeigt die Extremformen des Sterndreieckschalters in der Bauweise als Reihenschaltwalze.

Namen- und Sachverzeichnis

Die **fett** gedruckten Seitenzahlen verweisen auf Begriffsbestimmungen bzw.
-erläuterungen, die *kursiv* gedruckten Seitenzahlen verweisen auf Schaltregeln.

Abhängige Bedingungen 55, *57*
Abwicklung (Schaltwalze) 172
ACKERMANN 40
Addition 38 f., 54, 57 ff., 80 f.
Algebraische Logik = Logistik 12 f.,
  39, 45, 49, 53
Allpolige Schaltung(sform) **16**
Anreizschaltung = Impulsschaltung
  15, 29 f., 42, 101
Ansatzregel *57*
Antriebsvorrichtung **17**
Anzapfschalter = Wähler **18**, 170
— s. a. Drehwähler
Äquivalente Kontakte **27**, 34, 106,
  113, 130 f., *132*, *154*, 182
— Schaltfinger 182, 190, *200*
— Schaltungsformen **147** ff., 150, 152,
  154, 182 f.
— Walzenbeläge 153 f., 178 ff., 181 ff.
Äquivalenzgesetz = Tautologiesatz
  39, 45, *49*, 50, 54, 58, 74, 78, 87,
  137, 150
Arbeitsstellung **15**
Arbeitsstromschaltung **15**, 70, 144 f.
Assoziation, assoziatives Gesetz 45, 78
Aufgabe s. Schaltaufgabe
Aufwandskennzahl (Schaltwalzen)
  200, 217
Auslöseschaltung (Gefahrauslösung)
  122 ff., 129 ff.
Ausmittlung von Schaltungen 1 f., 5,
  7 ff., *11* ff., 106
Ausnützungsgrad (Schaltwalzen) =
  Füllfaktor **172** f., 195
Äußere Schaltung **5**
— Verknüpfung = Disjunktion 54
Axiale Schaltbrücken 168 ff., 173 ff.,
  189 ff.

BADER 12 f., 117
Bahnteilung = Fingerteilung **172**

Bedingung = Funktionsbedingung 13,
  36, 40, **53**, *57*, *84*
— = Leitbedingung 29, 53 ff., 57 ff.,
  62 ff., 70 ff., 74 ff., 84 ff.
— = Sperrbedingung 29, 73, 142
— abhängige = gleichzeitige 55, *57*
— inverse 28, 70 ff., *73*
— mittelbare **13**
— unabhängige 54 f., *57*
Bedingungsgleichung 55
Bedingungsoperand 54
Befehlsschalter **18** f.
Berandung (Masche) 143
Bezeichnungssystem 32, **34** ff.
BODA 10
Brücke = Schaltbrücke 6, **17**, 168 ff.,
  178 ff.
— axiale 168 ff., 173 ff., *186*, *188*,
  189 ff.
— gestreckte 6, 169
— Hebelbrücke = radiale Brücke
— Isolierbrücke 170, 173 ff., 181 ff.,
  186 f.
— Mehrfachbrücke 169
— radiale 169 f., 201 ff., *211*
— Ringbrücke = tangentiale Brücke
— tangentiale 169, 186, 201 ff., *211*
— Winkelbrücke 6, 169
Brückenkontakt = Quer- oder Ma-
  schenkontakt **23**, 113, 132 ff., *138*

CARNAP 40

Darstellung (Schaltungen) 30 ff.
Dauerschaltung **15**, 29
Determinante 155 ff.
Differenz (Leitbedingungen) 74 ff., 79,
  83 f., *84*
Disjunktion = äußere Verknüpfung 54
Distribution, distributives Gesetz
  45 f., 74, 78

Division (Leitbedingungen) 74 ff., 79,
    83 f., *84*, **133**
Drehschalter **6**, **168** ff., 171 ff.
Drehtaster 18
Drehwähler **60**, 163, 167
Drucktaster 18
Dualschaltung **70**, 163
Dunkelkammerschaltung 51 f.

Ebene Schaltung(sform) 143
EDLER 10, 32
Einseitige Schaltung(sform) **16**, 145
Elektromechanische Schaltung **14**, 19
Endschalter 88
Entwicklung von Schaltungen 9 f., 12
Ergebnis s. Resultat
Erlaubte (= zulässige) Verbindungen
    7 f., 99
Erweiterung **25** ff., **65** ff.
Erzwungene Reservehaltung 55 ff., 73
— (= formale) Vermaschung 26, **121**,
    124 ff., *127*

Fernschalter 18
Feste Schaltung **13**
Finger = Schaltfinger = Kontakt-
    finger 17, **168**
— äquivalente 182, 190, *200*
— Randfinger 176
Fingerfolge 153 f., 175 ff., 181 ff., *186*
Fingerleiste **168** f.
Fingerreihe 168, 172, 186 f., 191
Fingerteilung = Bahnteilung **172**
Flachbahnschalter **170**, 201
Formale (= erzwungene) Verma-
    schung 26, **121**, 124 ff., *127*
FRANKE **14**
Freiheitsgrade 6 ff., 11, 30, 74 ff., **83**,
    84, 90 ff., 95, 101, 106 ff., 116.
    179 ff., 183 ff.
Füllfaktor = Ausnützungsgrad
    (Schaltwalze) **172** f., 195
Funktion = Wirkungsweise 1, 8, 13,
    17, 19 f., 30, 36 f., 43, *68*, 147, 154 ff.
Funktionelle Vermaschung 26
Funktionsbedingungen s. Bedingungen
Funktionsgleichung **53**
Funktionsmatrix **149** ff., 183
Funktionsregel *68*

Gefahrauslösung 122 ff., 129 ff.
Gefahrenmeldung 14

Gekreuzte Schaltung(sform) **24** f.
Gestreckte Schaltbrücken **169**
Gleichlaufschaltung (Transforma-
    toren) 110 ff.
Gleichrichter-Ladestation 2 ff., 151 f.
Gleichung s. Leitwertgleichung
Gleichzeitige (= abhängige) Bedin-
    gungen 55, *57*

Haltekontakt 15, 18
Haltestrompfad 56
HANZAWA 12 f.
Hebelbrücke = radiale Brücke 169 f.,
    201 ff., *211*
Heizstufenschaltung 128 f., 177 ff.,
    193 f., 213 ff.
Heraushebung 45 f., *49*, 145 ff., *147*
HILBERT 40
Hilfskontakt 47, 56
Hilfsrelais 18, 43
Hilfsschaltgerät 18, 43
Hilfsschaltung 142 f.

Impulsschaltung = Anreizschaltung
    15, 29 f., 42, 101
Innere Schaltung **5** f., **168** ff.
Inverse = komplementäre = Kehr-
    Schaltung **28**, **70**, 142 ff., *145*
Inverse = komplementäre Kontakte,
    Variable 57 ff., *62*
Inversion = Umkehrung **28**, **70** ff.,
    73, 76, 79, 88 ff.
Inversionsoperator **35**, 50, 79
Irregulärer Komplex 43
Irreversible Schaltvorgänge **29**, 42 f.
Isolierbrücke 170, 173 ff., 181 ff.,
    186 f., *188*
Isolierplan 180 f.

Kehrform = Kehrschaltung = in-
    verse Schaltung **28**, **70**, 142 ff., *145*
Kombinationsklasse 164
Kombinationsschaltung 164 ff.
Kombinatorische Topologie 1, 12, **33**
Kommutation, kommutatives Gesetz
    45, *49*, 78
Komplementäre = inverse Schaltung
    28, **70**, 142 ff., *145*
Konjunktion = innere Verknüpfung
    55
Konstruktionsbedingungen 127
Kontaktbahn = Schaltbahn 16, 169 f.,
    172, 201

Kontakte 1, 5, 12, 14, 17, 20, 32
— äquivalente **27**, 34, 106, *113*, 130 f., *132*, *154*, 182
— Brückenkontakte = Maschenkontakte = Querkontakte **23**, 113, 132 ff., *138*
— Haltekontakte 15, 18
— Hilfskontakte 47, 56
— inverse = komplementäre 57 ff., *62*
— Längskontakte = Serienkontakte **23**, 132 ff., *138*
— Randkontakte 114, 127 ff., *132* ff.
— Schubkontakte 168
— selbststeuernde 15
Kontaktfinger = Schaltfinger, s. Finger
Kontaktfolge 5, 20, 106, 127 ff., *132* ff., *138*
Kontaktformen 168
Kontaktkombinationen, — verknüpfung 44 ff., *49*, *62*
Kontaktleitwert, -widerstand 12, 14, 17, 20, 32, 37 f.
Kontaktschaltung = elektromechanische Schaltung **14**, 19
Kontaktstelle 158
Kontaktverdoppelung 45
— s. a. äquivalente Finger
Kreuzschalter **18**
Kreuzung **24** f.
Kurzschlußregel *49*
Kürzung 11 f., **21** f., 27, 57 ff., *62*, 65 f., 86 ff., 90 ff., 93 ff., 132 ff., **134** f., *138*

Ladestation 2 ff., 151 f.
Längs-(= Serien-)Kontakt **23**, 132 ff., *138* f.
Lastschalter 170
Leerstellung 182, 187 f.
Leitbedingung s. Bedingungen
Leit-(= Wirk-)Bereich **80** ff., **136** f.
Leiter **17**, 107
Leiterpolygon 17, 23
Leiterstern 17, 23
Leitfähigkeit s. Kontaktleitwert
Leitung 17, 32
Leitungsregel *49*, *121*
Leitungsschar 114 ff., *121*
Leitungstabelle **111** f., 114 f., 117
Leitwert s. Kontaktleitfähigkeit

Leitwertfunktion 36 f., 44, 50, 53, 68, 147
Leitwertgleichung 12, 24, 37 f., 40, 46, 50 f., 53, 65
Leitwertmatrizen s. Matrix
Leitwertverhältnis 49, 54
LISCHKE 10, 12, 32
Logistik = algebraische Logik 12 f., **39**, 45, 49, 53
Lösungsverfahren (Schaltaufgaben) 1 f., 5, 7 ff., *11* ff., 106
Lusterwechselschaltung 117 ff.

Mannigfaltigkeit s. Freiheitsgrad
Masche **143**
— s. a. Vermaschung
Maschen-(= Brücken-, = Quer-) Kontakt **23**, 113, 132 ff., *138*
Maschinentelegraph 144 f.
Matrix (Leitwertmatrix), eindimensional 12, **82**, 117
— quadratische Funktionsmatrix **149** ff., 183
— — Verbindungsmatrix 24, 33, 148, **149** ff., 174 ff., 181, 183
Mehrdeutigkeit s. Freiheitsgrad
Mehrfachbrücke 170
Mehrpolige Schaltung(sform)en **16**, 147 ff.
Mehrreihige Nockenschalter 171
— Walzenschalter 6, 168 f., 186 f., *188*, 189 ff., *200*
Mehrteilige Schaltgeräte 170
— Schaltungen 142, *145* ff., 148
Meldeanlagen 14, 28, 47 ff., 125 f.
Mittelbare Verbindungen 7 f., 13, 149
— Verkettung **28**
Modulrechnung **66**
Mögliche Verbindung 7 f.
Multiplikation 39, 55, 58 ff., 81, 86

Nachsteuerschaltung 144 f.
NAKASIMA 12 f.
Natürliche Vermaschung 116, 124 ff., 141
Netzumschalter 203 f.
Nockenschalter 171
Normung 18 f.
Notschalter 122
Nullstellung 34, 171 *201*

Operand 47, 48 f., 50, 57, 86, 112, 148 f., 155
— inverser 58 ff., *62*
Operator 79
— Inversionsoperator **35**, 50, 79
Ordnung (Streckenkomplex) **92**

Paarige Verkettung 124
Paketschalter 6, **168**, **169** f., 201 ff., *211*
Parallellaufverriegelung 139 ff.
Parallelschaltung 37 ff., 44 f., *57*, 71 ff., *73*
PIESCH 128
PLECHL 13
Pol = Spannungspol **13**, 16, 138, 169
Polung 127 ff., *132*, 147, 149, 155 ff.
Potential 16, **32**
Potenzieren (Matrix) 150 f.

Quadrieren (Matrix) 150 f.
Quer-(Brücken-, Maschen-)Kontakt **23**, 113, 132 ff., *138*
Quotient (Leitbedingungen) 74 ff., 79, 83 f., *84*, **133** ff.

Radiale Schaltbrücke 169 ff., 201 ff., **211**
Randelement *132* f.
Randfinger 176
Randkontakt 114, 127 ff., *132* ff.
Rast 172
Rechenzeichen 35, 48, 79
Reduzierte Schaltungsform 158
Regelschaltung 14
Regulärer Komplex **42** f.
REICHENBERG 40
Reihen-Parallelschaltung 128
Reihenschalter 6, 168, **169** f., 201 ff., 208 ff., *211*
Reihenschaltung 37, 44 f., *57* f., 70 ff., *73*
Reihenwalzenschalter 170, 212 ff.
Resultat, arithmetisches **78** f.
— kürzestes **77** f.
— unbestimmtes 40, 74 ff., 82 ff., *84*, 90 ff.
— — s. a. Freiheitsgrade
Reversible Schaltvorgänge **29**, 42 f.
Richtungswechsel 23, 127
— s. a. Maschenkontakt
RIEDER 19

Ring-(Tangential-)Brücke 169 f., 186 ff., 201 ff., *211*
Ringsegment 169 f., 184 ff.
RITTER 13, 143
Rohrwalze 173
Ruhestellung 15
Ruhestromschaltung **15** f., 70, 144
RUSSEL 40

Schaltaufgabe 2, 6 ff., 13, 28 f., 53
Schalt-(= Kontakt-)Bahn 16, 169 f., **172**, 201
Schaltbedingung s. Bedingungen
Schaltbelag 6, 153 f., 174 f., 178 ff.
Schaltbereich **172**
Schaltbrücke s. Brücke
Schaltebene 170
Schaltelement 17
Schalter, Schaltgerät 1 ff., **17** f., 40, 168 ff.
— Anzapfschalter = Wähler 18, 170
— Befehlsschalter **18** f.
— Drehschalter 6, **168** ff., 171 ff.
— Endschalter 88
— Fernschalter 18
— Flachbahnschalter **170**, 201
— Hilfsschalter 18, 43
— Kreuzschalter 18
— mehrpolig 16
— mehrreihig 6, 168 f., 171, 186 f., *188*, 189 ff., *200*
— mehrteilig 170
— Netzumschalter 203 f.
— Nockenschalter 171
— Notschalter 122
— Reihenschalter 6, 168, 169 f., 201 ff., 208 ff., *211*
— **Reihenwalzenschalter 170, 212 ff.**
— Stellschalter 17
— Sterndreieckschalter 204 ff., 208 ff., 216 f.
— Steuerschalter 19
— Summenschalter 122
— Tastschalter 18
— Trafostufenschalter 110
— Umschalter 6, 18, 27 f., 192 ff.
— Verbraucherschaltgerät 19
— Wahlschalter 18, 170
— Walzenschalter 6, 153 f., **168** f., 173 ff., *186* ff., *188* ff., *200*
— Wendeschalter 192 ff.
Schalter-(= Kontakt-)Finger s. Finger

Schaltfolge **29** f.
Schaltfolgediagramm **42**
Schaltfolgeplan 41 ff., 92 ff.
Schaltgerät s. Schalter
Schalthebel 169
Schalt-(= Verbindungs-)Kette 154
Schaltplan 30 ff.
Schaltprogramm 169, 174, *186*, *188*, 212
— zyklisches 189 ff., *200*, 201 ff., *211*
Schaltregel 9 ff.
Schaltreihe (= Stellungskombination) **20**
Schaltschritt **29**, **172**
Schaltstellung 2 f., **20**, **29**, **172**
Schaltstück 1, 5 f., **17**
Schaltung **13**
— s. a. Schaltungsform
— allpolige **16**
— Anreizschaltung **15**, 29 f., 42, 101
— Arbeitsstromschaltung **15**, 70, 144 f.
— Auslöseschaltung 122 ff., 129 ff.
— äußere **5**
— Dauerschaltung **15**, 29
— Dualschaltung 70, 163
— Dunkelkammerschaltung 51 f.
— elektromechanische **14**, 19
— feste **13**
— Gleichlaufschaltung (Trafo) 110 ff.
— Heizstufenschaltung 128 f., 177 ff., 193 f., 213 ff.
— Hilfsschaltung 142 f.
— Impulsschaltung **15**, 29 f., 42, 101
— innere 5 f., 168 ff.
— inverse = Kehrschaltung **28**, **70**, 142 ff., *145*
— Kombinationsschaltung 164 ff.
— komplementäre s. inverse
— Kontaktschaltung **14**, 19
— Lusterwechselschaltung 117 ff.
— Nachsteuerschaltung 144 f.
— Parallelschaltung 37 ff., 44 f., *57*, 71 ff., *73*
— Reihen-Parallelschaltung 128 f., 177 ff., 193 f., 213 ff.
— Reihenschaltung 37, 44 f., *57* f., 70 ff., *73*
— Ruhestromschaltung **15** f., 70, 144
— Serienschaltung s. Reihenschaltung
— Sicherheitsschaltung **96**

Schaltung
— Variationsschaltung 164 ff.
— Wählerschaltung 60 ff., 69, 90 ff., 93 f.
— Wechselschaltung 15, 27, 40 ff., 63 ff., 72 f., 117
Schaltungsform 3 f., 106 ff.
— s. a. Schaltung
— äquivalente **147** ff., 150, 152, *154*, 182 f.
— ebene 143
— einseitige **16**, 145
— formal vermaschte 26, **121**, 124 ff., *127*
— funktionell vermaschte 26
— gekreuzte **24** f.
— mehrpolige **16**, 147 ff.
— mehrteilige 142, *145* ff., *147*
— mittelbar verkettete 28
— natürlich vermaschte 116, 124 ff., 141
— reduzierte 158
— Sperrform 116, 124 ff., *127*
— verkettete 5, 12 f., 21 ff., *27*, 89, **106** ff., *113*
— vermaschte 22 ff., 46, 105, 113 ff., **118**, *119*, 132 ff., *138*, 159 f.
— Vollform **65** f.
— Wahlformen s. äquivalente Schaltungsform und Freiheitsgrade
— wirkungsgleiche s. äquivalente
— zusammengesetzte 142, *145*
— zweiseitige **16**, 145
Schaltungstechnik 9 f., 12
Schaltvorgang **29**, 40 ff.
Schaltwalze 173, 212
— äquivalente 153 f., 168 ff., 181 ff.
Schaltwinkel **172**, 182, 187
Schaltzeichen 30
Schaltzustand 2 ff., **20**, 29 f., 62 f.
Schaltzyklus 30
Scheinbarer Strompfad 26
Schiffsschleuse 95 ff.
Schleifring 170, 191 ff., 194 ff., *200*
Schritteilung **172**
Schrittweise Vermaschung 132 f., *138*
Schrittwinkel 6, **172**, 187, *188*
Schubkontakt 168
Schütz 15, 18, 43, 55 f., 161 f.
Selbsthaltung 15, 18, 43, 55 f.
Selbstunterbrechung 15, 18, 60

Serien-(= Reihen-)Schaltung 37, 44 f.,
   57, 58, 70 ff., *73*
Skalar 49
Skelettwalze **173** f.
Spannungspol s. Pol, Polung
Spannungszustand 160 ff.
Sperrbedingung 29, 73, 142
Sperrbereich **80** ff.
Sperrform 116, *127*
Sperrkontakt 5, 26, **121**, 125, *127*, 137
Sperrstellung 181 f., 184, 186 f.
Stellschalter **17**
Stellung (= Schaltstellung) 2 f., 20,
   29, **172**
Stellungskombination (= -komplex)
   20, *68*, 164 ff.
Stellungsnummer 34, 63
Stellungstabelle 128
Stellungsvariation 20, 62 ff., *68*, 161 ff.
Sterndreieckschalter 204 ff., 208 ff.,
   216 f.
Steuerfunktion (= -wirkung) 30
Steuergerät (-schalter) **19**
Streckenkomplex 12, **32** f., 42, 92 ff.,
   177 ff., 182
Streckenpaar 43, 148
Stromkreis 30
Stromlaufplan 5, **31** f.
Strompfad 5, 11, **13**, 19 f., *62* ff.,
   69 f., 160, 165 f.
Stromquelle 17, 50, *53*, *73*
Stromrichtung 23, 127
Stufenschalter 171
Subtraktion 74 ff., 79, 82 f., *84*
Summe 38 f., 54, 57 ff., 80 f.
Summenschalter 122

Tabellenrechnung 80 ff., 84 ff.
Tangentialbrücke 169, 186, 201 ff., *211*
Taster **18**
Tautologiesatz s. Äquivalenzgesetz
Topologie 1, **12**, 33
Transformation *145*, 147 f., 183

Überdeckung (Leitbedingungen) 80
Umformung 72
Umkehranlasser 195 ff.
Umkehrbare Schaltvorgänge 29
Umkehrung (= Inversion) **28**, 70 ff.,
   73, 76, 79, 88 f., 142 ff., *145*
Umkehrungszeichen **35**, 50, 79
Umschalter 6, 18, 27 f., 192 ff.

Umschaltschütz 162
Unabhängige Bedingungen 54 f., *57*
Unbestimmtes Resultat 40, 74 ff.,
   82 ff., *84*, 90 ff.
— s. a. Freiheitsgrade
Unmittelbare Verbindungen 7 f., 13,
   149
Unterbrechungsloser Umschalter 21,
   35
Unterdeterminante 156 f.

Variable Leitwerte, Widerstände 17,
   19 f., 30 ff.
Variation (= Stellungsvariation) 20,
   62 ff., *68*, 161 f.
Variationsschaltung 164 ff.
Variationstabelle **63** ff., 70 ff., 80 ff.,
   84 ff., 95, 107 ff., 114 ff., 122 f.
Verbindungen, erlaubte 7 f., 99
— innere 5 f., 168 ff.
— Maschenverbindungen 23, 113,
   132 ff., *138*
— mittelbare 7 f., 13, 149
— mögliche 7 f.
— unmittelbare 7 f., 13, 149
— verbotene 7 f., 151
— verlangte 7 f.
— zulässige 7 f., 99
— (Schaltbrücken) 178 ff., 183 ff.,
   186, *188*, 194
Verbindungskomplex s. Streckenkomplex
Verbindungsmatrix **149** ff., 174 ff.,
   181, 183
Verbraucher 17, 50, *53*, *73*
Verbraucherschaltgerät 19
Verdoppelung von Kontakten 45
— — — s. a. äquivalente Finger
Vereinigung s. Zusammenlegung
Verfahren 1 f., 5, 7 ff., *11*, 106
Verjüngerung 45
Verkettung 5, 12 f., 21 ff., *27*, 89,
   106 ff., *113*, 116, 126
— mittelbare 28
Verknüpfung (Bedingungen) 53 ff.
— äußere 55
— innere 54
— (Kontakte) 44 ff., *49*, 62
Vermaschung **22** ff., 46, 105, 113 ff.,
   118, *119*, 132 ff., *138*, 159 f.
— erzwungene, formale **26** f., **121**,
   124 ff., *127*

Vermaschung, funktionelle **26**
— natürliche 116, 124 ff., 141
Verriegelung 88, 114 ff., 130, 142 f.
Vertauschung 45, *49*
Verteilung s. Heraushebung
Vollform **65** f.

Wächter 18
Wahlbereich **82**, **95**, 99 ff., 111, 135 ff.
Wählerschaltung 60 ff., 62 ff., 69,
90 ff., 93 f.
Wahlformen s. äquivalente Formen,
Freiheitsgrade
Wahlschalter (= Wähler) **18**, 170
Walzenanlasser 194 ff.
Walzenschalter 6, **168** f.
— einreihig 6, 153 f., 169, 173 ff., *186*
— mehrseitig 6, 168 f., 186 ff., 189 ff.,
188, 200
Wechselschaltung 15, 27, 40 ff., 63 ff.,
72 f., 117
Wendeschalter 181 f., 192
Whitehead 40
Widerstand (Kontakt) 12, 14, 17, 20,
32, 37 f.
Winkelbrücke 6, **169**

Wirkbereich (= Leitbereich) **80** ff.,
136 f.
Wirkschaltplan 6, **31**
Wirkungsbereich 78, *84*
Wirkungsgleiche Formen s. äqui-
valente Schaltungsformen
Wirkungsweise s. Funktion

Zerlegung von Matrizen 179 f., 189 ff.,
194 ff.
— — Schaltungen 145 ff., *147*
— — Variationstabellen 87, 90
Zulässige Verbindungen 7 f., 99
Zusammenlegung von Kontakten und
Leitungen s. Verkettung
— — Schaltbrücken 178 ff., 183 ff.,
186, 188, 194
Zusammensetzung von Schaltungen 9 f.
Zustand (= Schaltzustand) 2 ff., **20**,
29 f., 62 f.
Zustandswechsel (= Schaltvorgang)
29, 40 ff.
Zweiseitige Schaltung **16**, 145
Zwischenrelais 28
Zwischenschütz 28, 161 f.
Zyklische Verkettung 131, *132*
Zyklus (Schaltzyklus) 30